VOLUME 2

AMERICAN BEETLES

Polyphaga: Scarabaeoidea through Curculionoidea

VOLUME 2

AMERICAN BEETLES

Polyphaga: Scarabaeoidea through Curculionoidea

Edited by
the late Ross H. Arnett, Jr., Ph.D.
Michael C. Thomas, Ph.D.
Paul E. Skelley, Ph.D.
and
J. Howard Frank, D. Phil.

CRC Press
Taylor & Francis Group
Boca Raton London New York

CRC Press is an imprint of the
Taylor & Francis Group, an **informa** business

COVER FIGURES: Center - Coccinellidae, *Harmonia axyridus* (Palles) [Photo by Fred J. Santana]. Outer rim, clockwise from top: Ripiphoridae, *Macrosiagon cruentum* (Germar) [by Fred J. Santana]; Meloidae, *Lytta magister* Horn [by Charles L. Bellamy]; Carabidae, *Rhadine exilis* (Barr and Lawrence) [by James C. Cokendolpher]; Melyridae, *Malachius mirandus* (LeConte) [by Max E. Badgley]; Lampyridae, *Microphotus angustus* LeConte [by Arthur V. Evans].

CRC Press
Taylor & Francis Group
6000 Broken Sound Parkway NW, Suite 300
Boca Raton, FL 33487-2742

First issued in hardback 2017

CRC Press is an imprint of Taylor & Francis Group, an Informa business

Library of Congress Cataloging-in-Publication Data

American beetles / edited by Ross H. Arnett and Michael C. Thomas.
p. cm.
Contents: v. 1. Archostemata, Myxophaga, Adephaga, Polyphaga: Staphyliniformia.
Includes bibliographical references (p.).
ISBN 0-8493-0954-9 (alk. paper : v. 2))
1. Beetles—North America. I. Arnett, Ross H. II. Thomas, M. C. (Michael Charles).
1948–
QL581 .A43 2002
595.76'097—dc21 00-050809
CIP

Library of Congress Card Number 00-050809

ISBN 13: 978-1-138-42370-1 (hbk)
ISBN 13: 978-0-8493-0954-0 (pbk)

To Ross H. Arnett, Jr.
1919-1999

and

Mary Arnett
1919-2002

Preface

It has been nearly 40 years since Ross H. Arnett, Jr. published the first fascicle of *The Beetles of the United States: A Manual for Identification.* It quickly became an indispensable tool for professional and amateur coleopterists, general entomologists, and naturalists. Although there were four additional printings it has long been out of print and difficult to obtain. It was prepared to replace Bradley's *A Manual of the Genera of Beetles of America, North of Mexico*, which itself was some 30 years out of date in 1960. *American Beetles* is, in turn, designed to replace *The Beetles of the United States.* It is hoped that it will prove to be as useful as its predecessor.

Ironically, much of the preface to the original edition applies today as well as it did 40 years ago:

> Many genera have since been described and reported within the area concerned, and many families have been revised. Extensive changes have been made in the family classification of the beetles of the United States during this period.
>
> The aim of this series of fascicles is to provide a tool for the identification of adult beetles of the United States to family and genus with the aid of illustrations, keys, descriptions, and references to sources for keys and descriptions of the species of this area. All of the genera known to inhabit this area are included in the keys and lists of genera which follow.

The design and format of this work follow closely that of the original edition, but the way it was put together was quite different. Its predecessor was very much the work of one man, Ross H. Arnett, Jr. With a few exceptions (George Ball wrote the carabid treatment for both the 1960 edition and for this one), Dr. Arnett wrote the family treatments of *The Beetles of the United States.* Many specialists reviewed those chapters, but they were almost entirely Dr. Arnett's work.

When Dr. Arnett announced plans to prepare a work to replace *The Beetles of the United States*, coleopterists literally lined up to volunteer their time and expertise in preparing the family treatments. Ultimately, more than 60 coleopterists participated in the preparation of *American Beetles.* This has truly been a community project.

Due to the size of the ensuing work, *American Beetles* is being printed in two volumes. Volume 1 includes the introductory material, and family treatments for the Archostemata, Myxophaga, Adephaga, and Polyphaga: Staphyliniformia. The remainder of the Polyphaga and the keys to families appear here in Volume 2.

Sadly, although Dr. Arnett initiated this project and was instrumental in its planning, he did not live to see its fruition. He became seriously ill in late 1998 and died on July 16, 1999 at the age of 80. We hope he would be pleased with the outcome.

Michael C. Thomas, Ph.D.
Gainesville, Florida
April 3, 2002

Table of Contents

Preface vii

Acknowledgments xi

Contributors to Volume 2 xii

Series SCARABAEIFORMIA 1
Superfamily SCARABAEOIDEA 1
Family 23. LUCANIDAE 6
Family 24. DIPHYLLOSTOMATIDAE 10
Family 25. PASSALIDAE 12
Family 26. GLARESIDAE 15
Family 27. TROGIDAE 17
Family 28. PLEOCOMIDAE 20
Family 29. GEOTRUPIDAE 23
Family 30. OCHODAEIDAE 28
Family 31. HYBOSORIDAE 32
Family 32. CERATOCANTHIDAE 34
Family 33. GLAPHYRIDAE 37
Family 34. SCARABAEIDAE 39
Series ELATERIFORMIA 82
Superfamily SCIRTOIDEA 82
Family 35. EUCINETIDAE 82
Family 36. CLAMBIDAE 85
Family 37. SCIRTIDAE 87
Superfamily DASCILLOIDEA 90
Family 38. DASCILLIDAE 90
Family 39. RHIPICERIDAE 92
Superfamily BUPRESTOIDEA 95
Family 40. SCHIZOPODIDAE 95
Family 41. BUPRESTIDAE 98
Superfamily BYRRHOIDEA 113
Family 42 BYRRHIDAE 113
Family 43. ELMIDAE 117
Family 44. DRYOPIDAE 121
Family 45 LUTROCHIDAE 123
Family 46. LIMNICHIDAE 125
Family 47. HETEROCERIDAE 127
Family 48. PSEPHENIDAE 133
Family 49. PTILODACTYLIDAE 135
Family 50. CHELONARIIDAE 139
Family 51. EULICHADIDAE 142
Family 52. CALLIRHIPIDAE 144
Superfamily ELATEROIDEA 146
Family 53. ARTEMATOPODIDAE 146
Family 54. BRACHYPSECTRIDAE 148
Family 55. CEROPHYTIDAE 150
Family 56. EUCNEMIDAE 152
Family 57. THROSCIDAE 158
Family 58. ELATERIDAE 160
Family 59. LYCIDAE 174
Family 60. TELEGEUSIDAE 179

Family 61. PHENGODIDAE 181
Family 62. LAMPYRIDAE 187
Family 63. OMETHIDAE 197
Family 64. CANTHARIDAE 202
Series BOSTRICHIFORMIA 219
Family 65. JACOBSONIIDAE 219
Superfamily DERODONTOIDEA 221
Family 66. DERODONTIDAE 221
Superfamily BOSTRICHOIDEA 224
Family 67. NOSODENDRIDAE 224
Family 68. DERMESTIDAE 228
Family 69. BOSTRICHIDAE 233
Family 70. ANOBIIDAE 245
Series CUCUJIFORMIA 261
Superfamily LYMEXYLOIDEA 261
Family 71. LYMEXYLIDAE 261
Superfamily CLEROIDEA 263
Family 72. TROGOSSITIDAE 263
Family 73. CLERIDAE 267
Family 74. MELYRIDAE 281
Superfamily CUCUJOIDEA 305
Family 75. SPHINDIDAE 305
Family 76. BRACHYPTERIDAE 309
Family 77. NITIDULIDAE 311
Family 78. SMICRIPIDAE 316
Family 79. MONOTOMIDAE 319
Family 80. SILVANIDAE 322
Family 81. PASSANDRIDAE 327
Family 82. CUCUJIDAE 329
Family 83. LAEMOPHLOEIDAE 331
Family 84. PHALACRIDAE 335
Family 85. CRYPTOPHAGIDAE 338
Family 86. LANGURIIDAE 343
Family 87. EROTYLIDAE 348
Family 88. BYTURIDAE 354
Family 89. BIPHYLLIDAE 356
Family 90. BOTHRIDERIDAE 358
Family 91. CERYLONIDAE 363
Family 92. ENDOMYCHIDAE 366
Family 93. COCCINELLIDAE 371
Family 94. CORYLOPHIDAE 390
Family 95. LATRIDIIDAE 395
Superfamily TENEBRIONOIDEA 399
Family 96. MYCETOPHAGIDAE 399
Family 97. ARCHEOCRYPTICIDAE 401
Family 98. CIIDAE 403
Family 99. TETRATOMIDAE 413
Family 100. MELANDRYIDAE 417
Family 101. MORDELLIDAE 423
Family 102. RIPIPHORIDAE 431
Family 103. COLYDIIDAE 445
Family 104. MONOMMATIDAE 454
Family 105. ZOPHERIDAE 457
Family 106. TENEBRIONIDAE 463
Family 107. PROSTOMIDAE 510

Family 108. SYNCHROIDAE 512
Family 109. OEDEMERIDAE 514
Family 110. STENOTRACHELIDAE 520
Family 111. MELOIDAE 522
Family 112. MYCTERIDAE 530
Family 113. BORIDAE 534
Family 114. PYTHIDAE 537
Family 115. PYROCHROIDAE 540
Family 116. SALPINGIDAE 544
Family 117. ANTHICIDAE 549
Family 118. ADERIDAE 559
Family 119. SCRAPTIIDAE 564
Superfamily CHRYSOMELOIDEA 568
Family 120. CERAMBYCIDAE 568
Family 121. BRUCHIDAE 602
Family 122. MEGALOPODIDAE 609
Family 123. ORSODACNIDAE 613
Family 124. CHRYSOMELIDAE 617
Superfamily CURCULIONOIDEA 692
Family 125. NEMONYCHIDAE 692
Family 126. ANTHRIBIDAE 695
Family 127. BELIDAE 701
Family 128. ATTELABIDAE 703
Family 129. BRENTIDAE 711
Family 130. ITHYCERIDAE 720
Family 131. CURCULIONIDAE 722
Family Key 816
Taxonomic Index 836

Acknowledgments for Volume II

Originally, Ross Arnett was to have authored many of the family treatments, especially for those families with no specialists available. His death in 1999 left many families without an author. Several volunteers stepped forward, but Dan Young of the University of Wisconsin took responsibility for more than his fair share and got several of his enthusiastic graduate students involved in the project also. The members of the Editorial Board, listed in the Introduction, provided guidance, advice, and constructive criticism, but J. Howard Frank of the University of Florida has been outstanding in his unwavering demands for scholarship and proper English, and joined Paul E. Skelley and Michael C. Thomas, both of the Florida Department of Agriculture and Consumer Services, as an editor of Volume II. John Sulzycki of CRC Press has been more than helpful throughout some trying times.

Many of the excellent habitus drawings beginning the family treatments were done by Eileen R. Van Tassell of the University of Michigan for *The Beetles of the United States*, and for Volume 2 of *American Beetles* she produced excellent new ones for families 100, 108, and 119.

Authors of the family treatments often have acknowledgments in their respective chapters throughout the body of the text.

Ross Arnett's widow, Mary, was always his support staff throughout his long and productive career. After Ross' death, she helped by providing free and gracious access to Ross' files, and by her steady encouragement and quiet conviction that we would indeed be able to finish this, Ross Arnett's last big project. Unfortunately, Mary Arnett did not live to see Volume II published. She became ill in the fall of 2001 and died on January 3, 2002.

And I would like to again acknowledge my wife, Sheila, for her patience and forbearance during the long and sometimes difficult path that led to this volume.

Michael C. Thomas, Ph.D.
Gainesville, Florida
April 3, 2002

Contributors to Volume 2 of American Beetles

Authors

Rolf L. Aalbu, Ph.D.
Department of Entomology
California Academy of Sciences
Golden Gate Park
San Francisco, CA 94118-4599
106. Tenebrionidae.

Robert S. Anderson, Ph.D.
Canadian Museum of Nature/Entomology
P.O. Box 3443, Station D
Ottawa, ON K1P 6P4 CANADA
125. Nemonychidae; 127. Belidae; 129. Brentidae; 130. Ithyceridae; 131. Curculionidae.

Fred G. Andrews, Ph.D.
California Department of Food and Agriculture
Plant Pest Diagnostics Laboratory
3294 Meadowview Rd.
Sacramento, CA 95832-1448
95. Latridiidae.

Ross H. Arnett, Jr., Ph.D.
Senior Editor

C. L. Bellamy, D.Sc.
Entomology
Natural History Museum
900 Exposition Blvd
Los Angeles CA 90007, U.S.A.
40. Schizopodidae; 41. Buprestidae.

Marco A. Bologna, Ph.D.
Dipartimento di Biologia
Università degli studi "ROMA TRE"
Viale G. Marconi, 446
I-00146 ROMA
111. Meloidae.

Yves Bousquet, Ph.D.
Eastern Cereal and Oilseed Research Centre
Agriculture and Agri-Food Canada
Ottawa, ON K1A 0C6, CANADA
79. Monotomidae.

Stanley Bowestead, Ph.D.
Department of Entomology
The Manchester Museum, The University
Manchester M12 9PL England
94. Corylophidae.

Kirby W. Brown, Ph.D.
P.O. Box 1838
Paradise, CA 95967
106. Tenebrionidae.

J. Milton Campbell, Ph.D.
420 Everetts Lane
Hopkinsville, KY 42240
106. Tenebrionidae.

David C. Carlson, Ph.D.
4828 Dauntless Way
Fair Oaks, CA 95628
30. Ochodaeidae; 33. Glaphyridae.

Donald S. Chandler, Ph.D.
Department of Zoology
University of New Hampshire
Durham, NH 03824
117. Anthicidae; 118. Aderidae.

Shawn M. Clark, Ph.D.
Monte L. Bean Museum
Brigham Young University
Provo, UT 84602
122. Megalopodidae; 123. Orsodacnidae; 124. Chrysomelidae.

Arthur V. Evans, Ph.D.
Department of Entomology
National Museum of Natural History
Smithsonian Institution
c/o 1600 Nottoway Ave.
Richmond, VA 23227
34. Scarabaeidae: Melolonthinae.

Zachary H. Falin
Division of Entomology
Natural History Museum and Biodiversity Research Center
Snow Hall, 1460 Jayhawk Blvd.
Lawrence, KS 66045
102. Ripiphoridae

R. Wills Flowers, Ph.D.
Center for Biological Control
Florida A & M University
Tallahassee, FL 32307
124. Chrysomelidae.

J. Joseph Giersch
Department of Entomology
Montana State University
Bozeman, MT 59717
65. Jacobsoniidae

Arthur J. Gilbert
California Department of Food and Agriculture
2889 N. Larkin St., Suite 106
Fresno, CA 93727
124. Chrysomelidae.

Bruce D. Gill, Ph.D.
4032 Stonecrest Road
Woodlawn, ON K0A 3M0, CANADA
34. Scarabaeidae: Scarabaeinae.

Michael A. Goodrich, Ph.D.
Department of Biological Sciences
Eastern Illinois University
Charleston, IL 61920
88. Byturidae; 89. Biphyllidae

Robert D. Gordon, Ph.D.
Northern Plains Entomology
P.O.Box 65
Willow City, ND 58384
34. Scarabaeidae: Aphodiinae.

Dale H. Habeck, Ph.D.
Entomology and Nematology Department
University of Florida
Gainesville, FL 32611
76. Brachypteridae; 77. Nitidulidae.

Robert W. Hamilton, Ph.D.
Department of Biology
Loyola University Chicago
6525 North Sheridan Road
Chicago, IL 60626
128. Attelabidae.

Henry A. Hespenheide, Ph.D.
Department of Organismic Biology
University of California
Los Angeles, CA 90095-1606
131. Curculionidae: Conoderinae.

Frank T. Hovore, Ph.D.
14734 Sundance Place
Santa Clarita, CA 91387-1542
28. Pleocomidae.

Anne T. Howden, M.Sc.
Canadian Museum of Nature
P.O. Box 3443, Station D
Ottawa, ON K1P 6P4 Canada
131. Curculionidae: Entiminae.

Michael A. Ivie, Ph.D.
Department of Entomology
Montana State University
Bozeman, MT 59717
49. Ptilodactylidae; 50. Chelonariidae; 65. Jacobsoniidae; 67. Nosodendridae; 69. Bostrichidae; 90. Bothrideridae; 103. Colydiidae; 104. Monommatidae; 105. Zopheridae; Family Key.

John A. Jackman, Ph.D.
Department of Entomology
412 Heep Center
Texas A & M University
College Station, TX 77843
101. Mordellidae.

Mary L. Jameson, Ph.D.
W436 Nebraska Hall
Systematics Research Collections
University of Nebraska State Museum
Lincoln, NE 68588-0514
Scarabaeoidea; 24. Diphyllostomatidae; 26. Glaresidae; 27. Trogidae; 29. Geotrupidae; 31. Hybosoridae; 32. Ceratocanthidae; 34. Scarabaeidae.

Paul J. Johnson, Ph.D.
Insect Research Collection, Box 2207-A
South Dakota State University
Brookings, SD 57007
42. Byrrhidae; 58. Elateridae; 55. Cerophytidae; 57. Throscidae.

Kerry Katovich
Department of Entomology
University of Wisconsin-Madison
Madison, WI 53706
39. Rhipiceridae; 47. Heteroceridae.

John M. Kingsolver, Ph.D.
Florida State Collection of Arthropods
P. O. Box 147100
Gainesville, FL 32614
68. Dermestidae; 121. Bruchidae.

David G. Kissinger, Ph.D.
24414 University Avenue #40
Loma Linda, CA 92354
129. Brentidae: Apioninae.

Boris A. Korotyaev, Ph.D.
Zoological Institute
Russian Academy of Sciences
Universitetskaya nab 1.
St. Petersburg, 199034, Russia
131. Curculionidae: Ceutorhynchinae.

Nadine L. Kriska
Department of Entomology
University of Wisconsin-Madison
Madison, WI 53706
109. Oedemeridae.

John F. Lawrence, Ph.D.
12 Hartwig Road
Mothar Mountain
Gympie, QLD 4570 Australia
98. Ciidae.

Richard A.B. Leschen, Ph.D.
New Zealand Arthropods Collection
Landcare Research, Private Bag 92170
120 Mt. Albert Road, Mt. Albert
Auckland, New Zealand
66. Derodontidae; 72. Trogossitidae; 85. Cryptophagidae; 86. Languriidae; 92. Endomychidae; 94. Corylophidae.

James E. Lloyd, Ph.D.
Entomology and Nematology Department
University of Florida
Gainesville, FL 32611
62. Lampyridae.

Wenhua Lu
Wes Watkins Agricultural Research and Extension Center
Oklahoma State University
P.O. Box 128
Lane, OK 74555
101. Mordellidae.

Adriean J. Mayor, Ph.D.
Department of Entomology and Plant Pathology
205 Ellington Plant Sciences Bldg.
University of Tennessee
Knoxville, TN 37901
74. Melyridae.

Joseph V. McHugh, Ph.D.
Department of Entomology
University of Georgia
Athens, GA 30602
75. Sphindidae; 87. Erotylidae.

Richard S. Miller, Ph.D.
Department of Entomology
Montana State University
Bozeman, MT 59717
59. Lycidae; 60. Telegeusidae.

Jyrki Muona, Ph.D.
Division of Entomology
Finnish Museum of Natural History
P. O. Box 17, FIN-00014
University of Helsinki, Finland
56. Eucnemidae.

Gayle H. Nelson, Ph.D.
1308 N. W. Hawk Creek
Blue Springs, MO 64015-1787
40. Schizopodidae; 41. Buprestidae.

Sean T. O'Keefe, Ph.D.
Department of Biological and Environmental Sciences
Morehead State University
Morehead, KY 40351
38. Dascillidae; 61. Phengodidae.

Weston Opitz, Ph.D.
Department of Biology
Kansas Wesleyan University
100 Claflin Avenue
Salina, KS 67401
73. Cleridae.

T. Keith Philips, Ph.D.
Department of Biology
Western Kentucky University
Bowling Green, KY 42101-3576
65. Jacobsoniidae; 70. Anobiidae; 90. Bothrideridae.

John D. Pinto, Ph.D.
Department of Entomology
University of California
Riverside, California 92521
111. Meloidae.

Darren A. Pollock, Ph.D.
Department of Biology
Eastern New Mexico University
Portales, NM 88130
99. Tetratomidae; 100. Melandryidae; 112. Mycteridae; 113. Boridae; 114. Pythidae; 116. Salpingidae; 119. Scraptiidae.

Michele B. Price
Department of Entomology
University of Wisconsin-Madison
Madison, WI 53706
78. Smicripidae.

Robert J. Rabaglia, Ph.D.
Maryland Department of Agriculture
50 Harry S Truman Parkway
Annapolis, MD 21401
131. Curculionidae: Scolytinae.

Alistair S. Ramsdale
Department of Entomology
University of Wisconsin-Madison
Madison, WI 53706
63. Omethidae; 64. Cantharidae.

Brett C. Ratcliffe, Ph.D.
W436 Nebraska Hall
Systematics Research Collections
University of Nebraska State Museum
Lincoln, NE 68588-0514
23. Lucanidae; 24. Diphyllostomatidae; 34. Scarabaeidae.

Edward G. Riley
Department of Entomology
Texas A and M University
College Station, TX 77843
122. Megalopodidae; 123. Orsodacnidae; 124. Chrysomelidae.

Jack C. Schuster, Ph.D.
Universidad del Valle de Guatemala
Aptdo. 82
Guatemala, Guatemala
25. Passalidae.

William D. Shepard, Ph.D.
Department of Biology
California State University Sacramento
6000 J Street
Sacramento, CA 95819
43. Elmidae; 44. Dryopidae; 45. Lutrochidae; 46. Limnichidae; 48. Psephenidae; 51. Eulichadidae.

Paul E. Skelley, Ph.D.
Third Editor
Florida State Collection of Arthropods
Florida Department of Agriculture and Consumer Services
P. O. Box 147100
Gainesville, FL 32614-7100
34. Scarabaeidae: Aphodiinae; 85. Cryptophagidae; 86. Languriidae; 87. Erotylidae; 92. Endomychidae.

Andrew B. T. Smith
W436 Nebraska Hall
Systematics Research Collections
University of Nebraska State Museum
Lincoln, NE 68588-0514
34. Scarabaeidae.

Ronald E. Somerby, Ph.D.
Plant Pest Diagnostics Branch
California Department of Food and Agriculture
3294 Meadowview Road
Sacramento, CA 95832
106. Tenebrionidae.

Warren E. Steiner, Jr.
Department of Systematic Biology - Entomology
National Museum of Natural History
Smithsonian Institution
Washington, DC 20560
84. Phalacridae.

Margaret K. Thayer, Ph.D.
Field Museum of Natural History
1400 S. Lake Shore Drive
Chicago, IL 60605
98. Ciidae.

Donald B. Thomas, Ph.D.
USDA, ARS
2301 S. International Blvd.
Weslaco, TX 78596
106. Tenebrionidae.

Michael C. Thomas, Ph.D.
Second Editor
Florida State Collection of Arthropods
Florida Department of Agriculture and Consumer Services
P. O. Box 147100
Gainesville, FL 32614-7100
80. Silvanidae; 81. Passandridae; 82. Cucujidae; 83. Laemophloeidae; 91. Cerylonidae; 120. Cerambycidae.

Charles A. Triplehorn, Ph.D.
Museum of Biological Diversity
Ohio State University
1315 Kinnear Road
Columbus, OH 43212
106. Tenebrionidae.

Robert H. Turnbow, Jr., Ph.D.
ATZQ-DEL-PC
1404 Dilly Branch Road
Fort Rucker, AL 36362-5105
120. Cerambycidae.

Barry D. Valentine, Ph.D.
2359 Eastcleft Drive
Columbus, OH 43221
126. Anthribidae.

Natalia J. Vandenberg, Ph.D.
Systematic Entomology Laboratory
PSI, ARS, USDA
c/o National Museum of Natural History
Washington, DC 20560-0168
93. Coccinellidae.

Daniel K. Young, Ph.D.
Department of Entomology
University of Wisconsin
Madison, WI 53706
35. Eucinetidae; 36. Clambidae; 37. Scirtidae; 52. Callirhipidae; 53. Artemetopodidae; 54. Brachypsectridae; 71. Lymexylidae; 96. Mycetophagidae; 97. Archeocrypticidae; 99. Tetratomidae; 107. Prostomidae; 108. Synchroidae; 110. Stenotrachelidae; 115. Pyrochroidae.

Series SCARABAEIFORMIA Crowson 1960

(= Lamellicornia)

Superfamily SCARABAEOIDEA Latreille 1802

INTRODUCTION

by Mary Liz Jameson and Brett C. Ratcliffe

Common name: The scarabaeoid beetles

The superfamily Scarabaeoidea is a large, diverse, cosmopolitan group of beetles. Scarabaeoids are adapted to most habitats, and they are fungivores, herbivores, necrophages, coprophages, saprophages, and some are carnivores. They are widely distributed, even living in the Arctic in animal burrows. Some scarabs exhibit parental care and sociality. Some are myrmecophilous, termitophilous, or ectoparasitic. Many possess extravagant horns, others are able to roll into a compact ball, and still others are highly armored for inquiline life. Some are agricultural pests that may destroy crops while others are used in the biological control of dung and dung flies. Scarabaeoids are popular beetles due to their large size, bright colors, and interesting natural histories. Early Egyptians revered the scarab as a god, Jean Henri Fabre studied their behavior, and Charles Darwin used observations of scarabs in his theory of sexual selection.

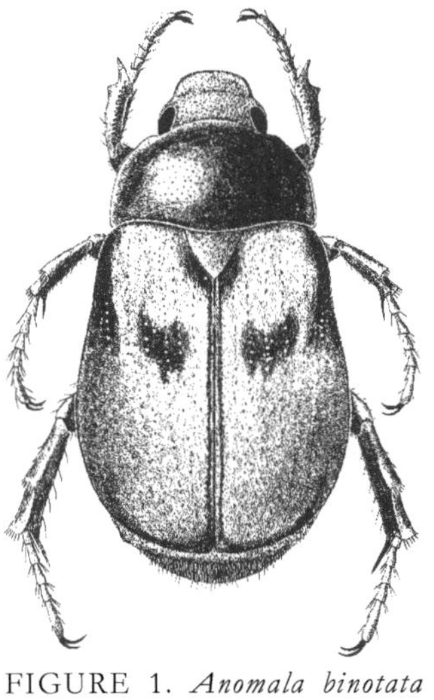

FIGURE 1. *Anomala binotata* (Gyllenhal) (Scarabaeidae) (Used by permission of University of Nebraska State Museum)

Description. Antennal club lamellate. Prothorax can be highly modified for burrowing, with large coxae (most with concealed trochantins and closed coxal cavities). Protibiae dentate with a single spur in most. Wing venation reduced and with a strong intrinsic spring mechanism for folding. Tergite 8 forming a true pygidium and not concealed by tergite 7. Four Malpighian tubules. Larvae scarabaeiform (cylindrical, C-shaped). Superfamily classification: Lawrence and Britton 1991; Lawrence and Newton 1995.

Status of the classification. The hierarchical level of families and subfamilies within the Scarabaeoidea is in disarray and remains unresolved. In the previous rendition of this work (Arnett's *The Beetles of the United States*, 1968), the Scarabaeoidea included three families: Passalidae, Lucanidae, and Scarabaeidae. This three-family system of classification was the "traditional" North American system and had several practical and conceptual advantages. First, it recognized the shared, derived characters that unite subfamilies within the family Scarabaeidae. Second, it provided a classification system that allowed easy retrieval of hierarchical information based on the fact that subfamilies were part of the family Scarabaeidae (*e.g.*, life history, morphology, larval type). Phylogenetic research indicates that the family Scarabaeidae (in the traditional sense) is not a monophyletic group. Therefore, we have chosen to follow the 12-family system established by Browne and Scholtz (1995, 1999) and Lawrence and Newton (1995). This system places emphasis on the differences that separate taxa rather than the similarities that unite them. Whereas families, subfamilies, and tribes in the staphylinoids and curculionoids are being combined because of shared characters (thus increasing efficient data retrieval), the scarabaeoids are being split into numerous families because of supposed differences (thus, in our view, decreasing information retrieval). The debate concerning scarabaeoid classification systems illustrates the weak phylogenetic foundation of the superfamily. This problem is the result of a number of factors including (1) lack of thorough study of all scarabaeoid taxa, (2) lack of diagnostic characters for all taxa, (3) lack of phylogenetic study of all taxa, (4) prevailing philosophies regarding categorical levels, and (5) emphasis in research on the less speciose groups of scarabaeoids and lack of research on the more speciose groups (such as the subfamilies of Scarabaeidae including the Melolonthinae, Rutelinae, Dynastinae, Aphodiinae, and Cetoniinae).

Acknowledgments. Mary Liz Jameson and Brett Ratcliffe thank our scientific illustrators, Mark Marcuson and Angie Fox, for their fine work. Al Newton, Margaret Thayer, Philip Harpootlian, and Andrew Smith are especially acknowledged for assistance with resolving authorship of some taxa and numerous grammatical corrections. For assistance with the keys and other aspects of the manuscript, we thank Andrew Smith, Federico Ocampo, William Warner, and Charles Brodel. Our development of the scarabaeoid chapters was supported by a NSF/PEET grant (DEB9712447) to B. C. Ratcliffe and M. L. Jameson. Parts of the Dynastinae treatment were supported by a NSF/BS&I grant (DEB9870202) to B. C. Ratcliffe.

Within the Scarabaeoidea there is a disparity in the knowledge between less speciose basal lineages and the more speciose groups of "higher" Scarabaeidae. For example, the family Trogidae includes approximately 300 species in four genera. Excellent revisionary, larval, and phylogenetic studies are available for this group (Baker 1968; Scholtz 1982, 1986, 1990, 1991, 1993; Scholtz and Peck 1990). Excellent monographs are also available for the approximately 600 species of Geotrupidae (Howden 1955, 1964, 1979, 1985a-b, 1992; Howden and Cooper 1977; Howden and Martínez 1978) and the Trogidae (Vaurie 1955), and these will provide the foundation for addressing relationships within this group. In comparison, the family Scarabaeidae (*sensu* Lawrence and Newton 1995) includes approximately 91% of the species (ca 27,800) of Scarabaeoidea. Within the Scarabaeidae, approximately 21,000 species are in the subfamilies Melolonthinae, Dynastinae, Rutelinae, and Cetoniinae (the "higher" scarabs). Only a few phylogenetic analyses have addressed relationships of pleurostict subtribes, genera, or species (Ratcliffe 1976; Ratcliffe and Deloya 1992; Jameson 1990, 1996, 1998; Jameson *et al.* 1994; Krell 1993), and only one analysis has been conducted to address tribal or subfamilial relationships (Browne and Scholtz 1999).

Historically, the superfamily Scarabaeoidea was divided into two generalized groups based on the position of the abdominal spiracles; the Laparosticti and Pleurosticti. Pleurostict scarabs were characterized by having most of the abdominal spiracles situated on the upper portion of the sternites (Ritcher 1969; Woodruff 1973) and included taxa whose adults feed on leaves, flowers and pollen, and whose larvae feed primarily on roots and decaying wood. Laparostict scarabs, on the other hand, were characterized by having most of the abdominal spiracles located on the pleural membrane between the tergites and sternites (Ritcher 1969) and included taxa whose adults and larvae feed on dung, carrion, hides, and feathers. The position of the spiracles, however, is not a consistent character (Ritcher 1969), and, in recent years, subfamilies and tribes that were once included in the Laparosticti have been raised to higher taxonomic status (family and subfamily, respectively).

The composition of the Scarabaeoidea remains a topic of debate. Lawrence and Newton (1995) proposed 13 families (12 found in the Nearctic, Belohinidae are Madagascan), and Scholtz and Browne (1996) and Browne and Scholtz (1995, 1998, 1999) proposed 13 families (all Nearctic, including Bolboceratidae; Belohinidae were not addressed). In this work we follow, with some hesitation, the system of Lawrence and Newton (1995) and treat the Scarabaeoidea as including 12 Nearctic families (eight or nine of which were previously considered subfamilies of the family Scarabaeidae, and one of which was previously considered a subfamily of the Lucanidae). Our reluctance to accept elevation of new families within the Scarabaeoidea stems from the fact that: 1) there have been no comprehensive taxonomic treatments of all higher categories of scarabaeoids (families and subfamilies) and, 2) there are few comprehensive, rigorous, phylogenetic analyses of higher scarabaeoid groups and, thus, a lack of synapomorphic characters that establish a basis for uniform familial and subfamilial levels. We *prefer* to see clades delimited by shared derived characters before the elevation of certain taxa to family level. Despite our reluctance to accept this classification system, we have little basis for disputing the validity of current taxonomic conclusions other than the fact that some of these taxonomic conclusions have been based on narrow taxonomic frameworks (only scarab taxa from certain geographic regions rather than all scarab groups) or based on few characters or suites of characters.

Underlying the classification problem is, of course, the fact that we are dealing with constructs that are 200 years old and that pre-date evolutionary theory. Linnaean classifications were based on overall morphological similarity rather than shared, derived characters. Thus, some groups within the scarabaeoids are not monophyletic lineages; instead, they are groups that were created historically because they superficially resembled each other. Our system of classification needs to convey information and concepts and allow for easy retrieval of information. Whether a certain taxon is classified at the level of family or subfamily may be trivial if we can continue to convey the needed information. We remain apprehensive that the trend of elevation to many families within the Scarabaeoidea will result, at least in the short term, in a net loss in retrievability of information.

Despite the considerable debate, phylogenetic analyses of scarabaeoid higher categories are on-going and their results bring us closer to understanding relationships of the groups. A preliminary "total evidence" phylogenetic analysis of 13 families of Scarabaeoidea (excluding Belohinidae, including Bolboceratidae) and most of the subfamilies was conducted using 134 adult and larval characters (Brown and Scholtz 1999). Results of this analysis showed that the superfamily Scarabaeoidea is comprised of three major lineages: the glaresid lineage that consists of only the family Glaresidae; the passalid lineage that consists of two major lines--a glaphyrid line (containing Glaphyridae, Passalidae, Lucanidae, Diphyllostomatidae, Trogidae, Bolboceratidae, and Pleocomidae), and a geotrupid line (containing Geotrupidae, Ochodaeidae, Ceratocanthidae, and Hybosoridae); and the scarab lineage (containing Aphodiinae, Scarabaeinae, Orphninae, Melolonthinae, Rutelinae, Dynastinae, and Cetoniinae).

The series Scarabaeiformia is comprised exclusively of the superfamily Scarabaeoidea. Monophyly of the group is well founded and undisputed (Lawrence and Britton 1991). The sister group for the Scarabaeoidea, however, is not resolved and continues to be debated. Two groups are considered: the Staphyliniformia and the Dascilloidea. The Scarabaeoidea and Staphyliniformia share characters of the wing venation and the abdomen that are not present in the dascilloids (Kukalová-Peck and Lawrence 1993). The Scarabaeoidea and Dascilloidea share similar larval characters (lack of urogomphi that are present in Staphyliniformia, cribriform spiracles, separate galea and lacinia) and adult characters (form of the ommatidium, male genitalia, mouthparts) (Scholtz *et al.* 1994). Lawrence and Newton (1982) argued that similarities in the Dascilloidea and Scarabaeoidea are attributable to either plesiomorphic or convergent characters that are associated with soil-dwelling habits.

Distribution. The superfamily Scarabaeoidea is one of the largest superfamilies in the Coleoptera and includes approximately 2,200 genera and about 31,000 species worldwide (Dalla Torre 1912-1913; Endrödi 1985; Hanski and Cambefort 1991; Krikken 1984; Lawrence 1982; Machatschke 1972; Scholtz 1982). While some of the smaller groups are well known worldwide (*e.g.,* Geotrupidae and Trogidae), some other groups (*e.g.,* Scarabaeidae that comprises 91% of the Scarabaeoidea) cannot be identified to even genus-level with reliability.

In the Nearctic region, the taxonomy of most scarabaeoids is now fairly well known although there remain a few areas of uncertainty. For example, the phylogenetic position of both the Pleocomidae and the Hopliini needs to be addressed. In this work, there are numerous changes in the author and/or date of many genera and even some higher categories since Arnett (1968). These are the result of greater scrutiny of the original literature rather than accepting at face value the often erroneus or incomplete information provided in older catalogs and faunal works.

There are approximately 170 genera and 2,000 species in the United States, Canada, and Nearctic Mexico. Smith (2001) provided a checklist of Nearctic scarabaeoids. Regional works: Blatchley 1910; Loding 1945; Saylor 1948; Edwards 1949; Helgesen and Post 1967; Hatch 1971; Woodruff 1973; Kirk and Balsbaugh 1975; Shook 1978; Lago *et al.* 1979; Ratcliffe 1991; Downie and Arnett 1996; Morón *et al.* 1997; Harpootlian 2001.

Summary of the Families and Subfamilies of Scarabaeoidea of the United States, Canada, and Nearctic Mexico

23. Lucanidae
    - Lucaninae
    - Nicaginae
    - Syndesinae
24. Diphyllostomatidae
25. Passalidae
26. Glaresidae
27. Trogidae
28. Pleocomidae
29. Geotrupidae
    - Bolboceratinae
    - Geotrupinae
30. Ochodaeidae
    - Ochodaeinae
    - Chaetocanthinae
31. Hybosoridae
32. Ceratocanthidae
33. Glaphyridae
34. Scarabaeidae
    - Aphodiinae
    - Scarabaeinae
    - Melolonthinae
    - Rutelinae
    - Dynastinae
    - Cetoniinae

Bibliography

ARNETT, R. H., Jr. 1968. The beetles of the United States. A Manual for Identification. The American Entomological Institute, Ann Arbor, MI, 1112 pp.

BAKER, C. W. 1968. Larval taxonomy of the Troginae in North America with notes on biologies and life histories (Coleoptera: Scarabaeidae). United States National Museum Bulletin, 279: 1-79.

BLATCHLEY, W. S. 1910. An illustrated descriptive catalogue of the Coleoptera or beetles known to occur in Indiana. Indiana Department of Geology and Natural Resources Bulletin, 1: 1-1386.

BROWNE, D. J. and C. H. SCHOLTZ. 1995. Phylogeny of the families of Scarabaeoidea (Coleoptera) based on characters of the hindwing articulation, hindwing base and wing venation. Systematic Entomology, 20: 145-173.

BROWNE, J. and C. H. SCHOLTZ. 1998. Evolution of the scarab hindwing articulation and wing base: A contribution toward the phylogeny of the Scarabaeidae (Scarabaeoidea: Coleoptera). Systematic Entomology, 23: 307-326.

BROWNE, J. and C. H. SCHOLTZ. 1999. A phylogeny of the families of Scarabaeoidea (Coleoptera). Systematic Entomology, 24: 51-84.

DALLA TORRE, K. W. von. 1912-1913. Scarabaeidae: Melolonthinae. Coleopterorum Catalogus, pars 45, 47, 49, 50: 1-450. W. Junk. Berlin.

DOWNIE, N. M. and R. H. ARNETT, Jr. 1996. The beetles of Northeastern North America, volumes 1 and 2. The Sandhill Crane Press. Gainesville, FL, 1721 pp.

EDWARDS, J. G. 1949. Coleoptera or beetles east of the Great Plains. Edwards Brothers. Ann Arbor, MI, 181 pp.

ENDRÖDI, S. 1985. The Dynastinae of the World. W. Junk. London, 800 pp.

HANSKI, I. and Y. CAMBEFORT. 1991. Dung beetle ecology. Princeton University Press. Princeton, NJ, 481 pp.

HARPOOTLIAN, P. J. 2001. Scarab beetles (Coleoptera: Scarabaeidae) of South Carolina. Biota of South Carolina, Volume 2. Clemson University. Clemson, SC, 157 pp.

HATCH, M. H. 1971. The beetles of the Pacific Northwest, part 5. University of Washington Publication in Biology, 16: 1-662.

HELGESEN, R. G. and R. L. POST. 1967. Saprophagous Scarabaeidae (Coleoptera) of North Dakota. North Dakota Insects, Publication Number 7: 1-60.

HOWDEN, H. F. 1955. Biology and taxonomy of North American beetles of the subfamily Geotrupinae with revisions of the genera *Bolbocerosoma, Eucanthus, Geotrupes* and *Peltotrupes* (Scarabaeidae). Proceedings of the United States National Museum, 104: 151-319.

HOWDEN, H. F. 1964. A Catalog of the Coleoptera of America North of Mexico. Family Scarabaeidae. Subfamily: Geotrupinae. United States Department of Agriculture, Agriculture Handbook 529-34a, 17 pp.

HOWDEN, H. F. 1979. A revision of the Australian genus *Blackburnium* Boucomont (Coleoptera: Scarabaeidae: Geotrupinae). Australian Journal of Zoology, Supplementary Series, 72: 1-88.

HOWDEN, H. F. 1985a. A revision of the South American genus *Parathyreus* Howden and Martínez (Coleoptera: Scarabaeidae: Geotrupinae). Coleopterists Bulletin, 39: 161-173.

HOWDEN, H. F. 1985b. A revision of the South American species in the genus *Neoathyreus* Howden and Martínez (Coleoptera, Scarabaeidae, Geotrupinae). Contributions of the American Entomological Institute, 21: 1-95.

HOWDEN, H. F. 1992. A revision of the Australian beetle genera *Eucanthus* Westwood, *Bolbobaineus* Howden and Cooper, *Australobolbus* Howden and Cooper and *Gilletus* Boucomont (Scarabaeidae: Geotrupinae). Invertebrate Taxonomy, 6: 605-717.

HOWDEN, H. F. and J. B. COOPER. 1977. The generic classification of the Bolboceratini of the Australian Region, with descriptions of four new genera (Scarabaeidae: Geotrupinae). Australian Journal of Zoology, Supplementary Series, 50: 1-50

HOWDEN, H. F. and A. MARTÍNEZ. 1978. A review of the New World genus *Athyreus* MacLeay (Scarabaeidae, Geotrupinae, Athyreini). Contributions of the American Entomological Institute, 15: 1-70.

JAMESON, M. L. 1990. Revision, phylogeny and biogeography of the genera *Parabyrsopolis* Ohaus and *Viridimicus* (new genus) (Coleoptera: Scarabaeidae: Rutelinae). Coleopterists Bulletin, 44: 377-422.

JAMESON, M. L. 1996. Revision and phylogeny of the Neotropical genus *Cnemida* (Coleoptera: Scarabaeidae: Rutelinae). Insecta Mundi, 10: 285-315.

JAMESON, M. L. 1998. Phylogenetic analysis of the subtribe Rutelina and revision of the *Rutela* generic groups (Coleoptera: Scarabaeidae: Rutelinae: Rutelini). Bulletin of the University of Nebraska State Museum, 14: 1-184.

JAMESON, M. L., B. C. RATCLIFFE and M. A. MORÓN. 1994. A synopsis of the Neotropical genus *Calomacraspis* Bates with a key to larvae of the American genera of Rutelini (Coleoptera: Scarabaeidae: Rutelinae). Annals of the Entomological Society of America, 87: 43-58.

KIRK, V. M. and E. U. BALSBAUGH, Jr. 1975. A list of the beetles of South Dakota. South Dakota State University, Agricultural Experiment Station, Technical Bulletin Number 42, 137 pp.

KRELL, F.-T. 1993. Phylogenetisch-systematische revision des Genus *Temnorhynchus* Hope, 1837 (Coleoptera: Scarabaeoidea: Melolonthidae: Dynastinae: Pentodontini). 1. Teil: Phylogenetische Analyse, mit Anmerkunge zur phylogenetisch-sytematischen Methodologie. Beitrage zur Entomologie, 43: 237-318.

KRIKKEN, J. 1984. A new key to the suprageneric taxa in the beetle family Cetoniidae, with annotated lists of the known genera. Zoologische Verhandelingen, No. 210: 1-75.

KUKALOVÁ-PECK, J. and J. F. LAWRENCE. 1993. Evolution of the hind wing in Coleoptera. Canadian Entomologist, 125: 181-258.

LAGO, P. K., R. L. POST and C. Y. OSETO. 1979. The Phytophagous Scarabaeidae and Troginae (Coleoptera) or North Dakota. North Dakota Insects, Publication No. 12. Schafer-Post Series, Bismarck, North Dakota, 131 pp.

LAWRENCE, J. F. 1982. Coleoptera, Pp. 482-553. *In:* S. P. Parker, ed. Synopsis and Classification of Living Organisms. Vol. 2. McGraw-Hill. New York.

LAWRENCE, J. F. and E. B. BRITTON. 1991. Coleoptera. Pp. 543-683. *In:* The Insects of Australia, 2nd edition. Volume 1. Melbourne University Press. Carlton.

LAWRENCE, J. F. and A. F. NEWTON, Jr. 1982. Evolution and classification of beetles. Annual Review of Entomology and Systematics, 13: 261-290.

LAWRENCE, J. F. and A. F. NEWTON, Jr. 1995. Families and subfamilies of Coleoptera (with selected genera, notes, references and data on family-group names). Pp. 779-1006. *In:* J. Pakaluk and S. A. Slipínski, eds. Biology, Phylogeny, and Classification of Coleoptera. Papers Celebrating the 80th Birthday of Roy. A. Crowson. Muzeum i Instytut Zoologii PAN. Warsaw, 1092 pp.

LODING, H. P. 1945. Catalogue of the beetles of Alabama. Alabama Geological Survey Monograph, 11: 1-172.

MACHATSCHKE, J. W. 1972. Scarabaeoidea: Melolonthidae, Rutelinae. Coleopterorum Catalogus Supplementa, 66: 1-361.

MORÓN, M. A., B. C. RATCLIFFE, and C. DELOYA. 1997. Atlas de los escarabajos de México. Coleoptera: Lamellicornia. Vol. 1. Familia Melolonthidae. Subfamilias Rutelinae, Dynastinae, Cetoniinae, Trichiinae, Valginae y Melolonthinae. Sociedad Mexicana de Entomología, Mexico 280 pp.

RATCLIFFE, B. C. 1976. A revision of the genus *Strategus* (Coleoptera: Scarabaeidae). Bulletin of the University of Nebraska State Museum, 10: 93-204.

RATCLIFFE, B. C. 1991. The scarab beetles of Nebraska. Bulletin of the University of Nebraska State Museum, 12: 1-333.

RATCLIFFE, B. C. and A. C. DELOYA. 1992. The biogeography and phylogeny of *Hologymnetis* (Coleoptera: Scarabaeidae: Cetoniinae) with a revision of the genus. Coleopterists Bulletin, 46: 161-202.

RITCHER, P. O. 1969. Spiracles of adult Scarabaeoidea (Coleoptera) and their phylogenetic significance. I. The abdominal spiracles. Annals of the Entomological Society of America, 62: 869-880.

SAYLOR, L. W. 1948. Contributions toward a knowledge of the insect fauna of lower California. No. 10. Coleoptera, Scarabaeidae. Proceedings of the California Academy of Sciences, Series 4, 24: 337-374.

SCHOLTZ, C. H. 1982. Catalogue of the world Trogidae (Coleoptera: Scarabaeoidea). Republic of South Africa, Department of Agriculture and Fisheries, Entomology Memoire, 54: 1-27.

SCHOLTZ, C. H. 1986. Phylogeny and systematics of the Trogidae (Coleoptera: Scarabaeoidea). Systematic Entomology, 11: 355-363.

SCHOLTZ, C. H. 1990. Phylogenetic trends in the Scarabaeoidea (Coleoptera). Journal of Natural History, 24: 1027-1066.

SCHOLTZ, C. H. 1991. Descriptions of larvae of Australian *Omorgus* Erichson, with implications for the phylogeny of the Trogidae (Insecta: Coleoptera). Invertebrate Taxonomy, 5: 827-835.

SCHOLTZ, C. H. 1993. Descriptions of larvae of African Trogidae (Coleoptera), with implications for the phylogeny of the family. African Entomology, 1: 1-13.

SCHOLTZ, C. H., D. J. BROWNE and J. KUKALOVÁ-PECK 1994. Glaresidae, archeopteryx of the Scarabaeoidea (Coleoptera). Systematic Entomology, 19: 259-277.

SCHOLTZ, C. H. and D. J. BROWNE. 1996. Polyphyly in the Geotrupidae (Coleoptera: Scarabaeoidea): a case for a new family Bolboceratidae. Journal of Natural History, 30: 597-614.

SCHOLTZ, C. H. and S. PECK. 1990. Description of a *Polynoncus* Burmeister larva, with implications for phylogeny of the Trogidae (Coleoptera: Scarabaeoidea). Systematic Entomology, 15: 383-389.

SHOOK, G. A. 1978. Records of some scarabs from Idaho (Coleoptera: Scarabaeidae). Coleopterists Bulletin, 32: 52.

SMITH, A. B. T. 2001. Checklist of the Scarabaeoidea of the Nearctic Realm (Includes Canada, the continental United States, and the following states of northern Mexico: Baja California, Baja California Sur, Chihuahua, Coahuila de Zaragoza, Durango, Nuevo Leon, Sinaloa, Sonora, Tamaulipas, and Zacatecas). URL: http://www-museum.unl.edu/research/entomology/nearctic.htm.

VAURIE, P. 1955. A revision of the genus *Trox* in North America. Bulletin of the American Museum of Natural History, 106: 1-89.

WOODRUFF, R. E. 1973. The scarab beetles of Florida (Coleoptera: Scarabaeidae). Part. I. The Laparosticti (subfamilies: Scarabaeinae, Aphodiinae, Hybosorinae, Ochodaeinae, Geotrupinae, Acanthocerinae). Arthropods of Florida and Neighboring Land Areas, 8: 1-220.

23. LUCANIDAE Latreille 1804

by Brett C. Ratcliffe

Family common name: The stag beetles

Stag beetles range in size from less than 1 cm to 9 cm. The largest Nearctic species attains a length of 6 cm. Most males possess greatly enlarged, curving mandibles that are sometimes used in combat with male opponents during fights to establish dominance. Female lucanids lack enlarged mandibles. In most, male development of the mandibles is allometric, that is, the size of the mandibles is proportional to the size of the body. Those males with the largest mandibles are referred to as "male majors" and those with the smallest mandibles are called "male minors."

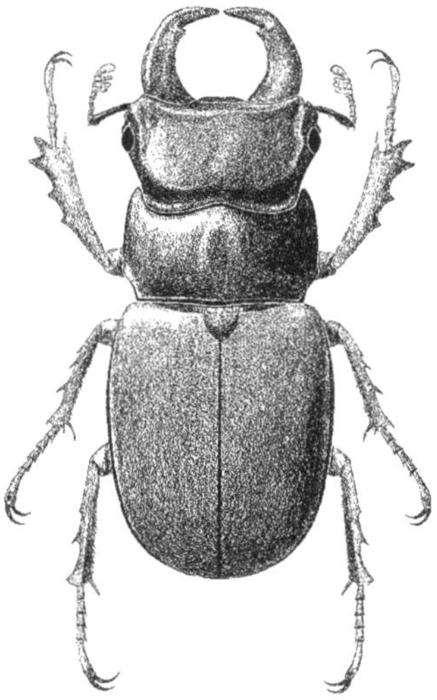

FIGURE 23.1. *Lucanus capreolus* (Linnaeus) (Used by permission of University of Nebraska State Museum)

Description. Length 8.0-60.0 mm. Shape usually weakly convex, subdepressed, or cylindrical, elongate. Color testaceous to reddish brown to black.

Head prognathus, not deflexed. Antennae geniculate or straight, 10-segmented, with 3-7 segmented club (all antennomeres unopposable and tomentose); first antennomere often subequal to remaining antennomeres. Eyes with eucone or acone ommatidia; eye canthus present or absent. Clypeus and labrum fused to frons. Mandibles produced beyond apex of labrum, prominent (males often with large, curved, elongate mandibles). Maxillae with 4-segmented palpi; labium with 3-segmented palpi.

Pronotum weakly convex, base narrower than elytral base, lacking tubercles, ridges, horns, or sulci. Elytra weakly convex, with or without impressed striae. Scutellum exposed, triangular or parabolic. Pygidium concealed by elytra or only weakly exposed. Legs with coxae transverse, mesocoxae separated; protibiae dentate on outer margin, apex with one spur; meso- and metatibia with ridges, apex with 2 spurs; tarsi 5-5-5; claws equal in size, simple; empodium present, extending weakly beyond fifth tarsomere or extending nearly one half claw length, with 2 to several setae. Profemora with tomentose patch anteriorly.

Abdomen with 5 visible sternites; 8 functional abdominal spiracles situated in pleural membrane. Wings well developed, with M-Cu loop and two, apical, detached veins. Male genitalia trilobed. References: Didier and Seguy 1953; Scholtz 1990.

Larvae are scarabaeiform (c-shaped, subcylindrical). Color creamy-white or yellowish (except at caudal end which may be darkened by accumulated feces). Cranium heavily sclerotized, lightly pigmented. Antennae 3-4 segmented, last segment greatly reduced in size. Stemmata absent (present in *Platycerus*). Frontoclypeal suture present. Labrum at apex rounded or weakly lobed. Epipharynx rounded or lobed, with symmetrical tormae. Maxilla with galea and lacinia distinctly separate; maxillary stridulatory teeth absent (present in *Platycerus*); maxillary palpus 4-segmented. Mandibles elongate, asymmetrical. Abdominal segments 3-7 with 2 annuli, each with 1 or more transverse rows of short setae. Spiracles cribriform. Anal opening Y-shaped or longitudinal, surrounded by 2 fleshy lobes. Legs 4-segmented. Stridulatory apparatus on meso- and metathoracic legs present; claws present. References: Ritcher 1966; Scholtz 1990.

Habits and habitats. Lucanids are usually associated with decaying wood and logs in coniferous and deciduous forest habitats. Adults of some species are attracted to lights at night and some feed at sap flows from fluxing trees. Adults of some smaller species have been observed feeding on flowers. The eggs are customarily laid in crevices in bark or logs, and the larvae feed on decaying wood. The larvae resemble those of Scarabaeidae, but in lucanids the anal opening is longitudinal or Y-shaped transverse or Y-shaped in a few. References: Ratcliffe 1991.

Status of the classification. The family Lucanidae has long been considered one of the most primitive groups in the Scarabaeoidea (Crowson 1967; Howden 1982; Ritcher 1966), and scarabaeoid classifications and evolutionary hypotheses have generally regarded the Lucanidae as basal to all scarabaeoids (Howden 1982; Iablokoff-Khnzorian 1977; Lawrence and Newton 1995). However, based on comparison with "primitive" scarabaeoid groups, Scholtz *et al.* (1994) hypothesized that the scarabaeoid family Glaresidae, rather than the Lucanidae, is the most primitive scarabaeoid. According to this hypothesis, the Lucanidae is a member of a clade including the Passalidae, Diphyllostomatidae, Glaphyridae, Trogidae, Pleocomidae, and Bolboceratinae (Geotrupidae).

Prior to the taxonomic elevation of the genus *Diphyllostoma* to the family Diphyllostomatidae (Holloway 1972), the Lucanidae was hypothesized to be most closely related to the Passalidae (Howden 1982). Based on shared characters, it is now thought that the Lucanidae is most closely related to the Diphyllostomatidae (Caveney 1986; Browne and Scholtz 1995).

The world Lucanidae (about 800 species) have been treated in checklists by Benesh (1960) and Maes (1992) and in illustrated

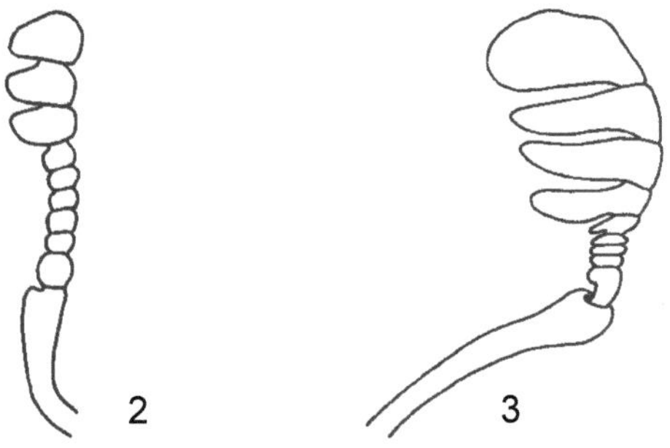

FIGURES 2.23-3.23. Right antenna dorsal view of: 2. *Ceruchus piceus* and 3. *Platycerus* sp.

catalogs by Didier and Seguy (1953) and Mizunuma and Nagai (1994). The latter is spectacular for its colored plates of the world fauna. Benesh (1960) recognized eight subfamilies, four of which occurred in the United States. The classification of the North American species had been relatively stable until Howden and Lawrence (1974) proposed a significant rearranging of genera within subfamilies based partly on the work of Holloway (1960, 1968, 1969). This newer classification is followed here, with three subfamilies now recognized as occurring in North America: Nicaginae, Syndesinae, and Lucaninae.

Distribution. The world fauna consists of about 800 species (Mizunuma and Nagai 1994) with more species found in Asia than in other areas. In North America there are three subfamilies with eight genera and 24 species. Keys to adults: Benesh 1946; Blatchley 1910; Howden and Lawrence 1974; Ratcliffe 1991. Keys to larvae: Ritcher 1966; Smith 2001. World catalog: Krajcik 2001. Regional works: Edwards 1949; Hatch 1971; Kirk and Balsbaugh 1975; Ratcliffe 1991. Biology: Milne 1933; Hoffman 1937.

Key to the Nearctic Subfamilies, Tribes, and Genera of Lucanidae

1. Antenna straight, not geniculate (Fig. 2); body form oval or cylindrical; prosternal process narrow, anterior coxae nearly contiguous 2

— Antenna geniculate (Fig. 3); body form elongate, weakly flattened; prosternal process broad, anterior coxae distinctly separated (Lucaninae) . 4

2(1). Form short, oval, convex; elytra coarsely punctate with conspicuous bristles; antenna with club lamellate (Nicaginae) .... *Nicagus*

— Form elongate, cylindrical; elytra striate and nearly glabrous; antenna with club only weakly lamellate (Syndesinae) 3

3(2). Head of male with long, median horn; female with median tubercle; mandibles in both sexes small and inconspicuous (Sinodendrini) . *Sinodendron*

— Head in both sexes lacking median horn or tubercle; mandibles large, conspicuous, especially in male (Ceruchini) .... *Ceruchus*

4(1). Eye nearly entire, canthus absent or very short; antenna weakly geniculate, scape lacking groove at apex; body length mostly less than 15 mm (Platycerini) 5

— Eye divided by canthus, canthus more than 1/4 as long as eye; antenna strongly geniculate, scape with apical groove; body length more than 15 mm 7

5(4). Body robust, obovate, strongly convex; posterior tibia stout, expanded at apex, triangular in cross section; apical spurs spatulate ..... *Platyceropsis*

— Body distinctly depressed; posterior tibia narrow, not expanded at apex, not triangular in cross section; apical spurs slender, not spatulate 6

6(5). Anterior margin of head semicircularly excised; clypeus mostly concealed; mandibles subequal to length of head; club of antenna with 4 segments in male, 3 segments in female .... *Platycerus*

— Anterior margin of head lacking excision on front; clypeus distinctly produced; mandibles about half length of head; club of antenna 3-segmented in both sexes .... *Platyceroides*

7(4). Elytra nearly smooth; pronotum with lateral margin strongly arcuate or angulate; posterior tibia with 0, 2, or 3 spines along outer edge (Lucanini) .... *Lucanus*

— Elytra striate-punctate; pronotum with lateral margins subparallel; posterior tibia with 1 spine on outer edge (Dorcini) .... *Dorcus*

Classification of the Nearctic Genera

Lucanidae Latreille 1804

Lucaninae Latreille 1804

Characteristics: Eye partly or completely divided by canthus. Antenna geniculate. Body form elongate, weakly flattened. Pronotal process broad between procoxae, coxae distinctly separated. Internal sac of aedeagus permanently everted.

Most of the lucanids of the world are found in this subfamily although many of the tribes are poorly or inconsistently characterized (Howden and Lawrence 1974). The classification of the North American genera seems now to be stable.

Lucanini Latreille 1804

Characteristics: Antenna strongly geniculate, scape with apical groove. Elytra nearly smooth or with minute and irregular punctation. Pronotum with lateral edges arcuate or angulate. Metatibia lacking spines or with 2 or 3 spines along outer edge. Most with body length greater than 25 mm.

This tribe consists of two genera in the New World. One genus, *Cantharolethrus*, is found from Mexico to South America. Keys to U.S. species: Fuchs 1882; Dillon and Dillon 1961. Keys to larvae: Ritcher (1966) was unable to distinguish between the three U.S. species he examined.

Lucanus Scopoli 1763
Hexaphyllus Mulsant 1839
Pseudolucanus Hope 1845
The genus *Lucanus* (Fig. 1) contains about 50 species distributed in Asia, Europe, and North America (Benesh 1960). Most of the species occur in Asia while five species occur in the United States and northern Mexico. Three species are restricted to the eastern United States while the other two are found in the Southwest and northern Mexico. Keys to species: Dillon and Dillon 1961. Biology: Milne 1933.

Dorcini Parry 1864

Characteristics: Antenna strongly geniculate, scape with apical groove. Elytra striate-punctate. Pronotum with lateral edges subparallel. Metatibia with 1 spine on outer edge. Body length more than 15 mm.

There is only one genus of Dorcini represented in North America: *Dorcus*. The tribe Dorcini has six genera, five of which are endemic to Asia.

Dorcus MacLeay 1819
The genus *Dorcus* contains about 30 species; most occur in Asia and India although two are also found in southern Europe, one in Mexico, and two in the United States (Benesh 1937). The two U.S. species are found in the eastern half of the country, and one species occurs in Quebec. Key to species: Downie and Arnett 1996.

Platycerini Mulsant 1842

Characteristics: Antenna weakly geniculate, scape lacking apical groove. Eye nearly entire, eye canthus absent or very short. Elytra striate in most. Most with body length less than 15 mm.

The tribe Platycerini consists of three genera in North America, with *Platycerus* also being found in Asia, Europe, and North Africa. Key to genera: Benesh 1946; Howden and Lawrence 1974.

Platyceroides Benesh 1946
This genus contains seven species, all of which are found in Washington, Oregon, and northern California. Key to species: Benesh 1946.

Platyceropsis Benesh 1946
This monobasic genus contains one species, *P. keeni* (Casey), which occurs from British Columbia to northern California.

Platycerus Geoffroy 1762
Systenocerus Weise 1883
Systenus Sharp and Muir 1912
Eight species are found in this genus, and five of these occur in the United States. Four of the U.S. species are found in the Pacific Northwest while the fifth species is found in the eastern United States and Canada. Key to species: Benesh 1946. Larval description: Ritcher 1966. Biology: Hoffman 1937.

Nicaginae LeConte 1860

Characteristics: Antenna not geniculate, instead straight, with strongly lamellate club. Eye entire (lacking canthus). Pronotum with lateral margins strongly crenulate. Elytra coarsely punctate, with conspicuous bristles. Prosternal process narrow so that anterior coxae almost contiguous. Aedeagus with internal sac eversible. Body form short, oval, convex.

According to Howden and Lawrence (1974), the Nicaginae contain two genera: *Ceratognathus* from Australia and New Zealand and *Nicagus* from North America and Japan. Previously, these genera were included in the subfamily Aesalinae.

Nicagus LeConte 1860
One species, *N. obscurus* LeConte, occurs in the eastern half of the United States.

Syndesinae MacLeay 1819

Characteristics: Antenna straight, not geniculate; club weakly lamellate. Eye entire (lacking canthus). Elytra striate, nearly glabrous. Prosternal process narrow, anterior coxae nearly contiguous. Aedeagus with internal sac eversible. Body form elongate, cylindrical.

This subfamily is composed of three tribes: Syndesini with the genera *Syndesus* (from Australia, Tasmania, New Caledonia, and Africa) and *Psilodon* (from South America); Sinodendrini with the genus *Sinodendron* (from North America); and Ceruchini with the single Holarctic genus *Ceruchus*.

Sinodendrini Mulsant 1842

Characteristics: Head of male with long, median horn; female with median tubercle. Mandibles in both sexes small and inconspicuous.

Sinodendron Hellwig 1894
Ligniperda Fabricius 1801
This genus contains two species and one, *S. rugosum* Mannerheim, is found in the Pacific Northwest. Generic overview: Hatch 1928.

Ceruchini Jacquelin du Val 1857

Characteristics: Head lacking median horn or tubercle in both sexes. Mandibles large, conspicuous (especially in male).

Ceruchus MacLeay 1819
Seven species occur in this Holarctic genus (Benesh 1960), and three species are found in North America. One species occurs in southeastern Canada and the northeastern quadrant of the United States while the other two species are found from California to

Vancouver Island. Larval description: Ritcher 1966. Biology: Hoffman 1937.

BIBLIOGRAPHY

BENESH, B. 1937. Some notes on boreal American Dorcinae (Coleoptera: Lucanidae). Transactions of the American Entomological Society, 63: 1-15.

BENESH, B. 1946. A systematic revision of the Holarctic genus *Platycerus* Geoffroy. Transactions of the American Entomological Society, 72: 139-202.

BENESH, B. 1960. Coleopterorum Catalogus Supplementa, pars 8: Lucanidae. W. Junk. Gravenhage, Netherlands.

BLACKWELDER, R. E. and R. H. ARNETT, Jr. 1974. Checklist of the beetles of Canada, United States, Mexico, Central America and the West Indies. Volume 1, Part 3. The scarab beetles, ant-loving beetles, clown beetles, and related groups (red version). The Biological Research Institute of America. Latham, NY, 120 pp.

BLATCHLEY, W. S. 1910. An illustrated descriptive catalogue of the Coleoptera or beetles known to occur in Indiana. Indiana Department of Geology and Natural Resources Bulletin, 1: 1-1386.

BROWN, D. J. and C. H. SCHOLTZ. 1995. Phylogeny of the families of the Scarabaeoidea (Coleoptera) based on characters of the hindwing articulation, hindwing base and wing venation. Systematic Entomology, 21: 145-173.

CAVENEY, S. 1986. The phylogenetic significance of ommatidium structure in the compound eyes of polyphagan beetles. Canadian Journal of Zoology, 64: 1787-1819.

CROWSON, R. A. 1967. The natural classification of the families of Coleoptera. E.W. Classey. Middlesex, 187 pp.

DIDIER, R. and E. SÉGUY. 1953. Catalogue illustré des lucanides du globe. Texte. Encyclopédie Entomologique (series A), 27: 1-223.

DILLON, E. S. and L. S. DILLON. 1961. A Manual of Common Beetles of Eastern North America. Row, Peterson and Co. Evanston, IL, 884 pp.

DOWNIE, N. M. and R. H. ARNETT, Jr. 1996. The Beetles of Northeastern North America, Vol. 1. The Sandhill Crane Press. Gainesville, FL, 880 pp.

EDWARDS, J. G. 1949. Coleoptera or Beetles East of the Great Plains. Edwards Brothers. Ann Arbor, MI, 181 pp.

FUCHS, C. 1882. Synopsis of the Lucanidae of the U.S. Bulletin of the Brooklyn Entomological Society, 5: 49-60.

HATCH, M. H. 1928. The species of *Sinodendron*. Pan-Pacific Entomologist, 4: 175-176.

HATCH, M. H. 1971. The beetles of the Pacific Northwest, part 5. University of Washington Publication in Biology, 16: 1-662.

HOFFMAN, C. H. 1937. Biological notes on *Pseudolucanus placidus* Say, *Platycerus quercus* Weber and *Ceruchus piceus* Weber (Lucanidae-Coleoptera). Entomological News, 48: 281-284.

HOLLOWAY, B. A. 1960. Taxonomy and phylogeny in the Lucanidae (Insecta: Coleoptera). Records of the Dominion Museum, 3: 321-365.

HOLLOWAY, B. A. 1968. The relationship of *Syndesus* MacLeay and *Sinodendron* Schneider (Coleoptera: Lucanidae). New Zealand Journal of Science, 11: 264-269.

HOLLOWAY, B. A. 1969. Further studies on generic relationships in Lucanidae (Insecta: Coleoptera) with special reference to the ocular canthus. New Zealand Journal of Science, 12: 958-977.

HOLLOWAY, B. A. 1972. The systematic position of the genus *Diphyllostoma* Fall (Coleoptera: Scarabaeoidea). New Zealand Journal of Science, 15: 31-38.

HOWDEN, H. F. 1982. Larval and adult characters of *Frickius* Germain, its relationship to the Geotrupini, and a phylogeny of some major taxa in the Scarabaeoidea (Insecta: Coleoptera). Canadian Journal of Zoology, 10: 2713-2724.

HOWDEN, H. F. and J. F. LAWRENCE. 1974. The New World Aesalinae, with notes on the North American lucanid subfamilies (Coleoptera, Lucanidae). Canadian Journal of Zoology, 52: 1505-1510.

IABLOKOFF-KHNZORIAN, S. M. 1977. Über die Phylogenie der Lamellicornia. Entomologische Abhandlungen der Staatlichen Museum für Tierkunde in Dresden, 41: 135-200.

KIRK, V. M. and E. U. BALSBAUGH, Jr. 1975. A list of the beetles of South Dakota. South Dakota State University, Agricultural Experiment Station, Technical Bulletin, Number 42, 137 pp.

KRAJCIK, M. 2001. Lucanidae of the World. Catalogue – Part 1. Checklist of the stag beetles of the world. M. Krajcik. Most, Czech Republic, 108 pp.

LAWRENCE, J. F. and A. F. NEWTON, Jr. 1995. Families and subfamilies of Coleoptera (with selected genera, notes, and references and data on family-group names). Pp. 779-1006. *In:* J. Pakaluk and S. A. Slipinski, eds. Biology, Phylogeny, and Classification of Coleoptera. Papers Celebrating the 80th Birthday of Roy A. Crowson. Muzeum i Instytut Zoologii PAN. Warsaw, Poland.

MAES, J.-M. 1992. Lista de los Lucanidae (Coleoptera) del mundo. Revista Nicaraguense de Entomologia, No. 22: 1-121.

MILNE, L. J. 1933. Notes on *Pseudolucanus placidus* (Say) (Lucanidae, Coleoptera). Canadian Entomologist, 65: 106-114.

MIZUNUMA, T. and S. NAGAI. 1994. The Lucanid Beetles of the World. Mushi-sha. Tokyo, 337 pp.

RATCLIFFE, B. C. 1991. The Lucanidae and Passalidae (Insecta: Coleoptera) of Nebraska. Great Plains Research, 1: 249-282.

RITCHER, P. O. 1966. White Grubs and Their Allies: A Study of North American Scarabaeoid Larvae. Oregon State University Press. Corvallis, Oregon, 219 pp.

SCHOLTZ, C. H. 1990. Phylogenetic trends in the Scarabaeoidea (Coleoptera). Journal of Natural History, 24: 1027-1066.

SCHOLTZ, C. H., D. J. BROWNE and J. KUKALOVÁ-PECK. 1994. Glaresidae, archaeopteryx of the Scarabaeoidea (Coleoptera). Systematic Entomology, 19: 259-277.

SMITH, A. B. T. 2001. Checklist of the Scarabaeoidea of the Nearctic Realm (Includes Canada, the continental United States, and the following states of northern Mexico: Baja California, Baja California Sur, Chihuahua, Coahuila de Zaragoza, Durango, Nuevo Leon, Sinaloa, Sonora, Tamaulipas, and Zacatecas). URL: http://www-museum.unl.edu/research/entomology/nearctic.htm.

24. DIPHYLLOSTOMATIDAE Holloway 1972

by Mary Liz Jameson and Brett C. Ratcliffe

Family common name: The false stag beetles

The family Diphyllostomatidae includes three species in the genus *Diphyllostoma*. The group is endemic to the west coast of California in the United States. The natural history of the group is poorly known, and larvae have not been described.

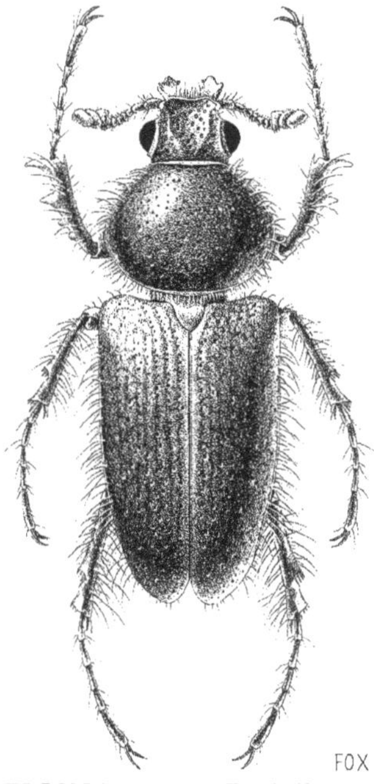

FIGURE 1.24 *Diphyllostoma fimbriata* (Fall) (Used by permission of University of Nebraska State Museum)

Description. Length 5.0-9.0 mm. Shape elongate-oval. Color brown to reddish-brown, without metallic reflections.

Head prognathous, not deflexed. Antennae straight (not geniculate), with 10 antennomeres including a 3-segmented, unopposable club (all antennomeres tomentose). Eyes with acone ommatidia, lacking canthus. Clypeus lacking tubercle or horn. Labrum concealed beneath clypeus. Mandibles prominent, quadrate or rounded at apex, produced beyond apex of labrum. Maxillae with 4-segmented palpi; labium with 4-segmented palpi.

Pronotum weakly convex, base narrower than elytral base, lacking tubercles, ridges, horns, or sulci. Elytra elongate, weakly convex, with weakly impressed, punctate striae. Pygidium concealed by elytra. Scutellum exposed, parabolic. Legs with protrochantin exposed, procoxae subconical, mesocoxae virtually contiguous; protibiae serrate on outer margin, apex lacking articulated spur; meso- and metatibia with ridges, apices with 2 spurs; tarsi 5-5-5; claws equal in size, simple; empodium present, extending to apex of fifth tarsomere, with 2 setae.

Abdomen with 7 visible sternites; 7 functional abdominal spiracles situated in pleural membrane. Wings well developed in male (vestigial in female), M-Cu loop and one apical detached vein present. Male genitalia trilobed. References: Holloway 1972; Scholtz 1990.

Larvae are not known but are probably of the scarabaeoid form and probably live in the soil.

Habits and habitats. Females differ from males in having greatly reduced eyes and vestigial wings. Life history information is scant. The adults are diurnal.

Status of the classification. The genus *Diphyllostoma* was originally placed in the family Lucanidae by Fall (1901) based on the 10-segmented antenna and lack of an eye canthus (typical characteristics of the family Lucanidae). However, several significant characters of *Diphyllostoma* are not found in other members of the family Lucanidae or other scarabaeoids. These include abdomen with 7 ventrites, exposed second abdominal segment, exposed protrochantin, and protibial spur lacking. Holloway (1972) proposed the family Diphyllostomatidae for the genus *Diphyllostoma* because of these unique characters.

Based on comparative studies, Holloway (1972) suggested the Diphyllostomatidae may be most closely related to the family Geotrupidae. Browne and Scholtz (1995, 1996, 1999) and Scholtz (1990) hypothesized that the family Diphyllostomatidae is the sister group to the Lucanidae. Diphyllostomatids and lucanids are the apical clade in a clade composed of the families Passalidae, Trogidae, Pleocomidae, Bolboceratidae, and Glaphyridae (Browne and Scholtz 1999).

Distribution. The genus *Diphyllostoma* includes three species that are found only in California in the western United States.

Classification of the Nearctic Genera

Diphyllostomatidae Holloway 1972

Diphyllostoma Fall 1901
Phyllostoma Fall 1901

This genus (Fig. 1) of fairly rare beetles includes only three species, all of which are found in California. Key: Linsley 1932; Fall 1932. Morphology: Holloway 1972.

Bibliography

BROWNE, D. J. and C. H. SCHOLTZ. 1995. Phylogeny of the families of the Scarabaeoidea (Coleoptera) based on characters of the hindwing articulation, hindwing base and wing venation. Systematic Entomology, 21: 145-173.

BROWNE, D. J. and C. H. SCHOLTZ. 1996. The morphology of the hind wing articulation and wing base of the Scarabaeoidea (Coleoptera) with some phylogenetic implications. Bonner Zoologische Monographien, No. 40: 1-200.

BROWNE, D. J. and C. H. SCHOLTZ. 1999. A phylogeny of the families of Scarabaeoidea (Coleoptera). Systematic Entomology, 24: 51-84.

FALL, H. C. 1901. Two new species of Lucanidae from California. Canadian Entomologist, 33: 289-292.

FALL, H. C. 1932. *Diphyllostoma*: a third species. Pan-Pacific Entomologist, 8: 159-161.

HOLLOWAY, B. A. 1972. The systematic position of the genus *Diphyllostoma* Fall (Coleoptera: Scarabaeoidea). New Zealand Journal of Science, 15: 31-38.

LINSLEY, E. G. 1932. The lucanid genus *Diphyllostoma*. Pan-Pacific Entomologist, 8 :109-111.

SCHOLTZ, C. H. 1990. Phylogenetic trends in the Scarabaeoidea (Coleoptera). Journal of Natural History, 24: 1027-1066.

25. PASSALIDAE Leach 1815

by Jack C. Schuster

Family common name: The bess beetles

The bess beetle family has a limited distribution in the Nearctic Region; most species occur in tropical regions rather than temperate regions. Members of this family in the United States are typically large beetles (20 to 43 mm long) with a sublamellate antennal club. The form of body (elongate and dorsoventrally depressed) and form of the mentum (deeply emarginate apically) help to distinguish this family from other scarabaeoids. Adults and larvae live together in subsocial groups in rotting logs.

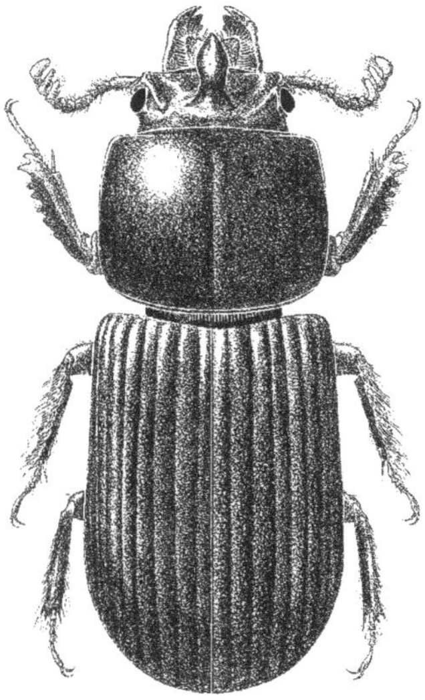

FIGURE 1.25 *Odontotaenius disjunctus* (Illiger)

Description. Length 20.0-43.0 mm. Shape elongate-cylindrical and depressed. Color black (tenerals orange to deep maroon); ventral surface with or without erect, moderately dense, yellow setae.

Head prognathous, narrower than thorax, often with dorsomedian horn. Antennae with 10 antennomeres including a 3-segmented club that is not opposable and not geniculate (but is capable of being rolled together); club antennomeres tomentose; antennae inserted under a prominent frontal margin. Eyes divided in half by canthus, with exocone ommatidium. Clypeus reduced, separated from frons by suture, or vertical and hidden beneath frons. Labrum distinct, prominent, projecting beyond apex of front of head, clypeal apex deeply emarginate, bisinuate, or truncate. Mandibles projecting beyond apex of labrum, large, curved, toothed, blunt. Maxillae with 4-segmented palpi; galea with apical hook; submentum large, prominent; mentum large, emarginate at apex; labium with 3-segmented palpi.

Pronotum broader than head, quadrate, surface smooth with median, longitudinal groove. Elytra elongate, sides parallel, apices rounded, with striae well developed. Pygidium concealed by elytra. Scutellum triangular, small (exposed only in groove between pronotum and elytra). Legs with transverse coxae, mesocoxae closed; protibiae with several external teeth on outer margin, apex with 1 spur; meso- and metatibiae with ridges, apices with 2 spurs; tarsal formula 5-5-5; claws equal in size, simple; empodium present, not extending beyond fifth tarsomere, with 2 setae.

Abdomen with 5 visible sternites; 7 functional abdominal spiracles situated in pleural membrane. Wings well developed without M-Cu loop, with 1 apical detached vein. Male genitalia trilobed. Female genitalia with paraprocts, proctiger, and styli absent; valvifers and coxite present. References: Scholtz 1990; Sharp and Muir 1912; Tanner 1927; Williams 1938.

Larvae are elongate, subcylindrical, slightly curved (not C-shaped). Color creamy white or blue-white (except at caudal end which may be darkened by accumulated feces). Cranium lightly sclerotized. Antennae 2-segmented with protruding fleshy base, short. Lateral ocelli lacking. Frontoclypeal suture distinct. Labrum rounded, setiferous at apex; maxilla with galea and lacinia distinctly separate, stridulatory area present, palpi 2-segmented (apparently 3-segmented). Abdominal segments without annuli. Spiracles cribriform with C-shaped peritremes. Venter of last abdominal segment with 2 anal lobes. Pro- and mesothoracic legs 4-segmented, with long, curved claw; metathoracic leg unsegmented, reduced to a stub with several apical teeth that rub against stridulatory area on coxa of mesothoracic leg. References: Bøving and Craighead 1930-1931; Krause and Ryan 1953; Ritcher 1966; Scholtz 1990. Keys to larvae: Schuster 1992; Schuster and Reyes-Castillo 1981.

Habits and habitats. Passalid adults live in well decayed logs and stumps with their larvae and subsocial family groups. All stages are found in galleries in wood that are excavated by the adults. Eggs are usually placed together in a "nest" of frass. In many species, recently oviposited eggs are red; as they mature, they change to brown, then to green. Adults and larvae communicate by stridulating and can produce 14 different calls. Adults care for larvae and prepare food by chewing it and presumably mixing it with saliva. Adults and larvae need to feed on adult feces that are predigested by microflora (essentially an external rumen). Biology: Reyes-Castillo 1970; Reyes-Castillo and Halffter 1984; Schuster 1975a, 1975b, 1983b; Gray 1946; Schuster and Schuster 1997; Ratcliffe 1991.

Status of the classification. Taxonomy of the species in the United States, Mexico and Central America is well known, although new species are still being found. Taxonomy of passalids in other regions of the world needs further study. Monophyly of the family is supported by larval and adult characteristics (see Scholtz 1990; Browne and Scholtz 1995). According to the phylogenetic analysis of Browne and Scholtz (1996), the family Passalidae is the basal member of a clade that includes the Diphyllostomatidae, Lucanidae, Glaphyridae, Trogidae, Bolboceratinae (Geotrupidae), and Pleocomidae. World catalog:

Acknowledgments: Jack Schuster thanks Enio Cano and Brett Ratcliffe for reviewing the passalid chapter.

Hincks and Dibb 1935. North American catalog: Smith 2001. Keys to Nearctic genera: Reyes-Castillo 1970; Schuster 1983a. Keys to genera of larvae: Schuster 1992. Regional works: Blatchley 1910; Edwards 1949; Hatch 1971; Ratcliffe 1991; Downie and Arnett 1996.

Distribution. There are over 500 described species, nearly all of which are tropical. Four extant species in two genera are recorded from the United States, and 90 species occur in Mexico. The presence of species in Texas, often cited, is highly doubtful, although one species, *Ptichopus angulatus* (Percheron), the only passalid obligatorially inhabiting leaf-cutter ant (*Atta* sp.) detritus chambers, is found on the Mexican side of the Rio Grande. The only confirmed U.S. species other than *Odontotaenius disjunctus* (Illiger) and *Odontotaenius floridanus* Schuster are *Passalus punctatostriatus* (Percheron) and *P. punctiger* LePeletier and Serville. *Passalus punctatostriatus* and *P. punctiger* were collected at the turn of the century in Arizona and have not been recorded since. *Odontotaenius disjunctus* is distributed from Nebraska, southern Manitoba and east Texas throughout the eastern U.S. and Canadian deciduous forests, south to the middle of Florida and north to Massachusetts and southern Ontario. *Odontotaenius floridanus* is endemic to certain sandhill areas of central Florida. A fossil passalid beetle, *Passalus indormitus* Cockerell, is known from Oligocene deposits of Oregon (Reyes-Castillo 1977). It is very similar to *P. punctiger.*

Key to The Genera of The Family Passalidae From The United States, Canada, and Northernmost Mexico

1. Anterior margin of labrum bisinuate; antennal lamellae curved; mandibles with 2 apical teeth; eyes reduced, canthus produced posteriorly more than radius of eye .... *Ptichopus*

— Anterior margin of labrum straight or concave; antennal lamellae straight, lying in one plane; mandibles with 3 apical teeth; eyes not reduced, canthus not produced posteriorly more than radius of eye 2

2(1). Frons with few to many large, disc-shaped punctures; frontoclypeal suture not visible; width of antepenultimate antennomere greater than 3/4 width of penultimate lamella .... *Passalus*

— Frons smooth, lacking disc-shaped punctures; frontoclypeal suture visible; width of antepenultimate antennomere 1/2 to 2/3 width of penultimate lamella .... *Odontotaenius*

Classification of the Genera

Passalidae Leach 1815

Passalinae Leach 1815

Proculini Kaup 1868

Characteristics: Frontal horn pedunculate or greatly raised, without lateral extensions. Frontoclypeal suture visible dorsally. Eastern U.S. and Canada.

Odontotaenius Kuwert 1896
Two species in North America: *O. disjunctus* (Illiger) (Fig. 1) is distributed in the eastern United States and Canada. It is commonly called the "horned passalus" or "bess-beetle." *Odontotaenius floridanus* Schuster is restricted to Florida (see Schuster 1994).

Passalini Leach 1815

Characteristics: Frontal horn absent or not pedunculate, with lateral extensions in basal half. Frontoclypeal suture not visible dorsally. Arizona and possibly Texas.

Passalus Fabricius 1792
Two species are recorded from Arizona: *P. punctiger* (LePeletier and Serville) and *P. punctatostriatus* Percheron. Two species are recorded from Texas, but these are probably erroneous.

Ptichopus Kaup 1869
One species, *P. angulatus* (Percheron) is distributed from Mexico to South America.

Bibliography

BLATCHLEY, W. S. 1910. An illustrated descriptive catalogue of the Coleoptera or beetles known to occur in Indiana. Indiana Department of Geology and Natural Resources Bulletin, 1: 1-1386.

BROWNE, D. J. and C. H. SCHOLTZ. 1995. Phylogeny of the families of the Scarabaeoidea (Coleoptera) based on characters of the hindwing articulation, hindwing base and wing venation. Systematic Entomology, 21: 145-173.

BROWNE, D. J. and C. H. SCHOLTZ. 1996. The morphology of the hind wing articulation and wing base of the Scarabaeoidea (Coleoptera) with some phylogenetic implications. Bonner Zoologische Monographien, No. 40: 1-200.

BØVING, A. G. and F. C. CRAIGHEAD. 1930-1931. An illustrated synopsis of the principal larval forms of the Coleoptera. Reprint edition, Brooklyn Entomological Society. Merrick, NY, 351 pp.

DOWNIE, N. M. and R. H. ARNETT, Jr. 1996. The Beetles of Northeastern North America, Volumes 1 and 2. The Sandhill Crane Press. Gainesville, FL, 1721 pp.

EDWARDS, J. G. 1949. Coleoptera or Beetles East of the Great Plains. Edwards Brothers. Ann Arbor, MI, 181 pp.

GRAY, I. E. 1946. Observations on the life history of the horned passalus. American Midlands Naturalist, 35: 728-746.

HATCH, M. H. 1971. The beetles of the Pacific Northwest, part 5. University of Washington Publication in Biology, 16: 1-662.

HINCKS, W. D. and J. R. DIBB. 1935. Passalidae, Coleopterorum Catalogus, 142: 1-118.

KRAUSE, J. B. and M. T. RYAN. 1953. Annals of the Entomological Society of America, 47: 1-20, 4 pls.

RATCLIFFE, B. C. 1991. The Lucanidae and Passalidae (Insecta: Coleoptera) of Nebraska. Great Plains Research, 1: 249-282.

REYES-CASTILLO, P. 1970. Coleoptera: Passalidae; morfología y división en grandes grupos; géneros americanos. Folia Entomologica Mexicana, 20-22: 1-240.

REYES-CASTILLO, P. 1977. Systematic interpretation of the Oligocene fossil *Passalus indormitus* (Coleoptera: Passalidae). Annals of the Entomological Society of America, 70: 652-654.

REYES-CASTILLO, P. and G. HALFFTER. 1984. La estructura social de los Passalidae (Coleoptera: Lamellicornia). Folia Entomologica Mexicana, 61: 49-72.

RITCHER, P. O. 1966. White Grubs and their Allies. Oregon State University Press. Corvallis, OR.

SCHOLTZ, C. H. 1990. Phylogenetic trends in the Scarabaeoidea (Coleoptera). Journal of Natural History, 24: 1027-1066.

SCHUSTER, J. C. 1975a. A comparative study of copulation in Passalidae (Coleoptera): New positions for beetles. Coleopterists Bulletin, 29: 75-81.

SCHUSTER, J. C. 1975b. Comparative behavior, acoustical signals and ecology of New World Passalidae (Coleoptera). Ph.D. Thesis. University of Florida. Gainesville, FL, 127 pp.

SCHUSTER, J. C. 1983a. The Passalidae of the United States. Coleopterists Bulletin, 37: 302-305.

SCHUSTER, J. C. 1983b. Acoustical signals of passalid beetles: complex repertoires. Florida Entomologist, 66: 486-496.

SCHUSTER, J. C. 1992. Passalidae: state of larval taxonomy with description of New World species. Florida Entomologist, 75: 358-369.

SCHUSTER, J. C. 1994. *Odontotaenius floridanus* new species (Coleoptera: Passalidae): a second U.S. passalid beetle. Florida Entomologist, 77: 474-479.

SCHUSTER, J. C. and P. REYES-CASTILLO. 1981. New World genera of Passalidae (Coleoptera): a revision of larvae. Anales de la Escuela Nacional de Ciencias Biólogicas, Mexico, 25: 79-116.

SCHUSTER, J. C. and L. B. SCHUSTER. 1997. The evolution of social behavior in Passalidae. Pp. 260-269. *In:* J. Choe and B. Crespi, eds. The Evolution of Social Behavior in Insects and Arachnids. Cambridge University Press. Cambridge.

SHARP, D. and F. MUIR. 1912. The comparative anatomy of the male genital tube in Coleoptera. Transactions of the Entomological Society of London, (1912): 477-642.

SMITH, A. B. T. 2001. Checklist of the Scarabaeoidea of the Nearctic Realm (Includes Canada, the continental United States, and the following states of northern Mexico: Baja California, Baja California Sur, Chihuahua, Coahuila de Zaragoza, Durango, Nuevo Leon, Sinaloa, Sonora, Tamaulipas, and Zacatecas). URL: http://www-museum.unl.edu/research/entomology/nearctic.htm.

TANNER, V. M. 1927. The female genitalia of Coleoptera. Transactions of the American Entomological Society, 53: 3-50.

WILLIAMS, I. W. 1938. The comparative morphology of the mouthparts of the order Coleoptera treated from the standpoint of phylogeny. Journal of the New York Entomological Society, 46: 245-289.

26. GLARESIDAE Semenov-Tian-Shanskii and Medvedev 1932

by Mary Liz Jameson

Family common name: The enigmatic scarab beetles

The beetle family Glaresidae contains one genus, *Glaresis*, that is found on all major continents except Australia. The family includes about 50 species worldwide, most of which inhabit arid and sandy regions. Members of the family are small (2.5-6.0 mm long) and light brown to dark brown. Adults are collected at lights. Larvae and biology of species in the family are not known.

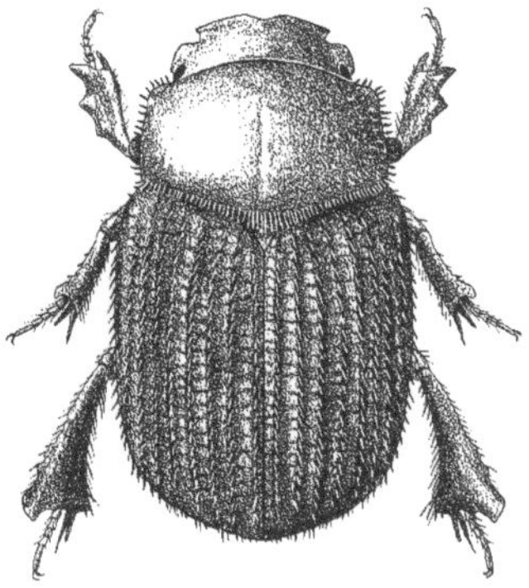

FIGURE 1.26. *Glaresis dakotensis* Gordon (Used by permission of University of Nebraska State Museum)

Description. Length 2.5-6.0 mm. Shape oblong-oval, convex. Color tan to dark brown; dorsal surface with moderately dense, short setae.

Head deflexed. Antennae 10-segmented with 3-segmented, opposable club; first antennomere of club often hollowed to receive club antenomeres 2 and 3; second and third antennomeres of club tomentose. Eyes divided in half by canthus, with eucone ommatidia. Clypeus lacking tubercle or horn. Labrum truncate, projecting weakly beyond apex of clypeus. Mandibles toothed and projecting weakly beyond apex of clypeus; maxillae with 4-segmented palpi; labium with 4-segmented palpi.

Pronotum short, broad, convex. Elytra convex with 10 distinct costae, intercostae with or without distinct punctures. Pygidium concealed by elytra. Scutellum exposed. Legs with coxae transverse; protibia fossorial, outer margin toothed, apex with one spur; meso- and metatibia with 2 apical spurs; metafemora and metatibiae enlarged to cover abdomen in retracted position; tarsi 5-5-5; claws equal in size, simple; empodium absent.

Abdomen with 5 free sternites; 8 functional abdominal spiracles situated in pleural membrane. Wings well developed, M-Cu loop present, without apical detached veins. Male genitalia trilobed. References: Cooper 1983; Scholtz 1990.

Larvae are not known but are probably of the scarabaeoid form.

Habits and habitats. Adults are found in dry, sandy areas and are attracted to lights. Adults stridulate weakly when handled (Scholtz *et al.* 1987). Based on the form of the lacinia, Scholtz *et al.* (1987) hypothesized that *Glaresis* species feed on subterranean fungi. Efforts to establish a laboratory culture from adults have not been successful (Scholtz *et al.* 1987).

Status of the classification. The genus *Glaresis* was traditionally placed in the family Trogidae (or the subfamily Troginae in the family Scarabaeidae) based on the trilobed male genitalia and five visible abdominal sternites. *Glaresis* has long been compared with primitive scarabaeoids (Ochodaeinae, Geotrupidae, Lucanidae, Hybosorinae) based on shared primitive characters. *Glaresis* shares no demonstrable, derived characters with any group and, as such, the family Glaresidae was established for the genus *Glaresis* (Scholtz *et al.* 1987). Based on plesiomorphic characters and comparison with "primitive" scarabaeoid groups, Scholtz *et al.* (1994) argued that the glaresids are the most primitive living scarabaeoids and that the family Glaresidae is the sister group to all other Scarabaeoidea (*e.g.*, Browne and Scholtz 1995, 1999; Scholtz *et al.* 1994).

Distribution. About 50 species are distributed on all major continents except Australia (Scholtz 1982), and about 15 species are found in Nearctic North America. North American catalog: Smith 2001. Regional works: Hatch 1971; Ratcliffe 1991; Downie and Arnett 1996.

Classification of the Genera of Glaresidae

Glaresidae Semenov-Tian-Shanskii and Medvedev 1932

The genus *Glaresis* (Fig. 1) is the sole member of the family Glaresidae. The tribe Glaresini was created by Semenov-Tian-Shanskii and Medvedev (1932) to accommodate the genus *Glaresis* and elevated to family status by Scholtz *et al.* (1987). Catalog: Scholtz 1982. Key to species of the United States: Gordon 1970, 1974; Warner 1995.

Glaresis Erichson 1848
  Eoglaresis Semenov-Tian-Shanskii and Medvedev 1932
  Afroglaresis Petrovitz 1968

Fifteen species in the genus occur from the western Great Plains to the west coast, and from Manitoba to northern Mexico and Baja California, Mexico.

Bibliography

BROWNE, D. J. and C. H. SCHOLTZ. 1995. Phylogeny of the families of the Scarabaeoidea (Coleoptera) based on characters of the hindwing articulation, hindwing base and wing venation. Systematic Entomology, 21: 145-173.

BROWNE, D. J. and C. H. SCHOLTZ. 1999. A phylogeny of the families of Scarabaeoidea (Coleoptera). Systematic Entomology, 24: 51-84.

COOPER, J. B. 1983. A review of the Nearctic genera of the family Scarabaeidae (exclusive of the subfamilies Scarabaeinae and Geotrupinae) (Coleoptera), with an evaluation of computer generated keys. Doctoral Thesis, Department of Biology, Carleton University. Ottawa, Ontario, Canada, 1121 pp.

DOWNIE, N. M. and R. H. ARNETT, Jr. 1996. The Beetles of Northeastern North America, Volumes 1 and 2. The Sandhill Crane Press. Gainesville, FL, 1721 pp.

GORDON, R. D. 1970. A review of the genus *Glaresis* in the United States and Canada (Coleoptera: Scarabaeidae). Transactions of the American Entomological Society, 96: 499-517.

GORDON, R. D. 1974. Additional notes on the genus *Glaresis* (Coleoptera: Scarabaeidae). Proceedings of the Biological Society of Washington, 87: 91-94.

HATCH, M. H. 1971. The beetles of the Pacific Northwest, part 5. University of Washington Publication in Biology 16: 1-662.

RATCLIFFE, B. C. 1991. The scarab beetles of Nebraska. Bulletin of the University of Nebraska State Museum, 12: 1-333.

SCHOLTZ, C. H. 1982. Catalogue of the world Trogidae (Coleoptera: Scarabaeoidea). Republic of South Africa, Department of Agriculture and Fisheries, Entomology Memoire, 54: 1-27.

SCHOLTZ, C. H. 1990. Phylogenetic trends in the Scarabaeoidea (Coleoptera). Journal of Natural History, 24: 1027-1066.

SCHOLTZ, C. H., D. J. BROWNE and J. KUKALOVÁ-PECK 1994. Glaresidae, archeopteryx of the Scarabaeoidea (Coleoptera). Systematic Entomology, 19: 259-277.

SCHOLTZ, C. H., D. D'HOTMAN and A. NEL. 1987. Glaresidae, a new family of Scarabaeoidea (Coleoptera) to accommodate the genus *Glaresis* Erichson. Systematic Entomology, 12: 345-354.

SEMENOV-TIAN-SHANSKII, A. and S. MEDVEDEV. 1932. Revisio synoptica specierum palaearcticarum novae tribus Glaresini (Coleoptera: Scarabaeidae). Pp. 337-342. *In:* Livre Centenaire 1932. Société Entomologique de France. Paris.

SMITH, A. B. T. 2001. Checklist of the Scarabaeoidea of the Nearctic Realm (Includes Canada, the continental United States, and the following states of northern Mexico: Baja California, Baja California Sur, Chihuahua, Coahuila de Zaragoza, Durango, Nuevo Leon, Sinaloa, Sonora, Tamaulipas, and Zacatecas). URL: http://www-museum.unl.edu/research/entomology/nearctic.htm.

WARNER, W. B. 1995. Two new *Glaresis* from the desert Southwest, with notes on the identity of *Glaresis mendica* Horn (Coleoptera: Scarabaeidae: Glaresinae). Insecta Mundi, 9: 267-271.

27. TROGIDAE MacLeay 1819

by Mary Liz Jameson

Family common name: The hide beetles

The family Trogidae is a small group (about 300 species worldwide) occurring on all major continents. Adults of the family are easily recognized by their overall warty, brown to gray to black, dirt-encrusted appearance, and their flat abdomen. The family includes three genera, two of which are present in North America. The genus *Trox* is widespread in the Holarctic and Ethiopian regions, and the genus *Omorgus* occurs primarily in arid regions in the southern continents. Adults and larvae can be found on the dry remains of dead animals (they are usually among the last of the succession of insects that invade carcasses) or in the nests of birds and mammals where they feed on hair, feathers, and skin.

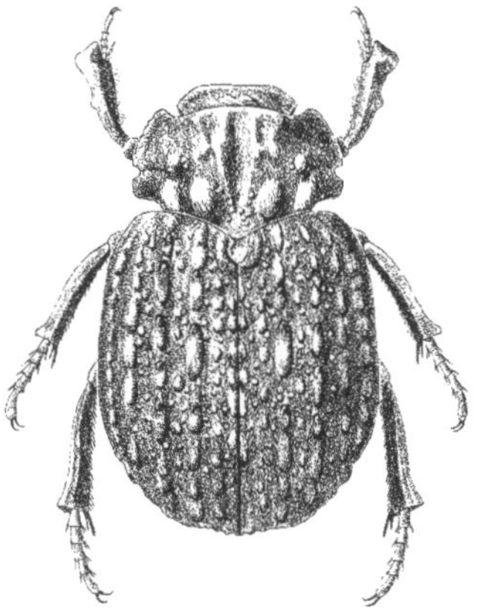

FIGURE 1.27. *Omorgus scabrosus* (Palisot de Beauvois) (Used by permission of University of Nebraska State Museum)

Description. Length 2.5-20.0 mm. Shape oblong-oval, convex. Color brown or gray to black, often with short, moderately dense, gray or brown setae; dorsal surface often greasy or encrusted with dirt.

Head deflexed. Antennae 10-segmented with 3-segmented, opposable club (all antennomeres tomentose); basal antennomere robust. Eyes with variable ommatidia (eucone, duocone, and exocone), not divided by canthus. Clypeus lacking tubercle or horn. Labrum truncate, not projecting beyond apex of clypeus. Mandibles with mandibular brush and prostheca, projecting weakly beyond apex of clypeus. Maxillae with 4-segmented palpi; labium with 4-segmented palpi.

Pronotum short, broad, convex; sculptured with ridges, depressions, or tubercles; with or without setae. Elytra convex with striae impressed and intervals ridged or tuberculate. Pygidium concealed by elytra. Scutellum exposed; shape hastate or oval. Legs with coxae transverse, mesocoxae contiguous or nearly so; protibia more or less slender, outer margin weakly toothed, apex with one spur; meso- and metatibia with 2 apical spurs; profemora enlarged (concealing or partially concealing head when deflexed); meso- and metafemora not enlarged; tarsi 5-5-5; claws equal in size, simple; empodium absent.

Abdomen with 5 free sternites; 7 or 8 functional abdominal spiracles situated in pleural membrane. Wings well developed, M-Cu loop present, with 1 or 2 apical detached veins. Male genitalia trilobed. References: Cooper 1983; Scholtz 1986, 1990.

Larvae are scarabaeiform (C-shaped, cylindrical). Color creamy-white or yellow (except at caudal end which may be darkened by accumulated feces). Cranium heavily sclerotized, brown to black. Antennae 3-segmented with apical segment much reduced. Lateral ocelli present. Frontoclypeal suture distinct or faint. Labrum bilobed. Epipharynx with an oval pedium often surrounded by phobae; heli absent, tormae united. Maxilla with galea and lacinia distinctly separate; maxillary stridulatory area with a row or patch of minute teeth; maxillary palp 4-segmented. Abdominal segments 1 to 6 with 3 annuli, each with one or more transverse rows of short, stiff setae. Spiracles with closing apparatus; cribriform or biforous. Ventral surface of last abdominal segment with bare, fleshy lobes surrounding anus. Legs 4-segmented, well developed, lacking stridulatory apparatus, each with a long, curved claw that has 2 setae at its base. References: Ritcher 1966, Baker 1968, Scholtz 1990.

Habits and habitats. Trogids are most diverse in temperate and subtropical regions and are most common in drier habitats. Adults and larvae are among the last scavengers that visit the dry remains of dead animals where they feed on feathers, fur, and skin. They also feed on organic matter found in nests of mammals and birds (*i.e.,* feces, feathers, and fur). Adults of many species are attracted to lights at night. The life histories of many species remain poorly known because of specialized associations with bird nests and mammal nests. Much biological information could be gathered by collecting from burrows and nests. When disturbed or frightened, adults feign death and remain motionless. This, in combination with their dirt-encrusted appearance, enables them to evade potential predators that might be scavenging at a carcass. Because organic debris and soil often adhere to the surface of these beetles, cleaning is necessary in order to see important characters such as sculpturing and setae. Adults stridulate by rubbing a plectrum (located on the penultimate abdominal segment) against a file (located on the internal margin of the elytra) (Lawrence and Britton 1994). Larvae of carcass-feeding species live in short, vertical burrows beneath the carcass (Baker 1968). Larvae do not stridulate.

Status of the classification. The Trogidae are considered either as a family within the Scarabaeoidea or as a subfamily of the family Scarabaeidae. This work follows Scholtz (1986) and Lawrence and Newton (1995) who treat the Trogidae as a family. Monophyly of the Trogidae is suggested by the fact that all larvae share well developed, lateral ocelli (unique in the Scarabaeoidea). The group is generally regarded as among the primitive groups of scarabaeoids (*e.g.,* Crowson 1954, 1981) based on the trilobate

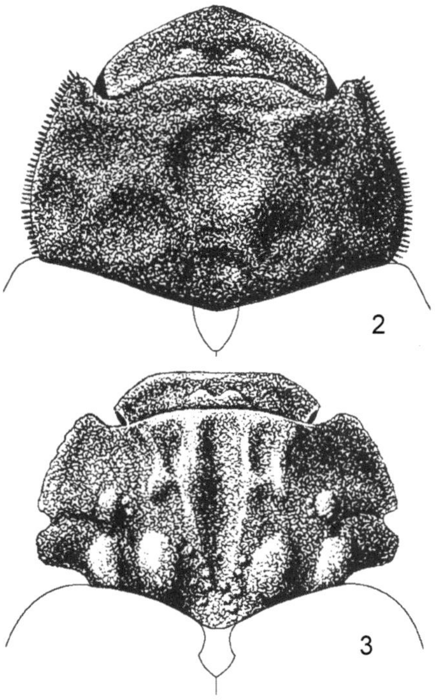

FIGURES 2.27-3.27. Dorsal view of pronotum and base of elytra showing: 2. *Trox* sp., scutellum oval, not narrowed at base; 3. *Omorgus* sp., scutellum hastate, distinctly narrowed at base (Used by permission of University of Nebraska State Museum).

form of the male genitalia. According to phylogenetic analyses, the family Trogidae is basal in a clade that includes the Pleocomidae and Bolboceratinae (Geotrupidae) (Browne and Scholtz 1995) or the Passalidae, Lucanidae, and Diphyllostomatidae (Browne and Scholtz 1999).

Distribution. About 51 species of trogids are known in the Nearctic region. In North America 43 species in 2 genera are known (Scholtz 1982). North American catalog: Smith 2001. Regional works: Blatchley 1910; Edwards 1949; Vaurie 1955; Hatch 1971; Kirk and Balsbaugh 1975; Lago *et al.* 1979; Ratcliffe 1991; Downie and Arnett 1996; Harpootlian 2001.

Key to the Genera of the Family Trogidae From the United States, Canada, and Nearctic Mexico

1. Scutellum oval, not narrowed at base (Fig. 2); base of pronotum not constricted; posterior tibia with lateral teeth or spines .... *Trox*

— Scutellum hastate, distinctly narrowed at base (Fig. 3); base of pronotum sharply constricted; posterior tibia without lateral teeth or spines............ .... *Omorgus*

Classification of the Genera of Trogidae

Trogidae MacLeay 1819

Omorgus Erichson 1847
  Chesas Burmeister 1876
  Lagopelas Burmeister 1876
  Megalotrox De Borre 1886

Sixteen species of *Omorgus* (Fig. 1) are generally distributed in southern Canada, the United States, and Mexico. Key: Vaurie 1955. Catalog: Scholtz 1982.

Trox Fabricius 1775
  Pseudatrox Robinson 1948

Twenty five species are found in southern Canada, the United States, and Mexico. Keys: Vaurie 1955; Howden and Vaurie 1957. Catalog: Scholtz 1982.

Bibliography

BAKER, C. W. 1968. Larval taxonomy of the Troginae in North America with notes on biologies and life histories (Coleoptera: Scarabaeidae). United States National Museum Bulletin, 279: 1-79.

BLATCHLEY, W. S. 1910. An illustrated descriptive catalogue of the Coleoptera or beetles known to occur in Indiana. Indiana Department of Geology and Natural Resources Bulletin, 1: 1-1386.

BROWNE, D. J. and C. H. SCHOLTZ. 1995. Phylogeny of the families of the Scarabaeoidea (Coleoptera) based on characters of the hindwing articulation, hindwing base and wing venation. Systematic Entomology, 21: 145-173.

BROWNE, D. J. and C. H. SCHOLTZ. 1999. A phylogeny of the families of Scarabaeoidea (Coleoptera). Systematic Entomology, 24: 51-84.

COOPER, J. B. 1983. A review of the Nearctic genera of the family Scarabaeidae (exclusive of the subfamilies Scarabaeinae and Geotrupinae) (Coleoptera), with an evaluation of computer generated keys. Doctoral Thesis, Department of Biology, Carleton University. Ottawa, Ontario, Canada, 1121 pp.

CROWSON, R. A. 1954 (reprint 1967). The Natural Classification of the Families of Coleoptera. E. W. Classey. Middlesex, England, 214 pp.

CROWSON, R. A. 1981. The Biology of Coleoptera. Academic Press. New York, 802 pp.

DOWNIE, N. M. and R. H. ARNETT, Jr. 1996. The Beetles of Northeastern North America, Volumes 1 and 2. The Sandhill Crane Press. Gainesville, FL, 1721 pp.

EDWARDS, J. G. 1949. Coleoptera or Beetles East of the Great Plains. Edwards Brothers. Ann Arbor, MI, 181 pp.

HARPOOTLIAN, P. J. 2001. Scarab beetles (Coleoptera: Scarabaeidae) of South Carolina. Biota of South Carolina, Volume 2. Clemson University. Clemson, SC, 157 pp.

HATCH, M. H. 1971. The beetles of the Pacific Northwest, part 5. University of Washington Publication in Biology, 16: 1-662.

HOWDEN, H. F. and P. VAURIE. 1957. Two new species of *Trox* from Florida (Coleoptera, Scarabaeidae). American Museum Novitates, 1818: 1-6.

KIRK, V. M. and E. U. BALSBAUGH, Jr. 1975. A list of the beetles of South Dakota. South Dakota State University, Agricultural Experiment Station, Technical Bulletin, Number 42, 137 pp.

LAGO, P. K., R. L. POST and C. Y. OSETO. 1979. The phytophagous Scarabaeidae and Troginae (Coleoptera) of North Dakota. North Dakota Insects Publication No. 12, Schafer-Post Series, Bismarck, North Dakota, 131 p.

LAWRENCE, J. F. and E. B. BRITTON. 1994. Australian Beetles. Melbourne University Press, 192 pp.

LAWRENCE, J. F. and A. F. NEWTON, Jr. 1995. Families and subfamilies of Coleoptera (with selected genera, notes, and references and data on family-group names). Pp. 779-1006. *In:* J. Pakaluk and S. A. Slipinski, eds. Biology, Phylogeny, and Classification of Coleoptera. Papers Celebrating the 80th Birthday of Roy A. Crowson. Muzeum i Instytut Zoologii PAN. Warsaw, Poland.

RATCLIFFE, B. C. 1991. The scarab beetles of Nebraska. Bulletin of the University of Nebraska State Museum, 12: 1-333.

RITCHER, P. O. 1966. White Grubs and their Allies. Oregon State University Press. Corvallis, OR.

SCHOLTZ, C. H. 1982. Catalogue of the world Trogidae (Coleoptera: Scarabaeoidea). Republic of South Africa, Department of Agriculture and Fisheries, Entomology Memoire, 54: 1-27.

SCHOLTZ, C. H. 1986. Phylogeny and systematics of the Trogidae (Coleoptera: Scarabaeoidea). Systematic Entomology, 11: 355-363.

SCHOLTZ, C. H. 1990. Phylogenetic trends in the Scarabaeoidea (Coleoptera). Journal of Natural History, 24: 1027-1066.

SMITH, A. B. T. 2001. Checklist of the Scarabaeoidea of the Nearctic Realm (Includes Canada, the continental United States, and the following states of northern Mexico: Baja California, Baja California Sur, Chihuahua, Coahuila de Zaragoza, Durango, Nuevo Leon, Sinaloa, Sonora, Tamaulipas, and Zacatecas). URL: http://www-museum.unl.edu/research/entomology/nearctic.htm.

VAURIE, P. 1955. A revision of the genus *Trox* (Coleoptera: Scarabaeidae) in North America. Bulletin of the American Museum of Natural History, 106: 1-89.

28. PLEOCOMIDAE LeConte 1861

by Frank T. Hovore

Family common name: The rain beetles

This family contains a single genus, *Pleocoma*, and includes 26 species that are generally distributed from southern Washington southward to northern Baja California, Mexico. As the generic name implies ("*pleos*" from the Greek, meaning full or abundant, and "*kome,*" Greek for hair), adult rain beetles are densely pubescent ventrally, on the appendages, and at the margins of the elytra and thorax.

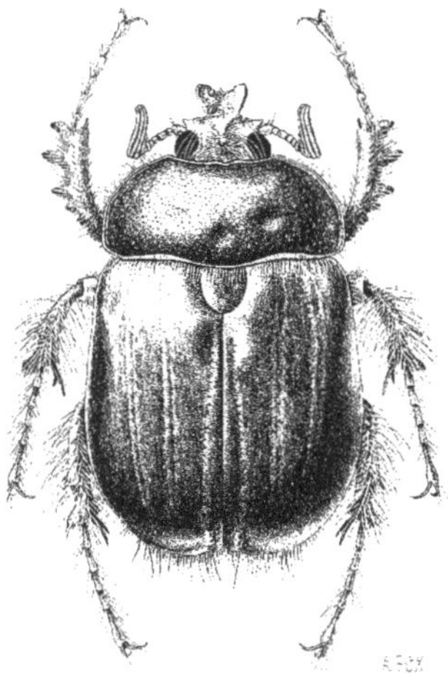

FIGURE 1.28. *Pleocoma fimbriata* LeConte

Description. Length: males 16.5-29.0 mm; females 19.5-44.5 mm. Body form robust, broadly oval in outline, strongly convex dorsally, dorsum glossy, venter densely clothed with long, fine hairs. Color variable, reddish-brown to piceous-black with setae variably colored (golden, reddish, chocolate-brown, black).

Head not deflexed or retractile. Clypeal process deeply bifurcated, outer angles produced and acute in males, broadly bilobed in females; vertex with a conical tubercle or erect horn medially. Antennae 11-segmented, scape subconical, antennomere 2 short and moniliform, antennomere 3 elongated and subcylindrical (angulated anteriorly in some), antennomeres 4-11 variable (moniliform, angulated anteriorly, or variously produced to form lamellae); antennal club of male elongate and comprised of 4 to 8 full lamellae; antennal club of female with 4-8 lamellae shorter and stouter than those of male. Eyes with exocone or duocone ommatidia, partially divided by canthus. Mouthparts partially fused and reduced; labrum connate with clypeus. Mandibles non-functional; esophageal opening closed by a membrane. Maxillary palpi 4-segmented, elongate; labial palpi shorter, 4-segmented.

Pronotum broad, evenly convex or depressed anteromedially, widest at or behind middle, lacking tubercles, horns or sulci, but some species with a low, transverse median ridge; mesothoracic spiracles primitive, slipper-shaped, with 2 intersegmentalia on each side. Scutellum exposed, subtriangular, narrowly to broadly rounded apically, lightly to densely pubescent. Elytra convex with variable sculpturing (costal striae lacking, feebly indicated, coriaceous, and/or strongly indicated); lateral margins rounded to sutural angle, sutural margins contiguous to apices. Pygidium exposed in both sexes. Legs with procoxae large, conical and prominent, procoxal cavities open; mesocoxae contiguous, prominent; protibiae strongly toothed on outer margin; meso- and metatibiae strongly ridged externally; tarsi simple and subcylindrical, elongate, 5-5-5, tarsomeres 1-4 subequal in length, tarsomere 5 longer than preceding 2 together; empodium present, produced beyond apex of tarsomere 5, with 2-3 setae.

Abdomen with 8 functional spiracles in pleural membrane. Male genitalia simple, trilobed, internal sac small, unarmed, setose. Female genitalia unsclerotized or with a few small, separated sclerites, ovipositor with styli, ovariole numbers 14-25 per ovary. Wings with M-Cu loop and 2 apical detached veins. Karyotype 9+Xyp. References: Ritcher 1969a, 1969b; Stemwedel 1973; Yadav *et al.* 1979; Ritcher and Baker 1974; Scholtz 1990; Browne and Scholtz 1995; Hovore 1977a, 1977b.

Larvae are scarabaeiform (C-shaped, cylindrical). Color creamy white except at caudal end which may be darkened by accumulated feces. Cranium heavily sclerotized, glossy, yellowish or reddish-brown. Mandibles piceous. Antennae 3-segmented, terminal segment minute, penultimate segment with apical sensory area. Frontoclypeal suture distinct. Galea and lacinia distinctly separate. Epipharynx with plegmatia and with prominent chaetoparia and acanthoparia; haptomerum with a longitudinal group of heli. Hypopharynx without oncyli. Maxillary palpi 4-segmented. Leg with apical claw bearing 2 basal setae; trochanters and femora of meso- and metathoracic legs with stridulatory organs. Spiracles cribriform, lacking closing apparatus, concavities of respiratory plates oriented ventrad. Terga of abdominal segments 3-7 each with 4 dorsal annulets. Anal opening V or Y-shaped, not surrounded by fleshy lobes. References: Ritcher 1947, 1966; Scholtz 1990.

Habits and habitats. *Pleocoma* larvae feed externally upon roots and often are found deep within the soil beneath their host plants. Although the duration of the larval stage is not known for most species, some species have nine or more instars and require from 8-13 years to reach adulthood. Pupation occurs in late summer in a simple, elongate cell. After pupation, both sexes dig to the surface and emerge more or less synchronously. Some species emerge at the onset of fall or winter rains while others are active during mid-winter or early spring. Above-ground activity of adults closely corresponds to rainfall or snowmelt, depending upon the species, elevation, and specific weather conditions. Most species are active during or immediately following precipitation. Because of the precipitation-oriented timing of adult activity, the common name for all members of the genus *Pleocoma* is "rain beetles."

Both sexes possess strongly toothed protibiae, and most species also have the clypeus and ocular canthi modified for digging through well-consolidated soils. Only male *Pleocoma* species are fully winged and capable of flight. Although the adults of most species have crepuscular flights, some fly in late morning and others fly in the night during rain. Males may be strongly attracted to light, particularly early in the flight season. Females generally are much larger than males, more heavy-bodied, and have the hind wings reduced to vestigial stubs. Females release pheromones that attract flying males, often in large numbers. Mating takes place either at the soil surface or within the female's larval burrow. Mated females return to the bottom of their burrow and wait for their eggs to mature (a process that may require several months) before depositing the eggs in a spiral pattern at the lower end of the burrow. Adult *Pleocoma* lack functional mouthparts or digestive tract, so the period of adult activity is relatively brief, dependent upon timing with conspecifics, temperature, and amount of precipitation during the emergence season. References: Hovore 1972, 1979; Fellin 1975, 1981.

Status of the classification. The genus *Pleocoma* has had a turbulent classification history. Previous authors placed *Pleocoma* in the subfamily Geotrupidae, the subfamily Melolonthinae, in its own subfamily (Pleocominae), or, in the current usage, its own family. The genera *Acoma* and *Benedictia* (currently in the subfamily Melolonthinae) were often treated in conjunction with *Pleocoma (e.g.,* Lawrence and Newton 1995). Browne and Scholtz (1995) consider the Pleocomidae to be a sister group to the Bolboceratinae (Geotrupidae) based primarily upon a series of minor structural synapomorphies. There are many basic structural dissimilarities between *Pleocoma* and the bolboceratines, however, and their respective larval characters, biologies, and geographical distributions are completely discordant, suggesting that any such putative relationship must have had an ancient point of divergence. Clearly, *Pleocoma* is a monophyletic and taxonomically isolated genus, and the relationship of the Pleocomidae to other scarabaeoids remains to be fully resolved.

The various forms of *Pleocoma* have been treated as species or subspecies with most taxa differing from one another by quantitative characters. Modern collecting methods have revealed considerable intraspecific variation in some taxa, resulting in synonymies and status changes. The only "revision" for the group was by Davis (1935). This work is out-dated and contains fewer than half of the present valid taxa. Linsley (1946) provided a provisional key to species. Currently there are 26 described species and 6 subspecies, and several additional new taxa await description.

Distribution. The genus *Pleocoma* is found from southern Washington through most of montane Oregon, southward through the Sierra Nevada and coast ranges of California, and into extreme northern Baja California Norte. The putative record of *Pleocoma* from Alaska is not considered valid. Regional work: Hatch 1971.

Classification of the Genera of Pleocomidae

Pleocomidae LeConte 1861

Pleocoma LeConte 1856
The genus *Pleocoma* (Fig. 1) includes 26 species and 6 subspecies that are restricted to regions of the west coast (from southern Washington to Baja California Norte). Key: Davis 1935. North American catalog: Smith 2001.

Bibliography

BROWNE, D. J. and C. H. SCHOLTZ. 1995. Phylogeny of the families of Scarabaeoidea based on characters of the hindwing articulation, hindwing base and wing venation. Systematic Entomology, 20: 145-173.

DAVIS, A. C. 1935. A revision of the genus *Pleocoma*. Bulletin of the Southern California Academy of Science, 33: 123-130, 34:4-36.

FELLIN, D. G. 1975. Feeding habits of *Pleocoma* larvae in coniferous forests of western Oregon. Northwest Scientist, 49: 71-86.

FELLIN, D. G. 1981. *Pleocoma* spp. in western Oregon coniferous forests: observations on adult flight habits and on egg and larval biology. Pan-Pacific Entomologist, 57: 461-484.

HATCH, M. H. 1971. The beetles of the Pacific Northwest, part 5. University of Washington Publication in Biology, 16: 1-662.

HOVORE, F. T. 1972. Three new sympatric *Pleocoma* from the southern Sierra Nevada mountains of California. Bulletin of the Southern California Academy of Science, 71: 69-80.

HOVORE, F. T. 1977a. New synonymy and status changes in the genus *Pleocoma* LeConte. Coleopterists Bulletin, 31: 229-238.

HOVORE, F. T. 1977b. A review of the taxonomic and distributional relationships of *Pleocoma hoppingi* Fall and *Pleocoma rubiginosa* Hovore. Coleopterists Bulletin, 31: 319-327.

HOVORE, F. T. 1979. Rain beetles: small things wet and wonderful. Terra Magazine, 17: 10-14.

LAWRENCE, J. F. and A. F. NEWTON, Jr. 1995. Families and subfamilies of Coleoptera (with selected genera, notes, and references and data on family-group names). Pp. 779-1006. *In:* J. Pakaluk and S. A. Slipinski, eds. Biology, Phylogeny, and Classification of Coleoptera. Papers Celebrating the 80th Birthday of Roy A. Crowson. Muzeum i Instytut Zoologii PAN. Warsaw, Poland.

LINSLEY, E. G. 1946. A preliminary key to the species of *Pleocoma*. Pan-Pacific Entomologist, 22: 61-65.

RITCHER, P. O. 1947. Description of the larva of *Pleocoma hirticollis vandykei* Linsley. Pan-Pacific Entomologist, 23: 11-20.

RITCHER, P. O. 1966. White Grubs and their Allies. A study of North American Scarabaeoid Larvae. Oregon State University Press. Corvallis, OR, 219 pp.

RITCHER, P. O. 1969a. Spiracles of adult Scarabaeoidea and their phylogenetic significance. I. The abdominal spiracles. Annals of the Entomological Society of America, 62: 869-880.

RITCHER, P. O. 1969b. Spiracles of adult Scarabaeoidea and their phylogenetic significance. II. Thoracic spiracles and adjacent sclerites. Annals of the Entomological Society of America, 62: 1388-1398.

RITCHER, P. O. and C. W. BAKER. 1974. Ovariole numbers in Scarabaeoidea. Proceedings of the Entomological Society of Washington, 76: 480-494.

SCHOLTZ, C. H. 1990. Phylogenetic trends in the Scarabaeoidea. Journal of Natural History, 24: 1027-1066.

SMITH, A. B. T. 2001. Checklist of the Scarabaeoidea of the Nearctic Realm (Includes Canada, the continental United States, and the following states of northern Mexico: Baja California, Baja California Sur, Chihuahua, Coahuila de Zaragoza, Durango, Nuevo Leon, Sinaloa, Sonora, Tamaulipas, and Zacatecas). URL: http://www-museum.unl.edu/research/entomology/nearctic.htm.

STEMWEDEL, T. A. 1973. The digestive, reproductive and nervous systems of *Pleocoma linsleyi* Hovore. Unpublished Thesis, California Polytechnic University, Pomona, 30 pp.

YADAV, J. S., R. K. PILLAI and KARAMJEET. 1979. Chromosome numbers of Scarabaeidae. Coleopterists Bulletin, 33: 309-318.

29. GEOTRUPIDAE Latreille 1802

by Mary Liz Jameson

Family common name: The earth-boring scarab beetles

As the name implies, the geotrupids are burrowers in the soil ("*geos*" from Greek meaning earth and "*trypetes*" from Greek meaning borer). In Europe, geotrupids are referred to as dor beetles. Adults of most species provision earthen burrows with dead leaves, cow dung, horse dung, or humus for their larvae.

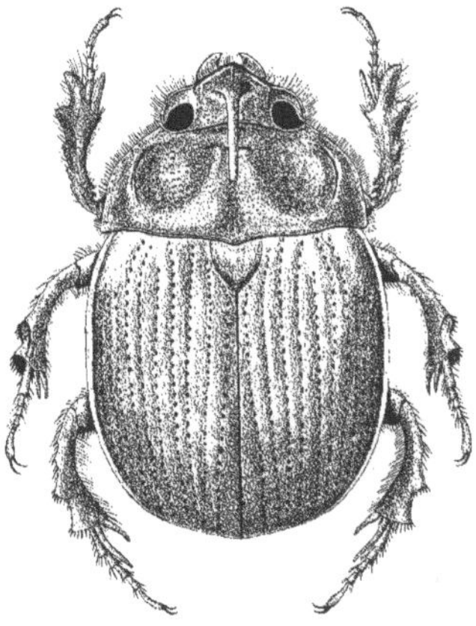

FIGURE 1.29. *Bolboceras filicornis* (Say) (Used by permission of University of Nebraska State Museum)

Description. Length 5.0-45.0 mm. Shape oval or round. Color yellowish, brown, orange-brown, reddish-brown, purple, brown, or black (with or without metallic reflections).

Head not deflexed. Antennae 11-segmented with 3-segmented, opposable club (all antennomeres tomentose). Eyes with eucone or exocone ommatidia, completely or partially divided by canthus. Clypeus often with tubercle or horn. Labrum truncate, prominent, produced beyond apex of clypeus. Mandibles produced beyond apex of labrum, prominent. Maxillae with 4-segmented palpi; labium with 3-segmented (Lethrinae) or 4-segmented (remaining taxa) palpi.

Pronotum convex with base wider than or subequal to elytral base and with or without tubercles, ridges, horns, or sulci. Elytra convex, with or without striae. Pygidium concealed by elytra. Scutellum exposed, triangular. Legs with coxae transverse, mesocoxae separated or contiguous; protibiae serrate on outer margin, apex with 1 spur; meso- and metatibia with ridges, apex with 2 spurs; tarsi 5-5-5; claws equal in size, simple; empodium present, extending beyond fifth tarsomere, with 2 setae.

Abdomen with 6 free sternites; 7 functional abdominal spiracles situated in pleural membrane (spiracles 1-7) and vestigial spiracle in pleural membrane (spiracle 8) [Bolboceratinae] or with 8 functional spiracles situated in the pleural membrane (spiracles 1-7) and the 8th pair in the tergite [Geotrupinae]. Wings well developed, M-Cu loop and two apical detached veins present. Male genitalia variable. References: Howden 1955; Scholtz 1990; Scholtz and Browne 1996.

Larvae are scarabaeiform (C-shaped, cylindrical). Color creamy-white or yellow (except at caudal end which may be darkened by accumulated feces). Cranium heavily sclerotized, brown to dark brown. Antennae 3-segmented, penultimate segment bearing 1 or more distal sense organs, last segment reduced in diameter. Lateral ocelli absent. Frontoclypeal suture absent (Geotrupinae and Bolboceratinae) or present (Taurocerastinae). Labrum at apex with 3 weak lobes or rounded. Epipharynx in most trilobed with symmetrical tormae. Maxilla with galea and lacinia distinctly separate; maxillary stridulatory area with teeth; maxillary palpi 4-segmented. Abdominal segments 3 to 7 with 2 annuli, each with one or more transverse rows of short setae. Spiracles cribriform (Geotrupinae and Lethrinae) or biforous (Bolboceratinae and Taurocerastinae). Venter of last abdominal segment V-shaped or Y-shaped, surrounded by fleshy lobes in some taxa. Legs 4-segmented (some Bolboceratinae) or pro- and mesothoracic legs 3-segmented and metathoracic leg reduced in size and 2-segmented (Geotrupinae and Taurocerastinae); stridulatory apparatus on meso- and metathoracic legs present (some Geotrupinae, some Bolboceratinae, some Taurocerastinae) or absent (some Geotrupinae, some Bolboceratinae, and Lethrinae); claws absent (Geotrupinae, Taurocerastinae, and some Bolboceratinae) or present (*Eucanthus* and *Bolbocerosoma*). References: Ritcher 1966; Scholtz 1990; Scholtz and Browne 1996.

Habits and habitats. Life histories of the geotrupids are diverse, and food habits vary from saprophagous to coprophagous and mycetophagous, and some adults apparently do not feed. Adults of most species are secretive, living most of their life in burrows. Although adults do not tend larvae, adults provision food for larvae in brood burrows. There is overlapping of generations in some species. For example, in the genus *Bolboceras*, eggs, larvae, pupae, and adults have been observed together in a single branching burrow. Adults dig vertical burrows (15-200 cm in depth) and provision larval cells with dead leaves, cow dung, horse dung, or humus. Burrows of some species extend to a depth of 3.0 meters. In restricted habitats, some species are semi-colonial. Geotrupids are not of economic importance, although their burrowing has occasionally caused damage in lawns. Adults of many geotrupids are nocturnal and are frequently attracted to lights at night. Some species are attracted to fermenting malt and molasses baits. Most adults and larvae stridulate. The natural history and behavior of many species, especially the Bolboceratinae, are poorly known. References: Howden 1955; Woodruff 1973.

Status of the classification. There is considerable debate concerning the classification of the Geotrupidae. The diversity in structure in both adults and larvae has led to differences of opinion regarding classification, evolution, and monophyly of the group and the genera assigned to it. There is evidence that the group (as defined in this work) includes two distinct lineages: the Bolboceratinae and Athyreinae forming one lineage, and the Geotrupinae, Taurocerastinae, and Lethrinae forming the other.

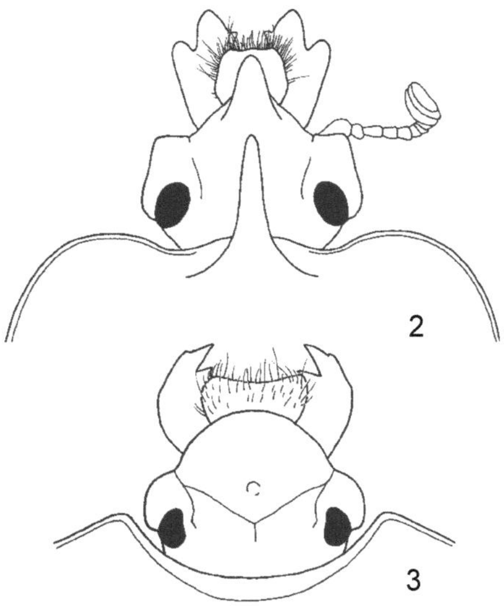

FIGURES 2.29-3.29. Dorsal view of head and apex of pronotum showing: 2. *Ceratophyus gopherinus* Cartwright, apex of mandibles deeply bidentate and pronotum with central horn; 3. *Geotrupes splendidus* (Fabricius) apex of mandibles rounded and pronotum without central horn.

Scholtz and Browne (1996) proposed the family Bolboceratidae for the former lineage. In addition to the possible division of the group into two families, the geotrupids are often considered as a subfamily of the family Scarabaeidae. In this volume, we follow Lawrence and Newton (1995) and consider the group a family. There is ample evidence that the group is not monophyletic, but the taxa that should be included, characters that support the groupings, and the ranking of the groupings are still debated. Past workers (*e.g.*, Davis 1935; Ritcher 1947) have also included the genus *Pleocoma* (treated in the family Pleocomidae in this work) in the Geotrupidae based on the 11-segmented antenna. However, *Pleocoma* species differ from the geotrupids based on the open procoxal cavities (closed in geotrupids) and 4 to 7-segmented club (3-segmented in geotrupids).

Howden (1982) hypothesized that the Geotrupinae (Geotrupini, Athyreini, Bolboceratini, and Lethrini) form a monophyletic lineage that is most closely related to the Pleocominae (= Pleocomidae) based on characters such as the form of the antennal club, adult provisioning, and diet. Alternatively, it was hypothesized (Browne and Scholtz 1995, 1999; Scholtz and Browne 1996; Scholtz and Chown 1995) that the geotrupids are a polyphyletic group: the Bolboceratidae (including Athyreinae) is the sister taxon of Pleocomidae in a clade including the Glaphyridae, Trogidae, Passalidae, Lucanidae, and Diphyllostomatidae while the family Geotrupidae (including Geotrupinae, Taurocerastinae, and Lethrinae) is part of a separate clade that includes the Ochodaeidae, Hybosoridae, and Ceratocanthidae. The placement of the unusual South American genera *Taurocerastes* and *Frickius* has also been a source of much debate. These genera were placed in the Geotrupini (*e.g.*, Howden 1982; Howden and Peck 1987) or in their own group, the Taurocerastinae (*e.g.*, Zunino 1984a; Browne and Scholtz 1995).

Taxonomy of the world Geotrupidae is well established, primarily due to the prolific work of Howden (*e.g.*, Howden 1955, 1964, 1974, 1980, 1984). Taxonomy of the North American geotrupids is treated in Howden (1955, 1964).

Distribution. The family Geotrupidae includes 68 genera and about 620 species worldwide (Scholtz and Browne 1996). The subfamily Geotrupinae (including Geotrupini, Taurocerastini, and Lethrini) is distributed in the Holarctic region. In the New World, the subfamily Geotrupinae occurs from Canada to El Salvador. The subfamily Bolboceratinae (including Bolboceratini and Athyreini) is best represented in Australia, Africa, and South America. In the New World, the subfamily Bolboceratinae is distributed from Canada to Central America.

Twelve genera and 28 species of geotrupids occur in the United States, Canada, and Nearctic Mexico. Keys to genera and species: Howden 1955; Olson *et al.* 1954; Howden 1964. Reference: Scholtz and Browne 1996; Zunino 1984b. Biology: Howden 1955. Catalog of the U.S. species: Howden 1984; Smith 2001. Regional works: Blatchley 1910; Edwards 1949; Helgesen and Post 1967; Hatch 1971; Woodruff 1973; Ratcliffe 1991; Downie and Arnett 1996; Harpootlian 2001. Larvae: Ritcher 1966.

Key to the Subfamilies and Genera of the United States, Canada, and Nearctic Mexico

1. Antennal club large (about as long as antennomeres 1 through 8), round; mentum not noticeably emarginate at apex (Bolboceratinae) 2
— Antennal club small (about half as long antennomeres 1 through 8), elongate; mentum deeply emarginate at apex (Geotrupinae) 9

2(1). Mesocoxal separation greater than width of labrum .... *Neoathyreus*
— Mesocoxal separation half or less than half width of labrum 3

3(2). Eyes entirely divided by canthus; dorsal color in some variegated 4
— Eyes only partially divided by canthus; dorsal color uniform brown to black 5

4(3). Color brownish orange with discrete areas of black or dark brown; mesocoxae narrowly separated by slender projection of mesosternal plate .... *Bolbocerosoma*
— Color uniformly brown to black; mesocoxae contiguous, not separated by projection of mesosternal plate .... *Bolboceras*

5(3). Elytra each with 5 striae between suture and humeral umbone; humeral angle of elytron not broadly rounded, margin almost always produced into tubercle .... *Eucanthus*
— Elytra each with 7 striae between suture and humeral umbone; humeral angle of elytron broadly rounded, margin never produced into tubercle 6

6(5). Mesocoxae nearly contiguous, intercoxal process less than 0.3 mm wide and linear ... *Bolbelasmus*

— Mesocoxae well separated, intercoxal process more than 0.3 mm wide and never linear 7

7(6). Base of elytra margined; pronotum without postapical carina 8

— Base of elytra not margined; pronotum with postapical carina extending almost to side margins .... *Bolborhombus*

8(7). Apex of middle and hind tibiae truncate; prosternal spine (behind anterior coxae) transverse .... *Bolbocerastes*

— Apex of middle and hind tibiae deeply emarginate; prosternal spine lacking .... *Bradycinetulus*

9(1). Mandible with apex deeply bidentate (Fig. 2) .... *Ceratophyus*

— Mandible with apex rounded, at most with weak apical tooth (Fig. 3) 10

10(9). Elytra fused, surface roughly granulate; wings vestigial .... *Mycotrupes*

— Elytra not fused, or if fused not roughly granulate; wings fully developed 11

11(10). Posterior tibia on outer edge with apex cariniform; lateral margin of elytra not widely flared at base .... *Geotrupes*

— Posterior tibia on outer edge with apex eroded, not cariniform; lateral margin of elytra widely flared at base .... *Peltotrupes*

Classification of the Subfamilies and Genera

Geotrupidae Latreille 1802

Bolboceratinae Mulsant 1842

Characteristics: Antennal club large, about as long as antennomeres 1-8, round, convex on both sides. Mesocoxae separated.

This subfamily contains 13 genera and over 300 species and occurs on all continents except Antarctica. Nearctic species belong to eight genera, over half of which belong to *Bolbocerosoma* and *Bolboceras*. Adults feed on fungi, and some are attracted to fermenting malt. Larval food consists of a brood ball of fine humus in *Eucanthus, Bolboceras*, and *Bolbocerosoma*. References: Howden 1955, 1964; Woodruff 1973. Biology: Woodruff 1973, Howden 1955. Larvae: Ritcher 1966; Howden 1964.

Bolbelasmus Boucomont 1911
 Kolbeus Boucomont 1911
Four species occur in California, Arizona, Texas, and Nearctic Mexico. Keys: Cartwright 1953; Howden 1964.

Bolboceras Kirby 1819
 Odonteus Samouelle 1819
 Odontaeus Dejean 1821
The genus (Fig. 1) includes ten species that are generally distributed from southern Canada to the southern United States. A proposal is currently pending with the International Commission of Zoological Nomenclature to conserve the generic name *Bolboceras* Kirby because the senior synonym, *Odonteus* Samouelle, has not been used in the primary literature for over 70 years (Jameson and Howden in press). Keys: Wallis 1928; Howden 1964.

Bolbocerastes Cartwright 1953
Four species are distributed from Kansas and Oklahoma to the west coast and south to Mexico. Keys: Cartwright 1953; Howden 1964.

Bolbocerosoma Schaeffer 1906
Includes twelve species that are widely distributed east of the Rocky Mountains and southward into Mexico. Keys: Howden 1955, 1964. (Volume 1, Color Figure 10).

Bolborhombus Cartwright 1953
Three species occur in Texas, New Mexico, Arizona, and Nearctic Mexico. Keys: Cartwright 1953; Howden 1964.

Bradycinetulus Cockerell 1906
 Amechanus Horn 1870
 Bradycinetus Horn 1871
Three species occur in the mid-United States and southeastern United States. Key: Cartwright 1953.

Eucanthus Westwood 1848
Five species are widely distributed from southern Canada to Mexico. Key: Howden 1964.

Neoathyreus Howden and Martínez 1963
The genus *Neoathyreus* is the only representative of the tribe Athyreini in the Nearctic region. Two species, *N. mixtus* Howden and *N. fissicornis* (Harold), are found in northeastern Mexico. Reference: Howden 1964; Howden and Gill 1984.

Geotrupinae Latreille 1802

Characteristics: Antennal club small, about half as long as the antennomeres 1-8, elongate, not convex on the sides. Mesocoxae almost contiguous.

This subfamily contains eight genera distributed throughout the northern hemisphere. They are absent from South America, South Africa, and Australia. In the United States, species in the subfamily occur mostly east of the Rocky Mountains with only *Certatophyus gopherinus* Cartwright found in California. References: Howden 1955, 1964, 1974; Woodruff 1973; Ritcher and Duff 1971. Larvae: Ritcher 1966; Howden 1955, 1964.

Ceratophyus Fischer von Waldheim 1823
One species, *C. gopherinus* Cartwright (Fig. 2), occurs in California. Larvae feed on leaf litter and twigs. Reference: Zunino 1973.

Geotrupes Latreille 1796
 subgenus *Geotrupes* Latreille 1796
 subgenus *Anoplotrupes* Jekel 1865
 subgenus *Cnemotrupes* Jekel 1865
  Onychotrupes Jekel 1865
 subgenus *Geohowdenius* Zunino 1984
 subgenus *Megatrupes* Zunino 1984

The genus includes ten species that are generally distributed from southern Canada to Texas. Adults feed on carrion, fungi, leaf litter, and dung. *Geotrupes stercorarius* (L.) was introduced to North America from Europe. Keys: Howden 1955, 1964.

Mycotrupes LeConte 1866

Five species in the genus are distributed in Georgia, Florida, and South Carolina. Adults are wingless and isolated in sandy areas. Keys: Olson *et al.* 1954; Woodruff 1973.

Peltotrupes Blanchard 1888

The genus includes two species, *P. profundus* (Howden) and *P. youngi* Howden, that live in the sandy ridges of northern and peninsular Florida. Larvae feed on leaves, twigs, bark, and male pine cones. Key: Howden 1955.

Bibliography

BLATCHLEY, W. S. 1910. An illustrated descriptive catalogue of the Coleoptera or beetles known to occur in Indiana. Indiana Department of Geology and Natural Resources Bulletin, 1: 1-1386.

BROWNE, D. J. and C. H. SCHOLTZ. 1995. Phylogeny of the families of the Scarabaeoidea (Coleoptera) based on characters of the hindwing articulation, hindwing base and wing venation. Systematic Entomology, 21: 145-173.

BROWNE, D. J. and C. H. SCHOLTZ. 1999. A phylogeny of the families of Scarabaeoidea (Coleoptera). Systematic Entomology, 24: 51-84.

CARTWRIGHT, O. L. 1953. The beetles of the genus *Bradycinetulus* and closely related genera in the United States (Coleoptera: Scarabaeidae). Proceedings of the United States National Museum, 103: 95-120.

DAVIS, A. C. 1935. A revision of the genus *Pleocoma*. Bulletin of the Southern California Academy of Science, 33: 123-130.

DOWNIE, N. M. and R. H. ARNETT, Jr. 1996. The Beetles of Northeastern North America, Volumes 1 and 2. The Sandhill Crane Press, Gainesville, FL, 1721 pp.

EDWARDS, J. G. 1949. Coleoptera or Beetles East of the Great Plains. Edwards Brothers. Ann Arbor, MI, 181 pp.

HARPOOTLIAN, P. J. 2001. Scarab beetles (Coleoptera: Scarabaeidae) of South Carolina. Biota of South Carolina, Volume 2. Clemson University, Clemson, SC, 157 pp.

HATCH, M. H. 1971. The beetles of the Pacific Northwest, part 5. University of Washington Publication in Biology, 16: 1-662.

HELGESEN, R. S., Jr. and R. L. POST. 1967. Saprophagous Scarabaeidae (Coleoptera) of North Dakota. North Dakota Insects, Publ. No. 7: 1-60.

HOWDEN, H. F. 1955. Biology and taxonomy of North American beetles of the subfamily Geotrupinae with revisions of the genera *Bolbocerosoma, Eucanthus, Geotrupes* and *Peltotrupes* (Scarabaeidae). Proceedings of the United States National Museum, 104: 151-319.

HOWDEN, H. F. 1964. The Geotrupinae of North America and Central America. Memoirs of the Entomological Society of Canada, 39: 1-91.

HOWDEN, H. F. 1974. Additional records and descriptions of North and Central American Geotrupinae (Coleoptera, Scarabaeidae). Canadian Journal of Zoology, 52: 567-573.

HOWDEN, H. F. 1980. Key to the Geotrupini of Mexico and Central America, with the description of a new species (Scarabaeidae, Geotrupinae). Canadian Journal of Zoology, 58: 1959-1963.

HOWDEN, H. F. 1982. Larval and adult characters of *Frickius* Germain, its relationship to the Geotrupini, and a phylogeny of some major taxa in the Scarabaeoidea (Insecta: Coleoptera). Canadian Journal of Zoology, 60: 2713-2724.

HOWDEN, H. F. 1984. A Catalog of the Coleoptera of America North of Mexico. Family Scarabaeidae. Subfamily: Geotrupinae. United States Department of Agriculture, Agriculture Handbook 529-34a, 17 pp.

HOWDEN, H. F. and B. Gill. 1984. Two new species of *Neoathyreus* Howden and Martínez from Costa Rica with distribution notes on other Athyreini from Mexico and Central America (Coleoptera: Scarabaeidae). Canadian Entomologist, 116: 1637-1641.

HOWDEN, H. F and S. B. PECK. 1987. Adult habits, larval morphology, and phylogenetic placement of *Taurocerastes patagonicus* Philippi (Scarabaeidae: Geotrupinae). Canadian Journal of Zoology, 65: 329-332.

JAMESON, M. L. and H. F. HOWDEN. In press. *Bolboceras* Kirby, 1819 and *Odonteus* Samouelle, 1819: proposed conservation of generic name. Bulletin of Zoological Nomenclature.

LAWRENCE, J. F. and A. F. NEWTON, Jr. 1995. Families and subfamilies of Coleoptera (with selected genera, notes, and references and data on family-group names). Pp. 779-1006. *In:* J. Pakaluk and S. A. Slipinski, eds. Biology, Phylogeny, and Classification of Coleoptera. Papers Celebrating the 80th Birthday of Roy A. Crowson. Muzeum i Instytut Zoologii PAN. Warsaw, Poland.

OLSON, A. L., T. H. HUBBELL and H. F. HOWDEN. 1954. The beetles of the genus *Mycotrupes* (Coleoptera: Scarabaeidae: Geotrupidae). Miscellaneous Publications of the Museum of Zoology, University of Michigan, 84: 1-59.

RATCLIFFE, B. C. 1991. The scarab beetles of Nebraska. Bulletin of the University of Nebraska State Museum, 12: 1-333.

RITCHER, P. O. 1947. Larvae of Geotrupinae with keys to tribes and genera (Coleoptera: Scarabaeidae). Kentucky Agricultural Experiment Station, Bulletin 506: 1-27.

RITCHER, P. O. 1966. White Grubs and Their Allies: A Study of North American Scarabaeoid Larvae. Oregon State University Press. Corvallis, OR, 219 pp.

RITCHER, P.O. and R.W. DUFF. 1971. A description of the larva of *Ceratophyus gopherinus* Cartwright, with a revised key to the larvae of North American Geotrupini and notes on the biology (Coleoptera: Scarabaeidae). Pan-Pacific Entomologist, 47: 158-163.

SCHOLTZ, C. H. 1990. Phylogenetic trends in the Scarabaeoidea (Coleoptera). Journal of Natural History, 24: 1027-1066.

SCHOLTZ, C. H. and D. J. BROWNE. 1996. Polyphyly in the Geotrupidae (Coleoptera: Scarabaeoidea): a case for a new family. Journal of Natural History, 30: 597-614.

SCHOLTZ, C. H. and S. L. CHOWN. 1995. The evolution of habitat use and diet in the Scarabaeoidea: a phylogenetic approach. Pp. 355-374. *In:* J. Pakaluk and S. A. Slipinski, eds. Biology, Phylogeny, and Classification of Coleoptera. Papers Celebrating the 80th Birthday of Roy A. Crowson. Muzeum i Instytut Zoologii PAN. Warsaw, Poland.

SMITH, A. B. T. 2001. Checklist of the Scarabaeoidea of the Nearctic Realm (Includes Canada, the continental United States, and the following states of northern Mexico: Baja California, Baja California Sur, Chihuahua, Coahuila de Zaragoza, Durango, Nuevo Leon, Sinaloa, Sonora, Tamaulipas, and Zacatecas). URL: http://www-museum.unl.edu/research/entomology/nearctic.htm.

WALLIS, J. B. 1928. Revision of the genus *Odontaeus*, Dej. (Scarabaeidae, Coleoptera). Canadian Entomologist, 60: 119-128, 151-156, 168, 176.

WOODRUFF, R. E. 1973. The scarab beetles of Florida (Coleoptera: Scarabaeidae). Part I. The Laparosticti (subfamilies Scarabaeinae, Aphodiinae, Hybosorinae, Ochodaeinae, Geotrupinae, Acanthocerinae). Arthropods of Florida and Neighboring Land Areas, 8: 1-220.

ZUNINO, M. 1973. II genere *Ceratophyus* Fisch. (Coleoptera, Scarabaeoidea). Bollettino del Museo di Zoologia dell'Università de Torino, 2: 9-40.

ZUNINO, M. 1984a. Analisi sistematica e zoogeografica della sottofamiglia Taurocerastinae Germain (Coleoptera, Scarabaeoidea: Geotrupidae). Bollettino del Museo Regionale di Scienze Naturali, Torino, 2: 445-464.

ZUNINO, M. 1984b. Sistematica generica dei Geotrupinae (Coleoptera, Scarabaeoidea: Geotrupidae), filogenesi della sottofamiglia e considerazioni biogeografiche. Bollettino del Museo Regionale di Scienze Naturali, Torino, 2: 9-162.

30. OCHODAEIDAE Mulsant and Rey 1870

by David C. Carlson

Family common name: The sand-loving scarab beetles

The family Ochodaeidae is relatively small and widely distributed. Adults are small, mostly brown, non-metallic beetles that are predominately active at night. They are most often collected at lights, sometimes in large numbers. Adults of a few species are active during the day. Many species prefer sandy areas, and many stridulate. Woodruff (1973) suggested that adults may spend the daylight hours in subterranean burrows, and that they might feed on fungi. Little else is known about the habits of adult or immature stages.

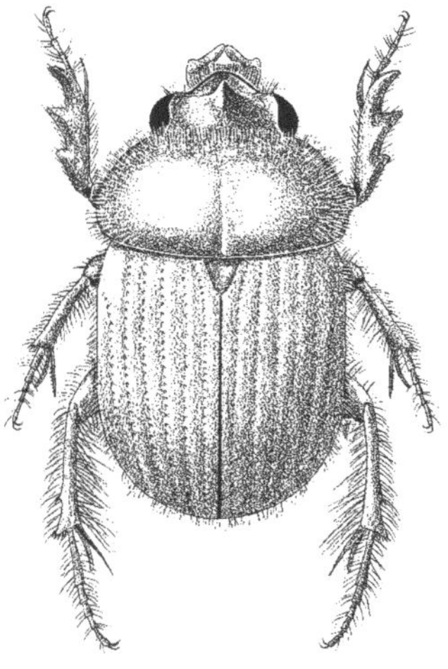

FIGURE 1.30. *Ochodaeus mandibularis* Linell

Description. Length 3.0-10.0 mm. Shape elongate and convex. Color yellowish, brown, reddish-brown, brown, or black; infrequently bicolorous.

Head not deflexed. Antennae 9 or 10-segmented, with 3-segmented, opposable, club (all antennomeres tomentose). Eyes with eucone ommatidia, not divided by canthus. Clypeus simple or with tubercle(s) on anterior margin. Labrum produced beyond apex of clypeus, often bilobed and emarginate, prominent. Mandibles produced beyond apex of labrum, prominent. Maxillae with 4 or 5-segmented palpi; labium with 3 or 4-segmented palpi.

Pronotum convex, subquadrate; most punctate and setose; without tubercles, ridges, horns, or sulci. Elytra convex, with or without striae, often punctate or granulate and setose, some smooth. Scutellum exposed, triangular. Pygidium exposed or concealed by elytra. Legs with procoxae conical or transverse; meso- and metacoxae transverse, mesocoxae separated or contiguous; protibia dentate on outer margin, apex with 1 spur; meso- and metatibia with 2 apical spurs; 1 mesotibial spur pectinate/crenulate, pro- and metatibial spur crenulate/pectinate in some; tarsi 5-5-5; claws equal in size, simple; empodium absent.

Abdomen with 6 visible sternites; stridulatory peg present in some; 8 functional abdominal spiracles with spiracles 1-6 situated in pleural membrane and spiracles 7-8 situated in tergites; tergite, pleurite, and sternite of female 9th abdominal segment visible as distinct sclerites. Wings well developed, M-Cu loop and two apical detached veins present. Male genitalia with divided basal piece, symmetrical parameres, partially sclerotized membranous median lobe and large internal sac; internal sac armed with spines, hooks, and toothed sclerites in many. Female genitalic hemisternites with styli present. Six ovarioles per ovary. References: Browne and Scholtz 1995; Carlson 1975; Carlson and Ritcher 1974; Ritcher 1969a, 1969b; Ritcher and Baker 1974; Scholtz 1990; Scholtz *et al.* 1988.

Larvae are scarabaeiform (C-shaped, cylindrical). Color whitish (except at caudal end which may be darkened by accumulated feces). Cranium sclerotized, yellow-brown to red-brown. Antennae 3 or 4-segmented, penultimate and apical segment with sense organs. Ocelli absent. Frontoclypeal suture absent. Labrum trilobed. Epipharynx with complete, symmetrical zygum; tormae fused and symmetrical. Maxilla with galea and lacinia distinctly separate. Maxillary palpi 4-segmented; maxillary and mandibular stridulatory areas present. Abdominal segments 1-7 with 3 dorsal lobes, anterior 2 lobes with transverse row of setae. Spiracles cribriform, inconspicuous. Anal opening Y-shaped, surrounded by fleshy lobes. Legs well developed, 4-segmented, with well-developed claws, stridulatory apparatus lacking. References: Carlson and Ritcher 1974; Medvedev 1960; Scholtz 1990.

Habits and habitats. Little is known about the biology of Ochodaeidae. There are few recorded observations of adult or larval habits except that adults of most species are nocturnal and are attracted to light, sometimes in large numbers. Adults of a few species are diurnally active and have been collected infrequently with sweep nets, malaise traps, or seining flumes. Adults of one species have been found associated with detritus deposits of harvester ants. Adults of another species were found to have basidiomycete spores in the midgut and hindgut. References: Arrow 1912; Carlson 1975; Carlson and Ritcher 1974; Deloya 1988.

Status of the classification. The ochodaeids have long been recognized as a distinct group. Their status has vacillated between the subfamily and family levels, but more recent works have tended to favor familial status (Scholtz 1990; Scholtz and Evans 1987; Scholtz and Chown 1995; Scholtz *et al.* 1988). Two subfamilies, Chaetocanthinae and Ochodaeinae are currently recognized (Lawrence and Newton 1995; Scholtz *et al.* 1988).

The phylogenetic position of the Ochodaeidae within the Scarabaeoidea has been discussed by numerous authors. Current views consider the ochodaeids to be an "intermediate" scarabaeoid family (Browne and Scholtz 1995; Scholtz 1990). A close relationship with the Hybosoridae was suggested by Carlson and Ritcher (1974). Recent studies suggest a close relationship between the Ochodaeidae, Hybosoridae, and Ceratocanthidae and consider the Ochodaeidae to be the sister group to the Hybosoridae plus

FIGURES 2.30-3.30. Spur at apex of metatibia of: 2. *Pseudochodaeus estriatus* (Schaeffer), pectinate; 3. *Ochodaeus* sp., not pectinate.

Ceratocanthidae (d'Hotman and Scholtz 1990; Scholtz *et al.* 1988; Browne and Scholtz 1995). The presence of a pectinate/crenulate mesotibial spur in the Ochodaeidae is unique among the Scarabaeoidea and, according to Scholtz (1990), establishes the monophyly of the group. Within the Ochodaeidae, d'Hotman and Scholtz (1990) consider *Ochodaeus* and *Codocera* to be the most primitive based on the structure of the male genitalia, and *Chaetocanthus* and *Namibiotalpa* are considered the most derived.

Taxonomy of the world Ochodaeidae is fairly well established for the Chaetocanthinae and the tribe Enodognathini (Ochodaeinae) (Scholtz and Evans 1987; Scholtz *et al.* 1988). The tribe Ochodaeini, however, is in a state of flux due to Nikolayev's (1995) description of two new genera. Nikolayev (1995) did not fully assign species to the new genera, and it is likely that more of the North American species currently included in *Ochodaeus* will be assigned to these genera. Nikolayev (1995) based these genera on characters that are well-recognized as separating groups (Fall 1909; Horn 1876; Carlson 1975).

Distribution. The family Ochodaeidae includes ten extant and two fossil genera and about 80 species worldwide (Arrow 1912; Scholtz and Evans 1987; Scholtz *et al.* 1988; Nikolayev 1995). In the Nearctic region, the family includes four genera and 35 species. North American catalog: Smith 2001. Regional works: Blatchely 1910; Edwards 1949; Woodruff 1973; Kirk and Balsbaugh 1975; Ratcliffe 1991; Downie and Arnett 1996.

Key to the Nearctic Subfamilies and Genera of the Family Ochodaeidae of the United States, Canada, and Nearctic Mexico

1. Metatibial spur not crenulate or pectinate (Fig. 3); metatibia cylindrical or flattened; abdominal stridulatory peg present in most (Ochodaeinae) 2

— Metatibial spur crenulate or pectinate (Fig. 2); metatibia flattened; abdominal stridulatory peg absent (Chaetocanthinae)........ *Pseudochodaeus*

2(1). Sutural angle of elytra dentiform; propygidium with pair of tubercles at midline that interlock with dentiform elytral apices .... *Parochodaeus*

— Sutural angle of elytra not dentiform; propygidium without tubercles at midline 3

3(2). Propygidium with longitudinal groove at midline that interlocks with medial edge of elytron. .... *Neochodaeus*

— Propygidum with distal margin modified into transverse ridge that interlocks with distal elytral margin .... *Ochodaeus*

Classification of the Subfamilies and Genera

Ochodaeidae Mulsant and Rey 1870

Ochodaeinae Mulsant and Rey 1870

Characteristics: Antennae 10-segmented. Pronotal surface punctate or granulate. Mesotibial spur crenulate or pectinate. Abdominal stridulatory peg present on sternum.

This subfamily contains six genera and about 70 species worldwide. Its members occur on all continents except Australia and Antarctica. North American species belong to three genera. Until Nikolayev (1995) described the genera *Neochodaeus* and *Parochodaeus*, *Ochodaeus* was the only genus from this subfamily represented in the Nearctic. Little is known about adult or larval habits. References: Horn 1876; Fall 1909; Arrow 1912; Howden 1968; Carlson 1975; Scholtz *et al.* 1988; Nikolayev 1995.

Neochodaeus Nikolayev 1995
Two species are distributed in the southwestern and southeastern United States and Mexico. References: Carlson 1975; Nikolayev 1995.

Ochodaeus Dejean 1821
This genus (Fig. 1) includes about 26 Nearctic and Neotropical species distributed from southern Canada to South America. References: Carlson 1975. Keys: Fall 1909; Woodruff 1973; Ratcliffe 1991.

Parochodaeus Nikolayev 1995
Six species are found in the southwestern and midwestern United States and Mexico. References: Carlson 1975; Nikolayev 1995.

Chaetocanthinae Scholtz 1988

Characteristics: Antennae 9 (*Chaetocanthus* only) or 10-segmented. Pronotal surface granulate in most. Meso- and metatibial spurs pectinate/crenulate. Abdominal stridulatory peg absent from sternum.

This subfamily contains four genera, three of which are found only in Africa. The fourth genus (*Pseudochodaeus*) is limited to the west coast of the United States. Little is known about the adult or larval habits. References: Carlson and Ritcher 1974; Scholtz *et al.* 1988.

Pseudochodaeus Carlson and Ritcher 1974
The genus *Pseudochodaeus* is monobasic. The single species, *P. estriatus* (Schaeffer), is distributed at mid-elevations from southern Oregon to central California. Reference: Carlson and Ritcher 1974.

Bibliography

ARROW, G. J. 1912. Scarabaeidae: Pachypodinae, Pleocominae, Aclopinae, Glaphyrinae, Ochodaeinae, Orphninae, Idiostominae, Hybosorinae, Dynamopinae, Acanthocerinae, Troginae. Coleopterorum Catalogus, 19: 1-66.

BLATCHLEY, W. S. 1910. An illustrated descriptive catalogue of the Coleoptera or beetles known to occur in Indiana. Indiana Department of Geology and Natural Resources Bulletin, 1: 1-1386.

BROWNE, D. J. and C. H. SCHOLTZ. 1995. Phylogeny of the families of the Scarabaeoidea (Coleoptera) based on characters of the hindwing articulation, hindwing base and wing venation. Systematic Entomology, 20: 145-173.

CARLSON, D. C. 1975. Taxonomic characters of the genus *Ochodaeus* Serville with descriptions of two new species in the *O. pectoralis* LeConte species complex (Coleoptera: Scarabaeidae). Bulletin of the Southern California Academy of Sciences, 74: 49-65.

CARLSON, D. C. and P. O. RITCHER. 1974. A new genus of Ochodaeinae and a description of the larva of *Pseudochodaeus estriatus* (Schaeffer) (Coleoptera: Scarabaeidae). Pan Pacific Entomologist, 50: 99-110.

DELOYA, C. 1988. Coleópteros lamelicornios asociados a depósitos de detritos de *Atta mexicana* (Smith) (Hymenoptera): Formicidae) in el sur del estado de Morelos, Mexico. Folia Entomologica Mexicana, 75: 77-91.

DOWNIE, N. M. and R. H. ARNETT, Jr. 1996. The Beetles of Northeastern North America, Volumes 1 and 2. The Sandhill Crane Press. Gainesville, FL, 1721 pp.

EDWARDS, J. G. 1949. Coleoptera or Beetles East of the Great Plains. Edwards Brothers. Ann Arbor, MI, 181 pp.

FALL, H. F. 1909. A short synopsis of the species of *Ochodaeus* inhabiting the United States. Journal of the New York Entomological Society, 17: 30-38.

HORN, G. H. 1876. Revision of the United States species of *Ochodaeus* and other genera of Scarabaeidae. Transactions of the American Entomological Society, 5: 177-198.

d'HOTMAN, D. and C. H. SCHOLTZ. 1990. Phylogenetic significance of the structure of the external male genitalia in the Scarabaeoidea (Coleoptera). Republic of South Africa, Department of Agricultural Development, Entomology Memoir, No. 77: 1-51.

HOWDEN, H. F. 1968. Canadian *Ochodaeus*, with a description of a new species (Coleoptera: Scarabaeidae). Canadian Entomologist, 100: 1118-1120.

KIRK, V. M. and E. U. BALSBAUGH, Jr. 1975. A list of the beetles of South Dakota. South Dakota State University, Agricultural Experiment Station, Technical Bulletin, Number 42, 137 pp.

LAWRENCE, J. F. and A. F. NEWTON, Jr. 1995. Families and subfamilies of Coleoptera (with selected genera, notes, and references and data on family-group names), Pp. 779-1006 *In:* J. Pakaluk and S. A. Slipinski, eds. Biology, Phylogeny, and Classification of Coleoptera. Papers Celebrating the 80th Birthday of Roy A. Crowson. Muzeum i Instytut Zoologii PAN. Warsaw, Poland.

MEDVEDEV, S. I. 1960. Descriptions of the larva of eight species of lamellicorn beetles from the Ukraine and Central Asia. Zoologicheskii Zhurnal, 39: 381-393.

NIKOLAYEV, G. V. 1995. New data on the systematics of the subfamily Ochodaeinae (Coleoptera, Scarabaeidae). Zoologicheskii Zhurnal, 74: 72-82.

RATCLIFFE, B. C. 1991. The scarab beetles of Nebraska. Bulletin of the University of Nebraska State Museum, 12: 1-333.

RITCHER, P. O. 1969a. Spiracles of adult Scarabaeoidea (Coleoptera) and their phylogenetic significance. I. The abdominal spiracles. Annals of the Entomological Society of America, 62: 869-880.

RITCHER, P. O. 1969b. Spiracles of adult Scarabaeoidea (Coleoptera) and their phylogenetic significance. II. The thoracic spiracles and adjacent sclerites. Annals of the Entomological Society of America, 62: 1388-1398.

RITCHER P. O. and C. W. BAKER. 1974. Ovariole numbers in Scarabaeoidea (Coleoptera: Lucanidae, Passalidae, Scarabaeidae). Proceedings of the Entomological Society of Washington, 76: 480-494.

SCHOLTZ, C. H. 1990. Phylogenetic trends in the Scarabaeoidea (Coleoptera). Journal of Natural History, 24: 1027-1066.

SCHOLTZ, C. H. and S. L. CHOWN. 1995. The evolution of habitat use and diet in the Scarabaeoidea: a phylogenetic approach. Pp. 355-374. *In:* J. Pakaluk and S. A. Slipinski, eds. Biology, Phylogeny, and Classification of Coleoptera. Papers Celebrating the 80th Birthday of Roy A. Crowson. Muzeum i Instytut Zoologii PAN. Warsaw, Poland.

SCHOLTZ, C. H. and A. V. EVANS. 1987. A revision of the African Ochodaeidae (Coleoptera: Scarabaeoidea). Journal of the Entomological Society of Southern Africa, 50: 399-426.

SCHOLTZ, C. H., D. d'HOTMAN, A. V. EVANS and A. NEL. 1988. Phylogeny and systematics of the Ochodaeidae (Insecta: Coleoptera: Scarabaeoidea). Journal of the Entomological Society of Southern Africa, 51: 207-240.

SMITH, A. B. T. 2001. Checklist of the Scarabaeoidea of the Nearctic Realm (Includes Canada, the continental United States, and the following states of northern Mexico: Baja California, Baja California Sur, Chihuahua, Coahuila de Zaragoza, Durango, Nuevo Leon, Sinaloa, Sonora, Tamaulipas, and Zacatecas). URL: http://www.museum.unl.edu/research/entomology/nearctic.htm.

WOODRUFF, R. E. 1973. The scarab beetles of Florida (Coleoptera: Scarabaeidae). Part I. The Laparosticti (subfamilies Scarabaeinae, Aphodiinae, Hybosorinae, Ochodaeinae, Geotrupinae, Acanthocerinae). Arthropods of Florida and Neighboring Land Areas, 8: 1-220.

31. HYBOSORIDAE Erichson 1847

by Mary Liz Jameson

Family common name: The scavenger scarab beetles

Members of the Hybosoridae are distinguished from other scarabaeoids by their prominent mandibles and labrum and by their 10-segmented antenna with a 3-segmented club in which the basal antennomere of the club is hollowed out to receive the penultimate and ultimate antennomeres (Fig. 2). Other than adults being attracted to lights, little is known about their biology. In the Nearctic region, only two genera occur, one of which was adventive.

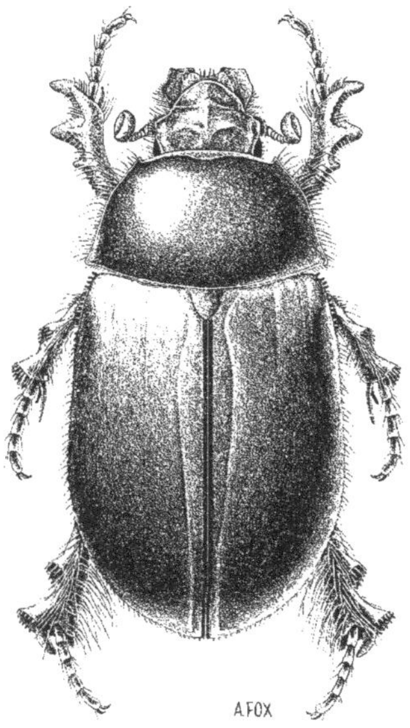

FIGURE 1.31. *Pachyplectrus laevis* (LeConte)

Description. Length 5.0-7.0 mm. Shape oval, dorsal surface convex. Color light brown to black, glossy.

Head not deflexed. Antennae 10-segmented with 3-segmented, opposable club (last 2 antennomeres tomentose), first antennomere of club hollowed to receive club antennomeres 2 and 3 (Fig. 2), basal antennomere expanded apically. Eyes with eucone ommatidia, divided by reduced canthus. Clypeus generally lacking tubercle or horn (*Pachyplectrus laevis* LeConte has a small tubercle on the frontoclypeal suture). Labrum truncate, produced beyond apex of clypeus, prominent. Mandibles produced beyond apex of labrum, prominent, external edge rounded, apex pointed. Maxillae with 4-segmented palpi; labium with 4-segmented palpi.

Pronotum convex, base wider than elytral base. Elytra convex, surface polished, glabrous. Pygidium concealed by elytra. Scutellum exposed, triangular. Legs with anterior coxae conical, contiguous; mesocoxae contiguous; protibiae tridentate on outer margin, apex with one spur; meso- and metatibia with prominent, oblique ridge, apex with 2 spurs; tarsi 5-5-5; claws equal in size, simple; empodium short, not extending beyond fifth tarsomere, with 2 setae.

Abdomen with 6 free sternites (first sternite obscured by hind coxae except at lateral edges); 8 functional abdominal spiracles, situated in pleural membrane (spiracles 1-7) and in tergite (spiracle 8). Wings well developed, with 2 apical, detached veins and M-Cu loop present. Male genitalia variable. References: Cooper 1983, Scholtz 1990.

Larvae are scarabaeiform (C-shaped, cylindrical). Color creamy-white or yellow (except at caudal end which may be darkened by accumulated feces). Cranium heavily sclerotized, brown to dark brown. Antennae 3 or 4-segmented (if 3-segmented, then third and fourth segments fused). Frontoclypeal suture distinct. Labrum at apex with 3 truncate lobes. Epipharynx with row of setae on each chaetoparia, a blunt tooth in the haptomeral region, and united tormae. Maxilla with galea and lacinia distinctly separate; maxillary stridulatory area consisting of a row of conical teeth; maxillary palp 3- or 4-segmented. Abdominal segments 1-6 with 3 annuli, each with 1 or more transverse rows of short setae. Spiracles cribriform, with closing apparatus. Venter of last abdominal segment with raster consisting of 2 curved rows of many short setae that converge caudally. Legs 4-segmented, well developed, with stridulatory apparatus on pro- and mesothoracic legs, each with a well-developed claw. References: Ritcher 1966; Scholtz 1990.

Habits and habitats. Little life history information is known for hybosorids. Adults feed on both invertebrate and vertebrate carrion in the early stages of decomposition; some species are found in dung, and others are attracted to lights at night. Adults of *Hybosorus illigeri* have been reported from turf in golf courses. Adults of *Hybosorus* are known to stridulate. Larvae stridulate by rubbing the front legs against the anterior margin of the epipharynx, a trait unique to the Hybosoridae (Paulian 1939). Larvae have been collected in decomposing plant material.

Status of the classification. The Hybosoridae are considered as a family within the Scarabaeoidea or as a subfamily of the family Scarabaeidae. In this volume, we follow Gardner (1935), Paulian (1939), and Lawrence and Newton (1995) and consider the group a family. Hybosorid larvae are distinct, and this is the primary reason that the group is treated as a family. Larvae possess prothoracic and mesothoracic stridulatory structures and three truncate lobes at the apex of the labrum, both of which are unique to this group. In adults, the form of the antenna (10-segmented with a 3-segmented club, the first club antennomere hollowed to receive club antennomeres two and three) is unique. The family Hybosoridae is hypothesized to be intermediate between the Ochodaeidae and Ceratocanthidae (Scholtz *et al.* 1988) or intermediate between the Trogidae and Ceratocanthidae (Howden and Gill 1988). Phylogenetic analyses of Browne and Scholtz (1995, 1999) hypothesize that the Hybosoridae and Ceratocanthidae are sister taxa. Aside from revisions of some genera and catalogs, little systematics work has been conducted on this group.

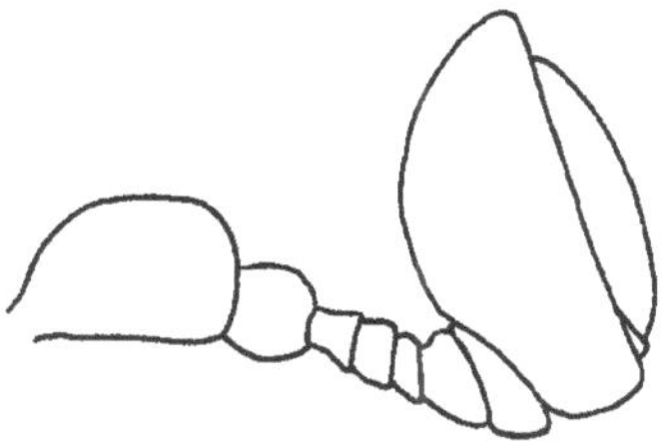

FIGURE 2.31. Right antenna, dorsal view of *Hybosorus illigeri* Reiche.

Distribution. The family Hybosoridae contains 33 genera world-wide and about 210 species (Allsopp 1984; Martínez 1994). Species are widely distributed in the tropics. Two genera and two species occur in the Nearctic region. *Phaeochrous emarginatus* Castelnau, a species that occurs in Asia, Australia, and India, was once reported from California, but the species is not established in California and the record was probably in error (Art Evans, personal communication 1998). Regional works: Hatch 1971; Woodruff 1973; Ratcliffe 1991; Downie and Arnett 1996; Harpootlian 2001.

KEY TO THE GENERA OF HYBOSORIDAE OF THE UNITED STATES, CANADA, AND NEARCTIC MEXICO

1. Mandibles narrow, falciform; frontoclypeal suture without tubercle .... *Hybosorus*
— Mandibles wide, outer edge angular; frontoclypeal suture with tubercle .... *Pachyplectrus*

CLASSIFICATION OF THE NEARCTIC GENERA

Hybosoridae Erichson 1847

Hybosorus MacLeay 1819
Hybosorus illigeri Reiche was inadvertently brought to the United States, apparently arriving before the 1840s from Europe. Its distribution now includes the southern United States, Caribbean, and Mexico. Adults are attracted to lights and have been reported from turf in golf courses. Larvae are not described. Key: Kuijten 1983.

Pachyplectrus LeConte 1874
The genus includes only *P. laevis* (LeConte) (Fig. 1). Specimens are rare and are distributed in Arizona and southern California. Adults have been taken at lights, on the crests of sand dunes, and in the burrows of kangaroo rats (Hardy 1977). Larvae remain unknown. Reference: Woodruff 1973.

BIBLIOGRAPHY

ALLSOPP, P. G. 1984. Checklist of the Hybosorinae (Coleoptera: Scarabaeidae). Coleopterists Bulletin, 38: 105-117.

BROWNE, D. J. and C. H. SCHOLTZ. 1995. Phylogeny of the families of Scarabaeoidea (Coleoptera) based on characters of the hindwing articulation, hindwing base and wing venation. Systematic Entomology, 20: 145-173.

BROWNE, J. and C. H. SCHOLTZ. 1999. A phylogeny of the families of Scarabaeoidea (Coleoptera). Systematic Entomology, 24: 51-84.

COOPER, J. B. 1983. A review of the Nearctic genera of the family Scarabaeidae (exclusive of the subfamilies Scarabaeinae and Geotrupinae) (Coleoptera), with an evaluation of computer generated keys. Doctoral Thesis, Department of Biology, Carleton University. Ottawa, Ontario, Canada, 1121 pp.

DOWNIE, N. M. and R. H. ARNETT, Jr. 1996. The Beetles of Northeastern North America, Volumes 1 and 2. The Sandhill Crane Press. Gainesville, FL, 1721 pp.

GARDNER, J. C. M. 1935. Immature stages of Indian Coleoptera (16) (Scarabaeidae). Indian Forest Records Entomology (New Series), 1: 1-33.

HARDY, A.R. 1977. Observations on some rare Scarabaeidae mainly from California. Coleopterists Bulletin, 31: 91-92.

HARPOOTLIAN, P. J. 2001. Scarab beetles (Coleoptera: Scarabaeidae) of South Carolina. Biota of South Carolina, Volume 2. Clemson University. Clemson, SC, 157 pp.

HATCH, M. H. 1971. The beetles of the Pacific Northwest, part 5. University of Washington Publication in Biology, 16: 1-662.

HOWDEN, H. F. and B. D. Gill. 1988. *Xenocanthus*, a new genus of inquiline Scarabaeidae from southeastern Venezuela (Coleoptera). Canadian Journal of Zoology, 66: 2071-2076.

KUIJTEN, P. J. 1983. Revision of the genus *Hybosorus* MacLeay (Coleoptera: Scarabaeidae, Hybosorinae). Zoologische Verhandelingen, 203: 1-49.

LAWRENCE, J. F. and A. F. NEWTON, Jr. 1995. Families and subfamilies of Coleoptera (with selected genera, notes, and references and data on family-group names). Pp. 779-1006. *In:* J. Pakaluk and S. A. Slipinski, eds. Biology, Phylogeny, and Classification of Coleoptera. Papers Celebrating the 80th Birthday of Roy A. Crowson. Muzeum i Instytut Zoologii PAN. Warsaw, Poland.

MARTÍNEZ, A. 1994. Notas sobre Hybosorinae (Coleoptera: Scarabaeidae), II. Elytron, 8: 223-239.

PAULIAN, R. 1939. Les caractères larvaires des Geotrupidae (Col.) et leur importance pour la position systématique du groupe. Bulletin de la Société Zoologique de France, 64: 351-360.

RATCLIFFE, B. C. 1991. The scarab beetles of Nebraska. Bulletin of the University of Nebraska State Museum, 12:1-333.

RITCHER, P. O. 1966. White Grubs and Their Allies: A Study of North American Scarabaeoid Larvae. Oregon State University Press. Corvallis, OR, 219 pp.

SCHOLTZ, C. H. 1990. Phylogenetic trends in the Scarabaeoidea (Coleoptera). Journal of Natural History, 24: 1027-1066.

SCHOLTZ, C. H., D. d'HOTMAN, A. V. EVANS and A. NEL. 1988. Phylogeny and systematics of the Ochodaeidae (Insecta: Coleoptera: Scarabaeidae). Journal of the Entomological Society of Southern Africa, 51: 207-240.

WOODRUFF, R. E. 1973. The scarab beetles of Florida (Coleoptera: Scarabaeidae). Part I. The Laparosticti (subfamilies Scarabaeinae, Aphodiinae, Hybosorinae, Ochodaeinae, Geotrupinae, Acanthocerinae). Arthropods of Florida and Neighboring Land Areas, 8: 1-220.

32. CERATOCANTHIDAE Martínez 1968

by Mary Liz Jameson

Family common name: The pill scarab beetles

Family synonym: Acanthoceridae Lacordaire 1856

Members of the Ceratocanthidae are distinguished from other scarabaeoids by the ability of the adult to form a nearly compact sphere. When disturbed, adults deflect the head, pronotum, and legs, thus forming a tight ball.

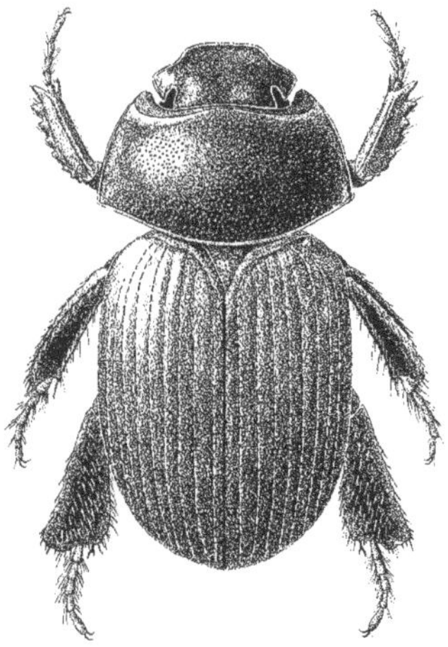

FIGURE 1.32. *Germarostes aphodioides* (Illiger) (Used by permission of University of Nebraska State Museum)

Description. Length 2.0-9.0 mm. Shape nearly spherical when head and pronotum deflexed. Color black, greenish black, or purplish, often with metallic luster.

Head deflexed. Antenna 9 or 10-segmented with 3-segmented, opposable club (all antennomeres tomentose); basal antennomere triangular and enlarged. Eyes partially divided by canthus, with eucone ommatidia. Clypeus lacking tubercle or horn. Labrum truncate, partially exposed beyond apex of clypeus. Mandibles partially exposed beyond apex of clypeus. Maxillae with 4-segmented palpi. Labium with 4-segmented palpi.

Pronotum broad, compressed laterally. Elytra convex, surface highly polished and glabrous. Pygidium concealed by elytra. Scutellum exposed, triangular. Legs with anterior coxae conical, prominent; mesocoxae transverse; tibiae (especially meso- and metatibae) horizontally flattened and broad (concealing sternites when contracted in spherical form), external surface striated; protibia with outer margin serrately toothed, apex with one spur; meso- and metatibia with 2 apical spurs; tarsi 5-5-5; claws equal in size, simple; empodium absent.

Abdomen with 5 free sternites; 8 functional abdominal spiracles situated in pleural membrane (spiracles 1-7) and in tergite (spiracle 8). Wings well developed, M-Cu loop reduced or absent, with 1 apical detached vein. Male genitalia variable. References: Cooper 1983; Scholtz 1990.

Larvae are scarabaeiform (C-shaped, cylindrical). Color creamy-white or yellow (except at caudal end which may be darkened by accumulated feces). Cranium heavily sclerotized, yellow-brown to dark brown. Antenna 4-segmented. Frontoclypeal suture distinct. Labrum with apical margin serrate, palpi 1-2 segmented. Epipharynx with dextral, beak-like process. Maxilla with galea and lacinia separate; maxillary stridulatory area with a row of conical teeth; maxillary palp 4-segmented. Abdominal segments 1-6 with 3 annuli, each with one or more transverse rows of short setae. Spiracles cribriform. Venter of last abdominal segment with transverse palidium of spatulate setae. Legs 4-segmented, well developed, with stridulatory apparatus on all legs or on meso- and metathoracic legs, each with a well-developed claw. References: Ritcher 1966; Scholtz 1990.

Habits and habitats. Adult ceratocanthids can be collected on the bark and branches of dead trees and vines, on fungi, in the burrows of passalid beetles, and occasionally at lights. Adults have also been found in association with termites and ants. When disturbed, these beetles are able to deflex their head and pronotum, thus concealing the entire ventral side. When contracted in this manner, they resemble spherical seeds. This behavior probably allows them to evade potential predators. This trait occurs in a lesser degree in some Hybosoridae. Adults probably feed on fungi (Nel and Scholtz 1990) or on rotting wood (Ohaus 1909). Larvae have been collected under bark (Ritcher 1966), reared from frass in passalid burrows (*Germarostes*, Woodruff 1973), and reared from wet tree holes (*Ceratocanthus*, Choate 1987). Adults and larvae of at least some species stridulate.

Status of the classification. The Ceratocanthidae are considered a family within the Scarabaeoidea or a subfamily of the family Scarabaeidae. In this volume, we follow Lawrence and Newton (1995) and consider the group a family. The group was previously referred to as the Acanthoceridae (Lacordaire 1856), a junior homonym that required replacement. The family name Ceratocanthidae has been erroneously attributed to Cartwright and Gordon (1971). Martínez (1968) first used the name at the family group level and should be credited as the author.

Based on phylogenetic analyses and character data, the family Ceratocanthidae is hypothesized to be the sister group to the Hybosoridae (Browne and Scholtz 1995, 1999; Lawrence and Britton 1994). However, Cooper (1983) postulated that the Ceratocanthidae are most closely related to the Trogidae. Aside from the work of Paulian (1982) for the South American Ceratocanthidae, little systematics work has been conducted on genera in the group. Howden and Gill (2000) provided a key to the New World genera, thus creating an excellent foundation for future studies.

Distribution. The family is widely distributed in the tropics. No ceratocanthids are known from Europe, and only three are known from Australia. In North and South America, the

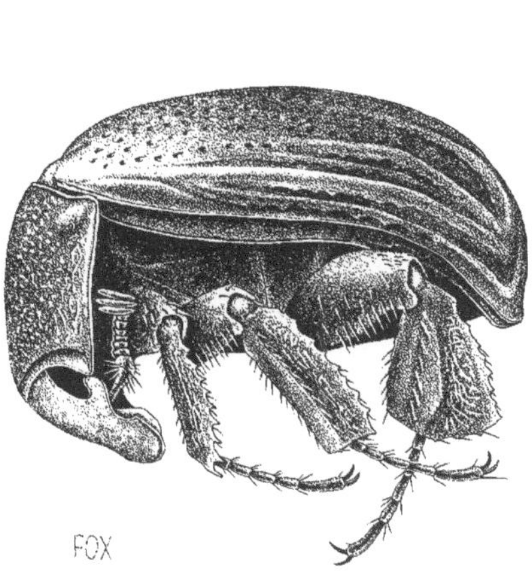

FIGURE 2.32. *Germarostes aphodioides* (Illiger) lateral view.

group includes 11 genera and about 155 species (Howden and Gill 2000). Two genera and three species occur in the Nearctic region. North American catalog: Smith 2001. Regional works: Woodruff 1973; Ratcliffe 1991; Downie and Arnett 1996; Harpootlian 2001.

Key to the Genera of the United States, Canada, and Nearctic Mexico

1. Middle and hind tibiae thickened at apical edge; posterior angles of pronotum nearly right-angled .... *Germarostes*

— Middle and hind tibiae flat, blade-like; posterior angles of pronotum rounded ...... *Ceratocanthus*

Classification of the Nearctic Genera

Ceratocanthidae Martínez 1968

Ceratocanthus White 1842

Acanthocerus MacLeay 1819

Sphaeromorphus Germar 1843

One species, *C. aeneus* (MacLeay), occurs in the southeastern United States. Specimens have been collected by beating dead vegetation; they are rare in collections (Howden and Gill 2000). Choate (1987) discussed the biology of this species.

Germarostes Paulian 1982

Two species of *Germarostes*, *G. aphodioides* (Illiger) (Figs. 1-2) and *G. globosus* (Say), occur in the Nearctic region. Previously, these species were placed in the genus *Cloeotus* but were removed by Paulian (1982). The two species occur from southern Ontario to Indiana, south to Florida and Texas, and west to Nebraska. Specimens can be collected by beating dead vines, from fungi and carrion, at lights at night, and in flight intercept traps. Larvae have been collected from the frass of passalid burrows in logs. Key: Woodruff 1973.

Bibliography

BROWNE, D. J. and C. H. SCHOLTZ. 1995. Phylogeny of the families of Scarabaeoidea (Coleoptera) based on characters of the hindwing articulation, hindwing base and wing venation. Systematic Entomology, 20: 145-173.

BROWNE, J. and C. H. SCHOLTZ. 1999. A phylogeny of the families of Scarabaeoidea (Coleoptera). Systematic Entomology, 24: 51-84.

CARTWRIGHT, O. L. and R. D. GORDON. 1971. Coleoptera: Scarabaeidae. Insects of Micronesia, 17: 257-296.

CHOATE, P. M. 1987. Biology of *Ceratocanthus aeneus* (Coleoptera: Scarabaeidae: Ceratocanthinae). Florida Entomologist, 70: 301-305.

COOPER, J. B. 1983. A review of the Nearctic genera of the family Scarabaeidae (exclusive of the subfamilies Scarabaeinae and Geotrupinae) (Coleoptera), with an evaluation of computer generated keys. Doctoral Thesis, Department of Biology, Carleton University. Ottawa, Ontario, Canada, 1121 pp.

DOWNIE, N. M. and R. H. ARNETT, Jr. 1996. The Beetles of Northeastern North America, Vol. 1. The Sandhill Crane Press. Gainesville, FL, 880 pp.

HARPOOTLIAN, P. J. 2001. Scarab beetles (Coleoptera: Scarabaeidae) of South Carolina. Biota of South Carolina, Volume 2. Clemson University, Clemson, SC, 157 pp.

HOWDEN, H. and B. D. GILL. 2000. Tribes of New World Ceratocanthinae, with keys to genera and descriptions of new species (Coleoptera: Scarabaeidae). Sociobiology, 35: 281-329.

LACORDAIRE, J. H. 1856. Histoire Naturelle des Insectes. Genera des coléoptères ou exposé méthodique et critique de tous les genres proposés jusqu'ici dans cet ordre d'insectes. Volume 3. Librairie Encyclopédique de Roret, Paris, 594 pp.

LAWRENCE, J. F. and E. B. BRITTON. 1994. Australian Beetles. Melbourne University Press, 192 pp.

LAWRENCE, J. F. and A. F. NEWTON, Jr. 1995. Families and subfamilies of Coleoptera (with selected genera, notes, and references and data on family-group names). Pp. 779-1006. *In:* J. Pakaluk and S. A. Slipinski, eds. Biology, Phylogeny, and Classification of Coleoptera. Papers Celebrating the 80th Birthday of Roy A. Crowson. Muzeum i Instytut Zoologii PAN. Warsaw, Poland.

MARTÍNEZ, A. 1968. Insectos nuevos o poco conocidos XIII. Ceratocanthini nom. nov. para Acanthocerini (Coleoptera, Scarabaeidae, Troginae). Revista de la Sociedad Entomológica Argentina, 30: 9-16.

NEL, A. and C. H. SCHOLTZ. 1990. Comparative morphology of the mouthparts of adult Scarabaeoidea (Coleoptera). Entomology Memoires of the Republic of South Africa Department of Agricultural Development, 80: 1-84.

OHAUS, F. 1909. Beiträge zur Kenntnis unserer einheimischen Rosskäfer. Deutsche Entomologische Zeitschrift, 109: 105-111.

PAULIAN, R. 1982. Révision des Cératocanthides (Coleoptera Scarabaeoidea) d'Amérique du Sud. Mémoires du Muséum National d'Histoire Naturelle, Série A, Zoologie, 124: 1-110.

RATCLIFFE, B. C. 1991. The scarab beetles of Nebraska. Bulletin of the University of Nebraska State Museum, 12:1-333.

RITCHER, P. O. 1966. White Grubs and Their Allies: A Study of North American Scarabaeoid Larvae. Oregon State University Press. Corvallis, OR, 219 pp.

SCHOLTZ, C. H. 1990. Phylogenetic trends in the Scarabaeoidea (Coleoptera). Journal of Natural History, 24: 1027-1066.

SMITH, A. B. T. 2001. Checklist of the Scarabaeoidea of the Nearctic Realm (Includes Canada, the continental United States, and the following states of northern Mexico: Baja California, Baja California Sur, Chihuahua, Coahuila de Zaragoza, Durango, Nuevo Leon, Sinaloa, Sonora, Tamaulipas, and Zacatecas). URL: http://www-museum.unl.edu/research/entomology/nearctic.htm.

WOODRUFF, R. E. 1973. The scarab beetles of Florida (Coleoptera: Scarabaeidae). Part I. The Laparosticti (subfamilies Scarabaeinae, Aphodiinae, Hybosorinae, Ochodaeinae, Geotrupinae, Acanthocerinae). Arthropods of Florida and Neighboring Land Areas, 8: 1-220.

33. GLAPHYRIDAE MacLeay 1819

by David C. Carlson

Family common name: The bumble bee scarab beetles

Glaphyrid beetles are active fliers during the day. Adults of many species are brightly colored and hairy and often possess markings and coloration resembling bees and bumble bees. They are strong fliers and are often observed hovering near flowers or foliage or flying over sandy areas.

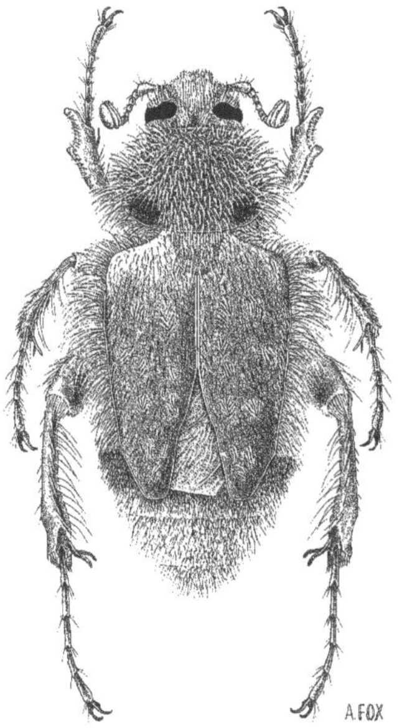

FIGURE 1.33 *Lichnanthe rathvoni* LeConte

Description. Length 6.0-20.0 mm. Shape elongate. Color testaceous to black, often with metallic reflections; setae dense, moderately long, color variable (white, yellow, orange, red, brown, or black).

Head deflexed. Antennae 9 or 10-segmented with 3-segmented, opposable club (all antennomeres tomentose). Eyes with eucone ommatidia, completely or partially divided by a canthus. Clypeus in most simple, anterior margin with or without teeth. Labrum emarginate, truncate or rounded, produced beyond apex of clypeus, prominent. Mandibles produced beyond apex of labrum, prominent. Maxillae filiform, plumose or truncate, with 4 or 5-segmented palpi. Labium with 4-segmented palpi.

Pronotum convex, most subquadrate, often densely punctate and setose, without tubercles, ridges, horns, or sulci. Elytra elongate, often thin and dehiscent at apex, without striae, often setose. Pygidium visible beyond elytra in most. Scutellum exposed, U-shaped or triangular. Legs with procoxae conical or transverse, meso- and metacoxae transverse; mesocoxae separated or contiguous; protibiae dentate on outer margin, apex with one spur; meso- and metatibia generally simple but some with apical modifications (spines or emarginations), apex with 2 spurs; tarsi 5-5-5, foretarsi modified medially (pectinate) in some Old World genera; claws equal in size with 1 tooth; empodium exposed beyond fifth tarsomere, dorso-ventrally flattened, with 2 setae.

Abdomen with 6 free sternites, most with 8 pairs of spiracles; spiracles 1-6 or 1-7 situated in pleural membrane, spiracles 7 or 8 in tergites (Glayphyrinae and some Lichniinae) or spiracles 1-8 in pleural membrane (some Lichniinae). Wings well developed, M-Cu loop present with 1 apical, detached vein. Male genitalia with well sclerotized, strongly arched basal piece, basal piece large relative to parameres; internal sac variable. Ovary with 6 ovarioles. References: Chapin 1938; Ritcher 1969; Ritcher and Baker 1974; Carlson 1980; d'Hotman and Scholtz 1990; Scholtz 1990; Browne and Scholtz 1995.

Larvae are scarabaeiform (C-shaped, cylindrical). Color bluish-white to yellow (prepupae) (except at caudal end which may be darkened by accumulated feces). Head capsule heavily sclerotized, reddish-brown (*Lichnanthe* with conspicuous median, circular depression on frons). Antennae 4-segmented, third segment with small sensory pits. Stemmata present or absent. Frontoclypeal suture present. Labrum trilobed. Epipharynx with asymmetrical tormae not fused. Maxilla with galea and lacinia separate; maxillary palpi 4-segmented; labial palpi 2-segmented; maxillary and mandibular stridulatory areas present. Abdominal segments 1 to 8 with 3 dorsal annuli. Spiracles cribriform. Anal slit transverse, located caudally on dorsum of last abdominal segment. Legs well developed, 4-segmented, lacking stridulatory organs; claws present. References: Ritcher 1966; Scholtz 1990.

Habits and habitats. Except for a few species, life histories of the glaphyrids are poorly documented. Adults are often brightly colored, densely setose, active diurnally, and strong fliers. Many species have colored setal bands on the abdomen and resemble various Hymenoptera (bumble bees and metallic bees). They have been observed frequenting flowers and foliage. Larvae are free living in sandy areas (riparian and coastal dunes) where they feed on decaying leaf litter and detritus that is layered in the sand. Larvae of *Lichnanthe vulpina* (Hentz) may be a pest of cranberry bogs in the northeastern United States. References: Ritcher 1966; Westcott 1976; Carlson 1977, 1980.

Status of the classification. The uniqueness of the glaphyrids has been recognized for a very long time, and the genera included in the group have changed little. However, the status of the group has been the subject of debate. Superfamily status was proposed by Machatschke (1959) but was not generally accepted. Workers have vacillated between using subfamily or family status, and familial status is now generally accepted (d'Hotman and Scholtz 1990; Scholtz 1990; Browne and Scholtz 1995).

The phylogenetic position of the Glaphyridae within the Scarabaeoidea has been discussed by numerous authors and is currently considered by most to be among the intermediate scarabaeoid families (d'Hotman and Scholtz 1990; Scholtz 1990; Browne and Scholtz 1995). Browne and Scholtz (1995) consider the Glaphyridae to be a monophyletic sister group of the trogid subgroup (Trogidae, Bolboceratinae [Geotrupidae] and Pleocomidae) based on characters of wing articulation.

Taxonomy of the world Glaphyridae is not well-established. Comprehensive taxonomic treatments are available for *Lichnanthe* (Carlson 1980), *Anthypna* (Endrödi 1952), and *Pygopleurus* (Petrovitz 1958). Most other genera have not been reviewed comprehensively. The taxonomy and nomenclature of the group were discussed by Chapin (1938) and Machatschke (1959). The extreme color polymorphism exhibited by many species has resulted in a proliferation of form, variety, or color morph names for some species, many of which are synonyms. The most recent world catalog for the family was Arrow (1912).

Distribution. The family Glaphyridae includes eight genera and about 80 species worldwide (Arrow 1912; Chapin 1938; Yawata 1942). The subfamily Lichniinae (including three genera) is limited to western South America, and the subfamily Glaphyrinae (including five genera) is widely distributed in the Holarctic region. One genus and eight species of glaphyrids occur in the United States, Canada, and Nearctic Mexico. Key to genera and subgenera: Chapin 1938. Key to U.S. species: Carlson 1980. Regional works: Edwards 1949; Hatch 1971; Downie and Arnett 1996. North American catalog: Smith 2001. Larvae: Ritcher 1966.

Classification of the Nearctic Genera

Glaphyridae MacLeay 1819

Glaphyrinae MacLeay 1819

Characteristics: Antennae 10-segmented, maxillary palpi truncate. This subfamily consists of five genera with about 68 species and is widely distributed in the Holarctic region. North American species belong to a single genus, *Lichnanthe* Burmeister. Adults are active diurnal fliers and frequent foliage and flowers. Adults of many species resemble various Hymenoptera (bumble bees and metallic bees). References: Chapin 1938; Westcott 1976; Carlson 1977, 1980. Larvae: Ritcher 1966.

Lichnanthe Burmeister 1844

This genus (Fig. 1) includes eight extant species and one fossil species that are Nearctic in distribution and primarily restricted to the continental United States. Species are distributed in the east coast states and western states. Key: Carlson 1980. (Volume 2, Color Figure 22)

Bibliography

ARROW, G. J. 1912. Scarabaeidae: Pachypodinae, Pleocominae, Aclopinae, Glaphyrinae, Ochodaeinae, Orphninae, Idiostominae, Hybosorinae, Dynamopinae, Acanthocerinae, Troginae. Coleopterorum Catalogus, 19: 1-66.

BROWNE, D. J. and C. H. SCHOLTZ. 1995. Phylogeny of the families of the Scarabaeoidea (Coleoptera) based on characters of the hindwing articulation, hindwing base and wing venation. Systematic Entomology, 21: 145-173.

CARLSON, D. C. 1977. Taxonomic revision of *Lichnanthe* Burmeister with studies on the biology of *L. rathvoni* (LeConte) (Coleoptera: Scarabaeidae). Ph.D. Thesis, Oregon State University. Corvallis, OR.

CARLSON, D. C. 1980. Taxonomic revision of *Lichnanthe* Burmeister (Coleoptera: Scarabaeidae). Coleopterists Bulletin, 34: 177-208.

CHAPIN, E. A. 1938. The nomenclature and taxonomy of the genera of the scarabaeid subfamily Glaphyrinae. Proceedings of the Biological Society of Washington, 51: 79-86.

DOWNIE, N. M. and R. H. ARNETT, Jr. 1996. The Beetles of Northeastern North America, Volumes 1 and 2. The Sandhill Crane Press. Gainesville, FL, 1721 pp.

EDWARDS, J. G. 1949. Coleoptera or Beetles East of the Great Plains. Edwards Brothers. Ann Arbor, MI, 181 pp.

ENDRÖDI, S. 1952. Monographie der Gattung *Anthypna* Latr. Folia Entomologica Hungarica (New Series), 5: 1-40.

HATCH, M. H. 1971. The beetles of the Pacific Northwest, part 5. University of Washington Publication in Biology, 16: 1-662.

d'HOTMAN, D. and C. H. SCHOLTZ. 1990. Comparative morphology of the male genitalia of derived groups of Scarabaeoidea (Coleoptera). Elytron, 4: 3-39.

MACHATSCHKE, J. W. 1959. Untersuchungen über die verwantschaftlichen Beziehungen der Gattungen der bisherigen Glaphyrinae (Coleoptera: Lamellicornia). Beiträge zur Entomologie, 9: 528-545.

PETROVITZ, R. 1958. Das Subgenus *Pygopleurus* Motschulsky der Gattung *Amphicoma* Latreille (Col., Scarabaeidae, Glaphyrinae). Entomologisk Tidskrift, 78: 38-63.

RITCHER, P. O. 1966. White Grubs and Their Allies: A Study of North American Scarabaeoid Larvae. Oregon State University Press. Corvallis, OR, 219 pp.

RITCHER, P. O. 1969. Spiracles of adult Scarabaeoidea (Coleoptera) and their phylogenetic significance. I. The abdominal spiracles. Annals of the Entomological Society of America, 62: 869-880.

RITCHER P. O. and C. W. BAKER. 1974. Ovariole numbers in Scarabaeoidea (Coleoptera: Lucanidae, Passalidae, Scarabaeidae). Proceedings of the Entomological Society of Washington, 76: 480-494.

SCHOLTZ, C. H. 1990. Phylogenetic trends in the Scarabaeoidea (Coleoptera). Journal of Natural History, 24: 1,027-1,066.

SMITH, A. B. T. 2001. Checklist of the Scarabaeoidea of the Nearctic Realm (Includes Canada, the continental United States, and the following states of northern Mexico: Baja California, Baja California Sur, Chihuahua, Coahuila de Zaragoza, Durango, Nuevo Leon, Sinaloa, Sonora, Tamaulipas, and Zacatecas). URL: http://www-museum.unl.edu/research/entomology/nearctic.htm.

WESTCOTT, R. L. 1976. Observations on the biology and ethology of *Lichnanthe rathvoni* LeConte (Coleoptera: Scarabaeidae) with emphasis on mating. University of Idaho, Department of Entomology Anniversary Publication, No. 11: 85-90.

YAWATA, H. 1942. Notes on the Glaphyrinae of Japan with description of a new genus and two new species. Transactions of the Kansai Entomological Society, 12: 33-37.

34. SCARABAEIDAE Latreille 1802

by Brett C. Ratcliffe, Mary Liz Jameson and Andrew B. T. Smith

Family common name: The scarab beetles

Scarab beetles comprise a speciose group, and they are a conspicuous component of the beetle fauna in the Nearctic region. Adults of many scarab beetles are noticeable due to their relatively large size, bright colors, often elaborate ornamentation, and interesting life histories. The family includes the goliath beetle from Africa [*Goliathus goliathus* (L.)], known as one of heaviest insects (up to 100 grams). It also includes the elephant beetle [*Megasoma elephas* (Fabricius)] and hercules beetle [*Dynastes hercules* (L.)], both from the American tropics, that are known for their large size (up to 160 mm for the hercules beetle) and highly developed horns in the males. The group includes over 27,800 species, an intriguing array of life histories, and many interesting adaptations.

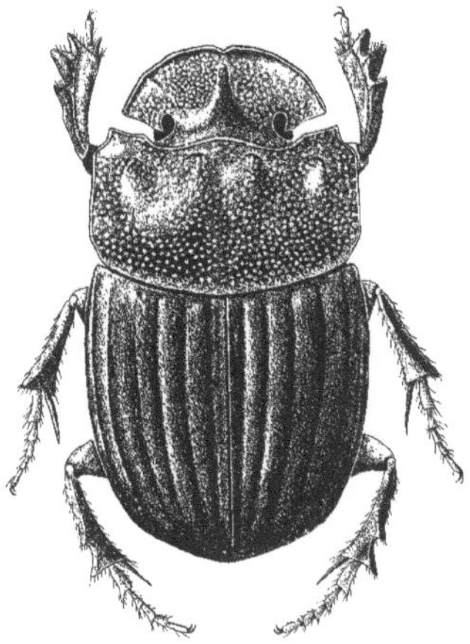

FIGURE 1.34. *Copris fricator* (Fabricius) (Used by permission of University of Nebraska State Museum)

Description. Length 2.0-60.0 mm (Nearctic species). Shape variable; ovate, obovate, quadrate, cylindrical. Color variable, with or without metallic reflections or metallic coloration; with or without vestiture.

Head weakly deflexed or not deflexed. Antennae 10-segmented (some 7-12 segmented) with 3 to 5-segmented, opposable club; club with apical antennomeres nearly glabrous [Melolonthinae, Dynastinae, Rutelinae, Cetoniinae] or with all antennomeres tomentose [Aphodiinae, Scarabaeinae]. Eyes with eucone ommatidia, partially divided by a canthus. Clypeus with or without tubercle or horn. Labrum distinct in most, produced beyond apex of clypeus or not. Mandibles variable, produced beyond apex of labrum or not. Maxillae with 4-segmented palpi. Labium with 3-segmented palpi.

Pronotum variable, with or without horns or tubercles. Elytra convex or flattened, with or without striae. Pygidium concealed by elytra [Aphodiinae, Scarabaeinae] or exposed [Scarabaeinae, Melolonthinae, Dynastinae, Rutelinae, Cetoniinae]. Scutellum exposed or not; shape triangular or parabolic. Legs with coxae transverse or conical; protibiae tridentate, bidentate, or serrate on outer margin, apex with one spur; meso- and metatibia slender or robust, apex with 1 or 2 spurs; tarsi 5-5-5, anterior tarsi absent in some Scarabaeinae; claws variable, equal in size or not, simple or toothed; empodium present, extending beyond fifth tarsomere, with 2-5 setae or with setae absent.

Abdomen with 6 free sternites; 7 functional abdominal spiracles situated in pleural membrane [Aphodiinae, Scarabaeinae] or in pleural membrane, in sternites and in tergite [Melolonthinae, Dynastinae, Rutelinae, Cetoniinae]. Wings well developed, M-Cu loop and 1 apical, detached vein present. Male genitalia variable, bilobed, or fused. References: Cooper 1983; Scholtz 1990.

Larvae are scarabaeiform (C-shaped, cylindrical), some hump-backed [Scarabaeinae]. Color creamy-white or yellow (except at caudal end which may be darkened by accumulated feces). Cranium heavily sclerotized, testaceous to brown to black. Antennae 4-segmented, last segment bearing 1 or more sensory spots. Ocelli absent, present [some Dynastinae, Cetoniinae], or with distinct pigmented spots. Frontoclypeal suture present. Labrum at apex variable, rounded or lobed. Epipharynx rounded or lobed, asymmetrical. Maxilla with galea and lacinia distinctly separate [Aphodiinae, Scarabaeinae], fused proximally and free distally [Melolonthinae], or fused to form mala [Dynastinae, Rutelinae, Cetoniinae]; maxillary stridulatory area present in most; maxillary palpus 4-segmented. Abdomen with segments 9 and 10 fused dorsally [Cetoniinae], segments 1-6 or 7 mostly with 3 annuli, each with 1 or more transverse rows of setae. Spiracles cribriform. Venter of last abdominal segment with or without fleshy lobes, apex with or without palidia, anal opening with transverse or Y-shaped slit. Legs 2-segmented [Scarabaeinae] or 4-segmented [Aphodiinae, Melolonthinae, Dynastinae, Rutelinae, Cetoniinae], lacking stridulatory process; claws present or absent with 1-2 setae. References: Ritcher 1966; Scholtz 1990.

Habits and habitats. Life histories of scarab beetles are incredibly diverse and include adults that feed on dung, carrion, fungi, vegetation, pollen, fruits, compost, or roots. Some scarab beetles live in the nests of ants (myrmecophiles), in the nests of termites (termitophiles), or in the nests of rodents or birds. Some species of dung beetles (Scarabaeinae) care for their larvae or the larval brood ball (*e.g.*, species of *Canthon* and *Copris*). Adults of some scarab beetles are diurnal and can be observed on flowers or vegetation (*e.g., Euphoria, Cotinis* [Cetoniinae]) while many species are nocturnal and attracted to lights at night (*e.g., Dynastes* [Dynastinae], *Chrysina* [Rutelinae], *Polyphylla* [Melolonthinae]). Adults and larvae of a few species of scarabs are economically important and may cause considerable damage due to defoliation or root-feeding (*e.g., Popillia japonica* Newman [Rutelinae]). Many scarabs are beneficial because they pollinate plants, recycle plant material, and are valuable dung recyclers. In 1968, for example, several species of dung beetles were introduced into Australia to aid in the decomposition of cattle feces (Waterhouse 1974). Domesticated cattle produce feces different from native Australian herbivores. Native Australian dung beetles were not adapted to feed on the dung of domesticated cattle. As a result, cattle feces accumulated at such a rate that grasses and forbes were smothered and killed. Dung beetles were introduced from Africa, and these beetles readily fed on cattle dung, enriched the soil with the dung, and allowed the grasses and forbes to thrive. The dung beetles also reduce breeding sites for pest flies.

"Dung beetle" is a common name applied to beetles in the subfamilies Scarabaeinae and Aphodiinae. Species in these groups often have specific ecological requirements. For example, *Dialytes* spp. and *Aphotaenius carolinus* (Van Dyke) (both Aphodiinae) are specialists on deer dung. Some dung beetles, the so-called "tumble bugs" (*e.g., Canthon pilularius* (L.) [Scarabaeinae]), form a ball of dung and roll it away from potential competition at a dung pat. Other dung beetles make a ball of dung under the dung pile (*e.g., Copris fricator* (Fabricius), *Phanaeus vindex* (MacLeay), *Onthophagus hecate* (Panzer), *Onthophagus subaeneus* (Palisot de Beauvois) [all Scarabaeinae]). Scarab beetles that live in the nests of vertebrates include: *Onthophagus polyphemi* Hubbard (Scarabaeinae) that lives in the nest of gopher tortoises in Florida and South Carolina; *Ataenius sciurus* Cartwright (Aphodiinae) that lives in the nest of tree squirrels in Florida, and many species of *Aphodius* (Aphodiinae) that live in the nests of prairie dogs and pocket gophers. Although most dung beetles feed on dung, some, such as *Onthophagus striatulus* (Palisot de Beauvois) (Scarabaeinae), defy their common name and feed on fungi.

Some scarabs, such as *Euparia castanea* LePeletier and Serville (Aphodiinae) and *Cremastocheilus* spp. (Cetoniinae) are inquilines in the nests of ants. Species in the genus *Valgus* (Cetoniinae: Valgini) are inquilines in the nests of termites. The life history of these beetles, their adaptations, and their body form is intimately intertwined with their hosts.

Most species in the subfamilies Melolonthinae, Dynastinae, Rutelinae, and Cetoniinae feed on plant products. Larvae of many dynastines and rutelines feed on rotting wood. Larvae of many melolonthines (*e.g., Phyllophaga* species), rutelines (*e.g., Anomala* species), and dynastines (*e.g., Cyclocephala* species) feed on grass roots. Some of these larvae may be lawn pests (*e.g., Popillia japonica* Newman [Rutelinae], *Cyclocephala borealis* Arrow [Dynastinae], *Amphimallon majalis* Razoumowski, *Plectris aliena* Chapin, and *Phyllophaga* species [all Melolonthinae]). As adults, most species in these subfamilies feed on leaves or fruits. Adults of *Phyllophaga falsa* (LeConte) (Melolonthinae) occasionally defoliate pine trees. Some adults in these subfamilies are also attracted to sap flows.

The larvae of most scarab beetles develop similarly. Eggs are deposited by the adult female in suitable soil, dung, compost, or other organic material. After hatching, the C-shaped grubs feed and grow, molting twice. In areas with cold winters, larvae overwinter below the frost line. As temperatures rise in the spring, larvae become active and feed until pupation. Emergence of the adults from pupal cells often occurs in response to environmental cues such as rainfall or temperature. After emergence, adults mate and begin the cycle anew. The biology and behavior of many species of scarabs are not known, and much remains to be studied. References: Woodruff 1973; Ratcliffe 1991.

Status of the classification. The past thirty years have seen many changes and debates in the classification of the Scarabaeidae. In the "traditional" North American system, the category Scarabaeidae has been treated as including the all scarabaeoid families except the Passalidae and Lucanidae. Old World scarab workers have tended to split the Scarabaeidae into several families. While the debate continues, we follow Lawrence and Newton (1995) and consider the family Scarabaeidae to include the subfamilies Aphodiinae, Scarabaeinae, Melolonthinae, Dynastinae, Rutelinae, and Cetoniinae. Several smaller subfamilies that are not present in the Nearctic region are also included in the Scarabaeidae: Orphninae, Phaenomeridinae, Pachypodinae, Allidiostomatinae, Dynamopodinae, Aclopinae, and Euchirinae. No phylogenetic analyses have addressed the relationships of all of these taxa. However, most hypotheses generally consider the Aphodiinae and Scarabaeinae as the sister group to the Melolonthinae, Dynastinae, Rutelinae, and Cetoniinae. The former Trichiinae and Valginae are here considered tribes of the Cetoniinae.

The family Scarabaeidae is sometimes referred to as the family Melolonthidae, especially by some of the Latin American workers. In this usage, the family includes the subfamilies Melolonthinae, Euchirinae, Phaenomeridinae, Dynastinae, Cetoniinae, Glaphyrinae, and Systellopodinae (Endrödi 1966) whereas the Scarabaeidae refers to everything else except Passalidae, Lucanidae, and Trogidae. This classification is not in wide use today and is incorrect. The family group names Rutelinae and Dynastinae were established by MacLeay in 1819, and the family group name Melolonthinae was established by Samouelle in 1819. However, the family group name Cetoniinae was established a few years earlier in 1815 by Leach. Thus, the family group name Cetoniidae has priority over Melolonthidae. Therefore, if one wants to consider all of these subfamilies in the same family (exclusive of Scarabaeinae, which was established by Latreille in 1802), then the valid name would be Cetoniidae! Accordingly, the family name Scarabaeidae (including Melolonthinae, Scarabaeinae, Dynastinae, Cetoniinae, etc.) is the correct family group name for these taxa and *not* Melolonthidae.

Classification of the world Scarabaeidae is variably known. The classification of the world Dynastinae is fairly well established due to the work of Endrödi (1985). Most Melolonthinae, Rutelinae, and Cetoniinae are so poorly known taxonomically that many New World genera cannot be reliably identified. Classification of the Scarabaeinae (Hanski and Cambefort 1991) and Aphodiinae are fairly well established (Dellacasa 1987, 1988a, 1988b, 1991, 1995). The taxonomy of the North American scarab beetles is relatively stable although no one volume is available for identification. Regional works (see listing) are sometimes the best sources for identification of Nearctic scarab beetles.

Distribution. The family Scarabaeidae includes about 91% of all scarabaeoids and includes about 27,800 species worldwide. Within the Scarabaeidae, the Aphodiinae and Scarabaeinae include approximately 6,850 species worldwide (about 22% of scarabaeoids and 25% of Scarabaeidae). The subfamilies

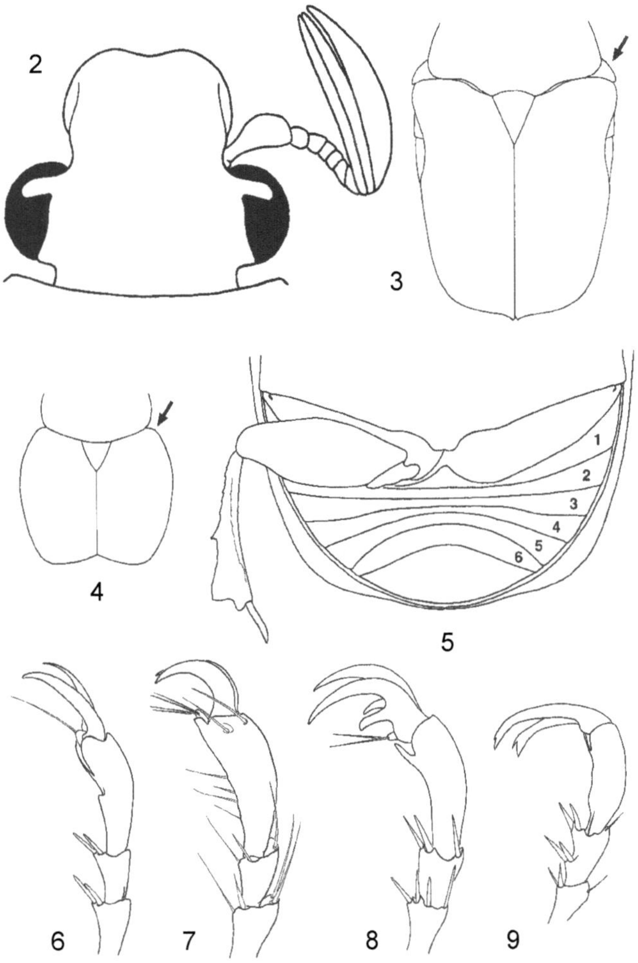

FIGURES 2.34-9.34. 2. Head and antenna (dorsal view) of *Euphoria* sp. showing clypeal sides constricted and with antennal insertion visible; 3. Cetoniini mesepimeron visible from above (arrow); 4. Trichiini mesepimeron not visible from above; 5. Abdomen and posterior leg of *Copris* sp. (Scarabaeinae). 6-9. Claws of posterior tarsi; 6. *Anomala* sp. (Rutelinae) (claws simple and unequal in length); 7. *Xyloryctes jamaicensis* (Drury) (Dynastinae) (claws simple and equal in length); 8. *Polyphylla* sp. (Melolonthinae); 9. *Dichelonyx* sp. (Melolonthinae) (claws cleft or toothed and equal in length) (Used by permission of University of Nebraska State Museum).

Orphninae, Melolonthinae, Dynastinae, Rutelinae, and Cetoniinae include approximately 20,950 species (about 69% of scarabaeoids and 75% of Scarabaeidae).

About 125 genera and 1,700 species of scarab beetles occur in the United States, Canada, and Nearctic Mexico. Keys to genera and species: Helgesen and Post 1967; Woodruff 1973; Ratcliffe 1991; Harpootlian 2001. Biology: Ritcher 1958. Catalog of the U.S. species: Blackwelder and Arnett 1974; Smith 2001. Regional works: Blatchley 1910; Loding 1945; Saylor 1948b; Edwards 1949; Helgesen and Post 1967; Hatch 1971; Woodruff 1973; Kirk and Balsbaugh 1975; Shook 1978; Lago *et al.* 1979; Ratcliffe 1991; Downie and Arnett 1996; Morón *et al.* 1997; Harpootlian 2001. Larvae: Ritcher 1966. Reference: Cooper 1983.

Classification of the Scarabaeidae of the United States, Canada, and Nearctic Mexico

The family Scarabaeidae is among the largest beetle families and is relatively well known. Despite the popularity of the group, there is no comprehensive treatment on the family as a whole. A comprehensive survey of this group is badly needed. The following subfamilies and tribes occur in the Nearctic region.

I. APHODIINAE
- 1. Aegialiini
- 2. Aphodiini
- 3. Didactyliini
- 4. Eupariini
- 5. Psammodiini

II. SCARABAEINAE
- 6. Canthonini
- 7. Coprini
- 8. Dichotomiini
- 9. Oniticellini
- 10. Onitini
- 11. Onthophagini
- 12. Phanaeini
- 13. Sisyphini

III. MELOLONTHINAE
- 14. Hopliini
- 15. Oncerini
- 16. Podolasiini
- 17. Sericini
- 18. Chasmatopterini
- 19. Melolonthini
- 20. Diplotaxini
- 21. Macrodactylini
- 22. Pachydemini
- 23. *Incertae Sedis* (*Acoma*)

IV. RUTELINAE
- 24. Anomalini
- 25. Rutelini

V. DYNASTINAE
- 26. Cyclocephalini
- 27. Pentodontini
- 28. Oryctini
- 29. Phileurini
- 30. Dynastini

VI. CETONIINAE
- 31. Gymnetini
- 32. Cetoniini
- 33. Cremastocheilini
- 34. Osmodermini
- 35. Trichiini
- 36. Valgini

Key to the Subfamilies of the United States, Canada, and Nearctic Mexico

1. Pygidium completely covered (or nearly so) by apex of elytra; length 1.5-13.0 mm I. Aphodiinae, p. 42
— Pygidium completely exposed; length longer than 5.0 mm 2

2(1). Antennal insertion visible from above; clypeus with sides constricted medially just before eyes (Fig. 2) .... VI. Cetoniinae, p. 68
— Antennal insertion not visible from above; clypeus with sides not constricted (*e.g.* Fig. 1, Superfamily Scarabaeoidea) 3

3(2). Abdominal sternites distinctly narrowed at midline (Fig. 5); length of all sternites shorter than length of metasternum; scutellum hidden in most .... II. Scarabaeinae, p. 48
— Abdominal sternites normal, not narrowed at midline; length of all sternites longer than length of metasternum; scutellum visible in most 4

4(3). Claws of both middle and posterior tarsi unequal in length and independently movable (Fig. 6) (exception: all legs in *Leptohoplia* with only one claw or with one claw greatly reduced); tarsomere 5 at apex with ventral, median, longitudinal cleft .... IV. Rutelinae, p. 60
— Claws of both middle and posterior tarsi equal in length and not independently movable (Figs. 7-9) (exception: posterior tarsi in *Hoplia* with only one claw); tarsomere 5 at apex lacking ventral, median, longitudinal cleft, instead with 2 parallel clefts either side of middle on ventral side..... 5

5(4). Claws of middle and posterior tarsi simple (Fig. 7); base of pronotum and elytra subequal in width; apex of posterior tibia always with 2 spurs; mandibles often exposed in dorsal view .... V. Dynastinae, p. 64
— Claws of middle and posterior tarsi cleft, toothed (Figs. 8-9), or simple (if simple, base of pronotum much narrower than base of elytra); apex of posterior tibia with 1-2 spurs or spurs absent; mandibles hidden in dorsal view .... III. Melolonthinae, p. 51

I. Aphodiinae Leach 1815

by Paul E. Skelley and Robert D. Gordon

Common name: The aphodiine dung beetles

Characteristics: Clypeus dilated to cover mouthparts (except Aegialiini and *Annegialia* [Eupariini]), most medially emarginate. Mandibles hardened, most concealed by clypeus. Antennae with 9 segments, club 3-segmented. Mesocoxae contiguous or nearly so. Metatibiae with 2 apical spurs. Elytra nearly or entirely covering pygidium. Abdomen with 6 visible sternites. Tarsi with distinct claws, few with claws reduced.

Marco Dellacasa (1987, 1988a, 1988b, 1991, 1995) reviewed the current classification scheme for this subfamily on a world basis, provided a computer generated catalog of all taxa of Aphodiinae (or Aphodiidae) (1987), and gave a bibliography of authors and publications for all taxa (1988a). Additions and corrections to these works are in Dellacasa 1988b, 1991, 1995. These publications are the only comprehensive references on the subfamily.

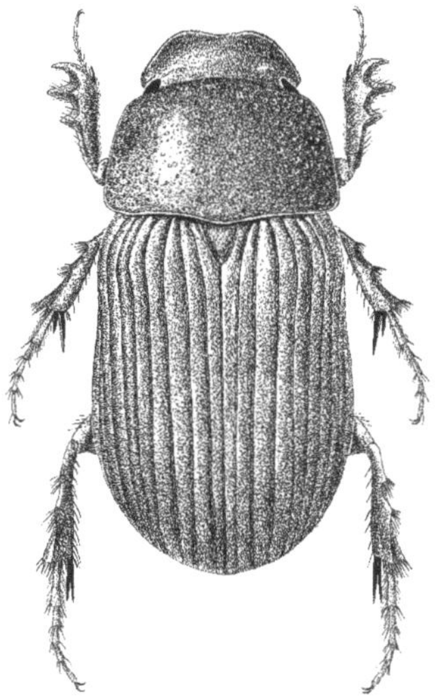

FIGURE 10.34. *Aphodius concavus* Say (Used by permission of University of Nebraska State Museum).

In the Nearctic region, the subfamily includes 26 genera and over 350 species. Keys to specific groups or revisionary studies are cited under the appropriate taxa. Regional works: Blatchley 1910 (Indiana); Helgesen and Post 1967 (North Dakota); Hatch 1971 (Pacific Northwest); Woodruff 1973 (Florida); Ratcliffe 1991 (Nebraska); Downie and Arnett 1996 (northeastern North America); Harpootlian 2001 (South Carolina). Faunal studies for other regions are in progress (see especially Dellacasa *et al.* in press). Larvae: Jerath 1960; Ritcher 1966. Biology: Jerath and Ritcher 1959.

Key to the Genera from the United States, Canada, and Nearctic Mexico

1. Apical spurs of posterior tibiae separated; basal tarsomere articulated between spurs (Figs. 29-31) 2
— Apical spurs of posterior tibiae adjoining basally so basal tarsomere not articulated between spurs (Figs. 32-41) 3

2(1). Labrum and mandibles clearly visible beyond clypeus (Fig. 11) (Aegialiini) 5
— Labrum and mandibles concealed beneath clypeus (as in Fig. 15) (Aphodiini) 6

3(1). Abdominal sternites mostly separated by transverse band of short, longitudinal carinae (Fig. 23); pygidium with transverse ridge or carina dividing it into anterior and posterior halves, posterior half frequently eroded; middle and posterior tibia with variably arranged setae (Figs. 33-41) 4
— Abdominal sternites not separated by band of short, longitudinal carinae; pygidium lacking transverse ridge, surface entire; middle and posterior tibia with 4-5 setae on lateral edge, often with weak

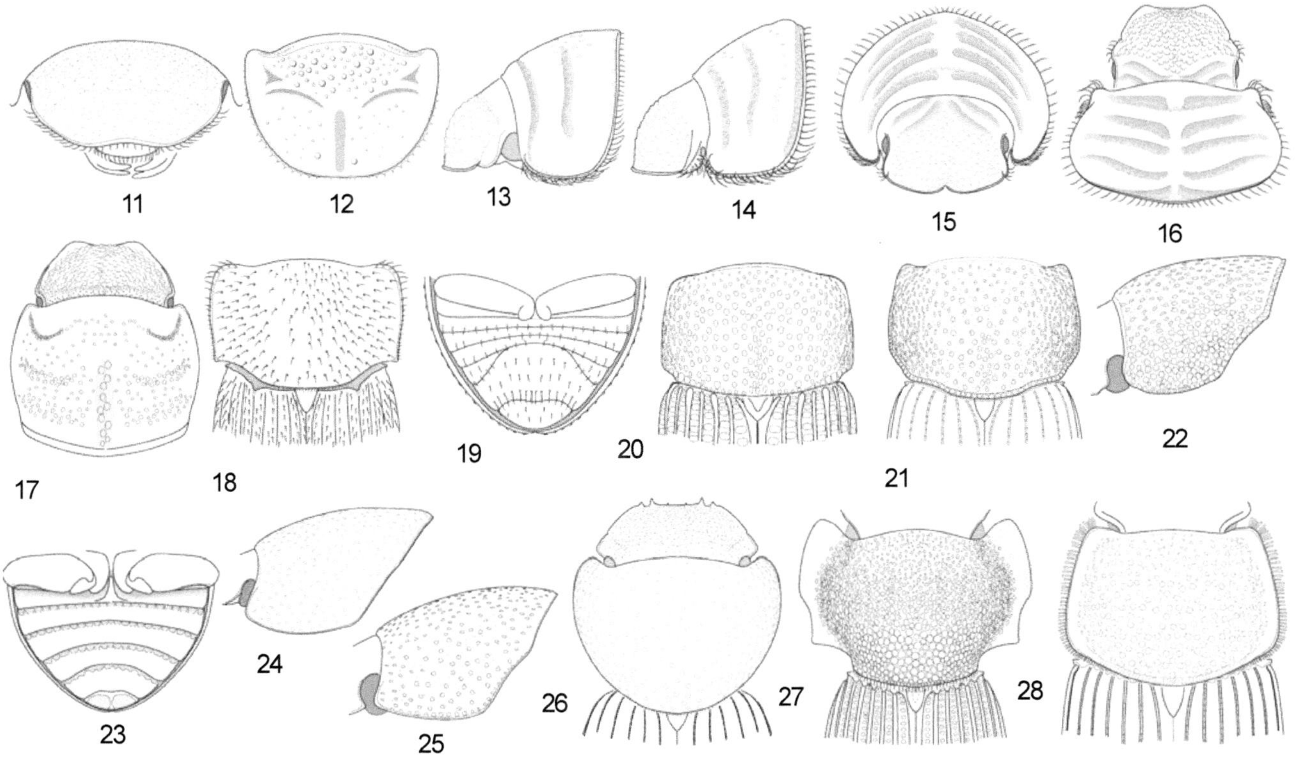

FIGURES 11.34-28.34. 11. *Aegialia montana* Brown, anterior view of head; 12. *Annegialia ataeniformis* Howden, pronotum (after Howden 1971a); 13. *Leiopsammodius malkini* (Cartwright), lateral view of head and pronotum; 14. *Geopsammodius relictillus* (Deyrup and Woodruff), lateral view of head and pronotum; 15. *Psammodius basalis* (Mulsant and Rey), anterior view of head and pronotum; 16. *P. basalis*, dorsal view of head and pronotum; 17. *Pleurophorus caecus* (Creutzer), dorsal view of head and pronotum; 18. *Euparia castanea* LePeletier and Serville, dorsal view of pronotum and elytral base; 19. *Micraegialia pusillus* (Horn), abdominal sternites; 20. *Oxyomus silvestris* (Scopoli), dorsal pronotum and elytral base; 21. *Dialytellus dialytoides* (Fall), dorsal view of pronotum and elytral base; 22. *D. dialytoides*, lateral view of pronotum; 23. *Aphotaenius carolinus* (Van Dyke), abdominal sternites; 24. *Aphodius erraticus* (Linnaeus), lateral view of pronotum; 25. *Aphodius badipes* Melsheimer, lateral view of pronotum; 26. *Aphodius pyriformis* Brown, pronotum and elytral base; 27. *Euparixia duncani* Brown, dorsal view of pronotum and elytral base; 28. *Martineziella dutertrei* (Chalumeau), pronotum and elytral base.

transverse carina (Fig. 32) (Didactyliini) .... *Didactylia*

4(3). Pronotum with transverse furrows separated by swellings (Figs. 13-17), furrows can be weak with only traces of impressions remaining, at least visible near eyes; head granular (Figs. 13-17); posterior tarsomeres of many triangularly expanded, especially basal tarsomere (Figs. 35-40) (Psammodiini) 11

— Pronotum without transverse swellings (except *Annegialia*, Fig. 12), some with weak transverse impressions (Figs. 18, 27-28); head lacking granules in most; posterior tarsomeres elongate, in a few triangularly expanded (Figs. 33-34, 41) (Eupariini) 20

5(2). Intervals of elytron with short setae; abdominal sternites narrowed medially, 5th sternite obliterated medially by 6th sternite (Fig. 19) .... *Micraegialia*

— Intervals of elytra without short setae; abdominal sternites not narrowed medially, 5th sternite not obliterated medially by 6th sternite ..... *Aegialia*

6(2). Lateral elytral margin disappears beneath carinate interval midway to apex, carinate interval becomes lateral margin at apex; other intervals not carinate .... *Hornietus*

— Lateral elytral margin complete from base to apex; intervals carinate or not 7

7(6). Anterior tibia with 4 teeth, apical tooth parallel to apical spur of tibia, lateral 3 teeth feeble, can be barely visible (Fig. 42) .... *Dialytes*

— Anterior tibia with 3 or fewer teeth, lacking apical tooth 8

8(7). Elytral striae each consisting of single row of large punctures; punctures of elytral striae wider than intervals; pronotum with longitudinal depression at base (Fig. 20) .... *Oxyomus*

— Elytral striae variable, each composed of distinctly impressed line with punctures single, doubled, or absent; elytral intervals wider than punctures of striae; pronotal disc lacking distinct depression 9

9(8). Posterior tarsomeres short, triangular; posterior tibia broadly, evenly expanded to strongly oblique apex (Fig. 29) .... *Xeropsamobeus*

— Posterior tarsomeres cylindrical, elongate in most; posterior tibia slender or often abruptly expanded to transverse apex (Figs. 30-31) 10

10(9). Posterior angle of pronotum emarginate (Figs. 21-22); basal pronotal margin much narrower than anterior margin and strongly defined with row of large punctures; pronotum not noticeably wider at apex; clypeal teeth absent .... *Dialytellus*

— Posterior angle of pronotum entire in most, rarely with posterior angle obsolete (Figs. 24-26); if angle obsolete then basal margin not strongly defined and pronotum distinctly wider at apex than at base (Fig. 26); clypeus with or without teeth .... *Aphodius*

11(4). Posterior tibia with weak transverse carina or row of tubercles (Figs. 35-36) 12

— Posterior tibia lacking transverse carina or row of tubercles (Figs. 37-40) 13

12(11). Apical clypeal angle with 1 small, sharp, clearly upturned tooth on each side of apical emargination .... *Odontopsammodius*

— Clypeus rounded or angled at each side of median emargination, without teeth .... *Tesarius*

13(11). Pronotal sculpture with at least 5 transverse ridges, ridges either continuous or broken into discrete granules, most with as many transverse furrows (Figs. 15-16) 14

— Pronotal sculpture with at most 3 weak, transverse ridges and furrows (Figs. 13-14, 17) 17

14(13). Posterior tarsomeres with at least basal tarsomere broad, strongly asymmetrically widened apically (*e.g.*, Figs. 35-36) 15

— Posterior tarsomeres elongate, basal tarsomere subcylindrical, slightly symmetrically or asymmetrically widened apically (Fig. 37) 16

15(14). Vertex of head with 1 or 2 pairs of oblique ridges arranged in chevron, ridges either continuous or broken into discrete tubercles (Figs. 15-16) .... *Psammodius*

— Vertex of head without oblique ridges .... *Neopsammodius*

16(14). Elytra with scale-like setae on raised intervals .... *Trichiorhyssemus*

— Elytra lacking setae.............................. *Rhyssemus*

17(13). Posterior tarsus short, tarsomeres 1-4 triangular (*e.g.*, Figs. 35-36); body short, robust, visibly dilated posteriorly; elytral base not margined.. 18

— Posterior tarsus elongate or somewhat shortened, all tarsomeres elongate or with basal 1 or 2 tarsomeres asymmetrically widened apically (Figs. 38-40); body slender, elongate, either suboval or subparallel, slightly dilated posteriorly; elytral base margined 19

18(17). Eye normal (Fig. 13); frontal suture distinct in most; hindwing normally developed; tarsal claw corneous .... *Leiopsammodius*

— Eye small, poorly developed (Fig. 14); frontal suture lacking, vertex lacking transverse ridge; hindwing vestigial; tarsal claws greatly reduced, setaceous .... *Geopsammodius*

19(17). Posterior tarsus elongate, at least as long as tibia; first tarsomere subequal in length to long spur,

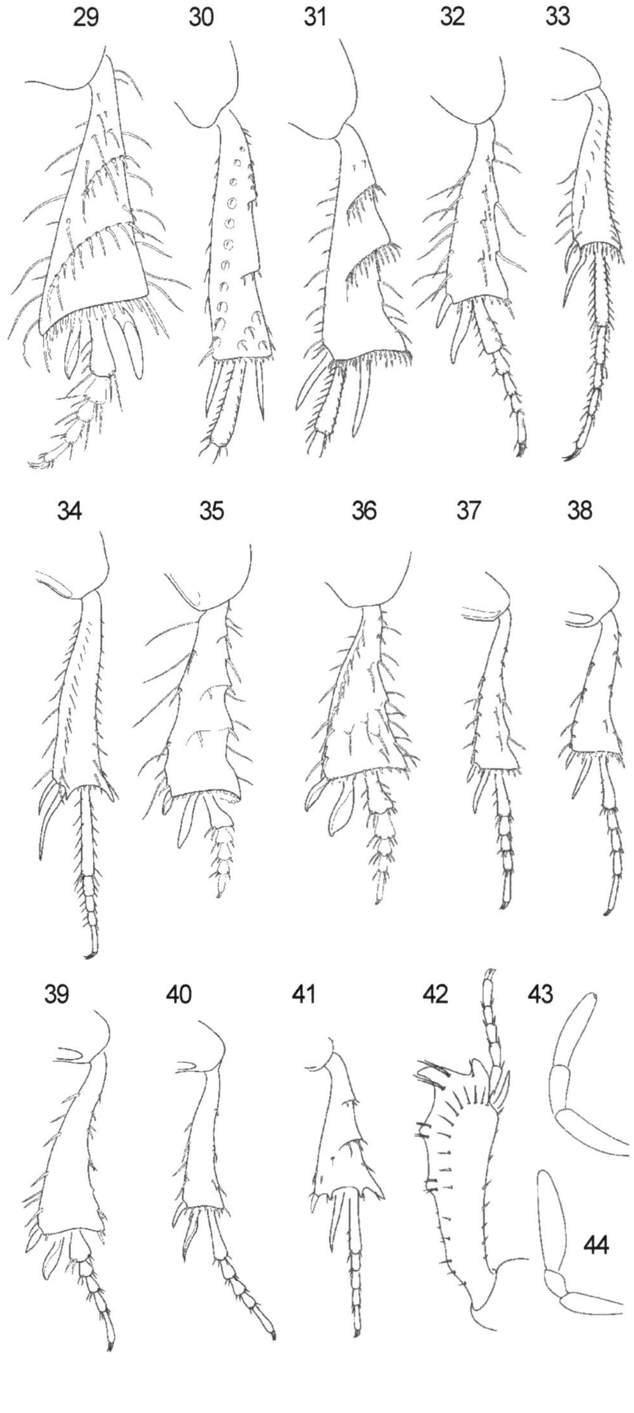

FIGURES 29.34-44.34. 29-41. Posterior tibiae and tarsi. 29. *Xeropsamobeus desertus* (Van Dyke); 30. *Dialytellus dialytoides* (Fall); 31. *Aphodius erraticus* (Linnaeus); 32. *Didactylia parcus* (Horn); 33. *Pseudataenius contortus* Cartwright; 34. *Ataenius rhyticephalus* (Chevrolat); 35. *Tesarius oregonensis* (Cartwright); 36. *Odontopsammodius cruentus* (Harold); 37. *Rhyssemus californicus* Horn; 38. *Pleurophorus caesus* (Creutzer); 39. *Platytomus micros* (Bates); 40. *Platytomus caelicollis* (Cartwright); 41. *Aphotaenius carolinus* (Van Dyke). 42. *Dialytes striatulus* (Say), anterior view of protibia; 43. *Pseudataenius contortus* Cartwright, maxillary palp; 44. *Ataenius rhyticephalus* (Chevrolat), maxillary palp.

cylindrical, slightly widened apically (Fig. 38)... .... *Pleurophorus*
— Posterior tarsus weakly to strongly shortened, visibly shorter than tibia; first tarsomere shorter than long spur, weakly to visibly widened at apex and slightly asymmetrical (Figs. 39-40) ... *Platytomus*

20(4). Pronotal disc with a T-shaped groove (Fig. 12) .... *Annegialia*
— Pronotal disc lacking groves or depressions 21

21(20). Elytral base notched (Fig. 18) .... *Euparia*
— Elytral base not notched (Figs. 27-28) 22

22(21). Pronotum constricted basally and flared in front (Fig. 27); middle coxa touching base of elytra .... *Euparixia*
— Pronotum not constricted basally, sides parallel or evenly rounded, most not flared; middle coxa separated from elytral base 23

23(22). Posterior tibial apex with 4 distinct, spinose processes (Fig. 41); clypeus with 2 spiniform teeth .... *Aphotaenius*
— Posterior tibial apex with 2 or fewer processes and fringed with many small spinules (Figs. 33-34); clypeus with or without spiniform teeth........ 24

24(23). Lateral pronotal margin flared for entire length and fringed with long, flattened setae (Fig. 28) .... *Martineziella*
— Lateral pronotal margin not flared for entire length, with or without flattened setae, without the above combination of characters 25

25(24). Mesosternum not carinate; terminal maxillary palpomere slender (Fig. 43); abdominal sternites with weakly developed, transverse band of longitudinal carinae; middle and posterior tarsus longer than tibia .... *Pseudataenius*
— Mesosternum carinate; terminal maxillary palpomere thickened at middle (Fig.44); abdominal sternites separated by transverse band of short, longitudinal carinae, most strongly developed (*e.g.*, Fig. 23); middle and posterior tarsus subequal to or shorter than tibia .... *Ataenius*

Classification of the Tribes and Genera of Aphodiinae

Aphodiinae Leach 1815

Aegialiini Laporte 1840

Characteristics: Head flat or slightly convex, surface granulate or punctate. Mandibles not covered by clypeus. Pronotum lacking ridges or grooves. Elytral base not bordered. Pygidium smooth, unmodified. Femora lacking grooves on anterior or posterior margin. Metatibia variable, broadly dilated at apex or not. Metatibial spurs separated by metatarsus. Metatarsomere 1 variable, broadly triangular or not.

Six of the nine genera and subgenera of the world Aegialiini are represented in North America. The tribe is composed of cold-adapted species with no truly tropical representatives. Only two species are known from the Southern Hemisphere: one from Tasmania and one from Argentina. Members of this tribe for which biological information is known are all detritivores, usually associated with sandy conditions, riparian, or littoral habitats.

Aegialia Latreille 1807
subgenus *Aegialia* Latreille 1807
subgenus *Caelius* Lewis 1895
Leptaegialia Brown 1931
subgenus *Psammoporus* Thomson 1863
Dimalia Mulsant and Rey 1869
subgenus *Rhysothorax* Bedel 1911
Anomalaegialia Brown 1931

The genus *Aegialia* (Fig. 11) is comprised of 41 species worldwide. In North America, the genus includes 31 species in four subgenera that are distributed from northeastern Canada to Alaska and south to South Carolina and southern California. Gordon and Cartwright (1988) revised the North American Aegialiini presenting keys and illustrations to all taxa as well as known habitat data. Gordon (1990) and Gordon and Rust (1997) described two additional species of *Aegialia.* The subgenera *Leptaegialia* and *Anomalaegialia* were placed as junior synonyms of *Caelius* and *Rhysothorax*, respectively, by M. Dellacasa (1987). Biology: Rust and Hanks 1982.

Micraegialia Brown 1931
The genus *Micraegialia* includes only *M. pusilla* (Horn), which is found in Manitoba, Minnesota, North Dakota, Washington, and British Columbia. This genus was described and illustrated in Gordon and Cartwright (1988).

Aphodiini Leach 1815

Characteristics: Head convex, surface granulate or punctate. Mandibles covered by clypeus. Pronotum lacking ridges or grooves. Elytral base not bordered. Pygidium smooth, unmodified. Femora not grooved at anterior or posterior margin. Metatibia broadly dilated at apex, may be slender, with 1-2 transverse carina in most. Metatibial spurs separated by metatarsus. Metatarsomere 1 slender, may be weakly triangular.

Six genera occur in North America according to the current classification, and some of these, especially *Aphodius*, are distributed worldwide. Keys to taxa: Horn 1887; Schmidt 1922; Helgesen and Post 1967; Hatch 1971; Woodruff 1973; Ratcliffe 1991. Keys to larvae: Jerath 1960.

Aphodius Illiger 1798
In North America this genus (Fig. 10) contains about 200 species with new species being discovered frequently. *Aphodius* is a tremendously large genus of more than 900 species worldwide. It has been variously divided into numerous subgenera. Poole and Gentili (1996) presented an extensive list of 138 subgeneric names associated with *Aphodius*. Dellacasa (1987, 1988a, 1988b, 1991, 1995) provided recent subgeneric placement of *Aphodius* species for the world. To date, no subgeneric classification has proved

satisfactory for North America. Reclassification efforts in progress will soon redefine most of our concepts, hence no subgeneric listing is included here. Keys to species of various North American groups can be found in Horn (1887); Brown (1927, 1928a, 1928b, 1929a); Fall (1932); Saylor (1940c); Cartwright (1972); Woodruff (1973); Gordon (1976, 1977, 1983); Ratcliffe (1991); Skelley and Gordon (2001).

As expected from the large number of species, *Aphodius* species occupy numerous niches. Many common species are generalist dung feeders. Many other species are highly specialized to feed on a specific type of dung or dung in specific situations (*e.g.*, animal burrows). Other species are detritivores, feeding on decaying plant materials. Gordon (1983) presented a key to the eastern species and discussed their feeding preferences. Skelley and Gordon (2001) readdress Gordon's (1983) ideas on their feeding preferences.

Dialytellus Brown 1929
Dialytellus contains only two species, both of which are endemic to North America. *Dialytellus dialytoides* (Fall) and *D. humeralis* (LeConte) are found in southeastern Canada, Massachusetts, Michigan, Maryland, North Carolina, and Virginia. Dellacasa and Gordon (1997) transferred *Dialytellus* from Eupariini to Aphodiini. Members of this genus feed on deer dung in shaded conditions but will also utilize sheep dung. Downie and Arnett (1996) diagnosed the species.

Dialytes Harold 1869
There are four species of *Dialytes*, all found in North America, *D. criddlei* Brown, *D. striatulus* (Say), *D. truncatus* (Melsheimer), and *D. ulkei* (Horn). Species are distributed from southeastern Canada to Virginia, west to Manitoba and Oklahoma. Stebnicka (1994) transferred *Dialytes* from Eupariini to Aphodiini and transferred all Old World members to other genera. Members of this genus feed on deer dung in shaded conditions and have also been found in sheep dung. Key: Downie and Arnett 1996.

Hornietus Stebnicka 2000
The genus *Hornietus* includes only one species, *H. ventralis* (Horn). It is known from eastern North America from the District of Columbia to Indiana and south to Kentucky and Alabama. *Hornietus ventralis* is a rare species that has been associated with logs containing passalid burrows and frass, but extensive efforts to survey this niche have been unsuccessful. The natural history of the species is unknown. This species was recently transferred from the genus *Saprosites* by Stebnicka (2000).

Oxyomus Laporte 1840
Xenoheptaulacus Hinton 1934
The genus *Oxyomus* is distributed in the Old World and consists of 27 species. One adventive species from the Old World, *O. silvestris* (Scopoli), has been recorded in New York, Pennsylvania, and Washington. It is known to live in dung of domestic animals and adults are attracted to lights.

Xeropsamobeus Saylor 1937
The genus is endemic to western North America and includes two species, *X. desertus* (Van Dyke) and *X. ambiguus* (Fall), that occur in California and New Mexico. Dellacasa and Gordon (1994) cite *Xeropsamobeus* as a subgenus of *Aphodius*. Because it traditionally has been considered a valid genus and because the current reclassification studies mentioned under *Aphodius* indicate it is valid, we choose to list it here with generic ranking. Members of *Xeropsamobeus* are detritivores and psammophilic (sand-loving).

Didactyliini Pittino 1984

Characteristics: Head slightly convex, surface punctate. Mandibles covered by clypeus. Pronotum lacking ridges or grooves. Elytral base not bordered. Pygidium smooth, unmodified. Femora lacking grooves on anterior or posterior margin. Metatibia dilated at apex, some with weak transverse carina. Metatibial spurs not separated by metatarsus. Metatarsomere 1 elongate, may be weakly triangular.

Pittino (1984) established the subtribe Didactyliina in the tribe Eupariini, and it was given tribal rank by Dellacasa (1987).

Didactylia d'Orbigny 1896
Worldwide, the genus *Didactylia* includes 31 species. Three species, *D. parcus* (Horn), *D. knausii* (Fall), and an undescribed species occur in the eastern and central United States. The taxonomy of American *Didactylia* is uncertain and is currently being studied (for pertinent comments see Woodruff 1973). The species are psammophiles and detritivores. Keys: Brown 1929b; Woodruff 1973.

Eupariini LePeletier and Serville 1828

Characteristics: Head slightly to strongly convex, surface variable, granulate to smooth with fine punctures. Mandibles covered by clypeus. Pronotum lacking ridges or grooves. Elytral base bordered in most. Pygidium modified, with transverse ridge and longitudinal grove at base (elytral locking mechanism). Femora with grooves on anterior or posterior margin, may be weakly developed. Metatibia slender, apically dilated in a few. Metatibial spurs not separated by metatarsus. Metatarsomere 1 slender in most.

Many genera and species worldwide are assigned to the Eupariini, but the genus *Ataenius* is by far the most significant taxon in terms of number of species, especially in the Western Hemisphere.

Annegialia Howden 1971
The genus *Annegialia* includes only *A. ataeniformis* Howden from Utah. *Annegialia* was described by Howden (1971a) in the Aegialiini and transferred to the Eupariini by Gordon and Cartwright (1988).

Aphotaenius Cartwright 1952
Five New World species are included in the genus and one species, *A. carolinus* (VanDyke), is widespread in the eastern United

States. Cartwright (1974) presented a description and other information. Woodruff (1973) illustrated an adult *A. carolinus*. *Aphotaenius* species are known to feed on deer dung in forested areas.

Ataenius Harold 1867
The genus *Ataenius* includes hundreds of species worldwide and is in need of taxonomic study. North of Mexico, the genus consists of about 65 species, occurring mostly in the eastern and southern United States. Cartwright (1974) revised the genera *Aphotaenius, Ataenius,* and *Pseudataenius*, providing keys to all taxa. Chalumeau (1992) transferred *Ataenius simulator* (Harold) to the genus *Parataenius* Balthasar. Additional species may be transferred to *Ataeniopsis* Petrovitz, *Phalangochaeta* Martínez, or other genera. Because of the amount of taxonomic work in progress on the Eupariini, the traditional classification (*sensu* Cartwright 1974) within *Ataenius* is maintained here. The majority of the known species are detritivores, while some species specialize on various types of dung. Many adults are attracted to lights.

Euparia LePeletier and Serville 1828
Worldwide, the genus *Euparia* includes 18 species. One species, *E. castanea* LePeletier and Serville, occurs in the southeastern U.S.: Florida, Alabama, and Louisiana. Woodruff (1973) provided a description and illustration. *Euparia castanea* lives primarily with native fire ants, *Solenopsis* species. Wojcik *et al.* (1977, 1978) and Wojcik and Habeck (1977) discussed the biology of the species.

Euparixia Brown 1927
Five species of *Euparixia* occur in the New World, and two species occur in the Nearctic region: *E. duncani* Brown in southern Arizona and *E. moseri* Woodruff and Cartwright in eastern Texas and Louisiana. The host of *E. duncani* is unknown, but *E. moseri* and several other members of *Euparixia* live in fungal gardens of leaf cutter ants (*Atta* sp.). Woodruff and Cartwright (1967) discussed this association (for additional information see Hinton 1936).

Martineziella Chalumeau 1986
Martinezia Chalumeau 1983
The genus *Martineziella* includes six American species. One adventive species, *M. dutertrei* (Chalumeau), known from South America and the West Indies, is widespread in the southeastern United States. It lives with imported fire ants (*Solenopsis* sp.). Woodruff (1973) provided a detailed description and illustrations. Wojcik *et al.* (1977, 1978), Wojcik and Habeck(1977), Vander Meer and Wojcik (1982), and Lanciani and Wojcik (1991) provided information on its biology in the United States.

Pseudataenius Brown 1927
The genus contains only three species that are endemic to the New World: *P. contortus* Cartwright, *P. socialis* (Horn), and *P. waltherhorni* (Balthasar). These species are known from New Jersey to Nebraska south to Florida and Texas and are primarily detritivores. Key: Cartwright 1974.

Psammodiini Mulsant 1842

Characteristics: Head strongly convex, surface with tubercles. Mandibles covered by clypeus. Pronotum with ridges or grooves, may be reduced to weak impressions. Elytral base bordered in most, may be weakly developed. Pygidium modified, with transverse ridge and longitudinal groove at base (elytral locking mechanism). Femora with grooves on anterior or posterior margin, may be weakly developed. Metatibia slender or broadly dilated. Metatibial spurs not separated by metatarsus. Metatarsomere 1 triangular or slender.

Cartwright (1955) revised North American "*Psammodius*" in the broad sense. Rakovic (1981, 1984, 1986, 1990) reclassified the tribe Psammodiini. Pittino and Mariani (1986) divided Psammodiini into two subtribes, Psammodiina for *Psammodius, Leiopsammodius, Tesarius* and several Old World genera, and Rhyssemina for *Platytomus, Pleurophorus, Rhyssemus, Trichiorhyssemus*, and several Old World genera. Marco Dellacasa (1988b) elevated Psammodiini to subfamily rank and the two subtribes to tribal status. Gordon and Pittino (1992) summarized the status of American taxa and provided a key to American genera. They treated Psammodiini as a tribe and disregarded the subtribe concept. All members of the Psammodiini are detritivores. A number of species are adventive, probably being transported in ships' ballast.

Geopsammodius Gordon and Pittino 1992
The genus *Geopsammodius* includes three species: *G. hydropicus* (Horn) from the Atlantic coast states; *G. relictillus* (Deyrup and Woodruff) from Central Florida; and *G. sabinae* Lavalette from French Guiana. Deyrup and Woodruff (1991) illustrated *G. relictillus*. Several undescribed species are known from the southeastern U.S., west to Texas.

Leiopsammodius Rakovic 1981
Thirty one species of *Leiopsammodius* are known worldwide with 15 occurring in the New World. There are four described and one undetermined species in the southeastern U.S.: *L. malkini* (Cartwright), *L. acei* Harpootlian *et al.*, *L. ocmulgeei* Harpootlian *et al.*, and *L. deyrupi* Harpootlian *et al.* Woodruff (1973) illustrated *L. malkini*, and Harpootlian *et al.* (2000) reviewed the genus in North America.

Neopsammodius Rakovic 1986
This genus includes nine species that are distributed in the western U.S. and Mexico. Five species occur in the western states and Nearctic region: *N. blandus* (Fall), *N. interruptus* (Say), *N. mimeticus* (Fall), *N. quinqueplicatus* (Horn), and *N. werneri* (Cartwright). In Cartwright's (1955) key, couplets 7-13 correspond to *Neopsammodius*.

Odontopsammodius Gordon and Pittino 1992
Odontopsammodius is comprised of 14 species in the New World. Three species occur in the southeastern United States: *O. armaticeps*

(Fall), *O. bidens* (Horn), and *O. cruentus* (Harold). In Cartwright's (1955) key, couplets 27-34 correspond to *Odontopsammodius*.

Platytomus Mulsant 1842
Of the 24 species in the genus, six are widely distributed in the United States: *P. atlanticus* (Cartwright), *P. caelicollis* (Cartwright), *P. longulus* (Cartwright), *P. micros* (Bates), *P. notialis* (Cartwright), and *P. tibialis* (Fabricius). Cartwright (1948) provided a key to Nearctic *Platyomus* species using the traditional generic name *Pleurophorus* for members of this group. Pittino and Mariani (1986) transferred all of the North American species, except *P. caesus*, to *Platytomus*.

Pleurophorus Mulsant 1842
The genus *Pleurophorus* contains 32 species that are distributed worldwide. Only one species, *P. caesus* (Creutzer), occurs in the United States, and it is widely distributed. Of the species included in *Pleurophorus* by Cartwright (1948), only *P. caesus* remains (see the genus *Platytomus*).

Psammodius Fallén 1807
Psammobius Heer 1841
The genus *Psammodius* includes approximately 19 Old World species and four New World species. Three species, *P. basalis* (Mulsant and Rey), *P. laevipennis* Costa, and *P. pierotti* Pittino occur in the northeastern United States. Couplets 2-4 in Cartwright's (1955) key correspond to the current definition of *Psammodius* as defined by Gordon and Pittino (1992).

Rhyssemus Mulsant 1842
Boucardius Petrovitz 1967
The genus *Rhyssemus* is a large (approximately 168 species), worldwide genus. The six species in the Nearctic region are widely distributed: *R. brownwoodi* Gordon and Cartwright, *R. californicus* Horn, *R. germanus* (L.), *R. neglectus* Brown, *R. scaber* Haldeman, and *R. sonatus* LeConte. Notes on the biology of *R. scaber* were given by Steiner (1980). Key: Gordon and Cartwright 1980.

Tesarius Rakovic 1981
The genus *Tesarius* includes five species, one of which is Australian and four of which are North American. *Tesarius caelatus* (LeConte), *T. oregonensis* (Cartwright), *T. mcclayi* (Cartwright), and *T. doyeni* (Cartwright) occur in the western United States. Couplets 22-23 in Cartwright's (1955) key correspond to species of *Tesarius*, with the exception of *T. doyeni*, which was described later. *Tesarius caelatus* is adventive to other continents, probably in ships' ballast (Skelley 2000).

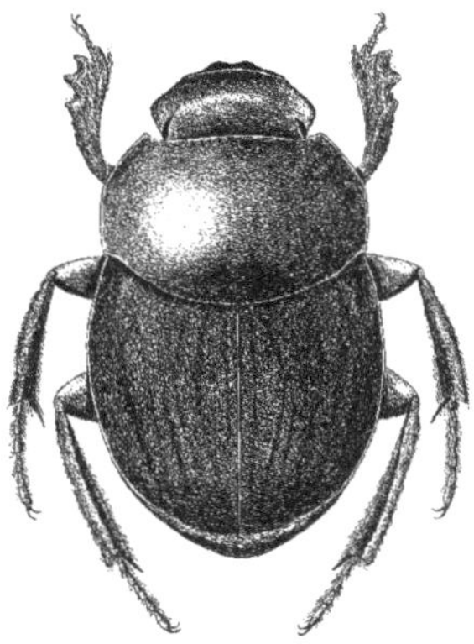

FIGURE 45.34. *Canthon pilularius* (Linnaeus) (Used by permission of University of Nebraska State Museum).

Trichiorhyssemus Clouet 1901
Trichiorhyssemus includes 33 species worldwide with only three known in the Americas. One species, *T. riparius* (Horn), occurs in the United States: Arizona, California, Nevada, New Mexico. Two species, *T. alternatus* Hinton and *T. cristatellus* (Bates), occur in Mexico. Key: Gordon and Cartwright 1980.

II. Scarabaeinae Latreille 1802

by Bruce D. Gill

Common name: The dung beetles

Characteristics: Clypeus expanded, covering mouthparts. Mandibles lamelliform, mostly membranous, with only outer margin hardened. Antennae with 8 or 9 antennomeres, club with 3. Middle coxae widely separated. Posterior tibiae nearly always with single apical spur. Elytra exposing pygidium, 6 visible, fused abdominal sternites. Anterior tarsi may be absent either in females or in both sexes.

This subfamily was divided by Cambefort (1991) into 12 tribes, eight of which are represented in the Nearctic (the Eucraniini and Eurysternini are strictly Neotropical, whereas the Gymnopleurini and Scarabaeini are limited to the Old World). Montreuil (1998) proposed significant changes to the classification of the Coprini and Dichotomiini by eliminating the Dichotomiini and transferring the genera to the Ateuchini and Coprini.

Members of this subfamily are commonly called dung beetles. Although many species feed on mammalian dung, others specialize to varying degrees upon the dung of other vertebrates and invertebrates, as well as on carrion, mushrooms, rotting fruit, and other decomposing plant material. The World fauna includes slightly over 5,000 described species in 234 genera, with close to 1,800 of these species belonging to the genus *Onthophagus*. In the Nearctic region, there are 17 genera and about 150 species. Biology: Halffter and Matthews 1966, Halffter and Edmonds 1982, Hanski and Cambefort 1991. Regional works: Woodruff 1973; Ratcliffe 1991; Downie and Arnett 1996; Harpootlian 2001. Keys to larvae: Ritcher 1966; Edmonds and Halffter 1978.

Key to the Tribes and Genera from the United States, Canada, and Nearctic Mexico

1. Middle and posterior tibiae slender, curved, only slightly expanded apically 2
— Middle and posterior tibiae strongly and often abruptly expanded apically 7

2(1). Body compressed laterally, sides of prothorax flattened, vertical (Sisyphini) .... *Sisyphus*
— Body not compressed laterally, sides of prothorax rounded (Canthonini) 3

3(2). Scutellum visible .... *Malagoniella*
— Scutellum hidden 4

4(3). Posterior edge of vertex rounded, not margined.. .... *Pseudocanthon*
— Posterior edge of vertex sharply margined, carinate 5

5(4). Anterior tarsi absent; elytral disc with humeral carina and apical tubercles .... *Deltochilum*
— Anterior tarsi present; elytral disc lacking carina and apical tubercles... 6

6(5). Posterior tibia with 2 apical spurs . *Melanocanthon*
— Posterior tibia with 1 apical spur .... *Canthon*

7(1). Scutellum visible although can be small 8
— Scutellum completely hidden 10

8(7). Pronotum with pair of basal pits near midline; antenna with 9 antennomeres (Onitini)........ *Onitis*
— Pronotum lacking basal pits; antenna with 8 antennomeres (Oniticellini).............................. 9

9(8). Sides of head abruptly notched at junction of clypeus and gena .... *Euoniticellus*
— Sides of head evenly rounded from clypeus to gena .... *Liatongus*

10(7). Basal antennomere of club cup-shaped, enclosing last 2 antennomeres (Phanaeini) 11
— Basal antennomere of club similar to other club antennomeres, not cup-shaped 12

11(10). Anterior margin of clypeus sharply bidentate medially .... *Coprophanaeus*
— Anterior margin of clypeus rounded or slightly emarginate .... *Phanaeus*

12(10). Third labial palpomere small or inconspicuous (Onthophagini) .... *Onthophagus*
— Third labial palpomere distinct 13

13(12). Elytron with 9 discal striae (Coprini) .... *Copris*
— Elytron with 7 or 8 discal striae (Dichotomiini) .. 14

14(13). First posterior tarsomere triangular 15
— First posterior tarsomere slender or broadly rectangular... 16

15(14). Abdominal sternites greatly compressed longitudinally along midline .... *Ontherus*
— Abdominal sternites not compressed along midline .... *Dichotomius*

16(14). Mesosternum very short; basal posterior tarsomere elongate, scarcely enlarged at apex .... *Canthidium*
— Mesosternum distinct; basal posterior tarsomere gradually enlarged towards apex......... *Ateuchus*

Classification of the Tribes and Genera of Scarabaeinae

Canthonini Péringuey 1901

Characteristics: Body size small to large (length 2-25 mm), shape round or oval. Head and pronotum lacking horns or carinae. Antenna with 9 segments. Middle and posterior tibiae slender, curved, only slightly expanded apically. Sexual dimorphism minimal.

The New World has 340 species in 27 genera, which equates to a third of the World fauna. Adults of most species are ball-rollers, and they shape carrion or dung into balls that are rolled away and buried at a distance from the food source. Key to genera: Halffter and Martínez 1977. Biology: Gordon and Cartwright 1974; Halffter and Halffter 1989.

Canthon Hoffmansegg 1817
Glaphyrocanthon Martínez 1948
Geocanthon Pereira and Martínez 1956
Boreocanthon Halffter 1958
Canthon (Fig. 45) is the most species-rich genus of the New World Canthonini with nearly 150 species distributed from southern Canada to Argentina. Twenty-one species are found in the Nearctic region. Keys to adults: Halffter 1958, 1961. Biology: Matthews 1963; Favila and Díaz 1996. Larval description: Ritcher 1966.

Deltochilum Eschscholtz 1822
Deltochilum is a New World genus with over 80 described species. Four species occur in northern Mexico and the southeastern United States, ranging as far north as Kentucky. Many of the species feed upon carrion. Key to adults: Howden 1966. Biology: Howden and Ritcher 1952. Larval description: Ritcher 1966.

Malagoniella Martínez 1961
There are nine species in the genus, eight of which are restricted to South America. *Malagoniella astyanax yucateca* (Harold) is widely distributed in Central America and reaches southern Texas. Adults are nocturnally active and frequently attracted to ultraviolet light. Key to adults: Martínez 1961; Halffter and Martínez 1966.

Melanocanthon Halffter 1958
Four species comprise this genus, all of which are restricted to the United States east of the Rocky Mountains. Key to adults: Halffter 1958.

Pseudocanthon Bates 1887
Opiocanthon Paulian 1947
Eight species occur in the Neotropics, and only *Pseudocanthon perplexus* (LeConte) extends into the eastern United States, ranging as far north as Arizona, Texas, Illinois and North Carolina. Reference: Bates 1887.

Coprini Leach 1815

Characteristics: Body 8-20 mm in length, elongate, robust, convex, generally black in color. Head or pronotum often with horns or tubercles, especially in males. Elytra with 9 discal striae. Scutellum hidden. Antenna with 9 segments. Middle and posterior tibia strongly expanded apically.

The tribe Coprini is a relatively small tribe with ten genera and slightly under 400 species worldwide. *Copris* is the only representative of the tribe in the New World.

Copris Geoffroy 1762
A widespread genus, *Copris* (Fig. 1) has nearly 200 species worldwide. Of the 28 species recorded from the New World, 14 are found from southern Canada to northern Mexico. Keys to adults: Matthews 1961; Matthews and Halffter 1968; Warner 1990. Keys to larvae: Ritcher 1966. Biology: Matthews 1961; Anduaga and Halffter 1991.

Dichotomiini Halffter and Matthews 1966

Characteristics: Body small and compact or large and robust (length 4-30 mm). Elytra with 7 or 8 discal striae. Scutellum hidden. Antenna with 9 segments. Middle and posterior tibia strongly expanded apically.

The tribe Dichotomiini is a diverse tribe with approximately 30 genera and 750 species worldwide. The New World has a large proportion of this diversity with close to 600 species in 23 genera.

Ateuchus Weber 1801
Choeridium LePeletier and Serville 1828
There are about 81 species in this New World genus. Five species are known from the eastern United States and northern Mexico. Key to adults: Balthasar 1939; Kohlmann 1984, Génier 2000. Larval description: Ritcher 1966. Phylogeny: Kohlmann and Halffter 1988.

Canthidium Erichson 1847
Canthidium is a Neotropical genus with over 150 described species and many more awaiting description. Several undescribed species occur in northern Mexico, at least one of which reaches southern Arizona. There is no modern taxonomic treatment for this group. The only revision (Harold 1867) treated just 38 species.

Dichotomius Hope 1838
Brachycopris Haldeman 1845
Pinotus Erichson 1847
Dichotomius is a New World genus with about 150 described species. Three or four species are found in northern Mexico, Arizona, and the eastern United States as far north as New York. The genus is in need of revision as the last complete study (Harold 1869c) is out of date. Taxonomic notes: Martínez 1951; Woodruff 1973.

Ontherus Erichson 1847
Ontherus is a Neotropical genus with 58 species, of which only *Ontherus mexicanus* Harold ranges into northeastern Mexico. Key to adults and phylogeny: Génier 1996.

Oniticellini d'Orbigny 1916

Characteristics: Body 7-14 mm in length, elongate, most flattened. Pronotum without basal pits near midline. Scutellum visible. Antenna with 8 segments.

Most of the 14 genera and 165 species in the tribe are found in Africa and Asia although *Anoplodrepanus* is restricted to the West Indies (Simonis 1981). Key to adults: Janssens 1953. Biogeography: Zunino 1982.

Euoniticellus Janssens 1953
Euoniticellus intermedius (Reiche) is an African species introduced into California and Texas (Fincher 1990), and it has spread into northern Mexico. There is a possibility that a native Cuban species may occur in Florida (Woodruff 1973). Key to adults: Janssens 1953. Biology and larva: Blume 1984.

Liatongus Reitter 1893
Two native species are found in California and northwestern Mexico. An African species, *L. militaris* (Laporte), has been introduced into the southern United States and may be established (Fincher 1990). Key to adults: Janssens 1953. Biology: Anduaga and Halffter 1993.

Onitini Laporte 1840

Characteristics: Body 10-20 mm in length, oblong, robust. Pronotum with pit on each side of midline near base. Scutellum small but visible. Elytron with prominent lateral carina. Antenna with 9 segments. Anterior tarsi lacking in both sexes. Middle and posterior tibiae strongly expanded at apex.

The tribe Onitini is an Old World group with 18 genera and about 200 species.

Onitis Fabricius 1798
Onitis is the largest genus in the tribe with over 150 described species in Africa, Asia, and Europe. Two African species have been released in the southern United States, although only *O. alexis* Klug is known to be established in California (Fincher 1990). Key to adults: Janssens 1937.

Onthophagini Lacordaire 1856

Characteristics: Body 2-12 mm in length, oval, convex. Head and/or pronotum often with horns or carinae. Scutellum hidden. Elytra with 7 discal striae. Third labial palpomere small or inconspicuous. Antenna with 9 segments. Tarsi and tarsal claws present on all legs. Middle and posterior tibiae broadly expanded at apex.

Sexual dimorphism is well developed in most *Onthophagus* species. Males typically have large horns on the head and/or pronotum (females with only carinae or rudimentary horns) or males have elongated protibiae with a long setal brush at the apex. The world fauna includes over 2,200 species in 34 genera.

Onthophagus Latreille 1807
Chalcoderus Erichson 1848
Monapus Erichson 1848
Psilax Erichson 1848
Gonocyphus Lansberg 1885
Tauronthophagus Shipp 1895
Digitonthophagus Balthasar 1959

Onthophagus is the largest genus of Scarabaeinae with representatives in nearly all parts of the world. Approximately 130 species are known from the New World, 40 of which occur from southern Canada to northern Mexico. An introduced species, *Onthophagus gazella* (Fabricius), has spread rapidly in the southern Nearctic realm. Adults commonly feed on mammalian dung, although some species feed on carrion, fungi, rotting plant material, or live in association with ground-burrowing mammals or tortoises. Keys to adults: Howden and Cartwright 1963; Zunino and Halffter 1988; additional species described by Howden 1973. Key to larvae: Ritcher 1966. Adventive species: Hoebeke and Beucke 1997. Biology: Hunter *et al.* 1996; Lobo and Halffter 1994.

Phanaeini Kolbe 1905

Characteristics: Body length 12-25 mm, robust, often with metallic coloration. Antenna with 9 segments, basal antennomere of club cup-shaped, enclosing distal 2 antennomeres. Tarsal claws absent. Anterior tarsi lacking in males, present or absent in females.

This tribe is restricted to the New World with most of the 9 genera and 150 species found in the Neotropics. Many species exhibit strong sexual dimorphism with the males possessing long horns on the head and/or pronotum. Zunino (1985) considered the Phanaeini to be a subtribe of the Onitini. Key to genera: Edmonds 1972, 1994.

Coprophanaeus d'Olsoufieff 1924

This Neotropical genus includes 32 species. Several species occur in northern Mexico, one of which (*C. pluto* [Harold]) reaches southern Arizona and Texas. Adults typically feed on carrion. Key to adults: d'Olsoufieff 1924.

Phanaeus MacLeay 1819
Lonchophorus Germar 1824
Onthurgus Gistel 1857
Paleocopris Pierce 1946

Nearly half of the 50 recognized species of *Phanaeus* are recorded from Mexico. Six species occur in northern Mexico and the eastern United States, extending as far north as Arizona in the west and Vermont in the east. Adults generally feed on mammalian dung. Key to adults: Edmonds 1994. Biology: Fincher 1972; Halffter and Lopez 1972; Halffter *et al.* 1974.

Sisyphini Mulsant 1842

Characteristics: Body length 6-10 mm, oval, compressed laterally with sides of pronotum and metasternum vertical. Color reddish brown to gray, surface matte, with setae on dorsum. Scutellum hidden. Antenna with 8 segments. Middle coxae very widely separated, forming lateral margin of body. Middle and posterior tibiae elongate, curved, barely enlarged at apex.

This tribe includes three genera of ball-rolling dung beetles. Most of the 60 species in the tribe are found in Africa and Asia, with only a few representatives in Europe and the New World.

Sisyphus Latreille 1807

Only two of the 30 species in the genus are native to the New World, and of these only *S. submonticolus* Howden reaches the Nearctic (Durango, Mexico). One African species (*S. rubrus* Paschalidis) has been introduced to the southern United States but is not known to be established (Fincher 1990). Key to adults: Haaf 1955; Howden 1965.

III. Melolonthinae Samouelle 1819

by Arthur V. Evans

Common names: The May beetles, June beetles, and chafers

Characteristics. Adults 3-58 mm in length. Dorsal surface often conspicuously setose or scaled; color reddish brown or black (*i.e.*

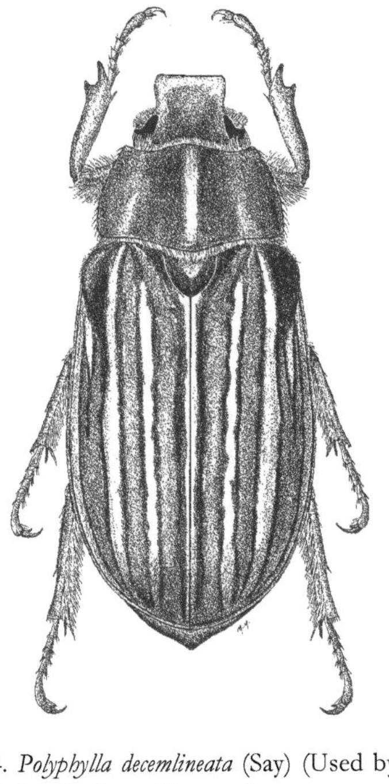

FIGURE 46.34. *Polyphylla decemlineata* (Say) (Used by permission of University of Nebraska State Museum).

Diplotaxis, Phyllophaga, Serica), some with metallic blue or green luster (*Dichelonyx*) or distinctly marked with patches of scales (*Hoplia, Polyphylla*). Head unarmed (except in *Chaunocolus*), most with mandibles well developed, sclerotized, completely concealed from above, or nearly so. Antennal insertions not visible from above; antennae 7 to 10-segmented, antennal lamellae folding tightly into a 3 to 7-segmented club, club oval to elongate, glabrous or with only a few setae. Labrum located below clypeus or on apical clypeal margin (Oncerini, Sericini), transverse, narrowed, or conical.

Thorax with pronotum unarmed; scutellum exposed. Elytral margins straight, without notch posterior to humerus. Mesepimeron covered by base of elytra (slightly exposed in *Gymnopyge*). Legs with claws simple; claws cleft, toothed, serrate or pectinate; metatarsal claws paired in most, equal in thickness and length or single (*Hoplia*).

Abdomen with 7 or fewer pairs of abdominal spiracles, most with posterior abdominal spiracles located in sternites, tergites or pleural membrane (see *Acoma* and *Podolasia*), with 1 pair exposed beneath edge of elytra; 5 or 6 visible sternites fused, most with sutures visible at least laterally, or completely effaced (Chasmatopterini), the sixth sternite (when present) can be partially or completely retracted within fifth (Diplotaxini); trace of suture between fifth sternite and propygidium evident in most, entirely absent in Diplotaxini; pygidium exposed. Sexual dimorphism not strongly developed; most males with abdomen less convex, with longer tarsi and antennal club than females, some genera with males having specialized front claws (*Hypotrichia*) or with protibial and metatibial spurs absent (some male Macrodactylini).

Ritcher (1966) characterized the soil dwelling, C-shaped larvae of North American melolonthines as follows: Head with mandibles lacking stridulatory area, or indistinct, with patch of minute granules; scissorial area of mandible with distal, blade-like portion which is separated from proximal tooth by scissorial notch. Galea and lacinia of maxilla fused proximally but separated distally; rarely galea and lacinia separated but tightly fitted together; lacinia with longitudinal row of 3 unci; maxillary stridulatory area without anterior process. Antennae 4-segmented; last antennal segment with single, elliptical, dorsal sensory spot. Thorax with all legs well developed, 4-segmented; each claw bearing 2 setae. Abdomen with anal opening angulate or Y-shaped in most. Lower anal lip of most with sagittal cleft or grooves. Keys to larvae: Bøving 1936, 1937, 1942b, 1942c; Ritcher 1949, 1966 (see also Bøving 1942a; King 1984).

Habits and habitats. Adults and larvae of the Melolonthinae are generally phytophagous, although some adults apparently do not feed. Adults and larvae of some genera may be of considerable economic importance (*i.e., Amphimallon, Diplotaxis, Phyllophaga, Polyphylla, Maladera, Serica*), damaging a wide variety of crops, pastures, and turf. Adult *Phyllophaga* may be so abundant locally that deciduous trees may be completely defoliated by their nocturnal feeding activities (Ritcher 1966). Many adults of flower or pollen-feeding species are diurnal (*Chnaunanthus, Gymnopyge, Hoplia, Macrodactylus, Oncerus*), but almost all melolonthines are crepuscular or nocturnal. Primarily nocturnal species in the genera *Diplotaxis, Phyllophaga, Polyphylla*, and *Serica* are usually encountered in numbers flying about lights or copulating, feeding, or resting on the foliage of host plants.

Status of the classification. Recognition of taxa as tribes or subfamilies has not been applied consistently (see discussion under Chasmatopterini and Oncerini). Moreover, the tribal classification is in a relative state of confusion due to the lack of definition and inconsistent use of characters (Hardy 1978b; Ratcliffe 1991). This state of affairs is further exacerbated by the fact that descriptions of the majority of genera and species published before 1940 are largely inadequate and seldom accompanied by illustrations, making reliable determinations difficult without examination of type specimens. Early workers rarely consulted types, resulting in numerous synonymies and incorrect placements. The present work is not designed to resolve these issues but is intended to make North American melolonthine genera recognizable. Not until the completion of a comparative analysis of all genera, including the larvae, will the higher classification of the Melolonthinae achieve some stability and begin to reflect the phylogenetic relationships of the group.

The North American melolonthines have not been considered in their entirety for more than 140 years, although many genera have subsequently been revised. Early attempts at establishing schemes of higher classification for the melolonthines were based primarily on taxa found outside of the Nearctic region. LeConte (1856) recognized the following tribes in the United States under the family Melolonthidae: Macrophyllae (*Phobetus*); Melolonthinae (*Polyphylla, Thyce*); Rhizotrogi (*Phyllophaga sensu latu*); Diplotaxes (*Diplotaxis*); Sericae (*Serica*); Macrodactyli (*Macrodactylus*); Dichelonychae (*Dichelonyx*); Lasipodes (*Podolasia, Oncerus*); Hopliae (*Hoplia*); Glaphyri (*Lichnanthe*). LeConte (1861) later modified his classification as follows: Macrophyllini, Melolonthini, Rhizotrogi, Diplotaxini, Sericini, Sericoidini (*Hypotrichia*), Macrodactylini, Dichelonychini, Oncerini (for *Lasipodes*), Hopliini, and Glaphyrini.

Harold (1869a) recognized the following tribes with North American taxa under the family Scarabaeidae: Glaphyrini (LeConte's Glaphyrini and Lasipodes, *Chnaunanthus*); Melolonthini (LeConte's Hopliini, Sericini, Dichelonychini, Macrodactylini, Diplotaxini, Rhizotrogini, Melolonthinae, Macrophyllini). The catalog of Dalla Torre (1912, 1913) listed the following tribes with Nearctic representatives under the subfamily Melolonthinae: Chasmatopterini (including LeConte's Oncerini); Sericini; Liparetrini (*Plectrodes, Hypotrichia*); Melolonthini (LeConte's Melolonthini, Rhizotrogini, Diplotaxini); Pachydemini (LeConte's Macrophyllini); Macrodactylini (LeConte's Macrodactylini, Dichelonychini); Hopliini. These tribes were not formally characterized by Dalla Torre, leading to considerable confusion of the higher classification of the subfamily. With the subsequent addition of hundreds of genera and species since its publication,

Acknowledgments. Art Evans thanks Paul Skelley, Paul Lago, James Hogue, Juan Limón and William Warner for checking the melolonthine keys and adding information and Rosser Garrison for assistance with the illustrations.

there is little basis for much of the systematic content of Dalla Torre's catalog although it remains a valuable bibliographic tool. The catalogues of Leng (1920), Leng and Mutchler (1927, 1933) and Blackwelder (1944) followed the classification of Dalla Torre. Saylor (1937b, 1938) removed the tribes Chasmatopterini and Oncerini from the Melolonthinae and considered them as separate subfamilies. On the basis of characters of known larvae, Ritcher (1966) recognized the following tribes: Sericini, Diplotaxini, Dichelonycini, Macrodactylini, Hopliini, Pachydemini, Melolonthini, and Plectrini. Hatch (1971) recognized the Diplotaxini and Dichelonycini. Arnett (1968) followed the system of Dalla Torre, with the exception of combining Saylor's Oncerinae and Chasmatopterinae within the Chasmatopterini of the Melolonthinae. Morón *et al.* (1997) presented the following tribal and subtribal classification for the melolonthines of Mexico: Melolonthini (Diplotaxina, Melolonthina, Rhizotrogina), Macrodactylini (Ceraspina, Macrodactylina), Hopliini (Hopliina), Sericini (Sericina), Chasmatopterini, and Pachydemini.

Useful faunal surveys, systematic catalogs, biogeographical analyses, or conservation notes on Nearctic Melolonthinae include: CANADA: Bousquet 1991; MEXICO: Morón 1996; Morón *et al.* 1997; Baja California peninsula: Saylor 1948b; Chiapas: Morón *et al.* 1985; Durango: Morón 1981, Morón and Deloya 1991; Hidalgo: Morón 1994; Jalisco: Morón *et al.* 1988; Morelos: Deloya *et al.* 1995; Mexico: Morón and Zaragoza 1976; Veracruz: Morón 1979; UNITED STATES: Arnett 1968; Alaska: Bousquet 1991; Colorado: Zimmerman *et al.* 1991; Florida: Woodruff and Deyrup 1994; Indiana: Blatchley 1910; Nebraska: Ratcliffe 1991; North Dakota: Lago *et al.* 1979; South Carolina: Harpootlian 2001; Northeast: Downie and Arnett 1996; Pacific Northwest: Hatch 1971; western sand dune systems: Andrews *et al.* 1979. Catalogs: Loding 1945; Edwards 1949; Kirk and Balsbaugh 1975; Shook 1978.

Distribution. The cosmopolitan subfamily Melolonthinae is one of the largest and most diverse subfamilies in the Scarabaeoidea, with approximately 750 genera and nearly 11,000 species worldwide (Houston and Weir 1992). There are nine tribes, 30 genera, and more than 1,000 species in North America. The greatest diversity of North American Melolonthinae appears to be in the southwestern United States and Mexico.

Key to the Tribes and Genera of the United States, Canada, and Mexico

Genera included in this key are from Nearctic *and* Neotropical Mexico.

1. Metatarsus with two claws 2
— Metatarsus with single claw (Hopliini) .... *Hoplia*

2(1). Labrum not readily visible or located beneath clypeus (Figs. 47-48) 3
— Labrum distinct, located on apical margin of clypeus (Fig. 49) (Oncerini) 10

3(2). Labrum indistinct (in the Sericini, labrum located on apical margin of clypeus, suture barely indicated by differences in sculpturing on dorsal surface and by faintly indicated notches on margin) (Fig. 47) 4
— Labrum visible, located below the clypeus (Fig. 48) 5

4(3). Metatibial spurs subcontiguous, located below tarsal articulation (Fig. 77) (Podolasiini) 12
— Metatibial spurs set apart, below and above tarsal articulation (Fig. 78) (Sericini) 13

5 (3). Sutures between abdominal sternites distinct, at least laterally 6
— Sutures between abdominal sternites with sutures obliterated (Chasmatopterini) 11

6(5). Anterior pronotal margin with membranous border (Fig. 58); if border absent, then metacoxae widely separated (Fig. 68) 7
— Anterior pronotal margin without membranous border (Fig. 59); metacoxae contiguous or subcontiguous (Fig. 69) (Melolonthini) 21

7(6). Fifth abdominal sternite and propygidium separated by partial or complete suture (Fig. 86) 8
— Fifth abdominal sternite and propygidium not separated by suture (Fig. 87) (Diplotaxini) . *Diplotaxis*

8(7). Metatibial spurs set apart, located above and below tarsal articulation, separated by tarsal articulation (Fig. 78) 9
— Metatibial spurs absent, or with 1 or 2 subcontiguous spurs located below tarsal articulation (Fig. 77) (Macrodactylini) 15

9(8). Claws bifid or toothed (Figs. 81-82); labrum distinct, transverse, not fused to labrum (Fig. 48) (Pachydemini) 38
— Claws simple (Fig. 79); labrum reduced, conical, fused to labrum .... *Acoma (incertae sedis)*

10(2). Lateral margins of clypeus parallel; metatibial spurs apical, subcontiguous .... *Oncerus*
— Lateral margins of clypeus convergent anteriorly; metatibial spurs subapical, subcontiguous .... *Nefoncerus*

11(5). Base of clypeus armed with erect horn; pronotum abruptly declivous anteriorly .... *Chaunocolus*
— Base of clypeus lacking horn; pronotum simply convex .... *Chnaunanthus*

12(4). Eye canthus obsolete or nearly so (Fig. 52); eyes separated ventrally by less than width of 1 eye (Fig. 54) .... *Podolasia*
— Eye canthus present (Fig. 53); eyes separated ventrally by more than width of 1 eye (Fig. 55) .... *Podostena*

13(4). Antennae with 10 antennomeres .... *Maladera*
— Antennae with 8 or 9 antennomeres 14

14(13). Elytra without apical, membranous border .. *Serica*
— Elytra with apical, membranous border .... *Nipponoserica*

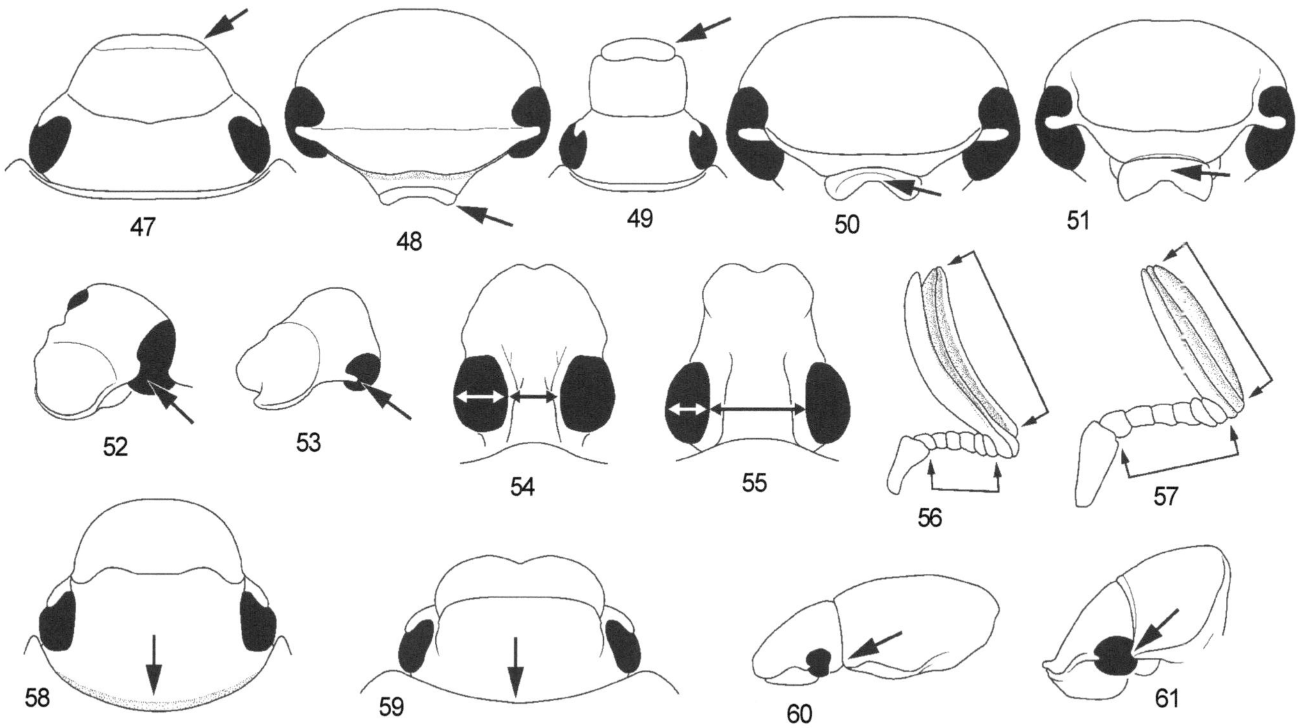

FIGURES 47.34-61.34. 47-55. Heads. 47. Dorsal view with labrum indistinct, located on apical margin of clypeus; 48. Frontal view with labrum located beneath clypeus; 49. Dorsal view with labrum distinct, located on apical margin of clypeus; 50. Frontal view with labrum and labral notch almost approaching clypeus; 51. Frontal view with labrum and labral notch distinctly separated from clypeus; 52. Oblique view with eye canthus nearly obsolete (after Howden 1997); 53. Oblique view with eye canthus present (after Howden 1997); 54. Ventral view with eyes separated by less than one eye width (after Howden 1997); 55. Ventral with eyes separated by more than one eye width (after Howden 1997). 56-57. Antennae. 56. Dorsal view with club twice the length of pedicel; 57. Dorsal view with antennal club subequal to length of pedicel. 58-59. Heads. 58. Dorsal view with anterior pronotal margin membranous; 59. Dorsal view with anterior pronotal margin lacking membranous border. 60-61. Head and pronotum, lateral view. 60. Anterior pronotal angle not reaching eyes; 61. Anterior, pronotal angle partially covering eyes.

15(8). Anterior pronotal angles never reaching eyes (Fig. 60) 16
— Anterior pronotal angles partially covering eyes (Fig. 61) 17

16(15). Apical margin of terminal tarsomeres entire (Fig. 75) .... *Macrodactylus*
— Apical margin of terminal tarsomere deeply notched (Fig. 76) .... *Dichelonyx*

17(15). Metacoxae contiguous or subcontiguous (Fig. 69) 18
— Metacoxae widely separated (Fig. 68) ..... *Ceraspis*

18(17). Claws cleft (Fig. 80) 19
— Claws bifid (Fig. 81) .... *Isonychus*

19(18). Claws distinctly cleft in lateral view 20
— Claws not cleft in lateral view .... *Coenonycha*

20(19). Antennae with 8 or 9 antennomeres .. *Gymnopyge*
— Antennae with 10 antennomeres .... *Plectris*

21 (6). Metepisternum narrow (Fig. 64) 22
— Metepisternum wide (Fig. 65) 32

22(21). Labrum narrow, less than half width of clypeus 23
— Labrum wide, more than half width of clypeus . 24

23(22). Clypeal suture carinate .... *Fossocarus*
— Clypeal suture not carinate .... *Cronocarus*

24(22). Labral notch almost approaching clypeus (Fig. 50) (orient specimen to achieve narrowest distance between notch and clypeus) (*Phyllophaga sensu lato*) 25
— Labral notch distinctly separated from clypeus (Fig. 51) .... *Amphimallon*

25(24). Outer surface of metatibiae with complete, oblique, transverse carina or dentiform processes (Figs. 72, 74) 26
— Outer surface of metatibiae with incomplete carina (Fig. 73) .... *Phyllophaga* (*Cnemarachis*)

26(25). Outer surface of mesotibiae with complete, oblique, transverse carina (Fig. 72) 27
— Outer surface of mesotibiae with dentiform processes (Fig. 74) .... *Phyllophaga* (*Triodonyx*)

27(26). Claws simple, bifid, or toothed (Figs. 79, 81-82) 28
— Claws serrate or pectinate (Figs. 84-85) .... *Phyllophaga* (*Listrochelus*)

28(27). All claws bifid or toothed 29
— Pro- and mesotarsal claws simple, metatarsal claws toothed .... *Phyllophaga* (*Chirodines*)

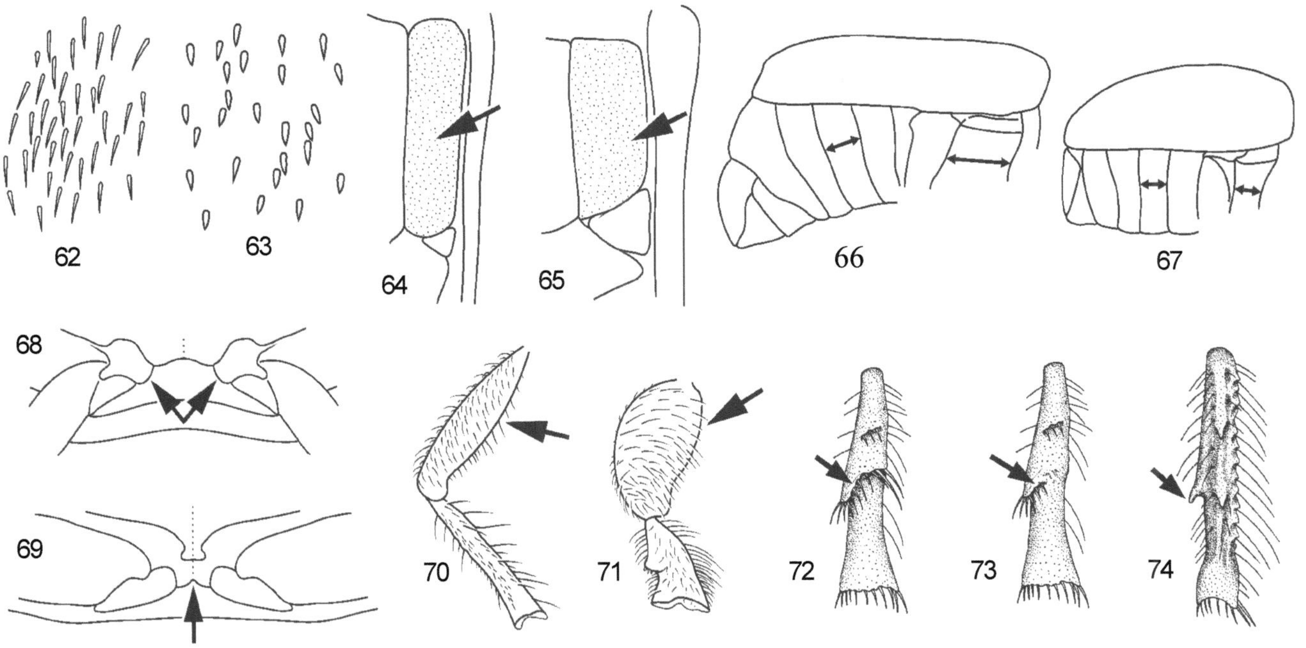

FIGURES 62.34-74.34. 62-63. Elytral vestiture. 62. Pubescent, or with hairs; 63. With scales or scale-like hairs. 64-65. Metathoraces, ventral view of left side. 64. Metepisternum narrow (after Lacroix 1989); 65. Metepisternum wide (after Lacroix 1989). 66-67. Lateral views of abdomen. 66. Metasternum longer than second abdominal sternite; 67. Metasternum subequal in length to second abdominal sternite. 68-69. Ventral views of metathorax and first abdominal sternite. 68. Metacoxae widely separated; 69. Metacoxae contiguous. 70-71. Hind legs, ventral view (after Britton 1957). 70. Metafemur slender; 71. Metafemur greatly enlarged. 72-74. Metatibiae, outer surface (after Morón 1986). 72. With complete, oblique, transverse carina; 73. With incomplete, transverse carina; 74. With dentiform process.

29(28). Claws bifid (Fig. 81) 30
— Claws toothed (Fig. 82) 31

30(29). Tarsi moderately to densely pilose beneath, less so in females; color light testaceous in most .... *Phyllophaga* (*Chlaenobia*)
— Tarsi sparsely setose beneath, never pilose; color variable .... *Phyllophaga* (*Phytalus*)

31(29). Metasternum longer than second abdominal sternite (Fig. 66) .... *Phyllophaga* (*Phyllophaga*)
— Metasternum subequal in length to second abdominal sternite (Fig. 67) .... *Phyllophaga* (*Tostegoptera*)

32(21). Antennal club with 5 (female) or 7 (male) segments .... *Polyphylla*
— Antennal club with 3 segments 33

33(32). Ventral abdominal sutures mostly distinct, imbricate, or with suture between sternites 4 and 5 obliterated only at midline 34
— Ventral abdominal sutures never imbricate, often obliterated medially 35

34(33). Outer tarsal claw on each leg with acute basal tooth; metafemora slender, less than or subequal in width to second abdominal sternite (Fig. 70) .... *Plectrodes*
— Inner tarsal claw of anterior and middle legs with large, oval, basal lobe; metafemora greatly enlarged, greater in width than second abdominal sternite (Fig. 71) .... *Hypotrichia*

35(33). Metafemora of males slender, less than or subequal in width to second abdominal sternite (Fig. 70). 36
— Metafemora of males enlarged, greater in width than second abdominal sternite (Fig. 71) 37

36(35). Antennal club of male twice length of pedicel (Fig. 56) .... *Dinacoma*
— Antennal club of male subequal in length to pedicel (Fig. 57) .... *Parathyce*

37(35). Dorsally pubescent, without scales or scale-like hairs (Fig. 62) .... *Hypothyce*
— Dorsally with at least some scales or scale-like hairs (Fig. 63) .... *Thyce*

38(9). Claws bifid (Fig. 81) .... *Phobetus*
— Claws tridentate (Fig. 83) .... *Benedictia*

Classification of the Tribes and Genera

The published distribution of each genus is indicated by country and state or province. The retrieval of generic provincial and state records was greatly facilitated by Bousquet (1991) and Morón *et al.* (1997). The inclusion of state and provincial records will readily identify range extensions and assist in compiling North American distributions.

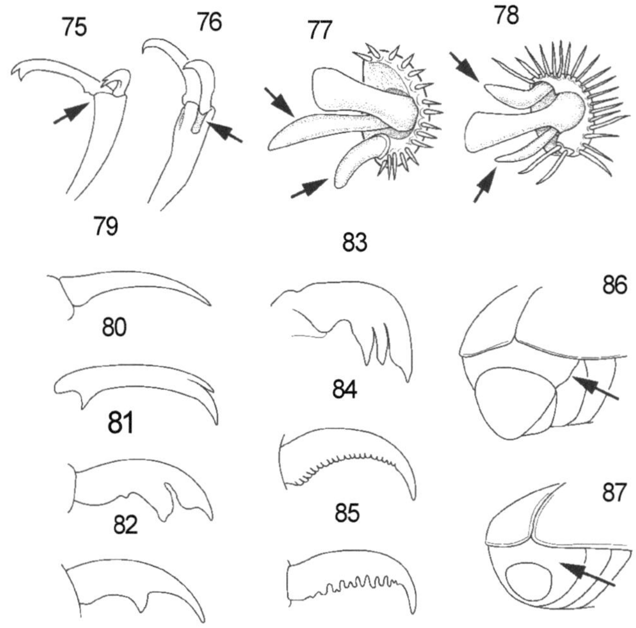

FIGURES 75.34-87.34. 75-76. Apices of terminal tarsomere, ventral view. 75. With apical margin without notched; 76. With apical margin deeply notched. 77-78. Apices of metatibia (after Britton 1957). 77. Metatibial spurs contiguous, located below tarsal articulation; 78. Metatibial spurs set apart, located above and below tarsal articulation. 79-85. Tarsal claws. 79. Simple; 80. Cleft (after Hardy 1977); 81. Bifid; 82. Toothed (after Morón 1986); 83. Tridentate; 84. Serrate (after Morón 1986); 85. Pectinate (after Morón 1986). 86-87. Caudal view of pygidium and abdomen. 86.With fifth abdominal sternite and propygidium separated by suture; 87. With fifth abdominal sternite and propygidium not separated by suture.

Melolonthinae Samouelle 1819

Hopliini Burmeister 1855

Characteristics: Antennae 9-segmented, club 3-segmented. Labrum located below clypeus, distinct. Fifth abdominal sternite and propygidium separated by suture. Protibiae with spurs, meso- and metatibiae without apical spurs. Opposing pro- and mesotarsal claws subequal, metatarsi with single claw.

The tribe Hopliini is predominately Afrotropical, with representatives in the Oriental and Palearctic regions. *Hoplia* is the sole representative of the tribe in the New World. The tribe has been considered as a separate subfamily. Saylor (1935) described the genus *Leptohoplia* and placed it within the Hopliini. The genus was later transferred to the Anomalini of the subfamily Rutelinae by Howden and Hardy (1971).

Hoplia Illiger 1803
This genus has 24 species found from CANADA: Alberta, British Columbia, Manitoba, New Brunswick, Nova Scotia, Ontario, Quebec; MEXICO: Baja California, Hidalgo, Oaxaca, Puebla, Tabasco, Veracruz; USA: generally distributed but not Alaska, Arkansas, Kansas, Louisiana, Oklahoma. Keys: Boyer 1940; Hardy 1977. Larvae: Ritcher 1966.

Oncerini LeConte 1861

Characteristics: Antennae 9-segmented, club 3-segmented. Labrum coplanar with clypeus. Fifth abdominal sternite and propygidium separated by suture. Protibiae without apical spurs, meso- and metatibiae with 2 apical spurs. Metatibial spurs subcontiguous, inserted below tarsal articulation; opposing tarsal claws similar, cleft apically.

Saylor (1938) placed *Oncerus* and *Nefoncerus* in the subfamily Oncerinae on the basis of the placement of the abdominal spiracles (entirely within the pleural membrane) and the apical attachment of the labrum to the clypeus. Both genera were included in the Chasmatopterini (Arnett 1968). The subfamily is restricted to the Nearctic region where there are two monotypic genera.

Nefoncerus Saylor 1938
One species, *N. convergens* (Horn), is found in MEXICO: Baja California Sur. Discussion: Saylor 1938; Morón *et al.* 1997.

Oncerus LeConte 1856
One species, *O. floralis* LeConte, is found in MEXICO: Baja California; USA: California. Discussion: Saylor 1938; Morón *et al.* 1997.

Podolasiini Howden 1997

Characteristics: Antennae 8 or 9-segmented, club 3-segmented. Labrum not visible, located below and fused to clypeus. Fifth abdominal sternite and propygidium separated by suture. Protibiae without apical spurs, meso- and metatibiae with 2 apical spurs. Metatibial spurs subcontiguous and located below tarsal articulation; opposing tarsal claws equal, simple.

Podolasia (as *Lasiopus*) and *Oncerus* were grouped in the Lasiopodes by LeConte (1856) and later the Oncerini (LeConte 1861). Harold (1869b) noted the preoccupation of *Lasiopus* and proposed the new name *Podolasia*. Harold (1869a) placed *Podolasia* and *Oncerus* in the tribe Glaphyrini. Dalla Torre (1912) listed both of these genera under the Chasmatopterini, but inexplicably listed LeConte's tribe Lasipodes as a synonym of the Pachydemini. Leng (1920) followed the classification of Dalla Torre. Both of these genera were later placed in the subfamily Oncerinae by Saylor (1938), but *Podolasia* was removed from the Oncerinae and placed *incertae sedis* by Howden (1954). Ritcher (1969) concurred with Howden, noting that the abdomen of *Podolasia* has seven pairs of functional spiracles and a vestigial eighth pair. All of the abdominal spiracles are located in the pleural membrane. Howden (1997) erected the tribe Podolasiini for *Podolasia* and included *Podostena*. The females are apparently flightless (Howden 1954, 1997).

Podolasia Harold 1869
Lasiopus LeConte 1856
Nine species occur from MEXICO: Baja California, Baja California Sur, Chihuahua, Durango, San Luis Potosí; USA: New Mexico, Texas, Utah. Keys: Howden 1954, 1997.

Podostena Howden 1997
Four species are found in MEXICO: Baja California Sur; USA: Texas. Key: Howden 1997.

Sericini Kirby 1837

Characteristics: Antennae 9 or 10-segmented, club 3-segmented. Labrum not easily visible, fused to apical margin of clypeus. Fifth abdominal sternite and propygidium separated by suture. Protibiae with apical spurs, meso- and metatibiae with 2 spurs. Metatibial spurs separated, inserted above and below tarsal articulation. Posterior femora broadly flattened. Opposing tarsal claws similar, cleft apically.

This large, cosmopolitan tribe contains many genera in all zoogeographical regions but primarily in the Afrotropical and Oriental regions. *Serica* is the only genus native to North America. Species of adult sercines are best identified by the configuration of the male genitalia. In California, adult *Serica* are known to damage the foliage of various fruit trees. The Asiatic garden beetle, *Maladera castanea* (Arrow), originally known from Japan, was first recorded from North America in New Jersey in 1921. Adult Asiatic garden beetles feed on the foliage of many plants, while the larvae damage the roots of lawns, strawberries, vegetable seedlings, and flowers. *Nipponoserica peregrina* (Chapin), another Japanese species, was described from specimens collected, in part, from Long Island, New York in 1937. Descriptions: Dawson 1919a, 1919b, 1920, 1921, 1922, 1932, 1933, 1947, 1952.

Maladera Mulsant and Rey 1870
Aserica Lewis 1895
Autoserica Brenske 1897
One species, *M. castanea* (Arrow), is found in the USA: Connecticut, Delaware, Georgia, Maryland, Massachusetts, Michigan, New Jersey, New York, North Carolina, Pennsylvania, Rhode Island, South Carolina, Vermont, Virginia, West Virginia (also Japan, Taiwan, and Korea). Description: Dawson 1967. Larvae: Ritcher 1966. Nomenclature: Pope 1961.

Nipponoserica Nomura 1973
One species, *N. peregrina* (Chapin), is found in USA: Connecticut, Maryland, Massachusetts, New Jersey, New York, Pennsylvania, Virginia, West Virginia (also Japan). Descriptions: Chapin 1938, Dawson 1952. (See Nomura 1976 for *Serica peregrina* transferred to *Nipponoserica*).

Serica MacLeay 1819
Stilbolemma Harris 1827
Camptorhina Kirby 1837
There are 99 species in CANADA: widely distributed, except Newfoundland and Yukon; USA: widely distributed, except Alaska: MEXICO: Baja California, Baja California Sur, Coahuila (also Afrotropical, Palearctic, Oriental regions). Keys: Dawson 1919-1967; Gordon 1975; Hardy and Andrews 1987; Ratcliffe 1991. Larvae: Ritcher 1966. Biology: Evans 1985. (See Nikolayev 1979 for additional nomenclatural commentary).

Chasmatopterini Lacordaire 1856

Characteristics: Antennae 9-segmented, club 3-segmented. Labrum located beneath and not fused to clypeus. Abdomen with sternites entirely fused, sutures indistinct except for apical sternite; fifth abdominal sternite and propygidium separated by suture. Protibiae without apical spurs, meso- and metatibiae with 2 apical spurs. Metatibial spurs subcontiguous, located below tarsal articulation. Opposing tarsal claws similar, cleft apically.

Two genera inhabit the Nearctic region and a third genus is found in the western Palearctic. Saylor (1937b) stated that the Chasmatopterini did not belong in the Melolonthinae and elevated the tribe to the rank of subfamily (see also Baraud and Branco 1990) on the basis that the abdominal spiracles in *Chnaunanthus* and the Palearctic genus *Chasmatopterus* are located in the pleural membrane. In erecting *Chaunocolus*, Saylor (1937a) related it to *Oncerus*, *Podolasia*, and *Chnaunanthus* but did not include it in his treatment of the Chasmatopterini. *Oncerus* and *Nefoncerus* became the basis for the subfamily Oncerinae (Saylor 1938). Arnett (1971) and Morón *et al.* (1997) recognized the Chasmatopterini as including all of the aforementioned genera. Ritcher (1969) found that the abdominal spiracles of *Chnaunanthus* were not entirely in the pleural membrane, bringing into question Saylor's assertion that the tribe could not be considered as part of the Melolonthinae.

Chaunocolus Saylor 1937
One species, *C. cornutus* Saylor, is found in MEXICO: Baja California Sur. Discussion: Saylor 1937b; Morón *et al.* 1997.

Chnaunanthus Burmeister 1844
Acratus Horn 1867
Pseudacratus Dalle Torre 1912
Four species occur in MEXICO: Baja California; Baja California Sur, Guanajuato, Michoacan, Mexico, Morelos, Nayarit, Oaxaca, Puebla; USA: Arizona, California, Oregon, Utah Key: Saylor 1937b. Discussion: Morón *et al.* 1997.

Melolonthini Samouelle 1819

Characteristics: Antennae 9 or 10-segmented, club 3, 5 or 7-segmented. Labrum located below clypeus, distinct in most. Fifth abdominal sternite and propygidium separated by suture. Protibiae with apical spurs, meso- and metatibiae with 2 spurs. Metatibial spurs contiguous, located below tarsal articulation. Opposing tarsal claws equal (protarsal claws unequal in *Hypotrichia*), simple, cleft, toothed, serrate or pectinate.

The tribe Melolonthini is the largest tribe of the subfamily. Its species are generally distributed throughout the world and include many of economic importance. The European chafer, *Amphimallon majalis* (Razoumowski), was inadvertently introduced from the western Palearctic. The larvae may cause damage to pastures, lawns, grain, and legumes (Ritcher 1966). The larvae of *Phyllophaga* and *Polyphylla* are also considered to be serious crop and turf pests as a result of their root-feeding activities, while the nocturnal feeding activities of some adult *Phyllophaga* may result in the defoliation of deciduous trees in the eastern United States.

LeConte's (1861) placement of *Hypotrichia* in the Sericoidini resulted in its subsequent listing under the Liparetrini by Dalla Torre (1912). Leng (1920) retained *Hypotrichia* in the Liparetrini but placed *Plectrodes* in the Melolonthini. Howden (1968a) placed *Hypotrichia* in the Melolonthini and added the new genus *Hypothyce*. Hardy (1974a) later added *Parathyce* and *Polylamina* to the tribe, but Coca-Abia (2000) synonymized *Polylamina* with *Polyphylla*. *Fossocarus* Howden and *Gronocarus* Schaeffer, previously placed in the Pachydemini, are placed here in the Melolonthini on the basis of the contiguous metatibial spurs located beneath the tarsal articulation and the partially obliterated abdominal sutures.

Amphimallon LePeletier and Serville 1828
One species, *A. majalis* (Razoumowski), is found in CANADA: Ontario, Quebec; USA: Connecticut, New York, West Virginia. Key: Baraud 1992. Larvae: Böving 1942c; Ritcher 1966. Discussion: Butt 1944; Gyrisco *et al.* 1954.

Dinacoma Casey 1889
Two species occur in the USA: California. Key: Hardy 1974a.

Fossocarus Howden 1961
One species, *F. creoleorum* Howden, is from the USA: Texas. Discussion: Howden 1961, 1971a.

Gronocarus Schaeffer 1927
Two species are from the USA: Alabama, Florida. Discussion: Howden 1961; Woodruff and Deyrup 1994.

Hypothyce Howden 1968
Two species occur in the USA: Georgia, Texas. Key: Hardy 1974a. Larvae: Ritcher 1973. Biology: Barfield and Gibson 1975.

Hypotrichia LeConte 1861
One species, *H. spissipes* LeConte, is from the USA: Florida. Discussion: Hardy 1974a; Woodruff and Deyrup 1994.

Parathyce Hardy 1974
Six species occur in MEXICO: Baja California; USA: California Key: Hardy 1974a. Larvae: Erwin 1970. Discussion: Morón *et al.* 1997.

Plectrodes Horn 1867
One species, *P. pubescens* Horn, occurs in the USA: California. Discussion: Hardy 1974a.

Phyllophaga Harris 1827
- subgenus *Phytalus* Erichson 1847. Key: Saylor 1939. Discussion: Frey 1975, Morón and Rivera 1992.
- subgenus *Chlaenobia* Blanchard 1850. Key: Chapin 1935. Discussion: Morón 1992.
- subgenus *Listrochelus* Blanchard 1851. Key: Saylor 1940a.
- subgenus *Chirodines* Bates 1888. Key: Morón 1991.
- subgenus *Cnemarachis* Saylor 1942. Key: Saylor 1942.
  - *Abcrana* Saylor 1942
  - *Clemora* Saylor 1942
- subgenus *Phyllophaga* Harris 1827. Keys: Luginbill and Painter 1953, Morón 1986, Woodruff and Beck 1989.
  - *Ancyclonycha* Dejean 1833
  - *Lachnosterna* Hope 1837
  - *Trichesthes* Erichson 1847
  - *Endrosa* LeConte 1856
  - *Gynnis* LeConte 1856
- subgenus *Tostegoptera* Blanchard 1851
  - *Eugastra* LeConte 1856
- subgenus *Triodonyx* Saylor 1942. Key: Warner and Morón 1992.

There are 400+ species generally distributed in North America except in: Alaska, Nunavut, Northwest Territories, and Yukon. Keys: Reinhard 1950; Luginbill and Painter 1953; Butler and Werner 1961; Morón 1986, 1993; Woodruff and Beck 1989; Ratcliffe 1991. Larvae: Bøving 1942a, 1942b; Ritcher 1949, 1966; Rosander and Werner 1970. Bibliography: Pike *et al.* 1976; Woodruff and Beck 1989. Discussion: Saylor 1940b; Wolcott 1948; Sanderson 1951; Frey 1975; Garcia-Vidal 1975, 1978, 1984, 1987; Chalumeau and Gruner 1976; Chalumeau 1983; King 1984; Woodruff and Deyrup 1994.

Polyphylla Harris 1841
- *Macranoxia* Crotch 1873
- *Polylamina* Hardy 1974

There are 31 species of *Polyphylla* (Fig. 46) distributed from CANADA: Alberta, British Columbia, Manitoba, Ontario, Quebec, Saskatchewan; MEXICO: Aguascalientes, Baja California, Chihuahua, Coahuila, Durango, Hidalgo, Jalisco, Morelos, Oaxaca, Puebla, Sonora; USA: widely distributed, except Ohio, Michigan, West Virginia. Keys: Hardy 1981; Young 1988. Larvae: Ritcher 1966. Discussion: Woodruff and Deyrup 1994; Morón *et al.* 1997.

Thyce LeConte 1856
Two species occur in MEXICO: Chihuahua; USA: California, New Mexico, Texas. Key: Hardy 1974a. Discussion: Morón *et al.* 1997.

Diplotaxini Burmeister 1855

Characteristics: Antennae 9 or 10-segmented, club 3-segmented. Labrum located below clypeus. Abdominal sutures distinct; fifth abdominal sternite and propygidium not separated by suture; sixth abdominal sternite nearly completely retracted beneath fifth sternite. Protibiae with apical spurs, meso- and metatibiae with 2 spurs; metatibial spurs contiguous, located below tarsal articulation. Opposing tarsal claws equal, toothed, or cleft apically.

The tribe Diplotaxini is found in the Afrotropical, Neotropical, and Oriental regions with one genus found in the Nearctic region. Adults are mostly nocturnal, feeding on the foliage of various plants. The larvae of some species may damage seedlings of commercially grown trees. *Diplotaxis* species occur from Canada to Panama and reach their greatest diversity in the southwestern United States and the central highlands of Mexico (Vaurie 1958).

Diplotaxis Kirby 1837
Alobus LeConte 1856
Orsonyx LeConte 1856
Diazus LeConte 1860
This genus includes 212 species from CANADA: generally distributed except Newfoundland, Northwest Territories, and Yukon; MEXICO: widely distributed; USA: widely distributed. Keys: Vaurie 1958, 1960, 1963; Ratcliffe 1991. Larvae: Ritcher 1966. Biology: Evans 1985. Discussion: Delgado-Castillo 1990; McCleve 1993; Woodruff and Deyrup 1994; Morón *et al.* 1997.

Macrodactylini Burmeister 1855

Characteristics: Antennae 7 to 10-segmented, club 3-segmented. Labrum located below clypeus. Fifth abdominal sternite and propygidium separated by suture. Protibiae with or without apical spurs, meso- and metatibiae with or without spurs; if present, paired metatibial spurs contiguous, located below tarsal articulation. Opposing tarsal claws equal or subequal, simple, cleft apically, or bifid.

The tribe Macrodactylini is primarily Neotropical, with species also occurring in the Oriental region. One species of *Plectris* has become established in Australia. Seven genera are known from North America, including one introduced species of the genus *Plectris*. The diurnal adults of *Gymnopyge, Macrodactylus*, and some *Dichelonyx* feed on flowers or pollen, while the nocturnal *Coenonycha* and most *Dichelonyx* feed on foliage. Ritcher (1966) provided a key to the tribe based on larvae. Both Ritcher (1966, 1969) and Hatch (1971) placed *Coenonycha* and *Dichelonyx* in the Dichelonycini.

Ceraspis LePeletier and Serville 1828
Faula Blanchard 1850
There are four species from MEXICO: Chiapas, Guerrero, Oaxaca, Tabasco, and Veracruz; Saylor (1935) reported a specimen from Arizona, but this record is undoubtedly erroneous. Key: Frey 1962. Biology: Capistran and Aquino 1992. Discussion: Delgado-Castillo *et al.* 1987; Morón *et al.* 1997.

Coenonycha Horn 1876
This genus has 34 species from MEXICO: Baja California; USA: Arizona, California, Nevada, Oregon, Utah, and Washington. Keys: Cazier and McClay 1943; Evans and Smith 1986; Evans and D'Hotman 1988. Larvae: Ritcher 1966. Biology: Tilden and Mansfield 1944; Evans 1985. Discussion: Morón *et al.* 1997.

Dichelonyx Harris 1827
Dichelonycha Kirby 1837
There are 29 species from CANADA: generally distributed except Newfoundland and Yukon; MEXICO: Baja California, Baja California Sur; USA: generally distributed. Keys: Saylor 1945a; Brown 1946. Larvae: Ritcher 1966. Biology: Evans 1985. Discussion: Morón *et al.* 1997.

Gymnopyge Linell 1895
There are four species from the USA: Arizona, California, Oregon, and Utah. Discussion: Linell 1896; Cazier 1939.

Isonychus Mannerheim 1829
This genus contains nine species from MEXICO: Durango, Guerrero, Sinaloa, Jalisco, Mexico, Michoacan, Morelos; Nayarit, and Oaxaca; USA: Arizona. Discussion: Howden 1959; Morón *et al.* 1997.

Macrodactylus Dejean 1821
Stenothorax Harris 1827
Macrodactylus has 27 species from CANADA: Ontario, Quebec; MEXICO: widely distributed; USA: Arizona, eastern and southern states. Keys: Horn 1876; Carrillo and Gibson 1960; Arce-Perez and Morón 2000. Larvae: Ritcher 1966. Discussion: Morón *et al.* 1997.

Plectris LePeletier and Serville 1828
One species, *P. aliena* Chapin, is found in the USA: Alabama, Georgia, Florida, South Carolina (also Paraguay). Key: Frey 1967. Larvae: Bøving 1936; Ritcher 1966. Discussion: Chapin 1934.

Pachydemini Reitter 1902

Characteristics: Antennae 10-segmented, club 3 to 5-segmented. Labrum located below clypeus. Fifth abdominal sternite and propygidium separated by suture. Protibiae with apical spurs, meso- and metatibiae with 2 spurs; metatibial spurs separated, located on either side of tarsal articulation. Opposing tarsal claws equal, toothed, or cleft apically.

Sanderson (1939) described the genus *Benedictia* and placed it in the Pleocominae (Pleocomidae). Hardy (1978b) transferred *Benedictia* to the Pachydemini and commented that this tribe had become a dumping ground for genera that are difficult to place elsewhere, attributing this situation to the lack of a formal definition of the tribal concept by Dalla Torre (1913). *Fossocarus*, and *Gronocarus* were provisionally placed in the Pachydemini at the time of their description, but they are treated as members of the Melolonthini in this work.

Benedictia Sanderson 1939
One species, *B. pilosa* Sanderson, is from MEXICO: Chihuahua, Coahuila; USA: Texas. Discussion: Sanderson 1939; Hardy 1978b.

Phobetus LeConte 1856
Eleven species are found in CANADA: British Columbia; MEXICO: Baja California; USA: California, Oregon, Washington. Key: Hardy 1978a. Larvae: Ritcher 1966. Biology: Evans 1985. Discussion: Morón *et al.* 1997.

Incertae Sedis

Acoma Casey 1889
There are 24 species from MEXICO: Baja California, Baja California Sur, Chihuahua, Durango, Sonora; USA: Arizona, California, New Mexico, and Texas. Keys: Cazier 1953; Howden 1958, 1962.

Characteristics: Antennae 9- or 10-segmented, club 3- to 5- or 7-segmented. Labrum located below and fused to clypeus. Abdomen with sutures distinct; fifth abdominal sternite and propygidium separated by suture. Protibiae with apical spurs, meso- and metatibiae with 2 apical spurs; 2 metatibial spurs separated by tarsal articulation. Opposing claws equal, simple.

Casey (1889) and Davis (1935) placed *Acoma* next to *Podolasia*. Arrow (1909, 1912) and Leng (1920) placed *Acoma* under the Pleocominae (Pleocomidae). Blackwelder (1944) placed *Acoma* under the Chasmatopterini in the subfamily Melolonthinae, whereas Cazier (1953) kept it in the Pleocominae (Pleocomidae). Females are believed to be apterous (Van Dyke 1928; Cazier 1953). Howden (1958) stated that the phylogenetic placement would remain in doubt until the morphology of the female became known. Ritcher (1969) recorded the functional abdominal spiracles on segments 1-5, all of which are situated in the pleural membrane. Spiracles corresponding with abdominal segments 6-8 are vestigial. Spiracles associated with segments 6 and 8 are located in the membrane, while the spiracles on segment 7 are located in the sternite.

IV. Rutelinae MacLeay 1819

by Mary Liz Jameson

Common name: The shining leaf chafers

Characteristics: Form elongate oval. Labrum produced weakly beyond apex of clypeus (except in *Anomalacra* [Anomalini]). Antenna with 8-10 segments, antennal club with 3 segments. Anterior coxae transverse. Scutellum exposed. Tarsal claws on all legs independently movable, claws unequal in length or size and frequently weakly split at apex, 1 claw of each pair greatly reduced (1 claw lacking on all legs in *Leptohoplia* [Anomalini]). Onychium laterally flattened. Pygidium exposed beyond apices of elytra.

The subfamily Rutelinae is composed of approximately 200 genera and 4,100 species that are distributed worldwide (Machatschke 1972), although many taxa remain to be described.

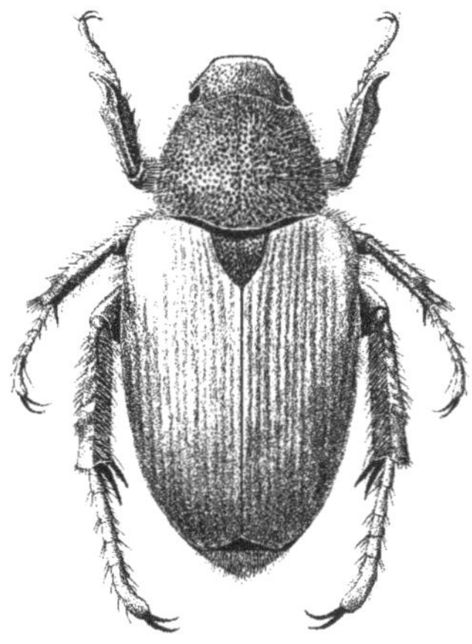

FIGURE 88.34. *Strigoderma arboricola* (Fabricius) (Used by permission of University of Nebraska State Museum).

Adult rutelines are phytophagous and feed on leaves, flowers, or flower parts. Larvae feed on roots, compost, and decaying vegetation. Some taxa, such as *Popillia japonica* Newman and *Anomala* species (both Anomalini), are agricultural pests. The common name of the subfamily, the shining leaf chafers, reflects the fact that many members of the subfamily are brightly colored, beautifully patterned, and often brilliantly metallic leaf-feeding beetles. Others in the subfamily, such as the genus *Anomala*, are small, obscure beetles. The subfamily is divided into six tribes, two of which occur in the United States. The tribe Spodochlamyini is found only in Central and South America; the tribe Anoplognathini occurs in the Australia and Central and South America; the tribe Adoretini is distributed throughout the Old World; the tribe Geniatini is distributed in Central and South America. The two remaining tribes, Rutelini and Anomalini, are the most speciose ruteline tribes and both occur in the Nearctic region. The tribe Rutelini is widely distributed but is most speciose in the Neotropics. The tribe Anomalini is widely distributed and is most speciose in the Old World. In the Nearctic region, the subfamily includes 14 genera and 95 species. Keys to genera and species: Casey 1915; Cooper 1983; Jameson 1990, 1998. Keys to larvae: Ritcher 1966; Jameson *et al.* 1994, Jameson 1998. United States catalog: Hardy 1991; Smith 2001. Regional works: Blatchley 1910; Loding 1945; Saylor 1948b; Edwards 1949; Hatch 1971; Kirk and Balsbaugh 1975; Lago *et al.* 1979; Ratcliffe 1991; Downie and Arnett 1996; Morón *et al.* 1997; Harpootlian 2001. World catalog: Machatschke 1972.

Key to the Tribes and Genera from the United States, Canada, and Nearctic Mexico

1. Lateral margin of elytra with membranous border (Fig. 89) (Anomalini) 2
— Lateral margin of elytra lacking membranous border (Fig. 90) (Rutelini) 6

2(1). Protibial spur absent; lacinia reduced, with 2 or fewer teeth ..... *Leptohoplia*
— Protibial spur present; lacinia not reduced, with more than 2 teeth 3

3(2). Base of pronotum tri-emarginate (Fig. 91); pygidium and abdominal sternites at sides with patches of dense, white setae .... *Popillia*

— Base of pronotum rounded posteriorly (Figs. 92-93); pygidium and abdominal sternites at sides without patches of dense, white setae 4

4(3). Clypeus parabolic; apex of labrum thin, narrowly exposed beneath clypeus apex..... *Anomalacra*

— Clypeus rounded or quadrate; apex of labrum moderately thick, broadly exposed beneath clypeal apex 5

5(4). Mesepimeron narrowly exposed in front of elytra (Fig. 93); base of pronotum narrower than base of elytra .... *Strigoderma*

— Mesepimeron not exposed in front of elytra (Fig. 92); base of pronotum subequal to width of elytra at base .... *Anomala*

6(1). Frontoclypeal suture complete, separating frons from clypeus (obsolete medially in *Paracotalpa deserta* Saylor) 7

— Frontoclypeal suture obsolete or lacking at middle 13

7(6). Mandible on external edge rounded 9

— Mandible on external edge with acute, apical tooth 8

8(7). Apex of mentum weakly emarginate; posterior border of metafemur in male without medial spine . ... *Homoiosternus*

— Apex of mentum with U-shaped emargination; posterior border of metafemur in male with medial spine .... *Plesiosternus*

9(7). Antenna with 10 antennomeres 10

— Antenna with 8 or 9 antennomeres .. *Parachrysina*

10(9). Terminal maxillary palpomere as long or longer than antennal club .... *Pseudocotalpa*

— Terminal maxillary palpomere shorter than antennal club (nearly as long as antennal club in *Cotalpa subcribrata* Wickham) 11

11(10). Apex of mentum sinuate or bisinuate or emarginate (Figs. 94-96) 12

— Apex of the mentum notched (Fig. 97) .... *Parabyrsopolis*

12(11). Pronotum glabrous; clypeus subrectangular, anterior angles narrowly rounded; large claw of at least metatarsus cleft in males, all claws simple in females .... *Cotalpa*

— Pronotum setose (at least in places); clypeus semicircular (widest at base), angles broadly rounded (*P. deserta* with subrectangular clypeus); claws of both sexes simple .... *Paracotalpa*

13(11). Frontoclypeal suture raised laterally; mandible on external edge with 1 acute, apical tooth....... 14

— Frontoclypeal suture not raised laterally; mandible on external edge rounded on bidentate 15

14(13). Antennal club 2 times longer than antennomeres 2-7; ventral apex of metatarsomere 4 with acute projection and 4 setae .... *Parastasia*

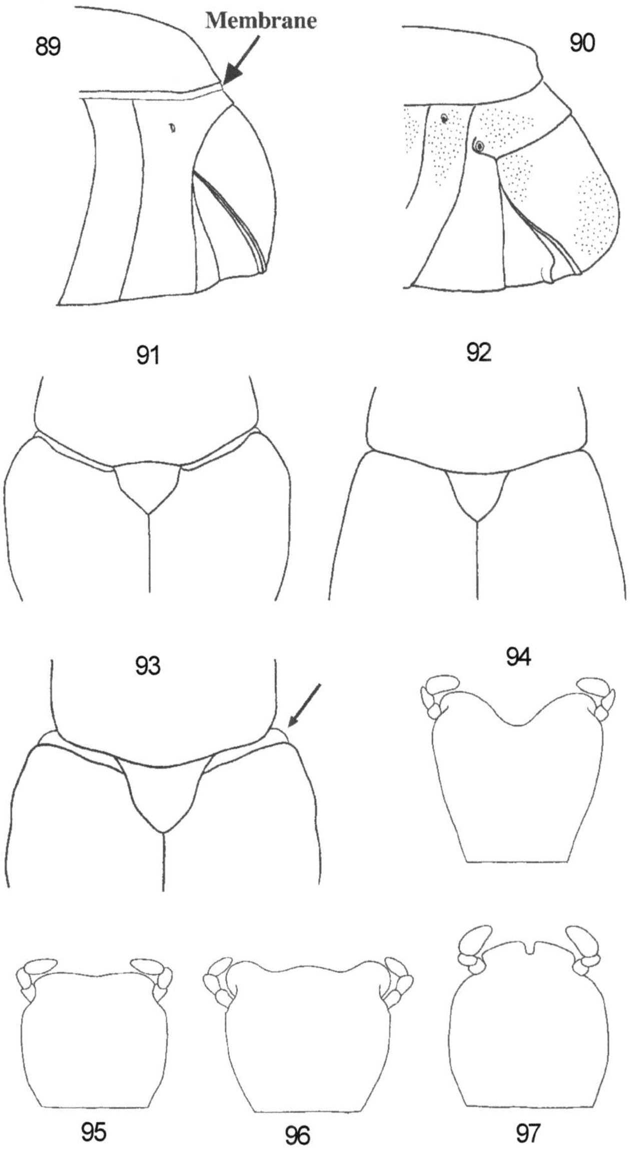

FIGURES 89.34-97.34. 89-90. Apex of pygidium and elytra. 89. Anomalini, margin of elytra with membranous border; 90. Rutelini, margin of elytra without membranous border. 91-93. Dorsal view of pronotum and elytra showing form of pronotal base. 91. *Popillia japonica* Newman; 92. *Anomala* sp.; 93. *Strigoderma* sp. 94-97. Mentum, showing apical margins. 94. *Paracotalpa* sp., deeply sinuate; 95. *Cotalpa* sp., weakly sinuate; 96. *Cotalpa consobrina* Horn, bisinuate; 97. *Parabyrsopolis chihuahuae* (Bates), notched.

— Antennal club subequal in length to antennomeres 1-7; ventral apex of metatarsomere 4 truncate and with 4 spines .... *Rutelisca*

15(13). Base of scutellum depressed below plane of elytron; elytron with marginal bead, epipleuron shelf-like 17

— Base of scutellum planar, extending anteriorly beneath pronotum; elytron without marginal bead, epipleuron rounded 16

16(15). Apex of metatibia with 10-18 spinules; mandibles rounded on external edge .... *Calomacraspis*
— Apex of metatibia with 0-3 spinules; mandibles bidentate on external edge .... *Rutela*

17(16). Apex of metatibia with 6-20 spinules ...... *Chrysina*
— Apex of metatibia with 0-1 spinules 18

18(17). Posterior tibia apicolaterally with small, articulated bristle; color rufous .... *Ectinoplectron*
— Posterior tibia apicolaterally lacking small, articulated bristle; color testaceous, castaneous, or black .... *Pelidnota*

Classification of the Tribes and Genera

Rutelinae MacLeay 1819

Anomalini Mulsant 1842

Characteristics: Antennae with 9 segments. Protibiae bidentate (rarely unidentate or tridentate), inner protibial spur subapical (lacking in *Leptohoplia*). Elytra with membranous border at lateral margin. Terminal spiracle not positioned in pleural suture.

The tribe Anomalini includes one of the largest genera in the Animal Kingdom: the genus *Anomala,* which includes approximately 1,000 species worldwide. Adult anomalines feed primarily on flowers and floral parts. Larvae feed primarily on plant roots. One adventive member of the tribe in North America, the Japnaese beetle (*Popillia japonica* Newman), causes economic damage to agricultural crops and ornamental plants. Because of potential damage to crop species, life histories of anomalines have been fairly well studied in the United States. Key to genera: Cooper 1983; Potts 1974, 1977a, 1977b. Key to larvae: Ritcher 1966.

Anomala Samouelle 1819
Phyllopertha Stephens 1830
Spilota Burmeister 1844
Callistethus Blanchard 1851
Pachystethus Blanchard 1851
Blitopertha Reitter 1903
Exomala Reitter 1903
Anomalepta Casey 1915
Anomalopus Casey 1915
Hemispilota Casey 1915
Oliganomala Casey 1915
Paranomala Casey 1915
Rhombonalia Casey 1915
Anomalopides Strand 1928

The genus *Anomala* (Fig. 1, Superfamily Scarabaeoidea) contains a heterogeneous assemblage of species and is in serious need of taxonomic study. In the United States and Canada 48 species are widely distributed. *Anomala marginata* (Fabricius) is considered a member of the genus *Callistethus* by some authors, and *A. orientalis* (Waterhouse) is placed in the genus *Exomala* by some authors. Adults feed on foliage and flowers, including the florets of grasses, and are diurnal and nocturnal. Larvae are subterranean root feeders and can be found in fields of oats, wheat, and corn, but they are also found under logs, rocks, and dried cow dung. Key to adults: Potts 1974, 1977a, 1977b. Larvae: Ritcher 1966.

Anomalacra Casey 1915
Anomalacra is a monotypic genus that includes only *A. clypealis* Schaeffer, and it occurs in Arizona and northern Mexico. Key: Potts 1974.

Leptohoplia Saylor 1935
Leptohoplia is a monotypic genus that includes only *L. testaceipennis* Saylor. The species occurs in the Colorado Desert of California and bordering regions in Arizona. In some specimens, one posterior claw may be so reduced as to appear absent. The genus *Hoplia* (subfamily Melolonthinae, tribe Hopliini) possesses only one posterior claw and, based on this character, *Leptohoplia* was previously placed in the Melolonthinae. Taxonomy and biology: Howden and Hardy 1971; Potts 1974.

Popillia Dejean 1821
The genus *Popillia* includes over 300 species in Asia and Africa. The Japanese beetle, *P. japonica* Newman, arrived from Asia into New Jersey in the roots of nursery stock in 1916. The species has expanded its range from the eastern regions (including Ontario) to as far west as Nebraska. Adults are severe pests of fruits and vegetables, field and forage crops, and ornamental plants. Larvae feed on various plant roots including ornamentals, grasses (including turfgrass), and vegetables. Larvae: Ritcher 1966.

Strigoderma Burmeister 1844
Alamona Casey 1915
Strigodermella Casey 1915

Five species occur in the United States and Canada, and an additional four species occur in Nearctic Mexico. Adults of *S. arboricola* (Fabricius) (Fig. 88) may cause damage to rose blossoms and are commonly encountered on many flowering plants. Keys to adults: Bader 1992. Larvae: Ritcher 1966.

Rutelini MacLeay 1819

Characteristics: Antennae with 10 segments (8 or 9 in *Parachrysina*). Protibiae tridentate, inner protibial spur apical. Elytral margin entirely chitinous. Terminal spiracle positioned in pleural suture.

The tribe Rutelini is distributed worldwide but is most speciose in the Neotropics. A wide array of morphological forms is exhibited by members of the tribe including taxa with enlarged, horn-like mandibles (*Fruhstorferia* from Asia), backward-projecting thoracic horns (*Peperonota* from Asia), enlarged hind femora (*Heterosternus* and *Chrysina* from the New World), and strikingly-colored, metallic silver and gold beetles (*Chrysina* from the New World). Keys to adults: Cooper 1983; Jameson 1990. Keys to larvae: Ritcher 1966; Jameson *et al.* 1994; Jameson 1998.

Calomacraspis Bates 1888

Calomacraspis splendens (Burmeister) is the only species of four in the genus that is distributed in the Nearctic region; it extends from southern Mexico to the southern limit of Nearctic Mexico. Adults are bright metallic green, diurnal, and feed on foliage, pollen, and floral parts (Jameson *et al.* 1994). Key to adults, biology, and larvae: Jameson *et al.* 1994.

Chrysina Kirby 1828
Plusiotis Burmeister 1844
Plusiotina Casey 1915
Pelidnotopsis Ohaus 1915

The genus *Chrysina* is well-known for its beautiful, metallic colors and relatively large size. Species in the genus are found from the southern United States to northern South America, and the genus includes 95 species (Hawks in press). Six species in this large genus occur in Arizona, New Mexico, Texas, and northern Mexico. *Chrysina gloriosa* (LeConte), a beautiful green and silver ruteline, occurs in Arizona, New Mexico, and Texas and feeds on juniper foliage. In some species, males possess greatly enlarged hind femora (such as *C. erubescens* Bates from northern Mexico). *Chrysina plusiotina* (Ohaus) is endemic to the Sierra Madre Oriental in Mexico and is rare in collections. Based on morphological and molecular data, Hawks *et al.* (in press) synonymized the genera *Plusiotis* and *Pelidnotopsis* with *Chrysina.* Key to adults: Cazier 1951; Morón 1990; Hawks and Bruyea 1999. Biology: Young 1957; Morón *et al.* 1997. Catalog: Hawks in press. Larvae: Ritcher 1966. (Volume 2, Color Figure 23)

Cotalpa Burmeister 1844
Ciocotalpa Saylor 1940 (subgenus)

The genus includes six species that are widely distributed in the United States. The goldsmith beetle, *Cotalpa lanigera* (L.), is found in sandy areas and feeds on the foliage of various trees including aspen, cottonwood, and willow. Key to adults: Saylor 1940d; Young 2002. Larvae: Ritcher 1966.

Ectinoplectron Ohaus 1915

Ectinoplectron includes only *E. oryctoides* (Ohaus) (= *Pelidnota howdeni* Hardy). This species is found only in northwestern Mexico. References: Morón 1990; Morón *et al.* 1997.

Homoiosternus Ohaus 1934

Homoiosternus includes two species and one of these, *H. beckeri* Ohaus, occurs in the Nearctic region. This species is found only in oak and pine/oak forests in the southern Sierra Madre Occidental mountain range of Mexico. Reference: Morón 1983a; Morón *et al.* 1997; Delgado and Blackaller-Bages 1997.

Parabyrsopolis Ohaus 1915
Byrsopolis Burmeister 1844
Parareoda Casey 1915

The genus *Parabyrsopolis* includes only *P. chihuahuae* (Bates). This species occurs in the Huachuca and Patagonia mountains of Arizona in the United States and in the Sierra Madre Occidental mountains in Mexico. The species is found in pine-oak forests and is attracted to lights at night. Reference: Jameson 1990.

Parachrysina Bates 1888

One species, *P. borealis* Jameson, occurs in the Sierra Madre Oriental mountains in Nuevo Leon, Mexico. The genus is unique in the Rutelini for its 8- or 9-segmented antenna. Key to adults: Jameson 1991.

Paracotalpa Ohaus 1915
Pocalta Casey 1915

Species in the genus are commonly called "little bears" due to their dense, long hair. The genus includes four species that are distributed west of the Rocky Mountains in the United States: *P. ursina* (Horn), *P. deserta* Saylor, *P. granicollis* (Haldeman), and *P. puncticollis* (LeConte). Larvae of *P. ursina* have been recorded feeding on sage brush (*Artemisia* sp.) (Ritcher 1966). Key to adults: Saylor 1940d. Larvae: Ritcher 1966.

Parastasia Westwood 1842
Barymorpha Guérin-Méneville 1843
Polymoechus LeConte 1856
Urleta Westwood 1875
Echmatophorus Waterhouse 1895
Ohkubous Sawada 1938

One species, *P. brevipes* (LeConte), is distributed in the eastern United States. The remaining species of *Parastasia* are distributed in Asia. Ratcliffe (1991) synonymized *P. conicicollis* (Casey) with *P. brevipes*. Adults are collected from lights at night. Key to adults: Kuijten 1992. Larva: Ritcher 1966.

Pelidnota MacLeay 1819
Aglycoptera Sharp 1885
Pelidnotidia Casey 1915

The genus *Pelidnota* includes about 100 species and is most speciose in South America. Six species are distributed in Nearctic North America. The spotted pelidnota, *P. punctata* (L.), occurs east of the Rocky Mountains. Adults feed on the foliage and fruits of grapes, and larvae feed on decaying roots and tree stumps. Key to adults: Hardy 1975. Larvae: Ritcher 1966.

Plesiosternus Morón 1983

Two species in the genus *Plesiosternus* are known, and *P. setosus* Morón occurs in the Nearctic region. The species is distributed from the Mexican states of Tamaulipas to Hidalgo and occurs in pine-oak forests. Reference: Morón 1983a; Morón and Howden 1992.

Pseudocotalpa Hardy 1971

The genus includes three species (*P. giulianii* Hardy, *P. andrewsi* Hardy, *P. sonorica* Hardy) that are distributed in sandy regions in California, Nevada, and Sonora, Mexico. Key to adults: Hardy 1974b. Biology: Hardy 1976, 1986.

Rutela Latreille 1802
Diabasis Hoffmansegg 1817
One species, *R. formosa* Burmeister, occurs in Florida, Georgia, the Bahamas, and Cuba. Keys, biology, larvae: Jameson 1998.

Rutelisca Bates 1888
This genus includes two species, one of which (*R. durangoana* Ohaus) is found in the Nearctic region. This species occurs in the pine/oak forests of the Sierra Madre Occidental mountains of Mexico. Key to adults: Jameson 2000. Larvae: Morón and Deloya 1991.

V. Dynastinae MacLeay 1819

by Brett C. Ratcliffe

Common names: The rhinoceros beetles and their hornless kin

Characteristics: Labrum hidden beneath clypeus. Clypeus with apex bidentate, rounded, acuminate, or truncate. Mandibles variable: Cyclocephalini and U.S. Phileurini lack teeth on lateral edge of mandibles, other tribes with teeth or lobes. Antennae 9-10 segmented, club small in most and with 3 segments. Scutellum normal, never enlarged. Procoxae transverse. Tarsal claws all subequal in size except for males of most Cyclocephalini and some Pentodontini with protarsal claws enlarged. Male genitalia diagnostic in nearly all species.

The sexes of most species are dimorphic (except Phileurini, some Cyclocephalini, and some Pentodontini) with males having either horns or enlarged tubercles or enlarged protarsi. The subfamily is divided into eight tribes, and five occur in the Nearctic region. Members of the Agaocephalini are Neotropical, while those of the Hexodontini are restricted to Madagascar and the Oryctoderini are found in the Australian and Oriental regions. In the United States, Canada, and northern Mexico there are 18 genera and about 62 species. Keys to adults: Endrödi 1985. Keys to larvae: Ritcher 1966. United States catalog: Hardy 1991; Smith 2001. Regional works: Blatchley 1910; Loding 1945; Saylor 1948a, 1948b; Edwards 1949; Hatch 1971; Kirk and Balsbaugh 1975; Shook 1978; Lago *et al.* 1979; Gordon and Anderson 1981; Ratcliffe 1991; Downie and Arnett 1996; Morón *et al.* 1997; Harpootlian 2001.

Key to the Tribes and Genera from the United States, Canada, and Nearctic Mexico

1. Head and pronotum lacking tubercles, carina, horn, or fovea; most males with anterior tarsomeres enlarged (Cyclocephalini) 5
— Head and pronotum with tubercles, carina, horn, or fovea; males with anterior tarsomeres slender or, if enlarged, with distinct transverse carina on head 2

2(1). Mentum greatly enlarged to cover bases of labial palps; clypeus sharply acuminate (Phileurini) . 7
— Mentum not enlarged, bases of labial palps visible; clypeal apex acuminate or not, but not sharply pointed 3

3(2). Apex of posterior tibia truncate or crenulate, upper apical angle not prolonged into large tooth (Fig. 99) (Pentodontini) 9
— Apex of posterior tibia crenulate or with triangular teeth or with prolonged tooth on upper apical angle (Fig. 100) 4

4(3). Mandibles concealed beneath clypeus or exposed; if exposed, then with 1-2 broadly rounded lobes or teeth (basal and subapical); males with median declivity on pronotum, females with or without declivity (Oryctini) 17
— Mandibles exposed at sides of clypeus and with 2 narrowly elongate teeth; neither males nor females with pronotal declivity (Dynastini) 18

5(1). Apex of clypeus angularly parabolic; mentum with apex deeply emarginate .... *Ancognatha*
— Apex of clypeus rounded or truncate; mentum with apex entire or weakly emarginate 6

6(5). Clypeus trapezoidal; color black .... *Dyscinetus*
— Clypeus rounded at apex; color testaceous or dark brown in some .... *Cyclocephala*

7(2). Apex of posterior tibia truncate, without teeth (dorsal angle also not spiniformly produced) .... *Archophileurus*
— Apex of posterior tibia with dorsal angle spiniformly produced into a tooth 8

8(7). Tubercles or horns of frons located just either side of middle, removed from lateral margin of head; pronotum without fovea in front of longitudinal furrow .... *Hemiphileurus*
— Tubercles or horns of frons located on lateral margin of head; pronotum with subapical fovea behind tubercle and in front of longitudinal furrow .... *Phileurus*

9(3). Antenna with 9 antennomeres . *Aphonides* (in part)
— Antenna with 10 antennomeres 10

10(9). Small, black species, most less than 15 mm in length; head with 2 small, transverse tubercles; pronotum lacking tubercles or fovea; posterior tibia strongly broadened to apex .... *Euetheola*
— Larger, reddish brown or black species generally greater than 15 mm in length; head with or without carina, tubercles, or horn; pronotum with or without tubercles or fovea; posterior tibia not strongly broadened apically 11

11(10). Clypeal apex acute and with subapical margins cariniform .... *Cxygrylius*
— Clypeal apex bidentate, truncate, or rounded; subapical margins not cariniform 12

12(11). Clypeus with high, transverse carina *immediately* behind apex 13
— Clypeus lacking high, transverse carina *immediately* behind apex 14

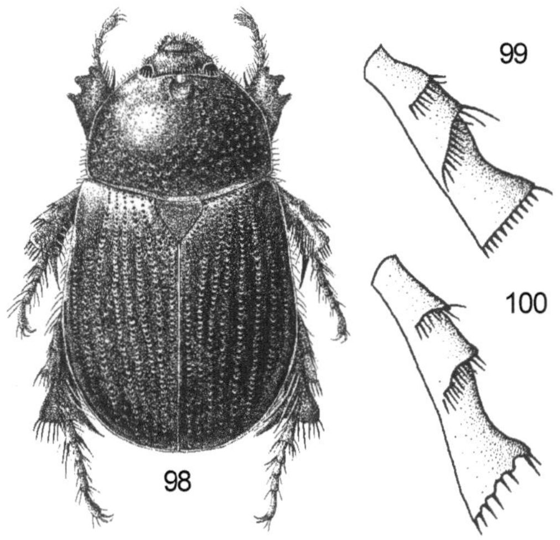

FIGURES 98.34-100.34. 98. *Tomarus gibbosus* (DeGeer). 99-100. Posterior tibia. 99. Pentodontini, apex truncate or crenulate; 100. Oryctini, apex crenulate or with triangular teeth (All used by permission of University of Nebraska State Museum).

13(12). Transverse carina just behind apex of clypeus tridentate. Pronotum lacking subapical tubercle. Protibia with apex truncate. Size mostly less than 17 mm ..... *Aphonus*

— Transverse carina just behind apex of clypeus bidentate or entire. Pronotum with subapical tubercle or not. Protibia with apex rounded. Size mostly greater than 17 mm ..... *Orizabus*

14(12). Frontoclypeal region with tubercles or carina; posterior tarsus shorter than femur 15

— Frontoclypeal region lacking tubercles or carina; posterior tarsus as long as or longer than femur ..... *Coscinocephalus*

15(14). Elytron with sutural stria and 3 feebly indicated discal striae; clypeus strongly convex on dorsal surface ..... *Gillaspytes*

— Elytron with sutural striae as well as 5-7 clearly indicated, punctate striae; clypeus not strongly convex on dorsal surface 16

16(15). Frontoclypeal region with 2 tubercles; apex of clypeus with 2 small, reflexed teeth ... *Tomarus*

— Frontoclypeal region with strong, arcuate, transverse carina ..... *Aphonides* (in part)

17(4). Anterior tibia with 3 teeth; frons in male with horn, female with a single tubercle; pronotum in females simply convex ..... *Xyloryctes*

— Anterior tibia with 4 teeth; frons in both sexes with 2 tubercles; pronotum in females with subapical fovea ..... *Strategus*

18(4). Clypeal apex broadly bidentate; elytra dark brown to black, some with conspicuous setae; pronotum at middle in males with tubercle, short horn, or simply rounded; base of horn or tubercle (if present) lacking 2 small teeth; anterior angles of pronotum often produced into small tubercle or horn ..... *Megasoma*

— Clypeal apex narrowly bidentate or emarginate; elytra greenish gray and mottled with dark spots, glabrous; pronotum at middle in males produced into long, forward projecting horn, horn with brush of tawny setae beneath; base of horn with 2 teeth; anterior angles of pronotum lacking horns or tubercles ..... *Dynastes*

Classification of the Tribes and Genera

Dynastinae MacLeay 1819

Cyclocephalini Laporte 1840

Characteristics: Head and pronotum lacking tubercles, horns, carinae, or foveae in both sexes. Club of antennae longer in some males of *Cyclocephala*. Mandibles without teeth or lobes on lateral edge. Legs with all tarsi cylindrical, never subtriangular. Anterior tarsi enlarged in males of most species. Prosternal process prominent in most. Onychium at apex with two setae.

This tribe of 13 genera is restricted to the New World except for one monobasic genus, *Ruteloryctes*, in Africa. In the U.S., there are three genera. Most species are Neotropical in distribution. Keys to species: Saylor 1945b; Endrödi 1985; Ratcliffe 1991.

Ancognatha Erichson 1847
One species, *A. manca* (LeConte), is found in Arizona, New Mexico, and northern Mexico.

Cyclocephala Dejean 1821
- *Mononidia* Casey 1915
- *Diaptalia* Casey 1915
- *Stigmalia* Casey 1915
- *Ochrosidia* Casey 1915
- *Spilosota* Casey 1915
- *Halotosia* Casey 1915
- *Homochromina* Casey 1915
- *Aclinidia* Casey 1915
- *Plagiosalia* Casey 1915
- *Isocoryna* Casey 1915
- *Graphalia* Casey 1915
- *Dichromina* Casey 1915
- *Aspidotites* Höhne 1922
- *Aspidodella* Prell 1936
- *Albridarollia* Bolivar, Jiminez and Martínez 1963
- *Paraclinidia* Martínez 1965

This is a large genus with more than 250 species with most of the species occurring in Central and South America. Fifteen species occur in the United States with about four more species in northern Mexico. The U.S. species are generally distributed except for the Pacific Northwest and extreme north where they are absent. Keys to adults: Saylor 1945b; Endrödi 1985. Key to larvae: Ritcher 1966. (Volume 1, Color Figure 6).

Dyscinetus Harold 1869
Chalepus MacLeay 1819
Palechus Casey 1915
There are 15 species in this genus, most of which are found in Central and South America. Two species, *D. morator* (Fabricius) and *D. picipes* Burmeister, occur in the southern and midwestern United States. Keys to adults: Saylor 1945b; Endrödi 1985. Larval description: Ritcher 1966.

Pentodontini Mulsant 1842

Characteristics: Head and pronotum with carina, tubercles, or fovea. Club of antenna small in most. Mandibles with 1-3 lateral lobes or teeth in most. Apex of posterior tibiae generally truncate or finely crenulate (Fig. 99), not toothed. Prosternal process long, columnar. Onychium at apex with two setae in most.

The Pentodontini is the largest tribe of Dynastinae, and its species occur everywhere except for the polar regions. There are 25 genera in the New World, and eight occur in the United States. In the last version of this work (Arnett 1968), the tribe Pentodontini was not recognized. Keys to genera: Ratcliffe 1981; Endrödi 1985. Keys to species: Saylor 1946a-b, 1948a (as Oryctini); Endrödi 1985.

Aphonides Rivers 1889
Anoplognatho Rivers 1889
One species, *A. dunnianus* (Rivers), is found in Arizona, New Mexico, Texas, and northern Mexico. This species has 9 or 10-segmented antennae.

Aphonus LeConte 1856
The genus *Aphonus* contains eight species. Six species are generally distributed from southeastern Canada, through the eastern United States, and to the southern U.S. and west to Texas. Key to adults: Gill and Howden 1985. Key to larvae: Ritcher 1966.

Coscinocephalus Prell 1936
Anoplocephalus Schaeffer 1906
Two species occur in northern Mexico with one, *C. cribrifrons* Schaeffer, reaching southern Arizona. This genus was transferred from the tribe Cyclocephalini to the tribe Pentodontini by Morón and Ratcliffe (1996). Key to adults and larval description: Morón and Ratcliffe 1996.

Euetheola Bates 1888
Euetheola is composed of four species. The genus contains two species in the Nearctic region, *E. subglabra* (Schaeffer) in Arizona and Mexico and *E. humilis* (Burmeister) in the southeastern U.S. and northern Mexico. Key to adults: Endrödi 1985. Larval description: Ritcher 1966.

Gillaspytes Howden 1980
One rare species, *G. janzeni* Howden, is found in Nearctic Tamaulipas state and also in Veracruz, Mexico.

Orizabus Fairmaire 1878
Pseudaphonus Casey 1915
Aztecalius Casey 1915
Eight species of *Orizabus* occur from the central U.S. to Central America. Three species occur in the southwestern and southcentral U.S. Key to adults: Endrödi 1985. Larval description: Ritcher 1966 (as *Cheiroplatys*).

Oxygrylius Casey 1915
One species, *O. ruginasus* (LeConte), occurs in the southwestern U.S. and northern Mexico.

Tomarus Erichson 1847
Ligyrus Burmeister 1847
Ligyrodes Casey 1915
Euligyrus Casey 1915
Grylius Casey 1915
Anagrylius Casey 1915
Ligyrellus Casey 1915
The genus *Tomarus* (Fig. 98) contains 25 species that occur from Canada through South America. There are four species, some generally distributed, in the Nearctic region.

Both *Tomarus* Erichson and *Ligyrus* Burmeister were described in 1847. Ever since the 1850s, *Ligyrus* has been used as the senior name while *Tomarus* has been used as a subgenus. The forward in Burmeister's *Handbuch der Entomologie* (volume 5) is dated February 1847, and it was received in the library of the Entomologischen Vereine zu Stettin in September 1847. But, the paper following Erichson's in the Archiv für Naturgeschichte is dated January 1847, and it was received in the library in Stettin in April 1847. *Tomarus* has priority although Lacordaire stated in 1859, for reasons known only to him, that *Tomarus* was described after *Ligyrus*. The confusion over the correct name for this genus seems to stem from this point.

Key to adults (as *Ligyrus*): Endrödi 1985. Larval description: Ritcher 1966.

Oryctini Mulsant 1842

Characteristics: Head and pronotum with tubercles or horns (especially males) and pronotum of most with fovea (especially females). Club of antenna small. Mandibles with lateral lobes or teeth. Apex of posterior tibia strongly crenulate or with teeth (including apical tooth)(Fig. 100). Prosternal process prominent, columnar. Onychium with 3 or more setae at apex.

The tribe Oryctini is world wide in distribution. There are 13 genera in the New World, and two of these are found in the United States. Key to genera and species: Endrödi 1985.

Strategus Kirby 1828
Anastrategus Casey 1915
Strategodes Casey 1915
Strategopsis Chapin 1932
The genus *Strategus* contains 31 extant species and one fossil species that occur from the southcentral U. S. through South

America. Five species are found in the southern U.S. with one species reaching central Kansas. Key to adults: Ratcliffe 1976a. Key to larvae: Ritcher 1966. Biology: Ratcliffe 1976a.

Xyloryctes Hope 1837
Ten mostly Mexican species occur in the genus, and two species are found in the eastern and southwestern U.S. Key to adults: Endrödi 1985. Larval description: Ritcher 1966.

Phileurini Burmeister 1847

Characteristics: Head with tubercles or short horns, pronotum with median sulcus and most with apical tubercle. Club of antenna small. Mandibles exposed, narrowly subtriangular, lacking teeth. Mentum large, covering base of labial palps. Clypeus with apex acute. Apex of posterior tibia truncate or with teeth. Apex of basal tarsomere on posterior leg with long spine.

Phileurines are found in all regions of the world although most species are found in the tropics. There are 21 genera in the New World, and three genera occur in the United States. Key to genera: Endrödi 1985.

Archophileurus Kolbe 1910
Amblyophileurus Kolbe 1910
Periphileurus Kolbe 1910
Anisophileurus Prell 1912
There are about 25 primarily Neotropical species of *Archophileurus*, and one species, *A. cribrosus* (LeConte), occurs in Texas.

Hemiphileurus Kolbe 1910
Epiphileurus Kolbe 1910
This genus is composed of about 40 species, all of which are Neotropical. One species, *H. illatus* (LeConte), is distributed northward into the southwestern U.S.

Phileurus Latreille 1807
Two species, *P. valgus* (Olivier) and *P. truncatus* (Palisot de Beauvois), occur in the southern United States and Mexico. The 19 species in the genus are mostly Neotropical in distribution. Key to adults: Endrödi 1985. Larval description: Ritcher 1966.

Dynastini MacLeay 1819

Characteristics: Head and pronotum with horn (males) or head with tubercle (females) in most species. Prosternal process flattened, subtriangular, generally adpressed to prosternum, most short. Onychium at apex multisetose.

The tribe Dynastini is comprised of three genera in the New World, and two are found in the United States and northern Mexico. Members of the Dynastini are among the largest insects on Earth, and the males of some species possess very large horns.

Dynastes MacLeay 1819
Theogenes Burmeister 1847
Six species are found in the New World, and three species are found in the U.S. and Mexico: *D. tityus* (L.) in the southeastern U.S., *D. granti* Horn in the southwestern U.S., and *D. hyllus* Chevrolat in Mexico (extending northward as far as Tamaulipas state). Key to adults: Endrödi 1985. Photographic synopsis of adults: Lachaume 1985. Key to larvae: Ritcher 1966. Biology: Glaser 1976.

Megasoma Kirby 1825
Megalosoma Burmeister 1841
Lyphontes Bruch 1910
Megasominus Casey 1915
Seven species are found in the southwestern U.S. and northern Mexico. Key to adults: Hardy 1972; Endrödi 1985. Photographic synopsis of adults: Lachaume 1985.

VI. Cetoniinae Leach 1815

by Brett C. Ratcliffe

Common name: The flower chafers

Characteristics: Mandibles weakly developed, hidden by clypeus. Labrum membranous in most, hidden. Antenna 10-segmented, club with 3 segments (these can be elongated); antennal insertion visible from above (Fig. 2) on side of clypeus. Eye canthus long, narrow. Elytron with distinct post-humeral emargination (Gymnetini, Cetoniini, Cremastocheilini) (Fig. 3) that reveals the mesepimeron at the base of elytron or lacking post-humeral emargination (Trichiini and Valgini) (Fig. 4). Pygidium exposed. Propygidium rigidly connected to fifth visible sternite. Procoxae protruding conically downward. Posterior coxae contiguous or nearly so in Trichiini. Tarsal claws simple and subequal in size.

The subfamily Cetoniinae was redefined by Krikken (1984) to include twelve tribes (citing Trichiini and Valgini as tribes). Previously, Schenkling (1921) had recognized seven tribes in the *Coleopterorum Catalogus* (not including the Trichiini and Valgini). Of the twelve tribes recognized by Krikken, five occur in North America. The tribes not occurring in North America are: Xiphoscelidini (Africa and Madagascar), Stenotarsiini (Madagascar), Schizorhinini (Australia, Oriental), Goliathini (Africa, Oriental, Madagascar, two genera and three species in southern Mexico), Diplognathini (Africa, Oriental), Phaedimini (Oriental), and Taenioderini (Oriental, Australia, Palearctic).

In the United States, Canada, and northern Mexico, there are 18 genera and about 105 species. Keys to adults: Casey 1915; Goodrich 1966; Howden 1968b, 1971b; Krikken 1976, 1984 (to subtribes); Hardy 1988; Harpootlian 2001. North American catalog: Smith 2001. Regional works: Blatchley 1910; Lago *et al.* 1979; Ratcliffe 1991; Morón *et al.* 1997; Harpootlian 2001. World Catalog: Schenkling 1921; Krikken 1984 (genera); Krajcík 1998, 1999. Larvae: Ritcher 1966.

Key to the Tribes and Genera from the United States, Canada, and Nearctic Mexico

1. Posthumeral elytral emargination present; mesepimeron visible from above (Fig. 3) 2
— Posthumeral elytral emargination lacking; mesepimeron not visible from above (Fig. 4) ... 4

2(1). Pronotum with basomedian lobe enlarged, covering scutellum (Fig. 102) (Gymnetini) 5
— Pronotum without enlarged basomedian lobe; scutellum visible (Fig. 101) 3

3(2). Labium not enlarged or cup shaped, less than half as wide as head (Cetoniini) 8
— Labium cup shaped, one half to three quarters as wide as head (Cremastocheilini) 11

4(1). Posterior coxae contiguous or nearly so; protibia with 1-3 teeth (Fig. 103) (Trichiini) 15
— Posterior coxae widely separate; protibia with 3-5 teeth (Fig. 104) (Valgini) .... *Valgus*

5(2). Apex of clypeus with median, upturned, lobiform process; vertex and frons with ridge-like, longitudinal process, often developed into distinct horn (Fig. 102) .... *Cotinis*
— Apex of clypeus subtruncate, rounded, or weakly emarginate; vertex and frons never armed...... 6

6(5). Dorsal surface shiny black, with or without cream colored band on lateral edge of pronotum, traversing elytra just past middle; elytral band occasionally reduced to spots .... *Gymnetina*
— Dorsal surface velutinous, either black with yellow markings or dusky brown, grey, yellowish green, or greyish green 7

7(6). Dorsal surface black with variable yellow markings on pronotum and elytra; prosternal process (in front of forecoxae) long..................... *Gymnetis*
— Dorsal surface a monochrome brown, grey, yellowish green, or greyish green; prosternal process absent .... *Hologymnetis*

8(3). Clypeus with apex quadridentate or bidentate; pronotum on posterior margin entire or weakly emarginate in front of scutellum ... *Stephanucha*
— Clypeus with apex subtruncate to rounded, lacking teeth; pronotum on posterior margin distinctly emarginate in front of scutellum in most 9

9(8). Apex of elytra spiniformly produced at suture .... *Protaetia*
— Apex of elytra rounded or angulate, not spiniformly produced 10

10(9). Prosternal process (in front of forecoxae) well-developed, almost as produced as coxae; body with dorsal surface conspicuously flattened; anterior tibia with 2 apical teeth very close together .... *Chlorixanthe*
— Prosternal process weakly developed, not noticeable in most; body of most with dorsal surface convex, not conspicuously flattened; anterior tibia with 3 teeth subequally spaced .. *Euphoria*

11(3). Pronotum with posterolateral angles produced, either acute or knob-like; trichomes present on basal or apical margin of pronotum 12
— Pronotum with posterolateral angle not produced, obtusely angulate or rounded in most; trichomes on pronotum absent 13

12(11). Pronotum with anterior angles acute; apex on sides with deep groove containing dense mat of setae .... *Cremastocheilus*
— Pronotum with apex entire, lateral grooves absent .... *Centrochilus*

13(11). Tarsomeres with numerous, small, longitudinal carinae; dorsal surface of pronotum and elytra smooth .... *Lissomelas*
— Tarsomeres smooth, punctate, or irregularly punctate; lacking numerous longitudinal carinae; dorsal surface punctate in most.......................... 14

14(13). Antennal scape concave, surface smooth and glossy; elytra glossy .... *Psilocnemis*
— Antennal scape flat or convex, surface finely to heavily punctate or rugose; elytra partly opaque or, if glossy, with chalky, white marks .... *Genuchinus*

15(4). Protibia with basomedial angle and adjacent femoral part acute; clypeus in males with horn-like projection; eye canthus angulate on anterior edge .... *Inca*
— Protibia lacking acute basomedial angle; clypeus in males without horn............................. 16

16(15). Body dorsally unicolorous brown to black, without markings or chalky, white marks .... *Osmoderma*
— Body dorsally distinctly bicolored or with chalky, white marks on either elytra, pygidium, metasternum or all 3 (Fig. 103) 17

17(16). Disc of pronotum with impression in form of triangle .... *Trigonopeltastes*
— Disc of pronotum without triangle-shaped impression 18

18(17). Disc of elytra with setae .... *Trichiotinus*
— Disc of elytra lacking setae .... *Gnorimella*

Classification of the Tribes and Genera

Cetoniinae Leach 1815

Gymnetini Kirby 1827

Characteristics: Pronotum with basomedian lobe strongly expanded, apex rounded, covering most of scutellum. Mesepimeron distinct. Posthumeral emargination of elytra distinct. Surface velutinous in most. Mesometasternal protrusion developed.

This tribe consists of approximately 29 genera, and 25 of these are restricted to the New World. In the U.S. there are four genera. Most of the species in the tribe are Neotropical in distribution. The taxa in this tribe are in great need of study, and Ratcliffe (in prep.) is working on this. Key to species: Casey 1915.

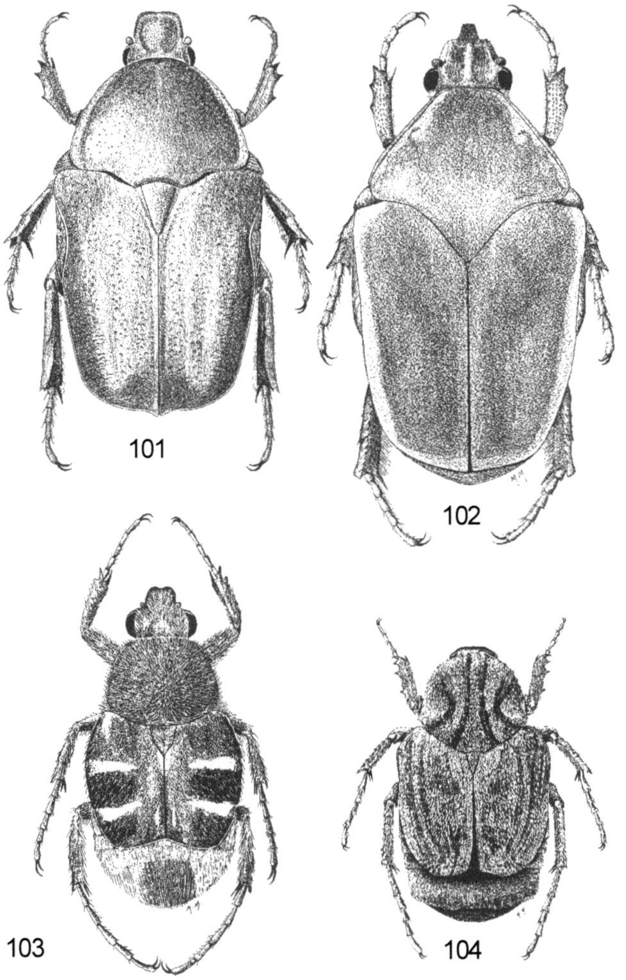

FIGURES 101.34-104.34. 101. *Euphoria fulgida* (Fabricius); 102. *Cotinis nitida* (Linnaeus); 103. *Trichiotinus assimilis* (Kirby); 104. *Valgus seticollis* (Palisot de Beauvois) (All used by permission of University of Nebraska State Museum).

Cotinis Burmeister 1842
subgenus *Cotinis* Burmeister 1842
Latemnis Thomson 1880
Cotinorrhina Schoch 1895
subgenus *Criniflavia* Goodrich 1966
subgenus *Liberocera* Deloya and Ratcliffe 1988

Eight species of *Cotinis* (Fig. 102) occur in the southern half of the United States and northern Mexico. Key: Goodrich 1966; Deloya and Ratcliffe 1988. Larvae: Ritcher 1966.

Gymnetina Casey 1915
One species, *G. cretacea* (LeConte), is found in Arizona.

Gymnetis MacLeay 1819
Paragymnetis Schürhoff 1937
Gymnetoides Martínez 1949

One species, *G. sallei* Schaum, is found in the southwestern U.S. and Mexico. Larvae: Ritcher 1966.

Hologymnetis Martínez 1949
Two species, *H. argenteola* (Bates) and *H. cinerea* (Gory and Percheron), are found in southern Arizona and northern Mexico. Key: Ratcliffe and Deloya 1992. Larvae: Micó *et al.* 2001.

Cetoniini Leach 1815

Characteristics: Pronotum lacking basomedian lobe, scutellum visible. Mesepimeron distinct. Posthumeral emargination of elytra distinct. Elytra glabrous, setose, velutinous, or with cretaceous patches. Mesometasternal protrusion developed.

The large tribe Cetoniini consists of about 107 genera. These beetles are widely distributed around the world although not as commonly in the Neotropics, Madagascar, or Australasia. The African and Neotropical taxa are in need of revision. There are five genera in the United States, Canada, and northern Mexico (*Protaetia* is adventive in Florida). Keys: Casey 1915; Hardy 1988; Ratcliffe 1991. Larvae: Ritcher 1966.

Chlorixanthe Bates 1889
Two species occur in the genus, and one species, *C. propinqua* (Gory and Percheron) (= *C. chapini* Cartwright), is found in Mexico and the southwestern U.S. Key: Hardy 1988.

Euphoria Burmeister 1842
Erirhipis Burmeister 1842
Euryomia Lacordaire 1856
Euphoriaspis Casey 1915
Erirhipidia Casey 1915
Euphorhipis Casey 1915
Haplophoria Casey 1915
Isorhipina Casey 1915
Rhipiphorina Casey 1915

The genus *Euphoria* (Fig. 101) is badly in need of revision, and even some of the U.S. species cannot be reliably identified. There are approximately 73 species in the genus, and about 20 species are broadly distributed in the United States. Key: Casey 1915 (not reliable due to its typological concept). Larvae: Ritcher 1966; Ratcliffe 1976b; Micó *et. al* 2000.

Protaetia Burmeister 1842
Protaetia is a huge genus with about 250 species that occur in the Oriental, Australian, and Palearctic regions. One species, *P. fusca* (Herbst), widely distributed in Asia, has become established in southeastern Florida (Thomas 1998) and Hawaii (Miksic 1987). Key: Miksic 1987 (for all Old World species).

Stephanucha Burmeister 1842
Anatropis Casey 1915

Stephanucha has six species that occur in the eastern United States west to Colorado, New Mexico, and Arizona and with two species in Mexico. Key: Casey 1915. Generic commentary: Hardy 1988. Immature stages and biology: Skelley 1991. (Volume 1, Color Figure 5).

Cremastocheilini Burmeister 1842

Characteristics: Labium cup-shaped, one half to three fourths as wide as head. Maxillary galea and lacinia dentate in most. Antenna with scape conspicuously enlarged into flattened, triangular shape. Middle coxae never separated by a broad protrusion. Body form of most subquadrate, flattened dorsally. Color black or brown.

The tribe Cremastocheilini contains about 51 genera. About 60% of the genera are found in Africa while none occur in Madagascar or Australia. In the New World there are ten genera, and five occur in North America. Most species are myrmecophilous. Keys: Potts 1945; Howden 1971b; Krikken 1976. Larvae: Ritcher 1966.

Centrochilus Krikken 1976
One species, *C. howdeni* Krikken, is known from Durango state in Mexico.

Cremastocheilus Knoch 1801
subgenus *Cremastocheilus* Knoch 1801
subgenus *Myrmecotonus* Mann 1914
Myrmeceicon Mann 1914
subgenus *Macropodina* Casey 1915
subgenus *Trinodia* Casey 1915
subgenus *Anatrinodia* Casey 1915
The genus *Cremastocheilus* is restricted to North America and includes about 35 species. Most of these species are uncommonly encountered. The genus is in need of a modern synopsis. Keys: Potts 1945; Alpert 1994. Larvae: Ritcher 1966; Ratcliffe 1977.

Genuchinus Westwood 1873
Nine species are placed in this genus, and two species are found in Arizona and California. The remainder are primarily South American in distribution. Key: Krikken 1981.

Lissomelas Bates 1889
Only one Neotropical species is known, *L. flohri* Bates, and its distribution extends northward to southern Arizona.

Psilocnemis Burmeister 1842
One species is known, *P. leucostictica* Burmeister. It is found from Maryland to North Carolina in the United States.

Trichiini Fleming 1821

Characteristics: Procoxae projecting conically. Pronotum evenly convex or with weakly impressed midline or pre-discal impression. Mandibles either weakly or strongly sclerotized. Galea of maxilla penicillate in most. Protibia with 1-3 teeth. Dorsum uniformly dark brown or black or bicolored or with cretaceous spots or bands or metallic.

Until relatively recently, the trichiines have been considered as a subfamily (frequently as a family in Europe). The tribe Trichiini consists of 43 genera that are found nearly worldwide except Australia and Madagascar. Thirteen genera are found in the New World with five genera occurring in North America. Key: Howden 1968b; Morón and Krikken 1990.

Gnorimella Casey 1915
Only one species is known, *G. maculosa* (Knoch), and it occurs in the eastern United States and Canada. Larvae: Ritcher 1966.

Inca LePeletier and Serville 1828
The genus contains six species, one of which (*Inca clathratus sommeri* Westwood) is found in Mexico's Tamaulipas state in the Sierra Madre Oriental. Distribution: Morón *et al.* 1997. Biology: Boos and Ratcliffe 1985. Larvae: Morón 1983b.

Osmoderma LePeletier and Serville 1828
Gymnodi Kirby 1827 (*nomen oblitum*)
Gymnodus Kirby 1837 (unjustified emendation; *nomen oblitum*)
This genus contains nine species, three of which occur in North America. Keys: Hoffmann 1939; Howden 1968b. Larvae: Ritcher 1966; Ratcliffe 1977.

Trichiotinus Casey 1915
Eight species of *Trichiotinus* (Fig. 103) are found throughout the United States (except the far west) and southern Canada. Keys: Hoffmann 1935; Howden 1968b. Larvae: Ritcher 1966.

Trigonopeltastes Burmeister 1840
Archimedii Kirby 1827 (*nomen oblitum*)
Euclidii Kirby 1827 (*nomen oblitum*)
Roplisa Casey 1909
This genus has 15 species distributed in Mexico and Central America. Six species are found in the Nearctic realm, and two species, *T. delta* (Forster) and *T. floridana* (Casey), are found in the southeastern United States. Key: Howden 1968b. Larvae: Ritcher 1966. (Volume 1, Color Figure 12).

Valgini Mulsant 1842

Characteristics: Antennal insertion visible from above. Pronotum narrower than elytra and most with 2 distinct, parallel, longitudinal ridges. Posthumeral elytral emargination absent. Mesepimeron not visible in dorsal view. Propygidium and pygidium exposed. Metacoxae widely separated. Protibia with 5 teeth. Body size small, generally less than 10 mm.

The tribe Valgini consists of 33 genera found nearly worldwide. One genus, *Valgus*, is found in the Nearctic region. Key: Krikken 1978. Overview: Arrow 1944.

Valgus Scriba 1790
Acanthurus Kirby 1827
Homovalgus Kolbe 1897
The genus *Valgus* (Fig. 104) contains about 11 species in the Oriental, Palearctic, and Nearctic regions, and five species, including an adventive one from Europe, are found in the eastern United States and California. Most species seem to be

termitophilous. *Valgus canaliculatus* (Olivier) adults are found on flowers nectar-feeding in late spring and again in late summer, and their larvae and pupae are found in termite galleries of softwood logs. Keys: Casey 1915; Cazier 1937. Larvae: Ritcher 1966.

Bibliography

ALPERT, G. D. 1994. A comparative study of the symbiotic relationships between beetles of the genus *Cremastocheilus* (Coleoptera: Scarabaeidae) and their host ants (Hymenoptera: Formicidae). Sociobiology, 25: 1-276.

ANDREWS, F. G., A. R. HARDY and D. GIULIANI. 1979. The coleopterous fauna of selected California sand dunes. Report to the BLM on the contract CA-960-1285-1288-DEOO. California Department of Food and Agriculture, Sacramento, CA, 142 pp.

ANDUAGA, S. and G. HALFFTER. 1991. Escarabajos asociados a madrigueras de roedores (Coleoptera: Scarabaeidae: Scarabaeinae). Folia Entomológica Mexicana, 81: 185-197.

ANDUAGA, S. and G. HALFFTER. 1993. Nidificación y alimentación en *Liatongus rhinocerulus* (Bates) (Coleoptera: Scarabaeidae: Scarabaeinae). Acta Zoologica Mexicana (N.S.), 57: 1-14.

ARCE-PEREZ, R. and M. A. MORÓN. 2000. Taxonomia y distribucion de las especies de *Macrodactylus* Latreille (Coleoptera: Melolonthidae) en Mexico y Estados Unidos de America. Acta Zoologica Mexicana (n. s.), 79: 123-239.

ARNETT, R. H., Jr. 1968. The Beetles of the United States. A Manual for Identification. The American Entomological Institute. Ann Arbor, MI, 1112 pp.

ARROW, G. J. 1909. On the characters and relationships of the less-known groups of lamellicorn Coleoptera, with descriptions of new species of Hybosorinae, etc. Transactions of the Entomological Society of London, 4: 479-507.

ARROW, G. J. 1912. Scarabaeidae: Pachypodiinae, Pleocominae, Aclopinae, Glaphyrinae, Ochodaeinae, Orphninae, Idiostominae, Hybosorinae, Dynamopinae, Acanthocerinae, Troginae. Coleopterorum Catalogus, Pars 43. W Junk. Berlin, 66 pp.

ARROW, G. J. 1944. The beetles of the lamellicorn subfamily Valginae, with a synopsis of the genera and descriptions of some new species. Transactions of the Royal Entomological Society of London, 94: 225-246.

BADER, A. M. 1992. A review of the North and Central American *Strigoderma* (Coleoptera: Scarabaeidae). Transaction of the Entomological Society of America, 118: 269-355.

BALTHASAR, V. 1939. Neue *Choeridium*-Arten (Ins. Col.). 6. Beitrag zur Kenntnis der Scarabaeiden der neotropischen Region. Senckenbergiana, 21: 44-66.

BARAUD, J. 1992. Coleopteres Scarabaeoidea d'Europe. Fauna de France. 78. Société Linnéene de Lyon, 856 pp.

BARAUD, J. and T. BRANCO. 1990. Revision des *Chasmatopterus* Latreille, 1825 (Coleoptera: Melolonthidae). Coleopterological Monographs, 1: 1-55.

BARFIELD, C. S. and W. M. GIBSON. 1975. Observations on the life history of *Hypothyce mixta* Howden (Coleoptera: Scarabaeidae). Coleopterists Bulletin, 29: 251-256.

BATES, H. W. 1887. Biologia Centrali-Americana. Coleoptera, Vol. 2 Pt. 2, pp. 1-381.

BLACKWELDER, R. E. 1944. Checklist of the coleopterous insects of Mexico, Central America, the West Indies, and South America. Part 2. Bulletin of the United States National Museum, 185: 189-341.

BLACKWELDER, R. E. and R. H. ARNETT, Jr. 1974. Checklist of the beetles of Canada, United States, Mexico, Central America and the West Indies. Volume 1, Part 3. The scarab beetles, ant-loving beetles, clown beetles, and related groups (red version). The Biological Research Institute of America. Latham, NY, 120 pp.

BLATCHLEY, W. S. 1910. An illustrated descriptive catalogue of the Coleoptera or beetles known to occur in Indiana. Indiana Department of Geology and Natural Resources Bulletin, 1: 1-1386.

BLUME, R. R. 1984. *Euoniticellus intermedius* (Coleoptera: Scarabaeidae): Description of adults and immatures and biology of adults. Environmental Entomology, 13: 1064-1068.

BOOS, J. and B. C. RATCLIFFE. 1985. A new subspecies of *Inca clathrataus* (Olivier) from Trinidad, West Indies, and range extensions for *Inca clathrata sommeri* Westwood (Coleoptera: Scarabaeidae: Trichiinae). Coleopterists Bulletin, 39: 381-389.

BOUSQUET, Y. 1991. Checklist of beetles of Canada and Alaska. Research Branch, Agriculture Canada Publication 1861/E, 430 pp.

BØVING, A. G. 1936. Description of the larva of *Plectris aliena* Chapin and explanation of new terms applied to the epipharynx and raster. Proceedings of the Entomological Society of Washington, 38: 169-185.

BØVING, A. G. 1937. Keys to the larvae of four groups and forty-three species of the genus *Phyllophaga*. United States Department of Agriculture, Bureau of Entomology and Plant Quarantine, E-417: 1-8.

BØVING, A. G. 1942a. Descriptions of the larvae of some West Indian melolonthinae beetles and a key to the known larvae of the tribe. Proceedings of the United States National Museum, 92: 167-176.

BØVING, A. G. 1942b. A classification of larvae and adults of the genus *Phyllophaga*. Memoirs of the Entomological Society of Washington, 2: 1-96.

BØVING, A. G. 1942c. Description of the third-stage larva of *Amphimallon majalis* (Razoumowsky). Proceedings of the Entomological Society of Washington, 44: 111-121.

BOYER, L. B. 1940. A revision of the species of *Hoplia* occurring in America North of Mexico (Coleoptera: Scarabaeidae). Microentomology, 5: 1-31.

BRITTON, E. B. 1957. A revison of the Australian chafers (Coleptera: Scarabaeidae: Melolonthinae). Volume 1. British Museum (Natural History). London. 185 pp.

BROWN, W. J. 1927. A revision of the species of Horn's series I-b. (Col.). Canadian Entomologist, 59: 162-167.

BROWN, W. J. 1928a. The subgenus *Platyderides* in North America (Coleoptera). Canadian Entomologist, 60: 35-40.

BROWN, W. J. 1928b. The subgenus *Platyderides* in North America (Coleoptera). Canadian Entomologist, 60: 10-21.

BROWN, W. J. 1929a. Revision of the species of the subgenus *Diapterna* (Coleoptera). Canadian Entomologist, 61: 224-231.

BROWN, W. J. 1929b. Studies in the Scarabaeidae, (II). Canadian Entomologist, 61: 86-93.

BROWN, W. J. 1946. Notes on some species of *Canthon* and *Dichelonyx* (Coleoptera, Scarabaeidae). Canadian Entomologist, 78: 101-109.

BUTLER, G. D. and F. G. WERNER. 1961. Distribution and host plants of May beetles in Arizona. Arizona Agricultural Experimental Station Technical Bulletin, 147: 1-19.

BUTT, F. H. 1944. External morphology of *Amphimallon majalis* (Razoumowski). Cornell University Agricultural Experimental Station Memoirs, 266: 1-118.

CAMBEFORT, Y. 1991. From saprophagy to coprophagy. Pp. 22-35. *In:* I. Hanski and Y. Cambefort, eds. Dung Beetle Ecology. Princeton University Press. Princeton, NJ, 481 pp.

CAPISTRAN, F. and O. AQUINO. 1992. Observaciones sobre biologia de *Ceraspis pilatei* Harold (Coleoptera: Melolonthidae) en el area de Catemaco, Veracruz, Mexico. Folia Entomologica Mexicana, 85: 119-122.

CARRILLO, J. S. and W. W. GIBSON. 1960. Repaso de las especies Mexicanas del genero *Macrodactylus* (Coleoptera, Scarabaeidae), con observaciones biologicas de algunas especies. Secretaria de Agricultura y Ganaderia, Mexico, Folleto Técnico Numero 39: 1-102.

CARTWRIGHT, O. L. 1948. The American species of *Pleurophorus* (Coleoptera: Scarabaeidae). Transactions of the American Entomological Society, 74: 131-145.

CARTWRIGHT, O. L. 1955. Scarab beetles of the genus *Psammodius* in the Western Hemisphere. Proceedings of the United States National Museum, 104: 413-462.

CARTWRIGHT, O. L. 1972. A key to the *crassulus* group of *Aphodius* with descriptions of new species from Texas and Maryland (Coleoptera: Scarabaeidae: Aphodiidae). Proceedings of the Biological Society of Washington, 85: 57-62.

CARTWRIGHT, O. L. 1974. *Ataenius*, *Aphotaenius*, and *Pseudataenius* of the United States and Canada. Smithsonian Contributions to Zoology, 154: 1-106.

CASEY, T. L. 1889. Coleopterological notices. I. Annals of the New York Academy of Sciences, 5: 39-198.

CASEY, T. L. 1915. A review of the American species of Rutelinae, Dynastinae, and Cetoniinae. Memoirs of the Coleoptera, 6: 1-394.

CAZIER, M. 1937. A new species of *Valgus* and a new generic record for Mexico. Pan-Pacific Entomologist, 13: 190-192.

CAZIER, M. A. 1939. A new California scarabaeid, with notes (Coleoptera-Scarabaeidae). Bulletin of the Southern California Academy of Sciences, 38: 17-19.

CAZIER, M. A. 1951. The genera *Chrysina* and *Plusiotis* of north central Mexico (Coleoptera, Scarabaeidae). American Museum Novitates, 1516: 1-8.

CAZIER, M. A. 1953. A review of the scarab genus *Acoma* (Coleoptera; Scarabaeidae). American Museum Novitates, 1624: 1-13.

CAZIER, M. A. and A. T. McCLAY. 1943. A revision of the genus *Coenonycha* (Coleoptera, Scarabaeidae). American Museum Novitates, 1239: 1- 27.

CHALUMEAU, F. 1983. Les Coléoptères Scarabaeides des Petites Antilles (Guadeloupe à Martinique). Encyclopédie Entomologique, Paris, 296 pp.

CHALUMEAU, F. 1992. Eupariini du nouveau monde: un mise au point (Coleoptera, Scarabaeidae) (1re partie). Nouvelle Revue d'Entomologie, 9: 189-206.

CHALUMEAU, F. and L. GRUNER. 1976. Scarabaeoidea des Antilles Françaises 2e partie: Melolonthinae et Rutelinae (Coleoptera). Annales de la Société Entomologique de France (N.S.), 12: 83-112.

CHAPIN, E. A. 1934. An apparently new scarab beetle (Coleoptera) now established at Charleston, South Carolina. Proceedings of the Biological Society of Washington, 47: 33-36.

CHAPIN, E. A. 1935. Review of the genus *Chlaenobia* Blanchard (Coleoptera: Scarabaeidae). Smithsonian Miscellaneous Collections, 94: 1-20.

CHAPIN, E.A. 1938. Three Japanese beetles of the genus *Serica* MacLeay. Journal of the Washington Academy of Sciences, 28: 66-68.

COCA-ABIA, M. M. 2000. *Polylamina* Hardy, a junior synonym of *Polyphylla* Harris (Coleoptera: Scarabaeidae, Melolonthinae, Melolonthini). Coleopterists Bulletin, 54: 11-22.

COOPER, J. B. 1983. A review of the Nearctic genera of the family Scarabaeidae (exclusive of the subfamilies Scarabaeinae and Geotrupinae) (Coleoptera), with an evaluation of computer generated keys. Doctoral Thesis, Department of Biology, Carleton University. Ottawa, Ontario, Canada, 1121 pp.

DALLA TORRE, K. W. von. 1912. Scarabaeidae; Melolonthidae I-III. Coleopterorum Catalogus, vol. 20, pars 45, 47, 49. pp. 1-290.

DALLA TORRE, K. W. von. 1913. Scarabaeidae; Melolonthidae IV. Coleopterorum Catalogus, vol. 20, pars 50. pp. 291-450.

DAVIS, A. C. 1935. A revision of the genus *Pleocoma*. Bulletin of the Southern California Academy of Sciences, 33: 123-130.

DAWSON, R. W. 1919a. New species of *Serica* (Scarabaeidae). I. Journal of the New York Entomological Society, 27: 32-39.

DAWSON, R. W. 1919b. New species of *Serica* (Scarabaeidae). II. Journal of the New York Entomological Society, 27: 223-225.

DAWSON, R. W. 1920. New species of *Serica* (Scarabaeidae). III. Journal of the New York Entomological Society, 28: 208-212.

DAWSON, R. W. 1921. New species of *Serica* (Scarabaeidae). IV. Journal of the New York Entomological Society, 29: 160-168.

DAWSON, R. W. 1922. New species of *Serica* (Scarabaeidae). V. Journal of the New York Entomological Society, 30: 154-169.

DAWSON, R. W. 1932. New species of *Serica* (Scarabaeidae). VI. Journal of the New York Entomological Society, 40: 529-548.

DAWSON, R. W. 1933. New species of *Serica* (Scarabaeidae). VII. Journal of the New York Entomological Society, 41: 435-440.

DAWSON, R. W. 1947. New species of *Serica* (Scarabaeidae). VIII. Journal of the New York Entomological Society, 55: 223-235.

DAWSON, R. W. 1952. New species of *Serica* (Scarabaeidae). IX. Journal of the New York Entomological Society, 60: 65-77.

DAWSON, R. W. 1967. New species of *Serica* (Scarabaeidae). X. Journal of the New York Entomological Society, 75: 161-178.

DELGADO, L. and J. BLACKALLER-BAGES. 1997. A new Mexican species of *Homoiosternus* (Coleoptera: Melolonthidae; Rutelinae). Journal of the New York Entomological Society, 105: 170-179.

DELGADO-CASTILLO, L. 1990. Dos nuevas especies Mexicanas de *Diplotaxis* del grupo "puberea" (Coleoptera: Melolonthidae; Melolonthinae). Folia Entomologica Mexicana, 78: 61-70.

DELGADO-CASTILLO, L., C. DELOYA and M. A. MORÓN. 1987. Descripción del macho de *Ceraspis velutina* (Bates) y nuevos registros de *C. centralis* (Sharp) para Mexico (Coleoptera; Melolonthidae: Macrodactylini). Folia Entomologica Mexicana, 71: 47-54.

DELLACASA, M. 1987. Contribution to a world-wide catalogue of Aegialiidae, Aphodiidae, Aulonocnemidae, Termitotrogidae (Coleoptera Scarabaeoidea). Memorie della Società Entomologica Italiana, 66: 1-455.

DELLACASA, M. 1988a. Contribution to a world-wide catalogue of Aegialiidae, Aphodiidae, Aulonocnemidae, Termitotrogidae (Coleoptera Scarabaeoidea) Part II. Memorie della Società Entomologica Italiana, 67: 1-229.

DELLACASA, M. 1988b. Contribution to a world-wide catalogue of Aegialiidae, Aphodiidae, Aulonocnemidae, Termitotrogidae (Coleoptera Scarabaeoidea). Addenda et corrigenda (First note). Memorie della Società Entomologica Italiana, 67: 291-316.

DELLACASA, M. 1991. Contribution to a world-wide catalogue of Aegialiidae, Aphodiidae, Aulonocnemidae, Termitotrogidae (Coleoptera Scarabaeoidea). Addenda et corrigenda (Second note). Memorie della Società Entomologica Italiana, 70: 3-57.

DELLACASA, M. 1995. Contribution to a world-wide catalogue of Aegialiidae, Aphodiidae, Aulonocnemidae, Termitotrogidae (Coleoptera Scarabaeoidea). Addenda et corrigenda (Third note). Memorie della Società Entomologica Italiana, 74: 159-232.

DELLACASA, G., P. BORDAT and M. DELLACASA. 2001. A revisional essay of world genus-group taxa of Aphodiinae (Coleoptera: Scarabaeoidea: Aphodiidae). Memorie della Società Entomologica Italiana, 79: 1-482.

DELLACASA, G. and R. D. GORDON. 1994. North American genus-group taxa of Aphodiini and their type species (Coleoptera Aphodiidae). Frustula Entomologica, Nova Serie (Pisa), 17: 157-174.

DELLACASA, G. and R. D. GORDON. 1997. Nearctic and Neotropical genus-group taxa of Aphodiini and their type-species (Coleoptera: Aphodiidae). Annali del Museo Civico di Storia Naturale "G. Doria" (Genova), 91: 355-382.

DELOYA, C., M. A. MORÓN and J. M. LOBO. 1995. Coleoptera Lamellicornia (MacLeay, 1819) del sur del estado de Morelos, Mexico. Acta Zoologica Mexicana (n.s.), No. 65: 1-42.

DELOYA, C. and B. C. RATCLIFFE. 1988. Los especies de *Cotinis* Burmeister en México (Coleoptera: Melolonthidae: Cetoniinae). Acta Zoologica Mexicana (N.S.), No. 28: 1-52.

DEYRUP, M. and R.E. WOODRUFF. 1991. A new flightless *Psammodius* from Florida's inland dunes (Coleoptera: Scarabaeidae). Coleopterists Bulletin, 45: 75-80.

DOWNIE, N. M. and R. H. ARNETT, Jr. 1996. The Beetles of Northeastern North America, Vol. 1. Sandhill Crane Press. Gainesville, FL, 880 pp.

EDMONDS, W. D. 1972. Comparative skeletal morphology, systematics and evolution of the phanaeine dung beetles (Coleoptera: Scarabaeidae). University of Kansas Science Bulletin, 49: 731-874.

EDMONDS, W. D. 1994. Revision of *Phanaeus* MacLeay, a New World genus of scarabaeine dung beetles (Coleoptera: Scarabaeidae, Scarabaeinae). Natural History Museum of Los Angeles County Contributions in Science, 443: 1-105.

EDMONDS, W. D. and G. HALFFTER. 1978. Taxonomic review of immature dung beetles of the subfamily Scarabaeinae (Coleoptera: Scarabaeidae). Systematic Entomology, 3: 307-331.

EDWARDS, J. G. 1949. Coleoptera or Beetles East of the Great Plains. Edwards Brothers. Ann Arbor, MI, 181 pp.

ENDRÖDI, S. 1966. Monographie der Dynastinae (Coleoptera, Lamellicornia) I. Teil. Entomologische Abhandlungen Museum für Tierkunde, 33: 1-460.

ENDRÖDI, S. 1985. The Dynastinae of the World. W. Junk. London, 800 pp.

ERWIN, T. L. 1970. A description of the larva of *Thyce harfordi* Casey (Scarabaeidae: Melolonthini). Psyche, 77: 50-53.

EVANS, A. V. 1985. New host plant associations for adult scarabs (Coleoptera: Scarabaeidae: Melolonthinae) from Arizona and California. Coleopterists Bulletin, 39: 86-88.

EVANS, A. V. and D. d'HOTMAN. 1988. *Coenonycha* Horn in Baja California, Mexico (Coleoptera, Melolonthidae: Melolonthinae). Coleopterists Bulletin, 42: 205-210.

EVANS, A. V. and K. A. SMITH. 1986. Four new species of *Coenonycha* Horn from California and Nevada with an illustrated key to all the species in the genus (Coleoptera: Scarabaeidae). Coleopterists Bulletin, 40: 81-92.

FALL, H. C. 1932. New North American Scarabaeidae, with remarks on known species. Journal of the New York Entomological Society, 40: 183-204.

FAVILA, M. E. and A. DÍAZ. 1996. *Canthon cyanellus cyanellus* LeConte (Coleoptera: Scarabaeidae) makes a nest in the field with several dung balls. Coleopterists Bulletin, 50: 52-60.

FINCHER, G. T. 1972. Notes on the biology of *Phanaeus vindex* (Coleoptera: Scarabaeidae). Journal of the Georgia Entomological Society, 7: 128-133.

FINCHER, G. T. 1990. Biological control of dung-breeding flies: pests of pastured cattle in the United States. Pp. 137-151. *In:* D. A. Rutz and R. S. Patterson, eds. Biocontrol of Arthropods Affecting Livestock and Poultry. Westview Studies in Insect Biology. Boulder, CO, 316 pp.

FREY, G. 1962. Revision der Gattung *Ceraspis* Serv. nebst Bechreibung einer dazuggohorigen neuen Gattung (Coleoptera: Melolonthidae). Entomologischen Arbeiten aus dem Museum G. Frey, 13: 1-66.

FREY, G. 1967. Die Gattung *Plectris* (Philochlaenia) (Coleoptera, Melolonthinae). Entomologischen Arbeiten aus dem Museum G. Frey, 18: 1-136.

FREY, G. 1975. Bestimmungstabelle der sudamerikanichen Arten der Gattung *Phyllophaga* Harris und ihrer Untergattung *Phytalus* Er. (Col. Melolonthinae). Entomologischen Arbeiten aus dem Museum G. Frey, 26: 201-226.

GARCIA-VIDAL, M. 1975. Compilación sobre los prinicipales aspectos relativos al genero *Phyllophaga* Harris. Revista Agrotecnia de Cuba, 7: 17-22.

GARCIA-VIDAL, M. 1978. El genero *Phyllophaga* Harris, 1826 (Coleoptera: Scarabaeidae) en Cuba. I. Descripción de cinco nuevas especies. Poeyana (Instituto de Zoologia, Academia Ciencias de Cuba), 182: 1-14.

GARCIA-VIDAL, M. 1984. The genus *Phyllophaga* Harris, 1826 (Coleoptera: Scarabaeidae) in Cuba. II. Description of eleven new species and illustrations of female genitalia of twelve other *Phyllophaga*. Pan-Pacific Entomologist, 60: 313-325.

GARCIA-VIDAL, M. 1987. The genus *Phyllophaga* Harris, 1826 (Coleoptera: Scarabaeidae) in Cuba. IV. Description of six new species. Journal of the New York Entomological Society, 95: 302-306.

GÉNIER, F. 1996. A revision of the Neotropical genus *Ontherus* Erichson (Coleoptera: Scarabeidae (*sic*), Scarabaeinae). Memoirs of the Entomological Society of Canada, 170: 1-169.

GÉNIER, F. 2000. A new North American *Ateuchus* Weber (Coleoptera: Scarabaeidae, Scarabaeinae, Scarabaeini). Coleopterists Bulletin, 54: 341-346.

GILL, B. D. and H. F. HOWDEN. 1985. A review of the North American genus *Aphonus* LeConte (Coleoptera: Scarabaeidae: Dynastinae). Coleopterists Bulletin, 39: 119-129.

GLASER, J. D. 1976. The biology of *Dynastes tityus* (Linn.) in Maryland (Coleoptera: Scarabaeidae). Coleopterists Bulletin, 30: 133-138.

GOODRICH, M. A. 1966. A revision of the genus *Cotinis* (Coleoptera: Scarabaeidae). Annals of the Entomological Society of America, 59: 550-568.

GORDON, R. D. 1975. A new *Serica* from Nevada (Coleoptera: Scarabaeidae). Coleopterists Bulletin, 29: 173-176.

GORDON, R. D. 1976. Studies on the genus *Aphodius* of the United States and Canada (Coleoptera: Scarabaeidae): IV. A taxonomic revision of Horn's group A. Proceedings of the Entomological Society of Washington, 78: 458-478.

GORDON, R. D. 1977. Studies on the genus *Aphodius* of the United States and Canada (Coleoptera: Scarabaeidae): V. An evaluation of Horn's group F. Proceedings of the Entomological Society of Washington, 79: 157-167.

GORDON, R. D. 1983. Studies on the genus *Aphodius* of the United States and Canada (Coleoptera: Scarabaeidae): VII. Food and habitat; distribution; key to eastern species. Proceedings of the Entomological Society of Washington, 85: 633-652.

GORDON, R. D. 1990. A new species of *Aegialia* (*Aegialia*) (Coleoptera: Scarabaeidae: Aphodiinae) from southern Arizona and new data on North American *Aegialia*. Coleopterists Bulletin, 44: 271-273.

GORDON, R. D. and D. M. ANDERSON. 1981. The species of Scarabaeidae (Coleoptera) associated with sugarcane in south Florida. Florida Entomologist, 64: 119-138.

GORDON, R. D. and O. L. CARTWRIGHT. 1974. Survey of food preferences of some North American Canthonini (Coleoptera: Scarabaeidae). Entomological News, 85: 181-185.

GORDON, R. D. and O. L. CARTWRIGHT. 1980. The Western Hemisphere species of *Rhyssemus* and *Trichiorhyssemus* (Coleoptera: Scarabaeidae). Smithsonian Contributions to Zoology, 317: 1-29.

GORDON, R. D. and O. L. CARTWRIGHT. 1988. North American representatives of the tribe Aegialiini (Coleoptera: Scarabaeidae). Smithsonian Contributions to Zoology, 461: 1-37.

GORDON, R. D. and R. PITTINO. 1992. Current status of the American genera and species of Psammodiini (Coleoptera: Scarabaeidae: Aphodiinae). Coleopterists Bulletin, 46: 260-273.

GORDON, R. D. and R. W. RUST. 1997. A new southern Nevada species of *Aegialia* (*Aegialia*) (Coleoptera: Scarabaeidae: Aphodiinae). Pan-Pacific Entomologist, 73: 168-171.

GYRISCO, G. G., W. H. WHITCOMB, R. H. BURRAGE, C. LOGOTHETIS and H. H. SCHWARDT. 1954. Biology of the European chafer, *Amphimallon majalis* (Razoumowsky) (Scarabaeidae). Cornell Agricultural Experimental Station Memoir, 3281: 1-35.

HAAF, E. 1955. Über die Gattung *Sisyphus* Latr. (Col. Scarab.). Entomologische Arbeiten aus dem Museum G. Frey, 6: 341-381.

HALFFTER, G. 1958. Dos nuevos géneros de Canthonini (Col., Scarabaeidae). Ciencia (México), 17: 207-212.

HALFFTER, G. 1961. Monografía de las especies norteamericanos del género *Canthon* Hoffsg. (Coleopt., Scarab.). Ciencia (México), 20: 225-320.

HALFFTER, G. and W. D. EDMONDS. 1982. The Nesting Behavior of Dung Beetles (Scarabaeinae). An Ecological and Evolutive Approach. Instituto de Ecología, México, D.F., 176 pp.

HALFFTER, G. and V. HALFFTER. 1989. Behavioral evolution of the non-rolling roller beetles (Coleoptera: Scarabaeidae; Scarabaeinae). Acta Zoologica Mexicana (N.S.), 32: 1-53.

HALFFTER, G., V. HALFFTER and Y. LOPEZ G. 1974. *Phanaeus* behavior: food transportation and bisexual cooperation. Environmental Entomology, 3: 341-345.

HALFFTER, G. and Y. LOPEZ G. 1972. Nidificación y comportamiento sexual en *Phanaeus* (Coleoptera: Scarabaeidae). Folia Entomológica Mexicana, 23-24: 105-108.

HALFFTER, G. and A. MARTÍNEZ. 1966. Revisión monográphica de los Canthonina americanos (Coleoptera, Scarabaeidae) (1a. parte). Revista de la Sociedad Mexicana de Historia Natural, 27: 89-177.

HALFFTER, G. and A. MARTÍNEZ. 1977. Revisión monográphica de los Canthonina americanos, IV parte. Clave para géneros y subgéneros. Folia Entomológica Mexicana, 38: 29-107.

HALFFTER, G. and E. G. MATTHEWS. 1966. The natural history of dung beetles of the subfamily Scarabaeinae (Co-

leoptera: Scarabaeidae). Folia Entomológica Mexicana, 12-14: 1-312.
HANSKI, I. and Y. CAMBEFORT (eds.). 1991. Dung Beetle Ecology. Princeton University Press. Princeton, NJ, 481 pp.
HARDY, A. R. 1972. A brief revision of the North and Central American species of *Megasoma* (Coleoptera: Scarabaeidae). Canadian Entomologist, 104: 765-772.
HARDY, A. R. 1974a. Revisions of *Thyce* LeConte and related genera (Coleoptera: Scarabaeidae). State of California Department of Food and Agriculture Occasional Papers in Entomology, No. 20: 1-47.
HARDY, A. R. 1974b. Two new species of *Pseudocotalpa* Hardy (Coleoptera: Scarabaeidae). Pan-Pacific Entomologist, 50: 243-247.
HARDY, A. R. 1975. A revision of the genus *Pelidnota* of American north of Panama (Coleoptera: Scarabaeidae; Rutelinae). University of California Publication in Entomology, 78: 1-43.
HARDY, A. R. 1976. Observations on the mating behavior of *Pseudocotalpa giulianii* Hardy. Coleopterists Bulletin, 30: 301-302.
HARDY, A. R. 1977. A revision of the *Hoplia* of the Nearctic realm (Coleoptera: Scarabaeidae). California Department of Food and Agriculture Occasional Papers in Entomology, No. 23: 1-48.
HARDY, A. R. 1978a. Three new Pachydemini and a key to the species of the genus *Phobetus* (Coleoptera: Scarabaeidae). Coleopterists Bulletin, 32: 47-52.
HARDY, A. R. 1978b. Placement of the genus *Benedictia* Sanderson (Coleoptera: Scarabaeidae). Coleopterists Bulletin, 32: 67-70.
HARDY, A. R. 1981. *Polyphylla* Harris in Baja California (Coleoptera: Scarabaeidae). Coleopterists Bulletin, 35: 299-302.
HARDY, A. R. 1986. Studies on the Coleoptera of western sand dunes. 2. Notes on four Scarabaeidae from the Algodones dune system. Coleopterists Bulletin, 40: 127-139.
HARDY, A. R. 1988. Studies in the Euphoriina of the Americas (Coleoptera: Scarabaeidae). I. Introduction and generic concepts. Coleopterists Bulletin, 42: 1-9.
HARDY, A. 1991. A catalog of the Coleoptera of America north of Mexico. Family: Scarabaeidae. Subfamilies: Rutelinae and Dynastinae. United States Department of Agriculture, Agriculture Handbook 529-34b, 56 pp.
HARDY, A. R. and F. G. ANDREWS. 1987. Studies in the Coleoptera of western sand dunes. 3. Remarks upon *Serica* (Scarabaeidae) from some Nevada sand dunes with descriptions of two new taxa. Coleopterists Bulletin, 41: 173-179.
HAROLD, E. von. 1867. Zur Kenntniss der Gattung *Canthidium* und ihrer nächsten Verwandten. Coleopterologische Hefte, 1: 1-61.
HAROLD, E. von. 1869a. Tom. IV. Scarabaeidae. Pp. 979-1346. *In:* M. Gemminger and E. von Harold, eds. Catalogus Coleopterorum husque descriptorum synonymicus et systematicus.
HAROLD, E. von. 1869b. Abanderungen vergebener Namen. Coleopterologische Hefte, No. 5: 112-125.
HAROLD, E. von. 1869c. Révision des espèces qui rentrent dans le genre *Pinotus* Erichs. Abielle, 6: 123-144.
HARPOOTLIAN, P. J. 2001. Scarab beetles (Coleoptera: Scarabaeidae) of South Carolina. Biota of South Carolina, Volume 2. Clemson University. Clemson, SC, 157 pp.
HARPOOTLIAN, P. J., R. D. GORDON and R. E. WOODRUFF. 2000. Review of the genus *Leipsammodius* Rakovic (Coleoptera: Scarabaeidae: Aphodiinae: Psammodiini) in America north of Mexico with descriptions of three new species. Coleopterists Bulletin, 54: 292-299.
HATCH, M. H. 1971. The Beetles of the Pacific Northwest. Part V: Rhipiceroidea, Sternoxi, Phytophaga, Rhynchophora and Lamellicornia. University of Washington Publications in Biology, 16: 1-662.
HAWKS, D. C. In Press. Taxonomic and nomenclatural changes in *Chrysina* and a synonymic checklist of species (Scarabaeidae: Rutelinae). Consortium Coleopterorum.
HAWKS, D. C., C. S. BABCOCK and J. M. HERATY. In Press. Phylogenetic relationships of *Chrysina* Kirby (Coleoptera: Scarabaeidae: Rutelinae), and a comparative method for building character state trees. Cladistics.
HAWKS, D. C. and G. P. BRUYEA. 1999. *Chrysina cavei*, a new species from Honduras (Coleoptera: Scarabaeidae: Rutelinae). Occasional Papers of the Consortium Coleopterorum, 3: 15-20.
HELGESEN, R. G. and R. L. POST. 1967. Saprophagous Scarabaeidae (Coleoptera) of North Dakota. North Dakota Insects, Publication 7: 1-60.
HINTON, H. E. 1936. Studies in the Mexican and Central American Eupariini. University of California Publications in Entomology, 6: 273-276.
HOEBEKE, E. R. and K. BEUCKE. 1997. Adventive *Onthophagus* (Coleoptera: Scarabaeidae) in North America: geographic ranges, diagnoses, and new distributional records. Entomological News, 108: 345-362.
HOFFMANN, C. H. 1935. The biology and taxonomy of the genus *Trichiotinus* (Scarabaeidae-Coleoptera). Entomologica Americana, 15: 133-214.
HOFFMANN, C. H. 1939. The biology and taxonomy of the Nearctic species of *Osmoderma*. Annals of the Entomological Society of America, 32: 510-525.
HORN, G. H. 1876. Revision of the United States species of *Ochodaeus* and other genera of Scarabaeidae. Transactions of the American Entomological Society, 5: 177-197.
HORN, G. H. 1887. A monograph of the Aphodiini inhabiting the United States. Transaction of the American Entomological Society, 14: 1-110.
HOUSTON, W. W. K. and T. A. WIER. 1992. Melolonthinae. Pp. 174-358. *In:* W. W. K. Houston, ed. Zoological catalogue of Australia. Vol. 9. Coleoptera: Scarabaeoidea. Australian Government Publishing Service, Canberra.
HOWDEN, H. F. 1954. Review of the species of *Podolasia* Harold (Coleoptera: Scarabaeidae). American Museum Novitates, 1661: 1-11.
HOWDEN, H. F. 1958. Species of *Acoma* Casey having a three-segmented antennal club (Coleoptera: Scarabaeidae). Canadian Entomologist, 90: 377-401.

HOWDEN, H. F. 1959. A new species of *Isonychus* Mannerheim from southern Arizona (Coleoptera : Scarabaeidae). Canadian Entomologist, 91: 59-61.

HOWDEN, H. F. 1961. New species and a new genus of Melolonthinae from the southeastern United States (Coleoptera: Scarabaeidae). Canadian Entomologist, 93: 807-812.

HOWDEN, H. F. 1962. New species and synonymy in *Acoma* (Coleoptera: Scarabaeidae). Canadian Entomologist, 94: 1147-1155.

HOWDEN, H. F. 1965. A second New World species of *Sisyphus* Latreille (Coleoptera: Scarabaeidae). Canadian Entomologist, 97: 842-844.

HOWDEN, H. F. 1966. Notes on the Canthonini of the "Biologia Centrali-Americana" and descriptions of new species (Coleoptera: Scarabaeidae). Canadian Entomologist, 98: 725-741.

HOWDEN, H. F. 1968a. Generic relationships of *Thyce*, *Plectrodes*, *Dinacoma*, and *Hypotrichia*, with a description of a new genus and species from eastern Texas (Coleoptera: Scarabaeidae: Melolonthini). Canadian Entomologist, 100: 542-548.

HOWDEN, H. F. 1968b. A review of the Trichiinae of North and Central America (Coleoptera: Scarabaeidae). Memoirs of the Entomological Society of Canada, 54: 1-77.

HOWDEN, H. F. 1971a. Five unusual genera of New World Scarabaeidae (Coleoptera). Canadian Entomologist, 103: 1463-1471.

HOWDEN, H. F. 1971b. Key to the New World Cremastocheilini, with notes and descriptions of a new genus. Proceedings of the Entomological Society of Washington, 73: 224-230.

HOWDEN, H. F. 1973. Four new species of *Onthophagus* from Mexico and the United States (Coleoptera; Scarabaeidae). Proceedings of the Entomological Society of Washington, 75: 329-337.

HOWDEN, H. F. 1997. Podolasiini Howden, new tribe, and a revision of the included genera, *Podolasia* Harold and *Podostena* Howden, new genus (Coleoptera: Scarabaeidae: Melolonthinae). Coleopterists Bulletin, 51: 223-255.

HOWDEN, H. F. and O. L. CARTWRIGHT. 1963. Scarab beetles of the genus *Onthophagus* Latreille north of Mexico. Proceedings of the United States National Museum, 114: 1-135.

HOWDEN H. F. and A. R. HARDY. 1971. Generic placement and adult behavior of the genus *Leptohoplia* Saylor. Proceedings of the Entomological Society of Washington, 73: 337-341.

HOWDEN, H. F. and P. O. RITCHER. 1952. Biology of *Deltochilum gibbosum* (Fab.) with a description of the larva. Coleopterists Bulletin, 6: 53-57.

HUNTER, J. S., III, G. T. FINCHER and D. C. SHEPPARD. 1996. Observations on the life history of *Onthophagus depressus* (Coleoptera: Scarabaeidae). Journal of Entomological Science, 31: 63-71.

JAMESON, M. L. 1990. Revision, phylogeny and biogeography of the genera *Parabyrsopolis* Ohaus and *Viridimicus* (new genus) (Coleoptera: Scarabaeidae: Rutelinae). Coleopterists Bulletin, 44: 377-422.

JAMESON, M. L. 1991. A new species of *Parachrysina* Bates (Coleoptera: Scarabaeidae: Rutelinae) from Nuevo León, Mexico. Coleopterists Bulletin, 45: 232-238.

JAMESON, M. L. 1998 (1997). Phylogenetic analysis of the subtribe Rutelina and revision of the *Rutela* generic groups (Coleoptera: Scarabaeidae: Rutelinae: Rutelini). Bulletin of the University of Nebraska State Museum, 14: 1-184.

JAMESON, M. L. 2000. Synopsis of the Mexican and Guatemalan genera *Rutelisca* Bates and *Metapachylus* Bates (Coleoptera: Scarabaeidae: Rutelinae) with comments on classification of the subtribe Rutelina. Proceedings of the Entomological Society of Washington, 102: 332-349.

JAMESON, M. L., B. C. RATCLIFFE and M. A. MORÓN. 1994. A synopsis of the Neotropical genus *Calomacraspis* Bates with a key to larvae of the American genera of Rutelini (Coleoptera: Scarabaeidae: Rutelinae). Annals of the Entomological Society of America, 87: 43-58.

JANSSENS, A. 1937. Révision des Onitides (Coleoptera Lamellicornia). Mémoires du Musée Royal d'Histoire Naturelle de Belgique, Deuxième Série, Fascicle 11, 200 pp

JANSSENS, A. 1953. Oniticellini (Coleoptera Lamellicornia). *In:* Exploration du Parc National de l'Upemba, Mission G.F. De Witte. Institut des Parcs Nationaux du Congo Belge. Fascicle 11: 1-118.

JERATH, M. L. 1960. Notes on larvae of nine genera of Aphodiinae in the United States. Proceedings of the United States National Museum, 3: 43-94.

JERATH, M. L. and P. O. RITCHER. 1959. Biology of Aphodiinae with special reference to Oregon (Coleoptera: Scarabaeidae). Pan-Pacific Entomologist, 35: 169-175.

KING, A. B. S. 1984. Biology and identification of white grubs (*Phyllophaga*) of economic importance in Central America. Tropical Pest Management, 30: 36-50.

KIRK, V. M. and E. U. BALSBAUGH, Jr. 1975. A list of the beetles of South Dakota. South Dakota State University, Agricultural Experiment Station, Technical Bulletin, 42: 1-137.

KOHLMANN, B. 1984. Biosistemática de las especies norteamericanas del género *Ateuchus* (Coleoptera: Scarabaeidae: Scarabaeinae). Folia Entomológica Mexicana, 60: 3-81.

KOHLMANN, B. and G. HALFFTER. 1988. Cladistic and biogeographical analysis of *Ateuchus* (Coleoptera: Scarabaeidae) of Mexico and the United States. Folia Entomológica Mexicana, 74: 109-130.

KRAJCÍK, M. 1998. Cetoniidae of the world. Catalogue-Part 1. Published by the author, Most, Czech Republic, 96 pp., indices separately paginated.

KRAJCÍK, M. 1999. Cetoniidae of the world. Catalogue-Part 2. M. Krajcík. Most, Czech Republic, 72 pp., index i-xxiii.

KRIKKEN, J. 1976. New genera of New World Cremastocheilini, with revisional notes (Coleoptera: Cetoniidae). Zoologische Mededelingen, 49: 307-315.

KRIKKEN, J. 1978. Valgine beetles: a preliminary review of the genera, with descriptions of two novelties. Zoologische Mededelingen, 53: 153-164.

KRIKKEN, J. 1981. Taxonomic review of the New World genus *Genuchinus* Westwood (Coleoptera: Cetoniidae). Proceedings of the Koninklijke Nederlandse Akademie van Wetenschappen (Series c), 84: 403-417.

KRIKKEN, J. 1984. A new key to the suprageneric taxa in the beetle family Cetoniidae, with annotated lists of the known genera. Zoologische Verhandelingen, 210: 1-75.

KUIJTEN, P. J. 1992. A revision of the genus *Parastasia* in the Indo-Australian region (Coleoptera: Scarabaeidae: Rutelinae). Zoologische Verhandelingen, 274: 1-207.

LACHAUME, G. 1985. Dynastini. 1: *Dynastes-Megasoma-Golofa.* Les Coléoptères du Monde, vol. 5. Sciences Nat, Venette (France), 85 pp., 29 plates.

LACORDAIRE, J. T. 1859. Histoire Naturelle des Insectes. Genera des Coléoptères. Volume 5. Librairie Encyclopédique de Roret, Paris, 750 pp.

LACROIX, M. 1989. Insectes Coléoptères Melolonthidae (1re partie). Fauna de Madagascar 73: 1-302.

LAGO, P. K., R. L. POST and C. Y. OSETO. 1979. The phytophagous Scarabaeidae and Troginae (Coleoptera) of North Dakota. North Dakota Insects, Publication No. 12. Schafer-Post Series, Bismarck, North Dakota, 131 pp.

LANCIANI, C. A. and D. P. WOJCIK. 1991. A high density of *Martineziella dutertrei* Chalumeau (Coleoptera: Scarabaeidae) in fire ant foraging areas on a pond edge. Coleopterists Bulletin, 45: 278-279.

LATREILLE, P. A. 1802. Histoire Naturelle, Générale et Particulière des Crustacés et des Insectes. Familles Naturelles des Genres. Volume 3. F. Dufart. Paris, 387 pp.

LAWRENCE, J. F. and A. F. NEWTON, Jr. 1995. Families and subfamilies of Coleoptera (with selected genera, notes, and references and data on family-group names). Pp. 779-1006. *In:* J. Pakaluk and S. A. Slipinski, eds. Biology, Phylogeny, and Classification of Coleoptera. Papers Celebrating the 80th Birthday of Roy A. Crowson. Muzeum i Instytut Zoologii PAN. Warsaw, Poland.

LEACH W. E. 1815. Entomology. Pp. 57-172. *In:* Brewster, ed. Edinburgh Encyclopedia. Volume 9 (1). Edinburgh.

LECONTE, J. L. 1856. Synopsis of the Melolonthidae. Journal of the Academy of Sciences of Philadelphia, (2) 3: 225-228.

LECONTE, J. L. 1861. Classification of the Coleoptera of North America. Part 1. Smithsonian Institution, Washington, DC, 214 pp.

LENG, C. W. 1920. Catalogue of the Coleoptera of America, North of Mexico. John D. Sherman. Mount Vernon, NY, 470 pp.

LENG, C. W. and A. J. MUTCHLER. 1927. Supplement 1919 to 1924 (inclusive) to Catalogue of the Coleoptera of America, North of Mexico. John D. Sherman. Mount Vernon, NY, 78 pp.

LENG, C. W. and A. J. MUTCHLER. 1933. Second and third supplements 1925 to 1932 (inclusive) to Catalogue of the Coleoptera of America, North of Mexico. John D. Sherman. Mount Vernon, NY, 112 pp.

LINELL, L. M. 1896. New species of North American Coleoptera of the family Scarabaeidae. Proceedings of the United States National Museum, 18: 721-731.

LOBO, J. M. and G. HALFFTER. 1994. Relaciones entre escarabajos (Coleoptera: Scarabaeidae) y nidos de tuza (Rodentia: Geomyidae): implicaciones biológicas y biogeográphicas. Acta Zoologica Mexicana (N.S.), 62: 1-9.

LÖDING, H. P. 1945. Catalogue of the beetles of Alabama. Alabama Geological Survey Monograph, 11: 1-172.

LUGINBILL, P. and H. R. PAINTER. 1953. May beetles of the United States and Canada. United States Department of Agriculture Technical Bulletin, 1060: 1-102.

MACHATSCHKE, J. W. 1972. Scarabaeoidea: Melolonthidae, Rutelinae. Coleopterorum Catalogus Supplementa, 66: 1-361.

MACLEAY, W. S. 1819. Horae Entomologicae, or Essays on the Annulose Animals. Volume 1, Part 1. I. S. Bagster. London, 160 pp.

MARTÍNEZ, A. 1951. Notas coleopterológicas V. La invalidez del nombre genérico *Pinotus* Erichson y dos nuevas sinonimias (Col. Scarab.). Anales de la Sociedad Científica Argentina, 152: 138-142.

MARTÍNEZ, A. 1961. Scarabaeoidea Neotropica VIII. Notas sobre el complejo *Megathopa* Eschscholtz, 1822 (Col. Scarab. Scarabaeinae). Neotropica, 7: 81-86.

MATTHEWS, E. G. 1961. A revision of the genus *Copris* Müller of the Western Hemisphere (Coleoptera, Scarabaeidae). Entomologica Americana (N.S.), 41: 1-137.

MATTHEWS, E. G. 1963. Observations on the ball-rolling behavior of *Canthon pilularius* (L.) (Coleoptera, Scarabaeidae). Psyche, 70: 75-93.

MATTHEWS, E. G. and G. HALFFTER. 1968. New data on American *Copris* with discussion of a fossil species (Coleopt., Scarab.). Ciencia (México), 26: 147-162.

McCLEVE, S. 1993. Three new species of flightless *Diplotaxis* from Oaxaca, Mexico (Coleoptera: Scarabaeidae: Melolonthinae). Coleopterists Bulletin, 47: 43-50.

MICÓ, E., W. E. HALL and B. C. RATCLIFFE. 2001. Descriptions of the larvae of *Hoplopyga singularis* (Gory and Percheron) and *Hologymnetis cinerea* (Gory and Percheron) with a revised key to the larvae of New World Gymnetini (Coleoptera: Scarabaeidae: Cetoniinae). Coleopterists Bulletin, 55: 205-217.

MICÓ, E., A. B. T. SMITH and M. A. MORÓN. 2000. New larval descriptions for two species of *Euphoria* Burmeister (Coleoptera: Scarabaeidae: Cetoniinae: Cetoniini: Euphoriina) with a key to the known larvae and a review of the larval biology. Annals of the Entomological Society of America, 93:795-801.

MIKSIC, R. 1987. Monographie der Cetoniinae der Paläarktischen und Orientalishcen Region. Coleoptera: Lamellicornia. Band 4. Graficki Zavod Hrvatske. Zagreb, 608 pp.

MONTREUIL, O. 1998. Analyse phylogénétique et paraphylie des Coprini et Dichotomiini (Coleoptera: Scarabaeidae). Scénarioi biogéographique. Annales de la Société Entomologique de France (N.S.) 34: 135-148.

MORÓN, M. A. 1979. Fauna de coleopteros lamelicornios de la Estacion de Biologia Tropical "Los Tuxtlas," Veracruz, UNAM, Mexico. Anales del Instituto de Biologia (UNAM, Mexico), Series Zoologica, 50: 375-454.

MORÓN, M. A. 1981. Fauna de coleopteros Melolonthidae de la Reserva de la Biosfera "La Michilia," Durango, Mexico. Folia Entomologica Mexicana, 50: 3-69.

MORÓN, M. A. 1983a. A revision of the subtribe Heterosternina (Coleoptera, Melolonthidae, Rutelinae). Folia Entomológica Mexicana, 55: 31-101.

MORÓN, M. A. 1983b. Los estados inmaduros de *Inca clathrata sommeri* Westwood (Coleoptera, Melolonthidae, Trichiinae): con observaciones sobre el crecimiento alometrico del imago. Folia Entomologica Mexicana, No. 56: 31-51.

MORÓN, M. A. 1986. El género *Phyllophaga* en Mexico. Morfología, distribución y sistemática supraespecifica. (Insecta: Coleoptera). Instituto de Ecologia. Mexico, D.F., 341 pp.

MORÓN, M. A. 1990. The Beetles of the World. 10: Rutelini 1. Sciences Nat, Venette, France, 145 pp.

MORÓN, M. A. 1991. Revision del subgenero *Chirodines* de *Phyllophaga* (Coleoptera: Melolonthidae, Melolonthinae). Giornale Italiano di Entomologia, 5: 397-407.

MORÓN, M. A. 1992. Nuevas especies mexicanas del subgenero *Chlaenobia* de *Phyllophaga* (Coleoptera: Melolonthidae). Giornale Italiano di Entomologia, 6: 15-31.

MORÓN, M. A. 1993. La especies de *Phyllophaga* (Coleoptera: Melolonthidae) del estado de Veracruz, Mexico. Diversidad, distribucion e importancia. Pp. 55-82. *In:* M A. Morón, ed. Diversidad y manejo de plagas subterraneas. Publicación Especial de la Sociedad Mexicana de Entomología e Instituto de Ecología, Xalapa, Veracruz, Mexico.

MORÓN, M. A. 1994. Fauna de Coleoptera Lamellicornia en las montanas del noreste de Hidalgo, Mexico. Acta Zoologica Mexicana (n.s.), 63: 7-59.

MORÓN, M. A. 1996. Melolonthidae (Coleoptera). Pp. 287-307. *In:* J. L. Llorente-Bousquets, A. N. Garcia-Aldrete, y E. Gonzalez-Soriano, eds. Biodiversidad, taxonomía y biogeografía de artropódos de Mexico. Hacia una sintesis de su conocimiento. Universidad Nacional Autónoma de México. Mexico, 660 pp.

MORÓN, M. A. and C. DELOYA. 1991. Los coleópteros lamelicornios de la Reserva de la Biosphera "La Michilia," Durango, Mexico. Folia Entomológica Mexicana, 81: 209-283.

MORÓN, M. A., C. DELOYA and L. DELGADO. 1988. Fauna de Coleopteros Melolonthidae, Scarabaeidae y Trogidae de la region de Chamela, Jalisco, Mexico. Folia Entomologica Mexicana, 77: 313-378.

MORÓN, M. A. and H. F. HOWDEN. 1992. A second species of *Plesiosternus* Morón with notes on other Heterosternina (Coleoptera: Scarabaeidae: Rutelinae). Coleopterists Bulletin, 46: 15-19.

MORÓN, M. A. and J. KRIKKEN. 1990. A mesoamerican genus of Trichiinae (Coleoptera: Scarabaeoidea). Folia Entomologica Mexicana, 78: 71-84.

MORÓN, M. A. and B. C. RATCLIFFE. 1996. New tribal placement of the genus *Coscinocephalus* Prell, 1936, with description of the larva, pupa and adult of a new species from Mexico (Coleoptera: Scarabaeoidea: Dynastinae). Journal of the New York Entomological Society, 104: 48-61.

MORÓN, M. A., B. C. RATCLIFFE and C. DELOYA. 1997. Atlas de los Escarabajos de México. Coleoptera: Lamellicornia. Vol. 1. Familia Melolonthidae. Subfamilias Rutelinae, Dynastinae, Cetoniinae, Trichiinae, Valginae y Melolonthinae. Sociedad Mexicana de Entomología, Mexico, 280 pp.

MORÓN, M. A. and L. E. RIVERA. 1992. Dos especies nuevas de *Phyllophaga* (*Phytalus*) de la Sierra de Manantlan, Jalisco, Mexico (Coleoptera: Melolonthidae). Anales del Instituto de Biologia (UNAM, Mexico), Series Zoologica, 63: 79-87.

MORÓN, M. A., F. J. VILLALOBOS and C. DELOYA. 1985. Fauna de coleópteros Lamelicornios de Boca del Chajul, Chiapas, Mexico. Folia Entomologica Mexicana, 66: 57-118.

MORÓN, M. A. and S. ZARAGOZA. 1976. Coleopteros Melolonthidae y Scarabaeidae de Villa de Allende, Estado de Mexico. Anales del Instituto de Biologia (UNAM, Mexico), Series Zoologica, 47: 83-118.

NIKOLAYEV, G. V. 1979. Neue Gattungen und Untergattungen der Blatthornkäfer (Coleoptera, Scarabaeidae). Reichenbachia, 17: 189-191.

NOMURA, S. 1976. On the Sericini of Japan (Scarabaeoidea), II. Toho Gakuho, 26:167-205.

d'OLSOUFIEFF, G. 1924. Les phanaeides (Coleoptera-Lamellicornia). Insecta, 13: 5-172.

PIKE, K. S., R. L. RIVERS, C. Y. OSETO and Z. B. MAYO. 1976. A world bibliography of the genus *Phyllophaga*. University of Nebraska, The Agricultural Experiment Station, 21 pp.

PITTINO, R. 1984. Insects of Saudi Arabia. Coleoptera Scarabaeoidea: a revision of the Family Aphodiidae. Fauna of Saudi Arabia, 6: 267-360.

PITTINO, R. and G. MARIANI. 1986. A revision of the Old World species of the genus *Diastictus* Mulsant and its allies (*Platytomus* Muls., *Pleurophorus* Muls., *Afrodiastictus* n. gen., *Bordatius* n. gen.) (Coleoptera, Aphodiidae, Psammodiini). Giornale Italiano di Entomologia, 3: 1-165.

POOLE, R. W. and P. GENTILI. 1996. Nomina Insecta Nearctica. Volume I: Coleoptera; Strepsiptera. Entomological Information Services. Rockville, MD, 827 pp.

POPE, R.D. 1961. *Aserica*, *Autoserica*, *Neoserica*, or *Maladera*? (Coleoptera: Melolonthidae). Annals and Magazine of Natural History, [1960] (series 13) 3: 545-550.

POTTS, R. W. 1945. A key to the species of Cremastocheilini of North America and Mexico (Coleoptera, Scarabaeidae). Bulletin of the Brooklyn Entomological Society, 40: 72-78.

POTTS, R. W. L. 1974. Revision of the Scarabaeidae: Anomalinae. 1. The genera occurring in the United States and Canada (Coleoptera). Pan-Pacific Entomologist, 50: 148-154.

POTTS, R. W. L. 1977a. Revision of the Scarabaeidae: Anomalinae. 2. An annotated checklist of *Anomala* for the United States and Canada. Pan-Pacific Entomologist, 53: 34-42.

POTTS, R. W. L. 1977b. Revision of the Scarabaeidae: Anomalinae. 3. A key to the species of *Anomala* of America north of Mexico. Pan-Pacific Entomologist, 53: 129-134.

RAKOVIC, M. 1981. Revision of species of the tribe Psammodiini from the Australian Region (Coleoptera, Scarabaeidae, Aphodiinae). I. Key to genera and revision of the genera *Aphodopsammobius* Endrödi, *Psammodius* Fallen, *Diastictus*

Mulsant, *Phycochus* Broun and *Tesarius g. n.* Annotationes Zoologicae et Botanicae (Bratislava), 139: 1-37.
RAKOVIC, M. 1984. A review of the genus *Tesarius* Rakovic (Coleoptera, Scarabaeidae, Aphodiinae). Acta Entomologica Bohemoslovaca, 81: 448-452.
RAKOVIC, M. 1986. A review of the genus *Neopsammodius gen. n.* (Coleoptera, Scarabaeidae, Aphodiinae). Acta Entomologica Bohemoslovaca, 83: 197-201.
RAKOVIC, M. 1990. Review of the genus *Leiopsammodius* Rakovic on the world basis with a key to species from the Western Hemisphere and description of a new species (Coleoptera, Scarabaeidae, Aphodiinae). Annotationes Zoologicae et Botanicae (Bratislava), 197: 1-18.
RATCLIFFE, B. C. 1976a. A revision of the genus *Strategus* (Coleoptera: Scarabaeidae). Bulletin of the University of Nebraska State Museum, 10: 93-204.
RATCLIFFE, B. C. 1976b. Notes on the biology of *Euphoriaspis hirtipes* (Horn) and descriptions of the larva and pupa (Coleoptera: Scarabaeidae). Coleopterists Bulletin, 30: 217-225.
RATCLIFFE, B. C. 1977. Descriptions of the larva and pupa of *Osmoderma subplanata* (Casey) and *Cremastocheilus wheeleri* LeConte (Coleoptera: Scarabaeidae). Journal of the Kansas Entomological Society, 50: 363-370.
RATCLIFFE, B. C. 1981. *Barutus hartmanni*, a new genus and species from Panama with a key to the genera of New World Pentodontini (Coleoptera: Scarabaeidae: Dynastinae). Coleopterists Bulletin, 35: 463-472.
RATCLIFFE, B. C. 1991. The scarab beetles of Nebraska. Bulletin of the University of Nebraska State Museum, 12: 1-333.
RATCLIFFE, B. C. and A. C. DELOYA. 1992. The biogeography and phylogeny of *Hologymnetis* (Coleoptera: Scarabaeidae: Cetoniinae) with a revision of the genus. Coleopterists Bulletin, 46: 161-202.
REINHARD, H. J. 1950. The *Phyllophaga* of Texas (Col. Scarabaeidae). Journal of the Kansas Entomological Society, 23: 27-51.
RITCHER, P. O. 1949. Larvae of Melolonthinae with keys to tribes, genera and species (Coleoptera: Scarabaeidae). Kentucky Experimental Station Bulletin, 537: 1-36.
RITCHER, P. O. 1958. Biology of Scarabaeidae. Annual Review of Entomology, 3: 311-334.
RITCHER, P. O. 1966. White Grubs and Their Allies. A Study of North American Scarabaeoid Larvae. Oregon State University Press. Corvallis, OR, 219 pp.
RITCHER, P. O. 1969. Spiracles of adult Scarabaeoidea (Coleoptera) and their phylogenetic significance. I. The abdominal spiracles. Annals of the Entomological Society of America 62: 869-880.
RITCHER, P. O. 1973. A description of the larvae of *Hypothyce mixta* Howden (Coleoptera: Scarabaeidae Melolonthini). Coleopterists Bulletin 27: 113-116.
ROSANDER, R. W. and F. G. WERNER. 1970. Larvae of some Arizona species of *Phyllophaga* (Coleoptera: Scarabaeidae). Annals of the Entomological Society of America, 63: 1,136-1,142.
RUST, R. W. and L. M. HANKS. 1982. Notes on the biology of *Aegialia hardyi* Gordon and Cartwright (Coleoptera: Scarabaeidae). Pan-Pacific Entomologist, 58:319-325.
SAMOUELLE, G. 1819. The Entomologist's Useful Compendium; or an Introduction to the knowledge of British Insects. Thomas Boys. London, 496 pp.
SANDERSON, M. J. 1939. A new genus of Scarabaeidae with descriptions and notes on *Phyllophaga*. Journal of the Kansas Entomological Society, 12: 1-15.
SANDERSON, M. J. 1951. The *Phyllophaga* of Hispaniola. Museum of Comparative Zoology (Harvard) Bulletin, 105: 249-283.
SAYLOR, L. W. 1935. A new genus and two new species of Coleoptera from California. Pan-Pacific Entomologist, 11: 132-134.
SAYLOR, L. W. 1937a. New scarab genera from lower and southern California (Coleoptera). Bulletin of the Southern California Academy of Sciences, 36: 35-37.
SAYLOR, L. W. 1937b. The beetles of the subfamily Chasmatopterinae in the New World. Journal of the Washington Academy of Sciences, 27: 531-535.
SAYLOR, L. W. 1938. Revision of the subfamily Oncerinae with description of a new genus (Coleoptera: Scarabaeidae). Proceedings of the Entomological Society of Washington, 40: 99-103.
SAYLOR, L. W. 1939. Revision of the beetles of the melolonthine subgenus *Phytalus* of the United States. Proceedings of the United States National Museum, 86: 157-167.
SAYLOR, L. W. 1940a. Revision of the scarabaeid beetles of the *Phyllophaga* subgenus *Listrochelus* of the United States, with discussion of related subgenera. Proceedings of the United States National Museum, 89: 59-130.
SAYLOR, L. W. 1940b. Ten new West Indian scarab beetles of the genus *Phyllophaga*, with two new names. Journal of the Washington Academy of Sciences, 30: 305-314.
SAYLOR, L. W. 1940c. Synopsis of the *cadaverinus* group of the genus *Aphodius* with descriptions of three new species (Coleoptera: Scarabaeidae). Proceedings of the Biological Society of Washington, 53: 99-104.
SAYLOR, L. W. 1940d. Synoptic review of the beetle genera *Cotalpa* and *Paracotalpa* of the United States, with description of a new subgenus. Proceedings of the Entomological Society of Washington, 42: 190-200.
SAYLOR, L. W. 1942. Notes on beetles related to *Phyllophaga* Harris, with descriptions of new genera and subgenera. Proceedings of the United States National Museum, 92: 157-165.
SAYLOR, L. W. 1945a. Revision of the scarab beetles of the genus *Dichelonyx*. Bulletin of the Brooklyn Entomological Society, 40: 137-157.
SAYLOR, L. W. 1945b. Synoptic revision of the United States scarab beetles of the subfamily Dynastinae, No. 1: tribe Cyclocephalini. Journal of the Washington Academy of Sciences, 35: 378-386.
SAYLOR, L. W. 1946a. Synoptic revision of the United States scarab beetles of the subfamily Dynastinae, No. 2: tribe

Oryctini (part). Journal of the Washington Academy of Sciences, 36: 16-21.
SAYLOR, L. W. 1946b. Synoptic revision of the United States scarab beetles of the subfamily Dynastinae, No. 3: tribe Oryctini (part). Journal of the Washington Academy of Sciences, 36: 41-45.
SAYLOR, L. W. 1948a. Synoptic revision of the United States scarab beetles of the subfamily Dynastinae, No. 4: tribe Oryctini (part). Journal of the Washington Academy of Sciences, 38: 176-183.
SAYLOR, L. W. 1948b. Contributions toward a knowledge of the insect fauna of lower California. No. 10. Coleoptera, Scarabaeidae. Proceedings of the California Academy of Sciences, Series 4, 24: 337-374.
SCHENKLING, S. 1921. Scarabaeidae: Cetoniinae. Coleopterorum Catalogus, 72: 1-431.
SCHMIDT, A. 1922. Coleoptera Aphodiinae. Das Tierreich, Lieferung 45. Walter de Gruyer. Berlin, 716 pp.
SCHOLTZ, C. H. 1990. Phylogenetic trends in the Scarabaeoidea (Coleoptera). Journal of Natural History, 24: 1027-1066.
SHOOK, G. A. 1978. Records of some scarabs from Idaho (Coleoptera: Scarabaeidae). Coleopterists Bulletin, 32: 52.
SIMONIS, A. 1981. *Anoplodrepanus*, nuovo genere di Oniticellini (Coleoptera, Scarabaeidae). Bollettino del Museo di Zoologia dell'Università di Torino, 7: 87-94.
SKELLEY, P. E. 1991. Observations on the biology of *Stephanucha thoracica* Casey (Coleoptera: Scarabaeidae: Cetoniinae). Coleopterists Bulletin, 45: 176-188.
SKELLEY, P. E. 2000. New record for *Tesarius* Rakovic in South America (Scarabaeidae: Aphodiinae: Psammodiini). Insecta Mundi, 14: 240.
SKELLEY, P. E. and R. D. GORDON. 2001. Scarab beetles from pocket gopher burrows in the southeastern United States (Coleoptera: Scarabaeidae). Insecta Mundi, (in press).
SMITH, A. B. T. 2001. Checklist of the Scarabaeoidea of the Nearctic Realm (Includes Canada, the continental United States, and the following states of northern Mexico: Baja California, Baja California Sur, Chihuahua, Coahuila de Zaragoza, Durango, Nuevo Leon, Sinaloa, Sonora, Tamaulipas, and Zacatecas). URL: http://www-museum.unl.edu/research/entomology/nearctic.htm.
STEBNICKA, Z. 1994. The status of some taxa of Aphodiinae with descriptions of new genus and species. Acta Zoologica Cracoviensia, 37: 71-80.
STEBNICKA, Z. 2000. A new genus for Nearctic *Pleurophorus ventralis* Horn, 1887, with phylogenetic inferences (Coleoptera: Scarabaeidae: Aphodiinae: Aphodiini). Acta Zoologica Cracoviensia, 43: 287-291.
STEINER, W. E., Jr. 1980. Notes on the habitats and distribution of *Rhyssemus scaber* (Coleoptera: Scarabaeidae). Coleopterists Bulletin, 34: 377-378.
THOMAS, M. 1998. A flower beetle, *Euphoria sepulcralis* (Fabricius), in Florida (Coleoptera: Scarabaeidae). Florida Department of Agriculture and Consumer Services, Entomology Circular No. 386: 1-2.
TILDEN, J. W. and G. S. MANSFIELD. 1944. Notes on three species of *Ceononycha* Horn (Coleoptera, Scarabaeidae). Pan-Pacific Entomologist, 20: 115-117.
VANDER MEER, R. K. and D. P. WOJCIK. 1982. Chemical mimicry in the myrmecophilous beetle *Myrmecaphodius exavaticollis*. Science, 218(4574):806-808.
VAN DYKE, E. C. 1928. Notes and descriptions of new species of Scarabaeidae from western North America. Pan-Pacific Entomologist, 4: 151-162.
VAURIE, P. 1958. A revision of the genus *Diplotaxis* (Coleoptera, Scarabaeidae, Melolonthinae). Part 1. Bulletin of the American Museum of Natural History, 115: 269-369.
VAURIE, P. 1960. A revision of the genus *Diplotaxis* (Coleoptera, Scarabaeidae, Melolonthinae). Part 2. Bulletin of the American Museum of Natural History, 120: 167-433.
VAURIE, P. 1963. Key to *Diplotaxis* of Baja California. Pan-Pacific Entomologist, 39: 67-73.
WARNER, W. B. 1990. Two new North American *Copris* Müller, with notes on other species (Coleoptera: Scarabaeidae). Pan-Pacific Entomologist, 66: 232-240.
WARNER, W. B. and M. A. MORÓN. 1992. A revision of the *Phyllophaga* subgenus *Triodonyx* Saylor (Coleoptera: Scarabaeidae). Journal of the Kansas Entomological Society, 65: 321-340.
WATERHOUSE, D.F. 1974. The biological control of dung. Scientific American, 230: 100-109.
WOJCIK, D. P., W. A. BANKS, D. M. HICKS and J. W. SUMMERLIN. 1977. Fire ant myrmecophiles: new hosts and distribution of *Myrmecaphodius excavaticollis* (Blanchard) and *Euparia castanea* Serville (Coleoptera: Scarabaeidae). Coleopterists Bulletin, 31: 329-334.
WOJCIK, D. P., W. A. BANKS and D. H. HABECK. 1978. Fire ant myrmecophiles: flight periods of *Myrmecaphodius excavaticollis* (Blanchard) and *Euparia castanea* Serville (Coleoptera: Scarabaeidae). Coleopterists Bulletin, 32: 59-64.
WOJCIK, D. P. and D. H. HABECK. 1977. Fire ant myrmecophiles: breeding period and ovariole number in *Myrmecaphodius excavaticollis* (Blanchard) and *Euparia castanea* Serville (Coleoptera: Scarabaeidae). Coleopterists Bulletin, 31: 335-338.
WOLCOTT, G. N. 1948. The insects of Puerto Rico. Coleoptera. Journal of Agriculture of the University of Puerto Rico, 32: 225-416. (actual date of publication footnoted as 1950, p. 416. See Woodruff and Beck 1989).
WOODRUFF, R. E. 1973. The scarab beetles of Florida (Coleoptera: Scarabaeidae). Part I. The Laparosticti (subfamilies Scarabaeinae, Aphodiinae, Hybosorinae, Ochodaeinae, Geotrupinae, Acanthocerinae). Arthropods of Florida and Neighboring Land Areas, 8: 1-220.
WOODRUFF, R. E. and B. M. BECK. 1989. The scarab beetles of Florida (Coleoptera: Scarabaeidae). Part II. The May or June beetles (genus *Phyllophaga*). Arthropods of Florida and Neighboring Land Areas, 13: 1-225.
WOODRUFF, R. E. and O. L. CARTWRIGHT. 1967. A review of the genus *Euparixia* with descriptions of a new species from leaf-cutting ants in Louisiana (Coleoptera: Scarabaeidae).

Proceedings of the United States National Museum, 123: 1-21.

WOODRUFF, R. E. and M. DEYRUP. 1994. Order Coleoptera. Family Scarabaeidae. Pp. 368-455. *In:* M. Deyrup and R. Franz, eds. Rare and Endangered Biota of Florida. Volume IV. Invertebrates. University Press of Florida. Gainesville, FL, 798 pp.

YOUNG, F. N. 1957. Notes on the habits of *Plusiotis gloriosa* LeConte (Scarabaeidae). Coleopterists Bulletin, 11: 67-70.

YOUNG, R. M. 1988. A monograph of the genus *Polyphylla* Harris in America North of Mexico. Bulletin of the University of Nebraska State Museum, 11: 1-115.

YOUNG, R. M. 2002. A new *Cotalpa* Burmeister taken on post oak in eastern Texas with notes and a key to species in the genus (Scarabaeidae: Rutelinae). Coleopterists Bulletin, In Press.

ZIMMERMAN, R. J., W. L. JACOBS and B. C. KONDRATIEFF. 1991. Plant feeding scarabs (Coleoptera: Scarabaeidae) of Colorado. Unpublished report, Colorado State University, 34 pp.

ZUNINO, M. 1982. Aspectos taxonómicos y biogeográficos del poblamiento americano de Oniticellini (Coleoptera, Scarabaeidae). Acta del VIII Congreso Latinoamericano de Zoología, Mérida, Venezuela, 2: 1083-1089.

ZUNINO, M. 1985. Las relaciones taxonómicas de los Phanaeina (Coleoptera, Scarabaeinae) y sus implicaciones biogeográphicas. Folia Entomológica Mexicana, 64: 101-115.

ZUNINO, M. and G. HALFFTER. 1988. Análisis taxonómico, ecológico y biogeográfico de un grupo americano de *Onthophagus* (Coleoptera: Scarabaeidae). Museo Regionale di Scienze Naturali - Torino. Monografia 9: 1-211.

Series ELATERIFORMIA Crowson 1960

Superfamily SCIRTOIDEA Fleming 1821

35. EUCINETIDAE Lacordaire 1857

by Daniel K. Young

Family common name: The plate-thigh beetles

The compact, elliptical body with greatly enlarged metathoracic coxal plates and saltatorial behavior characterize this small mycophagous family.

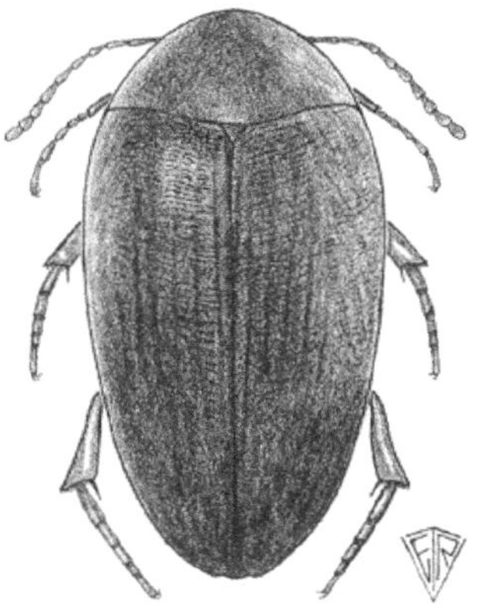

FIGURE 1.35. *Eucinetus terminalis* LeConte

Description: (after Lawrence 1982, Lawrence and Britton 1994, Vit 1999) Elongate-oval, convex; length 0.8-4.0 mm; color brown or black; vestiture consisting of evenly distributed, decumbent setae.

Head small, narrowed anteriorly, moderately (*Tohlezkus*) to strongly deflexed, resting on the prothoracic coxae; surface punctate. Antennae 11-segmented, slightly serrate to gradually expanded distally. Labrum moderately-sized; mandibles small, curved, apices acute; maxillary palpi 4-segmented, filiform; labial palpi 3-segmented, filiform. Mouthparts in several taxa modified, adapted for piercing-sucking function. Eyes lateral, moderate in size, entire.

Pronotum broader than head, very short, broad, narrowed in front; prosternum very short; prothoracic coxal cavities open behind. Scutellum of moderate size, triangular. Mesosternum short; metasternum broad. Legs short; prothoracic coxae conical and projecting (e.g., *Eucinetus*, *Nycteus*) to strongly transverse (e.g., *Euscaphurus*); mesothoracic coxae conical; metathoracic coxae dilated into large oblique plates, concealing much of metathoracic legs and part of first abdominal ventrite; femora slender; mesothoracic and metathoracic tibiae each bearing two large spurs (North American genera) and distinct rings of dark apical spurs; tarsal formula 5-5-5, tarsi slender, simple, decreasing in size apically; mesothoracic and metathoracic tarsomeres 1-4 bearing rings of apical spines; claws small, simple. Elytra entire, apically rounded, surface punctate, some taxa (e.g., *Eucinetus*) with transverse strigose lines; epipleura narrow and complete to elytral apices (*Tohlezkus*) or short.

Abdomen with five or six visible ventrites, 1-2 connate in several taxa; surface rugulose. Male genitalia consisting of a basal piece and parameres that are fused basally. Female genitalia with the valvifers absent, coxite and stylus present, proctiger large (Tanner 1927).

Larvae (after Lawrence 1991) oblong, slightly flattened, head broad, with five stemmata, bilaterally, antennae 3-segmented; mandibles each with a large, asperate molar surface, maxillae each with separate galea and lacinia; abdomen 10-segmented, urogomphi lacking; spiracles annular.

Habits and habitats. These beetles live in detritus or under fungus-covered bark of trees. Larvae are mycophagous on a variety of fungi including spores of slime molds and fruiting bodies of Basidiomycetes (Agaricaceae, Boletaceae, Coniophoraceae) (Bruns 1984, Weiss and West 1921, Wheeler and Hoebeke 1984).

Status of the classification. Historically, eucinetids have usually been considered a part of Dascilloidea, but several synapomorphies (larvae and adults) link Eucinetidae with Decliniidae (eastern Russia), Clambidae, and Scirtidae to form a well defined superfamily Scirtoidea (Crowson 1955, 1960, Lawrence and Newton 1982, 1995, Vit 1977).

Distribution. World-wide, approximately 37 species in nine genera are known: *Bisaya* (Central Asia), *Eucilodes* (Eurasia), *Eucinetus* (widely distributed), *Eucinetella* (China), *Euscaphurus* (western North America), *Jentozkus* (Neotropical), *Nycteus* (widely distributed), *Subulistomella* (Japan), and *Tohlezkus* (Turkey, Finland and eastern North America). Four genera and 11 species are known from the United States and Canada (Hatch 1961, Lawrence 1982, Lawrence and Newton 1995).

Key to the Genera of America north of Mexico

1. Metathoracic episterna present; epipleura developed anteriorly (Fig. 8); head strongly deflexed, not visible dorsally (Figs. 1-2, 5). 2
— Distinct metathoracic episterna lacking; epipleura narrow, complete to apex of elytra (Fig. 9); head partially visible from above .... *Tohlezkus*

2(1). Fifth tarsomere shorter than preceding; prothoracic coxae long and conical; six visible abdominal ventrites 3

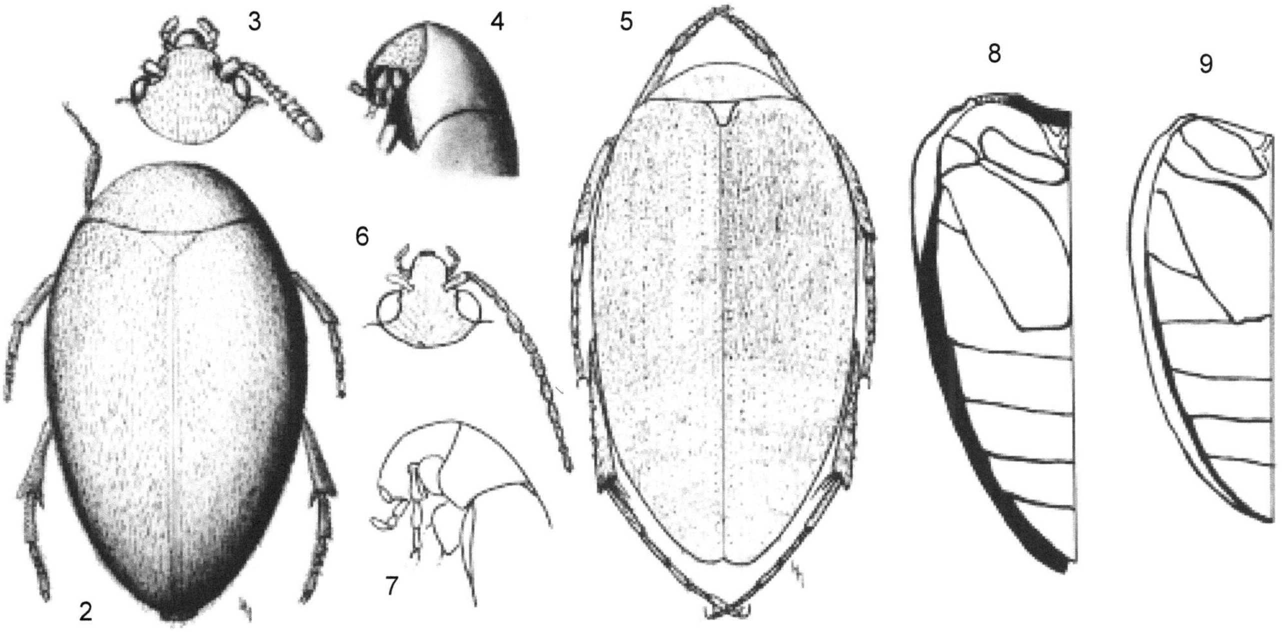

FIGURES 2.35-9.35. 2-4. *Euscaphurus saltator* Casey (from Hatch 1961), 2. habitus, 3. head, anterior view, 4. head and prothorax, lateral view; 5-7. *Nycteus infumatus* (LeConte) (from Hatch 1961), 5. habitus, 6. head, anterior view, 7. head and prothorax, lateral view; 8. *Euscaphurus saltator*, right venter (after Vit 1977); 9. *Tohlezkus inexpectus* Vit, right venter (after Vit 1977).

— Fifth tarsomere of meso- and metathoracic legs longer than fourth, prothoracic coxae strongly transverse; five visible abdominal ventrites (male with rudimentary trace of sixth) ..... *Euscaphurus*

3(2). Elytra densely punctate, lacking transverse strigose lines (Fig. 5) .... *Nycteus*

— Elytra distinctly substriate, with transverse strigose lines (Fig. 1). .... *Eucinetus*

Classification of the Genera of America north of Mexico

Eucinetidae Lacordaire 1857

Eucinetus Germar 1818, 3 spp., *E. morio* LeConte, Georgia, Illinois, Indiana, Virginia; *E. strigosus* LeConte, Georgia, Ohio, Pennsylvania; *E. terminalis* LeConte, British Columbia, Manitoba, Idaho, Illinois, Indiana, Michigan, New York, Oregon, Washington, Vermont. (Key to species in Downie and Arnett 1996)

Hamaxobium Duftschmid 1825

Euscaphurus Casey 1885, 2 spp., *E. saltator* Casey, Oregon, Washington; *E. spinipes* Vit, California. (Key to species in Vit 1996).

Nycteus Latreille 1829, 5 spp., distributed as follows: *N. falsus* Vit, Arizona, Mexico; *N. infumatus* (LeConte), British Columbia, Oregon, Washington; *N. oviformis* (LeConte), Illinois, Michigan, New York, Ohio, Pennsylvania, Virginia; *N. punctulatus* (LeConte), British Columbia, Massachusetts, Michigan, Oregon, Virginia; *N. testaceus* (LeConte), British Columbia, Quebec, Pennsylvania, Washington.

Tohlezkus Vit 1977, 1 sp., *T. inexpectus* Vit, North Carolina, Tennessee.

[*Jentozkus* Vit 1977, not in America north of Mexico]

Bibliography

BRUNS, T. D. 1984. Insect mycophagy in the Boletales: fungivore diversity and the mushroom habitat. Pp. 91-129. *In:* Q. D. Wheeler and M. Blackwell, eds., Fungus/insect relationships: perspectives in ecology and evolution. Columbia University Press. New York.

CROWSON, R. A. 1955. The Natural Classification of the Families of Coleoptera. Nathaniel Lloyd. London, 187 pp.

CROWSON, R. A. 1960. The phylogeny of Coleoptera. Annual Review of Entomology, 5: 111-134.

DOWNIE, N. M. and R. H. ARNETT, Jr. 1996. The beetles of Northeastern North America. Volume I: Introduction; Suborders Archostemata, Adephaga, and Polyphaga thru Superfamily Cantharoidea. The Sandhill Crane Press. Gainesville. 880 pp.

HATCH, M. H. 1961. The beetles of the Pacific Northwest. Part III: Pselaphidae and Diversicornia I. University of Washington Publications in Biology, Volume 16. University of Washington Press. Seattle ix + 503 pp.

LAWRENCE, J. F. 1982. Coleoptera. Pp. 482-553. *In:* S. B. Parker, ed., Synopsis and Classification of Living Organisms. Volume 2. McGraw-Hill. New York.

LAWRENCE, J. F. 1991. Eucinetidae. Pp. 364-365. *In:* F. W. Stehr, ed., Immature Insects. Volume 2. Kendall/Hunt. Dubuque, IA.

LAWRENCE, J. F. and E. B. BRITTON. 1994. Australian Beetles. Melbourne University Press. Carlton, Victoria, x+192 pp.

LAWRENCE, J. F. and A. F. NEWTON, Jr. 1982. Evolution and classification of beetles. Annual Review of Ecology and Systematics, 13: 261-290.

LAWRENCE, J. F. and A. F. NEWTON, Jr. 1995. Families and subfamilies of Coleoptera (with selected genera, notes, references and data on family-group names). Pp. 779-1006. *In:* J. Pakaluk and S. A. Slipinski, eds., Biology, Phylogeny, and Classification of Coleoptera: Papers Celebrating the 80th Birthday of Roy A. Crowson. Muzeum i Institut Zoologii, Polska Academia Nauk. Warsaw.

TANNER, V. M., 1927. A preliminary study of the genitalia of female Coleoptera. Transactions of the American Entomological Society 53: 5-50.

VIT, S. 1977. Contribution à la connaissance des Eucinetidae (Coleoptera). Revue Suisse de Zoologie, 84: 917-935.

VIT, S. 1996. Deux espèces nouvelles d'Eucinetidae d'Amérique du nord particulièrement intéressantes (Coleoptera: Eucinetidae). Elytron, 9: 125-137.

VIT, S. 1999. Sur les genres *Nycteus* Latreille, 1829 et *Eucinetus* Germar, 1818 (Coleoptera, Eucinetidae). Mitteilungen der Schweizerischen Entomologischen Gesellschaft, 72: 387-394.

WEISS, H. B. and E. WEST 1921. Additional fungous insects and their hosts. Proceedings of the Biological Society of Washington, 34: 59-62.

WHEELER, Q. D. and R. E. HOEBEKE. 1984. A review of mycophagy in the Eucinetoidea (Coleoptera), with notes on an association of the eucinetid beetle, *Eucinetus oviformis*, with a Coniophoraceae fungus (Basidiomycetes: Aphyllophorales). Proceedings of the Entomological Society of Washington, 86: 274-277.

36. CLAMBIDAE Jacquelin du Val 1857

by Daniel K. Young

Family common name: The minute beetles

Clambids are minute, oval and convex beetles, most of which are capable of rolling into a ball, with the head strongly deflexed. The partially or completely divided compound eyes (*Clambus, Loricaster*), the expanded metathoracic coxal plates concealing the metathoracic femora, and the hair-like fringe on the margins of the metathoracic wings (*Calyptomerus, Clambus*) separate Clambidae from other families.

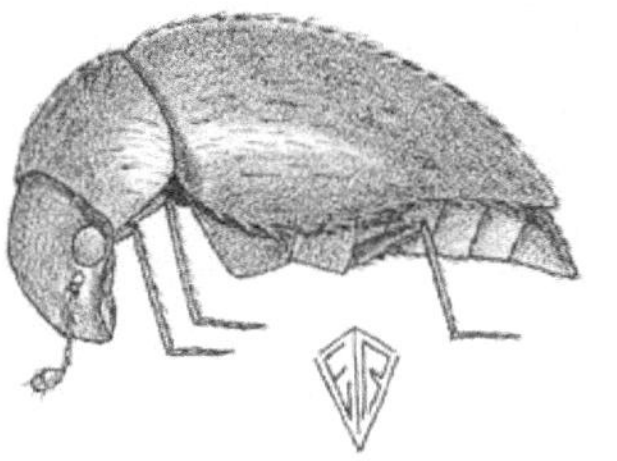

FIGURE 1.36. *Clambus pubescens* Redtenbacher

Description: (after Crowson 1979, Lawrence 1982, Lawrence and Britton 1994) Shape broadly oval, convex, compact; length 0.7-2.0 mm; color testaceous to piceous; body glabrous, or, in most species covered with moderately long, stout, and dense, decumbent (rarely erect) setae.

Head flattened, broadly and irregularly oval, strongly deflexed; surface finely punctate or rugulose. Antennae 8- (*Loricaster*) or 10-segmented (*Calyptomerus, Clambus*), scape large, the last two antennomeres expanded into an abrupt club. Frontoclypeal region forming a sharp ridge, concealing mouthparts, giving rise to antennal insertions, and partially to entirely dividing compound eyes. Mouthparts reduced, palpi awl-shaped.

Pronotum short, broader than head, sides explanate and slightly overlapping elytral bases, surface convex, finely margined laterally, evenly arcuate; surface smooth or rugulose. Trochantins of prothoracic legs moderately developed; prothoracic coxae conical, moderately prominent, contiguous; mesothoracic coxae separate; metathoracic coxae large, continguous, forming expanded plates that partially or completely cover metathoracic legs and at least part of the first abdominal ventrite. Femora swollen, tibiae and tarsi slender; tibiae without apical spurs; tarsal formula 4-4-4, tarsomeres subequal; claws simple. Scutellum elongate-triangular (*Calyptomerus, Clambus*) or very small (*Loricaster*). Elytra entire, slightly broader than thorax at widest part; surface smooth or rugulose; epipleural fold present or absent. Metathoracic wings with reduced venation, margins fringed with hair-like setae of moderate length (*Calyptomerus, Clambus*) or secondarily wingless (*Loricaster*).

Abdomen with five visible ventrites, the sutures distinct, first visible ventrite longer than the remaining ventrites. Male genitalia with penis thin, partly flattened, in most species bearing a dorsally projecting, curved, apical hook; parameres fused for two-thirds their length (completely fused in several taxa), apices unequal, each with two stout apical spines; basal piece present or not (Sharp and Muir 1912).

Larvae (after Crowson 1979, Lawrence 1991) similar to Eucinetidae: elongate, somewhat flattened, stemmata variable, many taxa with 5-6, bilaterally, reduced or lacking in some species; antennae 3-segmented; mandibles each with well developed, asperate mola; each maxilla with undivided mala (North American species). The abdomen lacks urogomphi.

Habits and habitats. These beetles live in decaying plant material; they have often been observed flying at dusk above the forest floor. Most appear to be mycophagous, especially on spores of Myxomycetes and Ascomycetes (Lawrence 1991).

Status of the classification. The family Clambidae is well defined, and support has been demonstrated for the monophyly of Scirtoidea (= Decliniidae + Eucinetidae+Clambidae + Scirtidae) (Lawrence *et al.* 1995, Lawrence and Newton 1995).

Distribution. There are five genera and about 70 described species in the world: *Calyptomerus* (western North America and Eurasia, with one adventive species in Australia and South Africa), *Acalyptomerus* (Africa, Asia, Central and South America), *Clambus* (widely distributed), *Loricaster* (western North America, Europe, Japan),and *Sphaerothorax* (Australia, Chile, New Zealand) (Crowson 1979, Hatch 1929, Lawrence and Newton 1995). Three genera and 12 species are known from North America north of Mexico (Endrödy-Younga 1981, Grigarick and Schuster 1961).

Key to the Genera of America north of Mexico

1. Antennae 10-segmented; scutellum well developed (Fig. 2) 2
— Antennae 8-segmented; scutellum very small, largely concealed by base of pronotum .... *Loricaster*

2(1). Eyes divided by lateral margin of head; antennal insertions close to eyes (Fig. 2) .... *Clambus*
— Eyes not divided; antennal insertions well in front of eye .... *Calyptomerus*

Classification of the Genera of America north of Mexico

Clambidae Jacquelin du Val 1857

Clambus Fischer von Waldheim 1820, 10 spp., widely distributed.
Sternuchus LeConte 1850

Calyptomerus Redtenbacher 1849, 1 sp., *C. oblongulus* Mannerheim, British Columbia, Alaska, Arizona, Colorado.
Comazus Fairmaire 1854

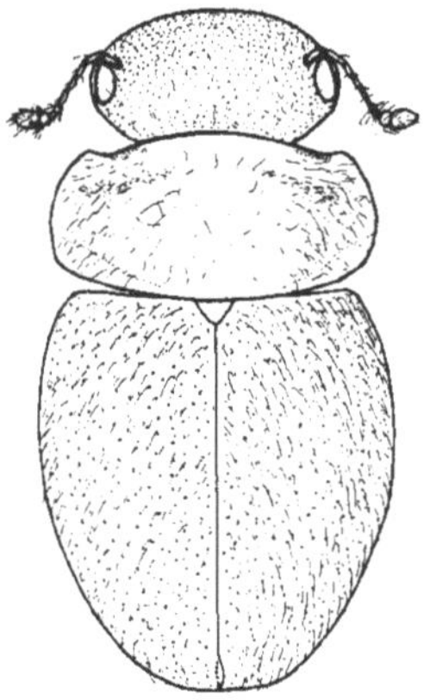

FIGURE 2.36. *Clambus vulneratus* LeConte (From Hatch 1957)

Loricaster Mulsant and Rey 1861, 1 sp., *L. rotundus* Grigarick and Schuster, California.

BIBLIOGRAPHY

CROWSON, R. A. 1979. Observations on Clambidae (Coleoptera), with descriptions of a new genus and species and of several larvae. Revue Suisse de Zoologie, 86: 611-623.

ENDRÖDY-YOUNGA, S. 1981. The American species of the familia Clambidae (Coleoptera: Eucinetoidea). Entomologia Generalis, 7: 33-67.

GRIGARICK, A. A. and R. O. SCHUSTER. 1961. A new species of *Loricaster* from California (Coleoptera: Clambidae). Pan-Pacific Entomologist, 37: 161-164.

HATCH, M. H. 1929. The genera and subgenera of Leiodidae and Clambidae. Journal of the New York Entomological Society, 37: 1-6.

LAWRENCE, J. F. 1982. Coleoptera, Pp. 482-553. *In:* S. B. Parker, ed. Synopsis and Classification of Living Organisms. Volume 2. McGraw-Hill. NY.

HATCH, M. H. 1957. The beetles of the Pacific Northwest. Part II. Staphyliniformia. University of Washington Publications in Biology, 16: 384 pp.

LAWRENCE, J. F. 1991. Clambidae, Pp. 365-366. *In:* F. W. Stehr, ed. Immature Insects. Volume 2. Kendall/Hunt. Dubuque, IA.

LAWRENCE, J. F. and E. B. BRITTON 1994. Australian Beetles. Melbourne University Press. Carlton, Victoria, x + 192pp.

LAWRENCE, J. F. and A. F. NEWTON, Jr. 1995. Families and subfamilies of Coleoptera (with selected genera, notes, references and data on family-group names), Pp. 779-1006. *In:* J. Pakaluk and S. A. Slipinski, eds. Biology, Phylogeny, and Classification of Coleoptera: Papers Celebrating the 80th Birthday of Roy A. Crowson. Muzeum i Institut Zoologii, Polska Academia Nauk. Warsaw.

LAWRENCE, J. F., N. B. NIKITSKY and A. G. KIREJTSHUK. 1995. Phylogenetic position of Decliniidae (Coleoptera: Scirtoidea) and comments on the classification of Elateriformia (*sensu lato*), Pp. 375-410. *In:* J. Pakaluk and S. A. Slipinski, eds. Biology, Phylogeny, and Classification of Coleoptera: Papers Celebrating the 80th Birthday of Roy A. Crowson. Muzeum i Institut Zoologii, Polska Academia Nauk, Warsaw.

SHARP, D. and F. A. G. MUIR. 1912. The comparative anatomy of the male genital tube in Coleoptera. Transactions of the Entomological Society of London, 1912: 477-642, pls. 42-78.

37. SCIRTIDAE Fleming 1821

by Daniel K. Young

Family common name: The marsh beetles

Family synonyms: Cyphonidae Stephens 1829; Elodidae Shuckard 1840; Helodidae Agassiz 1846

The strongly deflexed head, conical prothoracic coxae, broad, somewhat flattened, oval shape, and the short, broad prothorax serve to separate the adults of this family. The multisegmented antennae of the larvae separate this group not only from all other beetle families, but from all other holometabolous insects as well.

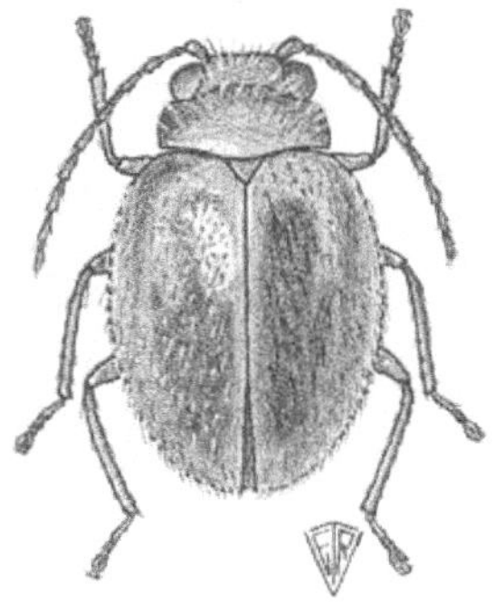

FIGURE 1.37. *Prionocyphon limbatus* LeConte

Description: Shape oval to more or less elongate, somewhat flattened; length 1.5 to 12 mm; color pale yellowish to black, in some spotted; vestiture consisting of deciduous, moderately long, suberect pubescence.

Head rather large, strongly deflexed, oval; surface punctate, with a pair of prominent genal ridges that rest against the prothoracic coxae when head is in deflexed position. Antennae with 11 antennomeres, filiform to subserrate, large, flattened, the first antennomere very large, inserted between the eyes, above bases of mandibles, beneath a very fine frontal margin. Labrum small, quadrate, anterior border arcuate; mandibles moderate, curved, apices acute or bifid; maxillary palpi with four awl-shaped palpomeres; labial palpi with three small, slender palpomeres. Eyes lateral, large, bulging, rounded.

Pronotum short, broad, sometimes partly covering the head, explanate, anterior margin broadly emarginate, lateral margin nearly straight, posterior margin sinuate; surface shallowly punctate; prosternum very short; prothoracic coxal cavities open behind externally and internally. Scutellum moderate, triangular. Mesosternum short; metasternum short and broad. Legs with the prothoracic coxae transverse, conical, narrowly separated; mesothoracic coxae subconical, slightly separate to contiguous, metathoracic coxae often with strongly dilated plates internally which partly cover the first abdominal sternite, coxae slightly separated; trochanters moderate, triangular; femora sometimes (*Scirtes, Ora*) greatly swollen; tibiae sulcate, spurs moderate; tarsal formula 5-5-5, the fourth tarsomere bilobed, larger than the third; claws simple. Elytra entire, apically rounded, surface punctate; epipleural fold broad, complete. Metathoracic wings with two anal veins (Forbes 1922); folding pattern with area B very large, and first A area enlarged, reaching the margin broadly, divided by subordinate folds; area C short, D broad on the costa (Forbes 1926).

Abdomen with five visible sternites, the sutures entire or with first two visible sternites connate. Male genitalia highly complex; penis stout, with two slender apical lobes; parameres apparently fused to the basal piece, basal piece large, trough-like, apically deeply emarginate (Sharp and Muir 1912, Young and Stribling 1990). Female genitalia with valvifers forming a long baculum; coxite partly divided into two segments; stylus small, apical; typically with complex internal armature (= prehensor) of uncertain homology (Young and Stribling 1990).

Larvae (Hannappel and Paulus 1987, LeSage 1991, Stribling and Young 1990) campodeiform, somewhat fusiform, slightly depressed; length 4 to 15 mm; vestiture of many small setae and a few longer setae on all segments; color creamy-yellowish to piceous. Head smaller than prothorax, depressed, wider than long, prognathous. Antennae long, multisegmented (unique among the Holometabola), in most individuals longer than head and thorax. Mouthparts highly complex, adapted for filter-feeding. Epipharynx with complex setae, brushes and sensoria and taxonomically distinctive epipharyngeal teeth; hypopharynx likewise complex, with taxonomically significant comb-like hypopharyngeal armature. Mandibles stout, broad at base, with a striate mola, complex prostheca, often with a dense fringe of setae along the mesal margin; maxilla with small cardo, distinct stipes, lacinia with inner fringe of setae, and galea with complex setae and comb-hairs, four-segmented palpi, the third segment bearing taxonomically distinctive sensory organs; labium with a broad, transverse submentum, a conspicuous, wide, disk-like combined mentum and ligula, palpi two-segmented, inconspicuous. Stemmata lateral, variable, one to three pairs. Thorax with five-segmented legs, including claw-like tarsungulus. Abdomen typically with eight dorsally visible segments, the ninth may be present, each segment depressed. Functional spiracles on the eighth segment only; many species with five anal papillae which may be branched.

Habits and habitats. Larvae are aquatic, although they can withstand significant fluctuations in water levels, and they have occasionally been found in moist wood and damp decaying leaves. Scirtid larvae frequent both stagnant and flowing water including ground pools in the forest floor and a diversity of marsh types (hence the common name, "marsh beetles"). Some marsh inhabiting species develop in microhabitats possessing high iron content, such as iron seepages. Larvae are also found in sphagnum bogs and lentic microhabitats within rivers and streams, such as marginal pools and backwaters. Perhaps the most habitat-specialized are the species associated with phytotelmata, or plant-held pools. This group includes such species as those developing in leaf axils of bromeliads and tree holes (*e.g.,* all known *Sacodes* spp. as well as some species of *Cyphon* and *Prionocyphon*) (Beerbower 1943, Benick 1924, LeSage 1991, Stribling and Young 1990).

Status of the classification. The Nearctic and Neotropical species of this family are badly in need of revision. Genitalic characters appear to be fairly reliable for both sexes. Unfortunately, the external anatomy of many species, especially *Cyphon* spp., is rather homogeneous, thereby necessitating the dissection of virtually every specimen for species level identifications (Hannappel and Paulus 1987, Klausnitzer 1976, 1980, Yoshitomi 1997).

Distribution. Worldwide there are about 30 genera and more than 600 species (Lawrence 1982), with the greatest diversity in the Northern and Southern Temperate regions. Presently, 50 species in eight genera are known from North America north of Mexico, with undoubtedly numerous undescribed species, especially in *Cyphon.*

Key to the Genera of America north of Mexico

1. Metathoracic femora greatly enlarged, broadened, saltatorial; spurs of metathoracic tibiae long, one being at least one-half as long as first tarsomere 7
— Metathoracic femora much like those of prothoracic and mesothoracic legs, not greatly enlarged for jumping; spurs of metathoracic tibiae small 2

2(1). Meso- and metasternal processes in contact between mesothoracic coxae, separating them 5
— Mesosternal process short, narrow, not contacting metasternum, mesothoracic coxae contiguous in about apical one-half 3

3(2). First tarsomere of metathoracic tarsi flattened above, finely margined laterally, second with part of hind margin prolonged, hiding basal section of third; labial palpi with third palpomere arising from side (about mid-point) of second 4
— First tarsomere of metathoracic tarsi rounded above, not laterally margined, second not produced posteriorly, not laterally margined, second not produced posteriorly, not hiding part of third, labial palpi with third palpomere arising from apex of second .... *Sarabandus*

4(3). Body elongate-oval, width less than half the length .... *Elodes*
— Body ovate, width greater than half the length .... *Sacodes*

5(2). First antennomere large, fully twice as broad as any of those following, expanded anteriorly; second arising from posterior apical angle of first and from under a slight margin; third very small, one-half as long as second; broadly ovate species .... *Prionocyphon*
— Antennae not as above, body shape various 6

6(5). Labial palpi with third palpomere arising from side (about mid-point) of second .... *Microcara*
— Labial palpi with third palpomere arising from end of second .... *Cyphon*

7(1). Metathoracic coxae meeting along full length of median line, hind margin conjointly forming a subquadrate plate which is not on same plane as intercoxal process of abdomen .... *Scirtes*
— Metathoracic coxae touching each other only anteriorly, arcuately diverging posteriorly where they are about on a plane with the abdominal process which separates them .... *Ora*

Classification of the Nearctic Genera

Scirtidae Fleming 1821

Elodes Latreille 1796, 7 spp.: *E. angusta* Hatch, Oregon, Washington; *E. apicalis* LeConte, California, Nevada, Oregon; *E. aquatica* Blaisdell, California; *E. emarginata* Hatch, Oregon, Washington; *E. impressa* Hatch, Oregon, Washington; *E. maculicollis* Horn, Quebec, New Hampshire, New York, Pennsylvania, Tennessee; *E. nunenmacheri* Wolcott, California (partial key to species in Hatch 1961).

Helodes Duval 1861
Hemicyphon LeConte 1866

Sacodes LeConte 1853, 3 spp., eastern U.S. and Canada; *S. fuscipennis* (Guérin), "Amer. bor.", known only from type; *S. pulchella* (Guérin), widespread in eastern United States; *S. thoracica* (Guérin), widespread in eastern United States (key to species in Horn 1880 — second half of his key to "*Helodes*").

Flavohelodes Klausnitzer 1980 (see also Yoshitomi 1997)

Microcara Thomson 1859, 1 sp., *M. explanata* (LeConte), Manitoba, Ontario, Michigan, Minnesota, New Hampshire, New York, North Dakota.

Cyphon Paykull 1799, 27 spp., generally distributed.

Sarabandus Leech 1955, 1 sp., *S. robustus* (LeConte), Georgia, Massachusetts, New Jersey, New York, North Carolina, Virginia.

Prionocyphon Redtenbacher 1858, 2 spp.: *P. discoideus* (Say), widespread in eastern United States; *P. limbatus* LeConte, widespread in eastern United States (key to species in Horn 1880).

Ora Clark 1865, 3 spp.: *O. hyacintha* Blatchley, Florida; *O. texana* Champion, Florida, Georgia; *O. troberti* Guérin, Florida, Texas.

Scirtes Illiger 1807, 6 spp., eastern United States, Arizona, California, and Texas: *S. californicus* Motschulsky, California; *S. orbiculatus* (Fabricius), widespread in eastern United States; *S. ovalis* Blatchley, Florida, North Carolina; *S. piceolus* Blatchley, Florida; *S. plagiatus*

Schaeffer, Arizona, Texas; *S. tibialis* Guérin, widespread in eastern United States.

REFERENCES

BEERBOWER, F. V. 1943. Life history of *Scirtes orbiculatus* Fabricius (Coleoptera: Helodidae). Annals of the Entomological Society of America, 36: 672-680.

BENICK, L. 1924. Zur Biologie der Käferfamilie Helodidae. Mitteilungen der Geographischen Gesellschaft und des Naturhistorischen Museums in Lübeck, (2)29: 47-75.

FORBES, W. T. M. 1922. The wing-venation of the Coleoptera. Annals of the Entomological Society of America, 15: 328-345.

FORBES, W. T. M. 1926. The wing folding patterns of the Coleoptera. Journal of the New York Entomological Society, 34: 42-115.

HANNAPPEL, U. and H. F. PAULUS. 1987. Arbeiten zu einem phylogenetischen System der Helodidae (Coleoptera)-Feinstructuruntersuchungen an europäischen Larven. Zoologische Beiträge N. F., 31: 77-150.

HATCH, M. H. 1961. The beetles of the Pacific Northwest. Part III: Pselaphidae and Diversicornia I. University of Washington Publications in Biology, 16: 503 pp.

KLAUSNITZER, B. 1980. Bemerkungen zur Kenntnis Helodidae Bulgariens mit Beschreibung einer neuen Art der gattung *Helodes* Latreille (Col., Helodidae) (Beitrag zue Kenntnis Helodidae). Entomologische Blätter, 76: 58-64.

HORN, G. H. 1880. Synopsis of the Dascyllidae of the United States. Transactions of the American Entomological Society, 8: 76-114.

LAWRENCE, J. F. 1982. Coleoptera. Pp. 482-553. *In*: S. P. Parker, ed. Synopsis and Classification of Living Organisms. Volume 2. McGraw-Hill. New York.

LESAGE, L. 1991. Helodidae. Pp. 366-369. *In*: F. W. Stehr, ed. Immature Insects, Volume 2. Kendall/Hunt. Dubuque, IA.

SHARP, D. and F. MUIR. 1912. The comparative anatomy of the male genital tube in Coleoptera. Transactions of the Entomological Society of London, 1912: 477-642.

STRIBLING, J. B. and D. K. YOUNG. 1990. Descriptions of the larva and pupa of *Flavohelodes thoracica* (Guérin-Méneville) with notes on a phytotelma association (Coleoptera: Scirtidae). Proceedings of the Entomological Society of Washington, 92: 765-770.

YOSHITOMI, H. 1997. A revision of the Japanese species of the genera *Elodes* and *Sacodes* (Coleoptera: Scirtidae). Elytra, 25: 349-417.

YOUNG, D. K. and J. B. STRIBLING. 1990. Systematics of the North American *Cyphon collaris* species complex with the description of a new species (Coleoptera: Scirtidae). Proceedings of the Entomological Society of Washington, 92: 194-204.

Superfamily DASCILLOIDEA Guérin-Méneville 1843

Fossipedes Mulsant and Rey 1865; Dascillida Meixner 1935

38. DASCILLIDAE Guérin-Méneville 1843

by Sean T. O'Keefe

Family common name: The soft-bodied plant beetles

Family synonyms: Atopidae LaPorte 1834

These beetles somewhat resemble the elateroids, but the prosternal keel is not as well developed as in that group, and the frontal ridge is absent.

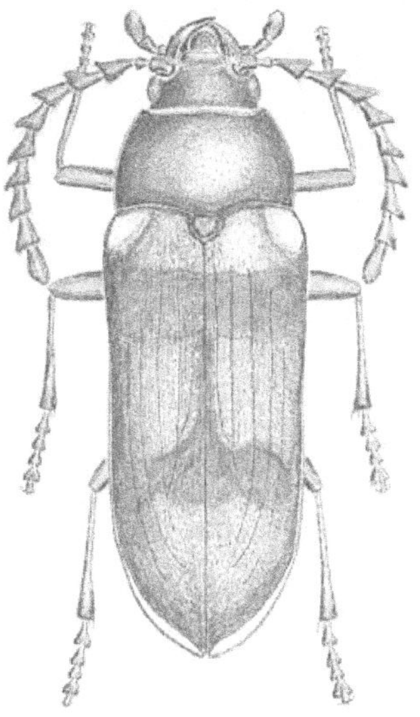

FIGURE 1.38. *Dascillus davidsoni* LeConte

Description: Elongate, slightly convex; length 8-18 mm; brownish or mottled grey-brown; vestiture dense, short, decumbent, covers entire body.

Head broadly triangular, slightly deflexed; surface finely punctate. Antennae with 11 antennomeres, about half body length, antennomere I short, club-shaped, antennomere II small, sub-triangular, antennomeres III-X subserrate to serrate, antennomere XI elongate, slightly ovate; antennae inserted on frons between eyes, just above mandibles, bases widely separated. Clypeus flat, transverse, distinctly separated from frons by impressed line; labrum nearly as long as broad; mandibles moderate in size, stout, curved, apices subacute; maxillary palpi with 4 palpomeres, palpomeres II-III subtriangular, palpomere IV enlarged, ovoid; submentum trapezoidal, narrow, mentum quadrate, membranous; ligula large, membranous, frequently bilobate; labial palpi with 3 palpomeres, palpomere III enlarged apically. Eyes lateral, round, finely faceted, moderate in size (*Dascillus*) or large (*Anorus*).

Thorax with pronotum distinctly (*Dascillus*) or slightly (*Anorus*) broader than head, moderately convex, subrectangular, with evenly arcuate, carinate lateral borders, posteriorly sinuate; surface finely punctate; pleural region moderate in size; prosternum broad in front of coxae with prominent, elevated keel between coxae (*Dascillus*) or without keel (*Anorus*); procoxal cavities open posteriorly; mesosternum short; metasternum broad; protrochantin exposed; procoxae large, transverse, separate (*Dascillus*) or contiguous (*Anorus*); mesocoxae subconical, narrowly separate; metacoxae transverse; trochanters moderate, triangular; femora slender; tibiae slender, with prominent apical spurs; tarsal formula 5-5-5, tarsomeres 1-4 bilobed beneath, claws simple. Scutellum subpentagonal. Elytra entire, apically rounded, striae punctate, intervals rugose-punctate; epipleural fold broad, entire, abruptly narrowed at basal third in some. Wings with five anals, wedge-cell distinct, closed, first r-m fine, extending from the small, subtriangular radial cell across to M which is long but weak and broken. Folding pattern with area B represented by a slender fold, concave on the upper side and extending nearly to the base of the wing, partially separated from area C by some crumpling, but functioning as a unit with it; median fold slender; first anal fold practically lost; area S imperfectly developed, but with area D large on the costa.

Abdomen with five visible sternites, the sutures entire. Male genitalia of the trilobed type; penis complex, with two parts, a dorsal piece broad, apically rounded, and a smaller ventral part with two broad basal struts; parameres large, arcuate; pars basalis smaller, complete. Female genitalia with the valvifers with a long bacula, coxite with a large bacula at the base; stylus small.

Larvae have only been described for *Dascillus*, those of *Anorus* are still unknown. Larvae elongate, subcylindrical, slightly flattened dorsoventrally, 10-35 mm in length, lightly sclerotized, surface smooth. Head large, protracted, declined, transverse, epicranial stem very short or absent; median endocarina absent; antennae long, with 3 antennomeres, antennomere III reduced; frontoclypeal suture absent; labrum partly fused to clypeus; mandibles asymmetrical, robust, bidentate; prostheca absent or reduced to a membranous area; maxilla with three palpomeres. Prothorax slightly larger than meso- or metathorax; thoracic and abdominal terga 1-9 well developed; meso- and metaterga and abdominal terga 1-8 each with transverse carina anteriorly; lega well developed, 5 segmented; abdomen with 10 visible segments; segments 1-8 subequal in length, segment 9 slightly shorter, segment 10 reduced (see Bøving and Craighead 1931, plate 63 and Lawrence 1991 for further details and figures).

Habits and habitats. Little has been recorded on the biology of dascillids. Larvae of *Dascillus* are found in moist soil or under rocks. Adults are often found on grass stems in spring.

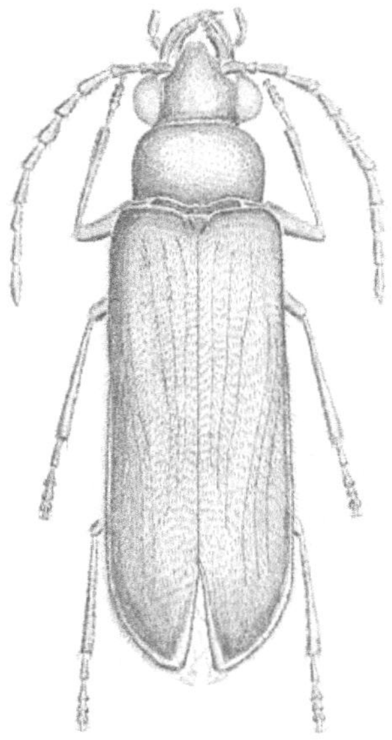

FIGURE 2.38. *Anorus piceus* LeConte.

Status of the classification. Crowson (1971) redefined the Dascilloidea and recognized three families, Dascillidae, Karumiidae, and Rhipiceridae. Lawrence and Newton (1995) recognized only two families, Dascillidae and Rhipiceridae, placing the Karumiidae as a subfamily in the Dascillidae. The composition of the family has changed considerably since the first edition of this book. Arnett (1963) included 14 genera in Dascillidae. Subsequently, *Macropogon* Motschulsky, *Eurypogon* Motschulsky, and *Allopogonia* Cockerell have been split off into the Artematopidae; *Anchytarsus* Guérin-Méneville, *Odontonyx* Guérin-Méneville, *Anchycteis* Horn, and *Araeopidius* Cockerell have been split off as the Ptilodactylidae; *Stenocolus* LeConte assigned to the Eulichadidae; and *Dicranopselaphus* Guérin-Méneville, *Ectopria* LeConte, *Acneus* Horn, and *Eubrianax* Kiesenwetter have been placed in the Psephenidae; which leaves only two genera, *Dascillus* Latreille and *Anorus* LeConte as representatives of Dascillidae in North America.

Distribution. There are 15 genera and 80 species (Lawrence 1991), generally distributed; there are five species in North America (Poole and Gentili 1996), limited to California and Arizona.

Key to the Nearctic genera

1. Larger (10-18 mm); mottled grey-brown; procoxae separated by prosternum (Dascillinae) . *Dascillus*

— Smaller (8-10 mm); brown; procoxae prominent and contiguous (Karumiinae) .... *Anorus*

Classification of the Nearctic Genera

Dascillidae Guérin-Méneville 1843

Dascillinae Guérin-Méneville 1843

Dascillus Latreille 1796, 2 spp., California. (Key to species: Horn 1880).

Atopa Paykull 1799

Petalon Laporte 1840

Karumiinae Escalera 1913

Anorus LeConte 1859, 3 spp., Arizona and California. (Key to species: Blaisdell 1934).

Bibliography

ARNETT, R. H., Jr. 1963. The Beetles of the United States: A Manual for Identification. Catholic University Press. Washington, DC. xi + 1112 pp.

BLAISDELL, F. E. 1934. Rare North American Coleoptera. Transactions of the American Entomological Society, 60: 317-326.

BØVING, A. G. and F. C. CRAIGHEAD. 1931. An illustrated synopsis of the principal larval forms of the Order Coleoptera. Entomologica Americana (New Series), 11: 1-351.

CROWSON, R. A. 1971. Observations on the superfamily Dascilloidea (Coleoptera: Polyphaga), with the inclusion of Karumiidae and Rhipiceridae. Zoological Journal of the Linnean Society of London, 50: 11-19.

HORN, G. H. 1880. Synopsis of the Dascillidae of the United States. Transactions of the American Entomological Society, 8: 76-114.

LAWRENCE, J. F. 1991. Dascillidae (Dascilloidea) (including Karumiidae). Pp. 369-370. *In:* F. W. Stehr, ed. Immature Insects, Volume 2. Kendall/Hunt. Dubuque, IA.

LAWRENCE, J. F. and A. F. NEWTON, Jr. 1995. Families and subfamilies of Coleoptera (with selected genera, notes, references and data on family-group names). Pp. 779-1006. *In:* J. Pakaluk and S. A. Slipinski, eds. Biology, Phylogeny, and Classification of Coleoptera: Papers celebrating the 80th birthday of Roy A. Crowson. Muzeum I Instytut Zoologii PAN, Warsaw.

POOLE, R. W. and P. GENTILI. 1996. Nomina Insecta Nearctica: A Check List of the Insects of North America, Volume 1: Coleoptera, Strepsiptera. Entomological Information Services. Rockville, MD. 827 pp.

39. RHIPICERIDAE Latreille 1834

by Kerry Katovich

Family common name: The cedar beetles or cicada parasite beetles

Family synonyms: Sandalidae Craighead 1921; Rhipidoceridae Lacordaire 1857

In the Nearctic region, Rhipiceridae are represented by the single genus, *Sandalus*. The sexually dimorphic, flabellate antennae of the males, arising from prominent tubercles at the bases of the prominent mandibles, large eyes, absence of an articulating prosternal process, and the presence of membranous lobes on tarsomeres 1-4 will distinguish Rhipiceridae from other families. Adults are uncommonly collected but, when found, commonly occur in aggregations. Species of *Sandalus* undergo hypermetamorphosis and are ectoparasitoids of nymphal cicadas.

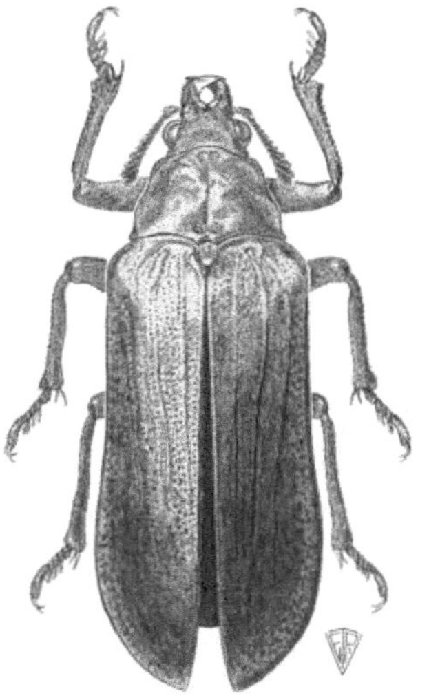

FIGURE 1.39. *Sandalus petrophya* Knoch

Description: (adapted in part from Rings 1942, Crowson 1955, Arnett 1971, Lawrence *et al.* 1999) [The following description is based on Nearctic *Sandalus*, and is not intended to be reflective of all rhipicerid genera]. Body elongate, moderately convex. Males 17-20 mm long, body dark brown to black, vestiture fine, setae short, dense, some males brown to dark brown, antennae orange or cedar-colored (suggesting common name), thorax darker than elytra; females 22-25 mm long, similar in color and vestiture to males; body more robust than in males.

Head elongate-quadrate, weakly deflexed, inserted slightly into the prothorax. Eyes relatively large, entire, finely facetted, lacking interfacetal setae. Antennae reaching at least middle of prothorax, but not beyond middle of elytra. Antennae with 11 antennomeres in both sexes; males flabellate, or females serrate. Antennae inserted into prominent antennal tubercles directly above the bases of mandibles; scape three times the length of pedicle. Males with antennomeres 3-11 8-10 times broader than long. Females with antennomeres 10-11 flattened, fused, all antennomeres 2-4 times broader than long. Labrum small, bilobed, appearing fused to head. Clypeus small and triangular. Mandibles generally large, well-sclerotized, visible in ventral view; apex strongly and abruptly curved mesally, unidentate, truncate or rounded. Dorsal surface smooth, flattened; ventral surface with a deep, concave depression at the innermost median half. Adoral face of mandible with a single tooth. Maxillae with cardo small; stipes often covered by lacinia; galea well-developed, apically densely setose or spinose; lacinia reduced. Maxillary palpi 4-segmented; palpomere small, remaining segments subequal in length and setose. Gula distinct, U-shaped, forming the ventral exoskeleton of the head; gular sutures distinct and separated. Labium divided into a subtriangular, plate-like mentum and reduced submentum; distal margin of mentum with small, rectangular palpiger, 3-segmented labial palpi, and mesal, conical, setose ligula.

Pronotum usually campanulate, slightly wider than long, widest posteriorly; dorsal surface finely to coarsely punctate; lateral carinae incomplete or apparently absent. Base of pronotum distinctly narrower than elytral bases, posterior margin sinuate, disk with mesal longitudinal groove or line.

Prosternum anterad coxae short, broad, flat to moderately convex. Prosternal process short, narrowed apically, flattened or weakly concave, apex acute or narrowly rounded. Procoxal cavities strongly transverse, contiguous medially, or narrowly separated, externally and internally open. Procoxae projecting below prosternum; protrochantin partially exposed.

Scutellum well-developed, subcordate. Mesosternal process extending at least to middle of mesocoxal cavity. Mesocoxae conical and projecting, coxal cavities moderately to widely separated, circular to slightly transverse, open laterally, not or partially closed by metepisterna.

Metasternum longer than first visible abdominal ventrite, moderately to strongly convex, with a transverse groove. Metacoxae not enlarged, horizontally oriented, contiguous or narrowly separated, extending laterally to meet elytra or sides of body.

Legs with prothoracic femora and tibiae narrow; femora with disto-ventral grooves; lateral margins of tibiae simple and rounded to carinate, lacking teeth, with paired spurs. Tarsal formula 5-5-5; tarsomeres 1-4 each with bilobed pulvillus or setose pads beneath. Fifth tarsomere elongate, with neither pulvillus nor setose pads. Pretarsus with ungitractor plate oval, deeply cleft distally; empodium club-shaped, trisetose; claws simple, paired.

Elytra entire, longer than wide, dorsal surfaces costate, punctate, punctures in rows. Elytral and sutural striae distinct to indistinct; elytral apices meeting or almost meeting along suture; elytral epipleurae complete, prominent posteriorly, not or gradually narrowing, lateral margin of each straight or weakly sinuate. Metathoracic wings well-developed, folding pattern dascilloid-like. Radial cell well-developed, elongate; inner posterior angle formed at base of radial cell right-angled or obtuse. Distal area of wing with one or more sclerotizations or pigment patches, or with one or more distinct veins; medial area with 5-6 free veins; oblongum cell absent; wedge cell well-developed or reduced, apex acute; anal lobe absent; posterior wing margins without fringes of long setae.

Abdomen with five visible ventrites, sutures entire. Functional spiracles located on ventrites 5-8. Spiracles on ventrites 5-7 located in pleural membrane. Propygidium (tergites 7 and 8) more or less horizontal. Tergite 9 in male slightly to moderately emarginate. Tergite 10 in male well-developed, free.

Male genitalia with aedeagus trilobate, symmetrical; parameres individually articulated to phallobase or base of penis; penis with basal and ventral lobes. Female genitalia of the ovipositor type; valvifers very long, slender; coxites long, slender, setiferous, bearing small, apical styli.

Larvae (adapted in part from Craighead 1921 and Elzinga 1977) of Rhipiceridae are believed to be hypermetamorphic, with a triungulin and ectoparasitoid form, based on the only available larval description, that of *Sandalus niger* Knoch (Craighead 1921). Triungulin less than 1 mm long; body elongate, parallel-sided, somewhat flattened, vestiture of scattered setae. Antennae 2-segmented. Mandibles flattened, apically rounded. Both maxillary and labial palpi 1-segmented, long and tusk-like. Legs long, 5-segmented; tarsungulus long and narrow, bearing a single seta. Paired egg bursters present on abdominal terga 3-7. Tergum 9 lacking urogomphi. Abdominal segment 10 well-developed, posteriorly oriented, with four lobes around the anus. Spiracles annular.

Mature larva 15-35 mm long. Body elongate, fusiform; abdomen physogastric, lightly sclerotized. Cuticle covered with scattered, short hairs. Head prognathous, slightly flattened. Antennae short, 1-segmented, conical. Labro-clypeal region fused to head capsule, labroclypeal and frontoclypeal sutures absent. Mandibles symmetrical, lightly sclerotized, apically acute, lacking accessory ventral process or mola. Each maxilla without distinct cardo; stipes elongate; palpi 2-segmented, fixed; galea falciform, fixed; lacinia rounded. Labrum with prementum, mentum and submentum fused into single plate; ligula present; labial palpi 1-segmented. Legs short, stout, widely separated, 3- to 4-segmented; tarsungulus with single seta. Abdominal segments gradually enlarged to segments 5 or 6, then reduced posteriorly. Segment 9 much smaller than segment 8. Tergum 9 with pair of short, upturned, fixed urogomphi. Segment 10 reduced, posteroventrally oriented. Spiracles annular-biforous, with long accessory tubes.

Habits and habitats. Adults of Nearctic *Sandalus* are active primarily from late September to early October. When encountered, they are often found aggregating in large numbers during the day. Neotropical and Nearctic species are known to be attracted to lights. Craighead (1921) described the larva of *Sandalus niger* Knoch, which was discovered within the exuviae of a nymphal cicada. From this evidence, he speculated that *S. niger* was an ectoparasitoid of cicada nymphs, an idea supported by Hicks (1942) and Young (1956), both of whom observed large aggregations of male and female *S. niger* in the vicinity of nymphal cicadas. Young (1956) suggested that the emergence of *S. niger* might be connected with the emergence of the periodical cicada *Magicicada septendecim* Linnaeus. Hicks (1942) noted that the cicada present was *Tibicen linnei* (Smith). Elzinga (1977) confirmed the relationship with cicadas by recording the life history of *S. niger* in Kansas. Adult emergence from soil occurred in late fall. Males flew frequently during warm afternoons, primarily around trees where a female was present. Adults apparently did not feed. Mating took place on the surface of the tree trunk. Large numbers of males were often seen attempting to mate with a single female, indicating a pheromone was used. After mating, the female inserted the tubular ovipositor into holes or crevices under bark, probable oviposition sites of cicadas. Sites previously occupied by the cicada, *Tibicen pruinosa* (Say), seemed to be preferred for oviposition, which lasted up to 15 hours, with many eggs deposited. Rings (1942) recorded 16,846 eggs from a single female *S. niger*. Eggs were easily washed off in rain storms, possibly along with those of cicadas. Egg eclosion occurred in the following spring, producing a triungulin. No further observations were made beyond this stage.

Status of the Classification. The status and classification of this family have historically been problematic. Horn (1881) considered Rhipiceridae closely related to Dascillidae, and included the Nearctic genus *Zenoa* along with *Sandalus*. Sharp and Muir (1912) placed Rhipiceridae in the Byrrhoidea, based on male genitalic morphology; however, they used *Callirhipis* (Callirhipidae) as their exemplar of Rhiphiceridae. Leng (1920) cataloged the family under Elateroidea. Forbes (1922) placed Rhipiceridae in his dascilloid group, based on wing venation. Forbes (1926, 1942) later noted that the wing-venation and wing folding of *Callirhipis* and *Zenoa* differed significantly from that of *Rhipicera* and *Sandalus*, and suggested the latter two resembled *Dascillus* (Dascillidae). Stickney's (1923) work on head capsule morphology suggested placement of Rhipiceridae in Elateroidea. Emden's (1924, 1931) discussion of Sandalidae, dealt mainly with genera currently placed in Callirhipidae. Pic (1925) cataloged the known genera and species of the world, including genera and species currently placed in Callirhipidae. Tanner (1927) studied the female genitalia, and Williams (1938) studied the mouthpart morphology. Both authors placed Rhipiceridae into what they termed the cantharoid series, close to Elateridae and Buprestidae. Craighead (1921) suggested that Rhipiceridae was comprised of *Zenoa* and Sandalidae was comprised of *Sandalus*. Crowson (1955) included both Callirhipidae and Sandalidae in the newly erected Rhipiceroidea, but later (Crowson 1971, 1973) separated the two families into different superfamilies, placing Rhipiceridae into Dascilloidea and Callirhipidae into Artematopoidea. Rhipiceridae is currently placed in the series Elateriformia and superfamily Dascilloidea (Lawrence and Newton 1995). No formal revision of the North American species has been undertaken, and a complete species key to *Sandalus* is unavailable.

The fossil record of Rhipiceridae is relatively young. Examples are known from the Paleocene and Eocene (Ponomarenko 1995). Crowson (1971) suggested a phylogeny with Rhipiceridae branching off as a member of Dascilloidea, sister group to Scarabaeoidea, in the mid-Jurassic (approx. 125 my. BP).

Distribution. This family occurs in all major zoogeographic regions. It is absent from the Hawaiian Islands and New Zealand. There are 57 described species in six genera worldwide (Lawrence *et al.* 1999). *Sandalus* Knoch (25 spp.) occurs in North and South America, Africa, Southeast Asia, China, India and Japan. *Rhipicera* Latreille (15 spp.) occurs in Australia, New Caledonia and Brazil. *Chamoerhipis* Latreille (6 spp.) occurs in Africa. *Ptiocerus* Castelnau (4 spp.) is restricted to South Africa. *Arrhaphipterus* Kraatz (6 spp.) occurs in southeastern Europe, North Africa and central Asia. *Polymerius* Philippi (monotypic, *P. marmoratus* Philippi) is restricted to Chile.

The Nearctic species are poorly known at present, and their distributions are poorly understood. *Sandalus californicus* LeConte occurs in California, Idaho and Nevada. *Sandalus cribricollis* Van Dyke is known only from California. *Sandalus niger* Knoch occurs from southern Ontario to the northeastern and midwestern United States, south into Texas. *Sandalus petrophya* Knoch occurs from the northcentral and northeastern United States south to Florida. *Sandalus porosus* LeConte occurs in Florida, Texas, New Mexico, Arizona and Colorado.

Classification of the Genera of America North of Mexico

Rhipiceridae Latreille 1834

Sandalus Knoch 1801

Five species widely distributed in the United States (see distribution section for species ranges). Partial key to species: Horn (1881), separating four species. *Sandalus cribicollis* is absent from the key; original description: Van Dyke (1923).

Bibliography

ARNETT, R. H. Jr. 1971. The beetles of the United States (a manual for identification). American Entomological Institute. Ann Arbor, MI, xii + 1112 pp.

CRAIGHEAD, F. C. 1921. Larva of the North American beetle *Sandalus niger* Knoch. Proceedings of the Entomological Society of Washington, 23: 44-48.

CROWSON, R. A. 1955. The natural classification of the families of Coleoptera. Nathaniel Lloyd. London, 187 pp.

CROWSON, R. A. 1971. Observations on the superfamily Dascilloidea (Coleoptera: Polyphaga), with the inclusion of Karumiidae and Rhipiceridae. Zoology Journal of the Linnean Society, 50: 11-19.

CROWSON, R. A. 1973. On a new superfamily Artematopoidea of polyphagan beetles, with the definition of two new fossil genera from the Baltic amber. Journal of Natural History, 7: 225-238.

ELZINGA, R. J. 1977. Observations on *Sandalus niger* Knoch (Coleoptera: Sandalidae) with a description of the triungulin larva. Journal of the Kansas Entomological Society, 50: 324-328.

EMDEN, F. I. VAN 1924. Zur Kenntnis der Sandalidae. II. u. III. Entomologische Blätter, 20: 86-99.

EMDEN, F. I. VAN. 1931. Zur Kenntnis der Sandalidae. XI-XVI. Entomologische Blätter, 27: 49-59, 107-116, 145-152.

FORBES, W. T. M. 1922. The wing venation of Coleoptera. Annals of the Entomological Society of America, 15: 328-352.

FORBES, W. T. M. 1926. The wing folding patterns of the Coleoptera. Journal of the New York Entomological Society, 34: 42-68, 91-139.

FORBES, W. T. M. 1942. The wing of the Schizopini (Coleoptera: Dascillidae). Entomological News, 53: 101-102.

HICKS, S. D. 1942. Observations on the adult of *Sandalus niger* Knoch in southern Ontario (Coleoptera: Rhipiceridae). Canadian Entomologist, 74: 78-79.

HORN, G. H. 1881. Notes on Elateridae, Cebrionidae, Rhipiceridae, and Dascyllidae. Transactions of the American Entomological Society, 9: 85-86.

LAWRENCE, J. F. and A. F. NEWTON, Jr. 1995. Families and subfamilies of Coleoptera (with selected genera, notes, references and data on family-group names). Pp. 841-842. *In:* J. Pakaluk and S.A Slipinski, eds. Biology, phylogeny, and classification of Coleoptera: Papers celebrating the 80th Birthday of R. A. Crowson. Volume 2. Muzeum I Instytut Zoologii PAN. Warsaw.

LAWRENCE, J. F., A. M. HASTINGS, M. J. DALLWITZ, T. A. PAINE and E. J. ZURCHER. 1999. Beetles of the world, a key and information system for families and subfamilies. Version 1.0 for Windows®. CSIRO Entomology, Canberra. Australia. CD-ROM.

LENG, C. W. 1920. Catalogue of the Coleoptera of America, north of Mexico. Mount Vernon, New York. 470 pp.

PIC, M. 1925. Pars 81. Rhipiceridae. *In*: S. Schenkling, ed. Coleopterorum Catalogus. W. Junk. Berlin, 13 pp.

PONOMARENKO, A. G. 1995. The geological history of beetles, Pp. 155-172 *In:* J. Pakaluk and S. A. Slipinski, eds. Biology, phylogeny, and classification of Coleoptera: Papers celebrating the 80th birthday of R. A. Crowson. Volume 1. Muzeum I Instytut Zoologii PAN. Warsaw.

RINGS, R. W. 1942. The external anatomy of *Sandalus niger* Knoch (Coleoptera: Rhipiceridae). Annals of the Entomological Society of America, 35: 411-425.

SHARP, D. and F. R. S. MUIR. 1912. The comparative anatomy of the male genital tube in Coleoptera, P. 545, pl. LXVI *In:* E.C. Zimmerman, ed. Male genital tube in Coleoptera. Entomological Society of America. Maryland.

STICKNEY, F. S. 1923. The head capsule of Coleoptera. Illinois Biological Monographs, 8: 1-104.

TANNER, V. M. 1927. A preliminary study of the genitalia of female Coleoptera. Transactions of the American Entomological Society, 53: 5-50.

VAN DYKE, E. C. 1923. New species of Coleoptera from California. Bulletin of the Brooklyn Entomological Society, 18: 37-38.

WILLIAMS, I. W. 1938. The comparative morphology of the mouthparts of the order Coleoptera treated from the standpoint of phylogeny. Journal of the New York Entomological Society, 46: 245-289.

YOUNG, F. N. 1956. Unusual abundance of *Sandalus* in southern Indiana. Coleopterists Bulletin, 9: 74.

Superfamily BUPRESTOIDEA Leach 1815

40. SCHIZOPODIDAE LeConte 1861

by G. H. Nelson and C. L. Bellamy

Family common name: The schizopodid or false jewel beetles

These beetles show some similarities to the Buprestidae, but they can be distinguished from them by the wide metepisternum and deeply bilobed fourth tarsomere of the adults. The schizopodines are a small group from western North America. Adults possess a primitive wing venation (Forbes 1942) compared to buprestids in general and yet the genital structures of both sexes are highly derived. In addition, the few known or putative larvae indicate primitive locomotory structures which suggest external root-feeding as opposed to most buprestids which develop within wood, stems or leaves of various host plants. A relationship to the second major buprestoid lineage, the Old World julodines, is supported by this groups' external plant-feeding habits, although no larvae of that group have any locomotory structures. The wing venation in the julodines is more typically buprestid and whereas the female ovipositor is highly modified for inserting the eggs into the soil, the male genitalia are also typically buprestid.

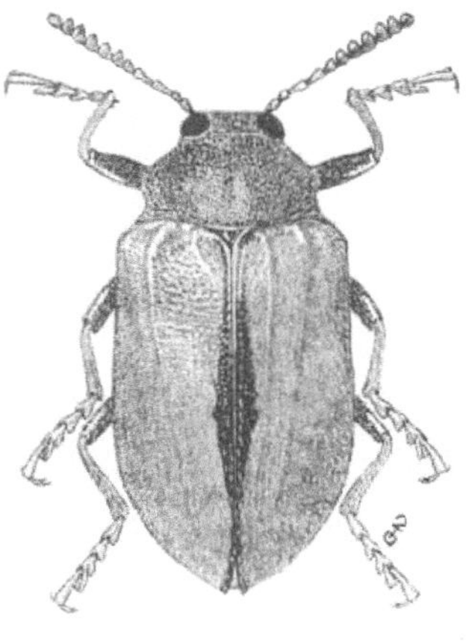

FIGURE 1.40. *Schizopus laetus* LeConte, male (from Nelson and Bellamy 1991, with permission of Taylor & Francis Ltd.)

Description: The body of adult schizopodids is somewhat to very stout and strongly convex. In color they vary from light brown in some species, green and coppery in other species to the dimorphic *Schizopus laetus* LeConte in which the male is green to blue with orange elytra, tibiae and tarsi, and the female is entirely green or blue to purple. The surface sculpture varies from coarse in the tribe Schizopodini to smooth and finely punctate in Dystaxiini. In size they vary from 6.2 mm to 18.0 mm in length.

Head with front either flat or slightly convex; eyes slightly emarginate near middle of anterior margin; antennae in the tribe Schizopodini have 11 antennomeres, each thick and transversely triangular with sensory pores diffuse on both surfaces. In the tribe Dystaxiini the antennae have 12 antennomeres that are flattened and elongate triangular with sensory pores in part concentrated in pits.

Thorax roundly trapezoidal; mesosternum forming cavity for prosternal process; metepisternum wide; femora unarmed; tarsi with five tarsomeres, fourth deeply bilobed.

Abdomen with five visible sterna in the female, six in the male with entire sterna five and six deeply emarginate; male aedeagus with basal piece and parameres fused into an asymmetrical tubular tegmen, median lobe also asymmetrical; female genitalia with relatively short paraprocts and valvifers and with sclerotized coxites plate-like. The metathoracic wings are characterized by media extending basally, the radial cell being situated more apically, and the radio-medial cross-vein much longer than is characteristic of the Buprestidae. The wings also have a normal hinge system at the end of the radial cell that is lacking in even the most complexly folded buprestid wings.

Larvae have a cylindrical body, sides nearly parallel; head with three well-developed unequal ocelli on each side posterior to and slightly dorsal to base of antennae; thoracic segments legless; abdominal segments 1-8 each with a pair of well-developed prolegs terminating in hoof-shaped structures; mesothoracic, metathoracic, and abdominal segments 1-7 each with a pair of ventral glands, each with a protracted duct (see Rees 1941).

The important studies regarding wing folding patterns and wing venation have been used by various authors in altering classification studies and proposals (Forbes 1922, 1926, 1942; Good 1925). The work of Sharp and Muir (1912) compared male genitalia and the study by Tanner (1927) detailed female genitalic structure.

Habits and habitats. Almost nothing is known of the life history. A short discussion of eggs was given by Nelson *et al.* (1996). First instar larvae of *Schizopus s. sallei* Horn, hatched from eggs laid in the laboratory, were described by Rees (1941). Only one other larva has been collected, in California, San Diego Co., 4 mi. west of Warner Springs, 16 March 1972. It was found on a hilltop in a mud cell about 4 inches below the soil surface under a clump of grass. The larvae are possibly external root-feeders. Adults of *Schizopus* are either found feeding on the blossoms of desert plants (*S. laetus*), or clinging to the dried leaves of grasses (*S. sallei*). *Dystaxia* adults have been collected on the foliage of *Quercus* spp., occasionally on *Juniperus*, with *D. murrayi* sometimes found on *Chrysothamnus*. Adults of *Glyptoscelimorpha* occur on the foliage of *Juniperus* spp.

Status of the classification. This family was recently revised by Nelson and Bellamy (1991), where cladistic analysis supported four major buprestoid lineages, with the schizopodids and julodines branching outside of the apparent true buprestid line. The genus *Dystaxiella* is now considered a subgenus of *Glyptoscelimorpha*. Schizopodidae, one of the major buprestoid lineages, was first defined as a distinct family by LeConte (1861). Subsequent authors have either accepted or ignored this status, although the three major buprestid authorities of the last century, Kerremans, Théry and Obenberger retained it as a buprestid tribe in the system of Lacordaire (1857). However, the concepts of Holynski (1993) did not accept this familial status and most recently Lawrence and Newton (1995) listed this group again as a subfamily, stating that there was no indication of what group, outside of Buprestoidea, the schizopodid or -podine line was related to. In an interesting paper on the dryopoid affinities of buprestids, Crowson (1982) suggested that a possible candidate for analysing buprestid relationships was the luctrochid genus *Lara*; indeed these beetles bear a remarkable resemblance to *Schizopus* in general morphology. This was the out-group used in the analysis by Nelson and Bellamy (1991), but the lack of a defined sister-group still allows conjecture. A recent paper by Kolibac (2001) supports a rank higher than subfamily for the schizopodids, because they are more distant from the other buprestid subfamilies than those are from one another, both from morphological and biological perspectives; of course we agree. The nomenclature question regarding the name *Schizopus* was resolved by the ICZN (1993) ruling.

Distribution: Seven species are now known from California, southern Nevada, southwestern Arizona, and northern Baja California.

Key to Nearctic Genera

1. Surface sculpture coarse; antennae with 11 antennomeres, serrate antennomeres transversely triangular; pronotal disk with distinct impressions (tribe Schizopodini) .... *Schizopus*

— Surface sculpture of fine punctures; antennae with 12 antennomeres, serrate antennomeres elongate triangular; pronotal disk without impressions (tribe Dystaxiini) 2

2(1). Robust oval; vestiture a mixture of hairlike and elongate flattened setae; terminal antennomere elongate slender; clypeus deeply emarginate; tarsal claws distinctly bifid .... *Dystaxia*

— Ovate-cylindrical; vestiture of elongate, lanceolate shaped, flattened setae; terminal antennomere small, oval; clypeus shallowly emarginate; tarsal claws simple or with small internal tooth .... *Glyptoscelimorpha*

Classification of the Nearctic Genera

Schizopodidae LeConte 1859

Schizopodinae Crowson 1955

Schizopodini LeConte 1866

Schizopodina Holynski 1993

Schizopus LeConte 1858, 2 spp., 1 ssp., Arizona, California, and Nevada (key to spp., Nelson and Bellamy 1991; other information, ICZN 1993, Obenberger 1939).

Yermoella Obenberger 1939

Dystaxiini Théry 1929

Dystaxiina Holynski 1993

Dystaxia LeConte 1866, 2 spp., California (key to spp., Nelson and Bellamy 1991).

Glyploscelimorpha Horn 1893, 3 spp., 1 ssp., California (key to spp., Nelson and Bellamy 1991).

subgenus *Dystaxiella* Knull 1940

Bibliography

CROWSON, R. 1982. On the dryopoid affinities of Buprestidae. Coleopterists Bulletin, 36: 22-25.

FORBES, W. T. M. 1922. The wing-venation of Coleoptera. Annals of the Entomological Society of America, 15: 328-352.

FORBES, W. T. M. 1926. The wing folding patterns of Coleoptera. Journal of the New York Entomological Society, 34: 42-68.

FORBES, W. T. M. 1942. The wing of the Schizopini. Entomological News, 53: 101-102.

GOOD, H. G. 1925. Wing venation of the Buprestidae. Annals of the Entomological Society of America, 18:251-276.

HOLYNSKI, R. 1993. A reassessment of the internal classification of the Buprestidae Leach (Coleoptera). Crystal, series Zoologica, 1: 1-42.

ICZN. 1993. Opinion 1727. *Schizopus* LeConte, 1858 (Insecta, Coleoptera): placed on the Official List of Generic Names. Bulletin of Zoological Nomenclature, 50: 174-175.

KOLIBAC, J. 2001. Classification and phylogeny of the Buprestoidea (Insecta: Coleoptera). Acta Musei Moraviae, Scientiae biologicae (Brno), 85: 113-184 (2000).

LACORDAIRE, J. T. 1857. Histoire naturelle des insectes. Genera des coléoptères ou exposé méthodique de critique de tous les genres proposés jusqu'ici dans cet ordre d'insectes, vol. 4, 1-554 pp. (Contenant les familles des buprestides, throscides, eucnémides, élatérides, cébrionides, cérophytides, rhipicérides, dascyllides, malacodermes, clérides, lyméxylones, cupésides, ptiniores, bostrichides et cissides.)

LAWRENCE, J. F. and A. F. NEWTON, Jr. 1995. Families and subfamilies of Coleoptera (with selected genera, notes, references and data on family-group names). Pp. 779-1006. *In:* J. Pakaluk and S.A. Slipinski, eds. Biology, Phylogeny, and Classification of Coleoptera. Papers Celebrating the 80th Birthday of Roy A. Crowson. Muzeum I Instytut Zoologii PAN. Warsaw.

LECONTE, J. L. 1861. Descriptions of some genera and species of Coleoptera from the vicinity of the southern boundary of the United States of America. Arcana Naturae, 1: 121-128.

NELSON, G. H. and C. L. BELLAMY. 1991. A revision and phylogenetic re-evaluation of the family Schizopodidae (Coleoptera, Buprestidae). Journal of Natural History, 25: 985-1026.

NELSON, G. H., R. L. WESTCOTT and T. C. MACRAE. 1996. Miscellaneous notes on Buprestidae and Schizopodidae occurring in the United States and Canada, including descriptions of previously unknown sexes of six *Agrilus* Curtis. Coleopterists Bulletin, 50: 183-191.

OBENBERGER, J. 1939. Über eine neue Buprestiden-Gattung und Art aus den Sammlungen des Deutschen Entomologischen Instituts nebst eingen Bemerkunger über die Buprestiden-Tribus Schizopini sensu Kerr. Arbeiten über Morphologische und Taxonomische Entomologie aus Berlin-Dahlem, 6: 254-265.

REES, B. E. 1941. First-instar larvae of *Buprestis rusticorum* (Kby.) and *Schizopus sallei* Horn, with notes on the classification of *Schizopus*. Proceedings of the Entomological Society of Washington 43: 210-222.

SHARP, D. and F. MUIR. 1912. The comparative anatomy of the male genital tube in Coleoptera. Transactions of the Royal Entomological Society, London 1912: 477-642.

TANNER, V. M. 1927. A preliminary study of the genitalia of female Coleoptera. Transactions of the American Entomological Society, 53: 5-50.

41. BUPRESTIDAE Leach 1815

by C. L. Bellamy and G. H. Nelson

Family common name: The metallic wood boring or jewel beetles

Adults of these hard bodied, generally glossy, iridescent-colored beetles are easily recognized by the hypognathous head, mostly serrate antennae, transverse metasternal suture and connate first two abdominal sterna.

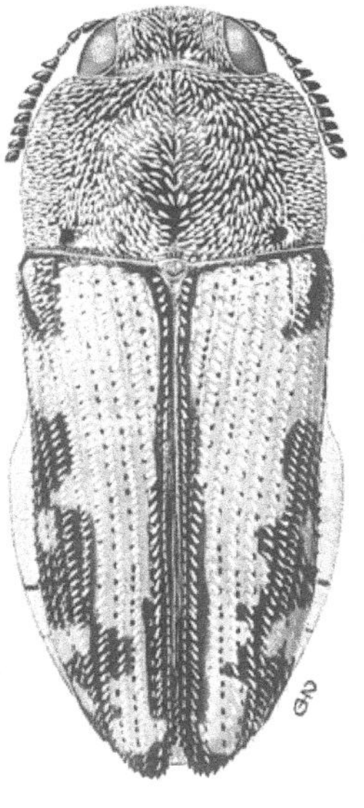

FIGURE 1.41. *Acmaeoderoides knulli* Nelson (from Nelson 1968)

Description: Shape cylindrical to flattened, elongate-ovoid, generally convex above, or cuneiform; size 3 to 100 mm or more in length, mostly less than 20 mm; color various, often bright iridescent hues, or dark-colored with patterned or irregular pigmented maculae; vestiture absent or variously covered with setae and sometimes broad scale-like setae.

Head greatly deflexed, resting on the prosternum, retracted into the prothorax, but mostly as broad as the anterior portion of pronotum; surface punctate or rugose punctate. Antennae with 11 antennomeres, mostly serrate, some males flabellate or pectinate; inserted some distance from the eyes and mandibles, on the front within distinct frontoclypeal cavities. Anteclypeus sometimes visible, labrum small, distinct, often bilobed and setose distally; mandibles small, stout basally, curved, the apices acute; maxillary palpi with four palpomeres, filiform; mentum quadrate to triangular; ligula mostly not prominent; labial palpi with three palpomeres, filiform. Eyes lateral, moderate to large, greatly elongate-oval to reniform, inner margins sometimes strongly converging.

Pronotum slightly broader than the head; shape irregularly quadrate, sometimes narrowed in front; lateral margins mostly carinate, carinae sometimes incomplete from posterior angle; surface punctate to rugose; hypomeron broad; prosternum long and broad, produced posteriorly between the coxae and inserted in the mesosternal cavity or the cavity is composed of the short mesosternal lobes laterally and the metatsternum distally; procoxal cavities open behind. Metasternum mostly with the vestige of a transverse suture near the posterior coxal plates. Legs with the trochantins of the fore- and middle legs exposed; anterior coxae small, oval, separate; middle coxae small, flat, almost quadrate, separate; hind coxae large, transverse, with thick plates; trochanters small, triangular; femora subparallel to fusiform; tibiae slender, sometimes dentate or spinose, the apical spurs moderate; tarsal formula 5-5-5, tarsi slender, some of the tarsomeres bilobed, distal tarsomeres each with ventral pulvillus; claws simple, appendiculate or bifid. Scutellum triangular to cordifrom, moderate to small. Elytra entire except in *Hesperorhipis*, apically rounded and often with one or more apical spines, rarely expose the pygidium; striae punctate or carinate; intervals smooth or rugose; margins, especially apical portion serrate to serrulate; epipleural fold indistinctly separate or with fine carina separating it from disc, broad basally. Wing venation with 2A with three branches; wedge cell, when present, acute apically with only one vein coming from it. Folding pattern of the wing with area A and B normally reduced to slender crumples, area C about half the length of the wing, frequently fused more or less completely with area D which is either open to the costa or reduced to a slender crumple; area H always well marked and reaching the margin for the full width, very slender and nearly longitudinal in the more typical forms; commonly with one chevron-like apical fold, but may be two, or absent; anal lobe highly variable, but never free.

Abdomen with five visible sterna, the first and second connate; sutures shallow, sometimes partly obsolete laterally; surface smooth, punctate, or rugose. Male genitalia of a modified trilobed type; the median lobe a flat, dorsal plate, nearly parallel-sided, the apex acute to transverse, grooved deeply ventrally; parameres sometimes highly modified from swollen to having projecting lateral lobes and sensory setae distally, parameres surround most of the median lobe, fused to the pars basalis; pars basalis fused, forming a basal plate. Female genitalia with the valvifers reduced to a large and strong baculum; coxite dorsally with a baculum, the membranous part fan-shaped; stylus greatly reduced, proctiger large with two bacculi which extend around to the ventral surface where they meet and articulate with the basal parts of the valvifers.

Larvae with the segments flattened, or oval, deeply notched and tapering behind; thorax enlarged, body long, slender, subcylindrical (flat-headed wood borers), or enlarged, club-like head and thorax may or may not be enlarged (leaf miners); rarely with the mid-abdominal segments the widest, length 5 to 50 mm or more; vestiture mostly absent; color cream to near white, sometimes with yellow, orange, or brown pigment spots. Head small, depressed, more or less retracted into the prothorax. An-

Acknowledgments. The authors of the chapters on Schizopodidae and Buprestidae thank Svatopluk Bílý, Jerry Davidson, Henry Hespenheide, Ted MacRae and Rick Westcott for comments and contributions which improved the text.

tennae three-segmented. Labrum arcuate, free; mandibles stout, toothed, spoon- shaped; maxillae with two-segmented palpi and a lobe-like mala; labium small, with ligula prominent, spatulate; labial palpi small, one-segmented, or absent. Stemmata absent. Thorax without legs; dorsally with a distinct V-shaped groove. Abdomen ten-segmented, mostly with two fleshy lobes apically; some with tenth segment sometimes with a pair of sharp, sclerotized, toothed forks or forceps laterally. Spiracles cribriform, mostly crescent-shaped, on the mesothorax and abdominal segments one to eight. The only key to larvae was that by Burke (1917) and is not repeated here. The eggs may be enormous; in the collection of exotic beetles at the Academy of Natural Sciences in Philadelphia are a few eggs glued to a piece of cardboard. These eggs measure 3.5 by 5 mm and likely belong to Asian or African species of either *Sternocera* or *Julodis*. Otherwise, little has been reported yet on the eggs of this family.

The important studies regarding wing folding patterns and wing venation have been used by various authors in altering classification studies and proposals (Forbes 1922, 1926, 1942; Good 1925). The work of Sharp and Muir (1912) compared male genitalia and the study by Tanner (1927) detailed female genitalic structure.

Habits and habitats. The larvae burrow through roots and logs, from within the bark to within the cambium layers, or are leaf and stem miners of herbaceous and woody plants, including grasses. Most of the wood boring species attack dying trees or dying/dead branches on healthy trees, only a few bore into green wood. Some produce galls on alder, roses, blue beech, ironwood, and hazelnut; a few live in pine cones or herbaceous plants. Some adults are active and very strong flyers (*e.g. Gyascutus* and *Chrysobothris*) and some make a loud buzzing noise as they fly, while others are slow and even approach being clumsy (*e.g. Chalcophora*, *Texania*, and *Buprestis*). The adults feed on foliage of their hosts or visit flowers to feed on energy-rich pollen or nectar except for some species of *Chrysobothris* and *Agrilus* which feed on fungus. With the often hirsute body and flower visiting habit, many species, particularly in the large genus *Acmaeodera*, serve as potential pollinators.

A number of recent papers on distribution and biology have helped fill in many gaps of understanding about many Nearctic species. These include works by Barr (1971), Bellamy (1982), Nelson (1959, 1960, 1962, 1965, 1967, 1968a, 1987), Nelson and MacRae (1990), Nelson and Westcott (1976), Nelson *et al.* (1981), Walters (1975, 1976), Walters and Bellamy (1982, 1990), Westcott (1990, 1991) and Westcott *et al.* (1979, 1989). State or regional synopses have been given by Barr (1971) for the Pacific Northwest, Bright (1987) for Alaska and Canada, Cazier (1951) for northcentral Mexico, Knull (1925) for Pennsylvania, MacRae (1991) for Missouri, Vogt (1949) for the lower Rio Grande valley, Texas and Wellso *et al.* (1976) for Michigan. The recent checklist by Davies (1991) listed the species known from Canada and Alaska.

Status of the classification. The classification of the family is becoming better understood, although a complete modern phylogenetic perspective is lacking. Some of the genera are fairly well known, but the classification of genera and their placement in higher taxa is still being contested. In terms of our phylogenetic understanding of the buprestids, it seems to be now generally agreed that there are five major lineages: schizopodines, julodines, polycestines, buprestines and agrilines. Some prefer that these five groups should be assumed to be the subfamiles of Buprestidae, but there are compelling data to argue for familial status for one or two of these groups as well. This five lineage concept is rather stark in contrast to the last major subfamily scheme proposed by Cobos (1980) where he outlined 13 subfamilies and suggested a 14th, which altered things substantially from the revision of LeConte (1860). Few attempts to apply real phylogenetic analysis techniques to the family, as a whole or groups of taxa within, have been conducted, but these are emerging slowly and will undoubtedly increase as we try to make sense of such a large group of beetles. There are approximately 15,000 species placed within about 450 genera. The last global accounting of the family came in the six buprestid parts of the Coleopterorum Catalogus by Obenberger (1926, 1930, 1934a, 1934b, 1936, 1937) with a summary of the higher categories provided by Bellamy (1985). The most recent higher system has been proposed by Holynski (1993) in which he suggested four subfamilies, 12 tribes and about 64 subtribes, but many of the proposals first brought forward there are untested, or intuitively unsupportable, so further refinement is necessary. This classification was the first attempt to organize the entire family since that by Kerremans (1893). Here, to some extent, we use the subtribal system of Holynski, although some changes are proposed that differ from his original scheme. The newest attempt at predicting the natural classification by Kolibac (2001) suggested some very different placements at the higher levels in a three-lineage, six-subfamily scheme, but considering the relatively small sample of

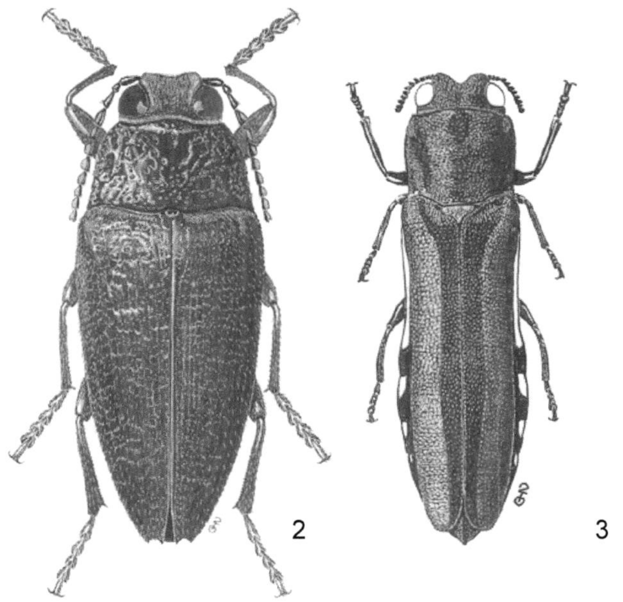

FIGURES 2.41-3.41. 2. *Gyascutus* (*Stictocera*) *caelatus* (LeConte) (from Nelson and Bellamy 1996, with permission of Taylor & Francis Ltd.); 3. *Agrilus* (*Engyaulus*) *pulchellus* Bland (from Nelson and Westcott 1991).

taxa used in his phylogenetic analysis, the resolution gained in his study is not robust enough for us to completely accept his conclusions. The most recent assessment of the classification was presented for the Nearctic region by Nelson (1982). A catalogue of the North American fauna by Nelson is nearly complete, while a complete world catalogue and bibliography by Bellamy is still in preparation.

The julodines are a large group composed of six genera and many species distributed from the southern Palaearctic and Oriental to the Cape region of South Africa. The largest two genera, *Julodis* and *Sternocera*, are found as far west as Pakistan and Southeast Asia, respectively. Interestingly, they are not present on Madagascar. The bodies of these insects are nearly cylindrical, tapering to the posterior end, from about 1 to 7 or 8 cm in length. This group is arguably supported at the family level too.

The remaining three main lineages comprise the majority of buprestid taxa and are virtually cosmopolitan. The polycestines and buprestines generally have very typical larvae, each with a strongly dorsoventrally flattened thoracic region, or ambulatory plate (Burke 1917), a feature that spawned the dubious moniker flat-headed wood-borers. The typical adult body is said to be bullet-shaped, but rather they are mostly flattened dorsoventrally and tapering towards the rear end. They are very active flyers, especially during the warmth of the day and are quick to escape both predators and collectors. Many species are found feeding on the foliage of their larval host plants and many other species visit flowers to feed and to rendezvous with others looking to mate. The range of host plants is rather broad and includes many families of gymnosperms and dicot angiosperms. Many species seem to be generalist feeders as larvae, often known to inhabit several plant hosts, sometimes developing in a variety of dead wood. Both groups are known from every biogeographical region and most every habitat.

The main differences between the polycestines and buprestines are found in larvae and adults. The polycestine larvae, as far as is known, possess a single central line or groove on the dorsal first thoracic segment, while in the buprestines this line is either "Y" or "V" shaped. In the adults, the main dividing characteristic is found in the structure of the sternal cavity. The polycestines have the prosternal process received distally by lateral lobes of the mesosternum, whereas in the buprestines and agrilines the sternal cavity involves the anteromedial portion of the metasternum.

POLYCESTINE TRIBES. Six tribes contain Neartic genera: Mastogeniini, Acmaeoderini, Polyctesini, Thrincopygini, Polycestini, and Tyndarini. In the past, Thrincopygini, Acmaeoderini and Mastogeniini were recognized as subfamilies. All but Acmaeoderini contain a single subtribe in the Nearctic region. The placement of Mastogeniini is still somewhat a matter of conjecture without any known larvae, but the most recent discussion by Bellamy (1996b) indicates a polycestine placement.

BUPRESTINE TRIBES. Eight tribes are included in this lineage: Chalcophorini, Hippomelanini, Psilopterini, Dicercini, Buprestini, Melanophilini, Anthaxiini and Chrysobothrini. The Chalcophorini, once accorded subfamily rank, contain two subtribes for Nearctic species: Chalcophorina and Agaeocerina, originally proposed at tribal rank. The Hippomelanini were originally proposed at subtribe level and is monotypic; this taxon is endemic to North America (including Mexico). The Psilopterini contain the nominate subtribe and Sphenopterina, once accorded subfamily rank. The Dicercini contain the nominate subtribe and Phrixiina. The Melanophilini are monotypic. The Buprestini have Nearctic taxa placed in two subtribes: Buprestina and Trachykelina. The Anthaxiini are composed of the nominate subtribe and Xenorhipina. The Chrysobothrini have two subtribes: the nominate taxon and Actenodina.

The agriline lineage is comparable in size and taxonomic diversity to the polycestine/buprestine lineage and perhaps exceeds it. These beetles are generally smaller, often very small, mostly subcylindrical or flattened and cuneiform or wedge-shaped. The larvae differ significantly in that they lack the proventriculus of the alimentary canal, indicating a very different way of handling food. The larvae also share a modification to the caudal segment, always in one way or another, bilobed, bifurcate and sometimes with this paired terminus sclerotized. The adults are most always collected from the foliage of their host plants and most seem to be very host specific. In fact, within this group, there are some very pronounced coevolutionary trends apparent with entire species-groups utilizing only one genus of host plant. The agrilines are also known from every biogeographical region and most every habitat, with leaf-mining taken precedence over wood-boring forms in the moist tropical areas as the larvae are much less exposed to the threat of fungal infection. *Agrilus* is one of the largest genera in the animal kingdom with more than 2,700 described species. There are no known associations in *Agrilus* with gymnosperms. Other members of this group utilize monocotyledonous plants as larval hosts (i.e. Poaceae and Cyperaceae)

AGRILINE TRIBES. The Agrilinae are composed of four tribes: Coraebini, Agrilini, Aphanisticini and Trachyini. The Coraebini are monotypic with *Eupristocerus* in the nominate subtribe. The coraebines are the most generically diverse group in the family, but with most of this diversity found in the Old World. The Agrilini contain two subtribes: Rhaeboscelina and Agrilina. The Aphanisticini are added to the Nearctic fauna with the recent discovery of an Asian species in Texas and Florida (Wellso and Jackman 1995, Peck and Thomas 1998). The tribe Trachyini contains mostly small, often cuneiform, leaf- or stem-mining beetles, with all four subtribes present in the Nearctic fauna: Trachyina (for the adventive *Trachys troglodytiformis* Obenberger), Brachyina, Leiopleurina and Pachyscelina. It is now thought from preliminary studies of larvae that this grouping is highly artificial and that leaf-mining may have evolved independently several times in the family and may represent a repeated evolutionary outcome at the distal end of several otherwise widely separated lineages (S. Bílý, pers. comm.).

HIGHER TAXON ADDITIONS AND CHANGES. The basic structure of the higher classification has evolved from the general system of Lacordaire (1857), with more specific changes from LeConte and Horn (1883), Kerremans (1893), Nelson (1982) and Holynski (1993). The tribes or subtribes that have more recently been defined and which contain, partly, Nearctic genera are: Agaeocerina Nelson 1982 (originally as tribe of Buprestini); Hippomelanina Holynski 1993 (originally as a subtribe); Trachykelina Holynski 1988 (originally in Anthaxiini); Rhaeboscelidina Cobos 1976; Brachydina Cobos 1979; Pachyschelina Cobos 1979; and Leiopleurina Holynski 1993. Higher level recombinations or synonymies are: Chalcophorinae Lacordaire as a synonym of Buprestinae by Toyama (1987) and Dicercini Kerremans as a synonym of Psilopterina Lacordaire by Holynski (1993). The placement of Polycestina Lacordaire under Buprestini was refuted by Bellamy (1996b) and later by Volkovitsh and Hawkeswood (1999). Bílý (2000) reviewed the Anthaxiini and Melanophilini, proposing a subtribal structure which differed from that proposed earlier by Holynski (1988). The Coraebini have been phylogenetically reviewed by Kuban *et al.* (2001).

GENERIC ADDITIONS, CHANGES AND PLACEMENTS. Recently described generic taxa for the Nearctic region include: *Acmaeoderopsis* Barr 1974; *Anambodera* Barr 1974; *Barrellus* Nelson and Bellamy 1996; *Beerellus* Nelson 1982; *Lepismadora* Velten 1987; and *Squamodera* Nelson 1996. Previously described genera recently added to the regional fauna are *Aphanisticus* Latreille 1810; *Leiopleura* Deyrolle 1864; *Micrasta* Kerremans 1893; *Sphaerobothris* Semenov-Tian-Shanskij and Rikhter 1934; and *Sphenoptera* Dejean 1833. Recent status changes or subgeneric elevations are: *Paratyndaris* Fisher 1919 was reduced to a subgenus of *Tyndaris* Thomson 1857 by Cobos (1980), which we do not support; *Texania* Casey 1909 was validated with the recognition of *Chalcophorella* Kerremans 1903 as a strictly Palaearctic genus by Obenberger (1942a); *Nanularia* Casey 1909 and *Ampheremus* Fall 1917 were recognized as valid genera by Bellamy (1987); *Gyascutus* LeConte 1859, with *Stictocera* Casey 1909 as its subgenus, and *Prasinalia* Casey 1909 were elevated from subgenera of *Hippomelas* Laporte and Gory 1837 by Nelson and Bellamy (1996); *Lampetis* Dejean 1833, subgenus *Spinthoptera* Casey 1909 contain the Nearctic species formerly placed in *Psiloptera* Solier 1833 according to Kurosawa (1993); *Spectralia* Casey 1909 was validated with the recognition of *Cinyra* Laporte and Gory 1837 as a Neotropical taxon; *Cypriacis* Casey 1909 was recognized as distinct from *Buprestis* L. 1758 by Kurosawa (1988); *Phaenops* Dejean 1833 and *Xenomelanophila* Sloop 1837 were recognized as distinct from *Melanophila* Eschscholtz 1829 by Cobos (1987); and *Agrilaxia* Kerremans 1903 was again recognized at the genus level by Bright (1987) and later by Bílý and Bellamy (1999). The placement of *Acmaeoderoides* Van Dyke 1942 in Ptosimina by Holynski (1993) was refuted by Bellamy and Westcott (1996). The placement of *Chrysophana* LeConte 1859 in Bubastina Obenberger by Holynski (1993) was refuted by Bellamy (1996c). The placement of Ptosimina under Acmaeoderini follows proposals by Volkovitsh and Hawkeswood (1999). Kuban *et al.* (2001) removed *Lepismadora* to the Agrilini, where it awaits subtribal placement.

Three recent rulings by the ICZN affected buprestid nomenclature. A very unstable situation was corrected (ICZN 1994) when type species were designated for *Buprestis* and *Chrysobothris*, Eschscholtz 1829. The fixation of the name *Poecilonota* Eschscholtz 1829 by type species designation (ICZN 1996a) reduced *Descarpentriesiola* Lerault 1983 to junior synonymy. A controversy between specialists from Europe and North America was solved with the designation of type species for *Melanophila* and *Phaenops* (ICZN 1996b).

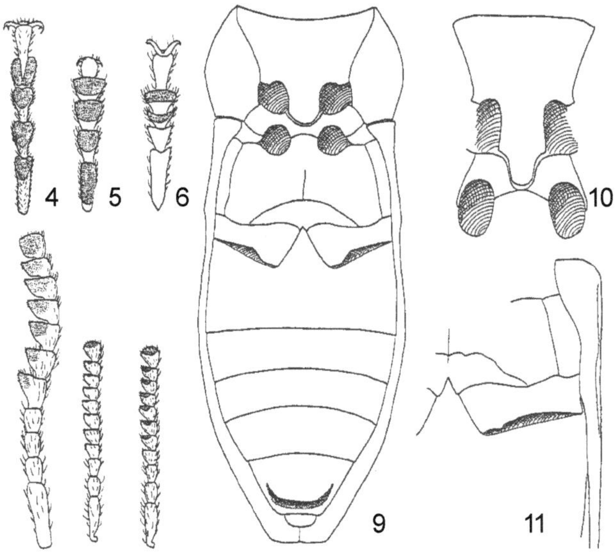

FIGURES 4.41 - 11.41. 4, *Schizopus laetus* LeConte, metatarsus of male, ventral view; 5, *Beerellus taxodii* Nelson, metatarsus of female, ventral view; 6, *Polycesta angulosa* Duval, metatarsus of female, ventral view; 7, *Acmaeodera gibbula* LeConte, antenna; 8, *Ptosima gibbicollis* (Say), antenna, dorsal (left), ventral (right); 9, *Thrincopyge ambiens* (LeConte), thorax and abdomen, ventral view; 10, *Chalcophora georgiana* (LeConte), thoracic sternal areas; 11, *Polycesta angulosa* Duval, metasternal area, female. (All from Nelson 1982)

Distribution. There are slightly more than 14,600 valid species known from the majority of the world's temperate and tropical biogeographical zones, making this family the eighth largest beetle family. There are 762 species and 26 subspecies currently listed for North America, north of Mexico, a substantial increase on those listed by Chamberlin (1926). The state of understanding for Mexican taxa does not allow a clear accounting for species found in the southern extent of the Nearctic area, although a general introduction and species estimates were given by Hespenheide (1996).

KEY TO NEARCTIC GENERA

1. Sternal cavity for reception of prosternal process formed entirely by the mesosternum (Fig. 9) (subfamily Polycestinae) 2
— Sternal cavity for reception of prosternal process attaining or formed in part by metasternum (Fig. 10) 12

Polycestinae

2(1). Metacoxal plates distinctly dilated medially; last visible abdominal sternum with deep groove around apical half (Fig. 9) (Thrincopygini, Thrincopygina) .... *Thrincopyge*

— Metacoxal plates not distinctly dilated medially (Fig. 11); last visible abdominal sternum without deep groove around apical half 3

3(2). Lobulated antennomeres with sensory pores diffuse, without vestiges of fossae or depressions on either surface (Fig. 7) (Acmaeoderini) 4

— Lobulated antennomeres with sensory pores in part concentrated in fossae on one or both surfaces, at least on apical segments (Fig. 8) 8

4(3). Scutellum visible; elytra free; epipleuron more or less hypertrophied toward base (Acmaeoderoidina) .... *Acmaeoderoides*

— Scutellum not visible; elytra fused; epipleuron not hypertrophied toward base (Acmaeoderina).... 5

5(4). Ventral surface clothed by a dense tomentum largely obscuring the surface........ *Squamodera*

— Ventral surface variously clothed otherwise...... 6

6(5). Pronotum without or with only faint indication of margin; front angles of pronotum in side view are rounded; suture between abdominal sterna one and two readily visible .... *Anambodera*

— Pronotum distinctly margined at least in part; front angles of pronotum in side view are angled; suture between abdominal sterna one and two obliterated or faint 7

7(6). Abdominal sterna three to five in females clothed differently than rest of body, consisting of dense long recurved hairlike setae .... *Acmaeoderopsis*

— Abdominal sterna three to five in females not distinctively clothed .... *Acmaeodera*

8(3). Articulating base of pronotum with row of rasplike grooves (Fig. 12) (Acmaeoderini, Ptosimina) .... *Ptosima*

— Articulating base of pronotum without row of rasplike grooves 9

9(8). Elytral apices not rounded, provided with special armiture (Fig. 14); epipleuron completely covering metepisternum (Fig. 15) (Tyndarini, Tyndarina) .... *Paratyndaris*

— Elytral apices rounded or more or less serrate (Fig. 13); epipleuron not completely covering metepisternum (Fig. 16) 10

10(9). Tarsomeres 1-4 with plantulae below (Fig. 5) (Polyctesini: Polyctesina) 11

— Only tarsomere 4 and in some 3 with plantulae below (Fig. 6) (Polycestini) .... *Polycesta*

11(10). Epipleural lobe rounded, covering all but anteroinferior angle of metepisternum (Fig. 17) .... *Beerellus*

— Epipleural lobe truncate leaving most of metepisternum exposed (Fig. 16)....................... .... *Chrysophana*

12(1). Metacoxal plates dilated medially or not, but only slightly longer medially than laterally, with anterior margin mostly sinuate (Fig. 20) 13

— Metacoxal plates distinctly dilated medially, mostly cut off laterally by prolongation of abdomen, with anterior margin rather straight, posterior margin oblique (Fig. 18) (Buprestinae)........................ 16

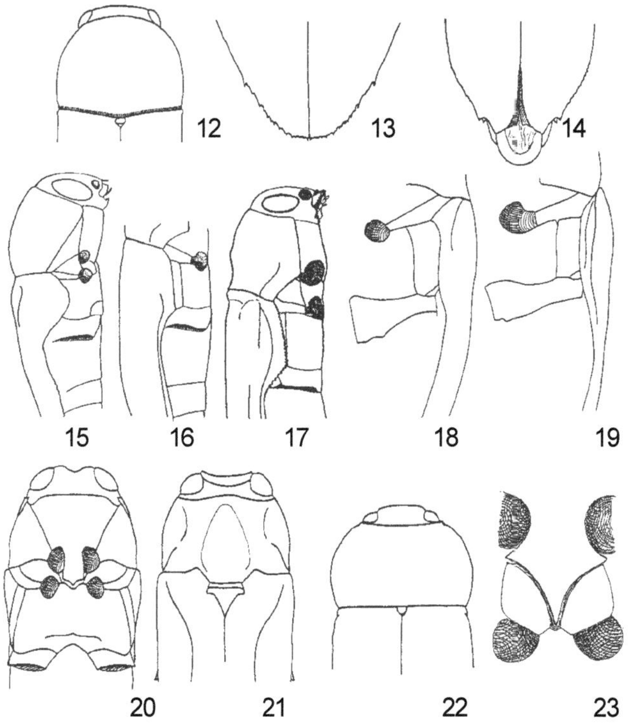

FIGURES 12.41 - 23.41. 12, *Ptosima gibbicollis* (Say), head and pronotum, dorsal view; 13, *Polycesta elata* LeConte, elytral apices; 14, *Paratyndaris olneyae* (Skinner), elytral apices; 15, *P. olneyae*, lateral view; 16, *Chrysophana placida* (LeConte), lateral view; 17, *Beerellus taxodii* Nelson, lateral view; 18, *Dicerca hesperoborealis* Hatch and Beer, meso-metasternal area; 19, *Buprestis maculativentris* Say, meso-metasternal area; 20, *Agrilus cavifrons* Waterhouse, ventral view; 21, *A. cavifrons*, dorsal view; 22, *Mastogenius robustus* Schaeffer, dorsal view; 23, *Chrysobothris octocola* LeConte, thoracic sternal areas. (Figures 12 - 22 from Nelson 1982).

13(12). Thorax truncate at base (Fig. 22) (Mastogeniini, Mastogeniina) 14

— Thorax lobed at base (Fig. 21) (Agrilinae) 44

14(13). Prosternum with distinctly limited antennal cavities, and carinae on each side of middle .... *Trigonogya*

— Prosternum without distinctly limited antennal cavities, and without carinae on each side of middle 15

15(14). Eyes parallel, size less than 1.5 mm, subcylindrical .... *Micrasta*

— Eyes converging slightly above, size more than 1.5 mm, body flattened above and below .... *Mastogenius*

Buprestinae

16(12). Prosternum obtusely angulate behind coxae (Fig. 10); frons not contracted by insertion of antennae 17

— Prosternum acutely angulate behind coxae (Fig. 23); frons contracted by insertion of antennae (Chrysobothrini) 42

17(16). Antennae with sensory foveae absent (sensory pores dispersed) or placed on ventral surface of antennomeres 18

— Antennae with sensory foveae placed on distal surface of antennomeres 33

18(17). Elytral epipleuron with denticle wedging between mesepimeron and rounded posterolateral angle of metepisternum (Fig. 18) 19

— Elytral epipleuron with lower margin more or less straight, without denticle (Fig. 19) (Buprestini, Buprestina) 31

19(18). Body subcylindrical in cross section; sensory pores diffuse on both surfaces of antennae; protrochanter mostly with sharp tooth (Fig. 30); antennomere 11 mostly with terminal notch (Fig. 31) (Hippomelanini, Hippomelanina) 20

— Body oval in cross section; sensory pores of antennae otherwise; protrochanter without sharp tooth; antennomere 11 without terminal notch. 25

20(19). Antennomeres 4-10 compact, triangular, not flattened (Fig. 28); inner margin of eyes parallel (Fig. 24); elytral apices rounded to slightly emarginate (Fig. 25) 24

— Antennomeres 4-10 elongate, subserrate or, in part parallel-sided, strongly flattened (Fig. 39); inner margin of eyes converging above (Fig. 40); elytral spices moderately emarginate and bidentate (Fig. 1) 21

21(20). Surface sculpture irregular, with conspicuous raised callosities; protibia arcuate (Fig. 29); antennomere 11 of male with strong terminal notch (Fig. 31) .... *Gyascutus*

— Surface sculpture uniform, without conspicuous raised callosities; protibia weakly arcuate or straight (Fig. 32); antennomere 11 of male mostly without strong terminal notch (Fig. 34) 22

22(21). Epipleuron with small marginal tooth near metacoxa (Fig. 37); tarsomere 1 of male about 2X as long as 5; metacoxa of male with acute tooth along inner margin (Fig. 35) .... *Prasinalia*

— Epipleuron not toothed (Fig. 38); tarsomere 1 subequal to 5 or shorter; metacoxa of male without acute tooth along inner margin (Fig. 36).. 23

23(22). Protrochanter without distinct tooth (Fig. 33); antennomeres 4-10 of males parallel-sided (Fig. 39) .... *Hippomelas*

— Protrochanter with distinct acute tooth (Figs. 41, 45); antennomeres 4-10 of male triangular with roundly truncate margin (Fig. 42) .... *Barrellus*

24(20). Lateral margin of pronotum carinate in posterior half; epipleuron carinate basally; hind margin of abdominal sterna 2-4 notched near sides (Fig. 26) .... *Nanularia*

— Lateral margin of pronotum without carina; epipleuron without sublateral carina; hind margin of abdominal sterna 2-4 entire (Fig. 27) .... *Ampheremus*

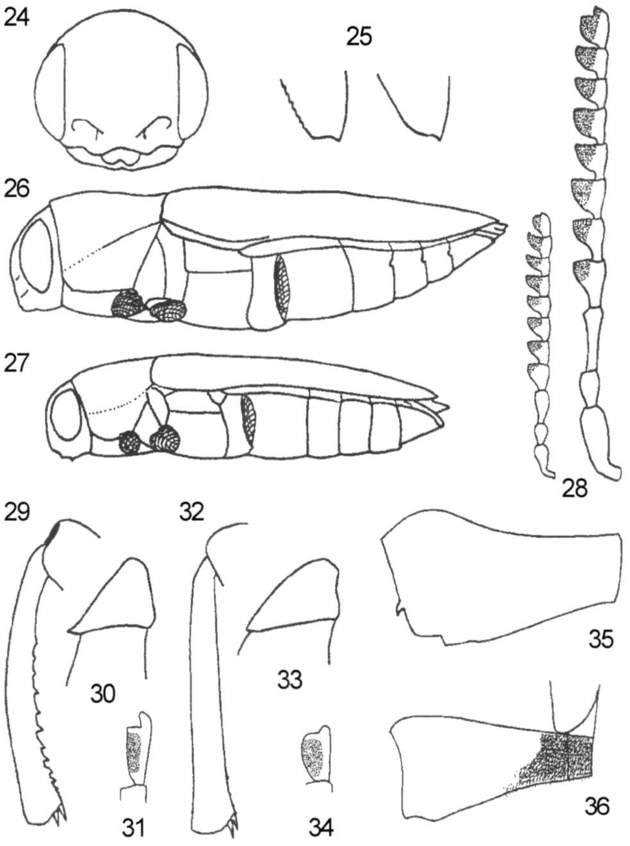

FIGURES 24.41 - 36.41. 24, *Nanularia*, head, anterior view; 25, *Nanularia*, left, *Ampheremus*, right, elytral apices; 26, *Nanularia*, lateral view; 27, *Ampheremus*, lateral view; 28, *Ampheremus*, left, *Nanularia*, right, antennae; Figs. 29 - 31, *Gyascutus* (*s. str.*) *planicosta* (LeConte), 29, protibia; 30, protrochanter; 31, antennomere 11; Figs. 32 - 34, *Hippomelas sphenicus* (LeConte), 32, protibia; 33, protrochanter; 34, antennomere 11; 35, *Prasinalia cuneata* (Horn), metacoxa of male; 36, *Gyascutus* (*Stictocera*) *caelatus* (LeConte), metacoxa of male. (From Nelson and Bellamy 1996, with permission of Taylor & Francis Ltd.).

25(19). Metacoxal plates slightly dilated medially, hind margin weakly oblique (Fig. 43); antennae mostly not extending beyond anterior third of pronotum when laid alongside, outer antennomeres transverse (Fig. 46) (Chalcophorini, Agaeocerina) .... *Agaeocera*

— Metacoxal plates strongly dilated medially, hind margin strongly oblique (Fig. 18); antennae mostly extending beyond anterior third of pronotum when laid alongside, outer antennomeres mostly elongate triangular (Fig. 47) 26

26(25). Terminal segment of maxillary palpi slender (Fig. 45) 27

— Terminal segment of maxillary palpi broadened (Fig. 44) 29

27(26). Elytral apex sharply bidentate; last visible abdominal sternum with thin rectangular lobe filling apical emargination between sharp lateral teeth (Dicercini, Phrixiina) .... *Spectralia*

— Elytral apex rounded and/or unidentate; last visible abdominal sternum not as above (Chalcophorini, Chalcophorina) 28

28(27). Pronotum unisulcate; apical third of elytral margins strongly serrate .... *Texania*
— Pronotum bisulcate, midline more or less costuliform; apical third of elytral margins entire or finely serrate .... *Chalcophora*

29(26). Metatarsomere 1 longer than 2; outer antennomeres regularly triangular (Fig. 47) (Dicercini, Dicercina) 30
— Metatarsomere 1 subequal in length to 2; outer antennomeres somewhat to distinctly truncate along external margin (Fig. 48) (Psilopterini, Psilopterina) .... *Lampetis*

30(29). Scutellum rounded; pronotum variably longitudinally sulcate in midline .... *Dicerca*
— Scutellum broader than long; pronotum with median longitudinal ridge or smooth line .... *Poecilonota*

31(18). Prosternal process not or only slightly widened behind front coxae 32
— Prosternal process strongly widened behind front coxae (Fig. 50) .... *Juniperella*

32(31). Elytra not striate but regularly costate; prosternum impressed and punctate along middle.............. .... *Cypriacis*
— Elytra striate; prosternum convex or flattened along middle .... *Buprestis*

33(17). Scutellum invisible; antennal grooves closed; epistoma with broad lateral lobes (Buprestini, Trachykelina) .... *Trachykele*
— Scutellum visible; antennal grooves open anteriorly; epistoma without distinct lateral lobes . 34

34(33). Scutellum transverse anteriorly, acuminate posteriorly; protibia with one apical spur (Buprestini, Sphenopterina) .... *Sphenoptera*
— Scutellum small, not as above; protibia with two spurs 35

35(34). Mentum coriaceous in front; prothorax sinuate at posterior margin; punctation of pronotum simple (Melanophilini, Melanophilina) 36
— Mentum entirely corneous; prothorax commonly truncate; punctation of pronotum variable (Anthaxini) 38

36(35). With mesothoracic pits next to lateral margin of middle coxal cavities (Fig. 51); flattened; glabrous .... *Melanophila*
— Without mesothoracic pits................................ 37

37(36). Flattened; glabrous; apices of elytra acute; head and pronotum with smooth facets .... *Xenomelanophila*
— More convex; elytra with fine, short hairlike setae; apices of elytra not acute; head and pronotum without smooth facets .... *Phaenops*

38(35). Pronotum truncate at base; antennae serrate in both sexes (Anthaxiina) 39

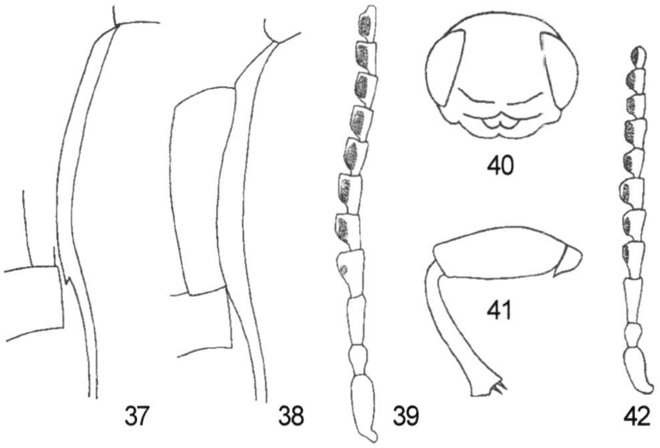

FIGURES 37.41 - 42.41. 37, *Prasinalia cuneata* (Horn), left epipleuron; 38, *Hippomelas sphenicus* (LeConte), left epipleuron; 39, *H. aeneocupreus* Kerremans, antenna of male; 40, *H. saginatus* (Mannerheim), head, anterior view; Figs. 41 - 42, *Barrellus femoratus* (Knull), 41, front leg of male; 42, antenna of male. (From Nelson and Bellamy 1996, with permission of Taylor & Francis Ltd.)

— Pronotum mostly sinuate at base; antennae of male pectinate (Xenorhipina).................................. 40

39(38). Body elongate, slender, *Agrilus*-like; pronotum with posteromedial depression; vestiture completely lacking; pygidium with margin serrate . *Agrilaxia*
— Body short, more robust, not *Agrilus*-like; pronotum without posteromedial depression; often with vestiture on head, elytra; pygidial margin entire .... *Anthaxia*

40(38). Pronotum broadly evenly rounded at sides without lateral margins .... *Trichinorhipis*
— Pronotum quadrate with lateral margins 41

41(40). Posterior coxal plates scarcely narrowed laterally (Fig. 52) .... *Xenorhipis*
— Posterior coxal plates triangular, hind margin strongly oblique (Fig. 53) .... *Hesperorhipis*

42(16). Tarsomere 3 prolonged on each side into a long, divergent spine that extends beyond tarsomere 4 (Fig. 54) (Actenodina)...................... *Actenodes*
— Tarsomere 3 truncate at apex, not extending beyond tarsomere 4 (Fig. 55) (Chrysobothrina) . 43

43(42). Eyes close together on vertex, space between them equal or less than width of eye; elytra with distinct foveae and more or less prominent carinae along main veins; tooth on front femur right or acute angled .... *Chrysobothris*
— Eyes widely separated on vertex, space between them almost twice or more than width of eye; elytra with distinct carinae and inconspicuous foveae; tooth on front femur short, obtuse .... *Sphaerobothris*

Agrilinae

44(13). Tarsi elongate, at least half as long as tibia; legs neither flattened nor tightly folding (Agrilini) 45

— If tarsi longer than 2/5 of tibiae, legs flat, tightly folding against body 48

45(44). Pronotum with entire marginal carina and one submarginal carina (Fig. 56) (Agrilini, Agrilina) .... *Agrilus*

— Pronotum without submarginal carina, some without any lateral carina (Fig. 57) 46

46(45). Frons with deep longitudinal groove; pronotum without lateral carinae; body covered by squamose setae; ovipositor with ventral pair of opposing setal brushes (Agrilini) .... *Lepismadera*

— Frons without deep longitudinal groove; pronotum with lateral carinae; body sparsely setose; ovipositor without ventral brushes 47

47(46). Antennae in repose received in sulci in hypomera ventral to pronotal marginal carinae; anterior prosternal margin arcuately produced medially (Agrilini, Rhaeboscelina) .... *Paragrilus*

— Antennae free in repose; anterior prosternal margin feebly bilobed (Coraebini) .... *Eupristocerus*

48(44). Propleura without sulci for antennae; if femora without sulci for tibiae, then either supraantennal pits absent, or propleural suture double (Aphanisticini, Aphanisticina) .... *Aphanisticus*

— Propleura with deep sulci for antennae; or femora without sulci for tibiae; supraantennal pits present, and propleural suture simple (Trachyini) 49

49(48). Tibiae not markedly flattened 50

— Tibiae strongly flattened; scutellum large, triangular (Pachyschelina) .... *Pachyschelus*

50(49). Propleura without antennal sulci (Trachyina) .... *Trachys*

— Propleura with deep sulci for antennae (Fig. 58) 51

51(50). Prosternal process rounded or truncated apically (Leiopleurina) .... *Leiopleura*

— Prosternal process pointed at apex (Fig. 58) (Brachyina) 52

52(51). Body broad, ovate, less than 2.0X longer than wide; elytra with sublateral carina extending from humerus to near apex; prosternal process sulcate (Fig. 58) .... *Brachys*

— Body narrow, elongate, at least 2.4X longer than wide; elytra without sublateral carina; prosternal process not sulcate .... *Taphrocerus*

Classification of the Nearctic Genera

Buprestidae Leach 1815

Polycestinae Lacordaire 1857

Polycestini Kerremans 1902

Polycestina Théry 1928

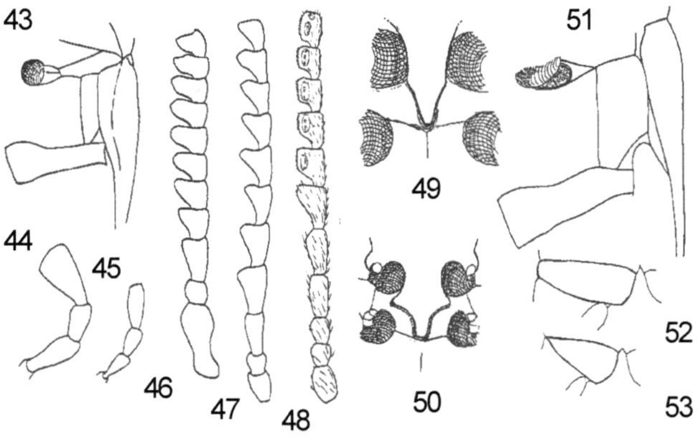

FIGURES 43.41 - 53.41. 43, *Agaeocera g. gentilis* (Horn), metasternal area of male; 44, *Dicerca querci* Knull, maxillary palpus; 45, *Buprestis* (*s. str.*) *maculativentris* Say, maxillary palpus; 46, *Agaeocera g. gentilis*, antenna of male; 47, *Poecilonota thureura* (Say), antenna of female; ig. 48, *Lampetis* (*Spinthoptera*) *webbii* (LeConte), antenna of female; 49, *Buprestis* (*s. str.*) *rufipes* (Oliver), prosternal area of male; 50, *Juniperella mirabilis* Knull, prosternal area of female; 51, *Melanophila acuminata* (DeGeer), meso-metasternal area; 52, *Xenorhipis brendeli* LeConte, metacoxa of female; 53, *Hesperorhipis mirabilis* Knull, metacoxa of female. (Figures 43 - 48, 51 from Nelson 1982)

Polycesta Solier 1833, 12 spp., Pennsylvania, Alabama, Arkansas, Florida, Missouri, Oklahoma, Texas, Arizona, and California (key to spp., Barr 1949; notes, Cobos 1981).

Subgenus *Arizonica* Cobos 1981
Subgenus *Nelsonella* Cobos 1981
Subgenus *Tularensia* Nelson 1997

Polyctesini Cobos 1955

Polyctesina Cobos 1955

Chrysophana LeConte 1859, 2 spp., Arizona, California, Colorado, Idaho, Oregon, Washington and British Columbia (key to spp., Barr in Hatch 1971).

Beerellus Nelson 1982, 1 sp. *B. taxodii* Nelson 1982, from *Taxodium*, Georgia.

Thrincopygini LeConte 1861

Thrincopygina LeConte 1861

Thrincopyge LeConte 1858, 2 spp. in *Dasylirion* and *Nolina* (Agavaceae) Texas, New Mexico, Arizona (key to spp., Nelson 1980).

Acmaeoderini Kerremans 1893

Acmaeoderina Kerremans 1893

Acmaeodera Eschscholtz 1829, 144 spp., widely distributed, mostly in southwestern United States, many visit flowers (key to spp., Fall 1899).

Squamodera Nelson 1996, 4 spp., Arizona, California, and Nevada (key to spp., Nelson 1996).

Acmaeoderopsis Barr 1974, 12 spp., Arizona, California, New Mexico, and Texas.

Anambodera Barr 1974, 6 spp., Arizona, California, Idaho, Nevada, Oregon, and Washington.

Ptosimina Kerremans 1902

Ptosima Serville in Dejean 1833, 4 spp., eastern and midwestern United States to Texas (key to spp., Nelson 1978; notes, Cobos 1980)

Acmaeoderoidina Cobos 1955

Acmaeoderoides Van Dyke 1942, 11 spp., Arizona, California, and Texas (key to spp., Nelson 1968b, 1970, 1999).

Tyndarini Cobos 1955

Tyndarina Cobos 1959

Paratyndaris Fisher 1919, 15 spp., Arizona, California, Florida, New Mexico, Oklahoma, and Texas (key to spp., as *Ancylotela*, Barr 1972; as subgenus of *Tyndaris* Thomson 1857 by Cobos 1980; revision in preparation, Nelson and Bellamy).
  Ancylotela Auctorum
  subgenus *Barberia* Cobos 1980, of *Tyndaris* (unavailable)
  subgenus *Knulliella* Cobos 1980, of *Tyndaris*
  subgenus *Tucsonia* Cobos 1980, of *Tyndaris* (unavailable)

Mastogeniini LeConte and Horn 1883

Mastogeniina LeConte and Horn 1883

Mastogenius Solier 1850, 5 spp., Connecticut, New Jersey, Indiana, Florida, Texas, and Arizona (key to spp., Nelson 1985; notes, Bellamy 1991).
  Haplostethus LeConte 1859

Trigonogya Schaeffer 1919, 1 sp., *T. reticulaticollis* (Schaeffer 1904), Texas.

Micrasta Kerremans 1893, 1 sp. *M. oakleyi* Fisher 1935, described from Puerto Rico, has been collected in Florida (Nelson *et al.* 1996).

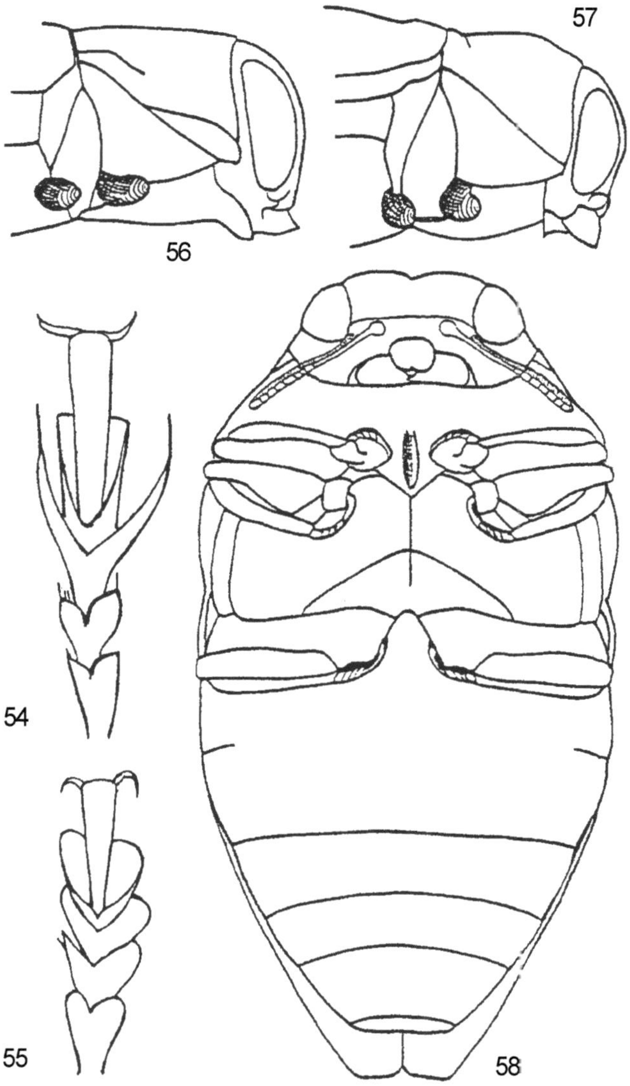

FIGURES 54.41 - 58.41. 54, *Actenodes calcarata* (Chevrolat), protarsus; 55, *Chrysobothris octocola* LeConte, protarsus; 56, *Agrilus cavifrons* Waterhouse, lateral view; 57, *Eupristocerus cogitans* (Weber), lateral view; 58, *Brachys floccosus* Mannerheim, ventral view. (Figures 56.41 - 58.41 from Nelson 1982)

Buprestinae Leach 1815

Chalcophorini Lacordaire 1857

Chalcophorina Lacordaire 1857

Chalcophora Solier 1833, 5 spp., generally distributed in coniferous forests throughout North America (Vol. 1, Color Fig 9).

Texania Casey 1909, 3 spp., New York, Pennsylvania, Indiana, Southeastern United States, Louisiana, and Texas (key to spp., as *Chalcophorella*, Casey 1909; review, Obenberger 1942a; notes, Nelson 1982).
  Chalcophorella auctorum

Agaeocerina Nelson 1982

Agaeocera Waterhouse 1882, 2 spp., Arizona, California, New Mexico, and Texas.
Anataxis Casey 1909

Hippomelanini Holynski 1993

Hippomelanina Holynski 1993

Hippomelas Laporte and Gory 1837, 4 spp., Arizona, New Mexico and Texas. (key to spp., Nelson and Bellamy 1996).

Prasinalia Casey 1909, 2 spp., Arizona and California (key to spp., Nelson and Bellamy 1996).

Gyascutus LeConte 1859, 12 spp., Arizona, California, Idaho, Nevada, New Mexico, Oregon, Texas, and Utah (key to spp., Nelson 2000).
subgenus *Stictocera* Casey 1909

Barrellus Nelson and Bellamy 1996, 1 sp., *B. femoratus* (Knull 1941), California.

Nanularia Casey 1909, 7 spp., Arizona, California, Idaho, Nevada, Texas, and Utah (key to spp., Bellamy 1987)

Ampheremus Fall 1917, 1 sp., *A. cylindricollis* Fall (1917). Arizona, California, Nevada, and New Mexico.

Psilopterini Lacordaire 1857

Psilopterina Lacordaire 1857

Lampetis Dejean 1833, 3 spp., Arizona, Colorado, Kansas, Louisiana, New Mexico, Oklahoma, and Texas (key to spp., Nelson 1986; notes, Kurosawa 1993).
Psiloptera auctorum
subgenus *Spinthoptera* Casey 1909

Sphenopterina Lacordaire 1857

Sphenoptera Dejean 1833, 1 sp., *S. jugoslavica* Obenberger 1926, introduced from eastern Europe to control diffuse knapweed, *Centaurea diffusa*, in the Pacific Northwest (notes, Nelson 1982; Rees *et al.* 1996).
subgenus *Chilostetha* Jakovlev 1889

Dicercini Kerremans 1893

Dicercina Kerremans 1893

Dicerca Eschscholtz 1829, 24 spp., occurs widely from Alaska, most Canadian Provinces and U.S. states (key to spp., Nelson 1975).
Stenuris Kirby 1837

Poecilonota Eschscholtz 1829, 9 spp., widely distributed (key to spp., Evans 1957).
Analampis Dejean 1836
Polydora Gistel 1848
Descarpentriesiola Lerault 1983

Phrixiina Cobos 1975

Spectralia Casey 1909, 6 spp., Arizona, California, northeastern United States to North Carolina, and Texas (key to spp., Chamberlin 1920).
Cinyra auctorum

Buprestini Leach 1815

Buprestina Leach 1815

Buprestis Linnaeus 1758, 18 spp., generally distributed; bore into such trees as beech, cottonwood, maple, and oak, as well as various conifers (key to spp., Helfer 1941).
Ancylocheira Eschscholtz 1829
Anoplis Kirby 1837
Gymnota Gistel 1834
subgenus *Stereosa* Casey 1909
subgenus *Knulliobuprestis* Kurosawa 1988

Cypriacis Casey 1909, 8 spp., generally distributed, hosts as in *Buprestis* (key to spp., Helfer 1941).
subgenus *Nelsonocheira* Kurosawa 1988

Juniperella Knull 1947, 1 sp., *J. mirabilis* Knull 1947, California.

Trachykelina Holynski 1988

Trachykele Marseul 1865, 6 spp., Arizona, Virginia, North Carolina, Georgia, Texas, New Mexico, California, Oregon, Washington, and British Columbia (notes, Fall 1906; key to spp., Burke 1920) (Vol. 1, Color Fig. 1).

Melanophilini Bedel 1921

Melanophilina Bedel 1921

Melanophila Eschscholtz 1829, 5 spp., generally distributed (key to spp., Sloop 1937; Cobos 1987).
Apatura Laporte and Gory 1838 (part)
Oxypteris Kirby 1837

Phaenops Dejean 1833, 15 spp., generally distributed (key to spp., Sloop 1937).

Xenomelanophila Sloop 1937, 1 sp., *X. miranda* (LeConte 1854), Arizona, Colorado, New Mexico, Oregon, Texas, and Utah (notes, Cobos 1987) (Vol. 2, Color Fig. 32).

Anthaxiini Gory and Laporte 1839

Xenorhipina Cobos 1986

Xenorhipis LeConte 1866, 3 spp., Connecticut, New York, Ohio, Pennsylvania, Illinois, and Texas (key to spp., Horn 1882; Obenberger 1939).
Lamesis Westwood 1883

Hesperorhipis Fall 1930, 4 spp., Arizona and California.

Trichinorhipis Barr 1948, 1 sp., *T. knulli* Barr 1948, California.

Anthaxiina Gory and Laporte 1839

Anthaxia Eschscholtz 1829, 36 spp., generally distributed (key to spp., Horn 1882; Obenberger 1942b; Cobos 1958; world catalog, Bílý 1997).
subgenus *Haplanthaxia* Reitter 1911 (key to spp., Cobos 1958)
subgenus *Melanthaxia* Rikhter 1945 (key to spp., Cobos 1958; revision of Nearctic spp., in prep., Bílý)

Agrilaxia Kerremans 1903, 2 spp., widely distributed. (key to spp., Cobos 1971; as valid genus, Bílý and Bellamy 1999).

Chrysobothrini Gory and Laporte 1838

Chrysobothrina Gory and Laporte 1838

Chrysobothris Eschscholtz 1829, 134 spp., generally distributed (key to spp., Fisher 1942).
Amblis Gistel 1834
Odonotomus Kirby 1837
Enocys Gistel 1856
Knowltonia Fisher 1935
Ceratobothris Pochon 1972

Sphaerobothris Semenov-Tian-Shanskij and Rikhter 1934, 2 spp., New Mexico to Arizona, California and Texas (key to spp., Bellamy and Volkovitsh 1997).

Actenodina Kerremans 1893

Actenodes Lacordaire 1857, 9 spp., widely distributed (key to spp., Nelson 1979).

Agrilinae Laporte 1835

Coraebini Bedel 1921

Coraebina Bedel 1921

Eupristocerus Deyrolle 1864, 1 sp., *E. cogitans* (Weber 1801), Eastern United States.
Coraebus LeConte 1859, not Laporte and Gory 1839

Agrilini Laporte 1835

Lepismadora Velten 1987, 1 sp. *L. algodones* Velten (1987), California (Velten and Bellamy 1987); moved to Agrilini, Kuban *et al.* (2001).

Agrilina Laporte 1835

Agrilus Curtis 1825, 171 spp., generally distributed; a few species occur on raspberries and blackberries; most species are wood borers (key to spp., Fisher 1928; notes, Bellamy 1996a). At least four immigrant species from the eastern or western Palaearctic region are established in the Nearctic fauna and one species, *A. hyperici* (Creutzer 1789), was introduced to control St. John's Wort, *Hypericum perforatum* L. Per. in the Pacific Northwest (notes, Rees, *et al.* 1996)
subgenus *Engyaulus* Waterhouse 1889 (key to spp. Nelson and Westcott 1991)

Rhaeboscelina Cobos 1976

Paragrilus Saunders 1871, 2 spp., eastern United States, Florida, and Texas (Cobos 1976).
Rhaeboscelis LeConte 1863, not Chevrolat 1837
Clinocera Deyrolle 1864, not Meigen 1803

Aphanisticini Jacquelin du Val 1863

Aphanisticina Jacquelin du Val 1863

Aphanisticus Latreille 1829, 1 ssp., *A. cochinchinae seminulum* Obenberger 1929, has recently been recorded from southern Texas (Wellso and Jackman 1995) and Florida (Peck and Thomas 1998).

Trachyini Gory and Laporte 1839

Trachyina Gory and Laporte 1839

Trachys Fabricius 1801, 1 sp., *T. troglodytiformis* Obenberger 1918 [= *T. pygmaea* (F. 1787)], naturalized on hollyhock (*Althaea rosea* (L.) Cav.) in New Jersey (Linsley 1949, Weiss 1954, Hespenheide 1968).

Pachyschelina Bøving and Craighead 1931

Pachyschelus Solier 1833, 5 spp., eastern and southern United States, Arizona, New Mexico, and Texas; leaf miners in *Croton* spp. and herbaceous spp. of Fabaceae (key to spp., Nicolay and Weiss 1920).
Metonius Say 1836

Leiopleurina Holynski 1993

Leiopleura Deyrolle 1864, 1 sp., *L. otero* (Fisher 1935), described from Cuba, has been collected in the Florida Keys (notes, Nelson *et al.* 1981).
Leiopleurella Fisher 1922
Enbrachys Fisher 1935

Brachyina Cobos 1979

Brachys Dejean 1833, 12 spp., eastern United States, Arizona, Colorado, New Mexico and Texas; species are leaf miners in various hardwoods, especially *Quercus* spp. (key to spp., Nicolay and Weiss 1923).

Taphrocerus Solier 1833, 13 spp., eastern United States to Texas, Arizona, California, and Washington; the larvae mine in *Scirpus* spp. and other Cyperaceae (key to spp., Obenberger 1934).

Bibliography

BARR, W. F. 1949. A revision of the species of the genus *Polycesta* Occurring in the United States (Coleoptera, Buprestidae). American Museum Novitates, 1432: 1-42.

BARR, W. F. 1971. Family Buprestidae. Pp. 55-89. *In:* M. H. Hatch, ed. Beetles of the Pacific Northwest, Part V. University of Washington Press. Seattle.

BARR, W. F. 1972. New species of *Ancylotela* from Mexico and the United States with a key to the known species. Journal of the Kansas Entomological Society, 45: 92-110.

BARR, W. F. 1974. New genera and species of North American Buprestidae (Coleoptera). Occasional Papers, Biological Society of Nevada, 39: 1-13.

BELLAMY, C. L. 1982. Observations on the biology and distribution of several species of Buprestidae (Coleoptera). Coleopterists Bulletin, 36: 358-361.

BELLAMY, C. L. 1985. A catalogue of the higher taxa of the family Buprestidae (Coleoptera). Navorsinge van die Nasionale Museum, Bloemfontein, 4: 405-472.

BELLAMY, C. L. 1987. A revision of the genera *Nanularia* and *Ampheremus* (Coleoptera, Buprestidae). Contributions in Science, Los Angeles County Museum of Natural History, 387: 1-20.

BELLAMY, C. L. 1991. Studies in the Mastogeniinae (Coleoptera: Buprestidae) III. New species, combinations and a world catalogue. Giornale Italiano di Entomologia, 5: 109-128.

BELLAMY, C. L. 1996a. Comments on the genus *Agrilus* Curtis, 1825: Where do we go now and do we go together? (Coleoptera: Buprestidae: Agrilinae). Elytron, 9(1995): 77-86.

BELLAMY, C. L. 1996b. A new genus and species of Buprestidae (Coleoptera) from the Namibian Richtersveld, with comments on the relationships of the subtribe Mastogeniina LeConte and Horn. African Entomology, 4: 137-142.

BELLAMY, C. L. 1996c. Further consideration of the subtribe Thomassetiina Bellamy: a new species, new records and placement in the contemporary classification (Coleoptera: Buprestidae). Annals of the Transvaal Museum, 36: 215-222.

BELLAMY, C. L. 2001. An annotated summary of the higher classification of the superfamily Buprestoidea (Coleoptera). Folia Heyrovskyana, Supplement, (in press).

BELLAMY, C. L. and R. L. WESTCOTT. 1996. The phylogenetic placement of two new genera and species of Buprestidae (Coleoptera) from Mexico. Journal of Natural History, 30: 229-245.

BELLAMY, C. L. and M. G. VOLKOVITSH. 1997. The chrysobothrine genus *Sphaerobothris* Semenov-Tian-Shankij and Rikhter, new status, with new combinations from North American and North Africa (Coleoptera: Buprestidae). Coleopterists Bulletin, 51: 59-70.

BÍLÝ, S. 1997. World catalogue of the genus *Anthaxia* Eschscholtz, 1829 (Coleoptera: Buprestidae). Folia Heyrovskyana, Supplementum 2, 190 pp.

BÍLÝ, S. 2000. A new concept of Anthaxiini (Coleoptera: Buprestidae). Folia Heyrovskyana, 8: 109-114.

BÍLÝ, S. and C. L. Bellamy. 1999. The genus *Agrilaxia* and description of a new genus from Africa (Coleoptera: Buprestidae). Folia Heyrovskyana, 7: 91-98.

BØVING, A. G. and F. C. CRAIGHEAD. 1931. An illustrated synopsis of the principal larval forms of the order Coleoptera. Entomologica Americana (n.s.), 41: 351 pp., 125 pl.

BRIGHT, D. 1987. Coleoptera. Buprestidae. The metallic wood-boring beetles of Canada and Alaska. Minister of Supply and Services. Ottawa, Canada. 335 pp.

BURKE, H. E. 1917. Flat-headed borers affecting forest trees in the United States. United States Department of Agriculture Bulletin No. 437, pp. 1-8.

BURKE, H. E. 1920. Some notes on the genus *Trachykele* with a description of a new species (Buprestidae, Coleoptera). Proceedings of the Entomological Society of Washington, 22: 18-170.

CASEY, T. L. 1909. Studies in the American Buprestidae. Proceedings of the Washington Academy of Science, 11: 47-178.

CAZIER, M. A. 1951. The Buprestidae of North Central Mexico (Coleoptera). American Museum Novitates, 1526 :1-56.

CHAMBERLIN, W. J. 1920. Description of one new buprestid with notes on other little known species (Coleop.). Entomological News, 31: 241-244.

CHAMBERLIN, W. J. 1926. Catalogue of the Buprestidae of North America north of Mexico. W. J. Chamberlin. Corvallis, Oregon. 289 pp. + 1 page index.

COBOS, A. 1958. Ensayo monográfico sobre las *Anthaxia* Eschs. (Coleoptera, Buprestidae) de América. Segunda parte: Subgénero *Haplanthaxia* Reitter y *Anthaxia* s. str. Archivos de Instituto de Aclimatación, 7: 69-126, 9 pl.

COBOS, A. 1971. Ensayo monográfico sobre las *Anthaxia* Eschs. (Coleoptera, Buprestidae) de América. Tercera parte: Subgénero *Agrilaxia* Kerremans. Archivos de Instituto de Aclimatación, 16: 5-235, 16 pl.

COBOS, A. 1976. Estudio sobre *Rhaeboscelis* Chevrolat, 1837 y géneros afines (Col. Buprestidae). EOS, 50(1974): 19-40.

COBOS, A. 1979. Revisión de la subfamilia Trachyinae a niveles supraespecíficos (Coleoptera, Buprestidae). Acta Entomologica Bohemoslovaca, 76: 414-430.

COBOS, A. 1980. Ensayo sobre los géneros de la subfamilia Polycestinae (Coleoptera, Buprestidae) (Parte I). EOS, 54(1978): 15-94.

COBOS, A. 1981. Ensayo sobre los géneros de la subfamilia Polycestinae (Coleoptera, Buprestidae) (Parte II). EOS, 55-56(1979-1980): 23-94.

COBOS, A. 1987. Ensayo monográfico sobre las *Melanophila* Eschscholtz sensu lato (Coleoptera: Buprestidae). 1 Parte: *Melanophila* Eschs., sensu novo; *Trachypteris* Kirby; *Xenomelanophila* Sloop. EOS, 62(1986): 45-90.

DAVIES, A. 1991. Family Buprestidae, Pp. 160-168. *In:* Y. Bousquet, ed. Checklist of Beetles of Canada and Alaska. Agriculture Canada. Ottawa.

EVANS, D. 1957. Revision of *Poecilonota* of America North of Mexico (Coleoptera: Buprestidae). Annals of the Entomological Society of America, 50: 21-37.

FALL, H. C. 1899. Synopsis of the species of *Acmaeodera* of America North of Mexico. Journal of the New York Entomological Society, 7: 1-37.

FALL, H. C. 1906. On the genus *Trachykele*, with notes and descriptions of other North American Buprestidae. Entomological News, 17: 160-168.

FISHER, W. S. 1928. A revision of the North American species of the buprestid beetles belonging to the genus *Agrilus*. United States National Museum Bulletin, 145: 1-347.

FISHER, W. S. 1942. A revision of the North American species of buprestid beetles belonging to the tribe Chrysobothrini. United States Department of Agriculture, Miscellaneous Publications, 470: 1-274.

FORBES, W. T. M. 1922. The wing-venation of the Coleoptera. Annals of the Entomological Society of America, 15: 328-345.

FORBES, W. T. M. 1926. The wing folding patterns of the Coleoptera. Journal of the New York Entomological Society, 34: 42-139.

FORBES, W. T. M. 1942. The wing of *Mastogenius*. Journal of the New York Entomological Society, 50: 193-194.

GOOD, H. G. 1925. Wing venation of the Buprestidae. Annals of the Entomological Society of America, 18: 251-276.

HELFER, J. R. 1941. A revision of the genus *Buprestis* of North America, North of Mexico (Coleoptera, Buprestidae). Entomologica Americana (N.S.), 21: 123-185.

HESPENHEIDE, H. A. 1968. An immigrant *Agrilus* (Coleoptera: Buprestidae). Entomological News, 79: 77-80.

HESPENHEIDE, H. A. 1996. Chapter 26, Buprestidae (Coleoptera). Pp. 411-421. *In:* J. Llorente B., A. N. García A., E. González S., eds. Biodiversidad, taxonomía y biogeografía de artrópodos de México: Hacia una síntesis de su conocimiento, Universidad Nacional Autónoma de México, Instituto de Biología, xii + 660 pp.

HOLYNSKI, R. 1988. Remarks on the general classification of Buprestidae Leach as applied to Maoraxiina Hol. Folia Entomologica Hungarica, 49: 49-54.

HOLYNSKI, R. 1993. A reassessment of the internal classification of the Buprestidae Leach (Coleoptera). Crystal, series Zoologica, No. 1, pp. 1-42.

HORN, G. H. 1882. Revision of the species of some genera of Buprestidae. Transactions of the American Entomological Society, 10: 101-112.

ICZN. 1994. Opinion 1784. *Buprestis* Linnaeus, 1758 and *Chrysobothris* Eschscholtz, 1829 (Insecta, Coleoptera): conserved by the designation of *Buprestis octoguttata* Linnaeus, 1758 as the type species of *Buprestis*, and *Chrysobothris* and *Dicerca* Eschscholtz, 1829: conserved as the correct original spellings. Bulletin of Zoological Nomenclature, 51: 280-282.

ICZN. 1996a. Opinion 1825. *Poecilonota* Eschscholtz, 1829, *Palmar* Schaefer, 1949 and *Scintillatrix* Obenberger, 1956 (Insecta, Coleoptera): conserved by the designation of *Buprestis variolosa* Paykull, [1799] as the type species of *Poecilonota* and *B. rutilans* Fabricius, [1777] as the type species of *Scintillatrix*. Bulletin of Zoological Nomenclature, 53: 57-59.

ICZN. 1996b. Opinion 1826. *Melanophila* Eschscholtz, 1829 and *Phaenops* Dejean, 1833 (Insecta, Coleoptera): conserved by the designation of *Buprestis acuminata* De Geer, 1774 as the type species of *Melanophila*. Bulletin of Zoological Nomenclature, 53: 60-61.

KERREMANS, C. 1893. Essai de groupement des Buprestides. Annales de la Société Entomologique de Belgique, 37: 94-122, 3 fig.

KERREMANS, C. 1904 -1914. Monographie des buprestides, 7 volumes, Bruxelles.

KNULL, J. N. 1925. The Buprestidae of Pennsylvania (Coleoptera). The Ohio State University Studies, Contributions from the Department of Zoology and Entomology, No. 87, vol. II, no. II , 77 pp.

KOLIBAC, J. 2001. Classification and phylogeny of the Buprestoidea (Insecta: Coleoptera). Acta Musei Moraviae, Scientiae Biologicae (Brno), 85(2000): 113-184.

KUBAN, V., K. MAJER and J. KOLIBAC. 2001. Classification of the tribe Coraebini Bedel, 1921 (Coleoptera, Buprestidae, Agrilinae). Acta Musei Moraviae, Scientiae Biologicae (Brno), 85(2000): 185-287.

KUROSAWA, Y. 1988. Reorganization of *Buprestis* and its allies (Coleoptera, Buprestidae). Kontyû, 56: 261-279.

KUROSAWA, Y. 1993. Reorganization of the genus *Psiloptera* (Coleoptera, Buprestidae). Japanese Journal of Entomology, 61: 577-583.

LACORDAIRE, J. T. 1857. Histoire naturelle des insectes. Genera des Coléoptères ou exposé méthodique de critique de tous les genres proposés jusqu'ici dans cet ordre d'insectes, vol. 4, 1-554 pp., illus. (Contenant les familles des buprestides, throscides, eucnémides, élatérides, cébrionides, cérophytides, rhipicérides, dascyllides, malacodermes, clérides, lyméxylones, cupésides, ptiniores, bostrichides et cissides.)

LAWRENCE, J. F. and A. F. NEWTON, Jr. 1995. Families and subfamilies of Coleoptera (with selected genera, notes, references and data on family-group names). Pp. 779-1006. *In:* J. Pakaluk and S.A. Slipinski, eds. Biology, Phylogeny, and Classification of Coleoptera. Papers Celebrating the 80th Birthday of Roy A. Crowson. Muzeum I Instytut Zoologii PAN, Warsaw.

LECONTE, J. L. 1860. Revision of the Buprestidae of the United States. Transactions of the American Philosophical Society, n. ser., 11(1859): 187-258.

LECONTE, J. L. and G. H. HORN, 1883. Classification of the Coleoptera of North America. Smithsonian Miscellaneous Collection, 26(507): 1-567.

LINSLEY, E. G. 1949. A hollyhock leaf miner new to North America. Journal of the Entomological Society of America, 41: 990.

MACRAE, T. C. 1991. The Buprestidae (Coleoptera) of Missouri. Insecta Mundi, 5: 101-126.

NELSON, G. H. 1959. Notes on the Buprestidae. Bulletin of the Brooklyn Entomological Society, 4: 21-24.

NELSON, G. H. 1960. Notes on the Buprestidae and Schizopodidae. Bulletin of the Brooklyn Entomological Society, 55: 70-74.

NELSON, G. H. 1962. Notes on the Buprestidae: part III. Bulletin of the Brooklyn Entomological Society, 57: 56-60.

NELSON, G. H. 1965. Notes on the Buprestidae: part IV, with a new synonym in *Chrysobothris*. Bulletin of the Brooklyn Entomological Society, 59/60: 37-41.

NELSON, G. H. 1967. Notes on the Buprestidae (Coleoptera): part V, with descriptions of previously unknown sexes. Coleopterists Bulletin, 21: 23-27.

NELSON, G. H. 1968a. Notes on Buprestidae (Coleoptera): part VI . Coleopterists Bulletin, 22: 28-30.

NELSON, G. H. 1968b. A revision of the genus *Acmaeoderoides* (Coleoptera: Buprestidae). Proceedings of the California Academy of Sciences, 4th series, 36: 125-146.

NELSON, G. H. 1970. Tribal placement of *Acmaeoderoides* and other notes (Buprestidae). Coleopterists Bulletin, 24: 30-31.

NELSON, G. H. 1975. A revision of the genus *Dicerca* in North America (Coleoptera: Buprestidae). Entomologische Arbeiten aus dem Museum G. Frey, 26: 87-180.

NELSON, G. H. 1978. A review of the genus *Ptosima* in North America (Coleoptera: Buprestidae). Coleopterists Bulletin 32: 327-336.

NELSON G. H. 1979. A new species of *Actenodes* from the United States with a key to the species (Coleoptera: Buprestidae). Coleopterists Bulletin, 33: 87-91.

NELSON, G. H. 1980. A review of the genus *Thrincopyge* LeConte (Coleoptera: Buprestidae). Pan-Pacific Entomologist, 56: 297-310.

NELSON, G. H. 1982. A new tribe, genus and species of North American Buprestidae with consideration of subfamilial and tribal categories. Coleopterists Bulletin, 35(1981): 431-450.

NELSON, G. H. 1985. Clarification of the taxonomic status in various genera of the family Buprestidae (Coleoptera). Coleopterists Bulletin, 39: 133-146.

NELSON, G. H. 1986. A review of the genus *Psiloptera* subgenus *Lampetis* Solier in the United States (Coleoptera: Buprestidae). Coleopterists Bulletin, 40: 272-284.

NELSON, G. H. 1987. Additional notes on the biology and distribution of Buprestidae in North America, II. Coleopterists Bulletin, 41: 57-65.

NELSON, G. H. 1996. A new genus, *Squamodera* Nelson, for the *vanduzeei* group of *Acmaeodera* Eschscholtz from western North America. Coleopterists Bulletin, 50: 167-175.

NELSON, G. H. 1997. Designations of type species for subgeneric names proposed by Cobos in the genus *Polycesta* Solier (Coleoptera:Buprestidae). Coleopterists Bulletin, 51: 318.

NELSON, G. H. 1999. A new species and other notes on the genus *Acmaeoderoides* Van Dyke (Coleoptera: Buprestidae). Coleopterists Bulletin, 53: 371-376.

NELSON, G. H. 2000. A revision of the subtribe Hippomelanina, part II: *Gyascutus* (*Gyascutus*) LeConte (Coleoptera: Buprestidae). Journal of Natural History, 34: 2251-2292.

NELSON, G. H. and C. L. BELLAMY. 1996. A revision of the subtribe Hippomelanina: *Hippomelas* Laporte and Gory, *Prasinalia* Casey, *Gyascutus* (*Stictocera*) Casey, and *Barrellus*, gen. nov. (Coleoptera: Buprestidae). Journal of Natural History, 30: 861-911.

NELSON, G. H. and T. C. MACRAE. 1990. Additional notes on the biology and distribution of Buprestidae (Coleoptera) in North America, part III. Coleopterists Bulletin, 44: 349-354.

NELSON, G. H., D. S. VERITY and R. L. WESTCOTT. 1981. Additional notes on the biology and distribution of Buprestidae (Coleoptera) of North America. Coleopterists Bulletin, 35: 129-151.

NELSON, G. H. and R. L. WESTCOTT. 1976. Notes on the distribution, synonymy, and biology of Buprestidae (Coleoptera) of North America. Coleopterists Bulletin, 30: 273-284.

NELSON, G. H. and R. L. WESTCOTT. 1991. Review of the *pulchellus* group of *Agrilus* with descriptions of new species (Coleoptera: Buprestidae). Coleopterists Bulletin, 45: 121-142.

NELSON, G. H., R. L. WESTCOTT and T. C. MacRAE. 1996. Miscellaneous notes on Buprestidae and Schizopodidae occurring in the United States and Canada, including descriptions of previously unknown sexes of six *Agrilus* Curtis. Coleopterists Bulletin, 50: 183-191.

NICOLAY, A. S. and H. B. WEISS. 1920. The group Traches in North America, part I. The genera *Pachyschelus* and *Taphrocerus*. Journal of the New York Entomological Society, 28: 136-150.

NICOLAY, A. S. and H. B. WEISS. 1923. The group Traches in North America, part II. The genus *Brachys*. Journal of the New York Entomological Society, 31: 59-76.

OBENBERGER, J. 1926. *In:* W. Junk and S. Schlenkling, eds. Coleopterorum Catalogus. Buprestidae 1. 84: 1-212; 1930. Buprestidae 2. 111: 213-568; 1934a. Buprestidae 3. 132: 569-781; 1934b. Buprestidae 4. 143: 782-934; 1936. Buprestidae 5. 152: 935-1246; 1937. Buprestidae 6. 157: 1247-1714.

OBENBERGER, J. 1934. Monographie du genre *Taphrocerus* Sol. Col. Bupr.). Acta Entomologica Musei Nationalis Pragae, 12: 5-62.

OBENBERGER, J. 1939. The *Xenorhipis* Group of the Family Buprestidae with a description of a new North American species (Col. Bupr.). Revise skupiny rodu krascu z pribuzenstva r. *Xenorhipis*, spolu s popisem noveho severoamerickeho druhu (Col. Bupr.). Vestnik Ceskoslovenske Zoologicke Spolecnosti, 6-7 (1938-39): 330-339.

OBENBERGER, J. 1942a. De genere *Chalcophorella* Kerr. et generibus vicinis (Col. Bupr.). O rodu *Chalcophorella* Kerr. a rodech pribuznych (Col. Bupr.). Acta Societatis Entomologicae Bohemiae, 39: 2-12.

OBENBERGER, J. 1942b. Tri nove druhy rodu *Anthaxia* ze severni Ameriky (Col. Bupr.). De tribus generis Anthaxiae Americae borealis speciebus novis. Acta Societatis Entomologicae Bohemiae, 39: 127-129.

PECK, S. B. and M. C. THOMAS. 1998. A distributional checklist of the beetles (Coleoptera) of Florida. Arthropods of Florida and Neighboring Land Areas, 16: i-viii + 1-180.

PETERSON, A. 1960. Larvae of insects. Part II, Coleoptera, Diptera, Neuroptera, Siphonaptera, Mecoptera, Trichoptera. Columbus, Ohio, 416 pp.

REES, N. E., P. C. QUIMBY, Jr., G. L. PIPER, E. M. COOMBS, C. E. TURNER, N. R. SPENCER and L. V. KNUTSON, eds. 1996. Biological control of weeds in the West. Western Society of Weed Science, in cooperation with USDA Agricultural Research Service, Montana Department of Agriculture, Montana State University. Color World Printers, Bozeman, Montana.

SCHAEFFER, C. 1919. Miscellaneous coleopterological notes and descriptions. Journal of the New York Entmological Society, 26(1918): 211-214.

SHARP, D. and F. A. G. MUIR. 1912. The comparative anatomy of the male genital tube in Coleoptera. Transactions of the Entomological Society of London, 3: 477-642.

SLOOP, K. D. 1937. A revision of the North American buprestid beetles belonging to the genus *Melanophila* (Coleoptera, Buprestidae). University of California Publications in Entomology, 7: 1-20.

TANNER, V. M. 1927. A preliminary study of the genitalia of female Coleoptera. Transactions of the American Entomological Society, 53: 5-50.

TOYAMA, M. 1987. The systematic positions of some buprestid genera (Coleoptera, Buprestidae). Elytra, 15(1/2): 1-11.

VELTEN, R. K. and C. L. BELLAMY. 1987. A new genus and species of Coroebini Bedel from Southern California with a discussion of its relationships in the tribe (Coleoptera, Buprestidae). Coleopterists Bulletin, 41: 185-192.

VOGT, G. B. 1949. A biologically annotated list of the Buprestidae of the Lower Rio Grande Valley, Texas. Annals of the Entomological Society of America, 42: 191-202.

VOLKOVITSH, M. G. and T. J. HAWKESWOOD. 1999. The larva of *Prospheres aurantiopicta* (Laporte & Gory) with comments on the larval characteristics of polycestoid taxa (Insecta, Coleoptera, Buprestidae). Mauritiana (Altenburg), 17: 295-314.

WALTERS, G. C., Jr. 1975. Notes on the distribution and biology of certain Buprestidae (Coleoptera): part I. Coleopterists Bulletin, 29: 69-70.

WALTERS, G. C., Jr. 1978. Notes on the distribution and biology of certain Buprestidae and Cerambycidae (Coleoptera): part II. Coleopterists Bulletin, 32: 355-356.

WALTERS, G. C., Jr. and C. L. BELLAMY. 1982. Notes on the distribution and biology of certain Buprestidae (Coleoptera): part III. Coleopterists Bulletin, 36: 218-220.

WALTERS, G. C., Jr. and C. L. BELLAMY. 1990. Notes on the distribution and biology of certain southwestern Buprestidae (Coleoptera): part iv. Coleopterists Bulletin, 44: 113-115.

WEISS, H. B. 1954. *Trachys pygmaea* (Fab.) the hollyhock leaf miner in New Jersey. Entomological News, 65: 230-232.

WELLSO, S. G. and J. A. JACKMAN. 1995. The first record of *Aphanisticus cochinchinae seminulum* Obenberger (Coleoptera: Buprestidae), a potential sugarcane pest in the Western Hemisphere. Coleopterists Bulletin, 49: 287-288.

WELLSO, S. G., G. V. MANLEY, and J. A. JACKMAN. 1976. Keys and notes on the Buprestidae (Coleoptera) of Michigan. Great Lakes Entomologist, 9: 1-22.

WESTCOTT, R. L. 1990. Notes on taxonomy, ecology and distribution for some species of *Chrysobothris* Eschscholtz occuring in the United States (Coleoptera: Buprestidae). Coleopterists Bulletin, 44: 323-343.

WESTCOTT, R. L. 1991. Distributional, biological, and taxonomic notes on North American Buprestidae. Insecta Mundi, 4(1990): 73-89.

WESTCOTT, R. L., T. ATKINSON, H. A. HESPENHEIDE and G. H. NELSON. 1989. New country and state records, and other notes for Mexican Buprestidae (Coleoptera). Insecta Mundi, 3: 217-232.

WESTCOTT, R. L., W. F. BARR, G. H. NELSON and D. S. VERITY. 1979. Distributional and biological notes on North and Central American species of *Acmaeodera* (Coleoptera: Buprestidae). Coleopterists Bulletin, 33: 169-181.

Superfamily BYRRHOIDEA Latreille 1804

42. BYRRHIDAE Latreille 1804

by Paul J. Johnson

Family common name: The pill or moss beetles

Family synonyms: Birrhii Latreille 1825; Brachymera Thomson 1859; Byrrhida Meixner 1935; Syncalyptidae Moursy 1961

The very convex and compact shape, deflexed head, transverse front coxae, retractile appendages, and association with mosses are distinctive of these beetles.

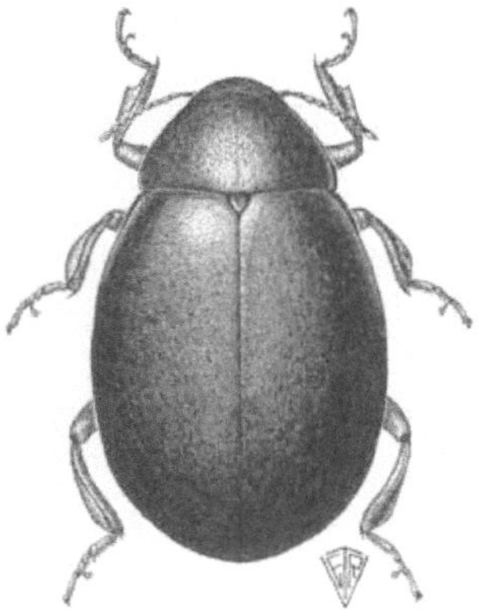

FIGURE 1.42. *Amphicyrta dentipes* Erichson

Description: Body form oval to elongate oval, strongly convex, length approximately 1-10 mm. Integument glabrous or covered with clavate bristles, erect setae, or with a dense coat of decumbent setae, giving the integument a velvety appearance. Color variable, mostly of gray, brown or black, or integument iridescent green and copper, or castaneous with a pronounced luster.

Head hypognathous, narrower than prothorax, frons convex. Eyes oval or slightly emarginate, situated on sides; partly or completely hidden when head retracted into prothorax. Clypeus reduced to obsolescent. Labrum freely articulating, mostly emarginate. Antenna 11-segmented, clavate, capitate, or subfiliform. Mandibles with a variable number of apical teeth, with a deep notch at middle, a blade-like molar area, and a basal brush of setae. Maxillary palpus 3-segmented, labial palpus 2-segmented, terminal palpomere pear-shaped or securiform.

Thorax compact, prothorax tightly fitting agains mesothorax. Pronotum convex, lateral margin finely to strongly carinate, posterior margin smooth or with peg-like crenulae. Hypomeron large, subtriangular, rarely with cural depressions. Prosternum V- or T-shaped on disc, relatively broad between coxae, intercoxal process received into a deep cavity on mesosternum; anterior coxal cavities broadly open behind. Scutellum small to minute, ovoid to subtriangular. Elytra entire, strongly convex, surface finely to coarsely punctate, striate to smooth, epipleural fold variously formed, extending to end of elytron, or mostly shorter; venter near apex with broad laminar flange. Mesosternum short, broad in front, narrow behind. Metathorax alate, brachypterous, or apterous. Wing in many nearly veinless, apical field long, in many subequal in length to veined basal portion; radial cell short, oblique; medial area veins reduced, wedge cell absent. Metasternum much broader and longer than pro- and mesosternum, mostly with a median longitudinal suture, transverse suture lacking. Legs with anterior coxae transverse; middle coxae globular to flat, slightly transverse, moderately to widely separated; hind coxae mostly transverse, approximate medially, nearly attaining the elytral epipleura laterally; trochanters triangular, large; femora mostly somewhat flattened; tibiae slender or stout, in many flattened and expanded apically, densely covered with setae or spines; tarsal formula 4-4-4 or 5-5-5, tarsomeres mostly increasingly larger from first to 3rd, 4th small, 5th long, 3rd in some lobate, remaining tarsomeres simple or with pubescent pads beneath; claws simple.

Abdomen with 5 ventrites; ventrites 1-2 connate. Male genitalia of the trilobed type; median lobe with apex flattened, pointed or hook-shaped, and short basal struts, parameres well developed, fused basally, asymmetrical, or lacking; most with basal piece longer than wide, rectangular, trapezoidal, or subcylindrical. Female genitalia with gonocoxites large; styli apical to subapical, short; baculi absent.

Larvae moderately fusiform and dorsoventrally arcuate to narrowly elateriform; mature length approximately 2-15 mm; terga with rows of setae. Head exserted, hypognathous, with lyriform or V-shaped epicranial suture surrounding frons; stemmata 4 or 6 each side; antennae short, 3-segmented; clypeus separate from frons, transverse; labrum freely articulating; mandibles symmetrical, single to multidentate, robust, with a large brush of setae at the base of the cutting edge, mola absent; maxilla with palpiform galea, 1-2 segmented; lacinia falciform or truncate, with spiniform mesal margin, palpus 4-segmented; cardo bipartite, fused stipes, setiferous. Labium with ligula, palpus 2-segmented. Thorax with the pronotum as long as the meso- and metathorax combined, legs short, 5-segmented, with bisetose tarsungulus. Abdomen 10-segmented; 9th segment in some transversely carinate, or operculate; 10th segment short, rarely with strong ventral hooks; urogomphi absent. Spiracles biforous, present on mesothorax and abdominal segments 1-8; closing apparatus absent.

Habits and habitats. Byrrhid hosts and habitats for North American species were studied by Johnson (1987). Most of the North American species are obligate moss feeders as both adults and larvae. Adults of all bryophagous taxa are surface grazers on host mosses and rarely on leafy liverworts. Larvae of most of these species burrow through gametophyte layers and underlying friable substrates. Larvae of *Cytilus* and *Arctobyrrhus* do not burrow but are surface grazers. Some *Cytilus* and *Byrrhus* species are occasionally reported as incidentally feeding on conifer seedlings, grasses, and clover grown in moist areas with abundant mosses. *Amphicyrta* species, adults and larvae, feed on succulent leaves and stems of forest and meadow herbs, and deciduous shrubs. They have been reported damaging various commercial and home garden vegetables and flowers, and are sometimes abundant in weedy lawns of northern California.

Adults of most species are found active on their hosts under dim light or at night, or under cover of wood fragments, leaves, edges of stones, and other objects during bright daytime hours. Species of *Morychus* and *Cytilus* are active in open mossy areas on sunny days. Alate species fly during spring and early summer and are often found on beaches as washup or windblown drift. Most species retract their legs and antennae into ventral recesses and become motionless when disturbed, leading to the common name of pill beetle.

Byrrhids inhabiting mesic coniferous forests express high fidelity to narrowly defined bryophyte and vascular plant communities. Byrrhids of open sites are adapted to relatively short-lived ruderal microhabitats dominated by the mosses *Ceratodon purpureus* (L.) Brid., *Polytrichum juniperinum* Wild. or *P. piliferum* Schreb., and *Carex* spp. on moist but thin or scarified soil, or mineral substrates (Johnson 1987, Berman 1990). Due to habitat fidelity, contemporary and subfossil byrrhids are used extensively for interpreting Pleistocene climates and biotic patterns.

Relationships of byrrhids with other organisms are poorly known. The fungus *Phaulomyces simplocariae* Kesel (Ascomycetes: Laboulbeniales) was found parasitizing *Simplocaria semistriata* (F.) by Kesel (1994), and species representing nine genera of byrrhids were found infected with the common insect pathogen *Beauveria bassiana* (Balsamo) Vuillemin (Ascomycetes: Clavicipitaceae). An undescribed species of *Anaphes* (Hymenoptera: Mymaridae) was reported (Schauff 1984) as reared from the eggs of *Lioligus nitidus* (Motschulsky) and *Lioon simplicipes* (Mannerheim). Undetermined Braconidae (Hymenoptera) were found in the larvae of *L. simplicipes* and *Eusomalia lecontei* (Wickham), specimens of a *Spilogona* sp (Diptera: Muscidae) were reared from larvae of *Byrrhus eximius* LeConte, and a nematode was found in the midgut of an adult *L. simplicipes* (Johnson 1987). An undescribed species of "*Pedilophorus*" from Australia was found to transmit a tymovirus among vascular plants in laboratory experiments (Guy and Gibbs 1981).

Status of the Classification. Dalla Torre (1911) provided the latest world catalogue. Johnson (1991) presented the most recent key to adults of species from Canada and the United States. Taxonomic revisions of holarctic genera are needed. El Moursy (1961) suggested that Syncalyptinae should be regarded as a family, Syncalyptidae. However, all subsequent authors retained the subfamily status of this group. Lawrence and Newton (1995) placed the family in an expanded Byrrhoidea that includes the former Dryopoidea and Psephenoidea; further relationship elucidation is needed. Considerable work remains for immature stages (Lawrence 1991). Johnson (1987) gave descriptions and a key to known larvae from North America. Fossil taxa are recorded from Oligocene strata (Carpenter 1992).

Distribution. About 290 species are known from all areas. Of these, 35 species are known from nearctic North America. Byrrhids are most common and diverse in the western mountains, the northeastern mountains, and the Great Lakes region. Many of the North American byrrhids show a close and probably geologically recent relationship to Eurasian byrrhids. Regional endemism is expressed in the far western regions with *Sierraclava* restricted to the Californian region; *Amphicyrta* to California and western Oregon; and *Lioligus*, *Lioon*, *Exomella*, *Listemus*, and *Eusomalia* to mesic coniferous forests in Oregon, Washington, Idaho, Montana, British Columbia and Alaska. Two species, one each of *Simplocaria* and *Chaetophora*, are adventitious from Europe (Johnson 1990).

Key to the Nearctic Genera

1. Antennae short, clavate or capitate; palps with ultimate segment fusiform or pyriform 2
— Antennae long, filiform, compressed; palps with ultimate segment securiform; appendages free, not fitting into recesses; tarsomere 3 with large fleshy lobe .... *Amphicyrta*

2(1). Antennae capitate; body short (0.9-2.6 mm), shallowly to moderately convex; appendages closely retractile into deep recesses; dorsum with clavate or truncate bristles 3
— Antennae clavate; body variable in length (1.9-8.7 mm) moderately to strongly convex, ovate to elongate-oval; appendages not or partially fitting into shallow recesses; dorsum with decumbent to erect fine setae 5

3(2). Length 2.4-3.2 mm; ovoid; dorsum with appressed scale-like setae and erect bristles 4
— Length 0.9-2.1 mm; ovate; dorsum with bristles only .... *Chaetophora*

4(3). Body ovate to oval, shallowly to moderately convex, lateral margins arcuate in silhouette; striae on elytral disc shallowly impressed, or serially punctate .... *Curimopsis*
— Body elongate, strongly convex, parallel-sided; elytral striae deeply sulciform .... *Sierraclava*

5(2). Frontal margin beaded, reflexed and carinate in a few species; mesosternum strongly reduced except for median fossa; palps with ultimate segment fusiform to narrowly pyriform; metacoxae small, distant from elytral margin; elytra in many connate 6
— Frontal margin rounded or truncate, not reflexed, beaded or carinate; mesosternum distinct laterad of median fossa; palps with ultimate segment compressed or cylindrically pyriform; metacoxae flattened, nearly reaching elytral margin; elytra not connate 9

6(5). Elytra connate, metathorax apterous; mesocoxae widely separated 7
— Elytra separate and metathoracic wings present; integument rufopiceous to piceous, a few species with submetallic reflections; mesocoxae moderately separated .... *Simplocaria*

7(6). Pubescence simple, slender, fine; epipleura flat; integument shining to metallic 8
— Pubescence stout, recurved; epipleura deeply emarginate to receive femoral knee; integument brunneous .... *Exomella*

8(7). Epipleura broad, extending length of elytron; dorsal integument piceous to olivaceous, glossy.. .... *Lioon*
— Epipleura narrow, attenuating and terminating before second ventrite; dorsal integument viridescent .... *Lioligus*

9(5). Frontal margin truncate and thickened; ventrite 1 without crural depressions 10
— Frontal margin obtusely rounded; ventrite 1 with distinct crural depressons or fossae 11

10(9). Form subparallel, moderately convex dorsally; piceous, without metallic patches or reflections; pubescence long, decumbent, with erect black setae .... *Arctobyrrhus*
— Form ovoid to subparallel, strongly convex dorsally; piceous, frequently with cupreous or viridescent patches or intervals; pubescence short, appressed, without erect black setae .... *Cytilus*

11(9). Tarsomere 4 with membranous ventral lobe; dorsal integument glossy, viridescent and/or cupreous 12
— Tarsomere 4 simple; dorsal integument opaque or glossy, piceous to black 14

12(11). Form ovate; aeneous to viridescent; punctures obsolescent or coarse and forming rugose patches on elytra; pubescence minute or moderately long and patchy, setae mixed cinereous and rufobrunneous 13
— Form subparallel; piceous with viridescent reflections to cupreous; punctures fine to moderate; pubescence moderately long, evenly distributed .... *Morychus*

13(12). Length 6.5-10.0 mm; dorsal punctures fine; subglabrous, pubescence minute; aeneous to viridescent, with elytra bearing alternating aeneous and viridaeneous vittae ...... *Eusomalia*
— Length 3.2-4.0 mm; punctures coarse; pubescence evident, moderately long .... *Listemus*

14(11). Form oval to subparallel, strongly convex dorsally; pubescence simple, appressed .... *Byrrhus*
— Form ovate, shallowly convex dorsally; pubescence including erect, bristle-like setae ..... *Porcinolus*

Classification of the Nearctic Genera

Byrrhidae Latreille 1804

Amphicyrtinae LeConte 1861

Amphicyrta Erichson, 1843, 2 spp., *A. dentipes* Erichson and *A. chrysomelina* Erichson, from California and Oregon.
Eucyphus Mannerheim 1852

Byrrhinae Latreille 1804

Byrrhini Latreille 1804

Arctobyrrhus Münster 1902, 1 sp., *A. subcanus* (LeConte), boreal and transcontinental.
Tylicus Casey 1912 (synonymy by Johnson 1998)

Byrrhus Linnaeus 1758, 7 spp., widely distributed in western and northeastern montane, boreal, subarctic, and Great Lakes regions.

Cytilus Erichson 1847, 2 spp., *C. alternatus* (Say) and *C. mimicus* Casey; boreal and montane.

Porcinolus Mulsant 1869, 1 sp., *P. undatus* (Melsheimer), from the eastern Rocky Mountain foothills and eastward.

Pedilophorini Casey 1912

Eusomalia Casey 1912, 1 sp., *E. lecontei* (Wickham), northern Idaho, northeastern Washington, northwestern Montana, and southeastern British Columbia.

Listemus Casey 1912, 2 spp., *L. acuminatus* (Mannerheim) from northern California to southeastern Alaska, and *L. kootenai* Johnson from northeastern Washington, northern Idaho, and northwestern Montana.

Morychus Erichson 1847, 2 spp.; *M. oblongus* (LeConte) and *M. aeneolus* (LeConte); montane western U.S., montane and boreal Canada.

Simplocariini Mulsant and Rey 1869

Exomella Casey 1912, 2 spp., *E. pleuralis* (Casey) from coastal Oregon and British Columbia, and *E. merickeli* Johnson from northern Idaho.
Exoma Casey 1908

Lioligus Casey 1912, 2 spp., *L. nitidus* (Motschulsky) from coastal California to Alaska, and *L. pallidus* Casey from northern Idaho.

Lioon Casey 1912, 2 spp., *L. simplicipes* (Mannerheim) from coastal northern California to southeastern Alaska, and *L. nezperce* Johnson from northern Idaho and northwestern Montana.

Simplocaria Stephens 1830, 3 spp., two holarctic species (*S. elongata* J. Sahlberg, *S. metallica* (Stürm)) in montane, boreal and subarctic regions, and one adventitious species (*S. semistriata* (F.)) in boreal U.S. and southern Canada.

Syncalyptinae Mulsant and Rey 1869

Chaetophora Kirby and Spence 1823, 1 sp., *C. spinosa* (Rossi), adventitious in northeast, north-central, and northwestern U.S. and adjacent Canada.

Syncalypta Stephens 1830

Curimopsis Ganglbauer 1902, 5 spp., transcontinental in montane, boreal and subarctic Canada, montane in western and northeastern U.S.

Sierraclava Johnson 1982, 1 sp., *S. cooperi* Johnson, southern California and northern Baja California.

Bibliography

BERMAN, D. I. 1990. Current habitats of pill beetle *Morychus viridis* (Coleoptera, Byrrhidae) and reconstruction of natural environment of Pleistocene in northeast USSR. Doklady Akademii Nauk SSSR, 310: 1021-1023.

CARPENTER, F.M. 1992. Superclass Hexapoda. Treatise on Invertebrate Paleontology, Part R, Arthropoda 4, 4: 279-655.

DALLA TORRE, K. W. von. 1911. Fam. Byrrhidae. Coleopterorum Catalogus, 14(33): 12-33.

EL MOURSY, A. A. 1961. A tentative classification of and a key to the North American genera of the family Byrrhidae (new sense) and family Syncalyptidae (new status). Coleopterists Bulletin, 15: 9-15.

GUY, P. and A. GIBBS 1981. A tymovirus of *Cardamine* sp. from alpine Australia. Australasian Plant Pathology, 10: 12-13.

JOHNSON, P. J. 1987. Larval taxonomy, biology, and biogeography of the genera of North American Byrrhidae (Insecta: Coleoptera). M.S. Thesis, University of Idaho, Moscow, 268 pp. [unpublished]

JOHNSON, P. J. 1990. Notes on the naturalization of two European Byrrhidae (Coleoptera) in North America. Journal of the New York Entomological Society, 98: 434-440.

JOHNSON, P. J. 1991. Taxonomic notes, new records, and a key to the adults of North American Byrrhidae (Coleoptera). Proceedings of the Entomological Society of Washington, 93: 322-332.

JOHNSON, P. J. 1998. *Arctobyrrhus* Münster, a senior synonym of *Tylicus* Casey (Coleoptera: Byrrhidae). Annales Zoologici, 47: 175-178.

KESEL, A. DE. 1994. *Phaulomyces simplocariae* sp. nov. (Ascomycetes, Laboulbeniales) from *Simplocaria semistriata* (Coleoptera, Byrrhidae). Mycotaxon, 50: 191-198.

LAWRENCE, J. F. 1991. Byrrhidae. Pp. 384-386. *In:* F. W. Stehr, ed. Immature Insects. Kendall/Hunt. Dubuque, IA.

LAWRENCE, J. F. and A. F. NEWTON, Jr. 1995. Families and subfamilies of Coleoptera (with selected genera, notes, references and data on family-group names). Pp. 779-1006. *In:* J. Pakaluk and S.A. Slipinski, eds. Biology, Phylogeny and Classification of Coleoptera: Papers Celebrating the 80th Birthday of Roy A. Crowson. Muzeum i Instytut Zoologii PAN, Warsaw.

SCHAUFF, M. E. 1984. The holarctic genera of Mymaridae (Hymenoptera: Chalcidoidea). Memoirs of the Entomological Society of Washington 12, 67 pp.

43. ELMIDAE Curtis 1830

by William D. Shepard

Family common name: The riffle beetles

Family synonyms: Limniidae Stephens 1828, Helminthidae Ganglbauer 1904

The combination of aquatic or semiaquatic habits with the slender antennae, hairless eyes, and non-swimming legs with large claws serve to distinguish adults of this family.

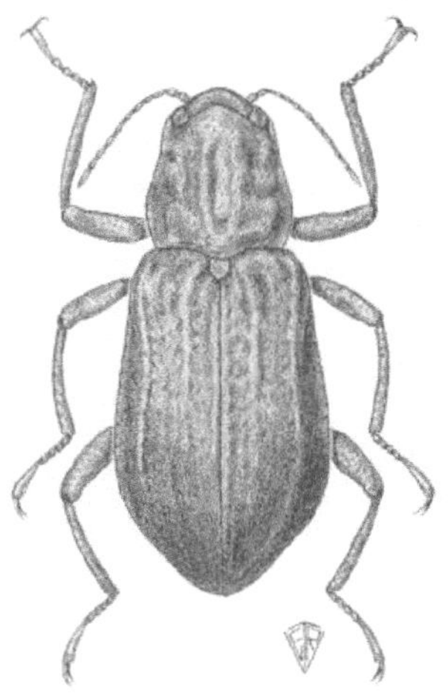

FIGURE 1.43 *Stenelmis crenata* (Say)

Description: Body ovate to elongate, parallel-sided to fusiform, somewhat depressed; 1-8 mm in length; color piceous to black, often with red or yellow maculae or vittae, appendages testaceous; setation generally fine and depressed, plastron covering much of venter and legs.

Head deflexed, may be deeply inserted in the prothorax; surface rugose. Antennae 7-11 segmented, filiform to clavate; inserted between the eyes, well separated from the mandibles. Clypeus mostly distinct. Labrum distinct, moderate, curved, thin. Mandible bi- or tridentate; prostheca large, membranous, setose. Maxillary palpi 3-4 segmented, segments slender; labium with the mentum trapezoidal, ligula large, labial palpi 3-segmented, segments slender. Eyes lateral, size small, rounded.

Pronotum broader that the head; irregularly quadrate, produced in front; borders laterally crenulate or smooth; disc smooth, punctate, sulcate, rugose, variously carinate, or a combination of these; pleural region broad; prosternum prominent, long, broad, extended anteriorly, prosternal process narrow to broad; procoxal cavities open. Mesosternum short, sulcate to receive prosternal process; metasternum longer, often with longitudinal or transverse impressed line. Legs with anterior and middle coxae rounded to globose and without the trochantin exposed or transverse and with the trochantin exposed, the coxae separate; hind coxae globular to transverse, separate; trochanters moderate, triangular; femora slender; tibiae slender, apical spurs absent; tarsal formula 5-5-5, segments filiform; claws slender to stout, some with ventral tooth. Scutellum small; subovate, triangular or pentagonal. Elytra entire, apically rounded; surface rugose, punctures diffuse or striate, in many carinate, or some combination of these; humeri usually distinct; epipleural fold narrow, entire. Wing with venation in basal half; costa, subcosta and radius crowded anteriorly; cubitus complete; 4 anal veins; may be reduced or absent.

Abdomen with 5 visible sternites, the sutures may be obscure; sides of 4th or 5th sternites may be upturned to clasp epipleuron. Male genitalia elongate, narrow, well-sclerotized; penis elongate, narrow to broad, apex rounded to acute; parameres well developed, apices blunt to acute, bases may be explanate; basal piece tubular to open dorsally. Female genitalia with elongate styli and coxites; bacula long and narrow, often sinuate.

Eggs are undescribed.

Larva elongate, convex, cylindrical, or depressed, or onisciform with lateral projections; 3-16 mm long; well sclerotized; surface with tubercles and setae. Head exserted, prognathous, epicranial suture distinct or absent; antennae 3-segmented, short, 2nd segment longest; stemmata clustered to form two ocelli. Labrum transverse; mandibles tridentate, without mola, with basal plumose processes; maxillae with 4-segmented palps, galea and lacinia separate; labium with palp 2-3 segmented; postmentum undivided. Thorax with prothorax as long as meso- and metathorax; with or without sternae, 1-3 thoracic pleura per segment; 5-segmented legs. Abdomen 9-segmented; sterna distinct; with 5-8 abdominal pleura; last abdominal segment slightly to deeply emarginate, bearing retractile anal tracheal gills below an operculum with a pair of internal hooks; apex of abdomen variously modified, but without urogomphi. Spiracles on mesothorax and abdominal segments 1-8; biforous.

Pupa elongate, exarate; dorsally and laterally with numerous elongate setae. Prothorax with anterior (and sometimes posterior) angles produced into elongate filiform spines. Abdomen with 1 pair of sclerotized elongate urogomphi; gin-traps absent.

Almost all of the derscriptive morphology has concerned the adult stage, particularly in regard to describing and naming new taxa. However, some papers have emphasized other stages. Larval morphology has been described by LeSage and Harper (1977) and reviewed by Brown (1991). Pupal morphology has been described by LeSage and Harper (1976) and Steedman (1983).

Habits and habitats. The larvae are aquatic; the adults of Larainae are riparian, those of Elminae are aquatic and seldom leave the water. Pupation is in cells under streamside substrates, and most pupation is associated with postflood periods. Life cycles are usually univoltine; in cold waters or with poor food they may be longer. While the larval stadium is about one year, adults (in captivity) have lived several years. Adults and larvae can be found year-around. There are 5-7 larval instars depending on the size of the adults. Most species live in rapid, cool, and highly oxygenated streams while some genera and species inhabit lakes, caves or warm springs. Most species are detritivorous or algivorous.

Life cycles have been described for species in both subfamilies (LeSage and Harper 1976, Steedman 1985, White 1978a), and Brown (1987) summarized much of the rest of the ecological literature.

Status of the classification. Worldwide there are still many new taxa being described, including additional tribes. The taxa of North America and Europe are relatively well known and stable. The US has 26 genera and 99 species. Differentiation of North American genera has largely been the result of work by Sanderson (1953, 1954). Brown (1972b, 1983) provided identification keys and distributional information. Generic level treatments are available for several genera: *Atractelmis* Chandler (Shepard and Barr 1991), *Heterelmis* Sharp (Brown 1981), *Huleechius* Brown (Brown 1981), *Neocylloepus* Brown (Brown 1979), *Optioservus* Sanderson (White 1978b), *Ordobrevia* Sanderson (Shepard 1982b), and *Stenelmis* Dufour (Sanderson 1938). Brown and White (1978) provide useful information on difficult-to-identify species.

Distribution. Elmids occur on all continents except Antarctica. In the United States there is a mix of broadly distributed species, such as *Stenelmis occidentalis* Schmude and Brown and *Microcylloepus pusillus* (LeConte), and species with highly restricted distributions, such as *Dubiraphia brunnescens* (Fall) and *Microcylloepus formicoideus* Shepard. Distributional information is summarized by Brown (1972b, 1983). There are some regional lists, such as Mingo (1979), Finni and Skinner (1975), and Shepard (1993). Shepard (1992a) also describes the elmids of Death Valley.

Key to the Genera of the United States

1\. Rather soft-bodied; pubescent, but without plastron; procoxae transverse and with trochantin exposed; riparian, usually not under water, agile fliers 2

— Hard-bodied; with plastron on various ventral parts; procoxae rounded and trochantin concealed; aquatic, typically slow-moving, clinging to substrate 3

2(1). Less than 4 mm long; antennae clubbed; pronotum with sublateral sulci ..... *Phanocerus*

— More than 5 mm long; antennae not clubbed; pronotum without sulci ..... *Lara*

3(1). Hind coxae globular and subequal to others; posterior margin of prosternal process almost as wide as head 4

— Hind coxae transverse and larger than others; posterior margin of prosternal process much narrower than head 5

4(3). Black; elytra with sublateral carinae; antennae 7-segmented, apically enlarged; pronotum without transverse impressions; tarsal claw lacking basal tooth ..... *Macronychus*

— Conspicuously colored with black and yellow or orange; elytra and pronotum without sublateral carinae; antennae 11-segmented, filiform; pronotum with oblique transverse impressions at apical third; tarsal claw with basal tooth ..... *Ancyronyx*

5(3). Antennae 8-segmented, apical segment enlarged ..... *Zaitzevia*

— Antennae with 10-11 segments, mostly filiform .. 6

6(5). Anterior tibia with fringe of tomentum 8

— Anterior tibia without fringe of tomentum 7

7(6). Elytron with accessory stria, sutural stria confluent with 2nd stria at about 5th puncture; granules of head and legs elongate ..... *Ordobrevia*

— Elytron without accessory stria; granules of head and legs round ..... *Stenelmis*

8(6). Lateral margin of 4th or 5th abdominal sternite upturned as a prominent lobe or tooth which clasps epipleuron; epipleuron widened to receive tooth then narrowing abruptly toward apex 14

— Lateral margin of abdominal sternites not upturned as a prominent lobe or tooth; epipleuron in many tapering uniformly to apex 9

9(8). Pronotum with sublateral carinae 10

— Pronotum smooth, without sublateral carinae ... 13

10(9). Epipleuron extending to middle of 5th abdominal segment; usually black, rarely black with red spots; prosternum projecting beneath head; 2.5-2.6 mm long ..... *Rhizelmis*

— Epipleuron ending at base of 5th abdominal segment; mostly black with red spots, rarely all black; prosternum not projecting beneath head; less than 2.3 mm long 11

11(10). Pronotal carinae forked at base ..... *Cleptelmis*

— Pronotal carinae not forked 12

12(11). Sides of pronotum converging anteriorly from base; body spindle-shaped; black, each elytron with a broad humeral and an oblique, narrow, subapical red spot; tarsi and claws prominent ... *Atractelmis*

— Sides of pronotum parallel or divergent anteriorly at base, strongly convergent apically; hump-backed; elytra black to red, uniformly colored or with basal half red, with or without broad apical spots; tarsi and claws not prominent ..... *Ampumixis*

13(9). Maxillary palpi 3-segmented; markings, if present, transverse ..... *Narpus*

— Maxillary palpi 4-segmented; markings, if present, longitudinal ..... *Dubiraphia*

14(8). Tooth that clasps epipleuron arising from lateral margin of 5th abdominal sternite 15

— Tooth that clasps epipleuron arising from posterolateral margin of 4th abdominal sternite 22

15(14). Elytron basally with a short accessory stria between sutural and 2nd major stria ..... *Macrelmis*

— Elytron without an accessory stria 16

16(15). Elytron with 1 sublateral carina; pronotum without oblique sculpturing 17

— Elytron with 2 sublateral carinae; rarely only 1 in *Microcylloepus* which has oblique sculturing on posterior half of pronotum 18

17(16). Posterior half of pronotum divided by a conspicuous median longitudinal impression, with a transverse impression slightly anterior to middle; brown to black .... *Neocylloepus*

— Pronotum undivided except by transverse impression at anterior two-fifths; testaceous *Neoelmis*

18(16). Pronotal hypomeron with plastron belt extending from coxa to lateral margin.......... *Hexacylloepus*

— Hypomeron with or without plastron, but if present it does not reach margin 19

19(18). Prosternal process broad and truncate apically; pronotum without median longitudinal impression, usually with transverse impression at middle; pronotal hypomeron with plastron near coxa; body short and wide .... *Heterelmis*

— Prosternal process relatively narrow, elongate with apex tapering or rounded; pronotum with median longitudinal impression; hypomeron without plastron; body elongate 20

20(19) Pronotum with a transverse impression at anterior two-thirds; epipleuron without plastron; small, less than 2.3 mm long .... *Microcylloepus*

— Pronotum without such a transverse impression; epipleuron with plaston 21

21(20). Gula distinctly narrower than submentum or mentum .... *Huleechius*

— Gula not distinctly narrower than submentum or mentum .... *Cylloepus*

22(14). Pronotum with sublateral carinae extending from base to anterior margin .... *Oulimnius*

— Pronotum with sublateral carinae absent or not extending beyond middle 23

23(22). Pronotum without or with only a trace of carinae 24

— Pronotum with carinae in basal half 25

24(23). Body elongate, spindle-shaped; pronotal surface smooth and shiny; each elytron with 2 oblique yellowish spots, third interval flat; legs long, claws prominent and recurved; 2-2.6 mm long.. .... *Gonielmis*

— Body short and broad; pronotal surface granulate, covered with plastron; elytra black, third interval strongly carinate; legs short, claws not prominent; less than 2 mm long .... *Xenelmis*

25(23). Body elongate; tarsi and claws long and prominent .... *Promoresia*

— Body plump; tarsi and claws not conspicuously enlarged 26

26(25). Convex, giving a rather hump-backed appearance, with sutural intervals slightly raised; with 3rd or 4th elytral stria converging and merging with 2nd or 3rd stria at about apical third; major striae entire, extending to elytral apex; antennae with 10-11 segments, last 3 somewhat enlarged; apex of 5th abdominal sternite usually somewhat truncate or emarginate; tarsal claws relatively slender; inhabiting cold water .... *Heterlimnius*

— Less convex; sutural interval usually not raised; elytral striae not ordinarily merging as described above, either being entire or becoming obsolete in posterior portion of elytra; antennae 11-segmented, the last 3 less enlarged; apex of 5th abdominal sternite usually evenly rounded; claws somewhat larger and more curved; inhabiting warmer waters .... *Optioservus*

Classification of the Genera of the United States

Elmidae Curtis 1830

Elminae Curtis 1830

Ancyronychini Ganglbauer 1904

Ancyronyx Erichson 1847, 1 sp., *A. variegatus* (Germar 1824), eastern states.

Elmini Curtis 1830

Ampumixis Sanderson 1954, 1 sp., *A. dispar* (Fall 1925), Washington, Oregon, California.

Atractelmis Chandler 1954, 1 sp., *A. wawona* Chandler 1954, California, Oregon, Idaho.

Cleptelmis Sanderson 1954, 1 sp., *C. addenda* (Fall 1907), western states.

Cylloepus Erichson 1847, 2 spp., Arizona.

Dubiraphia Sanderson 1954, 11 spp., generally distributed.

Gonielmis Sanderson 1954, 1 sp., *G. dietrichi* (Musgrave 1933), Tennessee, Florida to Louisiana.

Heterelmis Sharp 1882, 5 spp., California, southwestern states.

Heterlimnius Hinton 1935, 2 spp., western states.

Hexacylloepus Hinton 1940, 1 sp., *H. ferrugineus* (Horn 1870), New Mexico, Texas, Oklahoma.

Huleechius Brown 1981, 1 sp., *H. marroni* Brown 1981, Arizona.

Macrelmis Motschulsky 1859, 3 spp., Texas, New Mexico, Arizona.

Microcylloepus Hinton 1935, 6 spp., generally distributed.

Narpus Casey, 1893 3 spp., Washington, Oregon, California, Nevada, Wyoming, Utah.

Neocylloepus Brown 1970, 1 sp., *N. boeseli* Brown 1970, Texas, Arizona.

Neoelmis Musgrave 1935, 1 sp., *N. caesa* (LeConte 1874), Texas, Oklahoma.

Optioservus Sanderson 1954, 13 spp., generally distributed.

Ordobrevia Sanderson 1953, 1 species, *O. nubifera* (Fall 1901), Washington, Oregon, California.

Oulimnius des Gozis 1886, 2 spp., eastern states.

Promoresia Sanderson 1954, 2 spp., eastern states.

Rhizelmis Chandler 1954, 1 sp., *R. nigra* Chandler 1954, California.

Stenelmis Dufour 1835, 33 spp., generally distributed.

Xenelmis Hinton 1936, 1 sp., *X. sandersoni* Brown 1985, Arizona.

Macronychini Mulsant and Rey 1872

Macronychus Mueller 1806, 1 sp., *M. glabratus* Say 1825, eastern states.

Zaitzevia Champion 1923, 2 spp., western states.

Larainae LeConte 1861

Lara LeConte 1852, 2 spp., western states.

Phanocerus Sharp 1882, 1 sp., *P. clavicornis* Sharp 1882, Texas.

Bibliography

BROWN, H. P. 1970. *Neocylloepus*, a new genus from Texas and Central America. Coleopterists Bulletin, 24: 1-28.

BROWN, H. P. 1972a. Synopsis of the genus *Heterelmis* Sharp in the United States, with description of a new species from Arizona (Coleoptera: Dryopoidea: Elmidae). Entomological News, 83: 229-238.

BROWN, H. P. 1972b. Aquatic Dryopoid Beetles (Coleoptera) of the United States. Biota of Freshwater Ecosystems Identification Manual No. 6. Water Pollution Conference Series, United States Environmental Protection Agency, Washington, District of Columbia. 82 pp.

BROWN, H. P. 1981. *Huleechius*, a new genus of riffle beetle from Mexico and Arizona (Coleoptera, Dryopoidea, Elmidae). Pan-Pacific Entomologist, 57: 228-244.

BROWN, H. P. 1983. A Catalog of the Coleoptera of America North of Mexico. Family: Elmidae. United States Department of Agriculture. Agriculture Handbook No. 529-50. 23 pp.

BROWN, H. P. 1984. Neotropical dryopoids, III. Major nomenclatural changes affecting *Elsianus* Sharp and *Macrelmis* Motschulsky, with checklist of species (Coleoptera: Elmidae: Elminae). Coleopterists Bulletin, 38: 121-129.

BROWN, H. P. 1987. Biology of riffle beetles. Annual Review of Entomology, 32: 253-273.

BROWN, H. P. 1991. Elmidae (Dryopoidea). Pp. 404-407. *In*: F. W. Stehr, ed. Immature Insects. Volume 2. Kendall/Hunt. Dubuque, Iowa. 975 pp.

BROWN, H. P. and D. S., WHITE. 1978. Notes on separation and identification of North American riffle beetles (Elmidae). Entomological News, 89:1-13.

FINNI, G. R. and B. A. SKINNER. 1975. The Elmidae and Dryopidae (Coleoptera: Dryopoidea) of Indiana. Journal of the Kansas Entomological Society, 48: 388-395.

LESAGE, L. and P. P. HARPER. 1976. Cycles biologique d' Elmidae (Coléoptères) de Ruisseaux des Laurentides, Québec. Annales de Limnologie, 12: 139-174.

LESAGE, L. and P. P. HARPER. 1976. Descriptions de nymphes d' Elmidae néarctiques (Coléoptères). Canadian Journal of Zoology, 54: 65-73.

LESAGE, L. and P. P. HARPER. 1977. Description de cinq espèces de larves d' Elmidae néarctiques. Annales de la Société Entomologie du Québec, 22: 18-32.

MINGO, T. M. 1979. Distribution of aquatic Dryopoidea (Coleoptera) in Maine. Entomological News, 90: 177-185.

SANDERSON, M. W. 1938. A monographic revision of the North American species of *Stenelmis*. University of Kansas Science Bulletin, 25: 635-717.

SANDERSON, M. W. 1953. A revision of the Nearctic genera of Elmidae. Journal of the Kansas Entomological Society, 26: 148-163.

SANDERSON, M. W. 1954. A revision of the Nearctic genera of Elmidae. Part II. Journal of the Kansas Entomological Society, 27: 1-13.

SHEPARD, W. D. 1992a. Riffle beetles (Coleoptera: Elmidae) of Death Valley National Monument, California. Great Basin Naturalist, 52: 378-381.

SHEPARD. W. D. 1992b. A redescription of *Ordobrevia nubifera* (Fall) (Coleoptera: Elmidae). Pan-Pacific Entomologist, 68: 140-143.

SHEPARD, W. D. 1993. An annotated checklist of the aquatic and semiaquatic dryopoid Coleoptera of California. Pan-Pacific Entomologist, 69: 1-11.

SHEPARD, W. D. and C. B. BARR. 1991. Description of the larva of *Atractelmis* (Coleoptera: Elmidae) and new information on the morphology, distribution, and habitat of *Atractelmis wawona* Chandler. Pan-Pacific Entomologist, 67: 195-199.

STEEDMAN, R. J. 1983. The pupa of the elmid beetle *Lara avara* (Coleoptera: Dryopoidea: Elmidae). Aquatic Insects, 5: 17-19.

STEEDMAN, R. J. and N. H. ANDERSON. 1985. Life history and ecological role of the xylophagous aquatic beetle, *Lara avara* LeConte (Coleoptera: Elmidae). Freshwater Biology, 15: 535-546.

WHITE, D. S. 1978a. Life cycle of the riffle beetle *Stenelmis sexlineata*. Annals of the Entomological Society of America, 71: 121-125.

WHITE, D. S. 1978b. A revision of the Nearctic *Optioservus* (Coleoptera: Elmidae), with descriptions of new species. Systematic Entomology, 3: 59-74.

44. DRYOPIDAE Billberg 1820

by William D. Shepard

Family common name: The long-toed water beetles

Family synonym: Parnidae Leach 1817

The short antennae with most segments broader than long, and the aquatic association serve to separate adults of this family.

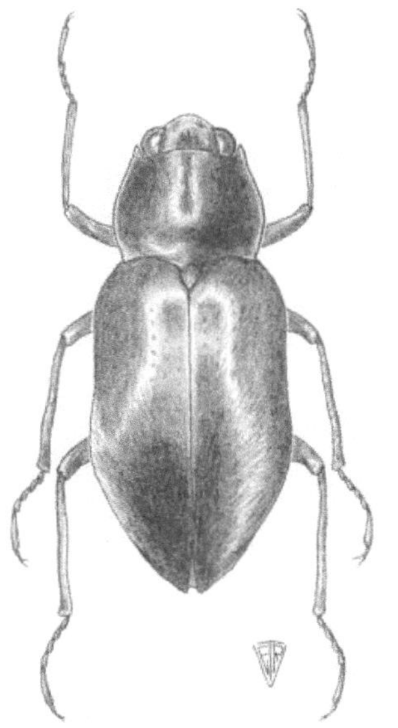

FIGURE 1.44. *Helichus lithophilus* (Germar)

Description: Elongate, oval, convex; 1-8 mm long; color dull silver gray, brown, or piceous; pubescence inconspicuous or conspicuously hairy.

Head deflexed, inserted into prothorax; surface rugose-punctate. Eyes lateral, small, rounded; rarely absent. Antennae 8-11 segmented, clavate, 2nd segment mostly ear-like and covering other segments, 3rd small, remaining segments laterally produced, forming a loose pectinate to lamellate club; inserted under a prominent frontal ridge. Labrum narrow, arcuate; mandibles small, stout, flattened, curved, apices denticulate; maxillae with the lacinia acute, internally setiferous; galea lobate, apically tufted; maxillary palpi 4-segmented, the apical segment large, fusiform; labium with the gula quadrate, the gular sutures distinct; mentum transverse, lobed laterally; labial palpi 3-segmented, the apical segment large, fusiform.

Pronotum larger than the head, ovate to rectangular; borders laterally arcuate, margined, anteriorly very broadly emarginate, posteriorly sinuate; surface rugose-punctate; pleural region long, narrow; prosternum quadrate, long and broad in front of the coxae, forming a broad lobe, prosternal process inserted into a mesosternal groove; procoxal cavities broadly open behind. Mesosternum narrow; mesocoxal cavities closed behind. Metasternum long and broad. Legs with anterior coxae rounded, small, separate; middle coxae rounded, small, separate; hind coxae transverse, separate; trochanters moderate, triangular; femora slender; tibiae slender, apical spurs small; tarsi slender, the apical spurs small; tarsal formula 5-5-5, segments filiform; claws simple. Scutellum small and triangular, or large and pentagonal. Elytra entire; striae punctate or smooth; epipleural fold narrow, entire. Wings with 1 or 2 anal veins; folding pattern normal, large anal fold present, area C and D large, normal, fold A well marked; may be vestigial.

Abdomen with 5 visible sternites, the sutures may be interrupted at the center; surface finely rugose. Male genitalia with the penis short, acute, parameres compressed; basal piece narrow. Female genitalia forming an elongate blade-like ovipositor; styli small or absent.

Eggs are undescribed.

Larva elongate, subcylindrical, tapering caudally; length 5-12 mm. Head exserted, concealed from dorsal view; epicranial suture usually distinct. Antennae 3-segmented, the distal segment may be very small. Frons, clypeus, and labrum distinct. Mandibles short, tridentate, bidentate, or apically acute, with prostheca modified or absent. Maxilla with inconspicuous cardo, stipes, galea, lacinia, and 4-segmented palpi; in few fused with labium. Labium with fused submentum, mentum, ligula, and small 2-segmented palpi. Stemmata one pair, or reduced to pigment spots, or absent, or rarely, five pairs present. Legs 4-5 segmented; apical, claw-like tarsunguli. Abdomen 9-segmented, the 8th and 9th segments often elongate; operculum present, retractable gills absent, pleura absent or not well developed; the first 5 abdominal sternites with a ventral fold. Spiracles mostly present, annular, perineustic, or bilabiate; on the mesothorax and abdominal segments 1-8, or on the 8th segment only.

Pupa elongate, exarate; dorsum and sides with elongate setae. Abdomen tapering; dorsally with gin-traps on intersegments 2-6; apically with a single, yellow-brown spicula.

Virtually all the morphological descriptions refer to adults. Larvae of North American species have been described by Barr and Spangler (1992), Hinton (1955), and Ulrich (1986), and reviewed by Brown (1991). Pupae have been described by Ulrich (1986) and Hinton (1939).

Habits and habitats. Worldwide most of the species are forest litter inhabitants. In the United States these beetles are aquatic mostly as adults, a few being riparian; larvae are generally terrestrial. The adults are covered with a plastron on truly aquatic species. They do not swim, but crawl about upon the stream substrates or on leaf packs and log jams. Riparian adults are on emergent vegetation in still waters. The larvae are terrestrial. Females lay eggs in the vicinity of the water, or on the stems of aquatic plants. Pupation occurs in streamside dirt. The little that is known about the life cycle indicates that it is univoltine. Adults overwinter in aquatic species. Adults and larvae are herbivorous.

Status of the classification. Worldwide this family is poorly known, and a revision is currently under way. In the United States the taxa are well known and relatively stable. There are five genera and 13 species in North America. Brown (1970, 1972) provided keys for all North American species except *Stygoparnus comalensis* Barr and Spangler. However, in Brown's works the genus *Postelichus* Nelson was still considered a part of *Helichus* Erichson. Most of the North American species have been reviewed by Hinton (1937) and Musgrave (1935). Nelson (1981) discussed problems with species recognition, and later (Nelson 1989) he split several species away from *Helichus* Erichson into the new genus *Postelichus* Nelson.

Distribution. Dryopids occur on all continents except Australia and Antarctica. They are especially diverse in the tropics. The distribution of the North American is described by Brown (1972, 1983), except for the regionally restricted *Stygoparnus comalensis* Barr and Spangler.

Key to the Genera of the United States

1. Eyes vestigial; antennae with 8 segments .... *Stygoparnus*
— Eyes normal; antennae with 11 segments 2

2(1). Pronotum on each side with a conspicuous, complete sublateral longitudinal sulcus .... *Dryops*
— Pronotum without such a sublateral sulcus 3

3(2). Second antennomere not enlarged; antennae pubescent; bases of antennae very close together; 3rd and 4th palpomeres of maxillary palp very elongate; without plastron but extremely pubescent .... *Pelonomus*
— Second antennomere enlarged, heavily sclerotized, forming a shield beneath which remaining antennomeres may be retracted; bases of antennae widely separate; 3rd and, in some, the 4th palpomere of maxillary palp short; parts of body and legs with plastron 4

4(3). Pubescence of last abdominal sternite different from that of preceding sternites .... *Helichus*
— All abdominal sternites similarly pubescent.......... .... *Postelichus*

Classification of the Genera of the United States

Dryopidae Billberg 1820

Pelonomus Erichson 1847, 1 species, *P. obscurus* LeConte 1852, southeastern United States. Riparian.
Oberonus Casey 1893

Dryops Olivier 1791, 1 sp., *D. arizonensis* Schaeffer 1905, Arizona. Riparian.
Parnus Fabricius 1792

Helichus Erichson 1847, 7 spp., generally distributed.
Dryops Leach 1817, not Olivier 1791

Postelichus Nelson 1989, 3 spp. Texas, California and Arizona.

Stygoparnus Barr and Spangler 1992, 1 species, *S. comalensis* Barr and Spangler 1992, Texas. Subterranean, found in spring orifices on the eastern side of the Edwards Aquifer of Central Texas.

Bibliography

BARR, C. B. and P. J. SPANGLER. 1992. A new genus and species of stygobiontic dryoid beetle, *Stygoparnus comalensis* (Coleoptera: Dryopidae) from Comal Springs, Texas. Proceedings of the Biological Society of Washington, 105: 40-54.

BROWN, H. P. 1970. A key to the dryopid genera of the New World. Entomological News, 81: 171-175.

BROWN, H. P. 1972. Aquatic dryopoid beetles (Coleoptera) of the United States. Biota of Freshwater Ecosystems Manual No. 6. Water Pollution Conference Series, United States Environmental Protection Agency. Washington, DC. 82 pp.

BROWN, H. P. 1983. A catalog of the Coleoptera of America North of Mexico. Family: Dryopidae. United States Department of Agriculture. Agriculture Handbook No. 529-49. 8 pp.

BROWN, H. P. 1991. Dryopidae. Pp. 399-401. *In*: F. W. Stehr, ed. Immature Insects. Kendall/Hunt. Dubuque, Iowa. 975 pp.

HINTON, H. E. 1937. *Helichus immsi*, sp. n., and notes on other North American species of the genus. Annals of the Entomological Society of America, 30: 317-322.

HINTON, H. E. 1939. An inquiry into the natural classification of the Dryopoidea, based partly on a study of their internal anatomy (Col.). Transactions of the Royal Entomological Society of London, 89: 133-184.

HINTON, H. E. 1955. On the respiratory adaptations, biology, and taxonomy of the Psephenidae, with notes on some related families (Coleoptera). Proceedings of the Zoolgocial Society of London, 125: 543-568.

MUSGRAVE, P. N. 1935. A synopsis of the genus *Helichus* Erichson in the United States and Canada, with descriptions of a new species (Coleoptera: Dryopidae). Proceedings of the Entomological Society of Washington, 37: 137-145.

NELSON, H. G. 1981. Notes on Nearctic *Helichus* (Coleptera: Dryopidae). Pan-Pacific Entomologist, 57: 226-227.

NELSON, H. G. 1989. *Postelichus*, a new genus of Nearctic Dryopidae (Coleoptera). Coleopterists Bulletin, 43: 19-24.

ULRICH, G. W. 1986. The larvae and pupae of *Helichus striatus* LeConte and *Helichus productus* LeConte (Coleoptera: Dryopidae). Coleopterists Bulletin, 40: 325-334.

45. LUTROCHIDAE Kasap and Crowson 1975

by William D. Shepard

Family common name: The travertine beetles

The general body shape, the dense pubescence, the short antennae with the first two antennomeres long and broad, and the apically elongate mandibles serve to separate adults of this family.

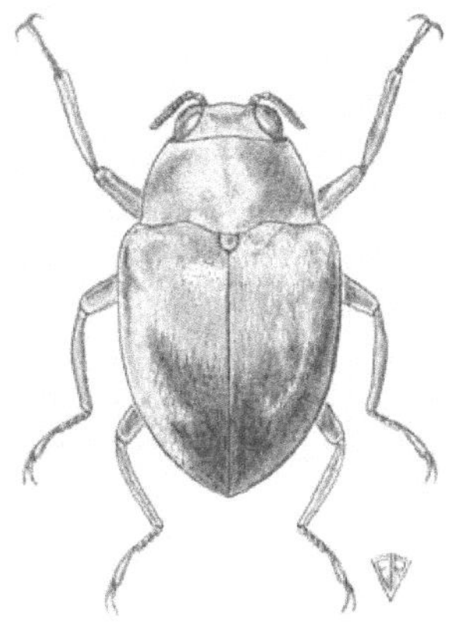

FIGURE 1.45. *Lutrochus luteus* LeConte

Description: Body ovate and strongly convex; 2-6 mm long; color yellowish; pubescence and punctation dense.

Head broad; hypognathous; vertex evenly convex. Antennae 11-segmented, short, not reaching apex of opened mandible; first 2 antennomeres broad and with conspicuous setae; remaining 9 antennomeres short, closely appressed, somewhat clavate, with shorter and straighter setae; length of antennomeres 3-11 subequal to 1-2. Eyes not prominent; densely setose; coarsely faceted. Labrum retractile; apical margin straight to arcuate. Mandibles large, conspicuous, strongly curved and elongate; tridentate; prostheca malleate. Maxillae with palpifer and 4-segmented palpi; galea with dense setal brush; lacinia shorter than galea and with elongate setae. Labium setose; with transverse mentum and submentum; ligula long, transverse, heavily setose; palpi 3-segmented.

Pronotum wider than head; sides convergent from base; disc broadly convex; surface densely punctate and pubescent. Prosternum distinctly transverse, prosternal process half as long as whole prosternum. Mesosternum with groove to receive prosternal process; mesoepisternum and mesoepimeron excavated to receive legs. Metasternum with faint longitudinal line and faint transverse groove at posterior two-thirds. Legs with pro- and metacoxae strongly transverse, coxal cavities open; middle coxae globose; pro- and mesocoxae with trochantin prominent; femora with grooves to receive tibiae; tibiae slender; tarsal formula 5-5-5, segments 1-4 short, 5th elongate, all segments glabrous; claws simple. Elytra densely punctate and setose. Wings with 4 anal veins; venation restricted to basal half.

Abdomen with 3-5 visible sternites, densely setose; 1st sternite deeply incised by coxal cavities and excavate to receive legs. Male genitalia with basal piece elongate, curved ventrally; parameres fused ventrally to each other and basal piece; penis slender and elongate, with acute tip; fibula present, bifurcate at base. Female genitalia an inverted V-shaped ovipositor; apex (fused coxites ?) blunt-tipped; baculae explanate at base.

Eggs are undescribed.

Larva elongate, elateriform; tapering caudally; length 4-10 mm. Head large, prognathous. Antennae 3-segmented, basal segment longest, apical segment subequal to sensory peg. Eyes a loose cluster of 5 stemmata, 6th stemma directed ventrally at base of mandible. Clypeus and labrum transverse. Mandibles strong, with 3 apical teeth; mola absent; mesal basal surface with patch of parallel bristles. Maxilla without palpifer, palp 4-segmented; galea separate from and larger than lacinia; stipes elongate, cardo membranous or absent. Labium with postmentum undivided; palps 2-segmented, with palpifer. Prothorax as long as meso- and metathorax together. Legs well developed; short; 5-segmented. Abdomen 9-segmented; pleura on segments 1-2 or 1-4; segments 4 or 5 to 8 ring-like; segment 9 with operculum covering retractile trachael gills, 2 hooks on dorsum. Spiracles present only in last instar; present on mesothorax and abdominal segments 1-8; biforous.

Pupa exarate, glabrous; anterior prothorax with 1 pair of elongate, sclerotized filiform projections; abdomen tapers to 1 pair of sclerotized urogomphi, posterior borders of segments 1-7 sclerotized.

As with many of the aquatic byrrhoids, most of the descriptive morphology concerns just the adults. However, the larvae are well described by Brown (1991), and Costa *et al.* (1996) describe larvae and pupae.

Habits and habitats. These beetles are aquatic as adults and larvae, and found on submerged and emergent substrates in rapids and other flowing areas of streams. Pupae are found under cover just above the water level. The life cycle is univoltine with the larvae overwintering. Adults and larvae feed on algae and waterlogged wood. In the United States lutrochids are common where travertine is being deposited in streams and near springs. Adults are quick to fly when captured.

Status of the classification. This family is found only in the New World. There is currently only the single genus *Lutrochus*, in which 12 species have been described. However, the genus needs revision, which will result in several northern (United States, Mexico and Belize) species being put into a new genus. There are several undescribed species.

Distribution. The three United States species are from Arizona, Texas and surrounding areas, and in the eastern and midwestern United States (Brown 1972, Brown and Murvosh

1970, Brown and Stoaks 1970). Other species extend southward to Bolivia and Argentina.

Classification of the Genera of the United States

Lutrochidae Kasap and Crowson 1975

Lutrochus Erichson 1847, 3 spp., widely distributed in eastern states, Texas, Arizona.

Bibliography

BROWN, H. P. 1972. Aquatic dryopoid beetles (Coleoptera) of the United States. Biota of Freshwater Ecosystems Identification Manual No. 6. Water Pollution Conference Series, United States Environmental Protection Agency. Washington, DC. 82 pp.

BROWN, H. P. 1991. Lutrochidae (Dryopoidea). Pp. 397-399. *In:* F. W. Stehr, ed. Immature Insects. Kendall/Hunt. Dubuque, IA. 975 pp.

BROWN, H. P. and C. M. MURVOSH. 1970. *Lutrochus arizonicus* new species, with notes on ecology and behavior. Annals of the Entomological Society of America, 63: 1030-1035.

BROWN, H. P. and R. D. STOAKS. 1970. Distribution records for *Lutrochus luteus* LeConte. Entomological News, 81: 7-8.

COSTA, C., S. IDE, S. A. VANIN and É. P. TEIXEIRA. 1996. Larvae of Neotropical Coleoptera. XXIII: *Lutrochus germari* Grouvelle, description of immatures, redescription of adult and bionomics (Dryopoidea, Lutrochidae). Revista Brasileira de Entomologia, 40: 47-56.

46. LIMNICHIDAE Erichson 1846

by William D. Shepard

Family common name: The minute marsh-loving beetles

The small size, dense and in many colorful pubescence, exposed anterior trochantins, and riparian habits serve to separate this family.

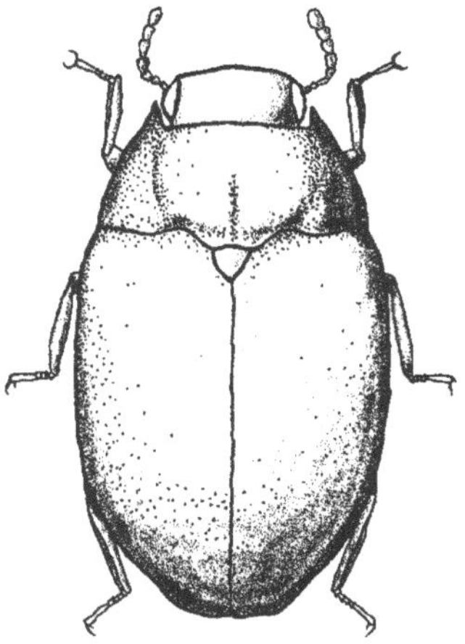

FIGURE 1.46. *Limnichites nebulosus* (LeConte)

Description: Oval to elongate oval, convex; 1-2 mm in length; brownish to piceous, in some metallic; pubescence of dense, fine, golden or grayish setae, flattened scale-like setae may be present.

Head small, inserted into the prothorax; surface finely punctate. Antennae 11-segmented, short, clavate, most antennomeres broad; inserted at the sides of the front near the eyes and the base of the mandibles; 2-7 segmented club. Clypeus distinct, the suture faint; labrum transverse, concealed; mandibles hidden; maxillary palpi 4-segmented, slender; labial palpi 2-3 segmented, slender. Eyes lateral, small, rounded, setose; may be hidden from dorsal view.

Pronotum subquadrate; sides strongly convergent; borders margined; surface punctate; pleural region narrow; prosternum broad, flat to longitudinally sulcate; prosternal process acutely to obtusely pointed; procoxal cavities open behind. Mesosternum short; metasternum broad. Trochantin of the fore legs well developed; anterior coxae transverse; middle coxae rounded; hind coxae transverse, mostly contiguous; trochanters moderate, triangular; femora slender; tibiae slender; tarsal formula 5-5-5, segments filiform; claws simple. Scutellum small, triangular. Elytra entire, convex; surface punctate; epipleural fold narrow.

Abdomen with 5 visible sternites. Male genitalia well sclerotized; basal piece elongate, frequently open dorsally; parameres in many fused with basal piece, elongate, in many extending beyond penis; penis slender, in many blunt apically; fibula absent. Female genitalia elongate, well-sclerotized, of the "ovipositor-type", the styli absent.

Eggs are undescribed.

Larva small in size (less than 5 mm), elongate, subcylindrical, terminating bluntly; pale brown in color; posterior margins of thoracic and first 8 abdominal segments with longitudinally striated borders. Head prognathus; partially retracted into prothorax. Antennae short, 3-segmented; 3rd segment with apical spine. Eyes a dorsolateral quadrangle of stemmata behind antennae; ventrally directed stemmata present or absent. Labrum and clypeus transverse. Mandible apically bidentate; excavate medially; without mola or basal process. Maxillae with 4-segmented palps; galea and lacinia stubby, with apical clusters of spines. Labium transverse; palpi 2-segmented. Thorax little wider than head; legs short, 5-segmented. Abdomen cylindrical, parallel-sided; rounded and decurved apically; segments 1-7 with broad membranous sterna, sternum 8 narrower, sternum 9 operculaform; gills lacking. Larvae are described by Brown (1991).

Pupae are undescribed.

Habits and habitats. The adults are generally riparian, on streamside plants, emergent vegetation and wood or on windrows of debris. *Throscinus* adults are intertidal along the Pacific and Gulf coasts. Larvae live in damp soil or humus near streams or other water bodies. The life cycle is probably univoltine. Most species are thought to be herbivorous.

Status of the classification. New World limnichids have been excellently revised by Wooldridge (1975, 1976, 1977, 1978, 1979, 1981a, 1981b, 1986). Casey (1889) was important in establishing genera in the Limnichinae. Taxa from the United States are now well known and stable. In contrast, Old World limnichids are poorly known, including those from Europe!

Distribution. Worldwide there are approximately 40 genera and 225 species arranged in 4 or 5 subfamilies. They occur on all continents except Antartica. There are 6 genera and 28 species in the United States.

KEY TO THE GENERA OF THE UNITED STATES

1. Body elongate; eyes large and dorsally prominent, venter not grooved for reception of the legs ... . *Throscinus*
— Body oval; eyes mostly small and not prominent; venter grooved for reception of legs 2

2(1). Pronotum with a deep excavation on each side of the head to receive antennae; size about 0.8-1.1 mm .... *Physemus*
— Pronotum not excavated; size variable, but mostly larger 3

3(2). Elytral setae of two types - dense, short decumbent setae and scattered, long, upright setae .. .. *Limnichoderus*

— Elytral setae all alike, either short and recumbent or long and upright 4

4(3). Eyes prominent, visible from above .... *Limnichites*
— Eyes vertical, flattened, not visible from above . 5

5(4). Prosternal process with a median longitudinal sulcus .... *Eulimnichus*
— Prosternal process without a median longitudinal sulcus .... *Lichminus*

Classification of the Genera of the United States

Limnichidae Erichson 1846

Cephalobyrrhinae Champion 1925

Throscinus LeConte 1874, 3 spp., Texas and California. Intertidal on mud flats.

Limnichinae Erichson 1846

Bothriophorini Mulsant and Rey 1868

Physemus LeConte 1854, 1 species, *P. minutus* LeConte, 1854, California, Arizona and Texas.
Ditaphrus Casey 1886

Limnichini Erichson 1846

Eulimnichus Casey 1889, 12 spp., generally distributed.

Lichminus Casey 1889, 1 species, *L. tenuicornis* (Casey 1889), California, Oregon and Washington.

Limnichites Casey 1889, 6 spp., generally distributed.

Limnichoderus Casey 1889, 5 spp., southern United States.

Bibliography

BROWN, H. P. 1991. Limnichidae (Dryopoidea). Pp. 401-402. *In*: F. W. Stehr, ed. Immature Insects. Volume 2. Kendall/Hunt. Dubuque, IA. 975 pp.

CASEY, T. L. 1889. II.--Coleopterological notices. I. With an appendix on the termitophilus Staphylinidae of Panama. Annals of the New York Academy of Sciences, 5: 39-198.

WOOLDRIDGE, D. P. 1975. A key to the New World genera of the beetle family Limnichidae. Entomological News, 86: 1-4.

WOOLDRIDGE, D. P. 1976. New World Limnichinae I. A revision of the genus *Physemus*. Coleopterists Bulletin, 30: 177-182.

WOOLDRIDGE, D. P. 1977. New World Limnichinae III: a revision of *Limnichites* Casey. Great Lakes Entomologist, 10: 179-189.

WOOLDRIDGE, D. P. 1978. New World Limnichinae IV: *Eulimnichus* Casey. A. Synonymies, lectotype designations and redescriptions. Great Lakes Entomologist, 11: 163-173.

WOOLDRIDGE, D. P. 1979. New World Limnichinae IV. *Eulimnichus* Casey. B. Descriptions of new species. Great Lakes Entomologist, 12: 1-11.

WOOLDRIDGE, D. P. 1981a. New World Limnichinae VI. A revision of *Limnichoderus* Casey (Coleoptera: Dryopoidea: Limnichidae). Journal of the Kansas Entomological Society, 54: 171-191.

WOOLDRIDGE, D. P. 1981b. Three new species of *Throscinus* LeConte, with notes on other species (Coleoptera: Limnichidae: Cephalobyrrhinae). Coleopterists Bulletin, 35: 217-222.

WOOLDRIDGE, D. P. 1986. A Catalog of the Coleoptera of America North of Mexico. Family: Limnichidae. United States Department of Agriculture. Agriculture Handbook No. 529-48. 8 pp.

47. HETEROCERIDAE MacLeay 1825

by Kerry Katovich

Family common name: The variegated mud-loving beetles

Family synonyms: Acanthapoda Latreille 1829, Heterocérites Castelnau 1840, Heterocerida Heer 1841, Heteroceri Redtenbacher 1849, Heterocérides Lacordaire 1854, Heterocerini Schiødte 1866, Spinipedes Mulsant and Rey 1872, Heterocériens Lameere 1900

Heteroceridae are a widespread and morphologically uniform family. They are easily recognized by the large mandibles, well developed comb of flattened spines on the dorsal margin of the prothoracic and mesothoracic tibiae, and the first visible ventrite with paired, arching stridulatory files. Adult and larval heterocerids are associated with riparian or moist sand and mud habitats.

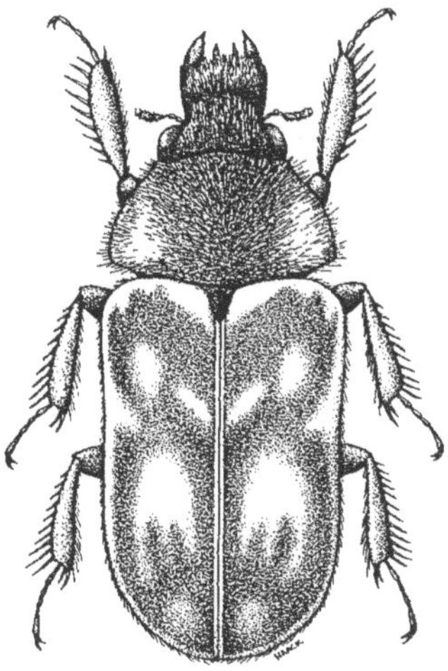

FIGURE 1.47 *Lanternarius brunneus* (Melsheimer) (from Clark and Ratcliffe 1989, used with permission)

Description: (adapted in part from Crowson 1955, Arnett 1971, Pacheco 1964, Lawrence *et al.* 1999) Body elongate, slightly flattened to moderately convex, minute to medium-sized (1-8 mm long). Dorsal surface with distinct hairs or setae. Color variable, light-brown or blackish; elytral disk often variegated, occasionally unicolorous; variegations (when present) usually with undulating yellowish or whitish bands or spots.

Head not or slightly deflexed, visible or partially so from above. Eyes well developed, not or slightly protuberant, finely facetted and lacking interfacetal setae. Antennae short, when fully extended, not reaching middle of prothorax; antennae with 9 -11 antennomeres; first antennomere less than three times the length of 2nd antennomere; antennomeres 6-7 compact, serrate (Augyliini, Heterocerini, Tropicini) or clubbed (Micilini and Elythomerini). Labrum visible, free, membranous or separated by suture line; apex of labrum subtruncate, convex, narrowly rounded or acute, occasionally abruptly produced into an elongate process. Frontoclypeal suture impressed, straight, weakly curved or angulate. Mandibles moderate to well developed; some genera with hypermandibulate males; mandibular apex gradually, moderately or strongly curved mesally, unidentate, bidentate, bilobed, truncate, or rounded; incisor edge of mandible with single, double, or multiple teeth present; mandible with well-developed mola; prostheca well-developed or reduced, if present, bearing a row of tooth-like prolongations, and divided by a notch medially. Maxilla with distinct galea and lacinia; maxillary palpi with apical palpomere cylindrical to fusiform. Ligula deeply emarginate or bilobed. Labial palpi slender, 3-segmented. Maxillary and labial lengths correlated with mandibular length.

Pronotum slightly wider than long, widest medially; lateral sides moderately to strongly curved, weakly or not explanate. Base of prothorax equal to or slightly narrower than base of elytra. Lateral pronotal carinae absent or incomplete; carinae, if present, not raised. Prosternum broadly lobed anteriorly, forming a "chin-piece". Prosternal process gradually expanded, or narrowed then expanded; process strongly elevated and curved dorsally behind coxae, slightly to extensively covering mesosternum, overall forming an elevated keel, separating the mesothoracic coxae. Procoxae slightly projecting below prosternum. Procoxal cavities strongly transverse, broadly open behind externally and internally; protrochantin partially exposed. Scutellum well-developed, triangular. Mesosternum narrow, not divided by a longitudinal suture. Mesocoxal cavities circular to transverse, open laterally, not closed by metepisterna, moderately to widely separated. Mesocoxae not projecting. Metacoxal cavities transverse, contiguous, or narrowly separated. Metacoxae not projecting.

Legs fossorial, with mesothoracic and metathoracic femora similar in size, somewhat swollen. Prothoracic and mesothoracic tibiae enlarged and armed with a row of well developed spines; metathoracic tibiae similar, spines less developed. Tarsal formula 4-4-4; tarsomeres filiform, ventrally with long, sparse hairs; pretarsus with long, slender claws.

Elytra entire, convex, posteriorly rounded; surface irregularly punctate, not punctate, or punctures and striae indistinct. Metathoracic wings well developed. Folding pattern dryopoid-like (*sensu* Crowson 1955). Radial cell of hind wing highly reduced or absent. Medial area of hind wing with three or fewer free veins; oblongum cell absent; wedge cell and anal lobe absent. Posterior edge of wing without fringe of long hairs.

Abdomen with five visible ventrites; sutures complete, occasionally indistinct medially; first visible ventrite with stridulatory file present, arching from antero-lateral corner toward meson; ninth abdominal sternum of the male invaginated, usually well sclerotized; modified into two elongate arms, with the anterior ends approximated, forming a single, anterior arm; ninth abdominal sternum with three distinct forms: U-shaped (Micilini), inverted Y-shaped (Tropicini), and inverted V-shaped (Elythomerinae, Augyliini and Heterocerini); ninth abdominal sternum of females not visible.

Male genitalia (terminology following Pacheco 1964) divided into two pieces (appearing single in Tropicini): the tegmen (phallobase and parameres) and the aedeagus (dorsal plate, anteriorly produced median strut and an internal sac). The phallobase may be divided into a basal piece (formed by two fused or separated, transverse, sclerotized bands) and median plate; the lateral portion of the median plate is designated as the lateral arms of the phallobase. Parameres present or absent, if present, membranous, arising from the basal piece of the phallobase. Internal sac generally well sclerotized dorsally. Female genitalia completely membranous, lacking visible coxites and styli.

Larval body elongate, campedeiform, subcylindrical, widest in thoracic region, tapering posteriorly; 2-10 mm long; dorsal body surface darkly pigmented, smooth; vestiture of short and long hairs. Head prognathous, broad, and protracted. Five stemmata on each side. Antennae very short, 3-segmented, with large, bulbous sensorium on 2nd segment longer than reduced 3rd segment. Labrum large and free. Frontoclypeal suture present. Mandibles symmetrical, somewhat flattened, bidentate. Ventral mouthparts retracted. Legs well developed, 5-segmented; tarsungulus with single seta. Abdominal tergum 9 simple, without urogomphi. Operculum absent. Abdominal segment 10 well-developed, its sternum forming a conical pygopod; tergum of 10th segment reduced; anal region posteriorly or posteroventrally oriented. Spiracles cribiform, appearing annular-uniforous.

Habits and habitats. Natural history information is available for relatively few species, however, Heteroceridae appear to be homogeneous in adult and larval habitats. In general, adults are most commonly associated with riparian habitats, although they are strong flyers and can be found some distance from water. Adults feed in shallow, horizontal tunnels excavated in mud or moist organic sand. Adult stomach contents suggest this family feeds on algae, plankton, and organic material (Silvey 1935). Kaufmann and Stansly (1979) described four distinct tunnel types in *Neoheterocerus pallidus* (Say): feeding galleries, pupal cells, egg chambers, and hibernacula. The tunnels of several individuals may intersect, hence these beetles are often gregarious as adults and larvae. Claycomb (1919) noted that several species may co-exist in a given habitat, and their tunnel systems may intersect. Tunnel construction has been documented in detail (Kaufmann and Stansly 1979, Clark and Ratcliffe 1989). Mating apparently occurs within the tunnel system.

Eggs are small, oblong, whitish to pale yellow, turning opaque as they mature. They are deposited together in large numbers, averaging 40-60 eggs within an egg chamber. The concentration of eggs is believed to be related to egg guarding behavior, as demonstrated in female *Lapsus tristis* (Mannerheim) (Folkerts 1989). When confronted by an intruder within the egg chamber, *L. tristis* raised its body, opened and closed its mandibles, and waved its antennae. This defense is likely directed against *Ellipes gurneyi* Gunther (Orthoptera: Tridactylidae), which often lives in close association and frequently use the same tunnel systems. Females apparently produce two broods; in *N. pallidus* a third brood is reabsorbed in the egg stage to provide a fat reserve for the winter months or adverse whether conditions. Adults may pass these periods in hibernacula. Diapause in northern latitudes usually occurs from October to April. The egg stage lasts approximately 3 days.

Larvae radiate out from the parental tunnel systems, forming their own tunnel systems or using existing tunnels. Larval food is similar to that of the adult. The larval stage is approximately seven days, with approximately four instars. Pupation ranges from three to six days.

The ecology of Heteroceridae is apparently closely tied to the rapidly changing riparian habitat. Frequent fluctuations in water level may cause heavy mortality in eggs, pupae, and larvae; adults are not usually affected (Kaufmann and Stansly 1979). For this reason, population levels may fluctuate dramatically in a given region. This may explain why heterocerids have fairly rapid egg to adult development, a long adult life stage, and overlapping generations. The presence of a breeding generation at any given time may enable survival of local populations.

Status of the classification. Lameere (1900) placed Heteroceridae in his Canthandiformia. Kolbe (1901) placed Heteroceridae under the superfamily Dascilloidea in the suborder Heterophaga. Ganglbauer (1904) placed Heteroceridae under the series Diversicornia. Based on wing folding and venation, Forbes (1926) placed the family under the superfamily Dryopoidea in the series Dryopiformia. Bøving and Craighead (1931) associated the family with Dascilloidea based on larval characters. Jeannel and Paulian (1944) placed the family under the division Cucujoidea, section Cucujaria. Based on adult and larval characters, Crowson (1955) placed Heteroceridae with the superfamily Dryopoidea in the series Dascilliformia. This position was supported by Sanderson (1953), who indicated that larval and adult Heteroceridae show a close relationship to the families Dryopidae, Limnichidae, and Elmidae. Later, Crowson (1981) suppressed Dascilliformia, placing Heteroceridae under Dryopoidea in the series Elateriformia. Currently, Lawrence and Newton (1995) placed Heteroceridae under the superfamily Byrrhoidea in the series Elateriformia; they further elevated Pacheco's (1964) Elythomerini and Heterocerini to subfamily status.

LeConte addressed Heteroceridae in the first true treatment of this family for North America north of Mexico in a series of works (1863-1866), discussing 15 species and describing three new species. Horn (1890) revised *Heterocerus* for "Boreal America" providing useful descriptions and illustrations of 11 species. Zaitzev (1910) provided the most recent catalogue of world Heteroceridae, listing three genera with 133 species, 27 of which were recorded from the New World. Leng (1920), Leng and Mutchler (1927, 1933) and Blackwelder (1939) catalogued the Heteroceridae of North America north of Mexico, listing 16 species under the genus *Heterocerus*. Blackwelder (1944-1957) catalogued the Heteroceridae of Mexico, Central America, the West Indies, and South America, listing 13 species under the genus *Heterocerus*. Pacheco (1964) monographed the New World Heteroceridae, dividing the family into five tribes and 20 genera. Hatch (1965) provided keys to ten species of *Heterocerus* in the Northwest. Downie and Arnett (1996) provided keys to *Augyles*, *Heterocerus*, and *Tropicus* species of northeastern North America.

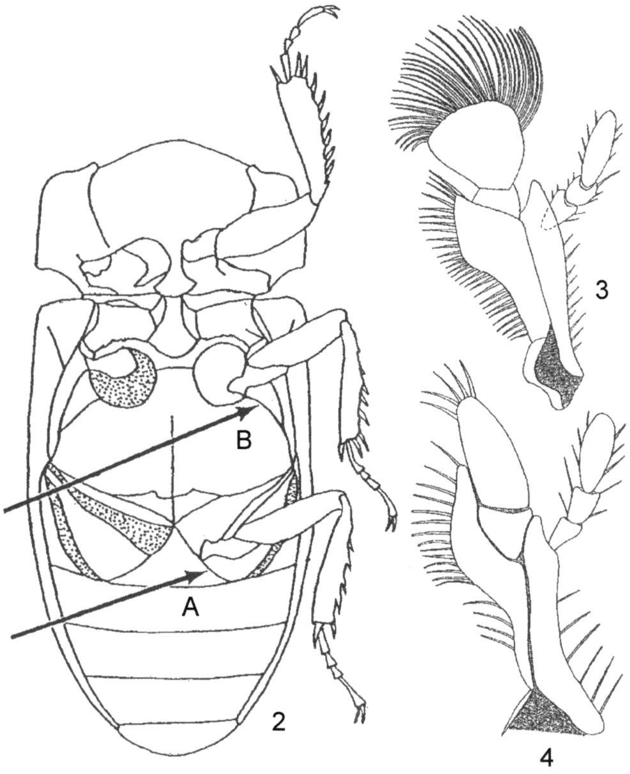

FIGURES 2.47-4.47. 2. Generalized Heteroceridae, ventral view; A. post-metacoxal line, B. post-mesocoxal line; 3. *Centuriatus auromicans* (Kiesenwetter), maxilla, dorsal view; 4. *Explorator canadensis* (Fall), maxilla, dorsal view.

Pacheco (1978) provided the most recent cataloge of Heteroceridae north of Mexico.

Charpentier (1967) noted that breaking up *Heterocerus* into numerous genera based on genitalic characters (*sensu* Pacheco 1964) was unnecessary. Miller (1988) stated that in Pacheco's genera, a high degree of variation exists in the male genitalia. Consequently, Miller retained *Heterocerus* (Pacheco's *Lapsus*, *Lanternarius*, *Dampfius*, *Neoheterocerus*, *Peditatus*, and *Efflagitatus*), *Augyles* (Pacheco's *Centuiatus*, *Microaugyles*, and *Explorator*), and *Tropicus* for North America, north of Mexico.

Currently there is no consensus on the tribal or generic level classification of the family. Recent work by several authors has shown some disparity in classification, some recognizing Pacheco's classification (Kaufmann and Stańsly 1979, Folkerts 1989), others retaining a more traditional classification (Hatch 1965, Downie and Arnett 1996). Pacheco's (1964) review of the Nearctic, Caribbean, and South American faunas remains the only work of its kind and scope. Heteroceridae has been reviewed for the Notogean and Ethiopian regions (Charpentier 1965, 1967). However, no formal phylogenetic revision for the world has been conducted. Lacking any competing classification schemes based on a complete analysis of New World taxa, Pacheco's work is followed in this chapter. Further work at the generic and tribal level is clearly needed to resolve this issue.

The fossil record of Heteroceridae is rather limited. Ponomarenko (1985) described a new genus, *Heterocerites*, from the lower cretaceous of Mongolia. Clark and Ratcliffe (1989) discussed the potential significance of trace fossils i.e., tunneling in paleoecological work.

Distribution.: Heteroceridae are found in all major zoogeographic regions, and consists of approximately 21 genera (20 extant) (Pacheco 1964) and 300 species. Elythomerini (monotypic: *Elythomerus* Waterhouse) are restricted to Australia. Micilini (monotypic: *Micilus*) are Palearctic. Heterocerini and Augyliini are distributed worldwide (absent from the Hawaiian Islands). Tropicini occur in North and South America. Three tribes (10 genera and 34 species) are discussed for North America north of Mexico below. Distributional information for genera is provided in the generic treatment below.

Keys to the North American Genera
(After Pacheco 1964)

In this key to genera of North America, only males can be identified beyond *Neoheterocerus*.

1. Post-metacoxal lines present (Fig. 2A) (Augyliini) 3
— Post-metacoxal lines absent 2

2(1). Antennae 11-segmented; usually medium to large in size (Heterocerini) 5
— Antennae 9-segmented; small in size, uniform in color, often with a whithish elytral margin (Tropicini) .... *Tropicus*

3(1). Antennae10-segmented .... *Microaugyles*
— Antennae 11-segmented 4

4(3). Galea broad, flattened, appearing as a golden brush of stout setae (Fig. 3) .... *Centuriatus*
— Galea not as above, galea with 5 well developed, white setae (Fig. 4) .... *Explorator*

5(2). Post-mesocoxal lines absent........ *Neoheterocerus*
— Post-mesocoxal lines present (Fig. 2B) 6

6(5). Male genitalia without membranous parameres; median plate (Fig. 5A) and basal piece of phallobase (Fig. 5C) separated by a sulcate line (Fig. 5B, 6) .... *Peditatus*
— Male genitalia with membranous parameres present; median plate and basal piece or pieces of phallobase clearly differentiated or continuous, not separated by a sulcate line 7

7(6). Male genitalia with median plate and basal piece of phallobase continuous (Fig. 7), basal piece often produced posteriorly; parameres relatively large and flexible; lateral lobes of phallobase arising from a central point, distally flexed inward .... *Dampfius*
— Male genitalia without the above combination of characters 8

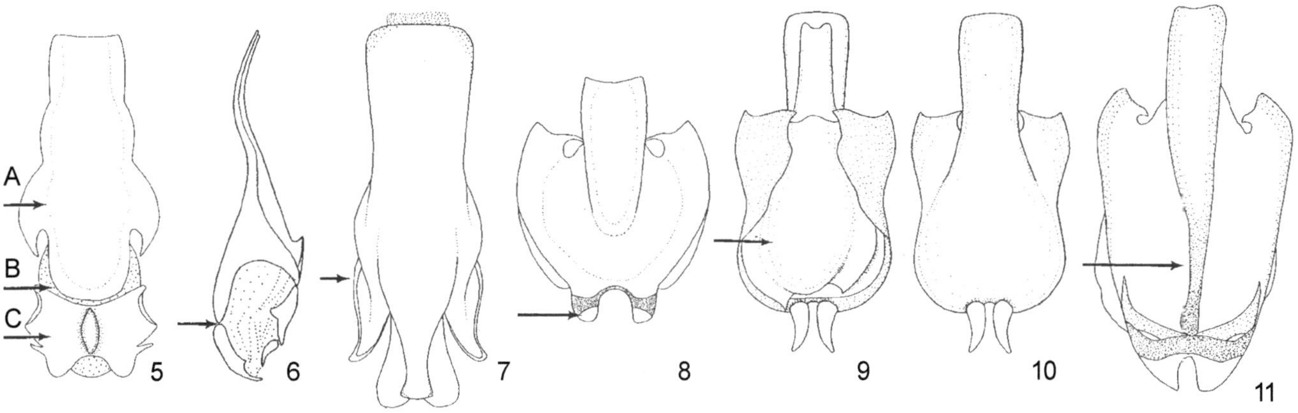

FIGURES 5.47-11.47. *Peditatus schwarzi* (Horn), male genitalia, ventral view; A. median plate of phallobase, B. transverse sulcate line, C. basal piece of phallobase; 6. *P. schwarzi*, arrow highlights sulcate line of male genitalia, lateral view; 7. *Dampfius collaris* (Kiesenwetter), male genitalia, arrow highlights lateral lobe of phallobase, ventral view; 8. *Lapsus tristis* (Mannerheim), arrow highlights male parameres, ventral view; 9. *Efflagitatus selanderi* Pacheco, male genitalia, arrow highlights dorsal plate of aedeagus, dorsal view; 10. *E. selanderi*, male genitalia, ventral view; 11. *Lanternarius brunneus* (Melsheimer), male genitalia, arrow highlights tapering medial plate of phallobase, ventral view.

8(7). Male genitalia with basal piece of phallobase about one half the width of phallobase; parameres rounded and relatively small (Fig. 8) .... *Lapsus*
— Male genitalia not as above 9

9(8). Male genitalia with parameres usually very small (Fig. 9, 10); dorsal plate of aedeagus usually elongate, distorted, and composed of several differentiated areas which vary in the degree of sclerotization .... *Efflagitatus*
— Male genitalia with parameres about as broad as long; median plate of phallobase usually tapering posteriorly (Fig. 11) .... *Lanternarius*

Classification of the Genera of America North of Mexico

Number of species, relevant keys, and distributional information are provided for North America north of Mexico. Traditional generic classification is noted where appropriate. Additional information pertinent to each genus is provided. Distributional information for the species should not be considered complete, and is only reflective of previous, limited works on this family.

Heteroceridae MacLeay 1825

Heterocerinae MacLeay 1825

Augyliini Pacheco 1964

Augyliini are distinguished by the following characters: post-mesocoxal and post-metacoxal lines present (Fig. 2A, 2B); antennae10-11 segmented. Five genera are currently placed under this tribe by Pacheco (1964). Three genera occur north of Mexico (see below); the genera *Taenheterocerus* (Kiesenwetter) and *Augyles* Schiødte are Palearctic and European, respectively. *Augyles* is considered present in North America under the traditional classification scheme *sensu* Miller (1988).

Explorator Pacheco 1964
Monotypic, *E. canadensis* (Fall), known from Alaska to southern Canada. A second species may be present in this genus. *Explorator* is distinguished from other Augyliini by the epipleural line well marked, and galea with five erect, white setae (Fig. 4).
Augyles Schiødte 1866, in part, *sensu* Miller 1988

Centuriatus Pacheco 1964
Two species: *C. auromicans* (Kiesenwetter), southern Canada south to Texas; *C. compactus* (Fall), southern Canada south to the north-central United States. Key to species in Pacheco 1964. *Centuriatus* is distinguished from other Augyliini by the squamiform hairs on its dorsal surface, epipleural line well marked, compact body, and galea flat, short, and tufted with wrinkled, golden setae (Fig. 3).
Augyles Schiødte 1866, in part, *sensu* Miller 1988

Microaugyles Pacheco 1964
Two species: *M. moleculus* (Fall), southern Canada to central United States; *M. mundulus* (Fall), western United States. Key to species in Pacheco 1964. *Microaugyles* is distinguished from other Augyliini by its unusually small size, poorly sclerotized male genitalia, prostheca small, with 12-17 setae, galea finger-like, with 4-6 long, curved, different sized setae, antennae10-segmented, and post-metacoxal lines curving gently.
Augyles Schiødte 1866, in part, *sensu* Miller 1988

Heterocerini Pacheco 1964

Heterocerini are distinguished by the following characters: antennae11-segmented, and post-metacoxal lines absent. Eleven genera are currently placed under this tribe (Pacheco 1964). Six genera are present north of Mexico (see below), with five Neotropical genera: *Culmus* (Mexico), *Olmedous* (Mexico), *Gradus* (Chile), *Erus* (southern Brazil to northeastern Argentina), and *Filiolus* (Cuba). Three species of Heterocerini remain *incertae sedis* (Pacheco 1978).

Lanternarius Pacheco 1964
Approximately six species, widespread across Canada and the United states. Key to species in Pacheco 1964. *Lanternarius* is distinguished by the usually black and shining male genitalia, lobate, usually triangular parameres, basal piece and lateral arms of phallobase heavily sclerotized (Fig. 11), three zig-zag bands on each elytron (these bands forming spots, occasionally vague on older specimens), galea finger-like, with few erect setae, and with oblique pronotal borders.

Heterocerus Fabricius 1792, in part, *sensu* Miller 1988

Neoheterocerus Pacheco 1964
Approximately 13 species, North America to Central Mexico, eleven species north of Mexico, widespread across the United States and Canada. Key to species in Pacheco 1964. *Neoheterocerus* is distinguished from other Heterocerini by the absence of the post-mesocoxal line, the medium to large body size, elytra with zig-zag banding, galea finger-like, with few erect white setae, labrum of hypermandibulate males abruptly produced into an elongate process, and epipleural line absent. Members of this genus will fall into Miller's *Heterocerus* "gnatho group". The natural history of *N. pallidus* was discussed by Kaufmann and Stansly (1979).

Heterocerus Fabricius 1792, in part, *sensu* Miller 1988

Efflagitatus Pacheco 1964
Eight species, primarily South American; one species, *E. selanderi* Pacheco, from Florida. Key to species in Pacheco 1964 and 1969 (modification of 1964 key). *Efflagitatus* is distinguished by small, lobular, usually approximate parameres, aedeagus usually elongate and distorted in appearance (Figs. 9, 10), galea usually short, with a row or tuft of golden colored setae.

Heterocerus Fabricius 1792, in part, *sensu* Miller 1988

Dampfius Pacheco 1964
Approximately six species, widespread, Canada to Texas. Key to species in Pacheco 1964. *Dampfius* is distinguished by the basal piece of the phallobase produced at the posterior end (Fig. 7), and prostheca small, with 15-30 setae. Species of *Dampfius* are morphologically similar and separated by male genitalic characters. Miller (1988) synonymized this genus with *Heterocerus*. The synonymy was based on the non-acceptance of Pacheco's generic designation. All of the members of *Dampfius* fall into Miller's *Heterocerus* "undatus group" (key to species Miller, 1988). The biology of *D. collaris* (Kiesenwetter) was discussed by Folkerts (1989).

Heterocerus Fabricius 1792, in part, *sensu* Miller 1988

Peditatus Pacheco 1964
Three species, two north of Mexico. *Peditatus schwarzi* (Horn), occurs from the north- central United States to the east coast, south to Texas. *Peditatus texanus* Pacheco 1964, is found in Texas. Key to species in Pacheco 1964. *Peditatus* is distinguished by the tegmen divided ventrally into two sclerotized parts by a sulcus (Fig. 5, 6), galea finger-like, with 4-7 short setae, post-mesocoxal lines strongly marked, and border anterior to stridulatory ridge absent.

Heterocerus Fabricius 1792, in part, *sensu* Miller 1988

Lapsus Pacheco 1964
Monotypic: *Lapsus tristis* (Mannerheim). This species is widespread across Canada and Alaska south to Florida and California. *Lapsus* is distinguished by the triangular tegmen, small, rounded parameres, and rounded, concave dorsal plate of the aedeagus (Fig. 8). Miller (personel communication) stated that *Lapsus* is synonymous with the European *Heterocerus fenestratus* Thunberg. Members of this genus fall into Miller's *Heterocerus* "mollinus group". Natural history information for *L. tristis* and *H. fenestratus* is unknown.

Heterocerus Fabricius 1792, in part, *sensu* Miller 1988

Tropicini Pacheco 1964

Tropicini are distinguished by the following characters: antennae 9-segmented, post-mesocoxal and post-metacoxal lines absent, color pattern simple, mandibles of males usually with dorsal ridge developed into a wide laminar process, and epipleural line absent. This tribe is represented by a single genus *Tropicus* Pacheco 1964. Twenty-seven species are distributed from North to South America.

Tropicus Pacheco 1964.
Twenty-seven primarily Neotropical species. Two species occur North of Mexico: *T. pusillus* (Say), northern United States to Panama, and *T. minutus* (Fall), Texas. No complete key to species is available. Pacheco (1964) provided a key to 11 species. Numerous species have subsequently been added or synonymized. Bameul (1995) provided the most recent species list for the world. *Tropicus* is distinguished by characters given in the tribal diagnosis.

Bibliography

ARNETT, R. H. Jr. 1971. Heteroceridae. Pp. 465-466. *In:* The beetles of the United States (a manual for identification). The American Entomological Institute. Ann Arbor, Michigan.

BAMEUL, F. 1995. Un nouveau *Tropicus* Pacheco de la Guadeloupe (Coleoptera: Heteroceridae). Bulletin de la Société Entomologique de France, 100: 475-480.

BLACKWELDER, R. E. 1939. Fourth Supplement to the Catalogue of the Coleoptera of America, north of Mexico. Mount Vernon. New York. 146 pp.

BLACKWELDER, R. E. 1944. Checklist of the Coleopterous Insects of Mexico, Central America, the West Indies, and South America. United States National Museum, Bulletin no. 185, pt. 2, pp. 189-341.

BØVING, A. G. and F. C. GRAIGHEAD. 1931. An illustrated Synopsis of the principal larval forms of the order Coleoptera. Entomologia Americana, 11: 1-351.

CHARPENTIER, R. 1965. A monograph of the family Heteroceridae (Coleoptera) of the Ethiopian region. Pp. 215-343. *In:* B. Hanström, P. Brinck, and G. Rudebeck, eds. South African Animal Life. Results of the Lund University Expedition in 1950-1951. Vol. XI. Almqvist and Wiksell. Stockholm.

CHARPENTIER, R. 1967. A monograph of the family Heteroceridae (Coleoptera) of the Notogean region. Arkiv för Zoologi, 20(11): 205-241.

CLARK, G. R., II, and B. RATCLIFFE. 1989. Observations on the tunnel morphology of *Heterocerus brunneus* Melsheimer (Coleoptera: Heteroceridae) and its paleoecological significance. Journal of Paleontology, 63: 228-232.

CLAYCOMB, G. B. 1919. Notes on the habits of *Heterocerus* beetles. Canadian Entomologist, 51: 25.

CROWSON, R. A. 1955. The natural classification of the families of Coleoptera. Nathaniel Lloyd. London. 187 pp.

CROWSON, R. A. 1981. The biology of the Coleoptera. Academic Press. New York. xii + 802 pp.

DOWNIE, N. M. and R. H. ARNETT, Jr. 1996. Heteroceridae, Pp. 723-727 *In:* The beetles of northeastern North America. Volume 1. The Sandhill Crane Press. Gainesville, Florida.

FOLKERTS, G.W. 1989. Egg guarding and its significance in the heterocerid beetle, *Dampfius collaris* (Kies.). Journal of Insect Behavior, 2: 139-141.

FORBES, W. T. M. 1926. The wing folding patterns of the Coleoptera. Journal of the New York Entomological Society, 34: 42-139, pls. VII-XVIII.

GANGLBAUER, L. 1904. Die Käfer von Mitteleuropa. Band 4. C. Gerold's Sohn. Vienna. 286 pp.

HATCH, M. H. 1965. Family Heteroceridae. Pp. 12-14. *In:* The beetles of the Pacific Northwest, part IV: Macrodactyles, Palpicornes, and Heteromera. University of Washington Press. Seattle.

HORN, G. H. 1890. The species of *Heterocerus* of Boreal America. Transactions of the American Entomological Society, 17: 1-16.

JEANNEL, R. and R. PAULIAN. 1944. Morphologie abdominale des coléoptères et systématique de l'ordre. Revue Francaise d'Entomologie, 11: 65-110.

KAUFMANN, T. and P. STANSLY. 1979. Bionomics of *Neoheterocerus pallidus* Say (Coleoptera: Heteroceridae) in Oklahoma. Journal of the Kansas Entomological Society, 52: 565-577.

KOLBE, H. J. 1901. Vergleichend-morphologische Untersuchungen an Coleopteren nebst Grundlagen zu einem System und zur Systematik derselben. Archiv für Naturgeschichte, 67 (Beiheft): 89-150, pls. II-III.

LAMEERE, A. A. L. 1900. Notes pour la classification des coléoptères. Annales de la Société Entomologique de Belgique, 44: 355, 377.

LAWRENCE, J. F. and A. F. NEWTON, Jr. 1995. Families and subfamilies of Coleoptera (with selected genera, notes, references and data on family-group names). P. 845. *In:* J. Pakaluk and S. A. Slipinski, eds. Biology, phylogeny, and classification of Coleoptera: Papers Celebrating the 80th Birthday of R. A. Crowson. Volume 2. Muzeum I Instytut Zoologii PAN. Warsaw.

LAWRENCE, J. F., A. M. HASTINGS, M. J. DALLWITZ, T. A. PAINE and E. J. ZURCHER. 1999. Beetles of the world, a key and information system for families and subfamilies. Version 1.0 for Windows®. CSIRO Entomology. Canberra, Australia. CD-ROM.

LECONTE, J. L. 1863-1866. New species of North American Coleoptera. Prepared for the Smithsonian Institution. Smithsonian Miscellaneous Collection, no. 167, pt. 1, 177 pp.

LENG, C. W. 1920. Catalogue of the Coleoptera of America, north of Mexico. Mount Vernon. New York. 470 pp.

LENG, C. W. and A. J. MUTCHLER. 1927. Supplement to the Catalogue of the Coleoptera of America, north of Mexico. Mount Vernon. New York. 78 pp.

LENG, C. W. and A. J. MUTCHLER. 1933. Second and Third Supplements to the Catalogue of the Coleoptera of America, north of Mexico. Mount Vernon. New York. 112 pp.

MILLER, W. V. 1988. Two new species of *Heterocerus* from North America, with notes on related species (Coleoptera: Heteroceridae). Coleopterists Bulletin, 42: 313-320.

PACHECO, M. F. 1964. Sistemática, filogenia y distribución de los heterocéridos de America. (Coleoptera: Heteroceridae). Escuela Nacional de Agricultura, Colegio de Post Graduados, Chapingo, Mexico. xii + 155 pp., 55 pls.

PACHECO, M. F. 1969. A new species of Heterocerini (Coleoptera: Heteroceridae). Florida Entomologist, 52: 37-39.

PACHECO, M. F. 1978. A catalog of the Coleoptera of America north of Mexico, family Heteroceridae. United States Department of Agriculture. Handbook 529-47, 8 pp.

PONOMARENKO, A. G. 1985. Beetles from the Jurassic of Siberia and West Mongolia. Pp. 47-87. *In:* A. P. Rasnitsyn, ed. Jurassic insects from the Siberia and West Mongolia. Trudy Paleontologicheskogo Instituta. 211 pp.

SANDERSON, M. W. 1953. A revision of the nearctic genera of Elmidae (Coleoptera). Journal of the Kansas Entomological Society, 26: 148-163.

SILVEY, J. R. G. 1935. An investigation of the burrowing inner-beach insects of some fresh water lakes. Papers of the Michigan Academy of Science, Arts and Letters, 21: 655-696.

ZAITZEV, P. 1910. Pars 17. Dryopidae, Cyathoceridae, Georyssidae, Heteroceridae. *In:* S. Schenkling, ed. Coleopterorum Catalogus. W. Junk. Berlin. 68 pp.

48. PSEPHENIDAE Lacordaire 1854

by William D. Shepard

Family common name: The water penny beetles

The soft body, serrate to pectinate antennae, anterior coxae with exposed trochantin, and stream association distinguish this family.

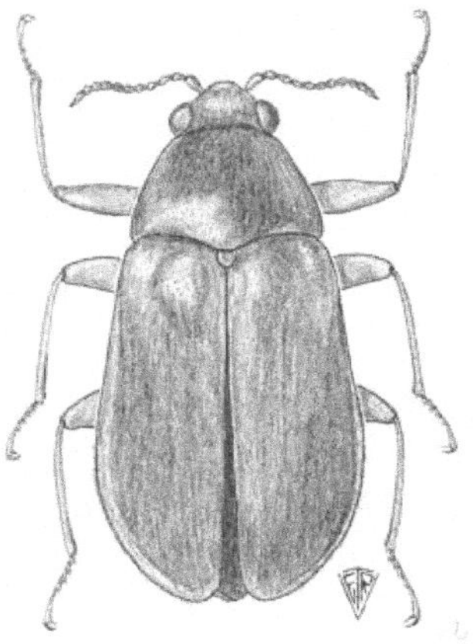

FIGURE 1.48. *Psephenus herricki* (DeKay)

Description: Oval to ovate, depressed; 3-7 mm long; soft bodied; color black to brown, may have red or yellow maculae; pubescence very short, fine, dense, sub-erect.

Head slightly deflexed, slightly to strongly retracted into the thorax, frons strongly declivous or narrow and elongate; surface rugose-punctate. Antennae 11-segmented, moniliform, serrate, or flabellate; inserted beneath frontal ridges well above the bases of the mandibles. Clypeus rostrate, prolonged between the antennae; labrum transverse, short; mandibles hidden from view; maxillary palpi 4-segmented, slender, segment 2 as long as 3 and 4 combined, segment 1 small; gula broader anteriorly, sutures distinct, separate; mentum trapezoidal; ligula hidden; labial palpi 3-segmented, very short, stout. Eyes lateral, size moderate to very large, rounded.

Pronotum twice as wide as long, broader than the head at the base, sometimes laterally explanate; laterally feebly margined; posterior border smooth or serrulate; surface finely rugose-punctate; pleural region broad; prosternum carinate, prosternal process fits into a mesosternal groove, or process lacking; procoxal cavities open behind. Mesosternum short; metasternum with a medial transverse impressed line. Legs with the anterior trochantin very large; anterior coxae large, globular to transverse, separate; middle coxae rounded to globular, separate; hind coxae transverse, contiguous to separate; trochanters large, triangular; femora swollen; tibiae slender, apical spurs very short; tarsal formula 5-5-5, segment 5 elongate; claws simple, toothed, apically bifurcate, or with a fleshy lobe below. Scutellum small to large, triangular to pentagonal; base smooth to serrulate. Elytra entire, bases smooth to serrulate; apices rounded, surface costate or vaguely sulcate; epipleural fold moderate, entire. Wings with stigma present; subcosta and radial viens anteriorly crowded; M_4 and cubitus fused; 4 anal veins.

Abdomen with 5-7 visible sternites. Male genitalia with basal piece rectangular, open dorsally; parameres fused in basal half ventrally; penis membranous to sclerotized, as long as parameres; fibula present or absent. Female genitalia symmetrical; coxites short and stout, each with a stylus.

Eggs (*Psephenus*) orange or yellow, in sheet-like masses on submerged rocks.

Larva onisciform, flattened, disk-like (water penny), almost as broad as long, with protruding projections concealing head, legs and gills; 6-10 mm long; color brownish to blackish. Head small, reflexed, epicranial suture sometimes present. Antennae long, narrow, 3-segmented. Clypeus membranous; labrum prominent, truncate; mandibles short, the apices blunt, sometimes with setae or prostheca on the mesal margin; maxilla small; labium with fused submentum, mentum, and ligula, with small 2-segmented palpi. Five pairs of stemmata closely grouped. Thorax large, legs 4-segmented, with apical claw-like tarsunguli. Abdomen either without an operculum on segment 9 and with exposed tracheal gills (Psepheninae and Eubrianacinae), or with tracheal gills below an operculum (Eubriinae). Larvae are described by Blackwelder (1930) and Brown (1991).

Pupae exarate, below last larval exuvia, weakly sclerotized. Prothorax and abdomen with laterally-projecting processes fitting into pockets on larval exuvia. Abdominal spiracles 1-7 functional.

Habits and habitats. Adults, which occur only in the summer months, are generally riparian, except females may lay eggs on rocks below the water. Larvae are aquatic, found creeping over stones in swiftly flowing water or on submerged wood. There are probably 6 larval instars. Pupation is usually under riparian substrates; some underwater pupation has been found in *Eubrianax* and *Psephenus*. The dorsal part of the exuvia of the last larval instar is free of the pupa except for the last three segments. The life cycle is semivoltine with the larvae the overwintering stage. All species are probably algivorous and/or detritivorous. Murvosh (1971) has described the ecology of *Psephenus*.

Status of the classification. The Psephenidae, Elmidae, and Dryopidae in the older classifications formed the family known as Parnidae. Currently there are four subfamilies: Eubrianacinae, Eubriinae, Psepheninae, and Psephenoidinae. The Eubriinae and Eubriianacinae are poorly known, yet particularly diverse in the Orient where they are currently being extensively studied. Numerous undescribed species are known in the Psepheninae from Central and South America.

Distribution. The family is found on all continents except Antarctica. The Eubrianacinae are circumPacific in distribution; the Psepheninae are mainly found in the New World; the Psephenoidinae are restricted to Asia and Africa; and the Eubriinae are globally distributed. There are 6 genera and 16 species in the United States. Their distributions are given in Barr and Spangler (1994), Brigham (1981), Brown (1983), Brown and Murvosh (1974), and Fender (1962).

Key to the Genera of the United States

1. Posterior margin of pronotum crenulate, serrulate, or finely beaded; males with at least the anterior claw on each tarsus forked at apex 2

— Posterior margin of pronotum smooth 4

2(1). Prosternum depressed between coxae; antennomere 3 at least as long as either first 2 antennomeres (females) or next 3 antennomeres combined (males); male with flabellate antennae and tarsal claws toothed at base; female larger than male, with serrate antennae and tarsal claws not toothed at base (western states) ...... *Acneus*

— Prosternum not depressed between coxae; tarsal claws of both sexes each with basal tooth; antennomere 3 as long as first 2 antennomeres but not as long as antennomeres 4-6, antennae of male not flabellate (eastern states) 3

3(2). Tarsi slender and not dilated, tarsomere 4not prolonged beneath 5; tibial length equal to tarsal length .... *Ectopria*

— Tarsi slightly dilated, tarsomeres 2-4 feebly emarginate, tarsomere 4 slightly prolonged beneath 5; tibial length approximately 1.25 times tarsal length .... *Dicranopselaphus*

4(1). Head usually hidden beneath broadly expanded pronotum; base of claws with a membranous appendage nearly reaching tip of claw; male antennae pectinate, female antennae serrate .... *Eubrianax*

— Head visible from above; base of claws without membranous appendage; male antennae subserrate to serrate, female antennae moniliform .... *Psephenus*

Classification of the Genera of the United States

Psephenidae Lacordaire 1854

Eubrianacinae Jacobson 1913

Eubrianax Kiesenwetter 1874, 1 sp., *E. edwardsii* (LeConte 1874), California, Oregon.

Eubriinae Lacordaire 1857

Acneus Horn 1880, 4 spp., California, Oregon.

Dicranopselaphus Guérin-Méneville 1861, 1 sp., *D. variegatus* Horn 1880, New York to Alabama.
Alabameubria Brown 1980

Ectopria LeConte 1853, 3 spp., Michigan east to New York, south to Alabama and west to Oklahoma.

Psepheninae Lacordaire 1854

Psephenus Haldeman 1853, 7 spp., Eastern United States to Kansas and Oklahoma (above fall line), Texas, New Mexico, Arizona, Idaho, Oregon, California.
Fluvicola DeKay 1844, not Swainson 1827
Eurypalpus LeConte 1852, not Dejean 1836

Bibliography

BARR, C. B. and P. J. SPANGLER. 1994. Two new synonymies: *Alabamuebria* Brown, a junior synonym of *Dicranopselaphus* and *alabameubria starki*, a synonym of *Dicranopselaphus variegatus* (Coleoptera: Psephenidae). Entomological News, 105: 299-302.

BLACKWELDER, R. E. 1930. The larva of *Eubrianax edwardsi* (LeC.) (Coleoptera: Psephenidae). Pan-Pacific Entomologist, 6: 139-142.

BRIGHAM, W. U. 1981. *Ectopria leechi*, a new false water penny from the United States (Coleoptera: Psephenidae). Pan-Pacific Entomologist, 57: 313-320.

BROWN, H. P. 1983. A Catalog of the Coleoptera of America North of Mexico. Family: Psephenidae. United States Department of Agriculture. Agriculture Handbook No. 529-49. 8 pp.

BROWN, H. P. 1991. Psephenidae (Dryopoidea). Pp. 395-397. *In*: F. W. Stehr, ed. Immature Insects. Volume 2. Kendall/Hunt. Dubuque, Iowa. 975 pp.

BROWN, H. P. and C. M. MURVOSH. 1974. A revision of the genus *Psephenus* of the United States and Canada (Coleoptera, Dryopoidea, Psephenidae). Transactions of the American Entomological Society, 100: 289-340.

FENDER, K. M. 1962. A review of *Acneus* (Coleoptera: Dascillidae). Northwest Science, 36: 44-49.

MURVOSH, C. M. 1971. Ecology of the water penny beetle *Psephenus herricki* (DeKay). Ecological Monographs, 41: 79-96.

49. PTILODACTYLIDAE Laporte 1836

by Michael A. Ivie

Family common name: The ptilodactylid beetles

Adult ptilodactylids are rather generalized elateriforms, best recognized in North America by the form of the pronotum, which is anteriorly narrowed and rounded, has complete, incomplete or obsolete lateral margins, and a crenulate base. Also diagnostic are the serrate to ramose antennae and cordate scutellum with a single median notch on the basal margin (in all but one case).

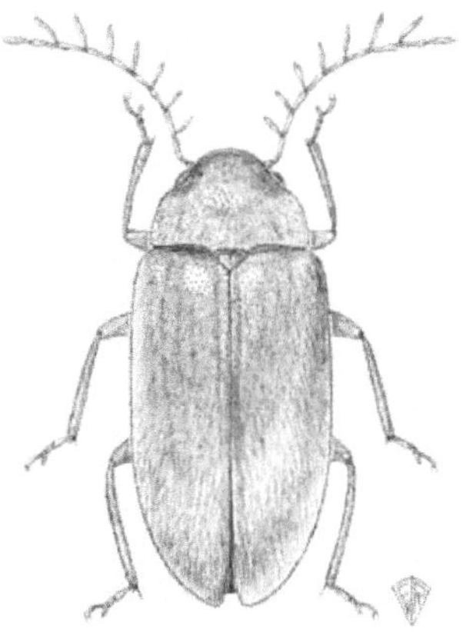

FIGURE 1.49 *Ptilodactyla serricollis* (Say)

Description: (North American species) Elongate-oval to elongate; 2-16 mm. Surface glaberous to variably clothed with dense to sparse, fine setae which in some species impart a color pattern.

Head strongly deflexed; eyes lateral, large, bulging, round; frontoclypeal suture distinct or not; clypeus transverse when distinct; labrum transverse, basal membrane visible. Antennae with eleven antennomeres, filiform, serrate or pectinate, most species sexually dimorphic; many males with articulated basal rami on antennomeres three or four to ten; inserted between inner margin of eye and anterior articulation of mandible. Mandibles with thin cutting blade interiorly along long axis; maxillae and labium often modified into setose lobes and/or with enlarged palps.

Pronotum broad at base, narrowed and rounded anteriorly, covering the head from dorsal view; lateral borders obsolete to completely margined; posterior border crenulate; pleuron large, extending behind procoxa as blunt to acute process; prosternum moderate to short before coxae, in some angularly deflexed anteriorly, in some laterally projecting to cover trochantin; intercoxal process complete and narrow, slightly to moderately expanded posteriorly; procoxal cavities broadly open behind; trochantins hidden or exposed, depending on the subfamily. Mesosternum short, with a narrow process between the coxae; metasternum moderate, broad. Pro- and mesocoxae conical; metacoxae contiguous, triangularly transverse and excavate posteriorly to produce ventral plates that are wide medially and narrowed laterally; pro- and mesotrochanters triangular, metatrochanters offset and oval to elongate-acute; femora and tibiae slender, with two prominent apical spurs; tarsi slender, 5-5-5, simple or with membranous lobe on third hiding minute fourth; claws simple, toothed or pectinate. Scutellum moderate, triangular, with a medial notch in basal margin, or basally smooth or crenulate. Elytra entire, apically rounded; striae punctuate or not, never ribbed; epipleuron moderate, entire, in many somewhat broader in apical 1/4. Metathoracic wing well developed, with elongate, obliquely closed radial cell.

Abdomen with five visible sternites, the sutures entire, the fifth visible sternite distinctly emarginate. Male genitalia very diverse (Stribling 1986b, Costa *et al.* 1999), symmetrical, trilobate, parameres individually articulated to phallobase, not specifically characterized for family. Female genitalia and ovipositor of selected species illustrated by Stribling (1986a, b) and discussed by Costa *et al.* (1999), not characterized for family.

Larvae 3 to 15 mm in length; yellowish to dark reddish brown; elongate, subcylindrical, straight to somewhat ventrally curved; abdominal apex strongly downturned in Ptilodactylinae; long, scattered conspicuous setae on surface. Head protracted, prognathous; frontal arms V-shaped, continuous, epicranial stem short or absent; three, five or six stemmata on each side. Antennae relatively long, three-segmented, segment 3 short. Mandibles tridentate, without mola, bearing various articulated processes, setae or hair brushes; maxilla and labium somewhat to highly retracted, maxilla with large cardo and stipes, four-segmented palpi, articulated galea which is two-segmented in some species; lacinia articulated; labium with submentum, mentum, and short broad ligula bearing two-segmented palpi.

Prothorax longer than meso- or metathorax; legs five-segmented, tarsungulus with single seta. Abdomen ten-segmented, ninth tergite concave or convex, with or without small urogomphi; ninth sternite simple; tenth segment mostly with hooks on pair of anal lobes; aquatic and semiaquatic taxa with three to several papilla-like gills in anal membrane. Biforous spiracles on mesothorax and abdominal segments 1-8 of mature larvae, all but A8 non-functional in *Araeopidius*. Described pupae with abdominal gin-traps.

See Süss and Puppin (1976), Stribling (1986a, b) Lawrence (1991), Beutel (1995) Lawrence *et al.* (1999a, b) and Costa *et al.* (1999) for further discussion of the morphology of this family.

Habits and habitats. Known ptilodactylid larvae feed on decaying vegetation or wood in damp or aquatic situations. As far as is known, all Ptilodactylinae are terrestrial, but only members of the genus *Ptilodactyla* has been described. Ptilodactyline larvae can be taken in numbers with Berlese funnels from moist leaf litter and rotten logs. Described larval Anchytarsinae (including both of the North American species) are truly aquatic, living in or on submerged wood. *Areopidius* larvae (Areopidiinae) live at the margins of streams, and although they possess gills, they apparently actually enter the water only by accident. Cladotominae larvae have been taken from litter along the margins of streams [the Japanese *Paralichus pectinata* (Kiesenwetter 1874)], or on the forest floor (the Australian *Austrolichus monteithi* Lawrence and Stribling 1992) (Lawrence and Stribling 1992, Lawrence *et al.* 1999b). The habits and larva of the North American *Paralichus trivittus* (Germar 1824) is undescribed.

Nothing has been reported on the larvae of the Aploglossinae.

Pupation, as far as is known, takes place terrestrially for all members of the family (reported for *Anchytarsis* and *Ptilodactyla*) (Lawrence 1991).

The adults are most commonly taken at lights, but can be beaten from vegetation, and, in the case of the terrestrial groups, occasionally by Berlese funnel from the larval habitat. Adult Ptilodactylinae have the maxillae modified into spores brushes to feed on molds (Stribling and Seymour 1988), and are often taken by beating vegetation with surface microfungi. Feeding habits of adults of other groups are unreported.

Status of the classification. This medium-sized (495 described species, many more to be described) and diverse family has a convoluted and difficult taxonomic history, many of its members having been confused with the Dascillidae, Eulichadidae and Cneoglossidae over the years (Costa *et al.* 1999). With the exception of an increasingly clear need to determine the relationship with the Chelonariidae, the ptilodactylids as defined here seem well supported as a family of Byrrhoidea (Costa *et al.* 1999). Although the current form of the family was documented by Crowson as early as 1955, and successively clarified by Lawrence (1982), Stribling (1986a,b), Lawrence (1991), Lawrence and Stribling (1992), Lawrence and Newton (1995), Lawrence *et al.* (1999a, b) and Costa *et al.* (1999), as recently as 1996 revanchist elements have returned to a structure for this group not based on phylogenetic relationships (Campbell 1991, Downie and Arnett 1996). Except as noted regarding the Chelonariidae, membership of the family is now well established (Lawrence *et al.* 1999a). However, subfamily, tribe and generic groups are still in need of clarification. Stribling (1986b) provided the first world-wide phylogenetic framework for this family, based provisionally on adult characters, with extensive reorganization of the membership not only of the family itself, but also of the genera and subfamilies. Although many of the changes he made remain unpublished in the sense of the International Code of Zoological Nomenclature, his work has been extensively cited in regards to this family (Lawrence and Stribling 1992, Lawrence and Newton 1995, Lawrence *et al.* 1999a).

Stribling recognized 25 genera in six subfamilies on the basis of hypothesized monophyly. Lawrence and Stribling (1992) added a genus and discussed the impact of larval characters on the classification. Lawrence and Newton (1995) rearranged some of Stribling's subfamilies, and recognized a total of five, four of which are represented in the Nearctic. These changes were documented with a limited set of adult and larval characters, but with no actual analysis. Costa *et al.* (1999) recognized six subfamilies, but these were, in fact, simply the six representative genera they redescribed and used as character sources in their investigation of the relationships of the Cneoglossidae.

Stribling's Epilichadinae, (validated by Lawrence and Stribling 1992) included four genera (*Areopidius* Cockerell, *Anchycteis* Horn, *Epilichas* Horn and *Byrrocryptus* Broun) but was supported only by convergent and reversed synapomorphies. Lawrence and Newton (1995) divided that subfamily into the Areopidiinae, including only the monotypic typical genus, and removed the others to the Anchytarsinae. Thus, the monotypic subfamily Areopidiinae is now defined only on the basis of an autapomorphy.

Stribling's Anchytarsinae was limited to the typical genus (*Anchytarsus* Guérin-Méneville) based on the synapomorpies of the 4 included species (Stribling 1986a, b). Lawrence and Newton's (1995) changes rendered it an admittedly artificial assemblage that included three of Stribling's Areopidiinae (*Anchycteis*, *Epilichas* and *Byrrocryptus*), plus *Daemon* Laporte and *Pseudoepilichus* Armstrong and Nakane from Stribling's Ptilodactylinae, the genus *Ocotoglossa* Guérin-Méneville from what Stribling treated as an unnamed monogeneric subfamily, and unspecified additional genera under the category of "etc." These later include *Therius* Guérin-Méneville and a still-undescribed genus from the Ptilodactylinae, as well as a genus treated as a synonym of *Octoglossa* by Stribling (1986b). Within this subfamily, they proposed a monophyletic group made up of *Anchytarsus*, *Anchycteis*, *Byrrocryptus*, and *Epilichas*, based on larval synapomorphies.

The removal of those genera from Stribling's somewhat weakly supported Ptilodactylinae left a group of genera that was hypothesized to be supported as monophyletic based on the concealment of the protrochantin by the posterior lateral extensions of the prosternum. Seven described genera (*Falsotherius* Pic, *Pherocladus* Fairmaire, *Lachnodactyla* Champion, *Ptilodactyla* Illiger, *Stirophora* Champion, *Chelonariomorphus* Pic, and *Lomechon* Wasmann [moved from Silphidae by Newton 1998]) and 2 still-undescribed genera remain here. These numbers reflect several unpublished generic synonymies proposed by Stribling (1986b).

The fourth subfamily represented in North American, the Cladotominae, included 4 described genera (*Cladotoma* Westwood, *Paralichus* White, *Pseudocladotoma* Pic and *Hovactyla* Fairmaire). The monophyly proposed for this group by Stribling (1986b) was supported by the discovery and addition of *Austrolichus* Lawrence and Striblilng 1992.

The last subfamily is the Aploglossinae, with the genera *Aploglossa* Guérin-Méneville and *Bradytoma* Guérin-Méneville.

Stribling listed 4 genera in *incertae sedis*. Of these, Lawrence and Newton (1995) removed *Brounia* Sharp to the Chelonariidae, and Lawrence *et al.* (1999a) removed *Podabrocephus* Pic to its own family. This leaves only *Falsoptilodactyla* Pic and *Valoka* Delève in this limbo status.

Larvae of seven ptilodactylid genera have been described (Lawrence and Stribling 1992), but show great promise in providing data to improve the classification of this interesting group. For instance, all described larvae of the Anchytarsinae *sensu* Lawrence and Newton are aquatic, and all known Ptilodactylinae larvae are terrestrial. However, an adult of the Malagasy genus *Daemon*, moved from Stribling's Ptilodactylinae to the

Anchytarsinae by Lawrence and Newton (1995), has been taken under bark with a terrestrial larva very similar to Neotropical *Ptilodactyla* (Ivie collection, pers. observation).

The many undescribed species and larvae, need for suprageneric work, and lack of current workers studying the species, makes this an excellent group for a young coleopterist to work with.

Distribution. The nearly 500 described species occur almost world-wide, absent only in the Chilean subregion, the Palearctic (except Japan and an introduced species in Italy, Süss and Puppin 1976), Antarctica and the more remote oceanic islands. The Anchytarsinae occur in the Nearctic, Neotropical, Australia/New Zealand, Oriental and Malagasy regions. The Cladotominae in Australia, South America, Eurasia, North America, the Greater Sundas and Madagascar. Aploglossinae are limited to the Neotropics, while the Araeopidinae are found in Western North America. The very large Ptilodactylinae occurs throughout the range of the family, excepting New Zealand.

Key to the Nearctic Genera

1. Prothoracic coxal cavities open laterally, trochantins exposed by gap between prosternum and posteroventral portion of hypomeron (Fig. 2) 2
— Prothoracic coxal cavities closed laterally, trochantins concealed by posterior lateral extension of prosternum that meets the hypomeron (Fig. 3) 5

2(1). Lateral margin of pronotum acute, complete; tarsal claws pectinate .... *Paralichus*
— Lateral margin of pronotum obsolete; tarsal claws simple or toothed 3

3(2). Head strongly deflexed, mentum received against vertical portion of prosternum which is set off by carina from horizontal portion; elytral setae giving mottled, sometimes transversely banded, appearance .... *Araeopidius*
— Head not strongly deflexed, mentum not received by prosternum; prosternum flat and horizontal in front of coxae; elytra without color pattern 4

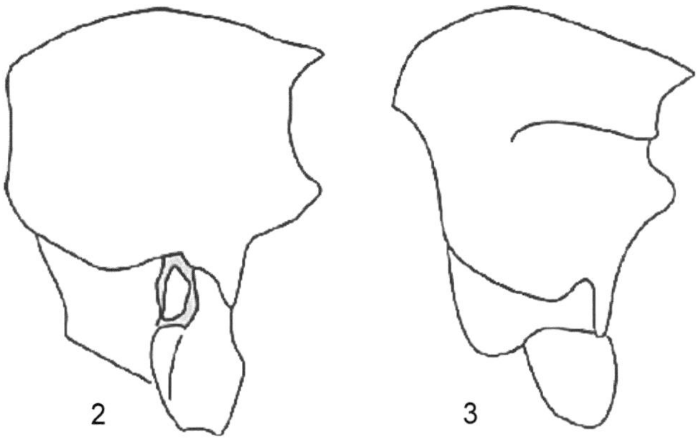

FIGURES 2.49-3.49. Generalized prothorax, lateral view, 2. *Anchytarsus* sp.; 3. *Ptilodactyla* sp.

4(3). Scutellum cordate, with a single median indentation in basal margin; West Coast .... *Anchycteis*
— Scutellum with basal margin simple or crenulate; eastern US .... *Anchytarsus*

5(1). Apical maxillary palpomere enlarged; in males elongate, largely membranous, in females somewhat elongate, mostly sclerotized; Arizona, Texas and Florida .... *Lachnodactyla*
— Apical maxillary palpomere simple, securiform; widespread .... *Ptilodactyla*

Classification of the Nearctic Genera

Ptilodactylidae Laporte 1836

Anchytarsinae Champion 1897

Anchycteis Horn 1880, a monotypic genus with *A. velutina* Horn 1880, is known from northern California to southwestern Oregon (illustrated by Hatch 1961).

Anchytarsus Guérin-Méneville 1843, has 3 described species ranging through North, Central and South America. In an uncommon twist, more species of larvae are known than adults. Stribling (1986a) revised and illustrated the species. One species, *A. bicolor* (Melsheimer 1846), occurs from Ontario and Quebec to Mississippi and Georgia.

Cladotominae Pic 1914

Paralichus White 1859, a genus of 10 currently recognized species from the Oriental region and eastern North America. At least 2 undescribed Neotropical species occur in Mexico and El Salvador (Stribling 1986b). *Paralichus trivittus* (Germar 1824) is a rare species occurring in the eastern United States, with records from Ohio, Pennsylvania, Georgia, and possibly Indiana (Downie and Arnett 1996, Florida State Collection of Arthropods).

Odontonyx Guérin-Méneville 1843, Stribling 1986b (propose synonymy), Lawrence and Stribling 1992 (validate synonymy)

Areopidiinae Lawrence 1991

Araeopidius Cockerell 1906, a monotypic genus known from *A. monachus* LeConte, ranging from British Columbia and Montana to Oregon and California. (Illustration in Hatch 1961.)

Ptilodactylinae Laporte 1836

Lachnodactyla Champion 1897, with 4 described and at least 2 undescribed species occurring from Columbia to the USA (Stribling 1986b). Three of these occur in the USA: *L. arizonica* Schaeffer 1906 from Arizona; *L. texana* Schaeffer 1906 from Texas (and Mexico), and *Lachnodactyla* sp. from Florida (Stribling 1986b, Peck and Thomas 1998).

Ptilodactyla Illiger 1807, a very large and widespread genus with 362 currently recognized species (including those from 3 exotic genera proposed as synonyms by Stribling 1986b), representing nearly three-quarters of the family world-wide. In North America there are nine described species, entirely based upon variation in the male genitalia (Johnson and Freytag 1982) ranging throughout the eastern forests from Manitoba to Prince Edward Island, south to Texas and Florida. Within this range there remain several states and provinces with no published records. The small number of collections and specimens studied by Johnson and Freytag (1982), coupled with their limited character source, indicates a need for a reexamination of this genus in North America. Judging solely on the genitalic characters used, several apparently undescribed species await recognition. Distributional records have been added by LeSage (1991) and Peck and Thomas (1998).

The described species are: *P. acuta* Johnson and Freytag 1978, known from Kentucky, Virginia, Georgia and Florida; *P. angustata* Horn 1880, Missouri to Pennsylvania, south to Texas and Florida; *P. carinata* Johnson and Freytag 1978, reported from New York and Quebec to Texas and Florida; *P. equilobata* Chapin 1927, Texas; *P. exotica* Chapin 1927, Illinois, Indiana, Pennsylvania, New York, Maryland, and the District of Columbia; *P. hyperglotta* Johnson and Freytag 1982, Texas; *P. isoloba* Johnson and Freytag 1982, New York, Pennsylvania, Ohio, Maryland and Kentucky; *P. nanoderma* Johnson and Freytag 1982, Illinois, Indiana, Ohio (Montana State University collection), Missouri and Florida; *P. serricollis* (Say 1823), Manitoba to Prince Edward Island, south to Texas and Florida (LeSage 1991 adds a report for British Columbia. This record is far out of the normal range for the group, and needs confirmation).

BIBLIOGRAPHY

BEUTEL, R. G. 1995. Phylogenetic analysis of Elatermiformia (Coleoptera: Polyphaga) based on larval characters. Journal of Zoological Systematics and Evolutionary Research, 33: 145-171.

CAMBPELL, J. M. 1991. Family Dascillidae. P. 143. *In*: Y. Bosuquet, ed. Checklist of the Beetles of Canada and Alaska. Agriculture Canada Publication 1861/E.

COSTA, C., S. A. VANIN and S. IDE. 1999. Systematics and bionomics of Cneoglossidae with a cladistic analysis of Byrrhoidea sensu Lawrence & Newton (1995) (Coleoptera, Elateriformia). Arquivos de Zoologia (São Paulo), 35: 231-300.

CROWSON, R. A. 1955. The natural classification of the families of Coleoptera. Nathaniel Lloyd. London, 187 pp.

DOWNIE, N. M. and R. H. ARNETT, Jr. 1996. The Beetles of Northeastern North America. Vol. 1. The Sandhill Crane Press. Gainesville, FL

HATCH, M. H. 1961. The beetles of the Pacific Northwest. Part III: Pselaphidae and Diversicornia I. University of Washington Publications in Biology, 16: 1-503

JOHNSON, V. and P. H. FREYTAG. 1982. A review of the species of Ptilodactyla in the United States with descriptions of three new species (Coleoptera: Ptilodactylidae). Entomological News, 93: 129-135.

LAWRENCE, J. F. 1982. Coleoptera. Pp. 482-553. *In*: S. P. Parker, ed. Synopsis and Classification of Living Organisms. Vol 2. McGraw-Hill. New York.

LAWRENCE, J. F. 1991. Ptilodactylidae (Dryopoidea), Pp. 391-394. *In*: F. W. Stehr, ed. Immature Insects. Vol. 2. Kendall/Hunt. Dubuque, IA.

LAWRENCE, J. F., A. M. HASTINGS, M. J. DALLWITZ, T. A. PAINE and E. J. ZURCHER. 1999a. Beetles of the World: A Key and Information System for Families and Subfamilies. CD-ROM, Version 1.0 for MS-Windows. CSIRO Publishing. Melbourne.

LAWRENCE, J. F., A. M. HASTINGS, M. J. DALLWITZ, T. A. PAINE and E. J. ZURCHER 1999b. Beetle Larvae of the World: Descriptions, Illustrations, Identification, and Information Retrieval for Families and Sub-families. CD-ROM, Version 1.1 for MS-Windows. CSIRO Publishing. Melbourne.

LAWRENCE. J. F. and A. F. NEWTON, Jr. 1995. Families and subfamilies of Coleoptera (with selected genera, notes, references and data on family-group names), Pp. 779-1006. *In*: J. Pakaluk and S. A. Slipinski, eds. Biology, phylogeny and classification of Coleoptera: Papers celebrating the 80th birthday of Roy A. Crowson. Muzeum i Instytut Zoologii PAN. Warsaw.

LAWRENCE, J. F. and J. B. STRIBLING 1992. A new genus of Ptilodactylidae (Coleoptera: Elateriformia) from North Queensland, with description of the presumed larva. Journal of the Australian Entomological Society, 31: 19-27.

LESAGE, L. 1991. Ptilodactylidae. P. 168. *In*: Y. Bousquet, ed. Checklist of the beetles of Canada and Alaska. Agriculture Canada Publication 1861/E.Canada Communications Group - Publishing. Ottawa.

NEWTON, A. F, Jr. 1998. Phylogenetic problems, current classification and generic catalog of world Leiodidae (including Cholevidae), Pp. 41-178. *In*: P. M. Giachino and S. B. Peck, eds. Phylogeny and evolution of subterranean and endogean Cholevidae (=Leiodidae Cholevinae). Proceedings of XX International Congress of Entomology. Museo Regionale di Scienze Naturali. Torino.

PECK, S. B. and M. C. THOMAS. 1998. A distributional checklist of the beetles (Coleoptera) of Florida. Arthropods of Florida and Neighboring Land Areas, 16: i-viii + 1-180.

STRIBLING, J. B. 1986a. Revision of *Anchytarsus* (Coleoptera: Dryopoidea) and a key to the New World genera of Ptilodactylidae. Annals of the Entomological Society of America, 79: 219-234.

STRIBLING, J. B. 1986b. World generic revision of Ptilodactylidae (Coleoptera: Dryopoidea). Ohio State University (PhD Thesis). Columbus.

STRIBLING, J. B. and R. L. SEYMOUR. 1988. Evidence of mycophagy in Ptilodactylidae (Coleoptera: Dryopoidea) with notes on phylogenetic implications. Coleopterists Bulletin, 42: 152-154.

SÜSS, L. and O. PUPPIN 1976. Osservationi sulla morfologia e sulla biologia di *Ptilodactyla exotica* Chapin (Coleoptera, Ptilodactylidae) nelle serre della Lombardia e contributo bibliografico allo studio del grupo. Bolletino di Zoologia Agraria e di Bachicoltura, (2) 13: 143-165.

50. CHELONARIIDAE Blanchard 1845

by Michael A. Ivie

Family common name: The chelonariid beetles

The single U.S. chelonariid species is a oval, compact, glossy beetle with sparse patches of white scale-like setae on the elytra. A sharp carina running around the entire anterior portion of the pronotum from hind angle to hind angle; basally crenulate pronotum and scutellum; pro- and mesosternal medial excavations that receive the basal segments of the antennae; lobed third tarsomere that obscures the small fourth; and the ability to retract the head and appendages into a compact, seed-like form with the head not at all visible from above will distinguish this group in the New World.

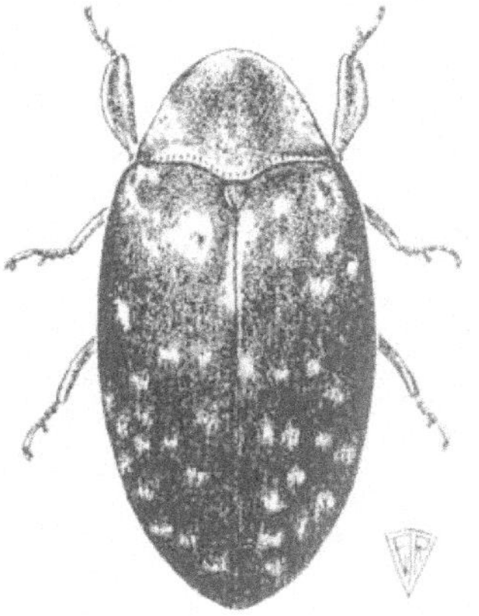

FIGURE 1.50. *Chelonarum lecontei* Thomson

Description: (New World species) Broad, oval, moderately convex, strongly contractile; length 2.5 to 10 mm; color black, red, brownish, or pale, rarely with contrasting markings; vestiture from scattered inconspicuous setae to contrasting patches of scale-like setae dorsally.

Head at rest retracted into the prosternum, exposing only the eyes and antennae; front distinctly margined, acute medially. Eyes lateral, large, bulging, round. Antennal insertions hidden below the frontal margin above the mandibles, antennae with eleven antennomeres, at rest held with scape hidden under front, pedicle and third antennomere visible, but received in median pro- and mesosternal grooves, antennomeres 4-11 exposed on mesosternum (Fig. 2). Scape short with bulbous apex, pedicle and antennomere three longer, both flattened and margined externally, the fourth very small, 5-11 filliform to subserrate. Clypeus not distinct, labrum hidden; mandibles short, broad, curved, the apices blunt; maxillary palpi with four palpomeres; labial palpi with three palpomeres. Gular sutures widely separate.

Pronotum much broader than the head; posteriorly broadly lobed; crenulate on posterior margin; with prominent ridge margining disc continuously from posterior angle to posterior angle, ridge evenly arcuate or faintly bisinuate to front; disc with a shallow, broad depression inside marginal ridge; hypomeron broad, excavate to receive prolegs. Prosternum broad, vertical anteriorly on either side of antennal cavity to receive head; intercoxal area deeply depressed to receive antennae, intercoxal process long, extending deep into mesosternal groove, with ventrally projecting digitiform antero-medial process that rests between apices of pedicles and bases of third antennomeres in retracted position; procoxal cavities open behind. Scutellum small, triangular, crenulate anteriorly. Mesosternum broad but short, deeply excavate medially to receive prosternal process and antennae; mesocoxal cavities open laterally. Metasternum broad, long and somewhat bulging, postcoxal lines vague to distinct.

Legs contractile, received in deep margined cavities involving head, prosternum, hypomeron and base of mesostermum (prolegs), lateral portion of mesothorax and base of epipleuron (mesolegs) and metacoxae and first ventrite (metalegs); anterior coxae flat, sub-oval; middle coxae small, flat, somewhat transverse, widely separate, hind coxae transverse, excavate posteriorly, with ventral flange; trochanters moderate, triangular; femora broad and flattened, excavate to receive tibia; tibiae flattened, laterally margined, excavate to receive tarsi; tarsal formula 5-5-5, first and second tarsomere pubescent beneath; third broadly lobed, fourth small and somewhat obscured; claws broad at base, often forming a tooth in apical half.

Elytra entire, closely fitting abdomen, rounded apically, irregularly punctate or with a faint indication of striae, especially in the postero-sutural area, rarely with striae-like raised trichomes behind humeri; the epipleuron narrow, entire, deeply grooved to apex, receiving lateral edge of abdomen in contracted state. Hind wing well-developed, similar to other families of this group; folding pattern with normal transverse folds (Forbes 1926); radial cell elongate, well-developed, acute at inner posterior angle; 5 free veins in median area.

Abdomen with five ventrites, sutures entire; surface punctate; distinct to vague postcoxal line on first ventrite. Male genitalia are not well characterized for family as a whole, but are described to be of the trilobe type, with penis short, stout, paired struts on anterior edge; parameres individually articulated to phallobase (Sharp and Muir 1912, Lawrence *et al.* 1999). Spangler (1980) illustrates the genitalia of *C. lecontei*, which differs in some details from the type described by Sharp and Muir (1912). The female genitalia are not described.

Larvae (Bøving and Craighead 1931, Spangler 1980, 1991) subcylindrical, integument sclerotized; post-cephalic segments each with a pair of increasingly large sublateral carinae which bear a row of stout filaments; head hypognathous, ocellus single; antenna short, 3-segmented; mandible broad, multidentate, lacking mola; gula present; legs normal; abdomen 9-segmented with operculum on segment 9; a pair of blunt projections on last tergite; nine

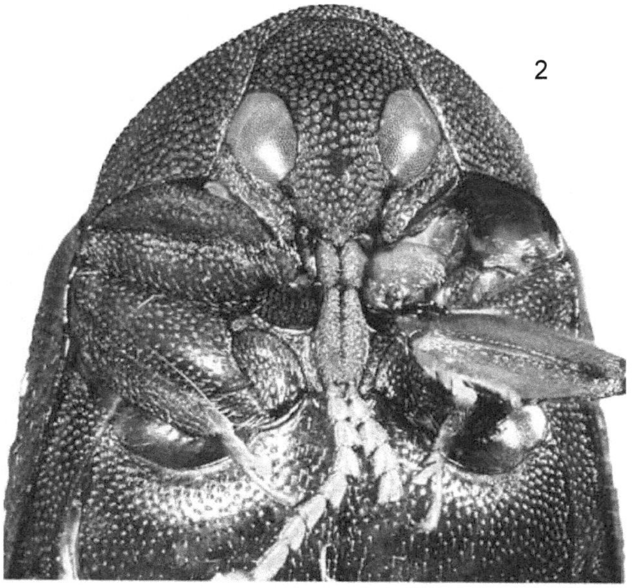

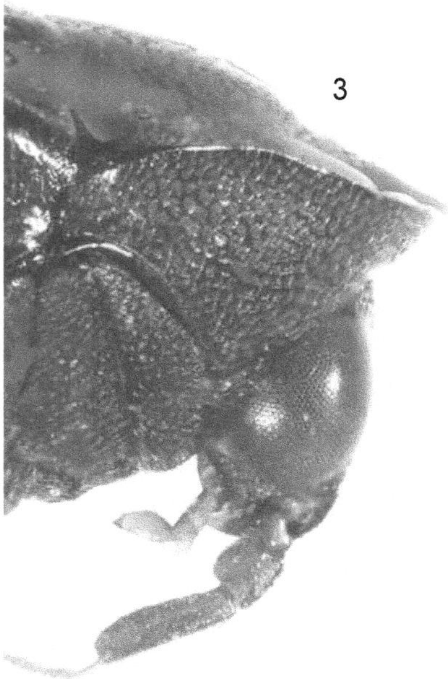

FIGURES 2.50-3.50. *Chelonarum lecontei* Thomson, head and thorax. 2. ventral view; 3. lateral view.

pairs of projecting spiracles sublateral on terga; spiracles of sinuous type.

Habits and habitats. The larvae are detritus feeders in moist to dry litter on the ground, at the bases of trees, orchids or other plants, in refuse heaps of ants or under bark. Previous suggestions that the larvae were aquatic has been shown to be incorrect (Brown 1972, Spangler 1980, 1991, Ivie pers. obs.). Some authors have suggested that the larvae of *Chelonarium* are associated with ants (Lenko 1967, Janzen 1974, Spangler 1980, 1991) either as inquilines or refuse feeders. Repeated collections of the larva of the West Indian *C. punctatum* Fabricius show that at least some species occur in dry litter on the forest floor and under bark without ants (Ivie pers. obs). The reported association with ants may be simply one of proximity, and requires further study. However, adults of some South American species have trichome-like structures on the elytron, which may indicate a relationship with ants in some particular species. The specific habits of the North American species are unreported, but its similarity to *C. punctatum* suggests that Berlese extractions of leaf litter in the areas where it occurs might be productive. The adults are usually taken at light, but may also be taken in Malaise traps or by beating vegetation, especially at night.

Status of the classification. The relationships of this family are unclear, but may involve the Ptilodactylidae (Hinton 1939, Lawrence and Newton 1995). Approximately 300 described species of this family are currently placed in 3 well-defined genera: *Chelonarium* Fabricius, *Pseudochelonarium* Sharp, and *Brounia* Sharp. There are 206 species described from the New World (Blackwelder 1947), all in *Chelonarium*, 167 of which were described by A. Méquignon in 8 papers totaling a mere 83 pages (*e.g.*, Méquignon 1934). The species are poorly documented, and in desperate need of revision.

Chelonariomorphus Pic was removed to the Ptilodactylidae by Stribling (1986), and *Brounia* added from that family by Lawrence and Newton (1995).

Distribution. Throughout the warmer parts of the Americas, Eurasia, Africa, the East Indies, northeastern Australia, and in New Zealand. The vast majority (Dalla Torre 1991, Blackwelder 1947, Costa *et al.* 1988) of the species belong to *Chelonarium*, which is most abundant in the Neotropics, but occurs throughout the familial range, with the exception of New Zealand. *Pseudochelonarium* occurs from India to South East Asia and New Guinea (Paulus 1969). The single species of *Brounia*, *B. thoracica* Sharp, is a rarely collected New Zealand endemic (Klimaszewski and Watt 1997).

Key to the World Genera

1. Pronotum margined anteriorly and laterally with a sharp, raised carina (Fig. 3); disc explainate laterally and anteriorly; widespread ..... *Chelonarium*
— Pronotum rounded anteriorly and laterally; without marginal carina; Old World 2

2(1). Antenna serrate (females) or pectinate (males); hypomeron not excavate, set off by low carina that meets anterior margin at back of eye; body elongate, nearly parallel-sided; New Zealand .... *Brounia*
— Antenna simple; hypomeron excavated to receive forelegs (as in *Chelonarium*); body ovoid; India to South East Asia and New Guinea .... *Pseudochelonarium*

Classification of the Nearctic Species

Chelonariidae Blanchard 1845

Chelonarium Fabricius 1801, 1 sp., C. *lecontei* Thomson 1867, Oklahoma and Virginia, south to Texas and Florida (Brown 1975, Coats and Selander 1979, Spangler 1980).

Bibliography

BLACKWELDER, R. M. 1947. Checklist of the coleopterous insects of Mexico, Central America, the West Indies, and South America, Part 5. United States National Museum Bulletin 185: 765-925.

BØVING, A. G. and F. C. CRAIGHEAD. 1931. An illustrated synopsis of the principal larval forms of the order of Coleoptera. Entomologica Americana (new series), 11: 1-351.

BROWN, H. P. 1972. Aquatic dryopoid beetles (Coleoptera) of the United States. Biota of freshwater ecosystems identification manual no. 6. U.S. Environmental Protection Agency. Washington, DC, 92 pp.

BROWN, H. P. 1975. A distributional checklist of North American genera of aquatic dryopoid and dasciloid beetles. Coleopterists Bulletin, 29: 149-160.

COATS, J. R. and R. B. SELANDER. 1979. Notes on *Chelonarium lecontei* Thomson (Coleoptera: Chelonariidae), including description of an unusual righting behavior. Coleopterists Bulletin, 33: 57-59.

COSTA, C., S. A.VANIN and S. A. CASARI-CHEN. (1988). Larvas de Coleoptera do Brasil. Museu de Zoologia, Universidade de Sao Paulo. Sao Paulo. vi + 282 pp.

DALLA TORRE, K. W. VON. 1911. Byrrhidae: Chelonariinae, Coleopterorum Catalogus, 14(33): 11-12.

FORBES, W. T. M. 1926. The wing folding patterns of the Coleoptera. Journal of the New York Entomological Society, 34: 42-68, 91-138.

HINTON, H. E., 1939. An inquiry into the natural classification of the Dryopoidea, based partly on a study of their internal anatomy. Transactions of the Royal Entomological Society of London, 89: 133-184.

JANZEN, D. H. 1974. Epiphytic myrmecophytes in Sarawak: mutualism through the feeding of plants by ants. Biotropica, 6: 237-259.

KLIMASZEWSKI, J. and J. C. WATT. 1997. Coleoptera: family group review and keys to identification. Fauna of New Zealand, 37: 1-199.

LAWRENCE, J. F., A. M. HASTINGS, M. J. DALLWITZ, T. A. PAINE and E. J. ZURCHER. 1999. Beetles of the World: A Key and Information System for Families and Subfamilies. CD-ROM, Version 1.0 for MS-Windows. CSIRO Publishing. Melbourne.

LAWRENCE. J. F. and A. F. NEWTON, Jr. 1995. Families and subfamilies of Coleoptera (with selected genera, notes, references and data on family-group names), Pp. 779-1006. *In*: J. Pakaluk and S. A. Slipinski, eds., Biology, phylogeny and classification of Coleoptera: Papers celebrating the 80th birthday of Roy A. Crowson. Muzeum i Instytut Zoologii PAN, Warsaw.

LENKO, K. 1967. *Chelonarium semivestitum* inquilino da formiga *Campanotus rufipes* (Coleoptera: Chelonariidae). Studia Entomologia, 10: 433-434

MÉQUIGNON, A. 1934. Les *Chelonarium* de l'Amerique continentale. Annales de la Société Entomologique de France, 103: 199-256.

PAULUS, H. F. 1969. *Pseudochelonarium* (*Neochelonarium*) *kalimantanense* n. sp. aus Borneo, mit Bemerkungen zum System der Chelonariidae (Col. Dryopoidea). Zeitschrift der Arbeitsgemeinschaft Österreichen Entomologen, 21: 105-109.

SHARP, D. and F. MUIR. 1912. Comparative anatomy of the male genital tube in Coleoptera. Transactions of the Entomological Society of London, (1912): 477-642, pls. XLII-LXXVIII.

SPANGLER, P. J. 1980. Chelonariid larvae: aquatic or not? (Coleoptera: Chelonariidae). Coleopterists Bulletin, 34: 105-114.

SPANGLER, P. J. 1991. Chelonariidae, Pp. 394-395. *In*: F. W. Stehr, ed., Immature Insects. Vol. 2. Kendall/Hunt. Dubuque, IA.

STRIBLING, J. B. 1986. Revision of *Anchytarsus* (Coleoptera: Dryopoidea) and a key to the New World genera of Ptilodactylidae. Annals of the Entomological Society of America, 79: 219-234.

WILLIAMS, I. W. 1938. The comparative morphology of the mouthparts of the order Coleoptera treated from the standpoint of phylogeny. Journal of the New York Entomological Society, 46: 245-289.

51. EULICHADIDAE Crowson 1973

by William D. Shepard

Family common name: The forest stream beetles

Family synonym: Lichadidae Forbes 1926

The elateriform shape and the scutellum covered with white setae serve to separate this family.

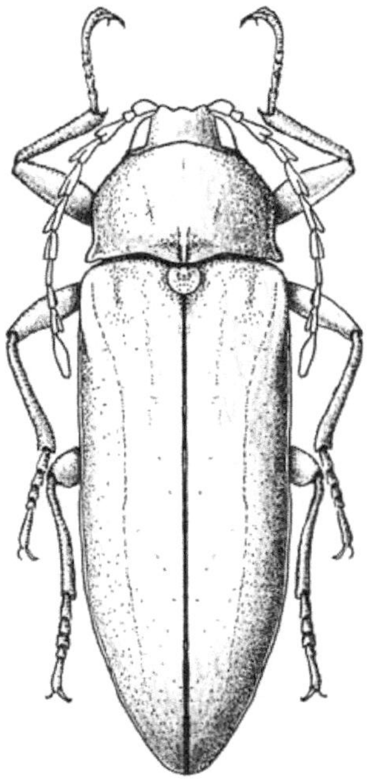

FIGURE 1.51. *Stenocolus scutellaris* LeConte

Description: Body elongate; sides parallel in the middle; brown in color; length 10-15 mm; setae short, brown and decumbent.

Head prognathus; frontal suture distinct; anterior part of clypeus membranous. Eyes moderately prominent, rounded and slightly covered by pronotal angles. Antennae 11-segmented; antennomere 1 conical, slightly arcuate, antennomere 2 small and rounded; serrate in males, subserrate in females. Labrum transverse, rounded apically. Mandibles prominent, acutely margined above, suddenly arcuate apically. Maxillae with last palpal segment elongate, obliquely truncate apically. Labium with mentum transverse, semicircular apically; ligula moderately prominent, emarginate; terminal palpal segment feebly triangular.

Pronotum twice as wide as long, sides convergent anteriorly, margined; posterior angles acute; based bisinuate; disc finely punctured. Prosternum projects posteriorly; mesosternum concave medially; metasternum long. Scutellum round, covered with dense white setae. Elytra complete, with shallow punctures and finely punctulate between punctures. Procoxae oval, with a large trochantin; metacoxae moderately dilated and slightly separate. Legs with tibiae having short spurs; tarsi 5-segmented, long and slender, segments 1-4 slightly pubescent beneath, segment 5 as long as segments 2-4; claws simple; bisetose onychium present.

Abdomen 5-segmented; segments 1-4 subequal, 5th longer.

Eggs are undescribed.

Larva elongate, parallel-sided, straight; length 15-50 mm. Head prognathus and protracted, slightly flattened; stemmata of one large lens on each side. Antennae 3-segmented, 3rd reduced. Frons, clypeus and labrum distinct; labrum upturned. Mandibles with 3 or 4 teeth, mola absent; mesal surface with with an articulated setose process. Maxilla with transverse cardo, elongate stipes; palp 4-segmented; rounded articulated galea and lacinia. Labium partially fused to base of mentum; palps 2-segmented. Thorax with legs well-developed. Abdomen subcylindrical, sides parallel; sternae 1-7 each with a pair of coarse tracheal gills; tergum 9 with pair of fixed upturned urogomphi; anal slit Y-shaped between 2 prominent fleshy lobes. Spiracles biforous; on meso- and metathorax and abdominal segments 1-8; metathoracic spiracles nonfunctional. Larvae are described by Costa and Vanin (1998) and Lawrence (1991).

Pupae are undescribed.

Habitats and habitats. Adults are short lived and present only for a few weeks each summer. They are riparian and found near the larval habitats. Larvae live in streams below moderate to large sized rocks where large particulate detritus (mainly leaves) collects (Shepard 1993). Larvae can be very numerous in streams. The life cycle is probably semivoltine due to the size of the larvae. Larvae are the overwintering stage.

Status of classification. The single American genus of this family usually has been placed within either the Dascillidae or the Ptilodactylidae. Eulichadidae are currently composed of *Stenocolus* from North America and *Eulichas* from Asia (Jäch 1995). *Stenocolus* has a single species and *Eulichas* has 20 species.

Distribution. *Stenocolus scutellaris* LeConte occurs widely across northern California.

CLASSIFICATION OF THE GENERA OF THE UNITED STATES.

Eulichadidae Crowson 1973

Stenocolus LeConte 1853, 1 species, *S. scutellaris* LeConte 1853, California.

Bibliography

COSTA, C. and S. A. VANIN. 1998. Eulichadidae: Description of the larva of *Eulichas dudgeoni* Jäch, with comparative notes on larvae of other *Eulichas* spp., and on *Stenocolus scutellaris* LeConte (Coleoptera). Water Beetles of China, 2: 327-336.

JÄCH, M. A. 1995. Eulichadidae: Synopsis of the species of the genus *Eulichas* Jacobson from China, Laos and Vietnam (Coleoptera). Water Beetles of China, 1: 359-388.

LAWRENCE, J. F. 1991. Eulichadidae (Dryopoidea). Pp. 390-391. *In*: F. W. Stehr (ed.). Immature Insects. Volume 2. Kendall/Hunt. Dubuque, IA. 975 pp.

SHEPARD, W. D. 1993. An annotated checklist of the aquatic and semiaquatic dryopoid Coleoptera of California. Pan-Pacific Entomologist, 69: 1-11.

52. CALLIRHIPIDAE Emden 1924

by Daniel K. Young

Family common name: The callirhipid beetles

The sexually dimorphic, pectinate to flabellate antennae arising on a frontal prominence, large eyes, pronotum with well-developed basal interlocking mechanism, and absence of an articulating prosternal process serve to separate members of this small family [monotypic in North America: *Zenoa picea* (Beauvois)].

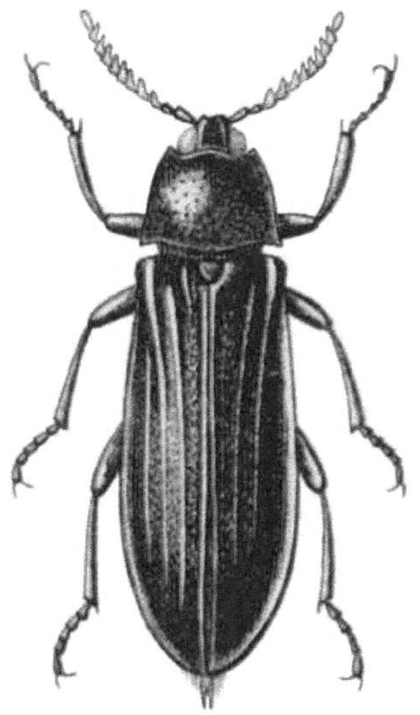

FIGURE 1.52. *Zenoa picea* (Beauvois) (after Lacordaire 1857)

Description: Elongate, somewhat convex, length 9-23 mm; color brownish to black, glabrous or with vestiture of fine setae.

Head elongate-quadrate, somewhat deflexed, inserted slightly into the prothorax. Antennae 11-segmented, flagellum pectinate to flabellate, more highly modified in the male, inserted on frontal prominence between the eyes and above the bases of the mandibles. Clypeus small, triangular, not distinctly separated from frons, labrum with base recessed and mostly concealed beneath edge of clypeus; mandibles stout, apical half sharply bent, apices acute with 2-3 apical teeth; maxillae each with cardo small, stipes covered by lacinia, lacinia modified into a large, triangular plate, galea heavily sclerotized, covered with long, dense setae, palpifer distinct, maxillary palpi filiform, 4-segmented, slightly enlarged apically; labium with gula distinct, gular sutures distinct, well separated, submentum greatly reduced, mentum triangular, ligula sclerotized, elongate, covered with long, dense setae; labial palpi filiform, 3-segmented. Eyes prominent, bulging, entire.

Pronotum narrowed anteriorly, posteriorly wider, but not quite as wide as base of elytra; lateral margins incomplete, posteriorly sinuate with basal interlocking mechanism; surface finely punctate; prosternum short and broad anterad the coxae, narrowly produced between the coxae, but not reaching the mesosternum; prothoracic coxal cavities broadly open behind. Scutellum of moderate size, subtriangular. Mesosternum short, broad, produced between the coxae, reaching the metasternum. Metasternum long and broad, produced anteriorly, meeting mesosternum between the coxae. Legs with prothoracic and mesothoracic trochantins visible, elongate, narrow; prothoracic coxae each prominent, conical, broad at base, slightly separate; mesothoracic coxae prominent, subtransverse; metathoracic coxae transverse, contiguous; trochanters elongate, triangular; femora slender; tibiae slender, the apical spurs prominent; tarsal formula 5-5-5, tarsi simple, empodium with several setae. Elytra entire, apically rounded, surface vaguely costate, the intervals punctate; epipleural fold indistinct; epipleura narrow. Metathoracic wings each with an elongate, closed radial cell.

Abdomen with five visible ventrites, first three connate; surface rugopunctate. Male genitalia with penis moderately stout, apically flattened, acute; parameres articulate, slender, curved; basal piece short, broad (Sharp and Muir 1912). Female genitalia with valvifers long, slender; coxite long, slender, setiferous; styli moderate, apical (Tanner 1927).

Larva elateriform, subcylindrical, heavily sclerotized; length 10-50 mm; body with a few short setae; color testaceous to dark reddish-brown. Head exserted, prognathous, epicranial stem long, frontal arms V-shaped. Antennae small, 1-segmented. Clypeus and labrum distinct; mandibles short, robust, tridentate (but typically appearing 5-dentate); mola absent; ventral mouthparts retracted; maxillae each with transverse cardo, elongate stipes, 4-segmented palpi, truncate mala with notched adoral angle; labium with mentum and submentum fused, ligula forming an apically truncate, 4-dentate plate; palpi 2-segmented. Stemmata absent. Prothorax subequal to meso- and metathorax combined, with 5-segmented legs. Abdomen 9-segmented, the ninth tergum produced into a cap-like, dorsally hinged operculum; segments 2-7 with deeply impressed, paired depressions. Spiracles biforous with accessory tubes, those of abdominal segments 3-8 reduced in size in some. Urogomphi absent.

Habits and habitats. Larvae of *Zenoa picea* are found in old, rotten wood, usually in the presence of white-rot fungi. Adults are apparently short-lived, but are active at night and have been found at lights.

Status of the classification. The position of this group has become somewhat more certain than when they were placed in Rhipiceridae, or as a separate family along with Rhipiceridae in a superfamily Rhipiceroidea (Crowson 1955). In most recent analyses, based upon larvae and adults, the family Callirhipidae falls near Eulichadidae; these, in turn show relationships to Byrrhoidea via Ptilodactylidae and Chelonariidae as well as to basal Elateroidea such as Artematopodidae and Brachypsectridae (Lawrence *et al.* 1995, Lawrence and Newton 1982, 1995).

Several authors have attributed authorship of Callirhipidae to Lacordaire (1857), but Lawrence and Newton (1995) found no use of the family name prior to Emden (1924). They also pointed out that Zenoidae LeConte (1866) would appear to have strict priority as a family group name, but noted that more research would be required and that, in any case, conservation of the more well-known name Callirhipidae would be a more propitious solution.

Distribution. There are eight genera and about 16 species, generally distributed; *Zenoa picea* is the only species known to inhabit the United States. It occurs in the southeastern states, northwest into southern Illinois.

Classification of the Nearctic Genera

Callirhipidae Emden 1924

Zenoa Say 1835, 1 sp., *Z. picea* (Beauvois), southeastern United States, northwest to southern Illinois.

Bibliography

CROWSON, R. A. 1955. The Natural Classification of the Families of Coleoptera. Nathaniel Lloyd. London, 187 pp.

LACORDAIRE, T. 1857. Histoire naturelle des Insectes. Genera des Coléoptères. Tome IV. Roret. Paris. 579 pp. [Atlas pl.43]

LAWRENCE, J. F. and A. F. NEWTON, Jr. 1982. Evolution and classification of beetles. Annual Review of Ecology and Systematics, 13: 261-290.

LAWRENCE, J. F. and A. F. NEWTON, Jr. 1995. Families and subfamilies of Coleoptera (with selected genera, notes, references and data on family-group names). Pp. 779-1006. *In:* J. Pakaluk and S. A. Slipinski, eds., Biology, Phylogeny, and Classification of Coleoptera: Papers Celebrating the 80th Birthday of Roy A. Crowson. Muzeum i Institut Zoologii, Polska Academia Nauk. Warsaw.

LAWRENCE, J. F., N. B. NIKITSKY and A. G. KIREJTSHUK. 1995. Phylogenetic position of Decliniidae (Coleoptera: Scirtoidea) and comments on the classification of Elateriformia (*sensu lato*). Pp. 375-410. *In:* J. Pakaluk and S. A. Slipinski, eds., Biology, Phylogeny, and Classification of Coleoptera: Papers Celebrating the 80th Birthday of Roy A. Crowson. Muzeum i Institut Zoologii, Polska Academia Nauk. Warsaw.

SHARP, D. and F. A. G. MUIR. 1912. The comparative anatomy of the male genital tube in Coleoptera. Transactions of the Entomological Society of London, 1912: 477-642, pls 42-78.

TANNER, V. M., 1927. A preliminary study of the genitalia of female Coleoptera. Transactions of the American Entomological Society, 53: 5-50.

Superfamily ELATEROIDEA Leach 1815

53. ARTEMATOPODIDAE Lacordaire 1857

by Daniel K. Young

Family common name: The artematopodid beetles

Family synonyms: Artematopidae; including Eurypogonidae Bøving and Craighead 1931

These beetles somewhat resemble the elaterids, but the prosternal keel is not as well developed as in that group, and the frontal ridge is absent. A curious feature of artematopodids is the tongue-like process associated with the apicoventral aspect of the elytra.

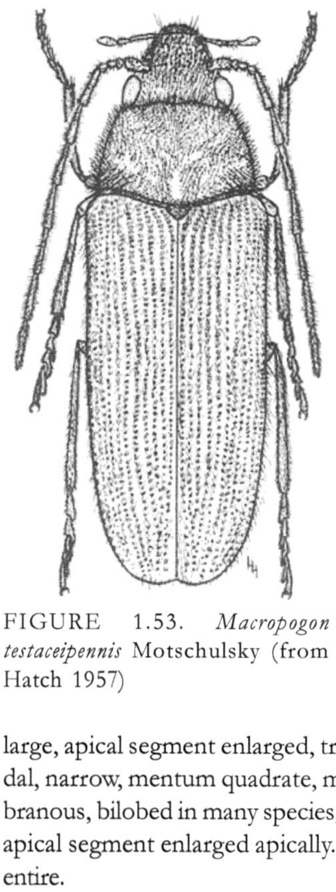

FIGURE 1.53. *Macropogon testaceipennis* Motschulsky (from Hatch 1957)

Description: (after Crowson 1973, Lawrence 1982, 1995) Ovate to elongate, moderately to strongly convex; length 3-10 mm; color brownish to black, vestiture consisting of moderately long setae.

Head slightly to strongly deflexed, inserted deeply into the prothorax; surface coarsely to finely punctate. Antennae 11-segmented, subserrate to serrate, inserted on the front between the eyes well above the mandibles, close together in some species. Clypeus distinctly separated by an impressed line; labrum more or less membranous, partially or entirely concealed beneath the clypeus; mandibles moderately-sized, stout, curved, apices subacute; maxillary palpi 4-segmented, large, apical segment enlarged, triangular; submentum trapezoidal, narrow, mentum quadrate, membranous; ligula large, membranous, bilobed in many species; labial palpi 3-segmented, large, apical segment enlarged apically. Eyes lateral, of moderate size, entire.

Pronotum broader than the head; convex, quadrate, with evenly arcuate lateral margins, base slightly sinuate but lacking crenulations and interlocking mechanism; borders margined; surface punctate; prosternum broad anterad the coxae, commonly with elevated keel between coxae; prothoracic coxal cavities open behind. Scutellum oval or triangular. Mesosternum short; metasternum broad. Prothoracic trochantins exposed; prothoracic coxae large transverse, separate; mesothoracic coxae subconical, narrowly separate; metathoracic coxae transverse; trochanters of moderate size, triangular; femora slender; tibiae slender, with prominent apical spurs; tarsal formula 5-5-5, tarsomeres 2-4 or 3-4 lobed beneath, claws simple or pectinate, onychium with two bristles in some taxa. Elytra entire, apically rounded, striate-punctate, or vaguely costate, intervals rugopunctate; epipleural fold broad, entire, sometimes abruptly narrowed along basal third; elytra with ventral (underside) tongue-like process near apex; metathoracic wings each with spirally rolled apex, well-developed radial sector, and a remnant of the first radiomedial crossvein.

Abdomen with 4-5 visible, connate ventrites. Male genitalia of the trilobed type; penis complex, with a broad, apically rounded dorsal piece and a smaller ventral part produced into two broad basal struts; parameres large, arcuate; basal piece smaller, complete (Sharp and Muir 1912). Female genitalia with the valvifers with a long baculum; coxite with a large baculum at the base; stylus small (Tanner 1927).

Larva (after Cooper 1991, Costa *et al.* 1985) elateriform, elongate, subcylindrical to slightly flattened, well sclerotized, especially dorsally. Head with the epicranial stem short to absent, frontal arms V-shaped; antennae short, 3-segmented, sensorium ventrolateral, half or more as long as cylindrical apical segment. Labrum free, deeply excavate mesally; mandibles lacking a mola, with 3-4 apical teeth; maxillae with slightly developed articulating area, distinct galea and lacinia, 4-segmented palpi; labium with short, broad ligula bearing two pairs of short, subapical setae and 2-segmented palpi. Thorax with 5-segmented, stout, spinous legs; tarsunguli bisetose. Abdomen with segments 1-8 simple, ninth weakly to crenulately margined; lacking urogomphi, tenth segment oriented ventrally, enclosing a protrusible anal papilla; spiracles biforous.

Habits and habitats. Adults have been found by sweeping and beating forest understory foliage; little is known of their habits. Larvae of several species have been consistently associated with moss mats growing on granitic boulders, where they tunnel into the rhizoids. Those of *Macropogon rufipes* Horn and *Eurypogon harrisii* (Westwood) have been associated with the moss *Paraleucobryum longifolium* (Hedw.)Loeske; those of *M. sequoiae* Hopping have been associated with several moss genera (Cooper 1991, Lawrence 1982).

Status of the classification. Formerly included in a vaguely defined Dascillidae, artematopodid taxa have also been associated with Byrrhoidea (= Dryopoidea) near Callirhipidae, or with Eulichadidae, Callirhipidae, and Brachypsectridae in a separate superfamily, Artematopodoidea. The taxon is thought to represent a relatively basal group of Elateroidea (Crowson 1973, Lawrence 1982, Lawrence and Newton 1995).

Distribution. This is a small family of approximately 60 species arranged in two subfamilies and seven genera: Electribiinae (*Electribius*, Mexico, Central America, Baltic amber of northern Europe); Artematopodinae (*Allopogonia*, California; *Artematopus*, Central and South America; *Carcinognathus*, South America; *Ctesibius*, Mexico and Central America; *Eurypogon*, North America, Italy, Japan; *Macropogon*, North America, Asia)(Crowson 1975, Lawrence 1982, 1995, Lawrence and Newton 1995). Three genera and 8-9 species are known from the United States and Canada (Cooper 1991).

Key to the Genera of America north of Mexico

1. Prosternum prolonged, meeting mesosternum and limited on each side by elevated line, divergent anteriorly 2
— Prosternum neither prolonged nor meeting mesosternum in front, convex, without raised lines .... *Allopogonia*

2(1). Antennomeres 2-4 very short, together not longer than fifth (Fig. 1) .... *Macropogon*
— Antennomeres 2-3 short, together equal to length of fourth .... *Eurypogon*

Classification of Genera of America north of Mexico

Artematopodidae Lacordaire 1857

Artematopodinae Lacordaire 1857
(including Ctesibiinae Crowson 1973; Eurypogoninae Bøving and Craighead 1931; Macropoginae LeConte 1861)

Allopogonia Cockerell 1906, 1 sp., *A. villosus* (Horn), California.
Allopogon Horn 1880, not *Allopogon* Schiner 1866 [Diptera]

Eurypogon Motschulsky 1859, 4 spp., *E. californicus* Horn, California; *E. confusus* Fall, California; *E. harrisii* (Westwood), Indiana, Michigan; *E. niger* (Melsheimer), Quebec, Indiana, Kansas, New York, Pennsylvania, Virginia. (partial keys to species in Horn 1880, and Downie and Arnett 1996)
Dasypogon LeConte 1861
Sclerodes Westwood 1862

Macropogon Motschulsky 1859, 3 spp., *M. piceus* LeConte, Alberta, British Columbia, New Brunswick, Quebec, Illinois, Michigan, Montana, New Hampshire, Oregon; *M. sequoiae* Hopping, California; *M. testaceipennis* Motschulsky, British Columbia, California, Oregon, Washington. (key to species in Hopping 1936)

Bibliography

COOPER, K. W. 1991. Artematopidae. Pp. 407-409. *In:* F. W. Stehr, ed. Immature Insects. Volume 2. Kendall/Hunt. Dubuque, IA.

COSTA, C., S. A. CASARI-CHEN, and S. A. VANIN. 1985. Larvae of Neotropical Coleoptera. XII. Artematopoidea, Artematopidae. Revista Brasileira de Entomologia, 29: 309-314.

CROWSON, R. A. 1973. On a new superfamily Artematopoidea of polyphagan beetles, with the definition of two new fossil genera from the Baltic Amber. Journal of Natural History, 7: 225-238.

CROWSON, R. A. 1975. The evolutionary history of Coleoptera, as documented by fossil and comparative evidence, Pp. 47-90. *In:* Atti del X Congresso Nazionale Italiano di Entomologia. Coppini. Firenza.

DOWNIE, N. M. and R. H. ARNETT, Jr. 1996. The Beetles of Northeastern North America. Volume I: Introduction; Suborders Archostemata, Adephaga, and Polyphaga thru Superfamily Cantharoidea. The Sandhill Crane Press. Gainesville, FL, 880 pp.

HATCH, M. H. 1957. The beetles of the Pacific Northwest. Part II. Staphyliniformia. University of Washington Publications in Biology, 16: 384 pp.

HOPPING, R. 1936. A revision of the genus *Macropogon*. Pan-Pacific Entomologist, 12: 45- 48.

HORN, G H. 1880. Synopsis of the Dascyllidae of the United States. Transactions of the American Entomological Society, 8: 76-115, 1 plate.

LAWRENCE, J. F. 1982. Coleoptera. Pp. 482-553. *In:* S. B. Parker, ed. Synopsis and Classification of Living Organisms. Volume 2. McGraw-Hill. NY.

LAWRENCE, J. F. 1995. *Electribius* Crowson: alive and well in Mesoamerica, with notes on Ctesibius Champion and the classification of Artematopodidae. Pp. 411-432. *In:* J. Pakaluk and S. A. Slipinski, eds. Biology, Phylogeny, and Classification of Coleoptera: Papers Celebrating the 80th Birthday of Roy A. Crowson. Muzeum i Institut Zoologii, Polska Academia Nauk. Warsaw.

LAWRENCE, J. F. and A. F. NEWTON, Jr. 1995. Families and subfamilies of Coleoptera (with selected genera, notes, references and data on family-group names). Pp. 779-1006. *In:* J. Pakaluk and S. A. Slipinski, eds. Biology, Phylogeny, and Classification of Coleoptera: Papers Celebrating the 80th Birthday of Roy A. Crowson. Muzeum i Institut Zoologii, Polska Academia Nauk. Warsaw.

SHARP, D. and F. A. G. MUIR. 1912. The comparative anatomy of the male genital tube in Coleoptera. Transactions of the Entomological Society of London, 1912: 477-642, pls 42-78.

TANNER, V. M., 1927. A preliminary study of the genitalia of female Coleoptera. Transactions of the American Entomological Society, 53: 5-50.

54. BRACHYPSECTRIDAE Bøving and Craighead 1931

by Daniel K. Young

Family common name: The Texas beetles

The acute, carinate hind pronotal angles and elongate prosternum forming a "chin piece" beneath the cranium give adults of this aberrant beetle a somewhat elaterid-like appearance, but the transverse prothoracic coxae, lack of a functional clicking mechanism, and five free abdominal ventrites (the first of which is keeled) readily separate members of this small family from elaterids.

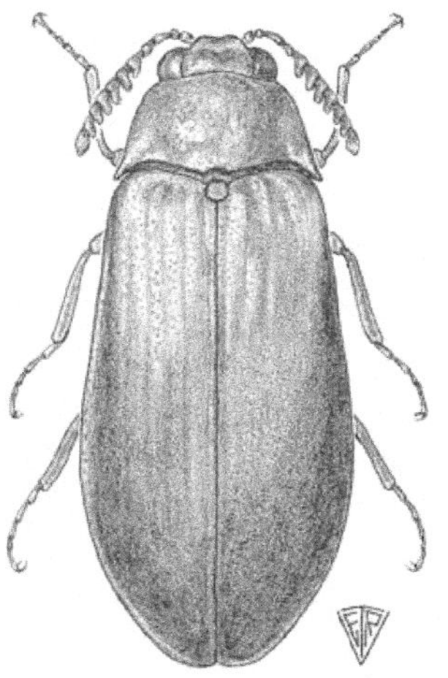

FIGURE. 1.54. *Brachypsectra fulva* LeConte

Description: (after Lawrence 1982, in part) Oblong, somewhat depressed, broad; length 4-7 mm; color yellowish-brown; vestiture consisting of fine, sparse pubescence.

Head deflexed, short and broad, surface coarsely punctate with large central depression. Antennae 11-segmented, antennomeres 5-11 serrate to weakly pectinate, inserted on the frons which is narrowed, with the clypeal area broader beyond. Clypeal suture indistinct; labrum small, oval; mandibles short, stout, with acute apices. Maxillary palpus with apical segment elongate, apex acute. Labial palpus with apical segment acute. Eyes large, prominent, oval.

Pronotum short, broader than long, narrowed in front with apical angles acutely produced, lateral margins fine but complete; with a weakly developed interlocking mechanism basally. Prosternum large, nearly as long as broad, forming a "chin piece" beneath head, intercoxal process well developed, articulating with mesosternal fossa; prothoracic coxal cavities open behind. Scutellum oval, shield-shaped. Mesosternum reduced; metasternum long, well developed. Legs with trochantins slightly exposed; prothoracic coxae transverse; mesothoracic coxae oval, metathoracic coxae slightly grooved for the reception of the femora. Tibial spurs absent; tarsal formula 5-5-5, segments slender; claws simple. Elytra entire, broad, weakly striate; epipleural fold entire. Metathoracic wings short, lacking a closed anal cell.

Abdomen with five visible ventrites, the first ventrite keeled between the metathoracic coxae. Parameres of male genitalia with excurved apices.

Larva broad, flattened, with pectinate branched lateral thoracic and abdominal processes, a vestiture of highly modified, scale-like setae, and with a prominent spine-like ninth abdominal tergum (Bøving and Craighead 1931, Lawrence 1991).

Habits and habitats. Larvae are slow moving, ambush predators and occur beneath bark, under pieces of wood, or in rock crevices. When a spider or other small arthropod walks onto the flat, sprawling, motionless body, the flexible anterior end and spiniform ninth tergum are abruptly raised, pinning the prey between the spine and the perforate, falciform mandibles. Pupation occurs in a delicate, wide-laced, silken cocoon. Adults have been found under bark; they are generally quite rare and are probably short-lived (Fleenor and Taber 1999, Lawrence 1982, 1991).

Status of the classification. LeConte included this species in Rhipiceridae when he described it (1874), but in LeConte and Horn (1883) it was placed as a tribe Brachypsectrini of Dascillidae (Dascilloidea). It is thought to represent a fairly basal lineage of a broadly conceived Elateroidea (i.e., including the artematopodoid and cantharoid lineages) (Barber 1905, Blair 1930, Crowson 1973, Kasap and Crowson 1975, Lawrence and Newton 1995).

Distribution. *Brachypsectra fulva* LeConte is known from southwestern North America; larvae can be locally rather common at times. Other species occur in India and Asia, an undescribed species is also known form northern Australia, and a Miocene record (larva) has been reported from Dominican amber (Lawrence and Britton 1994, Poinar 1992).

Classification of the Species of the United States

Brachypsectridae Bøving and Craighead 1931

Brachypsectra LeConte 1874, 1 sp., *B. fulva* LeConte, Arizona, Colorado, New Mexico, Texas, Utah, and California.

Bibliography

BARBER, H.S. 1905. Notes on *Phengodes* in the vicinity of Washington, D.C. Proceedings of the Entomological Society of Washington, 7: 196-197.

BLAIR, K. G. 1930. *Brachypsectra* LeC. -- The solution of an entomological enigma. Transactions of the Royal Entomological Society of London, 78: 45-50.

BØVING, A. G. and F. C. CRAIGHEAD. 1931. An illustrated synopsis of the principal larval forms of the order Coleoptera. Entomologica Americana (N.S.), 11 (1930): 1-351 (larvae). Pp. 13, 46, pls. 74, 75.

CROWSON, R. A. 1973. On a new superfamily Artematopoidea of polyphagan beetles, with the definition of two new fossil genera from the baltic amber. Journal of Natural History, 7: 225-238.

FLEENOR, S. B. and S. W. TABER. 1999. Review of *Brachypsectra* LeConte with a new record of the Texas Beetle (*B. fulva* LeConte; Coleoptera: Brachypsectridae). Coleopterists Bulletin, 53: 359-364.

KASAP, H. and R. A. CROWSON. 1975. A comparative anatomical study of Elateriformia and Dascilloidea (Coleoptera). Transactions of the Royal Entomological Society of London, 126: 441-495.

LAWRENCE, J. F. 1982. Coleoptera. Pp. 482-553. *In:* S. B. Parker, ed., Synopsis and Classification of Living Organisms. Volume 2. McGraw-Hill. NY.

LAWRENCE, J. F. 1991. Brachypsectridae. Pp. 421-422. *In:* F. W. Stehr, ed, Immature Insects. Volume 2. Kendall/Hunt. Dubuque, IA.

LAWRENCE, J. F. and E. B. BRITTON. 1994. Australian Beetles. Melbourne University Press. Carlton, Victoria, x+192 pp.

LAWRENCE, J. F. and A. F. NEWTON, Jr. 1995. Families and subfamilies of Coleoptera (with selected genera, notes, references and data on family-group names). Pp. 779-1006. *In:* J. Pakaluk and S. A. Slipinski, eds., Biology, Phylogeny, and Classification of Coleoptera: Papers Celebrating the 80th Birthday of Roy A. Crowson. Muzeum i Institut Zoologii, Polska Academia Nauk. Warsaw.

LECONTE, J. L. 1874. Descriptions of new Coleoptera chiefly from the Pacific slope of North America. Transactions of the American Entomological Society, 5: 43-72.

LECONTE, J. L. and G. H. HORN. 1883. Classification of the Coleoptera of North America. Smithsonian Miscellaneous Collections 507: xxxviii + 567 pp.

POINAR, G. O., Jr. 1992. Life in Amber. Stanford University Press. Stanford, CA, xiii + 350 pp.

55. CEROPHYTIDAE Latreille 1834

by Paul J. Johnson

Family common name: The rare click beetles

The strongly serrate to pectinate antennae, short body, deflexed head with protruding frons, small metacoxae lacking plates, and long trochanters distinguish adult cerophytids.

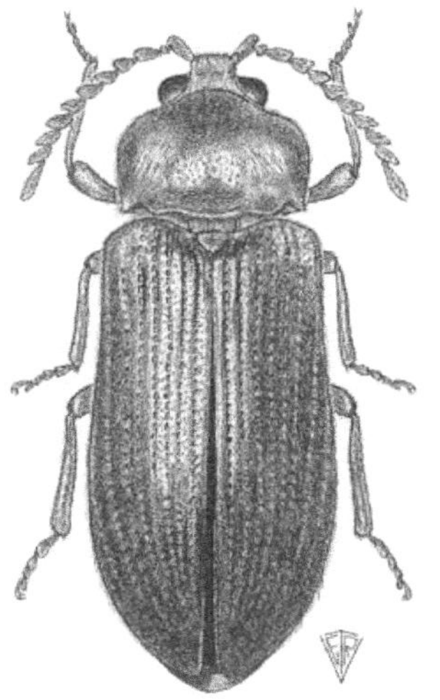

FIGURE 1.55. *Cerophytum pulsator* (Haldeman)

Description: Body elongate, shallowly convex; length approximately 5-9 mm; color red-yellow or red-brown to black; vestiture moderately dense, dark.

Head hypognathous, narrower than prothorax, deeply inserted into prothorax. Eyes large, bulging, coarsely faceted. Antenna 11-segmented, sexually dimorphic, strongly serrate (female) to pectinate (male); insertions moderately close on lateral portions of elevated frons; antennomere 1 long, flagellar attachment eccentric; antennomere 2 small. Labrum articulate, transverse, visible. Mandibles slender, arcuate, acute at apex. Maxilla with palpus 4-segmented, apical palpomere subsecuriform; labium with mentum small, narrowed anteriorly, palpus 3-segmented, apical palpomere subsecuriform.

Thorax with prothorax loosely articulate with mesothorax; pronotum transverse, hind angles briefly extended, lateral margins rounded; prosternum with a broadly rounded chin piece, posterior median process broadened behind coxae, with narrow ventral process fitting into mesosternal cavity; hypomeron not delimited laterally by pronotal carina; procoxal cavities open posteriorly. Scutellum moderately small, subtriangular. Mesosternum short, fused to metasternum; median with deep, narrow cavity to receive prosternal process; mesocoxae widely separated. Elytra entire, rounded at apex; striae with deep window punctures; intervals flat to shallowly convex, finely punctured; epipleura broad basally, narrow apically. Metasternum quadrate; metacoxae approximate medially, widening laterally, not laminate, shallowly impressed. Metathoracic wing with radial cell triangular, long radial sector, medial field with wedge cell absent. Legs not retractable (mutually or into cavities); meso- and metatrochanters elongate; femora fusiform, profemur excavate to receive tibia; tibiae slender, spinose; tarsi 5-5-5, tarsomere 2 with ventral setose pad, tarsomere 3 shallowly emarginate and with short ventral spongy pads, tarsomere 4 deeply emarginate and with large ventral lobate pad; claw pectinate.

Abdomen with 5 ventrites; ventrites 1-4 connate. Aedeagus trilobed; penis subparallel, tapering apically, apex narrowly obtuse; parameres separate, each with large membranous apical lobe; basal piece reduced, forming a concave plate wrapped around penial base. Female gonocoxites lightly sclerotized, elongate, narrowly subtriangular; styli narrowly cylindrical, attached eccentrically; bacula long slender, divergent.

Larvae are lightly sclerotized, grub-like; densely pubescent. Head small, prognathous, reduced, lacking epicranial suture; antenna 3-segmented, with large sensorium; single pair stemmata; mandibles reduced, flattened, styliform; maxilla with styliform mala, partially divided; palpus 3-segmented on large palpiger; labium a 5-toothed plate fused to head capsule, palpus absent. Prothorax with ventral supporting rods. Legs short, 5-segmented; front leg enlarged, with darkly sclerotized cleft tarsungulus. Mesothorax, metathorax, and abdominal segments 1-8 with distinctive pleural lobes, and tergal glands at posterior margin. Tergite 9 short, lacking urogomphi. Tergite 10 small, posteroventral. Spiracles biforous, on mesothorax and abdominal segments 1-8; closing apparatus present.

Habits and habitats. Little is known about cerophytid ecology. Adults are usually captured in traps or netted in late afternoon flights, by sweeping, or at light traps, in lowland and mid-elevational forested environments. Specimens have also been found in leaf litter and under ground debris (Steiner 2000). Larvae are associated with various angiosperm trees, apparently feeding in the laminated rotted xylem (Buysson 1910, Horion 1953, Mamaev 1978), but remain poorly known (Lawrence 1991). Cerophytid adults have an ability to "jump," apparently as do species of Elateridae, Throscidae, and Eucnemidae.

Status of the classification. There is only one extant genus, *Cerophytum*. North American species were last revised by Horn (1886). The taxonomy of *Cerophytum* species seems stable and identification of described world species is possible with Soares and Perrachi (1964) and Golbach (1983). Undescribed species are known. The most recent phylogenetic position is as the sister-group to Eucnemidae+ Throscidae+Elateridae (Lawrence and Newton 1995). Two fossil species of cerophytids are known from the Upper Cretaceous of Russia (Carpenter 1992).

Distribution. There are 10 described species of *Cerophytum*, occurring in the nearctic, Europe, and the neotropics, with undescribed species known from the latter region. Two species are known in the nearctic fauna. *Cerophytum pulsator* (Haldeman) is recorded from southern Indiana to Pennsylvania and south to

Florida (Peck and Thomas 1998; Steiner 2000), while *C. convexicolle* (LeConte) is only known from California and southwestern Oregon.

Classification of the Nearctic Species

Cerophytidae Latreille 1834

Cerophytum Latreille 1809; 2 spp., *C. pulsator* (Haldeman), eastern U.S.; *C. convexicolle* LeConte, California and Oregon.

Chorea Haldeman 1846

Bibliography

BUYSSON, H. du 1910. Matériaux pour servir à l'histoire des insectes de l'Aulne. Annales de la Société Entomologique de France, 79: 105-128.

CARPENTER, F. M. 1992. Superclass Hexapoda. Treatise on Invertebrate Paleontology, Part R, Arthropoda 4, 4: 279-655.

GOLBACH, R. 1983. Primera cita de la familia Cerophytidae (Coleoptera) para Paraguay, Bolivia y la Argentina. Acta Zoologica Lilloana, 37: 131-139.

HORION, A. 1953. Faunistik der mitteleuropäischen Käfer, Band III: Malacodermata, Sternoxia (Elateridae bis Throscidae). Horion. Munich.

HORN, G.H. 1886. A monograph of the species of the subfamilies Eucneminae, Cerophytinae, and Perothopinae inhabiting the United States. Transactions of the American Entomological Society, 13: 5-58.

LAWRENCE, J. F. 1991. Cerophytidae (Elateroidea). Pp. 409-410. *In:* F. W. Stehr, ed. Immature Insects. Kendall/Hunt. Dubuque, IA.

LAWRENCE, J. F. and A. F. NEWTON, Jr. 1995. Families and subfamilies of Coleoptera (with selected genera, notes, references and data on family-group names). Pp. 779-1006. *In:* J. Pakaluk and S.A. Slipinski, eds., Biology, Phylogeny and Classification of Coleoptera: papers Celebrating the 80th Birthday of Roy A. Crowson. Muzeum i Instytut Zoologii PAN. Warsaw.

MAMAEV, B. M. 1978. Morphology of the larvae of *Cerophytum elateroides* Latr. and the phylogenetic ties of the Cerophytidae (Coleoptera) family. Doklady Akademia Nauk S.S.S.R., Zoology, 238: 107-1008.

PECK, S. B. and M. C. THOMAS. 1998. A distributional checklist of the beetles (Coleoptera) of Florida. Arthropods of Florida and Neighboring Land Areas, 16: i-vii + 1-180.

SOARES, B. A. M. and A. L. PERRACHI. 1964. Sobre a presença de cerofitidas no Brasil, com a descrição de duas nova especies (Insecta: Coleoptera, Cerophytidae). Anais do Segundo Congresso Latino-Americano de Zoologia, 1962, 1: 127-143.

STEINER, W. E., Jr. 2000. Records and habitats of the "rare click beetle," *Cerophytum pulsator* (Haldeman), in Virginia and Maryland (Coleoptera: Cerophytidae). Banisteria, 15: 43-45.

56. EUCNEMIDAE Eschscholtz 1829

by Jyrki Muona

Family common name: The false click beetles

Family synonyms: Eucnemididae Lacordaire 1857, Melasidae Leach 1817, Phylloceridae Reitter 1905, Perothopidae Lacordaire 1857

These beetles resemble Elateridae but the strongly deflexed head, the membranous, mostly hidden labrum and the connate abdominal segments serve to distinguish this family.

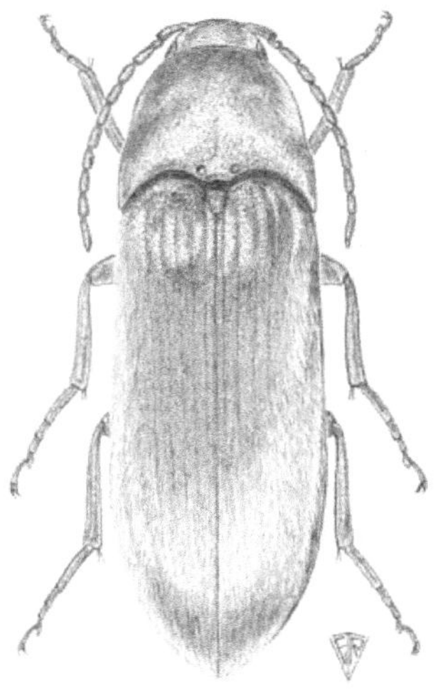

FIGURE 1.56. *Onichodon orchesoides* Newman

Description: Shape elongate, convex; size 2-30 mm in length, usually under 10 mm; color brownish to piceous, some with pale markings; vestiture mostly of short, sparse, depressed setae.

Head strongly deflexed, resting against the prosternum, retracted into the prothorax to the eyes; surface punctate, sometimes rugose punctate. Antennae with 11 antennomeres, moniliform, filiform, serrate, pectinate or flabellate, may be received in grooves along notopleural suture or hypomera; antennae inserted well forward on the front, close to the eyes and mandibles. Labrum membranous, hidden, although slightly visible in *Melasis*; mandibles moderate, curved, may be outwardly flattened, apices acute, bifid, or trifid; maxillary palpi with four palpomeres, the apical palpomere enlarged, triangular, or acute; labial palpi with three palpomeres, slender. Eyes lateral, moderate, slightly convex, round, or oval.

Pronotum broader than the head, convex, quadrate, or arcuate. anteriorly, posterior angles produced, acute; borders mostly finely margined laterally; surface punctate; hypomera wide; prosternum long in front of coxae, prosternal process produced between the coxae and fitted into a mesosternal fossa; procoxal cavities open behind. Mesosternum short; mesocoxal cavities open behind. Metasternum long and broad. Legs without exposed trochantins, middle trochantins exposed in *Perothops*; anterior coxae rounded, not prominent, separate; middle coxae rounded, flat, separate; hind coxae transverse, with femoral plates in some greatly expanded internally; trochanters of the fore and middle pair small, interstitial, hind pair triangular; femora swollen; tibiae slender, apical spurs obscure or moderate, prominent in some; tarsi 5-5-5, the tarsomeres slender, the penultimate may be somewhat lobed beneath, males frequently with sex-comb on first tarsomere; claws simple, dentate, or pectinate. Scutellum moderate, oval, obovate, or triangular. Elytra entire, rounded apically; striae punctate, intervals smooth; epipleural fold indistinctly separated, broad basally. Wing venation and folding pattern as in Elateridae.

Abdomen with five visible sterna, the sutures entire; surface punctate, microrugose in some. Male genitalia of the trilobed type but highly variable and in some greatly modified; penis elongate, apex acute, with two basal struts; parameres variable but mostly large, basally fused, distally spatulate; pars basalis mostly small, forming a round sclerite ventrally.

Female ovipositor with the valvifers reduced to a long slender baculum; coxite partly divided ventrally; stylus small. Male and female genitalia described in detail in Muona (1993b).

Larvae elongate, subcylindrical, either soft and whitish as buprestid larvae or hard, yellow or brown as many elaterid larvae; in some the prothoracic region is enlarged; length 7 to 25 mm or more; a few setae or none on the body, color white, yellowish or brown. Head specialized, prognathous, depressed, the lateral margins may be strongly serrate; epicranial suture absent Antennae vestigial or reduced to very small, two-segmented rudiments. Clypeus and labrum absent; mandibles with curved, dentate outer margins projecting cephalolaterad; maxillae, if present, palpiform, projecting beyond the margin of the head, or vestigial; labium reduced to very small projection ending in two reduced palpi. Thorax without legs, or legs reduced to either setae or very small, three-segmented appendages; prothorax may be twice as wide as the other segments, or it may be smaller than the other segments and taper toward the head. Abdomen nine-segmented; may have segments one to seven or two to eight subdivided into two annulets, a large one with the spiracle and a smaller one without; ninth segment may be disc-shaped, rounded, or depressed in the center; very small depressions or distinct; small asperities may occur on the ninth segment. Spiracles with conspicuous rings, annular-biforous on the mesothorax and abdominal segments one to eight. Urogomphi mostly either absent or very small. Descriptions of some immatures can be found in Bøving and Craighead (1930), Calder *et al.* (1993), Ford and Spilman (1979), Mamaev (1976), Mosher (1918), and Otto and Young (1998).

Habits and habitats. The larvae of most of the species develop in rotten wood. A few Melasinae species prefer coniferous trees (*Melasis*, *Hylis*), but most species are found in white-rot wood of deciduous trees. The larvae most likely feed on the fungus rather than the wood, using extraoral digestion The Pale-

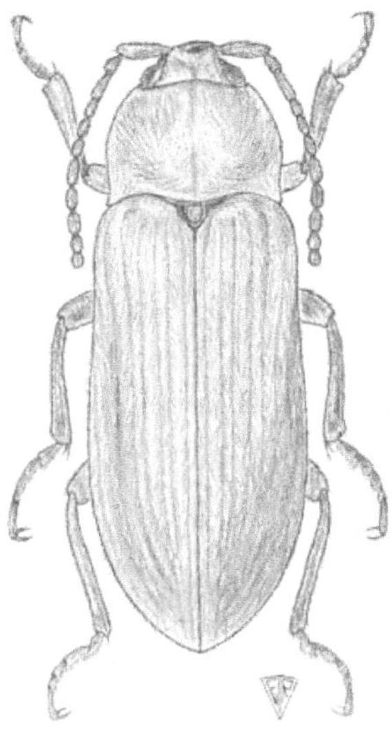

FIGURE 2.56. *Perothops mucida* (Say).

arctic *Phyllocerus* species develop in soil, and this may be the case with *Anelastes* and *Perothops* as well. Adult eucnemids are excellent fliers. Additional biological information can be found in Burke (1917), Dury (1904), Ford and Spilman (1979), Horn (1886), Kirk (1922), Knull (1946, 1947), Muona (1993b), Otto and Young (1998), Van Horn (1909), and in the regional studies cited below.

Status of the classification. The results of the recent comprehensive work of Muona (1993b, 1995, 2000) are included in this family section. Additional systematic information can be found in Bonvouloir (1871, 1872, 1875), Brown (1940), Cabos (1979), Eschscholtz (1829), Fall (1925, 1928, 1933), Guérin-Méneville (1843), Hopping (1926), Horn (1871, 1872, 1878, 1886, 1891), Knull (1946, 1947, 1948), Lawrence (1988), Melsheimer (1846), Muona (1987, 1995), and Van Dyke (1932, 1943, 1945, 1953). Muona (1993a) described fossil Eucnemidae from Baltic amber. Schenkling (1928) provides a catalog to taxa.

Distribution. There are approximately 1,300 described species, occurring in all areas; of these 85 species are known from America north of Mexico, 18 of these are described in Muona (2000). Regional studies can be found in Beaulieu (1919-1922, Canada), Blatchley (1910, Indiana), Dury (1888, Ohio), Hatch (1971, northwestern US), Horn (1890, Central America), LeConte (1852, 1866, 1878, US), and Muona (1996, northeastern US).

Key to the Nearctic Genera

1. Protibiae with two well developed apical spurs .. 2
— Protibiae mostly with one apical spur, rarely with none 5

2(1). Claws serrate .... *Perothops*
— Claws simple 3

3(2). Antennomere 11 asymmetrical, apically excavated, metacoxal plates abruptly narrowing in medial third .... *Anelastes*
— Antennomere 11 entire, metacoxal plates evenly narrowing laterad 4

4(3). Antennomeres 4-11 about equal in size, clypeus trilobed apically .... *Schizophilus*
— Antennomeres 9-11 enlarged, clypeus evenly convex apically .... *Palaeoxenus*

5(1). Antennomeres 9-11 enlarged, hypomera unmodified .... *Euryptychus*
— Antennomeres 9-11 not enlarged, or if so, hypomera with lateral antennal grooves 6

6(5). Hypomera with basally closed lateral antennal grooves, metatibiae with sharp angle between lateral and caudal surfaces, median lobe of aedeagus always free 7
— Hypomera simple or with notosternal or basally open lateral antennal grooves, metatibiae with rounded angle between lateral and caudal surfaces, median lobe of aedeagus mostly fused to lateral lobes 11

7(6). Antennomeres 9-11 enlarged .... *Proutianus*
— Antennomeres 8 and 9 about equal in size 8

8(7). Abdomen with tarsal grooves .... *Dendrocharis*
— Abdomen without tarsal grooves 9

9(8). Median wall of antennal grooves with dense fringe of excretory hairs, hypomera with deep sensory pits 10
— Median wall of antennal grooves and hypomera simple 36

10(9). Dorsal surface with well developed vestiture .... *Idiotarsus*
— Dorsal surface practically naked .... *Eucnemis*

11(6). Meso- and metatibiae without spines or spine combs on lateral surfaces, hypomera simple or with notosternal antennal grooves 12
— Meso- and metatibiae with spines or spine combs on lateral surfaces, hypomera simple or with lateral antennal grooves 25

12(11). Hypomera about parallel-sided 13
— Hypomera triangular in shape 14

13(12). Tibiae strongly flattened, wide .... *Melasis*
— Tibiae slender .... *Isoriphis*

14(11). Lateral pronotal ridge minutely serrate, hypomera often with notosternal antennal grooves 15
— Lateral pronotal ridge smooth, hypomera simple . 22

15(14). Antennomeres 4-10 biserrate or bipectinate .... *Sarpedon*
— Antennomeres never biserrate or bipectinate .. 16

16(15). Frons with week lateral keels .... *Dirrhagofarsus*
— Frons without lateral keels 17

17(16). Elytral apices produced in lateral view 35
— Elytral apices evenly curved in lateral view 18

18(16). Antennomeres 2 and 3 combined at most as long as 4 .... *Entomophthalmus*
— Antennomeres 2 and 3 combined always distinctly longer than 4 19

19(18). Metacoxal plates about parallel-sided 20
— Metacoxal plates distinctly, mostly strongly narrowing laterad 21

20(19). Form cylindrical, notosternal antennal grooves parallel sided, always well-defined at least laterad .... *Rhagomicrus*

— Form "top-heavy", notosternal antennal grooves widening caudad, mostly poorly delimited .... *Adelothyreus*

21(19). Frons between antennal sockets at least half as wide as distance between eyes, usually distinctly more .... *Golbachia*

— Frons between antennal sockets less than half as wide as distance between eyes, usually distinctly less .... *Microrhagus*

22(14). Metacoxal plates strongly narrowing laterad, body elongated, flattened 23

— Metacoxal plates parallel, body cylindrical 24

23(22). Antennomeres 8-11 abruptly longer than previous ones .... *Epiphanis*

— Antennomeres 5-10 about equal in length .... *Hylis*

24(22). Pronotum with median groove and conspicuous lateral impressions .... *Hylochares*

— Pronotum at most with median groove .. *Xylophilus*

25(11). Mandibles slender, hypomera and male protarsomere 1 unmodified, median lobe of adeagus notched apically .... *Hemiopsida*

— Mandibles stout with secondary basal tooth, hypomera often with lateral antennal grooves, male protarsomere 1 nearly always with sex comb, median lobe of aedeagus narrowly and deeply bifurcate apically 26

26(25). Antennae variable, but antennomeres 4 and 5 always about equal to 6 and 7, protibiae with simple apex 27

— Antennomeres 6-11 flattened, enlarged, protibiae with expanded apex, conspicuously so in males .... *Nematodes*

27(26). Elytral epipleura smooth, grooved basally 28

— Elytral epipleura punctate, not grooved basally 29

28(27). Elytral epipleura grooved at least for basal third, antennomeres 3-11 elongated, at most feebly serrate .... *Fornax*

— Elytral epipleura grooved only basally, antennomeres 3-10 pectinate or deeply and sharply serrate .... *Diphytaxis*

29(27). Claws with basal tooth 30

— Claws simple 31

30(29). Tarsomere 4 very wide, more than twice as wide as tarsomere 5, antennae elongated, antennomeres 9-11 extending beyond pronotum even in large females .... *Serrifornax*

— Tarsomere 4 less than twice as wide as 5, antennae stouter, antennomeres 9-11 extending beyond pronotum only in exceptionally small males .... *Onichodon*

31(29). Antennal grooves shallow, punctate, sharply defined medially by ridge, widening caudad, widest at base, antennomeres 3-10 pectinate or sharply serrate, flattened, pronotum bell-shaped .... *Deltometopus*

— Antennal grooves not shallow *and* sharply defined medially *and* widest at base 32

32(31). Black or deep dark brown western species, antennal grooves poorly delimited medially, always shallow and dullish .... *Asiocnemis*

— If antennal grooves not deep, glossy and sharply defined medially by ridge, then eastern species 33

33(32). Black, with yellowish antennae, less than 4.0 mm long .... *Thambus*

— All black, more than 4.0 mm long 34

34(33). Pronotum wider than long, body covered with unicolored, inconspicuous vestiture, antennal grooves medially defined by ridge.... *Isarthrus*

— If pronotum wider than long, then either with medially not defined, poorly developed antennal grooves or dorsum with conspicuous bicolored vestiture .... *Dromaeolus*

35(17). Elytral lateral keels extend to apex ..... *Protofarsus*

— Elytral lateral keels vanish caudally .... *Arrhipis*

36(9). Antennomeres flabellate or strongly serrate .... *Vitellius*

— Antennomeres feebly serrate, robust .... *Stethon*

Classification of the Nearctic Genera

Eucnemidae Eschscholtz 1829

Perothopinae Lacordaire 1857

Perothops Laporte 1838, 3 spp., Pennsylvania to Florida, and California. Key to spp., Cobos 1964.

Phyllocerinae Reitter 1905

Anelastini Reitter 1921

Anelastes Kirby 1818, 3 spp., generally distributed, absent from the Northeast.
Silenus Latreille 1834

Pseudomeninae Muona 1993

Schizophilini Muona 1993

Schizophilus Bonvouloir 1870, 1 sp., *S. subrufus* (Randall 1938), "Canada", Maryland, Michigan, New Jersey, New York, Oklahoma, Pennsylvania, Tennessee, Virginia, Wisconsin. Immature description and biology, Otto and Young 1998.

Palacoxeninae Muona 1993

Palaeoxenus Horn 1891, 1 sp., *P. dohrni* Horn 1878, southern California. Additional information, Fenyes 1903.
Cryptostoma Latreille 1825

Melasinae Fleming 1821

Melasini *sensu stricto*

Melasina *sensu stricto*

Melasis Olivier 1795, 4 spp., British Columbia, generally distributed in United States. Additional information, Chamberlin 1920, Hopping 1926.

Isorhipis Boisduval and Lacordaire 1835, 4 spp., generally distributed. Biology, Knull 1946, 1947.
Tharops Laporte 1835, not Rafinesque 1815; Huebner 1819.

Hylocharini Du Val 1859

Hylochares Latreille 1834, 1 sp., *H. nigricornis* (Say 1923), Illinois, Indiana, Iowa, Massachusetts, Ohio, Pennsylvania.

Xylobiini Reitter 1911

Xylophilus Mannerheim 1823, 2 spp., British Columbia, Ontario, Quebec, California, Kentucky, Maryland, Nevada, New Hampshire, New York, West Virginia, Wisconsin.
Xyloecus Dejean 1833
Xylobius Latreille 1934

Epiphanini Muona 1993

Epiphanis Eschscholtz 1829, 1 sp., *E. cornutus* Eschscholtz 1829, generally distributed.
Ceratotaxia Sharp 1908; Muona 1983
Prosopotropis Abeille de Perrin 1898

Hylis des Gozis 1886, 4 spp., generally distributed.
Hypocaelus auct., not Dejean 1833
Hypohylis Reitter 1902
Elatocoelus Hyslop 1924
Hylus Van Dyke 1943 (in error for *Hylis* Gozis)

Dirhagini Reitter 1911

Dirrhagofarsus Fleutiaux 1935, 1 sp., *D. lewisi* (Fleutiaux 1900), Georgia, Maryland, Pennsylvania, West Virginia. Additional information, Ford and Spilman 1979.

Microrhagus Dejean 1833, 4 spp., generally distributed.
Dirhagus Latreille 1834
Aulacostenus Motschulsky 1869
Aulacosternus Motschulsky 1869, not Marsham 1835
Arhagus Méquignon 1925
Emyirhagus Olexa 1975

Golbachia Cobos 1955, 2 spp., Alabama, Florida, Georgia, Louisiana, Missisippi, North Carolina, South Carolina.

Adelothyreus Chevrolat 1867, 3 spp., Florida, Georgia, Lousiana, Missisippi.

Entomophthalmus Bonvouloir 1870, 1 sp., *E. rufiolus* (LeConte 1866), Ontario, Quebec, generally distributed in United States.

Rhagomicrus Fleuteaux 1902, 4 spp., Ontario, eastern and southern United States.

Sarpedon Bonvouloir 1871, 1 sp., *S. scabrosus* Bonvouloir 1875, British Columbia, Ontario, generally distributed in United States.

Protofarsus Muona 2000, 1 sp., *P. caribicus* Muona 2000, Florida.

Arrhipis Bonvouloir 1870, 1 sp., *A. lanieri* (Guérin 1838), Florida. Additional information, Muona 1991.
Nematodinus Lea 1919

Eucneminae Eschscholtz 1829

Proutianini Muona 1993

Proutianus Muona 1993, 1 sp., *P. americanus* (Horn 1872), Vancouver, British Columbia, Oregon, Washington.
Phaenocerus auct., not Bonvouloir 1871

Dendrocharini Fleutiaux 1920

Dendrocharis Guérin-Méneville 1843, 2 spp., Louisiana, Missisippi, Georgia, Florida.
Galba LeConte 1853, not Latreille 1829

Mesogenini Muona 1993

Vitellius Bonvouloir 1971, 1 sp., *V. texanus* Knull 1957, Texas, Lower Rio Grande Valley. Additional information, Knull 1957.

Stethon LeConte 1866, 1 sp., *S. pectorosus* LeConte 1866, Ontario, Quebec, generally distributed in United States.

Eucnemini *sensu stricto*

Idiotarsus Bonvouloir 1871, 1 sp., *I. errans* (Horn 1886), Florida, Georgia, Louisana, Mississippi.

Eucnemis Ahrens 1812, 1 sp., *E. americana* Horn 1886, Ontario, generally distributed in United States.

Macraulacinae Fleutiaux 1922

Echtrogasterini Cobos 1964

Hemiopsida MacLeay 1872, 1 sp., *H. robusta* (Van Dyke 1943), California.

Euryptychini Mamaev 1976

Euryptychus LeConte 1852, 3 spp., eastern and southern United States, one species in Arizona and New Mexico.

Macraulacini *sensu stricto*

Diphytaxis Horn 1890, 1 sp., *D. excavata* Horn 1890, Texas.

Onichodon Newman 1838, 5 spp., Ontario, Quebec, generally distributed in United States.

Fornax Laporte 1835, 2 spp., Quebec, southern and eastern United States. Additional information, Blanchard 1889.

Serrifornax Fleutiaux 1926, 1 sp., *S. infelix* (Bonvouloir 1886), Florida.

Isarthrus LeConte 1852, 2 spp., Ontario, Quebec, generally distributed in United States, absent from the west.

Asiocnemis Mamaev 1976, 5 species, British Columbia, Western Territory, generally distributed in western United States.

Dromaeolus Kiesenwetter 1863, 9 spp., generally distributed, absent from western regions. Additional information, Blanchard 1904.

Thambus Bonvouloir 1871, 1 sp., *T. horni* Muona 2000, Ontario, Illinois, Indiana, Kansas, Missisippi, New York, Ohio, Texas, Virginia.

Deltometopus Bonvouloir 1870, 2 spp., generally distributed. Immature stages, Mosher 1919.

Nematodini Leiter 1976

Nematodes Berthold 1827, 4 spp., Ontario, Quebec, generally distributed in United States, absent from west.

Bibliography

BEAULIEU, G. 1919-1922. Monographie des melasides du Canada (1-17). Naturaliste Canadien 46: 73-83 to 49: 139-142.

BLANCHARD, F. 1889. Notes on *Fornax calceatus* Say, and *F. Hornii* Bonv., and on *Corymbites divaricatus* Lec., and *C. crassus* Lec. Entomologica Americana, 8: 139 - 140.

BLANCHARD, F. 1904. A new Californian species of *Dromaeolus* Kies. Entomological News, 15: 187-188.

BLATCHLEY, W. S. 1910. An illustrated descriptive catalogue of the Coleoptera or beetles (exclusive of the Rhynchophora) known to occur in Indiana. 1386 pp. Indiana.

BONVOULOIR, H. A. de. 1871. Monographie de la famille des eucnémides, 1er partie. Annales de la Société Entomologique de France, 40 (suppl.): 1-288, pls. 1-21.

BONVOULOIR, H. A. de. 1872. Monographie de la famille des eucnémides, 2me partie. Annales de la Société Entomologique de France, 40 (suppl.): 289-560, pls. 22-36.

BONVOULOIR, H. A. de. 1875. Monographie de la famille des eucnémides, 3me et 4me parties. Annales de la Société Entomologique de France, 40 (suppl.): 561-907, pls. 37-42.

BØVING, A. G. and F. C. CRAIGHEAD. 1930. An illustrated synopsis of the principal larval forms of Coleoptera. Entomologica Americana (new series), 11: 1-351.

BROWN, W. J. 1940. Some new and poorly known species of Coleoptera. Canadian Entomologist, 72: 182-188.

BURKE, H. E. 1917. Flat-headed borers etc. USDA Bulletin 437.

CALDER, A. A., J. F. LAWRENCE and J. W. H. TRUEMAN. 1993. *Austrelater*, gen. nov. (Coleoptera: Elateridae), with a description of the larva and comments on elaterid relationships. Invertebrate Taxonomy, 7: 1349-1394.

CHAMBERLIN, W. J. 1920. Notes on two little-known woodboring beetles. *Chrysobothris sylvania* Fall and *Melasis rufipennis* Horn (Buprestidae, Eucnemidae). Journal of the New York Entomological Society, 28: 151 - 157.

COBOS, A. 1964. Materiales para el estudio de la familia Eucnemidae. Primera parte. Eos, 40: 289-435.

COBOS, A. 1979. Géneros y especies nuevas de Eucnemini y Fornaxini. Revista Brasileira de Biologia, 39: 67-81.

DURY, C. 1888. Elateridae in the vicinity of Cincinnati, Ohio. Entomological News, 163 -164.

DURY, C. 1904. Notes on Coleoptera. Entomological News, 14: 52 -53.

ESCHSCHOLTZ, J. F. 1829. Zoologisches Atlas I. Reimer. Berlin.

FALL, H. C. 1925. New species of Coleoptera or (sic) recent discovery. Bulletin of the Brooklyn Entomological Society, 20: 180-183.

FALL, H. C. 1928. Miscellaneous notes and descriptions (Coleoptera). Entomological News, 23: 236-240.

FALL, H. C. 1933. New Coleoptera XVI. Canadian Entomologist, 65: 229-234.

FENYES, A. 1903. *Palaeoxenus* (*Cryptostoma*) *dohrnii* Horn. Entomological News, 11: 220.

FORD, E. J. and T. J. SPILMAN. 1979. Biology and immature stages of *Dirrhagofarsus lewisi*, a species new to the United States (Coleoptera, Eucnemidae). Coleopterists Bulletin, 33: 75 - 83.

GUÉRIN-MÉNEVILLE, F. E. 1843. Revue critique de la tribu des eucnémides. Annales de la Société Entomologique de France, (2) 1: 163-199, pls. 5-6.

HATCH, M. 1971. The Beetles of the Pacific Northwest. Part. V. Rhipiceroidea, Sternoxi, Rhyncophora and Lamellicornia. University of Washington Press. Seattle. 682 pp.

HOPPING, G. R. 1926. New *Melasis* with the key to the known species (Coleoptera). Canadian Entomologist, 58: 225 - 228.

HORN, G. H. 1871. Descriptions of new Coleoptera of the United States, with notes on known species. Transactions of the American Entomological Society, 3: 325-344.

HORN, G. H. 1872. Descriptions of some new North American Coleoptera. Transactions of the American Entomological Society, 4: 143-152.

HORN, G. H. 1878. Contribution to the coleopterology of the United States II. Transactions of the American Entomological Society, 7: 51-60.

HORN, G. H. 1886. A monograph of the species of the subfamilies Eucneminae, Cerophytinae and Perothopinae inhabiting the United States. Transactions of the American Entomological Society, 20: 5 - 58.

HORN, G. H. 1890. Insecta. Coleoptera. Throscidae and Eucnemidae. Biologia Centrali-Americana, 3: 193-257.

HORN, G. H. 1891. New species and miscellaneous notes. Transactions of the American Entomological Society, 18: 32-48.

KIRK, H. B. 1922. Biological notes on Elateridae and Melasidae (Col.). Entomological News, 32: 236 - 240.

KNULL, J. M. 1946. A new species of *Dirhagus* with notes on other Eucnemidae (Coleoptera). Annals of the Entomological Society of America, 39: 246-247.

KNULL, J. N. 1947. New Elateridae with notes on Eucnemidae. Entomological News, 57: 177 - 181.

KNULL, J. N. 1948. New species of *Agrilus* with Notes (Buprestidae and Eucnemidae). Entomological News, 58: 72-75.

KNULL, J. N. 1957. A new species of *Vitellius* from the United States (Coleoptera:Eucnemidae). Ohio Journal of Science, 57: 9-10.

LAWRENCE, J. F. 1988. Rhinorhipidae, a new beetle family from Australia, with comments on the phylogeny of Elateriformia. Invertebrate Taxonomy, 2: 1-53.

LECONTE, J. L. 1852. Synopsis of the eucnemides of temperate North America. Proceedings of the Academy of Natural Sciences of Philadelphia, 6: 45-49.

LECONTE, J. L. 1866. Additions to the Coleopterous fauna of the United States 1. Proceedings of the Academy of Natural Sciences of Philadelphia, 18: 361-394

LECONTE, J. L. 1878. Additional descriptions of new species. Proceedings of the American Philosophical Society, 100: 373-434.

MAMAEV, B. M. 1976. Morphological types of xylophagous beetle larvae (Coleoptera, Eucnemidae) and their evolutionary importance. *In*: B. M. Mamaev, ed., Evolutionary morphology of woodboring larvae. Moscow. 202 pp. (in Russian).

MELSHEIMER, F. E. 1846. Descriptions of new species of Coleoptera of the United States. Proceedings of the Academy of Natural Sciences of Philadelphia, 2: 134-160.

MOSHER, E. 1919. Notes on the immature stages of *Deltometopus rufipes* Melsh. (Coleoptera, Eucnemidae). Annals of the Entomological Society of America, 12: 49 - 53, 1 pl.

MUONA, J. 1987. The generic names of the beetle family Eucnemidae (Coleoptera). Entomologica Scandinavica, 18:79-92.

MUONA, J. 1991. The American species of the genus *Arrhipis* Bonvouloir (Coleoptera, Eucnemidae). Revista Brasileira de Biologia, 35: 135-146.

MUONA, J. 1993a. Eucnemidae and Throscidae from Baltic Amber. Entomologische Blätter zur Biologie und Systematik der Käfer, 89: 15-45.

MUONA, J. 1993b. Review of the phylogeny, classification and biology of the family Eucnemidae (Coleoptera). Entomologica Scandinavica Supplement, 44: 133 pp.

MUONA, J. 1995. The phylogeny of Elateroidea (Coleoptera) or which tree is best today? Cladistics, 11: 317-341.

MUONA, J. 1996. Eucnemidae. Pp. 828-837. *In*: R. H. Arnett, Jr. and N. M. Downie. The Beetles of Northeastern North America, II. Sandhill Crane Press. Gainesville, FL.

MUONA, J. 2000. Revision of the Nearctic Eucnemidae. Acta Zoologica Fennica, 212: 1-106.

OTTO, R. L. and D. K. YOUNG. 1998. Description of the larva of *Schizophilus subrufus* (Randall) (Coleoptera: Eucnemidae: Pseudomeninae), with notes on its natural history. Coleopterists Bulletin, 52: 306-312.

SCHENKLING, S. 1928. Melasidae. Coleopterorum Catalogus, 11 (pars 96): 1-110.

VAN DYKE, E. C. 1932. Miscellaneous studies in the Elateridae and related families of Coleoptera. Proceedings of the California Academy of Sciences, 20 : 291 - 299.

VAN DYKE, E. C. 1943. New species of West American Coleoptera. Pan-Pacific Entomologist, 19: 41-52.

VAN DYKE, E. C. 1945 New species of North American Coleoptera. Pan-Pacific Entomologist, 21: 101-109.

VAN DYKE, E. C. 1953. New Coleoptera from Western North America (Carabidae, Melasidae, Buprestidae, Curculionidae). Pan-Pacific Entomologist, 29 : 102-108.

VAN HORN, R. W. 1909. Notes on some of the Eucnemidae of the western states. (Coleoptera, Eucnemidae). Proceedings of the Entomological Society of Washington, 11: 54 - 62.

57. THROSCIDAE Laporte 1840

by Paul J. Johnson

Family common name: The throscid beetles

Family synonyms: Throscites Laporte 1840; Trixagidae Gistel 1856

The small size, oblong silhouette, depressed dorsum, deflexed head, pronotum tightly fitted against the elytral bases, and capitate antennae distinguish most adult throscids.

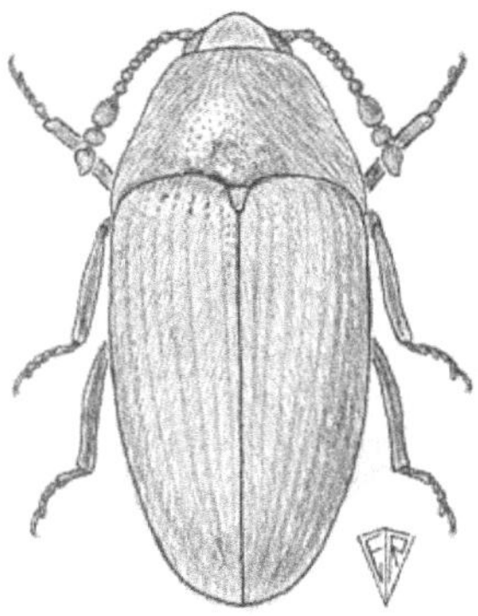

FIGURE 1.57. *Trixagus carinicollis* (Schaeffer)

Description: Body oblong to elongate, compact; length approximately 1.0-5.0 mm; color red-brown to black; vestiture fine, suberect.

Head hypognathous, narrower than prothorax, deeply inserted into prothorax when reposed. Eyes oval, vertically oriented, coarsely faceted, deeply emarginate, often nearly divided by frontal canthus. Antenna 11-segmented, narrowly clavate to loosely capitate, received into prosternal grooves, insertions distant on frons between eyes, and beneath carinate ridge. Labrum articulate, visible. Mandible stout. Maxilla with palpus 4-segmented, apical segment triangular; labium with mentum small, narrowed anteriorly, palpus 3-segmented.

Prothorax compact; pronotum trapezoidal, fitting closely to mesothorax and bases of elytra; hind angles briefly extended, lateral margins fine, obsolete anteriorly; prosternum truncate to briefly lobed anteriorly, extended and narrowed posterior as intercoxal process fitting into mesosternal cavity; pronotosternal suture grooved to accept antenna; hypomeron subtriangular, with deep antennal cavities posteriorly; procoxal cavities open posteriorly. Scutellum small, triangular. Mesothoracic wing short, apical field subequal in length to main portion; venation reduced, radial cell open, wedge cell absent; medial and anal field veins greatly reduced to absent. Mesosternum short, deeply excavated medially; mesocoxae widely separated. Elytra entire, rounded at apex; striae shallow, punctate; intervals flat, finely to moderately punctured; epipleura broad basally, narrow apically. Metasternum large, quadrate, with or without oblique tarsal grooves; metacoxae with broad ventral lamina, shallowly excavate posteriorly to receive femur. Legs retractile; trochanters short, triangular; femora fusiform; tibiae slender, spinose; tarsi 5-5-5, filiform or with tarsomere 4 with ventral membranous lobe, claw simple.

Abdomen with 5 ventrites, all connate. Aedeagus trilobed; penis acute at apex; parameres attenuate, densely setiferous laterally; basal piece longer than parameres, proportionately large, deeply emarginate posteromedially. Female gonocoxite lightly sclerotized, narrow, elongate; stylus clavate, apical; bacula long, slender.

Larvae are lightly sclerotized, grublike. Head small, prognathous, reduced; antenna short, sensorium as long as segment 3; labrum fused to head capsule, forming an inarticulate plate; mandibles narrowly elongate, flattened, fused to head capsule; maxilla reduced to a 3-segmented palpus; labium elongate, palpus 2-segmented. Legs short, 5-segmented, tarsungulus present. Tergite 9 with pair of small urogomphi. Tergite 10 obsolescent. Spiracles biforous, on mesothorax and abdominal segments 1-8; closing apparatus present.

Habits and habitats. Little is known about throscids. Adults are usually captured in traps or netted in late afternoon flights, at light traps, in flowers, recovered from duff and litter samples, near mammal burrows, in rotted wood, grain storage facilities, and on various foliage. Adults may be generalist pollen and mold feeders. They share an ability to click and "jump" with the Elateridae and Eucnemidae, by rapidly articulating the prothorax and mesothorax against a substrate. Adults are reported to possess four fat-body mycetomes containing three types of bacterial symbionts. Larvae have been found in blocky red-rotted portions of oak logs, in fungusy soil samples, and in grass tussocks. Larvae of the European *Trixagus dermestoides* (L.) were found in a variety of soils, always near tree stumps, and under the litter. They are liquid feeders on outer portions of ectotrophic mycorrhizal roots (Burakowski 1975).

Status of the Classification. Bonvouloir (1859) conducted the last worldwide study of the family, while Horn (1885, 1890) and Blanchard (1917) reviewed the North and Central American species. Schenkling (1928) compiled the current world catalog. With the removal of the Lissominae and Thylacosterninae to Elateridae (Lawrence and Newton 1995), the throscids have become a more clearly defined group. However, the family is in need of general study at all levels, and the position of certain exotic genera remains problematic. In North America, the most diverse genus, *Aulonothroscus*, is in need of revisionary study. Yensen (1975a) revised *Pactopus*, then (Yensen 1975b, 1980) revised the nearctic *Trixagus* and provided a key to the New World species. As noted by Burakowski (1975) and Becker (1991) larvae are poorly known. Fossil throscids are known from the Oligocene Baltic amber (Carpenter 1992).

Diversity and Distribution. There are 152 described species worldwide, with 20 species in the nearctic fauna. Most throscids occur in lowland and mid-elevational forested environments. Only a few species are transcontinental in North America. Most of our species occur east of the Great Plains, but *Pactopus horni* (LeConte) is known disjunctively from northwestern regions and the Black Hills of South Dakota (Johnson 1998).

Key to the Nearctic Genera

1. Abdominal ventrites with simple crural depressions; tarsomere 4 lobed; antenna capitate 2
— Abdominal ventrites with deep oblique tarsal sulcus; tarsus simple; antenna clavate .... *Pactopus*

2(1). Metasternum with a deep oblique tarsal sulcus .... *Aulonothroscus*
— Metasternum without sulcus .... *Trixagus*

Classification of the Nearctic Genera

Throscidae Laporte 1840

Throscinae Laporte 1840

Aulonothroscus Horn 1890; 13 spp., generally distributed in the U.S., and southernmost Canada. Key to species: Blanchard 1917.
Throscites Iablokov-Khnzorian 1962 (fossil, Oligocene)

Trixagus Kugelann 1794; 6 spp., generally distributed in the U.S., southernmost Canada. Key to species: Yensen 1980.
Throscus Latreille 1796
Palaeothroscus Iablokov-Khnzorian 1962 (fossil, Oligocene)

Potergini Cobos 1961

Pactopus LeConte 1868; 1 sp., *P. horni* LeConte; northwestern states, northern California, western South Dakota, and western Canada.

Bibliography

BECKER, E. C. 1991. Throscidae (Elateroidea) (= Trixagidae). Pp. 418-419. *In:* F. W. Stehr, ed. Immature Insects. Kendall/ Hunt. Dubuque, IA.

BLANCHARD, F. 1917. Revision of the Throscidae of North America (Coleoptera). Transactions of the American Entomological Society, 53: 1-26.

BONVOULOIR, H. de 1859. Essai monographique sur la famille des throscides. E. Deyrolle. Paris.

BURAKOWSKI, B. 1975. Development, distribution and habits of *Trixagus dermestoides* (L.), with notes on the Throscidae and Lissomidae (Coleoptera, Elateroidea). Annales Zoologici, 32: 375-405.

CARPENTER, F. M. 1992. Superclass Hexapoda.. Treatise on Invertebrate Paleontology, Part R, Arthropoda 4, 4: 279-655.

HORN, G. H. 1885. Synopsis of the Throscidae of the United States. Transactions of the American Entomological Society, 12: 198-208.

HORN, G. H. 1890. Fam. Throscidae. Biologia Centrali-Americana, 3(1): 193-209.

JOHNSON, P. J. 1998. New distribution records for three Throscidae (Coleoptera) in South Dakota. Prairie Naturalist, 29: 51-52.

LAWRENCE, J. F. and A. F. NEWTON, Jr. 1995. Families and subfamilies of Coleoptera (with selected genera, notes, references and data on family-group names). Pp. 779-1006. *In:* J. Pakaluk and S.A. Slipinski, eds. Biology, Phylogeny and Classification of Coleoptera: Papers Celebrating the 80th Birthday of Roy A. Crowson. Muzeum i Instytut Zoologii PAN, Warsaw.

SCHENKLING, S. 1928. Fam. Throscidae. Coleopterorum Catalogus, 11(101): 1-26.

YENSEN, E. 1975a. A review of the genus *Pactopus* LeConte (Coleoptera: Throscidae). Coleopterists Bulletin, 29: 87-91.

YENSEN, E. 1975b. A revision of the North American species of *Trixagus* Kugelann (Coleoptera: Throscidae). Transactions of the American Entomological Society, 101: 125-166.

YENSEN, E. 1980. A new species of *Trixagus* from Panama and a key to New World *Trixagus* Coleoptera: Throscidae). Coleopterists Bulletin, 34: 257-261.

58. ELATERIDAE Leach 1815

by Paul J. Johnson

Family common name: The click beetles

Family synonyms: Cebrionates Latreille 1802; Elaterides Leach 1815; Lissomidae Laporte 1835; Tetralobites Laporte 1840; Agriotites Laporte 1840; Phyllophoridae Hope 1842; Pangauridae Gistel 1856; Prosternidae Gistel 1856; Synaptidae Gistel 1856; Ampedidae Gistel 1856; Campylidae Candèze 1857; Agrypnides Candèze 1857; Hemirhipides Candèze 1857; Physodactylides Lacordaire 1857; Ludiides Lacordaire 1857; Monocrepidiites Candèze 1859; Athoites Candèze 1859; Chalcolepidiides Lacordaire 1857; Oxynopterides Candèze 1857; Melanactides Candèze 1857; Pomachilites Candèze 1859; Melanotites Candèze 1859; Dicrepidiites Candèze 1859; Physorhinites Candèze 1859; Cardiophorites Candèze 1859; Plastoceridae LeConte 1861; Pyrophorites Candèze 1863; Crepidomenites Candèze 1863; Dimites Candèze 1863; Cardiorhinites Candèze 1863; Adrastites Candèze 1863; Drapetini LeConte 1863; Alaites Candèze 1874; Hypodesites Candèze 1891; Octocryptites Candèze 1892; Hemicrepidiini Champion 1894; Dicronychidae Schwarz 1897; Protelateridae Schwarz 1902; Hypnoidini Schwarz 1906; Semiotina Jakobson 1913; Pityobini Hyslop 1917; Pachyderinae Fleutiaux 1919; Esthesopinae Fleutiaux 1919; Subprotelaterinae Fleutiaux 1920; Thylacosterninae Fleutiaux 1920; Campsosterninae Fleutiaux 1927; Senodoniinae Schenkling 1927; Cavicoxumidae Pic 1928; Ctenicerinae Fleutiaux 1936; Hemiopinae Fleutiaux 1941; Toxognathinae Fleutiaux 1941; Rostricephalinae Fleutiaux 1947; Negastriinae Nakane and Kishii 1956; Athoomorphinae Laurent 1966; Eudichronychinae Girard 1971; Campyloxeninae Costa 1975; Aplastinae Stibick 1979; Cebriognathinae Paulus 1981.

Click beetles are generally recognized by possessing a combination of an elongate, narrow body form, with a disproportionately large and freely articulating prothorax, long slender legs and serrate antennae. Most species produce an audible clicking or snapping noise and can forcibly articulate themselves by rapidly moving the prosternal intercoxal process from a locked position into a deep mesosternal cavity.

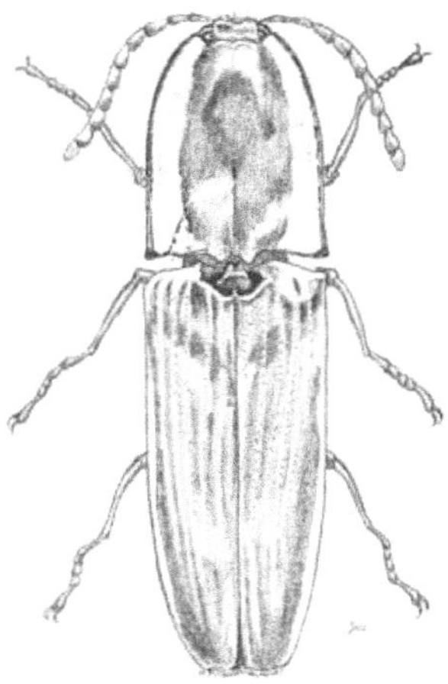

FIGURE 1.58. *Chalocolepidius tartarus* Fall.

Description: Body form ovoid to narrowly elongate, depressed to strongly convex; length ca. 1-60 mm; color somber browns and blacks, some iridescent, many brightly marked with red and yellow; most with pubescence hair-like, moderately long, decumbent, to appressed and scale-like.

Head hypognathous or prognathous, narrower than prothorax in most. Eyes lateral, moderate to large, hemispherical. Antenna serrate to pectinate; insertion beneath strong, often carinate ridge. Frons convex, depressed, to concave; frontal margin smooth and flat to strongly and transversely carinate. Clypeus absent. Labrum freely articulating. Mandible shallowly to strongly arcuate, apex beveled and truncate to emarginate, most with incisor strongly dentate. Maxilla densely setose medially; palpus 4-segmented. Labium with small quadrate mentum, ligula reduced; palpus 3-segmented.

Thorax freely articulating between prothorax and mesothorax. Prothorax subcylindrical to dorsally depressed. Pronotum with lateral edges deplanate and strongly carinate, to rounded and ecarinate; pronotal hind angles extended in most, dorsally ecarinate, unicarinate, or bicarinate. Hypomeron large, subtriangular, often with antennal or crural cavities or impressions. Prosternum with anterior margin truncate, or extended and deflexed to form a lobe, or chin piece, covering ventral mouthparts; posterior intercoxal portion extended and fitting into a mesosternal fossa. Mesosternum reduced, with deep median fossa to receive prosternal intercoxal process. Scutellum triangular, pentagonal, cordate, ovoid, to elongate. Elytra elongate, narrow, separately or conjointly rounded at apices, some attenuate and spinose, or emarginate and bispinose; with 9 striae, rarely with short stria 10, each stria serially punctate to sulciform, often obsolescent, subparallel or apically coalescent. Mesocoxae globular, narrowly to moderately separated. Metasternum large, subquadrate. Metacoxae nearly approximate, transverse, excavate posteriorly, most with a ventral plate or lamina. Metathorax alate and wings functional in most, frequently brachypterous to micropterous, rarely apterous; venation variable; radial cell large, elongate; medial field with wedge cell in most; apical field with 0-3 sclerotized patches. Leg long and slender, often received in cural depression or fossa; femur subfusiform, flattened ventrally; tibia slender, often spinose, with no or two apical spurs; tarsal formula 5-5-5, tarsomeres filiform to broadened or emarginate, most with spiniform setae, often with ventral setose pads or membranous lobes; pretarsal claws slender, to strongly toothed, cleft, or pectinate, with or without ventral and ectal setae.

Abdomen with 5 ventrites, 1-4 connate. Male aedeagus trilobed; most with penis simple, attenuate; parameres slender, attenuate, to sagittate or hooked apically, variously setiferous. Female gonocoxites dorsoventrally compressed; styli reduced, peg-like; bacculum slender, short to long, narrow to spatulate.

Larvae are elongate, narrow, subcylindrical to flattened; mature length approximately 5-60 mm. Integument darkly sclerotized to desclerotized and nearly membranous; most are sparsely setose. Head wedge-shaped, prognathous, darkly sclerotized; cranial sutures lyriform. Clypeus and labrum obsolete; frontal margin typically a projecting 1 or 3-dentate nasale. Antenna 3-segmented; 1-6 apical sensoria on segment 2. Stemmata 0-6 each side. Mandible robust, apically dentate, most with retinaculum, without mola (mandible elongate, deeply cleft, and strongly toothed dorsally in Cardiophorinae). Maxilla and labium well developed; palpi 3-4 segmented. Legs long, 5-segmented, tarsungulus elongate. Abdomen 10-segmented; segments 1-8 similar, often with patches of longitudinal striations anterolaterally; tergite 9 variously modified, entire to emarginate, often with tubercules, spines, pits, simple or tuberculate urogomphi, or depressed; sternite 9 transversely carinate in some, or with lateral hooks. Spiracles biforous, on mesothorax and abdominal segments 1-8; ecdysial scar adnate to or separate from atrium margin; lacking closing apparatus.

Habits and habitats. Adults of most species of click beetles are relatively common and are often collected in various traps, on vegetation, or at lights. They are generally active in afternoon and evening hours when temperatures exceed ca. 16°C (60°F), many species are nocturnal, and most fly well. In cool temperate climates most species overwinter as larvae or post-eclosion adults. Some are predaceous on soft-bodied sternorrhynchous insects and their "honey-dew" exudates, but many feed on overripe and decaying fruit, nectar, pollen, floral parts, ascomycete fungus fruiting bodies, and at extra-floral nectaries, glandular trichomes, and phloem weeps. Cohorts of adults are usually abundant for only a few weeks, and there are seasonal or phenologically related diversity shifts.

Larvae are found in soil, forest duff, or decaying plant materials, especially wood. Most are opportunistic predators. Wood-inhabiting larvae are predaceous on small and immature invertebrates or saprophagous on decay organisms such as myxomycetes, while soil dwelling species are generally predaceous or omnivorous. Some soil dwelling species (esp. *Agriotes*, *Limonius*, some *Melanotus* sp. and *Selatosomus* sp.) are herbivorous on sprouting seeds and roots of seedlings of grasses, and occasionally on roots and tubers of vegetables. Larvae are liquid feeders and possess extra-oral digestion. There are generally 3-5 larval instars taking 1-3 years for development, depending on food quality and availability. Reports of extended larval life-spans are based on inappropriate food offerings and abnormal rearing conditions. The pupal stage typically lasts 1-3 weeks, depending on ambient temperatures. Larvae of pest species are called wireworms.

Most members of the tribe Pyrophorini are bioluminescent. Adults of these beetles have paired light organs on the thorax, with an additional median spot ventrally at the base of the abdomen that is visible when the beetle is flying. Flash patterns may be involved with mate attraction. Embryos and larvae of these species are luminous through intersegmental membranes.

Interrelationships of click beetles with other organisms are poorly studied. Thomas (1929, 1931, 1932, 1940) discussed the parasites of larvae, predators, diseases, and general biology of click beetles. Balduf (1935) briefly discussed predation by larvae. Zacharuk (1963) examined feeding preferences of various species and summarized much of the literature on the subject. Numerous brightly colored click beetles with aposematic-like contrasting bright red or yellow against black color patterns are involved with mimicry complexes, probably as mimics. Such complexes often simultaneously involve moths (Lepidoptera), wasps (Hymenoptera), and beetles of the families Cerambycidae, Lycidae, Lampyridae, Cantharidae, and Cleridae. Possible color and gait mimicry combinations with ants (Hymenoptera) are suspected for species of *Aeolus* and *Conoderus*.

Eight adventive species are known to have established populations in various states and provinces in the Nearctic region: *Agriotes lineatus* (Linnaeus), *A. obscurus* (Linnaeus) and *A. sputator* (Linnaeus) (Becker 1956), and *Melanotus dichrous* Erichson (Becker 1973), are all from Europe; *Melanotus cete* Candèze (Johnson 1998) is from northeastern Asia; *Conoderus amplicollis* (Gyllenhall) (Cockerham and Deen 1936) and *Conoderus falli* (Lane) (Dobrovsky 1953, Lane 1953) are from Neotropical areas, and *C. exsul* (Sharp) (Graves 1938) is from the South Pacific.

Status of the classification. Elateridae are the largest family of the series Elateriformia and superfamily Elateroidea. The family was recently enlarged by inclusion of Lissomidae, Cebrionidae, Dicronychidae, and Subprotelaterinae and Thylacosterninae (ex Eucnemidae) (Lawrence and Newton 1995). Anischiinae (ex Cerophytidae) were placed into the Elateridae by these authors, but were segregated to Anischiidae by Lawrence *et al.* (1999).

Taxonomically, the family Elateridae is the ninth most diverse family of beetles. The last family-level revision of these beetles in the Nearctic region was by LeConte (1853). The Dalla Torre (1911, 1913), Schenkling (1925, 1927a-c) and Hetschko (1933) catalogs are the most recent for the world fauna, with the Leng (1925), Blackwelder (1944, 1957), and Bousquet (1991) catalogs for our region. Globally, Schwarz (1906, 1907a-c), Nakane and Kishii (1956), and Stibick (1979) provided the most recent keys to subfamilies and tribes. No general key to the genera has been published since Schwarz (1906). Important regional faunal studies that have keys to species are Dietrich (1945) for northeastern states and provinces, Dogger (1959) for the western Great Lakes region, Brooks (1960) for central Canada, Lane (1971) and Lanchester (1971) for the northwestern states and provinces, Smith and Enns (1977, 1978) for the central U.S., and Downie and Arnett (1996) for much of the region east of the Great Plains. In the Nearctic region several genera were taxonomically revised during the past 50 years, but most of the fauna remains in need of serious study. The most diverse genera in the Nearctic, such as *Ampedus*, "*Ctenicera*", *Limonius*, *Cardiophorus* and *Dalopius*, and many medium sized genera, remain in need of revisionary study. Keys to species are noted respectively with each taxon in the classification below. Useful regional keys to larvae were provided by Dietrich (1945), Glen (1950), Dogger (1959), and Dogger in Becker

(1991). Taxonomic and phylogenetic concepts in Elateridae are changing following important prior and contemporary studies by Hyslop (1917, 1921), Becker (1958), Crowson (1961), Zacharuk (1962), Dajoz (1964), Gurjeva (1969, 1974), Hayek (1973, 1979, 1990), Dolin (1976), Stibick (1979), Lawrence (1987), Calder *et al.* (1993), Muona (1995), and Lawrence *et al.* (1999). Significant reviews of family composition and internal relationships at the world level are underway by several workers. Major modifications in family group taxa should be expected in coming years. Considerable faunal work remains for the Nearctic species. Phylogenetic study is lacking for most taxa. The earliest fossils are known from the Upper Jurassic (Carpenter 1992).

Distribution. There are nearly 10,000 described species worldwide, with about 965 valid species described from the Nearctic region. An estimated 75-100 undescribed species from the region await study, particularly in the genera *Cardiophorus* and *Dalopius*.

Click beetles are found throughout the Nearctic region and are found in all but aquatic and the most severe arctic and alpine habitats. Forests and woodlands of mixed successional stages, and forest/meadow ecotonal areas, harbor the greatest local diversity of click beetles. Biogeographically, the Nearctic click beetle fauna is largely shared with that of the Palearctic region to form a Holarctic fauna, with only a small proportion of the fauna shared with Neotropical source areas, particularly Antillean (south Florida) and Mexican (Texas to California) biotas.

Key to the Nearctic Genera

S. Wells assisted with the Negastriinae, and H. Douglas assisted with Cardiophorinae. Wells, Douglas, M.C. Thomas, J. LaBonte, and R. Westcott tested portions of the key and provided valuable suggestions.

1. Claws bearing one or more setae (Fig. 4) emergent from the integument in basal half, and on ectal and ventral surfaces (do not include empodial setae) 2
— Claws lacking setae, or setae indistinct 17

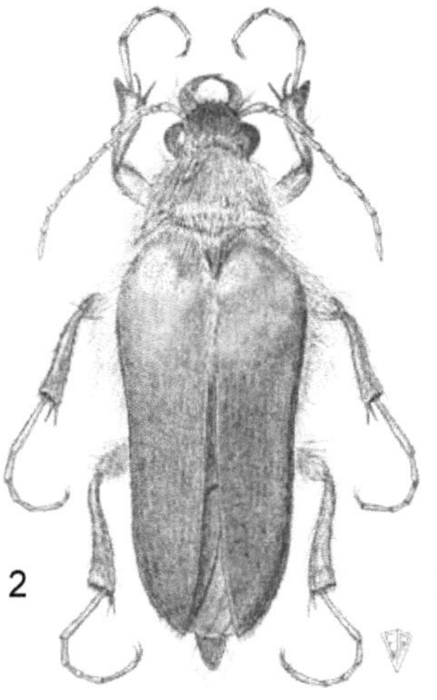

2

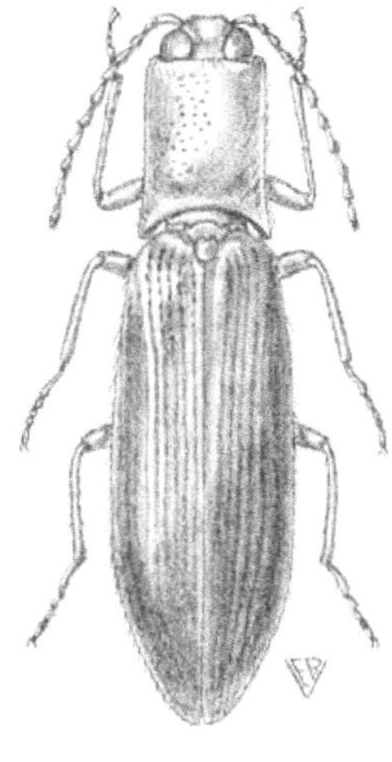

3

FIGURES 2.58-3.58. 2. *Scaptolenus lecontei* (Sallé); 3. *Athous cucullatus* (Say).

2(1). Body finely pubescent, setae narrow, recumbent to erect; elytral striae parallel to subparallel throughout length; tibial spurs present 3
— Body with scale-like setae; elytral striae convergent and coalescent apically, when distinct; tibial spurs absent 11

3(2). Tarsomere 4 with broad and emarginate membranous ventral lobe or a dense setal pad; elytral striae moderately to deeply engraved, sulciform to striatopunctate 4
— Tarsomere 4 ventrally, at most, with moderately-dense arrangement of erect setae; elytral striae fine and thinly engraved to serially arranged shallow punctures 7

4(3). Femur of front leg with anteroventral angle carinate (Fig. 5); hypomeron with punctures moderately large, deep .... *Aeolus*
— Femur of front leg with anteroventral angle rounded; hypomeron with punctures fine to small, shallow 5

5(4). Pronotum with hind angles slender, subparallel to divergent, weakly to moderately carinate dorsally, disc with fine to moderate punctures of subequal or variable sizes 6
— Pronotum with hind angles stout, strongly carinate dorsally, disc with coarse, deep punctures and interspaces with numerous fine punctures .... *Heteroderes*

6(5). Frontal margin complete, carinate or cariniform (Fig. 15); posterior margin of hypomeron broadly emarginate laterally .... *Conoderus*
— Frontal margin incomplete, ecarinate medially; posterior margin of hypomeron narrowly emarginate laterally .... *Deronocus*

7(3). Hypomeron (Fig. 6) with medial margin thickened and polished 8
— Hypomeron (Fig. 7) with medial margin deeply sulcate .... *Lanelater*

8(7). Body castaneous to rufopiceous; pronotum with yellowish to amber luminescence organs near hind angles 9
— Body and appendages black to piceous; pronotum concolorous and lacking luminescence organs .... *Anthracalaus*

9(8). Elytra conjointly rounded at apex; pronotum with lateral margins narrow; antennomeres 2 and 3 subequal in length, together subequal in length to antennomere 4, antennomere 3 subtriangular 10
— Elytra attenuate and spinose at apex; pronotum with lateral margins explanate; antennomere 3 slightly larger than antennomere 2, more or less triangular, antennomeres 2+3 shorter than antennomere 4 .... *Ignelater*

10(9). Pronotal punctures large and umbilicate; aedeagus with median lobe abruptly narrowed near apex; lateral lobes with large, acute subapical spines .... *Vesperelater*
— Pronotal punctures moderate, simple; aedeagus with median lobe not constricted, bearing numerous minute scales and long spines .... *Deilelater*

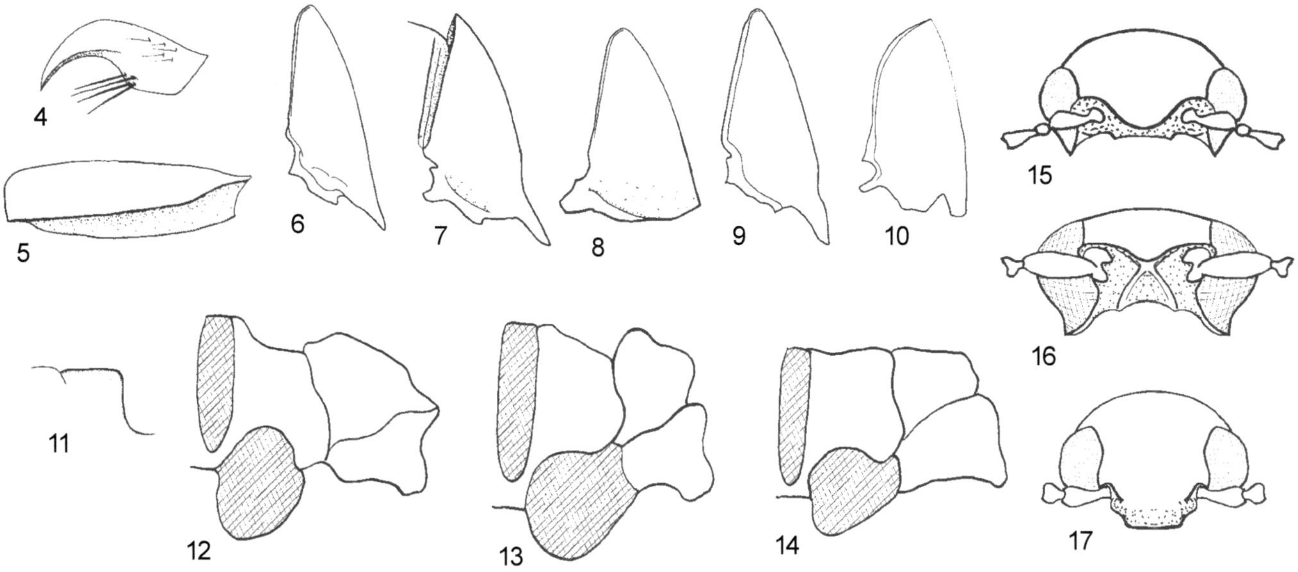

FIGURES 4.58-17.58. Body parts. 4. Tarsal claw; 5. Profemur, anteroventral edge; 6-10. Left hypomeron, ventral view; 11. Mesosternum, lateral view; 12-14, Mesothorax, ventral view, coxal cavities shaded; 15-17. Head, anterior view.

11(2). Smaller, body <4.8 mm in length 12
— Larger, body >5.0 mm in length 13

12(11). Pronotum strongly constricted at anterior angles; elytral intervals finely punctured .... *Rismethus*
— Pronotum evenly arcuate to anterior angles; elytral intervals coarsely granulate .... *Meristhus*

13(11). Hypomeron (Fig. 7) with medial margin deeply sulcate for more than half its length 14
— Hypomeron with medial margin evenly margined, not sulcate 15

14(13). Mesepimeron forming part of the border (Fig. 13) for the mesocoxal cavity .... *Lacon*
— Mesepimeron not forming part of the border (Fig. 14) for the mesocoxal cavity .... *Agrypnus*

15(13). Antennae moderately to strongly serrate, occasionally pectinate in males; pronotal median smooth or planar at posterior declivity or margin 16
— Antennae pectinate in both males and females; pronotal median with a strongly compressed tubercle at posterior declivity .... *Pherhimius*

16(15). Body elongate, narrow; pubescence short, fine, black and with or without intermixed pallid scales; meso-metasternal suture between mesocoxae sulciform .... *Alaus*
— Body broad, elongate, elliptical, to ovoid; pubescence of short, flattened, often brightly or metallic colored scales; meso-metasternal suture evanescent to obsolete.................... *Chalcolepidius*

17(1). Vestiture of broader, flattened scale-like hairs; hypomeron with medial margin deeply sulcate anteriorly 18
— Vestiture of slender hairlike setae; hypomeron with medial margin not deeply sulcate, though often flared anteriorly 20

18(17). Pronotum evenly convex on disc; dorsum with stout erect setae 19
— Pronotum deeply and broadly impressed on midline; scale-like setae appressed ...... *Danosoma*

19(18). Pronotum strongly constricted at anterior angles; elytral intervals finely punctured .... *Rismethus*
— Pronotum evenly arcuate to anterior angles; elytral intervals coarsely granulate .... *Meristhus*

20(17). Claws pectinate or serrate 21
— Claws simple, with a single tooth, or cleft, not pectinate or serrate.. 23

21(20). Nasale (Fig. 16) with 2 oblique, converging carinae forming an inverted V-shape, often with a short vertical carina connecting base of "V" with frontal margin .... *Glyphonyx*
— Nasale simple or thickened, lacking carinae..... 22

22(21). Scutellum subpentagonal to subtriangular, truncate anteriorly; hypomeron with medial margin strongly flared and reflexed anteriorly .... *Melanotus*
— Scutellum cordate, deeply emarginate anteriorly, attenuate posteriorly; hypomeron with medial margin simple throughout length .... *Aptopus*

23(20). Tarsomeres 3 or 4, 2-3, 1-3, or 1-4 with membranous ventral pads or projecting lobes 24
— Tarsomeres filiform or with ventral setal pads, without membranous pads or lobes 37

24(23). Hypomeron (Fig. 8) with hind margin more or less straight, shallowly arcuate or shallowly sinuate; pronotal hind angle little or not projecting from hypomeron 25
— Hypomeron (Fig. 9) with hind margin moderately to deeply emarginate; pronotal hind angle distinctly projecting from hypomeron 30

25(24). Antenna 11-segmented, serrate in both sexes . 26

— Antenna 12-segmented, pectinate in males, serrate in females .... *Pityobius*

26(25). Pronotosternal suture closed anteriorly; mesal margin of hypomeron not excavate 27
— Pronotosternal suture open anteriorly, mesal margin of hypomeron excavated 28

27(26). Mesocoxae widely separated, metasternal process wide .... *Athous*
— Mesocoxae approximate, metasternal process narrow .... *Denticollis*

28(26). Tarsomere 4 simple, not lobed 29
— Tarsomere 4 with a large membranous ventral lobe .... *Drapetes*

29(28). Tarsomere 3 with a short setose ventral lobe .... *Micrathous*
— Tarsomere 3 simple, lacking ventral lobe .... *Elathous*

30(24). Frontal margin elevated and projecting over labrum 31
— Frontal margin depressed medially to plane of labrum .... *Hemicrepidius*

31(30). Tarsomeres 3 or 4 with membranous pads or lobes 32
— Tarsomeres 2 and 3 with membranous pads or lobes 35

32(31). Pronotosternal suture closed anteriorly 33
— Pronotosternal suture excavate anteriorly 34

33(32). Claw cleft or strongly dentate at midlength; hypomeron (Fig. 10) with V-shaped incisure adjacent to pronotal hind angle .... *Esthesopus*
— Claw slender throughout; hypomeron (Fig. 9) with broad emargination separated from pronotal hind angle .... *Aeolus*

34(32). Hypomeron with mesal margin bearing a narrow, flattened bead that is reflexed and shallowly channeled anteriorly; metacoxal lamina subquadrate and angulate over trochanter .... *Anchastus*
— Hypomeron with mesal margin bearing a broad, flat to shallowly excavate bead that is broadly reflexed and shallowly channeled anteriorly; metacoxal lamina forming a broad, elongate lobe posteriorly .... *Physorhinus*

35(31). Nasale flat to shallowly convex; male with serrate antennae 36
— Nasale with two dorsally converging carinae (Fig. 16); male with pectinate antennae .. *Dicrepidius*

36(35). Dorsal sculpture, especially head and pronotum, coarsely to rugosely punctured .... *Blauta*
— Dorsal sculpture fine to moderately coarse .... *Dipropus*

37(23). Scutellum with anterior margin shallowly and emarginately arcuate, to deeply incised; prosternum with intercoxal process short, obtuse at apex 38
— Scutellum with anterior margin broadly arcuate to lobate, or subtruncate; prosternum with intercoxal process long, narrow, attenuate 39

38(37). Pronotum with a fine lateral carina or shallow arcuate groove extending onto hypomeron .... *Cardiophorus*
— Pronotum with a short and obtuse lateral carina at hind angle base, or absent .... *Horistonotus*

39(37). Supra-antennal carinae and frontal carina contiguous and forming a complete arcuate margin (Fig. 15) from eye to eye; frontal margin often extended shelf-like over clypeal remnant 40
— Supra-antennal carinae strong, widely separated, though occasionally extending transversely onto lateral portion of frons 52

40(39). Pronotosternal suture grooved or shallowly channeled anteriorly, with either mesal margin of hypomeron or prosternum excavate at anterior end 41
— Pronotosternal suture closed and mesal hypomeral margin carinate throughout length 46

41(40). Frontal margin and supra-antennal carina coincident across frons 42
— Frontal margin depressed, forming a frontoclypeal area; supra-antennal carinae (Fig. 17) arcuate anterad, extending along lateral margin of frontoclypeal area 44

42(41). Frontal margin finely carinate, not protruding over clypeus 43
— Frontal margin strongly arcuate, often protruding over clypeus .... *Ampedus*

43(42). Body shorter, <10.0 mm; pronotum with lateral margins more or less straight .... *Agriotella*
— Body longer, >14.0 mm; pronotum with lateral margins strongly sinuate .... *Leptoschema*

44(41). Lateral pronotal carina more or less straight from hind angle, joins anterior margin dorsad of pronotosternal margin 45
— Lateral pronotal carina joins anterior margin at pronotosternal margin .... *Agriotes*

45(44). Metacoxal plates distinctly longer medially, narrowing laterally .... *Dalopius*
— Metacoxal plates more or less similar in length throughout width .... *Sericus*

46(40). Antennomere 2 distinctly shorter and slightly narrower than antennomere 3 47
— Antennomeres 2 and 3 subequal in length and diameter .... *Megapenthes*

47(46). Prosternum with intercoxal process gradually declivous or arcuate between coxae 48
— Prosternum with intercoxal process with strongly to acutely angulate declivity immediately posterior of coxae .... *Diplostethus*

48(47). Mesosternal fossa with sides horizontal, anterior portion strongly declivous (Fig. 11) 49
— Mesosternal fossa with sides gradually declivous throughout length 50

49(48). Mesosternal fossa with sides distinctly divergent .... *Orthostethus*
— Mesosternal fossa with sides subparallel .... *Parallelostethus*

50(48). Metacoxal plates broadly rounded and sinuate near midwidth 51
— Metacoxal plates sharply angulate near midwidth .... *Elater*

51(50). Hypomeron with posterior margin more or less straight to sinuate from coxal cavity to pronotal hind angle .... *Dolerosomus*
— Hypomeron with posterior margin deeply emarginate near pronotal hind angle ... *Neotrichophorus*

52(39). Pronotum with posterior margin crenulate 53
— Pronotum with posterior margin with smooth edge 54

53(52). Prosternum with anterior lobe large and planar with prosternal disc; integument with pallid to silvery, long, hair-like setae often forming patterns dorsally .... *Oistus*
— Prosternum with anterior lobe short and slightly deflexed; integument subglabrous with pallid, short setae .... *Perissarthron*

54(52). Mesosternum gradually declivous anteriorly; prosternum with anterior lobe short and deflexed, or absent 55
— Mesosternum (Fig. 11) sharply declivous anteriorly; prosternum with anterior lobe large and projecting anteriorly .... *Melanactes*

55(54). Prosternum with anterior lobe extended, margin arcuate, typically deflexed 56
— Prosternum more or less truncate to broadly rounded anteriorly, lacking distinct anterior lobe 94

56(55). Mesocoxal cavity (Fig. 12) surrounded by mesosternum and metasternum, mesepimeron and mesepisternum not reaching mesocoxal cavity; frontal margin carinate 57
— Mesepisternum and mesepimeron forming part of mesocoxal cavity (Fig. 13); frontal margin ecarinate 64

57(56). Elytra with distinct striae at least in basal half .. 58
— Elytra with striae obsolescent to obsolete throughout .... *Paradonus*

58(57). Pronotum glossy and finely punctured on disc 59
— Pronotum dull to weakly glossy, granulose, strigate, or rugose on disc 61

59(58). Tarsal claws evenly arcuate 60
— Tarsal claws with a broad dentate flange in basal third .... *Oedostethus*

60(59). Antennomere 3 ca. 2X length of antennomere 2. . ... *Fleutiauxellus*
— Antennomere 3 subequal in length to antennomere 2 .... *Neohypdonus*

61(58). Hypomeron with mesal margin a flattened bead 62
— Hypomeron with mesal margin simple 63

62(61). Pronotum with anterior half more coarsely granulose or rugose than posterior half .... *Zorochros*
— Pronotum evenly granulose throughout ... *Migiwa*

63(61). Pronotum and elytra moderately to strongly convex; posterior margin of ventrite 4 with distinct fringe of setae .... *Negastrius*
— Pronotum and elytra depressed to shallowly convex; ventrite 4 lacking setal fringe .... *Microhypnus*

64(56). Mesepisternum, and in some the mesepimeron, not reaching mesocoxal cavity (Fig. 14) 65
— Mesepisternum and mesepimeron reaching mesocoxal cavity (Fig. 13) 72

65(64). Scutellum depressed; antennomeres petiolate 66
— Scutellum convex; antennomeres broad at base . ... *Desolakerrus*

66(65). Metasternum subequal in length to ventrites 2+3; metathorax macropterous to brachypterous . 67
— Metasternum slightly longer than ventrite 2; metathorax micropterous .... *Hypolithus*

67(66). Pronotum finely punctured, occasionally with intermixed larger punctures 68
— Pronotum evenly coarsely punctured, basal median impression with numerous fine punctures posteriorly .... *Ascoliocerus*

68(67). Pronotum dull, coarsely and densely punctured 69
— Pronotum glossy, finely punctured 70

69(68). Pronotum wider than long, strongly convex; ventrites with thick setae .... *Hypnoidus* (part)
— Pronotum longer than wide depressed; ventrites with fine hair-like setae .... *Ligmargus*

70(68). Pronotum longer than wide; hind angles long, slender, strongly divergent; scutellum with fine hair-like setae 71
— Pronotum wider than long; hind angles short, subparallel to slightly divergent; scutellum with thickened coarse setae .... *Hypnoidus* (part)

71(70). Maxillary palpus with terminal palpomere gradually broadening apically, widest at midlength .... *Berninelsonius*
— Maxillary palpus with terminal palpomere broadest at base, narrowing apically .... *Margaiostus*

72(64). Body elongate, moderately convex to subcylindrical; supra-antennal ridges slightly transverse to confluent with frontal margin, often conjoined medially 73
— Body short to moderately elongate, depressed to shallowly convex; supra-antennal ridges obtuse, briefly transverse at most, widely separated on frons 78

73(72). Prosternal process planar to shallowly convex; pronotal length/width ratio >0.9 74
— Prosternal process strongly arcuate; pronotal length/width ratio < 0.8 .... *Eanus*

74(73). Antennae slightly to moderately serrate; scutellum pentagonal to rounded 75
— Antennae strongly serrate to pectinate; scutellum subtriangular .... *Ctenicera* (*sensu stricto*)

75(74). Pronotum with distinct posteromedian impression; antennae moderately serrate 76
— Pronotum evenly convex; antennae narrowly serrate, antennomeres 3-10 longer than wide 77

76(75). Pronotum moderately convex, lateral margins subparallel; integument dull, brunneous 90
— Pronotum strongly convex, lateral margins convergent anteriorly from midlength; integument piceous, with submetallic reflections .... *Actenicerus*

77(75). Punctures of head, pronotum, and hypomeron umbilicate; dorsal carina of hind angle obsolescent .... *Liotrichus*
— Punctures of head, pronotum, and hypomeron simple; dorsal carina of hind angle distinct. .... *Ctenicera* (part, *sensu lato*)

78(72). Pronotal hind angle convex, with strong dorsal carina; elytron evenly brown to black 79
— Pronotal hind angle deplanate, with short, fine to obsolescent dorsal carina; elytron flavous to brunneous, often with piceous maculae 86

79(78). Body moderately convex; aedeagus with parameres hooked apically 80
— Body depressed to shallowly convex; aedeagus with parameres obtuse to subtruncate apically 84

80(79). Pronotosternal sutures simple, closed anteriorly; elytral intervals finely punctured, transversely wrinkled in some 81
— Pronotosternal sutures excavate anteriorly; elytral intervals coarsely punctured .... *Hadromorphus*

81(80). Pronotal hind angle broadly rounded to subtruncate at apex; prosternal process more or less plana. 82
— Pronotal hind angle attenuate, narrowly rounded at apex; prosternal process reflexed posteriorly.. .... *Hypoganus*

82(81). Pubescence of elytron with setae minute to moderately long, pallid, length < width of elytral interval; parameres hooked or sagittate at apex ... 83
— Pubescence of elytron with setae long, cinereous, length subequal to interval width; parameres slightly expanded at apex .... *Setasomus*

83(82). Metacoxa deeply excavated to receive femur; prosternum with large anterior lobe .... *Selatosomus*
— Metacoxa shallowly exavated; prosternum with short, transverse anterior lobe .... *Anthracopteryx*

84(79). Hypomeron with mesal margin moderately to densely punctured; pronotum wider than long. 85
— Hypomeron with mesal margin impunctate anteriorly, sparsely punctured posteriorly; pronotum longer than wide .... *Ctenicera* (part, *sensu lato*)

85(84). Antennae moderately serrate, antennomeres with ventroapical angle obtuse; elytral intervals convex .... *Ctenicera* (part, *sensu lato*)
— Antennae strongly serrate, antennomeres with ventroapical angle acute; elytral intervals flat.. .... *Neopristilophus*

86(78). Claw slender, subangular at base 87
— Claw with large dentate flange in basal half .... *Oxygonus*

87(86). Hypomeron with mesal margin thickened, reflexed, often shallowly excavate anteriorly 88
— Hypomeron with mesal margin simple 89

88(87). Pronotal hind angle ecarinate; antennomeres 4-10 obtuse ventroapically .... *Ctenicera* (part, *sensu lato*)
— Pronotal hind angle with dorsal carina; antennomeres 4-10 subangulate ventroapically .... *Pseudanostirus*

89(87). Pronotal hind angle ecarinate; antennae serrate (female) to subpectinate (male) .... *Ctenicera* (part, *sensu lato*)
— Pronotal hind angle with dorsal carina; antennae serrate in both sexes .... *Frosternon*

90(76). Pronotosternal sutures closed anteriorly.......... 91
— Pronotosternal sutures excavate anteriorly .... *Limonius* (part)

91(90). Pronotum moderately to strongly convex, declivous posteriorly; hind angle convex dorsally, with distinct carina 92
— Pronotum shallowly convex, deplanate posteriorly; hind angle flat dorsally, ecarinate 93

92(91). Hypomeron with mesal margin thickened to form a flat bead .... *Limonius* (part)
— Hypomeron with mesal margin without bead .... *Proludius*

93(91). Frons shallowly convex to depressed; pronotal hind angle broadly rounded to subtruncate at apex, dorsum ecarinate .... *Corymbitodes*
— Frons shallowly to moderately concave; pronotal hind angle narrowly rounded to subacute at apex, dorsal carina short to obsolescent ... *Euplastius*

94(55). Pronotum with long basal incisures; mandibles in closed position covered by broad labrum..... 95
— Pronotal incisures very short or absent; mandibles visible in closed position 96

95(94). Pronotum glossy, minutely punctured .. *Oestodes*
— Pronotum dull, coarsely punctured ....*Bladus*

96(94). Prosternum short, antecoxal portion short, narrow; tibial spurs long, distinct 97
— Prosternum long, antecoxal portion as long as or longer than width; tibial spurs short, indistinct. 98

97(96). Anterior tibia slender throughtout; antennae strongly serrate .... *Selonodon*
— Anterior tibia expanded at midlength and at apex, often sharply dentate (Fig. 2); appearing emarginate apically; antennae slightly serrate .... *Scaptolenus*

98(96). Antennae narrowly to strongly serrate in both sexes 99
— Antennae pectinate (males) with rami subequal in length or longer than segment, or strongly serrate to subpectinate (females) 100

99(98). Scutellum elongate, subrectangular, often constricted at midlength .... *Aplastus*
— Scutellum narrow, subcordate, narrowly attenuate posteriorly .... *Aphricus*

100(98). Antennae 11-segmented .... *Octinodes*
— Antennae 12-segmented .... *Euthysanius*

Classification of the Nearctic Genera

The following classification should be regarded as tentative and subject to significant change. Synonyms listed do not include names applied to extra-territorial taxa.

Elateridae Leach 1815

Cebrioninae Latreille 1802

Cebrionini Latreille 1802

Scaptolenus LeConte 1853; 5 spp., Texas and Arizona (Fig. 2). Key to species: Horn (1881).

Selonodon Latreille 1834; 25 spp., southeastern US, west to Arizona and Utah. Key to species: Galley (1999).
Anachilus LeConte 1861
Selenedon, misspelling

Aplastini Stibick 1979 (1861)

Aplastus LeConte 1859; 15 spp., southwestern states. Key to species: Van Dyke (1932).
Anamesus LeConte 1866

Euthysanius LeConte 1853; 9 spp., California. Key to species: Van Dyke (1932).

Octinodes Candèze 1863; 12 spp., southwestern states to western Texas. Key to species: Van Dyke (1932).
Plastocerus LeConte, 1853

Elaterinae Leach 1815

Elaterini Leach 1815

Diplostethus Schwarz 1907; 2 spp., *D. opacicollis* (Schaeffer), *D. peninsularis* (Champion), southwestern states. Key to species: Roache (1961).

Dolerosomus Motschulsky 1859; 3 spp., generally distributed. Keys to species: Roache (1961), Lane (1971).

Elater Linnaeus 1758; 6 spp., eastern states and west-coastal states. Key to species: Roache (1961).
Steatoderus Dejean 1833
Ludius Berthold 1827

Neotrichophorus Jakobson 1913; 5 spp., southern states to Arizona and Nevada. Key to species: Roache (1961).

Orthostethus Lacordaire 1857; 2 spp., *O. caviceps* (Schaeffer), *O. infuscatus* (Germar); southern states to Arizona. Key to species: Roache (1961).
Crigmus LeConte 1853.

Parallelostethus Schwarz 1907; 1 sp., *P. attenuatus* (Say), eastern states and provinces.

Sericus Eschscholtz 1829; 5 spp., boreal and montane. Key to species: Roache (1961).
Sericosomus Dejean 1833
Atractopterus LeConte 1853

Megapenthini Gurjeva 1973

Megapenthes Kiesenwetter 1858; 29 spp., generally distributed. Key to species: Becker (1971).
Dolopiosomus Motschulsky 1859

Agriotini Laporte 1840

Agriotina Laporte 1840

Agriotes Eschscholtz 1829; 37 spp., generally distributed. Key to species: Becker (1956).
Cataphagus Stephens 1830
Pedetes Kirby 1837

Dalopius Eschscholtz 1829; 54 spp., generally distributed. Key to species: Brown (1934).
Dolopius, misspelling

Pomachiliina Candèze 1859

Agriotella Brown 1933; 5 spp., northern and western states, Canada. Key to species: Brown (1933b).
Betarmon, auctorum

Leptoschema Horn 1884; 2 spp., *L. protractum* (Horn) and *L. praelontactum* Stibick, California. Key to species: Stibick (1970).

Synaptina Gistel 1856

Glyphonyx Candèze 1863; 17 spp.; eastern and southern states and provinces. Key to species: Smith and Balsbaugh (1984).

Ampedini Gistel 1856

Ampedina Gistel 1856

Ampedus Dejean 1833; 71 spp., generally distributed. Keys to species: LeConte (1884), Van Dyke (1932) (*cordifer* group), Dietrich (1945) (northeastern states), Brooks (1960) (cent. Canada), Lane (1971) (Pac. NW) [*Ectamenogonus* Buysson 1893 is sometimes used for those species with bicarinate pronotal hind angles.]
Elater Eschscholtz 1829

Dicrepidiina Candèze 1859

Blauta LeConte 1853; 2 spp., *B. cribraria* (Germar), *B. falli* Brown, southeastern states. Key to species: Brown (1936e).

Dicrepidius Eschscholtz 1829; 3 spp., southeastern states to California. Key to species: Candèze (1859).
Discrepidius, misspelling

Dipropus Germar 1839; 8 spp., southeastern states to California. Key to species: Candèze (1859).
Ischiodontus Candèze 1859
Tricrepidius Motschulsky 1859

Melanotina Candèze 1859 (1856)

Melanotus Eschscholtz 1829; 51 spp., generally distributed. Key to species: Quate and Thompson (1961).
Cratonychus Dejean 1833
Ctenonychus Melsheimer 1845

Physorhinina Candèze 1859

Anchastus LeConte 1853; 17 spp., eastern and southern states to California and Oregon. Key to species: Van Dyke (1932).
Brachycrepis LeConte 1853
Crepidotritus LeConte 1861
Elatrinus Horn 1871

Physorhinus Germar 1840; 3 spp., southwestern states to Texas, Florida. Key to species: Schaaf (1970).

Cardiophorinae Candèze 1859

Aphricus LeConte 1853; 7 spp., southwestern states to Texas. Key to species: Knull (1957).

Aptopus Eschscholtz 1829; 6 spp., southwestern states to Texas.
Eniconyx Horn, 1884

Cardiophorus Eschscholtz 1829; 82 spp., generally distributed. Keys to species: Blanchard (1889); Lanchester (1971).

Esthesopus Eschscholtz 1829; 9 spp., generally distributed. Key to species: Horn (1884).

Horistonotus Candèze 1860; 17 spp., generally distributed. Keys to species: Horn (1884), Wells (2000).

Lissominae Laporte 1835

Lissomini Laporte 1835

Drapetes Dejean 1821; 7 spp., east and south to Arizona. Keys to species: Schaeffer (1916), Blanchard (1917).

Oestodini Hyslop 1917

Oestodes LeConte 1853; 2 spp., *O. tenuicollis* (Randall), Maine to New York; *O. puncticollis* Horn, North Dakota and Manitoba.

Bladus LeConte 1861; 1 sp., *B. quadricollis* (Say), southern Midwest.

Agrypninae Candèze 1851

Agrypnini Candèze 1851

Agrypnus Eschscholtz 1829; 3 spp., southern, midwestern, southwestern states. Key to species: Arnett (1952).
Colaulon Arnett 1952

Danosoma Thomson 1859; 2 spp., *D. brevicornis* (LeConte), *D. obtecta* (Say); boreal and montane. Keys to species: Arnett (1952), Lane (1971).

Lacon Laporte 1838; 13 spp., generally distributed. Key to species: Arnett (1952).
Lepidotus Stephens 1830
Aulacon Arnett 1952
Diphyaulon Arnett 1952
Zalepia Arnett 1952
Lepidelater Smith 1969

Meristhus Candèze 1857; 1 spp., *M. cristatus* Horn; Texas and Arizona.
Rhaciaspis Arnett 1952

Rismethus Fleutiaux 1947; 2 spp., *R. scobinula* (Candèze) and *R. squamiger* (Champion), Florida to New Mexico. Key to species: Champion (1894).

Oophorini Gistel 1856

Aeolus Eschscholtz 1829; 14 spp., generally distributed, mostly in east and south. Key to species: LeConte (1884).
Oophorus Eschscholtz 1833

Conoderus Eschscholtz 1829; 27 spp; generally distributed, mostly in east and south. Key to species: Van Dyke (1932).
Monocrepidius Eschscholtz 1829
Conoderes of Van Dyke 1932, misspelling

Deronocus Johnson 1997; 1 sp., *D. sleeperi* (Becker), southern California, northern Baja California.
Deroconus Johnson 1995

Heteroderes Latreille 1834; 5 spp., southwestern states to Florida [note: separation of *Heteroderes* from *Conoderus* is not always reliable]. Key to species: Van Dyke (1932)

Pseudomelanactini Arnett 1967

Lanelater Arnett 1952; 4 spp., southeastern states to New York and Arizona. Key to species: Spilman (1985).

Anthracalaus Fairmaire 1888; 1 sp., *A. agrypnoides* (Van Dyke 1932); Arizona.
Pseudomelanactes Mathieu 1961

Pyrophorini Candèze 1863

Deilelater Costa 1975; 3 spp., Florida to Arizona. Key to species: Costa (1983).
Pyrophorus, auctorum

Ignelater Costa 1975; 1 sp.; *I. havaniensis* (Laporte), Florida.
Stilpnus Laporte 1840
Pyrophorus, auctorum

Vesperelater Costa 1975; 1 sp., *V. arizonicus* (Hyslop), Arizona.
Pyrophorus, auctorum

Hemirhipini Candèze 1857

Alaus Eschscholtz 1829; 5 spp., generally distributed (Volume 1, Color Figure 15). Key to species: Casari (1996).

Chalcolepidius Eschscholtz, 1829; 11 spp., Arizona to southeastern states (Fig. 1). Key to species: Fall (1898), Casey (1907).

Pherhimius Fleutiaux, 1942; 1 sp., *P. fascicularis* (F.), southeastern states. Key to species: Casari (1986).
Hemirhipus, auctorum

Prosterninae Gistel 1856

Athoini Candèze 1859

Athous Eschscholtz 1829; 40 spp., generally distributed (Fig. 3). Keys to species: Becker (1974, 1979).

Denticollis Piller and Mitterpacher 1783; 2 spp., *D. varians* (Germar), *D. denticornis* (Kirby), boreal. Key to species: Becker (1952).
Lepturoides Herbst 1784
Campylus Fischer von Waldheim 1824

Elathous Reitter 1890; 6 spp., eastern and west-coastal states. Keys to species: Van Dyke (1932) (U.S.); Lane 1971 (Pacific Northwest).

Euplastius Schwarz 1903; 5 spp., southeastern states. Key to species: Becker (1961).

Hemicrepidius Germar, 1839; 16 spp., generally distributed. Key to species: Horn (1880)
Asaphes Kirby 1837
Pedetes, auctorum

Limonius Eschscholtz 1829; 56 spp., generally distributed. Key to species: Van Dyke (1932).
Pheletes Kiesenwetter 1858
Gambrinus LeConte 1853

Micrathous Lane 1971; 1 sp., *M. brevis* (Van Dyke); Oregon and California.

Pityobiini Hyslop 1917

Pityobius LeConte 1853; 2 spp., *P. anguinus*, east of Great Plains; *P. murrayi*, California and Oregon.
Calocerus LeConte 1853

Oxynopterini Candèze 1857

Melanactes LeConte 1853; 6 spp., eastern and California. Key to species: Mathieu (1961).

Oistus Candèze 1857; 1 sp., *O. edmundsi* Lane, California and Oregon.

Perissarthron Hyslop 1917; 1 sp., *P. trapezium* LeConte, Texas and Oklahoma.

Prosternini Gistel 1856

Actenicerus Kiesenwetter 1858; 2 spp., *A. cuprascens* (LeConte) and *A. viridis* (Say) [not *A. siaelandicus* (= *sjaelandicus*),auctorum]; wetlands in boreal forests. Key to species: Van Dyke (1932).
Malloea Arnett 1955

Anthracopteryx Horn 1891; 1 sp., *A. hiemalis* Horn; Colorado and Wyoming.

Corymbitodes Buysson 1904; 2 spp., *C. lobatus* (Eschscholtz), *C. pygmaeus* (Van Dyke); montane, boreal and northeastern. Key to species: Van Dyke (1932), Lane (1971).

Ctenicera Latreille 1829; strictly only *C. kendalli* (Kirby) is attributable to this genus from North America. All other species cataloged in this genus are undergoing taxonomic review and generic transfer.

Cleniocerus Stephens 1829
Ctenicerus Stephens 1830
Ludius Eschscholtz 1829
Corymbites Latreille 1834

Eanus LeConte 1861; 8 spp., wetlands, northern and montane. Key to species: Brown (1930, 1936e).

Paranomus Kiesenwetter 1858

Hadromorphus Motschulsky 1859; 4 spp. [*inflatus* group], generally distributed. Key to species: Brown (1936c).

Eanoides Kishii 1966

Hypoganus Kiesenwetter 1858; 3 spp., eastern, northern, and western states and provinces. Key to Species: Van Dyke (1932).

Paroedostethus Van Dyke,1932

Liotrichus Kiesenwetter 1858; 7 spp. [*volitans* group]; boreal and montane forests. Key to species: Brown (1935c), Gurjeva (1987).

Neopristilophus Buysson 1894; 4 spp. [*cribrosus* group]; far-western states and provinces, northeastern. Key to species: Brown (1935c).

Oxygonus LeConte 1863; 4 spp.; eastern states and provinces, and California and Oregon. Key to species: Roache (1963).

Proludius Lane 1971; 1 sp., *P. iaculus* (LeConte), California and Oregon.

Prosternon Latreille 1834; 6 spp. [*fallax* group], boreal and montane forests, holarctic. Key to species: Brown (1936b).

Pseudanostirus Dolin 1964; 14 spp. [*triundulatus* and *propolus* groups], boreal and montane, holarctic. Key to species: Brown (1936b, 1936d).

Selatosomus Stephens, 1830; 23 spp. [*cruciatus*, *edwardsi*, *aeripennis* and *semivittatus* groups], generally distributed. Keys to species: Brown (1935a-b, 1936a), Tarnawski (1995).

Aphotistus Kirby 1837
Diacanthus Latreille 1834
Pristilophus Latreille 1834

Setasomus Gurjeva, 1985; 3 spp. [*nitidulus* group], northern and eastern montane. Key to species: Brown (1936a)

Hypnoidini Schwarz 1906 (1860)

Ascoliocerus Mequignon 1930; 1 sp., *A. sanborni* (Horn); holarctic.

Berninelsonius Leseigneur 1970; 1 sp., *B. hyperboreus* (Gyllenhal); holarctic.

Desolakkerus Stibick 1978; 1 sp., *D. pallidus* (Becker); southern Nevada.

Hypnoidus Dillwyn 1829; 7 spp.; northern and montane. Key to species: Stibick (1978).

Hypolithus Eschscholtz 1829; 1 sp., *H. littoralis* Eschscholtz; coastal Alaska, British Columbia [N Pacific, trans-Beringian].

Cryptohypnus Eschscholtz 1830

Ligmargus Stibick 1976; 3 spp.; western and northeastern montane. Key to species: Stibick (1976).

Margaiostus Stibick 1978; 3 spp; transcontinental. Key to species: Stibick (1978).

Negastriinae Nakane and Kishii 1956

Fleutiauxellus Mequignon 1930; 2 spp., *F. extricatus* (Fall), *F. manki* (Fall); western montane.

Microhypnus Kishii 1976; 1 sp., *M. striatulus* (LeConte), northern and montane.

Migiwa Kishii 1966; 1 sp., *M. dubius* (Horn); northern Great Plains.

Negastrius Thomson 1859; 13 spp., generally distributed. Key to species: Stibick (1990), eastern spp.; Wells (1996).

Neohypdonus Stibick, 1971; 7 spp., generally distributed, mostly montane and boreal. Key to species: Stibick (1990), eastern spp.; Wells (1991).

Oedostethus LeConte 1853; 1 sp., *O. femoralis* LeConte; transcontinental.

Paradonus Stibick 1971; 6 spp; generally distributed. Keys to species: Horn (1891); Stibick (1990), eastern spp. only.

Zorochros Thomson 1859; 5 spp; generally distributed, mostly western. Key to species: Horn (1891).

Zorochrus, misspelling

Bibliography

ARNETT, R.H., Jr. 1952. A review of the Nearctic Adelocerina (Coleoptera: Elateridae, Pyrophorinae, Pyrophorini). Wasmann Journal of Biology, 10: 103-126.

BALDUF, W.V. 1935. Bionomics of entomophagous Coleoptera. John S. Swift Co. Saint Louis, Missouri.

BECKER, E.C. 1952. The nearctic species of *Denticollis*. Proceedings of the Entomological Society of Washington, 54: 105-114.

BECKER, E.C. 1956. Revision of the nearctic species of *Agriotes* (Coleoptera: Elateridae). Canadian Entomologist, 88 (Supplement 1), 101 pp.

BECKER, E.C. 1958. The phyletic significance of the female internal organs of reproduction in the Elateridae. Proceedings of the 10th International Congress of Entomology, 1: 201-205

BECKER, E.C. 1961. The North American elaterid types of Otto Schwarz, with a revision of *Ctenicera bivittata* (Melsheimer) and allies (Coleoptera). Canadian Entomologist, 93: 161-181.

BECKER, E.C. 1971. Five new species of *Megapenthes* from the southwestern United States, with a key to the nearctic species (Coleoptera: Elateridae). Canadian Entomologist, 103: 145-167.

BECKER, E.C. 1973. A European species of *Melanotus* now established at Baltimore, Maryland (Coleoptera: Elateridae). Proceedings of the Entomological Society of Washington, 75:454-458.

BECKER, E.C. 1974. Revision of the nearctic species of *Athous* (Coleoptera: Elateridae) east of the Rocky Mountains. Canadian Entomologist, 106:711-758.

BECKER, E.C. 1979. Review of the western nearctic species of *Athous* (Coleoptera: Elateridae), with a key to the species north of Panama. Canadian Entomologist, 111: 569-614.

BECKER, E.C. 1991. Elateridae (Elateroidea), Pp. 409-410. *In*: F. W. Stehr, ed. Immature Insects. Kendall/Hunt. Dubuque, IA.

BLACKWELDER, R.E. 1944. Checklist of the coleopterous insects of Mexico, Central America, The West Indies, and South America, part 2. United States National Museum Bulletin 185, pp. 189-341.

BLACKWELDER, R.E. 1957. Checklist of the coleopterous insects of Mexico, Central America, The West Indies, and South America, part 6. United States National Museum Bulletin 185, pp. vii+ 927-1492.

BLANCHARD, F. 1889. Revision of the species of *Cardiophorus* Esch. of America north of Mexico. Transactions of the American Entomological Society, 16: 1-27.

BLANCHARD, F. 1917. Revision of the North American Throscidae. Transactions of the American Entomological Society, 43: 1-26.

BOUSQUET, Y. 1991. Checklist of beetles of Canada and Alaska. Agriculture Canada Publication 1861/E, vi + 430 pp.

BROOKS, A.R. 1960. Adult Elateridae of southern Alberta, Saskatchewan and Manitoba (Coleoptera). Canadian Entomologist, supplement 20.

BROWN, W.J. 1930. A revision of the North America species of *Eanus*. Canadian Entomologist, 62: 161-166.

BROWN, W.J. 1933a. Studies in the Elateridae, I (Coleop.). Canadian Entomologist, 65: 133-141.

BROWN, W.J. 1933b. Studies in the Elateridae, II. Canadian Entomologist, 65: 173-182.

BROWN, W.J. 1934. The American species of *Dalopius* Esch. (Elateridae, Coleop.). Canadian Entomologist, 66: 30-39, 66-72, 87-96, 102-110.

BROWN, W.J. 1935a. American species of *Ludius*; the *cruciatus* and *edwardsi* groups (Coleop.). Canadian Entomologist, 67: 1-8.

BROWN, W.J. 1935b. American species of *Ludius*; the *aeripennis* group (Coleop.). Canadian Entomologist, 67: 125-135.

BROWN, W.J. 1935c. American species of *Ludius*; the *cribrosus* and *volitans* groups. Canadian Entomologist, 67: 213-221.

BROWN, W.J. 1936a. American species of *Ludius*; the *semivittatus* and *nitidulus* groups. Canadian Entomologist, 68: 11-20.

BROWN, W.J. 1936b. American species of *Ludius*; the *fallax* and *triundulatus* groups. Canadian Entomologist, 68: 99-107.

BROWN, W.J. 1936c. American species of *Ludius*; the *inflatus* group. Canadian Entomologist, 68: 133-136.

BROWN, W.J. 1936d. American species of *Ludius*; the *propolus* group. Canadian Entomologist, 68: 177-187.

BROWN, W.J. 1936e. Notes on some species of Elateridae (Coleoptera). Canadian Entomologist, 68: 246-252.

CALDER, A.A., J.F. LAWRENCE and J.W.H. TRUEMAN. 1993. *Austrelater*, gen. nov. (Coleoptera: Elateridae), with a description of the larva and comments on elaterid relationships. Invertebrate Taxonomy, 7: 1349-1394.

CANDÈZE, E.C.A. 1859. Monographie des Élatérides, tome deuxieme. Mémoires de la Société Royale des Sciences de Liège, 14: 1-543, pl. i-vi.

CARPENTER, F.M. 1992. Superclass Hexapoda. Treatise on Invertebrate Paleontology, Part R, Arthropoda 4, 4: 279-655.

CASARI, S.A. 1986. Revision of *Pherhimius* Fleutiaux, 1942, with establishment of the *Saltamartinus*, new genus (Coleoptera, Elateridae, Pyrophorinae, Hemirhipini). Papéis Avulsos de Zoología, São Paulo, 39: 379-403.

CASARI, S.A. 1996. Systematics and phylogenetic analysis of *Alaus* Eschscholtz, 1829 (Coleoptera, Elateridae). Revista Brasileira Entomologia, 40: 249-298.

CASEY, T.L. 1907. Notes on *Chalcolepidius* and the Zopherini. Canadian Entomologist, 39: 29-46.

CHAMPION, G.C. 1894. Serricornia, Elateridae. Biologia Centrali-Americana, Insecta, Coleoptera, 3(1): v-viii, 258-296.

COCKERHAM, K.L. and O.T. DEEN. 1936. Notes on life history, habits and distribution of *Heteroderes laurentii* Guér. Journal of Economic Entomology, 29: 288-296.

COSTA, C. 1983. Revisão do gênero *Deilelater* Costa, 1975 (Elateridae, Pyrophorinae), com a descrição de nova espécie. Revista Brasileira Entomologia, 27: 115-124.

CROWSON, R.A. 1961. On some new characters of classificatory importance in adults of Elateridae (Coleoptera). Entomologist's Monthly Magazine, 96: 158-161.

DAJOZ, R. 1964. Anatomie et importance taxonomique des voies génitales femelles d'origine ectodermique chez les Elateridae (Insectes, Coléoptères). Cahiers des Naturalistes, Bulletin Naturalistes Parisiens, 20: 55-72.

DALLA TORRE, K.W. von 1911. Fam. Cebrionidae. Coleopterorum Catalogus, 11(25): 1-18.
DALLA TORRE, K.W. von 1913. Fam. Cebrionidae. Genera Insectorum, fasc. 127. 16p, 1 pl.
DIETRICH, H. 1945. The Elateridae of New York State. Cornell University Agricultural Experiment Station Memoir 269, 79 pp.
DOBROVSKY, T.M. 1953. Another wireworm of Irish potatoes. Journal of Economic Entomology, 46(6): 1115.
DOGGER, J.R. 1959. The Elateridae of Wisconsin, I. A list of the species found in Wisconsin and keys to the identification of genera of adults and larvae. Wisconsin Academy of Science, Arts and Letters, 48: 103-120.
DOLIN, V. 1976. Wing venation in click beetles and its significance for the taxonomy of the family. Zoologicheskii Zhurnal, 54: 1618-1633.
DOWNIE, N.M. and R.H. ARNETT, Jr. 1996. The beetles of northeastern North America, vol. 1: Introduction; Suborders Archostemata, Adephaga, and Polyphaga, thru Superfamily Cantharoidea. Sandhill Crane Press. Gainesville, Florida.
FALL, H.C. 1898. A new *Chalcolepidius*. Entomological News, 9: 238-239
GALLEY, K.E.M. 1999. Revision of the genus *Selonodon* latreille (Coleoptera: Cebrionidae). Occasional Papers of the Florida Collection of Arthropods, vol. 10, 49 pp.
GLEN, R. 1950. Larvae of the elaterid beetles of the tribe Lepturoidini (Coleoptera: Elateridae). Smithsonian Miscellaneous Collection, 111: 1-246.
GRAVES, H.W. 1938. A Hawaiian elaterid beetle introduced into California. Pan-Pacific Entomologist, 14: 91.
GURJEVA, E.L. 1969. Some trends in the evolution of click beetles (Coleoptera, Elateridae). Entomological Review, 48: 154-159.
GURJEVA, E.L. 1974. Thoracic structure of click beetles (Coleoptera, Elateridae) and the significance of the structural characters for the family. Entomological Review, 53: 67-79.
GURJEVA, E.L. 1987. A review of the genus *Liotrichus* Ksw. (Coleoptera, Elateridae) of the world. Entomological Review, 65: 76-85.
HAYEK, C.M.F. von. 1973. A reclassification of the subfamily Agrypninae (Coleoptera: Elateridae). Bulletin of the British Museum (Natural History), Entomology (Supplement 20), 309 pp.
HAYEK, C.M.F. von. 1979. Additions and corrections to 'A reclassification of the subfamily Agrypninae (Coleoptera: Elateridae). Bulletin of the British Museum (Natural History), Entomology, 38(5):183-261.
HAYEK, C.M.F. von. 1990. A reclassification of the *Melanotus* group of genera (Coleoptera: Elateridae). Bulletin of the British Museum (Natural History), Entomology, 59: 37-115.
HETSCHKO, A. 1933. Fam. Cavicoxumidae. Coleopterorum Catalogus, 127: 1.
HORN, G.H. 1871. Descriptions of new species of Elateridae of the United States. Transactions of the American Entomological Society, 3: 299-324.
HORN, G.H. 1880. Notes on the species of *Asaphes* of boreal America. Transactions of the American Entomological Society, 8: 69-75.
HORN, G.H. 1881. Notes on Elateridae, Cebrionidae, Rhipiceridae and Dascyllidae. Transactions of the American Entomological Society, 9: 76-90.
HORN, G.H. 1884. A study of some genera of Elateridae. Transactions of the American Entomological Society, 12: 33-52.
HORN, G.H. 1891. A monograph of the species of *Cryptophypnus* of boreal America. Transactions of the American Entomological Society, 18: 1-31, pl. I.
HYSLOP, J.A. 1917. The phylogeny of the Elateridae based on larval characters. Annales of the Entomological Society of America, 10: 241-263.
HYSLOP, J.A. 1921. Genotypes of the elaterid beetles of the world. Proceedings of the U.S. National Museum, 58: 621-680.
JOHNSON, P.J. 1998. *Melanotus cete* Candèze, a second adventitious species of *Melanotus* Eschscholtz in North America. Pan-Pacific Entomologist, 74: 118-119.
KNULL, J.F. 1957. Three new species of *Aphricus* with a note on *Ctenicera* (Coleoptera: Elateridae). Ohio Journal of Science, 57: 200-202.
LANCHESTER, H.P. 1971. Subfamily Cardiophorinae, Pp. 35-48. *In*: M. H. Hatch. Beetles of the Pacific Northwest, vol. 5. University of Washington Press. Seattle.
LANE, M.C. 1953. Distribution of the wireworm *Conoderus vagus* Cand. U.S. Bureau of Entomology and Plant Quarantine, Cooperative Economic Insect Report, 3(29): 536.
LANE, M.C. 1971. Family Elateridae [except Cardiophorinae], Pp. 6-35. *In*: M.H. Hatch. Beetles of the Pacific Northwest, vol. 5. University of Washington Press. Seattle.
LAWRENCE, J.F. 1987. Rhinorhipidae, a new beetle family from Australia with notes on the phylogeny of the Elateriformia. Invertebrate Taxonomy, 2: 1-53.
LAWRENCE, J.F. and A.F. NEWTON, Jr. 1995. Families and subfamilies of Coleoptera (with selected genera, notes, references and data on family-group names), Pp. 779-1006. *In*: J. Pakaluk and S.A. Slipinski, eds., Biology, Phylogeny and Classification of Coleoptera: papers Celebrating the 80th Birthday of Roy A. Crowson. Muzeum i Instytut Zoologii PAN, Warzawa.
LAWRENCE, J.F., A.M. HASTINGS, M.J. DALLWITZ, T.A. PAINE and E.J. ZURCHER. 1999. Beetles of the World: A key and information system for families and subfamilies. CD-ROM and manual. CSIRO Publishing. Collingwood, Victoria.
LECONTE, J.L. 1853. Revision of the Elateridae of the United States. Transactions of the American Philosophical Society, 10: 405- 508.
LECONTE, J.L. 1884. Short studies of North American Coleoptera (no. 2). Transactions of the American Entomological Society, 12: 1-32.
LENG, C.W. 1925. Catalogue of the Coleoptera of America north of Mexico. John D. Sherman, Jr. Mount Vernon, NY, 470 pp.
MATHIEU, J.M. 1961. Revision of the genus *Melanactes*, with a proposed new genus (Coleoptera, Elateridae). American Midland Naturalist, 65: 459-480.

MUONA, J. 1995. The phylogeny of Elateroidea (Coleoptera), or which tree is best today? Cladistics, 11: 317-341.

NAKANE, T. and T. KISHII 1956. On the subfamilies of Elateridae from Japan (Coleoptera). Kontyû, 24: 201-206.

ROACHE, L. 1961. A revision of the North American elaterid beetles of the tribe Elaterini (Coleoptera: Elateridae). Transactions of the American Entomological Society, 86: 275-324, pl. xv-xvii.

ROACHE, L. 1963. A revision of the genus *Oxygonus* LeConte with a description of one new species (Coleoptera: Elateridae) (Tribe: Agriotini). Coleopterists Bulletin, 17: 101-108.

QUATE, L.W. and S.E. THOMPSON. 1961. Revision of click beetles of genus *Melanotus* in America north of Mexico (Coleoptera: Elateridae). Proceedings of the U.S. National Museum, 121: 1-83, 1 pl.

SCHAAF, D. 1970. Revision of the genus *Physorhinus* (Coleoptera, Elateridae) in North, Central, and South America, Part I: Introduction and key to the species. Entomological News, 81: 225-232

SCHAEFFER, C. 1916. New species of Throscidae (Col.). Bulletin of the Brooklyn Entomological Society, 11: 62-63.

SCHENKLING, S. 1925. Fam. Elateridae I. Coleopterorum Catalogus, 11(80): 1-263.

SCHENKLING, S. 1927a. Fam. Elateridae II. Coleopterorum Catalogus, 11(88): 265-636.

SCHENKLING, S. 1927b. Fam. Plastoceridae. Coleopterorum Catalogus, 11(93): 3-7.

SCHENKLING, S. 1927c. Fam. Dicronychidae. Coleopterorum Catalogus, 11(93): 8-11.

SCHWARZ, O. 1906. Coleoptera, Fam. Elateridae. *In*: P. Wytsman, Genera Insectorum, fasc. 46B: 1-224.

SCHWARZ, O. 1907a. Coleoptera, Fam. Elateridae. *In*: P. Wytsman, Genera Insectorum, fasc. 46C: 225- 370, 6 pl.

SCHWARZ, O. 1907b. Coleoptera, Fam. Plastoceridae. *In*: P. Wytsman, Genera Insectorum, fasc. 50: 1-10, 1 pl.

SCHWARZ, O. 1907c. Coleoptera, Fam. Dicronychidae. *In*: P. Wytsman, Genera Insectorum, fasc. 51: 1-5, 1 pl.

SMITH, J.W. and E.U. BALSBAUGH, Jr. 1984. A taxonomic revision of the nearctic species of *Glyphonyx* (Coleoptera: Elateridae) with notes on *G. quadraticollis* Champion. North Dakota Insects Schafer-Post Series, no. 16, 85 pp.

SMITH, J.W. and W.R. ENNS 1977. The click beetle subfamilies Agrypninae, Pyrophorinae, and Melanotinae (Coleoptera: Elateridae) in Missouri - Part I. Journal of the Kansas Entomological Society, 50: 436-468.

SMITH, J.W. and W.R. ENNS 1978. The click beetle subfamilies Agrypninae, Pyrophorinae, and Melanotinae (Coleoptera: Elateridae) in Missouri - Part II. Journal of the Kansas Entomological Society, 51: 42-74.

SPILMAN, T.J. 1985. Review of *Lanelater* of North America (Coleoptera, Elateridae). Insecta Mundi, 1: 1-12.

STIBICK, J.N.L. 1970. A revision of the genus *Leptoschema* of California with a key to related genera (Coleoptera: Elateridae). Pan-Pacific Entomologist, 46: 153-166.

STIBICK, J.N.L. 1971. The generic classification of the Negastriinae (Coleoptera: Elateridae). Pacific Insects, 13: 371-390.

STIBICK, J.N.L. 1976. A revision of the Hypnoidinae of the world (Col. Elateridae); Part I, Introduction, phylogeny, biogeography. The Hypnoidinae of North and South America. The genera *Bernineisonius* and *Ligmargus*. Eos, 51: 143-223.

STIBICK, J.N.L. 1978. A revision of the Hypnoidinae of the world (Col. Elateridae); Part II. The Hypnoidinae of North and South America. The genera *Ascoliocerus, Desolakkerus, Margaiostus, Hypolithus* and *Hypnoidus*. Eos, 52: 309-386.

STIBICK, J.N.L. 1979. Classification of the Elateridae (Coleoptera): relationships and classification of the subfamilies and tribes. Pacific Insects, 20: 145-186.

STIBICK, J.N.L. 1990. North American Negastriinae (Coleoptera, Elateridae): The Negastriinae of the eastern United States and adjacent Canada. Insecta Mundi, 4: 99-131.

TARNAWSKI, D. 1995. A revision of the genus *Selatosomus* Stephens, 1830 (Coleoptera: Elateridae: Athoinae: Ctenicerini). Genus (Supplement), 183 pp.

THOMAS, C.A. 1929. The parasites of wireworms (Coleop.: Elateridae). Entomological News, 40: 287-293.

THOMAS, C.A. 1931. The predatory enemies of Elateridae (Coleoptera). Entomological News, 42: 137-140, 158-167.

THOMAS, C.A. 1932. The diseases of Elateridae (Coleoptera). Entomological News, 43: 149-155.

THOMAS, C.A. 1940. The biology and control of wireworms. Pennsylvania Agricultural Experiment Station Bulletin 392, 90 pp.

VAN DYKE, E.C. 1932. Miscellaneous studies in the Elateridae and related families of Coleoptera. Proceedings of the California Academy of Sciences, 4th Series, 20(9): 291-465.

WELLS, S.A. 1991. Two new species of *Neohypdonus* (Coleoptera: Elateridae) from North America with a key to nearctic species. Entomological News, 102: 73-78.

WELLS, S.A. 1996. Studies on nearctic *Negastrius* (Coleoptera: Elateridae). Great Basin Naturalist, 56: 308-318.

WELLS, S.A. 2000. Two new species of *Horistonotus* Candeze (Coleoptera: Elateridae), new synonymies, and a key to the species of the United States and Canada. Proceedings of the Entomological Society of Washington, 102(2): 412-420.

ZACHARUK, R.Y. 1962. Some new larval characters for the classification of Elateridae (Coleoptera) into major groups. Proceedings of the Royal Entomological Society of London, (B)31: 29-32.

ZACHARUK, R.Y. 1963. Comparative food preferences of soil-sand-, and wood-inhabiting wireworms (Coleoptera, Elateridae). Bulletin of Entomological Research, 54: 161-165.

59. LYCIDAE Laporte 1836

by Richard S. Miller

Family common name: The net-winged beetles

The relatively soft body with seven (female) or eight (male) visible abdominal sternites, the head partially covered by the pronotum, the distinctly separated mesocoxae (Western Hemsiphere), and the interstitial (not offset) trochanters (Fig. 29, Volume 1, Introduction) of the front and middle legs characterize adults of this distinctive family. Larvae can always be recognized by their unique longitudinally divided mandibles giving an appearance of two pair of non-opposing hypognathous mandibles.

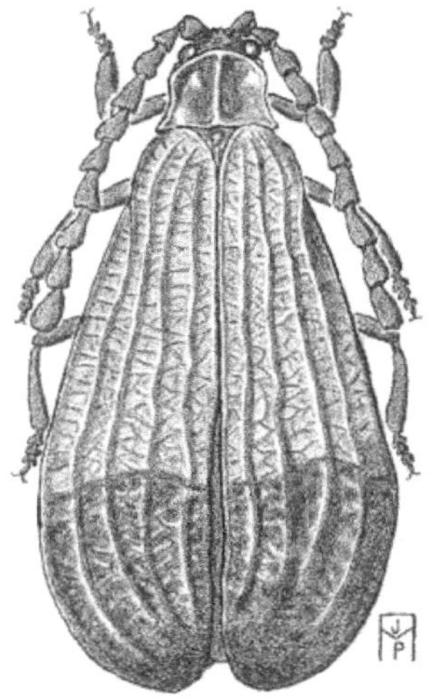

FIGURE 1.59. *Calopteron terminale* (Say)

Description: Elongate with elytra either nearly parallel or expanded beyond lateral margins of abdomen; size 2 to 22 mm in length, paedomorphic females may reach to 80 mm; variously colored, usually black, yellow, red, or a combination of these colors, vestiture usually sparse.

Head partly covered by pronotum, triangular or elongate (some Leptolycini); mouthparts ventral. Antennae with eleven (occasionally 10) antennomeres; filiform, serrate, pectinate, to flabellate; inserted between eyes or slightly below them, approximate, often under antennal prominence. Labrum distinct in most, free or occasionally fused to epistome; mandibles without teeth, straight and not meeting or slightly arcuate to falciform and crossing each other distally; maxillary palpi with three to four palpomeres, apical palpomere variously modified from accuminate, securiform, to cultriform; labium with gular region short and broad, submentum small, mentum small, labial palpi with one to three palpomeres, apical palpomere acuminate to expanded. Eyes lateral, size small to very large, bulging, oval, never incised by a canthus.

Pronotum broader than head; disk with carinae or none: borders explanate, surface rugose; hypomeron concave; prosternum short, transverse, intercoxal process short; procoxal cavities open. Mesosternum short, front margin not deeply excavate; mesocoxal cavities open. Metasternum long; metepisternum not sinuate on inner side. Legs gressorial; with a triangular trochantin on front coxae; front coxae conical, prominent; middle coxae separate or contiguous (*Duliticola*), oblique; hind coxae contiguous and transverse. Trochanters usually long, interstitial and not or little offset. Femora swollen, little to strongly compressed. Tibia swollen, little to strongly compressed, with or without apical spurs. Tarsal formula 5-5-5; tarsomeres one to four ventrally with apical plantar pads of dense pubescence, third and fourth cordate in most, but not emarginate, fifth tarsomere arising dorsally and sub-basally on fourth, claws usually simple. Scutellum quadrate, rarely triangular. Elytra parallel to expanded laterally, often with four to ten discal costae, but sometimes acostate; disk with regular reticulation in uniform cells, irregular reticulation, or none. Wing venation and folding elateriod.

Abdomen of male with eight visible sternites, of female with seven, sternites not fused, dorsal and ventral surfaces equally sclerotized, not membranous. Male genitalia trilobed; penis long, narrow, variously modified, some with penis membranous except for a few struts; parameres very short and fused to base of penis, or long and surrounding penis; pars basalis large and scoop-shaped, some with pars basalis reduced to a narrow ring.

Larvae elongate and parallel-sided or fusiform, slightly to moderately depressed; lightly to heavily sclerotized on dorsal and ventral surfaces; size 2 to 80 mm in length; vestiture sparse; color black, brown, red, yellow, or a combination of these. Head small, rounded or quadrate to pentagonal, widened anteriorly and somewhat depressed; epicranial stem, frontal arms and median endocarina absent; lateral ecdysial line present or absent; a pair of stemmata, one on each side of head, or none; antennae short and robust, prominent, two-segmented, the apical segment dome-shaped with sensorium at apex; mandibles symmetrical, narrow and falcate, inserted proximally at base, diverging distally, each divided longitudinally from base to apex, all appearing as 4 mandibles, without mola or accessory ventral process; labrum absent; maxillae with cardo and stipes distinct or fused, palps three-segmented; Labial mentum and submentum fused, labial palpi two-segmented; gula region very short and indistinct. Thoracic and abdominal tergites with margins entire or provided with processes; legs 5-segmented, with claw-like tarsungulus, each. Abdomen nine-segmented, the tenth, if present, very small, lightly sclerotized and not visible dorsally; usually without urogomphi; spiracles biforous. See: Lawrence (1991), Miller (1997), Bocák and Matsuda (in press).

Habits and habitats. Although there are anecdotal accounts that larvae are carnivores (Crowson 1955, Arnett 1968, Miller 1988), probably most, if not all, feed on myxomycetes or metabolic products of fungi (Lawrence 1982). Larvae are found in rotten fallen timber, but also on occasion in leaf litter, under bark, and *Lycus* larvae may be seen moving over open ground in the early evening. Adults are found on leaves and flowers and eat

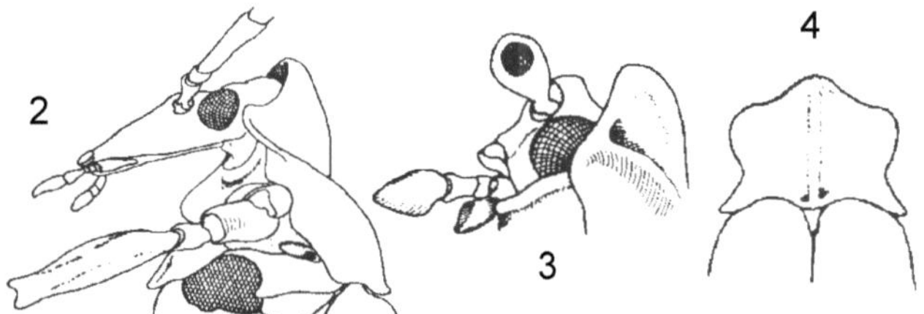

FIGURES 2.59-4.59. 2. *Lycus loripes* Chevrolat, head and prothorax, lateral view; 3. *Calopteron reticulatum* (Fabricius), head and prothorax, part, lateral view; 4. *Caenia dimidiata* (Fabricius), pronotum, dorsal view.

nectar and honeydew. Numerous species aggregate in one of their life stages and adults putatively serve as models in mimetic complexes. Both larvae and adults are distasteful and are often aposematically colored. A few Asian species with adult females, that differ little externally from larvae, are called trilobite larvae; their biology is reviewed in Mjöberg (1925) and Wong (1996). There are also a number of other species with females unknown and male morphology suggestive of species with flightless females (Bocák and Bocáková 1989, Miller 1991).

Status of the classification. This family has been completely reviewed for North America north of Mexico by Green (1949-1954). Bocák and Bocáková (1990) reviewed the then current super-generic classification of the family for the world, but continuing study has resulted in some significant changes and more are expected. The last world catalog (Kleine 1933) is woefully out of date, especially for North America. Several specialists currently are working on relationships within the family as well as describing new taxa and revising genera. The subfamilial classification used here follows Lawrence and Newton (1995) as modified by Miller (1997).

The family's relationship with other members of the superfamily is still uncertain and even the monophyly of the superfamily has been questioned. The traditional classification has been two superfamilies: Elaterioidea and Cantharoidea (Crowson 1955, 1972), but Lawrence (1996) placed the Cantharoidea within the Elateroidea and questioned the monophyly of the former. Beutel (1995, 1997), using internal structure within the larval head reaffirmed a monophyletic Cantharoidea. However, absence of the studied character states in Lycidae yields equivocal results in its placement.

Distribution. The family is cosmopolitan and, although it is most numerous in the tropics, it is also found throughout temperate regions into austral and boreal wooded areas. In deserts of these regions lycids are found in or near the more mesic drainages. There are more than 3500 (Lawrence 1991) described species worldwide. Seventy-six species are recorded north of Mexico, but several species remain undescribed. The key that follows (modified from Green 1949-1954) is based on the fauna of North America north of Mexico and extra-limital species may key incorrectly.

Key to the Nearctic Genera

1. Antennomere 2 expanded distally; antennae not or little compressed; body nearly linear in dorsal view; elytra not broadly expanded beyond lateral margins of abdomen (Erotinae) 2
— Antennomere 2 not expanded distally, parallel-sided and either longer than wide or markedly transverse; body linear, fan-shaped, or oval in dorsal view; elytra broadly or not expanded beyond lateral margins of abdomen 7

2(1). Pronotum with two more or less distinct longitudinal carinae enclosing a central cell 3
— Pronotum with single longitudinal carina, this either entire, apical, or absent 6

3(2). Longitudinal carinae of pronotum entire, confluent at base and at apex and enclosing central diamond-shaped cell 4
— Longitudinal carinae of pronotum not entire, central cell not diamond-shaped 5

4(3). Elytron with four raised costae; reticulation in ten rows of cells .... *Dictyoptera*
— Elytron with three raised costae; reticulation in eight rows of cells .... *Benibotarus*

5(3). Frons normally deflexed, not produced and transversely carinate; antennal sockets approximate .... *Eros*
— Frons acutely produced, transversely carinate between antennal sockets, latter widely separated .... *Platycis*

6(2). Longitudinal carina of pronotum entire ... *Lopheros*
— Longitudinal carina of pronotum apical and short. .... *Eropterus*

7(1). Pronotum without carinae, often with median longitudinal impression and pronotum with folds extending obliquely from lateral margin near basal fourth onto disk; elytra without reticulation; peritremes of anterior thoracic spiracles not exposed (Calochrominae)................................... 8
— Pronotum with at least a short carina on disk; antennomere 2 short, transverse.................. 12

8(7). Head rostrate; mandibles small, slender, and nearly straight, their tips distant when mandibles are closed 9
— Head not rostrate; mandibles well developed, strongly arcuate, their tips approximate or overlapping when closed 10

9(8). Median longitudinal line of pronotum linear apically, becoming broadly impressed, not sharply limited, widest near middle, narrowing to base; tibial spurs free and similar in shape; anterior coxae subcontiguous; dorsal pubescence short and decumbent, not sexually dimorphic; hypomera concave and widely explanate .... *Lygistopterus*
— Median longitudinal impression of pronotum linear, striaform throughout; tibial spurs in part rigidly attached and dissimilar; anterior coxae distinctly separated, dorsal pubescence long and erect,

sexually dimorphic; hypomera narrow, nearly flat, not broadly explanate .... *Lucaina*

10(8). Tibial spurs rigidly attached and dissimilar in shape; anterior coxae distinctly separated, elytral pubescence sparse, confined mostly to costae, intervals subglabrous; lateral elevations of pronotum parallel to each other .... *Caloptognatha*

— Tibial spurs free, similar in shape; anterior coxae contiguous or nearly so; elytra uniformly pubescent; lateral elevations of pronotum oblique 11

11(10). Median longitudinal line of pronotum linear apically, becoming broadly impressed, not sharply limited, widest near middle, narrowing to base, mandibles small, their tips slightly overlap when mandibles closed .... *Adoceta*

— Median longitudinal impression of pronotum linear, striaform throughout, sometimes partially obliterated .... *Calochromus*

12(7). Pronotum with complete median carina from front to posterior margins (Calopterini) 13

— Median pronotal carina not complete 16

13(12). Apical maxillary palpomere large and strongly transverse (Fig. 3); antennae serrate in both sexes (Fig. 1); anterior margin of pronotum arcuate; elytra broadly expanded beyond lateral margins of abdomen, each with four discal costae .... *Calopteron*

— Apical maxillary palpomere longer than wide; antennae flabellate in male, serrate in female; anterior margin of pronotum produced into subtriangular median lobe; lateral margins sinuate (Fig 4) 14

14(13). Apical labial palpomere transverse, broadly triangular, apically minutely digitate; mesosternum with entire median longitudinal impression; elytra broadly expanded beyond lateral margins of abdomen, each with four discal costae, intervals with irregular double row of cells; antennomere 3 of male simple .... *Caenia*

— Apical labial palpomere longer than wide; mesosternum not impressed; elytra each with three or four discal costae, intervals each with single row of cells; antennomere 3 of male with short process 15

15(14). Elytra expanded beyond lateral margins of abdomen, broadened posteriorly, each elytron with three discal costae; antennal branches of male arising apically .... *Idiopteron*

— Elytra subparallel or feebly broader posteriorly, each with four discal costae, these sometimes in part more or less abbreviated; antennal branches of male arising basally .... *Leptoceletes*

16(12). Peritremes of anterior thoracic spiracles prominently exposed; head rostrate (Fig. 2); pronotum without oblique lateral elevations; elytra broadly expanded beyond lateral margins of abdomen, costate and irregularly reticulate with raised lines, fourth discal costa prominently cariniform at humerus (Lycinae) 17

— Peritremes of anterior thoracic spiracles not exposed; elytra not broadly expanded beyond lateral margins of abdomen, body shape nearly linear; anterior coxae contiguous or nearly so (Platerodinae) 18

17(16). Prosternal crest carinate; mesosternum protruberant; inner angle of metacoxae somewhat prominent posteriorly; trochanters of male unmodified .... *Lycus*

— Prosternum and mesosternum unmodified; inner angle of metacoxae not produced; trochanters of male spinose .... *Lyconotus*

18(16). Antennae short and rather stout, less than half as long as body in male; middle antennomeres subquadrate; body smaller, less than 4 mm in length; genitalia of male with parameres .... *Falsocalleros*

— Antennae more long and slender, greater than half length of body in male, middle antennomeres elongate; body larger, usually over 4 mm in length; male genitalia without parameres .... *Plateros*

Classification of the Nearctic Genera

Lycidae Laporte 1836

Erotinae LeConte 1881

(Platerodini of Green 1951, in part)

Benibotarus Kôno 1932, a Holarctic genus with a single North American representative, *Benibotarus thoracicus* (Randall), found in northern United States and southern Canada from British Columbia through Ontario and Maine south to California and Florida. Bocák and Bocáková (1987) transferred this species from *Dictyoptera* to the current placement.

Dictyoptera Latreille 1829, a Holarctic genus with 4 species occurring in North America north of Mexico in coniferous or mixed coniferous forests. They are generally distributed from Alaska to Newfoundland south to California and Florida. Kasantsev (1990) has subdivided the genus into subgenera.

Dicytopterus Mulsant 1838 (see Pope 1977)

Eropterus Green 1951, this Holarctic genus, most numerous in species in the Eastern Palearctic, has 4 species in North America from Ontario to Maine south to Florida and Texas.

Eros Newman 1838, a former taxonomic suitcase, is apparently monotypic with *Eros humeralis* (Fabricius) found in Ontario south to Georgia and west to Ohio.

Lopheros LeConte 1881, two species of this Holarctic genus are known from Ontario and Quebec to New Hampshire south to Michigan and North Carolina.

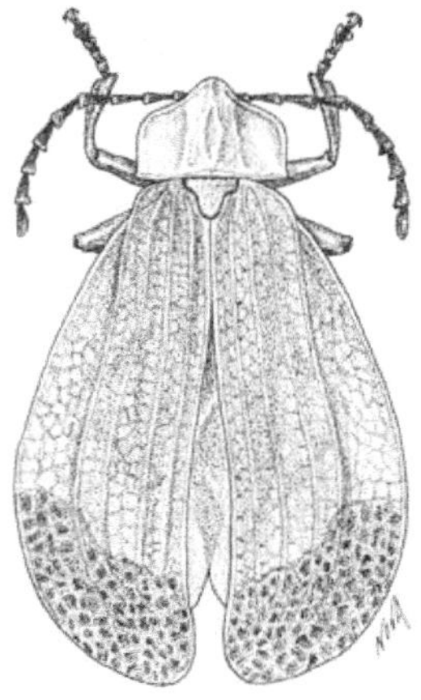

FIGURE 5.59 *Lycus arizonensis* Green

Platycis Thomson 1864, a Holarctic genus with one species, *Platycis sculptilis* (Say), in Ontario south to Florida, west to Illinois and Tennessee. McCabe and Johnson (1979) discussed its biology.

Erotides Waterhouse 1879

Calochrominae Lacordaire 1857

(=Lygistopterini of Green 1950)

Adoceta Bourgeois 1882, Green (1950) described 2 species from Arizona that he placed in this primarily African genus.

Calochromus Guérin 1833, 5 spp. are found north of Mexico: 4 from British Columbia to South Dakota south to California and New Mexico and the other, *C. perfacetus* (Say), from Michigan to Massachusetts south to Florida and New Mexico.

Caloptognatha Green 1954, *Caloptognatha beameri* Green, the only member of the genus, is found in Arizona.

Lucaina Dugés 1878, a genus restricted to the New World with 3 described species; 2 are found in southwestern Texas, Arizona, Nevada, and California.

Lycaina Horn 1885 (in error)

Lygistopterus Dejean 1833, found in both the Holarctic and Neotropical regions, only *L. rubripennis* LeConte is represented and ranges from Colorado to New Mexico, and Arizona.

Platerodinae Kleine 1928

Bocák and Bocáková (1990), following Green (1951, 1953), treated this as a tribe of Erotinae, but Miller (1997) demonstrated that the tribe shares apomorphies with more derived Lycidae, elevated the tribe to subfamily, and placed it there.

Falsocalleros Pic 1933, 2 spp. described, *F. moleculus* (Green) is known from Texas.

Pseudoplateros Green 1951, Bocáková 2001 (synonymy)

Plateros Bourgeois 1879, 31 spp., generally distributed throughout southern Canada and the United States.

Melaneros Bocák and Bocáková 1992, not Fairmaire 1877; corrected by Bocáková (2001)

Neoceletes Leng and Shoemaker 1915

Lycinae Laporte 1836

Lyconotus Green 1949, 1 sp., *L. lateralis* Melsheimer 1846, found in Eastern North America from Maine to Florida and west to Illinois, and Texas. The second species from Mexico was described by Zaragoza (1995), when he elevated *Lyconotus* to generic status.

Lycus Fabricius 1787, 11 spp. known north of Mexico from Arizona, Colorado, New Mexico, and Texas. *Lycus, sensu lato*, has an amphitropical distribution ranging into the temperate regions, but is not found in Australia. The aggregations of *Lycus loripes* Chevrolat were studied by Eisner and Kafatos (1962).

Rhyncheros LeConte 1881

subgenus *Neolycus* Bourgeois 1883

subgenus *Lycostomus* Motschulsky 1861

Calopterini Kleine 1933

Caenia Newman 1838, 12 spp. are known in North and South America with 2 found north of Mexico: *C. amplicornis* LeConte known only from Arizona and *C. dimidiata* (F.) from Quebec south to Missouri and Florida.

Caeniella Cockerell 1906 was listed as a synonym without comment by Green (1952). It is an unnecessary replacement name, based on *Coenia* Robineau-Desvoidy (Diptera: Ephydridae), but a single letter difference does not constitute homonym (1999 ICZN, Article 56.2).

Calopteron Castelnau 1838, a genus of 149 described species found throughout the New World, 3 are found north of Mexico from Manitoba to Ontario south to Texas and Florida. Young and Fisher (1972), McCabe and Johnson (1980), and Miller (1988) discussed the immature stages.

Digrapha Newman 1838

Calopteron Guérin-Méneville 1838 (see Bocák 1998)

Idiopteron Bourgeois 1905, 1 sp., *I. rufulum* (Gorham 1880). This record is based on a single specimen, the type of the synonym *Calopteron tricarinatum* LeConte (Green 1952) from Arizona. The species is found in southern Mexico and Central America, so the locality of the specimen is likely to be mislabeled. Green (1952) reported, and I have seen, no further records for Arizona.

Leptoceletes Green 1952, 6 spp. are recognized in the world, with 2 found only in Texas and a third is found from Nova Scotia south to Florida, west to South Dakota and Arkansas.

Celetes Newman 1838, not Schönherr 1836 (see Miller in press)

Bibliography

ARNETT, R. A., Jr. 1968. The Beetles of the United States: A Manual for Identification. American Entomological Institute. Ann Arbor, MI, xii + 1112 pp.

BEUTEL, R. G. 1995. Phylogenetic analysis of Elateriformia (Coleoptera: Polyphaga) based on larval characters. Journal of Zoologic Systematic Evolutionary Research, 33: 145-171.

BEUTEL, R. G. 1997. Uber Phylogenese und Evolution der Coleoptera (Insecta), insbesondere der Adephaga.

Abhandlungen des Naturwissenschaftlichen Vereins in Hamburg, (NF) 31: 1-164.
BOCÁKOVÁ, M. 2001. A revision and phylogenetic analysis of the subfamily Platerodinae (Coleoptera, Lycidae). European Journal of Entomology, 98: 53-85.
BOCÁK, L. 1998. Nomenclatural notes on taxa of the family Lycidae described by Guérin-Méneville (Insecta: Coleoptera). Annales Zoologici, 48: 245-251.
BOCÁK, L. and M. BOCÁKOVÁ. 1987. Notes on the taxonomy of some European species of the family Lycidae (Coleoptera). Acta Entomologica Bohemoslavaca, 84: 111-121.
BOCÁK, L. and M. BOCÁKOVÁ. 1989. New tribe Lyropaeini, with description of a new species of *Lyropaeus* (Coleoptera, Lycidae). Polskie Pismo Entomologiczne, 58: 717-723.
BOCÁK, L. and M. BOCÁKOVÁ. 1990. Revision of the supergeneric classification of the family Lycidae (Coleoptera). Polskie Pismo Entomologiczne, 59: 623-676.
BOCÁK, L and K. MATSUDA. In Press. Review of the immature stages of the family Lycidae (Coleoptera: Lycidae). Journal of Natural History.
CROWSON, R. A. 1955. The Natural Classification of the Families of Coleoptera. Nathaniel Lloyd, London (seen as 1967 reprint, E. W. Classey. Hampton, 187 pp.)
CROWSON, R. A. 1972. A review of the classification of Cantharoidea (Coleoptera), with the definition of two new families, Cneoglossidae and Omethidae. Revista de la Universidad de Madrid, 21(82): 35-77.
EISNER, T. E. amd F. C. KAFATOS. 1962. Defense mechanisms of Arthropods X. A pheromone promoting aggregation in an aposematic distasteful insect. Psyche 69: 53-61.
GREEN, J. W. 1949. The Lycidae of the United States and Canada, I. The tribe Lycini (Coleoptera). Transactions of the American Entomological Society, 75: 53-70.
GREEN, J. W. 1950. The Lycidae of the United States and Canada, II. The tribe Lygistopterini (Coleoptera). Transactions of the American Entomological Society, 76: 13-25.
GREEN, J. W. 1951. The Lycidae of the United States and Canada, III. The tribe Platerodini (in part) (Coleoptera). Transactions of the American Entomological Society, 77: 1-20.
GREEN, J. W. 1952. The Lycidae of the United States and Canada, IV. The tribe Calopterini (Coleoptera). Transactions of the American Entomological Society, 78: 1-19.
GREEN, J. W. 1953. The Lycidae of the United States and Canada, V. *Plateros*. Transactions of the American Entomological Society, 78: 149-181.
GREEN, J. W. 1954. An addition to the lycid fauna of the United States (Coleoptera). Coleopterists Bulletin, 8: 55-57.
INTERNATIONAL COMMISSION ON ZOOLOGICAL NOMENCLATURE. 1999. International Code of Zoological Nomenclature. Fourth Edition. International Trust for Zoological Nomenclature, London. xxix + 306 pp.
KASANTSEV, S. V. 1990. [Two new subgenera and a new species of the genus *Dictyoptera* (Coleoptera, Lycidae) from East Asia]. Vestnik Zoologii, 1: 10-15.
LAWRENCE, J. F. 1982. Coleoptera. Pp. 482-553. *In*: S. Parker, ed. Synopsis and Classification of Living Organisms, volume 2. McGraw-Hill. New York, NY.
LAWRENCE, J. F. 1991. Lycidae (Cantharoidea). Pp. 423-424. *In*: F. W. Stehr, ed. Immature Insects, volume 2. Kendall / Hunt. Dubuque, IA. xvi + 975 pp.
LAWRENCE, J. F. 1996. Rhinorhipidae, a new beetle family from Australia, with comments on the phylogeny of the Elateriformia. Invertebrate Taxonomy, 2: 1-53.
LAWRENCE, J. F. and NEWTON, A. F., Jr. 1995. Families and subfamilies of Coleoptera (with selected genera, notes, references and data on family-group names), Pp. 779-1006. *In*: J. Pakaluk and S. A. Slipinski, eds. Biology, Phylogeny, and Classification of Coleoptera: Papers Celebrating the 80th Birthday of Roy A. Crowson. Muzeum I Instytut Zoologii PAN. Warsaw. vi + 559-1092 pp.
McCABE, T. L. and L. M. JOHNSON. 1979. The biology of *Platycis sculptilis* (Say) (Coleoptera: Lycidae). Coleopterists Bulletin, 33: 297-302.
McCABE, T. L. and L. M. JOHNSON. 1980. Larva of *Calopteron terminale* (Say) with additional notes on adult behavior (Coleoptera: Lycidae). Journal of the New York Entomological Society, 87: 283-288.
MILLER, R. S. 1988. Behaviour of *Calopteron reticulatum* (F.) larvae (Coleoptera: Lycidae). Ohio Journal of Science, 88: 119-120.
MILLER, R. S. 1991. A revision of the Leptolycini (Coleoptera:Lycidae) with a discussion of paedomorphosis. Ph.D Dissertation, Ohio State University. Columbus, OH. 403 pp.
MILLER, R. S. 1997. Immature stages of *Plateros floralis* (Melsheimer) and discussion of phylogenetic relationships (Coleoptera: Lycidae). Coleopterists Bulletin, 51: 1-12.
MILLER, R. S. Submitted. *Leoptoceletes* Green (1952) a replacement for *Celetes* Newman (1838) [not *Celetes* Schönherr (1836)] (Coleoptera: Lycidae, Curculionidae). Coleopterists Bulletin.
MJÖBERG, E. 1925. The mystery of the so-called "trilobite larvae" or "Perty's larvae" definitely solved. Psyche, 32: 119-154
POPE, R. D. 1977. *In*: Kloet and Hincks. A Checklist of British Insects. Second Editon (Completely Revised). Part 3. Coleoptera and Strepsiptera. Royal Entomological Society of London. London. xiv + 105 pp.
WONG, A. T. C. 1996. A new species of neotenous beetle, *Duliticola hosiseni* (Insecta: Coleoptera: Cantharoidea: Lycidae) from peninsular Malaysia and Singapore. Raffles Bulletin of Zoology, 44: 173-187.
YOUNG, D. K. and R. L FISCHER. 1972. The pupation of *Calopteron terminale* (Say) (Coleoptera: Lycidae). Coleopterists Bulletin 26: 17-18.
ZARAGOZA CABALLERO, S. 1995. Cantharoidea (Coleoptera) de Mexico. II. Lycinae de Veracruz. Folia Entomologica Mexicana, 95: 23-84.

60. TELEGEUSIDAE Leng 1920

by Richard S. Miller

Family common name: The long-lipped beetles

The elongate and narrow body with apical maxillary palpomeres elongate and compressed, filiform to serrate antennae, variously shortened elytra, and submarginally impressed pronotal disk serve to distinguish this family.

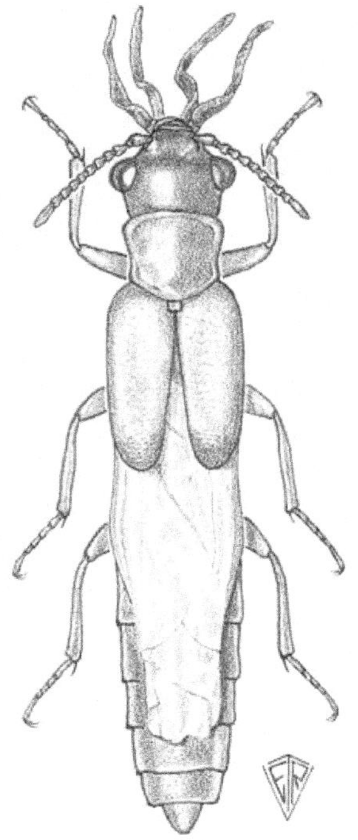

FIGURE 1.60. *Telegeusis nubifer* Martin

Description: Elongate; size 2.5 to 8 mm in length; color testaceous to black; vestiture sparse, fine, erect, setae usually same color as underlying sclerite.

Head prognathus; surface finely rugose to smooth. Antennae with 11 antennomeres, filiform to serrate, stout, variously reaching the anterior to the posterior pronotal margins; insertions widely separated, close to the mandibular bases. Clypeus not apparent; labrum short and visible, bi- or trilobed; mandibles stout, moderate, channeled and falcate, curved, apices acute; maxillary palpi with four palpomeres, last palpomere elongate, longer than basal palpomeres combined; labial palpi with one to three palpomeres; gular tentorial pits present, anteriorly placed and separate. Eyes lateral, separated in dorsal view by at least their own diameter; size small to moderate; shape oval. Gena elongate, length ½ times or more eye diameter.

Pronotum shorter than the head, broader than long, quadrate; borders smooth; surface smooth to finely rugose, lateral submarginal impressions vary from entire length to a single puncture on each side; prosternum short; procoxal cavities open behind. Mesosternum small, mesocoxal cavities open behind. Metasternum very long, two or more times as long as wide. Anterior and middle coxae conical, prominent, narrowly separate; hind coxae transverse, slightly narrower externally. Trochanters triangular, large; femora slender; tibiae slender, spurs large and serrate or not; tarsal formula 5-5-5, tarsomeres slender, the first elongate; claws normal. Elytra variously reduced, exposing one to several abdominal segments. Hindwing with venation reduced to a few longitudinal veins; wing folds reduced to anal fold and two simple apical folds.

Abdomen with seven or eight visible sterna; sternites not fused. Male genitalia of the trilobed type; penis long, slender; parameres nearly as long as and surrounding the penis; pars basalis a distinct and shallow concavity. The female genitalia are undescribed.

The larvae are unknown. Crowson (1972) speculated that the larva or female attributed by Barber (1908) to *Astraptor*, was a *Telegeusis*. It is, however, probably a member of Mastinocerini (Phengodidae).

Habits and habitats. These beetles are rare, and the immature stages probably live in soil. Adult males are occasionally numerous at lights and females are probably flightless, but unknown.

Status of the classification. *Telegeusis* was originally placed in the Drilidae (Horn, 1895). Following Barber (1913), who suggested that it is most closely related to *Atractocerus*, Leng (1920) placed both genera in Telegeusidae in his catalog, but Barber (1952) placed the genus in Lymexylidae. King (1955) argued placement of *Telegeusis* in his new suborder Aplicale based on shared "highly modified, flabellate maxillae" and reduced wing venation, but *Telegeusis* do not share the former state and Selander (1959) demonstrated the fallacy of the latter argument. However, Crowson (1955) had already removed *Atractocerus*, placed Telegeusidae in the Cantharoidea and, later (Crowson 1972), he suggested that telegeusids are closely related to phengodids. *Chaetocoelus* LeConte (Melyridae) is superficially similar to *Telegeusis*, but may be recognized by the lack of elongate maxillary palps and presence of protrusible vesicles on the pronotal margins.

Distribution. There are only eight described species, all restricted to the New World. Known *Pseudotelegeusis* range from Ecuador to northern Panama and *Telegeusis* from southern Panama to southwestern USA.

Key to the World Genera

1. Labial palps three-segmented, last palpomere elongate and similar to last maxillary palpomere .... *Telegeusis*

— Labial palps one-segmented, palpomere short .... *Pseudotelegeusis*

Classification of the Nearctic Genera

Telegeusis Horn, 1895, 6 spp., 2 in the southwestern United States (key to species: Barber 1952; Allen and Hutton 1970; Zaragoza 1975).

Pseudotelegeusis Wittmer 1976, 2 spp., Neotropical (diagnoses of species: Wittmer 1976).

Bibliography

ALLEN, R. T. and R. S. HUTTON. 1970. A new species of Telegeusidae (Leng) from Panama (Coleoptera: Cantharoidea). Coleopterists Bulletin, 23: 109-112.

BARBER, H. S. 1907 (1908). The glow-worm *Astraptor*. Proceedings of the Entomological Society of Washington, 9: 41-43.

BARBER, H. S. 1913. Observations on the life history of *Micromalthus debilis* Lec. (Coleoptera). Proceedings of the Entomological Society of Washington, 15: 31-38.

BARBER, H. S. 1952. Notes on *Telegeusis* and some relatives (Coleoptera: Lymexylidae). Pan-Pacific Entomologist, 28: 163-170.

CROWSON, R. A. 1955. The natural classification of the families of Coleoptera. Nathaniel Lloyd. London. 187 pp.

CROWSON, R. A. 1972. A review of the classification of Cantharoidea (Coleoptera), with the definition of two new families, Cneoglossidae and Omethidae. EOS Revista de la Universidad de Madrid, 21: 35-77.

HORN, G. H. 1895. Coleoptera of Baja California (Supplement I). Proceedings of the California Academy of Sciences, (2) 5: 242-243.

KING, E. W. 1955. The phylogenetic position of *Atractocerus* Palis. Coleopterists Bulletin, 9: 65-74.

LENG, C. W. 1920. Catalogue of the Coleoptera of America, north of Mexico. John D. Sherman, Jr. Mount Vernon, NY. x + 470 pp.

SELANDER, R. B. 1959. Additional remarks on wing structure in *Atractocerus*. Coleopterists Bulletin, 13: 1-3.

WITTMER, W. 1976. Eine neue Gattung der Familie Telegeusidae (Col.). Mitteilungen der Schweizerischen Entomologischen Gesellschaft Lausanne, 49: 293-296.

ZARAGOZA CABALLERO, S. 1975. Una nueva especie de *Telegeusis* Horn (Coleoptera: Telegeusidae) de Chamela, Jalisco, México. Anales Instituto de Biología Nacional Autónoma de México (Zoología), 46: 63-68.

61. PHENGODIDAE LeConte 1861

by Sean T. O'Keefe

Family common name: The glowworms

The adult males are easily identified by their short, leathery, often narrowed elytra, 12-segmented, bipectinate antennae, and large, strongly convex eyes. The adult females are larviform and possess bioluminescent organs.

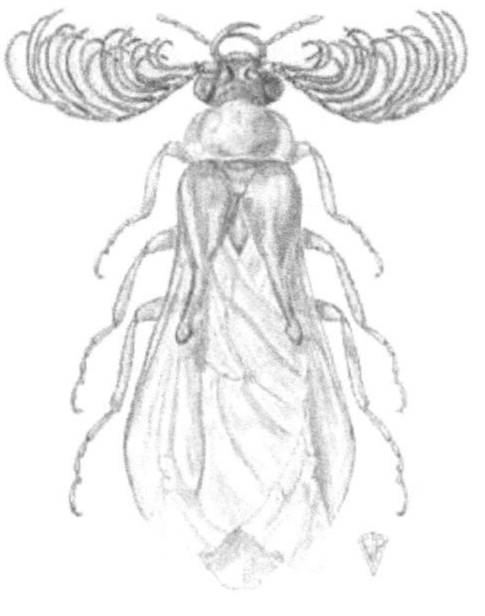

FIGURE 1.61. *Phengodes frontalis* LeConte

Description: (Based on Nearctic phengodids) Males. Shape elongate, dorso-ventrally compressed; 6-30 mm in length; head dark brown or black in most species, pronotum yellowish, elytra blackish gray or light to dark brown, body light yellow brown, some with red or yellow (only in *Distremocephalus* Wittmer are head, pronotum, and elytra concolorous); vestiture short, fairly dense, erect.

Head, including eyes, slightly narrower than pronotum in Phengodini (Figs. 1-2), as wide as or slightly wider than pronotum in Mastinocerini (Figs. 3-6); prognathous; surface glabrous in *Zarhipis* LeConte and *Stenophrixothrix* Wittmer, finely rugose in *Paraptorthodius* Schaeffer, *Phengodes* Illiger, and *Distremocephalus*, and coarsely rugose in *Cenophengus* LeConte. Vertex with transverse groove behind antennal insertions in *Phengodes*; flat, with triangular depression in *Zarhipis*; flat to slightly convex in *Stenophrixothrix* and *Cenophengus*; slightly to broadly depressed in *Distremocephalus* and *Paraptorthodius*. Frons flat, with median, longitudinal depression, slightly declivitous in *Phengodes*; flat, narrowed, slightly declivitous in *Stenophrixothrix*, *Distremocephalus*, and *Cenophengus*; sharply angled ventrally in *Zarhipis* and *Paraptorthodius*. Antenna with 12 antennomeres, antennomere I large, cylindrical to globose, antennomeres II-III short, transverse to submoniliform, antennomeres IV-XI elongate, 2-3 times as long as wide, cylindrical, bi-pectinate; pectines 5-6 times as long as antennomere in Phengodini, 2-4 times as long as antennomere in *Stenophrixothrix*, *Distremocephalus*, and *Cenophengus*, and 1.5 times as long as antennomere in *Paraptorthodius*; pectines cylindrical, densely covered with long setae in Phengodini, *Stenophrixothrix*, and *Distremocephalus*, pectines flattened, densely covered with short setae in *Cenophengus* and *Paraptorthodius*; antennomere XII greatly elongate, 4-6 times as long as wide. Antennal insertions widely separated (*i.e.*, greater than length of antennomere I) in Phengodini; moderately separated (*i.e.*, separated by length of antennomere I) in *Paraptorthodius*; and narrowly separated (*i.e.*, separated by less than length of antennomere I) in *Stenophrixothrix*, *Distremocephalus*, and *Cenophengus*. Antennal insertions exposed, except in *Phengodes* where they are covered by a prominent frontal ridge, and *Cenophengus* where they are covered by a small ridge. Labrum bilobed, distinct in Phengodini, small in Mastinocerini. Mandibles slender, sharply curved, apices acute; prominent in Phengodini, smaller in most Mastinocerini, reduced in *Cenophengus*. Maxillary palpi four-segmented, long, filiform; in Phengodini, maxillary palpomere I short, II very elongate, nearly as long as III + IV combined, III elongate and cylindrical, IV only moderately expanded at apex, nearly as long as or slightly longer than mandible; in Mastinocerini, maxillary palpomeres I-III short, IV distinctly broadly expanded at apex, longer than II+III combined, distinctly longer than mandible. Labial palp three-segmented, filiform, two-thirds length of maxillary palp in Phengodini; short, less than half length of maxillary palp in Mastinocerini. Gular area large, broad, convex, well developed between and behind eyes in Phengodini; moderate sized, flat, only moderately developed between eyes in Mastinocerini, except in *Stenophrixothrix* where it is narrowed, barely visible between eyes. In Phengodini, posterior tentorial pits combined, forming a single, deep pit immediately posterior to labium. In Mastinocerini, posterior tentorial pits separated, appearing as two small pits immediately posterior to labium, except in *Stenophrixothrix* where they are difficult to see. Eyes very large, strongly convex, circular to oval in outline, lateral in position on head, widely separated dorsally, widely to moderately separated ventrally, except in *Stenophrixothrix* where they nearly meet ventrally.

Pronotum in Phengodini (Figs. 1-2) and *Stenophrixothrix* (Fig. 3) subrectangular to trapezoidal, nearly twice as broad as long, broadly raised longitudinally at midline, anterior angles rounded, posterior angles acuminate, glabrous with moderately dense, long setation; in *Distremocephalus* (Fig. 4) hexagonal, slightly wider than long, finely rugose; in *Paraptorthodius* (Fig. 5), pentagonal, slightly longer than wide, flat, glabrous; in *Cenophengus* (Fig. 6), subrectangular, distinctly longer than wide, coarsely rugose. Lateral margins sharp in *Phengodes*, *Zarhipis*, *Stenophrixothrix*, and *Distremocephalus*, rounded in *Cenophengus* and *Paraptorthodius*. Prosternum small, triangular in Phengodini; broader, transverse in *Cenophengus* and *Paraptorthodius*; procoxal cavities open. Mesosternum very short, triangular in Phengodini, very small or mem-

Acknowledgments: I would like to thank Alan Gillogly, Alistair Ramsdale, and Edward Riley for comments improving this chapter, and to David Kavanaugh and Edward Riley for providing specimens.

FIGURE 2.61. *Zarhipis integripennis* (LeConte) (from Hatch 1961).

branous in Mastinocerini. Metasternum broader than long, strongly convex in Phengodini and *Stenophrixothrix*; longer than broad, flatter in *Cenophengus*, *Distremocephalus*, and *Paraptorthodius*. Procoxae elongate-ovoid, prominent, separate at base; mesocoxae similar in size and shape to procoxae; metacoxae transverse, broadest externally; trochanters small, triangular; femora normal; tibia slender, apical spurs absent or very small. Tarsal formula 5-5-5; tarsomeres I-III, V elongate, slender, IV short in all North American genera, except *Zarhipis*, with I, V elongate, slender and II-IV short; tarsomeres III-IV bilobed only in *Zarhipis*, IV bilobed on pro- and mesotarsomeres in *Phengodes*; pro- and meso-tarsomere I of *Distremocephalus* with distinct setal comb. Scutellum large, triangular, apex broadly pointed in Phengodini; transverse in Mastinocerini. Elytra leathery, without striae, extending less than half distance to apex of abdomen, distinctly narrowed to acuminate apically in *Phengodes*; extending half distance to apex of abdomen, gradually tapered in Mastinocerini; extending three-fourths distance to apex of abdomen, parallel-sided in *Zarhipis*; epipleural fold broad basally. Wings large, well developed, *i.e.* radial bar (RA), radial cell, RP, r4, medial bar (MP_{1+2}), CuA_{1+2}, CuA_{3+4} + AA_{1+2}, AA_{3+4}, and AP all present (e.g. *Nephromma* Wittmer, *Adendrocera* Wittmer, *Cenophengus*) to fairly reduced, *i.e.* radial bar and medial bar present, but radial cell and other veins indistinct (e.g. *Howdenia* Wittmer, *Spangleriella* Wittmer, *Steneuryopa* Wittmer), venation similar to that of *Pterotus* LeConte (Lampyridae) as illustrated in Kukalova-Peck and Lawrence (1993) (illustration of wing venation of *Adendrocera* in Wittmer 1976; *Cenophengus* in Zaragoza Caballero 1991; *Euryognathus* in Wittmer 1976; *Howdenia* in Wittmer 1988; *Nephromma* in Wittmer 1976; *Penicillophores* in Paulus, 1975; *Spangleriella* in Wittmer 1988; *Steneuryopa* in Wittmer 1986); the folding pattern is undescribed.

Abdomen elongate, dorsoventrally compressed with 7-8 visible sternites, sternites rectangular, twice as wide as long, slightly decreasing in size; 7th sternite trapezoidal, with deep emargination for genital capsule in Phengodini and *Distremocephalus*, biemarginate in *Cenophengus* and *Paraptorthodius*; 8th sternite elongate, slender; 2-5 tergites visible; sutures complete; lateral explanations distinct. Bioluminescent organs occur in *Phengodes*, *Stenophrixothrix*, and *Zarhipis* posterior to the metacoxae as two large, translucent areas.

Male genitalia trilobed with large, lateral parameres widely separated from slender, elongate median lobe in Phengodini (e.g. *Phengodes*, Zaragoza Caballero 1989, Zaragoza Caballero and Wittmer 1986), with elongate parameres closely adhered to slender median lobe in Mastinocerini (e.g. *Cenophengus*, Zaragoza Caballero 1986a, 1988, 1991; *Howdenia*, Wittmer 1988; *Mastinowittmerus*, Zaragoza Caballero 1984a; and *Taxinomastinocerus*, Wittmer 1988).

Adult females larviform, differing only slightly in external appearance from mature larvae in absence of larval setae from sterna and from terga of abdominal segments IX and X, presence of gonopore beneath large, transverse fold on abdominal segment IX, and presence of compound eyes.

Larvae of most species luminescent, with light orange spots. Shape elongate, somewhat depressed, slightly fusiform; mature larvae 15 to 65 mm long; body smooth, moderately sclerotized, shiny, with setae on legs only; color distinctly pigmented, cream, yellow, orange, red, or brown to near black. Head small, prognathous, somewhat depressed, one-third to one-half width of prothorax; epicranial sutures absent; frons, clypeus and labrum fused together. Antennae composed of three segments. Mandibles securiform, with a closed channel; retinaculum and mola absent; maxilla with distinct cardo, stipes, 3- or 4-segmented palpus, with a small mala; labium with small submentum, distinct prementum and ligula, and short 2-segmented palpus. One pair of stemmata on each side of head. Prothorax narrow in front, longer than wide, longer than the meso- or metathorax; legs short, 4-segmented, with spiniform setae and claw-like tarsunguli. Abdomen distinctly 10-segmented, segments I-VIII subequal in size; IX smaller; X reduced, possibly serving as a proleg. Spiracles normal, ovoid, located on parascutal areas above epipleurae on mesothorax and abdominal segments I-VIII. Urogomphi absent. Luminescent organs present as spots or bands on thoracic nota and abdominal terga.

Eisner *et al.* (1998) examined the larval gut and found it to consist of a slender esophagus, a muscular proventriculus, a long midgut, and a narrow hindgut. Tiemann (1967) and LeSage (1991) reviewed egg, larval, and pupal structure, and Eisner *et al.* (1998) provided photographs of the larval mouthparts.

Habits and habitats. Phengodid beetles receive their common name, glowworms, from the presence of bioluminescent organs on the larviform females and larvae. Within the Phengodidae, bioluminescent organs are found in the Phengodinae genera *Brasilocerus*, *Cenophengus*, *Euryopa*, *Mastinocerus*, *Mastinomorphus*, *Phengodes*, *Phrixothrix*, *Pseudophengodes*, *Stenophrixothrix*, *Taxinomastinocerus*, and *Zarhipis*, and the Rhagophthalminae genera *Cydistus* Bourgeois, *Dioptoma* Pascoe, *Diplocladon* Gorham, *Falsophrixothrix* Pic, and *Rhagophthalmus* Motschulsky (Herring 1987, Viviani and Bechara 1993, 1997). In the females and larvae, the bioluminescent organs are distributed along the body as transverse bands and lateral spots in *Phengodes* and *Zarhipis*, and as two lines of lateral organs and a head organ in *Mastinocerus*, *Phrixothrix*, and *Stenophrixothrix* (Viviani and Bechara 1993). In most species, the bioluminescent organs glow yellow-green, but in *Phrixothrix* the head organ glows red. Buck (1948) and Bassot (1974) studied the histology of the bioluminescent organs, Halverson *et al.* (1973) studied the physiology, and Viviani and Bechara (1993) studied the biochemistry and biophysics of phengodid bioluminescence. Viviani and Bechara (1993, 1997) found the bioluminescence systems of phengodids to be basically the same as those found in lampyrids and elaterids, and that the emission peaks for larvae are species-specific and range from 540 – 580 nm (yellow-green) for the thoracic and abdominal organs and 565 – 620 nm (red) for the head organ. For South American adult mastinocerine males, the emission peaks range from 549 to 580 nm.

Lloyd (1971, 1979) and Tiemann (1967) suggested that males are attracted to females at a distance by pheromones, and at close range by the female's bioluminescent organs. Tiemann (1967) described the courting and mating behavior of *Zarhipis*. The males of *Zarhipis* often seek females in the early evening in desert regions. In *Phengodes* and *Zarhipis* the males are weakly luminescent, but in *Phrixothrix*, the males are brightly luminescent, and possibly use the luminescence to communicate with the female (Tiemann 1970).

Phengodid adult females and larvae feed on millipedes (Tiemann 1967, 1970; Miller 1997, Eisner *et al.* 1998). Tiemann (1967), Miller (1997), and Eisner *et al.* (1998) studied the feeding behavior of *Zarhipis* and *Phengodes*. The feeding behavior is similar for these two groups (Eisner *et al.* 1998). Phengodid larvae are able to feed on millipedes and avoid the millipede's defensive compounds by a sequence of steps that begin with the phengodid mounting the millipede and quickly coiling around the millipede's head. The phengodid then bites the millipede in the intersegmental membrane just behind and underneath the head. After which, the millipede goes limp. Tiemann (1967) reports that phengodids sever the ventral nerve cord, but Eisner *et al.* (1998) argue that the phengodids actually inject gastric fluid (as witnessed by Tiemann 1970). The phengodid uncoils from the millipede, pulls it underground or into the leaf layer, and sucks out the contents of the head capsule. From the inside-out the phengodid devours the millipede's internal organs, except for the defensive glands. Tiemann (1967) reported that adult females did not feed (at least when provided millipedes in captivity). Tiemann (1970) briefly described the biology of the South American railroad worm, *Phrixothrix*.

Zarhipis occur in desert areas where the males are often attracted to blacklights. Adult females have been collected under loose bark (Miller 1997). Larvae are usually found on soil underneath leaves or rotting logs. Larvae are nocturnal and more active on humid nights (Viviani and Bechara 1997). Wing (1984) reported collecting up to 5 larvae per square meter from a flooded Florida field.

Status of the classification. Phengodids are closely related to Drilidae and Lampyridae of the Cantharoidea (Crowson 1972). Originally, LeConte (1881), in his revision of the Lampyridae of the U.S., included phengodids as a tribe of Lampyridae. Leng (1920), in his catalog of North American beetles, listed Phengodidae as a family, and recognized four tribes: Pterotini, Phengodini, Mastinocerini, and Omethini. Pic (1927), in his world catalog of the family, recognized three tribes: Pterotini, Phengodini, and Mastinocerini. Crowson (1972) removed the Omethini and placed it as its own family, Omethidae. Green (1948) removed the Pterotini and placed it into the Lampyridae. In the first edition of this book (Arnett 1963), two tribes were recognized: Phengodini with two genera, *Phengodes* and *Zarhipis*; and Mastinocerini, with four genera, *Cenophengus*, *Euryopa* Gorham, *Mastinocerus* Solier, and *Paraptorthodius*. During the 1970's and 1980's Walter Wittmer and Santiago Zaragoza Caballero greatly increased the number of known genera and species. Wittmer (1976) provided a comprehensive monograph of the family, proposing many new genera, organizing the taxonomy, and providing keys to the genera. Regarding the phengodid fauna of the United States, the two important changes he made were to place the North American species of *Mastinocerus* into a new genus, *Distremocephalus*, and to place the single species of *Euryopa* into *Stenophrixothrix*. Zaragoza Caballero (1984b) provided an updated world catalog, and recognized four tribes: Pterotini (now removed), Phengodini, Mastinocerini, and Penicillophorini. The checklist of U.S. species by Poole and Gentili (1996) failed to incorporate these updated changes, as well as those proposed by Linsdale (1964) in his revision of *Zarhipis*. In 1972, Crowson placed the five Asian genera of Rhagophthalminae into Phengodidae. Lawrence and Newton (1995) recognized two subfamilies, the Asian Rhagophthalminae and the New-World Phengodinae. However, Lawrence *et al.* (1999) elevated the Rhagophthalminae to family rank. To date, no cladistic analysis has been published on the phylogenetic relationships of Phengodidae.

Phengodinae are currently divided into 3 tribes (Paulus 1975, Wittmer 1975): the Phengodini, with 4 genera (*Phengodes*, 28 species, throughout North, Central, and South America; *Pseudophengodes* Pic, 22 species, South America; *Microphengodes* Wittmer, 2 species, South America; and *Zarhipis*, 3 species, North America); Penicillophorini, with 3 genera (*Penicillophores*, 1 species, Colombia; *Acladocera* Wittmer, 1 species, Dominican Republic; and *Adendrocera*, 1 species, Guatemala); and Mastinocerini, with 25 genera (*Brasilocerus* Wittmer, 9 species, South America; *Cenophengus*, 18 species, North America; *Cephalophrixothorax* Wittmer, 3 species, northern South America; *Decamastinocerus* Wittmer, 1 species, Venezuela; *Distremocephalus*, 11 species, North America; *Eurymastinocerus* Wittmer, 5 species, Mexico and Central America; *Euryopa*, 11 species, Central and South America; *Euryognathus*, 2 species, Paraguay; *Howdenia*, 9 species, South America; *Mastinocerus*, 26 species, Guatemala and South America; *Mastinomorphus* Wittmer, 12 species, South America; *Mastinowittmerus* Zaragoza Caballero, 2 species, Mexico; *Neophengus* Wittmer, 3 species, Chile; *Nephromma*, 2 species, Brazil; *Oxymastinocerus* Wittmer, 8 species, South America; *Paramastinocerus*, 1 species, Brazil; *Paraptorthodius* Schaeffer, 2 species, North America; *Phrixothrix*, 18 species, South America; *Pseudomastinocerus* Wittmer, 7 species, Central and northern South America; *Ptorthodiellus* Wittmer, 2 species, Venezuela; *Ptorthodius*, 4 species, northern South America; *Spangleriella*, 1 species, Venezuela; *Steneuropa*, 1 species, Costa Rica; *Stenophrixothrix*, 20 species, Mexico, Central and South America; and *Taximastinocerus* Wittmer, 17 species, Central and South America).

Distribution. Pic (1927) listed 47 species of Phengodidae (excluding Pterotini and subspecies), Blackwelder (1945) and Leng

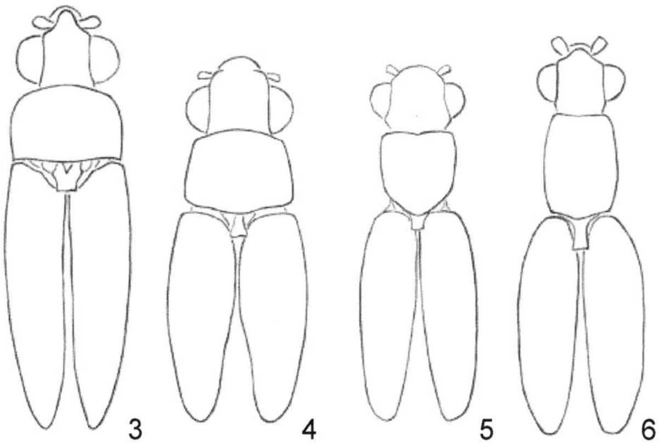

FIGURES 3.61-6.61. 3. *Stenophrixothrix fusca* (Gorham); 4. *Distremocephalus* sp., Texas; 5. *Paraptorthodius mirabilis* (Schaeffer); 6. *Cenophengus* sp., Texas.

(1920) combined, listed 87 species (excluding Pterotini, Omethini, and subspecies), Zaragoza Caballero (1984b) listed 181 species (excluding Pterotini and subspecies). Currently, there are over 250 species of Phengodinae in North, Central, and South America. From the United States, 23 species are known.

Key to Nearctic genera

1. Larger species (over 10 mm in length) (Phengodini) 2
— Smaller species (under 10 mm in length) (Mastinocerini) 4

2. Elytra truncate, narrowed distally; frons impressed between antennal bases (*Phengodes*) 3
— Elytra nearly covering entire abdomen; frons not impressed between antennal bases ..... *Zarhipis*

3. Galea short, apex with few scattered setae .... *Phengodes* (*s. str.*)
— Galae elongate, apex with numerous long setae .... *Phengodes* (*Phengodella*)

4. Pronotum distinctly wider than long (Figs. 3-4) ... 5
— Pronotum as wide as long or distinctly longer than wide (Figs. 5-6) 6

5. Eyes nearly meeting on ventral side of head; vertex longer than wide (Fig. 3), impunctate; antennal pectinations long, hair-like .... *Stenophrixothrix*
— Eyes widely separated on ventral side of head; vertex as wide as long (Fig. 4), punctate; antennal pectinations shorter, flattened .... *Distremocephalus*

6. Pronotum nearly as wide as long, pentagonal (Fig. 5); vertex and pronotum impunctate .... *Paraptorthodius*
— Pronotum distinctly longer than wide, rectangular (Fig. 6); vertex and pronotum coarsely punctate .... *Cenophengus*

Classification of the Nearctic genera

Phengodidae LeConte 1861

Phengodinae LeConte 1861

Phengodini LeConte 1861

Phengodes Illiger 1807
subgenus *Phengodes* (*s. str.*)
Of the 10 species of *Phengodes* (*s. str.*), 7 occur in the southwestern (Arizona), central, and eastern United States, and 3 others occur in Mexico. See Wittmer (1975, 1977) for revision and key to the species of the United States.

subgenus *Phengodes* (*Phengodella*) Wittmer 1975
Of the 18 species of *Phengodella*, one occurs in California and two others occur in Texas; the remaining 15 species are found in Mexico, Central America, Venezuela, or Ecuador. See Wittmer (1975) for revision and key to the species of the United States.

Zarhipis LeConte 1881
Zarhipis contains 3 species, restricted to the Pacific coastal and southwestern United States (Washington, Oregon, California, Arizona, Nevada, and New Mexico), and northern Baja California, Mexico. See Fall (1923) and Linsdale (1964) for revision and key to species.

Mastinocerini LeConte 1881

Cenophengus LeConte 1881
Cenophengus contains 18 species, mostly found in the United States or Mexico, but also occurring in Colombia and Chile. Four species occur in the United States (California, New Mexico, and Texas). See Wittmer (1976) for key to the species of the United States.

Distremocephalus Wittmer 1975
Distremocephalus contains 11 species found in the United States or Mexico. Four species occur in the southwestern United States (Texas, New Mexico, Arizona, and California). See Zaragoza Caballero (1986b) for review and discussion of the species.

[*Mastinocerus* Solier 1849, does not occur in North America.]

Stenophrixothrix Wittmer 1963
Stenophrixothrix ranges from the United States to Brazil, and contains 20 species. Only a single species, *S. fusca* (Gorham 1881), occurs in the United States (Arizona). See Gorham (1881) and Wittmer (1963,1976) for description of *Stenophrixothrix*.

[*Euryopa* Gorham 1881, does not occur in the United States (one species is reported from Mexico).]

Paraptorthodius Schaeffer 1904

Paraptorthodius contains 2 species, one, *P. mirabilis* Schaeffer, 1904, from the United States (Arizona, California, and Texas), and one from Mexico. See Wittmer (1976) for redescription of *Paraptorthodius*.

References

ARNETT, R. H. 1963. The Beetles of the United States: A manual for identification. Catholic University Press. Washington, DC, xii + 1112 pp.

BASSOT, J. M. 1974. Les cellules lumineuses du coleoptere Phengodes. Pp. 79-96. *In:* L. Arvy, ed. Recherches Biologiques Contemporareus. Vagner, Nancy.

BLACKWELDER, R. E. 1945. Checklist of the Coleopterous Insects of Mexico, Central America, the West Indies, and South America, Part 3. Bulletin of the United States National Museum, 185: 1-362.

BUCK, J. B. 1948. The anatomy and physiology of the light organ in fireflies. Annals of the New York Academy of Science, 49: 397-482.

CROWSON, R. A. 1972. A review of the classification of Cantharoidea (Coleoptera), with the definition of two new families, Cneoglossidae and Omethidae. Revista de la Universidad de Madrid, 21(82): 35-77.

EISNER, T., M. EISNER, A. B. ATTYGALLE, M. DEYRUP and J. MEINWALD. 1998. Rendering the inedible edible: Circumvention of a millipede's chemical defense by a predaceous beetle larva (Phengodidae). Proceedings of the National Academy of Sciences, 95: 1108-1113.

FALL, H. C., 1923. *Zarhipis* LeConte (Coleoptera). Canadian Entomologist, 55: 109-112.

GORHAM, H. S. 1881. Fam. Lymexylonidae, Pp. 106-112. *In:* F. D. Goodman and O. Salvin, eds. Biologia Centrali-Americana, 3(2).

GREENE, J. W. 1948. Two new species of Lampyridae from southeastern Florida, with a generic revision of the Nearctic fauna. Transactions of the American Entomological Society, 74: 61-73.

HALVERSON, C. R., J. F. CASE, J. BUCK and D. TIEMANN. 1973. Control of luminescence in phengodid beetles. Journal of Insect Physiology, 19: 1327-1339.

HATCH, M. H. 1961. The beetles of the Pacific Northwest. Part III: Pselaphidae and Diversicornia I. University of Washington Publications in Biology, 16: 503 pp.

HERRING, P. J. 1987. Systematic distribution of bioluminescence in living organisms. Journal of Bioluminescence and Chemiluminescence, 3: 147-163.

KUKALOVA-PECK, J. and J. F. LAWRENCE. 1993. Evolution of the hind wing in Coleoptera. The Canadian Entomologist, 125: 181-258.

LAWRENCE, J. F., A. M. HASTINGS, M. J. DALLWITZ, T. A. PAINE and E. J. ZURCHER. 1999. Beetles of the World: A Key and Identification System for Families and Subfamilies. CD-ROM, Version 1.0 for MS-Windows. CSIRO Publishing. Melbourne.

LAWRENCE, J. F. and A. F. NEWTON, Jr. 1995. Families and subfamilies of Coleoptera (with selected genera, notes, references and data on family-group names). Pp. 779-1006. *In:* J. Pakaluk and S. A. Slipinski, eds. Biology, Phylogeny, and Classification of Coleoptera: Papers Celebrating the 80th Birthday of Roy A. Crowson. Muzeum I Instytut Zoologii PAN, Warsaw.

LECONTE, J. L. 1861. Classification of the Coleoptera of North America. Prepared for the Smithsonian Institution. Part 1. Smithsonian Miscellaneous Collections, 3: 1-208.

LECONTE, J. L. 1881. Synopsis of the Lampyridae of the United States. Transactions of the American Entomological Society, 9: 15-72.

LENG, C. W. 1920. Catalog of the Coleoptera of America, North of Mexico. Cosmos Press. Cambridge, MA, 470 pp.

LESAGE, L. 1991. Phengodidae (Cantharoidea) (including Rhagophthalmidae). Pp. 424-426. *In:* F. W. Stehr, ed. Immature Insects, Volume 2. Kendall/Hunt, Dubuque, IA.

LINSDALE, D. D. 1964. A revision of the genus *Zarhipis* LeConte (Coleoptera: Phengodidae). Wasmann Journal of Biology, 22: 225-260.

LLOYD, J. E. 1971. Bioluminescent communication in insects. Annual Review of Entomology, 16: 97-122.

LLOYD, J. E. 1979. Sexual selection in luminescent beetles. Pp. 293-342. *In:* M. S. Blum and N. A. Blum, eds. Sexual Selection and Reproductive Competition in Insects. Academic Press, New York, NY.

MILLER, R. S. 1997. Female *Phengodes* feeding and an associated risk (Coleoptera: Phengodidae). Entomological News, 108: 213-214.

PAULUS, H. F. 1975. *Penicillophorus ctenotarsus* n. gen. et n. sp. aus Kolumbien, mit einer Beschreibung einer neuen Tribus Penicillophorini der Phengodidae (Col., Polyphaga, Cantharoidea). Zeitschrift der Arbeitsgemeinschaft Österreivhischer Entomologen, 25 (1973): 69-80.

PIC, M. 1927. Phengodidae and Karumiidae, part 94. *In:* S. Shenkling, ed. Coleopterorum Catalogus, W. Junk. Berlin. 8 pp.

POOLE, R. W. and P. GENTILI. 1996. Nomina Insecta Nearctica: A Check List of the Insects of North America, Volume 1: Coleoptera, Strepsiptera. Entomological Information Services, Rockville, MD. 827 pp.

TIEMANN, D. 1967. Observation on the natural history of the western banded glow-worm *Zarhipis integripennis*. Proceedings of the California Academy of Sciences, 35: 235-264.

TIEMANN, D. 1970. Nature's toy train, the railroad worm. National Geographic, 138: 58-67.

VIVIANI, R. V. and E. J. H. BECHARA. 1993. Biophysical and biochemical aspects of phengodid bioluminescence. Photochemistry and Photobiology, 58: 615-622.

VIVIANI, R. V. and E. J. H. BECHARA. 1997. Bioluminescence and biological aspects of Brazilian railroad-worms (Coleoptera: Phengodidae). Annals of the Entomological Society of America, 90: 389-398.

WING, S. 1984. A spate of glow-worms (Coleoptera: Phengodidae). Entomological News, 95: 55-57.

WITTMER, W. 1963. Zur Kenntnis der Phengodidae (Coleoptera). Mitteilungen der Schweizerischen Entomologischen Gesellschaft, 36: 73-99.

WITTMER, W. 1975. The genus *Phengodes* in the United States (Coleoptera: Phengodidae). Coleopterists Bulletin, 29: 231-250.

WITTMER, W. 1976. Arbeiten zur einer Revision der Familie Phengodidae (Coleoptera). Entomologische Arbeiten aus dem Museum G. Frey, 27: 415-524.

WITTMER, W. 1977. The family Phengodidae in the United States. Coleopterists Bulletin, 31: 282.

WITTMER, W. 1986. Ein Beitrag zur Kenntnis der Phengodidae (Coleoptera) (38. Beitrag zur Kenntnis der Neotropischen Fauna). Anales del Instituto de Biologia Universidad Nacional Autonoma de Mexico, 56: 159-176.

WITTMER, W. 1988. Die Familie Phengodidae (Coleoptera) vom Cerro de la Neblina, Venezuela. Mitteilungen Entomologische Gesellschaft, Basel, 38: 72-94.

ZARAGOZA CABALLERO, S. 1984a. Descripcion de un genero e una nueva especie de Mastinocerini (Coleoptera: Phengodidae). Anales del Instituto de Biologia Universidad Nacional Autonoma de Mexico, 55: 203-208.

ZARAGOZA CABALLERO, S. 1984b. Catalogo de la familia Phengodidae. Anales del Instituto de Biologia Universidad Nacional Autonoma de Mexico, 55: 307-324.

ZARAGOZA CABALLERO, S. 1986a. Descripcion de dos especies nuevas de Cenophengus LeConte (Coleoptera: Phengodidae; Mastinocerini). Anales del Instituto de Biologia Universidad Nacional Autonoma de Mexico, 56: 933-938.

ZARAGOZA CABALLERO, S. 1986b. El genero *Distremocephalus* Wittmer en Mexico (Coleoptera: Phengodidae). Anales del Instituto de Biologia Universidad Nacional Autonoma de Mexico, 56: 189-202.

ZARAGOZA CABALLERO, S. 1988. Una especie nueva de *Cenophengus* de Mexico (Coleoptera: Phengodidae). Anales del Instituto de Biologia Universidad Nacional Autonoma de Mexico, 58: 651-654.

ZARAGOZA CABALLERO, S. 1989. La familia Phengodidae en "Los Tuxtlas" Veracruz, Mexico. Anales del Instituto de Biologia Universidad Nacional Autonoma de Mexico, 59: 77-98.

ZARAGOZA CABALLERO, S. 1991. Descripcion de una especie nueva de *Cenophengus* (Coleoptera: Phengodidae; Mastinocerini) de Iguala, Guerrero, Mexico. Anales del Instituto de Biologia Universidad Nacional Autonoma de Mexico, 62: 109-114.

ZARAGOZA CABALLERO, S. and W. WITTMER. 1986. Nuevas especies de *Phengodes* Illiger (Coleoptera: Phengodidae) de Mexico. Anales del Instituto de Biologia Universidad Nacional Autonoma de Mexico, 56: 177-188.

62. LAMPYRIDAE Latreille 1817

by James E. Lloyd

Family common name: The fireflies, lightningbugs, and glowworms

Family synonyms: Malacodermidae Latreille 1806, in part; Telophoridae Leach 1815, in part

Adult American fireflies will usually be recognized by their flashes in the night; by their lantern, in males commonly appearing as two pale ventrites beneath the posterior abdomen; and by their large, flattened and flanged pronota which typically cover the eyes like a sun-shade, unless the head and neck are extended as when walking or climbing. Common lanternless species active in daylight typically have marginless, black elytra, many with ridges (costae), and pronota marked with red or orange. The integument of adult fireflies is soft and pliable, and with handling sometimes releases droplets of pale blood, especially from the elytra, which in some species is known to be poisonous when ingested by some predators — even lethal in some instances. Odor is often plant-like, and taste, bitter, astringent. Larvae are most often found and recognized by the intermittent glows they emit, especially on damp nights.

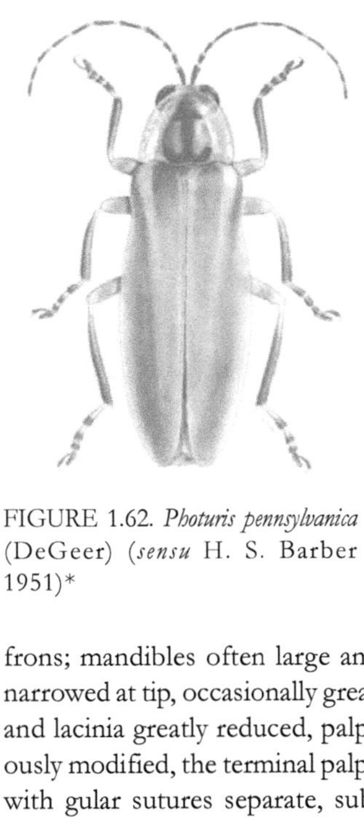

FIGURE 1.62. *Photuris pennsylvanica* (DeGeer) (*sensu* H. S. Barber 1951)*

Description: Shape, elongate or elongate oval, rarely oval; size, 4-18 mm.; color, variously, black, brown, tawny, charcoal, pearl, or olivaceous, with areas of pink, red, yellow, or salmon, especially on the pronotum, with color splashes below. Head deflexed/hypognathus; surface smooth or punctate, elongate. Antennae, with 8-13 but usually 11 antennomeres (in NA forms), filiform, serrate, sometimes with branches, and inserted above the mandibles. Labrum frequently indistinct, occasionally connate (fused at base) with frons; mandibles often large and curved, sometimes abruptly narrowed at tip, occasionally greatly reduced; maxillae with galea and lacinia greatly reduced, palpi with 4 palpomeres, and variously modified, the terminal palpomere usually enlarged; labium with gular sutures separate, submentum partly membranous, and mentum reduced, ligula small, 3-segmented labial palpi, terminal segment usually enlarged, sometimes lunate (rather, mitten-shaped). Eyes lateral, often large and sometimes bulging and nearly or actually meeting below.

Pronotum distinctive, covering head when at rest; shield-like and variously semicircular, sub-pentagonal, sub-triangular, sub-trapezoidal, etc., with explanate margins and raised disk, occasionally with a midline groove, rarely a carinula; surface smooth, punctate, or granulate, median and lateral vittae often present and diagnostic (i.e. vittagrams); hypomera, large and not reaching anterior margin (i.e. "open") in most American species (see Key). Bright colors (apparently pterodine pigments) are often present in the pronotal and neck region, and are potentially of some taxonomic use. Prosternum short, prominent, procoxal cavities open, procoxae conical, prominent, and contiguous; mesosternum short, mesocoxae prominent, conical, contiguous; metasternum longer, broad, mesal margins of metepisterna straight or nearly so, not curved or sigmoid; metacoxal cavities open, metacoxae transverse and contiguous. Trochanters quadrate, interstitial, angularly joined to their femora. Femora flattened, somewhat swollen. Tibia stout, somewhat flattened, apical spurs present or absent; tarsal formula 5-5-5, slender, penultimate tarsomere with pads (pulvilli); claws simple or variously forked (bifid), occasionally with a basal protuberance. Scutellum evident and triangular. Elytra with explanate margins, sometimes with longitudinal ridges (costae), and sides variously nearly parallel, somewhat or considerably widest in middle, or tapering and narrowing in posterior third or so. Elytral bead and margin often pale in color, contrasting with disk. Elytral vestiture sometimes provides useful characters for species identification, e.g., *Pyractomena*.

Abdomen of male with 7 visible ventrites, corresponding to abdominal segments 2-8 (except 6 in Luciolinae which may possibly be represented in North America), excluding the small genital segment ("clasper") which is occasionally absent. In some genera (e.g., *Phausis*) the distal abdominal dorsites (more or less equal tergites) at the sides extend over the pleural areas and are broadly visible from below (i.e. are lobed, foliate). The first abdominal segment is dorsal only, except in *Photinus granulatus* and

Acknowledgments: I am indebted to Marc Branham, Steven Wing, and Alistair Ramsdale for providing information and making helpful comments and technical corrections in the manuscript; to Jade Williams and Jennifer LaRosa, students who first used the key and made suggestions; to Laura Line who made the carbon-dust drawings and some of the line drawings; and to Seth Ambler and Mike Sanford for computer graphic and layout assistance. Florida Agr. Exp. Station Journal Series No. R-08641.

* For more than a century most *Photuris* spp. in NA were lumped under the name *P. pennsylvanica*, but Barber's work on flashing behavior led to recognition of several new and synonomized species, and to further research on the use of flash patterns in firefly taxonomy.

Phosphaenus hemipterus. Light organs are often present on ventral surface of the abdomen of one or both sexes, and in males of American lightningbug fireflies (*Aspisoma*, *Photinus*, *Photuris*, *Pyractomena*, *Micronaspis*, *Bicellonycha*) they completely occupy visible ventrites 5 and 6. Male genitalia consist of 2 lateral lobes, a median lobe, and a basal piece, and provide taxonomically useful characters variously at the generic, species, and/or species-group levels. In *Photuris* the basal piece has a pair of lateral filaments (Fig. 16), which remain outside the female during copulation and may have a proprioreceptive function.

Noteworthy sexual dimorphism: females of several species (e.g. *Pleotomus* spp., *Pleotomodes* spp., *Lucidota luteicollis* LeConte, and several in *Photinus*) have shortened elytra and wings, and females or both males and females of *Pyropyga nigricans* (G. A. Olivier) from some sites have reduced wings and elytra (Lloyd 1999). Brachyptery is not known to occur in *Photuris*, *Ellychnia*, or *Pyractomena*. Females of some genera, such as *Microphotus* and *Phausis* are larva-like to some degree, but whether they are "larviform" is problematic (see Cicero 1988). The lanterns of females of flashing lampyrids differ topographically from those of their males. Those of *Photuris* females occupy the same two segments, but each has a nonluminous frame; those of *Pyractomena* occupy the 4 remote corners of the 2 ventrites; and the lantern of *Photinus* is a median, sub-semicircular plate on the (visible) 5th ventrite. Known *Phausis* females glow from pairs of lanterns that are visible dorsally along the abdomen, in addition to their retained larval lanterns.

Larva: Full-grown larvae of North American species are 12-25 mm in length, well sclerotized, onisciform, and dorsoventrally flattened, or fusiform, or cylindrical and elongate (Figs. 2-7). Color varies among species, and is black, slate gray, brown, reddish-brown, or tan, and some have pink areas between sclerites. As examples, those of *Photuris congener* LeConte are brick red, those of *Pyractomena borealis* (Randall) vary from brick red to dark brown and are sometimes variegated, and those of *Ellychnia corrusca* (L.) are charcoal with pink pleural areas. Larvae have 12 visible dorsal plates (terga). The head is elongate and sub-cylindrical with prognathous mouthparts, curved, falcate and channeled mandibles, and retractable 3-segmented antennae, each bearing a globular accessory structure on the third joint and having stemmata near the base.

Though usually retracted within the prothorax except while walking, during feeding the head and neck are extended and inserted into the flesh of stunned (poisoned) prey (*Pyractomena* larvae in captivity have been found dead with greatly extended, limp, and perhaps even stretched necks, perhaps having been "strangled" by the powerful foot muscles of reviving snails.). Clypeus and labrum are not distinct; maxillae with a small cardo, large stipes, 3- or 4-segmented palpi, and 2-segmented mala; labium with mentum and submentum fused, ligula inconspicuous, and 2-segmented palpi.

Thorax with an elongate prothorax narrowing anteriorally, and as long as the meso- and metathorax combined; the meso- and metathoraces are sub-quadrate, and slightly wider than long. Legs are 5-segmented with apical, claw-like tarsunguli. The abdomen is 10-segmented, with the ultimate dorsite concealed, and with paired lanterns ventrally evident on the epipleura of segment 8. An unusual eversible, 8-fingered, membraneous structure sometimes with many delicate teeth over its surface, occurs at the tail. This pygopodium is extruded and used for grooming (washing) after eating. Functional annular biforous spiracles are located in the mesipleura and the epipleura of abdominal segments 1-8 (Archangelsky and Branham 1998 and 2001; Branham and Archangelsky 2000; Labella and Lloyd 1991; Peterson 1951).

Habits and habitats: Comparatively few studies have been made on the juvenile biology of fireflies (Archangelsky and Branham 1998; Buschman 1984a, 1984b, 1987; Lloyd 1997a; Wing 1989, 1991). In some species eggs are deposited a few at a time over several days or weeks. This is probably the case in *Photinus pyralis* (L.), and in several *Photuris* species which acquire nutritional substances as well as defensive compounds for their eggs from their predation on other fireflies (Eisner *et al.* 1997). In others, eggs may be laid together at one time, as in *Photinus collustrans* LeConte in which the brachypterous females live and die in their burrows (Wing 1989). The eggs of some North American species glow very dimly briefly after deposition, from luminous material that is placed on the eggs in the egg canal; in some the light organs of developing embryos may later begin to glow.

As far as known, larvae of all lampyrid species are luminescent, and emit glows of varying duration, nearly always from a pair of relatively simple lanterns at the tail (abdominal segment 8). Glowing larvae are most often seen in damp situations, for examples, herbaceous soggy ground along roadside verges, around ponds and marshes, at seeps of earthen dams, and along streams. Old fields, cattail marshes, and low wet pastures are especially rich habitats in the northeast and northern midwestern states. *Photuris* larvae are the emitters of the most commonly seen glows, and on damp and rainy nights especially, *Pyractomena* larvae sometimes glow several feet above ground on woody vegetation. Glowing larvae of *Micronaspis floridana* Green are found at the landward edges of the black needle-rush zone of saltwater marshes around the coastline of Florida. Larvae of some *Pyractomena* species are found on emergent vegetation of ponds and marshes. *Photinus* larvae are subterranean and seldom seen, and *Ellychnia* and *Lucidota* inhabit rotting logs, and glow when their logs are pulled apart and crumbled. In western states and Florida a few species occur in very dry habitats; those of *Pleotomodes* are inquilines of various ant species (Sivinski, *et. al* 1998), but adaptations of *Lucidota luteicollis* LeConte for Florida's dry scrub and sandhill habitats remain unknown. There has long been interest and speculation about the function of larval luminescence (Sivinski 1981); experimental evidence indicates that the light has aposematic value. As far as known, all lampyrid larvae are predaceous, with *Pyractomena* and *Photinus* possibly specializing on ("favoring") gastropod mollusks and earthworms respectively; *Photuris* larvae are omnivorous, predators of soft-bodied prey, and scavengers of dead insects and fallen berries. *Pyractomena lucifera* (Melsheimer) hunts snails underwater and brings them to the surface to eat (Buschman 1984a).

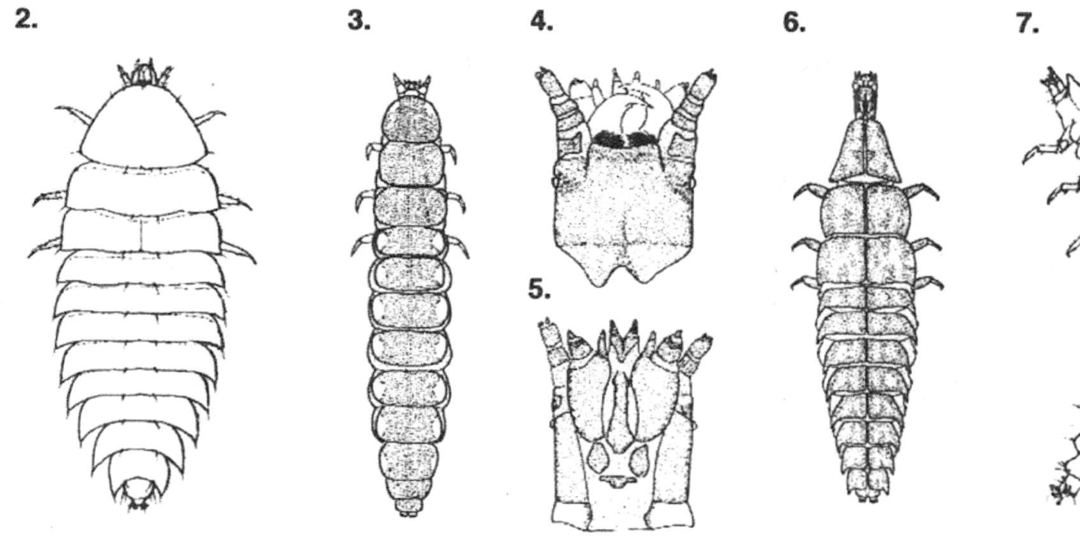

4. head, dorsal view, 5. head, ventral view (from Archangelsky and Branham 2001, with permission). 6. *Pyractomena borealis* (Randall) larva, dorsal view (from Archangelsky and Branham 1998, with permission). 7. *Lucidota atra* (G. A. Olivier) (not *Photinus*, see Branham and Archangelsky 2000) larva, lateral view (modified from Peterson 1951).

Lampyridae larvae and other luminous worm-form insects in the US are referred to by the term "glowworm," including Phengodidae larvae and females, and larval fungus gnats (Mycetophilidae: Diptera); the lampyrid *Lampyris noctiluca* (L.) in Britain and northern Europe has been called the glowworm for centuries.

Pupation occurs in earthen underground or surface cells in most species, and takes from one to three weeks depending upon ambient temperature and species. *Pyractomena* and apparently other Cratomorphini pupate on herbaceous and woody vegetation, which may keep them above rising flood water. *Pyactomena borealis* (Randall) in Florida often pupates next to vines, twig stubs, and other pupae, and in bark crevices, strongly favoring a southern exposure which provides increased warmth from insolation, accelerating winter development and eclosion, and mating follows immediately (Lloyd 1997a). In the northeast *Ellychnia corrusca* pupates in dead logs in autumn, then ecloses; during winter adults are sometimes found on tree trunks, though mating does not occur until spring.

Lampyrids whose adult males emit bioluminescence in "neurologically precise" flashing patterns (lightningbug fireflies, e.g. *Photinus*, *Pyractomena*, *Photuris*; Figs. 1, 25-27, 31-32) are common east of the Rocky Mountains but are presently known only from scattered western localities, in Arizona, Colorado, Idaho, Montana, Oregon, and British Columbia (Cicero 1982; Fender 1961b). Lampyrids in which only the females emit light, and males, with rare exception, are lightless with huge eyes (glowworm fireflies, a behavioral appellation with strong taxonomic/generic correlation, e.g., *Phausis*, *Microphotus*, *Pleotomus*; Figs. 22, 28), occur across North America and are well represented in western states — females of these species are typically brachypterous, apterous or "larviform," and apparently most live in burrows, but as noted, those of *Pleotomodes* (Fig. 33) live in ant nests (Fender 1961a,b; Sivinski et al 1998). Also common in the west, but well represented in the eastern and middle states, are several nonluminescent species (daytime dark fireflies). Some of these are known to use pheromones for sexual communication (e.g., *Lucidota atra* (G. A. Olivier) (Fig. 35), *Photinus indictus* (LeConte), *Pyropyga nigricans* (Say) (Fig. 34)), and probably most others do also, excepting perhaps *Photinus cookii* Green. Recent phylogenetic analyses using cladistic methods bring new insight into the evolution of sexual communication in fireflies, and also the evolution of bioluminescence in Coleoptera (Branham and Wentzel 2001, and in prep.).

In eastern and middle North America, males of flashing species often fly in conspicuous thousands over meadows and fields, beginning with *Photinus pyralis* (L.) (Fig. 31) shortly after sunset, and a little later various *Photuris* species continue well into the night during their peak mating season. Noteworthy conspicuous flashers include *P. tremulans* Barber, *P. versicolor* (Fabricius), *P. v. quadrifulgens* Barber, and *P. lucicrescens* Barber. These fireflies also fly at higher altitudes, over forest canopies and especially over the crowns of trees adjacent to and within grassland. *Photuris lucicrescens* seems to be most common along streams. Although populations of *Pyractomena angulata* (Say) (Fig. 26) are occasionally seen in roadside marshes, more often and on any late spring or summer evening individuals or a few males will be seen emitting their distinctive amber flicker as they weave around ecotonal shrubs and high boughs of trees. The flashing of some species in the *Photuris frontalis* group (Fig. 25) is often eye-catching, because males fly a meter or so above the ground emitting continuous trains of short flashes at short intervals, and at high densities they emit their flashes in synchrony.

In the sexual signaling of most Nearctic lightningbug firefly species, males emit species-typical flash patterns which their females answer with flashes. In general, the bioluminescence of *Photinus* is yellow, that of *Photuris* is green, and *Pyractomena*, orange-yellow or amber, but there are exceptions, a few being quite remarkable. Generic diagnosis in the field via color will often err because judgement is confused, apparently by variations in the dark adaptation of the observer's eyes and the color of background illumination. When firefly luminescence appears white it is apparently because dim light has not adequately stimulated human cone (color) receptors. As far as known, fireflies cannot discriminate color, but are differentially sensitive to different wavelengths — e.g., twilight *Photinus* spp. to yellow, the later active *Photuris* spp. to green.

Male flash patterns show considerable variation among species, and sometimes a range of distinct patterns within species. Several flash patterns consist of single flashes or groups of short flashes of various timing characteristics; these are repeated at fairly regular but temperature dependent (advertising) intervals. Other variations in species-typical flash patterns include intensity modulations within single pulses of light, resulting in flickers, crescendos, and bimodal twinkles. Flash patterns can often be used to identify the species of flying males at a distance, and in some cases, and in particular in the genus *Photuris*, flash patterns provide the only reliable character for identification (Barber 1951; McDermott 1914; Lloyd 1969a, 1990, 2001). Female response flashes are typically single flashes, with various timing characteristics in response-delay and duration, but in a few species, especially those in the *Photinus ardens* group and probably some *Photuris* species, female responses are more complex (Lloyd 1966a).

Males of several *Photuris* species use two or more distinctively different flash patterns. For examples, males of *Photuris tremulans* Barber emit a short flash and a flicker pattern, the latter explaining Barber's choice of epithet; and, in some populations during early evening, males of *Photuris pennsylvanica* (DeGeer) (sensu Barber 1951) emit a short flash, but within a half hour all have switched to the short-long (dot-dash) flash pattern described by Barber (1951). When flickering *P. tremulans* males and short-flashing *P. pennsylvanica* males receive a flashed answer (female or penlight) they switch (default) to their other (species-identifying) pattern, *P. tremulans* to the short flash and *P. pennsylvanica* to the dot-dash pattern, and continue their approaches toward respondents (Lloyd 1997b). During the approach of male fireflies to a female, after they have landed, patterns sometimes become more variable, and without specific knowledge, may become unreliable for identification.

Many fireflies emit light in contexts other than sexual communication, but little is known about these flashes, or whether they have special significance in the lives of fireflies. Some flashing in *Photuris* females and those of the *Photinus ardens* group clearly provides illumination at landing and takeoff, and others may be aposematic. A listing of situational emissions presently records 21 "varieties," such as "caught in spider-web", "walking through grass", and "grasped by wolf spider." Finally, *Photuris* females of many species take perches in the activity spaces of other species and flash deceptive, female-mimicking responses to passing males, attract them, and eat them. *Photuris* females (*fairchildi*?) were found to use defensive compounds from their prey for their own and their eggs' protection. In several cases, the "supernumerary" flash patterns of *Photuris* males closely match, except for color, those of their females' prey.

Status of the classification: In the last three decades, following the extensive and outstanding revisionary work by J. W. Green (1948-1961), and cataloging and organizational work and study by McDermott (1964, 1966), there has been extensive field study and collection of the lightningbug genera *Photinus*, *Photuris*, and *Pyractomena*, and a revisionary study of *Photuris* is nearly complete. About 40 new species will be named in the mentioned genera, raising the total count of United States Lampyridae to about 150 species, but it is expected that there are probably 50 additional species yet to be discovered throughout North America (Lloyd 2001 in press). DNA studies have recently been initiated and may establish methods for the DNA analysis from single legs of the several thousand flash-behavior-voucher specimens now archived.

Recent phylogenetic study has made a few significant changes in the composition of the family Lampyridae: the Matheteini (*Matheteus* and *Ginglymocladus*) have been removed to Omethidae (Crowson 1972), and the Pterotini (*Pterotus*) placed in limbo as *incertae sedis* (Branham and Wentzel 2001). It would appear that revisionary studies are needed for *Ellychnia* and *Phausis* accompanied by focused field collection and associated ecological/behavioral observation. Authors and dates for higher lampyrid categories are not given here; though these names occasionally appear in restricted usage, their authors and dates are absent from even highly structured and formal taxonomic treatments. I retain some reservation about the author of the family name itself.

Distribution: Fireflies occur throughout the world on all continents except Antarctica, with tropical America having more than half of the described 2000 species (McDermott 1966). Probably two or three times this many remain to be described, with thousands of unworked and undetermined specimens presently archived in museums and private collections, but in nature extinction rate may be especially high in this family because of habitat loss in tropical regions throughout the world. Crowson (1972) noted that "the family has endemic representatives in all the main zoogeographical regions except New Zealand; one or two endemic species are recorded from islands of an apparent oceanic type, such as Palau and Yap in Micronesia." Eighty to 85 genera have been described worldwide; 16 genera are known with certainty to occur in America north of the Rio Grande, and three others could be present. West coast species are daytime or glow-worm fireflies, with only occasional unsubstantiated reports of flashing (lightningbug) species, except for recent reports from British Columbia. A European genus (*Phosphaenus*) was found in Nova Scotia in the 1950s, presumably having arrived in ballast around the beginning of the 1800s with waifs of a number of plant and insect species. (And representatives of two Asian genera, *Luciola* and *Lamprigera*, were introduced into Hawaii for the biocontrol of snails in the 1950s, but there have been no subsequent reports of their survival or occurrence there).

Key to Males of Nearctic Genera

1. Pronotum with explanate lateral borders absent or very narrow (Fig. 8, pm); hypomera closed, i.e. reaching anterior margin of pronotum, or nearly so 2
— Pronotum with lateral borders explanate (i.e., flanged, flared out each side of disk; e.g., Fig. 9, pm); hypomera open anteriorally, i.e. not reaching anterior margin 4

2(1). All antennomeres without branches 3
— All except terminal antennomeres of flagellum with one (elongate, strap-like) branch; West Coast (CA, OR); superficially resembling phengodids .... *Pterotus*
— Antennomeres 3-10 each with two branches; eastern U.S. .... *Pollaclasis*

3(2). Locality in California; posterolateral corners of pronotum projecting and ridged, and hind margin with a sub-lateral pit (Fig. 8, SLP); length <5 mm .... *Brachylampis*
— Locality in southcentral Texas (see Classification); pronotum corners not projecting, ridged, etc., and length probably >6mm .... near *Luciola*

4(1). Antennae robust with antennomeres 3-10 each with two leaf-like branches that are fused at their bases (Fig. 22) .... *Pleotomus*
— Antennae not as above, instead unbranched, usually filiform or serrate, rarely compact, robust. 5

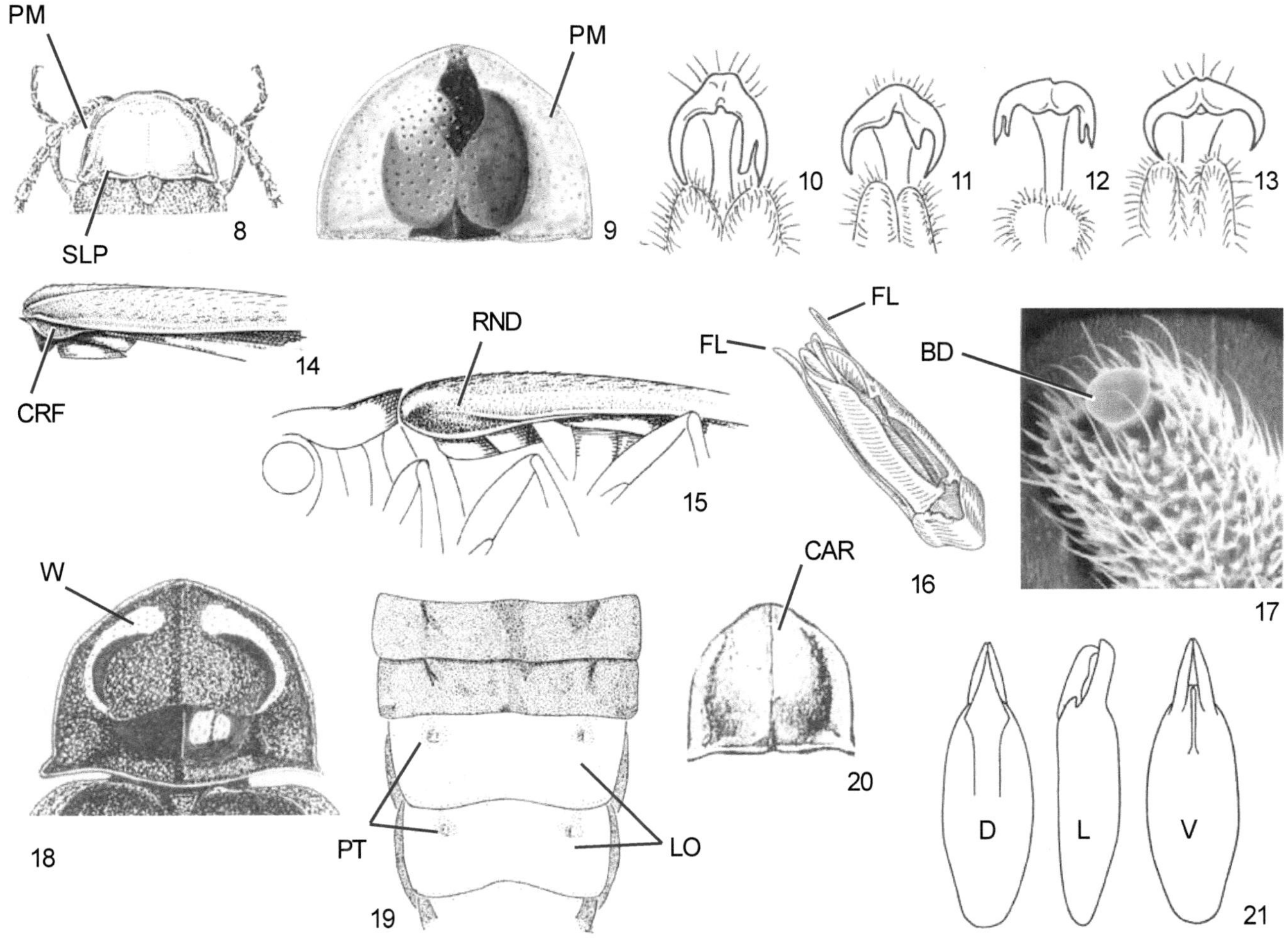

FIGURES 8.62-21.62. 8. Pronotum, etc. of *Brachylampis* (PM, pronotal margin, SLP, sub-lateral pit; from VanDyke 1939). 9. Pronotum of *Photuris* sp (PM, pronotal margin). 10. Sub-symmetrical fork of anterior tarsal claw of *Photuris* males. 11. Asymmetrical fork of anterior tarsal claw of *Micronaspis* males. 12. Forked anterior and posterior tarsal claws of *Bicellonycha*. 13. Simple tarsal claws of *Photinus* spp. and those of many other genera. 14. Folded and creased humeral area of left elytral epipleuron (CRF). 15. Rounded humeral area at base of left elytral epipleuron (RND). 16. Genitalia of *Photuris* spp. with filaments (FL) attached to the basal piece. 17. Bead on terminal antennomere of *Phausis reticulata* (Say) antenna. 18. Pronotum of *Phausis reticulata* (Say) with windows (W) over eyes. 19. Posterior ventrites of *Photinus concisus* Lloyd, showing two ventrites with lantern (LO) and pits (PT). 20. Pronotum of *Pyractomena* with mid-line carinula (CAR). 21. Genitalia of *Pyropyga nigricans* (Say), dorsal, lateral, and ventral views, showing general form found in genus, from Green 1961.

5(4). Anterior claws of fore and mid (pro and meso) tarsi bifid (2-pronged) (Figs. 10-12) 6
— Anterior claws of fore and mid tarsi simple, not bifid (e.g., Fig. 13) 8

6(5). Posterior claw of all tarsi bifid (Fig. 12); locality in Arizona; pygidium with pink or red (habitus, Fig. 23) .... *Bicellonycha*
— Posterior claw of all tarsi simple not bifid; locality not in Arizona; pygidium without pink or red, instead pale, dusky, or dark 7

7(6). Pronotal disk contoured and transparent over dorsolateral surfaces of eyes (Fig. 24); two prongs of anterior claw of fore-tarsi strongly asymmetrical, one basal and one apical (Fig. 11); epipleurae folded and creased basally (Fig. 14, CRF); basal piece of genitalia without lateral filaments (only known localities on Florida coastline); habitus Fig. 24) .... *Micronaspis*
— Pronotal disk not as above, instead rounded but not conforming to shape of eyes and opaque Figs. 1, 25, 27); two prongs of anterior claw of fore-tarsi nearly symmetrical, both apical as in Fig. 10; epipleura rounded basally (i.e. at humerus, Fig. 15, RND); basal piece of genitalia with lateral filament (Fig. 16, FL) .... *Photuris*

8(5). Terminal antennomere with a tiny sub-spherical, smooth, glabrous, sometimes vitreous (glass-like) bead (Fig. 17, BD, sometimes collapsed).......... 9
— Terminal article of antennae without such a bead 11

9(8). Antennae with 10 or fewer antennomeres (excluding terminal bead); abdominal spiracles ventral (occurrence almost exclusively western and southwestern US, i.e. AZ, CA, NM, UT, CO, wTX) .... *Microphotus*
— Antennae with 11 antennomeres (i.e., excluding terminal bead); abdominal spiracles dorsal (broad US occurrence, except AZ, sCA, CO, NV, UT).... 10

10(9). Pronotum with two anterior somewhat transparent (glassy) windows over eyes (one rare California exception) (Figs. 18, 28); eyes touching or nearly so below; head deeply concave above; pygidium not predominately yellow .... *Phausis*
— Pronotum without windows; eyes distant below; head rounded above; pygidium predominately yellow (only reported occurrence southeastern Arizona) .... *Paraphausis*

11(8). Body outline broadly oval; pronotum distinctly transverse and sub-triangular; epipleurae very wide (habitus neotropical in appearance, Figs. 29, 30) 12
— Body outline normal, i.e., narrowly oval or elongate; pronotum not distinctly transverse; epipleurae normal, not exceptionally wide (except in *Lamprohiza* which has pronotal windows) 13

12(11). Light organ (yellow or white enamel-like cuticle) occupying two posterior ventrites of abdomen; antennae filiform (Fig. 29) — (locality records, excepting airport quarantine interceptions, only southern-most points of USA, Florida Keys and near Brownsville, Texas) .... *Aspisoma*
— Light organ absent; antennae strongly serrate (Fig. 30). —(localities broadly arcing around Gulf states) .... *Tenaspis*

13(11). Light organ (yellow or pale, commonly enamel-like cuticle) present on 2 or 3 ventrites of abdomen 14
— Light organ as described not present on 2 or 3 ventrites of abdomen, instead either completely absent or merely a small, median spot on one ventrite... 16

14(13). Light organ occupying entire surfaces of the two ventrites (Fig.19, LO); two pits present on each light-organ-bearing ventrite (Fig. 19, PT); pronotum without transparent ("glassy") spots ("windows") over eyes 15
— Light organ not occupying entire surfaces of the two ventrites, instead margin of each lantern-bearing ventrite with nonluminous border ("frame"); pits not present on light-organ-bearing ventrites; pronotum with transparent (glassy, tear-shaped) windows over eyes .... *Lamprohiza*

15(14). Median line of pronotum with a distinct though low keel (i.e., a ridge, carina, carinula) along most or all of its length (Figs. 20, CAR, 26); pronotum sub-pentagonal .... *Pyractomena*
— Median line of pronotum without such a keel — instead flat, or with a shallow, often discontinuous groove (Figs. 31, 32); pronotum not distinctly sub-pentagonal (or if so then elytra shortened leaving most of abdomen exposed; rare exotic, only eastern maritime Canada, see *Phosphaenus*) .... *Photinus* (part)

16(13). Eyes large, approximate (actually or nearly touching) beneath head; tip of mandibles abruptly narrowing (habitus Fig. 33) .... *Pleotomodes*
— Eyes small and well separated; tip of mandibles not abruptly narrowing 17

17(16). Second antennal article very short, wider than long, about one-quarter as long as third; first antennal article shorter than third 18
— Second antennal article short but not wider than long, at least one-third as long as third; first antennal article longer than third 20

18(17). Antennae compressed, flattened in cross-section, and serrate or sub-serrate; elytra usually covering all or most of abdomen (except *P. nigricans* in a few and isolated localities; abdomen not lobed; localities occurring across North America).... 19
— Antennae not compressed, not serrate or sub-serrate; elytra very short, leaving at least 5 abdominal dorsites exposed and all strongly lobed; locality only maritime Canada .... *Phosphaenus*

19(18). Antennae delicate, only feebly compressed and serrate; pronotum sub-semicircular or hexagonal in outline, usually not more narrowly rounded in front; genitalia of general form in Fig. 21 (habitus of one common species Fig. 34) .... *Pyropyga*
— Antennae coarse and distinctive, strongly compressed and serrate; pronotum sub-pentagonal or -triangular in outline, its anterior margin usually more narrowly rounded than other rounded angles of outline; genitalia variable, not resem-

bling form in Fig. 21 (habitus of the most common species Fig. 35) .... *Lucidota*

20(17). Margins and sutural beads of elytra yellow or pale, contrasting with dark color of elytral disks; lateral margins of pronotal flanges always without dark (black, charcoal) stripe .... *Photinus* (part)

— Margins and beads of elytra concolorous with elytral disks; lateral margins of pronotal flanges with dark stripe in most and most common species (Fig. 36) .... *Ellychnia*

Classification of the Nearctic Genera

Lampyridae Latreille 1817

Lampyrinae

Lampyrini

Microphotus LeConte 1866, 7 spp., Arizona, California, New Mexico, and Texas (key to spp., Green 1959).

Paraphausis Green 1949, 1 sp., Arizona.

Pleotomini

Pleotomodes Green 1948 (Fig. 33), 3 spp., Florida and Arizona (tribal position and removal from synonymy with *Lampyris*, Geisthardt 1986).

Pleotomus LeConte 1861 (Fig. 22), 3 spp., Arizona and northeastward to Maryland. (Key to spp., LeConte 1881 [The three species are legitimate and warrant distinction.])

Lamprocerini

Tenaspis LeConte 1881 (Fig. 30), 1 sp., *T. angularis* (Gorham 1880), broadly around the Gulf states.

Cratomorphini

Aspisoma Laporte 1833 (Fig. 29), 1 sp., *A. ignitum* (Linnaeus 1767), southernmost Florida and Texas.

Micronaspis Green 1948 (Fig. 24), 1 sp., *M. floridana* Green 1948, Florida peninsular coastline.

Pyractomena LeConte 1845 (Fig. 26), 16 US spp., 2 to be named, all but sw NA, with only scattered localities w of Kansas (description and biology of juveniles, Archangelsky and Branham 1998; key to spp., Green 1957). This generic name is sometimes misspelled, often as *Pyractonema*, which is a South American genus.

Photinini

Ellychnia Blanchard 1845 (Fig. 36), 12 spp.+, but in need of revisionary field work, widely distributed in US (key to western species, Fender 1970).

Lamprohiza Motschulsky 1853, 1 sp. [two specimens voucher LeConte's 19th century IL and MD records, but none since; other reports apparently from misidentified native *Phausis reticulata*, spurious reports and museum IDs to the contrary]

Lucidota Laporte 1833 (Fig. 35), 3 spp., widely distributed US (description and homologies of lampyrid juveniles, Branham and Archangelsky 2000; key to spp., LeConte 1881; additional species descriptions, LeConte 1878).

Phausis LeConte 1851 (Fig. 28), 7+ spp., but in need of revisionary field work, widely distributed except sw US (key to spp., Fender 1961a).

Phosphaenus Fourcroy 1785, 1 sp. [a single specimen from 1950s in Nova Scotia, presumably arriving in ballast around 1800]

Photinus Laporte 1833 (Figs. 31-32), 34 spp., 13 in prep., widely distributed in NA, but only scattered or riparian localities w of Texas and Kansas (key to species, Green 1956; key update, Lloyd 1966b, 1968, 1969b.)

Pyropyga Motschulsky 1852 (Fig. 34), 4 spp., widely distributed in US (description and comparison of juveniles, Archangelsky and Branham 2001; key to spp., Green 1961).

Photurinae

Bicellonycha Motschulsky 1853 (Fig. 23), 1 sp., *P. wickershamorum* Cicero 1982, Arizona.

Photuris LeConte 1851 (Figs. 1, 25, and 27), 22 spp., 28 in prep., eastern US west to Colorado, sw Texas (keys to spp., Barber 1951 and McDermott 1967, should not be relied upon).

Luciolinae

Luciolini

Luciola Laporte 1833 (= near *Luciola*) [1 sp. possibly, in southcentral Texas but actual occurrence equivocal, one specimen reported but as possibly mislabeled].

Ototetrinae

Brachylampis Van Dyke 1939, 2 spp. California (key to species, Van Dyke 1939).

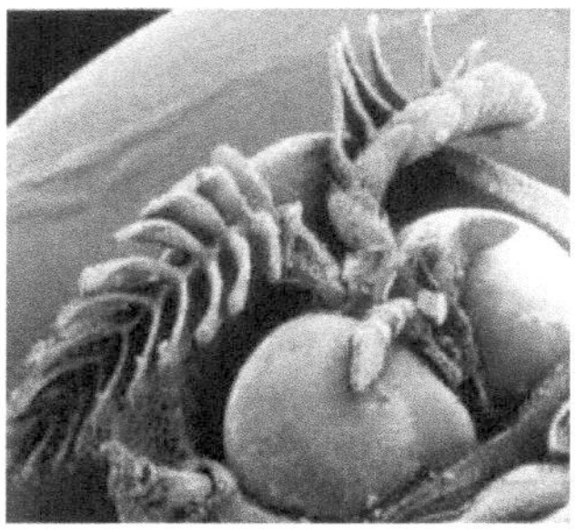
22

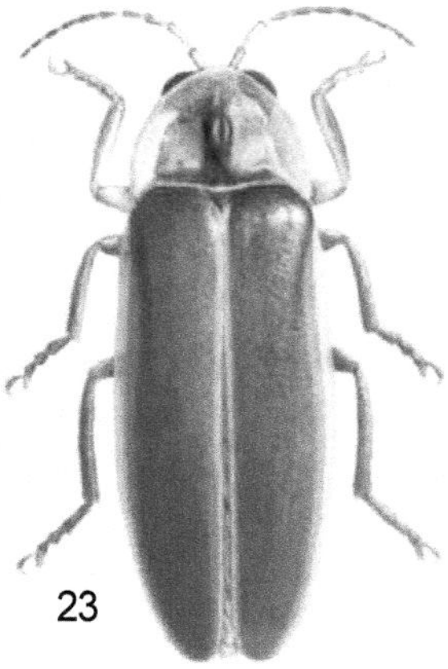
23

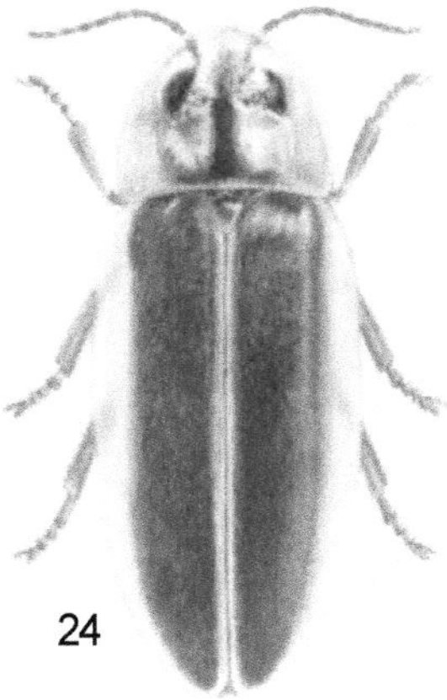
24

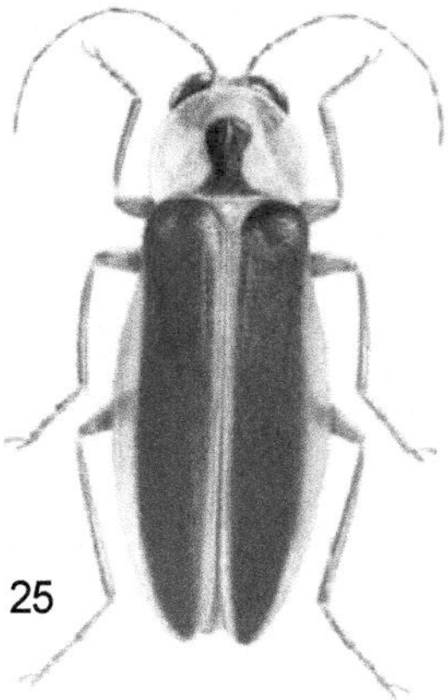
25

26

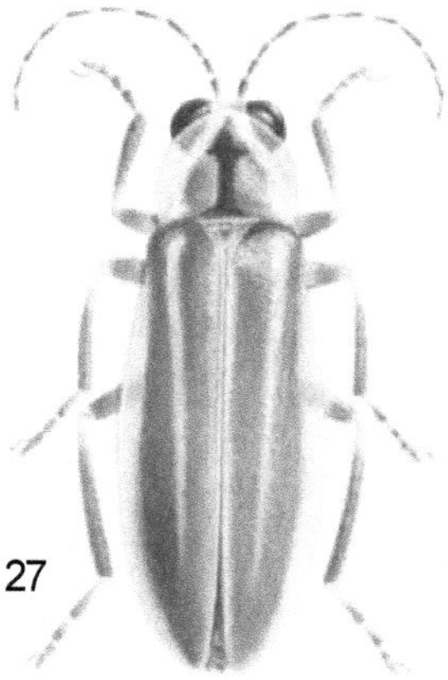
27

28

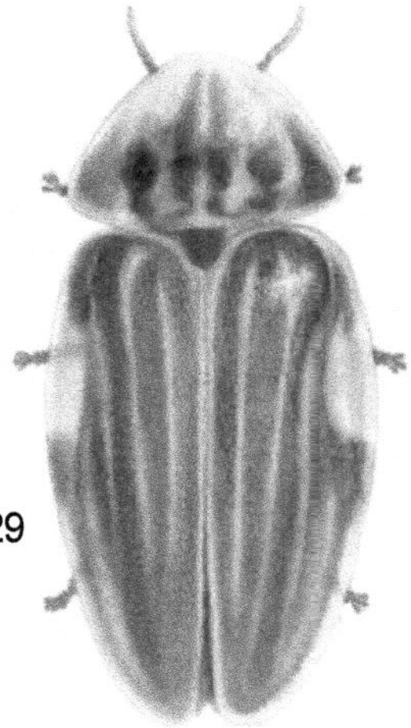
29

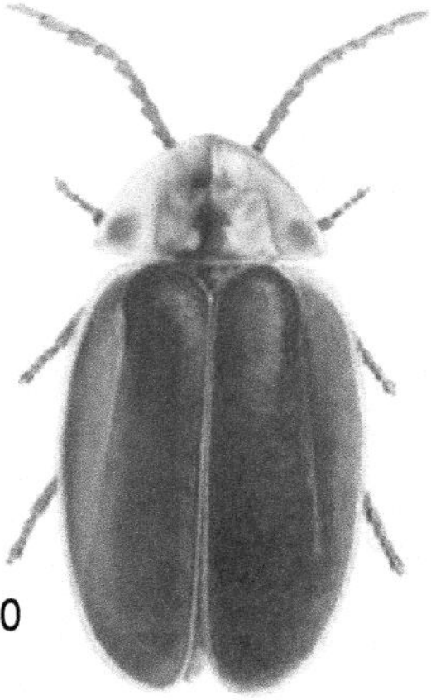
30

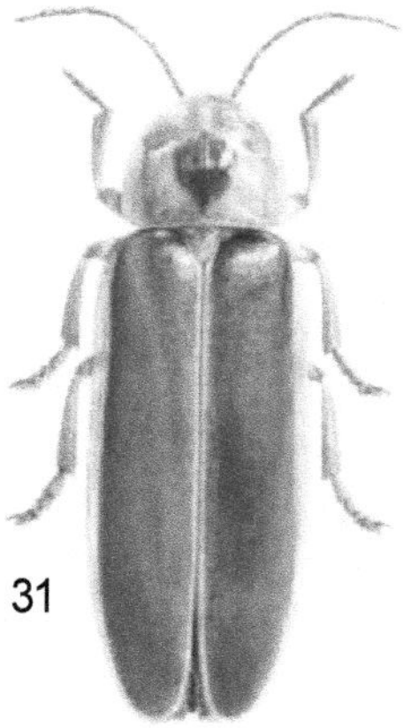
31

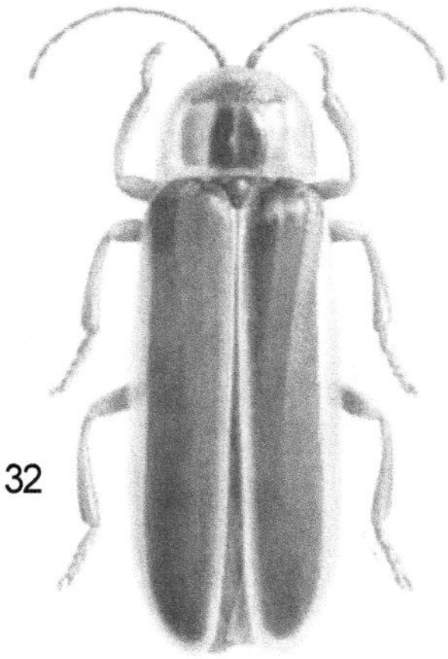
32

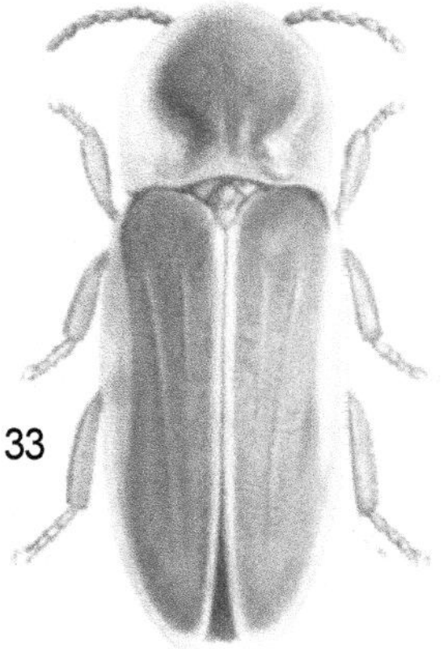
33

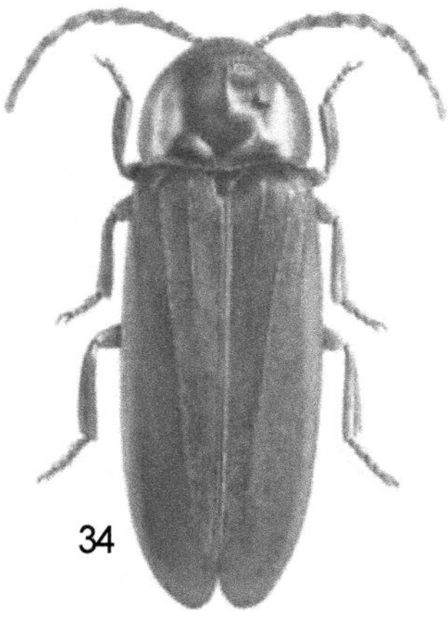
34

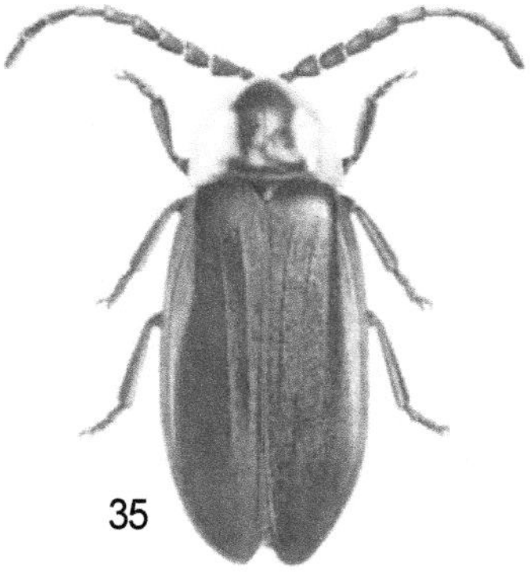
35

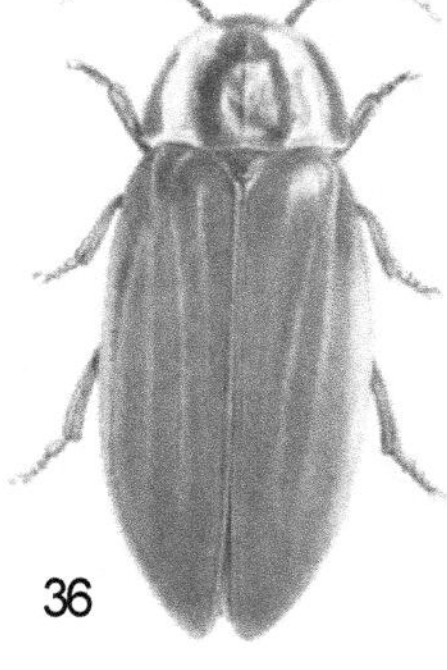
36

Cyphonocerinae

Pollaclasis Newman 1838, 1 sp., *P. bifaria* (Say) 1835, widely scattered localities in eastern half of US.

Incertae sedis

Pterotus LeConte 1859, 2 spp., western U.S. (northern California and southern Oregon).

Bibliography

ARCHANGELSKY, M. and M. A. BRANHAM. 1998. Description of the preimaginal stages of *Pyractomena borealis* (Randall, 1838) (Coleoptera: Lampyridae) and notes on its biology. Proceedings of the Entomological Society of Washington, 100:421-430.

ARCHANGELSKY, M. and M. A. BRANHAM. 2001. Description of last instar and pupa of *Pyropyga nigricans* (Coleoptera: Lampyridae, Photinini) and comparison with larvae of other Photinini genera. Canadian Entomologist, 133: 155-164.

BARBER, H. S. 1951. North American fireflies of the genus *Photuris*. Smithsonian Miscellaneous Collection, vol. 117, no. 1, pub. 4051, 58 pp.

BRANHAM, M. A. and ARCHANGELSKY, M. 2000. Description of the last larval instar and pupa of *Lucidota atra* (G. A. Olivier, 1790) (Coleoptera: Lampyridae), with a discussion of abdominal segmentation homology across life stages. Proceedings of the Entomological Society of Washington, 102: 869-877.

BRANHAM, M.A. and J. W. WENTZEL. 2001. The evolution of bioluminescence in cantharoids (Coleoptera: Elateroidea). Florida Entomologist (in press).

BRANHAM, M. A. and J. W. WENTZEL. Sexual communication in fireflies. (Coleoptera: Lampyridae). In prep.

BUSCHMAN, L.L. 1984a. Biology of the firefly *Pyractomena lucifera* (Coleoptera: Lampyridae). Florida Entomologist, 67: 529-542.

BUSCHMAN, L. L. 1984b. Larval biology and ecology of *Photuris* fireflies (Lampyridae: Coleoptera) in northcentral Florida. Journal of the Kansas Entomological Society, 57: 7-16.

BUSCHMAN, L.L. 1987. Larval development and its photoperiodic control in the firefly *Pyractomena lucifera* (Coleoptera: Lampyridae). Annals of the Entomological Society of America, 81: 82-90.

CICERO, J. M. 1982. The genus *Bicellonycha* in the United States with descriptions of a new species and subspecies (Coleoptera: Lampyridae, Photurinae). Coleopterists Bulletin, 36: 270-278.

CICERO, J. M. 1988. Ontophylogenetics of cantharoid larviforms (Coleoptera: Cantharoidea). Coleopterists Bulletin, 42: 105-151.

CROWSON, P. 1972. A revision of the classification of the Cantharoidea. Revista Universidad Madrid XXI (82): 35-74.

EISNER, T., M. A. GOETZ, D. E. HILL, S. R. SMEDLEY and J. MEINWALD. 1997. Firefly "femmes fatales" acquire defensive steroids (lucibufagins) from their firefly prey. Proceedings of the National Academy of Sciences. 94:9723-9728.

FENDER, K. M. 1961a. The genus *Phausis* in America north of Mexico. (Coleoptera-Lampyridae). Northwest Science, 40: 83-95.

FENDER, K. M. 1961b. Lampyroid section, Lampyridae. Pp. 34-43. *In:* M. H. Hatch. The beetles of the Pacific northwest. University of Washington Press. Seattle, WA.

FENDER, K. M. 1970. *Ellychnia* of western North America. (Coleoptera: Lampyridae). Northwest Science, 44: 31-43.

GEISTHARDT, M. 1986. *Pleotomodes* Green 1948 a valid genus, not a junior synonym of *Lampyris* Muller 1764 (Coleoptera: Lampyridae: Lampyrinae). Coleopterists Bulletin, 40: 297-300.

GREEN, J. W. 1948. Two new species of Lampyridae from southern Florida, with a generic revision of the Nearctic fauna. Transactions of the American Entomological Society, 74: 41-60.

GREEN, J. W. 1949. A new genus and new species of American Lampyrini, and other notes. Transactions of the American Entomological Society, 75: 1-6.

GREEN, J. W. 1956. Revision of the Nearctic species of *Photinus* (Lampyridae: Coleoptera). Proceedings of the California Academy of Sciences ser. 4, 28: 561-613.

GREEN, J. W. 1957. Revision of the Nearctic species of *Pyractomena* (Coleoptera: Lampyridae). Wasmann Journal of Biology, 15: 237-284.

GREEN, J. W. 1959. Revision of the species of *Microphotus*, with an emendation of the Lampyrini (Lampyridae). Coleopterists Bulletin, 13: 80-96.

GREEN, J. W. 1961. Revision of the species of *Pyropyga* (Lampyridae). Coleopterists Bulletin, 15:65-74.

LABELLA, D. and J. E. LLOYD. 1991. Lampyridae (Cantharoidea). Pp. 427-428. *In:* F. W. Stehr, ed., Immature insects, Vol. 2. Kimball Hunt. Dubuque, IA.

LeConte, J. L. 1878. Additional descriptions of new species. Pp. 405. *In*: H. G. Hubbard and E. A. Schwarz. Coleoptera of Florida and Michigan. Proceeding of the American Philosophical Society, 17: 353-669.

LeConte, J. L. 1881. Synopsis of the Lampyridae of the United States. Transactions of the American Entomological Society, 9: 15-72.

FIGURES 22.62-36.62. 22. Ventral view of head and pronotum of *Pleotomus pallens*. LeConte, showing biramous antennae, and also huge glowworm-type eyes, with open hypomeron at lower left. 23-36. Carbon-dust habitus drawings of various fireflies, by Laura Line. 23. *Bicellonycha wickershamorum*. Cicero. 24. *Micronaspis floridana* Green. 25. *Photuris frontalis* LeConte. 26. *Pyractomena angulata* (Say). 27. *Photuris lloydi* McDermott. 28. *Phausis reticulata* (Say). 29. *Aspisoma ignitum* (L.). 30.*Tenaspis angularis* LeConte. 31. *Photinus pyralis* (L.). 32. *Photinus macdermotti* Lloyd. 33. *Pleotomodes knulli* Green. 34. *Pyropyga nigricans* (Say). 35. *Lucidota atra* (G. A. Olivier). 36. *Ellychnia corrusca* (L.).

LLOYD, J. E. 1966a. Studies on the flash communication system in *Photinus* fireflies. Miscellaneous Publications No. 130, Museum of Zoology, University of Michigan, Ann Arbor.
LLOYD, J. E. 1966b. Two cryptic new firefly species in the genus *Photinus* (Coleoptera: Lampyridae). Coleopterists Bulletin, 20: 43-46.
LLOYD, J. E. 1968. A new *Photinus* firefly, with notes on mating behavior and a possible case of character displacement (Coleoptera: Lampyridae). Coleopterists Bulletin, 22: 1-10.
LLOYD, J. E. 1969a. Flashes of *Photuris* fireflies: their value and use in recognizing species. Florida Entomologist, 52: 29-35.
LLOYD, J. E. 1969b. Flashes, behavior and additional species of Nearctic *Photinus* fireflies (Coleoptera: Lampyridae). Coleopterists Bulletin, 23: 29-40.
LLOYD, J. E. 1990. Firefly semiosystematics and predation: a history. Florida Entomologist, 73: 51-66.
LLOYD, J. E. 1997a. On research and entomological education, and a different light in the lives of fireflies (Coleoptera: Lampyridae; *Pyractomena*). Florida Entomologist, 80: 120-131.
LLOYD, J. E. 1997b. Firefly mating ecology, selection, and evolution. *In:* J. C. Choe and B. J. Crespi, eds., Mating Systems in Insects and Arachnids. Cambridge University Press. New York.
LLOYD, J. E. 1999. On research and entomological education. III: Firefly brachyptery and wing "polymorphism" at Pitkin Marsh and watery retreats near summer camps (Coleoptera: Lampyridae: *Pyropyga*). Florida Entomologist, 82: 165-179.
LLOYD, J. E. 2001. On research and entomological education V: A species (c)oncept for fireflyers, at the bench and in old fields, and back to the Wisconsian glacier. Florida Entomologist (in press).
McDERMOTT, F. A. 1914. The ecologic relations of the photogenic function among insects. Zeitschrift fur Wissinschafliche Insektenbiologie, 10: 305-307.
McDERMOTT, F. A. 1964. The taxonomy of the Lampyridae. Transactions of the American Entomological Society, 90: 1-72.
McDERMOTT, F. A. 1966. Coleopterorum Catalogus Supplementa Edita, Pars 9 Lampyridae. W. O. Steele, Gravenhague: W. Junks.
PETERSON. A. 1951. Larvae of insects: An Introduction to Nearctic Species. Part II: Coleoptera, Diptera, Neuroptera, Siphonoptera, Mecoptera, Trichoptera. Privately published. Edward Bros. Columbus, OH.
SIVINSKI, J. M. 1981. The nature and possible functions of luminescence in Coleoptera larvae. Coleopterists Bulletin, 35: 167-179.
SIVINSKI, J. M., J. E. LLOYD, S. N. BESHERS, L. R. DAVIS, R. G. SIVINSKI, S. R. WING, R. T. SULLIVAN, P. E. CUSHING and E. PETERSON. 1998. A natural history of *Pleotomodes needhami* Green (Coleoptera: Lampyridae): a firefly symbiont of ants. Coleopterists Bulletin, 52: 23-30.
VAN DYKE, E. C. 1939. New species and subspecies of west American Coleoptera. Pan-Pacific Entomologist, 15: 15-20.
WING, S. R. 1989. Energetic costs of mating in a flightless female firefly, *Photinus collustrans* (Coleoptera: Lampyridae). Journal of Insect Behavior, 2: 841-847.
WING, S. R. 1991. Timing of *Photinus collustrans* reproductive activity: finding a mate in time (Coleoptera: Lampyridae). Coleopterists Bulletin, 45: 57-74.

63. OMETHIDAE LeConte 1861

by Alistair S. Ramsdale

Family common names: The false soldier beetles (Omethinae), the false firefly beetles (Matheteinae)

Omethidae are a small and poorly known family of soft-bodied beetles of uncertain affinity. Omethids are rarely collected, but have usually been found on foliage. Omethidae are best distinguished from other Elateriformia by the following combination of characters: antennae 11-segmented, filiform, serrate, or pectinate; antennal insertions widely separated; labrum visible and well sclerotized; prothoracic coxal cavities broadly open; prothoracic trochantins exposed; prothoracic coxae conical, projecting well below prosternum; mesothoracic coxae contiguous; tarsal formula 5-5-5; tarsomeres 3 and 4 with ventral bifid lobes; abdomen with 7-8 freely moveable ventrites; abdomen without luminous organs; abdominal tergites without paired lateral pores.

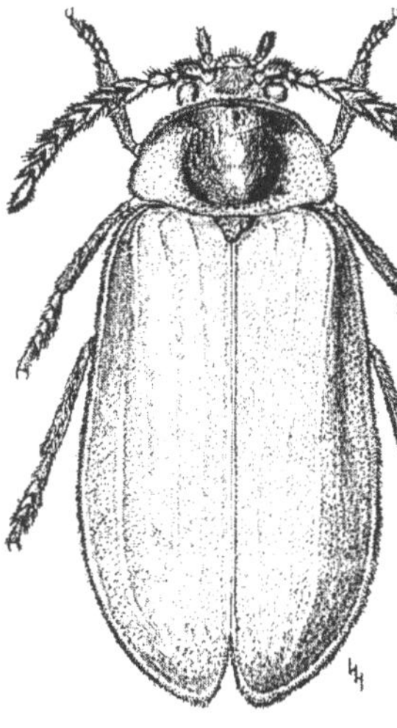

FIGURE 1.63. *Matheteus theveneti* LeConte (modified from Fender 1962)

Description: Length 3 -12 mm. Body elongate, narrow or broadly ovate, moderately flattened to slightly convex. Soft-bodied, with cuticle, especially on elytra and abdomen, lightly sclerotized, flexible. Coloration brown to black in most, many with head, pronotum, and/or elytra marked with or entirely red or yellow. Body typically with moderately dense, short, depressed pubescence.

Head exposed or partially concealed by pronotum, in some deeply excavated between eyes (male *Troglomethes*). Compound eyes entire, somewhat protruding in most. Antennae with 11 antennomeres, filiform (most Omethinae), serrate (Driloniinae), or pectinate with antennomeres 4 - 10 uniramose (Matheteinae), some filiform with antennomeres 4 and 5 enlarged and excavated (male *Blatchleya*); antennal insertions widely separated. Frontoclypeal and clypeolabral sutures present. Labrum free, well sclerotized and visible; labrum elongate with anterior margin bilobed (Driloniinae) or short with anterior margin entire (Omethinae, Matheteinae). Mandibles with inner margin unidentate in most, stout (Matheteinae) or more slender (Omethinae). Maxillae each with distinct galea and lacinia, with apical lobes densely setose in most; maxillary palpi 4-segmented; apical palpomere cylindrical to fusiform, in some apically glabrous (Matheteinae), rarely greatly elongated and flattened (*Drilonius* subgenus *Palpodrilonius*). Labium with ligula undivided or finely cleft; apical labial palpomere cylindrical to fusiform.

Acknowledgments: I thank J. F. Lawrence for generously sharing his knowledge of omethids and for reviewing the manuscript, R. L. Brett for arranging a loan of the omethid holdings of the California Academy of Sciences, including seven primary types, on short notice, M. A. Branham for providing additional specimen data, and J. H. Frank, P. E. Skelley and D. K. Young for reviewing the manuscript.

Pronotum transverse to subquadrate, lateral margins entire, posterior margin without acute angles, about as wide or narrower than elytra; pronotal margins in many strongly reflexed. Prothoracic coxae conical, projecting well below prosternum, attached externally, contiguous; prothoracic coxal cavities widely open; prothoracic trochantin exposed. Mesothoracic coxae contiguous, prominent in most, cavities open laterally. Scutellum well developed and visible. Metasternum with lateral margins straight (Driloniinae) or sinuate (Omethinae, Matheteinae); metathoracic coxae flattened, contiguous or narrowly separated. Legs more or less elongate, slender; tibiae with paired apical spurs. Tarsal formula 5-5-5; tarsomeres 3 and 4 with deeply bifid ventral membranous lobes; tarsomeres 1 and 2 sometimes with finely setose ventral pads (Matheteinae); tarsal claws in most with a reduced basal tooth, in some more or less appendiculate.

Elytra soft, usually parallel-sided to somewhat widened apically, some with lateral margins broadly explanate (*Matheteus*), most with elytra about as long as the abdomen, in some elongate and extending well beyond apex of abdomen (*Malthomethes*); elytra loosely fitted to pterothorax and abdomen, most without prominent costae (Omethinae), in some weakly costate (*Matheteus*), some with distinct longitudinal costae and/or reticulate cells (Driloniinae). Metathoracic wings fully developed, rarely reduced (female *Troglomethes oregonensis* Wittmer), with 5 anal veins in the main group and a closed anal cell.

Abdomen with 7 or 8 freely moveable ventrites (visible sternites) with complete sutures, some with dorsal paratergites (Driloniinae, Matheteinae), without luminous organs; abdominal tergites without paired glandular pores. Male genitalia (aedeagus) of trilobed type, symmetrical; basal piece (phallobase) strongly sclerotized in most; lateral lobes (parameres) well sclerotized, fused (e.g., *Symphyomethes*) or separated (e.g., *Omethes*), often apically curved or hooked; median lobe sclerotized, ventral, free, exposed.

Elements of this description were adapted from Crowson (1972) and Lawrence *et al.* (1999). The following habitus illustrations of Nearctic Omethidae are available: *Blatchleya gracilis*

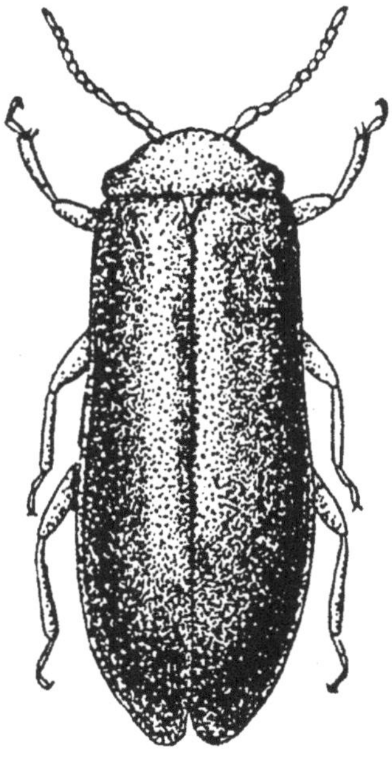

FIGURE 2.63. *Blatchleya gracilis* (Blatchley) (after Blatchley 1910)

(Blatchley) in Blatchley (1910) (Fig. 2), *Ginglymocladus discoidea* Van Dyke in Green (1948) and *Matheteus theveneti* LeConte in Fender (1962) (Fig. 1). Photographs of *Matheteus theveneti* LeConte are present in Lawrence *et al.* (1999).

Habits and habitats. Almost nothing is known about the ecology of omethids. Adults are rarely encountered. Adults of *Blatchleya, Troglomethes, Ginglymocladus* and *Matheteus* have been most frequently collected from foliage during the day. Adults of *Symphyomethes* have been extracted from forest floor debris. Adults emerge in the spring and summer and are probably all short-lived. All species are presumably univoltine. Nothing is known about their feeding habits. The larvae are unknown.

Status of the classification. The family Omethidae was erected by Crowson (1972) for taxa formerly included in Cantharidae, Drilidae, and Lampyridae. Omethidae are composed of three distinct subfamilies: Omethinae (formerly in Cantharidae), Matheteinae (formerly in Lampyridae), and Driloniinae (formerly in Drilidae). While differing considerably from one another, Crowson (1972) argued that these three subfamilies appeared to be more closely related to each other than to any other family of the former Cantharoidea (*sensu* Crowson 1972).

While the three subfamilies appear to be distinct natural groups, the family as a whole has never been adequately defined. No synapomorphies have been identified, and as a result the monophyly of Omethidae is uncertain (J. F. Lawrence, personal communication).

Crowson (1972) suggested that there might be a close affinity between Cantharidae, particularly the relictual subfamily Dysmorphocerinae, and Omethidae. This must remain a tentative hypothesis, as larvae of Omethidae are undescribed and considering the derived placement of Dysmorphocerinae in the phylogeny presented by Brancucci (1980). Omethidae are currently placed in the superfamily Elateroidea (*sensu lato*) of Lawrence (1987). Relationships within this lineage have not been satisfactorily resolved.

The immature stages of Omethidae, which would provide important insight into their phylogenetic relationships, remain unknown (Lawrence 1991). The females of several genera of Omethinae are also unknown. No fossil Omethidae are known.

Distribution. Omethidae are a small family with eight genera and 32 species described worldwide. In the New World, there are seven genera and 10 species. Omethinae are composed of five genera and eight species, with a single species occurring in Japan and the remainder restricted to North America. Matheteinae, with two genera and three species, are endemic to western North America. Driloniinae, composed of the single genus *Drilonius* Kiesenwetter, with 21 described species, are exclusively Asian, occurring from Japan to Indonesia (Nakane 1950, Wittmer 1944, 1948, 1956, 1957, 1995). No world catalog of the family exists, and omethids are poorly represented in museum collections.

Key to the New World Subfamilies and Genera

Many genera of Omethidae are currently based only on characters exhibited by males and the females of several genera are unknown. It is therefore not possible to reliably separate unassociated females in a key of this geographic scope.

1. Antennae filiform in most, some with antennomeres 4 and 5 strongly modified; abdomen without distinct paratergites; widely distributed; (Omethinae) 2
— Antennae pectinate (Fig. 3); abdomen with distinct paratergites (Fig. 4); California and Oregon: (Matheteinae) 6

Omethinae

2(1). Antennomeres 4 and 5 enlarged and excavated; eastern United States .... *Blatchleya*
— Antennae simple, without enlarged or excavated antennomeres; widely distributed 3

3(2). Head deeply excavated between eyes (Figs. 5, 6); California and Oregon .... *Troglomethes*
— Head not excavated as above, some with shallow impressions; widely distributed 4

4(3). Head covered by pronotum at least to posterior margins of eyes (Fig. 7); widely distributed 5
— Head more protruded, eyes not adjacent to anterior margin of pronotum (Fig. 8); California and Oregon .... *Malthomethes*

5(4). Head strongly rugose; elytra with finely impressed longitudinal costae; eastern United States .... *Omethes*
— Head smooth or weakly punctate; elytra without any traces of longitudinal costae; California .... *Symphyomethes*

Matheteinae

6(1). Elytra black or black with pale margins antennal branches narrowly attached to flagellomeres; bases of antennal branches recessed in semimembranous articulations; epipleura narrow .... *Ginglymocladus*
— Elytra red; antennal branches broadly attached to flagellomeres; bases of antennal branches not as above; epipleura very wide .... *Matheteus*

Classification of the New World Genera

The classification presented herein follows that of Crowson (1972), Lawrence and Newton (1995), and Lawrence *et al.* (1999). Within each subfamily, the genera are listed alphabetically. For

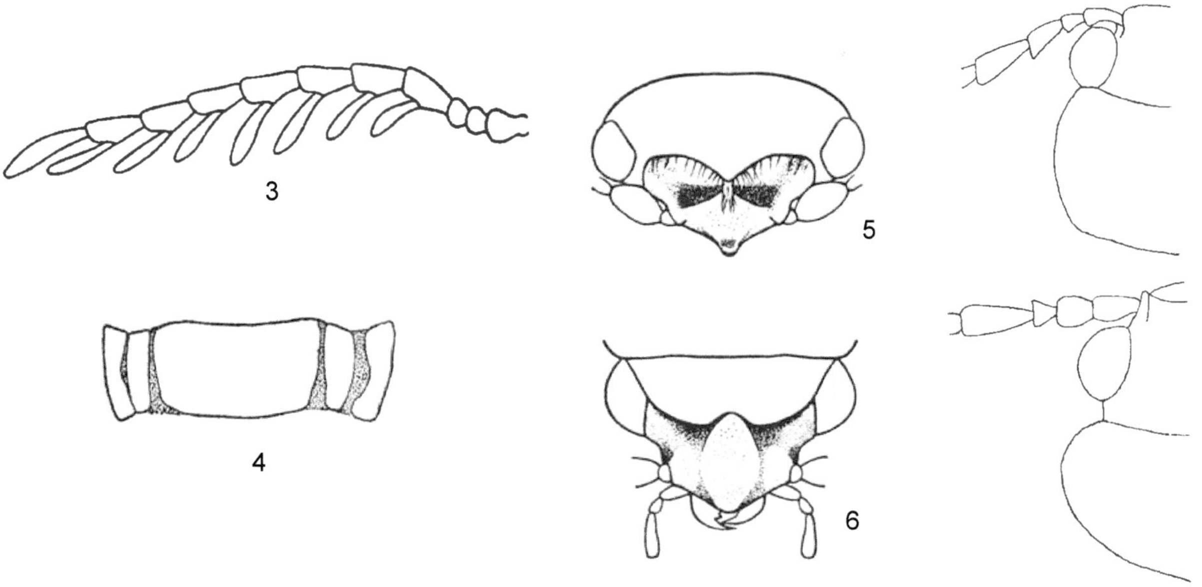

FIGURES 3.63-8.63. 3. *Ginglymocladus discoidea* VanDyke, antennae, lateral view (after Green 1948); 4. *Drilonius striatulus* Kiesenwetter, abdominal segment, dorsal view (after Crowson 1972); 5. *Troglomethes leechi* Wittmer, male head, frontal view (after Wittmer 1970); 6. *Troglomethes oregonensis* Wittmer, male head, frontal view (after Wittmer 1970); 7. *Omethes marginatus* LeConte, half of head, dorsal view (modified from Fender 1975); 8. *Malthomethes oregonus* Fender, half of head, dorsal view (after Fender 1975).

each genus, numbers of species and references are given on a worldwide basis. The ecological information presented is for adults only.

Omethidae LeConte 1861

Omethinae LeConte 1861

Omethini LeConte 1861 (not Blatchley 1910)

LeConte (1861) initially placed *Omethes*, the only genus known to him, in the subfamily Telephorinae (= Cantharidae) of his broadly defined Lampyridae. In his final revision of the Lampyridae *sensu lato* (= Cantharoidea *sensu* Crowson 1972) of North America, LeConte (1881) excluded *Omethes* and expressed uncertainty over its proper placement. Arnett (1963), Blatchley (1910), Downie and Arnett (1996) and Wittmer (1970) retained Omethini as a tribe of Cantharidae. Crowson (1972) elevated the group to subfamilial status and placed it in his new family Omethidae.

Lawrence and Newton (1995) attributed the earliest usage of a suprageneric name for this group to LeConte (1861), in contrast to Crowson (1972) who attributed it to Blatchley (1910).

Blatchleya Knab 1910
Blanchardia Blatchley 1910
One species, *Blatchleya gracilis* (Blatchley), known from Indiana (Crawford, Lawrence, Perry and Tippecanoe counties) and Ohio (Hocking County, **NEW STATE RECORD**). Blatchley (1910) reported collecting *B. gracilis* in Indiana by sweeping herbaceous vegetation alongside roads in May and June. The structure formed by the modified antennomeres 4 and 5 of the male probably serves a copulatory function similar to that of the analogous structures present on the antennae of male *Collops* (Melyridae) and *Meloe* (Meloidae). The antennae of the female are unmodified. Additional information: Knab (1910).

Malthomethes Fender 1975
One species, *Malthomethes oregonus* Fender, known from northern California (Humboldt County, **NEW STATE RECORD**) (K. M. Fender, unpublished note) and western Oregon (Coos, Curry and Lane counties). In Oregon, this species has been collected in a flight intercept trap in a young stand of Douglas Fir, *Pseudotsuga menziesii* (Mirbel) Franco, in July (Fender 1975). The female of *M. oregonus* is unknown. An undescribed species of *Malthomethes* from central California was recently collected by the author along the western slope of the southern Coast Range in Monterey County. The striking similarity of body form shared by *Malthomethes* and malthinine Cantharidae is a remarkable example of convergent evolution.

Omethes LeConte 1861
Elianus Lewis 1895
Omethes is currently composed of one Nearctic and one Palaearctic species. *Omethes marginatus* LeConte is known from Connecticut (locality unknown), Maryland (Baltimore County, **NEW STATE RECORD**), New Jersey (Warren County, **NEW STATE RECORD**), Ohio (near Cincinatti) (Blatchley 1910), and Pennsylvania (Northampton County, **NEW STATE RECORD**), where it has been collected from May to June. The three new state

records of *O. marginatus* were provided by M. A. Branham (personal communication). Female *O. marginatus* are fully winged and generally similar to the male. *Omethes rugiceps* (Lewis) occurs in Japan (Lewis 1895, Wittmer 1970).

Symphyomethes Wittmer 1970
Two species: endemic to California. Species descriptions: Wittmer (1970). The holotype, and possibly the only known specimen, of *Symphyomethes blandulus* Wittmer is known from coastal Mendocino County where it was collected from Coast Redwood, *Sequoia sempervirens* (D. Don) Endlicher, litter in July. *Symphyomethes californicus* Wittmer is known from Calaveras and Nevada counties and has been collected in May. The females of both species are undescribed.

Troglomethes Wittmer 1970
Two species: endemic to California and Oregon. Species descriptions: Wittmer (1970). *Troglomethes leechi* Wittmer is widely distributed in the Sierra Nevada Mountains of California (El Dorado, Fresno, Mariposa, Shasta, and Tulare counties), with adults active from late May to July. The female of *T. leechi* has fully developed metathoracic wings (Fender 1975). *Troglomethes oregonensis* Wittmer is known only from western Oregon (Benton, Clatsop, Columbia, Lane and Tillamook counties), with adults active in June and July. Fender (1975) described the female of *T. oregonensis*, which has reduced metathoracic wings. Fender (1975) reported that he and Wittmer collected a series of both sexes of *T. oregonensis* by sweeping herbaceous vegetation alongside a river. The cranial excavations of the males may be the site of pores for pheromone secretion and probably serve a role in courtship behavior (Wittmer 1970). Males of the two species can be readily separated by the different structure of these excavations (Figs. 5, 6). Females of both species lack cranial excavations.

Matheteinae LeConte 1881

Matheteini LeConte 1881

LeConte (1874) initially suggested that his genus *Mathetues* was probably most closely related members of the tribe Lampyrini (= Lampyridae *sensu stricto*, in part). LeConte (1881) then placed *Matheteus*, along with *Pollaclasis* Newman, in a new tribe, Mathetei (= Matheteini), of the family Lampyridae. VanDyke (1918), following his description of the genus *Ginglymocladus*, suggested that it was probably most closely related to *Matheteus*. Green (1948), recognizing the distinctiveness of the group, formally combined *Matheteus* and *Gingylomocladus* in a newly elevated subfamily Matheteinae of Lampyridae. Green (1948) also noted that *Pollaclasis* should not have been associated with this group and transferred it to the subfamily Amydetinae (Lampyridae). Matheteinae was retained as a subfamily of Lampyridae (Arnett 1963, Fender 1962, McDermott 1964) until Crowson (1972) transferred it to his new family Omethidae.

Ginglymocladus VanDyke 1918
Two species: endemic to California. Species descriptions: VanDyke (1918). *Ginglymocladus discoidea* VanDyke is known from coastal northern California (Butte, Del Norte, Humboldt, Marin, Mendocino counties), where it has been collected from late March to May. In the original description, it was stated that *G. discoidea* was collected "crawling over herbage" (VanDyke 1918: 4). *Ginglymocladus luteicollis* VanDyke is known from Sequoia National Park in the Sierra Nevada Mountains (Tulare County), where it has been collected from May to June. The holotype of *G. luteicollis* is of questionable provenance, but was "probably" collected in "the middle Sierras" in Tuolumne County (VanDyke 1918: 5). The females of both species are undescribed.

Matheteus LeConte 1874
One species, *Matheteus theveneti* LeConte, occurs from central California (El Dorado, Fresno, Humboldt, Madera, Mariposa, Santa Cruz, Tulare and Tuolumne counties) to southwestern Oregon (Coos County). Adults have been collected from March to July. Fender (1969) provided the only information available about the ecology of *M. theveneti*. Fender reported that, in Oregon, the adult activity period lasts from 10 days to two weeks, usually in late April. Fender suggested that the remarkable red coloration of this species, which would typically be interpreted as aposematic, might actually be an adaptation for crypsis among the fallen leaves of Salmonberry, *Rubus spectabilis* Pursh. The female of *M. theveneti* was briefly described as "winged and similar to the male" by Crowson (1972: 59).

Appendix

Complete label data for the new state records are as follows:

Blatchleya gracilis (Blatchley).
OHIO, Hocking Co., 2-June, col. D. J. Knull. (California Academy of Sciences).

Malthomethes oregonus Fender.
CALIFORNIA, Humboldt Co., Arcata, 21-July-1969, col. W. G. Goodman. (Bohart Museum of Entomology, University of California, Davis).

Omethes marginatus LeConte.
MARYLAND, Baltimore Co., Baltimore, May-1902, col. E. C. VanDyke. (California Academy of Sciences); NEW JERSEY, Warren Co., Phillipsburg, 8-June-1918, col. J. W. Green. (California Academy of Sciences); PENNSYLVANIA, Northampton Co., Wind Gap, 25-June-1931, col. J. W. Green. (California Academy of Sciences).

Bibliography

ARNETT, R. H., Jr. 1963. The beetles of the United States (a Manual for Identification). Catholic University of America Press. Washington, DC. xi + 1112 pp.

BLATCHLEY, W. S. 1910. An illustrated descriptive catalogue of the Coleoptera or beetles (exclusive of the Rhyncophora) known to occur in Indiana. The Nature Publishing. Indianapolis. 1386 pp.

BRANCUCCI, M. 1980. Morphologie Comparée, Évolution et Systématique des Cantharidae (Insecta: Coleoptera). Entomologica Basiliensia, 5: 215-388.

CROWSON, R. A. 1972. A review of the classification of Cantharoidea (Coleoptera), with the definition of two new families, Cneoglossidae and Omethidae. Revista de la Universidad de Madrid, 21: 35-77.

DOWNIE, N. M. and R. H. ARNETT, Jr. 1996. The beetles of northeastern North America, Volume I. Introduction, suborders Archostemata, Adephaga, and Polyphaga thru superfamily Cantharoidea. Sandhill Crane Press. Gainesville, FL. xiv + 880 pp.

FENDER, K. M. 1962. Family Lampyridae. Pp. 35-44. *In:* M. H. Hatch. Beetles of the Pacific Northwest. Part III: Pselaphidae and Diversicornia I. University of Washington Press. Seattle. 503 pp.

FENDER, K. M. 1969. Notes on *Matheteus theveneti* LeConte (Coleoptera: Lampyridae). Coleopterists Bulletin, 23: 52.

FENDER, K. M. 1975. Notes and descriptions of some North American Omethinae (Coleoptera: Omethidae). Pan-Pacific Entomologist, 51: 298-302.

GREEN, J. W. 1948. Two new species of Lampyridae from southeastern Florida, with a generic revision of the Nearctic fauna. Transactions of the American Entomological Society, 74: 61-73.

KNAB, F. 1910. Book review: "The Coleoptera or Beetles of Indiana". Science, 32: 838-840.

LAWRENCE, J. F. 1987. Rhinorhipidae, a new beetle family from Australia, with comments on the phylogeny of the Elateriformia. Invertebrate Taxonomy, 2: 1-53.

LAWRENCE, J. F. 1991. Omethidae (Cantharoidea). Pp. 428-429. *In:* F. W. Stehr, ed. Immature Insects. Vol. 2. Kendall/Hunt. Dubuque, IA.

LAWRENCE, J. F., A. M. HASTINGS, M. J. DALLWITZ, T. A. PAINE and E. J. ZURCHER. 1999. Beetles of the World: a key and information system for families and subfamilies. CD-ROM, Version 1.1 for MS-Windows. CSIRO Publishing. Melbourne.

LAWRENCE, J. F. and A. F. NEWTON JR. 1995. Families and subfamilies of Coleoptera (with selected genera, notes, references and data on family-group names). Pp. 779-1006. *In:* J. Pakaluk and S. A. Slipinski, eds. Biology, phylogeny, and classification of Coleoptera: Papers celebrating the 80th birthday of R. A. Crowson. Muzeum i Instytut Zoologii PAN. Warsaw.

LECONTE, J. L. 1861. Classification of the Coleoptera of North America. Prepared for the Smithsonian Institution. Part 1. Smithsonian Miscellaneous Collections, 3: 1-208.

LECONTE, J. L. 1874. Descriptions of new Coleoptera, chiefly from the Pacific Coast of the North America. Transactions of the American Entomological Society, 5: 43-72.

LECONTE, J. L. 1881. Synopsis of the Lampyridae of the United States. Transactions of the American Entomological Society, 9: 15-72.

LEWIS, G. 1895. On the Dascillidae and malacoderm Coleoptera of Japan. Annals and Magazine of Natural History, 6: 98-122.

McDERMOTT, F. A. 1964. The taxonomy of the Lampyridae (Coleoptera). Transactions of the American Entomological Society, 90: 1-72.

NAKANE, T. 1950. New or little known Coleoptera from Japan and its adjacent regions. IV* -Drilidae-. Mushi, 21: 29-31.

VANDYKE, E. C. 1918. Some new beetles in the families Cantharidae (Lampyridae), Ptinidae, and Scarabaeidae, from western North America, with notes upon others. Bulletin of the Brooklyn Entomological Society, 12: 1-15.

WITTMER, W. 1944. Catalogue des Drilidae E. Oliv. (Coleoptera-Malacodermata). Revista de la Sociedad Entomologica Argentina, 12: 203-221.

WITTMER, W. 1948. Supplement au Catalogue des Drilidae (1) E. Oliv. (Col. -- Malacodermata). Revista de la Sociedad Entomologica Argentina, 14: 115-116.

WITTMER, W. 1956. Drilidae, Cantharidae, Prionoceridae von Ceylon (Col.). (17. Beitrag zur Kenntis der indo-malayischen Malacodermata). Verhandlungen der Naturforschenden Gesellschaft in Basel, 67: 49-52.

WITTMER, W. 1957. 18. Beitrag zur Kenntnis der indo-malayischen Malacodermata (Col.) (2. Teil)*. Annals and Magazine of Natural History, 10: 567-578.

WITTMER, W. 1970. On some Cantharidae (Coleoptera) of the United States. Coleopterists Bulletin, 24: 42-46.

WITTMER, W. 1995. Lampyridae und Omethidae aus dem indo-malaiischen Faunengebiete (Coleoptera). Mitteilungen der Entomologischen Gesellschaft Basel, 45: 106-117.

64. CANTHARIDAE Imhoff 1856

by Alistair S. Ramsdale

Family common name: The soldier beetles

Family synonyms: Telephoridae Leach 1815, in part; including Chauliognathidae LeConte 1861

Cantharidae are a diverse group of soft-bodied, often aposematically colored, terrestrial beetles. They are often abundant on foliage and flowers, where they feed on other insects, nectar and pollen. Cantharidae are best distinguished from other families of Coleoptera by the following combination of characters: head not completely concealed beneath pronotum; antennae 11-segmented, filiform in most, can be serrate, pectinate or flabellate; labrum membranous and often concealed beneath clypeus; mesothoracic coxae contiguous or nearly so; tarsal formula 5-5-5; abdomen with 7 or 8 ventrites; and abdominal tergites 1-8 with paired lateral glandular pores.

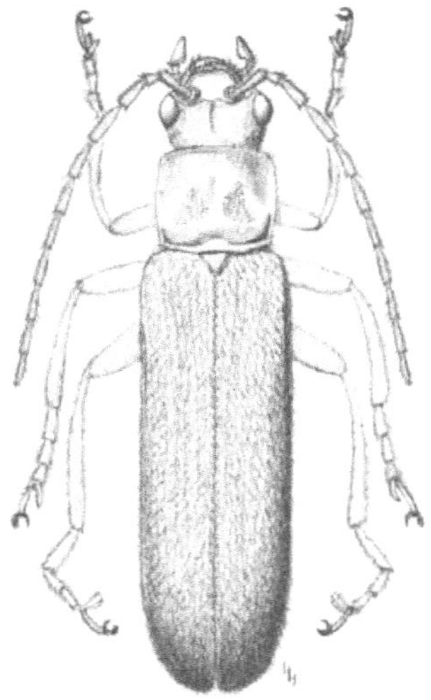

FIGURE 1.64. *Cyrtomoptera divisa* (LeConte) (after Fender 1962)

Description: Length 1.2 to 18 mm (to 28 mm in exotic species). Body elongate, more or less parallel-sided, most moderately flattened. Soft-bodied, with cuticle, especially on elytra and abdomen, lightly sclerotized, flexible. Coloration highly variable, from entirely black to predominantly red or orange, with aposematic coloration throughout body and elytra (e.g., most *Chauliognathus*) or restricted to the pronotum (e.g., many *Podabrus*, *Rhagonycha*, *Silis*). Body in most with moderately dense, short, depressed pubescence.

Head not completely concealed by pronotum, at least partially visible from above; head strongly prognathous (e.g., *Chauliognathus*, *Podabrus*) to somewhat deflexed (e.g., *Silis*), rarely hypognathous (*Ichthyurus*), someare abruptly constricted behind eyes (Podabrini). Compound eyes more or less entire, moderately to strongly protruding, often sexually dimorphic (smaller in females); ocelli absent. Antennae elongate, pubescent, with eleven antennomeres, filiform in most, infrequently serrate (e.g., most *Tytthonyx,* some *Ditemnus*), rarely pectinate or flabellate (*Tytthonyx* subgenus *Thinalmus*); flagellomeres in some with longitudinal sensory grooves (e.g., *Polemius*); antennal insertions exposed from above, most with insertions moderately to widely separated, rarely contiguous (*Ichthyurus*). Frontoclypeal (epistomal) suture absent in most (present in Chauliognathini); anterior margin of clypeus varied, entire to distinctly emarginate. Labrum free, membranous, often concealed beneath clypeus or apparently absent. Mandibles elongate in most, curved, with an acute apex; mola and prostheca absent; most with mandibles simple, falciform (e.g., *Podabrus*), with inner margin unidentate in some (e.g., *Frostia*, most *Chauliognathus*), uncommonly with inner margin serrate (e.g., some *Malthodes*). Maxillae with densely setose galea and lacinia in most, some with lacinia reduced or absent, some with apical lobes fused; maxillary palpi 4-segmented, apical palpomere securiform to cultriform, some ovate and terminating in an acute point (Malthininae). Labium with ligula undivided, finely cleft, or deeply bilobed; labial palpi 3-segmented, apical palpomere slightly to strongly expanded. Gular sutures distinctly separated (e.g., Cantharini), confluent (e.g., Podabrini) or apparently absent.

Pronotum subquadrate (e.g., most *Rhagonycha*) to transverse (e.g., most *Silis*), rarely narrowly elongate (few *Podabrus*); pronotum laterally margined, lateral margins entire, some are slightly to strongly explanate; lateral margins of some males shallowly incised (*Discodon*, most *Polemius*) or strongly excavated and modified into protuberances (*Ditemnus*, *Silis*); posterior margin of pronotum as wide or narrower than base of elytra. Prothoracic coxae conical, prominent, projecting well below prosternum, attached externally, contiguous or nearly so; prothoracic coxal cavities absent (appearing broadly open); prothoracic trochantin exposed. Mesothoracic coxae contiguous or nearly so, prominent in most, cavities open laterally. Scutellum well developed and visible. Metasternum with lateral margins sinuate; metathoracic coxae flattened, contiguous or narrowly separated. Most with legs elongate, slender, rarely with male mesofemora strongly dilated (some Ichthyurini); paired tibial spurs present in most,

Acknowledgments: I thank D. K. Young for his encouragement and for patiently commenting on several versions of the manuscript, S. Kazantsev for many informative discussions about cantharid systematics, for sharing some of his preliminary thoughts on the Nearctic Cantharini and for commenting on the manuscript, M. Brancucci for permitting me to use many of the figures from his dissertation, M.C. Thomas for giving me the opportunity to contribute to this book, P. J. Johnson for the extended loan of his virtually synoptic collection of North American Cantharidae (much of which came from the private collection of K. M. Fender), K. Katovich for producing most of the original figures, J. H. Frank, N. L. Kriska, and P. E. Skelley for commenting on the manuscript, and J. P. Gruber for testing the keys.

(absent in Chauliognathini). Tarsal formula 5-5-5; fourth tarsomere expanded and ventrally bilobed; claw structure highly variable, often sexually dimorphic (e.g., most Cantharinae and Silinae); claws simple, toothed, lobed or cleft.

Elytra soft, more or less parallel-sided, rarely somewhat expanded apically; elytra loosely fitted to pterothorax and abdomen; elytra either complete and normally elongate or abbreviated (brachelytrous) exposing multiple abdominal tergites and/or metathoracic wings (e.g., Malthininae, Ichthyurini); most with elytra lacking prominent costae; vestiture and degree of sculpturing highly varied. Metathoracic wings almost always well developed.

Abdomen with 7 (females, some males) or 8 (most males) freely moveable ventrites (visible sternites) with complete sutures; tergites 1-8 with paired lateral glandular pores. Males with caudal abdominal segment(s) often variously modified: ventrite 8 reduced to a narrow process or apparently absent and more or less concealed by deeply emarginate ventrite 7 (Silini); or ventrite 8 small, asymmetrical, with margins more or less entire (Chauliognathini); or caudal abdominal segment with a pair of dissimilar asymmetrical ventral lamellae (*Trypherus*); or caudal abdominal segments modified into an elaborate copulatory apparatus comprised of a modified, elongate-narrow and frequently bifurcated ventrite and some with variously modified tergite and/ or additional accessory processes (*Malthodes*); or (both sexes) with caudal abdominal segment posteriorly bifurcated and produced into a pair of elongate processes (*Ichthyurus*).

Male genitalia (aedeagus) of trilobed type; most consisting of a large and well sclerotized tubular tegmen which envelops the inner, often membranous, cylindrical median lobe. Tegmen composed of the phallobase (basal piece) and lateral lobes (parameres). Phallobase varied; often reduced, most membranous, with two lateral sclerites (most Cantharinae, Silinae), some are enlarged, strongly sclerotized, and produced ventrally (most Malthininae), or strongly sclerotized and fused to lateral lobes (Chauliognathinae). Lateral lobes in most large and strongly sclerotized (e.g., Cantharinae, Silinae), often fused, in some forming dorsal and ventral plates (e.g., many Silinae), and/or diverging apically into elongate paired processes. Median lobe (penis) membranous in most (e.g., Silinae, Cantharinae), sclerotized in some (Chauliognathinae); median lobe in most with eversible internal sac, highly variable in structure and often ornamented with asperities and/or elongate sclerotized processes; internal sac permanently invaginated in some (e.g., Malthininae). Additional sclerotized genitalic processes arising from interior of tegmen, termed basophyses and laterophyses, often present. Male genitalia (especially tegmen) of Chauliognathinae distinctively asymmetrical; tegmen produced into two dissimilar elongate processes, of which one is flexible and the other fixed.

Female genitalia (ovipositor): Coxites well developed in most, enlarged in some (e.g., Chauliognathinae); most with coxites terminating in a setose stylus (reduced or absent in Silinae); some with stylus enlarged and more or less fused to coxites (Malthininae, Tytthonyxini). Proctiger typically more or less triangular, usually well developed and sclerotized, sometimes membranous (e.g., Cantharinae) or reduced (Chauliognathinae, some Silini, Malthininae). Paraprocts well developed in most, some emarginate or divided (e.g., some Malthininae), expanded and fused to valvifers (Silini), reduced (Chauliognathinae) or enlarged (Dysmorphocerinae). Valvifers well developed and sclerotized in most (e.g., Cantharinae), reduced and/or fused to the paraprocts and coxites in some (Silini, Chauliognathinae). Additional structures, termed cupuliform sclerites and supplemental sclerotized piece present in some.

Brancucci (1980) contains a highly detailed and well illustrated treatise on the comparative anatomy of adult Cantharidae. Lawrence *et al.* (1999b) includes excellent descriptions of adults at the familial and subfamilial or tribal levels and habitus illustrations or photographs of numerous genera. The best sources of habitus illustrations of members of the Nearctic fauna are Dillon and Dillon (1961), Fender (1962) and White (1983).

Larva campodeiform, with a characteristic dense vestiture of hydrophobic setae. Head strongly prognathous; epicranial suture absent; with a single large stemmata on each side. Antennae 3-segmented, segment 2 truncate with sensorium terminal (Cantharinae, Silinae, Chauliognathinae), or segment 2 strongly emarginate with sensorium elongate and inserted subapically (Malthininae). Frontoclypeal suture indistinct or absent; clypeolabral suture absent. Labrum fused to head capsule forming a nasale; structure of nasale varied. Mandibles falciform, with more or less well developed longitudinal channel, channel broadly open in most, some nearly closed (e.g., *Malthinus*), retinaculum present in some, mola and prostheca absent, mesal margin with or without fringe of setae. Maxillae and labium contained in an anterior subfacial sinus. Maxillae with cardo small, stipes well developed, most with a 1-segmented palpiform mala (galea), mala absent in some (e.g., *Chauliognathus*); maxillary palpi 3- or 4-segmented. Labium comprised of prementum and postmentum, labial palpi 2-segmented, ligula absent. All thoracic terga and abdominal tergites 1-8 with paired lateral glandular pores (in some also present on tergite 9). Legs 5-segmented, tarsungulus present, tarsungulus with 4 or more setae. Spiracles biforous. Urogomphi absent; tenth abdominal segment rarely with sclerotized anal hooks (some *Malthinus*).

Crowson (1972) briefly described a larva of *Neoontelus* (Dysmorphocerinae) (Wittmer 1979), from New Zealand, which differs substantially from those described above, having a vestiture of scales instead of setae, dorsal surface with asperities, stemmata reduced, antennal sensorium extremely elongate-narrow, mandibles extremely slender, meso- and metatergum and abdominal tergites 1-8 with double paired glandular pores, abdominal tergites 1-8 with lateral outgrowths, and abdominal sternite 9 with ventral glandular pores.

Habits and habitats. Most adult cantharids frequent living vegetation and are often abundant in a wide variety of open and forested habitats. Some species are commonly encountered while they visit flowers. In arid regions, cantharids often prefer more riparian situations where they can be found on herbs, shrubs, and trees in close proximity to water. Adults of most species emerge in the spring or summer and are fairly short lived. Most species appear to be predominantly active diurnally, although there is usually a period of inactivity during the hottest part of the day in extremely warm weather. Some adults are attracted to lights at night.

Adult cantharids whose feeding habits are known prey on other insects or feed on nectar or pollen. Cantharini and Podabrini are predominantly fluid feeders, feeding on liquids with high nutrient quality, which they obtain by preying on other insects or by visiting flowers richly supplied with nectar (e.g., Apiaceae, Asteraceae). Adults of *Dichelotarsus* and *Podabrus* are primarily predators of aphids (Hemiptera: Aphididae) and other small, soft-bodied, foliage frequenting insects, but will also feed on nectar. Pollen forms a substantial part of the diet of Chauliognathini. Adults of *Chauliognathus* apparently feed exclusively on pollen and nectar. *Chauliognathus* species are very common on the flowers of a wide variety of plants and are important native pollinators. As a result of these feeding preferences, cantharids are among the most active fliers in the order Coleoptera. Nothing conclusive is known about the feeding habits of many groups (e.g., Silinae, Malthininae, Ichthyurini).

There is little information available about cantharid-plant associations, and it is suspected that specificity is minimal and that, in most cases, the association of adults with particular plants or habitat types is determined by larval requirements.

Many adult cantharids are highly active and spend a great deal of time exposed to predators on the surfaces of vegetation. They have therefore evolved an effective system of chemical defense. Adult and larval cantharids possess paired tergal glands which secrete repugnatorial compounds that serve to reduce their palatability to predators. Cantharids produce various defensive compounds, and it is suspected that these compounds are, in part, synthesized from dietary material (Dettner 1987). Adults of many species possess aposematic coloration, and some species are involved in mimicry complexes with other Cantharidae and members of many other families of Coleoptera, including Elateridae, Lycidae, Lampyridae, Omethidae, Cleridae, Melandryidae, Oedemeridae, Stenotrachelidae, Pyrochroidae, Cerambycidae, and Chrysomelidae.

Adults of some species of *Podabrus*, *Rhagonycha* and *Chauliognathus* are known to be subject to infections by fungal entomopathogens (Zygomycota: Entomopthorales) (Wheeler 1988). Infected beetles are typically found dead, attached to leaves by their mandibles, with their wings partially spread open (Fender 1969c).

Larvae are predominantly geophilic, preferring microhabitats with high relative humidity, and are common in leaf litter, loose soil, beneath rocks or debris, and in association with decaying wood. The dense larval hydrophobic vestiture is probably an adaptation for improved survival in areas prone to flooding. Some larvae will openly wander, probably foraging, exposed on the surface of the ground. Larvae of some species appear to be more active nocturnally.

Cantharid larvae are thought to be predominantly fluid-feeding predators. Larvae have been observed preying upon various arthropods, primarily the eggs and larvae of holometabolous insects. Larvae of *Chauliognathus* are known to be voracious predators and have even been observed preying on adult Chrysomelidae (Balduf 1935) and the larvae of cicindeline Carabidae (Schultz 1994). The eggs of Orthoptera are also commonly reported as food items. When artificially kept in close proximity (e.g., in rearing containers), larvae will sometimes engage in cannibalism. Larvae of some other genera (e.g., *Silis*) are probably at least facultatively phytophagous.

Overwintering occurs in the larval stage, and some larvae are known to remain active during winter even at extremely low temperatures. Pupation typically takes place in earthen cells below the surface of the soil. There can be as many as 10 instars. The first two or three instars of some Cantharinae are non-feeding and have reduced antennae, mouthparts and legs. Whereas all larvae are presumably protected by repugnatorial compounds, only those of *Chauliognathus* are known to exhibit aposematic coloration.

Our knowledge of cantharid ecology is extremely poor. This is further complicated by the fact that many of the historical published accounts of their natural history were based on ambiguously or unreliably determined species. Additionally, many of the published accounts on larvae were based on material not reared to adulthood, and can therefore not be reliably associated with any particular genus, despite the claims of the authors. Much additional study in this area is needed.

Status of the classification. For much of the Nineteenth Century, Cantharidae (typically known by its junior synonym Telephoridae) were included in the Malacodermata, a large and ill-defined assemblage of superficially similar soft-bodied beetles. LeConte (1851, 1881) treated cantharids as a subfamily (Telephorinae) of a broadly defined Lampyridae, and twice revised the known Nearctic fauna. More recently, Cantharidae were included in the superfamily Cantharoidea, as defined by Crowson (1955, 1972). Cantharoidea were composed of the following families: Cantharidae, Cneoglossidae, Drilidae, Lampyridae Lycidae, Omalisidae, Omethidae, Phengodidae, Plastoceridae and Telegeusidae. Crowson (1972) suggested that Cantharidae might be most closely related to Omethidae (q.v. Chapter 63, Omethidae, Status of the classification).

Lawrence (1987) treated the Cantharoidea as a single clade but suppressed the taxon, placing all the former Cantharoidea in the Elateroidea along with the families Artematopodidae, Brachypsectridae, Cerophytidae, Elateridae, Eucnemidae, and Throscidae. Synapomorphies for this assemblage include various modifications of the larval mouthparts to adapt for fluid feeding and extra-oral digestion (Lawrence and Newton 1982). Lawrence and Newton (1995), the classification adopted for this volume, maintained this broad sense of the Elateroidea. A recent phylogenetic analysis based exclusively on larvae suggested that Cantharidae may belong in a clade with Artematopodidae and Elateridae (Beutel 1995). At this stage, relationships within the Elateroidea (*sensu lato*) have not been satisfactorily elucidated, and the monophyly of the former Cantharoidea is uncertain (Beutel 1995, Lawrence 1987).

The family Cantharidae is almost certainly a monophyletic taxon, supported by the following putative synapomorphies:

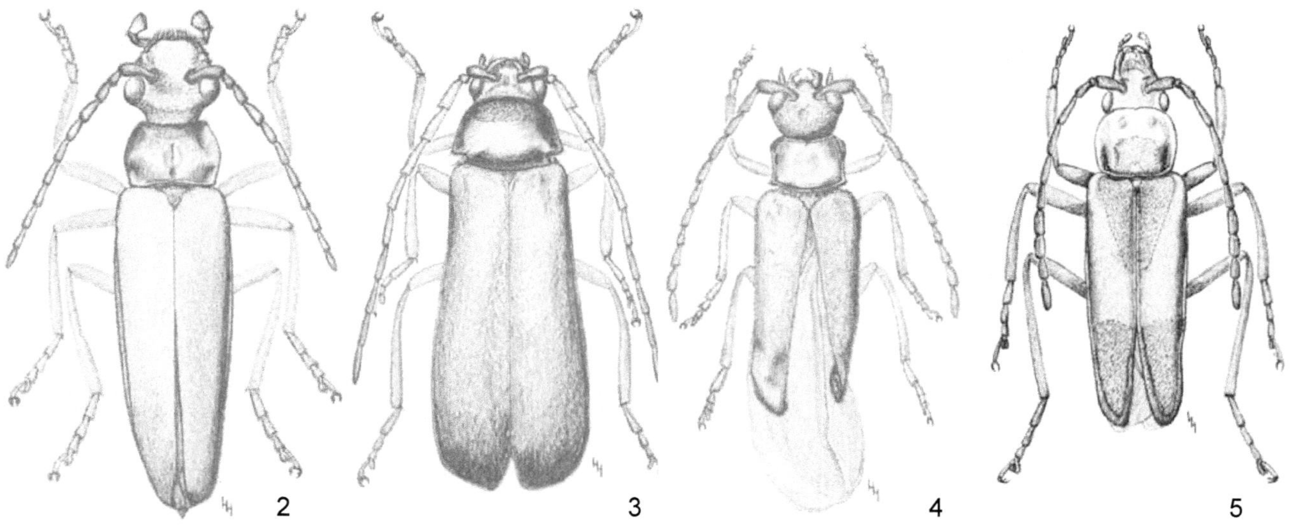

FIGURES 2.64-5.64. 2. *Podabrus pruinosus* LeConte, dorsal habitus); 3. *Silis difficilis* LeConte, dorsal habitus; 4. *Frostia malkini* Fender, dorsal habitus; 5. *Chauliognathus basalis* LeConte, dorsal habitus (all modified from Fender 1962).

labrum membranous (adult), abdominal tergites with paired repugnatorial glands (larval and adult), and hydrophobic vestiture (larval) (Brancucci 1980).

There have been substantial advances in our knowledge of the systematics of Cantharidae since the first edition of this book (Arnett 1963). The monumental body of taxonomic work produced by W. Wittmer during the latter half of the past century forms the basis for our understanding of the diversity of the family at the world level. A phylogenetic analysis and higher classification of the family was presented by Brancucci (1980). The family is now understood to consist of five subfamilies, four of which are represented in North America north of Mexico. The subfamily Cantharinae is a particularly problematic group, currently defined with only plesiomorphic characters. The affinities of the enigmatic genus *Tytthonyx* remain uncertain. The subfamily Chauliognathinae, treated as a distinct family by Miskimen (1961a), is thought to be the sister group of the remaining subfamilies (Brancucci 1980). The subfamily Dysmorphocerinae, previously thought to have a representative in Arizona (Brancucci 1980), does not actually occur in the United States. The single species in question, placed in the genus *Plectonotum* by Schaeffer (1908), must be transferred to the Silinae and will almost certainly necessitate the erection of a new genus for its reception.

Extensive study of the North American cantharid fauna was conducted during the last century by two American workers, K. M. Fender and J. W. Green, whose work epitomizes the tremendous contribution made to North American beetle systematics by avocational coleopterists. Despite this, several elements of the North American fauna are poorly known and in need of revision, including the species of *Cantharis*, *Rhagonycha*, *Dichelotarsus*, *Podabrus*, *Discodon*, *Polemius* and *Malthodes*, and the generic classification of Cantharini, Podabrini and Silini. Research in progress indicates that many new species (in various genera) await description, and that substantial changes may have be made to the generic classification presented in this summary.

Very little research has been conducted on the immature stages of these beetles, and the larvae of most North American genera are undescribed. There has never been any serious attempt to include characters from the immature stages in a phylogenetic analysis of the family. The only larval keys of any usefulness for North American taxa are those of Fitton (1975), which separates the genera present in Britain, and Bøving and Craighead (1930). Additional important references on the immature stages are Lawrence *et al.* (1999a), LeSage (1991) and Verhoeff (1917, 1923).

Many genera require the extraction and examination of male genitalia for species-level determination, making identification of females only possible when they are found in association with males. Cantharids should therefore, when possible, be collected in small series to help ensure that both sexes are present in a sample.

The earliest cantharids known have been found in Baltic amber from the early Tertiary (Eocene/Oligocene) (Crowson 1972, Ponomarenko 1995). There is, however, a good possibility that the absence of Cantharidae and related families from older deposits is an artifact of the poor fossilization potential of their lightly sclerotized cuticle and is not a reflection of their true evolutionary history (Miller 1991). The disjunct austral distribution of Dysmorphocerinae (Australia, New Zealand, South America and southern Africa) strongly indicates that its origin antedated the breakup of Gondwana.

Distribution. Cantharidae are the most speciose and widely distributed family of the former Cantharoidea (Crowson 1972). There were 5083 described species in 137 genera, present in all major zoogeographic regions, at the time of the compilation of the world catalog (Delkeskamp 1977, 1978). In America north of Mexico, there are currently 473 species in 23 genera.

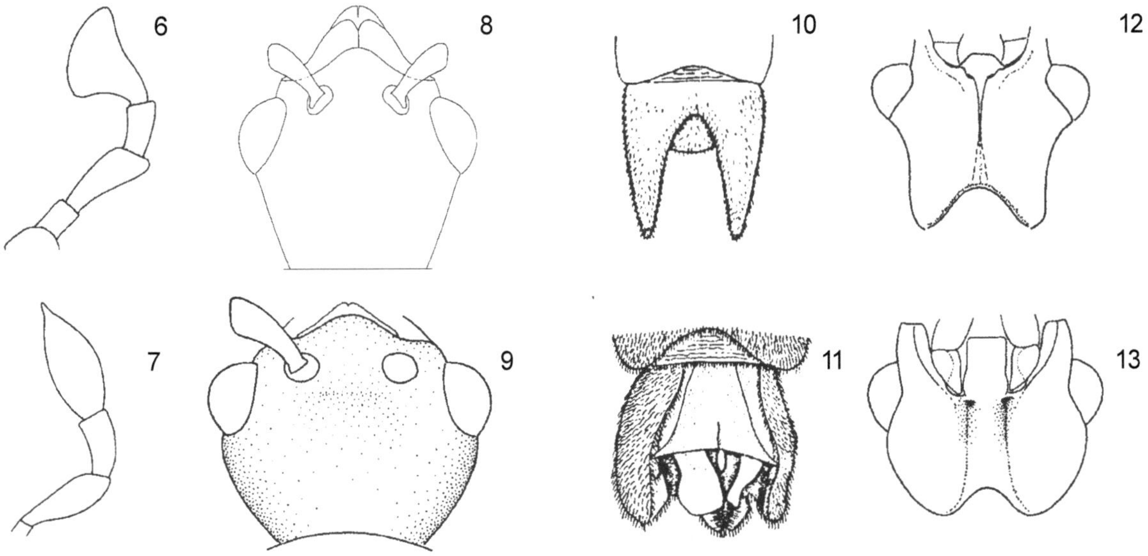

FIGURES 6.64-13.64. 6. *Cantharis (Cantharis) fusca* Linnaeus, maxillary palpus (modified from Brancucci 1980); 7. *Malthinus luteus* (Pic), maxillary palpus (modified from Brancucci 1980); 8. *Chauliognathus marginatus* (Fabricius), head, dorsal view (modified from Brancucci 1980); 9. *Silis nitidula* (Fabricius), head, dorsal view (modified from Brancucci 1980); 10. *Ichthyurus arizonensis* Fender, female caudal abdominal segments, dorsal view (after Fender 1960); 11. *Trypherus blaisdelli* Fender, male caudal abdominal ventrites, ventral view (after Fender 1960); 12. *Podabrus alpinus* (Paykull), head, ventral view (modified from Brancucci 1980); 13. *Rhagonycha fulva* (Scopoli), head, ventral view (modified from Brancucci 1980).

Extensive information about the distribution of the world fauna is contained in the recent Coleopterorum Catalogus (Delkeskamp 1977, 1978). A useful checklist of the fauna of Alaska and Canada (by province) was presented by McNamara (1991). Miskimen (1956) produced a faunal list of the Cantharidae of Ohio. Fender (1962) is an excellent, though somewhat out of date, source of information on the distribution and taxonomy of species present in the Pacific Northwest. Peck and Thomas (1998) included a provisional checklist of the Cantharidae of Florida. Downie and Arnett (1996) provide some useful distribution records for the Cantharidae of northeastern North America, but use of the keys presented therein is not recommended, as they contain numerous errors, were based on out of date literature, and are not accompanied by (the often necessary) illustrations of male genitalia.

KEY TO THE SUBFAMILIES, TRIBES, GENERA, AND SUBGENERA OF AMERICA NORTH OF MEXICO

1. Terminal maxillary palpomere bilaterally symmetrical, in most securiform or cultriform (Fig. 6); elytra varied; metathoracic wings in most concealed; size varied 2

— Terminal maxillary palpomere radially symmetrical, ovate and terminating in an acute point (Fig. 7); elytra more or less abbreviated; metathoracic wings in most exposed; size small 1.2 to 5mm 8

2(1). Frontoclypeal suture present (Fig. 8); mentum elongate in most; tibial spurs absent; male eighth (caudal) ventrite and genitalia asymmetrical (Chauliognathinae: Chauliognathini) 29

— Frontoclypeal suture absent (Fig. 9); mentum not elongate; tibial spurs present in most; male eighth (caudal) ventrite and genitalia varied 3

3(2). Elytra strongly abbreviated; metathoracic wings largely exposed; caudal abdominal segments often irregularly modified (Figs. 10, 11); male genitalia asymmetrical (Chauliognathinae: Ichthyurini) 30

— Elytra almost always normally elongate, very rarely abbreviated; caudal abdominal segments varied (Figs. 14, 15, 16, 17, 18); male genitalia symmetrical 4

4(3). Gular sutures confluent (Fig. 12); head distinctly constricted behind eyes (Fig. 2) (Cantharinae: Podabrini) 19

— Gular sutures widely separated (Fig. 13); head not as above (Fig. 1) 5

5(4). Posterior margin of seventh abdominal ventrite deeply invaginated, appearing longitudinally divided or nearly so (Figs. 14, 15, 16); **or** eighth (caudal) abdominal ventrite small and subtriangular (Fig. 17) (males) 6

— Posterior margin of seventh (caudal) abdominal ventrite never deeply invaginated or appearing longitudinally divided; **and** seventh (caudal) abdominal ventrite normally broad (Fig. 18), not small and subtriangular (females) 7

6(5). Posterior margin of seventh abdominal ventrite deeply invaginated, appearing longitudinally divided or nearly so (Figs. 14, 15, 16); lateral pronotal margins almost always incised or otherwise irregularly modified (Figs. 19, 20, 21, 22, 23) (Silinae: Silini) 22

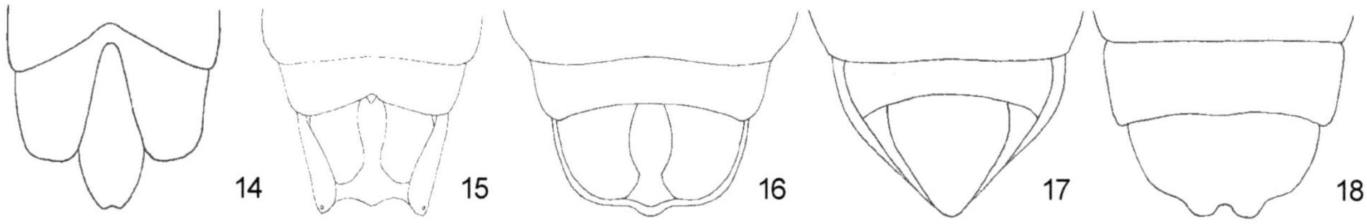

FIGURES 14.64-18.64. 14. *Silis percomis* (Say), male caudal abdominal ventrites, ventral view; 15. "*Plectonotum*" *excisum* Schaeffer, male caudal abdominal ventrites, ventral view; 16. *Polemius laticornis* (Say), male caudal abdominal ventrites, ventral view; 17. *Cantharis* (*Cantharis*) *rotundicollis* Say, male caudal abdominal ventrites, ventral view; 18. *Cantharis* (*Cantharis*) *rotundicollis* Say, female caudal abdominal ventrites, ventral view.

— Posterior margin of seventh abdominal ventrite entire, not as above; eighth (caudal) abdominal ventrite small and subtriangular (Fig. 17); lateral pronotal margins entire (Figs. 1, 2), never as above (Cantharinae: Cantharini) 10

7(5). Paired lateral glandular pores on abdominal tergites large and protruding (openings often concealed beneath elytra) (Fig. 24) .... (Silinae: Silini)*

— Paired lateral glandular pores on abdominal tergites not as above, small and inconspicuous (openings often concealed beneath elytra) (Cantharinae: Cantharini) 10

8(1). Antennae strongly serrate, pectinate or flabellate (Silinae, Tytthonyxini) 26

— Antennae filiform, rarely slightly subserrate (Malthininae) 9

9(8). Mandibles with a prominent tooth on the inner margin (Fig. 25); gular sutures confluent (Fig. 12); frons more or less convex (Malthinini) 27

— Mandibles simple, serrate, or with a tooth on the inflexed outer margin (Fig. 26); gular sutures almost always separated (Fig. 13); frons more or less concave (Malthodini) 28

Cantharinae: Cantharini

10(6,7). Third tarsomere simple, insertion of fourth tarsomere apical (Fig. 27) .... ***Rhagonycha***

— Third tarsomere emarginate, insertion of fourth tarsomere pre-apical (Fig. 28) 11

11(10). Mandibles with a prominent tooth; California to British Columbia .... *Cyrtomoptera*

— Mandibles simple; widely distributed 12

12(11). Both claws of all tarsi with an elongate basal tooth, making them appear broadly cleft (Fig. 29) .... *Rhaxonycha*

— All tarsal claws not as above 13

13(12). Anterior claws of all tarsi invariably with a short acute basal tooth (Fig. 30) .... "*Ancistronycha*"

— Anterior claws of all tarsi not exactly as above, i.e. anterior claws of some tarsi simple, or with a blunt (Fig. 31) or elongate (Fig. 32) basal tooth 14

14(13). Eighth (caudal) abdominal ventrite small and subtriangular (Fig. 17) (males) 15

— Seventh (caudal) abdominal ventrite normally broad, not small and subtriangular (Fig. 18) ...(females)*

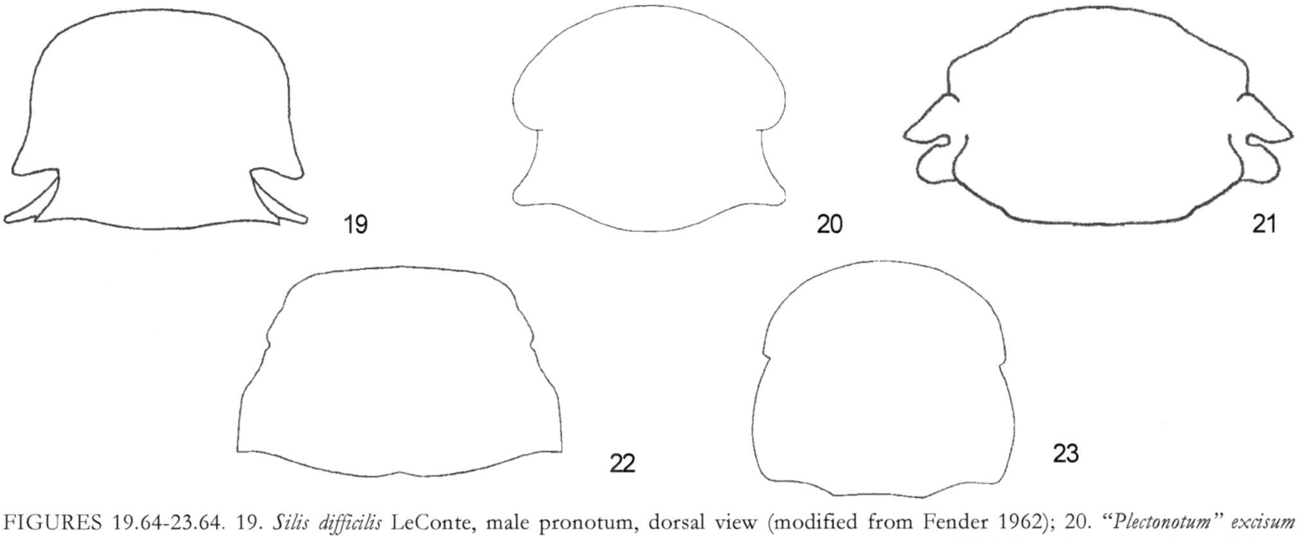

FIGURES 19.64-23.64. 19. *Silis difficilis* LeConte, male pronotum, dorsal view (modified from Fender 1962); 20. "*Plectonotum*" *excisum* Schaeffer, male ponotum, dorsal view; 21. *Ditemnus latilobus* Blatchley, male ponotum, dorsal view (modified from Green 1966); 22. *Polemius laticornis* (Say), male pronotum, dorsal view; 23. *Discodon planicolle* (LeConte), male pronotum, dorsal view.

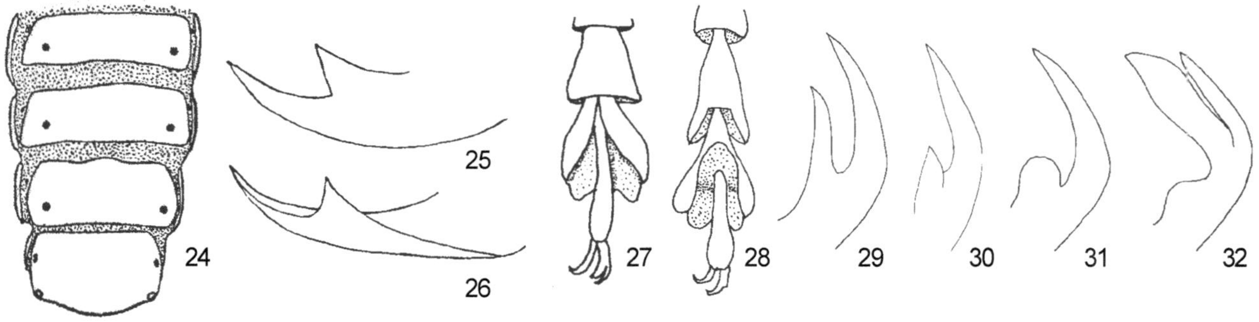

FIGURES 24.64-32.64. 24. *Crudosils ruficollis* (Fabricius), abdominal tergites, dorsal view (after Crowson 1972); 25. *Malthinus* sp., mandible, dorsal view (after Fender 1951); 26. *Frostia* sp., mandible, dorsal view (after Fender 1951); 27. *Rhagonycha recta* (Melsheimer), tarsomeres 3-5 (after Fender 1962); 28. *Cantharis (Cantharis) transmarina* (Motschulsky), tarsomeres 3-5 (after Fender 1962); 29. *Rhaxonycha carolina* (Fabricius), male anterior protarsal claw; 30. "*Ancistronycha*" *bilineata* (Say), male anterior protarsal claw (modified from McKey-Fender 1950); 31. *Cantharis (Cantharis) aneba* McKey-Fender, male anterior metatarsal claw (after Fender 1962); 32. *Cultellunguis larvalis* (LeConte), male anterior protarsal claw (after Fender 1962).

15(14). Anterior protarsal claws cleft, with one lobe elongate and cultriform and the other simple (Fig. 32), all other tarsal claws simple; California to Washington .... *Cultellunguis*
— Anterior protarsal claws not as above; widely distributed 16

16(15). Both claws of all tarsi simple; California 17
— At least one tarsal claw with a basal tooth; widely distributed 18

17(16). Pronotum longer than wide, narrower than head; male genitalia with laterophyses fused .... *Podistra (Pseudoabsidia)*
— Pronotum wider than long and wider than head; male genitalia with laterophyses separated .... *Cantharis (Cyrtomoptila)*

18(16). Anterior meso- and metatarsal claws simple; male genitalia with laterophyses fused; New Mexico .... *Podistra (Hemipodistra)*
— Anterior meso- and/or metatarsal claws typically with a basal tooth; male genitalia with laterophyses separated; widely distributed .... *Cantharis (Cantharis)*

Cantharinae: Podabrini

19(4). Eighth (caudal) abdominal ventrite small and subtriangular (Fig. 17); (males) 20
— Seventh (caudal) abdominal ventrite normally broad, not small and subtriangular (Fig. 18) ...(females)*

20(19). Male genitalia with laterophyses absent (Fig. 33) .... *Dichelotarsus*

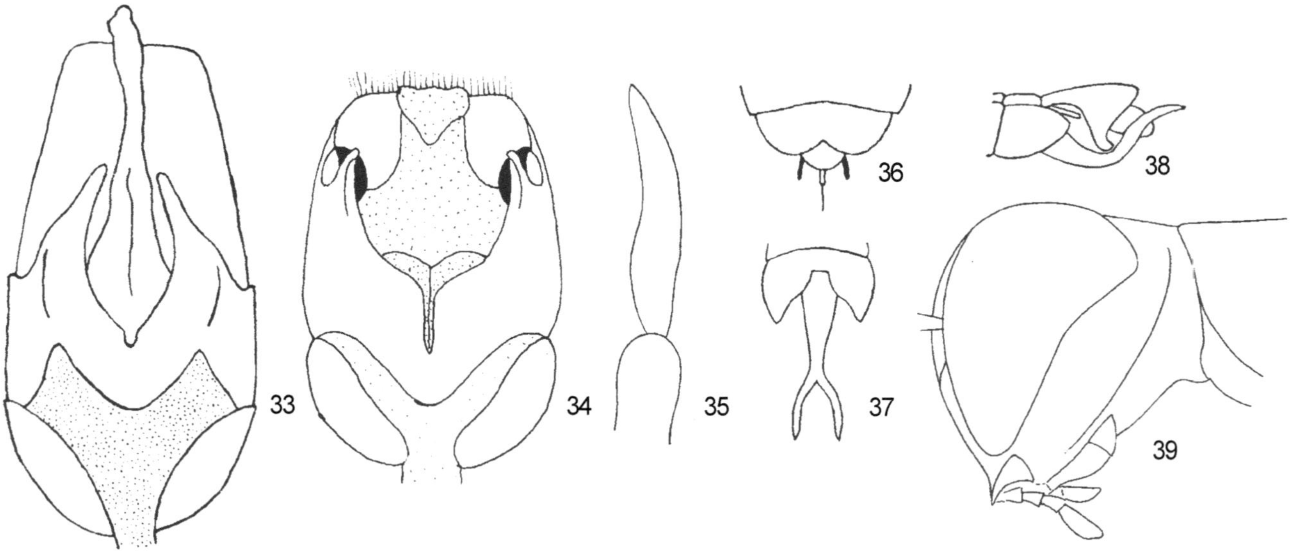

FIGURES 33.64-39.64. 33. *Dichelotarsus piniphilus* (Eschscholtz), male genitalia, ventral view (modified from Fender 1953); 34. *Podabrus sierrae* Fall, male genitalia, ventral view (after Fender 1979); 35. *Podabrus (Hatchiana) arizonensis* (Fender), apical antennomere (after Fender 1966b); 36. *Frostia laticollis* (LeConte), male caudal abdominal ventrites, ventral view (after Fender 1951); 37. *Malthodes forficatus* Fender, male caudal abdominal ventrites, ventral view (after Fender 1951); 38. *Malthodes forficatus* Fender, male caudal abdominal segments, lateral view (after Fender 1951); 39. *Ichthyurus lateralis* Westwood, head, lateral view (modified from Brancucci 1980).

— Male genitalia with laterophyses present (Fig. 34) 21

21(20). Apical antennal flagellomere distorted and tapered to an acute point (Fig. 35); Arizona .... *Podabrus (Hatchiana)*

— Apical antennal flagellomere normal, with apex rounded, not as above; widely distributed .... *Podabrus (Podabrus)*

Silinae: Silini

22(6). Seventh ventrite with invaginated posterior margin widely open posteriorly ("V-shaped"), eighth ventrite elongate but wide (Fig. 14); lateral margins of pronotum excavated and modified into angular processes in the region of the hind angles (Fig. 19) .... *Silis*

— Seventh ventrite with invaginated posterior margin narrowly open posteriorly ("U-shaped"), eighth ventrite reduced to a narrow process or apparently absent (Figs. 15, 16); lateral margins of pronotum varied (Figs. 20, 21, 22, 23) 23

23(22). Glandular pores on caudal visible abdominal tergite on sub-tuberculate protuberances (Fig. 15); caudal visible abdominal tergite strongly produced (Fig. 15); pronotum as in Fig. 20; Arizona .... *"Plectonotum"*

— Glandular pores on caudal visible abdominal tergite not as above; caudal visible abdominal tergite not produced as above (Fig. 16); pronotum varied (Figs. 21, 22, 23); widely distributed 24

24(23). Lateral margins of pronotum excavated and modified into angular processes (Fig. 21)... *Ditemnus*

— Lateral margins of pronotum shallowly incised in most (Fig. 22, 23), rarely entire, not modified as above 25

25(24). Anterior metatarsal claws apically cleft; most with anterior mesotarsal claws also apically cleft; Texas to Arizona .... *Discodon*

— Anterior meso- and metatarsal claws not apically cleft; all anterior tarsal claws with a broad basal tooth; widely distributed .... *Polemius*

Silinae: Tytthonyxini

26(8). Antennae of both sexes serrate; eastern U.S. to Texas .... *Tytthonyx (Tytthonyx)*

— Antennae of males pectinate, of females serrate; Arizona .... *Tytthonyx (Thinalmus)*

Malthininae: Malthinini

27(9). Elytra with coarse, striate punctures ..... *Malthinus*

— Elytra with fine, confused punctures ..... *Caccodes*

Malthininae: Malthodini

28(9). Inflexed outer margin of mandibles with a retinaculum (Fig. 26); pronotum with a moderately deep median depression; caudal abdominal segments of male never modified as below; elongate sclerotized processes of the lateral lobes of the male genitalia prominent, extending to or beyond the caudal abdominal ventrite (Fig. 36); California to Oregon .... *Frostia*

— Mandibles not as above, simple or finely serrate; pronotum without or with only a feeble median depression; caudal abdominal segment of male often modified into an elaborate copulatory apparatus of varied structure and degree of complexity; copulatory apparatus in most composed of a strongly modified, elongate, narrow and in some apically bifurcated caudal ventrite (Fig. 37), some also with a variously modified caudal tergite and/or additional accessory processes (Fig. 38); male genitalia without elongate sclerotized processes or with processes not prominent and not extending to or beyond caudal abdominal ventrite; widely distributed .... *Malthodes*

Chauliognathinae: Chauliognathini

29(2). Elytra abbreviated; Florida to Oklahoma ..... *Belotus*

— Elytra normally elongate (Fig. 5); widely distributed .... *Chauliognathus*

Chauliognathinae: Ichthyurini

30(3). Antennal bases contiguous; eyes very large (Fig. 39); caudal abdominal segment posteriorly bifurcated and produced into a pair of elongate processes (Fig. 10); Arizona .... *Ichthyurus*

— Antennal bases distinctly separated; eyes smaller; caudal abdominal segment of male with a pair of dissimilar asymmetrical ventral lamellae (Fig. 11), never modified as above in either sex; widely distributed (not in Arizona)................. *Trypherus*

* Many genera of Cantharini, Podabrini and Silini are currently based only on characters exhibited by males, and it is therefore not possible to reliably separate unassociated females in a key of this geographic scope.

Classification of the Genera of America North of Mexico

The subfamilial and tribal classification presented herein largely follows that of Brancucci (1980). Within each tribe, the genera are listed alphabetically. Only subgenera with representatives in America north of Mexico are listed. In the generic summaries, the numbers of species and the keys referenced (unless indicated otherwise) are for America north of Mexico only. The generic synonymies presented do not include unjustified emendations. Additional information provided includes, for each genus, all the species descriptions published after or not included in the most recent key, distribution records, ecological data, and other pertinent information. The biological information presented is for adults only, unless otherwise specified. No attempt has been made, with a few exceptions, to include lists of plant associations or prey records available from the literature and specimen label data, as they would be woefully incomplete and sometimes misleading. Oftentimes such plant associations do not represent feeding records and are, in fact, of no particular significance.

Cantharidae Imhoff 1856

Cantharinae Imhoff 1856 (not Thomson 1864)

Males of this subfamily can be distinguished by the following combination of characters: frontoclypeal suture absent; apical maxillary palpomere bilaterally symmetrical; elytra almost always normally elongate; seventh abdominal ventrite with posterior margin entire; and eighth (caudal) abdominal ventrite small and subtriangular. Cantharinae are widely distributed in the Holarctic and Oriental Regions, but are absent from Australia, South America and sub-Saharan Africa.

Cantharini Imhoff 1856

Male Cantharini can be distinguished by the following combination of characters: frontoclypeal suture absent; apical maxillary palpomere bilaterally symmetrical; gular sutures widely separated; elytra almost always normally elongate; seventh abdominal ventrite with posterior margin entire; and eighth (caudal) abdominal ventrite small and subtriangular.

As many elements of tribe are Holarctic in distribution, the key to the Palaearctic genera presented by Wittmer (1972), while somewhat outdated, is a useful reference.

Revisionary work on the Nearctic Cantharini is in progress, and it likely that the classification will undergo substantial changes (S. Kasantsev, personal communication). The classification presented herein largely follows that of McKey-Fender (1950), the most recent treatment of this tribe for the Nearctic region.

Cantharis Linnaeus 1758
Telephorus Schaeffer 1766
Dicranodes Motschulsky 1860
Oripa Motschulsky 1860
Absidiella Wittmer 1972
subgenus *Cantharis* (*sensu stricto*) Linnaeus 1758
subgenus *Cyrtomoptila* Motschulsky 1860

Twenty-two species in two subgenera. Subgenus *Cantharis* (*sensu stricto*) with 21 species, widely distributed in our area. Subgenus *Cyrtomoptila*, with 1 species, *Cantharis* (*Cyrtomoptila*) *sierrae* McKey-Fender, from the San Bernardino Mountains of southern California. Partial key to species: McKey-Fender (1950), as *Cantharis* (*sensu stricto*), only for the species with "Color predominantly pale or dusky". Key to species present in the Pacific Northwest: Fender (1962), as *Cantharis* (*sensu stricto*). Additional nominal species not present in those keys have been described by Blatchley (1910), Fall (1936), LeConte (1851), McKey-Fender (1950) and Skinner (1906). Additional taxonomic information: McKey-Fender (1951), Fender (1972c) and Wittmer (1972). This Holarctic genus was known to many earlier North American workers by its longstanding junior synonym *Telephorus*. *Cantharis* has a long and confusing taxonomic and nomenclatural history. North American workers retained a broad definition of the genus long after it was abandoned by European workers. At one time, nearly all of the Nearctic species in what is now the tribe Cantharini were placed in this genus. Many former subgenera of *Cantharis* have been elevated to generic status (e.g., *Ancistronycha*, *Cultellunguis*, *Cyrtomoptera*, *Rhagonycha* and *Rhaxonycha*). Wittmer and Kasantsev (1997) recently summarized the many taxonomic changes to *Cantharis*. The status and classification of the genus and its subgenera presented herein follows Švihla (1999). The 22 nominal Nearctic species currently placed in *Cantharis* almost certainly do not constitute a monophyletic group, and are otherwise poorly known and in need of revision. Two adventive, possibly introduced, European species, *C. livida* Linnaeus and *C. rufa* Linnaeus, are present in our fauna (Dahlgren 1979, McKey-Fender 1950). Both were first recorded from the Atlantic Coast, but *C. rufa* has been recently found along the southern shore of Lake Superior in Wisconsin (personal observation). *Cantharis* species are found in both forested and more open habitats, where they can be collected on foliage and flowers.

Cultellunguis McKey-Fender 1950
Nine species: California, Nevada, Oregon and Washington. Key to species: McKey-Fender (1950), as *Cantharis*, subgenus *Cultellunguis*. Additional information: Fender (1968b). *Cultellunguis* is endemic to North America. One species, *C. americana* (Pic), has been reported to be strongly associated with Oregon White Oak, *Quercus garryana* Douglas ex Hooker, in Oregon.

Cyrtomoptera Motschulsky 1860
Two species: California to British Columbia. Key to species: McKey-Fender (1950), as *Cantharis*, subgenus *Cyrtomoptera*. Additional taxonomic information: McKey-Fender (1944). *Cyrtomoptera* is endemic to North America. Nothing is known about their habits and habitats.

Podistra Motschulsky 1839
subgenus *Pseudoabsidia* Wittmer 1969
subgenus *Hemipodistra* Ganglbauer 1922

Two species, in two subgenera. Subgenus *Pseudoabsidia*, with one species, *Podistra (Pseudoabsidia) sanbernardensis* (Fender), from the San Bernardino Mountains of southern California Subgenus *Hemipodistra*, with one species, *Podistra (Hemipodistra) insipida* (Fall), from New Mexico. Delkeskamp's (1977) report of *P. insipida* from California is erroneous, reflecting an early published misidentification of the then undescribed *P. sanbernardensis* by McKey-Fender (1950), as discussed by Fender (1972c). *Podistra* is a Holarctic genus. The genus has been recently redefined by Kasantsev (1998) based on the Palaearctic species. Therein our two subgenera were reduced from generic status. *Podistra* can be definitively separated from *Cantharis* and related genera by the structure of the male genitalia. In *Podistra* the laterophyses are fused, in *Cantharis* they are distinctly separated (S. Kasantsev, personal communication). The Nearctic species were last addressed by Fender (1972c) and the current placement of these species needs to be reviewed (S. Kasantsev, personal communication). Additional information: Wittmer (1972). The female of *P. insipida* has greatly abbreviated elytra and metathoracic wings. Both Nearctic species are rarely collected, and nothing is known about their habits and habitats.

Rhagonycha Eschscholtz 1830
Spartiolepta Bedel and Bourgeois 1901
Forty-six species: widely distributed, most in eastern North America. Partial key to species: Green (1940) as *Cantharis*, Division I. Key to species present in the Pacific Northwest: Fender (1962). Additional species not included in those keys described by Fender (1972b) and Miskimen (1956). Additional taxonomic information: Fender (1971b). *Rhagonycha* is principally a Holarctic genus, although in North America it extends at least as far south as Veracruz, Mexico (personal observation). European workers have recognized that the structure of the internal sac of the male genitalia is key to recognizing what are otherwise indistinguishable or highly variable species (e.g., Dahlgren 1979). Green's (1940) revision illustrates only the tegmen of the aedeagus and, as a result, many of his species probably represent species complexes. One adventive species, *Rhagonycha fulva* (Scopoli), native to western Europe, has become established in British Columbia (Fender 1962) and possibly Texas (Green 1940). Adults of *Rhagonycha* are often present in large numbers on a wide variety of foliage and flowers. Members of the genus are reported to feed on other insects and on nectar (Jackson and Crowson 1969).

Rhaxonycha Motschulsky 1860
Two species: eastern North America, west to Oklahoma. Key to species: McKey-Fender (1950), as *Cantharis*, *carolina* group. Additional taxonomic information: McKey-Fender (1941). *Rhaxonycha* is endemic to North America. Adults of *Rhaxonycha* are found on foliage and flowers. *Rhaxonycha carolina* (Fabricius) is a generalist predator, feeding on foliage-frequenting insects including aphids (Balduf 1935) and adult Miridae; although it probably also feeds on nectar and/or pollen. *Rhaxonycha bilobata* (McKey-Fender) is attracted to lights.

[*Ancistronycha* Märkel 1852]
Three species: eastern North America, west to Oklahoma. Key to species: McKey-Fender (1950) as *Cantharis*, subgenus *Ancistronycha*. These three species, formerly in *Cantharis*, were transferred to *Ancistronycha* by McKey-Fender (1950). This generic placement was almost certainly erroneous (S. Kasantsev, personal communication), and *Ancistronycha* is otherwise restricted to the Palaearctic Region. Adults of these species are commonly collected on foliage and flowers and are also attracted to lights. In the Midwest, *A. bilineata* (Say) is often the first cantharid to emerge as an adult in spring.

Podabrini LeConte 1881

Members of this Holarctic tribe can be recognized by the following combination of characters: head distinctly constricted behind eyes; frontoclypeal suture absent; apical maxillary palpomere bilaterally symmetrical; gular sutures confluent; elytra almost always normally elongate.

Dichelotarsus Motschulsky 1860
Anolisus Mulsant 1862
For information see *Podabrus*, discussed below.

Podabrus Westwood 1838
Podabrus Fischer von Waldheim 1821 (*nomen nudum*).
Brachynotus Kirby 1837
Malthacus Kirby 1837
Malthaeus Motschulsky 1859
subgenus *Podabrus* (*sensu stricto*) Westwood 1838
subgenus *Hatchiana* Fender 1966
In *Podabrus* and *Dichelotarus* combined, there are 107 nominal species widely distributed in our area. There are two subgenera in the New World, *Podabrus* (*sensu stricto*), with 106 widely distributed species, and *Hatchiana*, with a single species, *Podabrus* (*Hatchiana*) *arizonensis* (Fender), known from Arizona. Partial key to species: Fall (1927). Key to groups of male Podabrini: Fender (1949a). Key to species present in the Pacific Northwest: Fender (1962). Additional species not present in Fall's (1927) key have been described by Brown (1940), Fender (1943, 1946, 1948a 1948b, 1949b, 1953, 1962, 1966b, 1979), Green (1947, 1948, 1949) and Miskimen (1956). Additional notes on species are presented in Fender (1945a, 1945b, 1949b, 1969a) and Miskimen (1958). *Dichelotarsus*, long considered a subgenus of *Podabrus* and ignored by North American workers, was returned to generic status by Kasantsev (1992) based on differences in structure of the aedeagus. This change will necessitate the transfer of many Nearctic species of *Podabrus* to *Dichelotarsus* in the future. For example, the type species of *Dichelotarsus*, *D. piniphilus* (Eschscholtz), is a Nearctic species. Most or all of the species in Fender's (1949a) group I will remain in *Podabrus*, many of the species in groups II through VIII will likely be placed in *Dichelotarsus* (Kasantsev 1992). Fender's genus *Hatchiana* was reduced to subgeneric status by Kasantsev (1992). Both *Dichelotarsus* and *Podabrus* are also present in the Palaearctic Region. There are two species of *Podabrus* known from Mexico and two nominal species from elsewhere in the Neotropics (Delkeskamp 1977). The Nearctic species of both genera are in need of taxonomic revision. The importance of the internal sac of the male genitalia for species recognition was discussed by Fender and McCarley (1968). Ecological information about many species in both genera was presented by Fender (1973). Species of *Dichelotarsus* and *Podabrus* are commonly found on the foliage and flowers of a wide variety of plants. Many species, especially those present in montane regions of the western U.S., tend to inhabit only foliage in close proximity to streams or other bodies of water. Some species are attracted to lights. Species of both genera are predators of aphids and other small soft-bodied insects. Some species are important native natural enemies of insects considered to be harmful to agriculture (Fender 1973). Some species are also known to feed on nectar. *Podabrus rugosulus* LeConte has been observed consuming discarded droplets of aphid honeydew (A.H. Williams, personal communication). The peculiar brachelytrous *P. brevipennis* LeConte, known only from high elevation (13,000 ft.) in Colorado, is discussed by Fender (1971a).

Silinae Mulsant 1862

Silinae, as currently defined, contains two distinct tribes, but is not, as a whole, easily characterized by readily observable characters. Synapomorphies are presented in Brancucci (1980). The bulk of the diversity lies is in the virtually cosmopolitan tribe Silini (q.v.). The tribe Tytthonyxini includes only the aberrant New World genus *Tytthonyx* (q.v.).

Silini Mulsant 1862

Males of this tribe can be readily distinguished by the following combination of characters: frontoclypeal suture absent; apical maxillary palpomere bilaterally symmetrical; lateral pronotal margins often incised or otherwise irregularly modified; elytra normally elongate; and seventh abdominal ventrite deeply invaginated, appearing longitudinally divided or nearly so. Without associated males, female Silini and Cantharinae can sometimes be difficult to distinguish. Both sexes of Silini posses large and protruding lateral glandular pores on the abdominal tergites, the presence of which is probably the most reliable method for recognizing unassociated females. Silini are more or less cosmopolitan in distribution, and are most diverse in tropical areas. They are notably absent from New Zealand. The classification of Nearctic genera is currently under revision (Ramsdale, in preparation). The classification used herein largely follows that of Champion (1915), as modified by Green (1966).

Discodon Gorham 1881
Five nominal species: Texas to Arizona. No key is currently available. Descriptions by LeConte (1858), Gorham (1881) and Schaeffer (1908, 1909). Additional information: Champion (1915) and Fender (1949b). As currently defined, *Discodon* is extremely diverse in the Neotropics, with 339 nominal species (Delkeskamp 1977). *Discodon* is endemic to the New World. The species of *Discodon* of North America north of Mexico are currently being revised (Ramsdale, in preparation). Two species, *D. bipunctatum* Schaeffer and *D. planicolle* (LeConte), are regularly collected, the other species are uncommon. Several species are attracted to lights.

Ditemnus LeConte 1861
Twelve species: widely distributed, most in Texas and Arizona. Key to species: Green (1966). Long treated as a subgenus or junior synonym of *Silis* due to superficial similarities (e.g., coloration, pronotal armature). *Ditemnus* is herein returned to generic status, as advocated by Green (1966). *Ditemnus* is composed of three distinct species groups, each of which may eventually have to be elevated to generic status. *Ditemnus* is endemic to the New World. The complete function of the elaborate pronotal armature of males in the genera *Ditemnus* and *Silis* remains unknown. It has been hypothesized that the armature provides increased surface area for the evaporation of pheromones that are secreted from pores located therein (Brancucci 1980). It also serves a role in courtship behavior, as female *Ditemnus latilobus* Blatchley have been observed with their mouthparts clasped onto the male armature for prolonged periods both before and after copulation (personal observation). *Ditemnus* species are found on foliage and have been collected at lights.

Polemius LeConte 1851
Silogaster Pic 1913
Fourteen nominal species: widely distributed, most in the Southwest. Partial key to species in eastern North America: LeConte (1881). Species descriptions by Blatchley (1928), Brown (1940), Fall (1907), Horn (1894), LeConte (1881, 1884), and Schaeffer (1908, 1909). Additional information: Champion (1915) and Gorham (1881). *Polemius* is a poorly defined, probably paraphyletic genus, in need of revision. *Polemius* is endemic to the New World and most of the species are Neotropical. The species of *Polemius* of North America north of Mexico are currently being revised (Ramsdale in preparation). Most species of *Polemius* are infrequently collected.

Silis Charpentier 1825
Hoosilis Pic 1923
Sixty-eight species: widely distributed, most in western North America. Partial key to species: Green (1966). Additional species not in that key have been described by Fender (1971c, 1972a). The center of diversity of this genus is in montane regions of western North America, particularly California, but additional species are present in the Palaearctic Region. Additional information: Kasantsev 1994. The aberrant *Silis spinigera* LeConte, currently placed herein, should probably be transferred to a new monobasic genus (Green 1966). *Silis* species are collected from foliage, flowers, and at lights. Many sympatric species of *Silis* and *Pedilus* (Pyrochroidae) strongly resemble one another and are undoubtedly involved in a mimicry complex. For a discussion of the function of the pronotal armature, see *Ditemnus*, above.

[*Plectonotum* Gorham 1891]
A single uncommonly collected species, *Plectonotum excisum* Schaeffer, which is apparently restricted to the Chiricahua and Huachuca Mountains of southeastern Arizona, was misplaced in the genus *Plectonotum* of the subfamily Dysmorphocerinae. *Plectonotum excisum* must be transferred to the Silinae, where it will be the type of a new genus with several additional undescribed species in Arizona and Mexico (Ramsdale, in preparation). Description: Schaeffer (1908). Additional information: Champion (1915), Green (1966), Gorham (1885, 1891), and Wittmer (1949).

Silinae *incertae sedis*

Tytthonyxini Arnett 1963

This monotypic tribe contains only the genus *Tytthonyx*, which is restricted to the New World. There has been much difficulty attempting to discern the true phylogenetic affinities of this genus. Species of *Tytthonyx* share significant characters with members of both the subfamilies Silinae and Malthininae. Tytthonyxini were retained in the Silinae *incertae sedis*, by Brancucci (1980). Species of *Tytthonyx* can be distinguished by the following combination of characters: antennae strongly serrate, pectinate, or flabellate; apical maxillary palpomere radially symmetrical and acutely pointed; elytra abbreviated.

Tytthonyx LeConte 1851
Spathizus LeConte 1884
subgenus *Tytthonyx* (*sensu stricto*) LeConte 1851
subgenus *Thinalmus* Gorham 1881

Five species in two subgenera. Subgenus *Tytthonyx* (*sensu stricto*) with 4 species, from the eastern U.S. to Texas. Subgenus *Thinalmus*, with one species, *Tytthonyx* (*Thinalmus*) *bicolor* (LeConte), from Arizona. No key is currently available. Descriptions by Blatchley (1920, 1924), LeConte (1881), and Schaeffer (1904). Additional information: Wittmer (1970, 1991). *Tytthonyx* is endemic to the New World, and also occurs in Mexico, Central America and the West Indies. Nothing is known about their ecology.

Malthininae Kiesenwetter 1852

Malthinines are characterized by their small size (1.2 – 5mm in length), radially symmetrical and acutely pointed apical maxillary palpomeres, and more or less abbreviated elytra. The superficially similar brachelytrous Ichthyurini or Tytthonyxini can be distinguished from malthinines by their bilaterally symmetrical apical maxillary palpomeres or distinctly serrate, pectinate, or flabellate antennae, respectively. Malthininae are widely distributed, but are most diverse in the Holarctic and Oriental Regions, with fewer species in the Neotropical (occurring south to Brazil) and Afrotropical Regions. They are absent from Australia and only one (possibly adventive) species occurs in New Zealand (Wittmer 1979). Brancucci (1980) includes a phylogenetic analysis and revised classification of the genera of this subfamily.

Malthinini Kiesenwetter 1852

Malthinini can be distinguished by the following combination of characters: antennae filiform to slightly subserrate; mandibles with a prominent tooth; apical maxillary palpomere radially symmetrical and acutely pointed; gular sutures confluent; elytra more or less abbreviated.

Malthinus Latreille 1806
Malachidius Motschulsky 1859
Ymnis Gozis 1886
Progeutes Abeille de Perrin 1894
Projeutes Pic 1911
Proceutes Sahlberg 1913

Three species: eastern North America, west to Texas. Key to species: Wittmer (1980). Most of the Nearctic and Mexican species formerly placed in this genus were transferred to *Caccodes* by Wittmer (1986). *Malthinus* is most diverse in the Palaearctic and Oriental zoogeographic regions. Like all malthinines, *Malthinus* species are small and secretive, and very little is known about their habits. They are found on foliage in moist or boggy situations, usually in close proximity to a body of water.

Caccodes Sharp 1885

Fourteen species: most from Texas to Arizona, one in the southeastern U.S. Key to species: Fender (1972d) as *Malthinus* with "elytra more finely, confusedly punctate." Additional information in Fender (1951, 1963, 1966b) and Wittmer (1980, 1986). *Caccodes* is present in the eastern Palaearctic and the Afrotropical, Neotropical and Oriental Regions. *Caccodes oceaniae* (Bourgeois), a species widely distributed on Pacific islands, is the only species of cantharid found in Hawaii. Biological information on *Caccodes* is limited (q.v. *Malthinus*). Some adults are attracted to lights.

Malthodini Bøving and Craighead 1930 (not Brancucci 1980)

Malthodini can be distinguished by the following combination of characters: antennae filiform; mandibles simple, serrate, or with a tooth on the inflexed outer margin; apical maxillary palpomere radially symmetrical and acutely pointed; gular sutures almost always separated; elytra more or less abbreviated.

Malthodes Kiesenwetter 1852
Hapaloderus Motschulsky 1852
Apodistrus Reitter 1882
Malthinellus Seidlitz 1889
Podistrella Seidlitz 1889
Podistrina Seidlitz 1889
Maltharcus Weise 1892
Paramalthodes Weise 1893
Dalthomes Fiori 1905
Malthodinus Fiori 1905
Mesomalthodes Fiori 1905
Promalthodes Fiori 1905

One-hundred and twenty-six species: widely distributed, most in western North America Partial key to species: Fender (1951). Additional species not included in that key described by Fender (1962, 1963, 1964a, 1966a, 1968a, 1969b, 1972d). Partial key to species present in the Pacific Northwest: Fender (1962). *Malthodes* is an extremely diverse genus present in the Holarctic, Afrotropical, Oriental, and Neotropical Regions. The Nearctic fauna of this, our most species rich cantharid genus, is very poorly known. Males possess remarkably diversified and often elaborately modified terminal copulatory structures. Habitat preferences are summarized in Fender (1951) and are similar to those of *Malthinus* (q.v.). Some species of *Malthodes* seem to be strongly associated with conifers, others with deciduous trees, and others with herbaceous vegetation. Like most malthinines, they are uncommonly collected, but some species may be locally abundant.

Frostia Fender 1951

Stenomaltypus Wittmer 1978

Five species: California and Oregon. Key to species: Fender (1951). All of our species are in the Holarctic subgenus *Frostia* (*sensu stricto*). The other subgenus, *Protomaltypus* Wittmer, is found in China. Nothing is known about their habits and habitats.

Chauliognathinae LeConte 1861 (not Champion 1914)

Chauliognathinae is composed of two distinct tribes, which should, perhaps, each be accorded subfamilial status. This subfamily is not easily characterized, and the only strong character uniting its members is the distinctly asymmetrical male genitalia. The distributions of the two tribes are dissimilar, and are discussed below. Magis and Wittmer (1974) revised the classification of this subfamily and refuted its elevation to familial status. Miskimen (1961b) provided an examination of the biogeography of the subfamily, which remains somewhat useful, although many of his conclusions are highly questionable in light of modern theories.

Chauliognathini LeConte 1861

Chauliognathini can be distinguished by the following combination of characters: frontoclypeal suture present; apical maxillary palpomeres bilaterally symmetrical; and tibial spurs absent. Chauliognathini are distributed throughout the New World and are also present in Australia and New Guinea.

Belotus Gorham 1881

Two species: Florida to Oklahoma. *Belotus* is endemic to the New World, and most of the species are Neotropical. Key to species: Brancucci (1979). The elongate structure of the head indicates that *Belotus* are probably pollen-feeders, like their close relatives *Chauliognathus*, although nothing has been published on their habits.

Chauliognathus Hentz 1830

Callianthia Dejean 1833

Cantharodema Castelnau 1840

Selenurus Fairmaire 1877

Xenismus Waterhouse 1878

Haplous Fairemaire 1885

Xeniscus Pic 1955

Nineteen species: widely distributed. Key to species: Fender (1964b, 1964c). Additional information: Miskimen (1966). The center of diversity of *Chauliognathus* is in the Neotropics, with 346 nominal species (Delkeskamp 1977). Additional species are present in Australia and New Guinea, indicating that the genus originated prior to the breakup of Gondwana. Species of *Chauliognathus* are the most commonly encountered and well known cantharids in much of North America. They visit the flowers of a great variety of plants where they can often be found in large numbers. Nearctic *Chauliognathus* are apparently strictly phytophagous as adults, feeding on pollen and nectar. The elongate prognathous form of the head, accessory protrusible labral lip and maxillary tongue are all adaptations to more successfully exploit this resource (Crowson 1981, Miskimen 1961a). Reports of adult Nearctic *Chauliognathus* as predators are, almost certainly, the result of the misinterpretation of larval feeding data (e.g., Clausen 1940, Fender 1962). Champion (1914), in his revision of the Mexican and Central American species, regarded the genus as a whole to be predatory. Miskimen (1961a) reported that some Australian species were omnivorous, but that the New World species were presumably all phytophagous. The conspicuous nature of many species of *Chauliognathus* has led to them be chosen as the subjects of recent research into color polymorphism, dispersal, genetics, and sexual selection.

Ichthyurini Champion 1915

Ichthyurini are best distinguished by the following characters: frontoclypeal suture absent, apical maxillary palpomere bilaterally symmetrical, elytra abbreviated, and caudal abdominal ventrite (and tergite in some) strongly modified. Members of this tribe are present in all zoogeographic regions except Australia. Miskimen (1961a) considered the members of this tribe to be "essentially carnivorous."

Ichthyurus Westwood 1848

Biurus Motschulsky 1853

Diurus Gemminger and Harold 1869

Dissacurus Gemminger 1872

One species: *Ichthyurus arizonensis* Fender, occurs in southeastern Arizona. Taxonomic information: Fender (1948b, 1960). Most of the species of this highly aberrant, somewhat wasp-like, genus occur in the Oriental and Afrotropical Regions. A few species are known from the Neotropics. *Ichthyurus arizonensis* is known only from the type series (of two females) collected in the Huachuca Mountains. Both male and, to a lesser degree, female *Ichthyurus* have a modified caudal abdominal segment which is posteriorly bifurcated and produced into a pair of elongate processes. Nothing is known about their ecology.

Trypherus LeConte 1851

Molorchus Say 1824

Lygerus Kiesenwetter 1852

Five species: eastern North America, west to Oklahoma. Key to species: Brancucci (1985). Additional information: Fender (1960). *Trypherus* has a typical eastern Asian - eastern North American disjunct Arcto-Tertiary distribution. Additional species are found in coastal regions of eastern Asia, with most of the diversity in Taiwan and Japan. Species of *Trypherus* are not uncommonly found on herbaceous foliage in the understory of deciduous forests.

Bibliography

ARNETT, R. H., Jr. 1963. The beetles of the United States (a Manual for Identification). Catholic University of America Press. Washington, DC. xi + 1112 pp.

BALDUF, W. V. 1935. The bionomics of entomophagous Coleoptera. John S. Swift. New York, NY. 220 pp.

BEUTEL, R. G. 1995. Phylogenetic analysis of Elateriformia (Coleoptera: Polyphaga) based on larval characters. Journal for Zoological Systematics and Evolutionary Research, 33: 145-171.

BLATCHLEY, W. S. 1910. An illustrated descriptive catalogue of the Coleoptera or beetles (exclusive of the Rhyncophora) known to occur in Indiana. The Nature Publishing. Indianapolis. 1386 pp.

BLATCHLEY, W. S. 1920. Notes on some Coleoptera taken in the vicinity of Dunedin, Florida, in the Spring of 1920, with descriptions of new species. Canadian Entomologist, 52: 259-264.

BLATCHLEY, W. S. 1924. New Coleoptera from southern Florida with notes on other interesting species. Canadian Entomologist, 56: 164-170.

BLATCHLEY, W. S. 1928. Notes on some Florida Coleoptera with descriptions of new species. Canadian Entomologist, 60: 60-73.

BØVING, A.G. and F. C. CRAIGHEAD. 1930. An illustrated synopsis of the principal larval forms of the order Coleoptera. Entomologica Americana (n. s.), 11: 1-351.

BRANCUCCI, M. 1979. Révision du genre *Belotus* Gorham (Col. Cantharidae). Entomologica Basiliensia, 4: 361-430.

BRANCUCCI, M. 1980. Morphologie Comparée, Évolution et Systématique des Cantharidae (Insecta: Coleoptera). Entomologica Basiliensia, 5: 215-388.

BRANCUCCI, M. 1985. Revision du genre *Trypherus* LeConte (Coleoptera, Cantharidae). Entomologica Basiliensia, 10: 251-322

BROWN, W. J. 1940. Some new species of Cantharidae and Chrysomelidae (Coleoptera). Canadian Entomologist, 72: 161-166.

CLAUSEN, C. P. 1940. Entomophagous Insects. McGraw-Hill. New York. 688 pp.

CHAMPION, G. C. 1914. Revision of the Mexican and Central American Chauliognathinae (Fam. Telephoridae), based on the genital armature of the males. Transactions of the Entomological Society of London: 128-168.

CHAMPION, G. C. 1915. Revision of the Mexican and Central American Telephorinae (Fam. Telephoridae), with descriptions of new species. Transactions of the Entomological Society of London: 16-146.

CROWSON, R. A. 1955. The natural classification of the families of Coleoptera. Nathaniel Lloyd. London. 187 pp.

CROWSON, R. A. 1972. A review of the classification of Cantharoidea (Coleoptera), with the definition of two new families, Cneoglossidae and Omethidae. Revista de la Universidad de Madrid, 21: 35-77.

CROWSON, R. A. 1981. The biology of the Coleoptera. Academic Press. London. xii + 802 pp.

DAHLGREN, G. 1979. Cantharidae (ohne Tribus Malthinini). Pp. 19-39. *In:* H. Freude, K. W. Harde and G. A. Lohse, eds. Die Käfer Mitteleuropas, Band 6, Diversicornia. Goecke & Evers. Krefeld. 367 pp.

DELKESKAMP, K. 1977. Cantharidae. *In:* J. A. Wilcox, ed. Coleopterorum Catalogus Supplementa Pars 165, Fasc. 1. W. Junk. The Hague. 485 pp.

DELKESKAMP, K. 1978. Cantharidae, Corrigenda et Addenda. *In:* J. A. Wilcox, ed. Coleopterorum Catalogus Supplementa Pars 165, Fasc. 2. W. Junk. The Hague. Pp. 487-556.

DETTNER, K. 1987. Chemosytematics and evolution of beetle chemical defenses. Annual Review of Entomology, 32: 17-48.

DILLON, E. S. and L. S. DILLON. 1961. A manual of common beetles of eastern North America. Row, Peterson. Evanston, IL. 768 pp.

DOWNIE, N. M. and R. H. ARNETT JR. 1996. The beetles of northeastern North America, Volume I. Introduction, suborders Archostemata, Adephaga, and Polyphaga thru superfamily Cantharoidea. Sandhill Crane Press. Gainesville, FL. xiv + 880 pp.

FALL, H. C. 1907. Descriptions of new species. *In:* H. C. Fall and D. A. Cockrell. The Coleoptera of New Mexico. Transactions of the American Entomological Society, 33: 145-272.

FALL, H. C. 1927. A review of the North American species of *Podabrus*. Entomologica Americana, 8: 65-103.

FALL, H. C. 1936. On certain species of *Cantharis* (*Telephorus*). Pan-Pacific Entomologist, 12: 179-183.

FENDER, K. M. 1943. Studies in the Cantharidae (Coleoptera). Pan-Pacific Entomologist, 19: 63-69.

FENDER, K. M. 1945a. Studies in the Cantharidae II (Coleoptera). Canadian Entomologist, 77: 37-39.

FENDER, K. M. 1945b. Notes on the species of *Podabrus* of Oregon and Washington (Coleoptera, Cantharidae). Pan-Pacific Entomologist, 21: 77-80.

FENDER, K. M. 1946. A new species of *Podabrus* (Coleoptera-Cantharidae). Bulletin of the Brooklyn Entomological Society, 41: 12.

FENDER, K. M. 1948a. The *clavicollis-corneus* group of *Podabrus* (Coleoptera: Cantharidae). Pan-Pacific Entomologist, 24: 131-140.

FENDER, K. M. 1948b. Some new or little known Cantharidae (Coleoptera). Wasmann Collector, 7: 117-123.

FENDER, K. M. 1949a. Studies in the Cantharidae III (Coleoptera). Pan-Pacific Entomologist, 25: 29-31.

FENDER, K. M. 1949b. Studies in the Cantharidae IV (Coleoptera). Pan-Pacific Entomologist, 25: 185-189.

FENDER, K. M. 1951. The Malthini of North America (Coleoptera-Cantharidae). American Midland Naturalist, 46: 513-629.

FENDER, K. M. 1953. New species of *Podabrus* from western North America (Coleoptera: Cantharidae). Pan-Pacific Entomologist, 29: 170-175.

FENDER, K. M. 1960. The Ichthyurini of North America (Coleoptera: Cantharidae). Pan-Pacific Entomologist, 36: 105-113.

FENDER, K. M. 1962. Family Cantharidae. Pp. 44-68. *In:* M. H. Hatch. Beetles of the Pacific Northwest. Part III: Pselaphidae and Diversicornia I. University of Washington Press. Seattle. 503 pp.

FENDER, K. M. 1963. Some undescribed North American Malthini (Coleoptera-Cantharidae). Pan-Pacific Entomologist, 35: 75-79.

FENDER, K. M. 1964a. New and little known species of *Malthodes* (Coleoptera: Cantharidae). Northwest Science, 38: 18-24.

FENDER, K. M. 1964b. The Chauliognathini of America north of Mexico (Coleoptera-Cantharidae). Part 1. Northwest Science, 38: 52-64.

FENDER, K. M. 1964c. The Chauliognathini of America north of Mexico (Coleoptera-Cantharidae). Part 2. Northwest Science, 38: 95-106.

FENDER, K. M. 1966a. Three new North American Malthini (Coleoptera-Cantharidae). Pan-Pacific Entomologist, 42: 219-222.

FENDER, K. M. 1966b. A new North American genus of Soldier Beetles (Coleoptera: Cantharidae). Pan-Pacific Entomologist, 42: 321-323.

FENDER, K. M. 1968a. New *Malthodes* from western North America (Coleoptera: Cantharidae). Northwest Science, 42: 108-111.

FENDER, K. M. 1968b. A new subspecies of *Cantharis hatchi* McKey-Fender (Coleoptera: Cantharidae). Pan-Pacific Entomologist, 44: 300-301.

FENDER, K. M. 1969a. On *Podabrus parvicollis* Motschulsky (Coleoptera: Cantharidae). Coleopterists Bulletin, 23: 40-41.

FENDER, K. M. 1969b. New California *Malthodes* and distribution notes on other species in California (Coleoptera: Cantharidae). Pan-Pacific Entomologist, 45: 119-126.

FENDER, K. M. 1969c. Violent deaths of *Podabrus* (Coleoptera: Cantharidae). Coleopterists Bulletin, 23: 106-107.

FENDER, K. M. 1971a. Redefinition of *Podabrus brevipennis* LeConte and recognition of a new subspecies (Coleoptera: Cantharidae). Coleopterists Bulletin, 25: 31-34.

FENDER, K. M. 1971b. A new California species of *Silis* (Coleoptera: Cantharidae). Pan-Pacific Entomologist, 47: 24-25.

FENDER, K. M. 1971c. The genus *Rhagonycha* Eschscholtz in North America (Coleoptera: Cantharidae). Coleopterists Bulletin, 25: 86-87.

FENDER, K. M. 1972a. Two new California species of *Silis* (Coleoptera: Cantharidae). Pan-Pacific Entomologist, 48: 215-217.

FENDER, K. M. 1972b. A new species of *Rhagonycha* from Alaska (Coleoptera: Cantharidae). Pan-Pacific Entomologist, 48: 242-243.

FENDER, K. M. 1972c. The genera *Podistra* and *Pseudoabsidia* in North America (Coleoptera: Cantharidae). Coleopterists Bulletin, 26: 13-16.

FENDER, K. M. 1972d. Some new and little known species of Malthini from the southwestern United States (Coleoptera: Cantharidae). Coleopterists Bulletin, 26: 43-52.

FENDER, K. M. 1973. Ecological notes on *Podabrus* (Coleoptera: Cantharidae). Coleopterists Bulletin, 27: 11-17.

FENDER, K. M. 1979. A new California species of *Podabrus* (Coleoptera: Cantharidae). Pan-Pacific Entomologist, 55: 155-158.

FENDER, K. M. and L. J. McCARLEY. 1968. The internal sac of the aedeagus of *Podabrus* (Coleoptera: Cantharidae). Coleopterists Bulletin, 22: 20-27.

FITTON, M. G. 1975. The larvae of British genera of Cantharidae (Coleoptera). Journal of Entomology, 44: 243-254.

GREEN, J. W. 1940. Taxonomic studies in *Cantharis* (Coleoptera: Cantharidae). Entomologica Americana, 20: 159-217.

GREEN, J. W. 1947. New eastern American species of *Podabrus* (Coleoptera: Cantharidae). Transactions of the American Entomological Society, 73: 63-67.

GREEN, J. W. 1948. New eastern American species of *Podabrus* II (Coleoptera: Cantharidae). Transactions of the American Entomological Society, 74: 75-82.

GREEN, J. W. 1949. A new genus and new species of American Lampyrini, and other notes (Coleoptera: Lampyridae, Cantharidae). Transactions of the American Entomological Society, 75: 1-6.

GREEN, J. W. 1966. Revision of the Nearctic species of *Silis* (Cantharidae: Coleoptera). Proceedings of the California Academy of Sciences, 32: 447-513.

GORHAM, H. S. 1881. Fam. Telephoridae. Pp. 65-106. *In:* F. D. Goodman and O. Salvin, eds. Biologia Centrali-Americana, 3(2).

GORHAM, H. S. 1885. Malacodermata. Pp. 289-312. *In:* F. D. Goodman and O. Salvin, eds. Biologia Centrali-Americana Supplement, 3(2).

GORHAM, H. S. 1891. Fam. Telephoridae. Pp. 49-52. *In:* E. Whymper. Supplementary Appendix to the Travels Amongst the Great Andes of the Equator. J. Murray. London. 143 pp.

HORN, G. W. 1894. The Coleoptera of Baja California. Proceedings of the California Academy of Sciences, 2: 302-446.

JACKSON, J. G. and R. A. CROWSON. 1969. A comparative anatomical study of the digestive, excretory and central nervous systems of *Malachius viridis* F. (Col., Melyridae) and *Rhagonycha usta* Gemm. (Col., Cantharidae), with observations on their diet and taxonomy. Entomologist's Monthly Magazine, 105: 93-97.

KASANTSEV, S. 1992. Contribution to the knowledge of Palaearctic Cantharidae (Coleoptera). Notes on *Dichelotarsus* Motschulsky. Entomologica Basiliensia, 15: 267-277.

KASANTSEV, S. 1994. The Palaearctic species of the genus *Silis* Charpentier, 1825 with the description of *Crudosilis* gen. nov. (Coleoptera: Cantharidae). Elytron, 8: 93-115.

KASANTSEV, S. 1998. On *Podistra* complex (Coleoptera, Cantharidae) of Russia and adjacent territories. Zoologicheskii Zhurnal, 77: 689-695.

LAWRENCE, J. F. 1987. Rhinorhipidae, a new beetle family from Australia, with comments on the phylogeny of the Elateriformia. Invertebrate Taxonomy, 2: 1-53.

LAWRENCE, J. F., A. M. HASTINGS, M. J. DALLWITZ, T. A. PAINE and E. J. ZURCHER. 1999a. Beetle larvae of the World: descriptions, illustrations, identification, and information retrieval for families and sub-families. CD-ROM, Version 1.1 for MS-Windows. CSIRO Publishing. Melbourne.

LAWRENCE, J. F., A. M. HASTINGS, M. J. DALLWITZ, T. A. PAINE and E. J. ZURCHER. 1999b. Beetles of the World: a key and information system for families and subfamilies. CD-ROM, Version 1.1 for MS-Windows. CSIRO Publishing. Melbourne.

LAWRENCE, J. F. and A. F. NEWTON, Jr. 1982. Evolution and classification of beetles. Annual Review of Ecology and Systematics, 13: 261-290.

LAWRENCE, J. F. and A. F. NEWTON, Jr. 1995. Families and subfamilies of Coleoptera (with selected genera, notes, references and data on family-group names). Pp. 779-1006. *In*: J. Pakaluk and S. A. Slipinski eds. Biology, phylogeny, and classification of Coleoptera: Papers celebrating the 80th birthday of R. A. Crowson. Muzeum i Instytut Zoologii PAN. Warsaw.

LECONTE, J. L. 1851. Synopsis of the Lampyrides of temperate North America. Proceedings of the Philadelphia Academy of Natural Sciences, 5: 331-347.

LECONTE, J. L. 1858. Catalogue of the Coleoptera of the regions adjacent to the boundary line between the United States and Mexico. Journal of the Philadelphia Academy of Natural Sciences (ser. 2), 4: 9-42.

LECONTE, J. L. 1881. Synopsis of the Lampyridae of the United States. Transactions of the American Entomological Society, 9: 15-72.

LECONTE, J. L. 1884. Short studies of North American Coleoptera (No.2). Transactions of the American Entomological Society, 12: 1-32.

LESAGE, L. 1991. Cantharidae (Cantharoidea). Pp. 429-431. *In:* F. W. Stehr, ed. Immature Insects. Volume 2. Kendall/Hunt. Dubuque, IA.

MAGIS, N. and W. WITTMER. 1974. Nouvelle répartition des genres de la sous-famille des Chauliognathinae (Coleoptera, Cantharoidea: Cantharidae). Bulletin de la Société Royale des Sciences de Liège, 43: 78-95.

McKEY-FENDER, D. 1941. Notes on *Cantharis* (Coleoptera: Cantharidae). Pan-Pacific Entomologist, 17: 126-129.

McKEY-FENDER, D. 1944. Notes on *Cantharis* II (Coleoptera: Cantharidae). Pan-Pacific Entomologist, 20: 20-22.

McKEY-FENDER, D. 1950. Notes on *Cantharis* III. Pan-Pacific Entomologist, 26: 25-33, 61-79.

McKEY-FENDER, D. 1951. A new *Cantharis* (Coleoptera, Cantharidae). American Midland Naturalist, 46: 630-633.

McNAMARA, J. 1991. Family Cantharidae soldier beetles. Pp. 192-195. *In:* Y. Bousquet, ed. Checklist of the Beetles of Canada and Alaska. Publication 1861/E. Research Branch, Agriculture Canada. Ottawa.

MILLER, R. S. 1991. A revision of the Leptolycinae (Coleoptera: Lycidae) with a discussion of paedomorphosis. Ph.D. Dissertation. Ohio State University. xv. + 403 pp. [unpublished]

MISKIMEN, G. W. 1956. A faunal list of the Cantharidae (Coleoptera) of Ohio with descriptions of new species. Ohio Journal of Science, 53: 129-134.

MISKIMEN, G. W. 1958. The problem of color variation in *Podabrus* (Coleoptera: Cantharidae). Proceedings of the Entomological Society of Washington, 60: 265-266.

MISKIMEN, G. W. 1961a. A new family of beetles found in the Cantharoidea. Coleopterists Bulletin, 15: 17-25

MISKIMEN, G. W. 1961b. Zoogeography of the coleopterous family Chauliognathidae. Systematic Zoology, 10: 140-153.

MISKIMEN, G. W. 1966. Zoogeography and systematics of North and Central American Chauliognathini (Coleoptera: Cantharidae). Ph.D. Dissertation. University of Florida. 162 pp. [unpublished]

PECK, S.B. and M. C. THOMAS. 1998. A distributional checklist of the beetles (Coleoptera) of Florida. Arthropods of Florida and Neighboring Land Areas, 16: vii + 180.

PONOMARENKO, A. G. 1995. The geological history of beetles. Pp. 155-171. *In:* J. Pakaluk and S. A. Slipinski, eds. Biology, phylogeny, and classification of Coleoptera: Papers celebrating the 80th birthday of R. A. Crowson. Muzeum i Instytut Zoologii PAN. Warsaw.

SCHAEFFER, C. 1904. New genera and species of Coleoptera. Journal of the New York Entomological Society, 12: 197-236.

SCHAEFFER, C. 1908. List of the Lampyridae from the Huachuca Mountains, Arizona, with descriptions of new species. Journal of the New York Entomological Society, 16: 61-67.

SCHAEFFER, C. 1909. New Coleoptera chiefly from Arizona. Science Bulletin of the Museum of the Brooklyn Institute of Arts and Sciences, 1: 375-386.

SCHULTZ, T. D. 1994. Predation by larval soldier beetles (Coleoptera: Cantharidae) on the eggs and larvae of *Pseudoxycheila tarsalis* (Coleoptera: Cicindelidae). Entomological News, 105: 14-16.

SKINNER, H. 1906. A new *Cantharis*. Entomological News, 17: 217.

ŠVIHLA, V. 1999. Contribution to the knowledge of the genus *Cantharis* L. and related genera from Turkey and adjacent regions (Coleoptera, Cantharidae). Entomologica Basiliensia, 21: 135-170.

VERHOEFF, K. W. 1917. Zur Entwicklung, Morphologie und Biologie der Vorlarven und Larven der Cantharidae. Archiv für Naturgeschichte (A), 83: 102-140.

VERHOEFF, K. W. 1923. Zur Kenntnis der Cantharidenn-Larven. 2. Beitrag. Archiv für Naturgeschichte (A), 89: 110-137.

WHEELER, A. G. 1988. "Violent deaths" of soldier beetles (Coleoptera: Cantharidae) revisited: new records of the fungal pathogen *Eryniopsis lampyridarum* (Zygomycetes: Entomophthoraceae). Coleopterists Bulletin, 43: 233-236.

WHITE, R. E. 1983. A field guide to the beetles of North America. The Peterson Field Guide Series. Houghton Mifflin. New York. xii + 368 pp.

WITTMER, W. 1949. 7 Beitrag zur Kenntnis der neotropischen Malacodermata. Revista de la Sociedad Entomologica Argentina, 14: 215-222.

WITTMER, W. 1970. On some Cantharidae (Coleoptera) of the United States. Coleopterists Bulletin, 24: 42-46.

WITTMER, W. 1972. 55. Beitrag zur Kenntnis der palaearktischen Cantharidae und Malachiidae (Col.). Entomologische Arbeiten aus dem Museum G. Frey (Tutzing bei München), 23: 122-141.

WITTMER, W. 1979. Zur Kenntnis der Cantharidae (Col.) Neuseelands. Entomologica Basiliensia, 4: 275-325.

WITTMER, W. 1980. Remarks on a few species of *Malthinus* Latreille (Coleoptera: Cantharidae: Malthinini). Coleopterists Bulletin, 34: 271-279.

WITTMER, W. 1986. Zur Kenntnis der Malthininae Mexicos (Coleoptera: Cantharidae). Mitteilungen der Schweizerischen Entomologischen Gesellschaft, 59: 197-220.

WITTMER, W. 1991. Zur Kenntnis der Gattung *Tytthonyx* LeConte, 1851, Subgenus *Thinalmus* Gorham, 1881, Beitrag 1. (Coleoptera: Cantharidae, Subfam. Silinae, Tribus Tytthonyxini). Mitteilungen der Schweizerischen Entomologischen Gesellschaft, 64: 115-126.

WITTMER, W. and S. KASANTSEV. 1997. On the classification of the genus *Cantharis* Linné (Coleoptera: Cantharidae). Entomologica Basiliensia, 20: 367-372.

Series BOSTRICHIFORMIA Forbes 1926

65. JACOBSONIIDAE Heller 1926

by T. Keith Philips, Michael A. Ivie and J. Joseph Giersch

Family common name: The jacobsoniid beetles

Family synonyms: Sarothriidae Crowson 1955

North American members of this family can be recognized by the minute (less than 1 mm), narrowly elongate (about 4 times as long as wide), yellowish brown body, an elongate prothorax narrowed posteriorly, lack of a visible scutellum, and the metasternum as long as or longer than all five ventrites combined. This family was only recently reported from North America (Peck and Thomas 1998, Arnett 2000).

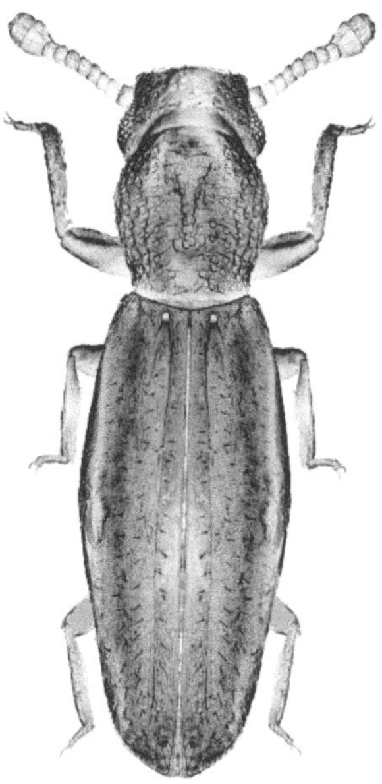

FIGURE 1.65. *Derolathrus* sp.

Description. Size generally minute, 0.7 to 2.1 mm (up to 4.0 mm in some species); shape narrowly elongate, cylindrical; yellowish-brown to brownish-gray, with a silvery tomentum in some; vestiture composed of few setae, which are either stout, broad, scale-like, or elongate, longitudinally aligned in some.

Head oval, deflexed, abruptly constricted behind eyes, forming short temples; surface smooth to microrugose; eyes lateral, somewhat bulging, small, rounded. Clypeus fused to frons; labrum small, deflexed; mandibles small, slender, curved, apices acute; apical maxillary palpomere aciculate and narrower than penultimate palpomere; gula narrow, gular sutures absent; labrum with mentum quadrate; labial palpi not visible. Antennae with 10-11 antennomeres; moniliform, with 1-or 2-segmented club, insertions concealed or raised on process. Trichomes sometimes present, inserted on top of lateral borders of head in well-defined pits, or laterally and insertion hidden.

Prothorax about as narrow as head, narrower than elytra, quadrate, twice as long as broad, borders not margined, sometimes narrowed posteriorly; pleural region broad; prosternum narrow in front of coxae; procoxal cavities closed or open posteriorly, trochantins concealed. Mesosternum short, fused to metasternum; metasternum very long, from 1 to 3 times as long as abdomen. Legs with anterior coxae globular, very narrowly separate; middle coxae globular, separate; hind coxae quadrate; trochanters somewhat triangular; femora slender, short; tibiae slender, short, spinose, apical spurs present; tarsomeres 3-3-3, short, first minute, third somewhat elongate; claws moderate, simple; scutellum not visible; elytra entire, apices truncate, but projecting beyond abdomen; elytral striae absent, surface microrugose; elytra ventrally closely appressed to sides of body in species examined, but epipleural fold absent, wings reduced, venation staphylinoid.

Abdomen with five very short, visible sterna, sutures entire, nonfunctional, surface rugose; apparently functional spiracles present on abdominal segment VIII. Male genitalia modified, with tubular tegmen enclosing penis with possibly a basal connection between these structures; female genitalia with paraprocts, proctigers, valvifers, coxites, and styli; baculi absent.

Larvae elongate, slightly flattened, and approximately parallel-sided. All terga are evenly sclerotized. The head lacks stemmata, an epicranial stem and median endocarina. Frontal arms are lyriform and widely separated at the base. Antennae are short and broad. Mandibles are unidentate with serrate incisors, and have slender prostheca and tuberculate mola. The maxillary palpi are 3-segmented with distinct galea (with a tuft of serrate hairs) and lacinia. The labial palpi are 2-segmented and the ligula is bilobed. The legs are short and the tarsungulus has a single seta. The abdomen (A9) has two short, fixed urogomphi and A10 is circular in shape and forms a ring around the ventrally oriented anus. All spiracles are annular.

Habits and habitats. Jacobsoniids adults and larvae have been found in leaf litter, rotten wood, fungi, bat guano and under bark (Lawrence 1991, Klimaszewski and Watt 1997, Lawrence *et al.* 1999a). The most widely distributed group, the Derolathrinae Sen Gupta 1979, are currently all lumped under the generic name *Derolathrus* Sharp 1908 and are very tiny (0.7- 0.9 mm). They are probably much more common and widely distributed than currently known, and specimens are most often taken with specialized techniques like Berlese extractions of litter or flight intercept traps. The Florida species were both taken by the latter method. Species of *Sarothrias* Grouvelle 1918 appear to be modified for myrmecophily.

Status of the classification. The relationships and status of this small family are still unclear. Löbl and Burckhardt (1988) found no synapomorphy to support this group of 3 disparate genera, considering it an amalgamation of groups that share odd symplesiomorphies. It has been placed variously in the Cucujiformia Merophysidae by Sen Gupta (1979); Bostrichiformia

Dermestoidea by Crowson (1955, 1959, 1960, 1981), who noted similarities to the Cucujoidea and Staphylinoidea in adult structures; Derodontoidea by Lawrence and Newton (1982), Lawrence (1991); and the Bostrichoidea by Lawrence and Britton (1991) and Klimaszewski and Watt (1997). Ivie (1985) placed it as the Jacobsoniiodea in the Cucujiformia on the basis of the hylocetoid metendosternite, but because of the presence of the symplesiomorphic condition of spiracles on A8 and the dermestoid aedeagus, it lacked two important synapomorphies of the rest of the series, and was considered weakly supported (at best) as the sister-group of the remaining Cucujiformia. Lawrence and Newton (1995), and Lawrence *et al.* (1999b) placed it *incertae sedis* in the Bostrichiformia [Derodontoidea and Bostrichoidea (including Dermestoidea)], where it remains today. At least *Sarothrias* lacks the bostrichoid synapomorphy of modified cryptonephridy (Ivie pers. obs.), but so do the Nosodendridae (including Orphilinae, see Chapter 67). This question obviously requires more study.

Löbl and Burckhardt (1988) combined the described members of the Derolathrinae into the single genus *Derolathrus*. The two Florida species, the first record of this family in the contiguous United States (Peck and Thomas 1998), are tentatively placed in *Derolathrus*, but this may soon change.

Distribution. The Jacobsoniidae are widespread. Two Florida species (Dade and Alachua Counties) of the Derolathrinae were reported by Peck and Thomas (1998). *Derolathrus s.l.* is also reported from seven described species from Brazil, Fiji, Guadeloupe, Hawaii, Madera, Mauritius, and Sri Lanka (Löbl and Burckardt 1988), and unstudied members from central and southern Africa, Australia (Queensland, A.C.T., S.A., Christmas Island, Lord Howe Island, Norfolk Island), Brazil, the Canary Islands, New Zealand, Panama, Solomon Islands, Reunion and the Virgin Islands (Ivie collection) (Lawrence 1991, Lawrence and Britton 1991, Klimaszewski and Watt 1997). The only other currently recognized genera in the family are the monotypic *Saphophagus* Sharp 1886 from New Zealand and *Sarothrias* Grouvelle 1918, with ten species distributed around the Indian and southwest Pacific Oceans, including records from Borneo, Fiji, Guam, India, Moluccas, New Britain, New Caledonia, New Guinea, Queensland (Australia), the Seychelles, Sri Lanka, Sumatra, and Tioman Island (west Malaysia) (Löbl and Burckardt 1988, Slipinski and Löbl 1995).

Classification of the Nearctic Genera

Jacobsoniidae Heller 1926

Derolathrus Sharp 1908, *in* Sharp and Scott 1908, 2 undescribed species, Florida (Dade and Alachua Co.); from flight intercept traps.

Lathridiomorphus Franz 1969
Gomya Dajoz 1973
Lathridiomimus (*sic*) Dajoz 1977

Bibliography

ARNETT, R. H., Jr. 2000. A handbook of the insects of America north of Mexico. Second edition. CRC Press LLC. Boca Raton, FL. 1001 pp.

CROWSON, R. A. 1955. The natural classification of the families of Coleoptera. Nathaniel Lloyd. London, 187 pp.

CROWSON, R. A. 1959. Studies on the Dermestoidea (Coleoptera), with special reference to the New Zealand fauna. Transactions of the Royal Entomological Society of London, 111: 81-94.

CROWSON, R. A. 1960. The phylogeny of Coleoptera. Annual Review of Entomology, 5: 111-134.

CROWSON, R. A. 1981. The Biology of the Coleoptera. Academic Press. London, xii + 802 pp.

IVIE, M. A. 1985. Phylogenetic studies in the Bostrichiformia (Coleoptera). Ohio State University (PhD. Thesis), Columbus, 137 pp.

KLIMASZEWSKI, J. and J. C. WATT. 1997. Coleoptera: family group review and keys to identification. Fauna of New Zealand, 37: 1-199.

LAWRENCE, J. F. 1991. Jacobsoniidae (Derodontoidea), Pp. 433-434. *In*: F. W. Stehr, ed., Immature Insects. Vol. 2. Kendall/Hunt. Dubuque, Iowa.

LAWRENCE, J. F. and E. B. BRITTON 1991. Coleoptera (beetles). Pp. 543-683. *In*: C.S.I.R.O. Division of Entomology, ed., The Insects of Australia: a Textbook for Students and Research Workers, Second Edition. Melbourne University Press. Carlton, Victoria.

LAWRENCE, J. F., A. M. HASTINGS, M. J. DALLWITZ, T. A. PAINE and E. J. ZURCHER. 1999a. Beetles of the World: A Key and Information System for Families and Subfamilies. CD-ROM, Version 1.0 for MS-Windows. CSIRO Publishing. Melbourne.

LAWRENCE, J. F., A. M. HASTINGS, M. J. DALLWITZ, T. A. PAINE and E. J. ZURCHER 1999b. Beetle Larvae of the World: Descriptions, Illustrations, Identification, and Information Retrieval for Families and Sub-families. CD-ROM, Version 1.1 for MS-Windows. CSIRO Publishing. Melbourne.

LAWRENCE, J. F. and A. F. NEWTON, Jr. 1982. Evolution and classification of beetles. Annual Review of Ecology and Systematics, 13: 261-290.

LAWRENCE. J. F. and A. F. NEWTON, Jr. 1995. Families and subfamilies of Coleoptera (with selected genera, notes, references and data on family-group names). Pp 779-1006. *In*: J. Pakaluk and S. A. Slipinski, eds., Biology, phylogeny and classification of Coleoptera: Papers celebrating the 80th birthday of Roy A. Crowson. Muzeum i Instytut Zoologii PAN, Warsaw.

LÖBL, I. and D. BURCKHARDT. 1988. Revision der Gattung *Sarothrias* mit Bemerkungen zur Familie Jacobsoniidae (Coleoptera). Stuttgarter Beiträge zur Naturkunde. Serie A (Biologie), 422: 1-23.

PECK, S. B. and M. C. THOMAS. 1998. A distributional checklist of the beetles (Coleoptera) of Florida. Arthropods of Florida and Neighbouring Land Areas, 16: i-viii + 1-180.

SEN GUPTA, T. 1979. A new subfamily of Merophysiidae (Clavicornia: Coleoptera) and descriptions of two new species of *Gomya* Dajoz and its larva. Revue suisse Zoologie, 86: 691-698.

SLIPINSKI, S. A. and I. LÖBL 1995. New species of *Sarothrias* (Coleoptera, Jacobsoniidae). Mitteilungen der Schweizerischen Entomologischen Gesellschaft, 68: 49-53.

Superfamily DERODONTOIDEA LeConte 1861

66. DERODONTIDAE LeConte 1861

by Richard A.B. Leschen

Family common name: The tooth-necked fungus beetles

Family synonyms: Laricobiidae Ganglbauer 1899; Peltasticidae LeConte 1861

The presence of paired ocelli, prothorax mostly with dentate lateral carinae, open mesocoxal cavities that contact the mesepimeron, excavate hind coxae, and rugose body forms of most adults is characteristic of the family.

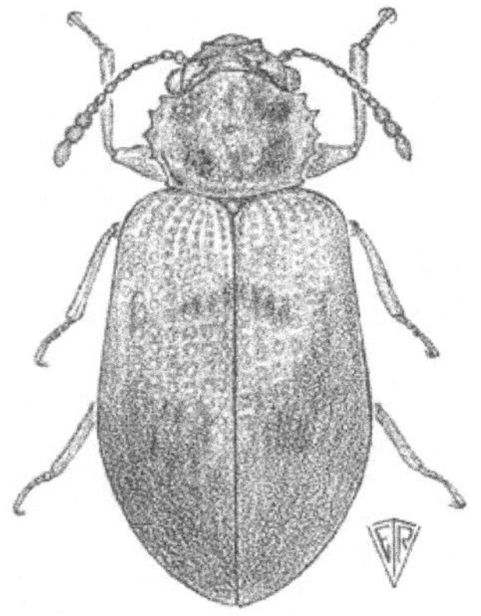

FIGURE 1.66. *Derodontus esotericus* Lawrence

Description (modified from Lawrence and Hlavac 1979): Size 2.0 mm to 3.2 mm, elongate, dorsally convex and ventrally flattened; vestiture of short fine hairs or longer erect or decumbent hairs. Color brownish (to black in *Laricobius*), sometimes mottled.

Head with 2 ocelli; compound eyes relatively large and entire; antennae 11-segmented with 3-segmented club, insertions concealed or exposed in dorsal view; frontoclypeal suture absent; frons and vertex often with a complex system of pores and canals; labrum well-developed. Mandible with well developed mola and prostheca (except *Laricobius*); maxilla with brush-like galea and lacinia bearing 1 or 2 spines at apex; palp 4-segmented, apical segment subulate; labial palps 3-segmented.

Prothorax relatively small with lateral carinae often explanate, and dentate; prosternum relatively short in front of coxae, with narrow intercoxal process; procoxae conical and projecting below sternal process, trochantin exposed or concealed; cavities open internally; externally open or closed. Mesocoxae approximate to almost contiguous; cavities broadly open laterally, bordered both by mesepimeron and metepisternum; perimeter of each cavity sometimes with several deep pores. Elytra seriate or striate (except *Peltastica*), with 10 or 11 punctate rows or striae; scutellary striole present or absent. Metasternum with long median suture and with transverse suture which may be interrupted or incomplete; anterolateral portion of metepisternum with elytral interlocking device; metacoxae somewhat motile, approximate, extending laterally to rim of elytra, with weakly developed plates. Metendosternite with stalk short, median process well-developed, anterior tendons approximate, lateral arms narrow and curved. Hindwing with 5 to 6 major veins; oblongum cell of hind wing absent; medial fleck of hind wing absent; wedge cell of hind wing well developed; anal lobe present. Legs relatively short trochanter attached obliquely to femur; tibial apex with 2 small spurs (except *Laricobius* which has an apical comb); tarsi 5-5-5, tarsomeres simple (except *Laricobius* with tarsomeres 1-3 lobed below); claws simple.

Abdomen with 5 free ventrites (except *Laricobius* with ventrites 3 and 4 connate); ventrite 1 not much longer than 2, intercoxal process of abdomen acute; longitudinal ridges sometimes present on ventrites 3-5, lateral pores and canals present in *Laricobius*; segments 8-10 enclosed; segment 8 with functional spiracles; segment 9 well-developed with 2 hemitergites, which may be fused at midline; tergite 10 well-developed and free.

Aedeagus simple and trilobed; parameres individually articulated to phallobase or base of penis; penis without dorsal and ventral lobes, anterior edge with paired struts. Ovipositor short, with 1-segmented coxites and short styli (long in *Derodontus*). Malpighian tubules free, 6 in number.

Larva (modified from Lawrence 1991) with dorsal surfaces granulate or tuberculate, mola asperate, galea and lacinia distinct, spiracles annular-biforous. Body elongate and more or less parallel sided or slightly fusiform, slightly flattened; dorsal surfaces granulate and tuberculate, sometimes darkly pigmented, vestiture of short simple setae. Head protracted and prognathous, moderately broad and slightly flattened. Epicranial stem very short or absent, frontal arms lyriform and contiguous at base. Median endocarina absent. Stemmata 6 on each side (5 in some *Nothoderodontus*). Antennae 3-segmented. Frontoclypeal suture absent or poorly developed; labrum free. Mandibles symmetrical, apex bidentate with accessory ventral process, incisor edge serrate or simple (*Laricobius*). Mola well-developed (poorly developed in *Laricobius*), asperate; prostheca absent (*Laricobius*) or present and consisting of fixed acute hyaline process with a brush of hairs. Ventral mouthparts usually retracted. Maxilla with transverse cardo, elongate stipes, palp 3-segmented with well-developed articulating area (absent in *Laricobius*), galea fixed narrowly or truncate, lacinea fixed and falciform, bifid or trifid at apex, galea and lacinia reduced in *Laricobius*. Labium usually free to base of mentum, more or less connate with stipites in *Laricobius*, ligula broad, not bilobed, labial palps 2-segmented and widely separated.

Hypopharyngeal sclerome as a transverse bar. Hypostomal rods absent. Ventral epicranial ridges present (absent in *Laricobius*). Gula wider than long. Legs well developed, 5-segmented; tarsungulus with 2 setae lying side by side; coxae moderately separated. Thoracic and abdominal terga 1-9 often with paired pigmented plates. Pleural lobes well-developed in most. Abdominal terga 1-8 with 1 or more pairs of setiferous tubercles or a median forked process (absent in *Laricobius*). Sternum A9 simple. Segment A10 circular, posteriorly oriented, cylindrical, in some sclerotized dorsally. Spiracles annular-biforous, borne on tubular cone-like process.

Habits and habitats. The biology of this small group of beetles is remarkably diverse (Lawrence and Hlavac 1979). Larvae and adults typically occur together in the same habitats and are easy to collect, especially in cool seasons and in temperate regions in the northern and southern hemispheres (Crowson 1980, Lawtrence 1982). *Peltastica* is present at sap flows where it feeds on fungi and other suspended nutrients while *Laricobius* are predators on Adelgidae (Hemiptera) (Franz 1958), with one species (*L. erichsoni* Rosenhauer) that was introduced to North America for biocontrol of *Adelges piceae* (Ratzeburg) in the Northwest and Northeast (Clark and Brown 1958, 1960; Hatch 1962). Predatory behavior is thought be derived in the family (Leschen 2000). *Derodontus* are found exclusively on fungi and are abundant in cooler seasons (Shepard 1976, Lawrence and Hlavac 1979). Host patterns have been described in detail for some of the North American species (Lawrence and Hlavac 1979, Leschen 1994) and an additional host for the western species *D. unidentatus* Lawrence is *Lentinellus ursinus* (Fr.:Fr.) Kühner (Agaricales). The south temperate genus *Nothoderodontus* is associated with sooty molds (Lawrence 1985), although some specimens have been collected in other habitats. Larvae have been described and keyed by several authors (Bøving and Craighead 1931, Fukuda 1963, Lawrence *et al.* 1999b).

Status of the classification. Derodontidae are an isolated and primitive group in Polyphaga (Crowson 1959) placed in their own superfamily Derodontoidea in the series Bostrichiformia by Lawrence and Newton (1995) and can be keyed in Lawrence *et al.* (1999a). The family may be sister taxon to Nosodendridae (Beutel 1996) and there are three subfamilies that contain the four genera. Older works of importance are Ganglbauer (1899), LeConte (1861), and Mulsant and Rey (1863-1864).

Distribution. Since the publication of Lawrence and Hlavac (1979), a few species have been described by Nitisky (1993a, 1993b) and Háva and Jelínek (1999) or verified taxonomically (Poggi 1995). There are 22 species from the Holarctic (*Laricobius*, 8 spp.; *Peltastica*, 2 spp.; *Derodontus*, 8 spp.) and south temperate regions of Australia, Chile, New Zealand (*Nothoderodontus*, 6 spp.) and nine species are present in North America.

Key to the Genera of the United States

1. Antennal insertions concealed in dorsal view; elytra with raised tubercles .... *Peltastica*
— Antennal insertions visible in dorsal view; elytra without raised tubercles 2

2(1). Dorsal surface with a vestiture of setae; tarsi 4-4-4 with tarsomeres 1-3 strongly lobed below .... *Laricobius*
— Dorsal surface without a vestiture of setae; tarsi 5-5-5 with tarsomeres 1-3 filiform .... *Derodontus*

Classification of the Nearctic Genera

Derodontidae LeConte 1861

Peltasticinae LeConte 1861

Peltastica Mannerheim 1852, 1 sp., *P. tuberculata* Mannerheim, 1852. Alaska, British Columbia, California, Idaho, Oregon.

Derodontinae LeConte 1861

Derodontus LeConte 1861, 4 spp., Alabama, Alaska, Arkansas, British Columbia, California, District of Columbia, Illinois, Indiana, Iowa, Kansas, Louisiana, Maryland, Massachusetts, Missouri, New Hampshire, Michigan, New York, Oklahoma, Ontario, Pennsylvania, South Carolina, Texas, Virginia.

Laricobiinae Mulsant and Rey 1863-64

Laricobius Rosenhauer 1846, 4 spp., British Columbia, District of Columbia, California, Idaho, Massachusetts, Michigan, New Brunswick, Newfoundland, New York, Nova Scotia, Oregon, Washington.

Bibliography

BEUTEL, R. G. 1996. Study of the larva of *Nosodendron fasciculare* (Olivier 1790) (Coleoptera, Nosodendridae) with implications for the phylogeny of Bostrichiformia. Journal of Zoological Systematics and Evolutionary Research, 34:121-134.

BØVING, A. G. and F. C. CRAIGHEAD. 1931. An illustrated synopsis of the principal larval forms of the Order Coleoptera. Entomologica Americana (New Series), 11: 1-351.

CLARK, R. C. and N. R. BROWN. 1958. Studies of predators of the Balsam Wooly Aphid, *Adelges piceae* (Ratz.) (Homoptera: Adelgidae) V. *Laricobius erichsonii* Rosen. (Coleoptera: Derodontidae), an introduced predator in eastern Canada. Canadian Entomologist, 62: 237-240.

CLARK, R.C. and N.R. BROWN. 1960. Studies of predators of the balsam wooly aphid, *Adelges piceae* (Ratz.) (Homoptera: Adelgidae) VII. *Laricobius rubidus* Lec. (Coleoptera: Derodontidae), a predator of *Pineus strobi* (Htg.) (Homoptera: Adelgidae). Canadian Entomologist, 62: 237-240.

CROWSON, R. A. 1959. Studies on the Dermestoidea (Coleoptera), with special reference to the New Zealand fauna. Transactions of the Royal Entomological Society of London, 111: 81-94.

CROWSON, R. A. 1980. On amphipolar distribution patterns in some cool climate groups of Coleoptera. Entomologia Generalis, 6: 281-292.

FRANZ, J. M. 1958. Studies on *Laricobius erichsonii* Rosenh. (Coleoptera: Derodontidae): a predator on chermesids. Entomophaga, 3: 109-196.

FUKUDA, A. 1963. Studies on the larva of *Peltastica reitteri* Lewis with comments on the classification of Derodontidae based on larval characters (Coleoptera, Derodontidae). Kontyû, 31: 189-193.

GANGLBAUER, L. 1899. Die Käfer Mitteleuropas; die Käfer der österreichisch-ungarischen Monarchie, Deutschlands, der Schweiz, sowie des französchen und italienischen Alpengebietes. Carl Gerold's Sohn. Vienna. iii + 1046 pp.

HATCH, M. H. 1962. The Beetles of the Pacific Northwest. Part III: Pselaphidae and Diversicornia I. University of Washington Press. Seattle. ix + 503 pp.

HÁVA, J. and J. JELÍNEK. 1999. A new species of the genus *Laricobius* (Coleoptera: Derodontidae) from China. Folia Heyrovskyana, 7: 115-118.

JELINEK, J. and J. HAVA. 2001. A new species of *Laricobius* (Coleoptera: Derodontidae) from Nepal. Revue Suisse de Zoologie, 108: 149-152.

LAWRENCE, J. F. 1982. Coleoptera. Pp. 482-553. *In:* S. P. Parker, ed., Synopsis and Classification of Living Organisms, Vol. 2. McGraw-Hill. New York.

LAWRENCE, J. F. 1991. Derodontidae (including Laricobiidae, Peltasticidae). Pp. 431-432. *In:* F.W. Stehr, Immature Insects, Vol. 2. Kendall/Hunt, Dubuque, IA.

LAWRENCE, J. F., A. M. HASTINGS, M. J. DALLWITZ, T. A. PAINE and E. J. ZURCHER. 1999a. Beetles of the World: A Key and Information System for Families and Subfamilies. CD-ROM, Version 1.0 for MS-DOS. CSIRO Publishing, East Melbourne. Victoria.

LAWRENCE, J. F., A. M. HASTINGS, M. J. DALLWITZ, T. A. PAINE and E. J. ZURCHER. 1999b. Beetle Larvae of the World: Descriptions, Illustrations, Identification, and Information Retrieval for Families and Sub-families CD-ROM, Version 1.0 for MS-DOS. CSIRO Publishing, East Melbourne. Victoria.

LAWRENCE, J. F. and T. HLAVAC. 1979. Review of the Derodontidae (Coleoptera: Polyphaga) with new species from North America and Chile. Coleopterists Bulletin, 33: 369-414.

LAWRENCE, J. F. 1985. The genus *Nothoderodontus* (Coleoptera: Derodontidae) with new species from Australia, New Zealand and Chile. Pp. 68-83. *In:* G. E. Ball, ed. Taxonomy, Phylogeny and Zoogeography of Beetles and Ants. W. Junk. Dordrecht.

LAWRENCE, J .F. and A. F. NEWTON, Jr. 1995. Families and subfamilies of Coleoptera (with selected genera, notes, references and data on family-group names). Pp. 779-1006. *In:* J. Pakaluk and S. A. Slipinski, eds., Biology, Phylogeny, and Classification of Coleoptera: Papers Celebrating the 80th Birthday of Roy A. Crowson. Muzeum Instytut Zoologii PAN: Warsaw.

LECONTE, J. L. 1861. Classification of the Coleoptera of North America, Part I. Smithsonian Miscellaneous Collections, 3: i-xxv + 1- 286.

LESCHEN, R.A.B. 1994. Fungal host use in two species of *Derodontus* (Coleoptera: Derodontidae). Coleopterists Bulletin, 48: 126-130.

LESCHEN, R.A.B. 2000. Beetles feeding on bugs (Coleoptera, Hemiptera): repeated shifts from mycophagous ancestors. Invertebrate Taxonomy, 14: 917-929.

MULSANT, E. and C. REY. 1863-64. Histoire Naturelle des Coléoptères de France. Angusticolles, Diversipalpes. Magnin, Blanchard & Co. Paris. 134 + 24 pp., 2 pls.

NIKITSKY, N. B. 1993a. Beetles of the family Derodontidae (Coleoptera, Polyphaga) in the eastern Palaearctic. Byulleten' Moskovsogo Obshchestva Ispytatelei Prirody, Otdel Biologicheskii, 92: 31-36 [in Russian].

NIKITSKY, N. B. 1993b. New species of Derodontidae and Biphyllidae (Coleoptera) from the Caucasus. Zoologicheskiy Zhurnal, 72: 152-156 [in Russian; reprinted in Entomological Review, 72: 52-55].

POGGI, R. 1995. Note su *Derodontus raffrayi* Grouvelle e *D. macularis* (Fuss) (Coleoptera: Derodontidae). Bollettino della Società Entomologica Italiana, 126: 249-254.

SHEPARD, W. D. 1976. Records and notes concerning *Derodontus maculatus* (Mels.). Southwestern Entomologist, 1: 168-170.

Superfamily BOSTRICHOIDEA Latreille 1802

67. NOSODENDRIDAE Erichson 1846

by Michael A. Ivie

Family common name: The nosodendrid beetles

The black, oval form, with the ability to retract appendages, the flattened prolegs held with the tibia anterior to the femur at rest, and the antennal club protected in cavities between the prolegs and hypomeron will serve to separate this family.

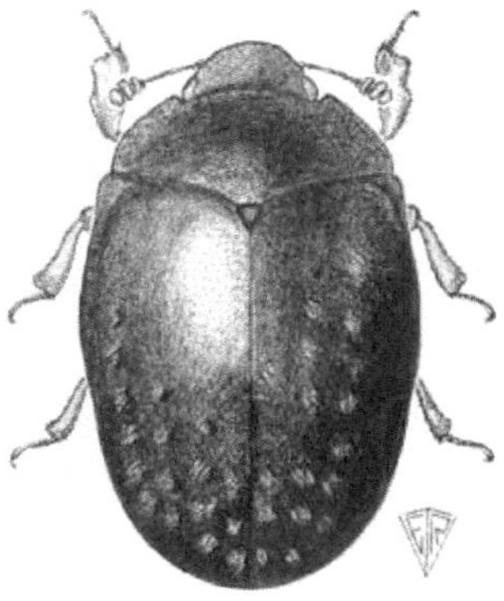

FIGURE 1.67. *Nosodendron californicum* Horn

Description: (see Figs. 1-3). Shape oval, convex, compact, size 2.3 to 9 mm in length; black; vestiture absent, with short erect or recumbent setae arising from distinct punctures or tufts of stout to scale-like setae on elytra, appendages retractile.

Head prominent or covered by the pronotum, prognathous or hypognathous; surface densely to sparely punctate. Eyes lateral, rounded, complete or slightly emarginate near antennal insertion. Antennae 11 segmented, distinct three-segmented club, inserted between eye and anterior corner of mandible. Gena with antennal groove along underside of eye. Labrum sclerotized, free, transverse, very narrowly to obviously visible. Mandibles short, broad, slightly curved, the apices acute, uni- or bidentate; mola well developed. Maxilla with galea and lacinia separate; galea narrow, elongate with dense bundle of long apical setae; lacinia elongate, with apical hooks, with or without mediad setose lobe; maxillary palpus four-segmented, cylindrical; mentum wide, heavily sclerotized, either expanded anteriorly to fit against venter of mandibles, enclosing maxillae and other parts of labium, or normal; labial palpus with three cylindrical palpomeres; ligula large, bilobate and densely setose.

Prothorax widest at base, broadly to narrowly rounded anteriorly, lateral margins acute; hypomeron excavate to receive antennal club and prolegs; prosternum short to very short, not extending anterior to procoxae, not contacting mouthparts in retracted position. Mesoscutellum visible, triangular. Mesosternum short, medially notched to receive prosternal process or anteriorly rounded; lateral portions of mesothorax depressed to receive legs in retracted position. Metasternum broad, apically truncate.

Tarsi 5-5-5. Procoxal cavities broadly open behind; procoxae countersunk, strongly transverse, with plate-like surfaces exposed behind and surrounding visible ball socket of trochanter obvious in retracted position; trochanters large, offset triangular; femur broad, flattened, excavate and sharply margined anteriorly to receive and cover edge of tibia, which is held anterior to femur in repose; tibia broad, flattened, arcuate to emarginate and armed with small to large spines on edge held away from femur in repose, excavate to receive and hide tarsus in repose; tarsi simple, with stout setae below on tarsomeres 1-4. Mesocoxae countersunk into mesothorax; cavities open laterally; legs similar to prolegs, but less broad, with tibia held posterior to femur in repose, tibia without spines on shaft and tarsi free in repose or received in grooves on elytral epipleuron. Metacoxae transverse, contiguous or narrowly separated; reaching epipleuron; excavate behind to receive femur; femur excavate to receive tibia, tarsi free or received in grooves on first ventrite.

Elytra entire, surface smooth and shining or microrugose; densely, confusedly punctate, or sparsely striate punctuate with a hexagonal-quadrangular meshwork (Reichardt 1976); epipleuron broad basally, obsolete beyond metacoxae; with or without depression to receive legs and/or tarsi. Hind wing well developed, with normal transverse folds and short, well developed radial cell (Forbes 1922, 1926).

Abdomen with five ventrites, all free; sutures impressed or not; first variously excavate to receive hind tibia, second with or without grooves to receive metatarsi; anterior portion of ventrites variously plain or sculptured; lateral margins of ventrites surrounded by fine to elevated carina; when elevated, this carina margining deep groove, widened on last ventrite and narrowly interupted medially; internally with unique dorso-posteriorly directed plates on anterior margins of ventrites 2-5 (Ivie 1985). Malpighian tubules free, not cryptonephridic.

Male genitalia of the trilobed type, rather short and broad; parameres and penis more or less equally developed; penis usually somewhat expanded at middle and with well developed apophyses; parameres individually articulated; phallobase relatively short (Reichardt 1976, Beal 1985). Females with paraproct, coxites and stylus well developed but variable; bacculi long and sclerotized or very short and complex (Reichardt 1976).

Larvae of 2 distinct types, but they share the presence of a mandibular mola, and a granulate dorsal surface with a transverse line on the anterior portion of each segment starting with the mesothorax (Beal 1985 considers this line to correspond to the antecostal suture). This line is straight and made up of clearly independent granules in *Nosodendron*, but is a scalloped carina in *Orphilus*. At high magnification this carina can be seen to be coalesced granules, especially mediolaterally on the abdomen. The

2

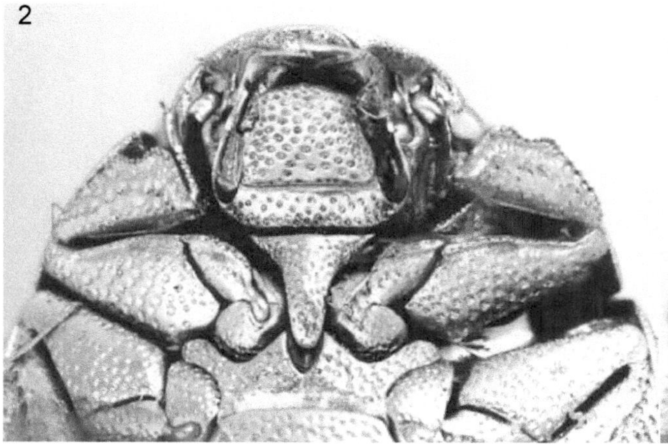

3

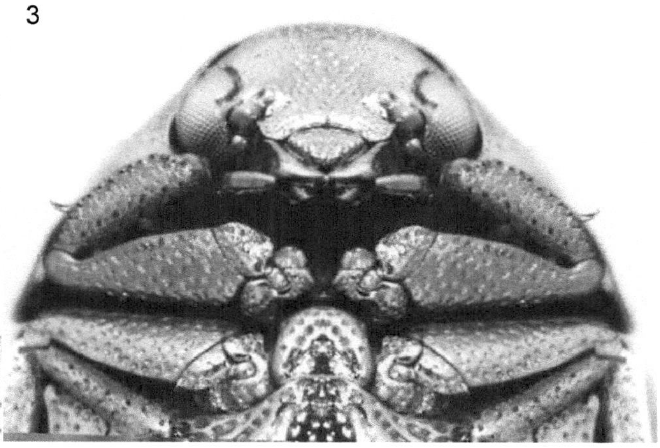

FIGURE 2.67-3.67. 2. *Nosodendron californicum* Horn, ventral view of head and thorax; 3. *Orphilus ater* Erichson, ventral view of head and thorax.

larva of *Nosodendron* has been well and repeatedly described (Bøving and Craighead 1931, Peterson 1960, Lawrence 1991, Beutel 1996, Lawrence *et al.* 1999b and others cited therein). Those of *Orphilus* have been described less extensively by Paulian (1942), Beal (1985, 1991), Beutel (1996) and Lawrence *et al.* (1999b).

Habits and habitats. Nosodendrids are most likely fungus feeders, living in association with fermenting sap flows or fungus infested wood. The larvae and adults of North American *Nosodendron* have been taken repeatedly in slime fluxes, which are rich in yeasts, other fungi, bacteria and fermentation products. The mandible, maxilla and ligula of larval *Nosodendron* are typical of fungal feeders. *Nosodendron unicolor* Say is recorded from sap flows on hardwood tress, including elm (*Ulmus* sp.), poplar (*Populus* sp.), willow (*Salix* sp.) and alder (*Alnus* sp.) (Hayes and Chu 1946, Peterson 1960), while *N. californicum* Horn is only reported from conifers, being reported from large grand fir [*Abies grandis* (Douglas) Lindley], white fir [*A. concolor* (Gord. And Glend.)], and Douglas fir [*Pseudotsuga menziesii* (Mirbel) Franco] (Sokoloff 1959, Osborne and Kulhavy 1975, Zack *et al.* 1979, Kulhavy 1980). Although there are reports of predation on dipteran larvae (Hayes and Chu 1946, Costa *et al.* 1986, etc.), this may simply involve facultative feeding on soft animals that co-occur in the slime flux habitat (Beutel 1996, Lawrence *et al.* 1999a). A rearing experiment to test this would make an excellent student project.

Adult *Orphilus* are commonly found on a wide variety of flowers with dense flower clusters, such as *Achillea, Spiraea, Ceanothus, Hydrangea, Eriogonum, Eriodictyon, Fallugia,* and various Umbelliferae. The known larvae have been found in dead wood where they probably feed on fungi. Beal (1985) reported larvae of *O. subnitidus* LeConte found in association with the fungus *Trametes sepium* on dead branches of madron (*Arbutus menzeisii* Pursh) from northern California. Paulian (1942) reported larval *O. niger* in oak (*Quercus* sp.) from Algeria.

Status of the classification. The members of this group have historically been assigned to 2 families: the Nosodendridae *s. str.* (i.e., *Nosodendron*) and the Dermestidae (i.e., *Orphilus*, Orphilinae). The Nosodendridae in that sense have been placed variously in the Dascilloidea, Dermestoidea, Byrrhoidea, Derodontoidea, and near the Histeroidea. Since Crowson's placement in 1955, and his clarification in 1959 (Crowson 1955, 1959), they have consistently been included, along with the Dermestidae, in either a Dermestoidea or Bostrichoidea in all major classifications except Hatch (1961), Lawrence (1991) and Beutel (1996). Beutel's (1996) placement of *Nosodendron* in the Derodontoidea was affected by the combination of plesiomorphic and highly autapomorphic characters of the adults and larvae of *Nosodendron*, associated with living in a functionally liquid environment. This situation lead to confusion of polarization, homology, and homoplasy, as well as at least one case of mistaken character state assumption. His placement was rejected by Lawrence *et al.* (1999a), who returned *Nosodendron* to the Bostrichoidea alongside the Dermestidae.

Ivie (1985) discovered that the recognition of only a single genus in this family left the Dermestidae a paraphyletic group, as the Orphilinae lack any exclusive synapomorphy with the Dermestidae *s. str.* (Ivie 1985). This required the transfer of the Orphilinae from the Dermestidae to the Nosodendridae in order to render both families monophyletic (Ivie 1985). Although the nomenclatural acts in that dissertation remain unpublished in the sense of the International Code of Zoological Nomenclature, its widespread citation (Lawrence and Newton 1995, Beutel 1996, Lawrence *et al.* 1999a) made the existence of the data known.

The problematic status of the Orphilinae has been discussed as far back as Crowson (1955), and the similarity of *Orphilus* to *Nosodendron* was noted by Beal (1985), among others. In a list of synapomorphies of the Dermestidae, *Orphilus* is an exception in every case: they lack the spinulate (haustate) larval setae, mouth parts received by the prosternum and the short metacoxa that is separated from the elytra by the metepisternum; and have a mola in the larval mandible, etc. (Ivie 1985). The continued placement of *Orphilus* in the Dermestidae has been supported solely by a combination of symplesiomorphies or incorrectly assumed character states (Lawrence and Newton 1995, Beutel 1996, Lawrence *et al.* 1999a). Lawrence and Newton (1995) simply cited unspecified "features in common" with the Dermestidae, apparently referring to the many symplesiomorphies they share. In the only actual phylogenetic analysis to contest the Orphilinae as Nosodendridae, Beutel (1996) stated his reason to reject the sister-group status of [*Orphilus* + *Nosodendron*] was that the larva of *Orphilus* belonged in the Bostrichoidea while *Nosodendron* did not. This argument is rendered a *non-sequitur* when the Nosodendridae are returned to that superfamily, as done by Lawrence *et al.* (1999a). The only synapomorphy Beutel (1996) specifically stated to support the Dermestidae including *Orphilus* was the dorsal surface of the larva covered in abundant setae, a character state that is shared with the Endecatominae (Bostrichidae), and some Anobiidae. In the phylogenetic reconstruction of Ivie (1985), *Endecatomus* emerged as the basal clade of the sister-lineage to the [Nosodendridae + Dermestidae], clearly rendering Beutel's character state suspect as symplesiomorphic. Other arguments for placing *Orphilus* in the Dermestidae, such as retention of adult ocelli (shared with staphiliniforms and derodontids), a campodeiform/eruciform larval type with a hypognathous head (shared with many basal groups), etc. are also symplesiomorphies. Lastly, some dermestid characters have been attributed to *Orphilus* that they simply do not possess, most importantly modifed cryptonephridism (Ivie 1985 vs. Beutel 1996). Most critical to the classification of the group, *Orphilus* and *Nosodendron* share an extensive set of complex and unrelated but concordant synapomorphies, including the form of antennal protection; unique structure of the procoxae, femora and tibiae; loss of modified cryptonephridism, larval sculpture, and unique internal plates on the anterior margin of the ventrites (Ivie 1985). Nevertheless, following Lawrence and Newton (1995), *Orphilus* is also treated in the Dermestidae in this volume (see Chapter 68).

There are 62 described species, 58 in *Nosodendron*, 4 in *Orphilus*. Reichart (1976) cataloged the world species of *Nosodendron*, and since then, ten African, Malagasy and Asian species have been described (Endrödy-Younga 1991, Hava 2000). The species are relatively well described, and most are easily identified. The New World species are illustrated and keyed by Reichart (1976), although there are remaining undescribed Neotropical species. Identification of specimens from the Ethiopian region is covered by Endrödy-Younga (1991), the Oriental region by Hava (2000), the Europe by Paulus (1979), the Russian Far East by Lafer (1992), and remaining Old World species can be identified with Champion (1923). The Holarctic and Neotropical faunae are apparently not closely related, the Nearctic species being rather distinct from those of the Neotropics (Reichart 1976). Beal (1985) covers the Orphilinae (see below, and Chapter 68. Dermestidae, in this volume).

Distribution. The Nosodendrinae are mainly a tropical group but the species occur throughout the Americas (Canada to Argentina), in Europe, the Kurils, Japan, South and South East Asia, Indonesia, the Philippines, New Guinea, New Caledonia, Fiji, Australia, New Zealand, South Africa, Madagascar and the Comoros. The Orphilinae are Holarctic, known from described species in both Eurasia and North America. North American distribution records for both groups have been added by Campbell (1991), Downie and Arnett (1996), Peck and Thomas (1998).

Key to World Genera

1. Head with median ocellus; venter of body convex; first ventrite without tarsal grooves; head hypognathous, reflexed against procoxae; mentum normal, not expanded; abdomen ventrally margined with a sharp marginal carina and associated groove running from coxa to coxa, setting off a false pygidium on last ventrite .... *Orphilus*

— Head without ocellus; venter of body flat; first ventrite with tarsal grooves; head prognathous; mentum expanded and plate-like, covering mouthparts in retracted position; abdomen without marginal carina .... *Nosodendron*

Classification of the Nearctic Species

Nosodendridae Erichson 1846

Nosodendrinae Erichson 1846

Nosodendron Latreille 1804, is a rather uniform genus of 58 described species with two North American members: *N. californicum* Horn 1874 from California to British Columbia, east to Idaho; *N. unicolor* Say 1824 in the east from Wisconsin and Ontario through Kansas and Texas to Florida. Reichart (1976) keys the American species. Oehme-Leonhardt (1954) synonymized *N. californicum* Horn with the European *N. fasciculare* (Olivier 1790) and Reichardt (1976) provided some support for this action, but then did not follow it. It remains for a future reviser to reevaluate this question.

Orphilinae LeConte 1861

Orphilus Erichson 1846, contains four described species with three very similar extant and one fossil species. Two of the extant species and the fossil are from North America: *Orphilus ater* Erichson 1846 occurs from Wisconsin and Ontario and Quebec to Arkansas and Florida; *Orphilus subnitidus* LeConte 1861 from British Columbia to Montana, east to western Nebraska and south to California and New Mexico. The fossil species, *Orphilus dubius* Wickham 1912, is from the Miocene of Florissant, Colorado. The remaining species is *Orphilus niger* (Rossi 1790) from Europe, Asia Minor, Middle Asia and the Mediterranean region. Beal (1985) illustrates and discriminates the extant species. The generic identity of the fossil species needs verification. It is not known if the East Indian distribution for *Orphilus* cited by Lawrence *et al.* (1999a) refers to *O. niger* or an undescribed species.

BIBLIOGRAPHY

BEAL, R. S. 1985. Review of the Nearctic species of *Orphilus* (Coleoptera: Dermestidae) with description of the larva of *O. subnitidus* LeConte. Coleopterists Bulletin, 39: 265-271.

BEAL, R. S. 1991. Dermestidae (Bostrichoidea). Pp. 434-439. *In:* F. W. Stehr, ed., Immature Insects. Vol. 2. Kendall/Hunt. Dubuque, IA.

BEUTEL, R. G. 1996. Study of the larva of *Nosodendron fasciculare* (Olivier 1790) (Coleoptera: Nosodendridae) with implications for the phylogeny of Bostrichiformia. Journal of Zoological Systematics and Evolutionary Research, 34: 121-134.

BØVING, A. G. and F. C. CRAIGHEAD. 1931. An illustrated synopsis of the principal larval forms of the order of Coleoptera. Entomologica Americana (new series), 11: 1-351.

CAMPBELL, J. M. 1991. Nosodendridae. P. 196. *In:* Y. Bousquet, ed., Checklist of the beetles of Canada and Alaska. Agriculture Canada Publication 1861/E. Canada Communications Group - Publishing. Ottawa.

CHAMPION, G. C. 1923. A revision of the eastern species of Nosodendridae (Coleoptera). Annals and Magazine of Natural History, 9: 578-591.

COSTA, C., S. A. CASARI-CHEN and S. A. VANIN. 1986. Larvae of Neotropical Coleoptera, XVI. Nosodendridae. Revista Brasileira de Entomologia, 30: 291-297.

CROWSON, R. A. 1955. The natural classification of the families of Coleoptera. Nathaniel Lloyd. London, 187 pp.

CROWSON, R. A. 1959. Studies on the Dermestoidea (Coleoptera), with special reference to the New Zealand fauna. Transactions of the Royal Entomological Society of London, 111: 81-94.

DOWNIE, N. M. and R. H. ARNETT, Jr. 1996. The Beetles of Northeastern North America. Vol. 1. The Sandhill Crane Press. Gainesville, FL.

ENDRÖDY-YOUNGA, S. 1991. Insectes Coléoptères Nosodendridae. Faune de Madagascar, 76: 1-29. Muséum National d'Histoire Naturelle, Paris.

FORBES, W. T. M. 1922. The wing-venation of the Coleoptera. Annals of the Entomological Society of America, 15: 328-345, plates 29-35.

FORBES, W. T. M. 1926. The wing folding patterns of the Coleoptera. Journal of the New York Entomological Society, 34: 42-68, 91-138.

HATCH, M. H. 1961. The beetles of the Pacific Northwest. Part III: Pselaphidae and Diversicornia I. University of Washington Publications in Biology, 16: 1-503.

HAVA, J. 2000. Distributional notes on some Oriental Nosodendridae (Coleoptera), with descriptions of five new species. Acta Musei Moraviae Scientiae Biologicae, 85: 57-65.

HAYES, W. F. and H. F. CHU. 1946. The larvae of the genus *Nosodendron* Latr. (Coleoptera: Nosodendridae). Annals of the Entomological Society of America, 39: 69-79.

IVIE, M. A. 1985. Phylogenetic studies in the Bostrichiformia (Coleoptera). Ohio State University (PhD. Thesis), Columbus, 137 pp.

KULHAVY, D.L. 1980. Characterization of grand fir colonized by *Nosodendron californicum* Horn (Coleoptera: Nosodendridae). Coleopterists Bulletin, 34: 171-173.

LAFER, G. Sh. 1992. 48. Cem. Nosodendridae. Pp. 44-46. *In:* P. A. LER, ed., Key to the Insects of the Russian Far East. Vol. III. Part 2. Dal'nauka, Vladivostok.

LAWRENCE, J. F. 1991. Nosodendridae (Derodontoidea). Pp. 432-433. *In:* F. W. Stehr, ed., Immature Insects. Vol. 2. Kendall/Hunt. Dubuque, IA.

LAWRENCE, J. F., A. M. HASTINGS, M. J. DALLWITZ, T. A. PAINE and E. J. ZURCHER. 1999a. Beetles of the World: A Key and Information System for Families and Subfamilies. CD-ROM, Version 1.0 for MS-Windows. CSIRO Publishing. Melbourne.

LAWRENCE, J. F., A. M. HASTINGS, M. J. DALLWITZ, T. A. PAINE and E. J. ZURCHER. 1999b. Beetle Larvae of the World: Descriptions, Illustrations, Identification, and Information Retrieval for Families and Sub-families. CD-ROM, Version 1.1 for MS-Windows. CSIRO Publishing: Melbourne.

LAWRENCE. J. F. and A. F. NEWTON, Jr. 1995. Families and subfamilies of Coleoptera (with selected genera, notes, references and data on family-group names). Pp 779-1006. *In:* J. Pakaluk and S. A. Slipinski, eds., Biology, phylogeny and classification of Coleoptera: Papers celebrating the 80th birthday of Roy A. Crowson. Muzeum i Instytut Zoologii PAN, Warsaw.

OEHME-LEONHARDT, B.G. 1954. A new Philippine species of *Nosodendron* Philippine Journal of Science, 82: 253-261.

OSBORNE, H.L. and D.L. KULHAVY. 1975. Notes on *Nosodendron californicum* Horn on slime fluxes of grand fir, *Abies grandis* (Douglas) Lindley, in northern Idaho (Coleoptera: Nosodendridae). Coleopterists Bulletin, 29: 71-73.

PAULIAN, R. 1942. The larvae of the sub-family Orphilinae and their bearing on the systematic status of the family Dermestidae (Col.). Annals of the Entomological Society of America, 25: 393-396.

PAULUS, H. F. 1979. 46. Familie: Nosodendridae. Pp. 327-328. *In:* H. Freude, K. W. Harde and G. A. Lohse, eds. Die Käfer Mitteleuropas. Band 6. Diversicornia. Goecke & Evers, Krefeld.

PECK, S. B., and M. C. THOMAS. 1998. A distributional checklist of the beetles (Coleoptera) of Florida. Arthropods of Florida and Neighboring Land Areas 16: i-viii + 1-180.

PETERSON, A. 1960. Larvae of insects. An introduction to Nearctic species. Part II. Coleoptera, Diptera, Neuroptera, Siphonaptera, Mecoptera, Trichoptera. Columbus, OH. 416 pp.

REICHARDT, H. 1976. Monograph of the New World Nosodendridae and notes on the Old World forms (Coleoptera). Papéis Avulsos de Zoologia, 29: 185-219.

SOKOLOFF, A. 1959. The habitat-niche of American Nosodendridae. Coleopterists Bulletin, 13: 97-98.

ZACK, R.S., E.J. DAVIS and K.F. RAFFA. 1979. A new host record and notes on *Nosodendron californicum* Horn. Coleopterists Bulletin, 33: 74.

68. DERMESTIDAE Gyllenhal 1808

by John M. Kingsolver

Family common name: The skin beetles

Family synonyms: Megatoillidae Leach 1815, Thorictidae Wollaston 1854.

The oval or obovate, compact shape, the clubbed antennae which are received into a shallow or deep antennal fossa, and the structure of the hind coxae which are excavated for the reception of the femora distinguish most adults of this family. A dense covering of setae or scales and the presence of a median ocellus facilitate the recognition of many of the members.

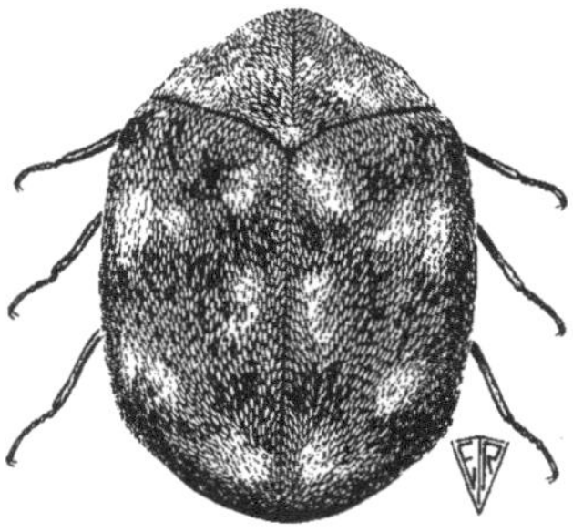

FIGURE 1.68. *Anthrenus verbasci* (Linnaeus)

Description: Shape compact, subparallel to obovate or nearly round, most are strongly convex; size 1 to 12 mm in length; color dark, some with white, yellow, brown, or reddish patterns of scales or setae; vestiture mostly consisting of a moderately dense covering of setae or scales.

Head small, deflexed, capable of being retracted into prothorax to the eyes in most. Antenna inserted in front of eyes, composed of five to eleven antennomeres, mostly clubbed (filiform in *Thylodrias*); antennal club in many of three antennomeres but in some of one to eight antennomeres, club compact or loosely joined, symmetrical, serrate, pectinate, or flabellate. Labrum distinct; mandibles small, curved, stout, the apices blunt; maxillary palpi with four palpomeres, these small, slender; mentum quadrate, ligula simple, labial palpi with three-palpomeres, short. Eyes mostly present, lateral, moderate, bulging, rounded. Median ocellus frequently present.

Pronotum narrowed in front, transverse, convex, the hind angles acute; lateral margins explanate or with fine carina at least on posterior half; pleural regions (hypomera) mostly concave or with a distinct and sharply circumscribed excavation for the reception of the antennae; prosternum usually produced posteriorly and lobed in front; procoxal cavities open behind. Mesosternum subquadrate or elongate, entire or sulcate; sulcus when present on anterior half only or completely dividing mesosternum. Metasternum subquadrate or shorter than wide. Legs mostly retractile; trochantin of fore and middle leg exposed; anterior coxa conical, prominent; middle coxa oval; hind coxa transverse, mostly with groove for reception of femur; trochanter large, triangular, fore and middle pair interstitial; femur swollen; hind femur usually with ventral groove for partial reception of the tibia; tibia slender, spinose, the apical spurs distinct; tarsal formula 5-5-5, tarsomeres one to four short; claws simple. Scutellum mostly visible, small to moderate, triangular. Elytra entire, apically rounded; not or very indistinctly striate; epipleural fold generally distinct, sometimes feebly developed or absent. Wing venation variable; wedge-cell open by lack of fusion of 2A and 3A. Folding pattern variable, incomplete, area A not well developed, replaced by a small adjustment fold between B and G only, the anal lobe free, or otherwise modified (see Forbes 1922, 1926).

Abdomen with five visible sterna (seven in *Thylodrias* sp.), the sutures entire, first sternum entire except in male of *Thylodrias* sp. where broadly and completely divided at middle. Male genitalia of the trilobed type; penis long, slender, apically blunt; parameres long, slender, apically rounded, pars basalis cap-shaped, short and broad. Female genitalia with the paraprocts and valvifers reduced to baculi; coxite two-segmented; stylus apical.

Larvae elongate and subcylindrical, obovate and subcylindrical, or ovate and somewhat flattened; 2 to 15 mm. in length; body densely covered with long or short spinulate setae, in some also with hastate setae, ramous setae, or club-shaped setae. Head hypognathous, subglobular; epicranial and frontal sutures present; frons triangular. Antenna with three segments, second with an accessory papilla. Clypeus present; labro-epipharyngeal margin with series of hooked or spatula-shaped setae, middle setae dissimilar or similar to lateral setal series; mandible distally rounded, in some dentate or pointed, with or without retinaculum and basal setal brush, with a basal membranous lobe or penicillus; maxillae with palpi three- or four-segmented; galea setiferous; lacinia with an elongate, curved, apically bifid spur (two in *Thylodrias*); labium with two-segmented palpi and bilobed ligula. Stemmata up to six pairs. Thorax with four-segmented legs with apical claw-like tarsunguli. Abdomen with nine or 10 segments, the last consisting of a sclerotized tergum only. Spiracles annuliform and inconspicuous on mesothorax and abdominal segments one to eight. Urogomphi prominent dorsally on the ninth segment of *Dermestes* spp., otherwise absent. (See Bøving and Craighead 1931, Peterson 1951, Rees 1943.)

Habits and habitats. Most dermestids for which the habits are known are scavengers, feeding on dried animal or plant materials of high protein content. Members of the genus *Dermestes* are commonly found in animal carcasses in the third or butyric state of decomposition. Many of the smaller species occur in bee and wasp nests, feeding on old pollen stores or on dried insect remains. Others more commonly are found in bird nests feeding on old feathers and other organic debris or in mammal nests

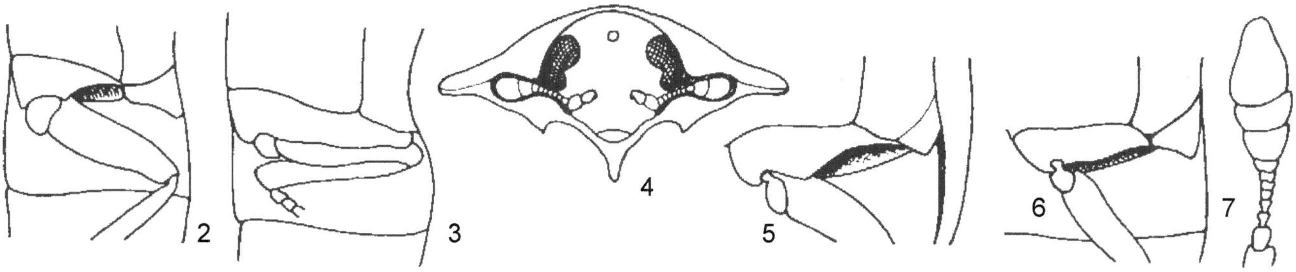

FIGURES 2.68-7.68. 2, *Dermestes carinus* Germar, thorax and abdomen, part, lateral view; 3, *Orphilus niger*, (Rossi), thorax and abdomen, part, lateral view; 4, *Anthrenus scrophulariae* (Linnaeus), head and prothorax, part, anterior view; 5, *Attagenus pellio* Linnaeus, metacoxal plate, ventral view; 6, *Trogoderma versicolor* Creutz, metacoxal plate, ventral view; 7, *Attagenus pellio* Linnaeus, antenna.

feeding on hair. Species of *Thaumaglossa* are found only in mantid egg cases; *Orphinus* sp. is sometimes found in mantid egg cases. *Apsectus* spp. live in spider webs, feeding on webbing and on dried spider eggs. A few species of *Trogoderma* appear primarily to be predators, feeding on wasp and bee larvae and on spider eggs. Other *Trogoderma* spp. are capable of subsisting on dried cereal products alone and, like the khapra beetle, are notorious pests of granaries (Hinton 1945, Kingsolver 1963). Adults of most of the smaller species of Dermestidae are commonly taken on flowers, where they feed on pollen and nectar.

Status of the classification. Because of the great economic importance of this family, the taxonomy is in fairly sound condition. A few of the non-economic genera are in need of revision. (See Jayne 1982, Beal 1959b, and Crowson 1959 for suprageneric revisions; catalogs include Dalla Torre 1911, Hetschko 1926, 1930, Mroczkowski 1968.)

Distribution. There are approximately 700 species (1945) known from all areas, 123 species occur in the United States (1961).

Key to the Nearctic Genera

1. Antenna with apical club, abdomen with five visible sterna; elytra present in both sexes 2

— Antenna without apical club; abdomen with seven visible sterna; female larviform (without elytra or hind wings) (Thylodriini) .... *Thylodrias*

2(1). Compound eyes present; legs more or less retractile; hind femur received in groove in coxa; large or small beetles 3

— Compound eyes absent; legs not retractile; hind coxa not grooved for reception of femur, minute beetles (less than 2 mm long) with general appearance of a *Dermestes* (Thorictini) .... *Thorictodes*

3(2). Dorsal surface covered with long or short setae or scales; front legs in repose not covering antennae in antennal fossae; metacoxal plate not extending to sides of body (Fig. 2) 4

— Dorsal surface glabrous; front legs in repose completely covering antennae within antennal fossae; metacoxal plate extending to sides of body (Fig. 3) (Orphilini) .... *Orphilus*

4(3). Pronotum without sublateral carinae; small or large beetles with suberect or subrecumbent setae or scales 5

— Pronotum with sublateral carinae extending from base nearly to anterior margin, mesal side of each carina depressed or somewhat sulcate; small (less than 2.5 mm long), strongly convex beetles covered with moderately long, erect setae (Trinodini) .... *Apsectus*

5(4). Head with median ocellus (Fig. 4; except in a few rare species less than 4 mm in length) species less than 5.5 mm in length; procoxae not contiguous at apices 6

— Head without an ocellus; species 5.5 to 12 mm in length; procoxae large and contiguous at apices (Dermestini) .... *Dermestes*

6(5). First hind tarsomere much shorter than second; metacoxal plate bearing a distinct tooth or distinctly broadened lateral to insertion of femur (Fig. 5) (Attagenini) 7

— First hind tarsomere as long as or longer than second; metacoxal plate with margins subparallel or gradually narrowed laterally (Fig. 6; Anthrenini)........... 8

7(6). Antennomeres of antennal club compact (Fig. 7); in male, length of ultimate antennomere greatly exceeds combined length of two preceding .... *Attagenus*

— Antennomeres of antennal club loosely joined; in male, length of ultimate shorter than combined length of two preceding, or all three greatly elongate, penultimate twice as long as wide............ .... *Novelsis*

8(6). Vestiture of setae (some of which may be slightly ensiform but never scale-like).......................... 9

— Vestiture of flat, conspicuously colored scales (Fig. 1).. *Anthrenus*

9(8). Antennal fossa partially or completely closed behind, or margined behind by a knifelike or threadlike carina; mesosternum deeply divided by sulcus at posterior margin 10

— Antennal fossa broadly open behind; posterior margin of fossa with or without a medial tumescence but never with a distinct carina; mesosternum with median sulcus becoming very shallow or obliterated posteriorly 15

10(9). Pronotum with small, short, diagonal, impunctate area on either side of basal lobe; male antennae with eleventh antennomere immensely enlarged, sub-triangular, approximately twice as wide as length of preceding combined; female with elev-

enth antennomere small, about as long as length of ninth and tenth combined ...... *Thaumaglossa*
— Pronotum without impunctate areas on either side of basal lobe; antennal club composed of two or more antennomeres; male antenna with terminal segments never as wide as length of preceding segment combined 11

11(10). Club of antenna composed of two antennomeres 12
— Club of antenna composed of three or more antennomeres 13

12(11). Club of antenna elongate or oval in outline; penultimate antennomere longer than terminal .... *Cryptorhopalum*
— Club of antenna nearly circular in outline; penultimate antennomere equal to or shorter than terminal .... *Orphinus*

13(11). Club of antenna composed of three large, subtriangular antennomeres, length of the club in male more than twice length of preceding eight antennomeres combined .... *Ctsias*
— Club of antenna composed of three to eight antennomeres, length of last three in male shorter than length of preceding eight combined; antennomeres of club symmetrical, pectinate, or flabellate, but if club of three or four antennomeres then these always more or less symmetrical.... 14

14(9). Antennal fossa at least partly enclosed behind by knifelike carina; metepisternum without transverse stria near anterior margin ...... *Trogoderma*
— Antennae fossa appearing broadly open behind but posterior margin marked by low, threadlike carina; metepisternum with transverse stria close to anterior margin .... *Reesa*

15(9). Antenna with 11 antennomeres 16
— Antenna with 9 antennomeres .... *Dearthrus*

16(15). Antenna of male with antennomere 10 at least one-fifth narrower than antennomere 11 and antennomere 9 at least one-fifth narrower antennomere 10; antenna of female with antennomere 9 five-sixths as wide as terminal antennomere .... *Pseudohadrotoma*
— Antenna of male with antennomere 10 as wide as or slightly wider than antennomere 11, and antennomere 9 as wide as or not less than four-fifths as wide as terminal antennomere; antenna of female with antennomere 9 subequal in width to terminal antennomere .... *Megatoma*

Classification of the Nearctic Genera

Dermestidae Latreille 1807

Dermestini Latreille 1807

Dermestes Linnaeus 1758, 15 spp., generally distributed (keys to spp., Casey 1900, Lepesme 1949, Scoggins and Tauber 1949, Rees 1947, larvae).

Thorictini Beal 1959

Thorictodes Reitter 1875, 1 sp., *T. heydeni* Reitter 1875, known in the United States from Texas and Nebraska (see Anderson 1949).
Thaumaphrastus Blaisdell 1927

Thylodrias Motschulsky 1839, 1 sp., *T. contractus* Motschulsky 1839, generally distributed (see Barber 1947).
Thelydrias Gemminger and Harold 1868
Ignotus Slosson 1903
Hospitopterus Pic 1921

Orphilini LeConte 1861

Orphilus Erichson 1846, 4 spp., generally distributed (see Beal 1985b).

Trinodini Casey 1900

Apsectus LeConte 1854, 2 spp., *A. hispidus* (Melsheimer 1844), Pennsylvania to southern Texas, and *A. araneorum* Beal 1959, Arizona (see Beal 1959).

Attagenini Castelnau 1840

Attagenus Latreille 1802, 9 spp., generally distributed (see Beal 1970, Halstead 1981).
Megatoma Kugelann 1792, not Herbst 1792.

Novelsis Casey 1900, 10 spp., Texas and states west of the Rocky Mountains (key to spp., Beal 1954a, 1970, 1984; see Kingsolver and Fales 1974).
Paranovelsis Casey 1900
Genattus Sharp 1902

Anthrenini LeConte 1861

Anthrenus Geoffroy 1792, 18 spp., generally distributed (see Beal 1998, Hoebeke *et al.* 1985, Kingsolver 1969).
subgenus *Nathrenus* Casey 1900
subgenus *Helocerus* Mulsant and Rey 1868
subgenus *Florilinus* Mulsant and Rey 1868

Thaumaglossa Redtenbacher 1868, 2 spp., *T. americana* Jayne 1882, Texas, and *T. libochoras* Beal 1952, Arizona (see Beal 1952).
Axinocerus Jayne 1882

Cryptorhopalum Guérin 1838, 22 spp., generally distributed (see Beal 1975, 1979, 1985a, 1995).

Orphinus Motschulsky 1858, 1 sp., *O. fulvipes* (Guérin 1838), Florida.
Aethriosia Reitter 1908

Ctesias Stephens 1830, 1 sp., *Ctesias dusmae* Beal 1960, California.
Megatoma Dejean 1821, not Herbst 1792
Tiresias Stephens 1932

Trogoderma Berthold 1827, 15 spp., generally distributed (keys to spp., Beal 1954b, 1956).

Macroprion Hope 1840
Eurhopalus Solier 1849
Psacus Pascoe 1866
Asidora Mulsant and Rey 1868
Globicornis Guérin 1868, not Latreille 1829
Acolpus Jayne 1882
Eucnocerus Sharp 1902
Entomotrogus Ganglbauer 1904

Dearthrus LeConte 1961, 2 spp., *D. longulus* LeConte 1861, eastern United States, and *D. stebbinsi* Beal 1954, California and Oregon (see Beal 1954c).

Megatoma Herbst 1792, 9 spp., western United States (key to spp., Beal 1967).

Pseudohadrotoma Kalik 1949, 4 spp., western United States (key to spp., Beal 1967).

Reesa Beal 21967, 1 sp., *R. vespulae* (Milliron 1939), widespread but spotty distribution.

Bibliography

ANDERSON, W. H. 1949. Larval description and transfer of *Thaumaphrastus karanisensis* from Colydiidae to a new subfamily of Dermestidae (Coleoptera). Bulletin of the Brooklyn Entomological Society, 64: 121-129.

BARBER, H. S. 1947. On the odd, or tissue-paper, beetle supposed to be *Thylodrias contractus* (Coleoptera: Dermestidae). Annals of the Entomological Society of America, 40: 344-349.

BEAL, R. S., Jr. 1952. Description of a new Arizona *Thaumaglossa* (Coleoptera: Dermestidae). Pan-Pacific Entomologist, 28: 171-172.

BEAL, R. S., Jr. 1954a. A revision of the species included in the genus *Novelsis*. Transactions of the American Entomological Society, 80: 73-90.

BEAL, R. S., Jr. 1954b. Biology and taxonomy of the Nearctic species of *Trogoderma* (Coleoptera: Dermestidae). University of California Publications in Entomology, 10: 35-102.

BEAL, R. S., Jr. 1954c. Classification of the Dermestid genus *Dearthrus* with description of a new western species (Coleoptera). Pan-Pacific Entomologist, 30: 231-234.

BEAL, R. S., Jr. 1956. Synopsis of the economic species of *Trogoderma* occurring in the United States with description of a new species (Coleoptera: Dermestidae). Annals of the Entomological Society of America, 49: 559-566.

BEAL, R. S., Jr. 1959a. Notes on the biology and systematics of the dermestid beetle genus *Apsectus* with description of two new species. Annals of the Entomological Society of America, 52: 132-137.

BEAL, R. S., Jr. 1959b. A key to the Nearctic genera of Dermestidae. Coleopterists Bulletin, 13: 99-101

BEAL, R. S., Jr. 1967. A revisionary study of the North American dermestid beetles formerly included in the genus *Perimegatoma* (Coleoptera). Entomological Society America, Miscellaneous Publications, 5: 281-312.

BEAL, R. S., Jr. 1970. A taxonomic and biological study of species of Attagenini (Coleoptera: Dermestidae) in the United States and Canada. Entomologica Americana, 45: 141-235.

BEAL, R. S., Jr. 1975. Description of adult and larval stages of a new species of *Cryptorhopalum* from Arizona and New Mexico (Coleoptera; Dermestidae). Proceedings of the Entomological Society of Washington, 77: 228-233.

BEAL, R. S., Jr. 1979. Systematics of the species of *Cryptorhopalum* (Coleoptera: Dermestidae) occurring in California. Contributions to Science, Natural History Museum of Los Angeles County, 306: 1-22.

BEAL, R. S., Jr. 1984. New sand-dune-inhabiting *Novelsis* (Coleoptera: Dermestidae) from California and Nevada. Proceedings of the Entomological Society of Washington, 86: 630-634.

BEAL, R. S., Jr. 1985a. A taxonomic revision of the Nearctic species of *Cryptorhopalum* (Dermestidae: Coleoptera). Transactions of the American Entomological Society, 111: 171-221.

BEAL, R. S., Jr. 1985b. Review of Nearctic species of *Orphilus* (Coleoptera: Dermestidae) with description of the larva of *O. subnitidus* LeConte. Coleopterists Bulletin, 39: 265-271.

BEAL, R. S., Jr. 1995. Taxonomic notes and distributional data on Nearctic species of *Cryptorhopalum* (Coleoptera: Dermestidae) with descriptions of a new southwestern U.S. species. Entomological News, 106: 169-177.

BEAL, R. S., Jr. 1998. Taxonomy and biology of Nearctic species of *Anthrenus* (Coleoptera: Dermestidae). Transactions of the American Entomological Society, 124: 271-332.

BØVING, A. G. and F.C. CRAIGHEAD, 1931. Illustrated synopsis of the principal larval forms of the order Coleoptera. Entomologica Americana (new series), 11: 1-352.

CASEY, T. L. 1900. Review of the American Corylophidae, Cryptophagidae, Tritomidae, and Dermestidae and other studies. Journal of the New York Entomological Society, 8: 138-165.

CROWSON, R. A. 1959. Studies on the Dermestoidea (Coleoptera), with special reference to the New Zealand fauna. Transactions of the Royal Entomological Society of London, 111: 81-94.

DALLA TORRE, K. W. VON 1911. Dermestidae, *In*: Coleopterorum Catalogus, 14(33): 39-96.

FORBES, W. T. M. 1922. The wing-venation of Coleoptera. Annals of the Entomological Society of America, 15: 328-352.

FORBES, W. T. M. 1926. The wing folding patterns of Coleoptera. Journal of the New York Entomological Society, 34: 42-68.

FORD, E. J. and J. M. KINGSOLVER 1966. Description and biological notes on the larvae of *Phradonoma tricolor* (Coleoptera: Dermestidae). Coleopterists Bulletin, 20: 27-29.

HALSTEAD, D. G. H. 1981. Taxonomic notes on some *Attagenus* spp. associated with stored products, including a new black species from Africa (Coleoptera: Dermestidae). Journal of Stored Product Research, 17: 91-99.

HETSCHKO, A. 1926. Thorictidae, Coleopterorum Catalogus 15(83): 1-11.

HETSCHKO, A. 1930. Thorictidae, Coleopterorum Catalogus supplement 15(109): 123.

HINTON, H. E. 1945. Monograph of Stored Product Beetles, Vol. I. London, (Pp. 234-401).

HOEBEKE, E., A. WHEELER and R. S. BEAL, Jr. 1985. *Anthrenus pimpinellae* F., a palearctic dermestid established in eastern North America (Coleoptera: Dermestidae). Journal of the New York Entomological Society, 93: 1216-1222.

JAYNE, H. F. 1882. Revision of the Dermestidae of the United States. Proceedings of the American Philosophical Society, 20: 343-377.

KINGSOLVER, J. M. 1963. Pictorial key for identification of the larvae of dermestid genera (Coleoptera). U. S. Department of Agriculture Cooperative Economic Insect Report, 13: 384-385.

KINGSOLVER, J. M. 1969. *Anthrenus coloratus* Reitter, a dermestid new to North America (Coleoptera). U.S. Department of Agriculture Cooperative Economic Insect Report, 19: 61-63.

KINGSOLVER, J. M. and J. H. FALES 1974. *Novelsis aequalis* (Sharp), a potential household insect in the Eastern United States. U.S. Department of Agriculture Cooperative Economic Insect Report, 24: 818-820.

LEPESME, P. 1949. Revision des *Dermestes* (Col. Dermestidae). Annales de la Societe Entomologique de France, 115: 37-68.

MROCZKOWSKI, M. 1968. Distribution of the Dermestidae (Coleoptera) of the world with a catalogue of all known species. Annales Zoologici, 26: 1-191.

PETERSON, A. 1951. Larvae of insects, Part. 2. Columbus OH. 416 pp.

REES, B. E. 1943. Classification of the Dermestidae (larder, hide, and carpet beetles) based on larval characters, with a key to the North American genera. U. S. Department of Agriculture, Miscellaneous Publications No. 511, 18 pp., 5 figs.

REES, B.E. 1947. Taxonomy of the larvae of some North American species of the genus *Dermestes* (Coleoptera: Dermestidae). Proceedings of the Entomological Society of Washington, 49: 1-14.

SCOGGIN, J. K. and O. E. TAUBER 1949. The bionomics of *Dermestes maculatus* DeG. Iowa State Journal of Science, 23: 363-373.

69. BOSTRICHIDAE Latreille 1802

by Michael A. Ivie

Family common name: The bostrichid beetles

Family synonyms: Apatidae Billberg 1820; Lyctidae Billberg 1820; Endecatomidae LeConte 1861; Psoidae Blanchard 1851; Bostrychidae *auctorum*.

This family as a whole is difficult to diagnose based on external features of dried adults. Most of them can be recognized by the distinctive hooded pronotum, and though the smaller species are often mistaken for Scolytinae, may be distinguished by their straight antennae with free club segments. The Psoinae, Polycaoninae and Lyctinae lack the hooded pronotum. The North American Psoinae are colorful red, yellow and green/blue species that resemble Cleridae, but lack their expanded maxilary palps. The Lyctinae are often confused with Colydiidae, from which they differ by the 5-5-5 tarsus.

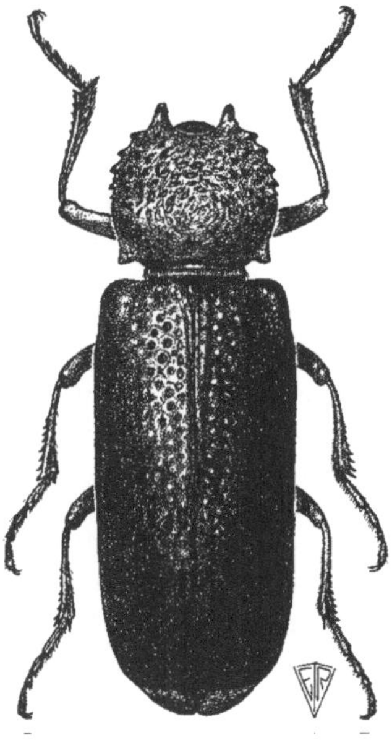

FIGURE 1.69 *Apatides fortis* (LeConte)

Description: Shape elongate, cylindrical or depressed, length 1.5 to 50 mm, most 2 to 20 mm in length; color black, various shades of yellow and red-browns, rarely red or yellow with metallic green/blue markings; vestiture absent or variably sparse to dense or patchy fine to scale-like setae.

Head somewhat hypognathous, inserted into the prothorax and usually not visible from above, some clearly prognathous and obvious from above; surface smooth, punctate or rugose. Antennae with eight to eleven antennomeres, straight, not geniculate, with club of two, three or four segments; rarely the antennomeres of the club strongly transverse, nearly lamellate, inserted laterally near the eyes, above the base of the mandibles or in front of the eyes under pronounced frontal angles. Labrum small, transverse; mandibles heavy, curved, the apices often blunt, sometimes bifid; maxillary palpi with four palpomeres, small, slender; gula small, the gular sutures distinct, mentum trapezoidal, labial palpi with three palpomeres, small, slender. Eyes prominent, bulging, small, oval.

Pronotum large cowled or flat; shape subquadrate or anteriorly rounded; borders not margined (except Lyctinae and some Xyloperthini, in which case margins either complete or basal, smooth or denticulate); cowled species with anterior dorsal surface rough, rasp-like, some with curved horns or hooks; pleural region broad; some flattened types (Lyctinae) with a median depression, or a median fovea; prosternum long to moderate in front of the coxae; intercoxal process wide to absent; procoxal cavities open or closed behind. Mesosternum moderate. Metasternum broad, long. Legs with the trochantins hidden; anterior coxae small, cylindrical to rounded, middle coxae round, nearly contiguous; hind coxae transverse, nearly contiguous to widely separate; trochanter obliquely or squarely attached to femur; femora broad to slender; tibiae slender, smooth, with single apical spur, tarsal formula 5-5-5, rarely 4-4-4 (*Psoa*), first tarsomere frequently very small, tarsomeres slender, simple. Scutellum small, triangular, oval, or quadrate. Elytra highly variable, usually coarsely punctate, with vague costae present, more rarely impunctate and plane; often with a distinct apical declivity variously modified with spines; epipleural fold obscure. Hind wing well developed, in fully-veined species with characteristic bostrichoid venation, a short r3 attached to the RP directly from the rp-mp2 cross-vein, a well-developed R cell; folding with normal transverse folds (Forbes 1926).

Abdomen usually with five ventrites, rarely six when the morphological segment II is visible at the base on each side; all segments free, sutures entire. Malphigian tubules exhibiting bostrichiform modified cryptonephridy (Sani 1964), hind gut with sclerotized ovoid rod that holds the "rectal pad" of Poll (1932) sensu Crowson (1981) (Ivie 1985). Male genitalia bostrichoid, large and flattened horizontally with two basal struts distally curved, apically acute; parameres fused at the base or free; or trilobed. Median lobe long and stout, parameres long, stout, apically acute, convergent, freely articulated (Sharp and Muir 1912). Female genitalia with the paraprocts reduced to baculi; valvifers reduced; coxite membranous, setiferous; styli subapical (Tanner 1927).

Larvae C-shaped; length 3 to 20 mm rarely to 60 mm; vestiture absent, with a few short setae on the ventral surface or dense and fine; color cream to dull white. Head retracted into the prothorax or not (*Endecatomus* and Anobiinae), prognathous,

Acknowledgments. Thanks to Charles Triplehorn, who supervised the dissertation on which much of this chapter is based. Thanks also to J. Joseph Giersch, Richard S. Miller, Katharine Marske and T. Keith Philips helped with literature, reviews, and illustrations, as well as discussion of various problems involved with this group. Stephen Wood donated a large and very valuable collection of bostrichids that made the testing of keys possible.

elongate and flattened or subglobular. Antennae two- or three-segmented, simple. Clypeus and labrum distinct; mandibles symmetrical, gouge-shaped, with or without mola; maxillae with cardo, stipes and two- or three-segmented palpi; labial palpi one- or two-segmented. Stemmata single or absent.

Thorax enlarged; meso- and metathorax sometimes with two plicae; prothorax of Bostrichinae with an oblique sclerotized rod imbedded in the tergum above the spiracle; legs five-segmented, with slender tarsungulus.

Abdomen ten-segmented, with three or less distinct plicae dorsally on each segment; apical segments simple, ninth enlarged, tenth small, with anal pads. Spiracles annular or annular-uniforous; present on prothorax and abdominal segments one to eight, last very large in Lyctinae. For larval descriptions see Bøving and Craighead (1931), Gardner (1933), Beeson and Bhatia (1937), Anderson (1939), Peterson (1960), Lawrence (1991) and Costa *et al.* (1999).

Habits and habitats: [For Anobiinae, see Chapter 70] The larvae of most groups are wood borers, with a few species found in stored grain and tubers (*Rhyzopertha, Prostephanus*), fungi (Endecatominae, Crowson 1961) and even books (*Rhyzopertha*, Hoffman 1933). Most wood borers receive their actual nutrition from the starch content in the wood they consume (Beeson and Bhatia 1937, Gerberg 1957 and references therein), making the switch to stored products obvious. Several species of this family are of great economic importance, including the larger and lesser grain borers (*Prostephanus* and *Rhyzopertha*), the powder-post beetles (*Lyctus* and *Trogoxylon*) and the bamboo powder-post beetles (*Dinoderus* spp.). Several other species are pests of living wild and cultivated trees and vines, (*Dinapate, Psoa, Melalgus* and Polycaon; Gardner 1933, Fisher 1950, Cooper 1986, Olson 1991), and in some cases the female may cause the death of wood utilized by the larvae (*Apate,* see Wolcott 1950). The larvae and adults of those that live in wood infest dead and dry wood of angiosperms, seasoned sapwood of commerce, pith of vines, and dried roots of herbaceous plants. The species infesting seasoned lumber and its products are now subject to wide distribution over the world. Beeson and Bhatia (1937) provide the best overview of larval and adult biology, covering many of the exotic species established or intercepted in North America. Fisher (1950) and Gerberg (1957) review the literature on North American species. Endosymbiosis involving bacteria in mycetomes of the posterior mid-gut are well-known in this group (Crowson 1981) and the modified cryptonephridism allows them to inhabit very low moisture environments (Sani 1964, Crowson 1981). See also Wright (1960).

Status of the classification: Due to the pioneering work of Pierre Lesne, this is one of the best know and most completely described families of beetles in the world. The membership of the family is virtually unchanged from his concepts as laid out in a brilliant series of revisions, faunal treatments, descriptions and catalog between 1894 and 1941 (Berland 1951). Anglophone authors continued to split the Bostrichidae and Lyctinae into the 1980s, but this is no longer standard usage (Downie and Arnett 1996). The early history of that debate is covered by Gerberg (1957), but was put to rest for most workers by Crowson (1961), who followed Lesne (1938 and others) by including the Lyctinae among the Bostrichidae.

The higher classification of the family was revised by Ivie (1985), who found the bostrichids to have a sister-group relationship with the [Dermestidae + Nosodendridae] (see Chapter 67). Relationships of the bostrichid clade resolve to be [Endecatominae [Dysidinae [[Polycaoninae + Bostrichinae] [Psoinae [[Lyctinae + Dinoderinae] [Euderinae + Anobiinae]]]]]]. This classification is followed here except for the treatment of the Anobiinae as an independent family, following Lawrence and Newton (1995) (see Chapter 70 below), and the return of the Endecatominae to this family from full-family status. This later is based on the argument that since *Endecatomus* seems to be the sister-group of the rest of the Bostrichidae (sensu Ivie 1985), it can be included in the Bostrichidae without negatively affecting the monophyly of either group. A family of only 4 species worldwide (Endecatomidae) is perhaps not the best use of the family-level taxon when the phylogenetic position is not in significant question.

The North American revisions of Fisher (1950) and Gerberg (1957) are still quite adequate for most species identifications. Because of the proclivity of bostrichid species to be moved around in wood, emerging far from their normal homes, subsequent regional treatments within North America are mostly useful for illustrations and additional distributions records (Hatch 1961, Downie and Arnett 1996, Beiriger and Sites 1996). Excellent faunal treatments are available for many extralimital regions, including Lesne (1901b and its supplements, see Lesne 1938) and Cymorek (1969) for the Palearctic; Lesne (1924) for tropical Africa, Lesne (1939) for Central America, Vrydagh (1951) for Congo (Zaire), Vrydagh (1958, 1959a, b) for Australia, Vrydagh (1959c) for Southern Africa, and Binda and Joly (1991) for Venezuela.

Distribution: There are about 550 species in 99 genera known from all inhabited regions of the world. They are most speciose in moderately xeric regions, and seem limited to absent in true rain forests. Seventy-three species occur in North America, four other exotic species are possibly established there, and an additional 34 more species have been intercepted at various ports or points inside North America but are apparently not established. The genera of these commonly intercepted species are included in the keys and classification because of the need for North American workers to recognize these elements, which are commonly found in wooden items from extralimital regions. Lumber, crates, pallets, furniture and artistic carvings are very commonly the sources of such specimens, which can be considered cosmopolitan in terms of where they may turn up.

Key to the World Subfamilies of Bostrichidae

1. Pronotum with explanate lateral margins with row of projecting straight or recurved setae; elytra with unisteose microtubercles arranged in an irregular, reticulate pattern, cuticle smooth between (Fig. 2, 34) (Endecatominae) .... *Endecatomus*

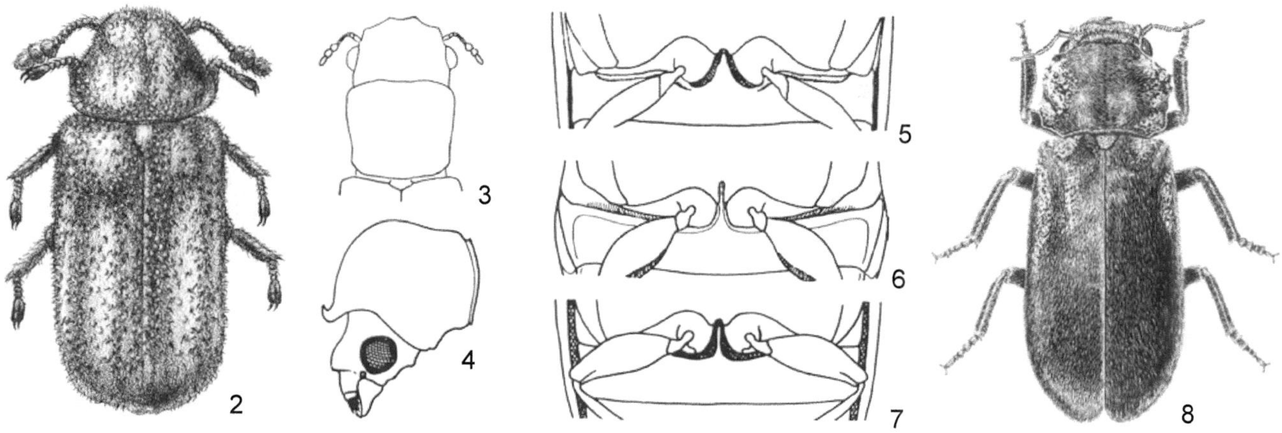

FIGURES 2.69-8.69. 2. *Endecatomus dorsalis* Mellié, habitus; 3. *Melalgus* sp, head and pronotum; 4. *Apatides fortis* (LeConte), head and pronotum, lateral view; 5. *Bostrichus capucinus* (Linnaeus), hind coxae, ventral view; 6. *Sinoxylon conigeriim* Gerstacker, hind coxae, ventral view; 7. *Tetropriocera longicornis* (Olivier), hind coxae, ventral view; 8. *Dysides obscurus* Perty, habitus.

— Pronotum with or without distinct lateral margins, but margin not explanate; elytra smooth, tuberculate, punctate, or rugose, not as above 2

2(1). Gular sutures confluent 3
— Gular sutures separate, some narrowly so 5

3(2). Head easily visible from above (Fig. 3), prosternal process long, reaching mesosternum and slightly expanded at apex; antennae 9-11 (usually 11) segmented (Polycaoninae) 9
— Head more or less covered by pronotum (Fig. 4); prosternal process short, acute, not extending between procoxae, antennae 8-10 (usually 10) segmented 4

4(3). Protibia with 1 apical spine; prothorax in dorsal view evenly rounded in front, first tarsomere subequal to second (Dinoderinae) 10
— Protibia with 2 apical spines; prothorax in dorsal view flattened or excavate in front (Fig. 1); first tarsomere very much shorter than second (Bostrichinae) 13

5(2). Hind trochanter attached squarely to femur; first tarsomere almost always longer than second . 6
— Hind trochanter obliquely attached to femur (Figs. 5); first tarsomere shorter than second, or tarsi 4-segmented 7

6(5). Hind coxa usually excavate to receive hind femur, with at least a flat, posterior face and carina along ventro-posterior margin, if not, antennal insertions approximate and antennae without pectinate 3-segmented club; worldwide (Anobiinae) .... (treated as Anobiidae, see Chapter 70)
— Hind coxa conical at trochanter insertion, not excavate to receive hind femur; antennae 11-segmented with a 3-segmented pectinate club, each ramus longer than antenna; New Zealand .... [Euderiinae]

7(5). Pronotum with complete lateral margins; antennae 11-segmented, usually with a 2, rarely 3-segmented club; procoxae widely separated, procoxal cavities open or closed externally (Lyctinae) 35
— Pronotum without lateral margins, or with only basal portion margined; antennae with 11 or less segments, club with 3 or 4 segments; procoxal cavities open externally 8

8(7). First ventrite with a postcoxal carina complete across its width, connected to wide intercoxal process (Figs. 8, 9) (Dysidinae) .... *Dysides*
— First visible sternite of abdomen (or second if first visible is II, narrowly visible at lateral edges in some genera) usually without a postcoxal carina, with intercoxal process lamiform or absent (Fig. 10) (Psoinae) .... *Psoa*

Key to the World Genera of Polycaoninae

9(3). Frons laterally expanded in male; dorsal surface of mandible with depression delimited laterally by ridge that fits against labrum and extending above it when closed; lateral profile of head not evenly arcuate .... *Polycaon*
— Frons not greatly dimorphic in extent of lateral margins (Fig. 3); mandible without dorsal depression, closing under labrum; lateral profile of head evenly arcuate .... *Melalgus*

Key to the New World Genera of Dinoderinae

10(4). Elytra with apical declivity truncate, margined, bituberculate or concave; pronotum narrowed anteriorly, the apical row of dentations V-shaped, the anterior-most pair markedly larger than the rest and adjacent at base .... *Prostephanus*
— Elytra evenly rounded and convex from disk to apex, not distinctly set off or bituberculate; pronotum evenly rounded and semicircular anteriorly, the denticles arranged in a U-shape, the median pair not differentiated (Figs. 12, 13) 11

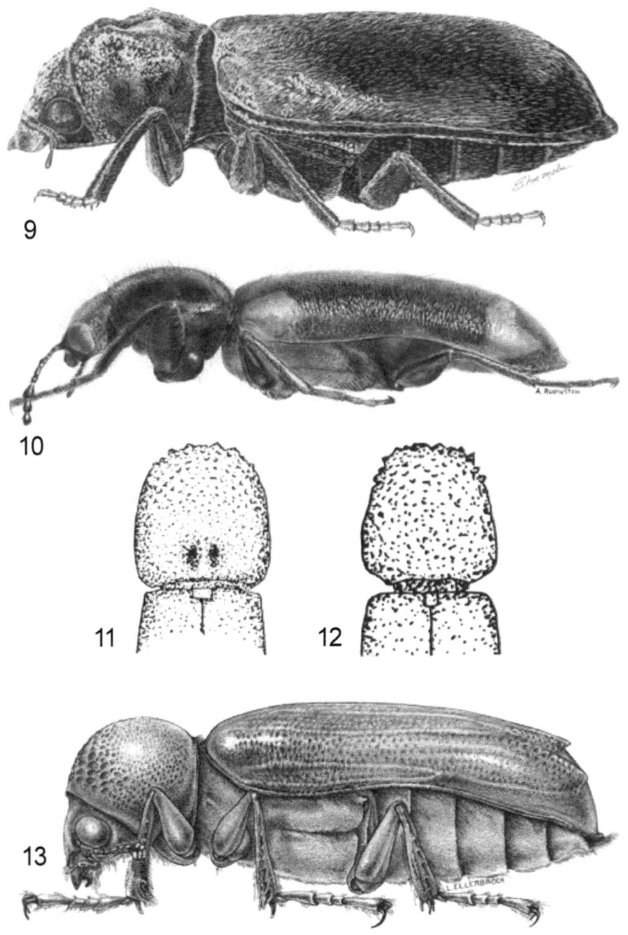

FIGURES 9.69-13.69. 9. *Dysides obscurus* Perty, lateral habitus; 10. *Psoa quadrisignata* (Horn), lateral habitus; 11. *Dinoderus* sp., head and pronotum; 12. *Rhyzopertha dominica* (Fabricius), head and pronotum; 13. *Dinapate wrightii* (Horn), lateral habitus.

11(10). Scutellum transversely rectangular (Fig. 11), pronotal disk punctate posteriolaterally; second antennomere distinctly shorter than first .... *Dinoderus*

— Scutellum nearly square (Fig. 12), pronotal disk punctate or tuberculate posteriolaterally; second antennomere subequal in length to first 12

12(11). Pronotum with tubercles on apical half low, blade-like and connected at bases into semicircular ridges (Fig. 12); posterior half with anteriorly directed, low, blunt, imbricate tubercles; elytra without tubercles, variably smooth to slightly rugose between punctures .... *Rhyzopertha*

— Pronotum with sharp, elevated, separate tubercles on entire disk, or with obtuse tubercles on basal half; elytra usually at least pustulate on declivity .... *Stephanopachys*

Key to the World Tribes of Bostrichinae

13(4). Intercoxal process of first ventrite always visible, with a ventral face; in cross section, this process is inverted U- or T-shaped; metepisternum usually broadly truncate behind, with posterior margin of metepimeron widely separated from metasternum (Figs. 5, 6) 14

— Intercoxal process of first ventrite a lamina or reduced to absent, without a ventral face; in cross section, this process is acute ventrally or process is below the level of the coxae and not visible; metepisternum usually narrowed posteriorly, often acute, metepimeron nearly contacting metasternum posteriorly (Fig. 7) 16

14(13). Mandibles long, pointed; almost always crossing in a shearing manner; without longitudinal striae on exterior face 15

— Both mandibles short, blunt, meeting along midline, the cutting edges directly opposable; each with 2 fine longitudinal striae on external face near tip (Sinoxylini) .... *Sinoxylon*

15(14). Anterior edge of pronotum curled under, with setal fringe of margin hidden behind this rolled edge; less than 25 mm (Bostrichini) 17

— Anterior edge of pronotum only slightly rounded, not curled under, marginal setal fringe insertion exposed (Fig. 13); greater than 30 mm (Dinapatini) .... *Dinapate*

16(13). Intercoxal process of first ventrite declivous, usually not visible except at base; females with a transverse line of setae in the middle of last ventrite, the apical margin densely fringed in long, spatulate setae; ovipositor wide and short. Old World and Chile, Old World species established in South America and the West Indies..... .... [Apatini]

— Intercoxal process of first ventrite plane, visible for entire length (Fig. 7); last ventrite of female not as above, ovipositor normal (Xyloperthini).. 24

Key to the Nearctic Genera of Bostrichini

17(15). First ventrite with postcoxal carina extending to tip of intercoxal process, giving intercoxal process a marginal V-shaped carina (Fig. 6)......... 18

— First ventrite with postcoxal carina not extending to intercoxal process; intercoxal process not margined (Fig. 5; not established in North America) .... [*Bostrichus*]

18(17). Margins of buccal cavities strongly dentate below eyes (not established in North America) .... [*Bostrychoplites*]

— Margins of buccal cavities not dentate below eyes 19

19(18). Antennomeres of antennal club with longitudinal grooves on each surface (Fig. 15) .. *Amphicerus*

— Antennomeres of antennal club without longitudinal grooves but with round depressions on each surface or none (Fig. 16) 20

20(19). Anterior angles of clypeus broadly rounded (Fig. 17) or elytra with dentate to spinose elytral apices .... *Lichenophanes*

— Anterior angles of clypeus obtuse, acute, or rectangular 21

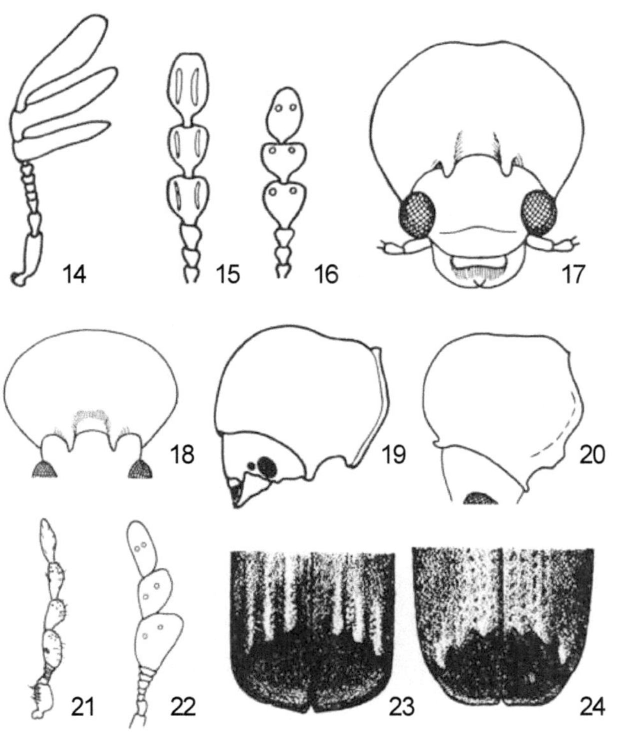

FIGURES 14.69-24.69. 14. *Sinoxylon conigerum* Gerstacker, antenna; 15. *Amphicerus* sp., antenna; 16. *Lichenophanes* sp., antenna; 17. *Lichenophanes bicornis* (Weber), head, anterior view; 18. *Heterobostryciiiis hamatipeiinis* (Lesne), pronotum, anterior view; 19. *Micropate labialis* Lesne, head and pronotum, lateral view; 20. *Tetrapriocera longicornis* (Olivier), pronotum, lateral view; 21. *Tetrapriocera longicornis* (Olivier), antenna; 22. *Xylobiops basilaris* (Say), antenna; 23. *Xylemeira torquata* (Fabricius), apex of elytra, dorsal view; 24. *Xylobiops basilaris* (Say), apex of elytra, dorsal view.

21(20). Pronotum transversely depressed behind anterior margin, anterior margins armed with prominent ventrally directed spines or obtuse angles (Fig. 18) .... *Heterobostrychus*
— Pronotum not transversely depressed behind anterior margin, armature of anterior margin variable 22

22(21). Frons of head transversely depressed or flattened behind eyes (Fig 4) 23
— Frons of head evenly convex behind eyes (Fig. 19) .... *Micrapate*

23(22). Frons of head transversely, perpendicularly elevated behind clypeus (Fig. 4); large, 9-20 mm in length .... *Apatides*
— Frons of head not transversely, perpendicularly elevated behind clypeus, sometimes with short, more or less distinct elevation at middle behind clypeus (not established in North America) .... [*Bostrychopsis*]

Key to Nearctic Genera of Xyloperthini

24(16). Pronotum without carina along lateral margins, or carina only indistinctly indicated at posterior angles (Fig. 20) 25
— Pronotum with very distinct carina along lateral margins posteriorly, sharply bent at posterior angles and slightly prolonged along base 33

25(24). Antennal club composed of four antennomeres (Fig. 21) .... *Tetrapriocera*
— Antennal club composed of three antennomeres (Fig. 22) 26

26(27). Antennae with eight or nine antennomeres 27
— Antennae with ten antennomeres (Fig. 22) 28

27(26). Antennae with nine antennomeres .... *Scobicia*
— Antennae with eight antennomeres (not established in North America) .... [*Octodesmus*]

28(26). Antennomeres of antennal club with sensory depressions on each surface (Fig. 22) 29
— Antennomeres of antennal club without sensory depressions on either surface (not established in North America) .... [*Xyloperthella*]

29(28). Apical declivity of elytra trituberculate on each side (Figs. 23, 24) 30
— Apical declivity of elytra bituberculate on each side 32

30(29). Lateral submargin on apical declivity of elytra very strongly elevated and acutely, angularly separated at sutural angles; frons of head convex in both sexes; antennae densely pubescent (not established in North America) .... [*Xyloprista*]
— Lateral submargin on apical declivity of elytra not strongly elevated or angularly separated at sutural margins 31

31(30). Tubercles along anterior margin of apical declivity of elytra costiform, cylindrical, all obtuse at apices (Fig. 23); antennae densely clothed with long setae; female with front of head flattened or concave, and anterior margin of pronotum with long setae .... *Xylomeira*
— Tubercles along anterior margin of apical declivity of elytra subtriangular, all or at least one pair spinose at apices (Fig. 24); antennae without or sparsely clothed with short setae; female with frons of head convex, and anterior margin of pronotum without long setae .... *Xylobiops*

32(29). First two antennomeres of antennal club elongate, first subtriangular; female with frons of head flattened or concave, and last visible abdominal sternum not emarginate at apex .... *Dendrobiella*
— First two antennomeres of antennal club surrounded or oval; female with frons of head convex, and last visible abdominal sternum very deeply emarginate at apex .... *Xyloblaptus*

33(24). Lateral margins of elytra very deeply, irregularly emarginate near apices (not established in North America) .... [*Xylion*]
— Lateral margins of elytra not emarginate near apices 34

34(33). First two antennomeres of antennal club elongate; frons of head of female with erect setae (not established in North America) .... [*Xylothrips*]

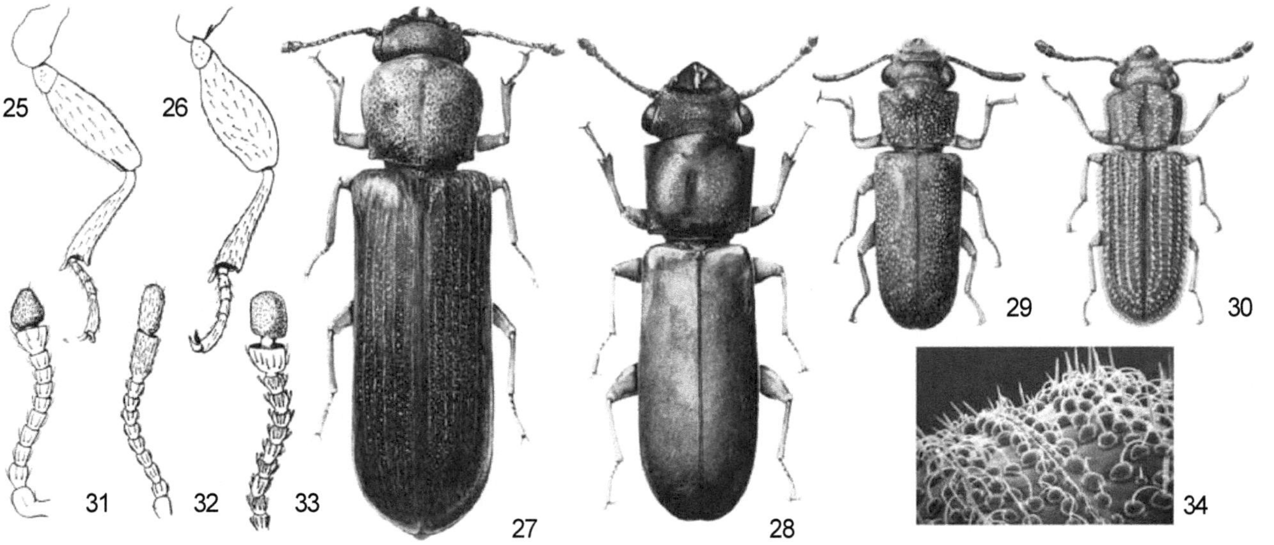

FIGURES 25.69-34.69. 25. *Lyctus planicollis* LeConte, proleg; 26. *Trogoxylon parallelopipedum* (Melsheimer), leg; 27. *Lyctus planicollis* LeConte, habitus; 28. *Trogoxylon parallelopipedum* (Melsheimer), habitus; 29. *Lyctus planicollis* LeConte, antenna; 30. *Lyctoxylon japonum* Reitter, antenna; 31. *Minthea rugicollis* (Walker), antenna; 32. *Lyctoxylon japonum* Reitter, habitus; 33. *Minthea rugicollis* (Walker),habitus; 34. *Endecatomus dorsalis* Mellié, detail of elytral sculpture. (Figures 25-33 from Gerberg 1957)

— First two antennomeres of antennal club transverse; frons of head of female without erect setae (not established in North America) .... [*Xylopsocus*]

Key to the Nearctic Genera of Lyctinae

35 (7). Metathoracic femur slender, not ellipsoidal or subglobose (Fig. 25, 27, 33); punctation and pubescence of elytra seriated, impressed (Lyctini) 36

— Metathoracic femur compressed, subglobose, or ellipsoidal; punctation and pubescence of elytra nonseriate (Fig. 26, 28) (Trogoxylini) .... *Trogoxylon*

36(35). Antennal club with apical antennomere ovoid becoming attenuated toward apex, usually longer than penultimate antennomere (Fig. 29); dorsal pubescence composed of fine, appressed, or thick, curvate setae (Fig. 27).................... *Lyctus*

— Antennal club with one or both antennomeres greatly elongate (Figs. 30, 31); dorsal pubescence of erect setae or semierect, thick setae........ 37

37(36). Antennal club with both antennomeres elongate and longer than broad (Fig. 30); dorsal pubescence composed of irregular thick, semierect setae (Fig. 32) .... *Lyctoxylon*

— Antennal club with apical antennomeres elongate (Fig. 31); elytra with lines of wide, flattened, erect, scale-like setae (Fig. 33)........................ *Minthea*

Classification of the Nearctic Genera

Bostrichidae Latreille 1802

I. Subfamily Endecatominae LeConte 1861

A monogeneric group, originally described in the Ciidae, and sometimes treated as an independent family (Lesne 1921, Ivie 1985, Lawrence and Newton 1995, Lawrence *et al.* 1999a,b). Ivie (1985) found support to place this group as the sister-lineage to the rest of the Bostrichidae (including Anobiidae). As such, it can either be included here or treated as its own family. Ivie (1985) argued for the later based on the plesiomorphic ability of the larva to emerge from its host and walk on the surface, something unknown in other Bostrichidae. However, the arguments against yet another tiny family have outweighed that idea in this treatment. The larva was described by Crowson (1961).

Diagnosis: The lateral margins of the pronotum are explanate and bear a fringe of stiff projecting setae, and the dorsal surface is covered in microtubercles, each bearing a stout seta (Figs. 2, 34).

Endectomus Mellié 1847, is an isolated group with four Holarctic species, two of which occur in eastern North America. *Endecatomus rugosus* (Randall 1838) is widespread from central Manitoba, down the Mississippi River corridor to Central Texas, east to Maine and Florida; *E. dorsalis* Mellié 1848 is known from Missouri, southern Illinois, Oklahoma, Alabama and Texas. *Endectomus* is under revision by Ivie.

Hendecatomus Bach 1852

Dictyalotus Redtenbacher 1847

II. Subfamily Dysidinae Lesne 1921

A small group of two genera and 3 species from Indo-Malaysia and South America, revised by Lesne (1894). The larva is described by Gardner (1933). It is relatively plesiomorphic in most characters, and forms the sister-group to the remaining subfamilies together (Ivie 1985).

Diagnosis: (Figs. 8, 9) Pronotum semicowled, distinctly tuberculate, and with lateral pronotal margins obsolete; with head visible from above; and first ventrite with wide intercoxal process.

Dysides Perty 1832. One species, *D. obscurus* Perty 1832, has been intercepted at Hoboken, New Jersey, but is not established in North America (Fisher 1950).

III. Subfamily Polycaoninae Lesne 1896

Formerly treated by Anglophones as part of the Psoinae, this group appears to be the sister-group of the large Bostrichinae (Ivie 1985), validating the views of Lesne and Vrydagh who treated it at equal rank with that group throughout their works. It contains 2 genera occurring throughout the warmer Americas and southeast Asia.

Diagnosis: The members of this subfamily have a cylindrical, unmargined pronotum; completely exposed head (Fig. 3), and wide intercoxal process on the prosternum, combined with a characteristic globular procoxa. The procoxae are flattened ventrally, and the trochanters and femora are laterally directed, compared to the more conical coxae of the related subfamilies.

Melalgus Dejean 1835, is a moderately large genus with species throughout the warmer Americas and southeast Asia. In North America there are four valid species, ranging from British Columbia south through the Pacific states, Arizona, Texas and Oklahoma. Several of the names in this group have been misapplied, and this genus is under revision by Ivie.

Heterarthron Dejean 1836
Exopioides Guérin-Méneville 1884
Polycaon Horn 1878

Polycaon Laporte 1836. Two species occur in western North America and Chile. *Polycaon stoutii* (LeConte 1853) occurs from extreme southern Vancouver Island and mainland British Columbia south through the Pacific states, inland to Brewster Co., Texas, and south to Chiapas, Mexico. It is intercepted from ornamental items such as grape stumps in many areas, including records for at least 5 eastern states where it does not seem to be established.

Exops Curtis 1839
Alloeocnemis LeConte 1853
Psoa Erichson 1834, not Herbst 1797

IV. Subfamily Bostrichinae Latreille 1802

With 61 genera in 6 tribes (Ivie 1985), the Bostrichinae are the largest of the subfamilies (except Anobiinae, treated here as a family, see Chapter 70). Native species of Bostrichinae occur in all major biotic regions except New Zealand. Several species are spread by commerce and have become nearly pan-tropical. These are the stereotypical bostrichids with the facies normally associated with this family.

Diagnosis: Most species of this subfamily can be recognized by the flattened or excavate front of the cowled pronotum that has a pair of small to large anteriorly or ventrally directed horns or angulate processes on each side; two protibial spines; and a short first tarsomere.

1. Tribe Dinapatini Lesne 1909

This monogeneric tribe contains the largest known bostrichid, ranging over 60 mm in length. Both known species are palm trunk specialists, and are unusual in attacking living trees.

Diagnosis: The large size, flat front edge of the pronotum with an exposed setal fringe, and short wide antennal club will diagnose this tribe.

Dinapate Horn 1886. There are two species of *Dinapate*, one in Mexico and *D. wrightii* Horn (Fig. 13) native to desert oases of the Californias (Cooper 1986). In its native state, *Dinapate wrightii* is a specialist on the California Fan Palm (*Washingtonia filifera* (Linden) Wendl.), a prized ornamental species in xeric urban landscapes. It has also been reported from the date palm, *Phoenix dactylifera* L. (Olson 1991). Larvae of *D. wrightii* have been moved in transplanted trees to Arizona (Olson 1991), Nevada and Utah, where it has become established as a periodic pest.

2. Tribe Apatini Billberg 1820

This tribe includes the genera *Apate* Fabricius 1775, *Phonapate* Lesne 1909 and *Xylomedes* Lesne 1902 from Africa and Asia, as well as *Chilenius* Lesne from the desert regions of Chile. The Apatini are not known to be established in North America, but two African species of *Apate* have become established in the Greater Antilles (Cuba, Jamaica, Hispaniola, Puerto Rico and the Virgin Islands), and should be watched for in the southern states. Revised by Lesne (1909, 1935), with *Chilenius* moved here by Ivie (1985).

Diagnosis: Mostly large beetles, with the first ventrite usually lacking a postcoxal line and the intercoxal process declivous, and often not visible between the hind coxae. The last ventrite of the female has a unique setal line; the broad hind margin bears a fringe of modified, clubbed setae, and the ovipositor is very broad and short.

3. Tribe Bostrichini Latreille 1802

A tribe of 18 genera, this is the second most diverse of the groups in the family. It occurs on all inhabited continents and on many islands, and some members are widely distributed through trade. This is the stereotypical bostrichid, and they occur in more mesic forest environments than the Xyloperthini. Last revised by Lesne (1899).

Diagnosis: The planar, completely margined intercoxal process of the first ventrite, long, scissors-like mandibles, and turned under front edge of the pronotum will distinguish this group.

Amphicerus LeConte 1861, contains 15 species from Eurasia and the Americas. Four species are generally distributed in North America as far north as Canada's Northwest Territories, excepting the Pacific Northwest. One additional European species has been intercepted at US ports, but is not established in North America (Fisher 1950).

Schistoceros Lesne 1899

Apatides Casey 1898, is a monotypic genus described for *A. fortis* (LeConte 1866), which is widespread in Mexico and is known from all the US border states, north to Utah and Oklahoma.

[*Bostrichus* Müller 1764. is a monotypic genus for *Bostrichus capucinus* (Linnaeus 1758). This European species has been intercepted numerous times in oak and cork at US ports, but is not yet established in North America (Lesne 1899, Fisher 1950).

Bostrychus Geoffroy 1762, is invalid as published in a work rejected by the International Commission for Zoological Nomenclature.]

[*Bostrychoplites* Lesne 1899, is an Africa genus with 15 described species. One of these, *B. cornutus* (Olivier 1790), is commonly intercepted from wood carvings and other artifacts from Africa. Not established in North America (Lesne 1899, 1924; Fisher 1950).]

[*Bostrychopsis* Lesne 1899. The 24 species of this genus occur in Asia, Africa and South America. Two species have been intercepted at various ports, but none are known to be established in North America. Teixeira (1992) revised the New World species.]

Heterobostrychus Lesne 1899. Three of the six species of this African and Asian genus have been intercepted at various ports. One of them, *H. aequalis* (Waterhouse 1884) is now established in Florida (Woodruff 1967), and another, *H. brunneus* (Murray 1867) is established in California's San Joaquin Valley (San Joaquin Co. Commissioner of Agriculture collection, Stockton). The third, *H. hamatipennis* Lesne 1895 may or may not be established in Florida (Peck and Thomas 1998). The genus was last revised by Lesne (1899).

Lichenophanes Lesne 1899. With 41 species, this is among of the largest genera in the family. It occurs in Eurasia, Africa and the Americas, being relatively common (compared to most bostrichids) in more mesic areas. It is imperfectly separated from the next genus, and needs revision. Eight species are known in North America ranging through California, Arizona, New Mexico and the eastern forests as far north as Montana, Manitoba and Quebec.

Micrapate Casey 1898, is a large genus that includes the smallest species in the tribe. It contains 40 described species from Africa, Eurasia and the Americas, but its greatest diversity is in Mexico, where numerous species remain undescribed. Three rarely collected species occur north of Mexico, recorded from Arizona, Texas, the District of Columbia, Maryland, Virginia and Florida. A single specimen from Nevada (Ivie collection), may represent a new, fourth species. Fisher (1950) records five species intercepted at various ports, but the difficulty of identifying this group probably means this is an underestimate. Of all genera of Bostrichidae, this genus is most in need of a revision. Lesne (1939) has the most recent treatment for Mexican species likely to be discovered across the border.

Bostrychulus Lesne 1899

4. Tribe Xyloperthini Lesne 1921

This large tribe of small species is particularly characteristic of xeric regions of the world, although some species do inhabit moist forests. There are 32 recognized genera occurring worldwide. They dominate the Bostrichinae fauna of Australia, which has only one native member belonging to another another tribe. Several species have been carried by commerce to remote oceanic islands and new continents, where they are often more common than in their homelands.

Diagnosis: The lamiform intercoxal process of the first ventrite, which is visible only as a narrow carina, and does not have a ventral face, coupled with mandibles that cross at the tips and a posteriorly narrowed metepimeron, is typical of the tribe.

Dendrobiella Casey 1998, is a genus of 5 Neotropical species, two of which extend across the border into California, Arizona, and Texas (Fisher 1950).

[*Octodesmus* Lesne 1901. One of the three Asian species, *O. episternalis* Lesne 1901, was intercepted at Boston, but not established (Lesne 1901a, Fisher 1950).]

Scobicia Lesne 1901. Six of the 11 species in this genus occur in North America, with one eastern species ranging from Ontario and New York to Georgia, four in Texas, Arizona, Colorado and California, and the lead cable borer, *S. declivis* (LeConte 1857), on the Pacific coast from British Columbia to California. Two exotic species have been intercepted at various ports (Fisher 1950).

Xylopertha Casey 1898, not Guérin-Méneville 1845

Tetrapriocera Horn 1878, contains 4 described species from the Neotropics and Marquises. One, *T. longicornis* (Olivier 1795), is common at lights throughout the Caribbean Basin and occurs as far north as Florida, but is also intercepted at ports farther north.

[*Xylion* Lesne 1901. One of the seven African species, *X. securifer* Lesne 1901, has been intercepted at various North American ports (Fisher 1950).]

Xylobiops Casey 1898. The five species recognized in this genus range from Venezuela to Canada Four of them occur in North America, ranging through eastern North America from Ontario to Florida, west to Montana, Nebraska and Texas, as well as Arizona and California (Fisher 1950).

Xyloblaptus Lesne 1901. Two species of *Xyloblaptus* occur in Texas, New Mexico, Arizona and California. The third species occurs from Mexico south to Costa Rica (Lesne 1939, Fisher 1950).

Xylomeira Lesne 1901, is a monotypic Neotropical genus. *Xylomeira tridens* (Fabricius 1792) reaches as far north as Florida and Texas (Fisher 1950).

[*Xyloperthella* Fisher 1950 contains 4 described species from Africa, one of which, *X. picea* (Olivier), has become established in South America and Jamaica, and should be watched for at North American ports. Another species, *X. crinitarsis* (Imhoff 1843), has been intercepted at various North American ports (Lesne 1924, Fisher 1950).

Xylopertha Lesne 1901, not Guérin-Méneville 1845]

[*Xyloprista* Lesne 1901. Of the four described species in this genus, three are Neotropical, and one is from India. Two of the Neotropical species have been intercepted at North American ports (Fisher 1950).]

Xylopsocus Lesne 1901. An Asian-Indian Ocean genus with 17 described species is represented in North America by *X. capucinus* (Fabricius 1781), which is established in Florida (Woodruff *et al.* 1978). Another species, *X. castanoptera* (Fairmaire 1850), has been intercepted at various ports (Fisher 1950).

[*Xylothrips* Lesne 1901. Two of the four described members of this originally Asian-Pacific genus are among the most commonly intercepted species of the family, showing up world-wide in packing materials, pallets and other wooden items. Neither is known to be established in North America (Fisher 1950).]

5. Tribe Sinoxylini Lesne 1898

This tribe of 6 genera is centered in Africa, but extends into Eurasia and through the Malay Archipelago to Australia, with one genus in temperate South America.

Diagnosis: The unique mandibles characterize this tribe. They are symmetrical and meet with their cutting edges against each other on the midline. Each mandible has two fine longitudinal striae on the external face near the tips (Lesne 1906, Ivie 1985).

Sinoxylon Duftschmid 1825. Five of the over 50 species of this genus have been intercepted at various places in North America. Two may be established: *S. ceratoniae* in California (Anonymous 1956) and *S. conigerum* Gerstäcker 1855 in Florida (Peck and Thomas 1998). One of these has become established in Venezuela (Fisher 1950, Binda and Joly 1991). Lesne (1906) last revised the genus, but this is now woefully out of date.

V. Subfamily Psoinae Blanchard 1851

This subfamily is known from five unusual genera with 9 often brightly colored species from India, South Africa, China, the Western Palearctic, and the Pacific Coast of North America. All are atypical bostrichids that resemble clerids more than other bostrichids.

Diagnosis: The first ventrite has the intercoxal process reduced or absent and lacks a postcoxal carina; the pronotum is rounded and narrowed behind, the procoxae are narrowly conical; and the head has a distinct neck (Ivie 1985).

Psoa Herbst 1797 is known from two species from British Columbia, Washington, Oregon, and California, and two from the Western Palearctic (Fisher 1950). See Tilden (1953) for observations on biology.

Acrepis LeConte 1852

VI. Subfamily Dinoderinae Thomson 1863

The six described genera in this subfamily are originally Holarctic, Oriental and Ethiopian in distribution. Representatives of this group are now commonly found throughout the world, spread through commerce in bamboo, grain and occasionally logs.

Diagnosis: The Dinoderinae are recognized by an evenly rounded, cowled prothorax, with no forward or ventrally directed horns (there is usually a crown of small dorsally directed teeth); a single protibial spine; and a relatively long first tarsomere which is subequal to the second.

Dinoderus Stephens 1830, is a taxonomically difficult genus with 26 currently recognized species distributed in two subgenera. Commonly known as the bamboo powder-post beetles, none are native to the New World, but several are now established there, and even more are intercepted each year. There is no current revision for the genus, although Vrydagh (1956) revised the nominal subgenus (22 of the species), recognizing the four *Dinoderastes* Lesne 1914 as a distinct genus. Spilman (1982) recombined the two groups as subgenera of *Dinoderus*, and reviewed the species established (three species) or commonly intercepted in North America (three species). The established species are recorded from Illinois and the District of Columbia to Texas and Florida, as well as California (Spilman 1982).

Prostephanus Lesne 1898 has four of its five described species known from North America. They occur generally throughout the eastern forests and through the Mexican border states (Fisher 1950, Wright 1986). One species, *P. punctatus* (Say) is the larger grain beetle of international commerce. Several undescribed species from Mexico and Central America will be included in a revision in progress by Ivie.

Rhyzopertha Stephens 1830. *Rhyzopertha dominica* (Fabricius 1792) is the lesser grain borer of international commerce and is virtually cosmopolitan in grain stores.

Rhizopertha Guérin-Méneville 1845

Stephanopachys Waterhouse 1888, is a Holarctic genus associated with conifers. The genus has 15 currently recognized names, of which Fisher (1950) recorded 9 from North America. They are generally distributed from the Northwest Territories to the Mexican border.

VII. Subfamily Lyctinae Billberg 1820

This group has been known as the powder-post beetles, and treated as an independent family in many North American treatments. The genera are placed in two tribes.

Diagnosis: The complete lateral margins of the pronotum and completely exposed head distinguish this family.

6. Tribe Lyctini Billberg 1820

Gerberg (1957) included six genera in this tribe.

Diagnosis: The seriate elytral punctation and/or pubescense and linear femur will distinguish this tribe.

Lyctoxylon Reitter 1878. One of the three Asian species, *L. japonum* Reitter 1878, has been repeatedly intercepted at various places in North America (Gerberg 1957).

Lyctus Fabricius 1792, is a large cosmopolitan genus of 25 species, several of which are widely spread through commerce. Five species are native to North America, three are established exotics, and one more has been intercepted. Wright (1960) reviews the biology of one of the North American species.

Minthea Pascoe 1866, is known from 8 described species from the Old World and Neotropics. Three of these have been intercepted in the eastern United States (Gerberg 1957, Vrydagh 1959b, Lawrence 1980).

7. Tribe Trogoxylini Lesne 1921

Gerber (1957) included six genera in this tribe.

Diagnosis: The globose femur and confused elytral punctation are diagnostic for this tribe.

Trogoxylon LeConte 1861, is another nearly cosmopolitan genus of 13 species, two of which are apparently native to North America. One of the exotic species is now established here, and two more have been intercepted (Gerberg 1957).

VIII. Subfamily Euderiinae Lesne 1934

This monogeneric tribe is known from two species (one undescribed) from New Zealand. It appears to be the sister-group of the Anobiinae.

Diagnosis: The squarely attached trochanters, three-segment pectinate antennal club of the male, and flat hind coxa that is not grooved to receive the femur will distinguish this unique animal (Crowson 1961, Klimaszewski and Watt 1997).

Bibliography

ANDERSON, W. H. 1939. A key to the larval Bostrichidae in the United States National Museum (Coleoptera). Journal of the Washington Academy of Sciences, 29: 382-391.

ANONYMOUS. 1956. U. S. Department of Agriculture Cooperative Economic Insect Report 6: 20.

BEESON, C. F. C. and B. M. BHATIA. 1937. On the biology of the Bostrychidae (Coleopt.). Indian Forest Records (N.S.) Entomology, 2: 223-323.

BEIRIGER, R.L. and R. W. SITES. 1996. The Bostrichidae (Coleoptera) of Missouri. Journal of the Kansas Entomological Society, 69: 45-68.

BERLAND, L. 1951. Pierre Lesne (1871-1949). Annales de la Société Entomologique de France, 120:1-15.

BINDA, F. and L. J. JOLY. 1991. Los Bostrichidae (Coleotpera) de Venezuela. Boletín de Entomología de Venezuela N.S., 6: 83-133.

BØVING, A. G. and F. C. CRAIGHEAD. 1931. An illustrated synopsis of the principal larval forms of the order of Coleoptera. Entomologica Americana (new series), 11:1-351.

COOPER, K. W. 1986. A lectotype for *Dinapate wrighti* Horn, the giant palm-borer, and description of a new species of *Dinapate* from eastern Mexico (Coleoptera: Bostrichidae). Transactions of the San Diego Society of Natural History, 21: 81-87.

COSTA, C., S. A. VANIN and S. IDE. 1999. Systematics and bionomics of Cneoglossidae with a cladistic analysis of Byrrhoidea sensu Lawrence & Newton (1995) (Coleoptera, Elateriformia). Arquivos de Zoologia (São Paulo), 35: 231-300.

CROWSON, R. A. 1961. Considerations on the genera *Endecatomus* Mellié and *Euderia* Broun (Coleoptera: Bostrychidae), with descriptions of their larvae. Proceedings of the Royal Entomological Society of London (B), 30: 113-120.

CROWSON, R. A. 1981. The Biology of the Coleoptera. Academic Press. London, xii + 802 pp.

CYMOREK, S. 1969. 67. Familie: Bostrychidae (Bohrkäfer), Pp. 13-25. *In*: H. Freude, K. W. Harde and G. A. Lohse, eds., Die Käfer Mitteleuropas. Band 8. Teredilia, Heteromera, Lamellicornia. Goecke & Evers, Krefeld.

DOWNIE, N. M. and R. H. ARNETT, Jr. 1996. The Beetles of Northeastern North America. Vol. 2. The Sandhill Crane Press. Gainesville, FL.

FISHER, W. S. 1950. A revision of the North American species of beetles belonging to the family Bostrichidae. United States Department of Agriculture Miscellaneous Publications 1-157.

FORBES, W. T. M. 1926. The wing folding patterns of the Coleoptera. Journal of the New York Entomological Society, 34(42): 98-99

GARDNER, J. C. M. 1933. Immature stages of Indian Coleoptera (13) (Bostrychidae). Indian Forest Records (Entomology Series), 18(9): 1-19.

GERBERG, E. J. 1957. A revision of the New World species of powder-post beetles belonging to the family Lycidae. United States Department of Agriculture, Technical Bulletin No. 1157, 55 pp., 14 pls.

HATCH, M. H. 1961. The beetles of the Pacific Northwest. Part III: Pselaphidae and Diversicornia I. University of Washington Publications in Biology, 16: 1-503.

HOFFMAN, W. A. 1933. *Rhizopertha dominica* F. is a library pest. Journal of Economic Entomology, 26: 293-294.

IVIE, M. A. 1985. Phylogenetic studies in the Bostrichiformia (Coleoptera). Ohio State University (Ph.D. Thesis), Columbus, 137 pp.

KLIMASZEWSKI, J. and J. C. WATT. 1997. Coleoptera: family group review and keys to identification. Fauna of New Zealand, 37: 1-199.

LAWRENCE, J. F. 1980. A new genus of Indo-Australian Gempylodini with notes on the constitution of the Colydiidae (Coleoptera). Journal of the Australian Entomological Society, 19: 293-310.

LAWRENCE, J. F. 1991. Bostrichidae (Bostrichoidea), Pp. 439-441. *In*: F. W. Stehr, ed., Immature Insects. Vol. 2. Kendall/ Hunt. Dubuque, IA.

LAWRENCE, J. F., A. M. HASTINGS, M. J. DALLWITZ, T. A. PAINE and E. J. ZURCHER. 1999a. Beetles of the World: A Key and Information System for Families and Subfamilies. CD-ROM, Version 1.0 for MS-Windows. CSIRO Publishing. Melbourne.

LAWRENCE, J. F., A. M. HASTINGS, M. J. DALLWITZ, T. A. PAINE and E. J. ZURCHER. 1999b. Beetle Larvae of the World: Descriptions, Illustrations, Identification, and Information Retrieval for Families and Sub-families. CD-ROM, Version 1.1 for MS-Windows. CSIRO Publishing: Melbourne.

LAWRENCE, J. F. and A. F. NEWTON, Jr. 1995. Families and subfamilies of Coleoptera (with selected genera, notes, references and data on family-group names). Pp. 779-1006. *In*: J. Pakaluk and S. A. Slipinski, eds., Biology, phylogeny and classification of Coleoptera: Papers celebrating the 80th birthday of Roy A. Crowson. Muzeum i Instytut Zoologii PAN, Warsaw.

LESNE, P. 1894. Le genere *Dysides* Pert. (*Apoleon* Gorh. in part.) Annales de la Société Entomologique de France, 63: 18-21.

LESNE, P. 1896. Révision des Coléoptères de la famille des Bostrychides. 1er Memoire. Bostrychides Procéphalides. Annales de la Société Entomologique de France, 65: 95-127, pls. 8-9.

LESNE, P. 1899. Révision des Coléoptères de la famille des Bostrychides. 3e Mémoire. Bostrychinae *sens. strict.* - I. Les *Bostrychus*. Annales de la Société Entomologique de France, [1898] 67: 438-621.

LESNE, P. 1901a. Révision des Coléoptères de la famille des Bostrychides. 4e Mémoire. Bostrychinae *sens. strict.* - II. Les *Xylopertha*. Annales de la Société Entomologique de France, 69(1900): 473-639.

LESNE, P. 1901b. Synopsis des Bostrychides paléarctiques. L'Abeille, 30: 73-136.

LESNE, P. 1906. Révision des Coléoptères de la famille des Bostrychides. 5e Mémoire. Sinoxylinae. Annales de la Société Entomologique de France, 75: 445-561.

LESNE, P. 1909. Révision des Coléoptères de la famille des Bostrychides. 6e Mémoire. Dinapatinae et Apatinae. Annales de la Société Entomologique de France, 78: 471-574, pls. 13-17.

LESNE, P. 1921. Classification des Coléoptères xylophages de la famille des Bostrychides. Compte Rendu. Association Française pour l'Avancement des Sciences, Congress de Strasbourg, 1920: 285-289.

LESNE, P. 1924. Les Coléoptères Bostrychides de l'Afrique tropical française. Encyclopédie Entomologique 3: 1-301. Librairie Paul LeChevalier, Paris.

LESNE, P. 1935. Un type primitif de bostrychides le genre *Chilenius*. Annales des Sciences Naturelles. Series 10, Zoologie, 18: 21-38.

LESNE, P. 1938. Pars 161. Bostrychidae. *In*: S. Schenkling, ed., Coleopterorum Catalogus. W. Junk, Berlin. 84 pp.

LESNE, P. 1939. Contribution a l'étude des Bostrychides de l'Amerique Centrale. Revue française d'entomologie, 6: 91-123.

OLSON, C.A. 1991. *Dinapate wrighti* Horn, the giant palm borer (Coleoptera: Bostrichidae), reported from Arizona infesting *Phoenix dactylifera* L. Coleopterists Bulletin, 45: 272-273.

PECK, S. B. and M. C. THOMAS. 1998. A distributional checklist of the beetles (Coleoptera) of Florida. Arthropods of Florida and Neighboring Land Areas, 16: i-viii + 1-180.

PETERSON, A. 1960. Larvae of insects. An introduction to Nearctic species. Part II. Coleoptera, Diptera, Neuroptera, Siphonaptera, Mecoptera, Trichoptera. Columbus, OH. 416 pp.

POLL, M. 1932. Contribution à l'étude des tubes de Malpighi des Coléoptères, leur utilité en phylogénése. Recueil de l'Institute Zoologique Torley-Rousseau, 4: 47-80.

SANI, R. S. 1964. Histology and physiology of the cryptonephridial system of insects. Transactions of the Royal Entomological Society of London, (A) 45: 160-168.

SHARP, D. and F. MUIR 1912. The comparative anatomy of the male genital tube in Coleoptera. Transactions of the Entomological Society of London, 1912: 533-534.

SPILMAN, T.J. 1982. False powder-post beetles of the genus *Dinoderus* in North America (Coleoptera, Bostrichidae). Coleopterists Bulletin, 36: 193-196.

TANNER, V. M. 1927. A preliminary study of the genitalia of female Coleoptera. Transactions of the American Entomological Society, 53: 5-50.

TEIXEIRA, É. P. 1992. Revisão das espécies Neotropicais do gênero Bostrychopsis Lesne, 1898 (Coleoptera, Bostrichidae). Revista brasiliera de Entomologia, 36: 607-646.

TILDEN, J. W., 1953. Observations on the biology of *Psoa maculata* LeConte (Psoidae). Coleopterists Bulletin, 7: 9-12.

VRYDAGH, J.-M. 1951. Faune entomologique des bois au Congo belge. Les insects bostrychides. Bulletin Agricole du Congo Belge, 42: 65-90.

VRYDAGH, J.-M. 1956. Contribution à l'étude des Bostrychidae. 7. Les genre Dinoderus Stephens 1830. Mémoires de la Société Royale d'Entomologie de Belgique, 27 [1955]: 495-513.

VRYDAGH, J.-M. 1958. Contribution à l'étude des Bostrychidae. 11. Les Bostrychidae de l'Australie, de la Tasmanie et la Nouvelle-Zélande. Bulletin et Annales de la Société Royale d'Entomologie de Belgique, 94: 35-64.

VRYDAGH, J.-M. 1959a. Contribution à l'étude des Bostrychidae. 18. Additions à l'étude des Bostrychidae de l'Australie. Bulletin et Annales de la Société Royale d'Entomologie de Belgique, 95: 274-285.

VRYDAGH, J.-M. 1959b. Contribution à l'étude des Bostrychidae. 19. Nouvelles additions et corrections à l'étude des Bostrychidae de l'Australie. Bulletin et Annales de la Société Royale d'Entomologie de Belgique, 95: 42-46.

VRYDAGH, J.-M. 1959c. Coleoptera Bostrychidae. South African animal life, Uppsala., 6: 97-124.

WOLCOT, G. N. 1950. Coleoptera. The Insects of Puerto Rico. Journal of Agriculture of the University of Puerto Rico, 32: 225-416.

WOODRUFF, R. E. 1967. An Oriental wood borer *Heterobostrychus aequalis* (Waterhouse) recently established in Florida (Coleoptera: Bostrichidae). Florida Department of Agriculture and Consumer Services, Entomology Circular 58: 1-2.

WOODRUFF, R. E., E. J. GERBERG and T. J. SPILMAN 1978. A false powder-post beetle new to the United States. Florida Department of Agriculture and Consumer Services, Entomology Circular 195: 1-2.

WRIGHT, C. G., 1960. Biology of the Southern *Lyctus* Beetle, *Lyctus planicollis*. Annals of the Entomological Society of America, 53: 285-292.

WRIGHT, V. 1986. An updated and annotated bibliography of *Prostephanus truncatus* (Horn) (Coleoptera: Bostrichidae): a pest of stored grain. Kansas State University Food and Grain Institute Postharvest Grain System R & D Research Report 24: 1-20.

70. ANOBIIDAE Fleming 1821

by T. Keith Philips

Family common name: The death watch and spider beetles

Family synomyms: Ptinidae Latreille 1802; Gnostidae LeConte and Horn 1883; Ectrephidae Sharp 1912

The short, contractile appendages and strongly deflexed head within a hood-like prothorax, or the spider-like appearance with greatly convex elytra, long legs and antennae, and moderately deflexed head will help separate adults of this family.

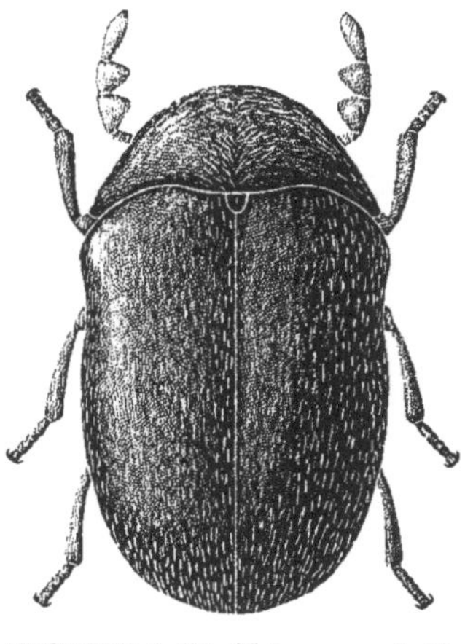

FIGURE 1.70. *Tricorynus similis* (LeConte)

Description: Shape strongly convex, elongate and cylindrical to oval or globular; length 1.1 to 9.0 mm; color tan, brown or piceous, some with lighter patches of setae or scales in various patterns; vestiture fine, mostly recumbent or erect but some with both present, in distinct tufts in some, occasionally appearing absent.

Head deflexed, inserted into prothorax, in some covered by pronotum; surface smooth, punctate, or rugose. Antennae mostly with 11 antennomeres, occasionally with 10, rarely three, with a 1-2 segmented club in some, more frequently with last three elongate, especially in male; either filiform, serrate, pectinate, or rarely flabellate; inserted on frons in front of eyes, narrowly to widely separated at base. Labrum small, transverse and in some emarginate anteriorly; mandibles small, curved, apices acute, bidentate apically in anobiids *sensu stricto* (i.e., excluding ptinines), medially in ptinines (second tooth absent in *Gnostus*); maxillary palp with four palpomeres, these short and slender, although some with apical palpomere expanded and truncate; gula distinct, gular sutures distinct and separate, mentum mostly quadrate; labial palp with three palpomeres, these small and slender, occasionally truncate or emarginate. Eyes lateral, small to large, bulging or flat, variously rounded, reduced in some.

Pronotum as wide or wider than head, broadly oval to subquadrate, borders mostly completely margined laterally, but margin may be present only at base or completely absent; surface smooth, punctate, or rugose; pleural region broad; prosternum very short, in some deeply excavated with mesosternum in front for head; procoxal cavities externally open, internally closed. Mesosternum short; mesocoxae moderately separated to separated by more than one coxal width, laterally coxae closed by sternum. Metasternum broad, convex.

Elytra entire, rarely connate, striae punctate or absent, intervals smooth, punctate, or rugose, epipleural fold variable, mostly obscure. Scutellum triangular, small, hidden in some. Wing venation frequently reduced. Radius anterior (RA+) divided into two branches beyond (Anobiidae *sensu stricto*) or before (Ptininae) middle of wing, but not extending and not forming a radial cell. Some with one or two cross-veins (r-r) from distal end of radius. Median spur in some nearly reaching wing margin. Proximal to media posterior (MP_{1+2}) veins absent or up to four main branches present. AP_3 absent although pigmentation occasionally visible at wing base. A wedge cell and/or jugal lobe is present in some. In ptinines, wings are atrophied to various degrees in some taxa, or completely absent in the most derived, globular-shaped species. In others, only females lack the ability to fly.

Legs short to long, with trochantins partially exposed to completely hidden on fore and middle legs, procoxae small, globular to conical, most contiguous; mesocoxae small, subconical, nearly contiguous; hind coxae small, nearly contiguous, transverse, can be completely fused to metasternum, most excavated to various degrees for reception of femora; trochanters short to very long, interstitial on all legs, squarely attached to bases of femora; femora slender to swollen; tibiae slender, some spinose, spurs obscure; tarsal formula 5-5-5, tarsomeres slender in some, one to four decreasing in length; claws simple, small.

Abdomen with five (apparently four in some, *Gibbium* with four, *Gnostus* with three) ventrites, strongly reduced in some, sutures entire or in some nearly obsolete, especially basally and at middle; 2 to 3 connate basally although rarely all five appearing fused or second and third fused, sculpture variable. Male genitalia trilobed; median lobe symmetrical or asymmetrical, curved, tubular, in some flattened apically or expanded, strongly curved at base dorsally where it articulates with base of parameres; parameres asymmetrical, or symmetrical, elongate; basal piece (pars basalis) forms a cap over base of parameres or (in ptinines) a small piece at paramere base ventrally. Female genitalia with valvifers large, articulating closely with coxite; coxite large, partly membranous, with two deep folds; styli large; proctiger small.

Acknowledgments. I thank the Annals of the Entomological Society of America (via the Copyright Clearance Center) and the Ohio Biological Survey for permission to use the illustrations drawn by Richard E. White. Also, appreciation goes to Michael A. Ivie, Norman F. Johnson, Brian J. Armitage, and Xavier Bellés for their help and encouragement with my ptinine research.

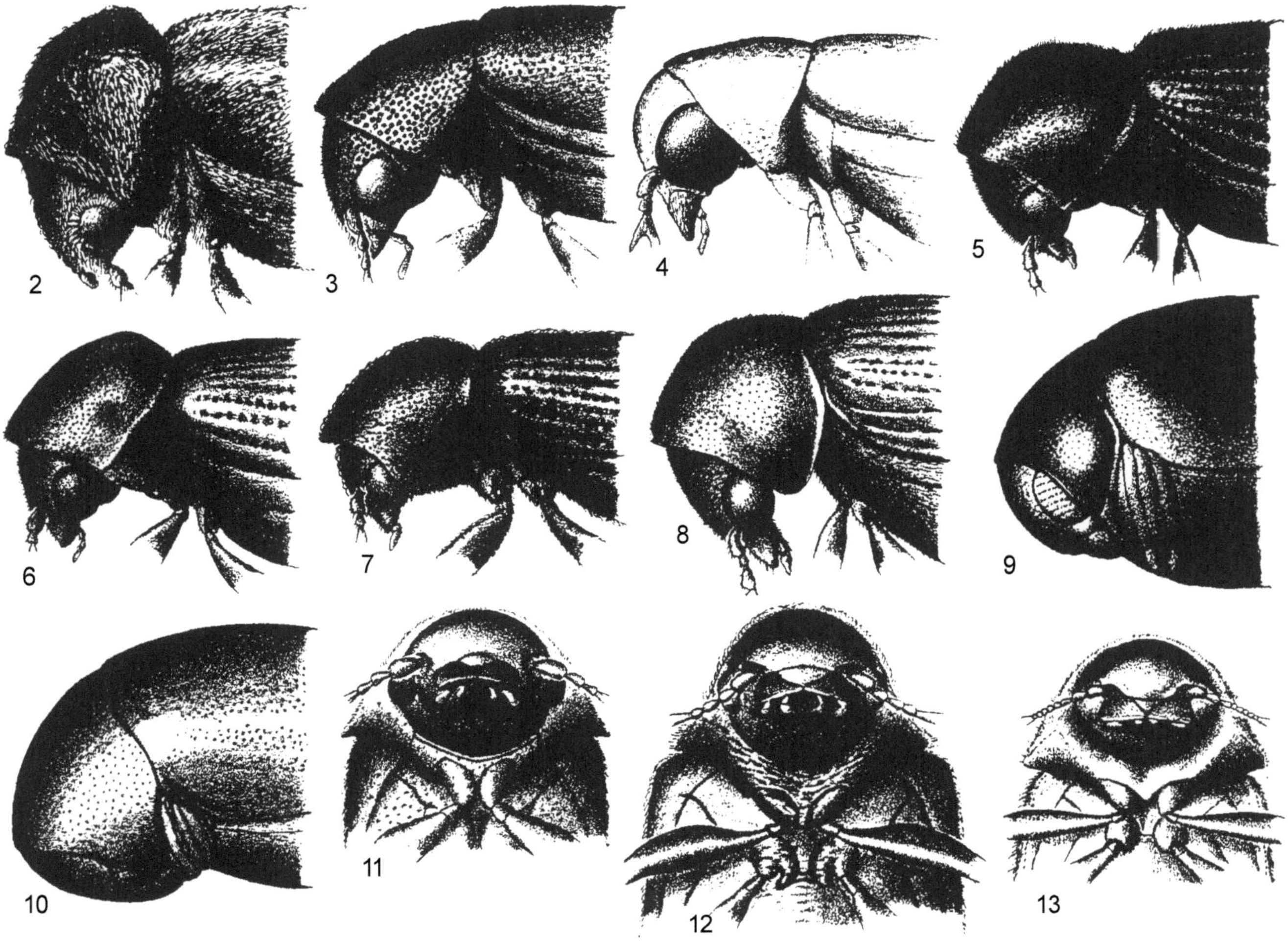

FIGURES 2.70-13.70. Anterior half of body, lateral view. 2. *Hedobia granosa* LeConte; 3. *Ernobius mollis* (L.); 4. *Xeranobium laticeps* Fall; 5. *Oligomerus sericans* (Melsheimer); 6. *Hemicoelus carinatus* (Say); 7. *Platybregmus canadensis* Fisher; 8. *Xyletinus peltatus* (Harris); 9. *Calymmaderus bahiensis* (Pic); 10. *Tricorynus estriatus* (Horn); anterior half and median portion of body, ventral view. 11. *Ernobius mollis* (L.); 12, *Utobium elegans* (Horn); 13. *Stegobium paniceum* (L.). (Illustrations from White 1971, reproduced with permission from the Entomological Society of America.)

Various papers describe aspects of adult morphology: Forbes (1922, wing venation); Forbes (1926, wing folding); Sharp and Muir (1912, male genitalia); Tanner (1927, female genitalia); Stickney (1923, head capsule); Williams (1938, mouthparts); Smith (1964, structure and development in flightless ptinines); Lawrence and Reichardt (1969, evolution related to myrmecophily); Philips (2000, evolution related to feeding habits, myrmecophily, and wing loss).

Larvae are C-shaped; up to 10 mm in length; vestiture may be composed of many short to long setae, scattered over body; color white or nearly so. Head exserted or nearly so, rounded and hypognathous, stemmata absent or one per side. Antennae with one or two segments, small, or vestigial. Fronto-clypeal suture distinct, labrum free (not fused); mandibles short, stout, uni- or bidentate, with irregular cutting edges; mola, retinaculum, or protheca absent, subbasal pseudomola rarely present; maxilla with galea and lacinia; distinct cardo, stipes, and three or four segmented palpi; galea setiferous or spiny; lacinia small or absent, acute; labium with ligula and two-segmented palpi; hypopharyngeal sclerome absent or rarely present; hypostomal rods absent; ventral epicranial ridges absent; one pair of ocelli or ocelli absent. Thorax with 5-segmented legs (absent in a few), each with two or more claw-like apical tarsunguli. Abdomen ten-segmented, ninth and tenth may be greatly reduced; segments in some with two or three variable dorsal plicae; mostly with rows of scattered asperites that are curved, straight, or hooked. Spiracles annular or annular-uniforous, located on prothorax and abdominal segments one to eight; a crescent or U-shaped sclerome present around cephalic end of the longitudinal anal slit; urogomphi absent.

Larval works relevant for species found in North America are Böving and Craighead (1931) and Peterson (1951). Bøving (1954) covers the Anobiidae *sensu stricto*, while those on ptinines are Manton (1945, stored product pest species), Hall and Howe (1953, a revised key to stored product pest species), Bøving (1956, description of *Ptinus californicus* Pic), and Andrews (1967, de-

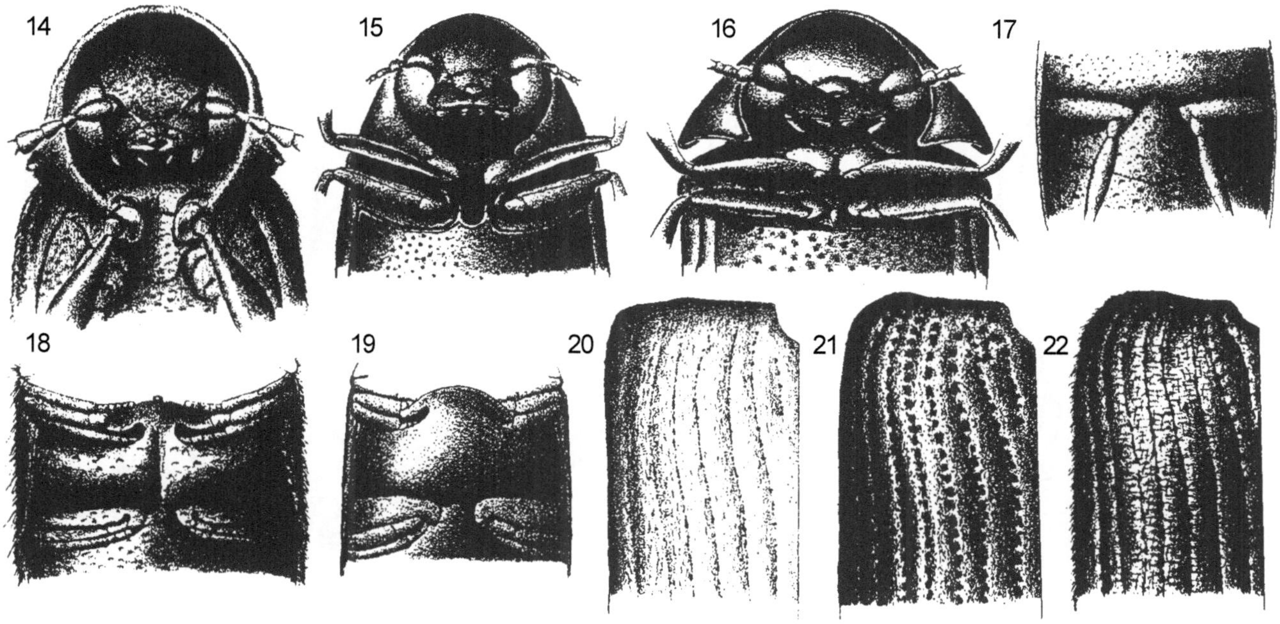

FIGURES 14.70-22.70. Anterior half and median portion of body, ventral view, and anterior half of elytra, dorsal view. 14. *Priobium sericum* (Say); 15. *Calymmaderus punctulatus* (LeConte); 16. *Tricorynus herbarius* (Gorham); 17. *Gastrallus marginipennis* LeConte; 18. *Protheca hispida* LeConte; 19. *Cryptoramorphus floridanus* White; 20. *Euvrilletta occidentalis* Fall; 21. *Oligomerus obtusus* LeConte; 22. *Stegobium paniceum* (L.). (Illustrations from White 1971, reproduced with permission from the Entomological Society of America.)

scription of *Ptinus latefasciatus* Gorham). Other larval works can be found cited in Lawrence (1991a and b).

Habits and Habitats. Larvae of Anobiidae *sensu stricto*, bore in bark, dry wood, twigs, seeds, woody fruits, galls, and fungi, or more rarely in the young stems or shoots of growing trees. Several species of *Ernobius* Thomson attack pine cones. One species, *Anobium punctatum* (DeG.) (adventive from Europe) causes great damage to furniture, the woodwork of houses, book bindings, and similar products. This species is also known as the "death-watch beetle" because of the habit in mating season of striking their vertex against the anterior margin of the pronotum, producing an audible tick, repeated in periods of seven or eleven strokes. This gave rise to ancient superstitions of being a portent of death; this habit of ticking is not confined to just *A. punctatum*. *Xestobium rufovillosum* (DeG.) in Europe also is known as the "death-watch" beetle (adventive from the United States). *Stegobium paniceum* (L.), the drug store beetle, and *Lasioderma serricorne* (Fab.), the cigarette beetle, harass the home-owner by breeding in stored products, tobacco, spices, and even cayenne pepper. *Ozognathus cornutus* LeConte (with large horns coming from the base of the mandible and extending back over the head in the males, absent in the females) are found in pine tree blossoms in Mexico and in oak galls in California. Members of this family have symbiotic yeast-like organisms in special organs called mycetomes which occur at the beginning of the mid-gut.

While some members of the Ptininae appear to be wood borers (Bellés 1980), most feed on accumulated dried animal or plant material (e.g. Howe 1959). Some species are common bird or mammal nest inhabitants. There are also species found in solitary bee nests, where they feed upon pollen stores (Linsley and MacSwaine 1942). Other genera, such as *Niptus* Boieldieu (Aalbu and Andrews 1992), some *Ptinus* Linnaeus (Andrews 1967), or the West Indian *Lachnoniptus* Philips (Philips 1998a) commonly breed in dung while at least one species of *Pitnus* Gorham (from the West Indies) is a leaf miner (Philips *et al.* 1998). One species of *Ptinus* (from Australia) is reported to feed on arthropod remains and spider eggs in nests of a salticid spider (Hickman 1974) while *Stereocaulophilus* Bellé (from the Canary Islands) is thought to graze on lichens (Bellés 1994). When in houses and other buildings, the larvae live in many sorts of dried vegetable or animal substances, such as stored flour, wool, and similar products (Hinton, 1941, Howe 1959). Most myrmecophilous species (the ectrephines) are found in Australia (Lawrence and Reichardt 1969). New World symphiles include *Coleoaethes* Philips, *Fabrasia* Martinez and Viana, and *Gnostus* Westwood and several are associated with the ant genera *Camponotus* and *Crematogaster* (Lawrence and Reichardt 1966, Philips 1998b). Some ptinines pupate within cocoons formed from their peritrophic membrane (Tristam 1977). Of note is the unusual biology of *Ptinus latro* Fabricius, a parthenogenic triploid which reproduces by gynogenesis (Woodroffe 1958). Mating with a male is still necessary for successful reproduction and males of three species have been found capable of inducing *P. latro* females to lay viable eggs. For a more detailed dicussion on the biology of spider beetles, see Howe (1959) and references within.

Status of the classification. Spider beetles have recently been considered a subfamily of the anobiids by Lawrence and Newton (1995). The anobiids *sensu stricto*, and the ptinines should probably be considered as subfamilies of the Bostrichidae based on adult evidence (Philips 2000) although larval (or molecular) evidence has yet to be critically examined (but see Bøving 1956 for comments on validity of the Ptinidae based on larvae). The ge-

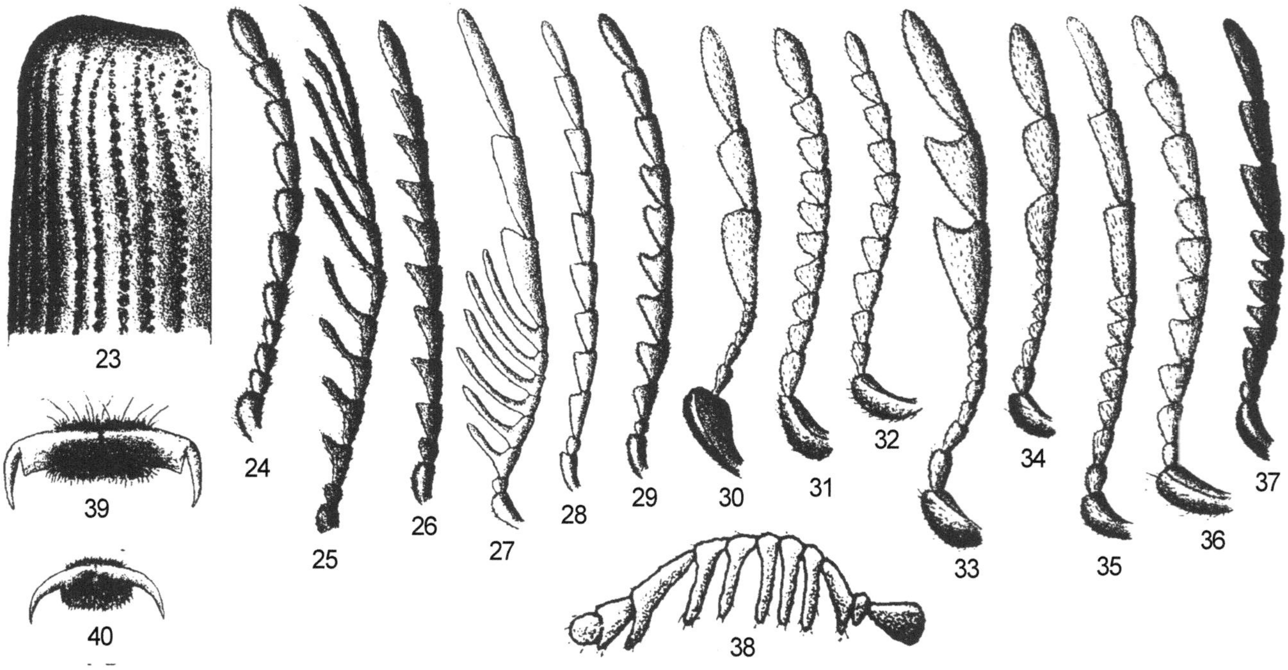

FIGURES 23.70-40.70. Anterior half of elytra, dorsal view. 23. *Hemicoelus carinatus* (Say); antennae and tarsal claws. 24. *Neohedobia texana* Fisher; 25. *Eucrada humeralis* (Melsheimer), male; 26. *Eucrada humeralis* (Melsheimer) female; 27. *Xeranobium laticeps* Fall, male; 28. *Actenobius pleuralis* (Casey); 29. *Euceratocerus hornii* LeConte; 30. *Tricorynus estriatus* (Horn); 31. *Mesocoelopus collaris* Mulsant and Rey; 32. *Lasioderma serricorne* (Fabricius); 33. *Megorama ingens* Fall; 34. *Xyletomerus histricus* Fall; 35. *Euvrilletta xyletinoides* Fall; 36. *Euvrilletta peltatus* (Harris); 37. *Vrilletta decorata* Van Dyke; 38. *Calytheca elongata* White; 39. *Trichodesma gibbosa* (Say); 40. *Hadrobregmus quadrulus* (LeConte). (Illustrations from White 1971, reproduced with permission from the Entomological Society of America.)

neric classification is fairly stable but many genera need revision and, although fairly well known in Canada, the United States and Europe, there are still many undescribed genera and species elsewhere. *Gnostus* and the Australian ectrephines, once considered separate familes, are now properly recognized as aberrant ptinines (Lawrence and Reichardt 1966, 1969).

Distribution. There are approximately 220 genera (nearly 70 are ptinines) and more than 2,200 species of Anobiidae known worldwide. Of these, 464 anobiid species occur in Mexico, United States and Canada, of which 69 species are ptinines.

Comprehesive species accounts with keys include Fall (1905, North America); Downie and Arnett (1996, Northeastern North American), and Hatch (1961, Pacific Northwestern North American). Catalogs and checklists include White (1982, North American Anobiidae, *sensu stricto*), Spillman (1975, North and Central America and the West Indies), Blackwelder (1945, Latin America and the West Indies), and McNamara (1991, Canada and Alaska). Although dated, more complete world lists are those of Pic (1912a, 1912b).

Key to the Genera of Mexico, the United States, and Canada

1. Antennal insertions close together between the eyes and separated by a ridge no wider than the length of the scape, the first three ventrites mostly fused, coxal plates absent 64

— Antennal insertions far apart just in front of each eye and separated by more than half the total width of the frons, the first two ventrites mostly fused and coxal plates nearly always present 2

2(1). Pronotum at side rounded throughout, without even a partial margin at base (Fig. 2) 3

— Pronotum with a complete lateral margin (Figs. 3, 5, 6, 8-10), or (infrequently) with a margin only basally (Figs. 4, 7) (Note: in some species, a lateral margin appears to be absent if the prothorax is fully retracted since the margin is vertical and adjacent to the mesothorax and front tibiae) .. 7

3(2). Last three antennomeres broad and elongate, their combined length nearly as great to greater than that of preceding segments (Note: antennae may be hidden in sternal grooves) 4

— Last three antennomeres not enlarged, their combined length no greater than that of four preceding segments (excluding pectinate portion of apical antennomere in *Eucrada* males) (Figs. 24-26) (Note: sternal grooves for antennae lacking) 5

4(3). Body contractile, head in retraction with mandibles attaining metasternum (Figs. 9, 49) .... *Calymmaderus*

— Body less contractile, head in retraction with mandibles distant from metasternum (similar to Fig. 2) .... *Xarifa*

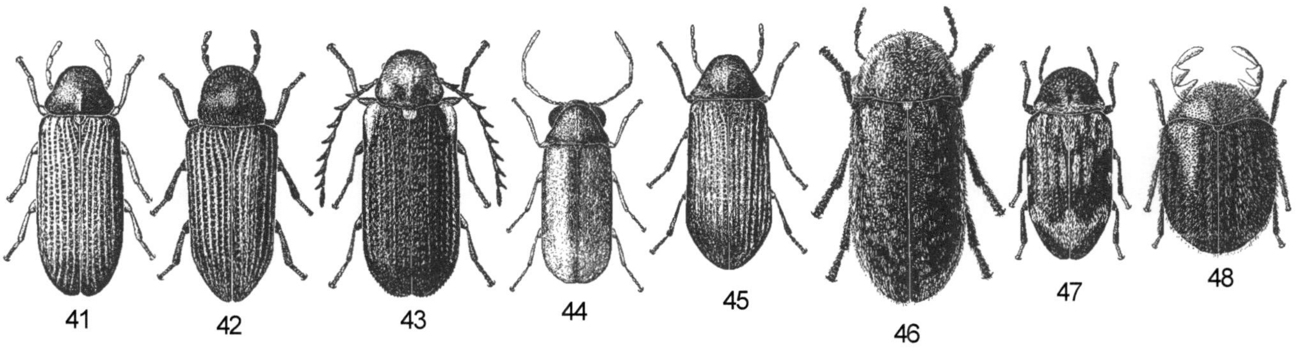

FIGURES 41.70-49.70. 41. *Hemicoelus carinatus* (Say), female; 42. *Platybregmus canadensis* Fisher, female; 43. *Eucrada humeralis* (Melsheimer), female; 44. *Ernobius granulatus* LeConte, female; 45. *Oligomerus obtusus* LeConte; 46. *Xestobium rufovillosum* (DeGeer); 47. *Hadrobregmus notatus* (Say); 48. *Byrrhodes intermedius* (LeConte). (Illustrations from White 1962, reproduced with permission from the Ohio Biological Survey.)

5(3). Elytra confusedly granulate-punctate, in some tuberculate, with no evidence of striae (Fig. 2) .... *Hedobia*

— Elytra rather obscurely, closely striate punctate, at least at sides 6

6(5). Antennomeres 3 and 4 nearly spherical, antennomeres 5 through 10 subtriangular (Fig. 24); Texas and Mexico .... *Neohedobia*

— Antennomeres 3 through 10 strongly serrate (females, Fig. 26) or pectinate (males, Fig. 25); Canada and northeastern U.S. (female habitus, Fig. 43) . ... *Eucrada*

7(2). Body in repose (i.e., with head and prothorax fully retracted) with head not markedly reflexed, mandibles distant from metasternum (Figs. 3-7) 8

— Body in repose with head markedly reflexed, mandibles near or reaching metasternum, in lateral view vertex of head mostly angled backwards and opisthognathus or nearly so (Figs. 8-10). 42

8(7). Pygidium exposed beyond elytra and vertical; pronotum at base distinctly indented on each side .... *Ptilineurus*

— Pygidium not exposed; pronotum at base not indented 9

9(8). Protibia with large tooth at apex, often most easily visible when apex is viewed head-on, margin finely toothed in most; pronotum mostly asperate anteriorly at middle, and with scattered protruberances, especially on anterior half .. 10

— Protibia rarely toothed at apex, margin above never toothed; pronotum anteriorly never asperate .. 11

10(9). Antennae strongly pectinate (males) or serrate (females); widespread (Fig. 64) .... *Ptilinus*

— Antennae filiform, with the last three antennomeres elongated and longer than remaining antennomeres combined; Baja California .... *Ptilinobium*

11(9). Antennomeres 3 and 4 nearly spherical, 5-10 subtriangular (Fig. 24); Texas and Mexico .... *Neohedobia*

— Antennae not as above; localities various 12

12(11). Pronotal lateral margin only at extreme base to basal 3/4 at most (Fig. 4); first ventrite suture weak; hind tarsus 0.9-1.7 times as long as tibia; antennomeres 4 through 8 nearly always serrate to pectinate .... *Xeranobium*

— Without above combination of characters 13

13(12). Prothorax with ventral surface convex or flat, not excavate (Figs. 11, 12) 14

— Prothorax with ventral surface concave to mostly excavate beneath, more or less enclosing head (Figs. 13, 14) 21

14(13). Procoxae conical, prominent, touching (Fig. 11)... 15

— Procoxae oval, less prominent, separated by a prosternal process (Fig. 12) 18

15(14). Antennomeres 3 through 8 serrate; prosternum before coxae longer than coxal diameter; tarsi longer than tibiae .... *Paralobium*

— Antennomeres 3 through 8 filiform to subtriangular; prosternal length before coxae 1/3 to 2/3 coxal diameter; tarsi shorter than tibiae 16

16(15). Lateral margin of pronotum distinct only basally, absent or indistinct anteriorly .... *Episernus*

— Lateral margin of pronotum distinct throughout 17

17(16). Prosternum before coxae about 2/3 coxal diameter, procoxae very narrowly separated... *Xestobium*

— Prosternum before coxae 1/3-1/2 coxal diameter, procoxae touching (Fig. 11, see also Figs. 3 and 44) .... *Ernobius*

18(14). Length 1.0-3.2 mm; dorsal surface nearly glabrous or with unicolorous pubescence not forming a pattern 19

— Length 5.0-7.5 mm; dorsal surface with light and dark patches of pubescence forming an irregular pattern 20

19(18). Antennae with 10 antennomeres; length 1.0-1.4 mm; pubescence of dorsal surface short and sparse, barely detectable; mandibles of male without horns .... *Microzogus*

— Antennae with 11 antennomeres; length 1.3-3.3 mm; pubescence of dorsal surface short and sparse to moderate in length; mandibles of male mostly with horns .... *Ozognathus*

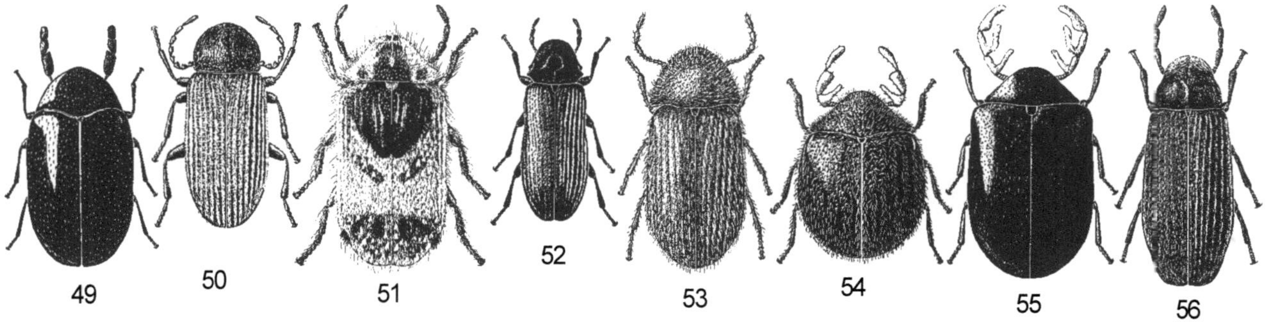

FIGURES 49.70-56.70. 49. *Calymmaderus nitidus* (LeConte); 50. *Priobium sericeus* (Say); 51. *Trichodesma gibbosa* (Say); 52. *Microbregma emarginatum* (Duftschmid); 53. *Stegobium paniceum* (L.); 54. *Caenocara oculata* (Say), male; 55. *Dorcatoma falli* White; 56. *Anobium punctatum* DeGeer, female. (Illustrations from White 1962, reproduced with permission from the Ohio Biological Survey.)

20(18). Antennomeres 4 through 8 subequal in length; pubescence of dorsal surface predominantly golden, dull yellowish or whitish (Fig. 46) .... *Xestobium*

— Antennomeres 4 through 8 unequal, antennomeres 5 and 7 broader and distinctly longer than others; pubescence white, brown, black, and (frequently) orange (see also Fig. 12) .... *Utobium*

21(13). Very small, length 1.0-1.3 mm; dorsal surface glossy, pubescence minute .... *Microzogus*

— Larger, length 1.9-9.0 mm; dorsal surface rarely glossy, pubescence distinct 22

22(21). Procoxae touching or very narrowly separated (as in Fig. 11) 23

— Procoxae distinctly to widely separated (Figs. 13, 14) 26

23(22). Combined length of last 3 antennomeres nearly as long as, to much longer than all preceding antennomeres (Fig. 35) 24

— Combined length of last 3 antennomeres not longer than preceding 4 or 5 antennomeres (Figs. 28, 29) 25

24(23). Mesocoxae touching; elytral striae indistinct (Fig. 20); tarsomere 3 not emarginate (Fig. 60) .... *Euvrilletta*

— Mesocoxae very narrowly separated, nearly touching; elytral striae distinct (Fig. 21); tarsomere 3 emarginate (barely visible in Fig. 45) .... *Oligomerus*

25(23). Antennae strongly serrate in female (Fig. 29), moderately pectinate in male (similar to Fig. 25 and habitus, Fig. 65) .... *Euceratocerus*

— Antennae moderately serrate in both sexes (Fig. 28) .... *Actenobius*

26(22). Combined length of antennomeres 9-11 not longer than preceding 5 antennomeres 27

— Combined length of antennomeres 9-11 much longer than preceding 5 antennomeres 28

27(26). Elytral punctures deep, nearly square, and forming distinct, even rows; abdominal sutures distinct at sides, feeble at center (Fig. 50, see also Fig. 14) .... *Priobium*

— Elytral punctures shallow, subcircular, not forming rows; abdominal sutures distinct throughout.... .... *Colposternus*

28(26). Prosternal intercoxal piece V-shaped (Fig. 13); punctures of elytral striae obscure, elongate and longitudinally closely spaced (Figs. 22, 53) .... *Stegobium*

— Prosternal intercoxal piece parallel-sided apex not pointed; punctures of elytral striae not as above 29

29(28). First ventrite suture feeble, curving posteriorly at middle (Fig. 17) 30

— First ventrite suture distinct, straight, or not curving as above 31

30(29). Lateral margin of pronotum present only in basal half; prosternum with anterior margin distinctly impressed between coxae; Colorado to California .... *Gastrallus*

— Lateral margin of pronotum present nearly throughout; prosternal anterior margin not depressed; Florida .... *Falsogastrallus*

31(29). Tarsal claws each with a broad, basal tooth (Figs. 39, 51) .... *Trichodesma*

— Tarsal claws each without a basal tooth (Fig. 40) 32

32(31). Antennomeres 4 through 8 serrate or pectinate; last abdominal segment conical, tip prominent and granulate .... *Ctenobium*

— Antennomeres 4 through 8 filiform, cylindrical, or subtriangular; last abdominal segment not conical, prominent, or granulate 33

33(32). Pubescence of upper surface and legs with intermixed, long, erect hairs 34

— Pubescence of upper surface and legs completely recumbent or with short, erect hairs 35

34(33). Metasternum excavate anteriorly; punctures at apical 1/3 of elytra confused .... *Anobiopsis*

— Metasternum not excavate anteriorly; punctures at apical 1/3 of elytra forming distinct striae .... *Nicobium*

35(33). Abdominal ventrite sutures 2, 3, and 4 distinct at sides and weak to nearly absent at middle.... 36

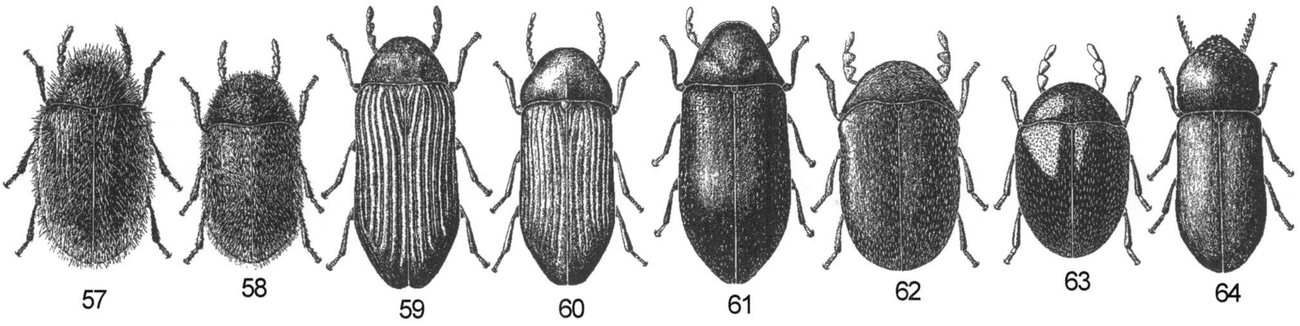

FIGURES 57.70-64.70. 57. *Stagetus profundus* (LeConte); 58. *Protheca hispida* LeConte; 59. *Vrilletta laurentina* Fall, female; 60. *Euvrilleta peltatus* (Harris); 61. *Petalium bistriatum* (Say), male; 62. *Tricorynus similis* (LeConte); 63. *Tricorynus rotundus* White; 64. *Ptilinus ruficornis* Say, female. (Illustrations from White 1962, reproduced with permission from the Ohio Biological Survey.)

— Abdominal ventrite sutures 2, 3, and 4 evenly distinct throughout 37

36(35). Punctures of abdomen large, distinct, ring-shaped (habitus, Fig. 47) .... *Hadrobregmus*

— Punctures of ventral surface small, obscured by granulation, not ring-shaped ..... *Desmatogaster*

37(35). Metasternum deeply excavate anteriorly (habitus, Fig. 56) .... *Anobium*

— Metasternum not excavate anteriorly, convex, flat, or slightly concave 38

38(37). Third ventrite half longer than ventrite 4; pronotum concave at center, produced laterally near base (Fig. 52) .... *Microbregma*

— Third ventrite about equal in length to ventrite 4; pronotum convex to flat at center 39

39(38). Lateral margin of pronotum present only at base (Fig. 7) 40

— Lateral margin of pronotum present throughout (Figs. 5, 6) 41

40(39). Prosternum longitudinally carinate at center (habitus, Fig. 42) (see also Fig. 7) .... *Platybregmus*

— Prosternum not carinate at center (Fig. 41) .... *Hemicoelus*

41(39). Procoxae touching, narrowly separated, or separated by up to half coxal diameter; elytral striae of punctures without distinct margins (Fig. 22), or of fine lines; pronotum stouter (Figs. 5, 45) .... *Oligomerus*

— Procoxae separated by half to entire coxal diameter; elytral striae of punctures with more or less distinct margins (Fig. 23); pronotum less stout (Figs. 6, 41) .... *Hemicoelus*

42(7). First ventrite depressed or grooved for reception of hind legs, posterior margin frequently distinctly carinate or appearing raised above second ventrite (Note: legs often tucked into groove and carina most easily visible near coxae)........... 43

— First ventrite not grooved as above................. 58

43(42). During body retraction procoxae visible (Figs. 9, 15) 44

— During body retraction procoxae concealed (Figs. 10, 16) 47

44(43). Metasternum produced into a broad rounded lobe, grooved laterally for mesotarsi, and which nearly conceals mandibles during retraction; second ventrite approximately as long as third through fifth combined (Fig. 61) .... *Petalium*

— Metasternum and second ventrite not as above . 45

45(44). Pubescence moderately dense, bristling; elytra with distinct, sharply impressed striae throughout (Fig. 57) .... *Stagetus*

— Pubescence absent to fine and appressed; elytra not striate (Fig. 9) or striate only at sides (see also Fig. 15) 46

46(45). Antennomeres 2-8 pectinate, apex of antennomere 9 acutely strongly produced (Fig. 38) .... *Calytheca*

— Antennomeres 2-8 triangular, apex of antennomere 9 not strongly produced (Figs. 9, 15, 49)........... .... *Calymmaderus*

47(43). When viewed front the front, eyes incised or notched to or beyond middle; body nearly spherical (Note: if head is fully retracted and eye canthus is large, only the dorsal half of the eye will be visible) (Fig. 54) .... *Caenocara*

— Eyes not as deeply incised; body broadly to narrowly oval 48

48(47). Each elytron with 10 striae, distinct from base to apex and not appearing as rows of punctures 49

— Elytra not striate as above or lacking striae 50

49(48). Metasternum produced anteriorly into a broad lobe visible in retraction and mostly with a linear pit of various depths at center (Fig. 48) ...... *Byrrhodes*

— Metasternum produced anteriorly into a narrow hook-like process concealed in retraction and at center longitudinally grooved throughout........ .... *Striatheca*

50(48). Metasternum anteriorly with distinct tarsal grooves (e.g., Figs. 18-19) 51

— Metasternum anteriorly lacking tarsal grooves (e.g., Fig. 16) 56

51(50). Elytra at sides notched for hind legs (Fig. 18) ... 52

— Elytra at sides not notched for hind legs (Fig. 19) 54

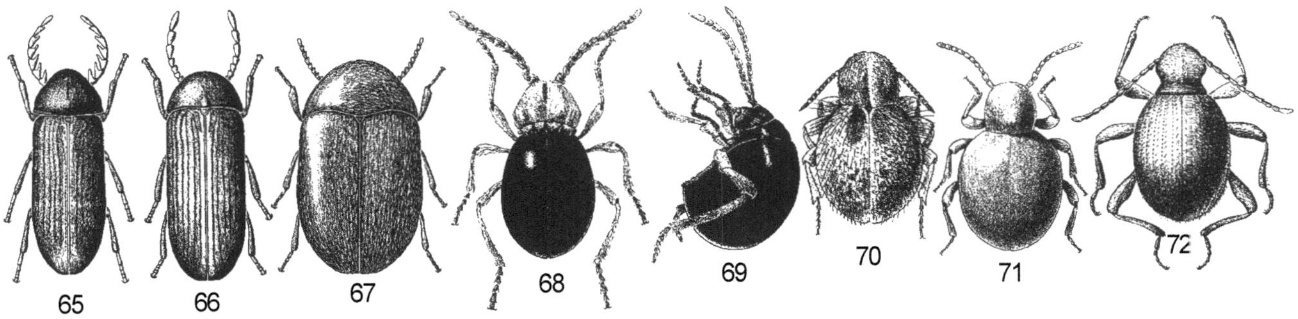

FIGURES 65.70-71.70. 65. *Euceratocerus gibbifrons* White, female; 66. *Xyletinus brevis* White; 67. *Lasioderma serricorne* (Fabricius); 68. *Mezium americanum* Laporte; 69. *Gibbium aequinoctiale* Boieldieu; 70. *Trigonogenius globulus* Solier; 71. *Sphaericus gibboides* (Boieldieu); 72. *Pseudeurostus hilleri* Reitter. (Figs. 65-67 from White 1962, reproduced with permission from the Ohio Biological Survey.)

52(51). Pubescence appressed; metepisternum parallel; length 3.0-3.5 mm .... *Stichtoptychus*
— Pubescence bristling; metepisternum narrowly triangular; length 1.6-2.4 mm 53

53(52). Antennae with 11 antennomeres; elytra mostly faintly punctate-striate, punctures small and not distinctly arranged in rows (Fig. 58)..... *Protheca*
— Antennae with 9 antennomeres; elytra obviously punctate-striate, punctures large and distinctly arranged in rows .... *Sculptotheca*

54(51). Elytra at sides not striate; anterior metasternal lobe broad, 1/3 or more of total width (Fig. 19) .... *Cryptoramorphus*
— Elytra at sides with 2 or 3 striae; anterior metasternal lobe narrow, 1/4 or less of total width........... 55

55(54). Eyes not or slightly notched; prosternum produced posteriorly into 2 slender parallel-sided processes below or posterior of procoxae and attaining or nearly attaining coxal apex (Note: this character is impossible to see unless head and prothorax are pulled away from the body); length 1.75-3.4 mm; antennae 10- segmented (Fig. 55) .... *Dorcatoma*
— Eyes distinctly notched; prosternum simple or produced into 2 broad, triangular processes up to half length of coxae; length 1.3-2.8 mm; antennae 8, 9, or 10- segmented (Fig. 48)... *Byrrhodes*

56(50). Mesosternum posteriorly not produced into a hook; metepisterna very narrow anteriorly (1/3 the width seen in Fig. 16), abruptly wider near posterior apex... *Cryptorama*
— Mesosternum posteriorly produced into a hook or small knob which is mostly visible with head retracted just below mandibles (Fig. 16); metepisterna as in Fig. 16 57

57(56). Antennae with 10 antennomeres, clubbed (Fig. 30); elytra laterally at apex and/or anterior tibiae nearly always striate (Figs. 10, 16, 62, 63)......... .... *Tricorynus*
— Antennae with 11 antennomeres, serrate (Fig. 31); neither elytral nor tibial striae present .... *Mesocoelopus*

58(42). Metasternum abruptly declivous anteriorly, declivity starting distinctly posterior of mesocoxae and may be posteriorly carinate; head deeply excavate beneath 59
— Metasternum not abruptly declivous anteriorly; head mostly not deeply excavate beneath (see also Fig. 8) 60

59(58). Antennae serrate, last 3 antennomeres not elongate (Fig. 32); metasternal declivity transversely carinate; length 1.8-3.0 mm (Fig. 67) .... *Lasioderma*
— Antennae not serrate, last 3 antennomeres elongate (Fig. 33); metasternal declivity not or partially carinate; length 3.2-8.4 mm ...... *Megorama*

60(58). Mesocoxae contiguous; antennae as in Fig. 34 ... . *Xyletomerus*
— Mesocoxae narrowly or distinctly separated.... 61

61(60). Mesocoxae distinctly separated, concealed in repose with only the trochanter visible .... *Neosothes*
— Mesocoxae narrowly separated, visible in repose 62

62(61). Antennomeres 4 through 10 strongly serrate to pectinate, at least as serrate as in Fig. 37; elytra mostly black with light longitudinal vittae or spots, intervals strongly convex (Fig. 59)............. *Vrilletta*
— Antennomeres 4-10 feebly to moderately serrate (Figs. 35, 36); elytra mostly brown, intervals less convex 63

63(62). Elongate bodied, about 2.2 or more times longer than wide; eyes large, separated by 2.0-3.5 times their width when viewed from the front (Fig. 60) .... *Euvrilletta*
— Stout bodied, about 2.0 times longer than wide; eyes small, separated by 4.0 or more times their width when viewed from the front (Fig. 66) .... *Xyletinus*

64(1). Ventrites about 1/3 as wide as elytra when viewed ventrally; elytra smooth and shiny although in some with dense setae at base and posteriorly with a few more scattered erect setae 65
— Ventrites at least half the width of the elytra when viewed ventrally; elytra mostly sparsely to densely covered with both erect and appressed or recumbent setae 66

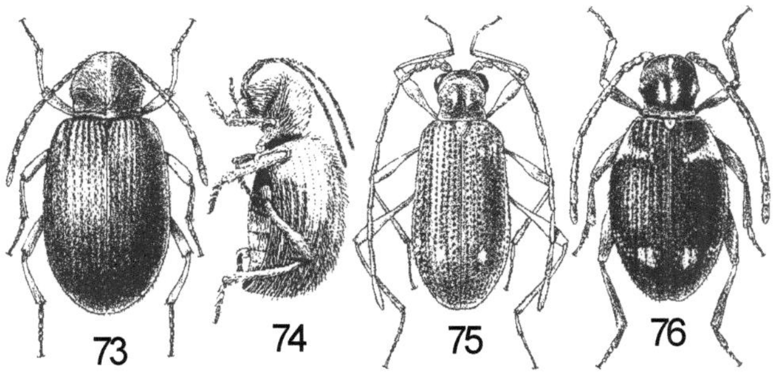

FIGURES 73.70-76.70. 73. *Ptinus tectus* Boieldieu; 74. *Niptus hololeucus* Falderman; 75. *Ptinus fur* L., male; 76. *Ptinus fur* L., female.

65(64). Head and thorax glabrous; abdomen with 4 ventrites; metatrochanter 2/3 the length of the femur (Fig. 69) .... *Gibbium*
— Head and thorax densely setose; abdomen with 5 ventrites; metatrochanter at most 1/3 the length of the femur (Fig. 68) .... *Mezium*

66(64). Trichomes present and visible as small clumps of reddish brown setae on the pronotum; antennae with 3 antennomeres; Florida .... *Gnostus*
— Trichomes absent; antennae with at least 9 antennomeres; widespread 66

67(66). Elytral surface smooth but very slightly granular, scattered small punctures difficult to see; elytral densely covered with recumbent scattered scale-like setae which obscures surface; ventrites slightly more than half width of elytra when viewed ventrally; introduced from Europe (Fig. 71) .... *Sphaericus*
— Elytral surface punctate; if scales present, surface of elytra not obscured and punctures easily visible and oriented in longitudinal rows; ventrites nearly equal in width to the elytra when viewed ventrally 68

68(67). Antennae with 9 antennomeres; small and black, 0.9-1.4 mm in length .... *Pitnus*
— Antennae with 11 antennomeres; medium to large in length, 1.6 mm or larger, most 2.5-4.5 mm . 69

69(68). Raised portion of frons between antennal fossae equal to about half or more of the length of the antennal scape 70
— Raised portion of frons between antennal fossae equal to less than 1/4 the length of the antennal scape 73

70(69). Pronotal width greater than length, about 1/10 to 1/5 wider than long (Fig. 70) .... *Trigonogenius*
— Pronotal width less than length, about 1/10 to 1/5 longer than wide 71

71(70). Elytral punctures distinctly visible among erect or suberect and appressed setae; punctures generally large, mostly at least 2 times the width of the elytral setae at setal base; metafemora not parallel-sided but very gradually increasing in width from base to near or past middle 72
— Elytral punctures hidden from view by a dense recumbent layer of setae (in addition to erect setae), if setae are abraded, punctures very small, about the same diameter of the elytral setae at setal base; metafemora parallel-sided from near base to just past middle (for more than half their total length); adventive from Europe (Fig. 74) (*N. hololeucus* Falderman) .... *Niptus*

72(71). Appressed or recumbent interpuncture elytral setae scattered and frequently 4 or 5 setae in any given area in a nearly transverse row and oriented towards a median point in the middle of the row and forming a distinct longitudinal setal carinae; adventive from Europe (*T. unicolor* P.and M.) .... *Tipnus*
— Appressed or recumbent interpuncture elytral setae scattered or aligned, but never forming a distinct longitudinal setal carinae; southwestern United States .... *Niptus*

73(69). In ventral view, hind trochanters long, exceeding margin of elytra when projected perpendicular to longitudinal axis of the body (Fig. 72) .... *Pseudeurostus*
— In ventral view, hind trochanters short, not reaching margin of elytra when projected perpendicular to longitudinal axis of the body 74

74(73). Prosternum expanded anteriorly and ventrally, capable of concealing mouthparts when head is in a retracted position; Mexico ..... *Prosternoptinus*
— Prosternum not expanded, mouthparts easily visible even if head is retracted 75

75(74). Punctures in anterior 1/5 to 1/3 of pronotum irregular, elongate, large and distinctly visible; elytra with white recumbent setae forming distinct maculations consisting of a transverse "W" at basal 1/3 and 4 spots at apical 1/3; first and second ventrite sutures absent or at most faintly impressed at middle; Texas and Mexico .... *Niptinus*
— Punctures in anterior 1/5 to 1/3 of pronotum round or ovoid, not large or distinctly visible in most, often obscured by setae; elytral setal maculations, if present, highly variable; most with first and second ventrite sutures complete and distinctly impressed at middle; widespread (Figs. 73, 75, 76) .... *Ptinus*

Classification of the Nearctic Genera

Anobiidae Fleming 1821, *sensu lato*

Ptininae Latreille 1802
(incl. Gibbiinae Pic 1912)

(Bellés 1982, suprageneric classification; Hinton 1941, key economically important Ptininae; Papp 1962, North American Ptininae; Papp and Okumura 1959, Californian Ptininae; Spilman 1991, spider beetle food pests)

Gibbium Scopoli 1777, 2 spp., *G. psylloides* (Czempinski 1778) and *G. aequinoctiale* Boieldieu, both probably adventive, generally distributed, sometimes pests. The latter appears to be common while the former is extremely rare in North America (Bellés and Halstead 1985, species differentiation; Bellés 1985a and 1985b, biology).

Scotias Czempinski 1778

Gnostus Westwood 1855, 1 sp., *G. floridanus* Blatchley 1930, Florida; myrmecophile (Thomas *et al.* 1992, observations; Lawrence and Reichardt 1966, revision).

Mezium Curtis 1828, 2 spp., *M. affine* Boieldieu 1856, and *M. americanum* Laporte 1935, both introduced and widely distributed cosmopolitan pests, although the former is more common (Bellés 1985a and 1985b, biology).

Niptinus Fall 1905, 4 spp., Texas and Mexico (Philips 1999, revision).

Niptus Boieldieu 1856, 9 spp., one adventive (*N. hololeucus* Falderman), Canada and northern United States to Texas, New Mexico, Arizona, Utah, Nevada, southern California, and Mexico (Aalbu and Andrews 1992, revision).

Pitnus Gorham 1883, 7 spp., Mexico including Baja California, Arizona, and Florida (Bellés 1992, revision; Philips *et al.* 1998, biology).

Pseudeurostus Heyden 1906, 2 spp., *P. hilleri* (Reitter 1877), New Brunswick to Ontario, Alberta, British Columbia (introduced), and *P. kelleri* Brown 1959, Utah to Montana (Brown 1959, revision; Aalbu and Andrews 1992, discussion of genus validity and see Philips 2000).

Eurostus Mulsant and Rey 1868, not Dallas 1851, not Dumeril 1853

Prosternoptinus Bellé 1985, 1 sp. undescribed, Mexico (Philips 1997, revision).

Ptinus Linnaeus 1767, 38 spp., generally distributed (Fall 1905 and Papp 1962, key to North American species; Papp and Okumura 1959, key to California species).

subgenus *Gynopterus* Mulsant and Rey 1868
subgenus *Ptinus, sensu stricto*

Sphaericus Wollaston 1854, 1 sp., *S. gibboides* (Boieldieu 1856), adventive, California to British Columbia.

Tipnus Thomson 1863, not DuVal 1860, not Boieldieu 1856, 1 sp., *T. unicolor* Piller and Mitterpacher 1882, Nova Scotia and New Brunswick (adventive).

Epauloecus Mulsant and Rey 1868
Epailaeus Pic 1912

Trigonogenius Solier 1849, 1 sp., adventive, California to British Columbia and Texas.

Anobiidae Flemming 1821, *sensu stricto*

(White 1971, evolution of body form; White 1973b, new neotropical genera; White 1962, Ohio Anobiidae; White 1974b, type species for World genera; White 1979a, nomenclatural changes). Author names are excluded for tribes as their authors are unclear at this time.

Eucradinae LeConte 1861
(= Hedobiinae Mulsant and Rey 1868)

Eucradini

Eucrada LeConte 1861, 2 spp., Iowa, Michigan, Kentucky, Indiana, Ohio, Pennsylvania, Quebec, Ontario, South Carolina, British Columbia, and California; bore between bark and wood of oak.

Hedobiini

Hedobia Dejean 1821, 3 spp., California; wood borer.

Neohedobia Fisher 1919, 1 sp., *N. texana* Fisher 1919, Texas.

Dryophilinae LeConte 1861

Ptilineurini

Ptilineurus Reitter 1901, 1 sp., *P. marmoratus* (Reitter 1877), Virginia; wood borer.

Ernobiinae Pic 1912

Ernobiini

Episernus Thomson 1863, 2 spp., California, Utah, and Colorado. One species from limbs of *Pinus flexis* James.

Amphibolus Mulsant and Rey 1863 (preoccupied Klug 1830)
Claudius Des Gozis 1882 (replacement name for *Amphibolus*)

Ernobius Thomson 1859, 30 spp., generally distributed; usually in conifer cones and under bark.(Ruckes 1957, key to California spp.)

Conophoribium Chevrolat 1861
Liozoum Mulsant and Rey 1863
Philoxylon LeConte 1862
Conopheribium (*sic*) Chevrolat 1864
Liozum (*sic*) Mulsant and Rey 1864
Conoploribium (*sic*) Pic 1912
Liozuom (*sic*) Gardiner 1953

Paralobium Fall 1905, 1 sp., *P. mundum* Fall 1905, California.

Ozognathini

Microzogus Fall 1905, 1 sp., *M. insolens* Fall 1905, Florida.

Ozognathus LeConte 1861, 3 spp., New Jersey, Virginia, Kentucky, Florida, Texas, Arizona, California, and Colorado; one species recorded infesting dead flower stalks, twigs, bark, and cynipid galls on oak in California.
Durangoum Pic 1903 (as subgenus)
Micranobium Gorham 1883
Microanobium (*sic*) Bøving 1954

Xarifa Fall 1905, 2 spp., California, southern California Islands. Reared from dead willow (White 1973c, key).

Xestobiini

Utobium Fall 1905, 4 spp., Western States through to Quebec and New York. Recorded from lodgepole pine (White 1966, key; White 1976a, key modification).

Xestobium Motschulsky 1845, 6 spp., Quebec and New England to Indiana, Illinois, British Columbia, and California; one species adventive, the death-watch beetle; European species occurs in old houses (see notes under Habits and habitats) (White 1975b, key).
Cnecus Thomson 1859
Hyperisus Mulsant and Rey 1863
Xesfobium (*sic*) Zaitsev 1956
Zestobium (*sic*) Little 1957
Hyperius (*sic*) Hatch 1961
Hyerisus (*sic*) Espanol 1964

Anobiinae Fleming 1821

Anobiini

Anobium Fabricius 1775, 1 sp., *A. punctatum* (DeGeer 1774), probably adventive and widely distributed. Commonly known as the furniture beetle, it feeds on many hard- and softwoods.
Byrrhus Mueller 1776 (preoccupied Linnaeus 1767)
Amobium (*sic*) Leach 1815
Anabium (*sic*) Ozer 1957
Anbium (*sic*) Chujo 1964
Hemicoelinum Espanol 1971 (as subgenus)

Hemicoelus LeConte 1861, 7 spp., northeastern North America west to Manitoba and Nebraska, south to North Carolina, and California north to Alaska; borer of many hard- and softwoods (White 1976a, key).
Cacotemnus LeConte 1861
Cacoteminus (sic) White 1974
Hadrobregmus of authors, not Thomson
Hemicoelum (emendation) Espanol 1971

Microbregma Seidlitz 1889, 1 sp. (2 subspp.), *M. emarginatum* (Duftschmid 1825), Massachusetts to British Columbia and California; bark borer.
Microbregmum (sic) Bøving 1954

Platybregmus Fisher 1934, 1 sp., *P. canadensis* Fisher 1934, Ontario to New Jersey; recorded from maple and basswood.

Colposternini

Colposternus Fall 1905, 1 sp., *C. tenuilineatus* (Horn 1894), Oregon, and California.

Euceratocerini

Actenobius Fall 1905, 1 sp., *A. pleuralis* (Casey 1898) California; recorded from live oak.
Actenobium (*sic*) Fall 1905

Ctenobium LeConte 1865, 1 sp., *C. antennatum* LeConte 1865, Virginia.

Euceratocerus LeConte 1874, 3 spp., Texas, Ohio, Kentucky, and South Carolina; one species from under redbud bark (White 1966, key; White 1974a, key modification).

Xeranobium Fall 1905, 13 spp., southwestern Texas, Arizona, Nevada, California, Oregon and Utah; larvae of two species in iodine bush, *Allenrolfea occidentalis* (S.Walton) Kuntze.

Gastrallini

Gastrallus Jacquelin du Val 1860, 2 sp., New Mexico, Colorado and California; found on oak and from stem of *Clematis ligusticifolia* Nutt.
Gastragallus (*sic*) Austin 1880

Falsogastrallus Pic 1914, 1 sp., *F. librinocens* (Fisher 1938), Florida; damaging books.
Neogastrallus Fisher 1938
Gastrallomimum Pic 1939

Hadrobregmini

Desmatogaster Knutson 1963, 1 sp., *D. subconnata* (Fall 1920), Manitoba, Quebec, and Maine; hosts listed as spruce and aspen.

Hadrobregmus Thomson 1859, 8 spp., widely distributed; various hard- and softwoods have been recorded as hosts (Knutson 1963, key).
Coelostethus LeConte 1861
Dendrobium Mulsant and Rey 1864 (as subgenus of *Anobium*)
Habrobregmus (*sic*) De Leon 1952
Hadrobrogmus (*sic*) Simeone 1962
Hadrobresmus (*sic*) Simeone 1962

Hadrobegmus (*sic*) Espanol 1970
Allobregmus Espanol 1970 (as subgenus)
Megabregmus Espanol 1970 (as subgenus)

Priobium Motschulsky 1845, 4 spp., broad distribution including Mexico, 1 sp. introduced from Old World; one species recorded from various hardwoods (White 1975a, key to North and Central America spp.; Ford 1984, modified key).
Trypopitys Redtenbacher 1849
Trypopithys (*sic*) Baudi di Selve 1873
Tripopitys (*sic*) Horn 1894
Tripopithys (*sic*) Pic 1931
Trpopitys (*sic*) Brimley 1938
Prioblum (*sic*) Zaitsev 1956

Nicobiini

Anobiopsis Fall 1905, 1 sp., *A. sericans* Fall 1905, southern California; in dead branches of live oak.

Nicobium LeConte 1861, 1 sp., *N. castaneum* (Olivier 1790), southeastern U.S. to Texas; found in various wood products and yellow pine.
Neobium Mulsant and Rey 1864 (as subgenus of *Anobium*)

Trichodesma LeConte 1861, 10 spp., widely distributed; hosts include various hardwoods.
Ptinodes LeConte 1861
Microtrichodesma Pic 1931

Stegobiini

Oligomerus Redtenbacher 1849, 14 spp., widely distributed; one species from *Chrysothamnus nauseosus* (Pall.) Britt. and two more from beech, walnut, oak, and chestnut (White 1976a, key).
Oligomerinus Portevin 1931

Stegobium Motschulsky 1860, 1 sp., *S. paniceum* (Linnaeus 1758 [*Dermestes*]), generally distributed; the drug-store beetle occurs in a wide variety of stored products.
Sitodrepa Thomson 1863
Artobium Mulsant and Rey 1864
Autobium (*sic*) Reitter 1901
Litropeda (*sic*) Stebbing 1914
Sidrotepa (*sic*) Bosq 1934
Sitoprepa (*sic*) Seabra 1939

Uncertain Tribal Placement

Ptilinobium White 1976, 1 sp., *P. aberrans* White 1976, Baja California (White 1976, discussion of placement and relationships).

Ptilininae Shuckard 1840

Ptilinini

Ptilinus Mueller 1764, 9 spp., including one of uncertain status, widely distributed, larvae mine in hardwood.

Xyletininae Gistel 1856

Xyletinini

Euvrilletta Fall 1905, 20 spp. (5 Mexican), Texas, Arizona, Nevada, and California and Mexico; adults of one species associated with Christmas berry, *Heteromeles arbutifolia* (Ait.) M. J. Roem (White 1973d, key North American spp.; White 1985, revised key North American spp.).
Oligomerodes Fall 1905
Oligomeroides (*sic*) Pic 1905
Euvrilleta (*sic*) Van Dyke 1946

Vrilletta LeConte 1874, 12 spp., including 2 of uncertain status, California, Oregon, Washington, and Ontario, Quebec, and Pennsylvania; recorded in oak, linden, and alder (White 1980a, revision).
Pseudoxyletinus Pic 1903 (as subgenus)
Vrilleta (*sic*) Pic 1905
Urilletta (*sic*) Zoological Record, 1924
Vriletta (*sic*) Böving 1927
Vriletta (*sic*) Van Dyke 1946

Xyletinus Latreille 1809, 18 spp., including one of uncertain status, generally distributed; recorded from various woods (White 1973d, key North American spp.; White 1977, key to Mexican spp.)
Xylotinus (*sic*) Sturm 1826
Xiletinus (*sic*) Stephens 1829
Xytelinus (*sic*) Griffith and Pidgeon 1832
Calypterus Mulsant and Godart 1859 (described as genus but listed as subgenus by Pic 1912)
Notiomimus Wollaston 1861
Xeronthobius Morawitz 1863
Sternoplus Mulsant and Rey 1864
Xyletomimus Reitter 1901 (as subgenus)
Ryletinus (*sic*) Zaitsev 1956
Xylentinus (*sic*) Santoro 1957

Xyletomerus Fall 1905, 2 spp., California; one species recorded from manzanita (*Arbutus*). The species are possible synonyms: see White 1976a: 169 (White 1976a, key)
Xyletomeridius Espanol 1968 (as subgenus)

Lasiodermini

Lasioderma Stephens 1835, 5 spp., generally distributed, includes the cigarette beetle, *L. serricorne* (Fabricius), which breeds in various stored products, particularly tobacco (White 1975a, key to Central and South American spp.; White 1990, biology).
Pseudochina Jacquelin du Val 1860
Hypora Mulsant and Rey 1864 (as subgenus)

Pseudochinus (*sic*) Zoological Record 1865
Tasioderma (*sic*) Chenu 1884
Lasiderma (*sic*) Schilsky 1899
Ladioderma (*sic*) Loeding 1945
Lassioderna (*sic*) Mukerji 1955

Megorama Fall 1905, 4 spp., California and Kentucky; one species from live oak.

Dorcatominae C. G. Thomson 1859

Calymmaderini

Calytheca White 1973, 1 sp., *C. elongata* White 1973, Sinaloa, Mexico (White 1983, key Neotropical spp.).

Calymmaderus Solier 1849, 17 spp. including one of uncertain status, widely distributed; dead grapevines recorded as a host (White 1983, key Neotropical spp.).
Eupactus LeConte 1861
Calymmoderus (sic) Gemminger and Harold 1869
Eutheca Kiesenwetter 1877
Thaptor Gorham 1883
Lioolius Gorham 1883
Byrrhocerus Brèthes 1919
Calymaderus (*sic*) Pic 1923
Nevermannia Fisher 1927 (preoccupied Enderlein 1921)
Nevermannus Fisher 1927 (replacement name for *Nevermannia* Fisher)
Colymmaderus (*sic*) Van Dyke 1936
Thapter (*sic*) Arnett 1962
Eupactidius Espanol 1969 (as subgenus)

Cryptoramorphini

Cryptoramorphus White 1966, 1 sp., *C. floridanus* White 1966, Florida and South Carolina (Note: possibly a synonym of *Pseudodorcatoma* Pic 1905, see White 1979a: 215).
Crytoramorphus (*sic*) Espanol 1967

Dorcatomini

Byrrhodes LeConte 1878, 11 spp., eastern U.S. and Michigan, Ohio, Texas, and California; in hard tree fungi (White 1966, key [as *Eutylistus*]; White 1973a, key).
Priotoma Gorham 1886
Eutylistus Fall 1905
Fachus Lucas 1920 (unjustified new name)
Notiotheca Brèthes 1923
Eutylistes (*sic*) Böving 1927
Byrrhoides (*sic*) White 1966

Caenocara Thomson 1859, 14 spp. including one of uncertain status, eastern U.S. and Texas, New Mexico, Arizona, Nevada, and California; immatures of most or perhaps all species live in puffballs.
Tylistis LeConte 1861
Enneatoma Mulsant and Rey 1864
Coenocara (*sic*) LeConte 1865
Cyphanobium Broun 1893

Dorcatoma Herbst 1792, 5 spp., widely distributed; feed on woody fungi (White 1966, key).
Dorkatoma Herbst 1792 (emendation)
Serrocerus Kugelann 1793
Dorcatorna (*sic*) Westwood 1838
Dorcotoma (*sic*) Lea 1924
Nerrocerus (*sic*) Wu 1937
Doratoma (*sic*) Hayashi 1951
Dorcadoma (*sic*) Horion 1961

Petaliini

Petalium LeConte 1861, 23 spp., widely distributed; recorded from several species of hardwoods.
Rhadine Baudi di Selve 1873 (preoccupied by LeConte 1846)
Synanobium Schilsky 1898
Radine (*sic*) Schilsky 1899
Fossanobium Pic 1903 (as subgenus)
Rahdine (*sic*) White 1962
Metapetalium Espanol 1970 (as subgenus)

Prothecini

Protheca LeConte 1865, 2 spp., New Jersey to Florida, west to Michigan, Texas and Mexico; one species from tuliptree (*Liriodendron*) (White 1979b, revision).
Photheca (sic) Espanol 1972
Picatoma Lepesme 1947

Sculptotheca Schilsky 1900 (as subgenus), 2 spp., *S. puberula* (LeConte 1895), Massachusetts to Florida and Georgia, west to Wisconsin; and one Mexican species.

Stagetus Wollaston 1861, 2 spp., *S. profundus* (LeConte 1865), Maryland to Oregon and California ; *S. grossus* White 1976, Arizona (White 1975a, key American spp.).
Theca Mulsant and Rey 1860 (preoccupied Morris 1845)
Anomotheca Schilsky 1899
Thecha (*sic*) Espanol 1968
Anotheca (*sic*) Espanol 1969

Stichtoptychus Fall 1905, 3 sp., Mexico and Texas (White 1980b, revision of New World spp.).
Stichtoplychus (*sic*) Arnett 1962

Striatheca White 1973, 2 spp., Mississippi, Georgia, Florida, and Sinaloa, Mexico.

Mesocoelopodinae Mulsant and Rey 1864
(= Tricoryninae White 1971)

Mesocoelopini

Mesocoelopus Jacquelin du Val 1860, 1 sp., *M. collaris* Mulsant and Rey 1864, Delaware, Virginia, Pennsylvania, and Ohio; in dead vines of English ivy, *Hedera helix* (L.) .

Tricorynini

Cryptorama Fall 1905, 21 spp., Mexico, New Jersey to Florida through to Oklahoma, and Texas; one species known from mesquite wood and twigs (White 1984, revision).

Peridorcatoma Lepesme 1947
Crytorama (*sic*) Espanol 1967

Tricorynus Waterhouse 1849, 106 spp., generally distributed; larvae bore in seeds, also in stems, galls, fungi, and wood (White 1965, revision North American species; White 1967, revision Mexican species; White 1981a, key modification; White 1981b, key tropical spp.).

Catorama Guérin-Méneville 1850
Hemiptychus LeConte 1865
Cathorama Gemminger and Harold 1869 (emendation)
Xylotheca Reitter 1897
Hemitychus (*sic*) Bøving 1954

Uncertain Tribal Placement

Neosothes White 1967, 3 spp. Baja California and Sinaloa, Mexico (White 1975a, key American spp.).

Bibliography

AALBU, R. L. and F. G. ANDREWS. 1992. Revision of the spider beetle genus *Niptus* in North America, including new cave and pholeophile species (Coleoptera: Ptinidae). Pan-Pacific Entomologist, 68: 73-96.

ANDREWS, R. W. 1967. Notes on the biology of two Central American *Ptinus* (Coleoptera: Ptinidae) with a description of a new species. Psyche, 74: 191-202.

BELLÉS, X. 1980. *Ptinus* (*Pseudoptinus*) *lichenum* Marsham, ptinido perforador de Madera (Col. Ptinidae). Boletín de la Estación Central de Ecología, 9: 89-91.

BELLÉS, X. 1982. Idees sobre la classificació supragenèrica de la familia Ptinidae (Col.). II Sessió Conjunta d'Entomologia, I.C.H.N.-S.C.L. Barcelona, 1981: Pp. 61-65.

BELLÉS, X. 1985a. Sistemática, filogenia y biogeografía de la subfamilia Gibbiinae (Coleoptera: Ptinidae). Treballs del Museu de Zoologia, 3: 1-94.

BELLÉS, X. 1985b. Hàbitats i hàbits d'alimentació dels Gibbiinae (Coleoptera: Ptinidae). Butlletí Institució Catalana d'Història Natural, 50: 263-267.

BELLÉS, X. 1992. Sistemática, historia natural y biogeografía del género *Pitnus* Gorham, 1880 (Coleoptera, Ptinidae). Eos, 68: 167-192.

BELLÉS, X. 1994. *Stereocaulophilus volcanius* gen. n., sp. n. (Coleoptera: Ptinidae) from Lanzarote (Canary Islands). Elytron, 8: 43-47.

BELLÉS, X. and D. G. H. HALSTEAD, 1985. Identification and geographical distribution of *Gibbium aequinoctiale* Boieldieu and *Gibbium psylloides* (Czenpinski) (Coleoptera: Ptinidae). Journal of Stored Products Research, 21: 151-155.

BLACKWELDER, R. E. 1945. Checklist of the coleopterous insects of Mexico, Central America, the West Indies, and South America, Part 3. United States National Museum Bulletin 185: 343-550.

BØVING, A. G. 1954. Mature larvae of the beetle-family Anobiidae. Danske Biologiske Meddelelser, 22: 1-298, 50 pls.

BØVING, A. G. 1956. A description of the mature larva of *Ptinus californicus* Pic. Entomologiske Meddelelser, 27: 229-24.

BØVING, A. G. and F. C. CRAIGHEAD, 1931. An illustrated synopsis of the principal larval forms of the order of Coleoptera. Entomologica Americana (new series), 11:1-351.

BROWN, W. J. 1959. *Niptus* Boield. and allied genera in North America (Coleoptera: Ptinidae). Canadian Entomologist, 91: 627-633.

DOWNIE, N. M. and R. H. ARNETT, Jr. 1996. The beetles of Northeastern North America, Vol. II. The Sandhill Crane Press. Gainesville, Florida, 1721 pp.

FALL, H. C. 1905. Revision of the Ptinidae [including Anobiidae] of Boreal America. Transactions of the American Entomological Society, 31: 97-296, pl. 7.

FORBES, W. T. M. 1922. The wing venation of the Coleoptera. Annals of the Entomological Society of America, 15: 328-352, 7 pls.

FORBES, W. T. M. 1926. The wing folding patterns of the Coleoptera. Journal of the New York Entomological Society, 34: 42-68, 91-138.

FORD, E. J. 1984. *Priobium carpini* (Herbst), an Old World anobiid now established in Maryland (Coleoptera). Coleopterists Bulletin, 38: 249-250.

HALL, D. W. and R. W. HOWE. 1953. Revised key to larvae of the Ptinidae associated with stored products in Britain. Bulletin of Entomological Research, 44: 85-96.

HATCH M. H. 1961. The beetles of the Pacific Northwest, Part III, Pselaphidae and Diversicornia I. University of Washington Publications in Biology, Vol. 16. Seattle, Washington.

HICKMAN, V. V. 1974. Notes on the biology of *Ptinus exulans* Erichson (Coleoptera: Ptinidae). Journal of the Entomological Society of Australia, 8: 7-14.

HINTON, H. E. 1941. The Ptinidae of economic importance. Bulletin of Entomological Research, 31: 331-381

HOWE, R. W. 1959. Studies on beetles of the family Ptinidae. XVII, conclusions and additional remarks. Bulletin of Entomological Research, 50: 287-326.

KNUTSON, L. V. 1963. Revision of the genus *Hadrobregmus* of North America (Coleoptera: Anobiidae). Proceedings of the Entomological Society of Washington, 63: 177-95.

LAWRENCE, J. F. 1991a. Anobiidae (Bostrichoidea). Pp. 441-444. *In:* F. W. Stehr, ed. Immature Insects, Vol. 2. Kendall/Hunt. Dubuque, IA.

LAWRENCE, J. F. 1991b. Ptinidae (Bostrichoidea). Pp. 444-445. *In:* F. W. Stehr, ed. Immature Insects, Vol. 2. Kendall/Hunt. Dubuque, IA.

LAWRENCE. J. F. and A. F. NEWTON, Jr. 1995. Families and subfamilies of Coleoptera (with selected genera, notes, references and data on family-group names). Pp. 779-1006. *In:* J. Pakaluk and S. A. Slipinski, eds., Biology, phylogeny and classification of Coleoptera: Papers celebrating the 80th birthday of Roy A. Crowson. Muzeum i Instytut Zoologii PAN, Warsaw.

LAWRENCE, J. F. and H. REICHARDT. 1966. Revision of the genera *Gnostus* and *Fabrasia* (Coleptera: Ptinidae). Psyche, 73: 30-45.

LAWRENCE, J. F. and H. REICHARDT. 1969. The myrmecophilous Ptinidae (Coleoptera), with a key to Australian species. Bulletin of the Museum of Compative Zoology, 138: 1-27.

LINSLEY, E. G. and J. W. MacSWAINE. 1942. The Bionomics of *Ptinus californicus*, a depredator in the nests of bees. Bulletin of the Society of the California Academy of Science, 40: 126-137.

MANTON, S. M. 1945. The larvae of the Ptinidae associated with stored products. Bulletin of Entomological Research, 35: 341-366.

McNAMARA, J. 1991. Family Anobiidae, Family Ptinidae. Pp. 201-206. *In:* Y. Bousquet, ed. Checklist of the beetles of Canada and Alaska, Agriculture Canada Publication 1861/E.

PAPP, C. S. 1962. An illustrated and descriptive catalogue of the Ptinidae of North America. Deutsche Entomologishe Zeitschrift, 5: 367-423.

PAPP, C. S. and G. T. OKUMURA, 1959. A preliminary study of the Ptinidae of California. Bulletin of the Department of Agriculture, California 49: 228-248.

PETERSON, A. 1951. Larvae of Insects, Part II. Edwards Brothers. Ann Arbour, MI, 416 pp.

PHILIPS, T. K. 1997. Systematics of the New World Ptininae (Coleoptera: Anobiidae). Ph.D. Dissertation. Ohio State University, Columbus, 355 pp.

PHILIPS. T. K. 1998a. A new genus and species of spider beetle from the Virgin Islands: *Lachnoniptus lindae* (Coleoptera: Anobiidae: Ptininae). Florida Entomologist, 81: 112-117.

PHILIPS. T. K. 1998b. A new genus and species of putatively myrmecophilous ptinine: *Coleoaethes tetralobus* (Coleoptera: Anobiidae: Ptininae). Pan-Pacific Entomologist, 74: 113-117.

PHILIPS. T. K. 1999. Revision of the New World spider beetle genus *Niptinus* Fall (Coleoptera: Anobiidae: Ptininae). Journal of the Kansas Entomological Society, 71: 137-158.

PHILIPS, T. K. 2000. Phylogenetic analysis of the New World Ptininae (Coleoptera: Bostrichoidea). Systematic Entomology, 25: 235-262.

PHILIPS, T. K., M. A. IVIE and L. L. IVIE. 1998. Leaf-mining and grazing in spider beetles (Coleoptera: Anobiidae: Ptininae): an unreported mode of larval and adult feeding in the Bostrichoidea. Proceedings of the Entomological Society of Washington 100: 147-153.

PIC, M., 1912a. Ptinidae, Coleopterorum Catalogus, 10(41): 1-46.

PIC, M., 1912b. Anobiidae, Coleopterorum Catalogus, 10(48): 1-92.

RUCKES, H., Jr. 1957. A synopsis of the California death-watch beetles of the genus *Ernobius* Thomson, with descriptions of two new species which attack pine (*Pinus*) cones (Col. Anobiidae). Pan-Pacific Entomologist, 33: 157-161.

SHARP, D. and F. MUIR. 1912. Comparative anatomy of the male genital tube in Coleoptera. Transactions of the Entomological Society of London, (1912): 477-642, pls. XLII-LXXVIII.

SMITH, D. S. 1964. The structure and development of flightless Coleoptera: a light and electron microscopic study of the wings, thoracic exoskeleton and rudimentary flight musculature. Journal of Morphology, 114: 107-184.

SPILMAN, T. J. 1975. Ptinidae, Anobiidae. *In:* R.E. Blackwelder and R. H. Arnett, Jr., eds. Checklist of the beetles of Canada, United States, Mexico, Central America, and the West Indies, Vol. 1(5) The darkling beetles, ladybird beetles and related groups (red version). Biological Research Institute of America. Rensselaerville, New York.

SPILMAN, T. J. 1991. Spider Beetles (Ptinidae, Coleoptera). Pp. 137-147, 567-572. *In:* J. R. Gorham, ed., Insect and mite pests in food, an illustrated key, Vol. 1 and 2. United States Department of Agriculture, Agricultural Research Service and United States Department of Health and Human Services, Agricultural Handbook No. 655.

STICKNEY, F. S. 1923. The head-capsule of Coleoptera. Illinois Biological Monographs, 8(1): 1-105.

TANNER, V. M. 1927. A preliminary study of the genitalia of female Coleoptera. Transactions of the American Entomological Society, 53: 5-50, pl. XV.

THOMAS, M. C., P. E. SKELLEY and R. W. LUNDGREN. 1992. New records for *Gnostus floridanus* (Coleoptera: Ptinidae) and observations on its behavior. Florida Entomologist, 75: 287-289.

TRISTAM, J. N. 1977. Normal and cocoon-forming peritrophic membrane in larvae of the beetle *Gibbium psylloides*. Journal of Insect Physiology, 23: 79-87.

WHITE, R. E. 1962. The Anobiidae of Ohio. Bulletin of the Ohio Biological Survey, (new series), 1(4): 58 pp.

WHITE, R. E. 1965. A revision of the genus *Tricorynus* of North America (Coleoptera: Anobiidae). Miscellaneous Publications of the Entomological Society of America, 4: 283-368.

WHITE, R. E. 1966. Six new Anobiidae from North America with keys (Coleoptera). Proceedings of the Entomological Society of Washington, 68: 228-236.

WHITE, R. E. 1967. The *Tricorynus* of Mexico (Coleoptera: Anobiidae). Transactions of the American Entomological Society, 93: 1-40.

WHITE, R. E. 1971. Key to North American genera of Anobiidae, with phylogenetic and synonymic notes (Coleoptera). Annals of the Entomological Society of America, 64: 179-191.

WHITE, R. E. 1973a. A new genus, two new species, and a species key for *Byrrhodes*. Proceedings of the Entomological Society of Washington, 75: 48-54.

WHITE, R. E. 1973b. Neotropical Anobiidae: new genera and species, and taxonomic notes (Coleoptera). Annals of the Entomological Society of America, 66: 843-848.

WHITE, R. E. 1973c. Taxonomic notes on North American Anobiidae with a new species (Coleoptera). Proceedings of the Entomological Society of Washington, 76: 459-463.

WHITE, R. E. 1973d. New North American *Euvrilletta* and *Xyletinus* with keys to species (Coleoptera: Anobiidae). Journal of the Washington Academy of Science, 63: 76-81.

WHITE, R. E. 1974a. Taxonomic notes on North American Anobiidae with a new species (Coleoptera). Proceedings of the Entomological Society of Washington, 76: 459-463.

WHITE, R. E. 1974b. Type-species for World genera of Anobiidae.Transactions of the American Entomological Society, 99: 415-475.

WHITE, R. E. 1975a. Sixteen new Neotropical Anobiidae with a new genus and keys (Coleoptera). Proceedings of the Entomological Society of Washington, 77: 169-188.

WHITE, R. E. 1975b. North American *Xestobium* (Anobiidae) with a new species. Coleopterists Bulletin, 29: 83-86.

WHITE, R. E. 1976a. Eight new North American species of Anobiidae with keys and notes (Coleoptera). Proceedings of the Entomological Society of Washington, 78: 154-170.

WHITE, R. E. 1976b. A new genus and four new species of North American Anobiidae with notes. Coleopterists Bulletin, 30: 337-342.

WHITE, R. E. 1977. Ten new North American species of *Xyletinus* (Anobiidae: Coleoptera). Proceedings of the Entomological Society of Washington, 79: 521-537.

WHITE, R. E. 1979a. New synonymy and new combinations primarily in American taxa (Coleoptera: Anobiidae). Proceedings of the Entomological Society of Washington, 81: 211-218.

WHITE, R. E. 1979b. The genus *Protheca* of the Americas (Coleoptera: Anobiidae). United States Department of Agriculture Technical Bulletin No. 1605, 24 pp.

WHITE, R. E. 1980a. Review of *Vrilletta*, with two new species and a key (Coleoptera: Anobiidae). Journal of the Washington Academy of Science, 70: 144-148.

WHITE, R. E. 1980b. A taxonomic study of the New World genus *Stichtoptychus* Fall (Coleoptera: Anobiidae). United States Department of Agriculture Technical Bulletin No. 1602, 35 pp.

WHITE, R. E. 1981a. Three new species of Anobiidae from Southwestern United States and Northwestern Mexico (Coleoptera). Proceedings of the Entomological Society of Washington, 83: 472-478.

WHITE, R. E. 1981b. A key to tropical species of *Tricorynus*, with taxonomic changes (Coleoptera: Anobiidae). Proceedings of the Entomological Society of Washington, 83: 772-784.

WHITE, R. E. 1982. A catalog of the Coleoptera of America North of Mexico, Family: Anobiidae. United States Department of Agriculture, Agricultural Research Service, Agricultural Handbook 529-70.

WHITE, R. E. 1983. Keys to Neotropical species of *Calymmaderus* Solier and species of *Calytheca* White, with taxonomic notes (Coleoptera: Anobiidae). Proceedings of the Entomological Society of Washington, 85: 229-250.

WHITE, R. E. 1984. A revision of the American genus *Cryptorama* (Coleoptera: Anobiidae). Transactions of the American Entomological Society, 110: 77-127.

WHITE, R. E. 1985. North American *Euvrilletta* (Coleoptera: Anobiidae) - Transferal of taxa from *Xyletinus*, two new species, and a key. Coleopterists Bulletin, 39: 185-193.

WHITE, R. E. 1990. *Lasioderma haemorrhoidale* (Ill.) now established in California, with biological data on Lasioderma species (Coleoptera: Anobiidae). Coleopterists Bulletin, 44: 344-348.

WILLIAMS, I. W. 1938. The comparative morphology of the mouthparts of the order Coleoptera treated from the standpoint of phylogeny. Journal of the New York Entomological Society, 46: 245-289.

WOODROFFE, G. E. 1958. The mode of reproduction of *Ptinus clavipes* Panzer form *mobilis* Moore (=*P. latro* Auct.) (Coleoptera: Ptinidae). Proceedings of the Royal Entomological Society of London (A), 33:25-30.

Series CUCUJIFORMIA Lameere 1938

Superfamily LYMEXYLOIDEA Fleming 1821

71. LYMEXYLIDAE Fleming 1821

by Daniel K. Young

Family common name: The ship-timber beetles

Family synonym: Lymexylonidae; including Atractoceridae Laporte 1840

The filiform to serrate (rarely flabellate) and fairly short antennae, highly modified maxillary palpal organ of the male (rarely in female or both sexes), cylindrical and projecting prothoracic coxae, and elongate, cylindrical body serve to distinguish this small family.

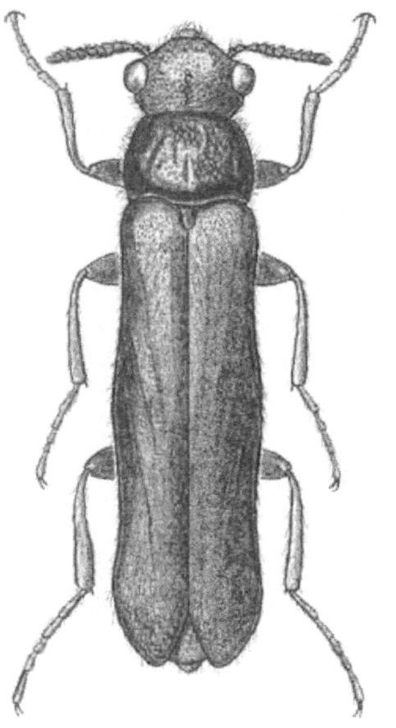

FIGURE 1.71. *Elateroides lugubris* (Say)

Description: (after Wheeler 1986, in part) Elongate, parallel-sided, subcylindrical to slightly flattened; length 5-40 mm; vestiture consisting of fine setae.

Head short, typically narrowed behind eyes forming slight neck; surface punctate, with (*Elateroides*) or without epicranial pit; antennae 11-segmented, short, filiform to serrate, often sexually dimorphic; inserted at sides of the head. Clypeus distinct; labrum of moderate size, slightly emarginate; mandibles moderately-sized, curved, apices subacute. Maxillary palpi 4-segmented, simple in most females, and with apical segment modified into a complex flabellate or plumose organ in the male. Gular sutures distinct and separate; mentum small, quadrate, ligula small, coriaceous; labial palpi 3-segmented, simple. Eyes rather large, protruding, entire, dorsally contiguous in some species.

Pronotum much longer than broad, elongate oval, lateral margins complete or incomplete, surface punctate; prosternum short; prothoracic coxal cavities widely open behind. Scutellum small, triangular. Mesosternum small, flat; mesothoracic episterna reaching the coxal cavities; mesothoracic coxal cavities open behind. Metasternum broad, strongly convex. Legs slender, moderately long; prothoracic coxae cylindrical to conical, projecting, contiguous or separate, trochantins exposed; mesothoracic coxae projecting, conical, contiguous; metathoracic coxae transverse, conical, prominent, contiguous; tibiae with apical spurs; tarsal formula 5-5-5, tarsi slender, claws simple, usually with multisetose empodium. Elytra loosely fitted to pterothorax and abdomen, often weakly carinate, but lacking distinct striae, entire or shortened and exposing 1-3 tergites; punctation confused, epipleurae reduced or absent. Metathoracic wings with fairly complete venation, radial cell short or absent, anal cell present or not.

Abdomen with 5-7 visible ventrites, that of male with one more visible ventrite than female (*Elateroides*), or sexes with same number of ventrites, sutures distinct. Aedeagus trilobed with articulated parameres or with a ring-like tegmen fused basally to the median lobe, parameres sometimes fused to tegmen or reduced to one (Sharp and Muir 1912, Wheeler 1986). Female genitalia with valvifers large with a baculum on inner margin; coxite symmetrical, partially segmented, stylus present, near the end of the coxite; proctiger vestigial (Tanner 1927). Malpighian tubules cryptonephridic.

Larvae (after Bøving and Craighead 1931, Wheeler 1986, 1991) elongate, subcylindrical, orthosomatic, sclerotized, with definite lateral folds; length 12-20 mm (up to 50 mm in many tropical species); vestiture consisting of a few setae and spines on some of the abdominal segments and the legs; color nearly white, yellow, to deep brown. Head large, somewhat hypognathous, somewhat retracted into the hood-like prothorax, epicranial suture short, mesal. Mandibles triangular, heavily sclerotized with transversely grooved or asperate mola, apices blunt. Each maxilla with cardo, stipes, and partially divided mala, mala bearing parallel rows of elongate setae; maxillary palpi 3-segmented; labium with submentum, mentum, and broad ligula; labial palpi 2-segmented. Prothorax large, distinctly larger than mesothorax or metathorax, hood-shaped, partially covering head, often asperate. Legs short, stout, 5-segmented, with pointed, sclerotized tarsunguli, tarsunguli bi- to multisetose. Abdomen with nine segments visible dorsally, tenth segment reduced, ventrally oriented; ninth segment in the form of an elongate, sclerotized spine (*Elateroides lugubris*), or heavily sclerotized, enlarged, cylindrical, terminating in an oblique, obtuse manner with a strongly serrate apical rim (*Melittomma sericeum*). Spiracles annular-elliptical, sometimes annular-multiforous.

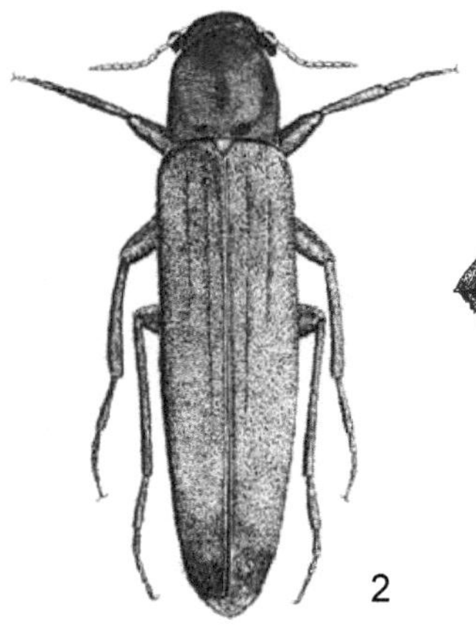

2

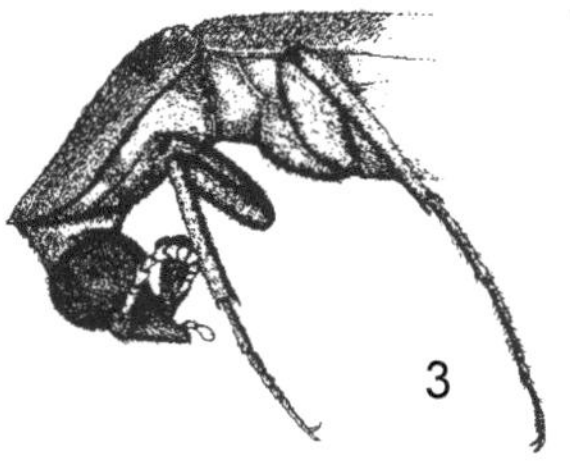

3

FIGURES 2.71-3.71. 2. *Melittomma sericeum* (Harris); 3. *Melittomma sericeum* (Harris), left lateral view of head and thorax (both courtesy of American Museum of Natural History).

Habits and habitats. Adults are found in decaying wood, running about on tree trunks, or under bark. They are probably short-lived and thus are not commonly encountered. The larvae are wood borers in heart and sapwood of oak, elm, and, historically, American chestnut (*Melittomma*), and softer-wooded species like poplar, birch, etc. (*Elateroides*). They are believed to form symbiotic associations with ambrosia fungi (Ascomycetes, *Ascoides* spp.) that grow on the walls of their tunnels. Adult females carry the fungal spores in mycangia near the end of the ovipositor and deposit them in a sticky matrix with their eggs. When the eggs eclose, the first instar larvae carry the fungal spores into the wood on their bodies. One northern European species was formerly very destructive to ship timbers (Batra and Franke-Grosmann 1961, Franke-Grosmann 1967, Wheeler 1986, 1991).

Status of the classification. Historically, authors associated Lymexylidae most commonly with Cleroidea or Cucujoidea. The family is viewed as forming a monofamilial superfamily, basal in the series Cucujiformia. An important synapomorphy is the presence of cryptonephridial Malpighian tubules (Wheeler 1986).

Distribution. Roughly 50 species are placed in seven genera and three subfamilies: Hylecoetinae (*Elateroides* (=*Hylecoetus*), Holarctic), Lymexylinae (*Atractocerus*, Pantropical; *Lymexylon*, Eurasia), Melittommatinae (*Australymexylon*, Australia; *Melittomma*, widely distributed; *Melittommopsis*, South America; *Promelittomma*, Madagascar, Seychelles) (Schenkling 1915, Wheeler 1986). Two species are confirmed from the United States and Canada: *Elateroides lugubris* (Say), and *Melittomma sericeum* (Harris). A third species, *Atractocerus gracilicornis* Schenkling, was described from "Californie" but it is unclear whether the reference was to California (United States) or Baja California (Mexico) (Wheeler 1986). In considering the possibility that the reference was to the United States, *Atracocerus* is included in the key.

Key to the Genera of America North of Mexico

1. Head with mesal epicranial pit; eyes round, small, entire (Fig. 1) .... *Elateroides*

— Head without epicranial pit; eyes (Fig. 3) moderate to large, emarginate anteriorly 2

2(1). Elytra very short, brachelytrous, not extending to metathoracic coxae .... *Atractocerus*

— Elytra covering abdomen, or nearly so (Fig. 2) .... *Melittomma*

Classification of the Genera of America north of Mexico

Lymexylidae Flemming 1821

Hylecoetinae Gistel 1856

Elateroides Schaeffer 1766, 1 sp., *E. lugubris* (Say), eastern United States; the sapwood timberworm, occurs commonly in fairly fresh *Populus* logs.
Hylecoetus Latreille 1806
Xylecoethus Gyllenhal 1827
Hyloceotus Melsheimer 1853
Hyloecotus Lacordaire 1857
Hyloecotus Chenu 1860
Hylecerus Jacquelin du Val 1863

Melittommatinae Wheeler 1986
(= Melittomminae)

Melittomma Murray 1867, 1 sp., *M. sericeum* (Harris), eastern United States, the chestnut timberworm, occurs primarily in *Quercus* logs.
Neohylecoetus Kurosawa 1956.

Bibliography

BATRA, L. R. and H. FRANKE-GROSMANN. 1961 Contributions to our knowledge of ambrosia fungi I. *Ascoidea hylecoeti* sp. nov. (Ascomycetes). American Journal of Botany, 48: 453-456.

BØVING, A. G. and F. C. CRAIGHEAD. 1931. An illustrated synopsis of the principal larval forms of the order Coleoptera. Entomologica Americana (N.S.), 11 (1930): 1-351.

FRANKE-GROSMANN, H. 1967. Ectosymbiosis in wood-inhabiting insects. Pp. 141-205. *In:* S. M. Henry, ed., Symbiosis. Volume 2. Academic Press. NY.

SCHENKLING, S., 1915. Lymexylonidae. *In:* S. Schenkling, ed., Coleopterorum Catalogus, Pars 10(64). W. Junk. Berlin, 13 pp. (2nd pagination).

SHARP, D. and F. A. G. MUIR. 1912. The comparative anatomy of the male genital tube in Coleoptera. Transactions of the Entomological Society of London, (1912): 477-642, pls 42-78.

TANNER, V. M., 1927. A preliminary study of the genitalia of female Coleoptera. Transactions of the American Entomological Society, 53: 5-50.

WHEELER, Q. D. 1986. Revision of the genera of Lymexylidae (Coleoptera: Cucujiformia). Bulletin of the American Museum of Natural History, 183: 113-210.

WHEELER, Q. D. 1991. Lymexylidae. Pp. 446-447. *In:* F. W. Stehr, ed., Immature Insects. Volume 2. Kendall/Hunt. Dubuque, IA.

Superfamily CLEROIDEA Latreille 1802

72. TROGOSSITIDAE Latreille 1802

by Richard A.B. Leschen

Family common name: The bark-gnawing beetles, cadelles

Family synonyms: Lophocateridae Crowson 1964; Peltidae Latreille 1807; Nemosomidae Leach 1815; Ostomatidae Reitter 1882; Temnochilidae Léveillé 1889; Ostomidae Ganglbauer 1899; Trogositidae, misspelling

The variable body shape (ovate to parallel sided and dorsoventrally flattened), relatively rigid body, exposed protrochantin and lateral carina of prothorax characterize adults of this heterogenous group of Cleroidea.

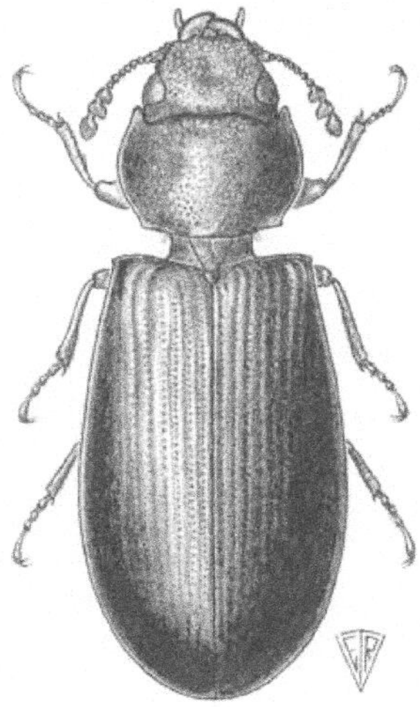

FIGURE 1.72. *Tenebioides corticalis* (Melsheimer).

Description: Body length 1.9 - 22.4 mm (up to 50 mm in foreign species). Body form elongate and parallel-sided, cylindrical in some, flattened, to moderately flattened or highly convex, oval, or round. Body color various, many dark, light, or red brown, some black, green or blue. Body setae absent or present and long and decumbent, in few erect, or short, and may be modified into short or elongate scales (*Calitys*).

Head wider than long, mostly retracted into thorax (or not in some Egoliinae and Trogossitinae), without a neck or line on vertex, without stridulatory files and subgenal spines; antennal grooves mostly present and well developed, or reduced and rarely absent; gular sutures mostly present and widely separate. Eyes mostly well developed and in many protuberant; finely or strongly faceted, mostly with ocular setae. Clypeus more or less quadrate, emarginate in some, mostly without a strongly impressed frontoclypeal suture (present in Lophocaterinae). Antennae mostly 11 segmented, but 8, 9, and 10 occur in some species; antennal club 3-segmented (1- or 2-segmented in some foriegn taxa), flattened in Trogossitinae; insertions hidden in dorsal view. Mandible with well developed mola (absent in Trogossitinae) and incisor apex; prostheca well developed, absent or reduced. Lacinia and galea brushlike, lacinia with apical spines (or lateral spines in Trogossitinae) or absent (Rentoniinae); terminal palpus variable in length and width, in few securiform. Labium with mentum mostly wider than long, without transverse ridge or line and middle process; labial palpus 2-segmented with basal palpomere mostly shorter in length than apical palpomere. Corporotentorium narrow; or incomplete or absent; without anterior median tendon (present in Rentoniinae).

Prothorax widest mostly near base or at middle, in few anteriorly; with well-developed lateral carina (weakly developed in *Cylidrella*); sides mostly strongly to moderately explanate (not in cylindrical Trogossitinae) and dentate or smooth with or without a bead. Pronotal disc irregular in some and with depressions. Anterior angles present (absent in some species) and produced anteriorly; process angulate or rounded. Prosternum in front of procoxae about equal in length to prosternal process; prosternal process variable mainly strongly and abruptly expanded at apex, parallel-sided, and in few carinate. Notosternal suture present (absent in some Larinotinae and Rentoniinae). Procoxae transverse and not projecting far below sternum (except Rentoniinae); trochantin present and mostly visible (absent or reduced in Rentoniinae); procoxal cavities externally open or closed; internally open (closed in Rentoniinae). Postcoxal projection of hypomeron long and often meeting prosternal process, but may be short, or very short or absent.

Mesosternum with or without procoxal rests; ventral surface horizontal throughout and at least partially visible in dorsal view (hidden in dorsal view in Rentoniinae). Mesocoxal cavities narrowly to widely separated; circular to transverse; open laterally and contacting mesepimeron (closed laterally in a some foreign taxa). Mesometasternal junction typically monocondylic. Metasternum mostly flat (convex in cylindrical forms); without axillary space (present in *Thymalus*) or postcoxal lines; discrimen present or absent; precoxal lines or groove present or absent. Metendosternite mostly with a narrow stalk (broader in some); arms well developed; with lamellae present, reduced or absent; anterior process mostly absent; anterior tendons mostly present (in few absent, especially in some flightless species), approximate or distant. Abdomen with 5 freely articulated ventrites (1 and 2 connate in *Thymalus*); of equal length (Rentoniinae with ventrite 1 longer); intercoxal process acute and narrow. Elytra completely covering abdomen; terminal 3 tergites exposed in *Cylidrella* by shortened elytra; punctation random or seriate; epipleuron complete, indistinct (*Cylidrella*), or present in basal half (Rentoniinae).

Wings (based on Lawrene *et al.* 1999a) well developed (absent or rediuced in some foreign taxa); radial cell of hind wing well developed or highly reduced or absent; medial bar of hind wing not crossed by fold; free veins in medial area of hind wing 4, 3 or fewer; oblongum cell and medial fleck absent; wedge cell of hind wing well developed, reduced or absent; anal lobe absent or present. Legs relatively long; tibia with outer margin smooth or spinose and with apical spines; tarsi 5-5-5, in few 4-4-4; tarsal claws typically extended (retracted in *Calitys* and toothed in Decamerinae); empodium with 2 setae (reduced in Rentoniinae).

Male with anterior edge of sternite 8 with or without median strut; anterior edge of sternite 9 with or without median strut (spiculum gastrale); terga 9 and 10 variable but mostly completely fused. Aedeagus of the cleroid type; anterior edge of tegmen mostly with single strut and opposing paired struts (opposing pair may be absent); parameres fused or articulated to phallobase.

Larva (modified from Foster and Lawrence 1991) with ventral mouthparts retracted, epicranial stem absent, and gular and paragular regions as sclerotized as the epicranium. Body elongate, subcylindrical or slightly flattened, enlarged thoracic region in *Ostoma* and other related genera; colour cream or white, but may be purple or pink; mostly lightly sclerotized, except for head, protergum, and part of T9 in some; paired sclerotized plates present on meso- and metaterga and on terga A6-8 in some; dorsal surfaces smooth; vestiture consisting of short and long scattered setae. Head protracted and prognathous, slightly elongate and parallel-sided in Trogossitinae, transverse with strongly rounded sides in Peltinae with intermediate conditions in *Calitys* and Lophocaterinae. Epicranial stem absent (short in some Peltinae); frontal arms v-shaped. Median endocarina extending between frontal arms in most species, a y-shaped endocarina co-occurring with short epicranial stem and frontal arms in most Peltinae; paired endocarinae located between frontal arms in *Thymalus*. Stemmata mostly 2 or 5 on each side, absent, 3, 4, or 6 in some. Antennae 3-segmented, short in Peltinae. Frontoclypeal suture absent, present in some Peltinae; labrum free. Mandibles subtriangular, apex bidentate lacking mola or ventral accessory process, inner edge of base with one to several hyaline processes forming lacinia mobilis. Ventral mouthparts retracted. Maxilla with transverse cardo, elongate stipes, maxillary articulating area mostly absent, mala simple and rounded or truncate, palp 3-segmented (2-segmented in Rentoniinae). Labium with mentum and submentum fused; postmentum elongate in Trogossitinae; labial palps mostly 2-segmented (1-segmented in Rentoniinae), ligula mostly absent. Hypostomal rods mostly subparallel and extending to posterior edge of head, delimiting sclerotized paragular plates; hypostomal rods absent in some Peltinae. Ventral epicranial ridges present in Trogossitinae and *Calitys*. Gula longer than wide, except Peltinae. Thorax and Abdomen. Legs well developed, 5-segmented; tarsungulus with 1 seta; coxae widely separated. Protergum mostly with a sclerotized plate in *Thymalus, Calitys,* and Trogossitinae; meso- and metaterga often with a single or two plates. Prothorax enlarged in *Ostoma* and related genera. Ventral regions with various slcerotizations in Trogossitinae. Abdominal terga 1-6 or sometime 7 with paired ampullae. Paired, coiled tubuler glands often present on abdominal terga 1-8. Tergum A9 mostly with sclerotized basal plate (absent in most Peltinae and Larinotinae) and with paired urogomphi (with accessory processes in *Calitys* and *Thymalus*; minute and projecting ventrally in Larinotinae). In Lophocaterinae, basal plate transversely divided and median, posteriorly projecting tooth present between urogomphi. Sternum A9 well-developed; segment A10 ventral, small and circular. Spiracles annular-biforous, or annular.

Habits and habitats. Most Trogossitinae (and possibly Egoliinae) are predatory and can be found under bark and in galleries of wood boring beetles. The cadelle, *Tenebroides mauritanicus* (Linnaeus), is a widespread species that may feed on other stored product insects or on damaged grain as a minor pest. Some trogossitines may also feed on fungi as do Protopeltinae, some New Zealand Rentoniinae, Peltinae, Lophocaterinae, and possibly Larinotinae which are found in leaf litter and in association with fungi. Decamerinae feed on pollen and are associated with flowers, especially of trees and shrubs. Lophocaterinae have larvae that may be predators (Crowson 1964) and adults and larvae are typically found under bark but those of *Eronyxa* are associated with flowers: one species, *Eronyxa expansus* Van Dyke, may be a predator on the incense cedar scale in California (Tait *et al.* 1990, Leschen 2000). The widespread *Lophocateres pusillus* (Klug) may feed on live or dead insects and detritus and is considered a minor stored grain pest (Halstead 1968). Larval and adult peltines and *Calitys* feed on fruiting bodies of various Polyporaceae (Barron 1971, 1996, Foster and Lawrence 1991) and *Thymalus marginicollis* Chevrolat is found commonly on the fruiting bodies of *Trametes cervinus* in the midwestern United States. Larval descriptions are available (Bøving and Craighead 1931, Costa *et al.* 1988, Foster and Lawrence 1991) and are keyed to higher taxa in Lawrence *et al.* (1999b). Adults can be keyed to higher taxa using Lawrence (1999a) and adult morphology, metamorphosis, and descriptions are available in Breed (1903) and Lawrence (1982).

Status of the classification. The classification of the group is far from settled (Crowson 1966, 1970; Lawrence and Newton 1995), and it is not clear whether the family is monophyletic within Cleroidea. Recently the genus *Eronyxa* was transferred from Decamerinae to Lophocaterinae based on larval characters (Tait *et al.* 1990) and Slipinski (1992) added a new subfamily (Larinotinae) and briefly reviewed the status of each of the subfamilies. A partial phylogeny was included in a detailed morphological study by Kolibac (1999).

Distribution. Approximately 600 species are known from all areas with 59 species present in the United States (Barron 1971, 1975, 1996; Léveillé 1919).

Key to the Genera of the United States
(modified from Barron 1971 and Slipinski 1992)

1. Procoxal cavities externally open 2
— Procoxal cavities externally closed 6

2(1). Protibia with a single spur; antennal club asymmetrical (Lophocaterinae, part) 3
— Protibia with two unequal spurs; antennal club symmetrical 4

3(2). Antenna 7-segmented .... *Lycoptis*
— Antenna 11-segmented 13

4(2). Mesocoxae distinctly separated; sternal process between them not much longer than wide; lacinia with strong apical hook but no spur on mesal surface (Peltinae) 5
— Mesocoxae not distinctly separated; sternal process narrow; lacinia without strong apical hook, spur present on mesal surface (Lophocaterinae, part) .... *Eronyxa*

5(4). Body dorsoventrally compressed; dorsal surfaces with a weak vesititure of setae; elytra with well-developed costae .... *Ostoma*
— Body convex; dorsal surfaces with a well developed vesititure of setae; elytra without well-developed costae .... *Thymalus*

6(1). Antennal club symmetrical, dorsal surfaces of elytra with costae and tubercles (Calitinae) ..... *Calitys*
— Antennal club asymmetrical, dorsal surfaces of elytra smooth (Trogossitinae) 7

7(6). Anterior prothoracic angles completely absent, pronotum with poorly developed lateral carina and width subequal to head 10
— Anterior prothoracic angles present and well developed, pronotum well developed lateral carina and width greater than head 8

8(7). All tibiae on outer margin with spines; antennae very short, present to level of anterior 1/4 of prosternum .... *Airora*
— All tibiae lacking spines, present on outer margin of protibiae in some species; antennae elongate, present to level of middle of prosternum or more 9

9(8). Pronotum with lateral margins behind middle angularly deflexed; labium at apex deeply triangularly emarginate; submentum of male with median pit .... *Temnoscheila*
— Pronotum with lateral margins behind middle not angularly deflexed; labium at apex shallowly emarginate; submentum of male without median pit .... *Tenebroides*

10(7). Vertex of head with a median groove 11
— Vertex of head without a median groove 12

11(10). Elytra short with apex of abdomen exposed; pronotum without raised lateral bead . *Cylidrella*
— Elytra elongate and covering the abdomen; pronotum with raised lateral bead ... *Nemosoma*

12(10). Elytra with subhumeral impression; pronotum with distinctly raised lateral bead .... *Corticotomus*
— Elytra without subhumeral impression; pronotum without distinctly raised lateral bead .... *Euschaefferia*

13(3). Elytra without well-developed costae .... *Lophocateres*
— Elytra without well-developed costae .... *Grynocharis*

Classification Of The Nearctic Genera

Trogossitidae Latreille 1802

[Protopeltinae Crowson 1966, not in North America]

[Larinotinae Slipinski 1992, not in North America]

Peltinae Kirby 1837

Ostoma Laicharting 1781, 3 spp., Alaska, Alberta, Arizona, British Columbia, Colorado, California, Idaho, Indiana, Labrador, Maine, Manitoba, Michigan, Minnesota, New Hampshire, Nevada, New Brunswick, New Mexico, Nevada, New Hampshire, New Mexico, New York, Northwest Territories, Nova Scotia, Ontario, Oregon, Pennsylvania, Quebec, Saskatchewan, South Dakota, Utah, Vermont, Washington, Wyoming, Yukon.
Gaurambe Thomson 1859

Thymalus Latreille 1802, 1 sp., *T. marginicollis* Chevrolat 1842, widespread in eastern North America.
Peltis Kugelann 1791
Thymalops Iablokoff-Khnzorian 1962

[Rentoniinae Crowson 1966, not present in North America]

[Decamerinae Crowson 1964, not present in North America]

Lophocaterinae Crowson 1964

Eronyxa Reitter 1876, 4 spp., British Columbia, California, Idaho, Oregon, Nevada, Washington.
Ostomodes Reitter 1877

Grynocharis Thomson 1859, 2 spp., British Columbia, California, Idaho, Iowa, Maine, Manitoba, Massachusetts, New Hampshire, New York, Nevada, Ohio, Ontario, Oregon, Pennsylvania, Quebec, Vermont, Virginia, Washington, Wisconsin.

Lophocateres Olliff 1883, 1 sp. (adventive), *L. pusillus* (Klug), Florida, Georgia, South Carolina, Texas.

Lycoptis Casey 1890, 1 sp., *L. americana* Casey 1890, Arkansas, District of Columbia, Florida, Georgia, Maryland, North Carolina, Oklahoma, South Carolina.

Calitinae Reitter 1922

Calitys Thomson 1859, 2 spp., Alabama, Alaska, Alberta, Arizona, British Columbia, California, Colorado, Idaho, Manitoba, Maryland, Minnesota, Montana, Nevada, New Hampshire, New

Mexico, New York, Northwest Territories, Ontario, Oregon, Quebec, Pennsylvania, South Dakota, Washington, Wisconsin, Yukon.

Peltis Kugelann 1791
Nosodes LeConte 1861
Peltidea Motschulsky 1858

[Egoliinae Lacordaire 1854, not present in North America]

Trogossitinae Latreille 1802

Airora Reitter 1876, 3 spp., widespread.
Alindria LeConte 1861

Corticotomus Sharp 1891, 6 spp., Arizona, California, Florida, New Mexico, Oregon, Texas, and eastern United States.
Parafilumis Casey 1916

Cylidrella Sharp 1891, 1 sp., *C. championi* Wickham 1916, Arizona, Colorado.

Euschaefferia Leng 1920, 2 sp., California, North Carolina, Texas.
Stenodema Schaeffer 1918
Pseudocotomus Van Dyke 1944

Nemosoma Latreille, 1804 5 spp., Arizona, California, Oregon.
Pseudalindria Fall 1910

Temnoscheila Westwood 1830, 10 spp., Florida, southern United States.
Temnochila Erichson 1844

Tenebroides Piller and Mitterpacher 1783, 21 spp., widespread.
Trogosita Olivier 1790
Trogossita Fabricius 1792

Bibliography

BARRON, J. R. 1971. A revision of the Trogositidae of America North of Mexico (Coleoptera: Cleroidea). Memoirs of the Entomological Society of Canada, 75: 1-143.

BARRON, J. R. 1975. A review of the genus *Lycoptis* Casey (Coleoptera: Trogositidae). Canadian Entomologist, 107: 1117-1122.

BARRON, J. R. 1996. Review of Nearctic species of *Ostoma* (Coleoptera: Cleroidea, Trogositidae). Annals of the Entomological Society of America, 89: 193-202.

BØVING, A. G. and F. C. CRAIGHEAD. 1931. An illustrated synopsis of the principal larval forms of the order Coleoptera. Entomologica Americana (New Series), 11: 1-351.

BREED, R. S. 1903. The changes which occur in the muscles of a beetle, *Thymalus marginicollis* Chevr., during metamorphosis. Bulletin of the Museum of Comparative Zoology, 40: 317-382 + 7 plts.

COSTA, C., S. A. VANIN and S. A. CASARI-CHEN. 1988. Larvas de Coleoptera do Brasil. Museu de Zoologia, Universidade de São Paulo. São Paulo, Brazil.

CROWSON, R. A. 1964. A review of the classification of Cleroidea (Coleoptera), with descriptions of two new genera of Peltidae and of several new larval types. Transactions of the Royal Entomological Society of London, 116: 275- 327.

CROWSON, R. A. 1966. Further observations on Peltidae (Coleoptera: Cleroidea), with definitions of a new subfamily and of four new genera. Proceedings of the Royal Entomological Society of London, 35: 119-127.

CROWSON, R. A. 1970. Further observations on Cleroidea (Coleoptera). Proceedings of the Royal Entomological Society of London, 39: 125-131.

FOSTER, D. E. and J.F. LAWRENCE. 1991. Trogossitidae (Cleroidea) (= Trogositidae; including Lophocateridae, Ostomidae, Peltidae, Temnochilidae). Pp. 448-450. *In:* F.W. Stehr, ed. Immature Insects, Vol. 2. Kendall/Hunt, Dubuque, IA.

HALSTEAD, D. G. H. 1968. Some observations on the biology of *Lophocateres pusillus* (Klug) (Coleoptera: Trogossitidae). Journal of Stored Products Research, 4: 197-202.

KOLIBAC, J. 1999. Comparative morphology of mandible, epipharynx, and alimentary canal in larval and adult Cleroidea (Coleoptera). Acta Musei Scientiae Biologicae (Brno), 84: 11-69.

LAWRENCE, J. F. 1982. Coleoptera. Pp. 482-553 *In:* S. P. Parker, ed. Synopsis and Classification of Living Organisms, Vol. 2. McGraw-Hill. New York.

LAWRENCE, J. F., A. HASTINGS, M.J. DALLWITZ, T.A. PAINE and E.J. ZURCHER. 1999a. Beetles of the World: A Key and Information System for Families and Subfamilies. CD-ROM, Version 1.0 for MS-DOS. CSIRO Publishing. East Melbourne, Victoria.

LAWRENCE, J. F., A.M. HASTINGS, M.J. DALLWITZ, T.A. PAINE and E.J. ZURCHER. 1999b. Beetle Larvae of the World: Descriptions, Illustrations, Identification, and Information Retrieval for Families and Sub-families CD-ROM, Version 1.0 for MS-DOS. CSIRO Publishing. East Melbourne, Victoria.

LAWRENCE, J. F. and A. F. NEWTON, Jr. 1995. Families and subfamilies of Coleoptera (with selected genera, notes, references and data on family-group names). Pp. 779-1006. *In:* Biology, Phylogeny, and Classification of Coleoptera: Papers Celebrating the 80th Birthday of Roy A. Crowson. J. Pakaluk and S. A. Slipinski, eds. Muzeum Instytut Zoologii PAN: Warsaw.

LESCHEN, R. A. B. 2000. Beetles feeding on bugs (Coleoptera, Hemiptera): repeated shifts from mycophagous ancestors. Invertebrate Taxonomy, 14: 917-929

LÉVEILLÉ, A. 1910. Temnochilidae, pars 11. Pp. 1-40. *In:* W. Junk, ed. Coleopterorum catalogus. Berlin. Vol. 15.

SLIPINSKI, A. S. 1992. Larinnotinae - a new subfamily of Trogossitidae (Coleoptera), with notes on the constitution of Trogossitidae and related families of Cleroidea. Revue Suisse de Zoologie, 99: 439-463.

TAIT, S. M., D. L. DAHLSTEN, R. J. GILL and J. T DOYEN. 1990. Life history of the incense cedar scale, *Xylococculus macrocarpae* (Homoptera: Margarodidae), on incense cedar in California with a description of the larva of one of its common predators, *Eronyxa expansus* Van Dyke (Coleoptera: Trogositidae). Hilgardia, 58: 1-19.

73. CLERIDAE Latreille 1804

by Weston Opitz

Family common name: The checkered beetles

Family synonyms: Corynetidae Lacordaire 1857; Korynetidae Chapin 1924

The adults of Cleridae are generally distinguished from other beetles by having bright or contrasting colors, abundant setal vestiture, large apical palpomeres in many, and lobed tarsomeres.

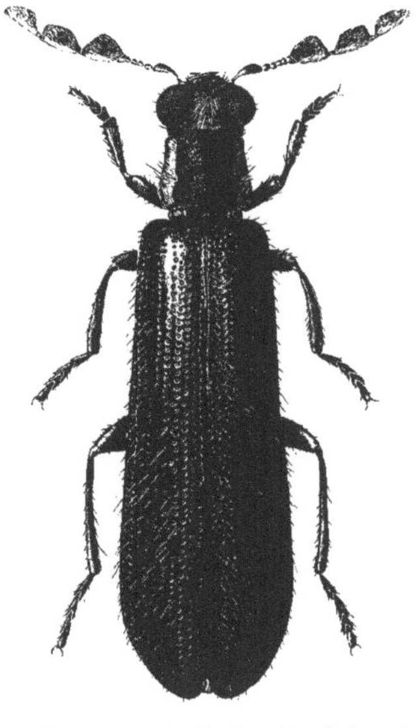

FIGURE 1.73. *Pyticeroides laticornis* (Say)

Description: Shape broad or narrow, some suboval, elongate, or convex; elytral humeral angle squared; 2 to 24 mm in length; color various, many brightly colored with red or yellow; vestiture dense in most, long or moderately elongated erect setae; most are predators on wood-infesting insects, some are anthophilic, and a very few are saprophagous.

Head prominent, strongly deflexed, as wide or wider than prothorax in most, may be inserted into prothorax; most with surface finely punctate. Antennae with 9 to 11 antennomeres, most clubbed or capitate, some serrate, or pectinate, rarely filiform; inserted on sides of frons proximal to ocular notch (Fig. 115) or distad to ocular noch (Fig. 114); clypeus distinct in most, very narrow, transverse; labrum distinct, corneous, transverse, incised (Fig. 76, # 3); mandibles large, arcuate, with one or more internal teeth, apices acute; maxillary palpi with four palpomeres, terminal palpomere cylindrical (Fig. 56) to securiform (Fig. 78, #1); labial palpi with three palpomeres, terminal palpomere cylindrical or securiform (Fig. 78); mentum quadrate, ligula membranous or coriaceous, without paraglossae; gular sutures approximate; eyes lateral, moderate, coarsely (Fig. 19) or finely granulate (Fig. 20), bulging in most, deeply incised (Fig. 77) or nearly entire (Fig. 53).

Pronotum as wide as or narrower than head; subquadrate or elongate, moderately convex, some constricted near front and hind margins; with or without distinct lateral ridges; surface of most finely punctate; pleural region large; prosternum short in front of coxae; procoxal cavities open or closed behind; mesosternum short; metasternum moderate, broad, epimera scarcely visible. Legs with trochantins some exposed; anterior coxae conical, rarely transverse, prominent in most, contiguous, or slightly separate; middle coxae rounded, not very prominent in most; hind coxae transverse; trochanters moderate, triangular; femora variously swollen; profemur spinose in some; tibiae slender, anterior or posterior face carinate or not, apical spurs small and variable in numbers; tarsal formula 5-5-5, fourth metatarsomere minute in some (Fig. 87), tarsomeres one to three or four of most lobed beneath (Figs. 9), when not lobed tarsomeres densely setose beneath (Fig. 8); pretarsal claws simple or bifid (Fig. 28, # 2) and variously denticulated at base, most with distinct denticle (Figs. 28); mesoscutellum small, triangular; elytra entirely covering abdomen, some abbreviated apically (Fig. 40), apex of most rounded, toothed in few, or truncated; wings reduced or absent in some, venation generally as in Figure 113; folding pattern with apical folding variable in amount.

Abdomen with five or six ventrites, sutures entire; surface micropunctate; venter of terminal segments strongly modified in some males; male genitalia with aedeagus elongate, slender, and moderately sclerotized, with two basal struts; parameres fused to phallobase, forming a sheath, apically variously emarginated; with long basal struts; phallus elongated, comprised of two plates, edges of plates denticulated or not; female genitalia with paraprocts with long bacilli; valvifers with long bacilli; coxite almost divided transversely into two segments; stylus very small.

Larvae elongate, orthosomatic; 2.0 to 20.0 mm, mostly 9.0 to 13.0 mm in length; vestiture moderate setae which may be numerous on all segments; color mottled, pink, red, yellow, orange, blue, or brown, with parts black or shades of brown, with a shiny luster; head exserted, most prognathous, mouthparts protracted; most dorsally depressed, ventrally convex; laterally parallel or divergent; epicranium in most separating frons or epicranial suture single; antennae three-segmented; clypeus and labrum distinct; labrum rectangular, with a series of setae; mandibles well developed, subtriangular, most simple with apices acute or blunt; a small retinaculum may be present; mola, prostheca, and penicillus absent; maxillae with cardo prominent, stipes present, maxillary palpi three-segmented, adjacent to a flat setiferous mala; labium with gula elongate, rectangular; labial palpi two-segmented; one to five pair of stemmata or stemmata absent; thorax with four-segmented legs with apical tarsunguli; legs widely separated; abdomen ten-segmented; ampullae may be present; tenth segment reduced and may be modified into a five-lobed organ of locomotion; spiracles annular or biforous on abdominal segments one to eight; urogomphi branched or unbranched, or absent.

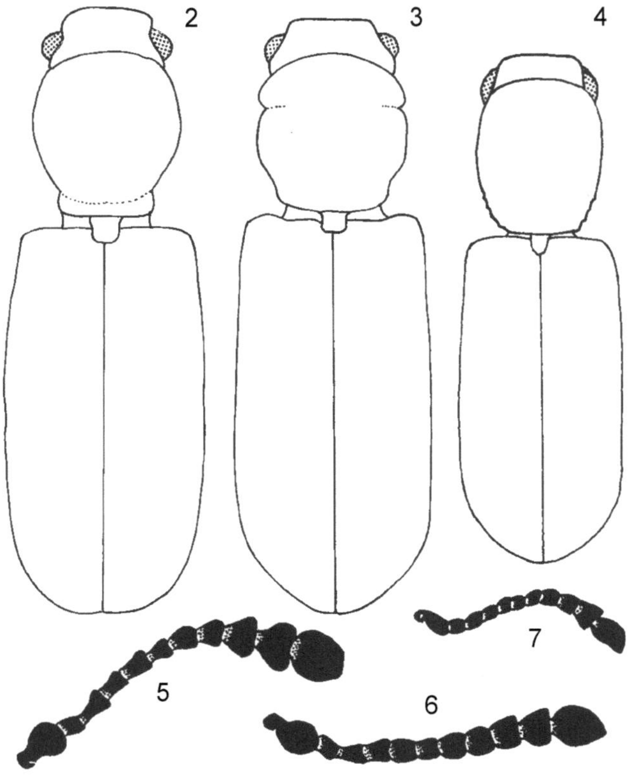

FIGURES 2.73 - 7.73. Body outlines. 2. *Thaneroclerus buquet* (Lefebvré); 3. *Zenodosus sanguineus* Say; 4. *Ababa tantilla* (LeConte); 5-7. Antennae. 5. *T. buquet*; 6. *A. tantilla*; 7. *Z. sanguineus*.

Habits and habitats. Most members of this family are predacious, as larvae and adults, on other insects. They generally are associated with woody plants and can be found on or under bark, in the tunnels of wood- and cone-borers, in galls or on plant foliage and dead twigs. Several species of *Enoclerus* and *Thanasimus* are believed to be important in natural control of bark beetles in coniferous forests. Adults of some other species, especially in the subfamily Clerinae, may be found on flowers where they feed on pollen. Many of the Tillinae adults are attracted to lights. The larvae of *Trichodes* spp. are known to complete their development on grasshopper egg pods or on animal or pollen contents of the cells of various bees and aculeate wasps. Larvae of an *Aulicus* sp. feed on eggs of lubber grasshoppers and as an adult on noctuid caterpillars. Some *Necrobia* spp. in addition to being predacious may also feed on stored food products of animal and vegetable origin.

Status of the classification. The family has received attention in America north of Mexico, especially by George H. Horn, Charles Schaeffer, Edwin C. Van Dyke, Albert B. Wolcott, Edward A. Chapin, and William F. Barr, and more recently by Jonathan R. Mawdsley, Jacques Rifkind, and Weston Opitz (formerly Ginter Ekis). From a world viewpoint, much remains to be done with clarification of generic concepts and zoogeographic relationships at supraspecific levels.

The higher classification of the Cleridae has undergone considerable categorical oscilations. Most recently a suprageneric classification based on an extensive analysis of morphology was proposed by Kolibac (1992, 1997). Assignment of categorical rank to supraspecific discontinuities are largely subjective. They become less so when higher classifications are based on gestalt-level analyses, i.e., involve criteria of morphology (internal and external), DNA, natural history, zoogeography, etc. Until such level of information is attained I consider it advisable to follow a system of classification that emphasizes kinship rather than evolutionary divergence. Hence, in this treatise the higher classification of Crowson (1964) is followed.

Distribution. There are 3,366 described species (Corporaal 1950), occurring in all areas, of which 291 described and undescribed species (Barr, personal communication) occur in America north of Mexico.

Key to Genera

1. Protarsomeres 1-4 compacted; tarsomeres very narrowly transverse (Fig. 8); tarsomere venter apparently rigid (Fig. 10) densely setose, without pulvilli; eyes not prominently incised; cranial (Fig. 26) and pronotal (Fig. 49) punctations long-oval (Thaneroclerinae) 2
— Protarsomeres 1-4 not compacted; tarsomeres not very narrowly transverse; tarsomere venter apparently pliable (Fig. 11), with fine setae and distinct pulvilli (Fig. 9); eyes deeply incised; cranial punctations spheroid 4

2(1). Anterior coxal cavities open behind (Fig. 13); body color red .... *Zenodosus*
— Anterior coxal cavities closed behind; body color various shades of pale brown to reddish brown 3

3(2). Pronotum subglobose (Fig. 2), strongly constricted at base, margins not minutely serrate ; tibial spur formula 2-2-2; first abdominal ventrite nearly twice length of other ventrites (Fig. 12); about 5.0 mm .... *Thaneroclerus*

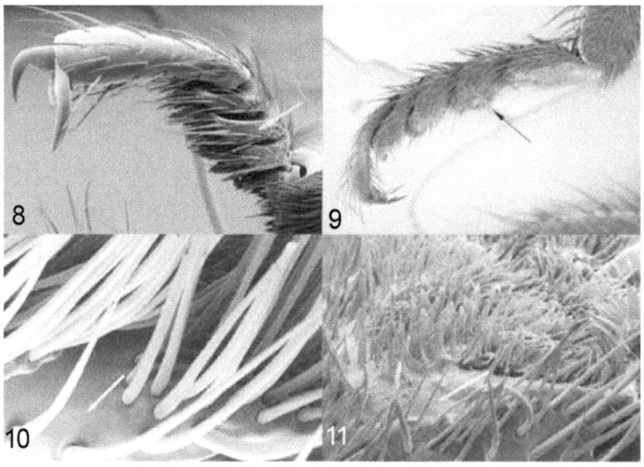

FIGURES 8.73-11.73. Tarsal structures. 8. *Thaneroclerus buquet* (Lefebvre) protarsus; 9. *Cymatodera bicolor* (Say) protarsus; 10. *T. buquet* venter of third protarsomere; 11. *C. bicolor* venter of second protarsomere.

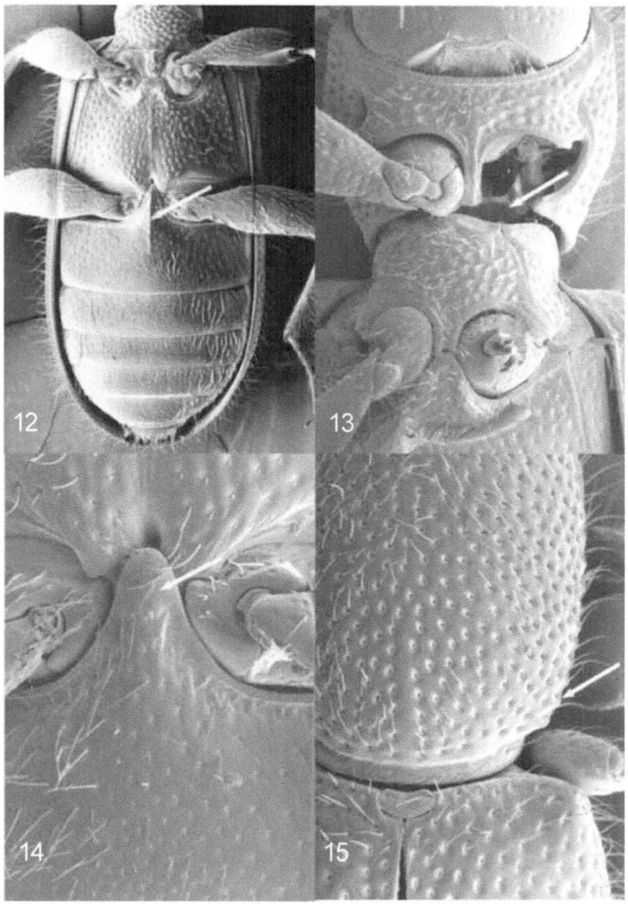

FIGURES 12.73-15.73. Thaneroclerinae external structures. 12. *Thaneroclerus buquet* (Lefebvre) hindbody, venter; 13. *Zenodosus sanguineus* Say forebody, venter; 14. *Z. sanguineus* abdomen, venter; 15. *Ababa tantilla* (LeConte) forebody, dorsum.

— Pronotum subrectangulate (Fig. 4), broadly rounded behind middle, margins minutely serrate near base (Fig. 15); tibial spur formula 1-2-2; first abdominal ventrite not twice the length of other ventrites; less than 2.0 mm .... *Ababa*

4(1). Anterior coxal cavities closed behind (Tillinae).. 5
— Anterior coxal cavities open behind (Figs. 21, 54) 12

5(4). Frons with prominent pair of horns arising from in front of eyes; elytra without striated punctations .... *Bostrichoclerus*
— Frons without horns; elytra with striated punctations 6

6(5). Last antennomere flattened, as long or much longer than combined length of preceding four antennomeres (Figs. 30, 31); metatarsus with three pulvilli, first reduced .... *Monophylla*
— Last antennomere circular in cross section; much shorter, at most as long as combined length of preceding three antennomeres; metatarsus with four pulvilli 7

7(6). Antenna boldly serrate (Fig. 36) 8
— Antenna feebly serrate (Fig. 33) 9

8(7). About 10 mm; flagellar antennomeres 4-10 trigonal (Fig. 36) .... *Bogcia*
— About 4 mm; most or all flagellar antennomeres trigonal or narrowly transverse (Fig. 42, 43) .... *Callotillus*

9(7). Basal denticle of tarsal claws digitiform (Fig. 28) .. .. *Araeodontia*
— Basal denticle of tarsal claws trigonal (Fig. 29) . 10

10(9). Last antennomere as long or longer than preceding two antennomeres (Figs. 34, 35) .... *Lecontella*
— Last antennomere only slightly longer than preceding antennomere 11

11(10). Eyes coarsely granulated (Fig. 19); body length about 4.0-20.0 mm .... *Cymatodera*
— Eyes finely granulated (Fig. 20); body length less than 6.0 mm .... *Cymatoderella*

12(4). Eyes feebly emarginated (Fig. 53) (Hydnocerinae) 13
— Eyes deeply emarginated (Fig. 77) 15

13(12). Basal denticle of tarsal claws well developed, narrowly trigonal (Fig. 44) .... *Phyllobaenus*
— Basal denticle of tarsal claw poorly developed, not narrowly trigonal (Fig. 41) 14

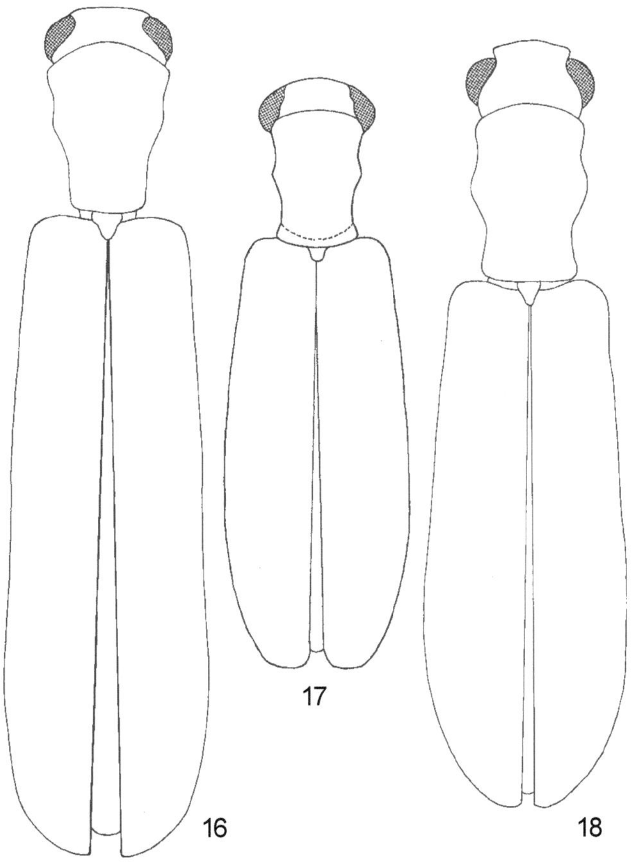

FIGURES 16.73-18.73. Body outlines. 16. *Lecontella gnara* Wolcott; 17. *Bogcia disjuncta* Barr; 18. *Cymatodera oblita* Horn.

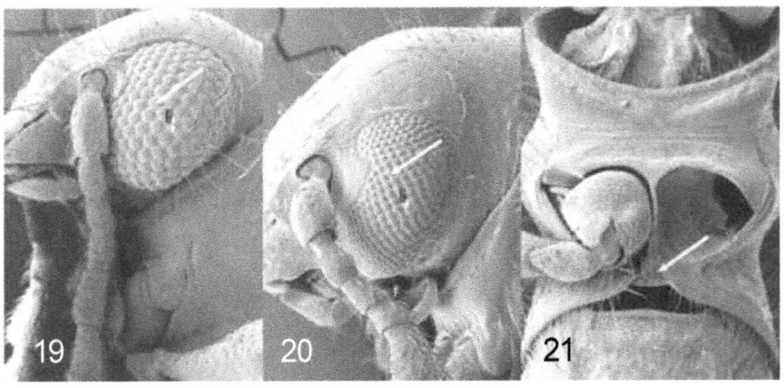

FIGURES 19.73-21.73. Tillinae structures. 19. *Cymatodera aegra* Wolcott forebody, lateral view; 20. *Cymatoderella collaris* (Spinola) forebody, lateral view; 21. *Cymatodera bicolor* (Say) forebody, ventral view.

14(13). Third antennomere about twice as long as wide (Fig. 47); head angular-trigonal when viewed from above (Fig. 45) tarsal claw devoid of basal denticle (Fig. 52) .... *Isohydnocera*
— Third antennomere short, about as long as wide (Fig. 48); head broad trigonal when viewed from above (Fig. 46); tarsal claw with feebly developed basal denticle (Fig. 41) .... *Wolcottia*

15(12). Fourth tarsomere about equal in size to third tarsomere (Fig. 50) (Clerinae)......................... 16
— Fourth tarsomere small, indistinctly set at base of tarsomere five (Fig. 87) 23

16(15). Antenna serrate (Fig. 63)................................... 17
— Antenna capitate (Fig. 68) or only partially capitate and serrate (Fig. 64)...................................... 18

17(16). Femora notably thickened at distal half; eyes coarsely granulated .... *Priocera*
— Femora not notably thickened in distal half; eyes finely granulated .... *Perilypus*

18(16). Terminal maxillary palpomere trigonal (as in Fig. 55) or subtrigonal (as in Fig. 78) 19
— Terminal maxillary palpomere cylindrical (Fig. 56) 21

19(18). Eyes coarsely granulated .... *Opilo*
— Eyes finely granulated 20

20(19). Last three antennomeres united into compact club, rendering antenna capitate (Fig. 68)... *Trichodes*
— Last three(or more) antennomeres loosely united into club, rendering antenna clavate (Fig. 64)... .... *Aulicus*

21(18). Antennal club loose (Fig. 62); base of elytra serrially punctate, scabrous .... *Thanasimus*
— Antennal club compact (Fig. 67); base of elytra not serially punctate, not scabrous 22

22(21). Elytra unicolored, bicolored, or tricolored; with pubescent markings; pronotum subspheroid, side margins prominently rounded (Fig. 70); anterior transverse depression evident laterally; pronotum varies in color .... *Enoclerus*
— Elytra unicolorous, black, some with a purpurescent sheen, without pubescent markings; pronotum subquadrate, side magins not prominently rounded (Fig. 59); anterior transverse depression not evident laterally; color of pronotum consistently yellow to red behind anterior angles and piceous in remainder (Fig. 120)........ *Placopterus*

23(15). Antennomeres 9-11 forming loose club that is as long or longer than antennomeres 3-8 combined (Fig. 94)... 24
— Antennomere 9-11 form club that is not as long as combined length of antennomeres 3-8 (Fig. 108) 33

24(23). Antennal insertion distad from ocular notch (Fig. 114); pronotum with two discal and two paralateral trichobothria (Figs. 79, 86) (Epiphloeinae) 25
— Antennal insertion proximal to ocular notch (Fig. 115); pronotum without trichobothria (Enopliinae) 27

25(24). Funicular antennomeres not densely setose, antennal club comprised of three antennomeres whose combined length are not longer than length of remainder of antenna (Fig. 95) .... *Madoniella*

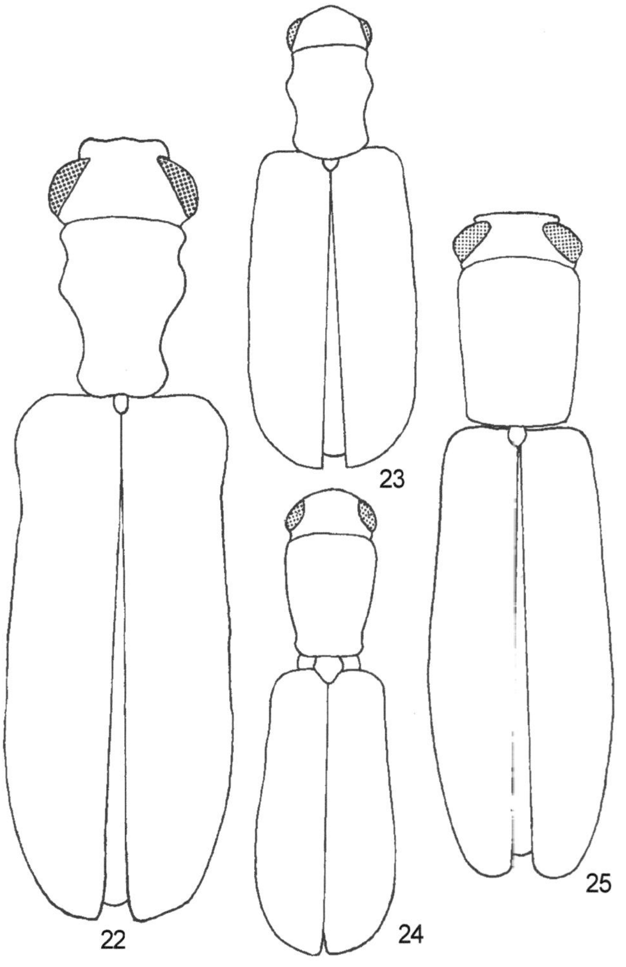

FIGURES 22.73-25.73. Body outlines. 22. *Araeodontia peninsularis* (Schaeffer); 23. *Cymatoderella collaris* (Spinola); 24. *Callotillus elegans* Erichson; 25. *Monophylla terminata* (Say).

FIGURES 26.73-29.73. Thaneroclerinae and Tillinae structures. 26. *Thaneroclerus buquet* (Lefebvre) epicranium; 27. *Cymatodera bicolor* (Say) protarsus; 28. *Araeodontia peninsularis* (Schaeffer) protarsus; 29. *Lecontella gnara* Wolcott protarsus.

— Funicular antennomeres densely setose; antennal club comprised of three antennomeres whose combined length is longer than length of remainder of antenna (Fig. 89) 26

26(25). Pronotum subquadrate (Fig. 79); antenna (Fig. 1) with 9 antennomeres .... *Pyticeroides*

— Pronotum narrow-elongated (Fig. 74); antenna (Fig. 94) with 10 antennomeres .... *Ichnea*

27(24). Antenna with 10 antennomeres 28

— Antenna with 11 antennomeres 29

28(27). Eyes coarsely granulated; pronotum abruptly expanded at middle (Fig. 83); elytral punctation large, serrate .... *Cregya*

— Eyes finely granulated; pronotum not abruptly expanded at middle (Fig. 81); elytral punctations small, not serrate .... *Pleonides*

29(28). Pronotum abruptly expanded at middle (Fig. 84) 30

— Pronotum not abruptly expanded at middle (Fig. 85) 31

30(29). Terminal maxillary and labial palpomeres notably trigonal (Fig. 78) .... *Pelonium*

— Terminal maxillary and labial palpomeres not trigonal .... *Boschella*

31(29). Anterior margin of protibia serrulate (Fig. 119) .... *Chariessa*

— Anterior margin of protibia not serrulate 32

32(31). Elytral punctations shallow, render elytral disc subscabrous; epipleuron in lateral position, most about 10 mm (males smaller); hind body piceous or brunneous .... *Neorthopleura*

— Elytral punctations deep, well defined; epipleuron in ventral position; about 4.0 mm; hind body castaneous .... *Parapelonides*

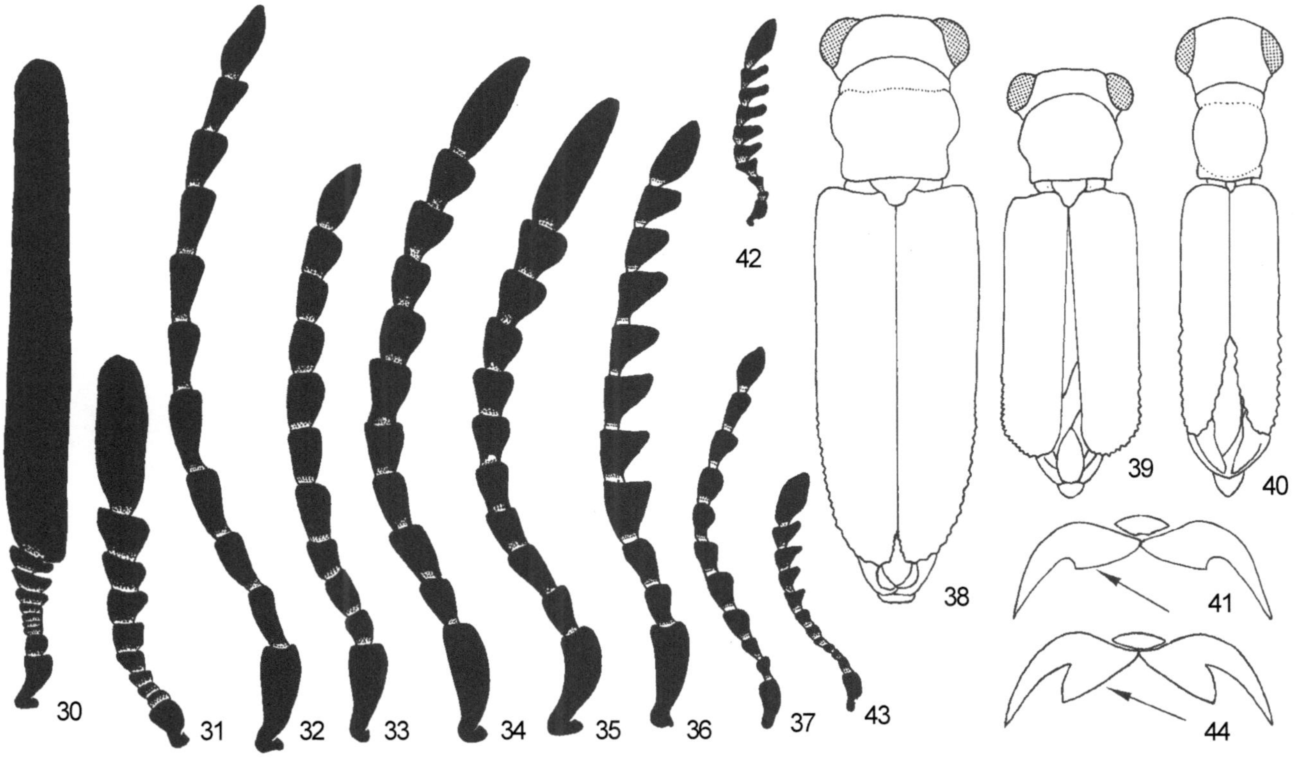

FIGURES 30.73-44.73. 30-37. Antennae. 30. *Monophylla terminata* (Say) male; 31. *M. terminata* female; 32. *Cymatodera oblita* Horn; 33. *Araeodontia peninsularis* (Schaeffer) male; 34. *Lecontella gnara* Wolcott female; 35. *L. gnara* male; 36. *Bogica disjuncta* Barr; 37. *Cymatoderella collaris* (Spinola); 38-40. Body outlines. 38. *Wolcottia pedalis* (LeConte); 39. *Phyllobaenus pallipennis* (Say); 40. *Isohydnocera tabida* (LeConte); 41. *W. pedalis* protarsal claws; 42. *Callotillus elegans* Erichson antenna, male; 43. *C. elegans* antennae, female; 44. *P. pallidipennis* protarsal claws.

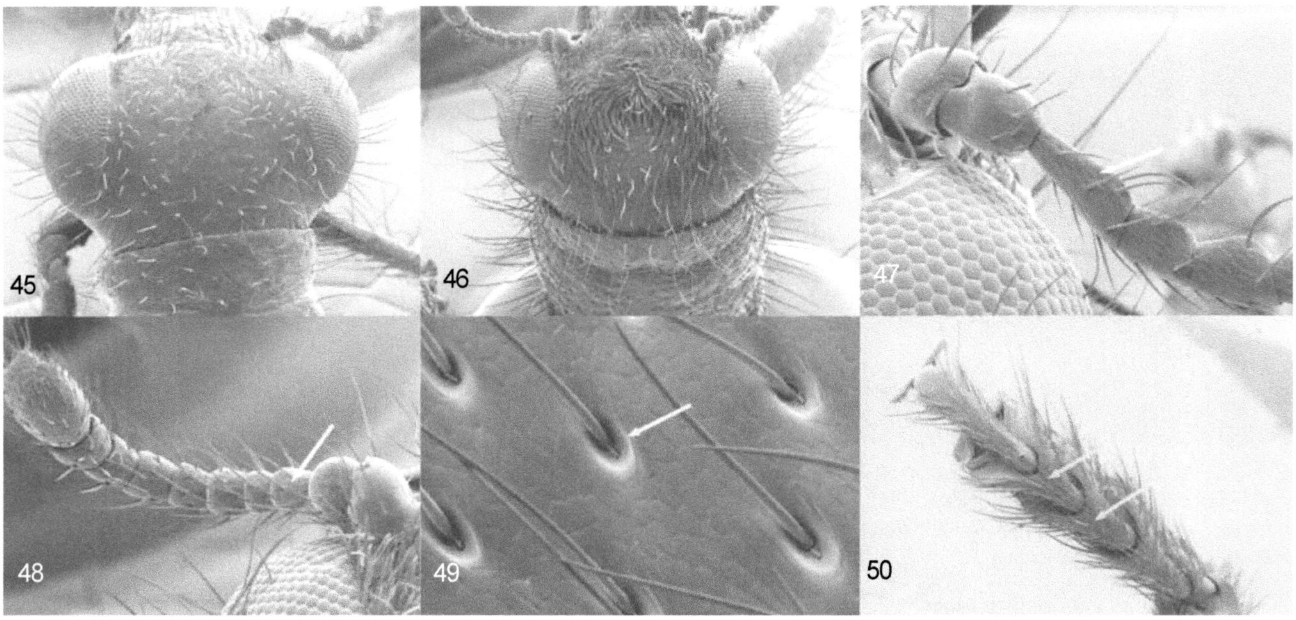

FIGURES 45.73-50.73. Thaneroclerinae, Clerinae and Hydnocerinae structures; *Isohydnocera tabida* (LeConte) forebody, dorsal view; 46. *Wolcottia sobrina* (Fall) forebody, dorsal view; 47. *I. tabida* antenna; 48. *W. sobrina* antenna; 49. *Thaneroclerus buquet* (Lefebvre) dorsum of pronotum; 50. *Enoclerus ichneumoneus* Fabricius protarsus.

33(23). Pronotal side-margin carina entire (Fig. 116); body form broad-oval (Korynetinae) 34
— Pronotal side-margin not entire (Fig. 117); body form narrow-rectangulate (Tarsosteninae) 36

34(33). Intestitial spaces of elytra polished, surface smooth and glossy .... *Lebasiella*
— Interstitial spaces of elytra gritty, surface nearly subrugose 35

35(43). Middle antennomere of antennal club narrowly transverse (Fig. 108) .... *Necrobia*
— Middle antennomere of antennal club trigonal (Fig. 109) .... *Loedelia*

36(33). Middle pale elytral fascia faintly elevated and devoid of asetiferous punctations; about 6.0 mm . ... *Paratillus*
— Middle pale elytral fascia not elevated and with asetiferous punctations; about 3.0 mm .... *Tarsostenus*

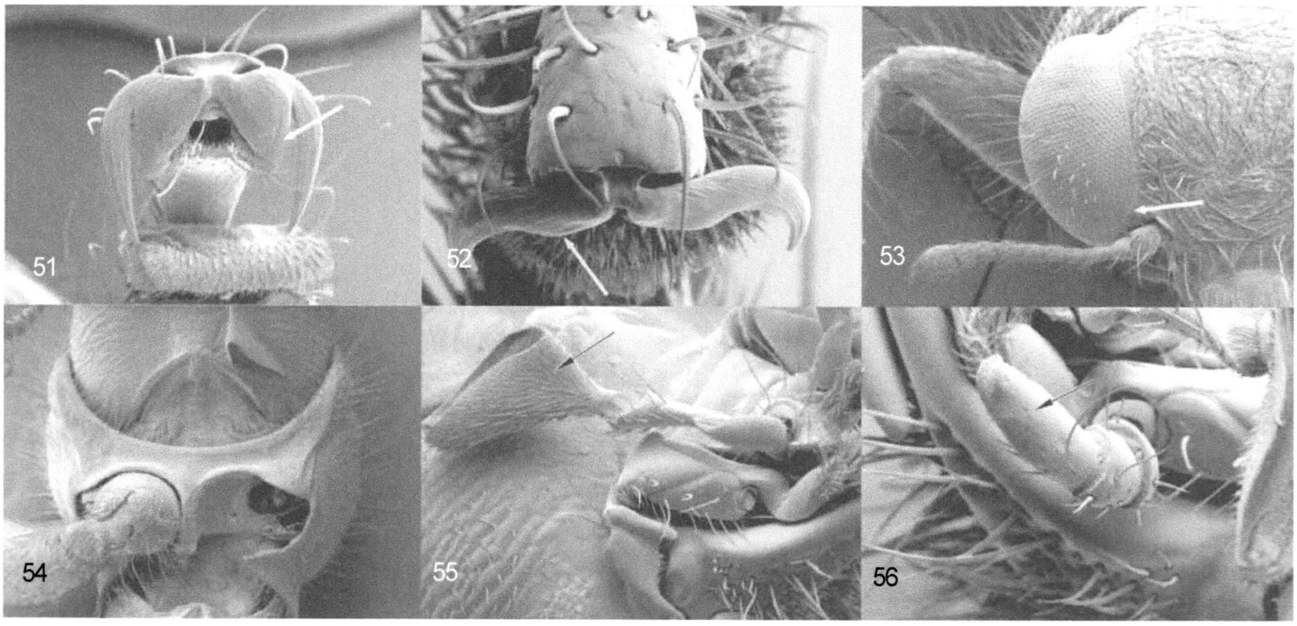

FIGURES 51.73-56.73. Thaneroclerinae, Clerinae and Hydnocerinae structures; 51. *Phyllobaeneus humeralis* (Say) protarsal claws; 52. *Isohydnocera tabida* (LeConte) protarsal claws; 53. *Phyllobaeneus humeralis* (Say) forebody; 54. *Enoclerus ichneumoneus* Fabricius forebody, ventral; 55. *E. ichneumoneus* head, ventral; 56. *E. ichneumoneus* mouthparts, ventral.

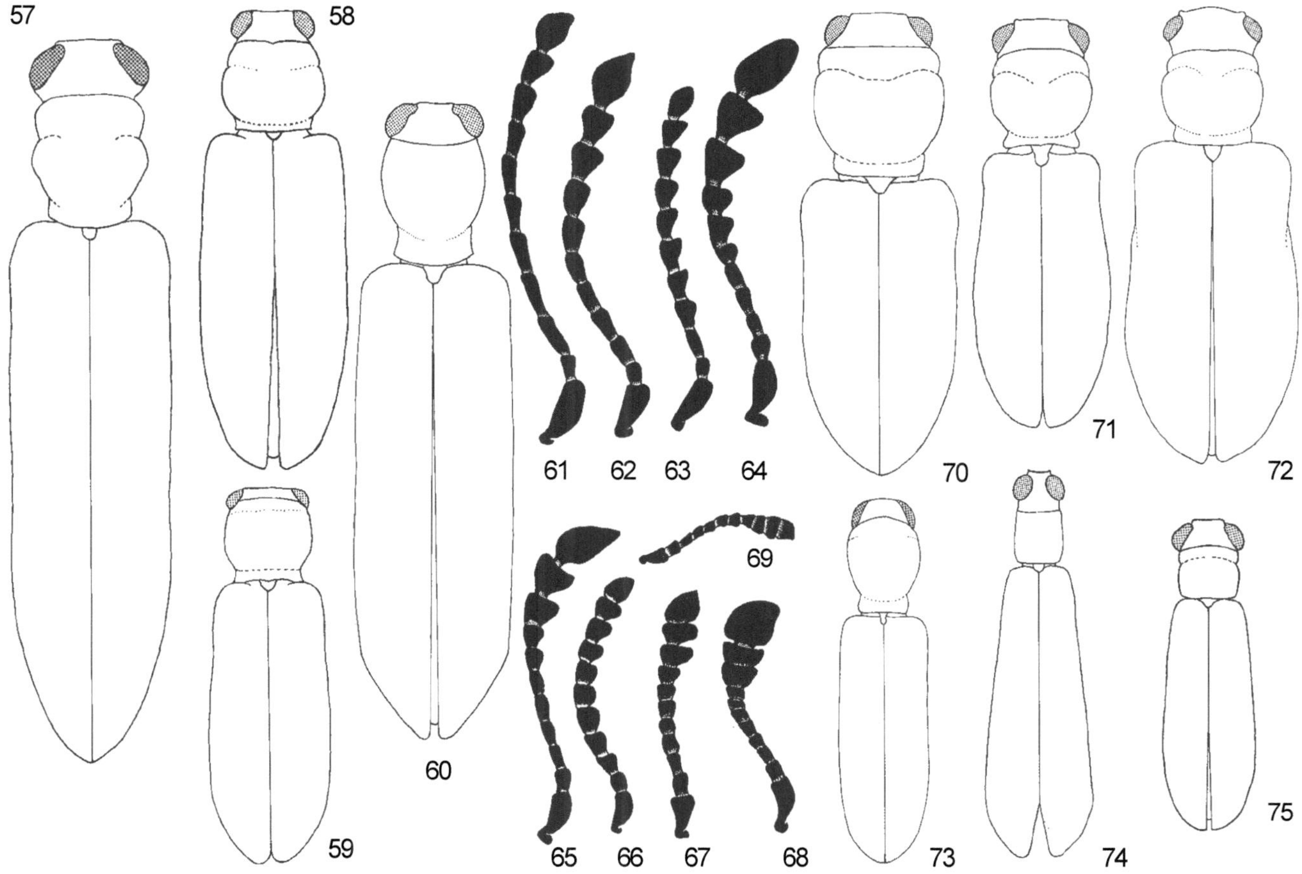

FIGURES 57.73-75.73. 57-60. Body outlines. 57. *Opilo mollis* (Linnaeus); 58. *Perilypus ornaticollis* (LeConte); 59. *Placopterus thoracicus* (Olivier); 60. *Trichodes ornatus* (Say). 61-69. Antenna. 61. *O. mollis*; 62. *Thanasimus dubius* (Fabricius); 63. *Priocera castanea* (Newman); 64. *Aulicus antennatus* Schaeffer male; 65. *A. antennatus* female; 66. *P. ornaticollis*; 67. *Enoclerus lunatus* (Klug); 68. *Trichodes ornatus* (Say); 69. *Placopterus thoracicus* (Olivier). 70-75. Body outlines. 70. *E. lunatus*; 71. *T. dubius*; 72. *A. antennatus*; 73. *P. castanea*; 74. *Ichnea atterima* (Klug); 75. *Madoniella dislocatus* (Spinola).

Classification of the Genera

Cleridae Latreille 1804

Thaneroclerinae Chapin 1921
(Corporaal 1939, Kolibac 1992)

Thaneroclerus Chapin 1924. Type species: *Clerus buquet* Lefebvré 1835. One cosmopolitan species known in North America. Habitus (Fig. 2). Antenna (Fig. 5). Notes: Conveniently distinguished from the superficially similar specimens of *Zenodosus* by its "duller" red-brown body color, closed procoxal cavities, proportionally longer first visible abdominal ventrite (Fig. 12), and carinate intermetacoxal process (Fig. 12). Corporaal (1939) reported *T. buquet* to be a predator of insects found in groceries, spices, drugs, and tobacco. The immature stages have been described most recently by Foster (1976a) while characteristics of adult internal organs were published by Ekis and Gupta (1971).

Thaneroclerus auctorum
Thaneclerus Desmarest 1860
Thanatereroclerus Gemminger and Harold 1869

Zenodosus Wolcott 1910. Type species: *Clerus sanguineus* Say 1835. One species from east of the Rocky Mountains. Habitus (Fig. 3). Antenna (Fig. 7). Notes: The bright red color of the elytra will separate the members of this genus from those of *Thaneroclerus*; moreover, the procoxal cavities are open (Fig. 13) and the intermetacoxal process is not carinate (Fig. 14). Knull (1951) reports that these insects are diurnal and found under bark of dead trees infested with lignicolous boring insects. Mawdsley (personal communication) collected overwintering adults under the bark of live birch trees. The mature larva of *C. sanguineus* is described in detail by Foster (1976a).

Ababa Casey 1897. Type species: *Ababa crinita* Casey 1897 (junior synonym of *A. tantilla* LeConte). One species from east of the Rocky Mountains. Habitus (Fig. 4). Antenna (Fig. 6). Notes: Within the subfamily, these reddish-brown beetles are most conveniently identified by their small size with most examples being in the 2.0 mm range. Also, unlike other members of this subfamily, the pronotal punctations are rounded and the pronotal side margins are minutely serrate behind the middle (Fig. 15).

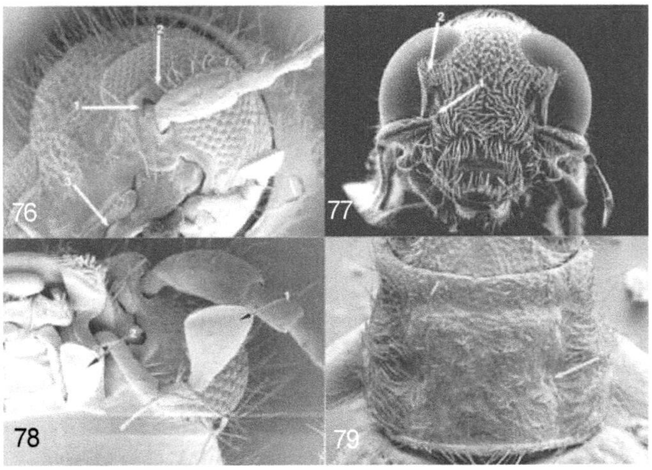

FIGURES 76.73-79.73. Enopliinae and Epiphloeinae structures. 76. *Pelonium leucophaeum* (Klug) head; 77. *Epiphloeus duodecimmaculatus* (Klug) head; 78. *P. leucophaeum* mouthparts; 79. *Pyticeroides laticornis* (Say) forebody.

Specimens have been reared in Texas from old bracket fungi, *Fomes ignarius* (Linnaeus) (Knull 1951).

Prionodera Wolcott 1910
Cleridopsis Champion 1913
Prionostichaeus Wolcott 1911 (replacement name)
Wolcottella Lucas 1920 (replacement name)

Tillinae Leach 1815
(Foster 1976b)

Bostrichoclerus Van Dyke 1938. Type species: *Bostrichoclerus bicornis* Van Dyke 1938. One species from Angel de La Guardia Island, Gulf of California, and southeastern California. Notes: The most distinctive characteristic of these beetles is the prominent horn arising from in front of each eye proximal to the antennal insertion. This group of clerids is associated with bark.

Monophylla Spinola 1841. Type species: *Monophylla megatoma* Spinola 1841. Three species distributed within eastern, southwestern, and western United States. Habitus (Fig. 25). Antenna (male Fig. 30; female Fig. 31). Notes: All species of this genus are characterised by sexual dimorphism in antennal development. Specimens are easily recognized as belonging to this genus by their elongate, spatulate terminal antennomere which is particularly long in the male. *Monophylla terminata* (Say) adults are known to feed on bark beetles in conifers, and on lignicolous borers in hardwoods. According to Knull (1951) specimens of *M. terminata* were collected from "dead bitter-sweet infested with small cerambycid larvae; wild grape infested with *Phymatodes amoenus* (Say) [Cerambydae]; persimmon with *Sinoxylon* [Bostrichidae], *Lyctus* [Bostrichidae], *Agrilus* [Buprestidae], *Chramesus* [Curculionidae] and other wood borers; sassafras with Anobiidae; ash lumber with *Lyctus;* hackberry with *Scolytus muticus* Say [Curculionidae]; feeding on white pine weevil larvae, also oak, honey locust, and mesquite" (bracketed comments are mine). Foster (1976b) reared *M. californica* Fall and *M. terminata* larvae from mesquite.

Macrotelus Klug 1842
Elasmocerus LeConte 1849

Bogcia Barr 1978. Type species: *Bogcia disjuncta* Barr 1978. One species from Texas. Habitus (Fig. 17). Antenna (Fig. 37). Notes: Members of this genus are superficially very similar to specimens of *Cymatodera* from which they are easily separated by the strongly serrate condition of the antenna (compare Figs. 32, 36). Also, in *Bogcia* specimens the protarsus has the subdivisions of the claw in close proximity (compare Figs. 121, 122) whereas in *Cymatodera* the claw subdivisions are distinctly separated (Fig. 27). *Bogcia obliquefasciata* (Schaeffer) specimens have been collected from *Acacia flexicaulis*.

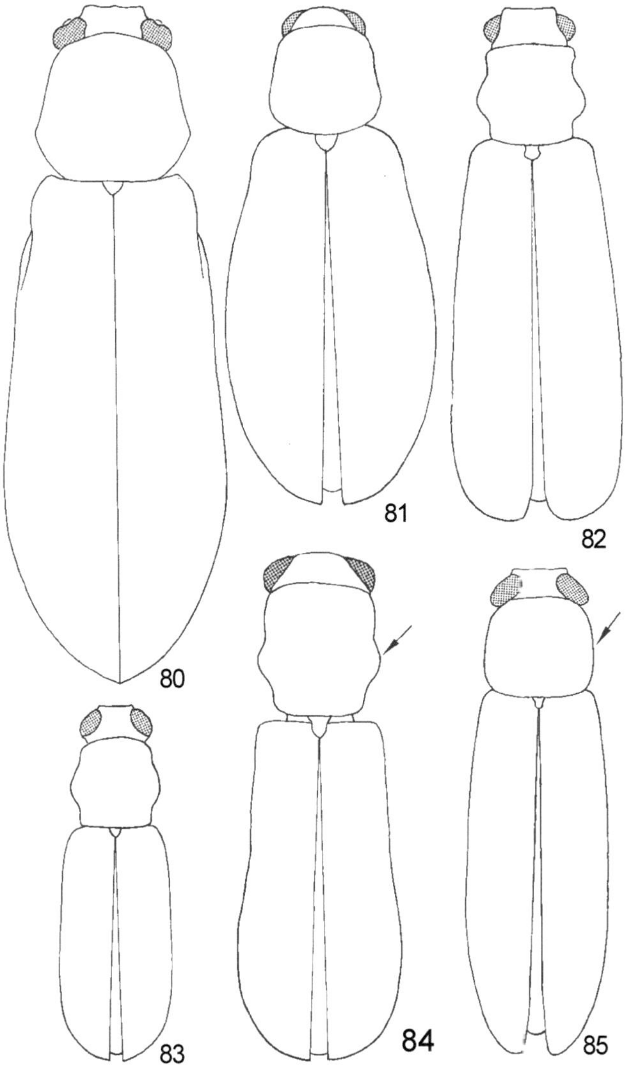

FIGURES 80.73-85.73. Body outlines. 80. *Chariessa pilosa* (Forster); 81. *Pelonides quadripunctatus* (Say); 82. *Boschella fasciata* (LeConte); 83. *Cregya oculata* (Say); 84. *Pelonium leucophaeum* (Klug); 85. *Neorthopleura thoracica* (Say).

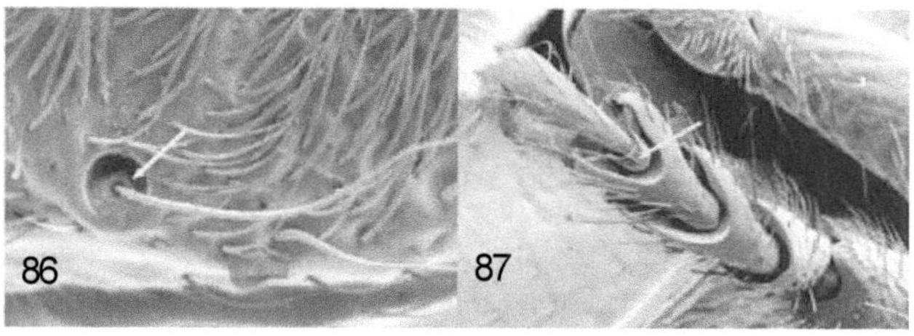

FIGURES 86.73-87.73. Epiphloeinae and Enopliinae structures. 86. *Pyticeroides laticornis* (Say) prothorax; 87. *Pelonium leucophaeum* (Klug) metatarsus.

Callotillus Wolcott 1911. Type species: *Callotillus eburneocinctus* Wolcott, 1911. Two species from Florida and southwestern United States. Habitus (Fig. 24). Antenna (male Fig. 42; female Fig. 43). Notes: These colorful little clerids (about 3 mm) are the only tillines of North America with the elytral surface adorned with a pale transverse, blister-like, swollen fascia. The male antenna has a flabellate appearance. Specimens of this genus have been reared from *Prosopis* sp. infested with the bostrichid *Xylobiops texanus*(Horn) and the lycid *Trogoxylon parallelopipedum* (Melsheimer), and from *Conocarpus erecta* (Jacq.) (Foster and Barr 1972).

Araeodontia Barr 1952. Type species: *Araeodontia penninsularis* Schaeffer 1904. Three species from southwestern United States. Habitus (Fig. 22). Antenna (Fig. 33). Notes: The tripartite condition of the tarsal claws (as shown in Fig. 28) will distinguish the members of this genus from all other North American Tillinae.

Lecontella Wolcott and Chapin 1918. Type species: *Cymatodera cancellata* LeConte 1854. Three species from eastern, central, and southwestern United States. Habitus (Fig. 16). Antenna (male Fig. 35; female Fig. 34). Notes: The members of this genus are very similar in appearance to the members of *Cymatodera*. However, the *Lecontella* beetles are easily recognized by the disproportionate elongation of the last antennomere most pronounced in the males. The pronounced serrate condition of large elytral punctations is an additional characteristic that helps recognize the members of this genus. The alimentary canal of *L. gnara* (LeConte) was illustrated and described by Ekis and Gupta (1971). Knull (1951) records *L. cancellata* (LeConte) from, "bee's nests in decayed log and nest of mud dauber wasp (*Sceliphron cementarium* Dru.)." Foster (1976b) described the larvae of *L. brunnea* (Spinola).

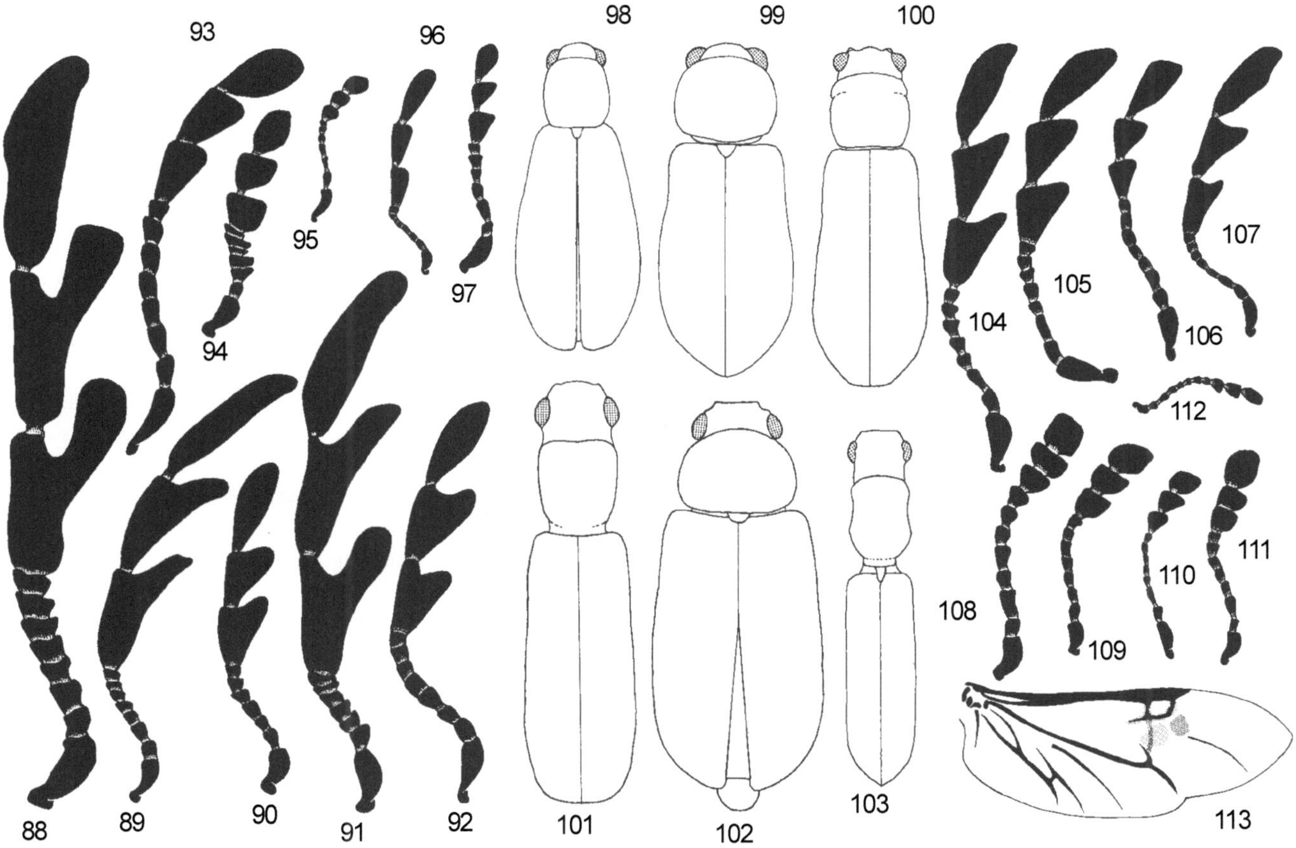

FIGURES 88.73-113.73. 88-97. Antennae. 88. *Chariessa pilosa* (Foster) male; 89. *C. pilosa* female; 90. *Pelonides quadripunctatus* (Say) female; 91. *P. quadripunctatus* male; 92. *Pelonium laeucophaeum* (Klug) male; 93. *P. leucophaeum* female; 94. *Ichnea atterima* (Klug); 95. *Madoniella dislocatus* (Spinola); 96. *Parapelonides nigrescens* (Schaeffer) female; 97. *Cregya oculata* (Say). 98-103. Body outlines. 98. *P. nigrescens*; 99. *Lebasiella pallipes* (Klug); 100. *Loedelia maculicollis* (LeConte); 101. *Tarsostennus univittatus* (Rossi); 102. *Necrobia rufipes* (DeGeer); 103. *Paratillus carus* (Newman). 104-112. Antennae. 104. *Neorthopleura thoracica* (Say) male; 105. *N. thoracica* female; 106. *Boschella fasciata* (LeConte) female; 107. *P. nigrescens* male; 108. *N. rufipes*; 109. *L. maculicollis*; 110. *L. pallipes*; 111. *P. carus*; 112. *T. univittatus*; 113. *Enoclerus lecontei* (Wolcott) metathoracic wing.

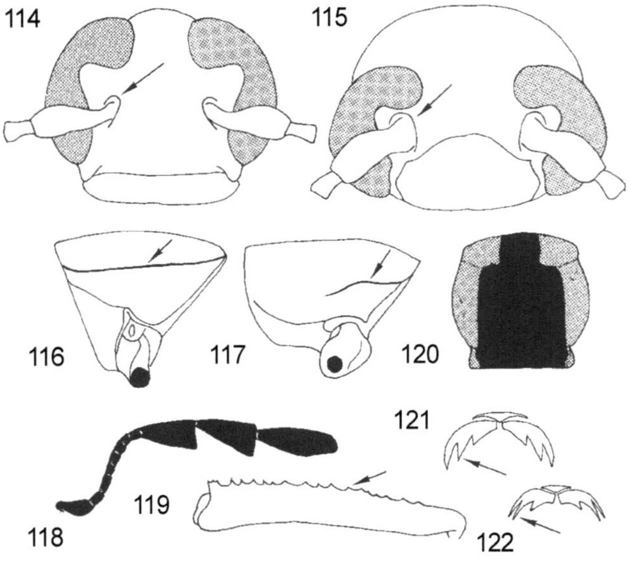

FIGURES 114.73-122.73. External anatomy of Cleridae. 114. *Pyticeroides laticornis* (Say) head, frontal view; 115. *Pelonium laeucophaeum* (Klug) head, frontal view; 116. *Necrobia rufipes* (DeGeer) prothorax, lateral view; 117. *Paratillus carus* (Newman) prothorax, lateral view; 118. *Boschella fasciata* (LeConte) male antenna; 119. *Chariessa pilosa* (Foster) protibia, lateral view; 120. *Placopterus thoracicus* (Olivier) pronotum; 121. *Cymatodera horni* Wolcott protarsal claws; 122. *Bogcia oaxacae* Barr protarsal claws.

Cymatodera Grey 1832. Type species: *Cymatodera hopei* Grey 1832. Sixty species widely distributed, particularly in southwestern United States. Habitus (Fig. 18). Antenna (Fig. 32). Notes: This is a very speciose genus whose members may be generally characterized as small (4.0 mm) to large (20.0 mm), dark or light brown, cylindrical beetles with long 11-segmented subfiliform antennae whose tarsal claws are bipartite (Fig. 27). Various species have a transverse pale midelytral fascia. Four species of *Cymatodera* were included in a treatise of the digestive system of the Cleridae by Ekis and Gupta (1971). Published reports suggest that adults of *Cymatodera* favor immature stages of prey species such as the larvae of cynipoid wasps and other grubs dissected from galls, fruit tree caterpillars, and larvae of bark beetles, cerambycids, and buprestids. Among the plants associated with these beetles one may include cedars, larch, junipers, acacias, apple and walnut, oaks, and sequoias. McSwain (1945) provided a thorough treatment of the life cycle of *C. ovipennis* (Say) including descriptions of the larva and pupa.

Cymatoderella Barr 1962. Type species: *Tillus collaris* Spinola 1844. Two species from the southwestern, midwestern and southern portions of the United States. Habitus (Fig. 23). Antenna (Fig. 37). Notes: The comparatively fine granulations of the eyes will distinguish members of this genus from smaller members of the superficially similar specimens of *Cymatodera*.

Hydnocerinae Spinola 1844
(Kolibac 1998)

Phyllobaenus Dejean 1837. Type species: *Clerus humeralis* Say 1823. Forty-nine species generally distributed in North America. Habitus (Fig. 39). Antenna (Fig. 53). Notes: The great majority of clerids collected that belong to this subfamily belong to this genus. In general, they are about 4mm, with a soft, squat hind body and forebody accentuated by protruding eyes and short "stubby" antennae. The eyes are only faintly incised and the elytra of many are shorter than the abdomen, tapered to the rear, and minutely serrulate along the posterior lateral margins. The subquadrate pronotum distinguishes *Phyllobaenus* specimens from the elongated pronotum of *Isohydnocera* specimens, while the distinctly formed tarsal denticle (Fig. 51) will distiguish them from specimens of *Wolcottia*. These little clerids are commonly collected by sweeping high grass, or by beating intermixture of live and dead branches with or without foliage. They have been reared from insect galls and from the bolls of cotton. Among their plant associations we may include wild grape, river birch, black and white oak, hickory, sumac, pine, blackberry, cotton, elm, and desert shrubs. They are known to prey on small wood borers, immature weevils, and hymenopterus larvae.

Hydnocera Newman 1838

Isohydnocera Chapin 1917. Type species: *Hydnocera curtipennis* Newman 1840. Ten species found east of the Rocky Mountains to southern California. Habitus (Fig. 40). Antenna (Fig. 47). Notes: These are small (5.0 mm), narrow- bodied, "spindly-legged," checkered beetles that are easily distinguished from other members of the subfamily by the comparative length of the third antennomere (Fig. 47, 48). Among isohydnocerans the third antennomere is considerable longer than broad (Fig. 47) which is not the case in members of *Phyllobaenus* or *Wolcottia* (Fig. 48). Viewed from above, the head of these beetles is seen as a highly triangular structure. These beetles are also characterized by the abbreviated length of the elytra whose apices are distinctly serrulate at the margin. *Isohydnocera curtipennis* Newman are reported from galls of *Euura salacis-nodus* Walsh [Hymenoptera: Tenthredinidae] on willow and *Gnorimoschema gallaesolidaginis* Riley [Lepidoptera: Gelechiidae] on golden rod, and *I. tabida* (LeConte) from plant stems infested with mordelid larvae (Knull 1951). Sabrosky (1934) described the parasitoid habits of *I. curtipennis* noting that the immatures of this species probably consumed the contents of a lepidopterous pupae found in a large gall of the solidago gallmoth. Several species have been taken by beating dogwood, witch hazel, oaks, and wild grape.

Wolcottia Chapin 1917. Type species: *Hydnocera pedalis* LeConte 1866. Two species from eastern and central United States and in particular Arizona. Habitus (Fig. 38). Antenna (Fig. 48). Notes: Distinguishable from specimens of *Phyllobaenus* by the lack of a well-developed tarsal denticle (compare Figs. 41, 44) and from those of *Isohydnocera* by the nearly equilateral condition of the third antennomere (Fig. 48). May be collected by sweeping grass and weeds, and by beating oaks.

Clerinae Latreille 1802

Priocera Kirby 1818. Type species: *Priocera variegata* Kirby 1818. Three species from southwest, eastern, central, and southern United States. Habitus (Fig. 73). Antenna (Fig. 63). Notes: These clerines are characterised by the extensive thickness of the femora, coarsely granulated eyes, serrate antenna, and comparatively elongate terminal palpomere of the maxillary palpus. These nocturnal checkered beetles are found during day time hidden under bark and in the tunnels of lignicolous insects. They have been associated with various species of pine and oak infested with *Melittomma* [Lymexilidae] larvae and *Dendroctonus frontalis* Zimmerman [Curculionidae], respectively.

Perilypus Spinola 1841. Type species: *Perilypus carbonarius* Spinola 1841. Two species, one from Arizona and the other from southern and midwestern states (Ohio to Texas). Habitus (Fig. 58). Antenna (Fig. 66). Notes: These clerids are easily recognized by the broadly serrate condition of the antenna. Beating oak branches onto a sheet is the preferred method for collecting them. The genus *Perilypus* was reviewed by Ekis (1977).

Opilo Latreille 1802. Type species: *Notoxus mollis* Fabricius 1794. Two cosmopolitan species. Habitus (Fig. 57). Antenna (Fig. 61). Notes: The triangular shape of the terminal maxillary palpomere will distinguish these beetles from all other North American clerines except those of *Aulicus*, which differ by having a distinct antennal club. In other North American clerines the terminal maxillary palpomere is digitiform, in *Trichodes* specimens this palpomere is only slightly expanded at the apex.

Trichodes Herbst 1792. Type species. *Clerus alvearius* Fabricius 1792. Ten species widely distributed in North America. Habitus (Fig. 60). Antenna (Fig. 68). Notes: Trichodeans are large (about 10.0 mm), very hairy, narrow-bodied beetles that exhibit a mixture of color patterns of red, orange, and yellow, and metallic blue-black in most. Their most outstanding distinguishing characteristic is the extent of development of the maxillary galea. The narrowly trigonal terminal maxillary palpomere distinguishes these beetles from those of *Aulicus* and *Enoclerus* in which the palpomere is cylindrical and the antennal club less compact. Adults of *Trichodes* are found most often on a great variety of flowers where they mate, feed, and oviposit. The larvae of some species are pernicious in bee nests. Females of *T. ornatus* Say are known to oviposit directly into bee nests (Mayer *et al.* 1976) while larvae of *T. oregonensis* Barr feed on the eggs in grasshopper egg pods (Foster 1976c). The North American species of *Trichodes* were reviewed by Foster (1976c).

Aulicus Spinola 1841. Type species: *Aulicus nero* Spinola 1844. Ten species from the southcentral and western United States. Habitus (Fig. 72). Antenna (male Fig. 64; female Fig. 65). Notes: Members of this genus range from about 7.0 mm (*A. dentipes* Schaeffer) to 20.0 mm (*A. edwardsii* Horn). *Aulicus* beetles are robust with elytral coloration showing combinations of metallic blue with red, orange, and yellow. The tarsal claws lack a basal denticle, a characteristic they share with *Trichodes* and *Opilo*. However, only in *Aulicus* and *Opilo* is the terminal maxillary palpomere strongly trigonal. In specimens of *Aulicus* the antennae are strongly clubbed (Fig. 64) while in *Opilo* the antennae are subfiliform (Fig. 61). *Aulicus* beetles are very active on flowers, foliage of shrubs and trees, or found scurrying on the soil surface. Linsley (1933, 1936) reported adult feeding on caterpillars of Noctuidae whereas the larvae feed on the egg masses of lubber grasshoppers. Most species of *Aulicus* are very active during the oviposition period of grasshoppers. The genus *Aulicus* was reviewed by Linsley (1936) and Barr and Foster (1979).

Serriger Spinola 1841
Xenoclerus Schenkling 1902

Thanasimus Latreille 1806. Type species: *Attelabus formicarius* Linnaeus1785. Five species widely distributed. Habitus (Fig. 71). Antenna (Fig. 62). Notes: The most convenient characteristic that distinguishes the members of this genus is the subseriate distribution of large punctations at elytral basal half. The punctations diminish abruptly at about midelytron. There is a general color pattern of the dorsum that involves a dark red forebody, dark red elytral base, and elytral disc with alternating regions of black disrupted by angular fascia of pale setae. These checkered beetles are most commonly associated with conifers in montane forests and tend to show a more northerly distribution. Many have been collected on dying spruce, pine, and elm trees ladened with bark beetles. Considerable research is in progress to investigate the pheromonal interactions of these clerids with bark-beetle prey species and the chemical volatiles of host trees (Reeve 2000). *Thanasimus formicarius* (Linnaeus), an Old World species, was introduced into North America to control the bark beetle *Dendroctonus frontalis* Zimmermann [Curculionidae] in 1892 and during the late 1900s.

Enoclerus Gahan 1910. Type species: *Clerus sexguttatus* Fabricius 1775. Thirty-six species widely distributed. Habitus (Fig. 70). Antenna (Fig. 67). Notes: This, the most speciose clerine genus in North America is closely allied to *Placopterus* and *Thanasimus*. *Enoclerus* beetles may be generally distinguished from those of *Placopterus* by having a more rounded forebody with the sides of the pronotum prominently rounded in most. To distinguish specimens of *Enoclerus* from *Thanasimus* one can use differences in development of the pronotal anterior transverse depression and in the texture of the elytral disc. In *Enoclerus* the anterior transverse depression is weakly impressed whereas in *Thanasimus* it is particularly prominent at the side margins. In *Thanasimus* the antennal club is very loose (compare Figs. 62, 67) and there is a coarse quality of the elytral base. Both characteristics are contrary to what one finds in most *Enoclerus* specimens in which the antennal club is compact and the disc of the elytral base notably smooth. Adults and larvae of *Enoclerus* are ferocious predators of bark beetles, weevils, and wood borers, and have been associated with such hardwoods as pine, spruce, elm, cedar, white oak, hickory, sweet gum, maple, peach trees, and various xerophilous plants of the American southwest. The beating-sheet technique involving admixtures of live and dead hardwood branches is a very productive collecting method for *Enoclerus* specimens, although Knull (1951) reports an abundance of *E. rosmarus* (Say) on flowers and weeds. Foster's (1971) field of study of *E. opifex* (Gorham) and *E. abdominalis* (Chevrolat) exemplifies the best of much needed biological observations to elucidate the fascinating natural history of these beetles. Schmid (1970) and Rice (1969) reported in considerable detail the natural history of *E. sphegeus* (Fabricius) and *E. barri* Knull, respectively.

Placopterus Wolcott 1910. Type species: *Poecilochroa plumbea* Gorham 1886. Three species widely distributed east of the Rocky Mountains. Habitus (Fig. 59). Antenna (Fig. 69). Notes: Specimens of this genus are most easily confused with those of *Enoclerus* from which they differ by the campaniform shape of the pronotum and generally less convex and less spheroid forebody (compare Figs. 59, 70). Further, the pronotal color of North American placopterans is quite distinctive; the pronotal sides are yellow whereas the central portion of the disc is vertically black (Fig. 120). These clerids have been associated with bark beetles and borers in branches of deciduous trees. Foster and Barr (1972) report adult emergence from the wasp cells of *Pemphredon lethifer* Wagner and *Trypoxylon*. Knull (1951) indicates emergence from red maple, *Acer rubrum* (Linnaeus), infested with *Melasis pectinicornis* Melsheimer [Eucnemidae].

Poecilochroa Chevrolat 1876
Ploeoptervus auctorum
Ploeopterus auctorum

Epiphloeinae Kuwert 1893
(Opitz 1997)

Madoniella Pic 1935. Type species: *Madoniella minor* Pic, 1935. Until recently (Opitz 1997) these species have been under the long standing nominal genus *Phlogistosternus*. Three species from east of the Rocky Mountains and in particular from the more southern midwestern states. Habitus (Fig. 75). Antenna (Fig. 95). Notes: These are dark, short-rectangulate clerids (4.0 mm) may be distinguished from specimens of the other two epiphloeine genera by structural and pubescence differences of the funicular articles. In *Madoniella* specimens the funicular articles are not significantly widened and they show very few hairs. Further, the antennal club antennomeres are very short compared to the composite length of the funicular articles. In specimens of *Ichnea* and *Pyticeroides* the body form is long-rectangulate, the funicular articles are densely setose and widened, and the club antennomeres are as long or longer than the composite length of the funicular articles. Elytral color is diagnostic as well; the elytral disc is predominantly dark-brown with a subbasal and subapical pale fascia that is contiguous along the sutural margin. *Madoniella dislocatus* (Spinola) is commonly found in the burrows of many wood inhabiting insects. These beetles seem to prey most predominantly on bark beetles associated with oak, juniper, cerdar, hickory, pine, hackberry, black spruce, butternut, redbud, and other hardwoods. For a thorough discourse about the plant and prey species associated with North American *Madoniella* see Blackman and Stage (1918), Bøving and Champlain (1920), and Knull (1951).

Phlogistosternus Wolcot 1944 (replacement name)
Phyllobaenus Spinola 1844

Pyticeroides Kuwert 1894. Type species: *Pyticeroides arrogans* Kuwert (1894). One species from eastern and central North America. Habitus (Fig. 1). Antenna (Fig. 1). Notes: From the other epiphloeine genus, *Ichnea*, with a long-rectangulate body form, *Pyticeroides* specimens can be distinguished by the subquadrate form of the pronotum (compare pronota in Figs. 1, 74). Adults of *P. laticornis* (Say) can be found running over bark beetle infested twigs of juniper, hickory, red cedar, cypress, mulberry, hackberry, black cherry, and elm. Bøving and Champlain (1920) and Knull (1951) provided a detail list of plants and insects with which *P. laticornis* is associated. The largely neotropical genus *Pyticeroides* is currently under revisionary study (Opitz, in manuscript).

Ichnea Laporte 1836. Type species: *Ichnea lycoides* Laporte 1836. One species from Arizona. Habitus (Fig. 74). Antenna (Fig. 94). Notes: This narrowly wedge-shaped clerid is easily identified by its entirely black coloration (in some specimens from near or south of Texas the lateral aspects of the pronotum may be faintly testaceous).

Ellipotoma Wolcott 1910, not Spinola 1844
Neichnea Wolcott and Chapin 1918 (replacement name)

Enopliinae Gistel 1856

Chariessa Perty 1830. Type species: *Chariessa ramicornis* Perty 1830. Four species widely distributed in North America. Habitus (Fig. 80). Antenna (male Fig. 88; female Fig. 89). Notes: Members of this genus are large (about 10.0-15.0 mm), oval in body form, and attractive in coloration. They are easily distinguished from members of *Pelonium,* the only other enopliine clerids in which the terminal labial and maxillary palpomeres are strongly triangular, by characteristics of the eyes, elytral punctations, and elytral disc interstitial spaces. In *Chariessa* specimens the eyes are finely granulated, the elytral punctations very small and the interstitial spaces are gritty. In *Pelonium* specimens the eyes are coarsely granulated, the elytral punctures are large, and the interstitial spaces of the elytral disc are polished. According to Knull (1951) members of *C. pilosa* (Forster) are recorded from infested oak, hickory, walnut, pear, sassafras, linden, grape, birch, chestnut, elm, and coniferous trees where it is predaceous on wood-boring larvae. These beetles are known to feed on *Scolytus multistriatus* (Marsham) [Curculionidae], *Xylotrechus* [Cerambycidae], and *Magdalis* [Curculionidae].

Brachymorphus Chelvrolat 1835

Pelonium Spinola 1844. Type species: *Pelonium lampyroides* Spinola 1844. Four species widespread in the United States. Habitus (Fig. 84). Antenna (male Fig 92; Female Fig. 93). Notes: These clerids are large (about 10 mm) and rectangulate, with an integumental coloration reminiscent of mossy dead wood often seen on live trees along the banks of rivers and flood plains; a niche from which they may be collected by the beating sheet technique. Bøving and Champlain (1920) report that specimens of *P. leucophaeum* Klug have been reared from sections of small trees and branches of cypress (*Taxodium*) containing the immature stages of long horned wood borers.

Cregya LeConte 1861. Type species: *Clerus oculatus* Say 1835. Three species from eastern, central, and southern United States. Habitus (Fig. 83). Antenna (Fig. 97). Notes: These short-rectangulate, small–sized clerids (5 mm) are the only members of the *Enopliinae* that have the antenna comprised of 10 antennomeres and the pronotal side margins conspicuously expanded at the middle. Adults of *C. oculata* (Say) have been collected from pitch pine infested with scolytine weevils and *Pogonocherus* [Cerambycidae], oak and sumac infested with *Leiopus* [Cerambycidae], and chestnut infested with *Euderces picipes* (Fabricius) and *Ecyrus dasycerus* (Say) [Cerambycidae] (Knull 1951). Larvae of the same species were taken from the pitch pine attacked by *Pityophthorus* [Curculionidae] and *Pogonocherus*. The cells of the latter genus were used by these clerids as pupal chambers which they line with frothy material and seal with cemented frass.

Boschella Barr 1980. Type species: *Enoplium fasciatum* LeConte 1852. One species from the western coast of the United States. Habitus (Fig. 82). Antenna (male Fig. 118; female Fig. 106). Among the enopliines only the members (about 7 mm) of this genus have two spurs at the apex of the metatibiae. Linsley and MacSwain (1946) reported a larva of *B. fasciatum* reared from a bee cell of *Hoplitis productus* (Cresson). These beetles are also associated with plants *Umbellularia californica* (Hook. and Arn.) Nutt. and a willow (*Salix*).

Parapelonides Barr 1980. Type species: *Enoplium nigrescens* Schaeffer 1904. One species from southcentral United States. Habitus (Fig. 98). Antenna (male Fig. 107; female Fig. 96). These are small (4 mm) clerids superficially similar to specimens to *Cregya* from which they most conspicuously differ by the uniformly curvate side margins of the pronotum (compare Figs. 83, 98).

Pelonides Kuwert 1894. Type species: *Enoplium quadripunctatum* Say 1823. Three species from central and southwestern United States. Habitus (Fig. 81). Antenna (male Fig. 91; female Fig. 90). Notes: These subovate colorful little clerids most commonly show a red dorsal facies with various combination of black spots and blotches. They range from about 4.0 to 10.0 mm. They have been collected primarily on flowers with records from flowers of *Crataegus* and *Prosopis juliflora* (Swartz) DC.

Neorthopleura Barr 1976. Type species: *Enoplium murina* Klug 1842. Three species from eastern, central, and southwestern United States. Habitus (Fig. 85). Antenna (male Fig. 104; female Fig. 105). Notes: These somewhat cylindrical enlongate clerids vary greatly in size. They range from 3.0 mm to about 12.0 mm, with males falling into the lower end of the scale. Typically, they have a dark, red pronotum and dark brown elytra; some specimens show a narrow, faintly visible, pale fascia at the middle of the elytral disc. These are the most likely clerids to emerge from a firewood pile of assorted hardwoods. They are often observed to be quite active at night scurrying on trees infested by various wood borers. They have been associated with dead oak and chestnut infested with *Chrysobothris* [Buprestidae]; chestnut infested with *Euderces picipes* (Fabricius) and *Ecyrus dasycerus* (Say) [both Cerambycidae], oak infested with *Neoclytus confusus* Van Dyke, *Tilliclytus geminatus* (Haldeman), and *Phymatodes aereus* (Newnan) [all Cerambycidae]; ash infested with *Obrium* [Cerambycidae]; and hickory, grape, locust, and persimmon infested with an assortment of lignicolous borers. They are known to be associated with weevils in recently felled and dying hardwoods.

Orthopleura auctorum

Tarsostenninae Jacquelin du Val 1861

Paratillus Gorham 1876. Type species: *Clerus carus* Newman 1840. One cosmopolitan species. Habitus (Fig. 103). Antenna (Fig. 111). Notes: These small (4.0 mm) narrow-rectangulate clerids are readily identified by the extraordinary length of their forebody. Also, the pale midelytral blister-like fascia is devoid of asetiferous punctations. The combined length of the head and pronotum is more than half the length of the elytra (Fig. 103). These attractive little beetles are most often encountered by quarantine station entomologists who find them in wood products infested with lyctids and anobiids.

Tarsostenus Spinola 1844. Type species; *Clerus univittatus* Rossi 1792. One cosmopolitan species. Habitus (Fig. 101). Antenna (Fig. 112). Notes: These very small (2.0-3.0 mm), narrow-rectangulate insects are known as ferocious predators of powder post beetles of the genera *Lyctus* and *Xylobius*. It is quite often intercepted by quarantine inspection of lumber products.

Korynetinae Laporte 1836: 34

Lebasiella Spinola 1844. Type species: *Lebasiella erythrodera* Spinola 1844. Two species east of the Rocky Mountains of the United States. Habitus (Fig. 99). Antenna (Fig. 110). Notes: This small (4.0 mm), somewhat oval checkered beetle, unlike any other member of the korynetines, has very polished and glossy elytra; mainly a consequence of the smooth surface of the elytral interstitial spaces. Specimens have been collected by beating oak branches.

Loedelia Lucas 1920. Type species: *Necrobiodes mexicanus* Gahan 1910. Two species from southwestern United States. Habitus (Fig. 100). Antenna (Fig. 109). Notes: This genus is distinguished from superficially similar specimens of *Necrobia* by the triangular shape of the middle antennomere of the antennal club. Specimens have been collected on various hardwoods including *Quercus*, *Ceanotus*, and *Prosopis*.

Necrobioides Gahan 1919, not Faimaire 1882

Necrobia Olivier 1795: 76. Type species: *Dermestes violaceus* Linnaeus 1758. Three cosmopolitan species. Habitus (Fig. 102). Antenna (Fig. 108). Notes: These beetles are about 5.0 mm long and have an ovoid body form. The ninth and tenth antennomeres are distinctly narrowly transverse. In specimens of *Lebasiella* and *Leodelia* these antennomeres are more trigonal. Entirely, or predominantly, blue-black checkered beetles found in the vicinity of carrion will probably belong to this genus. Members of this genus are saprophagous and predacious.

Bibliography

BARR, W. F. and D. E. FOSTER. 1979. Revision of the genus *Aulicus* (Coleoptera: Cleridae). Melanderia, 33: 1-31.

BLACKMAN, M. W. and H. H. STAGE. 1918. Notes on insects bred from the bark and wood of the American larch. Technical Publication No. 10, New York State College of Forestry at State University, 18: 95.

BØVING, A. G. and A. B. CHAMPLAIN. 1920. Larvae of North American beetles of the family Cleridae. Proceedings of the United States National Museum, 57: 575-649.

CORPORAAL, J. B. 1939. Revision of the Thaneroclerinae (Cleridae, Col.). Bijdragen tot de Dierkunde, 27: 347-359, pls. 9-11.

CORPORAAL, J. B. 1950. Cleridae. *In:* W. D. Hinks, ed., Coleopterorum Catalogus Supplemeta, pars 23 (editio secunda). W. Junk, 's-Gravenhage, 373 pp.

CROWSON, R. A. 1964. A review of the classification of Cleroidea (Coleoptera), with descriptions of two new genera of Peltidae and of several new larval types. Transactions of the Royal Entomological Society of London, 116: 275-327.

EKIS, G. 1977. Classification, phylogeny, and zoogeography of the genus *Perylipus* (Coleoptera; Cleridae). Smithsonian Contribution to Zoology, Number 227: 138 pp.

EKIS, G. and A. P. GUPTA. 1971. Digestive system of Cleridae (Coleoptera). International Journal of Morphology and Embryology, 1: 51-86.

FOSTER, D. E. 1971. Observations on the biologies of *Enoclerus abdominalis* (Chevrolat) and *Enoclerus opifex* (Gorham) (Coleoptera: Cleridae). Coleopterists Bulletin, 25: 127-129.

FOSTER, D. E. 1976a. North American Thaneroclerinae larvae (Coleoptera: Cleridae). Coleopterists Bulletin, 30: 75-80.

FOSTER, D. E. 1976b. A review of North American Tillini larvae (Coleoptera: Cleridae) University of Idaho, Department of Entomology Anniversary Publication, 19: 133-138.

FOSTER, D. E. 1976c. Revision of North American *Trichodes* (Herbst) (Coleoptera: Cleridae). Special Publications of the Museum of Texas Tech University, 11: 86 pp.

FOSTER, D. E. and W. F. BARR. 1972. Notes on the distribution and bionomics of some North American Cleridae. Journal of the Kansas Entomological Society, 45: 122-125.

KNULL, J. N. 1951. The checkered beetles of Ohio (Coleoptera: Cleridae). Ohio Biological Survey Bulletin, 8(42): 268-350.

KOLIBAC, J. 1992. Revision of Thanerocleridae n. stat. (Coleoptera: Cleroidea). Mitteilungen der Schweizerischen Entomologischen Gesellschaft, 65: 303-340.

KOLIBAC, J. 1997. Classification of the subfamilies of Cleridae (Coleoptera, Cleroidea). Acta Musei Moraviae Scientiae Naturales, 81: 307-361.

KOLIBAC, J. 1998. Classification of the subfamily Hydnocerinae Spinola, 1844 (Coleoptera: Cleridae). Acta Musei Moraviae Scientiae Biologicae, 83: 127-210.

LINSEY, E. G. 1933. A new California clerid beetle. Pan-Pacific Entomologist, 9: 95.

LINSEY, E. G. 1936. Studies in the genus *Aulicus*. University of California Publication of Entomology, 6: 249-262.

LINSEY, E. G. and J. W. MACSWAIN. 1946. Longevity of *Trichodes* and *Pelonium* larvae. Pan-Pacific Entomologist, 22: 18.

MACSWAIN, J.W. 1945. Notes on the habits of the predator *Cymatodera ovipennis* Say with description of the pupa (Coleoptera, Cleridae). Pan-Pacific Entomologist, 21: 97-100.

MAYER, D., C. JOHANSEN and H. DAVIS. 1976. Checkered flower beetle: A destructive predator of the alfalfa leafcutting bee. Cooperative Extension Service, College of Agriculture, Washington State University, Pullman. EM 4107, 2 pp.

OPITZ, W. 1997. Classification, natural history, and evolution of the Epiphloeinae (Coleoptera: Cleridae). Part I. The genera of Epiphloeinae. Insecta Mundi, 11: 51-96.

REEVE, J. D. 2000. Complex emergence patterns in a bark beetle predator. Agricultural and Forest Entomology, 2: 233-240.

RICE, R. E. 1969. Bionomics of *Enoclerus barri* (Coleoptera: Cleridae). Canadian Entomologist, 101: 382-386

SABROSKY, C. W. 1934. Notes on the larvae and larval habit of *Isohydnocera curtipennis* (Newman) (Coleoptera, Cleridae). Journal of the Kansas Entomological Society, 7: 65-68.

SCHMID, J. M. 1970. *Enoclerus sphegeus* (Coleoptera; Cleridae), a predator of *Dendroctonus ponderosae* (Coleoptera; Scolytidae) in the Black Hills. Canadian Entomologist, 102: 969-977.

74. MELYRIDAE Leach 1815

by Adriean J. Mayor

Family common name: The soft-winged flower beetles

Family synonyms: Attalomimidae Majer 1994; Carphuridae Champion 1923; Dasytidae Laporte 1840; Gietellidae Constantin and Menier 1987; Malachiidae Leach 1817; Mauroniscidae Majer 1994; Rhadalidae LeConte 1862.

The Melyridae can be distinguished from other Cleroidea by the presence of a distinct clypeus; the conical, prominent, nearly contiguous front coxae, which are open behind, and have distinctly exposed trochantins; and by confused elytral punctation which is not organized into striae or costae (except some *Melyris*).

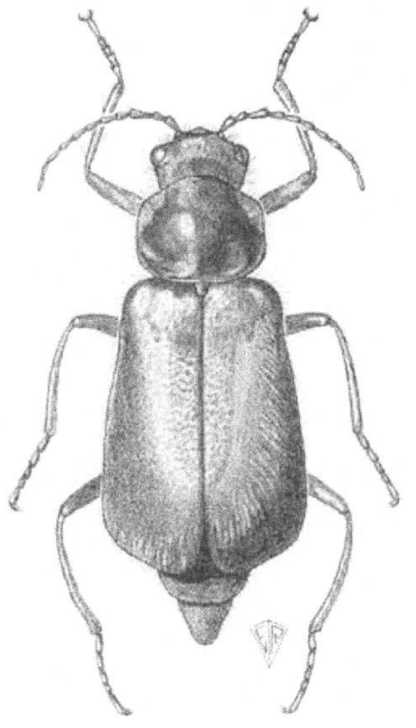

FIGURE 1.74. *Malachius aeneus* (Linnaeus)

Description: Form variable, graciliform to robust, lateral profile somewhat flattened to strongly convex; size <1.0 to 20 mm (most are 2 to 10 mm) in length; often brightly colored with red and blue; most with vestiture moderate, often dual with both decumbent and erect setae.

Head (Stickney 1923) large, nearly as broad as prothorax, somewhat deflexed, short to elongate; surface finely to coarsely punctate. Antennae 9- (*Ablechrus* in North America), 10- (*Cerallus*, not in North America) or 11-segmented; apparently 10-segmented in some (*Collops* and allied genera), antennomere II very small, hidden in distal end of I; antennomeres variable, serrate, pectinate, or filiform, in some with terminal antennomeres enlarged to form a loose club; inserted on front above base of mandibles. Clypeus distinct, trapezoidal, transverse and band-like or large and membranous (Malachiinae); epistomal suture present; labrum distinct, anteriorly arcuate; most with mandibles moderate in size, curved, apices acute or bifid; maxillary palpi (Williams 1938) 4-segmented, palpomeres slender, terminal palpomere conical in most, less commonly triangular (Rhadalinae); mentum small, quadrate; ligula slender. Eyes lateral, moderate, bulging, round.

Pronotum variable in shape, typically quadrate, less commonly elongate, somewhat broader than head, more or less explanate, surface often punctate, pleural region narrow, prosternum short, procoxal cavities open behind; pronotum often with lateral eversible vesicles (Malachiinae). Mesosternum short, transverse, intercoxal process acuminate. Metasternum broad, often with lateral eversible vesicles (Malachiinae). Legs with trochantin exposed on fore legs; anterior coxae conical, prominent, nearly contiguous; middle coxae conical, prominent, nearly contiguous; hind coxae transverse, contiguous; trochanters moderate, triangular; femora swollen or slender; tibiae slender, apical spurs obscure; tarsal formula 5-5-5, rarely 4-4-4 (*Anthriboclerus* in the Rhadalinae and *Attalomimus*, in the Malachiinae, not in North America), some males with 4-5-5 (*Troglops*, *Collops*, *Trophimus* and *Temnopsophus* in North America); tarsomeres short, not lobed (except Gietellinae, and some Malachiinae, not in North America); claws simple, toothed, or with ungual appendages beneath. Scutellum small, oval, or triangular. Elytra of most species entire, typically elongate, apically rounded, some are truncate, exposing three or more abdominal tergites (some Malachiinae in North America); most with surface rugose punctate; epipleural fold indistinct; epipleura entire. Wing venation with anal region degenerate; wedge-cell open (Forbes 1922); folding pattern with anal lobe free, or absent; area D well developed, not chitinized to nearly chitinized; area H reaching inner margin at a point, or not at all; area B well marked, but varying from not a principal fold to large, dominating folding of wing (Forbes 1926).

Abdomen with five, six or seven (Carphurini) visible sternites, occasionally with lateral eversible vesicles (Carphurini); sutures entire or first two abdominal sternites connate (Rhadalinae); surface micro-rugose or smooth. Male genitalia with penis tubular and slightly curved; parameres fused to pars basalis, apically slightly emarginate; pars basalis forming a ring-piece (Sharp and Muir 1912). Female genitalia with paraprocts reduced to a baculum; valvifers reduced to a long baculum which articulates with a short baculum of the coxite; coxite vaguely bipartate; stylus small, apical (Tanner 1927).

Larvae elongate, subcylindrical or somewhat depressed, laterally subparallel except head and caudal segments which are narrower, size <1 to 20 mm in length, most are 5 mm or less; vestiture composed of few to many short setae with some long setae on various segments; color pale. Head large, two-thirds width of thorax, exerted, prognathous, depressed, quadrate with a distinct Y-shaped epicranial suture; with one to five stemmata on each side. Clypeus transverse, labrum distinct; mandibles robust, with a modified prostheca, no molar areas, apices bidentate; maxillae short, cardo transverse, stipes distinct, with 3- palpomeres, galea lobe-like, setiferous, lacinia small; labium with a distinct

Acknowledgments: I thank Michael Thomas for suggesting this project, Paul Skelley for preparing the illustrations, and Jonathan Mawdsley and Doug LeDoux for helpful comments on the manuscript.

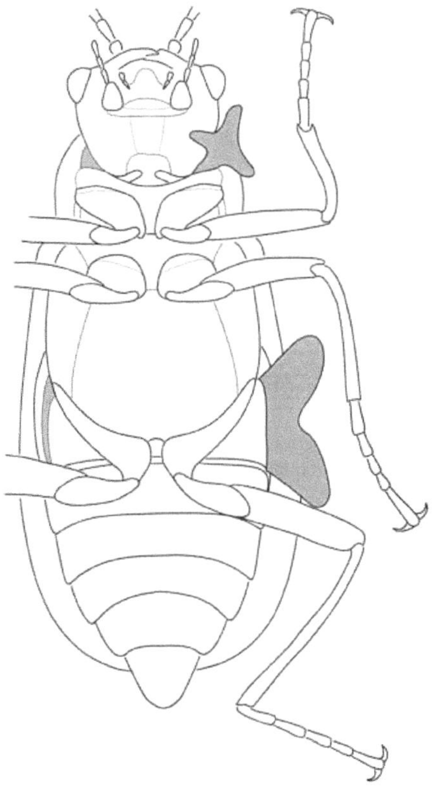

FIGURE 2.74. *Malachius* sp., eversible vesicles exposed, ventral view, female.

submentum and mentum, ligula small, with 2 palpomeres. Thoracic segments subequal; legs 4-segmented with apical, claw-like tarsunguli. Abdomen nine-segmented. Spiracles annular, inconspicuous, located on mesothorax and abdominal segments one to eight. Urogomphi caudally projecting, dorsally curved.

Habits and Habitats. Most adult Melyridae are, at least potentially, polyphagous, feeding on both plant and animal material. Many feed preferentially on pollen of various cone bearing and/or flowering plants, gathering in rather large, inconspicuous aggregations. Others, particularly members of the subfamily Malachiinae, are omnivorous scavengers and/or predators, feeding primarily on small arthropods, pollen and nectar. Fall (1901) provides notes on the habits and distribution of species known from southern California. Moore (1937) provides information on phenology and host plant associations for adults of all species known from San Diego County, California. The larvae are poorly known. Larval treatments are primarily descriptive (Urban 1912, 1914; Bøving and Craighead 1931; Balduf 1935; Schmidt 1944, 1945-1948; Peterson 1951; Moore 1956; Fiori 1959; Evers 1960; Medvedev and Galatz 1969; Foster and Antonelli 1973; Moore and Legner 1977). However, these references suggest that larvae are scavengers or predators, feeding primarily on detritus, fungi and small arthropods. Dix (1990) found larvae of *Malachius ulkei* Horn 1872 under bark of Siberian elm, feeding on eggs of spring cankerworm, *Paleacrita vernata* Peck. According to White (1983) larvae occur in leaf litter or under bark of dead trees, feeding on bark beetles, their larvae or eggs.

The presence of *Collops* in agricultural environments and the potential of its species as predators of agricultural pests has long been recognized (Dahms and Kagan 1938; Knowlton and Meier 1942; Knowlton 1944; Nielson and Henderson 1959). More recent studies suggest that various species of *Collops* are important predators in alfalfa, sorghum and cotton agroecosystems (Orphanides *et. al.* 1971; Rakickas and Watson 1974; Hussain 1975; and Lopez and Teetles 1976). The biologies of *Collops vittatus* (Say 1823) (Walker 1957; Nielson and Henderson 1959), *Collops balteatus* LeConte 1852 (Walker 1957) and *Collops georgianus* Fall 1910 (King 1985) have been studied and described in detail.

Status of the classification. In spite of its large size, the Melyridae have, until recently, been poorly studied particularly with respect to higher classification. At the beginning of the century and before, taxonomic confusion was caused by the descriptive efforts of several prolific writers. The first author to create some order out of this confusion was Crowson (1955) who, in his major work on the higher classification of Coleoptera and in subsequent works (Crowson 1964, 1970), defined the superfamily Cleroidea, and defined and provided keys to the subfamilies of Melyridae, the largest included family. Crowson's classification was based largely on characters of larval anatomy. He considered the Melyridae to be divisible into five subfamilies: Rhadalinae (Haplocneminae), Melyrinae, Malachiinae, Prionocerinae, and Dasytinae. These subfamilies have all, at one time or another, been given family rank (see Greiner 1937, Pic 1926, 1927, 1937) and are considered families by many recent authors (e.g., Constantin 1965, 1983; Liberti 1982, 1984; Wittmer 1995, 1999; Majer 1994a, 1996 and 1997).

Arnett (1962) included the Phloiophilinae in the family Melyridae. Crowson (1970) raised the Phloiophilinae to the rank of family where it remains today. Peacock (1987) reviewed the subfamily Rhadalinae, described new taxa, and provided a list, keys, descriptions and synonymy of the included genera. Majer (1987, 1989a, 1989b, 1990) revised the taxonomy of the family, described new taxa, divided the family into subfamilies, supertribes and tribes and discussed in detail the supposed phylogeny of the group. Constantin and Menier (1987) described the subfamily Gietellinae for the remarkable apterous species, *Gietella fortunata* Constantin and Menier 1987, found on the Canary Islands. Most recently, Majer (1994a, 1994b, 1995) has proposed that the "melyrid lineage" is comprised of eight families: Acanthocnemidae, Gietellidae, Mauroniscidae, Prionoceridae, Dasytidae, Melyridae, Malachiidae, and Attalomimidae.

Majer (1990) described a new tribe Listrini in the subfamily Dasytinae, including in the tribe 21 North American genera (see classification below), the Afrotropical genus *Calosotis* Redtenbacher 1867, the Palaearctic genus *Danacaeina* Reitter 1887, and two new Oriental genera, *Picolistrus* Majer 1990 and *Sinolistrus* Majer 1990. The species of *Listrus* Motschulsky 1859 were listed in the Coleopterorum Catalogus by Pic (1937) under the name *Amecocerus* Solier 1849 without explanation. Majer (1990) indicates that the species of *Amecocerus*, found primarily in Chile, are distinct and not related to the North American species of *Listrus* or other genera included by him in the tribe Listrini (Majer 1990) or the subfamily Listrinae (Majer 1994a).

Majer (1994a) described a new family, the Mauroniscidae, for the genera *Mauroniscus* Bourgeois 1911 from South America and *Mecomycter* Horn 1882 from North America. These taxa had previously been included in the melyrid subfamily Dasytinae. In the same paper Majer listed the genus *Hoppingiana* Blaisdell 1924 and the North American species of *Dasytes* Paykull 1798 as belonging to the subfamily Danaceinae Thompson 1859 of the family Dasytidae. The family Mauroniscidae was revised by Majer (1995), and in this revision he described the taxa *Amecomycter* Majer 1995 with species from South America, *Mectemycor* Majer 1995 with species from California, and *Scuromanius* Majer 1995 with species from South America, Mexico, and the United States. These changes have not been critically evaluated, and in this work the North American genera and species treated by Majer in the family Mauroniscidae and the subfamily Danaceinae are listed as members of the subfamily Dasytinae.

The genus *Dasyrhadus* Fall 1910 was transferred from the Rhadalinae to the subfamily Dasytinae by Crowson (1964) and this arrangement was followed by Peacock (1987). However, *Dasyrhadus* shares some key characters (setose eyes and triangular shape of apical maxillary palpomeres) with other Rhadaline genera and is considered by Majer (1994a) to belong in the subfamily Rhadalinae where it is treated here.

In reviewing the literature treating the taxonomy of the Melyridae of North America north of Mexico, all relevant species catalogs (Leng 1920; Leng and Mutchler 1927; 1933; Greiner, 1937; Pic 1926; 1929; 1937), all taxonomic studies published subsequent to the catalogs, and all papers by Blaisdell on the Melyridae are cited in this chapter. The Melyridae from Mexico and Central America are cataloged by Blackwelder (1945; 1957). Fossil Melyridae, from the Miocene fossil beds of Florissant, Colorado, described and cataloged by Wickham (1912, 1913, 1914, 1916, 1917, 1920) and Mawdsley (1999) are not included in this chapter.

The characters used to define genera, particularly in the subfamily Dasytinae, have not been critically evaluated and some, such as differences in vestiture, the degree of development of the elytral epipleura, the presence or absence and even degree of development of the ungual appendages, have been used to define genera. One result is that 12 of the 35 North American genera of Dasytinae are monotypic. A critical evaluation of this subfamilies classification is expected to result in a consolidation of genera, and eventually a much more stable classification. A character of the first abdominal sternite: raised into a keel between hind coxae in members of the subfamilies Dasytinae, Rhadalinae, and Melyrinae; or without a keel between hind coxae in members of the subfamily Malachiinae has been used by previous authors (Crowson 1955; Arnett 1962) to differentiate these subfamilies. This character has been a source of much confusion over the years, and because specimens must be dissected to properly see this character it is not used in the present work. The presence and number of spines (spinules of Blaisdell 1938a and Arnett 1962) on the protibiae in some Listrini (Figs. 15, 20, and 21) is apparently influenced by sex, and in some genera (*Trichochrous Emmenotarsus* and *Listropsis*) females have significantly more protibial spines than males. When present, the spines which are located on the external (dorsal) surface are most easily seen in profile, looking at the anterior or posterior face of the protibia. See Howell (1987) for a detailed discussion of protibial spines. There is at least one character, the shape of the tibial spurs on the pro- and mesotibiae, mentioned by both Casey (1895) and Blaisdell (1938a), but not used in defining genera. This character has several states: 1) pro- and mesotibial spurs absent or obscure in both sexes; 2) pro- and mesotibiae with two acute spurs in both sexes, and 3) pro- and mesotibiae with one acute spur and one broad, spoon shaped spur in males, two acute spurs in females. Using this character and others, some suspected relationships are alluded to in the section on "Classification of the North American Genera". The eversible vesicles (Fig. 2) present in all members of the subfamily Malachiinae are typically retracted into the body, and not everted as shown in figure 2. Though difficult to see, the presence of these vesicles is always marked by creases and/or folds of the integument.

Distribution. The family Melyridae, the largest in the superfamily Cleroidea, is a moderately large family with about 300 genera found worldwide. As Crowson (1955) points out, members of the family Melyridae are easily recognized from general facies, but are not easily defined using rigorous criteria. The group is morphologically diverse, its species partitioned among the subfamilies Dasytinae, Gietellinae (not in North America), Malachiinae, Melyrinae, and Rhadalinae. The Malachiinae is the largest, most diverse and probably the most advanced of the subfamilies. Lawrence (1982) reported that the family Melyridae contained about 200 genera and 5,000 species. Based on a tabulation of genera and species described since 1982, the Melyridae (including Attalomimidae, Carphuridae, Dasytidae, Gietellidae, Malachiidae, Mauroniscidae, and Rhadalidae) contains more than 300 genera and over 6,000 species. The family occurs throughout the world, but is particularly abundant and diverse in the dry temperate regions. In North America north of Mexico, 520 species are grouped into 58 genera which in turn comprise four subfamilies. In the Classification of the North American Genera (see below), North America refers to the continent, which extends from Panama north through Canada. The status and generic assignments of the species of Melyridae are poorly catalogued, and all of the genera treated here are in need of revision.

Key to the Genera of America North of Mexico

Many of the characters used in the key are sex linked, and in most cases identifications will be easier to make using males. Some of the genera of Melyridae may be difficult to identify using the key. As an aid to identification, a short diagnosis of most genera is included along with remarks on distribution and biology. In the following: the key to the genera of the subfamily Dasytinae is modified from Blaisdell (1938a); the key to the genera of the subfamily Rhadalinae is modified from Peacock (1987); and the key to the genera of the subfamily Malachiinae is modified from Wittmer (1962). Figures 3-14, and 18-19 are redrawn from Blaisdell (1938a) and, with figures 15-17 and 20-22 should help with identification of genera in the subfamily Dasytinae.

1. Body with lateral eversible vesicles (Fig. 2); number and arrangement of vesicles variable, but always present below anterior pronotal angles (Malachiinae) 40
— Body without eversible vesicles 2

2(1). Apical maxillary palpomere broadened, securiform or triangular (Rhadalinae) 37
— Apical maxillary palpomere not securiform or triangular, more or less conical 3

3(2). Tarsal claws with ungual appendages (as in Figs. 3-6); tarsomere I as long as or longer than II on all legs; (Dasytinae, in part) 6
— Tarsal claws without ungual appendages; tarsomere I variable 4

4(3). Tarsomere I slightly shorter than II on all legs (Melyrinae) 5
— Tarsomere I as long as or longer than II on all legs (Dasytinae, in part) 34

5(4). Size small, < 3 mm; brown to black, occasionally elytra with pale red basal and preapical spots .... *Melyrodes*
— Size larger, > 7 mm; reddish-orange, with head and elytra dark metallic blue (intercepted at North American ports) .... *Melyris*

6(3). Elytral epipleurae more or less distinctly defined at least toward base (as in Figs. 7-8); body form variable, most not elongate and sub-cylindrical ... 7
— Elytral epipleurae inflexed basally and apically, visible externally only in median one-third (Fig. 9); most with body form elongate and subcylindrical .... *Dolichosoma*

7(6). Male head and mandibles of moderate size or smaller 8
— Male head large; mandibles stout and longer in male .... *Pristoscelis*

8(7). Hind angles of pronotum produced (as in Figs. 10-13); body with pubescence decumbent, often flattened 9
— Hind angles of pronotum not produced (as in Fig. 14); body with pubescence variable 15

9(8). Pronotum with distinct, excavated lateral submarginal lines (as in Figs. 10-13) 11
— Pronotum lacking distinct excavated lateral submarginal lines, or if present, obscure and restricted to basal half of pronotum 10

10(9). Pronotum lacking distinct excavated lateral submarginal lines; pygidium truncate .... *Listromimus*
— Pronotum with excavated lateral submarginal lines obscure, restricted to basal half of pronotum; pygidium with a distinct V-shaped emargination .... *Listrimorpha*

11(9). Body graciliforn to cuneiform, elytra with lateral margins diverging to about apical one-third then converging to apical angles 12
— Body more robust, elytra with lateral margins parallel to at least apical one-third 14

12(11). Tarsal claws with ungual appendages asymmetrical, inner (anterior) appendage long, attached to claw throughout its length, outer (posterior) appendage rudimentary or absent (as in Figs. 5-6) 13
— Tarsal claws with ungual appendages symmetrical, as long as claws (as in Fig. 3) .... *Vecturoides*

13(12). Elytral epipleurae broad, horizontal, wide to near elytral apex .... *Vectura*
— Elytral epipleurae narrow, wider toward base, obsolete beyond middle of second abdominal sternite (as in Fig. 8) .... *Leptovectura*

14(11). Elytral epipleurae broad, horizontal, wide to near apex (Fig. 7) .... *Enallonyx*
— Elytral epipleurae narrow, wider toward base, obsolete beyond first abdominal sternite (Fig. 8) .... *Amphivectura*

15(8). Pronotum with distinct, excavated lateral submarginal lines (occasionally obscured by pronotal vestiture), or with extremely dense, rugose punctures laterally 16
— Pronotum lacking excavated submarginal lines, and without dense, rugose punctures laterally 19

16(15). Pronotum more or less constricted at sides behind the apex, lateral margins sinuate (Fig. 4) .... *Eschatocrepis*
— Pronotum not constricted at sides behind apex, lateral margins evenly rounded, not sinuate 17

17(16). Eyes setose; pronotum with extremely dense, rugose punctures laterally .... *Hoppingiana*
— Eyes glabrous; pronotum variable most lack dense, rugose puntures laterally 18

18(17). Tarsal claws with ungual appendages fully as long as claws (as in Fig. 3) .... *Dasytellus* and *Paradasytes*
— Tarsal claws with ungual appendages short, not as long as claws (Fig. 4) .... *Dasytes*

19(15). Tarsal claws with ungual appendages very short and rudimentary .. *Dasytastes* and *Neadasytes*
— Tarsal claws with ungual appendages well developed, half as long as claws or longer 20

20(19). Female protibiae with spines on external (dorsal) margin (as in Figs. 15 and 20) 21
— Female protibiae lacking spines on external (dorsal) margin (Fig. 22), or with spines few in number (Figs. 21) 31

21(20). Elytral epipleurae broad, horizontal, wide to near elytral apex (as in Fig. 7) 22
— Elytral epipleurae narrow, wider toward base (as in Fig. 8) 23

22(21). Apical pronotal angles anteriorly prominent; color black, legs pale or black .... *Eudasytes*
— Apical pronotal angles not prominent anteriorly; elytra and abdomen more or less pale rufous .... *Asydates*

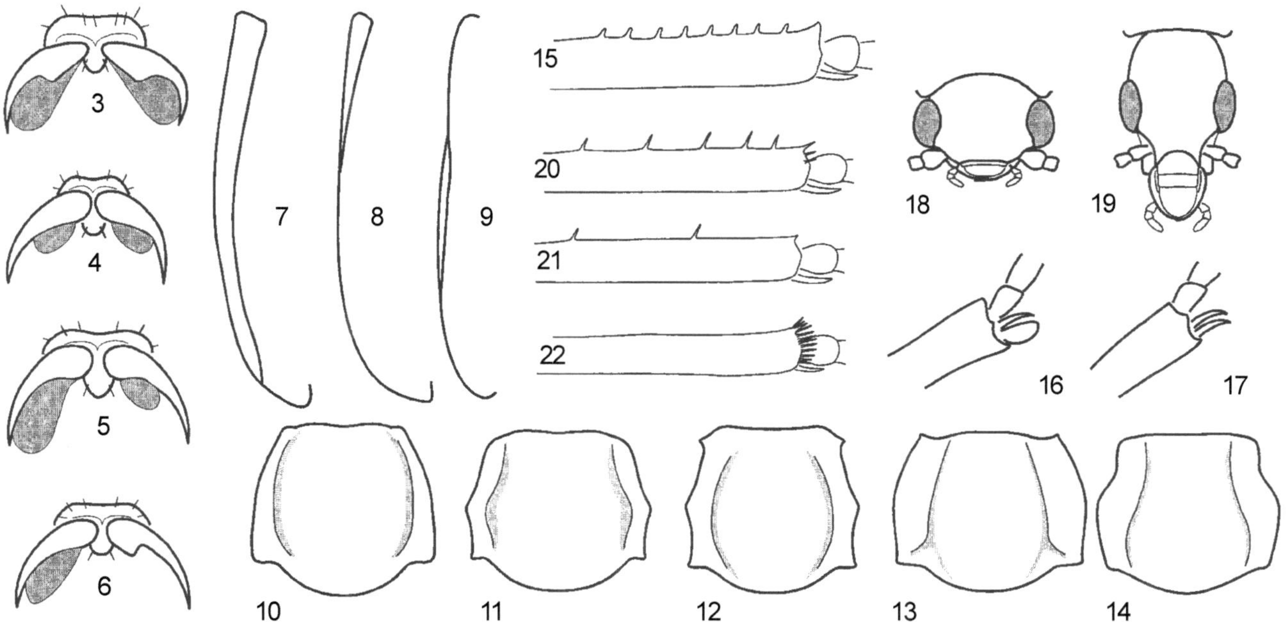

FIGURES 3.74-22.74. 3-6. Ungual appendages. 3. *Trichochrous oregonensis* (LeConte); 4. *Dasytes* sp.; 5. *Enallonyx sculptilis* (LeConte); 6. *Vectura longiceps* Casey. 7-9. Elytral epipleura. 7. *Enallonyx denudatus* (Casey); 8. *Amphivectura monticola* (Blaisdell); 9. *Dolichosoma foveicolle* (Kirby). 10-14. Pronotum, dorsal view. 10. *Vecturoides serrulata* (Blaisdell), female; 11. *Vecturoides pseudonycha* Fall; 12. *Vectura longiceps* Casey; 13. *Amphivectura monticola* (Blaisdell); 14. *Eschatocrepis constrictus* (LeConte). 15. *Byturosomus fucscus* (LeConte), protibia, lateral view, female. 16-17. Protibial spurs, lateral view, male. 16. *Eudasytes* sp.; 17. *Asydates* sp. 18-19. Head, frontal view. 18. *Hoppingiana hudsonicus* (LeConte); 19. *Mecomycter omalinus* Horn. 20-22. Protibia, lateral view, female. 20. *Trichochrous* sp.; 21. *Listropsis armatulus* Blaisdell; 22. *Listrus* sp.

23(21). Antennae short, not surpassing base of pronotum when directed posteriorly; antennomeres variable in shape, pilosity of most with short dense setae 24

— Antennae long, surpassing base of pronotum when directed posteriorly; antennomeres strongly serrate, clothed with long, sparse, flexible setae .... *Sydatopsis*

24(23). Lateral margins of pronotum not serrate 25

— Lateral margins of pronotum more or less serrate, especially anteriorly .... *Cradytes*

25(24). Pro- and mesotibiae of male with one acute spur and one broad, blunt, spoon shaped spur (as in Fig. 16) 26

— Pro- and mesotibae of both sexes with two acute spurs (as in Fig. 17) .... *Eutricholistra*

26(25). Protibiae straight, dorsal surface lacking conspicuous spines (as in Figs. 20 and 21); procoxae moderate in size, not flattened or excavated on their outer face 27

— Protibiae more or less arcuate (male), dorsal surface with numerous spines (Fig. 15); procoxae large, flattened, slightly excavated on their outer face .... *Byturosomus*

27(26). Body clothed with one kind of pubescence only, decumbent or sub-erect 28

— Body clothed with dual pubescence, decumbent and erect setae 29

28(27). Body clothed with more or less closely decumbent pubescence, without trace of intermixed erect setae; marginal pronotal fringe short, regular .... *Trichochrous*

— Body clothed with moderately long sub-erect blackish pubescence; marginal pronotal fringe rather short, composed of even, stiff black setae .... *Trichochronellus*

29(27). Body clothed with short cinereous or blackish subdecumbent pubescence, intermixed abundantly with long, erect, black or cinereous hispid setae; pronotal fringe long, erect, bristling, not regular and close-set 30

— Body clothed with dense sub-decumbent, cinereous pubescence, intermixed at least toward sides of elytra with short, sparse, erect cinereous setae which frequently seem to be serial in arrangement; pronotal fringe short, even, and close-set .... *Eutrichopleurus*

30(29). Fifth visible abdominal sternite of male truncate or sinuato-truncate at apex, disc simple, unmodified .... *Emmenotarsus*

— Fifth visible abdominal sternite of male sinuate at apex, disc with a spherical impression .... *Listropsis* (in part)

31(20). Tarsal claws with ungual appendages rounded apically 32

— Tarsal claws with ungual appendages acutely pointed apically, leaving nearly half of inner claw free .... *Sydates*

32(31). Pronotal epimeron with large rounded impression near inner edge anteriorly; pronotal margins smooth, not serrulate .... *Adasytes*
— Pronotal epimeron lacking rounded impression near inner edge anteriorly; pronotal margins serrulate or not 33

33(32). Lateral margins of pronotum minutely serrulate; protibiae lacking spines (as in Fig. 22); elytral vestiture simple, uniform in color or variegated .... *Listrus*
— Lateral margins of pronotum not serrulate; protibiae occasionally with a few spines (as in Fig. 21); elytral vestiture dual, sparse, uniform in color, not variegated .... *Listropsis* (in part)

34(4). Pubescence decumbent, short, inconspicuous; head strongly rostrate; pronotum elongate, lateral margins sinuate .... *Mectemycor*
— Pubescence dual, with both decumbent and erect setae conspicuous; head rostrate or not; pronotum variable, but if elongate, then lateral margins only slightly sinuate 35

35(34). Head rostrate (Fig. 19), pronotum elongate, without distinct lateral margins .... *Mecomycter*
— Head shorter, not rostrate; pronotum more or less transverse, lateral margins distinct 36

36(35). Pronotum with lateral margins more or less angulate medially .... *Scuromanius*
— Pronotum with lateral margins evenly rounded, not angulate medially .... *Pseudasydates*

37(2). Eyes setose 38
— Eyes glabrous .... *Semijulistus*

38(37). Elytra with sublateral carina in basal half, forming a double elytral edge 39
— Elytra lacking sublateral carina in basal half, elytra with a single edge .... *Dasyrhadus*

39(38). Body form elongate; disc of pronotum with punctures strongly rimmed or tuberculate; elytral carina close and subparallel to costal margin, forming an abrupt, narrow horizontal shelf .... *Rhadalus*
— Body form very short and broad; disc of pronotum with small simple punctures, tuberculate punctures near sides; elytral carina oblique to costal margin and less prominent, so scarcely forming a shelf .... [*Eucymbolus*]

40(1). Abdomen with 7 visible abdominal sternites; male protarsomere I with a pectinate comb of stout teeth on anterior margin (Carphurini) 41
— Abdomen with 6 visible abdominal sternites; male protarsomere I without comb of teeth on anterior margin (Malachiini) 42

41(40). Abdomen of males with visible sternites 5 and 6 excavated apically .... *Chaetocoelus*
— Abdomen of both sexes with visible sternites 5 and 6 unmodified .... [*Carphuroides*]

42(40). Antennae apparently 10- or 11-segmented 43
— Antennae 9-segmented .... *Ablechrus*

43(42). Antennae 11-segmented, second antennomere II easily visible 44
— Antennae apparently 10-segmented; antennomere II small, hidden in apex of I .... *Collops*

44(43). Protarsi of male 4-segmented 45
— Protarsi 5-segmented in both sexes 47

45(44). Head sulcate in male (introduced from Europe) .... *Troglops*
— Head simple in both sexes 46

46(45). Head rostrate; male abdominal sternites III-V modified, V sulcate .... *Trophimus*
— Head short; abdominal sternites III-V unmodified in both sexes .... *Temnopsophus*

47(44). Male protarsomere I expanded ventrally into a rounded lobe; male protibiae with 3-10 stout spicules ventro-anteriorly near the apex . *Nodopus*
— Male protarsi and protibiae not modified as described above 48

48(47). Male protarsomere II with dorsal lobe 49
— Protarsomere II simple, both sexes lacking a dorsal lobe 54

49(48). Elytra strongly abbreviated 50
— Elytra covering abdomen completely or nearly so 51

50(49). Head sulcate in male; size < 2 mm .... *Attalusinus*
— Head simple in both sexes; size larger > 2.5 mm .. .. *Endeodes*

51(49). Lobe on protarsomere II large, extending over III, most with a comb of stout black setae apically. .... *Attalus*
— Lobe on protarsomere II small, indistinct, not extending over III, most without a setal comb, occasionally elongate, but if so, arcuate and acicular in most 52

52(51). Male frons with a small knob-like projection .... *Condylops* (*Neocondylops*)
— Frons simple, lacking a knob-like projection in both sexes 53

53(52). Elytral apex of male prolonged, and with a small, cup-shaped process extending upward from the tip of each elytron; visible abdominal sternites IV and V simple in both sexes .... *Ebaeus*
— Elytral apex simple in both sexes; visible abdominal sternites IV and/or V sulcate in male .... *Tanaops* (in part)

54(48). Visible abdominal sternites IV and/or V sulcate in male; most with head flattened in front, moderately to strongly rostrate .... *Tanaops* (in part)
— Visible abdominal sternites IV and V unmodified in both sexes; head not flattened, not elongate 55

55(54). Pronotum elongate, nearly as long as or longer than wide 56
— Pronotum transverse, much wider than long..... 57

56(55). Antennae subfiliform; female wings reduced to small lobes, elytra inflated posteriorly .... *Charopus*

— Antennae subserrate to subfiliform or pectinate; female wings not reduced, elytra not inflated posteriorly .... *Microlipus* (in part)

57(55). Epistomal region of male head transversely sulcate (introduced from Europe) .... *Malachius* (*Malachius*)

— Epistomal region of head unmodified in both sexes 58

58(57). Elytral apex of male modified to form a variously shaped organ 59

— Elytral apex unmodified in both sexes 60

59(58). Elytral apex of male prolonged, and with a small cup-shaped process extending upward from the tip of each elytron .... *Hypebaeus*

— Elytral apex of male infolded horizontally, and with a variously shaped process below the plane of the elytra 61

60(58). Antennae of male subserrate to subfiliform, antennomere VII elongate 2.00 to 3.25 as long as wide .... *Microlipus* (in part)

— Antennae of male serrate to pectinate, antennomere VII shorter, 0.37 to 1.80 as long as wide .... *Malachius* (*Clanoptilus*) (in part)

61(59). Pronotum widest apically, lateral margins slightly convergent basally (intercepted at Mobile, Alabama) .... *Carapheles*

— Pronotum widest near middle or basally, lateral margins nearly parallel 62

62(61). Small, size < 2.5 mm; elytra with one or rarely two long, fine white setae near humeri at right angles to the posthumeral margin .... *Anthocomus*

— Larger, size > 3.0 mm; elytra with three to five long, fine, white setae near humeri at right angles to the posthumeral margin .... *Malachius* (*Clanoptilus*) (in part)

Classification of the North American Genera

Malachiinae Leach 1817

The diagnostic characters of the adults in this subfamily are the following: soft bodied, integument more or less flexible; body with lateral eversible vesicles; position of vesicles variable, but always present laterally at anterior pronotal angles; apical palpomere of maxillae variable, conical or triangular; first two visible abdominal sternites separated by a distinct suture, not connate; median lobe of aedeagus simple, without a dorsal appendage or 'lever'; tarsomere one as long as or longer than two on all legs; tarsal claws variable, most with ungual appendages. The Malachiinae is the largest, and probably most advanced subfamily of Melyridae. The adults are active, usually floricolous, and often brightly colored. In many species there is a marked sexual dimorphism, the males in particular possessing remarkable modifications of various body parts. These modifications are used during courtship, which often involves prolonged bouts of repetitive, stereotypic, chemotactic displays preceding copulation. The morphology and function of these courtship organs has been described in detail for a group of middle European species in which the organs occur on the head or on the elytral apices (Matthes 1962).

This subfamily is treated by Majer (1994a) as the family Malachiidae. Adults of the genus *Attalomimus* Wittmer 1976, usually treated in the subfamily Malachiinae, have a 4-4-4 tarsal formula and Majer (1994a) described the monobasic family Attalomimidae to include this unusual Neotropical genus. References: Leach 1817; Horn 1872; Marshall 1948 (key to genera), 1952 (species of North Central Mexico),1954a (key to world genera).

Carphurini Champion 1923

Adult members of this tribe can be recognized by the following characters: body with eight small pairs of eversible vesicles; one, present laterally at the anterior pronotal angles is bilobed; the second located on the metepimeron, is weakly bilobed; and three through eight, located laterally on abdominal segments one through six, are simple; male protarsomere I with a pectinate comb of stout, black teeth on anterior margin; abdomen with seven visible abdominal sternites, and with 3 or more tergites extending beyond elytral apices. The Carphurini is a relatively small specious tribe, with about 10 genera and more than 400 species, the genus *Carphurus* accounting for more than 300 species. The tribe is worldwide in distribution, but particularly well represented in the tropical Australasian fauna and rare in dry temperate areas. Only two genera occur in North America. References: Champion, 1923; Evers 1968.

Chaetocoelus LeConte 1880
The single species of *Chaetocoelus*, *C. setosus* LeConte, was described from Columbus, Texas and is also known from Maryland, Florida and Indiana. The diagnostic characters used to recognize members of this genus are the following: tarsomeres simple, not lobed; male abdominal sternites V and VI sulcate (Wittmer 1999). The female of *C. setosus* is brachelytrous and brachypterous, however, females of some undescribed species are alate with fully developed elytra. Four undescribed species have been found in recent years, one in southeastern Arizona, one in eastern Texas, one in Baja California Sur, and one in the mountains of southern Sonora, Mexico. References: LeConte 1880; Mayor 1982; Wittmer 1999.

[*Carphuroides* Champion 1923]
This genus does not occur north of Mexico. There are two species of *Carphuroides* known from North America; *C. californicus* Wittmer 1999, was described from Ceralbo Island in the Gulf of California, and *C. atratulus* (Gorham 1886), is known from Panama, Guatemala and Guerrero, Mexico. *Carphuroides* is a large genus with more than 50 species, and cannot be easily differentiated from *Chaetocoelus*. The only character useful in separating the two genera are male abdominal sternites V and VI, which are

sulcate in *Chaetocoelus* and unmodified in *Carphuroides*. References: Champion 1923; Marshall 1951; Wittmer 1999.

Malachiini Leach 1817

This tribe is well represented in America north of Mexico, where 186 species are grouped into 19 genera. Adult members of the tribe Malachiini can be recognized by the following characters: body with two large pairs of eversible vesicles; one, located laterally at the apical angles of the pronotum, is trilobed; the second, located between the metasternum and first abdominal segment, is bi- or trilobed; male protarsomere I simple in most, all without a pectinate comb of stout, black teeth on anterior margin; abdomen with six visible abdominal sternites. References: LeConte 1862; LeConte and Horn 1883; Blaisdell 1938a.

Condylops Redtenbacher 1850
This genus, with about 79 species, is widely distributed in southern Africa, with some species in the Palaearctic, Asia, the Indo-Malayan region, and one in North America. References: Redtenbacher 1850; Wittmer 1983, 1985.

Subgenus *Neocondylops* Wittmer 1987
The single North American species of *Condylops*, *C.* (*Neocondylops*) *obrieni* Wittmer 1987, was described from Florida. The diagnostic characters used to recognize members of this subgenus are the following: size small < 2 mm; male frons with a small knob-like projection; male protarsomere II lobed, partially covering III, with a comb of stout teeth apically; male pygidium deeply excavated. *Condylops obrieni* can be common in late winter and early spring on the leaves of saw palmetto and on the foliage of oaks, *Quercus* sp. and other trees. *C. obrieni* is not closely related to its Old World congeners. Reference: Wittmer 1987.

Troglops Erichson 1840
There are two species of *Troglops* in North America, *T. montanus* Evers 1993 described from Mexico, and *T. cephalotes* Olivier 1790, introduced from Europe and reported from Bedford, Massachusetts. *Troglops cephalotes* has not been reported since originally found, and may not be established in North America. The 4-segmented protarsi and excavated head of males will differentiate *Troglops* from other North American genera. References: Erichson 1840; Wittmer 1975; Evers 1993.

Nodopus Marshall 1951
There are two species in this small genus, *N. cariceps* Marshall 1951, described from Mississippi and *N. tibialis* Mayor and Wittmer 1981 described from Georgia. The diagnostic characters used to recognize members of this genus are the following: pubescence inconspicuous, sparse, decumbent, with few erect setae on elytra; male protibiae with 3-10 or more short, stout, black spicules near apex; male protarsomere I produced ventrally into a rounded lobe with 3-5 stout spicules near apex of lobe. In both species of *Nodopus* the male head is sulcate, however, based on similarities in the structure of the male protibiae and protarsi, an undescribed species from Louisiana with the male head and elytra unmodified belongs here. References: Marshall 1951; Mayor and Wittmer 1981.

Temnopsophus Horn 1872
Olistherarthrus Champion 1922
Afrocolotes Wittmer 1960
There are two species in this small genus, *T. bimaculatus* Horn 1872, described from Louisiana is widely distributed in the east and midwest, and *T. impressus* Schwartz 1878 is endemic to Florida. The diagnostic characters used to recognize members of this genus are the following: pubescence inconspicuous, sparse, pale, decumbent; body form ant-like, pronotum longer than wide and narrowed basally; elytra narrowed basally, expanded and inflated posteriorly; both sexes brachypterous; male protarsi 4-segmented; male maxillary palpi with apical palpomeres greatly enlarged and distorted. The remarkable species of this genus appear to be closely related to the African species of *Sphinginopalpus* Pic 1903. Evers (1989) treated the African genera *Olistherarthrus* Champion (sic. *Olisterarthrus*) and *Afrocolotes* Wittmer as synonyms of *Temnopsophus* (sic. *Temnosophus*). References: Horn 1872; Schwarz 1878; Pic 1903; Champion 1922; Hatch 1927; Wittmer 1960; Downie 1972; Evers 1989.

Hypebaeus Kiesenwetter 1863
Pseudebaeus Horn 1872
Ebaeus Erichson 1840 (in part); LeConte 1852 (in part)
Species of the genus *Hypebaeus* are found in all regions of the world, with four species widely distributed in North America. The diagnostic characters used to recognize members of this genus are the following: size small, length about 2 mm; pubescence inconspicuous, sparse, pale, decumbent; head simple in both sexes, epistomal region not sulcate; tarsi 5-segmented, unmodified in both sexes; elytral apex of male prolonged, and with a small cup-shaped process extending upward from the tip of each elytron. References: LeConte 1852b; Kiesenwetter 1863; Horn 1872 (key to species, as *Pseudebaeus*); Marshall 1955 (as *Pseudebaeus*).

Ebaeus Erichson 1840
This is a large genus found in all regions of the world. There are two species in North America, *E. mexicanus* Wittmer 1966, described from Mexico, and *E. viridescens* Wittmer 1968, known from Boulder, Colorado in the United States. The main character used to distinguish *Hypebaeus* (male protarsi simple) from *Ebaeus* (male protarsomere II prolonged as a free lobe over III) seems straightforward. However, all of the North American species of *Hypebaeus* have male protarsomere II more or less prolonged over III, and Platcher (1985) reported similar findings for some European *Hypebaeus*. The differences seem to be a matter of degree rather than structure. Platcher concluded that the two genera are closely related, but did not change their status. It appears that the North American species of *Ebaeus* and *Hypebaeus* may be congeneric. References: Wittmer 1966, 1968.

Charopus Erichson 1840

There are 20 species in this primarily Palaearctic genus, with two species, *C. moerens* LeConte 1859 and *C. longicollis* Motschulsky 1859, from North America, their collective ranges extending from British Columbia south to California. The diagnostic characters used to recognize adult members of this genus are the following: pubescence inconspicuous, sparse, pale, decumbent; head simple in both sexes; pronotum elongate, laterally sinuate; male elytral apices appendiculate; female brachypterus, elytra conspicuously divergent and inflated posteriorly; protarsi 5- segmented, tarsomeres unmodified in both sexes. The North American species of *Charopus* are commonly found in riparian or coastal areas associated with the flowers of numerous grasses. Several undescribed species have been found in California in recent years. References: LeConte 1859c; Motschulsky 1859.

Attalus Erichson 1840
 subgenus *Acletus* LeConte 1852
 Scalopterus Motschulsky 1859

The genus *Attalus* is very large, with hundreds of species found in all geographic regions of the world. There are 125 species in North America, with 54 species north of Mexico generally distributed, but most numerous in the southwestern United States. The diagnostic characters used to recognize adult North American members of this genus are the following: pubescence variable, often dual, with both decumbent and erect setae; head simple in both sexes; pronotum quadrate, wider than head; elytra lacking tactile setae near humeri, apex unmodified in both sexes; protarsi 5- segmented in both sexes; male protarsomere II prolonged as a lobe over III, most with lobe bearing an apical comb of stout black setae. *Attalus foveiventris* Fall 1917, *A. intermedius* Marshall 1953, and *A. santarosae* Marshall 1951 have the male protarsi elongate and acicular, and visible abdominal sternite IV and/or V sulcate, and belong in the genus *Tanaops*. *Attalus australis* Blatchley 1922 has 9-segmented antennae, and belongs in the genus *Ablechrus*. There are numerous undescribed species in the deserts of the southwestern United States and Mexico. References: LeConte 1852b; Fall 1917 (key to some species); Brown 1944; Marshall 1946, 1948, 1951 (key to some species), 1953, 1954b, 1955, 1957; Evers 1995.

Attalusinus Leng 1918

This genus contains 21 species, 17 are from Africa, one species each from Saudi Arabia and Jordan, and two species, *A. submarginatus* (LeConte 1852), and *A. mexicanus* Marshall 1955, live in the deserts of southwestern North America, their collective ranges extending from southern California to Arizona and Sonora, Mexico. The diagnostic characters used to recognize adult North American members of this genus are the following: body size small, length <2 mm; male head broadly, transversely impressed; pronotum saddle shaped, broadest in front of middle; elytra abbreviated, exposing three or more abdominal segments; protarsi 5-segmented in both sexes; male protarsomere II prolonged over III, and with a comb of black setae apically. Adults have been collected from desert shrubs, including *Olneya*, *Acacia*, *Larrea*, *Cercidium* and *Brickellia* (*A. submarginatus*), and *Atriplex* and *Cnidoscolus* (*A. mexicanus*). References: Leng 1918; Marshall 1948, 1955; Wittmer 1985, 1991.

Endeodes LeConte 1859; 1884
 Atelestus: LeConte 1852 (not Erichson 1840)

There are eight species in the genus *Endeodes*. Three species are from Sonora, Mexico, two species are from Baja California, and three species, *E. basalis* (LeConte 1852), *E. collaris* (LeConte 1852), and *E. insularis* Blackwelder 1932, occur north of Mexico. Their collective ranges extend from southern California north to British Columbia. The diagnostic characters used to recognize members of this genus are the following: pubescence dual, with both decumbent and erect setae; head simple in both sexes; elytra abbreviated, exposing three or more abdominal tergites in most species; apterous; male protarsomere II lobed, prolonged over III, with an apical border of stout black teeth. All included species, with the exception of *E. terminalis* Marshall 1957, have abbreviated elytra and are apterous. Species of *Endeodes* are known only from the seashore of Pacific North America where they occur in intertidal habitats, some under debris and dried seaweed and others on reefs and rocky shores exposed at low tide. References: LeConte 1852b, 1859a, 1884; Blackwelder 1932; Moore 1954, 1956, 1957, 1964 (keys to adults and larvae), 1971; Marshall 1957; Moore and Legner 1975 (revision and tabular key to species); Moore and Mayor 1976; Moore and Andrews 1985.

Tanaops LeConte 1859
 Cephalistes Motschulsky 1859

There are 24 species in the genus *Tanaops* generally distributed in western North America, their collective ranges extending from British Columbia in Canada south through Washington, Oregon, California and Arizona to Sonora, Mexico. There are 12 species endemic to California. The diagnostic characters used to recognize adult members of this genus are the following: pubescence conspicuous, dual, with both decumbent and erect setae; head variable, but typically rostrate; male protarsomere II with lobe small, indistinct, most with lobe not extending over III, less commonly elongate, arcuate and usually acicular, rarely unmodified (one undescribed species from the deserts of southern California), apex always lacking a comb of stout, black setae; male abdominal sternites IV and/or V sulcate. Adults of many species are associated with various buckwheats, *Eriogonum* spp., where they can be found in very large aggregations feeding on nectar and pollen. The genus is not known to occur east of Wyoming, and *T. terramariae* Evers 1993 described from Maryland is, based on an illustration in Evers paper, a species of *Attalus* near *A. melanopterus* (Erichson 1840). There are several undescribed species known from southern California, Arizona, and Baja California, Mexico. References: LeConte 1859b; Fall 1917 (key to some species); Marshall 1936, 1944, 1946 (key to some species), 1951, 1953, 1954b, 1955; Malkin 1948; Evers 1993.

Trophimus Horn 1870
There are two species in this small North American genus, *T. aeneipennis* Horn 1870, described from Colorado, and *T. mexicanus* Marshall 1952, described from Mexico. The diagnostic characters used to recognize adult members of this genus are the following: pubescence conspicuous, dual, with both decumbent and erect setae; head rostrate; male protarsi 4- segmented; male abdominal sternites with segment III prolonged over IV, segments IV and V sulcate. Females collected without males are easily mistaken for *Tanaops* species. References: Horn 1870; Marshall 1951.

Ablechrus Waterhouse 1877
Pseudattalus Champion 1914
The genus *Ablechrus*, with 38 species, is primarily Neotropical in distribution. There are 11 species known from North America, with two species in the United States, *A. granularis* (Erichson) 1840, widely distributed in the eastern United States, and *A. texensis* Marshall 1955, from Texas. The remaining 25 species are generally distributed in South America, and the islands of the Caribbean. The characters used to recognize adults of the genus *Ablechrus* are the following: size small, length typically <2 mm; antennae 9 segmented; male protarsomere II lobed, prolonged over III or III and IV. Wittmer (1961) treated *Tucumanius* Pic 1903 as a synonym of *Ablechrus*, but subsequently reinstated the genus for those species in which the penultimate male abdominal sternite is broadened and provided with grooves and setal tufts (Wittmer 1985). References: Waterhouse 1877; Marshall 1955 (as *Pseudattalus*); Wittmer 1961, 1976, 1979, 1984.

Microlipus LeConte 1852
There are 28 species in the genus *Microlipus*, their collective ranges including the Palaearctic, Middle East, Asia and North America, where seven species are distributed from British Columbia south to California (Evers 1988). The diagnostic characters of the adults of this genus are the following: pubescence variable, typically single decumbent, less commonly dual, with both decumbent and erect setae; head unmodified in both sexes; antennae subserrate to subfiliform, rarely pectinate; most with pronotum elongate, posterior angles truncate; elytra with two to three long, fine, white tactile setae near humeri, at right angles to posthumeral margin; elytral apex typically unmodified in both sexes, rarely forming a short, indistinct broadly rounded lobe (males of *M. productus* Fall 1917); protarsi 5- segmented, unmodified in both sexes. Male protarsomere II is prolonged in an apically pectinate lobe over III in two species, *Microlipus laevicollis* Horn 1872 and *Microlipus falli* Hopping 1925, and these species belong in the genus *Attalus*, near *A. nigrellus* (LeConte 1852). Many of the species placed by Evers (1988) in *Microlipus* will eventually be assigned to other genera, leaving the species from North America and a few from Asia in *Microlipus*. References: LeConte 1852b; Fall 1917; Marshall 1946, 1948 (as *Anthocomus*); Evers 1988.

Malachius (*Malachius*)Fabricius 1775
The subgenus *Malachius* is represented by one species, *M. aeneus* (Linnaeus 1758), adventive from Europe, and widely distributed in the United States and Canada. The diagnostic characters of the adults of this subgenus are the following: size large, length >6 mm; pubescence dual, with both decumbent and erect setae; male head with epistomal region sulcate; pronotum quadrate, wider than head, angles broadly rounded; elytral apex unmodified in both sexes; protarsi 5-segmented, unmodified in both sexes. Arnett (1962) reported that *M. aeneus* caused considerable unrecognized damage to developing wheat in the Midwest. This suggestion had its origin in a European account of feeding damage to anthers and pistils of wheat by adults of *M. aeneus* affecting crop yields (Frauenfeld 1866). In North America *M. aeneus* is common in the northeast and northwest, but is conspicuously absent from the central United States and Canada. It was first reported as occurring in New England by LeConte (1852b) and Dow (1914) and was probably introduced at Boston or Quebec. Leech (1947) first reported the species from the northwest, however, collection records indicate it was present in British Columbia as early as 1928. Barber (1949), citing the Frauenfeld paper, suggested that it had advanced through the wheat belt from New England, and should be studied for its potential to damage wheat crop yields. However, this view is not supported by the disjunct distribution of *M. aeneus*. Its absence from the central United States and Canada suggests that a second, much later western introduction took place, probably at Seattle or Vancouver. Additionally, there are no reports of *M. aeneus* causing damage to wheat in North America. References: Fabricius 1775; Downie 1950; Mawdsley 1993.

Malachius (*Clanoptilus*) Motschulsky 1854
Clanoptilus Motschulsky 1854
Hapalorhinus LeConte 1859
Anthocomus Erichson 1840 (in part)
The subgenus *Clanoptilus* (Volume 2, Color Figure 25) is represented by twenty-six species, generally distributed in the western United States and Canada. Only one species *M. ulkei* Horn 1872 occurs in the eastern United States. There are four undescribed species known from California. The diagnostic characters of the adults of this genus are the following: pubescence variable, often dual, with both decumbent and erect setae, less commonly single, with only pale decumbent setae; head unmodified in both sexes; antennae serrate to pectinate; pronotum quadrate, wider than head, angles broadly rounded; elytra with three to five long, fine, white tactile setae near humeri, at right angles to posthumeral margin; elytral apex of male often modified to form a variously shaped organ; protarsi 5- segmented, unmodified in both sexes. The species of *Malachius* (sensu lato) were arranged by Evers (1985) into nine genera including *Clanoptilus* Motschulsky 1854. Evers concept of *Clanoptilus*, includes many of the North American species placed by Marshall (1948) in *Anthocomus*. *Clanoptilus*, as defined by Evers (1985) and including some species in which the male elytra are unmodified, is here treated conservatively as a subgenus of *Malachius*. References: Motschulsky 1854; LeConte 1859b; Horn 1874; Fall 1901 (key to some species); Marshall 1946, 1948 (key to some species, as *Anthocomus*), 1951 (as *Anthocomus*),

1953 (as *Anthocomus*), 1954b (as *Anthocomus*); Evers 1985 (taxonomy).

Anthocomus Erichson 1840
There are eight species of *Anthocomus* in North America, with six species distributed north of Mexico. Only two species, *A. pristinus* (Fall 1901) from California, and *A. bipunctatus* (Harrer 1784) adventive from Europe and widely distributed in the eastern United States and Canada, belong in *Anthocomus*. The diagnostic characters of the adults of this genus are the following: pubescence inconspicuous, pale, decumbent, with few erect setae; head unmodified in both sexes; pronotum quadrate, wider than head, angles broadly rounded; elytra with one or two long, fine, white tactile setae near humeri, at right angles to posthumeral margin; elytral apex of male modified to form a variously shaped organ; protarsi 5- segmented, unmodified in both sexes. Marshall (1948) included 36 species in *Anthocomus*. This classification has not been followed by subsequent authors (Arnett 1962, 1975), and in this work 30 species are treated in *Charopus* (2 species), *Microlipus* (6 species) and *Malachius* (*Clanoptilus*) (22 species). Three species, *A. erichsoni* LeConte 1852, *A. flavilabris* (Say 1825), both widely distributed in the east, and *A. ventralis* Horn 1872, described from Texas, have the male protibiae with 3-10 or more short, stout, black spicules near apex, and male protarsomere I produced ventrally into a rounded lobe with 3-5 stout spicules near apex of lobe, and these species belong in the genus *Nodopus*. Male protarsomere II is prolonged in an apically pectinate lobe over III in *A. nigrinus* (Fall 1901), and this species belongs in the genus *Attalus*, near *A. nigrellus* (LeConte 1852). The genus *Anthocomus* is not known to occur in the tropics, and *A. fuscescens* Gorham 1886 (Panama) and *A. viridescens* Champion 1914 (Mexico) are probably misplaced. References: LeConte 1852b (in part); French 1942-43, 1944; Marshall 1948.

Cerapheles Mulsant in Mulsant and Rey 1867
The inclusion of this small European genus of six species is based on a single female, probably *C. terminatus* (Menetres 1832), found in the collection of the Florida State Collection of Arthropods in Gainesville and bearing the following label data: Mobile, Ala.; VI-12-1957; B. K. Dozier/ In Warehouse State Docks. No report in the literature on the occurrence of this genus in North America could be found, and it is probable that this record represents only an interception. *Cerapheles*, with the pronotum widest apically, is distinguished from *Malachius* (*Clanoptilus*) and *Anthocomus*, both with the pronotum widest at the middle. Reference: Mulsant and Rey 1867.

Collops Erichson 1840
There are 70 species of *Collops*, their combined ranges including northern South America (three species) and North America (67 species), with 45 species found in all geographic regions north of Mexico. The characters used to recognize adults of the this genus are the following: pubescence variable, but often dual, with both decumbent and erect setae, less commonly single, with only pale decumbent setae; antennae apparently 10 segmented, antennomere II small, hidden in apex of I; male protarsi 4-segmented; male antennae with antennomere 3 enlarged, most are excavated and appendiculate. Several undescribed species are known from California, and the deserts of the southwest. An additional seven species of *Collops* described from eastern Asia (Wittmer 1992), all with five protarsomeres in both sexes, probably belong in other allied genera. References: LeConte 1852b; Horn 1870; Fall 1912 (key to some species), 1913; Marshall 1951, 1953, 1954c, 1955; Russell 1966; King 1987, 1988; Evers 1993, 1994, 1995.

Rhadalinae LeConte 1862

The diagnostic characters of the adults of this subfamily are the following: hard bodied, integument more or less inflexible; body lacking eversible vesicles; apical palpomeres of maxillae securiform or triangular (except in some *Aplocnemus* Stephens 1830, not in North America); first two visible abdominal sternites connate (freely articulated in *Dasyrhadus* Fall 1910); median lobe of aedeagus with a dorsal appendage or 'lever' (absent in *Indiodasytes* Pic 1916 and *Dasyrhadus* Fall 1910); tarsomere I as long as or longer than II on all legs; tarsal claws variable, most with ungual appendages. There are sixteen extant genera in the subfamily Rhadalinae from all biogeographic regions except Australasia. Majer (1998) described the fossil genus *Aploceble* Majer 1998 from Baltic amber. The Rhadalinae are treated as a subfamily of the family Dasytidae by Majer (1994a). For a review of the subfamily see Peacock (1987). References: Stephens 1830; LeConte 1862 (as Rhadalini); LeConte and Horn 1883 (as Rhadalini); Casey 1895; Blaisdell 1938a.

Rhadalus LeConte 1852
Cymbolus Gorham 1886
There are eight species in the genus *Rhadalus* known from Brazil (one species), and in North America from Guatemala (two species), Mexico (three species), and the United States, where two species, *R. lecontei* Casey 1895, and *R. testaceous* LeConte 1852, are found in the Colorado and Sonoran deserts of southern California and Arizona. Adult members of this genus can be recognized by the following combination of characters: body elongate, convex, light to dark brown in color; eyes setose; sublateral elytral carina present, carina parallel to costal margin; pronotal punctures rimmed or tuberculate, disc often with raised, shiny impunctate patches. References: LeConte 1852a; Gorham 1886; Peacock 1987.

Semijulistus Schilsky 1894
Celsus Lewis 1895
Eurelymis Casey 1895
Eight species of *Semijulistus* are known from Turkestan (one species), Japan (two species), and North America, where there are five species, their collective ranges including Canada, Colorado, Utah, California and Arizona. Two species, *S. bicoloripes* Pic 1928 and *S. rubrithorax* Pic 1928 are reported by Pic (1928) from North America without definite locality. Adult members of the genus *Semijulistus* can be recognized by the following combination of

characters: eyes glabrous; antennae short, apical five or six antennomeres broadly serrate, appearing clubbed; pronotum broader than long, lateral margins serrate; elytra lacking sublateral carina, epipleura narrow, evanescent at level of second abdominal segment; female elytra with a characteristic oval, raised, shiny, unpunctured area on apical third. References: Schilsky 1894; Lewis 1895; Casey 1895 (key to some species as *Eurelymis*); Pic 1928; Peacock 1987.

[*Eucymbolus* Champion 1913]
This genus does not occur north of Mexico. The genus *Eucymbolus* was transferred from the Melyrinae to the subfamily Rhadalinae (as Haplocneminae) by Crowson (1964). The genus is known from a single specimen, *E. cyaneus* Champion 1913, from Guatemala, and according to Peacock (1987) may represent an aberrant *Rhadalus*. Reference: Champion 1913.

Dasyrhadus Fall 1910
There are two species of *Dasyrhadus*, *D. impressicollis* Fall 1910, and *D. longior* Fall 1910, both described from California. Crowson (1964) moved *Dasyrhadus* from the Rhadalinae to the subfamily Dasytinae based on specimens of *D. impressicollis* Fall in the British Museum. However, the setose eyes and triangular palpomeres of the maxillae mentioned by Fall (1910), suggest a relationship to the Rhadalinae. Majer (1994a) includes *Dasyrhadus* in the subfamily Rhadalinae of the family Dasytidae. The species of *Dasyrhadus* are easily recognized by the short, transverse pronotum which is strongly impressed around the margins.

Melyrinae Leach 1815

The diagnostic characters of the adults of this subfamily are the following: hard bodied, integument more or less inflexible; body lacking eversible vesicles; apical palpomere of maxillae variable, conical or triangular; first two visible abdominal sternites separated by a distinct suture, not connate; median lobe of aedeagus simple, without a dorsal appendage or 'lever'; tarsomere I shorter than II on all legs; tarsal claws toothed, lacking ungual appendages. Majer (1987) included 12 genera in the subfamily Melyrinae as follows: *Cerralus* and *Anthrodromius* (eastern Europe, Central Asia and Asia Minor); *Chalchas*, *Astylus*, *Arthrobrachus*, *Astylomorphus* and *Microzygia* (Neotropical); *Melyris* and *Falsomelyris* (Europe, Asia Minor and Africa); *Pseudozygia* (East Africa); *Procerallus* (India); and *Melyrodes* (New World). This subfamily is treated by Majer (1994a) as the family Melyridae. References: Leach 1815.

Melyrodes Gorham 1882
Alymeris Casey 1895
Melyris of authors, not Fabricius 1775
There are eight species of *Melyrodes* distributed from Bolivia (one species), north through Panama (two species), Costa Rica (one species) and Guatemala to Mexico (one species) and the United States, where three species, *M. basalis* (LeConte 1852), *M. cribratus* (LeConte 1852), and *M. floridana* (Casey 1895), occur in the southeastern states. The diagnostic characters of the adults of this genus are the following: size small < 3 mm; color brown to black, occasionally elytra with pale red basal and preapical spots; body punctures, especially on elytra, coarse and dense; pronotal lateral margins serrulate; elytral epipleurae wide and nearly equal in width from base to sutural angles, with the lower edge strongly serrulate. References: Gorham 1882; Casey 1895 (key to species, as *Alymeris*).

Melyris Fabricius 1775
The genus *Melyris* is large with more than 100 species found in Europe, Asia Minor, and Africa. Adults and larvae of *M. oblonga* Fabricius 1775 have been intercepted at North American ports of entry. Although the current status of this species in North America is unclear, specimen labels at the United States National Museum suggest that a small population may be established in New Jersey. Adults of *M. oblonga* are easily distinguished from species of *Melyrodes* by the much larger body size, > 7 mm, and by their coloration, which is reddish-orange, with the head and elytra dark metallic blue.

Dasytinae Laporte 1840

The diagnostic characters of the adults of this subfamily are the following: hard bodied, integument more or less inflexible; body lacking eversible vesicles; apical palpomere of maxillae variable, most are conical; first two visible abdominal sternites separated by a distinct suture, not connate; median lobe of aedeagus simple, without a dorsal appendage or 'lever'; tarsomere I as long as or longer than II on all legs; tarsal claws variable, most with ungual appendages. The subfamily Dasytinae is large with more than 50 genera from all biogeographic regions of the world This subfamily, along with the subfamilies Rhadalinae, Gietellinae, Danaceinae, Chaetomalachiinae, and Listrinae, is treated by Majer (1994a) in the family Dasytidae. References: Laporte 1840; Majer 1994a.

Dasytini Laporte 1840

Previous authors (Casey 1895; LeConte 1862; 1866; LeConte and Horn 1883; Pic 1937; Blaisdell 1938a) have treated most of the North American genera of the subfamily Dasytinae as belonging in the tribe Dasytini. Majer (1990) redefined the Dasytini to include 11 genera as follows: *Dasytes* (Palaearctic species only), *Enicopus*, *Divales*, *Psilothrix*, *Allotarsus*, *Dolichophron*, *Hauseria*, *Allotarsodasytes*, *Trochantodon*, *Graellsinus* (Europe, North Africa and Asia Minor) and *Dolichosoma* (Europe and North America). The characters used by Majer to distinguish the Dasytini from other groups of Dasytinae are the following: pronotum and elytra lacking a distinct marginal fringe of setae; tarsomere IV shorter than III on all legs; tarsal claws with a basal tooth, ungual appendages variable, most with appendages present. This definition excludes the genera *Hoppingiana, Dasytes* (North American species), *Dasytellus, Dasytastes, Eschatocrepis, Vectura, Vecturoides, Leptovectura, Amphivectura* and *Enallonyx* from the Dasytini. However, these genera have not yet been assigned to other tribes and are here treated in the tribe Dasytini. The genera *Mectemycor, Mecomycter*, and *Scuromanius*, assigned by Majer (1995) to the family Mauroniscidae are here treated in the tribe Dasytini. These genera will eventually be assigned to other tribes. References: Laporte 1840; LeConte 1862, 1866; LeConte and Horn 1883; Casey 1895; Pic 1937; Blaisdell 1938a; Majer 1990.

Dolichosoma Stephens 1830
This is a small Holarctic genus with two species known from North America, *D. foveicolle* (Kirby 1837), described from Nebraska, and *D. tenuiforme* Horn 1880, described from Texas. The diagnostic characters of the adults of this genus are the following: body extremely elongate, subcylindircal; elytral epipleurae inflexed basally and apically, visible externally only in median third (Fig. 9); pronotum with submarginal line of large contiguous punctures in basal half; tarsal claws and ungual appendages asymmetrical. References: Stephans 1830; Majer 1990.

The following genera (*Hoppingiana, Dasytes, Dasytellus,* and *Dasytastes*) share similarities which suggests a closer relationship than indicated by previous authors. *Paradasytes*, and *Neadasytes* placed by Majer (1990) in the tribe Listrini possess characters suggesting a close relationship to *Dasytellus* and *Dasytastes*, respectively.

Hoppingiana Blaisdell 1924
There are two species in the genus *Hoppingiana*, *H. hudsonicus* (LeConte 1866) described from British Columbia and *H. nitida* Hatch 1962 described from Oregon. The diagnostic characters of the adults of this genus are the following: size > 3 mm; pubescence dual, with both decumbent and erect setae; head (Fig. 18) short, eyes setose; pronotum with submarginal excavated lines marked by a series of large contiguous punctures, the lines delimiting a lateral coarsely sculptured area; elytral epipleurae short, restricted to basal one-third of elytra; tarsal claws with ungual appendages symmetrical, as long as claws (as in Fig. 3). Majer (1994a) places the genus *Hoppingiana* in the subfamily Danaceinae of the family Dasytidae. References: Blaisdell 1924a, 1934c; Majer 1994a (taxonomy).

Dasytes Paykull 1798
The primarily palaearctic genus *Dasytes* is very large, with hundreds of species. There are 22 North American species of *Dasytes*, their collective ranges including Arizona, California, Nevada, Colorado, Oregon, and British Columbia. The diagnostic characters of the adults of this genus are the following: size > 2 mm; pubescence variable either single, with pale decumbent setae or dual with some short, dark erect setae; eyes glabrous; pronotum with submarginal excavated lines often delimiting a lateral roughly sculptured area; elytral epipleurae wide at base, obsolete near middle of elytra; tarsal claws with ungual appendages symmetrical, shorter than claws (as in Fig. 4). Majer (1994a) indicates that most of the North American "*Dasytes*" belong in the subfamily Danaceinae of the family Dasytidae. Presumably, this would require a genus name change, which has not yet been proposed. References: Paykull 1798; Blaisdell 1906, 1921a (key to species), 1925a, 1926.

Dasytellus Casey 1895
There are eight species of *Dasytellus* widely distributed in western North America. The diagnostic characters of the adults of this genus are the following: size small < 2 mm; pubescence single, with sparse pale decumbent setae; eyes glabrous; pronotum with submarginal excavated lines, area lateral to the lines not more coarsely sculptured than disc; elytral epipleurae wide at base, obsolete near middle of elytra; tarsal claws with ungual appendages symmetrical, most with appendage as long as claws (as in Fig. 3). Reference: Casey 1895 (key to some species).

Dasytastes Casey 1895
The eleven species included in the genus *Dasytastes* are described from Baja California, California and Nevada. The diagnostic characters of the adults of this genus are the following: size small < 2 mm; pubescence single, with sparse pale decumbent setae; eyes glabrous; pronotum lacking submarginal excavated lines; elytral epipleurae wide at base, obsolete near middle of elytra; tarsal claws with ungual appendages symmetrical, shorter than claws (as in Fig. 4). References: Casey 1895 (key to some species); Blaisdell 1923; 1924d.

The following genera (*Eschatocrepis, Vectura, Vecturoides, Leptovectura, Amphivectura,* and *Enallonyx*) share similiarities in body form and vestiture that suggests a closer relationship than indicated by previous authors. The diagnostic characters of the adults of this group are the following: body more or less dorsally planate or flattened; vestiture simple, setae pale, decumbent and adpressed to the body, hair-like to strongly squamous; pronotum often with apical and/or basal angles prominent (broadly rounded in *Eschatocrepis* less so in *Enallonyx*), submarginal excavated lines present and well developed in most species, lateral and/or basal margins more or less sinuate; tibial spurs on front and middle legs indistinct or absent. The characters used by authors to differentiate between *Vectura*, *Vecturoides* (described by Fall 1930 as a subgenus of *Vectura*), *Leptovectura*, and *Amphivectura* (see below) may not warrant recognition of these taxa as distinct. The genera *Listromimus* and *Listrimorpha* placed by Majer (1990) in the tribe Listrini share characters which suggests a close relationship to this group of genera.

Eschatocrepis LeConte 1862
The single species of *Eschatocrepis*, *E. constrictus* (LeConte 1852), occurs in Arizona and California. Howell (1985) reduced the species of *Eschatocrepis* to subspecies and described a new subspecies. He recognized *E. c. constrictus* (LeConte 1852); *E. c. nigripes* Blaisdell 1921; *E. c. desertus* Blaisdell 1931; and *E. c. riversidensis* Howell 1985. The diagnostic characters of adults of *Eschatocrepis*

are the following: body graciliform; pronotum (Fig. 14) constricted laterally before the apex, apical and basal angles broadly rounded, submarginal excavated lines well developed; tarsal claws with ungual appendages symmetrical, as long as claws, attached at base. References: LeConte 1862; Blaisdell 1921b, 1931b (key to species); Howell 1985 (taxonomy).

Vectura Casey 1895
Pseudallonyx Casey 1895
The four species in this small genus are found in Arizona, California, Colorado and Oregon. The diagnostic characters of the adults of this genus are the following: body cuneiform, widened apically, pronotum as in Fig. 12; elytral epipleurae wide, flat, horizontal, broad to near apex (as in Fig. 7); tarsal claws with ungual appendages asymmetrical, inner (anterior) appendage as long as claw, outer (posterior) appendage obsolete (as in Fig. 6). Fall (1930) suggested that *Pseudallonyx* should be suppressed as a synonym of *Vectura*, and this synonymy was recently confirmed by Mawdsley (1999b). Reference: Blaisdell 1934a (key to species).

Vecturoides Fall 1930
Vectura (*Vecturoides*) Fall 1930
Menovectura Blaisdell 1931
There are three species in this small genus, *V. pseudonycha* Fall 1930, and *V. serrulata* (Blaisdell 1931), both endemic to California, and *V. albicans* (Casey 1895), from Utah and recently transferred from *Vectura* by Mawdsley (1999b). The diagnostic characters of the adults of this genus are the following: body graciliform, only slightly widened apically, pronotum as in Figs. 10-11; elytral epipleurae wide, flat, horizontal, broad to near apex; tarsal claws with ungual appendages symmetrical, as long as claws (as in Fig. 3). References: Fall 1930; Blaisdell 1931a (as *Menovectura*), 1938a.

Leptovectura Casey 1895
The single species of *Leptovectura*, *L. adspersa* Casey 1895, was described from New Mexico. The diagnostic characters of the adults of this genus are the following: body graciliform, only slight widened apically; elytral epipleurae narrow, sharply defined in basal half, disappearing beyond middle of second abdominal sternite; tarsal claws with ungual appendages asymmetrical, inner (anterior) appendage as long as claw, outer (posterior) appendage obsolete (as in Fig. 6). Reference: Blaisdell 1938a.

Amphivectura Blaisdell 1938
The single species of *Amphivectura*, *A. monticola* (Blaisdell 1934), was described from British Columbia, and has been reported from Oregon and Washington by Hatch (1962) and from Colorado by Mawdsley (1999b). The diagnostic characters of the adults of this genus are the following: body moderately robust, sides parallel, not widened apically; pronotum (Fig. 13) strongly, sinuately constricted basally, basal angles dorsally prominant and sharply acuminate; elytral epipleurae (Fig. 8) wide basally, inconspicuous in apical half; tarsal claws with ungual appendages symmetrical, as long as claws (as in Fig. 3). References: Blaisdell 1934c (as *Hoppingiana*); 1938a.

Enallonyx Wolcott 1944
Allonyx LeConte 1862, not Jacquelin Du Val 1860
There are four species of *Enallonyx*; *E. cinerescens* (Fall 1930), *E. disjunctus* (Casey 1895), *E. denudatus* (Casey 1895) and *E. sculptilis* (LeConte 1859), all endemic to California. The diagnostic characters of the adults of this genus are the following: body robust, sides parallel, not widened apically; pronotum strongly, sinuately constricted basally, sides parallel; elytral epipleurae wide, flat, horizontal, broad to near apex (as in Fig. 7); tarsal claws with ungual appendages asymmetrical, inner (anterior) appendage as long as claw, outer (posterior) appendage restricted to base of claw (as in Fig. 5). Fall (1930) suggested that *E. disjunctus* and *E. sculptilus* were the same species, but did not formalize the synonymy. References: LeConte 1859b, 1862; Jaquelin duVal 1860; Casey 1895 (key to species, as *Allonyx*); Wolcott 1944.

The following genera (*Mectemycor*, *Mecomycter*, and *Scuromanius*) are treated by Majer (1995) as members of the family Mauroniscidae in which he also includes the Neotropical genera *Amecomyter* Majer 1995 and *Mauroniscus* Bourgeois 1911. The diagnostic characters of adults of this group of genera are the following: antennomeres cylindrical to slightly serrate, terminal antennomeres tending to form a loose club; mouthparts often highly modified, galea and lacinea narrow, elongate (all North American genera); tarsal claws unarmed, lacking ungual appendages. The North American genera are nearly unique among the Dasytinae in that the tarsal claws lack ungual appendages. Only one other North American genus in the subfamily Dasytinae, *Pseudasydates*, placed by Majer (1990) in the Tribe Listrini, has tarsal claws lacking ungual appendages.

Mectemycor Majer 1995
This small genus contains only three species; *M. linearis* (Fall 1930), *M. sericeus* Majer 1995, and *M. strangulatus* Majer 1995, all endemic to California. The diagnostic characters of the adults of this genus are the following: pubescence inconspicuous; head rostrate (as in Fig. 19); pronotum elongate, strongly constricted laterally. Reference: Majer 1995 (taxonomy, and key to species).

Mecomycter Horn 1882
This is a small genus with only three species, *M. omalinus* Horn 1882 from California, Kansas and Wyoming, *M. testaceous* Majer 1995 from Arizona, and *M. majeri* Howell 1997 from California. The diagnostic characters of the adults of this genus are the following: pubesence dual, with both decumbent and erect setae; head rostrate (as in Fig. 19); pronotum elongate, without lateral magin. Howell (1997) wrote a key which includes species placed by Majer (1995) in the genera *Mectemycor* (*M. linearis*) and *Scuromanius* (*S. facetus* and *S. liebeckei*). References: Horn 1882; Majer 1995 (taxonomy, and key to some species); Howell 1997.

Scuromanius Majer 1995
There are six species of *Scuromanius* known from Columbia (one species), Mexico (two species) and the United States, where there are three species, *S. facetus* (Casey 1895) from in Arizona, *S. wickhami*

Majer 1995 from California, and *S. liebecki* (Blaisdell 1929) from Texas. The diagnostic characters of the adults of this genus are the following: pubesence dual, with both decumbent and erect setae; head short, not rostrate; pronotum wider than long, sides more or less angulate, lateral margin distinct. Reference: Majer 1995 (taxonomy and key to species).

Listrini Majer 1990

The tribe Listrini proposed by Majer (1990) is comprised of 25 genera, their collective ranges including tropical Africa, *Calosotis* Redtenbacher 1867, the Palaearctic, *Danacaeina* Reitter 1887, Asia, *Picolistrus* Majer 1990 and *Sinolistrus* Majer 1990, and western North America where 21 genera are found in all geographic regions. The characters used by Majer to distinguish the Listrini from other groups of Dasytinae are the following: pronotum and elytra nearly always with a distinct marginal fringe of setae; tarsomeres III and IV equal in length on all legs; tarsal claws without a basal tooth, with ungual appendages as long as or nearly as long as the claws, and more or less attached to claws throughout their length. In North American there are 10 monobasic genera in the tribe Listrini. The inclusion of *Eutricholistra, Pseudasydates, Neadasytes* and *Paradasytes* is provisional (Majer 1990). *Neadasytes*, and *Paradasytes*, share characters suggesting a closer relationship to *Dasytellus* and *Dasytastes*, respectively than to other genera of Listrini. *Listromimus* and *Listrimorpha* share characters suggesting a closer relationship to *Vectura* and allied genera than to other genera of Listrini. Reference: Majer 1990.

Listrus Motschulsky 1859

Amecocerus: Pic 1937 (in part); Arnett 1962 (in part)

There are 102 species of *Listrus* in North America, with 86 species generally distributed in all geographic regions of the western United States and Canada. The diagnostic characters of the adults of this genus are the following: pubescence single, decumbent; eyes glabrous; pronotum with lateral margins serrulate, apical angles not produced; protibiae lacking spines on external (dorsal) margin in both sexes (as in Fig. 22); tarsal claws with ungual appendages symmetrical, as long as claws (as in Fig. 3). The tibial spurs are probably variable in *Listrus*, but there are at least some species in which the male protibiae have two acute spurs (as in Fig. 17), and the mesotibiae one acute and one broad, spoon shaped spur (as in Fig. 16). Variation in the expression of this character may help to clarify species relationships within this genus. Without explanation *Listrus* species were listed in the Coleopterorum Catalogus by Pic (1937) under the name *Amecocerus* Solier 1849. According to Majer (1990), who studied the South American species of *Amecocerus*, reinstated the genus *Listrus* and described the new tribe Listrini, *Listrus* and *Amecocerus* are distinct genra. References: Motschulsky 1859; Blaisdell 1921a (key to some species), 1921c, 1924c, 1925b, 1925d, 1927b, 1928, 1931a, 1934b, 1936, 1937, 1938b, 1939, 1940a; Mawdsley 1999a.

The following genera (*Pristoscelis*, *Asydates*, *Pseudasydates*, and *Eutricholistra*) are treated here as a related group. Males of these genera all have two acute spurs on the pro- and mesotibiae (as in Fig. 17). *Asydates*, *Pseudasydatyes* and *Eutricholistra* share the following unique character combination: protibiae with spines on external (dorsal) margin in both sexes (as in Fig,. 20), and males with two acute tibial spurs on the pro- and mesotibiae.

Pristoscelis LeConte 1862

There are five species of *Pristoscelis*, all described from California. The diagnostic characters of the adults of this genus are the following: pubescence conspicuous, dual, with both decumbent and erect setae; eyes glabrous; pronotum with lateral margins not serrulate, apical angles not produced; protibiae with few spines on external (dorsal) margin in both sexes (as in Fig. 21); tarsal claws with ungual appendages symmetrical, as long as claws (as in Fig. 3); inner (anterior) appendage detached from claw in apical one-third. The species of *Pristoscelis* are recognized by the large male head, which is elongate behind the eyes, wider than the pronotum, and by the large elongate male mandibles. Howell (1997) regarded *P. volki* Howell 1979 as invalid, resembling *Pristoscelis* due to a deformity, but did not change its status. References: LeConte 1862; Blaisdell 1924c; Howell 1979, 1997 (key to species).

Asydates Casey 1895

This genus contains three species, *A. explanatus* Casey 1895, *A. puncticeps* Blaisdell 1924, and *A. rufiventris* Casey 1895, all endemic to California. The diagnostic characters of the adults of this genus are the following: pubescence variable, simple, decumbent or dual, with both decumbent and erect setae; eyes glabrous; pronotum with lateral margins not serrulate, apical angles not produced; elytral epipleurae broad, horizontal, wide to near elytral apex (as in Fig. 7); tarsal claws with ungual appendages symmetrical, as long as claws (as in Fig. 3). References: Casey 1895 (key to some species); Blaisdell 1923, 1924d.

Pseudasydates Blaisdell 1938

The single species of *Pseudasydates*, *P. inyoensis* Blaisdell 1938, was described from California. This genus is very close to *Asydates*, differing only in the tarsal claws lacking ungual appendages. Reference: Blaisdell 1938a.

Eutricholistra Blaisdell 1924

This genus has only one species, *E. punctata* Blaisdell 1924, described from Arizona. The diagnostic characters of the adults of this genus are the following: pubescence dual, with both decumbent and erect setae; eyes finely setose; pronotum with lateral margins not serrulate, apical angles not produced; tarsal claws with ungual appendages symmetrical, about two-thirds as long as claws. Reference: Blaisdell 1924c.

The following genera (*Sydates, Adasytes, Cradytes, Eudasytes, Trichochrous, Eutrichopleurus, Emmenotarsus, Byturosomus, Trichochronellus, Listropsis*, and *Sydatopsis*) are treated here as a related group. The diagnostic characters of the adults of this group are the following: male pro- and mesotibiae with one broad, blunt, spoon-shaped spur and one acute spur (as in Fig. 16); tarsal claws with ungual appendages symmetrical, as long as claws (as in Fig. 3). Species of *Eutrichopleurus, Emmenotarsus, Byturosomus, Trichochronellus, Listropsis* and *Sydatopsis*, are differentiated from *Trichochrous* in large part on differences in vestiture. All except *Sydatopsis* have species once included in *Trichochrous*.

Sydates Casey 1895
There is only one species in the genus *Sydates, S. zonatus* Casey 1895, described from El Paso, Texas. Three species described by Pic (1910) from Argentina and Peru, were subsequently transferred to *Amecomycter* and *Mauroniscus* by Majer (1995). The diagnostic characters of the adults of this genus are the following: pubescence simple, with pale and black decumbent setae; pronotum with lateral margins not serrulate, apical angles not produced; protibiae lacking spines on external (dorsal) margin in both sexes (as in Fig. 22), or spines few in number, very slender (as in Fig. 21); tarsal claws with ungual appendages acutely pointed. The dark spot on the pronotal disc, as well as humeral and preapical dark spots on the elytra result from the presence of black setae, and are not denuded spots as stated by Casey (1895) in his description of *S. zonatus*. References: Pic 1910; Casey 1895; Majer 1995.

Adasytes Casey 1895
There is only one species in this genus, *A. laciniatus* Casey 1895, described from southern California. The diagnostic characters of the adults of this genus are the following: pubescence dual, with both decumbent and erect setae, pronotal fringe long, irregular; pronotum with lateral margins not serrulate, apical angles not produced; protibiae lacking spines on external (dorsal) margin in both sexes (as in Fig. 22), or spines few in number, very slender (as in Fig. 21); tarsal claws with ungual appendages symmetrical, as long as claws (as in Fig. 21). Reference: Blaisdell 1938a.

Cradytes Casey 1895
There are five species of *Cradytes*, all found in the arid southwest of Arizona, New Mexico, and Colorado. The diagnostic characters of the adults of this genus are the following: pubescence dual, with both decumbent and erect setae; pronotum with lateral margins conspicuously serrate in male, less so in female, apical angles not produced; protibiae with spines on external (dorsal) margin in both sexes (as in Fig. 20), spines somewhat obscured by pale decumbent setae. In the species key (Blaisdell 1924b) only four species are recognized. A checklist for the North American Beetle Fauna Project (Arnett 1975) lists an additional species, *C. prosternalis* (Fall 1907). The potential synonymy of *Cradytes* with *Trichochrous* is discussed by Fall (1907) following the description of *T. placatus* Fall 1907. The inclusion of *T. prosternalis* in *Cradytes* by Arnett is probably based on the species list in Fall and Cockerell (1907: 183) which treats *T. serricollis* (LeConte 1866) and *T. prosternalis* in the subgenus *Cradytes*. References: Fall 1907 (as a subgenus of *Trichochrous*); Fall and Cockerell 1907; Blaisdell 1924b (key to species); Arnett 1975 (checklist).

Eudasytes Casey 1895
There are nine described species of *Eudasytes* living primarily in the desert regions of California, Utah, Arizona, and Nevada. The diagnostic characters of the adults of this genus are the following: form sub-oval, robust; pubescence dual, with both decumbent and erect setae; eyes large, minutely setose; pronotum with lateral margins not serrulate, apical angles produced, prominent; protibia with spines on external (dorsal) margin in both sexes (as in Fig. 20); tarsal claws with ungual appendages symmetrical, as long as claws (as in Fig. 20). References: Blaisdell 1921b, 1924b (key to some species), 1937, 1940b.

Trichochrous Motschulsky 1859
There are 152 species of *Trichochrous* in North America, with 115 species found in all geographic regions of the western United States and Canada. The diagnostic characters of the adults of this genus are the following: pubescence simple, with abundant decumbent setae, no trace of erect setae; pronotum with lateral margins not serrulate, apical angles not produced; protibia with spines on external (dorsal) margin in both sexes (as in Fig. 20), spines often fewer in number or absent in male (as in Figs. 21-22). The tibial spurs are variable in *Trichochrous*, but there are at least some species in which the male pro- and mesotibiae have one acute and one broad, spoon shaped spur. In other species the male pro- and mesotibiae have two acute spurs. Variation in the expression of this character may help to clarify species relationships in *Trichochrous*. Blaisdell (1938a) defined the genus to include only those species with simple decumbent pubescence. Unfortunately, he did not reassign the numerous species with dual pubescence, composed of both decumbent and erect setae. In the key, these species will run to *Emmenotarsus* (erect setae black) or *Eutrichopleurus* (erect setae pale). References: Casey 1895 (key to some species); Fall 1907; Blaisdell 1923, 1924a, 1924b, 1924d, 1925c, 1926, 1927b, 1930, 1931a, 1931b, 1937, 1938a, 1940b, 1942; Tanner 1928.

Eutrichopleurus Blaisdell 1938
There are six species in the genus *Eutrichopleurus* found in Washington, Oregon, Idaho and Utah. The diagnostic characters of the adults of this genus are the following: pubescence pale, dual, with both decumbent and erect setae; pronotum with lateral margins not serrulate, apical angles not produced; protibia with spines on external (dorsal) margin in both sexes (as in Fig. 20). Blaisdell (1938a) removed *Eutrichopleurus* from *Trichochrous* primarily based on differences in vestiture: single, short, decumbent in *Trichochrous*; dual, with pale decumbent and erect setae in *Eutrichopleurus*. In the key to genera, some *Trichochrous* species will key out to *Eutrichopleurus* (see *Trichochrous* above). References: Blaisdell 1938a, 1940b; Hatch 1962 (key to species).

Emmenotarsus Motschulsky 1859

Eummenotarsus Arnett 1962 (incorrect subsequent spelling)

Emmenotarsis Arnett 1962 (incorrect subsequent spelling)

Only two species, *E. quadricollis* (LeConte 1859), and *E. faulkneri* Howell 1997, both endemic to California, have been described in this small genus. The diagnostic characters of the adults of this genus are the following: pubescence dual, with both decumbent and erect setae; pronotum with lateral margins not serrulate, apical angles not produced; protibia with spines on external (dorsal) margin in both sexes (as in Fig. 20), spines often fewer in number or absent in male (as in Figs. 21-22). Blaisdell (1938a) removed *Emmenotarsus* from *Trichochrous* primarily based on differences in vestiture: single, short, decumbent in *Trichochrous*; dual, with both pale decumbent and erect black setae in *Emmenotarsus*. In the key to genera, numerous species of *Trichochrous* will key out to *Emmenotarsus* (see *Trichochrous* above). Arnett (1962) used the names *Eummenotarsus* (key) and *Emmenotarsis* (text) for this genus. *Emmenotarsus* is the correct original spelling (Motschulsky 1859: 394), and *Eummenotarsus* and *Emmenotarsis* are considered incorrect subsequent spellings. References: Blaisdell 1938a; Howell 1997.

Byturosomus Motschulsky 1859

Byturosoma Blaisdell 1938 (unjustified emendation); Arnett 1962

The single species of *Byturosomus*, *B. fuscus* (LeConte 1852), was described from southern California. The diagnostic characters of the adults of this genus are the following: pubescence dual, with both pale decumbent and erect black setae; pronotum with lateral margins not serrulate, apical angles not produced; protibia with stout spines on dorsal margin in both sexes (Fig. 15). The robust body and stout arcuate (in male) protibiae of *Byturosomus* are distinctive. Without explanation Blaisdell (1938a) used the name *Byturosoma* for this genus, and this usage was followed by Arnett (1962). *Byturosomus* is the correct original spelling (Motschulsky 1859: 395), and *Byturosoma* is considered an unjustified emendation. References: Motschulsky 1859; LeConte 1866; Blaisdell 1938a.

Trichochronellus Blaisdell 1938

This genus has only one species, *T. stricticollis* (Casey 1895), described from California. The diagnostic characters of the adults of this genus are the following: pubescence simple, long, suberrect, black; pronotum with lateral margins not serrulate, apical angles not produced, lateral margins sinuate behind the apex; protibia with spines on external (dorsal) margin in both sexes (as in Fig. 20). Reference: Blaisdell 1938a.

Listropsis Blaisdell 1924

Trichochroides Blaisdell 1938

There are 12 species in the genus *Listropsis*, all from California. The diagnostic characters of the adults of this genus are the following: pubescence dual, with both decumbent and erect setae; pronotum with lateral margins not serrulate, apical angles not produced; protibia with spines on external (dorsal) margin in both sexes (as in Fig. 20), spines fewer in number or absent in males and some females (as in Figs. 21-22); males with fifth ventral abdominal segment emarginate apically, the disk variously modified. This genus was described for the single species *L. tinctus* Blaisdell 1924, and a second species, *L. armatulus* was added by Blaisdell in 1925. Howell (1987) transferred 10 species of *Trichochroides* Blaisdell 1938 to *Listropsis*, treating *Trichochroides* as a junior synonym. References: Blaisdell 1924a, 1924b (as *Trichochrous*), 1925b, 1927a (as *Trichochrous*), 1941 (key to some species, as *Trichochroides*), 1938a; Howell 1987 (taxonomy).

Sydatopsis Casey 1895

The single species of *Sydatopsis*, *S. longicornis* Casey 1895, was described from Baja California. The diagnostic characters of the adults of this genus are the following: pubescence dual, with both decumbent and erect setae; pronotum with lateral margins not serrulate, apical angles not produced; protibia with spines on external (dorsal) margin in both sexes (as in Fig. 20). *Sydatopsis* is easily recognized by the characters given in the key. References: Casey 1895; Blaisdell 1938a.

[*Holomallus* Gorham 1886]

This genus does not occur north of Mexico. *Holomallus* is a small North American genus with two species, *H. aurivillus* Gorham 1886, and *H. serripes* Champion 1914, both from Mexico. The diagnostic characters of the adults of this genus are the following: pubescence dual, with both decumbent and erect setae, pubescence dense, shaggy; antennae short, antennomeres VII-XI transverse, forming an elongate club; pronotum with lateral margins not serrulate, apical angles not produced; protibia with stout spines on external (dorsal) margin in both sexes (as in Fig. 15); tarsal claws with ungual appendages symmetrical, slender, minute. References: Gorham 1886; Champion 1914; Blaisdell 1938a.

Paradasytes Hatch 1962

The single species of *Paradasytes*, *P. barri* Hatch 1962, was described from Idaho. The distinguishing characteristics of this genus appear to be identical with those of *Dasytellus*. In *Paradasytes* the submarginal excavated lines on the pronotum are completely obscured by dense decumbent setae (as they are in some *Dasytellus*), and were not mentioned by Hatch. Reference: Hatch 1962.

Neadasytes Hatch 1962

There is only one species in this genus, *N. testaceous* Hatch 1962, described from Idaho. The distinguishing characteristics of this genus appear to be identical with those of *Dasytastes*. Reference: Hatch 1962.

Listromimus Casey 1895

The single species of *Listromimus*, *L. sericatus* Casey 1895, was described from Nevada. The diagnostic characters of the adults of this genus are the following: body graciliform, slightly widened apically; pronotum sinuate at base, without excavated lateral submarginal lines, lateral margins not serrulate; elytral epipleurae wide, flat, horizontal, broad to near apex; tarsal claws with ungual appendages symmetrical, as long as claws. References: Casey 1895; Blaisdell 1938a.

Listrimorpha Blaisdell 1921

The single species of *Listrimorpha*, *L. pallipes* Blaisdell 1921, was described from Nevada. The diagnostic characters of the adults of this genus are the following: body graciliform, slightly widened apically; pronotum sinuate at base, excavated lateral submarginal lines obscure, restricted to basal half of pronotum, lateral margins minutely serrulate; elytral epipleurae narrow basally, obsolete beyond middle of elytra; tarsal claws with ungual appendages symmetrical, as long as claws. Majer (1990) placed *Listromimus* and *Listrimorpha* in the tribe Listrini; however, the shape of the pronotum (sinuate at base) along with similarities in vestiture and other characters, suggests that these taxa are related to *Vectura* and allied genera. Reference: Blaisdell 1921a.

Bibliography

ARNETT, R. H., Jr. 1960-62. The beetles of the United States (a manual for identification). Parts I-VI. Part IV. 1962. Fascicle 67. Melyridae. Pp. 607-616. The Catholic University of America Press. Washington, DC, xi + 1112 pp.

ARNETT, R. H., Jr. 1975. Checklist of the beetles of North and Central America and the West Indies. Volume 4: The click beetles, fireflies, checkered beetles, and related groups. Family 74. Melyridae. Flora and Fauna Publications, Gainesville, Florida, 22 pp.

BALDUF, W. V. 1935. The bionomics of entomophagous Coleoptera. John Swift, New York, 220 pp.

BARBER, H. S. 1949. A note on Malachiidae. Coleopterists Bulletin, 3: 72.

BLACKWELDER, R. E. 1932. The genus *Endeodes* LeConte (Coleoptera: Melyridae). Pan-Pacific Entomologist, 8: 128-136.

BLACKWELDER, R. E. 1945. Checklist of the coleopterous insects of Mexico, Central America, the West Indies, and South America. Bulletin of the United States National Museum, 185(3): 342-550.

BLACKWELDER, R. E. 1957. Checklist of the coleopterous insects of Mexico, Central America, the West Indies, and South America. Bulletin of the United States National Museum, 185(6): 926-1492.

BLAISDELL, F. E., Sr. 1906. New Californian Coleoptera. Entomological News, 17: 71-80.

BLAISDELL, F. E., Sr. 1921a. New species of Melyridae, Chrysomelidae, and Tenebrionidae (Coleoptera) from the Pacific Coast, with notes on other species. Stanford University Publications, University Series I, Biological Sciences, 1: 133-231.

BLAISDELL, F. E., Sr. 1921b. Four new species of Melyridae (Coleoptera). Canadian Entomologist, 53: 15-19.

BLAISDELL, F. E., Sr. 1921c. Miscellaneous studies in the Coleoptera - No. 1. Canadian Entomologist, 53: 129-132.

BLAISDELL, F. E., Sr. 1923. Expedition of the California Academy of Sciences to the Gulf of California in 1921. The Melyridae (lesser flower beetles). Proceedings of the California Academy of Sciences Fourth Series, 12(19): 409-421.

BLAISDELL, F. E., Sr. 1924a. Two new species of Melyridae from California and one from British Columbia, including two new genera. Canadian Entomologist, 56: 1-5.

BLAISDELL, F. E., Sr. 1924b. Studies in the Melyridae (Coleoptera) number two. Transactions of the American Entomological Society, 49: 315-337.

BLAISDELL, F. E., Sr. 1924c. Studies in the Melyridae (Coleoptera) III. Pan-Pacific Entomologist, 1(1): 15-21.

BLAISDELL, F. E., Sr. 1924d. New Melyrids from southeastern California. Proceedings of the California Academy of Sciences Fourth Series, 13(17): 249-259.

BLAISDELL, F. E., Sr. 1925a. A new species of *Dasytes* from California (Melyridae: Coleoptera). Pan-Pacific Entomologist, 1(4): 184-185.

BLAISDELL, F. E., Sr. 1925b. Studies in the Melyridae (Coleoptera) number four. Transactions of the American Entomological Society, 50: 313-318.

BLAISDELL, F. E., Sr. 1925c. Expedition to Guadelupe Island, Mexico. Proceedings of the California Academy of Sciences Fourth Series, 14(14): 321-343.

BLAISDELL, F. E., Sr. 1925d. Coleoptera of the Pacific coast, notes and criticisms. Entomological News, 36: 79-85.

BLAISDELL, F. E., Sr. 1926. Studies in the Melyridae, No. V (Coleoptera). Canadian Entomologist, 58: 8-13.

BLAISDELL, F. E., Sr. 1927a. Miscellaneous studies in the Coleoptera No. 2. Pan-Pacific Entomologist, 3(4): 163-168.

BLAISDELL, F. E., Sr. 1927b. Studies in the Melyridae No. 6. Pan-Pacific Entomologist, 4(2): 49-53.

BLAISDELL, F. E., Sr. 1928. Studies in the Melyridae No. 7. Pan-Pacific Entomologist, 5(1): 35-42.

BLAISDELL, F. E., Sr. 1930. Studies in the Melyridae, number eight. Pan-Pacific Entomologist, 7(1): 17-19.

BLAISDELL, F. E., Sr. 1931a. Studies in the Melyridae No. 9. (Coleoptera). Canadian Entomologist, 63: 178-183.

BLAISDELL, F. E., Sr. 1931b. Studies in the Melyridae (Coleoptera) number ten. Transactions of the American Entomological Society, 57: 325-331, plate 29.

BLAISDELL, F. E., Sr. 1934a. A new species of *Vectura* from Southern California. Pan-Pacific Entomologist, 10(2): 71-73.

BLAISDELL, F. E., Sr. 1934b. Rare North American Coleoptera. Transactions of the American Entomological Society, 60: 317-326, plate 21.

BLAISDELL, F. E., Sr. 1934c. A new species of *Hoppingiana* from British Columbia (Coleoptera: Melyridae). Canadian Entomologist, 66: 150-152.

BLAISDELL, F. E., Sr. 1936. Studies in the Melyridae No. 11. Pan-Pacific Entomologist, 12(4): 184-190, figures 1-4.

BLAISDELL, F. E., Sr. 1937. Miscellaneous studies in the Coleoptera, No. 5 (Tenebrionidae and Melyridae). Transactions of the American Entomological Society, 63: 127-145.

BLAISDELL, F. E., Sr. 1938a. A generic synopsis and generic revision of the tribe Dasytini of North America, North of Panama. Transactions of the American Entomological Society, 64: 1-31, plates 1-2.

BLAISDELL, F. E., Sr. 1938b. A new species of *Listrus* from the Sequoia National Park, California (Coleoptera: Melyridae). Pan-Pacific Entomologist, 14(4): 165-167.

BLAISDELL, F. E., Sr. 1939. Studies in the relationships of the subfamilies and tribes of the Tenebrionidae based on the primary genital characters, also descriptions of new species (Coleoptera). Transactions of the American Entomological Society, 65: 43-60, plates 4-5.

BLAISDELL, F. E., Sr. 1940a. A new species of *Listrus* from Cajon Pass, California (Coleoptera, Melyridae). Canadian Entomologist, 72: 212-213.

BLAISDELL, F. E., Sr. 1940b. Studies in the Melyridae No. 12. Transactions of the American Entomological Society, 66: 319-324.

BLAISDELL, F. E., Sr. 1941. A monographic study of the species belonging to the melyrid genus *Trichochroides*. Transactions of the American Entomological Society, 66: 283-306.

BLAISDELL, F. E., Sr. 1942. Miscellaneous studies in the Coleoptera, No. 6 (Melyridae and Tenebrionidae). Transactions of the American Entomological Society, 68: 129-144.

BØVING, A. G. and C. F. CRAIGHEAD. 1931. An illustrated synopsis of the principal larval forms of the order Coleoptera. Melyridae, Pp. 8, 55, pls. 91, 92. Entomologica Americana, 11: 1-351.

BROWN, W. J. 1944. Some new and poorly known species of Coleoptera, II. Canadian Entomologist, 76: 4-10.

CASEY, T. L. 1895. Coleopterological notices VI. Melyridae, Pp. 456-606. Annals of the New York Academy of Sciences, 8: 435-838.

CHAMPION, G. C. 1913. Notes on various Central American Coleoptera, with descriptions of new genera and species. Transactions of the Entomological Society of London (1913): 58-169, plates 3-4.

CHAMPION, G. C. 1914. Revision of the Mexican and Central American Malachiidae and Melyridae, with descriptions of new species. Transactions of the Entomological Society of London (1914): 13-127, plate 2.

CHAMPION, G. C. 1922. The S. African species of *Dinometopus*, *Troglops*, *Chalicorus*, *Colotes*, *Helcogaster*, and the allied genera, with an account of their accessory male-characters [Coleoptera]. Annals and Magazine of Natural History, Ser. 9, 10: 309-358, plates 4-6.

CHAMPION, G. C. 1923. A revision of the Malayan and Indian species of the melyrid subfamily Carphurinae represented in the Hope collection at Oxford and in the British Museum in London [Coleoptera]. Annals and Magazine of Natural History, Ser. 9, 12: 1-54, plate 1.

CONSTANTIN, R. 1965. Notes sur quelques Malacodermes du Nord de l'Espagne. L'Entomologiste, 21: 87-94.

CONSTANTIN, R. 1983. Dasytides nouveaux de Crete et de Turquie (Coleoptera, Dasytidae). Entomologica Basiliensia, 8: 329-333.

CONSTANTIN, R. and J. J. MENIER. 1987. Étude d'un remarquable Melyridae aptère des Îles Canaries: *Gietella fortunata*, n. gen., n. sp., type d'une souse-famille nouvelle Gietellinae (Coleoptera, Cleroidea). Revue Francaise d'Entomolgie (Nouvelle Serie), 9: 53-63.

CROWSON, R. A. 1955. The Natural Classification of the Families of Coleoptera. London, 187 pp.

CROWSON, R. A. 1964. A review of the classification of Cleroidea (Coleoptera) with descriptions of two new genera of Peltidae and of several larval types. Transactions of the Royal Entomological Society of London, 116: 275-327.

CROWSON, R. A. 1970. Further observations on Cleroidea (Coleoptera). Proceedings of the Royal Entomological Society of London Series B Taxonomy, 39: 1-20.

DAHMS, R. G. and M. KAGAN. 1938. Egg predator of the chinch bug. Journal of Economic Entomology, 31: 779-780.

DIX, M. E. 1990. *Malachius ulkei* Horn (Coleoptera: Malachiidae): An egg predator of *Paleacrita vernata* Peck (Lepidoptera: Geometridae). Journal of the Kansas Entomological Society, 63: 648-652.

DOW, R. P. 1914. *Malachius aeneus* Linn. Journal of the New York Entomological Society, 22: 73.

DOWNIE, N. M. 1950. Notes on the distribution of *Bruchus brachialis* Fahraeus and *Malachius aeneus* (L.). Coleopterists Bulletin, 4(2): 20-21.

DOWNIE, N. M. 1972. An interesting record of *Temnopsophus bimaculatus* Horn (Coleoptera: Melyridae). Coleopterists Bulletin, 26: 154.

ERICHSON, W. F. 1840. Entomographien, untersuchungen in dem Gebiete der Entomologie, mit besonderer Benutzung der Königl. Sammlung zu Berlin, ed. I, vol. III, Die Malachien. Pp. 44-134. F. H. Morin. Berlin, 180 pp.

EVERS, A. M. J. 1960. Beobachtungen an *Axinotarsus pulicarius* F. Entomologische Blätter für Biologie und Systematik der Käfer, 56: 77-88.

EVERS, A. M. J. 1968. Carphurinae oder Carphuridae? Entomologische Blätter für Biologie und Systematik der Käfer, 64: 17-27.

EVERS, A. M. J. 1985. Aufteilung der paläaktischen arten des gattungskomplexes *Malachius* F. Entomologische Blätter für Biologie und Systematik der Käfer, 81: 1-40.

EVERS, A. M. J. 1988. Die gattungen *Microlipus* LeC. und *Wittmerius* gen. nov. Entomologische Blätter für Biologie und Systematik der Käfer, 84: 165-170.

EVERS, A. M. J. 1989. Phylogenese der Malachiidae (Col.) der welt kladistik und biogeographie aus der sicht der pangaea-theorie. Entomologische Blätter für Biologie und Systematik der Käfer, 85: 1-57.

EVERS, A. M. J. 1993. Neue Zentral-Amerikanische Malachiidae. Entomologische Blätter für Biologie und Systematik der Käfer, 89: 49-58.

EVERS, A. M. J. 1994. Zur phylogenie von *Laius* Guér., *Collops* Er. und der verwandten gattungen. Entomologische Blätter für Biologie und Systematik der Käfer, 90: 169-181.

EVERS, A. M. J. 1995. Neue Malachiidae aus California, Nevada und Arizona. Entomologische Blätter für Biologie und Systematik der Käfer, 91: 165-170.

FABRICIUS, J. C. 1775. Systema Entomologie. Flensburgi et Lipsiae, Korte, 832 pp.

FALL, H. C. 1901. List of the Coleoptera of Southern California, with notes on habits and distribution and descriptions of new species. Malachidae, Pp. 23-24, 123-128, 242-251. Occasional Papers of the California Academy of Sciences No. 8, 282 pp.

FALL, H. C. 1907. Descriptions of new species, Pp. 218-272. *In*: H. C. Fall and T. D. A. Cockerell. The Coleoptera of New Mexico. Transactions of the American Entomological Society, 33: 145-272.

FALL, H. C. 1910. Miscellaneous notes and descriptions of North American Coleoptera. Transactions of the American Entomological Society, 36: 89-197.

FALL, H. C. 1912. A review of the North American species of *Collops* (Col.). Journal of the New York Entomological Society, 20: 249-274.

FALL, H. C. 1913. A correction. Journal of the New York Entomological Society, 21: 71-72.

FALL, H. C. 1917. Short studies in the Malachiidae. Transactions of the American Entomological Society, 43: 67-88.

FALL, H. C. 1930. New Coleoptera XIV, with notes on known species. Canadian Entomologist, 62: 251-257.

FALL, H. C. and T. D. A. COCKERELL. 1907. The Coleoptera of New Mexico. Transactions of the American Entomological Society, 33: 145-272.

FIORI, G. 1959. Contribute alla conoscenza morphologia ed etologica Coleotteri, VII. Su alcune larve di Malachiidi. Studia Sassaresi, 7: 323-259.

FORBES, W. T. M. 1922. The wing-venation of the Coleoptera. Malachiidae (Melyridae), Pp. 338-339, figs. 57, 58. Annals of the Entomological Society of America, 15: 328-345, plates 29-35, 71 figures.

FORBES, W. T. M. 1926. The wing folding patterns of the Coleoptera. Melyridae, P. 109, figs. 122-124. Malachiidae, Pp. 108-109, figs. 125-126. Journal of the New York Entomological Society, 34: 42-68, 91-115, plates 7-18, 145 figures.

FOSTER, D. E. and A. L. ANTONELLI. 1973. Larval description and notes on the biology of *Anthocomus horni* (Coleoptera: Melyridae). Pan-Pacific Entomologist, 49: 56-59.

FRAUENFELD, G. R. 1866. Über die bisher eingelangten diessjaerigen berichte von landwirthschaftlichen insektenschaeden. Verhandlungen Zoologisch-Botanische Gesellschaft in Wien, 16: 641-644.

FRENCH, G. T. 1942-43. A new household insect. Report of the Virginia Department of Agriculture and Immigration, P. 59.

FRENCH, G. T. 1944. *Anthocomus bipunctatus* (Harrer) a new household pest. Journal of Economic Entomology, 37: 103.

GORHAM, H. S. 1882. Fam. Melyridae (part), Cleridae (part), Pp. 113-168. *In*: F. D. Godman and O. Salvin, eds., Biologia Centrali-Americana. Insecta, Coleoptera. Vol. III. Part 2. Porter. London.

GORHAM, H. S. 1886. Supplement (part), Pp. 313-360. *In*: F. D. Godman and O. Salvin, eds., Biologia Centrali-Americana. Insecta, Coleoptera. Vol. III. Part 2. Porter. London.

GREINER, J. 1937. Pars 159. Malachiidae. *In*: S. Schenkling, ed., Coleopterorum Catalogus. W. Junk, Berlin, 199 pp.

HATCH, M. H. 1927. New aberrations of *Temnopsophus* and *Pseudebaeus* (Malachiinae, Coleoptera). Annals of the Entomological Society of America, 20: 366-367.

HATCH, M. H. 1962. The beetles of the pacific northwest. Part III: Pselaphidae and Diversicornia I. Family Melyridae. Pp. 68-95, 400-407, plates 21-23. University of Washington Publications in Biology. Vol. 16, 503 pp.

HORN, G. H. 1870. Contributions to the coleopterology of the United States. Transactions of the American Entomological Society, 3: 69-97.

HORN, G. H. 1872. Synopsis of the Malachiidae of the United States. Transactions of the American Entomological Society, 4: 109-127.

HORN, G. H. 1874. Descriptions of new species of United States Coleoptera. Transactions of the American Entomological Society, 5: 20-43.

HORN, G. H. 1882. Notes on some little known genera and species of Coleoptera. Transactions of the American Entomological Society, 10: 113-126, plate 6.

HOWELL, C. D. 1979. Studies of Dasytidae No. 1: New Species of *Pristocelis* (Coleoptera). Pan-Pacific Entomologist, 55: 41-45.

HOWELL, C. D. 1985. Diverse populations and subspecies in *Eschatocrepis* (Coleoptera: Dasytidae). Entomological News, 96: 129-141.

HOWELL, C. D. 1987. Sex-influenced protibial spines and sysnonymy in Dasytidae (Coleoptera), study number three. Pan-Pacific Entomologist, 63: 50-51.

HOWELL, C. D. 1997. Three new species of North American Dasytinae, with two new generic keys. Journal of the Kansas Entomological Society, 70: 11-17.

HUSSAIN, M. 1975. Predators of the alfalfa weevil, *Hypera postica* in western Nevada - a greenhouse study (Coleoptera: Curculionidae). Journal of the New York Entomological Society, 83: 226-228.

JACQUELIN DU VAL, P. N. C. 1859-1863. Genera des Colopteres d' Europe, Manuel Entomologique. Volume III. (*Allonyx*, Cleridae, p. 196). Chez A. Deyrolle, Naturaliste. Paris, 463 pp.

KIESENWETTER, E. A. H., VON. 1863. Naturgeschichte der Insecten Deutschlands, Coleoptera, ed. IV. Nicolai. Berlin, Pp. 569-746.

KING, P. S. 1985. Natural history of *Collops georgianus* (Coleoptera: Melyridae). Annals of the Entomological Society of America, 78: 131-136.

KING, P. S. 1987. Macro- and microgeographic structure of a spatially subdivided beetle species in nature. Evolution, 41: 401-416.

KING, P. S. 1988. Distribution and genetic structure of two allopatric beetle (Coleoptera: Melyridae) species on rock outcrops in the southeast. Annals of the Entomological Society of America, 81: 890-898.

KNOWLTON, G. F. 1944. *Collops* Feeding. Journal of Economic Entomology, 37: 443.

KNOWLTON, G. F. and W. MEIER. 1942. *Collops bipunctatus* Say. Journal of Economic Entomology, 35: 108.

LAPORTE, F. L. N. de C. 1840. Histoire naturelle des Insectes Coléoptères. Tome Premier. *In*: Histoire naturelle des animaux Articulés, Annelidés, Crustacés, Arachnides, Myriapodes et Insectes. P. Duménil, Paris, 324 pp., 19 plates.

LAWRENCE, J. F. 1982. Coleoptera, Pp. 482-553. *In*: S. P. Parker, ed., Synopsis and Classification of Living Organisms. Volume 2. McGraw-Hill. New York, NY, 1232 pp.

LEACH, W. E. 1815. Entomology. *In*: D. Brewster, ed., Edinburgh Encyclopaedie, 9: 57-172.

LEACH, W. E. 1817. Insecta. *In*: D. Brewster, ed., Edinburgh Encyclopaedie, 12: 155-164.

LECONTE, J. L. 1852a. Descriptions of new species of Coleoptera from California. Annals of the Lyceum of Natural History of New York, 5: 125-216.

LECONTE, J. L. 1852b. Catalogue of the Melyrides of the United States, with descriptions of new species. Proceedings of the Academy of Natural Sciences of Philadelphia, 6: 163-171.

LECONTE, J. L. 1859a. Descriptions of some genera and species of Coleoptera from the vicinity of the southern boundary of the United States of America. Arcana Naturae, 1: 121-128, plates 12-13.

LECONTE, J. L. 1859b. Catalogue of the Coleoptera of Fort Tejon, California. Proceedings of the Academy of Natural Sciences of Philadelphia, 11: 69-90.

LECONTE, J. L. 1859c. Additions to the coleopterous fauna of northern California and Oregon. Proceedings of the Academy of natural Sciences of Philadelphia, 11: 281-292.

LECONTE, J. L. 1862. Classification of the Coleoptera of North America. Smithsonian Miscellaneous Collections, 3(136): 190-194.

LECONTE, J. L. 1866. Revision of the Dasytini of the United States. Proceedings of the Academy of Natural Sciences of Philadelphia, 19: 349-361.

LECONTE, J. L. 1880. Short Studies of North American Coleoptera. *Chaetocoelus* n. g. (Malachidae). Transactions of the American Entomological Society, 8: 194.

LECONTE, J. L. 1884. Description of the genus *Endeodes* (Malachidae). Reprint from James Thomson's Arcana Naturae. Paris 1859, vol. I, P. 122. Bulletin of the Brooklin Entomological Society, 6: 136.

LECONTE, J. L. and G. H. HORN. 1883. Classification of the Coleoptera of North America. Prepared for the Smithsonian Institution. Fam. XLV.-Malachiidae, Pp. 212-216. Smithsonian Miscellaneous Collections, Volume 26, part 4. (No. 507), 567 pp.

LEECH, H. B. 1947. Collecting in southern British Columbia: hilltop to lakeshore for beetles. Canadian Entomologist, 79: 105-108.

LENG, C. W. 1918. Notes on some changes in the list of Coleoptera. Journal of the New York Entomological Society, 26: 201-211.

LENG, C. W. 1920. Catalogue of the Coleoptera of America, north of Mexico. John D. Sherman, Jr. Mount Vernon, NY, 470 pp.

LENG, C. W. and A. J. MUTCHLER. 1927. Catalogue of the Coleoptera of America, north of Mexico. Supplement 1919 to 1924 (inclusive). John D. Sherman, Jr. Mount Vernon, NY, 78 pp.

LENG, C. W. and A. J. MUTCHLER. 1933. Catalogue of the Coleoptera of America, north of Mexico. Second and Third Supplements 1925 to 1932 (inclusive). John D. Sherman, Jr. Mount Vernon, NY, 112 pp.

LEWIS, G. 1895. On the Dascillidae and malacoderm Coleoptera of Japan. Annals and Magazine of Natural History (6), 16: 98-122, plate 6, figure 11.

LIBERTI, G. 1982. *Dasytes (Mesodasytes)* of Taiwan (Col., Dasytidae). Entomologica Basiliensia, 7: 373-380.

LIBERTI, G. 1984. *Dasytes (Mesodasytes)* of Taiwan (Col., Dasytidae) (Part 2). Entomologica Basiliensia, 9: 324-327.

LOPEZ, E. G. and G. L. TEETLES. 1976. Selected predators of aphids in grain sorghum and their relation to cotton. Journal of Economic Entomology, 69: 198-204.

MAJER, K. 1987. Comparative morphology and proposed major taxonomy of the Family Melyridae (Insecta, Coleoptera). Polskie Pismo Entomologiczne, 56: 719-859.

MAJER, K. 1989a. A revision of the tribe Danacaeomimini (Coleoptera, Melyridae, Dasytinae). Deutsche Entomologische Zeitschrift, 36: 1-31.

MAJER, K. 1989b. Generic classification of the tribe Chaetomalachiini (Coleoptera, Melyridae). Polskie Pismo Entomologiczne, 58: 745-774.

MAJER, K. 1990. A new tribe Listrini trib. n. including two new genera (Coleoptera, Melyridae). Acta Entomologica Bohemoslovaca, 87: 368-384.

MAJER, K. 1994a. A review of the classification of the Melyridae and related families (Coleoptera:, Cleroidea). Entomologica Basiliensia, 17: 319-390.

MAJER, K. 1994b. Gietellidae, full family status for Melyridae: Gitetellinae (Coleoptera: Cleroidea). Entomological Problems, 25: 65-72.

MAJER, K. 1995. Revision of the family Mauroniscidae (Insecta: Coleoptera: Cleroidea). Entomologische Abhandlungen Staatliches Museum für Tierkunde Dresden, 57: 57-89.

MAJER, K. 1996. Complemetary generic revision of the subfamily Chaetomalachiinae (Insecta: Coleoptera: Dasytidae). Annalen des Naturhistorischen Museums in Wien serie B Botanik und Zoologie, 98B: 435-500.

MAJER, K. 1997. A revision of the tribe Amauronioidini (Coleoptera, Dasytidae). Acta Musei Moraviae Scientiae Naturales, 81: 363-402.

MAJER, K. 1998. Rhadalinae from the Baltic amber (Coleoptera, Dasytidae). Deutsche Entomologische Zeitschrift, 45: 255-264.

MALKIN, B. 1948. A new *Attalus* from the western United States. Pan-Pacific Entomologist, 24: 207-208.

MARSHALL, M. Y. 1936. A new Melyrid of the genus *Tanaops* (Coleoptera). Bulletin of the Southern California Academy of Sciences, 35: 164-165.

MARSHALL, M. Y. 1944. A new *Tanaops* from New Mexico, with some notes on the genus (Coleoptera: Melyridae). Canadian Entomologists, 76: 164-166.

MARSHALL, M. Y. 1946. Studies in the Malachiidae. Canadian Entomologist, 78: 183-195.

MARSHALL, M. Y. 1948. Studies in the Malachiidae II. Entomologica Americana, 28: 113-144.

MARSHALL, M. Y. 1951. Studies in the Malachiidae III. Proceedings of the California Academy of Sciences, 27: 77-132.

MARSHALL, M. Y. 1952. The Malachiidae of North Central Mexico (Coleoptera). American Museum Novitates, No. 1584, 20 pp.

MARSHALL, M. Y. 1953. Studies in the Malachiidae IV. American Midland Naturalist, 49: 825-831.

MARSHALL, M. Y. 1954a. A key to the world genera of Malachiidae. Coleopterists Bulletin, 8: 69-82.

MARSHALL, M. Y. 1954b. Studies in the Malachiidae V. Bulletin of the Brooklin Entomological Society, 49: 66-80.

MARSHALL, M. Y. 1954c. Notes on the Malachiidae. Coleopterists Bulletin, 8: 28.

MARSHALL, M. Y. 1955. Studies in the Malachiidae VI. Coleopterists Bulletin, 9: 35-48.

MARSHALL, M. Y. 1957. Studies in the Malachiidae VII. Coleopterists Bulletin, 11: 13-16.

MATTHES, D. 1962. Excitatoren und paarungsverhalten Mitteleuropäischer Malachiiden (Coleopt., Malacodermata). Zeitschrift für Morpholgie und Ökologie der Tiere, 51: 375-546.

MAWDSLEY, J. R. 1993. Re-description and Biology of *Malachius aeneus* (Linnaeus) (Coleoptera: Melyridae). Young Entomologists Society Quarterly, 10: 35-45.

MAWDSLEY, J. R. 1999a. Redescription and notes on the biology of *Amecocerus senilis* (LeConte) (Coleoptera: Melyridae: Dasytinae). Journal of the New York Entomological Society, 107: 68-72.

MAWDSLEY, J. R. 1999b. Review of the extant and fossil Dasytinae (Coleoptera: Melyridae) of Colorado, U.S.A. Transactions of the American Entomological Society (Philadelphia), 125: 251-267.

MAYOR, A. J. 1982. A review of the genus *Chaetocoelus* LeConte (Coleoptera: Malachiidae). Proceedings of the Entomological Society of Washington, 84: 475-479.

MAYOR, A. and W. WITTMER. 1981. A review of the genus *Nodopus* with a description of a new species from southeastern North America (Coleoptera; Malachiidae). Coleopterists Bulletin, 35: 473-476.

MEDVEDEV L. N. and L. P. GALATAZ. 1969. On the larvae of *Simoderus reflexicollis* Gebl. (Coleoptera, Melyridae). Polskie Pismo Entomologiczne, 39: 331-338.

MOORE, I. 1937. A list of the beetles of San Diego County, California. Occasional Papers of the San Diego Society of Natural History, No. 2: 109 pp.

MOORE, I. M. 1954. Notes on *Endeodes* LeConte with a description of a new species from Baja California. Pan-Pacific Entomologist, 30: 195-198.

MOORE, I. 1956. Notes on some intertidal Coleoptera with descriptions of the early stages (Carabidae, Staphylinidae, Malachiidae). Transactions of the San Diego Society of Natural History, 12: 207-230, plates 14-17.

MOORE, I. 1957. A northern extension of range for *Endeodes basalis* LeConte (Coleoptera: Malachiidae). Pan-Pacific Entomologist, 33: 140.

MOORE, I. 1964. A new species of *Endeodes* from Sonora, Mexico (Coleoptera: Melyridae). Pan-Pacific Entomologist, 40: 57-58.

MOORE, I. 1971. The type locality of *Endeodes terminalis* Marshall (Coleoptera: Malachiidae). Pan-Pacific Entomologist, 47: 278.

MOORE, I. and F. G. ANDREWS. 1985. Extensions of range for some seashore and intertidal beetles of western North America (Coleoptera: Staphylinidae, Carabidae, Malachiidae and Rhizophagidae). Pan-Pacific Entomologist, 61: 221-223.

MOORE, I. and E. F. LEGNER. 1975. Revision of the genus *Endeodes* LeConte with a tabular key to the species (Coleoptera: Melyridae). Journal of the New York Entomological Society, 83: 70-81.

MOORE, I. and E. F. LEGNER. 1977. The developmental stages of *Endeodes* LeConte (Coleoptera: Melyridae). Proceedings of the Entomological Society of Washington, 79: 172-175.

MOORE, I. and A. J. MAYOR. 1976. Notes on *Endeodes* with new synonymy and extensions of range (Coleoptera: Malachiidae). Wasman Journal of Biology, 34: 179-184.

MOTSCHULSKY, V. 1854. Etudes Entomologiques. Volume 2. Helsingfors, Impremerie de la Société de Litérature Finnoise. H. Molander, 69 pp.

MOTSCHULSKY, V. 1859. Coléoptères nouveaux de la Californie. Bulletin de la Société Impériale des Naturalistes de Moscou, 32: 122-185, 357-410.

MULSANT, E. and C. REY. 1867. Histoire naturelles des Coléoptères de France. Volume 18. Vésiculifères. Paris, 311 pp.

NIELSON, M. W. and J. A. HENDERSON. 1959. Biology of *Collops vittatus* (Say) in Arizona and feeding habits of seven predators of the spotted alfalfa aphid. Journal of Economic Entomology, 53: 159-162.

ORPHANIDES, G. M., D. GONZALEZ, and B. R. BARTLETT. 1971. Identification and evaluation of pink bollworm predators. Journal of Economic Entomology, 64: 421-424.

PAYKULL, G., VON. 1798. Fauna Suecica; Insecta. (Col.). ed. I. Edman, Upsaliae, 358 pp.

PEACOCK, E. R. 1987. A review of the Rhadalinae (= Aplocneminae) (Coleoptera: Melyridae). Bulletin of the British Museum (Natural History) Entomology, 56: 129-170.

PETERSON, A. 1951. Larvae of insects. Coleoptera, Diptera, Neuroptera, Siphonaptera, Mecoptera, Tricoptera Part II. Melyridae, Pp. 57, 90 and 192, figures C3, N and C54. A. Edwards Brother. Ann Arbor, Michigan, 416 pp.

PIC, M. 1903. Malachides de l'Afrique autrale. Fin. L'Echange, 19: 164-195.

PIC, M. 1910. Contribution a l'etude des Dasytides de l'Amerique meridionale. Hors-texte de L'Echange, 26(303): 1-5.

PIC, M. 1926. Pars 87. Phloeophilidae (Pp. 1-2), Rhadalidae (P. 3), Prionoceridae (Pp. 4-11). *In*: S. Schenkling, ed., Coleopterorum Catalogus. W. Junk. Berlin, 11 pp.

PIC, M. 1928. Notes diverses, descriptions et diagnoses. Hors-texte de L'Echange, 44(431): 54.

PIC, M. 1929. Pars 103. Dasytidae I. Melyrinae (Astylini, Cerallini, Melyrini). *In*: S. Schenkling, ed., Coleopterorum Catalogus. W. Junk, Berlin, 32 pp.

PIC, M. 1937. Pars 155. Dasytidae II. Dasytinae (Danacalini, Henicopini, Dasytini). *In*: S. Schenkling, ed., Coleopterorum Catalogus. W. Junk, Berlin, 130 pp.

PLATCHER, J. 1985. Morphology and biology of the European species of the genera *Ebaeus* Er. and *Hypebaeus* Kiesw. (Coleoptera: Malachiidae). Zoologische Jahrbuecher Abteilung für Anatomie und Ontogenie der Tiere, 113: 391-425.

RAKICKAS, R. J. and T. F. WATSON. 1974. Population trends of *Lygus* spp. and selected predators in strip-cut alfalfa. Environmental Entomology, 3: 781-784.

REDTENBACHER, L. 1850. Über den Character der Ins. Faun. von Südpersien. Denkschriften der Kaiserlichen Akademie der Wissenschaften, Wien, 1: 42-53.

RUSSELL, L. 1966. A new species of *Collops* from Washington state (Coleoptera: Melyridae). Coleopterists Bulletin, 20: 8-10.

SCHILSKY, J. 1894. Beitrag zur Kenntnis der Dasytinen. Deutsche Entomologische Zeitschrift, 38: 225-236.

SCHMIDT, H. 1944. Bemerkung zur lebensweise der larve von *Malachius bipustulatus* L. (Malacodermata, Telephoridae). Anzeiger fur Schadlingskunde, 20: 10-12.

SCHMIDT, H. 1945-48. Biologische und morphologische untersuchungen an Malachiiden (Col. Malacodermata). Entomologische Blätter für Biologie und Systematik der Käfer, 41-44: 167-177.

SCHWARZ, E. A. 1878. The Coleoptera of Florida. Proceedings of the American Philosophical Society, 17: 353-469.

SHARP, D. and F. MUIR. 1912. The comparative anatomy of the male genital tube in Coleoptera. Family Malacodermidae, Pp. 535-541, plates 64-65, figures 136-146. Transactions of the Entomological Society of London, Part III: 477-642, plates 42-78.

STEPHENS, J. F. 1830. Illustrations of British Entomology, Mandibulata Vol. III. Baldwin and Cradock, London, 374 pp.

STICKNEY, F. S. 1923. The head-capsule of Coleoptera. Melyridae, P. 12, figs. 40, 188, 333, 478. Illinois Biological Monographs 8: 1-105.

TANNER, V. M. 1927. A preliminary study of the genitalia of female Coleoptera. Melyridae, P. 23, figs. 54-55. Transactions of the American Entomological Society, 53: 5-50, figures 1-222.

TANNER, V. M. 1928. The Coleoptera of Zion National Park. Annals of the Entomological Society of America, 21: 269-279.

URBAN, C. 1912. Zur lebensgeschichte des *Anthocomus rufus* Hbst (Col.). Entomologische Mitteilungen, 1(11): 331-337.

URBAN, C. 1914. Zur Naturgeschichte des *Malachius bipustulatus* L. (Col.). Entomologische Mitteilungen, 3: 4-10.

WALKER, J. K., Jr. 1957. A biological study of *Collops balteatus* Lec. and *Collops vittatus* (Say). Journal of Economic Entomology, 50: 395-399.

WATERHOUSE, C. O. 1877. Coleoptera. *In*: Günther. Account of the Zoological Collection made during a visit of H. M. S. Peterel to the Galapagos Islands. Proceedings of the Zoological Society of London (1877): 77-82.

WHITE, R. E. 1983. Beetles. Peterson Field Guides. Houghton Mifflin. New York, NY, 368 pp.

WICKHAM, H. F. 1912. A report on some recent collections of fossil Coleoptera from the Miocene shales of Florissant. Bulletin from the Laboratories of Natural History of the State University of Iowa, 6(3): 19-20, plates 1-8.

WICKHAM, H. F. 1913. Fossil Coleoptera from the Wilson Ranch near Florissant, Colorado. Bulletin from the Laboratories of Natural History of the State University of Iowa, New Series No. 57, 6(4): 3-29, plates 1-7.

WICKHAM, H. F. 1914. New Miocene Coleoptera from Florissant (*Collops* spp., Pp. 446-448, plate 5). Bulletin of the Museum of Comparative Zoology, 58: 423-494, plates 1-16.

WICKHAM, H. F. 1916. New Fossil Coleoptera from Florissant beds. Bulletin from the Laboratories of Natural History of the State University of Iowa, New Series No. 119, 7(3): 3-20, plates 1-4.

WICKHAM, H. F. 1917. New species of fossil beetles from Florissant, Colorado. Proceedings of the United States National Museum, 52(2189): 463-472, plates 37-39.

WICKHAM, H. F. 1920. Catalogue of the North American Coleoptera described as fossils. Pp. 349-365. *In*: C. W. Leng. Catalogue of the Coleoptera of America, North of Mexico. John D. Sherman, Jr. Mount Vernon NY, 470 pp.

WILLIAMS, I. W. 1938. The comparative morphology of the mouthparts of the order Coleoptera treated from the standpoint of phylogeny. Melyridae, P. 247, fig. 28. Journal of the New York Entomological Society 46: 245-289.

WITTMER, W. 1960. Results of the Lund University Expedition in 1950-1951. Coleoptera: Cantharidae, Malachiidae & Dasytidae. 12. Beitrag sur kenntnis der Malacodermata Afrikas. South African Animal Life, 7: 113-127.

WITTMER, W. 1961. *Pseudattalus* Champ. a synonym of *Ablechrus* Waterh. (Malachiidae). Coleopterists Bulletin, 15: 36-37.

WITTMER, W. 1962. Fascicle 67. Subfamily Malachiinae. P. 612. *In*: R. H. Arnett Jr. The Beetles of the United States (a manual for identification). Catholic University of America Press. Washington, DC, xi + 1112 pp.
WITTMER, W. 1966. 26. Beitrag zur kenntnis der neotropischen Malacodermata (Col.). Entomologischen Arbeiten aus dem Museum Frey, 17: 25-32.
WITTMER, W. 1968. A genus of Malachiidae new to the United States. Coleopterists Bulletin, 22: 18-19.
WITTMER, W. 1975. First record of the genus *Troglops* Er. (Coleoptera: Malachiidae) in the United States. Coleopterists Bulletin, 29: 250.
WITTMER, W. 1976. Zur kenntnis der Malachiidae der Antillen (Col.). Mitteilungen der Schweizerischen Entomololgischen Gesellschaft, 49: 281-291.
WITTMER, W. 1979. Die gattung *Ablechrus* Waterh. auf Puerto Rico. Entomologica Basiliensia, 4: 347-350.
WITTMER, W. 1983. 36. Beitrag zur kenntnis der Malachiidae der indo-malaiischen region. Entomologica Basiliensia, 8: 310-328.
WITTMER, W. 1984. Über die gattungen *Ablechrus* Waterh., und *Tucmanius* Pic (Coleoptera, Malachiidae) (37. Beitrag zur kenntnis der netropischen fauna). Entomologica Basiliensia, 9: 226-243.
WITTMER, W. 1985. Zur kenntnis der mit *Hedybius* Er. verwandten gattungen unter besonderer beruecksichtigung der fauna des südlichen Afrikas. Entomologischen Arbeiten aus dem Museum Frey, 33-34: 291-347.
WITTMER, W. 1987. *Condylops* Redtenbacher, A genus new to the new world, with the description of a new subgenus and species from Florida (Coleoptera: Melyridae: Malachiinae). Coleopterists Bulletin, 41: 89-93.
WITTMER, W. 1991. 34. Beitrag zur kenntnis der fauna Afrikas - Coleoptera Malachiidae und Cantharidae. Entomologica Basiliensia, 14: 183-248.
WITTMER, W. 1992. 78. Beitrag zur kenntnis der palaearktischen Malachiidae (Coleoptera). Mitteilungen der Entomologischen Gesellschaft Basel, 42: 11-34.
WITTMER, W. 1995. Zur kenntnis der familie Malachiidae (Coleoptera). 2. Beitrag. Entomologica Basiliensia, 18: 287-391.
WITTMER, W. 1999. Zur kenntnis der familie Malachiidae (Coleoptera). 3. Beitrag. Entomologica Basiliensia, 21: 171-252.
WOLCOTT, A. B. 1944. *Allonyx* LeConte, a homonym. Bulletin of the Brooklin Entomological Society, 39: 47.

Superfamily CUCUJOIDEA Latreille 1802

Clavicornia Ganglbauer 1899, Coccinelloidea Leng 1920, Cucujides Leach 1815, Trimera Leach 1815

75. SPHINDIDAE Jacquelin duVal 1861

by Joseph V. McHugh

Family common name: The cryptic slime mold beetles

Family synonyms: Aspidiphoridae Kiesenwetter 1877 (1859), Coniporidae Thomson 1859

Adults of this group can be distinguished from similar taxa by the following features: antenna with 10-11 antennomeres and abrupt, pubescent 2-3 segmented club; scape and pedicel abruptly and asymmetrically inflated for apical 2/3; mandible bearing large central tubercle and setose cavity on dorsal surface; coxae transverse; tarsi simple, formula 5-5-5 female and 5-5-4 male; elytral punctures seriate.

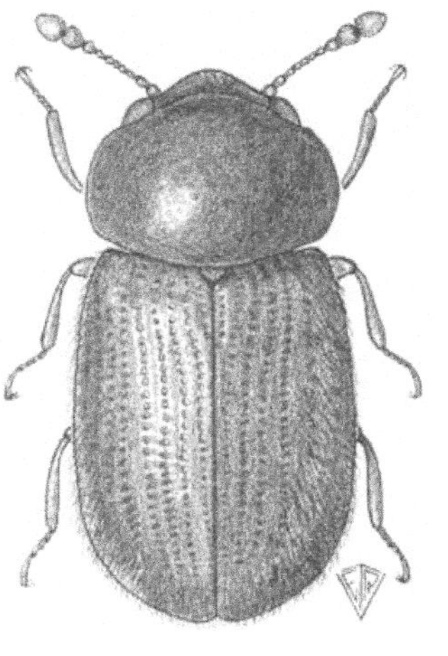

FIGURE 1.75. *Sphindus trinifer* Casey

Description: Shape elongate oval to broadly oval, convex, with head partially visible from above; length 1.5 to 3.5 mm.; color light brown to reddish black, elytra in some *Sphindus* and *Carinisphindus* bicolored; vestiture mostly sparse, short and suberect; rarely (in *Eurysphindus*) denser, longer and erect.

Head prolonged into short, broad muzzle; surface punctate, with one (Fig. 4) to many (Fig. 2) lateral longitudinal grooves. Antennae with 10-11 antennomeres, scape and pedicel asymmetrically inflated for apical 2/3 (Fig. 5); apical 2-3 antennomeres form abrupt, pubescent club about as long as stem; inserted near front margin of eye and base of mandible; stem often lies in groove on dorsal or ventral surface of head, club may rest in weak ventral concavities of head or hypomeron. Clypeus distinct, suture arcuate; labrum distinct, small; mandible large, prominent, apex with 1-3 well developed teeth, dorsal surface bearing large central tubercle and cavity (Fig. 3), prosthecal fringe well developed; mola prominent. Maxilla with slender lacinia, galea slender in most, or broad and truncate (in *Odontosphindus*), palpus 4-segmented, apical palpomere narrow, cylindrical. Labium with trapezoidal mentum; ligula coriaceous, wide; paraglossa small; labial palpus with 3 palpomeres, short, widely separate at bases, apical palpomere cylindrical. Eyes lateral, large, rounded; coarsely faceted (Fig. 4).

Pronotum oval to quadrate, broader than head, as wide as elytra at base; disk densely punctate, lateral margin acute with smooth, crenulate or acutely dentate edge; hypomeron flat or weakly concave for reception of antennal club; prosternum broad, posteriorly truncate to bilobed. Mesosternum narrow, slightly declivous to weakly concave (Fig. 6); metasternum long, coarsely punctate, weakly to moderately inflated. Legs slender; procoxa transverse, procoxal cavity broadly opened to closed posteriorly; mesotrochantins exposed; mesocoxae and metacoxae separate, transverse; trochanter small, triangular; tibial spurs absent except for distal transverse row of small spurs; tarsi 5-5-5 female, 5-5-4 male, shorter than tibia, tarsomeres slender, simple, apical tarsomere about length of others combined, claws simple. Scutellum quadrate to scutelliform, usually punctate. Elytron entire, convex, apically rounded; unicolorous in many, in some darkened apically, laterally, or by a transverse band; striae coarsely punctate; intervals finely punctate; scutellary striole present; epipleural fold narrow, not reaching apex. Wing with well formed jugal lobe, costa, radius, media, cubitus, medio-cubital crossvein and 1-3 anal veins; anal cell in *Odontosphindus* and *Carinisphindus* present.

Abdomen with five visible sterna, all sutures entire; ventrite I longest, most coarsely punctured; ventrites II-V with smaller sparser punctation, in some with proximal transverse band of enlarged punctures or irregular depressions. Pygidium uniformly punctate and setose, or with pair of large impunctate densely setulose binding patches (Fig. 7).

Male genitalia rotated 90 degrees when retracted, parameres fused together and joined to ring-shaped basal piece; median lobe broad, concave apically, housing apex of fused parameres. Female genitalia with membranous proctiger and paraprocts; coxite unilobed or bilobed, with short preapical stylus bearing 1-5 long setae.

Acknowledgments: A draft of this chapter was reviewed by E. Chiao, T. Kiselyova, C. Lewis and E. Tilgner. Figures 8-11 were drawn by T. Kiselyova. The author's work was supported on Hatch Grant #GEO 00787.

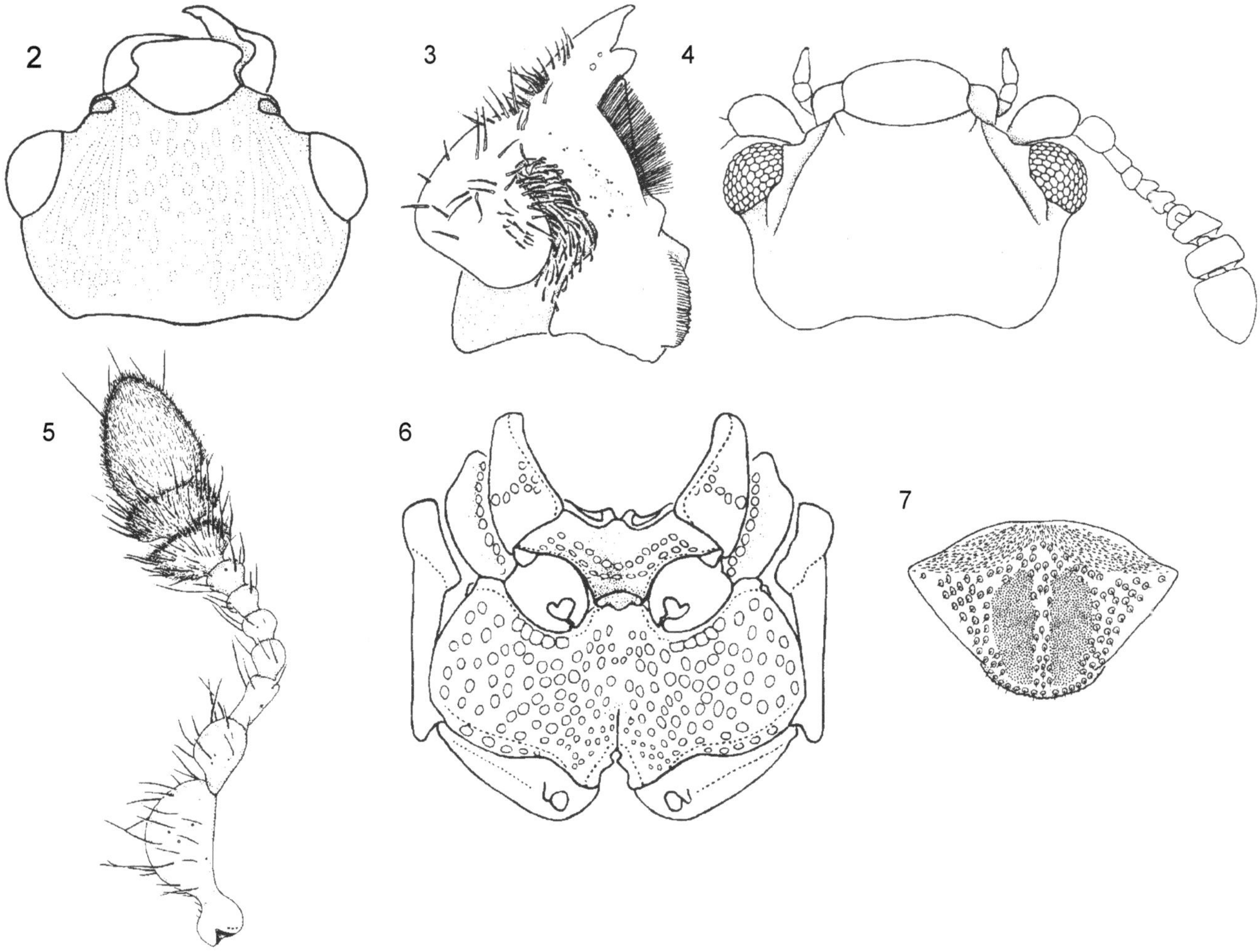

FIGURES 2.75-7.75. 2, *Carinisphindus isthmensis* McHugh, head, dorsal; 3, *Eurysphindus hirtus* LeConte, left mandible, dorsal; 4, *Eurysphindus hirtus* LeConte, head, dorsal; 5, *Eurysphindus hirtus* LeConte, right antenna, dorsal; 6, *Eurysphindus hirtus* LeConte, pterothorax, ventral; 7, *Carinisphindus isthmensis* McHugh, pygidium, dorsal.

Mature larvae 1.2-5 mm, elongate, parallel-sided to fusiform, weakly flattened; head dark with body pale yellow to bright white; thorax and abdomen in many with paired dark plates or maculae; integument smooth; vestiture sparse to moderately dense, setae moderately long to long, mostly simple, rarely (in *Eurysphindus*) with setal apices spiraled and weakly capitate. Head protracted, prognathous; epicranial stem short or absent; frontal arms lyriform, contiguous at base; stemmata 6 per side; antennae 3-segmented, antennomere 2 longest; frontoclypeal suture absent; labrum free; mandibles symmetrical or nearly so, unidentate to bidentate, narrow for distal 1/2; incisor edge serrate in some, or bearing several small denticles from broad subapical prominence (in *Sphindus*); prostheca narrow, hyaline, fixed; mola large, tuberculate or asperate; ventral mouthparts retracted; maxillary palpus 3-segmented; labial palpus 2-segmented, widely separated at bases; hypopharyngeal sclerome forming transverse bar; gula short, indistinct. Thoracic and abdominal terga I-VIII with paired darkened maculae or paired plates in many; legs 5-segmented; tarsungulus with 2 setae lying one distal to other; thoracic spiracles annular or annular uniforous; A9 simple or (in *Odontosphindus*) with pair of small upturned fixed urogomphi; segment A10 short, circular, posteriorly oriented; abdominal spiracles annular, in many positioned at end of small lobes.Generic level keys to larvae are provided by Sen Gupta and Crowson (1979), Burakowski and Slipinski (1987), and Chiao and McHugh (2000).

Burakowski and Slipinski (1987) provide the only descriptions of eggs and pupae for a sphindid, *Aspidiphorus orbiculatus* (Gyllenhal). Chiao and McHugh (2000) summarize the literature on immature stages.

Habits and habitats. All species are myxomycophagous. All life stages normally occur on or inside slime mold sporocarps, where adults and larvae feed on the spores and supporting structures. Sphindids are occasionally taken with flight intercept traps, ultraviolet light traps, or by Berlese funnel extraction from decaying plant matter. The best collecting approach is to remove myxomycete fruitings from tree stumps, logs and leaf litter and then to extract beetles from them using a mild heat source. Field-col-

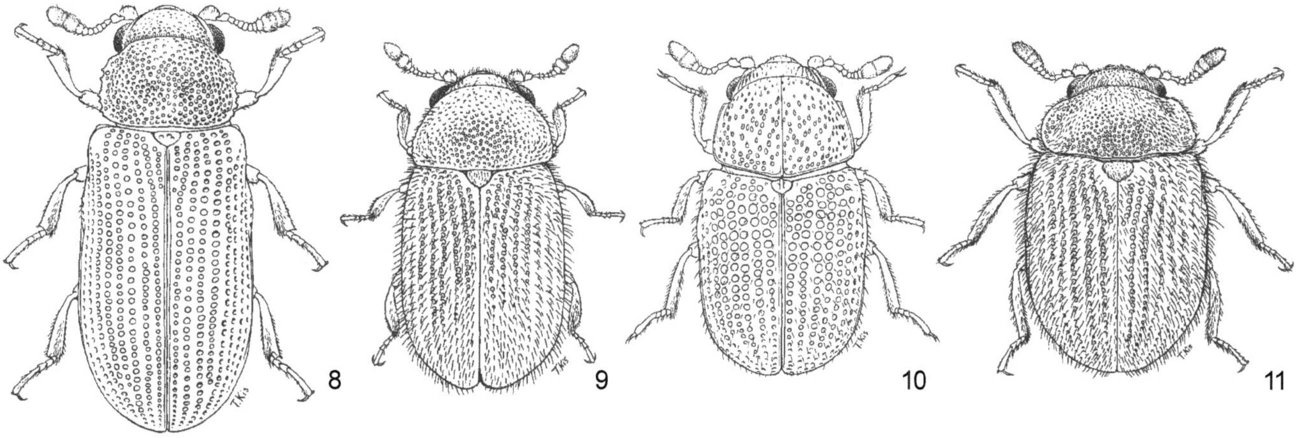

FIGURES 8.75-11.75. 8. *Odontosphindus clavicornis* Casey, habitus; 9. *Sphindus americanus* LeConte, habitus; 10. *Carinisphindus purpuricephalus* McHugh and Lewis, habitus; 11. *Eurysphindus comatulus* McHugh, habitus.

lected sphindids are easily reared in the lab if they are provided with sporocarps of an acceptable myxomycete. Host data for North American sphindids are provided by Lawrence and Newton (1980), Russell (1979) and Stephenson *et al.* (1994). Although sphindids show a broad range of acceptable hosts for adult feeding, they may show more selectivity for breeding (see McHugh and Kiselyova, in press). Burakowski and Slipinski (1987) describe the life history of the Palearctic species, *Aspidiphorus orbiculatus*. McHugh (1993b) provides information about parasitoids of Sphindidae.

Status of classification. Although its precise phylogenetic placement remains uncertain, it is clear that Sphindidae are a primitive family of Cucujoidea. Sen Gupta and Crowson (1979) proposed that strong affinities lie with Phloeostichidae, Silvanidae, Boganiidae and Protocucujidae. Based on mandibular morphology, Thomas (1984) agreed that Sphindidae forms a monophyletic group with the first three families and possibly the fourth. McHugh (1993a) and Chiao and McHugh (2000) used Protocucujidae as the outgroup for phylogenetic analyses of Sphindidae. McHugh (1993a) reviewed the classification of the family.

Distribution. This family occurs worldwide, with 61 species representing nine genera. In North America there are nine species representing four genera and two subfamilies. There are two small extralimital subfamilies, Protosphindinae Sen Gupta and Crowson (2 species, Chile) and Sphindiphorinae Sen Gupta and Crowson (1 species, South Africa).

Key to the Nearctic Genera

1. Adult with 11-segmented antenna; pronotal margin with six or seven acute teeth; mandible tridentate; length 2.8-3.5 mm; galea apex broad, truncate. Larva with urogomphi present, prominent, at least 2x as long as wide (Odontosphindinae) .... *Odontosphindus*

— Adult with 10-segmented antennae; pronotal margin smooth to crenulate; mandible with 1-2 teeth (Fig. 3); length 1.5-2.3 mm; galea apex narrow, slender. Larva with urogomphi absent or very small, length less than or equal to width (Sphindinae) 2

2(1). Adult body elongate-oval, parallel-sided; dorsal head punctures fusing to form 3 or more longitudinal grooves over eye (Fig. 2); pygidium with pair of large densely setulose impunctate binding patches (Fig. 7); procoxal cavities externally closed or nearly so by hypomera and prosternal process. Larva with antennomere II 2-2.5x length of antennomere I 3

— Adult body broadly oval, convex; head with single dorsal groove adjacent to eye (Fig. 4); pygidium evenly punctate and setose, lacking binding patches; procoxal cavities widely open externally. Larva with antennomere II greater than 3x length of antennomere I .... *Eurysphindus*

3(2). Adult pronotum and scutellum with dorsal median longitudinal carina; male metathoracic femur with acute posterior denticle; pronotal lateral margin smooth; wing with three anal veins and anal cell. Larva with abdominal tergite IX bearing small urogomphi; mandible bidentate, subapical tooth bearing a few weak serrations along mesal margin .... *Carinisphindus*

— Adult pronotum and scutellum lacking median longitudinal carina; posterior surface of male metathoracic femur smooth, lacking posterior denticle; pronotal lateral margin crenulate; wing with one anal vein, lacking anal cell. Larva with abdominal tergite IX lacking urogomphi; mandible with single tooth bearing serrations on mesal margin .... *Sphindus*

Classification of the Nearctic Genera

Sphindidae Jacquelin du Val 1861

Odontosphindinae Sen Gupta and Crowson 1979

Includes only the genus *Odontosphindus*. Distinguished from other North American Sphindidae by large size (2.8-3.5 mm); 11-

segmented antennae; tridentate mandible, and larvae with prominent (at least twice as long as wide) urogomphi.

Odontosphindus LeConte 1878, 2 spp., *O. denticollis* LeConte in Northeast from Quebec to North Carolina, extending westward to Ontario and Michigan, and *O. clavicornis* Casey in Northwest from British Columbia to northern California, extending eastward to Alberta and Montana. Casey (1898) provides an adult key for N.A. species. Larvae: *O. clavicornis* illustrated by Sen Gupta and Crowson (1979); *O. denticollis* described (as *O. clavicornis*) by Burakowski and Slipinski (1987). An additional species, *O. grandis* (Hampe), occurs in the Palearctic region.

Sphindinae Jacquelin du Val 1861

Recognized by the reduction to 10 antennomeres, bidentate or unidentate mandibles, and larvae with urogomphi highly reduced (length = width) or absent. The subfamily includes three extralimital genera: *Genisphindus* McHugh (5 species) from the Neotropics, *Notosphindus* McHugh and Wheeler (1 species) from Australia, and the widespread, Old World genus *Aspidiphorus* Ziegler *in* Dejean (18 species).

Sphindus Megerle *in* Dejean 1821, 4 spp., widely distributed. Casey (1898) provides adult key to 3 N.A. species. Larvae: *S. americanus* LeConte described by Chevrolat (1833) and figured by Bøving and Craighead (1931) and Lawrence (1991).
Coniophagus Mink 1853

Eurysphindus LeConte 1878, 2 spp., from Quebec to Florida, extending westward to Manitoba and Oklahoma. McHugh (1993a) provides species descriptions, distribution maps and a key to species. Larvae: both North American species described by McHugh and Kiselyova (2001). The genus includes five additional species from the Neotropics.

Carinisphindus McHugh 1990, 1 sp., *C. purpuricephalus* McHugh and Lewis (2000), in Florida. *Carinisphindus geminus* McHugh and Lewis occurs in Puerto Rico. McHugh (1990) gives a key to the adults of four species. Larvae: *C. purpuricephalus* is described by Chiao and McHugh (2000). The genus includes five additional species from the Caribbean and Central America.

Bibliography

BØVING, A. G. and F.C. CRAIGHEAD. 1931. An illustrated synopsis of the principal larval forms of the order Coleoptera. Entomologica Americana (N.S.), 11(1930): 1-351.

BURAKOWSKI, B. and S. A. SLIPINSKI. 1987. A new species of *Protosphindus* (Coleoptera: Sphindidae) from Chile with notes and descriptions of immature stages of related forms. Annali del Museo Civico di Storia Naturale Giacomo Doria, Genova, 86: 605-625.

CASEY, T. L. 1898. Studies in the Ptinidae, Cioidae and Sphindidae of America. Journal of the New York Entomological Society, 6: 61-93.

CHEVROLAT, L. A. A. 1833. Description du genre *Sphindus*. Revue Entomologique, 1(8).

CHIAO, E. and J. V. McHUGH. 2000. Larval Sphindidae (Coleoptera: Cucujoidea): Phylogenetic implications and new descriptions. Invertebrate Taxonomy, 14: 807-824.

LAWRENCE, J. F. 1991. Coleoptera. Sphindidae. *In:* F. W. Stehr, ed., Immature Insects. Vol. 2. Kendall/Hunt. Dubuque, IA. 975 pp.

LAWRENCE, J. F. and A. F. NEWTON, Jr. 1980. Coleoptera associated with the fruiting bodies of slime molds (Myxomycetes). Coleopterists Bulletin, 34: 129-143.

McHUGH, J. V. 1990. *Carinisphindus*, a new genus and three new species of neotropical Sphindidae (Coleoptera: Clavicornia). Coleopterists Bulletin, 44: 307-322.

McHUGH, J. V. 1993a. A revision of *Eurysphindus* LeConte (Coleoptera: Cucujoidea: Sphindidae) and a review of sphindid classification and phylogeny. Systematic Entomology, 18: 57-92.

McHUGH, J. V. 1993b. First records of parasitoids for slime mold beetles of the family Sphindidae (Coleoptera: Cucujoidea). Entomological News, 104: 136-138.

McHUGH, J. V. and C. N. LEWIS. 2000. Three new species of *Carinisphindus* (Coleoptera: Sphindidae) from Bahamas, Florida and Puerto Rico. Coleopterists Bulletin, 54: 143-153.

McHUGH, J. V. and T. G. KISELYOVA. 2001. First descriptions for larval stages of *Eurysphindus* (Cucujoidea: Sphindidae). Coleopterists Bulletin, [In press].

RUSSELL, L. K. 1979. Beetles associated with slime molds (Mycetozoa) in Oregon and California (Coleoptera: Leiodidae, Sphindidae, Lathridiidae). Pan-Pacific Entomologist, 55:1-9.

SEN GUPTA, T. and R. A. Crowson. 1979. The coleopteran family Sphindidae. Entomologist's Monthly Magazine, 113: 177-191.

STEPHENSON, S. L., Q. D. WHEELER, J. V. McHUGH and P. R. FRAISSINET. 1994. New North American associations of Coleoptera with Myxomycetes. Journal of Natural History, 28: 921-936.

THOMAS, M.C. 1984. A new species of apterous *Telephanus* (Coleoptera: Silvanidae) with a discussion of phylogenetic relationships of the Silvanidae. Coleopterists Bulletin, 38: 43-55.

76. BRACHYPTERIDAE Erichson 1845

by Dale H. Habeck

Family common name: The short-winged flower beetles

Family synonym: Cateretidae Ganglbauer 1899

The 5-5-5 tarsal formula, maxilla with galea and lacinia, weak antennal club and anterior coxal cavities open behind are characteristics that will identify members of this family.

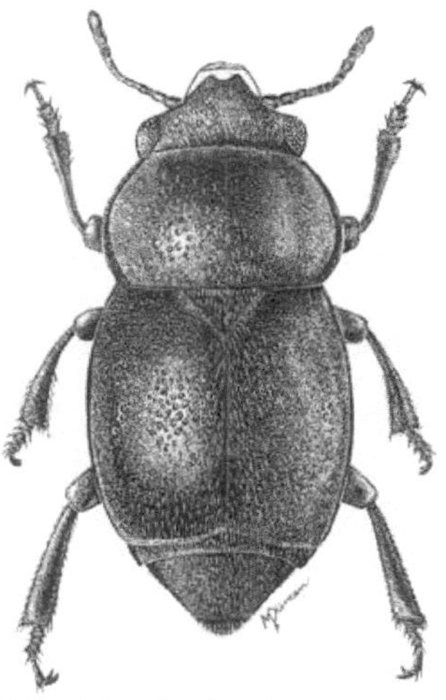

FIGURE 1.76. *Brachypterus urticae* (Fabricius)

Description: Shape elongate oval in most, some are depressed, size 1.5-6.0 mm long, color pale to piceous. Most with vestiture sparse, fine and short.

Head prognathous, much narrower than thorax. Antennae with eleven antennomeres, last 3 forming a feeble, indistinct club, inserted between eyes and base of mandibles. No ventral grooves to receive antennae. Clypeus distinct in some species, indistinguishable from frons in most. Labrum transverse, weakly to strongly emarginate. Mandible without setose prosthecal lobes, with or without bidentate apices. Maxilla with lacinia and galea. Four maxillary palpomeres; most with apical antennomere longer than palpomeres 2 and 3 combined. Mentum somewhat transverse. Labrum with 3 palpomeres; apical antennomere as long as palpomeres 1 and 2 combined. Eyes large, lateral, with large facets.

Pronotum slightly narrower than elytra. Scutellum large and usually triangular. Elytra with narrow epiplura on anterior half or less. Prosternal process narrow, subparallel and not extending beyond the coxae. Mesosternal process 2-5 times broader than prosternal process. Metasternum about as long as first 2 or 3 ventral abdominal segments. First ventral abdominal sternite as long or longer than 2 and 3 combined. Outer edge of meso- and metatibia without distinct carina. First 3 tarsomeres dilated, fourth very small and the fifth as long as the first 2 or 3 combined. Claws simple or dentate. Elytra exposing pygidium and at least 1 abdominal tergite. Pygidium free and broadly truncate or rounded. Female genitalia relatively little sclerotized, elongate, slender, and with well developed styli. Male genitalia with parameres distinctly articulated on phallobase; aedeagus in almost all asymmetrical.

Larva with front fused with clypeus. Stemmata on each side of head, usual number is 4 divided into 2 groups; 2 functional stemmata anterior and 2 posterior reduced to a small ocular spot or indistinct. Antenna with 3 segments, rather short and broad with a sensory appendage on antennomere 2. Labrum normally large and visible laterally. Epipharynx rather simple with short setae lateroanterior, and without a crest or longitudinal furrows. Mandible elongate with a single tooth distally. Prostheca absent. Mola distinct. Maxilla with 3 palpomeres, excluding the basal palpiger. Maxilla with a lacinia and galea. Labium with nonarticulated palpi. Legs relatively well developed. Femur and tibia about twice as long as wide. Spatulate adhesive hairs at base of tarsal claws. Tarsungulus with or without teeth basally. Abdominal spiracles in dorsolateral position, distinctly biforous, all having well developed air chambers with distinct openings outside the peritreme. Urogomphi absent. Abdominal tergites plain, without sclerotized plates or pigmented areas. Some larvae are illustrated by Bøving and Craighead (1931) and Lawrence (1991). Pupa exarate, without urogomphi.

Habits and habitats. Adults and larvae are phytophagous. Larvae develop in seed capsules of various plants. Adults feed on pollen and flower petals of the same plants or some others.

Status of the classification. Parsons (1943) last treated this group as Cateretinae: Nitidulidae. Since then, various authors have reviewed this family as Kateretidae, especially in Europe. Audisio (1979) provided characters to distinguish *Heterhelus* from *Kateretes*.

Distribution. This is a small family with 11 species in 7 genera known in the United States and Canada. Worldwide there are probably fewer than 100 species (Audisio 1993).

Key to Genera of America North of Mexico

1. Claws simple 2
— Claws distinctly dentate at base 3

2(1). Length under 3 mm 4
— Length over 3 mm 6

3(1). Posterior pronotal angles obtuse ... *Brachypterus*
— Posterior pronotal angles rectangular .... *Brachypterolus*

4(2). Elytra testaceous to piceous 5
— Elytra metallic greenish .... *Boreades*

5(4). Posterior corner of prothorax rounded .. *Kateretes*
— Posterior corner of prothorax sharply angled .... *Heterhelus*

6(2). Body convex .... *Amartus*
— Body depressed .... *Anthonaeus*

CLASSIFICATION OF THE GENERA OF AMERICA NORTH OF MEXICO

Brachypteridae Erichson 1845

Cateretidae Ganglbauer 1899

Amartus LeConte 1861, 2 spp., Oregon, California and Arizona.
Brachyleptus Motschulsky 1870, not Motschulsky 1845

Anthonaeus Horn 1879, 1 sp., *A. agavensis* (Crotch 1874), southern California.

Boreades Parson 1943, 1 sp., *B. abdominalis* (Erichson 1843), eastern Canada south to Georgia, west to Texas, Missouri, Nebraska, Iowa, and Manitoba.
Cercometes Leng 1920, not Reitter 1875

Brachypterolus Grouvelle 1913, 1 sp., *B. pulicarius* (Linnaeus 1758), Nova Scotia to Ontario, Wisconsin and Iowa, east to Delaware. Immigrant from Europe.
Heteromus DuVal 1858, not Bigot 1857

Brachypterus Kugelann 1794, 4 spp., generally distributed, except absent in the southeastern United States.
Virbius Des Gozis 1886
Heterobrachium Wollaston 1864
Parabrachypterus Hisamatsu 1976

Heterhelus DuVal 1858, 1 sp., *H. sericans* (LeConte 1869), western, distributed from British Columbia south to California, Utah and New Mexico, and east to North Carolina and New England.
Cateretes Heer 1841
Amartus Reitter 1875 (in part)
Pulion Parsons 1943

Kateretes Herbst 1793, 1 sp., *K. scissus* (Parsons 1943), Alberta and Massachusetts.
Cercus Latreille 1796 (in part)
Cateretes Illiger 1798 (in part)
Catheretes Gyllenhal 1808 (in part)
Anisocera Stephens 1832
Anomatocera Shuckard 1840
Pulion Des Gozis 1886

BIBLIOGRAPHY

AUDISIO, P. 1979. The nearctic species of the genus *Cateretes* (Coleoptera: Nitidulidae). Coleopterists Bulletin, 33: 48

AUDISIO, P. 1993. Coleoptera Nitidulidae-Kateridae. Fauna d'Italia, 32: 1-951.

BØVING, A.G. and F.C. CRAIGHEAD. 1931. An illustrated synopsis of the principal larval forms of the order of Coleoptera. Entomologica Americana (new series), 21: 1-351.

LAWRENCE, J.F. 1991. Nitidulidae (Cucujoidea) (including Brachypteridae, Cateretidae, Cybocephalidae, Smicripidae). Pp. 456-460. *In*: F. W. Stehr, ed., Immature Insects. Vol. 2. Kendall/Hunt. Dubuque, Iowa, 195 pp.

PARSONS, C.T. 1943. A revision of nearctic Nitidulidae (Coleoptera). Bulletin of the Museum of Comparative Zoology, 92: 121-273.

77. NITIDULIDAE Latreille 1802

by Dale H. Habeck

Family common name: The sap beetles

Family synonyms: Cybocephalidae Jacquelin du Val 1858

The transverse procoxal cavities, grooved metacoxae, dilated tarsomeres, small fourth tarsomere, and antennal club with three antennomeres distinguish these pentamerous beetles from other families.

FIGURE 1.77. *Glischrochilus fasciatus* (Olivier)

Description: Shape oval, some elongate, depressed; size 1.5 to 12 mm in length; color pale to piceous, often piceous with red or yellowish markings; most with vestiture sparse and fine, short.

Head prognathous, surface smooth, punctate, or rugose. Antennae with 11 antennomeres, apical three antennomeres forming a club, inserted between eyes and base of mandibles, often received in a groove on the underside of the head. Clypeus rarely distinct; labrum transverse, more or less bilobed, may be covered by clypeus; mandibles moderate to large, broad, a brush of setae on inner margin, apices often bidentate, inner tooth smaller and shorter than outer, or multidentate subapically, or only one mandible bidentate. Maxillae with lacinia present, galea absent, with four slender palpomeres. Labium with the mentum trapezoidal, anteriorly bisinuate; ligula variable; paraglossae large, reduced, or absent; labial palpi with three palpomeres. Eyes lateral, large, rounded, facets variable.

Pronotum with lateral borders explanate and laterally arcuate in most, with anterior border often broadly emarginate, posterior border straight or sinuate; most with borders finely margined; surface smooth, punctate or rugose; pleural region broad; prosternum produced posteriorly into a lamellate process which extends between the coxae and sometimes overlaps the mesosternum; procoxal cavities open behind only in Cryptarchinae. Mesosternum short, carinate in some; mesepisternum large; mesepimeron narrow; mesocoxal cavities closed; metasternum large, often divided by a suture medially from its posterior margin, often with a distinct "axillary piece"; metasternum broad; metacoxal cavities closed. Legs short, somewhat retractile; all coxae transverse, separate; trochanters small, triangular; femora sometimes canaliculate for reception of the tibiae; tibiae apically expanded, most with spurs obscure; tarsal formula 5-5-5, or 4-4-4 (Cybocephalinae); tarsomeres dilated, with pads of setae beneath, fourth minute, the fifth long; claws simple or toothed. Scutellum triangular in most, may be pentagonal, semicircular, or quadrangular. Elytra shortened, exposing part or all of pygidium and some with two or three terga exposed; epipleural fold moderate. Wings normal, venation reduced, with more or less normal median, cubitus, and some anal veins, lobed basally; folding pattern "extremely complex."

Abdomen with five visible sternites, sutures entire; most with surface microrugose. Male genitalia with penis short, moderately stout; parameres fused, with or without articulating lateral lobes; pars basalis scoop-shaped; genitalia segment well developed with a median strut; segment eight unmodified. Female genitalia with paraprocts dorsal on distal part of valvifer; valvifer present; coxite present, and stylus present.

Larvae elongate, 2-20 mm long, most less than 12 mm; parallel sided or fusiform, straight or curved ventrally, occasionally curved dorsally (*Pallodes*, *Pocadius*), subcylindrical to strongly flattened. Dorsal surface smooth to granulate or tuberculate, most are lightly pigmented except for head, protergum and ninth tergum. Some with single or paired plates on all exposed terga. Setae scattered and simple. Head prognathous. Epicranial stem absent in most, frontal arms lyriform or V-shaped. Median endocarina almost always absent. Two to 4 stemmata on each side. Antennae well developed with 3 segments. Frontal-clypeal suture absent (except in Cybocephalinae). Labrum almost always free but partly fused (in Cybocephalinae) and totally fused (in Meligethinae). Mandible symmetrical, uni- or bidentate; mola well developed. Most with prostheca complex consisting of a brush of simple or complex hairs or fringed membranes, absent (Cybocephalinae) or a simple membranous lobe (Meligethinae). Maxillary palps 3-segmented in most (2 in Cybocephalinae).

Protergum typically with pigmented plate; meso- and metaterga without plates. Abdominal segments 1-8 sometimes with paired lateral or dorsal processes. Legs 5 segmented, typically well developed. Spatulate adhesive hairs at the base of tarsal claws in Cybocephalinae and Meligethinae. Ninth tergum typically with paired fixed urogomphi often with a pair of pregomphi in front of urogomphi, without urogomphi on Meligethinae and Cybocephalinae. Segment 10 short, circular, ventrally or posterior-ventrally oriented. Spiracles bilabiate or annular-biforous at the end of tubes (except in Cybocephalinae). Additional information on larvae can be found in Bøving and Craighead (1931), Bøving and Rozen (1962), Connell (1977, 1984), Lawrence (1991), Okumura and Savage 1974; on pupae (Rozen, 1963).

Habits and habitats: The members are primarily saprophagous and mycetophagous. Although some live in flowers, the majority live in decaying fruits, fermenting plant juices, and in fungi. Some genera live in a particular fungus specific to that genus. *Epuraea depressa* Illiger, according to Dodge (Parsons 1943), breed in the nests of bumblebees, and the larvae are considered fungivorous. Cockerell has collected *Epuraea integra* Horn in the nest of *Bombus juxtus* [now *B. centralis* (Cresson), Apidae]. This habitat is similar to that of the Australian *Brachypeplus auritus* Murray of which the larvae and adult feed on the wax and honey of a wild bee. *Aethina tumida* Murray, the small hive beetle, can severely damage contents of beehives. *Amphotis* occurs in ant nests; *A. ulkei* LeConte is strictly myrmecophilous in the early spring but in the fall of the year is found in decaying fungi. *Nitidula* sp. and *Omosita* sp. breed in carrion. *Carpophilus hemipterus* Linnaeus is a minor pest of stored products and various species are involved in transmission of oak wilt fungus. In Europe the larvae of certain *Glischrochilus* sp., *Nitidula* sp., and *Pityophagus* sp. are predaceous on Scolytidae, and may have similar habits in the United States. *Cybocephalus* spp. larvae prey on coccids. Pupation takes place in the earth, which indicates that the pupae may hibernate also. According to Parsons (1943) in the tropics the life cycle seems to be continuous, although there must be aestivation over extremely dry seasons.

Status of the classification: For North America the group has been monographed by Parsons (1943). Connell (1984) provided an update on changes since Parsons' monograph. The subfamily Cateretinae is now considered a separate family, Brachypteridae (Lawrence and Newton 1995). The assignment of genera to tribes and subfamilies continues to change and appears to be far from stable. Kirejtshuk and Audisio (*in* Kirjtshuk 1986) divided the family into 7 subfamilies, with only Calonecrinae not represented in North America.

Distribution: There are about 2,800 species in about 172 genera (Audisio 1993) worldwide. About 165 species in 30 genera are known from the United States.

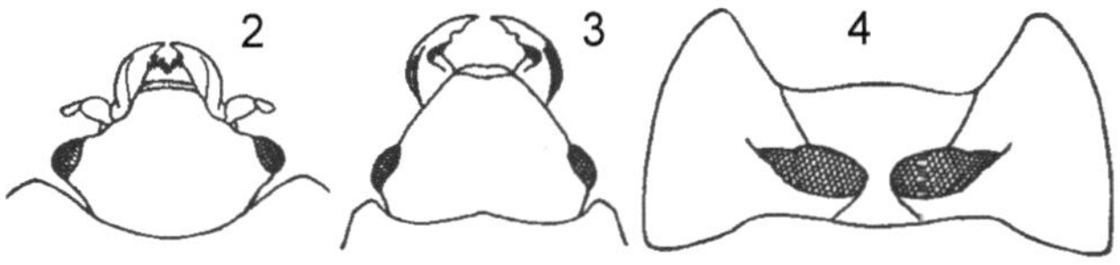

FIGURES 2.77-4.77. 2. *Stelidota geminata* (Say), head, dorsal view; 3. *Glischrochilus fasciatus* (Olivier) head, dorsal view; 4. *Prometopia sexmaculata* (Say), prosternum.

Key to the Genera of America North of Mexico

1. Tarsal formula 5-5-5 2
— Tarsal formula 4-4-4 (Cybocephalinae) .... *Cybocephalus*

2(1). Labrum and frons fused with clypeus, union marked only by a more or less distinct suture (Fig. 3) (Cryptarchinae) 28
— Labrum free, more or less visible (Fig. 2) 3

3(2). Body strongly flattened, frequently elongate or convex and filiform; sides explanate, often widely so; pygidium and preceding 2 tergites exposed (Cillaeinae) 6
— Body convex to flattened, elytra long or short; if pygidium and preceding 2 tergites exposed, the abdomen much more convex 4

4(3). Outer edge of middle and hind tibiae with a single row of small marginal spines (Fig. 8, Meligethinae) .... *Meligethes*
— Outer edge of middle and hind tibiae with two rows of small marginal spines 5

5(4). Elytra short and truncate apically, not covering pygidium and 1-2 preceding tergites, or, if only pygidium (partly or wholly) exposed, then metasternal axillary space virtually indistinct; terminal segment of labial palpi somewhat enlarged, hardly longer than or shorter than wide, widely truncate at apex; antennal grooves often very long, almost confluent posteriorly; elytra not longitudinally costate without long marginal hairs at sides, sutural striae, longitudinal carinae or longitudinal rows of hairs, setae or punctures (Carpophilinae) 8
— Elytra covering all of abdomen, or, at most, leaving pygidium wholly or partially exposed; if metasternal space absent or strongly reduced and terminal segment of labial palpi relatively short, then antennal grooves shorter and less convergent posteriorly; elytra longitudinally costate and with longitudinal rows of hairs and punctures, or with long marginal hairs at sides (Nitidulinae) 12

6(3). Body very elongate 7
— Body at most oblong (Fig. 5) .... *Colopterus*

7(6). Body convex, hypopygium and pygidium long and conical .... *Conotelus*
— Body depressed, hypopygium and pygidium short, depressed .... *Brachypeplus*

8(5). Elytra exposing only pygidium (or part of it) and, at most, posterior edge of penultimate abdominal segment 9
— Elytra exposing pygidium and at least penultimate abdominal segment 10

9(8). Prothorax margined at base, head horizontal .... *Amphicrossus*
— Prothorax not margined at base, head vertical 11

10(8). Elytra exposing two abdominal tergites (Fig. 6) .... *Carpophilus*
— Elytra exposing three abdominal tergites .... *Urophorus*

11(9). First 3 tarsomeres bilobed .... *Epuraea*
— All tarsomeres simple .... *Aphenolia*

12(5). Eyes with conspicuous setae .... *Aethina*
— Eyes without setae 13

13(12). Pronotum margined posteriorly, some with margin obsolete medially or laterally, head horizontal 14
— Pronotum not margined posteriorly, head vertical 22

14(13). Prosternum depressed and may be slightly expanded behind procoxae, not prolonged posteriorly (Fig. 4) 15
— Prosternum elevated behind, often prolonged posteriorly 24

15(14). Tarsi very distinctly dilated 16
— Tarsi not dilated, or only feebly so 17

16(15). Antennal grooves strongly convergent... *Nitidula*
— Antennal grooves parallel, passing directly backward .... *Stelidota*

17(15). Mentum broad, covering base of maxillae .... *Prometopia*
— Mentum not covering maxillae 18

18(17). Frons not lobed over antennae 19
— Frons lobed over antennae 20

19(18). Mandibles with apices slightly bifid .... *Phenolia*
— Mandibles with apices entire .... *Omosita*

20(18). Antennal grooves strongly convergent behind ... . *Soronia*
— Antennal grooves parallel 21

21(20). Elytra not, or merely apparently costate (Fig. 7).... .... *Lobiopa*
— Elytra distinctly costate .... *Amphotis*

22(13). Metatibiae strongly spined on outer edge .... *Thalycra*
— Metatibiae lacking distinct spines on outer edge. 23

23(22). Protarsi not dilated; body oval, pubescent .... *Pocadius*
— Protarsi dilated; body parallel and glabrous .... *Orthopeplus*

24(14). Mesosternum protuberant in front; mesocoxae widely separated 25
— Mesosternum small, oblique, not protuberant... 26

25(24). Labrum deeply bilobed; metatarsi longer than mesotarsi .... *Psilopyga*
— Labrum feebly bilobed; metatarsi and mesotarsi of equal length .... *Camptodes*

26(24). Metasternum not protuberant; mesocoxae narrowly separated 27
— Metasternum protuberant, widely separating mesocoxae; prosternum not prolonged apically .... *Cyllodes*

27(26). Metatarsi longer than others; body glabrous .... *Pallodes*
— Metatarsi equal to others; body pubescent .... *Cychramus*

28(2). Procoxal cavities open behind 29
— Procoxal cavities closed behind; pronotum not margined at base, body glabrous .... *Pityophagus*

29(28). Pronotum margined at base, slightly overlapping base of elytra; most with body pubescent .... *Cryptarcha*
— Pronotum not margined at base; body glabrous (Fig. 1) .... *Glischrochilus*

Classification of the Genera

Nitidulidae Latreille 1802

Cillaeinae Kirejtshuk and Audisio in Kirejtshuk 1986

Conotelus Erichson 1842, 4 spp., widely distributed except for Pacific northwest.

Brachypeplus Erichson 1842, 1 sp., *B. glaber* LeConte 1878, coastal from North Carolina to Texas.

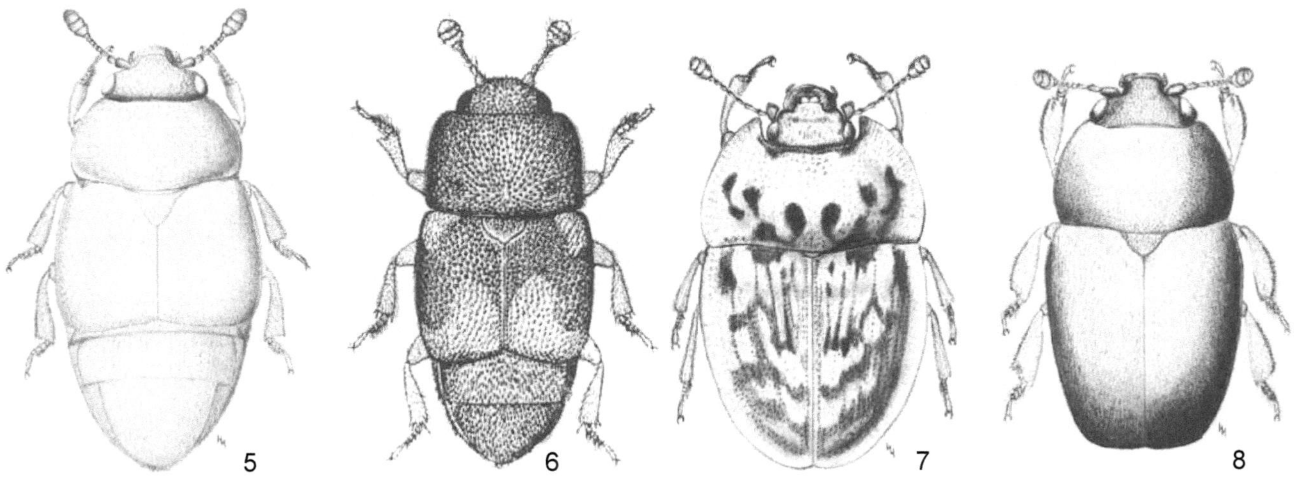

FIGURES 5.77-8.77. 5. *Colopterus truncatus* Randall; 6. *Carpophilus hemipterus* Linnaeus; 7. *Lobiopa setosa* Harold; 8. *Meligethes nigrescens* Stephen (all modified from Hatch 1961).

Nitidulopsis Walker 1856
Leropeplus Murray 1864
Liparopeplus Murray 1864
Onicotis Murray 1864
Selis Murray 1864
Tasmus Murray 1864

Colopterus Erichson 1842, 8 spp., generally distributed.
Colastus Erichson 1843
Cyllopodes Murray 1864

Carpophilinae Erichson 1842

Carpophilus Stephens 1830, 32 spp., generally distributed. Key to economic species adults: Connell 1977, 1991.
Ecnomorphus Motschulsky 1858
Tribrachys LeConte 1861
Endomerus Murray 1864
Nitops Murray 1864
Eidocolastus Murray 1864
Heterodontus Murray 1864
Myothorax Murray 1864
Stauroglossicus Murray 1864
Microxanthus Murray 1864
Idocolastus Gemminger and Harold 1868
Megagarpolus Reitter 1919
Loriarulus Kirejtshuk 1987

Urophorus Murray 1864, 1 sp., *U. humeralis* (Fabricius 1798). Worldwide, southern United States.
Anophorus Kerijtshuk 1990

Epuraea Erichson 1843, 32 spp., generally distributed.
Dadopora Thomson 1874
Epuraeanella Crotch 1874
Micruria Reitter 1874
Omosiphora Reitter 1875
Micrurula Reitter 1874
Apria Grouvelle 1919
Marinexa Kirejtshuk 1989
Polinexa Kirejtshuk 1989
Africaraea Kirejtshuk 1989
Haptoncus Murray 1864

Aphenolia Reitter 1884, 1 sp., *A. monogama* Crotch 1874. California and Oregon.

Amphicrossus Erichson 1843, 3 spp. Ontario to Florida, Arizona.
Lobostoma Fairmaire 1892
Rhacostoma Berg 1898

Nitidulinae Latreille 1802

Orthopeplus Horn 1879, 1 sp., *O. quadricollis* Horn 1879, Colorado, New Mexico, and Arizona.

Stelidota Erichson 1843, 4 spp., eastern United States to Arizona. Key to adults: Ford 1996. Key to larvae: Peng *et al.* 1990.

Omosita Erichson 1843, 2 spp., generally distributed.
Saprobia Ganglbauer 1899

Nitidula Fabricius 1775, 6 spp., widely distributed.

Aethina Erichson 1843, 1 sp., *A. tumida.* Murray 1867; adventive from Africa, widely distributed in honeybee hives.

Prometopia Erichson 1843, 2 spp., generally distributed.
Cerophorus Laporte 1840
Soronia Reitter 1873, not Erichson 1843

Lobiopa Erichson 1843, 7 spp., generally distributed.
Cerophorus Castelnau 1840
Soronia Reitter 1873, not Erichson 1843

Soronia Erichson 1843, 2 sp., widely distributed.

Phenolia Erichson 1843, 1 sp., *P. grossa* (Fabricius 1801), widely distributed east of the Rocky Mountains.

Amphotis Erichson 1843, 2 spp., Massachusetts to Florida, west to Texas.
Cerophorus Castelnau 1840
Soronia Horn 1879

Thalycra Erichson 1843, 16 spp., generally distributed. Key to spp.: Howden 1961.
Perthalycra Horn 1879
Pseudothalycra Howden 1961

Pocadius Erichson 1843, 4 spp., generally distributed.

Camptodes Erichson 1843, 3 spp., Texas and Arizona.
Eucamptodes Sharp 1890

Cychramus Kugelann 1794, 5 spp., New England to Florida, west to Texas, Missouri, and Michigan.
Campta Stephens 1830
Quadrifrons Blatchley 1916

Pallodes Erichson 1843, 3 spp., New England to Florida, west to Arizona, and north to Minnesota.
Neopallodes Reitter 1884

Cyllodes Erichson 1843, 1 sp., *C. biplagiatus* LeConte 1866, New England to New Jersey, west to Minnesota.
Strongylus Herbst 1792, not Mueller 1780
Volvoxis Kugelann 1794
Pseudocamptodes Grouvelle 1896

Psilopyga LeConte 1853, 4 spp., New England to Florida, west to Wisconsin, and Arizona.

Oxycnemus Erichson 1843
Eugoniopus Reitter 1884

Meligethinae C. G. Thomson 1859

Meligethes Stephens 1830, 6 spp., widely distributed. Key to spp.: Easton 1955.

Acanthogethes Reitter 1871
Odonthogethes Reitter 1871
Clypeogethes Scholz 1932
Idiogethes Kirejtshuk 1977
Astylogelhes Kirejtshuk 1979
Chromogethes Kirejtshuk 1989
Lariopsis Kirejtshuk 1989

Cryptarchinae C. G. Thomson 1859

Ipinae Erichson 1843
Pityophagini Fauconnet 1894

Cryptarcha Shuckard 1839, 5 spp., generally distributed.

Cryptarchus Heer 1841
Arhina Murray 1867
Lepiarcha Sharp 1891
Cryptarchula Ganglbauer 1899
Cryptarchina Iablokoff-Khnzorian 1966

Pityophagus Shuckard 1839, 3 spp., British Columbia to California and Arizona, also in Pennsylvania and District of Columbia.

Glischrochilus Reitter 1873, 9 spp., generally distributed.

Ips Fabricius 1776, not DeGeer 1775
Librodor Reitter 1884
Cryptarchips Reitter 1911
Cephalips Arrow 1931
Gymnoparomius Kirejtshuk 1987

Cybocephalinae Jacquelin du Val 1858

This subfamily is considered by many to be a separate family. It is not included in the Nitidulidae by Parsons (1943), but concrete reasons were not given for its exclusion.

Cybocephalus Erichson 1844, 4 spp., Indiana, Michigan, Georgia, Florida, Arizona, California, and New England.

Phantozomerus duVal 1854
Stegnomorpha Wollaston 1854
Acribus Waterhouse 1877

Bibliography

AUDISIO, P. 1993. Coleoptera Nitidulidae-Kateridae. Fauna d'Italia. 32: 1-951.

BØVING, A.G. and F.C. CRAIGHEAD. 1931. An illustrated synopsis of the principle larval forms of the order Coleoptera. Entomologica Americana (new series), 21: 1-351.

BØVING, A.G. and J.G. ROZEN. 1962. Anatomical and systematic study of the mature larvae of the Nitidulidae (Coleoptera). Entomologiske Meddelelelser 31: 265-299.

CONNELL, W.A. 1977. A key to *Carpophilus* sap beetles associated with stored foods in the United States. Cooperative Plant Pest Report, 2: 398-404.

CONNELL, W.A. 1984. Nearctic Nitidulidae - synonymy and additions since Parsons' revision (Coleoptera). Coleopterist Bulletin, 38: 160-164.

CONNELL, W.A. 1991. Chapter 8. Pp. 151-174. Sap beetles (Nitidulidae: Coleoptera). Insect and Mite Pests in Food, Vol. 3. United States Department of Agriculture Handbook 655.

EASTON, A.M. 1955. A review of the Nearctic species of the beetle genus *Meligethes* (Nitidulidae). Proceedings of the United States National Museum, 104 (3339): 87-103.

FORD, E.J. 1996. The genus *Stelidota* Erichson in North America: A new species from Florida, new synonymy and lectotype designations (Coleoptera: Nitidulidae). Coleopterist Bulletin, 50: 149-153.

HATCH, M. H. 1961. The beetles of the Pacific Northwest. Part III: Pselaphidae and Diversicornia I. University of Washington Publications in Biology, 16: 503 pp.

HOWDEN, H.F. 1961. A revision of the new world species of *Thalycra* Erichson, with a description of a new genus and notes on generic synonymy (Coleoptera: Nitidulidae). Canadian Entomologist, Supplement 25: 1-61.

KIREJTSHUK, A.G. 1986. On polyphyly of the Carpophilinae with description of a new subfamily, Cillacinae (Coleoptera: Nitidulidae). Coleopterists Bulletin, 40: 217-221.

LAWRENCE, J.F. 1991. Nitidulidae (Cucujoidea) (including Brachypteridae, Cateretidae, Cybocephalidae, Smicripidae). Pp. 456-460. *In*: F. W. Stehr, ed., Immature Insects, Vol. 2. Kendall/Hunt. Dubuque, Iowa, 975 pp.

LAWRENCE, J. F. and A. F. NEWTON, Jr. 1995. Families and subfamilies of Coleoptera (with selected genera, notes, references and data on family-group names). Pp. 779-1006. *In*: J. Pakaluk and S. A. Slipinski, eds. Biology, phylogeny, and classification of Coleoptera: Papers celebrating the 80th birthday of R. A. Crowson. Muzeum i Instytut Zoologii PAN. Warsaw.

OKUMURA, G.I. and I.E. SAVAGE. 1974. Nitidulid beetles most commonly found attacking dried fruits in California. National Pest Control Operators News, 34: 2-7.

PARSONS, C.T. 1943. A revision of nearctic Nitidulidae (Coleoptera). Bulletin of the Museum of Comparative Zoology, 92: 121-278.

PENG, C.R., R.N. WILLIAMS and J.R. GALFORD. 1990. Descriptions and key for identification of larvae of *Stelidota* Erichson (Coleoptera: Nitidulidae) found in America north of Mexico. Journal of the Kansas Entomological Society, 63: 626-633.

ROZEN, J.G. 1963. Preliminary systematic study of the pupae of the Nitidulidae (Coleoptera). American Museum Novitates, 2124: 1-13.

78. SMICRIPIDAE Horn 1879

by Michele B. Price

Family common name: The palmetto beetles

Smicripidae are an infrequently encountered group of small, elongate, parallel-sided, flattened, reddish-brown beetles. Smicripidae can be separated from other families by the following characters: head not narrowed behind the eyes; antennae 11-segmented, with a 3-segmented, capitate club; frontoclypeal suture distinctly impressed and curved; maxillae with single lobe; labial palpi 2-segmented; prothoracic coxal cavities externally open behind; elytra truncate exposing two abdominal tergites; abdomen with five ventrites, caudal ventrite equal in length to preceding four combined; tarsal formula 4-4-4 or 5-5-5. Smicripids are often found on decaying vegetation.

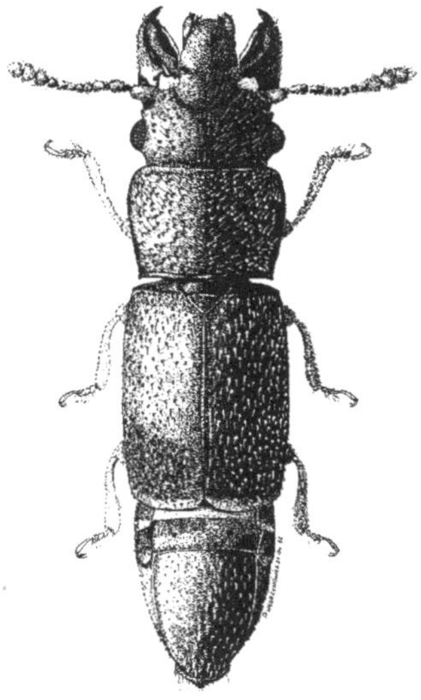

FIGURE 1.78. *Smicrips* sp. (from Lawrence *et al.* 1999b. Reproduced with permission from CSIRO, Australia.)

Description: (adapted, in part, from Audisio 1993, Bøving and Craighead 1931, Crowson 1955, Horn 1879, Kirejtshuk 1998, Lawrence 1991, Lawrence *et al.* 1999a, Lawrence *et al.* 1999b, and LeConte 1878). Body parallel-sided, elongate and flattened; length 1.0-1.7 mm; color mostly reddish-brown; pubescence fine, short, sparse, and pale in color.

Head prognathous, not narrowed behind eyes; eyes small, rounded, convex, strongly protuberant with fine facets; antennae 11-segmented, with a 3-segmented, capitate club; antennal insertions exposed from above, moderately to widely separated; labrum free, moderately prominent, transverse, apex strongly convex, narrowly rounded, or acute; clypeus wider than long, sides parallel, anterior edge concave to shallowly emarginate; frontoclypeal suture distinctly impressed, curved; mandibles short, broadly triangular; mandibular apex bidentate or multidentate; incisor edge with two or more teeth; mola well-developed; maxillae single-lobed, apex of lobe densely setose or spinose; apical maxillary palpomere cylindrical to fusiform, at least as wide as or longer than preapical palpomere; labial palpi 2-segmented; ligula prominent; mentum transverse, twice as wide as long, slightly narrowed in front, apex emarginate, slightly depressed mesally, concealing maxillae.

Pronotum as wide as or slightly wider than head, sides somewhat straight; anterior angles acute or obtuse, not produced forward; posterior angles acute or obtuse; posterior edge simple, more or less straight or evenly rounded, with narrow raised margin; prosternum slightly elevated and curved dorsally behind coxae, apex broadly rounded, angulate, or truncate; procoxae externally open behind, strongly transverse, moderately to widely separated, trochantin partly exposed; mesocoxae open laterally, circular to slightly transverse, moderately to widely separated; metacoxae widely separated; legs short, tibiae slender; tarsi pentamerous with apical tarsomere longer than preceding four together or tetramerous; claws simple.

Elytra scarcely wider than prothorax, truncate, exposing elongate pygidium, glabrous, irregularly punctate, or with five or fewer distinct striae; posterior edge of wings with fringe of long hairs, and with normal transverse folds present, radial cell reduced or absent, medial fleck absent, anal lobe absent.

Abdomen with five ventrites, last visible ventrite as long as preceding four combined; functional spiracles located on abdominal segments 5 and 6; male genitalia with deep longitudinal excision in tegmen, anterior edge of ninth ventrite with median strut, ninth tergite completely fused to tenth tergite, aedeagus asymmetrical, phallobase and articulated parameres secondarily segmented, anterior edge of phallobase with single strut, penis with single strut.

Larvae of Smicripidae resemble those of Laemophloeidae, Nitidulidae, and Phalacridae. Larvae are elongate, more or less parallel-sided; length less than 3.0 mm; dorsal and ventral surfaces lightly pigmented; vestiture consisting of fine setae. Head prognathus, narrower than prothorax, posterior edge of head capsule distinctly emarginate; epicranial stem absent; front arms present, V-shaped or U-shaped, bases distinctly separated; median endocarina absent; antennae 3-segmented; frontoclypeal suture absent; labrum free; epipharyngeal rods absent; mandibles symmetrical, apex bidentate, incisor edge with one or two subapical teeth; mola present, without brush at base; prostheca absent; ventral mouthparts strongly protracted; cardines absent; stipes wider than long; maxillae with single, fixed mala or single articulated mala; mala with single segment, apex simple, rounded or truncate, setose or spinose; labium consisting of prementum and postmentum; postmentum free or basally connate with maxillae; maxillary palpi 2-segmented; ligula absent; labial palpus one-segmented; hypopharyngeal sclerome tooth-like; hypostomal rods mostly absent, or long and parallel; ventral epicranial ridges absent; gula longer than wide. Prothoracic tergum with one or more sclerotized plates; mesothoracic and metathoracic terga without sclerotized plates; legs 5-segmented; 10 visible abdominal segments; ninth abdominal segment bearing strongly up-turned urogomphi, not forming a hinged plate; eighth abdominal segment not elongate; spiracles annular.

Habits and habitats. Little is known about the habits of these minute beetles. *Smicrips* adults and larvae occur in decaying flowers, leaf litter, and under bark (Lawrence 1982). In Florida, *Smicrips palmicola* LeConte has been collected in abundance on *Sabal palmetto* (Walter) Loddiges ex Sch. and Sch. (= *Chamaerops palmetto*) (LeConte 1878). In Panama, *Smicrips* adults and larvae were found among rotting flowers of *Pseudobombax* sp. (Bombacaceae) (Lawrence *et al.* 1999b). According to collection data, passive traps (barrier pitfalls and flight intercepts) prove to be a satisfactory method for capturing *Smicrips* adults in Florida.

Status of the classification. This small group has only recently been recognized as a separate family. Previously, they have been shifted between Nitidulidae and Monotomidae. Smicripidae is currently thought to be closely related to Nitidulidae.

LeConte (1878) described *Smicrips palmicola*, a new genus and species, from Florida, placing it in Monotomidae. Reitter (1876) had described this species from Cuba as *Tisiphone hypocoproides*, however, LeConte and Horn (1883) stated that the generic name was preoccupied.

Horn (1879) placed *Smicrips* in its own tribe, Smicripini, within Nitidulidae and provided a detailed description of the genus because its placement was still in question. He noted that *Smicrips* shared more similarities with Nitidulidae than with Monotomidae, such as, the single-lobed maxillae, 11-segmented antennae with a 3-segmented club, transverse procoxae and mesocoxae, and the procoxae open (however, it was mentioned that this character varied in Nitidulidae). LeConte and Horn (1883) maintained this position.

Sharp (1900) placed *Smicrips* (still under *Tisiphone*) in Monotomidae: Tisiphoninae. However, he was doubtful of the placement and suggested careful analysis of the genus to determine whether it should be placed in Monotomidae, Nitidulidae, or Cucujidae. Casey (1916) followed Sharp's placement of the genus and described a new species, *Tisiphone texana*. He explained that most of the characters and its general habitus harmonize well with Monotomidae. However, Casey also explained the differences between Tisiphoninae and Monotomidae, and suggested either disregarding the differences or establishing a separate family for *Tisiphone*. Leng (1920) placed the two species, *S. palmicola* and *S. texana*, in the subfamily Smicripinae, in Monotomidae. Hetschko (1930) placed *Smicrips* in Monotominae in the family Cucujidae.

Bøving and Craighead (1931) recognized *Smicrips* as a separate family, Smicripidae, based on larval characters. Parsons (1943) excluded *Smicrips* from the Nearctic Nitidulidae, mentioning its possible placement in Cucujidae. Crowson (1955) treated the group as its own family, but noted that based on adult characters, *Smicrips* should be treated as a subfamily of Nitidulidae. Arnett (1968) placed the group in Smicripinae of his Rhizophagidae (= Monotomidae). Kirejtshuk (1982) and Audisio (1993) likewise supported recognition of Smicripidae as a separate family based on adequately distinct morphological and ecological characters. Lawrence and Newton (1995) and Kirejtshuk (1998) maintained the family status of Smicripidae, however, stating that the group is sometimes treated as a subfamily of Nitidulidae.

Distribution. Smicripidae are composed of a single genus, *Smicrips*, with six species: *Smicrips chontalena* (Sharp) (Guatemala, Nicaragua), *S. distans* (Sharp) (Guatemala), *S. exilis* Murray (= *Tisiphone nitiduloides* Reitter) (Hispaniola, Guadeloupe, St. Vincent, Grenada), *S. mexicana* (Sharp) (Mexico), *S. palmicola* LeConte (= *T. hypocoproides* Reitter) (Cuba, Puerto Rico, U.S.A.), and *S. texana* (Casey) (U.S.A.) (Blackwelder 1944, Casey 1916, Hetschko 1930, Peck and Thomas 1998, Sharp 1900).

Classification of the United States Species

Smicripidae Horn 1879

Smicrips LeConte 1878
Tisiphone Reitter 1876, not Fitzinger 1826
Two species: *S. palmicola* LeConte 1878, California, Florida, and Georgia. *S. texana* (Casey 1916), Florida, Texas.

Bibliography

ARNETT, R. H., Jr. 1968. The beetles of the United States. (A manual for identification). The American Entomological Institute. Ann Arbor, Michigan, xii + 1112 pp.

AUDISIO, P. 1993. Coleoptera Nitidulidae-Kateretidae. Fauna d'Italia 32. Calderini. Bologna, xvi + 971 pp.

BLACKWELDER, R. E. 1944. Checklist of the coleopterous insects of Mexico, Central America, the West Indies, and South America. Part 3. Bulletin of the United States National Museum, 185: 343-550.

BØVING, A. G. and F. C. CRAIGHEAD. 1931. An illustrated synopsis of the principal larval forms of the order Coleoptera. Entomologia Americana (New Series), 11: 1-351.

CASEY, T. L. 1916. Some random studies among Clavicornia. Memoirs on the Coleoptera, 7: 35-292.

CROWSON, R. A. 1955. The natural classification of the families of Coleoptera. Nathaniel Lloyd. London, 187 pp.

HETSCHKO, A. 1930. Pars 109. Cucujidae, Thorictidae (Suppl.), Cossyphodidae (Suppl.). Pp. 1-122 *In:* S. Schenkling, ed. Coleopterorum Catalogus. W. Junk. Berlin.

HORN, G. H. 1879. Revision of the Nitidulidae of the United States. Transactions of the American Entomological Society, 7: 267-336.

KIREJTSHUK, A. G. 1982. Systematic position of the genus *Calonecrus* J. Thomas and notes on the phylogeny of the family Nitidulidae (Coleoptera). Entomologicheskoye Obozreniye, 61: 117-129. (In Russian, translation in Entomological Review, 61: 109-122).

KIREJTSHUK, A. G. 1998. Nitidulidae (Coleoptera) of the Himalayas and Northern Indochina. Part 1: subfamily Epuraeinae. Koeltz Scientific Books. Koeningstein (Theses Zoologicae, Vol. 28), 489 pp.

LAWRENCE, J. F. 1982. Coleoptera. Pp. 482-553. *In*: S. P. Parker, ed. Synopsis and Classification of Living Organisms. Volume 2. McGraw-Hill. New York.

LAWRENCE, J. F. 1991. Nitidulidae (Cucujoidea) (including Brachypteridae, Cateretidae, Cybocephalidae, and Smicripidae). Pp. 456-460. *In*: F. W. Stehr, ed. Immature Insects. Volume 2. Kendall Hunt. Dubuque, Iowa.

LAWRENCE, J. F., A. M. HASTINGS, M. J. DALLWITZ, T. A. PAINE and E. J. ZURCHER. 1999a. Beetle larvae of the World: Descriptions, illustrations, and information retrieval for families and sufamilies. CD-ROM, Version 1.1 for MS-Windows. CSIRO Publishing. Melbourne.

LAWRENCE, J. F., A. M. HASTINGS, M. J. DALLWITZ, T. A. PAINE and E. J. ZURCHER. 1999b. Beetles of the World: Descriptions, illustrations, and information retrieval for families and subfamilies. CD-ROM, Version 1.0 for MS-Windows. CSIRO Publishing. Melbourne.

LAWRENCE, J. F. and A. F. NEWTON, Jr. 1995. Families and subfamilies of Coleoptera (with selected genera, notes, references and data on family group names). Pp. 779-1006. *In*: J. Pakaluk and S. A. Slipinski, eds. Biology, Phylogeny, and Classification of the Coleoptera: Papers Celebrating the 80th Birthday of Roy A. Crowson. Muzeum i Instyut Zoologii PAN. Warsaw.

LECONTE, J. L. 1878. Additional descriptions of new species. Proceedings of the American Philosophical Society, 17: 373-434.

LECONTE, J. L. and G. H. HORN. 1883. Classification of the Coleoptera of North America. Smithsonian Miscellaneous Collections, 26 (507): xxxviii + 567 pp.

LENG, C. W. 1920. Catalogue of the Coleoptera of America, North of Mexico. Cosmos Press. Cambridge, Massachusetts, x + 470 pp.

PARSONS, C. T. 1943. A revision of the Nearctic Nitidulidae (Coleoptera). Bulletin of the Museum of Comparative Zoology, 92: 121-278.

PECK, S. B. and M. C. THOMAS. 1998. A distributional checklist of the beetles (Coleoptera) of Florida. Arthropods of Florida and Neighboring Land Areas, 16: 1-180.

REITTER, E. 1876. Revision der Monotomidae (sensu LeConte). Deutsche Entomologische Zeitschrift, 20: 295-301.

SHARP, D. 1900. Monotomidae. Pp. 563-579, pl. XVIII. *In*: F. D. Godman and O. Salvin, eds. Biologia Centrali-Americana. Insecta, Coleoptera. Vol. II. Part 1. Porter. London.

79. MONOTOMIDAE Laporte 1840

by Yves Bousquet

Family common name: The root-eating beetles

Family synonym: Rhizophagidae Redtenbacher 1845

This family of small beetles is usually considered to be closely related to the Nitidulidae on the basis of adult similarities. The one- or two-segmented antennal club on the ten-segmented antennae serves to separate adults of this family from those of Nitidulidae.

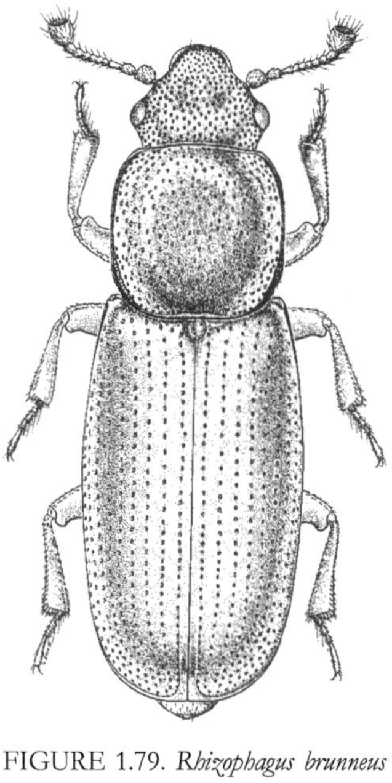

FIGURE 1.79. *Rhizophagus brunneus* Horn.

Description: Adult (see Lawrence 1982 and Sen Gupta 1988) shape subcylindrical to flattened, elongate; size 1.5 to 4.5 mm.; color dull; vestiture of short, sparse hairs or body glabrous. Head prognathous, constricted posteriorly in many groups. Antennae 10-segmented, with one- or two-segmented club; antennal grooves present only in a few groups. Frontoclypeal suture absent. Labrum very short, indistinct; mandibles short, curved, mostly with single apical tooth; maxilla with long and slender, finger-like galea, lacinia with setae along medial and apical margins, maxillary palpi four-segmented, the first segment short; labium with mentum in many elongate, labial palpi three-segmented. Eyes small, lateral, rounded, coarsely faceted.

Pronotum mostly subquadrate to elongate, with smooth or denticulate lateral margins; prosternum without apical prolongation, intercoxal process broadened toward apex; procoxae rounded with hidden trochantins in most groups, transverse with partly exposed trochantins in *Rhizophagus;* procoxal cavities closed behind. Mesocoxae narrowly to widely separated; mesocoxal cavities open laterally. Legs moderate; tibia broadened at apex, with distinct apical spurs; tarsal formula mostly 5-5-5 in females, 5-5-4 in males, in a few 5-5-5 or 4-4-4 in both sexes; tarsomeres not lobed, claws simple. Elytra truncate apically, exposing one (females) or two (males) tergites; striae distinct or not. Wing with single anal vein in many taxa; subcubital fleck present in a few genera; radial and anal cells absent.

Abdomen with five visible sternites; first sternite elongate, at least as long as next two sternites combined, sternites 2-4 short, subequal in length, sternite 5 elongate; coxal lines on first sternite present in many groups. Male genitalia with median lobe short, broad, flattened, with a pair of long, basal struts; parameres indistinct. Female genitalia with small paraprocts; valvifers small, chitinized; coxite two-segmented.

Larvae (see Lawrence 1991) with body subcylindrical to slightly flattened; surface with few to many tubercles. Head protracted and prognathous; frontoclypeal suture and epicranial suture absent. Two or four stemmata present in most groups, these absent in a few taxa. Antennae well developed, 3-segmented. Mandibles symmetrical, with serrate incisor lobe in some taxa, a slender, hyaline prostheca and a tuberculate or asperate mola. Maxilla with transverse cardo, 3-segmented palp, and falciform mala. Labial palps widely separated, 1- or 2-segmented. Gula markedly transverse. Ventral epicranial ridges present. Tergite IX with a pair of urogomphi, mostly branched. Legs well-developed, 5-segmented.

Habits and habitats. Most species of this family live under tree bark; some species occur in decaying vegetable matter and are often found in man-made habitats such as compost heaps and haystacks. A few species live in ants' nests. The larvae and adults of *Rhizophagus* have been listed as predators of xylophagous insects, such as scolytid larvae, although there are some indications that they also feed on fungi or fungal by-products. Information on the other groups of Monotomidae suggests that they feed on fungus.

Status of the classification. Crowson (1955) classified members of this family into four subfamilies: Lenacinae, Thioninae, Rhizophaginae and Monotominae. Lawrence and Newton (1995), following Sen Gupta (1988), recognized two subfamilies, Rhizophaginae and Monotominae which include lenacines and thionines. The entire group needs to be studied in detail. Sen Gupta (1988) published a review of the genera at the world level but since then nine new genera have been described.

Distribution. There are about 220 species described from all areas, of which 55 occur in the Nearctic Region.

KEY TO THE NEARCTIC GENERA

1. Fore coxa transverse; head with antennal grooves lateroventrally .... *Rhizophagus*
— Fore coxa more or less rounded; head without antennal grooves 2

2(1). Head distinctly longer than wide; anterior margin of clypeus markedly notched and flanked by 2 short, dentiform outgrowths .... *Thione*
— Head not longer than wide; anterior margin of clypeus rounded or slightly notched, without dentiform outgrowths 3

3(2). Antenna with one-segmented club 4

— Antenna with two-segmented club 7

4(3). Coxal bead on first visible abdominal sternite not triangularly produced; pronotum with pair of shallow depressions basally in most species, disc with large, subcontiguous punctures (except in *M. longicollis*), without impunctate median area .... *Monotoma*

— Coxal bead on first visible abdominal sternite triangularly produced; pronotum without depressions basally; disc with smaller, separated punctures, with impunctate median area 5

5(4). Head constricted basally, temples distinct; visible abdominal sternites 2-4 with 1-3 transverse rows of small, round punctures .... *Leptipsius*

— Head not constricted basally, temples indistinct; visible abdominal sternites 2-4 with one transverse row of large, oblong punctures 6

6(5). Pronotal disc distinctly depressed; pronotal disc and scutellum without microsculpture .... *Pycnotomina*

— Pronotum disc flat to convex; pronotal disc and scutellum with microsculpture .... *Bactridium*

7(3). Elytral disc with setigerous punctures irregularly placed, not forming distinct longitudinal rows .... *Phyconomus*

— Elytral disc with setigerous punctures arranged in longitudinal rows 8

8(7). Inflexed part of elytron (i.e., region from sixth stria to lateral margin of elytron) with 3 regular rows of setigerous punctures (lateral most row very close to epipleuron) 9

— Inflexed part of elytron with 4-5 rows of setigerous punctures (some of the intermediate rows are more or less regular in some species) 10

9(8). First visible abdominal sternite of male with median, oval plaque more microsculptured than adjacent area and bearing proportionally long setae .... *Aneurops*

— First visible abdominal sternite of male without median, oval plaque bearing long setae .... *Europs*

10(8). Pronotum markedly elongate (length pronotum/width pronotum > 1.25), with lateral margin not reaching anterior margin .... *Macreurops*

— Pronotum subquadrate to slightly elongate (length pronotum/width pronotum < 1.15), with lateral margin reaching anterior margin .... *Hesperobaenus*

Classification of the Nearctic Genera

Monotomidae Laporte 1840

Rhizophaginae Redtenbacher 1845

This subfamily includes a single genus, *Rhizophagus* Herbst; although some authors consider the monobasic taxon *Cyanostolus* Ganglbauer of Europe as a distinct genus, most authors treat it as a subgenus of *Rhizophagus*. The most distinctive character states of the subfamily are: body appearing glabrous except under high magnification; fore coxal cavities distinctly transverse with exposed trochantins; transverse suture on anterior part of gular region absent; antennal cavities well developed; wing with 3 anal veins; neck constriction absent.

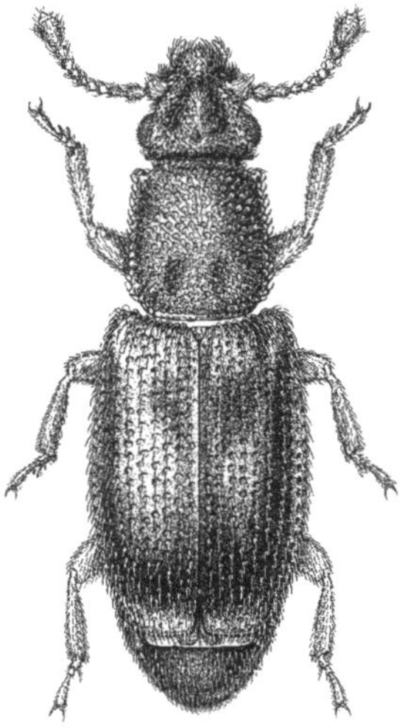

FIGURE 2.79. *Monotoma picipes* Herbst.

Rhizophagus Herbst 1793 (Fig. 1). This genus includes 14 species in North America. They are found over most of the temperate and boreal regions of the continent. The Nearctic species are classified in three subgenera: *Anomophagus* Reitter 1907 with three species, *Eurhizophagus* Méquignon 1909 with one species, and *Rhizophagus s. str.* (with *Syringobidia* Casey 1916 as a junior synonym) with the remaining species. Bousquet (1990) reviewed the Nearctic species and provided a key to the species.

Monotominae Laporte 1840

This subfamily includes 30 genera which are placed in four tribes: Monotomini (genus *Monotoma* Herbst), Thionini (genera *Thione* Sharp and *Shoguna* Lewis), Lenacini (genus *Lenax* Sharp of New Zealand), and Europini with the remaining genera. The main character states of the subfamily are: body with distinct setae in most groups; fore coxal cavities round with hidden trochantins; transverse suture on anterior part of gular region present; antennal cavities absent in most groups; wing with one anal vein in most groups; neck constriction distinct in most groups.

Monotomini Laporte 1840

The single genus included in this tribe is cosmopolitan. The species are collected in grass piles, compost heaps, even decaying seaweed and have been widely dispersed by man. A few species are apparently commensal in nests of ants.

Monotoma Herbst 1793 (Fig. 2). This genus contains about 15 species in North America. They are found all over the temperate region of the continent. Two of these species, *M. quadrifoveolata* Aubé 1837 and *M. myrmecophila* Bousquet and Laplante 2000, are classified in the subgenera *Monotomina* Nikitsky 1986 and *Gyrocecis* Thomson 1863 respectively whereas the rest of the species are placed in the nominotypical subgenus. The 12 Canadian species were reviewed by Bousquet and Laplante (2000).

Thionini Crowson 1955

Two genera belong to this tribe. Adults are subcylindrical and markedly elongate and superficially resemble those of the genus *Colydium* (Colydiidae). The genus *Shoguna* Lewis, which belongs here, occurs in southeastern Asia, New Guinea, New Britain, and Madagascar.

Thione Sharp 1899. This genus contains five described species, two from Australia and three from Central America, the West Indies, and Florida. The single Nearctic species, *T. championi* Sharp, was originally described from Guatemala and Panama (Sharp 1899) and subsequently reported from Guadeloupe (Hetschko 1930) and from Alachua and Orange Counties in Florida (Peck and Thomas 1998). A revision of the species is needed.

Europini Sen Gupta 1988

The bulk of the monotomine genera belong to this heterogenous tribe. Specimens of most of the 25 Nearctic species may be found under bark of dead trees.

Bactridium LeConte 1861. This genus includes eight species in North America. The species are found in the east, from southern Quebec and Ontario south to Texas, as well as in the west, in British Columbia and California. The genus is in need of taxonomic revision.

Pycnotomina Casey 1916. Only one species, *P. cavicolle* (Horn 1879), belongs to this genus. The species is found in eastern North America, from southern Quebec south to at least Kentucky.

Leptipsius Casey 1916. The genus currently contains two species. One, *L. dilutus* Casey 1916, is known from Illinois and the second one, *L. striatus* (LeConte 1858), from Idaho to California, east to Kansas. An undescribed species occurs in New Mexico and Arizona.

Hesperobaenus LeConte 1861. The genus includes six North American species which are found in eastern United States, from Ohio to Texas, and in western North America, from southern British Columbia to California and Arizona. The genus is in need of taxonomic revision.

Europs Wollaston 1854. Five North American species are currently included in the genus *Europs*. They are found in eastern United States, from New York to Florida and Texas, and in New Mexico, Arizona and California. The genus is in need of taxonomic revision.

Nomophloeus LeConte 1861

Aneurops Sharp 1900. The genus contains two described species, one from Guatemala and one previously reported only from Mexico that occurs also in Arizona, New Mexico, and Colorado.

Macreurops Casey 1916. The genus contains a single species, *M. longicollis* (Horn 1879), which is found from southwestern British Columbia south to California.

Phyconomus LeConte 1861. This genus includes a single species, *P. marinus* (LeConte 1858), which is found under seaweeds on beaches along the coast of California and Baja California. The other four species listed by Hetschko (1930) in this genus belong to other genera: *Mimema pallidum* Wollaston and *M. tricolor* Wollaston from South Africa belong to *Mimema* Wollaston, which is not a synonym of *Phyconomus* (personal observation) as reported by Reitter (1876); *Phyconomus luridipennis* Reitter from Mexico belongs to *Europs* (Sharp 1900) and *Phyconomus subtestaceus* Reitter from Mexico belongs to *Hesperobaenus* (personal observation).

Bibliography

BOUSQUET, Y. 1990. A review of the North American species of *Rhizophagus* Herbst and a revision of the Nearctic members of the subgenus *Anomophagus* Reitter (Coleoptera: Rhizophagidae). Canadian Entomologist, 122: 131-171

BOUSQUET, Y. and S. LAPLANTE. 2000. Taxonomic review of the Canadian species of the genus *Monotoma* Herbst (Coleoptera: Monotomidae). Proceedings of the Entomological Society of Ontario, 130 [1999]: 67-96.

CROWSON, R. A. 1955. The natural classification of the families of Coleoptera. Nathaniel Lloyd. London. 187 pp.

HETSCHKO, A. 1930. Fam. Cucujidae (Pars 109). *In:* W. Junk and S. Schenkling, eds. Coleopterorum Catalogus. Junk. Berlin. 122 pp.

LAWRENCE, J. F. 1982. Coleoptera. Pp. 482-553. *In:* S. P. Parker, ed. Synopsis and Classification of Living Organisms. Vol. 2. McGraw Hill. New York. 1232 pp.

LAWRENCE, J. F. 1991. Rhizophagidae (Cucujoidea) (including Monotomidae). Pp. 460-462 *In:* F.W. Stehr, ed. Immature Insects. Volume 2. Kendall/Hunt. Dubuque, IA.

LAWRENCE, J. F. and A. F. NEWTON, Jr. 1995. Families and subfamilies of Coleoptera (with selected genera, notes, references and data on family-group names). Pp. 779-1006. *In:* J. Pakaluk and S.A. Slipinski, eds. Biology, phylogeny, and classification of Coleoptera. Muzeum i Instytut Zoologii PAN, Warsaw.

PECK, S. B. and M. C. THOMAS. 1998. A distributional checklist of the beetles (Coleoptera) of Florida. Arthropods of Florida and Neighboring Land Areas, 16: viii + 180 pp.

REITTER, E. 1876. Revision der Monotomidae (*sensu* LeConte). Deutsche Entomologische Zeitschrift, 20: 295-301.

SEN GUPTA, T. 1988. Review of the genera of the family Rhizophagidae (Clavicornia: Coleoptera) of the world. Memoirs of the Zoological Survey of India, 17: 1-58, 24 pls.

SHARP, D. 1899. Fam. Cucujidae. Pp. 499-560. *In:* F. D. Godman and O. Salvin, eds. Biologia Centrali-Americana. Insecta. Coleoptera. Vol. II. Part 1. Taylor and Francis, London.

SHARP, D. 1900. Fam. Monotomidae. Pp. 563-579. *In:* F. D. Godman and O. Salvin, eds. Biologia Centrali-Americana. Insecta. Coleoptera. Vol. II. Part 1. Taylor and Francis, London.

80. SILVANIDAE Kirby 1837

by Michael C. Thomas

Common name: The silvanid flat bark beetles

This is a difficult family to characterize because it is so diverse. The presence of a mandibular mycangium, 5-5-5 tarsal formula, and head constricted behind eyes in most are diagnostic for the family as a whole. Its relationships within the Cucujoidea are unclear, but it is usually considered to be fairly primitive. Most species appear to be fungivorous, but several are important pests of stored grains and grain products.

FIGURE 1.80. *Telephanus velox* (Haldeman).

Description: Body elongate, parallel-sided to ovate; length, 2-15 mm; mostly brownish to blackish, some patterned; pubescence mostly conspicuous.

Head transverse, constricted behind eyes, temples present in many; with or without a frontoclypeal suture; surface mostly densely punctate; antennae of two types: elongate, filiform, with a very elongate scape and an inconspicuous club, or less elongate with a short scape and a conspicuous club of three antennomeres; labrum broadly rounded, small; mandibles broad, with two apical teeth, a subapical tooth, and a mola; dorsally with a large pit (mycangium) at the base (Brontinae) or remnants thereof (most silvanines); with a tusk-like process arising laterodorsally in some male Brontini; maxillary palpi with four palpomeres, apical longest, securiform in some, galea broadly rounded, lacinia narrow, both densely setose; mentum transverse, broadly emarginate; ligula quadrate or pentagonal; labial palpi with three palpomeres, apical longest in most, somewhat securiform in some; eyes lateral, more or less basal; size moderate to large, round or oval.

Pronotum transverse to elongate, mostly constricted basally; margined; with lateral spines or teeth in many, or prominent anterolateral tooth; surface punctate; prosternal process narrow to broad; procoxal cavities moderately to broadly separated, open posteriorly (Brontinae), or narrowly separated and closed posteriorly; mesocoxal cavities open laterally, broadly to narrowly separated; lateral sclerites of mesothorax in Silvaninae appearing fused; metacoxae transverse. Legs with trochantin concealed; anterior coxae round; middle coxae round; hind coxae transverse; trochanters normal, spined in males of some; femora slender to moderate, spined or otherwise modified in males of some; tibiae slender to moderate; apical spurs small, subequal; tarsi of two types: tarsomeres I and IV very short, tarsomeres II and III longer; lobed or bilobed in some Brontinae, or tarsomere I longer than II, III, or IV, IV very small, III lobed in some Silvaninae; tarsal formula 5-5-5 in both sexes; claws simple; scutellum moderate.

Elytra with epipleural fold broad, complete to apex; scutellary striole present or absent; metendosternite of normal hylecetoid type in genera with larger adults, reduced to lateral arms in smaller forms; wing venation well developed in larger forms, with six veins behind M_4Cu, radial recurrent, radial sector and radio-medial veins present, radial cell not closed; very reduced in smaller silvanines.

Abdomen with five sterna, sutures complete; intercoxal process mostly acute; abdominal sternum III with femoral lines in most; male genitalia either inverted (Brontinae) or normal (Silvaninae), parameres present and articulated; flagellum present in many, internal sac with well-developed armature in most. Female genitalia with proctiger membranous, paraprocts and valvifers reduced to bacula, coxite two-segmented, and stylus apical.

Larvae elongate, parallel-sided, somewhat flattened; mandible with acute retinaculum, and well-developed mola; mala acute apically, cardines bipartite; five or six pairs of stemmata; antennae either with three well-developed segments (Brontinae) or with segment III reduced and fused to segment II (Silvaninae); frontal suture somewhat lyriform; frontoclypeal suture absent; dorsal surface of thoracic and abdominal segments without asperities; spiracles annular; legs well-developed, two tarsungular setae, close together in Silvaninae, widely separated in Brontinae; long, whip-like urogomphi present in some Brontinae, absent in all Silvaninae (*Oryzaephilus* and *Nausibius* have tubercles apparently representing remnants of urogomphi); abdominal segment X located terminally, produced (see Thomas 1988; Sen Gupta and Pal 1996 present a useful review of adult and larval morphology).

Status of the classification. Like the other "cucujid" families treated here, this family was considered until recently a subfamily of the Cucujidae. Crowson (1955) removed the former subfamilies Silvaninae and Psammoecinae to the Silvanidae. Sen Gupta and Crowson (1966), Crowson and Sen Gupta (1969), and Crowson (1973) continued to treat the Silvanidae, including the subfamily Uleiotinae (=Brontinae), as a full family. Thomas

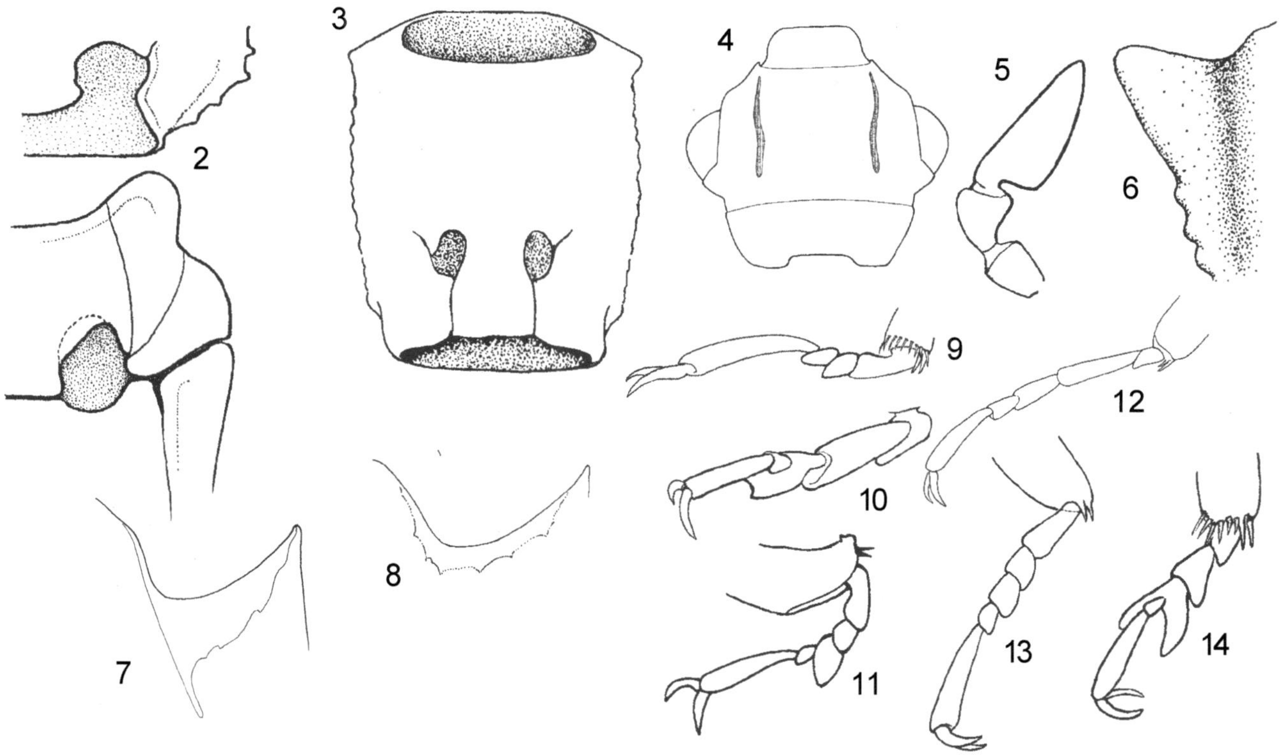

FIGURES 2.80 - 15.80. 2. *Uleiota dubius* (Fabricius), pro- and mesosterna; 3. *Cathartosilvanus imbellis* (LeConte), prosternum; 4. *Cryptamorpha desjardinsi* (Guérin-Méneville), head; 5. *Telephanus velox* (Haldeman), maxillary palpus; 6. *Pensus gilae* Casey; 7. *Cathartosilvanus imbellis*, femoral line; 8. *Silvanus muticus* Sharp, femoral line; 9. *Uleiota dubius*, hind tarsus; 10. *Telephanus velox*, hind tarsus; 11. *Cathartus quadricollis* (Guérin-Méneville), hind tarsus (major male); 12. *Dendrophagus cygnaei* Mannerhein, hind tarsus; 13. *Nausibius major* Zimmermann, hind tarsus; 14. *Cryptamorpha desjardinsi*, hind tarsus.

(1984) divided the Silvanidae into two subfamilies, Uleiotinae (=Brontinae) and Silvaninae, with the former composed of two tribes, Uleiotini (=Brontini) and Telephanini (=Psammoecini). Pal *et al.* (1985), retained Cryptamorphinae and Psammoecinae as subfamilies, in addition to Silvaninae and Uleiotinae (=Brontinae), which they, like Thomas (1984), considered to exhibit the most ancestral character states among the Silvanidae.

Hetschko (1930) listed 34 genera and 391 species worldwide that would now be assigned to Silvanidae. Genera and species described, and synonymies proposed since then (primarily by Nevermann, Halstead, Pal, and Sen Gupta), bring the world total to about 47 genera and approximately 470 species.

Habits and habitats. The biology and immature stages of the great majority of species and most genera are unknown, but the following generalizations can be made: Brontinae (Brontini) are found primarily under bark, where both adults and larvae probably feed on ascomycete and other fungi (Crowson and Ellis 1969; personal observations); Brontinae (Psammoecini) are found primarily on plants, especially withered, pendant leaves, and in plant debris, where they also probably feed predominantly on fungi; Silvaninae are found under bark (e.g., *Silvanus*, *Cathartosilvanus*) or in leaf litter or soil (e.g., *Silvanoprus*, *Monanus*, some *Ahasverus*), where they also seem to feed on fungi. Several Australian species are associated with ant nests, and two South American genera have been reported to milk honeydew from mealybugs. Some silvanines have been reported to be at least facultatively predacious.

Several species belonging to *Oryzaephilus*, *Nausibius*, *Cathartus*, and *Ahasverus* are important pests of stored grains, grain products, nuts, and spices (Halstead1993). Their taxonomy and biology are the best known of the Silvanidae.

Distribution. The family is worldwide in distribution, but is most abundant at both the generic and species level in the tropics. There are 47 genera and about 470 species in the world; 14 genera and 32 species are recorded from the United States, including several stored products species. Although the latter may not be established, they may be unintentionally introduced repeatedly.

Key to Nearctic genera

1. Antennae with an obvious club; scape not especially elongate; mandibles with dorsal mycangium reduced; aedeagus with parameres on dorsal aspect of median lobe; procoxal cavities closed posteriorly, usually broadly (Fig. 3) (Silvaninae)...... 5

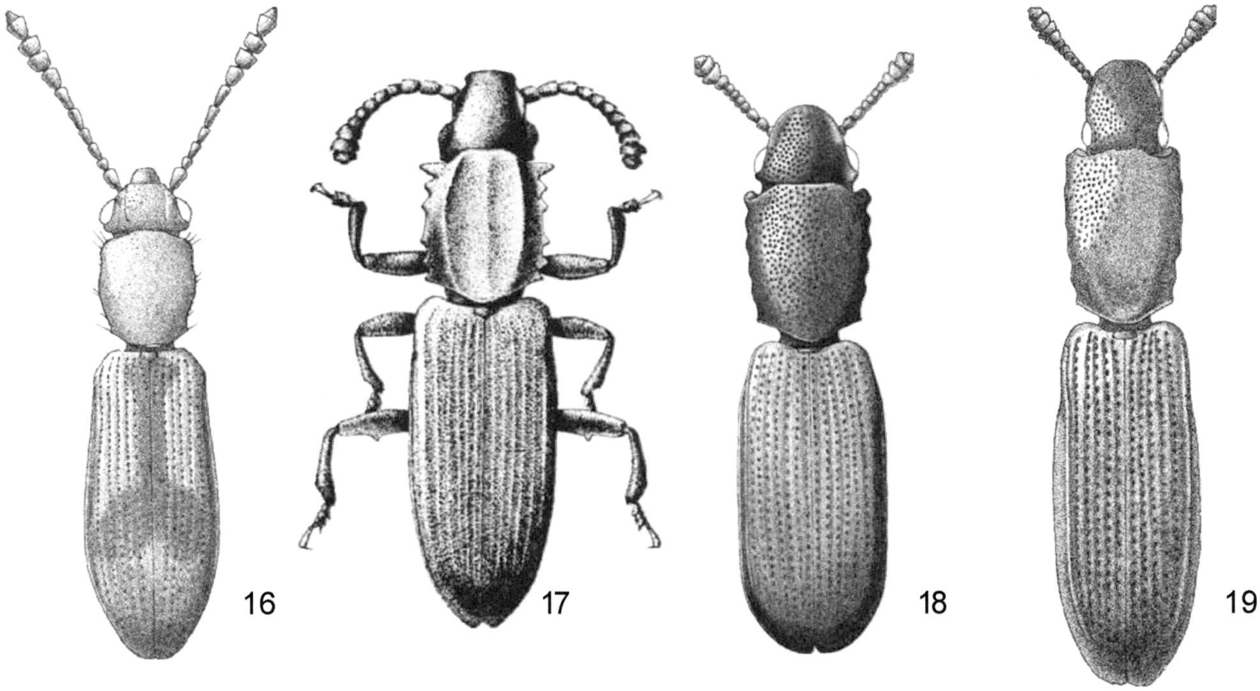

FIGURES 16.80-19.80. 16. *Cryptamorpha desjardinsi* (Guérin-Méneville); 17. *Oryzaephilus acuminatus* Halstead; 18. *Nausibius repandus* LeConte; 19. *Eunausibius salutaris* Parsons.

— Antennae elongate, filiform, without an obvious club; scape usually markedly elongate (Fig. 1); aedeagus with parameres on ventral aspect of median lobe; procoxal cavities open or closed posteriorly (Brontinae) 2

2(1). Procoxal cavities open posteriorly (Fig. 2); tarsomeres not lobed (Fig. 9, 12) (Brontini) 3
— Procoxal cavities closed posteriorly; tarsomeres distinctly lobed (Fig. 10, 14) (Psammoecini) 4

3(2). Tarsal formula 4-4-4 (Fig. 9); pronotum irregularly toothed laterally and with anterior angles strongly produced; males with curved mandibular processes .... *Uleiota*
— Tarsal formula 5-5-5, basal segment short but distinct (Fig. 12); pronotum simple laterally, anterior angles not produced; males without mandibular processes .... *Dendrophagus*

4(2). Head dorsally with longitudinal impressed lines; apical maxillary palpomeres not securiform; tarsomere III bilobed (Fig. 14) .... *Cryptamorpha*
— Head dorsally without impressed lines (Fig. 1); apical maxillary palpomeres securiform (Fig. 5); tarsomere III simply lobed (Fig. 10) .... *Telephanus*

5(1). At least tarsomere III lobed or incrassate (Fig. 11) 11
— No tarsomeres lobed (Fig. 13) 6

6(5). Lateral margins of pronotum simple; hind femora in males simple 9
— Lateral margins of pronotum dentate or undulating; hind femora in males toothed 7

7(6). Lateral margins of pronotum undulating (Fig. 18, 19) 8
— Lateral margins of pronotum dentate (Fig. 17) .... *Oryzaephilus*

8(7). Head ventrally with antennal cavity..... *Eunausibius*
— Head ventrally without antennal cavities .... *Nausibius*

9(7). Anterior angles of pronotum not produced anterolaterally (Fig. 3, 21); femoral lines open (Fig. 7) .... *Cathartosilvanus*
— Anterior angles of pronotum produced anterolaterally (Fig. 6); femoral lines closed (Fig. 8) 10

10(9). Anterior angles of pronotum acute, usually produced more laterally than anteriorly (Fig. 20).... *Silvanus*
— Anterior angles of pronotum obtuse, produced more anteriorly than laterally (Fig. 6) .... *Pensus*

11(5). Anterior angles of pronotum acute (Fig. 22); femoral lines closed (Fig. 8) .... *Silvanoprus*
— Anterior angles of pronotum lobed at most; femoral lines closed or open 12

12(11). Pronotal margins simple; femoral lines open (Fig. 7); elytra immaculate 13
— Pronotal margins denticulate (Fig. 23); femoral lines closed (Fig. 8); elytra maculate .... *Monanus*

13(12). Body elongate, parallel-sided (Fig. 24); antennomere XI at its broadest narrower than X; intercoxal process of sternum III pointed anteriorly . *Cathartus*
— Body ovate (Fig. 25); antennomere XI at its broadest equal in width to X; intercoxal process of sternum III rounded anteriorly .... *Ahasverus*

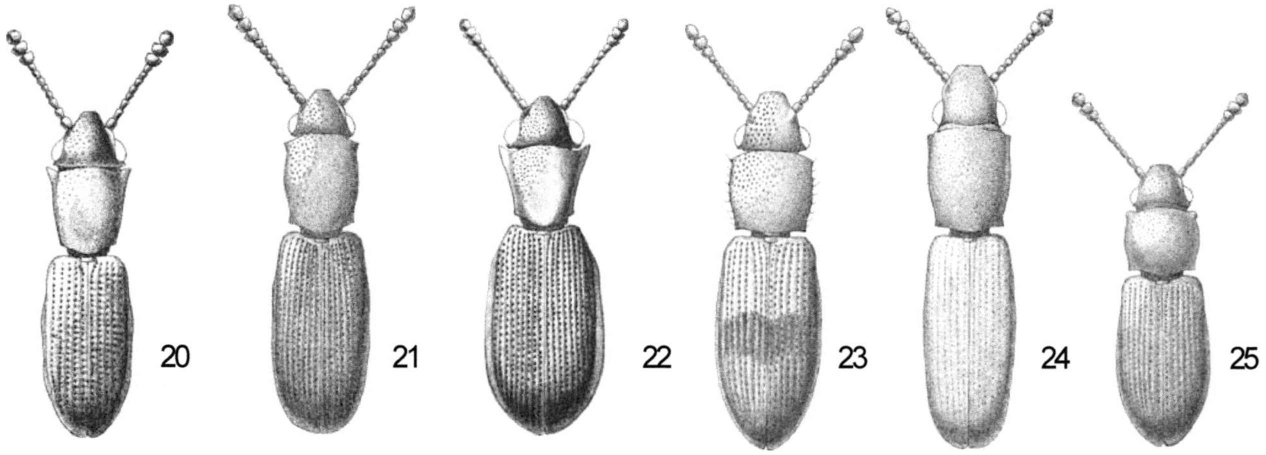

FIGURES 20.80 - 25.80. 20. *Silvanus planatus* Germar; 21. *Cathartosilvanus imbellis* (LeConte); 22. *Silvanoprus scuticollis* (Walker); 23. *Monanus concinnulus* (Walker); 24. *Catharus quadricollis* (Guérin-Méneville); 25. *Ahasverus rectus* (LeConte).

CLASSIFICATION OF THE NEARCTIC GENERA

Silvanidae Kirby 1837

Brontinae Erichson 1845

Diagnosis. Members of the Brontinae may be distinguished from members of the Silvaninae by the following combination of character states: mandible with large dorsal mycangium; aedeagus inverted with parameres located on the ventral aspect of the median lobe; antennae elongate and filiform, without a distinct club.

Brontini Erichson 1845

Diagnosis. Members of the Brontini may be distinguished from members of the Psammoecini by having the anterior coxal cavities wide open behind and the tarsi without lobed tarsomeres.

Uleiota Latreille 1796, 3 spp., generally distributed.
Brontes Fabricius 1801
Hyliota Reitter 1880

Dendrophagus Schönherr 1809, 1 sp., *D. cygnaei* Mannerheim, northern U.S.

Psammoecini Reitter 1879

Diagnosis. Members of the Psammoecini may be distinguished from members of the Brontini by having the anterior coxal cavities broadly closed behind and the tarsi usually with lobed tarsomeres.

Cryptamorpha Wollaston 1854, 1 sp., *C. desjardinsii* (Guérin-Méneville), Florida and Alabama, immigrant from tropical Asia.
Pseudophanus LeConte 1860

Telephanus Erichson 1846, 2 spp., eastern U.S. and Arizona.
Heterodromia Haldeman 1846

Silvaninae Kirby 1837

Diagnosis. Members of this subfamily can be distinguished by the following combination of character states: tarsomere I longer than II; mandibular mycangium present, but reduced in size compared to that found in Brontinae; antennae clubbed; anterior coxal cavities closed; aedeagus not inverted, parameres located on dorsal aspect of median lobe.

Nausibius Lentz 1857, 4 spp., including the stored products pest *N. clavicornis* (Kugelann); a Neotropical immigrant species, *N. sahlbergi* Grouvelle, established in Florida, and two native species, southeastern U.S. (key to spp., Thomas 1993)

Eunausibius Grouvelle 1912, 1 sp., *E. salutaris* Parsons, a Neotropical immigrant possibly established in Florida.

Oryzaephilus Ganglbauer 1899, 2 spp., *O. surinamensis* (L.), the saw-toothed grain beetle, and *O. mercator* (Fauvel), the merchant grain beetle, both generally distributed immigrant stored products pests from the Old World (key to spp., Halstead 1980).

Silvanus Latreille 1804, 9 spp., three of which are generally distributed native species, and six are immigrant species, two are generally distributed, three in Florida only, and one, *S. recticollis* Reitter, Florida, Georgia, and Oklahoma (key to spp., Halstead 1973).

Pensus Halstead 1973, 1 sp., *P. gilae* (Casey), Arizona.

Cathartosilvanus Grouvelle 1912, 2 spp., eastern U.S., Arizona, California (key to spp., Halstead 1980).

Silvanoprus Reitter 1911, 2 spp., both Old World immigrants occuring in the eastern and southeastern U.S.

Monanus Sharp 1879, 1 sp., *M. concinnulus* (Walker), an Asian immigrant possibly established in Florida.
Emporius Ganglbauer 1899

Cathartus Reiche 1854, 1 sp., *C. quadricollis* (Guérin-Méneville), the square-necked grain beetle, a generally distributed pest of corn, both in the field and in storage.

Ahasverus Gozis 1881, 3 spp., eastern U.S. to Texas, including *A. advena* (Waltl), the foreign grain beetle.

Bibliography

CROWSON, R. A. 1955. The natural classification of the families of Coleoptera. Nathaniel Lloyd. London. 187 pp.

CROWSON, R. A. 1973. Further observations on Phloeostichidae and Cavognathidae, with definitions of new genera from Australia and New Zealand. Coleopterists Bulletin, 27: 54-62.

CROWSON, R. A. and I. ELLIS. 1969. Observations on *Dendrophagus crenatus* (Paykull) (Cucujidae) and some comparisons with piestine Staphylinidae. (Coleoptera). Entomologists' Monthly Magazine, 104: 161-169.

CROWSON, R. A. and T. SEN GUPTA. 1969. The systematic position of Propalticidae and of *Carinophloeus* Lefkovitch (Coleoptera, Clavicornia) with descriptions of a new species of *Propalticus* and of its supposed larva. Proceedings of the Royal Entomological Society of London (B), 38: 132-140.

HALSTEAD, D. G. H. 1973. A revision of the genus *Silvanus* Latreille (s.l.) (Coleoptera: Silvanidae). Bulletin of the British Museum of Natural History (Entomology), 29: 39-112.

HALSTEAD, D. G. H. 1980. A revision of the genus *Oryzaephilus* Ganglbauer, including descriptions of related genera (Coleoptera: Silvanidae). Zoological Journal of the Linnean Society, 69: 271-374.

HALSTEAD, D. G. H. 1993. Keys for the identification of beetles associated with stored products-II. Laemophloeidae, Passandridae and Silvanidae. Journal of Stored Products Research, 29: 99-197.

HETSCHKO, A. 1930. Cucujidae. Coleoptorum Catalogus, 15(109): 1-93.

PAL, T. K., T. SEN GUPTA and R. A. CROWSON. 1985. Revision of *Uleiota* (Coleoptera: Silvanidae) from Indian and Sri Lanka and its systematic position. Oriental Insects, 18: 213-233 (1984).

SEN GUPTA, T. and R. A. CROWSON. 1966. A new family of cucujoid beetles, based on six Australian and one New Zealand genera. Annals and Magazine of Natural History, 13: 61-85.

SEN GUPTA, T. and T. K. PAL. 1996. Fauna of India and adjacent countries: Calvicornia [*sic*]: Coleoptera. Family Silvanidae. Zoological Survey of India. Calcutta. 262 pp.

THOMAS, M. C. 1984. A new species of apterous *Telephanus* (Coleoptera: Silvanidae) with a discussion of phylogenetic relationships of the Silvanidae. Coleopterists Bulletin, 38: 43-55.

THOMAS, M. C. 1988. Generic key to the known larvae of the Cucujidae, Passandridae, and Silvanidae of America North of Mexico (Coleoptera). Insecta Mundi, 2: 81-89.

THOMAS, M. C. 1993. The flat bark beetles of Florida (Laemophloeidae, Passandridae, Silvanidae). Arthropods of Florida and Neighboring Land Areas, 15: i-vii + 1-93.

81. PASSANDRIDAE Erichson 1845

by Michael C. Thomas

Family common name: The parasitic flat bark beetles

Family synonym: Scalidiidae Bøving 1921, Catogenidae Bøving and Craighead 1931

The adults of this unusual family tend to be elongate and parallel-sided and either strongly flattened dorsoventrally or almost cylindrical in cross section. The confluence of the gular sutures and the expanded genae are unique to this group of genera. They are typically heavily sculptured with longitudinal grooves and carinae and with heavy, often moniliform antennae. This group of genera has long been recognized as distinct, although usually as a subfamily of Cucujidae. However, Crowson (1955) recognized its distinctiveness and raised it to family rank. Thomas (1984) considered that Passandridae are more closely related to Laemophloeidae than to Cucujidae.

FIGURE 1.81. *Taphroscelidia linearis* (LeConte) (from Thomas 1993)

Description: Small to moderate in size, 3mm - 15mm; elongate, parallel-sided; dorsoventrally flattened or subcylindrical; mostly brown or black; nearly glabrous.

Head transverse to elongate, parallel-sided; mostly punctate dorsally, sculpture mostly rasp-like anteriorly; frons of many with median and admedian grooves, carinate laterally, mostly with a deep transverse groove basally; antennae inserted laterally behind mandibles, insertion concealed from above; antennae with 10 or 11 antennomeres, mostly moniliform, conspicuously pubescent; mostly without club, antennomere XI mostly asymmetrical and carinate; labrum small, semicircular, mostly concealed beneath clypeus; mandibles moderate in size, robust, with single apical tooth, and rounded retinaculum; molar area thickened, but without ridges; maxillary palps with four palpomeres, elongate, palpomere IV equal in length to first three combined; mentum transverse; ligula strongly V-shaped and heavily sclerotized; labial palps elongate, with three palpomeres, concealed in lateral grooves on ligula; genae produced anteriorly as rounded plates which conceal maxillae (Fig. 4); gular sutures confluent; eyes small to moderate, flat to moderately convex, round or oval; mostly situated basally.

Pronotum quadrate to elongate, strongly constricted basally in some; sublateral lines present or absent; margined laterally; anterior coxae globular, narrowly separated; procoxal cavities open (Fig. 5), trochantins hidden; mesocoxae globular, narrowly separated; mesocoxal cavities closed laterally by sterna (Fig. 6), trochantin hidden; hind coxae transverse, narrowly separated; tarsal formula 5-5-5 in both sexes, femora and tibiae stout; spurs large, unequal, anterior spur on proleg much longer than posterior spur and curved apically; tarsi not lobed; claws simple; scutellum small, semicircular; elytra with system of elytral cells or remnants thereof, and with lateral carina; scutellary striole absent; metendosternite unusual, with approximate anterior tendons and elongate lamina (Crowson 1938); hind wing venation well-developed, radial cell complete, medio-radial vein complete, radial recurrent absent, anal vein present, anal lobe absent (Wilson 1930).

Abdomen with five visible sterna, the sutures entire; intercoxal process acute; surface punctate, mostly with a subapical transverse marginal groove on sternum VII. Male genitalia with aedeagus inverted, with articulated parameres, and flagellum present; female genitalia very small, with proctiger membranous, paraprocts and valvifers reduced to bacula, coxite two-segmented, and stylus subapical.

Bøving and Craighead (1931) described and illustrated the larva of *Taphroscelidia linearis* (LeConte); Gravely (1916) described the larva of a species of *Passandra*. Body hypergastric, swollen; head and abdominal segment IX small; mouthparts reduced, articulations not evident on maxilla or labium; antennae short, inconspicuous, not articulated; legs short, stout; spiracles annular; urogomphi short, hook-like.

Habits and habitats. Larvae of these beetles are unusual because they seem to be exclusively ectoparasites on other wood-inhabiting insects, especially pupae of Cerambycidae. Dimmock (1884) and Fiske (1905) reported on the habits of the North American *Catogenus rufus* (Fabricius), and Gravely (1916) reported similar habits for *Hectarthrum* (=*Passandra*) *trigeminum* (Newman). Adult feeding habits are unknown.

Status of the classification. Slipinski (1987, 1989) and Burckhardt and Slipinski (1991, 1995) have completely revised the family. There are nine genera and 105 species worldwide. Two genera and three species occur in the Nearctic. The largest genus is the Old World *Ancistria* Erichson, with 32 species.

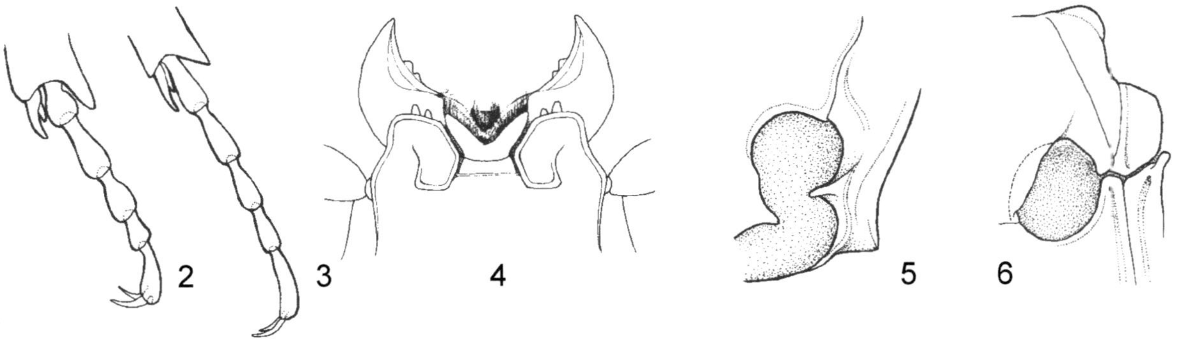

FIGURES 2.81 - 6.81. 2. hind tarsus and tibial spurs, *Catogenus rufus*; 3. same, *Taphroscelidia linearis*; 4. maxillae, *C. rufus*; 5. procoxal cavity, *C. rufus*; 6. mesocoxal cavity, *C. rufus* (all from Thomas 1993).

Distribution. Worldwide except for the western Palaearctic and New Zealand. The genus *Passandra* is found in all faunal regions in which the family occurs except for the Nearctic; *Ancistria* Erichson, *Nicolebertia* Burckhardt and Slipinski (=*Anisocerus* Westwood), *Passandrina* Reitter, and *Aulonosoma* Motschulsky (=*Laemotmetus* Gerstaecker) are restricted to the Old World; *Taphroscelidia* Crotch, *Passandrella* Grouvelle, *Scalidiopsis* Burckhardt and Slipinski, and *Catogenus* Westwood are restricted to the New World.

KEY TO NEARCTIC GENERA

1. Tibial spurs longer than tarsomere I of hind leg (Fig. 2); body dorsoventrally compressed .... *Catogenus*
— Tibial spurs shorter than tarsomere I of hind leg (Fig. 3); body subcylindrical .... *Taphroscelidia*

CLASSIFICATION OF THE NEARCTIC GENERA

Passandridae Erichson 1845

Catogenus Westwood 1835, 2 spp., eastern and southern North America, west to Arizona.
Scalidia Erichson 1846

Taphroscelidia Crotch 1873, 1 sp., *T. linearis* (LeConte), southeastern U.S. west to Texas, and Mexico, Central America, and the West Indies.
Scalidia auctorum, not *Scalidia* Erichson 1846
Syssitos Sharp 1899

BIBLIOGRAPHY

BØVING, A. G. and F. C. CRAIGHEAD. 1931. An illustrated synopsis of the principal larval forms of the order Coleoptera. Entomologica Americana (New Series), 11: 1-351.

BURCKHARDT, D. and S. A. SLIPINSKI. 1991. A review of the Passandridae of the world (Coleoptera: Cucujoidea). III. Genera *Anisocerus*, *Aulonosoma*, *Passandrella*, *Passandrina*, *Scalidiopsis* and *Taphroscelidia*. Revue Suisse Zoologie, 98: 453-497.

BURCKHARDT, D. and S. A. SLIPINSKI. 1995. A review of the Passandridae of the world (Coleoptera: Cucujoidea). IV. Revue Suisse Zoologie, 102: 995-1044.

CROWSON, R. A. 1938. The metendosternite in Coleoptera: a comparative study. Transactions of the Royal Entomological Society of London, 87: 397-415, 13 pls.

CROWSON, R. A. 1955. The natural classification of the families of Coleoptera. Nathaniel Lloyd. London. 187 pp.

DIMMOCK, G. 1884. Notes on *Catogenus rufus*. Psyche, 3: 341-342.

FISKE, W. F. 1905. *Catogenus rufus*, a coleopterous parasite. Proceedings of the Entomological Society of Washington, 7: 90-92.

GRAVELY, F. H. 1916. Some lignicolous beetle larvae from India and Borneo. Records of the Indian Museum, 12: 137-175.

SLIPINSKI, S. A. 1987. A review of the Passandridae of the world (Coleoptera, Cucujoidea). I - Genus *Passandra* Dalman. Annali del Museo di Civico Storia Naturale di Genova, 85: 553-603.

SLIPINSKI, S. A. 1989. A review of the Passandridae (Coleoptera, Cucujoidea) of the world. II. Genus: *Catogenus*. Polskie Pismo Entomologiczne, 59: 85-129.

THOMAS, M. C. 1984. A new Neotropical genus and species of rostrate Laemophloeinae (Coleoptera: Cucujidae), with discussion of the systematic position of the subfamily. Coleopterists Bulletin, 38: 67-83.

THOMAS, M. C. 1993. The flat bark beetles of Florida (Coleoptera: Silvanidae, Passandridae, Laemophloeidae). Arthropods of Florida and Neighboring Land Areas, 15: i-vii + 1-93.

WILSON, J. W. 1930. The genitalia and wing venation of the Cucujidae and related families. Annals of the Entomological Society of America, 23: 305-358.

82. CUCUJIDAE Latreille 1802

by Michael C. Thomas

Family common name: The flat bark beetles

Family synonym: Earophilidae Gistel 1856

These strikingly flat, often conspicuously colored beetles can usually be recognized by those characters alone. The extreme development of the temples in most, heteromerous tarsal formula in males and inverted male aedeagus (in Nearctic genera) are also diagnostic for the members of this family. They are considered to be rather primitive members of the Cucujoidea, but their exact affinities within the superfamily are unclear.

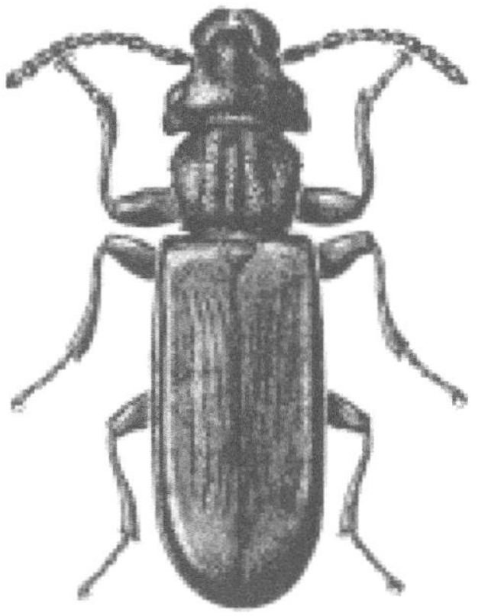

FIGURE 1.82. *Cucujus c. clavipes* Fabricius

Description: Small to moderate in size, 6mm - 25mm; elongate, parallel-sided, strongly dorsoventrally flattened; mostly brightly colored, red or yellow and black; pubescence inconspicuous to moderately pubescent.

Head large, transverse, distinctly triangular in most with well-developed temples (not *Pediacus*, although some exotic species have a denticle behind the eye); nearly impunctate to densely punctate; antennae with 11 antennomeres, filiform to nearly moniliform, without a distinct club, antennomere III longest; antennae inserted laterally, insertion hidden; labrum very small, transverse, fused to clypeus; mandibles large, robust, with two apical teeth and one subapical tooth (mandibles highly modified in males of *Palaestes*), mola not ridged; maxillary palpi with four palpomeres, terminal palpomere truncate apically; mentum transverse; ligula strongly V-shaped and membranous (except *Platisus*, where it is heavily sclerotized); labial palpi with three palpomeres, terminal palpomere truncate apically; eyes moderate, round, somewhat flattened to very convex.

Pronotum quadrate to very transverse, mostly smaller than head, without anterior angles except in *Palaestes*, laterally margined, dentate or denticulate, nearly impunctate to densely punctate, disc mostly with depressions; prosternum broad, prosternal process moderate to relatively narrow; procoxal cavities narrowly separated, open posteriorly, with an anterolateral extension exposing trochantins; mesocoxal cavities open laterally; metasternum elongate; trochantins visible on pro- and mesocoxae; anterior coxae globose, middle coxae globose, hind coxae transverse; trochanters normal; femora pedunculate; tibiae slender, apical spurs small, subequal; tarsal formula 5-5-5 females, 5-5-4 males; protarsi on males with tarsomeres somewhat expanded laterally compared to those of female; claws simple; scutellum quadrate to pentagonal; elytra confusedly punctate, not striate (vague longitudinal lines visible in some species do not appear to be surface structures); scutellary striole absent; epipleural fold complete; metendosternite of normal hylecetoid type (Crowson 1938); wing venation varies within family, but generally well-developed, with radial cell complete (not *Pediacus*), radial recurrent present (not *Pediacus*), medial radial vein complete or nearly so; six veins behind M_4 Cu, anal lobe present or absent (Wilson 1930).

Abdomen with five visible sterna, sutures entire; intercoxal process acute; surface nearly impunctate to densely punctate. Male genitalia of two types: one inverted, with parameres ventrad of the median lobe and with a flagellum (*Cucujus* and *Pediacus*), and one in which the aedeagus lays on its side in the abdomen and lacks a flagellum (*Platisus* and *Palaestes*). Female genitalia with proctiger membranous, paraprocts and valvifers reduced to bacula, coxite two-segmented, and stylus apical.

Known larvae dorsoventrally compressed, orthosomatic, well-sclerotized; vestiture consists of a few simple setae; color yellow to red-brown. Head protracted and prognathous, transverse, flattened and as wide or wider than prothorax; frontal arms lyriform; antennae long, with three antennomeres, antennomere II with sensorium; six pairs of stemmata; frontal suture absent; labrum free; mandibles broad, sclerotized, with three apical teeth, an acute prostheca, and a mola; ventral mouthparts retracted, maxillary articulating area well-developed, mala falciform; maxillary palpi with three palpomeres; labial palpi with two palpomeres. Legs long, five-segmented, and widely separated; abdomen with 10 segments, eighth normal to elongate; tergum IX bearing paired urogomphi supported on a short (*Cucujus*) to long (*Pediacus*) median stalk; sternum IX and tenth segment reduced and concealed beneath tergum IX (Bøving and Craighead 1931, Thomas 1988, Lawrence 1991).

Habits and habitats. Larvae and adults are found under bark of dead trees. Larvae of the North America species of *Cucujus* are reported to be predacious (Smith and Sears 1982), but otherwise there is little information of the biology of these beetles.

Status of the classification. After the removal of the Laemophloeinae, Silvaninae, and Passandrinae, as well as a number of true tenebrionoid genera, this family now is composed of only four genera: *Cucujus* Fabricius (12 spp.), *Pediacus* Shuckard

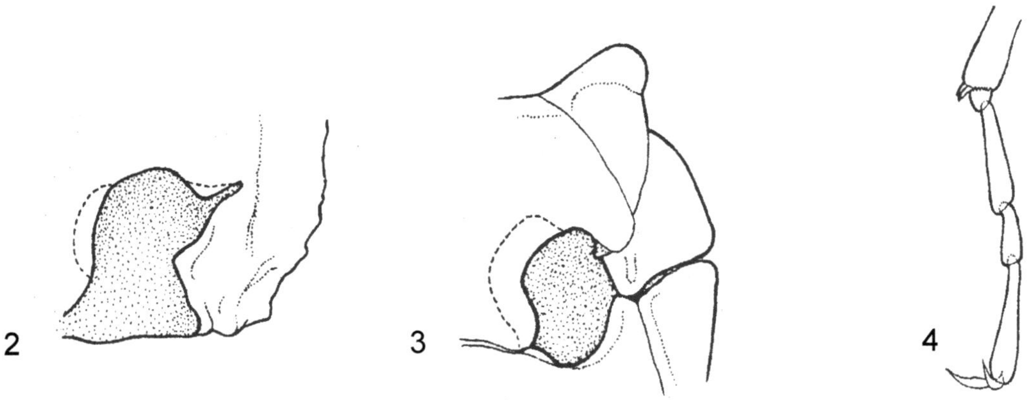

FIGURES 2.82-4.82. *Cucujus clavipes* Fabricius. 2. prothorax, ventral view; 3. mesothorax, ventral view; 4. hind tarsus (from Thomas 1993).

(16 spp.), *Palaestes* Perty (8 spp.), and *Platisus* Erichson (4 spp.). The literature consists primarily of scattered descriptions and the family is in need of revision.

Distribution. *Cucujus* and *Pediacus* are Holarctic in distribution, with *Pediacus* extending south into Central America. Both genera are represented in the Nearctic region. *Palaestes* is entirely Neotropical, and *Platisus* is restricted to Australia.

Key to the Nearctic Genera

1. Head widest across temples; body large (>10mm), brightly colored .... *Cucujus* Fabricius
— Temples absent, head widest across eyes; body small (<5mm), brown .... *Pediacus* Shuckard

Classification of the Nearctic Genera

Cucujidae Latreille 1802

Cucujus Fabricius 1775, 1 species, *C. clavipes* Fabricius, with 2 subspecies: *C. c. clavipes* Fabricius, eastern U.S. west to the Great Plains, and *C. c. puniceus* Mannerheim, Pacific coastal states.

Pediacus Shuckard 1839, 2 species, northern U.S. and Pacific Coast; both also occur in Europe.

Bibliography

BØVING, A. G. and F. C. CRAIGHEAD. 1931. An illustrated synopsis of the principal larval forms of the order Coleoptera. Entomologica Americana (New Series), 11: 1-351.

CROWSON, R. A. 1938. The metendosternite in Coleoptera: a comparative study. Transactions of the Royal Entomological Society, London, 87(17): 397-415.

LAWRENCE, J. F. 1991. Cucujidae. Pp. 463-488. *In:* F.W. Stehr, ed. Immature Insects, Vol. 2. Kendall/Hunt. Dubuque, IA.

SMITH, D. B. and M. K. SEARS. 1982. Mandibular structure and feeding habits of three morphologically similar coleopterous larvae: *Cucujus clavipes* (Cucujidae), *Dendroides canadensis* (Pyrochroidae), and *Pytho depressus* (Salpingidae). Canadian Entomologist, 114: 173-175.

THOMAS, M. C. 1988. Generic key to the known larvae of the Cucujidae, Passandridae, and Silvanidae of America North of Mexico (Coleoptera). Insecta Mundi, 2: 81-89.

THOMAS, M.C. 1993. The flat bark beetles of Florida (Laemophloeidae, Passandridae, Silvanidae). Arthropods of Florida and Neighboring Land Areas, 15: i-viii, 1-93.

WILSON, J. W. 1930. The genitalia and wing venation of the Cucujidae and related families. Annals of the Entomological Society of America, 23: 305-358.

83. LAEMOPHLOEIDAE Ganglbauer 1899

by Michael C. Thomas

Family common name: The lined flat bark beetles

This is a distinctive, well-defined family closely related to Phalacridae and the exotic Propalticidae. The mostly flattened body, sublateral lines on head and pronotum, elytral cells, and elongate antennae make the adults readily recognizable. Adults and larvae occur under bark, where they mostly feed on fungi; a few species are important stored products pests.

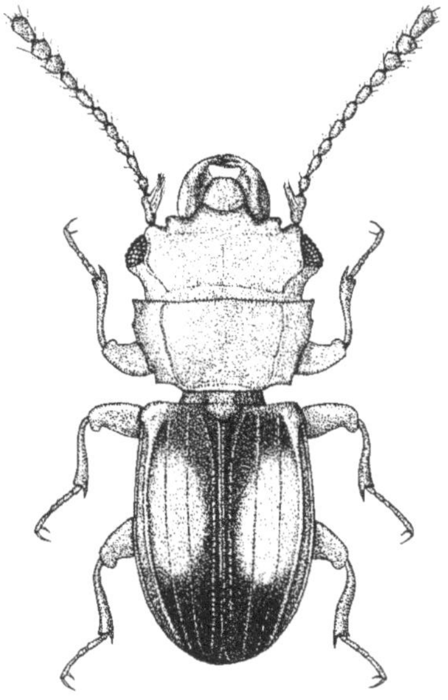

FIGURE 1.83. *Laemophloeus megacephalus* Grouvelle (from Thomas 1993)

Description: Shape ovate to elongate, mostly greatly flattened, a few sub-cylindrical; length, 1mm - 5mm (mostly 1.5mm - 3mm); color brown or black, some bicolored or maculate; pubescence mostly inconspicuous.

Head mostly transverse, broadest across eyes, rostrate in a few genera; surface sculpture punctate; disc of head mostly bordered by carinate or grooved sublateral lines; antennae elongate to very elongate, with 11 (rarely 10) antennomeres and a poorly to well-defined club of three (mostly) to six antennomeres; scape modified in males of some genera; form mostly filiform, rarely moniliform; insertion lateral, hidden; frontoclypeal suture present or absent; epistome with 3-5 emarginations; labrum large, rounded apically to deeply emarginate; mandibles large, stout to elongate, with three apical teeth, a subapical tooth in many, a membranous premolar prostheca, and a mola; many with a circular, internal structure medially; maxillary palpi with four palpomeres, distal one longest; galea obtuse, densely setose; lacinia narrow; uncinate, bifurcate apically, sparsely setose; mentum transverse; ligula pentagonal, slightly emarginate apically; labial palpi with three palpomeres, distal one longest; eyes lateral, oval or round, size variable, flat to hemispherical.

Pronotum quadrate to elongate, mostly constricted basally, margined laterally; surface sculpture of most punctate, disc bordered by carinate or grooved sublateral lines, anterior and posterior angles mostly well-marked; prosternum broad, mostly flattened; intercoxal process mostly broad; procoxal cavities open to closed, mostly widely separated; mesosternum short, broad; mesocoxal cavities mostly open laterally (closed in a few genera), broadly separated; metasternum elongate, with a complete or nearly complete median longitudinal line; metacoxal cavities moderately to broadly separated, intercoxal process broadly or narrowly rounded, or acuminate apically. Coxae with trochantins hidden, anterior coxae somewhat obliquely transverse, middle coxae globular; hind coxae transverse; trochanters normal; femora stout; tibiae stout; apical spurs small and subequal, or large and markedly unequal in size, especially on forelegs; tarsal formula in males mostly 5-5-4, 5-5-5 in some genera; in females 5-5-5; basal tarsomere mostly shortest, claws normal; scutellum transverse to triangular.

Elytra mostly with elytral cells and humeral carina; epipleural fold mostly moderate to broad and complete to posterolateral angle, reduced in some; scutellary striole absent; metendosternite reduced to a pair of lateral arms; wing venation very reduced.

Abdomen with five sterna visible, sutures entire. Male genitalia with aedeagus inverted; parameres reduced to tooth-like structures and fused to basal piece; armature of internal sac mostly well-developed; flagellum present in some; female genitalia with proctiger membranous, paraprocts and valvifers reduced to bacula, coxite two-segmented, and stylus apical.

Larvae with body elongate, dorsoventrally compressed, whitish, with head and abdominal segments VIII-IX pigmented; abdomen mostly with dorsal and ventral asperities arranged in incomplete, longitudinal ellipses; frontal arms divergent, not lyriform; mandibles with grinding mola, or with mola produced as acute tooth, or with complex structures in molar area; maxillary mala obtuse; maxillary articulating area reduced; cardines absent; abdominal segment VIII noticeably longer than VII; urogomphi present.

Habits and habitats. Little has been published on the biology of the non-economic laemophloeids, except that they are subcortical in habitat. They often have been reported to be predacious, although this is doubtful in many cases. Most appear to feed on fungus, especially ascomycetes. However, a few genera with subcylindrical adults are associated with scolytid burrows and may be predacious on bark beetles. Two species of *Cryptolestes* have been reported as predators of scale insects.

Several species, especially in the genus *Cryptolestes*, are important pests of stored grain and grain products. Larvae of some members of this genus are unique among insects in having paired

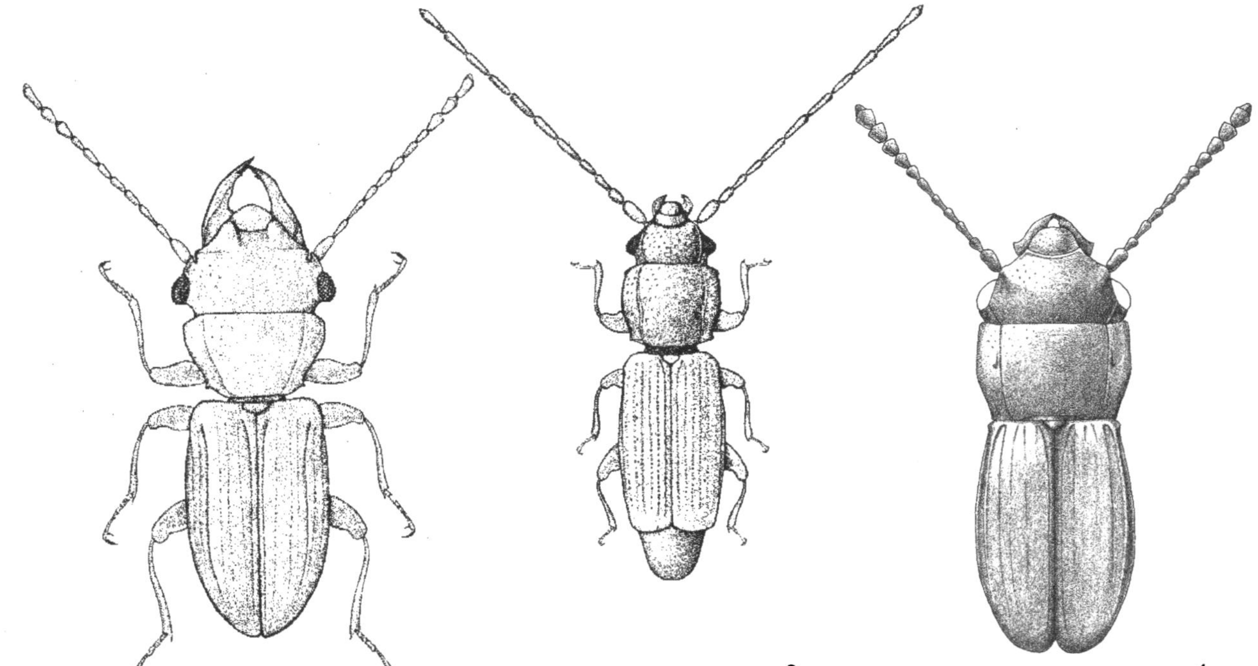

2 3 4

FIGURES 2.83-4.83. 2. *Charaphloeus bituberculatus* (Reitter); 3. *Placonotus modestus* (Say); 4. *Phloeolaemus chamaeropis* (Schwarz).

prosternal silk glands with which they spin a cocoon within which they pupate.

Status of the classification. This taxon has usually been treated as a subfamily (along with Cucujinae) of the family Cucujidae. However, it does not seem especially closely related to *Cucujus* Fabricius and its relatives, and has been increasingly accorded family status (see Thomas 1984, 1993, Lawrence and Newton 1995).

Although the family has been revised for the United States, Europe, and Africa (Casey 1884, Lefkovitch 1959, Lefkovitch 1962) and several genera have been revised for the New World, the family still needs much taxonomic study at all levels. For example, many Old World tropical species are still assigned to *Laemophloeus* Dejean, even though that genus is restricted to the New World and Palaearctic. Halstead (1993) provided keys to stored products species. Thomas (1988b) provided a larval key to some genera.

Distribution. This family is represented in all forested areas of the world, but reaches its greatest diversity in the tropics. There are about 400 species in the family in 37 genera. There are 52 species distributed among 13 genera in the United States.

KEY TO NEARCTIC GENERA

1\. Sublateral line of pronotum carinulate, obviously raised above the surface of the pronotum, some with associated inconspicuous medial groove; pedicel mostly longer and more massive than antennomere III; transverse epistomal groove absent, although epistome may be depressed ... 8

— Sublateral line of pronotum a groove, some with a weakly expressed associated ridge; or, disc of pronotum depressed but without obvious groove or carinulate line; pedicel mostly shorter or much shorter than antennomere III; transverse epistomal groove present in some ... 2

2(1). Intercoxal process of sternum III acuminate apically (Fig. 12) ... 3

— Intercoxal process of sternum III not acuminate apically ... 5

3(2). Tarsomere I shorter than penultimate tarsomere; labrum shallowly emarginate apically; elytra carinate laterally ... *Laemophloeus*

— Tarsomere I at least subequal to penultimate tarsomere; labrum semicircular, not emarginate anteriorly; elytra mostly not carinate laterally ... 4

4(3). Head rostrate ... *Metaxyphloeus*

— Head not rostrate ... *Charaphloeus*

5(2). Head with distinct transverse groove separating frons from epistome ... *Placonotus*

— Head without distinct transverse groove separating frons from epistome, although region of head corresponding to clypeus may be depressed below level of frons ... 6

6(5). Elytral epipleura well defined, complete to apical angle; head moderate in size; basal angles of epistome not foveate ... 7

— Elytral epipleura poorly defined, becoming obsolete at midpoint of elytra; head large, almost as large as pronotum; basal angles of epistome foveate ... *Parandrita*

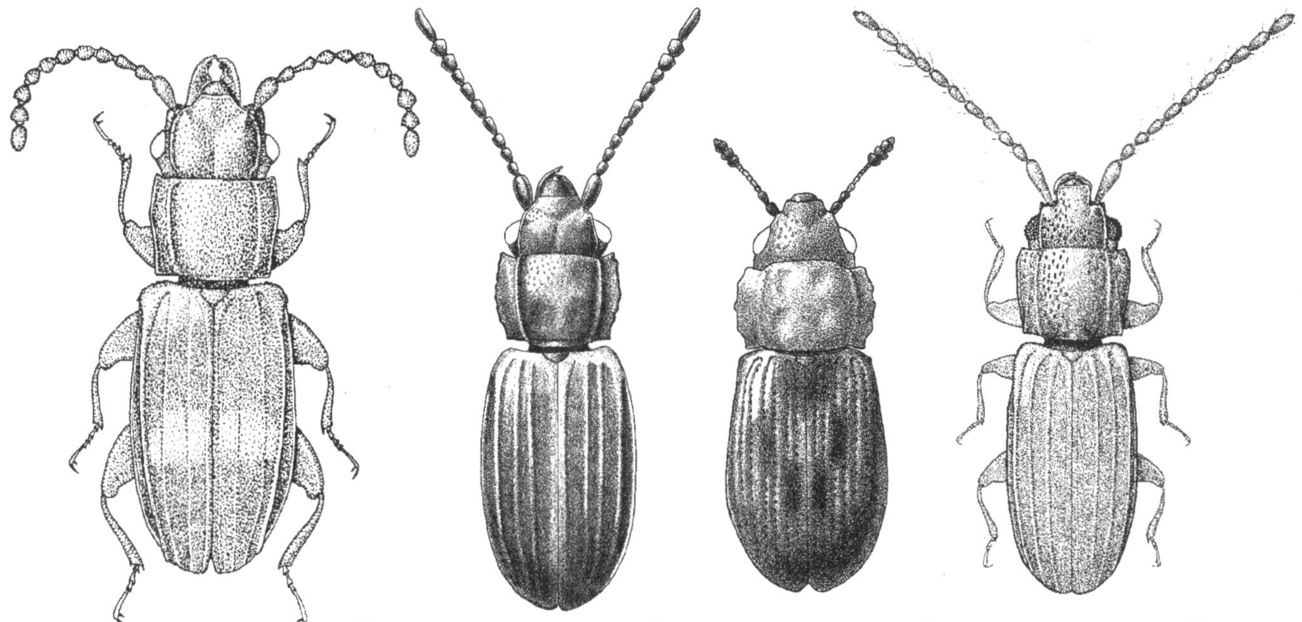

5 6 7 8

FIGURES 5.83-8.83. 5. *Deinophloeus impressifrons* (Schaeffer); 6. *Rhabdophloeus horni* (Casey); 7. *Lathropus pictus* (Schwarz); 8. *Cryptolestes puctatus* (LeConte)

7(6). Elytra mostly with only third cell present; sublateral line of pronotum composed of a distinct groove; anterior coxal cavities mostly wide open posteriorly; mandibles of males in some species expanded laterally .... *Phloeolaemus*

— Elytra with all cells present; sublateral line of pronotum represented by an abrupt change in elevation; anterior coxal cavities closed posteriorly; clypeal horns present in major males .... *Deinophloeus*

8(1). Lateral margins of pronotum slightly to strongly undulating; disc in some with paired anterior and posterior depressions, these often faint 9

— Lateral margins of pronotum evenly curved, or, at most, simply sinuate; disc without paired anterior and posterior depressions 10

9(8). Head with lateral lines; lateral margins of pronotum with irregular undulations; pronotum and elytra explanate laterally, elytra carinate laterally and with longitudinal discal costae; anterior coxal cavities open posteriorly; intercoxal process of sternum III narrowly rounded anteriorly .... *Rhabdophloeus*

— Head without obvious lateral lines; lateral margins of pronotum undulating at most; elytra without obvious elytral cells, not carinate; intercoxal process of sternum III broadly rounded anteriorly ... . *Lathropus*

10(8). Intercoxal process of sternum III broadly rounded anteriorly (Fig. 14) .... *Cryptolestes*

— Intercoxal process of sternum III narrowly rounded anteriorly (Fig. 13) 11

11(10). Epistome acuminate anteriorly .... *Narthecius*

— Epistome truncate or emarginate anteriorly 12

12(11). Pedicel attached laterally to scape, which is large and irregularly produced in males; pronotum quadrate .... *Dysmerus*

— Pedicel attached axially to scape, which is normal in both sexes; pronotum quadrate to elongate . ... *Leptophloeus*

Classification of the Nearctic Genera

Laemophloeidae Ganglbauer 1899

Lathropus Erichson 1846, 5 spp., eastern U.S. and Arizona (Fig. 7).

Rhabdophloeus Sharp 1899, 2 spp., Florida and Arizona (Fig. 6).

Cryptolestes Ganglbauer 1899, 11 spp., mostly eastern U.S. west to Arizona. One immigrant species, *C. curus* Lefkovitch from the Middle East, is associated with red date palm scale, *Phoenicoccus marlatti* Cockerell, in Arizona and California. Five cosmopolitan stored products species are generally distributed. (key to spp., Thomas 1988a) (Fig. 8).

Leptus Thomson 1863

Fractophloeus Kessel 1921

Leptophloeus Casey 1916, 1 sp., *L. angustulus* (LeConte), eastern U.S. to Oklahoma (Fig. 11).

Truncatophloeus Kessel 1921

Dysmerus Casey 1884, 1 sp., *D. basalis* Casey, southeastern U.S. (Fig. 10).

Brontophloeus Kessel 1921

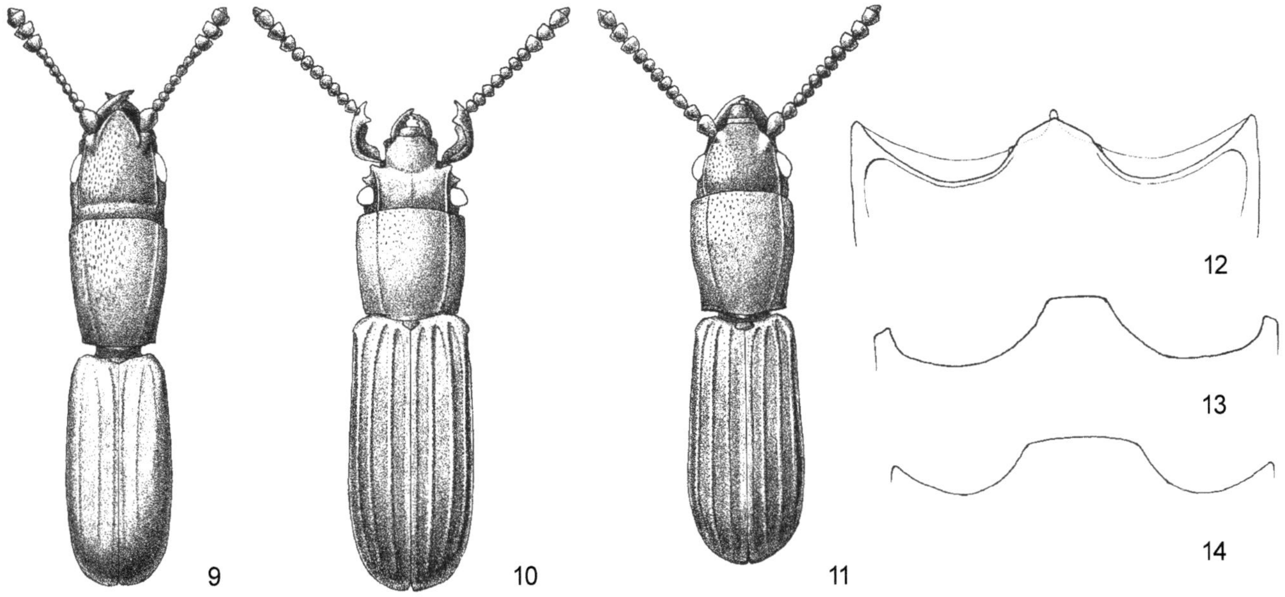

FIGURES 9.83-14.83. 9. *Narthecius grandiceps* (LeConte); 10. *Dysmerus basalis* Casey; 11. *Leptophloeus angustulus* (LeConte); 12-14. intercoxal process of sternum III. 12. *Laemophloeus* sp.; 13. *Leptophloeus* sp.; 14. *Cryptolestes* sp.

Narthecius LeConte 1861, 5 spp., generally distributed (Fig. 9).
Paraphloeus Sharp 1899

Laemophloeus Dejean 1836, 11 spp., generally distributed (Fig. 1).

Charaphloeus Casey 1916, 5 spp., eastern U.S., west to Arizona (Fig. 2).

Metaxyphloeus Thomas 1984, 1 sp., *M. texanus* (Schaeffer), southern Texas.

Placonotus Macleay 1871, 6 spp., generally distributed except for Pacific Coast. (key to spp., Thomas 1984) (Fig. 3).
Silvanophloeus Sharp 1899

Parandrita LeConte and Horn 1880, 2 spp., Florida and California.

Phloeolaemus Casey 1916, 3 spp., including the apparently immigrant *P. quinquearticulatus* (Grouvelle), southeastern U.S. and Arizona (Fig. 4).

Deinophloeus Sharp 1899, 1 sp., *D. impressifrons* (Schaeffer), Arizona (Fig. 5).

Bibliography

CASEY, T. L. 1884. Revision of the Cucujidae of America North of Mexico. Transactions of the American Entomological Society, 11: 69-112.

HALSTEAD, D. G. H. 1993. Keys for the identification of beetles associated with stored products-II. Laemophloeidae, Passandridae and Silvanidae. Journal of Stored Products Research, 29: 99-197.

LAWRENCE, J. F. and A. F. NEWTON, Jr. 1995. Families and subfamilies of Coleoptera (with selected genera, notes, and references and data on family-group names). Pp. 779-1006. *In:* J. Pakaluk and S.A. Slipinski, eds. Biology, Phylogeny, and Classification of Coleoptera. Papers Celebrating the 80th Birthday of Roy A. Crowson. Muzeum i Instytut Zoologii PAN, Warsaw.

LEFKOVITCH, L. P. 1959. A revision of European Laemophloeinae (Coleoptera: Cucujidae). Transactions of the Royal Entomological Society of London, 111: 95-118.

LEFKOVITCH, L. P. 1962. A revision of African Laemophloeinae (Coleoptera: Cucujidae). Bulletin of the British Museum of Natural History (Entomology), 12: 167-245.

THOMAS, M. C. 1984. A revision of the New World species of *Placonotus* Macleay (Coleoptera: Cucujidae: Laemophloeinae). Occasional Papers of the Florida State Collection of Arthropods, 3: i-vii, 1-28.

THOMAS, M. C. 1988a. A revision of the New World species of *Cryptolestes* Ganglbauer (Coleoptera: Cucujidae: Laemophloeinae). Insecta Mundi, 2: 43-65.

THOMAS, M. C. 1988b. Generic key to the known larvae of the Cucujidae, Passandridae, and Silvanidae of America North of Mexico (Coleoptera). Insecta Mundi, 2: 81-89.

THOMAS, M. C. 1993. The flat bark beetles of Florida (Laemophloeidae, Passandridae, Silvanidae). Arthropods of Florida and Neighboring Land Areas, 15: i-viii and 1-93.

84. PHALACRIDAE Leach 1815

by Warren E. Steiner, Jr.

Family common names: The shining flower beetles, shining mold beetles

The oval, convex shape, continuous body outline with wide pronotum, uncovered head, polished dorsal surface, clubbed antennae and the small fourth tarsomere to separate adults of this family.

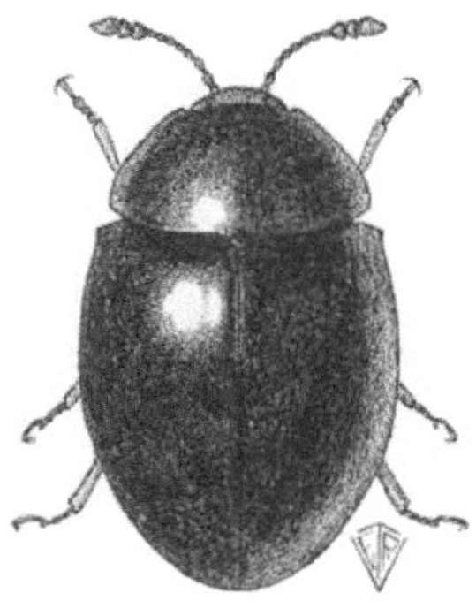

FIGURE 1.84. *Phalacrus politus* Melsheimer

Description: Body outline broadly oval to nearly circular; convex dorsally, flat to somewhat concave ventrally; size 1 to 3 mm long; color black to various brown hues or may be bicolored; vestiture absent or inconspicuous.

Head deflexed but not hidden from above; quadrate to oval; surface smooth. Antennae with 11 antennomeres, the apical three forming an elongate club with apical antennomere largest; insertion under or at sides of small frontal margin, above the base of the mandibles. Clypeus not separated; labrum small, transverse; mandibles small, curved, stout. Maxillae with lacinia and galea; maxillary palpi four-segmented, small. Labium with quadrate but variably shaped mentum. Labial palpi three-segmented, small. Eyes lateral, oval, not prominent.

Pronotum large, widest at base and nearly as wide as mid-body; head inserted in broad anterior emargination; lateral margins thin, finely beaded; pleural region broad; prosternum broad in front of the coxae; prosternal process flat, widened apically; procoxal cavities open behind. Mesosternum short, concave medially; metasternum broad, produced anteriorly between mesocoxae. Pro- and mesocoxae small, globular, separate; metacoxae transverse, flattened, nearly contiguous; trochanters transversely triangular, interstitial; femora swollen, flattened; tibiae slender, flattened with rows of spines and small or large apical spurs; tarsal formula 5-5-5, the fourth tarsomere small, obscure, tarsomeres one to three broad, tomentose beneath; claws moderate, with basal tooth or appendiculate beneath. Scutellum moderately large, triangular. Elytra entire, apically rounded; striae usually one or two in number near the suture, or absent; epipleural fold very broad basally, narrowing to absent at middle. Wing venation and folding pattern described as similar to Trogossitidae and Cucujidae.

Abdomen with five visible sternites, the sutures entire; surface smooth. Male genitalia with the penis broad and flattened, often with internal sclerites and spicules; tegmen with parameres fused into a single apical sclerite; basal piece generally much larger than apical, base ring-like with broad basal process. Female genitalia of variable sclerotization and structure.

Larvae (Steiner 1984, Steiner and Singh 1987, Lawrence 1991) elongate, slightly flattened, generally parallel-sided; size 2 to 6 mm in length; surfaces smooth with a few moderate setae; body color near white to lightly and variably pigmented with darker head and abdominal apex. Head prognathous, broad but not as wide as body, somewhat flattened; epicranial suture lyriform with stem very short or absent. Antennae short to moderately long, three-segmented. Labrum distinct; mandibles robust, with curved dentate apices, usually with distinct molar areas. Maxilla protracted, cardo not distinct, palpi three-segmented, stipes and mala fused. Labium exserted; ligula distinct, with two-segmented palpi. Stemmata five or six on each side. Thorax with well-developed five-segmented legs; tarsungulus claw-like. Abdomen nine-segmented, the ninth segment one-third the width of the eighth; urogomphi prominent, dorsally curved, with sharp apices. Spiracles annular-biforous on the mesothorax and abdominal segments one to eight, with those on segment eight often enlarged and oriented posteriorly.

Habits and habitats. The larvae of the genus *Olibrus* live in flower heads of Compositae and the pollen-feeding adults are often abundant on these plants. Members of other genera are specifically associated with various fungi including some that cause diseases of vascular plants, such as ergot (Steiner and Singh 1987) and smut on grasses; larvae and adults feed on spores or fungal tissue. Others graze on surface molds in hanging dead leaves and seed pods (White 1983, Steiner 1984).

Status of the classification. The group is poorly known, and badly in need of study; the most recent revisions are those of Casey (1890, 1916). Considered to be among the more primitive families of Cucujoidea, Phalacridae probably have a sister-group to be found in the Laemophloeidae or perhaps Nitidulidae. There are two subfamilies recognized (Lawrence and Newton 1995); all of the western hemisphere genera belong in the Phalacrinae. For accurate species identification, a study of the male genitalia is usually necessary.

Distribution. There are 504 species (Hetschko 1930) known from all areas; of these 122 (Blackwelder 1945) occur in the United States.

Key to the Nearctic Genera

1. Antennae inserted under frontal margin, their bases concealed; margin of clypeus evenly curved above antennal insertion; scutellum large 2
— Antennae inserted at sides of front, base visible from above; margin of clypeus above point of insertion more or less sinuate; scutellum small or moderate 3

2(1). Spurs of metatibiae very short, not projecting beyond terminal fringe of spinules; elytra with a single discal stria near suture; mandibles prominent, curved and pointed; body color black .... *Phalacrus*
— Spurs of metatibia long and distinct; elytra without striae; mandibles not prominent; body color brown .... *Phalacropsis*

3(1). Basal metatarsomere shorter than second 4
— Basal metatarsomere longer than second; elytra with two discal striae 9

4(3). Apex of prosternal process inflexed, without acute and free edge, devoid of spinose setae; metasternal process greatly developed, mesosternum before it reduced to fine bead 5
— Apex of prosternal process with acute free edge usually bearing a transverse series of stout setae; mesosternum more developed in front of metasternal process; elytra constantly with single discal striae, suture not beaded 6

5(4). Elytra with one or two discal striae, suture beaded; elytral surface very polished, punctures indistinct; second metatarsomere free .... *Olibrus*
— Elytra with one discal stria, suture not beaded; elytral surface less polished, with distinct rows of closely and regularly spaced punctures; second metatarsomere connate with first .... *Olibroporus*

6(4). Mesosternal plate not prolonged behind middle coxa, consisting of only a fine bead; second metatarsomere very long .... *Litolibrus*
— Mesosternal plate prolonged and sometimes greatly expanded behind middle coxa as a rounded to angulate lobe; second metatarsomere of variable length 7

7(6). Second metotarsomere very long, spongy-pubescent beneath in male, connate with first; tibial spurs variable in size; male genitalia with basal and apical pieces of tegmen separate, hinged 8
— Second metotarsomere of moderate length, not connate; posterior legs not different in sexes; tibial spurs small and inconspicuous; male genitalia with basal and apical pieces of tegmen fused .... *Stilbus*

8(7). Metatibial spurs long, very unequal; metatibiae, their spurs, and tarsi markedly modified or enlarged in some males; elytral surface with shallow crescentiform punctures; body outline rounded to broadly oval .... *Acylomus*
— Metatibial spurs small and inconspicuous; posterior legs not different between sexes; elytral surface, especially near suture, with rows of relatively distinct, rounded punctures; body outline more elongate-oval .... *Leptostilbus*

9(3). Prosternal process not inflexed, having broad apex with free acute edge, usually bearing terminal setae 10
— Prosternal process narrow and inflexed at apex, devoid of terminal setae 11

10(9). Mesosternal plate not developed behind middle coxae; first metatarsomere much longer than following two combined; elytra with three discal striae; prosternal process truncate, without stout setae .... *Ochrolitus*
— Mesosternal plate short and arcuate behind middle coxae; first metatarsomere subequal to following two combined; elytra with two discal striae; prosternal process with comb of stout setae on broad rounded apex .... *Gorginus*

11(9). Basal metatarsomere longer than or equal to all remaining tarsomeres together; elytral surface with scattered shallow crescentiform punctures .... *Litochrus*
— Basal metatarsomere shorter than all following tarsomeres together; elytral surface with transversely reticulate scratches .... *Litochropus*

Classification of the Nearctic Genera

Phalacridae Leach 1815

Phalacrus Paykull, 1798, 25 spp., widely distributed.

Phalacropsis Casey 1890, 1 sp., *P. dispar* (LeConte 1879), western mountain states.

Olibrus Erichson 1845, 28 spp., widely distributed.

[*Tinodemus* Guillebeau 1894 (not included in key above), 1 sp., *T. grouvellei* Guillebeau 1894, reported from Michigan but apparently not Nearctic.]

Olibroporus Casey 1890, 1 sp., *O. punctatus* Casey 1890, southeastern states.

Litolibrus Sharp 1888, 3 spp., Florida and Texas.

Acylomus Sharp 1888, 20 spp., widely distributed.

Stilbus Seidlitz 1872, 34 spp., widely distributed.
    Olistherus Seidlitz 1872
    Eustilbus Sharp 1888

Leptostilbus Casey 1916, 3 spp., southeastern states.

Litochrus Erichson 1845, 4 spp., southeastern states.

Gorginus Guillebeau 1894, 1 sp., *E. rubens* (LeConte 1856), southeastern states to Indiana.

Erythrolitus Casey 1916

Litochropus Casey 1893, 2 spp., southeastern states.

Ochrolitus Sharp 1889, 1 sp., *O. tristriatus* Casey 1893, Florida.

Bibliography

BLACKWELDER, R. E. 1945. Checklist of the Coleopterous Insects of Mexico, Central America, the West Indies, and South America. United States National Museum Bulletin 185: 343-550.CASEY, T. L. 1890. Phalacridae. Annals of the New York Academy of Sciences, 5: 89-144.

CASEY, T. L. 1916. Phalacridae. Memoirs on the Coleoptera, 7: 35-86.

HETSCHKO, A. 1930. Phalacridae. Coleopterorum Catalogus, 15(108): 1-48.

LAWRENCE, J. F. 1991. Phalacridae (Cucujoidea). Pp. 466-468. *In:* F. W. Stehr, ed. Immature Insects, Vol. 2. Kendall-Hunt. Dubuque, Iowa, 975 pp.

LAWRENCE, J. F. and A. F. NEWTON, Jr. 1995. Families and subfamilies of Coleoptera (with selected genera, notes, references and data on family-group names). Pp. 779-1006. *In:* J. Pakaluk and S. A. Slipinski, eds. Biology, Phylogeny, and Classification of Coleoptera. Papers Celebrating the 80th Birthday of Roy A. Crowson. Muzeum i Instytut Zoologii Polska Akademia Nauk. Warsaw.

STEINER, W. E. 1984. A review of the biology of phalacrid beetles (Coleoptera). Pp. 424-445. *In:* Q. Wheeler and M. Blackwell, eds. Fungus-Insect Relationships: Perspectives in Ecology and Evolution. Columbia University Press. New York, 514 pp.

STEINER, W. E. and B. P. SINGH. 1987. Redescription of an ergot beetle, *Acylomus pugetanus* Casey, with immature stages and biology (Coleoptera: Phalacridae). Proceedings of the Entomological Society of Washington, 89: 744-758.

WHITE, R. E. 1983. A Field Guide to the Beetles of North America. Houghton Mifflin. Boston, 385 pp.

85. CRYPTOPHAGIDAE Kirby 1837

by Richard A.B. Leschen and Paul E. Skelley

Family common name: The silken fungus beetles

Family synonyms: Hypocopridae Reitter 1879, Catopochrotidae Reitter 1889

The small size, antennal insertions exposed in dorsal view, mandible without a deep cavity or mycangium, tentorium without median tendon, prothorax with or without (Hypocoprini) well developed lateral carina, mesocoxal cavity closed laterally by the metasternum, abdominal ventrite I longer than remaining ventrites, epipleuron distinct in basal half and punctation of elytron random (or confused) define this group in the Cucujoidea.

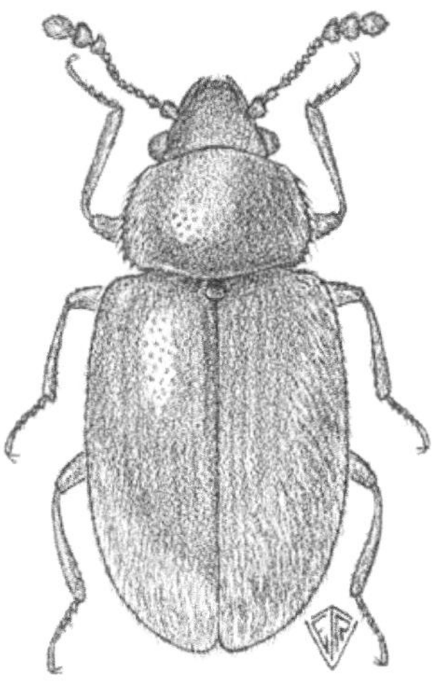

FIGURE 1.85. *Henoticus californicus* Mannerheim

Description: Body length 0.8 - 5.2 mm. Body form elongate and parallel-sided, moderately flattened or highly convex, oval, or round. Body color various, most dark, light, or red brown. Body setae long or short, decumbent, apressed, suberect, or erect. Glandular ducts present in various parts of the body (Atomariini, Cryptafricini, and Cryptophagini). Head longer than wide, retracted into thorax, with or without a short neck; with or without line on vertex, without stridulatory files (except for some species of *Atomaria*); with or without subgenal spines; frons without lateral tubercles (except in some foreign Atomariini); antennal grooves present in *Ephistemus*, anterior part of gula without a transverse groove, with or without gular suture. Eyes well developed and finely faceted, with or without ocular setae. Clypeus more or less quadrate, with or without frontoclypeal suture. Antennal club 3-segmented (1- or 2-segmented in some members of Caenoscelini and Atomariinae), funicle curved and conical or barrel-shaped. Mandible with well developed mola, incisor apex, and prostheca, dorsal tubercles present or absent, deep cavity or mycangia absent. Maxilla with brushlike galea and lacinia; apical spines and lateral setae present on lacinia; maxillary palpomeres more or less subequal to each other. Labium with mentum wider than long, with or without transverse ridge or line, with or without middle process; labial palpus 2-segmented with basal palpomere subequal to apical palpomere (Cryptophagini and Cryptosomatulini) or basal palpomere wider than apical palpomere (Atomariinae and Caenoscelini). Tentorium with anterior arms fused or separate, posterior part strap-like, without anterior median tendon.

Prothorax with well developed lateral carina (absent in Hypocoprini and present in basal half only in some foreign Atomariini); side with or without marginal teeth or processes or angularities. Pronotal disc with (Cryptophagini) or without basal pits; with depressions (in some Cryptosomatulini). Hypopleuron without antennal grooves (except some Caenoscelini and *Ephistemus*); with or without notch above procoxa. Prosternal process well developed (length less than anterior portion in some Atomariinae); vaulted or not (flat). Procoxae rounded and separate; procoxal cavities externally open or closed; internally open or closed. Mesosternum with or without procoxal rests. Mesepimeron with or without distinct pit, fused with mesosternum in *Amydropa*. Mesocoxal cavities closed laterally by sterna. Meso-metasternal articulation with either a double knob (most Cryptophaginae) or lateral processes (Atomariini) or unmodified. Metasternum elongate or short (in some flightless forms); with or without longitudinal line. Metendosternite with a broad stalk; most with basal plate transverse with well developed arms; with anterior tendons present (absent in some flightless forms), approximate or distant.

Elytra completely covering abdomen (tergite VII exposed in some species); punctation random (or confused) or arranged into ill-defined rows (*Henotiderus*); epipleuron distinct in basal half (distinct beyond basal half in some *Tisactia*, some *Antherophagus*, and some Cryptafricini). Wings with up to 5 major veins; cell formed by the merging of veins CuA_2 and CuA_{3+4}+AA_{1+2} (some Cryptophagini); with or without basal binding patch. Legs relatively long; length of trochanter about 1/4 that of femur; tibia parallel-sided or club-shaped, most with 2 apical spines (0 or 1 may be present); tarsi 5-5-5 in female, 5-5-4 (Cryptophaginae) or 5-5-5 in male.

Abdomen with 5 freely articulated ventrites. Ventrite I longer in length to remaining ventrites; with or without metasubcoxal lines. Intercoxal process moderately broad. Ventrite V with modified setae (Caenoscelini).

Male genitalia with spiculum gastrale with long anterior strut (most Cryptophaginae) or broad (most Atomariinae); asymmetrical in some foreign Atomariini. Orientation of aedeagus vertical or horizontal in body cavity, bilaterally symmetrical or not in some foreign Atomariini; parameres separate or fused (most Atomariinae), internal sac complex or not; dorsal arms narrow or joined at their apices or not, or broadly fused; flagellum present or not.

Larvae (modified from Lawrence 1991) with mala falciform, prostheca fixed and slender (apex acute, bifid, or serrate), 2

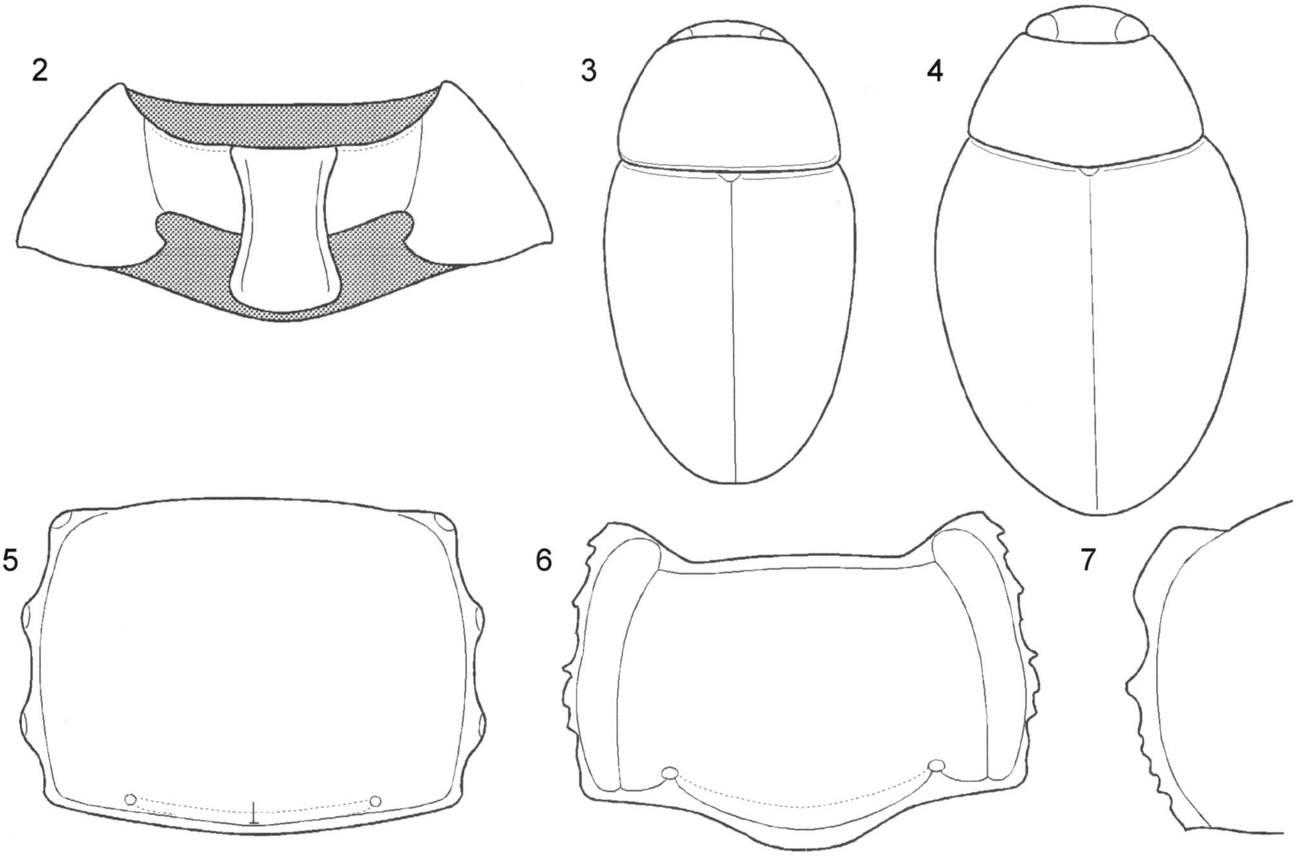

FIGURES 2.85-7.85. 2. *Ephistemus perminutus* Casey, prosternum; 3. *Tisactia subglabra* Casey, dorsal habitus; 4. *Ephistemus perminutus* Casey, dorsal habitus; 5. *Salebius octodentatus* (Malkin), pronotum; 6. *Henotiderus obesulum* Casey, pronotum; 7. *Cryptophagus* sp., left lateral pronotal margin

tarsungular setae, ventral epicranial ridges absent, 1-segmented labial palps (Atomariinae) or annular spiracles. Body elongate, more or less parallel-sided or slightly wider at middle; subcylindirical to slightly flattened; surfaces lightly pigmented, occasionally tergal plates yellow-brown; vestiture of simple setae. Head protracted and prognathous, slightly wider than long. Epicranial stem absent frontal arms lyriform, usually meeting at base. Median endocarina absent. Stemmata distinct or absent, 1, 2, or 5 on each side. Antennae well developed, 3-segmented. Frontoclypeal suture absent; labrum free. Mandibles symmetrical, apex bidentate often with serrate incisor edge and accessory ventral process; mola well-developed, tuberculate or asperate; prostheca fixed and slender consisting of acute, bifid, or serrate apex. Ventral mouthparts retracted. Maxilla with transverse cardo, elongate stipes, well-developed articulating area, 3-segmented palp and falciform mala. Labium free to base of mentum; ligula present; labial palps 1-segmented (Cryptophaginae) or 2-segmented (Atomariinae), separated. Hypopharyngeal sclerome usually toothlike. Hypostomal rods moderately long and diverging. Ventral epicranial ridges absent. Gula transverse. Thorax with legs well developed, 5-segmented; tarsungulus with 2 setae, either side by side or with one distal to the other; coxae moderately close together. Abdomen with tergum A9 usually with a pair of strongly upturned, fixed urogomphi; simple or sclerotized apically, sometimes reduced or absent. Sternum A9 well-developed, simple. Segment A10 well-developed, circular, posteriorly or posteroventrally oriented. Spiracles annular or annular-biforous, not on tubercles.

Habits and habitats. Cryptophagids typically are microphagous and occur in decaying habitats that promote fungal growth (Leschen 1996). Some species of *Atomaria*, *Cryptophagus*, and *Henoticus* are incidental stored products pests (Bousquet 1990, Hinton 1945). The most effective way to collect these beetles is by sifting leaf litter and rotting wood or using passive flight intercept traps in drier habitats. Some may be collected from their host fungi or with their social insect or mammal hosts. Most members feed on fungal hyphae, spores, and conidia while others are saprophagous. Adults of *Telmatophilus americanus* LeConte have been collected from the flowers of the aquatic plant *Sparganium*, while the adventive *T. typhae* (Fallen) is found on *Typha* sp. (Hoebeke and Wheeler 2000). Members of *Antherophagus* are phoretic on *Bombus* bees and are often found in the nests or at flowers (Leschen 1999). Members of *Myrmedophila* and *Hypocoprus* are present in the nests of mound-building ants (*Formica*). Various species of *Cryptophagus* and *Atomaria* may also occur incidentally in the nests of social insects and mammals. Other works on natural histories include Chavarria (1994), Crowson (1980), Dajoz (1988), Lawrence (1982), and Ljubarsky (1992a). Additional works illustrating and describing larvae include Bøving and Craighead (1931), Hinton and Stephens (1941), Lawrence *et al.* (1999b), and Peterson (1951).

Status of the classification. The classification of the family was studied in detail by Leschen (1996). Many taxa originally included in the family have been removed and placed into other families, mostly in Languriidae. The Cryptophagidae are probably monophyletic; however, the inclusion of Hypocoprinae is doubtful and the relationships to remaining members of the Cucujoidea is uncertain. While there are a few problems with higher classification, the most serious taxonomic problems concern the taxonomic status of *Cryptophagus* (see Woodroffe and Coombs 1961) and *Atomaria*, both of which contain new species and need revision in North America. Also there are undescribed species of *Henoticus*, *Henotiderus*, *Sternodea*, and *Tisactia* (Bousquet 1989, Leschen 1996). Additional information on identification and classification can be found in Bousquet (1989, 1990), Downie and Arnett (1996), Hatch (1962), Hinton (1945), Johnson (1971), Lawrence *et al.* (1999a), Leschen and Bowstead (2001), Ljubarsky (1992b), and Sharp (1900). There are various checklists or catalogs of importance; these include Blackwelder (1945), Crotch (1873), Lawrence and Newton (1995), Leng (1920), Pakaluk *et al.* (1995), and Schenkling (1923).

Distribution. There are approximately 600 described species (Leschen and Johnson in prep.) known from all areas, of which 145 species occur in the United States.

KEY TO THE NEARCTIC GENERA

1. Frontoclypeal suture absent 2
— Frontoclypeal suture present (Atomariini) 3

2(1). Lateral prothoracic carina absent (Hypocoprini) . 6
— Lateral prothoracic carina present (Cryptophaginae) 7

3(1). Prosternal process broad, vaulted and bearing two parallel lines (Fig. 2) 4
— Prosternal process narrow, flat and without parallel lines .... *Atomaria*

4(3). Body elongate and moderately convex (Fig. 3) .... *Tisactia*
— Body oval and highly convex (Fig. 4) 5

5(4) Antennal grooves present on the subgena and prosternum .... *Ephistemus*
— Antennal grooves absent .... *Curelius*

6(2). Head subparallel behind eyes, with conspicuous temples .... *Hypocoprus*
— Head abruptly narrowed behind eyes, without temples .... *Amydropa*

7(2). Pronotum with smooth lateral margin and lacking basal pits or groove (Caenoscelini) 8
— Pronotum (Figs. 5-7) with lateral margin serrate (can appear smooth) or lobed, and with basal pits or groove (Cryptophagini) 9

8(7). Body elongate; pronotum parallel-sided; prosternal process not vaulted; antennal club 3-segmented .... *Caenoscelis*
— Body oval; pronotum widest at base; prosternal process vaulted; antennal club 2-segmented .... *Sternodea*

9(7). Lateral prothoracic carina with three distinct lobes or undulations (Fig. 5) .... *Salebius*
— Lateral prothoracic carina without three lobes (two may be present) 10

10(9). Pronotum with sublateral line (Fig. 6) .. *Henotiderus*
— Pronotum without sublateral line 11

11(10). Pronotum with a distinct angularity (Fig. 7) or enlarged area in the anterior portion of the prothoracic carina 12
— Pronotum without a distinct angularity or widened area in the anterior portion of the prothoracic carina 14

12(11). Midlateral tooth absent on prothoracic carina; male with a distinct clypeal notch; body color golden-brown or yellow .... *Antherophagus*
— Midlateral tooth present on prothoracic carina; male without clypeal notch; body color variable .. 13

13(12). Color of body red-brown; vestiture of setae on elytron sparse .... *Myrmedophila*
— Color of body variable; setae of elytron well developed .... *Cryptophagus*

14(11). Tarsal formula 5-5-5 in male; tarsomeres II-III strongly lobed beneath .... *Telmatophilus*
— Tarsal formula 5-5-4 in male; tarsomeres not strongly lobed beneath 15

15(14). Body dorsoventrally compressed; mesosternal process greater in width than mesocoxa; anterior lateral angle of metasternum with a slight depression .... *Pteryngium*
— Body not dorsoventrally compressed; mesosternal process equal in width to mesocoxa; anterior lateral angle of metasternum with a pubescent pit .... *Henoticus*

CLASSIFICATION OF THE NEARCTIC GENERA

Cryptophagidae Kirby 1837

Cryptophaginae Kirby 1837

Review of North American genera by Bousquet (1989).

Cryptophagini Kirby 1837
Telmatophilides Jaquelin du Val 1859
Antherophagi LeConte 1861
Paramecosomina Reitter 1875
Spaniophaeni Casey 1900
Emphyli Casey 1900

Antherophagus Dejean 1821, 3 spp., Canada, New York, Pennsylvania, Indiana, Michigan, Wisconsin, Utah, and Pacific Northwest. Found in bumble bee nests or in flowers. Biology discussed by Chavarria (1994).

Cryptophagus Herbst 1792, 30 spp. (29 described, 1 undescribed), generally distributed. Woodroffe and Coombes (1961) revised the North American species. Biology and immatures of *C. acutangulus* Gyll. described by Hinton and Stephens (1941).
Mnionomus Wollaston 1864
Cryptophagistes Crotch 1873
Micrambinus Reitter in Heyden *et al.* 1906

Henoticus Thomson 1868, 3 spp., generally distributed.
Coniophthalma Kangas 1963
Glyptophorus Park 1929

Henotiderus Reitter 1877, 4 spp. (2 described, 2 undescribed), British Columbia, New York, Indiana, Iowa, Utah.
Crosimus Casey 1900
Henoticoides Hatch 1962

Myrmedophila Bousquet 1989, 1 sp., *M. americanus* (LeConte), Colorado, Utah, and California. Present in *Formica* ant nests.

Pteryngium Reitter 1887, 1 sp., *P. crenatum* (Fabricius), Indiana, Michigan, Oregon, Washington, and British Columbia.

Salebius Casey 1900, 5 spp. (4 described, 1 undescribed, see Dajoz (1988), California, Oregon, Washington and British Columbia.

Telmatophilus Heer 1841, 2 spp., *T. typhae* (Fallen) (adventive) and *T. americanus* LeConte, Indiana, Colorado, Oregon, Washington, New Brunswick, Nova Scotia, Prince Edward Is., and British Columbia; associated with aquatic plants.
Hydrophytophagus Shuckard 1839
Hydrophytophilus Erichson 1846

Caenoscelini Casey 1900

Sternodeini Casey 1900

Caenoscelis Thomson 1863, 4 spp., New York, Pennsylvania, Indiana, Kentucky, North Carolina, South Carolina, Colorado, California, Oregon, Washington, and British Columbia.
Macrodea Casey 1924

Sternodea Reitter 1875, no described U.S. species, but present in Florida and Mexico.

[Cryptosomatulini Crowson 1980. Not present in the United States
Picrotini Crowson 1980]

Atomariinae LeConte 1861

Hypocopriini Reitter 1879

Amydropa Reitter 1877b, 1 sp., *A. clarki* Leschen, Baja California.

Hypocoprus Motschulsky 1839, 1. sp., *H. tenuis* Casey, Rocky Mountains, in *Formica* ant nests or leaf litter.
Myrmecinomus Chaudoir 1845

Atomariini LeConte 1861

Ephistemini Casey 1900
Salltiini Crowson 1980

Atomaria Stephens 1829, 10 spp. (9 described, 1 undescribed), generally distributed.
Anchicera Thomson 1863
Agathengis Gozis 1886
Grobbenia Holdhaus 1903
Microum Wollaston 1854

Curelius Casey 1900, 1 sp., *C. japonicus* (Reitter 1877), southern United States.

Ephistemus Stephens 1829, 3 spp., New York to Indiana, Florida, Arizona, and Pacific Northwest.

Tisactia Casey 1900, 3 spp., (*T. subglabra* Casey, Indiana, and 2 undescribed).

[Cryptafricini Leschen 1996, not present in the United States.]

Bibliography

BLACKWELDER, R. E. 1945. Checklist of the the Coleopterous Insects of Mexico, Central America, the West Indies, and South America. Part 3. United States National Museum Bulletin, 185: 343-550.

BOUSQUET, Y. 1989. A review of the North American genera of Cryptophaginae (Coleoptera: Cryptophagidae). Coleopterists Bulletin, 43:1-17.

BOUSQUET, Y. 1990. Beetles associated with stored products in Canada: an identification guide. Research Branch, Agriculture Canada, Publication 1837, 220 pp.

BØVING, A. G. and F. C. CRAIGHEAD. 1931. An illustrated synopsis of the principal larval forms of the Order Coleoptera. Entomologica Americana (New Series), 11: 1-351.

CASEY, T. L. 1900. Review of the American Corylophidae, Cryptophagidae, Tritomidae and Dermestidae, with other studies. Journal of the New York Entomological Society, 8: 51-172.

CHAVARRIA, G. 1994. Phoresy on a neotropical bumblebee (Hymenoptera:Apidae) by *Antherophagus* (Coleoptera: Cryptophagidae). Psyche, 101: 109-111.

CROTCH, G. R. 1873. Check list of the Coleoptera of America, north of Mexico. Salem, MA.136 pp.

CROWSON, R. A. 1980. On amphipolar distribution patterns in some cool climate groups of Coleoptera. Entomologia Generalis, 6: 281-292.

DAJOZ, R. 1988. Les Coléoptères d'un polypore (Champignon, Basidiomycète) des montagnes rocheuses (États-Unis), et description de deux espèces nouvelles (Erotylidae et Cryptophagidae). Nouvelle Revue d'Entomologie (N. S.), 5: 209-215.

DOWNIE, N. M., and R. H. ARNETT, Jr. 1996. The beetles of Northeastern North America. Volume II: Polyphaga: Series Bostrichiformia through Curculionoidea. Sandhill Crane Press. Gainesville. x, 871-1721 Pp.

HATCH, M. H. 1962. The Beetles of the Pacific Northwest. Part III: Pselaphidae and Diversicornia I. University of Washington Press. Seattle, ix + 503 pp.

HINTON, H. E. 1945. A monograph of the beetles associated with stored products. Volume I. British Museum (Natural History). London, 443 pp.

HINTON, H. E. and F. L. STEPHENS. 1941. Notes on the biology and immature stages of *Cryptophagus acutangulus* Gyll. (Col., Cryptophagidae). Bulletin of Entomological Research, 32: 135-143.

HOEBEKE, E. R. and A. G. WHEELER. 2000. *Telmatophilus typhae* (Fallen) (Coleoptera: Cryptophagidae), a Palearctic cattail specialist established in the Canadian maritime provinces. Proceedings of the Entomological Society of Washington, 102: 398-402.

JOHNSON, C. 1971. The forgotten genus *Curelius* Casey in relation to *Ephistemus* Stephens, with keys to the Palaearctic species (Col., Cryptophagidae). Entomologische Blätter, 66: 159-162.

LAWRENCE, J. F. 1982. Coleoptera. Pp. 482-553. *In*: S. P. Parker, ed. Synopsis and Classification of Living Organisms, Vol. 2. McGraw-Hill. New York.

LAWRENCE, J. F. 1991. Cryptophagidae. Pp. 469-471. *In*: F. W. Stehr, ed.. Immature Insects, Vol. 2. Kendall/Hunt. Dubuque, IA.

LAWRENCE, J. F., A. M. HASTINGS, M. J. DALLWITZ, T. A. PAINE and E. J. ZURCHER. 1999a. Beetles of the World: Descriptions, illustrations, and information retrieval for families and subfamilies. CD-ROM, Version 1.0 for MS-DOS. CSIRO Publishing. Melbourne.

LAWRENCE, J. F., A. M. HASTINGS, M. J. DALLWITZ, T. A. PAINE and E. J. ZURCHER. 1999b. Beetle larvae of the World: Descriptions, illustrations, and information retrieval for families and subfamilies. CD-ROM, Version 1.1 for MS-DOS. CSIRO Publishing. Melbourne.

LAWRENCE, J. F. and A. F. NEWTON, Jr. 1995. Families and subfamilies of Coleoptera (with selected genera, notes, references and data on family-group names). Pp. 779-1006. *In*: J. Pakaluk and S.A. Slipinski, eds. Biology, Phylogeny and Classification of Coleoptera: Papers Celebrating the 80th Birthday of Roy A. Crowson. Muzeum i Instytut Zoologi PAN, Warsawa.

LENG, C. W. 1920. Catalogue of the Coleoptera of America, North of Mexico. Cosmos Press. Cambridge, MA. x + 470 pp.

LESCHEN, R.A.B. 1996. Phylogeny and revision of the genera of Cryptophagidae (Coleoptera: Cucujoidea). University of Kansas Science Bulletin 55:549-634.

LESCHEN, R.A.B. 1999. Origins of symbiosis: phylogenetic patterns of social insect inquilinism in Cryptophagidae (Coleoptera: Cucujoidea). Pp. 85-101. *In:* G. W. Byers, R. H. Hagen, and R. W. Brooks, eds. Entomological Contributions in Memory of Byron Alexander. University of Kansas Natural History Museum Special Publication, 24.

LESCHEN, R.A.B. and S. BOWSTEAD. 2001. Synonymical notes for Corylophidae and Cryptophagidae (Coleoptera: Cucujoidea). Coleopterists Bulletin, 55: 312-316

LESCHEN, R.A.B. and C. JOHNSON. World checklist of the species of Cryptophagidae (Coleoptera: Cucujoidea) [in prep.].

LJUBARSKY, G. Y. 1992a. Functional morphology of bulldozer structures and diversity of vital forms in the family Cryptophagidae (Coleoptera: Clavicornia). Russian Entomological Journal, 1: 3-16.

LJUBARSKY, G. Y. 1992b. On the taxonomical status of the genera *Myrmedophila* Bousquet, 1989 and *Catopochrotides* Kies. & Rchdt., 1936 (Coleoptera, Cryptophagidae). Zoologischeskii Zhurnal, 71: 595-598.

PAKALUK, J., S. A. SLIPINSKI and J. F. LAWRENCE. 1995. Current classification and family-group names in Cucujoidea (Coleoptera). Genus, 5: 223-268.

PETERSON, A. 1951. Larvae of insects: Coleoptera, Diptera, Neuroptera, Siphonaptera, Mecoptera, Trichoptera. Part II. Columbus, OH. 416 pp.

SCHENKLING, S. 1923. Cryptophagidae. *In*: W. Junk and S. Schenkling, eds. Coleopterorum Catalogus. Pars 76. W. Junk. Berlin, 92 pp.

SHARP, D. 1900. Cryptophagidae. *In*: F.Godman and O.Salvin, eds. Biologia Centrali-Americana. Insecta. Coleoptera. Vol II. Part 1 [1887-1905]: 579-624, pl. 18

WOODROFFE, G. E. and C. W. COOMBS. 1961. A revision of the North American *Cryptophagus* Herbst (Coleoptera: Cryptophagidae). Miscellaneous Publications of the Entomological Society of America, 2: 179-211.

86. LANGURIIDAE Wiedeman 1823

by Richard A. B. Leschen and Paul E. Skelley

Family common name: The lizard beetles

The antennal insertions concealed in dorsal view, mandible without a deep cavity or mycangium, tentorium without median tendon; prothorax with well developed lateral carinae, mesocoxal cavity closed laterally by the sternum, ventrite I equal to length of remaining ventrites; epipleuron distinct to apex, and punctation of elytron random or seriate define this group in the Cucujoidea.

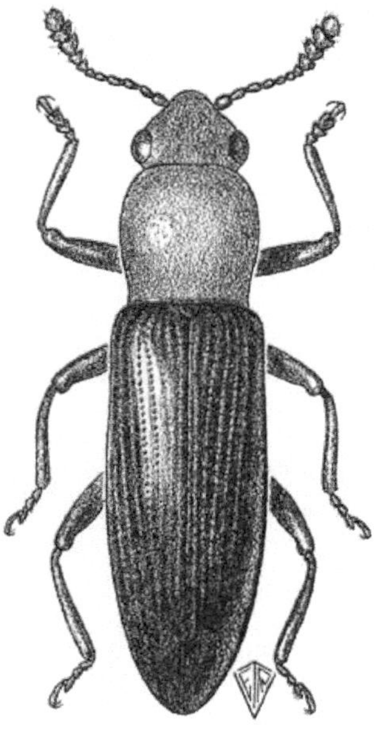

FIGURE 1.86 *Languria mozardi* Latreille

Description: Body length 2-20 mm. Body form elongate and parallel-sided, moderately flattened or relatively convex. Body usually glabrous; color various, adults of smaller species dark or light brown, some with elytral spots; adults of larger species blue, red or green. Body setae long or short or usually absent. Glandular ducts present mainly in head.

Head longer than wide, retracted into thorax, with or without line on vertex, usually with stridulatory file; subgenal spines present or absent; frons without lateral tubercles; antennal grooves absent; anterior part of gula sometimes with a transverse groove, gular suture present. Eyes well developed, with or without ocular setae. Clypeus more or less quadrate, with or without frontoclypeal suture. Antennal club 3-segmented, antennomeres symmetrical or not. Mandible with well-developed mola, incisor apex, and prostheca, dorsal tubercles present or absent, deep cavity or mycangia absent. Maxilla brushlike with well developed marginal setae and apical spines, maxillary palpomeres more or less subequal to each other. Labium with mentum wider than long, with or without transverse ridge or line, with or without middle process; labial palpus 2-segmented. Tentorium with anterior arms fused or separate, posterior part straplike, without anterior median tendon.

Prothorax with well developed lateral carina; side usually without well developed marginal processes or angularities. Pronotal disc with or without basal pits; with depressions. Procoxae rounded and separate; procoxal cavities externally open or closed; internally open or closed. Mesosternum with or without procoxal rests. Mesepimeron with or without distinct pit. Mesocoxal cavities closed laterally by sterna. Meso- metasternal articulation with either a double knob (Toraminae and most Cryptophilinae) or flattened, or button shaped. Metasternum with or without longitudinal line. Metendosternite with a broad stalk; most with basal plate transverse with well developed arms; with anterior tendons present (without in some flightless forms), approximate or distant.

Abdomen with 5 freely articulated ventrites. Ventrite I equal in length to remaining ventrites; with or without metasubcoxal lines. Intercoxal process moderately broad. Elytra completely covering abdomen; punctation confused or seriate; epipleuron distinct to apex. Legs relatively long; length of trochanter about 1/4 that of femur; tibia parallel-sided or club shaped, most with 2 apical spines; tarsi 5-5-5 in both sexes.

Male with spiculum gastrale with long anterior strut. Orientation of aedeagus vertical in body cavity; parameres separated and articulated, internal sac complex or not; dorsal arms narrow or joined at their apices or not, or broadly fused; flagellum present or not; tegmen with an elongate strut that is either fused or divided.

Larva (modified from Lawrence 1991) with mola asperate or tuberculate, mala falciform, prostheca fixed and broad at base, 2 tarsungular setae, granulate or tuberculate dorsal surface.

Body elongate, more or less parallel-sided or narrowed posteriorly; cylindrical to slightly flattened; dorsal surfaces usually granulated or tuberculate (smooth in Languriinae and *Leucohimatium*), lightly pigmented, sometimes with darker head and tergal plates; vestiture of simple setae, occasionally mixed with expanded or frayed setae. Head protracted and prognathous, wider than long, somewhat flattened. Epicranial stem usually absent (long in Languriinae); frontal arms usually lyriform (V-shaped in Languriinae) and contiguous at base (separated at base in some Loberini); epicranial suture sometimes indistinct. Median endocarina extending between frontal arms in some Xenoscelinae. Stemmata usually 5 or 6 on each side, occasionally 2 or absent. Antennae moderately long, 3-segmented. Frontoclypeal suture absent; labrum free. Mandibles symmetrical, apex bi- or tridentate, usually with accessory ventral process (absent in Languriinae) often with serrate incisor edge; mola well-developed, tuberculate or asperate, with asperities sometimes forming transverse rows, sometimes with hyaline lobe at base; prostheca mostly fixed, hyaline process broad at base and obtusely angulate at apex (broadly triangular), sometimes reduced. Ventral mouthparts retracted. Maxilla with transverse cardo, elongate stipes, well-developed articulating area, 3-segmented palp and falciform mala. Labium free to base of mentum; ligula short and broad; labial palps 2 segmented and widely separated. Hypopharyngeal sclerome tooth-like. Hypostomal rods moderately to very long and diverging. Ventral epicranial ridges absent except in Cryptophilinae and Toraminae.

Thoracic and abdominal terga of some with weak (Toraminae) or strong (Cryptophilinae) lateral processes. Legs well developed, 5-segmented; tarsungulus with 2 setae, either side by side or in some (Toraminae and Cryptophilinae) with 2 unequal setae lying with one distal to the other; coxae moderately widely separated. Tergum A9 well developed with pair of fixed urogomphi, which are strongly upturned (not in Cryptophilinae and Toraminae) and unpigmented or pigmented apex only; accessory setiferous tubercles or pregomphi present in front of urogomphi in many (absent in Languriinae). Segment A10 more or less circular and posteroventrally oriented (somewhat transverse in Languriinae). Spiracles biforous or annular-biforous, not on tubercles.

Habits and habitats. Some species are considered pests of stored grain (*Cryptophilus integer* [Heer], *Leucohimatium arundinaceum* [Forskal], *Pharaxonotha kirschi* Reitter) or herbaceous crops (*Languria mozardi* Latreille). In contrast, some species may be regarded as beneficial in pollinating African and New World cycads (*Pharaxonotha* spp.). Members of Languriinae (Languriini), and some Xenoscelinae are strictly phytophagous whereas most of the remaining taxa (some Xenoscelinae, Cryptophilinae, Setariolinae, and Cladoxenini [Languriinae]) are associated with decaying plant materials and are either saprophagous or mycophagous (spores and hyphae of microfungi) and pollen feeding. The larvae of Languriinae are stem borers of the composites and legumes. Languriines are collected on their host plants whereas the remaining taxa may be sifted from leaf litter, beaten from rotting flowers, or collected at lights. Other papers of interest are Leschen (1997) and Zablotny and Leschen (1996). Generalized works with larval information are Bøving and Craighead (1931), Lawrence (1991), Peterson (1951), and Rymer Roberts (1939, 1958), and Carlton *et al.* (2000).

Status of the classification. The classification of the family is under review by Leschen and Wegrzynowicz (1998). Many taxa now included in the family were transferred from Cryptophagidae in a series of papers by Sen Gupta (1967, 1968a, b, 1969) and Sen Gupta and Crowson (1969, 1971). There is some disagreement about the monophyly of the Languriidae because many members share characteristics in common with Erotylidae. There are problems with higher classification of the subfamily, including the monophyly of the subfamilies. Although the United States Languriinae have been described (Vaurie 1948), and a revision of the remaining taxa would be useful, the xenoscelines and toramines are poorly known in the Neotropics. Neotropical Languriinae were studied by Martins and Pereira (1965). Regional papers with keys include Downie and Arnett (1996), Hatch (1961), and Sharp (1900). Other works of interest to the systematics and identification of the Languriidae are Casey (1900), Lawrence (1982), Lawrence *et al.* (1999a, b), Lawrence and Newton (1995), and Pakaluk *et al.* (1995). In addition to Leschen and Wegrzynowicz (1998), catalogs on the Languriidae include Lawrence and Vaurie (1983) and Schenkling (1923, 1928).

Distribution. There are 1,037 described species (Leschen and Wegrzynowicz 1998) known from all areas, of which 33 species occur in the United States (Lawrence and Vaurie 1983).

Key to the Nearctic Genera

1. Antennal club asymmetrical (Figs. 2, 3), consisting of 4 or more antennomeres; body elongate, parallel-sided, cylindrical; size of most more than 5 mm; color dark black or brown, shining, many with metallic sheen and red markings; glabrous (Languriinae, "lizard beetles") 2
— Antennal club symmetrical, consisting of 3 antennomeres; body parallel-sided or not, slightly flattened; size of most less than 5 mm; color of most brown, shining or not; many pubescent 5

2. Ocular stria present, extending from near the antennal socket to the base of eyes (Figs. 3, 4) 3
— Ocular stria absent (Fig. 2) .... *Acropteroxys*

3(2). Elytral apices evenly rounded to sutural margin, or apices dentate; ocular stria deep, close to eyes (Fig. 3) 4
— Elytral apices terminating in acute tooth, inner third of elytral margin obliquely incised to suture which also terminates in tooth (Fig. 5); ocular stria shallow, distant from eyes, divided above eye (Fig. 4) .... *Langurites*

4(3). Elytral apices evenly rounded; thorax generally in great part red; elytra piceous, wholly or in part .... *Languria*
— Elytral apices dentate (Fig. 6); thorax and elytra of same piceous color .... *Dasydactylus*

5. Elytron with confused punctation, setae not in distinct rows 6
— Elytron with punctation in distinct rows, setae in rows 7

6. Procoxal cavities externally closed behind (Fig. 7); prothoracic width equal to elytral width .... *Cryptophilus*
— Procoxal cavities externally open behind (Fig. 8); prothoracic width less than elytral width .... *Toramus*

7. Body sparsely or not setose; form broad and not strongly parallel-sided 8
— Body densely setose; form elongate and parallel-sided .... *Hapalips*

8. Pronotum nearly parallel sided, with secondary basal line on each side (Fig. 9); size larger (4 mm); metasubcoxal lines absent on visible abdominal ventrite I .... *Pharaxonotha*
— Pronotum laterally arcuate, with basal pits (Fig. 10); size smaller (2 mm); metasubcoxal lines present on visible abdominal ventrite I .... *Loberus*

Classification of the Nearctic Genera

Languriidae Crotch 1873

Languriinae Crotch 1873

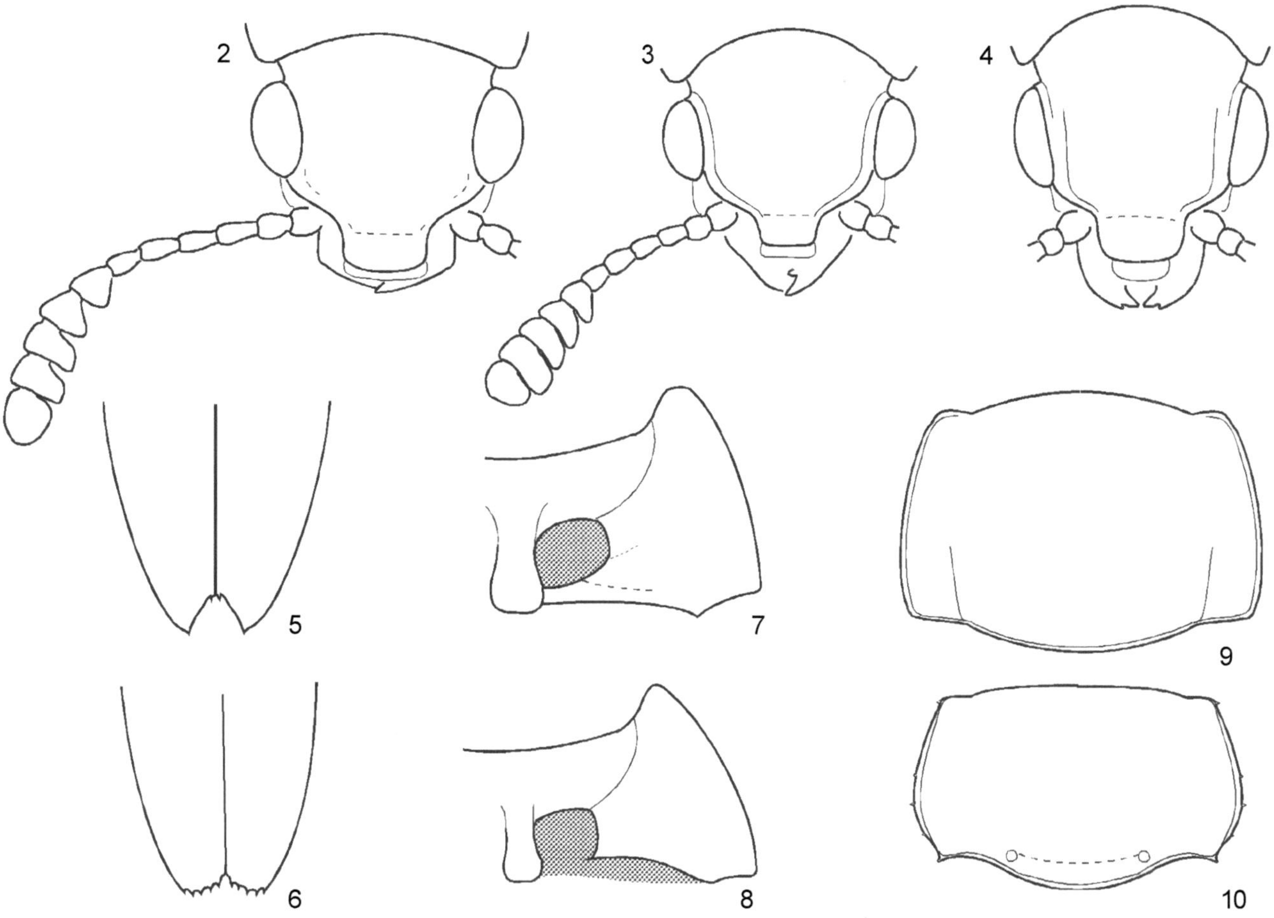

FIGURES 2.86 - 10.86. 2. *Acropteroxys gracilis* Newman, head and antenna; 3. *Dasydactylus cnici* Schaeffer, head and antenna; 4. *Langurites lineatus* Laporte, head; 5. *Langurites lineatus* Laporte, elytral apices; 6. *Dasydactylus cnici* Schaeffer, elytral apices; 7. *Cryptophilus fulminalis* Casey, prosternum; 8. *Toramus* sp., prosternum; 9. *Pharaxonotha floridana* (Casey), pronotum; 10. *Loberus* sp., pronotum

Species in the United States revised by Vaurie (1948); Neotropical species revised by Martins and Pereira (1965).

Languriini Crotch 1873

Acropteroxys Gorham 1887, 2 spp., widely distributed.

Dasydactylus Gorham 1887, 1 sp., *D. cnici* Schaeffer 1904, Texas.

Languria Latreille 1802, 14 spp., widely distributed.
Janessa Chevrolat *in* Dejean 1837
Languiria Hope *in* Gray 1831
Langura Rafinesque 1815

Langurites Motschulsky 1861, 1 sp., *L. lineatus* Laporte 1832, Florida and southern Arizona.

Cryptophilinae Casey 1900

Cryptophilini Casey 1900

Cryptophilus Reitter 1874, 3 spp., widely distributed.

Toraminae Sen Gupta 1967

Toramus Grouvelle 1916, 5 spp., widely distributed.
Itomarus Reitter 1919
Tomarus LeConte 1861

Xenoscelinae Ganglbauer 1899

Loberinae Bruce 1951

Pharaxonothini Crowson 1952

Pharaxonotha Reitter 1875, 2 spp., Illinois, Texas, and Florida. Larvae and adults of *Paraxonotha floridana* (Casey) feed on pollen of male cycad cones, *Zamia* spp. (coontie), in Florida. Hinton (1945) and Pratt and Scott (1962) report *P. kirschi* Reitter as a stored product pest. New World species revised by Pakaluk (1986).
Planismus Casey 1890

Xenoscelini Ganglbauer 1899

Hapalips Reitter 1877, 3 spp., *H. texanus* Schaeffer in Texas, and 2 undetermined species in southern Florida. They have been collected in palm sheaths.

Loberina Grouvelle 1902

Loberini Bruce 1951

Loberus LeConte 1861, 4 spp., Rhode Island, New Jersey, Indiana, Florida, Arkansas, Kansas, Louisiana, Michigan, and Texas. Carlton *et al.* (2000) describe the larva of *L. impressus* LeConte. (key to species, Sen Gupta 1967; review of related genera, Sen Gupta 1968a).

Glisonotha Motschulsky 1863

Bibliography

BØVING A.G. and F. C. CRAIGHEAD. 1930 [1930-1931]. An illustrated synopsis of the principal larval forms of the order Coleoptera. Entomologica Americana (New Series), 11: 1-351.

CARLTON, C. E., V. R. TOWNSEND, P. A. VAN ZANDT and S. MOPPER. 2000. Description of the larva of *Loberus impressus* LeConte with notes on its natural history (Coleoptera: Languriidae: Xenoscelinae). Annals of the Entomological Society of America, 93: 356-361.

CASEY, T. L. 1900. Review of the American Corylophidae, Cryptophagidae, Tritomidae and Dermestidae, with other studies. Journal of the New York Entomological Society, 8: 51-172.

DOWNIE, N.M. and R. H. ARNETT, Jr. 1996. The beetles of Northeastern North America. Volume II: Polyphaga: Series Bostrichiformia through Curculionoidea. The Sandhill Crane Press. Gainesville. x, 871-1721.

HATCH, M. H. 1961. The Beetles of the Pacific Northwest. Part III: Pselaphidae and Diversicornia I. University of Washington Publications in Biology, 16: XII + 503 pp., 66 pls.

HINTON, H. E. 1945. A Monograph of the Beetles Associated with Stored Products. Volume I. British Museum (Natural History). London, 443 pp.

LAWRENCE, J. F. 1982. Coleoptera. Pp. 482-553. *In:* S. P. Parker, ed. Synopsis and Classification of Living Organisms, Vol. 2. McGraw-Hill, New York.

LAWRENCE, J. F. 1991. Languriidae. Pp. 471-473. *In:* F.W. Stehr, ed. Immature Insects, Vol. 2. Kendall/Hunt. Dubuque, IA.

LAWRENCE, J. F., A. HASTINGS, M. J. DALLWITZ, T. A. PAINE and E. J. ZURCHER. 1999a. Beetles of the World: A Key and Information System for Families and Subfamilies. CD-ROM, Version 1.0 for MS-DOS. CSIRO Publishing, East Melbourne, Victoria.

LAWRENCE, J. F., A. HASTINGS, M. J. DALLWITZ, T. A. PAINE and E. J. ZURCHER. 1999b. Beetle Larvae of the World: Descriptions, Illustrations, Identification, and Information Retrieval for Families and Sub-families CD-ROM, Version 1.0 for MS-DOS. CSIRO. East Melbourne, Victoria.

LAWRENCE, J. F. and A. F. NEWTON, Jr. 1982. Evolution and classification of beetles. Annual Review of Ecology and Systematics, 13: 261-290.

LAWRENCE, J. F. and A. F. NEWTON, Jr. 1995. Families and subfamilies of Coleoptera (with selected genera, notes, references and data on family-group names). Pp. 779-1006. *In:* J. Pakaluk and S.A. Slipinski, eds. Biology, Phylogeny and Classification of Coleoptera: Papers Celebrating the 80th Birthday of Roy A. Crowson. Muzeum i Instytut Zoologii PAN, Warsaw.

LAWRENCE, J. F. and P. VAURIE. 1983. A Catalog of the Coleoptera of America North of Mexico. Family: Languriidae. United States Department of Agriculture, No. 529-92: 13 pp.

LESCHEN, R. A. B. 1997. The *Empocryptus*-group (Languriidae: Toraminae): relationships and a new genus associated with a lepidopteran cocoon. Coleopterists Bulletin, 51: 303-318.

LESCHEN, R. A. B. and P. WEGRZYNOWICZ. 1998. Generic catalogue and taxonomic status of Languriidae (Coleoptera: Cucujoidea). Annales Zoologici, 48: 221-243.

MARTINS, U. R. and PEREIRA, F. S. 1965. Revisão dos Languriinae Neotropicais (Coleoptera, Languriidae). Arquivos de Zoologia do Estado de São Paulo, 13: 139-309, 97 Figs., 4 maps.

PAKALUK, J. 1988. Review of the new world species of *Pharaxonotha* Reitter (Coleoptera: Languriidae). Revista de Biologia Tropical, 36: 447-451.

PAKALUK, J., S. A. SLIPINSKI and J. F. LAWRENCE. 1995. Current classification and family-group names in Cucujoidea (Coleoptera). Genus, 5: 223-268.

PETERSON, A. 1951. Larvae of insects. An introduction to Nearctic species. Part II. Coleoptera, Diptera, Neuroptera, Siphonaptera, Mecoptera, Trichoptera. A. Peterson. Columbus, OH. 416 pp.

PRATT, H.D. and H.G.SCOTT. 1962. A key to some beetles commonly found in stored foods (Coleoptera). Proceedings of the Entomological Society of Washington, 64: 43-50.

RYMER ROBERTS, A. W. 1939. On the taxonomy of Erotylidae (Coleoptera), with special reference to the morphological characters of the larvae. Transactions of the Royal Entomological Society of London, 88: 89-117.

RYMER ROBERTS, A.W. 1958. On the taxonomy of Erotylidae (Coleoptera), with special reference to the morphological characters of the larvae, II. Transactions of the Royal Entomological Society of London, 110: 245-284.

SCHENKLING, S. 1923. Cryptophagidae. *In:* W. Junk and S. Schenkling, eds. Coleopterorum Catalogus. Pars 76. W. Junk, Berlin, 92 pp.

SCHENKLING, S. 1928. Languriidae. *In:* W. Junk and S. Schenkling, eds. Coleopterorum Catalogus. Pars 100. W. Junk, Berlin, 40 pp.

SEN GUPTA, T. 1967. A new subfamily of Languriidae (Coleoptera) based on four genera, with a key to the species of

Toramus. Proceedings of the Royal Entomological Society of London (B), 36: 167-176.

SEN GUPTA, T. 1968a. Review of the genera of the tribe Loberini (Coleoptera: Languriidae). Breviora, 303: 1-27.

SEN GUPTA, T. 1968b. Revision of the genera of Cladoxenini (=Cladoxeninae Arrow) and Thallisellini trib. nov. of the family Languriidae (Coleoptera: Clavicornia). Journal of Natural History, 2: 463-475.

SEN GUPTA, T. 1969. On the taxonomy of Erotylidae [Insecta: Coleoptera: Clavicornia], with descriptions of two new larvae. Proceedings of the Zoological Society of Calcutta, 22: 97-107.

SEN GUPTA, T. and R. A. CROWSON. 1969. On a new family of Clavicornia (Coleoptera) and a new genus of Languriidae. Proceedings of the Royal Entomological Society of London (B), 38:125-131.

SEN GUPTA, T. and R. A. CROWSON. 1971. A review of classification of the family Languriidae (Coleoptera: Clavicornia) and the place of Languriidae in the natural system of Clavicornia. Memoirs of the Zoological Survey of India, 15: 1-42.

SHARP, D. 1900. Cryptophagidae. Pp. 579-626. *In*: F. D. Godman and O. Salvin, eds. Biologia Centrali-Americana. Insecta, Coleoptera II. Part 1. Dulau and Co., London. 717 pp.

VAURIE, P. 1948. A review of the North American Languriidae. Bulletin of the American Museum of Natural History, 92: 119-155.

ZABLOTNY, J. E. and R. A. B. LESCHEN. 1996. Two new genera of flightless Languriidae (Coleoptera: Cucujoidea) from the Australo-Pacific. Coleopterists Bulletin, 50: 382-390.

87. EROTYLIDAE Leach 1815

by Paul E. Skelley and Joseph V. McHugh

Common name: The pleasing fungus beetles

Family synonyms: Dacnidae LeConte and Horn 1883; Engidae MacLeay 1833; Engididae Latreille 1825

In many ways, adult erotylids resemble some fungus-associated tenebrionids. They may be similar in size, body-shape, habits, and coloration. The Erotylidae, however, have a 5-5-5 tarsal formula (although many appear tetramerous) and most have antennae bearing well developed clubs.

FIGURE 1.87 *Megalodacne fasciata* (Fabricius)

Description: Shape hemispherical to elongate-elliptical; some with inflated elytra giving a humped-back shape; size 3 to 22 mm in length; color mostly black with reddish to yellowish markings; vestiture mostly absent or short on the abdomen, tibiae, tarsi, and antennal stems.

Head inserted to eyes into prothorax, frons prolonged; surface smooth or punctate. Antennae with eleven antennomeres, short, most with an abrupt club of three antennomeres, club with four or five antennomeres in a few; third antennomere elongate, subequal to the fourth and fifth, or fourth to sixth combined; inserted on sides of frons at anterior margins of eyes. Labrum fused to frons; mandibles stout, arcuate, apices denticulate; maxillary palpi with four palpomeres, short, stout, apical palpomere fusiform to securiform. Gular sutures short or obscure; mentum strongly transverse, triangular, or subquadrate to pentagonal, some reduced and nearly obsolete; labial palpi with three palpomeres, short, stout, the apical palpomere fusiform to securiform. Eyes lateral, medium to small, rounded with margins entire.

Pronotum broader than head; shape subquadrate, excavated in front for reception of head, angles acute; borders strongly margined laterally; surface smooth or punctate; pleural region broad; prosternum broad in front of coxae with intercoxal process broad; procoxal cavities closed behind. Mesosternum subquadrate to transversely subrectangular; mesepimeron not reaching coxae; mesocoxal cavities closed behind. Metasternum somewhat wider than long in most. Legs with trochantins hidden; procoxae flat, globular, separate; mesocoxae globular separate; metacoxae triangular, separate; trochanters triangular, interstitial; femora weakly swollen; tibiae flattened, carinate, apically enlarged and with small apical spurs; tarsal formula 5-5-5, fourth tarsomere small in many, first three tarsomeres more or less broad and pubescent beneath; claws simple. Scutellum subcordate to pentagonal. Elytra entire, apically rounded; striae punctate, or surface smooth; intervals sparsely punctulate, a few smooth or confusedly punctate with large black impressions or small punctures; epipleural fold well developed. Wing has a dark, medial fleck (shagreened oval spot) near distal edge; two closed cells occur, an elongate wedge cell and a triangular radial cell; five veins nearly reach the hind edge of the wing; folding pattern with median area normal; anal lobe sessile.

Abdomen with five visible ventrites, basal and apical ventrites longer in most; sutures distinct; surface punctate or rugose punctate. Male genitalia with median lobe stout, curved, with a median, long, distally expanded, compressed basal strut; parameres very slender, short, setiferous; pars basalis elongate, hood-shaped, surrounding median lobe, with various basal struts. Female genitalia with paraprocts long and rather spatulate at both ends; valvifers large, with a strong baculum near ventral margin; coxite bipartite; stylus present or absent; proctiger small.

Larvae (see Lawrence 1991) elongate, subparallel to fusiform, subcylindrical to moderately flattened; length 3 to 25 mm; urogomphi present in most; many with dorsal and lateral surfaces bearing setiferous tubercles, granules, senti, or parascoli; integumental projections most distinct on terga AVII-VIII in many; color white to pale brown overall, some with darker brown tergal and pleural plates. Head exserted, hypognathous or nearly so, with five or six convex to conical stemmata per side; epicranial stem present or absent; frontal arms lyriform to V-shaped, contiguous basally; median endocarina absent in most. Antennae shorter than head capsule, three-segmented; sensorium conical, on apex of antennomere II, not exceeding length of antennomere III. Clypeus discrete or not; labrum distinct. Mandibles nearly symmetrical, robust, broad, apices bi- or tridentate; mola, ventral accessory process and prostheca absent in most (except in *Dacne*). Maxillolabial complex retracted; cardo transverse, divided in some; mala obtuse to falciform, apex entire or weakly cleft in most (deeply cleft in *Megalodacne*), bearing setae, spines, or unci; maxillary palpi three-segmented; articulating area large; mentum and submentum fused or not; ligula short; labial palpi two-segmented; hypopharyngeal bracon present or absent. Gula short, transverse.

Acknowledgment: The illustration of *Gibbifer californicus* (Lacordaire) was drawn by T.N. Kiselyova.

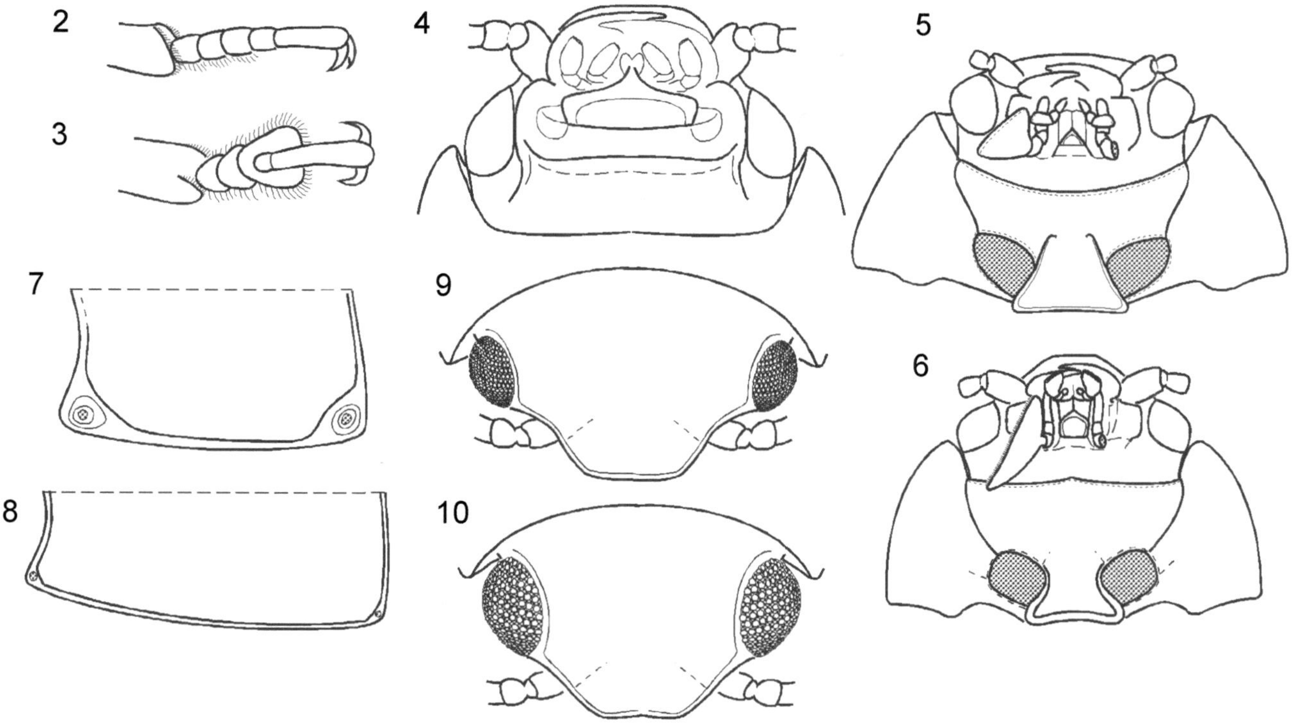

FIGURES 2.87-10.87. 2. *Megalodacne fasciata* (Fabricius), hind tarsus; 3. *Tritoma* sp., hind tarsus; 4. *Megalodance fasciata* (Fabricius) ventral view of head; 5. *Tritoma biguttata* (Say), ventral view of head and prosternum; 6. *Triplax thoracica* Say, ventral view of head and prosternum; 7. *Triplax californica* LeConte, left lateral view of pronotal margin; 8. *Tritoma* sp., left lateral view of pronotal margin; 9. *Pseudischyrus extricatus* (Casey), anterior view of head; 10. *Tritoma atriventris* LeConte, anterior view of head.

Thorax with moderately long to long, five-segmented legs; tarsunguli clawlike, with two setae side by side. Abdomen ten-segmented, AIX bearing urogomphi in most, urogomphi short to long, often acutely upturned and hook-like, long and thread-like in *Gibbifer*, replaced by a single, stout, medial projection at abdominal apex in *Microsternus*; AX small, surrounding the anus, some with spiny scoli, often bilobed, pseudopodia present in most. Spiracles annular-biforous to annular, on mesothorax and AI-VIII.

Habits and habitats. Larvae and adults feed on the fruiting bodies of the larger Basidiomycete fungi growing in decaying wood or in mycorrhizal associations with tree roots (Goodrich and Skelley 1994, Skelley *et al.* 1991). Adults deposit eggs on the fungus on which the larvae feed. Larvae of the supposed primitive erotylids (*Megalodacne*, *Dacne*, *Microsternus*) burrow in hard bracket fungi. Larvae of more derived erotylids are surface grazers on prostrate fungi or feed within mushrooms. Larvae that feed in rapidly decaying mushrooms are pale and maggot-like, with rapid development. McHugh (2001) provides a review of the literature on larval Erotylidae.

There are few accounts where erotylids are considered stored product pests (Hinton 1945), most of these are for taxa now considered members of the Languriidae. Boyle (1956) and Savary (1995) discuss the occasional pest status of *Dacne picta* Crotch (from Japan) in dried mushrooms.

Adults of some species (*e.g.*, *Megalodacne*) overwinter under bark or beneath logs in gregarious masses (Goodrich and Skelley 1991b, Navarrete-Heredia and Novelo-Gutierrez 2000). Many adults show nocturnal activity (*Ischyrus*, *Pseudischyrus*), while others appear to be active during the day (*Tritoma*, *Triplax*). Larvae of some are keyed or illustrated in Bøving and Craighead (1931), Peterson (1951), Lawrence (1991), and Skelley (1988).

Status of the classification. For North America, the family has been thoroughly revised (Boyle 1956); however, the affinities of the family are not discussed. Numerous studies on this family indicate a strong relationship with the Languriidae (Crowson 1955, Roberts 1958, Sen Gupta and Crowson 1971, Lawrence 1991, Lawrence and Britton 1994, Lawrence *et al.* 1999a, 1999b), and they may eventually be combined into a single family (Leschen pers. com.). There are many regional studies including keys or additional information; these include Downie and Arnett (1996), Goodrich and Skelley (1991b, 1993, 1995), Hatch (1962) and Skelley (1988).

Distribution. There are approximately 2500 described species (Alvarenga 1994, Boyle 1956, Chûjô and Chûjô 1988, 1989, 1990; Delkeskamp 1981) occurring in all areas, of which 49 occur in America north of Mexico (Boyle 1956; Goodrich and Skelley 1991a, 1997; Skelley 1993, 1994, 1997).

Key to the Nearctic Genera

1. Fourth tarsomere scarcely reduced, subequal to third and attached in normal manner to end of third (clearly 5-5-5, pentamerous) (Fig. 2); apical palpomere of maxillary palp cylindrical, not transverse or triangular; mentum strongly transverse (Fig. 4) 3
— Fourth tarsomere strongly reduced (pseudotetramerous), not more than half length or width of third, attached mediodorsally to third (Fig. 3); apical palpomere of maxillary palp strongly transverse or triangular; mentum not transverse (Figs. 5, 6) 2

2(1). Elytra non-striate, confusely punctate with large black punctures; prothorax at base little more than half as wide as greatest common elytral width; size large, 12 mm or more in length (Erotylinae) (Fig. 17) .... *Gibbifer*
— Elytra regularly striate-punctate; prothorax at base subequal to greatest common elytral width; size small, 8 mm or less in length (Tritominae) 5

3(1). Large forms, 9 mm or more in length; black, elytra bearing two reddish fasciae (Fig.1); body largely impunctate (Megalodacninae) .... *Megalodacne*
— Small forms, 6 mm or less in length; elytra without fasciae; body distinctly punctate (Dacninae) 4

4(3). Mesosternum extremely short and transverse, about ten times wider than long; prosternum bearing raised, triangular plateau which almost attains prosternal apex; elytron red with black spots (Fig. 14) .... *Microsternus*
— Mesosternum not more than three times as wide as long; prosternum without such plateau; elytra black or brown, many with orange spots .... *Dacne*

5(2). Pronotal angle pores large and conspicuous, umbilicate (Fig. 7); apical palpomere of maxillary palp bearing distinct brush at apex (Fig. 6) 6
— Pronotal angle pores small, inconspicuous, simple (Fig. 8); apical palpomere of maxillary palp without apical brush 7

6(5). Apical palpomere of maxillary palp strongly transverse; body without dorsal pubescence (Fig. 15) .... *Triplax*
— Apical palpomere of maxillary palp squared, not transverse; body strongly punctured and pubescent (Fig. 12) .... *Hirsutotriplax*

7(5). Pronotum and elytra bicolored, bearing complex black pattern on lighter background (Fig. 13) ... *Ischyrus*
— Pronotum unicolored, elytra unicolored or black with basal red spots 8

8(7). Prosternal lines long, either incurved anteriorly (Fig. 5) or meeting at prosternal apex; elytron immarginate basally; body broadly oval or egg-shaped 9
— Prosternal lines short, not or scarcely extending in front of inner coxal edges (as in Fig. 6); elytron margined basally in most; body elongate-elliptical. 10

9(8). Eyes coarsely faceted, relatively large and protuberant (Fig. 10) .... *Pseudischyrus*
— Eyes finely faceted, relatively small (Fig. 9) *Tritoma*

10(8). Head and pronotum reddish yellow to bright orange-red .... *Mycotretus*
— Head and pronotum entirely black ... *Haematochiton*

Classification of the Nearctic Genera

Erotylidae Leach 1815

Dacninae Gistel 1856

Dacne Latreille 1796, generally distributed, in fungi. Savary (1995) reports a U.S. interception of *Dacne picta* Crotch from Japan; Skelley (1997) provides a key and checklist to known species. Larvae are unknown for most US species, none has been formally described.

Engis Paykull 1800

subgenus *Dacne* Latreille 1796, 2 spp.; *D. californica* (Horn) western United States and Baja California, *D. quadrimaculata* (Say) (Fig. 11) eastern United States.

subgenus *Xenodacne* Boyle 1956, 3 spp., western US.; *D. cyclochilus* Boyle, *D. picea* LeConte, and *D. pubescens* Boyle.

Microsternus Lewis 1887, 1 sp., *M. ulkei* (Crotch) (Fig. 14), northeastern United States; details of its biology are unknown. It apparently feeds on hard bracket fungi in mature eastern deciduous forests (Goodrich 1994, Brown and Skelley 2001). No detailed larval description has been made. Dury (1878) briefly characterizes the larvae and comments on their habits. The larva is illustrated in Lawrence (1991) and Lawrence *et al.* (1993, 1999b).

Megalodacninae Sen Gupta 1969

Megalodacne Crotch 1873, 2 spp., eastern United States; *M. heros* (Say) and *M. fasciata* (Fabricius) (Fig. 1). Adults are frequently gregarious under bark in the day and nocturnally active. The heavily sclerotized larvae feed in bracket fungi and take 2-3 months to mature. The adult stage overwinters. Numerous references illustrate larvae of *Megalodacne* spp. Roberts (1958) describes the larvae of *M. fasciata* in detail. Skelley (1988) describes the larva of *M. heros*, and redescribes the larva of *M. fasciata*. McHugh *et al.* (1997) provide a detailed morphological study of *M. heros*. (Volume 2, Color Figure 24)

Tritominae Curtis 1834

Ischyrus Lacordaire 1842, 3 spp., eastern United States, and Arizona (many Neotropical species; Skelley 1998); Members are found on prostrate white polypore fungi on logs or under bark and are frequently collected at light. *Ischyrus quadripunctatus* (Olivier) (Fig. 13) is the only member whose larva is known. Its larva is illustrated in Lawrence (1991), Skelley (1998) and many other references. Skelley (1988) provides a description of the larva and pupa.

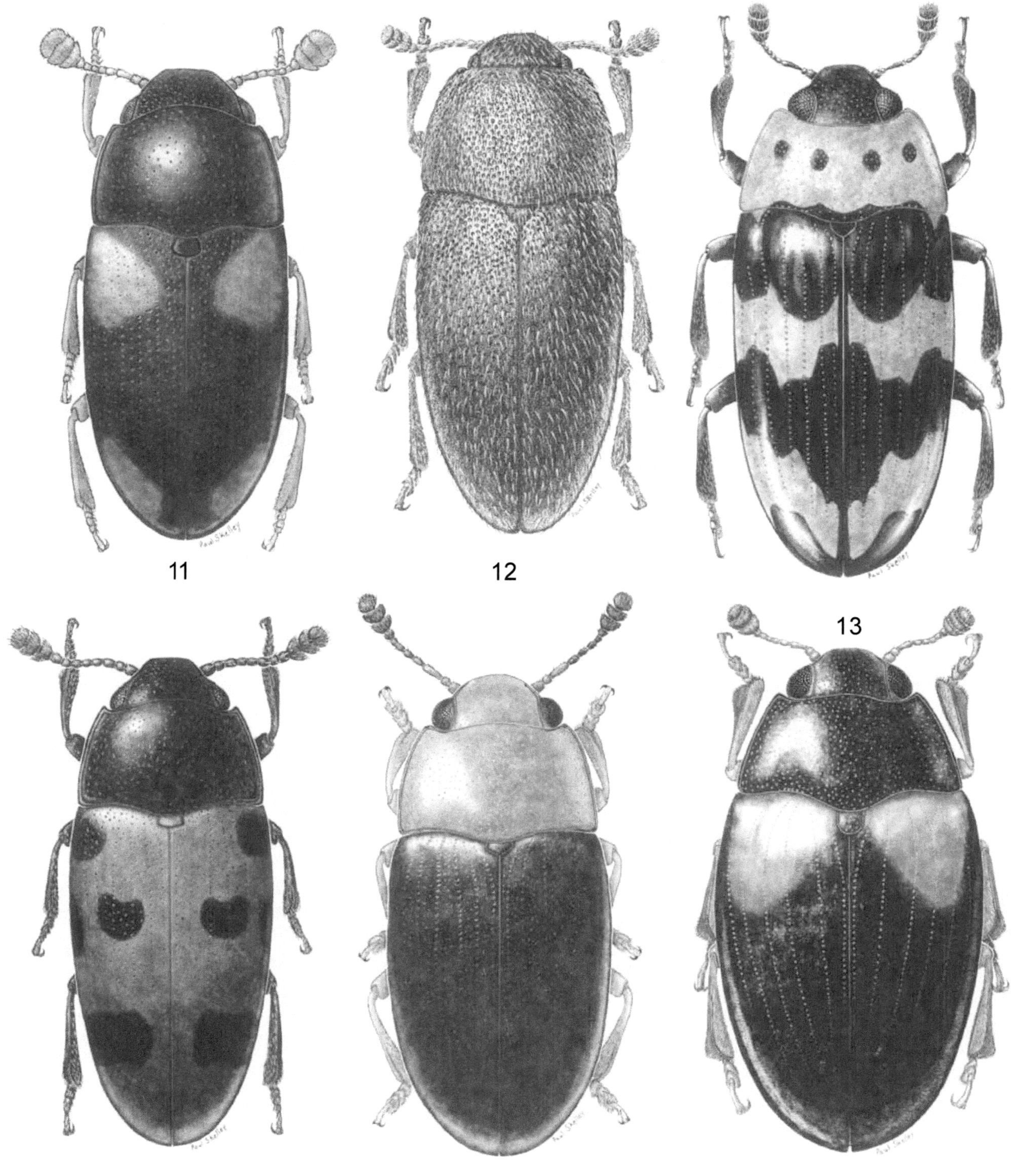

FIGURES 11.87-16.87. 11. *Dacne quadrimaculata* (Say); 12. *Hirsutotriplax mcclevei* Skelley; 13. *Ischyrus q. quadripunctatus* (Olivier); 14. *Microsternus ulkei* (Crotch); 15. *Triplax thoracicus* Say; 16. *Tritoma biguttata* (Say).

Triplax Herbst 1793, 18 spp. (Fig. 15) (Boyle 1962, Goodrich and Skelley 1997), generally distributed. Adults and larvae feed in soft bracket fungi, particularly *Inonotus* spp. and *Pleurotus* spp. Larvae are known for many species, but none has been formally described. Skelley (1988) provides larval diagnoses for *T. festiva* Lacordaire, *T. alachuae* Boyle, *T. thoracica* Say and *T. flavicollis* Lacordaire.

Hirsutotriplax Skelley 1993, 1 sp., *H. mcclevei* Skelley (Fig. 12), southeastern Arizona. Larvae and habits are unknown.

Tritoma Fabricius 1775, 11 spp., eastern United States; in fungi (Goodrich and Skelley 1994) (Fig. 16). Adults and larvae are found on various mushrooms or soft polypore fungi. The pale, maggot-like larvae mature rapidly, some taking only 2 weeks to reach the pupal stage. Roberts (1958) describes the larvae of *T. pulchra* Say in detail. Skelley (1988) provides a diagnosis for the larva of *T. sanguinipennis* (Say), and *T. atriventris* LeConte. No other species has a formal description of its larvae.

Cyrtotriplax Crotch 1873

Pseudischyrus Casey 1916, 3 spp., southeastern United States [California ?]; on mushrooms, often collected at light, probably nocturnal. Larvae are similar to *Tritoma* spp., but are undescribed.

Mycotretus Lacordaire 1842, 2 spp., *M. nigromanicatus* Boyle and an undetermined species, from southeastern Arizona; biology and larvae are unknown.

Haematochiton Gorham 1888, 2 spp., southeastern Arizona; biology and larvae are unknown. Label data indicate an association with pines at high elevations (Goodrich 1997).

Scaeother Gorham 1888

Erotylinae Latreille 1802

Gibbifer Voet 1778, 1 sp., *G. californicus* (Lacordaire) (Fig. 17), southwestern United States (Branham 1993); in fungi, or fungus-rotted wood. The large larvae of this species are covered with long scoli and graze the surface of fungi on dead wood. Larvae are illustrated in Lawrence (1991) and Peterson (1951). Roberts (1958) discusses larval characters. Larval habits and pupae are described by Graves (1965). *Gibbifer* is restricted to the New World, with most species occurring in the tropics. Observations and photographs of living adults indicate that all species in this genus have blue, pink, or lavender elytra. The elytral color fades to yellow-brown after death.

Cypherotylus Crotch 1873

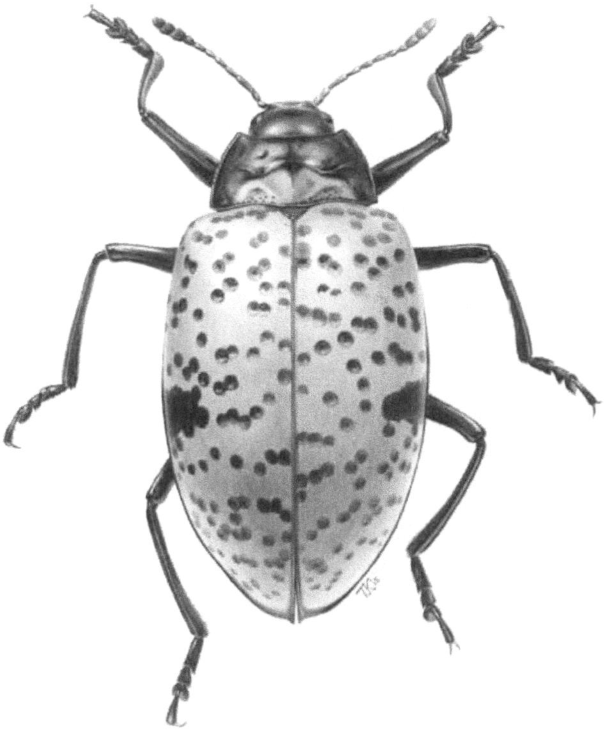

FIGURE 17.87. *Gibbifer californicus* (Lacordaire).

Bibliography

ALVARENGA, M. 1994. Catálogo dos Erotylidae (Coleoptera) Neotropical. Revista Brasiliera de Zoologia, 11: 1-175.

BØVING, A. G. and F. C. CRAIGHEAD. 1931. An illustrated synopsis of the principal larval forms of the order Coleoptera. Entomologica Americana (New Series), 11: 1-351, 125 pls.

BOYLE, W. W. 1956. A revision of the Erotylidae cf America north of Mexico (Coleoptera). Bulletin of the American Museum of Natural History, 110: 61-172, 141 figs., 8 pls.

BOYLE, W. W. 1962. A new species of *Triplax* from Arizona (Coleoptera: Erotylidae). Pan-Pacific Entomologist, 38: 29-30.

BRANHAM, M. A. 1993. Scientific Note. A new eastern record for *Cypherotylus californicus* Lacordaire in the Unites States (Coleoptera: Erotylidae). Coleopterists Bulletin, 47:81-82

BROWN, B. and P. E. SKELLEY. 2001. Scientific Note. New distribution records for *Microsternus ulkei* (Crotch) (Coleoptera: Erotylidae). Coleopterists Bulletin, (in press).

CHÛJÔ, M. and M. CHÛJÔ. 1988. A. catalog of the Erotylidae (Insecta, Coleoptera) from the Old World (excl. the Ethiopian Region). Esakia, 26: 139-185.

CHÛJÔ, M. and M. CHÛJÔ. 1989. A catalog of the Erotylidae (Insecta, Coleoptera) from the Old World (excl. the Ethiopian Region). Esakia, 28: 75-96.

CHÛJÔ, M. and M. CHÛJÔ. 1990. A catalog of the Erotyliae (Insecta; Coleoptera) from the Old World (excl. the Ethiopian Region) III. Esakia, 29: 1-67.

CROWSON, R. A. 1955. The natural classification of the families of Coleoptera. Nathaniel Lloyd. London, 187 pp

DELKESKAMP, K. 1981. Erotylidae von Africa und Madagascar. Coleopterorum Catalogus Supplementa, 34: 1-65

DOWNIE, N. M. and R. H. ARNETT, Jr. 1996. The beetles of Northeastern North America. Volume II: Polyphaga: Series Bostrichiformia through Curculionoidea. Sandhill Crane Press. Gainesville, FL. x, 871-1721.

DURY, C. 1878. Notes on several species of Coleoptera, with some accounts of habits, etc. Canadian Entomologist, 10: 210-211.

GOODRICH, M.A. 1994. *Microsternus ulkei* (Crotch) (Coleoptera: Erotylidae), a western extension of its range and first record from Illinois. Transactions of the Illinois State Academy of Science, 87: 171-174

GOODRICH, M.A. 1997. Scientific Note. New host records for *Haematochiton elateroides* Gorham. Coleopterists Bulletin, 51: 276.

GOODRICH, M. A. and P. E. SKELLEY. 1991a. New Synonymy in the genus *Tritoma* (Coleoptera: Erotylidae). Coleopterists Bulletin, 45: 31-36.

GOODRICH, M. A. and P. E. SKELLEY. 1991b. The Pleasing fungus beetles of Illinois (Coleoptera: Erotylidae). Part I. The Dacninae. Transactions of the Illinois State Academy of Science, 84: 155-172.

GOODRICH, M. A. and P. E. SKELLEY. 1993. The Pleasing fungus beetles of Illinois (Coleoptera: Erotylidae). Part II. Triplacinae - *Triplax* and *Ischyrus*. Transactions of the Illinois State Academy of Science, 84: 153-171.

GOODRICH, M. A. and P. E. SKELLEY. 1994. Fungal host records for species of *Tritoma* (Coleoptera: Erotylidae) of America north of Mexico. Entomological News, 105: 289-294.

GOODRICH, M.A. and P. E. SKELLEY. 1995. The Pleasing fungus beetles of Illinois (Coleoptera: Erotylidae). Part III. Triplacinae. The genus *Tritoma*. Transactions of the Illinois State Academy of Science, 88: 145-168.

GOODRICH, M. A. and P. E. SKELLEY. 1997. New synonymy in the genus *Triplax* Herbst (Coleoptera: Erotylidae), with notes on the biology of *Triplax californica* LeConte. Annales Zoologici, 47: 55-58.

GRAVES, R. C. 1965. Observations on the ecology, behavior and life cycle of the fungus-feeding beetle, *Cypherotylus californicus*, with a description of the pupa (Coleoptera: Erotylidae). Coleopterists Bulletin, 19: 117-122.

HATCH, M. H. 1962. The beetles of the Pacific Northwest. Part III: Pselaphidae and Diversicornia I. University of Washington Press. Seattle, WA. ix + 503pp.

HINTON, H. E. 1945. A Monograph of the Beetles Associated with Stored Products. Volume I. British Museum (Natural History), London. 443 pp.

LAWRENCE, J. F. 1991. Erotylidae (Cucujoidea) (Including Dacnidae). Pp. 473-475. *In:* F. W. Stehr, ed. Immature Insects, Volume 2. Kendall/Hunt. Dubuque, IA. xvi + 975 pp.

LAWRENCE, J. F., A.M. HASTINGS, M.J. DALLWITZ AND T.A. PAINE 1993. Beetle larvae of the world: Interactive identification and information retrieval for families and subfamilies. C.S.I.R.O. Information Services, East Melbourne, Victoria. CD-ROM disk and 52-page manual.

LAWRENCE, J. F. and E. B. BRITTON. 1994. Australian Beetles. Melbourne University Press. Melbourne. x + 192 pp.

LAWRENCE, J. F., A. HASTINGS, M. J. DALLWITZ, T. A. PAYNE and E. J. ZURCHER. 1999a. Beetles of the World: A key and information system for families and subfamilies. CD-ROM, Version 1.0 for MS-DOS. CSIRO Publishing, East Melbourne, Victoria.

LAWRENCE, J. F., A. HASTINGS, M. J. DALLWITZ, T. A. PAYNE and E. J. ZURCHER. 1999b. Beetle larvae of the World: Descriptions, illustrations, identification and information retrieval for families and subfamilies. CD-ROM, Version 1.0 for MS-DOS. CSIRO Publishing, East Melbourne, Victoria.

LAWRENCE, J. F. and A. F. NEWTON, Jr. 1995. Families and subfamilies of Coleoptera (with selected genera, notes, references and data on family-group names). Pp. 779-1006. *In:* J. Pakaluk and S.A. Slipinski, eds. Biology, Phylogeny and Classification of Coleoptera: Papers Celebrating the 80th Birthday of Roy A. Crowson. Muzeum i Instytut Zoologii PAN. Warsaw.

McHUGH, J. V. 2001. Description of immature stages for *Megischyrus* (Erotylidae: Triplacinae) and a review of literature on larval Erotylidae. Annales Zoologici, (in press).

McHUGH, J. V., C. J. MARSHALL and F. L. FAWCETT. 1997. A study of adult morphology in *Megalodacne heros* (Say). Transactions of the American Entomological Society, 123: 167-223.

NAVARRETE-HEREDIA, J. L. and R. NOVELO-GUTIERREZ. 2000. New distributional data and first record of gregarious behavior for *Aegithus melaspis* (Coleoptera: Erotylidae) from Mexico. Entomological News 111: 21-24.

PETERSON, A. 1951. Larvae of Insects: Coleoptera, Diptera, Neuroptera, Siphonaptera, Mecoptera, Trichoptera. Part II. Columbus, Ohio, 416 pp.

ROBERTS, A. W. RYMER. 1958. On the taxonomy of the Erotylidae (Coleoptera), with special reference to the morphological characters of the larvae. II. Transactions of the Royal Entomological Society of London, 110(8): 245-285.

SAVARY, W. E. 1995. *Dacne picta* Crotch: a recently introduced pest of stored, dried Shiitake mushrooms (Coleoptera: Erotylidae). Pan-Pacific Entomologist, 71: 87-91.

SEN GUPTA, T. and R. A. CROWSON. 1971. A review of classification of the family Languriidae (Coleoptera: Clavicornia) and the place of Languriidae in the natural system of Clavicornia. Memoirs of the Zoological Survey of India, 15: 1-42.

SKELLEY, P. E. 1988. The pleasing fungus beetles of Florida (Coleoptera: Erotylidae). Masters Thesis, University of Florida. Gainesville, FL. 172 pp., 53 figs. [unpublished]

SKELLEY, P. E. 1993. *Hirsutotriplax* Skelley, a new genus of Triplacinae, with a new species from Arizona (Coleoptera: Erotylidae). Coleopterists Bulletin, 47: 409-415.

SKELLEY, P. E. 1994. Scientific Note. New synonymy in the genus *Dacne* Latreille (Coleoptera: Erotylidae). Insecta Mundi, 8: 66.

SKELLEY, P. E. 1997. A new species of *Dacne* Latreille from Dominican amber, with a key and checklist to the known species of *Dacne* (Erotylidae: Dacnidae). Annales Zoologici, 47: 49-53.

SKELLEY, P. E. 1998. Revision of the genus *Ischyrus* Lacordaire (1842) of North and Central America (Coleoptera: Erotylidae: Tritominae). Occasional Papers of the Florida State Collection of Arthropods 9: vii + 134 pp.

SKELLEY, P. E., M. A. GOODRICH and R. A. B LESCHEN. 1991. Fungal host records for the Erotylidae (Coleoptera: Cucujoidea) of America north of Mexico. Entomological News, 102: 57-72.

88. BYTURIDAE Jacquelin du Val 1858

by Michael A. Goodrich

Family common name: The fruitworms

The plate-like lobes on the second and third tarsomeres, clavate antennae, closed procoxal cavities, and distinctive male genitalia serve to distinguish adults of this small family.

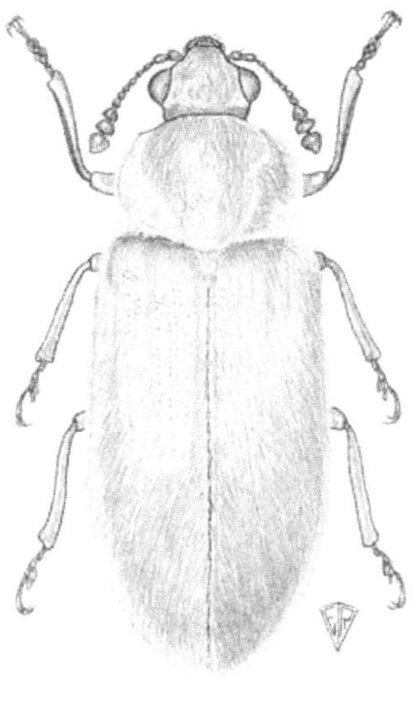

FIGURE 1.88. *Byturus unicolor* Say.

Description: Shape oblong, with a gently convex dorsum; adults of North American species 2.5-5.5 mm long; color ranges from ochraceous to piceous, some with pronotum darker than elytra; unpatterned in *Byturus,* some with elytral bands in *Xerasia;* vestiture fine and dense, with recumbent setae colored as body surface, white, or gray.

Head deflexed, inserted into prothorax to base of eyes; surface setigerously punctate. Antennae eleven-segmented, with a three-segmented club, inserted into depression between base of mandibles and eyes; antennomeres I and II each larger than antennomeres III-VIII, III to VIII progressively shorter and wider; antennomeres IX-XI forming a prominent club, oval in cross section. Labrum transverse, 2-5 times wider than long, margin entire, laterally curved to base; mandibles moderate in size, curved and apically bidentate or unidentate; maxillae well developed, with lacinia and galea both present; lacinia with medial and apical surface setose, a tiny tooth on the inner margin; galea larger than lacinia and bearing a setal brush apically; maxillary palpi four-segmented; labial palpi three-segmented; palpi variable in shape. Eyes variable in size, ovate, black to argenteous.

Pronotum wider than the head, as wide as the elytra behind, narrowed in front, sides arcuate; surface setigerously punctate as on head; pleural region broad; prosternum short between head and the coxae, with a narrow process extending between the procoxae; procoxae transverse with closed cavities and exposed trochantins. Mesosternum narrow; metasternum broad. Mesocoxae transverse; metacoxae narrowly transverse, slightly grooved to receive femora; trochanters triangular but interstitial; femora moderately swollen; tibiae slender; tarsal formula 5-5-5, the second and third tarsomeres broadly lobed beneath, the fourth tarsomere small; claws with a large basal tooth and sometimes with a bisetose empodium. Scutellum small. Elytra entire, parallel sided, tapering toward apex at meson; surface punctate, not striate; epipleural fold moderate, present on the basal one-half only. Wing venation with the anterior anal vein running into the subcubital fleck; folding pattern of the wings undescribed.

Abdomen with five visible sterna, the sutures entire; surface microrugose and finely pubescent. Male genitalia in Byturinae with aedeagus trilobate, long and slender; median lobe longer than tegmen; tegmen with parameres fused and bifid at apex both dorsally and ventrally, bearing macrotrichiae apically, basal struts well developed (Springer and Goodrich 1983, 1986, 1990; Goodrich and Springer 1988); basically similar to those of Biphyllidae (Goodrich and Springer 1992). Male genitalia in Platydascillinae (not in North America) are strikingly different from Byturinae in structure (Springer and Goodrich 1994; Goodrich and Springer 1995). Female genitalia in Byturinae have coxites divided into apical and basal sclerites and bear well developed apical styli, a valvifer supported by a sclerotized bacculum, and are generally similar to the ovipositor of Biphyllidae (Goodrich and Springer 1992). In Platydascillinae the coxites are undivided, flattened and bladelike; the structure otherwise quite variable within the subfamily (Springer and Goodrich 1994).

Mature larvae are 4-10 mm long. Body elongate, parallel-sided or slightly wider at middle; head protracted and prognathous. Generally similar to the larvae of Biphyllidae, but different in lacking an accessory ventral process on the mandible, ventral epicranial ridges, and modified 8th abdominal spiracles, and in having a projecting and setose hyaline process at the base of the mandibular mola. See Lawrence (1991) for a detailed description of the larvae.

Habits and habitats. Larvae of *Byturus unicolor* Say are known to develop in the fruit heads of species of *Rubus* and *Geum,* although adults may be found on a wide variety of other flowers. *Xerasia grisescens* (Jayne) has been reared from oak galls and from Monterey cypress. Adults are commonly taken beating oak branches.

Status of the classification. Barber (1942), Crowson (1967), and Springer and Goodrich (1983, 1986, 1990) agree that this group is closely allied to the Biphyllidae and therefore in the cucujoid series as here understood. The two North American species can be identified by the use of Springer and Goodrich (1983).

Distribution. There are 16 species known for the world (Goodrich and Springer 1995), with two of these found in the United States (Springer and Goodrich 1983). Species of the subfamily Byturinae (three genera and 10 species) are Holarctic in distribution, while the Platydascillinae (four genera and six species) are restricted to southeast Asia.

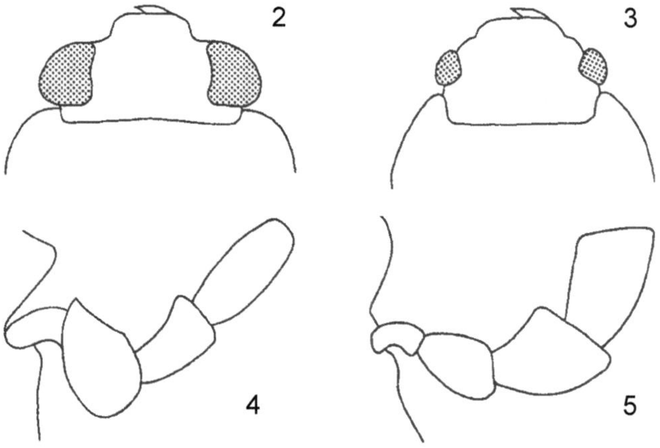

FIGURES 2.88-5.88. 2-3. Head, ventral aspect. 2, *Byturus unicolor* Say; 3, *Xerasia grisescens* (Jayne). 4-5. Maxillary palp. 4, *Byturus unicolor* Say; 5, *Xerasia grisescens* (Jayne).

Key to the Nearctic Genera

1. Eyes large, interocular distance on ventral side less than three transverse eye widths (Fig. 2); segment IV of maxillary palp 1.5 times length of segment III (Fig. 4); protibia of male with prominent tooth or swelling on medial surface at apical 1/3 or 1/4; protibial claws not cleft .... *Byturus*

— Eyes small, interocular distance on ventral side greater than four transverse eye widths (Fig. 3); segment IV of maxillary palp subequal to length of segment III (Fig. 5); protibia of male without prominent tooth or swelling; protibial claws cleft ...... *Xerasia*

Classification of the Nearctic Genera

There are two North American genera; both genera are Holarctic in distribution. The family is divided into two subfamilies, Byturinae and Platydascillinae (Springer and Goodrich 1994). Both North American genera belong to the subfamily Byturinae.

Byturidae Jacquelin du Val 1858

Byturus Latreille 1796, 1 sp., *B. unicolor* Say 1823, generally distributed in North America.

Horticola Lindemann 1865
Trixagus Seidlitz 1891 not Kugelann 1794
Terobyturus Ohta 1930

Xerasia Lewis 1895, 1 sp., *X. grisescens* (Jayne 1882), Arizona, California, Oregon and Washington.

Satorystia Reitter 1905
Byturellus Barber 1942

Bibliography

BARBER, H. S. 1942. Raspberry fruitworms and related species. United States Department of Agriculture Miscellaneous Publication Number 468: 32 pp.

CROWSON, R. A. 1967. The natural classification of the families of Coleoptera. E. W. Classey. Middlesex, England. 214 pp.

GOODRICH, M. A. and C. A. SPRINGER. 1988. A new species of *Xerasia* (Coleoptera: Byturidae) from the Middle East, with a key to the *Xerasia* of the world. Coleopterists Bulletin, 42: 345-351.

GOODRICH, M. A. and C. A. SPRINGER. 1992. A revision of the family Biphyllidae (Coleoptera) for America, north of Mexico. Coleopterists Bulletin, 46: 361-377.

GOODRICH, M. A. and C. A. SPRINGER. 1995. A new species of *Bisipinatus* Springer and Goodrich from Thailand (Coleoptera: Byturidae: Platydascillinae), with notes on the subfamily. Coleopterists Bulletin, 49: 183-190.

LAWRENCE, J.F. 1991. Coleoptera. Family Byturidae (Cucujoidea). *In:* F. W. Stehr, ed. Immature Insects, vol. 2. Kendall/Hunt. Dubuque, IA. 975 pp.

SPRINGER, C. A. and M. A. GOODRICH. 1983. A revision of the family Byturidae (Coleoptera) for North America. Coleopterists Bulletin, 37: 183-192.

SPRINGER, C. A. and M. A. GOODRICH. 1986. A revision of the family Byturidae (Coleoptera) in Europe. Coleopterists Bulletin, 40:335-352.

SPRINGER, C. A. and M. A. GOODRICH. 1990. A revision of the family Byturidae (Coleoptera) in Asia. Coleopterists Bulletin, 44: 461-483.

SPRINGER, C. A. and M. A. GOODRICH. 1994. A revision of the subfamily Platydascillinae (Coleoptera: Byturidae) from southeast Asia, with descriptions of two new genera and three new species. Coleopterists Bulletin, 48: 60-78.

89. BIPHYLLIDAE LeConte 1861

by Michael A. Goodrich

Family common name: The false skin beetles

Family synonym: Diphyllidae LeConte 1861

Adults of this family can be distinguished by the slender tarsal lobes on tarsomeres II and III, the closed procoxal cavities, the presence of lateral and/or femoral lines on the first abdominal sternite, and the structure of the male and female genitalia (Goodrich and Springer 1992).

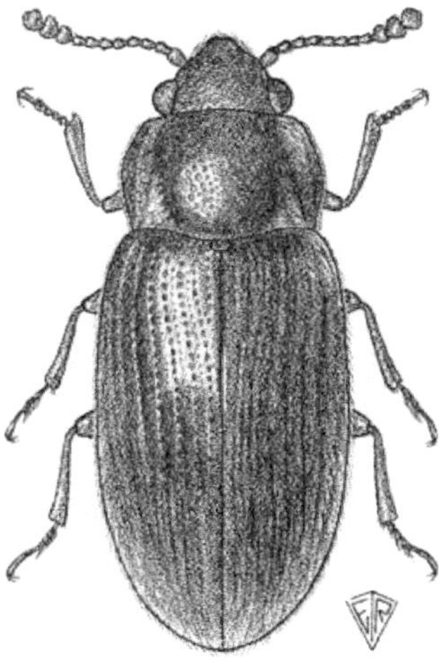

FIGURE 1.89. *Diplocoelus brunneus* LeConte

Description: Shape oval, somewhat convex; length, 2-4 mm; color testaceous; moderately to strongly pubescent, the hairs longer and more erect dorsally, shorter and recumbent ventrally.

Head prognathous, short; inserted into the prothorax to the base of the eyes; surface punctate. Antennae with 11-antennomeres, with a three-segmented club; inserted in front of the eyes; antennal grooves present between eyes and mandibular bases, each groove with a distinct lateral pore opening into a pregular pocket. Labrum transverse; mandibles moderate in size, curved and apically bidentate. Maxillae with laciniae elongate, three times as long as wide, apex rounded; long setae present on medial and apical margins; galeae wider than laciniae, twice as long as wide, with long setae as in laciniae; maxillary palpi four-segmented, slender. Labium with mentum transverse, trapezoidal; labial palpi three-segmented, slender. Eyes large and globose; coarsely faceted.

Pronotum transverse; procoxal cavities closed behind; trochantins concealed. Mesocoxal cavities open laterally; mesepimeron reaching middle coxal cavities. Exposed portion of scutellum transverse, diverging posteriorly. Legs with procoxae oval, transverse; mesocoxae rounded; metacoxae transverse, slightly grooved to receive femora; all coxae separate; femora slender to moderately robust; tibiae slender; tarsal formula 5-5-5, tarsomeres slender, the fourth small, the fifth as long or longer than all other tarsomeres taken together; second and third tarsomeres bearing slender pubescent lobes. Elytra distinctly punctate-striate. Abdomen with five visible sternites, the first sternite with distinct femoral and/or lateral lines.

Male genitalia similar to those of Byturidae, consisting of: a cylindrical tegmen with well-developed basal struts; parameres frequently divided by dorsal and ventral separations; and a relatively long and slender median lobe within the tegmen, also bearing struts. Female genitalia similar to Byturidae, but with a more elongate ovipositor, 3-6 times longer than wide. The ovipositor consists of: coxites with apical and basal sclerites; a subapical stylus with a long seta subequal in length to the stylus; a narrow vulvar lobe ventrad of the coxites; a membranous valvifer, supported ventrally for its entire length by two sclerotized baccular rods that curve laterally and anteriorly at base of valvifer to produce short extensions on the dorsal side (Goodrich and Springer 1992).

The larvae are similar to those of Byturidae, but differ in having an accessory process on the mandible, a basal hyaline mandibular lobe, well-developed ventral epicranial ridges, and enlarged, posteriorly oriented 8th spiracles, which may be raised on short tubes (Lawrence 1991).

Habits and habitats. These species live under bark of dead trees or fallen branches, where they apparently feed on fungi and their spores. Larvae of *Anchorius lineatus* Casey have been found under the fermenting bark of mesquite (*Prosopis*) in Arizona. Adults of *Anchorius* come strongly to light, but *Diplocoelus* spp. less strongly.

Status of the classification. This family has been placed within the Clavicornia (Crowson 1967) and is probably most closely related to the family Byturidae (Falcoz 1925; Goodrich and Springer 1992).

Distribution. There were 195 described species in 1934 (Schenkling 1934), mostly described from the world tropics, but found in all the major zoogeographic regions. A number of species have been described since 1934 for Africa, but since species level revisions have not been carried out in any zoogeographic region but the Nearctic, the number of species actually present worldwide is uncertain.

Key to the Nearctic Genera

1. Pronotum with 10 longitudinal ridges; ventral surface of first abdominal segment with strong lateral lines running from middle of hind coxal cavities to the posterior angles; femoral lines weak or lacking (Fig. 2) .... *Anchorius*

— Pronotal ridges weakly developed, restricted to lateral portions of prothorax or lacking; ventral surface of first abdominal segment without lateral lines as above; strong femoral lines running from ante-

2 3

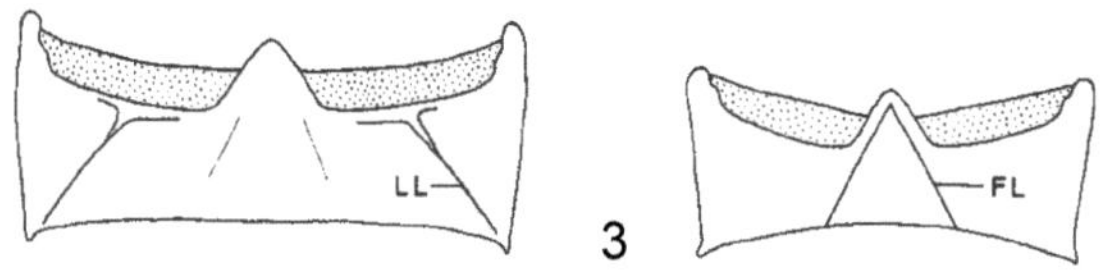

FIGURES 2.89-3.89. First abdominal sternite. 2. *Anchorius lineatus* Casey. 3. *Diplocoelus brunneus* LeConte (LL=lateral line; FL=femoral line).

rior intercoxal process posteriorly, forming a triangle (Fig. 3) .... *Diplocoelus*

Classification of the Nearctic Genera

Biphyllidae LeConte 1861

Anchorius Casey 1900, 1 sp., *A. lineatus* Casey 1900, Arizona and California.

Diplocoelus Guérin-Menéville 1838, 2 spp., widely distributed in North America east of the 100th Meridian.
Marginus LeConte 1862

Bibliography

CROWSON, R. A. 1967. The natural classification of the families of Coleoptera. E. W. Classey. Middlesex, England. 214 pp.

FALCOZ, L. 1925. Position systématique des genres *Diphyllus* Stephens and *Diplocoelus* Guérin. Encyclopedié de Entomologique, 1: 69-74.

GOODRICH, M. A. and C. A. SPRINGER. 1992. A revision of the family Biphyllidae (Coleoptera) for America, north of Mexico. Coleopterists Bulletin, 46: 361-377.

LAWRENCE, J. F. 1991. Coleoptera. Family Biphyllidae (Cucujoidea). *In:* F.W. Stehr, ed. Immature Insects, vol. 2. Kendall/Hunt. Dubuque, IA. 975 pp.

LECONTE, J. L. 1861. Classification of the Coleoptera of North America. Smithsonian Miscellaneous Collections, 3(136): 1-208.

SCHENKLING, S. 1934. Biphyllidae. Coleopterorum Catalogus 15(133): 1-7.

90. BOTHRIDERIDAE Erichson 1845

by T. Keith Philips and Michael A. Ivie

Family common name: The bothriderid beetles

Family synonyms: Anommatidae Ganglbauer 1899

The hetermeroid or greatly reduced trochanter, exposed antennal insertions, distinct antennal club, 4-4-4 tarsi, intercoxal process of the prosternum that is either obsolete, narrow or broad but expanded behind coxae, broad intercoxal process of ventrite I, five free ventrites and a body length at least 2.75 times the greatest width is the surest combination to confirm membership in this group. The one exception to this is a single introduced species that has 3-3-3 tarsi, lacks eyes and has nearly contiguous procoxal cavities. Most members of this family can be recognized by their various oblong or narrow shape, subcylindrical or slightly flattened body, antennae with a 1- to 2-segmented club, and superficial resemblance to some colydiids. They can be separated from the latter by the presence of exposed antennal insertions. The other group often confused with bothriderids are the Cerylonidae. The cerylonids never have a heteromeroid trochanter, are more ovoid, have a longer first sternite, and often a crenulate posterior face of the last ventrite.

FIGURE 1.90. *Bothrideres geminatus* (Say) (from Stephan 1989)

Description: Adult length 1.5-13 mm. Shape oblong to narrowly elongate, sub-cylindrical to moderately flattened; generally subglabrous, sometimes with decumbent and erect hairs, rarely with scales (*Dastarcus*). Antennae relatively short, 10-11 antennomeres, 1- to 2-segmented club in most, weakly 3-segmented in *Deretaphrus*; antennal insertions exposed. Frontoclypeal suture present in most. Front coxae globular to projecting, trochantins concealed, front coxal cavities externally open to closed, internally open; middle coxae narrowly to very widely separated, lateral closure closed or rarely open. Prosternal intercoxal process broad, except nearly contiguous in some, metasternal intercoxal process of ventrite I broad. Tibial apices expanded and spinous, some with enlarged and unequal tibial spurs; trochanterofemoral articulation strongly oblique, part of femur in contact with coxa (heteromeroid type), most with trochanter reduced and concealed by femoral base; tarsomeres 4-4-4, occasionally 3-3-3. Abdomen with five ventrites, none are connate. Aedeagus cucujoid or ring-type.

Free-living larvae (Teredinae) (modified from Lawrence *et al.* 1999b): length of mature larvae 3-4 mm. Shape elongate, approximately parallel-sided, moderate to strongly flattened; dorsally moderately pigmented and granulate-tuberculate; covered with long and short simple setae.

Head prognathous and protracted, broad and slightly flattened. Epicranial stem very short; frontal arms lyriform. Median endocarina absent. Five 5 stemmata on each side, some with an indistinct sixth present. Antennae 3-segmented, short. Frontoclypeal suture absent; labrum free. Ventral mouthparts retracted. Mandibles symmetrical, bidentate, with 2 or more teeth along edge of incisor, large accessory ventral process present. Mola well-developed and asperate; prostheca absent. Maxilla with 3-segmented palp, mala falciform, cardo transverse, stipes elongate, articulating area well-developed. Labium approximately free to base of mentum, ligula broad and short; labial palps 2-segmented, widely separated. Hypopharyngeal sclerome well-developed with characteristic pair of anterior horns. Hypostomal rods long, diverging. Ventral epicranial ridges absent. Gula transverse.

Thorax with moderately well-developed, widely separated legs; tarsugulus with 1 seta. Thoracic and abdominal terga (1-9) paramedially with two sharp tubercles, also smaller scattered tubercles and granules. Thoracic terga slightly expanded at sides; abdominal terga 1-8 even more laterally expanded and forming processes which are somewhat posteriorly directed.

Abdominal tergum A9 with well developed, widely separated urogomphi, each with a mesal and lateral accessory process at base; anterior edge (tergum A9) with a transverse row of internal processes visible under transmitted light. Sternum A9 simple, well-developed. Segment A10 oval, posteroventrally oriented. Spiracles annular biforous, with long tube-like accessory openings; those on A8 slightly larger than others.

First instar larva differs with a broader and flatter, smoother body; dorsal surface more heavily pigmented, some pleural and sternal sclerites pigmented. Vestiture includes modified setae with expanded tips (probably glandular). Thoracic and abdominal (1-8) lateral tergal processes more developed, forming flattened plates. Paramedian tubercles on thorax and abdomen reduced, except tergum A9. Urogomphi accessory basal processes reduced. Spiracles located at ends of longer tubes.

Ectoparasitic larva (Bothriderinae) (modified from Lawrence *et al.* 1999b, Roberts 1980): Length of mature larvae 3-15 mm; body elongate, subcylindrical but abdomen enlarged, slightly sclerotized, smooth; covered with short scattered setae.

Head prognathous and protracted, small, slightly flattened; epicranial stem absent, frontal arms sometimes v-shaped but usually indistinct or absent; median endocarina absent; stemmata lacking or a single pair. Antennae very short, usually 2-segmented, some only 1-segment, long sensorium on segment 1. Frontoclypeal suture absent, labrum usually free. Ventral mouthparts retracted. Mandibles symmetrical, uni- or bidentate and lacking mola and accessory ventral process. Maxilla with 2-segmented palp, palpifer sometimes distinct; usually with transverse or oblique cardo; stipes slightly elongate; mala narrowly rounded, blunt; in *Sosylus*, cardo, stipes, articulating areas indistinct, and palp and mala sometimes lacking, although palp sometimes visible as minute papilla. Labium usually free to mentum base, ligula longer than 2- segmented palps; in *Sosylus*, labium may not be subdivided, ligula and palps are absent, although the latter may be visible as papillae. Hypopharyngeal sclerome and hypostomal rods absent. Ventral epicranial ridges absent. Gula longer than wide.

Thorax short and narrower than abdomen; in *Sosylus* strongly narrowed between meso- and metathorax. Legs 5-segmented, usually reduced, widely separated; tarsungulus usually with 1 seta; *Sosylus* lacking legs.

Abdominal tergum A9 usually much shorter than A8, urogomphi absent or short, strongly upturned; in *Sosylus*, A9 not reduced and lacking urogomphi; sternum A9 well developed; segment A10 circular, orientation posterior or posteroventral. Spiracles annular, not raised on tubes, usually with strongly sclerotized peritreme; in *Sosylus*, absent on thorax and A8.

Triungulin larvae: (*Sosylus*) length: 1-1.5 mm. Body fusiform, elongate, very flat, covered with stout spines and more heavily sclerotized than later instars. Head strongly transverse, labrum fused to head capsule. Single, well developed stemmata on each side. Antennae 2-segmented, segment 2 much longer than segment 1, with long apical seta, and longer than sensorium. Mandibles narrow and falcate, curved and sickle-shaped, lacking mola. Ventral mouthparts retracted. Maxilla with 2-segmented palp, cardo longitudinally oblique, stipes elongate, with indistinct articulation area; mala broadly rounded. Labium with mentum and submentum fused, palps 2-segmented. Legs 5-segmented, well developed, tibiae long and narrow. Segment A9 slightly shorter than A8, lacking urogomphi, pair of long setae at apex. A10 circular and posteriorly oriented. Spiracles annular, present on abdominal segments 1-9 and mesothorax.

Habits and Habitats: This family includes members that are both ectoparasites and fungal feeders. Many species are found in the galleries and tunnels of wood-boring beetles where they are ectoparasites of larvae and pupae. Wood-borers attacked include species of anobiids, bostrichids, mycterids (Ivie, pers. obs.), cerambycids, platypodines, scolytines and other curculionids (Browne 1962, Lawrence 1985, Lawrence 1991, Lieu 1944, Piel 1938, Rasmussen 1967, Roberts 1968, Schedl 1962). More specifically, our North American *Bothrideres* species attack larvae and pupae of the buprestid *Chrysobothris*, the cerambycid *Elaphidionoides*, and a large desert weevil, *Cactophagus validus* LeConte, that feeds on cactus (Stephan 1989).

Species in the genus *Sosylus* prey upon platypodine larvae (Browne 1962) and are hypermetamorphic with active first instar triungulins and subsequent instars modified for an ectoparasitic lifestyle. There is one record of *Dastarcus* larvae parasiting carpenter bee larvae (*Xylocopa*) in Asia (Lieu 1944). *Dastarcus helophoroides* Fairmaire is currently being examined for use as a biological control agent of woodborers in Japan and China via mass rearing (Ogura *et al.* 1999). Some species (*Sosylus*) are known to construct a waxen scale-like pupal chamber while others (Bothriderini, *Dastarcus* and *Deretaphrus*) spin a silken cocoon (Lawrence 1991).

The non-parasitic groups include species of Teredinae, which probably feed on fungi, including those cultivated by ambrosia beetles, within tunnels in wood (Lawrence 1985). Members of the Anommatinae are found in the soil, leaf litter, and subterranean wood while the Xylariophilinae feed on the fruiting bodies of pyrenomycetous fungi (Pal and Lawrence 1986). None of these species are known to construct cocoons.

Status of the classification: The history of this group goes back to a time when what is now the Colydiidae, Cerylonidae and Bothrideridae were all in a single family (Hetschko 1930). Using larval characters, Craighead (1920) and Bøving and Craighead (1931) first proposed family status for the Bothrideridae, but it did not gain general acceptance for another 50 to 60 years, and the composite Colydiidae continued to be used. Crowson (1955) removed the Cerylonidae, but commented on the problems of both the Anommatinae and Bothriderinae remaining in the Colydiidae. Arnett (1973) rejected the split of the Colydiidae, and returned the Cerylonidae to that family. Sen Gupta and Crowson (1973) revised the Cerylonidae and added the Anommatini as a tribe of the Euxestinae. Dajoz (1977), who had treated the Anommatinae as Colydiidae (Dajoz 1968), later promoted the Anommatinae as a separate family. In most of these classifications, the remaining Bothrideridae were included as an awkward subfamily of the Colydiidae. Lawrence (1980) made a clear case for moving the Bothriderids from the Tenebrionoidea, where the Colydiidae belong, to the Cucujoidea, but did not act. After years of these authors and others discussing that this group did not fit in either the Colydiidae or the Cerylonidae, Lawrence (1985) seems to be the first to actually return the Bothrideridae to a full family status. Pal and Lawrence (1986) followed up quickly, defining the Bothrideridae as an independent family, with the four subfamilies Bothriderinae, Teredinae, Xylariophilinae and Anommatinae. They discussed problems diagnosing the group, and admitted that it still might not be monophyletic. Lawrence (1991) further supported the status of this family though, based on larval characters. Ivie and Slipinski (1990) listed all genera moved to the family dating back to Hetschko's (1930) Colydiidae. The placement of the Anommatinae has gone between the Bothrideridae (Pal and Lawrence 1986), Cerylonidae (Lawrence and Stephan 1975) and Anommatidae (Dajoz 1977) over the

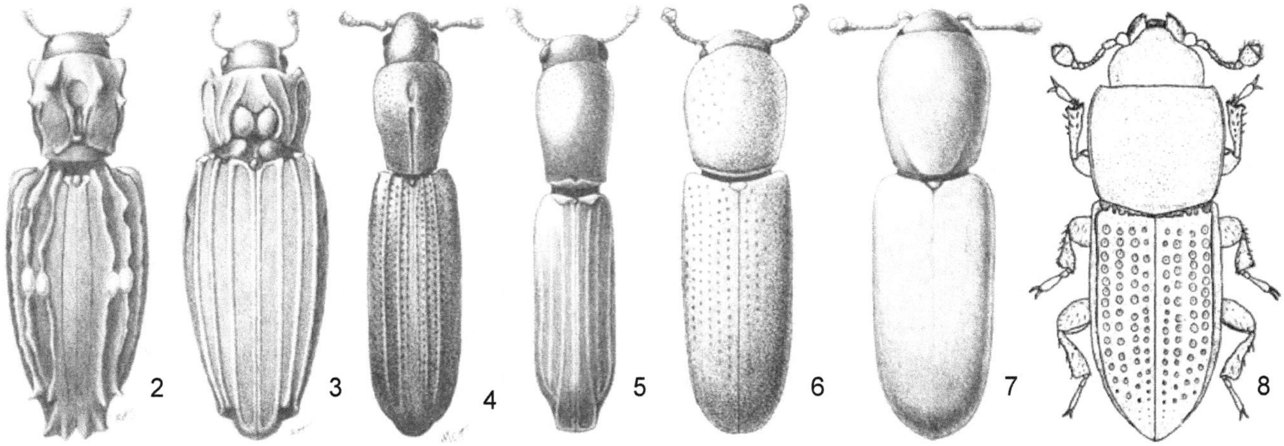

FIGURES 2.90-8.90. 2. *Lithophorus ornatus* Arrow; 3. *Prolyctus exaratus* (Melsheimer); 4. *Deretaphrus oregonensis* Horn; 5. *Sosylus costatus* LeConte; 6. *Oxylaemus americanus* LeConte; 7. *Rustleria obscura* Stephan; 8. *Anommatus duodecimstriatus* (Müller 1821) (Figures 2-7 from Stephan 1989; Figure 8 from Burakowski and Slipinski 1986, reproduced with permission)

years. Stephan (1989) in his treatment of the North American Bothrideridae excluded the Anommatinae.

Slipinski (1990) monographed and characterized the world Cerylonidae, compared them to related families, and placed the Anommatinae within the Bothrideridae. He pointed out the heterogeneous nature of the bothriderids as a whole. Of particular concern were several overlapping symplesiomorphies between the bothriderids and Euxestinae (Cerylonidae), especially *Metacerylon* Grouvelle 1906. However, since that time, most major works have continued to view the family as proposed by Pal and Lawrence (1986), but some still split the group between the Cerylonidae and Colydiidae (Downie and Arnett 1996).

Distribution. Worldwide, this group contains about 35 genera and about 300 species (Lawrence 1991). Our two native North American subfamilies, the Teredinae and Bothriderinae have representatives worldwide. The Teredinae Seidlitz 1888, with the tribes Teredini, Sosylopsini Dajoz 1980 and Sysolini Slipinski and Pal 1985, occur throughout most of the world. Slipinski and Pal (1985) give a key to these tribes. The Teredini encompasses four genera only one of which, *Oxylaemus*, occurs in the New world. The others are widespread in the Old World (Lawrence *et al.* 1999a). Sosylopsini includes a single genus with about 50 species from Africa, Australia, the Pacific and the Neotropics (Dajoz 1980). The Oriental Sosylini are also monogeneric, with a single small genus revised by Slipinski and Pal (1985).

The ectoparasitic Bothriderinae are divided into two tribes, the Deretaphrini Horn 1878 and Bothriderini. A key to these tribes is provided by Slipinski and Pal (1985). The Bothriderini are generally distributed around the globe, and include 21 genera which were reviewed by Slipinski *et al.* (1989). The Deretaphrini include 5 genera (Lawrence *et al.* 1999a), 2 of which reach North America. Heinze (1943) is the most comprehensive treatment of this tribe.

Of our introduced Anommatinae, one genus is widespread in the western Palearctic while the second is restricted to the Mediterranean region. One species, including our North American representative, is now widespread via human commerce. Dajoz (1977) provides the most recent treatment of this subfamily. The monogeneric Xylariophilinae Pal and Lawrence 1986, with three species, are found from India through to Australia and Melanesia (Lawrence and Newton 1995).

The bothriderids have been included in many regional treatments of colydiids, and taxonomic resources can be found therein. See Dajoz (1977, 1980) for the Palearctic and Malagasy faunas, Pope (1961) for Africa, and Stephan (1989) and Lawrence and Stephan (1975) for the Nearctic species. More regional works include Hatch (1961) and Downie and Arnett (1996) for the Northwest and Northeast Nearctic respectively. A key to the 4 world subfamilies is provided by Pal and Lawrence (1986).

Key to the Nearctic genera

1. Eyes absent, length less than 2 mm (Fig. 8) .... *Anommatus*
— Eyes present, length more than 2 mm 2

2(1). Procoxae contiguous or nearly so 3
— Procoxae separated by half or greater than width of one procoxa 4

3(2). Trochanters small and indistinct 6
— Trochanters large and distinct 7

4(2). Elytral costae interrupted or notched; each elytron with a pair of yellow callosities at or near middle (Fig. 2) .... *Lithophorus*
— Elytral costae uninterrupted and even; elytra lacking callosities 5

5(4). First ventrite with distinct costal lines which extend over half its length and usually nearly reach the posterior margin (Fig. 3) .... *Prolyctus*
— First ventrite with coxal lines at most vaguely indicated (Fig. 1) .... *Bothrideres*

6(3). Length greater than 9.0 mm; procoxae slightly separated (Fig. 4) .... *Deretaphrus*

— Length less than 6.0 mm; procoxae contiguous (Fig. 5) .... *Sosylus*

7(3). Elytral punctures in rows and coarse (Fig. 6) .... *Oxylaemus*
— Elytral punctures scattered and very fine (Fig. 7) . ... *Rustleria*

Classification of the Nearctic Genera

Bothrideridae Erichson 1845

Bothriderinae Erichson 1845

Bothriderini Erichson 1845

Bothrideres Erichson 1845, 8 spp., eastern U.S. to Colorado and Arizona, and extending to Guatemala.

Lithophorus Sharp 1894, 1 sp., *L. ornatus* Arrow 1909, southern Texas to Guatemala.

Prolyctus Zimmerman 1869, 1 sp., *P. exaratus* (Melsheimer 1846), western Pennsylvania to northern Alabama, west to Texas and Oklahoma; under dead elm bark and dead pine bark.
Machlotes Horn 1878

Deretaphrini Horn 1878

Deretaphrus Newman 1842, 1 sp., *D. oregonensis* Horn 1872, British Columbia, Idaho, Montana (Montana State Entomology Collection), Oregon, and Washington.
Sigerpes Germar 1848

Sosylus Erichson 1845, 3 spp., southern U.S., Baja California, and the Virgin Islands.
Pleuridium LeConte 1861
Pycnocephalus Kraatz 1895
Cephalopycnus Arrow 1909

Teredinae Seidlitz 1888

Oxylaemus Erichson 1845, 2 spp., northern California through to British Columbia, and New Jersey to Florida, west to eastern Oklahoma.
Redistes Casey 1924

Rustleria Stephan 1989, 1 sp., *R. obscura* Stephan 1989, southeast Arizona.

Anommatinae Ganglbauer 1899

Anommatus Wesmael 1835, 1 sp., *A. duodecimstriatus* (Müller 1821), recorded from Illinois, New York, Ohio, Wisconsin (Lawrence and Stephan 1975, Downie and Arnett 1996). Introduced from Europe (Cooper 1962, Peck 1972).

Bibliography

ARNETT, R. H. 1973. Beetles of the United States (A Manual for Identification). American Entomological Institute. Ann Arbor, MI. 1112 pp.

BØVING, A. G. and F. C. CRAIGHEAD. 1931. An illustrated synopsis of the principal larval forms of the order Coleoptera. Entomologica Americana (New Series), 1: 1-351.

BROWNE, F.G. 1962. *Sosylus spectabilis* Grouvelle (Coleoptera, Colydiidae), a predator and parasite of African ambrosia beetles. Report of the West African Timber Borer Research Unit 5 (1961-62): 91-96.

BURAKOWSKI, B. and S. A. SLIPINSKI. 1986. Gwozdnikowate – Colydiidae, Bothrideridae, Cerylidae, Anommatidae. *In*: Klucze do Oznaczania Owadów Polski. Czesc XIX Chrzaszcze – Coleoptera (Zeszyt 59). Polskie Towarzystwo Entomologiczne, Warszawa-Wroclaw. 86 pp.

COOPER, K.W. 1962. A remarkable anopthalmic immigrant to North America: *Anommatus duodecimstriatus* Müller (Coleoptera: Colydiidae). Entomological News, 73: 187-190.

CRAIGHEAD, F.C. 1920. Biology of some Coleoptera of the families Colydiidae and Bothrideridae. Proceedings of the Entomological Society of Washington, 22: 1-13.

CROWSON, R. A. 1955. The natural classification of the families of Coleoptera. Nathaniel Lloyd. London, 187 pp.

DAJOZ, R. 1968. Révision des Colydiidae anophthalmes de la faune paléarctique (Col.) IV. Etudes sur les genres *Anommatus* et *Langelandia*. Annales de la Société entomologique de France (N.S.), 4: 975-988.

DAJOZ, R. 1977. Coléoptères Colydiidae et Anommatidae Paléarctiques. Faune de l'Europe et du Bassin Méditerranéen. Masson, Paris. 8: i-vi, 1- 280.

DAJOZ, R. 1980. Insectes Coléoptères: Colydiidae et Cerylonidae. Faune de Madagascar, 54: 1- 256.

DOWNIE, N. M. and R. H. ARNETT, Jr. 1996. The Beetles of Northeastern North America. Vol. 2. The Sandhill Crane Press. Gainesville, FL.

HATCH, M. H. 1961. The beetles of the Pacific Northwest. Part III: Pselaphidae and Diversicornia I. University of Washington Publications in Biology, 16: 1-503.

HEINZE, E. 1943. Studien zur Kenntnis der Tribus Deretaphrini und deren Stellung im System (Colydiidae). Entomologische Blätter, 39: 85-93, 97-124.

HETSCHKO, A. 1930. Colydiidae. *In*: W. Junk and S. Schenkling, eds., Coleopterorum Catalogus, Pars 107. W. Junk, Berlin. 124 pp.

IVIE, M. A. and S. A. SLIPINSKI. 1990. Catalog of the genera of world Colydiidae (Coleoptera). Annales Zoologici, 43 (Suppl. 1): 1-32.

LAWRENCE, J. F. 1980. A new genus of Indo-Australian Gempylodini with notes on the constitution of the Colydiidae (Coleoptera). Journal of the Australian Entomological Society, 19: 293–310.

LAWRENCE, J. F. 1985. The genus *Teredolaemus* Sharp (Coleoptera: Bothrideridae) in Australia. Journal of the Australian Entomological Society, 24: 205-206.

LAWRENCE, J. F. 1991. Bothrideridae (Cucujoidea), Pp. 477-479. *In*: F. W. Stehr, ed., Immature Insects. Vol. 2. Kendall/Hunt. Dubuque, Iowa.

LAWRENCE, J. F., A. M. HASTINGS, M. J. DALLWITZ, T. A. PAINE and E. J. ZURCHER. 1999a. Beetles of the World: A Key and Information System for Families and Subfamilies. CD-ROM, Version 1.0 for MS-Windows. CSIRO Publishing. Melbourne.

LAWRENCE, J. F., A. M. HASTINGS, M. J. DALLWITZ, T. A. PAINE and E. J. ZURCHER 1999b. Beetle Larvae of the World: Descriptions, Illustrations, Identification, and Information Retrieval for Families and Sub-families. CD-ROM, Version 1.1 for MS-Windows. CSIRO Publishing: Melbourne.

LAWRENCE. J. F. and A. F. NEWTON, Jr. 1995. Families and subfamilies of Coleoptera (with selected genera, notes, references and data on family-group names). Pp 779-1006. *In*: J. Pakaluk and S. A. Slipinski, eds., Biology, phylogeny and classification of Coleoptera: Papers celebrating the 80th birthday of Roy A. Crowson. Muzeum I Instytut Zoologii PAN, Warsaw.

LAWRENCE, J. F. and K. STEPHAN. 1975. The North American Cerylonidae (Coleoptera: Clavicornia). Psyche, 82: 131-166.

LIEU, K.O.V. 1944. A preliminary note on the colydiid parasite of the willow-branch cerambycid. Indian Journal of Entomology, 6: 125-128.

OGURA, N., K. TABATA and W. WANG. 1999. Rearing of the colydiid beetle predator, *Dastarcus helophoroides*, on artificial diet. BioControl, 44: 291-299.

PAL, T.K. and J.F. LAWRENCE. 1986. A new genus and subfamily of mycophagous Bothrideridae (Coleoptera: Cucujoidea) from the Indo-Australian Region, with notes on related families. Journal of the Australian Entomological Society, 25: 185-210.

PECK, S. B. 1972. The eyeless european soil colydiid, *Anommatus duodecimstriatus*, in North America (Coleoptera: Colydiidae). Coleopterists Bulletin, 26: 19-20

PIEL, O. 1938. Note sur le parasitisme de *Dastracus helophoroides* Fairmaire (Coléoptère: Colydiidae). Notes d'entomologie Chinoise, 4: 1-15, pls. 1-2.

POPE, R. D. 1961. Exploration du Parc National de la Garamba. Mission H. de Saeger. Fascicule 25. Colydiidae (Coleoptera Clavicornia). Institut des Parcs Nationaux du Congo et du Ruanda-Urundi, Bruxelles. 115 pp.

RASMUSSEN, S. 1967. *Hylotrupes* (Col., Cerambycidae) in dead trees on Farön, a Swedish Island. Entomologiske Meddelelser, 35: 223-226.

ROBERTS, H. 1968. Notes on the biology of ambrosia beetles of the genus *Trachyostus* Schedl (Coleoptera: Platypodidae) in West Africa. Bulletin of Entomological Research 58: 325-352.

ROBERTS, H. 1980. Description of the developmental stages of *Sosylus* spp. (Coleoptera: Colydiidae) from New Guinea, parasites and predators of ambrosia beetles (Coleoptera: Platypodidae). Bulletin of Entomological Research, 70: 245-252.

SCHEDL, K.E. 1962. Forstentomologie Beiträge aus dem Kongo. Räuber und Kommensalen. Entomologische Abhandlungen und Berichte aus dem staatlichen Museum für Tierkunde in Dresden, 28: 37-84.

SEN GUPTA, T. and R. A. CROWSON. 1973. A review of the classification of Cerylonidae (Coleoptera, Clavicornia). Transactions of the Royal Entomological Society of London, 124: 365-446.

SLIPINSKI, S. A. 1990. A monograph of the world Cerylonidae (Coleoptera; Cucujoidea). Part I - Introduction and higher classification. Annali del Museo Civico di Storia Naturale "G. Doria", 88: 1-273.

SLIPINSKI, S. A. and T. PAL. 1985. Sysolini - new tribe of Bothriderinae (Coleoptera, Colydiidae), with a description of a new species of *Sysolus* from Viet-Nam. Polskie Pismo Entomologiczne, 55: 39-44.

SLIPINSKI, S. A., R. D. POPE and R. J. W. ALDRIDGE. 1989. A review of the world Bothriderini (Coleoptera: Bothrideridae). Polskie Pismo Entomologiczne, 59: 131-202.

STEPHAN, K. 1989. The Bothrideridae and Colydiidae of America north of Mexico (Coleoptera: Clavicornia and Heteromera). Occasional Papers of the Florida State Collection of Arthropods, 6: 1-65.

91. CERYLONIDAE Billberg 1820

by Michael C. Thomas

Family common name: The minute bark beetles

Family synonyms: Aculagnathidae Oke 1932; Dolosidae Dajoz 1963; Euxestidae Grouvelle 1908; Murmidiidae Jacquelin du Val 1858; Pleosomidae Fauvel 1891

These small to minute, mostly smooth and shining beetles are almost unique in the piercing-sucking mouthparts possessed by most larvae and some adults. They were traditionally included in the Colydiidae but are abundantly distinct from members of that family.

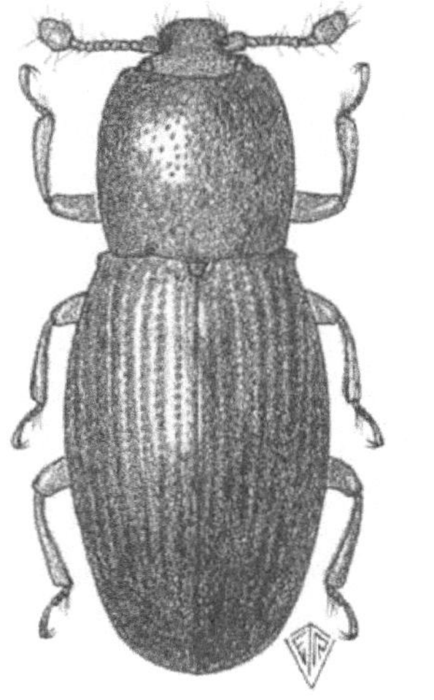

FIGURE 1.91. *Cerylon castaneus* Say

Description: Oblong-ovate to ovate, rather broad; length 1-5 mm (most less than 2 mm); mostly brownish or blackish; surface smooth and shining in most, pubescence inconspicuous or absent.

Head small, deeply inserted into prothorax in most; surface punctate, stridulatory file on vertex absent; antennae rather short, inserted laterally, insertion exposed; with less than 11 antennomeres in most; club mostly of less than three antennomeres; labrum unmodified or elongate and pointed in piercing forms; mandibles with two or three apical teeth, a prosthecal fringe, and a well-developed mola, or much modified and blade-like; lacinia and galea unmodified or elongate and blade-like in piercing forms; maxillary and labial palpi geniculate, terminal palpomere in many small or aciculate.

Pronotum transverse to quadrate, many with a pair of basal impressions; anterior angles or hypomera in some with antennal cavities; procoxal cavities: open or closed behind, moderately to broadly separated; trochantin hidden; mesocoxal cavities closed laterally by sterna; moderately to widely separated; metasternum in some with femoral lines; metendosternite with anterior tendons in almost all widely separated; anterior coxae small, rounded; middle coxae mostly small, rounded; hind coxae transverse; trochanters of normal clavicorn type or slightly heteromeroid; femora short, stout; tibiae short and rather stout, apical spurs mostly present; tarsal formula usually 4-4-4, rarely 3-3-3; tarsi mostly simple, rarely with tarsomere 1 lobed; claws simple.

Elytra with epipleural fold distinct and complete; mostly striate or with longitudinal lines of punctures; scutellary striole absent; wing without closed radial cell; if more than one anal vein present the front vein runs into the subcubital fleck.

Abdomen with 5 free sterna, the first visible one much longer than the second and in many with femoral lines. Male genitalia resting on one side in abdomen; tegmen complete or incomplete, with or without parameres, median lobe rarely with basal struts. Female with spiculum gastrale on abdominal segment 8; ovipositor short, in many with distinct valvifers, coxites, and styli.

Larva more or less flattened, elongate to oval; head inclined to strongly hypognathous; well-marked frontal sutures absent; two or three pairs of stemmata, if present; antennae three-segmented, short; sensory appendage of segment 2 longer than segment 3; mouthparts unmodified or modified for piercing; mandibles in most without distinct prostheca; prothorax in many large and covering head; legs with single tarsungular seta; body segments in many expanded laterally, clothed with modified setae in many; abdomen with segment 9 short, with or without urogomphi; pregomphal processes absent, pygopod short, simple; spiracles annular and not on stalks (adult and larval descriptions adapted from Sen Gupta and Crowson 1973).

Habits and habitats. Adults and larvae are found under bark and in leaf litter, where they apparently feed on fungi.

Status of the classification. Lawrence and Stephan (1975) revised the United States species, and the family has been revised at the generic level (Slipinski 1990; Sen Gupta and Crowson 1973) for the world.

Distribution. The family is represented throughout the world but is most diverse in the tropics. There are 10 genera and 19 species in the U.S., and about 52 genera and more than 300 described species worldwide (Slipinski 1990).

KEY TO THE NEARCTIC GENERA
(modified from Lawrence and Stephan 1975)

1. Frontoclypeal suture present; terminal maxillary palpomere at least twice as long as and more than half as wide as palpomere 3, which is shorter than or subequal to palpomere 2; last ventrite not or very finely crenulate 2

— Frontoclypeal suture absent; terminal maxillary palpomere aciculate, shorter than and less than half as wide as palpomere 3, which is longer than palpomere 2; apex of last ventrite always strongly crenulate (Ceryloninae) 7

2(1). Procoxal cavities closed behind; length at least 2 mm; body oval and highly convex, with smooth, even, lateral edges and no antennal cavities (Euxestinae) 3

— Procoxal cavities open behind; length less than 1.5 mm; without other characters in combination 4

3(2). Tarsi not lobed; penultimate antennomere asymmetrical, almost as wide as ultimate; length 2.2 - 2.6 mm .... *Hypodacne*

— Tarsomere 1 lobed; penultimate antennomere symmetrical, much narrower than ultimate; length 1.8 - 2.0 mm .... *Euxestus*

4(2). Prothorax without antennal cavities; lateral edges of pronotum serrate; metasternum and first ventrite without femoral lines; tarsi with 3 tarsomeres; body more elongate, more than 1.75X width across elytra (Ostomopsinae) .... *Ostomopsis*

— Prothorax with antennal cavities; lateral edges of pronotum smooth; metasternum and first ventrite with femoral lines; tarsi with 4 tarsomeres; body round or oval in outline, less than 1.75X width across elytra (Murmidiinae) 5

5(4). Antennal cavities of prothorax dorsal, easily visible from above; antennae with 10 antennomeres; body strongly convex and elytral punctation striate .... *Murmidius*

— Antennal cavities of prothorax ventral, not visible from above; antennae with less than 10 antennomeres; body flattened or elytral punctation confused 6

6(5). Body distinctly flattened; antennae with 9 antennomeres; elytral punctation seriate; mesosternum truncate anteriorly; metasternal suture present .... *Mychocerinus*

— Body strongly convex; antennae with 8 antennomeres; elytral punctation confused; mesosternum rounded anteriorly; metasternal suture absent ...... *Botrodus*

7(1). Prothorax with large, ventral antennal cavities situated laterally and formed primarily from the hypomera; metasternum and first ventrite with femoral lines; antennae with 8 antennomeres .... *Mychocerus*

— Prothorax without antennal cavities, or with smaller, mesal cavities formed entirely from the prosternum; metasternum and abdomen without femoral lines; antennae with 10 or 11 antennomeres 8

8(7). Procoxal cavities open behind or narrowly closed, the postcoxal bridge at apex less than a fourth as wide as cavity; intercoxal process of prosternum not widened behind; mesosternum concave; antennae with 11 antennomeres, club composed of two antennomeres; lateral margins of pronotum visible for their entire lengths from above .... *Philothermus*

— Procoxal cavities broadly closed behind, the postcoxal bridge more than half as wide as cavity; intercoxal process of prosternum strongly widened posteriorly; mesosternum flat or slightly convex; antennae with 10 antennomeres, club composed of one antennomere; lateral margins of pronotum not visible for their entire lengths from above...... *Cerylon*

Classification of the Nearctic Genera

Cerylonidae Billberg 1820

Euxestinae Grouvelle 1908

Diagnosis: Tarsal formula 4-4-4; anterior coxal cavities closed behind; wings with distinct subcubital fleck; abdominal segment 8 in female with short non-articulated spiculum gastrale.

Euxestus Wollaston 1858, 1 sp., *E. erithacus* Chevrolat, Florida, Neotropical (Thomas *et al.* 1995).
Tritomoidea Motschulsky 1859
Neoplotera Belon 1879

Hypodacne LeConte 1875, 1 sp., *H. punctata* LeConte, 1875, eastern North America from Ontario to Florida and west to Texas.
Pachyochthes Reitter 1875

Murmidiinae Jacquelin du Val 1858

Diagnosis: Frontoclypeal suture present; maxillary palps with palpomere 2 larger than 3; lacinia not spined or exceptionally elongate; last ventrite with hind margin inflexed and crenulate.

Botrodus Casey 1890, 1 sp., *B. estriatus* Casey 1890, Florida and Texas.

Murmidius Leach 1822, 1 sp., *M. ovalis* (Beck 1817), widespread U.S.; cosmopolitan in stored products.

Mychocerinus Slipinski 1990, 2 spp., eastern North America from Ontario to Florida and west to Arizona.
Mychocerus LeConte 1869, not Erichson 1845

Ostomopsinae Sen Gupta and Crowson 1973

Diagnosis: Frontoclypeal suture present; transverse line of vertex absent; antennal cavities absent; procoxal cavities open; tarsal formula 3-3-3; femoral lines absent.

Ostomopsis Scott 1922, 1 sp., *O. neotropicalis* Lawrence and Stephan 1975, Florida, Mexico, Panama.

Ceryloninae Billberg 1803

Diagnosis: Frontoclypeal suture absent; lacinia and galea long, slender, and blade-like or needle-like; maxillary palps aciculate; last visible sternum with distinctly crenulate margin.

Cerylon Latreille 1802, 5 spp., throughout North America.
Aphardion Gozis 1886

Mychocerus Erichson 1845, 2 spp., North Carolina, Tennessee, California.

Lapethus Casey 1890
Lytopeplus Sharp 1895
Brachylon Gorham 1898
Lapecautomus Sen Gupta and Crowson 1973
Decalapethus Dajoz 1978

Philothermus Aubé 1843, 4 spp., eastern North America from Ontario to Florida and west to Texas, California.

Cerylcautomus Sen Gupta and Crowson 1973
Pseudophilothermus Dajoz 1973
Kenyalon Dajoz 1974
Comalon Dajoz 1974
Pologlyptus Dajoz 1974
Caecodium Dajoz 1974
Neoglyptus Dajoz 1974
Neoglyptoides Dajoz 1976
Batufia Dajoz 1978
Madacerylon Dajoz 1980

BIBLIOGRAPHY

LAWRENCE, J. F. and K. STEPHAN. 1975. The North American Cerylonidae (Coleoptera: Clavicornia). Psyche, 82: 131-166.

SLIPINSKI, S.A. 1990. A monograph of the world Cerylonidae (Coleoptera: Cucujoidea) Part I - Introduction and higher classification. Annali del Museo Civico di Storia Naturale "Giacomo Doria", 33: 1-273.

SEN GUPTA, T. and R. A. CROWSON. 1973. A review of the classification of Cerylonidae (Coleoptera, Clavicornia). Transactions of the Royal Entomological Society of London, 124: 365-446.

THOMAS, M. C., P. E. SKELLEY and R. W. LUNDGREN. 1995. Two species of Cerylonidae new to Florida and the U.S. (Coleoptera). Insecta Mundi, 9: 46.

92. ENDOMYCHIDAE Leach 1815

by Paul E. Skelley and Richard A. B Leschen

Family common name: The handsome fungus beetles

Family synonyms: Mycetaeidae DuVal 1859; Merophysiidae Seidlitz 1872; Mychothenidae Sasaji 1978

The two longitudinal sulci or sublateral lines on the pronotum (lacking in some Anamorphinae), absence of internal antennal vessicles, presence of a frontoclypeal suture, 4-4-4 tarsal formula, and absence of subcoxal lines on abdominal ventrite I in most taxa will distinguish this family from other members of the cerylonid series of Cucujoidea.

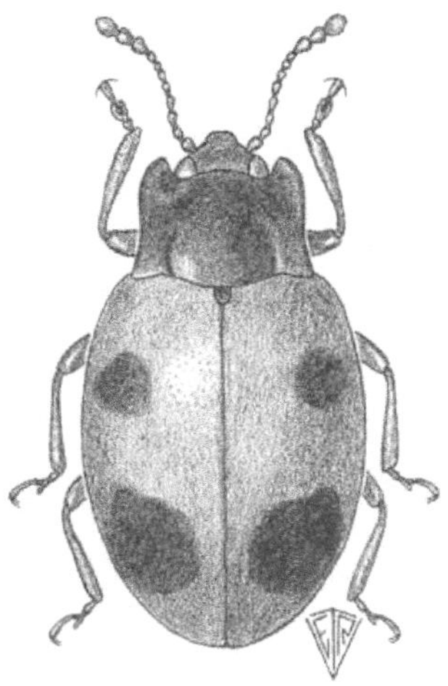

FIGURE 1.92. *Endomychus biguttatus* Say

Description: Oval to elongate-oval or round; size 1.0 to 10 mm, mostly 4 to 8 mm in length; color piceous with reddish or pale markings; vestiture fine or reduced, moderately dense, recumbent to suberect setae. Head slightly deflexed and prognathous; frontoclypeal suture present, vertex with or without a stridulatory file, labrum well-developed. Antennae with 11 antennomeres (in a few, 4, 5, or 8 to 10), with a club of one to three antennomeres (internal vesicles absent), which is enlarged or dimorphic in some; insertions exposed (concealed in Merophysiinae). Clypeus trapezoidal, narrow; labrum small, quadrate, apically arcuate; mandibles moderate, curved, the apices acute, dentate; maxillary palpi with four palpomeres, the apical palpomere securiform, oval, or triangular; gular region broad, mentum transverse, triangular, or rhomboidal; ligula membranous apically; labial palpi with three palpomeres, these short, slender, the apical one larger, cylindrical, or triangular.

Pronotum mostly much broader than the head; shape irregularly trapezoidal, explanate; without internal glandular ducts; borders margined; surface mostly with a transverse, subbasal groove and two longitudinal impressions or pair of basal pits with short sulci extending up to half the distance to the apex (absent in some Anamorphinae and *Eidoreus*); surface finely punctate; pleural region broad; anterior portion sometimes with a stridulatory membrane; prosternum moderate to long in front of coxae and with a narrow to moderate posterior process reaching the mesosternum; procoxal cavities open behind and closed internally. Mesosternum short; mesocoxal cavities laterally open or closed by the metasternum. Metasternum with subcoxal fovea in many species. Legs moderately long; trochantins not exposed; procoxae globose, or slightly transverse, not prominent, separate; mesocoxae rounded to globose, only slightly projecting, widely separated; metacoxae transverse, widely separated; trochanters small to moderate; tibiae slender, apical spurs obscure; tarsal formula 4-4-4 or 3-3-3, the third tarsomere may be minute, tarsomeres I or I-II may be broad and lobed slender; claws simple. Scutellum small to moderate, subtriangular, or arcuate. Elytra entire, apically rounded; surface punctate; epipleuron well developed; without internal glandular ducts, narrowing apically, mostly reaching apex. Abdomen with five or six ventrites, ventrite I mostly longer and mostly without subcoxal lines.

Larva (modified from Lawrence 1982, 1991): Body of various shapes, some are elongate or fusiform, subcylindrical, with dorsal scoli and verrucae; others are onisciform or flattened dorsally with roughened lateral and dorsal protuberances; others are cryptosomatic with numerous fan-shaped setae; size 2 to 10 mm, mostly 5 to 8 mm in length; vestiture various, mostly with setae. Head exserted, mostly prognathous, narrower than the thorax; epicranial suture, when present, U or V-shaped, without coronal suture. Antennae three-segmented, second segment greatly elongate, third inconspicuous. Clypeus transverse or absent; labrum lobe-like; mandibles mostly unidentate to tridentate, mola and membranous protheca present, the latter sometimes very large, concealing the molar area; maxillae with the cardo triangular, palpi three-segmented, stripes fused, mala setiferous or spiny; labium with fused submentum and mentum, ligula, and two-segmented palpi. Most with four stemmata on each side of head. Thorax with the prothorax longer and narrower than the meso- and metathorax; legs four-segmented, with apical tarsunguli. Abdomen ten-segmented with scoli, verrucae, roughened tubercles, or fan-shaped setae dorsally. Spiracles small, annular on the mesothorax and abdominal segments one to eight. Urogomphi present or absent on A9.

Habits and habitats. These beetles are typically mycophagous on spores and hyphae of microfungi (Anamorphinae, Eupsilobiinae, Merophysiinae, Holoparamecinae) or on large Basidiomycetes (remaining subfamilies). The most effective way to collect the smaller species is by sifting leaf litter and rotting wood, while other taxa may be collected from their host fungi under bark or on rotting wood. Some species of anamorphines are collected in flight interecept traps while other taxa may come to lights (*e.g.*, *Danae*) or can be collected in cantharidin-baited traps (*Xenomycetes*) (Young 1989). The genus *Lycoperdina* is a specialist on puffballs (Pakaluk 1984). Debris-cloaking behavior has been reported in Anamorphinae (Leschen and Carlton 1993). Some species reflex bleed (*Endomychus*, *Lycoperdina*, *Rhanidea*), while others apparently do not (*Bystus*, *Clemmus*). *Mycetaea subterranea* (Fab.) is an incidental stored product pest (Bousquet 1990). Information on immatures is scattered and includes Beutel *et al.* (2000), Bøving and Craighead (1931), Johnson (1986), Lawrence *et al.* (1999b), Leschen and Carlton (1993), Pakaluk (1984), and Burakowski and Slipinski (2000).

Status of the classification. The Endomychidae are part of the cerylonid series of Cucujoidea (see reviews by Slipinski 1990, Tomaszewska 2000b). The higher classification of the family is in need of revision because the inclusion of some taxa, such as the Eupsilobiinae amd Anamorphinae, suggest that the group is not monophyletic (Slipinski and Pakaluk 1991). The systematics of the family for the United States is relatively well known while there are numerous undescribed species in Mexico. There is one undescribed species of epipocine from northern United States and Canada (Bousquet and Leschen 2001, 2002), and other anamorphine and holoparamecine taxa that are undescribed from the southeastern United States. Catalogs containing information on the Endomycidae include Blackwelder (1945), Crotch (1873), Leng (1920), and Strohecker (1986). There are also many regional or specialty works with valuable information and keys to taxa, some of these include Downie and Arnett (1996), Hinton (1945), Lawrence *et al.* (1999a), Leschen and Carlton (2000), and Strohecker (1953). In addition to all of the references mentioned above several others are of systematic importance, and include Lawrence and Newton (1995), Pakaluk and Slipinski (1995), Pakaluk *et al.* (1995), Sasaji (1978, 1987, 1990), and Tomaszewska (2000a). Tomaszewska (2000b) reviews the subfamilial classification and presents a key to the subfamilies of the world. That classification is followed here.

Distribution. There are approximately 1300 described species (Strohecker 1986) known from all areas, of which 45 species occur in the United States.

Key to the Nearctic Genera

1. Antennae with 5 antennomeres; terminal 2 antennomeres greatly enlarged in both sexes, appearing fused (Fig. 2) .... *Trochoideus*
— Antennae with more than 5 antennomeres; club with 1-3 distinct antennomeres, enlarged only on males of some genera and not appearing fused (Fig. 3) 2

2(1). Antennal club consists of 1-2 antennomeres 3
— Antennal club consists of 3 antennomeres 4

3(2). Body elongate; pronotum with basal grooves .... *Holoparamecus*
— Body oval; pronotum lacking basal groves .... *Eidoreus*

4(2). Tarsi linear, with 3-4 tarsomeres; penultimate tarsomere exposed (Fig. 4) 8
— Tarsi pseudotrimerous, with 4 tarsomeres; tarsomere III minute and fused to tarsomere IV; tarsomere II lobed (Fig. 5) 5

5(4). Front margin of pronotum with stridulatory membrane (Fig. 6) 18
— Front margin of pronotum without membrane (Figs. 7, 8) 6

6(5). Elytra distinctly pubescent 20
— Elytra glabrous or minutely setose 7

7(6). Mesosternum with strong median carina .... *Xenomycetes*
— Mesosternum flat medially .... *Endomychus*

8(4). Form coccinelloid, strongly arched in long axis . 9
— Form long-oval to somewhat elongate 13

9(8). Pronotum without sulci 10
— Pronotum with lateral sulci 11

10(9). Antenna 9-segmented; dorsal surfaces with vesititure of short sparse fine hairs; tarsal claws basally dentate .... *Rhymbomicrus*
— Antenna 8-segmented; dorsal surfaces without vestiture of short sparse fine hairs; tarsal claws basally smooth .... *Micropsephodes*

11(9). Tarsal claws basally dentate .... *Anamorphus*
— Tarsal claws basally smooth 12

12(11). Tarsi with 3 tarsomeres; antennae 11-segmented .... *Clemmus*
— Tarsi with 4 tarsomeres; antennae 10-segmented .... *Bystus*

13(8). Pubescence long, rather dense, and suberect; form oval 14
— Pubescence, if present, fine, short, and decumbent; form more elongate 15

14(13). Side margins of pronotum with long, coarse setae; margin broad, raised side margin defined by a complete sublateral line .... *Mycetaea*
— Side margins of pronotum with fine setae; pronotum with deep grove near each lateral margin at anterior 3/4, lacking sublateral lines .... *Symbiotes*

15(13). Pronotum with groove on each side of mid-line in addition to short lateral sulci .... *Rhanidea*
— Pronotum without discal grooves 16

16(15). Base of pronotum with four distinct pits; body subglabrous .... *Hadromychus*
— Base of pronotum with two pits; body glabrous 17

17(16). Males without modified antennal club; size 2-3 mm .... *Stethorhanis*
— Males with swollen antennal club (Fig. 3); size 4-6 mm .... *Phymaphora*

18(5). Prosternum narrow, not prolonged behind; coxae contiguous or nearly so 19
— Prosternum broader, prolonged behind and spatulate; coxae well separated .... *Mycetina*

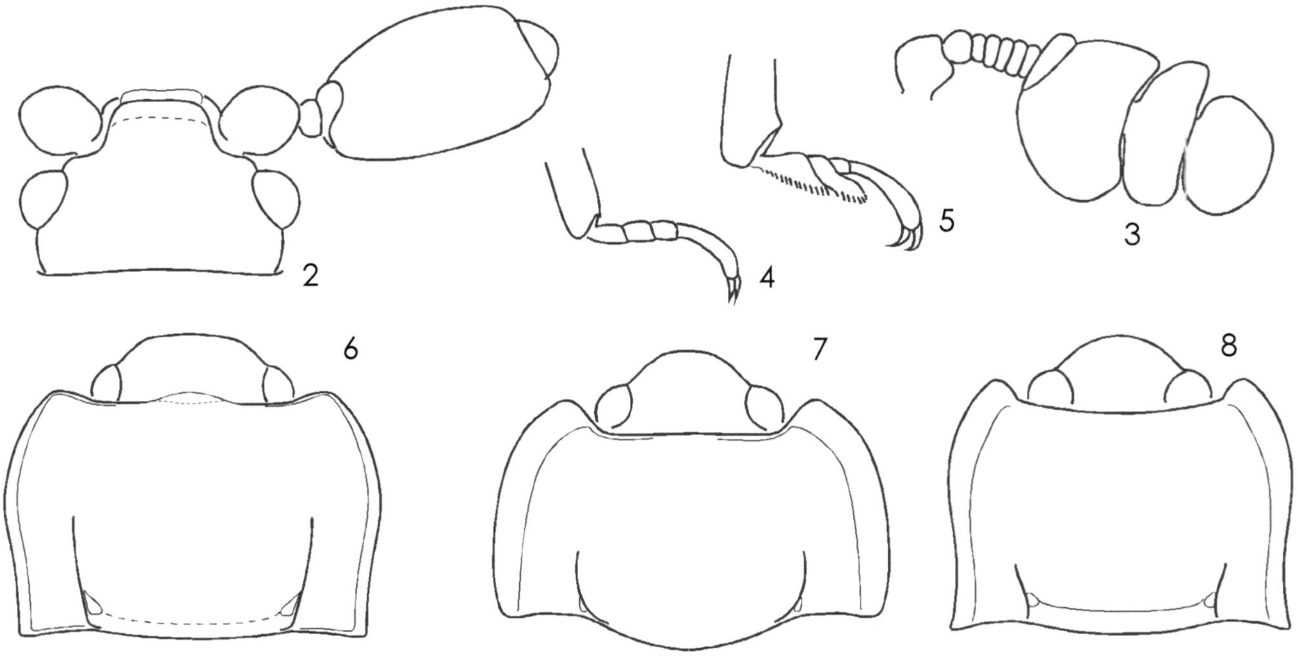

FIGURES 2.92-8.92. 2. *Trochoideus desjardinsi* Guérin, head and antenna; 3. *Phymaphora pulchella* Newman, male antenna; 4. *Phymaphora pulchella* Newman, hind tarsus; 5. *Lycoperdina ferruginea* LeConte, hind tarsus; 6. *Lycoperdina ferruginea* LeConte, head and pronotum; 7. *Stenotarsus hispidus* (Herbst), head and pronotum; 8. *Danae testacea* (Ziegler), head and pronotum.

19(18). Procoxae globular, distinctly separated .... *Aphorista*
— Procoxae subcylindrical, contiguous .... *Lycoperdina*

20(6). Pronotum with broad, raised side margins (Figs. 7, 8) 21
— Pronotum with narrow margins .... *Epipocus*

21(20). Short-oval, basal sulcus of pronotum feeble or absent (Fig. 7) .... *Stenotarsus*
— Long-oval, basal sulcus of pronotum distinct (Fig. 8) .... *Danae*

Classification of the Nearctic Genera

Endomychidae Leach 1815

Merophysiinae Seidlitz 1872

Holoparamecinae Seidlitz 1888

Holoparamecus Curtis 1833, 6 spp., widespread. Specimens have been collected in leaf litter or in flight intercept traps.
Calyptobium Aubé 1843

Leiestinae C. G. Thomson 1863
(review by Tomaszewska 2000a)

Phymaphora Newman 1838, 2 spp., *P. californica* Horn in western United States, *P. pulchella* Newman in eastern United States.

Rhanidea Strohecker 1953, 1 sp., *R. unicolor* (Ziegler), central and eastern United States.
Rhanis LeConte 1854, not Heyden 1837, not Koch 1846

Stethorhanis Blaisdell 1931, 2 spp., California and British Columbia.

Eupsilobiinae Casey 1895
(review Pakaluk and Slipinski 1990)

Eidoreus Sharp 1885, 1 sp., *E. politus* (Casey), Florida.
Eupsilobius Casey 1895
Pseudalexia Kolbe 1910

Endomychinae Leach 1815

Endomychus Panzer 1795, 2 spp., *E. biguttatus* Say in the eastern United States, *E. limbatus* (Horn) in the western United States. Larvae and biology of *E. biguttatus* are described by Leschen and Carlton (1988). A hymenopterous parasite of *E. biguttatus* is described by Leschen and Allen (1987).
Eudomychus Latreille 1796
Cyanauges Gorham and Lewis 1874
Cynauges Gemminger and Harold 1876
Caenomychus Lewis 1893

Stenotarsinae Chapuis 1876

Danae Reiche 1847, 1 sp., *D. testacea* (Ziegler), eastern United States.
Oediarthrus Gerstaecker 1858
Coniopoda Gorham 1873
Rhabduchus Gorham 1873

Stenotarsus Perty 1832, 2 spp., eastern United States.
Quirinus Thomson 1857
Systaechea Gorham 1890
Stenotarsoides Csiki 1900

Epipocinae Gorham 1873

Epipocus Germar 1943, 6 spp., widely distributed in eastern and southern United States (revision by Strohecker 1977).

Hadromychus Bousquet and Leschen 2002, 1 sp., *H. chandleri* Bousquet and Leschen, Nova Scotia, Ontario, Quebec, New Hampshire. An undescribed species is also known from Idaho and Montana.

[*Epopterus* Chevrolat 1844, not in North America. Review by Strohecker (1997)]

Lycoperdininae Redtenbacher 1844

Aphorista Gorham 1873, 3 spp., generally distributed throughout the United States. Hoebeke *et al.* (1987) comment on host associations and biology.

Lycoperdina Latreille 1907, 1 sp., *L. ferruginea* LeConte 1824, eastern United States and New Mexico. Larvae and biology are described by Pakaluk (1984).

Mycetina Mulsant 1846, 3 spp., generally distributed throughout the United States. Hoebeke *et al.* (1987) comment on host associations and biology.

Mycetaeinae Jacquelin du Val 1857

Mycetaea Stephens 1830, 1 sp., *M. subterranea* (Fabricius), widespread, adventive.

Anamorphinae Strohecker 1953

Anamorphus LeConte 1878, 2 spp., eastern United States.
Trichopsephus Arrow 1920

Bystus Guérin 1857, 1 sp., *B. ulkei* (Crotch), eastern United States. Larvae and biology of a tropical species described by Leschen and Carlton (1993).
Rhymbus Gerstaecker 1858

Clemmus Hampe 1850, 1 sp., *C. minor* (Crotch), eastern United States.
Clemnus Redtenbacher 1858
Clemmys Seidlitz 1888

Micropsephodes Champion 1913, 1 sp., *M. lundgreni* Leschen and Carlton (2000), southeastern United States.

Rhymbomicrus Casey 1916, 3 spp., District of Columbia, Oklahoma and Kansas. Revision by Pakaluk (1987).

Symbiotes Redtenbacher 1849, 3 spp., widely distributed (primarily eastern United States), includes one introduced species.
Microchondras Wollaston 1854
Eponomastus Buysson 1891

Pleganophorinae Jacquelin du Val 1858

Trochoideinae Chapuis 1876

Trochoideus Westwood 1833, 1 sp., *T. desjardinsi* Guérin, southern Florida, apparently with ants or termites, immigrant from southeast Asia (Skelley and Burgess 1995).

Xenomycetinae Strohecker *in* Arnett 1962

Xenomycetes Horn 1880, 2 spp., California, Nevada, and Washington. Larvae and hosts of *X. laversi* Hatch described by Johnson (1986). Habits of *X. morrisoni* Horn described by Young (1989).

Bibliography

BEUTEL, R.G., D. WEIDE and D. BERNHARD. 2000. Characters of the larval of *Mycetina cruciata* (Schaller) (Coleoptera: Endomychidae) and their phylogenetic implications. Annales Zoologici, 50: 7-14.

BLACKWELDER, R. E. 1945. Checklist of the the Coleopterous Insects of Mexico, Central America, the West Indies, and South America. Part 3. United States National Museum Bulletin, 185: 343-550.

BOUSQUET, Y. 1990. Beetles associated with stored products in Canada: an identification guide. Research Branch, Agriculture Canada Publication 1837. 220 pp.

BOUSQUET, Y. and R. A. B. LESCHEN. 2002. Description of a new genus and species of Endomychidae (Coleoptera: Cucujoidea) from northern North America. Coleopterists Bulletin, (in press).

BØVING, A. G. and F. C. CRAIGHEAD. 1931. An illustrated synopsis of the principal larval forms of the Order Coleoptera. Entomologica Americana (New Series), 11: 1-351.

BURAKOWSKI, B. and S.A. SLIPINSKI. 2000. The larvae of Leiestinae with notes on the phylogeny of Endomychidae (Coleoptera: Cucujoidea). Annales Zoologici, 50: 559-573.

CROTCH, G. R. 1873. Check list of the Coleoptera of America, north of Mexico. Salem, MA. 136 pp.

DOWNIE, N. M. and R. H. ARNETT, Jr. 1996. The beetles of Northeastern North America. Volume II: Polyphaga: Series

Bostrichiformia through Curculionoidea. Sandhill Crane Press. Gainesville, FL. x, 871-1721.

HINTON, H. E. 1945. A Monograph of the Beetles Associated with Stored Products. Volume I. British Museum (Natural History). London, 443 pp.

HOEBEKE, E. R., Q. D. WHEELER and R. L. GILBERTSON. 1987. Second Eucinetidae-Coniophoraceae association (Coleoptera; Basidiomycetes), with notes on the biology of *Eucinetus oviformis* LeConte (Eucinetidae) and on two species of Endomychidae. Proceedings of the Entomological Society of Washington, 89: 215-218.

JOHNSON, P. J. 1986. A description of the late-instar larva of *Xenomycetes laversi* Hatch (Coleoptera: Endomychidae) with notes on the species' host and distribution. Proceedings of the Entomological Society of Washington, 88: 666-672.

LAWRENCE, J. F. 1982. Coleoptera. Pp. 482-553. *In*: S. P. Parker, ed. Synopsis and Classification of Living Organisms, Vol. 2. McGraw-Hill. New York.

LAWRENCE, J. F. 1991. Chapter on Endomychidae. Pp. 482-485. *In*: F. W. Stehr, ed. Immature Insects, Vol. 2. Kendall/Hunt. Dubuque, IA.

LAWRENCE, J. F. and A. F. NEWTON, Jr. 1995. Families and subfamilies of Coleoptera (with selected genera, notes, references and data on family-group names). Pp. 779-1006. *In*: J. Pakaluk and S.A. Slipinski, eds. Biology, Phylogeny and Classification of Coleoptera: Papers Celebrating the 80th Birthday of Roy A. Crowson. Muzeum i Instytut Zoologi PAN, Warsaw.

LAWRENCE, J. F., A. HASTINGS, M. J. DALLWITZ, T. A. PAYNE and E. J. ZURCHER. 1999a. Beetles of the World: A key and information system for families and subfamilies. CD-ROM, Version 1.0 for MS-DOS. CSIRO Publishing. East Melbourne, Victoria.

LAWRENCE, J. F., A. HASTINGS, M. J. DALLWITZ, T. A. PAYNE and E. J. ZURCHER. 1999b. Beetle larvae of the World: Descriptions, illustrations, identification and information retrieval for families and subfamilies. CD-ROM, Version 1.0 for MS-DOS. CSIRO Publishing. East Melbourne, Victoria.

LENG, C. W. 1920. Catalogue of the Coleoptera of America, North of Mexico. Cosmos Press. Cambridge, MA. x + 470 pp.

LESCHEN, R. A. B. and R. T. ALLEN. 1987. Range extension and biology of *Endomychobius flavipes* (Hymenoptera: Pteromalidae). Entomological News, 98: 186-188.

LESCHEN, R. A. B. and C.E. CARLTON. 1988. Immature stages of *Endomychus biguttatus* Say (Coleoptera: Endomychidae) with observations on the alimentary canal. Journal of the Kansas Entomological Society, 61: 321-327.

LESCHEN, R. A. B. and C. E. CARLTON. 1993. Debris cloaking in Endomychidae: a new species from Peru (Coleoptera). Zoological Journal of the Linnaean Society, 109: 35-51.

LESCHEN, R.A.B. and C. E. CARLTON. 2000. A new species of *Micropsephodes* from southern United States (Coleoptera : Endomychidae : Anamorphinae). Coleopterists Bulletin, 54: 232-238.

PAKALUK, J. 1984. Natural history and evolution of *Lycoperdina ferruginea* (Coleoptera: Endomychidae) with descriptions of immature stages. Proceedings of the Entomological Society of Washington, 86: 312-325.

PAKALUK, J. 1987. Revision of the North American *Rhymbomicrus* Casey (Coleoptera: Endomychidae). Annals of the Entomological Society of America, 80: 456-461.

PAKALUK, J. and S. A. SLIPINSKI. 1990. Review of Eupsilobiinae (Coleoptera: Endomychidae) with descriptions of new genera and species from South America. Revue Suisse de Zoologie, 97: 705-728.

PAKALUK, J. and S. A. SLIPINSKI. 1995. Acrisomatinae, a new subfamily of Endomychidae for *Acrisoma* new genus based upon two new species from Mexico and Peru. Genus, 6: 327-335.

PAKALUK, J., S. A. SLIPINSKI and J. F. LAWRENCE. 1995. Current classification and family-group names in Cucujoidea (Coleoptera). Genus, 5: 223-268.

SASAJI, H. 1978. Notes on the Japanese Endomychidae, with an establishment of a new subfamily (Coleoptera). Memoirs of the Faculty of Education, Fukui University, Series II (Natural Science), 28: 1-31.

SASAJI, H. 1987. On the higher classification of the Endomychidae, and their related families (Coleoptera). Entomological Journal of Fukui, 1: 44-51.

SASAJI, H. 1990. The family Mychothenidae of Japan. Esakia (Special Issue), 1: 65-75.

SKELLEY, P. E. and G. R. BURGESS. 1995. *Trochoideus desjardinsi* Guérin found in Florida (Endomychidae: Trochoideinae). Coleopterists Bulletin, 49: 289-291.

SLIPINSKI, S. A. 1990. A monograph of the world Cerylonidae (Coleoptera: Cucujoidea), part 1, introduction and higher classification. Annali del Museo Civico di Storia Naturale "Giacomo Doria," 88: 1-273.

SLIPINSKI, S. A. and J. PAKALUK. 1992. Problems in the classification of the Cerylonid series of Cucujoidea (Coleoptera). Pp. 79-88. *In*: M. Zunino, X. Bellés and M. Blas, eds. Advances in Coleopterology. European Association of Coleopterology. Silvestrelli and Cappelletto. Torino.

STROHECKER, H. F. 1953. Coleoptera. Fam. Endomychidae. *In*: P. Wytsman, ed. Genera Insectororum. Fasc. 210. L. Desmet-Verteneuil, Brussels, 140 pp.

STROHECKER, H. F. 1977. Revision of the genus *Epipocus* (Coleoptera: Endomycidae). Transactions of the American Entomological Society, 103: 303-325.

STROHECKER, H. F. 1986. A Catalog of the Coleoptera of America North of Mexico. Family: Endomychidae. United States Department of Agriculture (Agriculture Handbook 529-98), Washington, DC. viii + 19 pp.

STROHECKER, H. F. 1997. An illustrated descriptive catalogue of the genera *Anidrytus* Gerstaecker and *Epopterus* Chevrolat, with descriptions of new species (Coleoptera: Endomychidae). Insecta Mundi, 11: 158-188.

TOMASZEWSKA, K. W. 2000a. A review and a phylogenetic analysis of the genera of Leiestinae (Coleoptera, Endomychidae). Mittelungen Museum Naturkunde Berlin, Deutsche Entomologische Zeitschrift, 47: 65-85.

TOMASZEWSKA, K. W. 2000b. Morphology, phylogeny and classification of adult Endomychidae (Coleoptera: Cucujoidea). Annales Zoologici, 50: 449-558.

YOUNG, D. K. 1989. Notes on the bionomics of *Xenomycetes morrisoni* Horn (Coleoptera: Endomychidae), another cantharidin-orienting fungus beetle. Pan-Pacific Entomologist, 65: 447-448.

93. COCCINELLIDAE Latreille 1807

by Natalia J. Vandenberg

Family common name: The lady beetles

Family synonyms: Epilachnidae Mulsant and Rey 1846; Securipalpes Mulsant 1846 (unavailable name, not based on genus).

Although officially called lady beetles, members of the family Coccinellidae are more commonly known as ladybugs (American) or ladybirds (Britain). The charismatic red and black dappled members of the tribe Coccinellini are easily recognized, even by young school children, but the family as a whole is somewhat difficult to characterize. Most species can be identified by the compact, rounded, body form with convex dorsum and flattened venter, clubbed antennae, and the presence of a postcoxal line on the first abdominal ventrite (lacking in *Paranaemia*, *Naemia*, and *Coleomegilla*). The tarsal formula of most species is 4-4-4 with the third tarsomere minute and tucked within the broad triangular second (cryptotetramerous or pseudotrimerous), only a few have the tarsomeres more equal (true tetramerous), some have tarsi reduced to 3-3-3 (true trimerous).

FIGURE 1.93. *Anatis mali* (Say)

Description: Shape (Figs. 2-4) rounded, varying from circular to elongate oval, and superhemispherical to somewhat depressed; size 0.8 to 11 mm (some exotic species up to 18 mm). Many are aposematically colored, red, orange or yellow with contrasting markings in black and or white, some less conspicuously colored, black, brown, ivory or gray, a few metallic blue, green, or violet; body glabrous to finely pubescent.

Head deeply inserted into prothorax in most, but exposed, except basally, in more elongate species (e.g. *Coleomegilla* spp.) (Fig. 2); form subquadrate with epicranium, frons, genae and clypeus fused; surface punctate. Antennae (Figs. 20-28) moderately short, with eight to eleven antennomeres; terminating in a compact or loose club of one to six antennomeres and smooth to serrate lateral margins; antennae inserted (Figs. 7, 8, 9, 10, 20) at the inner front margin of the eyes, or below the eyes; antennal insertion exposed or covered dorsally by lateral extension of the fronto-clypeal region. Labrum short, transverse; mandibles (Figs. 36-39) moderately robust, strongly arcuate, the apices simple, bifid, or dentate, mola generally with two teeth (reduced or absent in some Epilachninae and Sticholotidinae), membranous prostheca present; maxillary palpus (Figs. 29-35) with four palpomeres, the apical palpomere typically large, securiform (hatchet-shaped), cup-shaped, or barrel-shaped, in some species elongate conical or oval (many Sticholotidinae); gula quadrate, the sutures generally distinct; mentum (Figs. 17, 18) trapezoidal or triangular; ligula prominent, rectangular or oval; labial palpus with three palpomeres (two in Noviini), the apical palpomere oval or conical, distally truncate. Eyes lateral, moderate, somewhat bulging, reniform (Fig. 9), entire (Fig. 5), or deeply divided by eye canthus (Figs. 10, 11), finely to coarsely faceted, glabrous or hirsute.

Pronotum (Figs. 8, 10, 11, 12) broader than the head; transversely oval to quadrate, weakly to strongly convex, with deep to shallow anterior emargination; lateral margins generally explanate, with fine raised border at sides and less commonly at base; surface punctate, pleural region broad; prosternum (Figs. 15, 16) long, generally T-shaped, elevated in many species; intercoxal process bicarinate, laterally margined, or unmodified; procoxal cavities generally closed behind, open in a few. Mesosternum short, trapezoidal to subquadrate, anteriorly emarginate or truncate; metasternum long and broad.

Legs (Figs. 40-44) short relative to body width (Figs. 3, 4) with most or all of femur hidden by elytra in dorsal view (except *Hippodamia*, *Coleomegilla*, and allies; Fig. 2); trochantins of the fore and middle legs exposed; procoxae transverse, separate; mesocoxae round, separate; metacoxae transverse, widely separate; trochanters small, triangular; femora swollen; tibiae cylindrical (Fig. 40) or flattened, in some externally toothed and or grooved (Figs. 42-44), finely spinose, with or without single (Fig. 41) or paired (Fig. 40) minute apical spurs on middle and hind leg; tarsal formula generally 4-4-4 but apparently 3-3-3 (cryptotetramerous or pseudotrimerous) (Figs. 40, 43, 44) with the first and second tarsomeres apically dilated and spongy-pubescent beneath, the third minute, appearing as a basal annulation of the elongate fourth tarsomere; alternatively, formula 3-3-3 (true trimerous) through loss of or fusion of tarsomere three with four (some Scymnini, Noviini), exceptionally true tetramerous (4-4-4) with all tarsomeres more or less cylindrical and ventrally spinose

Acknowledgments: I thank my colleagues, J. Mawdsley, Smithsonian Department of Entomology, Washington, D.C. and A. Solis, A. Norrbom and E. Roberts, USDA Systematic Entomology Laboratory, for providing scientific and editorial reviews of this manuscript. I also thank P. Skelley for helpful suggestions on manuscript and plate preparation.

(*Lithophilus* and allies; soil-dwelling species not occurring in North America); claws (Figs. 45-48) simple, cleft, or with basal to median quadrate or triangular tooth. Scutellum small, triangular.

Elytra entire; surface shagreened to highly polished, finely to moderately punctate, non-striate, laterally explanate or steeply descending; epipleural fold entire (most species) to obsolete in apical third, with or without foveae to receive retracted tibiae and tarsi. Hindwing normally present and functional; wing venation reduced, nervature of the cantharoid type, veins feebly marked or absent in distal half; anal lobe generally present, but reduced or absent in some minute species. Abdomen with five to seven ventrites (= visible sternites), sutures entire; first ventrite composed of the fused 1-3 morphological sternites, normally bearing a postcoxal line of variable form (Figs. 49-54), line absent (Fig. 55) in *Paranaemia*, *Naemia* and *Coleomegilla*.

Male genitalia with a curved tubular sipho (= penis); basal end of sipho typically enlarged to form a siphonal capsule with an outer arm through which the ejaculatory duct is threaded, and in most cases an inner arm or apodeme which articulates with a median tegminal strut known as the trabes. The predistal portion of the sipho is embraced by the tegmen which consists of the following identifiable parts: a central projection termed the basal lobe (or incorrectly known as the median lobe), pair of lateral projections known as parameres or accessory lobes, a basal portion called the basal piece and the median strut or trabes. Note: the parameres of this and related taxa are considered by some to be secondary developments not homologous to the parameres of non-cucujoid lineages. The entire coccinellid tegmen, therefore, is often referred to as the phallobase, while in other groups the term phallobase corresponds to the tegmen minus the parameres.

Female genital segments (ovipositor) composed of a well developed proctiger (tenth tergite), the paired ninth pleurites, and the paired coxites also called "hemisternites" (ninth sternites); the shape of the coxites are important taxonomically and may be oval, tapered distally or paddle-shaped, generally heavily chitinized and bearing a small apical stylus. Receptaculum seminis (spermatheca) present, or exceptionally absent, typically with three branches, but weakly developed or obsolete in some: the proximal branch or nodulus which communicates with the sperm duct, middle branch or ramus bearing the spermathecal accessory gland, and distal branch or cornu. Sperm duct straight or coiled, of variable length and rigidity, generally bearing a sclerotized "infundibulum" of variable form and position; sperm duct attached to bursa distally, basally or dorsally; bursa variously shaped, thin or thick-walled, membranous, in many enclosing a sclerotized plate-, rod-, or bulb-like extension of the infundibulum (for additional readings on morphology and anatomy of the adult coccinellid, see Sasaji 1968 and Kovár 1996a).

Eggs cream-colored, yellow or reddish orange, darkening prior to eclosion; typically oval to fusiform, positioned on end in clusters (e.g. Coccinellini, some Chilocorini), or ellipsoidal laid flat in loose groupings (most Scymnini), in some taxa rounded or flattened, deposited individually near leaf vein, in cracks and crevices, or adjacent to/under prey (e.g. Platynaspini, Chilocorinae).

Larva elongate fusiform or oblong to broadly ovate, slightly to strongly flattened; color gray, blue, brown, or pink, many with contrasting markings of white, yellow, orange, or red, particularly in the later instars; body surfaces rugose to microtrichose with or without glabrous patches, armed with setae and setose processes of variable architecture (strumae, verrucae, parascoli, scoli or senti), in some taxa well covered with white waxy exudates. Head exserted, circular or oblong, hypo- to prognathous; frontal sutures forming a V or inverted omega shape, lacking in some; epicranial stem obsolete except in Epilachninae. Antennae of one to three segments, not over three times as long as wide. Labrum distinct, mandibles moderate, stout, triangular, or sickle-shaped, apically acute, bidentate, or multidentate; mola generally present but reduced or absent in Epilachninae; maxillae with fused cardo and stipes; maxillary palpi with two or three palpomeres; labium with fused submentum and ligula, labial palpi with one to two palpomeres. Three pair of stemmata generally present. Thorax with distinct armature and normally with tergal plates on each segment; legs elongate, each with four segments plus tarsungulus; apex of tibiae with clavate or flattened setae. Spiracles on the mesothorax and abdominal segments one to eight, annular. Abdomen ten-segmented; most abdominal segments with scattered setae or chalazae and transverse row of six setose processes visible in a dorsal view; the tenth segment provided with an anal organ or sucking disk (pygopod); pores of repugnatorial glands may occur on each lateral margin of the tergum in the coria between segments. Urogomphi absent. Rees et al. (1994) provide a key to genera and selected species of North America; LeSage (1991) provides family and tribal level diagnoses and a key to North American tribes; Pope (1979) describes the phenomenon of wax production by larvae in diverse tribes. Descriptions and keys to selected North American genera and species include Gordon and Vandenberg (1993: genus *Cycloneda*) and Gordon and Vandenberg (1995: genus *Coccinella*). Introduced species are described and illustrated in Gordon and Vandenberg (1991).

Pupa rounded or oval, attached to substrate at caudal end, exarate; generally exposed, but some enclosed within the last larval exuvium except for a narrow dorsolongitudinal strip (Chilocorini, Noviini, some Scymninae); most brightly colored or patterned. Phuoc and Stehr (1974) provide descriptions and a key to subfamilies and tribes of North America.

Habits and habitats. Adults and larvae of most species are predacious on aphids, psyllids, mealybugs, scales, or other small soft-bodied insects and mites. The larger, aposematically colored lady beetles can usually be found feeding amongst colonies of their prey or basking openly on vegetation. *Coleomegilla* species, which prefer a more humid microclimate, are frequently associated with aquatic vegetation, such as water lettuce. In agricultural plantings these beetles are often abundant inside the tightly whorled leaves of corn plants, feeding on wind blown pollen, aphids, mites, and the eggs or larvae of beetles and moths. Although most predacious lady beetles will consume pollen in the absence of prey, *Coleomegilla* is one of the few genera able to complete development on pollen alone. In North America, members of the subfamily Coccinellinae are typically aphidophagous.

In the United States, *Neoharmonia* species were assumed to be normal aphidophages until they were found feeding on the eggs and larvae of an exotic leaf beetle (Whitehead and Duffield 1982). In Central and South America the same and related species feed on native leaf beetles, suggesting that this is the more normal food preference for the genus. Many predacious coccinellids will feed occasionally on spider mites. The minute *Stethorus* species have specialized in this particular prey group. Many of the smaller predacious lady beetles are adept at squeezing into the tiny nooks and crannies provided by plant architecture, or crawling inside the folds of distorted leaves damaged by sucking insects and mites; they can even find their way inside of hollow thorns or insect galls if furnished with a tiny opening, often provided by an ant. Ants are inimical to most lady beetles, but some lady beetle taxa are either tolerated or ignored. The larvae of *Brachiacantha* species have been reported feeding on Homoptera housed within ant colonies, and the Australian *Scymnodes bellus* Blackburn has been recorded preying on the ants themselves under eucalyptus bark (Pope and Lawrence 1990). The blind larvae of *Ortalistes* are termitophiles and live in Central America. Although the entomophagous members of the lady beetle family can all be categorized as predators, the larvae of some minute scale feeding *Hyperaspis* species will complete their development by burrowing into the large egg sac attached to a single female scale and thus approach a parasitic mode of existence.

Lady beetles in the subfamily Epilachninae are exceptional in following a completely phytophagous diet, feeding primarily on the leaf parenchyma, particularly of plants in the families Solanaceae, Curcurbitaceae and Leguminosae. This lady beetle subfamily is distributed worldwide, but is most diverse in the tropics. The adventive Mexican bean beetle, *Epilachna varivestis* Mulsant, is a notorious pest of bean and alfalfa crops in the Eastern United States. A North American native, the squash lady beetle, *Epilachna borealis* (Fabricius), damages squash and related crop plants.

The Halyziini (= Psylloborini) of the subfamily Coccinellinae feed on powdery mildews (Ascomycetes: Erysiphales). Their greatest diversity is also in the tropics. Systematists argue whether this group of mycetophagous Coccinellinae represents a single or multiple phyletic lines. As with pollenivory, consumption of mold spores occurs among the entomophagous lady beetle taxa as well. *Neocalvia*, a genus of Coccinellinae restricted to the Neotropics, feeds on the larvae of the Halyziini.

Lady beetles have their share of natural enemies, but they are well protected against most birds, mammals, ants, and other generalist predators. Adult lady beetles are capable of releasing a bitter fluid from specialized glands at the tibio-femoral articulations which serve as a repellent. This renders most species unpalatable, and some are even highly toxic if ingested. Larvae are similarly protected by repugnatorial glands on the abdomen.

In North America many lady beetle species become dormant during the hot dry summer or the cold winter when prey are scarce. Some species migrate to the mountains and form large aggregations, while others remain in situ or fly only a short distance to form smaller clusters at the bases of prominent objects such as fence posts or rocky outcrops. The introduced multicolored Asian lady beetle, *Harmonia axyridis* (Pallas), annoys many home owners by moving indoors in large numbers toward the end of fall and remaining until spring.

Several predacious lady beetles have been used with great success as agents of biological control, particularly against scale insects (Drea and Gordon 1990). The vedalia beetle, *Rodolia cardinalis* (Mulsant) was successfully introduced from Australia into many parts of the world for control of a notorious citrus pest, the cottony cushion scale, *Icerya purchasi* Maskell. *Cryptognatha nodiceps* Marshall and *Rhyzobius pulchellus* Montrouzier were successfully employed against coconut scale in Fiji and New Hebrides respectively, and *Rhizobius lophanthae* Blaisdell has proven to be an asset in controlling several scale pests in tropical areas of the world. The attempt to release generalist aphidophagous predators for biological control of introduced aphids has been less effective and sometimes produced undesirable side effects such as the displacement of native lady beetle species, predation on non-target species, or, in the case of the multicolored Asian lady beetle, *H. axyridis,* the creation of a new public nuisance. Further valuable readings on the habits and habitats of the Coccinellidae include Hodek and Honek (1996), Majerus (1994), Kuznetsov (1997), and Klausnitzer and Klausnitzer (1997).

Status of the classification. The family Coccinellidae belongs to the cerylonid series (8 families, 38 subfamilies), section Clavicornia, of superfamily Cucujoidea. Its closest affinities are believed to be with Corylophidae and Endomychidae (Crowson 1955, Sasaji 1971a) or with Alexiidae (=Sphaerosomatidae) and Endomychidae (Slipinski and Pakaluk 1991). Various viewpoints on the systematic position of the Coccinellidae are summarized by Sasaji (1971a), Slipinski and Pakaluk (1991), Pakaluk *et al.* 1994, and Kovár (1996b)). Many contemporary works recognize six subfamilies of Coccinellidae (Sticholotidinae (= Sticholotinae), Scymninae, Coccidulinae, Chilocorinae, Coccinellinae, and Epilachninae) (Booth *et al.* 1990, Pakaluk *et al.* 1994, Lawrence and Newton 1995, Kuznetsov 1997), a system first proposed by Sasaji (1968), and based on a detailed morphological study of adult and larval characteristics. Other authors have built upon this classification through the addition of one or more subfamilies. There is no current consensus for the higher classification of the Coccinellidae, despite an attempt by Chazeau *et al.* (1989) and Fürsch (1996) to develop one, soliciting the input of the greater community of coccinellid specialists. The tribal level classification in the system developed by Sasaji (1968) has proven to be even less stable. Although coccinellid subfamilies are more or less worldwide in distribution, many proposed tribes are restricted to particular biogeographic regions where they fall outside the consideration of regional revisionists. This has resulted in a proliferation of alternative classifications which can not be easily reconciled.

Even a cursory review of the New World Coccinellidae suggests that generic and higher level taxa have not been rigorously defined, nor do they maintain a consistent hierarchical value throughout the family. Often the more derived members of a taxon have been stripped away and isolated under a separate name, leaving the parent group with a para- or polyphyletic assemblage of residual taxa. Thus, some broadly defined genera with many species and species groups exist alongside other narrowly defined genera with one or only a few species (e.g., *Hyperaspis* vs. *Hyperaspidius*, *Hippodamia* vs. *Ceratomegilla*, *Coleomegilla* vs. *Paranaemia* or *Naemia*). To achieve a balanced classification, either these generic sets should be reunited, or the larger paraphyletic genera should be split into multiple genera.

The Halyziini (= Psylloborini, see Pakaluk *et al.* 1994) of North America are a small and easily recognized group consisting of the single genus *Psyllobora*, but the world fauna is much more diverse. Various authors have classified these mycophagous coccinellines in multiple tribes (Fürsch 1996, Kovár 1996b), or combined some or all of them with their predacious relatives in the tribe Coccinellini (Kuznetsov 1997, Iablokoff-Khnzorian 1982). The single tribe classification (Sasaji 1968) is followed here as a matter of convenience and without an independent attempt at evaluating alternatives.

Pope (1988) revised the Australian coccinellid fauna and identified the classification of subfamilies Coccidulinae and Scymninae (*sensu* Sasaji 1968) as one problem area in the higher classification of the family. He suggested that a single subfamily with 5 tribes would more accurately portray the phylogeny of the cocciduline–scymnine lineage, but unfortunately he did not employ or elaborate upon this suggestion. Other authors sought to solve the same problem through diverse methods involving either (1) reshuffling of scymnine and cocciduline tribes within the two existing subfamilies, (2) segregating specialized members of this lineage in additional subfamilies (3) elevating existing tribes to the subfamily level, and thus effectively sidestepping the issue of their relationship to one another, or a combination of the above (for further discussion, see Kovár 1996b, Gordon 1994a). None of these methods has proven entirely satisfactory, but the effect on the classification of the North American fauna has been minimized due to the fact that many of the problematic taxa do not occur in this region, or occur only as isolated introductions of exotic species.

The tribe Azyini Mulsant was resurrected by Gordon (1980) for two closely related genera (*Azya* Mulsant and *Pseudoazya* Gordon) which were deemed sufficiently distinct from the rest of the neotropical fauna to justify their separation from Coccidulinae. Had exemplars from the Australian region been included in this study, it would have become evident that problematic allied taxa exist, in particular the probable sister genus *Bucolus* Mulsant (variously classified in Coccidulinae and Scymninae). The tribe Azyini is used here provisionally, but its subsequent elevation to subfamily status (Gordon 1994a), is not implemented as it does nothing to resolve existing problems. The classification of Scymninae and Coccidulinae is in need of serious study on a worldwide basis.

The classification presented here recognizes the split of Hyperaspidini (*sensu lato*) into Hyperaspidini and Brachiacanthini (Duverger 1989, as Hyperaspini and Brachiacanthadini), but does not employ Hyperaspidinae (= Hyperaspinae) as a subfamily level taxon (Duverger 1989). Although the separation of Hyperaspidinae from Scymninae has considerable merit, to achieve a holophyletic classification, the Hyperaspidinae would need to include other taxa remaining in Scymninae, such as Selvadiini and Diomini (in part). The Selvadiini share important derived antennal characters with *Hyperaspis,* and neotropical members of Diomini appear to be polyphyletic with respect to Selvadiini.

The tribe Scymnillini (Scymninae) shares many characteristics with members of Sticholotidini (Sticholotidinae), not only in external morphology (Sasaji 1971b), but in the genitalia of both sexes. Its placement is problematic. Kovár (1996b) identifies two major series within the Sticholotidinae, and it may be that this latter subfamily is polyphyletic. Gordon (1994b) uses the scymnilline genus *Zilus* Mulsant as an outgroup for his cladistic analysis of the West Indian Sticholotidini, remarking that they have many similarities. The correspondence between both internal and external character states of Scymnillini and Sticholotidini suggests that the similarities are not due merely to convergence.

An additional problem area is with the generic level classification of the subfamily Chilocorinae. Kovár (1995) briefly reviewed the New World classification in a work primarily focusing on palearctic members of the chilocorine genera *Brumus* Mulsant, *Exochomus* Redtenbacher, and *Brumoides* Chapin. He restricted use of these names to the Old World species, but provided no alternative placement for New World members.

The identification of Coccinellidae from America has been greatly facilitated by the publication of a comprehensive and well illustrated work (Gordon 1985) with keys and descriptions of 57 genera and 475 species. A few additional species have subsequently become established either by immigration or through biological control efforts, or represent earlier establishments which were overlooked until recent times (Vandenberg 1990, Gordon and Vandenberg 1991, Peck and Thomas 1998). Some native Coccinellini, once quite common, have become rare in the last decade, most likely due to competition with exotic species. Similarly, some species ranges reported here may no longer be accurate.

In the future, we can also expect changes in the number of recorded North American species due to a reassessment of the species versus subspecies or varietal status in some problematic groups, or the discovery of new species and species synonyms, particularly in the more minute, cryptically colored taxa. Changes in the higher classification of the Coccinellidae and a better understanding of their position within the group of related families can be anticipated, particularly as information is shared among specialists from around the world.

Distribution. Nearly 6,000 species occur worldwide in about 360 genera. Gordon (1985) documented 475 species in 57 genera from America north of Mexico. Additional records (Vandenberg 1990, Gordon and Vandenberg 1991, Peck and Thomas 1998) bring that figure up to 481 species and 60 genera.

Key to Nearctic Genera

1. Eye deeply divided by transverse projection (canthus) from inner ventral margin (head positioned vertically); canthus broad, band-like, expanded to cover basal antennomeres from frontal view; clypeus not projecting, with semicircular emar-

gination medially (Fig. 10); antenna of ten or fewer antennomeres with spindle-shaped flagellum (Fig. 21); mandible scythe-like with a single apical tooth (Fig.37) (Chilocorinae: Chilocorini) 2

— Eye usually weakly emarginate (Figs. 7, 8, 9) or with brief lobe-like or digitiform canthus (Fig. 13, 14); if canthus deeply dividing eye (some Sticholotidinae, Exoplectrini, Cryptognathini) then its form narrow and clypeus not as above (Fig.11); other characters variable 10

2(1). Postcoxal line of first abdominal ventrite merging with posterior margin of ventrite (Fig. 49) .... *Chilocorus*

— Postcoxal line of first abdominal ventrite not merging with posterior margin of ventrite (Figs. 50, 52, 53, 54) 3

3(2). Postcoxal line of first abdominal ventrite parallel and close to posterior margin of first ventrite (Fig. 50); antenna composed of 7 antennomeres .... *Halmus*

— Postcoxal line of first abdominal ventrite recurved apically, complete or not; antenna of 8-10 antennomeres 4

4(3). Postcoxal line of first abdominal ventrite complete (Figs. 53, 54) 5

— Postcoxal line of first abdominal ventrite incomplete (Fig. 52) 8

5(4). Tarsal claw simple, without basal tooth (Fig. 45); antenna composed of 8 antennomeres .... *Brumoides* (*sensu lato*)

— Tarsal claw with basal tooth (Fig. 47); antenna composed of 10 antennomers (exceptionally of 9 antennomeres but these species not occurring in North America) 6

6(5). Pronotum finely margined at base 7

- Pronotum not margined at base (North America natives) .... *Exochomus* (*sensu lato*; in part)

7(6). Postcoxal line of first abdominal ventrite reaching or directed toward the inner end of lateral line (Fig. 54) .... *Brumus* (*sensu* Kovár 1995)

— Postcoxal line of first abdominal ventrite reaching or directed toward midpoint of lateral line (Fig. 53) .... *Exochomus* (*sensu stricto*; Kovár 1995)

8(4). Elytron metallic blue without spots .... *Curinus*

— Elytron blackish with one or more red to yellow spots 9

9(8). Elytral margin not reflexed, with marginal bead; length less than 3.6 mm .... *Arawana*

— Elytral margin feebly reflexed, with or without marginal bead; length more than 5.0 mm .... *Axion*

10(1). Mandible with apex multidenticulate, bearing three or more large irregular teeth (Fig. 38); all tibiae with one or two apical spurs present (Figs. 40, 41) (North American fauna); dorsal surfaces pubescent; antenna inserted dorsally between eyes and distant from inner ocular margin (Fig. 9), long, loosely articulated, with 11 antennomeres and inner margin of club weakly serrated; eye bean-shaped without an abrupt notch or emargination (Fig. 9); length 3.5mm or greater (Epilachninae: Epilachnini) 11

— Mandible with bifid or single apex (Figs. 36, 37), a few with very weak subapical tooth (some Sticholotidinae), but then body size minute, length less than 3.0 mm; if with additional well developed apical teeth (Halyziini (=Psylloborini)) then teeth regular, comb-like (Fig. 39), all tibiae lacking apical spurs and dorsal surfaces glabrous; other characters variable 12

11(10). Anterior tibia slender, not angulate at outer margin; anterior tibia with single spur at apex (as in fig. 41); body length more than 6.0 mm .... *Epilachna*

— Anterior tibia relatively robust, with outer margin angulate at apical 1/4; anterior tibia with pair of spurs at apex (as in fig. 40); body length less than 5.0 mm .... *Subcoccinella*

12(10). Dorsal surfaces glabrous; distal maxillary palpomere broadly securiform (hatchet-shaped) with sides strongly divergent apically, base narrowly articulated with preceding palpomere (Fig. 29); antenna equal to 2/3 head width or longer (Fig. 20, right); femur not strongly flattened; tibia simple, without angulations (Fig. 40) (Coccinellinae) 13

— Dorsal surfaces glabrous or pubescent; distal maxillary palpomere barrel-shaped, oblong, oval or conical (tapered toward apex) (Figs. 31-34); if securiform, then base rather broadly articulated with previous palpomere (Figs. 30, 35) and antenna less than 2/3 head width (Fig. 20, left), or dorsal surfaces pubescent; femur sometimes strongly flattened; tibia simple or modified (Fig. 42-44) 14

13(12). Apex of mandible multidenticulate with small comb-like denticles (Fig. 39); eye bean-shaped without an abrupt emargination (Fig. 8); eye facets coarse, bead-like; spurs lacking on all tibiae; elytral ground color yellow or white with brown speckles or blotches; mycetophagous on powdery mildews (Halyziini (=Psylloborini)) .... *Psyllobora*

— Apex of mandible bifid (Fig. 36); eye circular or oval with an abrupt notch or digitiform emargination produced by eye canthus (Figs. 13, 14); eye facets fine, somewhat flattened; one or pair of spurs usually present on apex of middle and hind tibia (Figs. 40, 41), rarely absent (*Mulsantina, Neoharmonia, Harmonia, Aphidecta*); elytral color pattern variable; predacious on insects and mites (Coccinellini) 45

14(12). Distal maxillary palpomere elongate: conical, or parallel-sided with an acute apex (Figs. 33, 32) (mouthparts may be hidden from view, see couplet 15, below); mentum usually narrowly articulated with submentum (Fig. 17); length of body less than 3.0 mm (Sticholotidinae) 15

— Distal maxillary palpomere short and/or broad (although somewhat elongate in Scymnillini, Fig. 31): barrel-shaped, securiform, or with apex weakly convergent (Figs. 30, 31, 34, 35); mentum rather broadly articulated with submentum (Fig. 18); length of body variable 20

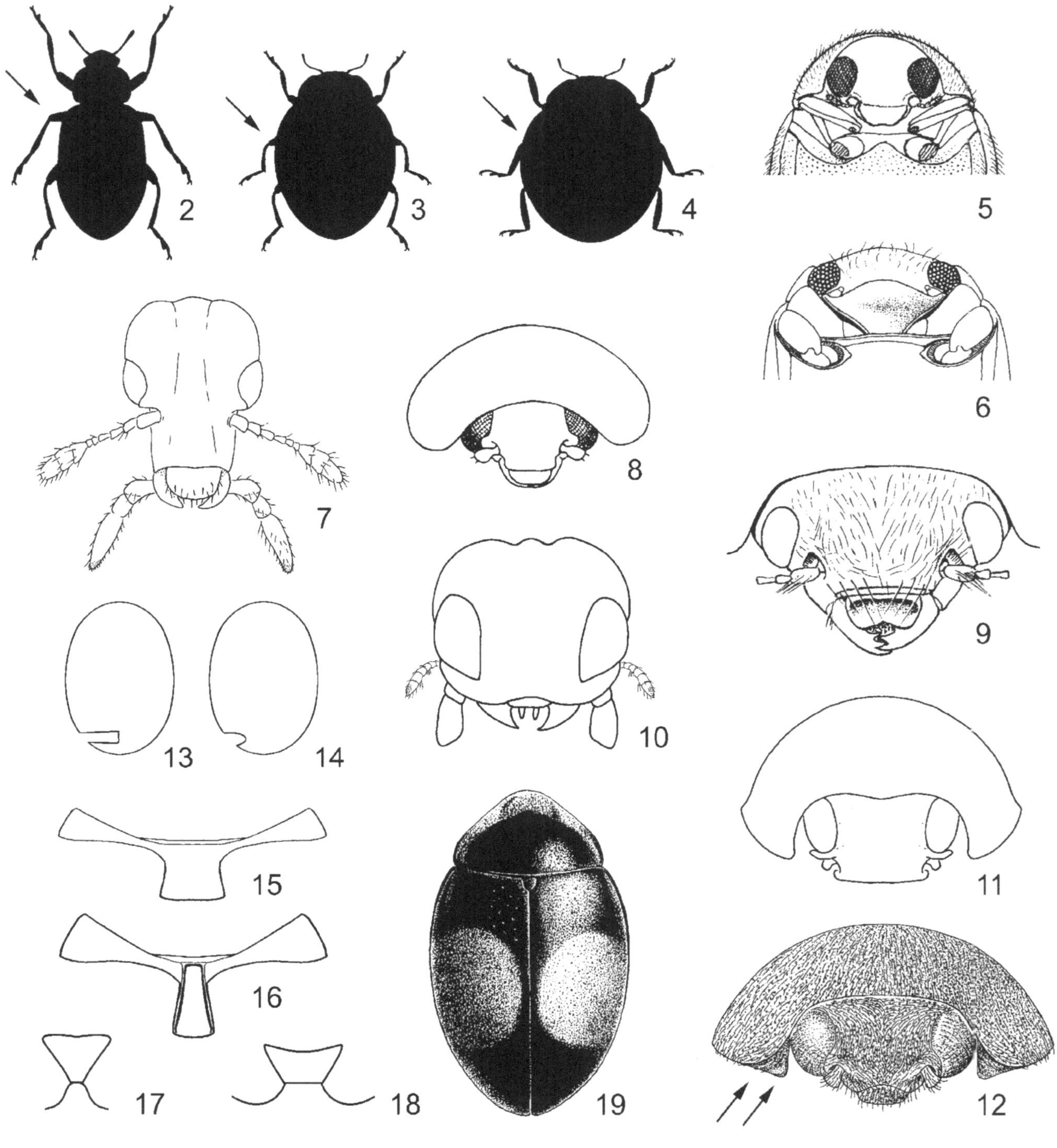

FIGURE 2.93-19.93. 2-4 silhouettes showing a range of coccinellid body types (arrow identifies apex of mesofemur); 5. *Nephaspis* sp., head, anterior part of thorax (ventral view of body); 6. *Delphaspis* sp., head, anterior part of thorax (ventral view of body); 7. *Gnathoweisea* sp., head; 8 *Psyllobora* sp. (Halyziini), head, pronotum; 9. *Epilachna* sp. (Epilachnini), head, anterior edge of pronotum; 10. *Chilocorus* sp. (Chilocorini), head; 11. *Cryptognatha* sp. (Cryptognathini), head, pronotum; 12. *Azya* sp. (Azyini), head, pronotum, arrows indicate inner and outer edge of thickened anterolateral margin; 13. left compound eye with elongate eye canthus (Scymnillini) (diagrammatic); 14. left compound eye with brief notch-like eye canthus (diagrammatic); 15. Scymnillini sp., prosternum (ventral view, diagrammatic); 16. *Scymnus* sp., prosternum (ventral view, diagrammatic); 17. mentum (above) showing narrow articulation with submentum (typical of Sticholotidinae, diagrammatic);18. mentum (above) showing broad articulation with submentum (typical of most Coccinellidae, diagrammatic); 19. *Nipus* sp. (dorsal view, appendages retracted). (Some figures modified from the works of Sasaji and Gordon)

15(14). Prosternum greatly expanded to conceal mouthparts (Fig. 6); antennal club composed of a single oblong or elongate-oval antennomere (Fig. 25); femur broad, flat, fitting into depressions on ventral surface (Serangiini) .... *Delphastus*
— Prosternum not expanded or with small lobe-like expansion that does not conceal mouthparts; antennal club composed of more than one antennomere, terminal antennomere not oblong or elongate-oval; femur not broad or flat; ventral surface without depressions for femora........ 16

16(15). Dorsal surface clothed in conspicuous long semirecumbent pubescence; head large, exposed, vertical; eye large, narrow, elongate (Cephaloscymnini) .... *Cephaloscymnus*
— Dorsal surface glabrous or with inconspicuous stubble-like pubescence; head small, concealed or exposed; eye small, round or oval (Microweiseini) 17

17(16). Head entirely concealed beneath pronotum (Fig. 19) .... *Nipus*
— Head exposed or only partially concealed 18

18(17). Head unusually long and narrow (Fig. 7) .... *Gnathoweisea*
— Head more or less transverse 19

19(18). Antenna composed of 10 antennomeres; length of antepenultimate antennomere subequal to penultimate antennomere .... *Microweisea*
— Antenna composed of 9 antennomeres; antepenultimate antennomere much shorter than penultimate antennomere .... *Coccidophilus*

20(14). Procoxa broad, obscuring lateral arm of prosternum; antenna of 8 antennomeres with weakly formed, spindle-shaped club (Fig. 26); eye densely pubescent; tarsi trimerous (Noviini. The two genera occurring in North America are distinguished primarily by larval characteristics; species specific colorpatterns are used to separate the adults, below)... 21
— Procoxa normal, not obscuring lateral arm of prosternum; antenna of 9-11antennomeres, club of various forms (Figss. 22-24, 27, 28); eye glabrous or with sparse to moderate pubescence laterally; tarsi trimerous or cryptotetramerous 22

21(20). Elytron reddish with a complicated pattern of dark marks (Fig. 56); apical one-third with a reddish mark entirely or partially enclosed by a darker border (color pattern of single introduced species: *Rodolia cardinalis* (Mulsant)); pronotum with outline of basal half not completely arcuate; posterior angles apparent .... *Rodolia*
— Elytron predominantly dark with a median red spot and reddish anterolateral border, these reddish areas sometimes confluent; apical one-third dark without additional marks (color pattern of single native species *Anovia virginalis* (Wickham)); pronotum with outline of basal half completely arcuate; posterior angles obsolete.......... *Anovia*

22(20). Pronotum with anterolateral margin thickened, with sharply defined inner and outer edge (Fig. 12); hypomeron with foveae to accommodate the antennal club and part of anterior leg (Azyini) .. 23
— Pronotum with anterolateral margin not thickened; hypomeron not as above 24

23(22). Prosternum with intercoxal process elevated, narrow.. *Azya*
— Prosternum with intercoxal process flat, not elevated .... *Pseudoazya*

24(22). Prosternum broadly rounded anteriorly (similar to Fig. 6), at least partially concealing mouthparts and antennae 25
— Prosternum not as above 27

25(24). Clypeus with anterior margin upturned; eye canthus long and narrow, nearly dividing eye (Cryptognathini) (Fig. 11)............ *Cryptognatha*
— Clypeus with anterior margin not upturned; eye canthus short to obsolete (Figs. 5, 13, 14) 26

26(25). Size 3.4 mm or greater; prosternal intercoxal process carinate; distal maxillary palpomere diverging toward apex (similar to Fig. 30) (Scymnini in part) .... *Cryptolaemus*
— Size less than 2.0 mm; distal maxillary palpomere weakly tapered toward apex (Fig. 34); prosternal intercoxal process not carinate (Stethorini)...... .... *Stethorus*

27(24). Eye canthus long and narrow, nearly dividing eye; basal antennomere greatly enlarged (Exoplectrini) .... *Exoplectra*
— Eye canthus extending half way across eye or less (Figs. 13, 14); basal antennomere variable 28

28(27). Antennae long, more than 2/3 head width, inserted laterally; flagellum and club well differentiated, terminal antennomere usually large and quadrate (Fig. 20, right) (Coccidulini) 29
— Antenna short, 2/3 head width or less (Fig. 20, left), inserted laterally or ventrally; flagellum and club merging gradually, not well differentiated, terminal antennomere reduced, often tapered (Figs. 22, 23, 24, 27, 28) 30

29(28). Dorsal pubescence uniform, decumbent............. .... *Coccidula*
— Dorsal pubescence of mostly decumbent hairs with some long, erect hairs scattered throughout.... .... *Rhyzobius*

30(28). Abdomen with 5 ventrites; prosternal intercoxal process very broad and flat, without carinae (Fig. 15); eye canthus extending about halfway across eye (Fig. 13) (Scymnillini) 31
— Abdomen with 6 or 7 ventrites; prosternal intercoxal process normal, with (Fig. 16) or without carinae; eye canthus extending distinctly less than halfway across eye (Fig. 14) 32

31(30). Elytron apparently glabrous or with only sparse hairs in evidence; often with metallic sheen ...... *Zilus*
— Elytron with dense, mostly erect pubescence; without metallic sheen .... *Zagloba*

32(30). Dorsal surfaces glabrous 33
— Dorsal surfaces pubescent 37

33(32). Anterior tibia with external tooth or spine (Fig. 42); eye emarginate; male of many species with cusp

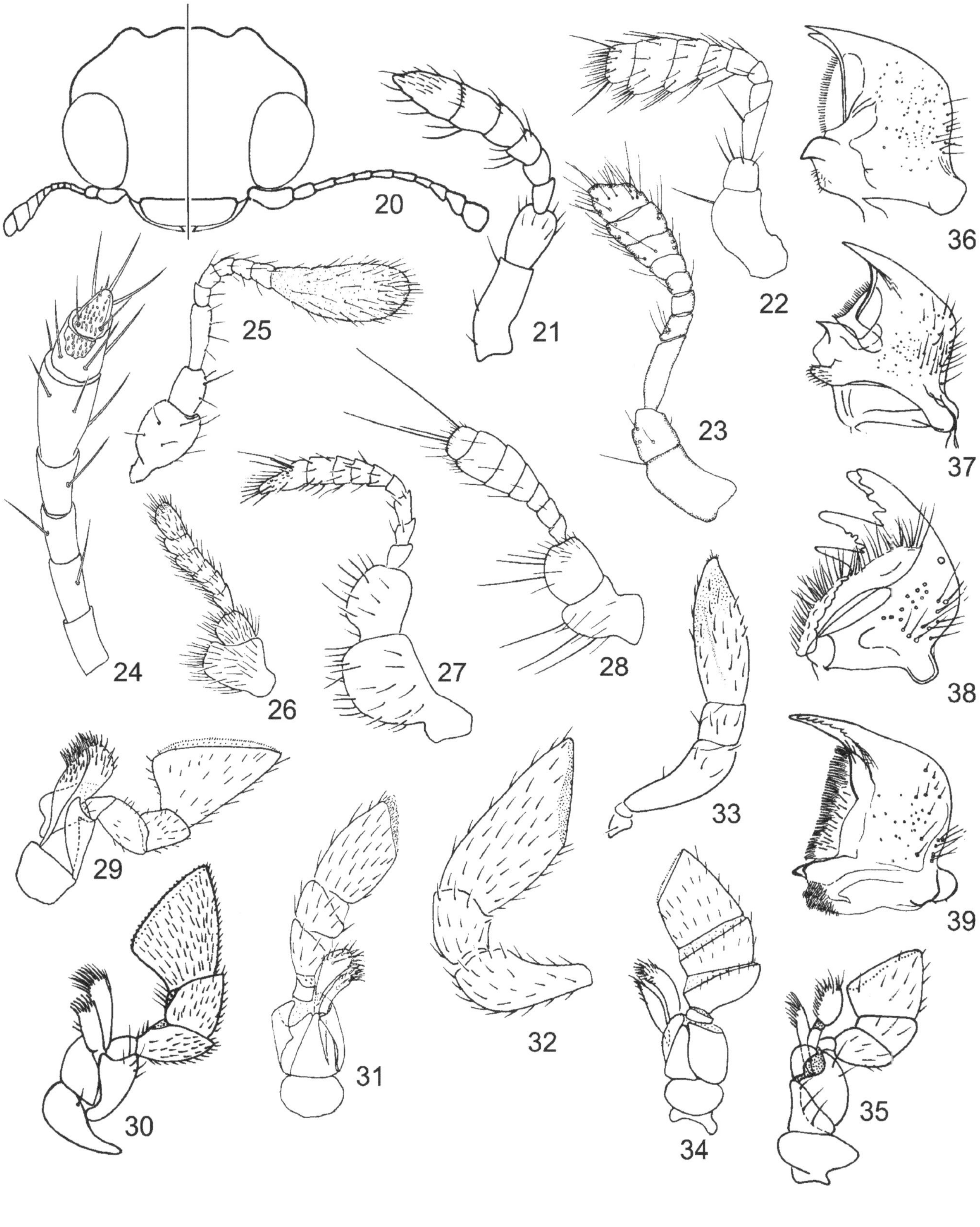
20
21
22
23
24
25
26
27
28
29
30
31
32
33
34
35
36
37
38
39

on abdominal ventrite 3; female genitalia with simple spermathecal capsule, infundibulum present (Brachiacanthini) .... *Brachiacantha*

— Anterior tibia without external tooth or spine; eye not emarginate; male without cusp on abdominal ventrite 3; female genitalia with retort-shaped appendage (= modified cornu?) on spermathecal capsule (Hyperaspidini, in part), infundibulum absent or unsclerotized 34

34(33). Epipleuron of elytron not excavated for reception of middle and hind femoral apices; tarsal claw simple (Fig. 45) .... *Hyperaspidius*

— Epipleuron of elytron excavated for reception of middle and hind femoral apices; tarsal claw toothed (Fig. 47) or simple (Fig. 45) 35

35(34). Epipleuron of elytron strongly slanting down and away from body; anterior tibia wide, rounded externally in basal one-third, subangulate externally at apical one-fourth (Fig. 44); elytron greenish black with red spot behind middle ...... *Thalassa*

— Epipleuron of elytron flat or only feebly inclined; anterior tibia slender throughout or with a lobe-like preapical expansion (Fig. 43); elytron not greenish black 36

36(35). Femur short, stout; tibia with a lobe-like preapical expansion (Fig. 43); elytron reddish brown, without maculation .... *Helesius*

— Femur slender; tibia slender, without a lobe-like preapical expansion; elytron usually black or brown with pale maculation, rarely immaculate. .... *Hyperaspis*

37(32). Head with mouthparts directed postero-ventrad in repose, concealing prosternum (Fig. 5); basal 2 antennomeres greatly enlarged relative to remaining antennomeres (Fig. 27) .... *Nephaspis*

— Head with mouthparts not directed postero-ventrad, not concealing prosternum; basal 2 antennomeres of normal size (Figs. 22, 23), or at least not greatly enlarged relative to club (Fig. 28) 38

38(37). Antenna very short, of 9 antennomeres, with two or more terminal setae much longer than last 3 antennomeres combined (Fig. 28); eye large, elongate .... *Pseudoscymnus*

— Antenna longer, of 10-11 antennomeres; terminal setae never longer than last 3 antennomeres combined (Figs. 22-24); eye smaller, rounded 39

39(38). Antennal club fusiform, symmetrical, with lower margin even; distal antennomere conical; last 2 antennomeres with concentration of shorter setae in membranous area on inner surface (Fig. 24); third antennomere subequal to remaining flagellomeres 40

— Antennal club oval, asymmetrical, with lower margin somewhat uneven; distal antennomere quadrate or rounded with concentration of shorter setae on distal or oblique outer face; third antennomere often elongate relative to remaining flagellomeres (Figs. 22, 23) 41

40(39). Head with clypeus more or less truncate in frontal view; postcoxal line of first abdominal ventrite not recurved at outer end (Fig. 50) (Selvadiini) . ... *Selvadius*

— Head with clypeus strongly arcuately emarginate in frontal view; postcoxal line of first abdominal ventrite recurved at outer end (Fig. 52) (Hyperaspidini in part) .... *Blaisdelliana*

41(39). Postcoxal line of abdomen reaching and joining posterior margin of ventrite; apex not recurved (Fig. 49); distal maxillary palpomere securiform, strongly expanded distally (as in fig. 30) (Diomini in part) 42

— Postcoxal line of abdomen not reaching posterior margin of ventrite; continuing parallel to margin (Fig. 50) or with apex recurved (Figs. 52, 54); distal maxillary palpomere roughly parallel-sided or barallel-shaped (as in Fig. 31), at most only weakly expanded distally (Scymnini in part)............. 43

42(41). Antenna with 11 antennomeres .... *Diomus*

— Antenna with 10 antennomeres (Fig. 23) .... *Decadiomus*

43(41). Prosternum with distinct carinae on intercoxal process; carinae often reaching anterior margin of prosternum (Fig. 16) .... *Scymnus*

— Prosternum lacking distinct carinae, or with only abbreviated ridges near coxal cavities 44

44(43). Postcoxal line complete, recurved to base of first abdominal ventrite (Fig. 54)...................... *Didion*

— Postcoxal line incomplete, not reaching base nor lateral margin of first abdominal ventrite; apex recurved (Fig. 52) or parallel to posterior margin (Fig. 50) .... *Nephus*

45(13). Tarsal claws each with a small median triangular tooth (Fig. 46); postcoxal line of abdomen not recurved toward anterior margin of ventrite (Fig. 49); specimens 6.0 to 10.0 mm in length; elytron vittate, or solid brown to beige in color.... *Myzia*

— Tarsal claws variable; if small triangular tooth present, then position of tooth more apical (Fig. 48) and postcoxal line of abdomen absent (Fig.

FIGURE 20.93-39.93. 20. diagram: variation in form/length of antenna: a shorter antenna with gradually formed club typical of scymnines (left), a longer antenna with more abruptly differentiated club typical of coccinellines and coccidulines (right); 21. Chilocorini, antenna; 22. *Scymnus* sp. (Scymnini), antenna; 23. *Decadiomus* sp. (Diomini) antenna; 24. *Brachiacantha* sp. (Brachiacanthini), antenna, apical part (rotated to expose inner surface) note specialized membranous sensory area of last two antennomeres, typical of Brachiacanthini, Hyperaspidini, Selvadiini (art: E. Roberts); 25. *Delphastus* sp., antenna; 26. *Rodolia* sp. (Noviini), antenna; 27. *Nephaspis* sp., antenna; 28. *Pseudoscymnus* sp., antenna; 29.Coccinellini (generalized), maxillary palpus; 30. *Rodolia* sp. (Coccidulinae; Noviini), maxillary palpus; 31. *Zagloba* (Scymnilini), maxillary palpus; 32. *Cephaloscymnus* (Sticholotidinae; Cephaloscymnini), maxillary palpus, apical part; 33. *Microweisea* (Sticholotidinae; Microweiseini), maxillary palpus, apical part; 34. *Stethorus* sp., (Stethorini; Scymninae), maxillary palpus; 35. *Hyperaspis* sp. (Hyperaspidini; Scymninae), maxillary palpus; 36. Coccinellini, mandible; 37. Chilocorini, mandible; 38. Epilachnini, mandible; 39. Halyziini, mandible.(Some figures modified from the works of Sasaji and Gordon; Figures 36-39 after Kovár 1996a).

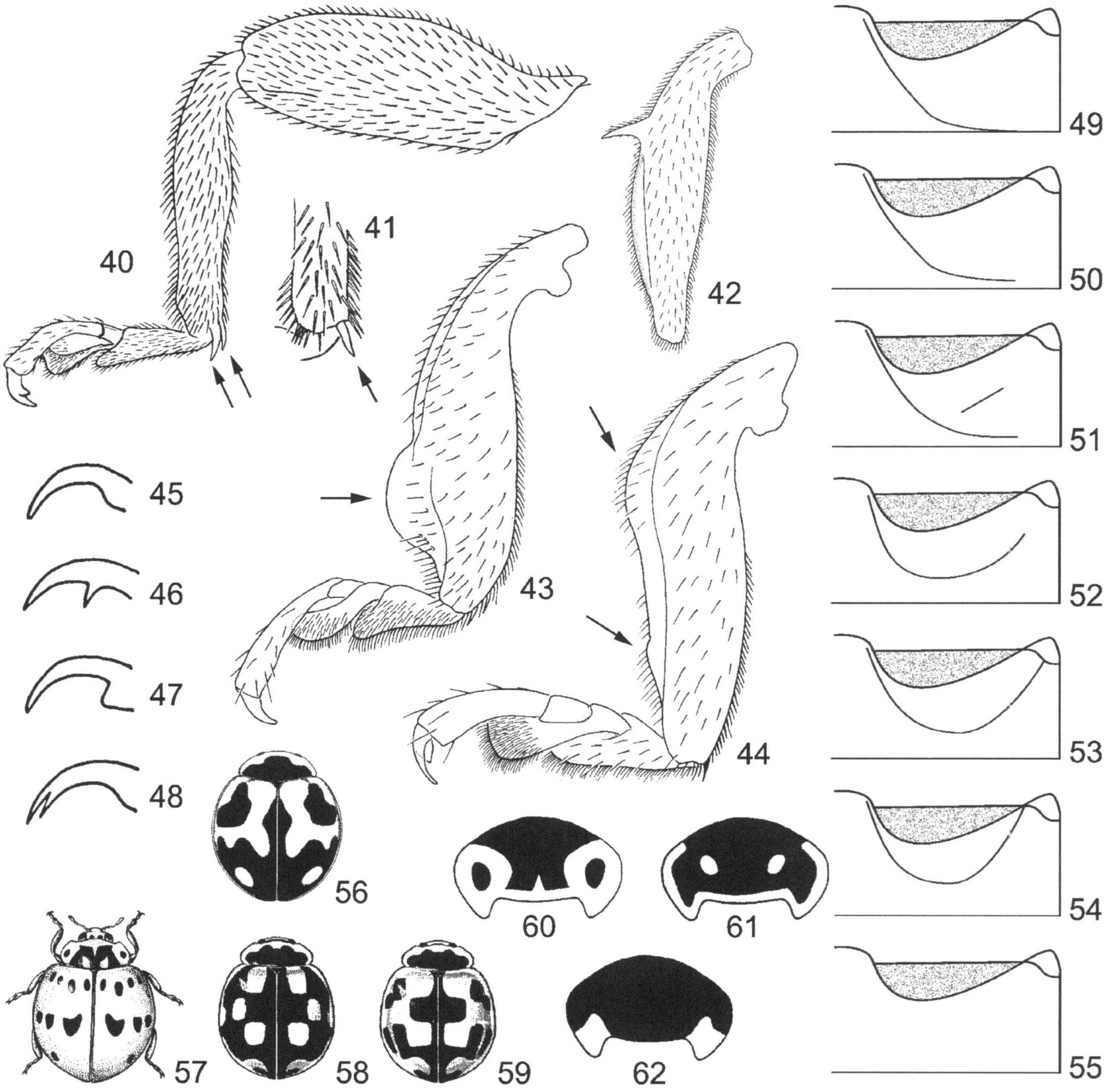

FIGURE 40.93-62.93. 40. *Myzia* sp. (Coccinellinae; Coccinellini), hind leg, arrows indicate position of paired tibial spurs; 41. *Anisosticta* sp. (Coccinellinae; Coccinellini), apex of hind tibia, arrow indicates position of single tibial spur; 42. *Brachiacantha* sp. (Brachiacanthini), front tibia; 43. *Helesius* sp. (Hyperaspidini), front leg, arrow shows position of preapical semi-circular expansion of tibia; 44. *Thalassa* sp. (Hyperaspidini), front leg, arrows show positions of rounded basal expansion (above) and subangulate preapical expansion (below); 45.-48. different configurations of the tarsal claw: 45. simple; 46. with triangular median tooth; 47. with subquadrate basal tooth; 48. with apical cleft; 49.-55. Left side of first abdominal ventrite showing different forms of postcoxal line(s): 49. postcoxal line merges with posterior margin of ventrite, not recurved at apex; 50. postcoxal line runs parallel to posterior margin of ventrite; 51. postcoxal line as above, but with oblique line present; 52. postcoxal line recurves at apex, incomplete; 53. postcoxal line reaches midpoint of lateral line; 54. postcoxal line reaches inner margin of lateral line; 55. postcoxal line obsolete; 56.-59. Dorsal color patterns: 56. *Rodolia cardinalis* (Mulsant); 57. *Olla v-nigrum* (Mulsant); 58. *Propylea quatuordecimpunctata* (L.), darker form; 59. as previous, lighter form; 60.-62. pronotal color patterns: 60. *Cycloneda* sp.; 61. *Cycloneda* sp. 62. *Coccinella* sp. (Some figures modified from the works of Gordon)

55) or recurved toward anterior margin of segment (Figs. 52-54) 46

46(45). Surface between pronotal punctures not shagreened; anterior margin of mesosternum with deep, broad, triangular emargination .... *Calvia*

— Surface between pronotal punctures shagreened; anterior margin of mesosternum truncate, with shallow emargination or with deep but narrow emargination 47

47(46). Prosternum strongly convex and thickened along midline, with anterior face forming a wide, hairy, triangular platform at middle; antenna with distal margin of ninth and tenth antennomeres protuberant on inner side; body form rhomboidal to oval (Fig. 1); 7.0 to 10.0 mm in length ...... *Anatis*

— Prosternum and antenna not as above; body form and size variable 48

48(47). Postcoxal line of first abdominal ventrite always present, joining or running parallel to posterior margin of ventrite (Figs. 49, 50); oblique dividing line often present (Fig. 51). Body form compact subcircular to slightly elongate oval (Figs. 4, 3); tips of femora hidden by or just visible beyond lateral margins of elytra (Figs. 4, 3) 57

— Postcoxal line of abdomen absent (Fig. 55) or recurved toward anterior margin of segment (Figs. 52-54); body form elongate oval to highly elongate oval or elliptical (Figs. 3, 2); tips of femora just visible to well extended beyond lateral margins of elytra (Figs. 3, 2) 49

49(48). Tarsal claw not toothed or cleft, simply widened basally (Fig. 45) 50

— Tarsal claw toothed or cleft (Figs. 47, 48) 53

50(49). Pronotal base with fine entire marginal bead; abdomen with postcoxal line obsolete (Fig. 55) 51

— Pronotal base without marginal bead; abdomen with postcoxal line distinct (Fig. 52) 52

51(50). Elytron with large black spots; metasternum with postcoxal line....................................... *Naemia*

— Elytron vittate; metasternum without postcoxal line .... *Paranaemia*

52(50). Apex of middle and hind tibia each with 2 spurs (Fig. 40); elytron with straight, regular vittae; epipleuron declivitous .... *Macronaemia*

— Apex of middle and hind tibia each with single spur (Fig. 41); elytron spotted, or with spots joined to form irregular and sinuous vittae; epipleuron horizontal .... *Anisosticta*

53(49). Tarsal claw cleft near apical 1/3 (Fig. 48) .... *Hippodamia*

— Tarsal claw with subquadrate basal tooth (Fig. 47) 54

54(53). Metasternum and first abdominal ventrite (Figs. 52-54) with distinct postcoxal lines 55

— Metasternum without postcoxal line; first abdominal ventrite without postcoxal lines (Fig. 55) or with a trace indication only .... *Coleomegilla*

55(54). Apex of middle and hind tibia each with pair of spurs (Fig. 40) 56

— Apex of middle and hind tibia each without spurs .... *Aphidecta*

56(55). Pronotal base arcuate, without marginal bead; body form slightly elongate oval (Fig. 3) .... *Adalia*

— Pronotal base sinuate, with marginal bead; body form highly elongate (Fig. 2) .... *Ceratomegilla*

57(48). Apex of middle and hind tibia without spurs..... 58

— Apex of middle and hind tibia with pair of spurs (Fig. 40) 60

58(57). Postcoxal area of first abdominal ventrite without an oblique dividing line (Fig. 50)...... *Mulsantina*

— Postcoxal area of first abdominal ventrite with an oblique dividing line (Fig. 51)........................ 59

59(58). Scutellum with apical angle much more acute and attenuated than basal angles............ *Harmonia*

— Scutellum with apical and basal angles similar .... *Neoharmonia*

60(57). Hind margin of mesepimeron with median triangular projection; pronotal hypomeron with a well defined fovea to accommodate the antennal club .... *Coelophora*

— Hind margin of mesepimeron straight or curved, without projection; pronotal hypomeron without a well defined fovea 61

61(60). Pronotum black with large, subtrapezoidal or triangular white spot on each anterolateral angle (Fig. 62); elytral ground color yellow to red with black bands or spots in many .... *Coccinella*

— Pronotal and elytral color pattern not as above 62

62(61). Pronotum black with white lateral border and discal spot in each lateral third (Fig. 61); spot may be connected anteriorly and laterally to form a complete or broken ring-shaped mark (Fig. 60); elytra without black markings (Fig. 63) .... *Cycloneda*

— Pronotum not as above; elytra with black markings 63

63(62). Distal antennomere elongate, oval; scutellum with base slightly longer than side; maculation on elytron typically forming a yellow and black "checkerboard" pattern (Figs. 58, 59) .. *Propylea*

— Distal antennomere short, robust, obtriangular; scutellum with side slightly longer than base. Elytron black with red spot or pale, ashen with minute dark spots not forming a "checkerboard" pattern (Fig. 57) .... *Olla*

Classification of the Nearctic Genera

Coccinellidae Latreille 1807

For most of the genera listed below, keys to the North American species, morphological and habitus illustrations and other useful information can be found in Gordon (1985). This citation will not be repeated for each entry. More recent papers are referenced below along with older but more complete works when appropriate.

Sticholotidinae Weise 1901

Pharini Casey 1899 (unavailable, preoccupied type genus)
Pharini Ganglbauer 1899 (unavailable, preoccupied type genus)
Sticholotini Weise 1901
Clanini Weise 1901 (unavailable, preoccupied type genus)
Coelopterini Della Beffa 1912
Sticholotidinae Gordon 1977 (emendation)

Diagnosis. North American members of this subfamily can be easily distinguished by the shape of the terminal maxillary palpomere which is distinctly elongate (Fig. 32, 33): conical, oval, or parallel-sided with an oblique apex (taxa from other parts of the world may have this palpomere shortened or distally expanded). Additional diagnostic characteristics include: mentum generally narrowly articulated with submentum (Fig. 17); middle coxal cavities broadly separated; size less than 3.0 mm; dorsal surfaces glabrous or hirsute.

Gordon (1977) discusses the taxonomy, phylogeny and zoogeography of the New World members. Kovár (1996b) provides a revised phylogeny which recognizes two phyletic series, each with a distinct form to the metendosternite and genitalia of both sexes. Gordon (1994b) contributes additional West Indian genera to the tribe Sticholotidini. Predominantly scale predators, but *Delphastus* (Serangiini) are predacious on whiteflies.

Microweiseini Leng 1920

Pharini Casey 1899 (unavailable, preoccupied type genus)
Microweisini Gordon 1985 (incorrect subsequent spelling)

Microweisea Cockerell 1903 (new name for *Epismilia* Cockerell 1900). New World, from southern Canada into South America; 5 species described from north of Mexico.

Smilia Weise 1891, not Germar 1833
Epismilia Cockerell 1900, not Fromental 1861 (new name for *Smilia* Weise)
Pseudoweisea Schwarz 1904 (name made available by accident)

Coccidophilus Brethes 1905. New World; 2 species described from north of Mexico; *C. atronitens* (Casey), California, Nevada, Arizona, Oregon, and Utah; *C. marginata* (LeConte), Maine, New York, Pennsylvania, New Jersey, and Michigan; 1 additional new species reported from Florida (Peck and Thomas 1998).

Cryptoweisea Gordon 1970

Gnathoweisea Gordon 1970. Known only from North America; 6 species, California, Nevada, Arizona, New Mexico, and Texas.

Nipus Casey 1899. Southwestern United States; 4 species, California, Utah, Wyoming, Arizona, and Colorado.

Serangiini Pope 1962

Serangiini Blackwelder 1945 (unavailable name, published without description).

Delphastus Casey 1899. New World; 3 species from north of Mexico: Rhode Island, Connecticut, New York, west to California, south to Texas and Florida. Gordon (1994c) revises, keys and illustrates members of the genus from the Western Hemisphere.

Oeneis LeConte 1852, not Mulsant 1850
Cryptognatha Crotch 1874 (in part), not Mulsant 1850
Lioscymnus Champion 1913

Cephaloscymnini Gordon 1985

Cephaloscymnus Crotch 1873. New World, most diverse in the tropics; 3 species from north of Mexico with scattered distributional data: Illinois, New Jersey, Maryland, Virginia, West Virginia, District of Columbia, Indiana, Tennessee, South Carolina, Texas, New Mexico, Arizona, California.

Scymninae Mulsant 1846

Scymniens Mulsant 1846
Scymninae Della Beffa 1912

Diagnosis. Antennae relatively short (Fig. 22-24, 27, 28), usually two-thirds head width or less (Fig. 20, left); middle coxal cavities broadly separated; size generally less than 3.0 mm (except *Cryptolaemus*, *Thalassa*). Mentum broadly articulated with submentum (Fig. 18). This tribe is difficult to characterize and probably polyphyletic; antennae are of at least two different types, exemplified by *Scymnus* (Fig. 22) and allies, on the one hand, and *Hyperaspis* and allies (Fig. 24), on the other; maxillary palpus (Fig. 31, 34, 35) with terminal palpomere parallel-sided (e.g., Scymnillini) to barrel-shaped (e.g., Scymnini), short and weakly convergent apically (Stethorini), or more or less securiform and apically expanded (Hyperaspidini, Brachiacanthini, Selvadiini, Diomini). Predacious on various Homoptera and mites.

Scymnillini Casey 1899

Zilini Gordon 1985 (unnecessary replacement name for Scymnillini)

Note: This tribe has many affinities with Sticholotidini (Sticholotidinae) and may be misclassified in Scymninae. The two included genera are predacious on whiteflies.

Zilus Mulsant 1850. Primarily neotropical with 4 species recorded from the United States; in the east, from Maryland to Florida and west to Wisconsin with disjunct localities in Louisiana; in the west from Idaho and Washington to California and Arizona.

Scymnus (*Zilus*) Mulsant 1850
Scymnillus Horn 1895
Scymnillodes Sicard 1922

Zagloba Casey 1899. New World tropical and temperate; 4 species from north of Mexico: Oregon, California, Arizona, Texas, and Florida.

Stethorini Dobzhansky 1924

Stethorus Weise 1885. Worldwide; 5 species widely distributed north of Mexico. Predators of spider mites. Gordon and Chapin (1983) treated the Western Hemisphere species.

Scymnini Mulsant 1846

Scymniaires Mulsant 1846
Scymnini Costa 1849

Nephaspis Casey 1899. 4 neotropical species, one of which, *N. oculatus* (Blatchley), established in scattered localities in the United States: Florida, Louisiana, Texas, Iowa, and Vermont.
Nephasis: Korschefsky 1931 (error)

Cryptolaemus Mulsant 1853. 1 species, *C. montrouzieri* Mulsant, introduced from Australia for biocontrol of *Planococcus citri* (Risso); established in Indiana, Missouri, Florida, and California.

Didion Casey 1899. Restricted to North America; 3 species, generally distributed.

Scymnus Kugelann 1794. Worldwide distribution. Gordon (1976b) revised the genus north of Mexico.
- subgenus *Scymnus* Kugelann 1794. 11 species, widely distributed north of Mexico.
- subgenus *Pullus* Mulsant 1846. 82 species, widely distributed north of Mexico.

Pseudoscymnus Chapin 1962 (replacement name needed; preoccupied by *Pseudoscymnus* Herre 1935). *Pseudoscymnus tsugae* McClure and Sasaji, imported from Japan to control woolly hemlock adelgid, *Adelges tsugae* Annand, has become established at release sites in Connecticut, Virginia, and New Jersey.
Clitostethus Kamiya 1961, not *Clitostethus* Weise 1885

Nephus Mulsant 1846. Worldwide (at least nominally); 5 subgenera are recognized in the New World fauna. Gordon (1976b, 1985) revised the genus from north of Mexico.
- subgenus *Nephus* Mulsant 1846. 1 species, *N. (N.) ornatus* LeConte, with 2 subspecies, United States and Canada
- subgenus *Sidis* Mulsant 1850 (as subgenus of *Scymnus*). 1 species, *N. (Sidis) binaevatus* (Mulsant), California.
- subgenus *Turboscymnus* Gordon 1976b. 1 species, *N. (Turboscymnus) georgei* (Weise), California, Oregon, Idaho, Alberta, Ontario, and Quebec (Paquin and Dupérré 2000).
- subgenus *Scymnobius* Casey 1899. 9 species, widely distributed in the United States, extending into southern Canada.
- subgenus *Depressoscymnus* Gordon 1976b. 1 species, *N. (Depressoscymnus) schwartzi* Gordon, Arizona.

Diomini Gordon 1999

Diomus Mulsant 1850. Worldwide; 18 species recorded from north of Mexico, generally distributed. The generic placement of some of these species may need to be reassessed. Primarily mealybug predators. Gordon (1999) revised the South American members of *Diomus* and related taxa.

Decadiomus Chapin 1933. Primarily Carribean; 1 species, *D. bahamicus* (Casey) reported in Florida (Peck and Thomas 1998).

Selvadiini Gordon 1985

Selvadius Casey 1899. A New World genus; 4 species occur north of Mexico: Texas, Arizona, California, and Colorado. Prey unknown; possibly scale insects.

Hyperaspidini Mulsant 1846

Hyperaspiens Mulsant 1846
Iperaspini Costa 1849 (= Hyperaspini)
Hyperaspini Casey 1899
Hyperaspides Crotch 1873
Hyperaspidae Berg 1874
Hyperaspites Chapuis 1876
Hyperaspidina Jacobson 1916
Hyperaspidini Wingo 1952 (emendation)

Blaisdelliana Gordon 1970. Monobasic genus; *B. sexualis* (Casey), California, Arkansas, and Utah.

Helesius Casey 1899. Only 3 known species; 2 species in North America: Montana, Colorado, and Texas; 1 species, Colombia.

Thalassa Mulsant 1850. Neotropical with 6 described species, 1 species, *T. montezumae* Mulsant, penetrating north of Mexico: Arizona, Texas, and Louisiana.

Hyperaspis Redtenbacher 1844. Worldwide; 94 species north of Mexico, generally distributed. Predators of various Homoptera.
Oxynchus LeConte 1850

Hyperaspidius Crotch 1873. New World; 26 species, generally distributed in the United States and southern Canada; undescribed species occur in Mexico and Central America. Predators of scale insects and mealybugs.

Brachiacanthini Mulsant 1850

Brachyacanthaires Mulsant 1850
Brachiacanthadini Duverger 1989:143
Brachyacanthadini Duverger 1989 (misspelling)
Brachiacanthini Pakaluk *et al.* 1994 (emendation)

Brachiacantha Dejean 1837. New World; 25 species north of Mexico, generally distributed. Predators of coccids in ant nests and possibly other Homoptera.

Brachyacantha Chevrolat 1842 (unjustified emendation).

Cryptognathini Mulsant 1850

Cryptognathaires Mulsant 1850
Pentiliaires Mulsant 1850
Oeneini Casey 1899 (genus preoccupied)
Cryptognathini Gordon 1971
Oeniini Gordon 1985 (error)

Cryptognatha Mulsant 1850. Neotropical; 1 species, *C. nodiceps* Marshall, introduced from Trinidad for biocontrol of *Aspidiotus destructor* Signoret, established in Florida.

Chilocorinae Mulsant 1846

Chilocoriens Mulsant 1846
Exochomaires Mulsant 1850
Chilocorinae Sasaji 1968
Clanini Pakaluk *et al.* 1994 (presumably based on *Clanis* Mulsant 1850, misspelling of *Cladis* Mulsant 1850; not Clanini Weise 1901, see entry under Sticholotidinae)

Diagnosis. North American members of this subfamily all belong to the tribe Chilocorini (tribes Telsimiini, Platynaspini and Aspidimerini occur in the Eastern Hemisphere), and can be readily identified by the following combination of character states: eye canthus deeply dividing eye (Fig. 10), broad, band-like, expanded to cover basal antennomeres from dorsal or frontal view; clypeus not projecting, with semicircular emargination medially; antenna (Fig. 21) of ten or fewer antennomeres, with spindle-shaped flagellum; mandible (Fig. 37) scythe-like with single apical tooth; tibia angulate externally in many species; dorsum apparently glabrous (North America natives), but may exhibit lateral pubescence at least on pronotum; only the introduced species *Exochomus metallicus* Korschefsky has the pronotum and elytron evenly covered with moderately long silky hair. Predominantly scale predators, but some species known to feed on mealybugs, aphids, adelgids and psyllids.

Chilocorini Mulsant 1846

Chilocoriens Mulsant 1846
Exochomaires Mulsant 1850
Chilocorini Costa 1849
Clanini Pakaluk *et al.* 1994 (presumably based on *Clanis* Mulsant 1850, misspelling of *Cladis* Mulsant 1850; not Clanini Weise 1901, see entry under Sticholotidinae)

Chapin (1965) revised the genera of the World. Kovár (1995) treated members of the genera, *Brumus*, *Brumoides* and *Exochomus*, but focused primarily on the palearctic fauna.

Brumoides Chapin 1965 (*sensu lato*, not *Brumoides sensu* Kovár 1995 who restricted use of this name to certain Old World species). 3 species, 1 with 3 subspecies, widely distributed in the United States but absent from the southeastern states, extending into Canada.

Brumus (of authors; not Mulsant 1850)

Brumus Mulsant 1850 (in part). 1 introduced palearctic species, *B. quadripustulatus* (L.), transferred from *Exochomus* by Kovár (1995).

Axion Mulsant 1850. 2 species; *A. plagiatum* (Olivier), Pacific Coast and southwestern states from Oregon to Louisiana; *A. tripustulatum* (De Geer), Pennsylvania south to Florida, west to Colorado and Texas.

Curinus Mulsant 1850. Neotropical; 1 adventive species, *Curinus coeruleus* (Mulsant), reported in Florida (Peck and Thomas 1998).

Arawana Leng 1908. New World; 1 species in North America, *A. arizonica* (Casey), Arizona.

Exochomus Redtenbacher 1843 (*sensu lato*). Worldwide; 9 species north of Mexico, generally distributed. The 7 native nearctic species which have a non-bordered pronotal base are excluded from *Exochomus sensu* Kovár (1995), but as no alternative placement is provided, they remain in *Exochomus* for the time being. Only the two introduced species, *E. flavipes* and *E. metallicus* with bordered pronotal base belong to *Exochomus sensu* Kovár (1995) (see also *Brumus*).

Halmus Mulsant 1850. 1 species, *H. chalybeus* (Boisduval), introduced from Australia, established in California.

Orcus Mulsant 1850

Chilocorus Leach 1815, in Brewster. Worldwide; 8 species in North America, generally distributed.

Coccidulinae Mulsant 1846

Cocciduliens Mulsant 1846
Trichosomides Mulsant 1846 (unavailable name, not based on genus)
Coccidulinae Sasaji 1968

Diagnosis. Members of this subfamily are difficult to characterize, but can usually be recognized by the following combination of character states: Dorsal surfaces conspicuously pubescent; body length 2.0 to 7.5 mm; antenna usually long (more than two-thirds head width) (Fig. 20, right), loosely articulated, with irregularly shaped club (externally serrate or papillate), but shorter, more compact in Noviini (Fig. 26) and Exoplectrini; meso- and metasternum narrowly articulated; maxillary palpus securiform (Fig. 30) to parallelsided; legs slender, simple, to flattened and highly modified. Predominantly scale predators.

Coccidulini Mulsant 1846

Coccidulini Costa 1849
Rhizobiares Mulsant 1846
Cocciduliens Mulsant 1846
Coccidulides Crotch 1873
Rhizobiides Crotch 1874
Rhizobiini Weise 1885
Rhizobiinae Della Beffa 1912
Coccidulina Jacobson 1916

Gordon (1994a) revised the South American genera and species.

Coccidula Kugelann 1798. Europe and North America; 1 species, *C. lepida* LeConte, in northern United States and southern Canada.
Strongylus Panzer 1813
Cacidula Curtis 1827
Cacicula Stephens 1828

Rhyzobius Stephens 1829. 2 species, southern United States.
Rhizobius Stephens 1832 (error)
Rhizobius Agassiz 1846 (unjustified emendation)
Lindorus Casey 1899
Rhizobiellus Oke 1951 (unnecessary replacement name)

Noviini Mulsant 1850

Noviares Mulsant 1850
Rodoliaires Mulsant 1850
Noviini Ganglbauer 1899

Rodolia Mulsant 1850. 1 species, *R. cardinalis* (Mulsant), South Carolina, Florida, Louisiana, Texas, New Mexico, Arizona, and California; introduced from Australia for biocontrol of *Icerya purchasi* Maskell.
Rodolia (*Macronovius*) Weise 1885

Anovia Casey 1920. New World; 1 species, *A. virginalis* (Wickham), occurring north of Mexico: Texas, New Mexico, Arizona, and Utah.

Exoplectrini Crotch 1874

Chnoodiens Mulsant 1850
Chnoodiaires Mulsant 1850
Siolaires Mulsant 1850
Exoplectrae Crotch 1874
Exoplectrides Gorham 1895
Exoplectrinae Weise 1904
Exoplectrini Casey 1908

Exoplectra Chevrolat 1837. Primarily neotropical; 1 species, *E. schaefferi* Gordon occurring north of Mexico: Arkansas.

Azyini Mulsant 1850

Azyaires Mulsant 1850
Azyae Crotch 1874
Azyini Schilder and Schilder 1928
Azyinae Gordon 1994

Gordon (1980) revised the neotropical members of this tribe.

Azya Mulsant, 1850. Neotropical; 1 adventive species, *A. orbigera orbigera* Mulsant 1850 (=*A. luteipes* Mulsant 1850, misidentification in Woodruff and Sailer 1977) established in Florida.

Pseudoazya Gordon 1980. Neotropical; 1 species, *P. trinitatis* (Marshall), released in the Miami, Florida area in 1938 and recovered in 1939, but no evidence to suggest survival in Florida after 1939.

Coccinellinae Latreille 1807

Coccinellides Leach 1815, in Brewster
Aphidiphages LaPorte 1840
Gymnosomides Mulsant 1846
Coccinelliti Costa 1849
Coccinellidae Crotch 1873
Coccinellides Aphidiphages Chapuis 1876
Coccinellidae Aphidiphages Weise 1885
Coccinellinae Ganglbauer 1899

Diagnosis. This subfamily contains some of the larger, more conspicuously colored members of the North American lady beetle fauna, and can be easily recognized by the following combination of character states: dorsal surfaces glabrous; body length 1.75 to 10.5 mm; terminal maxillary palpomere securiform (Fig. 29); antenna two-thirds head width or longer (Fig. 20, right), with 11 antennomeres (except in some neotropical species). Primarily predacious on aphids and other Homoptera, but occasionally specializing in other prey groups; all members of the Halyziini feed on powdery mildews.

Coccinellini Latreille1807

Coccinellinae Latreille 1807
Adoniates Mulsant 1846
Coccinellaires Mulsant 1846
Coccinellates Mulsant 1846
Coccinelliens Mulsant 1846
Hippodamiaires Mulsant 1846
Micraspaires Mulsant 1846
Mysiates Mulsant 1846
Hippodamiini Costa 1849
Micraspidarii Costa 1849
Cariaires Mulsant 1850
Alesiaires Mulsant 1850

Coelophoraires Mulsant 1850
Cydoniaires Mulsant 1850
Coccinellina Thomson 1866
Coccinellides Thomson 1866
Coccinellidae Berg 1874
Hippodamiidae Berg 1874
Tytthaspides Crotch 1874
Cariites Chapuis 1876
Coccinellites Chapuis 1876
Hippodamiites Chapuis 1876
Coccinellini Weise 1885
Synonychini Weise 1885
Halyziides Gorham 1892 (in part)
Synonychinae Della Beffa 1912
Anisostictini Jacobson 1916
Coccinellina Jacobson 1916
Synonychina Jacobson 1916
Hippodamiina Dobzhansky 1926

Paranaemia Casey 1899. Monobasic genus; 1 species, *P. vittigera* (Mannerheim), western United States, western Canada.

Naemia Mulsant 1850. North America through Central America and the Caribbean; currently treated as 1 species with 2 subspecies; in the United States *N. s. seriata* Melsheimer ranges from Rhode Island, south to Texas (coastal localities); *N. s. litigiosa* Mulsant is recorded from southern California and southern New Mexico. This genus is in need of revision.

Coleomegilla Timberlake 1920. Restricted to the New World, most diverse in the tropics. Gordon (1985) followed Timberlake (1943) in recognizing 3 subspecies of *C. maculata* (DeGeer) from north of Mexico: *C. m. lengi*, eastern United States; *C. m. strenua*, southwestern United States; and *C. m. fuscilabris* (Mulsant), South Carolina to Florida and west to Louisiana (coastal localities). Problems with the current species level classification are discussed in Krafsur and Obrycki (2000). This genus is in need of revision.

Megilla Mulsant 1850 (in part), not Fabricius 1805, not Erichson 1804

Ceratomegilla Crotch 1873. Monobasic; *C. ulkei* Crotch, Alaska, and arctic and subarctic Canada. Although separable from indigenous North American members of *Hippodamia*, *Ceratomegilla* is not so easily distinguished if the entire holarctic fauna is considered; generic limits in need of reassessment.

Ceratomagilla Malkin 1943 (in error)
Megilla Mulsant 1850 (in part), not Fabricius 1805, not Erichson 1804
Spiladelphia Tian-Shanskij and Dobzhansky 1923
Hippodamia (*Ceratomegilla*) Iablokoff-Khnzorian 1982

Hippodamia Dejean 1837. Primarily holarctic; 18 species occur north of Mexico (1 recently introduced), generally distributed. Chapin (1946) illustrates some of the variability of colorpatterns within the New World species.

Hemisphaerica Hope 1840
Adonia Mulsant 1846

Anisosticta Dejean 1837. Holarctic; 2 species in North America: *A. bitriangularis* (Say), Labrador to New Jersey, west to Alaska, California, and British Columbia; *A. borealis* Timberlake, Manitoba to Alaska.

Anisostica Malkin 1943 (error)

Macronaemia Casey 1899. Oriental and Nearctic; 1 species, *M. episcopalis* (Kirby) in North America: Ontario to New York, west to Yukon Territory and northern California.

Micronaemia Weise 1905

Aphidecta Weise 1899 (emendation). A monotypic, palearctic genus; *A. obliterata* (Linnaeus) was released in the United States and Canada for biocontrol of *Adelges piceae* (Ratzeburg); established in North Carolina.

Aphideita Weise 1893 (error)

Adalia Mulsant 1846 (addenda). Worldwide; 1 holarctic species, *A. bipunctata* (L.), widely distributed in the United States and Canada, as well as temperate parts of South America (Argentina, Chile).

Idalia Mulsant 1846, not Hübner 1819
Arrowella Brethes 1925

Coccinella Linnaeus 1758. Primarily holarctic; 12 species occur in the United States, generally distributed. Brown (1962) and Brown (1967) provide keys, illustrations and additional discussion of this genus in the United States and Mexico respectively.

Spilota Billberg 1820
Neococcinella Savoyskaya 1969
Dobzhanskia Iablokoff-Khnzorian 1970

Cycloneda Crotch 1871. Primarily neotropical; 3 species, generally distributed north of Mexico. Gordon and Vandenberg (1993) provide a larval key to the North American species.

Daulis Mulsant 1850 (not Erichson 1842)
Coccinellina Timberlake 1943

Harmonia Mulsant 1850. An exotic genus with 3 introduced species, now widely distributed in the United States. Gordon and Vandenberg (1991) provide a key and illustrations of the introduced species. (Volume 2, Color Figure 31)

Anatis Mulsant 1846. Holarctic and neotropical (species from the latter region previously placed in other genera); 4 species occur north of Mexico; most common in coniferous forests, woodland habitats, urban plantings of mature trees.

Myzia LeConte 1852 (in part)
Pelina Mulsant 1850
Palla Mulsant 1850, not Hübner 1819, not Billberg 1820
Neopalla Chapin 1955 (new name for *Pelina* Mulsant and *Palla* Mulsant)

FIGURE 63.93. *Cycloneda sanguinea* Linnaeus.

Myzia Mulsant 1846. Holarctic and neotropical; 3 species north of Mexico, generally distributed; arboreal.

Mysia Mulsant 1846
Neomysia Casey 1899

Calvia Mulsant 1850. Primarily Old World; 1 holarctic species, *C. quatuordecimguttata* (L.), with numerous color forms, northern United States and Canada.

Anisocalvia Crotch 1871
Eocaria Timberlake 1943

Propylea Mulsant 1846. Old World; 1 species, *P. quatuordecimpunctata* (L.) introduced into North America for control of various aphid species; established in eastern Canada southward along the eastern United States from Maine into New York, New Jersey, and Pennsylvania. Vandenberg and Gordon (1991) review and illustrate the known species.

Propylaea Mulsant 1846 (error)

Coelophora Mulsant 1850. Primarily Old World tropics, including Africa, Asia and Australia; 1 species, *C. inaequalis* (Fabricius 1775), introduced from Puerto Rico; eastern United States and Hawaii.

Olla Casey 1899. New World with 4 described species; 1 species, *O. v-nigrum* (Mulsant), generally distributed in the United States, complete range from southern Canada as far South as Argentina. Vandenberg's revision (1992) provides a key and illustrations of the known species.

Neoharmonia Crotch 1871. Only 1 species, *N. venusta* (Melsheimer), with 2 subspecies currently recognized, but these may well prove to be distinct; in North America, *N. v. venusta* widely distributed in the eastern half of the United States; *N. v. ampla* (Mulsant) recorded from Arizona, Texas and New Mexico. These and related taxa are well represented in the Neotropics as predators of chrysomelid eggs and larvae.

Agrabia Casey 1899
Neoharmonia Casey 1899, not Crotch 1871
Harmoniaspis Casey 1908

Mulsantina Weise 1906. Restricted to the New World; 4 species widely distributed in North America; primarily arboreal in habits. Chapin (1985) provides a key and illustrations of the known species (mostly North American). The diverse related fauna of South America has not been studied.

Cleis Mulsant 1850, not Guerin 1831
Pseudocleis Casey 1908

Halyziini Mulsant 1846

Halyziaires Mulsant 1846
Halyziates Mulsant 1846
Aliziarii Costa 1849 (= Halyziini)
Halyziides Gorham 1892 (in part)
Psylloborini Casey 1899
Halyziini Capra 1927

Psyllobora Dejean 1836. Worldwide, most diverse in the tropics; 6 species occur north of Mexico, generally distributed. Feed on powdery mildew.

Psyllobora (Psyllobora) Mulsant 1850
Thea Mulsant 1846

Epilachninae Ganglbauer 1899

Epilachniens Mulsant 1846

Diagnosis. North American members of this subfamily all belong to the tribe Epilachnini, and can be easily identified by the following combination of character states: Mandible (Fig. 38) with apex multidenticulate, bearing three or more large irregular teeth; all tibiae with one or two apical spurs present (North American fauna); dorsal surfaces pubescent; antenna inserted dorsally between eyes and distant from inner ocular margin (Fig. 9), long, loosely articulated, with 11 antennomeres and inner margin of club weakly serrated; eye bean-shaped without an abrupt notch or emargination; length 3.5 mm or greater. Gordon (1976a) revised the species of the Western Hemisphere.

Epilachnini Costa 1849

Epilachna Dejean 1837. Primarily neotropical; 3 species occur in the United States, but absent from western and northern central states. Two plant families serve as hosts for the North American species. *Epilachna borealis* (F.) and *E. tredecimnotata* (Latreille) feed on members of the Cucurbitaceae, *E. varivestis* Mulsant feeds on members of the Leguminosae.

Solanophila Weise 1898
Afissa Dieke 1947

Subcoccinella Huber 1842. Old World; 1 species, *S. vigintiquatuorpunctata* (L.), accidentally introduced from Europe where it is a pest of alfalfa; established in Illinois, Maryland, New Jersey, New York, Ohio, Pennsylvania, and West Virginia. Fortunately the established biotype feeds primarily on bouncing bet (*Saponaria officinalis* L.).

Bibliography

BOOTH, R. G., M. L. COX and R. B. MADGE. 1990. IIE guides to insects of importance to man. 3. Coleoptera. University Press. Cambridge, 384 pp.

BROWN, W. J. 1962. A revision of the forms of *Coccinella* L. occurring in America north of Mexico. Canadian Entomologist, 94: 785-808.

BROWN, W. J. 1967. The Mexican forms of *Coccinella* (Coleoptera: Coccinellidae). Canadian Entomologist, 99: 107-108.

CHAPIN, E. A. 1946. Review of the New World species of *Hippodamia* Dejean (Coleoptera: Coccinellidae). Smithsonian Miscellaneous Collections, 106(11): 1-39, pl. 1-22.

CHAPIN, E. A. 1965. The genera of the Chilocorini (Coleoptera: Coccinellidae). Bulletin of the Museum of Comparative Zoology, 133: 227-271.

CHAPIN, J. B. 1985. Revision of the genus *Mulsantina* Weise (Coleoptera: Coccinellidae). Annals of the Entomological Society of America, 78: 348-368.

CHAZEAU, J. H., H. FÜRSCH and H. SASAJI. 1989. Taxonomy of Coccinellids. Coccinella, 1: 6-8.

CROWSON, R. A. 1955. The natural classification of the families of Coleoptera. Nathaniel Lloyd. London, 187 pp.

DREA, J. J. and R. D. GORDON. 1990. Coccinellidae. Pp. 19-40. *In:* D. ROSEN, ed. The armored scale insects, their biology, natural enemies and control, vol. B. Elsevier Science. Amsterdam.

DUVERGER, C. 1989. Contribution à l'étude des Hyperaspinae. 1ère note (Coleoptera, Coccinellidae). Bulletin de la Société Linnéenne de Bordeaux, 17: 143-157.

FÜRSCH, H. 1996. Taxonomy of Coccinellids. Coccinella, 6: 28-30.

GORDON, R. D. 1976a. A revision of the Epilachninae of the Western Hemisphere (Coleoptera: Coccinellidae). Agricultural Research Service, United States Department of Agriculture, Technical Bulletin, No. 1493: 1-409.

GORDON, R. D. 1976b. The Scymnini (Coleoptera: Coccinellidae) of the United States and Canada: Key to genera and revision of *Scymnus*, *Nephus* and *Diomus*. Bulletin of the Buffalo Society of Natural Sciences, 28: 1-362.

GORDON, R. D. 1977. Classification and phylogeny of the New World Sticholotidinae (Coccinellidae). Coleopterists Bulletin, 31: 185-228.

GORDON, R. D. 1980. The tribe Azyini (Coleoptera: Coccinellidae): historical review and taxonomic revision. Transactions of the American Entomological Society (Philadelphia), 106: 1149-203.

GORDON, R. D. 1985. The Coccinellidae (Coleoptera) of America north of Mexico. Journal of the New York Entomological Society, 93: 1-912.

GORDON, R. D. 1994a. South American Coccinellidae (Coleoptera) Part III: definition of Exoplectrinae Crotch, Azyinae Mulsant, and Coccidulinae Crotch; a taxonomic revision of Coccidulini. Revista Brasileira de Entomologia, 38: 681-775.

GORDON, R. D. 1994b. West Indian Coccinellidae VI (Coleoptera): new genera and species of Sticholotidini and a cladistic analysis of included genera. Journal of the New York Entomological Society, 102: 232-241.

GORDON, R. D. 1994c. South American Coccinellidae (Coleoptera) Part III: taxonomic revision of the Western Hemisphere genus *Delphastus*. Frustula Entomologica, (1994) 17(30): 71-133.

GORDON, R. D. 1999. South American Coccinellidae Part IV: A systematic revision of the South American Diomini, new tribe. Annales Zoologici (Warsaw), 49 (Supplement 1): 1-219.

GORDON, R. D. and E. A. CHAPIN. 1983. A revision of the New World species of *Stethorus* Weise (Coleoptera: Coccinellidae). Proceedings of the Entomological Society of Washington, 84: 250-255.

GORDON, R. D. and N. J. VANDENBERG. 1991. Field guide to recently introduced species of Coccinellidae (Coleoptera) in North America, with a revised key to North American genera of Coccinellini. Proceedings of the Entomological Society of Washington, 93: 845-864.

GORDON, R. D. and N. J. VANDENBERG. 1993. Larval systematics of North American *Cycloneda* Crotch (Coleoptera: Coccinellidae). Entomologica Scandinavica, 24: 301-312.

GORDON, R. D. AND N. J. VANDENBERG. 1995. Larval systematics of North American *Coccinella* L. Entomologica Scandinavica, 26: 67-86.

HODEK, I. and A. HONEK. 1996. Ecology of Coccinellidae. Kluwer Academic Publishers. Netherlands, 464 pp.

IABLOKOFF-KHNZORIAN, S. M. 1982. Les Coccinelles, Coléoptères-Coccinellidae. Tribu Coccinellini des Régions Paléarctique et Orientale. Société Nouvelle des Éditions Boubée, Paris, 568 pp.

KLAUSNITZER, B. and H. KLAUSNITZER. 1997. Marienkäfer (Coccinellidae). Die Neue Brehm-Bücherei, 451 pp.

KOVÁR, I. 1995. Revision of the genera *Brumus* Muls. and *Exochomus* Redtb. (Coleoptera, Coccinellidae) of the Palaearctic Region. Part I. Acta Entomologica Musei Nationalis Pragae, 44(1995): 5-124.

KOVÁR, I. 1996a. Morphology and anatomy. Pp.1-18. *In:* I. Hodek and A. Honek, eds. Ecology of Coccinellidae. Kluwer Academic Publishers. Netherlands, 464 pp.

KOVÁR, I. 1996b. Phylogeny. Pp.19-31. *In:* I. Hodek and A. Honek, eds. Ecology of Coccinellidae. Kluwer Academic Publishers. Netherlands, 464 pp.

KRAFSUR, E. S. and J. J. OBRYCKI. 2000. *Coleomegilla maculata* Degeer (Coleoptera: Coccinellidae) is a species complex. Annals of the Entomological Society of America, 93: 1156-1163.

KUZNETSOV, V. N. 1997. Lady beetles of the Russian Far East. Memoir No. 1, Center for Systematic Entomology, Sandhill Crane Press. Gainesville, 248 pp.

LAWRENCE, J. F. and A. F. NEWTON, Jr. 1995. Families and subfamilies of Coleoptera (with selected genera, notes, references and data on family-group names). Pp. 779-1006. *In:* J. Pakaluk and S.A. Slipinski, eds. Biology, phylogeny and classification of Coleoptera; papers celebrating the 80th birthday of Roy A. Crowson. Museum I Instytut Zoologii PAN, Warsaw.

LESAGE, L. 1991. Coccinellidae (Cucujoidea), the lady beetles, ladybirds. Pp. 485-494. *In:* F. W. Stehr, ed. Immature Insects. Vol. 2. Kendall/Hunt. Dubuque, Iowa.

MAJERUS, M. E. N. 1994. Ladybirds. Harper Collins. London, 367 pp.

PAKALUK, J., S. A. SLIPINSKI and J. F. LAWRENCE. 1994. Current classification and family-group names in Cucujoidea (Coleoptera). Genus, 5: 223-268.

PAQUIN, P. and N. DUPÉRRÉ 2000. Biologie, répartition géographique et variation du patron élytral d'une coccinelle rarement trouvée en Amérique du Nord: *Nephus georgei* Weise (Coleoptera: Coccinellidae: Scymnini). Fabreries, 25: 7-14.

PECK, S. B. and M. C. THOMAS. 1998. A distributional checklist of the beetles (Coleoptera) of Florida. Arthropods of Florida and Neighboring Land Areas, 16: i-viii, 1-180.

PHUOC, D. T. and STEHR, F. W. 1974. Morphology and taxonomy of the known pupae of Coccinellidae (Coleoptera) of North America, with a discussion of phylogenetic relationships. Contributions of the American Entomological Institute, 10: 1-125.

POPE, R. D. 1979. Wax production by coccinellid larvae (Coleoptera). Systematic Entomology, 4: 171-196

POPE, R. D. 1988. A revision of the Australian Coccinellidae (Coleoptera). Part I. Subfamily Coccinellinae. Invertebrate Taxonomy, 2: 633-735.

POPE, R. D. and J. F. LAWRENCE. 1990. A review of *Scymnodes* Blackburn, with the description of a new Australian species and its larva (Coleoptera: Coccinellidae). Systematic Entomology, 15: 241-252.

REES, B. E., D. M. ANDERSON, D. BOUK, and R. D. GORDON. 1994. Larval key to genera and selected species of North American Coccinellidae (Coleoptera). Proceedings of the Entomological Society of Washington, 96: 387-412.

SASAJI, H. 1968. Phylogeny of the family Coccinellidae (Coleoptera). Etizenia, Occasional Publications of the Biological Laboratory, Fukui University, 35: 1-37.

SASAJI, H. 1971a. Fauna Japonica. Coccinellidae (Insecta: Coleoptera). Academic Press of Japan, Keigaku Publishing. Tokyo, ix +340pp, 16pls.

SASAJI, H. 1971b. Phylogenetic positions of some remarkable genera of the Coccinellidae (Coleoptera), with an attempt of the numerical method. Memoirs of the Faculty of Education, Fukui University, Series II (Natural Science), 21: 55-73.

SLIPINSKI, S. A. and J. PAKALUK. 1991. Problems in the classification of the Cerylonid series of Cucujoidea (Coleoptera). Pp.79-88. *In:* M. Zunino, X. Belles and M. Blas, eds. Advances in Coleopterology. European Association of Coleopterology. Barcelona.

TIMBERLAKE, P. H. 1943. The Coccinellidae or ladybeetles of the Koebele collection. Part 1. Bulletin of the Experiment Station of the Hawaiian Sugar Planters' Association, Entomological Series, 22: 1-67.

VANDENBERG, N. J. 1990. First North American records for *Harmonia quadripunctata* (Pont.) (Coleoptera: Coccinellidae); a lady beetle native to the palaearctic. Proceedings of the Entomological Society of Washington, 92: 407-410.

VANDENBERG, N. J. 1992. Revision of the New World lady beetles of the genus *Olla* and description of a new allied genus (Coleoptera: Coccinellidae). Annals of the Entomological Society of America, 85: 370-392.

VANDENBERG, N. J. and R. D. GORDON. 1991. Farewell to *Pania* Mulsant (Coleoptera; Coccinellidae); a new synonym of *Propylea* Mulsant. Coccinella, 3: 30-35.

WHITEHEAD, D. R. and R. M. DUFFIELD. 1982. A specialized predator prey association (Coleoptera: Coccinellidae, Chrysomelidae): failure of a chemical defense and possible practical application. Coleopterists Bulletin, 36(1): 96-97.

WOODRUFF, R. E. and R. I. SAILER. 1977. Establishment of the genus *Azya* in the United States. Florida Department of Agriculture and Consumer Services, Division of Plant Industry, Entomology Circular, 230: 1-2.

94. CORYLOPHIDAE LeConte 1852

by Stanley Bowestead and Richard A.B. Leschen

Family common name: The minute hooded beetles, minute fungus beetles, hooded beetles

Family synonym: Orthoperidae Thomson 1859

The small size, coccinelloid body form, head often covered by the pronotum, frontoclypeal suture absent, single maxillary lobe, antennae fairly elongate with distinct 3-segmented club, externally closed procoxal cavities, 4-4-4 tarsal formula, and exposed pygidium in most taxa define this group in the Cucujoidea.

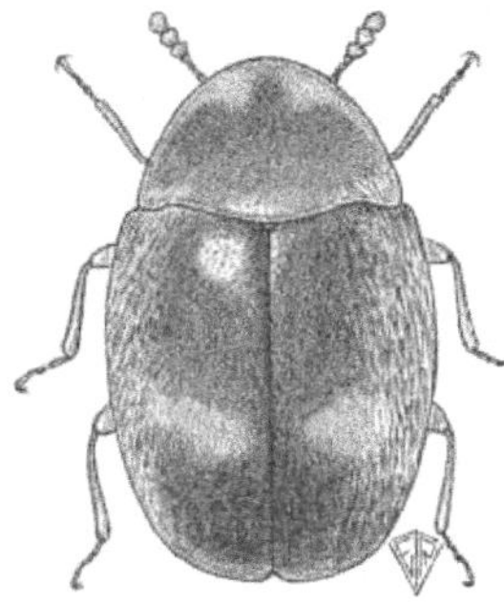

FIGURE 1.94. *Clypastraea lugabris* (LeConte).

Description: Color variable but unicolorous in most, from pale testaceous to pitchy; in some with paler pronotum or discal patches or areas, or with elaborate coccinellid-like patterns on the elytra (*Holopsis* in other faunas); antennae frequently pale with club in many slightly infuscate; legs in most unicolorous and pale.

Body form variable; in a few elongate and latridiid-like (Aenigmaticini); most elongate, oval and depressed, with head completely concealed beneath the pronotum (Parmulini, and Corylophini); acrotrichine-like with broad pronotum and tapering elytra (Sericoderini); globose with anterior margin of pronotum emarginate (*Orthoperus*); globose with variable anterior margin to pronotum (Rypobiinae); globose with head completely concealed beneath pronotum (Teplinini) (West Palaearctic) and Peltinodinae); length ranges from 0.56 mm (*Orthoperus*) to 2.3 mm (*Holopsis*, New Zealand). Punctation variable: punctures minute to strongly impressed; interspaces variable from smooth and glossy (*Holopsis*; some *Clypastraea*, *Arthrolips*; *Orthoperus*; *Gloeosoma*); with simple isodiametric cells (some *Clypastraea*, *Corylophus*, *Teplinus*, *Orthoperus*, *Rypobius*); with transverse wavy lines (some *Arthrolips*, *Microstagetus*, *Corylophus* (West Palaearctic), *Orthoperus*, *Rypobius*); with asperites (*Orthoperus*, some *Hoplicnema* and *Gloeosoma*).

Pronotum variable with the anterior margin emarginate, revealing the head (*Aenigmaticum*, *Foadia*, *Orthoperus*, most Rypobiini, and some Gloeosomatini) or with variable emargination (Gloeosomatini) or entire and completely covering the head (most Corylophinae, Peltinodinae).

Elytra mostly apically truncate, exposing the apex of the abdomen (Corylophinae) or more rounded apically in the more globose subfamilies (Rypobiinae, Orthoperinae and Peltinodinae); sutural stria always present but very variable in length, sutural interval variable in width: from entire with medially widened sutural interval with one row of punctures (*Aenigmaticum*) to long with sutural interval wide with 1, 2 or 3 rows of punctures providing good taxonomic characters (Parmulini, Sericoderini) and short with sutural interval narrow, with only 1 row of punctures where the punctures are traceable (these are the more globose subfamilies - Rypobiinae, Orthoperinae and Peltinodinae); epipleura incomplete, in most narrow and horizontal, but wider and inflexed in Sericoderinae; scutellum variable from triangular to strongly transverse, providing useful diagnostic characters at species level.

Head in most quadrate but very elongate in some Parmulini; eyes large and mostly prominent, but less prominent in the more elongate head forms; frontoclypeal suture absent; labrum large; antennae fairly elongate with 9, 10, or 11 antennomeres; scape and pedicel long; segments 3-8 forming narrower funicle, with 7 wider than its other antennomeres (a very strong family characteristic), 9-11 forming a distinct club with sensory vesicles, the whole antenna articulates into a z-shape when at rest. Antennae may have fused antennomeres (*Sericoderus*); vestigial antennomeres at the point of articulation (*Corylophus*), loss of the smaller antennomere 6 (*Gloeosoma*) or smaller antennomere 8 (*Arthrolips*), and reduced segmentation in the genera with minute species (*Teplinus* and *Orthoperus*).

Maxillary palpi composed of 3 large segments and a 4th narrow stem segment which can be vestigial or absent, plus stipes and cardo; lacinia mostly elongate with apical tooth-like setae; mandibles mostly simple but in the Rypobiinae are elongate and articulated with bifid hook-shaped tip; ligula large and mostly with apical glossae and paraglossae; labial palpi 1- or 2-segmented, bulbous, plus narrow stem segment; mentum mostly quadrate; submentum variable at tribal level; from shallow and wide to deep and narrow or deep and tapering, with corresponding changes in the emargination of the postgena.

Prosternum variable from entire (*Aenigmaticum, Foadia*), laterally emarginate with antennal slots (*Clypastraea*), with variable sinuate emargination (*Arthrolips*), or completely emarginate (*Sericoderus*). Interprocoxal area variable; narrow and widened basally with anterior projection (Peltinodinae), flat and triangular or rectangular (Corylophinae), with prosternal process (Rypobiinae), with anterior projections (Orthoperinae); procoxal cavities closed internally and variable from short and laterally truncate, to elongate and laterally obtusely angled. Procoxae correspondingly variable from short and almost round, to very elongate, all with

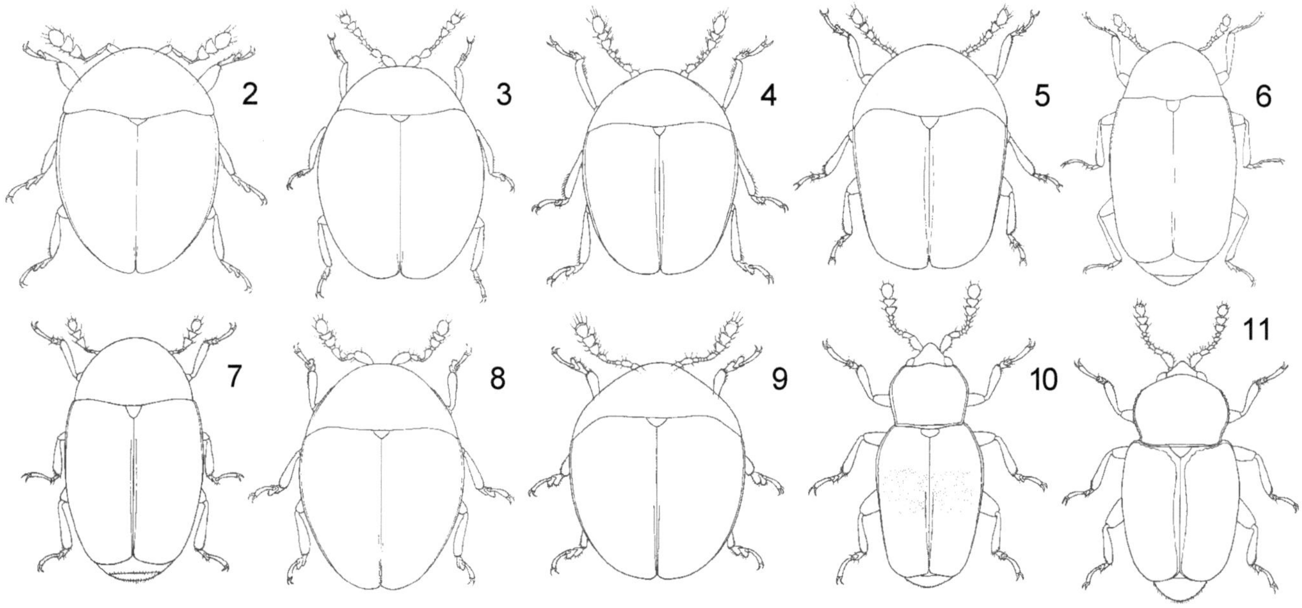

FIGURE 2.94-11.94. 2. *Holopsis carolinae* (Casey); 3. *Orthoperus brunneus* (Casey); 4. *Microstagetus parvulus* Wollaston; 5. *Sericoderus lateralis* (Gyllenhal); 6. *Clypastraea lugabris* (LeConte); 7. *Arthrolips decolor* LeConte; 8. *Rypobius marinus* LeConte; 9. *Gloeosoma sticticum* (Casey); 10. *Foadia maculata* Pakaluk; 11. *Aenigmaticum californicum* Casey.

lateral extensions that are disc-shaped apically to attach the coxae to the inner face of the pronotum.

Mesosternum with or without an angle between disc and collar; disc often providing strong characters at tribal level; mesocoxal cavities closed internally and variable from round to elongate; mesocoxae similarly variable. Metasternum wide, with or without femoral lines; males in some genera with various secondary sexual characters metacoxae widely separated, flat and forming an almost continuous surface with the disc of the metasternum and fitting into pocket-like emargination of basal sternite; metacoxal cavity narrow without internal closure. Mesepisternum and mesepimeron fused; metepisternum elongate and simple, apically curving around metacoxal emargination of metasternum.

Legs with femora broad and flattened with large trochanters; tibiae simple, with or without sexual modifications; tarsi 4-4-4 segmented; 1 and 2 large, 3 very small, 4 narrow and elongate.

Wings either micropterous and bud-like; brachypterous, extending only to the apex of the cubital vein; or fully winged and entirely or partially fringed with hairs; veins reduced, only long cubital and reduced costal and subcostal veins present.

Abdomen with 9 sternites; 1-3 fused and very wide; tergites 1-7 with spiracles (1-5 in *Orthoperus*), tergite 7 with narrow pleurite; sternite 9 and tergite 9 fused and often with sexual modifications.

Aedeagus of simple cucujoid form and composed of penis and tegmen. Tegmen encloses penis dorsally and fits against its ventral side; divided into a basal piece which is hinged to an elongate oval piece with or without a central tegminal strut; basal piece has characteristic paired oval apertures and apically has blunt parameres with setae. Penis mostly simply curved in lateral outline and symmetrical in ventral aspect; lateral aspect may also be sinuate; apex of the median lobe may be with or more often without setae; and with or more often without lateral extensions. Internal armature of the penis composed of flagellum that may be cone-shaped, hook-shaped or bulbous, lateral sclerites that may extend beyond the apex of the flagellum or be shorter than the apex of the flagellum, and winged sclerites that can be single or multiple. Sperm duct, which extends from the basal opening of the median lobe, is simple or with complex windings.

Spermatheca situated at base of abdomen and located against the large first sternite (differences among higher taxa are discussed in Bowestead 1999).

Larvae (modified from Lawrence 1991) with form variable from broadly ovate to strongly flattened or elongate; abdominal segments 1-7, 1 and 8, or 2 and 8 with paired lateral or dorsolateral glandular openings; mandible with or without prostheca; mala obtuse or falciform; labial palp 2-segmented with apical segment much longer than basal, or 1-segmented; 1 tarsungular seta present. Body variable from broadly ovate to strongly flattened (*Holopsis* and *Gleosoma*) and elongate and parallel-sided like Endomychidae and Latridiidae (*Arthrolips*, *Orthoperus, Sericoderus*); subcylindirical to slightly flattened; surfaces lightly to strongly pigmented; vestiture of simple or modified setae. Head protracted and prognathous, hidden in dorsal view in strongly flattened forms, elongate to strongly transverse. Epicranial stem absent frontal arms v-shaped (if present). Median endocarina absent. Stemmata absent or 2 on each side. Antennae moderately to very long, 2 or 3-segmented; segment 2 in some very long. Frontoclypeal suture absent; labrum free or fused to head capsule. Mandibles symmetrical, apex bi- or tridentate often without accessory ventral process; mola well-developed, tuberculate or asperate; prostheca absent or present and consisting of 2 hyaline processes. Ventral mouthparts usually protracted. Maxilla with cardo and stipes, indistinct articulating area; mala obtuse or falciform; palp 2- or 3-segmented, with apical segment much longer than basal. Labium mostly consisting of a single plate, palps 1- or 2-segmented with apical segment much longer than basal. Hypopharyngeal sclerome mostly complex. Hypostomal rods absent or long and diverging. Ventral epicranial ridges absent. Gula in some longer than wide, gular sutures mostly well-marked by parallel internally ridges. Thorax with legs well developed, 5-segmented; tarsungulus with 1 setae, mostly clavate; coxae moderately separated. Thoracic and abdominal terga in some broadly expanded at edges forming contiguous flat plates making a disc-like form. Abdominal segments 1-7, 1 and 8, or 2 and 8 with paired lateral or dorsolateral glandular openings. Tergum A9 simple, lacking urogomphi, in some with pigmented plate or macula. Sternum A9 short, simple. Segment A10 circular, ventrally oriented. Spiracles annular, dorsally placed on abdomen.

Habits and habitats. Adults and larvae feed on fungal spores and are present in a diversity of habitats (see review in Lawrence 1991 and a comprehensive list of special niches of Palaearctic species is provided in Bowestead 1999). Many species are present on the leaves and flowers of plants and may be found in nests of birds and caterpillars (*Orthoperus* and *Sericoderus*). Others can be sifted from leaf litter and grass heaps, while some may be found underneath bark. Though details about fungal host associations are fragmented, it appears that corylophids have diverse associations. Some species feed on various Hyphomycetes and Zygomycetes (*Orthoperus* and *Sericoderus*) while others have associations with Ascomycetes or Basidiomycetes (*Arthrolips*, *Holopsis*). Larvae have been described by Chandler (1983), Costa *et al.* (1988), Hinton (1941), Lawrence (1991).

Status of the classification. The family name is based on LeConte (1852) where he first used the prefix "Corylophi" when he described a section of the Coccinellidae of the United States. Wollaston (1854) formally designated the family name Corylophidae when working on the fauna of Madeira. The synonymous name Orthoperidae was used by Csiki (1910) in his catalogue.

North American taxa can be identified using Downie and Arnett (1996); Hatch (1957, 1962); Lawrence *et al.* (1999a, 1999b, larvae). Pakaluk (1985a, 1985b, 1987) and Pakaluk and Lawrence (1986) provided recent systematic treatments on specific groups and higher taxa, and Leschen and Bowestead (2001) made changes in the taxonomy of North American taxa. Other works of importance to the North American fauna include: Bøving and Craighead 1931, Bruce 1948, Casey 1889, 1900, Lawrence 1982, Matthews 1888, 1889, and Paulian 1950.

The family is a member of the cerylonid series of Cucujoidea (Slipinski and Pakaluk 1991, Lawrence and Newton 1995) and is sister taxon to the family Latridiidae (Bowestead 1999). The family is divided into 4 subfamilies, all of which are represented in the North American fauna. Phylogenetic relationships based on adult characters were presented by Bowestead (1999).

Corylophids have been referred to as 'minute fungus beetles' (Arnett 1973) or 'hooded beetles' (Kuschel 1990) but are perhaps more aptly thought of as 'minute hooded beetles' because of their size combined with distinctive habitus, where the head is concealed beneath the pronotum in the great majority of species.

Distribution. There are 284 species known from all areas, of which 61 occur in the United States.

Key to the genera of Corylophidae
(modified from Bowestead 1999)

1. Procoxal cavities open basally (Peltinodinae, Fig. 2) .... *Holopsis*
— Procoxal cavities closed basally 2

2(1). Interprocoxal area flat and triangular or rectangular (Corylophinae) 4
— Interprocoxal area not flat and triangular or rectangular 3

3(2). Mouthparts simple; antennae 9-segmented; metasternum with femoral lines, of males with small median keel (Fig. 3) .... *Orthoperus*
— Mouthparts modified with mandibles articulated and elongate, bifid and hook-shaped apically; antennae 10- or 11-segmented; metasternum without femoral lines, metasternum of males with slight longitudinal median impression (Rypobiinae) .. 9

4(2). Prosternum entire and cup-like; habitus latridiid-like with head projecting beyond anterior margin of pronotum (Aenigmaticini) 10
— Anterior margin of prosternum more or less emarginated, habitus oval with head concealed beneath hood-like and entire anterior margin of pronotum 5

5(4). Interprocoxal area flat and rectangular; procoxal cavities elongate laterally (Palaearctic) .... Teplinini
— Interprocoxal area flat and triangular; procoxal cavities short 6

6(5). Triangular interprocoxal area elongate. Angle between disc and collar more acute (Fig. 4) .... *Microstagetus*
— Triangular interprocoxal area transverse. Angle between disc and collar obtuse 7

7(6). Basal angles of pronotum acute and projecting beyond basal margin, forming widest point of habitus (Fig. 5) .... *Sericoderus*
— Basal angles of pronotum less projecting; elytra forming widest part of habitus (Fig. 6) (Parmulini) 8

8(7). Anterior margin of prosternum sharply emarginate laterally by antennal slots, leaving a truncate median plate; antennae 11-segmented (Figs. 1 and 6) .... *Clypastraea*
— Anterior margin of prosternum without antennal slots, with various degrees of emargination from sinuate to completely emarginate; antennae 10- or 11-segmented (Fig. 7) .... *Arthrolips*

9(3). Procoxal cavities shorter and laterally truncate; anterior margin of pronotum emarginate, revealing front of head, or entire with head completely covered; protibia simple or with a tooth-like enlargement of the inner margin in males (*Hoplicnema*, Neotropical); prosternal process angled towards head (*Rypobius* (Fig. 8)) or towards mesosternum (*Hoplicnema*, Neotropical) (Rypobiini) .... *Rypobius*

— Procoxal cavities more elongate and laterally obtusely angled; anterior margin of pronotum with variable emargination of anterior margin; protibia simple in females but with apical inner projection in males; prosternal process angled towards head or absent (*Catoptyx*, Oriental) (Fig. 9) .... *Gleosoma*

10(4). Basal sternite with femoral lines; antennae 10-segmented; sutural stria extending for only half length of suture (Fig. 10) .... *Foadia*

— Basal sternite without femoral lines; antennae 9-segmented; sutural stria entire, extending past scutellum along the basal margin of elytra (Fig. 11) .... *Aenigmaticum*

Classification Of The Nearctic Genera

Corylophidae LeConte 1852

Corylophi LeConte 1852
Orthopérites Jacquelin du Val 1857
Orthoperidae Thomson 1859

Peltinodinae Paulian 1950

Peltinoditae Paulian 1950
Corylophodini Paulian 1950

Holopsis Broun 1883, 9 spp., Atlantic Coast states to Texas and Oklahoma (Florida, North Carolina, Pennsylvania, Virginia), southern California.
Corylophodes Matthews 1885
Bathona Casey 1900

Corylophinae LeConte 1852

[Teplinini Pakaluk, Slipinski and Lawrence 1994, Palaearctic]

Aenigmaticini Casey 1900

Aenigmaticum Matthews 1888, 2 spp., Arizona, California, Florida.

Foadia Pakaluk 1985, 1 species, *F. maculata* Pakaluk, Florida.

Parmulini Poey 1854

Arthrolipinae Bøving and Craighead 1931

Clypastraea Haldeman 1842, 16 spp., Alaska, Alberta, Arizona, British Columbia, California, Colorado, Florida, Idaho, Iowa, Kentucky, Louisiana, Manitoba, Massachusetts, Michigan, Montana, New Jersey, New York, North Carolina, North Dakota, Ontario, Pennsylvania, Texas, Utah, Virginia
Aspidocha Gistel 1848
Sacium LeConte 1852
Parmulus Gundlach 1854
Molamba Casey 1900
Clypeastodes Casey 1900
Sacinops Paulian 1950
Aposacium Paulian 1950

Arthrolips Wollaston 1854, 8 spp., Arkansas, Arizona, California, Florida, Illinois, Indiana, Massachusetts, Michigan, Minnesota, Missouri, North Dakota, Oklahoma, Texas.
Lepadodes Matthews 1887
Sacina Broun 1893
Meioderus Matthews 1899
Meizoderus Matthews 1899
Sericoderistes Reitter 1913
Alloparmulus Paulian 1950
Pseudoparmulus Paulian 1950

Sericoderini Matthews 1888

Sericoderus Stephens 1829, 5 spp., widely distributed.
Gryphinus Redtenbacher 1849
Anisomeristes Matthews 1886

Corylophini LeConte 1852

Microstagetus Wollaston 1861, 1 species; *M. parvulus* Wollaston, Oklahoma.

Orthoperinae Jacquelin du Val 1857

Orthopérites Jacquelin du Val 1857

Orthoperus Stephens 1829, 16 spp., widely distributed.
Pithophilus Heer 1841
Microsphera Redtenbacher 1845
Eutrilia Casey 1900

Rypobiinae Paulian 1950

Rypobiini Paulian 1950

Rypobius LeConte 1852, 2 spp., Delaware, Florida, Maryland, New Jersey, New Hampshire, New York, Rhode Island, Texas, Virginia.

Gloeosomatini Bowestead 1999

Gloeosoma Wollaston 1854, 5 spp., eastern United States to Manitoba, Arkansas, Florida, Indiana, Iowa, Kansas, Ohio, Massachusetts, Michigan, Nebraska, New Hampshire, North Carolina, North Dakota, Ontario, Virginia, Wisconsin.

Moronillus Jacquelin du Val 1854
Lewisium Matthews 1899
Gronevus Casey 1900
Sahlberginus Bruce 1948

Bibliography

ARNETT, R. H., Jr. 1973. The beetles of the United States (A manual for identification).The Entomological Institute. Michigan.

BØVING, A. G. and F. C. CRAIGHEAD. 1931. An illustrated synopsis of the principal larval forms of the Order Coleoptera. Entomologica Americana (New Series), 11: 1-351.

BOWESTEAD, S. 1999. A revision of the Corylophidae (Coleoptera) of the West Palaearctic Region. Instrumentum Biodiversitatis, Museum d'Histoire Naturelle, Geneva. 3, 203 pp, 426 figs.

BRUCE, N. 1948. The Scandinavian species of the genus *Orthoperus* Steph. (Coleoptera), with a discussion on some species occurring outside Scandinavia and a description of a new genus. Opuscula Entomologica Supplementa, 9: 1-36, 4 pls.

CASEY, T. L. 1900. Review of the American Corylophidae, Cryptophagidae, Tritomidae and Dermestidae with other Studies. Journal of the New York Entomological Society, 8: 60-75.

CHANDLER, D. S. 1983. Larvae of the wrack Coleoptera in the families Corylophidae, Rhizophagidae, and Lathridiidae. Psyche, 90: 287-296.

COSTA, C., S. A. VANIN and S. A. CASARI-CHEN. 1988. Larvas de Coleoptera do Brasil. Museu de Zoologia, Universidade de São Paulo. 282 p +165 pls.

CSIKI, E. 1910. Coleopterorum Catalogus, Part 18: 5-28. W. Junk. Berlin.

DOWNIE, N. M. and R. H. ARNETT, Jr. 1996. The Beetles of Northeastern North America, 2 vols. The Sandhill Crane Press. Gainesville, FL.

HATCH, M. H. 1957. The beetles of the Pacific Northwest. Part II. Staphyliniformia. University of Washington Publications in Biology, 16: 384 pp. [pp. 284-286].

HATCH, M. H. 1962. The beetles of the Pacific Northwest. Part III: Pselaphidae and Diversicornia I. University of Washington Publications in Biology, 16: 503 pp.

HINTON, H. E. 1941. The immature stages of *Sericoderus lateralis* (Gyllenhal) (1827) (Coleoptera, Corylophidae). The Entomologist, 74: 198-202.

KUSCHEL, G. 1990. Beetles in a suburban environment: A New Zealand case study. The identity and status of Coleoptera in natural and modified habitats of Lynfield, Auckland (1974-1989). DSIR Plant Protection report 3.

LAWRENCE, J. F. 1982. Coleoptera. Pp. 482-553. *In*: S. P. Parker, ed. Synopsis and Classification of Living Organisms, Vol. 2. McGraw-Hill, New York.

LAWRENCE, J. F. 1991. Corylophidae (= Orthoperidae). Pp. 495-497. *In*: F.W. Stehr, ed., Immature Insects, Vol. 2. Kendall-Hunt. Dubuque, IA

LAWRENCE, J. F., A. HASTINGS, M. J. DALLWITZ, T. A. PAINE and E. J. ZURCHER. 1999a. Beetles of the World: A Key and Information System for Families and Subfamilies. CD-ROM, Version 1.0 for MS-DOS. CSIRO Publishing. East Melbourne, Victoria.

LAWRENCE, J. F., A. HASTINGS, M. J. DALLWITZ, T. A. PAINE and E. J. ZURCHER. 1999b. Beetle Larvae of the World: Descriptions, Illustrations, Identification, and Information Retrieval for Families and Sub-families CD-ROM, Version 1.0 for MS-DOS. CSIRO Publishing. East Melbourne, Victoria.

LAWRENCE, J. F. and A. F. NEWTON, Jr. 1995. Families and subfamilies of the Coleoptera. 2: 884-886. *In*: J. Pakaluk and S. A. Slipinski, eds. Biology, Phylogeny and Classification of the Coleoptera. 2 vols. Pp. 1092. Museum i Institut Zoologii PAN. Warsaw.

LECONTE, J. L. 1852. Remarks upon the Coccinellidae of the United States. Proceedings of the Academy of Natural Sciences of Philedelphia, 6: 129-145.

LESCHEN, R.A.B. and S. BOWESTEAD. 2001. Synonymical notes for Corylophidae and Cryptophagidae. Coleopterists Bulletin, 55: 312-316.

MATTHEWS, A. 1888. Biologia Centrali-Americana. Insecta, Coleoptera 2, pt. 1: 118-119. London.

MATTHEWS, A. 1899. A monograph of the Coleopterous Families Corylophidae and Sphaeriidae. O. E. Janson and Son. London. 220 pp., 8 pls.

PAKALUK, J. 1985a New genus and species of Corylophidae (Coleoptera) from Florida, with a description of its larva. Annals of the Entomological Society of America, 78: 406-409.

PAKALUK, J. 1985b. Phylogenetic position of *Hyplathrinus* (Coleoptera: Corylophidae). Entomological News, 96: 69-70.

PAKALUK. J. 1985c. Review of the New World genus *Aenigmaticum* Matthews (Coleoptera: Corylophidae). Coleopterists Bulletin, 39: 207-214.

PAKALUK, J. 1987. Revision and phylogeny of the neotropical genus *Hoplicnema* Matthews (Coleoptera: Corylophidae). Transactions of the American Entomological Society, 113: 73-116.

PAKALUK, J. and J. F. LAWRENCE. 1986. *Priamima*, a new genus of Corylophidae (Coleoptera) from Australia. Journal of the Australian Entomological Society, 25: 81-84.

PAULIAN, R. 1950. Les Corylophidae d'Afrique. Memoires de l'Institute Français d'Agfrique Noire, No. 12: 1-126.

SLIPINSKI, S. A. and J. PAKALUK. 1991. Problems in the classification of the cerylonid series of Cucujoidea (Coleoptera). Pp. 79-88. *In*: M. Zunino, X. Belles, and M. Blas, eds. Advances in Coleopterology. Silvestrelli and Cappelletto. Turin.

WOLLASTON, T. V. 1854. Pp. 470-482. *In*: Insecta Maderensia. Van Voorst. London.

WOLLASTON, T. V. 1861. Additions to the Madeiran Coleoptera. Annals and Magazine of Natural History (3), 8: 103-106.

95. LATRIDIIDAE Erichson 1842

by Fred G. Andrews

Family common name: The minute brown scavenger beetles

Family synonym: Lathridiidae LeConte 1861

The tarsi with three tarsomeres, small size, and elongate-oval shape serve to separate this family.

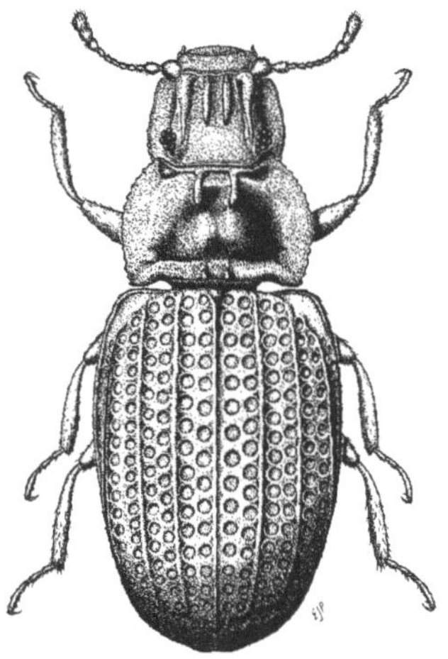

FIGURE 1.95. *Metophthalmus rudis* Fall

Description: Elongate oval, somewhat convex; size 1-3 mm long; color mostly brown to piceous; vestiture moderate, fairly long, semierect, recumbent or absent.

Head oval or elongate trapezoidal, slightly deflexed; surface punctate or impunctate; dorsal surface even or with median longitudinal depression. Antennae with 10 or 11-antennomeres, with two or three antennomeres forming a sensilla bearing club, antennae inserted on frons between eyes and slightly above base of mandibles. Mouthparts with mandibles moderate, curved, with a filamentous prostheca and well developed mola; maxillary palp with four palpomeres, first small and cylindrical, fourth acuminate; labium with mentum and prementum fused or divided, oblong; ligula indistinct; labial palp with two palpomeres, first small, cylindrical, second large, round. Eyes lateral, prominent or reduced, rounded or absent.

Pronotum wider than head, narrower than base of elytra, oval or trapezoidal, laterally margined or rounded; margins sigmoid, straight or sharply invaginated; dorsal surface punctate, or impunctate with ridges or depressions and laterally explanate; prosternum moderate in front of coxae, occasionally with setose fovea anterior to coxa; procoxal cavities open or closed behind. Mesosternum narrow, lightly punctate or impuncate with faintly impressed rugae. Metasternum broad, placate sculptured with recumbent setae or smooth with various combinations of fovea, rugae, punctation and setae. Legs with procoxae moderately conical or rounded, contiguous or separate; mesocoxae moderately conical or rounded, separate; metacoxae transverse, widely separate; trochanters triangular, or long and cylindrical; femora slightly swollen; tibiae slender, finely spinose, may have small apical spurs; tarsal formula 3-3-3, tarsomeres long, simple, first tarsomere may be ventrally pubescent; claws moderate, simple; pulvillus a single stiff hair. Elytra entire, apically rounded, suture may be connate; scutellum small, triangular; epipleural fold moderate on basal half, narrowed in apical half; usually regularly striate with striae setose and interstriae setose or glabrous, interstria may be raised or carinate. Wings present or absent; venation reduced to a short costal and cubital.

Abdomen with five or six visible sternites, sutures entire; sternites usually lightly setose and punctate; first sternite may have paired fovea or a transverse furrow between metacoxa and longitudinal rugae or oblique coxal lines behind coxa. Aedeagus a relatively large, simple, downwardly curved tube without accessory appendages. Female genitalia are undescribed.

Larvae elongate, subcylindrical; 1-3 mm in length; pearly white except for dark head; setation long and acicular, or long and short with tip expanded and jagged. Head protracted; antennae three-segmented, second and third antennomeres subequal and twice as long as first, or with second half again as long as third and three times the first and with a long acicular setae at tip. Mandibles with apices rounded and bearing two elongate setae, membranous medially and with large tuberculate mola. Maxillae with prominent stipes, three-segmented palp, mala obtuse, rarely with the galea and lacinia present. Labrum with two palpomeres. Eyes each consisting of four ocelli. Thoracic legs with five segments including single clawed tarsunguli. Abdomen 10-segmented. Spiracles inconspicuous, annular, on abdominal segments one to eight. Larvae described in Bøving and Craighead 1931, and Lawrence 1991.

Habits and habitats. Adults and larvae feed on the conidia of fungi and Myxomycetes. Most published records associate them with fungi in the classes Phycomycetes, Deuteromycetes and Ascomycetes. Species of *Metophthalmus*, *distemia*, *Cartodere*, *Dienerella* and *Corticaria* have been reared from the Deuteromycetes and Phycomycetes, *Mucor*, *Aspergillus*, *Penicillium*, *Botrytis* and *Helminthosporium* and *Alternaria* (Andrews 1976a-c, Hammad 1953, Hinton 1941, Chandler 1983). *Enicmus maculatus* LeConte has been reared from the ascomycete *Hypoxylon* (Lawrence 1977). Most *Enicmus* species and *Revelieria californica* Fall feed on the spores of Myxomycetes (Andrews 1976, Russell 1979, Lawrence and Newton 1980, Newton and Stephenson 1990). Members of the subfamily Latridiinae are generally associated with leaf litter in nature and can be collected by sifting or Berlese techniques. Adult Corticariinae can be collected by sweeping dead, low lying vegetation. They are most prevalent in wetter seasons of the year.

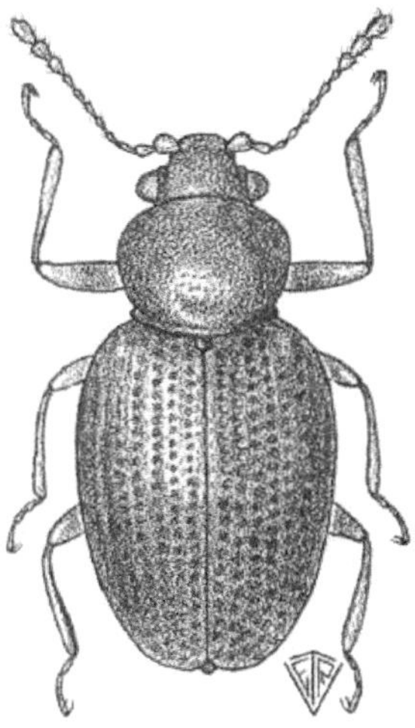

FIGURE 2. *Melanophthalma americana* Mannerheim

Several species, including *Aridius nodifer* (Westwood), *Latridius minutus* (Linnaeus), *Dienerella ruficollis* (Marsham), *D. filum* (Aubé), *D. arga* Reitter, *Adistemia watsoni* (Wollaston) and *Cartodere constricta* (Gyllenhal) are widely spread throughout the world in the stored products of man (Hinton 1941). Of these, only *A. nodifer* and *C. constricta* are know to occur in natural environments away from man. Most nearctic species in most genera are found in the western United States. Immature stages are relatively unknown. Larvae of selected species of *Adistemia*, *Dienerella*, *Lithostygnus* and *Corticaria* (Hinton 1941, Hammad 1953, Chandler 1983) have been described, but the larvae of most genera are unknown. Other important references are Balduff 1935 (bionomics), Bousquet 1991 (Canada), Fall 1926 (Alaska), Hatch 1962 (Pacific Northwest), Walkely 1952 (revision Latridiini, Washington).

Status of the classification. Included in the Cucjoidea by Crowson (1955) and Lawrence and Newton (1995); and the Cucujoidea, Cerylonid series by Slipinski and Pakaluk (1991). The genus *Holoparamecus* Curtis (Holoparamecinae) was moved to the Endomychidae by Lawrence and Newton (1995) and the genus *Dasycerus* Brong to the Staphylinidae by Newton and Thayer (1992). The group needs considerable study at the species level.

Distribution. About 1050 species are known from all areas, with 140 species recorded from the United States.

Key to the Nearctic Genera

1. Clypeus at a lower plane than frons and sharply delimited from frons by deep transverse suture (Latridiinae) 2
— Clypeus and frons on same plane, at most delimited from frons by fine transverse suture (Corticariini) 13

2(1). Procoxal cavities open behind .... *Akalyptoischion*
— Procoxal cavities closed behind 3

3(2). Dorsal and ventral surface smooth, evenly punctate, elytra with eleven poorly defined rows of punctate striae, labrum embracing clypeus .... *Revelieria*
— Dorsal and ventral surface sculptured, head, pronotum and ventral surface with impunctate areas 4

4(3). Elytra with 10 rows of punctate striae in posterior half .... *Thes*
— Elytra with eight or fewer rows of punctate striae 5

5(4). Metasterna and first abdominal sternite between coxa separated by a suture 6
— Metasterna and first abdominal sternite between coxa not separated by a suture .... *Adistemia*

6(5). Trochanters four times longer than wide .... *Eufallia*
— Trochanter at most twice as wide as long 7

7(6). Antennomere 3 widest proximally, trapezoidal in profile; head with paired longitudinal carina, tempora wanting (Fig. 1) .... *Metophthalmus*
— Antennomere 3 not widest proximally, head without paired longitudinal carina, with or without distinct tempora 8

8(7). Eyes small, with fewer than 20 facets ... *Dienerella*
— Eyes normally developed with more than 70 facets 9

9(8). Pronotum with paired longitudinal carina nearly extending entire length 10
— Pronotum without paired longitudinal carina extending entire length, may have short paired carina across or behind posterior transverse sulcus 12

10(9). Antennae 11-segmented with 2-segmented club .... *Cartodere*
— Antennae 11-segmented with 3-segmented club 11

11(10). Lateral margins of pronotum deeply incised behind .... *Aridius*
— Lateral margins of pronotum inwardly angulate, but not sharply .... *Stephostethus*

12(9). Prosternal process not keeled or raised above coxa .... *Latridius*
— Prosternal process keeled, elevated above coxa .... *Enicmus*

13(2). Eyes absent or reduced to a single facet .... *Fuchsina*
— Eyes well developed, distinct, many facets 14

14(13). Mentum and submentum distinct sclerites 15
— Mentum and submentum fused to form a single elongate sclerite 16

15(14). Antennae 11-segmented with 3-segmented club (Fig. 2) ...... *Melanophthalma* (*Melanophthalma*)
— Antennae 11-segmented with 2-segmented club .... *Melanophthalma* (*Cortilena*)

16(14). Antennae 10-segmented with 3-segmented club .... *Migneauxia*
— Antennae 11-segmented with 3-segmented club 17

17(16). Female with five visible abdominal sternites, prosternum with setose fovea anterior to prosternal coxa, male at most with a small incon-

spicuous calcareous spine at terminus of protibia .... *Corticaria*
— Female with six visible abdominal sternites, prosternum without setose fovea anterior to procoxa, male with large distinct spine a short distance from terminus of protibia 18

18(17). Second tarsomere inserted near middle of first tarsomere and not exceeding, or barely exceeding first segment, pronotum with angulate projection at posterior corners, pronotum with circular depression at midline and posterior one-third .... *Corticarina*
— Second tarsomere inserted near terminus of first tarsomere, pronotum without angulate projection at posterior corners, pronotum with oblique depression extending from midline to posterior corners .... *Cortinicara*

Classification of the Nearctic Genera

Latridiidae Erichson 1842

Latridiinae Erichson 1842

Adistemia Fall 1899, 1 sp., *A. watsoni* (Wollaston 1871), cosmopolitan.

Akalyptoischion Andrews 1976, 8 spp., California, southwestern United States (Andrews 1976c).

Aridius Motschulsky 1866, 2 spp., cosmopolitan.
Aridionomus Reitter 1911

Dienerella Reitter 1908, 9 spp., generally distributed.
(*Cartodere sensu* Thomson 1863)
Cartoderema Reitter 1908
Microgramme Walkley 1948

Cartodere Thomson 1859, 1 sp., *C. constrictus* (Gyllenhal 1827), cosmopolitan.
Coninomis Thomson 1863, type species by subsequent monotypy (*carinatus=constrictus*)

Enicmus Thomson 1859, 13 spp., generally distributed.

Eufallia Muttkowski 1910, 1 sp., *E. seminiveus* (Motschulsky 1866), Florida.
Belonia Fall 1899, not Kirby 1889

Latridius Herbst 1793, 9 spp., generally distributed.
Lathridius Illiger 1801 (emendation)
Conithassa Thomson 1859, not Thomson 1863

Metophthalmus Wollaston 1854, 9 spp., West Coast and southern United States (Andrews 1976b, 1998).
Metatypus Belon 1897

Revelieria Perris 1870, 1 sp., *R. californica* Fall 1899, California, Nevada.

Stephostethus LeConte 1878, 9 spp., generally distributed.
Enicmoderes Reitter 1911

Thes Semenov 1909, 2 spp., Pacific Coast and Ontario.
Lar Semenov 1904, not Gosse 1857

Corticariinae Curtis 1829

Corticaria Marshall 1902, 39 spp., generally distributed.

Corticarina Reitter 1880, 19 spp., generally distributed (Andrews 1985, 1992, 2001, Johnson 1972, 1990).

Cortinicara Johnson 1975, 1 sp. *C. gibbosus* (Gyllenhal 1827).

Melanophthalma Motschulsky 1866, 21 spp., generally distributed (Johnson 1972).
subgenus *Cortilena* Motschulsky 1867, 4 spp., generally distributed.
subgenus *Melanophthalma* Motschulsky 1866, 11 sp., generally distributed.

Fuchsina Fall 1899, 2 spp., California, Nevada, Oregon, and Arizona (Andrews 1976a).

Migneauxia Jacquelin du Val 1857, 1 sp. *Migneauxia crassiuscula* (Aubé 1850), western U.S. (Andrews 1989).

Bibliography

ANDREWS, F. G. 1976a. A new species of *Fuchsina* Fall with notes on some California Lathridiidae (Coleoptera). Pan-Pacific Entomologist, 52: 339-347.

ANDREWS, F. G. 1976b. A revision of the North American species of *Metophthalmus*. Coleopterists Bulletin, 30: 37-56.

ANDREWS, F. G. 1976c. *Akalyptoischion*, A new genus of Lathridiidae from Western North America (Coleoptera). Occasional Papers in Entomology, Number 22, 23 pp.

ANDREWS, F. G. 1985. *Corticarina scissa* (Leconte): a valid species (Coleoptera: Lathridiidae). Coleopterists Bulletin, 39: 147-149.

ANDREWS, F. G. 1989. Notes on *Migneauxia orientalis* Reitter, a genus and species not previously recorded from North America (Coleoptera: Lathridiidae: Corticariinae). Coleopterists Bulletin, 43: 173-175.

ANDREWS, F. G. 1992. Two new species of *Corticarina* from coastal California (Lathridiidae: Corticariini) with notes on J. L. LeConte Types. Coleopterst Bulletin, 46: 274-280.

ANDREWS, F. G. 1998. A new species of *Metophthalmus* Wollaston (Coleoptera: Latridiidae) from Texas. Coleopterist Bulletin, 52: 55-59.

ANDREWS, F. G. 2001. A new species of *Corticarina* from Arizona (Latridiidae: Corticariini). Pan-Pacific Entomologist, 77: 37-38.

BALDUF, W. V. 1935. Latridiidae. Pp. 167. *In:* The Bionomics of Entomophagus Coleoptera, John S. Swift. St. Louis, 220 pp.

BOUSQUET, Y. 1991. Lathridiidae. Pp. 239-241. *In:* Checklist of the beetles of Canada and Alaska. Canada Communications Group - Publishing. Ottawa, Canada.

BØVING, A. G. and F. C. CRAIGHEAD. 1931. Lathridiidae. *In:* An illustrated synopsis of the principal larval forms of the order Coleoptera. Entomologica Americana, 11: 8, 33, 38, 74 pp., pl. 25.

CHANDLER, D. S. 1983. Larvae of wrack Coleoptera in the families Corylophidae, Rhizophagidae, and Lathridiidae. Psyche, 90: 287-296.

CROWSON, R. A. 1955. The natural classification of the families of Coleoptera. Nathaniel Lloyd. London, 187pp.

FALL, H. C. 1899. Revision of the Lathridiidae of boreal America. Transactions of the American Entomological Society, 26: 101-190.

FALL, H. C. 1926. A list of the Coleoptera taken in Alaska and adjacent parts of the Yukon territory in the summer of 1924. Pan-Pacific Entomologist, 2: 191-208.

HAMMAD, S. M. 1953. The immature stages of *Metophthalmus serripennis* Broun (Coleoptera: Lathridiidae). Proceedings of the Royal Entomological Society London. Series A, 28:133-138.

HATCH, M. H. 1962. The Beetles of the Pacific Northwest. Part III: Pselaphidae and Diversicornia I. University Washington Press. Seattle, ix + 503 pp.

HETSCHKO, A. 1926. Lathridiidae. Coleopterum Catalogus. W. Junk. Berlin.85: 1-86.

HINTON, H. E. 1941. The Lathridiidae of economic importance. Bulletin of Entomological Research, 32: 191-247.

JOHNSON, C. 1972. Studies on the genera *Corticarina* Reitter and *Melanophthalma* Motschulsky. Nouvelle Revue Entomologie, 2: 185-199.

JOHNSON, C. 1990. New exotic species of Corticariinae (Col.: Latridiidae). Entomologist's Record, 102: 11-15.

LAWRENCE, J. F. 1977. Coleoptera associated with *Hypoxylon* Species (Ascomycetes: Xylariacae) on oak. Coleopterists Bulletin, 31: 309-312.

LAWRENCE, J. F. and A. F. NEWTON, Jr. 1980. Coleoptera associated with fruiting bodies of slime molds (Myxomycetes). Coleopterists Bulletin, 34: 129-144.

LAWRENCE, J. F. 1991. Lathridiidae. Pp. 497-498. *In:* F.W. Stehr, ed. Immature Insects, Vol. 2. Kendall/Hunt. Dubuque, IA.

LAWRENCE, J. F. and A. F. NEWTON, Jr. 1995. Families and subfamilies of Coleoptera (with selected genera, notes, references and data on family-group names). Pp. 779-1006. *In:* J. Pakaluk and S. A. Slipinski, eds. Biology, Phylogeny, and Classification of Coleoptera: Papers celebrating the 80th birthday of Roy A. Crowson. Muzeum I Instytut Zoologii PAN, Warsaw.

NEWTON, A. F., Jr. and M. K. THAYER. 1992. Current classification and family-group names in Staphyliniformia (Coleoptera). Fieldiana Zologogy, 67: 1-92.

NEWTON, A.F., Jr. and S. L. STEPHENSON. 1990. A beetle slime mold assemblage from northern India (Coleoptera: Myxomycetes). Oriental Insects, 24: 197-218.

RUSSELL, L. K. 1979. Coleoptera associated with slime molds (Mycetozoa) in Oregon and California (Coleoptera: Leiodidae, Sphindidae, Lathridiidae). Pan-Pacific Entomologist, 55: 1-9.

SLIPINSKI, S. A. and J. PAKALUK. 1991. Problems in the classification of the Cerylonid series of Cucujoidea (Coleoptera). Pp. 79-88. *In:* M. Zunino, X. Bellés and M. Blas, eds. Advances in Colopterology. European Association of Coleopterology, Barcelona.

WALKLEY, L. M. 1948. Notes on nomenclature in the Lathridiini. Proceedings of the Entomological Society of Washington, 50: 149-150.

WALKLEY, L. M. 1952. Revision of the Lathridiini of the state of Washington. Proceedings of the Entomological Society of Washington, 54: 217-235.

Superfamily TENEBRIONOIDEA Latreille 1802

96. MYCETOPHAGIDAE Leach 1815

by Daniel K. Young

Family common name: The hairy fungus beetles

The oblong to ovate, somewhat flattened, pubescent body, relatively large, coarsely faceted compound eyes, and clubbed antennae partially distinguish this family of primitive tenebrionoid beetles.

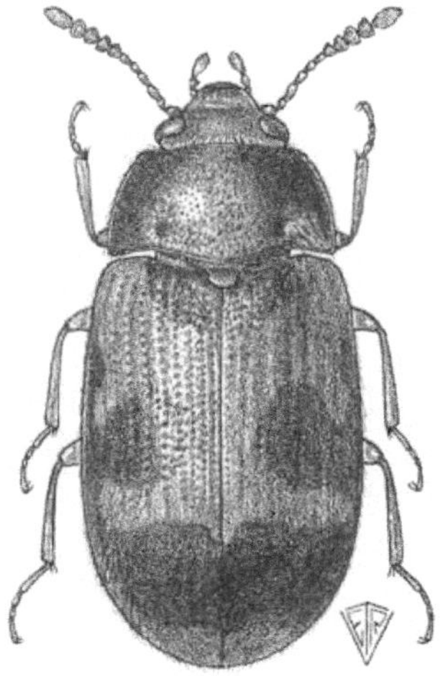

FIGURE 1.96. *Mycetophagus punctatus* Say

Description: (after Lawrence 1982) Shape obovate, broad, depressed; length 1.0 to 6.5 mm; color brownish to piceous, some with orange or reddish markings; vestiture consisting of short to moderately long, sparse to dense, suberect setae.

Head short, triangular, slightly deflexed; surface punctate or rugopunctate. Antennae with 11 antennomeres, clavate, antennomeres seven to 11 enlarged, or with two to three apical antennomeres forming a loose club; inserted under small frontal ridge. Clypeus distinctly separated by impressed line which may be incomplete, trapezoidal; labrum quadrate; mandibles curved, blunt; maxillae consisting of separate galeae and laciniae; maxillary palpi with four palpomeres, long, apex of each enlarged, palpi nearly as long as head; gula present, the gular sutures distinct, widely separate; labium with mentum trapezoidal; ligula corneous; paraglossae absent; labial palpi short, with three small, slender palpomeres. Eyes lateral, large, obovate, entire or slightly emarginate near the antennal insertions.

Pronotum broader than the head, as broad as elytra basally, narrowed and truncate anteriorly, posteriorly sinuate, laterally narrowing toward head, feebly arcuate; sides finely, distinctly margined; surface punctate; pleural region large, triangular; prosternum broad anterad coxae with elevated keel between the coxae; prothoracic coxal cavities open behind. Scutellum of moderate size, arcuate. Mesosternum short; metasternum broad. Legs with protrochantins visible; procoxae elongate, globular, separate; mesocoxae subconical, narrowly separated; metacoxae transverse, contiguous; trochanters small, triangular; femora swollen; tibiae slender with distinct apical spurs; tarsal formula 4-4-4 or 3-4-4 (males), segments slender, the first and fourth segments elongate; claws simple. Elytra entire, the apices rounded; striae punctate; intervals rugulose; epipleural fold broad basally, narrowing to absent apically. Metathoracic wings with radial cell present or absent, anal cell lacking, subcubital fleck entire.

Abdomen with five visible ventrites, the sutures entire. Male genitalia of the trilobed type; penis large, flattened, apically acute with a pair of basal, median struts; parameres large, flattened, surrounding the basal part of the median lobe; basal piece large, membranous above.

Larvae elongate, subcylindrical, or subdepressed; length 3 to 8 mm; vestiture sparse, consisting of a few moderately long setae on each segment; color light brown. Head exserted, prognathous, nearly as wide as prothorax; epicranial suture with stem short or absent, frontal arms lyriform, contiguous basally; antennae three-segmented with segment two much longer than segment one, sensorium sometimes dome-shaped. Labrum free; mandibles asymmetrical, stout, bidentate, left mola nearly vertical, right molar area oblique; maxilla with cardo, elongate stipes, well-developed articulating area, rounded, setiferous mala, and three-segmented palpi; labium with submentum and mentum fused, ligula and two-segmented palpi present (one-segmented in *Thrimolus*). Four or five stemmata present on each side of head. Thorax with five-segmented legs including apical, claw-like tarsungulus that bears two juxtaposed setae. Abdomen 10-segmented, the tenth segment reduced, transverse; urogomphi slightly up-curved, simple, caudally projecting. Spiracles annular, annular-biforous, or modified-annular (with short accessory tubes).

Habits and habitats. As the family name implies, mycetophagids feed largely upon fungi, and are most commonly found under fungus-grown bark, in shelf-fungi, and on moldy vegetative material. *Mycetophagus* spp. are not uncommon in decaying fruiting bodies of mushrooms and fleshy polypores, especially those that have begun to dehydrate. *Typhaea stercorea* (Linnaeus) can be abundant in stored products (Hinton 1945). *Berginus* spp. have been observed to feed on pollen; one Indian species is said to prey on scale insects. The Chilean *Filicivora chilensis* (Philippi and Philippi) is documented as feeding on fern spores (Leschen and Lawrence 1991).

Status of the classification. Parsons' (1975) posthumous revision of the North American species provides an adequate, albeit terse, treatment of our fauna. Although fairly widespread in both temperate and tropical regions, diversity is greatest in temperate areas.

Distribution. There are approximately 18 genera and 200 species arranged into three subfamilies (Hetschko 1930), known from all areas, with two subfamilies, five genera and 26 species in the United States (Parsons 1975).

Key to the Genera of America North of Mexico

1. Head across eyes nearly as wide as pronotum; antennal club two-segmented .... *Berginus*
— Head across eyes much narrower than pronotum; antennal club, if present, of three or more antennomeres 2
2(1). Basal angles of pronotum well defined 3
— Basal angles of pronotum broadly rounded; body very minute (1-1.2 mm) .... *Thrimolus*

3(2). Epipleural fold of elytra horizontal and flat 4
— Epipleural fold of elytra concave .... *Litargus*

4(3). Eyes transverse, sinuate anteriorly *Mycetophagus*
— Eyes more rounded, not sinuate............. *Typhaea*

Classification of the Genera of America North of Mexico

Mycetophagidae Leach 1815

Mycetophaginae Leach 1815

Mycetophagus Hellwig 1792, 15 spp., generally distributed.
Tritoma Geoffroy 1762, not binominal; Mueller 1765, not binominal; not Fabricius 1775
subgenus *Mycetophagus*, *sensu stricto*
subgenus *Ilendus* Casey 1900
subgenus *Parilendus* Casey 1900
subgenus *Gratusus* Casey 1900

Typhaea Curtis 1830, 1 sp., *T. stercorea* (Linnaeus 1758), widely distributed throughout the US.

Litargus Erichson 1846, 6 spp., widely distributed as follows: *L. balteatus* LeConte, widespread throughout the US and Canada; *L. didesmus* (Say), eastern US and Canada, west to the Dakotas and Texas; *L. grandis* Schaeffer, Arizona (known only from type locality in Huachuca Mountains); *L. nebulosus* LeConte, eastern US, west to Utah and New Mexico.; *L. sexpunctatus* (Say), eastern US and Canada, west to Nebraska and Texas; *L. tetraspilotus* LeConte, eastern US and Canada, west to Iowa and Texas. Casey (1900) recognized five subgenera for the Nearctic species of *Litargus*; these were rejected by Parsons (1975).
Tilargus Casey 1900
Alitargus Casey 1900
Litargellus Casey 1900
Paralitargus Casey 1900

Thrimolus Casey 1900, 1 sp., *T. minutus* Casey 1900. District of Columbia, Florida, Indiana, Ohio, Pennsylvania, and Texas.

Bergininae Leng 1920

Berginus Erichson 1846, 3 spp., *B. bahamicus* Casey, Florida, Bahamas, Cuba; *B. nigricolor* Champion, Central America to southern Texas, *B. pumilus* LeConte, California, south into Mexico. (The type is labeled as Pennsylvania; Parsons questioned the locality data).

Bibliography

CASEY, T. L. 1900. Review of the American Corylophidae, Cryptophagidae, Tritomidae, and Dermestidae, with other studies. Journal New York Entomological Society, 8: 51-172.
HETSCHKO, A. 1930. Coleopterorum catalogus. Phalacridae, Mycetophagidae, Tretothoracidae, Jacobsoniidae, Cavicoxumidae, Gnostidae, 15(108): 1-76. (2nd pagination)
HINTON, H. E. 1945. A monograph of the beetles associated with stored products. Volume 1. British Museum (Natural History). London. 443 pp.
LAWRENCE, J. F. 1982. Coleoptera. Pp. 482-553. *In:* S. P. Parker, ed., Synopsis and Classification of Living Organisms. McGraw-Hill. New York, NY.
LESCHEN, R. A. B. and J. F. LAWRENCE. 1991. Fern sporophagy in Coleoptera from the Juan Fernandez Islands, Chile, with descriptions of two new genera in Cryptophagidae and Mycetophagidae. Systematic Entomology, 16: 329-352.
PARSONS, C. T. 1975. Revision of Nearctic Mycetophagidae (Coleoptera). Coleopterists Bulletin, 29: 93-108.

97. ARCHEOCRYPTICIDAE Kaszab 1964

by Daniel K. Young

Family common name: The archaeocryptic beetles

The heteromerous, simple tarsi, simple tarsal claws, hard body, and closed prothoracic coxal cavities, distinguish these beetles from all but Tenebrionidae, a family with which they have long been associated. They differ from tenebrionids most notably in having lateral posterior extensions of the prothoracic intercoxal process that embrace the prothoracic coxae and partially close the coxal cavities. Archeocrypticids also have only the basal two abdominal sterna connate whereas tenebrionids generally have the basal three sterna connate.

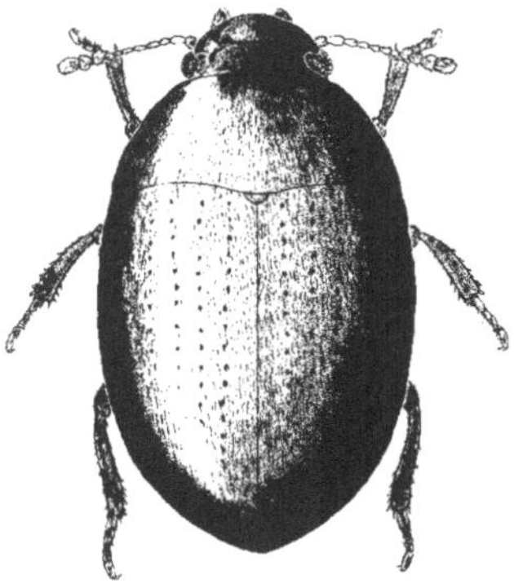

FIGURE 1.97. *Enneboeus caseyi* Kaszab (from Triplehorn and Wheeler 1979)

Description (after Lawrence 1994, Merkl 1988, in part): Elongate-oval to broadly oval, somewhat to strongly convex; length 1.5-3.7 mm; brown to black and finely pubescent. Head subglobular, prognathous, narrower than pronotum, with distinct frontoclypeal suture, surface finely punctate to coarsely punctured. Antennae with 11 antennomeres, terminal 2-7 antennomeres forming a club; insertions concealed by frontal ridges. Eyes of moderate size, protruding and coarsely faceted. Labrum distinct, transverse. Mandibles short, robust, bidentate, with well developed mola and pubescent prostheca. Maxillary palpi with 4 palpomeres, the fourth enlarged, usually securiform.

Pronotum as wide as elytra basally, basal margin slightly to conspicuously bisinuate, unmargined; laterally narrowed toward apex; sides narrowly to broadly margined. Prosternum prominent, with intercoxal process extending laterally well behind coxae. Prothoracic coxal cavities closed internally and externally. Scutellum visible, rounded posteriorly. Mesosternum with intercoxal process short, broad; mesothoracic coxal cavities open laterally. Metasternum transverse. Legs moderately long; trochantins of prothoracic legs often partially exposed by slit-like extension of coxal cavity; prothoracic coxae globose, always well separated; mesothoracic coxae rounded, well separated, metathoracic coxae transverse, contiguous; trochanters small, triangular, heteromeroid; femora slender to slightly swollen; tibiae slender, sometimes spinose along outer edge; tibial spurs prominent; metathoracic tibiae sometimes modified in males. Tarsal formula 5-5-4 (rarely 4-4-4); tarsomeres 1-2 or 1-3 densely setose ventrally; penultimate tarsomere often highly reduced. Elytra slightly longer than wide, strongly convex, with fine to coarse rows of punctures; elytral intervals smooth and shining or punctulate; epipleural fold broad. Metathoracic wings usually fully developed with short radial cell; medial fleck present, undivided.

Abdomen with five visible sterna; sterna one and two connate; apex of last visible sternum modified in some males. Male genitalia with normal heteromeroid orientation (tegmen dorsad penis); basal piece longer than parameres which are typically setose; penis with well developed, paired basal struts; ejaculatory duct with sclerotized seminal pump. Female genitalia elongate; coxites subequal in length to valvifers; apical styli long, narrow; copulatory bursa short, continuing anteriorly as long, tubular spermatheca with large reservoir attached by long narrow duct.

Larvae with body elongate, straight, slightly flattened, moderately to lightly sclerotized, sides subparallel; length 2-6 mm. Head exserted from prothorax; mouthparts prognathous. Epicranial stem short; frontal arms lyriform; endocarinae lacking. Labrum intact, symmetrical. Frontoclypeal suture distinct. Stemmata five on each side. Antennae elongate, 3-segmented, sensorium conical. Mandibles asymmetrical, mola present with tubercules forming transverse ridges separated by rows of deep pits; prostheca lacking; dorsal mandibular asperities present. Ventral mouthparts retracted; maxilla with cardo divided, well developed maxillary articulating area, mala present, entire, palpi 3-segmented. Labium composed of mentum, prementum; submentum fused with gula, ligula absent, palpi 2-segmented. Hypopharyngeal sclerome dentiform. Hypostomal rods short, subparallel or slightly diverging. Thorax bearing well developed, 5-segmented legs, tarsungulus bearing two setae, one distal to the other. Abdomen 10-segmented, ninth tergum bearing well developed, linear, posteriorly projecting, lightly pigmented urogomphi; ninth sternum simple; tenth segment transverse, posteriorly oriented. Spiracles annular-biforous.

Habits and habitats. Larvae and adults appear to be largely saprophagous and are generally found in leaf litter and other vegetative debris. They have been most commonly collected by extraction from litter using Berlese funnels. However, some Australian species have been associated with the fruiting bodies of Polyporaceae.

Status of classification. In the past 30 years, this group has become well defined; it is easily distinguished from all other tenebrionoid families on the basis of numerous autapomorphies, particularly in the adults (Kaszab 1981, 1984, Lawrence 1991, Watt 1974). Although larvae bear a striking similarity to those of many mycetophagids, the resemblance appears to be a mixed bag of plesiomorphies and homoplasies relating to microhabitat and trophic similarities (Lawrence 1991).

Distribution. There are approximately 10 genera and 50 species (Lawrence 1994), worldwide but most diverse pantropically. A single species, *Enneboeus caseyi* Kaszab, occurs in the southern United States, south into Mexico and Central America.

Classification of the Nearctic Species

Archeocrypticidae Kaszab 1964

Enneboeus Waterhouse 1878, 1 sp., *E. caseyi* Kaszab 1981, Florida, Kentucky, North Carolina, South Carolina, Texas, Mexico, Central America (Triplehorn and Wheeler 1979).

Uloporus Casey 1889

Bibliography

KASZAB, Z. 1981. Die Gattungen und Arten der Tribus Archeocrypticini (Coleoptera: Tenebrionidae). Folia Entomologica Hungarica, 42: 95-155.

KASZAB, Z. 1984. Revision der australischen Archeocrypticinen (Coleoptera: Tenebrionidae). Annales Historico-Naturales Musei Nationalis Hungarici, 76: 143-163.

LAWRENCE, J. F. 1991. Archeocrypticidae. Pp. 500-501. *In:* F. W. Stehr, ed., Immature Insects, Volume 2. Kendall/ Hunt. Dubuque.

LAWRENCE, J. F. 1994. Review of the Australian Archeocrypticidae (Coleoptera), with descriptions of a new genus and four new species. Invertebrate Taxonomy, 8:449-470.

MERKL, O. 1988. Novelties of *Sivacrypticus* Kaszab, 1964 and *Enneboeus* Waterhouse, 1878 (Coleoptera, Archeocrypticidae). Anneles Historico-Naturales Musei Nationalis Hungarica, 80: 71-78.

TRIPLEHORN, C. A. and Q. D. WHEELER. 1979. Systematic placement and distribution of *Uloporus ovalis* Casey (Coleoptera: Heteromera: Archeocrypticidae). Coleopterists Bulletin, 33: 245-250.

WATT, J. C. 1974. A revised subfamily classification of the Tenebrionidae (Coleoptera). New Zealand Journal of Zoology, 1: 381-452.

Color Figure 17
FAMILY: Staphylindae (Pselaphinae)
SPECIES: *Batrisodes gravesi* Chandler and Reddell
PHOTOGRAPHER: James C. Cokendolpher

Color Figure 18
FAMILY: Histeridae
SPECIES: *Spilodiscus gloveri* (Horn)
PHOTOGRAPHER: David M. Dennis

Color Figure 19
FAMILY: Carabidae
SPECIES: *Cicindela repanda* Dejean
PHOTOGRAPHER: Arthur V. Evans

Color Figure 20
FAMILY: Gyrinidae
SPECIES: *Dineutus sublineatus* (Chevrolat)
PHOTOGRAPHER: Arthur V. Evans

Color Figures 21
FAMILY: Silphidae
SPECIES: *Heterosilpha ramosa* (Say)
PHOTOGRAPHER: Arthur V. Evans

Color Figure 22
FAMILY: Glaphyridae
SPECIES: *Lichnanthe alpina* Carlson
PHOTOGRAPHER: Arthur V. Evans

Color Figure 23
FAMILY: Scarabaeidae
SPECIES: *Chrysina beyersi* (Skinner)
PHOTOGRAPHER: Arthur V. Evans

Color Figure 24
FAMILY: Erotylidae
SPECIES: *Megalodacne heros* (Say)
PHOTOGRAPHER: David M. Dennis

Color Figure 25
FAMILY: Melyridae
SPECIES: *Malachius* (*Clinoptilus*) *mirandus* (LeConte)
PHOTOGRAPHER: Max E. Badgley

Color Figure 26
FAMILY: Cerambycidae
SPECIES: *Crossidius coralinus coccineus* Casey
PHOTOGRAPHER: Robert H. Turnbow, Jr.

Color Figure 27
FAMILY: Cerambycidae
SPECIES: *Neoclytus mucronatus* (Fabricius)
PHOTOGRAPHER: David M. Dennis

Color Figure 28
FAMILY: Attelabidae
SPECIES: *Homoeolabis analis* (Illiger)
PHOTOGRAPHER: Fred J. Santana

Color Figure 29
FAMILY: Ripiphoridae
SPECIES: *Macrosiagon cruentum* (Germar)
PHOTOGRAPHER: Fred J. Santana

Color Figure 30
FAMILY: Curculionidae
SPECIES: *Sphenophorus aequalis* (Gyllenhal)
PHOTOGRAPHER: Arthur V. Evans

Color Figure 31
FAMILY: Coccinellidae
SPECIES: *Harmonia axyridus* (Palles)
PHOTOGRAPHER: Fred J. Santana

Color Figure 32
FAMILY: Buprestidae
SPECIES: *Xenomelanophila miranda* (LeConte)
PHOTOGRAPHER: Nathan M. Schiff

98. CIIDAE Leach in Samouelle 1819

by Margaret K. Thayer and John F. Lawrence

Family common name: The minute tree-fungus beetles

Family synonyms: Ciidae Leach in Samouelle 1819; Cissides Lacordaire 1857; Cioidae Gistel 1856 and Rosenhal 1856; Orophyidae Kiesenwetter 1877 [based on misspelling]; Octotemnidae Reitter 1878

Ciidae adults are small to minute, convex, usually oval or elongate beetles having 4-4-4 tarsi (3-3-3 in a few non-North American forms); antennae with 8-10 antennomeres (11 in one California species), the last 2 or 3 forming a club; eyes well-developed, oval, entire; base of pronotum not or slightly narrower than base of elytra; prosternal process narrowed apically or parallel-sided; procoxal cavities open internally, narrowly open or closed externally; mesocoxae not closed by sterna laterally; elytra without punctate striae; abdomen with 5 visible sternites, all articulated or (one California species) the first two connate. Larvae may be recognized by their subcylindrical form, more or less hypognathous head, absence of a mandibular mola (pseudomola sometimes present), head with long epicranial stem and V-shaped frontal arms, laciniar lobe of the maxilla placed dorsally, and reduced antennae with antennomere 3 absent or (one California species) present but shorter than sensorium.

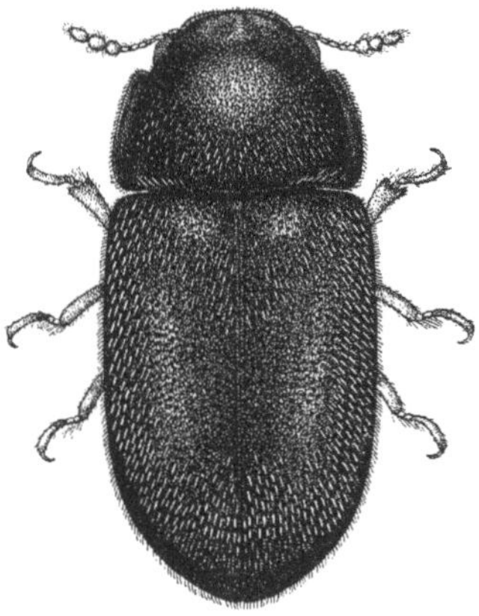

FIGURE 1.98. *Cis fuscipes* Mellié, female (after Lawrence 1974a; drawing by C. Green)

Description: Convex, generally oval to elongate (Figs. 1, 20, 22, 36-41, 42, 45); length 0.5-6 mm; brown to piceous, in many variable within species; glabrous or subglabrous, in some with vestiture of short erect dark bristle-like setae.

Head subglobular, without temples, at least slightly deflexed, partly hidden from dorsal view; eyes well-developed, entire, strongly protuberant. Frontoclypeal suture distinct, mostly strongly curved, in males of many elevated into a ridge or other protuberance. Antennae of 8-10 antennomeres (11 in Sphindociinae), the last 2 or 3 forming a club and each (except in Sphindociinae) bearing at least 4 tufts of sensilla (Fig. 5); antennal insertions well separated, visible or concealed in dorsal view. Clypeus varying in shape; labrum well-sclerotized, free, varying in shape; mandibles short, broad, curved, apices bidentate, basally with well-developed mola; maxillae with well-developed galea, lacinia either well-developed (Sphindociinae) or reduced (Ciinae); maxillary palpi of four palpomeres, apical one cylindrical to fusiform or subglobular, not or only slightly smaller than preapical palpomere. Labium with ligula reduced; labial palpi of 3 palpomeres, the apical one cylindrical to fusiform.

Pronotum transverse to nearly quadrate, as wide as the elytra at the base or nearly so, widest at middle or base, sides straight or curved, smooth (Ciinae) or crenulate (Sphindociinae), front margin sometimes with teeth or horns; lateral margins explanate or not, front and hind angles variable in shape; prosternal process narrowed apically or parallel-sided, varying in length, rarely absent; postcoxal process well-developed; procoxal cavities internally open, externally open or closed; trochantins partly exposed or concealed; procoxae externally slightly transverse or conical, contiguous or separated. Mesosternum transverse, in some very short; mesocoxae globose, coxal cavities circular to slightly transverse, contiguous or narrowly separated, not closed laterally by sterna; mesotrochantins hidden or exposed, joint between mesotrochanter and femur strongly oblique, at least partly (in most completely) separating base of femur from coxa; scutellum well-developed or reduced; meso- and metasternal processes meeting in a point or a complex joint; metasternum subquadrate, without coxal lines, with (Fig. 17) or without (Fig. 8) metasternal suture; metacoxae transverse, either contiguous or narrowly separated; tibiae slender to strongly dilated apically or at middle, externally spinose or not, apical spurs present or absent; tarsal formula 4-4-4 (3-3-3 in a few exotic forms), tarsomeres slender, preapical ones together shorter than apical one. Elytra entire, without punctate striae, though punctures occasionally in rows; epipleura very narrow, nearly reaching elytral apex. Wings present in most (occa-

Acknowledgments. We thank A. F. Newton and J. L. Navarrete-Heredia for comments on a draft of the chapter. The second author is grateful to the many institutions and individuals who provided specimens and research support for his earlier studies on the North American Ciidae (cited in Lawrence 1971, 1974a) and CSIRO Entomology for later support. He also thanks artists S. P. Kim (Fig. 2), A. Atkins (Fig. 3), and C. Greene (Figs. 1, 20, 22, 33, 36-42, and 45). We thank the University of California Press for permission to use Figs. 1, 20, 22, 36-42, and 45 and the Museum of Comparative Zoology, Harvard University, for permission to use Figs. 4-19, 21, 23-35, and 43-44. The first author heartily thanks the second author for stimulating and subsequently encouraging her interest in beetles over a quarter-century ago – through the seemingly tedious task of pointing and labeling thousands of Ciidae from around the world.

sionally absent), with reduced venation: radial cell absent, oblongum cell present, anal lobe absent, medial area with one vein (Ciinae) or three (Sphindociinae); medial fleck present.

Abdomen with five visible sternites, all free (Ciinae) or first two connate (Sphindociinae); first visible sternite without coxal lines, in males often with a pubescent median fovea. Sexes similar or males with horns, tubercles, or other structures on head and/or pronotum, their development often highly variable within species. Male sternite 9 with or without median strut, tergites 9 and 10 highly reduced or membranous; aedeagus of tenebrionoid type, symmetrical, median lobe membranous (Sphindociinae) or sclerotized (Ciinae), ventral tegmen without struts or accessory lobes, parameres fused to each other but articulated with phallobase (basal piece). Female genitalia incompletely known, mostly well-developed (similar to basal tenebrionid type of Tschinkel and Doyen 1980): basally with a pair of elongate valvifers each supported by a longitudinal baculus, coxites about as long as the valvifers, each divided transversely in 4 parts and with a transverse baculus supporting its basal part, styli moderately well-developed, inserted at apices of coxites; genitalia occasionally (e.g., *Octotemnus*) strongly shortened and reduced, with coxites of only 2 parts. Spiculum ventrale of females usually long, reaching junction of visible sternites 2 and 3 in *Sphindocis* and base of abdomen in *Plesiocis*, but only junction of visible sternites 3 and 4 in *Octotemnus*. Internal anatomy of Ciidae is little known; Klopfenstein and Graves (1992) described that of *Hadraule blaisdelli* Casey.

Larva: Subcylindrical, parallel-sided (Figs. 2-3); 1-7 mm long, but mostly under 3.5 mm; vestiture of only scattered long simple setae; lightly pigmented except abdominal apex and front of head. Head globular, exserted, moderately to strongly hypognathous, with V-shaped frontal arms (in North American taxa), and median endocarina (except in Sphindociinae) coinciding with long median epicranial suture. Usually 3-5 stemmata on each side, occasionally 2, 1, or none. Antennae very short, with 2 (Ciinae) or 3 (Sphindociinae) antennomeres, the second with a long sensory appendage, apical antennomere with a long apical seta. Labrum free, distinct frontoclypeal suture present; mandibles without true mola, sometimes with pseudomola bearing transverse ridges or with narrow hyaline process, usually asymmetrical; maxilla characteristic, with transverse cardo, elongate stipes, subapical short truncate lacinia placed dorsally, rounded or truncate galea, and palpi with 3 palpomeres. Labium with submentum, mentum, and short ligula (sometimes absent), sometimes basally connate with maxilla, labial palpi with 2 palpomeres. Gula transverse, ventral epicranial ridges present, hypostomal rods and hypopharyngeal sclerome absent. Legs short, broad, coxae relatively close together. Abdominal tergite 9 usually with pair of pointed upturned urogomphi, occasionally instead with 1 or more than 2 processes or (Fig. 3) a concave disc-like impression; sternite 9 simple (with row of fine asperities apically in Sphindociinae); sternite 10 transverse. Spiracles in Ciinae usually annular, rarely annular-uniforous, in Sphindociinae annular-biforous.

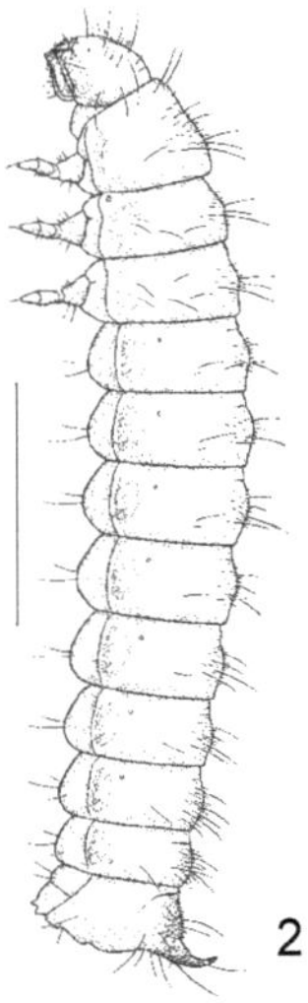

2

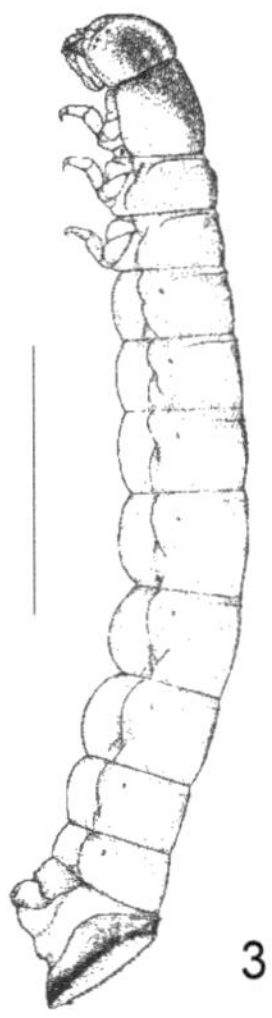

3

FIGURES 2.98-3.98. 2. *Cis vitulus* Mannerheim, larva (scale = 1 mm; drawing by S. P. Kim); 3. *Sphindocis denticollis* Fall, larva (scale = 2 mm; drawing by A. Atkins).

Pupa (based on the few known) adecticous, exarate, pale, with darkened urogomphus-like projections on the abdominal apex; spiracles present on first five abdominal segments (of *Porculus*: Costa *et al.* 1988; and *Plesiocis*: Lawrence unpublished).

Habits and habitats. Adults and larvae feed on fungus, almost all of them internal feeders (Lawrence 1989) on fruiting bodies of persistent wood-rotting Basidiomycetes, primarily those of the order Aphyllophorales (=Polyporales), the polypore or bracket fungi. Most feed on the context (non-spore-producing part) of such fungi after the spores have been shed and sometimes long after the fungus dies. Pupation usually occurs within the fungus. Ciids are occasionally found under bark of logs, in rotting wood (logs, vines, or branches), or in the galleries of bark beetles (Scolytidae), but are probably associated with fungi found there (Lawrence 1973: 177). Most species of Ciidae feed, and especially breed, on a relatively small number of fungus species. Paviour-Smith (1960) found that fungi in her study area fell into two clusters she called "habitat groups," which had almost completely distinct breeding ciid faunas. Lawrence (1971, 1973) extended this work to North America, showing that polypore hosts for a given ciid species usually fall within just one of the groups he referred to as "host preference groups," although a few species do feed on a wider variety of hosts. Similar patterns are found for at least some of these host preference groups in other parts of the world (Lawrence and Milner 1996). This limitation seems partly related to the hyphal structure of the fungi and/or the types of wood-rot they produce, which is a function of their metabolic capabilities (Lawrence 1973, Gilbertson 1984). Lawrence (1971, 1973) discussed general ciid biology as well as patterns of host preference; the generic and host preference group names used below for fungi follow his 1973 paper. Ackerman and Shenefelt (1973) provided additional host records for several species. Lawrence's (1973) *Coriolus* and *Phellinus* groups have particularly distinct ciid faunas associated with them (Lawrence and Milner 1996). Fossli and Andersen (1998) and Økland (1995) found

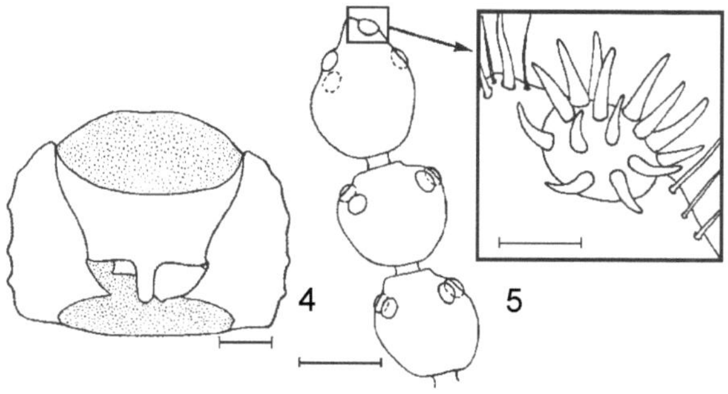

FIGURES 4.98-5.98. 4. *Sphindocis denticollis* Fall, male prothorax, ventral, right coxa removed (scale = 0.250 mm) (after Lawrence 1974b); 5. *Cis vitulus* Mannerheim, antennal club with positions of tufts of sensilla (scale = 0.076 mm); inset: detail of tuft of sensilla (scale = 0.018 mm) (after Lawrence 1971).

similar associations between several species groups of ciids and fungi in Norway, with slightly narrower host specificity than recorded by Lawrence (1973). Jonsell and Nordlander (1995) found that some ciid species in Sweden were strongly attracted by odor to one host fungus but not another, and Jonsson *et al.* (1997) found no evidence for long-distance attraction to fungi by beetle pheromones in the few species they studied. Guevara *et al.* (2000) provided strong evidence for use by at least some ciid species of fungal chemicals for detecting and recognizing host fungi. The affinity of ciids for dried fungi has resulted in their recently documented potential to become pests of dried fungi in commerce (Lohse and Reibnitz 1991, Madenjian *et al.* 1993) as well as fungal herbarium specimens (Lawrence 1971, J. L. Navarrete-Heredia, personal communication). Males of at least some species of *Ceracis* and *Cis* use the horns on the head and sometimes pronotum to fight with each other in tunnels within the polypores they inhabit (Eberhard 1979).

Status of the classification. The family's placement has varied considerably over the years, from Bostrichoidea or Cleroidea in some early works to, more recently, Cucujoidea (Crowson 1955) or Tenebrionoidea (Crowson 1960, Lawrence 1971, 1974b, 1991). On the basis of adult and larval characters, Ciidae now seem most reasonably placed within Tenebrionoidea; the family's exact position within the superfamily remains somewhat uncertain, but it appears to be a relatively basal group close to Tetratomidae and Mycetophagidae (Lawrence 1991). The last world catalog (Abdullah 1973), based largely on Dalla Torre (1911), unfortunately did not incorporate Lawrence's (1971) North American revision; the last world revision was that of Mellié (1848). The composition of the family has been relatively stable in the 20th century except for the removal of two Asian genera to Pterogeniidae (Crowson 1955) and the brief removal of *Sphindocis*, tentatively to Tetratomidae by Lawrence (1971; returned to Ciidae by Lawrence 1974a,b).

The North American ciid fauna has been revised (Lawrence 1967b, 1971, 1974b) and catalogued (Lawrence 1982) in recent decades and is reasonably well known, although several undescribed species have been discovered more recently, particularly from Florida and Texas (Lawrence 1982 and unpublished data). New genera and hundreds of new species remain to be described from other parts of the world, particularly the Neotropics and Asia (Lawrence 1982). The interactive keys of Lawrence *et al.* (1993 and 1999a, 1999b) include considerable information on larvae and adults, respectively, of both subfamilies.

Distribution. There are about 550 described species in 40 genera worldwide, of which 84 species in 13 genera occur in the United States and Canada (Lawrence 1982). Of these, 23 or 24 also occur in Mexico (Navarrete-Heredia and Burgos-Solorio 2000).

Key to the Genera of North America
(modified from Lawrence 1971, 1974b)

1. Pronotal lateral margins broadly coarsely crenulate (Fig. 4); antennae with 11 antennomeres, without tufts of sensilla on club; abdominal sternites 3 and 4 (first two visible) connate; mesotrochantin visible (Sphindociinae) .... *Sphindocis*

— Pronotal lateral margin smooth, occasionally very finely toothed; antennae with 8-10 antennomeres, each article of club with at least 4 tufts of sensilla (Fig. 5); abdominal sternites 3 and 4 free; mesotrochantin concealed (Ciinae) 2

2(1). Procoxae subconical, strongly projecting below intercoxal process, which does not extend to middle of coxae (Figs. 6-7); abdominal sternite 3 (first visible) of male with posteriorly projecting triangular flap partly concealing pubescent fovea (Fig. 9) 3

— Procoxae transverse or globular, not projecting below intercoxal process, which extends beyond middle of coxae (Figs. 10-16); abdominal sternite 3 of male simple or foveate (Fig. 18), but without triangular flap 4

3(2). Outer edges of all tibiae spinose for more than one third of their lengths (Fig. 19); antennae with 8 antennomeres; body form oval (Fig. 20); vestiture of minute setae, not visible under 10x magnification, plus a few scattered long fine setae; pronotal punctation finer and sparser, the punctures much smaller than those on elytra and separated by at least 1.5 diameters .... *Octotemnus*

— Outer edges of tibiae with spines at apex only (Fig. 21); antennae with 10 antennomeres; body form cylindrical; vestiture of long fine setae only; pronotal punctation coarser and denser, punctures only slightly smaller than those on elytra and mostly separated by less than 1 diameter .. .. *Rhopalodontus*

4(2). Prosternum in front of coxae almost twice as long as intercoxal process (Fig. 16); sides of pronotum subparallel or slightly diverging toward apex (Figs. 16, 22); body small, elongate, and extremely flattened (Fig. 22), pronotal plus elytral length mostly less than 1.4 mm; elytral length more than 1.6 x width; depth through metasternum mostly less than 0.65 x elytral width; antennae with 9 antennomeres, club of 2 or 3 antennomeres; elytral punctation of two sizes, distinctly seriate, micropunctures with short fine setae or bristles .... *Hadraule*

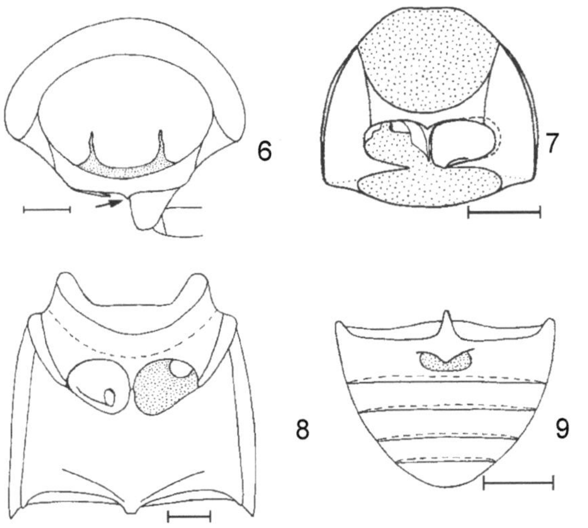

FIGURES 6.98-9.98. 6. *Octotemnus laevis* Casey, prothorax, anterior (scale = 0.125 mm), showing very short prosternal process (arrow); 7. *Rhopalodontus americanus* Lawrence, male prothorax, ventral (scale = 0.250 mm); 8. *Octotemnus laevis* Casey, meso- and metathorax, ventral (scale = 0.125 mm); 9. *Rhopalodontus americanus* Lawrence, abdomen of male, ventral (scale = 0.250 mm) (all after Lawrence 1971).

— Prosternum in front of coxae not or only slightly longer than intercoxal process; without other characters in combination 5

5(4). Outer apical angle of protibia expanded, rounded, and bearing several spines (Figs. 23-24) 6

— Outer apical angle of protibia not as above, usually produced and dentate (Figs. 25-27), blunt and angulate (Figs. 28-29), or narrowly rounded (Figs. 30-31); if expanded and rounded, then not spinose (Fig. 32) 9

6(5). Prosternum in front of coxae carinate and on same plane as intercoxal process (Fig. 33); lateral edges of pronotum visible from above for their entire lengths; anterior pronotal angles slightly produced (Fig. 33); antennae with 10 antennomeres .... *Strigocis*

— Prosternum in front of coxae concave or biconcave and on different plane from intercoxal process (Fig. 35); lateral edges of pronotum not visible from above for their entire lengths; anterior pronotal angles rounded or obtusely angulate, not produced (Fig. 35); antennae with 8-10 antennomeres 7

7(6). Intercoxal process of prosternum laminate, less than 0.15 x as wide as a procoxal cavity (Fig. 13); vestiture of very short fine setae; pronotum of male usually with tubercles, horns, or laminae at apex (Figs. 13, 36); antennae with 8-10 antennomeres; habitus as in Fig. 36 ...... *Ceracis*

— Intercoxal process of prosternum not laminate, at least 0.20 x as wide as a procoxal cavity (Fig. 15); vestiture of short stout bristles; pronotum of male always simple; antennae with 9-10 antennomeres 8

8(7). Pronotum very short and broad, length less than 0.73 x width; prosternum in front of coxae only half as long as intercoxal process (Fig. 15); metasternum at midline less than 0.40 x as long as broad; antennae with 10 antennomeres (in North American species) .... *Malacocis*

— Pronotum more elongate, length more than 0.73 x width; prosternum in front of coxae not or only slightly shorter than intercoxal process; metasternum at midline more than 0.40 x as long as broad; antennae with 9 or 10 antennomeres; habitus as in Fig. 37 .... *Sulcacis*

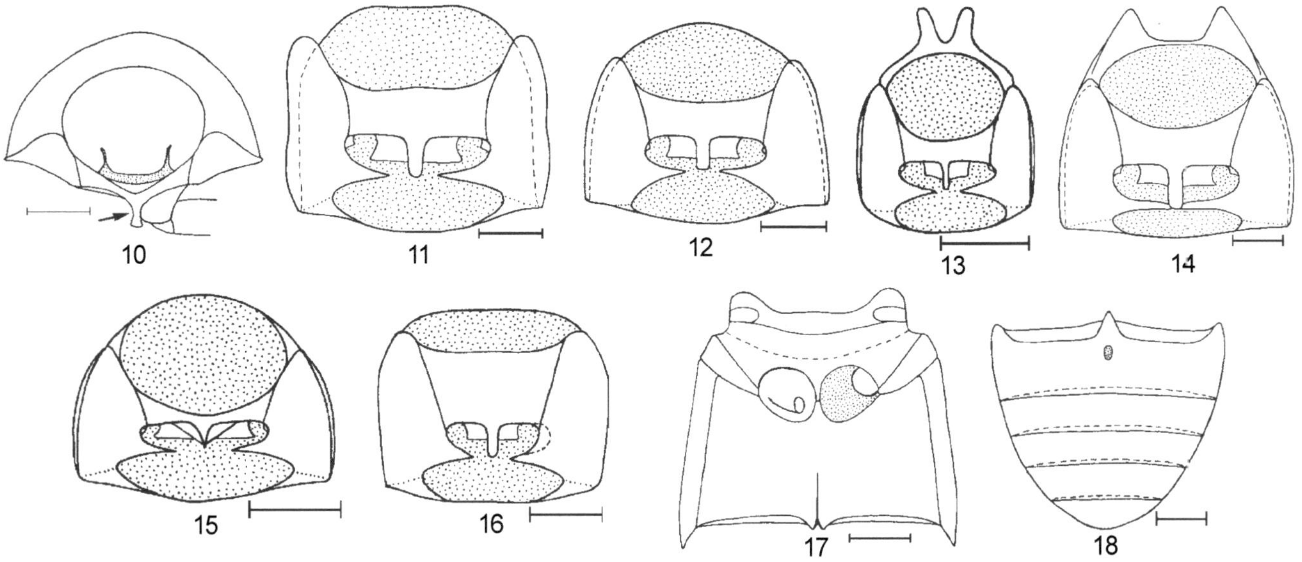

FIGURES 10.98-18.98. 10. *Cis fuscipes* Mellié, prothorax, anterior, showing long prosternal process (arrow); 11-16. Male prothorax, ventral: 11. *Orthocis punctatus* Mellié; 12. *Cis fuscipes* Mellié; 13. *Ceracis thoracicornis* (Ziegler); 14. *Cis vitulus* Mannerheim; 15. *Malacocis brevicollis* (Casey); 16. *Hadraule blaisdelli* (Casey) (scale = 0.125 mm); 17. *Cis vitulus* Mannerheim, meso- and metathorax, ventral; 18. *Cis tetracentrum* Gorham, abdomen of male, ventral (all after Lawrence 1971; scale = 0.250 mm unless noted otherwise).

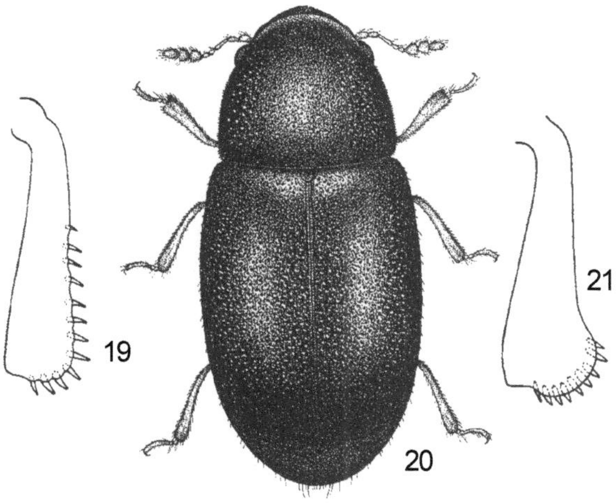

FIGURES 19.98-21.98. 19. *Octotemnus laevis* Casey, male right tibia, posterior (after Lawrence 1971); 20. *Octotemnus laevis* Casey, male (after Lawrence 1974a; drawing by C. Green); 21. *Rhopalodontus americanus* Lawrence, male right tibia, posterior (after Lawrence 1971).

9(5). Outer apical angle of protibia narrowly rounded (Fig. 30); elytral sutural apices meeting asymmetrically (Fig. 34); head and pronotum in both sexes without tubercles or horns, male sometimes with densely pubescent area on clypeus; elytral punctation single and uniform; vestiture of very short fine setae; body elongate and parallel-sided (Fig. 38) .... *Orthocis*

— Outer apical angle of protibia mostly produced and dentate (as in Figs. 25-27) or blunt and angulate (as in Figs. 28-29); if somewhat rounded, then without other characters in combination; elytral sutural apices meeting symmetrically 10

10(9). Antennae with 10 antennomeres; habitus varied (as in Figs. 1, 39-41) .... *Cis*

— Antennae with 9 antennomeres 11

11(10). Outer apical angle of protibia rounded (Fig. 31); body more elongate (Fig. 42), elytral length mostly more than 1.50 x width; apex of pronotum simple in both sexes .... *Dolichocis*

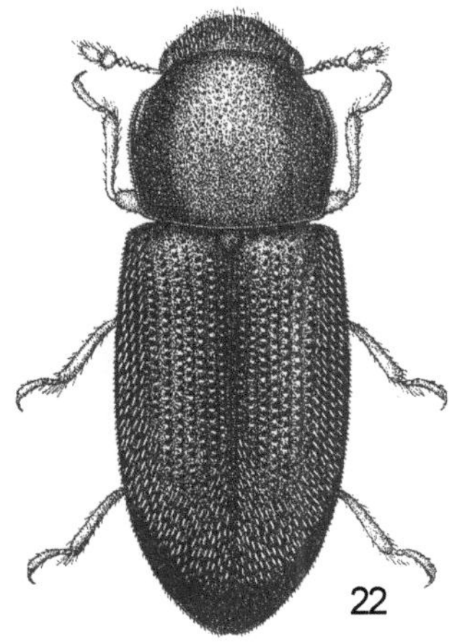

FIGURE 22.98. *Hadraule blaisdelli* (Casey), male (after Lawrence 1974a; drawing by C. Green).

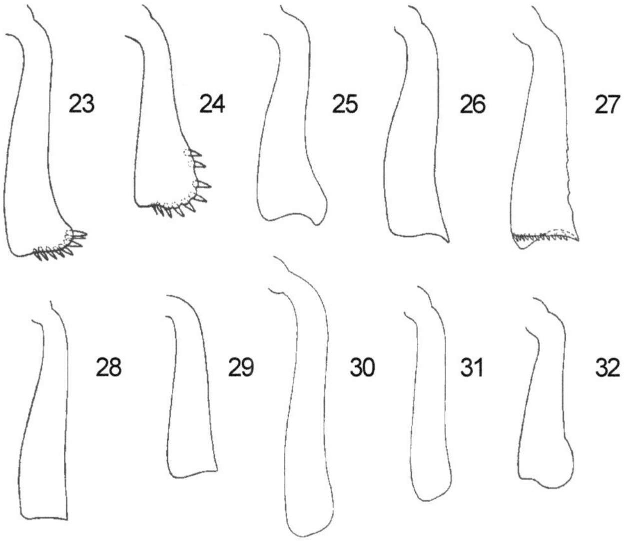

FIGURES 23.98-32.98. Male right tibia, posterior: 23. *Malacocis brevicollis* (Casey); 24. *Ceracis thoracicornis* (Ziegler); 25. *Cis robiniophilus* Lawrence; 26. *Cis cornelli* Lawrence; 27. *Cis levettei* Casey; 28. *Cis acritus* Lawrence; 29. *Cis stereophilus* Lawrence; 30. *Orthocis punctatus* (Mellié); 31. *Dolichocis manitoba* Dury; 32. *Cis cayensis* Lawrence (all after Lawrence 1971).

— Outer apical angle of protibia produced and dentate (as in Figs. 25-27); body shorter and broader, elytral length mostly less than 1.50 x width; apex of pronotum in male produced and emarginate, forming two horns or tubercles (Figs. 43-44). 12

12(11). Intercoxal process of prosternum less than 0.25 x as wide as a procoxal cavity; metasternal suture less than 0.25 x as long as median length of metasternum; frontoclypeal ridge in male bearing 4 sharp teeth (Fig. 43); habitus as in Fig. 45 .... *Plesiocis*

— Intercoxal process of prosternum more than 0.25 x as wide as a procoxal cavity; metasternal suture more than 0.25 x as long as median length of metasternum; frontoclypeal ridge in male bearing 2 subtriangular plates with a distinct notch between them (Fig. 44) .... *Ennearthron*

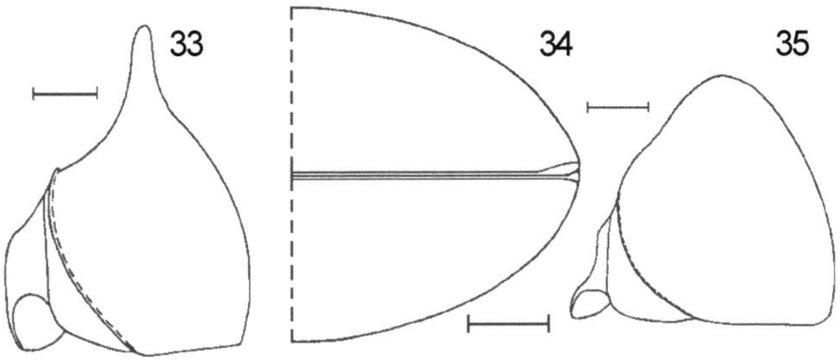

FIGURES 33.98-35.98. 33. *Strigocis opacicollis* Dury, male prothorax, left lateral (scale = 0.125 mm); 34. *Orthocis punctatus* (Mellié), elytral apices, posterodorsal (scale = 0.250 mm); 35. *Sulcacis lengi* Dury, male prothorax, left lateral (scale = 0.125 mm) (all after Lawrence 1971).

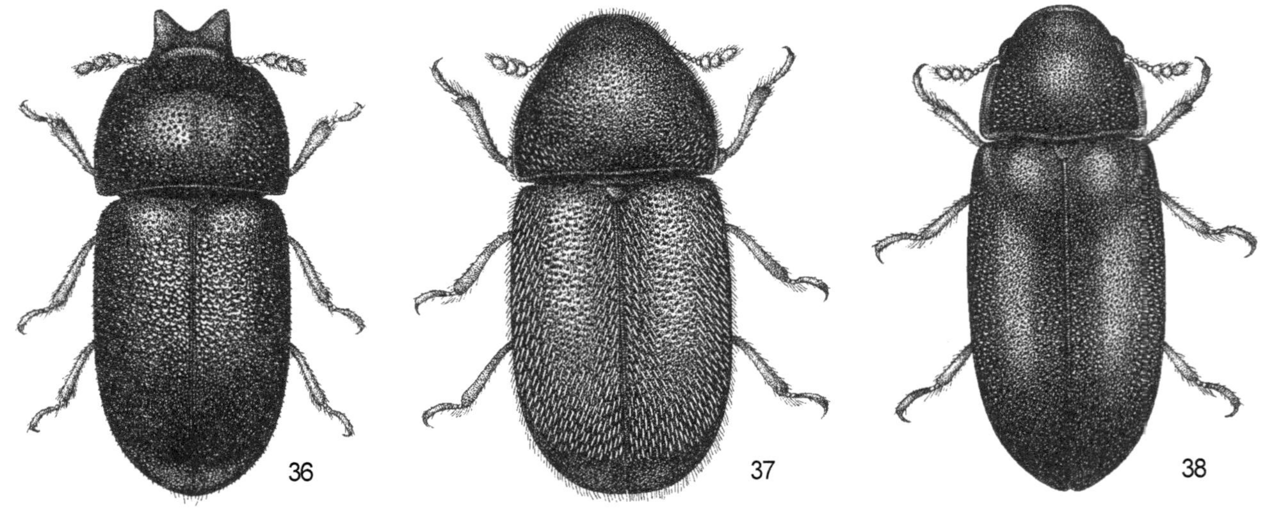

FIGURES 36.98-38.98. 36. *Ceracis californicus* (Casey), male; 37. *Sulcacis curtulus* (Casey), male; 38. *Orthocis punctatus* (Mellié), male (all after Lawrence 1974a; drawings by C. Green).

CLASSIFICATION OF THE NEARCTIC GENERA OF CIIDAE
(after Lawrence 1971, 1974a,b, 1982)

Ciidae Leach in Samouelle 1819

Sphindociinae Lawrence 1974

Antennae with 11 antennomeres, last 3 forming a club and lacking tufts of sensilla; pronotal lateral margins crenulate; mesotrochantins exposed; tibial apices not expanded, but with 2 small spurs; abdominal sternites 3 and 4 (first two visible) connate. Larva without endocarina on head; tergite 9 with concave disc-like impression (Fig. 3). A single monotypic genus, known only from California, breeding in polypore fruiting bodies. Larva, biology, and diagnoses: Lawrence (1974a,b, 1991), Lawrence *et al.* (1993). Lawrence (1971) excluded *Sphindocis* from Ciidae but later (1974a,b) recognized it as the sister group of the rest of the family.

Sphindocis Fall 1917
One species, *S. denticollis* Fall 1917, California; in *Trametes* and *Poria* fruiting bodies, pupating in the wood beneath them. Description, larva, biology: Lawrence (1974b, 1991), Lawrence *et al.* (1993).

Ciinae Leach in Samouelle 1819

Orophiina Thomson 1863
Orophyidae Kiesenwetter 1877 (based on misspelling)

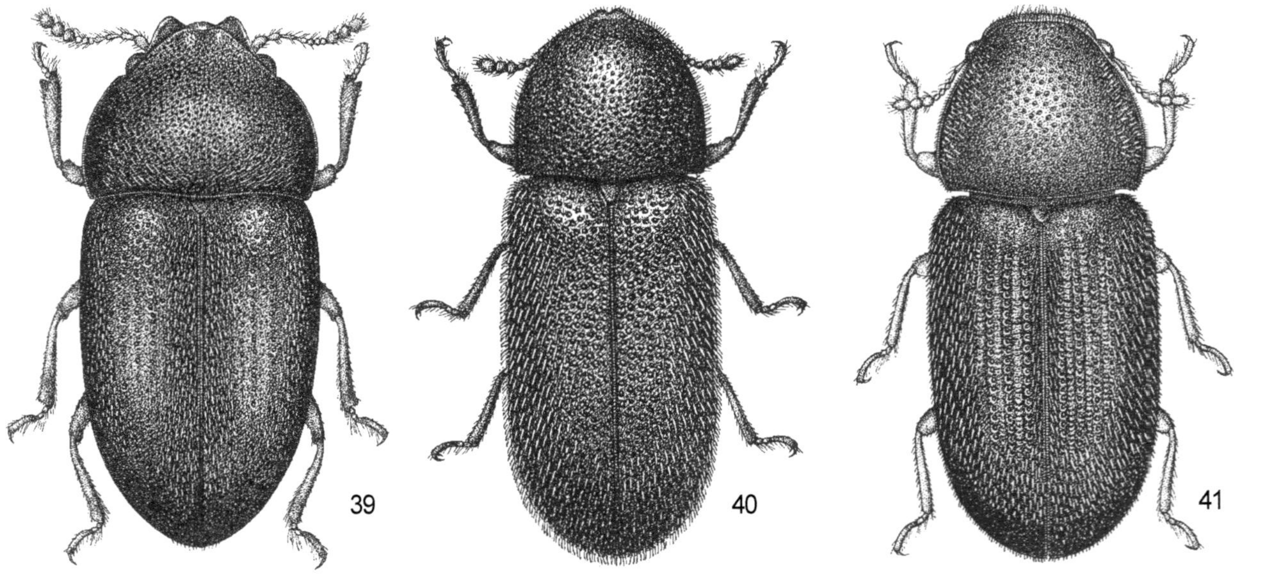

FIGURES 39.98-41.98. 39. *Cis biarmatus* Mannerheim, male; 40. *Cis hystriculus* Casey, male; 41. *Cis versicolor* Casey, male (all after Lawrence 1974a; drawings by C. Green).

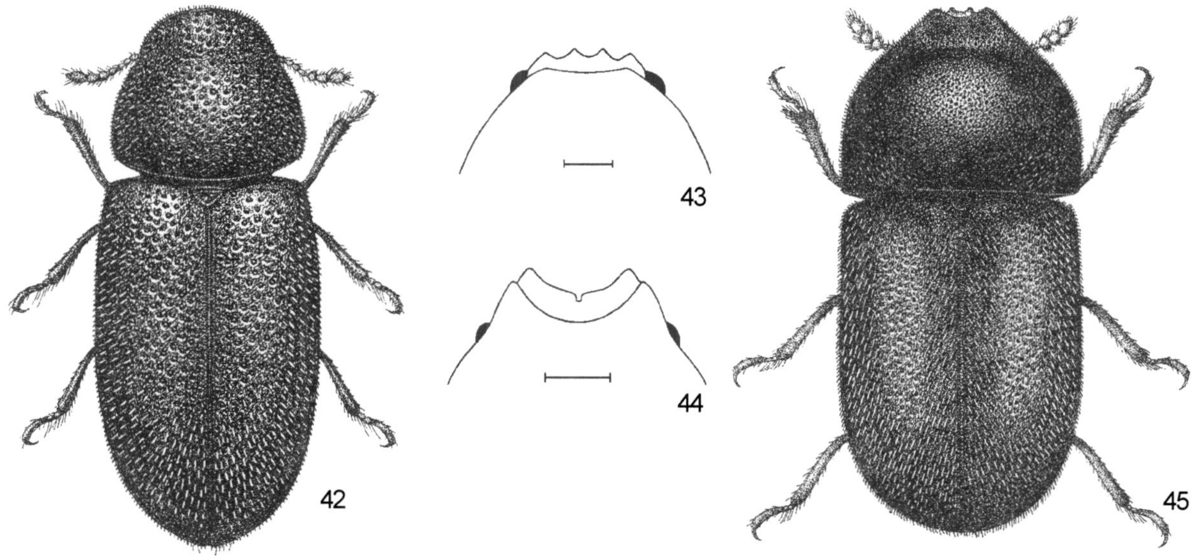

FIGURES 42.98-45.98. 42. *Dolichocis manitoba* Dury, male (after Lawrence 1974a; drawing by C. Green); 43. *Plesiocis cribrum* Casey, head and pronotal apex of male, dorsal (scale = 0.250 mm) (after Lawrence 1971); 44. *Ennearthron aurisquamosum* Lawrence, head and pronotal apex of male, dorsal (scale = 0.125 mm) (after Lawrence 1971); 45. *Plesiocis cribrum* Casey, male (after Lawrence 1974a; drawing by C. Green).

Octotemnidae Reitter 1878
Orophiini; Lawrence 1982

Antennae with 8-10 antennomeres, last 2 or 3 forming a club and each bearing at least 4 tufts of sensilla; pronotal lateral margins smooth; mesotrochantins concealed; tibial (especially protibial) apices variously expanded or modified, but without or with reduced spurs; abdominal sternite 3 (first visible) not connate with following one, in males often having a pubescent fovea or patch medially. Larva with moderately to very long epicranial stem underlain by a median endocarina; tergite 9 usually with a pair of upturned urogomphi (Fig. 2; occasionally one or more than two processes, very rarely with a concave disk). About 40 genera worldwide, 12 of them occurring in various parts of North America; the tribal classification needs additional study (Lawrence 1982) and none is used here. Most species are found in fruiting bodies of polypore fungi (Lawrence 1971, 1973). Revision and keys: Lawrence (1971); revised diagnosis: Lawrence (1974b). Biology: Lawrence (1971, 1973). Larvae: Lawrence (1974b, 1991), Lawrence *et al.* (1993), Rühm and Milewski (1996; species identification characters).

Ceracis Mellié 1848 (Fig. 36)
Nineteen species, generally distributed; on polypores of all host preference groups (most individual species with limited host ranges) and occasionally a few other fungi. Worldwide, there are 42 species, mostly New World but also in southern Africa, Japan, and the Indo-Pacific, with only a single introduced species in Europe. Revision and keys: Lawrence (1967b, 1971). Biology: Lawrence (1967b, 1971, 1973), Ackerman and Shenefelt (1973), Eberhard (1979; male use of horns). Larvae: Lawrence (1991), Bøving and Craighead (1931; as *Ennearthron* spp.).

Cis Latreille 1796 (Figs. 1, 39-41)
Eridaulus Thomson 1863
Macrocis Reitter 1878
Xestocis Casey 1898
Forty-three species, generally distributed; on polypore fungi (almost exclusively) of all four host preference groups, though most species have a host range limited to one such group. One widespread North American species, *C. fuscipes* Mellié 1848, is partly parthenogenetic (Lawrence 1967a). Two Asian species of this genus have recently been found infesting commercial dried mushrooms and polypores (Madenjian *et al.* 1993). There are about 350 described species, distributed worldwide except Africa. Revision and keys: Lawrence (1971). Biology: Ackerman and Shenefelt (1973), Lawrence (1973), Eberhard (1979; male use of horns). Larvae: Lawrence (1991), Bøving and Craighead (1931), Peterson (1951; as *Xestocis*).

Dolichocis Dury 1919 (Fig. 42)
Two species, *D. indistinctus* Hatch 1962 and *D. manitoba* Dury 1919, both Rocky Mountains and westward and transcontinental along Canada-United States boundary, the former also southeastern Arizona; in polypore fungi of the *Ganoderma* host preference group. Two additional species are known from Japan and Europe. Revision and keys: Lawrence (1971). Biology: Ackerman and Shenefelt (1973), Lawrence (1973).

Ennearthron Mellié 1847
Two species, *E. aurisquamosum* Lawrence 1971, Ohio, Kentucky, North Carolina; *E. spenceri* (Hatch 1962), British Columbia (probably introduced from Japan); known only from single records in *Bjerkandera adusta* (Fr.) P. Karst. and *Cryptoporus volvatus* (Pk.) Hub., respectively. The genus contains a total of 7-22 species across the Holarctic; its limits are somewhat uncertain (Lawrence

1971). Revision and keys: Lawrence (1971). Biology: Lawrence (1973). The larvae illustrated by Bøving and Craighead (1931) as *Ennearthron* are actually *Ceracis* (Lawrence 1991).

Hadraule Thomson 1863 (Fig. 22)
Diphyllocis of authors, in part, not Reitter 1885
Maphoca Casey 1900
Pityocis Peyerimhoff 1918
Three species, *H. blaisdelli* (Casey 1900), widely distributed and with broad host preferences; *H. elongatula* (Gyllenhal 1827), New Brunswick (probably adventive, also Palearctic) on *Piptoporus betulinus* (Bull. ex Fr.) P. Karst.; *H. explanata* Lawrence 1971, Maine, host unknown. These are the only species in the genus. Lawrence (1973) reported breeding records of *H. blaisdelli* from both *Ganoderma* and *Coriolus* host preference groups and non-breeding records from the *Phellinus* group. Klopfenstein and Graves (1989) found this species had a strong preference for and better adult survival on *Bjerkandera fumosa* (Fr.) P. Karst. (a relatively soft member of the *Ganoderma* group) compared to harder *Ganoderma*-group fungi and members of other groups. Lawrence (1971) cited *H. blaisdelli* as the only ciid known to be a pest in fungal herbaria. Revision and keys: Lawrence (1971). Biology: Lawrence (1973), Klopfenstein and Graves (1989). Internal anatomy: Klopfenstein and Graves (1992).

Malacocis Gorham 1886
Brachycis Casey 1898
One species, *M. brevicollis* (Casey 1898), eastern United States and Canada; in polypores of the *Phellinus* host preference group (and occasionally *Ganoderma* group, non-breeding). One other described and several undescribed species from Central America and the West Indies. Revision and keys: Lawrence (1971). Biology: Ackerman and Shenefelt (1973), Lawrence (1973). Larvae: Lawrence (1991).

Octotemnus Mellié 1847 (Fig. 20)
Orophius Redtenbacher 1849
One species, *O. laevis* Casey 1898, widely distributed except southwestern United States; almost exclusively in polypores of the *Coriolus* host preference group. The genus contains 16 additional species that are Palearctic and Indo-Pacific, one of which (*O. glabriculus* (Gyllenhal 1827)) may be the same as *O. laevis*. Revision and keys: Lawrence (1971). Biology: Lawrence (1971, 1973), Ackerman and Shenefelt (1973).

Orthocis Casey 1898 (Fig. 38)
Mellieicis Lohse 1964 (subgenus)
Five species, generally distributed; recorded from the polypore *Auricularia auricula* (Hook) Under. (one species), under bark, and in dead branches. The genus occurs worldwide, consisting of 38 described and 30 or more undescribed species. An Asian species of this genus has recently been found infesting commercial dried *Auricularia* (Madenjian *et al.* 1993). Revision and keys: Lawrence (1971). Biology: Lawrence (1973).

Plesiocis Casey 1898 (Fig. 45)
One species, *P. cribrum*, widespread except southeastern United States; mainly in the polypore *Cryptoporus volvatus* (Pk.) Hub. The genus is restricted to North America. Revision and keys: Lawrence (1971). Biology: Lawrence (1973).

Rhopalodontus Mellié 1847 (unjustified emendation, universally adopted and thus now valid)
Ropalodontus Mellié 1847 (original spelling)
One species, *R. americanus* Lawrence 1971, Wisconsin; apparently favoring polypores of the *Ganoderma* host preference group (Ackerman and Shenefelt 1973; Lawrence 1973). This mostly Holarctic genus includes 10 described species and a few undescribed ones from southeastern Asia. Revision and keys: Lawrence (1971). Biology: Lawrence (1971, 1973), Ackerman and Shenefelt (1973).

Strigocis Dury 1917
Two or three species, *S. opacicollis* Dury 1917, and *S. opalescens* (Casey 1898), eastern United States (also Mexico); *S. bilimeki* (Reitter 1878), Mexico and probably southern Arizona (Lawrence 1971); mostly in polypores of the *Coriolus* host preference group. The genus includes two other described species from Japan and Europe and a few undescribed ones from Central America. Revision and keys: Lawrence (1971). Biology: Lawrence (1973).

Sulcacis Dury 1917 (Fig. 37)
Entypus Redtenbacher 1848, not Dahlbom 1843
Two species, *S. curtulus* (Casey 1898), generally distributed; *S. lengi* Dury 1917, eastern United States to Texas; mostly in polypores of the *Coriolus* host preference group. The genus is Holarctic, containing a total of 7 species (Kawanabe 1997). Revision and keys: Lawrence (1971). Biology: Ackerman and Shenefelt (1973), Lawrence (1973).

Bibliography

ABDULLAH, M. 1973. The systematic position of Cisidae (Heteromera) including a catalogue of the world and comments on central European families of Cucujoidea (Coleoptera). Zoologische Beiträge (N.F.), 19: 189-246.

ACKERMAN, J. K. and R. D. SHENEFELT. 1973. Notes concerning Ciidae (Coleoptera) associated with macro-fruiting bodies of higher fungi (Basidiomycetes) in Wisconsin. Proceedings of the Entomological Society of Washington, 75: 55-62.

BØVING, A. G. and F. C. CRAIGHEAD. 1931. An illustrated synopsis of the principal larval forms of the order Coleoptera. Entomologica Americana (new series), 11: 1-351.

COSTA, C., S. A. VANIN and S. A. CASARI-CHEN. 1988. Larvas de Coleoptera do Brasil. Museu de Zoologia, Universidade de São Paulo. 4 + 1-282, estampas 1-165.

CROWSON, R. A. 1955. The natural classification of the families of Coleoptera. Nathaniel Lloyd. London. 187 pp.

CROWSON, R. A. 1960. The phylogeny of Coleoptera. Annual Review of Entomology, 5: 111-134.

DALLA TORRE, K. W. von. 1911. Cioidae. *In:* Coleopterorum Catalogus, Pars 30: 1-32. W. Junk. Berlin.

EBERHARD, W. G. 1979. The function of horns in *Podischnus agenor* (Dynastinae) and other beetles. Pp. 231-258. *In:* M. S. Blum and N. A. Blum, eds., Sexual selection and reproductive competition in insects. Academic Press. New York.

FOSSLI, T.-E. and J. ANDERSEN. 1998. Host preference of Cisidae (Coleoptera) on tree-inhabiting fungi in northern Norway. Entomologica Fennica, 9: 65-78.

GILBERTSON, R. 1984. Relationships between insects and wood-rotting Basidiomycetes. Pp. 130-165. *In:* Q. D. Wheeler and M. Blackwell, eds., Fungus-insect relationships; perspectives in ecology and evolution. Columbia University Press. New York.

GUEVARA, R., A. D. M. RAYNER, and S. E. REYNOLDS. 2000. Orientation of specialist and generalist fungivorous ciid beetles to host and non-host odours. Physiological Entomology, 25: 288-295.

JONSELL, M. and G. NORDLANDER. 1995. Field attraction of Coleoptera to odours of the wood-decaying polypores *Fomitopsis pinicola* and *Fomes fomentarius*. Annales Zoologici Fennici, 32: 391-402.

JONSSON, M., G. NORDLANDER and M. JONSELL. 1997. Pheromones affecting flying beetles colonizing the polypores *Fomes fomentarius* and *Fomitopsis pinicola*. Entomologica Fennica, 8: 161-165.

KAWANABE, M. 1997. A taxonomic study on the Japanese species of the genus *Sulcacis* (Coleoptera, Ciidae). Elytra, Tokyo, 25: 67-78.

KLOPFENSTEIN, P. C. and R. C. GRAVES. 1989. Feeding preference and adult survival of *Hadraule blaisdelli* (Coleoptera: Ciidae) on different host fungi (Polyporaceae). Entomological News, 100: 157-164.

KLOPFENSTEIN, P. C. and R. C. GRAVES. 1992. Morphology of the digestive and reproductive systems of adult *Hadraule blaisdelli* (Casey) (Coleoptera: Ciidae). Coleopterists Bulletin, 46: 344-356.

LAWRENCE, J. F. 1967a. Biology of the parthenogenetic fungus beetle *Cis fuscipes* Mellié (Coleoptera: Ciidae). Breviora, 258: 1-14.

LAWRENCE, J. F. 1967b. Delimitation of the genus *Ceracis* (Coleoptera: Ciidae) with a revision of North American species. Bulletin of the Museum of Comparative Zoology, 136: 91-143.

LAWRENCE, J. F. 1971. Revision of the North American Ciidae (Coleoptera). Bulletin of the Museum of Comparative Zoology, 142: 419-522.

LAWRENCE, J. F. 1973. Host preference in ciid beetles (Coleoptera: Ciidae) inhabiting the fruiting bodies of Basidiomycetes in North America. Bulletin of the Museum of Comparative Zoology, 145: 163-212.

LAWRENCE, J. F. 1974a. The ciid beetles of California (Coleoptera: Ciidae). Bulletin of the California Insect Survey, 17: 1-41.

LAWRENCE, J. F. 1974b. The larva of *Sphindocis denticollis* Fall and a new subfamily of Ciidae. (Coleoptera: Heteromera). Breviora, 424: 1-14.

LAWRENCE, J. F. 1982. A catalog of the Coleoptera of America north of Mexico. Family: Ciidae. United States Department of Agriculture, Agriculture Handbook, 529-105: x + 18 pp.

LAWRENCE, J. F. 1989. Mycophagy in the Coleoptera: Feeding strategies and morphological adaptations. Pp. 1-23. *In:* N. Wilding, N. M. Collins, P. M. Hammond and J. F. Webber, eds. Insect-fungus interactions. Academic Press. London.

LAWRENCE, J. F. 1991. Ciidae. Pp. 502-504. *In:* F. W. Stehr, ed. Immature Insects, Vol. 2. Kendall/Hunt. Dubuque, IA.

LAWRENCE, J. F., A. M. HASTINGS, M. J. DALLWITZ and T. A. PAINE. 1993. Beetle larvae of the World: Interactive identification and information retrieval for families and subfamilies. CD-ROM, Version 1.0 for MS-DOS. CSIRO Publishing. Melbourne.

LAWRENCE, J. F., A. M. HASTINGS, M. J. DALLWITZ, T. A. PAINE and E. J. ZURCHER. 1999a. Beetle larvae of the World: Descriptions, illustrations, and information retrieval for families and subfamilies. CD-ROM, Version 1.1 for MS-Windows. CSIRO Publishing. Melbourne.

LAWRENCE, J. F., A. M. HASTINGS, M. J. DALLWITZ, T. A. PAINE and E. J. ZURCHER. 1999b. Beetles of the World: Descriptions, illustrations, and information retrieval for families and subfamilies. CD-ROM, Version 1.0 for MS-Windows. CSIRO Publishing. Melbourne.

LAWRENCE, J. F. and R. J. MILNER. 1996. Associations between arthropods and fungi. Fungi of Australia, 1B: 137-202.

LOHSE, G. A. and J. REIBNITZ. 1991. *Cis multidentatus* (Pic), ein unbeachteter Vorratsschädling (Coleoptera, Cisidae). Acta Coleopterologica (Munich), 7: 103-105.

MADENJIAN, J. J., J. D. EIFERT and J. F. LAWRENCE. 1993. Ciidae: Newly recognized beetle pests of commercial dried mushrooms. Journal of Stored Products Research, 29: 45-48.

MELLIÉ, J. 1848. Monographie de l'ancien genre *Cis* des auteurs. Annales de la Société Entomologique de France, ser. 2, 6: 205-274, 313-396.

NAVARRETE-HEREDIA, J. L. and A. BURGOS-SOLORIO. 2000. Ciidae (Coleoptera), pp. 413-420. *In:* Llorente Bousquets, J. E., E. González Soriano, and N. Papavero, eds. Biodiversidad, taxonomía y biogeografía de artrópodos de México: hacia una sítesis de su conocimiento. Volumen II. Universidad Nacional Autónoma de México. Mexico, D. F. xvi + 676 pp.

ØKLAND, B. 1995. Insect fauna compared between six polypore species in a southern Norwegian spruce forest. Fauna norvegica, Series B, 42: 21-26.

PAVIOUR-SMITH, K. 1960. The fruiting-bodies of macrofungi as habitats for beetles of the family Ciidae (Coleoptera). Oikos, 11: 43-71.

PETERSON, A., 1951. Larvae of insects. An introduction to Nearctic species. Part II. Coleoptera, Diptera, Neuroptera, Siphonaptera, Mecoptera, Trichoptera. Columbus, OH, v + 1-416 pp.

RÜHM, W. and I. MILEWSKI. 1996. Zur Bedeutung der larvalen Bestimmungsmerkmale beim Studium von Abbausukzessionen: Bestimmungsmerksmale einiger Cisidenlarven (Cisidae, Col.). Entomologische Mitteilungen aus dem Zoologischen Museum Hamburg, 12: 55-62.

TSCHINKEL, W. R. and J. T. DOYEN. 1980. Comparative anatomy of the defensive glands, ovipositors and female genital tubes of tenebrionid beetles (Coleoptera). International Journal of Insect Morphology and Embryology, 9: 321-368.

99. TETRATOMIDAE Billberg 1820

by Daniel K. Young and Darren A. Pollock

Family common name: The polypore fungus beetles

The oblong to ovate, somewhat flattened, pubescent body, relatively large, emarginate compound eyes, prothoracic coxae separated by a prosternal process, and 5-5-4 tarsal formula with no lobed tarsomeres, partially distinguish this family of primitive tenebrionoid beetles.

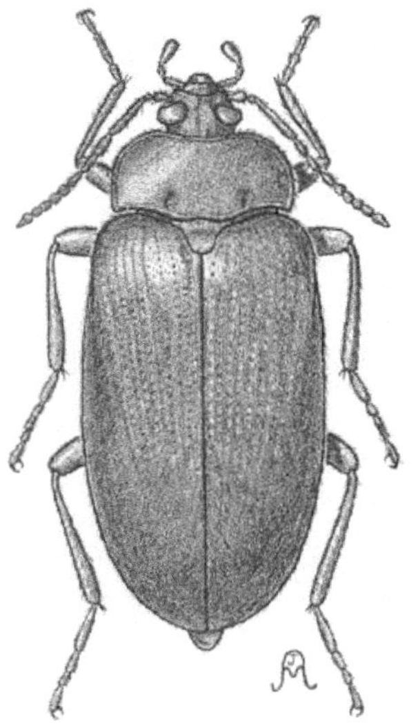

FIGURE 1.99. *Penthe pimelia* (Fabricius)

Description: (after Crowson 1955, 1964, Lawrence 1982) Shape oblong to elongate, strongly convex to slightly flattened; length 2 to 17 mm; color brownish to black, some with orange or reddish markings; subglabrous or with vestiture consisting of sparse to dense decumbent setae.

Head short, triangular, slightly deflexed; surface punctate or rugopunctate. Antennae with 11 antennomeres, clavate, or with apical 3-4 antennomeres forming a loose club; antennal insertions exposed or slightly concealed under small frontal ridge. Clypeus distinctly separated by impressed line which may be incomplete, trapezoidal; labrum quadrate; mandibles curved, blunt; maxillae reduced; maxillary palpi with four palpomeres, the first small, the remaining three conspicuous; gula small, the gular sutures sometimes indistinct, moderately separated; labium with the mentum transverse; ligula moderate, conspicuously setose; labial palpi with three palpomeres, the first minute, third the longest. Eyes lateral, large, obovate, weakly to conspicuously emarginate near the antennal insertions, narrowly (especially some Eustrophinae) to widely separated, dorsally.

Pronotum broader than head, nearly as broad as basal width of elytra, subquadrate or narrowed in front, with anterior margins rounded; posterior margin entire or sinuate, laterally narrowing toward head; sides distinctly, smoothly or crenulately margined; surface punctate; prosternum broad in front of coxae with moderately broad intercoxal process; prothoracic coxal cavities open behind externally and open or closed internally. Scutellum arcuate or subtriangular. Mesosternum short; metasternum broad. Legs with anterior trochantins usually visible; prothoracic coxae transverse or oval; mesothoracic coxae narrowly separated, with laterally open cavities; metathoracic coxae transverse, contiguous; trochanters small, triangular; tibiae slender with distinct apical spurs; tarsal formula 5-5-4, segments slender, the first segment elongate; claws simple. Elytra entire, the apices rounded; punctation usually confused; epipleural fold narrow and incomplete or (Eustrophinae) relatively wide and distinct. Metathoracic wings with 3-4 anal veins, subcubital fleck entire or absent.

Abdomen with five visible ventrites, the sutures entire or with the first two segments connate. Male genitalia sometimes inverted; basal piece sometimes with a pair of accessory lobes.

Larvae (after Bøving and Craighead 1931, Crowson 1964, Lawrence 1991, Nikitsky 1992) elongate, subcylindrical, or subdepressed, parallel-sided or slightly narrowed anteriorly and posteriorly; length 3 to 17 mm; vestiture sparse, consisting of a few moderately long and shorter setae on each segment; color creamy white to light brown, sometimes with darker tergal maculations. Head exserted, prognathous, somewhat flattened, nearly as wide as the prothorax; epicranial suture with stem short to moderately long (but absent in some Eustrophinae), frontal arms lyriform or forked, with straight outer and lyriform inner branches; median endocarina absent or (*Penthe*) coincidental with stem; antennae short, three-segmented. Labrum distinct, emarginate in Eustrophinae; frontoclypeal suture generally absent (present in Piseninae); mandibles symmetrical or weakly to strongly asymmetrical, stout, bidentate, molar area well developed (*Pisenus*), tuberculate (Hallomeninae), reduced (*Triphyllia elongatus*), or replaced by hyaline processes (Tetratominae) or a membranous lobe (*Penthe*); maxillae each with transverse cardo, elongate stipes, well developed maxillary articulating area, blunt mala, and three-segmented palpi; labium free to base of mentum, short ligula (long and narrow in *Penthe*), and two-segmented palpi. Five stemmata present on each side of head. Thorax without asperity patches or (*Penthe*) with meso- and metanotum and abdominal terga 1-8 bearing asperity patches; thoracic legs five-segmented, with bisetose tarsunguli. Abdomen 10-segmented; ninth tergum bearing paired, usually upcurved, bifid urogomphi; ninth sternum simple or (Hallomeninae) with a single, stout asperity on each side; tenth segment transversely oval, oriented posteroventrally. Spiracles annular-biforous, with moderately long accessory tubes.

Habits and habitats. (See also Crowson 1964, Lawrence 1982, 1991, Nikitsky and Lawrence 1992) Tetratomids feed largely upon the fruiting bodies of hymenomycete fungi, especially Polyporaceae and Tricholomataceae. Thus, they are most commonly found under fungus-grown bark and in softer shelf fungi, where adults tend to browse primarily on the surface while larvae bore into the tissues. Adults are commonly encountered at night on fungi; several species are also not uncommon in Lindgren funnel samples from forested habitats.

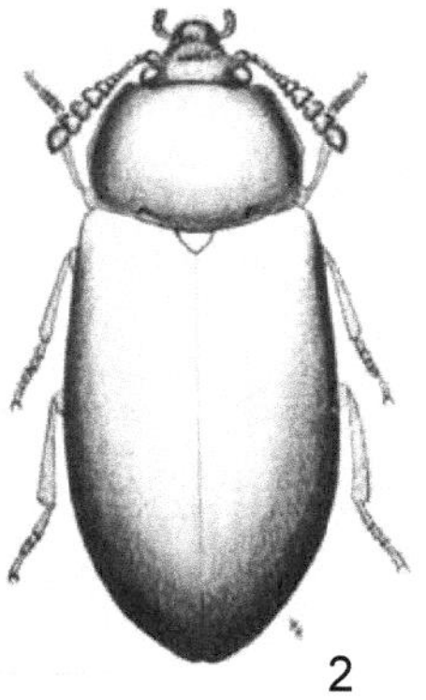
2

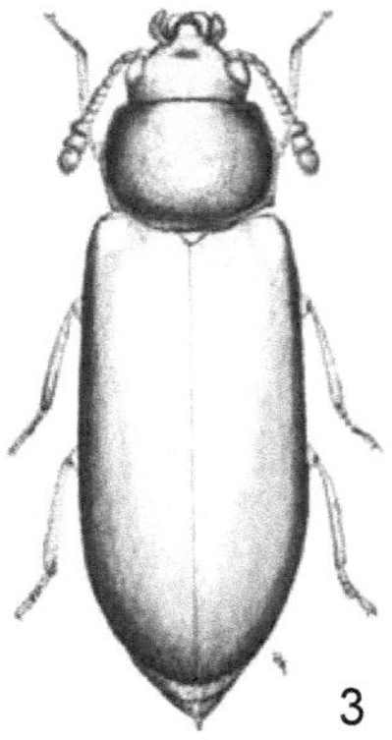
3

FIGURE 2.99-3.99. 2.*Tetratoma concolor* LeConte; 3.*Triphyllia elongatus* (LeConte)(from Hatch 1965).

Status of the classification. Various elements of this family have historically been associated with several lineages of Melandryidae (Crowson 1955, Lawrence and Newton 1995, Miyatake 1960, Nikitsky 1992); there may also be some close links to Mycetophagidae (Nikitsky 1998). The world generic classification recently proposed by Nikitsky (1998), based upon character sets from both larvae and adults, has done much to consolidate a working hypothesis for the group, but no formal cladistic analysis has yet been attempted.

Distribution. There are 13 genera and about 155 species arranged into five subfamilies (Nikitsky 1998), known from all major biogeographical regions except the Australian. All five subfamilies, 10 genera and 26 species have been recorded from the United States and Canada (Chantal 1985, Lawrence 1991, Nikitsky 1998).

KEY TO THE GENERA OF AMERICA NORTH OF MEXICO

1. Body broad (Fig. 1), relatively large (>9 mm in length); antennomere 3 as long as combined lengths of antennomeres 4-5, much longer than combined lengths of antennomeres 1-2; each elytron with 11+ rows of punctures (Penthinae) .... *Penthe*
— Body relatively narrow or oblong-oval, usually much less than 9 mm in length; antennomere 3 usually much shorter; each elytron with fewer than 11 rows of punctures 2

2(1). Apical 3-4 antennomeres abruptly dilated forming loose, parallel club (Figs. 2-3) 3
— Antennae filiform, serrate, moniliform, or but gradually, mildly clavate distally 5

3(2). Antennae with apical four antennomeres dilated (Fig. 2) (Tetratominae) .... *Tetratoma*
— Antennae with apical three antennomeres abruptly and strongly dilated, forming a loose club (Fig. 3) (Piseninae) 4

4(3). Pronotum broadly reflexo-explanate at sides, transversely truncate and scarcely perceptibly bisinuate at base, foveae obsolete (Fig. 3); abdominal segments decreasing uniformly and gradually in length, first scarcely visibly longer than second .... *Triphyllia*
— Pronotum not at all explanate at sides, base broadly, arcuately lobed at middle, foveae distinct though not very well developed; first visible ventrite as long as next two combined .... *Pisenus*

5(2). Body form oval; prosternum elongate anterad prothoracic coxae; distal antennomeres distinctly wider than antennomeres 3-4 (Figs. 5-7) (Eustrophinae) 6
— Body form elongate; prosternum short, sunken anterad prothoracic coxae; antennomeres 3-10 subequal in width (Fig. 4) (Hallomeninae) .... *Hallomenus*

6(5). Outer faces of mesothoracic and metathoracic tibiae with numerous oblique comb-like ridges (Fig. 8) 7
— Outer faces of mesothoracic and metathoracic tibiae smooth, without oblique ridges 8

7(6). Elytral punctures rather coarse, forming distinct striae; posterior margin of pronotum with distinct medial lobe; prothoracic episterna each with a distinct, transverse suture........... *Eustrophopsis*
— Elytral punctures fine, not forming obvious striae; posterior margin of pronotum with slight lobe only; prothoracic episterna without a transverse suture .... *Eustrophus*

8(6). Prosternal process narrowed distally, not extended posterad prothoracic coxae; elytral punctation coarse, forming distinct striae ...... *Synstrophus*
— Prosternal process widened distally, separating prothoracic coxae; elytral punctation fine, not forming striae (Holostrophini) 9

9(8). Eyes only indistinctly emarginate, distance between eyes greater than transverse ocular diameter; prosternal process at most slightly projecting behind posterior edge of prothoracic coxae .... *Pseudoholostrophus*
— Eyes distinctly emarginate, distance between eyes less than transverse ocular diameter; prosternal process longer, protruding behind posterior edge of prothoracic coxae .... *Holostrophus*

CLASSIFICATION OF THE GENERA OF AMERICA NORTH OF MEXICO

Tetratomidae Billberg 1820

Tetratominae Billberg 1820

This subfamily is represented solely by 22 Holarctic species of *Tetratoma* Fabricius. We follow Nikitsky (1998) in recognizing five subgenera: *Tetratoma sensu stricto*; *Tetratoma* (*Abstrulia*) Casey 1900; *Tetratoma* (*Incolia*) Casey 1900; *Tetratoma* (*Paratetratoma*) Nikitsky, 1998; *Tetratoma* (*Falsoxanthalia*) Pic 1934.

Diagnosis. Body relatively narrow to oblong-oval, usually much less than 9 mm in length; each elytron with fewer than 11 rows of punctures; apical four antennomeres abruptly dilated forming loose, parallel club.

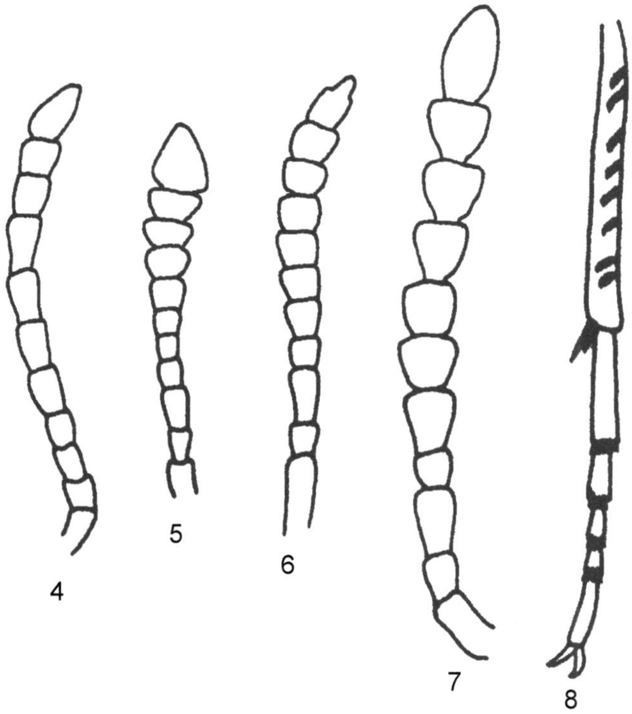

FIGURE 4.99-8.99. 4-7. Antenna, 4. *Hallomenus* sp.; 5. *Holostrophus bifasciatus* (Say); 6. *Eustrophus tomentosus* Say; 7. *Synstrophus repandus* (Horn). 8. *Eustrophopsis bicolor* (Fabricius), mesothoracic tibia and tarsus.

Tetratoma Fabricius 1790, 5 species, distributed as noted below (key to species in Casey 1900).

- subgenus *Tetratoma sensu stricto: T. truncorum* LeConte, northeastern and northcentral United States and adjacent Canada.
- subgenus *Abstrulia* Casey 1900: *T. tessellata* Melsheimer, northeastern and northcentral United States and adjacent Canada; *T. variegata* Casey, northeastern United States.
- subgenus *Incolia* Casey 1900: *T. concolor* LeConte, northwestern United States and adjacent Canada; *T. longipennis* (Casey), Indiana.

Piseninae Miyatake 1960

Piseninae consist of three, largely Holarctic, genera. Six species of *Pisenus* Casey (1900) are known from Russia, Korea, Taiwan, Japan, and North America. The monotypic, Chilean *Notopisenus boleti* Nikitsky and Lawrence (Nikitsky 1992) is the only non-Holarctic pisenine. The third genus, *Triphyllia* Reitter 1898, contains one Eurasian and one North American species.

Diagnosis. Body relatively elongate-narrow, much less than 9 mm in length; each elytron with fewer than 11 rows of punctures; apical three antennomeres abruptly dilated forming loose, subparallel club.

Pisenus Casey 1900, 2 species: *P. humeralis* (Kirby), northeastern and northcentral United States and adjacent Canada; *P. pubescens* Casey, Virginia (key to species in Casey 1900).

Triphyllia Reitter 1898, 1 species, *T. elongatus* (LeConte); Alaska to California.

Eupisenus Casey 1900

Penthinae Lacordaire 1859

Two genera are included in Penthinae: *Penthe* Newman, with nine Holarctic species (China, Himalayan region, Oriental region, Japan, and North America), and *Cyanopenthe* Nikitsky 1998, with two species from India, Bhutan, and Indochina.

Diagnosis. Body broad, relatively large (>9 mm in length); antennomere three as long as combined lengths of antennomeres 4-5, much longer than combined lengths of antennomeres 1-2; each elytron with more than 11 rows of punctures or, if elytral punctation confused (*Cyanopenthe*), then antennae terminating in an abrupt, 4-5 segmented, pectinate club.

Penthe Newman 1838, 2 species: *P. obliquata* (Fabricius), widespread throughout the eastern and central United States and adjacent Canada; *P. pimelia* (Fabricius), widespread throughout the eastern and central United States and adjacent Canada, southwest to Texas (key to species in Casey 1900).

Hallomeninae Mulsant 1856

This subfamily is represented by 13 species of *Hallomenus* Panzer (Holarctic, Mexico) and four Palearctic species of *Mycetoma* Dejean.

Diagnosis. Members of Hallomeninae may be distinguished by the following combination of external characters: eyes deeply emarginate, but margins removed from antennal fossae; antennomeres 3-10 moniliform or serrate, without distinct widening or distal club; apical maxillary palpomere elongate, fusiform or only very slightly expanded; prosternal process elongate, separating coxae, curved dorsally at apex; tibial spurs very short.

Hallomenus Panzer 1794, 6 species, widely distributed in North America, south into Mexico and Central America. *Hallomenus serricornis* LeConte is placed in subgenus *Xeuxes* Champion by Nikitsky (1998); the remaining North American species belong to *Hallomenus sensu stricto.*

Eustrophinae Gistel 1856

Eustrophines are widespread, and are found in North America, New World tropics, Eurasia, Africa, and Madagascar. Nikitsky (1998) divided the subfamily into the tribes Holostrophini and Eustrophini.

Diagnosis. The Eustrophinae can be distinguished by the following combination of characters: body shape elliptical, convex; eyes emarginate or not, reaching antennal fossae; pronotum with distinct lateral carinae; metepisternum divided into a short, triangular anterior piece and a longer, quadrangular posterior piece.

Eustrophini Gistel 1856

The tribe Eustrophini includes three genera: *Eustrophopsis* Champion, with approximately 55 widely distributed species, *Eustrophus* Illiger, with four Holarctic species, and *Synstrophus* Seidlitz, with five Holarctic species.

Eustrophopsis Champion 1889; 5 species, widely distributed (key to species in Horn 1888).
Eustrophinus Seidlitz 1898

Eustrophus Illiger 1807; 1 species, *E. tomentosus* Say, western and eastern North America, with a gap in the interior states.

Synstrophus Seidlitz 1898; 1 species, *S. repandus* (Horn), widely distributed in North America.

Holostrophini Nikitsky 1998

This tribe is divided into two genera: *Holostrophus* Horn, with 17 Holarctic species, and *Pseudoholostrophus* Nikitsky with four species, two each from China and North America.

Holostrophus Horn 1888; 1 species, *H. bifasciatus* (Say), eastern North America.

Pseudoholostrophus Nikitsky 1983; 2 species, *P.* (*Holostrophinus*) *discolor* (Horn), eastern North America, and *P.* (*Pseudoholostrophus*) *impressicollis* (LeConte), western North America (key to species in Nikitsky 1998).

Bibliography

BØVING, A. G. and F. C. CRAIGHEAD. 1931. An illustrated synopsis of the principal larval forms of the order Coleoptera. Entomologica Americana, 11: 1-351.

CASEY, T. L. 1900. Review of the American Corylophidae, Cryptophagidae, Tritomidae, and Dermestidae, with other studies. Journal of the New York Entomological Society, 8: 51-172.

CHANTAL, C. 1985. Les Tetratomidae (Coleoptera) du Quebec. Fabreries 11: 43-66.

CROWSON, R. A. 1955. The Natural Classification of the Families of Coleoptera. N. Lloyd. London.

CROWSON, R. A. 1964. Observations on British Tetratomidae (Col.), with a key to the larvae. Entomologist's Monthly Magazine, 99 (1963): 82-86.

HATCH, M. H. 1965. The beetles of the Pacific Northwest. Part IV. Macrodactyles, Palpicornes, and Heteromera. University of Washington Publications in Biology 16: 268 pp.

HORN, G. H. 1888. Miscellaneous coleopterous studies. Transactions of the American Entomological Society, 15: 26-48.

LAWRENCE, J. F. 1982. Coleoptera. Pp. 482-553. *In*: S. P. Parker, ed., Synopsis and classification of living organisms, Volume 2. McGraw-Hill. NY.

LAWRENCE, J. F. 1991. Tetratomidae. Pp. 504-505. *In*: F. W. Stehr, ed., Immature insects, Volume 2. Kendall/Hunt. Dubuque, IA.

LAWRENCE, J. F. and A. F. NEWTON, Jr. 1995. Families and subfamilies of Coleoptera (with selected genera, notes, references and data on family-group names). Pp. 779-1006. *In*: J. Pakaluk and S. A. Slipinski, eds., Biology, phylogeny and classification of Coleoptera: Papers celebrating the 80th birthday of Roy A. Crowson. Muzeum i Instytut Zoologii PAN. Warsaw.

MIYATAKE, M. 1960. The genus *Pisenus* Casey and some notes on the family Tetratomidae (Coleoptera). Transactions of the Shikoku Entomological Society, 6: 121-134.

NIKITSKY, N. B. 1992. News on the genus *Triphyllia* Reitter, 1898 (Coleoptera Tetratomidae). Elytron, 5(1991): 159-168.

NIKITSKY, N. B. 1998. Generic classification of the beetle family Tetratomidae (Coleoptera, Tenebrionoidea) of the world, with description of new taxa. Pensoft, Sofia. 80 pp.

100. MELANDRYIDAE Leach 1815

by Darren A. Pollock

Family common name: The false darkling beetles

Family synonym: Serropalpidae Latreille 1825

This family exhibits moderate structural diversity, and is difficult to characterize; most species have elongated first hind tarsomeres and distinct hind tibial spurs, some have posterior pronotal pits; the distal maxillary palpomere in most groups is large, either securiform or cultriform.

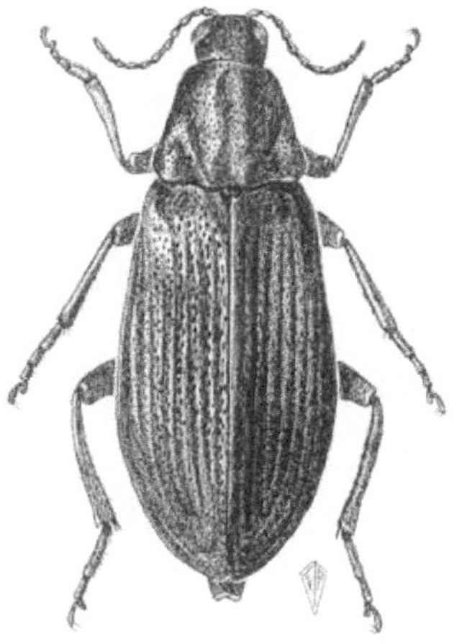

FIGURE 1.100. *Melandrya striata* Say

Description: Body variously shaped: slender, parallel-sided or tapered posteriorly, to broad, ovate to subcylindrical; total length about 2-20 mm; color of most taxa uniform brown to black; some with contrasting color pattern; dorsal vestiture present, often combination of depressed and erect setae.

Head very short (especially in Orchesiini) to short in length, slightly to moderately hypognathous, parallel-sided, without distinct constriction posterior of eyes; eyes narrowly (especially in Orchesiini) to widely separated on vertex; attaining posterior margin of head (some Orchesiini), or ending well anterior of posterior limit of visible portion of head; eyes coarsely to finely faceted, with or without intrafacetal setae, not, slightly, or distinctly emarginate anteriorly, around antennal insertions; antennae 11-segmented, moniliform to filiform and serrated, with or without indistinct to distinct club of 3-5 antennomeres; antennal insertions completely visible, frons not produced laterally; frontoclypeal area undefined, marked laterally only, or with distinct arcuate or straight suture; translucent strip of cuticle between clypeus and labrum present, short to relatively long; labrum transverse, although relatively narrowly so in some taxa; mandibles relatively short and stout, not significantly extended anterior of labrum; maxillary palpi well-developed, palpomeres variously shaped (Figs. 2-8), moderately serrated in Serropalpini; last maxillary palpomere triangular, securiform or cultriform, often very large; labial palpi short, last palpomere fusiform, pyriform to flattened and expanded; males of some genera (especially in Serropalpini and Melandryini) with setose pit on mentum.

Pronotum subquadrate to slightly elongate, to distinctly transverse, widest posterior of midlength in most taxa; pronotal disc evenly convex or with various grooves and depressions, laterally explanate in some genera (especially Melandryini); disc with posterolateral pits in most taxa; lateral margins of pronotal disc without, or with partial to distinct carina; posterior edge of pronotum tightly appressed against, or lying loosely over, base of elytra; prosternum in front of coxae shrunken and very short (especially in some Serropalpini) to relatively long; suture between prosternum and propleuron carinate and somewhat excavated in some genera; prosternal process short and indistinct; procoxae subglobular to elongate; procoxal cavities open or closed; protrochantin exposed in some genera; legs moderately to distinctly elongate, sexually dimorphic in some groups (especially *Osphya*); front legs generally smaller than middle, and especially hind legs; femora subcylindrical to distinctly flattened, parallel-sided to slightly enlarged at midlength; tibiae thin to moderately wide, parallel to slightly widened distally; dorsal surface in some groups (notably Serropalpini) with ornamentation consisting of transverse combs or longitudinal carinae; tibial spurs distinct (especially in Serropalpini and Orchesiini), relatively straight (males of *Enchodes* with ventral protibial spur distinctly bent), either subequal or distinctly unequal in length; tarsi 5-5-4 in both sexes; tarsomeres narrow, simple (*e.g.,* Orchesiini) or widened, with penultimate tarsomere expanded ventrally; tarsal claws narrow, simple, except for *Osphya*, with distinct basal tooth.

Elytra rounded to somewhat acute apically, covering all of abdomen, lateral margins parallel, tapered and narrowed posteriorly, or broadly ovate; dorsal surface flattened to convex, punctation arranged in distinct striae or scattered (*Melandrya* with longitudinal carinae on elytra); humeri distinct; epipleura indistinct in most groups, ranging from very narrow and short to narrow and extended to elytral apex, relatively wide and distinct in some Orchesiini; scutellum relatively small, or entirely hidden (*Lederia*), posteriorly rounded to triangular; mesosternum evenly convex, to elevated medially (especially in *Anisoxya*); intercoxal process various, from blunt to long and narrow, extended to midlength or to posterior extent of mesocoxae; intercoxal process variously narrowed, bluntly triangular to subparallel, distinctly notched in some groups, attaining anterior margin of metasternum or not; mesotrochantin exposed; mesocoxal cavities open laterally, partly closed by mesepimera; hind wing well developed; radial cell absent or distinct, longer than wide; wedge cell short to elongate; four terminal veins in medial region; metasternum variously convex, with discrimen from 1/3 to 1/2 length of sternum; anterior margin truncate, to slightly produced, in contact with mesocoxal process in some taxa.

Abdomen with five ventrites; V1 and at least V2 connate; aedeagus elongate (especially so in some Serropalpini) to moderately stout, with or without distinct medial struts on median lobe; distal part of apical portion of aedeagus with or without distinct lobes; ovipositor elongate, flexible, with rod-like bacula; coxites sparsely setose; styli cylindrical.

Larvae elongate, subparallel sided or slightly widened at middle, subcylindrical to slightly depressed; integument only slightly sclerotized, except for mouthparts and abdominal apex; some taxa with distinctly pigmented abdominal and thoracic tergites; integument smooth, with vestiture of scattered setae; head protracted and prognathous, relatively wide; epicranial stem in most taxa relatively long, very short or absent in Osphyinae and some Melandryinae; frontal arms U- or V-shaped, entirely absent in some taxa; median endocarina in most taxa coincident with epicranial stem, occasionally extended along frontal arms; stemmata in most taxa 5 per side, in some groups, reduced to 2 or 0; antennae in most groups relatively short, 3-segmented (in *Osphya*, appearing 2-segmented); frontoclypeal suture usually absent; labrum free; mandibles symmetrical, stout, without accessory processes; mola absent from most taxa, or represented by few teeth or tubercles; ventral mouthparts retracted; maxilla with transverse cardo, elongate stipes, distinct articulating area; maxillary palpi 3-segmented; mala rounded or cleft (Osphyinae) distally; labium with fused mentum and submentum; ligula when present, longer than 2-segmented labial palpi in most taxa; hypopharyngeal sclerome absent from most taxa; hypostomal rods short to moderately long, absent in some genera of Melandryinae; ventral epicranial ridges absent; thorax with distinct, 5-segmented legs; tarsungulus with 1 or 2 setae; abdomen with tergite 9 without or with (some Melandryinae) urogomphi; median process present in Osphyinae; some Melandryinae with urogomphal pit; sternite 9 simple; spiracles annular-biforous, in some groups annular, annular-uniforous, or annular biforous; accessory tubes of various lengths.

Habits and habitats. It is possible to divide the members of this family according to their dominant feeding habits, with two general categories: fungivory and xylophagy. Within Melandryinae, a combination of fungivory (*e.g.*, most Orchesiini) and xylophagy (Serropalpini and Melandryini) is present. However, it is likely that fungi comprise a significant portion of the diet of the seemingly xylophagous groups. Adult Melandryidae are primarily nocturnal, and can often be seen crawling about on fungi or dead logs at night. Lawrence (1991) provided a detailed discussion of the feeding habits of various groups of Melandryidae.

Status of the classification. Some authors (*e.g.*, Crowson 1955, Chantal 1985, Nikitsky 1998) have placed Hallomeninae and Eustrophinae in Tetratomidae; there is evidence for this, and this classification is followed in this treatment of Melandryidae. Several distinct larval types exist in Melandryidae, which may indicate the non-monophyly of Melandryidae (Lawrence 1991). Lawrence and Newton (1995) recognized four subfamilies, all of which are represented in North America. According to Lawrence and Newton (1995), Melandryinae are probably monophyletic, although the tribal classification (*e.g.*, Arnett 1963) seems inadequate and probably artificial. Unfortunately, there has been no comprehensive systematic study of Melandryidae, or of their relationships to other lower Tenebrionoidea. A comprehensive, phylogenetic study is needed of all lower tenebrionoids, in order to solidify the placement of the Hallomeninae and Eustrophinae in Tetratomidae rather than Melandryidae.

This family has historically comprised many taxa now placed in various other families (*e.g.*, Tetratomidae, Stenotrachelidae, Pythidae, Pyrochroidae, Scraptiidae, etc.). The last world catalogue was that of Csiki (1924). Even with the removal of Tetratomidae, Synchroidae and Scraptiidae, the tribal classification of Melandryidae presented in Arnett (1963) is likely artificial; as a convenience, these categories are repeated in the classification at the end of this chapter, but all are not used in the key to genera. Taxa included by Arnett (1963) as Melandryini, Hypulini, and Serropalpini are keyed together; consistent, diagnostic features for these tribes were not found. A detailed phylogenetic analysis of the entire family is needed to clarify the relationships among the tribes and subfamilies.

If all taxa not currently placed in Melandryidae (including Hallomeninae and Eustrophinae) are removed from Csiki's (1924) catalogue, there are about 290 species worldwide. A more current checklist (Pollock unpublished) lists about 430 species worldwide with 24 genera and about 60 species in the Nearctic Region.

Distribution. Melandryidae are distributed widely in both temperate and tropical areas of the world, with highest diversity in the tropics. In North America, the highest diversity occurs in forested regions of the eastern and western United States and Canada.

Key to Nearctic Genera

1. Metatibia shorter than either femur or first tarsomere; metatibial spurs at least 1/3 length of metatibia (Fig. 20) (Melandryinae: Orchesiini) 2
— Metatibia as long as either femur or first tarsomere; metatibial spurs less than 1/3 length of metatibia (Figs. 21, 22) 4

2(1). Body form elongate, wedge-shaped; antenna elongate, gradually widened distally (Fig. 9) .... *Orchesia*
— Body form rounded, sub-globose; antenna short, with distinct 3-4 segmented club (Fig. 10) 3

3(2). Scutellum visible .... *Microscapha*
— Scutellum not visible .... *Lederia*

4(1). Tarsal claw with large basal tooth (Osphyinae) .... *Osphya*
— Tarsal claw with at most a slight swelling basally (Melandryinae: excl. Orchesiini) 5

5(4). Pronotum with pair of depressions between medial area and lateral margins, either confined to basal region or extended further anteriorly 6
— Pronotum without depressions between medial area and lateral margins 11

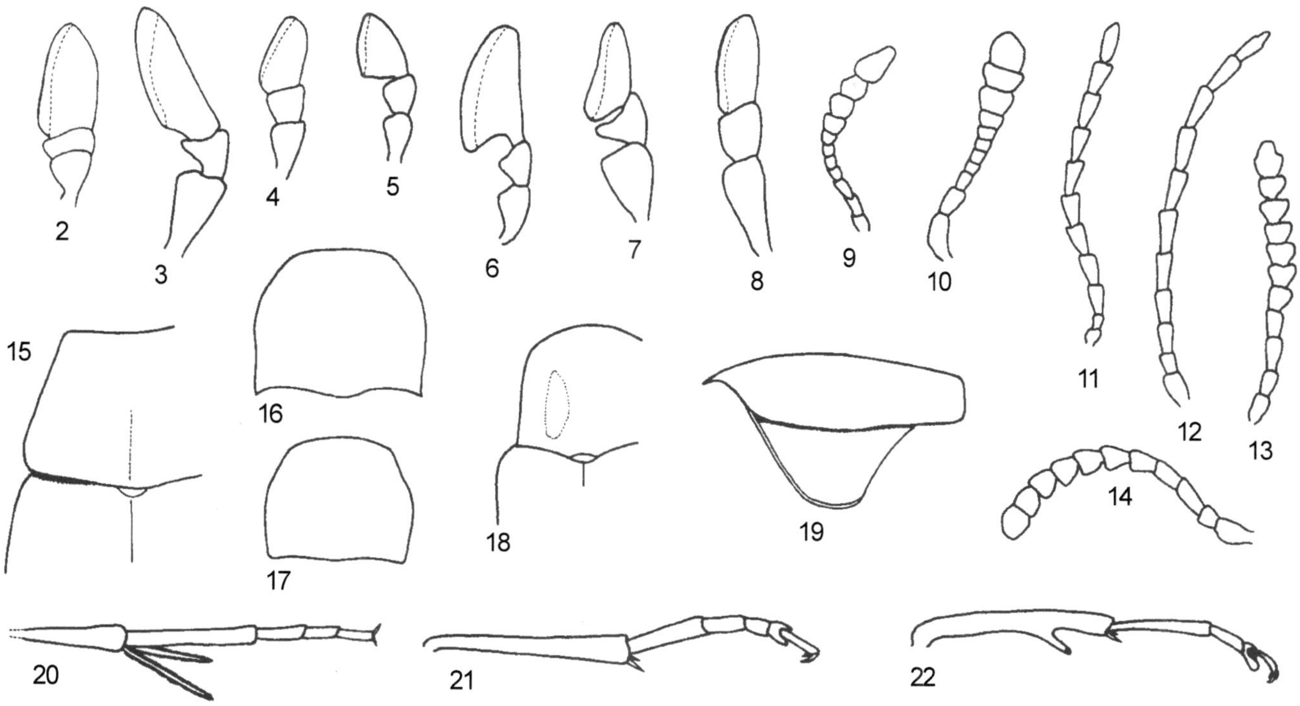

FIGURES 2.100-22.100. 2-8 Maxillary palpus, 2. *Orchesia castanea* (Melsheimer); 3. *Melandrya striata* Say; 4. *Phryganophilus collaris* LeConte; 5. *Scotochroa basalis* LeConte; 6. *Rushia longula* (LeConte); 7. *Serropalpus substriatus* Haldeman; 8. *Enchodes sericea* (Haldeman). 9-14. Antenna, 9. *Orchesia cultriformis* Laliberté; 10. *Lederia arctica* Horn; 11. *Scotochroides antennatus* Mank; 12. *Enchodes sericea* (Haldeman); 13. *Rushia longula* (LeConte); 14. *Xylita laevigata* Hellenius. 15-18. Pronotum, 15. *Melandrya striata* Say; 16. *Enchodes sericea* (Haldeman); 17. *Serropalpus substriatus* Haldeman; 18. *Hypulus simulator* Newman. 19. *Dircaea liturata* (LeConte), prothorax, lateral view. 20-22. Tibia and tarsus, 20. *Orchesia ornata* Horn, hind leg; 21. *Emmesa connectens* Newman, middle leg; 22. *Osphya varians* LeConte, hind leg, male.

6(5). Base of pronotum without posterior bead, lying rather loosely over base of elytra (Fig. 15) 7

— Base of pronotum with posterior bead, in relatively close contact with base of elytra (e.g., Fig. 18). 9

7(6). Elytra with at least seven longitudinal, raised intervals; pronotum distinctly narrowed anteriorly (Fig. 15) .... *Melandrya* (*s. str.*)

— Elytra without longitudinal, raised intervals; pronotum not narrowed anteriorly 8

8(7). Pronotum reddish, with two black spots .... *Phryganophilus*

— Pronotum unicolorous, without contrasting spots .... *Emmesa*

9(6). Pronotum elongate, lateral margins subparallel; pronotal depressions long, extended anteriorly beyond midlength (Fig. 18) .... *Hypulus*

— Pronotum wider, lateral margins arcuate; depressions shorter, confined to posterior third of pronotal length 10

10(9). Elytra uniformly colored, without contrasting pattern of light and dark .... *Zilora*

— Elytra with contrasting pattern of light and dark color .... *Prothalpia*

11(5). Procoxal cavities without distinct fissure 12

— Procoxal cavities with distinct fissure 18

12(11). Frontoclypeal suture distinct, deeply impressed. 13

— Frontoclypeal suture absent, or vaguely suggested only 14

13(12). Lateral margins of pronotum sinuate before hind angles; hind angles distinct, about square .... *Microtonus*

— Lateral margins of pronotum rounded before hind angles; hind angles rounded .... *Symphora*

14(12). Hind tibia ornament consisting of single longitudinal dorsal carina; male with elliptical to round pit on mentum; suture between prosternum and proepimeron distinct, ± carinate (Fig. 19) 15

— Hind tibia ornament absent; male without pit on mentum; suture between prosternum and proepimeron not carinate .... *Spilotus*

15(14). Prosternum sunken anterior of procoxae, very short; mesosternum elevated medially .... *Anisoxya*

— Prosternum visible anterior of procoxae, not sunken; mesosternum not distinctly elevated medially. 16

16(15). Apices of meso- and metatibiae obliquely truncated .... *Dircaea*

— Apices of meso- and metatibiae squarely truncated 17

17(16). Body small, mostly < 4 mm; elytra with contrasting light and dark color pattern (in North American species) .... *Abdera*
— Body larger, mostly >5 mm; elytra uniformly brown to piceous .... *Phloeotrya*

18(17). Antennomeres 5-10 short and triangular, subequal in length and maximum width (e.g., Figs. 13, 14) 19
— Antennomeres 5-10 elongate, filiform to triangular, distinctly longer than maximum width (e.g., Figs. 11,12) 22

19(18). Tarsomeres distinctly compressed laterally; elytral striae present; lateral pronotal carina absent .... *Amblyctis*
— Tarsomeres cylindrical, not compressed laterally; elytra without striae; lateral pronotal carina present 20

20(19). Pronotal disc flattened posterolaterally, slightly explanate; lateral margins of pronotal disc distinctly converging anteriorly 21
— Pronotal disc evenly rounded posterolaterally, not explanate; lateral margins subparallel, not or only slightly converging anteriorly .... *Rushia*

21(20). Apical maxillary palpomere short, approximately triangular (Fig. 5); posterior margin of pronotum distinctly bisinuate .... *Scotochroa*
— Apical maxillary palpomere elongate, cultriform; posterior margin of pronotum truncated to only slightly bisinuate .... *Xylita*

22(18). Antennomeres triangular, distinctly widened distally (Fig. 11); pronotal disc posterolaterally explanate; metatibial spurs short, subequal in length .... *Scotochroides*
— Antennomeres filiform, at most indistinctly widened distally; pronotal disc not explanate postero-laterally; metatibial spurs long, unequal in length. 23

23(22). Hind angles of pronotum distinct, subacute (Fig. 16); lateral pronotal carina absent; elytra without impressed striae, evenly tapered from humeri to apex .... *Enchodes*
— Hind angles of pronotum indistinct, rounded to nearly square (Fig. 17); lateral pronotal carina present posteriorly; elytra with distinctly impressed carinae, more parallel-sided .... *Serropalpus*

Classification of the Nearctic Genera

Melandryidae Leach 1815

Melandryinae Leach 1815

The melandryines are the most diverse subfamily of Melandryidae. Historically, they were often divided into several tribes: Orchesiini, Hypulini, Melandryini, and Serropalpini. Among these groups, the melandryines and especially the orchesiines are well defined, but convincing diagnostic characters for the hypulines and serropalpines are unknown. Although Melandryinae are probably monophyletic (Lawrence and Newton 1995), the relationships among the constituent tribes are unknown. Any attempt to provide a tribal classification of the melandryines before a detailed character analysis is performed will be less than satisfactory.

Orchesiini

The orchesiines are a well-defined group of melandryines, with variously widened or clubbed distal antennomeres, very long metatibial spurs, and pronounced jumping ability (*e.g.*, Sasaji 1995).
Orchesia Latreille 1807, 5 spp., widely distributed from Northwest Territories to Texas and Florida. Key to species: Laliberté (1966).

Microscapha LeConte 1860, 1 sp., *M. clavicornis* LeConte, eastern United States.

Lederia Reitter 1879, 1 sp., *L. arctica* (Horn), west coast of North America.
Crioscapha Horn 1893

Serropalpini

It is unclear whether this tribe is a natural group. In most treatments, it is defined on general elongate body shape, vertical orientation of head, modifications of maxillary palpi (Figs. 5-8), and size (mostly large) of tibial spurs.

Amblyctis LeConte 1879, 1 sp., *A. praeses* LeConte 1879, northern New York, Indiana, southern Quebec.
Dapsiloderus Fairmaire 1887, Nikitsky *et al.* 1998 (synonymy)

Xylita Paykull 1798, 3 spp., widely distributed.

Rushia Forel 1893, 3 spp., eastern and western North America, with gap in interior of continent.
Carebara LeConte 1862, not Westwood 1840

Spilotus LeConte 1862, 1 sp., *S. quadripustulatus* (Melsheimer), eastern North America from Quebec to Louisiana.

Scotochroa LeConte 1874, 3 spp., Canada and northern United States. Key to species: Mank (1939a).

Scotochroides Mank 1939, 1 sp., *S. antennatus* Mank, eastern Canada and United States.

Enchodes LeConte 1866, 1 sp., *E. sericea* (Haldeman), eastern United States, and across Canada to British Columbia.

Serropalpus Hellenius 1786, 3 spp., widespread, in areas of coniferous forest. Key to species: Mank (1939b). Life history: Hoebeke and McCabe (1977).

Dircaea Fabricius 1798, 2 spp., *D. liturata* (LeConte), eastern USA and across northern Canada; *D. horiei* Hatch, British Columbia.

Phloeotrya Stephens 1832, 4 spp., widely distributed, with general gap in interior of continent.

Abdera Stephens 1832, 2 spp., western North America.

Anisoxya Mulsant 1856, 1 sp., *A. glaucula* LeConte 1866, eastern North America, from Ontario to Louisiana.

Hypulini

Nearctic genera of Hypulini are characterized as having the base of the pronotum somewhat narrower than the base of the elytra, and the elytra with distinct humeri.

Zilora Mulsant 1856, 4 spp., relatively widespread in areas of coniferous forest, with slight gap in interior. Key to species: Mank (1938).

Hypulus Paykull 1798, 2 spp., *H. simulator* Newman, eastern North America, from Maritimes to Mississippi and *H. californicus* Van Dyke, California and Oregon. NOTE.—*H. californicus* Van Dyke is not congeneric with *H. simulator*, and is very similar to *Marolia* Mulsant (Palearctic) or *Nippomarolia* Miyatake (Japan).

Microtonus LeConte 1862, 1 sp., *M. sericans* LeConte, east-central United States, from Maine to Texas.

Symphora LeConte 1866, 2 spp., eastern North America, into southern Canada.

Melandryini

The generic classification within Melandryini is entirely unsettled. For example, the genera *Emmesa* and *Melandrya* are variously considered separate genera or are united. Also, Japanese workers use *Prothalpia* for species probably more accurately placed in *Emmesa*. Adults of Melandryini are characterized as having the pronotum laying rather loosely over the base of the elytra.

Prothalpia LeConte 1862, 2 spp., *P. undata* (LeConte), eastern North America, from Quebec and Newfoundland to North Carolina, west to Wisconsin; *P. holmbergi* (Mannerheim), west coast of North America, from Alaska to California.

Melandrya Fabricius 1801, 1 sp., *M. striata* Say, eastern United States, and across Canada from New Brunswick to British Columbia.

Emmesa Newman 1838, 6 spp., widespread (mostly north) but with general gap in interior of continent. Key to species: Hatch (1927).

Phryganophilus Sahlberg 1834, 1 sp., *P. collaris* LeConte, widespread in Canada, Pacific Northwest, Minnesota; absent from most of United States.

Osphyinae Mulsant 1856

This subfamily is represented in North America by the single genus *Osphya* Illiger. Adults of *Osphya* somewhat resemble Cantharidae; males of most species exhibit marked sexual dimorphism on the legs and/or abdominal ventrites.

The Osphyinae may be distinguished from other Melandryidae by the following combination of characters: antennae elongate, filiform; apical maxillary palpomere distinctly cultriform; apical labial palpomeres distinctly widened, securiform; prosternum and prosternal process very short; penultimate tarsomere with distinct lobe, extended under last tarsomere; base of tarsal claw with tooth.

Osphya Illiger 1807, 3 spp., *O. varians* (LeConte) widespread in eastern and central North America; *O. lutea* (Horn) and *O. essigi* Van Dyke, California. Key to species: Van Dyke (1928).

Melandryidae, *Incertae Sedis*

Polypria Chevrolat 1874, 1 sp., *P. cruxrufa* Chevrolat, known only from southern Texas (with a wider distribution in Mexico and Central America).

The genus *Polypria* was described originally as a member of Othniidae (now included in Salpingidae); Champion (1889) transferred the genus to his rather wide concept of Melandryidae, specifically in the group of genera now included in Mycteridae. Spilman (1952), who examined the male genitalia of *P. cruxrufa* Chevrolat "temporarily" transferred *Polypria* from Mycteridae to Melandryidae. It was included in Melandryidae by Arnett (1963). Lawrence and Newton (1995) placed the genus in Tenebrionoidea, *incertae sedis*. Because of its former placement in Melandryidae (Arnett 1963), *Polypria* is reluctantly included here in this volume. Structurally the genus does not agree with the concept of Melandryidae (or any other family of Tenebrionoidea) and is not included in the generic key above. Future discovery and study of the larval stages of *Polypria*, and more detailed examination of adults, will hopefully solidify its family placement.

Bibliography

ARNETT, R. H., Jr. 1963. The Beetles of the United States (a Manual for Identification). Catholic University of America Press. Washington, DC. 1112 pp.

BØVING, A. G. and F. C. CRAIGHEAD. 1931. An illustrated synopsis of the principal larval forms of the order Coleoptera. Entomologica Americana, 11: 1-351.

CHAMPION, G. C. 1889. Biologia Centrali-Americana, Insecta, Coleoptera (Lagriidae, Melandryidae, Pythidae, Oedemeridae), Volume 4, Part 2, 120 pp.

CHANTAL, C. 1985. Les Tetratomidae (Coleoptera) du Québec. Fabreries, 11: 43-66.

CROWSON, R. A. 1955. The Natural Classification of the Families of Coleoptera. N. Lloyd. London.

CSIKI, E. 1924. Serropalpidae. Coleopterorum Catalogus 17(77): 1-62.

HATCH, M. H. 1927. Concerning Melandryidae. Annals of the Entomological Society of America, 20: 363-367.

HOEBEKE, E. R. and T. L. MCCABE. 1977. The life history of *Serropalpus coxalis*, with a description of the larva and pupa (Coleoptera: Melandryidae). Coleopterists Bulletin, 31: 57-63.

HORN, G. H. 1888. Miscellaneous coleopterous studies. Transactions of the American Entomological Society, 15: 26-48.

LALIBERTÉ, F. 1966. Les Orchesiini du Québec. Le Naturaliste Canadien, 93: 689-700.

LAWRENCE, J. F. 1991. Melandryidae (Tenebrionoidea) (= Serropalpidae). Pp. 505-508. *In*: F. W. Stehr, ed. Immature Insects. Volume 2. Kendall/Hunt. Dubuque, IA. xvi + 975 pp.

LAWRENCE, J. F. and A. F. NEWTON, Jr. 1995. Families and subfamilies of Coleoptera (with selected genera, notes, references and data on family-group names). Pp. 779-1006. *In*: J. Pakaluk and S. A. Slipinski, eds. Biology, Phylogeny, and Classification of Coleoptera. Papers Celebrating the 80th Birthday of Roy A. Crowson. Volume 2. Muzeum i Instytut Zoologii PAN, Warsaw.

MANK, E. W. 1938. A revision of the genus *Zilora*. Psyche, 45: 101-104.

MANK, E. W. 1939a. *Scotochroa* and a closely allied new genus, *Scotochroides*, (Coleoptera, Melandyridae). Canadian Entomologist, 71: 181-183.

MANK, E. W. 1939b. A review of the genus *Serropalpus*, (Coleoptera, Melandryidae). Canadian Entomologist, 71: 237-239.

NIKITSKY, N. B. 1998. Generic classification of the beetle family Tetratomidae (Coleoptera, Tenebrionoidea) of the world, with description of new taxa. Pensoft. Sofia, 80 pp.

NIKITSKY, N. B., V. B. SEMENOV and M. M. DOLGIN. 1998. The beetles of the Prioksko-Terrasny Biosphere Reserve – Xylobiontes, Mycetobiontes, and Scarabaeidae (with the review of the Moscow region fauna of these groups). Supplement 1 (with remarks on nomenclature and systematics of some Melandryidae of the world fauna). Archives of the Zoological Museum of Moscow State University [Sbornik Trudov Zoologicheskogo Museya MGU] (suppl. 1), 55 pp., 4 pls. [in Russian]

PETERSON, A., 1951. Larvae of Insects: An Introduction to Nearctic Species. Part II. Coleoptera, Diptera, Neuroptera, Siphonaptera, Mecoptera, Trichoptera. A. Peterson. Columbus, OH (Pp. 54-55).

SASAJI, H. 1995. On the adaptive characteristics of the genus *Lederia* (Coleoptera, Melandryidae), with description of a new species from Japan. Special Bulletin of the Japanese Society of Coleopterology, 4: 425-431.

SPILMAN, T. J. 1952. The male genitalia of the Nearctic Salpingidae. Coleopterists Bulletin, 6: 9-13.

VAN DYKE, E. C. 1928. New species of heteromerous Coleoptera. Bulletin of the Brooklyn Entomological Society, 23: 251-261.

101. MORDELLIDAE Latreille 1802

by John A. Jackman and Wenhua Lu

Family common name: The tumbling flower beetles

These small, wedge-shaped beetles, with short antennae, fore bodies retractile, elytra entire, pygidium pointed and mostly long, and long hind legs, form a distinctive and easily recognized group.

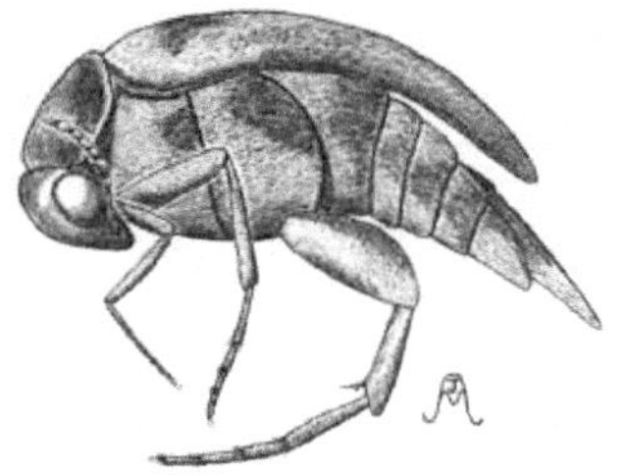

FIGURE 1.101 *Mordella marginata* Melsheimer

Description: Shape smooth, streamlined, hump-backed, pointed posteriorly, wedge-shaped or linear; size 1.5 to 15 mm, mostly less than 8 mm in length; color various, commonly black, brown, red, and yellow, often bicolored with head and pronotum differing from elytra; vestiture sparse to dense, coarse decumbent setae or scales of various colors, commonly white or greyish and frequently formed into distinctive patterns.

Head large, triangular in frontal view, lying flush to thorax, as wide as apex of thorax, sharply constricted behind eyes, hypognathous; surface smooth or rugose-punctate. Antennae with eleven antennomeres; mostly short, not extending beyond thorax in most species; commonly filiform, somewhat serrate or clavate, pectinate only in Ctenidiinae; inserted under frontal ridge, close to base of mandibles. Clypeus distinct, small; labrum prominent, entire; mandibles short, stout, somewhat curved, apices acute or bifid. Maxillary palpi long with four palpomeres, apical palpomere large, securiform, scalene, cultriform, or malleiform. Gular process present; mentum trapezoidal; ligula prominent, cordiform; labial palpi with three palpomeres, apical palpomere triangular. Eyes lateral, large, well developed, shape oval, round, to almost square, anterior margin behind antenna emarginate in some; eye facets fine to coarse.

Pronotum small, narrowed in front, as wide as elytra at base; shape irregular, anteriorly produced, broadest at base or just before, arcuately rounded at lateral margins, posteriorly broadly sinuate, somewhat explanate; borders margined; surface rugose-punctate; pleural region broad; prosternum very short; procoxal cavities open behind, confluent. Mesosternum short, carinate, pointed posteriorly; mesocoxal cavities separate. Metasternum large, moderate. Legs without trochantin; procoxae conical, large, contiguous; mesocoxae small, slightly separate; metacoxae flat, contiguous. Metafemora slender or enlarged for jumping. Metatibiae slender, with large apical spurs, metatibiae frequently expanded apically. Tarsal formula 5-5-4, tarsomeres compressed, slender. Protibiae of males in some with row of hairs on inner surface, enlarged or flattened. Metatibiae and metatarsi mostly with a few to many comb-like or serrate ridges. Claws simple, cleft, or pectinate. Scutellum variable in shape, typically triangular, semicircular, or quadrate with apex bifid. Elytra narrowed behind, exposing pygidium, surface rugose-punctate; epipleural fold not distinct, epipleura moderate. Wings well developed, folded and rolled in some; subcosta distinct at base of wing; radius closely parallel to subcosta with few heavy branches, R1 prominent, Rs missing at base and cross-vein r connected to the two veins closing a radial cell proximally; median vein obsolescent near base but distally curving to heavy cubitus forming a loop. Vein M4 +Cu extended from loop toward wing margin; cross vein r-m connected to Rs and M. Four weak anal veins besides jugal vein, base of first two anal veins united somewhat indistinctly with base of cubitus. Several thickened areas represent aberrant veins. Principal folds of wing interrupt distal branches of vein Rs and somewhat vein M.

Abdomen with five visible sterna, sutures distinct; pygidium pointed and mostly long; surface micro-rugose. Male genitalia with penis very elongate; parameres variously modified, often asymmetric, meeting at base; phallobase distinct from parameres. Female genitalia with baculum on paraprocts long; valvifers with a baculum; coxite with a baculum at base articulating with that of valvifer.

Larvae elongate, cylindrical, or shorter and more stout; size 5 to 16 mm, mostly less than 10 mm in length; vestiture a few short setae generally distributed; color white. Head exserted, hypognathous, sclerotized but not deeply pigmented; epicranial sutures absent; antennae vestigial, or two-segmented with vestigial third segment. Labrum and clypeus distinct, not prominent; mandibles short, stout, conical, with or without a molar area; maxillae with cardo, fused stipes and setiferous mala; maxillary palpi three-segmented; labium with small two-segmented palpi. Stemmata absent, or from 1-3 indistinct pairs. Thorax in some enlarged, protruding above head, legs without claws or pigment, only two or three indistinct segments. Abdomen with simple apical spine or paired urogomphi; fleshy dorsal extensions present in some; nine-segmented. Spiracles annular on lower part of mesothorax and on abdominal segments one to eight.

Habits and habitats. Adults are phytophagous, apparently feeding on pollen of many plants, especially in umbelliferous (Apiaceae) and composite (Asteraceae) flowers. Adults are commonly collected on flowers but there does not seem to be any association between adult food plants and larval host plants. Franciscolo (1985) describes a species with fungus feeding adults. There are probably more species associated with woodland habitats but grasslands also support some species. Adults are good fliers and are commonly collected in flight intercept traps (Jackman and Nelson 1995). When disturbed or captured the adults kick with their hind legs which makes them bounce or "tumble" erratically, hence the common name. This habit is probably accentuated by the wedge-shaped body. The most complete account of the life cycle of a North American mordellid is of *Mordellistena unicolor* LeConte 1862 (probably *sericans* Fall 1907) by Ping (1915). Plant stems are the only known oviposition site (Ping 1915).

A summary of known larval food plants is given by Ford and Jackman (1996). Larvae are primarily herbivores feeding in living herbaceous stems, decaying wood, and fungi (Lawrence 1991). Woody hosts are more commonly reported in Mordellini while herbaceous stems are typically used by the Mordellistenini. Pithy stems of composites (Asteraceae) and grasses (Poaceae) (Schwarz 1887) are most commonly reported for Mordellistenini. A few *Mordellistena* detected in agricultural quarantine inspections are reported to be leafminers of orchids (Ray 1937). Several *Mordellistena* have been found in the goldenrod stem-gall (*Solidago*: Asteraceae) formed by *Eurosta solidaginis* Fitch (Diptera: Tephritidae) (Liljeblad 1945, Abrahamson and Weis 1997). A study of mordellid larvae in stems (Rathcke 1976) showed an inverse correlation with other occupants which can best be interpreted as opportunistic rather than obligate carnivory because some develop in stems without any evidence of the presence of prey. Reports of carnivorous mordellid larvae appear in the literature (e.g., Hill 1922) but provide little evidence that they are true carnivores. Larvae of some species of *Mordellistena* are reported to bore in stems of hemp, *Cannabis* (McPartland 1996).

Status of the classification. Mordellidae are a relatively homogeneous group which is most easily recognized by the pointed last apparent abdominal tergite or pygidium, flattened hind femora and the spiny ridges on the tibia and tarsi of the hind legs. Franciscolo (1957, 1965, 1967) and Ermisch (1950) provide the most comprehensive keys to genera worldwide. Subsequent workers have built upon these generic keys but a comprehensive world key is not available. There has been a proliferation of genera with few species and there are now 113 genera worldwide. Mordellidae have two subfamilies: Ctenidiinae with one species, *Ctenidia mordelloides* Castelnau 1840, from South Africa, which is distinguished by pectinate antennae; and Mordellinae. Mordellinae consist of five tribes: Stenaliini with four genera; Reynoldsiellini with two genera; Mordellini with 58 genera; Conaliini with 9 genera; and Mordellistenini with 39 genera. Stenaliini and Reynoldsiellini are restricted to the Old World and the other tribes are cosmopolitan. The genera *Mordella* and *Mordellistena* have been the historical "default genera" to which many species have been assigned. The only phylogenetic studies of North American mordellids were studies of wing venation (Khalaf 1971a), a morphometric analysis of *Mordellistena* (Burne 1987), and male genitalia (Lu *et al.* 1997). A detailed study is needed for the world genera before a satisfactory understanding of phylogenetic relationships can be achieved.

A revision of the family in North America was published by Liljeblad (1945) but this work was actually completed in 1929. The works of Ray (1936, 1939, 1944, 1946a-c, 1947) and several other authors have contributed to the systematics of Mordellidae in North America. The latest catalog to North American mordellids was completed by Bright (1986) with subsequent additions and corrections by Jackman (1991) included 12 genera and 203 species. Jackman and Lu (2001) proposed new name combinations which included five genera new to North America bringing the number to 17 genera.

Some of the North American species have not been correctly assigned to genera as these genera are understood by European and Oriental authorities. Taxonomy of mordellids in North America has relied considerably on the ridges on the hind legs and the shape of the maxillary palpi. Each ridge consists of a row of short spines which appears more like a comb with an elevated ridge hidden beneath. However, the count of the ridges can vary between individuals in the same species and even from one side of a beetle to the other. Interpretation of what constitutes a "true" ridge rather than a rudimentary ridge is also judgmental. Franciscolo (1957) points out that the front legs are important to separate genera but this character has been ignored in North American mordellid studies. Palpi remain important characters to distinguish genera even though they vary between sexes. Ongoing studies of genitalia characters (Lu *et al.* 1997) and possibly host plants will contribute to the systematics of this group in North America.

The process of assigning the mordellids of America north of Mexico to genera as they are now understood is still under investigation. Consequently, additional changes in the generic placement of North American species are to be expected. The larger genera, especially *Mordellistena*, are still likely to have species that should be reassigned. The key to genera here will not work adequately for species that have not been evaluated with a modern understanding of genera in the family.

Distribution. There are about 1500 species, occurring in all areas of the world. The number of species recorded from the United States is about 205. A few more undescribed species will be discovered in North America, with additional study of biology and genitalia.

Key to Nearctic genera

The key presented below covers all the genera known to occur in America north of Mexico. The primary characters given in the key will separate these genera in the world fauna. The secondary characters in square brackets are specific for American species. The key has not been verified for all species in America north of Mexico. Some species now considered in *Mordellistena* will key out to *Mordellina, Falsomordellistena, Glipostenoda, Pseudotolida* and *Mordellochroa.*

1. No lateral ridges on hind tibia except subapical ridge, which may be elongated obliquely on lateral face of tibia almost to base, or short, at various angles to apical margin, a fine dorsal ridge may also be present (Figs. 3-7); hind tarsomeres with fine dorsal ridges, lateral ridges or no ridges at all 2

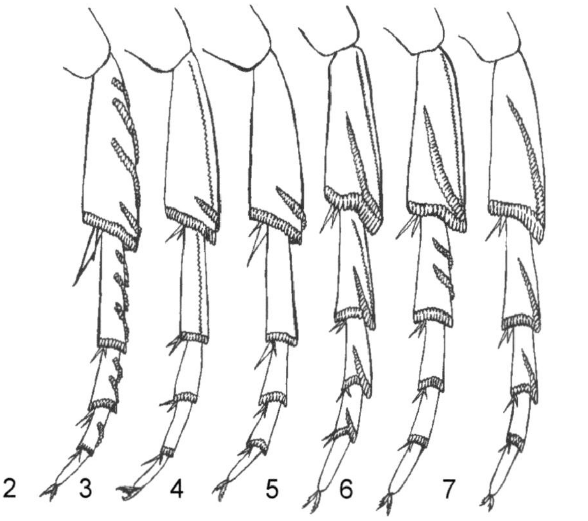

FIGURES 2.101-7.101. Ridges on the hind tibia and tarsus of mordellid genera. 2. *Mordellistena*; 3. *Tomoxia*; 4. *Mordella*; 5. *Glipodes*; 6. *Conalia*; 7. *Isotrilophus*.

— One or more lateral ridges on hind tibia in addition to subapical ridge, which is generally short and more or less parallel to apical margin (Fig. 2); hind tarsomeres with variable number of lateral ridges (Fig. 2) (Tribe Mordellistenini) 3

2(1). Subapical ridge of hind tibia extremely long, more or less parallel to dorsal margin (Figs. 5-7) (Tribe Conaliini) 8

— Subapical ridge of hind tibia short, mostly parallel to apical margin or slightly oblique (Figs. 3, 4) (Tribe Mordellini) 10

Tribe Mordellistenini

3(1). Penultimate tarsomeres of fore and middle legs transversely truncate at apex, or very moderately emarginate, neither dilated nor bilobed (Fig. 8, 9) 4

— Penultimate tarsomeres of fore and middle legs deeply emarginate, more or less bilobed and mostly strongly dilated (Figs. 10-13) 6

4(3). Apical palpomere of maxilla scalene, securiform, or cultriform, with sharp or in some with smooth angles (Figs. 14-17) in both sexes 5

— Apical palpomere of maxilla malleiform (Fig. 18) in males and scalene or securiform (Figs. 14, 15) in females [black elytra with humeral spots reddish to yellow; lateral ridges on hind tibia short] .... *Mordellochroa*

5(4). Eyes small, shape oval, not emarginate; facets fine, not exceeding 0.015 mm in diameter .... *Mordellistena*

— Eyes large, shape unusually broad, emarginate behind antenna; facets coarse, never smaller than 0.022 mm in diameter .... *Mordellina*

6(3). Apical palpomere of maxilla scalene or securiform (Figs. 14, 15) or broadly triangular (Fig. 16) in both sexes 7

— Apical palpomere of maxilla malleiform (Fig. 18) in males and scalene (Fig. 15) in females .... *Pseudotolida*

7(6). Eyes large, shape broad, often emarginate behind antenna; facets coarse [uniformly brown] .... *Glipostenoda*

— Eyes small, shape oval, never emarginate behind antenna; facets fine [elytra patterned] .... *Falsomordellistena*

Tribe Conaliini

8(2). Apical palpomere of maxilla forked (Fig. 19) in males; hind tibia with a subapical ridge extending along lateral face, plus a fine dorsal ridge; first, second and third tarsomeres of hind leg with one lateral ridge and occasionally also a fine dorsal ridge (Fig. 5) [uniformly brown] .... *Glipodes*

— Apical palpomere of maxilla not forked in males; ridges on hind tibia otherwise 9

9(8). Hind tibia with a fine dorsal ridge and long, oblique, subapical ridge, extending along entire outer face nearly to base; basal tarsomere of hind leg with two lateral ridges (Fig. 6) [uniformly brown] .... *Conalia*

— Hind tibia with elongate subapical ridge, extending from outer angle to just below base; first and second tarsomeres of hind leg with similar ridges (Fig. 7); pygidium short, nearly concealed by elytra in dorsal view [small under 3 mm; black; southern] .... *Isotrilophus*

Tribe Mordellini

10(2). Hind tibiae with a fine dorsal ridge (Fig. 3); basal tarsomere of hind leg also with fine dorsal ridge 11

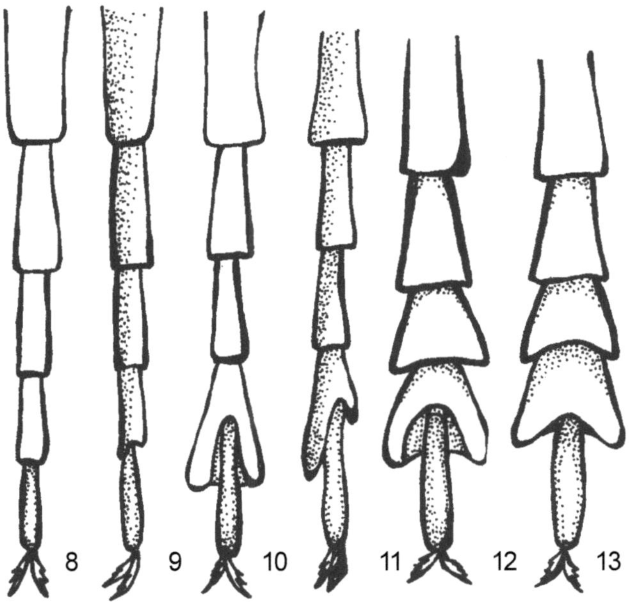

FIGURES 8.101-13.101. Anterior tarsi. 8. *Tomoxia* (dorsal view); 9. *Mordellistena* (lateral view); 10. *Pseudotolida* (dorsal view); 11. *Pseudotolida* (lateral view); 12. *Hoshihananomia* (dorsal view); 13. *Mordella* (dorsal view).

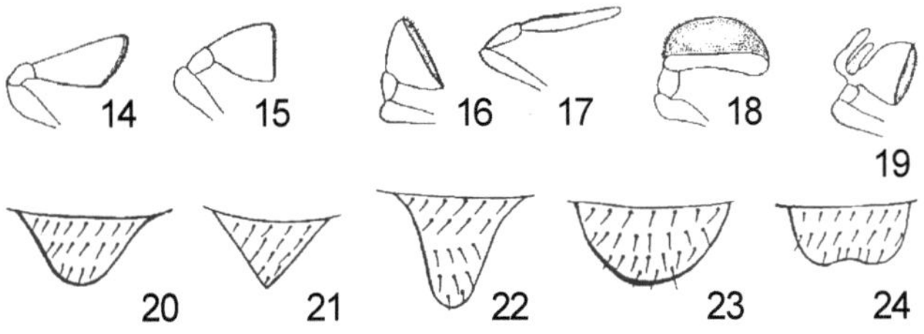

FIGURES 14.101-24.101. 14-19. Types of mordellid maxillary palpus. 14. Scalene; 15. Securiform; 16. Triangular of *Glipa*; 17. Cultriform; 18. Malleiform; 19. Branched of *Glipodes*; 20-24. Scutellum shapes. 20-22. Triangular; 23. Semicircular; 24. Quadrilateral.

— Hind tibiae without a distinct dorsal ridge (in some with small modified spinulae simulating a ridge) (Fig. 4) 15

11(10). Penultimate tarsomere of fore and middle legs truncate at apex, or only feebly emarginate (Figs 8, 9) 12

— Penultimate tarsomere of fore and middle legs more or less deeply emarginate, bilobed or dilated (Figs. 10-12) 13

12(11). Scutellum more or less quadrilateral, in some rectangular, or trapezoidal (Fig. 24), apical margin emarginate, or triangularly excised ..... *Tomoxia*

— Scutellum not quadrilateral nor trapezoidal, mostly triangular or more or less rounded at apex (Figs. 20-23) .... *Mordellaria*

13(11). Apical palpomere of maxilla broadly triangular (Fig. 16) in both sexes [large species with multiple colored patterns on elytra] .... *Glipa*

— Apical palpomere of maxilla scalene (Fig. 14), securiform (Fig. 15), or malleiform (Fig. 18) [small, bicolored, and patterned] 14

14. Apical palpomere of maxilla securiform (Fig. 15) in both sexes .... *Paramordellaria*

— Apical palpomere of maxilla malleiform (Fig. 18) in males and scalene (Fig. 14) in females [fulvous to black, with yellow spots or bands] .... *Tolidomordella*

15(10). Scutellum more or less quadrilateral, in some rectangular, or trapezoidal (Fig. 24), apical margin emarginate, or triangularly excised. [apical angles of scutellum acute, large species over 7 mm]... .... *Yakuhananomia*

— Scutellum never quadrilateral nor trapezoidal, mostly triangular or more or less rounded at apex (Figs. 20-23) 16

16(15). Penultimate tarsomere of fore and middle legs truncate at apex, or only feebly emarginate (Fig. 13) .... *Mordella*

— Penultimate tarsomere of fore and middle legs more or less deeply emarginate, bilobed or dilated (Fig. 12); maxillary palpi not broadly triangular (Fig. 16) [large, often spotted or striped]........................ .... *Hoshihananomia*

Classification of the Nearctic Genera

Mordellidae Latreille 1802

Mordellinae Fowler 1912

Characteristics: The family characters are sufficient to define the subfamily except that none have truly pectinate antennae. This subfamily, containing 112 genera, is worldwide in distribution.

Mordellini Smith 1882

Characteristics: No lateral ridges on hind tibia except subapical ridge, which is short, mostly parallel to apical margin or slightly oblique; hind tarsomeres with fine dorsal ridge, lateral ridges or no ridges at all. This tribe, containing 58 genera, is worldwide in distribution.

Tomoxia Costa 1854

Franciscolo (1982) recognized only four species in this genus and recommended that the approximately 40 other species from around the world be reevaluated. The squarish shape of the scutellum is a diagnostic character for genera in the *Tomoxia* group. However, several species in North America north of Mexico previously assigned to *Tomoxia* having the scutellum triangular or rounded in shape were reassigned to *Mordellaria* and *Paramordellaria*.

Only two species found north of Mexico remain in this genus. *Tomoxia inclusa* LeConte 1862, has been recorded from western to the northeastern U. S., and *T. lineella* LeConte 1862, recorded from the middle to the northeastern U. S. *Tomoxia inclusa* has been reared from *Tilia* sp. (Tiliaceae) (Brimley 1951). Keys to species: Ermisch (1949) for Africa; Liljeblad (1945) for North America which includes several species now assigned to other genera.

Mordellaria Ermisch 1950

This genus contains about 14 species and is found in Africa, Japan, Chile, Taiwan, and North America. Five species are now recognized in *Mordellaria* in North American (Jackman and Lu 2001): *Mordellaria borealis* (LeConte 1862); *M. fascifera* (LeConte 1878); *M. latipalpis* (Ray 1946b); *M. serval* (Say 1835); *M. undulata* (Melsheimer 1845).

Mordellaria borealis, *M. serval* and *M. undulata* are reported primarily from the northeastern to the midwest United States. *Morellaria latipalpis* is from Florida and *M. fascifera* is from Florida, Texas and the District of Columbia. Key to species: none; Liljeblad (1945) and Ray (1946b) separate species in North America.

Paramordellaria Ermisch 1968

This genus contains eight species, five from South America, two from North America, and one from the Ivory Coast. *Paramordellaria carinata* (Smith 1883b) has been recorded from the Gulf Coast States. *Paramordellaria triloba* (Say 1824) has been recorded from the northeastern United States to Alabama, Kansas, and Missouri. Key to species: none; Ermisch (1968) treats the genus worldwide.

Yakuhananomia Kôno 1935
This genus contains six species and is found in Japan, Africa, South America and North America. North of Mexico only *Yakuhananomia bidentata* (Say 1824) is reported from the northeastern U. S., west to Missouri and Iowa. Key to species: none.

Mordella Linnaeus 1758
This genus contains over 500 species and is distributed worldwide. There are 25 species north of Mexico. Many of these species are widely distributed and common on various flowers as adults, e.g., *Mordella marginata* Melsheimer 1845 and *M. atrata* Melsheimer 1845.

Mordella marginata Melsheimer has been reared from dead wood of *Quercus* (Fagaceae), *Carya* (Juglandaceae), and *Sapindus* (Sapindaceae) (Ford and Jackman 1996). It is also reported from the fungus, *Gloeophyllum* spp. (Weiss 1920). Key to species: several regional keys are available; Liljeblad (1945) for North America.

Tolidomordella Ermisch 1950
There are only two species in this genus from America north of Mexico: *Tolidomordella fenestrata* (Champion 1891) from Panama and *T. discoidea* (Melsheimer 1845) with two subspecies from America north of Mexico: *T. discoidea discoidea* (Melsheimer 1845) from northeastern United States to Iowa; and *T. discoidea flaviventris* (Smith 1883b), from Texas and Florida. There may be a dozen species in the Neotropics. Key to species: none; treated by Ermisch (1950).

Hoshihananomia Kôno 1935
This genus contains over 50 species and is cosmopolitan. Three species occur north of Mexico. *Hoshihananomia octopunctata* (Fabricius 1775) is widely distributed from Texas to the northeastern United States; *H. inflammata* (LeConte 1862) is known only from Florida; and *H. perlineata* (Fall 1907) only from Arizona and New Mexico.

Hoshihananomia octopunctata has been reported in bark of a dying beech tree, *Fagus* (Fagaceae) (Ford and Jackman 1996) and has been found in rotten oak logs, *Quercus* (Fagaceae) (Hubbard in Riley 1892). Key to species: regional keys for Japan and Hungary; Liljeblad (1945) separates the North American species but did not recognize the genus.

Glipa LeConte 1857
This genus contains about 100 species and is distributed worldwide, but the placement of many of the species needs study. Only two species are reported in America north of Mexico. *Glipa hilaris* (Say 1835) is widely distributed in the eastern United States to Texas. *Glipa hieroglyphica* Schwarz 1878, is recorded only from Florida and areas south. Key to species: regional keys for China (Fan and Yang 1993) and Japan.

Conaliini Ermisch 1956

Characteristics: No lateral ridges on hind tibia except subapical ridge, subapical ridge of hind tibia extremely long, more or less parallel to dorsal margin; hind tarsomeres with lateral ridges variable. This tribe, containing nine genera, is worldwide in distribution.

Glipodes LeConte 1862
The genus has four species and is found in South America and North America. Only one species, *Glipodes sericans* (Melsheimer 1845) is known from north of Mexico, where it occurs in the eastern United States, from Pennsylvania and Kansas, south into Mexico. Key to species: Franciscolo (1962, 1989).

Conalia Mulsant and Rey 1858
The genus contains three species from the Holarctic Region. Two species occur in America north of Mexico: *C. helva* (LeConte 1862), from the eastern United States, and *C. melanops* (Ray 1946b) which is recorded only from Quebec. Key to species: none.

Isotrilophus Liljeblad 1948
This genus is currently considered monotypic with a distribution from Iowa to Argentina. In America north of Mexico, *Isotrilophus erratica* (Smith 1883a) is found in the southern United States and occasionally northward to Iowa (Hendrickson 1933).

Mordellistenini Ermisch 1941

Characteristics: One or more lateral ridges on hind tibia in addition to the subapical ridge, subapical ridge is generally short and parallel to apical margin; hind tarsomeres with lateral ridges of variable number. There are 39 genera worldwide in distribution.

Mordellistena Costa 1854
Mordellistena is a very large genus found throughout the world and includes 131 species in America north of Mexico. This genus occurs throughout our region and many of these species are also widely distributed. Some of our North American species formerly assigned to this genus belong to other genera and have been reassigned (Jackman and Lu 2001) to *Falsomordellistena*, *Mordellina*, *Mordellochroa*, and *Pseudotolida*. The basis for these reassignments are the characters in the key to world genera by Franciscolo (1967). Some species currently in *Mordellistena* may also key out to these genera. While most of our species properly remain in *Mordellistena*, further investigation will certainly justify the reassignment of others.

Larval host plants for the genus north of Mexico have been summarized by Ford and Jackman (1996). Larval host plants include: *Actinomeris*, *Ambrosia*, *Aster*, *Cirsium*, *Elephantopus*, *Eupatorium*, *Iva*, *Solidago*, *Xanthium* (Asteraceae); *Desmodium* (Fabaceae); *Castanea* (Fagaceae); *Calycocarpum* (Menispermaceae); *Spartina*, *Uniola* (Poaceae) (Brodie 1892, Ford and Jackman 1996, Harrington

1895, Liljeblad 1945, Riley 1892, Schwarz and Barber [in Liljeblad 1945], Strong *et al.* 1984).

Mordellid larvae (mostly *Mordellistena*) have also been reported feeding in stems of the following plants: *Amaranthus* (Amaranthaceae); *Helianthus*, *Silphium* (Asteraceae); *Monarda* (Labiatae); *Ruellia* (Acanthaceae); and *Triodia* (Poaceae) (Ford and Jackman 1996, Rathcke 1976, Riley 1892). Key to species: none worldwide; many regional keys.

Falsomordellistena Ermisch 1941

This genus contains about 25 species and is found primarily in Japan, Taiwan, and China. Two species are now recognized in *Falsomordellistena* in North America (Jackman and Lu 2001): *F. hebraica* (LeConte 1862) and *F. pubescens* (Fabricius 1798). They are all generally distributed at least in the eastern United States. Key to species: none.

Mordellina Schilsky 1908

This genus contains about 45 species found in Africa, China, Japan, Taiwan and North America. Five species are now recognized in North America (Jackman and Lu 2001): *M. blatchleyi* (Liljeblad 1945), *M. pilosella* (Ray 1947), *M. pustulata* (Melsheimer 1845), *M. ustulata* (LeConte 1862), and *M. wickhami* (Liljeblad 1945).

The above five species are the only species of this genus now recognized in America north of Mexico. They are all generally distributed at least in the eastern United States. The species *M. ustulata* has been considered a subspecies of *Mordellistena andreae* (LeConte 1862) (Liljebald 1945). However, the penultimate segment of the anterior and middle tarsi are truncate in *M. ustulata* and emarginate in *M. andreae*. This character separates genera according to Franciscolo (1967).

Larval host plants for *M. pustulata* (Melsheimer) include: *Ambrosia, Chrysopsis, Cirsium, Elephantopus, Erigeron, Eupatorium, Helianthus, Vernonia*, and *Xanthium* (Asteraceae) (Ford and Jackman 1996, Riley 1892, Liljeblad 1945). *Mordellina ustulata* has been reared from the thorns of *Gleditisia* (Fabaceae) (Ford and Jackman 1996). Key to species: Africa (Ermisch 1968, Franciscolo 1967); Netherlands (Batten 1976).

Mordellochroa Emery 1876

This genus has at least six species and is found in Japan, Mongolia, Poland, Hungary, and probably Russia. One species *Mordellochroa scapularis* (Say 1824) is now recognized from America north of Mexico (Jackman and Lu 2001) and is found in Quebec, Ontario, Illinois eastward to Maine and British Columbia. Key to species: Palearctic (Batten 1977); Hungary (Kaszab 1979, Horák 1993).

Glipostenoda Ermisch 1950

This genus has about 45 species and is found in Africa, Japan, China, Taiwan, South America, North America, South Africa, and Queensland.

The only species now recognized in this genus in America north of Mexico is *Glipostenoda ambusta* (LeConte 1862) which is widely distributed. Several other species in America probably belong in this genus. Key to species: Africa (Franciscolo 1967; Ermisch 1968); East Asia (Ermisch 1953).

Pseudotolida Ermisch 1950

This genus has about 10 species and is found in Japan, China, Taiwan, Guatemala, and Brazil (Franciscolo 1982). Three species are now recognized in North America (Jackman and Lu 2001): *P. arida* (LeConte 1862), *P. knausi* (Liljeblad 1945), and *P. lutea* (Melsheimer 1845).

Pseudotolida arida and *P. lutea* are widely distributed except for the western slope of the Rocky Mountains and *P. knausi* is reported only from Utah. Key to species: Franciscolo (1982).

Bibliography

ABRAHAMSON, W. G. and A. E. WEIS. 1997. Evolutionary ecology across three trophic levels, goldenrods, gallmakers, and natural enemies. Monographs in Population Biology 29, Princeton University Press. 456 pp.

BATTEN, R. 1976. De Nederlandse soorten van de keverfamilie Mordellidae. Zoologische Bijdragen, 19: 3-37.

BATTEN, R. 1977. *Mordellochroa* species of the Western Palaearctic region (Coleoptera, Mordellidae). Entomologische Berichten, 37: 21-26.

BRIGHT, D. E. 1986. A catalog of the Coleoptera of America North of Mexico, Family: Mordellidae. USDA, ARS, Agriculture Handbook 529-125, 22 pp.

BRIMLEY, J. F. 1951. Mordellidae of Prince Edward County, Ontario (Coleoptera). Canadian Entomologist, 83: 278-279.

BRODIE, W. 1892. Canadian galls and their occupants - *Eurosta solidaginis*, Fitch. Canadian Entomologist, 24: 137-139.

BURNE, J. C. 1987. A morphometric analysis of *Mordellistena* Costa in Southwestern United States (Coleoptera: Mordellidae). Pan-Pacific Entomologist, 63: 224-235.

CASTELNAU, COMTE DE. [aka François Louis Nompar de Caumont de Laporte]. 1840. Historie naturelle des animaux articulés. Volume 1, 324 pp., Volume 2, 564 pp.

CHAMPION, G. C. 1891. Biologia Centrali-Americana, Insecta, Coleoptera (Mordellidae, Rhipidophoridae, Meloidae), 4(2): 257-368.

COSTA, A. 1854. Mordellidae. Fauna dell Regno di Napoli, 32 pp.

EMERY, M. C. 1876. Essai monographique sur les Mordellides de l'Europe et des contrées limitrophes. L'Abeille, 14: 1-128.

ERMISCH, K. 1941. Tribus Mordellistenini (Col. Mordell.) (8. Beitrag zur Kenntnis der Mordelliden). Mitteilungen der Münchener Entomologischen Gesellschaft, 31: 710-726.

ERMISCH, K. 1949. Neue Arten der Gattung *Tomoxia* und verwandter Gattungen aus Afrika. (17. Beitrag zur Kenntnis der Mordelliden). Entomologische Blätter, 41-44: 98-110.

ERMISCH, K. 1950. Die Gattungen der Mordelliden der Welt (16. Beitrag zur Kenntnis der Mordelliden). Entomologische Blätter, 46: 34-92.

ERMISCH, K. 1953. Über die Typen der von Boheman beschriebenen Mordelliden des Reichsmuseum Stockholm (23. Beitrag zur Kenntnis der Mordelliden). Arkiv för Zoologi, 6(19): 425-431.

ERMISCH, K. 1955. Mordellidae. *In:* A. D. Horion, ed. Faunistik der Mitteleuropäischen Käfer. Entomologische Arbeiten aus dem Museum G. Frey, 4: 269-321.

ERMISCH, K. 1968. Contributions à la connaissance de la faune entomologique de la Côte-D'Ivoire (J. Decelle, 1961-1964). 19. Coleoptera Mordellidae. (55. Beitrag zur Kenntnis der Mordelliden). Musee Royal de l-Afrique Centrale, Tervuren, Belgique Annales, Serie in-8E , Sciences Zoologiques, 165: 257-287.

FABRICIUS, J. C. 1775. Systema entomologiae sistens insectorum classes, ordines, genera, species, adjectis synonymis, locis, descriptionibus, observationibus. Korte. Flensburg, 832 pp.

FABRICIUS, J. C. 1798. Supplementum entomologiae systematicae. Proft et Storch. Copenhagen, 572 pp.

FALL, H. C. 1907. Descriptions of new species. *In:* H. C. Fall and T. D. A. Cockerell, eds. The Coleoptera of New Mexico. Transactions of the American Entomological Society, 33: 218-272.

FAN, X. and C.-K. YANG. 1993. Revision of the Genus *Glipa* of China (Coleoptera: Mordellidae). Memoirs of Beijing Natural History Museum, 53: 45-68.

FORD, E. J. and J. A. JACKMAN. 1996. New larval host plant associations of tumbling flower beetles (Coleoptera: Mordellidae) in North America. Coleopterists Bulletin, 50: 361-368.

FOWLER, W. W. 1912. Fauna of British India, including Ceylon and Burma, Coleoptera, General Introduction, and Cicindelidae and Paussidae. London, 529 pp, pl. i-xx.

FRANCISCOLO, M. 1957. A monograph of the South African genera and species of Mordellidae. Part 1. Morphology, subfamily Ctenidiinae and tribe Stenaliini. South African Animal Life, Results of the Lunds University Expedition in 1950-1951, 4: 207-291.

FRANCISCOLO, M. 1962. The genus *Glipodes* LeConte 1862 (Coleoptera: Mordellidae) with description of a new species from Venezuela and Costa Rica. Proceedings of the Royal Entomological Society of London (B), 31: 131-136.

FRANCISCOLO, M. 1965. A monograph of the South African genera and species of Mordellidae. Part 2. Tribe Mordellini. South African Animal Life, Results of the Lunds University Expedition in 1950-1951, 11: 244-458.

FRANCISCOLO, M. 1967. A monograph of the South African genera and species of Mordellidae. Part 3. Tribe Mordellistenini. South African Animal Life, Results of the Lunds University Expedition in 1950-1951, 13: 67-203.

FRANCISCOLO, M. 1982. Mordellidae (Coleoptera: Heteromera) from Rennell and Bellona Islands (Solomons). *In:* T. Wolff, ed. The Natural History of Rennell Island British Solomon Islands, University of Copenhagen, 8: 49-72.

FRANCISCOLO, M. 1985. About a new fungus-eating Mordellid-beetle from Ecuador (Col. Mordellidae). (52. Contribution to the knowledge of Mordellidae). Annali del Museo Civico di Storia Naturale, Genova, 85: 79-94.

FRANCISCOLO, M. 1989. A new species of *Glipodes* LeConte 1862 from Venezuela, the fourth in the genus (Coleoptera: Mordellidae: Conaliini). (58th Contribution to the knowledge of Mordellidae). Coleopterists Bulletin, 44: 105-111.

HARRINGTON, W. H. 1895. Occupants of galls of *Eurosta solidaginis*. Canadian Entomologist, 8: 197-199.

HENDRICKSON, G. O. 1933. Some Mordellidae (Coleoptera) of Iowa prairies. Bulletin of the Brooklyn Entomological Society, 28: 193.

HILL, G. F. 1922. A new species of *Mordellistena* (Coleoptera, Fam. Mordellidae) parasitic on termites. Proceedings of the Linnean Society of New South Wales, 3: 346-347.

HORÁK, J. 1993. Mordellidae, Scraptiidae. *In:* J. Jelínek, ed. Checklist of Czechoslovak Insects IV (Coleoptera). Seznam ...eskoslovenských broukç. Folia Heyrovskyana, Suppl. I, 111-112, 115.

JACKMAN, J. A. 1991. Notes on the nomenclature of Mordellidae of America north of Mexico. Coleopterists Bulletin, 45: 323-330.

JACKMAN, J. A. and W. LU. 2001. Nomenclatural changes for Mordellidae (Coleoptera) in North America. Insecta Mundi, (in press).

JACKMAN, J. A. and C. R. NELSON. 1995. Diversity and phenology of tumbling flower beetles (Coleoptera: Mordellidae) captured in a malaise trap. Entomological News, 106: 97-107.

KASZAB, Z. 1979. Felemás lábfejízes bogarak II. Heteromera II. Mordellidae. Magyarorszag Allatvilaga, Fauna Hungaria, Akademiai Kiado, Budapest, 9: 1-100.

KHALAF, K. T. 1971. Wing venation and phylogenetic relationships in Mordellidae (Coleoptera: Heteromera). Florida Entomologist, 53: 153-160.

KÔNO, H. 1935. Die Mordelliden Japans (Fuenfter Nachtrag). Transactions of the Sapporo Natural History Society, 14: 123-130.

LATREILLE, P. A. 1802. Histoire naturelle, générale et particulière des crustaces et des insectes. Paris, Vol. 3, 467 pp.

LAWRENCE, J. F. 1991. Mordellidae. *In:* F. W. Stehr, ed. Immature Insects. Kendall/Hunt. Dubuque, IA, 975 pp.

LECONTE, J. L. 1859. The Coleoptera of Kansas and eastern New Mexico. Smithsonian Contribution to Knowledge, 2: 1-58.

LECONTE, J. L. 1862. Synopsis of the Mordellidae of the United States. Proceedings of the Natural Sciences of Philadelphia, 14: 43-51.

LECONTE, J. L. 1878. Additional descriptions of new species. *In:* E. A. Schwarz. The Coleoptera of Florida. Proceedings of the American Philosophical Society, 17: 373-472.

LILJEBLAD, E. 1945. Monograph of the family Mordellidae (Coleoptera) of North America, north of Mexico. Miscellaneous Publications, Museum of Zoology, University of Michigan, No. 62, 229 pp.

LINNAEUS, C. 1758. Systema naturae per regna tria naturae secundum classes, ordines, genera, species, cum characteribus, differentiis, synonymis, locis. 10 ed. Laurentii Salvii. Stockholm, vol. 1, 823 pp.

LU, W., J. A. JACKMAN and P. W. JOHNSON. 1997. Male Genitalia and phylogenetic relationships in North American Mordellidae (Coleoptera). Annals of the Entomological Society of America, 90: 742-767.

McPARTLAND, J. M. 1996. Cannabis pests. Journal of International Hemp Association, 3: 51-55.

MELSHEIMER, F. E. 1845. Descriptions of new species of Coleoptera of the United States. Proceedings of the Academy of Natural Sciences of Philadelphia, 2: 213-223, 302-318.

MULSANT, E. and C. REY. 1858. Descriptions d'une espèce constitutant un genre nouveau dans la famille des Mordelliens. Annales des Science Physiques et Naturelles d'Agriculture et d'Industrie, Publiées par la Société Nationale d'Agriculture, etc. Lyon, 3: 313-315.

PING, C. 1915. Some inhabitants of the round gall of the goldenrod. Journal of Entomology and Zoology, Pomona College, Claremont, CA, 7: 161-179.

RATHCKE, B. J. 1976. Competition and coexistence within a guild of herbivorous insects. Ecology, 57: 76-87.

RAY, E. 1936. Studies on North American Mordellidae, I. (Coleoptera). Canadian Entomologist, 68: 124-129.

RAY, E. 1937. Two new beetles of the family Mordellidae from orchids. Proceedings of the United States National Museum, Washington, 84(3016): 239- 241.

RAY, E. 1939. A taxonomic study of neotropical beetles of the family Mordellidae, with descriptions of new species. Proceedings of the United States National Museum, Washington, 87(3075): 271-314.

RAY, E. 1944. New mordellid beetles from the Western Hemisphere. Zoological Series of Field Museum of Natural History, Chicago, 29: 117-133.

RAY, E. 1946a. Studies on North American Mordellidae, II (Coleoptera). Pan-Pacific Entomologist, 22: 41-50.

RAY, E. 1946b. Studies on North American Mordellidae, III (Coleoptera). Pan-Pacific Entomologist, 22: 90-99.

RAY, E. 1946c. Studies on North American Mordellidae, IV (Coleoptera). Pan-Pacific Entomologist, 22: 121-132.

RAY, E. 1947. Studies on North American Mordellidae, V (Coleoptera). Pan-Pacific Entomologist, 23: 121-131.

RILEY, C. V. 1892. Coleopterous larvae with so-called dorsal prolegs. Proceedings of the Entomological Society of Washington, 2: 319-325.

SAY, T. 1824. Descriptions of coleopterous insects collected in the late expedition to the Rocky Mountains, performed by order of Mr. Calhoun, Secretary of War, under command of Major Long. Journal of the Academy of Natural Sciences of Philadelphia, 3: 238-282, 298-331, 403-464; 4: 83-99.

SAY, T. 1835. Descriptions of North American coleopterous insects and observations on some already described. Boston Journal of Natural History, 1: 151-203.

SCHILSKY, J. 1908. XII. Beitrag zur deutschen Käferfauna. Deutsche Entomologische Zeitschrift, 604-608.

SCHWARZ, E. A. 1878. The Coleoptera of Florida. Proceedings of the American Philosophical Society, 17: 353-472.

SCHWARZ, E. A. 1887. On the insects found on *Uniola paniculata* found in southeastern Florida. Proceedings of the Entomological Society of Washington, 1: 104-108.

SMITH, J. B. 1882. A synopsis of the Mordellidae of the United States. Transactions of the American Entomological Society, 10: 73-100.

SMITH, J. B. 1883a. New Mordellidae and notes. Bulletin of the Brooklyn Entomological Society, 5: 80-81.

SMITH, J. B. 1883b. Mordellidae, notes and descriptions. Bulletin of the Brooklyn Entomological Society, 6: 3-5.

STRONG, D. R., J. H. LAWTON and T. R. E. SOUTHWOOD. 1984. Insects on plants, community patterns and mechanisms. Harvard University Press. Cambridge, MA, 313 pp.

WEISS, H. B. 1920. *Mordella marginata* Melsh., bred from fungus (Coleop.). Entomological News, 31: 67-68.

102. RIPIPHORIDAE Gemminger and Harold 1870 (1853)

by Zachary H. Falin

Family common name: The ripiphorid beetles

The ripiphorid beetles form a small group of special interest because of their remarkable biology: the larvae are phoretic in one subfamily, hypermetamorphic in at least two subfamilies and are parasitoids of other insects in all known cases.

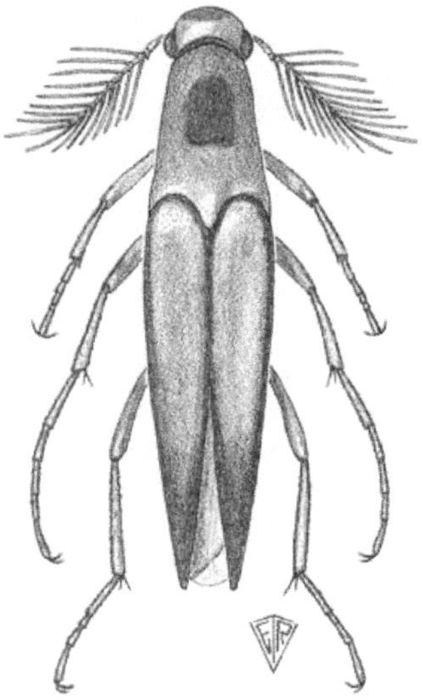

FIGURE 1.102. *Macrosiagon limbatum* (Fabricius)

Description: The four subfamilies represented in North America are morphologically quite diverse and difficult to characterize as a group. The most commonly encountered ripiphorid genus, *Macrosiagon* (Ripiphorinae: Macrosiagonini), resembles the Mordellidae in its humpbacked, streamlined appearance (Fig. 1). In North America it is easily separated from the mordellids by the bipectinate to biflabellate antennae of the males and unipectinate antennae of the females, the absence of an apical spine on the abdomen and the absence of preapical ridges on the hind tibiae. The next most commonly collected taxon, *Ripiphorus* (Ripiphorinae: Ripiphorini), along with the males of the rare, as yet undescribed species of *Pirhidius* (Ripidiinae) found in Florida, vaguely resembles strepsipterans or perhaps flies, although can easily be distinguished from the former by their 5-5-4 tarsal formula and the presence of pro- and mesotrochanters. The females of the Floridian ripidiine are currently unknown, though they are expected to be apterous and larviform in appearance. Species in *Trigonodera*, *Pelecotoma* (Pelecotominae), and *Toposcopus* (Ptilophorinae) are more generalized in appearance; they most closely resemble other tenebrionoid beetles in the Melandryidae and Scraptiidae though they are quickly separated from these taxa by the strongly unipectinate antennae of the males and the unipectinate or serrate antennae of the females and by their filiform apical maxillary palpomere.

Ripiphoridae as currently understood encompass many poorly defined, rare genera whose morphological characteristics tend to grade into each other, making comparative statements between them difficult if not impossible. Given the current lack of a modern, comprehensive morphological study of the family, this chapter will treat the adult morphology of Nearctic taxa only, drawing mainly from Rivnay (1929), Selander (1957) and personal observation.

Adult shape variable, commonly humpbacked, wedge-shaped; head prominent; antennae bipectinate to biflabellate in males, unipectinate in females; elytra strongly acuminate, revealing partially folded hind wings; most are black or red with red, orange, yellow or brown elytra, some with maculations; integument glossy with sparse or rarely relatively dense, decumbent pubescence; length 4.0-14.0 mm (*Macrosiagon*, Fig. 1, 2). Some taxa vaguely dipterous; head prominent; antennae biflabellate in males, unipectinate in females; elytra small, scale-like, revealing unfolded hind wings; black, some with brownish or yellowish abdomen, elytra and male antennae; integument glossy with variable, pale, suberect pubescence; length 2.5-11.0 mm (*Ripiphorus*, Fig. 3). A few taxa elongate, widest at humeri, tapering posteriorly or subparallel; head prominent, partially recessed in pronotum in one species; antennae strongly unipectinate in males, strongly serrate in females; elytra fully covering abdomen; brownish to blackish; short, pale decumbent pubescence; length 3.5-10.0 mm (*Trigonodera*, Fig. 4; *Pelecotoma*, Fig. 5; and *Toposcopus*, Fig. 6). One rare species with males strepsipteran-like; head prominent, eyes holoptic; antennae strongly uniflabellate; elytra reduced, leathery, not reaching the third abdominal tergite; brownish; integument glossy and granular with pale suberect pubescence; length 3.0 mm; females putatively larviform (*Pirhidius*, undescribed species, Fig. 7).

In the following section it is convenient to consider the following order of genera as least to most morphologically derived: *Trigonodera*, *Pelecotoma*, *Toposcopus*, *Macrosiagon*, *Ripiphorus* and *Pirhidius*. While the actual relationships between these taxa remain to be tested by cladistic analysis, assuming that *Trigonodera* represents the ground plan for the family makes comparison among the taxa easier. Descriptions are for the *Trigonodera*-like ground plan with exceptions to morphological characteristics in other taxa explicitly noted. Descriptions of the undescribed *Pirhidius* species are for the males only.

Head large, greatly deflexed; subglobular or anteroposteriorly compressed (*Ripiphorus*) or compressed and variously dorsoventrally elongate (*Macrosiagon*); distinct from pronotum or partially recessed into pronotum (*Trigonodera*); with distinct neck hidden in pronotum, becoming obsolete in *Ripiphorus* and *Pirhidius*; vertex scarcely to slightly raised above anterior pronotal margin or distinctly raised above margin (*Macrosiagon*, *Ripiphorus*); surface variously punctate and pubescent or glabrous in some *Macrosiagon* species. Antennae extremely variable; 11-segmented or 10-segmented (in some *Ripiphorus* females); strongly unipectinate in males and serrate to weakly unipectinate in females (Figs. 8-13) or bipectinate to biflabellate in males and unipectinate in females (*Macrosiagon*, Figs. 14, 15; *Ripiphorus*, Figs. 16, 17) or strongly uniflabellate in males (*Pirhidius*, Fig. 18); four basal antennomeres simple or three simple (*Pelecotoma*, *Pirhidius*) or two simple (*Toposcopus*, *Macrosiagon*, *Ripiphorus*), the third antennomere with an abbreviated process in *Toposcopus* females; basal antennomere elongate, slightly curved or short, irregularly cup-shaped (*Pirhidius*) or very short, annular (*Ripiphorus*); antennae inserted in front of eyes, approximately at midline or slightly below (*Toposcopus*) or slightly above (*Pirhidius*) or else dorsal to eyes (*Ripiphorus*). Eyes large, oval with slight excavations, nearly contiguous anteriorly, relatively coarsely faceted or smaller, reniform and distant anteriorly and more finely faceted (*Pelecotoma*) or divided, the dorsal and ventral lobes connected by a thin corneous strip (*Toposcopus*) or prominent, bulging laterally, oval to slightly emarginate, very finely faceted (*Macrosiagon*, *Ripiphorus*) or occupying most of the head, contiguous dorsally, ventrally and anteriorly between base of maxillary palpi and antennae, relatively coarsely faceted (*Pirhidius*).

Mouthparts entire, functional (except in *Pirhidius*, see below); clypeus distinct, membranous with labrum prominent, broad, apex slightly emarginate to broadly truncate or clypeus reduced, labrum attached under ventral margin of frons, less prominent, broad, apex slightly excavate (*Toposcopus*) or generally long and narrow, apex rounded to broadly truncate (*Macrosiagon*) or less prominent, broad and deeply bilobed (*Ripiphorus*). Mandibles stout, triangular, internal tooth and membranous setaceous patch present or thinner, more falciform, simple, internal tooth and membranous setaceous patch absent (*Toposcopus*, *Macrosiagon*, *Ripiphorus*). Maxillae with palpi four-segmented, distal palpomere always filiform though apex sometimes obliquely truncate; galea blade-like, external edge sclerotized, internal edge membranous and setose or thin and greatly elongate (though never longer than head), completely setose with a sclerotized tip (*Macrosiagon*) or short, fleshy and peg-like (*Ripiphorus*); lacinia free or broadly fused with galea (*Toposcopus*, *Macrosiagon*) or absent (*Ripiphorus*); stipes and subgalea separate or fused (*Toposcopus*, *Macrosiagon*, *Ripiphorus*). Labial palpi three-segmented except apparently two-segmented in *Pelecotoma*; ligula generally broad, flat and bilobed or truncate (*Toposcopus*) or elongate and narrow (*Macrosiagon*); articulation point of mentum to gular area broad, even (*Trigonodera*, *Toposcopus*) or petiolate (*Pelecotoma*, *Macrosiagon*, *Ripiphorus*). In the undescribed *Pirhidius* species the mouthparts are vestigial, consisting of two-segmented maxillary palpi with the basal segments fused.

Pronotum large, distinct, generally bell-shaped, narrowed anteriorly, as wide as elytra at base or nearly so; without sharp lateral margin except basally in *Trigonodera*, lateral suture absent except in *Pelecotoma*; lateral aspect of pronotal base not dorsoventrally compressed or dorsoventrally compressed such that the hypomeral area is nearly obsolete (*Macrosiagon*, *Ripiphorus*, *Pirhidius*); hind margin sinuate, slightly projecting with a shallow medial notch or expanded into a triangular lobe (*Macrosiagon*) or straight (*Ripiphorus*, *Pirhidius*); prosternum generally small, with a projection extending ventrally between procoxae or not (*Ripiphorus*, *Pirhidius*); procoxal cavities large, open behind, close except distant in *Ripiphorus*. Mesoscutellum visible or hidden by pronotum (*Macrosiagon*) or apparently absent or highly modified (*Ripiphorus*, *Pirhidius*); mesosternum moderate to short in length except extremely abbreviated in *Ripiphorus*; with anterior keel or without (*Toposcopus*, *Macrosiagon*, *Ripiphorus*, *Pirhidius*); mesocoxal cavities close except distant in *Ripiphorus*. Metascutum with medial furrow or without (*Ripiphorus*, *Pirhidius*); posterior half of metepimeron narrow or greatly expanded (*Macrosiagon*); metasternum large, greatly expanded in *Ripiphorus* and *Pirhidius*; longitudinal metasternal suture present or absent (*Macrosiagon*, *Ripiphorus*, *Pirhidius*).

Legs slender, long; trochantins absent. Procoxae large, conical, contiguous except distant in *Ripiphorus*. Mesocoxae oblique, conical, close except in *Ripiphorus*; not projecting except in *Pirhidius*. Metacoxae transverse, lamellate or irregularly rounded (*Ripiphorus*, *Pirhidius*); contiguous except nearly so in *Ripiphorus*; coxal apophysis well developed or absent (*Ripiphorus*, *Pirhidius*); dorsal margin rounded except straight in *Macrosiagon*; posterior margin well developed or rounded (*Ripiphorus*, *Pirhidius*); lateral face in same plane as thoracic pleuron or deflexed posteriorly, giving a pinched appearance to base of abdomen (*Ripiphorus*, *Pirhidius*). Trochanters small, triangular except elongate in *Pirhidius*; femora normal, slender. Tibiae with a ring of short apical spines or not (*Ripiphorus*, *Pirhidius*); never with preapical ridges as in Mordellidae; spur formula 2-2-2 (*Trigonodera*, *Toposcopus*, *Ripiphorus*) or 1-2-2 (*Macrosiagon*) or 1-1-1 (*Pelecotoma*) or 0-0-0 (*Pirhidius*); spurs simple, generally subequal, moderate to large except small in *Pelecotoma*. Tarsal formula 5-5-4; tarsomeres generally slender with slightly to moderately flared apices; protarsomeres dilated in some males of *Macrosiagon*; basal metatarsomere enlarged and obliquely truncate in some *Ripiphorus*; claws variously pectinate or bifid (*Macrosiagon*) or simple (*Pirhidius*).

Elytra without costae or striae; entire or acuminate (*Macrosiagon*) or scale-like (*Ripiphorus*) or reduced and leathery (*Pirhidius*); epipleural fold present or absent (*Ripiphorus*, *Pirhidius*). Hind wings large, functional; folded under those with entire elytra, partially folded apically in *Macrosiagon*, folded accordian-like in *Pirhidius* and unfolded in *Ripiphorus*; venation reduced in genera with modified elytra, apparently "mordelloid" (Rivnay 1929) or similar to that of Meloidae (Forbes 1926) in others; R-M crossvein present or absent (*Macrosiagon*, *Ripiphorus*, *Pirhidius*); cell 2dA3 present in *Trigonodera*.

Abdomen without apical spine; with five visible sternites (III-VII) (often difficult to see in *Macrosiagon*) or seven visible sternites (II-VIII) in *Pirhidius*; junction with metathorax abrupt, sternite II effectively hidden or junction more gradual, lateral aspects of sternite II partially or entirely visible (*Ripiphorus*, *Pirhidius*); well sclerotized and relatively inflexible or more thinly sclerotized and telescoping (*Macrosiagon*) or poorly sclerotized and tubular (*Pirhidius*); abdominal tip somewhat deflexed in *Macrosiagon*, significantly decurved in *Ripiphorus*, tergite VII forming a pygidium in the latter.

Male genitalia generally similar throughout taxa and considered "mordelloid" (as compared to *Tomoxia* in Sharp and Muir 1912) except *Toposcopus*, described separately. Male genitalia in the undescribed species of *Pirhidius* currently unstudied, though likely similar to that of *Ripidius quadriceps* Abeille 1872 as described in Besuchet (1956). Gonocoxae (= tegmen) large, fused, forming an asymmetrical dorsal trough for aedeagus, produced to the right in dorsal view with apex towards the top except towards the left in *Pelecotoma*; gonostyli (= parameres) distinct, well sclerotized; symmetrical or variously asymmetrical (*Macrosiagon*, *Ripiphorus*); variously attached to gonocoxae; ventral baculi paired or fused (*Macrosiagon*, *Ripiphorus*); aedeagus generally long, slender, distinct, arched, simple except with decurved apical hook in *Macrosiagon*. Male genitalia in *Toposcopus* of significantly different appearance, apparently more similar to *Anaspis* (Scraptiidae) according to Rivnay (1929); gonocoxae fused into a complex, symmetrical structure with paired anteriorly and posteriorly projecting flattened arms; putative gonostyli symmetrical, cylindrical, basally indistinct from gonocoxae; ventral baculi possibly forming a posteriorly projecting loop supporting apex of aedeagus; aedeagus a long, slender, simple sclerotized tube.

Female genitalia poorly characterized; gonostyli present or absent (*Macrosiagon*, *Ripiphorus*); genital tube membranous, gonocoxae, dorsal and ventral baculi distinct or becoming more sclerotized, gonocoxae and baculi becoming less distinct, more fused, tip becoming sclerotized and blade-like (*Macrosiagon*, *Ripiphorus*). The female genitalia of *Pirhidius* are unknown, but likely a simple, poorly sclerotized tube.

In addition to the larva of *Pelecotoma fennica* (Paykull 1799), only those of a few species in *Macrosiagon*, *Ripiphorus* and *Ripidius* are known. The first instar of the former differs greatly in appearance from the more well known triungulin type larvae of the Ripiphorinae and Ripidiinae and is described first. The larval stages of the Nearctic species *Pelecotoma flavipes* Melsheimer 1846 are currently undescribed though nearly identical to those of the European *Pelecotoma* species (Acciavatti and Falin unpublished data). Likewise, the first instar of the undescribed *Pirhidius* species is unknown but likely similar to those of *Ripidius*; the larva *R. quadriceps* Abeille 1872 is described below. The larval descriptions are taken directly from Linsley *et al.* (1952) and Besuchet (1956) as summarized in Selander (1991), and Švácha (1994).

First, endophagous instar of *Pelecotoma fennica* small, 0.5 mm; moderately sclerotized, unpigmented; campodeiform, somewhat flattened and narrow before feeding, greatly distended but not strongly crescentic after feeding; spiracles apparent only on abdominal segments I and VIII; little setation. Head prognathous, flat, spade-shaped; epicranial suture absent; two pairs of stemmata placed laterally; antennae three-segmented, apical antennomere with seta; antennae placed anteriorly, near mandibles. Mandibles falciform distally with a large, flat medial plate at base; maxillary palpi two-segmented; labium reduced. Thoracic terga and sterna more or less well defined. Legs moderately long, robust, five-segmented; pretarsus with single simple claw and a basal plate. Abdomen slender, elongate; 10-segmented, last segment forming an anal sucker; urogomphi absent; segment IX with two pairs of elongate setae. First ectophagous phase (second to fourth instars) reduced, unsclerotized, more or less glabrous; head large, antennae and mouthparts reduced except for robust mandibles; legs short, stout, indistinctly segmented; spiracles present on mesothorax and abdominal segments I-VIII; head and leg characteristics becoming more defined in later instars. Second ectophagous phase (fifth instar) more robust with short setae; cranium and mouthparts well-sclerotized and defined; three-segmented antennae; legs long, five-segmented.

First, triungulin instars of the Ripiphorinae and Ripidiinae small, 0.45-0.95 mm; heavily sclerotized, fusiform before feeding, crescentic with intersegmental membranes greatly expanded after feeding; spiracles on mesothorax and abdominal segments I-VIII (*Macrosiagon*) or segment VIII (*Ripiphorus*) or VII-VIII (*Ripidius*); setation somewhat reduced, particularly in *Macrosiagon* and *Ripiphorus*; lacking line of dehiscence on dorsal surface. Head prognathous, rounded in front; epicranial suture absent; stemmata placed laterally, behind antennae at posterior of head; five stemmata (*Macrosiagon*, *Ripiphorus*) or four (*Ripidius*); antennae three-segmented (*Macrosiagon*, *Ripiphorus*) or two-segmented (*Ripidius*), apical antennomere with long terminal seta; apical segment bifurcate in *Ripidius*. Labrum fused with frontoclypeus; mandibles falciform distally, smooth, with large basal lobe; maxillae with cardo fused with stipes; stipes elongate except vestigal in *Ripidius*; maxillary palpi two- or three-segmented; labium reduced. Thoracic nota well developed. Legs slender, long, five-segmented; tibiae greatly elongate in *Ripidius*; pretarsus apparently with none (*Macrosiagon*), one (*Ripiphorus*) or two (*Ripidius*) simple claws; pulvillae present. Abdominal tergites and sternites well sclerotized, 10-segmented, last segment tubular; subdivided in *Ripidius*; urogomphi absent; segment IX with pair of caudal setae, particularly developed in *Macrosiagon*. Ectophagous instars of *Macrosiagon* and *Ripiphorus* (second to sixth) crescentic, lightly sclerotized, stemmata and labial palpi absent, maxillary palpi reduced, antennae vestigial, legs reduced, thoracic and abdominal segments with large, conical horns. Endophagous phase of *Ripidius* larvae (second instar) greatly reduced, legs, spiracles, antennae, and mouthparts absent, then pseudoeruciform (third instar) with spiracles absent but mouthparts and legs distinct. Free living last phase (fourth instar) with spiracles, two-segmented antennae, and well-developed five-segmented legs.

Habits and habitats. Information about ripiphorid biology is drawn almost entirely from detailed studies of three species: *Ripiphorus smithi* Linsley and MacSwain 1950 (Ripiphorinae: Ripiphorini) (Linsley *et al.* 1952), *Ripidius quadriceps* Abeille 1872 (Ripidiinae) (Besuchet 1956), and *Pelecotoma fennica* (Paykull 1799) (Pelecotominae) (Švácha 1994). Much piecemeal and anecdotal information about ripiphorid biology is scattered in the literature, particularly for the Ripiphorinae, but the comparative value of this information in illuminating broad patterns in the family is severely limited without a much-needed comprehensive review. A general biological summary follows the detailed life histories below.

Details on the life histories of most species of the genus *Ripiphorus* are lacking and generally have only been cursorily noted in studies of the ground-nesting bees they parasitize. The host taxa include species of *Diadasia* (Apidae: Apinae: Emphorinae) (Barber 1939, Linsley and MacSwain 1950, 1951, 1957), *Dieunomia* (Halictidae: Nomiinae) (Pierce 1904), and species in various halictine genera (Halictidae), including *Halictus* (Fabre 1879, 1886, Dolphin 1979, Nye 1980), *Lasioglossum* (Batra 1965, Eickwort unpublished data), *Augochlora* (Barber 1939, Eickwort and Eickwort 1972), and *Augochlorella* (Mueller and Falin unpublished data).

Pierce (1904) and Barber (1939) have published some notes on the life history of *Ripiphorus*, though the definitive work on this genus was done by Linsley *et al.* (1952), in which parasitism of the bee *Diadasia consociata* Timberlake by *Ripiphorus smithi* Linsley and MacSwain 1950 is described. The host bee occurs in California, Nevada and Arizona, specializes on the pollen of the alkali mallow, *Malvella leprosa* (Ortega) Krapov. (=*Sida hederacea*) (Malvaceae), nests gregariously in hard-packed soil and overwinters as last-instar larvae.

Females of *Ripiphorus smithi* emerge over the course of the summer and after mating deposit approximately 6 to 15 eggs within the unopened flower buds of the alkali mallow. Dissected females contained over 850 eggs, and were observed to oviposit on many flower buds. Egg hatching is more or less synchronized with the opening of the buds and consequently with visitation by host bees. The first instar is a very active, well-sclerotized triungulin larva that holds itself erect on the apex of the abdomen in the presence of insect activity on the surface of the flower. This "searching" position is mirrored in *Metoecus* (Chapman 1870), *Macrosiagon* (Cros 1920) and *Ripidius* (Besuchet 1956), albeit on different substrates. Once a suitable host is contacted, the larva climbs onto the host, attaches itself in a variety of positions (Tomlin and Miller 1989) and is carried back to the host nest.

Once in the nest, the beetle enters a cell and waits on the pollen mass for a host egg to be laid and hatch. Shortly after the bee larva hatches, the beetle penetrates it and lives endoparasitically while the bee larva consumes its provisions. During this time, the beetle feeds, growing progressively more distended, though it does not molt. It then overwinters in the now mature bee larva and resumes feeding as the bee enters the prepupal stage the next spring. When this occurs, the larva expands enormously over a short period of time and then emerges from the mesothorax of the bee. It molts to the second instar while doing so; the first instar exuviae of the beetle plug the exit wound upon its emergence. The second instar, rather reduced and unsclerotized, curls around the host to the side opposite that of the exit wound and begins to feed by peristaltic motions visible within the head, thorax, and abdomen. The beetle undergoes four more molts (for a total of six instars) within approximately two weeks of emergence from the bee. Once the bee larva is consumed, the beetle pupates, and emerges as an adult about 12 days later, waiting an additional day or so to exit from the host cell. In the field, adults usually live a maximum of one to two days.

There are several important biological differences in the host taxa that affect various aspects of the life history of *Ripiphorus* species. For example, bees of the Halictinae tend to be pollen generalists, making the choice of ovipositional sites by beetles that parasitize these bees more varied than for those that parasitize the specialist genera *Diadasia* and *Dieunomia*. Also, halictine bees tend to overwinter as inseminated females, presumably requiring the first instar beetle larva to overwinter attached to the outside of a diapausing adult bee rather than as an endoparasite of a mature larva as in *Diadasia* and *Dieunomia* (though a first instar larva has yet to be observed attached to a diapausing adult bee).

Although the life histories of species of *Metoecus* (Ripiphoridae: Macrosiagonini), a relatively small genus occurring in tropical Asia and Europe, appear generally similar to those of *Ripiphorus*, they exclusively parasitize eusocial species of vespine wasps (Vespidae: Vespinae) (see Heitmans and Peeters 1996 and references therein). General aspects of the life history of *Metoecus paradoxus* (Linnaeus 1761) have been known for quite some time (Chapman 1897) and recent work has added detail (Okutani 1952, Hattori and Yamane 1975), but a comprehensive study of its bionomics has yet to be undertaken. In addition to host taxa, the life history of *Metoecus* differs from that of *Ripiphorus* in that eggs are laid in the fall, constituting the overwintering stage and further, are laid in crevices of weathered wood of the kind attractive to vespid wasps for paper-making rather than on flowers (Švácha 1994). Also, it appears that the triungulin *Metoecus* larvae wait until the fourth or fifth instar of the wasp larva before penetrating it and beginning to feed (Hattori and Yamane 1975), rather than penetrating it immediately upon hatching as in *Ripiphorus*.

Species of *Macrosiagon* (Ripiphorinae: Macrosiagonini) parasitize a wide variety of aculeate Hymenoptera, including genera within Eumeninae (Vespidae) (Chobaut 1891, Bequeart 1918, Snelling 1963), Crabronidae (Barber 1915, Williams 1928, Hook and Evans 1991), Sphecidae (Eberhard 1974, Callan 1981), Scoliidae (Jarvis 1929), Tiphiidae (Wolcott 1914, Rivnay 1929) and the bee genera *Exomalopsis* (Apidae: Apinae: Exomalopsini) (Raw 1977, Rozen 1997) and *Megalopta* (Halictidae: Halictinae: Augochlorini) (Falin *et al.* 2001). Like *Metoecus*, there are no definitive life history studies for this genus as there are for *Ripiphorus*, *Ripidius* and *Pelecotoma*, though certain details can be gleaned from the above references and the work of Chobaut (1895, 1906), Grandi (1936), Iwata (1939), and Callan (1977). Obviously with such a wide range of hosts only a generalized picture of the life history of this genus is currently possible. Unlike *Metoecus*, with which *Macrosiagon* shares close morphological and biological affinities, the eggs are laid on the undersides of leaves, in crevices in the soil, and on flowers of Asteraceae (= Compositae of authors). It is uncertain how the triungulin first instar larvae that are hatched on the ground locate suitable hosts. Similar to *Metoecus*, it appears that the first instar larva waits until the host larva is fairly well developed before penetrating it and feeding (Kifune 1956). Kifune also indicates that *Macrosiagon nasutum* (Thunberg 1784), at least, undergoes a total of six larval instars as do species in the other two genera of the Ripiphorinae. It is also unclear in what stage or stages species of *Macrosiagon* overwinter.

Besuchet (1956) presents an excellent study of the life history of *Ripidius quadriceps* Abeille 1872 (Ripidiinae); the following is taken from this work. Flightless, larviform females of *R. quadriceps* lay eggs in crevices in the soil occupied by roaches in the genus *Ectobius* (Blattidae: Ectobiinae). After hatching, the first instar triungulin larva, rather similar in form to those of the Ripiphorinae, seeks out a young roach nymph and attaches itself by inserting its head and a portion of its thorax into an intersegmental membrane in the ventral side of the roach's thorax. It remains as an ectoparasite for two to three weeks, feeding on the host's haemolymph. It eventually molts into a legless, worm-like second instar, during which it inserts itself fully into the host's body. It then migrates to the host's abdomen and becomes quiescent for eight to nine months, allowing the roach nymph to develop and gain in size. It then molts into a much larger, legged, less-reduced third instar, and then several months later, into a final fourth instar larva. Several days after molting to the fourth instar, the beetle emerges from the host nymph and pupates very shortly thereafter. The pupal stage lasts 5 to 19 days depending on the sex of the beetle, and the adults live three to four days after eclosion. *Ripidius pectinicornis* Thunberg 1806, parasitic on the german roach *Blattella germanica* Linnaeus, sometimes allows its host to reach adulthood before emerging (Sundevall 1831, Stamm 1936) and, like *R. quadriceps*, multiple larvae occasionally parasitize a single roach. The only other known hosts of the Ripidiinae are for Australian genera and include blattids in the Blattinae (*Cutilia* and *Platyzosteria*), Panchlorinae (*Onicosoma*), Pseudomopinae (*Elipsidion* and *Escala*) and Ectobiinae (*Balta*, *Choristima*, and *Choristimodes*) (Riek 1955).

Švácha's (1994) excellent paper on the European species *Pelecotoma fennica* (Paykull 1799) is nearly the only source of biological information on any members of the heterogeneous and likely non-monophyletic Pelecotominae. Adult individuals of *P. fennica* are found on dead or partially dead poplars and willows infested with *Ptilinus fuscus* Geoffroy (Coleoptera: Anobiidae). The females oviposit in damaged xylem cells at the openings of *Ptilinus* galleries and, after hatching, the mobile first instar larva enters the gallery seeking *Ptilinus* immatures. The first instar larva is quite different from the triungulin larva of the Ripiphorinae and Ripidiinae in that it is less sclerotized and campodeiform. Once a suitable host larva is found, it is penetrated by the parasite in a manner similar to that found in the Ripiphorinae. The first instar beetle overwinters inside the host, then emerges the next year without having molted. Upon emergence, the beetle attaches itself to the host and begins to feed. Within seven to eight days the beetle undergoes four molts (for a total of five instars) and, as in Ripiphorinae, consumes the host larva. The beetle pupates in the host's original pupal chamber, then emerges as an adult in the late spring and early summer. The Nearctic species *Pelecotoma flavipes* Melsheimer 1846 has been reared from *Ptilinus ruficornis* Say and so far as is known possesses a life history very similar to that of *P. fennica* (Acciavatti and Simeone 1976, Acciavatti and Falin unpublished data). The only other biological information on members of the Pelecotominae is evidence of the parasitism of *Prionoplus reticularis* White (Cerambycidae: Prioninae) by *Rhipistena cryptarthra* Broun 1904 in New Zealand (Hudson 1934). The same cerambycid may also host *Allocinops brookesi* Broun 1921 in New Zealand as well (Watt 1983).

Nothing is known of the biology of the three very rare genera in the Micholaeminae (*Ancholaemus*, *Clinops*, and *Micholaemus*) though they are likely closely related to *Pelecotoma*, *Allocinops* and *Rhipistena*, and if so they may also have coleopterous wood-boring larval hosts.

The taxa currently placed in the Hemirhipidiinae are also extremely rare and very little is known of their biology. Tôyama and Hatayama (1985) state that the host of *Nephrites kurosawai* Tôyama and Hatayama 1985 is a species of *Blepephaeus* (Cerambycidae: Lamiinae), though no further details are given. Švácha (1994) states that the host of *Nephrites japonicus* Tôyama and Hatayama 1985 is *Pterolophia annulata* (Chevrolat) (Cerambycidae: Lamiinae), though this information was not presented in Tôyama and Hatayama's original paper and must have been conveyed to Švácha by personal communication. A word of caution is necessary here: the composition of, and generic concepts in this subfamily are poorly enough understood that making biological comparisons between these species and others in the family should be done tentatively.

Lastly, nothing is known of the biology of *Ptilophorus* and *Toposcopus*, the two genera in the Ptilophorinae. Such information may help bridge the gap between the putatively basal, wood-boring larval parasitism of *Pelecotoma* and the relatively derived, hymenopterous and blattid parasitism of the Ripiphorinae and Ripidiinae.

In summary, all known ripiphorids are endoparasitic of immature stages of other insects at some point in their larval development and free-living as adults. All appear to emerge from their host before pupation, and all but the Ripidiinae completely consume their host upon emergence. The putatively basal pelecotomine genus *Pelecotoma* uses wood-boring anobiid larvae as hosts; the closely related genera *Allocinops* and *Rhipistena* appear to use cerambycid larvae as do the indeterminately placed species of *Nephrites*. Members of the Ripiphorinae attack ground-nesting bees in the case of *Ripiphorus*, social wasps in the case of *Metoecus*, and solitary wasps and rarely other bees in *Macrosiagon*, and all species in the Ripidiinae attack roaches. The first instar larva is the host-finding stage, modified from a campodeiform larval type in *Pelecotoma* to a highly derived (possibly independently) triungulin type in the Ripiphorinae and Ripidiinae. Larvae apparently undergo five instars in *Pelecotoma*, six in the Ripiphorinae, and four in the Ripidiinae; only the first instar is endoparasitic in the former two groups, the latter passes the second to fourth instars internally. Eggs are laid quite near the host taxa in *Pelecotoma*, in the general vicinity of the host in Ripidiinae, and distantly in the Ripiphorinae; the triungulin type larva appears to have arisen in the latter two taxa to facilitate the various forms of phoresy they exhibit. Ripiphorids most commonly overwinter as first instar endoparasites, though certain species of *Ripiphorus*, at least, appear to overwinter as first-instar ectoparasites and *Metoecus* species as eggs.

Adults tend to be fast moving and short lived. It is not known and difficult to guess what most adults eat; those of

Macrosiagon have greatly prolonged proboscis-like galeae and is likely that they suck nectar. The mouthparts of *Ripiphorus* are reduced though functional, and those of Ripidiinae are completely degenerate and non-functional. To date, no gut-content analysis has been performed on any ripiphorid.

Most ripiphorids appear to be diurnal, though some have been caught at lights, and the large, coarsely faceted eyes in taxa such as *Trigonodera* and the Ripidiinae suggest nocturnal activity, though none has specifically been noted in the literature. It is difficult to tell how rare many of these species are: some are certainly relatively more plentiful in collections than others. Adults are often extremely spatially and temporally localized, particularly those parasitizing strictly univoltine, gregariously nesting hosts and, combined with their short lifespan, are therefore rarely encountered while sometimes ephemerally abundant.

The fossil ripiphorid taxa have recently been cataloged and reviewed by Kaupp *et al.* (in press) and consist almost exclusively of inclusions of male ripidiines in Baltic (Eocene) and Dominican (mid-Tertiary) amber. One such ripidiine, described by Cockerell (1917) as *Myodites burmiticus* from Burmese (Upper Cretaceous) amber, is as fully derived as extant taxa and proves the rather surprising age of this lineage. Clearly, the major lineages of Ripiphoridae must have split at the very latest 100 million years ago and, at least in the case of the Ripidiinae, remained more or less in morphological stasis since. Other taxa in the fossil record include a *Trigonodera*-like individual in Baltic amber and a doubtful "*Macrosiagon*" species (= *Rhipiphorus sensu* Scudder) from Florissant, Colorado (Oligocene shale) (Scudder 1890). Many of the fossil taxa need to be re-examined and their taxonomy revised to current usage, and several fossil specimens remain to be described.

Status of the classification. The first ripiphorid to be described was the European species *Mordella paradoxa* Linnaeus 1761 (now placed in *Metoecus*). Fabricius (1792) transferred the 20 previously described ripiphorid species into *Ripiphorus*, created by Bosc d'Antic (1791) to contain *Ripiphorus subdipterus* Bosc 1791. *Ripiphorus* and subsequently described genera remained undifferentiated from the Mordellidae until Gerstaecker (1855) separated them into their own family while reiterating their general similarity to mordellids. Lacordaire (1859) concurred with Gerstaecker's treatment and Ripiphoridae were placed next to the Mordellidae in Gemminger and Harold's (1870) world catalog.

In contrast, Forbes (1926) suggested a relationship between Ripiphoridae and Meloidae based on similarities in wing venation. Bøving and Craighead (1931) similarly divorced Ripiphoridae from Mordellidae, instead placing them with the Meloidae in the superfamily Meloidea based on similarities between their respective triungulin larvae. Interestingly, White (1846), in describing the Australian ripiphorid genus *Sitarida*, originally placed it near *Sitaris*, a European meloid genus, based on adult similarities. Pascoe (1863) also placed his new genus *Goetymes* (subsequently synonymized under *Sitarida*) in the Meloidae (= Cantharidae *sensu* Pascoe) for the same reasons, going so far as to compare *Sitaris*, *Goetymes*, and the then-recognized ripiphorid genus *Ptilophorus* (= *Evaniocera sensu* Pascoe). The discovery of non-triungulin larvae in the putatively basal ripiphorid *Pelecotoma fennica* (Paykull 1799) by Švácha (1994) and in the putatively basal meloid genus *Electica* by Pinto *et al.* (1996) at the same time negates a close relationship between the two families based on triungulin-type larvae and illustrates another curious parallel in the evolution of their larval habits. Bologna and Pinto's (2001) phylogeny of the Meloidae, based primarily on larval characters, is noteworthy in proposing multiple origins of phoretic triungulin-type larvae in that family, a phenomenon that is also a distinct possibility in the Ripiphoridae. Although the current consensus is that similarities between adult ripiphorids and meloids are merely coincidental (Crowson 1955, Selander 1957) and their similar triungulin larvae are clearly independently derived, a familial-level phylogenetic analysis refuting a close relationship between the two would be required to convincingly settle the issue.

That Ripiphoridae belong within the heteromerous cucujoids of Crowson (1955, 1960) and the cucujoid superfamily Tenebrionoidea (= Heteromera) of Lawrence and Newton (1995) seems clear, though its particular sister-group relationship within this assemblage remains obscure. Crowson (1955) suggests close relationships with the scraptiids and mordellids based on metendosternite and pronotal morphology, though he points out that ripiphorids lack securiform apical maxillary palpomeres and pubescent tibial spurs that characterize the other groups. Later, Crowson (1960) suggests a common ripiphorid-mordellid ancestor whose larvae live in dead woody or herbaceous stems and then (Crowson 1966) posits a tenebrionoid lineage consisting of Melandryidae, Mordellidae + Ripiphoridae and Scraptiidae. Lawrence and Newton (1982) suggest a slightly different group of relationships, placing Tetratomidae, Melandryidae, Mordellidae and Ripiphoridae into a single assemblage, allying Scraptiidae with Anthicidae and Aderidae (= Euglenidae *sensu* Lawrence and Newton 1982) without suggesting specific sister-group relationships.

Selander (1957) treats the relationship between Ripiphoridae and Mordellidae as obvious and does not suggest synapomorphies uniting the families. In moving the South African genus *Ctenidia* from Ripiphoridae to its own subfamily in Mordellidae, Franciscolo (1952) makes the most detailed morphological comparison between the families to date. While valuable, this effort is fraught with taxonomic and morphological errors due, surely, to a limited comparative knowledge of ripiphorid morphology and his points should not be accepted uncritically. Later Franciscolo (1957) suggests the similarities in the elongate form of the galea and lacinia of *Ctenidia* and *Macrosiagon* as evidence of relationship between the two groups, repeating his previous mistake in believing the elongate mouthparts of *Macrosiagon* to be plesiomorphic for, and widespread in the Ripiphoridae (they are almost assuredly not the former and certainly not the latter). Franciscolo (1962) again mentions a ripiphorid-mordellid relationship in his treatment of the North American genus *Glipodes* and most recently (Franciscolo 2000) in the description of the curious genus *Ideorhipistena* that superficially resembles the New Zealand ripiphorid genus *Rhipistena*. The Mordellidae are almost certainly a well-defined monophyletic lineage, but the definition of the Ripiphoridae and what, exactly, unites that lineage or lineages with the mordellids remains an open question. Švácha (1994) recently calls this close relationship into question with his preliminary comparative observations on the larvae of *Pelecotoma fennica* (Paykull 1799) and various mordellids.

While a sister-group relationship between the ripiphorids and the mordellids is certainly possible, one should remember Crowson's (1960, p. 127) admonition that "The interrelationships of the Heteromeran families are complex and can hardly be usefully discussed in the present state of our knowledge." These relationships are today nearly as opaque as they were then; no inter-familial level phylogenetic analyses have been conducted on this group of beetles and concepts of their inter-relationships continue to be based on authority. To complicate matters, the apparent lack of a strong synapomorphy (other than the variously modified antennae and a few plesiomophic absences) to unite the Ripiphoridae, combined with the relatively disparate larval and adult morphologies found within the family, make the *a priori* belief in a monophyletic Ripiphoridae dubious at best. In fact, Iablokov-Khnzorian (1986) described a new tribe, Eorhipidiini, based on *Eorhipidius januschevi* Iablokov-Khnzorian 1986, a taxon whose males strongly resemble less-derived ripidiines in possessing full mandibles, complete maxillary palpi, less atrophied elytra and 11-segmented, filiform antennae. Should this species, along with a number of closely related, undescribed Neotropical taxa, be recognized as basal to the rest of the ripidiines, that subfamily's placement within the Ripiphoridae may have to be reconsidered given its plesiomorphic filiform antennae. Clearly, much work needs to be done to confidently discuss the relationships within this family as well as the relationship(s) it holds to other lineages within the Tenebrionoidea.

Ripiphoridae have often been considered a possible sister-group to the order Strepsiptera by those who believe the latter to be highly derived Coleoptera (Bøving and Craighead 1931, Arnett 1963, and see Whiting *et al.* 1997 for a general review of the "strepsipteran problem"). Crowson believed the Strepsiptera to be a family within the polyphagan Cucujiformia and drew parallels between them and the Ripidiinae in the similarity of the strepsipteran forewing and the reduced ripidiine elytron, the flabellate antennae in the males, the triungulin first instar larvae, the hypermetamorphosis and the parasitic lifestyle (Crowson 1955, 1960, 1981). Conversely, he also noted that basal adult male strepsipterans have a 5-5-5 tarsal formula (instead of 5-5-4 or less as in all Heteromera) and the larvae lack trochanters (instead of possessing them). Further, since the most basal strepsipterans appear more highly modified than the most derived ripiphorids, he argued that the strepsipteran lineage must be older than the putative ripiphorid-mordellid split and suggests a possible affinity with the Lymexylidae. While the former morphological characters are evidence against a ripiphorid-strepsipteran relationship, the latter does not in and of itself disprove relationship. Regardless, assuming *Pelecotoma* and its non-triungulin larvae to be a basal ripiphorid and assuming the monophyly of the family, the strepsipterans could hardly be considered the sister-group to the family but would rather have to be derived from within it. As it is, there are significant morphological and biological differences between Ripiphoridae and Strepsiptera in all their life stages (see references below). Also, the proposed similarities (reduced elytra not covering the hind wings, generalized triungulin type larvae, winged males and flightless females and other forms of paedogenesis, loss of larval legs in certain instars, and flabellate antennae) are all found with various frequencies in widely disparate lineages of beetles (Crowson 1981). These facts suggest that similarities shared by the two groups are coincidental or convergent.

Further, evidence is mounting that Strepsiptera should not be considered a sister-group to the Coleoptera, much less placed within the Tenebrionoidea. Kathirithamby (1989) convincingly re-interprets a number of supposed morphological synapomorphies shared with Coleoptera as do Pix *et al.* (1993) for the function of the haltere-like strepsipteran forewing and Whiting and Kathirithamby (1995) for hind wing venation. Kathirithamby *et al.* (1993) and Afzelius and Romano (1994) give evidence that strepsipteran sperm are morphologically different from those of *Pelecotoma* and the rest of Neuropteroidea while Buschbeck *et al.* (1999) report that the fine structure of the strepsipteran eye is fundamentally different from that of Coleoptera and indeed the rest of the extant insects. Neither Whiting *et al.*'s (1997) nor Wheeler *et al.*'s (2001) combined molecular and morphological phylogenetic analyses place the Strepsiptera next to the ripiphorid example; rather, both analyses removed them entirely from the Coleoptera and placed them instead as sister-group to the Diptera (though this result is not without controversy, see: Huelsenbeck 1997, 1998, Whiting 1998, Siddall and Whiting 1999, Rokas *et al.* 1999). While the final placement of the Strepsiptera remains uncertain, the preponderance of evidence does not support a close relationship between these insects and ripiphorids.

There has been considerable confusion regarding the correct spelling, formation and attribution of many family-group names in the Ripiphoridae (Selander 1957, Lawrence and Newton 1995), stemming in large part from a misspelling (Fabricius 1792) and continued subsequent misidentification of the correct type species and author of the genus *Ripiphorus*. Krell (1996) properly indicates the correct original spelling of *Ripiphorus* Bosc 1791 and *Ripidius* Thunberg 1806, that is, without an "h" following the first letter of the name. According to article 33.3 of the ICZN (1999), the correct original spelling of these names must be used. Krell also suggests, given article 35.4.1, that the family-group names based on these genera must be corrected to Ripiphoridae, Ripiphorinae, Ripiphorini, and Ripidiini. However, Selander (1957) points out and Lawrence and Newton (1995) reiterate that the family-group names currently based on *Ripiphorus* may not be valid. Indeed, because these names were based on a misidentified type species, family names such as Ptilophoridae, Myoditidae, or Ripidiidae may in fact be correct according to the zoological code. Similarly convoluted issues surround the proper formation of what I use here as the Hemirhipidiinae. For the purposes of this chapter, I will follow both Krell's directive in altering the affected family group names to reflect the correct original spellings of the type genera as well as Lawrence and Newton's recommendation in invoking article 40.2 to retain the long-established family group names currently in use. Clearly, an exhaustive taxonomic accounting of the family and potentially a subsequent petition or petitions to the ICZN will be required to satisfactorily resolve these issues. For expediency, I will follow current practice in attributing the family name to Gemminger and Harold 1870 (1853) while not attempting to correctly attribute the remaining family-group names.

There are currently six subfamilies recognized in the Ripiphoridae: Pelecotominae, Micholaeminae, Ptilophorinae, Hemirhipidiinae, Ripidiinae, and Ripiphorinae (Lawrence and Newton 1995). Brewer (1963) has recently re-diagnosed the Pelecotominae, Riek (1973) the Hemirhipidiinae (= Nephritinae *sensu* Riek) and Viana (1971) erected the Micholaeminae. Rivnay (1929) provides limited characterizations of the four subfamilies represented in the U.S. and Besuchet (1956) offers some insight to the characterizations of the Ripidiinae, Ripiphorinae, and Pelecotominae (*sensu* Csiki 1913). However, none of these treatments have been globally comprehensive or phylogenetic in nature, leaving the subfamilial status of a number of smaller, morphologically intermediate genera uncertain and the value of the natural groups encompassed by the subfamilies as currently defined questionable. The tribe Macrosiagonini (*Macrosiagon* + *Metoecus*) and the subfamily Ripidiinae (excluding the aforementioned tribe Eorhipidiini) may be the only higher taxa in the family that could, *a priori*, be considered monophyletic, though even this should be considered tentative without a supporting analysis.

Selander (1957) has come closest to a comprehensive comparative treatment of lineages within the Ripiphoridae and is currently the best reference for students of ripiphorid comparative morphology. He is the only author to suggest a plausible, though pre-cladistic, subfamilial-level phylogeny for the family. He hypothesized an early split of the Ripiphorinae off the ancestral lineage, with *Trigonodera* in the Pelecotominae representing the least-derived extant taxon. Then Ptilophorinae split from the ancestral lineage, leaving Hemirhipidiinae (= Nephritinae *sensu* Selander) and Ripidiinae as the most derived sister taxa (Micholaeminae post-dates Selander's work and does not conveniently fall into place as a cohesive clade). Preliminary analyses have confirmed Selander's suggestion that *Trigonodera* is the least derived genus, the Pelecotominae as the least derived subfamily, and the Ptilophorinae as the next-least derived lineage, though relationships among the remaining lineages continue to be obscure (Falin unpublished data).

Distribution. The last world catalog (Csiki 1913) listed 232 described species, though at least several hundred have been described since, unfortunately many of them dubiously so by less than exacting authors (see Franciscolo 1957 for a frank discussion of an identical problem in the Mordellidae). The relative rarity, frequency of large specific distributions, sexual dimorphisms, and considerable intraspecific variability associated with their parasitic lifestyle has caused many ripiphorid species to be described repeatedly under different names. However, careful comparative studies of fine morphological detail (e.g., male genitalia) are rare; such studies may reveal a greater species diversity than is externally apparent. Nonetheless, it is likely that synonymies will greatly outnumber new descriptions in future revisions, except perhaps in the relatively under collected and under studied Ripidiinae. There are currently 38 genera in the Ripiphoridae, excluding three described by Pic (*Dunbrodianus*, *Madrasiindus* and *Setosicornia*) that probably do not belong in the family but have yet to be officially removed and reassigned.

The larger genera in the family are world-wide or nearly so: *Macrosiagon* is found on every continent (except Antarctica), Australia and Madagascar as is *Ripiphorus*, excepting the latter two islands. *Trigonodera* likewise is found nearly everywhere except Europe. Other, smaller genera or groups of related genera have similarly wide, and seemingly relictual distributions. The three species of *Pelecotoma* are found in North America, Europe, and Japan, and related genera can be found in New Zealand (*Rhipistena*, *Allocinops*), the Galapagos and Brazil (*Ancholaemus*), Greece (*Scotoscopus*), and South Africa and Syria (*Clinops*). The described and undescribed taxa in the Eorhipidiini are found in Tadzhikistan, South Africa, and the Neotropics. These patterns, combined with the surprising age of fossil ripidiines (likely the most derived group in the family), point to an old divergence and radiation in the family.

Generic distributions logically tend to mirror those of their respective host taxa: species in *Macrosiagon* and *Ripiphorus* tend to be more common in temperate to arid localities as are their particular hymenopterous hosts and species in the Ripidiinae are nearly exclusively restricted to tropical or subtropical localities where their blattid hosts are most numerous. Likewise, taxa such as *Toposcopus* and *Ptilophorus* tend to be found in arid to temperate regions whereas certain taxa like *Micropelecotoides* are exclusive to the Asian tropics, but for unknown reasons.

There are 51 currently valid species representing six genera and four subfamilies found in the United States.

Key to the Genera of the United States

1. Elytra long, entire, covering abdomen and folded wings 2
— Elytra reduced, partially or entirely exposing wings 4

2(1). Eyes divided into two lobes connected by a narrow, non-faceted corneous strip (Ptilophorinae) .... *Toposcopus*
— Eyes shallowly emarginate (Pelecotominae) 3

3(2). Small insects, not over 6 mm.; eyes small not nearly contiguous anteriorly; first 3 antennomeres simple (Figs. 10, 11); claws with two minute teeth .... *Pelecotoma*
— Large insects, 6 mm or larger; eyes large, nearly contiguous anteriorly; first 4 antennomeres simple (Figs. 8, 9); claws strongly pectinate .... *Trigonodera*

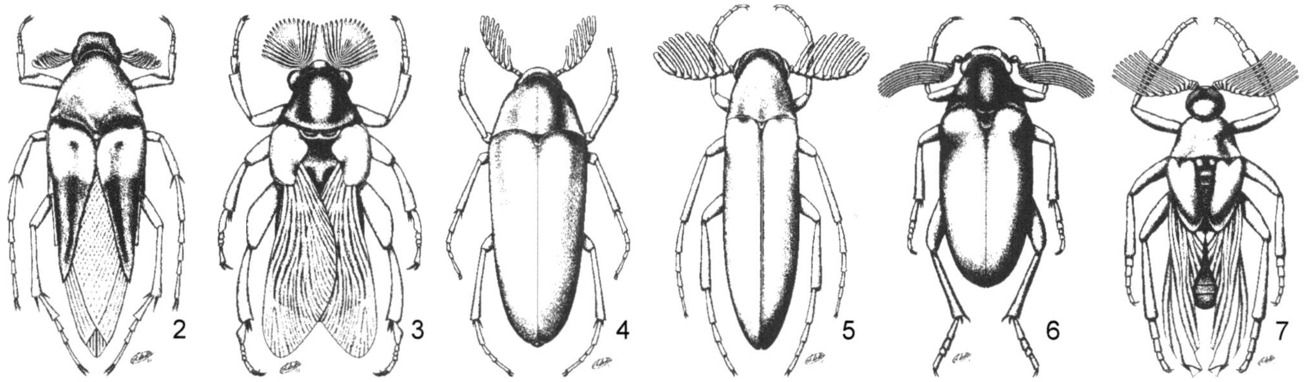

FIGURES 2.102-7.102. Dorsal habitus drawings of ripiphorid genera. 2. *Macrosiagon flavipenne* (LeConte), male; 3. *Ripiphorus solidaginis* (Pierce), male; 4. *Trigonodera schaefferi* Rivnay, male; 5. *Pelecotoma flavipes* Melsheimer, male; 6. *Toposcopus wrighti* LeConte, male; 7. *Pirhidius* sp., male.

4(1). Mouthparts reduced to 2-segmented maxillary palpi, basal pair fused; eyes holoptic; antennae uniflabellate (Fig. 18); females putatively larviform (Ripidiinae) .... *Pirhidius*

— Mouthparts entire, functional, eyes smaller, not holoptic; male antennae biflabellate or bipectinate (Figs. 14, 16), female antennae unipectinate (Figs. 15, 17); females not larviform (Ripiphorinae) 5

5(4). Elytra reduced to scale-like convex plates, not extending beyond thorax; antennae insert dorsal to eyes; claws pectinate (Ripiphorini) .... *Ripiphorus*

— Elytra longer, dehiscent; antennae insert anterior to eyes; claws bifid (Macrosiagonini)................. .... *Macrosiagon*

Classification of the Genera of the United States

Ripiphoridae Gemminger and Harold 1870

Given the uncertainty surrounding the proper formation and attribution of family-group names in the Ripiphoridae described above, I am purposely omitting authors to avoid inadvertently perpetuating or augmenting error. An exhaustive taxonomic review of the family is necessary before such information can be conclusively presented.

The family-group diagnoses provided are for convenience in dealing with the Nearctic taxa and are not definitive or universal; much work needs to be done to properly diagnose the higher taxa in Ripiphoridae, particularly the putatively basal lineages. The family description and the simple key couplets given above should be adequate to characterize and differentiate the Nearctic genera; they are not described below.

Pelecotominae

Diagnosis. Head subglobular, vertex little raised above anterior pronotal margin; eyes expanded or reniform but never divided; male antennae with robust, tubular pectinations, female antennae strongly serrate; at least three basal antennomeres simple; gonostyli of male genitalia more or less symmetrical, clearly articulated with but separate from gonocoxae; ventral baculi free.

Pelecotoma Fischer 1809 (Figs. 5, 10, 11)

Pelecotoma contains three species, two of which, *P. fennica* (Paykull 1799) and *P. septentrionalis* Kôno 1936, are Palaearctic. The single Nearctic species, *P. flavipes* Melsheimer 1846, occurs in the eastern United States and Canada, approximately from North Carolina to Ontario and Quebec, and west to Michigan. It is a parasitoid of *Ptilinus ruficornis* (Say) (Anobiidae) larvae and is most often

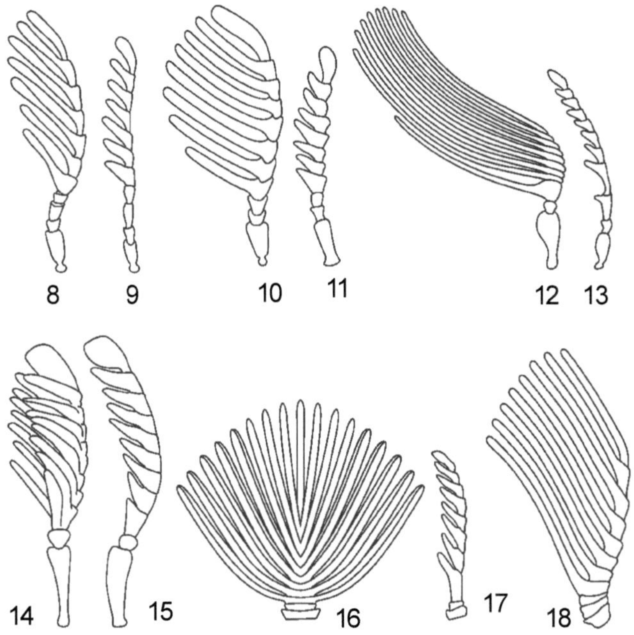

FIGURES 8.102-18.102. Male and female antennae. 8. *Trigonodera schaefferi* Rivnay, male; 9. *Trigonodera schaefferi*, female; 10. *Pelecotoma flavipes* Melsheimer, male; 11. *Pelecotoma flavipes*, female; 12. *Toposcopus wrighti* LeConte, male; 13. *Toposcopus wrighti*, female; 14. *Macrosiagon flavipenne* (LeConte), male; 15. *Macrosiagon flavipenne*, female; 16, *Ripiphorus solidaginis* (Pierce), male; 17. *Ripiphorus solidaginis*, female; 18. *Pirhidius* sp., male.

found on exposed, dead wood of sugar maple, beech, and oak that exhibits anobiid boring activity (Stephens 1968, Acciavatti and Simeone 1976, Acciavatti and Falin unpublished data). So far as is known, its biology closely resembles that of *P. fennica* as presented by Švácha (1994).

Trigonodera Dejean 1834 (Figs. 4, 8, 9)
Pelecotoides Laporte 1840
Pelecotomoides Gemminger and Harold 1870
Caspyria Fairmaire 1901
With approximately 71 current species found world-wide exclusive of Europe, *Trigonodera* is the second largest ripiphorid genus, though it is in serious need of revision. Although five species of *Trigonodera* have been recorded from Central America and Mexico and several more potentially remain to be described from this region, only *T. schaefferi* Rivnay 1929 has been found in the U.S. (Rivnay 1929). This species has been collected in southern Texas and Costa Rica at lights, by beating trees, and in malaise traps. Nothing else is known of the biology of this or any other species of *Trigonodera*.

Ptilophorinae

Diagnosis. Head subglobular, vertex slightly raised above anterior pronotal margin; eyes strongly divided into dorsal and ventral lobes connected by a thin corneous strip; male antennae with thin, hair-like pectinations, female antennae strongly serrate; two basal antennomeres simple, third antennomere with abbreviated process in female; gonostyli of male genitalia symmetrical, fused with gonocoxae; ventral baculi possibly drawn posteriorly, fused medially supporting apex of aedeagus.

Toposcopus LeConte 1868 (Figs. 6, 12, 13)
Toposcopus wrighti LeConte 1868, the only species in the genus and the sole representative of the Ptilophorinae in the New World, is apparently restricted to southern Arizona, New Mexico and Texas. Little is known of the biology of species in this subfamily; they are most frequently collected by sweeping vegetation. *Ptilophorus*, the other genus in the subfamily, contains 21 primarily African and Australian species and is considerably more morphologically diverse than LeConte had surmised in his description of *Toposcopus* (LeConte 1868). A revision of the group will likely reduce the generic name *Toposcopus* to junior synonymic status.

Ripiphorinae

Diagnosis. Head anteroposteriorly compressed, vertex distinctly raised above anterior pronotal margin; eyes oval or slightly excavate, lateral; male antennae bipectinate to biflabellate, female antennae unipectinate; two basal antennomeres simple; gonostyli of male genitalia asymmetrical, clearly articulated with but separate from gonocoxae; ventral baculi fused anteriorly.

Macrosiagonini

Diagnosis. Head dorsoventrally elongate; labrum long and narrow; maxillae with elongate galea; antennae inserted anterior to eyes; basal antennomere elongate; pronotum with posterior lobe; elytra acuminate; abdomen telescoping; tarsal claws bifid.

Macrosiagon Hentz 1830 (Figs. 1, 2, 14, 15, Vol. 2, Color Fig. 29)
Emenadia Laporte 1840
The genus *Macrosiagon*, with approximately 156 species, is world-wide in distribution and the largest genus in the family. Eleven species are found throughout North America. Distributions of individual species vary enormously in this genus: *M. fernaldum* Rivnay 1929 is restricted to limited areas of southern California (Rivnay 1929) while *M. flavipenne* (LeConte 1866) ranges from New York to California and as far south as Argentina (Brewer 1966). Rivnay's (1929) key still suffices to roughly distinguish species in this genus, although his use of varieties is of questionable utility. Because of their large distributions, relative rarity and remarkable intraspecific variability, a careful revision of this group would likely result in the synonymy of several Nearctic species and possibly the recognition of a few new species, particularly in the *M. pectinatum* (Fabricius 1775) complex. Adults are almost exclusively collected by sweeping flowers, particularly of Asteraceae (= Compositae of authors), though have occasionally been reared from the nests of their host.

Ripiphorini

Diagnosis. Head not dorsoventrally elongate; labrum broad and deeply bilobed; maxillae with galea reduced and peg-like; antennae inserted dorsal to eyes; basal antennomere short, annular; pronotum without posterior lobe; elytra scale-like; abdomen not telescoping, strongly decurved; tarsal claws pectinate

Ripiphorus Bosc 1791 (Figs. 3, 16, 17)
Myodes Latreille 1818
Myodites Latreille 1819
Dorthesia Say 1823
Rhipidophorus Gemminger and Harold 1870
Ripiphorus occurs worldwide except for Australia. Of the 67 species in the genus, 36 occur in North America, Central America and the Caribbean, 29 north of Mexico. However, the alpha-level taxonomy of the Nearctic *Ripiphorus* species is extremely poor and seriously hampers a proper understanding of the patterns and types of biological diversity found in the genus. Widespread synonymy in this group has been recognized for over 100 years (Horn 1892) though neither Rivnay (1929) in his revision of the family nor Linsley and MacSwain (1951) in their treatment of the Ripiphoridae of California have adequately addressed this problem. As no new Nearctic species have been described since 1951, one can use a combination of Rivnay's (1929) and Linsley and MacSwain's (1951) keys and the original descriptions to accurately identify a portion of the species, though many, particularly those allied with *R. fasciatus* (Say 1823), are impossible to identify with

confidence. A careful phylogenetic revision of the Nearctic species along with a comprehensive review of the known host associations is required to clarify the fascinating diversity and biology of these beetles. Unless reared from the nest of their hosts, adults are almost exclusively found sweeping vegetation and flowers in the vicinity of their host colonies.

Ripidiinae

Diagnosis. Males with head subglobular, vertex little raised above anterior pronotal margin; eyes holoptic; mouthparts reduced, non-functional; male antennae robustly uniflabellate; three basal antennomeres simple; gonostyli of male genitalia more or less symmetrical, membranous, closely articulated to the gonocoxae; ventral baculi absent.

Pirhidius Besuchet 1957 (Figs. 7, 18)
An undescribed species belonging to the currently monotypic genus *Pirhidius* has recently been discovered in northern and central Florida, and represents the single endemic Nearctic representative of the Ripidiinae. A full description of this species is in preparation by the author. The one described species in this genus, *P. beaumonti* Besuchet 1957, as well as a number of undescribed species belonging to the genus, is Neotropical in distribution. The females and putatively blattid hosts of this genus are not known. Males of the undescribed Floridian species have been most effectively collected using malaise traps; flight intercept traps have also been quite useful in collecting other Neotropical ripidiine taxa. Females are likely to be found in leaf litter or in loose bark of trees.

Ripidius pectinicornis Thunberg 1806, a second ripidiine species, is infrequently but consistently collected in and around the Honolulu International Airport on Oahu, Hawaii, in light and malaise traps though it is unclear whether it has established permanent populations there (Falin 2001). It has also rarely been found at other ports of entry in the continental U.S. The males of *R. pectinicornis* can be distinguished from those of *Pirhidius* by possessing the following characters: basal maxillary palpomeres free, presence of postocular ommatidia, mesoscutellum present, and metaprescutum poorly developed.

Bibliography

ACCIAVATTI, R. E. and J. B. SIMONE. 1976. The occurrence of anobiid species in a northern hardwood forest and the emergence, flight and host selection behavior of *Ptilinus ruficornis* (Say) (Coleoptera: Anobiidae). Material und Organismen, 3: 419-427.

AFZELIUS, B. A. and D. ROMANO. 1994. Characteristics of the flagellar axoneme in Neuroptera, Coleoptera and Strepsiptera. Journal of Morphology, 219: 15-20.

ARNETT, R. H., Jr. 1963. The beetles of the United States (a manual for identification). Catholic University of America Press. Washington, DC, xi + 1112 pp.

BARBER, H. S. 1915. *Macrosiagon flavipennis* in cocoon of *Bembex spinolae* (Coleoptera: Rhipiphoridae). Proceedings of the Entomological Society of Washington, 17: 187-188.

BARBER, H. S. 1939. A new parasitic beetle from California (Rhipiphoridae). Bulletin of the Brooklyn Entomological Society, 34: 17-20.

BATRA, S. W. T. 1965. Organisms associated with *Lassioglossum zypherum* (Hymenoptera: Halictidae). Journal of the Kansas Entomological Society, 38: 367-389.

BEQUART, J. C. 1918. A revision of the Vespidae of the Belgian Congo based on the collection of the American Museum Congo Expedition, with a list of Ethiopian diplopterous wasps. Bulletin of the American Museum of Natural History, 39: 1-384.

BESUCHET, C. 1956. Biologie, morphologie et systématique des *Rhipidius* (Col. Rhipiphoridae). Bulletin de la Société Entomologique Suisse, 29(2): 74-144.

BOLOGNA, M. A. and J. D. PINTO. 2001. Phylogenetic studies of Meloidae (Coleoptera), with emphasis on the evolution of phoresy. Systematic Entomology, 26: 33-72.

BOSC D'ANTIC, L. 1791. "Mémoire sur l'etablissement d'un nouveau genre d'insectes coléoptères". Pp. 327-328. *In*: Fourcroy, ed., La médecine eclaireé par les sciences physiques, ou journal des déscouvertes relatives aux différentes parties l'art de guérir, vol. 1. Bousson. Paris.

BØVING, A. G. and F. C. CRAIGHEAD. 1931. An illustrated synopsis of the principal larval forms of the order Coleoptera. Entomologica Americana, II (New Series), 351 pp., 125 pls.

BREWER, M. M. 1963. Contribución al conocimiento de las Ripiphoridae Argentinas (Coleoptera). Opera Lilloana, 11: 1-108.

BREWER, M. M. 1966. Subfamilia Ripiphorinae (Col. Ripiphoridae), revision de las especies argentinas del genero *Macrosiagon* Hentz. Revista de la Facultad de Ciencias Exactas, Fisicas, y Naturales, 27(1-2): 5-47.

BUSCHBECK, E., B. EHMER and R. HOY. 1999. Chunk versus point sampling: visual imaging in a small insect. Science, 286: 1178-1180.

CALLAN, E. 1977. *Macrosiagon diversiceps* (Coleoptera: Rhipiphoridae) reared from a sphecid wasp, with notes on other species. Australian Entomological Magazine, 4: 45-47.

CALLAN, E. 1981. Further records of *Macrosiagon* (Coleoptera: Rhipiphoridae) reared from eumenid and sphecid wasps in Australia. Australian Entomological Magazine, 7: 81-83.

CHAPMAN, T. A. 1870. Some facts towards a life history of *Rhipiphorus paradoxus*. Annals and Magazine of Natural History (Series 4), 5: 314-326.

CHAPMAN, T. A. 1897. Sketch of the life history of *Metoecus* (*Rhipiphorus*) *paradoxus*. Entomologist's Record and Journal of Variation, 9: 321-322.

CHOBAUT, A. 1891. Moeurs et métamorphoses de *Emenadia flabellata* F., insecte coléoptère de le famille des rhipiphorides. Mémoires de l'Academie de Vaucluse, 10: 83-94.

CHOBAUT, A. 1895. Étude sur les *Macrosiagon* Hentz (*Emenadia* Lap.-Cast.) de la région Méditerraneene I: Moeurs et métamorphoses des *Macrosiagon*. Abeille, 28: 181-183.

CHOBAUT, A. 1906. Le triongulinide de *Macrosiagon tricuspidata* Lepech. Bulletin de la Société Entomologique de France, (1906): 270-272.

COCKERELL, T. D. A. 1917. Fossil insects. Annals of the Entomological Society of America, 10: 1-22.

CROS, A. 1920. Contribution a l'étude des rhipiphorides Algériens. Bulletin de la Société d'Histoire Naturelle de l'Afrique du Nord, 11: 56-68, 70-75.

CROWSON, R. A. 1955. The natural classification of the families of Coleoptera. Nathaniel Lloyd. London, 187 pp.

CROWSON, R. A. 1960. The phylogeny of Coleoptera. Annual Review of Entomology, 5: 111-134.

CROWSON, R. A. 1966. Observations on the constitution and subfamilies of the family Melandryidae. Eos, 41: 507-513.

CROWSON, R. A. 1981. The biology of the Coleoptera. Academic Press. London, xii + 802 pp.

CSIKI, E. 1913. Rhipiphoridae, Pars 54, 29 pp. *In*: S. Schenkling, ed., Coleopterorum Catalogus auspiciis et auxilio, vol. 17. W. Junk. Berlin.

DOLPHIN, R. E. 1979. Associates of the native bee *Halictus* (*Seladonia*) *confusus* Smith (Hymenoptera: Halictidae). Proceedings of the Indiana Academy of Sciences, 88: 228-234.

EBERHARD, W. G. 1974. The natural history and behavior of the wasp *Trigonopsis cameronii* Kohl (Sphecidae). Transactions of the Royal Entomological Society of London, 125: 295-328.

EICKWORT, G. C. and K. R. EICKWORT. 1972. Aspects of the biology of Costa Rican bees, IV: *Augochlora* (*Oxystoglossa*) (Hymenoptera: Halictidae). Journal of the Kansas Entomological Society, 45: 18-45.

FABRE, J. H. 1879. Étude sur les moeurs et la parthénogenése des halictes. Annales des Sciences Naturelles, 6(9): 3-27.

FABRE, J. H. 1886. Pp. 220-222. *In*: Souvenirs Entomologique, études sur l'instinct et les moeurs des insects (Series 3).

FABRICIUS, J. C. 1792. Entomologia systematica emendata et aucta, vol. 1. Proft. Copenhagen, XX + 330 pp. (part 1), 538 pp. (part 2).

FALIN, Z. H. 2001. Notes on the occurrence of *Ripidius pectinicornis* Thunberg (Coleoptera: Rhipiphoridae) in the continental United States and Hawaii. Coleopterists Bulletin, 55: 194-197.

FALIN, Z. H., L. C. ARNESON and W. T. WCISLO. 2001. Night-flying sweat bees *Megalopta genalis* and *Me. ecuadoria* (Hymenoptera: Halictidae) as hosts of the parasitoid beetle *Macrosiagon gracilis* (Coleoptera: Rhipiphoridae). Journal of the Kansas Entomological Society, 73: 183-185.

FORBES, W. T. M. 1926. The wing folding patterns of Coleoptera. Journal of the New York Entomological Society, 34: 42-68, 91-139, 18 pls.

FRANCISCOLO, M. E. 1952. On the systematic position of the genus *Ctenidia* Castelnau, 1840 (Coleoptera). Contribution 34 to the knowledge of the Mordellidae. Proceedings of the Royal Entomological Society of London (Series B), 21: 155-163.

FRANCISCOLO, M. E. 1957. Coleoptera: Mordellidae, a monograph of South African genera and species. I. Morphology, subfamily Ctenidiinae and tribe Stenaliini. South African animal life. Hanström, Brink and Rudebeck (eds.), vol. 4: 207-291.

FRANCISCOLO, M. E. 1962. The genus *Glipodes* LeConte, 1862 (Coleoptera: Mordellidae) with description of a new species from Venezuela and Costa Rica. Proceedings of the Royal Entomological Society of London (Series B), 31: 131-136.

FRANCISCOLO, M. E. 2000. A new mordellid genus with rhipiphorid traits (Coleoptera: Mordellidae). Coleopterists Bulletin, 54: 395-402.

GEMMINGER, M. and B. HAROLD. 1870. Catalogus Coleopterorum, vol. 7. Sumptu E. H. Gummi. Munich, Pp. 1801-2180.

GERSTAECKER, A. 1855. Rhipiphoridum coleopterorum familiae dispositio systematica. Friderici Nicolai. Berolini. 36 pp., 1 pl.

GRANDI, G. 1936. Morfologia ed etologia comparate di insetti a regime specializzato, XII. *Macrosiagon ferrungineum flabellatum* (F.). Bollettino del Laboratorio de Entomologia del R. Instituto Superiore Agrario de Bologna, 9: 33-64.

HATTORI, T. and S. YAMANE. 1975. Notes on *Metoecus paradoxus* and *M. vespae* parasitic on the *Vespula* species in northern Japan (Coleoptera: Rhipiphoridae; Hymenoptera: Vespidae), I. New Entomologist, 24: 1-7.

HEITMANS, W. and T. PEETERS. 1996. *Metoecus paradoxus* in the Netherlands (Coleoptera: Rhipiphoridae). Entomologische Berichten, Amsterdam, 56: 109-117.

HOOK, A. W. and H. E. EVANS. 1991. Prey and parasites of *Cerceris fumipennis* (Hymenoptera: Sphecidae) from central Texas, with description of the larva of *Dasymutilla scaevola* (Hymenoptera: Mutillidae). Journal of the Kansas Entomological Society, 64: 257-264.

HORN, G. H. 1892. Random studies in North American Coleoptera. Transactions of the American Entomological Society, 19: 40-48.

HUDSON, G. V. 1934. New Zealand beetles and their larvae. Furgeson and Osborn. Wellington, 236 pp.

HUELSENBECK, J. P. 1997. Is the Felsenstein Zone a fly trap? Systematic Biology, 46: 69-74.

HUELSENBECK, J. P. 1998. Systematic bias in phylogenetic analysis: is the Strepsiptera problem solved? Systematic Biology, 47: 519-537.

IABLOKOV-KHNZORIAN, S. M. 1986. [A new genus and species of Rhipiphoridae from Tadzhikistan (Coleoptera, Rhipiphoridae)] *in Russian*. Doklady Akademii Nauk Armianskoi SSR, 82: 89-92.

ICZN. 1999. International Code of Zoological Nomenclature, Fourth Edition, adopted by the International Union of Biological Sciences. International Trust for Zoological Nomenclature. London, xxix + 306 pp.

IWATA, K. 1939. Biology of *Macrosiagon nasutum* (Thun.) with some biological notes on *Macrosiagon bipunctatum* (F.) in

Japan. Transactions of the Kansai Entomological Society, 9(1): 44-52.

JARVIS, E. 1929. Some notes on the economy of cockchafer beetles. Bulletin of the Bureau of Sugar Experiment Stations, Queensland, 20: 1-36.

KATHIRITHAMBY, J. 1989. Review of the order Strepsiptera. Systematic Entomology, 14: 41-92.

KATHIRITHAMBY, J., M. CARCUPINO and M. MAZZINI. 1993. Comparative spermatology of four species of Strepsiptera and comparison with a species of primitive Coleoptera (Rhipiphoridae). International Journal of Insect Morphology and Embryology, 22: 459-470.

KAUPP, A., Z. H. FALIN and P. NAGEL. (*in press*). An annotated catalogue of fossil Ripiphoridae, taxonomic notes, and the description of a new genus and species from Baltic amber (Coleoptera: Ripiphoridae: Ripidiinae). Mitteilungen des Geologisch-Palaeontologischen Instituts der Universität Hamburg.

KIFUNE, T. 1956. Notes on the bionomics of two Japanese species of the genus *Macrosiagon* (Coleoptera: Rhipiphoridae). Insect Ecology, 5(13): 158-164.

KRELL, F.-T. 1996. Ripiphoridae oder Rhipiphoridae? (Col., Tenebrionoidea). Entomologische Nachrichten und Berichte, 40: 47-48.

LACORDAIRE, J. T. 1859. Histoire naturelle des insectes. Genera des coléoptères, vol. 5. Paris, 750 pp., illus.

LAWRENCE, J. F. and A. F. NEWTON, Jr. 1982. Evolution and classification of beetles. Annual Review of Ecology and Systematics, 13: 261-290.

LAWRENCE, J. F. and A. F. NEWTON, Jr. 1995. Families and subfamilies of Coleoptera (with selected genera, notes, references and data on family-group names). Pp. 779-1006. *In*: J. Pakaluk and S. A. Slipinski, eds., Biology, Phylogeny, and Classification of Coleoptera: Papers celebrating the 80th birthday of Roy A. Crowson. Muzuem i Instyut Zoologii PAN. Warsaw.

LECONTE, J. L. 1868. New Coleoptera collected on the survey for the extension of the Union Pacific Railway, E. D. from Kansas to Fort Craig, New Mexico. Transactions of the American Entomological Society, 2: 49-59.

LINSLEY, E. G. and J. W. MACSWAIN. 1950. New western species of Rhipiphoridae (Coleoptera). Wasmann Journal of Biology, 8: 229-239.

LINSLEY, E. G. and J. W. MACSWAIN. 1951. The Rhipiphoridae of California. Bulletin of the California Insect Survey, 1: 79-87.

LINSLEY, E. G. and J. W. MACSWAIN. 1957. The nesting habits, flower relationships, and parasites of some North American species of *Diadasia* (Hymenoptera: Anthophoridae). Wasmann Journal of Biology, 15: 199-235.

LINSLEY, E. G., J. W. MACSWAIN and R. F. SMITH. 1952. The life history and development of *Rhipiphorus smithi* with notes on their phylogenetic significance. University of California Publications in Entomology, 9: 291-314.

NYE, W. P. 1980. Notes on the biology of *Halictus* (*Halictus*) *farinosus* Smith (Hymenoptera: Halictidae). U.S. Department of Agriculture, Science and Education Administration, Agricultural Research Results (Western Series), vol. 11, iv + 28 pp.

OKUTANI, T. 1952. Notes on the life history of *Metoecus paradoxus*, nest parasite of *Vespula lewisii*. Insect Ecology, 4: 71-74.

PASCOE, F. P. 1863. On some new or little known genera of Coleoptera. Journal of Entomology, 2: 26-56.

PIERCE, W. D. 1904. Some hypermetamorphic beetles and their hymenopterous hosts. Nebraska University Studies, 4(2): 153-190.

PINTO, J. D., M. A. BOLOGNA and J. K. BOUSEMAN. 1996. First-instar larvae, courtship and oviposition in *Electica*: amending the definition of the Meloidae (Coleoptera: Tenebrionoidea). Systematic Entomology, 21: 63-74.

PIX, W., G. NALBACK and J. ZEIL. 1993. Strepsipteran forewings are haltere-like organs of equilibrium. Naturwissenschaften, 80: 807-809.

RAW, A. 1977. The biology of two *Exomalopsis* species (Hymenoptera: Anthophoridae) with remarks on sociality in bees. Revista de Biología Tropical, 25: 1-11.

RIEK, E. F. 1955. The Australian rhipidiine parasites of cockroaches (Coleoptera: Rhipiphoridae). Australian Journal of Zoology, 3: 71-94.

RIEK, E. F. 1973. Rhipiphorid beetles of the subfamily Nephritinae (Coleoptera: Rhipiphoridae). Journal of the Australian Entomological Society, 12: 261-276.

RIVNAY, E. 1929. Revision of the Rhipiphoridae of North and Central America (Coleoptera). Memoirs of the American Entomological Society, 6: 68 pp.

ROKAS, A., J. KATHIRITHAMBY and P. W. H. HOLLAND. 1999. Intron insertion as a phylogenetic character: the engrailed homeobox of Strepsiptera does not indicate affinity with Diptera. Insect Molecular Biology, 8: 527-530.

ROZEN, J. G. 1997. New taxa of brachynomadine bees (Apidae: Nomadinae). American Museum Novitates, 3200, 26 pp.

SCUDDER, S. H. 1890. The fossil insects of North America, with notes on some European species, II. The Tertiary insects of North America. Report of the U.S. Geological Survey of the Territories, vol. 13, 734 pp., 28 pls., 1 map.

SELANDER, R. B. 1957. The systematic position of the genus *Nephrites* and the phylogenetic relationships of the higher groups of Rhipiphoridae (Coleoptera). Annals of the Entomological Society of America, 50: 88-103.

SELANDER, R. B. 1991. Rhipiphoridae (Tenebrionoidea). Pp. 509-512. *In*: F. W. Stehr, ed., Immature Insects, vol. 2. Kendall-Hunt. Dubuque, Iowa.

SHARP, D. and F. MUIR. 1912. The comparative anatomy of the male genital tube in Coleoptera. Transactions of the Entomological Society of London, 1912: 477-642.

SIDDALL, M. E. and M. F. WHITING. 1999. Long-branch abstractions. Cladistics, 15: 9-24.

SNELLING, R. R. 1963. A host of *Macrosiagon cruentum* (Germar) in Georgia. Pan-Pacific Entomologist, 39: 87-88.

STAMM, R. H. 1936. A new find of *Rhipidius pectinicornis* Thbg. (*Symbius blattarum* Sund.) (Col. Rhipiphor.) [continued; published posthumously by K. L Henrikson]. Entomologiske Meddelelser, 19: 286-297.

STEPHENS, K. 1968. Observation on *Pelecotoma flavipes* Melsheimer (Coleoptera: Rhipiphoridae). Coleopterists Bulletin, 22: 30.

SUNDEVALL, J. C. 1831. Beschreibung einer neuen Coleopteren Gattung, *Symbius blattarum*. Isis von Oken, 11 (6): 1222-1228.

ŠVÁCHA, P. 1994. Bionomics, behaviour, and immature stages of *Pelecotoma fennica* (Paykull) (Coleoptera: Rhipiphoridae). Journal of Natural History, 28: 585-618.

TOMLIN, A. D. and J. J. MILLER. 1989. Physical and behavioral factors governing the pattern and distribution of Rhipiphoridae (Coleoptera) attached to wings of Halictidae (Hymenoptera). Annals of the Entomological Society of America, 82: 785-791.

TÔYAMA, M. and T. HATAYAMA. 1985. Occurrence of the rhipiphorid subfamily Hemirhipidiinae in Japan and Taiwan (Coleoptera: Rhipiphoridae). Gekkan-Mushi, 176: 18-24.

VIANA, J. M. 1971. Micholaeminae, nueva subfamilia de Ripiphoridae y *Micholaemus gerstaeckeri*, nuevo genero y especie de la Republica Argentina. Revista de la Sociedad Entomologica Argentina, 33: 69-76.

WATT, J. C. 1983. The huhu *Prionoplus reticularis* (Cerambycidae) and other Coleoptera in *Acacia*. New Zealand Entomologist, 7: 364-365.

WHEELER, W. C., M. WHITING, Q. D. WHEELER and J. M. CARPENTER. 2001. The phylogeny of the extant hexapod orders. Cladistics, 17: 113-169.

WHITE, A. 1846. Descriptions of new or unfigured species of Coleoptera from Australia. Pp. 505-512. *In*: J. L. Stokes, Discoveries in Australia, vol. 1. T. and W. Boone. London.

WHITING, M. F. 1998. Phylogenetic position of the Strepsiptera: review of molecular and morphological evidence. International Journal of Insect Morphology and Embryology, 27: 53-60.

WHITING, M. F. and J. KATHIRITHAMBY. 1995. Strepsiptera do not share hind-wing venational synapomorphies with the Coleoptera. Journal of the New York Entomological Society, 103: 1-14.

WHITING, M. F., J. C. [*sic*: M.] CARPENTER, Q. D. WHEELER and W. C. WHEELER. 1997. The Strepsiptera Problem: Phylogeny of the holometabolous insect orders inferred from 18S and 28S ribosomal DNA sequences and morphology. Systematic Biology, 46: 1-68.

WILLIAMS, F. X. 1928. Studies in tropical wasps- their hosts and associates. Bulletin of the Experiment Station of the Hawaiian Sugar Planter's Association (Entomological Series), vol. 19: vi + 180 pp.

WOLCOTT, G. N. 1914. Notes on the life history and ecology of *Tiphia inornata* Say. Journal of Economic Entomology, 7: 382-389.

103. COLYDIIDAE Erichson 1845

by Michael A. Ivie

Family common name: The colydiid beetles

Family synonyms: Adimeridae Sharp 1894; Monoedidae Schaeffer 1911; Orthoceridae Reitter 1882

The Colydiidae are a very diverse group, difficult to diagnose. Most members share the combination of concealed antennal insertions; closed mesocoxal cavities; heteromeroid trochanters; at least a slightly clubbed antenna; either open procoxal cavities, closed procoxal cavities with an expanded, apically spined protibia, or the procoxal cavities closed by a lateral extension of the intercoxal process; and (at least in the North American fauna) 4-4-4 tarsi. However, if it has 4-4-4 tarsi and doesn't fit somewhere else, try this family.

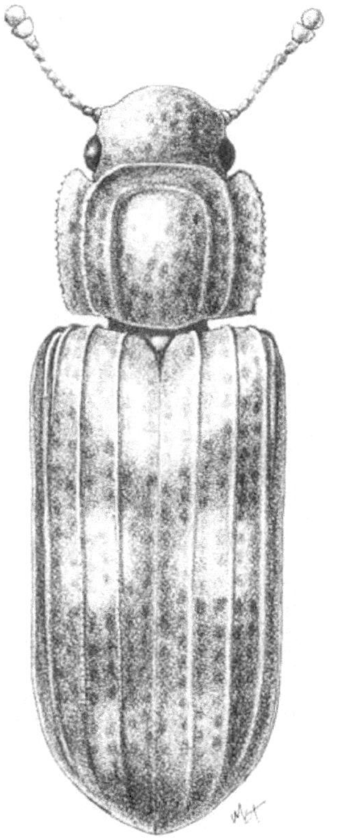

FIGURE 1.103. *Bitoma quadriguttata* (Say) (from Stephan 1989)

Description: (Adapted from Lawrence 1991, Slipinski and Lawrence 1997, 1999, and Lawrence *et al.* 1999a, b) Convex to strongly flattened, elongate cylindrical to depressed and parallel-sided, or broadly oval; 1.2 to15 mm in length; color usually brown to black, occasionally with yellow, red or green shades, sometimes maculate; glabrous or variously covered in sparse to dense setae, often modified into scales or bristles.

Head usually visible from above; eyes present or absent, entire or deeply emarginate; antenna with ten or eleven antennomeres, with a one, 2- or 3-segmented, weak to strong club; inserted under frontal margin; often with short to long antennal groove on gena below ventral margin of eye. Mouthparts highly variable, depending on food utilized. Maxillary palps 4-segmented, labial palps 3-segmented, last segments variable in shape.

Pronotum quadrate, elongate or transverse; lateral margins obsolete, finely to strongly carinate, explanate, smooth or denticulate to elaborately produced; disc sharply or indistinctly carinate, tuberculate or smooth; surface smooth, punctate, rugose, or microtuberculate; hypomeron simple or with antennal grooves or cavities; prosternum long and broad in front, intercoxal process very narrow to broad, occasionally expanded posteriorly, rarely very strongly so; procoxal cavities open or less commonly closed behind by mediad extension of the hypomeron or laterad extension of intercoxal process.

Mesoscutellum visible or not. Mesocoxal cavities narrowly to broadly separated, closed laterally. Hind wing may be reduced or absent; venation variable, in its most fully developed form with normal transverse folds, well developed medial fleck, radial and wedge cells; anal lobe absent. Elytra entire, covering pygidium; plain, costate, carinate, tuberculate, and/or with up to 11 punctate striae; epipleuron complete to absent.

Tarsal formula 4-4-4, rarely 3-3-3 because of fusion of basal 2 tarsomeres (can be seen in cleared specimens with compound microscopy); tarsomeres simple or with first strongly lobed, or with 1-3 expanded and/or setose below. Tibiae usually slender, rarely expanded apically and armed with stout spine, or carinate. Trochanters usually strongly oblique, of the heteromeroid type, rarely transverse.

Abdomen with five ventrites; no, two or three basal ventrites connate. Male genitalia tenebrionoid, symmetrical, inverted parameres apically glabrous.

Larvae elongate, parallel-sided, subcylindrical or flattened; straight or slightly curved; lightly pigmented except for head, tergum 9, and asperites; vestiture of simple, scattered setae. Head exserted, prognathous; somewhat flattened or subglobular, transverse to slightly elongate; epicranial stem usually short, occasionally long or absent; frontal arms lyriform or V-shaped. Five stemmata arranged in 2 groups with 3 anterior and 2 posterior or reduced to 3 (2 and 1) or none. Antennae 3-segmented and well developed or rarely very short and reduced to two segments. Prothorax sometimes enlarged. Dorsum of mesothorax metathorax and or anterior abdominal segment sometimes with rows or patches of asperites. Legs well developed, 5-segmented, tarsal claw with 2 setae. Tergum 9 variously ornamented with granules or tubercles and pair of urogomphi, occasionally reduced. Spiracles annular-biforous or apparently annular.

Acknowledgments. I want to thank S. Adam Slipinski for fostering and mentoring my interest in colydiids, and for his long discussions of the various problems with this group, as well as permission to use Figs. 15 and 18. He, John F. Lawrence and Karl Stephan have all provided extensive data, discussion and specimens critical to the production of this chapter. Albert Allen provided data on the *Pseudocorticus* sp. from his collection and Fig. 11. Michael C. Thomas provided Figs. 1-10, 12-14, 16-17 and 19-25. Richard S. Miller, J. Joseph Giersch, Paul E. Skelley, and Katharine Marske all provided help with the text, keys and/or figures.

Habits and habitats: Adult colydiids are usually taken under bark or at light, but may also be taken on bark, in the galleries of ambrosia beetles or other wood boring beetles, by beating dead branches and twigs, or by Berlese funnels from forest duff. Larvae are found in a variety of habitats, depending on the group involved. The ancestral food seems to be fungi, with several independent moves to predation on insect larvae and perhaps some to living plant tissue.

The larvae of the Adimerini are known only from a single specimen found in association with adults in the pith of a Florida milkweed, *Cynanchun scoparium* Nuttall (Asclepidaceae) (Craighead 1920). Since then, numerous additional collections of adults have been made in this habitat (Stephan 1989). It is not known if the larva was feeding on the plant or fungi therein, but hundreds of adult *Monoedus* have been taken in the West Indies by beating dead twigs and grasses infested with molds.

Among the Colydiini, *Colydium* and *Aulonium* have each been associated with both fungivory and predatory habits. *Aulonium* commonly occur in bark beetle galleries in association with both fungi and larvae, and one North American species has been directly observed to feed on scolytine larvae, but the same species has also been reared entirely on rotten plant material (Craighead 1920, Marshall 1978, Lawrence 1991). Podoler *et al.* (1990, and references therein) have found *Aulonium* to be important scolytine predators in Israeli pine plantations. In the case of *Colydium*, the evidence for predatory habits is entirely circumstantial, with larvae possibly occasionally dining on scolytine eggs or young larvae that happen to be on the fungi or rotting wood on which they are feeding (Lawrence 1991, Wegrzynowicz 1999).

Known larvae of Nematidiini are found in platypodine ambrosia beetle galleries, where they feed on the larvae (Beeson 1941, Roberts 1977, Lawrence *et al.* 1999a). The long, thin cylindrical body form is well adapted to the tunnels of their prey.

Rhagoderini are usually taken with pitfall traps in deserts along the Mexican border, south to the state of Puebla. The larvae of the isolated genus *Rhagodera* are unknown, and their discovery would solve one of the more interesting mysteries left among North American Coleopterology.

The huge tribe Synchitini are mostly associated with rotting wood and bark or fungus-ridden duff. Most Synchitini probably get their nutrition from fungi, both Ascomycetes and Basidiomycetes (Lawrence 1991). They may feed directly on fungal masses or fruiting bodies, or on rotting plant material containing fungi. Among the genera recorded from North America, *Bitoma*, *Synchita*, and *Namunaria* have specifically been associated with fungi (Lawrence 1991). Most other North American genera have some indirect reference to fungi associated with collection records, such as being found under bark of rotting trees, in white rot wood, etc. (Stephan 1989). *Colobicus parilis* occurs in commercial shipments and stores of sweet potatoes and other root and fruit crops, where they are implicated in the spread of a fungal disease (Hinton 1945), probably the true food source for the beetles. Some *Lasconotus* spp. are known scolytine predators, although early instars retain the fungivorous habit (Hackwell 1973).

Among the exotic tribes, the known Sarrotriini feed on lichens and mosses (Crowson 1984, Slipinski and Lawrence 1997). The cylindrical Gempylodini live in ambrosia beetle galleries, and at least some larvae seem to be at least facultatively predaceous, but have the mandibles of a fungivore (Lawrence 1980, Slipinski and Lawrence 1997, Lawrence *et al.* 1999a, b). Adult Acropini are surface active, with very large eyes. They are taken by beating hanging dead branches, twigs and leaves, or on the surface of tree trunks, and may be associated with microfungi or molds. Their larval biology is unknown, and no larvae have been described.

Status of the classification. The group recognized here is composed of the genera included in the Colydiinae of the Zopheridae by Slipinski and Lawrence (1999). The composition of this group has been changed almost beyond recognition in the last 35 years. Exactly half of the 36 genera included here by Arnett (1968) have been removed to other families. Of the seven Hawaiian species recorded by Ford (1968), only four remain in the family, the others departing for three different families. Removal of the Cerylonidae (Crowson 1955, Sen Gupta and Crowson 1973, Lawrence and Stephan 1975), Bothrideridae (Lawrence 1980, Pal and Lawrence 1986, Slipinski *et al.* 1989) and Pycnomerini (to the Zopheridae, Slipinski and Lawrence 1999) have been the greatest changes, but Ivie and Slipinski (1990), in their catalog of world genera, noted members of 11 other families that were removed from the colydiids since the last world catalogs (Hetchko 1926, 1930). What remains is a far more focused but still not demonstrably monophyletic family. Slipinski and Lawrence (1999) included this group as one of two subfamilies in their Zopheridae, but only the zopherine lineage (including the monommatines and pycnomerines) is well-supported as monophyletic.

The Colydiidae are arrayed in eight tribes, following the treatment of Slipinski and Burakowski (1988) as modified by Slipinski and Lawrence (1997), with nearly 140 recognized genera (Ivie and Slipinski 1990, Slipinski and Lawrence 1997). In the text of their 1999 work (p. 4), Slipinski and Lawrence indicated their intent to reduce the number of tribes to four, but did not follow this in their list of taxa studied. Therefore, the 1997 system is retained here. This tribal system is far from satisfactory, as it splits off seven small groups with obvious synapomorphies, leaving most genera and species behind in the Synchitini as a heterogeneous and unnatural amalgamation. Three of the tribes are monogeneric, four others together total only 19 genera, leaving the Synchitini with about 115 genera — over 80% of the total. A synopsis of the tribes follows.

Acropini Sharp 1894 include some of the most spectacular of the Coleoptera, with somewhat to greatly enlarged and swept-back eyes and a narrow mouth creating a characteristic heart-shaped head. The four genera are all Neotropical, and are under revision by Ivie and Slipinski. Slipinski and Lawrence (1997) suggested that with more study this tribe may be combined with the Synchitini.

Adimerini Sharp 1894 are Neotropical, with a monotypic genus in South America, and the large widespread genus *Monoedus* concentrated in the West Indies. The greatly enlarged first tarsomere often hides the small second and third, and the species have an indistinct 1-segmented antennal club.

Colydiini Erichson 1842 include four genera, two of which occur in North America. It is distributed throughout the New World except the far north and Chile. They share a long basal tarsomere, and elongate, cylindrical body with the Nematidiini and Gempylodini, but have sublateral carinae on the pronotum, protibiae that are widened and spinose apically, a 3-segmented antennal club, and in most genera the procoxal cavities are closed.

Gempylodini Sharp 1893 are a tropical group of six genera, ranging across Africa, Asia, the Indo-Pacific, Australia and the Neotropics. Another long, slender cylindrical group, with long first tarsomeres, their prosternal process is very narrow between the coxae, and expanded behind to broadly close the coxal cavities. The tribe was reviewed by Lawrence (1980) in his description and revision of *Pseudendestes* Lawrence 1980, with Slipinski and Pal (1997) following up with a review of *Aprostoma* Guérin-Méneville 1939. Slipinski is revising the remaining genera.

Nematidiini Sharp 1894 are a monogeneric tribe from the New World and Indo-Australian regions. As the name implies, they are long, thin and cylindrical. Nematidiini are characterized by a 2-segmented club on an 11-segmented antenna; dorsally visible mandibular bases; no supraorbital ridges; a broad, flat prosternal process; broadly closed procoxal cavities; a reduced, fine lateral margin on the pronotum; and long first tarsomeres. The single genus has only a few described species, and needs desperately to be revised on a world scale, the most recent work being that of Pal and Slipinski (1984) on the Old World forms.

Rhagoderini LeConte and Horn 1883 are a monogeneric tribe endemic to North America (including Mexico). They are fully wingless, have relatively narrow hind coxae, and an 11-segmented antenna with an indistinct 3-segmented club. The rhagoderine membership in this family is uncertain, and discovery of a larva would be important to settling it either here or in the Zopheridae (= Zopherinae of Slipinski and Lawrence 1999). The genus is in desperate need of a revision.

Rhopalocerini Reitter 1911 are another monogeneric tribe, this one occurring in Europe, Africa, Madagascar and the East Indies. They are characterized by the reduced first tarsomere, and a raised metasternal tooth in front of the metacoxa, as well as the nearly moniliform, scaled antennae with a 1-segmented club. The 17 species of *Rhopalocerus* Redtenbacher 1842 were revised by Slipinski and Burakowski (1988).

Sarrotriini Billberg 1820 include three Palearctic genera, characterized by a unique, apparently 10-segmented antenna in which the actual eleventh segment is enclosed in the tenth, and together appearing to be a weakly 1-segmented club. The reduced wings; scales on the antenna; broad, apically emarginate prosternal process; and open procoxal cavities help separate this group. Dajoz (1977) can be used to key members of this group.

Synchitini Erichson 1845 are virtually cosmopolitan, and contain the remaining 115 or so genera. Its members are mostly defined by not belonging to any of the other tribes. The need for phylogenetic work on this group cannot be overstated, including the need to evaluate at least the Acropini, Colydiini, and Adimerini as potential members of a clade that includes this group.

Five of these tribes and 26 genera occur in North America. Slipinski and Burakowski (1988) key the world tribes. Except for the Gempylodini, Sarrotriini and monogeneric tribes there are no world keys for most genera or species, but Burakowski and Slipinski (1986) provide an excellent guide to the Polish species, Dajoz (1977, 1980) covers the Palearctic and Malagasy faunae, Stephan (1989) the Nearctic, Hatch (1961) and Downie and Arnett (1996) treated the Northwest and Northeast Nearctic, Pope (1961) the African, and Slipinski and Lawrence (1997) the Austro-Pacific fauna. The Asian and Neotropical Synchitini species are the biggest problem, and much descriptive and phylogenetic work remains to be done on this group world-wide.

Larvae are known for only a fraction of the genera, but at least one is described for all tribes except Acropini and Rhagoderini, although that of the Sarrotriini is inadequate. More larval associations will provide critical data for resolving the relationships of this group. Works useful for larval (and occasionally pupal) identification include Craighead (1920), Bøving and Craighead (1931), Hayashi (1972), Nikitsky and Belov (1980), Lawrence (1980, 1991 and references therein), Burakowski and Slipinski (1986), Costa *et al.* (1988), Slipinski and Burakowski (1988), and Lawrence *et al.* (1999b).

Distribution. Colydiidae occur throughout forested areas of the world, including many oceanic islands. Although many species occur in the tropics, the temperate zones are also well represented. Temperate New Zealand is particularly well represented, with over 200 known species in 25 genera, many of them endemic. The 73 species of Colydiidae that have been reported in North America are known to occur in every state, province and territory of North America north of Mexico expect Alaska, Yukon, Nunavut, Nova Scotia, Prince Edward Island, the mainland portion of Newfoundland (Labrador), and Saint-Pierre et Miquelon.

Key to the North American and Hawaiian Genera

1\. Margin of frons arcuate, exposing lateral corner of mandibular base in frontal view; body long, thin, cylindrical; pronotum excavate laterally (Fig. 2) .... *Nematidium*

— Mandibular bases concealed by frons; body and pronotal shape variable 2

2(1). Apex of protibia expanded, armed with stout apical spine; procoxal cavities closed by mediad extension of hypomeron; antenna with distinct 3-segmented club 3

— Apex of protibia only rarely expanded and armed; procoxae open, narrowly closed or closed by laterad extension of intercoxal process; antenna with 1-, 2- or very weak 3-segmented club 5

3(2). First tarsomere at least 1.5 times as long as second; lateral margin of frons extending into eye as a canthus; eye emarginate (Colydiini) 4

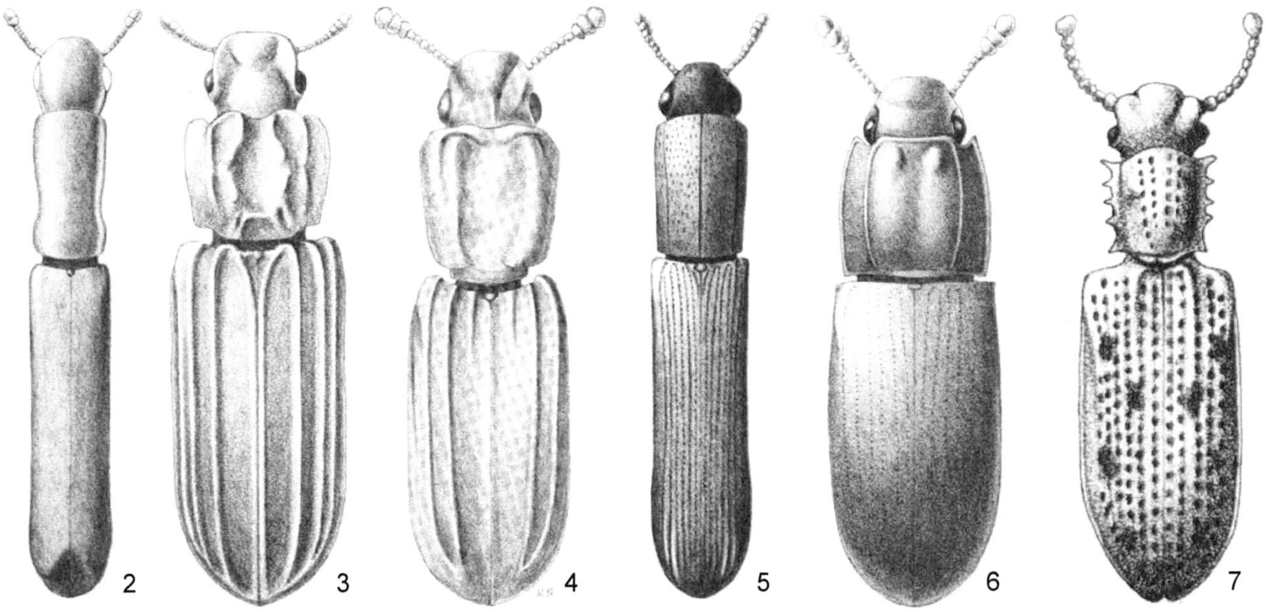

FIGURES 2.103-7.103. Dorsal habitus. 2. *Nematidium filiforme* LeConte; 3. *Lasconotus complex* LeConte; 4. *Lasconotus laqueatus* LeConte; 5. *Colydium lineola* Say; 6. *Aulonium parallelopipedum* (Say); 7. *Monoedus guttatus* LeConte (all from Stephan 1989).

— First tarsomere not or little longer than second; lateral margin of frons continuing above eye as supraorbital carina; eye round (Fig. 3, 4) (Synchitini, in part) .... *Lasconotus*

4(3). Body elongate cylindrical; elytra with alternate intervals strongly carinate, especially on declivity; pronotum without distinct, strong carinae except laterally; last ventrite with pair of long setae on hind margin (Fig. 5) .... *Colydium*

— Body broad or depressed-cylindrical; elytra with flat intervals, with at most a subhumeral groove; pronotum with strong sublateral carinae, disk variously ornamented by tubercles, striae or carinae; last ventrite without pair of long setae (Fig. 6) .. .. *Aulonium*

5(2). First tarsomere large, with greatly expanded lobe below that engulfs very small second and larger third tarsomeres; lateral margins of pronotum denticulate; antenna with 1-segmented club; length less than 3 mm; south Florida (Fig. 7) .... *Monoedus*

— First tarsomere normal, not lobed beneath; lateral margins of pronotum variable; antenna with 1-, 2- or 3-segmented club 6

6(5). Antennomeres 9-11 forming indistinct 3-segmented club; antennomeres 1-8 covered in dense scale-like setae; antennomere 11 asymmetrical; hind wings absent, metasternum very short relative to length of metafemur; body length 7 mm or greater; Mexican border states (Figs. 25, 26) .... *Rhagodera*

— Antenna with distinct 1- or 2-segmented club; antenna not densely-scaled, surface visible; other characters not in combination (Synchitini)...... 7

7(6). Antennal club clearly 2-segmented 8

— Antennal club 1-segmented (may appear to be 2 connate antennomeres) 20

8(7). Hypomeron with deep, margined pocket to receive antenna (Fig. 8) .... *Megataphrus*

— Hypomeron with at most shallow antennal depressions 9

9(8). Pronotum with obvious mid-lateral secretory pore (specimen MUST be clean); body elongate cylindrical; first antennomere of club cup-like, receiving second; pronotum never carinate 10

— Pronotum without visible mid-lateral secretory pores; antennomeres of club separate; body form variable 11

10(9). Pronotum with deep, wide, smooth-bottomed lateral-longitudinal canals associated with mid-lateral secretory pore; pronotum with pair of anterior pronotal horns extending above head; elytra glabrous, with upturned apical margin; Florida, Hawaii and Neotropics (Fig. 9) .... *Lobogestoria*

— Medio-lateral secretory pore without associated linear canal, laterally with enlarged microtubercles; pronotum without horns; elytra with scale-like setae and simple apex; Hawaii and Neotropics . ... *Neotrichus*

11(9). Procoxal cavities closed by lateral extension of apically expanded prosternal process (Fig. 10) .... *Namunaria*

— Procoxal cavities open, sometimes narrowly so... 12

12(11). Pronotum simple or rugose, elytra simple or tuberculate, without carinae 13

— Pronotum with longitudinal, straight or sinuate carinae or tuberculate ridges; elytra variably ornamented 16

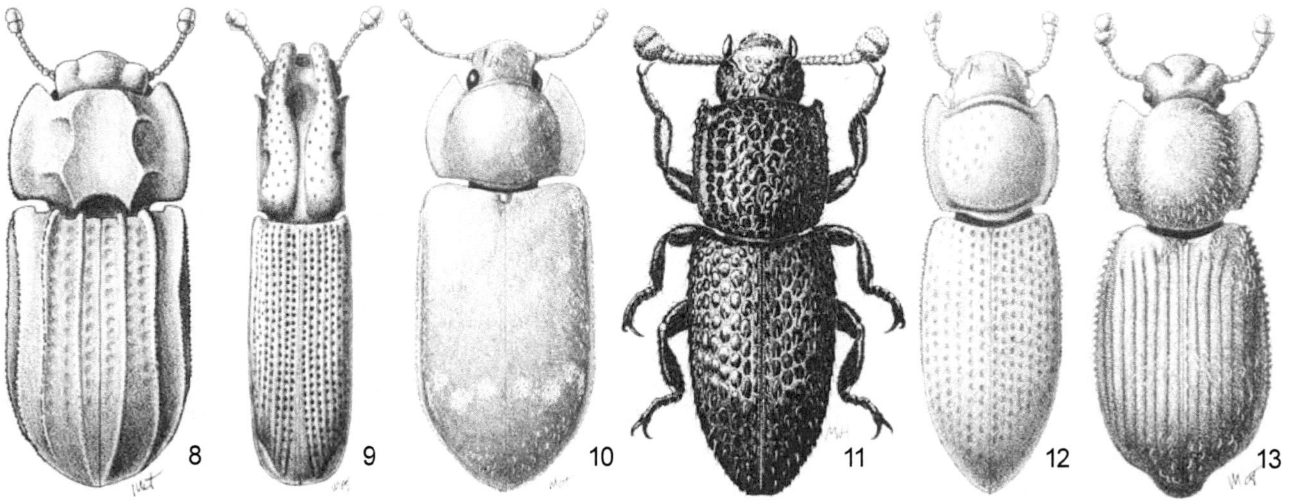

FIGURES 8.103-13.103. Dorsal habitus. 8. *Megataphrus tenuicornis* Casey; 9. *Lobogestoria gibbicollis* Reitter; 10. *Namunaria guttulata* (LeConte); 11. *Lyreus alleni* Ivie and Slipinski; 12. *Stephaniolus longus* (Stephan); 13. *Coxelus serratus* Horn (Figure 11 compliments A. Allen, others from Stephan 1989)

13(12). Elytra with hair-like setae 14
— Elytra with scale-like setae, scales narrow to wide 15

14(13). Eyeless; Alabama (Fig. 11) .... *Lyreus*
— Eyes well developed; Arizona (Fig. 12) .... *Stephaniolus*

15(13). Pronotum widest behind middle, metasternum normal, more than half as long as metafemur, winged; eyes large .... *Colobicus*
— Pronotum widest before middle; metasternum short, less than half length of metafemur; wingless, eyes small (Fig. 13) .... *Coxelus*

16(12). Lateral margins of frons continuous with supraorbital carinae or ending at front of eye; eye round; pronotal sculpture and antennal groove variable 17
— Lateral margins of frons extending into eye at mid-eye level to end as a canthus, not continuous with supra-ocular carina; eye strongly emarginate; median pair of pronotal carinae diverging medially, encircling mid-discal area, and forked behind this encirclement; antennal groove moderately long (Fig. 14) .... *Phloeonemus*

17(16). Antennal groove long, extending nearly to back of eye 18
— Antennal groove short, not reaching posterior half of eye 19

18(17). Elytra striate-punctate, with simple carinae on alternate intervals; length less than 4 mm; southern California (Fig. 15) .... *Micropriu*s
— Elytra striae interrupted by sinuate irregular blunt carinae; length 6 mm or more; northern California and southern Oregon (Fig. 16) .... *Denophoelus*

19(17). Pronotum with parallel pairs of longitudinal carinae; setation on elytra indistinct, narrow (Fig. 1) .... *Bitoma*
— Pronotum with rounded complex ridges; setation of elytra scale-like, forming light and dark maculations (Fig. 17) .... *Eudesma*

20(7). Pronotum with obvious mid-lateral secretory pores and deep, smooth bottomed linear canals (specimen MUST be clean); body elongate cylindrical; pronotum not carinate; Hawaii (Fig. 18) .... *Antilissus*
— Pronotum without visible mid-lateral secretory pores, often carinate, sometimes microtuberculate; body form variable; widespread 21

21(20). Elytra with scutellary striole, sutural (first) interstria diverging from suture near base (Fig. 19) .... *Acolobicus*
— Elytra lacking scutellary striole; sutural (first) interstria coincident with suture to base 22

22(21). Dorsal vestiture indistinct; elytra carinate or with elongate tubercles (Fig. 20) .... *Paha*
— Dorsum with obvious setae; elytra simple, with round tubercles, or irregular depression 23

23(22). Elytral setae hair-like; third antennomere 3-times as long as wide, as long as antennomeres 4-6 or 1-2 together (Fig. 21) .... *Endeitoma*
— Elytral setae scale-like; third antennomere less than 3-times as long as wide 24

24(23). Pronotum distinctly wider anteriorly than basally; antennae with scale-like setae; eyes densely scaled; antennae longer, reaching approximately to midway point of pronotum ...... *Pseudocorticus*
— Pronotum quadrate or wider at base than anteriorly; antennal setation not scale-like; eye scales difficult to see; antenna short, not reaching past anterior third of pronotum 25

25(24). Antennal groove long, curved, and reaching far behind eye (Fig. 22) .... *Eucicones*
— Antennal groove short and straight to obsolete 26

26(25). Elytra solid reddish-brown, not patterned (Fig. 33) .... *Synchita*
— Elytra patterned with light and dark maculations (Fig. 24) .... *Microsicus*

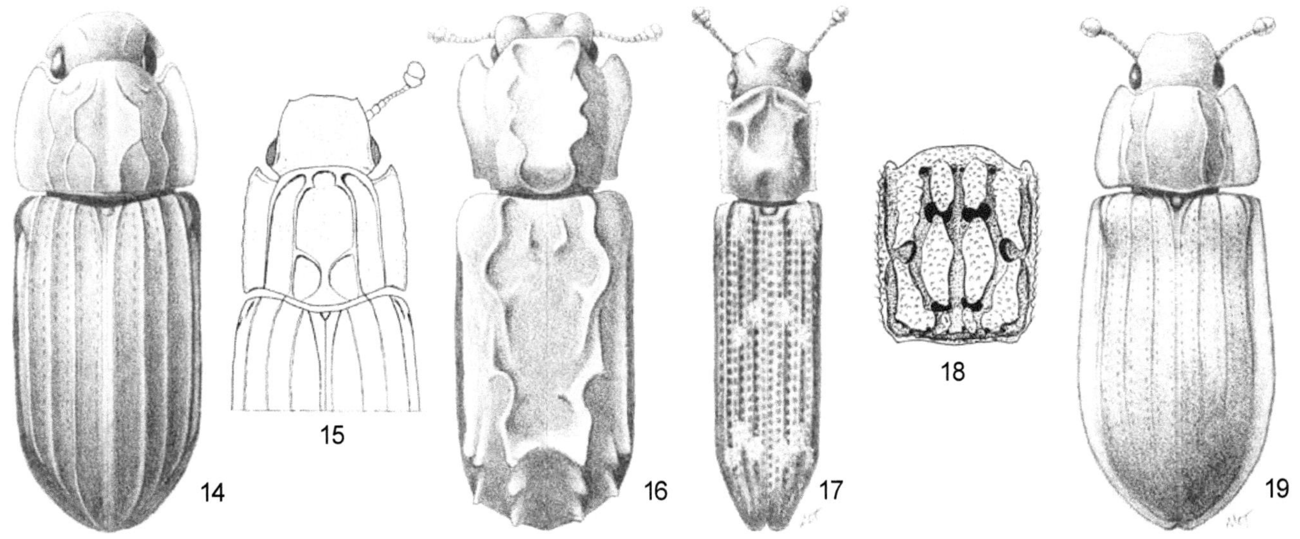

FIGURES 14.103-19.103. Dorsal views. 14. *Phloeonemus catenulatus* Horn, habitus; 15. *Microprius rufulus* (Motschulsky), head and pronotum; 16. *Denophoelus nosodemoides* (Horn), habitus; 17. *Eudesma undulata* Melsheimer, habitus; 18. *Antilissus aper* Sharp, pronotum; 19. *Acolobicus erichsoni* (Reitter), habitus (Figures 15 and 18 compliments S. A. Slipinski, others from Stephan 1989).

Classification of the North American and Hawaiian Genera

Colydiidae Erichson 1845

Except as noted below, the most recent North American key to species for each genus is Stephan (1989). Only synonymous names changed since that work or Ivie and Slipinski (1990) are listed.

Adimerini Sharp 1894

Monoedini Schaeffer 1911

Monoedus Horn 1882. *Monoedus guttatus* (Horn 1882) is the single representative of this mostly West Indian genus to reach our area. It occurs in Florida in the pith of *Cynanchum scoparium* Nuttall (Asclepidaceae) stems. This genus is under revision by Slipinski and Ivie.

Adimerus Sharp 1894

Colydiini Erichson 1845

Aulonium Erichson 1845. Five North American species are generally distributed throughout forested North America as far north as British Columbia, Michigan and Massachusetts. Ivie *et al.* (2001b) synonymized the North American species described mistakenly by Dajoz (1992) from Chile. The genus occurs widely in the Americas and Eurasia.

Colydium Fabricius 1792. The five North American species are generally distributed throughout forested North America from southern Canada to California and Florida. Wegrzynowicz (1999) revised the world fauna in a beautifully illustrated paper, reporting 32 species distributed across the Americas and Western Palearctic. He synonymized the Arizona species described by Dajoz (1992).

Rhagoderini LeConte and Horn 1883

Rhagodera Mannerheim 1843. Four described North American species are known from Arizona, southern California and Texas. This flightless genus forms many distinctive local populations. It is best represented in the deserts of northwest Mexico, but ranges as far south as Puebla.

Nematidiini Sharp 1894

Nematidium Erichson 1845. One species, *N. filiforme* LeConte 1863, of the Neotropical nominotypical subgenus, is known from the southeastern USA, as far north as Tennessee, west to Louisiana. The two subgenera were keyed and the old world members, belonging to the subgenus *Paleonematidium* Heinze 1954, were revised by Pal and Slipinski (1984). The genus was redescribed by Slipinski and Lawrence (1997), but is in great need of species-level revision.

Synchitini Erichson 1845

Acolobicus Sharp 1894. The validity of this genus is being evaluated as part of a study of New World Colydiidae genera by Slipinski and Ivie. What for now is known as *A. erichsoni* (Reitter 1837) is recorded only from South Carolina, Florida, and Mexico. It is part of a difficult group of Neotropical species that requires revision.

Antilissus Sharp 1879. This genus is known from one described species from Hawaii, *A. aper* Sharp 1879, and undescribed species

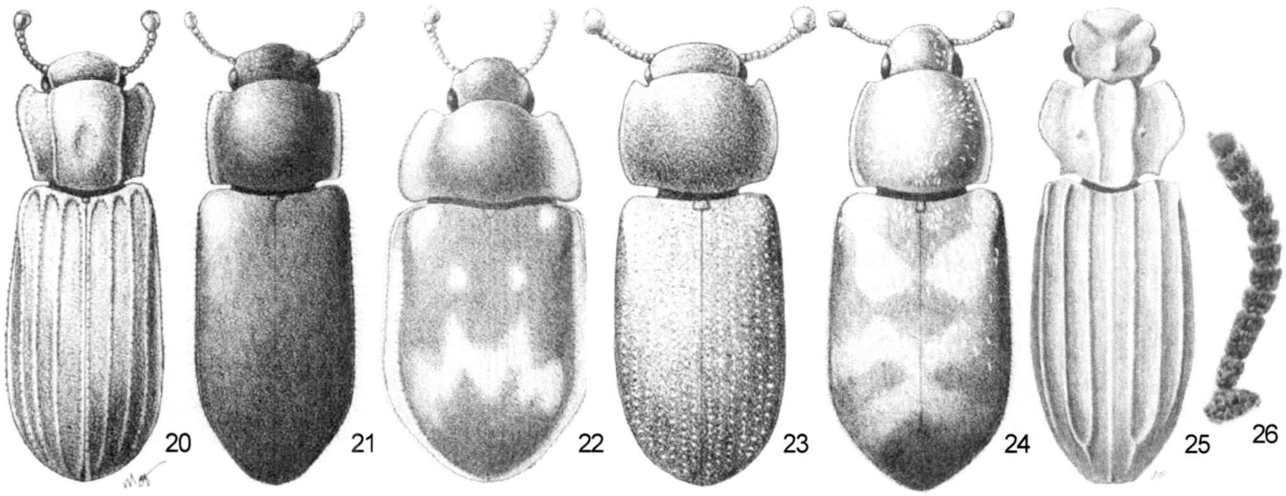

FIGURES 20.103-26.103. Dorsal habitus. 20. *Paha laticollis* (LeConte); 21. *Endeitoma granulata* (Say); 22. *Eucicones marginalis* (Melsheimer); 23. *Synchita fuliginosa* Melsheimer; 24. *Microsicus parvulus* (Guérin-Méneville); 25. *Rhagodera costata* LeConte. 26. *Rhagodera* sp., antenna. (Figures 20-25 from Stephan 1989).

from Indonesia and Australia (Ford 1968, Slipinski and Lawrence 1997).

Bitoma Herbst, 1793. This genus of over 100 species is in great need of revision. The 14 species recorded from North America and one from Hawaii are recorded from 37 states and 3 provinces, and probably occur in all 61. They include native and exotic species, and can be very abundant. Slipinski and Lawrence (1997) redescribed the genus, but its limits and the species placed here require extensive revisional work before a truly useful classification will be available.

Eulachus Erichson 1845. Lawrence and Slipinski (1997) returned this group to synonymy with *Bitoma*.

Colobicus Latreille 1807. *Colobicus parilis* Pascoe 1860 is an Old World native that has been moved widely through trade (Hinton 1945, Dajoz 1977, DeLoble and Tran 1993). It has recently been found to be established in Louisiana, and since it has been implicated in the spread of the fungal disease *Diplodia* (Coelomycetes) in sweet potatoes, may be of economic interest (see Ivie *et al.* 2001b). It also occurs in Hawaii (Ford 1968, Nishida 1992). This sizable genus occurs widely from Asia to Australia, and is in need of revision. Slipinski and Lawrence (1997) redescribed the genus.

Coxelus Dejean 1821. *Coxelus serratus* Horn 1885 from California is the sole New World member of this genus, which is widespread in the Old World.

Denophoelus Stephan 1989. *Denophoelus nosodermoides* (Horn 1878) from California and Oregon is the sole member of this genus.

Endeitoma Sharp 1894. This nearly cosmopolitan genus is represented in our region by two eastern species, ranging from Oklahoma and Delaware to Texas and Florida.

Eucicones Sharp 1894. *Eucicones marginalis* (Melsheimer 1846) is our only representative of this otherwise Neotropical genus. It ranges from southern Ontario to Florida, west as far as Kansas and Texas.

Eudesma LeConte 1863. The only species in this genus is *Eudesma undulata* (Melsheimer 1846), a rare species known only from Illinois, Indiana, Ohio and Pennsylvania. This genus is under study by M. Goodrich.

Eudesmula Cockerell 1906, unjustified emendation (see Ivie and Slipinski 1990).

Lasconotus Erichson 1845. This very difficult genus is one of the largest in the family, occurring virtually worldwide. There are 21 recognized species in North America, with records from coast to coast and from the Northwest Territories to the Mexican border. An additional undescribed species was reported from Colorado by Stephan (1989) under the name *Chrysopogonius* sp., a genus synonymized by Ivie and Slipinski (1990). Although that species is very distinctive from the other North American *Lasconotus*, on a world scale two genera cannot be maintained. In 1989, Stephan reported that this species would be described by John Kingsolver. Slipinski and Lawrence (1997) redescribed the genus, but a species-level revision is badly needed. *Lasconotus* is perhaps the best North American example of the problems with the current tribal structure. It has the pronotal structure of the classic Synchitini, but the tibia, antenna, procoxal cavities and biology of the Colydiini.

Chrysopogonius Hinton 1935, Ivie and Slipinski 1990 (synonymy).

Lobogestoria Reitter 1878. This monotypic genus is based on *Lobogestoria gibbicollis* Reitter 1878, a Neotropical species that reaches the southeastern USA, as far north as South Carolina.

Lyreus Aubé 1861. *Lyreus alleni* Ivie and Slipinski 2001 is a blind species known only from a sinkhole in Alabama. The other two known members of the genus occur in France and Sardinia (Ivie and Slipinski 2001).

Megataphrus Casey 1890. As currently defined, this genus contains three species from Oregon, California and Arizona. Other described species from the Neotropics will be moved into this genus by Slipinski and Ivie as part of our study of New World genera.

Microprius Fairmaire 1868. *Microprius rufulus* (Motschulsky 1863) is an Old World native established in California and the West Indies (see Ivie *et al.* 2001b).

Microsicus Sharp 1894. Three *Microsicus* species occur in North America, one in the west (Oregon, Idaho, California and Arizona), and the other 2 in the east, from New Jersey to Oklahoma, south to Texas and Florida. This genus is part of a worldwide complex that requires further study.

Namunaria Reitter 1882. Two species of this genus occur in our area, one each in the east (Ontario to North Carolina and Texas) and the west (British Columbia to California). Others occur in the Oriental and Australian regions. Slipinski and Lawrence (1997) redescribed the genus.

Neotrichus Sharp 1886. This nearly cosmopolitan genus is absent in Anglophonic North America, Europe and New Zealand, but does occur in Mexico near the northern border and in Hawaii, where *N. latiusculus* (Fairmaire 1881) is reported (Ford 1968, Nishida 1992, Slipinski and Lawrence 1997, Jamieson 1999).

Paha Dajoz 1984. The North American *P. laticollis* (LeConte 1863) is one of four described New World species that belong in this genus. It occurs from Oklahoma to New York, south to Texas and the Florida keys.

Phloeonemus Erichson 1845. In North America, two species of *Phloeonemus* are known from California, Arizona and Texas. This Neotropical genus is under revision by Ivie and Slipinski.

Pseudocorticus Hinton 1935. One specimen of this genus is known from Jeff Davis Co., Texas (Albert Allen collection). There is a single described species known from central Mexico, and the specific identity of the Texas specimen will require revisionary studies.

Stephaniolus Ivie, Slipinski and Wegrzynowicz 2001. A monotypic genus for *S. longus* (Stephan 1989) from the mountains of southeast Arizona.

Pseudotaphrus Stephan 1989, not Cossmann 1888 (Ivie *et al.* 2001a).

Synchita Hellwig 1792. *Synchita fuliginosa* Melsheimer 1846, widespread in eastern North America, is the only recognized North American species of this large and difficult genus. Schuh (1998) listed 43 described species in the nearly world-wide *Synchita*, and divided them into 5 species groups. *Synchita fuliginosa* was placed in the *S. humeralis*-species group.

Bibliography

ARNETT, R. H., Jr. 1968. Beetles of the United States (A Manual for Identification). American Entomological Institute. Ann Arbor, MI. 1112 pp.

BEESON, C. F. C. 1941. The Ecology and Control of Forest Insects. Forest Research Institute, Dehra Dun, and Imperial Forestry Institute. Oxford. 1007 pp.

BØVING, A. G. and F. C. CRAIGHEAD. 1931. An illustrated synopsis of the principal larval forms of the order Coleoptera. Entomologica Americana (New Series), 11: 1-351.

BURAKOWSKI, B. and S. A. SLIPINSKI. 1986. Gwczdnikowate – Colydiidae, Bothrideridae, Cerylidae, Anommatidae. *In*: Klucze do Oznaczania Owadów Polski. Czesc XIX Chrzaszcze – Coleoptera (Zeszyt 59). Polskie Towarzystwo Entomologiczne, Warszawa-Wroclaw. 86 pp.

COSTA, C., S. A. VANIN and S. A. CASARI-CHEN. (1988). Larvas de Coleoptera do Brasil, Museu de Zoologia, Universidade de São Paulo: São Paulo. vi + 282 pp.

CRAIGHEAD, F. C. 1920. Biology of some Coleoptera of the families Colydiidae and Bothrideridae. Proceedings of the Entomological Society of Washington, 22: 1-13.

CROWSON, R. A. 1955. The natural classification of the families of Coleoptera. Nathaniel Lloyd. London, 187 pp.

CROWSON, R. A. 1984. The associations of Coleoptera with Ascomycetes, Pp. 256–285. *In*: Q. D. Wheeler and M. Blackwell, eds., Insect/Fungus Relationships: Perspectives in Ecology and Evolution. Columbia University Press. New York.

DAJOZ, R. 1977. Coléoptères Colydiidae et Anommatidae Paléarctiques. Faune de l'Europe et du Bassin Méditerranéen. Masson, Paris. 8: i-vi, 1-280.

DAJOZ, R. 1980. Insectes Coléoptères: Colydiidae et Cerylonidae. Faune de Madagascar, 54: 1-256.

DAJOZ, R. 1992. Description de trois nouvelles espèces de Colydiidae de l'Arizona et de Californie (Coleoptera). Bulletin Mensuel de la Société Linnéenne de Lyon 61: 60-66.

DELOBLE, A. and M. TRAN. 1993. Les Coléoptères des denrées alimentaires entreposées dans les régions chaudes. Faune tropicale. Orstrom, Paris. 32: 1-425.

DOWNIE, N. M. and R. H. ARNETT, Jr. 1996. The Beetles of Northeastern North America. Vol. 2. The Sandhill Crane Press. Gainesville, FL. 891-1721 Pp.

FORD, E. J., Jr. 1968. Colydiid beetles of Hawaii, with the description of a new species. Pacific Insects, 10: 161-165.

HACKWELL, G. A. 1973. Biology of *Lasconotus subcostulatus* (Coleoptera: Colydiidae) with special reference to feeding behavior. Annals of the Entomological Society of America, 66: 62–65.

HATCH, M. H. 1961. The beetles of the Pacific Northwest. Part III: Pselaphidae and Diversicornia I. University of Washington Publications in Biology, 16: 1-503.

HAYASHI, N. 1972. On the larvae of some species of Colydiidae, Tetratomidae and Aderidae occurring in Japan (Coleoptera: Cucujoidea). Kontyû, 40: 100-111.

HETSCHKO, A., 1926. Pars 83. Thorictidae, Catapochrotidae, Monoedidae, Synteliidae, Cossyphodidae. *In*: S. Schenkling, ed., Coleopterorum Catalogus. W. Junk, Berlin, 15 pp.

HETSCHKO, A. 1930. Colydiidae. *In*: W. Junk and S. Schenkling, eds., Coleopterorum Catalogus, Pars 107. W. Junk, Berlin. 124 pp.

HINTON, H. E. 1945. A monograph of the beetles associated with stored products. Volume 1. British Museum of Natural History. London. 440 pp.

IVIE, M. A. and S. A. SLIPINSKI. 1990. Catalog of the genera of world Colydiidae (Coleoptera). Annales Zoologici, 43 (Suppl. 1): 1-32.

IVIE, M. A. and S. A. Slipinski. 2001. A new species of *Lyreus* Aubé from Alabama, first report of the genus from the New World (Zopheridae: Colydiinae: Synchitini). Coleopterist Bulletin 55: (in press).

IVIE, M. A., S. A. SLIPINISKI and P. WEGRZYNOWICZ. 2001a. Generic homonyms in the Colydiinae (Coleoptera: Zopheridae). Insecta Mundi 15: (in press).

IVIE, M. A., S. A. SLIPINISKI and P. WEGRZYNOWICZ. 2001b. New records and synonyms in the colydiinae. Insecta Mundi 15: (in press).

JAMIESON, D. W. 1999. New arthropod records for Kaua'i. Bishop Museum Occasional Papers, 59: 19-26.

LAWRENCE, J. F. 1980. A new genus of Indo-Australian Gempylodini with notes on the constitution of the Colydiidae (Coleoptera). Journal of the Australian Entomological Society, 19: 293–310.

LAWRENCE, J. F. 1991. Colydiidae (Tenebrionoidea), Pp. 512-514. *In*: F. W. Stehr, ed., Immature Insects. Vol. II. Kendall Hunt: Dubuque, Iowa.

LAWRENCE, J. F., A. M. HASTINGS, M. J. DALLWITZ, T. A. PAINE and E. J. ZURCHER. 1999a. Beetles of the World: A Key and Information System for Families and Subfamilies. CD-ROM, Version 1.0 for MS-Windows. CSIRO Publishing. Melbourne.

LAWRENCE, J. F., A. M. HASTINGS, M. J. DALLWITZ, T. A. PAINE and E. J. ZURCHER 1999b. Beetle Larvae of the World: Descriptions, Illustrations, Identification, and Information Retrieval for Families and Sub-families. CD-ROM, Version 1.1 for MS-Windows. CSIRO Publishing. Melbourne.

LAWRENCE. J. F. and A. F. NEWTON, Jr. 1995. Families and subfamilies of Coleoptera (with selected genera, notes, references and data on family-group names), Pp. 779-1006. *In*: J. Pakaluk and S. A. Slipinski, eds., Biology, phylogeny and classification of Coleoptera: Papers celebrating the 80th birthday of Roy A. Crowson. Muzeum i Instytut Zoologii PAN, Warsaw.

LAWRENCE, J. F. and K. STEPHAN. 1975. The North American Cerylonidae (Coleoptera: Clavicornia). Psyche, 82: 131-166.

MARSHALL, J. E. 1978. The larva of *Aulonium trisulcum* (Fourcroy) (Coleoptera: Colydiidae) and its association with elm bark beetles (Scolytus spp.). Entomologist's Gazette, 29: 59-69.

NIKITSKY, N. B. and V. V. BELOV. 1980. Larvae of cylindrical bark-beetles (Coleoptera, Colydiidae) of the European part of the USSR and Caucasus, with comments on taxonomy of the family. 1. Zoologicheskii Zhurnal, 59: 1040-1053.

NISHIDA, G. M., ed. 1992. Hawaiian terrestrial arthropod checklist. Bishop Museum Technical Report, 1: i-viii and 1-262.

PAL, T. K. and J. F. LAWRENCE. 1986. A new genus and subfamily of mycophagous Bothrideridae (Coleoptera: Cucujoidea) from the Indo-Australian Region, with notes on related families. Journal of the Australian Entomological Society, 25: 185–210.

PAL, T. K. and S. A. SLIPINSKI. 1984. Notes on the *Nematidium* Erichson (Coleoptera, Colydiidae) with description of new species. Polskie Pismo Entomologiczne, 53: 531–543.

PODOLER, H., Z. MENDEL and H. LIVNE. 1990. Studies on the biology of *Aulonium ruficorne* Olivier (Coleoptera: Colydiidae). Environmental Entomology, 19: 1010-1016.

POPE, R. D. 1961. Colydiidae (Coleoptera Clavicornia). Exploration du Parc National de la Garamba. Mission H. de Saeger. Bruxelles. Fascicule 25, 115 pp.

ROBERTS, H. 1977. The Platypodidae (Coleoptera) of Fiji (with descriptions of two new species). Journal of Natural History, 11: 555-578.

SCHUH, R. 1998. Revision of the *Synchita variegata* species group (Coleoptera: Zopheridae: Colydiinae). Annales Zoologici, 48: 313-324.

SEN GUPTA, T. and R. A. CROWSON. 1973. A review of the classification of Cerylonidae (Coleoptera, Clavicornia). Transactions of the Royal Entomological Society of London, 124: 365-446.

SLIPINSKI, S. A. and B. BURAKOWSKI. 1988. A review of the genus *Rhopalocerus* W. Redtenbacher of the world (Coleoptera, Colydiidae). Annales Zoologici, 42: 75-118.

SLIPINSKI, S.A. and J.F. LAWRENCE. 1997. Genera of Colydiinae (Coleoptera: Zopheridae) of the Australo-Pacific region. Annales Zoologici, 47: 341–440.

SLIPINSKI, S. A. and J. F. LAWRENCE. 1999. Phylogeny and classification of Zopheridae *sensu novo* (Coleoptera: Tenebrionoidea) with a review of the genera of Zopherinae (excluding Monommatini). Annales Zoologici, 49: 1-53.

SLIPINSKI, S.A. and T. PAL. 1997. A review of the genus *Aprostoma* Guérin-Méneville (Coleoptera: Colydiidae). Annales Zoologici, 47: 93-105.

SLIPINSKI, S. A., R. D. POPE and R. J. W. ALDRIDGE. 1989. A review of the world Bothriderini (Coleoptera: Bothrideridae). Polskie Pismo Entomologiczne, 59: 131-202.

STEPHAN, K. H. 1989. The Bothrideridae and Colydiidae of America north of Mexico (Coleoptera: Clavicornia and Heteromera). Occasional Papers of the Florida State Collection of Arthropods , 6: xii + 65 pp.

WEGRZYNOWICZ, P. 1999. A revision of the genus *Colydium* Fabricius, 1792 (Coleoptera: Zopheridae: Colydiinae). Annales Zoologici, 49: 265-328.

104. MONOMMATIDAE Blanchard 1845

by Michael A. Ivie

Family common name: The monommatid beetles

Family synonyms: Monommatini *sensu* Slipinski and Lawrence 1999, Monommidae *auctorum*

This group of beetles is now considered a tribe of the Zopheridae (Chapter 105). The smooth, oval, convex form; vertical eyes that nearly meet dorsally; 5-5-4 tarsi; open procoxal cavities; four connate ventrites; oval convex shape; and clubbed antennae that are received in margined fossae on the hypomera will distinguish the North American members of this tribe. On a world level, the 5-5-4 tarsi, four connate ventrites, open procoxal cavities and antennal cavity with a unique carinately margined groove connecting the main fossa and anterior margin will set this tribe apart.

FIGURE 1.104. *Hyporhagus gilensis opuntia* Horn 1872

Description (extra-limital exceptions in brackets [ ]): (Modified from Lawrence 1982, Lawrence *et al.* 1999a, Slipinski and Lawrence 1999) (Figs. 1-5) Shape oval [oblong or parallel-sided], convex [or flattened] dorsally, flattened ventrally; smooth and lacking obvious vestiture [to scaled and tuberculate] above, size 2.3 to 12 mm in length; color black [brown, red or yellow], vestiture absent [or moderate to densely setose] above.

Head horizontal, prominent; eyes large, shape vertical, extending from ventral surface of head to nearly contiguous dorsally [or absent from dorsal surface or reduced to single facet]; antennal insertion concealed under the frontal margins, eleven antennomeres, oval flattened club of two or three antennomeres, received in hypomeral fossa. Underside of head strongly armored, mouthparts and base of antenna received in grooves around mentum, on gena and underside of mandibles. Labrum short, transverse, sometimes retracted, anterior margin fringed with setae; mandibles short; maxillae palpi with four palpomeres which are stout, the apical palpomere slightly expanded and truncate; mentum narrow, pentagonal, often with pits and grooves; ligula corneous, somewhat prominent; labial palpi with three palpomeres, the apical palpomere awl-shaped. Pronotum narrowed in front [rarely parallel-sided], lateral margins distinct, evenly arcuate, posteriorly as wide as the elytra; surface punctate; hypomeron grooved at anterior margin to receive antennae, groove opening into wide fossa that receives curled end of antenna. Prosternum short [to long], broad, with complex to simple sculpture; broad intercoxal process fitting into emargination of the mesosternum; procoxal cavities open both internally and externally; mesosternum broad, short, emarginate to receive prosternal process; metasternum large; metepimeron distinct. Procoxae globular with long internal processes; mesocoxae flat, widely separated; metacoxae flat, transverse, widely separated. Prolegs received in cavities involving prosternum, hypomeron and mesepisternm, mesolegs in cavities involving mesepisternm, metasternum, metepisternum and elytral epipleuron. Trochanters strongly oblique; femora constricted at base, excavate to receive tibia in retracted condition; tibiae slender to wide, flattened, excavate to receive tarsi to variable extent on pro- and mesolegs; tarsal formula 5-5-4, tarsi slender, claws simple. Scutellum small, triangular [or not visible]. Elytra smooth [rarely tuberculate or carinate], apically rounded, entire; striae present [or absent], epipleural fold moderate [to very wide], extending to the apex. Hind wings present [to absent], with normal transverse folds (Forbes 1926), well developed radial cell, lacking anal cell and median fleck.

Abdomen with five visible sterna, the first elongate, intercoxal process broad. Aedeagus tenebrionoid, symmetrical, inverted; penis long, thin, tubular; parameres large, fused, articulated to phallobase (Sharp and Muir 1912). Female genitalia poorly described (Tanner 1927).

Larvae (modified from Lawrence 1991, Lawrence *et al.*. 1999b) subcylindrical, slightly depressed; pale to reddish; size up to 20 mm, usually less than 10 mm in length; lightly sclerotized except for head and ninth tergite. Head prognathous, broad, flattened; frontal arms lyre-shaped, epicranial stem absent; five stemmata on each side; antennae three-segmented; clypeus and labrum distinct; mandible short, stout, symmetrical, with two or more teeth; ventral mouthparts retracted; maxillae with transverse cardo, elongate stipes, truncate, apically cleft mala, three-segmented palpi; labium free, mentum and ligula present, palpi two-segmented.

Legs short, spinose, five-segmented. All or part of metanotum and abdominal tergum 1 to 6 with transverse, straight to curved, single or double row of asperites on each side of midline. Tergum nine with pair of pigmented urogomphi separated by sclerotized pit; sternum nine simple. Spiracles annular-biforous.

Habits and habitats: (see Chapter 105). Monommatini are associated with rotting vegetable matter. Large numbers of the

Acknowledgments: I thank J. Joseph Giersch for Figs. 2-5, and Katharine Marske for help with the manuscript.

larvae of at least three genera have been repeatedly found (in South America, the West Indies and Madagascar) between the bark and wood of trees with wet to relatively dry rotting cambium (Ivie, pers. obs.). A few species have been reported from other rotting vegetation, such as papaya, *Euphorbia* and *Yucca* stems (Lawrence 1991), but all of these seem to be isolated cases with the commonality that all the "hosts" were rotting, indicating that the true nutrition is from the products of rot, probably fungi.

Status of the classification. This group is in fact a tribe of the Zopheridae (see Chapter 105). Historically, it has been uniformly treated as a family since it was described in 1845. Doyen and Lawrence (1979) pointed out its relationship to the Zopheridae and Colydiidae, and Lawrence (1994) suggested it formed a monophyletic group with the Zopheridae and Pycnomerini. Slipinski and Lawrence (1999) tested this idea with a phylogenetic analysis, and discovered that this group does indeed form a well-supported, if highly autapomorphic, member of the Zopheridae. The monommatines are treated in their own chapter only because Slipinski and Lawrence's work appeared too late to be used in the classification laid out in Volume 1 of this work, and they are maintained as a separate chapter only for consistency's sake.

Although the monophyly and position of the monommatines is reasonably secure, the generic and species level work is far from complete. The world monommatines have been revised by Heinz Freude (1955a, b, 1957, 1958), but the work was very sparsely illustrated. Freude has subsequently published two major updates on the New World fauna (Freude 1962, 1976) and more recently provided an updated key to the New World genera and species (Freude 1993). This key (Freude 1993) of 74 couplets to a group of very similar-appearing species has no figures, which is an indication of the difficulty encountered in the taxonomy of this group. Even more recently, he has described an additional species from Texas (Freude 2000).

Distribution. There are approximately 300 species described in 15 genera world-wide. In the New World there are 75 species in six genera, all endemic to that region. Diversity is highest in Madagascar, which has six genera. Three genera and six (maybe seven) species are recorded from the United States.

Key to the Nearctic Genera

1. Groove around dorsal edge of eye following facets throughout 2
— Groove around dorsal edge of eye encompassing a narrow cuticular process directed away from facets .... *Spinhyporhagus*

2(1). Antennal club with two antennomeres; antenna in cavity not covered by proleg in retracted condition; posterior carina of antennal fossa recurved dorsally to meet lateral margin of pronotum (Figs. 2-3) .... *Aspathines*
— Antennal club with three antennomeres; antenna in cavity covered by proleg in retracted condition; posterior carina of antennal fossa ending on hypomeron, not recurved to meet lateral margin of pronotum (Figs. 4-5) .... *Hyporhagus*

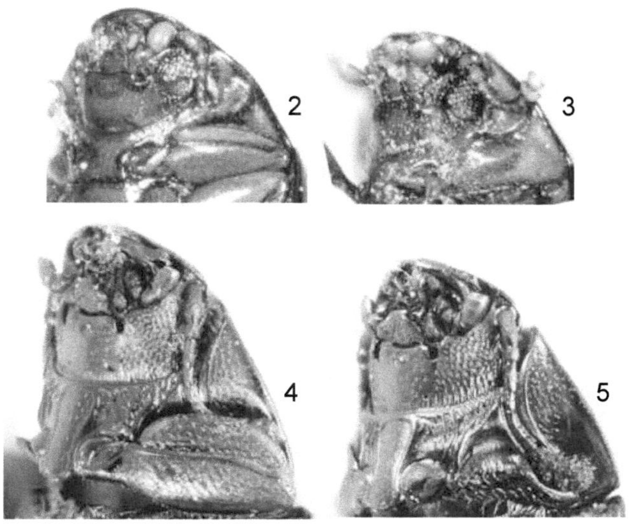

FIGURES 2.104-5.104. 2. *Aspathines aeneus* Thomson 1860, oblique ventral view of head and prothorax; 3. *A. aeneus*, oblique ventral view of head and prothorax with leg removed; 4. *Hyporhagus gilensis opuntia* Horn 1872, oblique ventral view of head and prothorax with leg in retracted position; 5. *H. g. opuntia*, oblique ventral view of head and prothorax with leg removed.

Classification of the Nearctic Genera

Monommatidae Blanchard 1845

Monommatini Blanchard 1845

Aspathines Champion 1888 is a currently monotypic genus for the Neotropical *A. aeneus* Thomson 1860, recorded from Florida and Mexico to Paraguay (Freude 1993). Although there are many distinct forms in the Greater Antilles and Neotropical mainland, Freude (1993) recognized only one species with three subspecies. The variation exhibited by this group cries out for a revisional treatment. The populations in southern Florida have been assigned to *A. aeneus ovatus* Champion 1888, which is also recorded from Mexico and Central America (Freude 1955b, 1993, Peck and Thomas 1998).

Hyporhagus Thomson 1860 is the largest genus in the New World, with 67 species and several subspecies recognized (Freude 1993). Freude (1993) reports 4 species from the southern United States: *H. opaculus* LeConte 1866 (three of the four subspecies in the US, occurring in all the states bordering Mexico); *H. punctulatus* Thomson 1860 (one of two subspecies occurs in Florida) (Peck and Thomas 1998) and Louisiana (E. G. Riley collection)); *H. gilensis* Horn 1872 (three of four subspecies, including *H. g. opun-*

tia Horn 1872 and *H. g. texanus* Linell 1899, occur in the border states, Nevada and Utah); *H. pseudogilensis* Freude 1955b (Texas and Arizona). He records a fifth species, *H. valdepunctatus* Thomson 1869 from simply "Nord America, ? Antillen" (Freude 1993). Thomson described it from "Amer. Merid. (Brésil?)" and Freude (1955) recorded it from the Antilles. Whether this species actually belongs to our fauna is unknown. *Hyporhagus leechi* Freude 1955, sometimes considered a California species, was described from Santa Catalina Island, Gulf of California. This is not the island of the same name in California's Channel Islands, but is a part of Mexico. Horn's (1872) treatment of the genus, although old, may still be of some use with this group.

Spinhyporhagus Freude 1984 is known from one species from central Brazil and *S. cuneispinatus* Freude 2000, was described from a single old specimen labeled simply "Texas." No specimens referable to this genus or species have been seen.

Bibliography

DOYEN, J. T. and J. F. LAWRENCE. 1979. Relationships and higher classification of some Tenebrionidae and Zopheridae (Coleoptera). Systematic Entomology, 4: 333-377.

FORBES, W. T. M. 1926. The wing folding patterns of the Coleoptera. Journal of the New York Entomological Society 34: 42-139, pls. 7-18.

FREUDE, H. 1955a. Die Monommiden der Welt: I. Teil: Die Monommiden der indo-australischen Region. Entomologische Arbeiten aus dem Museum G. Frey, 6: 1-73.

FREUDE, H. 1955b. Die Monommiden der Welt: II. Teil: Die Monommiden der amerikanischen Region. Entomologische Arbeiten aus dem Museum G. Frey, 6: 684-763.

FREUDE, H. 1957. Die Monommiden der Welt: III. Teil: Die Monommiden Madagaskars und der umliegenden Inseln. Entomologische Arbeiten aus dem Museum G. Frey, 8: 279-332, 560-608.

FREUDE, H. 1958. Die Monommidae der afrikanischen Region (Coleoptera). (IV. Teil der Monommiden der Welt mit Zusammenfassung der Ergebnisse). Annales du Musée Royal du Congo Belge, Tervuren (Série 8°: Sciences Zoologiques), 61: 1-115.

FREUDE, H. 1962. Nachtrag zur Monographie der Monommiden der amerikanishcen Region; Mitteilungen der Münchener Entomologischen Gesellschaft, 52: 111-114.

FREUDE, H. 1976. Monommidae aus dem Museu de Zoologia, Universidade de Sao Paulo, und der collection Alvarenga. Papéis Avulsos do Departamento de Zoología São Paulo, 30: 99-105.

FREUDE, H. 1993. Neue Monommidae und Epitragini (Tenebrioidae) des British Museum und eine Tabell der amerikanishen Monommidae. Spixiana, 16: 213-225.

FREUDE, H. 2000. Zur Monommatiden-Fauna der Afrotropis, Orientalis, Nearktis sowie Mittelamerikas mit Beschreibung von vier neuen Arten (Insecta, Coleoptera: Monommatidae). Mitteilungen aus dem Museum für Naturkunde in Berlin Zoologische Reihe, 76: 135-141.

HORN, G. H. 1872. Descriptions of some new North American Coleoptera. Transactions of the American Entomological Society, 4: 143-152.

LAWRENCE, J. F. 1982. Coleoptera. Pp. 482-553. *In*: S. P. Parker, ed., Synopsis and Classification of Living Organisms. Vol. 2. McGraw-Hill, New York.

LAWRENCE, J. F. 1991. Monommidae (Tenebrionoidea), Pp. 514-515. *In*: F. W. Stehr, ed., Immature Insects. Vol. II. Kendall Hunt. Dubuque, Iowa.

LAWRENCE, J. F. 1994. The larva of *Sirrhas variegatus*, sp. nov., with notes on the Perimylopidae, Ulodidae (stat. nov.), Zopheridae and Chalcodryidae (Coleoptera: Tenebrionoidea). Invertebrate Taxonomy, 8: 329–349.

LAWRENCE, J. F., A. M. HASTINGS, M. J. DALLWITZ, T. A. PAINE and E. J. ZURCHER. 1999a. Beetles of the World: A Key and Information System for Families and Subfamilies. CD-ROM, Version 1.0 for MS-Windows. CSIRO Publishing. Melbourne.

LAWRENCE, J. F., A. M. HASTINGS, M. J. DALLWITZ, T. A. PAINE and E. J. ZURCHER. 1999b. Beetle Larvae of the World: Descriptions, Illustrations, Identification, and Information Retrieval for Families and Sub-families. CD-ROM, Version 1.1 for MS-Windows. CSIRO Publishing: Melbourne.

PECK, S. B. and M. C. THOMAS. 1998. A distributional checklist of the beetles (Coleoptera) of Florida. Arthropods of Florida and Neighbouring Land Areas, 16: i-viii + 1-180.

SHARP, D. and F. A. G. MUIR. 1912. The comparative anatomy of the male genital tube in Coleoptera. Transactions of the Entomological Society of London, 1912: 477-642, pls. 42-78.

SLIPINSKI, S. A. and J. F. LAWRENCE. 1999. Phylogeny and classification of Zopheridae *sensu novo* (Coleoptera: Tenebrionoidea) with a review of the genera of Zopherinae (excluding Monommatini). Annales Zoologici, 49: 1-53.

TANNER, V. M. 1927. A preliminary study of the genitalia of female Coleoptera. Transactions of the American Entomological Society, 53: 5-50, pls. 2-15.

105. ZOPHERIDAE Solier 1834

by Michael A. Ivie

Family common name: The zopherid beetles

Family synonyms: Monommatidae Blanchard 1845, Monommidae *auctorum*, Pycnomerinae Erichson 1845

The four connate ventrites, antenna inserted under a frontal margin, and either 5-5-4 tarsi or 4-4-4 with metacoxae separated by greater than the width of a metacoxa, and the glabrous elytra will distinguish this family in North America.

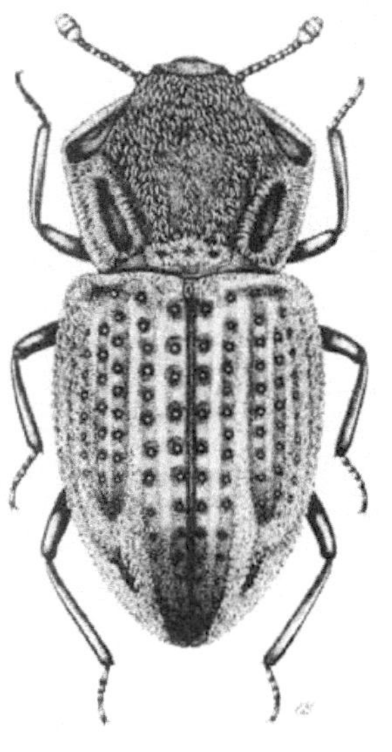

FIGURE 1.105. *Usechimorpha montanus* Doyen and Lawrence (from Doyen and Lawrence 1979)

Description: (Modified from Lawrence 1982, Lawrence *et al.* 1999a, Slipinski and Lawrence 1991) Elongate, flattened, parallel-sided to convex oval; glabrous to covered in setae or scales; smooth or tuberculate, carinae or pustulate, often with coating of secretory products; length 1.8 – 34 mm. Head deeply to weakly inserted into the prothorax. Eyes emarginate, round, reniform, vertical or reduced. Antennae 8-11 segmented, with weak to strong 1-3 segmented club; antennal insertions concealed. Maxillary palps variable from globular, subulate, rarely securiform or apically truncate. Submentum often with a median setose pit in males. Lateral edges of pronotum smooth or dentate, sometimes indistinct or absent. Hypomeron often with antennal grooves and/or fossae, or simple. Prosternum short to long in front of coxae, intercoxal process moderate to very broad, sometimes expanded at apex. Procoxae apparently globular, but with long internal extensions; procoxal cavities widely or narrowly open or closed externally, open internally. Mesocoxae moderately to very widely separated, cavities open or closed laterally. Elytral punctation sometime seriate; epipleura often broad and complete; sometimes narrow or indistinct. Hind wings often absent; when present with or without subcubital fleck. Tarsi 5-5-4 or 4-4-4; tarsomeres not lobed. Ventrites 1-3, 1-4, or 1-5 connate, rarely all free; aedeagus sometimes inverted.

Acknowledgments: For permission to use figures, I thank S. Adam Slipinski for Figs. 5-7, to Michael C. Thomas for Fig. 2, to Charles Triplehorn for Fig. 8. John Lawrence, Adam Slipinski, John Doyen and Charles Triplehorn have all patiently discussed problems with my understanding of this group over the years, and donated many specimens and reprints to make this chapter possible. Katharine Marske and J. Joseph Giersch helped with figures and the manuscript.

Larvae (modified from Dajoz 1977, Lawrence 1982, 1991a-c, Lawrence 1999b, Slipinski and Lawerence 1999) up to 45 mm in length. Elongate, subcylindrical to slightly flattened, usually lightly sclerotized except for part or all of head and A9 (may be just urogomphi or even just tips); rarely all visible tergites dark. Dorsal surface usually smooth, sometimes granulate, vestiture scattered simple setae.

Head protracted and prognathous, broad, slightly flattened; epicranial stem ranges from absent short to moderately long, frontal arms approximate, lyriform or V-shaped; endocarina absent or present beneath epicranial stem and/or Y-shaped and extending under or just mesad of frontal arms. Zero, three or five stemmata on each side; frontoclypeal suture usually absent, labrum free. Antenna 3-segmented, well developed, some very short. Mandibles symmetrical or slightly asymmetrical, robust, bidentate or tridentate, without accessory ventral process; mola present, reduced in some, form concave, tuberculate, coarsely ridged or a row of hyaline teeth. Ventral mouthparts retracted.

Maxilla with cardo transverse, often divided; stipes elongate with well developed articulating areas; mala truncate, apically cleft or complete, with one or more teeth at inner apical angle (?absent in Pycnomerini); palps 3-segmented. Labium more or less free to base of mentum; ligula present; sometimes longer than labial palps; labial palps 2-segmented and narrowly or broadly separated. Hypopharyngeal sclerome absent or tooth-like. Hypostomal rods absent or short to moderately long and diverging or subparallel. Gula transverse.

Legs well developed, 5-segmented, sometimes short and spinose, tarsungulus usually with 2 setae lying side by side; coxae narrowly to widely separated.

Usually mesotergum and always metatergum and first four to eight abdominal tergites with patches, rows or ridges of asperites or tubercles; infrequently all tergites bear discrete plates; Tergite IX smooth or tuberculate, with pair of upturned urogomphi, approximate or separated by a sclerotized pit, urogomphi simple or rarely with accessory process. Sternite IX simple. Segment X transverse, ventrally or posteroventrally oriented. Spiracles annular-biforous with long or occasionally very short accessory tubes, in the later case appearing annular.

Habits and habitats: Slipinski and Lawrence (1999) reviewed the biology of the Zopheridae. Most zopherids live in dead and rotten wood or other rotting plant material. The actual source of

FIGURE 2.105. *Pycnomerus haematodes* (Fabricius) (from Stephan 1989).

nutrition is less certain, but the association with fungi is rather strong. Adults of Usechini, Phellopsini and some Zopherini are known to feed on polypore fruiting bodies (Guppy 1951, Doyen and Lawrence 1979, Lawrence 1991c, Steiner 1992, Slipinksi and Lawrence 1999, Lawrence *et al.* 1999a,b). The larvae of *Phellopsis obcordata* (Kirby 1837) bore through soft wood, where they feed on white sheet fungi between the laminae of large rotting spruce (*Picea*) stumps in Montana (Ivie, pers. obs.). Larvae of Usechini also bore in rotting wood, in this case associated with white rot fungus (Lawrence 1991c).

Adult and larval Pycnomerini are associated with rotten plant material, most species being found in old rotten logs Species of one lineage of tiny *Pycnomerus* are found in the very rotten ribs of old palm fronds (Ivie and Slipinski 2001). Monommatini are also associated with the products of fungi. Large numbers of the larvae of at least three genera have been repeatedly found (in South America, the West Indies and Madagascar) between the bark and wood of trees with wet to relatively dry rotting cambium (Ivie, pers. obs.). A few species have been reported from other rotting vegetation, such as papaya, *Euphorbia* and *Yucca* stems (Lawrence 1991b), but all of these seem to be isolated cases with the commonality that all the "hosts" were rotting, indicating that the true nutrition is from the products of rot, probably fungi.

Adult Latometini are found in old dead wood, but their larval habitat is unreported (Slipinski and Lawrence 1999). In the Zopherini, the larva of *Phloeodes diabolicus* LeConte show modifications for a wood-boring habit, and were described from the rotting roots of cottonwood (*Populus* sp.) and mulberry (*Morus*) as well as a rotting oak log (*Quercus*) in California (Doyen 1976). This wood-boring trend is continued in *Zopherus*, where some taxa may have moved into sound dead wood. Doyen and Lawrence (1979) reported the larvae of *Zopherus* spp. are adapted to boring through sound wood. They described larvae of *Z. nodulosus* Solier from decaying pecan wood (*Carya*) in Texas. Triplehorn (1972) and Doyen and Lawrence (1979) reported *Zopherus granicollis* Horn reared by William Tyson, who chopped larvae from roots of *Pinus monophylla* Torrey and Frémont in the mountains of Death Valley National Monument. Adults of *Zopherus gracilis* Horn have repeatedly been found in association with emergence holes in sound trunks of living alligator juniper (*Juniperus deppeana* Steudel) in Arizona (Ivie pers. obs.). The investigation of actual food habits of the Zopheridae is a fertile area of unexplored research.

Adults of *Zopherus* are long-lived, tough and rather general in their feeding. *Zopherus chilensis* Gray is the Makech, a living broach decorated with jewels, gold and bric-a brac, and tethered to blouses by chains in Mexico (Triplehorn 1972). Adult *Z. gracilus* have been kept alive for more than a year on apple slices (Ivie, pers. obs.).

The loss or reduction of flight wings is common in this group. All tribes but Latometini have flightless members, and some (Usechini, Phellopsini) are totally flightless, while the vast majority of Zopherini cannot fly. Flightlessness seems to have arisen several times in the Pycnomerini, often on islands or mountaintops (Pope 1955, Ivie and Slipinski 1989).

Status of the classification: This family, as defined here, is equal to the Zopherinae of Slipinski and Lawrence (1999). It incorporates elements placed in 3 families in earlier works (*e.g.*, Arnett 1973). One tribe of the current Zopheridae is herein listed as the family Monommatidae in Chapter 104 because the classification for this book was set to follow Lawrence and Newton (1995) (before the appearance of Slipinski and Lawrence 1999), and could not be changed in time for publication of Volume 1 of American Beetles. This does not diminish the fact that the monommatines have been shown to be, and should be treated as, true zopherids (see Chapter 104 for further discussion).

The origin of the Zopheridae as a family dates to Bøving and Craighead (1931), who based their argument to remove the Zopherini and Nosodermini from the Tenebrionidae to a family of their own solely on larvae. Further refinement, additions from other Tenebrionidae tribes, and use of adult characters were made by Crowson (1955), Kamiya (1963), Watt (1974) and Doyen and Lawrence (1979). During all of this, Watt (1974) added the southern hemisphere Ulodinae (including the Merycidae), which has since moved in and out of the family numerous times (*c.f.* Lawrence 1991c vs. Lawrence 1994, etc.).

This movement of groups has continued until very recently. Doyen and Lawrence (1979) pointed out the close relationship of the Zopheridae, Colydiidae and Monommatidae, but it took 20 years for this observation to be tested. In a cladistic-based revision of the zopherids, Slipinski and Lawrence (1999) found support for a monophyletic [Ulodidae [Colydiidae [Zopheridae + Monommatidae + Pycnomerini (the later formerly a subfamily of the Colydiidae)]]], with a well-supported [Colydiidae [Zopheridae + Monommatidae + Pycnomerini]]. Although they could not convincingly demonstrate monophyly for the remaining Colydiidae, the [Zopheridae + Monommatidae + Pycnomerini] was very solidly supported by mulitiple non-homoplastic characters. They chose to recognize a separate Ulodidae, but to combine the Colydiidae and Zopheridae, recongnizing two subfamilies — the Colydiinae being the old colydiids minus the pycnomerines, and the Zopherinae being the old zopherids plus the old monommatids and pycnomerine colydiids.

Although they had very complete representation of the old zopherid and pycnomerine genera, and rather good representation of the monommatines, Slipinski and Lawrence had included only 8 of the 125 genera from the core Colydiidae in their analysis. Even with that limited Colydiidae data set, they still did not discover any unique, or even very convincing, characters to support the colydiid lineage as monophyletic, and as such, its sister-group relationship as a single lineage is in question. Therefore, in order to have a demonstrably monophyletic Zopheridae and a Colydiidae that can draw attention as a continuing phylogenetic problem, both are recognized as families here.

Even today, the adoption of the Zopheridae is not universal, and some recent workers continue to treat the old Zopheridae as Tenebrionidae (Downie and Arnett 1996), although from a phylogenetic standpoint, the debate is well settled that the two groups are not closely related (Watt 1974, Lawrence 1994, etc.). Further, some will undoubtedly object to the fact that progress leads to the submersion of the Monommatidae, a family recognized uniformly since 1845, and the Pycnomerinae, a subfamily that has always been in the Colydiidae, as tribes of a family with a checkered past. Nevertheless, progress in understanding phylogenetic relationships will produce these novel combinations as we move toward a more natural classification of the Coleoptera.

Because of the fractured nature of the history of this family, few regional faunal treatments have been produced, and the reader is referred to the generic revision of Slipinski and Lawrence (1999), where it will be found that except as noted in the classification below, few species-level treatments exist for the genera of Zopheridae. This is a fertile group for species-level revisions.

Distribution: With the addition of the Pycnomerini, this family is nearly cosmopolitan. The distribution of the various tribes is discussed below. In North America there are five tribes, nine genera and about 30 species.

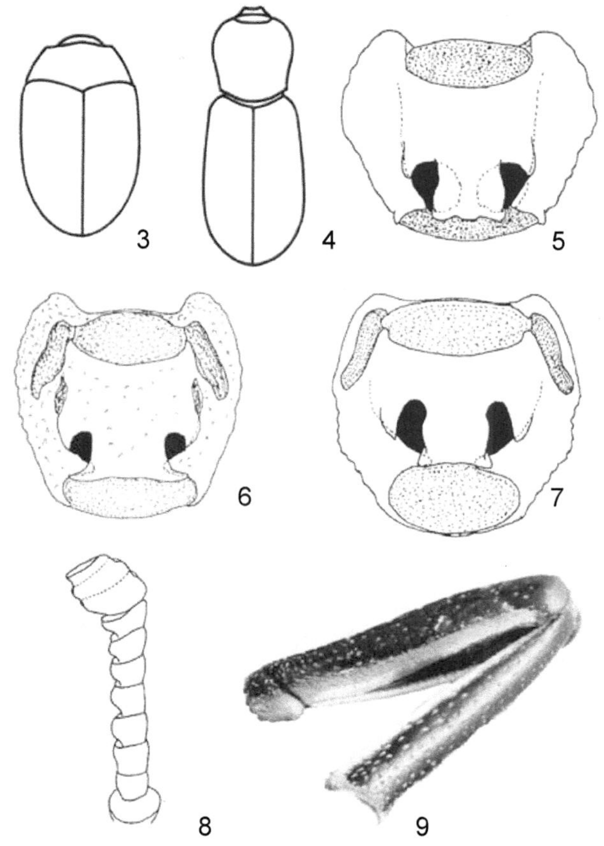

FIGURES 3.105-9.105. 3. Generalized Monommatini, dorsal view; 4. Generalized Zopherini/Phellopsini, dorsal view; 5. *Phellopsis obcordata* (Kirby), ventral view of prothorax; 6. *Phloeodes diabolicus* LeConte, ventral view of prothorax; 7. *Zopherus nodulosus haldemani* Horn, ventral view of prothorax; 8. *Zopherus nodulosus haldemani* Horn, antenna; 9. *Zopherus gralcilis* Horn, venter of mesofemur (Figures 5-7 from Slipinski and Lawrence 1999; Figure 8 from Triplehorn 1972.

Key to the Nearctic Genera
(modified from Doyen and Lawrence 1979 and Slipinski and Lawrence 1999)

1. Body broadly oval (Fig. 3); smooth dorsally; hypomeron with antennal cavity; eyes vertical, subcontiguous dorsally, separated by a distance less than eye diameter (eyes often concealed by head being withdrawn into prothorax) .... (Monommatini, see Chapter 104)

— Body elongate, parallel sided (Fig. 2) or constricted between pronotum and elytra (Fig. 4); often strongly punctate, tuberculate or costate above; hypomeron with or without antennal cavity; eyes variable, separated dorsally by distance far greater than eye diameter 2

2(1). Tarsi 4-4-4; elytra glabrous, reddish-brown to black, regularly striatopunctate (Fig. 2) (Pycnomerini) . ... *Pycnomerus*

— Tarsi 5-5-4; elytra variously setose, rarely striatopunctate, often tuberculate or confusedly costate 3

3(2). Pronotum with antennal cavities located above lateral margin (Fig. 1), on anterolateral portion of pronotum, visible from above; length 7 mm or less; Pacific coastal regions (Usechini) 4

— Pronotum with or without antennal cavities, but if present, confined to hypomeron below lateral margin (Figs. 5-7), not visible from above; length 10 mm or more; widespread 5

4(3). Procoxal cavities open behind .... *Usechimorpha*

— Procoxal cavities closed behind by widened apex of prosternal process .... *Usechus*

5(3). Scutellum visible; hypomeron without antennal fossa (Fig. 5); procoxal cavities open behind (Fig. 5); male submentum with setose pit (Phellopsini) .... *Phellopsis*

— Scutellum not visible (Fig 4); hypomeron with small to large antennal fossa (Fig. 6-7); procoxal cavities closed behind by widened apex of prosternal process; male submentum without setose pit (Zopherini) 6

6(5). Antenna appearing 9-segmented, club of 3 fused antennomeres (Fig. 8); front and hind angles of venter of femur fringed in golden setae (Fig. 9) .... *Zopherus*

— Antenna appearing 10-segmented, club of 2 fused antennomeres; femur uniformly clothed below .... *Phloeodes*

CLASSIFICATION OF THE WORLD TRIBES AND NEARCTIC GENERA

Zopheridae Solier 1834

Latometini Slipinski and Lawrence 1999

This is the sole tribe of zopherids *s.s.* that does not occur in North America. The genera *Latometus* Erichson 1842, *Orthocerodes* Slipinski and Lawrence 1999 and *Notorthocerus* Slipinski and Lawrence 1999 include small species (< 5.5 mm) that occur in Australia and Chile (Slipinski and Lawrence 1999).

Diagnosis: The latometines are unique among the zopherids in having all abdominal sterna free.

Usechini Horn 1867

Kamiya (1963) first associated this tribe with the Zopheridae, and Doyen and Lawrence (1979) sealed it with larval characters. Slipinski and Lawrence (1999) placed the tribe near the base of the zopherine lineage. The two genera occur in western North America and Japan.

Diagnosis: The dorsally visible antennal cavities on the anterolateral portion of the pronotum will distinguish this tribe.

Usechus Motschulsky 1845 contains five species, two of which, *U. lacerta* Motschulsky 1845 and *U. nucleatus* Casey 1889, occur in California, Oregon and Washington (key to spp., Blaisdell 1929). The other three occur in Japan (Boddy 1965, Doyen and Lawrence 1979, Saitô 1999).

Usechimorpha Blaisdell 1929. Two species, *U. barberi* Blaisdell 1929 and *U. montanus* Doyen 1979, occur on Vancouver Island (British Columbia), and in Oregon and California (Boddy 1965, Doyen and Lawrence 1979, Cannings and Cannings 1997).

Monommatini Blanchard 1845

Monommidae *auctorum* (see Chapter 104.)

Phellopsini Slipinski and Lawrence 1999

This is a monotypic tribe of moderately large, flightless beetles that seem to be associated with old growth forests (Egorov 1992, Steiner 1992). Slipinski and Lawrence (1999) found it closely related to the Monommatini.

Diagnosis: Its large size (10-22 mm), and elongate, rugose body with parallel sided elytra, combined with 11-segmented antennae and open procoxal cavities will distinguish this group.

Phellopsis LeConte 1862. Seven names currently recognized as species of this genus occur in the Holarctic region. *Phellopsis obcordata* (Kirby 1837) is transcontinental in old growth boreal forest from Alaska to Newfoundland. It ranges south in the Appalachians to Tennessee and North Carolina, and in the Western Cordillera to California, Idaho and Montana. This species has previously been treated as two, one western and one eastern until synonymized by Campbell 1991. There are two other validated names for western North American forms (*Phellopsis robustula* Casey 1907 and *Phellopsis montana* Casey 1907) but they are almost certainly synonyms. The other four named species occur in the Russian Far East, China and Japan, but these have not been critically examined either (Gebien 1937, Guppy 1951, Boddy 1965, Campbell 1991, Steiner 1992, Egorov 1992, Slipinski and Lawrence 1999). A revision of this genus would make and excellent graduate project.

Pseudonosoderma Heyden 1885

Pycnomerini Erichson 1845

Lawrence (1994) first pointed out the relationship of this tribe of then-colydiids to the Zopheridae, but it was not formally moved until the revision by Slipinski and Lawrence (1999). However, the classification of this group by Slipinski and Lawrence (1999) is quite novel from previous treatments (*c.f.* Ivie and Slipinski 1990). They combined the previously recognized group of Colydiidae, now reduced through synonymy to two genera (the nearly cosmopolitan genus *Pycnomerus,* and *Pycnomerodes* Broun 1886 from New Zealand and the Juan Fernandez Islands) with two Australian genera (*Cotulades* Pascoe 1860 and *Docalis* Pascoe 1860) originally moved to the Zopheridae by Watt (1974). A very solid phylogenetic analysis supports their move.

Diagnosis: In the New World, members of this group can be recognized as the only Zopheridae *s.s.* with a 4-4-4 tarsal formula. For a world diagnosis see Slipinski and Lawrence (1999).

Pycnomerus Erichson 1842. Stephan (1989) revised the North American species of this genus. Ivie and Slipinski (2001) added a species from Florida, bringing the total to six. Dajoz (1992) described an obvious synonym, which was synonymized by Ivie *et al.* (2002). The genus as a whole is very large with more than 70 species, many undescribed, and badly requires a revision. Extralimital faunal treatments include Pope (1955), Dajoz (1977, 1980), Slipinski (1984) and Ivie and Slipinski (1989).

Penthelispa Pascoe 1863
Endectus LeConte 1863
Pycnomorhus Motschulsky 1858
Dechomus Jacquelin du Val 1859

Zopherini Solier 1834

This tribe contains eight genera, widely distributed throughout the world. There are two genera in North America, and six that occur elsewhere: *Meralius* Casey 1907 (2 species from Cuba and Venezuela), *Noserinus* Casey 1907 (two species from Brazil), *Nosoderma* Solier 1841 (23 species from Mexico, see Doyen and Lawrence 1979 for a list of generic synonyms), *Scoriaderma* Fairmaire 1894 (3 species from Africa and the Comoro Islands), *Zopher*

Slipinski and Lawrence 1999 (monotypic from Malaysia), and *Zopherosis* White 1859 (1 species from Australia) (Slipinski and Lawrence 1999).

Diagnosis: The large size (length > 10 mm), reduced number of apparent antennomeres (9 or 10), at least a trace of an antennal fossa on the hypomeron, and the lack of a visible scutellum will diagnose this tribe in North America.

Nosodermini Casey 1907
Zopherosini Casey 1907

Zopherus Laporte 1840 was revised by Triplehorn (1972), who recognized a total of 19 species, all from the Americas, ranging from Venezuela north to the southwestern United States. The ten species that occur in the United States occur in the states bordering Mexico, Nevada, Utah and Colorado, ranging as far north as Napa Co., California, the Great Salt Lake Valley of Utah and Glade Park, Colorado (Canadian National Collection of Insects).

Zopherodes Casey 1907 (syn. Triplehorn 1972)

Phloeodes LeConte 1862. There are 14 names still considered valid in this genus, because there have been no published synonymies to reduce this total. According to Doyen and Lawrence (1979), there should be only eight valid species in this genus (listed by them as two *Phloeodes* and six *Noserosus* [sic]), but they list no names and synonyms, so their action was not validated. Thirteen of these names refer to California populations, centered around *P. diabolicus* LeConte 1851, *P. pustulosus* LeConte 1859 and *P. plicatus* (LeConte 1859) and the other to a Texas species, *P. emarginatus* (Horn 1878). The remaining 10 names are T. L. Casey species (Casey 1907), and many if not all of them are probably synonyms (*c.f.* Triplehorn 1972 who synonymized 21 of 23 Casey names in *Zopherus*). Garcia-Paris *et al.* (2000) recorded the group from mainland Mexico for the first time. Obviously, the genus is in desperate need of revision, and would make an excellent graduate project.

Noserus LeConte 1862 (syn. Slipinski and Lawrence 1999)
Ageonoma Pascoe 1866
Phoeodes Arnett 1962: 668 (*lapsus calami*)
Noserosus Doyen and Lawrence 1979: 345 (*lapsus calami*)

Bibliography

ARNETT, R. H. 1973. Beetles of the United States (A Manual for Identification). American Entomological Institute. Ann Arbor, MI. 1112 pp.

BLAISDELL, F. E. 1929. A revision of the beetles of the tenebrionid tribe Usechini, with descriptions of a new genus and new species. Proceedings of the United States National Museum, 75 (19): 1-14.

BODDY, D. W. 1965. Zopheridae. Pp. 77-79. *In*: M. H. Hatch, ed. The beetles of the Pacific Northwest. Part IV: Macrodactyles, Palpicornes and Heteromera. University of Washington Publications in Biology, 16 (4): 1-268.

BØVING, A. G. and F. C. CRAIGHEAD. 1931. An illustrated synopsis of the principal larval forms of the order of Coleoptera. Entomologica Americana (new series), 11:1-351.

CAMPBELL, J. M. 1991. Zopheridae. p. 252. *In*: Y. Bousquet, ed. Checklist of the beetles of Canada and Alaska. Agriculture Canada Publication 1861/E. Canada Communications Group - Publishing. Ottawa.

CANNINGS, R. A. and S. G. CANNINGS. 1997. Terrestrial arthropods of Brooks Peninsula. British Columbia Parks Occasional Paper, 5: 10.1-10.69.

CASEY, T. L. 1907. Notes on *Chalcolepidius* and the Zopherini. Canadian Entomologist, 39: 34-46.

CROWSON, R. A. 1955. The natural classification of the families of Coleoptera. Nathaniel Lloyd. London, 187 pp.

DAJOZ, R. 1977. Coléoptères Colydiidae et Anommatidae Paléarctiques. Faune de l'Europe et du Bassin Méditerranéen, 8: i-vi, 1-280.

DAJOZ, R. 1980. Insectes Coléoptères: Colydiidae et Cerylonidae. Faune de Madagascar, 54: 1-256.

DAJOZ, R. 1992. Description de trois nouvelles espèces de Colydiidae de l'Arizona et de Californie (Coleoptera). Bulletin Mensuel de la Société Linnéenne de Lyon, 61: 60-66.

DOYEN, J. T. 1976. Description of the larva of *Phloeodes diabolicus* LeConte (Coleoptera: Zopheridae). Coleopterists Bulletin, 30: 267-272.

DOYEN, J. T. and J. F. LAWRENCE. 1979. Relationships and higher classification of some Tenebrionidae and Zopheridae (Coleoptera). Systematic Entomology, 4: 333-377.

DOWNIE, N. M. and R. H. ARNETT, Jr. 1996. The Beetles of Northeastern North America. Vol. 1. The Sandhill Crane Press. Gainesville, FL.

EGOROV, A. B. 1992. 94. Cem. Zopheridae, Pp. 504-505. *In*: P. A. Ler, ed. Key to the Insects of the Russian Far East. Vol. III. Part 2. Dal'nauka, Vladivostok.

GARCIA-PARIS, M., G. PARRA-OLEA and M. COCA-ABIA. 2000. First records of the genus *Noserus* LeConte (Coleoptera: Zopheridae) in Mexico. Proceedings of the Entomological Society of Washington, 102: 473-474.

GEBIEN, H. 1937. Katalog der Tenebrioniden (Col.: Heteromera). Teil I. Pubblicazioni del Museo Entomologico 'Pietro Rossi', 2: 505-883.

GUPPY, R. 1951. Habitat of *Phellopsis porcata* LeC. Coleopterists Bulletin, 5: 28.

IVIE, M. A. and S. A. SLIPINSKI. 1989. The Pycnomerini (Coleoptera: Colydiidae) of the West Indies. Florida Entomologist, 72: 64-80.

IVIE, M. A. and S. A. SLIPINSKI. 1990. Catalog of the genera of world Colydiidae (Coleoptera). Annales Zoologici, 43 (Suppl. 1): 1-32.

IVIE, M. A. and S. A. SLIPINISKI. 2001. *Pycnomerus thrinax*, a new North American zopherid. Insecta Mundi, [2000] 14: 225-227.

IVIE, M. A., S. A. SLIPINISKI and P. WEGRZYNOWICZ. 2002. New records and synonyms in the Colydiinae and

Pycnomerini (Coleoptera: Zopheridae). Insecta Mundi, 15: (in press).

KAMIYA, H. 1963. On the systematic position of the genus *Usechus* Motschulsky, with a description of a new species from Japan Coleoptera). Mushi, 37: 19-26.

LAWRENCE, J. F. 1991a. Colydiidae (Tenebrionoidea), Pp. 512-514. *In*: F. W. Stehr, ed., Immature Insects. Vol. II. Kendall/ Hunt. Dubuque, Iowa.

LAWRENCE, J. F. 1991b. Monommidae (Tenebrionoidea), Pp. 514-515. *In*: F. W. Stehr, ed., Immature Insects. Vol. II. Kendall/Hunt. Dubuque, Iowa.

LAWRENCE, J. F. 1991c. Zopheridae (Tenebrionoidea) (including Merycidae), Pp. 518-519. *In*: F. W. Stehr, ed., Immature Insects. Vol. II. Kendall/Hunt. Dubuque, Iowa.

LAWRENCE, J. F. 1994. The larva of *Sirrhas variegatus*, sp. nov., with notes on the Perimylopidae, Ulodidae (stat. nov.), Zopheridae and Chalcodryidae (Coleoptera: Tenebrionoidea). Invertebrate Taxonomy, 8: 329–349.

LAWRENCE, J. F., A. M. HASTINGS, M. J. DALLWITZ, T. A. PAINE and E. J. ZURCHER. 1999a. Beetles of the World: A Key and Information System for Families and Subfamilies. CD-ROM, Version 1.0 for MS-Windows. CSIRO Publishing. Melbourne.

LAWRENCE, J. F., A. M. HASTINGS, M. J. DALLWITZ, T. A. PAINE and E. J. ZURCHER. 1999b. Beetle Larvae of the World: Descriptions, Illustrations, Identification, and Information Retrieval for Families and Sub-families. CD-ROM, Version 1.1 for MS-Windows. CSIRO Publishing: Melbourne.

LAWRENCE. J. F. and A. F. NEWTON, Jr. 1995. Families and subfamilies of Coleoptera (with selected genera, notes, references and data on family-group names), Pp 779-1006. *In*: J. Pakaluk and S. A. Slipinski, eds., Biology, phylogeny and classification of Coleoptera: Papers celebrating the 80th birthday of Roy A. Crowson. Muzeum i Instytut Zoologii PAN. Warsaw.

POPE, R. D. 1955. Los Insectos de las Islas Juan Fernandez. 25. Colydiidae (Coleoptera). Revista Chilena de Entomologia, 4: 153-158.

SAITÔ, M. 1999. Notes on the Japanese species of the genus *Usechus* (Coleoptera, Zopheridae). Elytra, 27: 103-111.

SLIPINSKI, S. A. 1984. Studies on the African Colydiidae (Coleoptera). Part II. Genera: *Afrorthocerus* Pope and *Pycnomerus* Erichson. Annales Zoologici, 38: 129-158.

SLIPINSKI, S. A. and J. F. LAWRENCE. 1999. Phylogeny and classification of Zopheridae *sensu novo* (Coleoptera: Tenebrionoidea) with a review of the genera of Zopherinae (excluding Monommatini). Annales Zoologici, 49: 1-53.

STEINER, W. E., Jr. 1992. "Ironclad Beetles", family Zopheridae, in Maryland: notes on the natural history and distribution of *Phellopsis obcordata* (Kirby). Maryland Naturalist, 35 (1991): 25-30.

STEPHAN, K. H. 1989. The Bothrideridae and Colydiidae of America north of Mexico (Coleoptera: Clavicornia and Heteromera). Occasional Papers of the Florida State Collection of Arthropods, 6: 1-65.

TRIPLEHORN, C. A. 1972. A review of the genus *Zopherus* of the World (Coleoptera: Tenebrionidae). Smithsonian Contributions to Zoology, 108: 1-24.

WATT, J. C. 1974. A revised subfamily classification of Tenebrionidae (Coleoptera). New Zealand Journal of Zoology, 1: 381-452.

106. TENEBRIONIDAE Latreille 1802

by Rolf L. Aalbu, Charles A. Triplehorn, J. Milton Campbell, Kirby W. Brown, Ronald E. Somerby and Donald B. Thomas

Family common name: The darkling beetles

Family synonyms: Alleculidae Laporte 1840; Blapsidae Leach 1815; Cossyphodidae Wasmann 1899; Diaperidae Latreille 1802; Helopidae Latreille 1825; Lagriidae Latreille 1825; Nilionidae Champion 1888; Petriidae Semenow 1893; Pimeliidae Leach 1815; Rhysopaussidae Wasmann 1896; Tentyriidae Eschscholtz 1829.

The Tenebrionidae are the most highly evolved and diverse family of Tenebrionoidea. Because of the many exceptions that exist in most characters, these beetles are often difficult to recognize as a family. However, by using a combination of adult and larval characters, they are relatively easily distinguished from other Tenebrionoidea. These include: (adults) typically hard bodied, antennae typically 11-segmented, inserted under lateral expansion of frons; closed procoxal cavities; abdomen with visible sternites 1-3 connate, 4 and 5 movable; tarsal formula 5-5-4, rarely other than (see below); (larvae) subcylindrical, head with frontoclypeal suture present; malar apex without cleft; annular or annular-multiforous spiracles, simple 9th sternum and lacking an endocarina, mandibular prostheca, hypostomal rods and ventral prolegs.

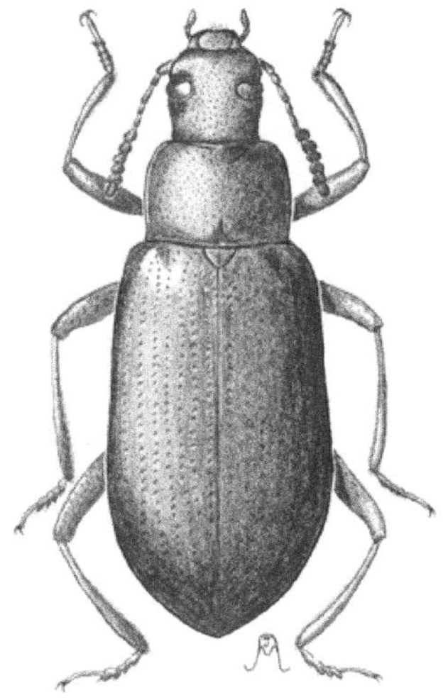

FIGURE 1.106 *Alobates pennsylvanica* (DeGeer)

Description: Highly variable in shape and size (slightly over 1 to over 80 mm), winged or apterous; eyes very rarely absent, often separated by epistomal canthus into upper and lower portions. Antennae inserted under lateral expansion of frons, 11-segmented, rarely 10-segmented (some Archaeoglenini, Hypophloeinae, Bolitophaginae and Opatrinae) or 9-segmented (some Archaeoglenini); variable in shape; apical antennomeres often bearing compound (or sometimes placoid) sensilla. Epistoma with clypeolabral membrane exposed or not. Labrum subquadrate, sometimes elongate or transverse. Mandibles bidentate or tridentate (Phrenapatinae) at apex. Mandibular mola with or without fine transverse ridges. Lacinia with or without apical hook or tooth.

Pronotum typically carinate or explanate laterally. Procoxal cavities closed behind externally (with rare exceptions: see Pimeliinae), open or closed internally; mesocoxal cavities with or without exposed trochantin; closed laterally by meso- and metasternite or by mesepimeron or rarely by trochantin (Phrenapatinae). Legs with penultimate tarsomere sometimes lobed (some Alleculinae and Lagriinae); anterior trochanters overlapped laterally by the femora (except Scaphidemini); tarsal claws sometimes pectinate (Alleculinae). Tarsal formula 5-5-4; rarely 4-4-4 (some Bolitophaginae, Diaperinae, Pimeliinae and Phrenapatinae) or 3-3-3 (some non-North American Phrenapatinae).

Elytra, if striate, with scutellary striole (except Phrenapatinae); typically with 9 striae (10 in Pimeliinae, Lagriinae and some Bolitophaginae). Abdomen with intersternal membrane of abdomen exposed (except most Pimeliinae and some Lagriinae and Bolitophaginae); hinge position lateral or medial (Pimeliinae and some Lagriinae); visible sternites 1-3 connate, 4 and 5 movable. Abdominal defensive glands present or absent (Zolodinae, Pimeliinae and some Lagriinae). Aedeagus typically not inverted, with tegmen dorsal (except Zolodinae and Pimeliinae and a few other genera).

Mature larvae 5 to 70 mm in length. Body typically elongate and cylindrical to slightly flattened, occasionally short and broad, fusiform, or strongly flattened; cuticle lightly to heavily sclerotized (occasionally head and 9th tergum more heavily sclerotized), bearing sparse to dense, short to long setae. Head with frontoclypeal suture present, endocarina absent; antennae 1 (Leochrini), 2 (Lagriinae) or 3-segmented, with segments slightly to highly reduced apically; antennal sensorium almost always flat, dome-like, typically C-shaped, forming an incomplete ring around terminal segment (or fused remnant in species with 2-segmented antenna). Labrum transverse with few to many setae. Mandibles stout, subtriangular, with 1 to 3 apical teeth and a sharp incisor edge; sometimes with dorsolateral or rarely ventrolateral surface with membranous elevation bearing numerous setae (Pimeliinae); without prostheca, malar apex simple (not cleft), hypostomal rods absent; gula well developed. Thorax with prothorax typically slightly larger than meso- or metathorax. Legs well developed, 5-segmented; prothoracic legs often much stouter, more sclerotized than other legs (especially in soil or sand dwelling species). Thoracic and abdominal spiracles annular or annular-multiforous. Abdomen without ventral prolegs; 9th sternum unmodified; 9th abdominal tergum typically terminal, extending onto ventral surface, its shape and setal and spicule patterns typically diagnostic at various levels, from simple, rounded to trian-

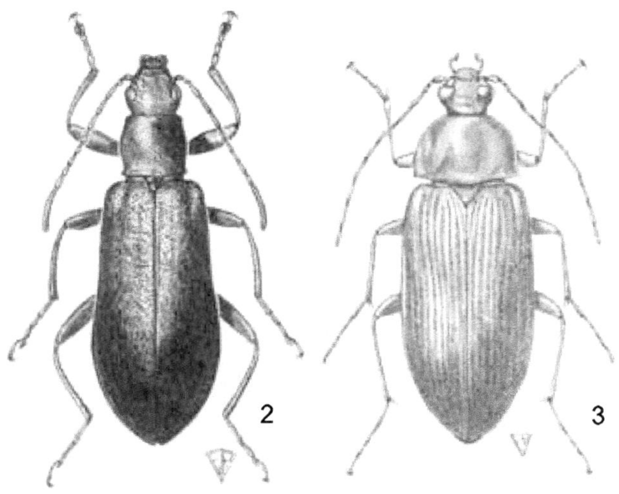

FIGURES 2.106-3.106. 2. *Arthromacra aenea* Say; 3. *Capnochroa fuliginosa* (Melsheimer) (both from Arnett 1973).

gular; urogomphi, concave plates or other armature present or absent. The larvae of many Tenebrionidae are referred to as "false wireworms." They differ from true wireworms (Elateridae) in having a less flattened head, a labrum which is articulated, no lateral stemmata and by the presence of a mandibular mola and annular or annular-multiforous spiracles (see Lawrence and Spilman 1991).

Habits and habitats. Most tenebrionids are heavily sclerotized, dark in color, long lived, mostly crepuscular or active at night (although some Old World forms are day active), on the ground, logs or tree trunks. Brachyptery and aptery are found in many species. Adaptation to arid environments is also highly developed.

Tenebrionid beetles present ideal biological models for studies of evolutionary systematics, biogeography and ecology. Reasons for this are the following: (1) they are a large and diverse group (perhaps fifth largest beetle family). (2) They are geographically and ecologically diverse: found from fungus on the top of rain forest trees to soil in caves and desert sand dunes; from sea level to some of the highest peaks (even flightless species, as *Eleodes vandykei* Blaisdell are commonly found at 13000' elevation in the White Mountains, California). (3) They are a relatively old group with fossils dating back to the Jurassic (some Eocene amber specimens have been placed in extant genera). (4) Some species are economically important to man both as pests or beneficials. (5) They are relatively easy to collect and maintain live for the purpose of behavioral, ecological or molecular studies or rearing. (6) They are relatively character rich in not only adult morphology but also in more evolutionarily conservative immature and internal features as well.

Habitats of tenebrionid beetles can be divided into two major groups: those that are associated with trees, usually found in forest or savannas and those that are associated with soil or sand, usually found in plains or deserts as well as in forests and savannas. Among tenebrionids that are associated with trees, different areas are exploited as larvae and adults. Those associated with dead, rotten wood often softened and chemically altered by the action of various fungi include most Phrenapatinae, Tenebrionini, Amarygmini, Ulomini, Alleculinae, Gnathidiini, Strongyliini and Coelometopini and some Diaperini, Alleculini, and Helopini. Those found under bark and associated cambium and subcortical spaces or in galleries of bark beetles and wood borers include many Coelometopini, Centronopini, Ulomini, Tenebrionini and some Diaperini, Alleculini, and Helopini. Some Diaperini are found in fermenting tree wounds. Large proportions of tenebrionids are also associated with fruiting bodies of Basidiomycetes (polypores). These include the Bolitophagini, some Diaperini and Alleculinae (some Bolitophagini being restricted to certain species of polypores). Surface grazers on lichens, algae and mosses growing on bark and rock surfaces include some Helopini, Amarygmini, and Alleculinae. Those found on leaves and flowers include some Alleculinae, Lagriinae, Epitragini and Eurymetopini.

Many tenebrionids are found on or in the soil, both as adults and larvae. These usually take shelter in leaf litter under rocks or logs. These include most Pimeliinae, Opatrinae, Apocryphini, some Goniaderini, Lagriinae, Scaurini, Blaptini, Eleodini, and some Alleculinae. Many Tenebrionidae have adapted to life on sand dunes, both on the surface and in the sand. These include many Pimeliinae (almost all tribes have sand adapted species) and Opatrinae and a number of Eleodini and Blaptini. Many tenebrionids adapted for life in very hot, arid regions have sealed subelytral cavities, which reduces transpiratory water loss. The Phaleriini and some Coniontini are found on sand at beaches. A few species have also become adapted to life in caves. These include members of the Stenosini, Cryptoglossini, Eurymetopini, Eleodini, Tenebrionini, and Hypophloeini.

Some species are associated with nests of vertebrates or insects. These include mammals, especially rodents [Eleodini, Melanimini, Cerenopini, many Pimeliinae and Opatrinae, some Alleculinae], birds [some Alphitobiini, Tenebrionini, Diaperini, and Alleculinae], bees [Triboliini], ants [a few Triboliini, many Pimeliinae and Opatrinae, Eulabini, some Alleculinae] and termites [many Amargymini, some Pimeliinae, Opatrinae and Alleculinae]. Predaceous tenebrionids exist mainly as larvae. These are found in bark beetle galleries [Hypophloeini], in nests of birds [Alphitobiini] or in palm fronds where they prey on other beetle larvae [Triboliini].

Species of Tenebrionini, Triboliini and several other tribes are pests of stored products (For a key to stored product pests in the United States see Spilman 1991). Some Eleodini are occasional pests of commercial crops. The larvae of these, called false wireworms. *Alphitobius diaperinus* (Panzer) may play a beneficial role in the control of fly populations in poultry houses (for details on this and other aspects of biology see Lawrence and Spilman 1991).

Status of the classification. Although much progress towards the higher classification of the Tenebrionidae has been made in the last few decades, cladistic placement of some tribes and subfamilies remains uncertain. The subfamily Tenebrioninae as delineated by Lawrence and Newton (1995) is paraphyletic with groups like the Alleculinae, Coelometopinae, and Diaperinae fall-

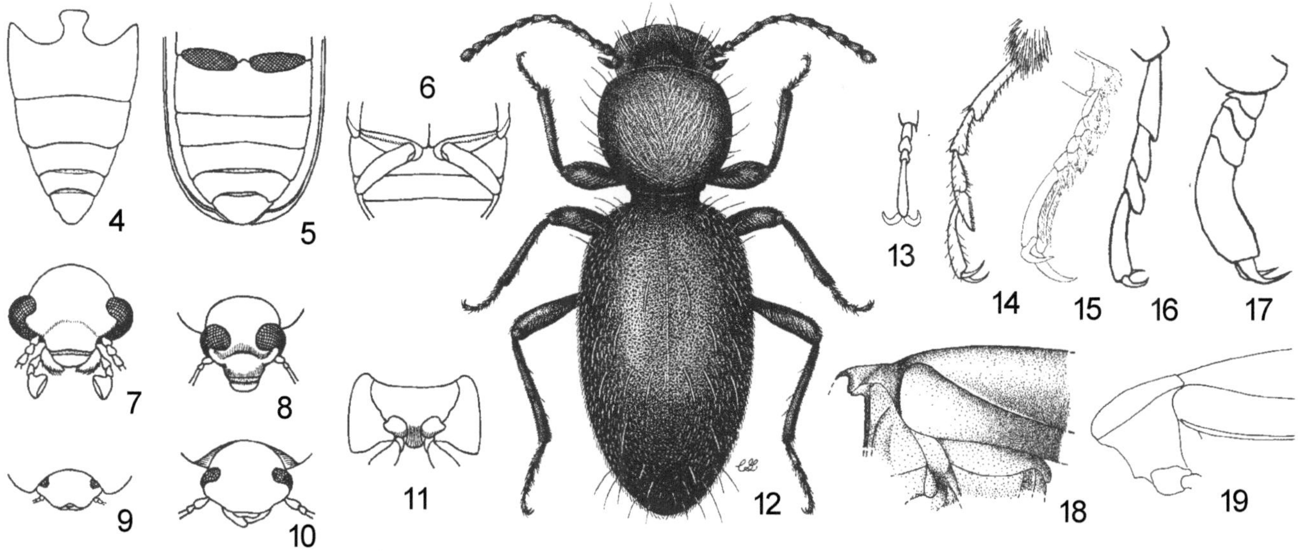

FIGURES 4.106-19.106. 4-5. Abdomen, ventral view, 4. *Eleodes dentipes* Eschscholtz; 5. *Uloma* sp.; 6. *Crypticus obsoletus* (Say), metacoxae. 7-10. Head, dorsal view, 7. *Talanus langurinus* (LeConte); 8. *Strongylium tenuicolle* Say; 9. *Ammodonus fossor* (LeConte); 10. *Amphidora littoralis* Eschscholtz. 11. *Tribolium castaneum* (Herbst), prosternum; 12. *Apocrypha setosa* Doyen and Kitayama, habitus. 13-17. Tarsus. 13. *Talanus langurinus* (LeConte); metatarsus; 14. *Arthromacra* sp., protarsus; 15. *Strongylium* sp.; 16. *Paratenetus* sp., metatarsus; 17. *Boletotherus* sp., metatarsus. 18-19. Pronotum and elytral base, lateral view, 18. *Eleodes* (*Blapylis*) *productus* Mannerheim; 19. *Blaps lethifera* Marshall. (Figures 4-11, 13, 14 from Arnett 1973; Figure 12 from Doyen and Kitayama 1980; Figure 15 from Doyen 1989).

ing within the "Tenebrioninae". We have adapted use of divisions to clarify these relationships. Classification follows Lawrence and Newton (1995) except for the following: (1) The Bolitophagini have in the past been placed in the Diaperinae. Recently the Bolitophagini were moved to the Tenebrioninae by Doyen *et al.* (1989b) without giving specific explanation. The Bolitophagini and Tenebrioninae share similar spermathecal configurations. The Bolitophagini also share many characters and life habits with the Diaperinae and were included in or basal to the "diaperine" clade in the "tenebrionoid" lineages proposed by Doyen and Tschinkel (1982). The Bolitophagini also have unique larval characters and are sometimes placed as a separate subfamily in keys to immature Tenebrionidae (see Skopin 1978). We have taken this approach here until further studies clearly illustrate the position of the Bolitophaginae. (2) The "Opatrine" tribes form a large monophyletic lineage within the tenebrionines. Indeed worldwide the opatrines are nearly as large a group as the rest of the Tenebrioninae with approximately 240 genera placed in 15 tribes. These tribes have been included in the past in the subfamily Opatrinae as defined by Koch (1956), Medvedev (1977) and other authors. The Opatrinae are delineated in part by groups defined by Medvedev (1977) and Aalbu and Triplehorn (1985) and expanded to include the Eleodini.

We have tried to keep the tribal descriptions relatively short. For this reason we have omitted internal characters such as structure of the tentorium, meso and metendosternite, female internal reproductive system and defensive gland structure.

Distribution. This family, the largest of the Tenebrionoidea is at present represented by approximately 19,000 described species worldwide in more than 2000 genera. In the previous version of this book, the Tenebrionidae were represented in the United States by 3 subfamilies, 45 tribes, 207 genera and 1303 species. Of the 207 genera, 9 have been transferred to other families, 27 have been placed as synonyms and 6 do not occur in the United States. At present the family is represented in the United States by 10 subfamilies, 44 tribes, 191 genera and 1184 species. Of the 191 genera, 13 are replacement names, 15 are new and 20 are transferred from other families.

I. Key to the Subfamilies and Tribes of the United States

1. Tarsal formula 4-4-4 2
— Tarsal formula 5-5-4 4

2(1). Abdomen without visible membrane along hind margin of visible sternites 3 and 4; pronotum and elytra costate (Anepsiini, *Anchomma*, in part) ... . Key I
— Abdomen with visible membrane along hind margin of visible sternites 3 and 4 (Fig. 4); pronotum and elytra not costate 3

3(2). Antennae with 4-segmented club (Myrmechixenini) .... *Myrmechixenus*
— Antennae with 7-segmented, pectinate club (Bolitophaginae, Bolitophagini, in part) ..... Key M

4(1). Abdomen without visible membrane along hind margin of visible sternites 3 and 4 5
— Abdomen with visible membrane along hind margin of visible sternites 3 and 4 (Fig. 4) 6

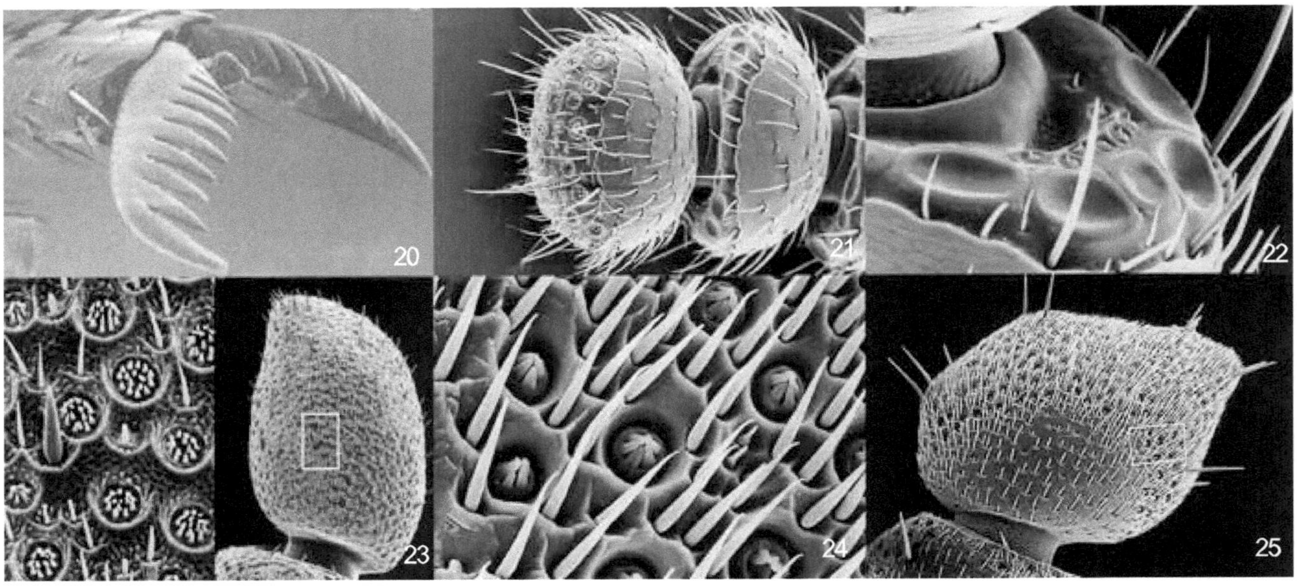

FIGURES 20.106-25.106. 20. *Stenochidus gracilis* LeConte, tarsal claws. 21. *Uloma longula* LeConte, terminal antennomeres; 22. Same, apex of penultimate antennomere; 23. *Coelocnemis magna* LeConte, terminal antennomere with magnified inset; 24. *Zophobas atratus* (Fabricius), magnification of inset in next Figure; 25. Same, terminal antennomere. (Figure 20 from Campbell 1984).

5(4). Meso- and metafemora with strong ventral tooth apically (Belopini) .... *Adelonia*

— Meso- and metafemora without strong ventral tooth apically (Pimeliinae, in part) .... Key II

6(4). Tarsal claws pectinate (Fig. 20) (Alleculinae, Alleculini) .... Key AA

— Tarsal claws simple 7

7(6). At least penultimate tarsomere ventrally truncate, broader than other tarsomeres; apical tarsomere arising before apex (Fig. 14); body slender, elongate, typically setose, sometimes metallic in color; apical antennomere often several times as long as preceding (Lagriini) .... Key A

— Penultimate tarsomere not ventrally truncate, not broader than other tarsomeres; apical tarsomere arising at apex; apical antennomere not several times as long as preceding 8

8(7). Middle and hind tibiae bearing a longitudinal, finely crenulate carina on outer (dorsal) margin; eyes typically more prominent than sides of frons, typically rounded, feebly or not emarginate (Diaperini) .... Key O

— Middle and hind tibiae lacking carina on outer margin; eyes less prominent than sides of frons, more or less transverse, always emarginate in front 9

9(8). Seventh abdominal tergite at least partially exposed as pygidium; mesocoxal cavities closed laterally by sternite, protibiae typically with serrate outer (dorsal) margins 10

— Seventh tergite concealed by elytra; mesocoxal cavities with epimeron separating meso and metasternite; protibiae with nonserrate outer (dorsal) margins 11

10(9). Body elongate, cylindrical; elytra truncate, pygidium completely exposed; antennae short, flattened, clavate, without placoid sensilla (Hypophloeini) .... *Corticeus*

— Body oblong-oval, convex; elytra abbreviated, seventh abdominal tergite partially exposed as pygidium; antennae slender, moniliform, bearing large placoid sensilla, often appearing as large white scales at apex of antennomeres 7 and 8 (Figs. 21, 22) (Ulomini) .... Key U

11(9). Abdomen and thorax globose, greatly narrowed at base to less than half width of pronotum; antennae very slender; epistoma with clypeolabral membrane clearly exposed (Fig. 12) (Apocryphini) .... *Apocrypha*

— Abdomen and thorax not narrowed at base to less than half width of pronotum; antennae variable; epistoma with clypeolabral membrane exposed or not 12

12(11). Anterior trochanters not overlapped laterally by the femora; coxal cavities (especially metacoxae) widely separated; intercoxal process of basal abdominal sternite short, truncate (Scaphidemini) .... *Scaphidema*

— Anterior trochanters overlapped laterally by the femora; coxal cavities variably separated, not widely separated in most 13

13(12). First tarsomere short, tarsi compressed (Fig. 17); genae sulcate, antennomeres 3 to 5 sometimes pectinate (Bolitophaginae, Bolitophagini, in part) .... Key M

— First tarsomere moderate or elongate, never very short, tarsi not compressed (Fig. 16); genae not sulcate 14

14(13). All tibiae more or less expanded apically, fossorial 15

— Protibiae only, or none of the tibiae, expanded apically 16

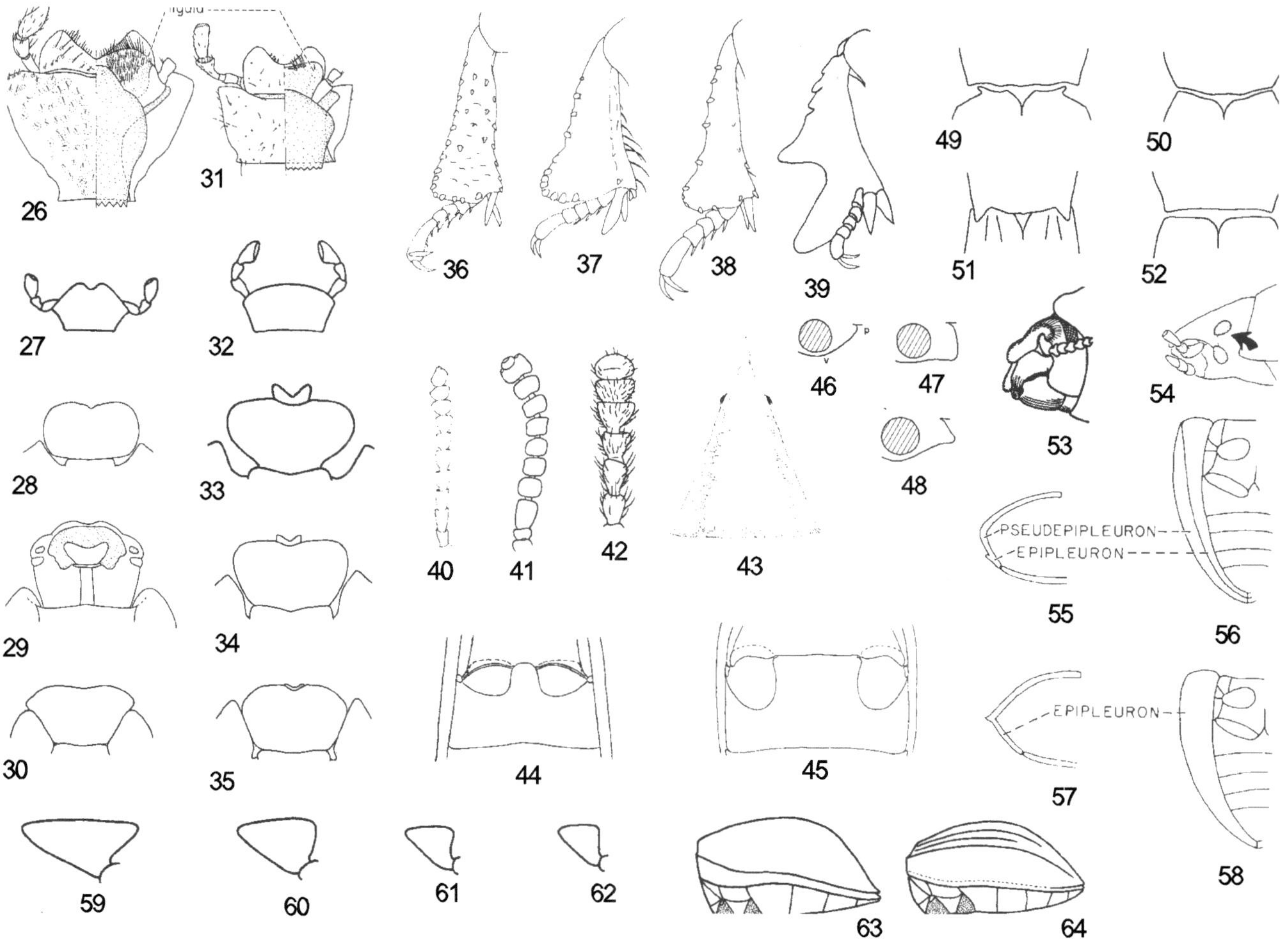

FIGURES 26.106-64.106. 26-35. Labium, 26. *Branchus floridanus* LeConte; 27. *Eurymetopon rufipes* Eschscholtz; 28. *Philolithus carinatus* (LeConte); 29. *Eusattus mexicanus* Champion; 30. *Astrotus* sp.; 31. *Eusattus muricatus* LeConte; 32. *Bothrotes canaliculatus* (Say); 33. *Asidopsis* sp.; 34. *Megasida* sp.; 35. *Zaleucus* sp. 36-39. Protibia and tarsus, anterior view, 36. *Anepsius delicatulus* LeConte; 37. *Anepsius montanus* Casey; 38. *Anepsius valens* Casey; 39. *Cheirodes californica* (Horn). 40-42. Antenna, 40. *Eusattus convexus* LeConte; 41. *Microschatia solieri* Brown and Doyen; 42. *Branchus obscurus* Horn. 43. Platynotini, stridulatory gula. 44-45. Metasternum and first abdominal sternite, 44. *Eulabis bicarinata* Eschscholtz; 45. *Cerenopus concolor* LeConte. 46-48. Prosternal process, lateral view. 46. *Stenosides* sp.; 47. *Astrotus* sp.; 48. *Stethasida* sp. 49-52. Pronotal-elytral junction, dorsal view, 49. *Glyptasida* sp.; 50. *Stenomorpha* sp.; 51. *Parasida* sp.; 52. *Megasida* sp.. 53-54. Head, lateral view, 53. *Astrotus contortus* (LeConte); 54. *Blapstinus discolor* Horn. 55-56. *Eusattus difficilis* LeConte, 55. X-section of elytron and first abdominal sternite; 56. Pterothorax, abdomen and eltyra, ventral view. 57-58. *Eusattus robustus* LeConte, 57. x-section of elytron and first abdominal sternite; 58. Pterothorax, abdomen and eltyra, ventral view. 59-62. Apical maxillary palpomere, 59. *Philolithus carinatus* (LeConte), male; 60. Same, female; 61. *Trichiasida* sp., male; 62. Same, female. 63-64. Elytron, pterothorax and abdomen, lateral view, 63. *Microschatia* sp.; 64. *Parasida* sp. (Figure 39 from Aalbu and Triplehorn 1985; Figures 27, 32, 53 from Arnett 1973; Figures 44, 45 from Berry 1973; Figures 28, 30, 33-35, 46-52, 59-64 from Brown 1971; Figure 41 from Brown and Doyen 1991; Figures 26, 31 from Doyen 1972; Figures 29, 40, 42, 55-58 from Doyen 1984; Figures 36-38 from Doyen 1987; Figure 43 from Koch 1956; Figures 54 from Spilman 1991).

15(14). Antennae moderately long, gradually broadened apically (Phaleriini) .... Key N
— Antennae short, apical 4 antennomeres abruptly dilated (Trachyscelini) .... *Trachyscelis*

16(14). Penultimate tarsomere weakly to strongly bilobed (Goniaderini) .... Key B
— Penultimate tarsomere simple 17

17(16). Procoxae subtransverse (Fig. 11); mesocoxae with trochantin hidden; first antennomere short .. 18
— Procoxae rounded; mesocoxae with trochantin visible; first antennomere typically longer than following antennomeres 19

18(17). Scutellum small; elytra without scutellar striae; anterior striae non-parallel (Penetini) .... Key C
— Scutellum large; elytra with scutellar striae; anterior striae parallel (Triboliini) .... Key W

19(17). Antennae with compound, stellate sensoria (visible at 50x or higher, at lower power appearing as minute circular white pits) on at least apical 3 antennomeres (Figs. 23, 24) 20
— Antennae with simple, setiform sensilla (Figs. 70, 71) 26

20(19). Head vertical in repose, nearly encompassed within pronotum; eyes very large, embracing antennal

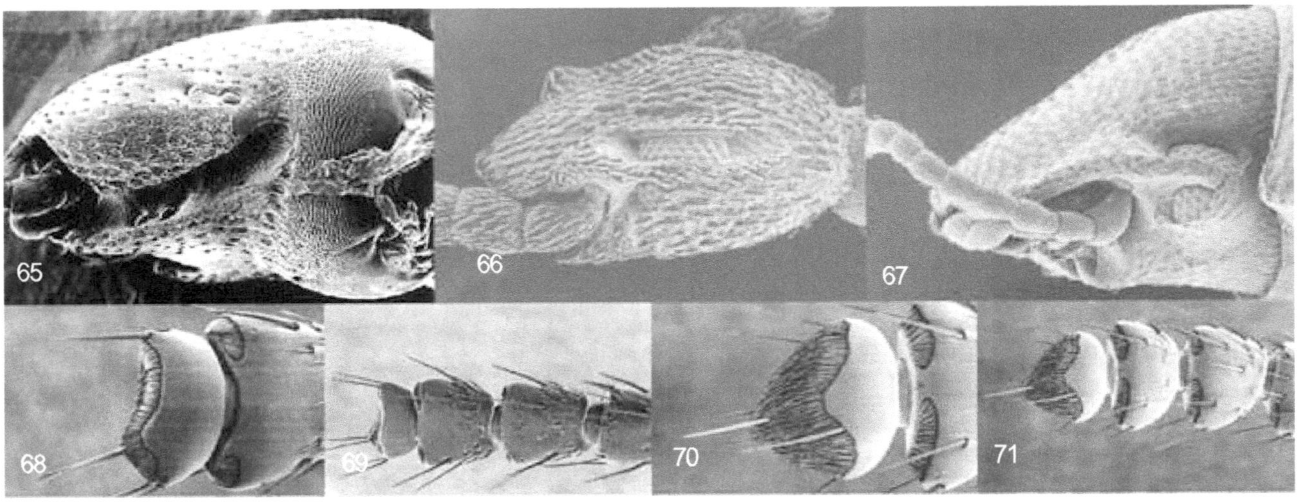

FIGURES 65.106-71.106. 65-67. Head, lateral view, 65. *Typhlusechus singularis* Linell; 66. *Araeoschizus andrewsi* Papp; 67. *Anepsius delicatulus* LeConte. 68-71. Terminal antennomeres, 68-69. *Asbolus verrucosus* LeConte; 70-71. *Cryptoglossa muricata* (LeConte).

insertion in genae; epistoma with clypeolabral membrane clearly exposed (Amarygmini) . Key Z

— Head normal, subhorizontal in repose; eyes variable; epistoma with clypeolabral membrane exposed or not 21

21(20). Antennae with compound, stellate sensoria on at least apical 5 antennomeres 22

— Antennae with compound, stellate sensoria on apical 3 or 4 antennomeres only 33

22(21). Antennae with compound, stellate sensoria on apical 5 or 6 antennomeres 23

— Antennae with compound, stellate sensoria on apical 7 or 8 antennomeres 25

23(22). Ventral surface of basal 3 or 4 tarsomeres covered by pads of dense, typically yellow pubescence (Fig. 9); ventral surface of tarsomeres flattened in most; head and eyes variable 24

— Ventral surface of basal 3 tarsomeres typically covered by stiff, dark colored setae, tarsomeres cylindrical; head widest anterior to eyes; eyes small, separated ventrally by distance greater than 3 times width of eye (Alphitobiini) .... Key X

24(23). Fourth tarsomere much smaller than 3rd, bearing only a few long, ventral setae (Fig. 13); body form cylindrical; epistoma with clypeolabral membrane clearly exposed (Fig. 7) (Talanini) .... *Talanus*

— Tarsomeres 3 and 4 subequal, most with pads of dense, yellowish pubescence (Fig. 15); body form variable; epistoma with clypeolabral membrane exposed or not (Centronopini and Coelometopini, in part) .... Key BB

25(22). Prosternal process declivous, flattened behind coxae; broadly rounded or truncate; mesosternal fossa very broad, shallow; epistoma with clypeolabral membrane clearly exposed (Fig. 8) (Strongyliini) .... *Strongylium*

— Prosternal process prominent, horizontal behind coxae; sharply acute and received in deep mesosternal fossa; epistoma with clypeolabral membrane exposed or (Centroncpini and Coelometopini, in part) .... Key BB

26(19). Eyes vertically elongate, subparallel, strongly inflated; epistoma with clypeolabral membrane clearly exposed (Helopini) .... Key V

— Eyes variable, not vertically elongate, subparallel or strongly inflated; epistoma with clypeolabral membrane rarely exposed 27

27(26). Metacoxae transverse, never oblique (Fig. 5); tarsi variable; epistoma with clypeolabral membrane exposed or not 28

— Metacoxae oblique (Fig. 6); tarsi spinous epistoma with clypeolabral membrane not exposed (Crypticini) .... Key P

28(27). Head strongly dilated and prominent at sides in front of eyes; anterior margin of clypeus with a deep triangular or arcuate notch (Fig. 9); epistoma with clypeolabral membrane not exposed (Opatrinae, in part) .... Key III

— Head not prominent laterally; anterior margin of clypeus without deep triangular or arcuate notch (Fig. 10); epistoma with clypeolabral membrane exposed or not 29

29(28). Epipleura narrow throughout length; pseudepipleura (Figs. 55, 56) very broad, gradually widening anteriorly, anterior aspect evenly rounded (Fig. 19); basal margin of pronotum truncate, closely embracing and overlapping basal aspect of elytra; males with pubescent pit at base of visible 2nd abdominal sternite; epistoma with clypeolabral membrane not exposed (Blaptini) . ... *Blaps*

— Epipleura gradually widening anteriorly (Fig. 18), sometimes obsolete anteriorly; pseudepipleura variable, anterior aspect angulate, never evenly rounded; basal margin variable, never closely embracing and overlapping basal aspect of elytra; males never with pubescent pit at base of 2nd visible abdominal sternite; epistoma with clypeolabral membrane exposed or not 30

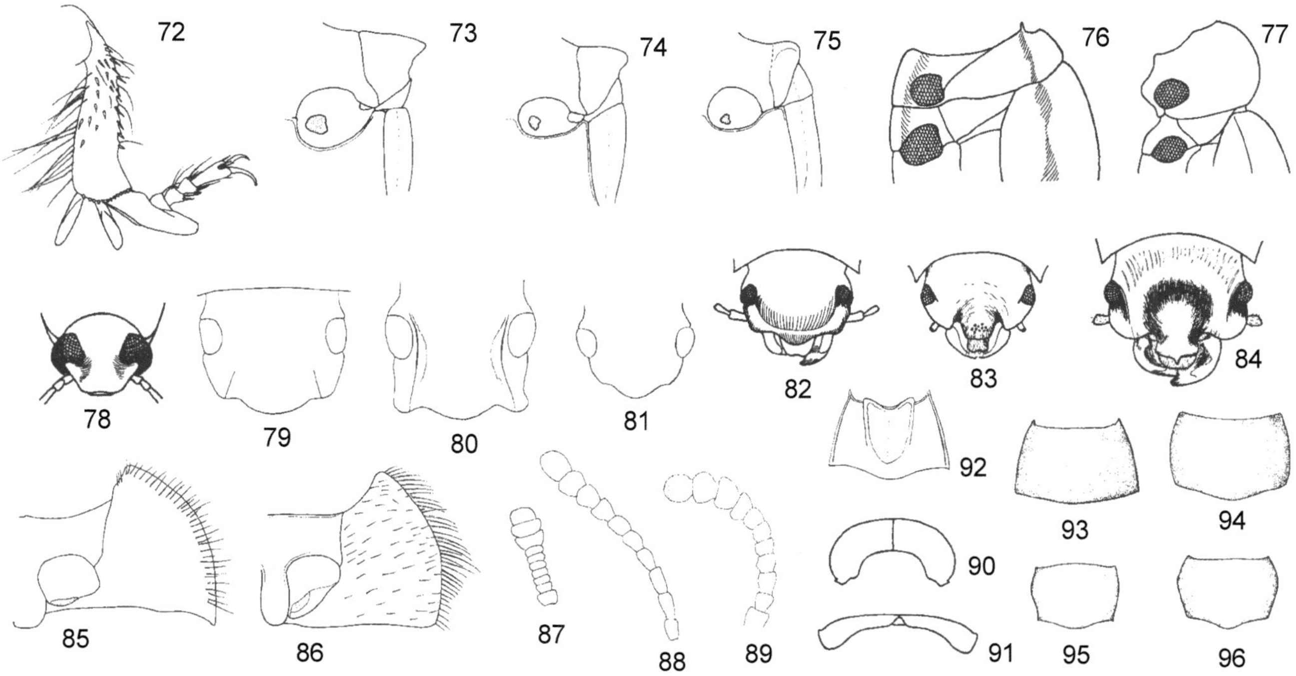

FIGURES 72.106-96.106. 72. *Coelus globosus* LeConte, protibia and tarsus. 73-75. Mesocoxae and mesothoracic structure, 73. *Batuliomorpha comata* Doyen; 74. *Anepsius valens* Casey; 75. *Batuliodes rotundicollis* (LeConte). 76-77. Meso- and metasternum, lateral view, 76. *Edrotes ventricosus* LeConte; 77. *Craniotus pubescens* LeConte. 78-84. Head, dorsal view, 78. *Alaephus pallidus* Horn; 79. *Metopoloba pruinosa pruinosa* (Horn); 80. *Polemiotus submetallicus* (LeConte); 81. *Schoenicus puberulus* LeConte; 82. *Stibia imperialis* Blaisdell; 83. *Trimytis pruinosa* LeConte; 84. *Triorophus rugiceps* LeConte. 85-86. Left hypomeron, ventral view, 85. *Eusattus minimus* Doyen; 86. *Conisattus rectus* Casey. 87-89. Antenna, 87. *Batuliomorpha comata* Doyen; 88. *Anepsius delicatulus* LeConte; 89. *Batulius setosus* LeConte. 90-91. Elytra, posterior view, 90. *Triorophus laevis* LeConte; 91. *Bothrotes canaliculatus* (Say). 92-96. Pronotum, 92. *Bothrotes canaliculatus arundinis* (LeConte); 93. *Trimytis pruinosa* LeConte; 94. *Chilometopon helopioides* Horn; 95. *Chilometopon brachystomum* Doyen; 96. *Chilometopon abnorme* (Horn). (Figures 76-78, 90, 91 from Arnett 1973; Figures 82-84 from Blaisdell 1936; Figures 72, 85, 86 from Doyen 1984; Figures 73-75, 87, 88 from Doyen 1987; Figure 89 from Doyen and Lawrence 1979; Figures 79, 80, 92 from Freude 1967; Figure 81 from Freude 1968; Figures 93-96 from MacLachlan and Olson 1990).

30(29). Epipleural fold forming only part of inflexed side of elytra (Figs. 55, 56); metacoxae moderately to broadly separate; epistoma with clypeolabral membrane not exposed 31

— Epipleural fold forming entire inflexed sides of elytra (Fig. 57, 58); metacoxae generally narrowly separated; epistoma with clypeolabral membrane exposed or not 33

31(30). Antennomeres 2-5 about as long as broad, or only slightly longer than broad; fronto-clypeal suture typically distinct; labrum mostly concealed; metacoxae variable 32

— Antennomeres 2-5 distinctly longer than broad; fronto-clypeal suture indistinct; labrum clearly exposed; metacoxae moderately separated (Fig. 4) (Eleodini, in part) .... Key R

32(31). Intercoxal process of abdomen 0.7 to 1.3 times wider than long; narrowing anteriorly, with anterior angles broadly rounded (Fig. 44); males and females lacking teeth on hind femur; size small (length, typically less than 1 cm); often pubescent (Eulabini) .... Key S

— Intercoxal process of abdomen 1.1 to 1.9 times wider than long; not widening or slightly widening anteriorly, with anterior angles more narrowly rounded (Fig. 45); males with one or more teeth on hind femur; size large (length more than 1 cm); never pubescent (Cerenopini) .... Key T

33(30). Body setose or with buccal process of subgenae narrow, acute, strongly produced; antennae never bearing compound sensilla; epistoma with clypeolabral membrane not exposed (Eleodini, in part) .... Key R

— Body glabrous, with buccal process of subgenae unmodified, not produced, never acute; antennae sometimes bearing compound sensilla (Figs. 24, 25); epistoma with clypeolabral membrane rarely exposed (Tenebrionini) .... Key Y

II. Key to the Tribes of Pimeliinae of the United States

1. Protibiae with single apical spur and submedian dorsal tooth; mandibles grooved externally throughout their length (Cnemodini) .... *Cnemodinus*

— Protibiae with 2 apical spurs not toothed dorsally near middle; most with mandibles not grooved externally 2

2(1). Body small (less than 4 mm), at least head and pronotum covered with erect and/or sessile appressed, scales; antennae shorter than width of head, last 3 antennomeres forming a club;

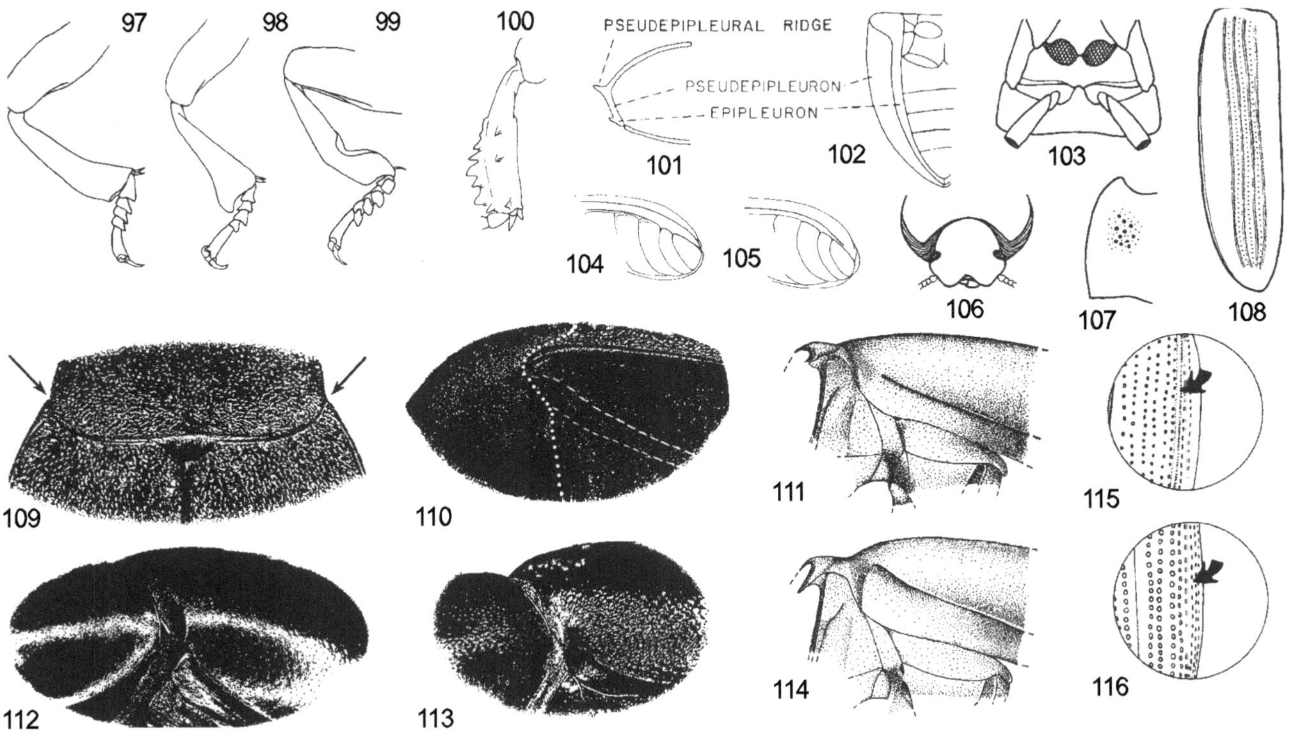

FIGURES 97.106-116.106. 97-99. Protibia and tarsus, anterior view. 97. *Trichoton sordidum* (LeConte), female; 98. *Ulus crassus* (LeConte); 99. *Trichoton sordidum* (LeConte), male. 100. *Uloma longula* LeConte, protibia, posterior view. 101. *Eusattus reticulatus* (Say), x-section of elytron and first abdominal sternite; 102. *Eusattus reticulatus* (Say), pterothorax, abdomen and eltyra, ventral view; 103. *Helops aereus* Germar, metasternum; 104. *Ulosonia* sp., apex of abdomen and elytron; 105. *Tharsus* sp., apex of abdomen and elytron; 106. *Lariversius tibialis* Blaisdell, head, dorsal view; 107. *Ulosonia tricorni* LaPorte, pronotum; 108. *Tribolium castaneum* (Herbst), left elytron, dorsal view; 109. *Eleodes (Promus) opacus* (Say), pronotum and elytral base, dorsal view. 110-114. Pronotum and elytral base, lateral view, 110. *Eleodes (Promus) opacus* (Say); 111. *Eleodes (Blapylis) tenebrosus* Horn; 112. *Eleodes (Steneleodes) giganteus* Mannerheim; 113. *Eleodes (Discogenia) marginatus* Eschscholtz; 114. *Eleodes (Blapylis) blanchardi* Blaisdell. 115-116. Lateral elytral margin, dorsal view, 115. *Coelopalorus foveicollis* (Blair); 116. *Palorus subdepressus* Wollaston. (Figures 103, 106-108 from Arnett 1973; Figures 97-99 from Aalbu and Triplehorn 1985; Figures 101, 102 from Doyen 1984; Figures 100, 104, 105 from Doyen 1985; Figures 115, 116 from Spilman 1991).

protibiae strongly expanded, covered with scales (Cnemeplatiini) .... Key D

— Body size variable, most are larger, head and pronotum without erect and/or sessile scales; antennae typically longer than head; protibiae, if expanded apically, never covered with scales 3

3(2). Mentum very large, concealing maxillae and usually ligula 4

— Mentum generally smaller in size, never concealing both ligula and maxillae 6

4(3). Eighth antennomere enlarged, much larger than tenth or seventh (Fig. 41) (Asidini, in part) .... Key G

— Antennae gradually enlarged in most, eighth antennomere never much larger than tenth or seventh (Fig. 40) 5

5(4). Mentum hexagonal, with apex more or less distinctly emarginate or sinuate (Fig. 27) (Eurymetopini)... .... Key L

— Mentum transversely trapezoidal, apex very broadly arcuate from side to side and not sinuate at middle, generally much more transverse (Fig. 32) (Epitragini) .... Key K

6(3). Size small (1.6 to 6 mm); eyes divided into upper and lower groups of facets 7

— Size typically larger; eyes rarely divided, f divided, then size much larger than 10 mm 8

7(6). Femora slender; protibiae never apically expanded; eyes with lower group of facets always positioned anteriorly under upper group (Fig. 66) (Stenosini) ....Key E

— Femora stout; protibiae apically expanded, triangular (Figs. 36-38); eyes with lower group of facets positioned centrally under upper group (Fig. 67) (Anepsiini, in part) .... Key I

8(6). Tarsi ventrally with dense short yellow setae; cuticle rough, densely punctate-tuberculate (Nyctoporini) .... *Nyctoporis*

— Tarsi ventrally spinose, rarely with long yellow setae, if so, cuticle smooth 9

9(8). Body and appendages elongate, slender; eyes large, weakly emarginate anteriorly; antennae slender, filiform, clearly longer than head and pronotum (Vacronini) .... Key J

— Body ovate or elongate-ovate, appendages stout; eyes vertically elongate; antennae shorter, not reaching base of pronotum 10

10(9). Labrum covered by expanded clypeus, scarcely visible (Cryptoglossini) .... Key F

— Labrum not covered by clypeus, mostly visible 11

11(10). Eighth antennomere enlarged, larger than seventh, much larger than reduced tenth (Fig. 41); mentum trapezoidal, filling at least three-quarters of buccal space, ligula small or not visible (Figs. 28, 33), cardo partly or entirely concealed (Asidini, in part) .... Key G

— Antennae gradually enlarged or with 3-segmented club, eighth antennomere never much larger than tenth or seventh; mentum lunate, filling half or less of buccal space (Fig. 29), ligula large (Fig. 26, 31) 12

12(11) Antenna with distinct, 3-segmented club (Fig. 42); elytra bearing numerous clavate setae (Branchini) .... *Branchus*

— Antenna gradually enlarged (Fig. 40); elytra glabrous or bearing slender, long setae (Coniontini) .... Key H

III. Key to the Tribes of Opatrinae of the United States

1. Head strongly dilated and prominent at sides in front of eyes; anterior margin of clypeus with a deep triangular or arcuate notch (Fig. 9); epistoma with clypeolabral membrane not exposed 2

— Head not prominent laterally; anterior margin of clypeus without deep triangular or arcuate notch (Fig. 10) (Eleodini, in part) .... Key R

2(1). Gula with stridulating surface, consisting of symmetrically arranged, slender, transverse ridges and fossae (Fig. 43); eyes not entirely divided by epistomal canthus (Platynotini) .... *Opatrinus*

— Gula simple, irregularly sculptured or smooth, without stridulating surface; eyes entirely divided or not 2

3(2). Eyes with erect scales between facets (Leichenini) .... *Leichenum*

— Eyes without erect scales between facets 3

4(3). Protibiae with 2 strongly produced dorsal extensions, one apical, one near midpoint of tibia (Fig. 39) (Melanimini) .... *Cheirodes*

— Protibiae unmodified, triangular or at most with one strongly produced dorsal extension 5

5(4). Eyes completely divided by epistomal canthus (Fig. 54) (Opatrini, in part) .... Key Q

— Eyes whole or emarginate, not completely divided by epistomal canthus 6

6(5). Elytra costate (Eleodini, in part) .... Key R

— Elytra not costate 6

7(6). Meso-metacoxae separated by at least width of coxae (Eleodini, in part) .... Key R

— Meso-metacoxae separated by much less than width of coxae (Opatrini, in part) .... Key Q

A. Key to the Genera of the Tribe Lagriini

1. Elytra without striae; coarsely punctate; eyes transverse, reniform, not prominent; head not constricted behind to form a neck ..... *Arthromacra*

— Elytra with distinct striae; eyes large, convex, prominent; head constricted behind forming a neck; punctures variable .... *Statira*

B. Key to the Genera of the Tribe Goniaderini

1. Eyes absent, northern California ...... *Eschatoporis*

— Eyes present 2

2(1). Body glabrous; thorax quadrate, margin simple, antenna with last 3 antennomeres enlarged .... *Prateus*

— Body pubescent; margin of thorax sometimes denticulate; antenna variable 3

3(2). Antenna gradually expanded; margin of thorax simple, often laterally expanded .... *Anaedus*

— Antenna with last 3 antennomeres abruptly enlarged; margin of thorax denticulate .... *Paratenetus*

C. Key to the Genera of the Tribe Penetini

1. Antennal club composed of 3 antennomeres .... *Clamoris*

— Antennal club composed of 2 antennomeres .... *Dioedus*

D. Key to the Genera of the Tribe Cnemeplatiini

1. Eyes well developed, composed of many facets . ... *Lepidocnemeplatia*

— Eyes reduced to single facet .... *Alaudes*

E. Key to the Genera of the Tribe Stenosini

1. Antennae very small, slender; elytra smooth, glabrous; eyes reduced, composed of 2 widely separated pair of facets (Fig. 65) .... *Typhlusechus*

— Antennae moderately long, thick; elytra costate; eyes larger, not composed of pair of facets, divided into upper and lower clusters .... *Araeoschizus*

F. Key to the Genera of the Tribe Cryptoglossini

1. Eyes with dorsal and ventral portions completely separated by epistomal canthus, epistomal canthus strongly inflated laterally .... *Schizillus*

— Eyes with dorsal and ventral portions narrowed at center, not completely divided; epistomal canthus slightly inflated laterally 2

2(1). Antennae in cross section flattened, narrowly oval, with size of sensory area on terminal antennomere reduced, truncate (Figs. 68, 69); eyes strongly narrowed at center to width of 3 facets .... *Asbolus*

— Antennae in cross section broadly oval, with size of sensory area subtriangular (Figs. 70, 71); eyes only slightly narrowed at center (eye facets at center 5 or more) .... *Cryptoglossa*

G. Key to the Genera of the Tribe Asidini

1. Mentum tightly contacting postgenal process exposing at most a minute space at basal corners; mentum sometimes contacting mandibles apically; gular pedestal weakly developed or not evident (Fig. 30); mesotrochantin small to absent 2
— Mentum not or but loosely contacting postgenal process (Fig. 34), or if tightly contacting postgenal process, a large space is exposed at basal corners; mentum never contacting mandibles apically; gular pedestal evident and well developed in most; mesotrochantin large 4

2(1). Ligula hidden; mesotrochantin absent or small, epipleura of elytra narrow from apex to base (Fig. 64), epipleural fold distinct only near apex and weak or absent over remaining length 3
— Ligula visible (Fig. 35); mesotrochantin present, epipleura narrow apically, widening toward elytral base (Fig. 63), most with epipleural fold distinct over entire length .... *Microschatia* (in part)

3(2). Lateral margin of mentum bent inward toward mandible leaving a very narrow space (Fig. 53); postgenal process often swollen; prosternal process often porrect behind coxae (viewed laterally), sloping precipitously toward mesosternum (Fig. 47); mesotrochantin absent; sculpture of elytra rough and uneven .... *Astrotus*
— Lateral margin of mentum not bent inward toward mandible, leaving a wider space near apex of moderately thickened postgenal process; prosternal process deflexed behind coxae (viewed laterally), sloping at an incline toward mesosternum (Fig. 46); mesotrochantin small but visible; elytra with fine anastomosing dorsal ridges .... *Stenosides*

4(1). Frons rounded laterally in front of eyes, width subequal to width between eyes; pronotum not globular, with lateral margins present, body variable 5
— Frons pointed and toothlike laterally, projecting well beyond eyes; pronotum globular, lacking lateral margins; body hirsute .... *Craniotus*

5(4). Epipleura of elytra narrow apically, widening toward elytral base (Fig. 63), most with epipleural fold distinct over entire length; prosternal process porrect and prominent behind coxae 6
— Epipleura of elytra remaining narrow from apex to base (Fig. 64), epipleural fold very distinct near cauda and often weak or absent over remaining length; prosternal process variable 7

6(5). Mentum large and in close contact with postgenal process, concealing cardo; antenna and tarsi thick and compact; hind angles of pronotum not or slightly prolonged backward over elytra .... *Microschatia* (in part)
— Mentum small and separated from postgenal process, revealing cardo; antennae and tarsi slender and elongate; hind angles of pronotum acutely prolonged backward over elytra .... *Litasida*

7(5). Ligula completely retracted and hidden under apical margin of mentum (Fig. 28), ligula very small; maxilla with apical palpomere extremely enlarged, rectitriangular in the female, scalene in the male with proximal angle acute and prolonged 8
— Ligula always at least partially exposed and projecting beyond apical margin of mentum (Fig. 34), ligula size variable; most with apical palpomere of maxilla only moderately enlarged, but if extremely enlarged then ligula is very large and tumid 10

8(7). Tarsi with short and inconspicuous spiniform setae beneath; elytra with distinct, though sometimes irregular carinae; pronotum variable 9
— Tarsi with long, fulvous, silky setae beneath; elytra may be deeply sulcate, but not with distinct, well separated carinae; pronotum disk smooth, sides often dilated and reflexed (wing-like) .... *Gonasida*

9(8). Width of pronotum at basal angles at least 1/5 greater than width of elytra at humeri; elytra each with marginal carina and 2 distinct, subequal discal carinae .... *Clyptasida*
— Width of pronotum at basal angles subequal to width of elytra at humeri; elytra each with marginal carina and only 1 distinct discal carina, a 2nd inner carina absent or poorly defined .... *Philolithus*

10(7). Hind angles of pronotum weakly or not at all overlapping elytra (Figs. 50, 52); elytral sculpture variable 11
— Hind angles of pronotum distinctly prolonged backward and overlapping elytra (Fig. 51); elytra each with 2 or 3 strongly carinate discal ridges, lateral margin always carinate .... *Parasida*

11(10). Base of pronotum convex or convexo-emarginate (Fig. 50); hind angles of pronotum obtuse to rounded, not prominent 12
— Base of pronotum truncate, concave (Fig. 52), or bisinuate (Fig. 49); hind angles of pronotum rectangular to acute, somewhat prominent in most. 14

12(11). Mentum small, not in contact with postgenal process which is typically thin; base of postgenal process not parallel to gular pedestal, forming a rounded or angled open space; body, maxillary palpi variable 13
— Mentum large, often in parallel contact with postgenal process which is thickened, base of postgenal process parallel to gular pedestal and often forming a quadrate open space; body glabrous; maxillary palpi weakly sexually dimorphic .... *Pycnomorpha*

13(12). Maxilla with apical palpomere greatly enlarged and sexually dimorphic, rectitriangular in the female (Fig. 59), larger and scalene with proximal angle prolonged in the male (Fig. 60); body rarely hirsute; ligula large and tumid .... *Stenomorpha* (in part)
— Maxilla with apical palpomere smaller, only moderately enlarged and often not notably sexually dimorphic, at most rectitriangular in the male (Fig. 61, 62); disc of pronotum and elytra with setae present, body often hirsute; ligula small, not tumid .... *Trichiasida* (in part)

14(11). Prosternal process (viewed laterally) deflexed or obtusely prominent behind coxae, not sharply margined posteriorly 15
— Prosternal process (viewed laterally) porrect and extending over mesosternum, broadly lobed and sharply margined posteriorly (Fig. 48) .... *Stethasida*

15(14). Body clothed with short, but numerous and clearly visible setae 16
— Body glabrous, or setae very inconspicuous ... 18

16(15). Lateral margin of elytra strongly carinate 17
— Lateral margin of elytra not carinate; body often hirsute .... *Trichiasida* (in part)

17(16). Disc of elytra flat .... *Asidina* (in part)
— Disc of elytra convex .... *Trichiasida* (in part)

18(15). Elytra lacking any discal carinae; lateral margin variable 19
— Elytra each with at least one distinct discal carina, lateral margin always carinate .. *Asidina* (in part)

19(18). Mentum small (Fig. 33), with wide space between it and postgenal process; width of mentum at base about half the width at middle....................... 20
— Mentum large (Fig. 34), often contacting postgenal process; width of mentum at base about three-quarters the width at middle .... *Megasida*

20(19). Elytra humeri prominent and raised .... *Asidopsis*
— Elytra humeri inconspicuous and deflexed .... *Stenomorpha* (in part)

H. Key to the Genera of the Tribe Coniontini

1. Protarsus cylindrical, apically truncate 2
— Protarsus with basal tarsomere bearing long spatulate process extending beyond 2nd tarsomere (Fig. 72) .... *Coelus*

2(1). Protibiae with outer margin expanded as prominent, lamellate process 2
— Protibiae subcylindrical to apex, sometimes briefly hooked at apex of outer margin .... *Coniontis*

3(2). Protarsus with basal tarsomere more than twice length of 2nd; pronotal edge glabrous or with few short, squamose setae; hypomeron mostly glabrous except for fringe just below pronotal edge (Fig. 85) .... *Eusattus*
— Protarsus with basal tarsomere subequal in length to 2nd; pronotal edge bearing fringe of long setae (Fig. 86); hypomeron densely setose throughout .... *Conisattus*

I. Key to the Genera of the Tribe Anepsiini

1. Tarsi 4-4-4; thorax and elytra costate, protibiae not apically expanded .... *Anchomma*
— Tarsi 5-5-4: thorax and elytra not costate, protibiae apically expanded 2

2(1). Middle coxal cavity with trochantin concealed; sternal lobes meeting laterad of coxal cavities (Fig. 75) .... *Batuliodes*
— Middle coxal cavity with trochantin exposed; sternal lobes separated by narrow space laterad of coxal cavities (Figs. 73, 74) 3

3(2). Metasternum about 1.0 to 1.5 times as long as mesocoxa; pronotum and elytra subglabrous or with short, appressed setae; apical antennomere longer than broad (Fig. 88) .... *Anepsius*
— Metasternum about twice as long as mesocoxa; pronotum and elytra fimbriate along lateral margins; apical antennomere subquadrate or broader than long 4

4(3). Posterior pronotal angles strongly angulate; antennal club composed of 4 antennomeres (Fig. 89); lateral fimbriation short, stiff; venter with few short setae .... *Batulius*
— Posterior pronotal angles obliterated or represented by minute tubercles; antennal club composed of 3 antennomeres (Fig. 87); lateral fimbriation long, flying; venter with numerous long setae .... *Batuliomorpha*

J. Key to the Genera of the Tribe Vacronini

1. Epistoma with anterior and lateral borders strongly reflexed (Fig. 78); pronotum with angulation between dorsum and hypomeron acute .... *Alaephus*
— Epistoma with anterior border not reflexed and lateral border only weakly reflexed; pronotum with angulation between dorsum and hypomeron obtuse or rounded .... *Eupsophulus*

K. Key to the Genera of the Tribe Epitragini

1. Lateral lobes of epistoma straight or convex, not prominent apically; eyes without furrows (Fig. 79) 2
— Lateral lobes of epistoma obliquely prominent apically; eyes with deep furrows over inner sides (Fig. 80) .... *Polemiotus*

2(1). Eyes not inflated (Fig. 79), scarcely projecting outward beyond lateral margin of head, not overlapping temporal ridge; prosternum and mesosternum variable, body variable 3
— Eyes inflated (Fig. 81), projecting outward beyond lateral margin of head, overlapping temporal ridge, prosternum rounded, deflexed, mesosternum not excavate, body glabrous .... *Schoenicus*

3(2). Pronotum widest at or posterior to midpoint, sometimes more or less parallel-sided posteriorly; most with base wider than anterior margin 4
— Pronotum widest anterior to midpoint, anterior margin subequal in width to base, hind angles rounded (Texas) .... *Conoecus*

4(3). Prosternum broad, obtuse, mesosternum at most shallowly excavate 5
— Prosternum narrow, acute, mesosternum deeply excavate 6

5(4). Body metallic black, typically glabrous, eyes flat or weakly convex (Texas) .... *Phegoneus*

— Body brownish, distinctly mottled with scattered setal patches, eyes moderately convex (Florida and Georgia) .... *Epitragodes* (in part)

6(4). Pronotum not narrowing posteriorly; body rarely mottled with scattered setal patches 7

— Pronotum narrowing posteriorly; body distinctly mottled with scattered setal patches; no supraorbital carina (Florida and Georgia) .... *Epitragodes* (in part)

7(6), Supraorbital carina distinct, body setae short and indistinct 8

— Supraorbital carina weak or absent; body setae variable 9

8(7). Temples immediately constricted behind eyes; elytra sub-acute at apex; females without pronotal elevations .... *Metopoloba*

— Temples not immediately constricted behind eyes; elytra broadly rounded at apex; females with distinct pronotal elevations (Fig. 92) .... *Bothrotes* (in part)

9(7). Fully alate, elytra not fused at suture; temples not immediately constricted behind eyes 10

— Brachypterous, elytra fused at suture; temples immediately constricted behind eyes .... *Epitragosoma*

10(9). Body moderately elongate; elytra subacute at apex, sometimes notably hirsute; temples narrow behind eyes; females without pronotal elevations 11

— Body stout; elytra broadly rounded at apex; temples broad behind eyes; females with distinct pronotal elevations (Fig. 92) .... *Bothrotes* (in part)

11(10). Elytra with white setae forming striate or mottled patterns; pronotum wider L:W = 2.3 ... *Pechalius*

— Elytra uniformly or indistinctly hirsute; pronotum longer L:W = 3.4 .... *Lobometopon*

L. KEY TO THE GENERA OF THE TRIBE EURYMETOPINI

1. Ventral length of thorax greater than visible length of abdomen; mesosternum often elevated, flat, abutting closely against and on the same plane as apex of flattened prosternal process (Fig. 76); broadly oval, setose species .... *Edrotes*

— Ventral length of thorax less than visible length of abdomen; mesosternum not elevated, its surface not continuous with surface of prosternal process (Fig. 77) body not broadly oval, setose or not 2

2(1). Elytra embracing sides of body, inflexed parts not wholly occupied by epipleural fold (Fig. 90) ... 3

— Elytra not embracing sides of body, inflexed parts wholly occupied by epipleural fold (Fig. 91) ... 9

3(2). Epistomal lobe very prominent, expanded, angulate at apex (Fig. 84), dorsally forming large elevated tubercle .... *Triorophus*

— Epistomal lobe not expanded, angulate at apex, without dorsal large elevated tubercle 4

4(3). Pronotum subcylindrical, not angulate laterally, pronotum with lateral carina present only in basal and apical fifths; antennae very long and slender, pedicel much longer than twice its maximum width .... *Eschatomoxys*

— Pronotum either angulate laterally throughout its length or with lateral carina present throughout its length; antennae with pedicel subequal to or less than twice its maximum width 5

5(4). Body strongly pubescent 6

— Body glabrous 8

6(5). Pronotum strongly inflated, globose, strongly constricted with elytra at base; epistomal apex slightly produced, apex broadly triangular with slight submarginal declination .... *Trichiotes*

— Pronotum only moderately inflated, not strongly constricted with elytra at base; epistomal apex strongly produced 7

7(6). Epistomal apex triangular; tarsi not pubescent..... .... *Oxygonodera*

— Epistomal apex laterally straight, with central strongly produced lobe, apex rounded; tarsi pubescent, with coarse, dense yellow setae .... *Triphalopsis*

8(5) Thorax with anterior and posterior margins approximately equal in width 9

— Thorax with posterior margins much wider than anterior margin .... *Auchmobius*

9(8). Epistomal lobe distinctly deflexed, line forming upper margin of deflexed portion continuous with sides of frons when viewed from above (Fig. 82); larger species, more than 5 mm in length .... *Stibia*

— Epistomal lobe not distinctly deflexed; smaller species, less than 5 mm in length .... *Micromes*

10(8). Protibiae produced dorsally at apex 11

— Protibiae not produced dorsally at apex 15

11(10). Head without distinct supra-orbital carina .... *Eurymetopon*

— Head with distinct supra-orbital carina 12

12(11). Epistomal lobe not produced, straight to concave .... *Metoponium*

— Epistomal lobe slightly produced, angulate, apex truncate 13

13(12). Body elongate .... *Telabis*

— Body broadly oval, wingless 14

14(13). Prosternum produced into a tumescent keel with an angular apex .... *Texaponium*

— Prosternum declivous behind coxae .. *Cryptadius*

15(10). Epistomal lobe moderately to strongly prolonged (Fig. 83) 16

— Epistomal lobe at most only slightly produced (Fig. 82) 17

16(15) Pronotum widest at base (Fig. 93); wings absent; lateral frontal margin reaching middle of eye .... *Trimytis*

— Pronotum widest at or near mid point (Figs. 94-96); wings present; lateral frontal margin reaching no more than one-third across eye sometimes barely affecting margin .... *Chilometopon*

17(15). Eyes small, not prominent, lateral margin straight, about half divided by epistomal canthus .... *Melanastus*
— Eyes large, prominent, lateral margin convex ... 18

18(17). Tarsi pubescent ventrally .... *Armalia*
— Tarsi with sparse, stiff, subspiniform setae ventrally 19

19(18). Body elongate, parallel .... *Hylocrinus*
— Body oval, convex .... *Steriphanus*

M. Key to the Genera of the Tribe Bolitophagini

1. Tarsal formula 4-4-4 (Subtribe Rhipidandrina)........ .... *Rhipidandrus*
— Tarsal formula 5-5-4 (Subtribe Bolitophagina)...... 2

2(1). Antennae 10-segmented .... *Bolitotherus*
— Antennae 11-segmented 3

3(2). Eyes not divided; apical antennomere free, eighth truncate 4
— Eyes divided; apical antennomere set partially within emarginate apex of eighth .... *Megeleates*

4(3). Pronotum tuberculate; elytra with coarse interrupted ridges .... *Bolitophagus*
— Pronotum punctate; elytra with fine continuous ridges .... *Eleates*

N. Key to the Genera of the Tribe Phaleriini

1. Antennae long, extending well beyond pronotal base; prosternal process deflexed .... *Phaleromela*
— Antennae short, not extending middle of pronotum; most with prosternal process not deflexed....... .... *Phaleria*

O. Key to the Genera of the Tribe Diaperini

1. Antennae with terminal 5-8 antennomeres distinctly enlarged as club 2
— Antennae gradually enlarged to apex, or subfiliform 5

2(1). Antennal club composed of 6-8 enlarged antennomeres; eyes emarginate anteriorly 3
— Antennal club composed of 5 antennomeres; eyes round, not at all emarginate .... *Pentaphyllus*

3(2). Ventral distance between eyes greater than width of oral fossa; most with epipleuron narrowed abruptly near anterior margin of 5th sternite ... 4
— Ventral distance between eyes less than width of oral fossa; epipleuron narrowing gradually to elytral apex .... *Ulomoides*

4(3). Body broadly oval, strongly convex; without frontal horns; body length greater than 4 mm .... *Diaperis*
— Body elongate oval, subcylindrical; males with frontal horns, females with tubercles; body length less than 5 mm .... *Neomida*

5(1). Mesocoxal cavities closed by meso- and metasternite 7
— Mesocoxal cavities bordered laterally by mesepimeron 6

6(5). Mesosternum concave between middle coxae .... *Platydema*
— Mesosternum projecting anterad as prominent, horizontal lobe .... *Liodema*

7(5). Epipleuron narrowed abruptly near anterior margin of 5th sternite; dorsum glabrous 8
— Epipleuron narrowed gradually to elytral apex; dorsum setose .... *Alphitophagus*

8(7). Posterior pronotal corners angulate 9
— Posterior pronotal corners broadly rounded .... *Doliodesmus*

9(8). Elytra with 8th interval simply convex or nearly flat; antennomeres 6-8 at least as wide as long 10
— Elytra with 8th interval sharply angulate, most are weakly carinate; antennomeres 6-8 longer than wide .... *Adelina*

10(9). Posterior pronotal border evenly convex; mandibles of males bearing dorsal horns 11
— Posterior pronotal border bisinuate; mandibular horns absent in both sexes 12

11(10). Pronotum longer than broad, anterior corners rounded .... *Iccius*
— Pronotum broader than long, anterior corners angulate .... *Gnatocerus*

12(10). Anterior pronotal corners angulate; pronotal disc evenly convex .... *Cynaeus*
— Anterior pronotal corners rounded; pronotal disc flat medially, declivous laterally, especially near the front margin .... *Sitophagus*

P. Key to the Genera of the Tribe Crypticini

1. Body elongate oval in dorsal view, more than twice as long as wide; dorsum brightly colored, with pronotum orange and elytra yellow and black .. .. *Poecilocrypticus*
— Body broadly oval in dorsal view, not more than twice as long as wide; most with dorsum dark and uniformly colored; if elytra maculate, then pronotum brown or black, not orange .... *Gondwanocrypticus*

Q. Key to the Genera of the Tribe Opatrini

1. Eyes completely divided by epistomal canthus (Fig. 54) 4
— Eyes whole or emarginate, not completely divided by epistomal canthus 2

2(1). Body covered by whitish-grey scales; maxilla with apical palpomere elongate-oval, acuminate .... *Ammodonus*
— Body glabrous; maxilla with apical palpomere triangular or securiform 3

3(2). Elytra without striae, but with dense confused granules .... *Ephalus*

— Elytra with punctate striae, fine granules on intervals only .... *Pseudephalus*

4(1). Scutellum triangular to subtriangular; metathoracic wings often well developed; most with protarsus of male distinctly dilated 5

— Scutellum very broad and short; metathoracic wings absent; protarsus of male not dilated 10

5(4). Base of pronotum bisinuate................................ 6

— Base of pronotum not bisinuate 8

6(5). Protibiae distinctly bent apically (Figs. 97, 99); vestiture of 2 types, erect brownish setae gradually broadening to apex and narrow flat pale setae .... *Trichoton*

— Protibiae straight; vestiture simple...................... 7

7(6). Protibiae produced dorsally at apex (Fig. 98); body laterally fimbriate .... *Ulus*

— Protibiae not produced dorsally at apex; body not laterally fimbriate .... *Blapstinus*

8(5). Basal pronotal margin evenly arcuate; body in dorsal view, narrow, convex; surface and sides densely fimbriate .... *Aconobius*

— Basal pronotal margin straight; surface and sides not densely fimbriate 9

9(8). Body, in dorsal view, broadly oval, strongly convex; pronotum widened behind; basal pronotal margin equal in width to elytral base; males with dense setal patches laterally to midline of metasternum .... *Cybotus*

— Body, in dorsal view, broadly oval, strongly convex; pronotum widened behind; basal pronotal margin equal in width to elytral base; males with dense setal patches laterally to midline of metasternum .... *Mecysmus*

10(4). Pronotum densely fimbriate laterally................. 11

— Pronotum not fimbriate laterally......................... 12

11(10). Protibiae broadly triangular, compressed; body stout, oblong-oval .... *Notibius*

— Protibiae narrow, non-fossorial; body narrow parallel-sided .... *Conibiosoma*

12(10). Elytral intervals elevated on disc...................... 13

— Elytral intervals not elevated on disc....... *Conibius*

13(12). Elytra, pronotum distinctly narrowed at base; basal width of pronotum narrower than basal width of elytra, widest anterior to middle; elytral intervals moderately convex with serial rows of asperate punctures or tubercles on crest ...... *Nocibiotes*

— Elytra subparallel, not narrowed at base; pronotum with lateral margins subparallel, basal width of pronotum equal to basal width of elytra; elytral intervals strongly convex, minutely punctured along crest .... *Tonibius*

R. Key to the Genera of the Tribe Eleodini

1. Pseudoepipleura narrow throughout most of length, slightly broadening at humeral angles, sometimes basally obsolete 2

— Pseudoepipleura of moderate width to broadly expanded, gradually broadening throughout length; never basally obsolete 6

2(1). Pseudoepipleural ridge present, distinct (Figs. 101, 102) 3

— Pseudoepipleural ridge absent........................... 5

3(2). Elytra each with 4 sharp discal costae (not including sutural costae which may or may not be present) .... *Trogloderus*

— Elytra without discal costae 4

4(3). Pronotum margin laterally acute to foliaceous; dorsal lateral sides of elytra angled to acutely angled .... *Embaphion*

— Pronotum margin laterally acute; dorsal lateral sides of elytra rounded .... *Neobaphion* (in part)

5(2). Pronotum laterally hirsute; sides of epistoma distinctly dilated, prominent (Fig. 106); profemora with basal 2/3 laterally expanded .... *Lariversius*

— Pronotum not hirsute; sides of epistoma not dilated, prominent; profemora without basal 2/3 laterally expanded .... *Neobaphion* (in part)

6(1). Pseudoepipleural fold occupying entire inflexed portion of elytra; buccal processes of subgenae strongly and acutely produced .. *Eleodimorpha*

— Pseudoepipleural fold occupying only part of inflexed portion of elytra; buccal processes of subgenae never strongly produced (*E. blanchardi* Blaisdell, subgenae weakly produced) 7

7(6). Front margin of profemora sinuate in apical 1/5 in both sexes; elytral disc somewhat flattened, laterally angled; female ovipositor triangular, depressed, dorsal surface not setose, apices of coxites strongly produced beyond the gonostyle, chitinous, more or less everted, convex dorsally, concave ventrally .... *Neobaphion* (in part)

— Not with above combination of characters (*Eleodes*) .... Key Ra

Ra. Key to Subgenera of *Eleodes*

Note: As in the previous version of this book, we have included a key to the subgenera of *Eleodes*. Please note that subgeneric classification is primarily based on female genitalia. Certain characters involving external morphology are often variable, making the construction of a key that will work for all specimens (and species) difficult. For instance, the basal protarsomere character is difficult to determine in old specimens on which the setae on plantar surfaces may be eroded. The following key is the result. It should work for approximately 85% of the species.

1. Basal dorsolateral margin of elytra angled to acutely angled (Fig. 112)... 2

— Basal dorsolateral margin of elytra rounded, never acutely angled (Fig. 113); dorsal setae minute to strongly hirsute... 5

2(1). Profemora armed in both sexes; form not oval .... subgenus *Eleodes* (in part)

— Profemora unarmed in both sexes; form elongate to oval, rounded to subdepressed 3

3(2). Males with a large tooth on inner surface of hind tibia; body brownish, elongate; conspicuously clothed with erect golden setae .... subgenus *Cratidus*

— Males without a large tooth; form oval, subdepressed 4

4(3). Dorsolateral margin of elytra projecting anteriorly to embrace pronotum (Figs. 109, 110); dorsal setae short, yellowish, recumbent; male basal 2 protarsomeres brush-like with fine pubescence beneath, densely clothed along entire plantar surface .... subgenus *Promus* (in part)

— Dorsolateral margin not projecting anteriorly; dorsal setae short, whitish, recumbent; male with plantar surface of protarsomere interrupted apically by a small tuft of fine pubescence, tarsomere 2 with a smaller tuft of fine pubescence (Texas) .... subgenus *Heteropromus*

5(1). If profemora unarmed or sinuate then elytra in both sexes with a definite caudal extension............ .... subgenus *Eleodes*

— If profemora armed only in the male, incipiently armed, sinuate or unarmed; elytra ovate to elongate, never with a caudal extension 6

6(5). Dorsum with conspicuous, setiferous vestiture, setae long, sometimes thin and flying; femora unarmed in both sexes 7

— Dorsum glabrous or with inconspicuous setal vestiture; male femora armed or unarmed........ 8

7(6). Basal protarsomere of male with at least a minute apical tuft of fine silken pubescence obscuring plantar groove to having a well developed pad covering its entire length; tufts or pads of pubescence may also be present on additional pro- and meso tarsomeres; basal protarsomere of female with plantar groove open, partially interrupted by coarse setae, or interrupted by a moderately projecting bundle of coarse setae, occasionally swollen and produced apically .... subgenus *Blapylis* (in part)

— Basal protarsomere with plantar grooves open in both sexes .... subgenus *Tricheleodes*

8(6). Basal protarsomere of male with plantar groove having a minute tuft of fine pubescence at apex, to having well developed silken pads; femora armed or unarmed 9

— Basal protarsomere of male with plantar groove open or interrupted by an incipient to well developed bundle of coarse setae (*Litheleodes granulatus* and *L. papillosus* key here, see couplet 10 below) 12

9(8). Humeral edge of elytral pseudoepipleuron sharply inset at a right angle; dorso-anterior margin of elytra may be abruptly stepped (Fig. 112), elytral surface often smooth, occasionally tuberculate; femora armed or unarmed 10

— Humeral edge of elytral pseudoepipleuron rounded, dorso-anterior margin smoothly contoured (Fig. 111); occasionally, if anterior width of pseudoepipleuron moderately to strongly expanded, anterior edge appears secondarily inset at a moderate (Fig. 114) to strong angle (Fig. 18); elytral surface muricately tuberculate to fully tuberculate; femora unarmed to occasionally sinuate; plantar surface of basal protarsomere in most females simple, uninterrupted, occasionally interrupted apically by a produced bundle of coarse setae........... subgenus *Blapylis* (in part)

10(9). Basal protarsomere of male with plantar groove having a minute to small tuft of fine pubescence at apex; basal protarsomere of female with planter surface simple, open, to having an interruped apical bundle of course setae; male femora spined to sinuate, sinuate in females; elytral surface smooth, weakly muricate tuberculate to lightly tuberculate..................... subgenus *Litheleodes*

— Basal protarsomere of male having plantar groove with a well developed pad being densely clothed with fine pubescence; male femora, elytral surface variable 11

11(10). Male with plantar surface of one or 2 protarsomeres with pads densely clothed with fine pubescence; female tarsi typically simple and unmodified; femora of males spined, if not, anterior dorsal surface of pseudoepipleuron strongly produced, embracing pronotum (Fig. 110); femora of female sinuate, occasionally unarmed; proximal end of tibia normal; elytral surface smooth, occasionally slightly muricate; punctures tend to form rows; prosternal process strongly enlarged in 4 species to moderately or little developed; body variable .... subgenus *Promus* (in part)

— Male with plantar surface of 3 protarsomeres having well developed pads densely clothed in fine pubescence; protarsomere of female with plantar surface interrupted by a thick bundle of coarse setae; male with proximal end of tibia constricted, female with proximal end of tibia constricted to weakly constricted; femora of male and female strongly sinuate; body elongate .... subgenus *Holeleodes*

12(8). Protibial spurs usually dissimilar in the sexes, with outer (dorsal) spur longer and stouter than inner ventral spur .... subgenus *Melaneleodes*

— Protibial spurs similar in the sexes 13

13(12). Humeral edge of elytral pseudoepipleuron sharply inset at a right angle, dorso-anterior margin of elytra may be abruptly stepped, most with smooth contour, if rounded, elytra never tuberculate to strongly tuberculate, elytral surface smooth, occasionally slightly muricate; femora sinuate to unarmed; plantar surface of basal protarsomere in female variable 14

— Humeral edge of elytral pseudoepipleuron rounded, dorso-anterior margin smoothly contoured; elytral surface moderately to strongly tuberculate (Fig. 113); femora unarmed to sinuate, one species with an incipient spine; most with plantar surface of basal protarsomere in female simple, uninterrupted, occasionally interrupted by a produced bundle of coarse setae in male, female or both sexes 16

14(13). Antennae long and slender, antennomere 3 at least five to six times as long as broad subgenus .... *Caverneleodes*

— Antennomere 3, less than five times as long as broad 15

15(14). Middle lobe of mentum small (1/3 anterior width of mentum or less) to absent; most with plantar surface of tarsi in males and females simple, uninterrupted; in male of *E. nigrinus*, slightly interrupted apically, in female of *E. nigrinus*, interrupted apically by a bundle of coarse setae; elytral surface smooth to slightly muricate, occasionally with punctae in rows; femora sinuate, slightly sinuate to unarmed .... subgenus *Metablapylis*

— Middle lobe of mentum moderate in size (3/4 anterior width of mentum), plantar surface of protarsomere in males and females interrupted apically by one, or a pair of small merging bundles of coarse setae; elytra smooth, some with slight costae; femora unarmed to sinuate .... subgenus *Steneleodes*

16(13). Middle lobe of mentum large, transversely oval; lateral lobes rudimentary, invisible without dissection; in male, planter surface of protarsomere interrupted apically by a bundle of coarse setae, or open; in female, interrupted apically by a bundle of coarse setae, or open; in male profemora armed or sinuate, sinuate in female; elytral surface tuberculate .... subgenus *Discogenia*

— Middle lobe of mentum moderate, lateral lobes visible without dissection; plantar surface of protarsomere in male and female simple, not interrupted; femora unarmed; elytral surface strongly tuberculate .... subgenus *Pseudeleodes*

S. Key to the Genera of the Tribe Eulabini

1. Elytra distinctly costate, irregularly margined at base; eyes narrowly reniform; metasternal epimera coarsely punctate 2

— Elytra distinctly costate, irregularly margined at base; eyes narrowly reniform; metasternal epimera coarsely punctate .... *Epantius*

2(1). Pronotum bicarinate; mentum tuberculate in the male, carinate in the female, middle lobe atrophic, lateral lobes broadly exposed; sutural costae of elytra always present; apical antennomere as wide as long, as wide or slightly wider than the eighth; tarsi slender .... *Eulabis*

— Pronotum not carinate, ventral edge of epipleuron horizontal apically; middle lobe of the mentum well developed, carinate in both sexes, lateral lobes small; elytron with a basal carina; apical 4 antennomeres transverse; tarsi more or less robust .... *Apsena*

T. Key to the Genera of the Tribe Cerenopini

1. Dorsal apical angle of protibiae extended and flattened 2

— Dorsal apical angle of protibiae not extended or flattened .... *Argoporis* (in part)

2(1). Clypeus deeply notched at middle; metasternum 1.28 - 1.53 times as long as mesocoxa .... *Cerenopus*

— Clypeus not deeply notched at middle; metasternum 0.83 - 1.00 times as long as mesocoxa .... *Argoporis* (in part)

U. Key to the Genera of the Tribe Ulomini

1. Eighth elytral interval not produced apically as a rounded crest supertending the enlarged punctures of the epipleural stria; protibia with outer (dorsal) margin coarsely serrate (Fig. 100) (except *U. carolynae*) .... *Uloma*

— Eighth elytral interval produced apically as a rounded crest supertending the enlarged punctures of the epipleural stria; protibia with outer (dorsal) margin not coarsely serrate ..... *Eutochia*

V. Key to the Genera of the Tribe Helopini

1. Mesosternum nearly flat, horizontal, at least as long as mesocoxal diameter; prosternum bent down behind procoxae and prolonged posteriorly..... .... *Tarpela*

— Mesosternum excavate, declivous in front, length variable; prosternum variable 2

2(1). Metasternum shorter than mexocoxal diameter (Fig. 103) .... *Helops*

— Metasternum very elongate, much longer than mesocoxal diameter .... *Nautes*

W. Key to the Genera of the Tribe Triboliini

1. Elytral intervals finely carinate (Fig. 108) .... *Tribolium*

— Elytral intervals flat or convex, sometimes with weak carina on seventh or eighth interval (Figs. 115, 116) 2

2(1). Elytral epipleuron ending abruptly well before elytral apex (Fig. 105) .... *Tharsus*

— Elytral epipleuron tapering gradually posteriorly, attaining elytral apex (Fig. 104) 3

3(2). Antenna with apical 4 antennomeres forming distinct, capitate club .... *Lyphia*

— Antenna gradually clavate 4

4(3). Pronotal disk with punctures of 2 distinct sizes (Fig. 107) 5

— Pronotal disk with punctures of single size 6

5(4). Distance between eye and cardo greater than width of cardo .... *Mycotrogus*

— Distance between eye and cardo much less than width of cardo .... *Hypogena*

6(4). Eye reniform, emarginated in dorsal third by epistomal canthus .... *Latheticus*

— Eye round, entirely ventrad of epistomal canthus 7

7(6). Elytra with interval 7 carinate (Fig. 115) .... *Coelopalorus*

— Elytra with interval 7 rounded or flat (Fig. 116) .... *Palorus*

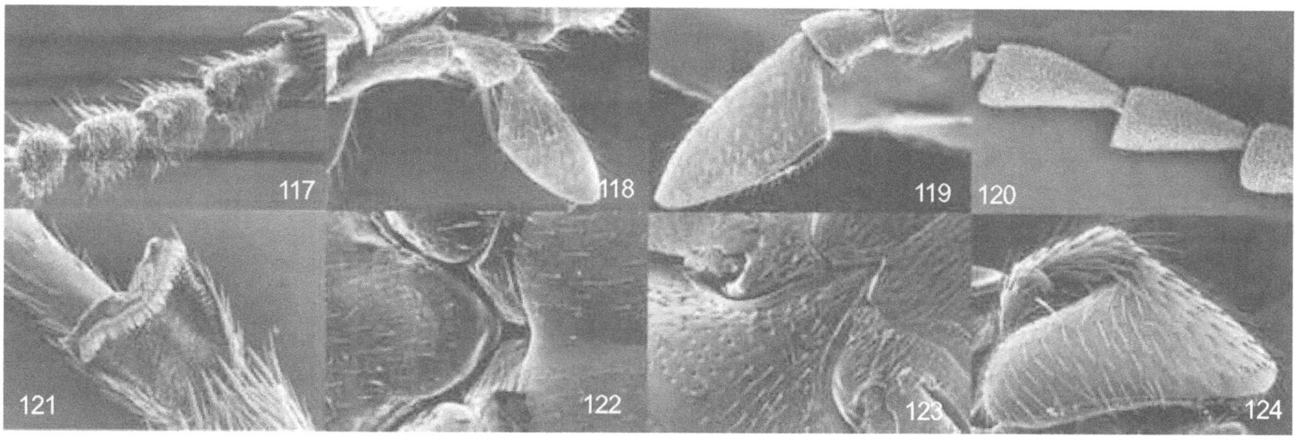

FIGURES 117.106-124.106. 117. *Andrimus murrayi* LeConte, male protarsomeres 1-4. 118-119. Apical maxillary palpomere. 118. *Andrimus murrayi* LeConte; 119. *Capnochroa fuliginosa* (Melsheimer). 120. *Pseudocistela pinguis* LeConte, antennomeres 7-8. 121. *Stenochidus gracilis* LeConte, lobes of protarsomere 4, ventral view. 122-123. Intercoxa process of abdomen, 122. *Stenochidus gracilis* LeConte; 123. *Lobopoda punctulata* (Melsheimer). 124. *Lobopoda punctulata* (Melsheimer), apical maxillary palpomere. (Figures 117-124 from Campbell 1984).

X. Key to the Genera of the Tribe Alphitobiini

1. Epipleuron gradually narrowing to elytral apex; prosternal process prominent, sub-horizontal behind coxae; mesosternum acutely concave . ... *Alphitobius*

— Epipleuron abruptly narrowing at anterior margin of visible sternite five, not reaching elytral apex; prosternal process declivous, declivous behind coxae; mesosternum obtusely concave .... *Metaclisa*

Y. Key to the Genera of the Tribe Tenebrionini

1. Tarsi with ventral pads of dense, pilose, yellowish setae; most with clypeolabral membrane exposed, at least medially 2

— Tarsi with stiff, sparse, most with dark colored setae ventrally; clypeolabral membrane concealed 3

2(1). Metasternum about twice length of mesocoxa..... .... *Zophobas*

— Metasternum about as long as mesocoxa .... *Rhinandrus*

3(1). Eye divided by epistomal canthus into dorsal and ventral lobes .... *Idiobates*

— Eye not divided by epistomal canthus 4

4(3). Elytra with confused punctation, without striae; epipleuron not reaching elytral apex .... *Bius*

— Elytra with distinct striae; epipleuron reaching elytral apex 5

5(4). Fifth visible sternite with very fine marginal groove .... *Neatus*

— Fifth visible sternite without marginal groove .... *Tenebrio*

Z. Key to the Genera of the Tribe Amarygmini

1. Elytra with humeral angles prominent; pronotum finely and shallowly punctate or impunctate; apterous .... *Cymatothes*

— Elytra with humeral angles obtusely rounded; pronotum coarsely, confluently punctured; winged .... *Meracantha*

AA. Key to the Genera of the Tribe Alleculini

1. Penultimate tarsomeres of anterior and intermediate tarsi with distinct membranous ventral lobes (Alleculina) (Fig. 121) 2

— All tarsomeres without lobes, but often densely pubescent ventrally (Fig. 117) 9

2(1). Wings greatly reduced or absent, much shorter than elytra; one species, Arizona .... *Phedius*

— Wings fully developed, when unfolded distinctly longer than elytra 3

3(2). Intercoxal process of abdomen broad (Fig. 122), with apex rounded; body black dorsally, most with distinct blue sheen, appearing glabrous dorsally; Arizona, California .... *Stenochidus*

— Intercoxal process of abdomen narrowly triangular (Fig. 123), with apex acute; color variable, most are testaceous to dark brown; body distinctly pubescent dorsally 4

4(3). Apical palpomeres of maxilla expanded, oblong-ovate (Fig. 124); prosternum (Fig. 132) short, nearly vertical; eyes oblique in position (Fig. 133) with inner margins distinctly closer to each other anteriorly than posteriorly; eastern North America to western Texas .... *Lobopoda*

— Apical palpomeres of maxilla triangular (Figs. 118, 119); prosternum (Fig. 131) longer, nearly horizontal; eyes more transverse (Fig. 134) with inner margins only slightly closer to each other anteriorly than posteriorly 5

5(4). Head almost completely covered by pronotum; pronotum distinctly wider than elytra at base; one species, southern Texas .... *Menoeceus*

— Head fully visible dorsally; pronotum subequal to or narrower than elytra at base 6

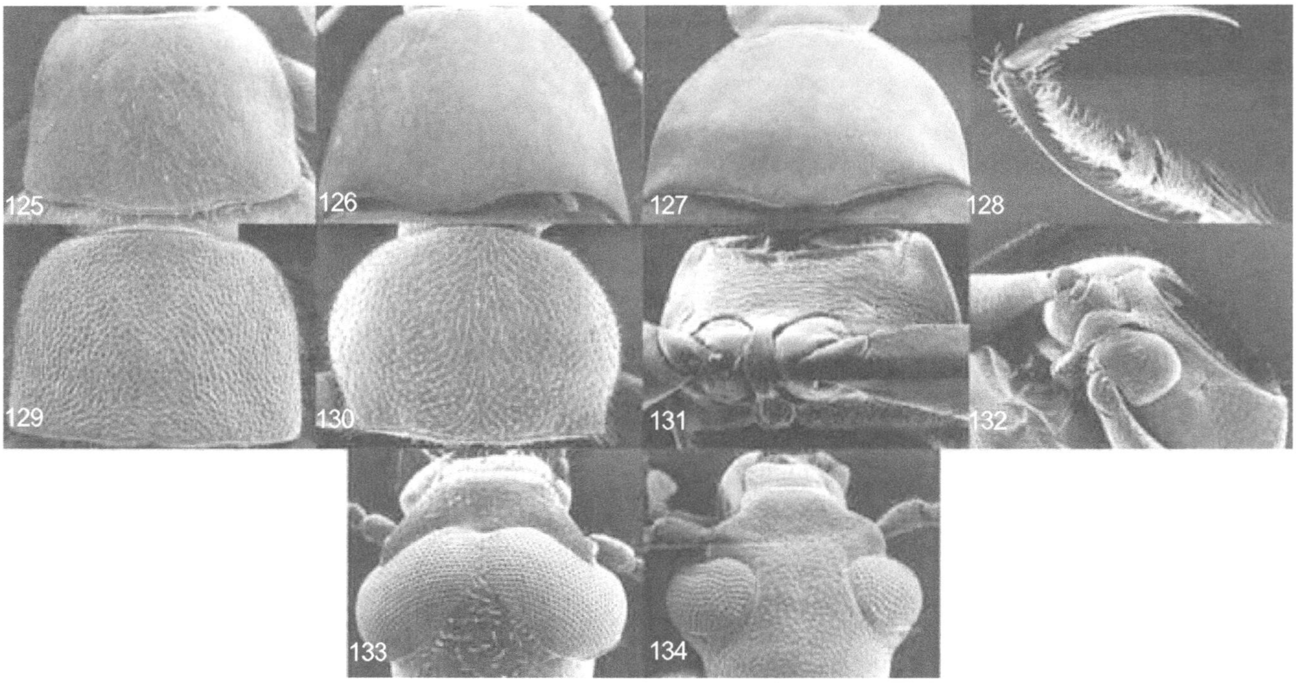

FIGURES 125.106-134.106. 125-127. Pronotum, 125. *Andrimus murrayi* LeConte; 126. *Androchirus erythropus* Kirby; 127. *Pseudocistela pinguis* LeConte. 128. *Onychomira floridensis* Campbell, protarsal claw. 129-130. Pronotum, 129. *Alethia liebecki* (Fall); 130. *Telesicles cordatus* Champion. 131-132. Prosternum, 131. *Stenochidus gracilis* LeConte; 132. *Lobopoda punctulata* (Melsheimer). 133-134. Head, male, dorsal view, 133. *Lobopoda punctulata* (Melsheimer); 134. *Andrimus murrayi* LeConte. (Figures 125-134 from Campbell 1984).

6(5). Basal angles of pronotum acute, strongly produced posteriorly; antenna of male with antennomeres 4-10 strongly compressed (females unknown); one species, Texas, New Mexico .... *Knausia*

— Basal angles of pronotum rectangular to obtuse, not produced posteriorly; antenna varying from almost filiform to obconical 7

7(6). Pronotum with sides widest near middle, distinctly narrowed basally (Fig. 130); one species, southwestern United States and one, Pacific Northwest .... *Telesicles*

— Pronotum (Fig. 129) with sides parallel basally to slightly narrowed from base to apex 8

8(7). Pronotum nearly quadrate (Fig. 129), conspicuously narrower than elytra at base; antenna elongate, filiform, antennomeres 4-10 each at least 2.5 times longer than wide; extreme southern Texas to southern Arizona .... *Alethia*

— Pronotum rarely quadrate, most with sides narrowed anteriorly from middle and forming continuous curve with anterior margin; width equal to, or slightly narrower than elytra at base; antenna shorter, antennomeres 4-10 each obconical, slightly widened from base to near apex, typically no more than 2 times longer than wide; many species United States, southern Canada .... *Hymenorus*

9(8). Intercoxal process of abdomen broadly triangular (Fig. 122); dorsal surface glabrous or coarsely setose with suberect setae (Lystronychina) . 10

— Intercoxal process of abdomen narrowly triangular (Fig. 123); dorsal surface finely, densely pubescent with setae subrecumbent (Gonoderina and Mycetocharina) 11

10(9). Wings fully developed; body coarsely pubescent dorsally; southern Texas and Arizona .... *Lystronychus*

— Wings reduced, distinctly shorter than elytra; body glabrous dorsally; one species, western Texas .... *Anamphidora*

11(9). Tarsal claws very elongate (Fig. 128), nearly as long as basal 2 tarsomeres of protarsus combined; length of row of teeth of tarsal claws reduced, restricted to basal third of claws; one species, Florida .... *Onychomira*

— Length of tarsal claws normal (Fig. 20), typically no longer than basal tarsomeres of protarsus; row of teeth of tarsal claws extending well beyond middle of claws 12

12(11). Antenna with antennomeres 3 and 4 generally subequal in length, if antennomere 3 shorter than 4 then elytra lacking rows of impressed striae and length less than 10 mm 17

— Antenna with antennomere 3 half or less as long as antennomere 4 in male, or two-thirds or less as long as 4 in female; elytral striae distinctly impressed; most with length greater than 10 mm (except some species of *Pseudocistela*) 13

13(12). Sides of pronotum (Fig. 125) subparallel or sinuate basally, basal angles rectangular..................... 14

— Sides of pronotum (Fig. 126) convergent anteriorly from base, basal angles slightly to strongly acute 15

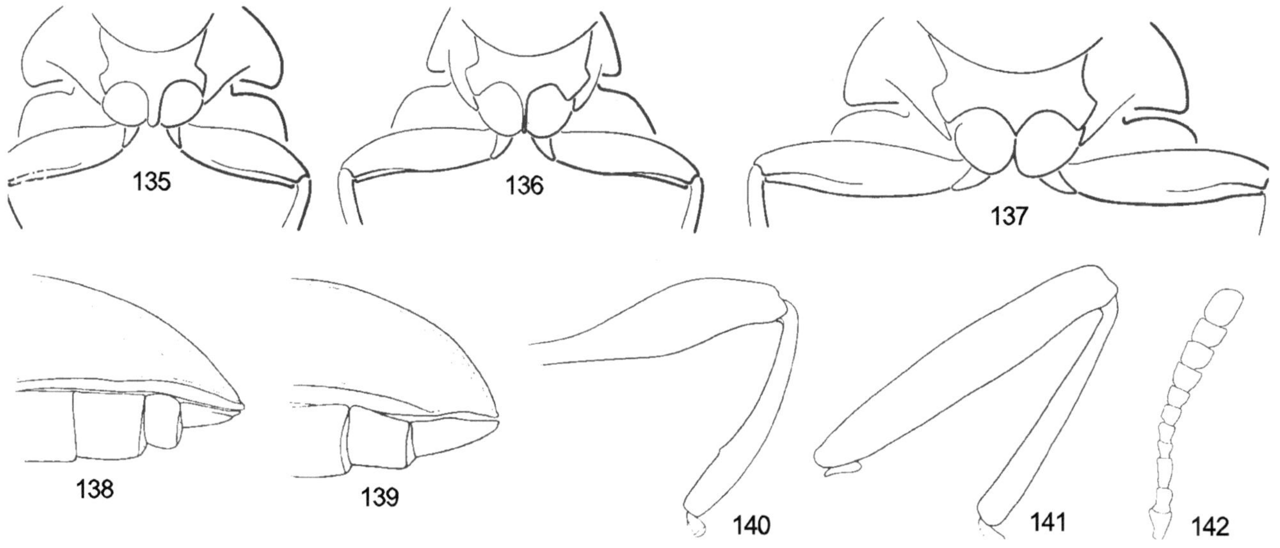

FIGURES 135.106-142.106. 135-137. Alleculini prosternal process, 135. broad; 136. narrow or lamellate; 137. lacking, with coxae contiguous. 138-139. Elytral apex, lateral view, 138. *Hiceaon* sp.; 139. *Haplandrus fulvipes* Herbst. 140-141. Mesothoracic leg, 140. *Merinus* sp.; 141. *Alobates* sp. 142. *Apsida* sp., antenna. (Figures 135-137 from Campbell 1984; Figures 138-142 from Doyen 1989).

14(13). Antenna with antennomeres 4-10 each narrowly elongate, either filiform or slightly widened apically; light to dark brown dorsally, occasionally with pronotum paler 16

— Antenna with antennomeres 4-10 each triangular (Fig. 120), strongly widened from base to apex; black dorsally, occasionally with pronotum reddish-orange .... *Pseudocistela* (in part)

15(13). Apical palpomere of maxillary palpus broadly triangular (Fig. 119); anterior portion of metasternum declivous, mesosternal intercoxal process strongly convex; size large, length 10.0-14.0 mm; one species, widely distributed in eastern North America .... *Capnochroa*

— Apical palpomere of maxillary palpus narrowly triangular to scapulate (Fig. 118); anterior portion of metasternum horizontal, mesosternal intercoxal process slightly convex; size typically smaller, length 6.5-10.5 mm; southeastern United States .... *Andrimus*

16(14). Posterior angles of pronotum strongly acute (Fig. 126) and projecting posteriorly along sides of humeral angles of elytra; antenna almost filiform; two species known from eastern North America .... *Androchirus*

— Posterior angles of pronotum slightly acute (Fig. 127), not projecting posteriorly along sides of humeral angles of elytra; antenna with antennomeres 4-10 each strongly widened apically; United States and southern Canada ... . *Pseudocistela* (in part)

17(12). Elytra with striae completely unimpressed or rarely, vaguely impressed and without distinct rows of strial punctures; with procoxae widely separated and without red humeral markings on elytra . 18

— Elytra of most with striae distinctly impressed and with distinct rows of strial punctures; if unimpressed, species with procoxae contiguous to narrowly separated or with distinct red humeral markings on elytra 19

18(17). Ventral surface of tarsi densely, finely pubescent; male with apical margin of 8th sternite deeply bilobed, extending beyond margin of last visible abdominal sternite; one species, southeastern United States .... *Tedinus*

— Ventral surface of tarsi sparsely, coarsely pubescent; male with apical margin of 8th sternite evenly convex, not lobed, not extending beyond margin of last visible sternite .... *Isomira*

19(17). Procoxae widely separated by broad prosternal process (Fig. 135) 20

— Procoxae contiguous (Fig. 137) or separated by thin lamella of prosternum (Fig. 136) .... *Mycetochara* (in part)

20(19). Elytra uniformly dark brown; pronotum markedly, evenly convex with basal foveae only slightly evident and coarsely, contiguously punctate .. .. *Hymenochara*

— Elytra normally with reddish markings on humeral angles; pronotum slightly and unevenly convex or with basal foveae deeply impressed and surface finely and sparsely to moderately coarsely punctate, but never with punctures contiguous .... *Mycetochara* (in part)

BB. Key to the Genera of the Tribes Centronopini and Coelometopini

1. Body small, (<8 mm), broadly oval; antennae with terminal 5 antennomeres expanded to form a loose club (Fig. 142), with antennomeres 8 and 9 twice as broad as long; elytra dark with red, gold, and green iridescent stripes and blotches .... *Apsida*

— Body medium to large, (typically > 10 mm), elongate; antennae without terminal 5 antennomeres expanded to form a loose club (antennomeres 8 and 9 never twice as broad as long); elytra without red, gold, and green iridescent stripes and blotches 2

2(1). Dorsal lobe of eye with marginal groove around apex 3
— Dorsal lobe of eye without groove around apex 11

3(2). Metasternal length between coxae equal to or less than mesocoxal diameter (Fig. 103) 4
— Metasternal length between coxae 1.5 to 2 times mesocoxal diameter 7

4(3). Tibiae with paired, narrow longitudinal lines of yellowish pubescence on apical half to two-thirds of inner surfaces 5
— Tibiae without lines of pubescence, or with apical one-eighth to one-quarter with faint setal lines or patch 6

5(4). Gena with deep, abrupt excavation bordering ventral lobe of eye anteroventrally .... *Oenopion*
— Gena often coarsely rugose but never with distinct excavation at apex of ventral lobe of eye .... *Coelocnemis* (in part)

6(4). Legs long, metafemur reaching at least to base of fifth abdominal sternite, frequently extending beyond apex of abdomen .... *Cibdelis* (in part)
— Legs shorter, metafemur reaching no farther than 3rd abdominal sternite .... *Polypleurus*

7(3). Epipleuron abruptly narrowing at about anterior margin of fifth abdominal sternite, disappearing before elytral apex (Fig. 139) . *Haplandrus* (in part)
— Epipleuron complete to apex of elytra, gradually narrowed posteriorly from fifth abdominal sternite (Fig. 138) 8

8(7). Fifth abdominal sternite with marginal groove or impressed line .... *Glyptotus*
— Fifth abdominal sternite without marginal groove or line 9

9(8). Femora strongly clavate; profemur 2.5 to 4 times thicker in middle than at base 10
— Femora subcylindrical; profemur no more than 2 times thicker in middle than at base 11

10(9). Pronotum finely and shallowly punctate (punctures much smaller than single eye facet); elytra punctate-striate .... *Merinus* (in part)
— Pronotum coarsely and deeply punctate (punctures as large as several eye facets combined); elytra reticulately rugose .... *Upis* (in part)

11(2, 9). All tibiae with paired, narrow longitudinal lines of yellowish pubescence on apical half to two-thirds of inner surface .... *Coelocnemis* (in part)
— Tibiae without lines of pubescence or with apical one-eighth to one-quarter with single faint setal lines or with setal patch 12

12(11). Femora clavate; profemur about twice as thick in middle as at base 13
— Femora subcylindrical; profemur less than 1.5 times thicker in middle than at base 15

13(12). Eyes with dorsal lobe equal to subequal to ventral lobe (Centronopini) 14
— Eyes with dorsal clearly much larger than ventral lobe .... *Polopinus* (in part)

14(13). Fifth visible abdominal sternite with marginal groove .... *Scotobaenus*
— Fifth visible abdominal sternite without marginal groove .... *Centronopus*

15(12). Metasternum length between coxae equa to or less than coxal diameter 16
— Metasternum length between coxae about 1.5 to 2 times mesocoxal diameter 17

16(15). Epipleuron gradually narrowing posteriorly, with subparallel margins throughout except near humerus and apex (Fig. 138), extending to elytral apex or nearly so .... *Polopinus* (in part)
— Epipleuron abruptly narrowed at about anterior margin of fifth abdominal sternite, not reaching elytral apex (Fig. 139) .... *Cibdelis* (in part)

17(15). Pronotum with base unmargined, at least medially .... *Haplandrus* (in part)
— Pronotal base with complete raised margin 18

18(17). Tarosmeres with 2 longitudinal rows of setae, separated by groove with few sparse setae .... *Iphthiminus*
— Basal 4 tarsomeres with ventral surface entirely covered by pads of dense pubescence 19

19(18). Elytra irregularly rugose; striae not discernible ... . *Upis* (in part)
— Elytra regularly striate (strial punctures sometimes very fine) 20

20(18). Fifth abdominal sternite with marginal groove .... *Xylopinus*
— Fifth abdominal sternite without marginal groove. 21

21(20). Meso- and metatibia arcuately curved (Fig. 141); pronotum with hind angles rounded .... *Merinus* (in part)
— Meso-and metatibia straight (Fig. 140); pronotum with hind angles sharp, right angled .... *Alobates*

Summary of the Subfamilies and Tribes of the United States

Note: A number of priority problems exist in the higher classification of the Tenebrionidae. For instance, the Lagriinae, Alleculinae and Coelometopinae all have earlier available names based on priority. These were discussed but not implemented recently by Lawrence and Newton (1995:796, 892-894). However at least two of these names have been long utilized and fall within the fourth edition of the International Code of Zoological Nomenclature (ICZN) summary of changes number 11 entitled "Measures empowering authors to act in the interest of preserving established usage" which states that: "An author will be required (without a ruling by the Commission) not to displace a name which has been used as valid by at least 10 authors in 25 publications during the past 50 years by an earlier synonym or homonym which has not been used as valid since 1899". Furthermore, the second paragraph of the preamble of the rules of nomenclature states "The object of the code is to promote stability and universality in the scientific names of animals and to ensure that the name of each taxon is unique and distinct. All its provisions are subservient to these ends and none restricts the freedom of taxonomic thought or action." For this reason we have decided to let these names stand at present. Some tribal names also present priority problems. These are discussed below for each case.

Family Tenebrionidae (Latreille 1802)

Division Lagriiformes

I. Subfamily Lagriinae Latreille 1825
(incl. Cossyphinae Latreille 1802)
1. Tribe Lagriini Latreille 1825
2. Tribe Goniaderini Lacordaire 1859
3. Tribe Belopini Reitter 1917
II. Subfamily Phrenapatinae Solier 1834
4. Tribe Penetini Lacordaire 1859

Division Pimeliiformes

III. Subfamily Pimeliinae Latreille 1802
(incl. Tentyridae Eschscholtz 1829)
5. Tribe Cnemeplatiini Csiki 1953
6. Tribe Stenosini Lacordaire 1859
1. Subtribe Araeoschizina Casey 1907
2. Subtribe Typhlusechina Casey 1907
7. Tribe Anepsiini Leconte 1862
8. Tribe Cryptoglossini Leconte 1862
9. Tribe Nyctoporini Lacordaire 1859
10. Tribe Coniontini Lacordaire 1859
11. Tribe Branchini Leconte 1862
12. Tribe Asidini Solier 1834
13. Tribe Vacronini Casey 1907
14. Tribe Cnemodini Casey 1907
15. Tribe Epitragini Lacordaire 1859
16. Tribe Eurymetopini Casey 1907*

Division Tenebrioniformes

IV. Subfamily Bolitophaginae Kirby 1837
17. Tribe Bolitophagini Kirby 1837
1. Subtribe Bolitophagina Kirby 1837 (*sensu* Doyen *et al.* 1989b)
2. Subtribe Rhipidandrina Leconte 1862 (*sensu* Doyen *et al.* 1989b)
V. Subfamily Diaperinae Latreille 1802
18. Tribe Phaleriini Mulsant 1854
19. Tribe Diaperini Latreille 1802
1. Subtribe Adelinina Leconte 1862
2. Subtribe Diaperina Latreille 1802
20. Tribe Scaphidemini Reitter 1922
21. Tribe Crypticini Mulsant 1854
22. Tribe Trachyscelini Mulsant 1854
23. Tribe Myrmechixenini Jacquelin Duval 1857
VI. Subfamily Hypophloeinae Billberg 1820
24. Tribe Hypophloeini Billberg 1820
VII. Subfamily Opatrinae Laporte De Castelnau 1840
25. Tribe Opatrini Laporte De Castelnau 1840
26. Tribe Melanimini Seidlitz 1894
27. Tribe Platynotini Mulsant 1853
28. Tribe Leichenini Mulsant 1854
29. Tribe Eleodini Blaisdell 1909**
VIII. Subfamily Tenebrioninae Latreille 1802
30. Tribe Eulabini Horn 1870
31. Tribe Cerenopini Horn 1870
32. Tribe Blaptini Leach 1815
33. Tribe Ulomini Mulsant 1854
34. Tribe Helopini Latreille 1802
35. Tribe Triboliini Mulsant 1854
36. Tribe Alphitobiini Reitter 1917
37. Tribe Tenebrionini Latreille 1802
38. Tribe Centronopini Doyen 1989
39. Tribe Amarygmini Lacordaire 1859
40. Tribe Apocryphini Lacordaire 1859
IX. Subfamily Alleculinae Laporte De Castelnau 1840
41. Tribe Alleculini Laporte De Castelnau 1840
1. Subtribe Alleculina Laporte De Castelnau 1840
2. Subtribe Gonoderina Seidlitz 1896
3. Subtribe Mycetocharina Mulsant 1856
4. Subtribe Lystronychina Lacordaire 1859
X. Subfamily Coelometopinae Lacordaire 1859
42. Tribe Coelometopini Lacordaire 1859
43. Tribe Talanini Champion 1887 ***
44. Tribe Strongyliini Lacordaire 1859

* LeConte (1862) used the tribal names Epiphysini, Gnathosini and Thinobatini (all Lacordaire 1859) while referring to North American Eurymetopini. However the first two are synonyms of the distantly related old world tribes Adesmiini and Tentyriini respectively (both also Lacordaire 1859) and the third is a distantly related South American tribe.

** Eleodini is preoccupied by Amphidorini (LeConte 1862). However Eleodini falls within the intentions of the preamble and/or ICZN(11) mentioned above.

*** Talanini is preoccupied by Dignamptini (LeConte and Horn 1883), but has been used by Gebien (1943:818) and perhaps incorrectly placed as a synonym by Doyen (1989:302).

Classification of Nearctic Genera

Tenebrionidae Latreille 1802

I. Subfamily Lagriinae Latreille 1825

Characteristics: Body small to large in size (2 to 45 mm); elongate, glabrous to pubescent, winged or rarely apterous. Eyes variable in size or rarely absent; antennae 11-segmented, filiform, gradually expanded to clubbed, the apical antennomere sometimes very long; bearing only simple, setiform sensoria on apical antennomere. Labrum often elongate. Mandible with mola longitudinally elongate with 3 transverse ridges (except Belopini). Procoxal cavities closed externally and internally; mesocoxal cavities with or without exposed trochantin; closed laterally by mesepimeron or sternum. Elytra, if striate, with 10 striae and scutellary striole. Intersternal membrane of abdomen exposed or concealed (Belopini); hinge position lateral or medial. Abdominal defensive glands variable or absent: if present of the pycnocerine, adeliine or lagriine type (see Tschinkel and Doyen 1980). Penultimate tarsomere sometimes lobed. Tarsal formula 5-5-4. Aedeagus with tegmen dorsal.

This subfamily contains eight tribes, generally distributed, three of which occur in the United States.

1. Tribe Lagriini Latreille 1825

Characteristics: Lagriinae; body elongate, cylindrical; winged. Head prominent, prognathous; antennae 11-segmented, filiform, the apical antennomere often very long; antennae inserted under the very small, oblique frontal ridges. Clypeus subcoriaceous; labrum prominent, emarginate; apical palpomere of maxillary palpi enlarged; mentum transverse, trapezoidal; palpi very small, somewhat flattened, the apical palpomere enlarged. Eyes lateral, small, transverse, emarginate, the facets coarse. Pronotum narrower than the elytra at the base, subcylindrical, borders unmargined. Procoxae globose, prominent; middle coxae conical, moderate; hind coxae transverse, moderate; femora slender; tibiae slender, spurs absent. Intersternal membrane of abdomen exposed.

This tribe contains 83 genera, of which two occur in the United States.

Arthromacra Kirby 1837, 3 spp., eastern United States. Key to species: Parsons (1975).

Statira Serville 1828, 15 spp., mainly southern United States. Key to species: Parsons (1975).

2. Tribe Goniaderini Lacodaire 1859

Characteristics: Lagriinae; adults moderate to small in size, elongate, winged. Body glabrous to pubescent, typically punctate; thorax with margins sometimes denticulate or laterally expanded. Antennae gradually expanded or clubbed. Eyes large or absent (*Eschatoporis*). Labrum subquadrate, Mandibles sometimes elongate, with long, highly asymmetrical molar lobes. Mentum small, clearly exposing cardo and stipes. Penultimate tarsomere weakly to strongly bilobed. Abdominal defensive glands absent. Intersternal membranes of abdomen exposed. Penultimate tarsomere weakly to strongly bilobed. Ovipositor reduced, with paraprocts subquadrate or oval and the basal coxite lobes fused; with large, digitate gonostyli and elongate, subdigitate apical coxite lobes.

This tribe contains 19 genera, of which four occur in the United States.

Anaedus Blanchard 1845, 5 spp., occurring in the United States. Key to species: none exist, see Schaeffer (1915). This genus is in need of revision.

Eschatoporis Blaisdell 1906, 1 sp., *E. nunenmacheri* Blaisdell 1906, northern California.

Paratenetus Spinola 1844, 5 spp., occurring in the United States. Key to species: none exist, see Spilman (1973). This genus is in need of revision.
Storthephora M klin 1875

Prataeus LeConte 1862, 1 sp., *P. fusculus* LeConte 1856, eastern United States.

3. Tribe Belopini Reitter 1917

Characteristics: Lagriinae; adults moderate to small in size, very elongate, parallel to subparallel, winged. Labrum elongate, asymmetrical. Mandible with mola coarsely striate. Mesocoxal cavities closed by mesepisternite. Intersternal membranes of abdomen concealed. Margin of 7th abdominal sternite dorsally grooved. For greater details see Doyen and Tschinkel 1982, and Doyen *et al.* 1989b.

This tribe contains eight genera, of which one occurs in the United States.

Adelonia Laporte 1840, 2 spp., California and Texas. Key to species: Doyen (1982).
Merotemnus Horn 1870
Rhacius Champion 1885

II. Subfamily Phrenapatinae Solier 1834

Characteristics: Body very small to large in size (slightly over 1 to over 30 mm); oval to elongate; winged or apterous; eyes entire, rarely reduced. Antennae 11-segmented (10- or 9-segmented in Archaeoglenini); often clubbed, first antennomere sometimes very elongate (Phrenapatini); bearing only simple, setiform sensoria on apical antennomeres. Mandibles moderately or strongly projecting (Phrenapatini), tridentate at apex. Procoxal cavities closed externally and internally; mesocoxal cavities with or without exposed trochantin, closed laterally by meso- and metasternite or by trochantin. Elytra with scutellary striole and 9 or 10 complete striae or estriate. Intersternal membrane of abdomen exposed; hinge position lateral. Abdominal defensive glands present. Penultimate tarsomere not lobed; tarsal claws not pectinate. Tarsal formula 5-5-4 (or 4-4-4 or 3-3-3 in Archaeoglenini: not North American). Aedeagus with tegmen dorsal.

This subfamily contains three tribes, generally distributed, one of which occurs in the United States.

4. Tribe Penetini Lacordaire 1859

Characteristics: Phrenapatinae; body small, oval, winged. Antennae with 11 antennomeres, first antennomere typically shorter than following three antennomeres: if longer, then club with more than three antennomeres; mandibles moderately projecting, most with apices sharply bent towards midline; galea shorter than lacinia. Scutellum small; elytra without scutellar striae, and anterior stria nonparallel. Procoxae subtransverse. Mesocoxal cavities not closed laterally by the sternite, mesotrochantin exposed (but often difficult to see).

This tribe consists of 22 genera, generally distributed, two of which occur in the United States.

Dioedus LeConte 1866, 1 sp., *D. punctatus* LeConte 1866, eastern United States.
Arrhabaeus Champion 1886
Brachycilibe Carter 1911
Tagalus Gebien 1913

Clamoris Gozis 1886, 1 sp., *C. americana* (Horn 1874), Oregon.
Phthora Dejean 1834
Phthora Mulsant 1854

III. Subfamily Pimeliinae Latreille 1802

Characteristics: Body very small to very large (slightly over 1 to over 60 mm), most often apterous with fused elytra; wings, if present, almost always without subcubital flecks. Antennae gradually enlarged or rarely clubbed; bearing only simple, setiform sensilla, often grouped into dense rings apically around last 3 antennomeres. Labrum enlarged, transverse to subquadrate to slightly longer than broad. Procoxal cavities closed externally and internally (with rare New Zealand exception). Elytra rarely striate, but if so then with 10 striae (rarely 8) and a scutellary striole. Hinge position between abdominal sternites 5-7 medial. Abdomen with intersternal membrane of visible abdominal sternites 3-5 concealed (except some old world Tribes). Abdominal defensive glands absent. Protibiae with 2 apical spurs rarely one (*Cnemodinus*); Tarsal formula 5-5-4, rarely 4-4-4. Tarsi grooved ventrally with ventral spines or setae, rarely pubescent; tarsal claws not pectinate. Aedeagus inverted from normal position, with tegmen ventral, very rarely dorsal (*Alaudes*). Female genitalia very diverse. Additional characters include internal and larval characters (see Doyen 1972, 1979, 1993, Watt 1974, 1992, Keleinikova 1971).

This subfamily is abundant worldwide except in Australia, where only a few species occur. The Pimeliinae are especially abundant in desert areas. This subfamily contains 37 tribes, generally distributed, 12 of which occur in the United States.

5. Tribe Cnemeplatiini Csiki 1953

Characteristics: Pimeliine; adults small elongate and parallel-sided or oval, brown in color, almost glabrous to densely pubescent, or clothed with erect and/or sessile scales. Antennae short, always shorter than width of head, 11-segmented, the last 3 antennomeres forming a club. Eyes typically convex, oval, not or scarcely encroached upon anteriorly by epistomal canthus, large or reduced to single facet (*Alaudes*). Antennae serrate-clubbed (*Lepidocnemeplatia*) or moniliform-clubbed (*Alaudes*); Labrum distinctly wider than long. Mandible with mola reduced, attenuate; right mandible with dorsolateral tooth absent. Mentum much broader than long, subequal in width to prementum; prementum membranous with basal sclerites, with base concealed beneath mentum. Maxilla with articulation of cardo concealed in socket overlain by subgena. Mesocoxal cavity closed by sternites. Legs, especially front legs, modified for digging. Front tibia broadly expanded at apex, bearing scales or spines. Tibial spurs long on all legs, but larger and blade-like on front legs. Front tarsi short, stout with apical tarsomere about twice as long as other tarsomeres; meso-metatarsi more elongate, basal metatarsomere longer than any other tarsomere; aedeagus with tegmen ventral or dorsal (*Alaudes*). Ovipositor rudimentary, proctiger, paraproct and sometimes coxites atrophied or lacking; most with spiculum absent, if present, (ratio to ovipositor length) short; gonostyle distinct.

This tribe is composed of 6 genera, in three subtribes. The tribe has a worldwide distribution with two genera occurring in the United States (for detailed description see Watt 1992).

Alaudes Horn 1870, 6 spp., restricted to California and Oregon. Key to species: Fall (1928). Revision in progress (Aalbu).

Lepidocnemeplatia Kaszab 1938, 1 sp., *L. sericea* (Horn 1870), widespread in the southwestern United States.

6. Tribe Stenosini Lacordaire 1859

Characteristics: Pimeliine; adults apterous, small to very small (1.6 to 6 mm) in size. Eyes divided into upper and lower groups of facets, with lower group of facets positioned anterior to upper group. Antennae moderate in size, compact, often scaly, with reduced eleventh antennomere. Labrum distinctly wider than long. Right mandible with dorsolateral tooth absent. Genae strongly, acutely produced around mentum; mentum subquadrate, small in size, never covering both the ligula and the maxillae; prementum with base concealed beneath mentum. Mesocoxal cavities closed by sternites externally. Mesotrochantin may be either visible or hidden. Tarsi grooved ventrally with ventral spines or setae. Ovipositor with coxite lobes 3 and 4 fused.

These beetles are often associated with ants, rodent nests or soil. Thirty-four genera, in five subtribes, are assigned to the tribe, which has a worldwide distribution. Two genera in two subtribes occur in the United States (see Aalbu and Andrews 1996).

1. Subtribe Araeoschizina Casey 1907

Characteristics: Stenosine; body slender, elongate. Eyes coarsely faceted. Antennae moderate to long, serrate-moniliform and perfoliate, usually scaly. Mentum subequal in width to prementum.

Prementum entirely membranous; Ovipositor rudimentary, proctiger, paraproct and sometimes coxites atrophied or lacking; spiculum (ratio to ovipositor length) short; gonostyle rudimentary or absent.

Only the one genus occurs in this subtribe.

Araeoschizus LeConte 1851, 32 spp., western United States. Key to species: Papp (1981), see also Dajoz (1984, 1989a, 1989b). Genus needs further study.

2. Subtribe Typhlusechina Casey 1907

Characteristics: Stenosine; adults very small (1.6 to 3 mm) elongate, flattened. Head large and subquadrate. Eyes composed of 2 pairs of ommatidia, lower pair located in the basal portion of the antennal fossae, each pair is separated by connecting furrow of a little longer than diameter of ommatidia. Antennae slightly longer than head, moniliform, set in genal fossae, last three antennomeres forming loosely joined, clavate club. Genae acutely produced around mentum. Mentum much broader than prementum; prementum membranous with basal sclerites. Thorax quadrate to subquadrate, basal margin broadly lobed to depressed in center with lateral acute elevations. Elytra as long as head and thorax combined, base of scutellar region depressed. Ovipositor with gonostyle distinct.

Only the one genus occurs in this subtribe.

Typhlusechus Linell 1897, 3 spp., southwestern United States southwest. Key to species: Aalbu and Andrews (1985).

7. Tribe Anepsiini LeConte 1862

Characteristics: Pimeliine; body small (2-6 mm long), apterous; most with lateral epistomal sutures distinct, medial suture obliterated. Eye completely divided by epistomal canthus and super-tended by low carina; antenna with 11-segments, clavate or with apical 3 or 4 antennomeres forming club; labrum transverse with long, slender tormae with medial processes directed obliquely posteriad; mandible with small, smooth mola remote from bidentate incisor lobe, with dorsolateral tooth absent; maxilla with galea densely setose. Mentum subquadrate, subequal in width to prementum; prementum entirely membranous, nearly or entirely concealed beneath mentum. Prosternal process unmargined, horizontal a short distance behind coxae, then abruptly declivous; mesosternum scarcely excavated. Mesocoxal cavity closed by mesepimeron or sternites (*Anchomma*); mesotrochantin exposed or concealed. Elytra constricted basally, collar-like, amplected into prothorax; scutellum angulate or rounded posteriorly, retracted into prothorax, not visible on elytral disc. Metasternum about one to 2 times length of mesocoxa. Femora stout; protibiae apically expanded (except *Anchomma*); tarsi short with few spinose setae beneath. Ovipositor short, thick, with paraproct and coxite subequal; coxites lobes 3 and 4 fused; gonostyles minute, rudimentary, papilliform, inserted dorsolaterally very near apex of coxite.

This tribe has assigned to it five genera, restricted to North America. All five genera occur in the United States (for detailed description see Doyen 1987 and 1993).

Anchomma LeConte 1858, 1 sp., *A. costatum* LeConte 1858, southern California.

Anepsius LeConte 1851, 4 spp., southwestern United States to North Dakota and Texas. Key to species: Doyen (1987).

Batulius LeConte 1851, 1 sp., *B. setosus* LeConte 1851, southern California, Arizona.

Batulinus Papp 1961

Batuliomorpha Doyen 1987, 2 spp., southern California and Arizona. Key to species: Doyen (1987).

Batuliodes Casey 1907, 4 spp., southwestern United States. Key to species: Doyen (1987).

Batulius Doyen and Lawrence 1987 (in part)

8. Tribe Cryptoglossini LeConte 1862

Characteristics: Pimeliinae; body moderate to large; surface glabrous to rugose-tuberculate. Eyes dorso-ventrally elongate, narrowed at center to completely divided (*Schizillus*). Labrum distinctly wider than long, almost entirely concealed by the dilated frons. Right mandible with dorsolateral tooth absent. Mentum subquadrate, much broader than prementum; prementum membranous with basal sclerites, nearly or entirely concealed beneath mentum. Maxilla with articulation of cardo exposed laterad of submentum. Mesocoxal cavity closed by mesepimeron. Ovipositor with coxites lobes 3 and 4 fused; gonostyle rudimentary or absent.

This tribe has assigned to it three genera, restricted to North America. All three genera occur in the United States (for detailed description see Aalbu 1985 and in press).

Asbolus LeConte 1851, 4 spp., southwestern United States. Key to species: Aalbu (1985).

Cryptoglossa Solier 1836, 4 spp., southwestern United States. Key to species: Aalbu (1985).

Centrioptera Mannerheim 1843

Oochila LeConte 1862

Amblycyphrus Motschulsky 1870

Schizillus Horn 1874, 2 spp., southwestern United States. Key to species: Aalbu (1985).

9. Tribe Nyctoporini Lacordaire 1859

Characteristics: Pimeliinae; body moderate in size, elongate, parallel, convex, rough, coarsely sculptured; apterous. Head longer than wide, often with a longitudinal keel medially. Eyes moderately small, coarsely faceted, lunate around episthomal canthus. Antennae moniliform, cylindrical, coarsely sculptured, the apical antennomeres only slightly wider. Labrum distinctly wider than long. Mentum much broader than long, much broader than prementum; prementum membranous with basal sclerites, nearly or entirely concealed beneath mentum; maxillae laterally exposed. Right mandible with dorsolateral tooth absent. Subgenae laterally prominent, acutely produced around mentum. Maxilla with articulation of cardo exposed laterad of submentum. Mesocoxal cavity closed by mesepimeron. Tarsi ventrally with dense short yellow setae. Ovipositor with coxites lobes 3 and 4 fused; gonostyle distinct.

This tribe includes only the following genus.

Nyctoporis Eschscholtz 1831, 4 spp., southwestern United States. Key to species: Casey (1907). This genus needs revision.
Emeax Pascoe 1866
Enneacoides Fairmaire 1881

10. Tribe Conionitini Lacordaire 1859

Characteristics: Pimeliinae; body stout, oval to ovate-elongate to subglobose, apterous. Labrum distinctly wider than long. Right mandible with dorsolateral tooth absent. Mentum small, trapezoidal, weakly emarginate anteriorly; subquadrate, much broader than prementum; prementum membranous with basal sclerites, with base concealed beneath mentum; submental sclerite present, but may be reduced; ligula large, sclerotized, projecting anteriorly beyond mentum; labial palps inserted dorsally on distinct palpifers; maxilla with articulation of cardo exposed laterad of submentum, lacinia dentiform; gular sutures separate. Antennae long and slender, eleventh antennomere almost as large as tenth. Epipleura of elytra gradually narrowing from humeri to apex. Intercoxal process of abdomen narrow, pointed or rounded. Mesocoxal cavity closed by mesepimeron. Ovipositor with coxite lobes 3 and 4 fused; gonostyle rudimentary or absent.

The tribe contains four genera, all of which occur in the United States.

Coniontis Eschscholtz 1829, 48 spp., western and central United States. Key to species: Casey (1908), also see Doyen (1977). This genus needs revision.
Coniontellus Casey 1890
Coniontides Casey 1908
Crypticomorpha Casey 1908
Brachyontis Casey 1908

Conisattus Casey 1895, 1 sp., *C. rectus* Casey 1895, northwestern United States.

Eusattus LeConte 1851, 14 spp., western and central United States. Key to species: Doyen (1984b).
Discodemus LeConte 1866
Conipinus LeConte 1866
Nesostes Casey 1908
Megasattus Casey 1908
Eusattodes Casey 1908
Sphaeriontis Casey 1908
Coelosattus Blaisdell 1927

Coelus Eschscholtz 1829, 4 spp., California. Key to species: Doyen (1976).
Coelomorpha Casey 1908
Pseudocoelus Casey 1908

11. Tribe Branchini LeConte 1862

Characteristics: Pimeliinae; body oval, moderately convex; apterous. Eyes transverse. Labrum distinctly wider than long. Right mandible with dorsolateral tooth absent. Mentum moderately small, trapezoidal, emarginate anteriorly; much broader than prementum; prementum membranous with basal sclerites, with base concealed beneath mentum; submental sclerite absent; ligula large, sclerotized, projecting anteriorly beyond mentum; labial palps inserted dorsally on distinct palpifers. Maxilla with articulation of cardo exposed laterad of submentum, lacinia dentiform; gular sutures confluent. Antennae serrate-clubbed, apical three antennomeres enlarged, eleventh larger than tenth. Epipleura of elytra abruptly narrowing behind humeri. Intercoxal process of abdomen moderately broad, truncate. Mesocoxal cavity closed by mesepimeron. Ovipositor with coxites lobes 3 and 4 fused; spiculum (ratio to ovipositor length) short; gonostyle rudimentary or absent.

The three genera included in this tribe occur in the Western Hemisphere only. One is found in the United States.

Branchus LeConte 1862, 2 spp., Florida and Texas. Key to species: Steiner (1991).

12. Tribe Asidini Solier 1834

Characteristics: Pimeliinae; body ovate to ovate-elongate; apterous. Head scarcely narrowed behind the eyes. Eyes transverse, reniform, and moderately finely granulate. Antennae serrate-clubbed, with 10th antennomere large and subquadrate, the eleventh small. Labrum distinctly wider than long. Right mandible with dorsolateral tooth absent. Mentum large, much broader than long, either filling entirely the gular cavity or inserted upon a very short and wide peduncle, thus usually leaving the base of the maxillae exposed; prementum distinctly narrower than mentum, membranous with basal sclerites, with base concealed beneath mentum. Maxillary and labial palpi with the apical palpomere large and securiform; frons transversely impressed; epistoma very short, not covering the base of the mandibles; labrum prominent. Mesocoxae with visible trochantin, but may be reduced and barely visible in a few genera; mesepimeron reaching mesocoxal cavities; mesocoxal cavities closed by mesepimeron or sternites (*Craniotus*); metasternum very short; metepisternum broad, metepimeron hidden. Hind coxae moderately separated. Legs moderate, apical spurs of the tibiae distinct; tarsi with setae ventrally, but not sulcate. Epipleural fold typically indistinct except at apex of elytra. Intercoxal process of abdomen obtuse; fourth and fifth visible abdominal sternite somewhat prolonged posteriorly at the sides, abdominal laterotergites extremely small. Ovipositor with coxites lobes 3 and 4 fused; spiculum (ratio to ovipositor length) short to moderate; gonostyle rudimentary or absent.

This tribe has assigned to it 46 genera, distributed in southern Europe, Africa, Madagascar, North and South America, 16 genera occur in the United States (see Brown, 1971b).

Asidina Casey 1912, 6 spp., southwestern United States. Key to species: Triplehorn and Brown (1971).

Asidopsis Casey 1912, 21 spp., southwestern and central western United States. Key to species: Casey (1912). This genus needs revision.

Astrotus LeConte 1858, 3 spp., Texas. Key to species: Casey (1912).

Craniotus LeConte 1851, 2 spp., California, Nevada, Arizona. Key to species: Tanner (1963). Revision in progress (Aalbu and S nchez Pi ero).

Glyptasida Casey 1912, 13 spp., south central western United States. Key to species: Casey (1912). This genus needs revision.

Gonasida Casey 1912, 7 spp., western United States. Key to species: Casey (1912) and Brown (1971a). This genus needs revision.

Litasida Casey 1912, 1 sp., *L. townsendi* Casey 1912, Arizona.

Megasida Casey 1912, 2 spp., southwestern United States. Key to species: Triplehorn (1967).

Microschatia Solier 1836, 7 spp., southwestern United States. Key to species: Brown and Doyen (1991).
Pycnonotida Casey 1912
Acroschatia Wilke 1922

Parasida Casey 1912, 2 spp., southwestern United States. Key to species: Casey (1912). This genus needs revision.

Philolithus Lacordaire 1859, 24 spp., western United States. Key to species: Casey (1912). This genus needs revision.
Pelecyphorus of Casey 1912, not Solier 1836

Pycnomorpha Motschulsky 1870, 2 spp., California. This genus is currently being revised (Brown).

Stenomorpha Solier 1836, 61 spp., western United States. Key to species: Casey (1912). This genus needs revision.
Euschides LeConte 1858

Stenosides Solier 1836, 7 spp., southwestern and central western United States. Key to species: Casey (1912). This genus needs revision.
Ologlyptus Lacordaire 1859
Pactostoma LeConte 1858

Stethasida Casey 1912, 18 spp., California and Arizona. Key to species: Casey (1912). This genus needs revision.

Trichiasida Casey 1912, 11 spp., southwestern United States, California. Key to species: Casey (1912). This genus needs revision.

13. Tribe Vacronini Casey 1907

Characteristics: Pimeliine; adults winged, moderate in size (4-15 mm), glabrous, with slender, elongate body and appendages. Head small, subquadrate. Eyes large, weakly emarginate anteriorly; antennae long, slender, filiform, clearly longer than head and pronotum. Clypeus transversely truncate with rounded angles, almost covering the mouthparts; labrum distinctly wider than long, transverse with long slender tormae with medial processes directed obliquely posteriad; mandibles slender, with small, smooth mola remote from bidentate incisor lobe; with dorsolateral tooth absent; maxilla with galea densely setose; lacinia bearing 2 to 4 blunt teeth in addition to fine setae; mentum small, exposing ligula and maxillae, subquadrate, subequal in width to prementum; prementum membranous with basal sclerites, entirely exposed anteriad mentum. Mesocoxal cavities closed partly by epimera, mesotrochantins exposed. Legs long and slender. Protibiae linear, not expanded apically, the apical spurs small, distinct, subequal. Tarsi with very short, sparse spinules beneath. Elytra confusedly punctate, with narrowly reflexed and acute margin; epipleural fold rather wide, flat, entire, occupying the entire inflexed side with the raised flat lower margin very narrow. Scutellum large, transverse, truncate and conspicuous. Abdomen with the fourth visible sternite similar to the third, a little shorter. Ovipositor with gonostyle distinct.

This tribe has assigned to it two genera, both which occur in the southwestern United States.

Eupsophulus Cockerell 1906, 2 spp., Arizona and California. Key to species: Spilman (1959).
Eupsophus Horn 1870

Alaephus Horn 1870, 11 spp., southwestern United States. Key to species: Fall (1907). This genus needs revision.
Vacronus Casey 1907

14. Tribe Cnemodini Casey 1907

Characteristics: Pimeliine; adults winged, moderate in size (6-9 mm), elongate, rounded. Frons with lateral lobes slightly dilated over the insertion of the antennae. Clypeus separated from frons by costate ridge, prominent, anterior bilobed, slightly spatulate, concealing labrum medially; labrum very wide, visible laterally. Antennae long, slender; apical antennomere oval, acuminate. Mandibles grooved externally throughout their length; right mandible with dorsolateral tooth present. Mentum feebly sinuate at apex, hexagonal, much broader than long, much broader than prementum; prementum entirely membranous, nearly or entirely concealed beneath mentum. Maxilla with articulation of cardo overlain by both submentum and subgena. Protibiae with single apical spur and submedian and apical dorsal extension; Mesocoxal cavity closed by sternites. Ovipositor with coxite lobes 3 and 4 fused; spiculum (ratio to ovipositor length) short; gonostyle rudimentary or absent.

This tribe includes only the following genus.

Cnemodinus Cockerell 1906, 3 spp., Utah, Arizona and California. Key to species: Casey (1907). This genus needs revision.
Cnemodus Horn 1870, not Herrich-Schaeffer 1850

15. Tribe Epitragini Lacordaire 1859

Characteristics: Pimeliinae; body oval to elongate. Mentum transversely trapezoidal, the apex very broadly arcuate from side to side, not sinuate at the middle, generally transverse; prementum entirely membranous and mostly concealed beneath mentum; epistoma lobed; labrum subquadrate; mandibles not grooved externally, dorsal tooth on basal half of mandible, right mandible with small dorsolateral tooth. Metasternum with transverse grooves parallel to the metacoxae; body winged or brachypterous; scutellum distinct; mesocoxal cavities closed by sternites. Elytra not embracing the sides of the body, the inflexed parts occupied wholly by the epipleural fold. Protibiae with 2 apical spurs, most lacking a median tooth; metacoxae narrowly separated. Abdominal process acute and broadly rounded between the metacoxae.

This tribe contains 33 genera of which the following 11 occur in the United States.

Bothrotes Casey 1907, 2 spp., eastern, central and southwestern United States. Key to species: Freude (1967).

Conoecus Horn 1885, 1 sp., *C. ovipennis* Horn 1885, Texas, Louisiana.

Epitragodes Casey 1890, 1 sp., *E. tomentosus* (LeConte 1866), southeastern United States.

Epitragosoma Brown and Triplehorn (in press), 1 sp., Texas.

Lobometopon Casey 1907, 3 spp., southwestern United States to Kansas. Key to species: Freude (1968).

Metopoloba Casey 1907, 1 sp., *M. pruinosa* (Horn 1870), southwestern United States.

Pechalius Casey 1907, 4 spp., southwestern United States. Key to species: Triplehorn (1974).

Phegoneus Casey 1907, 1 sp., *P. julichi* (Casey 1891), Texas.

Polemiotus Casey 1907, 1 sp., *P. submetallicus* (LeConte 1854), southwestern United States.

Schoenicus LeConte 1866, 1 sp., *S. puberulus* LeConte 1866, eastern United States.

16. Tribe Eurymetopini Casey 1907

Characteristics: Pimeliinae; body oval to elongate, parallel, glabrous or pubescent, winged or apterous. Eye variable in size, with or without distinct supraorbital carina. Frons with clypeus, epistomal lobe highly variable: unmodified or slightly to strongly produced, apex variable. Antennae filiform, serrate or serrate-clubbed, variable in length. Labrum subquadrate or slightly longer than wide. Mentum very large, concealing both the ligula and the maxillae, with apex more or less distinctly emarginate or sinuate, much broader than long, much broader than prementum; prementum entirely membranous, nearly or entirely concealed beneath mentum. Mandibles grooved or not, right mandible with dorsolateral tooth present or absent, large or small. Maxilla with articulation of cardo overlain by both submentum and subgena. Mesocoxal cavity closed by sternites. Anterior tibiae with 2 apical spurs. Tarsi with sparse, stiff, subspiniform setae or pubescent ventrally. Ovipositor with coxite lobes 3 and 4 fused; spiculum (ratio to ovipositor length) moderate or very long; gonostyle rudimentary or absent.

This tribe is almost entirely New World, with 46 genera, 20 of which occur in the United States.

Armalia Casey 1907, 2 spp., Texas. This genus is common in Mexico and Central America. Key to species (United States): Casey (1907). This genus needs revision.

Auchmobius LeConte 1851, 8 spp., California to British Columbia. Key to species: Blaisdell (1934). This genus needs revision.

Chilometopon Horn 1874, 5 spp., southwestern United States. Key to species: MacLachlan and Olson (1990).
Prometopion Casey 1907

Cryptadius LeConte 1851, 1 sp., *C. inflatus* Leconte 1851, California. Key to species: Thomas (1985).

Edrotes LeConte 1851, 4 spp., southwestern United States. Key to species: Doyen (1968).

Eschatomoxys Blaisdell 1935, 2 spp., California and Utah. Key to species: Sorenson and Stones (1959). Revision in progress (Aalbu and Thomas).

Eurymetopon Eschscholtz 1831, 2 spp., California. This genus is unknown to authors. It is most likely that *Metoponium* Casey 1907 is a synonym of this genus. See comments by Casey (1907:286, 289).

Hylocrinus Casey 1907, 16 spp., southwestern United States. Key to species: Casey (1907). This genus needs revision.

Melanastus Casey 1907, 24 spp., southwestern United States. Key to species: Casey (1907). This genus needs revision.

Metoponium Casey 1907, 52 spp., southwestern United States. Key to species: Casey (1907). This genus needs revision.

Micromes Casey 1907, 2 spp., California. Key to species: Casey (1907). Revision in progress (Aalbu).

Oxygonodera Casey 1907, 3 spp., Utah. Key to species: Casey (1907). Revision in progress (Aalbu).

Steriphanus Casey 1907, 16 spp., southwestern United States. Key to species: Casey (1907). *Steriphanus* is most likely a synonym of *Hylocrinus* Casey. This genus needs revision.

Stibia Horn 1870, 5 spp., southwestern United States. Key to species: Blaisdell (1936a). Revision in progress (Aalbu).
Eutriorophus Casey 1924

Telabis Casey 1890, 32 spp., southwestern United States. Key to species: Casey (1907). This genus needs revision.

Texaponium Thomas 1984, 1 sp., *T. triplehorni* (Berry 1974), Texas.

Trichiotes Casey 1907, 1 sp., *T. seriatus* Casey 1907, Texas.

Trimytis LeConte 1851, 2 spp., southwestern United States to Kansas. Key to species: MacLachlan and Olson (1990).

Triorophus LeConte 1851, 5 spp., southwestern United States. Key to species: Berry and Triplehorn (in progress).

Triphalopsis Blaisdell 1923, 1 sp., *T. californicus* Doyen 1982, California.

IV. Subfamily Bolitophaginae Kirby 1837

Characteristics: Body very small to moderate in size (1 to 20 mm). Labrum subquadrate or slightly longer than broad. Antennae 10- or 11-segmented, with apical antennomeres capitate or pectinate, with simple setiform sensilla. Procoxal cavities closed externally and internally; Elytra if striate, with 10 or less than 10 striae and a scutellary striole. Abdomen with intersternal membrane of visible abdominal sternites 3-5 exposed or hidden; hinge position lateral. Abdominal defensive glands present, of the bolitophagine type (see Tschinkel and Doyen 1980). Tarsal formula 4-4-4 (Rhipidandrina and some Bolitophagina) or 5-5-4; tarsal claws not pectinate. Aedeagus with tegmen dorsal. Ovipositor with fourth coxite lobe free, digitate.

This subfamily contains one tribe, which occurs in the United States.

17. Tribe Bolitophagini Kirby 1837

Characteristics: Bolitophagine; body oblong convex, surface roughly verrucose or with sharp elytral costae. Eyes large, oval, slightly to deeply emarginate; antennae 10- (*Bolitotherus*) or 11-segmented (7 apical antennomeres pectinate in *Rhipidandrus*), antennal base inserted in a subocular groove. Elytra estriate or with 9 or 10 striae. Epipleural fold narrow. Abdomen with intercoxal process triangular.

This tribe is divided into two subtribes, the Bolitophagina and Rhipidandrina. There are 21 genera in this tribe, widely distributed. Six genera occur in the United States.

Subtribe Bolitophagina Kirby 1837

Characteristics: Body oblong convex, surface roughly verrucose or with sharp elytral costae. Antennae 10- (*Bolitotherus*) or 11-segmented. Elytra estriate or with 9 striae. Tarsal formula 5-5-4, rarely 4-4-4.

Bolitophagus Illiger 1798, 1 sp., *B. corticola* Say 1825, northeastern United States.

Bolitotherus Candèze 1861, 1 sp., *B. cornutus* (Panzer 1794), eastern United States.
Phellidius Leconte 1862

Eleates Casey 1886, 3 spp., Maine, Oregon and California. Key to species: Casey (1890).

Megeleates Casey 1895, 1 sp., *M. sequoiarum* Casey 1895, California.

Subtribe Rhipidandrina LeConte 1862

Characteristics: Body oblong convex, surface roughly verrucose. Antennae 11-segmented; apical antennomeres pectinate. Elytra ten-striate. Tarsal formula 4-4-4.

Rhipidandrus LeConte 1862, 3 spp., eastern United States, Florida, Texas and Arizona. This genus needs revision.
Heptaphylla Friedenreich 1883
Cherostus Waterhouse 1894
Eutomus Lacordaire 1866, not Hope 1838

V. Subfamily Diaperinae Latreille 1802

Characteristics: Body small to moderate in size (2 to 15 mm). Antennae bearing simple, setiform or compound, stellate sensoria on apical antennomeres. Mandible with molar surface bearing fine transverse ridges. Lacinia without apical hook or tooth. Procoxal cavities closed externally and internally. Elytra with scutellary striole and 9 complete striae or estriate. Wings, if present, almost always without subcubital flecks. Abdomen with intersternal membrane of visible abdominal sternites 3-5 exposed; hinge position lateral. Abdominal defensive glands present, of the platydemine, phaleriine or diaperine type (see Tschinkel and Doyen 1980). Tarsal formula 5-5-4, rarely 4-4-4; tarsal claws not pectinate. Aedeagus with tegmen dorsal.

This subfamily contains nine tribes, generally distributed, six of which occur in the United States.

18. Tribe Phaleriini Mulsant 1854

Characteristics: Diaperine; adults small to moderate in size; body oval or rounded; usually winged. Mentum small; maxilla with apical palpomere simple; epistoma rounded or truncate; frons feebly dilated laterally. Eyes round (*Phaleromela*) or transverse and emarginate (*Phaleria*). Antennae short to moderately long, antennomeres gradually broader apically. Legs stout; mesotrochantin distinct; procoxae transverse; metacoxae narrowly separated; protibiae apically dilated; apical spurs of tibiae distinct; tarsi ventrally setose. Abdomen with intercoxal process triangular.

There are nine genera in this tribe, generally distributed. Two genera occur in the United States.

Phaleria Latreille 1802, 8 spp., ocean beaches. Key to species: Triplehorn and Watrous (1979) and Triplehorn (1991).
Empysara Pascoe 1866
Epiphaleria Lewis 1894
Phalerisida Kulzer 1959
Atahualpina Espa ol 1950

Phaleromela Reitter 1916, 4 spp., Oregon, California. Key to species: Triplehorn (1961).

19. Tribe Diaperini Latreille 1802

Characteristics: Diaperine; adults small to moderate (2-15 mm), broadly oval to elongate, flattened to convex; winged. Eyes large, transverse, anteriorly emarginate, or occasionally small, entire; antennae gradually enlarged or bearing distinct 5-8 segmented club; apical 5-7 antennomeres bearing large, stellate sensoria; labrum strongly transverse, basal membrane almost always exposed; mandible with mola finely, transversely striate, rarely flat or highly modified; lacinia finely setose, without uncus; palp subcylindrical or weakly triangular; meso- and metatibiae bearing fine, crenulate longitudinal ridge on outer (posterior) surface, or rarely smooth; apical spurs of tibia small. Procoxae transverse; mesocoxal cavities closed by mesepimeron or meso- and metasternites. Ovipositor with gonostyles terminal, moderate in size; coxite with basal lobe reduced, folded under 2nd lobe; or ovipositor highly modified (*Diaperis*). Aedeagus with median lobe free or occasionally adnate, or aedeagus highly modified.

This tribe contains 49 genera, generally distributed, with 13 genera placed in two subtribes, represented in the United States.

Subtribe Adelinina LeConte 1862

Characteristics: Flattened, elongate oval beetles, antennae gradually enlarged; eyes narrower than frons across epistomal canthus (except *Alphitophagus*); mesocoxal cavities closed by sternite; epipleuron abruptly narrowed near anterior margin of fifth abdominal sternite; aedeagus with clavae.

Adelina Dejean 1835, 3 spp., Maryland, Florida and Arizona. Key to species: Ardoin (1977).
Doliema Pascoe 1860
Schedarosus Reitter 1876
Sitophagus Horn 1870, not Mulsant 1854

Alphitophagus Stephens 1832, 1 sp., A. *bifaciatus* (Say 1824) adventive from Europe, generally distributed.
Phyletes Redtenbacher 1845
Phylethus Redtenbacher 1849

Cynaeus LeConte 1862, 2 spp., Arizona, California. *Cynaeus angustus* LeConte 1851 is widespread and a minor pest of stored cereal products.

Doliodesmus Spilman 1967, 1 sp., *D. charlesi* Spilman 1967, Arizona.

Gnatocerus Thunberg 1814, 4 spp., generally distributed. Key to species: Spilman (1991).
Cerandria Lucas 1849
Echocerus Horn 1870
Sicinus Champion 1886
Uloma Stephens 1832

Iccius Champion 1866, 1 sp., *I. cylindricus* Champion 1866, Texas and Arizona (see Triplehorn 1995).

Sitophagus Mulsant 1854, 2 spp., one cosmopolitan, Arizona. Key to species: Doyen (1984a).

Subtribe Diaperina Latreille 1802

Characteristics: Globular or short oval beetles. Antennae gradually enlarged or with distinct 5-8 segmented club; eyes exceeding epistomal canthi; mesocoxal cavities closed by mesepimeron (except *Pentaphyllus*); epipleuron gradually narrowed to elytral apex (except *Diaperis*, some *Neomida*); aedeagus without clavae.

Diaperis Muller 1764, 4 spp., eastern United States, South Dakota, Kansas, Texas, Arizona and California. Key to species: Triplehorn (1965).
Allophasia Pascoe 1871

Neomida Latreille 1829, 2 spp., eastern United States and Texas.
Hoplocephala Laporte and Brulle 1831
Arrhenoplita Kirby 1837
Hoplocephala Agassiz 1846
Evoplus LeConte 1866

Liodema Horn 1870, 1 sp., *L. laeve* (Haldeman 1848), Florida to Texas.

Ulomoides Blackburn 1888, 1 sp., *U. ocularis* (Casey 1891), Florida.
Palembus Casey 1891
Alphitobius Carter (in part, see Doyen *et al.* 1989b)
Tenebriomimus Kolbe 1901
Martianus Fairmaire 1893

Pentaphyllus Dejean 1821, 2 spp., eastern United States and California.
Iphicorynus Jacquelin du Val 1861

Platydema Laporte and Brull 1831, 19 spp., generally distributed. Key to species: Triplehorn (1965).
Typhobia Pascoe 1869
Neomida of Motschulsky 1873
Histeropsis Chevrolat 1878

20. Tribe Scaphidemini Reitter 1922

Characteristics: Diaperine. Similar to Diaperini except the coxae are very widely separated and the anterior trochanters are not overlapped laterally by the femora. Most of the reasons for separating this tribe from Diaperini (where it has traditionally been placed are based on internal structures and the larvae (see Doyen 1989).

Three genera compose this tribe. Only one species occurs in the United States.

Scaphidema Redtenbacher 1849, 1 sp., *S. aeneolum* (LeConte 1850), north central and eastern United States.
Nelites LeConte 1850

21. Tribe Crypticini Mulsant 1854

Characteristics: Diaperine. Body elongate-oval; small, winged. Maxilla and labium with apical palpomeres enlarged. Frons feebly expanded laterally, covering base of mandibles, rounded or truncate anteriorly. Eyes transverse, reniform, slightly emarginate. Antennae moderately long, slender, outer 5 antennomeres slightly broader. Prosternum produced posteriorly; mesosternum concave; mesepimera short; metepimera moderate; mesotrochantin distinct; procoxae oval, almost round; metacoxae narrowly separated; tibiae slender; apical spurs distinct, tarsi finely spinous ventrally, basal tarsomere very long. Epipleural fold occupying entire inflexed portion of elytra.

Eleven genera are included in this tribe, widely distributed. Two genera occur in the United States.

Poecilocrypticus Gebien 1928, 1 sp., *P. formicophilus* Gebien 1928, adventive into Alabama, Florida and Mississippi (see Steiner 1982).

Gondwanocrypticus Espa ol 1955, 1 spp., *G. obsoletus* (Say 1823), eastern United States.

22. Tribe Trachyscelini Mulsant 1854

Characteristics: Diaperine. Body rounded, convex, winged. Mentum small, gular sutures diverging; maxilla and labium with apical palpomere simple; epistoma truncate. Frons laterally dilated, covering base of mandibles. Eyes transverse, slightly emarginate. Antennae short, outer 5 antennomeres expanded, forming an abrupt compact club. Legs stout; mesotrochantin distinct; procoxae transverse; metacoxae narrowly separated; protibiae broadly dilated apicolaterally; apical tibial spurs distinct; tarsi ventrally setose. Abdomen with intercoxal process triangular.

This tribe contains three genera, all Old World except for the following species.

Trachyscelis Latreille 1809, 1 sp., *T. flavipes* Melsheimer 1846, Gulf States.

23. Tribe Mymechixenini Jacquelin Du Val 1857

Characteristics: Diaperine; body very small (slightly over 1 mm); antennae with 4-segmented club; antennal insertion hidden. Scutellum small, obscure. Procoxae large, prominent. Metacoxae widely separated. Tarsal formula 4-4-4.

This tribe contains only the single genus.

Myrmechixenis Chevrolat 1835, 1 sp., *M. lathridioides* Crotch 1873, southern United States.

VI. Subfamily Hypophloeinae Billberg 1820

Characteristics: Body very small to moderate in size (1 to 16 mm); slender elongate, cylindrical, sometimes flattened, usually winged; head recessed into prothorax as far as eyes; epistoma truncate, separated from labrum by a slender membranous band. Eyes large and emarginate, rarely absent. Antennae short, flattened, clavate, with club composed of from 1 to 8 antennomeres; bearing simple, setiform or compound, stellate sensoria on apical antennomeres. Lacinia without apical tooth. Procoxal cavities closed externally and internally; mesocoxal cavities without exposed trochantin; closed laterally by meso- and metasternite. Elytra with scutellary striole and 9 complete striae or estriate. Elytra truncate apically with a distinct epipleural fold and epipleural carina, the pygidium exposed. Abdomen with intersternite membrane of visible abdominal sternites 3-5 exposed; hinge position lateral. Abdominal sternites with lateral grooves. Abdominal defensive glands present. Tarsal formula 5-5-4; tarsal claws not pectinate. Aedeagus with tegmen dorsal.

This subfamily is divided into two tribes containing 25 genera. Only the tribe Hypophloeini, with one genus (*Corticeus*) occurs in the United States. Most of the species have an intimate association with bark beetles (Scolytidae) and are regarded as facultative predators of those beetles as well as feeding on fungus and frass.

24. Tribe Hypophloeini Billberg 1820

Characteristics: Hypophloeine; body small to moderate in size (2-16 mm), elongate, subcylindrical or flattened. Eyes large and emarginate, rarely absent. Antennae 10- or 11-segmented, with 7 or 8 apical antennomeres flattened, gradually clubbed; exposed portion of labrum transverse and subquadrate; Mandible with mola small to moderately large, triangular, transversely ridged. Prothorax with simple hind angles; elytra with apices truncate; metacoxae subcontiguous; apical 1 to 5 abdominal sternites with a broad, longitudinal impression on each side; pygidium well developed, slightly to strongly oblique, heavily sclerotized, convex to concave.

This tribe contains six genera, only one of which occurs in the United States.

Corticeus Piller and Mitterpacher 1783, 13 spp., widely distributed. Key to species: Triplehorn (1990).
Hypophloeus Fabricius 1790
Paraphloeus Seidlitz 1894

VII. Subfamily Opatrinae Laporte de Castelnau 1840

Characteristics: Body small to large (about 3 mm to 30 mm). Head often dialated and prominent at sides in front of eyes. Anterior margin of clypeus with deep triangular or arcuate notch, at least in plesiomorphic state. Labrum transverse with basal membrane exposed or concealed. Eyes emarginate, constricted at middle or completely divided by genae into upper and lower parts; sometimes with erect setae between facets. Antennae usually relatively short, sometimes clubbed, 11-segmented (rarely 10); bearing only simple, setiform sensoria on apical antennomeres. Gula simple or with stridulating surface (Platynotini and Oncotini). Mentum small to very large in size, often with anterior aspect tridentate. Trochantin of mesocoxae usually prominent (or absent: Leichenini). Procoxal cavities closed externally and internally. Metasternum long to very short. Mesocoxal cavities closed laterally by mesepimeron. Metacoxae close to widely separated. Pseudepimeron clearly visible on elytra, epipleuron poorly developed to externally absent, sometimes forming narrow band. Wings present or absent. Elytra with scutellary striole and 9 complete striae or estriate. Abdomen with intersternal membrane of visible abdominal sternites 3-5 exposed; hinge position lateral. Abdominal defensive glands present, of the opatrine-eleodine type (see Tschinkel and Doyen 1980). Apical spurs on tibiae short, well developed. Tarsal formula 5-5-4; tarsal claws not pectinate; protarsi in males often dilated. Aedeagus often highly modified with presence of at least one pair of lacinae (=clavae), occasionally with additional pairs; with tegmen dorsal.

This subfamily has a worldwide distribution as is especially abundant in desert areas. The Opatrinae contain 16 tribes, five of which occur in the United States.

25. Tribe Opatrini Laporte de Castelnau 1840

Characteristics: Opatrine. Body small to large in size, oval to elongate oval. Eyes constricted at middle or completely divided by genae into upper and lower parts, without erect scales between facets. Gula simple. Males with aedeagus with tegmen consisting of three parts; apical, basal and intermediate sclerite situated dorsally at base of parameres; without lacinae.

The Opatrini are worldwide in distribution, with 118 genera, 14 of which occur in the United States.

Aconobius Casey 1895, 3 spp., southwestern United States. Key to species: none exist, see Casey (1914). Revision in preparation.

Ammodonus Mulsant and Rey 1859, 2 spp., eastern United States and Arizona. Key to species: none exist, see Fall (1912).
Pseudonomus Fairmaire 1884
Scaptes Champion 1886

Blapstinus Sturm 1826, 29 spp., occurring in the United States. Key to species: Davis (1970). Revision in preparation.
Pedonoeces Waterhouse 1845
Tessaroma Boheman 1858
Lachnodere Mulsant and Rey 1859
Heteropus Laporte 1840
Aspidius Mulsant and Rey 1859
Lodinus Mulsant and Rey 1859

Conibiosoma Casey 1890, 1 sp., *C. elongatum* Horn 1870, southern California.

Conibius LeConte 1851, 5 spp., occurring in the United States. Key to species: Casey (1890). Revision in progress (Aalbu and Triplehorn).
Ooconibius Casey 1895
Euconibius Casey 1895

Cybotus Casey 1890, 1 sp., *C. estriatus* LeConte 1878, Florida.

Ephalus LeConte 1862, 1 sp., *E. latimanus* (LeConte 1854) eastern United States.

Mecysmus Horn 1870, 5 spp., western United States. Key to species: Casey (1890). This genus needs revision.

Nocibiotes Casey 1895, 3 spp., occurring in the United States. Key to species: Casey (1895). Revision in progress (Aalbu and Triplehorn).

Notibius LeConte 1851, 3 spp., western United States. Key to species: Casey (1890). Revision in progress (Aalbu and Triplehorn).

Pseudephalus Casey 1924, 1 sp., *P. brevicornis* Casey 1924, Alabama, Florida.

Tonibius Casey 1895, 1 sp., *T. sulcatus* LeConte 1851, California, Arizona.

Trichoton Hope 1840, 1 sp., *T. sordidum* (LeConte 1851), California, Arizona.

Ulus Horn 1870, 5 spp., Florida, Texas, western United States. Key to species: Casey (1890). This genus needs revision.

26. Tribe Melanimini Seidlitz 1894

Characteristics: Opatrine. Body oval, small, (less than 5 mm in North American species). Eyes appearing divided, actually deeply emarginate, joined posteriorly by thin line of facets, without erect scales between facets. Gula simple. Protibia with 2 strongly produced, dorsal extensions, one apical, one near midpoint of tibia. Wings well developed. Metathorax elongate; Pseudepipleuron extending to apex. First visible abdominal sternite short; process between coxae pointed. Aedeagus with tegmen consisting of 2 parts without intermediate dorsal sclerite between apical and basal parts; parameres fused; penis without lacinae. Anterior tarsi of male not swollen.

This tribe, whose species resemble aphodiine scarabs, is predominately Old World. Three genera are known, only one species occurring in North America.

Cheirodes Gen 1839, 1 sp., *C. californica* (Horn 1870), California, Arizona.

Ammidium Erichson 1843
Anemia Laporte 1840
Anaemia Horn 1870
Eremonomus Wollaston 1861
Pseudanemia Wollaston 1864
Histiaea Fairmaire 1892
Ammidaenemia Reitter 1904
Spinianemia Ardoin 1971
Anemiadena Ardoin 1971
Trichanemia Ardoin 1971

27. Tribe Platynotini Mulsant 1853

Characteristics: Opatrine. Body small to moderate in size, elongate oval, with or without wings. Eyes emarginate, without erect scales between facets, rarely divided. Gula with stridulating surface, consisting of symmetrically arranged, slender, transverse ridges and fossae. Fronto-clypeal suture indistinct; mentum with anterior aspect tridentate or cordate, maxilla with apical palpomere securiform, moderate in size; labrum prominent. Antenna short with first antennomere twice as long as wide or less. Metacoxae close to often widely separated. Metasternum short to long. Anterior tibia often dilated and ornate; femur and tibia occasionally spined. Male aedeagus with tegmen consisting of 2 sclerites, basale and apicale; apicale distinctly shorter than basale; penis with lacunae, parameres always divided at least apically. Coxite of female ovipositor usually with 4 lobes.

The Platynotini contains 39 genera. The tribe is restricted to the Ethiopian and Oriental biogeographic regions except for one genus which occurs in the New World. Species are absent from the west coast of North America.

Opatrinus Dejean 1821, 13 spp., New World, 3 spp. occurring in the United States. Key to species: Iwan (1995).

Hopatrinus Gemminger and Harold 1869
Zidalus Mulsant and Rey 1852
Zodinus Mulsant and Rey 1852

28. Tribe Leichenini Mulsant 1854

Characteristics: Opatrine. Body small, oval (approximately 5 mm in North American species). Body densely covered with scales and setae. Wings developed. Eyes globose, slightly emarginate, with erect setae between facets. Antennae short, with clearly defined 4-segmented club. Gula simple. Pronotum narrower than elytra; with anterior lateral angles reaching level of eyes. Trochantin absent. Aedeagus with tegmen consisting of one part (not divided into basal and apical parts); parameres functionally replaced by strongly sclerotized, hook-shaped apical parts of 9th abdominal sternite. Penis with lacinae.

The tribe contains two genera, one of which occurs in the United States.

Leichenum Blanchard 1845, 1 sp., *L. canaliculatum variegatum* (Klug 1833), southeastern United States (adventive).

29. Tribe Eleodini Blaisdell 1909

Characteristics: Opatrine; body oblong to ovate, rarely oval; highly variable in size; apterous; sometimes setose. Head prominent slightly narrowed behind eyes, not dilated at sides in front of eyes, except for *Lariversius* and *Trogloderus* which also have anterior margin of clypeus with a deep arcuate notch. Fronto-clypeal suture indistinct; labrum clearly exposed; mentum with anterior aspect tridentate or cordate, maxillae exposed; ligula partially concealed; maxilla with apical palpomere securiform, moderate in size; labrum prominent. Eyes transverse, reniform. Antenna rarely short (*Lariversius* and *Trogloderus*), usually long with first antennomere 2.5 to 4 times, rarely 6 times as long as wide. Antennomeres 2-5 usually longer than broad, outer 4 antennomeres globular, sometimes forming loose, 3-segmented club (*Lariversius*). Metasternum short; metepisternum narrow;

metepimeron distinct; mesotrochantin large. Legs long, profemora frequently toothed (sometimes the meso- and metafemora as well). Metacoxae widely separated. Elytra fused at suture, with pseudepipleura forming only part of inflexed sides of elytra (rarely entire inflexed sides of elytra [i.e., *Elediomorpha* and some *Eleodes* (*Cratidius*)], gradually widening basally (sometimes obsolete anteriorly); males never with pubescent pit at base of 2nd abdominal sternite. Male aedeagus with tegmen consisting of 2 sclerites, basale and apicale, apicale distinctly shorter than basale, penis with lacunae, parameres fused. Coxite of female ovipositor without lobes.

Six genera are included in this tribe, all of them North American; all are found in Central and Western United States. The larval stages of members of the large genus *Eleodes* are known as "false wireworms" and are of considerable economic importance on semiarid lands and irrigated cropping areas where they feed on underground portions of many commercially grown plants.

Eleodes Eschscholtz 1829, 129 spp., western United States. Key to species: Blaisdell (1909). Revision in progress (Triplehorn).

Amphidora Eschscholtz 1829.
Cratidus LeConte 1862
Trichoderulus Blaisdell 1923.
Eleodopsis Blaisdell 1939

Species assigned to subgenera as follows:

Subgenus *Melaneleodes* Blaisdell 1909, 11 spp., western United States.
Subgenus *Litheleodes* Blaisdell 1909, 8 spp., western United States.
Subgenus *Tricheleodes* Blaisdell 1909, 6 spp., western United States.
Subgenus *Pseudeleodes* Blaisdell 1909, 2 sp., California, Nevada.
Subgenus *Promus* LeConte 1862, 7 spp., western United States.
Subgenus *Heteropromus* Blaisdell 1909, 1 sp., *E. veterator* Horn 1874, Texas.
Subgenus *Eleodes, sensu stricto*, 33 spp., western United States.
Subgenus *Blapylis* Horn 1870: 45 spp., western United States.
Subgenus *Cratidus* LeConte 1862, 1 sp., *C. osculans* (LeConte 1851), California.
Subgenus *Metablapylis* Blaisdell 1909, 5 spp., western United States.
Subgenus *Steneleodes* Blaisdell 1909, 4 spp., western United States.
Subgenus *Discogenia* LeConte 1866, 2 spp., California.
Subgenus *Holeleodes* Blaisdell 1937, 1 sp., *E. beameri* Blaisdell 1937 (see Triplehorn and Doyen 1972).
Subgenus *Caverneleodes* Triplehorn 1975, 3 spp., Arizona and Texas.

Eleodimorpha Blaisdell 1909, 1 sp., *E. bolcan* Blaisdell 1909, California.

Embaphion Say 1824, 8 spp., southwestern United States to South Dakota. Key to species: Blaisdell (1909).

Lariversius Blaisdell 1947, 1 sp., *L. tibialis* Blaisdell 1947, California, Nevada.

Neobaphion Blaisdell 1925, 3 spp., southwestern United States. Key to species: Triplehorn and Aalbu (1985).

Trogloderus LeConte 1879, 1sp., *T. costatus* LeConte 1879, Arizona, Utah, California, Nevada.

VIII. Subfamily Tenebrioninae Latreille 1802

Characteristics: Body small to large (about 3 mm to 30 mm). Antennae filiform, serrate, incrassate or rarely capitate bearing simple, setiform or compound, stellate or placoid sensoria on apical 4 to 6 antennomeres. Labrum transverse or subquadrate with basal membrane exposed or concealed. Mandible with mola striate or not. Maxilla with galea finely setose or with uncus of one or 2 teeth. Procoxal cavities closed externally, open or closed internally. Mesocoxal cavities closed laterally by mesepimeron or sternum. Abdomen with intersternal membrane of visible abdominal sternites 3-5 exposed; hinge position lateral. Defensive reservoirs variable, often distinctive at tribal level (see Doyen 1989 and Tschinkel and Doyen 1980). Elytra with scutellary striole and 9 complete striae or estriate. Tarsal formula 5-5-4; tarsal claws not pectinate. Tarsi usually with ventral surface coarsely setose or spinose, occasionally with pads of pilose setae. Aedeagus with tegmen dorsal or rotated about 45° to 90° at rest, rarely inverted; median lobe freely extrusible or adnate to tegmen; sometimes with accessory lobes. Ovipositor usually with coxites clearly 4-lobed, occasionally with ovipositor shaft shortened and lobing reduced; lobes usually subequal in length; fourth lobe rarely free and digitate; paraprocts parallel to axis of ovipositor at rest.

This subfamily contains 14 tribes, generally distributed, 11 of which occur in the United States.

30. Tribe Eulabini Horn 1870

Characteristics: Tenebrionine; body small, oblong (length, usually less than 1 cm); often pubescent; head short; antennal joints 4 to 11 subequal in length, outer joints more or less transverse; last joint rounded at apex. Intercoxal process of abdomen 0.7 to 1.3 times wider than long; narrowing anteriorly, with anterior angles broadly rounded; Elytra usually finely costate (except *Epantius*, where costae are obsolete); epipleuron gradually widening anteriorly, with moderately large or subobsolete punctures; first abdominal sternite shorter than next 2 combined; 3rd and fourth sternite feebly emarginate at apex. Metafemora lacking teeth, never suddenly clavate.

The three genera composing this tribe are restricted to the west coast of the United States and northern Mexico.

Apsena LeConte 1862, 5 spp., California. Key to species: Blaisdell (1932).

Eulabis Eschscholtz 1829, 1 sp., *E. bicarinata* Eschscholtz 1829, California.

Epantius LeConte 1862, 1 sp., *E. obscurus* LeConte 1851, California.

31. Tribe Cerenopini Horn 1870

Characteristics: Tenebrionine; body moderate to large, elongate (length more than 1 cm); never pubescent; head much longer than wide; antennomeres 2 to 9 gradually diminishing in size and width, terminal antennomere acute at apex. Intercoxal process of abdomen 1.1 to 1.9 times wider than long; not widening or slightly widening anteriorly, with anterior angles more narrowly rounded; epipleuron gradually widening anteriorly, with minute punctures or granules; first abdominal sternite equal or longer than 2nd and 3rd combined; 3rd and fourth sternites very short and deeply emarginate at apex. Metafemora toothed in males, strongly clavate.

Two genera are included in this tribe, which is restricted to western United States and Mexico.

Argoporis Horn 1870, 5 spp., California, Arizona, New Mexico, and Texas. Key to species: Berry (1980).
Threnus Motschulsky 1870

Cerenopus LeConte 1851, 1 sp., *C. concolor* LeConte 1851, California, Arizona (see Berry 1975).

32. Tribe Blaptini Leach 1815

Characteristics: Tenebrionine; body large, oblong-elongate (length usually more than 2 cm). Head with antennomeres 2-5 moniliform; bearing only simple, setiform sensoria on apical antennomeres. Mouthparts, other aspects of head as in Eleodini; Pronotum with basal margin truncate, closely embracing and overlapping basal aspect of elytra. Elytra with epipleura narrow throughout its length; pseudepipleura very broad, gradually widening anteriorly, anterior aspect evenly rounded; males with pubescent pit at base of 2nd abdominal sternite.

Fifteen genera compose this tribe, all Old World. Two species are adventive in the United States.

Blaps Fabricius 1775, 2 spp., adventive, in the United States. Key to species: Triplehorn (1952).

33. Tribe Ulomini Mulsant 1854

Characteristics: Tenebrionine; body small to large (3-25 mm), elongate to broadly oval, flattened to subcylindrical, winged. Eyes moderately large, anteriorly emarginate; antennae gradually clavate, apical 5-8 antennomeres bearing large placoid sensoria; apical antennomere with compound sensoria; labrum transverse, constricted at base; clypeolabral membrane exposed or concealed; mandibles with molas estriate; right mola weakly convex, surface simple or bluntly lobed; left mola concave, complexly lobed; lacinia with 1-2 enlarged apical unci; mentum often sexually dimorphic, subtrapezoidal to cordate to oval, flat or weakly convex in females; broadly oval, enlarged and concave, frequently setose in males; ligula sclerotized; labial palp subcylindrical or enlarged. Procoxal cavities closed internally; protibia usually modified in males; mesocoxal cavities closed laterally by sternite or barely open (some *Uloma*); 7th abdominal tergite usually exposed as heavily sclerotized pygidium. Ovipositor with gonostyles lateral, small or moderate in size; coxite about as long as paraproct, with 4 subequal lobes (for more detailed description see Doyen 1985).

The Ulomini are worldwide in distribution, with approximately 40 genera, only 2 of which occur in the United States.

Eutochia LeConte 1862, 2 spp., eastern United States and Texas.
Aniara Lacordaire 1859
Delpygus LeConte 1866
Cenoscelis Wollaston 1867
Aniarus Gemminger and Harold 1870
Holaniara Fairmaire 1871

Uloma Dejean 1821, 5 spp., occurring in the United States. Key to species: Horn (1870).
Prioscelida White 1846
Melasia Mulsant 1856
Apterulomoides Kaszab 1982 (subgenus)

34. Helopini Latreille 1802

Characteristics: Tenebrionine; body small to moderate in size elongate to broadly oval; apterous or winged; antennae moniliform or gradually clavate; bearing only simple, setiform sensoria on apical antennomeres. Eyes, transverse, vertically elongate, subparallel, strongly inflated. Labrum transverse; mandibles with mola not planar; lacinia with single tooth; maxillary palp securiform. Mentum small, trapezoidal; ligula prominent. Procoxal cavities broadly closed. Mesocoxal cavities closed by mesepimeron. Tarsi pubescent ventrally.

The Helopini are worldwide in distribution, with 32 genera, three of which occur in the United States.

Helops Fabricius 1775, 40 spp., occurring in the United States. Key to species: Horn (1870). This genus needs revision.
Anteros Laporte de Castelnau 1840
Biomorphus Motschulsky 1872
Coscinopter Allard 1876 (subgenus)
Coscinoptilix Allard 1877 (subgenus)
Mesohelops Reitter 1904 (subgenus)
Stenotrichus LeConte 1862 (synonym)

Nautes Pascoe 1866, 1 sp., *N. viridimicans* Horn 1878, Florida.

Tarpela Bates 1870, 3 spp., eastern United States. Key to species: Triplehorn (1952). This genus needs revision.
Lamperos Allard 1876

35. Tribe Triboliini Mulsant 1854

Characteristics: Tenebrionine; adults small to moderate in size, elongate to broadly oval, flattened to subcylindrical. Eyes moderate in size, anteriorly emarginate or entire (*Palorus*); antennae gradually clavate or with abrupt club composed of apical 4 antennomeres (*Lyphia*), with simple, conical sensoria, or with compound (tenebrionoid) sensoria on inner and outer margins of apical 7-8 antennomeres. Labrum subquadrate (*Palorus*) to broadly transverse; clypeolabral membrane concealed; mandibles with molas similar with simple or striate surface; lacinial uncus present or absent; ligula membranous. Mesocoxal cavities closed laterally by sternite. Ovipositor with gonostyles lateral, small; coxite as long or longer than paraproct, with 4 subequal lobes, sometimes poorly defined.

The Triboliini are worldwide in distribution, with 12 genera including many stored product pests, seven genera occur in the United States.

Coelopalorus Blair 1930, 1 sp., *C. foveicollis* (Blair 1930), (adventive) has established in the United States. Key to species: Spilman (1991).

Hypogena Dejean 1834, 2 spp., California, Arizona and Texas.
Ulosonia Laporte 1840

Latheticus Waterhouse 1880, 2 spp., southwestern United States and cosmopolitan.

Lyphia Mulsant and Rey 1859, 1 sp., *L. tetraphylla* (Fairmaire 1856), cosmopolitan.
Lindia Blackburn 1883

Mycotrogus Horn 1870, 2 spp., Arizona and California. Key to species: Spilman (1963).

Palorus Mulsant 1854, 2 spp., both cosmopolitan. Key to species: Spilman (1991).
Caenocorse Thomson 1859
Schizophthalmotribolium Kaszab 1940
Eba Pascoe 1863
Platyotus Gerstaecker 1871
subgenus *Circornus* Fleischer 1900

Tharsus LeConte 1862, 1 sp., *T. seditiosus* LeConte 1862, southern United States.

Tribolium Macleay 1825, 7 spp., southwest and cosmopolitan. Key to species: Spilman (1991), see also Triplehorn (1978).
Margus Dejean 1833
Stene Stephens 1832
Tenebrioloma Gebien 1910
Lindia Blackburn 1885
Aphanotus LeConte 1862
subgenus *Leanum* Uyttenboogaart 1934
subgenus *Eusemostene* Gebien 1940

36. Tribe Alphitobiini Reitter 1917

Characteristics: Tenebrionine; adults small (about 4 mm to 7 mm). Eyes emarginate but never divided by epistomal canthus; antennae incrassate, bearing stellate, compound sensoria on apical 6 antennomeres; labrum about 2.5 to 3 times broader than long, with basal membrane concealed; epipharynx asymmetrical; mandibles with incisors bifid, molas striate or nonstriate; lacinia with uncus; palp narrowly triangular. Procoxal cavities closed externally and internally; mesocoxal cavities closed by epimeron or sternite; elytra 9-striate with scutellary striole. Tarsi clothed ventrally with spinose setae. Ovipositor as in Tenebrionini; aedeagus with median lobe adnate to tegmen, without accessory lobes.

The Alphitobiini are worldwide in distribution, with four genera including *Alphitobius diaperinus* (Panzer 1797), a common stored product pest. Two of these genera occur in the United States.

Alphitobius Stephens 1832, 2 spp., cosmopolitan (see Spilman 1991).
Heterophagus Redtenbacher 1845
Cryptops Solier 1851

Metaclisa Jacquelin du Val 1861, 1 sp., *M. marginalis* Horn 1870, California.
Amarantha Faust 1875

37. Tribe Tenebrionini Latreille 1802

Characteristics: Tenebrioninae; adults small to large (about 6 mm to 30 mm). Eyes moderate in size, weakly emarginate to entirely divided by epistomal canthus; antennae serrate-filiform, weakly incrassate, bearing simple, setiform sensilla or compound stellate sensoria on apical 4 antennomeres; labrum about twice as broad as long, with basal membrane concealed or exposed; epipharynx asymmetrical (as in Fig. 11; Doyen and Tschinkel 1982); mandibles with incisors bifid, molas striate or estriate; lacinia with or without (*Rhinandrus*) uncus, palp subcylindrical to broadly triangular. Procoxal cavities closed externally, open or closed internally; mesocoxal cavities closed laterally by mesepimeron; elytra 9-striate with scutellary striole or estriate. Tarsi clothed ventrally with spinose or pilose (*Zophobas, Rhinandrus*) setae. Ovipositor flexible with coxites and paraprocts subequal in length; coxites divided into 4 subequal lobes; fourth lobe not digitate.

There are approximately 40 genera worldwide; six of these, including *Tenebrio* occur in the United States.

Bius Mulsant 1854, 1 sp., *B. estriatus* LeConte 1851, California.

Idiobates Casey 1890, 1 sp., *I. castaneus* (Knoch 1801), eastern United States.

Neatus LeConte 1862, 1 sp., *N. tenebrioides* (Beauvois 1805), generally distributed.

Rhinandrus LeConte 1866, 1 sp., *R. sublaevis* Horn, Arizona.
Exerestus Bates 1870
Proderops Fairmaire 1873

Tenebrio Linnaeus 1758, 2 spp., cosmopolitan (Spilman 1991).
Menedrio Motschulsky 1872
Tenebrionellus Crotch 1873

Zophobas Blanchard 1845, 2 spp., Florida, Texas to Arizona.
Macrozophobas Pic 1913

38. Tribe Centronopini Doyen 1989

Characteristics: Tenebrioninae; adults moderate to large (about 10 mm to 20 mm), elongate, flattened beetles. Eyes moderate in size, strongly emarginated by epistomal canthus; antennae incrassate; apical 5 or 6 antennomeres bearing large, stellate sensoria, especially on inner apical margins; labrum about twice as broad as long, with basal membrane concealed; epipharynx symmetrical or nearly so; mandibles with incisors bifid, molas nonstriate; lacinia with uncus, palp weakly triangular. Procoxal cavities closed externally and internally; elytra 9-striate with short, sometimes poorly defined scutellary striole; mesocoxal cavities closed laterally by mesepimeron. Tarsi clothed ventrally with pads of dense, fine, yellowish pubescence. Ovipositor (Fig. 1) strongly sclerotized, slightly compressed in lateral plane with lobing of coxites sometimes obscured and gonostyli papilliform; paraprocts about twice as long as coxites. Aedeagus rotated about 45° to 60°; median lobe adnate to tegmen or nearly so.

Only three genera compose this New World tribe; two are found in the United States.

Centronopus Solier 1848, 2 spp., eastern United States to Texas (Spilman 1962).
Scotobates Horn 1875
Pachyurgus LeConte 1866
Menechides Motschulsky 1872
Pyres Champion 1885

Scotobaenus LeConte 1859, 4 spp., California. Key to species: Blaisdell (1933).

39. Tribe Amarygmini Lacordaire 1859

Characteristics: Tenebrionine; adults with body oval or elongate oval, convex, apterous or winged; head deflexed, nearly vertical in respose, inserted into prothorax to the eyes or beyond; eyes large, reniform, emarginate, sides of frons obliquely elevated; epistoma with a short membranous band between it and labrum; maxilla and labium with apical palpomeres securiform. Antennae long and slender, extending posteriorly well beyond base of pronotum, apical antennomeres slightly broader. Legs slender, tibia with apical spurs small, tarsi densely pubescent ventrally.

This tribe has more than 70 genera assigned to it. It is widely distributed but mostly Old World. Only two genera occur in the United States.

Cymatothes Dejean 1834, 2 spp., Florida, Louisiana and Texas.
Pyanisia Laporte de Castelnau 1840

Meracantha Kirby 1837, 1 sp., *M. contracta* (Beauvois 1805), central and eastern United States.
Falacer Laporte de Castelnau 1840
Physocoelus Dejean 1834

40. Tribe Apocryphini Lacordaire 1859

Characteristics: Tenebrionine; adults very small, pedunculate, with globular prothorax, constricted, petiolate mesothorax and long slender legs with clavate femora. Wings absent. Eyes nearly round or with slight anterior emargination. Antennae slender, filiform basally, becoming slightly clavate and serrate apically. Labrum transversely subrectilinear with medial processes of tormae transverse. Mandibles elongate. Prothorax globular, rounded, without defined angles or lateral carinae; procoxal cavities closed externally and internally. Mesocoxal cavities closed laterally by apposed sternite, trochantin concealed. Elytra oval, inflated, abruptly constricted basally; epipleuron narrow, extending from humerus to elytral apex. Ovipositor with gonostyles small, subapical.

This tribe contains five genera, one of which occurs in the United States.

Apocrypha Eschscholtz 1831, 3 spp., California. Key to species: Doyen and Kitayama 1980.
Compsomorphus Solier 1851

IX. Subfamily Alleculinae Laporte de Castelnau 1840

Characteristics: Body small to large (3-15 mm), elongate, convex; surface pubescent or glabrous, never coarsely sculptured. Head narrowed behind eyes; eyes small to large, transverse or oblique, distinctly emarginate on anterior margin; antennae elongate, filiform, obconical or rarely subserrate or pectinate, inserted under small, oblique frontal ridge (in North American species); bearing simple, setiform or compound, stellate sensoria on apical antennomeres. Labrum prominent, rectangular. Mandibles hidden under labrum, distinctly notched apically (in North American species). Maxillary palpi long, with apical palpomere enlarged, securiform to broadly triangular. Pronotum evenly convex; distinctly narrowed apically with anterior margin truncate to evenly convex, basal margin usually bisinuate. Procoxal cavities closed externally and internally. Procoxae globose or conical, separated by broad prosternal process (except some species of *Mycetochara*). Mesocoxal cavities closed laterally by mesepimeron. Abdomen with intersternal membrane of visible abdominal sternites 3-5 exposed; hinge position lateral. Abdominal defensive glands present. Male eighth and ninth sternite slightly emarginate apically to bilobed, often highly modified and function as secondary sexual structures. Elytra with scutellary striole and 9 complete striae or estriate. Legs with apical spurs of tibiae distinct; tarsi with at least penultimate tarsomere lobed ventrally (Alleculini) or not lobed. Tarsal formula 5-5-4; tarsal claws distinctly pectinate (reduced in *Onychomira*). Aedeagus with tegmen dorsal. The subfamily was revised by Seidlitz (1896) for the World, the North American species were revised by Casey (1891), and an illustrated key to the North American genera was published by Campbell (1984).

In North America, species occur throughout the contiguous United States and southern Canada, but the greatest diversity is found in the arid regions of the southwestern United States and in Florida. The subfamily is divided into two tribes, the Alleculini which are widely distributed in all faunal regions and the Omophlini which are known only from the Palearctic Region. The tribe Alleculini is divided into four subtribes. The largest subtribe, Alleculina, is distributed in all faunal regions; the subtribes Gonoderina and Mycetocharina are both Holarctic in distribution; and the subtribe Lystronychina is known only from the New World.

41. Tribe Alleculini Laporte de Castelnau 1840

Characteristics: Alleculine; body small to moderate in size. Eyes deeply emarginate anteriorly, with the basal antennomeres of the antennae inserted under a frontal ridge within this emargination; the hind coxae do not project over the first visible abdominal sternite; the intercoxal process of the first visible abdominal sternite distinct, either narrowly or broadly triangular; abdomen with only 5 visible abdominal sternite (a small sixth sternite is often visible in the male and rarely in the female), sides and base margined and acutely narrowed; the apical palpomere of the maxillary palpi expanded and much larger than the penultimate palpomere; the apex of the mandibles cleft.

There are approximately 140 genera worldwide, 19 genera and 159 species recorded from the United States and Canada. The tribe is in need of revision, but most species have been described.

1. Subtribe Alleculina Laporte de Castelnau 1840

Characteristics: North American species densely pubescent dorsally, usually with fine, subrecumbent setae. Antennae filiform to obconical, moderately coarsely setose. Intercoxal process of first visible abdominal sternite narrowly triangular. At least penultimate tarsomeres of the tarsi distinctly lobed ventrally (lobes obsolete on posterior tarsi of *Lobopoda*). The Mexican and Central American genus *Charisius* Champion was mistakenly reported from Florida.

Alethia Champion 1888, 3 spp., southwestern United States.

Hymenorus Mulsant 1852, 96 spp., widely distributed. Key to species: Fall (1931). This genus needs revision. Note: The original form is *Hymenophorus* Mulsant 1851. However *Hymenorus* Mulsant 1852 (an unjustified emendation) has been utilized by all published citations since then and clearly falls under ICZN(11) mentioned above.

Knausia Fall 1931, 1 sp., *K. crassicornis* Fall 1931, New Mexico.

Lobopoda Solier 1835, 7 spp., eastern United States to west Texas. Key to species: Campbell (1966).

Allecula of Say 1827, not Fabricius 1801

Menoeceus Champion 1888, 1 sp., *M. texanus* Champion 1888, Texas.

Phedius Champion 1888, 1 sp., *P. funereus* Schaeffer 1905, Arizona.

Stenochidus LeConte 1862, 3 spp., Nevada and California. Key to species: Marshall (1967).

Telesicles Champion 1888, 2 spp., *T. cordatus* Champion 1888, southwestern United States and *T. magnus* Hatch 1965, Washington.

2. Subtribe Gonoderina Seidlitz 1896

Characteristics: Body with surface pubescent dorsally, with fine, subrecumbent setae. Antennae obconical to subserrate, often with 3rd antennomere distinctly shorter than fourth antennomere; finely, densely pubescent. Intercoxal process of first visible abdominal sternite narrowly triangular. Tarsi never lobed ventrally.

Andrimus Casey 1891, 7 spp., southeastern United States. Key to species: Casey (1891, 5 spp. only). This genus needs revision.

Cteniopus of LeConte 1866, not Solier 1835

Androchirus LeConte 1862, 2 spp., *A. erythropus* (Kirby 1837) and *A. femoralis* (Olivier 1795), eastern United States and southeastern Canada (2 species in Casey 1891).

Capnochroa LeConte 1862, 1 sp., *C. fuliginosa* (Melsheimer 1846), eastern United States and southeastern Canada.

Isomira Mulsant 1856, 12 spp., widely distributed in United States and southern Canada. Key to species: Marshall (1970).

Onychomira Campbell 1984, 1 sp., *O. floridensis* Campbell 1984, Florida.

Pseudocistela Crotch 1873, 6 spp., widely distributed in United States and southern Canada. Key to species: Hopping (1933). Note: *Pseudocistela* Crotch 1873 is a junior synonym of *Chromatia* LeConte 1862. However, *Pseudocistela* falls within the intentions of the preamble and/or ICZN(11) mentioned above.

Cistela LeConte 1862, not Fabricius 1775, not Forester 1771
Chromatia LeConte 1862

Tedinus Casey 1891, 1 spp., *T. angustus* Casey 1891, Georgia.

3. Subtribe Mycetocharina Mulsant 1856

Characteristics: North American species pubescent dorsally, usually with fine, subrecumbent setae. Antennae obconical, moderately coarsely setose. Intercoxal process of first visible abdominal sternite narrowly triangular. None of tarsomeres lobed ventrally.

Hymenochara Campbell 1978, 2 spp., *H. rufipes* (J. E. LeConte 1824), eastern United States and *H. arizonensis* Campbell 1978, Arizona.

Mycetochara Berthold 1827, 10 spp., widely distributed in United States and southern Canada. Key to species: Campbell (1978).
Ernocharis Thomson 1859
Mycetophila Gyllenhal 1810, not Meigen 1803
Mycetocharis Gyllenhal 1827
Myacetochares Latreille 1829
Mycetocharus Stephens 1829
Pterna Seidlitz 1896
Stigmatoma LeConte 1862

4. Subtribe Lystronychina Lacordaire 1859

Characteristics: North American species glabrous dorsally, or with coarse, elongate, erect setae. Antennae filiform to slightly obconical, moderately finely pubescent. Intercoxal process of first visible sternite broadly triangular. None of tarsomeres lobed ventrally.

Anamphidora Casey 1924, 1 sp., A. *campbelli* Marshall 1967, Texas.

Lystronychus Latreille 1829, 3 spp., southwestern United States. Key to species: Marshall (1969). Note: The original spelling is *Lystronichus*. However *Lystronychus* has always been utilized in published references.

X. Subfamily Coelometopinae Lacordaire 1859

Characteristics: Body small to very large (about 5 mm to 45 mm) beetles of diverse shape and color. Winged or apterous. Antennae filiform, serrate, incrassate or weakly capitate; apical 5 to 8 antennomeres bearing stellate compound sensoria. Labrum transverse with basal membrane exposed or concealed. Mandible with mola finely striate, flat or occasionally coarsely ridged. Maxilla with galea finely setose or with uncus of 1 or 2 teeth. Procoxal cavities broadly closed both externally and internally. Mesocoxal cavities closed laterally by mesepimeron; elytra with scutellary striole and 9 complete striae or estriate. Abdomen with intersternal membrane of visible abdominal sternites 3-5 exposed; hinge position lateral. Defensive glands present, of the coelometopine, cnodalonine or catapeistus type (see Doyen 1989 and Tschinkel and Doyen 1980). Tarsal formula 5-5-4; tarsal claws not pectinate. Tarsi with ventral pads of fine, dense pubescence or with sparser, coarser setae; tibiae frequently with setose inner apical margins. Aedeagus with tegmen dorsal at rest, rotated about 60° to 90°, or occasionally inverted (rotated 180°); median lobe usually adnate to tegmen, rarely freely extrusible. Ovipositor (Tschinkel and Doyen 1980, Figs. 22, 23, 39) with coxites clearly 4-lobed; basal lobe usually elongate (often longer than three apical lobes combined); paraprocts rotated about 145° about articulation with coxite at rest, or, rarely, rotated about 60° to 90° (*Menephilus,* not North American).

This subfamily contains three tribes, all occurring in the United States.

41. Tribe Coelometopini Lacordaire 1859

Characteristics: Coelometopine; adults small to very large (5 to 45 mm), of diverse shape and color; apterous or winged. Eyes reniform, moderate in size, separated dorsally by much more than width of single eye lobe. Antennae incrassate or weakly capitate, with stellate (coelometopine) sensoria on apical 5 or 6 (rarely 7 or eight) antennomere. Labral membrane exposed or concealed. Tarsi with ventral surface almost always flattened, bearing pads of yellowish, usually pilose setae; inner margins of tibiae frequently pilose, especially near apices; tarsomeres three and 4 subequal.

This large tribe contains 45 genera, generally distributed, 11 of which occur in the United States.

Alobates Motschulsky 1872, 2 spp., eastern and southern United States. Key to species: Triplehorn (1952).
Nyctobates of Horn 1870, not Gu rin 1834.

Apsida Lacordaire 1859, 2 spp., eastern and southern United States. Key to species: Triplehorn (1952).

Cibdelis Mannerheim 1843, 5 spp., California. This genus needs revision.

Coelocnemis Mannerheim 1843, 7 spp., western United States. Key to species: Doyen (1973).

Glyptotus LeConte 1858, 1 spp., *G. cribratus* LeConte 1858, Florida and Texas.

Haplandrus LeConte 1862, 5 spp., eastern United States. This genus needs revision.

Iphthiminus Spilman 1973, 6 spp., western and northeastern United States. This genus needs revision.
Iphthimus Truqui 1857 (see Spilman 1973).

Merinus LeConte 1866, 1 spp., *M. laevis* (Olivier 1795) eastern United States.

Oenopion Champion 1885, 1 spp., *O. zopheroides* (Horn 1874), Texas.

Polopinus Casey 1924, 4 spp., Florida. Key to species: Kritsky (1989). (Volume 1, Cover Figure, center)

Polypleurus Eschscholtz 1831, 2 spp., Atlantic states.

Upis Fabricius 1792, 1 sp., *U. ceramboides* (Linneaus 1758) Northeastern United States.

Xylopinus LeConte 1862, 2 spp., eastern United States. Key to species: Triplehorn (1952).
Taenobates Motschulsky 1872

43. Tribe Talanini Champion 1887

Characteristics: Coelometopine; body elongate, cylindrical, winged. Eyes moderate to large, bulging, but separated dorsally by much more than width of dorsal eye lobe. Antennae incrassate, bearing stellate, compound sensoria on apical 5 or 6 antennomeres. Labral membrane exposed. Tarsi with ventral surface flattened, with pads of yellowish, pilose setae; 4th tarsomere of first 2 pairs of legs much smaller than preceding tarsomeres and bearing only few long setae. Inner tibial surface setose or pilose.

The tribe is monogeneric.

Talanus Jacquelin DuVal 1856, 4 spp., Florida, Louisiana and Texas. Key to species: Triplehorn (1968).
Dignamptus LeConte 1878

44. Tribe Strongyliini Lacordaire 1859

Characteristics: Coelometopine; body moderate in size, elongate, subcylindrical; winged. Eyes large, usually separated dorsally by less than width of single eye lobe, frequently contiguous or nearly so, especially in males. Antennae usually serrate, sometimes weakly incrassate, rarely pectinate; apical 7 or 8 antennomeres bearing stellate, compound sensoria. Labral membrane exposed. Prosternal process declivous, flattened behind coxae; broadly rounded or truncate; mesosternal fossa very broad, shallow. Tarsi nearly cylindrical, ventral surfaces usually covered by stiff, often dark colored setae; tibiae seldom bearing strips or patches of pile.

There are approximately 58 genera in this generally distributed, mainly tropical tribe, only one of which occurs in the United States.

Strongylium Kirby 1818, 10 spp., eastern United States, Texas, Arizona. Key to species: Triplehorn and Spilman (1973).
Anomaearthrum M klin 1864
Doichopterum Murray 1862
Eustrongylium Kolbe 1894
Stenochia Kirby 1818
Zurecheria Reitter 1908
Reminius Casey 1924

Bibliography

AALBU, R. L. (in press) The Pimeliine tribe Cryptoglossini: classification, biology and inferred phylogeny (Coleoptera: Tenebrionidae). Knull Series, Ohio Biological Survey.

AALBU, R. L. 1985. Systematics and Biology of the Cryptoglossini (Coleoptera: Tenebrionidae). Ph.D. dissertation, Ohio State University. 329 pp. [unpublished]

AALBU, R. L. and F. G. ANDREWS. 1985. New species, Relationships and notes on the biology of the Endogean Tentyriine genus *Typhlusechus* (Tenebrionidae: Stenosini). Occasional Papers in Entomology, Department of Food and Agriculture, State of California, 30: 1-28.

AALBU, R. L. and F. G. ANDREWS. 1996. Revision of the Neotropical genus *Discopleurus* Lacordaire (Tenebrionidae: Stenosini). Coleopterists Bulletin, 50: 14-38.

AALBU, R. L., T. J. SPILMAN and K. W. BROWN. 1995. The systematic status of *Amblycyphrus asperatus, Threnus niger, Pycnomorpha californica, Emmenastus rugosus*, and *Biomorphus tuberculatus* Motschulsky (Coleoptera: Tenebrionidae). Proceedings of the Entomological Society of Washington, 97: 481-488.

AALBU, R. L. and C. A. TRIPLEHORN. 1985. Redefinition of the Opatrine tribes in North America with notes on some apterous genera. (Coleoptera: Tenebrionidae: Tenebrioninae). Coleopterists Bulletin, 39: 272-280.

AGASSIZ, J. L. R. 1846. Nomenclator zoologicus. Faciculus XI. Continens Coleoptera. 170 pp.

ALLARD, E. 1876. R visions des helopides vrais de Lacordaire. L'Abeille, 14: 1-80.

ALLARD, E. 1877. R visions des helopides vrais de Lacordaire. Mitteilungen der Schweizerischen Entomologischen Gesellschaft, 5: 13-268.

ARDOIN, P. 1971. Contribution l' tude des esp ces africaines et malgaches du genre *Anemia* Laporte [Col. Tenebrionidae]. Annales de la Soci t Entomologique de France (n.s), 7: 357-422.

ARDOIN, P. 1977. Contribution l' tude des esp ces Am ricaines du genre *Doliema* Pascoe (Col. Tenebrionidae). Annales de la Soci t Entomologique de France (n.s), 13: 1-20.

ARNETT, R. H. 1973. The Beetles of the United States (A manual for identification). American Entomological Institute. Ann Arbor, MI. 1112 pp.

BARBER, H. S. 1913. Notes on Rhipidandri (Coleoptera). Proceedings of the Entomological Society of Washington, 15: 188-193.

BATES, H. W. 1870. Descriptions of new genera and species of Heteromera. Entomologist's Monthly Magazine, 6: 268-275.

BATES, H. W. 1873. Notes on Heteromera, and descriptions of new genera and species (No. 8). Entomologist's Monthly Magazine, 10: 14-17.

BEAUVOIS, A. M. F. J. PALISOT de. 1805. Insectes recueillis en Afrique et en Am rique. 1805-1821, pp. I-XVI and 1-276. Paris.

BERRY, R. L. 1973. The Cerenopini and Eulabini, two tribes previously included in the Scaurini (Coleoptera: Tenebrionidae). Annals of the Entomological Society of America, 66: 70-77.

BERRY, R. L. 1974. New species of *Cryptadius* from Texas and Sonora. (Coleoptera: Tenebrionidae). Proceedings of the Entomological Society of Washington 76: 172-177.

BERRY, R. L. 1975. A revision of the genus *Cerenopus* (Coleoptera: Tenebrionidae). Annals of the Entomological Society of America, 68: 925-934.

BERRY, R. L. 1980. A revision of the North American genus *Argoporis* (Coleoptera: Tenebrionidae: Cerenopini). Ohio Biological Survey Bulletin, 6: 109 pp.

BILLBERG, G. J. 1820. Enumeratio insectorum in museo Billberg. 138 pp. Stockholm.

BLACKBURN, T. 1888. Notes on Australian Coleoptera with descriptions of new species. Proceedings of the Linnean Society of New South Wales (Ser. 2), 3: 805-875.

BLAIR, X. G. 1930. The Indian species of *Palorus* Muls. (Coleoptera: Tenebrionidae) and some associated beetles. Indian Forest Records, Entomological Series, 14(5): 133-152.

BLAISDELL, F. E. 1906. New Californian Coleoptera. Entomological News, 17: 71-80.

BLAISDELL, F. E. 1909. A monographic revision of the Coleoptera belonging to the tenebrionid tribe Eleodiini inhabiting the United States, Lower California, and adjacent islands. Smithsonian Institution, United States National Museum Bulletin, 63: 1-524.

BLAISDELL, F. E. 1923. Expedition of the California Academy of Sciences to the Gulf of California in 1921. Proceedings of the California Academy of Sciences, 12: 201-288.

BLAISDELL, F. E. 1925. Studies in the Tenebrionidae, No. 2 (Coleoptera). Proceedings of the California Academy of Sciences, 14(16): 369-390.

BLAISDELL, F. E. 1927. Miscellaneous studies in the Coleoptera No. 2. Pan-Pacific Entomologist, 3: 163-168.

BLAISDELL, F. E. 1932. Studies in the tenebrionid tribe Scaurini, a monographic revision of the *Eulabes* (Coleoptera). Transactions of the American Entomological Society, 58: 35-101.

BLAISDELL, F. E. 1933. A monographic revision of the species of *Centronopus* inhabiting America north of Mexico. Transactions of the American Entomological Society, 59: 211-228.

BLAISDELL, F. E. 1934. Studies in the genus *Auchmobius* LeConte. Transactions of the American Entomological Society, 60: 223-267.

BLAISDELL, F. E. 1935. Two new species of *Eleodes* from the Pacific coast region. (Coleoptera, Tenebrionidae). Canadian Entomologist, 67: 28-31.

BLAISDELL, F.E. 1936a. Studies in the tenebrionid tribe Triorophini: A monographic revision of the species belonging to the genus *Stibia* (Coleoptera). Transactions of the American Entomological Society, 62: 57-105.

BLAISDELL, F. E. 1936b. Facts concerning the rearing of Tenebrionidae (Coleoptera). Entomological News, 67: 39-42.

BLAISDELL, F. E. 1937. Miscellaneous studies in the Coleoptera, No. 5 (Tenebrionidae and Melyridae). Transactions of the American Entomological Society, 68: 127-145.

BLAISDELL, F. E. 1939. Studies in the relationships of the subfamilies and tribes of the Tenebrionidae based on the primary genital characters, also descriptions of new species (Coleoptera). Transactions of the American Entomological Society, 65: 43-60.

BLAISDELL, F. E. 1947. A new genus and species of the coleopterous family Tenebrionidae. Pan-Pacific Entomologist, 23: 59-62.

BLANCHARD, C. E. 1845. Histoire des insectes, traitant de leurs moeurs et de leurs m tamorphoses en g n ral, et comprenant une nouvelle classification fond e sur leurs rapports naturels. Vol. 2., 524 pp. Paris.

BOHEMAN, C. H. 1858. Coleoptera. Species novas descripsit. *In*: Kongliga Svenska Fregatten Eugenies resa omkring Jorden ... Zoologi, I, Insecta, 112 pp.

BROWN, K. W. 1971a. A population approach to computer taxonomy with applications in the genus *Gonasida* (Coleoptera: Tenebrionidae). Ph.D. dissertation, University of California, Riverside, 317 pp. [unpublished]

BROWN, K. W. 1971b. Redefinition of the genera *Pelecyphorus* and *Philolithus* with a key to the genera of the tribe Asidini (Coleoptera: Tenebrionidae). Coleopterists Bulletin, 25: 17-30.

BROWN, K. W. and J. T. DOYEN. 1991. Review of the genus *Microschatia* (Solier) (Tenebrionidae: Coleoptera). Journal of the New York Entomological Society, 99: 539-582.

BROWN, K. W. and C. A. TRIPLEHORN. (in press). *Epitragosoma arenaria*, a new genus and species from Texas (Coleoptera: Tenebrionidae: Epitragini). Coleopterists Bulletin.

CAMPBELL, J. M. 1966. A revision of the genus *Lobopoda* (Coleoptera: Alluculidae) in North America and the West Indies. Illinois Biological Monographs, 37: 203 pp.

CAMPBELL, J. M. 1978. A review of the North American species of *Mycetochara* Berthold (Coleoptera: Alleculidae). Canadian Entomologist, 110: 921-948.

CAMPBELL, J. M. 1984. *Onychomira floridensis*, a new genus and species from Florida with a revised key to the genera of North American Alleculidae. Coleopterists Bulletin, 38: 288-300.

CAND ZE, E. C. A. 1861. Histoire des metamorphoses de quelques col opt res exotiques. M moires de la Soci t Royale de Sciences de Li ge, 16: 325-410.

CASEY, T. L. 1886. Descriptive notes of North American Coloeptera. I. Bulletin of the California Academy of Sciences, 2(6): 157-264.

CASEY, T. L. 1890. Coleopterological notices, II. Annals of the New York Academy of Sciences, 5: 307-504.

CASEY, T. L. 1891. Coleopterological notices, III. Annals of the New York Academy of Sciences, 6: 9-214.

CASEY, T. L. 1895. Coleopterological notices. VI. Annals of the New York Academy of Sciences, 8: 435-838.

CASEY, T. L. 1907. A revision of the American components of the tenebrionid subfamily Tentyriinae. Proceedings of the Washington Academy of Sciences, 9: 275-522.

CASEY, T. L. 1908. A revision of the tenebrionid subfamily Coniontinae. Proceedings of the Washington Academy of Sciences, 10: 51-166.

CASEY, T.L. 1912. A revision of the American genera of the tenebrionid tribe Asidini. Pp. 70-214. *In*: Memoirs on the Coleoptera III. Lancaster Press. Lancaster, PA.

CASEY, T. L. 1914. Miscellaneous notes and new species. Pp. 355-378. *In*: Memoirs on the Coleoptera V. Lancaster Press. Lancaster, PA.

CASEY, T. L. 1924. Additions to the known Coleoptera of North America. Pp. 1-347. *In*: Memoirs on the Coleoptera. XI. Lancaster Press. Lancaster, PA.

CHAMPION, G. C. 1884-1893. *In*: Godman and Salvin, eds., Biologia Centrali-Americana, Insecta, Coleoptera. Vol. 4, pt. 1: pp. 1-88 [1884]; pp. 89-136 [1885]; pp. 137-264 [1886]; pp. 265-353 [1887]; pp. 354-476 [1888]; pp. 477-524 [1892]; pp. 525-572 and 23 pl. [1893].

CHEVROLAT, L. A. 1835. Col opt res du Mexique, fasc. 5-8. Strasbourg.

CHEVROLAT, L. A. 1878. Diagnoses de diaperides du Mexique. Petites Nouv lles Entomologiques, 2: 194-195.

COCKERELL, T. D. A. 1906. Names of Coleoptera. Entomological News, 17: 349.

COOPER, K. W. 1933. On the American species of *Alobates* Mots. (Col. Tenebrionidae). Bulletin of the Brooklyn Entomological Society, 28: 105-107.

CROTCH, G. R. 1873. Check List of the Coleoptera of America, north of Mexico. Salem, Mass. 136 pp.

CSIKI, E. 1953. Ueber neue and bekannte Colopteren aus Ungarn und angr nzenden L nden. Annals Historico-naturales Musei Nationalis Hungarci, 3: 115-135.

DAJOZ, R. 1984. Une esp ce nouvelle du genre *Araeoschizus* LeConte (Col opt res, Tenebrionidae). Bulletin Mensuel de la Soci t Linn nne de Lyon, 53(7): 246-247.

DAJOZ, R. 1989a. Une nouvelle esp ce texane du genre *Araeoschizus* LeConte (Col. Tenebrionidae). Bulletin de la Soci t Entomologique de France, 93(5-6): 149-152.

DAJOZ, R. 1989b. Une nouvelle esp ce du genre *Araeoschizus* LeConte du Sud-Est de l'Arizona (tats-Unis) (Coleoptera. Tenebrionidae). Nouvelle Revue d'Entomologie (N.S), 6(1): 33-36.

DAJOZ, R. 1991. Description de deux esp ces nouvelles du genre *Araeoschizus*, de la larve, de la nymphe et notes sur leur biologie (Col. Tenebrionidae). Bulletin de la Soci t Entomologique de France, 96(2): 165-174.

DAVIS, J. C. 1970. Revision of the genus *Blapstinus* Sturm of America north of Mexico with notes on extralimital species (Coleoptera: Tenebrionidae). Thesis. The Ohio State Univ. 459 pp. [unpublished]

DEJEAN, P. F. M. A. 1821. Catalogue de la collection de Col opt res de M. Le Baron Dejean. 136 pp. Paris.

DEJEAN, P. F. M. A. 1833. Catalogue des Col opt res de la collection de M. Le Baron Dejean, livr. 1 and 2, pp. 1-176, Paris.

DEJEAN, P. F. M. A. 1834. Iconographie des Col opt res de la collection de M. Le Baron Dejean livr. 3, pp. 176-256, Paris.

DOYEN, J. T. 1968. The phylogenetic position of *Edrotes* and a new species of the genus (Coleoptera: Tenebrionidae). Pan-Pacific Entomologist, 44: 218-227.

DOYEN, J. T. 1972. Familial and subfamilial classification of the Tenebrionoidea (Coleoptera) and a revised generic classification of the Coniontini (Tentyriidae). Quaestiones Entomologicae, 8: 357-376.

DOYEN, J. T. 1976. Biology and systematics of the genus *Coelus* (Coleoptera: Tentyriidae). Journal of the Kansas Entomological Society, 49: 595-624.

DOYEN, J. T. 1977. Synonymy in Coniontini. Pan-Pacific Entomologist, 53: 1-7.

DOYEN, J. T. 1982. New species of Tenebrionidae from western North America (Coleoptera). Pan-Pacific Entomologist, 58: 81-91.

DOYEN, J. T. 1984a. Reconstitution of the Diaperini of North America, with new species of *Adelina* and *Sitophagus* (Coleoptera: Tenebrionidae). Proceedings of the Entomological Society of Washington, 86: 777-789.

DOYEN, J. T. 1984b. Systematics of *Eusattus* and *Conisattus* (Coleoptera; Tenebrionidae; Coniontini; Eusatti). Occasional Papers of the California Academy of Sciences, No. 141. 104 pp.

DOYEN, J. T. 1985. Reconstitution of the tribes Ulomini and Triboliini for North and Central America. Proceedings of the Entomological Society of Washington, 87: 512- 524.

DOYEN, J. T. 1987. Review of the tenebrionid tribe Anepsiini (Coleoptera). Proceedings of the California Academy of Sciences, 44(15): 343-371.

DOYEN, J. T. 1989a. Reconstitution of Coelometopini, Tenebrionini and related tribes of America north of Colombia. (Coleoptera: Tenebrionidae). Journal of the Entomological Society, 97: 277-304.

DOYEN, J. T. 1993. Cladistic relationships among pimeliine Tenebrionidae (Coleoptera). Journal of the New York Entomological Society, 101: 443-514.

DOYEN, J. T. and C. Y. KITAYAMA. 1980. Review of the North American species of *Apocrypha* Eschscholtz, with a description of the immature stages of *Apocrypha anthicoides* (Coleoptera: Tenebrionidae). Pan-Pacific Entomologist, 56: 121-136.

DOYEN, J. T. and J. F. LAWRENCE. 1979. Relationships and higher classification of some Tenebrionidae and Zopheridae (Coleoptera). Systematic Entomology, 4: 333-377.

DOYEN, J. T., E. G. MATHEWS and J. F. LAWRENCE. 1989. Classification and Annotated Checklist of the Australian Genera of Tenebrionidae (Coleoptera). Invertebrate Taxonomy, 3: 229-260.

DOYEN, J. T. and W. R. TSCHINKEL. 1982. Phenetic and cladistic relationships among Tenebrionid Beetles. Systematic Entomology, 7: 127-183.

ERICHSON, W. F. 1843. Beitrag zur Insecten-Fauna von Angola, in besonderer Beziehung zur geographischen Verbreitung der Insecten in Afrika. Archiv fur Naturgeschichte, 9: 199-267.

ESCHSCHOLTZ, J. F. 1829-1831. Zoologischer atlas, enthaltend Abbildungen und Beschreibungen neuer Thierarten, wahrend des Flottcapitains von Kotzebue zweiter Reise um die Welt, auf der Russich-Kaiserlichen Kriegsschlupp Predpriaetie in den Jahren 1823-1826. Erstes Heft, 17 pp; Zweites Heft, 13 pp; Drittes Heft, 18 pp; Viertes Heft, 19 pp.

ESPA OL, F. 1955. Los Crypticini pale rcticos (Col. Tenebrionidae). EOS, 31: 7-38.

ESPA OL, F. 1960. Un Nuevo tipo de Tenebri nido sabuli ola de las costas del Peru. Instituto de Biolog a Aplicada (Barcelona), 31: 113-117.

FABRICIUS, J. C. 1775. Systema Entomologiae sistens Insectorum classes, ordines, genera, species, adjectis synonymis, locis, descriptionsibus, observationibus. 832 pp.

FABRICIUS, J. C. 1790. Nova insectorum genera. Skriftes af naturhistorie Selskabet Kj berhaun, 1: 14-31, 213-228.

FABRICIUS, J. C. 1792. Entomologia systematica emendata et aucta, secundum classes, ordines, genera, species, adjectis synonymis, locis, observationibus, descriptionsibus. 1(2): 1-538. Copenhagen.

FABRICIUS, J. C. 1801. Systema Eleutheratorum secundum ordines, genera, species, adjectis synonymis, locis, observationibus, descriptionsibus 1: 1-506, 2: 1-687.

FAIRMAIRE, L. 1856. Larves et nymphes de col opt res qui ont t recueilles la Gaudeloupe. Bulletin de la Soci t Entomologique de France (Ser 3), 4: 534.

FAIRMAIRE, L. 1871. Notes les Col opt res recueillis par Charles Coquerel Madagascar et sur les c tes d'Afrique, part. 3. Annales de la Soci t Entomologique de France (Ser. 5), 1: 29-78.

FAIRMAIRE, L. 1873. Description d'un nouveau genre de la famille des t n brionides. Annales de la Soci t Entomologique de France (Ser. 5), 3: 393-394.

FAIRMAIRE, L. 1881. Col opt res des iles Viti (Fidgi). Annales de la Soci t Entomologique de France (Ser. 6), 1: 243-318.

FAIRMAIRE, L. 1883. Note sur quelques col opt res de Magellan et de Santa-Cruz. Annales de la Soci t Entomologique de France (Ser. 6), 3: 483-506.

FAIRMAIRE, L. 1892. Col op res d'Obock. Troisi me partie. Revue Fran aise d'Entomologie, 11: 77-126.

FAIRMAIRE, L. 1893. Col opt res des Iles Comores. Annales de la Soci t Entomologique de Belgique, 37: 521-555.

FALL, H. C. 1907. Descriptions of new species. *In*: H. C. Fall and T. D. A. Cockerell, The Coleoptera of New Mexico. Transactions of the American Entomological Society, 33: 218-272.

FALL, H. C. 1912. New Coleoptera chiefly from the Southwest. V. Canadian Entomologist, 43: 40-48.

FALL, H.C. 1928. *Alaudes*. Pan-Pacific Entomologist 4: 145-150.

FALL, H. C. 1931. The North American species of *Hymenorus* (Coleoptera: Alleculidae). Transactions of the American Entomological Society, 57: 161-247.

FAUST, J. 1875. Beitr ge zur Kenntnis der K fer des europ ischen und asiatischen Russlands mit Einschluss der K sten des Kaspichen Meeres. Horae Societatis Entomologique Rossicae. St. Petersburg, 11: 163-252.

FLEISCHER, A. 1900. Ueber sichtstabelle der Arten der Coleopteren-Gattung *Palorus* Duv. Wiener Entomologische Zeitung, Vienna, 19: 236-237.

FORSTER, J. R. 1771. Novae species insectorum, centuria I, 100 pp. London.

FREIEDENREICH, C. W. 1883. Pilzbewohnende K fer in der Provinz Santa Catharina (S dbrasilien). Stettiner Entomologische Zeitung, 44: 375-380.

FREUDE, H. 1967. Revision der Epitragini (Coleoptera: Tenebrionidae). Part. 1. Entomologische Arbeiten aus dem Museum Georg Frey, 18: 137-307.

FREUDE, H. 1968. Revision der Epitragini (Coleoptera: Tenebrionidae). Part. 2. Entomologische Arbeiten aus dem Museum Georg Frey, 19: 32-143.

GEBIEN, H. 1910. Ergebn. Exped. Kilimandjaro, 1(7): 386.

GEBIEN, H. 1928. ber einige Gruppen amerikanischer Tenebrioniden (Col. Heter.). Stettiner Entomologische Zeitung, 89: 97-164, 167-234.

GEBIEN, H. 1938-1944. Katalog der Tenebrioniden. Mitteilungen der M nchener Entomological Gesellschaft. Teil II: 28[1938], 283-428 (370-465); 29[1939], 443-474, 739-770 (446-529); 30[1940], 405-436, 755-786, 1061-1092, (530-625); 31[1941], 331-362, 803-834, 1131-1146 (626-705); 32[1942], 308-346 (706-744). Teil III: 32[1942], 729-760 (746-777); 33[1943], 339-430, 895-926 (778-841); 34[1944], 497-555 (842-899).

GEMMINGER, M. and E. HAROLD. 1869. Catalogus Coleopterorum. 6: 1609-1800. Munich.

GEN , J. 1839. De quibusdam insectis Sardiniae novis aut minus cognitis. Fasiculus II. Memorie della Reale Accademia delle Scienze di Torino, ser. E. 1: 43-84.

GERSTAECKER, C. E. A. 1871. Beitrag zur Insektenfauna von Zanzibar. III Coleoptera. Archiv fur Naturgeschichte, 37(1): 42-86.

GOZIS, M. des. 1886. Recherche de l'esp ce typique de quelques anciens genres. Rectifications synonymiques et notes diverses. 36 pp. Montlu on.

GU RIN-M NEVILLE, F. E. 1834. Mat riaux pour une classification des M lanosomes. Magasine Zoologique, 4: 1-39.

GYLLENHAL, L. 1810. Insecta Suecica descripta. Classis 1. Coleoptera sive Eleutherata. 1(2): 20-660. Scaris.

GYLLENHAL, L. 1827. Insecta Suecica descripta. Classis 1. Coleoptera sive Eleutherata. 1(4): 1-761. Lipsiae.

HALDEMAN, S. S. 1848. Descriptions of North American Coleoptera, chiefly in the cabinet of Journal. LeConte, M.D., with references to described species. Journal of the Academy of Natural Sciences of Philadelphia, 1: 95-110.

HERRICH-SCHAEFFER, G. A. W. 1850. Wanzenartigen Insecten. 9(4) p.184. [Hemiptera].

HOPE, F. W. 1838. The coleopterist's manual. Part 2. 168 pp. London.

HOPE, F. W. 1840. The coleopterist's manual. Part 3. 191 pp. London.

HOPPING, R. 1933. New Coleoptera from western Canada IV. Canadian Entomologist, 65: 281-286.

HORN, G. H. 1866. Descriptions of some new genera and species of Central American Coleoptera. Proceedings of the Academy of Natural Sciences of Philadelphia, 397-401.

HORN, G. H. 1870. Revision of the Tenebrionidae of America, north of Mexico. Transactions of the American Philosophical Society, 14: 253-404.

HORN, G. H. 1874. Description of new species of United States Coleoptera. Transactions of the American Entomological Society, 5: 20-43.

HORN, G. H. 1875. Synonymical notes and description of new species of North American Coleoptera. Transactions of the American Entomological Society, 5: 126-176.

HORN, G. H. 1876. Notes on the coleopterous fauna of Guadalupe Island. Transactions of the American Entomological Society, 5: 198-201.

HORN, G. H. 1878. Contributions to the coleopterology of the United States, No. 2. Transactions of the American Entomological Society, 7: 51-60.

HORN, G. H. 1885. Synonymical notes (No. 3). Entomologia Americana, 1: 108-113.

ILLIGER, J. C. W. 1798. Kugelann Verzeichniss der K fer Preussens. 510 pp. Halle.

IWAN, D. 1995. Revision of the genus *Opatrinus* Dejean, 1821 (Coleoptera: Tenebrionidae: Platynotini). Genus (International Journal of Invertebrate Taxonomy), 6(1): 90 pp.

JACQUELIN-DUVAL, P. N. C. 1856. Coleoptera. *In*: Sagra, Historie physique, politique et naturelle de l'Ile de Cuba. Animaux articules, insectes, 136 pp. Paris.

JACQUELIN-DUVAL, P. N. C. 1858-59. Genera des Col opt res d'Europe comprenant leur classification en familles naturelles, la description de tous les genres, les tableaux synoptiques destin s faciliter l'etude, le catalogue de toutes les esp ces ... 2: 129-232.

JACQUELIN-DUVAL, P. N. C. 1861. Manuel entomologique. Genera des Col opt res d'Europe. 3: 125-200.

KASZAB, Z. 1938. Die arten der Gattungen *Cnemeplatia* Costa und *Psylachnopus* Reitter (Col. Tenebr. Opatrinae). Entomologisk Tidskrift, Stockholm, 50: 77-83.

KASZAB, Z. 1940. Neue exotische Tenebrioniden (Coleop.). Annales Musei Nationalis Hungarici, Budapest (Zoology), 33: 172-174.

KASZAB, Z. 1982. Revision der australischen Uloma-Arten (Coleoptera, Tenebrionidae). Acta Zoologica Hungarica, 28: 233-291.

KELEJNIKOVA, S. I. 1971. Morphological peculiarities of first-stage larvae of Tenebrionidae (Coleoptera). Proceedings XIII International Congress of Entomology, [1968] 1: 154-155.

KIRBY, W. 1818. A century of insects, including several new genera described from his cabinet. Transactions Linnean Society London, 12: 375-453.

KIRBY, W. 1837. Insects. Coleoptera. *In*: Richardson, Fauna Boreali-Americana; or the zoology of the northern parts of British America...., 249 pp. Norwich.

KLUG, J. C. F. 1833. Berich ber eine auf Madagascar veranstaltete Sammlung von Insecten aus der Ordnung Coleoptera. Abhandlungen der Preussiche Akademie der Wissenschaften, 91-223.

KNOCH, A. W. 1801. Neue Beitr ge zur Insectenkunde. 1: 1-208. Leipzig.

KOCH, C. 1953. The Tenebrionidae of Southern Africa. XXIV (marked as XXVI). New Portuguese East African species collected by Dr. A.J. Barbosa, Revista da Facultade de Ciencas (Lisbon), 2,c, 3: 239-310.

KOCH, C. 1956. S.A. Tenebrionidae XXIX. Tenebrionidae (Coleoptera, Polyphaga). II. Opatrinae. First part: Platynotini Litoborini and Loensini. Exploration du Parc National de l'Upemba, 40. 1-472.

KOLBE, H. J. 1894. Coleopteren aus Afrika. Stettiner Entomologische Zeitung, 55: 361-397.

KOLBE, H.J. 1901. Allgemeine Zeitschrift f r Entomologie, 6: 342.

KRITSKY, G. 1989. A review of the genus *Polopinus* Casey with descriptions of two new species. Coleopterists Bulletin, 43: 127-133.

KULZER, 1959. Neue Tenebrioniden aus Sudamerika (Col.) 18. Beitrag zur Kenntnis der Tenebrioniden. Entomologische Arbieten aus dem Museum Georg Frey, 10: 523-567.

LACORDAIRE, T. 1859. Histoire Naturelle des Insectes. Genera des Col opt res. Tome Cinqui me. Premier Partie. (Heteromera), 750 pp.

LACORDAIRE, T. 1866. Genera des col opt res, 7: 1-620.

LAPORTE de CASTELNAU, F. L. and G. A. BRULL . 1831. Monographie du genre *Diaperis*. Annales des Science Naturelles, 23: 325-410.

LAPORTE de CASTELNAU, F. L. 1840. Histoire naturelle des insectes. Col opt res, 2: 1-563.

LATREILLE, P. A. 1802. Histoire naturelle, g n rale et particuli re des crustaces et des insectes, 3: 1-467. Paris.

LATREILLE, P. A. 1809. Genera crustaceorum et insectorum secundum ordinem naturalem in familias disposita, iconibus exemplisque plurimus explicata, 4: 1-399. Paris.

LATREILLE, P. A. 1817. Les crustac s, les arachnides et les insectes. *In:* G. Cuvier, ed. Le R gne Animal distribu dapr s son Organisation ... Vol. 3. Deterville. Paris, xxix + 653 pp.

LATREILLE, P. A. 1825. Encyclopedie methodique..., Entomologie, 10(1): 1-344.

LATREILLE, P. A. 1829. Les crustac s, les archnides et les insectes, distribu s en familles naturelles, ouvrage formant les tomes 4 et 5 de celiu de M. le Baron Cuvier sure le R gne animal (deuxi me dition), 1: 1-584. Paris.

LAWRENCE, J. F. and A. F. NEWTON, Jr. 1995. Families and subfamilies of Coleoptera (with selected genera, notes references and data on family-group names). *In*: J. Pakaluk and S. A. Slipinski, eds. Biology, Phylogeny, and Classification of Coleoptera, Papers Celebrating the 80th Birthday of Roy A. Crowson. Museum i Instytut Zoologii PAN. Warsaw, 1092 pp.

LAWRENCE, J. F. and T. J. SPILMAN. 1991. Tenebrionidae (Tenebrionoidea), (including Alleculidae, Cossyphodidae, Lagriidae, Nilionidae, Rhysopaussidae, Tenetyriidae). Pp. 520-528. *In:* F. W. Stehr, ed., Immature Insects, Vol. 2. Kendall/Hunt. Dubuque, IA.

LEACH, W. E. 1815. The zoological miscellany, 2: 1-154. London.

LECONTE, J. E. 1824. Description of some new species of North American insects. Annals of the Lyceum Natural History of New York, 1: 169-173.

LECONTE, J. L. 1850. General remarks upon the Coleoptera of Lake Superior. Pp. 201-242. *In*: Agassiz, Lake Superior..., Boston.

LECONTE, J. L. 1851. Descriptions of new species of Coleoptera from California. Annals of the Lyceum Natural History of New York, 5: 125-216.

LECONTE, J. L. 1854. Notice of some coleopterous insects, from the collections of the Mexican Boundary Commission. Proceedings of the Academy of Natural Sciences of Philadelphia, 7: 79-85.

LECONTE, J. L. 1858. Descriptions of new species of Coleoptera, chiefly collected by the United States and Mexican boundary commission, under Major W.H. Emory, U.S.A. Proceedings Academy of Natural Sciences of Philadelphia, 10: 59-89.

LECONTE, J. L. 1858. Catalog of the Coleoptera of the regions adjacent to the boundary line between the United States and Mexico. Journal of the Academy of Natural Sciences of Philadelphia, (2)4: 9-42.

LECONTE, J. L. 1859. The Coleoptera of Kansas and eastern New Mexico. Smithsonian Contributions to Knowledge, 2(6): 58 pp.

LECONTE, J. L. 1862. Classification of the Coleoptera of North America. Smithsonian Miscellaneous Collections, 136: 209-286.

LECONTE, J. L. 1866. New species of North American Coleoptera. Smithsonian Miscellaneous Collections, 167: 87-177.

LECONTE, J. L. 1878. Additional descriptions of new species. *In*: E. A. Schwarz, The Coleoptera of Florida. Proceedings of the American Philosophical Society, 17: 373-434.

LECONTE, J. L. 1879. New North American Coleoptera. North American Entomology, 1: 1-5.

LEWIS, G. 1894. On the Tenebrionidae of Japan. Annals and Magazine of Natural History (Ser. 6), 13: 377-400.

LINELL, M. L. 1897. A new, nearly blind genus of Tenebrionidae. Entomological News, 8: 154-156.

LINNAEUS, C. 1758. Systema Naturae per regna tria naturae secundum classes, ordines, genera, species, cum caracteribus, deferentilis, synnonymis, locis. Vol. 1, 823 pp. Stockholm.

LUCAS, H. 1849. L'histoire naturelle des animaux articul s de l'Alg rie. 2: 1-527, Paris.

MacLACHLAN, W. B. and C. A. OLSON. 1990. A revision of the Trimytini of America North of Mexico (Coleoptera: Tenebrionidae). Coleopterists Bulletin, 44: 69-82.

MacLEAY, W. S. 1825. Annulosa javanica or an attempt to illustrate the natural affinities and analogies of the insects collected in Java by T. Horsfield. 50 pp. London.

M KLIN, F. W. 1864. Monographie der Gattung *Strongylium* Kirby, Lacordaire und der damit zurn chst verwandten Formen. Acta Societatis Scientiarum Fennicae, 8: 217-518.

M KLIN, F. W. 1875. Neue *Statira*-Arten und einige mit der genannten Gattung verwandte Formen. Acta Societatis Scientiarum Fennicae, 10: 633-660.

M KLIN, F. W. 1878. N gra bidrag till k nedom af sl gtet Talanus Dejean Cat. Ofversigt af Finska Vetenskaps-Soci t tens F rhandlingar. Helsinki, 20: 95-103.

MANNERHEIM, C. G. 1843. Beitrag zur Kaferfauna der Aleutischen Inseln, der Insel Sitka und Neu-Californiens. Bulletin de la Soci t Imperiale des Naturalistes de Moscou, 16(2)175-314.

MARSHALL, J. D. 1967. A new species and new United States record of *Anamphidora* Casey (Coleoptera: Alleculidae). Entomological News, 78: 209-210.

MARSHALL, J. D. 1967. Synopsis of *Stenochidus* LeConte (Coleoptera, Alleculidae). Occasional Papers of the Biological Society of Nevada, No. 15, 3 pp.

MARSHALL, J. D. 1969 A review of *Lystronychus* Latreille in America north of Mexico (Coleoptera, Alleculidae). Occasional Papers of the Biological Society of Nevada, No. 21, 5 pp.

MARSHALL, J. D. 1970a. *Isomira* Mulsant in America north of Mexico (Coleoptera: Alleculidae): Redescriptions, new synonymies, and taxonomic notes on eastern North American species. Entomological News, 81: 41-49.

MARSHALL, J. D. 1970b. *Isomira* Mulsant in America north of Mexico (Coleoptera: Alleculidae): Species of eastern North America concluded. Coleopterists Bulletin, 24(3): 88-95.

MEDVEDEV, G. S. 1977. Taksonomicheskoye znacheniye antennal'nikh sensill zhukov-chernotelok (Coleoptera: Tenebrionidae). Akad. Nauk S.S.S.R. Trudi Vesesoyuznoivo Entomologicheskovo Obschestva, 58: 61-86.

MOTSCHULSKY, V. 1870. numeration des nouvelles esp ces de col opt res rapport s de ses voyages. Bulletin de la Soci t Imperiale des Naturalistes de Moscou, 43(1): 379-407.

MOTSCHULSKY, V. 1872. num ration des nouvelles esp ces des col opt res rapport s de ses voyages (Part 7). Bulletin de la Soci t Imperiale des Naturalistes de Moscou, 45(2): 22-55.

MOTSCHULSKY, V. 1873. num ration des nouvelles esp ces des col opt res rapport s de ses voyages (Part 12). Bulletin de la Soci t Imperiale des Naturalistes de Moscou, 46: 466-482.

M LLER, O. T. 1764. Fauna insectorum fridrichsdalina, sive methodica descriptio insectorum agri fridrichsdalensis, cum characteribus genericis et specificis, nominibus trivialibus, locis natalibus, iconibus allegatis novisque pluribus speciebus additis, 96 pp.

MULSANT, E. 1854. Histoire naturelle de col opt res de France, pt. 5, (Latig nes) 396 pp. Paris.

MULSANT, E. 1856. Histoire naturelle de col opt res de France (Barbipalpes) 116 pp. Paris.

MULSANT, E. and C. REY. 1852. Essai d'une division des derni rs m lanosomes. M moires de la Soci t Agriculturale de Lyon (Ser. 2), 2: 226-329.

MULSANT, E. and C. REY. 1853. Essai d'une division des derni rs m lanosomes. P dinites. Opatrinaires. Opuscules Entomologiques, 4: 69-107.

MULSANT, E. and C. REY. 1859. Essai d'une division des derni rs m lanosomes. M moires de la Soci t Agriculturale de Lyon (Ser. 3), 3: 129-201.

MURRAY, 1862. Transactions Linnean Society of London, 23: 4

OLIVIER, G. A. 1795. Entomologie, ou histoire naturelle des insectes, avec leurs caract res g n riques et sp cifiques, leur description, leur synonymie et leur figure enlumin e. Col opt res, 3: 1-557.

PANZER, G. W. F. 1794. Faunae Insectorum Americae borealis prodrumus. 8 pp. N rnberg.

PAPP, C. S. 1961. Checklist of Tenebrionidae of America north of the Panama Canal., Opuscules Entomologiques, 26: 97-140.

PAPP, C. S. 1981. Revision of the genus *Araeoschizus* LeConte (Coleoptera: Tenebrionidae). Entomologische Arbeiten aus dem Museum Georg Frey, 29: 273-420.

PARSONS, C. T. 1975. A key to Nearctic *Statira* and *Arthromacra* (Lagriidae). Coleopterists Bulletin, 29: 211-226.

PASCOE, F. P. 1860. Notices of new or little known genera and species of Coleoptera. Journal of Entomological Description and Geography, 1: 36-64.

PASCOE, F. P. 1863. Notices of new or little known genera and species of Coleoptera. Journal of Entomological Description and Geography, 2: 26-56.

PASCOE, F. P. 1866. Notices of new or little known genera and species of Coleoptera. Journal of Entomological Description and Geography, 2: 443-493.

PASCOE, F. P. 1869. Descriptions of new genera and species of Tenebrionidae from Australia and Tasmania (cont.). Annals and Magazine of Natural History, (4)3: 277-296.

PASCOE, F. P. 1871. Notes on Coleoptera, with descriptions of new genera and species (Part 1). Annals and Magazine of Natural History, (4)8: 345-361.

PIC, M. 1913. Nouveaux genres et nouvelles esp ces. M langes Exotiques-Entomologiques, 6: 1-7.

PILLER, M. and L. MITTERPACHER. 1783. Iter per Poseganam, Sclavoniae provinciam mensibus Junio et Julio 1782 susceptum. 147 pp. Budae.

REDTENBACHER, L. 1845. Die Gattungen der deutchen K fer-Fauna nach der analytischen Methode bearbeitet, nebst einem durz gefassten Leifaden, zum Studium dieses Zweiges der Entomologie, 177 pp.

REDTENBACHER, L. 1849. Fauna Austriaca: Die K fer, nach der analytischen Methode bearbeitet, 883 pp.

REICHARDT, A. 1936. Tabl. Analyt. Faune U.R.S.S. 19. Inst. Zool. Academy of Sciences Leningrad, pp. 24, 62.

REITTER, E. 1876. Neue Gattungen und Arten aus Familie der Cucujidae. Coleopterologische Hefte, 15: 42-43.

REITTER, E. 1904. Bestimmungstablellen europ ischer Coleoptera. -53 (bersicht der Subfamilien der Tenebrioniden) Verhandlungen des Naturforschenden Vereins, Br nn, XLII pp. 25-189.

REITTER, E. 1908. Zwei neue Coleopteren aus Adana in Kleinasien. Wiener Entomologische Zeitung, 27: 133-136.

REITTER, E. 1916. Fauna Germanica. Die K fer des Deutschen Reiches. Nach der analytischen Methode bearbeitet 5: 1-343.

SAY, T. 1824. Descriptions of coleopterous insects collected in the late expedition to the Rocky Mountains performed by order of Mr. Calhoun, Secretary of War, under the command of Major Long. Journal of the Academy of Natural Sciences, Philadelphia, 3: 139-216; 251-252.

SAY, T. 1825. Descriptions of coleopterous insects inhabiting the United States. Journal of the Academy of Natural Sciences, Philadelphia, 5: 160-204.

SAY, T. 1827. Descriptions of coleopterous insects inhabiting the United States (cont.). Journal of the Academy of Natural Sciences, Philadelphia, 5: 237-304.

SCHAEFFER, C. 1905. Some additional new genera and species of Coleoptera found within the limit of the United States. Science Bulletin of the Museum of the Brooklyn Institute of Arts and Science, 1(7): 141-179.

SCHAEFFER, C. 1915. New Coleoptera and miscellaneous notes, III. Journal of the New York Entomological Society, 23: 235-238.

SEIDLITZ, G.Von. 1894. Naturgeschichte der Insecten Deutschlands, Coleoptera, 5(1)401-608.

SEIDLITZ, G.Von. 1895. Naturgeschichte der Insecten Deutschlands, Coleoptera, 5(2)609-800.

SEIDLITZ, G.Von. 1896. Naturgeschichte der Insecten Deutschlands, Coleoptera, 5(2)1-304.

SERVILLE, J. G. (AUDINET-SERVILLE). 1825. *In*: Encyclop die m thodique, Entomologie, 10(1): 1-344. Paris.

SERVILLE, J. G. (AUDINET-SERVILLE). 1828. *In*: Encyclop die m thodique, Entomologie, 10(2): 345-832. Paris.

SKOPIN, N. G. 1978. 5.42. Tenebrionidae, Pp. 223-301. *In*: B. Klausnitzer, ed., Ordnung Coleoptera (Larven), W. Junk. The Hague, Netherlands.

SOLIER, A. J. 1834. Essai d'une division des col opt res h t rom res, et d'une monographie de la famille des collapt rides. Annales de la Soci t Entomologique de France, 3: 479-636.

SOLIER, A. J. 1835. Essai sur les collapt rides. Annales de la Soci t Entomologique de France, 4: 249-419.

SOLIER, A. J. 1836. Essai sur les collapt rides (Suite). Annales de la Soci t Entomologique de France, 5: 635-684.

SOLIER, A. J. 1848. Essai sur les collapt rides. Baudi et Truqui Studi Entomologici, 2: 146-370.

SOLIER, A. J. 1851. Orden III. Coleopteros. *In*: Gay, Historia fisica y politica de Chile, Vol. 5, 285 pp.

SORENSON, E. B. and R. C. STONES. 1959. Description of a new tenebrionid (Coleoptera) from Glen Canyon, Utah. Great Basin Naturalist, 19: 63-66.

SPILMAN, T. J. 1959. Notes on *Edrotes, Leichenum, Palorus, Eupsophulus, Adelium,* and *Strongylium* (Tenebrionidae). Coleopterists Bulletin, 13: 58-64.

SPILMAN, T. J. 1962. The new world genus *Centronopus*, with new generic synonymy and a new species (Coleoptera: Tenebrionidae). Transactions of the American Entomological Society, 88: 1-19.

SPILMAN, T. J. 1963. The American genus *Mycotrogus*: a synopsis, a new species from Cuba, and a note on a larva. Proceedings of the Entomological Society of Washington, 65: 21-30.

SPILMAN, T. J. 1967. A new North American ulomine genus and species, *Doliodesmus charlesi* (Coleoptera; Tenebrionidae). Pan-Pacific Entomologist, 43: 149-154.

SPILMAN, T. J. 1973. Nomenclatural problems in six genera of Tenebrionidae. Proceedings of the Entomological Society of Washington, 75: 39-44.

SPILMAN, T. J. 1991. 11, Darkling Beetles (Tenebrionidae, Coleoptera). *In*: J. R. Gorham, ed., Insect and Mite Pests in Food. United States Department of Agriculture, Agricultural Handbook 655: 185-214, 589-598.

SPINOLA, M. M. 1844. Essai monographique sur les cl rites, insectes col opt res. Ponthenenier. Genoa, 616 pp.

STEINER, W. E. 1982. *Poecilocrypticus formicophilus* Gebien, a South American beetle established in the United States (Coleoptera: Tenebrionidae). Proceedings of the Entomological Society of Washington, 84: 232-239.

STEINER, W. E. 1991. *Branchus whiteheadi*, new species, from southern Texas, with notes on the genus *Branchus* (Coleoptera: Tenebrionidae). Proceedings of the Entomological Society of Washington, 93: 425-432.

STEPHENS, J. F. 1829. A systematic catalog of British insects... 338 pp. London.

STEPHENS, J. F. 1832. Illustrations of British Entomology. Mandibulata. 5: 1-447. London.

STURM, J. 1826. Catalog meiner Insecten-Sammlung, pt. 1, 207 pp.

TANNER, V. M. 1961. A checklist of the species of *Eleodes* and descriptions of new species (Coleoptera: Tenebrionidae). Great Basin Naturalist, 21: 55-78.

TANNER, V. M. 1963. A new species of *Craniotus* (Coleoptera: Tenebrionidae). Great Basin Naturalist, 23: 167-170.

THOMAS, D. B. 1984. *Texaponium*, a new genus for *Cryptadius triplehorni* Berry (Coleoptera: Tenebrionidae). Proceedings of the Entomological Society of Washington, 86: 658-659.

THOMAS, D. B. 1985. A morphometric and revisionary study of the littoral beetle genus *Cryptadius* LeConte, 1852 (Tenebrionidae: Coleoptera). Pan-Pacific Entomologist, 61: 189-199.

THOMSON, C. G. 1859. Skandinaviens Coleoptera, synoptiskt bearbetade, 1, 290 pp.

TRIPLEHORN, C. A. 1952. The darkling beetles of Ohio (Coleoptera: Tenebrionidae). Thesis, Ohio State University, 159 pp. [unpublished]

TRIPLEHORN, C. A. 1961. The genus *Phaleromela* Reitter in North America (Tenebrionidae). Coleopterists Bulletin, 15: 125-127.

TRIPLEHORN, C. A. 1965. Revision of Diaperini of America north of Mexico with notes on extralimital species (Coleoptera: Tenebrionidae). Proceedings of the United States National Museum, 117: 349-458.

TRIPLEHORN, C. A. 1967. Notes on the species of *Megasida* Casey from the United States (Coleoptera: Tenebrionidae). Ohio Journal of Science, 67(1): 38-41.

TRIPLEHORN, C. A. 1968. A synopsis of the genus *Talanus* Jacquelin du Val in America north of Mexico, with descriptions of two new species. Coleopterists Bulletin, 22: 33-39.

TRIPLEHORN, C. A. 1974. A new species of *Pechalius* Casey for New Mexico and Arizona (Coleoptera: Tenebrionidae). Proceedings of the Entomological Society Washington, 76: 73-75.

TRIPLEHORN, C. A. 1975. A new subgenus of *Eleodes*, with three new cave-inhabiting species (Coleoptera: Tenebrionidae). Coleopterists Bulletin, 29: 39-43.

TRIPLEHORN, C. A. 1978. A new species of *Tribolium* from Arizona (Coleoptera: Tenebrionidae). Coleopterists Bulletin, 32: 73-75.

TRIPLEHORN, C. A. 1979. Two new species of *Corticeus* from Florida and the West Indies. Proceedings of the Entomological Society of Washington, 81: 46-50.

TRIPLEHORN, C. A. 1990. Review of the genus *Corticeus* (Coleoptera: Tenebrionidae) of America north of Mexico. Annals of the Entomological Society of America, 83: 287-306.

TRIPLEHORN, C. A. 1991. A review of the genus *Phaleria* Latreille from the Western Hemisphere (Coleoptera: Tenebrionidae: Phaleriinae). Coleopterists Bulletin, 45: 258-270.

TRIPLEHORN, C. A. 1995. First United States records for the genus *Iccius* Champion (Coleoptera: Tenebrionidae: Diaperini). Coleopterists Bulletin, 49: 397.

TRIPLEHORN, C. A. and R. L. AALBU. 1985. A review of the genus *Neobaphion* Blaisdell with description of a new species from Nevada (Coleoptera: Tenebrionidae: Eleodini). Proceedings of the Entomological Society of Washington, 87: 587-592.

TRIPLEHORN, C. A. and K. W. BROWN. 1971. A synopsis of the species of *Asidina* in the United States with description of a new species from Arizona (Coleoptera: Tenebrionidae). Coleopterists Bulletin, 25: 73-86.

TRIPLEHORN, C. A. and J. T. DOYEN. 1972. Synonymy and an emendation in *Eleodes* (Coleoptera: Tenebrionidae) Coleopterists Bulletin, 26: 79-80.

TRIPLEHORN, C. A. and T. J. SPILMAN. 1973. A review of *Strongylium* of America north of Mexico, with descriptions of two new species (Coleoptera, Tenebrionidae). Transactions American Entomological Society, 99: 1-27.

TRIPLEHORN, C. A. and L. E. WATROUS. 1979. A synopsis of the genus *Phaleria* in the United States and Baja California (Coleoptera: Tenebrionidae). Coleopterists Bulletin, 33: 275-295.

TSCHINKEL, W. R. and J. T. DOYEN. 1980. Comparative anatomy of the defensive glands, ovipositors and female genital tubes of tenebrionid beetles (Coleoptera). International Journal of Insect Morphology and Embryology, 9: 321-368.

UYTTENBOOGAART, D. L. 1934. Revision des genus *Tribolium* (Col. Ten.). Entomologische Bl tter, 30: 20-31.

WASMANN, E. 1896. Kritische Bemerkungen ber einige Myrmecophilen und Termitophilen. Wiener Entomologische Zeitung, 15: 32-36.

WATERHOUSE, G. R. 1845. Descriptions of some new genera and species of heteromerous Coleoptera. Annals and Magazine of Natural History, 16: 317-324.

WATERHOUSE, C. O. 1880. Description of a new genus and species of heteromerous Coleoptera. Annals and Magazine of Natural History, (5)5: 147-148.

WATERHOUSE, C. O. 1894. Coleoptera (partim). *In*: J. Walker. A visit to Damma Island, East Indian Archipelago. Annals and Magazine of Natural History, (6)14: 64-71.

WATT, J. C. 1974. A revised subfamily classification of Tenebrionidae (Coleoptera). New Zealand Journal of Zoology, 1: 381-452.

WATT, J. C. 1992. Relationships of *Actizeta* and Cnemeplatiini (Coleoptera: Tenebrionidae). Systematic Entomology, 17: 287-299.

WHITE, A. 1846. Insects of New Zealand. *In*: J. Richardson and J. E. Gray. The zoology of the voyage of the H.M.S. Erebus and Terror, 2: 1-24.

WILKE, S. 1922. Beitrage zur Systematik und geographischen Verbreitung ungeflugelter Tenebrioniden (Unterfam-Asidinae). Archiv f r Naturgeschichte, 12: 248-312.

WOLLASTON, T. V. 1861. Annals and Magazine of Natural History (Ser. 3), 7: 199.

WOLLASTON, T. V. 1864. Catalogue of the coleopterous insects of the Canaries in the collection of the British Museum, 648 pp. London.

WOLLASTON, T. V. 1867. Coleoptera Hesperidum, being an enumeration of the coleopterous insects of the Cape Verde Archipelago, 285 pp. London.

107. PROSTOMIDAE Horn 1888

by Daniel K. Young

Family common name: The jugular-horned beetles

The elongate, parallel-sided, flattened form, prognathous head with large, projecting mandibles and 4-4-4 tarsal formula characterize adults of the single North American prostomid species.

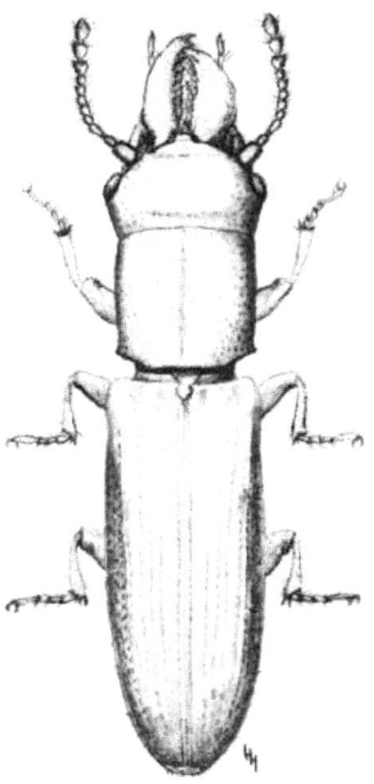

FIGURE 1.107. *Prostomis mandibularis* (Fabricius) (from Hatch 1961)

Description: (after Hatch 1961, in part) Elongate, narrow, somewhat flattened; length 5 to 10 mm; color yellowish- to reddish-brown; subglabrous.

Head easily visible from above, prognathous; genae strongly, acuminately produced anteriorly. Antennae with 11antennomeres, ending in a weakly 3-segmented club; insertions exposed. Frontoclypeal suture distinct; labrum small; mandibles large, strongly projecting; maxilla with the laciniae small; maxillary palpi with four palpomeres; gula well developed, gular sutures distinct; labium small and elongate, conspicuously setose; labial palpi with three elongate palpomeres. Eyes small, entire.

Pronotum slightly narrower than the head, as broad as basal width of elytra, quadrate, anteriorly truncate; sides rounded, without margins; surface shallowly, coarsely punctate; prosternum broad anterad coxae, with wide, well developed intercoxal process between the coxae; prothoracic coxal cavities closed behind externally and internally, trochantins concealed. Scutellum visible, of moderately small size. Mesosternum short, mesothoracic coxal cavities closed laterally; metasternum broad. Legs with prothoracic coxae small, rounded, widely separated; mesothoracic coxae small, rounded, widely separated; metathoracic coxae transverse, nearly contiguous; trochanters small, triangular; femora swollen; tibiae stout with well developed, short, stout apical spurs; tarsal formula 4-4-4, tarsomeres slender, first three tarsomeres subequal in length, fourth the longest; claws simple. Elytra entire, the apices rounded; striae punctate; epipleural fold narrow, complete. Metathoracic wings with subcubital fleck entire (Wilson 1930).

Abdomen with five visible ventrites, the first two ventrites connate. Male genitalia of inverted heteromeroid type; penis narrow, apically acute; parameres somewhat flattened, fused along much of their length, strongly tapered and narrowly separated distally (Wilson 1930).

Larvae (after Bøving and Craighead 1931, Young 1991) elongate, strongly flattened, widest slightly beyond middle; length 8 to 9 mm; vestiture sparse, a few short setae on each segment; color creamy white. Head exerted, prognathous, head capsule conspicuously asymmetrical with right side larger than left; epicranial suture poorly defined, stem short or absent and frontal arms lyriform; antennae elongate, three-segmented, second segment bearing a conical sensorium. Labrum distinct; frons and clypeus fused; mandibles asymmetrical, stout, tridentate; left mandible with prominent molar tooth; maxillae with cardo, maxillary articulating area, shallowly cleft mala with three-segmented palpi; labium with submentum and mentum fused, ligula and two-segmented palpi present. Stemmata absent. Thorax with five-segmented legs including tarsungulus; legs bearing numerous stout, spine-like setae. Abdomen visibly nine-segmented, the tenth segment partially fused to ninth; tergite nine extending ventrally, bearing lightly sclerotized, paired, short, fixed, urogomphi apically; ninth sternite partially enclosed by eighth sternite, bearing a transverse row of apical asperities. Spiracles annular-biforous, those of thorax placed apically on well-developed spiracular tubes.

Habits and habitats. Larvae and adults are typically found in dead logs within a characteristic clay-like material between layers of the decaying wood fibers.

Status of the classification. Historically treated as Cucujidae, *Prostomis* was excluded largely on the basis of characters provided by the larvae (Bøving and Craighead 1931). Wilson (1930) noted that the wing venation and male genital structure of *Prostomis* were consistent with those of Heteromera rather than Cucujoidea.

Distribution. Worldwide, there are two genera and about 20 species (Young 1991). *Prostomis* is known from western North America, eastern Asia, Africa, and the Pacific region. A few species of *Dryocora* are known from Australia, New Zealand, and Tasmania. Only *Prostomis mandibularis* (Fabricius) is found in North America (Hatch 1961).

Classification of the Genera of America north of Mexico

Prostomidae Horn 1888

Prostomis Latreille 1838, 1 sp., *P. mandibularis* (Fabricius 1838), British Columbia, Washington, Oregon, northern California to the eastern foothills of the Cascade Mountains.

Bibliography

BØVING, A. G. and F. C. CRAIGHEAD. 1931. An illustrated synopsis of the principal larval forms of the Order Coleoptera. Entomologica Americana (New Series), 11: 1-351.

HATCH, M. H. 1961. Beetles of the Pacific Northwest. Part III: Pselaphidae and Diversicornia I. Volume 16, University of Washington Publications in Biology. University of Washington Press. Seattle. 503 pp.

WILSON, J. W. 1930. The genitalia and wing venation of Cucujidae and related families. Annals of the Entomological Society of America, 23: 305-341.

YOUNG, D. K. 1991. Prostomidae. Pp. 515-516. *In:* F. W. Stehr, ed., Immature insects, Volume 2. Kendall/Hunt. Dubuque, IA.

Øø

108. SYNCHROIDAE Horn 1888

by Daniel K. Young

Family common name: The synchroa bark beetles

The elongate, narrow, tapered and somewhat flattened form of synchroids provides a superficial resemblance to Elateridae, from which they can easily be separated by the heteromerous tarsal formula, serrulate tibial spurs, and antennae that are inserted beneath a frontal ledge in the emargination of the compound eyes.

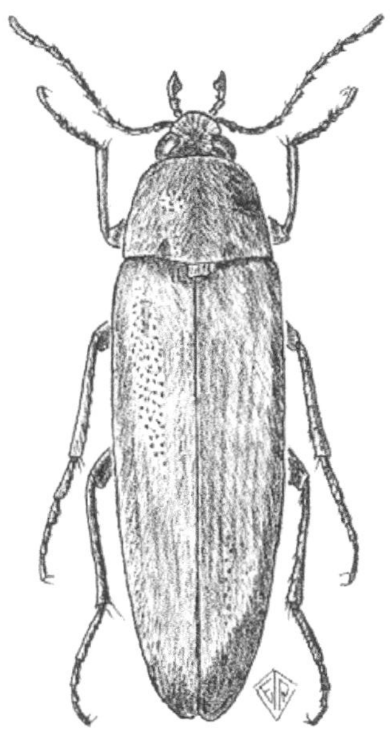

FIGURE 1.108. *Synchroa punctata* Newman.

Description: Elongate, narrow, tapered and somewhat flattened; length 7 to 13 mm; color brownish to piceous; vestiture consisting of decumbent setae.

Head easily visible from above, prognathous; surface punctate, conspicuously setose. Antennae with 11 antennomeres, filiform; inserted under small frontal ledge. Clypeus distinctly separated by impressed line which may be incomplete; labrum quadrate; mandibles strongly curved; maxilla with the lacinia small; maxillary palpi with four palpomeres, terminal segment slightly securiform; gula well developed, the gular sutures and posterior tentorial pits distinct, widely separate; labium small, conspicuously setose; labial palpi with three palpomeres, first two small, third a little larger and somewhat swollen. Eyes lateral, large, emarginate near frontal ledge into which antennae insert.

Pronotum very slightly broader than head, as broad as basal width of elytra, slightly narrowed in front, anteriorly truncate, posteriorly slightly sinuate, laterally narrowing toward head; sides distinctly margined; surface punctate; pleural region large, triangular; prosternum broad anterad coxae with well developed (*Synchroa*) or incomplete (*Mallodrya*) intercoxal process between coxae; prothoracic coxal cavities open behind externally, closed internally, and with an external fissure. Scutellum broader than long, trapezoidal. Mesosternum short, with a shallow mesal groove (*Synchroa*) to receive the prosternal process; metasternum broad. Legs with the anterior trochantin visible; prothoracic coxae oval (*Mallodrya*) or slightly projecting (*Synchroa*), separate (*Synchroa*) or nearly contiguous (*Mallodrya*); mesothoracic coxae oblong-oval, narrowly separated; metathoracic coxae transverse, nearly contiguous; trochanters small, triangular; tibiae slender with well developed, serrulate apical spurs; tarsal formula 5-5-4, tarsomeres slender, first tarsomere longest; claws simple. Elytra entire, the apices rounded; punctation confused; epipleural fold narrow, incomplete. Metathoracic wings with an entire subcubital fleck.

Abdomen with five visible ventrites, the first two connate. Male genitalia of inverted heteromeroid type; penis narrow, apically acute; parameres somewhat flattened, fused along much of their length, strongly tapered and narrowly separated distally.

Larvae (based on *Synchroa*, after Bøving and Craighead 1931, Young 1991) elongate, subcylindrical, subparallel; length 15 to 18 mm; vestiture sparse, consisting of a few moderate setae on each segment, dorsum of meso- and metathorax and abdominal segments 1-5 with asperity patches; color yellowish-white. Head exserted, prognathous, a little narrower than prothorax; epicranial suture with moderately elongate stem and lyriform frontal arms; antennae elongate, three-segmented with a small palpiform sensorium on second segment. Labrum distinct; frons and clypeus fused; mandibles asymmetrical, stout, bidentate with single subapical tooth; mola of right mandible more prominent than that of left; maxillae with cardo, small maxillary articulating area, subapically cleft mala with three-segmented palpi; labium with submentum and mentum fused, ligula and two-segmented palpi present. Five stemmata present on each lateral aspect of head, each side with two stemmatal groups: 3 anterior and 2 posterior. Thorax with five-segmented legs including tarsungulus; legs bearing numerous stout, spine-like setae; prosternum with mesal dentiform process. Abdomen nine-segmented, the tenth segment partially fused to ninth; tergite nine extending ventrally, with a single, shallow pit between bases of paired, fixed, upcurved urogomphi; ninth sternite with a single, stout asperity near each anterolateral margin. Spiracles annular-biforous.

Habits and habitats. Known synchroids feed largely upon fungal material and decaying wood, commonly beneath the bark of dead deciduous trees (Payne 1931). Adults are most active at night when they can be collected at light or as they run over the surface of logs. They have also been commonly collected at Malaise and Lindgren traps either unbaited or baited with ethanol.

Status of the classification. Historically treated as Melandryidae, *Synchroa* was excluded largely on the basis of characters provided by the larvae (Bøving and Craighead 1931). Viedma (1966) supported the move and Crowson (1966) noted similarities of synchroid larvae to those of Zopheridae and Stenotrachelidae.

Distribution. Two genera and eight species (Young 1991) are known to occur in Indonesia, Japan and eastern North America. Two species are found in eastern North America.

Key to the Nearctic Genera

1. Prothoracic coxae moderately prominent, nearly contiguous .... *Mallodrya*
— Prothoracic coxae not prominent, rather widely separated .... *Synchroa*

Classification of the Nearctic Genera

Synchroidae Horn 1888

Synchroa Newman 1838, 1 sp., *S. punctata* Newman 1838, Connecticut, Florida, Indiana, Maine, Michigan, New York, Ohio, Pennsylvania, Vermont, and West Virginia.

Mallodrya Horn 1888, 1 sp., *M. subaenea* Horn 1888, Ohio.

References

BØVING, A. G. and F. C. CRAIGHEAD. 1931. An illustrated synopsis of the principal larval forms of the Order Coleoptera. Entomologica Americana (N.S.), 11: 1-351.

CROWSON, R. A. 1966. Observations on the constitution and subfamilies of the family Melandryidae. Eos, 41: 507-513.

PAYNE, N. M. 1931. Food requirements for the pupation of two coleopterous larvae, *Synchroa punctata* Newn. and *Dendroides canadensis* LeC. (Melandryidae, Pyrochroidae). Entomological News, 42: 13-15.

VIEDMA, M. G. de. 1966. Contribución al conocimiento de las larvas de Melandryidae de Europa (Coleoptera). Eos, 41: 483-506.

YOUNG, D. K. 1991. Synchroidae. Pp. 516-518. *In*: F. W. Stehr, ed., Immature insects, Volume 2, Kendall/Hunt. Dubuque, IA.

109. OEDEMERIDAE Latreille 1810

by Nadine L. Kriska

Family common name: The false blister beetles

Family synonyms: Ascleridae Semenov 1894; Calopodidae Costa 1852; Ditylidae Mulsant 1858; Nacerdidae Mulsant 1858; Ganglbaueriidae Semenov 1894; Sparedridae Semenov 1894; Stenostomatidae Semenov 1894.

Oedemeridae are soft-bodied, elongate, variably colored beetles. They are distinguished from all other families by the following characters: pronotum broadened anteriorly, lacking margin on lateral borders, posteriorly narrower than the elytra; front coxal cavities open; tarsal formula heteromerous (5-5-4).

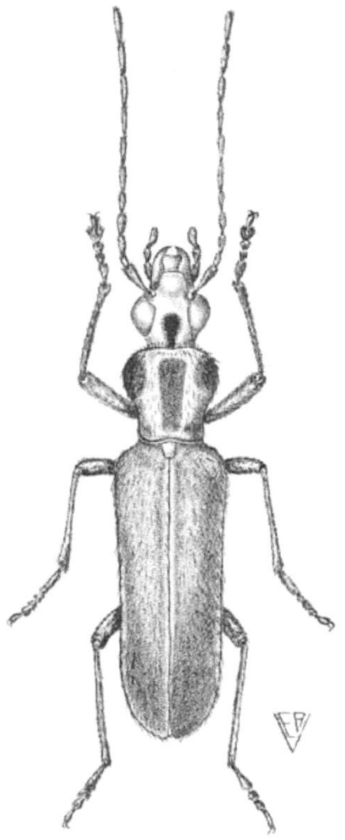

FIGURE 1.109. *Oxacis trimaculata* Champion

Description: (adapted, in part, from Arnett 1968 and Lawrence *et al.* 1999a) Body elongate, slightly flattened to moderately convex; 5-20 mm long; color variable, pale to piceous, in some satiny or metallic, often with yellow, orange, or red markings; dorsum with moderately fine to coarse, sparse to dense, recumbent to erect setae.

Head slightly deflexed, narrower than thorax, almost all longer than broad, surface smooth, finely punctate, or micro-rugose. Eyes entire, or anterior or mesal margin of eye deeply (Calopodinae) or weakly emarginate; posterior margin of eye not or barely emarginate; ommatidium of the acone type. Frontoclypeal suture absent, incomplete, or indistinctly impressed; suture (when present) straight, slightly curved, or angulate; anterior clypeal margin straight to convex; clypeus in some with distinct epistomal ridge; labrum at least partly visible, free, membranous, or separated by a suture; labrum distinctly longer than wide, apex subtruncate to slightly convex, slightly concave, or emarginate, moderately to heavily sclerotized. Mandibles short and broad to moderately elongate, moderately to strongly curved mesally; apices truncate, rounded, bidentate, or bilobed; each mandible with well-developed or reduced mola, and well-developed or reduced prostheca. Maxilla with distinct galea and lacinia; apex of galea or maxillary lobe densely setose or spinose; maxillary palpi 4-segmented; apical maxillary palpomere cylindrical to fusiform, or slightly expanded and truncate to subtriangular; labial palpi 3-segmented; apical labial palpomere as long as or longer than preapical palpomere, cylindrical to fusiform or slightly to strongly expanded apically. Ligula emarginate, in some deeply emarginate to bilobed. Antennae 11-segmented, few males with false segment 12; antennae filiform or serrate, inserted in deep emargination of eyes (Calopodinae) or anterad to eyes.

Thorax with pronotum slightly to moderately broader anteriorly or mesally than posteriorly; base of pronotum distinctly narrower than bases of elytra; laterally rounded, lacking lateral margins; anterior and posterior angles absent, rounded or right-angled, never produced or acute; surface punctate or rugose, rarely smooth. Prosternum anterad coxae flat to moderately convex; prosternal process incomplete, narrowed apically, not extending to mesosternum; prothoracic coxal cavities open, nearly contiguous. Scutellum well-developed or reduced, broadly to narrowly rounded, obtusely angulate, or acute posteriorly. Mesosternum short; mesothoracic coxal cavities open, contiguous. Metasternum broad, long, flat to moderately convex; metathoracic coxal cavities contiguous or narrowly separated.

Hind wings well-developed, reduced, or absent; wing venation reduced, M_4+Cu united apically, M_4 incomplete; radial cell well-developed, reduced, or absent; oblongum cell absent; wedge cell well developed or absent; anal lobe of hind wing absent.

Legs slender (North American males do not have expanded metathoracic femora); prothoracic coxae large, conical, prominent; mesothoracic coxae large, conical, not as prominent; metathoracic coxae transverse; apical tibial spurs mostly present, in some only one apical spur on prothoracic tibia; tarsal formula 5-5-4; penultimate (and rarely antepenultimate) tarsomeres cordate or bilobed, ventral surface spongy; claws simple or basally toothed.

Elytral surface glabrous, irregularly punctate, or with five or fewer distinct rows of punctures, microrugose, or subcostate; elytra entirely covering abdominal tergites or (non-North American species) exposing up to the first, second, or third tergite; elytral apices nearly or exactly meeting at suture, or each rounded or acute and separated by a broad gap.

Abdomen with five visible abdominal ventrites, sutures complete; surface smooth or micro-rugose; genital segment (ventrite

Acknowledgments. I thank D. K. Young and A. S. Ramsdale for reviewing the manuscript and offering useful suggestions, and to Paul Skelley, FSCA, for providing a loan of material and access to Ross Arnett's oedemerid collection and reprints.

9) mostly modified; tergite 9 (males) almost completely divided into two parts, tergite 10 (males) well developed and free, or partly fused to tergite 9. Aedeagus of male tenebrionoid, symmetrical; anterior edge of tegmen or phallobase without struts or accessory lobes; parameres fused to phallobase, separate from each other or partly to nearly entirely fused together. Female genitalia with rudimentary paraprocts on a long baculum; valvifers long; coxite indistinctly 2-segmented; stylus small, lateral on the coxite.

Mature larvae (adapted, in part, from Arnett 1968, Lawrence 1991, Lawrence *et al.* 1999b, and Rozen 1960) 10-40 mm long, mostly 25 mm or less. Body elongate, parallel-sided, subcylindrical, straight or slightly curved ventrally, very weakly sclerotized, color nearly white; surface smooth, vestiture of scattered, simple setae.

Head protracted, prognathous, broad, slightly flattened, in some asymmetrical. Stemmata mostly absent, some with two or five on each side. Antennae well developed, 3-segmented, segment 3 reduced. Frontoclypeal suture present (*Calopus*) or absent. Mandibles asymmetrical, large, bi- or tridentate, lacking accessory ventral condyle; molae large, transversely ridged, the left one parallel to long axis and produced apically, the right one strongly oblique; prostheca absent. Ventral mouthparts retracted. Maxilla with cardo transverse, in some distinctly divided, stipes elongate, palpi 3-segmented, and mala rounded or truncate (from fused galea and lacinia). Labium well developed, free to base of mentum; ligula well developed or reduced; labial palpi 2-segmented. Prehypopharynx mostly bearing an apically pilose, columnar projection situated immediately distad hypopharyngeal sclerome.

Thorax mostly with paired patches of asperities on all terga, absent in some, or present on meso- and metatergum only; meso- and metatergum and abdominal terga 1-2, 1-3, or 1-5 (*Calopus*) with asperity-bearing ampullae. Legs short, 5-segmented; tarsungulus with two setae lying side by side. Abdominal sterna 2-3, 3-4, or 2-4 mostly with paired, asperity-bearing ampullae; urogomphi mostly absent, or, if present (on tergite nine), small, lightly sclerotized, sharply pointed, immovable; *Calopus* with well-developed, upturned urogomphi with a single pit between them, sternite nine with 3-4 asperities on each side at base. Spiracles annular or annular-multiforous.

Habits and habitats. Most species in this family are uncommonly encountered. Many adults are occasionally attracted to light. Species found on islands can be particularly abundant and attracted to light in large numbers (Arnett 1968).

Oedemerids are most abundant near oceanic coastlines, particularly on islands (Hudson 1975). Adults are found on flowers and foliage, under driftwood, or in moist to wet, rotten logs (Arnett 1968, Hudson 1975). Many species are attracted to light, although usually not in large numbers.

Larvae occur in moist, decaying wood. Inland species seem particularly fond of conifers, while coastal species use various kinds of driftwood. Rozen (1960) provided detailed information regarding the larvae for each subfamily. Larval Calopodinae are typically found in rotting logs, stumps, and roots. Larval Nacerdinae are typically found in wet, rotting wood, often that which is buried or intermittently submerged in fresh or salt water. Larval Oedemerinae also use rotting wood and require a very moist environment as well. Members of this subfamily often inhabit wet and decaying driftwood, moist, rotten logs, stumps, or roots. Some species reside in dead roots, stems or branches of herbaceous plants, such as oak, chestnut, and mulberry.

Larvae of a single species, *Nacerdes melanura* (Linnaeus) (the wharf borer), are of minor economic importance and are closely monitored for potential damage to wharves, pilings, and other structural wood.

Oedemeridae have earned their common name, "false blister beetle," not only because some adults resemble slightly members of Meloidae (the true "blister beetle" family), but also because many species can cause blisters when pinched or squashed on the skin. Vaurie (1951) reported that in the Bimini Islands, Bahamas, several species, particularly those of *Oxacis* and *Hypascleris*, caused varying degrees of blistering on exposed areas of skin, such as the face, neck, forearms, and hands, areas commonly exposed to the beetles. Susceptibility to blistering varies, from no reaction, to mild, but painless blistering, to the extreme of painful blisters or sores taking weeks to heal (Arnett 2000). Chemical analysis of 21 oedemerid species from Japan identified the blistering agent as cantharidin; all 21 species contained this chemical (Kurosa 1977). Judging by the aposomatic coloration found throughout Oedemeridae, many species probably possess cantharidin to varying degrees. Since most adults are soft-bodied flower feeders, they probably use coloration and cantharidin to deter potential predators.

Status of the classification. Oedemeridae are a well-defined, monophyletic group within Coleoptera (Crowson 1967). Various classifications have recognized either two or three subfamilies. Crowson (1967) and Lawrence and Newton (1995) recognized Calopodinae (considered to be the most primitive oedemerid group) and Oedemerinae. Arnett (1951), Hudson (1975), Rozen (1960), and Svihla (1985) recognized Nacerdinae in addition to the other two subfamilies; these authors conducted various revisions for different portions of the world, and their data lend strong support for the recognition of three subfamilies.

Some discrepancy exists concerning tribal classification. Arnett (1951) recognized three tribes of Oedemerinae: Asclerini, Ditylini, and Oedmerini (not occurring in the New World). Rozen (1960) studied larval Oedemeridae systematically and noted a discrepancy between adult and larval classifications. He indicated that based on larvae, Asclerini and Ditylini are more closely related to Nacerdinae. Svihla (1985) in a revision of adult Oedemeridae of the Old World, placed Ditylini into Nacerdinae but maintained Asclerini in Oedemerinae.

The classification of North American Oedemeridae followed here recognizes three subfamilies: Calopodinae, Nacerdinae, and Oedemerinae, the latter with two tribes: Asclerini and Ditylini. An unpublished manuscript by Arnett on generic classification is not followed. Higher-level phylogenetic analyses, based on both adults and larvae, are needed.

Distribution. Oedemeridae are present in all major zoogeographic regions (Lawrence *et al.* 1999a). The family is composed

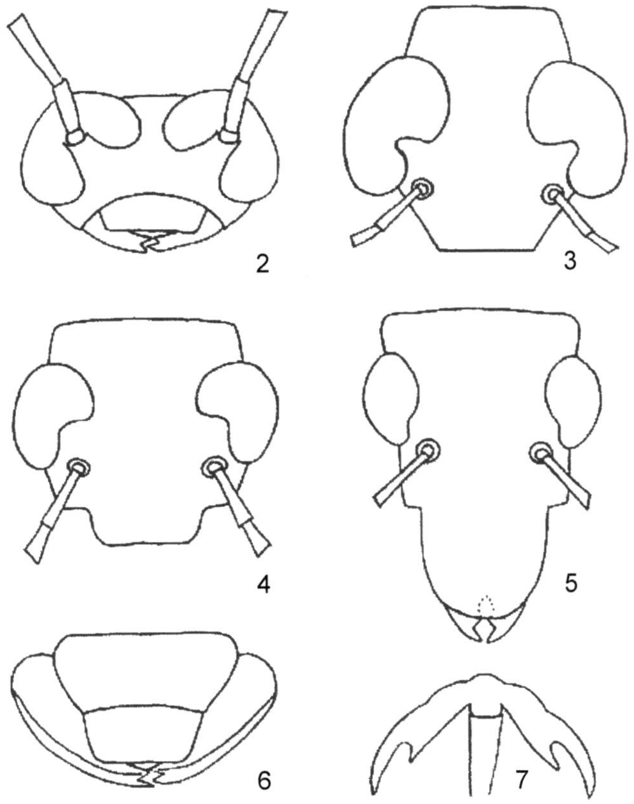

FIGURES 2.109-7.109. 2-5. Head, dorsal view. 2. *Calopus angustus* LeConte; 3. *Xanthochroa erythrocephala* (Germar); 4. *Nacerdes melanura* (Linnaeus); 5. *Rhinoplatia mortivallicola* Arnett; 6. *Asclera nigra* LeConte, mandibles, clypeus, and frons, anterodorsal view; 7. *Asclera nigra* LeConte, mesothoracic tarsal claws (basally toothed).

of about 1500 species (Britton 1970) in 115 genera world-wide. In North America north of Mexico, there are 87 species in 17 genera. Schenkling (1915) provided the most recent world catalogue of Oedemeridae, including general distributional information.

Key to the Nearctic Genera
(Adapted, in part, from Arnett 1968)

Some characters used in the following key, particularly those of the mouthparts, may be difficult to see in dried specimens. Soaking specimens in a weak ammonia solution for an hour, and then rinsing with water, or boiling them briefly in water should sufficiently relax the specimens and enable positioning of the mouthparts to be clearly visible.

1. Antennae set in deep emargination of eyes (Fig. 2) (Calopodinae) 2
— Antennae set in front of eyes, eyes emarginate or entire (Figs. 3-5) 3

2(1). Mandibles apically bifid (Fig. 2) .... *Calopus*
— Mandibles simple .... *Sparedrus*

3(1). Prothoracic tibiae each with single apical spur (Nacerdinae) 4
— Prothoracic tibiae each with two apical spurs (Oedemerinae) 5

4(3). Eyes large, each wider than frons (Fig. 3), moderately emarginate .... *Xanthochroa*
— Eyes small, each narrower than frons (Fig. 4), weakly emarginate.. *Nacerdes*

5(3). Body stout (vaguely "carabid-like"), black, some with fine, light gold vestiture (Ditylini) .... *Ditylus*
— Body slender (vaguely "cantharid-like"), variably colored, never entirely black; lacking light gold vestiture (Asclerini) 6

6(5). Head elongate; clypeus prolonged, rostrate (Fig. 5) .... *Rhinoplatia*
— Head elongate or short, but clypeus not prolonged or rostrate 7

7(6). Both mandibles bifid apically (Fig. 6) AND tarsal claws toothed basally (Fig. 7) .... *Asclera*
— Both mandibles simple, or one simple/one apically bifid, or both apically bifid; if both bific, then tarsal claws not toothed basally 8

8(7). Right mandible apically bifid or with small apical tooth, left mandible entire .... *Hypasclera*
— Both mandibles apically bifid or entire... 9

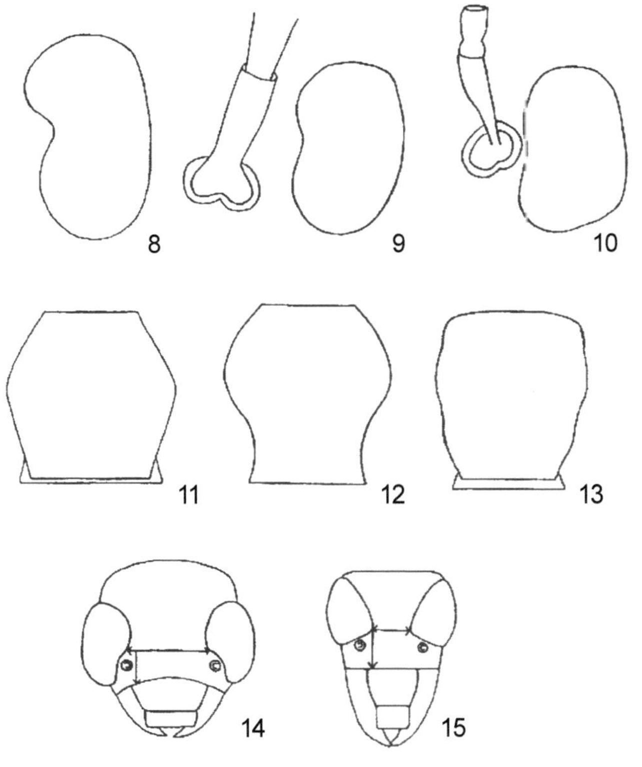

FIGURES 8.109-15.109. 8. *Eumecomera obscura* (Horn), eye, lateral view; 9-10. eye and antennal insertion, lateral view. 9. *Sissenes championi* Horn; 10. *Oxacis trimaculata* Champion; 11-13. pronotum, dorsal view. 11. *Sissenes championi* Horn; 12.*Oxacis trimaculata* Champion; 13. *Copidita quadrimaculata* (Motschulsky); 14-15. Head in dorsal view. 14. *Xanthochroina bicolor* (LeConte); *Oxacis trimaculata* Champion.

9(8). Eyes emarginate (Fig.8); body often metallic colored .... *Eumecomera*
— Eyes entire to weakly emarginate (Fig. 9); usually not metallic colored 10

10(9). Antennal insertion (Fig. 9) well separated from eye (exception: *Vasaces* with base of antenna approximate to eye); pronotum widest medially, more parallel-sided (Fig. 11); head elongate 11
— Antennal insertion approximate to eye (Fig. 10); pronotum widest subapically or anteriorly, narrowed at the base (Figs. 1, 12-13); head short or elongate 13

11(10). Eyes entire .... *Heliocis*
— Eyes weakly emarginate 12

12(11). Body slightly dorsoventrally flattened; dorsal surface with a satiny sheen .... *Sisenes*
— Body not dorsoventrally flattened; dorsal surface glossy, many with a metallic sheen ...... *Vasaces*

13(10). Both mandibles apically bifid; tarsal claws simple 14
— Both mandibles simple; tarsal claws simple or toothed 15

14(13). Smaller (5-12 mm long); length of maxillary palpi < length of head .... *Oxycopis*
— Large (8-22 mm long); length of maxillary palpi > length of head .... *Copidita*

15(13). Tarsal claws basally toothed .... *Paroxacis*
— Tarsal claws simple 16

16(15). Head very short, frons from anterior corner of eye to epistomal ridge shorter than distance between eyes (Fig. 14); mandibles small, curved, sharp .. .. *Xanthochroina*
— Head short to moderately long, frons from anterior corner of eye to epistomal ridge as long as or longer than distance between eyes (Fig. 15); mandibles mostly stout, blunt .... *Oxacis*

Classification of the Nearctic Genera

Oedemeridae Latreille 1810

Calopodinae Costa 1852

Calopus Fabricius 1775

One species, *C. angustus* LeConte, occurs in Canada from British Columbia to Quebec, and in the United States from Washington south to New Mexico and east to Pennsylvania. This species is uncommon throughout its range, particularly so in northeastern North America (Arnett 1968). Burke (1906) reported rearing an adult from a gallery in a live western cedar (*Thuja plicata* Don ex D. Don), and finding a small number of larvae, pupae, and fragments of adults in galleries in both dead and living branches of alpine fir (*Abies balsamea lasiocarpa* (Hook)). Burke (1906) indicated that larvae cause damage to living trees by entering through small wounds and working into the living tissues, keeping the wounds open and enlarged instead of allowing them to heal over.

Sparedrus Latreille 1829

Two species occurring in southern California and Texas, but are rarely encountered (Arnett 1951).

Nacerdinae Mulsant 1858

Nacerdes Falderman 1836

This genus has traditionally been attributed to Dejean (1834), but should be attributed to Falderman (1836) as *Nacerdes* Dejean is a *nomen nudum* (Neave 1939, Hudson 1975). One species, *N. melanura* (Linnaeus), is widespread throughout North America, and is common in coastal states and states containing major rivers. Known as the wharf borer, this Old World species is now cosmopolitan. The larvae feed on decomposing wood brought down by flooding or washed ashore. They are known to infest pilings and are capable of withstanding immersion by the tide (Arnett 1951). This species is monitored but is of minor economic importance (Rozen 1960).

Xanthochroa Schmidt 1846

Six species: four west coast (British Columbia to California) and two east coast (Maine to Florida, west to Louisiana). Key to species: Arnett (1951). Adults of *X. erythrocephala* (Germar) were found feeding on dogwood (*Cornus* sp.) and *X. californica* Horn is attracted to light (Arnett 1951).

Oedemerinae Costa 1852

Ditylini Mulsant 1858

Ditylus Fischer von Waldheim 1817

Three species: two occur in western North America, from Alaska south to California, and one occurs in the northeast from Ontario south to New York and west to Minnesota. Key to species: Arnett (1951). Arnett (1951) described the biology of *D. quadricollis* LeConte and stated that the life histories of its congeners are probably similar. Young (1990) reported some county records of *D. caerules* (Randall) from Upper and Lower Michigan, and he noted that several specimens were found in June and July in Malaise traps and among waveline beach drift. Larvae have been found in old, wet cedar logs; the larval stage may last up to three years. Adults overwinter in their pupal cell and are active from May through August. They have been observed running over swampy ground or under wet logs. Adults have also been observed on flowers of cultivated *Spiraea.*

Asclerini Semenov 1894

Asclera Stephens 1839

Six species: three occur in western sections of Canada and the United States (British Columbia south to California, east to Arizona), two occur in eastern Canada and United States (Ontario and Quebec south to Alabama, west to Nebraska), and one is widespread throughout North America. Key to species: Arnett (1951). Adults of this genus are found on flowers and foliage of various plants. Arnett (1951) provided food plant information for each species.

Copidita LeConte 1866
One species, *C. quadrimaculata* (Motschulsky), occurs in the western United States (Oregon, California, Arizona, Texas). Adults are attracted to light at night. A few specimens have been encountered diurnally on a beach (Arnett 1951).

Eumecomera Arnett 1951
Three species, all restricted to western North America, from the Rocky Mountain region west to the Pacific Coast. Key to species: Arnett (1951).

Heliocis Arnett 1951
One species, *H. repanda* (Horn), known from Maryland, Florida, Texas, and Arizona.

Hypasclera Kirsch 1866
Eight species, five restricted to the southeastern United States (Georgia and Florida), one occurring in California, one occurring from New York south to Florida and west to Texas, and one occurring from Virginia south to Florida, west to Texas and southern California. Key to species: Arnett (1951). Adults are active primarily in the summer, from mid-May through August; a few have been collected as late as December in southern states. Little is known of the biology of *Hypasclera*. Larvae of *H. dorsalis* (Melsheimer) appear abundant in soil and wood (Arnett 1951). Adults of *H. nesiotes* (Arnett) are attracted to light.
Alloxacis Horn 1896

Oxacis LeConte 1866
Twenty-nine species occur throughout the United States. Partial keys to species: Arnett (1951, 1960, 1963). This genus is in need of revision; no complete key to species exists. Adults have been found on flowers of a variety of plants. Arnett (1951) provided information on specific plant associations for several species.

Oxycopis Arnett 1951
Thirteen species occur throughout the United States. Partial key to species: Arnett (1951), treated as a subgenus of *Oxacis*. Adults of *O. mimetica* (Horn) have been found at light. Adults of most Nearctic species are found on flowers of a variety of plants. Arnett (1951) provided information on specific plant associations for several species.

Paroxacis Arnett 1951
Four species, all occurring in the southern United States. Two species occur in California and Arizona, one in Louisiana and Florida, and one in Texas, Louisiana, and Alabama. Key to species: Arnett (1951), treated as a subgenus of *Oxacis*.

Rhinoplatia Horn 1862
Two species occur in the western United States (California, Nevada, and Arizona). Key to species: Arnett (1951). Adults have been found on flowers (Arnett 1951).

Sisenes Champion 1889
One species, *S. championi* Horn, is known only from Arizona (Arnett 1951).

Vasaces Champion 1889
Four species, occurring in the southwestern United States (Arizona and Texas). Key to species: Arnett (1953). This genus is closely related to the previous genus, *Sisenes*.

Xanthochroina Ganglbauer 1881
One species, *X. bicolor* (LeConte), occurs in British Columbia south to California and east to Idaho, Wyoming, and Texas. Arnett (1951) treated it as a subgenus of *Oxacis*. Larvae have been found in wooden flooring and railroad ties (Arnett 1951). Several adults were found in the lower levels of a mine in California (Arnett 1951).

Bibliography

ARNETT, R. H., Jr. 1951. A revision of the Nearctic Oedemeridae (Coleoptera). American Midland Naturalist, 45: 257-391.

ARNETT, R. H., Jr. 1953. Beetles of the oedemerid genus *Vasaces* Champion. Proceedings of the United States National Museum, 103: 87-94.

ARNETT, R. H., Jr. 1960. Contribution towards a monograph of the Oedemeridae 13. The "Fragilis" complex of the genus *Oxacis*. Coleopterists Bulletin, 14: 33-44.

ARNETT, R. H., Jr. 1963. The phenogram, a method of description for studies on *Oxacis* (Coleoptera, Oedemeridae). Coleopterists Bulletin, 17: 6-18.

ARNETT, R. H., Jr. 1968. Oedemeridae (Latreille, 1810). The false blister beetles. The Beetles of the United States. The American Entomological Institute. Ann Arbor, MI, 1112 pp.

ARNETT, R. H., Jr. 2000. False Blister Beetles. University of Florida EENY-154, Featured Creatures web site: http://creatures.ifas.ufl.edu/urban/medical/false_blister_beetles.htm. [originally published as: Arnett, R. H., Jr. 1984. The false blister beetles of Florida (Coleoptera: Oedemeridae). Florida Department of Agriculture and Consumer Services, Entomology Circular No. 259. 4 pp]

BRITTON, E. B. 1970. Coleoptera (Beetles). Pp. 495-621. *In:* C.S.I.R.O., ed.. The Insects of Australia. Melbourne University Press. Australia.

BURKE, H. E. 1906. Notes on the larva of *Calopus angustus* Lec. Proceedings of the Entomological Society of Washington, 8: 64-66.

CROWSON, R. A. 1967. The natural classification of the families of Coleoptera. Classey.

DEJEAN, P. F. M. A. 1834. Catalogue des coléoptères de la collection de M. le comte Dejean, livre 3, pp. 177-256.

FALDERMAN, F. 1836. Bereicherung der Käferkunde des Russischen Reichs. Bulletin de la Société Impériale des Naturalistes de Moscou, 9:351-398.

HUDSON, L. 1975. A systematic revision of the New Zealand Oedemeridae (Coleoptera, Insecta). Journal of the Royal Society of New Zealand, 5: 227-274.

KUROSA, K. 1977. Poisonous beetles (Oedemeridae). Pp. 126-129. *In:* M. Sara, H. Takabasi, R. Kano, and H. Tanaka. Animals of medical importance in the Nansei Islands in Japan. Shinjuku Shodo, Tokyo. 410 pp.

LAWRENCE, J. F. 1991. Oedemeridae (Tenebrionoidea). Pp. 534-535. *In:* F. W. Stehr, ed. Immature Insects, Vol. 2. Kendall/Hunt. Dubuque, IA.

LAWRENCE, J. F., A. M. HASTINGS, M. J. DALLWITZ, T. A. PAINE and E. J. ZURCHER. 1999a. Beetles of the world, a key and information system for families and subfamilies. Version 1.0 for Windows. CSIRO Entomology. Canberra, Australia. CD-ROM.

LAWRENCE, J. F., A. M. HASTINGS, M. J. DALLWITZ, T. A. PAINE and E. J. ZURCHER. 1999b. Beetle larvae of the world, descriptions, illustrations, identification, and information retrieval for families and sub-families. Version 1.1 for Windows. CSIRO Entomology. Canberra, Australia. CD-ROM.

LAWRENCE, J. F. and A. F. NEWTON, Jr. 1995. Families and subfamilies of Coleoptera (with selected genera, notes, references and data on family-group names). *In:* J. Pakaluk and S. A. Slipinski., eds. Biology, Phylogeny, and Classification of Coleoptera. Papers celebrating the 80th birthday of Roy A. Crowson. Muzeum I Instytut Zoologii PAN. Warsaw, 558 pp.

NEAVE, S. A. 1939. Nomenclator Zoologicus. Zoological Society of London. London.

ROZEN, J. G. 1960. Phylogenetic-systematic study of larval Oedemeridae (Coleoptera). Miscellaneous Publications of the Entomological Society of America, 1: 35-68.

SCHENKLING, S. 1915. Pars 65. Oedemeridae. *In:* S. Schenkling, ed. Coleopterorum Catalogus. W. Junk. Berlin, 82 pp.

SVIHLA, V. 1985. Revision of the generic classification of the Old World Oedemeridae (Coleoptera). Sbornik Narodniho Muzea v Praze (Acta Musei Nationalis Pragae), 41B (1985): 141-238.

VAURIE, P. 1951. Blistering caused by oedemerid beetles. Coleopterists Bulletin, 5: 78-79.

YOUNG, D. K. 1990. The occurrence of *Ditylus caeruleus* in Michigan (Coleoptera: Oedemeridae), with observations on its range. Great Lakes Entomologist, 23: 39-41.

110. STENOTRACHELIDAE Thomson 1859

by Daniel K. Young

Family common name: The false longhorn beetles

Family synonym: Cephaloidae LeConte 1862

The 5-5-4 tarsal formula and distinctive pretarsal claws bearing a ventral lobe serve to distinguish this small heteromerous family.

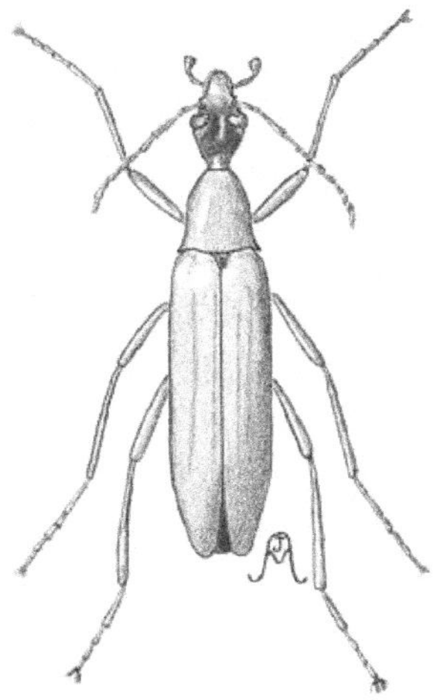

FIGURE 1.110. *Cephaloon tenuicorne* LeConte

Description: Body elongate, narrow and convex in many, somewhat fusiform; length 6-20 mm; vestiture consisting of very fine, short, decumbent setae.

Head elongate, diamond- or bell-shaped in many (especially *Cephaloon*), porrect, narrowed behind eyes, gradually (e.g., *Cephaloon*) or abruptly constricted behind (e.g., *Nematoplus*), forming a slender "neck"; surface smooth; antennae 11-segmented, filiform, or terminating in a weakly developed, 3-segmented club (some *Cephaloon*), inserted between the emarginate eyes and bases of mandibles under a small frontal ridge. Labrum prominent; mandibles moderately elongate, apices acute, subserrate internally with broad membrane from along the basal half; maxillary palpi 4-segmented, the first segment small, obscure, the apical segment triangular; labium with gular process developed; mentum small, nearly square; ligula membranous, broad, bilobed, prominent; labial palpi 3-segmented, slender.

Pronotum elongate and narrowed anteriorly from near the middle (*Anelpistus*, *Cephaloon*, *Nematoplus*), or elongate-quadrate (*Stenotrachelus*); entirely margined (*Anelpistus*, *Stenotrachelus*), weakly, incompletely margined (*Nematoplus*), or unmargined (*Cephaloon*); surface smooth (*Cephaloon*, *Nematoplus*), punctate (*Stenotrachelus*), or granulate-punctate (*Anelpistus*); pleural region broad; prosternum long to short in front of coxae, intercoxal process very narrow or incomplete; prothoracic coxal cavities open behind. Scutellum small, triangular. Mesosternum narrowed anteriorly; metasternum long. Legs long and slender; pro- and mesothoracic trochantins distinct; prothoracic coxae prominent, conical, confluent; mesothoracic coxae prominent, conical, confluent; metathoracic coxae obliquely transverse, prominent; femora slender; tibiae with two large apical spurs; tarsal formula 5-5-4, segments simple, not lobed or tomentose; claws simple (*Anelpistus*, *Nematoplus*, *Stenotrachelus*) or pectinate (*Cephaloon*) each with a subequal, membranous lobe beneath. Elytra gradually narrowed apically, entire, vaguely costate, minutely, confusedly punctate, never striate; epipleurae narrow, usually incomplete. Metathoracic wings well developed, radial and anal cells closed, subcubital fleck lacking.

Abdomen with five visible ventrites; sutures entire or with first two visible ventrites connate. Male genitalia of inverted heteromeroid type, with penis small, somewhat curved, tube-like, fused to the very large, trough-like basal piece, but with parameres distinct and freely articulating. Ninth segment commonly (*Cephaloon*) with triangular tergite and ventrite, filling emargination of eighth, segment eight bulbose, with apex forming two laterally triangular pieces.

Larvae elongate, subcylindrical to slightly flattened; head protracted, prognathous, cranium sometimes asymmetrical, with median epicranial suture distinct; mandibles strongly asymmetrical, with mola large, transversely ridged; maxillae each with transverse cardo, well developed articulating area, mala, and 3-segmented palpi; labium with well developed ligula, 2-segmented palpi. Abdomen with urogomphi dorsally upcurved and with sclerotized apices (Stenotrachelinae), with posteriorly projecting, straight, poorly sclerotized urogomphi (*Cephaloon*) or lacking urogomphi entirely (*Nematoplus*).

Habits and habitats. Little is known about the habits of adults of the Stenotrachelidae. They are sometimes found on flowers or in Malaise, flight intercept or other such traps; they are probably short-lived. Larvae feed and develop in decaying wood. Some, such as *Nematoplus*, are suggested to be associated with logs infested with brown rot fungi (wood in the red rot stage of decay). Larvae of *Cephaloon* are sometimes found in similar situations.

Status of the classification. Various elements of Stenotrachelidae have historically been associated with Melandryidae (Stenotrachelinae) and Pyrochroidae or "Pedilidae" (Nematoplinae). The pretarsal claw structure is an important synapomorphy for members of this family within the Tenebrionoidea (Abdullah 1965, Crowson 1955, Lawrence and Newton 1995, Mamaev 1973).

Distribution. This family contains about 20 species in three subfamilies and six genera: Stenotrachelinae (*Anelpistus*, Nearctic; *Scotodes*, Palearctic; *Stenocephaloon*, Japan; *Stenotrachelus*, Circumboreal), Nematoplinae (*Nematoplus*, Holarctic), Cephaloinae

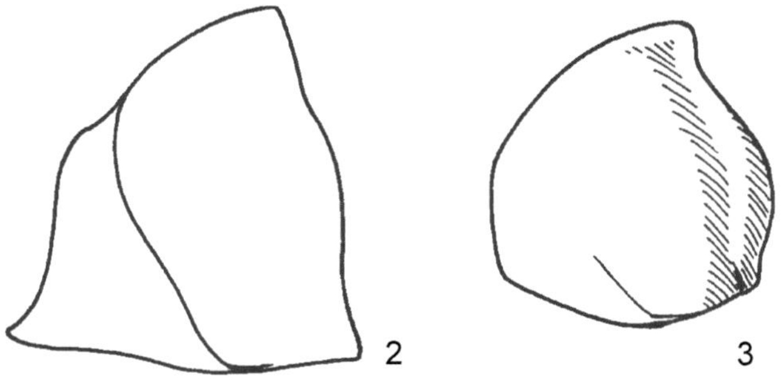

FIGURE 2.110-3.110. Prothorax, left dorsolateral view. 2. *Stenotrachelus aeneus* (Fabricius); 3. *Nematoplus collaris* LeConte.

(*Cephaloon*, Holarctic). The Japanese genus *Stolius* is sometimes added to Stenotrachelidae as a monogeneric Stoliinae (Arnett 1953, Borchman 1917, Lawrence and Newton 1995). In the United States and Canada, four genera and 10 species are known (Arnett 1953, Hopping and Hopping 1934, Lawrence and Newton 1995).

Key to the Genera of the United States

1. Prothorax with lateral margin entire, separating pronotum from hypomera (Fig. 2; Stenotrachelinae) 2
— Prothorax lacking complete lateral margin; pronotum and hypomera evenly rounded laterally, not separated (Fig. 3) 3

2(1). Head largely exposed; pronotum punctate, but without granulations .... *Stenotrachelus*
— Head inserted into prothorax nearly to eyes; pronotum densely granulate-punctate .... *Anelpistus*

3(1). Pronotum in dorsal view bell-shaped; lateral aspects of head gradually converging behind eyes (Fig. 1) (Cephaloinae) .... *Cephaloon*
— Pronotum in dorsal view hexagonal, head strongly, abruptly constricted behind eyes (Nematoplinae) .... *Nematoplus*

Classification of the Genera of America north of Mexico

Stenotrachelidae Thomson 1859

Stenotrachelinae Thomson 1859

Stenotrachelus Latreille 1825, 1 sp., *S. aeneus* (Fabricius), transcontinental: Canada and the northernmost United States.

Anelpistus Horn 1870, 2 spp., *A. americanus* Horn, New Hampshire; *A. canadensis* Mank, British Columbia (review of species, Mank 1942).

Nematoplinae LeConte 1862

Nematoplus LeConte 1855, 1 spp., *N. collaris* LeConte, northeastern United States and Canada, west to the Great Lakes region.

Cephaloinae LeConte 1862

Cephaloon Newman 1838, 6 spp., distributed as follows: *C. bicolor* Horn, British Columbia, California, Oregon, Washington; *C. lepturides* Newman, northeastern United States and Canada; *C. pacificum* Van Dyke, British Columbia, Oregon, Washington; *C. tenuicorne* Le Conte, Alberta, British Columbia, Idaho, Montana, Oregon, Washington; *C. ungulare* Le Conte, "Lake Superior," New Hampshire, New York, North Carolina; *C. vandykei* Hopping and Hopping, California (key, Arnett 1953).

Ichnodes Dejean 1834, *nomen nudum*
Cephalaon Motschulsky 1860 (misspelling)
Typitium Casey 1898
Sponidium Casey 1898
Drachylis Casey 1898
Ephamillus Semenov 1900

Bibliography

ABDULLAH, M. 1965. The genus *Nematoplus* LeConte (Coleoptera, Cephaloidae) and its affinities with Meloidae. Entomologist, 98: 54-59.

ARNETT, R. H., Jr. 1953. A review of the beetle family Cephaloidae. Proceedings of the United States National Museum, 103 (no. 3321): 155-161.

BORCHMAN, F. 1917. Cephaloidae. *In:* S. Schenkling, ed. Coleopterorum Catalogus, Pars 17(69): 206-208. W. Junk. Berlin.

CROWSON, R. A. 1955. The Natural Classification of the Families of Coleoptera. Nathaniel Lloyd. London, 187 pp.

HOPPING, R. and G. R. HOPPING. 1934. A revision of the genus *Cephaloon* Newm. Pan-Pacific Entomologist, 10: 64-70.

LAWRENCE, J. F. and A. F. NEWTON, Jr. 1995. Families and subfamilies of Coleoptera (with selected genera, notes, references and data on family-group names). Pp. 779-1006. *In:* J. Pakaluk and S. A. Slipinski, eds. Biology, Phylogeny, and Classification of Coleoptera: Papers Celebrating the 80th Birthday of Roy A. Crowson. Muzeum i Institut Zoologii, Polska Academia Nauk. Warsaw.

MAMAEV, B. M. 1973. The morphology of the larva of the beetle genus *Nematoplus* Lec. and the phylogenetic connections of some families of Heteromera (Coleoptera, Cucujoidea). Entomologicheskoe Obozrenie, 52: 586-598 [in Russian, translation in Entomological Review, 52: 388-395].

MANK, E. W. 1942. A review of the genus *Anelpistus* Horn (Coleoptera, Melandryidae). Canadian Entomologist, 74: 186-193.

111. MELOIDAE Gyllenhal 1810

by John D. Pinto and Marco A. Bologna

Family common name: The blister beetles

Family synonyms: Horiidae Latreille 1802; Lyttidae Wellman 1910; Tetraonycidae Bøving and Craighead 1931. See Bologna and Pinto (1997) and ICZN (1999) for precedence of the family name Meloidae and subfamily name Nemognathinae over Horiidae.

Adults of this distinctive family of terrestrial phytophagous beetles can be recognized by the following characters: soft bodied; generally rather elongate; head deflexed, with a narrow neck; pronotum not carinate at sides; tarsi heteromerous; legs long; tarsal claws each with a lower or ventral blade which is reduced to a tooth adnate to the claw itself in some groups; integument smooth to distinctly rugose.

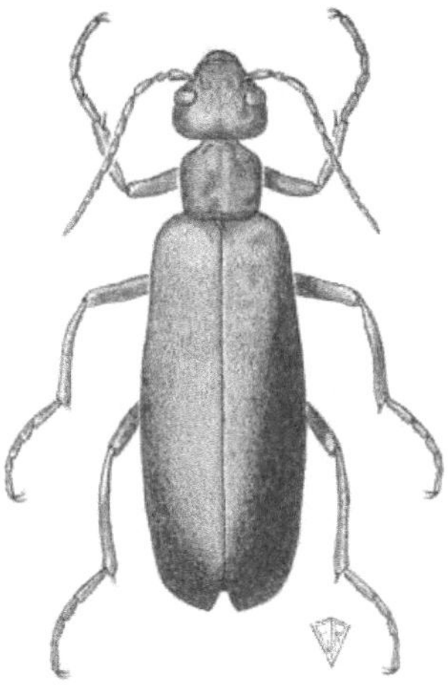

FIGURE 1.111. *Epicauta pensylvanica* (Degeer)

Description: Family heterogeneous in shape, most species elongate with long legs; length 3 to 30 mm, most 10 to 20 mm, occasionally to 70 mm in some tropical species; color variable, opaque or metallic; vestiture fine in most, often very sparse and short above, but in some (*Epicauta*) dense, covering surface of integument.

Head large, larger than thorax at its apex in most (extremely enlarged in the Horiini), deflexed, abruptly and markedly constricted at postocciput to form a narrow neck; frons mostly wide, convex, flat or rarely furrowed- surface smooth, punctate, or rugose. Antennae with 11 antennomeres, reduced in number (7- 10) in the Old World tribes Mylabrini and Cerocomini and in one species of the North American genus *Cordylospasta,* short to moderately long, mostly subfiliform or submoniliform, rarely serrate, often with antennomeres modified in male (extremely so in Cerocomini); insertion between eyes, above base of mandibles. Clypeus distinct; labrum mostly prominent, but distinctly shorter than mandibles in some (*e.g.,* Horiini). Mandibles more or less curved; apex mostly acute, rarely abruptly recurved at apical half (some Epicautini), in some blunt, extremely elongate in the Horiini. Maxillae unmodified, or, in the subfamily Nemognathinae, with galeae poorly sclerotized and penicillate apically or forming an elongate sucking tube which may be as long or longer than the body; palpi with 4 palpomeres (5 in the African tribe Morphozonitini), slender, the apical segment somewhat enlarged but never securiform, modified in male of some species. Labium with gular process large; mentum trapezoidal; ligula prominent; palpi with 3 palpomeres, slender, in some with last palpomere slightly enlarged but highly modified to form a sucking tube in *Leptopalpus,* an Old World genus of Nemognathini. Eyes lateral, oval or, more commonly, emarginate, extending considerably ventrally in a few genera (*e.g., Pseudozonitis*), completely absent in the Neotropical tetraonycine genus *Meloetyphlus.*

Pronotum generally narrower than base of elytra, narrowest at apex in most; sides not carinate or margined, surface smooth, punctate, or rugose; pleural region large; prosternum short; procoxal cavities large, confluent, open behind. Mesosternum triangular, more or less modified anteriorly in the Old World tribe Mylabrini; mesocoxal cavities confluent. Metasternum long or (in wingless forms) short. Legs with anterior coxae prominent, contiguous; middle coxae prominent, contiguous, hind coxae transverse, nearly contiguous; trochanters large, triangular, in some modified in male; profemora subcylindrical, or with an apicoventral excavation with silky pubescence (most Epicautini); tibiae mostly elongate, variously modified in male of several taxa, with two distinct apical spurs, one on the protibiae of males in some; hind tibial spurs often dissimilar, the external one commonly spoon-like and in some greatly enlarged; tarsal formula 5-5-4; tarsi slender, bilobed only in the Tetraonycinae and some tropical Eleticinae, most with narrow pads beneath segments; tarsi of male modified in several species of Meloinae; claws each subtended by a ventral blade which is rarely reduced to a filiform appendix or fused basally to the dorsal blade (claw) and forming a tooth, ventral blade microsetate in the Eleticinae, in some Lyttini, and in two species of North American *Zonitis* (Nemognathinae); dorsal blade often with a double or single row of teeth on ventral margin (particularly in the Nemognathinae). Elytra entire or shortened, rarely overlapping at suture; apex of elytra with a pit-like organ in males of several Eleticinae; epipleural fold small. Wings normal, rarely reduced, or absent, in rare cases (*Cordylospasta*) absent only in female; venation without radial vein and cell; the most primitive wing (in Eleticinae) includes a $2A_2$ vein which may be interrupted at base.

Abdomen soft, more sclerotized in the Eleticinae, with 6 visible sterna, or only 5 visible in several eleticines; sutures entire; only in the Eleticinae is sternum I excavated for reception of the hind coxae as occurs in most Tenebrionoidea; last visible sternum of male emarginate to almost completely divided (in Nemognathinae). Male genitalia with aedeagus elongate, either with 1-2 distal dorsal hooks and 1 ventral endophallic hook or with hooks absent (Nemognathinae); parameres fused at base only or entirely fused (Nemognathinae); phallobase large. Female genitalia short, lacking long membranous tube-like ovipositor (except in the primitive subfamily Eleticinae); gonocoxites variously shaped, gonostyli more or less elongate, borne apically on gonocoxites.

Larvae of Meloidae differ from those of most other Tenebrionoidea in lacking a mola and urogomphi. With 0-2 stemmata; antennae 3 -segmented; with or without a distinct epicranial suture; legs present. Larvae of Eleticinae apparently have typical holometabolic development and are the most similar to those of other tenebrionoids. In the other subfamilies larvae are parasitoids and development is hypermetamorphic with the various instars differing considerably in morphology and behavior. The first-instar larvae of hypermetamorphic blister beetles are heavily sclerotized, campodeiform, prognathous and highly mobile; they are often referred to as triungulins. This instar is followed by four grub-like, weakly sclerotized and hypognathous scarabaeiform instars referred to as feeding grubs. The feeding grub is followed by instar VI, the coarctate, mostly a heavily sclerotized diapausing larva with aborted mouthparts and legs. Instar VII, or the second grub, is morphologically similar to V but does not feed. The second grub is followed by a typical exarate pupa. Larvae of the Meloidae have been reviewed by Cros (1940), Bologna (1991), and Selander (1991b); MacSwain (1956) published a monograph on first-instar larval morphology with a classification; the first instar of the Eleticinae and its relationship to larvae in the other subfamilies is detailed by Pinto *et al.* (1996).

Habits and habitats. Adult meloids are phytophagous and feed on leaves and flowers of several families of plants, particularly Asteraceae, Leguminosae, and Solanaceae; only a few specialized species do not feed. Larvae are parasitoids except in the the Eleticinae where they probably are subcorticolous predators (Pinto *et al.* 1996). Larval hosts of the parasitoid subfamilies include the provisions and immature stages of wild bees and, in the Epicautini and Mylabrini, the eggs of grasshoppers; rare cases of parasitism on sphecid and masarid wasps also occur in certain non-Nearctic genera. Larvae of most Meloinae directly attain their food source; those of Nemognathinae, Tetraonycinae, and certain Meloinae, all bee parasitoids, reach their host nest by phoresy, attached to the adult bees visiting flowers. Larvae of several species of the Old World tropical genus *Cyaneolytta* are phoretic on carabid beetles, although their actual host remains unknown (Bologna *et al.* 1990). Similarities in parasitic habits and features of first-instar larval morphology and behavior in the Meloidae, Ripiphoridae and Strepsiptera are convergent.

The blood and softer parts of the body of Meloidae contain a vesicating substance called cantharidin. This defensive material is commonly released through reflex bleeding when adult beetles are disturbed. Cantharidin is produced by the male. It is transferred to the female during copulation and also is present in larvae (Carrel *et al.* 1993, Dettner 1997). It is used medicinally to a limited extent but can be dangerous, causing external blisters or internal hemorrhaging and death if ingested by man or animals. Alfalfa hay containing blister beetles is known to cause acute illness and death in horses (Schoeb and Panciera 1978). Certain species, particularly those of *Epicauta,* may cause localized economic damage to various crops.

Status of the classification. The most recent family classifications are by Bologna (1991) and Selander (1991a). Bologna's classification, followed here, recognizes four subfamilies. Selander treats three subfamilies, placing the tetraonycines in the Nemognathinae. A recent cladistic analysis supports four subfamilies (Bologna and Pinto 2001). Classification at the tribal level remains unsettled.

Most Nearctic taxa of Meloidae have been revised relatively recently. Exceptions are *Gnathium, Pyrota,* and the subgenus *Macrobasis* of *Epicauta.* Also the revisions of the nemognathine genera *Nemognatha, Zonitis, Rhyphonemognatha* and *Pseudozonitis* by Enns (1956) only address the United States species. The Mexican fauna has not been treated since Champion (1891-93). The West Indian fauna was reviewed by Selander and Bouseman (1960).

A key to the New World genera was recently published (Pinto and Bologna 1999). The key presented below treats the Nearctic genera but should function for all from North and Central America. The only genus included that may not occur in the Nearctic is *Denierota.*

Distribution. This family is widespread throughout the world except in New Zealand and the Antarctic. Diversity is greatest in arid and semiarid regions. There currently are 120 genera and about 2500 species assigned to the family. Approximately 410 species are known from the Nearctic. Of the 22 Nearctic genera, only five (*Meloe, Epicauta, Lytta, Nemognatha* and *Zonitis*) occur in the Old World. There also is relatively little generic overlap with the Neotropical Region (Pinto and Bologna 1999).

Key to the Nearctic Genera

1. Dorsal blade of tarsal claws with a single or double row of teeth along ventral margin (Fig. 13, 14) ... 2
— Dorsal blade of tarsal claws smooth along ventral margin, or only with a single large tooth (Fig. 15, 16) ... 9

2(1). Orange beetles with black fasciae or rows of spots on elytra. Head very broad, greatly bulged at temples; labrum narrow and very short, not extending beyond base of mandibles. Large beetles, body length 15-35 mm. Known from central Mexico south ... [*Cissites*]
— Color pattern not as above. Head variable, mostly not noticeably bulged at temples; labrum longer, extending to middle of mandibles or beyond. Body length variable but mostly less than 15 mm ... 3

3(2). Antennae relatively short, submoniliform, antennomeres IV and V only slightly longer than wide (Fig. 9). Black beetles with brick red elytra. Dorsal blade of tarsal claws with a single row of teeth ventrally ... *Tricrania*

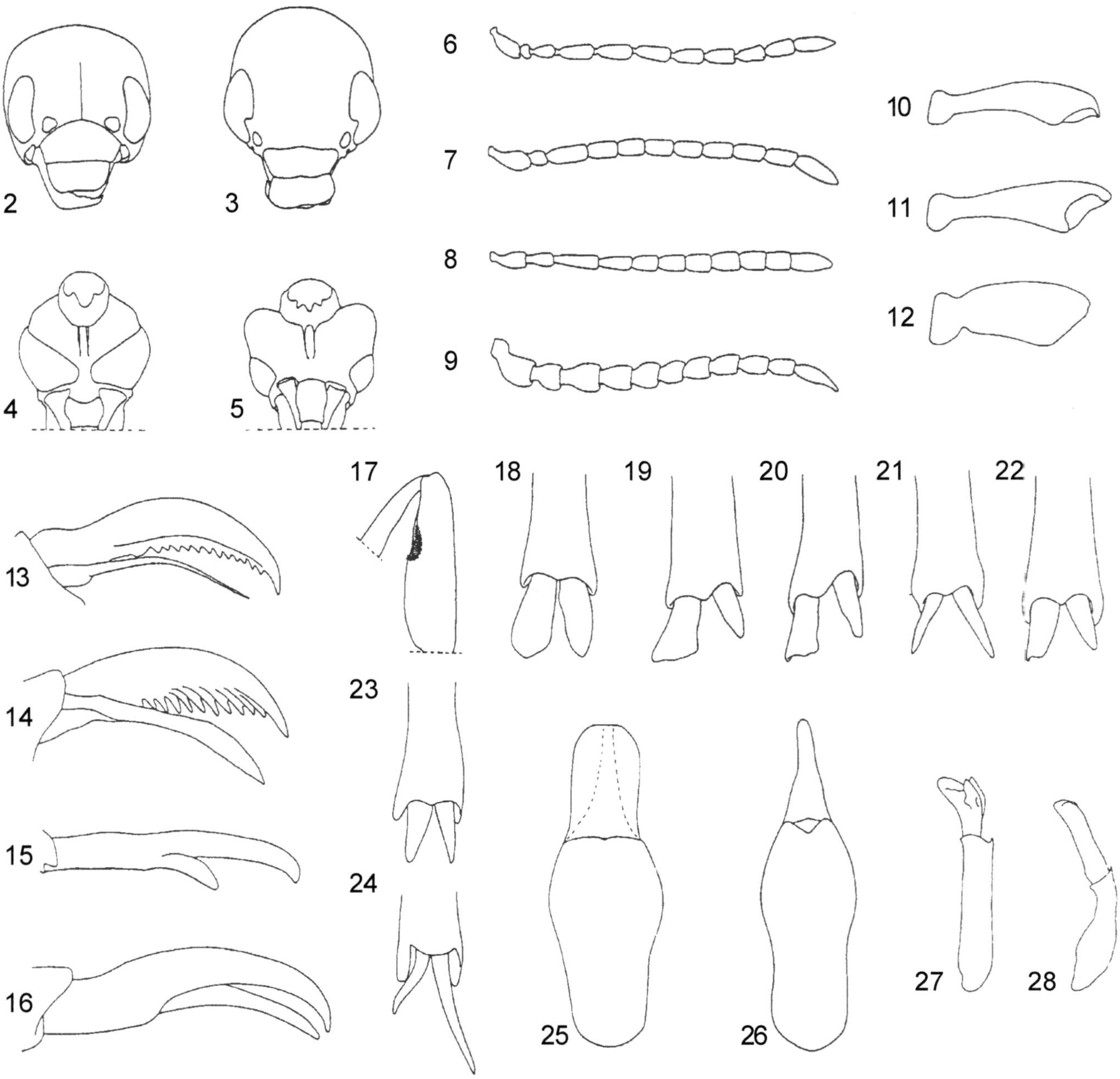

FIGURES 2.111-28.111. 2. *Phodaga marmorata* (Casey), head; 3. *Eupompha elegans* LeConte, head; 4. *Pseudozonitis vaurieae* Enns, head (underside); 5. *Zonitis dunniana* Casey, head (underside); 6. *Denierota kraatzi* Haag-Rutenberg, antenna; 7. *Eupompha elegans*, antenna; 8. *Gnathium minimum* (Say), antenna; 9. *Tricrania sanguinipennis* (Say), antenna; 10. *Pyrota palpalis* Champion, antennomere I (lateral at bottom); 11. *Pyrota quadrinervata* (Herrera and Mendoza), antennomere I (lateral at bottom); 12. *Lytta biguttata* LeConte, antennomere I (lateral at bottom); 13. *Tricrania sanguinipennis*, claw (medial view); 14. *Nemognatha nigripennis* LeConte, claw (medial view); 15. *Phodaga alticeps* LeConte, claw (medial view); 16. *Pyrota palpalis*, claw (lateral view, base of ventral blade behind dorsal blade); 17. *Epicauta oregona* Horn, forefemur; 18. *Zonitis dunniana*, hind tibial spurs (= hts); 19. *Nemognatha lurida* LeConte, hts; 20. *Nemognatha lutea* LeConte, hts; 21. *Nemognatha nigripennis*, hts; 22. *Nemognatha piazata* (Fabricius), hts; 23. *Rhyphonemognatha rufa*, hts; 24. *Linsleya convexa* (LeConte), hts; 25. *Pseudozonitis vaurieae*, male gonoforceps (ventral view, apex at top); 26. *Zonitis dunniana*, male gonoforceps (ventral view, apex at top); 27. *Zonitis dunniana*, aedeagus; 28. *Nemognatha lurida*, aedeagus.

— Antennae more elongate, antennomeres IV and V at least 1.4x as long as wide (*e.g.*, Fig. 8). Color pattern not as above. Dorsal blade of tarsal claws with one or two rows of teeth ventrally........... 4

4(3). Antennae subclavate with apical antennomeres wider than basal antennomeres (Fig. 8). Head elongate, varying from almost as long as wide to longer than wide; maxillary galeae elongate to form a distinct sucking tube. Small beetles, body length 6 mm or less .... *Gnathium*

— Antennae subfiliform, apical antennomeres not wider than basal antennomeres. Head mostly not as elongate, if so then maxillary galeae not modified to form a distinct sucking tube. Size variable but body length mostly greater than 6 mm 5

5(4). Maxillary galeae produced into a filiform sucking tube which is at least as long as mandibles 6

— Maxillary galeae short, lobiform or penicillate, not produced into a filiform sucking tube 7

6(5). Aedeagus of male genitalia with a pair of large apicoventral lobes (Fig. 27). Hind tibial spurs similar, spatulate, subequal in width (Fig. 18) (in part) .... *Zonitis*

— Aedeagus of male genitalia without a pair of large apicoventral lobes (Fig. 28). Hind tibial spurs variable, similar or not but both rarely spatulate (Fig. 19-22) [the 4 species of *Nemognatha* (*Pronemognatha*) are exceptions] .... *Nemognatha*

7(5). Aedeagus of male genitalia with a pair of large apicoventral lobes (Fig. 27). Hind tibial spurs mostly similar, spatulate, concave dorsally, subequal in width (Fig. 18) 8

— Aedeagus of male genitalia without a pair of large apicoventral lobes at apex (as in Fig. 28). Hind tibial spurs not spatulate (Fig. 23) .... *Rhyphonemognatha*

8(7). Fused gonostyli of male genitalia narrowing gradually or not at all apically (Fig. 25), distinctly recurved in lateral view. Eyes mostly large, protuberant, extensively produced on underside of head (Fig. 4). Antennae elongate, slender, antennomere X about one-fourth as wide as long .... *Pseudozonitis*

— Fused gonostyli of male genitalia narrowing considerably to apex (Fig. 26), not distinctly recurved in lateral view. Eyes smaller, not as protuberant, not as extensively produced on underside of head (Fig. 5). Antennae variable but commonly shorter and with antennomere X considerably wider (in part) .... *Zonitis*

9(1). Forefemora with apical half of ventral surface (that opposing tibia) slightly excavated, excavation with a patch of appressed, transversely directed silky pubescence (Fig. 17). .... *Epicauta*

— Forefemora without a ventroapical excavation and lacking a patch of transversely directed pubescence 10

10(9). Ventral blade of claws fused to dorsal blade and shorter, resulting in claws appearing toothed (Fig. 15) 11

— Ventral blade of claws free from dorsal blade and mostly subequal in length (Fig. 16), claws never appearing toothed 20

11(10). Head black with a distinct orange frontal macula. Elytra entirely black. Central California (in minor part) .... *Lytta*

— Head variable but never black with an orange frontal macula. Elytra variable in color 12

12(11). Claws with a curved blade-like spine (ungual spine) arising from base of lateral surface between dorsal and ventral blades. Small beetles (<10 mm in length) with elytra uniformly black or brown and pronotum red .... *Spastonyx*

— Claws without an ungual spine. Size and color variable, but never with combination of uniformly black or brown elytra and red pronotum 13

13(12). Hind wings absent. Elytra abbreviated 14

— Hind wings present. Elytra of normal length 16

14(13). Elytra greatly inflated, fused along suture .... *Cysteodemus*

— Elytra not inflated, entirely separate along suture 15

15(14). Head and elytra entirely black. Southwestern North America, but absent from Chihuahuan Desert ... . *Cordylospasta* (females)

— Head and elytra black but with red, orange or yellow coloration present at least on elytra. Restricted to Chihuahuan Desert of USA and Mexico .... *Megetra*

16(13). Antennal sockets on frons placed dorsomedial to lateral margins of clypeus (Fig. 2). Tarsomere I of foretarsi flanged in male .... *Phodaga*

— Antennal sockets on frons placed directly above lateral margin of clypeus (Fig. 3). Tarsomere I of foretarsi modified or not in male, but never flanged 17

17(16). Head, pronotum and elytra entirely dull black .... *Cordylospasta* (males)

— Head pronotum and elytra variously colored, but never entirely dull black - elytra mostly orange, yellow or red at least in part, or entirely metallic blue or green 18

18(17). Elytra with four distinct longitudinal costae on disk, costae distinct at least in basal half, disk never reticulate; elytra yellow with brown to black apical and post-median bands, and at least one pair of dark basal maculae .... *Pleuropasta*

— Elytra not as above, if with apical and median dark bands then lacking basal spots and without four distinct longitudinal costae, disk strongly reticulate or not 19

19(18). Elytra strongly reticulate, yellow or yellow brown at least in part, never metallic .... *Tegrodera*

— Elytra mostly not strongly reticulate, but if so then entirely metallic green or blue .... *Eupompha*

20(10). Tarsomeres I-IV of foreleg distinctly bilobed and padded; tarsomere IV wider than long. Pronotum transverse, much wider than long, at least as wide at base as in apical half .... *Tetraonyx*

— Tarsomeres I-IV of foreleg not distinctly bilobed, padded or not; tarsomere IV as long as wide or, more commonly, distinctly longer than wide. Pronotum variable but generally narrower at base 21

21(20). Hind wings absent. Elytra abbreviated. Metasternum reduced, apex of mesocoxae approaching base of metacoxae by a distance not exceeding 1/2 length of metacoxae 22

— Hind wings present. Elytra not obviously abbreviated. Metasternum well developed, mesocoxae separated from metacoxae by a distance at least subequal to length of metacoxae 24

22(21). Elytra strongly abbreviated, shorter than pronotum. Eyes very narrow, transverse, subperpendicular to front of head. Adults entirely subterranean, restricted to nests of solitary bees .... *Hornia*

— Elytra not as strongly abbreviated, distinctly longer than pronotum. Eyes broader, not perpendicular to front of head. Adults not restricted to nest of solitary bees, free living 23

23(22). Head entirely dark (mostly black or blue), without red coloration. Generally distributed........ *Meloe*

— Head not entirely dark, with at least a red macula on vertex. Restricted to central California .... *Lytta sublaevis* (Horn)

24(21). Antennomere III noticeably shorter than IV, V and VI, each at least 1.5x the length of III (Fig. 6). Male with a heavy tubercle at base of tarsomere I of middle tarsus and last visible tergum with an apically dilated median process. Central America, southern Mexico .... [*Denierota*]

— Antennomere III longer relative to IV, V and VI (as in Figs. 7, 8). Male without similar modifications... 25

25(24). Antennae subfiliform. Antennomere I elongate, gradually increasing in width apically, and mostly at least 2.5x as long as wide, with its lateral surface distinctly arcuate (Figs. 10, 11). Pronotum elongate, at least 1.1 x as long as wide, disk smooth and subglabrous. Mesepisterna not meeting at midline. Most species with head, pronotum and elytra with bold orange and black coloration .... *Pyrota*

— Antennae heavier. Antennomere I shorter, more robust, less than 2.5x as long as wide, lateral surface not distinctly arcuate (Fig. 12). Pronotum broader, if rarely about 1.1 x as long as wide then disk not smooth and subglabrous. Mesepisterna meeting at midline. Only a few species with bold orange and black coloration 26

26(25). Hind tibial spurs narrow and spiniform (Fig. 24). Antennomere I longer than III, VI-X more or less transverse, rarely slightly longer than wide...... .... *Linsleya*

— Hind tibial spurs dissimilar, outer spur obliquely truncate apically (as in Figs. 19, 20), if both spurs rarely narrow and spiniform then antenna with antennomere I shorter or subequal to III and VI-X distinctly longer than wide (as in Fig. 7) (major part) .... *Lytta*

Classification of the Nearctic Genera

Meloidae Gyllenhal 1810

Meloinae Gyllenhal 1810

Diagnosis. Tarsomeres I-IV of forelegs not bilobed, length of tarsomere IV subequal to its width or greater; tarsal claws of North American taxa with ventral margin of dorsal blade smooth or with only a single large tooth (Eupomphini, *Spastonyx* and a few *Lytta*); male genitalia with aedeagus having 1-2 dorsal hooks and 1 ventral (endophallic) hook, and parameres separated apically; maxillary galeae unmodified. Copulation linear and prolonged. Oviposition in excavated burrows in soil at sites not necessarily adjacent to hosts. First instar larvae of most genera not phoretic (exceptions in Nearctic are *Meloe* and *Spastonyx*).

Pyrotini MacSwain 1956

[*Denierota* Kaszab 1959, 2 spp., southern Mexico and Central America.]

Pyrota Dejean 1834, 40 spp., restricted to New World, about 30 in Nearctic; southwestern, central and eastern U.S., south into Mexico. (key to spp. of five species groups, Selander 1982a, b, 1983a, b).

Eupomphini LeConte 1862

The Eupomphini are the only tribe of Meloidae confined to the Nearctic.

Cordylospasta Horn 1875, 2 spp., western U.S. (key to spp., Pinto 1972)
Gynaecomeloe Wellman 1910
Brachyspasta Van Dyke 1928

Cysteodemus LeConte 1851, 2 spp., southwestern U.S., northern Mexico. (key to spp., Pinto 1984).

Eupompha LeConte 1858, 12 spp., southwestern U.S. and Mexico (key to spp., Pinto 1979; additional sp., Pinto 1983).
Calospasta LeConte 1862

Megetra LeConte 1859, 3 spp., southwestern U.S., northern and central Mexico. (key to spp., Selander 1965).

Phodaga LeConte 1858, 2 spp., southwestern U. S., northern Mexico. (key to spp., Pinto 1984).
Negalius Casey 1891

Pleuropasta Wellman 1909, 2 spp., southwestern U.S., northern Mexico. (key to spp., Pinto 1984).

Tegrodera LeConte 1851, 3 spp., southwestern U.S., northwestern Mexico. (key to spp., Pinto 1975).

Epicautini Denier 1935

Epicauta Dejean 1834, 173 spp., widely distributed in North and Central America, but primarily southwestern U.S. and Mexico. Two subgenera recognized, the nominate subgenus and *Macrobasis* LeConte 1862. (key to spp., Pinto 1991).

Causima Dejean 1836
Henous Haldeman 1852
Isopentra Mulsant 1858
Apterospasta LeConte 1862 (syn. of *Macrobasis)*
Nomaspis LeConte 1866
Pleuropompha LeConte 1866
Gnathospasta Horn 1875 (syn. of *Macrobasis)*
Anomalonychus Saylor 1940 (repl. name for *Anomalonyx* Denier, not Weise 1903)
Maculicauta Dillon 1952

Linsleya MacSwain 1951, 5 spp., restricted to western U.S. and northern Mexico. (key to spp., Selander 1955).

Lyttini Solier 1851

Lytta Fabricius 1775, 69 spp., widely distributed in Nearctic, but primarily western U.S. and Mexico. Four Nearctic subgenera, *Pomphopoea* LeConte 1862, *Poreospasta* Horn 1868, *Adicolytta* Selander 1960, and *Paralytta* 1960. The nominate subgenus is Eurasian. (key to spp., Selander 1960; additional sp., Pinto 1985).

Meloini Gyllenhal 1810

Meloe Linnaeus 1758, 22 spp., New World, all but one occurring in the Nearctic. Ranges from Alaska and northern Canada south into Mexico; also West Indies. Three Nearctic subgenera, the nominate subgenus, *Eurymeloe* Reitter 1911, and *Treiodous* Dugés 1869. (key to spp., Pinto and Selander 1970; additional sp., Pinto 1998).
Proscarabaeus Schrank 1781
Melittophagus Kirby 1818
Triungulinus Dufour 1828
Cnestocera Thomson 1859

Incertae Sedis

Spastonyx Selander 1954, 2 spp., southwestern U.S., northwestern Mexico. (key to spp., Selander 1954, as subgenus of *Calospasta).* This genus originally was considered a member of the Eupomphini and later was transferred to the Meloini by Pinto and Selander (1970). Its tribal placement is unclear, however (Bologna and Pinto 2001); larval and adult traits suggest relationship to the South American *Lyttomeloe* Denier 1921(Selander 1988).

Tetraonycinae Bøving and Craighead 1931

Diagnosis. Tarsomeres I-IV of forelegs mostly distinctly bilobed and padded beneath, tarsomere IV wider than long; tarsal claws with ventral margin smooth; male genitalia with aedeagus hooked or not and with parameres separated apically; maxillary galeae unmodified. Copulation dorsoventral and brief. Oviposition occurring on vegetation, or rarely (in Neotropical taxa) in nests of host bees. First instar larvae phoretic. Restricted to New World.

Tetraonycini Bøving and Craighead 1931

Tetraonyx Latreille 1805, 6 spp., Nearctic (most spp. Neotropical), southern U.S. and Mexico; also West Indies. (key to spp. in southwestern U.S., Werner *et al.* 1966; spp. cat., Selander 1983c). Two subgenera are recognized; all Nearctic species assigned to the nominate subgenus.
Iodema Pascoe 1862
Jodema Gemminger and Harold 1870 (unjustified emendation)

Nemognathinae Castelnau 1840

Diagnosis. Tarsomeres I-IV of forelegs not bilobed, length of tarsomere IV not wider than long; tarsal claws with ventral margin of dorsal blade with a single or double row of small teeth (*Hornia* an exception); male genitalia with aedeagus without hooks and parameres separated apically (Horiini: *Cissites*) or not (Nemognathini); maxillary galeae mostly modified, penicillate or prolonged to form a distinct sucking tube. Copulation dorsoventral and brief. Oviposition occurring on vegetation or, in *Cissites, Hornia* and *Tricrania,* in or adjacent to nests of host bees. First instar larvae phoretic.

Horiini Latreille 1802

[*Cissites* Latreille 1804, 2 spp., from central Mexico south into South America; also in West Indies. (key to spp., Enns 1958).]

Nemognathini Castelnau 1840

Gnathium Kirby 1818, 14 spp., from middle latitudes in U.S. south into Mexico. (key to spp., MacSwain 1952; additional sp., Dillon 1952; two spp. *nomina dubia* [see MacSwain 1952]).

Hornia Riley 1877, 3 spp., widely distributed (keys to spp., Linsley 1942; also see MacSwain 1958).
Leonidia Cockerell 1900 (repl. name for *Leonia* Dugés 1889, not Baird 1850).

Nemognatha Illiger 1807, 28 spp., widely distributed; also in West Indies. (key to U.S. spp., Enns 1956). Four subgenera in Nearctic, the nominate subgenus, *Meganemognatha* Enns 1956, *Pauronemognatha* Enns 1956, and *Pronemognatha* Enns 1956. (Volume 1, Color Figure 11)

Pseudozonitis Dillon 1952, 17 spp., southern United States, south into Mexico; also in West Indies. (key to U.S. spp., Enns 1956).

Rhyphonemognatha Enns 1956, 1 sp., *R. rufa* (LeConte), Illinois and Nebraska south to Texas and Arizona; three other species from southern Mexico and Central America. Also known from South America.

Tricrania LeConte 1860, 3 spp., widely distributed in U.S.
Tricranioides Wellman 1910

Zonitis Fabricius 1775, 16 spp., widely distributed. (key to U.S. spp., Enns 1956; additional spp., Pinto 2001). Two subgenera in the Nearctic, *Neozonitis* Enns 1956, and *Parazonitis* Enns 1956. The nominate subgenus is Palaearctic. Aksentjev's (1988) elevation of *Neozonitis* to genus is not adopted here.

Bibliography

AKSENTJEV, S. I. 1988. A catalogue of the genus-group taxa of the beetle family Meloidae (Coleoptera). Entomologicheskoe Obozrenie, 67: 569-582 [in Russian]. [English translation, 1989, Entomological Review, 68: 11-26].

BOLOGNA, M. A. 1991. Coleoptera Meloidae, Fauna d'Italia. XXVIII. Calderini. Bologna, xiv + 541 pp.

BOLOGNA, M. A. and J. D. PINTO. 1997. Case 2924. Meloidae Gyllenhal, 1810 and Nemognathinae Castelnau, 1840 (Insecta, Coleoptera): proposed precedence over Horiidae Latreille, 1802. Bulletin of Zoological Nomenclature, 54: 226-230.

BOLOGNA, M. A. and J. D. PINTO. 2001. Phylogenetic studies of Meloidae (Coleoptera), with emphasis on the evolution of phoresy. Systematic Entomology, 26: 33-72.

BOLOGNA, M. A., G. ALOISI and A. VIGNA TAGLIANTI. 1990. Phoretic association of some *Cyaneolytta* Péringuey 1909 with carabids, and morphology of first instar larvae in Meloini (Coleoptera Meloidae). Tropical Zoology, 3: 159-180.

CARREL, J. E., M. H. McCAIREL, A. J. SLAGLE, A. J. DOOM, J. P. BRILL and J. P. McCORMICK. 1993. Cantharidin production in a blister beetle. Experientia, 49: 171-174.

CHAMPION, G. C. 1891-93. Family Meloidae. Pp. 364-464. *In*: F. E. Godman and O. Salvin, eds. Biologia Centrali-Americana, Coleoptera. London. Vol. 4, pt. 2, pls. 17-21.

CROS, A. 1940. Essai di classification des Meloidae algériens. VI Congresso Internacional de Entomologia, Madrid, 1935: 312-338.

DETTNER, K. 1997. Inter- and intraspecific transfer of toxic insect compound cantharidin. Pp. 115-145. *In*: K. Dettner, G. Bauer and W. Völkl, eds. Vertical Food Web Interactions, Ecological Studies, Vol. 130 Springer. Berlin.

DILLON, L.S. 1952. The Meloidae (Coleoptera) of Texas. American Midland Naturalist, 48: 330-420.

ENNS, W. R. 1956. A revision of the genera *Nemognatha*, *Zonitis* and *Pseudozonitis* (Coleoptera, Meloidae) in America north of Mexico, with a proposed new genus. University of Kansas Science Bulletin, 37: 685-909.

ENNS, W. R. 1958. Distribution records of two American species of *Cissites* (Coleoptera - Meloidae). Coleopterists Bulletin, 12: 61-64.

ICZN. 1999. OPINION 1918. Meloidae Gyllenhal, 1810 and Nemognathinae Castelnau, 1840 (Insecta, Coleoptera): given precedence over Horiidae Latreille, 1802. Bulletin of Zoological Nomenclature, 56: 89-91.

LINSLEY, E. G. 1942. Systematics of the meloid genera *Hornia* and *Allendesalazaria* (Coleoptera). University of California Publications in Entomology, 7: 189-206.

MACSWAIN, J. W. 1952. A synopsis of the genus *Gnathium*, with description of new species (Coleoptera, Meloidae). Wasmann Journal of Biology, 10: 205-224.

MACSWAIN, J. W. 1956. A classification of the first instar larvae of the Meloidae (Coleoptera). University of California Publications in Entomology, 12: 1-182.

MACSWAIN, J. W. 1958. Taxonomic and biological observations on the genus *Hornia* (Coleoptera: Meloidae). Annals of the Entomological Society of America, 51: 390-396.

PINTO, J. D. 1972. A taxonomic revision of the genus *Cordylospasta* (Coleoptera: Meloidae) with an analysis of geographic variation in *C. opaca*. Canadian Entomologist, 104: 1161-1180.

PINTO, J. D. 1975. A taxonomic study of the genus *Tegrodera* (Coleoptera: Meloidae). Canadian Entomologist, 107: 45-66.

PINTO, J.D. 1979. A classification of the genus *Eupompha* (Coleoptera: Meloidae). Transactions of the American Entomological Society, 195: 391-459.

PINTO, J. D. 1983. A new *Eupompha* from Baja California with additional information on *E. decolorata* (Horn) (Coleoptera: Meloidae). Pan-Pacific Entomologist, 58(1982): 184-195.

PINTO, J. D. 1984. A taxonomic review of *Cysteodemus* LeConte, *Phodaga* LeConte and *Pleuropasta* Wellman (Coleoptera: Meloidae: Eupomphina) with a new generic synonymy. Proceedings of the Entomological Society of Washington, 86: 127-143.

PINTO, J. D. 1985. A new species of *Lytta* (Coleoptera: Meloidae) from Baja California with notes on relationships in the subgenus *Paralytta*. Coleopterists Bulletin, 39: 335-339.

PINTO, J.D. 1991. The taxonomy of North American *Epicauta* (Coleoptera: Meloidae), with a revision of the nominate subgenus and a survey of courtship behavior. University of California Publications in Entomology, 110: i-x, 1-372 + 40 pls.

PINTO, J. D. 1998. A new *Meloe* Linnaeus (Coleoptera: Meloidae, Meloinae) from southern California chaparral: a rare and endangered blister beetle or simply secretive? Coleopterists Bulletin, 52: 378-385.

PINTO, J. D. 2001. Two new species of *Zonitis* (Coleoptera: Meloidae) from southwestern North America, with comments on generic definitions in the Nemognathinae. Proceedings of the Entomological Society of Washington, 103: 319-324.

PINTO, J.D. and M.A. BOLOGNA. 1999. The New World genera of Meloidae (Coleoptera): a key and synopsis. Journal of Natural History, 33: 569-620.

PINTO, J. D. and R. B. SELANDER. 1970. The bionomics of blister beetles of the genus *Meloe* and a classification of the New World species. Illinois Biological Monographs, 42: 1-222.

PINTO, J. D., M. A. BOLOGNA and J. K. BOUSEMAN. 1996. First-instar larvae, courtship and oviposition in *Eletica*: amending the definition of the Meloidae (Coleoptera: Tenebrionoidea). Systematic Entomology, 21: 63-74.

SCHOEB, T. R. and R. J. PANCIERA. 1978. Blister beetle poisoning in horses. Journal of the American Veterinary Medical Association, 173: 75-77.

SELANDER, R. B. 1954. Notes on the tribe Calospastini, with description of a new subgenus and species of *Calospasta* (Meloidae). Coleopterists Bulletin, 8: 11-18.

SELANDER, R. B. 1955. The blister beetle genus *Linsleya* (Coleoptera, Meloidae). American Museum Novitates, 1730: 30 pp.

SELANDER, R. B. 1960. Bionomics, systematics and phylogeny of *Lytta*, a genus of blister beetles (Coleoptera: Meloidae). Illinois Biological Monographs, 28: i-iv, 1-295.

SELANDER, R. B. 1965. A taxonomic revision of the genus *Megetra* (Coleoptera: Meloidae) with ecological and behavioral notes. Canadian Entomologist, 97: 561-580.

SELANDER, R. B. 1982a. A revision of the genus *Pyrota*. I. The Mylabrina group (Coleoptera: Meloidae). Journal of the Kansas Entomological Society, 55: 665-717.

SELANDER, R. B. 1982b. A revision of the genus *Pyrota*. III. The Pacifica and Nobilis groups (Coleoptera, Meloidae). Transactions of the American Entomological Society, 108: 459-478.

SELANDER, R. B. 1983a. A revision of the genus *Pyrota*. II. The Postica group (Coleoptera: Meloidae). Proceedings of the Entomological Society of Washington, 85: 69-85.

SELANDER, R. B. 1983b. A revision of the genus *Pyrota*. IV. The Tenuicostatis group (Coleoptera, Meloidae). Journal of the Kansas Entomological Society, 56: 1-19.

SELANDER, R. B. 1983c. An annotated catalogue of blister beetles of the tribe Tetraonycini (Coleoptera, Meloidae). Transactions of the American Entomological Society, 109: 277-293.

SELANDER, R. B. 1988. Blister beetles of the genus *Lyttomeloe* (Coleoptera: Meloidae). Journal of the Kansas Entomological Society, 61: 80-101.

SELANDER, R. B. 1991a. On the nomenclature and classification of the Meloidae (Coleoptera). Insecta Mundi, 5: 65-94.

SELANDER, R. B. 1991b. Meloidae (Tenebrionoidea). Pp. 530-534. *In:* F.W. Stehr, ed. Immature Insects, Vol. 2. Kendall/Hunt. Dubuque, IA.

SELANDER, R. B. and J.K. BOUSEMAN. 1960. Meloid beetles (Coleoptera) of the West Indies. Proceedings of the United States National Museum, 111: 197-226.

WERNER, F. G., W. R. ENNS and F. H. PARKER. 1966. The Meloidae of Arizona. University of Arizona Agricultural Experiment Station Technical Bulletin, 75: 1-96.

112. MYCTERIDAE Blanchard 1845

by Darren A. Pollock

Family common name: The palm and flower beetles

Family synonym: Hemipeplidae Lacordaire 1854

The three subfamilies of Mycteridae are very diverse in appearance, and it is difficult to present a brief diagnosis of adults at the family level. Within the subfamilies Mycterinae and Hemipeplinae, there is moderate structural homogeneity; in Lacconotinae (at a world level) there is an amazing structural diversity. Among North American taxa, Mycteridae share the following features: antennomeres relatively short, ranging from submoniliform/triangular (*Hemipeplus*) to slightly elongate (*Lacconotus*) to serrated/subflabellate (*Mycterus*); prosternum sunken anterad of procoxae; elytra with apical binding patch on ventral surface, often visible dorsally as elliptical area of contrasting color; males with raised area and/or patch of setae on abdominal ventrite(s) (absent from Hemipeplinae) aedeagus with distinct, paired, parameres.

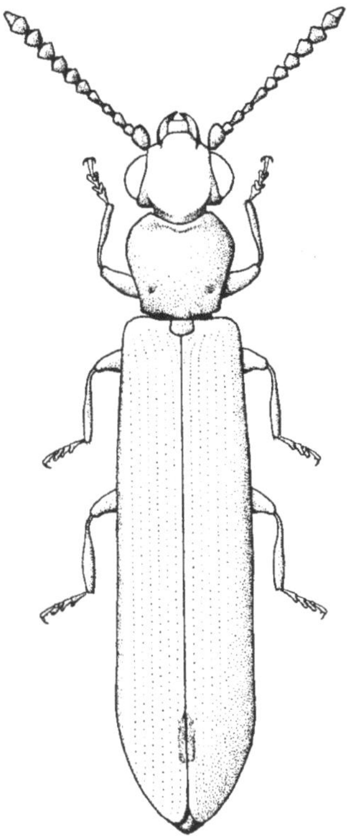

FIGURE 1.112. *Hemipeplus chaos* Thomas (from Thomas 1985)

Description: Body elongate, rather stout and convex (*Mycterus*), slightly ovate and somewhat depressed (*Lacconotus*) or distinctly parallel-sided and flattened (*Hemipeplus*); body from 2.5 to over 4 times longer than maximum width; dorsum with relatively dense vestiture of short, adpressed setae; total length approximately 2.5-9.0 mm; greatest (elytral) width about 0.5-3.0 mm.

Head short or elongate and variously rostrate (*Mycterus*), not, or distinctly (*Hemipeplus*) narrowed behind eyes; tempora present (some *Hemipeplus*) or absent; eyes distinct, relatively small and protuberant (*Lacconotus* and *Hemipeplus*) or larger and less convex (*Mycterus*), entire or only very slightly emarginate anteriorly; facets coarse to fine, without distinct interfacetal hairs; antennae short to moderately long; antennomeres submoniliform to filiform and distal antennomeres gradually widened, exhibiting sexual dimorphism in males of *Mycterus* (variously flattened and serrated); antennomere 11 constricted near midlength; antennal insertions not, or only very slightly concealed by lateral extension of frons; frontoclypeal suture indicated laterally only (*Hemipeplus* and *Lacconotus*); translucent strip of cuticle present over base of transverse labrum; anterior margin of labrum evenly and slightly convex to slightly emarginate; mandibles slightly asymmetrical, terebral teeth present or absent; molae present, with relatively smooth surface; ventral groove with row of microtrichia absent; prostheca finger-like; maxillary palpi moderately long, distal palpomere from slightly expanded, securiform, to nearly cultriform; gula distinct, flat, sutures subparallel (most species of *Mycterus* and *Hemipeplus*) to convergent anteriorly (*Lacconotus*); labial palpi of similar shape to maxillary palpi, last palpomere less expanded than last maxillary palpomere.

Prothorax subquadrate (*Lacconotus*), campanulate (*Mycterus*), or slightly cordate (*Hemipeplus*), widest near midlength, at hind angles, or distinctly anterior of midlength; disc pronouncedly to slightly flattened, to convex, with or without shallow paired depressions; lateral margins of pronotal disc smooth, carinae absent; anterior margin dorsally with raised, setose band (some *Hemipeplus* only); posterolateral area of disc with pair of small, deep pits; prosternum long anterior of coxae, somewhat flattened; intercoxal process short, triangular, somewhat sunken, not extended between coxae, or narrow, elongated, extended well between coxae (some *Hemipeplus*); coxae subglobular to projecting; coxal cavities open, except for some *Hemipeplus*; protrochantins not exposed.

Elytra elongate, parallel-sided to subovate, covering entire abdomen (except one species of *Hemipeplus*, exposing much of last visible tergite); disc with scattered punctation and vestiture, in most taxa short and inconspicuous, adpressed; apicoventral binding patches not, or only slightly visible dorsally through elytral integument; humeri distinct, basal elytral margins prolonged inwards to scutellum; epipleura various, either very narrow and visible only to V3 or V4, to wide and carinate dorsally, distinct to, and around elytral apex; scutellum relatively small, narrowly subtriangular, raised above level of elytra in some taxa (especially *Mycterus*); mesosternum flattened to slightly convex, anteriorly blunt or rounded; mesepisterna narrowly to relatively widely separated; mesosternal process narrow, extended between coxae and in contact with metasternum; mesotrochantin concealed; coxal cavities closed laterally, meso- and metasternum approximate lateral of coxa; hind wing long, functional; venation normal, or slightly reduced (*Hemipeplus*), with 3-4 veins reaching hind margin posterad of MP; metasternum flattened to convex, discrimen distinct, about half length of sternum; anterior margin variously produced, in contact with mesocoxal process; legs similar in size and shape on all thoracic segments; femora slightly

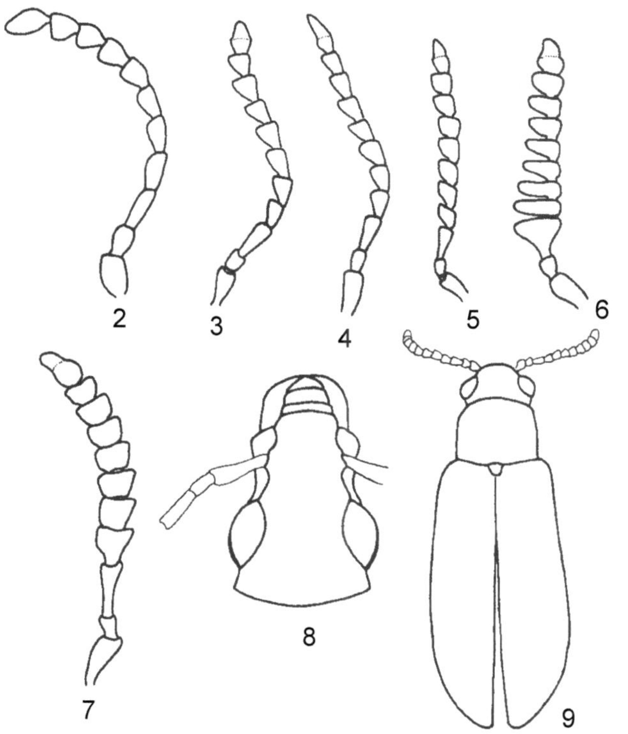

FIGURES 2.112-9.112. 2-7. Antenna. 2. *Lacconotus pallidus* Van Dyke; 3. *Mycterus marmoratus* Pollock; 4. *M. concolor* LeConte; 5. *M. canescens* Horn; 6. *M. scaber* Haldeman; 7. *M. quadricollis* Horn; 8. *Mycterus concolor* (female), outline of head; 9. *Laccontus punctatus* LeConte, outline of body.

to distinctly swollen distad of midlength; tibiae slender, relatively straight, spurs small and inconspicuous, equal in size; tarsi 5-5-4 in both sexes; penultimate tarsomere expanded laterally, forming fleshy ventral lobe; antepenultimate tarsomere slightly expanded in some species; ventral pubescence sparse to dense; tarsal claws with slight basal swelling, or short but distinct basal tooth.

Abdomen with five ventrites, V1 and V2 connate; males of *Lacconotus* and *Mycterus* with sex patch on V1, V2, or V1-V3, consisting of glabrous protuberance or tuft of setae; anterior process of V1 narrow and elongate (*Hemipeplus*) or bluntly triangular; male genitalia with aedeagus consisting of two distinct pieces, slender (*Hemipeplus* and *Mycterus*) to short and stout (*Lacconotus*); parameres with inner margins variously sinuate, without setae; median lobe subparallel to slightly tapered (*Hemipeplus* and *Mycterus*) or distinctly expanded basally (*Lacconotus*); female with ovipositor elongate, and flexible.

Larva (from Lawrence 1991) elongate, parallel-sided, distinctly flattened dorsoventrally; integument slightly sclerotized except for head and urogomphal plate; vestiture of scattered setae. Head protracted and prognathous; epicranial stem short or absent; frontal arms lyriform and contiguous at base; median endocarina absent in most groups, extending anterad of epicranial stem in Hemipeplinae; stemmata 2 or 5 on each side; antennae 3-segmented, in most groups with short, dome-like sensorium apically on antennomere 2; frontoclypeal suture absent; mandibles symmetrical or slightly asymmetrical; mola in most groups small, either smooth, tuberculate, or (in Hemipeplinae) distinct and transversely ridged; ventral mouthparts slightly retracted; maxilla with longitudinally oblique cardo, short and stout stipes; maxillary articulating area distinct; palp 3-segmented with rounded mala, cleft distally, with or without inner apical tooth; ligula distinct, longer than 1-segmented labial palpi (2-segmented in *Mycterus*); hypostomal rods distinct, elongate and divergent posteriorly. Thorax slightly narrower than abdomen; legs relatively short, 5-segmented; coxae widely separated; tarsungulus with 2 distal setae. Abdomen with terga and sterna 2-6 each with paired rows of asperities, forming incomplete, longitudinally oriented rings (note: these asperities absent in many non-Nearctic species of Lacconotinae); segment 8 slightly to distinctly longer than 7; tergite 9 distinctly sclerotized, forming hinged plate; urogomphi distinct, with or without pair of pits between them; median process present in some species; tergite 9 in some groups with distinct median suture, dividing it longitudinally; sternite 8 partly enclosing sternite 9; sternite 9 deeply emarginate posteriorly, forming U-shaped sclerite, enclosing segment 10; 1 to 3 asperities present at each end of sternite 9; spiracles annular-multiforous; peritreme with a perimeter of small accessory openings, or with cluster of openings on one side.

Habits and habitats. Little is known about the biology or natural history of Mycteridae. Larvae and adults of *Hemipeplus* are associated with grasses and palms (Thomas 1985), where they feed on unopened vegetative material. One species of *Hemipeplus* has been implicated as a pest of palms (Lepesme 1947). Adults of most species of *Mycterus* are mainly collected at flowers of *Ceanothus*, *Daucus*, *Yucca*, etc., where they can be very abundant. No larval habits are known for North American *Mycterus* species, although the larva of the Palearctic species *M. curculionoides* (Fabricius) was described by Crowson and de Viedma (1964) from under dead pine bark. Relatively few Lacconotinae have had adequate larval descriptions (Costa and Vanin 1977, Costa and Vanin 1984, Pollock 1995, Pollock *et al.* 2000). A western species of *Lacconotus* was collected under bark of dead poplar (Lawrence 1991); specific habits of adult *Lacconotus* remain unknown. Based on other published and unpublished data, it appears that most if not all larvae of Lacconotinae live under dead tree bark.

Status of the classification. As with many groups of Tenebrionoidea, the components of Mycteridae have only recently had a stable classification, brought about mainly by inclusion of larval characters in their systematics. What are now considered Mycteridae were placed in Melandryidae (*e.g.,* Leng 1920), Pythidae (Blair 1928), or Salpingidae (Arnett 1963). Crowson and de Viedma (1964) proposed a separate family Mycteridae, including *Hemipeplus*, the latter of which was on occasion treated as a separate family (*e.g.*, Crowson 1955). Analyses of world genera and species of Lacconotinae are underway (Pollock, unpublished). The relationships among the three subfamilies as well as the relationship of Mycteridae to other Tenebrionoidea are unknown, but it is likely that Mycteridae have affinities to the salpingid group of families (*sensu* Pollock 1994).

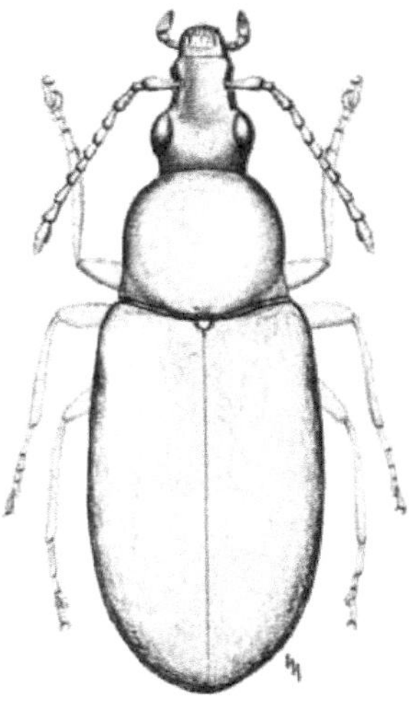

FIGURE 10.112. *Mycterus concolor* LeConte (modified from Hatch 1965).

Distribution. Mycteridae are worldwide in distribution, with approximately 30 genera and 160 species (Lawrence 1982). Many undescribed genera and species are known in collections (Pollock, unpublished). There are only three genera in North America: *Mycterus* (6 species), *Hemipeplus* (3 species), and *Lacconotus* (3 species). *Mycterus* is primarily northern and southern hemisphere temperate, with several Indian species. In North America, species are either western (4 species) or eastern (2 species), with a general gap in the interior of the continent. *Lacconotus* is restricted to the Nearctic region, with two western and one eastern species. It is possible that the two western species will require a new genus, distinct from the eastern *Lacconotus* (s. str.) *punctatus* LeConte (Pollock, unpublished). *Hemipeplus* is found in warm regions worldwide. In North America, the overwhelming majority of records for the three species are from Florida, with few collections in neighboring Gulf states. One species, *H. marginipennis* (LeConte), has been transported into extralimital regions through commerce in palmetto.

Key to the Nearctic Subfamilies and Genera

1. Body distinctly flattened dorsoventrally, very narrow and parallel-sided (Fig. 1); head narrowed distinctly behind eyes, often with tempora; males without patch of setae or tumidity on abdominal ventrite(s) (Hemipeplinae) .... *Hemipeplus*

— Body convex dorsally, elongate oval; head not narrowed distinctly behind eyes, without tempora; males with patch of setae and/or tumidity on one or more abdominal ventrites 2

2(1). Head with distinct rostrum, length between apices of mandibles and line drawn between anterior margins of eyes greater than inner width between eyes (Fig. 8, 10); antennae relatively elongate, antennomeres variously serrated and/or compressed (Figs. 3-7), exhibiting sexual dimorphism; epipleuron wide, carinate dorsally, distinct to elytral apex (Mycterinae) .... *Mycterus*

— Head without rostrum, length between apices of mandibles and line drawn between anterior margin of eyes less than inner width between eyes; antennae relatively short, antennomeres moniliform to slightly filiform (Figs. 2, 9), not exhibiting distinct sexual dimorphism; epipleuron narrow, not carinate dorsally, traceable only to ventrite 4 (Lacconotinae) .... *Lacconotus*

Classification of the Nearctic Genera

Mycteridae Blanchard 1845

Mycterinae Blanchard 1845

Lawrence and Newton (1995) listed *Mycterus* and *Mycteromimus* Champion (Seychelles) as the constituents of Mycterinae. However, analysis of male genitalic characters has revealed that the latter is in fact a lacconotine and not a mycterine (Pollock, unpublished). Adults of *Mycterus* are robust, elongate oval beetles, with a distinct rostrum, and serrated, sexually dimorphic antennomeres. Males have a sexual patch of setae, or merely a raised area, on one or more abdominal ventrites. The aedeagus is elongate, and relatively narrow, with elongate parameres (see Pollock 1993).

Mycterus Clairville 1798, 6 spp., widely distributed (with gap in interior of continent). Key to Nearctic species: Hopping (1935); Pollock (1993). Description of larva: Crowson and de Viedma (1964).

Lacconotinae LeConte 1862

The lacconotines are the most diverse group of Mycteridae, and consist of a pantropical assemblage and a southern hemisphere disjunct group. They are very diverse in body form, and structure; it is difficult to provide an adequate diagnosis for the group. Only a single genus is known from North America.

Lacconotus LeConte 1862, 3 spp., two in western (British Columbia, Colorado, California, Arizona, New Mexico) and one in eastern (Quebec, Pennsylvania, Ohio, Michigan) North America. There are numerous differences between the eastern *L. punctatus* LeConte and the two western species *L. pinicolus* Horn and *L. pallidus* Van Dyke; a new genus may be needed for these latter two species. Key to species: Van Dyke (1928). Description of larva Lawrence (1991).

Hemipeplinae Lacordaire 1854

The hemipeplines are the second most diverse subfamily of Mycteridae, although only two genera are recognized: *Hemipeplus* (widespread) and *Holopeplus* Arrow (West Indies). The body form is very narrow and elongate, and dorsoventrally flattened. These beetles are known from palms and grasses, in warmer areas worldwide.

Hemipeplus Latreille 1825, 3 spp., Georgia, Florida, Texas, Alabama, and adventive in palmetto in California. Key to Nearctic species: Thomas (1985). Key to New World species: Pollock (1999). Description of larva: Thomas and Woodruff (1986).

Ochrosanis Pascoe 1866

Nemicelus LeConte 1836

Bibliography

ARNETT, R. H., Jr. 1963. The beetles of the United States (a Manual for Identification). Catholic University of America Press., Washington, DC, 1112 pp.

BLAIR, K. G. 1928. Pars 99: Pythidae. *In:* W. Junk and S. Schenkling, eds. Coleopterorum Catalogus, 17: 1-56.

COSTA, C. and S. A. VANIN. 1977. Larvae of Neotropical Coleoptera. I: Mycteridae, Lacconotinae. Papéis Avulsos de Zoologia, 31: 163-168.

COSTA, C. and S. A. VANIN. 1984. Larvae of Neotropical Coleoptera. X: Mycteridae, Lacconotinae. Revista Brasileira de Zoologia, 2: 71-76.

CROWSON, R. A. 1955. The natural classification of the families of Coleoptera. N. Lloyd. London.

CROWSON, R. A. and M. G. de VIEDMA. 1964. Observations on the relationships of the genera *Circaeus* Yablok. and *Mycterus* Clairv. with a description of the presumed larva of *Mycterus*. Eos, 40: 99-107.

HATCH, M. H. 1965. The beetles of the Pacific Northwest. Part IV: Macrodactyles, Palpicornes, and Heteromera. University of Washington Publications in Biology, 16:viii +268 pp.

HOPPING, R. 1935. Revision of the genus *Mycterus* Clairv. (Coleoptera, Pythidae). Pan-Pacific Entomologist, 11: 75-78.

LAWRENCE, J. F. 1982. Coleoptera. Pp. 482-553. *In:* S. P. Parker, ed. Synopsis and Classification of Living Organisms. Volume 2. McGraw-Hill. New York.

LAWRENCE, J. F. 1991. Mycteridae (Tenebrionoidea) (including Hemipeplidae). Pp. 535-537. *In:* F. W. Stehr, ed. Immature Insects. Volume 2. Kendall/Hunt. Dubuque, IA. xvi + 975 pp.

LAWRENCE, J. F. and A. F. NEWTON, Jr. 1995. Families and subfamilies of Coleoptera (with selected genera, notes, references and data on family-group names). Pp. 779-1006. *In:* J. Pakaluk and S. A. Slipinski, eds). Biology, Phylogeny, and Classification of Coleoptera. Papers Celebrating the 80th Birthday of Roy A. Crowson. Volume 2. Muzeum i Instytut Zoologii PAN. Warsaw.

LENG, C. W. 1920. Catalogue of the Coleoptera of America, North of Mexico. Cosmos Press. Cambridge, MA. x + 470 pp.

LEPESME, P. 1947. Les Insectes des Palmiers. Lechavalier. Paris. 904 pp.

POLLOCK, D. A. 1993. A new species of *Mycterus* Claireville (Coleoptera: Mycteridae) from Florida, with a checklist of described species. Coleopterists Bulletin, 47: 309-314.

POLLOCK, D. A. 1994. Systematic position of Pilipalpinae (Coleoptera: Tenebrionoidea) and composition of Pyrochroidae. Canadian Entomologist, 126: 515-532.

POLLOCK, D. A. 1995. The Antillean *Physcius fasciatus* Pic (Coleoptera: Mycteridae: Lacconotinae): redescription of the adult and description of the larva. Coleopterists Bulletin, 49: 387-392.

POLLOCK, D. A. 1997. Review of the Afrotropical species of *Hemipeplus* Latreille (Coleoptera: Mycteridae: Hemipeplinae). Annales Zoologici, 47: 79-92.

POLLOCK, D. A. 1999. Review of the New World Hemipeplinae (Coleoptera: Mycteridae) with descriptions of new species. Entomologica Scandinavica, 30: 47-73.

POLLOCK, D. A., S. IDE and C. COSTA. 2000. Review of the Neotropical genus *Physiomorphus* Pic (Coleoptera: Mycteridae: Lacconotinae), with description of the larvae of three species. Journal of Natural History, 34: 2209-2239.

THOMAS, M. C. 1985. The species of *Hemipeplus* Latreille (Coleoptera: Mycteridae) in Florida, with a taxonomic history of the genus. Coleopterists Bulletin, 39: 365-375.

THOMAS, M. C. and R. E. WOODRUFF. 1986. Description of the larvae of two species of *Hemipeplus* Latreille (Coleoptera: Mycteridae). Insecta Mundi, 1: 121-124.

VAN DYKE, E. C. 1928. New species of heteromerous coleoptera. Bulletin of the Brooklyn Entomological Society, 23: 251-262.

113. BORIDAE C. G. Thomson 1859

by Darren A. Pollock

Family common name: The conifer bark beetles

Adults of these beetles may be distinguished by their concealed antennal insertions, antennae with abruptly widened, 3-segmented club, distinct lateral pronotal carinae, and lack of conspicuous dorsal vestiture.

FIGURE 1.113 *Boros unicolor* Say (from Berry 1972)

Description: Body elongate, about 3 times longer than maximum width, parallel-sided, convex dorsally; dorsum with distinct, coarse punctation without distinct vestiture; length 8-25 mm; width 2-8 mm; color uniformly brown.

Head slightly elongate, parallel-sided (*Lecontia*) behind eyes or abruptly narrowed (*Boros*) with distinct tempora; eyes slightly (*Lecontia*) to moderately (*Boros*) convex, slightly emarginate anteriorly; antennae relatively short, antennomeres 2-8 moniliform to slightly filiform; antennomeres 9-11 distinctly widened, forming club; antennomeres 9-10 with distal reniform, sensory patches; antennal insertions concealed dorsally, beneath laterally produced frons; frontoclypeal suture indicated laterally only; translucent strip of cuticle present over base of distinctly transverse labrum; mandibles moderate to large, protruding beyond labrum (*Lecontia*) or not (*Boros*), bidentate apically; maxillary palpi short, last palpomere slightly widened, rounded to truncate distally; gula moderately convex, gular sutures moderately long, subparallel; labial palpi short, similar in structure to maxillary palpi; ligula small, distinctly sclerotized.

Pronotum slightly transverse, widest near (*Boros*), or anterior (*Lecontia*) of midlength; disc slightly to distinctly flattened, with (*Lecontia*) or without (*Boros*) slight lateral concavities; lateral oblique groove present on each side in front of posterior margin, which is narrow and beaded; lateral pronotal carinae distinct along entire length, visible or concealed dorsally, extended to, and united with, oblique groove; prosternum anterior of coxae well developed, moderately convex; males of *Lecontia* with dense group of setae near anterior margin; intercoxal process evenly tapered posteriorly, extended to near end of coxal cavity, apex of process slightly curved dorsally; coxae relatively elongate, and projecting below intercoxal process; coxal cavity broadly open internally and externally; legs all similar in shape and size; femora moderately slender, broadest near midlength; tibiae slender, slightly widened distally; tibial spurs distinct, relatively short, more or less equal on each leg; tarsal formula 5-5-4 in both sexes; tarsomeres slender, setose ventrally, not lobed or expanded; tarsal claws simple.

Elytra elongate, from 2 to 3 times longer than maximum width, parallel-sided with broadly rounded combined apices; disc with scattered, coarse punctation, with (*Lecontia*) or without (*Boros*) suggestions of convex intervals; humeri distinct; epipleura distinct, prolonged inside humeri, extended posteriorly to about 4th ventrite; scutellum small, semicircular (*Boros*) to triangular (*Lecontia*), rounded posteriorly; mesosternum slightly to moderately convex, bluntly triangular anteriorly; mesepisterna separated by approximately distance between mesocoxae; intercoxal process broadly triangular, with distinct margins, reaching anterior margin of metasternum; mesotrochantin exposed; mesocoxal cavities open laterally, partly closed by mesepimera; hind wing well developed, radial cell triangular to distinctly wider than long; wedge cell narrow; medial fleck present; four terminal veins in medial region; metasternum distinctly convex, with discrimen from 1/3 to 1/2 length of sternum; anterior margin broadly rounded, in contact with mesocoxal process.

Abdomen with five ventrites, V1 and V2 connate; male genitalia with articulated lateral lobes, setose distally; basale and apicale subequal in length; edophallic armature absent; ovipositor flexible, elongate, with distinct bacula; coxites setose; styli elongate, subcylindrical.

Larvae elongate, parallel-sided, distinctly flattened, lightly sclerotized except for head capsule and abdominal segment 9; thoracic and abdominal sclerites with sinuate parabasal ridges near anterior margins; vestiture of scattered setae; head prognathous, evenly arcuate laterally; epicranial suture with short stem; anterior arms lyriform, complete to near antennal insertions; endocarinae absent; stemmata 5 (*Boros*) or entirely absent (*Lecontia*); antennae elongate, segment 2 with small, conical sensorium; mouthparts retracted, hypostomal rods distinct, posteriorly divergent; mandibles distinctly sclerotized, slightly asymmetrical, left mandible with molar tooth; apices tridentate, with several subapical teeth; cardo appearing 2-segmented from internal fold; maxillary articulating area distinct, pad-like; maxillary mala undivided; maxillary palp 3-segmented; labium with distally rounded ligula and 2-segmented palpi; hypopharyngeal sclerome trans-

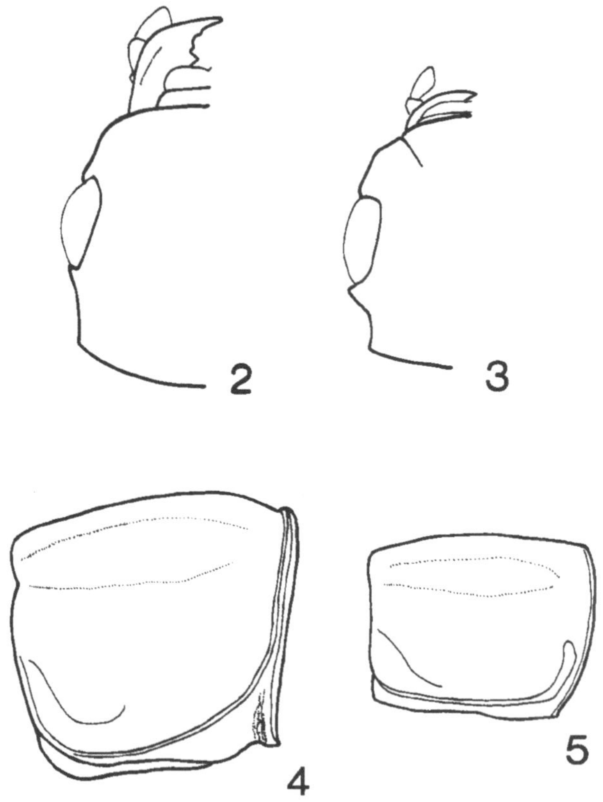

FIGURES 2.113-5.113. 2. *Lecontia discicollis* (LeConte), head; 3. *Boros unicolor* Say, head; 4. *L. discicollis*, pronotum, left lateral; 5. *B. unicolor*, pronotum, left lateral.

verse, rectangular; mentum wider than long with anterior margin slightly emarginate; thorax elongate, parallel-sided; posterior margin of pronotum (*Lecontia*) or anterior margin of mesonotum (*Boros*) with pair of posteriorly directed, flat processes medially; legs distinct, 5-segmented, all similar in size and shape, bearing fine, scattered setae; abdomen distinctly flattened; tergite 9 (urogomphal plate) hinged, distinctly sclerotized; plate divided ventrally by longitudinal suture (*Boros*) or wide sulcus (*Lecontia*); posterior margin of plate with 2 urogomphal pits between urogomphi; 9th sternite broadly U-shaped (*Boros*) or transversely rectangular (*Lecontia*), with series of small dentiform plates (*Boros*) or single asperity on each side (*Lecontia*); thoracic spiracles annular-biforous (*Boros*) or ovate with partially crenulated peritreme (*Lecontia*); abdominal spiracles annular or annular-biforous.

Habits and habitats. Immature stages of *Boros unicolor* Say and the Palearctic species *B. schneideri* (Panzer) are found in the subcortical region of dead (often standing or leaning) coniferous trees (personal observation; St. George 1931; Young 1991). Until recently, larvae of *Lecontia* were known only from very old specimens (Young 1985) with limited associated ecological data. However, it is now known that larvae of *L. discicollis* (LeConte) inhabit the subcortical region of fire-killed coniferous trees (Young *et al.* 1996). The figure of the larva of *L. discicollis* in Peterson (1951) represents *Priognathus monilicornis* Randall (family Pythidae).

Status of the classification. Spilman (1954) considered *Boros* and *Lecontia* to be synonymous; Arnett (1963) included the genera of Boridae in Salpingidae. Pollock (1994) considered Boridae to be a sister group to Pyrochroidae. Lawrence and Pollock (1994) examined the family, and divided it into the Borinae (*Boros* and *Lecontia*) and the Synercticinae (for the Australian genus *Synercticus* Newman).

Distribution. This small family is amphitropical, with representatives in both northern (2 genera, 3 species) and southern (1 genus, 1 species) hemisphere temperate regions. In North America, Boridae are found primarily in the boreal forest region. However, specimens of *Lecontia* are known from as far south as Arizona and northern Mexico. *Boros* is primarily an eastern insect, while *Lecontia* is known from across North America without significant longitudinal gaps.

Key to the Nearctic Genera

1. Head ± abruptly narrowed behind eyes, forming tempora (Fig. 3); pronotal disc evenly convex or only slightly flattened; elytra not striate; mandibles not distinctly protruding from beneath labrum; lateral pronotal carina not joining with posterior bead (Fig. 5) .... *Boros*

— Head ± parallel-sided behind eyes, without tempora (Fig. 2); pronotal disc distinctly flattened in middle and concave on each side; elytra substriate, with at least 5 indistinct striae; mandibles distinctly protruding from beneath labrum; lateral pronotal carina joining dorsally with posterior bead (Fig. 4) .... *Lecontia*

Classification of the Nearctic Genera

Boridae C. G. Thompson 1859

Borinae C. G. Thompson 1859

Boros Herbst 1797. One species, *B. unicolor* Say, relatively widespread in North America, predominantly in the east.

Lecontia Champion 1893. One species, *L. discicollis* (LeConte); relatively widespread in Canada, United States and in northern Mexico (Sierra Madre de Durango).

Crymodes LeConte 1850, not Guénée 1841

Bibliography

ARNETT, R. H., Jr. 1963. The beetles of the United States (a manual for identification). Catholic University of America Press. Washington, DC. 1112 pp.

LAWRENCE, J. F. and D. A. POLLOCK. 1994. Relationships of the Australian genus *Synercticus* Newman (Coleoptera: Boridae). Journal of the Australian Entomological Society, 33: 35-42.

PETERSON, A. 1951. Larvae of insects: an introduction to the Nearctic species. Part II. Coleoptera, Diptera, Neuroptera,

Siphonaptera, Mecoptera, Trichoptera. A. Peterson. Columbus, OH. 416 pp.
POLLOCK, D. A. 1994. Systematic position of Pilipalpinae (Coleoptera: Tenebrionoidea) and composition of Pyrochroidae. Canadian Entomologist, 126: 515-532.
ST. GEORGE, R. A. 1931. The larva of *Boros unicolor* Say and the systematic position of the family Boridae Herbst. Proceedings of the Entomological Society of Washington, 33: 103-113.
SPILMAN, T. J. 1954. Generic names of the Salpingidae and their type species (Coleoptera). Journal of the Washington Academy of Sciences, 44: 85-94.
YOUNG, D. K. 1985. The true larva of *Lecontia discicollis* and change in the systematic position of the genus (Coleoptera: Boridae). Great Lakes Entomologist, 18: 97-101.
YOUNG, D. K. 1991. Boridae (Tenebrionoidea). Pp. 537-539. *In:* F. W. Stehr, ed. Immature Insects. Volume 2. Kendall/Hunt. Dubuque, IA. xvi + 975 pp.
YOUNG, D. K., K. KATOVICH and M. SCHWENGEL 1996. The larval habitat of *Lecontia discicollis* (LeConte) (Boridae). Mola, 6: 2-3.

114. PYTHIDAE Solier 1834

by Darren A. Pollock

Family common name: The dead log beetles

Although relatively few taxa are included in Pythidae, it is difficult to characterize the family in the adult stage. Among known larvae, all taxa have a single urogomphal pit and a transverse row of tubercles across the ninth abdominal segment.

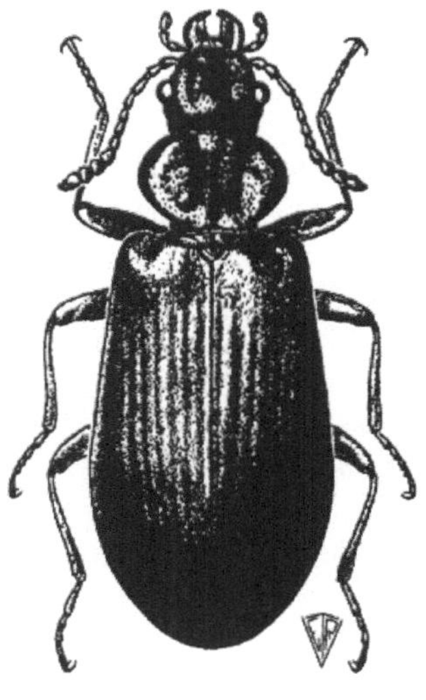

FIGURE 1.14. *Pytho niger* Kirby

Description: Body elongate, subcylindrical to depressed, about three times longer than maximum width; dorsum with distinct punctation of various depth; vestiture indistinct or consisting of few scattered setae only (*Trimitomerus*); total 6-22 mm; width 2-6 mm.

Head subquadrate, slightly elongated (*Pytho*, *Priognathus*) or short; lateral margins subparallel and not narrowed posteriorly of eyes (slight tempora developed in *Priognathus*); eyes small or large (*Trimitomerus*) and variously protuberant, separated widely from antennal insertions (*Pytho*, *Priognathus*), or slightly emarginate around antennal insertions (*Trimitomerus*, *Sphalma*); facets relatively fine, without interfacetal setae; frons depressed medially (except *Sphalma*, and only slightly in *Trimitomerus*); antennal insertions concealed dorsally only in *Trimitomerus* and especially *Sphalma*; frontoclypeal suture absent; antennae of three general types: moniliform (except antennomere 3 filiform), with very slightly developed 3-4 segmented club (*Pytho*, *Priognathus*); subfiliform with antennomeres 7-10 expanded laterally (*Sphalma*); or antennomeres 2-8 moniliform (male), filiform (female), with distinctly elongated antennomeres 9-11 (*Trimitomerus*); labrum relatively short, transverse, slightly to moderately concave anteriorly; mandibles relatively slender and elongate (rather stout in *Sphalma*), visible distinctly anterad of labrum (slightly in *Sphalma*); apices sharply (*Pytho*, *Priognathus*) or bluntly (*Sphalma*) bidentate, or unidentate (*Trimitomerus*); inner margins with distinct terebral teeth (indistinctly developed in *Sphalma*); mola indistinct or very large (*Sphalma*); maxillary palpi moderately elongate, terminal palpomere variously securiform (*Pytho*, *Sphalma*), or only slightly expanded distally (*Priognathus*, *Trimitomerus*); male mentum with pit with projecting setae (except *Trimitomerus*); labial palpi similar in shape to maxillary palpi.

Pronotum rounded or subquadrate, wider than long, widest near midlength; disc evenly and variously convex to slightly flattened, with vague suggestions of, or distinct paired depressions (*Pytho*); lateral margins of disc smooth (*Pytho*, *Priognathus*), with slightly developed carinae (*Trimitomerus*) or with very distinct border (*Sphalma*); prosternum anterior of coxae well developed, slightly to moderately convex; prosternal process short, acute, extended posteriorly to at most half length of coxae; coxae relatively elongate, projecting ventrad of intercoxal process (especially so in *Trimitomerus*); front coxal cavities broadly open internally and externally.

Elytra elongate, more or less parallel-sided (in *Trimitomerus*, tapered posteriorly, and in some specimens of *Pytho*, widened posteriorly), covering all of abdomen or revealing only posterior part of last visible segment; disc smooth, with scattered punctures, or with distinctly raised, longitudinal intervals (*Pytho*); integument translucent in *Trimitomerus*, more distinctly sclerotized in other genera; humeri distinct; epipleura variously developed, either quite short or extended to near apices of elytra; scutellum visible, relatively long and rounded posteriorly, or wide and triangular (*Sphalma*); mesosternum slightly to moderately convex, subtruncate anteriorly; mesepisterna narrowly to moderately widely separated anteriorly; intercoxal process triangular, variously acute, reaching, or not quite, anterior margin of metasternum; mesotrochantins exposed; mesocoxal cavities open laterally, closed partly by mesepimera; hind wing functional; radial cell indistinctly or distinctly closed; wedge cell present, narrow; four terminal veins in medial region; metasternum convex, with discrimen about half length of sternum; anterior margin variously triangular, in contact with mesosternum, or sunken below mesocoxae (*Trimitomerus*); legs similar in size and shape, relatively slender; femora only slightly widened toward midlength; tibiae straight or slightly curved, without stout spines; tibial spurs short and stout to moderately long and slender (*Trimitomerus*), slightly unequal in length; tarsal formula 5-5-4 in both sexes, tarsi slender and simple, without lobes or lateral expansion; ventral surfaces without distinct adhesive setae.

Abdomen elongate, with 5 ventrites; all ventrites free (*Pytho*, *Priognathus*) or ventrites 1 and 2 connate (*Trimitomerus* and *Sphalma*); male aedeagus of tenebrionoid type, divided into distinct basal and apical pieces; apical portion with distinct, elongate paired lateral lobes (very short in *Trimitomerus*, absent in *Sphalma*); lateral lobes setose basally and apically; median lobe narrowed distally, without discernible internal sac sclerites; ovipositor elongate, flex-

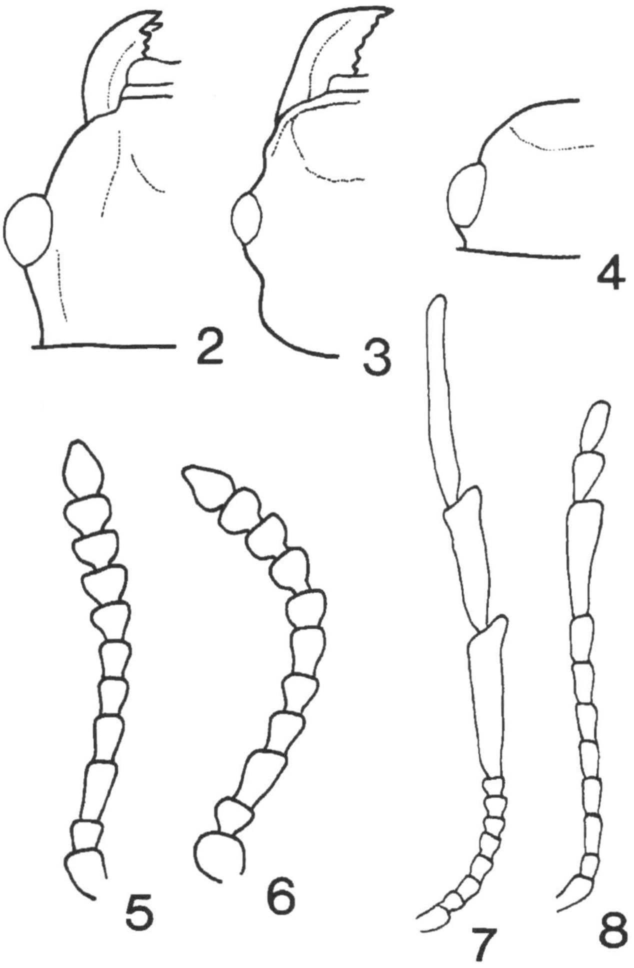

FIGURES 2.143-8.143. 2. *Pytho niger* Kirby, head; 3. *Priognathus monilicornis* (Randall), head; 4. *Sphalma quadricollis* Horn, head; 5. *P. niger*, antenna; 6. *P. monilicornis*, antenna; 7. *Trimitomerus riversii* Horn, male antenna; 8. *T. riversii*, female antenna.

ible, with 4 elongate bacula; coxites variously setose; styli moderately long, cylindrical.

Larvae moderately (*Sphalma*) to distinctly flattened (*Pytho*) or subcylindrical (*Priognathus*), sides subparallel, indistinctly sclerotized except for head and abdominal apex; body surface generally smooth, although with sclerotized ridges in some taxa; vestiture of scattered, elongate setae; head prognathous, frontal arms of epicranial suture lyriform, traceable to near antennal insertions, stem absent or very short; endocarinae absent; stemmata 5 per side; antennae relatively elongate, 3-segmented, with conical sensorium on antennomere 2; hypostomal rods present or absent; mandibles slightly to moderately asymmetrical, moderately stout and not extended anterior of labrum, to elongate and extended distinctly beyond labrum; base of mandibles with distinct mola, molar tooth larger on left mandible; mala cleft apically, with distinct uncus; thorax subparallel-sided, without distinct, sclerotized ridges or asperities, except in *Pytho*, with flat, triangular processes along midline; legs well-developed, 5-segmented, similar on all thoracic segments, with spinose setae; spiracles annular or annular-biforous; abdomen elongate, parallel-sided, variously flattened to almost cylindrical; tergites 1-8 without, or with distinct transverse parabasal ridges with posteriorly directed medial processes (*Pytho*); segments 1-9 subequal in size, segment 10 highly reduced, surrounding anus; tergite 9 distinctly sclerotized, forming urogomphal plate; urogomphi distinct, short (*Priognathus*), moderate (*Sphalma*) or relatively long (*Pytho*), separated by single urogomphal pit; inner margins of urogomphi with several accessory branches or teeth; distinct urogomphal lip present (*Pytho*) or not; dorsum of tergite 9 with transverse row of setiferous tubercles, with or without gap at middle; sternite 9 with double arch of asperities along anterior margin; spiracles annular or annular biforous.

Habits and habitats. Larvae of Pythidae live in the subcortical region of dead coniferous (*Pytho*) or deciduous (*Sphalma*) logs, or in the sapwood of red rotten conifer logs (*Priognathus*) (Pollock 1991; Young 1976, 1991). *Pytho* and *Priognathus* are primarily boreal insects, although the latter can be found as far south as Arizona, at high elevation. Adults of *Trimitomerus riversii* Horn have been collected repeatedly at ultraviolet lights, but the larval stage and natural history remain unknown. The food requirements of Pythidae are poorly known, although larvae of *Pytho* have been reared solely on cambium of conifers, and are apparently xylophagous (Andersen and Nilssen 1978; Pollock 1991). The mandibles of adult *Pytho*, *Priognathus* and *Trimitomerus* are long and slender and exhibit features indicative of a predacious habit, namely the very small mola and numerous terebral teeth. *Sphalma* adults, however, have a very large mola and reduced number of terebral teeth, indicating possibly xylophagous habits. Adults and larvae of *Pytho* have been used in studies of cold tolerance, and have been shown to possess relatively high supercooling points (*e.g.*, Ring and Tesar 1980; Zachariassen 1977).

Status of the classification. The definition of Pythidae as here presented, agrees quite closely with that of Crowson (1955). Arnett (1963) grouped Pythidae within Salpingidae, along with various other families including Mycteridae and Boridae. Recently, Pollock (1994) removed Pilipalpinae and Tydessinae from Pythidae and transferred these to Pyrochroidae. Pollock and Lawrence (1995) provided a key to world genera of Pythidae, tentatively including the aberrant Palearctic genus *Osphyoplesius* Winkler in the family.

Distribution. In the present restricted sense, Pythidae are represented in North America by four genera and eight species. *Sphalma quadricollis* Horn is known only from west of the Rocky Mountains, and *Trimitomerus riversii* Horn seems to be restricted to southern Arizona. *Pytho* and *Priognathus*, are transcontinental. *Pytho* reaches the treeline in the north, and extends into the south in areas of high elevation with coniferous forests. *Priognathus* has a similar distribution, but its range extends even further south, especially in the west, where it occurs in California and Arizona.

Key to the Nearctic Genera

1. Last 3 antennomeres elongate, distinctly modified, much longer than antennomeres 1-8 (male) (Fig.

7) or about 1/2 length of 1-8 (female) (Fig. 8); elytra pale, testaceous, almost translucent; known only from SW United States .... *Trimitomerus*

— Last 3 antennomeres not significantly different than preceding antennomeres, at most forming very slight club (Fig. 5-6); elytra darker, not translucent; widespread 2

2 (1). Antennal insertions concealed by slight canthus; lateral pronotal carina distinct, visible dorsally; mandible short, not distinctly extended anterior of labrum (Fig. 4) .... *Sphalma*

— Antennal insertions visible dorsally; lateral pronotal carina absent; mandibles distinctly extended beyond labrum (Fig. 2-3) 3

3 (2). Pronotal disc with distinct, paired depressions; elytra with variously convex, longitudinal ridges; epipleura short, indistinct posterior of thorax... .... *Pytho*

— Pronotal disc evenly convex or with only vague impressions; elytra without longitudinal ridges; epipleura distinct to first or second ventrite .... *Priognathus*

Classification of the Nearctic Genera

Pythidae Solier 1834

Pytho Latreille 1796, 4 spp., transcontinental in areas of coniferous forest. Key to world species: Pollock (1991); larvae and pupa: Pollock (1991).

Priognathus LeConte 1850, 1 sp., *P. monilicornis* (Randall 1838), transcontinental in areas of coniferous forest. Larva: Peterson (1951, as *Lecontia discicollis*).

Sphalma Horn 1888, 1 sp., *S. quadricollis* Horn 1888, is found in western North America, from British Columbia to California. Larva: Young (1976).

Trimitomerus Horn 1888, 1 sp., *T. riversii* Horn 1888, known from southern Arizona.

Bibliography

ANDERSEN, J. and A. C. NILSSEN. 1978. The food selection of *Pytho depressus* L. (Col., Pythidae). Norwegian Journal of Entomology, 25: 225-226.

ARNETT, R. H., Jr. 1963. The beetles of the United States (a manual for identification). Catholic University of America Press. Washington, DC. 1112 pp.

CROWSON, R. A. 1955. The natural classification of the families of Coleoptera. Nathaniel Lloyd. London. 187 pp.

PETERSON, A. 1951. Larvae of Insects: an Introduction to the Nearctic Species. Part II. Coleoptera, Diptera, Neuroptera, Siphonaptera, Mecoptera, Trichoptera. A. Peterson. Columbus, OH. 416 pp.

POLLOCK, D. A. 1991. Natural history, classification, reconstructed phylogeny, and geographic history of *Pytho* Latreille (Coleoptera: Heteromera: Pythidae). Memoirs of the Entomological Society of Canada, 154: 1-104.

POLLOCK, D. A. 1994. Systematic position of Pilipalpinae (Coleoptera: Tenebrionoidea) and composition of Pyrochroidae. Canadian Entomologist, 126: 515-532.

POLLOCK, D. A. and J. F. LAWRENCE. 1995. Review of *Anaplopus* Blackburn (Coleoptera: Pythidae), with comments on constituents and systematics of Pythidae. Pp. 449-472. *In:* J. Pakaluk and S. A. Slipinski, eds. Biology, phylogeny, and classification of Coleoptera: Papers celebrating the 80th birthday of Roy A. Crowson. Muzeum i Instytut Zoologii. Warsaw.

RING, R. A. and D. TESAR. 1980. Cold-hardiness of the arctic beetle, *Pytho americanus* Kirby. Coleoptera, Pythidae (Salpingidae). Journal of Insect Physiology, 26: 763-774.

YOUNG, D. K. 1991. Pythidae (Tenebrionoidea). Pp. 539-541. *In:* F. W. Stehr, ed. Immature insects. Volume 2. Kendall/Hunt. Dubuque, IA. xvi + 975 pp.

YOUNG, D. K. 1976. The systematic position of *Sphalma quadricollis* Horn (Coleoptera: Salpingidae: Pythini) as clarified by discovery of its larva. Coleopterists Bulletin, 30: 227-231.

ZACHARIASSEN, K. E. 1977. Effects of glycerol in freeze-tolerant *Pytho depressus* L. (Col., Pythidae). Norwegian Journal of Entomology, 24: 25-29.

115. PYROCHROIDAE Latreille 1807

by Daniel K. Young

Family common name: The fire-colored beetles

The heteromerous tarsal formula with at least the penultimate tarsomere lobed below, conical, projecting prothoracic coxae, with cavities that are externally and internally open behind, emarginate compound eyes, and a head that is abruptly constricted behind the eyes, forming a conspicuous "neck" characterize the more common North American pyrochroids.

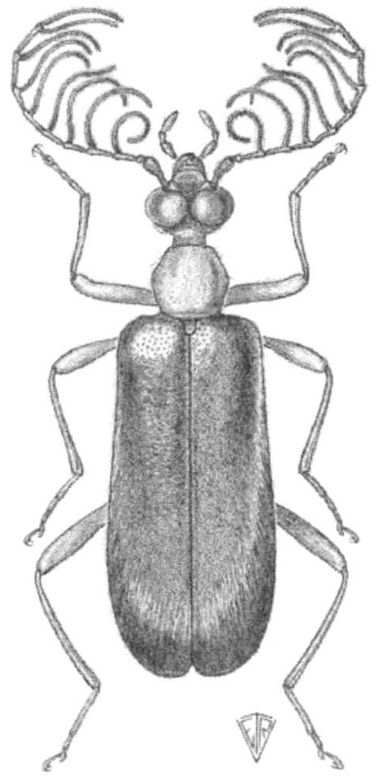

FIGURE 1.115. *Dendroides canadensis* Latreille

Description: Body slightly to moderately flattened; length 4 to 20 mm; color yellowish to black, commonly black with red or yellow thorax; vestiture usually consisting of dense, semierect hairs over surface of body.

Head deflexed, strongly constricted behind the eyes forming a broad neck; surface rugose; antennae with 11 antennomeres; antennae mostly filiform to pectinate (females) or serrate to plumose (males), rarely (*Anisotria*) filiform to submoniliform with last three antennomeres greatly elongated; inserted at side of the frons in emargination of the eyes. Labrum prominent; mandibles short, curved, apices emarginate or bi- to tridentate; maxillary palpi with four palpomeres, the apical palpomere cultriform to slightly securiform; labium with gular process large; mentum trapezoidal, narrowed in front; ligula large, prominent, bilobed; labial palpi with three palpomeres. Eyes moderate in size, emarginate, rarely (*Anisotria*) entire, setose.

Prothorax narrower than basal width of elytra, ovate, margins absent, surface finely punctate, pleural region broad; prosternum large, prosternal process acute; prothoracic coxal cavities open behind externally and internally, trochantins slightly exposed. Scutellum small, shield-shaped. Mesosternum pointed posteriorly, keeled; metasternum long and broad. Legs long, trochantins visible on the mesothoracic coxae; prothoracic coxae confluent, large, conical; mesothoracic coxae confluent, conical; metathoracic coxae transverse, slightly separate, with elongate internal keel; trochanters heteromeroid, angular; femora slender; tibiae with apical spurs small, slender; tarsal formula 5-5-4, penultimate tarsomere small, lobed beneath; claws simple to dentate. Elytra broader than abdomen, apices entire, surface rugulose; epipleural fold almost absent, except at the base. Metathoracic wings with radial cell open or closed, anal cell usually present, subcubital fleck poorly defined to absent; metathoracic wings rarely reduced (*Anisotria*) or absent (*Cononotus*).

Abdomen with five or six ventrites, the last visible ventrite emarginate in the male, sutures distinct. Male genitalia with the penis long, somewhat flattened, with two struts at the base; parameres fused proximally, separate and subparallel to divergent distally; basal piece well developed. Female genitalia with first and second valvifers reduced to rod-like oblique and ventral bacculi, coxites two-segmented, coxital stylus short, stout, setose.

Larva (after Doyen 1979, Mamaev 1976, Young 1991, Young and Pollock 1991) well sclerotized, smooth and shiny; shape depressed, orthosomatic; length 9 to 35 mm; vestiture sparse, consisting of a few setae on most segments; color dull white or yellow with the ninth segment usually more heavily sclerotized. Head exserted, prognathous, nearly width of the thorax, epicranial suture lyriform, stem short to absent, frontal arms complete to antennal insertions; frons and clypeus fused. Labrum distinct; mandibles well developed, asymmetrical, apices bi- to tridentate, curved with a distinct molar area on the right mandible, left mandible with a prominent molar tooth; maxilla with diagonally folded cardo, well developed maxillary articulating area, stipes with fused setiferous mala, bearing a distinct uncus, palpi three-segmented; labium with elongate submentum, subquadrate mentum, and well developed ligula; palpi two-segmented. Thorax elongate, sides subparallel, cervicosternum divided into three plates; legs five-segmented, including tarsungulus. Abdomen with nine visible segments, the eighth segment often nearly two times as long as the seventh; ninth tergum hinged, extending ventrally to form the entire terminal segment; paired, fixed, heavily sclerotized, simple or branched urogomphi present caudally or dorsocaudally on ninth tergum. Spiracles annular, on mesothoracic laterotergite and abdominal segments one to eight.

Habits and habitats. Larvae of Pyrochroinae (Young 1991: 541-544), and at least some *Pedilus* (Young 1991: 544-547), are associated with the somewhat cool, moist subcortical conditions beneath slightly loosened bark and to some extent decaying wood of dead deciduous and coniferous trees. Larvae of the most commonly encountered pyrochroine, *Dendroides canadensis* Latreille, are associated primarily with exposed regions of dead logs, while larvae of *D. concolor* (Newman), *Neopyrochroa flabellata* (Fabricius) and *N. sierraensis* Young are typically found beneath bark and in decaying wood of the undersides of logs, adjacent to, or partially

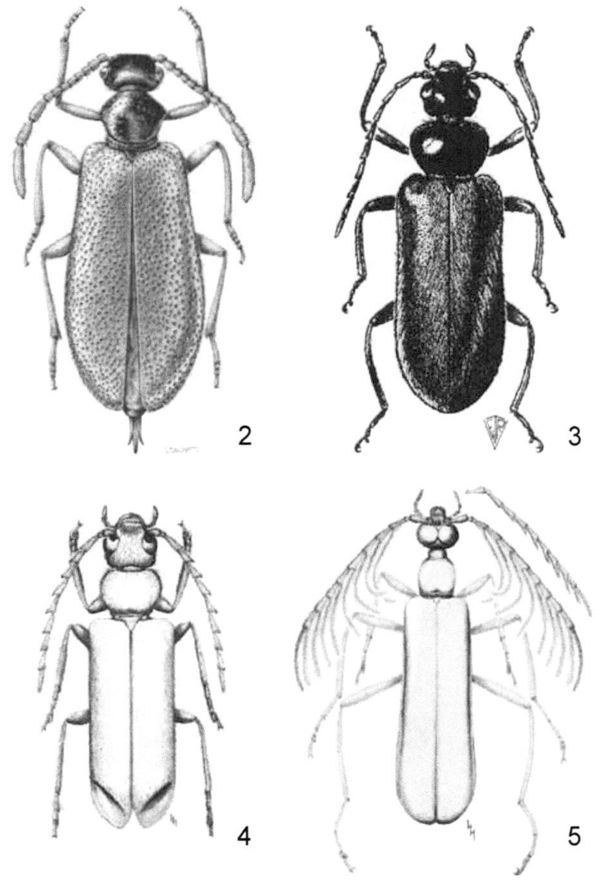

FIGURE 2.115-5.115. *Anisotria shooki* Young, male; 3. *Pedilus lugubris* (Say), female; 4. *Pedilus cavatus* Fall, male; 5. *Dendroides ephemeroides* (Mannerheim), male [inset, female antenna] (Fig. 2 from Young 1984; Figs. 4-5 from Hatch 1965].

buried in the soil. Larvae of *N. femoralis* (LeConte) are sometimes found beneath bark and in decomposing wood within standing dead logs, especially in riparian areas; those of *Schizotus cervicalis* Newman specialize in decaying woody sections of moss-covered logs in boreal-like areas (e.g., edges of bogs). While both woody and fungal materials are found in the gut, fungi are thought to play a key role in larval development. In crowded situations, larvae may sometimes become cannibalistic, but they are not normally predaceous (Young 1975). Larvae of most *Pedilus* are unknown; known larvae have been found within decaying vegetative materials on or in the soil. Larvae of *Cononotus bryanti* Van Dyke were taken in numbers from beneath dried cow dung (Doyen 1979). Adults of *Neopyrochroa* and *Dendroides* are often collected at lights or on forest vegetation while collecting at night with a head lamp; adults of both genera appear to be largely nocturnal. Adults of *Pedilus* are sometimes very common on shrubs and on flowers; those of *Cononotus* appear to be adapted to xeric habitats; they have been found under rocks and in rather dry, decaying vegetation. Adult males of *Schizotus*, *Neopyrochroa*, *Anisotria* and most *Pedilus* are commonly attracted in large numbers to cantharidin, or Spanishfly (Young 1975, 1984b, 1984c).

Status of the classification. Pyrochroidae are closely related to Boridae and perhaps Pythidae, Salpingidae and Trictenotomidae. With the removal of *Ischalia* (Young 1985), the return of *Pedilus* (Young 1984a, Young and Pollock 1991) and expansion of Pedilinae, the inclusion of Tydessinae (Peacock 1982, Pollock 1992, 1994) and, most recently, the convincing inclusion of Pilipalpinae (Pollock 1995), the size and scope of the family has grown significantly. The exact relationships of the aberrant genera *Cononotus* and *Agnathus* are still uncertain, as adults of both taxa exhibit numerous autapomorphies. Nonetheless, ties to Pyrochroidae seem reasonably well documented through what we have been able to ascertain from larvae (Mamaev 1976, Doyen 1979). The North American taxa and relationships have become reasonably well understood in the past 20 years (Doyen 1979, Young 1975, 1984a, 1985, Young and Pollock 1991).

Distribution. Approximately 200 species have been described from all major regions of the world (Blair 1914, 1928, Lawrence and Newton 1995, Pic 1911). The most speciose subfamilies, Pedilinae and Pyrochroinae are largely temperate in distribution while the 12 genera of Pilipalpinae illustrate a Southern Hemispheric distribution. Including *Cononotus*, there are 50 pyrochroid species in the United States and Canada (Young 1975, 1983, 1991).

KEY TO THE GENERA OF AMERICA NORTH OF MEXICO

1. Prothoracic coxae small; first two visible ventrites connate (Agnathinae) .... *Cononotus*

— Prothoracic coxae conical, projecting; visible ventrites freely articulated 2

2(1). Lateral aspects of pronotum margined about half their length; posterolateral angles of pronotum each with a small, deep pit (Tydessinae) .... *Tydessa*

— Lateral aspects of pronotum smooth, not margined; posterolateral angles of pronotum lacking pits 3

3(2). Eyes entire, setose; antennomeres 1-8 filiform to submoniliform, 9-11 greatly elongate, cylindrical (Fig. 2) .... *Anisotria*

— Eyes emarginate, glabrous; antennae variable, never with last three segments greatly elongate, cylindrical 4

4(3). Tarsal claws simple; antennae usually pectinate to flabellate (Figs. 1, 5); elytral apices normal (Pyrochroinae) 5

— Tarsal claws each bearing a small to prominent basal tooth; antennae subfiliform to flabellate, most commonly serrate (Figs. 3, 4); males with elytral apices often modified (conspicuously colored, indented, infolded, acuminate, etc.) (Fig. 4) .... *Pedilus*

5(4). Eyes separated dorsally by more than the dorsal width of an eye; eyes of male widely separated dorsally; males with cranial pits 6

— Eyes separated dorsally by less than the dorsal width of an eye (Figs. 1, 5); eyes of male nearly contiguous dorsally; males lacking cranial pits. .... *Dendroides*

6(5). Eyes large, minimal distance across eye (anterior to posterior margin at emargination) greater than width of genal area (between posterior margin of eye and point of cranial constriction forming "neck"); males with cranial pits interocular........ .... *Neopyrochroa*

— Eyes smaller, minimal distance across eye much less than width of genal area; males with cranial pits postocular .... *Schizotus*

Classification of the Genera of America north of Mexico

Pyrochroidae Latreille 1807

Tydessinae Nikitsky 1986

Tydessa Peacock 1982, 1 sp., *T. blaisdelli* Pollock, California, Nevada.

[Pilipalpinae Abdullah 1964, not in North America (see Pollock 1995)]

Pedilinae Lacordaire 1859

Anisotria Young 1984, 1 sp., *A. shooki* Young, Idaho, Montana, Washington.

Pedilus Fischer 1822, 30 spp., mostly western United States, also northeastern United States.
Corphyra Say 1835
Neopedilus Abdullah 1969

Pyrochroinae Latreille 1807

[*Pyrochroa* Geoffry 1762, not in North America.]

Neopyrochroa Blair 1914, 4 spp., two species endemic to California, two in eastern United States and southeastern Canada: *N. californica* (LeConte), California (Los Angeles area; known only from holotype female); *N. femoralis* (LeConte), southeastern Canada and eastern United States west to Kansas and Texas; *N. flabellata* (Fabricius), southeastern Canada and eastern United States west to Kansas and Texas; *N. sierraensis* Young, California (south-central Sierra Nevada mountains). Key to spp., Young 1975.

Schizotus Newman 1838, 2 spp., widely distributed across southern Canada and the northern United States: *S. cervicalis* Newman, transcontinental southern Canada and the northern United States; *S. fuscicollis* (Dejean), Alaska (and eastern Asia). Key to spp., Young 1975.

Dendroides Latreille, 1810, 6 spp., generally distributed: *D. canadensis* Latreille, eastern North America; *D. concolor* (Newman), northeastern and northcentral North America; *D. ephemeroides* (Mannerheim), Alaska south to northern California; *D. marginata* VanDyke, British Columbia, California, Oregon; *D. picipes* Horn, British Columbia, California, Oregon; *D. testacea* LeConte, transcontinental southern Canada and northern United States. Key to spp., Young 1975.

Pyrochroidae *incertae sedis*:

Agnathinae Lacordaire 1859

[*Agnathus* Germar 1818, not in North America.]

Cononotus LeConte 1851, 6 spp., southwestern US: *C. bryanti* Van Dyke, southern Arizona; *C. lanchesteri* Van Dyke, Idaho, Utah; *Cononotus macer* Horn, California; *C. punctatus* LeConte, California; *Cononotus sericans* LeConte, southern California; *C. substriatus* Van Dyke, California. Key to spp., Van Dyke 1939.

Bibliography

BLAIR, K. G. 1914. A revision of the family Pyrochroidae (Coleoptera). Annals and Magazine of Natural History, (8)13: 310-326.

BLAIR, K. G. 1928. Pyrochroidae. *In:* S. Schenkling, ed. Coleopterorum Catalogus, Pars 17(99). W. Junk. Berlin, 14 pp.

DOYEN, J. T. 1979. The larva and relationships of *Cononotus* LeConte (Coleoptera: Heteromera). Coleopterists Bulletin, 33: 33-39.

LAWRENCE, J. F. and A. F. NEWTON, Jr. 1995. Families and subfamilies of Coleoptera (with selected genera, notes, references and data on family-group names). Pp. 779-1006. *In:* J. Pakaluk and S. A. Slipinski, eds. Biology, phylogeny and classification of Coleoptera: Papers celebrating the 80th birthday of Roy A. Crowson. Muzeum i Instytut Zoologii PAN, Warszawa.

MAMAEV, B. M. 1976. Larval morphology of the genus *Agnathus* Germ. (Coleoptera, Pedilidae) and the position of the genus in the system of the Coleoptera. Entomologicheskoe Obozrenie, 55: 642-645. [In Russian; English translation in Entomological Review 55: 97-99.]

PEACOCK, E. R. 1982. *Tydessa* n. gen. from Japan (Coleoptera: Pyrochroidae). Entomologica Scandinavica, 13: 361-365.

PIC, M., 1911. Pedilidae. Pp. 11-27. *In:* S. Schenkling, ed. Coleopterorum Catalogus, Pars 17(26). W. Junk. Berlin (2nd pagination).

POLLOCK, D. A. 1992. A new species of *Tydessa* Peacock (Coleoptera: Pythidae: Pilipalpinae) from western North America. Pan-Pacific Entomologist, 68: 243-247.

POLLOCK, D. A. 1994. Systematic position of Pilipalpinae (Coleoptera: Tenebrionoidea) and composition of Pyrochroidae. Canadian Entomologist, 126: 515-532.

POLLOCK, D. A. 1995. Classification, reconstructed phylogeny and geographical history of genera of Pilipalpinae (Coleoptera: Tenebrionoidea: Pyrochroidae). Invertebrate Taxonomy, 9: 563-708.

VAN DYKE, E. C. 1939. New species and subspecies of west American Coleoptera. Pan-Pacific Entomologist, 15: 15-20.

YOUNG, D. K. 1975. A revision of the family Pyrochroidae (Coleoptera: Heteromera) for North America based on the larvae, pupae, and adults. Contributions of the American Entomological Institute, 11: 1-39.

YOUNG, D. K. 1983. A catalog of the Coleoptera of America North of Mexico. Family: Pyrochroidae. United States Department of Agriculture. Agricultural Handbook Number 529-120: 8 pp.

YOUNG, D. K. 1984a. *Anisotria shooki*, a new genus and species of Pedilinae (Coleoptera: Pyrochroidae), with a note on the systematic position of *Lithomacratria* Wickham and a key to the genera. Coleopterists Bulletin, 38: 201-208.

YOUNG, D. K. 1984b. Field studies of cantharidin orientation by *Neopyrochroa flabellata* (Coleoptera: Pyrochroidae). Great Lakes Entomologist, 17: 133-135.

YOUNG, D. K. 1984c. Field records and observations of insects associated with cantharidin. Great Lakes Entomologist, 17: 195-199.

YOUNG, D. K. 1985. Description of the larva of *Ischalia vancouverensis* Harrington (Coleoptera: Anthicidae: Ischaliinae), with observations on the systematic position of the genus. Coleopterists Bulletin, 39: 201-206.

YOUNG, D. K. 1991. Pyrochroidae, Pp. 541-544; Pedilidae, Pp. 544-547. *In:* F. W. Stehr, ed. Immature Insects, Volume 2. Kendall/Hunt. Dubuque.

YOUNG, D. K. and D. A. POLLOCK. 1991. Description of the mature larva of *Pedilus flabellatus* (Coleoptera: Pyrochroidae: Pedilinae), with phylogenetic implications of the discovery. Canadian Journal of Zoology, 69: 2234-2238.

116. SALPINGIDAE Leach 1815

by Darren A. Pollock

Family common name: The narrow-waisted bark beetles

Family synonyms: including Aegialitidae LeConte 1862; Dacoderidae LeConte 1862; Elacatidae Chapin 1923; Eurystethidae Seidlitz 1910; Inopeplidae Grouvelle 1808; Othniidae LeConte 1861; TretothoracidaeLea 1910

This family is very difficult to characterize, and diagnoses are presented only for the subfamilies of Salpingidae. It is entirely possible that this group is a composite and non-monophyletic.

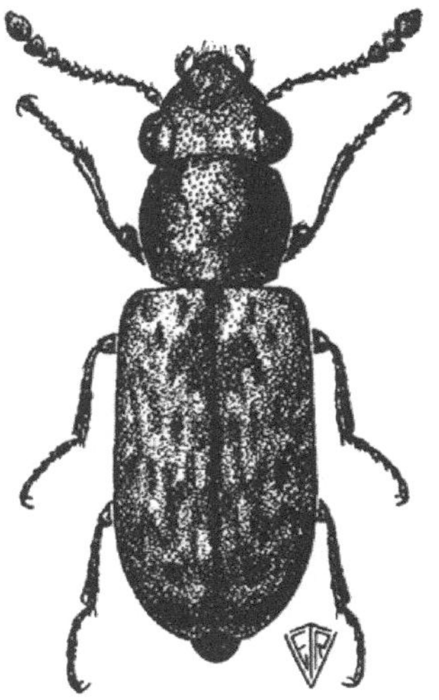

FIGURE 1.116. *Elacatis umbrosus* (LeConte)

Description: Body variously shaped, elongate to slightly ovate, distinctly depressed to subcylindrical; dorsum generally punctate, with or without distinct vestiture; length 1.5-7 mm; width 0.5-2 mm.

Head slightly to distinctly elongate and rostrate (especially in some Salpinginae, e.g., *Rhinosimus*), convex to flattened; head not, only slightly, or distinctly (*Dacoderus*) narrowed posterior of eyes; eyes absent (*Aglenus*), small or very large (especially Othniinae), facets coarse to fine; antennae of moderate length, either appearing 10-segmented (*Dacoderus*), or distinctly 11-segmented; antennomeres ranging from moniliform, subfiliform to filiform; antennal club, when present, slightly to distinctly developed, of 3-5 antennomeres; terminal antennomeres at least vaguely widened (except in Inopeplinae); antennal insertions exposed or partially concealed by lateral projection of frons; frontoclypeal area poorly differentiated in most taxa, indicated laterally only, in *Aglenus*, marked by deep pit in *Dacoderus*, and suture in *Inopeplus*; translucent strip of cuticle at apex of clypeus absent, short, or very long (*Aegialites*); labrum slightly to distinctly transverse; mandibles relatively short, mostly concealed by labrum; maxillary palpi relatively short, distal palpomere more or less parallel-sided, not expanded; gula moderately long, variously convex (especially in *Dacoderus*); labial palpi short, similar in structure to maxillary palpi.

Pronotum transverse (*Inopeplus*) to subquadrate to elongate (*Dacoderus*), basally narrower than elytra (except *Aglenus* and *Aegialites*); lateral margins slightly arcuate, widest near midlength, or distinctly narrowed posteriorly (especially *Inopeplus*); disc evenly convex, flattened, or slightly concave (*Dacoderus*); lateral margins of disc smooth, or distinctly carinate; prosternum well developed anterior of coxae; procoxae various, rounded to projecting, very narrowly to very widely separated; prosternal process narrow, knifelike to broad, truncate (especially in *Inopeplus* and *Aegialites*); front coxal cavities open or closed; protrochantins concealed.

Elytra elongate, nearly parallel-sided to ovate, concealing most of, or exposing most of abdomen (*Inopeplus*); disc convex, flat or slightly concave, with or without raised, longitudinal intervals; punctures indistinct (*Inopeplus*), scattered, or in distinct striae (Salpinginae); vestiture absent to distinct (*Elacatis*), mostly adpressed setae with few erect setae; humeri distinct (except *Dacoderus*), angulate in *Aglenus*; epipleura very short to distinct to apex, not prolonged inside humeri; scutellum visible, relatively small; mesosternum flat to slightly convex, with median keel in *Dacoderus*, anteriorly broadly rounded to truncate; mesepisterna widely separated; mesosternal process short, triangular to very wide, truncate (especially in *Aegialites*); mesotrochantin concealed; middle coxae narrowly to widely separated, cavities closed to partly closed by mesepimera; hind wing absent or present, functional; metasternum short (*Aegialites*) to relatively long, variously convex; anterior margin narrowly to very widely truncate, in contact with mesosternum (suture between meso- and metasternum obliterated in some *Inopeplus*); legs similar in size and shape on each thoracic segment; femora not, or slightly to moderately swollen near midlength; tibiae slender, relatively straight; tibial spurs small, inconspicuous; tarsi 5-5-4 in both sexes, stout to relatively slender, not lobed or laterally expanded; claw-bearing tarsomere longer than any other, longer than combined length of others in *Aegialites*; ventral surface of tarsomeres setose, not distinctly adhesive, except long, curled setae in *Aegialites*; tarsal claws relatively small to large (especially in *Aegialites*), simple, without teeth.

Abdomen with 5 ventrites, V1 and V2 connate (*Aegialites*), with indistinct suture; all ventrites free in other groups, with distinct sutures; male genitalia with bipartite tegmen, with or without parameral struts, most taxa with lateral lobes articulated to apical piece of tegmen; median lobe elongate, slender, without internal sac sclerites; ovipositor relatively elongate and flexible; bacula distinct.

Larvae moderately to distinctly flattened, sides subparallel, indistinctly sclerotized except for head and abdominal apex; body surface generally smooth, with vestiture of scattered, elongate setae; head prognathous, frontal arms of epicranial suture lyriform, traceable to near antennal insertions, stem absent; endocarinae

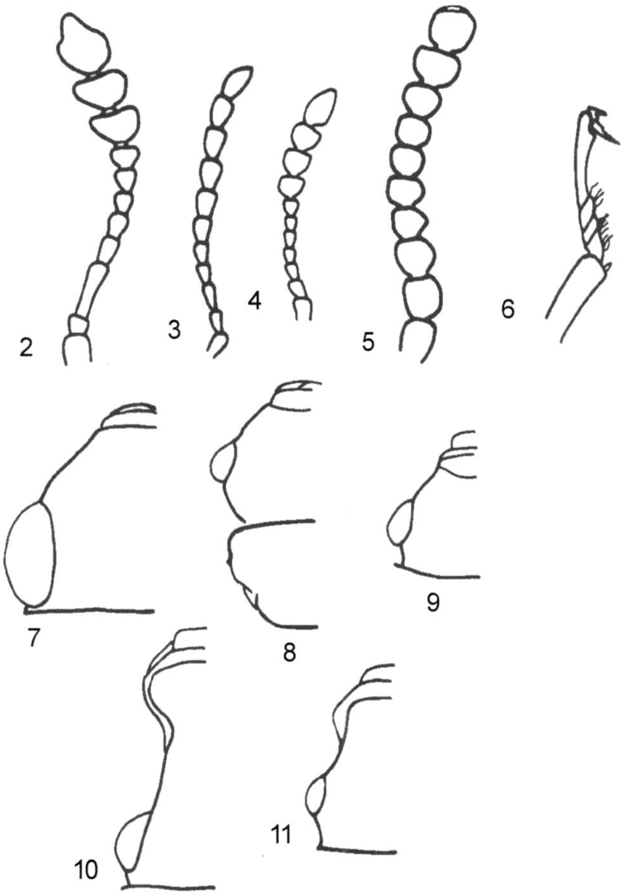

FIGURES 2.116-11.116. 2. *Elacatis* sp., antenna; 3. *Vincenzellus elongatus* (Mannerheim), antenna; 4. *Rhinosimus viridiaeneus* Randall, antenna; 5. *Dacoderus* sp., antenna; 6. *Aegialites californicus* (Motschulsky), hind tarsomere; 7. *Elacatis* sp., head; 8. *Inopeplus* sp., head and pronotum; 9. *Sphaeriestes virescens* LeConte, head; 10. *R. viridiaeneus*, head; 11. *V. elongatus*, head.

absent; stemmata present, 5 per side, or absent (*Aglenus*); antennae elongate, 3-segmented, with sensorium on antennomere 2; hypostomal rods distinct, parallel or divergent; mandibles moderately stout, basally with mola or hyaline lobe; mala undivided, with or without distinct uncus; thorax subparallel-sided, without distinct, sclerotized ridges or asperities; legs well-developed, 5-segmented, similar on all thoracic segments, with either fine or spinose setae; spiracles annular or annular-biforous, placed on laterotergite or spiracular tube or not; abdomen elongate, parallel-sided, variously flattened to almost cylindrical; segments 1-9 subequal in size, segment 10 highly reduced, surrounding anus; surface of all tergites (*Elacatis*) or tergite 9 only with or without distinct, sclerotized parabasal ridges; surface of tergite 9 smooth, or with small tubercles or callosities (*Aegialites*); tergite 9 distinctly sclerotized, possessing pair of distinct 2-branched urogomphi; sternite 9 with either 1 or 2 asperities near anterolateral margin, or with double arch of asperities (*Elacatis*); spiracles annular or annular biforous.

Habits and habitats. See individual subfamilies, below.

Status of the classification. The classification of Salpingidae has undergone many changes in the past century. In most European works, Salpingidae were restricted to what are now the Salpinginae. In contrast, the concept of Salpingidae presented in Arnett (1963), in an arrangement adopted from Spilman (1951), was very broad. Elements from Mycteridae, Boridae, and Pythidae, in addition to true salpingids, were included. The present, enlarged concept of Salpingidae has its roots in a paper by Lawrence (1977), which was one of the very few comprehensive analyses of Tenebrionoidea. In the latest classification of Lawrence and Newton (1995), seven subfamilies are included, although no detailed phylogenetic analyses have been done. In fact, the monophyly of the family is not established convincingly. Pollock (1994) included Salpingidae with Pythidae and Trictenotomidae in an unresolved trichotomy within the salpingid group of families.

Distribution. This family is worldwide in distribution, and is well represented in both tropical and temperate regions. There are about 45 genera and 300 species in the world, and 8 genera and about 20 species in Canada and the United States.

Key to Nearctic Subfamilies And Genera

1. Antennae appearing 10-segmented (Fig. 5); pronotal disc with pair of deep sulci, open laterally; eyes with ventral ridge (Dacoderinae) ...... *Dacoderus*

— Antennae 11 segmented; pronotal disc entire, without sulci; eyes without ventral ridge 2

2. Body distinctly flattened; pronotum abruptly narrowed basally (Fig. 8); elytra short, exposing most of abdominal tergites (Inopeplinae) .... *Inopeplus*

— Body at least slightly convex dorsally; pronotum not abruptly narrowed basally; elytra complete, concealing most of abdomen 3

3. Eyes absent (Agleninae) .... *Aglenus*

— Eyes present 4

4. All coxae widely separated, intercoxal processes broad, blunt; last tarsomere (Fig. 6) at least equal in length to all others combined (Fig. 12) (Aegialitinae) .... *Aegialites*

— At least procoxae nearly contiguous, intercoxal processes slender, narrowed; last tarsomere shorter than all other tarsomeres combined 5

5. Eyes large, convex, occupying almost entire length between antennal insertion and anterior margin of pronotum (Fig. 7); antennal club distinct, of 3 antennomeres (Fig. 2); pronotal disc with lateral carinae, present at least basally, often slightly toothed; dorsal vestiture distinct; elytral punctation confused, not forming distinct striae (Fig. 1) (Othniinae) .... *Elacatis*

— Eyes smaller, occupying about half length between antennal insertion and anterior margin of pronotum; antennal club of more than 3 antennomeres, but less distinct (Fig. 3-4); pronotal disc with smooth lateral margin; dorsal vestiture indistinct; elytral punctures forming distinct striae (Salpinginae) 6

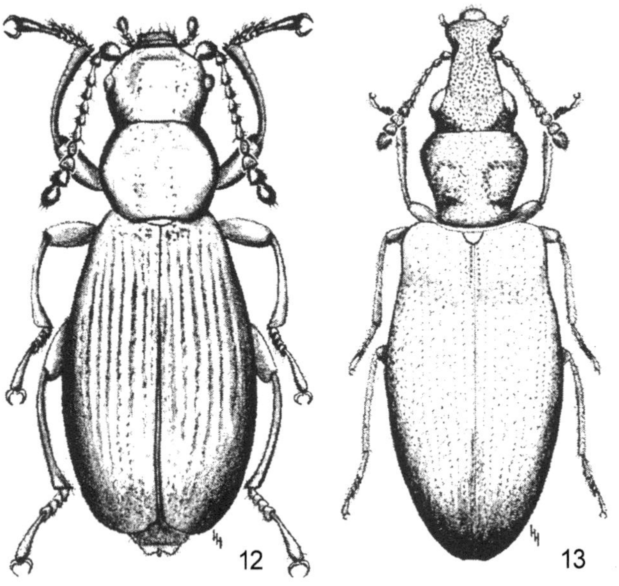

FIGURE 12.116-13.116. 12. *Aegialatis californicus* Motschulsky (after Hatch 1965); 13. *Rhinosimus viridiaeneus* Randall (after Hatch 1965).

6. Head (Fig. 9) without rostrum, distance from apex of clypeus to imaginary line between anterior margin of eyes much less than inner width between eyes .... *Sphaeriestes*
— Head (Fig. 10-11) with rostrum, distance from apex of clypeus to imaginary line between anterior margin of eyes much more than inner width between eyes 7

7. Antennae distinctly clavate (Fig. 4); rostrum (Fig. 10) with lateral margins deeply concave dorsally; supra-antennal margins anteriorly divergent before antennal insertions (Fig. 13) ..... *Rhinosimus*
— Antennae slightly clavate (Fig. 3); rostrum (Fig. 11) with lateral margins indistinctly concave dorsally; supra-antennal margins subparallel before antennal insertions .... *Vincenzellus*

Classification of the Nearctic Genera

Salpingidae Leach 1815

The present broad concept of Salpingidae (Lawrence and Newton 1995) includes taxa historically accorded family status (e.g., Othniidae, Dacoderidae, Inopeplidae, etc.). No phylogenetic analysis has been conducted on the subfamilies, and hence their relationships are unknown.

Othniinae LeConte 1861

This subfamily, often called false tiger beetles, has been treated as a family by many workers. *Elacatis* Pascoe 1861 was described for a species from Asia a year before *Othnius* LeConte 1862 was described for a North American species. The two names are synonyms, with *Elacatis* the valid name. However, the subfamily name Othniinae has priority over Elacatinae, and is the valid name for this group (Lawrence and Newton 1995). There are about 50 described world species, almost all of which are placed in the genus *Elacatis*. Most species are from tropical Asia, Africa, and Central America. Little is known of the habits of larval or adult othniines. However, larvae have been collected from under dead pine bark at high elevations in Arizona (Pollock, unpubl. notes), and adults are usually found on dead foliage (Young 1991a).

Diagnosis. Adults of Othniinae may be distinguished from other Salpingidae by the following combination of characters: body with distinct vestiture of decumbent setae, often in contrasting patterns of color; eyes large, convex, occupying almost entire length between antennal insertion and anterior margin of pronotum; pronotal disc with lateral carinae, present at least basally, often slightly toothed.

Elacatis Pascoe 1860, 5 spp., widely distributed, mainly in the western areas of North America.
Othnius LeConte 1861

Prostominiinae Grouvelle 1808

There are no representatives of this subfamily from North America north of Mexico, but several genera and species are known from Mexico and Central America. The subfamily is worldwide in distribution (Lawrence 1977). Members of this subfamily were included by Crowson (1955) in Othniidae. There is currently no key to genera of Prostominiinae, and very little is known about them other than the scattered species descriptions. As mentioned by Lawrence and Newton (1995: 900) prostominiines have been included historically in Cryptophagidae, Tenebrionidae, Prostomidae, and Colydiidae.

Agleninae Horn 1878

This subfamily contains a single species, which is widespread in distribution, presumably through repeated introductions from North America. This species was historically considered to be a member of Colydiidae, and has only recently been transferred to Salpingidae, in its own subfamily. *Aglenus brunneus* has been collected in caves, where it seems to be a scavenger (Young 1991b).

Diagnosis. Adults of the genus *Aglenus* may be distinguished from other Salpingidae by the following combination of characters: body small (less than 2 mm long), cylindrical; eyes absent; pronotum with complete lateral carina; humeri distinct, produced and angulate.

Aglenus Erichson 1845, 1 sp., *A. brunneus* (Gyllenhal 1808), widespread.

Inopeplinae Grouvelle 1908

There are four genera in this subfamily, with *Inopeplus* being by far the most diverse and widespread, with approximately 75 described species, most from tropical areas. Unpublished analyses (Slipinski and Lawrence) indicate that *Inopeplus* will likely be divided into several genera. This group has been given family rank by many authors, and historically was placed even in Staphylinidae (for example, by Blackwelder 1944) and Cucujidae (Arnett 1963). Larvae and adults are found under dead tree bark (see Young 1991c).

Diagnosis. Inopeplines may be distinguished from other Salpingidae by the following combination of characters: body very flat; antennae filiform; pronotum distinctly narrowed posteriorly; elytra short, exposing most of abdomen.

Inopeplus Smith 1851, 2 spp., *I. reclusa* LeConte and *I. immunda* Reitter, eastern and southern United States. Key to spp., Casey (1884).

Ino Laporte 1834, not Schrank 1803
Euryplatus Motschulsky 1860
Eleusinus Blackwelder 1943 (in Staphylinidae)
Pseudino Fairmaire 1869

Salpinginae Leach 1815

This subfamily is the most diverse and widespread of the Salpingidae (Blair 1928). The taxonomy of the many genera is in need of revision, and the relationships among genera are unknown. Representatives are small, often glabrous, with distinct rows of elytral punctures. Historically, genera in this subfamily have been divided into the Lissodemini (with denticulate pronotal margins) and Salpingini (with smooth pronotal margins). Young (1991b) presented information on larvae and biology of salpingids.

Diagnosis (North American species only). Adults of Salpinginae may be distinguished from other Salpingidae by the following combination of characters: body glossy, often slightly metallic, without distinct vestiture; elytral punctures in distinct striae; base of elytra slightly depressed; head produced into rostrum (except *Sphaeriestes*).

Sphaeriestes Stephens 1831, 3 spp., widely distributed.
Salpingus Gyllenhal 1810, of authors [not North American]

Rhinosimus Latreille 1805, 3 spp., widely distributed. Key to spp., Blair (1932). Larva: Howden and Howden (1981).

Vincenzellus Reitter 1911, 1 sp., *V. elongatus* Mannerheim 1852, Alaska.

Aegialitinae LeConte 1862

Eurystethinae Seidlitz 1916

This subfamily is composed of two genera, *Aegialites* Mannerheim and *Antarcticodomus* Brookes, which occur along the Pacific coast, and on Campbell and Auckland Island, respectively. These beetles live in the intertidal zone (Spilman 1967), and the widely separated coxae, and long tarsal claws of adults are adaptations to this habitat. The name Eurystethinae Seidlitz was an unnecessary replacement name for Aegialitinae (Spilman 1954).

Diagnosis. The Aegialitinae may be distinguished from other Salpingidae by the following combination of characters: coxae very widely separated; last tarsomere elongate, equal in length to all others combined; ventral surface of all but last tarsomere with long, curled hairs; tarsal claws distinct, long; V1 and V2 connate, suture obliterated.

Aegialites Mannerheim 1853, 3 spp., along Pacific coast, Alaska to California. Key to spp., Spilman (1967).
Eurystethes Seidlitz 1916.

Dacoderinae LeConte 1862

There are two described genera in Dacoderinae, *Dacoderus* LeConte (New World) and *Tretothorax* Lea (Australia). Little is known about their biology, although some of the peculiar skeletal features of *Dacoderus* possibly are indicative of association with ants. *Tretothorax cleistostoma* Lea was collected originally from ant nests. The species of Dacoderinae are currently under review.

Diagnosis. Dacoderinae may be distinguished from other Salpingidae by the following combination of characters: antennae distinctly moniliform, appearing 10-segmented; eyes coarsely faceted, produced, shelf-like; pronotum with pair of deep pits, open dorsally and laterally; elytra flat to slightly concave.

Dacoderus LeConte 1858, 2 spp., Texas, Arizona, and southern California. Key to spp., Horn (1893).

Bibliography

ARNETT, R. H., Jr. 1963. The beetles of the United States (a Manual for Identification). Catholic University of America Press. Washington, DC, 1112 pp.

BLACKWELDER, R. E. 1944. Checklist of the coleopterous insects of Mexico, Central America, the West Indies, and South America. Part 1. Smithsonian Institution, United States National Museum, Bulletin 185, 188 pp.

BLAIR, K. G. 1928. Pythidae, Pyrochroidae. Coleopterorum Catalogus, 17(99): 1-56.

BLAIR, K. G. 1932. The North American species of *Rhinosimus* (Col., Pythidae). Entomologists Monthly Magazine. 68: 253-255.

CASEY, T. L. 1884. Revision of the Cucujidae of America north of Mexico. Transactions of the American Entomological Society, 11: 69-112 + pls. 4-8.

CROWSON, R. A. 1955. The Natural Classification of the Families of Coleoptera. Nathaniel Lloyd. London, 187 pp.

HATCH, M. H. 1965. The beetles of the Pacific Northwest. Part IV. University of Washington Publications in Biology 16: 268 pp.

HORN, G. H. 1893. Miscellaneous Coleoptera studies. Transactions of the American Entomological Society, 20: 136-144.

HOWDEN, A. T. and H. F. HOWDEN 1981. The larva and adult biology of *Rhinosimus viridiaeneus* (Coleoptera: Salpingidae). Canadian Entomologist, 113: 1055-1060.

LAWRENCE, J. F. 1977. The family Pterogeniidae, with notes on the phylogeny of the Heteromera. Coleopterists Bulletin, 31: 25-56.

LAWRENCE, J. F. and A. F. NEWTON, Jr. 1995. Families and subfamilies of Coleoptera (with selected genera, notes, references and data on family-group names). Pp. 779-1006. *In:* J. Pakaluk and S.M. Slipinski, eds. Biology, Phylogeny, and Classification of Coleoptera. Papers Celebrating the 80th Birthday of Roy A. Crowson. Volume 2. Muzeum i Instytut Zoologii PAN. Warsaw.

POLLOCK, D. A. 1994. Systematic position of Pilipalpinae (Coleoptera: Tenebrionoidea) and composition of Pyrochroidae. Canadian Entomologist, 126: 515-532.

SPILMAN, T. J. 1951. The genera of Salpingidae (*olim* Pythidae, Coleoptera), with particular reference to the Nearctic forms. M.Sc. thesis, Cornell University.

SPILMAN, T. J. 1954. Generic names of the Salpingidae and their type species (Coleoptera). Journal of the Washington Academy of Sciences, 44: 85-94.

SPILMAN, T. J. 1967. The heteromerous intertidal beetles (Coleoptera: Salpingidae: Aegialitinae). Pacific Insects, 9: 1-21.

YOUNG, D. K. 1991a. Othniidae (Tenebrionoidea). Pp. 547-549. *In:* F. W. Stehr, ed. Immature Insects. Volume 2. Kendall Hunt. Dubuque. xvi + 975 pp.

YOUNG, D. K. 1991b. Salpingidae (Tenebrionoidea). Pp. 549-551. *In:* F. W. Stehr, ed. Immature Insects. Volume 2. Kendall/Hunt. Dubuque, IA, xvi + 975 pp.

YOUNG, D. K. 1991c. Inopeplidae (Tenebrionoidea). Pp. 551-552. *In:* F. W. Stehr, ed. Immature Insects. Volume 2. Kendall/Hunt. Dubuque, IA, xvi + 975 pp.

117. ANTHICIDAE Latreille 1819

by Donald S. Chandler

Family common name: The ant-like flower beetles

Family synonym: Notoxidae Stephens 1832, Ischaliidae Blair 1920

The Anthicidae are readily recognized by the abruptly constricted head forming a long to short neck, the pronotum widest in the anterior half and narrowing to the base, pronotal base narrower than elytral base and the first two visible abdominal sternites being freely articulated.

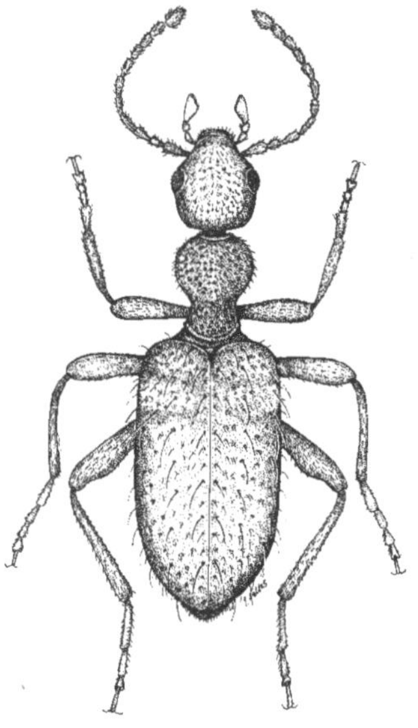

FIGURE 1.117. *Malporus formicarius* (LaFérte-Senéctère)

Description: Head deflexed, strongly constricted behind eyes to form neck one-half or less head width (barely constricted in *Lagrioida*); antennae with eleven antennomeres; antennae filiform, serrate, weakly clubbed (*Lagrioida*), while most are subclavate; antennal bases exposed, inserted at sides of the frons. Frontoclypeal suture mostly distinct; labrum large to small in size, transverse. Mandibles small, strongly curved; apices truncate or emarginate. Maxillary palpi with four segments, the last mostly securiform, enlarged and basally angulate in a few groups (*Pergetus* [Eurygeniinae], *Ischyropalpus* [Anthicinae]); second and third segments protruding mesally in a few groups, particularly Macratriinae and some Eurygeniinae. Labium with the gular region long and narrow, gular sutures present or absent; submentum indistinct from the gula; mentum greatly modified, ridged; ligula large and fleshy, only slightly emarginate; labial palpi with three segments, filiform, small. Eyes transversely oval, emarginate in some Eurygeniinae, with short setae originating between facets in many.

Pronotum mostly elongate to trapezoidal, globose in Notoxini, larger than head, widest in anterior third, narrowing in basal half (except Ischaliinae); lateral margins not bordered (except Ischaliinae). Prosternum narrow in front of contiguous procoxae; procoxae conical, prominent; procoxal cavities open posteriorly (closed in Ictystignini), closed internally by internal apodeme extending from lateral foveae above procoxae (foveae lacking in Lagrioidinae, Ischaliinae, and Steropinae). Thin, well defined sulcus in many originating at lateral foveae and extending posterodorsally to cross dorsum just anterior to base. Pronotal apex simple, dorsally flanged, or with distinct collar; prominent apicodorsal horn projecting over head in Notoxini.

Mesosternum broadly triangular, rounded to angulate anteriorly, with mesepisterna nearly meeting to meeting at middle of anterior margin; mesocoxal cavities narrowly separated by mesosternal extension in almost all groups; mesocoxae oval, not prominent; trochantin evident. Mesosternal-mesepisternal suture distinct (except in Macratriinae); lateral and posterior margins of mesosternum and mesepisterna greatly bowed or angled in some groups; fringes or patches of dense or long setae found on mesosternal or mesepisternal margins in some. Metasternum moderate; metacoxal cavities separate; metacoxae transverse, with short internal flange. Legs with femora slender; tibiae slender, apical spurs present (lacking in Ischaliinae and Tomoderinae); tarsal formula 5-5-4, tarsi slender, penultimate tarsomeres narrowly lobed beneath; claws simple to appendiculate. Scutellum very small, triangular (quadrate in some Lemodinae). Elytra typically entire (shortened in *Tanarthrus* and *Leptanthicus* [Anthicinae]); epipleural fold very narrow. Characterization of pubescence often based on three types on elytra, the first two types are typically present: tactile setae are sparse, forming ordered rows on elytra when clearly seen, typically suberect (angled at 40-60°) to erect (60-90°), may be short and difficult to see; setae are shorter and variably denser, lacking obvious organization, varying from appressed (tip touches elytra) to erect; undersetae are dense hairs mostly appressed to elytra, in many directed laterally (*Sapintus* [Anthicinae]). Flight wings with well developed postcubital patch.

Abdomen with five visible sterna, sterna free (basal two fused in Lagrioidinae). Males of many groups with pygidium (morphological tergum 8) at least slightly visible, completely covered by tergum 7 in other groups. Sternite 9 internal, mostly Y-shaped with long basal stalk and shorter arms, variously modified, a simple rod in Tomoderinae. Male genitalia with separate parameres and penis in Macratriinae, Eurygeniinae, and Notoxini; most with parameres fused to form tegmen in the other groups; phallobase separate in all groups except Copobaeninae, Lemodinae, and Tomoderinae. Internal sac in many with patches of spines of varing widths and lengths. Female genitalia with coxites very close together in the mid-ventral region and with a partial division on the ventral side; styli small, borne on unsegmented or incompletely 2 segmented coxites. Lawrence (1982); Werner and Chandler (1995); de Marzo (1996, male and female genitalia).

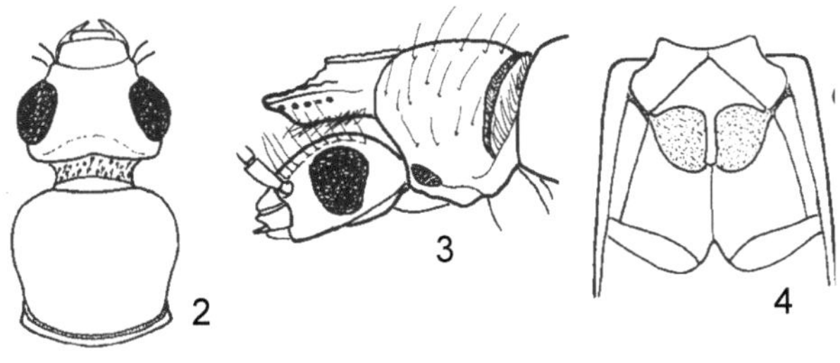

FIGURES 2.117-4.117. 2. Dorsal view head and pronotum, *Stereopalpus pruinosus* LeConte; 3. Left lateral view head and pronotum, *Notoxus monodon* Fabricius; 4. Ventral view thorax and portion of abdomen, *Anthicus ephippium* LaFérte-Senéctère.

Larva elongate, subcylindrical, subparallel; body lightly sclerotized, whitish; size 3 to 15 mm.; vestiture a few long setae on most of the segments. Head exserted from prothorax, prognathous; epicranial sutures with arms lyriform (U-shaped in Lagrioidinae and Ischaliinae), arms diverging anteriorly and reaching antennal insertions, stem lacking (*Pergetus* [Eurygeniinae]; Lagrioidinae) or short; frontoclypeal suture lacking (present in Lagrioidinae). Endocarina lacking (Ischaliinae, Lagrioidinae), single (Anthicinae), or paired (*Pergetus*). Antennae with three segments, second segments with conical to broad sensorium; antennal insertions exposed. Labrum small. Mandibles asymmetrical, the apices tridentate (*Pergetus*) or bidentate, with a distinct molar area, molar area of right mandible with patch of coarse serrulations or brush of spine-like setae; maxillae with cardo, fused stipes, setiferous entire mala (cleft in Lagrioidinae), and three-segmented palpi; labium with a distinct ligula and a pair of two-segmented palpi. One pair of stemmata, or lacking (*Pergetus*) (Lagrioidinae with 5 pairs). Thorax with four-segmented legs and claw-like tarsunguli. Abdomen nine-segmented; spiracles annular on the mesothorax and abdominal segments one to eight; segment nine with pair of fixed, heavily sclerotized urogomphi curved slightly upward (lacking in Ischaliinae), in many branched at bases, lacking urogomphal pits; ninth sternite lacking asperities. Bøving and Craighead (1931); Kitayama (1982); Young (1991a, 1991b); Costa *et al.* (1995). The larvae for two of the North American subfamilies, Macratriinae and Tomoderinae, are completely unknown.

Habits and habitats. Adults are omnivorous scavengers and opportunistic predators of small arthropods, but also may feed on pollen, plant exudates, or fungal hyphae and spores (Werner and Chandler 1995). A few species in crop ecosystems have been investigated as biological control agents, primarily through eating eggs or small larvae of pest species (McCutcheon and Webster 1996). In North America the only groups commonly collected on flowers are members of *Ischyropalpus* (Anthicinae) and adults of some genera of Eurygeniinae (Abdullah 1969), while Macratriinae, Microhoriini (Anthicinae), and species in several other genera are typically collected on vegetation during the day. The rest are mostly crawling about on the ground, usually in areas where there are exposed areas of soil with prostrate plants, pieces of debris, stones, or clumps of litter to hide beneath. Adults of *Amblyderus* (Weissmann and Kondratieff 1999) and *Mecynotarsus* (Chandler 1977, 2001) burrow into sandy soils and dunes during the day, while adults of *Tanarthrus* scurry about on saline mud flats like tiger beetles during the day (Peterman 1973) [all Anthicinae]. Diversity is highest in the Southwest and California, as the North American members of this family seem to prefer hot, dry habitats (except Tomoderinae and Macratriinae). Over half of the species in North America and Mexico may be collected at ultraviolet light (Wolda and Chandler 1996).

Adults of species in a number of genera, but particularly *Notoxus* [Anthicinae], are known to be attracted to cantharidin or meloid beetles that produce cantharidin (Young 1984), and this can be used to attract specimens into traps (Chandler 1976). The cantharidin is usually obtained initially by males, particularly in those groups where the males have secretory glands that concentrate the cantharidin in canals at the elytral apices, which are indicated by apical tubercles in many (Schütz and Dettner 1992). Cantharidin is believed to act as a feeding deterrent to predators of anthicids, and when the females feed upon the secretions of the males, they presumably take the cantharidin into their own bodies, and pass it on to the eggs and larvae that eventually develop as has been documented for pyrochroids (Eisner *et al.* 1996a, 1996b). Mesothoracic glands are found in the adults in some genera of Anthicinae, and these have been found to secrete toxic iridoids through a median mesothoracic pore, that appears to act as an effective feeding deterrent for predatory ants (Hemp and Dettner 1997). This may explain why certain species of Anthicinae may freely run in ant columns or search for food on leaves when a variety of ant species are present.

Larvae are found on the ground where they are omnivorous or mycetophagous on hyphae or spores (Hinton 1945; Kitayama 1982; Young 1991b), or are opportunistic predators (Davidson and Wood 1969). The larvae of one *Notoxus* species have been documented boring into sweet potato tubers (Cuthbert 1967), and there have been further reports of this behavior.

Status of the classification. Almost all North American genera have been revised in the last 30 years, though the genera of Eurygeniinae should be reexamined. Major papers have typically treated only regional faunas rather than monophyletic groups, in part because most genera occur on several or all continents, but also because generic limits and variation are poorly understood at the world level. The family also has grown in recent years through the addition of several groups from the dismembered Pedilidae. The higher classification needs a modern analysis, with inclusion of groups of the following three groups being questioned: Lagrioidinae (Costa *et al.* 1995), Ischaliinae (Young 1985; Nikitsky and Egorov 1992), and Afreminae (Werner and Chandler 1995). Last world catalogue: Pic (1911a [in part], 1911b).

Distribution. There are over 3,000 species in about 100 genera found in all biogeographic regions. In North America there are 32 genera with 231 species, with nearly half of these members of *Anthicus* or *Notoxus*.

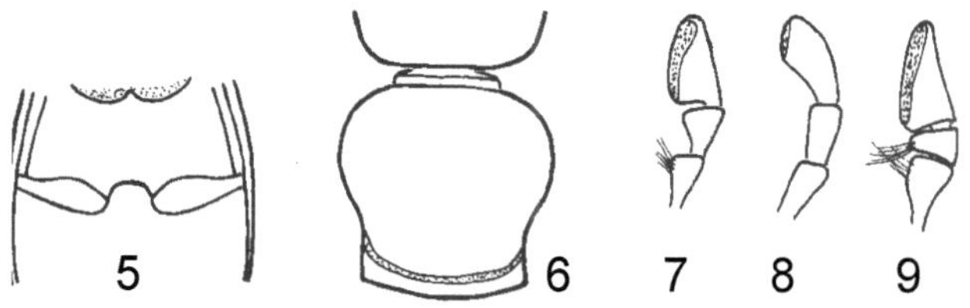

FIGURES 5.117-9.117. 5. Ventral view thorax and portion of abdomen, *Tomoderus constrictus* (Say); 6. Dorsal view pronotum, *Anthicus ephippium* LaFérte-Senéctère; 7. Dorsal view right maxillary palpus, *Retocomus wildii* (LeConte); 8. Dorsal view right maxillary palpus, *Mastoremus* sp.; 9. Dorsal view right maxillary palpus, *Pergetus campanulatus* (LeConte).

Key to the Subfamilies of Anthicidae

1. Elytra with prominent lateral and epipleural ridges extending to or near apices; pronotum with lateral margins parallel from middle to base, base strongly angulate at basolateral angles and at middle (Ischaliinae) .... *Ischalia*

— Elytra convex, lacking any strong ridges; pronotum with lateral margins converging in basal half, basal margin gently convex to subtruncate (Figs. 1, 2) 2

2(1). Neck width one third head width or more, neck exposed and rugosely striate (Fig. 2) (except *Thambospasta*, neck narrower); eyes with anterior margins lengthily flattened, or slightly sinuate to deeply emarginate; metacoxae contiguous to narrowly separated, intercoxal process of first visible sternite with margins diverging at about 45° or less (Key A) .... Eurygeniinae

— Neck width about one-fourth or less head width and neck smooth (Fig. 1), or head base closely articulated with pronotum and with pronotal horn (Fig. 3); eyes with anterior margins rounded to briefly flattened; metacoxae narrowly to widely separated, intercoxal process of first visible sternite with margins diverging at about 60-90° (Anthicinae, Fig. 4), broadly separated (Tomoderinae, Fig. 5), or much narrower (Macratriinae) 3

3(2). Pronotum lacking distinct apical collar or prominent rim, or lacking anterodorsal horn; lateral antebasal constriction distinct across dorsum of pronotum; metacoxae widely separated, intercoxal process of first visible sternite broadly subtruncate (Fig. 5) (Tomoderinae) .... *Tomoderus*

— Pronotum with distinct apical collar (Figs. 1, 6), or prominent anterodorsal horn present (Fig. 3), lateral antebasal constriction absent to strongly developed, but rarely as deeply and narowly defined; metacoxae narrowly to widely separated by angulate intercoxal process of first visible sternite (Fig. 4) 4

4(3). Elytra with thin elongate longitudinal sulcus along lateral angles; mesosternum and mesepisterna smoothly fused, lacking evidence of carina at line of fusion; lacking clypeofrontal suture (Macratriinae) .... *Macratria*

— Elytra lacking longitudinal sulcus at lateral angles; mesosternum and mesepisterna with distinct carinae marking their lines of fusion; clypeofrontal suture present (Key B) .... Anthicinae

Key A. Genera of Eurygeniinae

1. Pronotum with apical collar encircling neck; neck width at apical portion of constriction about one-fourth head width (cf. Fig. 6); western Texas (*Incertae sedis*) .... *Thambospasta*

— Pronotal apex with dorsally protruding flange over neck, flange not extending to or strongly narrowing laterally to venter; neck width at apical portion of constriction one-third head width or greater (Fig. 2) (Eurygeniini) 2

2(1). Eyes with margins near antennal insertions narrowly emarginate, depth of emarginations equivalent to apical width of first antennomeres 3

— Eyes with margins near antennal insertions slightly and broadly emarginate to nearly straight, clearly less than apical width of first antennomeres... 4

3(2). Elytra with erect (60-90°) tactile setae twice as long as suberect (40-60°) setae; neck narrowest at junction with head; fourth segments of maxillary palpi with bases strongly angulate at mesal margins (Fig. 7) .... *Retocomus*

— Elytra with suberect tactile setae about as long as setae; neck narrowing toward pronotum; fourth segments of maxillary palpi with margins straight from middle to base (Fig. 8) .... *Mastoremus*

4(2). Eleventh antennomeres four or more times longer than tenth antennomeres............... *Bactrocerus*

— Eleventh antennomeres no more than twice length of tenth antennomeres 5

5(4). Body covered with dense, appressed, white, silky pubescence, slightly clumped on dorsum; lacking raised setae on elytra; males with antennae elongate, antennomeres III-VII strongly serrate, antennomeres XI creased at middle to suggest two antennomeres, or articulated to form twelve antennomeres; female antennae much shorter and only feebly serrate .... *Leptoremus*

— Body with silky appressed setae only present on dorsum and lacking on venter, or lacking such setae; antennae of both sexes subequal in length, at most only slightly widened to apices 6

6(5). Fourth segment of maxillary palpus with base prominently angulate mesally (Fig. 9); elytra with erect tactile setae twice as long as suberect setae; pronotum with lateral margins broadly and shallowly constricted in basal half, disc with deep median longitudinal line...................... *Pergetus*

— Fourth segment of maxillary palpus not protruding mesally at base, about as wide or narrower than apex of third segment (Fig. 10); elytra with tactile setae as long as other setae, long and distinct only in *Rilettius*; prontoum with lateral margins slightly convex to slightly concave in basal half, median line on disc of pronotum shallow or lacking 7

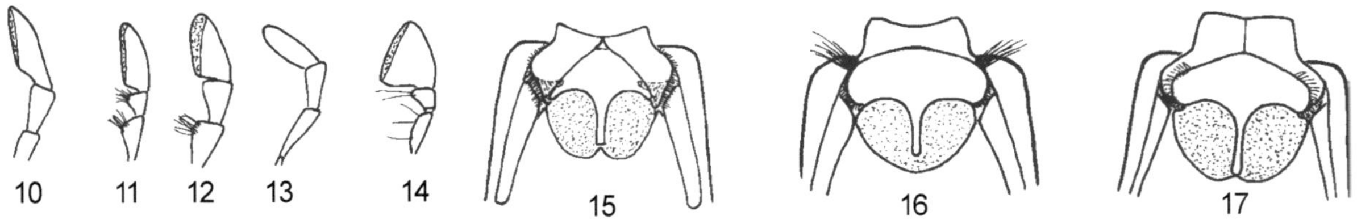

FIGURES 10.117-17.117. 10. Dorsal view right maxillary palpus, *Qadrius quietus* Abdullah; 11. Dorsal view right maxillary palpus, *Stereopalpus pruinosus* LeConte; 12. Dorsal view right maxillary palpus, *Duboisius arizonensis* (Champion); 13. Dorsal view right maxillary palpus, *Neoeurygenius portoricensis* Abdullah; 14. Dorsal view right maxillary palpus, *Ischyropalpus turgidicollis* (Casey); 15. Ventral view mesothorax, *Sapintus fulvipes* (LaFérte-Senéctère); 16. Ventral view mesothorax, *Formicilla munda* LeConte; 17. Ventral view mesothorax, *Acanthinus zeteki* Werner.

7(6). Elytra with erect (60-90°) tactile setae twice as long as suberect (40-60°) setae .... *Rilettius*

— Elytra with tactile setae suberect to decumbent (20-40°), about as long as setae, tactile setae and setae in many difficult to separate 8

8(7). Eyes small, separated by more than four times their width in dorsal view; clypeus short, with apex broadly rounded; labrum in dorsal view not visible or barely visible laterally .... *Qadrius*

— Eyes large, mostly separated by distance equal to their width in dorsal view; clypeus with apex broadly subtruncate to truncate; labrum short to elongate, clearly visible at middle in dorsal view (Fig. 2) 9

9(8). Pronotum with lateral margins broadly and shallowly concave in basal half; fourth segments of maxillary palpi elongate, margins nearly parallel for most of length (Fig. 11) .... *Stereopalpus*

— Pronotum with lateral margins straight to slightly convex in basal half; fourth segments of maxillary palpi shorter, margins not parallel for most of length (Fig. 12), or palps simple (Fig. 13) 10

10(9). Maxillary palpus with second segment triangular, width more than half length (Fig. 12); basolateral angles of head distinct, narrowly rounded .... *Duboisius*

— Maxillary palpus with second segment elongate, nearly three times as long as wide (Fig. 13); basolateral angles of head indistinct, gently curved 11

11(10). Elytra with clumps of appressed silken setae, setae giving mottled appearance to elytra.................. .... *Neoeurygenius*

— Elytra lacking clumps of appressed silken setae, all setae evenly dispersed .... *Eurygenius*

Key B. Genera of Anthicinae

1. Pronotum with prominent dorsoapical horn extending anteriorly, covering head; head closely articulated with pronotum, neck very short (Fig. 3) (Notoxini) 2

— Pronotum bluntly to smoothly rounded on anterior margin, head completely exposed; neck clearly exposed (Anthicini) (Fig. 1) 4

2(1). Head lacking erect setae on lateral margins of vertex; gula with tubercles in apical half .... *Mecynotarsus*

— Head with row of prominent setae along lateral margins of vertex; gula smooth 3

3(2). Lateral margins of horn varying from smooth to with multiple small teeth, never with 3-4 promiment, widely-spaced teeth on each side; pronotum with setose pits clearly visible just posterior to occipital articulations with head (Fig. 3); metatibiae longer than metatarsi .... *Notoxus*

— Lateral margins of horn with 3-4 prominent, widely-spaced teeth to each side; pronotum with pits posterior to articulations with head minute and nude; metatibiae mostly clearly shorter than metatarsi .... *Squamanotoxus*

4(1). Fourth segment of maxillary palpus angularly expanded medially, form close to that of isosceles triangle (Fig. 14).......................... *Ischyropalpus*

— Fourth segment of maxillary palpus smoothly curved on mesal margins, not angularly expanded (Fig. 1) 5

5(4). Posterolateral margins of mesepisterna with thick fringe of setae overlying obliquely impressed mesepimera, impression densely setose, with fovea from depressed portion of mesepimera projecting internally anterior to mesocoxae (Fig. 15); elytra with dense layer of appressed to decumbent (10-20°) undersetae mixed with or beneath setae, in many with suberect to erect tactile setae, undersetae in many directed obliquely .... *Sapintus*

— Mesepisterna in some with very short fringe of setae on posterolateral margins; narrow mesepimera forming nude oblique sulcus, lacking foveae (Figs. 4, 16); elytra lacking undersetae, with at most setae and tactile setae present 6

6(5). With at least most of large last tergite exposed by truncated elytra; body flattened, sides parallel; on saline mud flats or beaches of western North America... 7

— With at most tip of last tergite exposed (excluding small apical pygidium of males); body more robust 8

7(6). Antennomeres XI constricted near middle to give the appearance of an additional antennomere; most of large last tergite exposed by truncate elytra .... *Tanarthrus*

— Antennomeres XI not constricted; elytra abbreviated to expose three abdominal tergites .... *Leptanthicus*

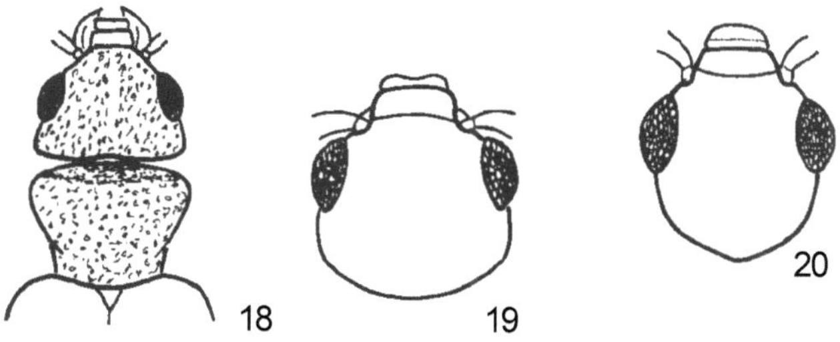

FIGURES 18.117-20.117. 18. Dorsal view head and pronotum, *Amblyderus scabriceps* (LeConte); 19. Dorsal view head, *Stricticomus tobias* (Marseul); 20. Dorsal view head, *Cyclodinus texanus* LaFérte-Senéctère.

8(6). Body with setae sparse and appressed, short (0.04 mm long), erect tactile setae short and inconspicuous; head and pronotum with reticulate microsculpture; elytra brown to dark brown, reddish across base .... *Omonadus*
— Most with distinct tactile setae or longer setae, if not then coloration different and lacking microreticulation on the head and pronotum .. 9

9(8). Lateral mesosternal margins straight, lacking fringe of setae (Fig. 4) 10
— Lateral mesosternal margins strongly curved, in many with short to long fringe of setae (Figs. 16, 17) 14

10(9). Pronotum with antebasal constriction deeply indicated on lateral margins, continuing very shallowly across dorsum; elytra with long erect tactile setae; large dark species with pale postbasal band on elytra (Fig. 1) .... *Malporus*
— Lateral pronotal margins either lacking distinct constriction, or when present antebasal constriction not continuing across dorsum; different color and setation patterns (Figs. 6, 18) 11

11(10). Pronotum widest very near anterior margin, abruptly declivous and flattened on anterior face; head strongly triangular (Fig. 18) .... *Amblyderus*
— Pronotum evenly convex on anterior margin, mostly widest in apical third; head subquadrate to oval 12

12(11). Pronotum with lateral margins nearly straight (Fig. 6); elytra with tactile setae mostly distinct........ .... *Anthicus*
— Pronotum with lateral margins broadly constricted in basal half; elytra with erect tactile setae very short, not obvious 13

13(12). Head with basal margin broadly rounded (Fig. 19); elytral setae dense, each elytron with light spot near apex .... *Stricticomus*
— Head narrowly pointed to angularly rounded at middle (Fig. 20); pubescence sparse and fine; color mostly light brown with obscure dark band across middle of elytra .... *Cyclodinus*

14(9). Pronotum with lateral margins straight, evenly converging in basal half 15
— Pronotum with lateral margins distinctly constricted in basal half 16

15(14). Head base with deep medial cleft; elytra lacking visible tactile setae, setae dense and appressed; northern plains area to Rocky Mountains .... *Euvacusus*
— Head base broadly subtruncate, lacking medial notch; elytra with distinct short to long tactile setae .... *Vacusus*

16(14). Elytra with tactile setae very short, not obvious; body covered with dense, short, appressed setae; body brown, elytra with light postbasal transverse band .... *Baulius*
— Elytra with visible short to long tactile setae; body sparsely to densely setose; coloration variable 17

17(16). Anterolateral margins of mesepisterna strongly compressed by expansion of mesosternum, with tuft of setae originating from mesepisterna, lacking prominent setae on mesosternal margins (Fig. 16); pronotal disc gently convex to base; body with prominent tactile setae .... *Formicilla*
— Lateral mesosternal margins with fringe of setae evenly arranged along nearly entire length (Fig. 17), lacking tuft of setae arising from mesepisterna; pronotum with constriction across disc in some species; length of tactile setae variable .... *Acanthinus*

Classification of the Nearctic Genera

Anthicidae Latreille 1819

The Anthicidae currently contain ten subfamilies, with five of these occurring in North America. Of the groups occurring in North America, the Ischaliinae are also eastern Eurasian and Oriental, while the rest are represented world-wide. Subfamilies not found in North America are the Steropinae (Eurasian), Afreminae (African), Copobaeninae (Chilean), Lemodinae, and Lagrioidinae (both found in Australia, New Zealand, and temperate South America). Biology: Werner and Chandler (1995); Key to genera: Uhmann (1976).

Ischaliinae Blair 1920

Diagnosis: This group is unique in the family by the presence of longitudinal lateral and epipleural ridges on the elytra, the pronotal base that is angulate at the basolateral margins and the middle, the lateral pronotal margins parallel from the middle to the base, the neck one-third the head width, and the lack of tibial spurs. This subfamily contains a single genus, *Ischalia* Pascoe, which was transferred from the Pyrochroidae to the Anthicidae by Young (1985) primarily on the basis of larval characters. It does not fit well with the rest of the Anthicidae, and Nikitsky and Egorov (1992) have placed this group as a separate family.

Ischalia Pascoe 1860
subgenus *Eupleurida* LeConte 1862, other subgenera Old World
Three North American species, northern United States and southern Canada. Revision and key: Young (1975). Larvae: Young (1985). Biology: Young (1985). Larvae and adults appear to feed

on fungal mycelia on rotting wood. A small genus that is also found in eastern Eurasia and the Oriental region.

Eurygeniinae LeConte 1862

Diagnosis: The Eurygeniinae are characterized by: intercoxal projection of the first visible abdominal sternite narrower, margins diverging at about 45° or less, mostly about 20°; metacoxae contiguous to narrowly separated; eyes mostly large and anterior margins very shallowly to deeply emarginate; neck width about one-third head width or greater, neck coarsely strigose (except *Thambospasta*); and with tibial spurs. There are three tribes, with the Eurygeniini found world-wide, and the two other tribes restricted to Chile (Mitraelabrini Abdullah, 1 genus) or Australia, Africa, and Brazil (Ictistygnini Abdullah, 4 genera). *Thambospasta* Werner, was recently transferred to the Anthicidae, and this distinctive group is here placed in the Eurygeniinae as *Incertae sedis*. Revision and key: Abdullah (1969).

Eurygeniini LeConte 1862

Diagnosis: The members of this tribe have the procoxal cavities open externally, and closed internally. The pronotal apex has a prominent dorsal flange extending over the cervix, with this flange less developed to lacking ventrally. Eighteen genera are placed here, with twelve of these occurring in North America. Abdullah revised or commented on all of the genera, but the status of several of the genera he created needs to be reevaluated.

Retocomus Casey 1895
Seventeen species, two occurring in southeastern states, the rest from Utah to California with fourteen species in California. Revision and key: Abdullah (1965).

Rilettius Abdullah 1964
Four species, Texas to Arizona and northern Mexico. Revision and key: Abdullah (1964c).

Pergetus Casey 1895
Two species, Pacific Northwest. Revision: Abdullah (1960). Larvae: Bøving and Craighead (1931). The type species, *P. campanulatus* (LeConte), is found in the Pacific Northwest, while *P. wilati* (Lacordaire) was taken from "North America."

Stereopalpus LaFérte-Senéctère 1846
    Stereopselaphus Gemminger and Harold 1870
Eleven species, North America. Revision and key: Abdullah (1964d).

Leptoremus Casey 1904
A single species, *Leptoremus argenteus* Casey, Arizona to southern California. Revision: Abdullah (1961).

Mastoremus Casey 1895
Three species, Arizona and Idaho. Key: Abdullah (1964a).

Bactrocerus LeConte 1866
A single North American species, *Bactrocerus concolor* LeConte, Southwest to Arizona and northern Mexico. Redescription: Abdullah (1963). Other species are in Mexico and Central America.

Eurygenius LaFérte-Senéctère 1849
Three North American species, Texas and California. Revision and key: Abdullah (1967). These are the northernmost representatives of a small Neotropical genus.

Duboisius Abdullah 1961
Five North American species, Texas to Arizona. Revision and key: Abdullah (1964b). Seven more species are found in northern and central Mexico.

Qadrius Abdullah 1964
    Quadrius Uhmann 1976
Two species, Arizona to Texas. Revision and key: Abdullah (1964c).

Neoeurygenius Abdullah 1963
A single North American species, *Neoeurygenius grahami* Abdullah, Arizona. Key: Abdullah (1964e). The other species in this genus occurs in Puerto Rico and the Virgin Islands.

Eurygeniinae, *Incertae sedis*

Thambospasta Werner 1974
A single species, *Thambospasta howdeni* Werner, western Texas. Description: Werner (1974). This genus was originally placed in the Meloidae, and was only recently clearly transferred to the Anthicidae (Aksentjev 1988) without more specific placement. It cannot be placed in any of the three present tribes. The pronotum has a collar at the apex, the eyes are large and emarginate, the neck constriction is narrow like in the Anthicini, and the antennae are elongate and filiform. Two other species have been seen from Mexico and Belize, and this genus appears to be very close to *Salimuzzamania* Abdullah (1968), which is based on *S. uniformis* (Champion) from Guatemala.

Macratriinae LeConte 1862

Diagnosis: The Macratriinae are characterized by: head lacking clypeofrontal suture; prominent apical rim on pronotum; head with neck width at constriction one-fourth or less head width, neck smooth; mesosternum completely fused to mesepisterna; elytra with a thin longitudinal sulci along lateral margins of elytral disc; metacoxae narrowly separated; and tibial spurs present. This group is based on the large, world-wide genus, *Macratria*.

Macratria Newman 1838
    Macrarthria Erichson 1840
    Macrarthrius LaFérte-Senéctère 1849
Three North American species, eastern North America to Texas and Arizona. Revision and key: Casey (1895). Members can be

very common on riparian vegetation. A large, worldwide group most diverse in the tropics.

Anthicinae Latreille 1819

The Anthicinae may be recognized by: metacoxae moderately to widely separated by strongly divergent intercoxal piece of first visible sternite; line of contact between mesosternum and mesepisterna indicated by distinct suture; lacking longitudinal sulci at lateral margins of elytral disc; neck width at constriction one-fourth or less head width; and with tibial spurs, or (Notoxini) neck very short and width of head base about one half head width. Five tribes are included, with both North American tribes having members world-wide, with the other three tribes (Endomiini, Formicomini, Microhoriini) restricted to the Old World.

Anthicini Latreille 1819

Diagnosis: Members of the Anthicini have the pronotal apex broadly and smoothly curved without any development into a large tubercle, lack pits on the pronotum posterior to the cervical articulations, and have a distinct rounded collar at the pronotal apex encircling the neck. Fifteen of the approximately 25 genera occur in North America.

Amblyderus LaFérte-Senéctère 1849
Inamblyderus Pic 1911
Seven North American species, coastal and interior dune areas. Revision and key: Chandler (1999). Biology: Weissmann and Kondratieff (1999). A moderate-sized world-wide genus.

Anthicus Paykull 1798
Nathicus Casey 1895
Forty-five North American species. Revision and key: Werner (1964). Larvae: Davidson and Wood (1969); Kitayama (1982). This very large genus occurs in all biogeographic areas.

Omonadus Mulsant and Rey 1866
Trapezicomus Pic 1894
Hemantus Casey 1895
Two widely introduced Old World species, now throughout North America. Revision and key: Werner (1964, as *Anthicus*). Until recently treated as a subgenus of *Anthicus* in North American literature. Larvae: Hinton (1945).

Tanarthrus LeConte 1851
Tanarthropsis Casey 1895 (subgenus)
Fifteen species, southwestern states and California to Mexico. Key and revision: Chandler (1975). Biology: Peterman (1973). Larvae: Kitayama (1982). Strongly associated with saline mud flats or beaches.

Leptanthicus Werner 1958
One species, *Leptanthicus staphyliniformis* Werner, Nevada and southern California. Strongly associated with saline mud flats.

Sapintus Casey 1895
Eleven North American species, eastern North America and the Southwest, occurring rarely in Pacific Northwest and Great Basin. Revision and key: Werner (1962, 1983). Larvae: Kitayama (1982; as *Anthicus vexator* Werner). Adults are associated with riparian areas. This is a large, world-wide genus.

Stricticomus Pic 1894
Sticticomus Pic 1894
One widely immigrant Old World species, *Stricticomus tobias* (Marseul), now throughout North America. Description: Werner (1961b). This is a moderate-sized Old World genus.

Baulius Casey 1895
A single species, *Baulius tenuis* (LeConte), southwestern United States and California. Description: Casey (1895).

Malporus Casey 1895
Four species, eastern North America and Arizona to northern Mexico. Revision and key: Chandler (1997). Larvae: Kitayama (1982). Other species occur in Japan.

Cyclodinus Mulsant and Rey 1866
Thicanus Casey 1895
Four North American species, western North America, marine beaches in eastern United States. Revision and key: Casey (1895). Larvae: Kitayama (1982). A large genus that has only a few species in North America, and is lacking from the Neotropics except for some Caribbean Islands. Often found on marine beaches or near inland saline lakes.

Euvacusus Casey 1904
A single species, *Euvacusus coloradanus* Casey, high northern plains into Rocky Mountains. Description: Casey (1904). A rarely collected species.

Vacusus Casey 1895
Five North American species, North America except for the Northeast. Revision and key: Werner (1961a). Larvae: Kitayama (1982). One species, *V. formicetorum* (Wasmann), is found in nests of ants in the genus *Formica*. This genus is most diverse in the Neotropics.

Acanthinus LaFérte-Senéctère 1849
Dilandius Casey 1895
Liobaulius Casey 1904
Seven North American species, southern and midwestern states to the Southwest. Revision and key: Werner (1966-1970). This is a large Neotropical and Australian genus that reaches its northernmost limit in the United States.

Formicilla LeConte 1851
A single North American species, *Formicilla munda* LeConte, southwestern United States to California and Mexico. Revision: Chan-

dler (1973). Larvae: Kitayama (1982). A small Neotropical genus with one species reaching the southwestern states.

Ischyropalpus LaFérte-Senéctère 1849
Lappus Casey 1895
Thirteen species in North America. Revision and key: Werner (1973). Biology: Landwehr (1977). Adults are often found on flowers where they feed on pollen, or on vegetation where they feed on small arthropods such as mites. A large New World genus that is most diverse in the Neotropical region.

Notoxini Stephens 1829

Diagnosis: The Notoxini are readily recognized by: presence of the apicodorsal horn on the pronotum that protrudes anteriorly to cover head; neck short, occipital margin of head articulating closely with pronotum; and pronotum with small to large pits posterior to articulations with head. Eight genera are known, with three of these occurring in North America. Two genera (*Plesionotoxus* Chandler and *Squamanotoxus* Chandler) are wholly or primarily Neotropical, and three others are confined to Africa and the Oriental region. The pronotal horn is typically used for burrowing into sand dunes or sandy soils.

Notoxus Geoffroy 1762
Monocerus Faldermann 1837
Ceratoderus Blanchard 1846
Forty-seven species in North America. Revision and key: Chandler (1982). Many species are attracted to cantharidin and drying meloid beetles (Chandler, 1976). Larvae: Böving and Craighead (1931). A large worldwide genus that is most diverse in Africa, western North America, and Mexico.

Squamanotoxus Chandler 2001
Three species occur from Florida to New Mexico in North America. Revision and key: Chandler (2001). This small New World genus of seven species is most diverse in the Caribbean area.

Mecynotarsus LaFérte-Senéctère 1847
Two species in North America. Revision and key: Chandler (1977). Larvae: Bøving and Craighead (1931). Often associated with sand dunes or sandy soils. A moderate-sized worldwide genus.

Tomoderinae Bonadona 1961

Diagnosis: Members of the Tomoderinae have: metacoxae widely separated by broadly subtruncate intercoxal projection of the first visible sternite; pronotal apex with only thin rim around cervical insertion, lacking distinct collar or flange; neck width at constriction one-fourth head width, neck smooth; pronotum strongly constricted laterally, most with constriction distinct across dorsum; and lacking tibial spurs. Eight genera are placed here, with only the worldwide *Tomoderus* found in North America, another is Neotropical (*Holcopyge* Champion), and the others are from Africa, Australia, and the Oriental region.

Tomoderus LaFérte-Senéctère 1849
Four North American species, eastern United States and Washington. Revision and key: Werner (1958). A large tropical genus that is poorly represented in the Nearctic region, and lacking from the Palearctic region.

Bibliography

ABDULLAH, M. 1960. A revision of the genus *Pergetus* (Coleoptera: Pedilidae). Canadian Journal of Zoology, 38: 875-878.

ABDULLAH, M. 1961. A revision of the genus *Leptoremus* (Coleoptera: Pedilidae). Annals of the Entomological Society of America, 54: 73-75.

ABDULLAH, M. 1963. A key to the genera of Eurygeniini, with redescription of the genus *Bactrocerus* and description of a new genus (Coleoptera. Anthicidae. Pedilinae). Annals and Magazine of Natural History (1962), 13: 595-600.

ABDULLAH, M. 1964a. *Mastoremus idahoensis*, a new species of Pedilinae (Coleoptera, Anthicidae) from southern Idaho. Entomological News, 75: 221-222.

ABDULLAH, M. 1964b. A revision of the genus *Duboisius* Abdullah (Coleoptera, Anthicidae, Pedilinae). Entomologisk Tidskrift, 85: 57-83.

ABDULLAH, M. 1964c. New Nearctic Eurygeniini (Coleoptera, Anthicidae, Pedilinae). Annals and Magazine of Natural History, series 13, 7: 81-94, pl. III.

ABDULLAH, M. 1964d. World species of the genus *Stereopalpus* and a proposed new oriental genus (Col., Anthicidae, Pedilinae). Opuscula Entomologica, 30: 25-78. [preprint distributed Dec. 23, 1964; journal published, Feb. 19, 1965].

ABDULLAH, M. 1964e. A new species of *Neoeurygenius* Abdullah (Coleoptera, Anthicidae, Pedilinae) from Arizona. Entomologisk Tidskrift, 85: 199-202.

ABDULLAH, M. 1965. A revision of the genus *Retocomus* Casey (Coleoptera, Anthicidae). Annales historico-naturales Musei Nationalis Hungarici, 57: 297-328.

ABDULLAH, M. 1967. Some phylogenetic conclusions on the Eurygeniinae (Coleoptera: Anthicidae), with a review of the North American species of *Eurygenius* including the description of a new species (*E. darlingtoni*) from Texas. Entomological News, 78: 180-188.

ABDULLAH, M. 1968. *Salimuzzamania uniformis* (Champion) gen. n. et sp. comb. nov. (Coleoptera: Anthicidae, Eurygeniinae) from Guatemala. Pakistan Journal of Scientific and Industrial Research, 11: 188-189.

ABDULLAH, M. 1969. The natural classification of the family Anthicidae with some ecological and ethological observations (Coleoptera). Deutsche Entomologische Zeitschrift, 16: 323-366.

AKSENTJEV, S. I. 1988. [in Russian] Catalogue of the genus group taxa of Meloidae (Coleoptera). Entomologochiskoye Obozreniye, 68: 569-582.

BØVING, A. G. and F. C. CRAIGHEAD. 1931 (1930). An illustrated synopsis of the principal larval forms of the order Coleoptera. Entomologica Americana, n.s. 11: 1-351.

CASEY, T. L. 1895. Coleopterological notices, VI. Annals of the New York Academy of Science, 8: 435-838.

CASEY, T. L. 1904. On some new Coleoptera, including five new genera. Canadian Entomologist, 36: 312-324.

CHANDLER, D. S. 1973. A revision of the New World genus *Formicilla* (Coleoptera: Anthicidae). Pan-Pacific Entomologist, 49: 31-42.

CHANDLER, D. S. 1975. A revision of *Tanarthrus* LeConte with a presentation of its mid-Cenozoic speciation (Coleoptera: Anthicidae). Transactions of the American Entomological Society, 101: 319-354.

CHANDLER, D. S. 1976. Use of cantharidin and meloid beetles to attract Anthicidae (Coleoptera). Pan-Pacific Entomologist, 52: 179-180.

CHANDLER, D. S. 1977. New *Mecynotarsus* with a key to the New World species (Coleoptera: Anthicidae). Coleopterists Bulletin, 31: 363-370.

CHANDLER, D. S. 1982. A revision of North American *Notoxus* with a cladistic analysis of the New World species (Coleoptera: Anthicidae). Entomography, 1: 333-438.

CHANDLER, D. S. 1997. Revision of the genus *Malporus* Casey (Coleoptera: Anthicidae: Anthicinae). Coleopterists Bulletin, 51: 265-275.

CHANDLER, D. S. 1999. Revision of the North American species of *Amblyderus* with a checklist of the world species (Coleoptera: Anthicidae). Transactions of the American Entomological Society, 125: 269-293.

CHANDLER, D. S. 2001. A new genus of Notoxini based on the *elegans* and *schenklingi* groups of *Mecynotarsus* (Coleoptera: Anthicidae). Transactions of the American Entomological Society, 127: 473-494.

COSTA, C., S. A. VANIN and S. IDE. 1995. Larvae of Neotropical Coleoptera XXII. Description of adults and immatures of *Lagrioida nortoni* sp. n., and bionomics (Coleoptera, Tenebrionoidea, Anthicidae). Iheringia, Série Zoologica, 78: 113-126.

CUTHBERT, F. P., Jr. 1967. Insects affecting sweet potatoes. Agriculture Handbook No. 329. Agriculture Research Service, U. S. D. A., 28 pp.

DAVIDSON, J. A. and F. E. WOOD. 1969. Description and biological notes on the larva of *Anthicus heroicus* Casey (Coleoptera: Anthicidae). Coleopterists Bulletin, 23: 5-8.

DE MARZO, L. 1996. Anatomia dei genitali interni e modalità di inseminazione in alcuni Coleotteri Anticidi. Entomologica (Bari), 30: 147-165.

EISNER, T., S. R. SMEDLEY, D. K. YOUNG, M. EISNER and B. ROACH. 1996a. Chemical basis of courtship in a beetle (*Neopyrochroa flabellata*): Cantharidin as precopulatory "enticing" agent. Proceedings of the National Academy of Sciences, USA, 93: 6494-6498.

EISNER, T., S. R. SMEDLEY, D. K. YOUNG, M. EISNER and B. ROACH. 1996b. Chemical basis of courtship in a beetle (*Neopyrochroa flabellata*): Cantharidin as "nuptial gift." Proceedings of the National Academy of Sciences, USA, 93: 6499-6503.

HEMP, C. and K. DETTNER. 1997. Morphology and chemistry of mesothoracic glands in anthicid beetles (Coleoptera: Anthicidae). Entomologia Generalis 22: 97-108.

HINTON, H. E. 1945. A monograph of the beetles associated with stored products. Vol. 1. British Museum (Natural History). London, viii + 372 pp.

KITAYAMA, C. Y. 1982. Biosystematics of anthicid larvae (Coleoptera: Anthicidae). Coleopterists Bulletin, 36: 76-95.

LANDWEHR, V. R. 1977. *Ischyropalpus nidituls* (Coleoptera: Anthicidae), a predator of mites associated with Monterey Pine. Annals of the Entomological Society of America, 70: 81-83.

LAWRENCE, J. F. 1982. Coleoptera. Pp. 482-553. *In:* S. P. Parker, ed. Synopsis and classification of living organisms, vol. 2. McGraw Hill. New York, 1232 pp.

McCUTCHEON, G. S. and D. M. WEBSTER. 1996. The hooded beetle: a predator of the tobacco budworm. Pp. 796-798. Cotton Insect Research and Control Conference: 1996 Beltwide Cotton Conference.

NIKITSKY, N. B. and A. B. EGOROV. 1992. Fam. Ischaliidae, stat. n. - false fire-red beetles. Pp. 497-498. *In:* P. A. Ler, ed. Keys to the insects of the far eastern USSR in six volumes. Vol. 3, Coleoptera or beetles, Part 2. Nauka. St. Petersburg [in Russian].

PETERMAN, R. M. 1973. Possible behavioral thermo-regulation in *Tanarthrus salinus* and *T. inyo* (Coleoptera: Anthicidae). Pan-Pacific Entomologist, 49: 67-73.

PIC, M. 1911a. Pedilidae; Pars 26: Scraptiidae, Pedilidae. Pp. 11-27. *In:* S. Schenkling, ed. Coleopterorum Catalogus. W. Junk. Berlin.

PIC, M. 1911b. Anthicidae; Pars 36. *In:* S. Schenkling, ed. Coleopterorum Catalogus. Berlin, W. Junk. 102 pp.

SCHÜTZ, C. and K. DETTNER. 1992. Cantharidin-secretion by elytral notches of male anthicid species (Col.: Anthicidae). Zeitschrift für Naturforschung, 47: 290-299.

UHMANN, G. 1976. Die Gattungen der Anthicidae und ihre systematische Anordnung. Entomologische Blätter, 72: 166-182.

WEISSMANN, M. J. and B. C. KONDRATIEFF. 1999. Two new species of *Amblyderus* (Coleoptera: Anthicidae) from Great Sand Dunes National Monument, Colorado. Entomological News, 110: 137-142.

WERNER, F. G. 1958. A revision of the Nearctic species of *Tomoderus* (Coleoptera: Anthicidae). Psyche, 64: 51-59.

WERNER, F. G. 1961a. A revision of the genus *Vacusus* Casey (Coleoptera: Anthicidae). Annals of the Entomological Society of America, 54: 798-909.

WERNER, F. G. 1961b. *Anthicus tobias* Marseul, another tramp species (Coleoptera: Anthicidae). Psyche, 68: 70-72.

WERNER, F. G. 1962. A revision of the Nearctic species of *Sapintus* (Coleoptera: Anthicidae). Annals of the Entomological Society of America, 55: 492-498.

WERNER, F. G. 1964. A revision of the North American species of *Anthicus*, s. str. (Coleoptera: Anthicidae). Miscellaneous Publications of the Entomological Society of America, 4: 195-242.

WERNER, F. G. 1966-1970. A revision of *Acanthinus* (Coleoptera Anthicidae). Annals of the Entomological Society of America. 1966: III, The Spinicollis-group, 59: 1267-1276. 1967: IV, 60: 255-273. 1967: VI, 60: 1217-1234. 1970: VII, 63: 111-128. 1970: IX, The Leporinus, Scitulus, Cuyabanus, and Myrmecops-groups, 63: 719-731. 1970: X, Albicinctus-group and conclusion, 63: 859-876.

WERNER, F. G. 1973. Revision of the Nearctic *Ischyropalpus* (Coleoptera: Anthicidae). Annals of the Entomological Society of America, 66: 1055-1064.

WERNER, F. G. 1974. A new genus of primitive Meloidae from West Texas (Coleoptera). Psyche, 81: 147-154.

WERNER, F. G. 1983. Neotropical *Sapintus*, with a general key to species (Coleoptera: Anthicidae). Proceedings of the Entomological Society of Washington, 85: 405-425.

WERNER, F. G. and D. S. CHANDLER. 1995. Anthicidae (Insecta: Coleoptera). Fauna of New Zealand, No. 34, 59 pp.

WOLDA, H. and D. S. CHANDLER. 1996. Diversity and seasonality of tropical Pselaphidae and Anthicidae (Coleoptera). Proceedings of the Koninklijke Nederlandse Akademie van Wetenschappen, Amsterdam, 99: 313-333.

YOUNG, D. K. 1975. A revision of the family Pyrochroidae (Coleoptera: Heteromera) for North America based on the larvae, pupae, and adults. Contributions of the American Entomological Institute, 11(3), 39 pp.

YOUNG, D. K. 1984. Cantharidin and insects: an historical review. Great Lakes Entomologist, 17: 187-194.

YOUNG, D. K. 1985. Description of the larva of *Ischalia vancouverensis* Harrington (Coleoptera: Anthicidae: Ischaliinae), with observations on the systematic position of the genus. Coleopterists Bulletin, 39: 201-206.

YOUNG, D. K. 1991a. Pedilidae (Tenebrionoidea). False ant-like flower beetles. Pp. 544-546. *In:* F. W. Stehr, ed Immature insects, vol. 2. Kendall/Hunt. Dubuque, IA, xvi + 975 pp.

YOUNG, D. K. 1991b. Anthicidae (Tenebrionoidea). Ant-like flower beetles. Pp. 552-554. *In:* F. W. Stehr, ed Immature insects, vol. 2. Kendall/Hunt. Dubuque, IA, xvi + 975 pp.

118. ADERIDAE Winkler 1927

by Donald S. Chandler

Family common name: The ant-like leaf beetles

Family synonyms: Xylophilidae Shuckard 1840; Euglenesidae (=Euglenidae) Seidlitz 1875; Hylophilidae Pic 1900

The emarginate, hairy, coarsely granulate eyes, transverse deflexed head, head abruptly constricted at the base, and the fused basal two abdominal sternites serve to quickly distinguish most members of this family.

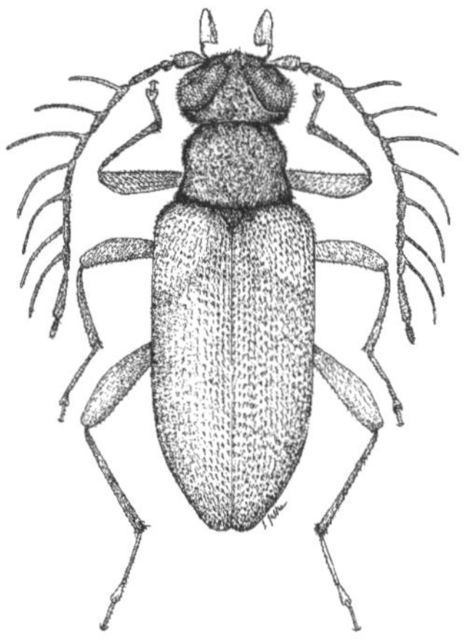

FIGURE 1.118. *Emelinus butleri* Werner

Description: Somewhat ant-like, oval to elongate; size 1-4 mm long; body with sparse distinct tactile setae in a few genera, erect to suberect when present; setae dense, varying from appressed to suberect; undersetae clearly present in some genera, short and appressed, same form as appressed setae in some genera; body mostly with no more than two types of pubescence.

Head strongly deflexed, transverse; abruptly constricted at base (except Phytobaenini); head mostly wider than pronotum, same width in Phytobaenini. Antennae with 11 antennomeres, mostly filiform, subserrate, or clavate, flabellate in males of Emelinini, antennal insertions exposed. Eyes mostly large and coarsely granulate, with anterior margins weakly to deeply emarginate, rounded anteriorly in a few genera, with conspicuous setae between facets; antennae inserted near the eyes or in the emargination of the eyes. Frontoclypeal suture distinct (except *Cnopus*); labrum distinct; mandibles small, curved, apices blunt, emarginate or denticulate. Maxillary palpi with four segments; apical segments large, with posterior mesal margins prominently and angularly protruding. Labium with the gular region short, indistinct, gular pits absent, gular sutures nearly absent, interrupted; submentum indistinct; mentum quadrate; ligula large and partly chitinized, evenly arcuate; labial palps with three segments, third segments large and flattened, oval and angulate on anterior margins, venters membranous and densely setose.

Pronotum trapezoidal to transversely oval, frequently narrowed at apex, base mostly narrower than elytra; lateral margins rounded; pleural region broad; prosternum narrow; procoxal cavities open posteriorly, open internally. Mesosternum short; mesocoxal cavities narrowly to moderately widely separated. Trochantins not evident; procoxae conical, prominent; mesocoxae obliquely transverse, not as prominent as the procoxae; metacoxae nearly contiguous to widely separated, transverse. Trochanters large; pro- and mesofemora slender, metafemora in many enlarged, modified with setal brushes or setose sulci in certain genera; tibiae lacking apical spurs; tarsal formula 5-5-4; first tarsomeres very long, particularly on middle and hind legs; penultimate tarsomeres very small so tarsi appear 4-4-3; antepenultimate tarsomeres lobed ventrally; claws simple. Scutellum triangular to quadrate, moderate to large. Elytra entire, punctation confused; epipleura narrow to lacking. Hind wings lacking radial and anal cells; anal venation reduced; subcubital flecks are entire.

Abdomen with five visible sterna; first two visible sterna fused, suture only partially visible in some groups. Male genitalia with aedeagus inverted (Abdullah 1974, Lawrence 1982).

Larva: Only three genera have had their larvae thoroughly described: *Escalerosia* from Japan (Hayashi 1972); *Megaxenus* from Australasia (Lawrence *et al.* 1990); and *Vanonus* from Mexico (Yanega and Leschen 1994). Larvae reach up to 6 mm; body elongate, subcylindrical to flattened; whitish except for darkened mouthparts and urogomphi; body with scattered, elongate hairs. Head transverse and flattened, exserted from prothorax, narrower than prothorax; epicranial suture with stem short or lacking, frontal arms abruptly divergent at apical third toward antennae; lacking median endocarina; labrum free, lacking frontoclypeal suture (present in *Vanonus*); lacking stemmata; mandibles asymmetrical (nearly symmetrical in *Vanonus*), left mandible with prominent molar tooth, both with well developed mola and fleshy hyaline lobe at base (lacking in *Megaxenus* and *Vanonus*); antennae with sensorium of second segments large, conical; maxillae with 1-segmented cardo, simple maxillary mala, and 3-segmented palps; lacking hypostomal rods. Ninth tergite with pair of fixed, upwardly curved urogomphi at apex (lacking in *Megaxenus*); ninth sternite lacking asperities; anal orifice oriented caudally; spiracles annular (Abdullah 1974, Young 1991).

Habits and habitats. Adults are found by sweeping and beating vegetation, and are usually found resting on the undersides of leaves of various shrubs and trees, particularly angiosperms with broad leaves. Oaks attract several species in North America and Mexico, while *Axylophilus yuccae* Casey was collected on *Yucca*. Báguena (1948) described the general behavior of adults on a number of cultivated tree crops in Africa. Larvae have been taken from rotting wood (Hayashi 1972), bee nests (Yanega and Leschen 1994), termite nests (Lawrence *et al.* 1990), leaf litter, and under bark (Young 1991).

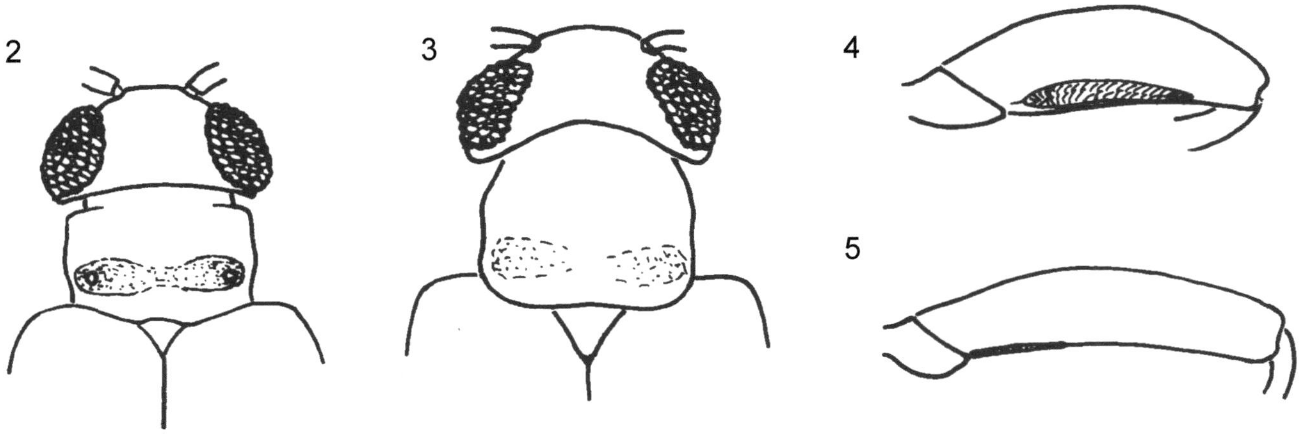

FIGURES 2.118-5.118. 2. Dorsal view head and pronotum, *Cnopus impressus* (LeConte); 3. Dorsal view head and pronotum, *Vanonus densus* (Casey); 4. Anterior view left metafemur, *Pseudariotus notatus* (LeConte); 5. Anterior view left metafemur, *Ariotus quercicola* (Schwarz).

Status of the classification. This family was recently revised for eastern North America (Werner 1990), but the fauna of the West remains largely undescribed. This family is most diverse in the southern tier of states where large, poorly known Neotropical genera reach their northern limits. The world classification is based on Báguena's (1948) massive work on the African fauna, and subsequently modified (Báguena 1962). However, Báguena lacked representatives of almost all North American genera, and his placement of these genera in his classification relied on published descriptions. As a result, the actual characters of the North American genera frequently lead to contradictions with his tribal and subtribal characterizations where they have been placed. The generic placements made here are based on Báguena's (1962) modified classification. Many aderid genera seem to be widely distributed, and it is quite probable that some of the North American genera may eventually be placed in synonymy with some Old World genera or vice versa. Báguena (1948) placed some North American genera as synonyms of Old World genera, but since he lacked representatives of these genera, those placements are not followed here but indicated in the discussion under the relevant genera.

The current bases for separating the tribes and genera badly need to be reexamined, and any statements on generic limits or higher categories should only be made after examining the world fauna. The last world catalogue was that of Pic (1910). Aderidae Winkler 1927 has been given priority over Euglenesidae (=Euglenidae) Seidlitz 1875 (Mroczkowski and Slipinski 1987).

Distribution. There are about fifty genera with approximately 1000 species distributed in all areas but concentrated in the tropics. In North America there are currently 11 genera with 48 species.

Key to the Nearctic Genera

1. Pronotum with broad transverse antebasal sulcus not reaching lateral margins, deepest in lateral portions of sulcus (Fig. 2); head lacking frontoclypeal suture; first metatarsomeres one-third or less length of metatibiae .... *Cnopus*

— Pronotum lacking transverse antebasal sulcus, vague lateral impressions may be present (Fig. 3); frontoclypeal suture present; first metatarsomeres one-half or more length of metatibiae 2

2(1). Eyes with margins closest to antennal insertions rounded .... *Gymnoganascus*

— Eyes with margins closest to antennal insertions weakly to deeply emarginate (Fig. 1) 3

3(2). Elytral disc with large flat-bottomed punctures as wide as eye facets; elytral setae suberect (40-60°), with dense appressed undersetae directed laterally .... *Elonus*

— Elytral disc with punctures smaller than eye facets; setae usually more depressed, undersetae never directed laterally 4

4(3). Setae elongate, decumbent to suberect (20-60°); lacking short appressed undersetae or setae; tactile setae erect at about 90°, often clearly visible along elytral flanks, short to as long as setae . 5

— Setae short, no longer than undersetae, in some genera difficult to separate from dense appressed undersetae; lacking distinct tactile setae 6

5(4). Pronotum with loose tuft of setae at basolateral angles; body elongate, elytra parallel through most of length; elytra with tactile setae easily overlooked, about two-thirds length of decumbent setae (20-30°); males with antennomeres lengthily flabellate (Fig. 1) .... *Emelinus*

— Pronotum lacking tuft of setae at basolateral angles; body usually more robust, elytral margins more curved in most species; tactile setae as long as subdecumbent to suberect setae (30-60°); males with simple antennomeres .... *Zonantes*

6(4). Metafemora simple, lacking ventral setal brush or elongate setose pit 7

— Metafemora with prominent setal brush or setose pit (Fig. 4) extending at least one-fifth of distance along ventral margins; brush is obscure, narrow and appearing as thin carina in some *Pseudariotus* (Fig.5) 9

7(6). Second and third antennomeres small, subequal in length, together about as long as fourth antennomeres (Fig. 6) .... *Aderus*

— Third antennomeres elongate (Fig. 7), subequal in length to fourth antennomeres 8

8(7). Head with frons swollen; with two sulci extending ventrally from antennal insertions and from ventral angle of emarginations of eyes to reach the frontoclypeal suture or the mouthparts; body small, less than 1.1 mm .... *Axylophilus*

— Head with frons flat; lacking sulci between antennae or eyes to mouthparts; body larger, greater than 1.4 mm .... *Ganascus*

9(6). Pronotum convex in basal third, or base vaguely impressed and lateral margins slightly constricted near middle .... *Ariotus*

— Pronotum with pair of oblique faint impressions near base (Fig. 3) 10

10(9). Elytra with short appressed setae; body usually dark brown, only one species from Texas with light bands on elytra .... *Vanonus*

— Elytra with setae decumbent to subdecumbent, with dense undersetae; elytra with light bands or spots .... *Pseudariotus*

6

7

FIGURES 6.118-7.118. 6. Dorsal view right antenna, *Aderus brunnipennis* (LeConte); 7. Dorsal view right antenna, *Ganascus ventricosus* (LeConte).

Classification of the Nearctic Genera

Revision and key: Werner (1990)

Phytobaenini Báguena 1948

Diagnosis. This tribe holds those genera with: head almost hypognathous, width of the head across the eyes no greater than the maximum pronotal width, head partially obscured by pronotum in dorsal view, antennae simple, elytra have both setae and undersetae, and metafemora lack setal brushes. This group is also found in the Neotropics, Australia, and the Oriental region.

Axylophilus Casey 1895

A single species, *A. yuccae* Casey; known only from Florida; adults taken on *Yucca* plants. Revision: Werner (1990).

Ganascus Casey 1895

Sandytes Casey 1895

Two species, *G. ventricosus* (LeConte) and *G. ptinoides* Schwarz, eastern United States. Revision and key: Werner (1990). Other Neotropical species have been seen.

Euglenesini Seidlitz 1875

Diagnosis. Members of this subtribe have: head visible and wider than the pronotum, elytral setae dense and raised, undersetae lacking, antennae inserted on ocular canthus, second antennomeres short and subglobular while third antennomeres more elongate and about as long as fourth antennomeres, and metafemora lack setal brushes. There are two subtribes, with the Pseudolotelina Báguena not yet recorded from North America, though undescribed taxa from Arizona have been seen that would be placed in this subtribe. Members are found on all continents, but only one genus is common in the New World.

Euglenesina Seidlitz 1875

Diagnosis. Members of this subtribe have setal brushes present on the metafemora of both sexes, while members of the Pseudolotelina have setal brushes present only on the metafemora of males.

Zonantes Casey 1895

Ten species, eastern North America to Arizona. Revision and key: Werner (1990). Báguena (1948) placed this group as a subgenus of *Syzeton* Blackburn. This is a large New World genus.

Emelinini Báguena 1948

Diagnosis. This tribe can only be separated from the Euglenesini by the flabellate antennae of the males, while the female antennae are simple and similar to those of females in the Euglenesini. The body lacks undersetae, the setae are suberect and simple, the head is wider than the pronotum and clearly visible, and the metafemora lack setal brushes. This tribe is based on two genera, one from the New World and the other from Africa.

Emelinus Casey 1895

Three species, eastern North America and Arizona. Revision and key: Werner (1956, 1990). Other species are found in Mexico and Central America.

Aderini Winkler 1927

Diagnosis. The Aderini are based on genera that have: head clearly visible in dorsal view, antennae simple, and elytra with both setae and undersetae, though both may be appressed and indistinguishable. There are four subtribes, all of these with representatives in North America.

Cnopina Báguena 1948

Diagnosis. The single genus in this subtribe, *Cnopus* Champion, has decumbent to appressed setae on the elytra, the metatarsomeres are one-third or less the length of the metatibiae, the metafemora lack setal brushes, and (not mentioned by Báguena) the pronotum has a broad antebasal sulcus that doesn't reach the lateral margins, while the head lacks a frontoclypeal suture.

Cnopus Champion 1893
Two species, *C. impressus* (LeConte) and *C. nucleus* (Fall), respectively southeastern United States, and Arizona to California. Revision: Werner (1990, eastern species). A small Neotropical genus.

Aderina Winkler 1927

Diagnosis. The subtribe Aderina holds those groups that have: first metatarsomeres one-half or more length of metatibiae, third antennomeres not elongate, about as long as second antennomeres, and metafemora lacking setal brushes. This group has representatives in all biogeographic regions.

Aderus Stephens 1829
Xylophilus Latreille 1825 (not *Xylophilus* Mannerheim 1823)
Hylophilus Berthold 1827 (not *Hylophilus* Temminck 1822)
Phomalus Casey 1895
Four species. Revision and key: Werner (1990). A world-wide genus, with one species, *P. populneus* (Panzer), immigrant from Europe.

Syzetoninina Báguena 1948

Diagnosis. Members of this subtribe have: third antennomeres elongate, longer than second antennomeres, first metatarsomeres one-half or more length of metatibiae, and metafemora with setal brushes or setose sulci. Represented in all biogeographic regions.

Vanonus Casey 1895
Tanilotes Casey 1895
Thirteen species, eastern North America. Revision and key: Werner (1990). Other species are found in Mexico, Central America, and Europe. Biology: Yanega and Leschen (1994). Báguena (1948) placed this genus as a synonym of *Pseudanidorus* Pic.

Ariotus Casey 1895
Scanylus Casey 1895
Four species, eastern North America and Arizona. Revision and key: Werner (1990). Other species are found in Mexico and Central America. Báguena (1948) treated this genus as a synonym of the world-wide group, *Syzeton* (subgenus *Syzeton*) Blackburn.

Pseudariotus Casey 1895
A single species, *P. notatus* (LeConte), eastern United States. Revision: Werner (1990). This is a Neotropical genus, which Báguena (1948) treated as a subgenus of *Syzeton* Blackburn.

Elonus Casey 1895
Six species, eastern North America to the Southwest and Arizona. Revision and key: Werner (1990, 1993). This distinctive genus is diverse in Mexico and Central America.

Olotelina Báguena 1948

Diagnosis. This subtribe is characterized by: third antennomeres elongate, longer than second antennomeres, first metatarsomeres one-half or more length of metatibiae, and metafemora lack setal brushes. Placement of *Gymnoganascus* here is the first definite record of this group for North America.

Gymnoganascus Werner 1990
A single species, *G. stephani* Werner, Kentucky and Texas to Mexico and Cuba. Description: Werner (1990). Members of this genus have also been seen from Panama.

Incertae sedis

Xylophilus Latreille 1825 (junior homonym of *Xylophilus* Mannerheim, 1823); senior available name is *Aderus* Stephens

A single North American species, *Xylophilus constrictus* Fall, is found in California and Arizona. It has not been moved from *Xylophilus* (=*Aderus*), and while it is not a member of this genus, it cannot be placed in any described genus. The pronotum has transverse antebasal and anteapical sulci, with neither being present in any of the known world genera. The body lacks tactile setae and undersetae, and the legs lack setal brushes.

Bibliography

ABDULLAH, M. 1974. A comparative study of the adults and larvae of Xylophilidae and 32 other families of Cucujoidea (Coleoptera). Entomologische Arbeiten der Museum G. Frey, 25: 281-315.

BÁGUENA, L. 1948. Estudio sobre los Aderidae (Coleópteros Heterómeros). Instituto de Estudios Africanos. Madrid, xiv + 547 pp.

BÁGUENA, L. 1962. Aderidae (Coleoptera Heteromeroidea). Exploration du Parc National de la Garamba. Mission H. de Saeger, Fasc. 26. Imprimerie Hayez. Brussels, 135 pp.

HAYASHI, N. 1972. On the larvae of some species of Colydiidae, Tetratomidae and Aderidae occurring in Japan (Coleoptera: Cucujoidea). Kôntyu, 40: 100-111.

LAWRENCE, J. F. 1982. Coleoptera. Pp. 482-553. *In:* S. P. Parker, ed. Synopsis and classification of living organisms, vol. 2. McGraw Hill. New York, 1232 pp.

LAWRENCE, J. R., D. H. KISTNER and J. M. PASTEELS. 1990. A new genus and three new species of termitophilous Aderidae (Coleoptera) from Australia, Papua New Guinea and the Philippines, with notes on their biology. Invertebrate Taxonomy, 4: 643-654.

MROCZKOWSKI, M. and S. A. SLIPINSKI. 1987. EUGLENIDAE Stein, 1878 (Protista, Flagellata) and EUGLENIDAE Seidlitz, 1875 (Insecta, Coleoptera): proposals to remove the homonymy, with conservation of ADERIDAE Winkler, 1927 (Insecta, Coleoptera). Bulletin of Zoological Nomenclature, 44: 230-232.

PIC, M. 1910. Hylophilidae, Pars 14. *In:* S. Schenkling, ed. Coleopterorum Catalogus, 25 pp.

WERNER, F. G. 1956. Two new species of *Emelinus* from Arizona (Coleoptera: Aderidae). Psyche, 63: 30-37.

WERNER, F. G. 1990. Revision of the Aderidae of eastern North America. Journal of the New York Entomological Society, 98: 187-232.

WERNER, F. G. 1993 (1992). The Nearctic species of *Elonus* (Coleoptera: Aderidae). Psyche, 99: 245-264.

YANEGA, D. and R. LESCHEN. 1994. Beetles associated with bee nests (Hymenoptera: Apidae) in Chiapas, Mexico, with descriptions of the immature stages of *Vanonus balteatus* Werner (Coleoptera: Aderidae, Endomychidae, Meloidae). Coleopterists Bulletin, 48: 355-360.

YOUNG, D. K. 1991. Euglenidae (Tenebrionoidea). Ant-like leaf beetles. Pp. 554-555. *In:* F. W. Stehr, ed. Immature insects, vol. 2. Kendall/Hunt. Dubuque, IA, xvi + 975 pp.

119. SCRAPTIIDAE Mulsant 1856

by Darren A. Pollock

Family common name: The false flower beetles

Family synonym: Anaspididae Mulsant 1856

This family is comprised of two subfamilies of quite different appearance, and therefore is difficult to characterize. The constituents of Scraptiidae have been placed historically among the Melandryidae, Mordellidae, and other families. Most scraptiines may be recognized by their soft bodies, deeply emarginate eyes, and conspicuously setose vestiture. Anaspidines have a very abruptly narrowed head, and most have distinct, transverse elytral strigae dorsally, without conspicuous dorsal vestiture.

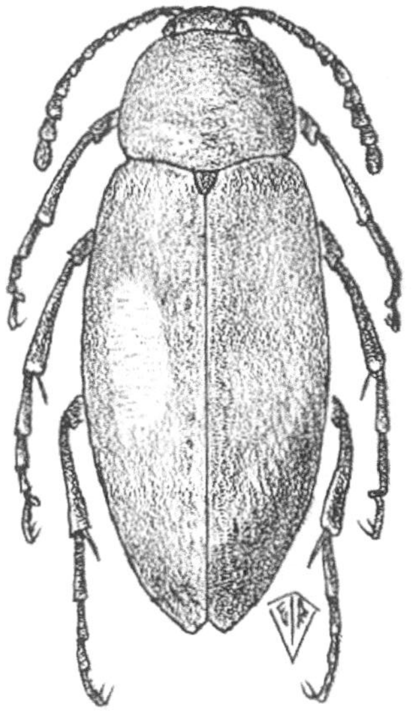

FIGURE 1.119. *Anaspis rufa* Champion

Description: Body elongate, about 2.0-2.8 times longer than maximum width, parallel-sided to slightly ovate; slightly flattened to moderately convex dorsally; dorsum with distinct punctation or other macrosculpturing, with (Scraptiinae) or without (Anaspidinae) distinct vestiture; length 1.0-13.5 mm; color of most taxa uniform, testaceous to black, other taxa with contrasting color patterns.

Head relatively short, slightly to distinctly declined ventrally; temples either shorter or longer (some Scraptiini) than length of eye, associated closely with anterior margin of pronotum, or not; eyes not, or moderately protuberant, with facets coarse (Scraptiinae) or fine (Anaspidinae); anterior margin of eye slightly to deeply emarginate near antennal insertion; antennae relatively short (most Anaspidinae) to moderately long (some Scraptiinae), 11-segmented; antennomeres moniliform, filiform, or flabellate (Scraptiinae: *Pectotoma*), with or without distinct club; antennal insertions exposed dorsally, moderately to widely separated; frontoclypeal suture indistinctly to distinctly impressed, straight to curved; labrum slightly to distinctly transverse, with subtruncate to emarginate apex; mandibles short and broad, apex moderately to distinctly curved mesally, with distinct mola and prostheca; maxilla with galea and lacinia distinct, apical palpomere of various shape, from slightly expanded to cultriform to broadly securiform; apex of galea densely setose (Scraptiinae) or not (Anaspidinae); labial palpi small, with apical palpomere simple, truncate, or broadly expanded, cultriform (some Scraptiinae), or deeply emarginate to crescent-shaped (some Anaspidinae).

Prothorax widest posteriorly, ratio of pronotal length to maximum width 0.5-1.1; disc of pronotum flattened to slightly and evenly convex, without distinct paired depressions; posterolateral area of disc with small punctiform depression, especially in Scraptiinae; sides of pronotum evenly arcuate; lateral margins of pronotum disc carinate in posterior half, smooth or only indistinctly carinate near anterior margin; posterolateral angles of pronotum rounded, slightly obtuse, to nearly right; prosternum anterior of coxae very short to short in length, flat or slightly sunken medially; procoxal process short, triangular, not extended between coxae; procoxae projecting, well below level of prosternum; procoxal cavities externally broadly open, contiguous medially; prothoracic protrochantin exposed or not.

Elytra elongate to subovate, covering entire abdomen; disc somewhat flattened to moderately convex, with uniform punctation and long, raised setae (Scraptiinae) or with distinct transverse reticulation and shorter, adpressed setae (Anaspidinae); epipleuron distinct anteriorly, evenly or abruptly narrowed, not visible to elytral apex; scutellum visible, bluntly rounded to triangular posteriorly; mesosternum flattened to slightly convex, anteriorly blunt or rounded; mesosternal process narrow, extended between coxae; mesocoxal cavities open laterally; hind wing long, functional; venation normal; radial cell present or absent; 3-4 free veins in medial area of wing; metasternum variously convex, discrimen distinct, about half length of sternum; legs similar in size and shape on all thoracic segments; femora slightly swollen near midlength, somewhat compressed laterally; tibiae slender, terminal spurs small to moderate in length; tarsi 5-5-4 in both sexes; penultimate tarsomere variously expanded laterally, forming fleshy ventral lobe (esp. pronounced in some Scraptiinae), absent from hind tarsomere of Anaspidinae and some Scraptiinae.

Abdomen with 5 free ventrites, ventrites entirely simple (Scraptiinae and most Anaspidinae), with 1-2 pairs of digitiform, articulated appendages on third ventrite of males (*Anaspis, sensu stricto*), or some ventrites medially with very long setae (*Larisia*); aedeagus tenebrionoid, symmetrical, with or without distinct parameres; medial lobe slender, with or without apical knob-like enlargement.

Larvae (Young 1991) subcylindrical, subparallel, lightly sclerotized except for mouthparts and abdominal apex; body surface smooth, with vestiture of fine setae. Head prognathus, exserted from or slightly retracted within prothorax; epicranial suture with short stem (Anaspidinae) or absent (Scraptiinae); frontal arms lyriform, extended to antennal insertions; endocarinae absent; labrum symmetrical, frons and clypeus fused (Anaspidinae) or separated by distinct suture (Scraptiinae); stemmata 1 on each side (Anaspidinae) or absent (Scraptiinae); antennal insertions exposed; antennae 3-segmented, segment 2 with small dome-like sensorium; retracted mouthparts with short (Anaspidinae) or no (Scraptiinae) hypostomal rods; mandibles distinctly sclerotized, symmetrical to slightly asymmetrical; mola distinct, with (Anaspidinae) or without (Scraptiinae) basal brush of stout spines; maxilla with single-segmented cardo, and distinct articulating area; mala simple (Scraptiinae) or cleft subapically (Anaspidinae); maxillary palpi 3-segmented; labium free to base of mentum, with short (Anaspidinae) or long (ligula) and 2-segmented palpi; hypopharyngeal sclerome transverse; gular sutures separated. Thorax elongate, sides subparallel; legs distinct, 5-segmented, of similar shape, each with fine setae. Abdomen subcylindrical, only indistinctly sclerotized; tergite 9 oriented ventrally, bearing urogomphi (Anaspidinae) or dorsally, with large, oblong, dehiscient process (Scraptiinae); sternite 9 lacking asperities, not enclosed by sternite 8; segment 10 reduced.

Habits and habitats. Adults of Scraptiinae and especially Anaspidinae are floricolous and may be taken in very large numbers from blossoms of Apiaceae and Rosaceae (Young 1991). Larvae of *Canifa* have been collected under bark of dead *Populus* sp. (Pollock, pers. observ.), and larvae of other taxa have been found associated with dead logs or lichens (Young 1991). Hatch (1965) reported that larvae of *Anaspis* occur under loose tree bark. Adults of an unidentified species of *Pentaria* were found commonly within unopened ears of corn in northern Texas (Pollock, pers. observ.).

Status of the classification. Lawrence and Newton's (1995) concept of Scraptiidae differs little from that of Crowson (1955), in that two subfamilies are recognized: Scraptiinae and Anaspidinae. The latter has been placed in its own family, or even as members of Mordellidae (e.g., Liljeblad 1945). Crowson (1955) recognized affinities of Scraptiidae for Melandryidae and Mordellidae (*i.e.*, "lower" tenebrionoids), but Lawrence and Newton (1995) placed Scraptiidae among the presumably more highly derived Tenebrionoidea. It is possible that a close relationship exists between Scraptiidae (especially Anaspidinae) and Anthicidae, at least based on larval characters (Young 1991). Females of the four genera of Anaspidini (*Anaspis*, *Larisia*, *Nassipa* and *Silaria*) are very difficult to differentiate; most keys to these genera have included the secondary male sexual characteristics, without mentioning females (e.g., Liljeblad 1945). Because of this inadequacy, these four genera are grouped under the single genus *Anaspis* in the key to Nearctic genera. Comments on classification are given below, for individual taxa.

Distribution. This family contains an estimated 30 genera and about 400 species world-wide (Lawrence 1982). In North America, there are 46 described species in 13 genera (Poole and Gentili 1996). Most of the genera (in North America, especially *Canifa*, *Scraptia*, and *Anaspis*) are widespread, while several are more or less restricted in distribution.

Key to the Nearctic Genera

1. Pronotum and elytra punctate, without distinct transverse strigae (Scraptiinae) 2
— Pronotum and at least base of elytra with distinct transverse strigae (Anaspidinae) 6

2(1). Apical maxillary palpomere securiform 3
— Apical maxillary palpomere cultriform (Fig. 2) .... *Canifa*

3(2). Penultimate tarsomeres on all legs lobed ventrally (Figs. 9, 11-12) 5
— Penultimate hind tarsomere not lobed ventrally (Fig. 10) 4

4(3). Eye attaining posterior margin of head; antennomeres in both sexes similar, not flabellate .... *Allopoda*
— Eye separated from posterior margin of head by nearly width of single eye; male with flabellate antennae (Fig. 6); female with filiform antennae (Fig. 7) .... *Pectotoma*

5(3). Apical labial palpomere narrow distally; emargination of eye rather shallow; combined length of second and third antennomere greater than length of fourth antennomere; dorsal vestiture depressed .... *Scraptia*
— Apical labial palpomere widened, triangular; emargination of eye deep; combined length of second and third antennomere less than length of fourth antennomere; dorsal vestiture erect to decumbent .... *Neoscraptia*

6(1). Fourth pro- and mesotarsomere relatively elongate, at least half length of third tarsomeres (Fig. 11) 7
— Fourth pro- and mesotarsomeres short, less than half length of, and often concealed between lobes of third tarsomeres (Fig. 12) .... *Anaspis* (*s. lat.*)

7(6). Body size usually < 2 mm; antennae short, antennomeres moniliform to only slightly filiform (Fig. 5), with antennomeres 6-11 gradually widened 8
— Body size usually > 2 mm; antennae more slender, antennomeres 2-4 at least slightly filiform; antennomeres 6-11 only indistinctly widened (*e.g.*, Fig. 8) 9

8(7). Transverse strigae present only on base of elytra; last labial palpomere slender, acuminate .... *Naucles*
— Transverse strigae present on entire elytra; last labial palpomere expanded, lunate .... *Sphingocephalus*

9(7). Antennae relatively elongate, antennomeres 3 and 4 filiform, subequal in length, each subequal in length to combined lengths of antennomeres 1 and 2 (Fig. 8) .... *Diclidia*
— Antennae short, antennomeres 3 and 4 stout, each much shorter than combined lengths of antennomeres 1 and 2 .... *Pentaria*

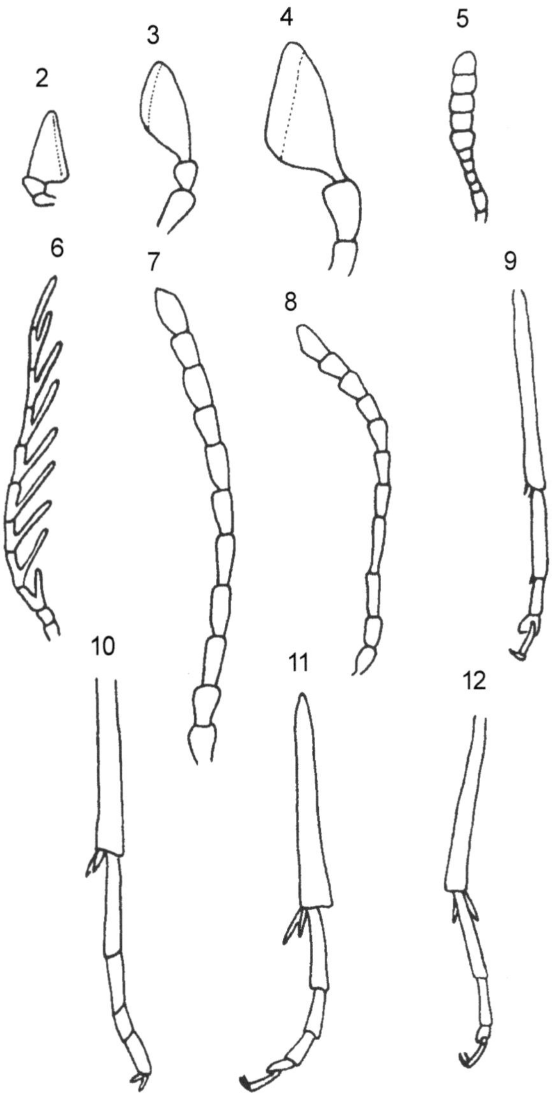

FIGURES 2.119-12.119. 2-4. Maxillary palpus. 2. *Canifa* sp.; 3. *Allopoda lutea* (Haldeman); 4. *Scraptia* sp.; 5-8. Antenna. 5. *Naucles* sp.; 6. *Pectotoma hoppingi* Hatch, male; 7. *P. hoppingi*, female; 8. *Diclidia* sp.; 9-10. Hind leg, tibia and tarsus. 9. *Scraptia* sp.; 10. *Allopoda lutea*; 11-12. Middle leg, tibia and tarsus. 11. *Diclidia* sp.; 12. *Anaspis rufa* Say.

CLASSIFICATION OF THE NEARCTIC GENERA

Scraptiidae Mulsant 1856

Scraptiinae Mulsant 1856

Taxa now placed among the Scraptiinae historically have been included mainly in Melandryidae. Franciscolo (1972) lists 10 genera in Scraptiinae ("Philum scraptioide").

Diagnosis. Members of Scraptiinae may be distinguished from other groups of Tenebrionoidea by the following combination of characters: body soft, integument rather flimsy in construction; head rounded behind eyes, not distinctly and sharply constricted; antennae filiform or pectinate, distal antennomeres not widened or forming club; pronotum and elytra punctate, not transversely strigate; elytra with distinct vestiture; known larvae with dehiscient appendage at abdominal apex.

Allopodini

Allopoda LeConte 1866, 3 spp., New York, New Jersey, Indiana, Maryland, Florida, Arizona, and California. This genus is usually placed among Scraptiinae, but is included in Anaspidinae by Lawrence and Newton (1995). Franciscolo (1964) placed *Allopoda* and the tribe Allopodini in Scraptiinae, but then in 1972 placed the genus and tribe in his "Philum pentarioide," in the subfamily Anaspidinae. *Allopoda* lacks the transverse strigae, that are generally characteristic of all Anaspidinae.

Scraptiini

Scraptia Latreille 1807, 2 spp., New York, Indiana, District of Columbia, Florida, and Arizona.

Neoscraptia Fender 1946, 1 sp., *N. testacea* Fender 1946, Pacific Northwest.

Canifa LeConte 1866, 4 spp., widespread in Canada and United States.

Pectotoma Hatch 1965, 1 sp., *P. hoppingi* Hatch 1965, Montana, Alberta, British Columbia, Arizona

Anaspidinae Mulsant 1856

The anaspidine scraptiids were treated as Mordellidae in many early classifications, *e.g.*, Liljeblad (1945), or as a separate family, *e.g.*, Bøving and Craighead (1931). There are approximately 20 recognized genera of Anaspidinae (Franciscolo 1972).

Diagnosis: Members of Anaspidinae may be distinguished by the following combination of characters: body oblong, distinctly tapered posteriorly, somewhat similar to that of Mordellidae; head distinctly and sharply constricted behind eyes; antennae relatively short, antennomeres gradually widened distally; elytra with distinct, transverse micro-reticulation; pubescence indistinct, adpressed.

Pentariini

Diclidia LeConte 1862, 9 spp., Ohio, Colorado, Texas, New Mexico, Arizona, and California (key to spp., Liljeblad 1945).

Pentaria Mulsant 1856, 6 spp., generally distributed (key to spp., Liljeblad 1945).

Anthobates LeConte 1850, not Gistel 1847
Anthobatula Strand 1929

Naucles Champion 1891, 3 spp., southwestern United States (key to spp., Liljeblad 1945).

Sphingocephalus Liljeblad 1945, 1 sp., *S. ovalis* Liljeblad 1945, Florida.

Anaspidini

Anaspis Geoffroy 1762, 13 spp., widely distributed. Franciscolo (1972) used a broad concept of *Anaspis*, including in it the three genera listed below, which were accorded full generic status by Liljeblad (1945) and Hatch (1965). However, in both treatments, the keys work only for males; females of *Anaspis* are virtually indistinguishable from females of these three genera. They are included within *Anaspis* in this work, although it is likely that they are deserving of at least subgeneric rank (key to spp., Liljeblad 1945).

Silaria Mulsant 1856
Larisia Emery 1876
Nassipa Emery 1876

Bibliography

BØVING, A. G. and F. C. CRAIGHEAD. 1931. An illustrated synopsis of the principal larval forms of the order Coleoptera. Entomologica Americana (N.S.), 11 (1930): 1-351.

CROWSON, R. A. 1955. The natural classification of the families of Coleoptera. N. Lloyd. London.

FRANCISCOLO, M. E. 1964. Nota preliminare sulla filogenia degli Scraptiidae (Coleoptera Heteromera). Atti dell'Accademia Nazionale Italiana di Entomologia, Bologna, 11: 175-181.

FRANCISCOLO, M. E. 1972. Su alcuni generi poco noti di Anaspidinae. Memorie della Società Entomologica Italiana, 51: 123-155.

HATCH, M. H. 1965. The Beetles of the Pacific Northwest. Part IV: Macrodactyles, Palpicornes, and Heteromera. University of Washington Press. Seattle. viii + 268 pp.

LAWRENCE, J. F. 1982. Coleoptera. Pp. 482-553. *In*: S. P. Parker, ed. Synopsis and Classification of Living Organisms. Volume 2. McGraw-Hill. New York.

LAWRENCE, J. F. and A. F. NEWTON, Jr. 1995. Families and subfamilies of Coleoptera (with selected genera, notes, references and data on family-group names). Pp. 779-1006. *In*: J. Pakaluk and S. M. Slipinski, eds. Biology, Phylogeny, and Classification of Coleoptera. Papers Celebrating the 80th Birthday of Roy A. Crowson. Volume 2. Muzeum i Instytut Zoologii PAN. Warsaw.

LILJEBLAD, E. 1945. Monograph of the family Mordellidae (Coleoptera) of North America, north of Mexico. Miscellaneous Publications, Museum of Zoology, University of Michigan, 62: 229 pp.

PIC, M. 1911. Scraptiidae, Coleopterorum Catalogus, 17(26): 1-10.

POOLE, R. W. and P. GENTILI, eds. 1996. Fauna Insecta Nearctica. Vol. 1: Coleoptera, Strepsiptera. Entomological Information Services. Rockville MD. 827 pp.

YOUNG, D. K. 1991. Scraptiidae (Tenebrionoidea). Pp. 555-556. *In*: F. W. Stehr, ed. Immature Insects. Volume 2. Kendall/Hunt. Dubuque, IA. xvi + 975 pp.

Superfamily CHRYSOMELOIDEA Latreille 1802

Cerambycoidea Leng 1920; Phytophagoidea Peyerimhoff 1933; Phytophaga LeConte and Horn 1883; Pseudotetramera Westwood 1839; Tetramera Leach 1815.

120. CERAMBYCIDAE Leach 1815

by Robert H. Turnbow, Jr. and Michael C. Thomas

Family common name: The longhorn beetles

Family synonyms: Disteniidae Thomson 1860; Hypocephalidae Imhoff 1856; Lamiidae Latreille 1825; Lepturidae Leach 1815; Prionidae Leach 1815; Spondylidae LeConte 1873; Totropiidae *auctorum*

The long antennae, absence of a beak, and characteristic tarsi serve to separate most adults of this very large family from other beetles.

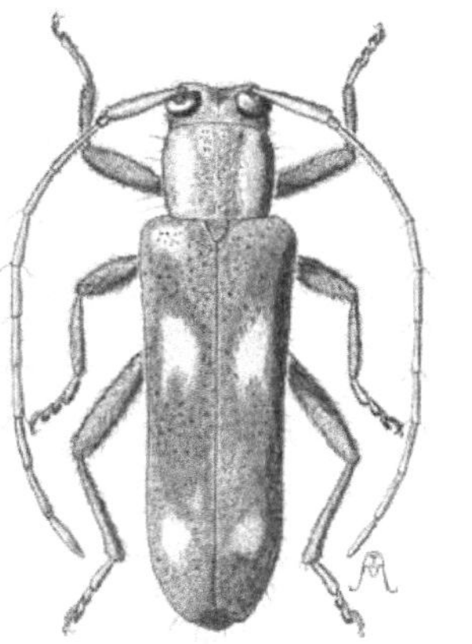

FIGURE 1.120. *Saperda cretata* Newman

Description: Shape extremely variable, most more or less elongate and cylindrical or depressed, may be short and depressed; size 3 to 150 mm or more in length; color generally brownish or with bright colors; vestiture variable, glabrous, hairy, or scaly.

Head long in most, slightly to strongly deflexed; surface smooth, punctate, or rugose. Antennae eleven-segmented, ten-segmented in a few, some with twelve to twenty-five or more antennomeres; most very long, longer than body, some several times length of body, may be shorter than body; insertion variable, most insert in emargination of eyes. Labrum prominent in most; mandibles variable, most stout, some very long, shape curved with apices acute in most, may be blunt or dentate; maxillar with four palpomeres, apical palpomere largest; gular sutures present or absent; mentum variable, transverse and entirely corneous to trapezoidal, and membranous at apex; ligula membranous or corneous; labium with three palpomeres, palpomeres moderate, filiform. Eyes lateral, most with deep emargination, may be divided; moderate in size.

Pronotum variable, quadrate, oval, or elongate; borders margined only in Prioninae; surface variable; pleural region broad; prosternum long in front of coxae, with broad prosternal process; procoxal cavities closed behind. Mesosternum short. Metasternum moderate or long. Legs with trochantins hidden in most; procoxae globular, separate; mesocoxae round, flat, separate; metacoxae transverse, somewhat separate; trochanters moderate, triangular; femora of most somewhat swollen; tibiae slender, apical spurs present in most, moderate; tarsal formula 5-5-5, most apparently 4-4-4 because of minute fourth tarsomere in cleft of bilobed third tarsomere, but may be distinctly 5-5-5; claws of most simple, a few with appendiculate or cleft claws. Scutellum moderate, most triangular. Elytra entire, but may be very short, exposing most or all of abdomen; surface smooth, punctate, or rugose, some are striate; epipleural fold distinct, entire, or absent beyond basal third. Wing venation with anal area more or less normal; folding pattern very similar to Erotylidae and related families; some are apterous.

Abdomen with five sternites, some with six; sutures entire in most; surface of most microrugose. Male genitalia with penis stout, curved, with very stout, uniform, compressed basal apodeme; parameres fused to pars basalis, some with separating suture, distally setiferous, stout, or moderate; pars basalis hood-shaped, with slender basal struts fused ventrally. Female genitalia with paraprocts present; valvifers some reduced to a baculum; coxite two-segmented; stylus present; proctiger present. Larvae elongate, or thick and robust, fleshly, cylindrical or somewhat depressed, thorax wider than abdomen of most, constricted between abdominal segments in most; size 15 to 80 or more mm in length; vestiture with fine setae; color near white. Head with sides rounded and broad behind or sides straight and convergent behind, somewhat depressed, deeply inserted into prothorax, but extensible; occipital foramen large; epistomal ridge conspicuous in most; epicranial sutures surrounding triangular frons, with a distinct coronal suture. Antennae three-segmented, inconspicuous, with large basal membrane into which antennae may be retracted. Clypeus and labrum small; mandibles quadrangular, or rounded, robust, with sharp cutting edge, gouge-like, dentate in a few; maxillae with distinct cardo, stipes, three-segmented palpi and lobe-like mala; labium with gula, submentum, mentum, ligula, and two-segmented palpi. One to five pairs of ocelli present. Thorax wider than head or abdomen, smooth, shining, granulated, recticulated, pinnately striated, velured, or asperate; legs very small, widely separated, five-segmented, with spine-like tarsus, or legs absent. Abdomen nine-segmented, telescoping, some with ambulatory ampullae, prominent, fleshy, on segments one to six or one to seven; a distinct lateral fold present on most abdominal segments; ninth segment with one to two sclerotized spines dorsally, and with three anal lobes. Spiracles elliptical, bilabiate, oval, or annular, on mesothorax, or between prothorax and mesothorax, and abdominal segments one to eight.

Habits and habitats. The adults feed on wood, roots, leaves, pollen, and are rarely carnivorous. The larvae bore into wood and roots. Some, such as *Hylotrupes* spp., are structural pests in wood-frame buildings. Eggs are laid in or under bark or in cracks in the wood. The females of the first group lay their eggs in crevices by means of long ovipositors or gnaw a slit which penetrates to the cambium, but are devoid of an ovipositor, the head being better adapted for gnawing. Larval tunnels are of three types: a) between the bark and wood, serving also as the pupal chamber, as in *Rhagium* spp.; b) between the bark and wood but with the pupal chamber in the wood, as in *Clytus* spp., *Callidium* spp., and others; c) between the bark and wood, but mostly in the wood, where the pupal chamber is located, as in *Monochamus* spp. Among some species, the position of the tunnel can be modified to suit conditions. The adults feed: a) on pollen and parts of flowers, as in Lepturini; b) on green parts of plants, leaves, needles, as in *Saperda* spp.; c) on bark of twigs, leaf-stems, even leaf ribs, as in *Monochamus* spp.; d) not at all during the adult stage, so far as is known. A host plant index for the North American fauna has been compiled by Linsley and Chemsak (1997).

The larvae of some Lamiinae live in the stems of herbaceous plants (e.g., *Oberea* spp., *Mecas* spp., *Dorcasta* spp.). There are among the Lamiinae some species in which the female, or sometimes both sexes together (as in *Oncideres* spp.), girdle the branches of various trees below the spot where the eggs have been deposited, the resulting girdle providing the larvae with suitable host material. The life histories of most species are unknown, but Solomon (1995) offers an account of many species attacking hardwood trees and shrubs.

Status of the classification. Our knowledge of the Cerambycidae of North America has been greatly expanded over the past 40 years. A monographic treatment of the family has been completed (Linsley 1961, 1962a, 1962b, 1963, 1964 and Linsley and Chemsak 1972, 1976, 1984 and 1995). To complete that series, Linsley and Chemsak (1997) compiled an extensive bibliography for the family in North America. Illustrated identification manuals have been produced for the Parandrinae, Spondylidinae, Aseminae and Prioninae (Chemsak 1996) and the fauna of the northeastern United States (Yanega 1996). In addition, a checklist of the Cerambycidae of the Western Hemisphere has been completed (Monné and Giesbert 1994). Although additional species have been described, synonymies have been recognized and changes have been implemented in the placement of the various genera within higher taxa since the appearance of these various publications, as references, they remain extremely useful as guides to the North American fauna. The longhorn beetles have always been popular with collectors because of their bright colors, highly adapted and varying morphologies and interesting habits. Taxonomically, however, the family is a difficult group, and the relationships within the family are not well understood. The position of the family near the Chrysomelidae and Curculionidae has not been challenged.

Distribution. More than 20,000 species have been described worldwide. In the Western Hemisphere, more than 8,700 species and subspecies have been described in almost 1500 genera. In America north of Mexico, more than 900 species are known in 300+ genera.

Keys to the Subfamilies, Tribes and Genera of the Cerambycidae of North America

I. Key to the Subfamilies of the Cerambycidae of North America (adapted from Chemsak 1996)

1. Mandibles scalpriform, clypeus oblique to frons; wings lacking a spur on the radio-medial crossvein (Disteniinae) .... *Distenia*
— Mandibles not as above; wings with a spur on the radio-medial crossvein 2

2(1). Tarsi distinctly pentamerous, without pubescent ventral pads, third tarsomere not dilated, not concealing fourth tarsomere (Fig. 2); antennae short, a few surpassing base of pronotum, sensory areas deeply impressed, second antennomere more than 1/2 as long as third, remaining antennomeres subequal in length 3
— Tarsi pseudotetramerous, padded beneath, third tarsomere dilated, concealing true fourth tarsomere (Fig. 3); antennae surpassing base of pronotum, most very long, sensory areas diffused, second antennomere in a few 1/2 as long as third, remaining antennomeres unequal in length 4

3(2). Pronotum with elevated lateral margin (Fig. 4); labrum fused with epistoma; protibiae without apical lamellae; third tarsomere entire or feebly emarginate; wings with radial cell open, cubito-anal crossvein present (Parandrinae) .... Key II
— Pronotum without lateral margin; labrum free; protibiae with broad terminal lamellae (Fig. 5); third tarsomere deeply bilobed; wings with radial cell closed, cubito-anal crossvein absent (Spondylidinae) .... Key IV

4(2). Head obliquely inclined anteriorly or subvertical, genal margin never directed posteriorly (Fig. 6); protibiae without mesial sinus; mesotibiae never notched or grooved externally; apical maxillary palpomere obtuse or truncate at apex 5
— Head vertical or retracted, genal margin always directed posteriorly (Fig. 7); protibiae with mesal sinus, mesotibiae of most notched or grooved externally (Fig. 8); apical maxillary palpomere pointed at apex (Lamiinae) .... Key VIII

5(4). Pronotum with elevated lateral margin; labrum fused with epistoma; inner lobe of maxillae obsolete; procoxae strongly transverse (Fig. 9); mesonotum without striated stridulatory area; wings with closed cell in anal sector, vein $2A_1$ absent (Prioninae) .... Key III

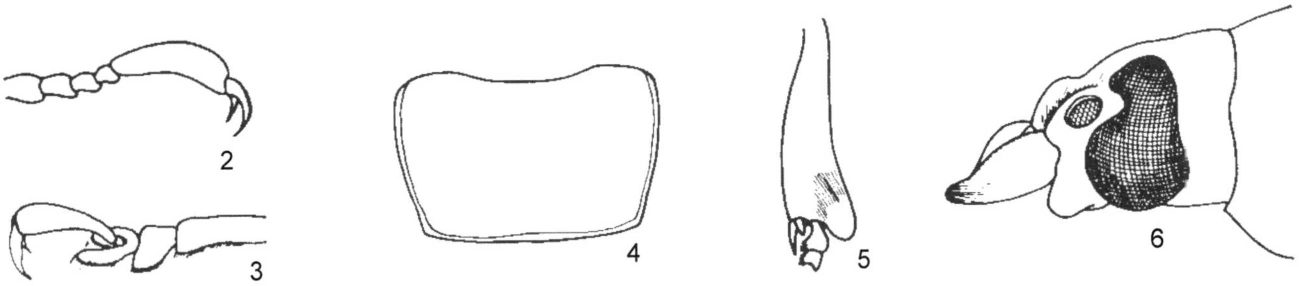

FIGURES 2.120-6.120. 2. *Parandra brunnea* Fabricius protarsus; 3. *Monochamus titillator* (Fabricius) mesotarsus; 4. *P. brunnea* pronotum, dorsal view; 5. *Spondylis uniformis* Mannerheim protibia, anterior view; 6. *Prionus californicus* Motschulsky head, lateral view.

— Pronotum without lateral margin; labrum free; inner lobe of maxillae well developed in most; procoxae transverse in a few, most are rounded; mesonotum with finely striated stridulatory area (absent in a few); wings without closed cell in anal sector except when vein 2A present 6

6(5). Stridulatory plate of mesonotum divided by a median vitta; wings of many with closed cell in anal sector, veins $2A_1$ and $2A_2$ present, 1A connected with $2A_{1+2}$ 7

— Stridulatory plate of mesonotum large (absent in a few), undivided; wings without closed cell in anal sector; vein 2A, or $2A_1$ and $2A_2$ absent (Cerambycinae) .... Key VII

7(6). Head short, not narrowed behind eyes; procoxae subglobular; second antennomere longer than broad, nearly 1/2 as long as third antennomere; submentum without intermaxillary process; mandibles without molar teeth, not fringed with pubescence (Aseminae) .... Key V

— Head elongate, narrowed behind eyes; procoxae conical (Fig. 10); second antennomere not longer than broad, much less than 1/2 as long as third antennomere; submentum with intermaxillary process; most mandible with molar tooth, densely fringed with pubescence on inner margin (Lepturinae) .... Key VI

II. Key to the Genera of the Parandrinae of the United States (from Chemsak 1996)

1. Tarsus with paronychium distinct and bearing a seta on each side of apex; eyes entire .... *Hesperandra*

— Tarsus with paronychium almost invisible, lacking setae; eyes emarginate internally ...... *Parandra*

III. Key to the Tribes, Genera and Subgenera of the Prioninae of North America (adapted from Chemsak 1996)

1. Eyes finely faceted (Solenopterini) 2

— Eyes coarsely faceted 3

2(1). Pronotum at base nearly as wide as base of elytra, sides obtusely toothed near base; disk without median impression .... *Sphenostethus*

— Pronotum at base much narrower than elytra; sides serrulate on margin; disk with median impression .... *Elateropsis*

3(1). Pronotum evenly rounded, crenulate or multispinose at sides 4

— Pronotum with 1-4 teeth at sides 5

4(3). Sides of pronotum retracted basally, posterior angles distinct; eyes at most feebly emarginate (Tribe Macrotomini) 6

— Sides of pronotum not retracted basally, posterior angles feeble; eyes deeply emarginate; third antennomere much longer than fourth (Callipogonini) .... *Ergates*

5(3). Episterna of metathorax parallel-sided (Prionini).. 11

— Episterna of metathorax narrowed posteriorly (Meroscelisini) .... *Tragosoma*

6(4). Third antennomere shorter than scape; scutellum concave 7

— Third antennomere longer than scape; scutellum very convex .... *Strongylaspis*

7(6). Antennal tubercles obtuse; mandibles nearly vertical, not carinate above, unidentate internally .... *Archodontes*

— Antennal tubercles acute; mandibles nearly horizontal, prolonged in male, more or less carinate above, most bidentate and pubescent internally 8

8(7). Antennae attaining at least apical third of elytra in males, middle of elytra in females; mandibles of males much longer than head, almost glabrous .... *Stenodontes*

— Antennae not surpassing middle of elytra in males, not attaining middle in females; mandibles of males at most but little longer than head, most pilose internally 9

9(8). Mandibles not strongly retracted at base, outer margins arcuate, inner margin with or without a tooth before apex 10

— Mandibles strongly retracted at base, outer margins tumid and subangulate; inner margin with a strong double tooth before apex.. *Neomallodon*

10(9) Metepisternum broad, inner margin straight or feebly convex; mandible without strong preapical tooth on inner margin .... *Mallodon*

— Metepisternum narrow, inner margin slightly concave in female, strongly concave in male; mandible with a strong preapical tooth on inner margin .... *Nothopleurus*

11(5). Antennae filiform, eleven-segmented, poriferous system in longitudinal grooves, separated by fine striae; elytra elongate, sides subparallel 12
— Antennae imbricated, twelve- to many-segmented, poriferous system not separated by regular longitudinal striae; elytra robust *Prionus* 13

12(11). Third antennomere distinctly longer than first and second antennomeres together, about as long as fourth and fifth antennomeres together; maxillary palpi elongate; posterior wings with two unconnected postcubital veins; abdomen of male with fifth sternite broadly emarginate, sixth sternite exposed .... *Derobrachus*
— Third antennomere about as long as first and second antennomeres together, distinctly shorter than fourth and fifth antennomeres together; maxillary palpi short; posterior wings with one postcubital vein; abdomen of male with fifth sternite broadly truncate, sixth sternite concealed in most .... *Orthosoma*

13(11). Poriferous system of antennae striolate; metatarsi with lobes of third tarsomere dentate or with a short spine at apex 14
— Poriferous system of antennae dull, not striolate; tarsi with lobes of third tarsomere triangular, spinose at apex; antennae with 12-14 antennomeres .... subgenus *Homaesthesis*

14(13). Antennae with 12-13 antennomeres .... subgenus *Prionus*
— Antennae with 15-30 antennomeres 15

15(14). Antennae with 15-20 antennomeres .... subgenus *Neopolyarthron*
— Antennae with 25-30 antennomeres .... subgenus *Antennalia*

IV. Key to the Genera of the Spondylidinae of North America (from Chemsak 1996)

1. Second antennomere about half as long as third, following antennomeres elongate, oval; protibiae with lamellae small but distinct; color black .... *Spondylis*
— Second antennomere nearly as long as third, following antennomeres transverse; protibiae with lamellae large, conspicuous; color reddish-brown to brown .... *Scaphinus*

V. Key to the Tribes and Genera of the Aseminae of North America (adapted from Chemsak 1996)

1. Procoxal cavities open behind, intercoxal process not dilated at apex; mesocoxal cavities open to epimeron; base of antennae not embraced by eyes (Tribe Asemini) 3
— Procoxal cavities closed behind, intercoxal process dilated at apex; mesocoxal cavities closed; base of antennae embraced by eyes (Tribe Atimiini) 5

2(1). Base of antennae embraced by eyes; elytra subparallel-sided or widest at base; wings present; integument with pubescent patterns formed by dense patches of scales or hairs (Tribe Atimiini) 6
— Base of antennae scarcely embraced by eyes; form ant-like, elytra elongate oval, widest behind middle; wings absent; integument black, without pubescent patterns formed by dense patches of scales or hairs (Tribe Michthisomini) .... *Michthisoma*

3(1). Eyes entire or shallowly emarginate 4
— Eyes deeply emarginate or completely divided; protibiae with two spurs 5

4(3). Protibiae with one spur .... *Arhopalus*
— Protibiae with two spurs .... *Asemum*

5(3). Eyes coarsely faceted, emarginate for nearly half their width but not divided .... *Megasemum*
— Eyes finely faceted, completely divided .... *Tetropium*

6(2). Elytra, and in most pronotum and abdomen, with denuded areas; apical maxillary palpomere cylindrical; procoxae widely separated by prosternum .... *Atimia*
— Elytra, pronotum, and abdomen without denuded areas; apical maxillary palpomere triangular; procoxae narrowly separated by prosternum ... . *Paratimia*

VI. Key to the Genera and Tribes of Lepturinae of North America (adapted from Linsley and Chemsak 1972, 1976)

1. Mandibles short, broad, without an internal pubescent fringe (Desmocerini) .... *Desmocerus*
— Mandibles slender, acute, with a pubescent fringe along inner margin 2

2(1). Elytra abbreviated, scarcely extending over base of abdomen; wings not folded; stridulatory plate of mesonotum entire; mentum transverse (Necydalini) 3
— Elytra entire, wings folded at apex; stridulatory plate of mesonotum divided; mentum trapezoidal (Lepturini) 4

3(2). Body slender, pronotum sparsely hairy; fifth antennomere shorter than third and fourth antennomeres together; palpi with ultimate antennomere expanded, campanuliform .... *Necydalis*
— Body robust, pronotum very hairy; fifth antennomere as long as third and fourth antennomeres together; ultimate palpomere oval .... *Ulochaetes*

4(2). Pronotum with acute lateral spines or distinct tubercles and/or eyes entire 5
— Pronotum with sides sinuate, rounded or at most angulate, without spines or tubercles; eyes notched or emarginate 26

5(4). Eyes coarsely faceted (nocturnal species) 6
— Eyes finely faceted (diurnal species) 8

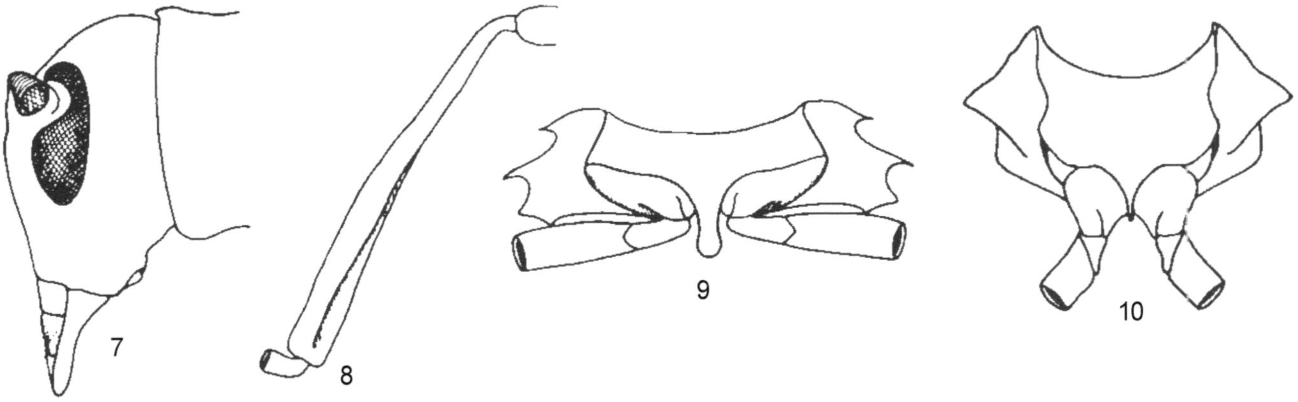

FIGURES 7.120-10.120. 7. *Monochamus titillator* (Fabricius) head, lateral view; 8. *M. titillator* protibia anterior view; 9. *Prionus californicus* Motschulsky prosternum, ventral view; 10. *Centrodera decolorata* (Harris) prosternum, ventral view.

6(5). Head with tempora absent or sharply convergent behind eyes; eyes deeply notched or emarginate 7

— Head with tempora inflated behind eyes; eyes shallowly emarginate .... *Xylosteus*

7(6). Head with tempora sharply convergent behind eyes; apical palpomeres broadly dilated .... *Leptorhabdium*

— Head with tempora absent; apical palpomeres cylindrical or slightly dilated .... *Centrodera*

8(5). Tibial spurs subterminal, inserted into an emargination near apex of tibiae 9

— Tibial spurs terminal, inserted at apices of tibiae 10

9(8). Antennomeres very short, stout, almost subequal in length from third antennomere, not extending beyond middle of elytra; tempora subparallel; eyes entire .... *Piodes*

— Antennomeres slender, elongate, extending well beyond middle of elytra; tempora convergent; eyes notched or emarginate .... *Stenocorus*

10(8). Eyes deeply notched or emarginate 11

— Eyes entire, not deeply emarginate or notched, at most very shallowly concave along inside margin 16

11(10). Front of head vertical, front and vertex meeting at an angle of nearly 90 degrees; front short, distance from tips of mandibles to forward edge of eyes shorter than width between outside edges of genae 12

— Front of head oblique, meeting vertex at an angle of more than 90 degrees; front long, distance from tips of mandibles to forward edges of eyes greater than width across genae 13

12(11). First antennomere shorter than third; hind tarsi with second tarsomere much longer than third; intercoxal process of prosternum narrow, straight .... *Encylops*

— First antennomere longer than third; hind tarsi with second tarsomere subequal to third; intercoxal process of prosternum broad, arcuate .... *Pyrotrichus*

13(11). Hind tarsi with third tarsomere cleft to base; basal margins of elytra not elevated around scutellum 14

— Hind tarsi with third tarsomere cleft to about middle; basal margins of elytra elevated around scutellum .... *Pachyta*

14(13). Pronotum transverse, sides strongly tuberculate or spined; eyes deeply emarginate 15

— Pronotum longer than broad, sides very feebly tuberculate; eyes shallowly notched dorsally .... *Evodinus*

15(14). Third antennomere longer than fourth; middle tibiae of males with a distinct spur or tubercle between long terminal spurs; integument shining or metallic .... *Anthophyllax*

— Third antennomere equal to fourth; tibiae of males normal; integument subopaque *Neanthophylax*

16(10). Prosternum with intercoxal process narrow, extending between coxae; antennae extending well beyond humeri; pronotal spines, if present, small, obtuse 17

— Prosternum with intercoxal process broad, extending over coxae, abruptly declivous behind; antennae very short; pronotal spines large, acute .... *Rhagium*

17(16). Mesosternum with intercoxal process protruding slightly or level with margins of coxae 18

— Mesosternum with intercoxal process lying well below margins of coxae 20

18(17). Head with tempora not inflated, convergent behind eyes 19

— Head with tempora inflated, subparallel, abruptly constricted at neck; body distinctly pubescent, most with small patches of whitish hairs .... *Comacmaeops*

19(18). Elytra coarsely, irregularly rugose, with eburneous fasciae; pronotum strongly, irregularly rugose .... *Sachalinobia*

— Elytra distinctly punctate, not rugose, fasciae absent, most are metallic; pronotum finely punctate .... *Gaurotes*

20(17). Pronotum with disk convex, not flattened behind nor elevated on each side into a broad, obtuse tubercle 21
— Pronotum with disk flattened behind and elevated on each side into a broad, obtuse tubercle, apical margin collared .... *Acmaeops*

21(20). Head with tempora inflated, parallel, abruptly constricted at neck 22
— Head with tempora not inflated, convergent, not abruptly constricted at neck 23

22(21). Pronotum with lateral tubercles; prosternum with intercoxal process extending behind coxae .... *Leptalia*
— Pronotum with sides rounded to angulate, not tuberculate; prosternum with intercoxal process not extending beyond coxae .... *Cortodera*

23(21). Elytra with basal margin not elevated around scutellum 24
— Elytra with basal margin strongly elevated around scutellum; integument bright metallic greenish or bluish .... *Pseudogaurotina*

24(23). Head with front longer than broad; outer antennomeres slender, without poriferous areas 25
— Head with front short; outer antennomeres thickened, most with small poriferous areas .... *Brachysomida*

25(24). Antennae short, not extending beyond middle of elytra, scape longer than third antennomere; metepisternum broad .... *Gnathacmaeops*
— Antennae elongate, extending beyond middle of elytra, scape shorter than third; metepisternum narrow .... *Metacmaeops*

26(4). Pronotum with hind lateral angles acute, most are expanded over humeri 27
— Pronotum with hind lateral angles rounded, not acute nor produced over humeri 47

27(26). Elytra elongate, narrow, strongly attenuated posteriorly and strongly constricted behind middle; most males with apex of abdomen strongly inflated at sides 28
— Elytra of most shorter, not strongly attenuated posteriorly nor strongly constricted behind middle; abdomen of males not inflated apically 31

28(27). Antennae with distinct sensory poriferous areas on outer antennomeres; males of most with apex of abdomen modified 29
— Antennae lacking poriferous areas on outer antennomeres; abdomen of males not apically inflated .... *Neobellamira*

29(28). Pronotum transversely impressed across base and apex in most, most with disk inflated 30
— Pronotum elongate, not impressed nor inflated; antennae with poriferous areas very distinct in most .... *Strangalia*

30(29). Pronotum deeply impressed at apex and base; front of head short .... *Bellamira*
— Pronotum shallowly impressed at base only, disk inflated; front of head moderately long .... *Analeptura*

31(27). Elytra broadly to narrowly rounded or transversely truncate at apices 32
— Elytra obliquely emarginate to acuminate at apices, outer angles produced in most 42

32(31). Pronotum campanuliform, base of most strongly impressed on each side of middle giving hind angles a toothed appearance, hind angles strongly produced over humeri or integument metallic 33
— Pronotum with sides sinuate, disk inflated or plane, base of most impressed across middle, hind angles not toothed-appearing 39

33(32). Eyes large, length about equal to distance between front of eyes to tips of mandibles; tempora hardly inflated 34
— Eyes small, length much less than distance between front of eyes and tips of mandibles; tempora strongly inflated .... *Charisalia*

34(33). Second antennomere at most as long as broad, never 1/3 as long as third antennomere 35
— Second antennomere about 1/3 as long as third antennomere .... *Alosternida*

35(34). Head with genae much longer than width of third antennomere 36
— Head with genae very short, eyes contiguous to base of mandibles or separated from base of mandibles by width of third antennomere .... *Grammoptera*

36(35). Mesosternum with intercoxal process narrow, deeply recessed 37
— Mesosternum with intercoxal process prominent, not or shallowly recessed; pronotum with punctures coarser than those of elytra; elytra of most expanded toward apices .... *Megachoriolaus*

37(36). Form slender, most with elytra expanded toward apices; pronotum with punctures smaller than those at base of elytra 38
— Form moderately robust, pronotum with punctures equal to or larger than those at base of elytra; antennae filiform, outer antennomeres not thickened .... *Choriolaus*

38(37). Elytra expanded toward apices, apices subtruncate, outer angle dentate; head with front measured from apices of antennal tubercles to anterior margin of labrum longer than width across genae; antennae thickened .... *Lychochoriolaus*
— Elytra parallel, not expanded toward apices, apices rounded; head with front from apex of antennal tubercles to anterior margin of labrum shorter than width across genae; antennae filiform, not thickened .... *Orthochoriolaus*

39(32). Head with front short, quadrate, tempora prominent, parallel; antennae stout, most are thickened 40
— Head with front moderately long, tempora not prominently produced, most convergent; antennae slender; body form short, stout .... *Judolia*

40(39). Antennae elongate, longer than body in males, extending at least to middle of elytra in females... 41

— Antennae short, subserrate, shorter than body in males, extending to a little behind humeri in females; form more elongate, elytra more than twice as long as broad .... *Pygoleptura*

41(40). Elytra about twice as long as broad, form robust, short; apices unarmed; antennae of males opaque .... *Dorcasina*

— Elytra 2-1/2 times as long as broad, form elongate, parallel, apices toothed at suture; antennomeres from fifth opaque .... *Idiopidonia*

42(31). Antennae with distinct, elongate, poriferous areas on outer antennomeres 43

— Antennae lacking poriferous areas or poriferous areas reduced to small round pits near apices of outer antennomeres 44

43(42). Pronotum deeply, transversely impressed at apex; antennomeres of most expanded apically; form robust, tapering .... *Typocerus*

— Pronotum not impressed at apex; antennae slender, antennomeres not expanded apically; form elongate, subparallel .... *Neoalosterna*

44(42). Elytra not strongly dehiscent at apex; pronotum transversely impressed at apex 45

— Elytra strongly dehiscent at apex; pronotum not transversely impressed at apex .... *Pseudostrangalia*

45(44). Tempora not inflated, most short and convergent; antennae slender, not subserrate, third and fourth antennomeres together much longer than fifth antennomere in males 46

— Tempora inflated, parallel; antennae subserrate, third and fourth antennomeres together as long as fifth antennomere in males .... *Pygoleptura*

46(45). Elytra about 2-1/2 times as long as broad, color variable; most with antennae lacking poriferous areas .... *Leptura*

— Elytra less than 2-1/2 times as long as broad, always with transverse yellow and black bands; antennae with small poriferous areas .... *Stenostrophia*

47(26). Intercoxal process of prosternum very narrow, not expanded at apex, coxal cavities widely open behind 48

— Intercoxal process of prosternum expanded at apex, coxal cavities closed or nearly closed behind 49

48(47). Pronotum barely impressed at apex and base, disk almost plane; apical palpomeres cylindrical .... *Trachysida*

— Pronotum deeply impressed at base and apex, disk convex; apical palpomeres dilated .... *Pidonia*

49(47). Eyes finely faceted; form small to moderate sized, not elongate and subparallel (diurnal species) 50

— Eyes coarsely faceted; most with elytra spined at inner angles at apex; most are large, elongate and subparallel (nocturnal species) .... *Ortholeptura*

50(49). Antennae slender, most subserrate but not thickened; pronotum impressed across disk at base; head with front short or only moderately long 51

— Antennae thickened toward apex; pronotum inflated dorsally, not impressed at base; head with front narrow, very long .... *Cyphonotida*

51(50). Pronotum deeply to shallowly transversely impressed behind apical margin 52

— Pronotum not impressed behind apical margin, disk very convex; base of elytra strongly protruding on each side of scutellum .... *Strangalepta*

52(51). Elytra rounded, truncate, or shallowly emarginate at apices, angles not strongly produced; antennae never 12-segmented 53

— Elytra strongly emarginate at apices, angles strongly produced; antennae 12-segmented or eleventh antennomere very strongly appendiculate, outer antennomeres subserrate .... *Stictoleptura*

53(52). Pronotum with disk strongly convex, basal and apical transverse impressions deep 54

— Pronotum with disk very feebly convex, almost plane, apical and basal impressions very shallow .... *Anastrangalia*

54(53). Pronotum with basal margin sinuate to straight, most with disk deeply impressed longitudinally; most with elytra not bilobed around scutellum; if bilobed, pronotum deeply impressed longitudinally 55

— Pronotum with basal margin broadly lobed at middle, disk not or barely impressed longitudinally; elytra bilobed around scutellum; most with body rather slender, subparallel .... *Xestoleptura*

55(54). Pronotum not or very shallowly impressed longitudinally, impression if present not extending length of disk 56

— Pronotum deeply impressed longitudinally, impression extending length of disk .... *Lepturopsis*

56(55). Prosternum with intercoxal process narrow, width much less than length of second antennomere 57

— Prosternum with intercoxal process broad, wider than length of second antennomere .... *Trigonarthris*

57(56). Pronotum with sides sinuate to rounded; elytra with apices truncate to emarginate-truncate or acuminate 58

— Pronotum with sides distinctly angulate; elytra with apices rounded .... *Acmaeopsoides*

58(57). Elytra about twice or less as long as broad, sides strongly tapering; most with pronotal disk strongly inflated 59

— Elytra more than 2-1/2 times as long as broad, subparallel; pronotum with disk not strongly inflated .... *Idiopidonia*

59(58). Elytra not strongly dehiscent at apex, sides not strongly sinuate behind middle; hind tibiae of males normal 60
— Elytra rather strongly dehiscent at apex, sides strongly sinuate behind middle; hind tibiae of most males modified .... *Brachyleptura*

60(59). Elytra with apices obliquely truncate, outer angles produced, surface always yellow and black banded .... *Strophiona*
— Elytra with outer apical angles not produced, surface never yellow and black banded .... *Cosmosalia*

VII. Key to the Tribes and Genera of Cerambycinae of North America (adapted from Linsley 1962b, 1963, 1964; Fragoso *et al.* 1987; Lingafelter 1998)

1. Eyes coarsely faceted (except *Stenosphenus*); most with coloration somber 2
— Eyes finely faceted; coloration of many bright, contrasting in most 77

2(1). Terminalia shortened, with gonopharsa when normally invaginated shorter or subequal to sum of sternites 6 + 7; females with abdominal sternite 8 transversely oblong, most bounded at apical margin by a brush of hairs; prothorax typically tumid and expanded; most with scutellum disproportionately developed 3
— Terminalia elongated, with gonopharsa when normally invaginated surpassing the length of sternites 6 + 7; females with distal borders of abdominal sternite 8 straight, most without brush of hairs at apical margin; prothorax generally not tumid; scutellum not disproportionately developed.. 6

3(2). Prothorax not tubercled laterally (Torneutina) .... *Chrotoma*
— Prothorax with one or more tubercles on each side (Bothriospilina) 4

4(3). Prothorax armed with a median lateral spine and an atemedian lateral tubercle .... *Chlorida*
— Prothorax armed with an acute median or postmedian lateral spine only 5

5(4). Prosternum protuberant, ridged between the coxae, vertical or concave behind .... *Gnaphalodes*
— Prosternum arcuate behind .... *Knulliana*

6(2). Pronotum broadly arcuate-emarginate at base, emargination filled with thin, corneous plate; second antennomere 2 or 3 times as long as broad, at least half as long as third antennomere; maxillary palpi very much elongated (Opsimini) 7
— Pronotum truncate, lobed, or slightly sinuate at base; second antennomere less than twice as long as broad, or if (in a few) twice as long as broad, less than half as long as third antennomere 8

7(6). Pronotum armed on each side with a single, median spine .... *Opsimus*
— Pronotum armed on each side with 2 spines, one median, one basal .... *Dicentrus*

8(6). Second antennomere half as long as third antennomere; mesothoracic epimera narrowed and acutely pointed internally; body parallel-sided, flattened (Smodicini) .... *Smodicum*
— Second antennomere less than half as long as third antennomere 9

9(8). Anterior coxae prominent, conical or subconical, their cavities narrowly separated or contiguous, widely open posteriorly, broadly angulate externally and exposing a large trochantin; ligula corneous, feebly emarginate; antennae very long, slender, most with second antennomere transverse; many with elytra abbreviated; eyes very deeply emarginate or divided (Methiini) 10
— Anterior coxae globose or subglobular, if (in a few) subconical, anterior coxal cavities closed posteriorly or rounded externally; ligula membranous or partly membranous, deeply emarginate or broadly bilobed 19

10(9). Elytra entire; anterior coxal cavities separated by a narrow prosternal process 11
— Elytra abbreviated, or if (in a few) nearly as long as abdomen, anterior coxal cavities contiguous, without prosternal process 15

11(10). Pronotum rounded, unarmed at sides 12
— Pronotum tuberculate at sides 14

12(11). Pronotum lobed at base; palpi broadly dilated; intercoxal process of prosternum short .... *Malacopterus*
— Pronotum constricted at base; palpi feebly dilated; intercoxal process of prosternum long 13

13(12). Intercoxal process of prosternum narrow but not laminiform, slightly wider at apex, not extending beyond coxae .... *Paranoplium*
— Intercoxal process of prosternum extremely narrow, laminiform, pointed at apex, extending beyond coxae .... *Oeme*

14(11). Pronotum subquadrate, with a lateral, obtuse, subbasal tubercle; elytra strongly costate; anterior coxae separated by prosternal process .... *Eudistenia*
— Pronotum elongate, with a small anterior and a large postmedian lateral tubercle; elytra without costae; anterior coxae contigous behind .... *Vandykea*

15(10). Elytra gradually narrowing apically, suture entire; abdomen much longer than metathorax; posterior tarsi cylindrical, first tarsomere at least twice as long as broad 16
— Elytra dehiscing suddenly from base, suture deeply arcuate; abdomen abbreviated, shorter than metathorax; posterior tarsi broad, flat, first tarsomere barely longer than broad .... *Coleomethia*

16(15). Eyes with dorsal and ventral lobes connected posteriorly by a line or by 1 or more rows of facets; second antennomere distinct but can be small; abdomen with sternites not imbricated at sides in female 17
— Eyes with dorsal and ventral lobes completely separated, not connected by line or row of facets; second antennomere obsolete; abdomen

with sternites imbricated at sides in female .... *Tessaropa*

17(16). Basal antennomeres not distinctly thickened, more slender than scape, third antennomere longer than scape; pronotum cylindrical or rounded at sides; posterior tarsi less than half as long as tibiae 18

— Basal antennomeres thickened, as wide as scape, third antennomere shorter than scape; pronotum armed with an obtuse lateral tubercle; posterior tarsi more than half as long as tibiae .... *Pseudomethia*

18(17). Pronotum longer than broad, subcylindrical, sides parallel or very feebly convex; femora clavate .... *Styloxus*

— Pronotum as wide as or wider than long, sides rounded; femora slender, not clavate .... *Methia*

19(9). Anterior coxae subconical, cavities closed behind; abdomen of female with basal sternite as long as remaining antennomeres together, sternites 2 to 5 greatly modified (Obriini) .... *Obrium*

— Anterior coxae globose, most with cavities open behind; abdomen of female normally segmented 20

20(19). Prosternum with intercoxal process not expanded behind coxae, or, if slightly expanded, anterior coxal cavities broadly angulate externally and intermediate coxal cavities open to epimera; antennomeres without apical spines (except in some *Osmidus*) 21

— Prosternum with intercoxal process expanded behind coxae, or, if not expanded, anterior coxal cavities rounded or very narrowly angulate externally and intermediate coxal cavities not open to epimera, or antennae spined at apex of 1 or more antennomeres beginning with third...... 39

21(20). Antennomeres 3 to 6 successively increasing in length, or with fifth antennomere longest and maxillary palpi twice as long as labial palpi; mesepimera small, narrow, not extended to coxae; most with pronotum elongate, widest behind middle, tapering anteriorly; many with body small, slender, very finely punctate and pubescent; antennae very slender, many very much longer than body in male; small, delicate forms (Graciliini) 22

— Antennomeres 3 to 6 successively decreasing in length, or subequal, or with fourth antennomere shorter than third or fifth; maxillary palpi at most but little longer than labial palpi, or if very much longer, fifth antennomere shorter than third; mesepimera large; medium to large, robust forms 24

22(21). Head broader across eyes than width of pronotum; antennae not ciliate, third antennomere shorter than scape; maxillary palpi less than twice as long as labial palpi 23

— Head narrower across eyes than width of pronotum; antennae ciliate, third antennomere not shorter than scape; maxillary palpi more than twice as long as labial palpi .... *Gracilia*

23(22). Antennae 11-segmented, antennomeres 3 to 11 gradually increasing in length .... *Hypexilis*

— Antennae 12-segmented, 1 or more of the subterminal antennomeres shorter than those preceding .... *Ferigracilia*

24(21). Anterior coxal cavities rounded externally; terminal palpomere feebly dilated; elytra without eburneous fasciae on costae; posterior femora slender; ligula corneous at base (Achrysonini) 25

— Anterior coxal cavities angulate externally, if rounded, terminal palpomere broadly dilated and posterior femora clavate or elytra with eburneous fasciae on costae (Hesperophanini) 26

25(24). Mesosternum with intercoxal process broad, apex more or less truncate; elytral apices obliquely truncate or subtruncate, outer angle spiniform .... *Achryson*

— Mesosternum with intercoxal process narrow, subtriangular, apex acuminate; elytral apices rounded.. *Geropa*

26(24). Elytra without ivory marks; pronotum rounded or tuberculate at sides, never spinose; femora always unarmed 27

— Elytra with ivory marks; most with pronotum bearing a lateral spine; femora in many spinose .... *Eburia*

27(26). Anterior coxal cavities rounded externally; femora clavate 28

— Anterior coxal cavities angulate externally; femora linear to feebly clavate.................................. 30

28(27). Maxillary palpi less than twice as long as labial palpi; pronotum wider than long, nearly as wide as base of elytra; front of head not canaliculated; elytral pubescence uniform 29

— Maxillary palpi more than twice as long as labial palpi; pronotum longer than wide, much narrower than base of elytra; front of head canaliculated; elytral pubescence with an intermixture of very long, scattered, coarse, straight, suberect hairs .... *Penichroa*

29(28). Antennae bisulcate; pronotum with dorsal callosities .... *Tylonotus*

— Antennae not sulcate; pronotum without dorsal callosities .... *Zamodes*

30(27). Frontal suture deep 31

— Frontal suture faint; elytra densely pubescent except for small round denuded spots.... *Osmidus*

31(30). Metepisterna broadened anteriorly................. 32

— Metepisterna parallel-sided 36

32(31). Maxillary and labial palpi short, subequal, terminal palpomeres cylindrical.................................. 33

— Maxillary and labial palpi unequal in length, terminal palpomeres expanded, elongate-triangular 34

33(32). Pronotum granulate-punctate; body clothed with recumbent cinereous pubescence intermixed with longer erect and suberect pale brownish hairs .... *Cacophrissus*

— Pronotum finely or distinctly punctate, not granulate-punctate; body thinly clothed with fine, pale, recumbent pubescence, not intermixed with longer erect or suberect hairs .... *Haplidus*

34(32). Antennomeres unarmed; elytra without costae, apices rounded, not spinose 35

— Antennomeres 3-5 spinose at apex; pronotum distinctly wider than long, sides obtusely angulate or tuberculate; elytra feebly costate, with a short, subsutural spine .... *Eucrossus*

35(34). Third antennomere distinctly shorter than fourth; pronotum densely punctate, without discal callosities...................................... *Hesperanoplium*

— Third antennomere distinctly longer than fourth; prontoum with polished discal callosities .... *Meganoplium*

36(31). Pronotum coarsely punctate, with dorsal callosities or tubercles; most with maxillary and labial palpi unequal in length 37

— Pronotum without dorsal callosities or tubercles; maxillary and labial palpi subequal in length . 38

37(36). Vestiture dense; elytra with numerous round, elevated, denuded areas .... *Brothylus*

— Vestiture thin, elytra without round, elevated, denuded areas .... *Xeranoplium*

38(36). Antennal tubercles sharply angulate or spined on inner side, especially in male; prothorax of male with a large tomentose depression on each side .... *Stromatium*

— Antennal tubercles depressed; prosternal process gradually sloped posteriorly, slightly dilated at apex; prothorax of male without large tomentose depression on each side .... *Hesperophanes*

39(20). Femora with a broad tooth beneath; mesepimera fused to mesepisterna; body small; femora strongly clavate; antennomeres without apical spines (Curiini) 40

— Femora without broad tooth beneath; mesepimera separated from mesepisterna by distinct suture 41

40(39). Form depressed, dull; anterior coxae nearly contiguous; abdomen with first ventrite as long as 2 following .... *Curius*

— Form cylindrical, shining; anterior coxae distinctly separated; abdomen with first ventrite as long as 3 following..................................... *Plectromerus*

41(39). Femora not clavate or pedunculate, or, if so, antennae distinctly spinose at apex of 1 or more antennomeres beginning with third, or anterior coxal cavities wide open behind, or intermediate coxal cavities open to epimera (Elaphidiini) .. 42

— Femora clavate and pedunculate in most; antennomeres without apical spines; anterior coxal cavities rounded externally, closed or very narrowly open behind; intermediate coxal cavities closed externally and most not at all angulate 75

42(41). Eyes coarsely granulated, facets very convex with eye surface appearing uneven; lower lobe of eye very large with respect to lateral aspect of head, occupying more than 50% (most much more) of the gena; noctural species 43

— Eyes finely granulated, facets not very convex with eye surface appearing smooth; lower lobe of eye small with respect to lateral aspect of head, occupying less than 50% of the gena; diurnal species .... *Stenosphenus*

43(42). Prosternal process narrow, forming a vertical plane between the coxae, apex not or barely expanded; acute lateral tubercles absent .... *Anoplocurius*

— Prosternal process widened, apex nearly always expanded; acute lateral pronotal tubercles present in some species 44

44(43). Spines present on mesal antennomeres (dentiform in some species but always with at least a small spine on third antennomere).......................... 45

— Spines absent from mesal antennomeres........ 72

45(44). Elytral humeri with small distinct tooth at epipleural margin; integument with distinct and demarcated dense patches of white or yellow pubescence on head, pronotum and elytra; some antennomeres weakly bispinose .... *Linsleyonides*

— Elytral humeri without a tooth at epipleural margin; pubescent patches, if present on head, pronotum or elytra diffuse around edges; spines on antennomeres variable 46

46(45). Pronotum with impunctate regions and longitudinally oriented rugosities; bands of dense pubescence on elytra......................... *Elaphidionopsis*

— Pronotum not as above 47

47(46). Procoxal cavities open posteriorly 48

— Procoxal cavities closed posteriorly................... .... *Psyrassa* (in part)

48(47). Mesotibial carina present 49

— Mesotibial carina absent 63

49(48). Lateral metafemoral apices spinose 50

— Lateral metafemoral apices rounded or dentiform 51

50(49). Prosternal process between coxae acutely declivous in most species; integument covered with dense pubescence which may be interrupted by glabrous regions on pronotum and elytra; body generally not shiny in appearance; antennomeres bispinose in most species; pronotum without strong posterior constriction .... *Elaphidion*

— Prosternal process between procoxae gradually declivous; integument sparsely covered with long, erect flying hairs and never with patches of dense pubescence; body shiny in appearance; antennomeres not strongly spinose laterally, spinose mesally only in most specimens; pronotum strongly constricted posteriorly, with the constriction having a slight anterior extension at the middle... *Stizocera*

51(49). Metafemora clavate .... *Anopliomorpha*

— Metafemora gradually enlarged or linear 52

52(51). Spine on antennomere 3 blunt 53

— Spine on antennomere 3 acute 54

53(52). Lateral projections into mesocoxae from mesosternal process present; elytral apicolateral spines present, but weak; body narrow and elongate, covered with short appressed pubescence .... *Aneflomorpha* (in part)
— Lateral projections into mesocoxae from mesosternal process absent; elytral apicolateral spines absent; body not narrow, very sparsely pubescent with scattered long hairs only on the elytra .... *Stenelaphus*

54(52). Antennomere 3 of males and females about 1/2 the length of the pronotum or less; antennae with dense fringe of pubescence ventrally in most species 55
— Antennomere 3 of males (and most females) about 2/3 the length of the pronotum; antennae without dense fringe of long pubescence ventrally 60

55(54). Profemora clavate in most specimens; median impunctate region on pronotum absent; pronotum with small patches of pubescence laterally on most specimens; femoral and elytral apices rounded and without spines .... *Anelaphus* (in part)
— Profemora gradually enlarged; median pronotal impunctate region present on most specimens; spination of femoral and elytral apices variable 56

56(55). Antennomeres 3+4 less than length of pronotum; length of most specimens greater than 18 mm, width greater than 8 mm; patches of dense pubescence present on anterolateral areas of pronotum .... *Eustromula*
— Antennomeres 3+4 approximately length of pronotum or longer; without patches of dense pubescence on anterolateral areas of pronotum 57

57(56). Antennomeres strongly widened at apices or appendiculate; length of most specimens greater than 15 mm, width variable 58
— Antennomeres not strongly widened at apices nor appendiculate; length of most specimens less than 15 mm, width of most specimens less than 6 mm 59

58(57). Elytra with strong apicolateral spines; length greater than 20 mm; peripheral pronotal calli present but reduced in most specimens .... *Aneflus*
— Elytra without apicolateral spines; length 10 to 20 mm; peripheral pronotal calli absent.. *Neaneflus*

59(57). Antennal tubercles without small tufts of pubescence; elytral apices rounded to suture and without spines .... *Gymnopsyra*
— Antennal tubercles with small tufts of pubescence; elytral apices truncate .... *Anelaphus* (in part)

60(54). Lateral projections into mesocoxae from mesosternal process present 61
— Lateral projections into mesocoxae from mesosternal process absent.............................. *Orwellion*

61(60). Apicolateral spines strong; lateral antennomere shape linear; body somewhat robust, not particularly narrow and elongate .... *Parelaphidion*
— Apicolateral spines weak or absent; antennomere shape variable; body narrow and elongate in most specimens 62

62(61). Antennomeres gradually widened at apices; pronotum moderately punctate and pubescent .... *Aneflomorpha* (in part)
— Antennomeres linear and not gradually widened at apices; pronotum sparsely punctate and glabrous or with only scattered long hairs .... *Psyrassa* (in part)

63(48). Antennomere 3 of males and most females about as long as pronotum or longer .... *Psyrassa* (in part)
— Antennomere 3 of males and most females less than length of pronotum 64

64(63). Antennomere 3 of males and females about 1/2 as long as pronotum or shorter; elytra glabrous 65
— Antennomere 3 of males and most females about 2/3 as long as pronotum 67

65(64). Metafemoral apices spinose; elytra glabrous with distinctly visible costae .... *Atylostagma*
— Metafemoral apices rounded, pubescence variable; elytra with costae not easily visible 66

66(65). Antennal tubercles without small tufts of pubescence; elytral apices rounded to suture and without spines .... *Gymnopsyra*
— Antennal tubercles with small tufts of pubescence; elytral apices truncate .... *Anelaphus* (in part)

67(64). Lateral projections into mesocoxae from mesosternal process present 68
— Lateral projections into mesocoxae from mesosternal process absent... 70

68(67). Mesofemoral apices spinose; elytral apices strongly spinose in most specimens; body not elongate and narrow 69
— Mesofemoral apices rounded; elytral apices weakly spinose at most; body elongate and narrow .... *Psyrassa* (in part)

69(68). Prosternal process abruptly declivous or vertical; posterior notch on metepisternum wide and deep; taxa not associated with *Eucalyptus* spp.; elytra with small shallow punctures not present in rows .... *Elaphidion*
— Prosternal process gradually declivous or rounded; posterior notch on metepisternum absent or weak; taxa associated primarily with *Eucalyptus* spp.; elytra with very large and deep punctures arranged in rows .... *Phoracantha*

70(67). Elytra and pronotum covered with conspicuous, dense, connected patches of white or yellowish pubescence; females with an additional dense coating of short, erect pubescence; body length variable, not exceeding 30 mm in most specimens 71
— Elytra and pronotum with inconspicuous appressed pubescence mixed with erect, sparse hairs, many areas appearing glabrous and free of pubescence in both sexes; body length greater than 30 mm in most specimens .... *Romulus*

71(70). Pronotum not broadly rounded or inflated, distinctly narrower than elytra at middle .... *Parelaphidion*
— Pronotum broadly rounded and inflated, as wide as elytra at middle .... *Enaphalodes*

72(44). Lateral projections into mesocoxae from mesosternal process present 73
— Lateral projections into mesocoxae from mesosternal process absent......... *Micranoplium*

73(72). Antennae 11-segmented; body length less than 20 mm; elytral sutural spines weak or absent 74
— Antennae 12-segmented; body length greater than 20 mm; elytral sutural spines strong .... *Aneflus* (in part)

74(73). Eleventh antennomere not strongly constricted; eyes when viewed ventrally situated on an extension from head, so that lower eye lobes are very widely separated; elytral sutural spines absent; pale in appearance .... *Curtomerus*
— Eleventh antennomere strongly constricted; eyes not as in *Curtomerus*; elytral sutural spines weak; body dark .... *Micraneflus*

75(41). Antennae elongate, slender in females, thickened at base in male (Ibidionini) 76
— Antennae depressed and serrate, very much shorter than body in female (Piezocerini).......... .... *Piezocera*

76(75). Eyes with dorsal lobes with four rows of facets, dorsal lobes well developed; males with antennomeres 3 to 6 enlarged and most without carinae; prothorax short; posterior tibiae without carinae .... *Heterachthes*
— Eyes with dorsal lobes connected at most by three rows of facets, dorsal lobes small; antennae strongly carinate; prothorax elongate .... *Neocompsa*

77(1). Anterior coxal cavities rounded or feebly angulate externally 78
— Anterior coxal cavities distinctly angulate externally 107

78(77). Elytral epipleura prolonged to apex; second antennomere 1/3 as long as third antennomere (Pteroplatini) 79
— Elytral epipleura attenuated before apex; second antennomere less than 1/3 as long as third antennomere 81

79(78). Scutellum not or scarcely longer than broad, rounded behind or obtusely triangular, apex not produced; body lyciform, pronotum and elytra flattened; smaller species, length 8-14 mm ... 80
— Scutellum elongate-triangular, apex acute, produced in most; body subcylindrical; larger species, length 14-19 mm .... *Parevander*

80(79). Antennae about half as long as body in males, shorter in female, antennomeres serrate, eleventh antennomere appendiculate; most with elytra distinctly costate.................. *Elytroleptus*
— Antennae about as long as body in male, shorter in female, antennomeres filiform or nearly so, eleventh antennomere not divided; elytra flat, not costate .... *Holopleura*

81(78). Metepisternum very large, subtriangular, broad in front, narrowly separated from coxae; elytra attenuated or subulate, apices separated (Rhinotragini) .... *Odontocera*
— Metepisternum in most only slightly broader in front, narrow, subparallel; elytra entire, not attenuated or abbreviated 82

82(81). Elytral epipleurae deeply sinuate near humeri, humeri prominent (Agallissini) 83
— Elytral epipleurae not deeply sinuate near humeri 85

83(82). Integument sparsely and mostly finely punctate; elytra distinctly, gradually narrowed from basal emargination 84
— Integument densely and very coarsely puntcate; elytra subparallel, narrowed over apical third; pubescence sparse, not condensed into a pattern; most with pronotum red and elytra banded with red .... *Zagymnus*

84(83). Pronotum truncate and shallowly emarginate at base, surface with polished longitudinal ridges separated by lines of dense coarse white pubescence; elytra without pale maculations, clothed with long coarse pale hairs, apices rounded to suture, suture subspiniform .... *Osmopleura*
— Pronotum feebly lobed at base, surface evenly convex, minutely punctate and obscurely finely pubescent except for median discal smooth area; elytra ornamented with conspicuous pale spots, sparsely pubescent, apices finely serrate .... *Agallissus*

85(82). Scutellum elongate, acutely pointed behind; terminal palpomere oval, squarely truncate at apex 86
— Scutellum small, rounded or broadly triangular; terminal palomere dilated 130

86(85). Anterior coxal cavities open behind 87
— Anterior coxal cavities closed behind; integument generally metallic blue, green, or coppery; posterior legs elongated, compressed; large species (Callichromatini) .... *Plinthocoelium*

87(86). Head declivous, antennae compressed, flabellate; pronotum without discal tubercles (Lissonotini) .... *Lissonotus*
— Without above combination of characters (Trachyderini part; Trachyderina) 88

88(87). Pronotum not lobed at base; stridulatory area of mesonotum large.. 89
— Pronotum broadly lobed at base; stridulatory area of mesonotum narrow 90

89(88). Metepisternum with episterna not excessively broad .... *Trachyderes*
— Metepisternum with episterna very broad, scent pores absent, pronotum angulate at sides behind middle; antennae shorter than body in both sexes .... *Megaderus*

90(88). Elytra with elevated ivory-colored costae 91

— Elytra without costae, or costae narrow, filiform 94

91(90). Mesosternum feebly convex, declivous anteriorly 92

— Mesosternum horizontal, sharply limited and vertical anteriorly 93

92(91). Body robust, elytra only twice as long as broad; mandibles truncate at apex; pronotum hairy, black; elytra each with 2 ivory costae .... *Perarthrus*

— Body elongate, elytra about 2-1/2 times as long as broad; mandibles acute at apex; pronotoum subglabrous, most bright red; elytra each with 2 ivory costae (although some are black) .... *Mannophorus*

93(91). Elytra each with 2 ivory costae; antennae 12-segmented, much longer than body in male, longer than body in female .... *Sphaenothecus*

— Elytra each with a single ivory costa; antennae 11-segmented in both sexes, barely longer than body in male, shorter than body in female .... *Lophalia*

94(90). Mandibles with apex acute 95

— Mandibles with apex emarginate-truncate, the edge chisel-like 104

95(94). Prosternum protuberant or ridged between the coxae, vertical or concave behind 96

— Prosternum arcuate behind 97

96(95). Form large, robust, 20 mm or more in length; legs moderate, posterior femora at most attaining elytral apices in male, falling short of apices in female; mesosternum not protuberant .... *Stenaspis*

— Form slender 15 mm or less in length; legs elongate, posterior femora exceeding elytral apices in both sexes; mesosternum protuberant between coxae .... *Rhodoleptus*

97(95). Elytra distinctly margined at sides; mostly large, robust forms 98

— Elytra not or obtusely margined at sides 99

98(97). Elytra metallic green or blue; prothoracic sculpture and punctation strongly sexually dimorphic .... *Callona*

— Elytra black with reddish or yellow markings, surface not metallic; prothoracic sculpture and punctation similar in both sexes .... *Crioproposus*

99(97). Pronotum armed with a lateral spine 100

— Pronotum subcylindrical or rounded, without lateral spine 102

100(99). Body pubescent 101

— Body glabrous .... *Purpuricenus*

101(100). Elytra sulcate or with distinct threadlike costae; genae moderate .... *Tragidion*

— Elytra without sulci, costae feeble or evanescent; genae prominent .... *Metaleptus*

102(99). Elytra concolorus or clouded or vittate, without transverse yellowish bands; antennal tubercles feebly elevated, vertex nearly flat 103

— Elytra with a pair of transverse yellowish bands; antennal tubercles prominent, vertex distinctly concave; body finely pubescent .... *Aethecerinus*

103(102). Pronotum without elevated apical margin, surface densely punctate, pubescent; elytra finely bicostulate, pale with dark vittae; venter very densely pubescent, punctation obscured .... *Ammanus*

— Pronotum with an elevated apical margin, surface sparsely punctate, thinly clothed with erect hairs; elytra not costulate, concolorous or clouded; venter thinly clothed with erect hairs, punctation not obscured .... *Batyle*

104(94). Eyes emarginate but not divided; pubescence fine 105

— Eyes divided; pubescence coarse, dirty-white, scutellum, suture and lateral margin of elytra clothed with yellowish pubescence .... *Schizax*

105(104). Mesosternum not protuberant, declivous in front 106

— Mesosternum protuberant, vertical in front; pronotum rounded at sides, narrowed in front .... *Plionoma*

106(105). Pronotum red, with round polished black dorsal callosities, sides rounded or subangulate, unarmed; antennae distinctly 12-segmented .... *Tylosis*

— Pronotum without round polished dorsal callosities, sides rounded or with a lateral tubercle or spine; antennae 11-segmented or with eleventh antennomere appendiculate .... *Crossidius*

107(77). Anterior coxae globose or subconical; eyes embracing antennal insertion 108

— Anterior coxae transverse; eyes not or slightly embracing antennal insertion 117

108(107). Elytra entire, neither abbreviated nor attenuated 109

— Elytra either greatly abbreviated or attenuated 114

109(108). Anterior coxal cavities closed posteriorly; mesosternum with intercoxal process parallel-sided; pronotum with sides angulate or trunculate; second antennomere much less than 1/2 as long as third; abdomen of female with first sternite as long as remaining antennomeres together, sternites 2 to 5 greatly modified; ultimate palpomere cylindrical (Hyboderini) 110

— Anterior coxal cavities open posteriorly; mesosternum with intercoxa process triangular; pronotum with sides rounded; second antennomere 1/2 as long as third; abdomen of female with first sternite shorter than remaining sternites together, sternites 2-5 unmodified; ultimate palpomere dilated (Eumichthini) 113

110(109). Pubescence long, sparse, depressed hairs few; punctation coarse 111

— Pubescence short, dense, appressed, flying hairs very sparse; punctation fine .... *Hybodera*

111(110). Intermediate coxae separated by at least their own diameters; metepisterna broad, parallel-sided; elytral punctures coarse, distinct throughout, elytral apices dehiscent, inner angle acute or dentate; posterior tibiae arcuate .... *Lampropterus*
— Intermediate coxae separated by much less than their own diameters; metepisterna attenuated apically; elytral punctures obsolete over apical 1/3, apices separately rounded; posterior tibiae straight 112

112(111). Anterior coxae distinctly separated; intercoxal process of mesosternum less than twice as wide as that of prosternum; antennae of male shorter than body .... *Pseudopilema*
— Anterior coxae narrowly separated; intercoxal process of mesosternum several times as wide as that of prosternum; antennae of male distinctly longer than body .... *Megobrium*

113(109). Third antennomere shorter than fourth; terminal palpomere narrowly triangular; first 2 protarsomeres tumid in male; integument nonmetallic .... *Eumichthus*
— Third antennomere longer than fourth; terminal palpomere broadly triangular; first 2 protarsomeres slender in both sexes; integument metallic .... *Poecilobrium*

114(108). Maxillary palpi very much longer than labial palpi; stridulatory plate present; wings with a single postcubital vein; abdomen of female with apical sternites modified, middle segments very hairy and setose 115
— Maxillary palpi short, subequal in length to labial palpi; apical sternites of female normal 116

115(114). Eyes divided or emarginate (Psebiini) .... *Nathriobrium*
— Eyes entire (Nathriini) .... *Nathrius*

116(114). Anterior coxal cavities open posteriorly; elytra less than 1/2 as long as abdomen, apices rounded or truncate, never attenuated or subulate; intercoxal process of mesosternum narrow, triangular, truncate behind; antennae longer than body in male; posterior wings without postcubital veins (Molorchini) .... *Molorchus*
— Anterior coxal cavities closed posteriorly; elytra more than 1/2 as long as abdomen, apices attenuated, subulate; intercoxal process of mesosternum broad, flat, emarginate behind; antennae shorter than body in both sexes; posterior wings with a single postcubital vein (Stenopterini) .... *Callimoxys*

117(107). Second antennomere short, transverse; form moderately large, elongate, somewhat depressed with contrasting coloration 118
— Second antennomere as long as broad, 1/3 as long as third, outer antennomeres without tufts of hair; legs robust, femora clavate (Callidiini) 120

118(117). Legs slender, femora not clavate; antennae with tufts of long hair at apex of antennomeres 3 to 7; pubescence pale blue or grayish, broken by black bands (Rosaliini) .... *Rosalia*
— Legs with femora moderately to strongly clavate; antennae densely hairy beneath, without tufts, 1 or more antennomeres beginning with the third armed with an inner apical spine; with dense patches of brightly colored pubescence (Dryobiini) 115

119(118). Posterior femora feebly clavate; antennal scape without an inner basal tooth, second antennomere scarcely longer than broad; pronotum rounded at sides .... *Dryobius*
— All of the femora strongly clavate; antennal scape with an inner basal tooth, second antennomere elongate; pronotum subangulate at sides .... *Ornithia*

120(117). Femora clavate; antennae 11-segmented in both sexes; eyes not embracing antennal insertion. 121
— Femora slender; terminal antennomere appendiculate or divided in male; third antennomere elongate; eyes feebly embracing antennal insertion; black, pronotum rufous, elytra with a bluish or greenish luster .... *Pronocera*

121(120). Mesonotum polished or finely tessellate, undivided, sides not pubescent, punctures scattered, if present 122
— Mesonotum punctate, pubescent at sides, with a small medial stridulatory surface 129

122(121). Anterior coxal cavities separated by a parallel-sided intercoxal process, process truncate, emarginate, or expanded posteriorly 123
— Anterior coxal cavities contiguous or with a narrow, abbreviated, acute, or feebly enlarged intercoxal process, in a few extending beyond middle of cavities 125

123(122). Intercoxal process of prosternum moderate or narrow; third antennomere shorter than or but little longer than fourth; tarsal claws simple 124
— Intercoxal process of prosternum very broad; third antennomere nearly twice as long as fourth; tarsal claws toothed .... *Hylotrupes*

124(123). Femora distinctly clavate; pronotum finely punctate, dorsal callosities feeble; antennae slender, filiform .... *Callidiellum*
— Femora feebly clavate; pronotum coarsely punctate, with prominent polished dorsal callosities; antennae stout, outer antennomeres flattened, expanded .... *Semanotus*

125(122). Mesonotum polished, without a stridulatory area, surface with scattered large punctures 126
— Mesonotum with a very finely tessellate or punctulate stridulatory area, without any coarse punctures (if coarsely punctate, body small, subcylindrical) 127

126(125). Mesosternum broad, emarginate; elytra with elevated yellowish costae more or less distinct basally .... *Meriellum*
— Mesosternum obtusely triangular; elytra without elevated yellowish costae................. *Callidium*

127(125). Posterior coxae prominent; metasternum with scent pores 129

— Posterior coxae not prominent; metasternum without scent pores; many with body small, subcylindrical .... *Phymatodes*

128(123). Elytra moderately convex, with elevated ivory-like ridges; pronotum dark .... *Physocnemum*

— Elytra flattened, without ivory-like ridges; pronotum red .... *Ropalopus*

129(121). Femora feebly clavate; antennae slender, third antennomere distinctly longer than fourth; pronotum with polished longitudinal callosities; posterior tarsi with first tarsomere slender, much more than twice as long as broad .... *Elatotrypes*

— Femora distinctly clavate; antennae robust, third antennomere subequal in length to fourth; pronotum without polished callosities; posterior tarsi with first tarsomere less than twice as long as broad .... *Xylocrius*

130(85). With scent pores on metathorax; integument reddish, with antennae, elytral apices and legs excepting femoral bases black; elytra each with a pair of eburneous markings beginning as spots at the base of the elytra and extending as irregular lines to the basal margin of the apical black apical marking (Heteropsini) .... *Heterops*

— Without scent pores on the metathorax; integument not colored as above; eburneous markings of elytra, if present, not as described above 131

131(130). Tibial spurs large; scutellum rounded or broadly triangular 132

— Tibial spurs small; scutellum rounded; legs long, slender, pedunculate and clavate 138

132(131). Metathoracic epimera not produced, episterna linear 134

— Metathoracic epimera produced over angles of first abdominal sternite and enclosing posterior coxae externally, most with wide episterna (Clytini) 140

133(132). Elytra gibbose at base, declivous at apex, without transverse ivory-like ridges (Anaglyptini) 134

— Elytra narrowed at middle in most, most with transverse raised ivory-like ridges (Tillomorphini)..... 136

134(133). Eyes oblique, emarginate 135

— Eyes rounded, entire; third antennomere not or feebly spinose; elytra strongly gibbose at base, apices rounded .... *Tilloclytus*

135(134). Third antennomere with a prominent spine at apex; pronotum gibbose; elytra strongly gibbose at base, apices emarginate-truncate .... *Cyrtophorus*

— Third antennomere not or feebly spinose; pronotum not gibbose; elytra feebly gibbose at base, apices rounded .... *Microclytus*

136(133). Antennomeres 3 to 6 or 7 distinctly enlarged in male 137

— Antennae slender in both sexes .... *Euderces*

137(136). Eyes emarginate, pointed behind; antennomeres 3 to 6 enlarged in male .... *Tetranodus*

— Eyes entire, oval; antennomeres 3 to 7 enlarged male .... *Pentanodes*

138(131). Antennae without poriferous system 139

— Antennae with poriferous system; elytra scarlet (Trachyderini part; Ancylocerina) .... *Ancylocera*

139(138). Antennae short, serrate (Cleomenini) .... *Dihammaphora*

— Antennae filiform (Rhopalophorini) .... *Rhopalophora*

140(132). Basal antennomere spined at apex 141

— Antennae without apical spines 144

141(140). Prosternum with intercoxal process arcuate or declivous; pronotum not excavated laterally at base 138

— Prosternum with intercoxal process prominent, apex vertical or concave; pronotum excavated laterally at base .... *Megacyllene*

142(141). Elytra not distinctly narrowed at middle, subsutural carina flattened or lacking, apices not spinose; posterior tarsi with first tarsomere cylindrical... 143

— Elytra narrowed at middle, subsutural carina prominent apically, apices with outer angle spiniform; posterior tarsi with first tarsomere compressed, arcuate; antennae slender .... *Tylcus*

143(142). Outer antennomeres tapering; pronotum black with transverse patches of bright yellow pubescence at sides; elytra black with narrow, transverse, bright yellow bands on basal 2/3, antemedian band deeply angulate, posterior third largely yellow pubescent, apices emarginate ..... *Glycobius*

— Outer antennomeres subserrate; pronotum densly clothed with white or yellowish pubescence, obscuring surface; elytra with 5 transverse bands of white or yellowish pubescence, antemedian band shallowly angulate, apices rounded; last abdominal tergite of male and last 2 or 3 tergites of female exposed beyond elytral apices .... *Megacheuma*

144(140). Front of head without carinae........................ 145

— Front of head with a bicarinate ridge or a V-shaped or Y-shaped carina; antennae short, filiform; form robust .... *Xylotrechus*

145(144). Pronotum without transverse carinae............ 146

— Pronotum with transverse carinae; legs elongate; form slender 153

146(145). Pronotum not lobed at middle, not excavated laterally at base; elytra subparallel, apices not spinose.. 147

— Pronotum lobed medially at base, broadly excavated laterally; elytra strongly tapering, with a prominent subsutural carina, apices acutely spinose.. *Placosternus*

147(146). Episterna of metathorax broad, not more than 3 times as long as wide 148

— Episterna of metathorax narrow, about 4 times as long as wide 150

148(147). Abdomen with intercoxal process broadly rounded; plane of face elevated above anterior margin of eyes; large, robust forms, length 19-26 mm .... *Calloides*
— Abdomen with intercoxal process acute; plane of face not elevated above anterior margin of eyes; smaller forms, length 7-14 mm 149

149(148). Antennae 11-segmented .... *Clytus*
— Antennae 12-segmented, antennomere12 abbreviated; pronotum of male with discal impressions; elytra flattened, dehiscent .... *Plesioclytus*

150(147). Pronotum broader than long, most with apex narrower than base; form robust; antennae only moderately slender; larger forms, length 9-16 mm 151
— Pronotum longer than broad, apex a little wider than base; form slender; metathoracic episterna narrow, parallel-sided, more than 5 times as long as broad; antennae very slender; smaller forms, length 6.5-8 mm .... *Clytoleptus*

151(150). Mesosternum with intercoxal process abruptly, subvertically declivous 152
— Mesosternum with intercoxal process gradually, feebly declivous 154

152(151). Elytra with undulant bands, apices emarginate-truncate .... *Ochraethes*
— Elytra without undulant bands, apices rounded or subtruncate .... *Triodoclytus*

153(145). Antennae 11-segmented .... *Neoclytus*
— Antennae 12-segmented .... *Euryscelis*

154(151). Elytra gradually narrowed posteriorly, subsutural carina lacking, apices narrowly feebly truncate; ventral surface sparsely pubescent .... *Sarosesthes*
— Elytra subparallel or slightly attenuated, with an obtuse nonprominent subsutural carina over apical half, apices feebly undulate-truncate; ventral surface densely pubescent .... *Tanyochraethes*

VIII. Key to the Tribes and Genera of the Lamiinae of North America (adapted from Linsley and Chemsak 1984, 1995)

1. Claws simple, divergent or divaricate 2
— Claws bifid, or appendiculate 68

2(1). Tarsal claws divergent 3
— Tarsal claws divaricate 14

3(2). Antennal scape with a distinct cicatrix at apex (feeble in some *Ataxia*) 4
— Antennal scape without a cicatrix at apex 5

4(3). Middle tibiae with an external sinus toward apex (Ataxiini) .... *Ataxia*
— Middle tibiae without an external sinus toward apex (Mesosini) .... *Synaphaeta*

5(3). Eyes completely divided 6
— Eyes entire or emarginate, lobes connected by at least a row of facets 8

6(5). Elytra bigibbose near base; size very small (2.3-3.5 mm), ant-like (Cyrtinini) .... *Cyrtinus*
— Elytra lacking gibbosities; many with wings lacking; form cylindrical to ovoid, size larger (Apomecynini part) 7

7(6). Metasternum very short; elytra with humeral angles lacking .... *Parmenonta*
— Metasternum of normal length; elytra with small humeral angles .... *Adetus*

8(5). Antennae with scape extending to basal margin of pronotum; form slender and elongate (Agapanthiini) 9
— Antennae with scape not extending to basal margin of pronotum; most with form robust 10

9(8). Eyes entire; antennae short, fringed .... *Spalacopsis*
— Eyes emarginate; antennae elongate, not fringed .... *Hippopsis*

10(8). Eyes with lower lobes rounded, not longer than broad; third antennomere straight; antennal tubercles small; form slender (Apomecynini part) 11
— Eyes with lower lobes longer than broad; third antennae of many arcuate; antennal tubercles prominent (Onciderini) 12

11(10). Pubescence gray and yellow-orange, that of the pronotum arranged in five yellowish, longitudinal bands separated by gray bands, each elytron with three yellowish longitudinal bands separated by gray bands; antennomeres moderately, densely fringed beneath, antennomeres from third narrowly white annulate at bases; punctures of elytra coarse, each bearing a long, erect, black seta; elytral apices emarginate .... *Dorcasta*
— Pubescence while similar in color, not arranged in distinct longitudinal bands but rather irregular patches; antennomeres not fringed; punctures of elytra not bearing setae; outer angles of elytral apices strongly dentate .... *Sybra*

12(10). Third antennomere straight; elytra basally asperate; form robust, moderate-sized to large 13
— Third antennomere curved; elytra not asperate basally; form small, slender .... *Cacostola*

13(12). Pronotum broader at apex than at base; males with antennal tubercles not prominently produced .... *Oncideres*
— Pronotum as broad at apex as base; males with antennal tubercles strongly produced into horns .... *Lochmaeocles*

14(2). Antennal scape strongly asperate (Dorcaschematini) .... *Dorcaschema*
— Antennal scape punctate 15

15(14). Antennal scape with a distinct cicatrix at apex (Lamiini) 16
— Antennal scape lacking a distinct cicatrix at apex 22

16(15). Pronotum not strongly tuberculate nor spined at sides 17

— Pronotum strongly tuberculate, most distinctly spined at sides 18

17(16). Prosternum with intercoxal process laterally expanded in coxae; elytra spined at apex .... *Neoptychodes*

— Prosternum with intercoxal process laterally linear; elytra rounded at apex .... *Hebestola*

18(16). Antennal scape with a distinctly delimited cicatrix; elytra densely punctate basally, most are asperate 19

— Antennal scape with a vague cicatrix; elytra very sparsely punctate at base .... *Plectrodera*

19(18). Pronotum with lateral tubercles small to moderate-sized; most elytra finely punctate at base, asperate but not rugose, apices rounded; front legs of male short 20

— Pronotum with lateral tubercles large, prominent; most elytra coarsely rugose at base, many with apices angled to spined; front legs of male elongate 21

20(19). Elytra coarsely, closely punctate at base, with two black spots at apical one-third; antennae at least 1-1/2 times longer than body .... *Microgoes*

— Elytra rather finely, separately punctate at base, lightly asperate; antennae never 1-1/2 times as long as body .... *Goes*

21(19). Mesosternal intercoxal process gradually inclined anteriorly; integument, gray, reddish-brown, brown, or black, if black, pubescence not condensed into distinct and well demarcated patches .... *Monochamus*

— Mesosternal intercoxal process abruptly inclined, tuberculate or truncate anteriorly; shiny black with distinct and well demarcated dense patches of whitish pubescence .... *Anoplophora*

22(15). Elytra lacking humeral angles 23

— Elytra with humeri normal 26

23(22). Form small, ovoid; front coxal cavities angulate externally; eyes coarsely to moderately coarsely faceted (Parmenini) 24

— Form large, convex; front coxal cavities rounded externally; eyes finely faceted (Moneilemini) ... . *Moneilema*

24(23). Pronotum laterally tuberculate; elytra moderately convex 25

— Pronotum laterally unarmed; elytra very strongly convex .... *Ipochus*

25(24). Elytra with apices strongly spined .... *Plectrura*

— Elytra with apices rounded .... *Parmenosoma*

26(22). Antennal scape clavate 27

— Antennal scape cylindrical or conical 30

27(26). Pronotum strongly, acutely tuberculate at sides; form depressed or tapering posteriorly (Acanthoderini) 28

— Pronotum with a vague, obtuse callus at sides; form globose (Anisocerini) .... *Thryallis*

28(27). Head with front elongate, quadrate; antennal scape shorter than third antennomere 29

— Head with front transverse; antennal scape longer than third antennomere .... *Peritapnia*

29(28). Pronotum with prominent dorsal calluses; middle coxal cavities closed to epimeron .... *Aegomorphus*

— Pronotum with dorsal calluses vague; middle coxal cavities open to epimeron .... *Oplosia*

30(26). Pubescence variable, long or short, not condensed into long tufts on pronotum and elytra 31

— Pubescence condensed into long tufts on pronotum and elytra (Desmiphorini part) .... *Desmiphora*

31(30). Middle tibiae with a distinct external sinus near apex 32

— Middle tibiae lacking an external sinus or sinus very vague 35

32(31). Antennal scape cylindrical to conical, seldom extending beyond middle of pronotum 33

— Antennal scape slender, linear, most extending beyond middle of pronotum (Acanthocinini) . 44

33(32). Episternum of metathorax broad anteriorly, strongly tapering posteriorly (Saperdini) .... *Saperda*

— Episternum of metathorax narrow, subparallel (Desmiphorini part) 34

34(33). Body and antennae with long erect hairs; basal gibbosities lacking on elytra .... *Eupogonius*

— Body and antennae lacking long erect hairs; basal gibbosities present on elytra .... *Psenocerus*

35(31). Body short and convex, long flying hairs present in many, if elongate and flying hairs absent, pronotum laterally unarmed (Pogonocherini). 36

— Body elongate and parallel-sided, flying hairs absent; pronotum tuberculate laterally (Desmiphorini part) 43

36(35). Femora not distinctly clavate 37

— Femora distinctly clavate 39

37(36). Pronotum without a lateral tubercle; head flat or convex between antennal tubercles 38

— Pronotum with a distinct lateral tubercle; head fully concave between antennal tubercles .... *Pygmaeopsis*

38(37). Body small, slightly robust; head convex between antennal tubercles; antennae not extending beyond elytral apices in either sex .... *Zaplous*

— Body elongate; head flat between antennal tubercles; antennae extending beyond elytral apices in both sexes .... *Lypsimena*

39(36). Pronotum with or without conical discal tubercles 40

— Pronotum with discal tubercles in the form of an elongated ridge .... *Callipogonius*

40(39). Antennal scape stout, fourth antennomere incurved, longer than third antennomere...... 41

— Antennal scape slender, fourth antennomere shorter than third antennomere .... *Poliaenus*

41(40). Pronotum with lateral tubercles; body clothed with long flying hairs 42
— Pronotum rounded at sides, without lateral tubercles; body without long flying hairs ... *Ecyrus*

42(41). Elytra with a large subbasal crest; pronotum with prominent discal tubercles and elongate, blunt lateral tubercles .... *Lophopogonius*
— Elytra without a large subbasal crest; pronotum with small or obsolete discal tubercles and short, acute lateral tubercles .... *Pogonocherus*

43(35). Pronotum with sides broadly tuberculate, antennae relatively short; body with gray and black appressed pubescence .... *Tigrinestola*
— Pronotum with acute lateral tubercles; antennae longer; body with concolorous pubescence .... *Estoloides*

44(32). Elytra with long or short erect setae or flying hairs 45
— Elytra with appressed pubescence or tufted tubercles 51

45(44). Pronotum with obtuse tubercles at sides 46
— Pronotum with acute tubercles or spines at sides 48

46(45). Elytra with apices rounded to emarginate 47
— Elytra with outer margins of apices produced into broad spines; pronotum with small obtuse lateral tubercles behind middle; form convex, broad .. .. *Alcidion*

47(46). Prothorax with intercoxal process angulate at sides; elytra with apices rounded; abdomen with last sternite not elongate in female .... *Glaucotes*
— Prothorax with intercoxal process not angulate at sides; elytra with apices truncate to emarginate; abdomen with last sternite elongate in female.. .... *Urographis*

48(45). Elytra with small tufted tubercles, at least on basal crests 49
— Elytra lacking tufted tubercles 50

49(48). Pronotum with sides broadly tuberculate behind middle; male with front tarsi broad, fringed laterally; female with last abdominal sternite greatly elongate .... *Trichocanonura*
— Pronotum with sides acutely tuberculate behind middle; male with front tarsi slender; female with last abdominal sternite moderately elongate .... *Eutrichillus*

50(48). Elytra with pubescence dense, appressed, suberect setae short; body cylindrical .. *Dectes*
— Elytra with pubescence very fine, short, not obscuring surface; body subdepressed ... *Valenus*

51(44). Pronotum with acutely spined tubercles at sides 52
— Pronotum with lateral tubercles rounded or obtuse 60

52(51). Pronotum with lateral spines a little behind middle 53
— Pronotum with lateral spines near base 56

53(52). Pronotum with basal transverse impression extending below tubercles onto sides 54
— Pronotum with basal impression extending only to bases of lateral tubercles, not onto sides .... *Sternidius*

54(53). Antennomeres three to five at most with several suberect hairs beneath, at most one-third longer than body; abdomen of female not elongate . 55
— Antennomeres three to five densely fimbriate beneath, at least 1-1/2 times as long as body; abdomen with last sternite elongate in female .... *Acanthocinus*

55(54). Pronotum with disk convex, not callused; antennal scape subequal in length to third antennomere; protibiae with short apical spurs .... *Sternidocinus*
— Pronotum with five obtuse discal calluses; antennal scape shorter than third antennomere; tibial spurs normal .... *Astylopsis*

56(52). Elytra with epipleura rounded 57
— Elytra with epipleura vertical, delimited by a lateral carina; most elytra with small round spots.. .... *Hyperplatys*

57(56). Pronotum with basal sulcus extending onto sides 58
— Pronotum with basal sulcus confined to disk, not extending beyond bases of lateral tubercles 59

58(57). Mesosternum with mesocoxal process as broad as prosternal process, about 1/6 as broad as coxal cavities; lateral tubercles placed at or near extreme base; pronotum lacking discal calluses... .... *Urgleptes*
— Mesosternal process much broader than prosternal process; lateral tubercles placed before base; pronotum with three discal calluses..... *Liopinus*

59(57). Mesosternum with intercoxal process about 2/3 as broad as coxal cavities, much broader than prosternal process; body ovoid ... *Nyssodrysina*
— Mesosternal process about 1/6 as broad as coxal cavities, only slightly broader than prosternal process; body small, subdepressed .. *Lepturges*

60(51). Prothorax with intercoxal process less than half as broad as coxal cavities 61
— Prothorax with intercoxal process more than half as broad as coxal cavities 63

61(60). Elytra with costae and small tufted tubercles; form small to moderate-sized 62
— Elytra lacking costae and tufted tubercles; antennae of males with sixth antennomere apically produced; form large, robust .... *Coenopoeus*

62(61). Pronotum with obtuse tumid tubercles at sides just before basal impression; disk with three calluses; mesosternal process about as broad as coxal cavities .... *Styloleptus*
— Pronotum with obtuse lateral tubercles at middle; disk with five calluses; mesosternal process about 2/3 as broad as coxal cavities. *Pseudastylopsis*

63(60). Pronotum with three discal calluses 64

— Pronotum with five or more discal calluses..... 65

64(63). Legs with numerous long flying hairs; prosternal process 2/3 as broad as coxal cavities, mesosternal process about as broad as coxal cavities, abruptly declivous anteriorly .... *Trichastylopsis*

— Legs lacking long flying hairs; prosternal process half as broad as coxal cavities, mesosternal process as broad as coxal cavities, arcuate anteriorly .... *Astylidius*

65(63). Form small to moderate-sized; pronotum with lateral tubercles small to moderate-sized; antennae and front tarsi of males not modified 66

— Form large, robust; pronotum with robust lateral tubercles; antennae of males with an apical spur on sixth sement; front tarsi of males broadened and strongly fringed .... *Lagocheirus*

66(65). Mesosternum with intercoxal process as broad as or slightly less broad than coxal cavities; prosternal process about 1/2 to 2/3 as broad as coxal cavities 67

— Mesosternal process broader than coxal cavities; prosternal process about 2/3 as broad as coxal cavities .... *Leptostylopsis*

67(66). Pronotum with discal calluses obtuse, shallow area between calluses densely punctate; mesosternal process about 4/5 as broad as coxal cavities .. .. *Astylopsis*

— Pronotum with discal calluses prominent, punctures between calluses not distinct; mesosternal process as broad as coxal cavities . *Leptostylus*

68(1). Eyes completely divided (Tetraopini) 69

— Eyes emarginate 71

69(68). Tarsal claws bifid; most with body broad, robust; elytra always with dark spots or chevrons .... *Tetraopes*

— Tarsal claws appendiculate; most with body more narrow, elongate; elytra without dark spots or chevrons 70

70(69). Form small, 4-6 mm, integument black, elytra pale orange-brown with apices black .... *Tetrops*

— Form larger, 7-12 mm, elytra and abdomen black, head and pronotum excepting mandibles and thoracic umbone reddish or orange .... *Phaea*

71(68). Abdomen with first four sternites equal or gradually decreasing in length; episternum of metathorax broad anteriorly, strongly tapering posteriory (Phytoeciini) 72

— Abdomen with sternites two to four shorter than others; episternum of metathorax narrow, gradually tapering posteriorly (Hemilophini) 73

72(71). Tarsal claws bifid .... *Mecas*

— Tarsal claws appendiculate .... *Oberea*

73(71). Elytra with post-humeral carinae well developed, sides steeply declivous below carina .... *Cathetopteron*

— Elytra lacking post-humeral carinae, sides less abrupt .... *Hemierana*

Classification of the Genera of America North of Mexico

Cerambycidae Leach 1815

Parandrinae Blanchard 1845

Hesperandra Arigony 1977, 1 sp., *H. polita* (Say 1835), eastern United States, west to Nebraska, Mexico and Central America.

Parandra Latreille 1804
Subgenus *Neandra* Lameere 1912, 2 spp., eastern United States to Texas, Arizona, southern California and Central America (key to spp. Chemsak 1996; biology Solomon 1995).

Prioninae Latreille 1804

Macrotomini Thomson 1860

Archodontes Lameere 1903, 1 sp., *A. melanopus* (Linnaeus 1767), and 3 subspp., southern United States (biology Solomon 1995).
Paramallus Casey 1912

Mallodon Lepeletier and Audinet-Serville 1830, 1 sp., *M. dasytomus* (Say 1823), southeastern United States to Arizona, and Mexico (biology Solomon 1995).

Neomallodon Linsley 1957, 1 sp., *N. arizonicus* (Casey 1912), Arizona.

Nothopleurus Lacordaire 1869, 2 spp., southwestern United States (key to spp. Chemsak 1996).

Stenodontes Audinet-Serville 1832, 1 sp., *S. chevrolati* Gahan 1890, southern Florida, Cuba, and the Bahamas.

Strongylaspis Thomson 1860, 1 sp., *S. corticarius* (Erichson in Schomburg 1848), Florida, Mexico, Central America, South America, and the Antilles.

Callipogonini Thomson 1860

Ergates Audinet-Serville 1832, 2 spp., Washington to California, Arizona, New Mexico, Colorado, Utah, Wyoming, and South Dakota (key to spp. Chemsak 1996).
Callergates Lameere 1904
Trichocnemis LeConte 1851

Solenopterini Lacordaire 1869

Elateropsis Chevrolat 1862, 1 sp., *E. scabrosus* Gahan 1890, southern Florida, West Indies.
Derancistrus Audinet-Serville 1832 (part)

Sphenostethus Haldeman 1845, 1 sp., *S. taslei* (Buquet 1841), eastern United States to Florida and Texas.
Hoplopteryx Westwood 1845
Derancistrus Audinet-Serville 1832 (part)

Prionini Latreille 1804

Derobrachus Audinet-Serville 1832, 4 spp., southeastern United States to Texas, New Mexico, Arizona, and southern California (key to spp. Chemsak 1996).

Orthosoma Audinet-Serville 1832, 1 sp., *O. brunneum* (Forster 1771), eastern and central United States.

Prionus Fabricius 1775, 16 spp., generally distributed.
subgenus *Prionus* Fabricius 1775, 5 spp., generally distributed (key to spp. Chemsak 1996; biology Solomon 1995).
Prionellus Casey 1924
subgenus *Neopolyarthron* Semenov 1899, 2 spp., eastern and central United States (key to spp. Chemsak 1996; biology Solomon 1995).
Riponus Casey 1912
subgenus *Antennalia* Casey 1912, 1 sp., *P. (Antennalia) fissicornis* Haldeman 1845, central United States
subgenus *Homaesthesis* LeConte 1862, 8 spp., central and western United States (key to spp. Chemsak 1996).
Prionina Casey 1912

Meroscelisini Thomson 1860

Tragosoma Audinet-Serville 1832, 3 spp., northeastern and western United States (key to spp. Chemsak 1996).

Spondylidinae Audinet-Serville 1832

Scaphinus LeConte 1851, 1 sp., *S. muticus* (Fabricius 1801), southeastern United States.

Spondylis Fabricius 1775, 1 sp., *S. upiformis* Mannerheim 1843, western North America and northern Mexico.

Aseminae Thomson 1860

Asemini Thomson 1860

Arhopalus Audinet-Serville 1834, 4 spp., generally distributed (key to spp. Chemsak 1996).
Criocephalum Dejean 1835
Criocephalus Mulsant 1839

Asemum Eschscholtz 1830, 4 spp., generally distributed (key to spp. Chemsak 1996).
Liasemum Casey 1912

Megasemum Kraatz 1879, 1 sp., *M. asperum* (LeConte 1854), western United States.
Nothorhina LeConte 1873
Notorhina Leng 1884

Tetropium Kirby in Richardson 1837, 6 spp., including an immigrant sp. from the Palearctic, *T.* fuscum (Fabricius 1787) and 1 subsp., generally distributed except gulf coastal plain of the southeastern United States (key to spp. Chemsak 1996; see also Smith and Hurley 2000).
Isarthron Dejean 1835 (Chemsak 1996, indicated that this name has priority over *Tetropium* but did not validate the change).
Criomorphus Mulsant 1839
Isarthrum Agassiz 1846

Atimiini LeConte 1873

Atimia Haldeman 1847, 6 spp. and 2 subspp., generally distributed (key to spp. Chemsak 1996).

Paratimia Fisher 1915, 1 sp., *P. conicola* Fisher 1915, California and Oregon.

Michthisomini LeConte 1873

Michthisoma LeConte 1850, 1 sp., *M. heterodoxum* LeConte 1850, North Carolina and Georgia.
Mychthisoma Gemminger and Harold 1873
Michthysoma LeConte 1873

Lepturinae Latreille 1804

Lepturini Latreille 1804

Acmaeops LeConte 1850, 2 spp., eastern and montane regions of North America (key to spp. Linsley and Chemsak 1972).
Acmaeops (Pseudodinoptera) Aurivillius 1912 (part)

Acmaeopsoides Linsley and Chemsak 1976, 1 sp., *A. rufula* (Haldeman 1847), northeastern North America.

Alosternida Podany 1961, 1 sp., *A. chalybea* (Haldeman 1847), eastern United States.

Analeptura Linsley and Chemsak 1976, 1 sp., *A. lineola* (Say 1824), eastern North America.

Anastrangalia Casey 1924, 3 spp., southern Canada and western and southwestern United States (key to spp. Linsley and Chemsak 1976).
Anoplodera Swaine and Hopping 1928 (part)

Anthophylax LeConte 1850, 4 spp., southeastern Canada and northeastern United States to southern Appalachians (key to spp. Linsley and Chemsak 1972).
   Anthophilax LeConte in Agassiz 1850

Bellamira LeConte 1873, 1 sp., *B. scalaris* (Say 1826), eastern North America.

Brachyleptura Casey 1913, 7 spp., eastern and western North America (key to spp. Linsley and Chemsak 1976).
   Anoplodera Swaine and Hopping 1928 (part)

Brachysomida Casey 1913, 5 spp., southern Canada and eastern and western United States and Canada to Rocky Mountains (key to spp. Linsley and Chemsak 1972).

Centrodera LeConte 1850, 11 spp. and 2 subspp., eastern North America and western United States (key to spp. and diagnoses of subspp., Linsley and Chemsak 1972).
   Rhamnusium Haldeman 1847
   Parapachyta Casey 1913
   Centrodera (*Apatophysis*) Gressitt 1947

Charisalia Casey 1913, *C. americana* (Haldeman 1847), eastern and central United States.

Choriolaus Bates 1885, 1 sp., *C. sabinoensis* (Knull 1954), Arizona
   Euryptera LeConte and Horn 1883 (part)

Comacmaeops Linsley and Chemsak 1972, 1 sp., *C. brunnea* (Knull 1962), Texas.

Cortodera Mulsant 1864, 20 spp. and 2 subspp., western North America (key to spp. and diagnoses of subspp., Linsley and Chemsak 1972).
   Acmaeops LeConte 1850 (part)
   Leptacmaeops Casey 1913
   Leptacmaeops (*Acmaeopsilla*) Casey 1913

Cosmosalia Casey 1913, 2 spp. and 1 subsp., boreal North America, western United States and northern Mexico (key to spp. and diagnosis of subsp., Linsley and Chemsak 1976).
   Anoplodera Swaine and Hopping 1928 (part)

Cyphonotida Casey 1913, 2 subspp of *C. rostrata* (Bates 1880) (for diagnoses of subspp. see Giesbert and Hovore 1998).
   Ophistomis Bates 1880 (part)

Dorcasina Casey 1913, 2 spp., Pacific coast states (key to spp. Linsley and Chemsak 1976).
   Anoplodera Swaine and Hopping 1928 (part)

Encyclops Newman 1838, 2 spp., eastern North America and California and Oregon (key to spp. Linsley and Chemsak 1972).

Evodinus LeConte 1850, 2 spp. and 1 subsp., eastern United States, western Canada, Alaska, northwestern United States and Wyoming (for diagnoses see Chemsak and Linsley 1972, Lewis 1976).
   Brachyta Fairmaire in Jacquelin du Val 1864

Gaurotes LeConte 1850, 2 spp., eastern North America (key to spp. Linsley and Chemsak 1972).

Gnathacmaeops Linsley and Chemsak 1972, 1 sp., *G. pratensis* (Laicharting 1784), boreal regions of North America.

Grammoptera Audinet-Serville 1835, 6 spp., generally distributed in boreal and montane habitats (key to spp. Linsley and Chemsak 1976).
   Parallelina Casey 1913

Idiopidonia Swaine and Hopping 1928, 1 sp., *I. pedalis* (LeConte 1861), eastern North America.

Judolia Mulsant 1863, 9 spp. and 3 subspp., generally distributed in America north of Mexico (key to spp. and diagnoses of subspp., Linsley and Chemsak 1976).
   Anoplodera Swaine and Hopping 1928 (part)

Leptalia LeConte 1873, 1 sp., *L. macilenta* (Mannerheim 1853), west coast of North America to Alaska.

Leptorhabdium Kraatz 1879, 1 sp., *L. pictum* (Haldeman 1847), eastern United States.

Leptura Linnaeus 1758, 14 spp., generally distributed.
   Stenura Haldeman 1847 (part)
   Strangalia LeConte 1850 (part)
   Strangalia (*Strangalia*) Aurivillius 1912 (part)
   Leptura (*Megaleptura*) Casey 1913
   Strangalia Casey 1924, not Audinet-Serville 1835
   Nakanea Ohbayashi 1963
   subgenus *Leptura* Linnaeus 1758, 11 spp. and 2 subspp., generally distributed (key to spp. and diagnoses of subspp., Linsley and Chemsak 1976).
   subgenus *Stenura* Dejean 1837, 3 spp., eastern United States west to Arizona (key to spp. Linsley and Chemsak 1976).

Lepturopsis Linsley and Chemsak 1976, 2 spp., eastern and western United States (key to spp. Linsley and Chemsak 1976).
   Anoplodera Swaine and Hopping 1928 (part)

Lychochoriolaus Linsley and Chemsak 1976, 1 sp., *L. lateralis* (Olivier 1795), southeastern United States to Texas and southern Mexico.
   Euryptera LeConte 1873 (part)
   Choriolaus Linsley and Chemsak 1971 (part)

Megachoriolaus Linsley 1970, 3 spp., Texas and Arizona (key to spp. Giesbert and Hovore 1998).
   Euryptera Swaine and Hopping 1928 (part)

Metacmaeops Linsley and Chemsak 1972, 1 sp., *M. vittata* (Swederus 1787), eastern United States.

Neanthophylax Linsley and Chemsak 1972, 4 spp. and 2 subspp., western North America to Rocky Mountains (key to spp. and diagnoses of subspp., Linsley and Chemsak 1972).

Neoalosterna Podany 1961, 2 spp., eastern and western North America (key to spp. Linsley and Chemsak 1976).

Neobellamira Swaine and Hopping 1928, 1 sp., *N. delicata* (LeConte 1874) and 1 subsp., California and Oregon (diagnosis of subsp., Linsley and Chemsak 1976).

Orthochoriolaus Linsley and Chemsak 1976, 1 sp., *O. chihuahuae* (Bates 1885), Arizona, Mexico.

Ortholeptura Casey 1913, 3 spp., Pacific coast states (key to spp. Linsley and Chemsak 1976).
    Anoplodera Swaine and Hopping 1928 (part)

Pachyta Dejean 1821, 2 spp.and 1 subsp., Canada, Alaska and western United States (diagnoses of spp. and subsp., Linsley and Chemsak 1972).
    Argaleus LeConte 1850
    Acmaeops Thomson 1866, not LeConte 1850
    Anthophylax Blessig 1873, not LeConte 1850
    Neopachyta Bedel 1906
    Pachyta (*Linsleyana*) Podany 1964

Pidonia Mulsant 1863, 6 spp., generally distributed in North America (key to subgenera, Linsley and Chemsak 1976).
    Grammoptera Thomson 1864, not Audinet-Serville 1835
    Pseudopidonia Pic 1900
    Hapalosalia Casey 1913
    Pidnota Bayer and Shenefelt 1969
    subgenus *Pidonia* Mulsant 1863, 4 spp., generally distributed in North America (key to spp. Linsley and Chemsak 1976).
    subgenus *Thesalia* Casey 1891, 1 sp., *P. gnathoides* (LeConte 1873), southwestern Canada and western United States.
    subgenus *Ceratopidonia* Linsley and Chemsak 1976, 1 sp., *P. quadrata* (Hopping 1931), southwestern Canada and northwestern United States.

Piodes LeConte 1850, 1 sp., *P. coriacea* LeConte 1850, northwestern United States.

Pseudogaurotina Plaviltstshikov 1958, 2 spp. and 1 subsp., southeastern Canada and northeastern United States and southwestern Canada and western United States (key to spp. and diagnosis of subsp., Linsley and Chemsak 1972).
    Gaurotes Lacordaire 1869 (part)

Pseudostrangalia Swaine and Hopping 1928, 1 sp., *P. cruentata* (Haldeman 1847), eastern North America to Texas.

Pygoleptura Linsley and Chemsak 1976, 3 spp. and 1 subsp., northern and western United States to Alaska (key to spp. and diagnosis of subsp., Linsley and Chemsak 1976).
    Anoplodera Swaine and Hoppings 1928 (part)

Pyrotrichus LeConte 1862, 1 sp., *P. vitticollis* LeConte 1862, Pacific coast of North America.

Rhagium Fabricius 1775, 1 sp., *R. inquisitor* (Linnaeus 1758), Holarctic.
    Stenocorus Geoffroy 1762
    Stencorus Lamarck 1817
    Hargium Samouelle 1819
    Allorhagium Kolbe 1884
    Harpium Reitter 1912

Sachalinobia Jacobson 1899, 1 sp., *S. rugipennis* (Newman 1844), eastern Canada and northeastern United States.
    Pseudopachyta Swaine and Hopping 1928

Stenocorus Fabricius 1775, 12 spp. generally distributed in North America (key to spp. Linsley and Chemsak 1972; see also Giesbert and Hovore 1998).
    Toxotus Dejean 1821
    Anisorus Mulsant 1863
    Minaderus Mulsant 1863
    Toxotochorus Reitter 1907
    Stenochorus Reitter 1912
    Stenocorus (*Eutoxotus*) Casey 1913
    Stenocorus (*Toxotopsis*) Casey 1913

Stenostrophia Casey 1913, 3 spp. and 2 subspp., western United States and British Columbia (key to spp. and diagnoses of subspp., Linsley and Chemsak 1976; biology Solomon 1995).
    Anoplodera Swaine and Hopping 1928 (part)

Stictoleptura Casey 1924, 1 sp., *S. canadensis* (Olivier 1795), and 3 subspp., northeastern United States and southeastern Canada and southwestern United States (diagnoses of subspp. Linsley and Chemsak 1976).
    Anoplodera Swaine and Hopping 1928 (part)

Strangalepta Casey 1913, 2 spp., eastern North America (key to spp. Linsley and Chemsak 1976).
    Anoplodera Swaine and Hopping 1928 (part)

Strangalia Audinet-Serville 1835, 10 spp. and 1 subsp., generally distributed (key to spp. and diagnosis of subsp., Linsley and Chemsak 1976).
    Ophistomis Thomson 1857
    Ophiostomis Gemminger and Harold 1872
    Strangalina Aurivillius 1912

Strophiona Casey 1913, 3 spp., generally distributed (key to spp. Linsley and Chemsak 1976).
Anoplodera Swaine and Hopping 1928 (part)

Trachysida Casey 1913, 2 spp. and 2 subspp., eastern and western North America (key to spp. and diagnoses of subspp., Linsley and Chemsak 1976).
Anoplodera Swaine and Hopping 1928 (part)

Trigonarthris Haldeman 1847, 3 spp., eastern North America (key to spp. Linsley and Chemsak 1976).
Leptura (*Cercolia*) Casey 1913
Anoplodera Swaine and Hopping 1928 (part)

Typocerus LeConte 1850, 15 spp. and 3 subspp., generally distributed (key to spp. and diagnoses of subspp., Linsley and Chemsak 1976).
Strangalia (*Typocerus*) Aurivillius 1912

Xestoleptura Casey 1913, 6 spp., generally distributed, but mostly western (key to spp. Linsley and Chemsak 1976).
Anoplodera Swaine and Hopping 1928 (part)

Xylosteus Frivaldsky 1838, 1 sp., *X. ornatus* LeConte 1873, Oregon, and California.

Desmocerini Blanchard 1845

Desmocerus Dejean 1821, 3 spp. and 4 subspp., eastern and western North America (key to spp. and diagnoses of subspp., Linsley and Chemsak 1972; biology Solomon 1995).

Necydalini Latreille 1825

Necydalis Linnaeus 1758, 7 spp. and 1 subspp., eastern North America and Pacific coast to northwestern Mexico (key to spp. and diagnosis of subsp., Linsley and Chemsak 1972).

Ulochaetes LeConte 1854, 1 sp., *U. leoninus* LeConte 1854, Pacific coast from British Columbia to California.

Disteniinae Thomson 1860

Distenia Lepeletier and Audinet-Serville in Latreille 1828, 1 sp., *D. undata* (Fabricius 1775), eastern United States.

Cerambycinae Latreille 1804

Smodicini Lacordaire 1869

Smodicum Haldeman 1847, 3 spp., eastern United States to Texas and Arizona; also northwestern Mexico (key to spp. Martins 1975).

Oemini Pascoe 1869

Eudistenia Fall 1907, 1 sp., *E. costipennis* Fall 1907, southern California.

Haplidoeme Chemsak and Linsley 1965, *H. schlingeri* Chemsak and Linsley 1965, California.

Malacopterus Audinet-Serville 1833, 1 sp., *M. tenellus* (Fabricius 1801), Texas to southern California.
Malacomacrus White 1853
Ganimus LeConte 1873

Oeme Newman 1840, 2 spp. and 1 subsp., generally distributed (key to spp. and diagnosis of subsp., Linsley 1962b).
Sclerocerus LeConte 1850

Paranoplium Casey 1924, 1 sp., *P. gracile* and 1 subsp. (LeConte 1881), California (diagnosis of subsp., Linsley 1962b).

Pseudomethia Linsley 1937, 1 sp., *P. arida* Linsley 1937, Arizona and California to northwestern Mexico.

Vandykea Linsley 1932, 1 sp., *V. tuberculata* Linsley 1932, northern Califomia.

Methiini Thomson 1860

Coleomethia Linsley 1940, 1 sp., *C. xanthocollis* (Knull 1935), Texas.

Methia Newman 1842 16 spp., southern United States (key to spp. Linsley 1962b; see also Philips and Ivie 1998).
Thia Newman 1840
Staphylinoeme Gounelle 1913

Styloxus LeConte 1873, 3 spp. and 1 subsp., southwestern United States and northern Mexico (key to spp. and diagnosis of subsp., Linsley 1962b; biology of *S. fulleri* (Horn), Solomon 1995).
Idoemea Horn 1880
Malthopia Casey 1912
Malthophia Casey 1913

Tessaropa Haldeman 1847, 1 sp., *T. tenuipes* (Haldeman 1846), eastern United States to Texas.
Dysphaga LeConte 1852
Tessaroptes Gemminger and Harold 1873

Opsimini LeConte 1873

Dicentrus LeConte 1880, 2 spp., Pacific coast of North America (key to spp. Linsley 1962b).
Paraopsimus Champlain and Knull 1926

Opsimus Mannerheim 1843, 1 sp., *O. quadrilineatus* Mannerheim 1843, Pacific coast of North America.

Achrysonini Lacordaire 1869

Achryson Audinet-Serville 1833, 1 sp., *A. surinamum* (Linnaeus 1767), Atlantic coast and southwestern United States to Argentina and West Indies.

Geropa Casey 1912, 1 sp., *G. concolor* (LeConte 1873), southern Texas to southern Mexico.

Eburiini Blanchard 1845

Eburia Lepeletier and Audinet-Serville in Lacordaire 1830, 12 spp., eastern to southwestern United States to South America (key to spp. Linsley 1962b).
Dissacanthus Hope 1835
Coeleburia Thomson 1860
Dissacantha Thomson 1864
Drymo Thomson 1864
Eleutho Thomson 1864
Pantomallus Lacordaire 1869

Hesperophanini Mulsant 1839

Brothylus LeConte 1859, 2 spp., western United States (key to spp. Linsley 1962b).

Cacophrissus Bates 1885, 1 sp., *C. pauper* Bates 1885, Arizona to Baja California and western Mexico.
Knullanoplium Linsley 1957

Eucrossus LeConte 1873, 1 sp., *E. villicornis* LeConte 1873, southwestern United States to Mexico.

Haplidus LeConte 1873, 2 spp., western United States (key to spp. Linsley 1962b).

Hesperanoplium Linsley 1957, 2 spp., California (key to spp. Linsley 1962b).

Hesperophanes Dejean 1835, 1 sp., *H. pubescens* (Haldeman 1847), eastern North America.
Anoplium Haldeman 1847
Trichoferus Wollaston 1854
Stromatium LeConte 1873, not Audinet-Serville 1834

Malobidion Schaeffer 1908, 1 sp., *M. brunneum* Schaeffer 1908, Arizona and Utah to Sonora.

Meganoplium Linsley 1940, 1 sp., *M. imbelle* (LeConte 1881), California.

Osmidus LeConte 1873, 1 sp., *O. guttatus* LeConte 1873, southwestern United States to Mexico.

Penichroa Stephens 1839, 1 sp., *P. fasciata* (Stephens 1831), an immigrant sp. from Europe, found in Pennsylvania and New York.
Exilia Mulsant 1863, not Conrad 1860
Liagrica Costa 1855

Stromatium Audinet-Serville 1834, 1 sp., *S. fulvum* (Villers 1789), an immigrant sp. from Europe reported from North America, not known to be established.

Tylonotus Haldeman 1847, 2 spp., eastern North America to Arizona (key to spp. Linsley 1962b).

Xeranoplium Linsley 1957, 1 sp., *X. tricallosum* (Knull 1938), Arizona and Colorado.

Zamodes LeConte 1873, 1 sp., *Z. obscurus* LeConte 1873, Pennsylvania.

Elaphidiini Thomson 1864

Aneflomorpha Casey 1912, 21 spp., generally distributed in the United States and into Central America (key to spp. Linsley 1963; biology of some species, Solomon 1995).
Anespyra Casey 1912

Aneflus LeConte 1873, 7 spp. and 2 subspp., southwestern United States (Texas to California) and northwestern Mexico (key to spp. and diagnoses of subspp, Linsley 1963; see also Lingafelter 1998).
Axestinus LeConte 1873
Protaneflus Linsley 1934

Anelaphus Linsley 1936, 22 spp., generally distributed except northwestern North America (keys to spp., Linsley 1963; see also Skiles 1985 and Lingafelter 1998; biology Solomon 1995).
Elaphidionoides Linsley 1957 (part)
Peranoplium Linsley 1957

Anopliomorpha Linsley 1936, 1 sp., *A. rinconium* (Casey 1924), Arizona to Baja California and western Mexico.

Anoplocurius Fisher 1920, 3 spp., Texas, Arizona and California to Baja California (key to spp. Linsley 1963).
Astromula Chemsak and Linsley 1965, 1 sp., *A. nitidum* Chemsak and Linsley 1965, California.

Atylostagma White 1853, 1 sp., *A. glabrum* Schaeffer 1909, Arizona to southern Mexico.

Curtomerus Stephens 1939, 2 spp., Florida and West Indies to South America (key to spp. Linsley 1963).
Cylindera Newman 1833
Lampromerus Thomson 1860
Cyrtomerus Gemminger and Harold 1872
Cylindrodera Gemminger and Harold 1872
Sotenus Sharp 1878

Elaphidion Audinet-Serville 1834, 8 spp., eastern and southeastern United States to Texas, West Indies into South America (key to spp. Linsley 1963).
Cyclopleurus Hope 1835
Elaphidium Agassiz 1846
Centrocerum Thomson 1864, not Chevrolat 1861

Elaphidionopsis Linsley 1936, 1 sp., *E. fasciatipennis* Linsley 1936, Texas and northern Mexico.

Enaphalodes Haldemam 1847, 8 spp., generally distributed (key to spp. Linsley 1963; biology Solomon 1995).
Romaleum White 1855
Thersalus Pascoe 1865
Hypermallus Lacordaire 1869

Eustromula Cockerell 1906, 1 sp., *E. validum* (LeConte 1858), California to Texas, Baja California and northern Mexico.
Eustroma LeConte 1873

Gymnopsyra Linsley 1937, 3 spp., Arizona to Texas (key to spp. Linsley 1963).

Linsleyonides Skiles 1985, 1 sp., *L. albomaculatus* (Champlain and Knull 1922), southern Florida and Cuba (key to spp. Skiles 1985).
Elaphidionoides Linsley 1963 (part)

Micraneflus Linsley 1957, 1 sp., *M. imbellis* (Casey 1914), Arizona, California and Baja California.

Micranoplium Linsley 1957, 1 sp., *M. unicolor* (Haldeman 1847), eastern United States.
Anoplium Haldeman 1847

Neaneflus Linsley 1957, 2 spp., southwestern United States and northern Mexico (key to spp. Linsley 1963).

Orwellion Skiles 1985, 2 spp., southwestern United States and northern Mexico (key to spp. Skiles 1985).
Elaphidionoides Linsley 1963 (part)

Parelaphidion Skiles 1985, 3 spp., eastern North America to Texas and Cuba (key to spp. Skiles 1985).
Elaphidionoides Linsley 1963 (part)

Phoracantha Newman 1840, 2 spp., immigrants from Australia, found in California.

Psyrassa Pascoe 1866, 5 spp., eastern United States, Texas to Panama (key to spp. Linsley 1963; biology Solomon 1995).
Stizocera Haldeman 1847, not Audinet-Serville 1834
Pseudibidion Casey 1912

Romulus Knull 1948, 1 sp., *R. globosus* Knull 1948, Florida.

Stenelaphus Linsley 1936, 1 sp., *S. alienus* (LeConte 1875), Arizona, New Mexico and Mexico.

Stenosphenus Haldeman 1847, 5 spp., eastern and southwestern United States to Guatemala (key to spp. and revision of genus Giesbert and Chemsak 1989).
Stenosphenopsis Linsley 1935

Stizocera Linsley 1961, 1 sp., *S. floridana* Linsley 1949, Florida.
Nesostizocera Linsley 1961

Piezocerini Lacordaire 1869

Piezocera Audinet-Serville 1834, 1 sp., *P. serraticollis* Linell 1897, Texas and northern Mexico.

Ibidionini Thomson 1860

Heterachthes Newman 1840, 5 spp., eastern North America to South America (key to spp. Linsley 1963).
Ibidion (*Heterachthon*) Thomson 1867
Heterachthon Aurivillius 1912
Heterachtes Aurivillius 1912
Brydaeon Thomson 1867

Neocompsa Martins 1965, 4 spp. and 1 subspp., Texas to Arizona and Mexico.

Curiini LeConte 1873

Curius Newman 1840, 1 sp., *C. dentatus* Newman 1840, southeastern United States.

Plectromerus Haldeman 1847, 1 sp., *P. dentipes* (Olivier 1790), southeastern United States and Cuba.

Graciliini Mulsant 1839

Gracilia Audinet-Serville 1834, 1 sp., *G. minuta* (Fabricius 1781), an immigrant from Europe, found in eastern United States.
Nothrus Haldeman 1847
Oesyophila Bedel 1894

Hypexilis Horn 1885, 1 sp., *H. pallida* Horn 1885, southern Texas to Mexico.

Perigracilia Linsley 1942, 1 sp., *P. delicata* Knull 1942, Arizona.

Obriini Mulsant 1839

Obrium Dejean 1825, 9 spp., generally distributed (key to spp. Linsley 1963).
Phyton Newman 1840
Diozodes Haldeman 1847

Hyboderini Linsley 1940

Hybodera LeConte 1873, 2 spp., western North America (key to spp. Linsley 1963).

Lampropterus Mulsant 1863, 2 spp., Pacific coast of North America (key to spp. Linsley 1963).
    Callimus Mulsant 1864
    Pilema LeConte 1873
    Callimellum Strand 1928

Megobrium LeConte 1873, 1 sp., *M. edwardsi* LeConte 1873, California.

Pseudopilema Linsley 1940, 1 sp., *P. hoppingi* (Van Dyke 1920), California.

Eumichthini Linsley 1940

Eumichthus LeConte 1873, 1 sp., *E. oedipus* LeConte 1873, Pacific coast of North America.

Poecilobrium Horn in LeConte and Horn 1883, 1 sp., *P. chalybeum* (LeConte 1873), British Columbia to California and Colorado to New Mexico.

Nathriini Linsley 1963

Nathrius Bréthes 1916, 1 sp., *N. brevipennis* (Mulsant 1839), an immigrant from Europe, found in California.
    Leptides Mulsant 1839
    Leptideella Strand 1936
    Deuteroleptidea Paclt 1946

Molorchini Mulsant 1863

Molorchus Fabricius 1792, 3 spp. and 6 subspp., eastern North America to Texas and California (key to spp. and diagnoses of subspp., Linsley 1963).
    Heliomanes Newman 1840
    Glaphyra Newman 1840
    Laphyra Newman 1842
    Caenoptera C. G. Thomson 1859
    Limomius Mulsant 1863
    Sinolus Mulsant 1863
    Conchopterus Fairmaire in Jacquelin du Val 1864

Psebiini Lacordaire 1869

Nathriobrium Hovore 1980, 1 sp., *N. methioides* Hovore 1980, Texas (for diagnosis see Hovore 1980).

Stenopterini Fairmaire 1864

Callimoxys Kraatz 1863, 1 sp., *C. sanguinicollis* (Olivier 1795), with two subspp., eastern and western United States: *C. s. sanguinicollis* (Olivier) eastern US and *C. s. fuscipennis* (LeConte) western US.

Rhinotragini Thomson 1860

Odontocera Audinet-Serville 1833, 1 sp., *O. aurocincta arizonensis* Linsley 1961, Arizona to Mexico.

Callichromatini Blanchard 1845

Plinthocoelium Schmidt 1924, 2 spp. and 1 subsp., Delaware to Florida to Texas and Arizona (key to spp. and diagnosis of subspp., Linsley 1964).
    Callichroma LeConte 1850
    Plinthocoalium Arnett 1962

Rosaliini Fairmaire 1864

Rosalia Audinet-Serville 1833, 1 sp., *R. funebris* Motschulsky 1845, western North America (Alaska to California and New Mexico).

Dryobiini Linsley 1964

Dryobius LeConte 1850, 1 sp., *D. sexnotatus* Linsley 1957, eastern North America (biology Solomon 1995).

Ornithia Thomson 1864, 1 sp., *O. m. mexicana* (Stürm 1843), Texas, eastern Mexico to Panama.

Callidiini Mulsant 1839

Callidiellum Linsley 1940, 3 spp., including an immigrant from Asia, *C. rufipenne* (Motschulsky 1860) in North Carolina and two native spp. in California and Arizona (key to native spp. Linsley 1964).

Callidium Mulsant 1839 not Fabricius 1775, 19 spp., generally distributed throughout North America (key to spp. Linsley 1964).
    Meridion des Gozis 1886
    Callidium (*Callidostola*) Reitter 1912

Elatotrypes Fisher 1919, 1 sp., *E. hoferi* Fisher 1919, Colorado, Idaho, Nebraska.

Hylotrupes Audinet-Serville 1834, 1 sp., *H. bajulus* (Linnaeus 1758), the Old House Borer; an immigrant from Europe, found in eastern North America.

Meriellum Linsley 1957, 1 sp., *M. proteus* (Kirby in Richardson 1837), northeastern and western (Alaska to Colorado) North America.
  Merium LeConte 1873, not Kirby 1837

Phymatodes Mulsant 1839, 26 spp. and 6 subspp., generally distributed in the United States and Canada (key to spp. and diagnoses of subspp., Linsley 1964).
  Poecilium Fairmaire in Jacquelin du Val 1864
  Phymatodes (*Melasmetus*) Reitter 1912
  Phymatodes (*Phymatoderus*) Reitter 1912
  Phymatodes (*Phymatodellus*) Reitter 1912
  Microcallidium Casey 1912
  Pseudopoecilium Planet 1924
  Paraphymatodes Plavilstshikov 1934

Physocnemum Haldeman 1847, 3 spp., eastern North America to Texas (key to spp. Linsley 1964; biology Solomon 1995).
  Dularius Thomson 1860

Pronocera Motschulsky 1859, 1 sp., P. *collaris* (Kirby in Richardson 1837), and 1 subsp., eastern North America and Alaska, Utah and Washington (diagnosis of subsp., Linsley 1964).
  Pronocerus Motschulsky 1875
  Pseudophymatodes Pic 1901
  Protocallidium Csiki 1904
  Potocallidium Reitter 1912
  Gonocallus LeConte 1873

Ropalopus Mulsant 1839, 1 sp., *R. sanguinicollis* (Horn 1860), northeastern United States and southern Canada.
  Rhopalopus Agassiz 1846
  Euryoptera Horn 1860

Semanotus Mulsant 1839, 5 spp., generally distributed in North America (key to spp. Linsley 1964).
  Sympiezocera Lucas 1851
  Xenodorum Marseul 1856
  Hylotrupes LeConte 1873 (part)
  Anocomis Casey 1912
  Hemicallidium Casey 1912
  Anacomis Leng 1920

Xylocrius LeConte 1873, 2 spp., western North America (key to spp. Linsley 1964; biology Solomon 1995).
  Hylocrius Lameere 1883

Clytini Mulsant 1839

Calloides LeConte 1873, 2 spp., western United States and eastern North America (key to spp. Linsley 1964).

Chlorophorus Chevrolat 1863, 1 sp., *C. annularis* (Fabricius 1787), an immigrant from Asia, found in the United States.

Clytoleptus Casey 1912, 1 sp., *C. albofasciatus* (Laporte and Gory 1835), eastern North America.

Clytus Laicharting 1784, 8 spp., generally distributed in boreal North America (key to spp. Linsley 1964; see also Hovore and Giesbert 1974).

Euryscelis Dejean 1835, 1 sp., *E. suturalis* (Olivier 1795), Florida and West Indies.

Glycobius LeConte 1873, 1 sp., *G. speciosus* (Say 1828), northeastern United States and southern Canada (biology Solomon 1995).
  Arhopalus Lacordaire 1869, not Audinet-Serville 1834
  Plagionotus LeConte and Horn 1883, not Mulsant 1842

Megacyllene Casey 1912, 10 spp., eastern, central and southwestern United States to South America (key to spp. Linsley 1964; biology Solomon 1995).
  Cyllene Newman 1840, Gray 1834)

Megacheuma Mickel 1919, 1 sp., *M. brevipennis* (LeConte 1873), and 2 subspp., western United States (for diagnosis of subsp., see Hovore 1979; biology Solomon 1995).

Neoclytus Thomson 1860, 26 spp., generally distributed (key to spp. Linsley 1964; biology Solomon 1995) (Volume 2, Color Figure 27).
  Rhopalopachys Chevrolat 1860
  Rhopalomerus Chevrolat 1860
  Plagithmysus Horn 1875

Ochraethes Chevrolat 1860, 1 sp., *O. citrinus* (Chevrolat 1860), southwestern United States to Mexico and Honduras.
  Ochroesthes Chevrolat 1860
  Ochraesthes Thomson 1860
  Ochrestes Lacordaire 1869
  Ochresthes Bates 1880

Placosternus Hopping 1937, 2 spp., Florida and southwestern United States and northern Mexico (key to spp. Linsley 1964).

Plesioclytus Giesbert 1993, 1 sp., *P. relictus* Giesbert 1993, Florida.

Sarosesthes Thomson 1866, 1 sp., *S. fulminans* (Fabricius 1775), eastern North America.
  Arhopalus LeConte 1873, not Audinet-Serville 1834

Tanyochraethes Chemsak and Linsley 1965, 1 sp., *T. tildeni* Chemsak and Linsley 1965, Texas (for sp. diagnosis see Chemsak and Linsley 1965).

Triodoclytus Casey 1912, 1 sp., *T. lanifer* (LeConte 1873), California and Oregon.
  Paraclytus Casey 1912, not Bates 1884
  Synclytus Lucas 1920

Tylcus Casey 1912, 1 sp., *T. hartwegi* (White 1855), southwestern United States through Mexico to Guatemala.

Xylotrechus Chevrolat 1860, 22 spp., generally distributed (key to spp. Linsley 1964; biology Solomon 1995) (Cover Figure, Volume 1, right side).

Anaglyptini Lacordaire 1869

Cyrtophorus LeConte 1850, 1 sp., *C. verrucosus* (Olivier 1795), eastern North America to Texas.

Microclytus LeConte 1873, 2 spp., northeastern North America (key to spp. Linsley 1964).

Tilloclytus Bates 1885, 1 sp., *T. geminatus* (Haldeman 1847), eastern North America.
  Tillomorpha Chevrolat 1860

Tillomorphini Lacordaire 1869

Euderces LeConte 1850, 6 spp., eastern and southwestern United States to Mexico (key to spp. Linsley 1964; revision of genus Giesbert and Chemsak 1997).
  Eplophorus Chevrolat 1861
  Apelocera Chevrolat 1862
  Apilocera Chevrolat 1962

Pentanodes Schaeffer 1904, 1 sp., *P. dietzi* Schaeffer 1904, Texas.

Tetranodus Linell 1896, 1 sp., *T. niveicollis* Linell 1897, Texas.
  Tetranodes Schaeffer 1904

Cleomenini Lacordaire 1869

Dihammaphora Chevrolat 1859, 1 sp., *D. dispar* Chevrolat 1859, Texas.

Rhopalophorini Blanchard 1845

Rhopalophora Audinet-Serville 1834, 7 spp., eastern and southwestern United States to Mexico and Honduras (key to spp. Linsley 1964; revision of genus Giesbert and Chemsak 1993).
  Rhopalophorus LeConte 1873
  Tinopus LeConte 1850

Heteropsini Lacordaire 1869

Heterops Blanchard 1842, 1 sp., *H. dimidiatus* (Chevrolat 1838), southern Florida, Bahamas and Cuba.

Agallissini LeConte 1873

Agallissus Dalman 1823, 1 sp., *A. lepturoides* (Chevrolat in Orbigny 1844), Texas and Mexico to Honduras.
  Aplectrus Chevrolat in Orbigny 1844
  Cryptopleura LeConte 1850
  Agalissus Thomson 1864

Osmopleura Linsley 1964, 1 sp., *O. chamaeropis* (Horn 1893), Florida and Georgia.

Zagymnus LeConte 1873, 1 sp., *Z. clerinus* (LeConte 1873), Florida and Georgia.

Holopleurini Chemsak and Linsley 1974

Holopleura LeConte 1873, 1 sp., *H. marginata* LeConte 1873, British Columbia to California.

Lissonotini Thomson 1860

Lissonotus Dalman in Schoenherr 1817, 1 sp., *L. flavocinctus puncticollis* Bates 1885, Texas to southern California, northern Mexico and Baja California.

Torneutini Thomson 1860

Torneutina Thomson 1860

Chrotoma Casey 1891, 1 sp., *C. dunniana* Casey 1891, Texas to southern California.

Bothriospilina Lane 1950

Chlorida Audinet-Serville 1834, 1 sp., *C. festiva* (Linnaeus 1758), southern Florida.
  Tetracanthus Hope 1835

Gnaphalodes Thomson 1860, 1 sp., *G. trachyderoides* Thomson 1860, Texas to Costa Rica.

Knulliana Linsley 1962, 1 sp., *K. cincta* (Drury 1773), and 3 subspp., eastern United States and Texas to Arizona (key to subspp. and diagnoses of subspp., Linsley 1962b).
  Cerasphorus Dejean 1835
  Chion Newman 1840, not Scopoli 1777

Trachyderini Dupont 1836

Ancylocerina Thomson 1864

Ancylocera Audinet-Serville 1834, 1 sp., *A. bicolor* (Olivier 1795), southeastern United States to Texas.

Trachyderina Dupont 1836

Aethecerinus Fall and Cockerell 1907, 3 spp., Florida and Texas to California (key to spp. Linsley 1962b).
Aethecerus Chevrolat 1862, not Wesmael 1844

Amannus LeConte 1858, 3 spp., southwestern United States to Baja California (key to spp. Linsley 1962b; biology Solomon 1995).

Batyle Thomson 1864, 3 spp. and 8 subspp., widely distributed in North America except Pacific coast (key to spp. and diagnoses of subspp., Linsley 1962b).
Batyleoma Casey 1912

Callona Waterhouse 1840, 1 sp., *C. rimosa* (Buquet 1840), southern Texas and northern Mexico (biology Solomon 1995).

Crioprosopus Audinet-Serville 1834, 2 spp., Texas and Arizona (key to spp. Linsley 1962b; biology Solomon 1995).

Crossidius LeConte 1851, 12 spp. and 37 subspp., western United States and northern Mexico (key to spp. and diagnoses of subspp., Linsley 1962b) (Volume 2, Color Figure 26).
Oxoplus LeConte 1862

Deltaspis Audinet-Serville 1834, 2 spp., Arizona to California and Baja California (key to spp. Beierl and Barchet-Beierl 1999).
Eudoxilus Lacordaire 1869

Elytroleptus Dugès 1879, 8 spp., eastern United States to Texas and Arizona (key to spp. Linsley 1962b).

Lophalia Casey 1912, 1 sp., *L. cyanicollis* (Dupont 1838), Texas and Arizona to southern Mexico.
Entomosterna LeConte 1873, not Chevrolat 1862

Mannophorus LeConte 1854, 2 spp., Arizona and Texas to northern Mexico.

Megaderus Dejean 1821, 1 sp., *M. bifasciatus* Dupont 1836, Texas to Guatemala.

Metaleptus Bates 1872, 1 sp., *M.* batesi Horn 1885, Arizona and southern California.

Parevander Aurivillus 1912, 1 sp., *P. hovorei* Giesbert 1984 (for diagnosis see Giesbert and Hovore 1984).
Evander Lacordaire 1869, not Thomson 1860

Perarthrus LeConte 1851, 2 spp., Arizona and southern California to Baja California (key to spp. Linsley 1962b).

Plionoma Casey 1912, 2 spp., southwestern United States and northern Mexico (key to spp. Linsley 1962b).
Sphaenothecus LeConte 1873, not Dupont 1838

Purpuricenus Dejean 1821, 6 spp., eastern United States to Texas and California (revision of genus in North America and key to spp. MacRae, 2000; biology Solomon 1995).
Acanthoptera Latreille 1829
Acanthopterus Guérin-Méneville 1844
Cyclodera White 1846
Philagathes Thomson 1864

Rhodoleptus Linsley 1962, 1 sp., *R. femoratus* (Schaeffer 1909), Arizona to western Mexico.

Schizax LeConte 1873, 1 sp., *S. senex* LeConte 1873, western Texas to southern California.

Sphaenothecus Dupont 1838, 1 sp., *Sphaenothecus bivittata* (Dupont 1838), southwestern United States to Honduras (revision of genus Chemsak and Noguera 1998).
Sphoenothecus Guérin-Méneville 1839
Sphenothecus White 1853
Taranomis Casey 1912
Ischnocnemis LeConte 1873

Stenaspsis Audinet-Serville 1834, 2 spp., southwestern United States to Honduras (key to spp. Linsley 1962b).
Smileceras LeConte 1850

Trachyderes Dalman in Schoenherr 1817, 1 sp. *T. mandibularis* with 3 subspp., Florida and Texas to southern California (revision of genus and key to spp. and subspp. Hüdepohl 1985).
Dendrobias Dupont in Audinet-Serville 1834
Dendrobias LeConte 1885

Tragidion Audinet-Serville 1834, 5 spp. and 2 subspp., eastern North America to southwestern United States (key to spp. and diagnoses of subspp., Linsley 1962).
Trichoscelis Bates 1885

Tylosis LeConte 1850, 4 spp., Texas to Arizona and Mexico (key to spp. Linsley 1962b).

Lamiinae Latreille 1825

Parmenini Mulsant 1839

Ipochus LeConte 1852, 1 sp., *I. fasciatus* LeConte 1852, California to Baja California.

Parmenosoma Schaeffer 1908, 1 sp., *P. griseum* Schaeffer 1908, Texas.

Plectrura Mannerheim 1852, 1 sp., *P. spinicauda* Mannerheim1852, Alaska to northern California.

Moneilemini Thomson 1864

Moneilema Say 1824, 6 spp., Texas and western North America to Mexico (key to spp. Linsley and Chemsak 1984).
Collapteryx Newman 1838
Monilema LeConte 1852
Collapterix Thomson 1857
Omoscylon Thomson 1867
Monoplesa Motschulsky 1875

Lamiini Latreille 1825

Anoplophora Hope 1839, 1 sp., *A. glabripennis* (Motschulsky 1853), an immigrant from Asia found on Long Island, New York and in Chicago, Illinois. As of this writing, State and Federal agricultural agencies are attempting eradication.

Goes LeConte 1852, 9 spp., eastern and central United States to western Texas (key to spp. Linsley and Chemsak 1984; biology Solomon 1995).

Hebestola Haldeman 1847, 1 sp., *H. nebulosa* Haldeman 1847, eastern United States.
Cacoplia LeConte 1852

Microgoes Casey 1913, 1 sp., *M. oculatus* (LeConte 1862), eastern North America.

Monochamus Megerle in Dejean 1821, 8 spp. and 6 subspp., generally distributed (key to spp. and diagnoses of subspp., Linsley and Chemsak 1984) .
Monachammus Gray in Griffith 1832
Monohammus Dejean 1837
Monohamus Guérin-Méneville 1844

Neoptychodes Dillon and Dillon 1941, 1 sp., *N. trilineatus* (Linnaeus 1771), southern United States to South America and West Indies (biology Solomon 1995).

Plectrodera Dejean 1837, 1 sp., *P. scalator* (Fabricius 1792), eastern and central United States (biology Solomon 1995).

Mesosini Thomson 1860

Synaphaeta Thomson 1864, 1 sp., *S. guexi* (LeConte 1852), California to British Columbia (biology Solomon 1995).
Synaphe Thomson 1864, not Huebner 1789
Mesosa LeConte 1852, not Latreille 1829

Dorcaschematini Thomson 1860

Dorcaschema Haldeman 1847, 4 spp., eastern United States to Texas (key to spp. Linsley and Chemsak 1984; biology Solomon 1995).
Hetoemis Haldeman 1847
Dorchaschema Leng and Hamilton 1896

Apomecynini Thomson 1860

Adetus LeConte 1852, 2 spp., Kansas to Arizona and Texas and northeastern Mexico (key to spp. Linsley and Chemsak 1984).
Polyopsia Haldeman 1847, not Mulsant 1839
Agennopsis Thomson 1857
Talaepora Fairmaire and Germain 1859
Stygnesis Pascoe 1866
Atimuropsis Thomson 1868
Pterichyta Thomson 1868
Tautoclines Thomson 1868
Pterichthya Bates 1880
Sicyobius Horn 1880
Pterichtya Aurivillius 1922

Dorcasta Pascoe 1858, 1 sp., *D. cinerea* (Horn 1860), south central United States to northeastern Mexico.
Aegilopsis Horn 1860

Parmenonta Thomson 1868, 2 spp., Florida and Texas (key to spp. Linsley and Chemsak 1984).

Sybra Pascoe 1865, 1 sp., *S. alternans* Wiedemann, immigrant from southeast Asia or Pacific islands, found in southern Florida.

Agapanthiini Mulsant 1839

Hippopsis Lepeletier and Audinet-Serville in Latreille 1825, 1 sp., *H. lemniscata* (Fabricius 1801), eastern United States to Texas and Mexico.

Spalacopsis Newman 1842, 5 spp., Florida to Louisiana and Texas (key to spp. Linsley and Chemsak 1984).
Eutheia Dejean 1835, not Stephens 1830
Spacalopsis LeConte 1852
Euthuorus Jacquelin du Val in Sagra 1857
Systene Pascoe 1858

Onciderini Thomson 1860

Cacostola Fairmaire and Germain 1859, 2 spp., Texas and Mexico (key to spp. Linsley and Chemsak 1984).
Aporataxia Hamilton in Leng and Hamilton 1896
Cylindrataxia Linsley 1934
Hyagniellus Breuning 1943
Paratucumaniella Breuning 1943

Lochmaeocles Bates 1880, 2 spp., Texas and Arizona to Mexico (key to spp. Linsley and Chemsak 1984).
Lochmaeodes Aurivillius 1923
Lochmalodes Arnett 1962

Oncideres Lepeletier and Audinet-Serville in Lacordaire 1830, 4 spp. and 1 subsp., eastern United States to Texas and Arizona (key to spp. and diagnosis of subsp., Linsley and Chemsak 1984; biology Solomon 1995).

Pteropliini Thomson 1860

Ataxia Haldeman 1847, 7 spp., eastern United States to Texas and Arizona (key to spp. Linsley and Chemsak 1984).
Stenidea Haldeman 1847
Stenosoma LeConte 1852, not Mulsant 1839
Proecha Thomson in Chevrolat 1862
Esthlogena Bates 1866, not Thomson 1864
Parysatis Thomson 1868
Parepectasis Bruch 1926

Pogonocherini Mulsant 1839

Callipogonius Linsley 1935, 1 sp., *C. cornutus* (Linsley 1930), Texas and eastern Mexico.

Ecyrus LeConte 1852, 3 spp. and 1 subsp., eastern and central United States south to Florida and west to Texas and Mexico (key to spp. and diagnosis of subsp., Linsley and Chemsak 1984).
Exocentrus Haldeman 1847, not Mulsant
Oebaceres Thomson 1868

Lophopogonius Linsley 1935, 1 sp., *L. crinitus* (LeConte 1873), Pacific coast of North America.

Lypsimena Haldeman 1847, 1 sp., *L. fuscata* Haldeman 1847, southeastern United States and California to Mexico and South America.
Alloeoscelis Bates 1885

Pogonocherus Megerle in Dejean 1821
Pogonochaerus Gemminger and Harold 1873
Pityphilus Mulsant 1863
Pityophilus Bedel 1889
subgenus *Eupogonocherus* Linsley 1935, 5 spp., generally distributed (key to spp. Linsley and Chemsak 1984).
subgenus *Pogonocherus sensu stricto*, 1 sp., *P. penicillatus* LeConte in Agassiz 1850, northern United States, Canada and Alaska.

Poliaenus Bates 1880, 6 spp. and 3 subspp., western North America to Baja California and Guatemala (key to spp. Linsley and Chemsak 1984).

Pygmaeopsis Schaeffer 1908, 1 sp., *P. viticola* Schaeffer 1908, Texas.

Zaplous LeConte 1878, 1 sp., *Z. annulatus* (Chevrolat 1862), southeastern United States and Cuba.

Desmiphorini Thomson 1860

Desmiphora Audinet-Serville 1835 (key to spp. Giesbert 1998)
Euchaestes Chevrolat 1861
Pyrracita Thomson 1868
Therchaetes Thomson 1868
Terchaetes Lacordaire 1872
Desmophora Gemminger and Harold 1873
Pyrrhacita Gemminger and Harold 1873
subgenus *Desmiphora sensu stricto*, 2 spp., southwestern United States to South America and West Indies.

Estoloides Breuning 1940
Estoloides sensu stricto, 1 sp., *E. aquilonius* Linsley and Chemsak 1984, Arizona.

Eupogonius LeConte 1852, 7 spp., eastern and central United States to Arizona (key to spp. Linsley and Chemsak 1984).
Phidola Chevrolat 1862
Eriopsilus Bates 1866

Psenocerus LeConte 1852, 1 sp., eastern North America to Texas.

Tigrinestola Breuning 1949 1 sp., *T. tigrina* (Skinner 1905), Arizona to Texas and northern Mexico.

Anisocerini Thomson 1860

Thryallis Thomson 1857, 1 sp., *T. undatus* (Chevrolat 1834), southern Texas to Central America (revision of genus, Chemsak and McCarty 1997).

Acanthoderini Thomson 1860

Aegomorphus Haldeman 1847, 5 spp., eastern and central United States to Arizona and Mexico (key to spp. Linsley and Chemsak 1984).
Acanthoderes Haldeman 1847, not Audinet-Serville 1835
Aethiopoctines Thomson 1868
Psapharochrus Casey 1913, not Thomson 1864
Aegoschema Knull 1946, not Aurivillius 1923

Oplosia Mulsant 1863, 1 sp., *O. nubila* (LeConte 1862), northeastern North America.
Hoplosia Fairmaire 1864
Lepargus Schiødte 1864

Peritapnia Horn 1894, 1 sp., *P. fabra* Horn 1894, Arizona to to Baja California and Mexico.

Acanthocinini Blanchard 1845

Acanthocinus Megerle 1821, 8 spp., generally distributed (key to spp. Linsley and Chemsak 1995).
Astynomus Haldeman 1847, not Stephens 1839
Aedilis LeConte 1852, not Audinet-Serville 1835
Canonura Casey 1913
Graphisurus Casey 1913, not Kirby 1837
Tylocerina Casey 1913
Neacanthocinus Dillon 1956

Alcidion Sturm 1 sp., *A. umbraticus* (Jacquelin du Val in Sagra 1857), southern Florida, Cuba and Puerto Rico.
Probatius White 1855
Hirsutographis Dillon 1956

Astylidius Casey 1913, 1 sp., *A. parvus* (LeConte 1873), eastern United States to Texas.
Astylopsis Casey 1913, 6 spp., eastern United States (key to spp. Linsley and Chemsak 1995; see also Schiefer, 2000).
Amniscus Haldeman 1847, not Dejean 1835)

Coenopoeus Horn 1880, 1 sp., *C. palmeri* (LeConte 1873), southwestern United States and northern Mexico.

Dectes LeConte 1852, 2 spp., eastern and southwestern United States to northwestern Mexico and Baja California (key to spp. Linsley and Chemsak 1995).

Eutrichillus Bates 1885, 4 spp., eastern North America to southwestern United States and Mexico (key to spp. Linsley and Chemsak 1995).
Ceratographis Gahan 1887
Lepturgoides Schaeffer 1905

Glaucotes Casey 1913, 1 sp., *G. yuccivorus* (Fall 1907), Arizona.

Hyperplatys Haldeman 1847, 4 spp., generally distributed (key to spp. Linsley and Chemsak 1995).

Lagocheirus Dejean 1835, 2 spp., Florida and Texas (key to spp. Linsley and Chemsak 1995).
Lagochirus Erichson 1847

Leptostylopsis Dillon 1956, 5 spp., southeastern United States and Texas (key to spp. Linsley and Chemsak 1995).

Leptostylus LeConte 1952, 3 spp., eastern North America and central United States to Texas and Arizona (key to spp. Linsley and Chemsak 1995).
Leptostylis Bradley 1930

Lepturges Bates 1863
Maculurges Dillon 1956
subgenus *Lepturges sensu stricto*, 9 spp., eastern North America to Arizona (key to spp. Linsley and Chemsak 1995).

Liopinus Linsley and Chemsak 1995, 10 spp., eastern and central United States to Arizona (key to spp. Linsley and Chemsak 1995).
Leiopus Haldeman 1847
Liopus LeConte 1852
Sternidius LeConte 1873

Nyssodrysina Casey 1913, 1 sp., *N. haldemani* (LeConte 1852), eastern United States to Panama.
Nyssodrys Bates 1864 (part)
Nyssodrystes Gilmour 1963
Cicanyssodrys Gilmour 1965

Pseudastylopsis Dillon 1956, 3 spp., western United States and Mexico to Guatemala (key to spp. Linsley and Chemsak 1995).
Leptostylus Horn 1880 (part)
Pseudostylopsis Arnett 1962

Sternidius LeConte 1873, 1 sp., *S. variegatus* (Haldeman 1847), eastern North America to North Dakota and Arizona.
Astyleiopus Dillon 1956

Sternidocinus Dillon 1956, 1 sp., *S. barbarus* (Van Dyke 1920), southern California.

Styloleptus Dillon 1956, 1 sp., *S. biustus* (LeConte 1852), eastern United States to Texas, the Bahamas and Cuba.

Trichastylopsis Dillon 1956, 1 sp., *T. albidus* (LeConte 1852), southwestern United States.

Trichocanonura Dillon 1956, 1 sp., *T. linearis* (Skinner 1905), Arizona and northern Mexico.

Urgleptes Dillon 1956, 5 spp., eastern North America to Texas (key to spp. Linsley and Chemsak 1995).
Lepturges Bates 1863 (part)

Urographis Horn 1880, 3 spp., eastern North America to Minnesota and Texas (key to spp. Linsley and Chemsak 1995).
Graphisurus LeConte 1852, not Kirby 1837

Valenus Casey 1892, 1 sp., *V. inornatus* Casey 1892, southwestern United States and northern Mexico.

Cyrtinini Thomson 1864

Cyrtinus LeConte 1852, 2 sp., eastern United States to western Texas (key to spp. Linsley and Chemsak 1995).

Saperdini Mulsant 1839

Saperda Fabricius 1775, 15 spp. and 2 subspp., generally distributed (key to spp. and diagnoses of subspp., Linsley and Chemsak 1995; biology Solomon 1995).

Anaerea Mulsant 1839
Compsidia Mulsant 1839
Amilia Mulsant 1863
Argalia Mulsant 1863

Phytoeciini Pascoe 1864

Mecas LeConte 1852 (key to spp. Linsley and Chemsak 1995)

subgenus *Dylobolus* Thomson 1868, 1 sp., *M. rotundicollis* (Thomson 1868), southwestern United States to Mexico and Costa Rica.

subgenus *Mecas sensu stricto*, 9 spp., generally distributed in North America and Mexico.

Oberea Mulsant 1839, 13 spp., eastern North America to Rocky Mountains (key to spp. Linsley and Chemsak 1995; biology Solomon 1995).

Isosceles Newman 1842
Amaurostoma Mueller 1906

Tetraopini Thomson 1860

Phaea Newman 1840, 2 spp., eastern and central United States to Texas (key to spp. in America north of Mexico Linsley and Chemsak 1995; revision of genus Chemsak 1999).

Oberopa Haldeman 1947
Tetrops LeConte 1852, not Stephens 1831
Lamprocleptes Thomson 1857

Tetraopes Dalman in Schoenherr 1817, 13 spp., generally distributed (key to spp. Linsley and Chemsak 1995).

Tetrops Kirby 1826, 1 sp., *T. praeusta* Linnaeus 1758, immigrant from Europe, found in the northeastern United States (characterization of sp. Yanega 1996).

Hemilophini Thomson 1868

Cathetopteron Hamilton in Leng and Hamilton 1896, 1 sp., *A. amoena* Hamilton 1896, southern Texas.

Hemierana Aurivillius 1923, 1 sp., *H. marginata* (Fabricius 1798), and 3 subspp., eastern United States to Kansas and Texas (diagnoses of subspp., Linsley and Chemsak 1995).

Amphionycha LeConte 1852, not Dejean 1835

BIBLIOGRAPHY

BEIERL, J. D. and E. J. BARCHET-BEIERL. 1999. A new species of *Deltaspis* Audinet-Serville (Coleoptera: Cerambycidae) from southern California. Occasional Papers of the Consortium Coleopterorum, 3: 5-7.

CHEMSAK, J. A. 1996. Illustrated revision of the Cerambycidae of North America. Volume I. Parandrinae, Spondylidinae, Aseminae, Prioninae. Wolfsgarden Books. Burbank, CA, 131 pp.

CHEMSAK, J. A. 1999. Revision of the genus *Phaea* Newman (Coleoptera: Cerambycidae). Occasional Papers of the Consortium Coleopterorum 3: 36-101.

CHEMSAK, J. A. and E. G. LINSLEY. 1965. New genera and species of North American Cerambycidae. Pan-Pacific Entomologist, 41: 141-153.

CHEMSAK, J. A. and J. D. McCARTY. 1997. Review of the genus *Thryallis* Thomson (Coleoptera: Cerambycidae). Coleopterists Bulletin, 51: 101-112.

CHEMSAK, J. A. and F. A. NOGUERA. 1998. Review of the genus *Sphaenothecus* Dupont (Coleoptera: Cerambycidae). Pan-Pacific Entomologist, 74: 12-26.

FRAGOSO, S. A., M. A. MONNÉ and C. A. C. SEABRA. 1987. Preliminary considerations on the higher classification of Cerambycinae (Coleoptera, Cerambycidae), with nomenclatural alterations. Revista Brasileira de Biologia, 47: 189-202.

GIESBERT, E. F. 1998. A review of the Genus *Desmiphora* Audinet-Serville (Coleoptera: Cerambycidae: Lamiinae: Desmiphorini) in North America, Mexico and Central America. Occasional Papers of the Consortium Coleopterorum, 2: 27-43.

GIESBERT, E. F. and J. A. CHEMSAK. 1989. The genus *Stenosphenus* Haldeman (Coleoptera: Cerambycidae). Pan-Pacific Entomologist, 65: 269-301.

GIESBERT, E. F. and J. A. CHEMSAK. 1993. A review of the Rhopalophorini (Coleoptera: Cerambycidae) of North and Central America. Insecta Mundi, 7: 27-64.

GIESBERT, E. F. and J. A. CHEMSAK. 1997. A review of the genus *Euderces* LeConte (Coleoptera: Cerambycidae; Tillomorphini). Proceedings of the California Academy of Sciences, 49: 211-286.

GIESBERT, E. F. and F. T. HOVORE. 1984. Two new purpuricenine longhorns (Coleoptera: Cerambycidae) from the Tamaulipan biotic province. Coleopterists Bulletin, 38: 59-65.

GIESBERT, E. F. and F. T. HOVORE. 1998. Descriptions and synonymies in the North American Lepturini (Coleoptera: Cerambycidae: Lepturinae). Occasional Papers of the Consortium Coleopterorum, 2: 16-23.

HOVORE, F. T. 1979. A new subspecies of *Megacheuma brevipennis* (LeConte) from southeastern California (Coleoptera, Cerambycidae). Coleopterists Bulletin, 33: 459-463.

HOVORE, F. T. 1980. A new genus and species of Cerambycidae from southern Texas (Coleoptera). Coleopterists Bulletin, 34: 115-119.

HOVORE, F. T. and E. F. GIESBERT. 1974. Two new species of Cerambycidae from southern California (Coleoptera). Pan-Pacific Entomologist, 50: 139-144.

HÜDEPOHL, K.-E. 1985. Revision de Trachyderini (Coleoptera, Cerambycidae, Cerambycinae). Entomologische Arbeiten aus dem Museum G. Frey, 33/34: 1-167.

LEWIS, A. 1976. A new species of *Evodinus* from Wyoming (Coleoptera, Cerambycidae). Pan-Pacific Entomologist, 52: 227-228.

LINGAFELTER, S. W. 1998. The genera of Elaphidiini Thomson 1864 (Coleoptera: Cerambycidae). Memoirs of the Entomological Society of Washington No. 20, 118 pp.

LINSLEY, E. G. 1961. The Cerambycidae of North America. Part I. Introduction. University of California Publications in Entomology, 18: 1-97, 35 pls.

LINSLEY, E. G. 1962a. The Cerambycidae of North America. Part II. Taxonomy and classification of the Parandrinae, Prioninae, Spondylinae, and Aseminae. University of California Publications in Entomology, 19: 1-102, 1 pl.

LINSLEY, E. G. 1962b. The Cerambycidae of North America. Part III. Taxonomy and classification of the subfamily Cerambycinae, tribes Opsimini through Megaderini. University of California Publications in Entomology, 20: 1-188.

LINSLEY, E. G. 1963. The Cerambycidae of North America. Part IV. Taxonomy and classification of the subfamily Cerambycinae, tribes Elaphidionini through Rhinotragini. University of California Publications in Entomology, 21: 1-165.

LINSLEY, E. G. 1964. The Cerambycidae of North America. Part V. Taxonomy and classification of the subfamily Cerambycinae, tribes Callichromatini through Ancylocerini. University of California Publications in Entomology, 22: 1-197.

LINSLEY, E. G. and J. A. CHEMSAK. 1972. Cerambycidae of North America. Part VI, No. 1. Taxonomy and classification of the subfamily Lepturinae. University of California Publications in Entomology, 69: 1-138, 2 pls.

LINSLEY, E. G. and J. A. CHEMSAK. 1976. Cerambycidae of North America. Part VI. No. 2: Taxonomy and classification of the subfamily Lepturinae. University of California Publications in Entomology, 80: 1-186.

LINSLEY, E. G. and J. A. CHEMSAK. 1984. The Cerambycidae of North America, Part VII, No. 1: Taxonomy and classification of the subfamily Lamiinae, tribes Parmenini through Acanthoderini. University of California Publications in Entomology, 102: 1-258.

LINSLEY, E. G. and J. A. CHEMSAK. 1995. The Cerambycidae of North America, Part VII, No. 2: Taxonomy and classification of the subfamily Lamiinae, tribes Acanthocinini through Hemilophini. University of California Publications in Entomology, 114: 1-292.

LINSLEY, E. G. and J. A. CHEMSAK. 1997. The Cerambycidae of North America, Part VIII: Bibliography, index, and host plant index. University of California in Publications Entomology, 117: 1-534.

MACRAE, T. C. 2000. Review of the genus *Purpuricenus* Dejean (Coleoptera: Cerambycidae) in North America. Pan-Pacific Entomologist, 76: 137-169.

MARTINS, U. R. 1975. A taxonomic revision of the world Smodicini (Coleoptera, Cerambycidae). Arquivos de Zoologia, 26: 319-359.

MONNÉ, M. A. and E. F. GIESBERT. 1994. Checklist of the Cerambycidae and Disteniidae (Coleoptera) of the Western Hemisphere. Wolfsgarden Books. Burbank, CA, 409 pp.

PHILIPS, T. K. and M. A. IVIE. 1998. The Methiini of the West Indies (Coleoptera: Cerambycidae) with notes on the circum-Caribbean species. Entomologica Scandinavica, 29: 57-87.

SCHIEFER, T. L. 2000. A new species of *Astylopsis* Casey (Coleoptera: Cerambycidae: Acanthocinini) from the southeastern United States. Coleopterists Bulletin, 54: 533-539.

SKILES, D. D. 1985. New genera and species of elaphidionine Cerambycidae (Coleoptera) from North America and the West Indies. Coleopterists Bulletin, 39: 305-320.

SMITH, G. and J. E. HURLEY. 2000. First North American record of the Palearctic species *Tetropium fuscum* (Fabricius) (Coleoptera: Cerambycidae). Coleopterists Bulletin, 54: 540.

SOLOMON, J. D. 1995. Guide to insect borers of North American broadleaf trees and shrubs. USDA Forest Service Agriculture Handbook 706. Washington, DC.

YANEGA, D. 1996. Field guide to northeastern longhorned beetles (Coleoptera: Cerambycidae). Illinois Natural History Survey Manual 6, 174 pp., 32 pls.

121. BRUCHIDAE Latreille 1802

by John M. Kingsolver

Family common name: The bean weevils

Family synonyms: Acanthoscelidae Schilsky 1905; Lariidae Bedel 1901; Mylabridae Heyden *et al.* 1883, not Lacordaire 1959; Spermophagidae *auctorum*

Adults of this family are easily recognized by their characteristic compact body shape, small head, rather large fully exposed pygidium, pseudotetramerous tarsi, and in life, in their ability to feign death by applying the head, antennae, and legs closely to the body and falling from their resting place. In size they vary from 1 to about 10 mm in length for the United States species, reaching 22 or more mm in the American tropics.

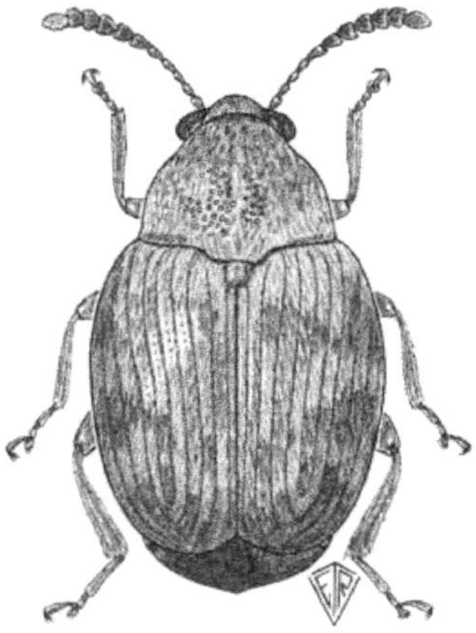

Figure 1.121. *Acanthoscelides obtectus* (Say).

Description: Mostly oval in shape, varying to oblong or to nearly quadrate; integument in most black or fuscous but in some ochreous, rufus, or flavous, or variously marked or mottled with one or more of these colors; surface more or less densely punctate and micropunctate throughout, in a few rugose-punctate in part; vestiture mostly of recumbent setae varying considerably in density, color, and pattern, according to the species.

Head, when at rest, closely applied to the pro- and mesosternum, mandibles then directed backward; mandibles elongate, acute, with a membranous flap on the inner margin between the cutting apex and a basal molar mass; the clypeus is separated from the frons by the frontoclypeal suture; eyes shallowly to deeply emarginate at antennal insertion; antennae with 11 antennomeres, compressed, apical two or three mostly expanded and subperfolate, serrate, pectinate or flabellate, rarely subfiliform; antennal insertion in front of and adjoining the emargination of the eye; head with a transverse furrow extending across the ventral surface and on either side behind the eyes, setting off the occiput as a neck, which is largely concealed from above when the head is extended, and exposed with the head at rest; gular sutures distinct behind, abruptly bent inward and joined in the middle behind the transverse furrow; gula at and before the furrow not separated from epicranium, continued forward between the buccal fissures to receive the mentum on the entire anterior margin.

Procoxae elongate, received in elongate coxal cavities, nearly closed behind by posterolateral pieces which are not joined at the middle; prosternum short, vertical or nearly so, with lateral wings definitely limited on either side and an intercoxal process partially or completely separating the coxae; elytra separately rounded at apex, and exposing the pygidium, disk punctate-striate, with 10 striae, some of which are abbreviated and joined at apex; procoxal and mesocoxal cavities in some species with a visible trochantin; pro- and mesofemora slender; metafemur usually compressed and more or less incrassate, in many flattened on ventral margin, with one or two longitudinal carinae, in many these are armed with serrations, teeth and denticles, especially near apex; metatibiae mostly longitudinally carinate, truncate and variously armed with fixed spines or teeth at apex, less frequently bicalcarate. Pygidium exposed behind the elytra, and in some males with an additional tergum visible between the pygidium and the last sternum.

Abdomen with five visible sterna, with the first, and in most the fifth longer than one of the three equal intermediate sterna. Male genitalia (Kingsolver 1970) with the apical portion of the tegmen (lateral lobes) well developed, frequently bilobate, connected by lateral commissures with a flattened or carinate ventral strut; internal sac well developed with sclerotizations on the internal surface characteristic of species; median lobe with characteristic apical structures, basal portion frequently spoon-shaped basally. Female genitalia with two segments supported by sclerotized struts and telescoping one into the other; copulatory bursa in many bearing on its internal surface sclerotizations characteristic of the species.

First stage larva has on the pronotum a transverse pectinate plate unknown in other coleopterous larvae. In this stage some species have functional legs while others are without legs. Full-grown larvae are curved, plump, soft, in part covered with minute hairs or setae, yellow to nearly white in color, and from less than 2 to 8 mm or more in length in the United States species. Head small, elongate, prognathous, deeply retracted into the prothorax; epicranial sutures, except for the coronal suture, inconspicuous or absent; antennae very small, two- or three-segmented; labial palpomeres in most transversely oval with apical setae; one or three pairs of ocelli if present, or reduced to pigmented spots. Thorax three-segmented (in some faintly indicated), increasing in diameter basally; mesothorax and metathorax each in most with two subequal dorsal plicae; mesothorax with a small lateral annular spiracle; legs rudimentary. Abdomen widest in basal half, ten-segmented, the first seven or eight each with two subequal dorsal

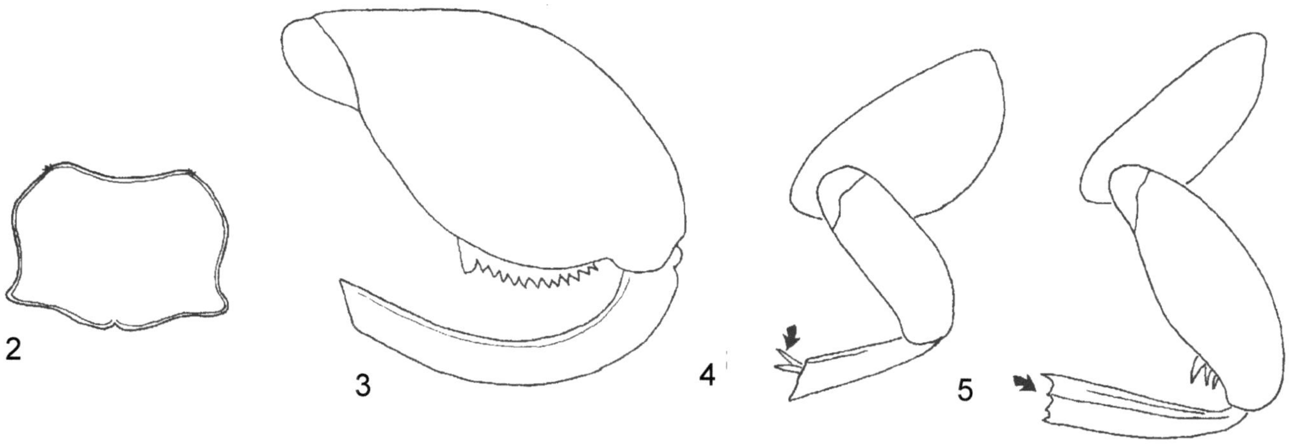

FIGURES 2.121-5.121. 2. *Caryobruchus* pronotum, dorsal view; 3. *Caryobruchus* hind leg, lateral view; 4. *Zabrotes subfasciatus* (Boheman), hind leg; 5. *Acanthoscelides* sp., hind leg..

plicae, the ninth segment simple, short, buttonlike, the 10th reduced to a small mammallate structure which includes the anus; lateral spiracles on segments one to eight.

Habits and habitats. Bruchids vary considerably in their habits but they depend entirely upon seeds of various types for their larval development, the larvae feeding within the seeds, or seed envelopes, and pupating within the cell or cavity excavated during their growth. Normally, oviposition takes place in the field when the seeds are nearly fully developed, but a few species deposit eggs on young fruits, while still other species wait until the seeds are fully matured and exposed since they will deposit eggs only on the seeds. In most instances the seeds attacked are large enough to support but a single adult and in such cases the seeds are rendered worthless for propagation. Some species have a single generation per year while others are continuous breeders and are therefore able to live in stored dried seeds. About one percent of the United States species are world-wide in distribution attacking legumes that are grown for food or forage. Among these are the most serious pests of these agricultural crops. Since these insects are definitely associated with plants it is interesting to note that perhaps 80 percent or more of the United States species live in the seeds or pods of Leguminosae, with Convolvulaceae, Palmaceae, and Malvaceae comprising most of the other host plant families. Some species are restricted to a single plant species or to one genus of plants, while a few are less specific and will develop normally in the fruits of several plant genera. More information can be found in Bottimer (1961).

Status of the classification. The family Bruchidae is apparently a monophyletic, chrysomeloid group springing from a common ancestor with the chrysomeloid subfamily Sagrinae. Some workers in chrysomeloid taxonomy treat the Bruchidae as a subfamily of the Chrysomelidae. While not denying that the two taxa are closely related, Kingsolver (1995) considered the following combination of characteristics sufficient for familial status for Bruchidae: Eye shallowly to deeply emarginate; elytral striae always present; pygidium always fully exposed; tarsal claws appendiculate; development of unique adaptations for ovipositing on seeds, or on a fruit or pod containing seeds; presence of an egg burster on the first dorsal pronotal segment of the first instar larva; larval development and transformation within the seed; the larval habit of cutting an exit from the seed prior to their return to the feeding chamber within the seed to pupate; male genitalia with median lobe and tegmen ring connected by membrane; base of median lobe modified into a pump to evert the internal sac during copulation; and lateral lobes always present. It has variously been placed also as a subfamily of the Chrysomelidae, or as being related to the Curculionidae, or the Anthribidae. The two latter associations have been discounted.

The family name Bruchidae was first used by Latreille (1810) based on the genus *Bruchus* Linnaeus with its type species *Dermestes pisorum* Linnaeus.

The first serious attempt to form species groups in American Bruchidae was by Horn (1873). He listed three genera: *Spermophagus* Schoenherr with one species - *robiniae* Fabricius, now in *Amblycerus*; *Caryoborus* Schoenherr with one species - *arthriticus* Fabricius, now a synonym of *Caryobruchus gleditsiae* Linnaeus; and *Bruchus* Linnaeus with 44 species. *Spermophagus* is now restricted to the Old World, *Caryoborus* is a Central and South American genus, and *Bruchus* is restricted to the Old World with three species introduced into America. Later, Horn (1885) added *Zabrotes* with five new species and one previously described.

The genus *Bruchus* as conceived by Horn was divided into nine groups, several of which were given generic names by Bridwell (1946) in his key to genera. Bridwell only listed the type species for each new genus but inexplicably did not assign other described species although he certainly was familiar with them. He also included *Acanthoscelides* Schilsky, *Callosobruchus* Pic, *Gibbobruchus* Fahraeus, *Kytorhinus* Fischer, and *Megacerus* Fahraeus in his key to genera. Blackwelder and Blackwelder (1948) moved the species formerly in "*Bruchus*", except those Bridwell designated as type species for his new genera in 1946, into *Acanthoscelides*. From this matrix, species names have been assigned to their current genera by Bottimer, Johnson, Leng, Kingsolver, Whitehead and Teran. Two of Bridwell's genera described in 1946, *Cercidiestes* and *Sparteus*,

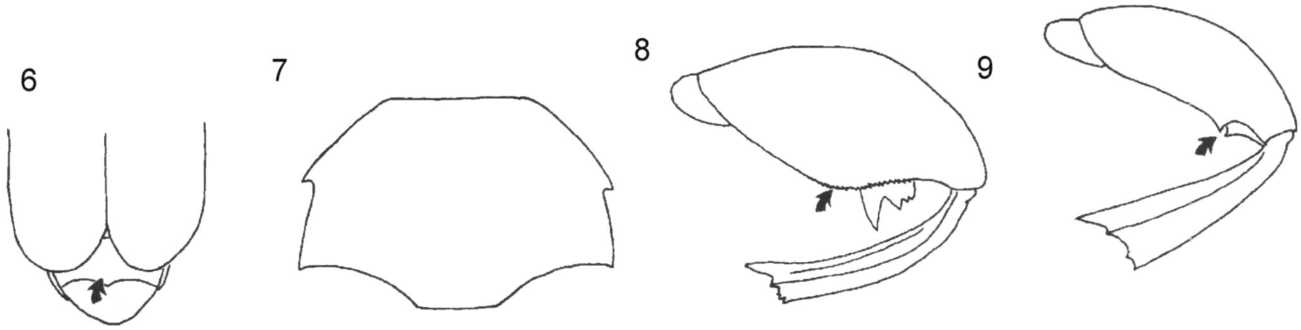

FIGURES 6.121-9.121. 6. *Kytorhinus prolixus* (Fall), terminal abdominal segments; 7. *Bruchus* sp. pronotum, dorsal view; 8. *Gibbobruchus* sp., hind leg; 9. *Callosobruchus* sp., hind leg.

have subsequently been synonymized. More information can be found in Bottimer (1968), Bøving (1927), Bridwell (1932), Bradley (1946), Pic (1913), and Pope (1956).

Distribution. Bruchidae are found naturally in all continents except Antarctica, and on New Zealand. Relatively few species are found on the Pacific Islands because of the difficulty of finding a suitable food plant for establishment of a colony.

In the Western Hemisphere, one species is known from Alaska and another from the extreme south- Tierra del Fuego, but the greatest diversity of species is in Central America and northern South America. In the United States, the greatest number of species is found in the southern states.

An estimated 1350 species are known throughout the world. New World species number about 760 of which 149 are known from the United States.

Key to Nearctic Bruchid Genera

1. Pronotum flat or slightly convex with at least basal and lateral margins delimited by distinct submarginal sulcus (Fig. 2); large species with metafemur enlarged, metatibia curved to fit closely to ventral margin of femur (Fig. 3); antenna serrate; eye shallowly emarginate (Pachymerinae) 2

— Pronotum convex, without submarginal sulcus except on anterolateral margin behind eye, other characters variable 3

2(1). Prosternum flat between procoxal apices; elytra uniformly red to black; metatibia lacking lateral carina .... *Caryobruchus*

— Prosternum short, triangular, acute, not separating apices of procoxae; elytra yellowish brown speckled with dark brown, metatibia with lateral carina; Hawaii, Florida .... *Caryedon*

3(1). Metatibia with two apical movable spurs (Fig. 4); metacoxal face broader than metafemur (Amblycerinae) 4

— Metatibia lacking movable apical spurs but may have fixed apical spines or denticles (Fig. 5); metacoxal face as wide as or narrower than metafemur ... 5

4(3). Eye feebly emarginate at antennal socket; tenth stria reaching nearly to apex of elytron; prosternum separating coxae at apex; metatibia lacking carinae; tibial spurs unequal in length .. .. *Amblycerus*

— Eye deeply emarginate; tenth stria abbreviated opposite metacoxa; procoxae not separated by prosternum; tibial spurs equal in length metatibia carinate (Fig. 5) .... *Zabrotes*

5(3). Pygidium and preceding tergal sclerite fully sclerotized and exposed beyond apices of elytra (Fig. 6); male antenna strongly pectinate, female antenna serrate; mesepimeron not strongly narrowed at ventral end, broadly reaching coxal cavity (Kytorhininae) .... *Kytorhinus*

— Pygidium only sclerotized and exposed beyond elytra; antenna of various shapes; mesepimeron either very narrow between mesepisternum and metepisternum, or reduced to triangular sclerite on dorsal margin of mesepisternum 6

6(5). Lateral pronotal margin with single large tooth projecting horizontally (Fig. 7); metafemur with tooth or angulation on lateroventral margin lying outside metatibia when legs are closed .... *Bruchus*

— Lateral pronotal margin without large tooth but may be set with small denticles; metafemur with or without external tooth, or angulation, or row of denticles on lateroventral margin 7

7(6). Ventrolateral margin of metafemur with an angulation, a tooth, or a row of denticles 8

— Ventrolateral margin of metafemur without angulation, tooth, or row of denticles 11

8(7). Ventrolateral margin of metafemur with row of denticles, ventromesal margin with several denticles (Fig. 8); pronotum gibbous; female pygidium with glabrous median spot .... *Gibbobruchus*

— Ventrolateral margin angulate, ventromesal margin with single denticle (Fig. 9); pronotum gibbous or not; female pygidium without glabrous spot ... 9

9(8). Lateral pronotal margin with distinct, arcuate carina .... *Stator*

— Lateral pronotal margin lacking distinct carina . 10

10(9). Hind legs red, or reddish brown with at most partial darker clouded areas, especially along ventral margin of hind femur .... *Callosobruchus*

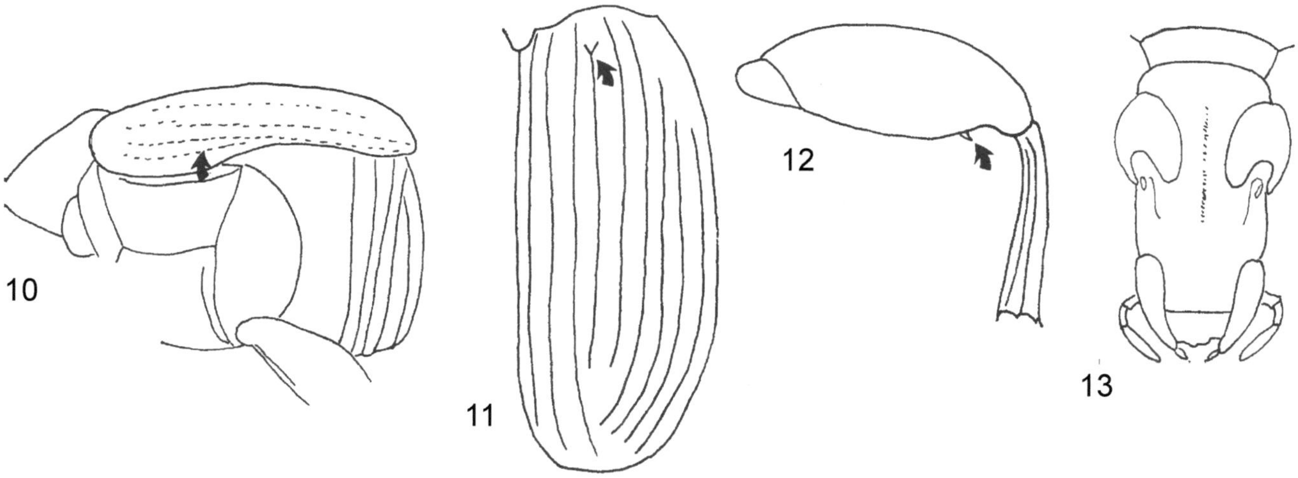

FIGURES 10.121-13.121. 10. *Megacerus* sp., body, lateral view; 11. *Meibomeus musculus* (Say), right elytron, dorsal; 12. *Sennius* sp., hind leg; 13. *Caryedes helvinus* (Motschulsky), head.

— Hind legs entirely black except tarsi yellow .... *Borowiecius*

11(7). Tenth elytral stria ending opposite metacoxa (Fig. 10) (except *Megacerus impiger*); frontal carina sharp; male eyes enlarged; male antenna pectinate, female antenna serrate; metafemur not enlarged, ventral face flat between ventral carinae; ventromesal carina smooth, or serrate .... *Megacerus*

— Tenth stria extending to apical margin; frontal carina various; eyes dimorphic or not; antenna various; metafemur usually moderately or strongly enlarged with ventral face taking varied forms 12

12(11). Front of head with V- or Y-shaped glabrous boss between dorsal margins of eyes; male metafemur with setose pocket on ventral margin; mucro short, usually not as long as lateral denticle..... .... *Mimosestes*

— Front of head otherwise; male metafemur lacking pocket; mucro variable 13

13(12). Fourth stria only abbreviated at base and ending in small denticle (Fig. 11); metatibia strongly bent at base; mucro minute; pecten of 5-7 minute denticles .... *Meibomeus*

— Fourth stria not abbreviated; with more than one stria denticulate at base, or stria not denticulate; metatibia usually slightly arcuate at base, sometimes with entire tibia arcuate; mucro of variable length 14

14(13). Scutellum 1.5 to 2 x as long as wide; female pygidium with 2 deep, polished sulci; male genitalia with H-shaped sclerite in armature of internal sac *Algarobius*

— Scutellum quadrate or transverse; female pygidium lacking polished sulci; male genitalia lacking H-shaped sclerite 15

15(14). Metafemur with single, sometimes minute, denticle on ventromesal margin, or without denticle or denticles (Fig. 12); mucro usually no longer than lateral denticle 16

— Metafemur with multiple denticles on ventromesal margin; mucro variable in length 20

16(15). Metafemoral denticle triangular, serrate on caudal margin; in seeds of malvaceous plants; east of 100th meridian .... *Althaeus*

— Metafemoral denticle usually small, not serrate, or absent; host plants various; generally distributed 17

17(16). Metatibia lacking lateral and ventral carinae; metafemoral denticle extremely minute or absent, elytral color uniform, without maculae; elytra together as wide as long; body length less than 2 mm. South Texas .... *Abutiloneus*

— Metatibia with at least one carina, usually 2-4; metafemur with one denticle 18

18(17). Pronotum moderately or strongly gibbous; southwestern U.S. .... *Neltumius*

— Pronotum uniformly convex, not gibbous; widely distributed 19

19(18). Body and appendages all black except 2 basal antennal segments sometimes reddish brown; metafemoral denticle minute, hidden among marginal setae .... *Bruchidius*

— Body and/or appendages, in part or wholly, red or reddish orange; metafemoral denticle easily visible (Fig. 12), sometimes as long as width of metatibia at its base .... *Sennius*

20(15). Head elongate; width across eyes equal to distance from top of eyes to end of clypeus (Fig. 13); third and fourth elytral striae ending in prominent single basal gibbosity; pronotum with prominent median and lateral gibbosities. Florida.............. .... *Caryedes*

— Head short (3:1); elytral striae usually ending in basal denticles, sometimes on slight swelling; pronotum not gibbous. Widely distributed 21

21(20). Antenna extremely long, that of male extending beyond apices of elytra, of female extending to first or second abdominal segment; mucro slender, curved, half as long as basitarsus; third and fourth striae prominently denticulate; hind femur with three minute, slender denticles; eye deeply emarginate .... *Stylantheus*
— Antenna not exceptionally long, never extending beyond middle of elytra; mucro various; striae denticulate or not; denticles of hind femur variously formed; emargination of eye shallow or deep 22

22(21). Elytral striae 3 and 4 prominently denticulate at base, denticles sometimes on transverse ridge or slight swelling (except *M. major* with denticles on striae 3 to 6); anal notch of female fifth sternum deep, usually laterally flanged; metafemur strongly swollen; tibia strongly arcuate; ventral valve of male genitalia broadly rounded or truncate; armature of internal sac usually including variant of forked sclerite .... *Merobruchus*
— Denticles of elytral striae various, sometimes absent, never on basal gibbosity; metafemur usually not strongly swollen; tibia various; female fifth sternum not notched or flanged; male genitalia of various forms, seldom with forked sclerite .... *Acanthoscelides*

Note: Some species of *Merobruchus* resemble certain species of *Acanthoscelides*, and the distinction between the two genera remains nebulous at various points. Unequivocal definitions of these two genera are not possible at this time.

Classification of the Nearctic Genera

Bruchidae Stephens 1929

Pachymerinae Bridwell 1929

Caryobruchus Bridwell 1929, 1 sp., *C. gleditsiae* Linnaeus 1763, Florida and Carolinas along Gulf Coast to Texas.

Caryedon Schoenherr 1823, 1 sp., *C. serratus* (Olivier 1790), Florida and Hawaii. Additional information in Kingsolver (1992).

Amblycerinae Bridwell 1932

Amblycerini Borowiec 1987

Amblycerus Thunberg 1815, 7 spp., Florida, Arizona, and eastern United States (Kingsolver 1996, Romero *et al.* 1996).
Spermophagus Leng 1920, not Schoenherr 1833

Spermophagini Borowiec 1987

Zabrotes Horn 1885, 17 spp., Connecticut, New Jersey, Illinois, District of Columbia, Georgia, Florida, Texas, Arizona, Colorado, Nevada, and California (Kingsolver 1990).

[*Spermophagus* Schoenherr is an Old World genus not present in the Western Hemisphere.]

Kytorhininae Bridwell 1932

Kytorhinus Fischer 1907, 1 sp., *K. prolixus* (Fall 1926), Alberta and South Dakota.

Bruchinae Pic 1913

Bruchidini Bridwell 1946

Bruchidius Schilsky 1905, 2 spp. Montana and eastern U.S. (adventive).
Sparteus Bridwell 1946

Callosobruchus Pic, 1902, 3 spp., generally distributed (adventive). Key to species, Kingsolver (1969).

Bruchini Bridwell 1946

Bruchus Linnaeus 1767, 3 spp., generally distributed, (adventive) (Kingsolver 1964b).
Mylabris Mueller 1764, of authors (invalid name)
Laria Scopoli 1763, of authors (invalid name)

Megacerini Bridwell 1946

Megacerus Fahraeus 1839, 9 spp., eastern and southern U.S. to California (revision by Teran and Kingsolver 1977)
subg. *Megacerus* Fahraeus 1839
subg. *Pachybruchus* Pic 1912
subg. *Serratibruchus* Teran and Kingsolver 1977

Acanthoscelidini Bridwell 1946

Abutiloneus Bridwell 1946, 1 sp., *A. idoneus* Bridwell 1946, Texas, Arizona (Kingsolver 1965).

Acanthoscelides Schilsky 1905, approx. 54 spp., generally distributed (Kingsolver 1968).

Algarobius Bridwell 1946, 2 spp., southwestern U.S. (Kingsolver 1986).

Althaeus Bridwell 1946, 3 spp., eastern United States (Kingsolver *et al.* 1989).

Borowiecius Anton 1994, 1 sp. *B. ademptus* (Sharp 1886), Asia, Alabama, D.C., North Carolina, Maryland (adventive, see Anton 1994)

Caryedes Hummel 1827, 2 spp., adventive, Florida (Kingsolver and Whitehead 1974).

Gibbobruchus Pic 1913, 3 spp., eastern and southwestern U.S. (Whitehead and Kingsolver 1975).

Meibomeus Bridwell 1946, 3 spp., eastern and southwestern U.S. (Whitehead and Kingsolver 1976).

Merobruchus Bridwell 1946, 8 spp., Florida, southwestern U.S. (Kingsolver 1988).

Mimosestes Bridwell 1946, 7 spp. Florida and southwestern U.S. (Kingsolver and Johnson 1979).
Cercidiestes Bridwell 1946

Neltumius Bridwell 1946, 3 spp., Texas, Arizona, Utah, and California (Kingsolver 1964a).
Sennius Bridwell 1946, 11 spp., eastern and southwestern U.S. (revision by Johnson and Kingsolver 1973).

Stator Bridwell 1946, 9 spp. Florida, Texas, Arizona, and southern California (Johnson and Kingsolver 1976).

Stylantheus Bridwell 1946, 1 sp., *S. macrocerus* (Horn 1873), New Jersey, District of Columbia, Maryland, and Central States.

Bibliography

ANTON, K.-W. 1994. The Bruchidae (Coleoptera) of Oman, with descriptions of a new genus and two new species. Fauna of Saudi Arabia, 14: 105-122.

BLACKWELDER, R. E. and R. M. BLACKWELDER. 1948. Fifth supplement 1939 to 1947 (inclusive) to the Leng Catalogue of Coleoptera of America, north of Mexico. John D. Sherman, Jr. Mount Vernon, N.Y. 87 pp.

BOTTIMER, L. J. 1961. New United States records in Bruchidae with notes on host plants and rearing procedures. Annals of the Entomological Society of America, 54: 291-298.

BOTTIMER, L. J. 1968. Notes on Bruchidae of America north of Mexico with a list of world genera. Canadian Entomologist, 100: 1009-1049.

BØVING, A. G. 1927. On the classification of the Mylabridae larvae. Proceedings of the Entomological Society of Washington, 29: 133-143.

BØVING, A. G. and F. C. CRAIGHEAD. 1931. Illustrated synopsis of the principal larval forms of the order Coleoptera. Entomologica Americana, 11: 1-352.

BRADLEY, J. C. 1946. Contributions to our knowledge of the Mylabridae, seu Bruchidae, with special reference to the fauna of northeastern America. Psyche, 53: 33-42.

BRIDWELL, J. C. 1932. The subfamilies of the Bruchidae. Proceedings of the Entomological Society of Washington, 29: 100-106.

BRIDWELL, J. C. 1946. The genera of beetles of the family Bruchidae in America north of Mexico. Journal of the Washington Academy of Sciences, 36: 52-57.

HORN, G. H. 1873. Revision of the Bruchidae of the United States. Transactions of the American Entomological Society, 4: 311-342.

HORN, G. H. 1885. Contributions to the coleopterology of the United States. Transactions of the American Entomological Society, 12: 128-162.

JOHNSON, C. D. and J. M. KINGSOLVER. 1973. A revision of the genus *Sennius* of North and Central America (Coleoptera: Bruchidae). ARS, U.S. Department of Agriculture, Technical Bulletin 1462. 135 pp.

JOHNSON, C. D. and J. M. KINGSOLVER. 1976. Systematics of *Stator* of North and Central America (Coleoptera: Bruchidae). ARS, U.S. Department of Agriculture, Technical Bulletin 1537. 101 pp.

KINGSOLVER, J. M. 1964a. The genus *Neltumius*. Coleopterists Bulletin, 18: 105-111.

KINGSOLVER, J. M. 1964b. A preliminary key to the species of the genus *Bruchus* (Bruchidae) commonly intercepted in U.S.D.A. Plant Quarantine interceptions. ARS, U.S. Department of Agriculture, Plant Quarantine Mimeo Document. 7 pp.

KINGSOLVER, J. M. 1965. On the genus *Abutiloneus* Bridwell. Coleopterists Bulletin, 19: 125-128.

KINGSOLVER, J. M. 1968. A review of the obtectus group in *Acanthoscelides* Schilsky, with designations of lectotypes (Coleoptera: Bruchidae: Bruchinae). Proceedings of the Entomological Society of Washington, 70: 4-9.

KINGSOLVER, J. M. 1969. A key to the species of *Callosobruchus* (Bruchidae) intercepted in USDA Plant Quarantine Inspections. ARS, U.S. Department of Agriculture Memo. No. 690, 14 pp.

KINGSOLVER, J. M. 1970. A study of male genitalia in Bruchidae (Coleoptera). Proceedings of the Entomological Society of Washington, 72: 370-386.

KINGSOLVER, J. M. 1986. A taxonomic study of the genus *Algarobius* (Coleoptera: Bruchidae). Entomography, 4: 109-136.

KINGSOLVER, J. M. 1988. Biosystematics of the genus *Merobruchus* of continental North America and the West Indies. ARS, U.S. Department of Agriculture, Technical Bulletin 1744. 63 pp.

KINGSOLVER, J. M. 1990. Biosystematics of the genus *Zabrotes* of America north of Mexico (Coleoptera: Bruchidae). Transactions of the American Entomological Society, 116: 135-163.

KINGSOLVER, J. M. 1992. *Caryedon serratus* (Olivier) new to continental United States (Coleoptera: Bruchidae). Insecta Mundi, 6: 22.

KINGSOLVER, J. M. 1995. On the family Bruchidae. Chrysomela, 30: 3.

KINGSOLVER, J. M. 1996. *Amblycerus schwarzi* Kingsolver (Coleoptera: Bruchidae) recorded new for North America. Insecta Mundi, 10: 24.

KINGSOLVER, J. M., T. J. GIBB and G. S. PFAFFENBERGER. 1989. Synopsis of the genus *Althaeus* Bridwell (Coleoptera: Bruchidae) with descriptions of two new species. Transactions of the American Entomological Society, 115: 57-82.

KINGSOLVER, J. M. and C. D. JOHNSON. 1979. Systematics of the genus *Mimosestes* (Coleoptera: Bruchidae). ARS, U.S. Department of Agriculture, Technical Bulletin 1590. 106 pp.

KINGSOLVER, J. M. and D. R. WHITEHEAD. 1974. Classification and comparative biology of the seed beetle genus *Caryedes* Hummel (Coleoptera: Bruchidae). Transactions of the American Entomological Society, 199: 341-436.

KINGSOLVER, J. M. and D. R. WHITEHEAD. 1976. The North and Central American species of *Meibomeus* (Coleoptera: Bruchidae: Bruchinae). U. S. Department of Agriculture Technical Bulletin 1523, 54 pp.

LATREILLE, P. A. 1810. Considerations generale sur l'ordre naturel des animaux conposant les classes des Crustaces, des Arachnides et des Insectes. C. F. Schoell, Paris. 430 pp.

PIC, M. 1913. Bruchidae. Coleopterorum Catalogus, 55: 1-75. Junk, Berlin.

POPE, R. D. 1956. The family name Bruchidae. Entomological Monthly Magazine, 92: 45-46.

ROMERO, N. J., C. D. JOHNSON and J. M. KINGSOLVER. 1996 Revision of the genus *Amblycerus* of the United States and Mexico (Coleoptera: Bruchidae: Amblycerinae). ARS, U.S. Department of Agriculture, Technical Bulletin 1845. 166 pp.

SHARP, D. and F. MUIR. 1912. The comparative anatomy of the male genital tube in Coleoptera. Transactions of the Royal Entomological Society, London 1912: 477-642.

TANNER, V. M. 1927. A preliminary study of the genitalia of female Coleoptera. Transactions of the American Entomological Society, 53: 5-50.

TERAN, A. L. and J. M. KINGSOLVER. 1977. Revision del genero *Megacerus* (Coleoptera: Bruchidae). Opera Lilloana, 25: 1-287.

WHITEHEAD, D. R. and J. M. KINGSOLVER. 1974. Classification and comparative biology of the seed beetle genus *Caryedes* Hummel (Coleoptera: Bruchidae). Transactions of the American Entomological Society, 101: 167-225.

WHITEHEAD, D. R. and J. M. KINGSOLVER. 1975. Biosystematics of the North and Central American species of *Gibbobruchus* (Coleoptera: Bruchidae: Bruchinae) Transactions of the American Entomological Society, 101: 167-225.

WILLIAMS, I. W., 1938. The comparative morphology of the mouthparts of the order Coleoptera treated from the standpoint of phylogeny. Journal of the New York Entomological Society, 46: 245-289.

122. MEGALOPODIDAE Latreille 1802

by Shawn M. Clark and Edward G. Riley

Family common name: The megalopodid leaf beetles

This family is an assemblage of three leaf beetle subfamilies that differ from one another in their outward appearance and habits. Collectively, the megalopodids have been placed basal to the chrysomelid lineage of the Chrysomeloidea. They share with most members of the cerambycid lineage the plesiomorphic characters of paired apical spurs on all tibiae and the elongate and separate anterior apodemes of the male genitalia. Their chief distinguishing feature is the non-cerambycid-like antennae that are not reflexed backwards, their insertion points not on tubercles and located low on the face above the base of the mandibles. All members of this family have a mesonotal stridulatory file, and most possess a large, membranous, bilobed ligula. The known larvae are stem borers, leaf miners, or consume pollen inside the male cones of Araucariaceae.

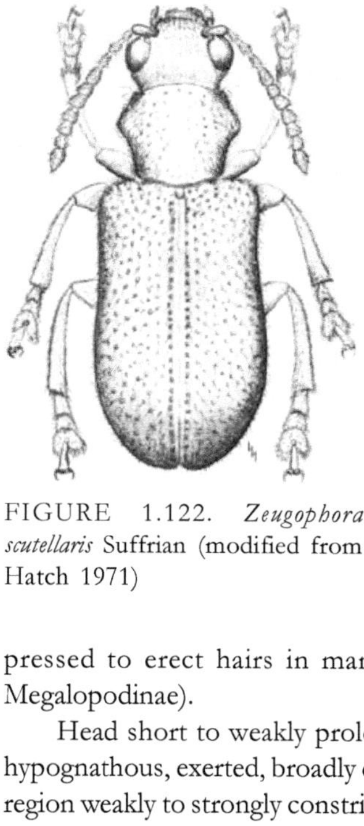

FIGURE 1.122. *Zeugophora scutellaris* Suffrian (modified from Hatch 1971)

Description: Shape variable, elongate in most species, parallel to subparallel-sided (North American species), subcylindrical to depressed; length from 3.3 to 4.5 mm (North American species), up to 15 mm in some Megalopodinae; color of most nonmetallic (North American species), a few weakly metallic, generally dull, black to brownish, dorsally unicolorous or bicolored with pattern of maculae; color of extralimital Megalopodinae bright, metallic or not, patterned; dorsum with sparse to dense vestiture of appressed to erect hairs in many species (glabrous in some Megalopodinae).

Head short to weakly prolonged, without a true rostrum, hypognathous, exerted, broadly exposed in dorsal view; occipital region weakly to strongly constricted; surface smooth to densely, deeply punctate; front without distinct grooves; antennal calli weak, well separated; gular sutures evident at hind margin of cranium, separate. Eyes large, lateral, protuberant, strongly protuberant and weakly stalked in some species, coarsely or finely faceted; interior margin of each eye with emargination ranging from broad and shallow to narrow, sharply angled and deep. Antenna with eleven freely articulated antennomeres, short to moderate in length, extending to point just beyond humerus (in Palophaginae long and filiform, reaching middle of elytra or just beyond in some species), antennomeres 5-11 subserrate; insertion lateral, low on face, between eye and base of mandible, internally delimited by elevated flange in some Megalopodinae. Mouthparts exposed behind, free from prosternum; labrum distinct; clypeus subquadrate to transverse, projecting slightly or deeply between base of mandibles, well delineated from frons; mandible long, apically narrow, evenly curved with acute, unidentate apex, or bidentate in Zeugophorinae (exceptionally long and blade-like in Megalopodinae with apices broadly overlapping in repose), without mola in most species, with distinct mola in Palophaginae. Maxilla with large galea and lacinia, each with dense brush of setae; palp 4-segmented, not asymmetrically dilated; palpomeres long and slender, moderate in length in Zeugophorinae; terminal palpomere cylindrical, with apex evenly rounded or angularly truncate. Mentum transverse, not large; ligula large, membranous, deeply bilobed in most species, small and not distinctly lobed in Zeugophorinae; labial palp 3-segmented.

Pronotum subequal to distinctly broader than head, nearly flat to moderately convex, laterally weakly arcuate or strongly angled in Zeugophorinae, with weak to strong basal constriction, narrowed anteriorly and somewhat conical in some Megalopodinae, anteriorly with a sharply to weakly defined transverse impression in many Megalopodinae; lateral borders unmargined in most genera, margined in *Palophagus* (Palophaginae); anterior angles with one or more sensorial setae; posterior angles with or without sensorial setae; dorsum smooth to densely, deeply punctate. Prosternum moderately long in front of coxae, accounting for nearly half total length of prothorax; prosternal process very narrow and depressed, or raised, lamellate, and hidden between protruding contiguous procoxae, not prolonged behind coxae; procoxal cavities closed behind by narrow strip of hypomeron which reaches prosternal process. Mesonotum with stridulatory file visible as very finely ridged, iridescent patch which may be single or in Palophaginae narrowly divided to form a pair of broad patches. Scutellum exposed, narrow, triangular to subtrapezoidal or rounded apically, flat. Mesosternum narrow and subparallel-sided to triangular, extending between mesocoxae; metasternum long and broad, flat or moderately convex, or sharply projected and conical behind mesocoxae in some Megalopodinae. Legs moderate to long; procoxae transverse, prominent, contiguous; mesocoxae subconical, small; metacoxae transverse; femora not or moderately and equally swollen, or frequently in Megalopodinae hind femora enlarged and with or without subapical spines or denticles on ventral margin; tibiae slender, straight or weakly curved in most species, strongly curved in many Megalopodinae; both sexes with paired apical spurs on all tibiae; tarsi 5-5-5, pseudotetramerous, with fourth tarsomere small and more or less hidden between lobes of third; tarsomeres 1-3 with expanded ventral pads which are densely covered with adhesive setae; pad of tarsomere 3 strongly bilobed; claws paired, symmetrical, simple or with broad basal tooth in Zeugophorinae; empodium present or absent.

Elytra entire, apically rounded; surface smooth, finely to strongly punctate, without costate or organized striae; margins entire, unmodified; epipleura narrow throughout, reaching apex or nearly so, visible from lateral view, appearing as thickened marginal bead in some Megalopodinae. Hind wing present and well developed, venation (Reid 1995; terminology from Kukalová-Peck and Lawrence 1993) complete to strongly reduced; two cubito-anal cells or one elongate basal cell present; "anal" veins with three or five apical branches; MP_{3+4} connected or almost connected to MP_{1+2} or not connected; AA_{3+4} present or absent; CuA_1 + MP_4/MP_{3+4} fused at base; CuA_1 connected to cubito-anal cell(s) or not; RP_{3+4} spur present.

Abdomen having five free ventrites, 1-4 of subequal length, 5 somewhat longer; mesal region of last ventrite slightly flattened or weakly impressed in some males, weakly impressed before posterior margin in females; apical margin of last ventrite in male transverse or weakly incised, without median lobe; apical margin of last ventrite of female either transverse, or weakly to strongly arcuate. Male genitalia with shaft of median lobe elongate, subcylindrical or depressed, with apex symmetrical and lacking fringe of subapical setae on lower margin; anterior apodemes fused proximally, distally separate, long and narrow; tegmen completely encircling median lobe, with long ventral arm extending anteriorly; dorsal part of tegmen expanded and fused; articulating parameres absent, or present as pair of articulating parameral sclerites in Palophaginae (Kuschel and May 1990). Female with internalized sternite 8 consisting of a separate distal blade and an anteriorly directed apodeme connected by a folded membrane; ovipositor telescopic, with sclerites elongate; spiculum gastrale absent; vaginal pouches absent (Reid 1995); valves present (hemisternites of Kuschel and May 1990). Rectum without specialized sclerotized plates (kotpresse) or spinose patches (Reid 1995).

Larva: Body elongate, parallel-sided, cylindrical and cerambycid-like (orthosomatic), or in Zeugophorinae dorsoventrally compressed with margins of each abdominal segment drawn outward, angled, and tipped with a seta on each side; length from three to five times greatest width; color mostly pale whitish except head capsule, mouthparts, and pronotal shield which are pigmented; dorsal and ventral ambulatory ampullae distinct, or in Zeugophorinae weakly developed. Setae of head capsule and body well developed. Head exerted, or in Zeugophorinae partially retracted into prothorax. Head capsule transverse, with frontal sutures long, straight, and meeting at hind margin of occiput; coronal (epicranial) suture absent; clypeus indistinct and not well delimited from front; six pairs of stemmata present, two of these fused in some species. Antenna short, three-segmented. Mandible bidentate apically; mola absent, or in Palophaginae strongly developed; maxilla with mala broad and apically spinose, with palp three-segmented; labium with mentum and submentum divided by distinct suture; labial palp two-segmented; ligula rectangular without setae. Prothoracic shield transverse. Egg bursters in first instar present on pronotum, meso- and metathorax, and A 1-6 or A1-7 in Zeugophorinae, on meso- and metathorax and A 1-8 in Megalopodinae (Cox 1988). Leg minute to small, reduced, or in Zeugophorinae absent, three-segmented when present, with or without apical claw; pulvillus absent. Abdominal segments A-1 to A-10 visible from above; A-9 weakly sclerotized, with tergite simple and lacking setiferous urogomphi; A-10 small; anal opening round or Y-shaped, terminal; spiracles biforous.

Habits and habitats. The Zeugophorinae (including the only Nearctic members of the family), as far as known, are folivorous as adults and larvae. The latter are true leaf miners, and they are legless with dorsoventrally flattened bodies and prognathous, retracted heads. Little is known of the food plants in North America, but some species utilize willows and poplars (Salicaceae). Larvae produce large, darkened, blotch-type mines in the leaves of their hosts and exit the mine to pupate in the soil. There appears to be only one generation per year in North America.

Members of the subfamily Megalopodinae should perhaps be called stem beetles, as the adults inhabit the stems rather than the foliage of their hosts, and their larvae are stem borers resembling the larvae of Cerambycidae. Adults have large, slicing mandibles which are used to clip off leaves and the terminal growth of the stems they inhabit and to construct oviposition sites in these stems. Adults clearly feed on the plant fluids which exude from chewed-off stems. Adults have been observed to expel fluids in short bursts from the tips of their abdomens, suggesting that the quantity of fluid intake is great. Adults also feed on solids, including the pith of stems (Eberhard and Marin 1996, Yu and Xingke 1994) and leaf buds (Yu and Xingke 1994). Larvae tunnel lengthwise in the stems of their hosts but exit the stem to pupate in the soil. The enlarged and modified hind legs of the Central American *Megalopus armatus* Lacordaire 1845 are used as weapons in male to male combat for feeding and breeding sites (Eberhard and Marin 1996). Some solanaceous plants are utilized in the Neotropics (Eberhard and Marin 1996, Jolivet and Hawkeswood 1995, Santos 1981), Anacardiaceae in South Africa (Schulze 1996), and Oleaceae in China (Yu and Xingke 1994).

As larvae, the Palophaginae are pollen feeders inside the male cones (strobili) of *Araucaria* and possibly *Agathis* (Araucariaceae). Their morphology is that associated with an internal mode of life, and their bodies resemble those of cerambycid larvae. Morphology of the adult mandibles suggests that adults are probably also pollen feeders, but timing would probably preclude their utilization of *Araucaria* pollen (Kuschel and May 1996a). As in the other subfamilies, their larvae exit the host material to pupate in the soil. Like at least some Megalopodinae, they are hypermetamorphic with an inactive ultimate larval instar. Adult specimens of Palophaginae are rare even though the larvae can be abundant in their host material (Kuschel and May 1996a). Efforts to rear adult specimens have met with minimal success (Kuschel and May 1990, 1996b).

Status of the classification. The Megalopodidae have often been treated as a subfamily of the Chrysomelidae, as have the Zeugophorinae which were also treated as part of the Orsodacninae in some older classifications. The classification used here follows Kuschel and May (1990) and recent phylogenetic studies, especially those of Reid (1995, 2000), where three distinctive groups of beetles are united as subfamilies of Megalopodidae. The Megalopodinae are the largest of the three subfamilies and are represented by about 25 genera (Seeno and Wilcox 1982). The Zeugophorinae contain what is probably only one wide-ranging genus (possibly two). The recently recognized Palophaginae contain three genera and four species. Their strict association with coniferous pollen and the classic Gondwanan distribution supports the contention that their morphology is perhaps the most plesiomorphic of extant chrysomeloids. Phylogenetic studies place the Megalopodidae basal to Chrysomelidae + Orsodacnidae, and they place Palophaginae as sister group to Zeugophorinae + Megalopodinae (Reid 1995, 2000). Potential relationships with basal members of the cerambycid lineage may need further investigation. Reid (2000) suggested that the family is possibly not monophyletic. A review of the world genera of Megalopodinae and Zeugophorinae is needed, including a detailed morphological study of *Zeugophora.*

Distribution. The Zeugophorinae are absent from the Neotropics but are otherwise widely distributed. The Megalopodinae are essentially circumtropical but absent from Australia, with the genera split nearly equally between the New and Old Worlds (Seeno and Wilcox 1982). The three genera of Palophaginae occur in Australia (*Cucujopsis* Crowson and *Palophagus* Kuschel) and southern Chile and Argentina (*Palophagoides* Kuschel). These beetles are known from only a few collections. These data and what is known of palophagine habits indicate a strictly south temperate or Gondwanan distribution for this subfamily. Only the Zeugophorinae are found in America north of Mexico.

Classification of Nearctic Genera

Megalopodidae Latreille 1802

Zeugophorinae Bøving and Craighead 1931

Characteristics. Dorsum distinctly punctate, with appressed to erect setae. Eye with broad and shallow internal emargination. Clypeus transverse, projecting slightly between bases of mandibles; mandible apically bidentate. Antenna short, reaching base of pronotum; antennomeres 5 and beyond subserrate. Mesonotal stridulatory file composed of a single broad patch. Pronotal margins without marginal bead, broadly angled near mid-length. Tarsal claws each with broad basal tooth. Larvae exhibiting leaf mining habits.

This subfamily, represented by what is probably a single genus containing roughly 55 described species, occurs nearly worldwide with the notable exception of the Neotropical Region.

Zeugophora Kunze 1818 [conserved name, ICZN 1986, Opinion 1382]

Zeugophora contains two subgenera with roughly 55 species and is distributed in the Holarctic, Afrotropical, Australian, and Oriental Regions.

subgenus *Zeugophora* Kunze 1818

Auchenia Thunberg 1792 [suppressed name, ICZN 1986, Opinion 1382]

Taraxis LeConte 1850

This subgenus is composed of about 15 species and is widespread in the Holarctic Region. Nine species are currently recognized in North America, but for the most part they are poorly characterized. A taxonomic revision is much needed. Some of our species are associated with *Populus* and *Salix* (Salicaceae). This host family is used in Europe, as well as *Betula* (Betulaceae), *Corylus* (Corylaceae), and *Juglans* (Juglandaceae) (Jolivet and Hawkeswood 1995). One North American species, *Z. scutellaris* Suffrian 1840, is adventive from Europe. Adults feed externally on leaves while larvae are leaf miners. Biology and descriptions of immature stages, Bøving and Craighead (1931), Grave (1917), Jolivet (1948), Lee (1998), Weiss and Nicolay (1919). Partial keys to species, Brisley (1928), Crowson (1946), Hatch (1971).

subgenus *Pedrillia* Westwood 1864 [Afrotropical, Australian, and Oriental]

Macrozeugophora Achard 1914

Pedrilliomorpha Pic 1917

Austrolema Oke 1932

Pedrillimorpha Papp 1946 [unavailable]

Pedrinella Papp 1946 [unavailable]

Pedrilonga Papp 1946 [unavailable]

Bibliography

BØVING, A. G. and F. C. CRAIGHEAD. 1931. An illustrated synopsis of the principal larval forms of the Order Coleoptera. Entomologica Americana (new series), 11: 1-351 (1930).

BRISLEY, H. R. 1928. A short review of the tribes Orsodacnini and Criocerini of the coleopterous family Chrysomelidae with special reference to species of western United States. Pan-Pacific Entomologist, 4: 54-60, 114-119.

COX, M. L. 1988. Egg bursters in the Chrysomelidae, with a review of their occurrence in the Chrysomeloidea and Curculionoidea (Coleoptera). Systematic Entomology, 13: 393-432.

CROWSON, R. A. 1946. A revision of the genera of the chrysomelid group Sagrinae (Coleoptera). Transactions of the Royal Entomological Society of London, 97: 75-115.

EBERHARD, W. G. and M. C. MARIN. 1996. Sexual behavior and the enlarged hind legs of *Megalopus armatus* (Coleoptera, Chrysomelidae, Megalopodinae). Journal of the Kansas Entomological Society, 69: 108.

GRAVE, B. H. 1917. *Zeugophora scutellaris* (Suffr.). Journal of Morphology, 30: 245-259.

HATCH, M. H. 1971. The beetles of the Pacific Northwest. Part V: Rhipiceroidea, Sternoxi, Phytophaga, Rhynchophora, and Lamellicornia. University of Washington Publications in Biology, 16:1-662.

JOLIVET, P. H. A. 1948. Les Orsodacnidae de la fauna française (Col., Orsodacnidae). Miscellanea Entomologica, 45: 33-46.

JOLIVET, P. H. A. and T. J. HAWKESWOOD. 1995. Host-plants of Chrysomelidae of the world. An essay about the relationships between the leaf-beetles and their food-plants. Backhuys. Leiden. [11] + 281 pp.

KUKALOVÁ-PECK, J. and J. F. LAWRENCE. 1993. Evolution of the hind wing in Coleoptera. Canadian Entomologist, 125: 181-258.

KUSCHEL, G. and B. M. MAY. 1990. Palophaginae, a new subfamily for leaf-beetles, feeding as adult and larva on Araucarian pollen in Australia (Coleoptera: Megalopodidae). Invertebrate Taxonomy, 3: 697-719.

KUSCHEL, G. and B. M. MAY. 1996a. Palophaginae, their systematic position and biology. Pp. 173-185. *In:* P. H. A. Jolivet and M. L. Cox, eds. Chrysomelidae Biology, vol. 3: General Studies. SPB Academic Publishing. Amsterdam. vi + 365 pp.

KUSCHEL, G. and B. M. MAY. 1996b. Discovery of Palophaginae (Coleoptera: Megalopodidae) on *Araucaria araucana* in Chile and Argentina. New Zealand Entomologist, 19: 1-13.

LEE, J. E. 1998. Leaf beetle larvae of Zeugophorinae from North America and Europe. Coleopterists Bulletin, 53: 118-125.

REID, C. A. M. 1995. A cladistic analysis of subfamilial relationships in the Chrysomelidae sensu lato (Chrysomeloidea). Pp. 559-631. *In:* J. Pakaluk and S. A. Slipinskí, eds. Biology, Phylogeny, and Classification of Coleoptera. Papers Celebrating the 80th Birthday of Roy A. Crowson. Muzeum i Instytut Zoologii PAN, Warsaw. Poland. vi + pp. 559-1092.

REID, C. A. M. 2000. Spilopyrinae Chapuis: a new subfamily in the Chrysomelidae and its systematic placement (Coleoptera). Invertebrate Taxonomy, 14: 837-862.

SANTOS, H. R. 1981. Biologia de *Agathomerus sellatus* (Germar, 1824) (Coleoptera, Chrysomelidae, Megalopodinae) broca do tomateiro. Revista brasileira de Entomologia, 15: 165-170.

SCHULZE, L. 1996. Life-history and description of early stages of *Sphondylia tomentosa* (Lacordaire). Coleoptera: Chrysomelidae: Megalopodinae). Pp. 187-199. *In:* P. H. A. Jolivet and M. L. Cox, eds. Chrysomelidae Biology, vol. 3: General Studies. SPB Academic. Amsterdam. vi + 365 pp.

SEENO, T. N. and J. A. WILCOX. 1982. Leaf beetle genera (Coleoptera: Chrysomelidae). Entomography, 1: 1-222.

WEISS, H. B. and A. S. NICOLAY. 1919. Notes on *Zeugophora scutellaris* Suffr., a European poplar leaf-miner, in New Jersey (Col.). Entomological News, 30: 124-127.

YU, P. and Y. XINGKE. 1994. Biological studies on *Temnaspis nankinea* (Pic) (Chrysomelidae: Megalopodidae). Pp. 527-531. *In:* P. H. A. Jolivet and M. L. Cox, eds. Novel Aspects of the Biology of Chrysomelidae. Kluwer Academic. The Netherlands. xxiii + 582 pp.

123. ORSODACNIDAE Thompson 1859

by Shawn M. Clark and Edward G. Riley

Family common name: The orsodacnid leaf beetles

Orsodacnidae are distinguished from other families of the Chrysomeloidea by the elongate and separate anterior apodemes of the male genitalia, in combination with the large, membranous and distinctly bilobed ligula and the absence of a mesonotal stridulatory file. Other more easily observed characters for the family include the generally short to medium length of the antennae which are inserted laterally between eye and base of mandible; the quadrate clypeus which projects between the base of the mandibles; the paired apical spurs on all tibiae; and the elongate, subparallel body form.

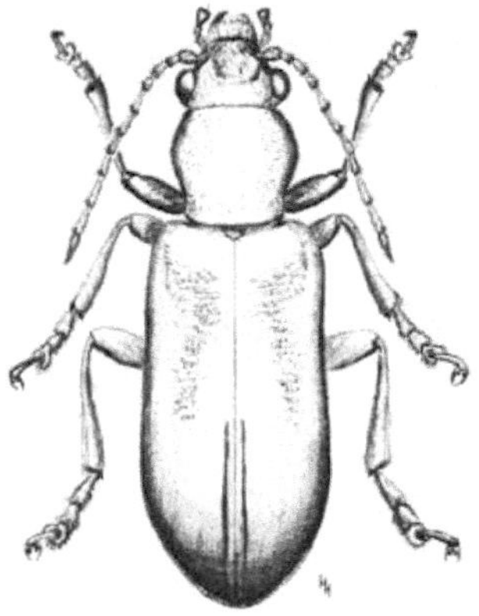

FIGURE 1.123. *Orsodacne atra* (Ahrens) (modified from Hatch 1971)

Description: Shape elongate, parallel to subparallel-sided, subcylindrical to depressed; length from 4.0 to 8.7 mm (in our species), up to 15 mm in Neotropical *Janbechynea*. Most species nonmetallic, a few weakly metallic, color generally dull, black to brownish, some individuals yellow, orange, or bright red, dorsally unicolorous or bicolored with pattern of maculae on elytra. Dorsum of most with sparse vestiture of suberect hairs; some with elytral vestiture dense and recumbent.

Head short, not rostrate, weakly constricted behind eyes, hypognathous, exerted and broadly exposed from dorsal view; surface smooth to densely, deeply punctate; front without distinct grooves, antennal calli separated by distinct quadrate depression; gular sutures evident at hind margin of cranium, separate. Eyes lateral, round, moderately protuberant. Antenna with eleven freely articulated antennomeres, filiform to subserrate, moderate in length, extending to point just beyond humerus, or in some species nearly to middle of elytra; insertion lateral between eye and base of mandible. Mouthparts exposed; labrum distinct; clypeus quadrate, projected between base of mandibles, well delineated from frons; mandible long, apically narrow, evenly curved with acute, uni- or bidentate apex; maxillary palp 4-segmented, palpomeres moderate in length, not asymmetrically dilated; terminal palpomere flattened, with apex evenly or angularly truncate; mentum transverse, not large; ligula large, membranous, and deeply lobed; labial palp 3-segmented, with palpomeres short.

Pronotum subequal to distinctly broader than head, nearly flat to moderately convex, laterally arcuate or weakly constricted in posterior half; borders unmargined or with distinct, reflexed marginal bead; anterior and posterior angles without sensorial setae; surface smooth to densely, deeply punctate; prosternum narrow in front of coxae; prosternal process convex, narrow, subparallel-sided between procoxae, not prolonged behind; procoxal cavities closed. Mesonotum lacking stridulatory file. Scutellum exposed, flat, broad at base, obtusely rounded apically. Mesosternum narrow and subparallel-sided, extending between mesocoxae; metasternum long and broad. Legs moderate to long; procoxae transverse, prominent; mesocoxae subconical, small; metacoxae transverse; trochanters of most small, triangular; femora moderately and equally swollen, without ventral teeth; tibiae slender, weakly curved in most species, those of both sexes with paired apical spurs on all tibiae; tarsi 5-5-5, pseudotetramerous, fourth tarsomere minute and more or less hidden between lobes of third; tarsomeres 1-3 with expanded ventral pads; tarsomere 3 with pad strongly bilobed; tarsal pads densely covered with simple or spatulate adhesive setae; claws paired, symmetrical, either bifid or simple.

Elytra entire and apically rounded; surface finely to strongly punctate, costate in some species, especially laterally, smooth between punctures; organized striae absent; margins entire, narrowly flattened in some species; epipleura well defined and complete to near apex, visible from lateral view. Hindwing present and well developed; venation (Reid 1995; terminology after Kukalová-Peck and Lawrence 1993): cubito-anal cell single, elongate; "anal" veins with three or four apical branches; MP_{3+4} connected or almost connected to MP_{1+2}; AA_{3+4} present; CuA_1 + MP_4/MP_{3+4} fused at base or reduced to one vein; CuA_1 connected to cubito-anal cell; RP_{3+4} spur present.

Abdomen with five free ventrites of subequal length; mesal region of last ventrite slightly flattened or impressed in some males, unmodified in female; apical margin of last ventrite of male either entire or weakly to deeply incised, that of female unmodified. Male genitalia with shaft of median lobe elongate, subcylindrical with symmetrical apex, without fringe of subapical setae on lower margin; anterior apodemes fused proximally, distally free, long and narrow. Tegmen completely encircling median lobe, with long ventral arm extending anteriorly; parameres present, dorsal in position, fused basally, separated apically. Female with internalized sternite 8 forming a continuous structure, not divided by a folded membrane into an apical blade and anteriorly directed apodeme; female ovipositor telescopic, with sclerites elongate; spiculum gastrale present; vaginal pouches absent

(Reid 1995); valves present, tipped with numerous long setae and each with short stylus (Cox and Windsor 1999a). Rectum without specialized sclerotized rectal plates (kotpresse) or spinose patches (Reid 1995).

Larva (known only from first instar): Body elongate, parallel-sided, depressed (orthosomatic), six times as long as wide (ca. 1.5 mm), mostly soft, white in coloration except head capsule, pronotal shield, and tergite of A-9 which are slightly sclerotized and yellowish or brownish; setae of head capsule and body well-developed. Head exerted, oval, flattened; frontal sutures long and straight, meeting near hind margin of occiput; coronal (epicranial) suture very short; clypeus indistinctly separated from front; one pair of stemmata present. Antenna short, two-segmented. Mandible bidentate apically, with three to four or six small subapical teeth along cutting edge, without mola but with penicillus; maxilla with mala broad and apically spinose, with palp three-segmented; labium with mentum broad and rectangular, submentum about twice as long as mentum, palp two-segmented and inconspicuous, ligula rectangular with pair of apical setae. Prothoracic shield transverse; egg bursters present on pronotum and meso- and metathorax, absent from abdominal segments. Leg four-segmented, with apical, claw-like tarsungulus, pulvillus absent. Abdominal segments A-1 to A-9 visible from above; A-9 with weakly sclerotized tergite bearing a pair of short, posteriorly-directed, setiferous urogomphi; A-10 small, ventral; anal opening Y-shaped, ventral; spiracles annular or biforous.

Habits and habitats. Adults of Orsodacninae feed on pollen and other floral parts of a wide range of mostly woody plants. Larval habits are unknown but are presumed to involve either internal or external root feeding. There is circumstantial evidence suggesting that first instar larvae hibernate beneath bud scales of oak (Cox 1981). Adults of some Aulacoscelidinae have been associated with cycads (Gymnospermae: Cycadales) upon which at least some of them feed (Monrós 1954; Cox and Windsor 1999a, 1999b), and some have been collected from flowers of various angiosperms. Angiosperm-type pollen has been recovered from the gut of one adult *Aulacoscelis* specimen (Crowson 1991). Feeding by at least one species of *Aulacoscelis* consists of damaging the upper surface of cycad leaflets and ingesting the juices from these perforations (Cox and Windsor 1999a). The size and build of these beetles suggests that they are unlikely to be involved in the pollination of cycads (Crowson 1991). Nothing is known about orsodacnine larval habits but a cryptic life is suspected (Cox and Windsor 1999a).

Status of the classification. Opinions regarding the rank, limits, and taxonomic affinities of this small group of beetles have varied considerably. The group has been treated as a subfamily of Chrysomelidae rather than a separate family. The Zeugophorinae, were once included here but are now classified in the Megalopodidae. Similarly, the eumolpine chrysomelid tribe Synetini was formerly classified here. Also, the inclusion of the Aulacoscelidinae is not universally accepted, the position of this group within the Chrysomeloidea has been somewhat controversial. Suzuki (1994) indicated that the aulacoscelidines should be classified near the Megalopodidae or the chrysomelid subfamily Sagrinae. Crowson and Crowson (1996) provided an argument relating Orsodacninae to the chrysomelid subfamily Galerucinae. Even so, the classification adopted here follows the phylogenetic studies of Reid (1995, 2000) and Schmitt (1996). In these analyses, Orsodacninae and Aulacoscelidinae consistently form a monophyletic group which, in Reid's study, as well as that of Farrell (1998), is recognized as the sister group to Chrysomelidae. The taxonomic history of the Aulacoscelidinae was reviewed by Cox and Windsor (1999a, 1999b). There still appear to be a few remaining undescribed species of *Aulacoscelis* in Mexico and Central America, and some of the described species are known only from the original type material. A complete revision of *Orsodacne* is needed. Catalogs, Clavareau (1913), Jolivet (1957).

Distribution. The subfamily Orsodacninae, consisting of a single genus, is strictly Holarctic in occurrence. The Aulacoscelidinae, consisting of two genera, occurs only in the New World, ranging from the southwestern United States to South America. With not more than 30 recognized species, the Orsodacnidae is the smallest of the chrysomeloid families.

Key to the Subfamilies and Genera of Orsodacnidae

1. Lateral margin of pronotum without marginal bead; tarsal claws bifid (Orsodacninae) ...... *Orsodacne*
— Lateral margin of pronotum with distinct, narrowly reflexed marginal bead; tarsal claws simple (Aulacoscelidinae) 2

2(1). Pronotum with short longitudinal fold on each side, extending from basal margin to beginning of disc; elytra with sparse erect setae .... *Aulacoscelis*
— Pronotum without such folds but with two basal grooves contiguous to basal margin; elytra uniformily covered with appressed pubescence .... *Janbechynea*

Classification of Nearctic Genera

Orsodacnidae Thompson 1859

Orsodacninae Thompson 1859

Characteristics. Body elongate, narrow, subcylindrical, moderately convex; maxilla with terminal palpomere flattened with apex evenly or angularly truncate; pronotum weakly convex with lateral margins arcuate anteriorly, constricted before base; elytron deeply, densely punctate, without plicae; tarsal claws bifid; hind margin of last ventrite of male unmodified.

This subfamily is composed of a single genus confined to the Holarctic Region.

Orsodacne Latreille 1802
Orsodanca Latreille 1804 [error]
Orsodachna Leach 1815 [error]

This genus contains ten species, including a few which are of questionable status. Only a single species, *O. atra* (Ahrens 1810), occurs in North America where it is rather widespread throughout most of Canada and all but the southern-most portions of the United States. This species is extremely variable in color pattern even within populations, this accounting for the numerous synonyms. Adults are among the first phytophagous beetles to become active in the spring, sometimes being abundant when snow still covers much of the ground. They have been associated with *Acer* (Aceraceae); *Rhus* (Anacardiaceae); *Alnus, Betula, Carpinus, Corylus* (Betulaceae); *Viburnum* (Caprifoliaceae); *Cornus* (Cornaceae); *Cercis* (Fabaceae); *Quercus* (Fagaceae); *Hamamelis* (Hamamelidaceae); *Iris* (Iridaceae); *Lindera* (Lauraceae); *Amelanchier, Crataegus, Malus, Prunus, Rubus, Spiraea* (Rosaceae); *Hepatica* (Ranunculaceae); *Galium* (Rubiaceae); *Salix* (Salicaceae); and *Verbascum* (Scrophulariaceae). They feed on flowers and will apparently accept practically any plant that happens to be in bloom. The larva of *O. atra* is unknown; the description given by Bøving and Craighead (1931) was based on misidentified material (Cox 1981). However, larvae of two Palearctic species have been described (Cox 1981, Mann and Crowson 1981). Only the first instar is known, the descriptions being of neonate larvae hatched from laboratory-oviposited eggs. The biology of the immature stages remains unknown, although larval morphology suggests a cryptic habit such as internal or external root feeding.

Aulacoscelidinae Chapuis 1874

Aulacoscelinae, *auctorum*

Characteristics. Body elongate, narrow, convex to depressed; maxilla with terminal palpomere narrow, cylindrical, with apex narrowly rounded; pronotum weakly convex to nearly flat, with lateral margins arcuate and having well-developed, narrowly reflexed marginal bead; elytron finely and sparsely punctate, in some species with complete or incomplete discal or lateral plicae which are only present, or best developed, in females; tarsal claws simple; hind margin of last ventrite of male weakly to strongly emarginate.

This strictly New World group is composed of two genera and 19 described species, nearly all of them Neotropical in distribution. Adults feed on the fronds of cycads and on pollen of various angiosperm flowers. Larval habits are unknown.

Aulacoscelis Duponchel and Chevrolat 1842
This genus, comprised of 15 described species, is mostly Neotropical in distribution, with most species inhabiting Mexico and Central America. However, two species occur in the United States, *A. candezei* Chapuis 1874 (Texas to California) and *A. vogti* Monrós 1959 (Texas). Adults have been found in the flowers of Asteraceae and *Hechtia* (Bromeliaceae) (Monrós 1954, 1959). Adults are reportedly associated with the foliage of cycads (Monrós 1954), and have been observed feeding on fronds of ornamental cycads in Costa Rica and on native cycads in Panama (Cox and Windsor 1999a). It is interesting to note that no native Cycadales occur in the southwestern United States. Larvae of the United States species are unknown; neonate larvae of a Central American species have recently been described from specimens hatched from laboratory-oviposited eggs (Cox and Windsor 1999a). Keys to species, Crowson (1946), Monrós (1954), Cox and Windsor (1999a); also see Monrós (1959) and Medvedev (1975).

Janbechynea Monrós 1953
This genus contains two subgenera and five described species, most of which are recorded from isolated localities in Mexico and South America. A species of the nominotypical subgenus has been associated with cycads (Monrós 1954).

subgenus *Janbechynea* Monrós 1954 [Neotropical]
subgenus *Bothroscelis* Monrós 1954

This subgenus is comprised of two described species, one occurring only in Mexico and the other, *J. fulvipes* (Jacoby 1888), occurring in Arizona as well as Mexico. Keys to species, Crowson (1946), Monrós (1954).

Bibliography

BØVING, A. G. and F. C. CRAIGHEAD. 1931. An illustrated synopsis of the principal larval forms of the Order Coleoptera. Entomologica Americana (new series), 11: 1-351 (1930).

CLAVAREAU, H. 1913. Chrysomelidae: 1. Sagrinae, 2. Donaciinae, 3. Orsodacninae, 4. Criocerinae. vol. 24 (Pars 51). *In:* W. Junk and S. Schenkling, eds. Coleopterorum catalogus. W. Junk. Berlin. 103 pp.

COX, M. L. 1981. Notes on the biology of *Orsodacne* Latreille with a subfamily key to the larvae of the British Chrysomelidae (Coleoptera). Entomologist's Gazette, 32: 123-135.

COX, M. L. and D. M. WINDSOR. 1999a. The first instar larva of *Aulacoscelis appendiculata* n. sp. (Coleoptera: Chrysomelidae: Aulacoscelinae) and its value in the placement of the Aulacoscelinae. Journal of Natural History, 33: 1049-1097.

COX, M. L. and D. M. WINDSOR. 1999b. The first instar larva of *Aulacoscelis* sp. and *Megascelis puella* Lacordaire (Coleoptera: Chrysomelidae: Aulacoscelinae, Megascelinae) and their value in the placement of the Aulacoscelinae and Megascelinae. Pp. 51-70. *In:* M. L. Cox, ed. Advances in Chrysomelidae Biology. Backhuys. Leiden, xii + 671 pp.

CROWSON, R. A. 1946. A revision of the genera of the chrysomelid group Sagrinae (Coleoptera). Transactions of the Royal Entomological Society of London, 97: 75-115.

CROWSON, R. A. 1991. The relations of Coleoptera to Cycadales. Pp. 13-28. *In:* M. Zunino, X. Bellés and M. Blas, eds. Advances in Coleopterology. Asociación Europea de Coleopterologia. Barcelona, 323 pp.

CROWSON, R. A. and E. A. CROWSON. 1996. The phylogenetic relations of Galerucinae-Alticinae. Pp. 97-118. *In:* P. H. A. Jolivet and M. L. Cox, eds. Chrysomelidae Biology, vol. 1: The Classification, Phylogeny and Genetics. Academic Publishing. Amsterdam, 443 pp.

FARRELL, B. D. 1998. "Inordinate Fondness" explained: why are there so many beetles? Science, 281 (no. 5376): 555-559.

HATCH, M. H. 1971. The beetles of the Pacific Northwest. Part V: Rhipiceroidea, Sternoxi, Phytophaga, Rhynchophora, and Lamellicornia. University of Washington Publications in Biology 16: 662 pp. (Chrysomelidae, Pp. 157-243)

JOLIVET, P. 1957. Chrysomelidae: Orsodacninae (Pars 51: Fasc. 3). *In:* W. D. Hincks, ed. Coleopterorum Catalogus, Supplementa. W. Junk. s-Gravenhage. 16 pp.

KUKALOVÁ-PECK, J. and J. F. LAWRENCE. 1993. Evolution of the hind wing in Coleoptera. Canadian Entomologist, 125: 181-258.

MANN, J. S. and R. A. CROWSON. 1981. The systematic positions of *Orsodacne* Latr. and *Syneta* Lac. (Coleoptera Chrysomelidae), in relation to characters of larvae, internal anatomy and tarsal vestiture. Journal of Natural History, 15: 727-749.

MEDVEDEV, L. N. 1975. Review of the chrysomelid subfamily Aulacoscelinae in entomological collections of the German Democratic Republic. Mitteilungen aus dem Zoologischen Museum in Berlin, 51: 37-40.

MONRÓS, F. 1954. Revision of the chrysomelid subfamily Aulacoscelinae. Bulletin of the Museum of Comparative Zoology, 112:321-360.

MONRÓS, F. 1959. Notas sobre Chrysomelidae (Coleoptera). Acta Zoológica Lilloana, 17: 1-24.

REID, C. A. M. 1995. A cladistic analysis of subfamilial relationships in the Chrysomelidae sensu lato (Chrysomeloidea). Pp. 559-631. *In:* J. Pakaluk and S. A. Slipinski, eds. Biology, Phylogeny, and Classification of Coleoptera. Papers Celebrating the 80th Birthday of Roy A. Crowson. Muzeum i Instytut Zoologii PAN. Warsaw, Poland.

REID, C. A. M. 2000. Spilopyrinae Chapuis: a new subfamily in the Chrysomelidae and its systematic placement (Coleoptera). Invertebrate Taxonomy, 14: 837-862.

SCHMITT, M. 1996. The phylogenetic system of the Chrysomelidae -history of ideas and present state of knowledge. Pp. 57-96. *In:* P. H. A. Jolivet and M. L. Cox, eds. Chrysomelidae Biology, vol. 1: The Classification, Phylogeny and Genetics. SPB Academic Publishing. Amsterdam, 443 pp.

SUZUKI, K. 1994. The systematic position of the subfamily Aulacoscelinae (Coleoptera: Chrysomelidae). Pp. 45-59. *In:* D. G. Furth, ed. Proceedings of the Third International Symposium on the Chrysomelidae Beijing, 1992. Backhuys Publishers. Leiden, The Netherlands.

124. CHRYSOMELIDAE Latreille 1802

by Edward G. Riley, Shawn M. Clark, R. Wills Flowers, and Arthur J. Gilbert

Common name: The leaf beetles

Family synonyms: Camptosomes, of authors; Cassididae Gyllenhal 1813; Chlamydidae Lacordaire 1848; Chrysomelidae Latreille 1802; Clytridae Lacordaire 1848 (Clythridae); Crioceridae Latreille 1807; Cryptocephalidae Gyllenhal 1813; Cryptostomes, of authors; Cyclica, of authors; Donaciidae Kirby 1837; Eumolpidae Hope 1840; Eupodes, of authors; Fulcidacidae Jacobson 1924; Galerucidae Latreille 1802; Alticidae Newman 1835 (Halticidae); Hispidae Gyllenhal 1813; Lamprosomatidae Lacordaire 1848 (Lamprosomidae); Megascelididae Chapuis 1874 (Megascelidae); Mesomphaliidae Hope 1840; Phytophaga (part), of authors; Sagridae Leach 1815; Synetidae Edwards 1953

The leaf beetle family Chrysomelidae is one of the largest beetle families and among the phytophagous families is second in number of species only to the Curculionidae. Chrysomelids are generally small to medium-sized beetles with most brightly colored or strikingly patterned. The defining character for distinguishing this family from other families of the Chrysomeloidea is the broadly fused anterior apodemes of the male genitalia. The resulting cap-like structure at the base of the median lobe is unique among the families of Phytophaga (Chrysomeloidea + Curculionoidea). Chrysomelids also lack a mesonotal stridulatory file and only a few possess an exposed, membranous ligula. Other less rigorous but more observable characters for the family include the generally short to medium-length antennae, antennal insertions not on prominences and not partly encircled by the eyes, general reduction of tibial spurs, highly variable but generally elongate-oval body form, leaf-feeding habits of adults, and larvae which are not true borers.

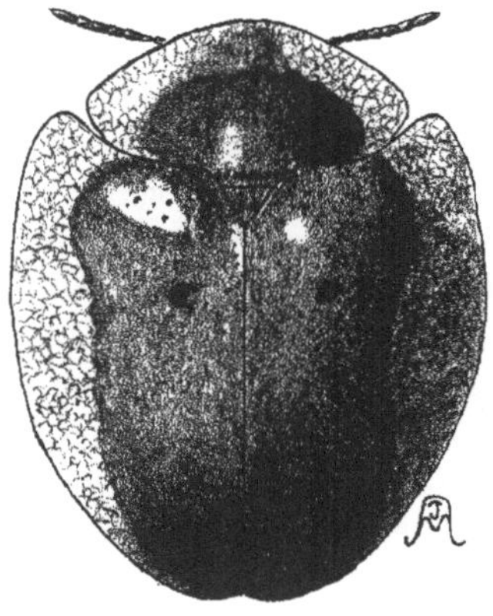

FIGURE 1.124. *Charidotella sexpunctata bicolor* (Fabricius)

Description: Shape highly variable, elongate-cylindric to oval-convex or depressed; length from 1 to 17 mm (in our species); colors various, commonly bright or metallic, often dorsally bicolored and formed into distinctive patterns, often with pale pronotum and dark elytra; dorsum usually glabrous, vestiture when present sparse to dense and consisting of simple hairs or narrow to moderately wide scales which can be erect, inclined, or appressed.

Head usually short, not or weakly rostrate, usually not constricted behind eyes but sometimes distinctly so, usually hypognathous but sometimes prognathous or opisthognathous, broadly exposed or completely hidden from dorsal view, sometimes deeply recessed into prothorax; surface smooth, punctate, or rugose, with or without various grooves, sometimes with distinct antennal calli; gular sutures separate. Mouthparts usually exposed, sometimes partly concealed by anterior margin of prosternum or proepisternal lobes; labrum distinct; clypeus variably delineated from frons. Eyes lateral, round or elliptical, sometimes emarginate on internal margins. Antenna usually with eleven freely articulated antennomeres, rarely 10-segmented or sometimes with a variable number of fused terminal antennomeres, usually moderate in length, usually filiform but may be short and serrate, clavate, or rarely capitate; insertion on flat portion of frons between eyes or between eye and the base of mandible. Mandible short, stout, curved with apex acute, blunt, or dentate; maxillary palp 4-segmented, palpomeres usually enlarged but not elongate or asymmetrically dilated; mentum usually transverse, not large; ligula usually coriaceous and entire, rarely membranous and lobed; labial palp 3-segmented, palpomeres short, apically acute.

Pronotum usually broader than head, highly variable in shape; borders usually margined, sometimes broadly explanate; one or more setae usually present in anterior and/or posterior angles; surface smooth, punctate, or rugose; prosternum various in front, frequently narrowed between coxae, not strongly prolonged behind procoxae; procoxal cavities open or closed. Mesonotum lacking stridulatory file. Scutellum usually exposed, broad, triangular to subtrapezoidal, sometimes inclined with apex elevated. Mesosternum moderate in length, narrow to moderately wide; metasternum usually long and broad. Legs usually short to moderate in length; trochantins not exposed; procoxae round to transverse, often conical and prominent; mesocoxae subconical, small; metacoxae transverse; trochanters small, trian-

Acknowledgments: We wish to thank the following people for their contributions to the content of this chapter: Andrzej O. Bienkowski provided helpful information pertaining to the synonymy of *Chrysolina*; Igor Lopatin provided helpful information pertaining to the synonymy of *Cryptocephalus*; Douglas G. LeDoux provided selected food plant data and reviewed a late draft of the manuscript.

gular; femora usually moderately swollen, but sometimes distinctly enlarged; tibiae slender, usually without apical spurs, these usually inconspicuous when present or distinct on metatibia only; tarsi usually 5-5-5, pseudotetramerous, fourth tarsomere minute and more or less hidden between lobes of third, less commonly tarsi acually 4-4-4 with the penultimate tarsomere lost; tarsomeres 1-3 usually lobed or with expanded ventral pads; pad of third tarsomere usually strongly bilobed, sometimes only weakly lobed; tarsal pads densely covered with spatulate or bifid adhesive setae; claws paired, rarely single or asymmetrical, usually simple or appendiculate, rarely pectinate.

Elytra usually entire and apically rounded; surface smooth to rugose, impunctate to strongly punctate, sometimes costate; striae present or absent, when fully developed usually 10 in number plus abbreviated subscutellar stria, sometimes striae reduced in number or irregular; margins entire or serrate, sometimes broadly explanate; epipleura usually well defined and complete to near apex, variably inclined. Hind wing venation (Reid 1995; terminology after Kukalova-Peck and Lawrence 1993): two cubito-anal cells or one elongate basal cell present; with from 2 to 5 apical branches of "anal" veins present; MP_{3+4} connected or almost connected to MP_{1+2} or not connected; AA_{3+4} present or absent; CuA_1 + MP_4/MP_{3+4} fused at base, free or reduced to one vein; CuA_1 connected to cubito-anal cell(s) or not; RP_{3+4} spur present; overall venation often reduced in small forms. Some species brachypterous, micropterous, or apterous.

Abdomen with five ventrites of variable size, their lengths sometimes markedly shortened at midline, usually free with sutures entire or first and second connate with suture partially obliterated; mesal region of last ventrite of some females with moderate to deep median fossa; apical margin of last ventrite of male entire or variously modified. Male genitalia with shaft of median lobe highly variable, short and stout to elongate, cylindric to flattened, usually symmetrical but may display great range of asymmetry, sometimes lower distal margin fringed with setae; anterior apodemes fused to form hood-like basal piece which is rarely incised apically. Tegmen variable, may completely encircle median lobe or more commonly reduced to ventral Y- or V-shaped yoke; parameres usually absent, but present when tegmen complete and fused to form dorsal sclerotized cap which is incised apically. Ovipositor telescopic with sclerites elongate, or not telescopic with broad sclerites in a rigid structure; spiculum gastrale (tignum, Konstantinov 1998) present or absent; vaginal pouches present or absent (Reid 1995); vaginal palpi present or absent (Konstantinov 1998).

Larval body form highly variable, cyphosomatic, eruciform, fusiform, orthosomatic, or weakly to strongly scarabaeiform, free or contained in case made from larval fecal material; length up to 13 mm. (our species). Body surface various, sometimes setiferous, covered with microspines or numerous fleshly lobes, or naked; lateral scoli sometimes present, the posterior pair on A-8 positioned dorsally and elongated to form specialized manipulator for holding exuviae and/or fecal material. Some taxa with eversible glands present in paired dorsolateral cones on meso- and metathorax and abdomen. Coloration usually white to cream in subterranean forms, highly variable in open-feeding forms, sometimes distinctly patterned. Head small, usually exserted, sometimes hidden beneath prothoracic shield, usually hypognathous; epicranial suture usually present, short; from one to six pairs of stemmata present, or rarely stemmata lacking. Antenna usually inconspicuous, one- to three-segmented. Clypeus distinct or indistinct, sometimes fused to frons. Labrum free or fused with clypeus. Mandible simple or palmate, with from three to five teeth on distal margin; mola usually absent; penicillus sometimes present. Maxilla with lightly sclerotized stipes, short mala, two- to four-segmented palp. Labium usually with submentum and mentum fused to form small, basal plate; ligula small; palp inconspicuous, one- or two-segmented. Prothoracic shield variably sclerotized, formed into a transverse plate or expanded and covering head. Thoracic egg bursters of first instar absent from prothorax, present or absent from meso- and metathorax. Leg three- or four-segmented, sometimes greatly elongated, with apical, claw-like tarsungulus; rarely legs vestigial. Abdomen usually with eight dorsally visible segments; dorsum of segments smooth or with transverse plicae; ambulatory lobes sometimes present ventrally; A-9 and 10 usually hidden from dorsal view, sometimes tube-like and retractile, A-10 sometimes bearing an eversible anal disc. Anal opening usually caudoventral, rarely dorsal. Spiracles annular, annular-biforous, or cribriform, usually present and well-developed on A-1 to 8, the eighth pair sometimes vestigial or absent; in Donaciinae, pairs on A-1 to 7 vestigial, and each spiracle on A-8 joined to a sclerotized spur.

Habits and habitats. Chrysomelids are phytophagous and are among the most diverse and conspicuous insect families on plants. The adults feed on living plant material, usually consuming leaves or sometimes various flower parts including pollen. Many chrysomelid larvae feed on leaves, but a great number are subterranean attacking roots and underground stems. A smaller but significant number are leaf miners or occupy concealed spaces between appressed leaves or other plant parts. The larvae of some, perhaps most, Cryptocephalinae consume dead plant materials including fallen leaves and the bark of living or dead twigs. Some members of this group occur in ants' nests where they have been recorded eating ant eggs and fecal material, and possibly the bodies of dead ants. The larvae of *Sagra* (subfamily Sagrinae, not found in North America) are gall-forming in the stems of semiwoody plants. The great majority of chrysomelid species feed on Angiospermae, both dicotyledons and monocotyledons; a small number on Coniferae, and a few on ferns and cycads. Most species are fairly specific in their food plant preferences, being either monophagous or more commonly oligophagous, consuming a few closely related plants. In many instances, related groups of chrysomelids feed on groups of related plants, and these patterns are frequently evident worldwide.

A fair number of leaf beetles are of some economic importance, either through direct consumption of valuable plants or as vectors of phytopathogenic bacteria and viruses. Others have proven to be beneficial as agents for the biological control of invasive weeds in various areas of the world.

Generally, terrestrial plants serve as hosts, but a few chrysomelids spend their lives on emergent portions of aquatic vegetation, and the larvae of Donaciinae are truly aquatic, living on the submerged stems and roots of their hosts. The adults of most Donaciinae live on the exposed parts of aquatic or semiaquatic plants. Most of the ventral surface of their bodies is covered with a dense pile of silver-colored pubescence which traps a layer of air for the brief periods they spend submerged. Eggs are deposited above or below the water and may be uncovered, contained in a gelatinous mass, or inserted into plant tissues. The larvae possess a pair of sclerotized spurs on the dorsum of the eighth abdominal tergite, and the only functional pair of spiracles is found on this segment, a single spiracle positioned within the base of each spur. Larvae bury these processes into plant tissues in such a manner as to effect their respiration by taking in oxygen from the plant vessels. Transformation to the pupal stage takes place under water in a parchment-like cocoon, probably secreted by two shortened Malpighian tubules. This cocoon is attached to a root and contains oxygen that has come from the tissue of the root, the larva having prepared the root earlier by making an incision into it. Some species of this subfamily are reported to only rarely leave the water as adults (probably all Haemoniini); adults of this group lack the expanded tarsal pads characteristic of the family. Except for one species found in brackish waters of the Baltic Sea, all Donaciinae are associated with freshwater communities.

The larvae of Chrysomelinae, most Criocerinae, and many of the Galerucinae and Hispinae are folivorous (leaf-eating), inhabiting the exposed surfaces of their host plants. In this predator rich space, a number of defensive strategies have evolved. Some chrysomeline larvae possess dorsal eversible glands stocked with defensive chemicals which are either synthesized independently or from precursors obtained from their food plants. Defensive use of fecal material is seen in several groups, especially in cassidoid Hispinae and the Criocerinae, but also in *Blepharida* (Alticini). While the latter two taxa simply deposit feces on the dorsum of their bodies, the larvae of most tortoise beetles are equipped with elongate scoli which arise from abdominal tergite 8, and with a protrusible anus which permits the construction of a maneuverable fecal "parasol" held above the body. This fecal shield can be elaborately shaped and may incorporate the exuviae. Females of most taxa with folivorous larvae place eggs directly on their food plants, usually in small groups and sometimes with a covering of fecal material added for protection. Some larvae bury themselves for pupation and may form a "cocoon;" others transform to naked pupae and remain attached to their food-plants. The larvae of the Nearctic galerucine genus *Ophraella* pupate on their host plants within mesh, basket-like cocoons.

The "casebearers," larvae of Cryptocephalinae and Lamprosomatinae, live in variably-shaped cases they construct of fecal material and sometimes bits of vegetable debris. The adult females of Cryptocephalinae capture each new egg with the hind tibiae and hold it in a pocket-like fossa on the last ventrite. A specialized set of sclerotized rectal plates, the "kotpresse," is used to compress bits of fecal material into platelets which are glued to the surface of the egg in succession to form an intricate encrustation. This egg coating serves the first instar larvae as a case which they enlarge as they grow by adding their own fecal material. In the Cryptocephalini with leaf-litter inhabiting larvae, and in some Clytrini, the females simply drop the completed eggs to the ground. In the Nearctic Chlamisini with folivorous larvae, the eggs are glued to the surface of the host plant. Larvae of casebearers pupate within their cases after cementing the open end to a substrate. Before pupation, they reverse their position within to face toward the posterior wall of the case. The adult, after eclosion, uses its mandibles to neatly cut a circular "cap" off the end of the case to free itself.

The leaf mining habit has evolved independently in a few chrysomelid groups. The larvae of many Hispinae are leaf miners, producing either blotch or trail-type mines in the leaves of their hosts. Larvae may exit one leaf and enter another before completing development; pupation takes place within the mine. Some flea beetle larvae (Alticini) are also leaf miners, and some mine stems as well; at least some of these exit the mine to pupate in the soil. The Nearctic genus *Monoxia* (Galerucini) includes some species which are also leaf miners as larvae.

The larvae of all Eumolpinae are subterranean, as are those of a great many Galerucinae. Soil-dwelling chrysomelid larvae are either free-living in soil, tunneling in roots, or mining under the skin of roots and underground stems. Some Eumolpinae oviposit on plant materials above ground, and the newly hatched larvae drop to the ground and enter the soil. In Galerucinae, some females oviposit in cracks in the soil surface, others on the base of the plant or on the soil surface near the host.

The length of the life cycle can vary greatly among taxa. Most Nearctic species are probably univoltine, but others may have multiple generations per warm season, and some probably cease reproduction only because of the onset of cold weather. Some *Chrysolina* species (Chrysomelinae) of the high arctic require up to three years to complete a generation. For most chrysomelids, the time spent in immature stages is generally short unless these stages overwinter or exhibit diapause. It is probable that most Nearctic species overwinter as adults, but some are known to overwinter as eggs or quiescent larvae. Oviparity is the rule; however, a few Chrysomelinae are viviparous or ovoviviparous, a characteristic common in arctic, subarctic, and high mountain populations.

The major food plant genera and citations to key studies on the biology of Nearctic Chrysomelidae are provided in the generic discussions which follow. These citations are by no means exhaustive but are intended to serve as a starting point for those interested in more detailed, taxon-specific biological data.

Status of the classification. Although there has been general agreement on major groupings within the Chrysomelidae, there has been little agreement on their rank and relationships or on the boundaries of the family. However, this era may be coming to an end. In recent years, important data have come to light that have helped to settle some questions. Some problematic groups (Megalopodidae, Zeugophorinae, Orsodacnidae) now appear to be rightfully excluded from Chrysomelidae, while others (*Syneta* and *Megascelis*) have probably found their proper subfamily placement. The recent phylogenetic analyses by Reid (1995, 2000) have produced a working hypothesis of relationships among subfamilies. This hypothesis will no doubt be tested as new data are recovered. The subfamilies recognized in those works are followed here, except the Bruchinae (commonly called bean weevils or seed beetles) are treated as a separate family (family no. 121 in this volume). In recent phylogenetic studies of the Chrysomeloidea (Farrell 1998, Reid 1995 and 2000, and Schmitt 1996), these beetles clearly fall within the Chrysomelidae (as sister group to Sagrinae in studies of the latter two authors). In spite of their ecological differences and specialized morphological adaptations, these beetles should be ranked as a chrysomelid subfamily rather than a separate chrysomeloid family. In this book, they are treated separately for convenience only.

The monophyly of the following chrysomelid groups is demonstrated and accepted by Reid (1995): Sagrinae + Bruchinae; Chrysomelinae + Galerucinae; and Eumolpinae + (Lamprosomatinae + Cryptocephalinae). The position of Hispinae is problematic, but a sister-group relationship with Criocerinae is favored. The Donaciinae are either sister to (Sagrinae + Bruchinae), or they belong with Criocerinae, or with (Hispinae + Criocerinae) as sister group to (Sagrinae + Bruchinae). The preferred position of the Spilopyrinae, a group recently elevated from obscurity to subfamily status, is basal to the clade (Eumolpinae + (Lamprosomatinae + Cryptocephalinae)) (Reid 2000). Support for some of the above groupings is weak. See Reid (1995, 2000) and Schmitt (1996) for detailed discussions.

A much different situation exists regarding the classification of genera within subfamilies. With the exception of the Donaciinae and part of the Hispinae, no entire subfamily has received a modern phylogenetic treatment. Most tribes, subtribes, etc., in use today are the products of many years of alpha taxonomy. Until broad-based phylogenetic studies are carried out within the larger subfamilies (especially Chrysomelinae, Eumolpinae, Galerucinae, and Hispinae), some of these higher categories should be considered tentative at best. Many appear to represent defensible natural groups, while others are obviously artificial. Nonetheless, until further phylogenetic studies appear, the majority of these groupings should be maintained for their utility. A thorough review of family-group names in the Chrysomelidae is needed.

Many species remain to be discovered and described in this enormous group of beetles, and concepts for certain genera will likely need refinement as this work progresses. An especially troublesome problem in the Chrysomelidae is the regional nature of many revisionary or monographic works, a problem inherent in any large, taxonomically difficult group. Some of the generic names applied to North American species will likely change as more global generic concepts evolve. Modern world catalogs are needed for most of the larger subfamilies.

Distribution. This is one of the largest beetle families with as many as 35,000 described species (Jolivet 1988) and perhaps up to 60,000 total species (Reid 1995). In the last comprehensive world catalog (Coleopterorum Catalogus (1913-1940) and later Supplementa for some groups (1958-1975)), there were 26,387 species recorded from all areas except the antarctic and most of the high arctic. By far the greatest diversity is found in the tropical regions where the taxonomy of the group remains poorly known. Presently, there are 195 genera and approximately 1,720 valid species and subspecies (excluding Bruchinae) accepted as occurring in America north of Mexico.

The keys presented below cover all genera known to occur in America north of Mexico. Also included are a few genera once thought to be part of our fauna but not known to exist in the region now. The keys should also work for the genera of the Baja California peninsula of Mexico, but not necessarily for the little-known fauna of temperate northern and central Mexico.

I. Key to North American Subfamilies of Chrysomelidae

1. Prothorax usually with well-formed lateral bead extending without interruption between anterolateral and posterolateral corners, sometimes with lateral margin strongly explanate and continuous with similarly explanate anterior margin; head usually not strongly constricted nor neck-like behind eyes, often inserted into prothorax to posterior margin of eyes; aedeagal tegmen (except rarely) Y-shaped, not forming complete ring around median lobe 2

— Lateral bead of prothorax usually absent or incomplete, sometimes present posteriorly but not extending to anterolateral corner; head constricted or neck-like behind eyes, usually not inserted into prothorax to level of eyes; aedeagal tegmen variable, forming complete ring around median lobe or not 7

2(1). Head opisthognathous, with front or vertex projecting strongly forward (Figs. 2, 24); tarsal formula 4-4-4; abdomen with ventrites 1 and 2 connate, although suture between them indicated by groove .... Hispinae

— Head usually normal, prognathous or hypognathous, with front or vertex not projecting strongly forward (if head opisthognathous, then hind femora greatly swollen); tarsal formula 5-5-5, pseudotetramerous, with penultimate tarsomere minute and usually hidden between lobes of tarsomere 3; abdomen with ventrites 1 and 2 usually free 3

3(2). Elytral epipleuron strongly angulate near basal third, excavated behind angulation for reception of metafemoral apex; prothorax, at each side of prosternal process, grooved for reception of antennae; body compact, hemispherical (Fig. 78); head deeply inserted into prothorax .... Lamprosomatinae

— Elytral epipleuron sometimes strongly curved, but without abrupt angulation near basal third, nor with excavation for reception of metafemoral apex; antennal grooves usually absent (present in chlamisine Cryptocephalinae); body variable in form 4

4(3). Abdomen with ventrites 2 to 4 usually strongly shortened mesally (Fig. 3); body subcylindrical, compact (Figs. 90-93); head deeply inserted into pro-

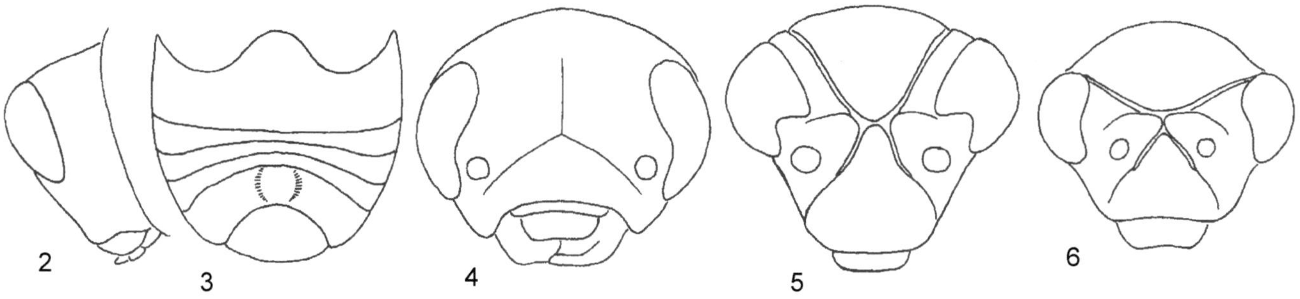

FIGURES 2.124-6.124. 2. Hispinae (diagrammatic), head, lateral view; 3. *Cryptocephalus* sp., female ventrites; 4. *Zygogramma exclamationis* (Fabricius), head, frontal view; 5. *Lema trilinea* R. White, head, frontal view; 6. *Oulema melanopus* (Linnaeus), head, frontal view.

thorax, vertically flattened; pygidium usually exposed, vertical; antennal insertions separated by much more than length of basal antennomere .. .. Cryptocephalinae

— Abdomen with intermediate ventrites not abnormally shortened; head usually less deeply inserted into prothorax; form usually not subcylindrical; abdomen without exposed, vertical pygidium, except in species with antennae separated by length of antennomere 1 or less 5

5(4). Antennal insertions widely separated, usually by distance much greater than length of basal antennomere 6

— Antennal insertions usually separated by distance less than length of basal antennomere .... Galerucinae

6(5). Front coxae transverse; ventral lobe of tarsomere 3 usually entire or weakly emarginate apically, less commonly bilobed; anteclypeus usually present (Fig. 4) .... Chrysomelinae

— Front coxae globular; ventral lobe of tarsomere 3 deeply bilobed; anteclypeus usually absent .... Eumolpinae (part)

7(1). Eyes entire, not emarginate near antennal insertions 8

— Eyes emarginate near antennal insertions 9

8(7). Tarsal claws bifid or appendiculate, each with broad to narrow tooth; head without deep median groove between antennae; aedeagal tegmen Y-shaped, not formed into complete ring around median lobe .... Eumolpinae (part)

— Tarsal claws simple; head usually with deep median groove between antennae; aedeagal tegmen formed into complete ring around median lobe, with sclerotized dorsal cap .... Donaciinae

9(7). Pygidium broadly exposed; hind femora greatly swollen, much larger than femora of front and middle legs, often with large ventral teeth; aedeagal tegmen forming complete ring around median lobe; tarsal claws divergent, usually with basal tooth, never connate; pronotum often with lateral carina in basal portion .... Bruchinae

— Abdomen usually without exposed pygidium; hind femora slightly to moderately larger than those of front and middle legs; metafemur without ventral teeth; aedeagal tegmen Y-shaped, not forming ring around median lobe; tarsal claws not toothed, often basally connate 10

10(9). Head with X-shaped groove between eyes (Figs. 5, 6); elytra usually not pubescent; abdomen without exposed pygidium .... Criocerinae

— Head without X-shaped groove between eyes; elytra pubescent; abdomen with exposed pygidium .... Eumolpinae (part)

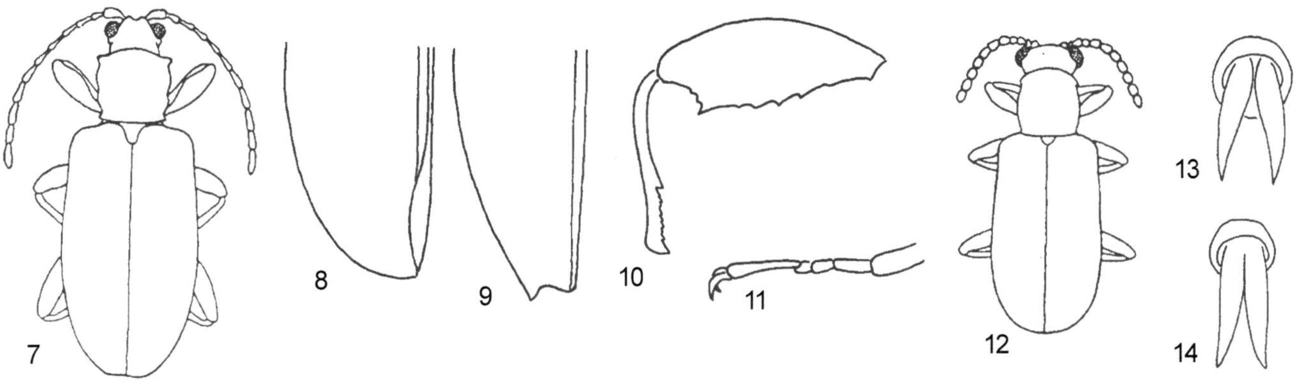

FIGURES 7.124-14.124. 7. *Donacia biimpressa* Melsheimer; 8. *Plateumaris* sp., apex left elytron, dorsal view; 9. *Neohaemonia nigricornis* (Kirby), apex left elytron, dorsal view; 10. *Poecilocera harrisii* (LeConte), metafemur and tibia; 11. *Neohaemonia* sp., tarsus; 12. *Crioceris asparagi* (Linnaeus); 13. *Crioceris* sp., tarsal claws; 14. *Lema* sp., tarsal claws.

II. Key to Tribes, Genera, and Subgenera of Donaciinae of America North of Mexico (modified from Askevold 1990)

1. Inner sutural bead of elytron before apex narrow, not expanded (Fig. 9); pronotal hypomeron usually with pubescent area above coxa 2
— Inner sutural bead of elytron expanded before apex, sinuate and joined to outer bead (Fig. 8); pronotal hypomeron glabrous (Tribe Plateumarini) 3

2(1). Dorsum testaceous; legs almost entirely testaceous; metafemur edentate, slender; tarsi with apical tarsomere elongate (Fig. 11), with ventral pads on tarsomeres 1-3 reduced; elytron usually with spine at outer apical angle (Fig. 9) (Tribe Haemoniini) .... *Neohaemonia*
— Most specimens dorsally metallic or dark brown, not testaceous; metafemur usually with 1 or 2 subapical teeth; apical tarsomere much shorter than preceding tarsomeres combined; tarsomeres 1-3 with ventral pads; elytron without apical spine (Tribe Donaciini) 4

Tribe Plateumarini

3(1). Apex of elytron broadly rounded; antennomere 3 shorter than 4; metatibia and metafemur not denticulate on underside, although metafemur usually with single large tooth............. *Plateumaris*
— Apex of elytron truncate or emarginate; antennomere 3 as long as or longer than 4; metatibia and metafemur distinctly denticulate on underside (Fig. 10) .... *Poecilocera*

Tribe Donaciini

4(2). Pronotal disc distinctly, finely pubescent; legs entirely rufous, at most with indistinct infuscation; metafemur without distinct subapical tooth below; elytral apex broadly rounded; subapical punctation of elytra close and entirely confused .... *Donaciella*
— Pronotal disc glabrous, or if pubescent then elytral apex truncate or elytra also pubescent; legs rufous to entirely metallic; metafemur of most specimens with one or more subapical teeth; elytral apex truncate, with obtuse to acute outer angle; subapical punctation of elytra more widely spaced and more or less regular 5

5(4). Legs usually with no more than dorsum of metafemur darkened; pronotal disc without coarse punctures; mesosternal process, especially in females, broad between mesocoxae, at least half as broad as diameter of mesocoxal cavity; ventral punctation obscured by dense pubescence; pygidium of most females more or less triangular, elongate .... *Donacia* (subgenus *Donacia*)
— Legs entirely dark to entirely red, but not with only dorsum of metafemur dark; pronotal disc with coarse punctures, usually forming coarse to moderate lateral rugae; mesosternal process no more than half diameter of mesocoxal cavity; ventral punctation coarse, often not obscured by pubescence; pygidium of females broadly rounded or emarginate, not elongate (Fig. 7) .... *Donacia* (subgenus *Donaciomima*)

III. Key to Tribes, Genera, and Subgenera of Criocerinae of America North of Mexico (modified from White 1993)

1. Tarsal claws connate, contiguous for much of their length (Fig. 14) (Tribe Lemiini) 2
— Tarsal claws separated, divergent from base to apex (Fig. 13) (Tribe Criocerini) 6

Tribe Lemiini

2(1). Ninth elytral stria with gap of 5-12 punctures; dorsum largely yellow, orange or red, and elytra nearly always with 1-4 black spots ...... *Neolema*
— Ninth elytral stria complete, or if with gap, then elytra dark throughout 3

3(2). Antennal tubercles touching or clearly more near to one another than are frons and vertex (Fig. 6); elytra entirely black or very dark, usually with metallic blue or green reflection; prothoracic constriction weaker and more basal 4
— Antennal tubercles separated by an upward extension of frons making frons and vertex more near to one another than are antennal tubercles (Fig. 5); elytra usually bicolored, with bands, stripes, or pale lateral margins, but rarely entirely dark; prothorax with strong, median constriction 5

4(3). Elytron with short row of subscutellar punctures. .... *Oulema* (subgenus *Oulema*)
— Elytron without row of subscutellar punctures..... .... *Oulema* (subgenus *Hapsidolemoides*)

5(3). Elytron with short row of subscutellar punctures. .... *Lema* (subgenus *Lema*)
— Elytron without row of subscutellar punctures..... .... *Lema* (subgenus *Quasilema*)

Tribe Criocerini

6(1). Prothorax strongly constricted at midlength; elytra entirely reddish-yellow, without pattern .... *Lilioceris*
— Prothorax feebly constricted near base, with sides arcuate (Fig. 12); elytra with dark spots or pattern .... *Crioceris*

IV. Key to Tribes, Genera, and Subgenera of Hispinae of America North of Mexico

1. Body narrow, margins of pronotum and elytra not explanate; antennae porrect, usually not retractable 2
— Body broad, margins of pronotum and elytra explanate; antennae usually capable of retraction 3

2(1). Dorsum smooth, punctation fine and weakly impressed, elytra without costae; tactile seta on anterior margin of pronotum arising posterior to eye, not from anterolateral corner of pronotum (Fig. 20); scutellum hastate, apex acute (Fig. 20) (Tribe Cephaloleiini) .... *Stenispa*

— Dorsum rough, punctation coarse and deeply impressed, elytra usually costate; tactile seta of anterior margin of pronotum usually arising from anterolateral corner of pronotum, not from behind eye; scutellum subquadrate, apex rounded to subtruncate (Tribe Chalepini) 6

3(1). Head broadly exposed; anterior corner of pronotum with seta (Fig. 25); tarsal claws singular (Tribe Hemisphaerotini) .... *Hemisphaerota*

— Head narrowly exposed or usually concealed from dorsal view (Figs. 22-24, 26); anterior corner of pronotum lacking seta and setal pore; tarsal claws in pairs 4

4(3). Head narrowly exposed (Fig. 26); vertex of head without apparent stridulatory file; tarsal claws appendiculate, each with broad basal tooth (Tribe Mesomphaliini) 20

— Head concealed by broadly expanded anterior margin of pronotum; vertex of head with stridulatory file; claws variable 5

5(4). Clypeus slanted posteriorly, short to moderately long; claws simple (Tribe Ischyrosonychini) 21

— Clypeus horizontal, long; claws variable (Tribe Cassidini) 22

Tribe Chalepini

6(2). Antenna appearing 8-segmented due to fusion of terminal antennomeres, apparent terminal antennomere enlarged and usually faintly annulate 7

— Antenna composed of 11 distinct antennomeres, terminal antennomere not especially enlarged nor annulate 14

7(6). Antenna short, capitate 8

— Antenna moderately long to long, filiform to clavate (Fig. 19) 9

8(7). Tarsomere 3 bilobed, lobes narrow and inconspicuous .... *Brachycoryna*

— Tarsomere 3 simple, not lobed .... *Stenopodius*

9(7). Protibia flattened, broadest at midlength, externally with sharp carina; antenna with abrupt club; elytra with complex network of irregular, anastomosed carinae .... *Octotoma*

— Protibia not flattened, broadest subapically, externally without distinct carina; antenna gradually enlarged distally; elytral punctation and sculpture usually regular, if irregular, then without complex, anastomosed carinae 10

10(9). Antennomere 3 as long as 1 and 2 combined; elytra conjointly emarginate at apex (Fig. 18) .... *Platocthispa*

— Antennomere 3 shorter than 1 and 2 combined; elytra usually conjointly rounded at apex 11

11(10). Vertex trisulcate, median sulcus distinct, lateral sulcus distinct and adjacent to eye; body form narrow .... *Glyphuroplata*

— Vertex with sulci variably developed, lateral sulcus when present distant from eye; body form usually broad 12

12(11). Elytral margin entire; lateral sulci of vertex poorly developed; narrow semitransparent cuticular fringe present along anterior margin of pronotum .... *Microrhopala* (part)

— Elytral margin weakly to strongly serrate; lateral sulci of vertex distinct; anterior margin of pronotum without distinct semitransparent cuticular fringe 13

13(12). Elytral margins gradually and evenly curved at apex (Fig. 15) without apical foliaceous expansion or large spines; elytral costae variably developed, sometimes weak or irregular .... *Microrhopala* (part)

— Elytral margins abruptly curved at apex (Fig. 16), sometimes with apical foliaceous expansion and large spines; elytral costae strong, well defined .... *Pentispa*

14(6). Elytron with 8 rows of punctures 15

— Elytron with 10 rows of punctures, sometimes narrowed to 8 rows at midlength 16

15(14). Body slender (Fig. 19); mesotibia strongly incurved (Fig. 21) .... *Anisostena*

— Body broader; mesotibia straight or nearly so .... *Sumitrosis*

16(14). Clypeus flat, wider than long, surface smooth or micropunctate; elytral costa on interval 6 disrupted at middle third of elytron, extended obliquely to join or nearly join costa of interval 2. .... *Baliosus*

— Clypeus not as above, surface scabrous; costa on interval 6 present or absent, broken at middle or entire, but never oblique 17

17(16). Clypeus longer than broad, elytron slightly narrowed at midlength, elytral punctation between intervals 4 and 8 reduced or confused at midlength .... *Chalepus* (part)

— Clypeus as wide as long, elytron not narrowed at midlength, punctation between elytral intervals 4 and 8 regular and complete 18

18(17). Elytron without short subscutellar row of punctures .... *Chalepus* (part)

— Elytron with short row of subscutellar punctures (at least with one extra puncture) 19

19(18). Elytra conjointly emarginate at apex (Fig. 17); elytral interval 4 flat, interval 2 strongly costate .... *Xenochalepus*

— Elytra conjointly rounded at apex (Fig. 15); elytral interval 4 with moderately developed costa comparable to that on intervals 2 and 6 .... *Odontota*

Tribe Mesomphaliini

20(4). Base of pronotum emarginate before posterolateral angle (Fig. 26); proepisternum next to head with raised flange defining short groove .... *Chelymorpha*

— Base of pronotum straight before posterolateral angle; proepisternum without raised flange next to head .... *Hilarocassis*

Tribe Ischyrosonychini

21(5). Body outline nearly continuous, with base of pronotum nearly as wide as base of elytra .... *Physonota* (subgenus *Physonota*)
— Body outline discontinuous, with base of pronotum distinctly narrower than base of elytra .... *Physonota* (subgenus *Eurypepla*)

Tribe Cassidini

22(5). Anterior margin of elytron crenulate (sometimes faintly) 23
— Anterior margin of elytron smooth 28

23(22). Proepisternum next to head with short groove bordered externally by short carina (Fig. 24) 24
— Proepisternum without groove next to head 25

24(23). Tarsal claws simple but appearing appendiculate due to projected flanks of terminal tarsomere at base of claws; elytral disc punctate-striate, smooth .... *Deloyala*
— Tarsal claws simple and appearing simple, flanks of terminal tarsomere not projected; elytral disc rugosely punctate or tuberculate .... *Parorectis*

25(23). Tarsal claws simple 26
— Tarsal claws appendiculate, each with broad basal tooth 27

26(25). Elytral profile weakly arched, evenly rounded; explanate margins of body opaque; apex of elytral epipleuron essentially glabrous; crenulations of anterior margin of elytron faint .... *Cassida* (subgenus *Cassida*) (part)
— Elytral profile strongly arched, angled; explanate margins of body transparent-reticulate; apex of elytral epipleuron with brush of erect setae; crenulations of anterior margin of elytron distinct .... *Psalidonota*

27(25). Body outline broadly rounded; elytral profile angled; antenna long and slender, usually porrect in repose .... *Plagiometriona*
— Body outline oblong, subparallel; elytral profile evenly rounded; antenna short and thick, usually retracted in repose .... *Gratiana*

28(22). Meso- and metatarsal claws distinctly pectinate, with pectens large and easily observed (Fig. 27) .... *Microctenochira*
— Meso- and metatarsal claws indistinctly pectinate or smooth; pectens, if present, minute and visible only with high magnification 29

29(28). All tarsal claws simple 30
— Protarsal claws appendiculate, claws of meso- and metatarsi variable 34

30(29). Frontoclypeal sulcus obsolete, base of clypeus poorly defined 31
— Frontoclypeal sulcus distinct, clearly delineating base of clypeus (Fig. 28) 32

31(30). Tarsal claws partly hidden between lobes of tarsomere 3 which extend beyond apex of last tarsomere .... *Cassida* (subgenus *Mionycha*)
— Tarsal claws protruding well beyond lobes of tarsomere 3 which do not reach apex of last tarsomere ..... *Cassida* (subgenus *Cassida*) (part)

32(30). Elytral margins at apices projected caudad, shelf-like; epipleura at apex horizontal, moderately wide; elytral margins impunctate 33
— Elytral margins at apices deflexed; epipleura at apex deflexed, narrow; elytral margins distinctly punctate .... *Floridocassis*

33(32). Edge of elytral margin thickened and weakly reflexed; mesotarsal claws of male distinctly asymmetrical, with outer claw enlarged; body usually larger, length 5.7-9.1 mm .... *Jonthonota*
— Edge of elytral margin not especially thickened nor reflexed; mesotarsal claws of male subequal; body smaller, length 5.4-6.7 mm .... *Opacinota*

34(29). Clypeus inflated, extended well beyond level of eyes; elytra vittate (Fig. 23); all claws of male and female with basal tooth .... *Agroiconota*
— Clypeus flat, level with eyes; elytra without vittae; male with at least outer claw of mesotarsus simple 35

35(34). Elytral margins at apices projected caudad, shelf-like; epipleura at apex horizontal, moderately wide 36
— Elytral margins at apices deflexed; epipleura at apex deflexed, narrow 37

36(35). Longitudinal frontal groove extended onto clypeus as distinct median groove; lateral grooves of clypeus distinct; elytral margin impunctate .... *Strongylocassis*
— Longitudinal frontal groove not extended onto clypeus; lateral grooves of clypeus indistinct; elytral margin deeply, irregularly punctate .... *Erepsocassis*

37(35). Clypeus slightly swollen with distinct median groove; lateral grooves of clypeus distinct and evenly impressed to base (Fig. 28) .. *Metrionella*
— Clypeus flat or depressed, without distinct median groove; lateral grooves of clypeus evident between base of mandible and eye, not extended to base 38

38(37). Coloration of living specimens always non-metallic; ventral surface of body black; trochanters pale; ventral surface of elytral disc immaculate .... *Charidotella* (subgenus *Chaerocassis*)
— Living specimens capable of attaining bright metallic coloration; venter of body pale or if largely black, then trochanters black, or if venter black and trochanters pale, then ventral surface of elytral disc with dark maculae .... *Charidotella* (subgenus *Charidotella*)

V. Key to Tribes, Subtribes, Genera, and Subgenera of Chrysomelinae of America North of Mexico

1. Procoxal cavities closed behind 2
— Procoxal cavities open behind 4

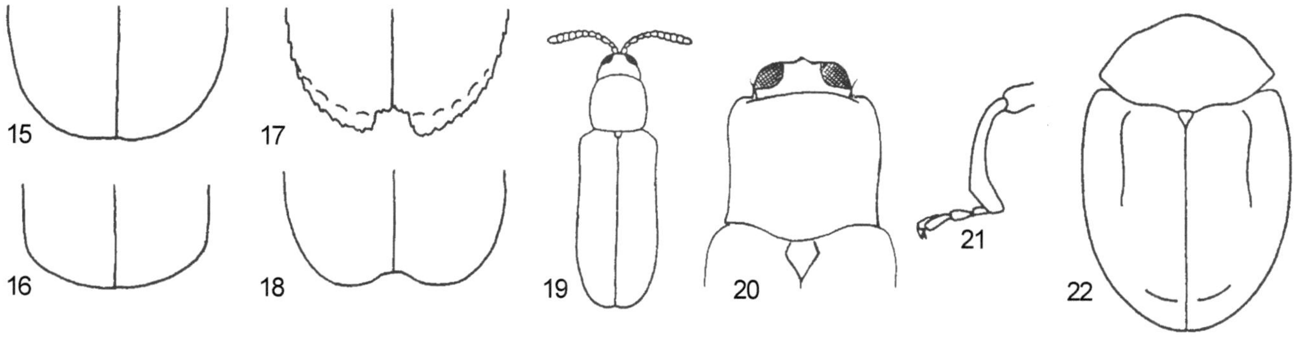

FIGURES 15.124-22.124. 15. *Microrhopala vittata* (Fabricius), apex of elytra; 16. *Pentispa suturalis* (Baly), apex of elytra; 17. *Xenochalepus* (*s. str.*) *omorgus* (Crotch), apex of elytra; 18. *Platocthispa lateritia* (J. Smith), apex of elytra; 19. *Anisostena nigrita* (Olivier); 20. *Stenispa metallica* (Fabricius), forebody, dorsal view; 21. *Anisostena* sp., mesotibia; 22. *Cassida* (*s. str.*) *rubiginosa* Müller, dorsal view.

2(1). Metasternum reduced, with mesal length less than that of prosternum; aedeagus with tegmen ring-like, encircling median lobe; species flightless, dark colored, tenebrionid-like (Fig. 29) (Tribe Timarchini) .... *Timarcha*

— Metasternum normal, longer than prosternum; aedeagal tegmen Y-shaped, not encircling median lobe (Tribe Chrysomelini) 3

3(2). Tarsomeres 1, 2, and especially 3 broad, with well-developed setose pads ventrally (Figs. 36, 37); procoxae separated by prosternum; pronotum glabrous except for single seta at each corner, without specialized pore near anterolateral angle (Subtribe Entomoscelina) 7

— All tarsomeres nearly cylindrical; tarsal pads absent or poorly developed; procoxae contiguous; pronotum sparsely pubescent throughout, with small, specialized, asetose pore near anterolateral corner (Chrysomelini: incertae sedis) .... *Cadiz*

4(1). Tibia armed with preapical tooth followed by coarsely setose emargination (Fig. 34), or meso- and metatibiae with weak preapical emargination with brush of setae (Fig. 35) (Subtribe Paropsina) 6

— Tibia sometimes angularly broadened at apex but without preapical tooth and emargination, or distinct brush of setae 5

5(4). Epipleuron apically setose; maxilla with terminal palpomere broadly truncate at tip, sub-quadrangular or distally dilated (Figs. 38-40); third tarsomere apically truncate or slightly emarginate (Subtribe Doryphorina) 8

— Epipleuron without apical setae; maxilla with terminal palpomere apically attenuate (Fig. 41); third tarsomere variable but often deeply bilobed (Subtribe Chrysomelina) 21

Subtribe Paropsina

6(3). Epipleuron slanted upwards from outer edge, not visible in lateral view; meso- and metatibiae with weak emargination and brush of setae (Fig. 35); each tarsal claw with small basal tooth; elytral punctation dense and confused; dorsum pale with irregular dark reticulation .... *Trachymela*

— Epipleuron slanted downwards from outer edge, visible in lateral view; tibia armed with preapical tooth followed by coarsely setose emargination (Fig. 34); each tarsal claw with broad basal tooth (Fig. 32); elytral punctation fine, with major punctures aligned in rows; dorsum usually pale, often with dark spots .... *Gonioctena*

Subtribe Entomoscelina

7(3). Elytral punctation confused; pronotum black with pale lateral margins; elytra red with suture and

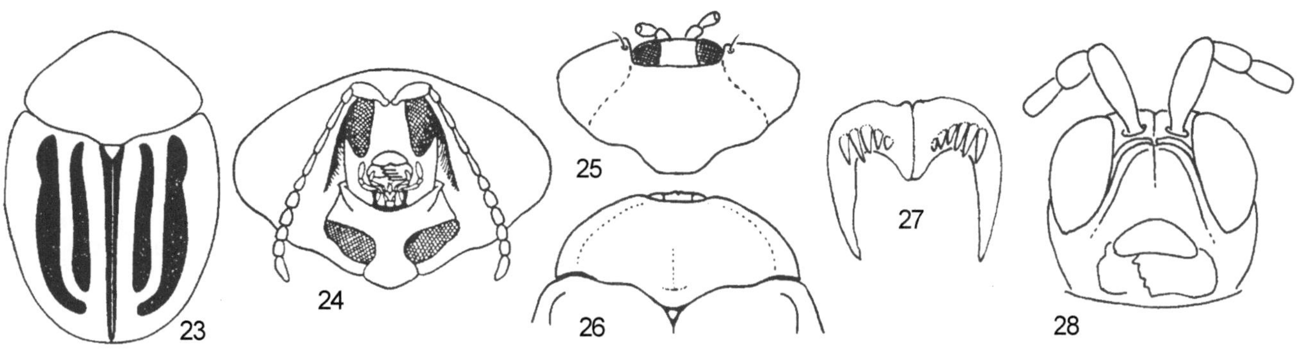

FIGURES 23.124-28.124. 23. *Agroiconota bivittata* (Say), dorsal view; 24. *Deloyala guttata* (Olivier), prothorax and head, ventral view; 25. *Hemisphaerota cyanea* (Say), head and pronotum, dorsal view; 26. *Chelymorpha* sp., forebody, dorsal view; 27. *Microctenochira bonvouloiri* (Boheman), mesotarsal claws; 28. *Metrionella bilimeki* (Spaeth), head, ventral view.

discal vitta black; length 6.5-10 mm .... *Entomoscelis*

— Punctures of each elytron arranged in four regular rows; pronotum entirely dark; elytra entirely dark, or predominantly dark with lateral margins, base, and basal portion of suture yellow; length 4.1-6.0 mm .... *Microtheca*

Subtribe Doryphorina

8(5). Tarsal claws connate, contiguous for much of length (Fig. 14); elytra bicolored, patterned with spots or vittae .... *Zygogramma*

— Tarsal claws separated and usually divergent from base to apex (Fig. 13); elytra variable, sometimes unicolorous 9

9(8). Terminal palpomere of maxilla shorter than preceding, parallel-sided or slightly narrowed apically (Figs. 38, 39) 10

— Terminal palpomere of maxilla at least as long as preceding, often dilated apically (Fig. 40) 11

10(9). Front femur strongly toothed in male (Fig. 31), normal in female; mesosternum between middle coxae forming blunt tubercle raised above level of prosternum .... *Labidomera*

— Front femur of male and female normal, without tooth; mesosternum not raised above level of prosternum .... *Leptinotarsa*

11(9). Pronotum not thickened at sides, without distinct longitudinal impressions; elytral disc partially pale, patterned with spots or vittae 12

— Sides of pronotum thickened, the thickened portion usually separated from disc by longitudinal impression; elytra entirely dark, entirely pale, or dark with pale lateral margins 15

12(11). Elytra pale with two or three dark brown to black vittae; sutural vitta with sides nearly straight; lateral vitta sometimes interrupted or deeply incised slightly behind middle; elytra without small, discal spots 13

— Elytral vittae, when present, irregular in form, pale brown, or accompanied by numerous spots . 14

13(12). Elytra with short subscutellar row of punctures in addition to long striate rows .... *Calligrapha* (subgenus *Bidensomela*)

— Elytra without short subscutellar row of punctures .... *Calligrapha* (subgenus *Calligramma*)

14(12). Elytra yellow with pale brown vittae; isolated elytral spots absent or few in number .... *Calligrapha* (subgenus *Graphicallo*)

— Elytral pattern including numerous isolated spots; darker elytral markings frequently black or dark brown, sometimes with metallic luster .... *Calligrapha* (subgenus *Calligrapha*)

15(11). Elytral punctures usually small, similar in size to those of pronotal disc, arranged in striate rows, sometimes obsolete; interstriae usually convex or variably elevated into ridges; flightless species occurring in treeless areas of Alaska and northwestern Canada 16

— Elytral punctures much larger than those of pronotal disc; elytra not costate 17

16(15). Apical abdominal ventrite transversely impressed and declivous before apex, very strongly so in males, not strongly but distinctly so in females; median lobe of aedeagus with large, heavily sclerotized, triangular area dorsally on each side near apex; legs and apical half of last abdominal ventrite usually reddish-yellow or reddish-brown, sometimes blackish .... *Chrysolina* (subgenus *Pleurosticha*)

— Apical abdominal ventrite with a strongly impressed marginal line but not declivous before apex, flattened or feebly depressed at middle in male, unmodified in female; aedeagus without such sclerotized areas; legs and abdomen entirely blackish .... *Chrysolina* (subgenus *Arctolina*)

17(15). Pronotum not sulcate although usually with feeble impression on each side at base; elytra strongly metallic, never with epipleura or lateral margins pale; elytral punctures not at all seriate .... *Chrysolina* (subgenus *Sphaeromela*)

— Pronotum with distinct sulcus on each side, at least basally, except in some specimens with blackish or bicolored elytra; elytral punctures often subseriate in part 18

18(17). Coarse punctures of each elytron arranged in subsutural row and four pairs of irregular rows; elytra metallic, without pale lateral margins .... *Chrysolina* (subgenus *Hypericia*)

— Elytral punctures rarely arranged in paired rows; elytral color variable 19

19(18). Pronotum with sulci feebly to deeply impressed basally, feebly impressed or obsolete at middle and apically; elytra entirely dark, or with pale lateral margins .... *Chrysolina* (subgenus *Chalcoidea*)

— Pronotum with submarginal sulci at least moderately impressed from base to apex; elytral color variable 20

20(19). Elytra entirely reddish-brown; species occurring from Newfoundland to Nova Scotia .... *Chrysolina* (subgenus *Chrysolina*)

— Elytral color uniformly dark, often with metallic luster; species occurring throughout much of North America, but absent from eastern Canada .... *Chrysolina* (subgenus *Pezocrosita*)

Subtribe Chrysomelina

21(5). Tarsal claws appendiculate, each with basal tooth (Fig. 32); dorsum entirely dark, metallic .... *Phratora*

— Tarsal claws simple; dorsum variable, sometimes bicolored 22

22(21). Larger elytral punctures essentially arranged in nine well-separated striae plus short subscutellar row, although some rows sometimes irregular or confused; striate punctures rarely only slightly larger than interstrial punctures and therefore distinguished only upon careful examination; head with coronal suture usually not discernible.......... 23

— Elytral punctures completely confused or arranged in very poorly defined rows; if substriate, rows closely spaced and therefore numbering more than nine; coronal suture visible, usually indicated by shallow mesal impression 26

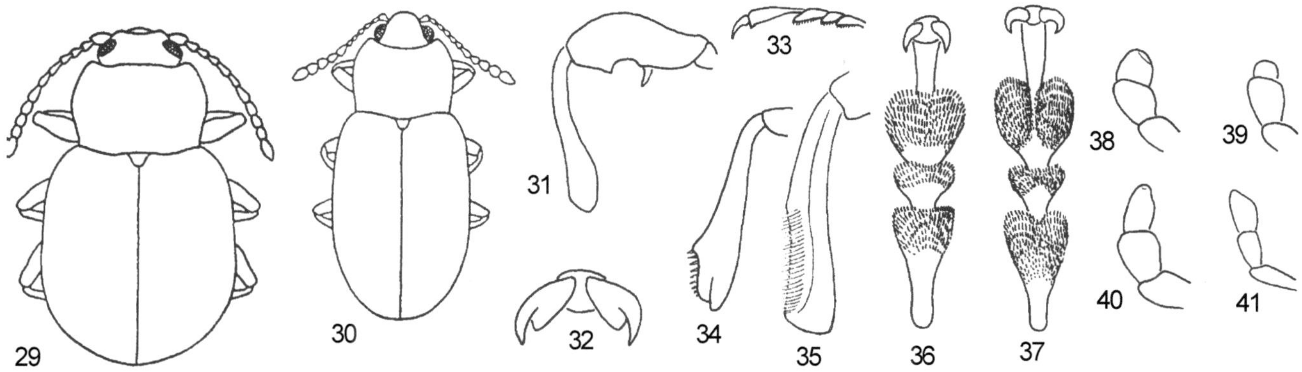

FIGURES 29.124-41.124. 29. *Timarcha intricata* Haldeman; 30. *Prasocuris* (*Hydrothassa*) sp.; 31. *Labidomera clivicollis* (Kirby), anterior leg of male; 32. *Phratora* sp., tarsal claws; 33. *Plagiodera* (*Plagiomorpha*) *arizonae* Crotch, tarsus, lateral view; 34. *Gonioctena* sp.; 35. *Trachymela sloanei* (Blackburn), mesotibia; 36. *Leptinotarsa peninsularis* (Horn), metatarsus, ventral view; 37. *Chrysomela crotchi* Brown, metatarsus, ventral view; 38. *Leptinotarsa collinsi* Wilcox, maxillary palp; 39. *Leptinotarsa decemlineata* (Say), maxillary palp; 40. *Calligrapha rowena* Knab, maxillary palp; 41. *Gastrophysa dissimilis* (Say), maxillary palp. (Figs. 31, 32, and 36-41 redrawn from Wilcox 1972).

23(22). Dorsum bicolored; lateral margins of pronotum, lateral margins of elytra, and narrow discal vitta on each elytron yellow 24

— Dorsum entirely dark, often with metallic luster 25

24(23). Basal margin of pronotum with fine elevated bead; elytra with sutural dark vitta not strongly widened around scutellum; body very elongate, with length at least 2.5 times width .... *Prasocuris* (subgenus *Prasocuris*)

— Basal margin of pronotum without bead; elytra with sutural dark vitta abruptly widened around scutellum; body broader, with length less than 2.5 times width (Fig. 30) .... *Prasocuris* (subgenus *Hydrothassa*)

25(23). Elytra with interstrial punctures much smaller than strial punctures ... *Phaedon* (subgenus *Phaedon*)

— Elytra with interstrial punctures nearly equal in size to strial punctures, distinguishable only upon close examination .... *Phaedon* (subgenus *Allophaedon*)

26(22). Pronotum with distinct longitudinal impression on each side; elytral disc usually bicolored or entirely pale, but sometimes entirely dark with weak metallic luster 27

— Pronotum without distinct longitudinal impressions; elytra usually entirely dark with strong metallic luster, but sometimes brown with pale lateral margins 28

27(26). Elytra immaculate, pale brown or brownish yellow; sides of pronotum not or scarcely paler than disc; apex of aedeagus angulate on each side, not attenuate or evenly rounded .... *Chrysomela* (subgenus *Chrysomela*)

— Elytra bicolored, entirely dark, or entirely pale; when elytra entirely pale, pronotum laterally pale also; apex of aedeagus broadly rounded or somewhat attenuate, lacking lateral angles 28

28(27). Elytron without traces of striae except for stria delimiting external ridge, punctures not arranged in series except near suture; elytra maculate in most .... *Chrysomela* (subgenus *Macrolina*)

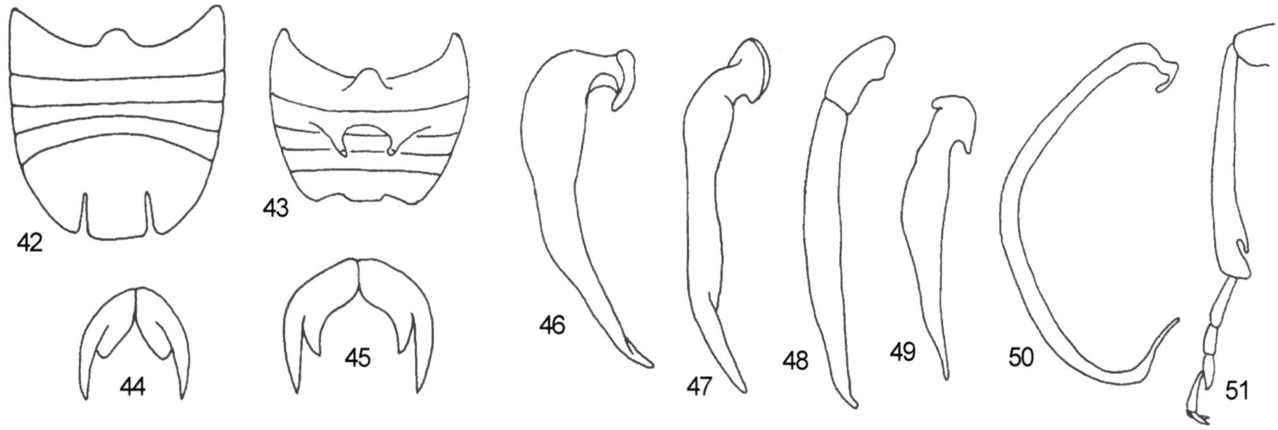

FIGURES 42.124-51.124. 42. *Eusattodera delta* Wilcox, male ventrites; 43. *Pseudoluperus tuberculatus* (Blake), male ventrites; 44. *Lygistus streptocephallus* Wilcox, tarsal claws; 45. *Triarius melanolomatus* (Blake), tarsal claws; 46. *Galeruca rudis* LeConte, median lobe of male genitalia, lateral view; 47. *Miraces aeneipennis* Jacoby, median lobe of male genitalia, lateral view; 48. *Pseudoluperus cyanellus* (Horn), median lobe of male genitalia lateral view; 49. *Pseudoluperus fulgidus* Wilcox, median lobe of male genitalia, lateral view; 50. *Erynephala maritima* (LeConte), median lobe of male genitalia, lateral view; 51. *Phyllecthris dorsalis* (Olivier), male mesotibia. (Figs. 42-50 redrawn from Wilcox 1965).

— Elytron distinctly substriate, most with discal punctures substriate in part (in a few confused as in *Macrolina*) and with third, fifth, and seventh intervals largely impunctate and weakly convex; elytral entirely blackish; arctic species .... *Chrysomela* (subgenus *Pachylina*)

29(27). Tibiae, especially of middle and hind legs, angularly expanded at apex; prosternum, between coxae, narrower than third antennomere; body strongly convex, hemispherical in cross section; sides of elytra nearly parallel .... *Gastrophysa*

— Tibiae not abnormally expanded at apex; prosternum, between coxae, much broader than third antennomere; body less convex, not hemispherical in cross section; sides of elytra strongly curved 30

30(29). Apical tarsomere without ventral tooth near base of each claw; elytral punctures comparatively sparse and fine, separated by much more than twice diameter of puncture .... *Plagiodera* (subgenus *Plagioschema*)

— Apical tarsomere with acute ventral tooth near base of each claw (Fig. 33); elytral punctures coarser, with most punctures separated by not more than twice their diameter 31

31(30). Ventral lobe of tarsomere 3 apically cleft; elytra with epipleuron horizontal, scarcely visible in lateral view; pronotal punctation uniform, not noticeably coarser in lateral areas than in mesal area .... *Plagiodera* (subgenus *Plagiodera*)

— Ventral lobe of tarsomere 3 truncate or feebly emarginate apically; elytra with epipleuron slanted downward from outer edge, clearly visible in lateral view; pronotum with lateral punctation coarser than mesal punctation .... *Plagiodera* (subgenus *Plagiomorpha*)

VI. Key to Tribes, Subtribes, and Sections of Galerucinae of America North of Mexico

Much of the classification within the Galerucinae is based on male characters, especially those found on the aedeagus. This is reflected in the following key to tribes, subtribes, and sections. However, with the intent of enabling identification of females and undissected males, an independent key to genera is provided (Key VII). It incorporates more superficial characters (color, etc.) that are easily visible in either sex. These characters are not the basis for classification, nor will they always allow for the proper determination of beetles from extralimital areas. But they will enable generic identification of species occurring north of Mexico. Thus, identification to genus can be accomplished without first determining the tribe, subtribe, and section.

1. Hind femur usually adapted for jumping, broad, with internal extensor apodeme (spring) (Fig. 54); if hind femora slender, then prosternum comparatively broad, forming small horizontally flattened area between front coxae; pronotum variable, but often with transverse prebasal groove in basal fourth (this groove not to be confused with fine line delimiting basal bead); inner wall of epipleuron usually with two elytron-to-body binding patches .... Tribe Alticini

— Hind femur not abnormally broad, without sclerotized internal extensor apodeme (spring); front coxae contiguous, or narrowly separated by keel-like prosternum; pronotum sometimes shallowly impressed near base, but never with well-defined, transverse groove in basal fourth; inner wall of epipleuron with single elytron-to-body binding patch 2

2(1). Median lobe of aedeagus with prominent basal spurs (Fig. 46) or with constriction slightly beyond base (Fig. 47); last ventrite of male abdomen without apical lobe; antennae usually inserted low on frons, beyond middle of eyes; larvae, where known, feeding on leaves 3

— Basal spurs of aedeagus small or absent (Fig. 48); if small spurs present, aedeagus not strongly constricted slightly beyond base (Fig. 49) male abdomen variable, sometimes with lobe at apex; antennae usually inserted higher, near middle of eyes; larvae, where known, feeding on roots (Tribe Luperini) 4

3(2). Last ventrite of male abdomen with median, apical, semicircular depression; abdominal apex sometimes emarginate behind impression; tarsal claws of Nearctic species either simple or bifid with narrow, sharply pointed appendage (Fig. 45); anterior and posterior tibiae usually without terminal spurs; larvae feeding on leaves .... Tribe Galerucini: Sections Atysites, Coelomerites, Galerucites, Schematizites

— Last ventrite of male abdomen without distinct impression, although sometimes flattened; tibiae usually without terminal spurs; tarsal claws of Nearctic species appendiculate, with broad, blunt lobe (Fig. 44); larvae unknown .... Tribe Metacyclini

4(2). Last ventrite of male abdomen with rectangular lobe (Figs. 42-43) (Subtribe Luperina) 5

— Last ventrite of male apically rounded or slightly truncate, without lobe (Subtribe Diabroticina). 7

5(4). Elytral epipleura extremely narrow, indistinct .... Section Phyllobroticites

— Elytral epipleura normal, well defined 6

6(5). Tarsomere 1 of hind leg distinctly longer than 2 and 3 combined; apical lobe of last ventrite of male abdomen large, nearly square (Fig. 42); aedeagal orifice covered by sclerotized plate . ... Section Monoleptites

— Tarsomere 1 of hind leg usually shorter than 2 and 3 combined; apical lobe of male abdomen usually much wider than long; aedeagal orifice variable, but usually without sclerotized covering .... Sections Exosomites, Scelidites

7(4). Mesotibia of male with deep, inner, subapical notch (Fig. 51) .... Section Phyllecthrites

— Mesotibia of male without subapical notch 8

8(7). Tarsal claws bifid, with narrow, sharply pointed inner lobe (Fig. 45) .... Section Diabroticites

— Tarsal claws appendiculate, with comparatively broad, blunt inner lobe (Fig. 44) 9

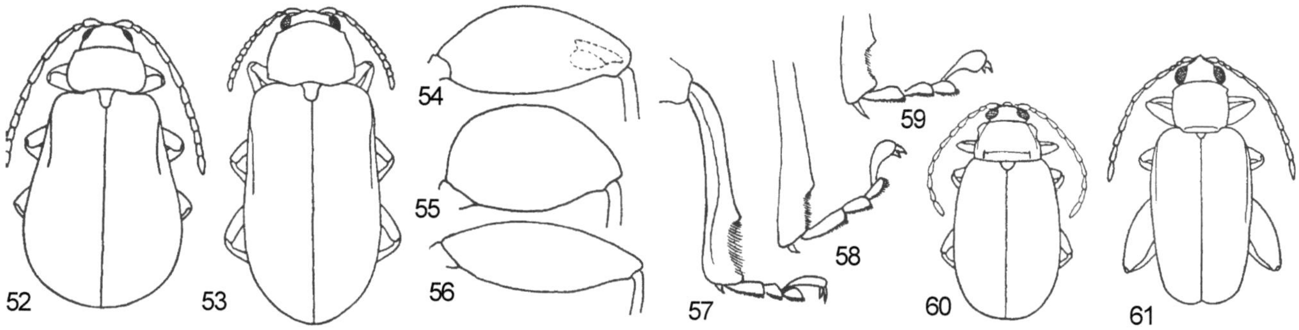

FIGURES 52.124-61.124. 52. *Monocesta coryli* (Say); 53. *Erynephala maritima* (LeConte); 54. *Asphaera lustrans* (Crotch), metafemora; 55. *Kuschelina gibbitarsa* (Say), metafemora; 56. *Luperaltica nigripalpis* (LeConte), metafemora; 57. *Blepharida rhois* (Forster), metatibia and tarsus; 58. *Asphaera lustrans* (Crotch), metatibia and tarsus; 59. *Kuschelina gibbitarsa* (Say), metatibia and tarsus; 60. *Parchicola tibialis* (Olivier); 61. *Glenidion flexicaulis* (Schaeffer).

9(8). Elytra entirely dark, with distinct transverse impression near basal third ... Section Trachyscelidites
— Elytra without transverse impression across basal third, those of Nearctic species yellow, orange, or red, with dark markings .... Section Cerotomites

VII. Key to Genera and Subgenera of Galerucinae of America North of Mexico

[Note: Key also includes one genus from the Baja California peninsula which is not recorded from America north of Mexico]

1. Hind femur usually adapted for jumping, broad, with internal, sclerotized extensor apodeme (spring) (Fig. 54); if hind femora slender, then prosternum comparatively broad, forming small, horizontally flattened area between front coxae; pronotum variable, but often with transverse prebasal groove in basal fourth (this groove not to be confused with fine line delimiting basal bead); inner wall of epipleuron usually with two elytron-to-body binding patches (Tribe Alticini) 58
— Hind femur not abnormally broad, without sclerotized internal extensor apodeme (spring); front coxae contiguous, or very narrowly separated by keel-like prosternum; pronotum sometimes shallowly impressed near base, but never with well-defined, transverse groove in basal fourth; inner wall of epipleuron with single elytron-to-body binding patch 2

2(1). Tarsal claws simple, or bifid with inner lobe comparatively narrow and pointed (Fig. 45) 3
— Tarsal claws appendiculate, with inner lobe comparatively broad and blunt (Fig. 44) 31

3(2). Procoxal cavities closed behind 4
— Procoxal cavities open behind 6

4(3). Tarsomere 1 about as long as 5; external surface of tibia with ridge extending most of length 5
— Tarsomere 1 about half as long as 5; tibia without ridge .... *Diorhabda*

5(4). Elytra densely pubescent; elytral disc brown with narrow, pale discal vittae .... *Neolochmaea*
— Elytral pubescence sparse, noticeable only upon close examination; elytra often with pale lateral margins, but without pale discal vittae .... *Galeruca*

6(3). Metatibia usually without apical spur; if apical spur present, elytra densely and finely pubescent; pronotum at base nearly as wide as elytra; median lobe of aedeagus with prominent basal spurs; antennae usually inserted low on frons, beyond middle of eyes 7
— Metatibia usually with apical spur; elytra glabrous or with a few scattered, erect hairs; elytra distinctly wider than pronotum at base; median lobe of aedeagus without prominent basal spurs; antennae usually inserted higher on frons, near middle of eyes 24

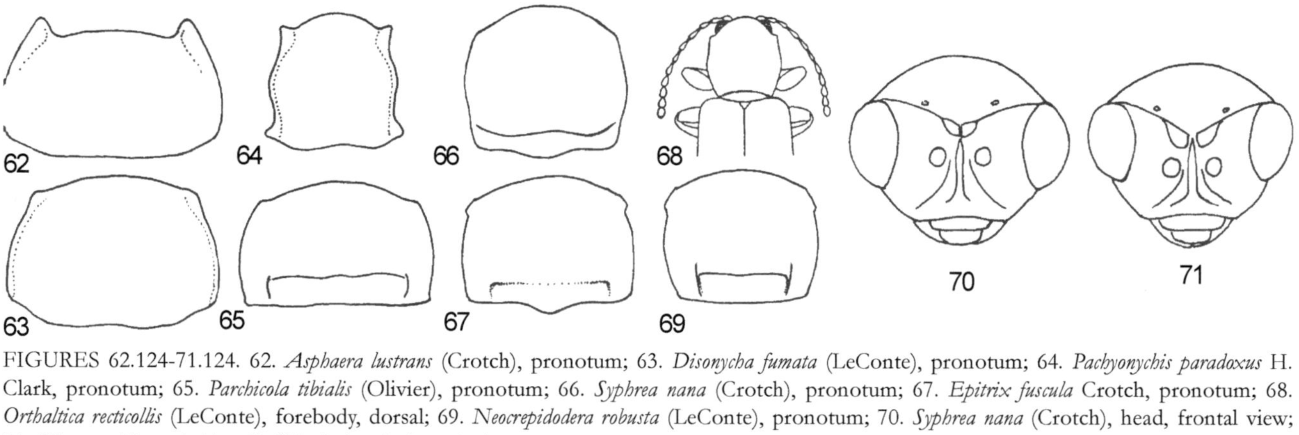

FIGURES 62.124-71.124. 62. *Asphaera lustrans* (Crotch), pronotum; 63. *Disonycha fumata* (LeConte), pronotum; 64. *Pachyonychis paradoxus* H. Clark, pronotum; 65. *Parchicola tibialis* (Olivier), pronotum; 66. *Syphrea nana* (Crotch), pronotum; 67. *Epitrix fuscula* Crotch, pronotum; 68. *Orthaltica recticollis* (LeConte), forebody, dorsal; 69. *Neocrepidodera robusta* (LeConte), pronotum; 70. *Syphrea nana* (Crotch), head, frontal view; 71. *Nesaecrepida asphaltina* (Suffrian), head, frontal view.

7(6). Size large, at least 13 mm long; elytra often with dark blotches, but without longitudinal vittae (Fig. 52) .... *Monocesta*
— Length less than 12 mm, or elytra with longitudinal vittae 8

8(7). Size small, less than 4 mm in length; elytra glabrous or with only a few scattered hairs; color black or dark metallic blue, green, or bronze; all tibiae without apical spurs .... *Miraces*
— Size larger, or combination of characters other than above; mesotibia of male often with broad, curved apical spur 9

9(8). Tarsal claws simple; female 10
— Tarsal claws bifid (Fig. 45); male or female 11

10(9). Length 6-8 mm .... *Erynephala*, females
— Length less than 6 mm ...... *Monoxia*, females (part)

11(9). Median lobe of aedeagus long, flat, C-shaped with apex strongly deflexed (Fig. 50); pronotal punctation conspicuously coarser than that of elytra (Fig. 53) .... *Erynephala*, males
— Median lobe of aedeagus not as above; pronotal punctation not conspicuously coarser than that of elytra 12

12(11). Pronotum very short, broad, with width equal to 2.5 times length at middle; median lobe of aedeagus short, apically truncate .... *Derospidea*
— Pronotum longer in proportion to width; apex of aedeagus pointed or rounded, not truncate . 13

13(12). Elytra unicolorous reddish brown, usually with faint but distinct green luster; antennae long, with antennomeres 3 and 4 equal in length; length 7.5 to 8 mm .... *Coraia*
— Elytra not colored as above, if with metallic luster then also with pale margin 14

14(13). Base of pronotum strongly sinuate; head and elytra black; pronotum red with median black spot; elytra glabrous .... *Ophraea*
— Base of pronotum not strongly sinuate; if elytra entirely black, then also densely pubescent 15

15(14). Antennomere 3 shorter than 4 .... *Trirhabda*
— Antennomere 3 longer than 4 16

16(15). Elytra testaceous, sometimes spotted or mottled with black; antennae short, not reaching beyond base of elytra; abdomen of male usually with deflexed pygidium .... *Monoxia* (part)
— Color variable, rarely mottled or spotted; antennae usually reaching nearly to middle of elytra; male abdomen without deflexed pygidium 17

17(16). Front coxae narrowly but distinctly separated by prosternum; middle coxae separated by distance subequal to half coxal width; pronotum polished and nearly impunctate, except in depressions; all tibiae lacking apical spurs in both male and female .... *Galerucella*
— Front coxae not separated by prosternum; middle coxae closely approximate but rarely in actual contact; male with broad, often curved, apical spur on middle tibia 18

18(17). Apical spurs lacking on hind tibiae of male; elytral pubescence even, not accentuating color pattern 19
— Male with apical spurs on middle and hind tibiae, also with tubercle on ventral surface at base of tarsomere 1 of front leg; median lobe of aedeagus symmetrical, strongly deflexed in apical third, tapering to acute point; elytra dark testaceous with two slightly darker vittae, one submarginal and one running longitudinally along middle of each elytron, also with three faint subsutural spots; surface of elytra densely pubescent, pale areas accentuated by arrangement of pale pubescence .... *Brucita*

19(18). Elytra with distinct dark vittae 20
— Elytra without distinct dark vittae 21

20(19). Elytra pale yellow (darker in overwintering form) with broad black vitta extending from humerus nearly to apex, also usually with short subscutellar vitta; elytra without darkened suture or subsutural vitta; median lobe of aedeagus short, thick, with apex asymmetrical and having point to right side .... *Xanthogaleruca*
— Elytra pale, each with subsutural vitta and at least two other vittae; median lobe of aedeagus long, slender, slightly curved, symmetrical .... *Ophraella* (part)

21(19). Outer margin of epipleuron (marginal bead between disc of elytron and epipleuron) sometimes becoming obscure near apex, but inner margin (next to body) always distinct to apex or to point where it joins outer margin; antennomere 4 distinctly longer than 2 22
— Inner margin of elytral epipleuron remaining distant from outer margin and ending rather abruptly before apex; antennomeres 2 and 4 about equal in length .... *Neogalerucella*

22(21). Elytral punctation relatively fine, with punctures similar in size to those of abdomen; pronotum pale with three dark markings, one mesal and one at each side lateral to depressed area .... *Pyrrhalta*
— Elytral punctures much larger than those of abdomen; dark pronotal markings sometimes present on sublateral tubercles or in sublateral depressions, but not lateral to depressions 23

23(22). Median lobe of aedeagus symmetrical; aedeagal orifice located very near apex, small, without weakly sclerotized area above it; body oval, usually strongly convex; hosts Asteraceae ... . *Ophraella* (part)
— Median lobe of aedeagus strongly asymmetrical, with apex curved to one side; aedeagal orifice large, with weakly sclerotized area above it; body usually more oblong, not as convex; hosts other than Asteraceae .... *Trichochmaea*

24(6). Elytra dark metallic blue or green, except for narrow pale lateral and apical margins; male with antennomeres 5, 6, and 7 modified, twice as wide as 4 or 8; elytra of male with large depression across suture at apical fourth .... *Paratriarius*
— Elytral color other than described above; male with antennae and elytra not modified 25

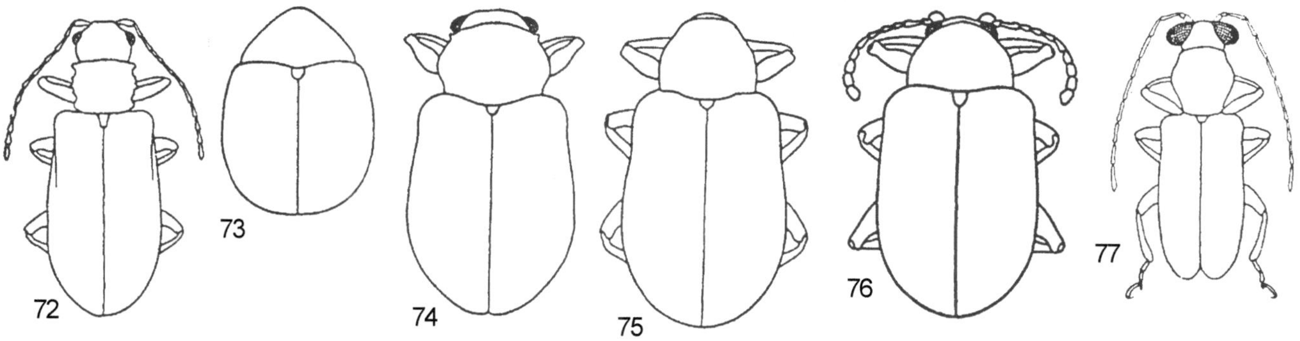

FIGURES 72.124-77.124. 72. *Syneta ferruginea* (Germar); 73. *Spintherophyta globosa* (Olivier); 74. *Colaspidea smaragdula* (LeConte); 75. *Glyptoscelis squamulata* Crotch; 76. *Eumolpus robustus* (Horn); 77. *Megascelis texana* Linell.

25(24). Last ventrite of male abdomen with prominent, rectangular, median apical lobe; apical palpomere of maxilla as long as, and usually nearly as wide as, penultimate palpomere; body more elongate, with sides more parallel .... *Triarius* (part)

— Last ventrite of male truncate or vaguely emarginate, without median apical lobe; maxilla with apical palpomere usually shorter, more slender at base than penultimate palpomere; body more oval, usually wider behind middle of elytra 26

26(25). Pronotum with two rather small, deep, circular impressions, one on each side of middle, separated from each other by distance greater than diameter of impression; pronotum pale with small black spot on each side; each elytron pale with suture and discal vitta black .... *Amphelasma*

— Pronotum evenly convex or with larger, shallow depressions; depressions, when present, usually separated by a distance less than their diameters; pronotum unicolorous pale or dark, without small spots 27

27(26). Antennomere 3 small, not more than 1.5 times as long as 2; antennomeres 2 and 3 together not as long as 4 28

— Antennomere 3 large, at least twice as long as 2, about as long as 4 29

28(27). Elytron with two or more distinct sinuate plicae, extending from behind humerus to beyond middle; pronotum deeply bifoveate .... *Diabrotica* (*virgifera* group)

— Elytron without plicae in post-humeral area; pronotum weakly bifoveate .... *Diabrotica* (*fucata* group)

29(27). Lateral margin of prothorax with six or more fine, evenly spaced setae (lost in rubbed specimens and often hard to see at low magnifications); elytra with erect or suberect discal hairs arranged in longitudinal rows; elytral color variable, but often with longitudinal vittae 30

— Lateral margin of prothorax with only large setae on anterior and posterior angles, or with one or two small setae placed near the large setae; elytra largely glabrous, sometimes with scattered hairs on margins or in apical third; elytra without longitudinal vittae .... *Paranapiacaba*

30(29). Elytra vittate and at least vaguely costate .... *Acalymma* (*gouldi* group)

— Elytra neither vittate nor costate .... *Acalymma* (*peregrinum* group)

31(2). Procoxal cavities closed behind 32

— Procoxal cavities open behind 34

32(31). Antennomere 3 distinctly longer than 4, both enlarged and otherwise modified in some males; elytra red or testaceous with black markings, without metallic luster .... *Cerotoma*

— Antennomere 4 distinctly longer than 3 33

33(32). Head and elytra entirely dark, sometimes with blue luster .... *Eusattodera* (part)

— Head or elytra at least partially pale; blue luster absent .... *Metrioidea* (part)

34(31). Elytron with epipleural fold well defined........... 35

— Elytral epipleura extremely narrow, indistinct .... *Phyllobrotica*

35(34). Prothorax with lateral marginal bead; male clypeus without unusual hole 36

— Prothorax without lateral marginal bead; male with deep hole in middle of clypeus, surrounded by funnel-shaped rim; head and pronotum yellow; elytra black with yellow suture and margins; sutural yellow vitta not reaching apex but bending away from suture on each side just before apex .... *Cyclotrypema*

36(35). Middle tibia with deep, inner, subapical notch (Fig. 51); male 37

— Middle tibia without subapical notch; male or female 38

37(36). Each antenna with only ten antennomeres; antennomere 3 nearly four times as long as 2... .... *Phyllecthris*, males

— Each antenna with eleven antennomeres; antennomere 3 only slightly longer than 2........ .... *Luperosoma*, males

38(36). Abdomen with prominent ventral appendages (Fig. 43); male 39

— Abdomen without ventral appendages; male or female 41

39(38). Elytron with distinct impression at posterolateral angle; distal portion of epipleuron bent upward at point of impression .... *Androlyperus*, males (part)

— Elytra lacking impression at posterolateral angle.. 40

40(39). Upper surface shining testaceous; elytra with suture and narrow lateral vitta dark .... *Pseudoluperus*, males (part)

— Elytra bright metallic blue or green; head and pronotum testaceous .... *Scelida*, males (part)

41(38). Tarsomere 1 of hind leg distinctly longer than 2 and 3 combined; antennomere 3 about as long as 2; antennomere 4 usually as long as 2 and 3 combined; last ventrite of male abdomen with large, nearly square apical lobe 42

— Tarsomere 1 of hind leg shorter than 2 and 3 combined; apical lobe of male abdomen, when present, usually much wider than long 43

42(41). Head and elytra entirely dark, sometimes with blue luster .... *Eusattodera* (part)

— Head or elytra at least partially pale; blue luster absent .... *Metrioidea* (part)

43(41). Body broadly oval, with length less than twice width; male without distinct lobe at abdominal apex; color entirely testaceous, except for black eyes, black elytra, and brown mandibles .... *Trachyscelida*

— Body more elongate, with length at least twice width across widest part of elytra 44

44(43). Antennae short, reaching only slightly beyond humerus; antennomeres 2, 3, and 4 about equal in length and width; body length 5-6 mm ... *Pteleon*

— Either antennae longer, reaching at least to middle of elytra, or body much smaller 45

45(44). Basal margin of pronotum with fine bead extending without interruption between posterior angles (this character often observable only under close examination, sometimes obscured by extremely coarse pronotal punctation) 46

— Basal bead of pronotum absent, or restricted to extreme lateral areas; pronotal punctation not extremely coarse 52

46(45). Abdomen of female greatly inflated, mostly not covered by elytra; eyes small, separated from each other by more than twice their maximum diameter .... *Metacycla*

— Abdomen of female not abnormally enlarged, mostly or entirely covered by elytra; eyes larger, separated from each other by less than twice their maximum diameter 47

47(46). Rectangular lobe at apex of male abdomen at least half as long as broad, large, not strongly impressed towards dorsum; terminal spurs of tibiae very large, several times longer than setae encircling tibial apex; length of antennomere 1 about equal to maximum diameter of eye .... *Triarius* (part)

— Rectangular lobe at apex of male abdomen either shorter, absent, or distinctly impressed towards dorsum; tibiae with terminal spurs smaller, often hidden among adjacent setae, sometimes absent; antennomere 1 usually distinctly shorter than maximum diameter of eye 48

48(47). Distance between antennal insertions equal to at least twice diameter of antennal fossa; frons with mesal elevation broad, nearly continuous with frontal tubercles, delimited behind by shallow, sometimes indistinct impression; elytra usually pale with darker vittae or entirely pale, rarely entirely dark with very faint metallic luster, never with obvious metallic luster...... *Synetocephalus*

— Distance between antennal insertions equal to less than twice diameter of antennal fossa; mesal elevation of frons narrower, usually separated from frontal tubercles by distinct sulci; transverse interocular sulcus usually distinct, abruptly delimiting frontal tubercles behind; elytral color variable, often with distinct metallic luster 49

49(48). Elytra not metallic, usually bicolored with sutural area pale; if elytral suture dark (or entire elytron dark), then eye distant from mandible, the genae being about as long as antennomere 2; pronotum pale; elytra sometimes modified, with odd lateral depression or tubercle 50

— Elytra with at least sutural area dark, often entirely dark, often with metallic luster; if elytra bicolored or lacking metallic luster, then pronotum dark or genae much shorter than antennomere 2; elytra not modified as described above 51

50(49). Front tibiae lacking terminal spurs; elytron of male with distinct impression at posterolateral angle; distal portion of epipleuron bent upward at point of impression; abdomen of male with ventral, rectangular lobe at apex; median lobe of aedeagus without prominent basal spurs .... *Androlyperus* (part)

— Front tibia armed with terminal spur; male elytron sometimes unmodified, sometimes with lateral impression or tubercle at mid-length, but never with modification at posterolateral angle; abdomen of male with apex slightly truncate at apex, nearly as in female; median lobe of aedeagus with prominent basal spurs .... *Malacorhinus*

51(49). Frons with mesal elevation narrow, usually forming angulate ridge; frontal tubercles often extending to and continuous with orbit, rarely delimited laterally by poorly defined impression; genae usually about as long as terminal palpomere of maxilla; vertex often with alutaceous microsculpture; head and elytra dark, often metallic; apex of aedeagus often truncate, emarginate, or asymmetrical, rarely symmetrical and pointed .... *Scelolyperus*

— If mesal frontal elevation narrow or acutely elevated, then frontal tubercles delimited laterally by distinct impression; genae usually not much more than half as long as terminal palpomere of maxilla; vertex not distinctly alutaceous; head and elytra variable in color, sometimes largely pale; apical portion of aedeagus usually pointed and symmetrical .... *Pseudoluperus* (part)

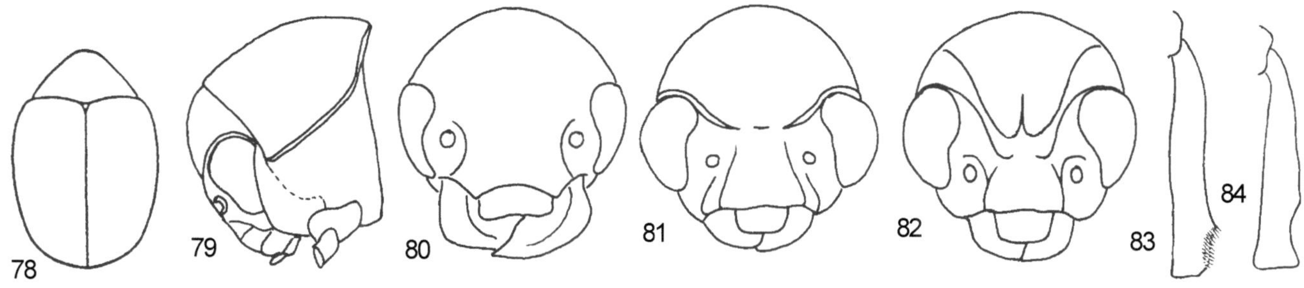

FIGURES 78.124-84.124. 78. *Oomorphus floridanus* (Horn); 79. *Paria sellata* (Horn), head and prothorax, lateral view; 80. *Metaparia viridimicans* (Horn), male head, frontal view; 81. *Paria scutellaris* (Notman), head, frontal view; 82. *Typophorus nigritus* (Fabricius), head, frontal view; 83. *Typophorus nigritus* (Fabricius), metatibia; 84. *Promecosoma arizonae* (Crotch), mesotibia. (Figs. 79 and 81 redrawn from Wilcox 1957).

52(45). Pronotum of most species with one or more large, conspicuous impressed areas, either at base or on each side near mid-length; if pronotal impressions weak or absent, then elytron with numerous longitudinal costae 53

— Pronotal impressions, when present, small (usually separated by more than their diameters) or very shallow; elytron sometimes with sublateral ridge, but never with numerous costae 56

53(52). Male with femora of front legs much larger than those of middle and hind legs; elytra entirely dark with metallic blue luster .... *Lygistus*

— Male with femora of front legs not larger than those of middle and hind legs; elytra not uniformly metallic blue 54

54(53). Elytra dark with pale sutural and lateral margins, or entirely pale; female 55

— Elytra pale with small dark spots; male or female .... *Neobrotica*

55(54). Metasternum dark brown or black .... *Luperosoma*, females

— Metasternum yellow .... *Phyllecthris*, females

56(52). Male with basitarsus of hind leg enlarged, much different in form than basitarsi of front and middle legs; elytral punctures very fine, those in pale vittae hardly visible .... *Keitheatus*

— Male with basitarsus of hind leg not conspicuously enlarged; elytral punctures coarser, easily seen 57

57(56). Pronotum and vertex entirely pale; elytra entirely dark with metallic luster .... *Scelida* (part)

— Pronotum and vertex entirely pale only in species with largely pale elytra .... *Pseudoluperus* (part)

Tribe Alticini

58(1). Terminal tarsomere of metatarsus moderately to strongly inflated (Figs. 58, 59) 59

— Terminal tarsomere of metatarsus weakly or not inflated 66

59(58). Elytra with punctures aligned in rows 60

— Elytra with punctation confused 62

60(59). Elytra glabrous or nearly so; pronotum with prebasal, transverse impression .... *Pachyonychus*

— Elytra pubescent; pronotum without basal impression 61

61(60). Procoxal cavities closed behind; vestiture of dorsum consisting of a mixture of both short, appressed hairs and long, erect hairs .... *Distigmoptera*

— Procoxal cavities open behind; vestiture of dorsum composed only of dense, appressed hairs .... *Pseudolampsis*

62(59). Pronotum as long as broad, lateral margin distinctly angled at middle, posterior angle strongly produced (Fig. 64); metatarsal claws simple .... *Pachyonychis*

— Pronotum distinctly broader than long, lateral margin evenly arcuate, posterior angle not prolonged; metatarsal claws each with broad, basal tooth 63

63(62). Terminal tarsomere of metatarsus moderately inflated (Fig. 58); metafemur moderately dilated (Fig. 54) 64

— Terminal tarsomere of metatarsus apically globose (Fig. 59); metafemur strongly dilated (Fig. 55) 65

64(63). Elytra dark brownish with pale spots . *Omophoita*

— Elytra uniformly dark metallic blue .... *Asphaera* (part)

65(63). Reflexed lateral margin of elytron narrow; body generally larger, length 4-8 mm, robust; eyes smaller, interocular space half width of head in most .... *Kuschelina*

— Reflexed lateral margin of elytron broad; body generally smaller, length 3-6 mm, depressed; eyes larger, interocular space less than half width of head in most .... *Capraita*

66(59). Base of pronotum without lateral furrows or transverse impression (note: a broad, shallow impression or vaguely expressed transverse groove may be present in some species or specimens) 67

— Base of pronotum with lateral furrow on each side and/or distinct transverse impression 91

67(66). Antenna 10-segmented .... *Psylliodes*

— Antenna 11-segmented 68

68(67). Articulated spur of metatibia with apex broad, bifid .... *Dibolia*

— Articulated spur of metatibia with apex narrow, tapered to single point 69

69(68). Metatibia with subapical angular dilation or tooth along outer margin, followed by weak emargination and brush of setae (Fig. 57-59) in most . 70

— Metatibia with outer subapical margin entire .. 74

70(69). Anterior angles of pronotum projected forward (Fig. 62); terminal tarsomere of metatarsus weakly inflated .... *Asphaera* (part)

— Anterior pronotal angles not projected forward; terminal tarsomere of metatarsus normal 71

71(70). Outer margin of metatibia with irregular series of teeth above angular tooth; inner margin of metatibia at apex extended to form broad, denticulate lobe .... *Pseudodibolia*

— Outer margin of metatibia without series of teeth above angular dilation or tooth; inner margin of metatibia at apex not modified as above 72

72(71). Tarsal claws simple; body small, length 1.1-3.5 mm, most with metallic green or bronze luster .... *Chaetocnema*

— Tarsal claws bifid, or if simple, then body large, black or brownish in a few, length 5.5-7.0 mm. 73

73(72). Tarsal claws simple .... *Euplectroscelis*

— Tarsal claws bifid .... *Blepharida*

74(69). Elytron with punctures aligned in rows, intervals may also be punctate 75

— Elytral punctation confused, indistinct, or lacking, not aligned in distinct rows 78

75(74). Procoxal cavities closed behind; body robust and strongly convex .... *Pseudorthygia*

— Procoxal cavities open behind; body not as above 76

76(75). Body ovate; base of pronotum as wide as base of elytra; male antenna with intermediate antennomeres 3-7 dilated, terminal antennomeres successively narrowed .... *Phydanis*

— Body narrower; base of pronotum narrower than base of elytra; male antenna not modified as above 77

77(78). Elytral epipleuron uniformly broad, subvertical; elytral apex subtruncate; hind femur long, nearly reaching elytral apex .... *Dysphenges*

— Elytral epipleuron narrow, gradually tapering towards apex; elytral apex rounded; hind femur short, not reaching elytral apex .... *Glyptina*

78(74). First tarsomere of metatarsus at least half as long as metatibia .... *Longitarsus*

— First tarsomere of metatarsus less than half as long as metatibia.. 79

79(78). Procoxal cavities closed behind 80

— Procoxal cavities open behind 81

80(79). Body robust, convex; elytron smooth with a few large punctures forming partial rows .... *Acrocyum*

— Body elongate, depressed; elytral punctation, if present, confused and uniformly distributed (note: some species with weak transverse impression before base of pronotum)..... *Systena*

81(79). Body round and strongly convex; head strongly deflexed, opisthognathous; elytral epipleuron slanted upward from margin, not visible from lateral view .... *Argopistes*

— Body not as above 82

82(81). Prosternal process narrow, depressed between coxae .... *Hemiphrynus*

— Prosternal process broad, not depressed between coxae 83

83(82). Antennal callus delineated behind by distinct groove which extends laterally to superior margin of eye (Fig. 71) 84

— Antennal callus not well delineated behind, or if so, then groove irregular, interrupted before reaching eye or not extended to eye 85

84(83). Body elongate, depressed above, sides subparallel; base of pronotum broadly flattened; males with apex of median lobe of aedeagus exserted from abdominal cavity in most .... *Luperaltica*

— Body oblong, convex above, sides rounded; base of pronotum convex; males with aedeagus normally hidden within abdomen .... *Aphthona*

85(83). Antennal calli small, most fused medially and poorly delineated behind; small species, length 1.8-3.6 mm .... *Phyllotreta*

— Antennal calli large, most flattened and separated by longitudinal median groove, well delineated behind or not; most are larger species 86

86(85). Posterior border of pronotum angled before posterolateral angle (Fig. 63); eyes vertically elliptical, inner margin weakly emarginate .... *Disonycha*

— Posterior border of pronotum not angled as above; eyes round, inner margin not emarginate 87

8786). Body above uniformly red to yellow with black scutellum, appendages dark; eyes small; genal space subequal to diameter of eye (note: weak transverse impression may be present before base of pronotum) .... *Strabala*

— Body not colored as above; eyes larger; genal space distinctly less than diameter of eye 88

88(87). Body long and narrow; elytron black with yellow vittae; fifth ventrite of male with large fossa (note: weak transverse impression may be present before base of pronotum) .... *Agasicles*

— Body broader; elytron immaculate; fifth ventrite of male without median fossa 89

89(88). Body smaller, length 2.4-2.7 mm; dorsum bicolored or uniformly metallic; legs, except hind femur in some, entirely yellow; basal margin of pronotum at middle flat or faintly impressed in most .... *Lupraea*

— Body larger, length of most 3-5 mm; body entirely metallic; legs metallic, rarely entirely yellow, some partially yellow; basal margin of pronotum

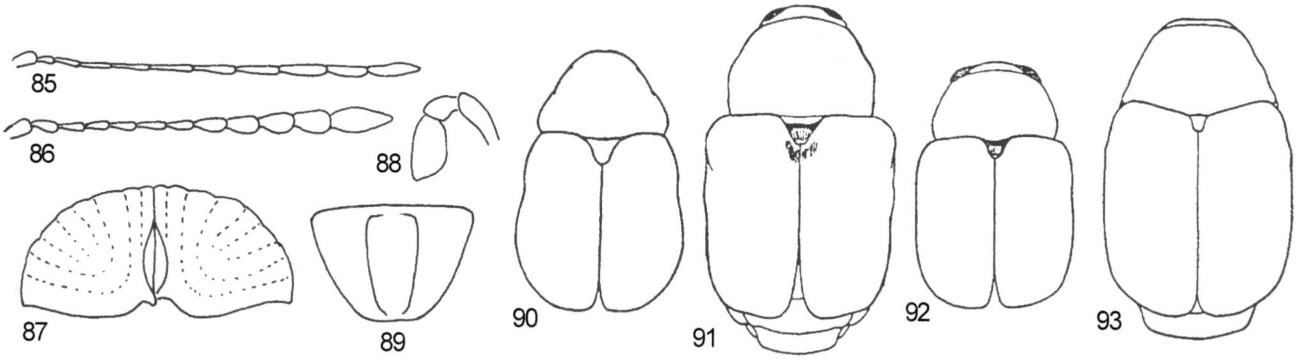

FIGURES 85.124-93.124. 85. *Colaspis* sp., antenna; 86. *Brachypnoea* sp., antenna; 87. *Promecosoma arizonae* (Crotch), apex of elytra, caudal view; 88. *Euphrytus intermedius* Jacoby, maxillary palp, male; 89. *Colaspis brunnea* (Fabricius), pygidium; 90. *Babia quadriguttata* (Olivier); 91. *Griburius scutellaris* (Fabricius); 92. *Pachybrachis othonus* (Say); 93. *Cryptocephalus venustus* (Fabricius).

at middle not noticeably flattened or impressed 90

90(91). Metafemur strongly swollen; body form broader, sides rounded; pronotum and elytra finely punctate; appendages of most metallic, partially pale in some .... *Altica* (part)

— Metafemur weakly swollen; body form narrow, sides subparallel; pronotum and elytra distinctly punctate; legs entirely pale (note: weak transverse impression may be present before base of pronotum) .... *Lysathia*

91(66). Elytron with major punctures aligned in rows, intervals in some irregularly punctate 92

— Elytron with punctation confused, or if poorly defined rows present, then majority of the punctation confused 104

92(91). Base of pronotum weakly sinuate on each side producing broad median lobe (Figs. 66, 67). 93

— Base of pronotum more or less straight or weakly sinuate but without median lobe 99

93(92). Elytra pubescent .... *Epitrix* (part)

— Elytra glabrous, or at most with few scattered hairs on disc or along apical margin 94

94(93). Pronotum without transverse impression, lateral furrow present on each side before base .... *Mantura*

— Pronotum with transverse impression 95

95(94). Tarsal claws simple; dorsum bronzed with elytral apices yellowish .... *Hippuriphila*

— Tarsal claws with broad basal tooth; color not as above 96

96(95). Procoxal cavities open behind; basal callus of elytron indistinct or lacking 97

— Procoxal cavities closed behind 98

97(96). Interocular sulcus steeply angled; pronotum essentially impunctate; elytra with rows of punctures regular to apex; median lobe of aedeagus evenly tapered to apex .... *Margaridisa*

— Interocular sulcus nearly horizontal; pronotum with fine punctation; elytra with rows of punctures becoming irregular towards apex; median lobe of aedeagus with short, angular flange on each side before apex .... *Hornaltica*

98(96). Body convex, shining black; basal callus of elytron distinct .... *Acallepitrix*

— Body depressed, pale brownish; basal callus of elytron indistinct .... *Epitrix* (part)

99(92). Hind femur long, reaching elytral apex; elytral epipleuron uniformly broad, subvertical .... *Glenidion*

— Hind femur short, not reaching elytral apex; elytral epipleuron of most not as above 100

100(99). Procoxal cavities open behind; elytra with sparse, erect hairs .... *Trichaltica*

— Procoxal cavities closed behind 101

101(100). Transverse impression at base of pronotum extended to basal corners of pronotum, not interrupted by lateral furrow (Fig. 68); lateral margin of pronotum serrate in most; elytra in most pubescent or with sparse erect hairs; metafemur without enlarged extensor apodeme (spring) .... *Orthaltica*

— Transverse impression of pronotum ending on each side in lateral furrow; pronotal margins entire; elytra glabrous 102

102(101). Lateral furrow on each side of pronotal base long, extended to point well in front of junction with transverse impression (Fig. 69) .... *Neocrepidodera*

— Lateral furrow on each side of pronotal base short, ending near junction of transverse impression (Figs. 65, 67) 103

103(102). Pronotal disc with coarse punctation; posterior pronotal space enclosed by basal grooves flat or weakly elevated .... *Crepidodera*

— Pronotal disc impunctate or nearly so; posterior pronotal space enclosed by basal grooves convex .... *Derocrepis*

104(91). Hind margin of pronotum weakly sinuate on each side, producing broad median lobe (Figs. 66, 69) 105
— Hind margin of pronotum more or less straight, not bisinuate as above 106

105(104). Base of pronotum on each side with distinct lateral furrow extending from transverse impression to pronotal base; antennal calli separated by frontal ridge (Fig. 71); punctation on basal portion of elytron fine and partly aligned in rows .... *Nesaecrepida*
— Base of pronotum with lateral furrows indistinct or lacking; antennal calli approximate, separated by narrow groove, not by frontal ridge (Fig. 70); punctation on basal portion of elytron coarse and entirely confused .... *Syphrea* (part)

106(104). Elytra with scattered erect hairs; pronotum with lateral furrow long, extending to point well in front of juncture with transverse impression .... *Hemiglyptus*
— Elytra glabrous; lateral furrows of pronotum short or obsolete 107

107(106). Transverse impression of pronotum curved basad to join hind margin or ending in lateral depression (Fig. 65) 108
— Transverse impression reaching or nearly reaching lateral margin 109

108(109). Supraorbital punctures one or few in number .... *Parchicola*
— Supraorbital punctures several in number, forming a dense cluster .... *Monomacra*

109(107). Pronotum convex; head partially retracted under anterior pronotal margin in repose; groove behind antennal callus distinct and continuous to superior margin of eye .... *Syphrea* (part)
— Pronotum depressed; head exserted; groove above antennal callus indistinct or interrupted before reaching eye 110

110(109). Elytron with faint to strong lateral plica which reaches apical declivity; body larger, length of most 5.0-8.0 mm, sides subparallel in most; transverse impression before base of pronotum distinct in most .... *Altica* (costate species group)
— Elytral plica absent in most, if present (in few) then weakly developed on apical declivity only; body smaller, length of most 2.5-5.0 mm, up to 8.0 mm in a few, sides rounded in most; transverse impression of pronotum variable, may be indistinct or absent (also see *Lysathia*, couplet 90) .... *Altica* (part)

VIII. Key to Genera of Eumolpinae of America North of Mexico

1. Pygidium with median longitudinal groove for locking elytra (Fig. 89) 2
— Pygidium without median longitudinal groove, or at most with trace of groove on basal third 16

2(1). Dorsum covered with scales (Tribe Adoxini, part) 3
— Dorsum glabrous (Tribe Eumolpini, part) 4

3(2). Protibia with ventral tooth; tarsal claws widely divergent, each with short basal tooth; lateral margin of pronotum usually dentate, sometimes undulate or entire, without marginal bead .... *Myochrous*
— Protibia without ventral tooth; tarsal claws weakly divergent, bifid or rarely simple; lateral pronotal margin entire, with marginal bead .... *Glyptoscelis* (part)

4(2). Anterior margin of prosternum and proepisterna distinctly excavated, or flattened with raised outer margin; head at rest (hypognathous position) with at least rear of mouthparts covered by prosternum 5
— Anterior margin of prosternum and proepisterna not as above, usually simple and meeting cervix well behind mouthparts 9

5(4). Anterior margin of proepisterna distinctly arcuate (Fig. 79) 6
— Anterior margin of proepisterna straight or slightly arcuate 7

6(5). Dorsum coarsely, densely punctate, lateral punctures often confluent; antenna filiform; body smaller, length 3.8-7 mm .... *Tymnes* (part)
— Dorsum not coarsely punctate; antenna subclavate; body larger, length 8-13 mm .... *Chrysochus*

7(5). Body large, length 15 mm or more; color metallic blue or green .... *Eumolpus*
— Body smaller, length less than 10 mm; color variable 8

8(7). Mandibles of male enlarged, much larger than those of female (Fig. 80) .... *Metaparia*
— Mandibles similar in both sexes, at most slightly larger in male .... *Spintherophyta*

9(4). Antenna distinctly expanded in apical half (Fig. 86) 10
— Antenna slender throughout or gradually expanded from base to apex (Fig. 85) 11

10(9). Body oval; elytra of male dorsally flattened in apical fourth, epipleuron almost horizontal; elytron of female with more than one strong, posthumeral costa .... *Chrysodinopsis*
— Body generally elongate-oval; elytra of male not flattened in apical fourth, epipleuron slanted downward; elytron of female with posthumeral costae lacking, or if present, then usually only weakly developed .... *Brachypnoea*

11(9). Outer margin of mesotibia of male with distinct, broad emargination at apex (Fig. 84); body depressed; elytra with apical notch at sutural angle (Fig. 87) .... *Promecosoma*
— Outer margin of mesotibia of male without emargination at apex, or at most weakly emarginate; elytra without apical notch 12

12(11). Prosternum broader, at least 0.8x width of procoxa; dorsum uniformly dark reddish brown overlain with strong greenish luster; elytra without costae .... *Rhabdopterus*
— Prosternum narrower, less than 0.75x width of procoxa; color of dorsum highly variable; elytra sometimes distinctly costate 13

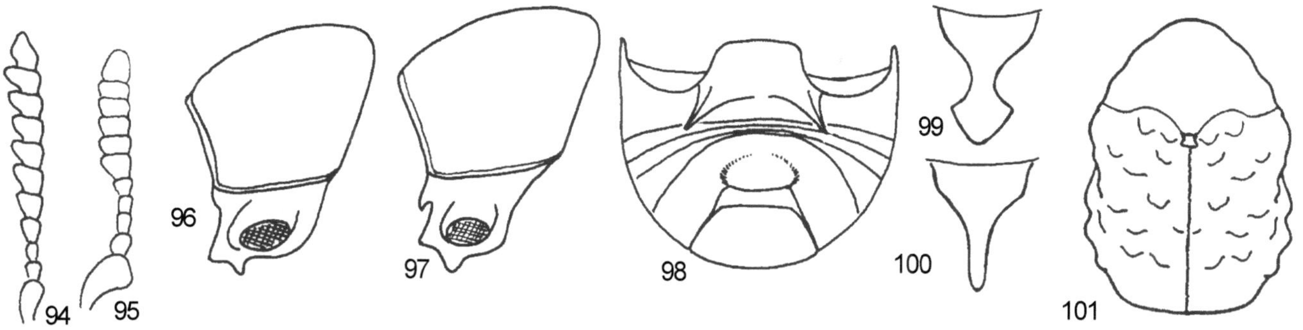

FIGURES 94.124-101.124. 94. *Anomoea laticlavia* (Forster), antenna; 95. *Exema* sp., antenna; 96. *Bassareus lituratus* (Fabricius), male prothorax, lateral view; 97. *Bassareus brunnipes* (Olivier), male prothorax, lateral view; 98. *Bassareus brunnipes* (Olivier), male ventrites; 99. *Diplacaspis prosternalis* (Schaeffer), prosternum; 100. *Neochlamisus* sp., prosternum; 101. *Exema dispar* Lacordaire.

13(12). Male maxilla with terminal palpomere truncate (Fig. 88) .... *Euphrytus*
— Terminal palpomere of maxilla rounded or acute in both sexes 14

14(13). Elytron with basal callus lacking or indistinct, postbasal depression absent or very shallow .... *Colaspis* (part)
— Elytron with basal callus and distinct postbasal depression 15

15(14). Base of elytron with distinct, complete marginal bead extending from humeral area to scutellum; lateral margins of ventrites 4 and 5 serrate; color metallic green above, yellowish below .... *Percolaspis*
— Base of elytron with marginal bead restricted to humeral area, or absent; lateral margin of ventrite 4 smooth or crenulate only at apex, lateral margin of ventrite 5 crenulate or smooth; color variable .... *Colaspis* (part)

16(1). Meso- and metatibiae with apical emargination on outer edge lined with dense brush of setae (Fig. 83) (Tribe Typophorini) 17
— Meso- and metatibiae unmodified, or never both modified as above 19

17(16). Anterior margin of proepisternum straight, head exposed .... *Metachroma*
— Anterior margin of proepisternum distinctly arcuate, partially covering head in repose (Fig. 79) 18

18(17). Ocular sulcus expanded above and behind eye (Fig. 82); body metallic, length 6.5 mm or more .... *Typophorus*
— Ocular sulcus not expanded above eye (Fig. 81); body non-metallic, length less than 4.5 mm .... *Paria*

19(16). Maximum width of head equal to or greater than basal width of pronotum (Figs. 72, 77); elytron densely punctate 20
— Maximum width of head less than basal width of pronotum; elytral punctation variable 22

20(19). Tarsal claws simple, very narrowly divergent; elytron without costae; dorsum metallic (Tribe Megascelidini) .... *Megascelis*
— Tarsal claws bifid, moderately to widely divergent; elytron usually with lateral costa; dorsum non-metallic (Tribe Synetini) 21

21(20). Elytral punctation, at least in part, serially arranged; elytron with sparse erect hairs .... *Syneta*
— Elytral punctation entirely confused; elytron densely clothed with long recumbent pubescence, without erect hairs .... *Thricolema*

22(19). Body covered with scales or dense pubescence (Tribe Adoxini, part) 23
— Body glabrous (Tribe Eumolpini, part) .. *Tymnes* (part)

23(22). Eye emarginate .... *Glyptoscelis* (part)
— Eye round 24

24(23). Anterior margin of proepisternum distinctly arcuate (Fig. 79) 25
— Anterior margin of proepisternum straight......... 26

25(24). Lateral margin of pronotum with marginal bead; dorsum metallic .... *Colaspidea*
— Lateral margin of pronotum without marginal bead; dorsum not metallic .... *Bromius*

26(24). Ocular sulcus present (Fig. 81) .... *Graphops*
— Ocular sulcus absent 27

27(26). Body covered with narrow recurved scales; lateral margin of pronotum serrate .... *Demotina*
— Body covered with simple hairs; pronotum unmargined 28

28(27). Prothorax distinctly wider than long, depressed; profemur with moderate to minute ventral tooth .... *Xanthonia*
— Prothorax very little wider than long, subcylindrical; profemur lacking ventral tooth .... *Fidia*

IX. Key to Genera of Lamprosomatinae of America North of Mexico

1. Tarsal claws simple; antenna long, surpassing prosternal process, with antennomere 8 distinctly narrower than 7 or 9; body small, length 1.5-2.0 mm. .... *Oomorphus*
— Tarsal claws appendiculate, each with broad tooth; antenna short, not surpassing prosternal process, with antennomere 8 nearly as wide as 7 or 9; body larger, length 4.5 mm. .... *Lamprosoma*

X. Key to Tribes, Subtribes, Genera, and Subgenera of Cryptocephalinae of America North of Mexico

1. Dorsal surface of body rough or tuberculate (Fig. 101); legs contractile, fitting into grooves or depressions; prothorax with antennal groove at each side of prosternal process; antenna short, subserrate (Fig. 95) (Tribe Chlamisini) 19
— Dorsal surface of body relatively smooth, not tuberculate although sometimes with coarse punctation; legs not contractile; prothorax without antennal groove at each side of prosternal process; antenna variable 2

2(1). Antennae short, serrate beyond fourth antennomere (Fig. 94) (Tribe Clytrini) 3
— Antennae long and filiform, usually reaching to or beyond elytral umbone (Tribe Cryptocephalini) 13

Tribe Clytrini
(modified from Moldenke 1970)

3(2). Prothorax with lateral antennal grooves (Subtribe Ischiopachina) .... *Ischiopachys*
— Prothorax without lateral antennal grooves 4

4(3). Tarsal claws simple 5
— Tarsal claws appendiculate or bifid (Subtribe Babiina) 10

5(4). Prosternum very narrow between procoxae; procoxae conical, contiguous; scutellum inclined relative to plane of elytra; elytra thin, pliable (Subtribe Clytrina) 6
— Prosternum evident, separating procoxae; procoxae globose; scutellum not distinctly inclined, on same plane as elytra or only slightly inclined; elytra rigid (Subtribe Megalostomina) 7

Subtribe Clytrina

6(5). Body smaller, length 4.0 to 7.5 mm; protibia of both sexes straight; elytra of male usually glabrous; dark color of elytra usually with metallic luster, including a median spot adjacent to or contiguous with scutellum .... *Smaragdina*
— Body larger, length 6.5 to 11 mm; protibia of male elongate and arcuate, that of female normal; elytra of males with long, regularly arranged setae; dark markings, if any, without metallic luster, never involving a median spot adjacent to scutellum . ... *Anomoea*

Subtribe Megalostomina

7(5). Eyes entire 8
— Eyes distinctly emarginate .... *Megalostomis*

8(7). Dorsum with metallic reflection; epipleuron at postbasal angle densely pubescent, otherwise elytra entirely glabrous .... *Coscinoptera*
— Dorsum without metallic reflection; epipleuron pubescent only if entire dorsum is pubescent ... 9

9(8). Elytra striate, glabrous; eyes not stalked posterolaterally .... *Coleorozena*
— Elytra with confused, dense punctation, usually obscured by dense pubescence; eyes stalked posterolaterally .... *Coleothorpa*

Subtribe Babiina

10(4). Postbasal epipleural lobes pronounced, strongly angulate; dorsum metallic with coarse punctation .... *Saxinis*
— Postbasal epipleural lobes weakly developed, not angulate; dorsum non-metallic with fine punctation 11

11(10). Smaller, length 2.5 to 5.5 mm; pygidium with transverse, subapical angle 12
— Larger, length 4.5 mm to 10.5 mm; pygidium flat, to slightly convex .... *Urodera*

12(11). Lateral margin of pronotum broad, explanate........ .... *Babia* (subgenus *Babia*)
— Lateral margin of pronotum narrow, not explanate .... *Babia* (subgenus *Archaebabia*)

Tribe Cryptocephalini

13(2). Claws simple, or if appendiculate, then prosternum longer than wide to subquadrate 14
— Claws appendiculate, each with broad, basal tooth; prosternum wider than long (Subtribe Monachulina) .... *Lexiphanes*

14(13). Pronotum margined at base, not crenulate (Subtribe Pachybrachina) 15
— Pronotum not margined at base, crenulate (Subtribe Cryptocephalina) 16

Subtribe Pachybrachina

15(14). Prosternum broad, as wide as long, flat anteriorly, depressed posteriorly along the lateral edges; female with sides of abdomen usually visible from above .... *Griburius*
— Prosternum narrower, longer than wide, weakly sulcate or shallowly depressed medially; female with sides of abdomen hidden from above .... *Pachybrachis*

Subtribe Cryptocephalina

16(14). Tarsal claws simple 17
— Tarsal claws appendiculate, each with broad basal angulation 18

17(16). Leading edge of prothorax laterally sinuous (Fig. 96) or toothed (Fig. 97); male with first ventrite mesally excavated and with large spine on each side (Fig. 98) .... *Bassareus*
— Leading edge of prothorax laterally straight; first ventrite of male without spines ... *Cryptocephalus*

18(16). Antennomeres 6 to 11 widened; body metallic or partially metallic in color .... *Diachus*
— Antennomeres 7 to 11 widened; body nonmetallic in color .... *Triachus*

Tribe Chlamisini
(modified from Karren 1972)

19(1). Prosternal process distally widened between mesocoxae (Fig. 99); metascutellum broadly exposed, tarsal claws bifid .... *Diplacaspis*

— Sides of prosternal process parallel, convergent or serrate (Fig.100), if slightly widened between mesocoxae, then metascutellum not exposed; tarsal claws simple or appendiculate............. 20

20(19). Mandibles enlarged in males, normal in females; prosternum strongly and abruptly constricted behind anterior margin; prosternal process more than half as long as prosternum .... *Pseudochlamys*

— Mandibles of males and females similar; prosternum gradually constricted behind anterior margin; prosternal process about half length of prosternum 21

21(20). Body generally larger, more or less metallic; frons in emargination of eye with yellow spot, or if without spot, then elytron with two velvety spots; posterolateral slope of pronotum usually with striations; internal sac of median lobe of male genitalia with ejaculatory guide asymmetrical, with sheath .... *Neochlamisus*

— Body size variable, but usually smaller; color variable, but usually not metallic; frons in emargination of eye usually without yellow spot, but if spot present or if yellow from face extends into emargination, then elytron without velvety spots; posterolateral slope of pronotum without striations; internal sac of median lobe of male genitalia with ejaculatory guide symmetrical, without sheath 22

22(21). Body small, length less than 3.5 mm; males with spine or spinulae medially on first ventrite; serration of elytral suture complete .... *Exema*

— Body usually larger, length from 2.5 to 6.0 mm; males without spines or spinulae on first ventrite; serration of elytral suture usually incomplete, absent from postscutellar area .... *Chlamisus*

Higher Classification of North American Chrysomelidae

CHRYSOMELIDAE Latreille 1802
- SAGRINAE Leach 1815 [subfamily]
- BRUCHINAE Latreille 1802 [see Family 121]
- DONACIINAE Kirby 1837
  - Tribe Plateumarini Askevold 1990
  - Tribe Donaciini Kirby 1837
  - Tribe Haemoniini Chen 1941
- CRIOCERINAE Latreille 1807
  - Tribe Criocerini Latreille 1807
  - Tribe Lemiini Heinze 1962
- HISPINAE Gyllenhal 1813
  - Tribe Cephaloleiini Chapuis 1875
  - Tribe Chalepini Weise 1910
  - Tribe Hemisphaerotini Monrós and Viana 1952
  - Tribe Mesomphaliini Hope 1840
  - Tribe Ischyrosonychini Chapuis 1875
  - Tribe Cassidini Gyllenhal 1813
- CHRYSOMELINAE Latreille 1802
  - Tribe Timarchini Motschulsky 1860
  - Tribe Chrysomelini Latreille 1802
    - Subtribe Entomoscelina Chapuis 1874
    - Subtribe Paropsina Motschulsky 1860
    - Subtribe Doryphorina Motschulsky 1860
    - Subtribe Chrysomelina Latreille 1802
    - *incertae sedis*
- GALERUCINAE Latreille 1802
  - Tribe Galerucini Latreille 1802
    - Section Coelomerites Chapuis 1875
    - Section Atysites Chapuis 1875
    - Section Schematizites Chapuis 1875
    - Section Galerucites Latreille 1802
  - Tribe Metacyclini Chapuis 1875
  - Tribe Hylaspini Chapuis 1875
  - Tribe Luperini Chapuis 1875
    - Subtribe Diabroticina Chapuis 1875
      - Section Diabroticites Chapuis 1875
      - Section Cerotomites Chapuis 1875
      - Section Phyllecthrites Horn 1892
      - Section Trachyscelidites Wilcox 1972
    - Subtribe Luperina Chapuis 1875
      - Section Scelidites Chapuis 1875
      - Section Phyllobroticites Chapuis 1875
      - Section Exosomites Wilcox 1973
      - Section Monoleptites Chapuis 1875
  - Tribe Alticini Newman 1835
- SPILOPYRINAE Chapuis 1874 [subfamily]
- EUMOLPINAE Hope 1840
  - Tribe Synetini Edwards 1953
  - Tribe Typophorini Chapuis 1874
    - Section Metachromites Chapuis 1874
    - Section Typophorites Chapuis 1874
  - Tribe Eumolpini Hope 1840
    - Section Iphimeites Chapuis 1874
    - Section Eumolpites Hope 1840
    - Section Corynodites Chapuis 1874
  - Tribe Megascelidini Chapuis 1874
  - Tribe Adoxini Baly 1865
    - Section Scelodontites Chapuis 1874
    - Section Leprotites Chapuis 1874
    - Section Adoxites Baly 1865
    - Section Myochroites Chapuis 1874
- LAMPROSOMATINAE Lacordaire 1848
  - Tribe Lamprosomatini Lacordaire 1848
- CRYPTOCEPHALINAE Gyllenhal 1813
  - Tribe Cryptocephalini Gyllenhal 1813
    - Subtribe Pachybrachina Chapuis 1874
    - Subtribe Monachulina Leng 1920
    - Subtribe Cryptocephalina Gyllenhal 1813
  - Tribe Clytrini Lacordaire 1848
    - Subtribe Clytrina Lacordaire 1848
    - Subtribe Megalostomina Chapuis 1874
    - Subtribe Babiina Chapuis 1874
    - Subtribe Ischiopachina Chapuis 1874
  - Tribe Chlamisini Gressitt 1946

Donaciinae Kirby 1837

Characteristics: Head prominent, prognathous, slightly narrowed behind the eyes; eyes entire, convex, prominent, moderate in size; antennae closely inserted on frons, filiform, extending to middle of elytra. Prothorax subquadrate, narrower than elytra, about as wide as head, without lateral marginal bead. Each elytron with ten rows of punctures. Procoxae conical, closely inserted; procoxal cavities closed. Abdomen with first ventrite as long as the remaining four together. Legs long; metafemora often enlarged and often toothed ventrally; protibiae and mesotibiae with apical articulated spur; tarsi 5-5-5, pseudotetramerous, with bifid setae present on tarsomere 3.

Larvae aquatic, attached to and feeding on the submerged stems or roots of their host plants.

This is a rather small group comprising about 165 species worldwide. Most species are Holarctic with a few representatives in Africa, Asia, and Australia. There are no species in South America. In the Nearctic Region, the group consists of three tribes and five genera. The subfamily is completely aquatic in the larval stage. Many species are microsympatric, sharing the same habitats and hosts. Adults of several species can often be collected intermixed at a single site. Except for the absence of a complete revision of the genus *Donacia*, the subfamily in America north of Mexico has been treated recently (Askevold 1987a, 1987b, 1988, 1990, 1991a, and 1991b). Key to world genera and subgenera, Askevold (1990). Larval descriptions and key to some larvae, Hoffman (1940). General larval biology and morphology, Bøving (1910).

Plateumarini Askevold 1990

Characteristics: Sutural bead of elytron expanded apically, with inner sutural bead sinuate far from apex and joined to outer bead; elytral apex rounded in most species; hypomeron of prothorax glabrous (or with a few scattered setae); mandible with apical teeth approximate, lacking serrulate occlusal edge; median lobe of aedeagus without basal angulation; aedeagal tegmen robust and tapered (Askevold 1990).

Plateumaris Thompson 1859
Donacocia Gistel 1857
Juliusina Reitter 1920
Euplateumaris Jablokoff-Khnzorian 1966
Juliusiana: Mohr 1966 [error]

This is principally a Holarctic genus with 17 Nearctic and 9 Palearctic species. It is generally distributed throughout the Nearctic Region, although the occurrence of species and number of collections is far greater in the northeastern portion of the United States and adjacent Canada. Host plants include *Acorus* (Araceae); *Carex, Eleocharis, Scirpus* (Cyperaceae); *Juncus* (Juncaceae); *Caltha*, and *Ranunculus* (Ranunculaceae). Key to species, Askevold (1991b). Larvae, Hoffman (1940), MacGillivray (1903).

Poecilocera Schaeffer 1919

This genus contains a single distinctive species, *P. harrisii* (LeConte 1851), recorded from the New England states of Connecticut, Massachusetts, New Hampshire, New Jersey, New York, and Vermont. Very little is known about the biology. Host records include *Carex* (Cyperaceae), and Askevold (1990) suggests *Scirpus* (Cyperaceae) as well. Treatment of the genus, Askevold (1990).

Donaciini Kirby 1937

Characteristics: Elytron with sutural margin before apex unmodified; elytral apex truncate in most species; pronotal hypomeron with broad pubescent area in most species; mandible with apical teeth slightly to markedly divergent, with occlusal edge long, serrulate; median lobe of male genitalia with basal angulation in most species; aedeagal tegmen thin and slender in most species (Askevold 1990).

Donacia Fabricius 1775
subgenus *Donacia* Fabricius 1775

Fifteen species occur worldwide, of which 10 are Nearctic. Nearctic species are distributed primarily in the eastern half of Canada and the United States, with one species extending into British Columbia and California. Principal host plants are all dicotyledons, including *Brasenia* (Cabombaceae); *Nuphar* and *Nymphaea* (Nymphaeaceae). Key to species, Marx (1957). List of Nearctic species, Askevold (1991a). Larvae, Hoffman (1940), Lawson (1991), MacGillivray (1903).

subgenus *Cyphogaster* Goecke 1934 [Oriental and Australian]

subgenus *Donaciomima* Medvedev 1973

Seventy-four species occur worldwide, of which 21 are Nearctic. This subgenus is distributed over most of Canada and the United States, but species are most abundant in the eastern and northern part of this range and are not recorded from the southwestern states of Arizona, Nevada, and New Mexico. Host plants are, with the exception of *Nymphaea* (Nymphaeaceae), monocotyledons including *Sagittaria* (Alismataceae); *Acorus, Peltandra* (Araceae); *Eleocharis, Scirpus* (Cyperaceae); *Eriocaulon* (Eriocaulaceae); *Pontederia* (Pontederiaceae); *Sparganium* (Sparganiaceae); and *Typha* (Typhaceae). Key to species, Marx (1957); key to species of the *subtilis* group, Askevold (1987a); list of Nearctic species, Askevold (1991a). Larvae, Hoffman (1940), MacGillivray (1903).

subgenus *Askevoldia* Kippenberg 1994 [Palaearctic]

Donaciella Reitter 1920

One North American species, *D. pubicollis* (Suffrian 1872), belongs to this genus which was recently elevated to generic level by Askevold (1990). Distribution includes the Canadian provinces of Alberta, Manitoba, Ontario, and Saskatchewan, as well as Illinois (northern), Indiana (northern), Michigan, Minnesota and Wisconsin in the United States. The recorded host is *Phragmites*

(Poaceae). Treatment of genus, Askevold (1990). Larva, Hoffman (1940).

Haemoniini Chen 1941

Characteristics: Dorsum testaceous; legs almost entirely testaceous; metafemur edentate, slender; tarsi without ventral pads, with apical tarsomere elongate; elytron apically with spine at outer angle in most species; pronotal hypomeron with pubescent area above coxa; mandible forming serrate occlusal edge, with apical teeth divergent; median lobe of aedeagus with basal angulation; aedeagal tegmen slender (Askevold 1990).

Neohaemonia Székessy 1941
This Nearctic genus is considered distinct from the Old World genus *Macroplea* Samouelle 1819, based on numerous adult characters. The Nearctic fauna consists of five species, including *N. flohri* (Jacoby 1884) from central Mexico and now possibly extinct (Askevold 1988). Distribution of the other species extends across the southern provinces of Canada and the northern third of the United States, with one species, *N. flagellata* Askevold 1988, extending into northern California. Most collections are from the eastern portion of this range. Both adults and larvae are completely aquatic, except for some adult activity, such as movement to and from over-wintering habitats in leaf litter. Specimens are uncommonly collected, probably because of their aquatic nature. Host plants are restricted to *Potamogeton* (Potamogetonaceae). Key to species, Askevold (1988). Larvae, Hoffman (1940) and MacGillivray (1903).

Criocerinae Latreille 1807

Characteristics: Body oblong, elongate; head and pronotum distinctly narrower than elytra; dorsum glabrous. Head prognathous, usually as wide as or wider than pronotum; frons with distinct X-shaped grooves that continue around eyes forming a distinct neck behind eyes; antennae clavate to slightly filiform, reaching beyond humeri; antennal insertions near lower margins of eyes, widely separated on frons; eyes emarginate. Pronotum lacking lateral marginal bead and usually strongly constricted at or near the middle. Procoxae conical; procoxal cavities closed. Elytra with punctures in distinct rows. First ventrite as long as next two combined; tergite 7 with stridulatory file; pygidium covered by elytra. Tarsi 5-5-5, pseudotetramerous, with bifid setae present on third tarsomere.

Larvae are mostly external feeders on the leaves, and rarely fruit, of their hosts. As a defensive mechanism they cover themselves with a slimy secretion of excrement.

This is one of the smaller subfamilies, with approximately 1,500 species (White 1993) recorded in three tribes and twenty genera worldwide (Seeno and Wilcox 1982). In America north of Mexico and in the Baja California peninsula, there are 46 species in five genera (not including three species described from British Columbia by White (1993) with identical label data; these are most certainly mis-labeled Neotropical specimens). The three genera of Lemiini are native, however, some species of *Lema* and *Oulema*, as well as both genera of the Criocerini, are introduced. The Criocerinae of North America feed on both monocotyledonous and dicotyledonous plants representing at least seven families. The subfamily was recently revised by White (1993) for America north of Mexico. Key to world genera and subgenera, Monrós (1960b). Biology, phylogeny, and evolution, Schmitt (1988).

Criocerini Latreille 1807

Characteristics: Tarsal claws separated at base, divergent.

Lilioceris Reitter 1912
This is an Old World genus best represented in Africa, China and Southeast Asia.

subgenus *Lilioceris* Reitter 1912
A single species belonging to this subgenus, *L. lilii* (Scopoli 1763), is adventive in North America. Its current Nearctic distribution includes the Canadian provinces of Ontario and Quebec, and adjacent United States. This distribution is likely to increase. Adults and larvae feed on *Fritillaria* and *Lilium* (Liliaceae).

subgenus *Bradyceris* Chûjô 1951 [Palearctic]

subgenus *Chujoita* Monrós 1960 [Oriental and Australian]

Crioceris Geoffroy 1762 [conserved name, ICZN 1970, 1994; Opinions 908 and 1754]
Crioceris: Müller 1764 [subsequent usage]

subgenus *Crioceris* Geoffroy 1762
Old World in original distribution, this subgenus is represented in America north of Mexico by two adventive species, *C. asparagi* (Linnaeus 1758), the asparagus beetle, and *C. duodecimpunctata* (Linnaeus 1758), the spotted asparagus beetle. Both species feed on asparagus, *Asparagus officinalis* L. (Liliaceae), and both are widely distributed in southern Canada and across the northern United States from Maine to California. Key to species, White (1993). Larva, Bøving and Craighead (1931), Lawson (1991), Peterson (1951), Sailsbury (1943). Life history, Fink (1913).

subgenus *Pseudolema* Jacoby 1903 [Oriental]

Lemiini Heinze 1962

Characteristic: Tarsal claws contiguous at base and for much of length.

Lema Fabricius 1798 [conserved name, ICZN 1970 Opinion 908]
This large genus includes hundreds of described species and probably many more undescribed. Most species are found in the tropical and subtropical areas of the world. Except for America north of Mexico, this genus is in need of revision. With further study,

some extralimital species presently listed in *Lema* will be found to belong in *Neolema* or *Oulema*. There are 16 species recorded from America north of Mexico and the Baja California peninsula of Mexico.

subgenus *Lema* Fabricius 1798
Sulcatolema Pic 1928
Although this is an Old World subgenus, *L. puncticollis* (Curtis 1830) was released in Alberta, Saskatchewan, and New Brunswick for the biological control of Canada thistle, *Cirsium arvense* (L.) Scop. (Asteraceae). Apparently, the introduced populations either died out naturally, or they were intentionally exterminated following concerns about herbivory to non-target species of *Cirsium*.

subgenus *Bradylema* 1901 [Old World]
Pseudolema Pic 1928, not Jacoby 1903 (Insecta)
Bradylemoides Heinze 1930
Microlema Pic 1932
Enoplolema Heinze 1943

subgenus *Quasilema* Monrós 1951
The Nearctic species were not assigned to subgenera by White (1993), but selected species were assigned to this subgenus by Balsbaugh and Hays (1972), Schmitt (1988), and Monrós (1960b). All our native *Lema* belong here, as well as most of the other New World species correctly assigned to this genus. *Lema trilinea* White 1981, the threelined potato beetle, and *L. trivittata trivittata* Say 1824, range from the southern portions of eastern Canada to northern Mexico; only the Pacific Northwest is excluded from this distribution. The remaining 13 species have various distributions from the southeastern United States west to Arizona and south into Mexico. Principal host plants are Solanaceae of the genera *Datura*, *Physalis*, and *Solanum*. *Lema* species from the Baja California peninsula have been reportedly collected on *Nolina* (Liliaceae). *Lema t. trivittata* is recorded as a non-commercial pest on potato, *Solanum tuberosum* L., and tomatillo, *Physalis philadelphica* Lam. Key to species, White (1993). Biology and larva, Kaufmann (1967); larvae, Kogan and Goeden (1970), Lawson (1991), Peterson (1951), Sailsbury (1943).

subgenus *Pachylema* Monrós 1951 [Neotropical]

Neolema Monrós 1951
This genus was originally established as a subgenus of *Lema* by Monrós (1951) for a single South American species. White (1993) elevated *Neolema* to genus, described new species, and transferred others from *Lema*. Together with two species found in the Baja California peninsula, the Nearctic fauna consists of at least nine species, with a collective distribution from Pennsylvania west to Missouri, southwest to Arizona and into northern Mexico. Primary host plants include *Commelina* and *Tradescantia* (Commelinaceae). One species from the Baja California peninsula has been collected on *Nolina* (Liliaceae). Key to species, White (1993). Biology and larvae, Green (1939), Kaufmann (1967), Peterson (1951), Sailsbury (1943).

Oulema Des Gozis 1886
Distributed worldwide, this genus is in need of revision except in America north of Mexico. Our native North American species have only recently been transferred to this genus. As refinements are made in our knowledge of the vast Neotropical criocerine fauna, it is likely that more New World species will be transferred. White (1993) did not assign the 18 Nearctic species to subgenera, but some native Nearctic species were assigned to the subgenus *Hapsidolemoides* by Monrós (1960b). The subgenera as proposed by Monrós (1951, 1960b) are provisionally accepted in the present work (but see under subgenus *Hapsidolemoides*). Key to species, White (1993). Biology, Wellso and Hoxie (1988).

subgenus *Oulema* Des Gozis 1886
Ulema: Bedel 1889 [error]
Hapsidolema Heinze 1927
Incisophthalma Heinze 1929
Xoidolema Heinze 1931
Conradsia Pic 1936
This is a large Old World subgenus with one adventive species, *O. melanopus* (Linnaeus 1758), occurring in North America. Known as the cereal leaf beetle, this species is established in most of the eastern United States and adjacent Canada west to Utah and Idaho. It is a serious pest of grain crops and has recently been reported to feed on ornamental turf grasses (Staines 1997b) Recorded North American hosts include *Avena*, *Hordeum*, *Secale*, *Triticum*, and *Zea* (Poaceae). Larva, Lawson (1991).

subgenus *Parhapsidolema* Monrós 1951 [Neotropical]

subgenus *Hapsidolemoides* Monrós 1951
All 17 of the native Nearctic species should be assigned to this New World subgenus. Eleven of these species always have the ninth elytral stria complete and with regularly spaced punctures, and thus best fit the diagnosis for *Hapsidolemoides*. These species occur in the United States east of the Rocky Mountains and in parts of adjacent Canada. Principal host plants include *Carduus*, *Cirsium*, *Senecio* (Asteraceae); and *Commelina* and *Tradescantia* (Commelinaceae). The morphology of the other six Nearctic *Oulema* species approaches that of the subgenus *Parhapsidolema* Monrós in that they sometimes have the ninth elytral stria interrupted or at least with irregularly spaced punctures at midlength. In general body form, these species are somewhat intermediate between the North American *Neolema* and *Oulema* (*Hapsidolemoides*). This group of species is distributed from the southeastern and central United States, west to Arizona, and south into Mexico. Food plants include *Commelina* and *Tradescantia* (Commelinaceae). Further study, especially of the Neotropical fauna, is needed to fully test the subgeneric concepts for *Oulema* set forth by Monrós (1951). Key to species, White (1993). Biology and larva (as *Lema simulans*), Kaufmann (1967).

subgenus *Gracilema* Chûjô 1964 [Oriental]

Hispinae Gyllenhal 1813
The hispines and tortoise beetles

Characteristics: Body narrow, parallel to broadly ovate, above nearly flat to highly convex, finely to deeply punctate; margins narrow to broadly explanate, frequently spined or serrate. Head broadly to narrowly exposed or hidden by explanate margin of pronotum, usually opisthognathous; frons and clypeus usually slanted posteriorly, sometimes clypeus horizontal; mouthparts sometimes partially hidden by anterior margin of prosternum; antenna with from 3 to 11 distinct antennomeres (8 or 11 in our species), inserted on frons between eyes, insertions narrowly separated, terminal antennomeres sometimes fused to form club. Pronotum with or without tactile setae in anterior and posterior angles. Elytral punctures generally arranged in 10 striae and short subscutellar row but frequently greatly modified or absent; intervals frequently costate in narrow-bodied forms. Ventrites 1 and 2 connate. Tarsi 4-4-4, the normal penultimate tarsomere lost; bifid tarsal setae present on tarsomeres 1-3.

Larvae highly variable in body form, living as leaf miners, between appressed leaves or other plant parts, or openly on foliage; exposed feeders usually with variably developed lateral scoli and fecal annex.

The union of the former Hispinae and Cassidinae poses a problem regarding the status of the tribes in both groups. Ranking the former subfamilies as tribes and reducing former tribes to subtribes is unsatisfactory since it appears that neither of the former subfamilies is monophyletic. Until a comprehensive analysis can be completed, it is probably best to maintain most of the former tribes at tribal rank. There are approximately 320 genera and 5,300 species worldwide. Six tribes, 33 genera, and 114 species are recorded from America north of Mexico. Relationships of cassidoid groups, Borowiec (1995); world catalog of cassidoid groups, Borowiec (1999).

Cephaloleiini Chapuis 1875

Imatidiini Chapuis 1875

Characteristics: Body elongate, subparallel to broadly ovate, depressed, with margins entire, narrow or explanate (subparallel in our species); punctation of dorsum usually fine. Head exposed in dorsal view; clypeus short, nearly vertical; antenna with 10 or 11 distinct antennomeres (11 in our species). Tactile setae of anterior pronotal margin arising from behind eyes, not from anterolateral corner of pronotum. Elytra usually with punctures aligned in striae, intervals not costate.

Larvae are broad and flattened with a concealed head and appendages and bear a strong resemblance to water pennies (larvae of *Psephenus* spp., Psephenidae). They are external feeders living between tightly appressed plant parts such as leaf sheaths, the bases of leaves and bracts, and the folds of young unopened leaves. Adults of some genera also reside in these microhabitats. Host plants are a diverse array of monocotyledonous families.

This group, as here constituted, contains both narrow-bodied forms and the broad, cassid-like species which were traditionally kept separate in the tortoise beetle tribe Imatidiini. The close relationship between these groups has been suggested by several authors, and they were recently united by Borowiec (1995). The tribe is widespread in the Neotropical Region and includes 17 genera, one of which is also represented in the Nearctic.

Stenispa Baly 1858
This genus is largely Neotropical with 21 species. Two species are generally distributed in the eastern United States west to South Dakota and Texas. Larvae and adults of *S. metallica* (Fabricius 1801) are on sedges of the genera *Carex* and *Scirpus* (Cyperaceae). Larval description, Ford and Cavey (1985). Key to species, Downie and Arnett (1996).

Chalepini Weise 1910

Octotomites Chapuis 1875
Uroplatini Weise 1910

Characteristics: Body narrow, parallel-sided to dilated posteriorly, above nearly flat to moderately convex from side to side, deeply punctate; margins frequently spined or serrate, not explanate or only slightly so at elytral apex. Head broadly exposed; clypeus moderately long to long, nearly vertical to strongly sloped posteriorly; antenna with from 3 to 11 distinct antennomeres (8 or 11 in our species), filiform to distinctly clubbed. Tactile setae usually arising from anterolateral corners of pronotum. Elytron with ground plan of ten punctate striae and short subscutellar row, in many genera these striae variably reduced, sometimes to as few as seven (eight in our genera), and subscutellar rows may be reduced or lacking; intervals frequently costate; elytral punctation and costae sometimes irregular. Larvae are leaf miners in both monocotyledonous and dicotyledonous plant families.

The concept for the tribe followed here is broader than that of most previous treatments. Combined with the traditional genera of the Chalepini are those genera previously assigned to the tribe Uroplatini. These two groups appear inseparable except for the arbitrary distinction in number of articulating antennomeres. It is likely that fusion of terminal antennomeres is an adaptive attribute and has arisen independently in several separate lines within this large group. Found throughout the New World, this group is comprised of fifty-seven genera, of which 13 are found in our region. The limits of several large genera are not well established. Key to larvae of the northeastern United States, Ford and Cavey (1985).

Anisostena Weise 1910
Twenty-nine species comprise this genus which is found throughout the Neotropical Region and warmer portions of the Nearctic Region. Key to subgenera, Staines (1993).

subgenus *Anisostena* Weise 1910
This subgenus is represented by fourteen species in America north of Mexico and is generally distributed throughout the United States and adjacent Canada. Food plants are various grasses including the genera *Bothriochloa*, *Panicum*, *Schizachyrium*, and *Tripsacum* (Poaceae). Immature stages and biological notes, Ford and Cavey (1982, 1985). Keys to species, Staines (1994a, 1994b, and 1994c).

subgenus *Neostena* Monrós and Viana 1947 [Neotropical]

subgenus *Apostena* Staines 1993 [Neotropical]

Sumitrosis Butte 1968
About 50 species comprise this genus which is found throughout the Neotropical Region, excluding the West Indies; five species occur in America north of Mexico. The four eastern species have a combined range throughout the eastern United States and adjacent Canada west to Texas and Wyoming. The one western species occurs in southern Arizona. Food plant genera include *Aster*, *Eupatorium*, *Solidago*, *Vernonia* (Asteraceae); *Cassia*, *Desmodium*, *Strophostyles*, *Robinia* (Fabaceae); and *Laportia* (Urticaceae). Immatures and biology, Ford and Cavey (1985), Wheeler and Snook (1986). Key to species, Butte (1969).

Odontota Chevrolat 1836
Nine species comprise this genus which occurs from northern South America to southeastern Canada. The locust leaf miner, *O. dorsalis* (Thunberg 1805), the best known member of the genus, and five other species occur in the eastern United States, and some range west to the Great Plains. A species from Arizona, *O. arizonica* (Uhmann 1938), seems more closely related to some Neotropical species assigned to *Xenochalepus* (*Neochalepus*) than to the eastern species. Food plant genera include *Amorpha*, *Amphicarpaea*, *Apios*, *Desmodium*, *Glycine*, and *Tephrosia* (Fabaceae). Numerous other plants are fed upon by adults only. Immature stages, Ford and Cavey (1985), Lawson (1991), Peterson (1951); biology, Chittenden (1902), Kogan and Kogan (1979), Wheeler and Snook (1986). Key to species, Butte (1968c).

Chalepus Thunberg 1805
  Anoplitis Kirby 1837
  Parachalepus Baly 1885
  Parachelepus: Seeno and Wilcox 1982 [error]
This is a large Neotropical genus which is in need of revision and reassessment of its limits. One of our species, *C. walshii* (Crotch 1873), maintains a position isolated from the remaining four species from the region. The combined range of our species is throughout most of the eastern United States west to Arizona. Food plants in America north of Mexico are grasses (Poaceae) in the genera *Elymus*, *Hystrix*, and *Panicum*. Immature stages, Ford and Cavey (1985). Key to species, Butte (1968b).

Xenochalepus Weise 1910
About 80 Neotropical species are assigned to this genus. Two subgenera are recognized, Staines and Riley (1994).

subgenus *Xenochalepus* Weise 1910
  Hemichalepus Spaeth 1937 [unavailable]
  Hemichalepus Uhmann 1957
About 15 species comprise this subgenus. Four species occur in America north of Mexico, three in the southwestern United States from Texas to Arizona, and one, *X. potomacus* Butte 1968, in the Atlantic states from Maryland to Georgia and also Missouri. Food plants are Fabaceae in the genera *Glycine*, *Phaseolus*, and *Robinia*. Larva, Bøving and Craighead (1931). Key to species, Butte (1968a).

subgenus *Neochalepus* Staines and Riley 1994 [Neotropical]

Baliosus Weise 1905
  Parabaliosus Monrós and Viana 1947
Two species occur in our area, *B. californicus* (Horn 1883) in Arizona, California, and Oregon, and *B. nervosus* (Panzer 1794) generally distributed in the eastern United States and adjacent Canada west to Colorado. About 40 additional species assigned to the genus occur in the Neotropical region, excluding the West Indies. Food plant genera in America north of Mexico include *Alnus* (Betulaceae), *Tilia* (Tiliaceae), *Quercus* (Fagaceae), and *Salix* (Salicaceae) in the east, and *Ceanothus* (Rhamnaceae) in the west. Larva, Ford and Cavey (1985); biology Hodson (1942).

Octotoma Dejean 1836
Two species groups of this New World genus are recognized, one in South America with four species and one in North America, including the West Indies, with seven species. One species, *O. plicatula* (Fabricius 1801), is restricted to and generally distributed in the eastern United States west to Kansas and Texas, and two other species occur in the southwestern states extending into Central America. Larval food plants in America north of Mexico include *Campsis radicans* (L.) Seem. (Bignoniaceae), *Lantana* (Verbenaceae), and *Perezia thurberi* Gray (Asteraceae); adults commonly feed on *Fraxinus* (Oleaceae). One or more species have been introduced to Australia, China, Hawaii and various other Pacific islands, Ghana, India, and the Republic of South Africa for the control of *Lantana camara*, an invasive tropical weed. Larva, Ford and Cavey (1985). Key to species, Staines (1989).

Stenopodius Horn 1883
Seven species comprise this genus with a combined range from the cape region of the Baja California peninsula to California east to Minnesota and Texas. Food plants are *Sphaeralcea* (Malvaceae). Key to species, Blaisdell (1939), but a revision is needed. Taxonomic note, Staines (1986a).

Brachycoryna Guérin-Méneville 1844
Six species occur in America north of Mexico and one of these extends south to northern South America. A seventh species is restricted to South America. Food plant genera include *Madia*

(Asteraceae), *Ceanothus* (Rhamnaceae), and *Malvastrum*, *Sida*, and other Malvaceae. Key to species, Staines (1986b).

Glyphuroplata Uhmann 1940
Glyphuroplata Uhmann 1937 [unavailable]
Four described species comprise this genus which is found throughout most of the eastern United States, west across the southern states to California, the Baja California peninsula, and as far south as Costa Rica. All of them can be found in the United States. Food plants are grasses (Poaceae) including the genera *Digitaria*, *Panicum*, and *Valota*. Key to species, Riley (1985b).

Platocthispa Uhmann 1940
Platocthispa Uhmann 1939 [unavailable]
Seven species, mostly from Mexico and Central America, comprise this genus. One species, *P. lateritia* (J. Smith 1886), is found in Arizona and Sonora (Staines 1997a).

Pentispa Chapuis 1875
Penthispa Weise 1910 [error]
About 25 Neotropical species are assigned to this genus. Three species extend into the southwestern United States, occurring from Texas to Arizona. The limits of this genus are not well understood. Food plants in America north of Mexico are *Verbesina*, *Baccharis*, and other Asteraceae. Immature stages and biology, Boldt and Staines (1993).

Microrhopala Chevrolat 1836
About 20 species comprise this genus and nine occur in America north of Mexico. Their combined distribution covers all of the United States and adjacent Canada south to Colombia. *Microrhopala vittata* (Fabricius 1798), the goldenrod leaf miner, is the type species of the genus, and, together with the closely related *M. laetula* (LeConte 1859), forms a distinct group, separate from the remaining species presently assigned to the genus. The latter species seem more closely related to *Pentispa* than to *Microrhopala s. str.* Food plants in America north of Mexico are Asteraceae of the genera *Aster*, *Boltonia*, *Encelia*, *Franseria*, *Haplopappus*, *Heterotheca*, *Helianthus*, *Pityopsis*, *Sericocarpus*, *Silphium*, *Solidago*, and probably others. Immature stages and biology, Ford and Cavey (1985), Lawson (1991), McCauley (1938), Peterson (1951). Key to species, Clark (1983).

Hemisphaerotini Monrós and Viana 1951

Characteristics: Body ovate, highly convex, margins explanate. Head broadly exposed, clypeus short and horizontal, vertex with stridulatory file, antenna 11-segmented and gradually enlarged to form a weak club. Pronotum with tactile seta present on tooth of anterior angle. Tarsal claws simple, single or paired (single in our species). Larvae on foliage and completely concealed by their fecal shield which consists of tightly-woven strands and resembles an inverted bird's nest. Hosts belong to various monocotyledonous plant families.

This tribe contains about 45 species in two genera and is found throughout the Neotropical Region and in the southeastern United States.

Hemisphaerota Chevrolat 1836
Porphyraspis Hope 1840
Emperochela Spaeth 1901
Eight or nine species occur in the West Indies (Cuba, Bahamas, and Hispaniola) and one, *H. cyanea* (Say 1824), occurs in the southeastern United States from South Carolina to Florida to Mississippi. This species is also found in Texas where it is thought to have been introduced (Jackman 1976). It breeds on both native and ornamental palms of the genera *Coccothrinax*, *Sabal*, *Serenoa*, and *Washingtonia* (Arecaceae). Biology and immature stages, Beshear (1969), Woodruff (1965).

Mesomphaliini Hope 1840

Stolaini Hincks 1952
Eugenysini Hincks 1952

Characteristics: Body ovate, moderately to highly convex, margins explanate. Head broadly to narrowly exposed; clypeus moderately long and slanted posteriorly; vertex without apparent stridulatory file; antenna with 11 antennomeres, usually enlarged distally. Pronotum without tactile setae. Tarsal claws paired and symmetrical, appendiculate. Larvae on foliage, with loose or poorly formed fecal shield. Lateral scoli and those on A-8 well developed. Hosts are in various dicotyledonous plant families.

This tribe is restricted to the New World and includes about 20 genera and approximately 670 described species, mostly South American. Two genera and four species occur in America north of Mexico.

Hilarocassis Spaeth 1913
Six species comprise this genus which is found throughout most of the Neotropical Region. The widespread species *Hilarocassis exclamationis* (Linnaeus 1767) has been reported from Arizona (Werner 1960). Food plants are *Ipomoea* and *Jacquemontia* (Convolvulaceae).

Chelymorpha Chevrolat 1836
Cyphomorpha Hope 1840
Chelimorpha Schaeffer 1925 [error]
This is a large genus of approximately 100 species, found mostly throughout the Neotropical Region. Three species occur in America north of Mexico. The argus tortoise beetle, *C. cassidea* (Fabricius 1776), is generally distributed throughout the United States and southern Canada; *C. phytophagica* Crotch 1873, occurs in the southwestern United States; and *C. cribraria* (Fabricius 1775), is a South American species recently established in Florida (Thomas 1994). Food plant genera in North America are *Calystegia*, *Convolvulus*, and *Ipomoea* (Convolvulaceae). Larva, Lawson (1991), Peterson (1951); biology and immature stages, Chittenden (1924a).

Ischyrosonychini Chapuis 1875

Physonotini Spaeth 1942
Asterizini Hincks 1952

Characteristics: Body ovate, moderately to highly convex, margins explanate. Head concealed by broadly explanate anterior margin of pronotum; clypeus short, slanted posteriorly; vertex with distinct stridulatory file; antenna with 11 antennomeres, weakly enlarged distally. Pronotum without tactile setae. Tarsal claws paired and symmetrical, simple. Bodies of living specimens usually brightly metallic, fading after death and desiccation. Larvae on foliage with or without loose, poorly-formed fecal shield or with feces deposited directly on dorsum. Lateral scoli and those on A-8 well developed, or those on A-8 absent. Hosts are in various dicotyledonous plant families.

This tribe is restricted to the New World and includes four genera and approximately 70 species; one genus and five species occur in America north of Mexico.

Physonota Boheman 1854
About 40 species comprise this genus which occurs from northern South America to southern Canada, with its greatest species diversity centered in Mexico. The genus needs revision, and nomenclature for some of the five species confirmed for America north of Mexico may eventually change. Two subgenera occur in our region.

subgenus *Physonota* Boheman 1854
Four species of the nominotypical subgenus are positively recorded from America north of Mexico: *P. arizonae* Schaeffer 1925, in Arizona; *P. alutacea* Boheman 1854, in south Texas; and two others found throughout much of the central and eastern United States and adjacent Canadian provinces. Food plant genera include *Franseria*, *Gaertneria*, *Helianthus* (Asteraceae); *Cordia* (Boraginaceae); and *Monarda* (Lamiaceae). Another species, *P. pacifica* Spaeth 1932, was described from "California" but remains unconfirmed for that state. It is confirmed for Sonora and Baja California, Mexico. Sanderson (1948) provided keys to adults, larvae, and pupae; larva, Peterson (1951).

subgenus *Eurypepla* Boheman 1854
The status of the five species and subspecies belonging to this subgenus needs review. The subgenus occurs in the Bahamas, the major islands of the West Indies, the Yucatán peninsula of Mexico, and one subspecies, *P.* (*E.*) *calochroma floridensis* (Blake 1966), is found in south Florida. Our subspecies breeds on *Cordia sebestena* L. (Boraginaceae) in both native hammock and ornamental plantings. Immature stages and biological notes, Woodruff (1976a).

subgenus *Platycycla* Boheman 1854 [Neotropical]

subgenus *Enagria* Spaeth 1913 [Neotropical]

subgenus *Physonaspis* Spaeth 1952 [Neotropical]

Cassidini Gyllenhal 1813

Aspidimorphini Chapuis 1875
Basiptini Chapuis 1875
Charidotini Spaeth 1942

Characteristics: Body ovate, moderately to highly convex, margins explanate. Head concealed by broadly explanate anterior margin of pronotum; clypeus long, horizontal; vertex with stridulatory file which is sometimes indistinct; antenna with 11 antennomeres, usually enlarged distally. Pronotum without tactile setae. Tarsal claws highly variable, paired, symmetrical or asymmetrical, simple, appendiculate, or pectinate. Bodies of living specimens in many genera brightly metallic, fading after death and desiccation. Larvae on foliage and usually with well-organized fecal shield. Lateral scoli and those on A-8 well developed. Hosts in various dicotyledonous plant families.

This is the largest tortoise beetle tribe and is found worldwide with approximately 90 genera and 1,700 species; 15 genera and 29 species occur in America north of Mexico. Key to North American genera and taxonomic notes, Riley (1986a).

[*Aspidimorpha* Hope 1840 (*Aspidomorpha*, of authors), intercepted in North America but not established.]

Cassida Linnaeus 1758
A heterogeneous assemblage of approximately 350 described species from the Old World remains grouped in this genus. Further work is needed to adequately define subgenera and species groups (see Borowiec 1990 and 1994). Presently, fourteen subgenera are recognized. Three of these, including two which were introduced, have been found in the Nearctic Region, but one is no longer extant.

subgenus *Cassida* Linnaeus 1758
The Afrotropical, Palearctic, and Oriental Regions are rich in species of this subgenus, the limits of which are not well established on a worldwide basis. Four species have been recorded in the Nearctic Region, but only *C. relicta* Spaeth 1927 is endemic to North America. *Cassida rubiginosa* Müller 1776 was unintentionally introduced and is now a biological control agent against introduced thistles in the northeastern United States and adjacent Canada. Known food plants for the subgenus in North America include the genera *Arctium*, *Carduus*, *Carthamus*, *Cirsium*, *Cynara*, and *Silybum* (Asteraceae); and *Spergula* and *Stellaria* (Caryophyllaceae). Key to Nearctic species, Riley (1986b). Larva, Peterson (1951); biology, Ward and Pienkowski (1978); history of *C. rubiginosa* in North America and biological notes, White (1996b).

subgenus *Pseudocassida* Desbrochers 1891 [Palearctic]

subgenus *Mionycha* Weise 1891
This Palearctic subgenus contains five species. One species, *C. azurea* Fabricius 1801, was intentionally introduced in Manitoba and Saskatchewan for the control of bladder campion, *Silene vulgaris* (Moench) Garcke (Caryophyllaceae); it reproduced in the field and may be established. Biology (misidentified as *Cassida hemisphaerica* Herbst), Maw (1976).

subgenus *Odontionycha* Weise 1891 [Palearctic]

subgenus *Crepidaspis* Spaeth 1912 [Oriental and Australian]
Taiwania Spaeth 1913
Cyclocassida Chen and Zia 1961
Yunocassis Chen and Zia 1961

subgenus *Lordicassis* Reitter 1926 [Palearctic]

subgenus *Lordiconia* Reitter 1926 [Palearctic]

subgenus *Tylocentra* Reitter 1926 [Palearctic]
Eremocassis Spaeth 1926

subgenus *Onychocassis* Spaeth 1926 [Palearctic]

[subgenus *Cassidulella* Strand 1928
Cassidula Weise 1889, not de Blainville 1830 (Echinodermata)
One species, *C. nobilis* Linnaeus 1758, was intercepted at Sitka, Alaska, in the mid-1800's and has not been seen since. This is a Palearctic subgenus containing six species.]

subgenus *Alledoya* Hincks 1950 [Palearctic]
Deloyala Redtenbacher 1858, not Chevrolat 1836 (Insecta)
Lasiocassis Gressitt 1952

subgenus *Mionychella* Spaeth 1952 [Palearctic]

subgenus *Dolichocassida* Günther 1958 [Palearctic]

subgenus *Cyrtonocassis* Chen and Zia 1962 [Oriental]

[*Coptocycla s. str.* Chevrolat 1836, not in America north of Mexico.]

[*Nuzonia* Spaeth 1912, not in America north of Mexico.]

Psalidonota Boheman 1854
Twenty species comprise this genus which is distributed throughout the Neotropical Region, includng the West Indies. Sometimes it is treated as a subgenus of *Coptocycla* Chevrolat 1836. One Mexican species, *P. texana* Schaeffer 1933, ranges into south Texas where it breeds on *Ehretia anacua* (Mier and Berl.) I. M. Johnst. (Boraginaceae).

Deloyala Chevrolat 1836
Chirida Chapuis 1875
This genus needs a thorough revision. Seven species are presently recognized with a collective range throughout the Neotropical Region and most of the United States and adjacent Canada, but absent from the Pacific Northwest. Four species are recognized in our region but some are of questionable status. Food plants include the genera *Calystegia*, *Convolvulus*, and *Ipomoea* (Convolvulaceae). Biology, Barrows (1979).

Parorectis Spaeth 1901
Orectis Spaeth 1901, not Lederer 1857 (Insecta)
Three described species comprise this genus which ranges from Costa Rica north to the southern tier of the United States. Two species occur in our area, *P. sublaevis* (Barber 1946) from Sonora, Mexico, and Arizona east to Texas, and *P. callosa* (Boheman 1854) from South Carolina to Texas. Both are on *Physalis* (Solanaceae).

Gratiana Spaeth 1913
The seven species belonging to this genus are found in South America. One of these, *G. pallidula* (Boheman 1854), the eggplant tortoise beetle, is widespread in the eastern United States from West Virginia south to Florida, west to California and Baja California Sur. It also occurs in mainland Mexico as far south as Yucatán. Food plants are *Solanum* (Solanaceae). Immature stages and biology, Rolston *et al.* (1965).

Plagiometriona Spaeth 1899
Parametriona Spaeth 1937
This heterogeneous group includes about 75 species which are concentrated in South America, with a few in Central America and Mexico, and one extending north to Canada. One species, *P. clavata* (Fabricius 1798), is generally distributed in the eastern United States and adjacent Canada, west across the southern states to Arizona and south to northern South America. Its food plant genera are *Datura*, *Solanum*, and *Physalis* (Solanaceae). Larva, Lawson (1991). Notes on biology, Woodruff (1975), Barrows (1979).

[*Metriona* Weise 1896, not in America north of Mexico.]

Agroiconota Spaeth 1913
This genus includes about 16 described species and is found throughout the Neotropical Region, being most diverse in South America. One species, *A. bivittata* (Say 1825), is widely distributed in the eastern United States west to Arizona and south into western Mexico. Food plants are *Convolvulus* and *Ipomoea* (Convolvulaceae).

Floridocassis Spaeth 1952
This genus is monotypic, with *F. repudiata* (Suffrian 1868) originally described from Cuba but also found in the southeastern coastal United States from North Carolina to Mississippi. It breeds on *Ipomoea sagittata* Poir. (Convolvulaceae), a plant usually found growing in low coastal areas.

Jonthonota Spaeth 1913
Three species comprise this genus which is generally distributed in the United States and adjacent southern Canada, with one species extending southward to the state of Mexico in Mexico. Food plants belong to the genera *Convolvulus* and *Ipomoea* (Convolvulaceae). Larva, Lawson (1991).

Opacinota E. Riley 1986
This genus is monotypic, with *O. bisignata* (Boheman 1855) found in the southeastern coastal United States from Maryland to Texas. It feeds on *Ipomoea pandurata* (L.) Mey. (Convolvulaceae). Taxonomic notes, Riley (1985a, 1986a).

Erepsocassis Spaeth 1936
This genus is monotypic, with *E. rubella* (Boheman 1862) occurring in the southeastern coastal United States from Virginia to Mississippi. It is uncommon, and the host is unknown although it is probably one of the Convolvulaceae. Taxonomic notes, Riley (1982).

Strongylocassis Hincks 1950 [conserved name, ICZN 1987, Opinion 1453]
Strongylaspis Spaeth 1936, not Thomson 1860 [Insecta]
This genus is monotypic, with *S. atripes* (LeConte 1859) generally distributed in the eastern United States from Connecticut to Georgia, west to Texas, Utah, and Manitoba. Food plant genera are *Convolvulus* and *Ipomoea* (Convolvulaceae). Taxonomic notes, Riley (1985a).

Metrionella Spaeth 1932
This genus of approximately ten species is spread throughout the Neotropical Region except for the West Indies. One species, *M. bilimeki* (Spaeth 1932), ranges through much of Central America and Mexico, including the Baja California peninsula, and reaches southeastern Arizona. It feeds on *Ipomoea* (Convolvulaceae).

Charidotella Weise 1896
Two of the four subgenera occur in our region. Key to subgenera and list of species, Borowiec (1989).

subgenus *Charidotella* Weise 1896
Three species are known from America north of Mexico: *Charidotella sexpunctata* (Fabricius 1781), the golden tortoise beetle, is transcontinental and extends southward into South America; *C. succinea* (Boheman 1855) is found from Arizona to Colorado and Texas, south to Panama; and *C. bifossulata* (Boheman 1855) is recorded from Central America, Mexico and Arizona. Many other species occur throughout the Neotropical Region. Food plant genera in America north of Mexico are *Calystegia*, *Convolvulus*, and *Ipomoea* (Convolvulaceae). Illustrations of immature stages, Orton and Chittenden (1917); biology, Barrows (1979).

subgenus *Philaspis* Spaeth 1913 [Neotropical]

subgenus *Metrionaspis* Spaeth 1942 [Neotropical]

subgenus *Chaerocassis* Spaeth 1952
Three species of this subgenus occur in America north of Mexico, one widespread in the eastern and central United States and adjacent Canada, one in central Florida, and one in the southwestern states and southward to Costa Rica. Two additional species are found in Mexico. Food plant genera in America north of Mexico are *Calystegia*, *Convolvulus*, and *Ipomoea* (Convolvulaceae).

Microctenochira Spaeth 1926
Ctenochira Chapuis 1875, not Förster 1855 [Insecta]
Euctenochira Hincks 1950
Approximately 100 species comprise this genus which is found throughout the Neotropical Region except for the West Indies. One species, *M. bonvouloiri* (Boheman 1862), ranges through much of Central America and Mexico, including the Baja California peninsula; it reaches extreme south Texas where it breeds on *Merremia dissecta* (Jacq.) Hallier f. (Convolvulaceae).

Chrysomelinae Latreille 1802

Characteristics: Body broadly oval to elongate, convex, most hemispherical. Head inserted into prothorax to eyes, only partly visible dorsally; eyes feebly emarginate; antennae somewhat enlarged apically, composed of 11 freely articulated antennomeres; antennal insertions widely separated, positioned between eye and mandible. Prothorax of most broad and somewhat convex, many emarginate in front; lateral margins well defined. Elytra convex, covering abdomen in most species; epipleura well defined. Procoxae transverse, most widely separated. Metafemora not abnormally enlarged, similar in size and form to front and middle femora. Tarsal formula 5-5-5, most pseudotetramerous; bifid tarsal setae absent.

Larvae feed externally on leaves. Unlike those of some other subfamilies, they are not equipped with fecal coverings. However, many are chemically protected from predation, and some are aposematically colored.

This subfamily contains 133 genera and is well represented in most parts of the world, especially in tropical areas. Following the classification of Daccordi (1994), it is composed of only two tribes. Both occur in America north of Mexico where they are represented by a total of 16 genera and 135 species. Zoogeography, Daccordi (1996).

Timarchini Motschulsky 1860

Characteristics: Species flightless, dark colored, tenebrionid-like. Procoxal cavities closed behind; metasternum reduced, with mesal length less than that of prosternum. Aedeagus with tegmen ring-like, encircling median lobe, with short dorsal cap.

As demonstrated by the form of the aedeagus, this group is quite distinct from the remainder of the Chrysomelinae. Unlike the Y-shaped tegmen found in other chrysomeline lineages (and in most other chrysomelid subfamilies), the ring-like tegmen of timarchines is reminiscent of that found in the basal subfamilies Donaciinae, Bruchinae, and extralimital Sagrinae. The tribe includes a single genus and has a relictual distribution, with most species restricted to the Mediterranean area, but with one subgenus occurring in western North America.

Timarcha Latreille 1829
subgenus *Americanotimarcha* Jolivet 1948
This strictly Nearctic subgenus contains two species distributed from British Columbia to Montana to California. They feed on Rosaceae, particularly *Fragaria* and *Rubus*, and on Ericaceae. Biology, Jolivet (1976). Keys to species, Jolivet (1948, 1989), Wilcox (1972a).

subgenus *Timarcha* Latreille 1829 [Palearctic]
Timarchostoma Motschulsky 1860

subgenus *Metallotimarcha* Motschulsky 1860 [Palearctic]

Chrysomelini Latreille 1802

Characteristics: Procoxal cavities variable, but if closed behind, metasternum normal, longer than prosternum. Aedeagal tegmen Y-shaped, not encircling median lobe.

This tribe, containing 132 genera, is nearly worldwide in distribution. The taxa treated below as subtribes are sometimes given full tribal rank. However, the classification of Daccordi (1994) is followed here.

Entomoscelina Chapuis 1874

Characteristics: Procoxal cavities closed behind; procoxae separated by prosternum. Tarsomeres 1, 2, and especially 3 broad, with well-developed setose pads ventrally; tarsal claws simple.

This subtribe, containing 27 genera, is nearly worldwide in distribution.

Entomoscelis Chevrolat 1836
This genus includes about twelve species, mostly from Eurasia. The single North American species, *E. americana* Brown 1942, occurs from Alaska to Wisconsin to Colorado. Although it has been reported to damage a few plants in other families, this insect, also known as the red turnip beetle, usually feeds on Brassicaceae, including *Armoracia*, *Brassica*, *Capsella*, *Erysimum*, *Lepidium*, *Raphanus*, and *Sisymbrium*. Biology, Gerber (1994).

Microtheca Stål 1860
This genus is composed of about eight South American species. Two of them, *M. ochroloma* Stål 1860, and *M. picea* (Guérin-Méneville 1829), are adventive in the southeastern United States. They feed principally on Brassicaceae, including *Brassica*, *Lepidium*, *Nasturtium*, and *Raphanus*. Key to species, Balsbaugh (1978).

Paropsina Motschulsky 1860

Gonioctenini Motschulsky 1860

Characteristics: Procoxal cavities open behind. Tibiae armed with preapical tooth followed by coarsely setose emargination, or meso- and metatibiae with weak emargination and dense brush of setae; tarsal claws appendiculate, each with basal tooth.

This subtribe contains about 20 genera. Although it is nearly worldwide in distribution, it is poorly represented in the Western Hemisphere.

Gonioctena Chevrolat 1836
This genus is most diverse in the Old World. However, as indicated below, one subgenus is represented in North America.

subgenus *Spartophila* Stephens 1834 [Palearctic]
Spartiophila Weise 1884
Amblis Gistel 1847

subgenus *Gonioctena* Chevrolat 1836
Phytodecta Kirby 1837
This subgenus is distributed throughout much of Europe and Asia. It also includes four species that occur in the northern part of the United States and in Canada, where they feed on *Populus* and *Salix* (Salicaceae). Keys to species, Brown (1942a), Wilcox (1972a). Larval descriptions, Lawson (1991), Takizawa (1989); biology, Mason and Lawson (1982).

subgenus *Goniomena* Motschulsky 1860 [Palearctic]

subgenus *Spartoxena* Motschulsky 1860 [Palearctic]
Phytodectella Cantonnet 1968

subgenus *Spartomena* Reitter 1912 [Palearctic]

subgenus *Ceryonops* Jacobson 1900 [Palearctic]

subgenus *Asiphytodecta* Chen 1935 [Oriental]

subgenus *Sinomela* Chen 1935 [Oriental]

subgenus *Platyphytodecta* Bechyné 1948 [Oriental]

subgenus *Brachyphytodecta* Bechyné 1948 [Palearctic and Oriental]

Trachymela Weise 1908
Approximately 120 species comprise this genus, all from the Australian Region (Weise 1916). Recently, one of these, *T. sloanei* (Blackburn 1897), has become established in southern California where it is a pest of red gum eucalyptus, *Eucalyptus camaldulensis* Dehnh. (Myrtaceae).

Doryphorina Motschulsky 1860

Zygogrammini Weise 1915
Chrysolinina Chen 1936

Characteristics: Maxilla with terminal palpomere broadly truncate at tip, subquadrangular or distally dilated. Epipleuron apically setose. Tibiae without preapical tooth or emargination; third tarsomere apically truncate or slightly emarginate.

With the exception of Australia where only an introduced species occurs, this subtribe naturally occurs almost worldwide. It includes 30 genera.

Chrysolina Motschulsky 1860 [conserved name, ICZN 1984, Opinion 1279]
This is a large genus, with hundreds of species occurring in the Palearctic Region. More than 70 subgenera are recognized. However, in contrast to the treatments of other chrysomelid genera in this work, only the subgenera known to occur in North America or those that have been reported from North America are listed below (see Daccordi 1994 for a listing of subgenera). Some of the characters that have previously been used to separate subgenera are difficult to interpret or otherwise do not work well for Nearctic species. Such characters are not used in the preceding key but are replaced by others (perhaps superficial) that are not widely used in Old World literature. Although the key may not work for extralimital species, it will enable the proper sorting of North American beetles. Also, note that the classification within the genus *Chrysolina* is currently being studied by European workers, and some subgeneric placements may change after their scrutiny.

subgenus *Chalcoidea* Motschulsky 1860
This subgenus is widely distributed in Europe and Asia, and there are four species that occur in North America, from Alaska to Newfoundland to Ohio to California. Nearctic species feed on *Achillea*, *Artemisia*, *Aster*, *Hymenoclea*, and *Tanacetum* (Asteraceae). Keys to species, Brown (1962), Jolivet (1992), Wilcox (1972a).

subgenus *Chrysolina* Motschulsky 1860
Stichosoma Motschulsky 1860
This is a Palearctic group, with one species, *C. staphylea* (Linnaeus 1758), introduced into eastern Canada. In the Old World, this species has been associated with *Achillea*, *Hymenoclea* (Asteraceae); *Melissa*, *Mentha*, *Ocimum* (Lamiaceae); *Plantago* (Plantaginaceae); and *Ranunculus* (Ranunculaceae). Larval description, Marshall (1979).

subgenus *Pleurosticha* Motschulsky 1860
Pelurosticha: Kimoto and Gressitt 1981 [error]
This subgenus includes one North American species, *C. cavigera* (J. Sahlberg 1885), occurring in Alaska and the Yukon, and also in Siberia. This species feeds on *Parrya* (Brassicaceae). Biology, Khruleva (1996).

subgenus *Hypericia* Bedel 1899
Two species of this Palearctic subgenus, *C. hyperici* (Forster 1771) and the well-known Klamathweed beetle, *C. quadrigemina* (Suffrian 1851), have been intentionally introduced into North America for the biological control of *Hypericum* (Clusiaceae). They are well established, both in western and eastern areas of the continent. Keys to species, Brown (1962), Wilcox (1972a). Biology and descriptions of immature stages, Marshall (1979), Paterson (1931), Wilson (1943).

subgenus *Sphaeromela* Bedel 1899
In an effort to control *Hypericum* (Clusiaceae), the European species *C. varians* (Schaller 1782), was introduced into western areas of Canada and the United States. However, it may not be established. Descriptions of immature stages, Marshall (1979), Paterson (1931).

[subgenus *Euchrysolina* Bechyné 1950; not in North America]
One Palearctic species, *C. virgata* (Motschulsky 1860), has been reported from Florida, but this probably represents an interception.

subgenus *Pezocrosita* Jacobson 1901
This is a Holarctic subgenus with five species native to North America. They feed on Lamiaceae, including *Pycnanthemum* and *Satureja*. Identification of species, Brown (1962), Jolivet (1992), LeSage (1989), Wilcox (1972a).

subgenus *Arctolina* Kontkanen 1959
This Holarctic subgenus includes two species, *C. caurina* Brown 1962 and *C. subsulcata* (Mannerheim 1853), that occur in Alaska and the Yukon. They feed on *Nordosmia*, *Senecio* (Asteraceae); *Parrya* (Brassicaceae); and *Oxytropis* (Fabaceae). Keys to species, Brown (1962), Jolivet (1992), Wilcox (1972a).

[subgenus *Fastuolina* Warchalowski 1991; not in North America]
One Palearctic species, *C. fastuosa* (Scopoli 1763), has been reported from Pennsylvania, but this record is extremely doubtful.

Calligrapha Chevrolat 1836
This genus of showy beetles contains more than 80 currently recognized species native to North and South America. Key to subgenera, Monrós (1955); but see also Clark and Cavey (1995).

subgenus *Bidensomela* Monrós 1955
Acalligrapha Monrós 1955
Coreopsomela Monrós 1955
This subgenus is composed of four species, all of which occur in America north of Mexico. They have been associated with *Ambrosia*, *Bidens*, *Coreopsis*, *Dahlia*, *Eupatorium*, and *Solidago* (Asteraceae). Keys to species, Clark and Cavey (1995), Wilcox (1972a). Descriptions of immature stages, Papp (1959).

subgenus *Calligramma* Monrós 1955
This subgenus is monotypic, with *C. cephalanthi* (Schwarz 1878) endemic to Florida. It is reported to feed on *Cephalanthus* (Rubiaceae).

subgenus *Calligrapha* Chevrolat 1836
Polyspila Hope 1840
Phyllis Gistel 1847
Boliographa Motschulsky 1860
Metallographa Motschulsky 1860
Most species of *Calligrapha* belong in this subgenus, 31 of them occurring north of Mexico. The host plants north of Mexico include *Heliopsis* (Asteraceae); *Alnus, Betula, Corylus, Ostrya* (Betulaceae); *Cornus* (Cornaceae); *Althaea, Malva, Sidalcea, Sphaeralcea* (Malvaceae); *Crataegus, Physocarpus, Prunus* (Rosaceae); *Populus, Salix* (Salicaceae); *Tilia* (Tiliaceae); *Ulmus* (Ulmaceae); and possibly *Illicium* (Illiciaceae). Keys to species, Brown (1945), Wilcox (1972a). Descriptions of immature stages and biology, Brown (1945), Lawson (1991), Peterson (1951), Wheeler and Hoebeke (1979).

subgenus *Graphicallo* Monrós 1955
This subgenus is distinguished by only a few weak characters, and it may eventually fall into synonymy with the subgenus *Calligrapha* or perhaps *Bidensomela*. It is composed of a single species, *C. lunata* (Fabricius 1787), which occurs from Alberta to New Brunswick to Louisiana to Colorado. It feeds on *Rosa* (Rosaceae).

Zygogramma Chevrolat 1836
This genus contains about 100 species distributed throughout much of North and South America.

subgenus *Zygogramma* Chevrolat 1836 [Neotropical]

subgenus *Tritaenia* Motschulsky 1860 [Neotropical]

subgenus *Zygospila* Achard 1923
This subgenus contains about 50 species distributed from Canada to Panama, with 13 of them occurring north of Mexico. Of these, *Z. suturalis* (Fabricius 1775) is an eastern species occurring from Manitoba to Maine to North Carolina to Colorado, and *Z. heterothecae* Linell 1896 is rather widespread in the southern half of the United States. The other eleven species occur mostly in the western half of the continent, but the distribution of one of them, *Z. exclamationis* (Fabricius 1798), the sunflower beetle, extends into eastern Canada. North of Mexico, species feed on *Ambrosia, Aster, Helianthus, Heterotheca, Iva, Parthenium, Simsia, Tithonia*, and *Viguiera* (Asteraceae). *Zygogramma exclamationis* is a serious pest of cultivated sunflowers (*Helianthus annuus* L.). Key to species, Wilcox (1972a). Larval descriptions, Lawson (1991). Life history and biology, Goeden and Ricker (1979), Piper (1978).

Leptinotarsa Chevrolat 1836 [conserved name, ICZN 1984, Opinion 1290]
Polygramma Chevrolat 1836 [suppressed name, ICZN 1984, Opinion 1290]
Myocoryna Dejean 1836 [*nomen nudum*]
Leptinotarsa: Stål 1858 [a subsequent usage]
Myocoryna Stål 1859
Thlibocoryna C. Riley 1875
This genus contains more than 40 species distributed throughout much of North and South America, with at least ten species occurring north of Mexico. Among these is the widespread and notorious Colorado potato beetle, *L. decemlineata* (Say 1824). Beyond this, *L. juncta* (Germar 1824), is found in the eastern United States, and the rest of the species north of Mexico occur in the southwestern United States, with the distributions of many extending southward into Mexico. Among the southwestern species is *L. collinsi* Wilcox 1972, from Arizona. Contrary to the opinion of Jacques (1988), this species belongs in this genus and not in *Calligrapha*. Host plants for species north of Mexico include *Hymenoclea, Tithonia* (Asteraceae); *Atropa, Datura, Hyoscyamus, Lycium, Lycopersicon, Nicotiana, Physalis, Solanum* (Solanaceae); and *Kallstroemia* (Zygophyllaceae). Keys to species, Jacques (1988), Wilcox (1972a). Larval descriptions, Lawson (1991), Peterson (1951). Biology, Hsiao (1988), Jacques (1988).

Labidomera Chevrolat 1836
Paropsimena Motschulsky 1860
This genus is composed of four species. One of them, *L. clivicollis* (Kirby 1837), is known as the milkweed leaf beetle and is rather widespread in North America, occurring from Manitoba to Quebec south to Florida to western Texas, and into northern Mexico. The other species occur in Mexico and Central America. North of Mexico, the beetles feed on *Asclepias, Cynanchum*, and *Sarcostemma* (Asclepiadaceae). Key to species, Daccordi and LeSage (1999). Larval descriptions, Lawson (1991), Peterson (1951). Biology, Dickinson (1996), Palmer (1985).

Chrysomelina Latreille 1802

Characteristics: Maxilla with terminal palpomere apically attenuate. Epipleuron without apical setae. Front coxal cavities open behind. Tibiae without preapical tooth or emargination, although some are angularly broadened at apex; third tarsomere variable, deeply bilobed in many.

This subtribe, including 54 genera, is nearly worldwide in distribution.

Gastrophysa Chevrolat 1836
subgenus *Gastrophysa* Chevrolat 1836
Gastroeidea Hope 1840
Gastroidea: Gemminger and Harold 1874 [error]
This essentially Holarctic subgenus is composed of nine species, five of which occur in the Western Hemisphere (four north of Mexico and one in Mexico). North of Mexico, the beetles feed on *Polygonum, Rheum*, and *Rumex* (Polygonaceae). Keys to species, Jolivet (1951b), Wilcox (1972a). Larval descriptions and biology, Bøving and Craighead (1931), Girault (1908), Goe (1918), Jolivet (1951a), Lawson (1950, 1991), Paterson (1931).

subgenus *Exiguipenna* Jolivet 1951 [Palearctic]

Phaedon Megerle von Muhlfeld 1823
subgenus *Phaedon* Megerle von Muhlfeld 1823
Alitene Gistel 1857
Emmetrus Motschulsky 1860
Phaeton: Kuster 1846 [error]
This Holarctic and Neotropical subgenus is represented by seven species in America north of Mexico and is generally distributed throughout most of the region. Although these beetles have been reported from a variety of plants, their true hosts are probably *Bidens*, *Cosmos*, *Tagetes* (Asteraceae); *Armoracia*, *Barbarea*, *Brassica*, *Lepidium*, *Nasturtium* (Brassicaceae); and *Veronica* (Scrophulariaceae). Some of the beetle species live in aquatic or semi-aquatic habitats. Key to species, Balsbaugh (1983). Descriptions of immature stages, Lawson (1991), Paterson (1931); biology, Chittenden (1907).

subgenus *Orthosticha* Motschulsky 1860 [Neotropical]

subgenus *Hemiphaedon* Jacobson 1901 [Palearctic]

subgenus *Paraphaedon* Sharp 1910 [Palearctic]

subgenus *Allophaedon* Kontkanen 1933
This Nearctic subgenus is composed of a single species, *P. prasinellus* (LeConte 1861), which occurs from British Columbia to Indiana to Texas to California. Of the three plants mentioned by Balsbaugh (1983), tansy mustard (*Descurainia*, Brassicaceae) is a likely host. Species identification, Balsbaugh (1983).

Prasocuris Latreille 1802
subgenus *Prasocuris* Latreille 1802
Helodes Dejean 1836, not Paykull 1799
Hellodes: Redtenbacher 1845 [error]
This is essentially a Palearctic subgenus; however, one species, *P. phellandrii* (Linnaeus 1758), occurs in wetlands from Alaska to Quebec to Ohio to Colorado. It was likely introduced from Europe. In North America, it has been associated with *Sium* (Apiaceae) and *Caltha* (Ranunculaceae). Descriptions of immature stages, Paterson (1931).

subgenus *Hydrothassa* Thomson 1859
Eremosis Des Gozis 1882
Agrostithassa Jacobson 1912
This is a Holarctic subgenus, with four species occurring in North America. They are associated with semi-aquatic species of *Caltha* and *Ranunculus* (Ranunculaceae). Keys to species, Schaeffer (1928), Wilcox (1972a).

Phratora Chevrolat 1836
subgenus *Phratora* Chevrolat 1836
Phyllodecta Kirby 1837
Seven species of this Holarctic subgenus occur in North America. They feed on *Betula* (Betulaceae) and on *Populus* and *Salix* (Salicaceae). Keys to species, Brown (1951), Wilcox (1972a). Biology and descriptions of immature stages, Lawson (1991), Lindquist and Davis (1971).

subgenus *Chaeroceroides* Strand 1935 [Palearctic]
Chaetocera Weise 1884, not Agassiz 1846 [Insecta]
Chaeroceta Mroczkowski 1990

subgenus *Alonsina* Warchalowski 1995 [Palearctic]

Chrysomela Linnaeus 1758 [conserved name, ICZN 1984, Opinion 1279]
subgenus *Chrysomela* Linnaeus 1758
Gymnota Gistel 1847
Lina Dejean 1836
Melasoma Stephens 1831
Melosoma: Bedel 1899 [error]
Two species, *C. crotchi* Brown 1956 and *C. invicta* Brown 1956, of this Holarctic subgenus occur in North America. They feed on *Populus* (Salicaceae). Keys to species, Brown (1956), Wilcox (1972a). Biology, Smereka (1965).

subgenus *Macrolina* Motschulsky
Microdera Stephens 1839, not Eschshcoltz 1831 [Insecta]
Macrolina Motschulsky 1860
Strickerus Lucas 1920
Microlina Lopatin 1977 [unavailable]
This is essentially a Holarctic subgenus, with the distributions of a few species extending southward into tropical areas. Represented by 14 species, it is generally distributed north of Mexico, occurring even in Arctic areas. It feeds on *Alnus* (Betulaceae) and on *Populus* and *Salix* (Salicaceae). Keys to species, Brown (1956), Wilcox (1972a). Larval descriptions, Lawson (1991), Peterson (1951). Biology, Brown (1956), Lowe (1898).

subgenus *Pachylina* Medvedev and Chernov 1969
A single American species, *C. blaisdelli* Van Dyke 1938, from the Arctic of Canada and Alaska is in this subgenus. Other species are present in the Palearctic.

Plagiodera Chevrolat 1836
This nearly worldwide genus is often distinguished from its relatives by the nature of tarsomere 3, the apex of the ventral lobe being truncate or slightly emarginate, rather than deeply bilobed. This condition holds true for three of the four species that occur north of Mexico. However, contrary to published reports, this is not the condition in the type species, *Chrysomela armoraciae* Fabricius 1775 (= *P. versicolora* (Laicharding 1781)). In this species, the tarsal lobes are deeply incised, as in true *Chrysomela*. Key to subgenera, Daccordi (1986).

subgenus *Plagiodera* Chevrolat 1836
This is an Old World group with one species, *P. versicolora* (Laicharding 1781), being adventive in North America. It now occurs from Manitoba to New Brunswick to North Carolina.

Also known as the imported willow leaf beetle, it usually feeds on *Salix* (Salicaceae) but can also be found associated with *Populus* (Salicaceae). Descriptions of immature stages, Hood (1940), Lawson (1991), Peterson (1951), Weiss and Dickerson (1917). Biology, Wade (1994), Wade and Breden (1986).

subgenus *Linamorpha* Motschulsky 1860 [Neotropical]

subgenus *Plagiomorpha* Motschulsky 1860
Melasomida Schaeffer 1920
Pseudolina Schaeffer 1919, not Jacoby 1896 [Insecta]
This is a strictly North American subgenus, with two species occurring north of Mexico, from California to Colorado to Texas, and with three more distributed from Mexico to Guatemala. In the United States, *P. arizonae* Crotch 1874 and *P. californica* (Rogers 1856) feed on *Populus* or *Salix* (Salicaceae). Key to species, Wilcox (1972a).

subgenus *Plagiosterna* Motschulsky 1860 [Palearctic]

subgenus *Pseudoparopsis* Blackburn 1899 [Australian and Oriental]

subgenus *Paraplagiomorpha* Daccordi 1986 [Neotropical]

subgenus *Plagioschema* Daccordi 1986
This is essentially a Neotropical group, but one species, *P. thymaloides* Stål 1860, occurs in south Texas where larvae and adults feed on *Xylosma flexuosa* (H. E. K.) Hemsl. (Flacourtiaceae).

Chrysomelini: *incertae sedis*

Cadiz Andrews and Gilbert 1992
In contrast to the opinion of Daccordi (1994) who excluded this genus from the Chrysomelinae, its placement within the subfamily is here reaffirmed. Although its affinities to other chrysomelines is not clear, it is obviously not closely related to *Timarcha*, and, accordingly, is here placed within the tribe Chrysomelini.

Cadiz includes a single species, *C. hardyi* Andrews and Gilbert 1992, that feeds on *Coldenia* (Boraginaceae) in southern California sand dunes. The absence of expanded tarsal pads is probably an adaptation for a life spent partially in loose sand. The larva is known to be folivorous but remains undescribed. Biology and adult description, Andrews and Gilbert (1992).

Galerucinae Latreille 1802

Characteristics: Body oval or oblong. Head exposed, easily visible from above, inserted into prothorax but without neck-like constriction at base; frontal tubercles present behind antennal insertions, delimited posteriorly by more or less distinct, transverse, interocular groove; eyes of most entire; antennae of most shorter than body, filiform or clavate, not capitate, composed of 11 (only 10 in a few) freely articulated antennomeres; some antennomeres occasionally enlarged and modified in male; antennal insertions on frons between eyes, narrowly separated from each other. Pronotum truncate or emarginate in front; lateral bead present in most. Tarsi 5-5-5, pseudotetramerous, the fourth tarsomere being very small; tarsomere 3 with ventral, setose, bilobed pad; tarsal setae not bifid.

Adults feed on leaves or flower parts including pollen. Larvae usually feed on leaves or roots.

In many previous classifications, this group of beetles has been treated as two separate subfamilies, the Galerucinae and the Alticinae, and this view is still held by many chrysomelid workers. Lingafelter and Konstantinov (2000) provide a brief review of the old and recent history of this question. Their introductory comments highlight an emerging trend of merging the two groups into a single subfamily. Support for this single-subfamily view comes from the results of their cladistic analysis of adult characters which demonstrates that the alticines are a well-supported monophyletic group within a larger Galerucinae. Farrell's (1998) analysis of both DNA sequence data and morphological data conflicts with their conclusions in that both groups are recognized as monophyletic, but this study was not designed to address this specific question. Based on larval morphology, Bøving and Craighead (1931) treated the groups as a single family and failed to find satisfactory characters to segregate Alticinae apart from other Galerucinae. Reid (1995, 2000) treated the two groups as a single subfamily in his assessments of phylogenetic relationships among chrysomelid subfamilies. In keeping with this trend, the two groups are treated as one in the present work. Together, they form by far the largest leaf beetle subfamily, with roughly 1,000 genera and from 13,000 to 15,000 species described worldwide.

Within the group formerly recognized as the Alticinae, no satisfactory classification exists. The systems proposed by early workers are no longer useful. Although modern workers have grouped a few closely related genera under tribal and subtribal names, they have not classified the majority of genera into any tribal arrangement. No doubt, future phylogenetic investigations will provide a well-supported internal classification for these beetles. Until then, it seems best to treat the entire group as a single taxon, the Alticini.

In contrast, modern workers (Seeno and Wilcox 1982; Wilcox 1965) have recognized an internal classification for the galerucines, exclusive of the Alticini. As expected, the subfamily has been divided into tribes, and these have in turn been divided into subtribes. Beyond this, subtribes are subdivided into sections ("-ites" endings). Unfortunately, there are a few oddities in this system. For instance, the subtribal grouping is occasionally bypassed, the tribes being divided directly into sections. The groups recognized as tribes and subtribes seem to be quite natural and fairly well characterized, at least in the New World. However, the sections vary in their value. Some of them are quite distinct, but others are hardly, if at all, separable.

Reid (1995) treated all of the group formerly recognized as Galerucinae as a single tribe, the Galerucini. If his classification were strictly followed, the rankings of lesser groups within this single tribe would need to be adjusted downward. Although a revision of the galerucine classification is sorely needed, it is far beyond the scope of this book. At present, it seems best to keep the tribal classification intact, unchanged in rank from when alticines were excluded from the Galerucinae.

Galerucini Latreille 1802

Characteristics: Antennae generally inserted low on frons, beyond middle of eyes. Hind femur slender, without internal extensor apodeme (spring); most with anterior and posterior tibiae lacking terminal spurs; tarsal claws of Nearctic species either simple or bifid with narrow, sharply pointed appendage. Last ventrite of male abdomen with median, apical, semicircular depression, some with emargination behind impression; abdomen without apical lobe in either sex. Aedeagus with prominent basal spurs or with constriction slightly beyond base. Larvae feeding on leaves.

In the early period of chrysomelid taxonomy, characters were used to separate the sections now included in this tribe. Since then, numerous species have been added to each of these sections. Although these additional species appear to be closely related to those already included, they do not always possess characters previously thought to be diagnostic for the sections. Taxonomic revision is needed to demonstrate that these sections are truly monophyletic and hopefully to elucidate characters that will easily enable their separation. If future revision does not accomplish these goals, these three sections might best be treated as synonyms.

Coelomerites Chapuis 1875

Monocesta H. Clark 1865
This New World genus contains about 30 described species, most of which occur in tropical areas. A single species, *M. coryli* (Say 1824), occurs in the United States from Kansas to Pennsylvania to Florida to Oklahoma. Known as the larger elm leaf beetle, the normal host of this species is *Ulmus* (Ulmaceae). Life history and larval descriptions, Anderson and Papp (1961), Bøving (1929), Bøving and Craighead (1931), Peterson (1951).

Coraia H. Clark 1865
As presently constituted, this genus contains four Mexican and Central American species. The distribution of one of them, *C. subcyanescens* (Schaeffer 1906), extends northward into southern Texas where adults and larvae feed on *Karwinskia* (Rhamnaceae). Species identification, Blake (1931a).

Derospidea Blake 1931
This genus contains three species. One of them, *D. brevicollis* (LeConte 1865), occurs from Kansas to Massachusetts to Florida to Texas and in Mexico. The other two are essentially Mexican, but the distribution of *D. ornata* (Schaeffer 1905) extends into southern Texas. North of Mexico, species feed on *Citrus* or *Zanthoxylum* (Rutaceae). Key to species, Wilcox (1965). Larval description, Bøving (1929).

Trirhabda LeConte 1865
  Trirrhabda: Gemminger and Harold 1876 [error]
  Trihabda: Crotch 1873 [error]
This North and Central American genus contains about 30 species, 24 of which occur north of Mexico. Our species feed on *Acamptopappus*, *Artemisia*, *Baccharis*, *Bahia*, *Bigelowia*, *Brickellia*, *Chrysothamnus*, *Cirsium*, *Encelia*, *Franseria*, *Gutierrezia*, *Haplopappus*, *Helianthus*, *Hymenoclea*, *Parthenium*, and *Solidago* (Asteraceae); and *Covillea* and *Eriodictyon* (Hydrophyllaceae). Keys to species, Blake (1931a), Hatch (1971), Wilcox (1965). Larval description, Bøving (1929), Lawson (1991). Biology, Boldt (1989), Eckberg and Cranshaw (1994), Messina (1982), O'Brien and Atsatt (1982), Palmer (1986), Palmer and Haseler (1992), Redak *et al.* (1995).

Miraces Jacoby 1888
  Halticidea Horn 1893
This New World genus contains three described mainland species, all of which occur in the United States, the distribution of one of them extending as far south as Guatemala. It also contains three West Indian species, as well as unnamed species from Mexico and Central America. In Arizona, the genus has been associated with *Rhamnus* (Rhamnaceae), and, in southern Texas, adults of *M. aeneipennis* Jacoby 1880, are repeatedly found on *Condalia hookeri* M. C. Johnst. (Rhamnaceae). In south Florida, the adults of *M. placida* (Horn 1893) have been taken on *Eugenia* (Myrtaceae). Key to species, Wilcox (1965).

Atysites Chapuis 1875

Diorhabda Weise 1883
  Prophyllis Reitter 1912
  Radymna Reitter 1912
This genus contains 15 Old World species. One of them, *D. elongata* (Brulle 1832), has received USDA approval as a biological control agent against *Tamarix* (Tamaricaceae). It will likely be released in southwestern states in the near future.

Galerucella Crotch 1873
  Hydrogaleruca Laboissiére 1922
This genus is nearly worldwide in distribution. However, it is notably absent from the Neotropical Region, many of the species previously classified in *Galerucella* properly belonging in the genus *Yingaresca* Bechyné 1956 or in other genera. *Galerucella* is represented in the New World by a single species, *G. nymphaeae* (Linnaeus 1758), that is distributed in wetlands throughout much of the United States and Canada, as well as in the Palearctic Region. In North America, it feeds commonly on *Nuphar* (Nymphaeaceae) and *Polygonum* (Polygonaceae). It has also been reported from plants such as *Sagittaria* (Alismataceae), *Brasenia* (Cabombaceae), *Myrica* (Myricaceae), *Nymphaea* (Nymphaeaceae), *Rumex* (Polygonaceae), and *Potamogeton* (Potamogetonaceae). Descriptions of immature stages, Bøving (1929), Lawson (1991), MacGillivray (1903), Paterson (1931).

Neogalerucella Chûjô 1962
This group of beetles is quite distinct and deserves the full generic status attributed to it by some recent European authors. In North America, it contains two native species, *N. quebecensis* Brown 1938 and *N. stefanssoni* Brown 1938, distributed from Alaska to Nova Scotia to Michigan. They feed on *Potentilla* and *Rubus* (Rosaceae). Beyond this, two European species, *N. calmariensis* (Linnaeus 1767) and *N. pusilla* (Duftschmid 1835), have been introduced into wetlands in many areas of the United States for the biological control of *Lythrum* (Lythraceae). They are apparently well established. Keys to species, Manguin, *et al.* (1993), Wilcox (1965). Descriptions of immature stages, Paterson (1931).

Tricholochmaea Laboissiére 1932
This is a mostly Holarctic genus, but it is also represented in Indonesia. More or less thirteen species occur in America north of Mexico, the number varying slightly depending on the definitions of species and subspecies. Nearctic species feed on *Alnus* (Betulaceae); *Kalmia*, *Rhododendron*, *Vaccinium* (Ericaceae); *Prunus*, *Spiraea* (Rosaceae); *Populus*, *Salix* (Salicaceae); and *Ribes* (Saxifragaceae). Identification of adults, Wilcox (1965), Brown (1969). Biology and descriptions of immature stages, Cushman (1916), Hartzell (1917), Herrick (1916), Woods (1924).

Pyrrhalta Joannis 1866
Although this is a Palearctic genus, one species, *P. viburni* (Paykull 1799), has been accidentally introduced and become established in North America (Maine, New York, Nova Scotia, Ontario, Quebec). It feeds on *Viburnum* (Caprifoliaceae). Biology, Becker (1979). Descriptions of immature stages, Bøving (1929), Paterson (1931).

Xanthogaleruca Laboissiére 1934
This is a Palearctic genus, but one adventive species, *X. luteola* (Müller 1766), is widespread throughout much of Canada and the United States. Also known as the elm leaf beetle, it feeds on *Ulmus* (Ulmaceae). Larval descriptions, Bøving (1929), Bøving and Craighead (1931), Lawson (1991), Peterson (1951). Biology, Britton (1907).

[*Lochmaea* Weise 1883, not in America north of Mexico.]

Schematizites Chapuis 1875

Brucita Wilcox 1965
This genus is presently monotypic, with *B. marmorata* (Jacoby 1886) occurring from south Texas to Guatemala. This species feeds on *Ehretia anacua* (Mier and Berland.) I. M. Johnst. (Boraginaceae). Additional Mexican or Central American species, either undescribed or currently catalogued in the genus *Yingaresca* Bechyné 1956, may eventually be classified here.

Erynephala Blake 1936
This New World genus contains six described species, four of them occurring north of Mexico. One of these, *E. puncticollis* (Say 1824) is an inland species and is rather widespread in the western half of North America. The distributions of our other three species are limited to areas along either the Atlantic or the Pacific Coast. North of Mexico, species feed on *Sesuvium* (Aizoaceae); *Amaranthus* (Amaranthaceae); *Batis* (Batidaceae); *Atriplex*, *Beta*, *Chenopodium*, *Salicornia*, *Salsola*, *Spinacia*, and *Suaeda* (Chenopodiaceae). Species identification of adults, Blake (1936 1970a), Wilcox (1965). Biology and descriptions of immature stages, Bøving (1929), Chittenden (1920).

Monoxia LeConte 1865
As currently constituted, this New World genus contains 18 described species, all but two of them occurring in the United States. However, this is a heterogeneous group, and some of the species might be better classified in *Yingaresca* Bechyné 1956, or in some still undescribed genus. Also, several species are yet to be described. Most of the species north of Mexico are confined to the western half of the United States or to Florida. They have been associated with *Artemisia*, *Chrysothamnus*, *Encelia*, *Grindelia*, *Gutierrezia*, *Haplopappus*, *Hemizonia*, *Iva*, *Solidago* (Asteraceae); *Batis* (Batidaceae); *Lepidium* (Brassicaceae); *Atriplex*, *Beta*, *Chenopodium*, *Salicornia*, *Suaeda* (Chenopodiaceae); *Lycium* and *Solanum* (Solanaceae). The larvae are leaf miners, or they feed on flowers or fruit. Keys to species, Blake (1939), Wilcox (1965). Larval descriptions, Bøving (1929), Bøving and Craighead (1931), Lawson (1991).

Ophraella Wilcox 1965
This genus is generally distributed in the United States and Canada, occurring even in Arctic areas. It contains 14 species, all Nearctic, plus one other species, questionably classified in *Ophraella*, that occurs in Mexico and Guatemala. Nearctic species feed on *Ambrosia*, *Artemisia*, *Aster*, *Chrysopsis*, *Eupatorium*, *Franseria*, *Helianthus*, *Iva*, *Liatris*, *Parthenium*, *Ratibida*, *Solidago*, and *Xanthium* (Asteraceae). Identification of adults, Futuyma (1990, 1991), LeSage (1986b). Biology and descriptions of immature stages, Bøving (1929), Futuyma (1990), Futuyma and McCafferty (1990), LeSage (1986b), Palmer and Goeden (1991), Welch (1978). Phylogeny, Funk *et al.* (1995), Futuyma and McCafferty (1990).

Ophraea Jacoby 1886
This genus contains twelve species native to Mexico and Central America. The distribution of one of them, *O. rugosa* Jacoby 1886, extends northward into Arizona. This species has been collected from the leaves of *Beloperone* (Acanthaceae).

Neolochmaea Laboissiére 1939
This Neotropical genus contains about twelve species distributed throughout much of Central America, South America, and the West Indies. One of them, *N. dilatipennis* (Jacoby 1886), is adventive in Florida. It feeds on *Borreria* (Rubiaceae). Species redescription, Moura (1998), White (1979). Key to species, Moura (1998).

Galerucites Latreille 1802

Characteristics: Elytra of most with pale lateral margins, but without pale discal vittae; elytral pubescence sparse, noticeable only upon close examination. Procoxal cavities closed behind. External surface of tibia with ridge extending most of length.

Galeruca Geoffroy 1762 [conserved name, ICZN 1994, Opinion 1754]
subgenus *Galeruca* Geoffroy 1762
Galeruca: Müller 1764 [a subsequent usage]
Adimonia Laicharting 1781
Galleruca Fabricius 1792 [error]
Although this subgenus is mostly Holarctic in distribution, it also occurs in Mauritius. Five species occur in North America where their host plants include *Arabis, Brassica, Dentaria, Lepidium, Raphanus* (Brassicaceae); *Lupinus* (Fabaceae); and *Phlox* (Polemoniaceae). Keys to species, Blake (1945), Wilcox (1965). Biology, Davis (1907).

subgenus *Emarhopa* Weise 1886 [Palearctic]

subgenus *Haptoscelis* Weise 1886 [Palearctic]

subgenus *Galerima* Reitter 1903 [Palearctic]

subgenus *Galerotoma* Reitter 1903 [Palearctic]

subgenus *Fassatia* Havelka 1954 [Palearctic]

Metacyclini Chapuis 1875

Exorini Wilcox 1965

Characteristics: Hind femur slender, without internal extensor apodeme (spring); most with tibiae lacking terminal spurs; tarsal claws of Nearctic species appendiculate, with broad, blunt lobe. Last ventrite of male abdomen without apical lobe, without distinct impression, although some are flattened. Median lobe of aedeagus with prominent basal spurs. Larvae unknown.

This tribe, containing 37 described genera, occurs not only in the New World, but also in Africa, Madagascar, India, and southeast Asia. Most species are tropical.

[*Metacycla* Baly 1861, probably not north of Mexico]
Old specimens of *M. insolita* (LeConte 1861) are labeled from San Diego, California. However, this species has not been found there in recent years. It occurs in Baja California Sur where it feeds on *Ambrosia ambrosioides* (Cav.) Payne and *Hymenoclea monogyra* J. Torr. and Gray ex A. Gray (Asteraceae).

Malacorhinus Jacoby 1887
This genus includes 24 described species, plus several undescribed species. Although they basically occur from Mexico south to Bolivia, the distributions of three of them extend into the United States (Arizona, California, and Texas). North of Mexico, they have been associated with *Acacia* (Fabaceae) and *Salix* (Salicaceae). Keys to species, Wilcox (1951, 1965).

Hylaspini Chapuis 1875

Sermylini Chapuis 1875

Characteristics: Hind femur slender, without internal extensor apodeme (spring); tibiae with terminal spurs on middle and hind tibiae of both male and female; tarsal claws appendiculate, with broad, blunt lobe. Last ventrite of male abdomen with short, rounded apical lobe. Median lobe of aedeagus without basal spurs. Larvae feeding on leaves.

This Old World tribe is probably not in North America, but two species have been reported from the United States.

[*Sermylassa* Reitter 1912, not established in North America.]
The records of *Sermylassa halensis* (Linnaeus 1767) from North America are at best interceptions.

[*Agelastica* Chevrolat 1836, not established in North America.]
The records of *Agelastica alni* (Linnaeus 1758) from North America are at best interceptions.

Luperini Chapuis 1875

Characteristics: Antennae of most inserted high on frons, near middle of eyes. Hind femur not abnormally broad, without sclerotized internal extensor apodeme (spring). Basal spurs of aedeagus small or absent; if small spurs present, aedeagus not strongly constricted slightly beyond base. Larvae, where known, feeding on roots.

Diabroticina Chapuis 1875

Characteristics: Last ventrite of male abdomen apically rounded or slightly truncate, without lobe.

This is a strictly New World group, with no representatives native to other areas.

Diabroticites Chapuis 1875

Characteristics: Tarsal claws bifid, with narrow, sharply pointed inner lobe; mesotibia of male without subapical notch.

This section contains 16 described genera, mostly Neotropical in distribution. Key to genera and list of species, Smith and Lawrence (1967).

Diabrotica Chevrolat 1836

This genus includes more than 350 described species, eight of which occur north of Mexico. Some species are commonly referred to as cucumber beetles, since they feed on Cucurbitaceae, including *Citrullus, Cucumis, Cucurbita,* and *Ibervillea.* Also, the larvae of several species are known as corn rootworms and are extremely serious pests of cultivated corn (*Zea*, Poaceae). Beyond this, the beetles are quite polyphagous, having been reported to eat even fungi. They have been found on a great variety of plants. Although some of these reports involve larval feeding on roots, many others involve only adult feeding in flowers. Flower-associated leaf beetles are often not choosy as to the plants attacked, and they can sometimes be found on almost anything that happens to be in bloom. Some notorius pests species are the northern corn rootworm, *D. barberi* Smith and Lawrence 1967; the western corn rootworm, *D. virgifera virgifera* LeConte 1868; the Mexican corn rootworm, *D. virgifera zeae* Krysan and Smith 1980; the southern corn rootworm or spotted cucumber beetle, *D. undecimpunctata howardi* Barber 1947; and the banded cucumber beetle, *D. balteata* LeConte 1865. Keys to species, Krysan and Smith (1987), Krysan *et al.* (1980, 1983), Wilcox (1965). Larval descriptions, Bøving (1927), Bøving and Craighead (1931), Lawson (1991), Peterson (1951).

Paranapiacaba Bechyné 1958
This genus includes more than 60 described species. However, only two of them, *P. connexa* (LeConte 1865) and *P. tricincta* (Say 1824), occur north of Mexico, from Colorado to Kansas to Texas to Arizona. They have been associated with *Citrullus*, *Cucumis*, and *Cucurbita* (Cucurbitaceae), and with *Zea* and perennial grasses (Poaceae). Key to species, Wilcox (1965).

Amphelasma Barber 1947
This genus contains eleven described species. For the most part, they occur from Mexico south to Venezuela. But the distribution of one of them, *A. cavum* (Say 1835), extends into Arizona. In Mexico, this species is apparently associated with *Salvia* (Lamiaceae).

Acalymma Barber 1947
This genus contains more than 70 described species, six of which occur north of Mexico. Five of the six species have vittate, more or less costate elytra, and belong to the *gouldi* group of *Acalymma*. Feeding on cucurbits, including *Citrullus*, *Cucumis*, *Cucurbita*, *Echinocystis*, *Marah*, *Sechium*, and *Sicyos*, they are commonly known as striped cucumber beetles. These species have also been associated with numerous plants in a variety of other families, but at least some of these associations involve flowers rather than leaves, and the plants almost certainly are not suitable hosts for the root-feeding larvae. The sixth species is *A. peregrinum* (Jacoby 1892), originally described from Mexico but later reported from south Texas associated with *Physalis* (Solanaceae). This species and also an undescribed species from Arizona are neither vittate nor costate, and they belong to the *peregrinum* group. Although no taxonomic change is made here, this group is quite distinct and probably deserves at least subgeneric status. Keys to species, Munroe and Smith (1980), Wilcox (1965). Biology and descriptions of immature stages, Balduf (1922), Bøving (1927), Gould (1944), Lawson (1991) and Peterson (1951).

Paratriarius Schaeffer 1906
Chanchamayia Bechyné 1956
This genus includes more than 50 species. However, only one of them, *P. dorsatus* (Say 1824), occurs north of Mexico (Oklahoma to Ohio to Arkansas). This species has been associated with *Commelina* and *Tradescantia* (Commelinaceae).

Cerotomites Chapuis 1875

Characteristics: Tarsal claws appendiculate, with comparatively broad, blunt inner lobe; mesotibia of male without subapical notch.

This section contains eleven genera, and is distributed throughout much of North and South America.

Neobrotica Jacoby 1887
This genus includes more than 60 described species, but only two of them occur north of Mexico, *N. pluristica* Fall 1910 in Arizona and *N. septemmaculata* Blake 1966 in Texas. The Arizona species has been found on *Chilopsis* (Bignoniaceae). Key to species, Blake (1966b).

[*Metrobrotica* Bechyné 1958, not in North America.]

Cyclotrypema Blake 1966
This genus is monotypic, with *C. furcata* (Olivier 1808) found in south Texas and nearby areas of Mexico. Although it has been reported from *Solanum* (Solanaceae) and *Lippia* (Verbenaceae), the true adult food plant is *Calyptocarpus vialis* Less. (Asteraceae).

Cerotoma Chevrolat 1836
Andrector Horn 1872
Cerotana: Bowditch 1913 [error]
Ceratoma: Dozier 1922 [error]
This genus includes 18 species, three of which occur north of Mexico. One of them, *C. trifurcata* (Forster 1771), is commonly known as the bean leaf beetle and is an important pest in the eastern half of North America. The other two United States species occur in Arizona, Florida, and Texas, and southward into the West Indies, Mexico, and Central America. Hosts of species occurring north of Mexico include *Amphicarpaea*, *Arachis*, *Cajanus*, *Desmodium*, *Glycine*, *Lespedeza*, *Phaseolus*, *Strophostyles*, *Vigna*, and *Wisteria* (Fabaceae). Key to species, Wilcox (1965). Larval description, Bøving (1931), Peterson (1951); biology, Isley (1930).

Phyllecthrites Horn 1893

Characteristics: Tarsal claws of Nearctic species appendiculate, each with a comparatively broad, blunt inner lobe; mesotibia of male with deep, inner, subapical notch.

This section contains twelve genera, most occurring in Neotropical areas. Key to genera, Blake (1966a).

Phyllecthris Dejean 1836

Phyllechthrus Agassiz 1846 [emendation]

Phyllecthrus Horn 1893 [error]

This genus includes only three species, all of which occur north of Mexico (Montana to New York to Georgia to Texas). They feed on *Amorpha*, *Amphicarpaea*, *Desmodium*, *Lespedeza*, *Robinia*, and *Tephrosia* (Fabaceae). Keys to species, Blake (1958), Wilcox (1965).

Luperosoma Jacoby 1891

Deuterobrotica Bechyné 1958

This genus includes 13 described species, three of which occur north of Mexico (Arizona to Kansas to Louisiana to Texas). In the United States, they have been associated with *Helianthus* (Asteraceae), *Psoralea* (Fabaceae), and *Larrea* (Zygophyllaceae). In Baja California Sur, *L. nigricolle* Blake 1966, occurs on *Aeschynomene vigil* Brandegee (Fabaceae). Keys to species, Blake (1958), Wilcox (1965).

Trachyscelidites Wilcox 1972

Characteristics: Elytra entirely dark, with distinct transverse impression near basal third. Tarsal claws appendiculate, with comparatively broad, blunt inner lobe; mesotibia of male without subapical notch.

This section contains a single genus, mostly Neotropical in distribution.

Trachyscelida Horn 1893

Racenisa Bechyné 1958

As currently recognized, this genus contains seven Neotropical species, the distribution of one of them, *T. bicolor* (LeConte 1884), extending northward into Arizona. This species is a secondary homonym of *T. bicolor* (Bechyné 1958). Although no taxonomic change is made here, a replacement name is needed for Bechyné's Bolivian species, if the two nominate species are indeed separate species.

Luperina Chapuis 1875

Characteristics: Tarsal claws of most appendiculate, each with comparatively broad basal tooth, less commonly bifid. Last ventrite of male abdomen with rectangular lobe.

This subtribe is distributed nearly worldwide. Although the immature stages of one species were illustrated (Wilcox 1965) and the larva of another described (Bøving 1927), the larval biology is essentially unknown for New World genera of this group.

Scelidites Chapuis 1875

Androlyperini Leng 1920

Characteristics: Epipleura normal, well defined. Tarsomere 1 of hind leg shorter than 2 and 3 combined in most. Apical lobe of male abdomen much wider than long in most. Aedeagal orifice variable, most without sclerotized covering.

This group of beetles, comprised of 30 genera, occurs in most major geographical areas, but it is notably absent from Australia and South America. Key to North America genera, Clark (1999a).

Synetocephalus Fall 1910

Synetocephala: Weise 1924 [error]

This genus contains ten described species (plus several that are undescribed), all of which occur in our area, from Washington to Wyoming to Arizona to the Baja California peninsula. Specimens have been associated with *Atriplex* (Chenopodiaceae); *Aesculus* (Hippocastanaceae); *Eriogonum* (Polygonaceae); *Ceanothus* (Rhamnaceae); *Adenostoma*, *Prunus*, *Purshia* (Rosaceae); and *Salix* (Salicaceae). Keys to species, Blake (1942), Wilcox (1965).

Pseudoluperus Beller and Hatch 1932

As recognized here, this genus contains eleven described species, eight occurring in western areas of Canada and the United States, one confined to the Baja California peninsula, one occurring in central Mexico, and one occurring in Central America. There are also numerous undescribed species, mostly in Mexico. However, note that this group is a heterogeneous assemblage of probably unrelated species. Future study will almost certainly require that several new genera be described. North of Mexico, beetles have been associated with *Yucca* (Agavaceae); *Heracleum*, *Pteryxia* (Apiaceae); *Achillea*, *Artemisia*, *Balsamorhiza*, *Chaenactis*, *Cirsium*, *Crepis*, *Encelia*, *Senecio*, *Tetradymia* (Asteraceae); *Thelypodium* (Brassicaceae); *Opuntia* (Cactaceae); *Astragalus*, *Lupinus*, *Melilotus* (Fabaceae); *Geranium* (Geraniaceae); *Sphaeralcea* (Malvaceae); *Syringa* (Oleaceae); *Oenothera* (Onagraceae); *Agropyron* (Poaceae); *Phlox* (Polemoniaceae); *Eriogonum* (Polygonaceae); *Adenostoma*, *Cercocarpus*, *Purshia*, *Rosa* (Rosaceae); *Salix*, *Populus* (Salicaceae); and *Penstemon* (Scrophulariaceae). Adults of at least some species are frequently found on inflorescences rather than leaves. Such flower-associated leaf beetles are often not very choosy and will accept practically any plant that happens to be in bloom. Key to species, Wilcox (1965).

Keitheatus Wilcox 1965

This genus contains only two described species, *K. histrio* (Horn 1895) from the Baja California peninsula and *K. blakeae* (White 1944) from the Big Bend area of Texas south to Durango and Nuevo Leon in Mexico. In Texas, the host is *Condalia* (Rhamnaceae).

Lygistus Wilcox 1965

This genus contains a single described species, *L. streptophallus* Wilcox 1965, which occurs in Arizona. It has been associated with *Bouteloua* and *Muhlenbergia* (Poaceae).

Triarius Jacoby 1887

This genus includes six described species, plus at least one undescribed species, all of which occur in the United States, some of which also occur in Mexico. Species of *Triarius* have been collected from *Dasylirion, Nolina, Yucca* (Agavaceae); *Baccharis, Erigeron* (Asteraceae); *Xerophyllum* (Liliaceae); *Sphaeralcea* (Malvaceae); *Pinus* (Pinaceae); *Condalia* (Rhamnaceae); *Adenostoma* and *Fallugia* (Rosaceae). However, some of these associations are with flowers rather than foliage. Flower associated leaf beetles are often not very specific as to which plants they will accept. Key to species, Wilcox (1965).

Scelida Chapuis 1875

Some species in this genus have odd ventral appendages on the male abdomen. Also, the median lobe of the aedeagus in most is characteristically broadened or laterally angulate just before the apex. This genus contains nine described species, and at least five undescribed species, which occur in Mexico and Central America. Two species *S. flaviceps* (Horn 1893) and *S. nigricornis* (Jacoby 1888) range into Arizona. Specimens have been collected from *Baccharis* (Asteraceae). Key to species, Wilcox (1965).

Androlyperus Crotch 1873

Malacamerus Wilcox 1951

This genus contains six described species, five of which occur north of Mexico (Arizona, California, and Utah). Two of these and one additional species are found in Baja California (norte), Mexico. In the United States, beetles have been associated with *Encelia, Eriophyllum* (Asteraceae); *Dalea* (Fabaceae); *Quercus* (Fagaceae); *Salvia* (Lamiaceae); *Mentzelia* (Loasaceae); *Sphaeralcea* (Malvaceae); *Clarkia, Oenothera* (Onagraceae); *Eriogonum* (Polygonaceae); and *Larrea* (Zygophyllaceae). Keys to species, Clark (2001), Wilcox (1951, 1965).

Scelolyperus Crotch 1874

Eugalera Brancsik 1899

This genus contains 27 described Nearctic species, plus two described Palearctic species, plus one species from Baja California (norte), Mexico. Although four of the Nearctic species occur in eastern areas of the United States or Canada, all of the others are found in western areas, and the genus is rare or entirely absent throughout much of the central plains. In North America, beetles have been associated with *Apocynum* (Apocynaceae); *Madia* (Asteraceae); *Betula, Carpinus* (Betulaceae); *Erysimum* (Brassicaceae); *Ribes* (Grossulariaceae); *Hamamelis* (Hamamelidaceae); *Hemerocallis* (Liliaceae); *Leptodactylon, Phlox* (Polemoniaceae); *Ceanothus* (Rhamnaceae); *Adenostoma, Rosa* (Rosaceae); *Ribes* (Grossulariaceae); and *Salix* (Salicaceae). They have also been reported from many other plants in a variety of families, but at least some of these associations probably involve mere resting sites rather than true hosts. Keys to species, Clark (1996), Wilcox (1965). Immature stages, Wilcox (1965).

Phyllobroticites Chapuis 1875

Characteristics: Elytral epipleura extremely narrow, indistinct. Tarsal claws appendiculate, each with broad basal tooth.

This section contains 13 genera, but only one of them occurs in the New World.

Phyllobrotica Chevrolat 1836

Stachysivora Farrell and Mitter 1990 [unavailable]

This genus contains about 30 species. For the most part, it is Holarctic in distribution, with 17 described species (plus at least two that are undescribed) occurring in temperate North America. In North America, hosts of *Phyllobrotica* are *Physostegia, Scutellaria,* and *Stachys* (Lamiaceae). Identification of species, Abdullah and Abdullah (1968), Blake (1956), Hatch (1971), Riley (1979), Wilcox (1965). Larva, Bøving (1927). Biology and phylogeny, Farrell and Mitter (1990).

Exosomites Wilcox 1973

This is essentially an Old World group, with only a single genus occurring in the Western Hemisphere. Minimal differences separate the section from the Scelidites, and future taxonomic revision, including a thorough study of the Old World genera, may show that these two should be joined.

Pteleon Jacoby 1888

A single species occurs in Arizona, California, Texas, Utah, and nearby areas of Mexico, and two additional species are found only in Mexico. Adults are found on the flowers of *Opuntia* (Cactaceae), *Sphaeralcea* (Malvaceae), and other plants.

[*Cneorane* Baly 1867, not in America north of Mexico.]

Monoleptites Chapuis 1875

Characteristics: Elytral epipleura well defined, not extremely narrow. Tarsomere 1 of hind leg distinctly longer than 2 and 3 combined; tarsal claws appendiculate, each with broad basal tooth. Male abdomen with apical lobe of last ventrite large, nearly square. Aedeagal orifice covered by sclerotized plate.

This is a large group containing about 40 genera and more than 1,000 species worldwide. Unfortunately, many of the genera are very poorly characterized. Often, type species are quite different, but genera run together when other species, with intermediate characters, are considered. In older classifications, the condition of the procoxal cavities (open versus closed behind) was used. Whereas this character is very useful in other galerucine groups, it has little value in the Monoleptites. Sometimes the procoxal cavities vary even among beetles that clearly belong to the same species. A thorough reassessment of the genera of this section is needed.

Eusattodera Schaeffer 1906

As currently recognized, this genus is comprised of six species, all of which occur in the United States. The only eastern species is *E. thoracica* (Melsheimer 1847), which has been found in Georgia, Kansas, and New York. The others occur in Arizona, Colorado, New Mexico, or southward into Mexico. Beetles have been found on *Pinus* (Pinaceae) and *Prunus* (Rosaceae). Key to species, Wilcox (1965).

[*Luperodes* Motschulsky 1858, not in America north of Mexico.]

Metrioidea Fairmaire 1881

As currently recognized, this genus contains about 20 species distributed in Fiji and the islands of southeast Asia, as well as in North and South America. Eleven of them occur in America north of Mexico, most in southwestern states, but some also in eastern areas. They have been associated with *Rhus* (Anacardiaceae); *Dahlia*, *Helianthus* (Asteraceae); *Abelmoschus*, *Althaea*, *Callirhoe*, *Gossypium*, *Hibiscus* (Malvaceae); *Zea* (Poaceae); and *Anemone* (Ranunculaceae). Adults are often found on blooms. Keys to species, Blake (1942), Wilcox (1965).

[Section Luperites Chapuis 1875, not in North America; although species now included in various American genera were formerly classified in *Luperus* Müller 1764, that genus, as well as the entire section Luperites, is restricted to the Old World.]

Alticini Newman 1835
The flea beetles

Characteristics: Antennae usually inserted high on frons, near middle of eyes; metafemur usually distinctly swollen, with sclerotized extensor apodeme (spring) internally near distal apex (usually not visible without dissection or rarely absent, see *Orthaltica*); metatibia with strong, articulated, apical spur; last ventrite of male with a more or less developed median lobe, sometimes inflexed; aedeagus without prominent basal spurs. Larvae folivorous, leaf or stem miners, or subterranean root feeders living free in soil, tunneling in roots, or mining under skin of roots and underground stems. Almost all the beetles of this tribe are capable of jumping.

A satisfactory internal classification for flea beetle genera does not exist. While some groupings of genera seem reasonably distinct, e.g., the "Monoplatini," "Oedionychina," and "Disonychina," a comprehensive overview of the group is needed. The last such work, Chapuis (1875), divided the "Halticides" into 19 groups and was largely adopted by Horn (1889) who produced the last comprehensive treatment of the North American Alticini. The arrangement in the checklist of world chrysomelid genera by Seeno and Wilcox (1982) roughly follows the order used in the world catalog by Heikertinger and Csiki (1939-1940) but with extensive additions. The presentation of genera in the present work is only slightly modified from that of Seeno and Wilcox; the nameless subdivisions indicated in that work are omitted here.

The great size and taxonomic complexity of the Alticini is one of the major factors hindering research into the classification and phylogeny of the group. Approximately 500 genera and between 8,000 and 10,000 species are recognized worldwide (Scherer 1988); the vast majority are tropical and poorly known. The Neotropical fauna is especially large, and here the morphological limits of many genera are poorly understood, including those for some familiar and long-standing genera. Changes to the nomenclature of some Nearctic genera are expected as work continues on the Neotropical fauna. Forty-seven genera are here accepted as occurring in America north of Mexico, represented by about 470 described species. One additional genus, endemic to the Baja California peninsula, is also included in the present treatment.

Blepharida Chevrolat 1836

This genus is found in both the New and Old World and contains about 60 species (Konstantinov and Vandenberg 1996). Taxonomic notes, Furth (1992).

subgenus *Blepharida* Chevrolat 1836
Metacolaspis Horn 1895
Blepharonycha Fall 1927

A single species of this New World subgenus, *B. rhois* (Forster 1771), occurs in America north of Mexico, ranging throughout the eastern United States and adjacent Canada to Arizona and southern California. Larvae and adults feed on the foliage of *Rhus* and *Schinus*, and adults are also recorded on *Cotinis obovatus* Raf. (all Anacardiaceae). Two additional species occur in the cape region of the Baja California peninsula where both adults and larvae of one of them feed on *Bursera* (Burseraceae). Key to species: Furth (1998). Larva, Bøving and Craighead (1931), Lawson (1991), Peterson (1951); biology, Frost (1972, 1973).

subgenus *Calotheca* Heyden 1887 [Afrotropical and Oriental]
Eutheca Baly 1878, not Kiesenwetter 1877 (Insecta)
Blepharidula Weise 1916

subgenus *Blepharidella* Weise 1910 [Afrotropical]

subgenus *Blepharidina* Bechyné 1968 [Afrotropical]

Acrocyum Jacoby 1885

The few described species comprising this genus are found in Central America, Mexico, or the West Indies. One Mexican species, *A. sallei* Jacoby 1885, is also recorded from Texas.

Euplectroscelis Crotch 1873

This genus is monotypic, with *E. xanti* Crotch 1873, endemic to the cape region of the Baja California peninsula. Adults and larvae are on *Bursera* (Burseraceae).

Pseudorthygia Csiki 1940
Orthygia Jacoby 1891, not Mörch 1853 (Mollusca)

One species, *P. nigritarsis* (Jacoby 1891), occurs in the Chisos Mountains, Texas, and in Mexico; other species are found in Mexico as far south as Chiapas.

Phydanis Horn 1889

One species, *P. bicolor* Horn 1889, occurs in Texas; a few other species are found in Mexico as far south as Guerrero. Larva, Bøving and Craighead (1931).

Luperaltica Crotch 1873
As currently cataloged, four species comprise this genus, all occurring in America north of Mexico; two are widely distributed in the eastern United States west to South Dakota and Texas, and two others are found in western Texas. Other species are found in Mexico, but these, if described, are currently placed under other generic names. In the eastern United States, adults feed on pollen of summer and fall-blooming flowers, especially *Eupatorium*, *Helianthus*, and *Solidago* (Asteraceae). Key to species, Wilcox (1953).

Lupraea Jacoby 1885
Palaeothona Jacoby 1885
Trachymetopa Weise 1899
Luprea: Scherer 1962 [error]
Aphthonia: Crotch 1873 (part) [error, *Aphthona* intended]
One species, *L. picta* (Say), is generally distributed in the eastern United States west to Kansas and New Mexico; an additional species is recorded from Arizona. Many Neotropical species are assigned to this genus, but the limits of the genus are not well established. Some Mexican species are congeneric with our species. Adults feed on *Quercus* (Fagaceae). Taxonomic notes, Blake (1950a).

Phyllotreta Chevrolat 1836
Orchestris Crotch 1873, not Kirby 1837 (Insecta)
Tanygaster Blatchley 1921
About 150 species comprise this genus which is nearly worldwide with the greatest species diversity in the north temperate zone. The genus is generally distributed in America north of Mexico where it is represented by 48 species; a few of these are adventive from the Palearctic Region. Known food plants in the Nearctic Region are Brassicaceae and Capparaceae, including *Alyssum*, *Arabis*, *Armoracia*, *Barbarea*, *Brassica*, *Cakile*, *Cardamine* (=*Dentaria*), *Descurania*, *Lepidium*, *Nasturtium*, *Radicula*, and *Rorippa* (Brassicaceae); and *Cleome* and *Polanisia* (Capparaceae). The Nearctic species were treated by Chittenden (1927), but his work is outdated and has been revised, in part, by Smith (1979, 1985). Key to maculate North American species, Smith (1985). The immaculate species need revision, most are poorly known and not recognizable using Chittenden's treatment. Larva and biology, Chittenden (1917), Peterson (1951).

Aphthona Chevrolat 1836
Aphtona: Allard 1860 [error]
Ectonia Weise 1922
Cerataltica Crotch 1873
Properly defined, this genus is restricted to the Old World, excepting six species introduced to North America for the biological control of leafy spurge, *Euphorbia esula* L., and other *Euphorbia* species. These beetles are found from Ontario and North Dakota west to British Columbia and Oregon. Keys to species, LeSage and Paquin (1996) and Konstantinov and Vandenberg (1996). Revision of Palaearctic species, Konstantinov (1998). Native North American species previously assigned to *Aphthona* belong elsewhere. Larvae of *Aphthona* are subterranean on roots. Biology, Jackson (1997).

Glyptina LeConte 1859
Aphthonia: Crotch 1873 (part) [error]
Balophila: Crotch 1873 [error, *Batophila* intended]
As presently cataloged, this genus appears to be almost entirely Nearctic, but this is misleading. Many Neotropical species belong here, including some of those presently assigned to *Aphthona*. A thorough study is needed to firmly establish the limits of this genus in the Neotropical Region. The Neotropical genus *Centralaphthona* Bechyné 1960, is probably a synonym. Most Nearctic species previously assigned to *Aphthona* have been reassigned to *Glyptina*. The regularity and development of serial punctation on the elytra of Nearctic species varies considerably among species, but minimally there are partially formed rows present. These can be faint or partially obscured by interstitial punctation. *Cerataltica* Crotch 1873, based on *Sphaeroderma insolita* Melsheimer 1847, is still listed in synonymy with *Aphthona*; however, this genus is much closer to *Glyptina* and yet does not agree entirely. *Cerataltica* should probably stand as a separate genus (A. Konsantinov, personal communication).

Food plant genera for North American *Glyptina* species are *Croton*, *Euphorbia*, *Sebastiania*, *Stillingia* (Euphorbiaceae), and one species eats *Geranium* (Geraniaceae). The Nearctic species need a modern revision, as there appear to be numerous unnamed species. Incomplete keys to species are found in Balsbaugh and Hays (1972), Downie and Arnett (1996), Horn (1889), and Wilcox (1954).

Pseudodibolia Jacoby 1891
According to Scherer (1983), this genus includes three described species and is known from Mexico, Central America, and the Lesser Antilles. An additional species, *P. opima* (LeConte 1878), is found in North America ranging from Florida to Maryland west to Missouri and Texas. Although this genus is very close to *Heikertingerella* Csiki 1940, the serrate outer margin of the metatibia is distinctive. A recorded food plant in America north of Mexico is *Ruellia caroliniensis* (J. F. Gmel.) Steud. (Acanthaceae).

Longitarsus Latreille 1829
Thyamis Stephens 1831
Teinodactyla Chevrolat 1836
Inopelonia Broun 1893
Testergus Weise 1893
Apterius Blatchley 1921
Truncatus Palii 1970
This genus is nearly worldwide and contains about 500 species, of which half occur in the Palearctic Region (Konstantinov and Vandenberg 1996). It is generally distributed in southern Canada and the United States and represented in this region by 46 described species. At least 10 North American species are introduced from the Palearctic Region, including *L. jacobaeae* (Waterhouse 1858) which was deliberately introduced for control of *Senecio jacobaea* L. (Asteraceae). The native North American species are in

need of revision and there appear to be numerous undescribed species present. Until genitalia of the type material can be studied, there are few Nearctic species names which can be applied with certainty. Applying species names with the keys in Blatchley (1921), Horn (1889), and various faunal works should be avoided. Confirmed food plants in North America include *Ambrosia*, *Cacalia*, *Eupatorium*, *Liatris*, *Senecio* (Asteraceae); *Echium*, *Heliotropium*, *Onosmodium* (Boraginaceae); *Mentha*, *Salvia* (Lamiaceae); *Lantana*, *Phyla*, and *Verbena* (Verbenaceae). Some of the adventive species probably feed on *Convolvulus* (Convolvulaceae) and *Plantago* (Plantaginaceae). Larvae feed on roots. Taxonomy of adventive species, LeSage (1988c); also see LeSage (1988a, 1988b), Westcott *et al.* (1985). Larva, Bøving and Craighead (1931); biology, Gentner (1926).

Systena Chevrolat 1836

About 90 described species comprise this genus which is distributed throughout the New World with most species found in the tropics. Nineteen species are recorded from America north of Mexico with a combined range throughout the United States and southern Canada. Many different plants are utilized by adults, and some species appear to have a wide host range. Some attack common garden plants and as larvae occasionally damage various field crops. Other plants preferred by adults include *Ambrosia*, *Baccharis*, *Borrichia* (Asteraceae); *Cornus*, *Nyssa* (Cornaceae); *Quercus* (Fagaceae); *Polygonum* (Polygonaceae); and *Taxodium* (Taxodiaceae). Larvae feed on roots. For taxonomic notes, see Blake (1935), but a revision is needed, as several North American species are poorly known. Larva and biology, Underhill (1928), Lee *et al.* (1998). Adult morphology, Lingafelter *et al.* (1998). The larva illustrated by Lawson (1991) as *Systena blanda* (Melsheimer 1847) is not a *Systena* and probably does not belong to the Chrysomelidae.

Dysphenges Horn 1894

Disphenges: Furth 1985 [error]

Sysphenges: Furth 1989 [error]

This genus includes a single named species, *Dysphenges elongatulus* Horn 1894, originally described from the Baja California peninsula and later reported from Texas (Wilcox 1975a). There appear to be several other species in mainland Mexico and more than one from Baja California Sur. An unidentified species, not *D. elongatulus*, is found in Arizona, and another possibly different species is known from Missouri and Texas. This genus is associated with *Mimosa* (Fabaceae) in the cape region of the Baja California peninsula.

Glenidion H. Clark 1860

Ptinomorpha Harold 1875

Sangaria Harold 1876

A few Neotropical species comprise this genus. One species, *G. flexicaulis* (Schaeffer 1905), is found in extreme south Texas on *Acacia smalli* Isely (Fabaceae).

Hemiphrynus Horn 1889

The type species, *H. intermedia* Jacoby 1884, is found from western Texas to Arizona and Sonora, Mexico. A second species, presently undetermined, occurs in New Mexico. Several additional species are found in Mexico; most are probably undescribed.

Nesaecrepida Blake 1964

Two species belonging to this genus are found in America north of Mexico (Louisiana, Texas) and south into Mexico, and one of these, *N asphaltina* (Suffrian 1868), and one additional described species are found in the West Indies. In the United States, *N. asphaltina* feeds on *Caperonia palustris* (L.) A. St.-Hil. (Euphorbiaceae) growing in disturbed habitats including agricultural fields. A short series of specimens examined from Puerto Rico are labeled as having been taken on this same plant. *Nesaecrepida infuscata* (Schaeffer 1906) has been collected in Texas from two species of *Mimosa* and from *Neptunia* (both Fabaceae).

Syphrea Baly 1876

Micraltica LeConte and Horn 1883

Syphraea: Csiki 1939 [error]

This genus includes more than 100 species and is found throughout the Neotropical Region (Scherer 1983). Our species need to be revised. Four species, three named and one undetermined, are found in America north of Mexico. *Syphrea nana* (Crotch 1873) is the best known and ranges from Florida to South Carolina west to Missouri and Texas. The other species inhabit the southern tier of states from Florida to southern California. The morphological limits of this genus are not well established, and even the few United States species form a heterogeneous assemblage. *Syphrea burgessi* (Crotch), recorded once from extreme south Florida, probably belongs in another genus. As adults, our species of *Syphrea* feed on *Bernardia*, *Croton*, *Crotonopsis*, and *Argythamnia* (=*Ditaxis*) (Euphorbiaceae).

Lysathia Bechyné 1959

This genus is recorded from throughout the Neotropical Region. A single species, *L. ludoviciana* (Fall 1910), occurs in the southeastern United States from Georgia and Florida west to Texas and also in the West Indies. Larvae are folivorous on *Myriophyllum* (Haloragaceae) and *Ludwigia* (Onagraceae); adults feed on these plants as well as *Oenothera* (Onagraceae). Biology, Habeck and Wilkerson (1980), Campbell and Clark (1983).

Altica Geoffroy 1762 [conserved name, ICZN 1994, Opinion 1754]

Altica: Müller 1764 [a subsequent usage]

Haltica Illiger 1807 [unjustified emendation]

Graptodera Chevrolat 1836

Rybakowia Jacobson 1892

Megaltica: Hatch 1971 [error, *Macrohaltica* intended]

About 300 species are recognized in this genus which is found nearly worldwide (Konstantinov and Vandenberg 1996); it is generally distributed in America north of Mexico and represented by about 70 described species. Members of the costate species group (eight species) feed on foliage of trees and shrubs in the genera *Alnus* and *Betula* (Betulaceae) and in the genera *Populus* and *Salix* (Salicaceae). The distribution of this group is primarily northern and western, with one species reaching as far north as central Alaska. The greater part of the genus in America north of Mexico feeds on a diverse array of plants belonging to at least ten families: genera include *Silybum* (Asteraceae); *Cornus* (Cornaceae); *Kalmia*, *Vaccinium* (Ericaceae); *Ribes* (Grossulariaceae); *Lagerstroemia* (Lythraceae); *Calylophus*, *Epilobium*, *Gaura*, *Fuchsia*, *Ludwigia*, *Oenothera* (Onagraceae); *Fallugia*, *Fragaria*, *Potentilla*, *Prunus*, *Rosa* (Rosaceae); *Heuchera* (Saxifragaceae); *Ulmus* (Ulmaceae); *Parthenocissus* and *Vitis* (Vitaceae). Larvae are folivorous. Key to species of costate group, LeSage (1995). The non-costate species need to be revised; therefore, some names may be misapplied in keys to *Altica* (*s. lat.*) in Horn (1889), Wilcox (1954), and Balsbaugh and Hays (1972). Biology and larvae, Barstow and Gittens (1971, 1973), DeSwarte and Balsbaugh (1973), Lawson (1991), Peterson (1951), Woods (1918).

[*Macrohaltica* Bechyné 1959, not in America north of Mexico.]

Strabala Chevrolat 1836
About 30 species comprise this genus; two subgenera are recognized (Bechyné and Špringlová de Bechyné 1975).

subgenus *Strabala* Chevrolat 1836
This subgenus contains eight species and occurs throughout most of the Neotropical Region. One species, *S. rufa* (Illiger 1807), is widely distributed in eastern North America, ranging from New Hampshire to Ontario, south to Florida, and west to Missouri and Texas. Three Neotropical species range into the southern limits of our area; two reach south Texas, and one Cuban species has been found in south Florida. *Strabala rufa* feeds on *Diodia* (Rubiaceae). Key to species, Blake (1953).

[subgenus *Isostrabala* Bechyné and Špringlová de Bechyné 1975 Neotropical]

Orthaltica Crotch 1873
Leptotrix Horn 1889
Leptotrichaltica Heikertinger 1925
Serraticollis B. White 1942
The placement and synonymy of this genus have been the focus of considerable debate, largely because this, and a few apparently closely related Old World genera, lack the extensor apodeme (spring) found in the metatibia of most Alticini. The latest installment in this debate was provided by Furth and Suzuki (1994) who consider *Orthaltica* a member of the Galerucinae (exclusive of Alticini). The revised synonymy suggested in that work is followed here; however, the genus is retained in the Alticini following the work by Samuelson (1994a, 1996) on elytron-to-body binding patches. A single binding patch was found in the diverse array of non-alticine Galerucinae studied by Samuelson, whereas most Alticini were found to have two separate patches. *Orthaltica* has two separate patches and also possesses a transverse, prebasal impression on the pronotum and a broad prosternal intercoxal process, both common conditions in Alticini but uncommon in non-alticine Galerucinae.

Four species are found in our area, generally distributed in the United States and southeastern Canada. An additional species was described from the cape region of the Baja California peninsula, and other species, possibly undescribed, are present in mainland Mexico. Adults feed on Anacardiaceae in the genera *Cryptocarpus*, *Rhus*, and *Toxicodendron*. Key to species, Scherer (1974), and see Andrews and Gilbert (1993).

Epitrix Foudras 1860
Epithrix: Bedel 1897 [error]
Epitrix is found nearly worldwide and represented by more than 100 species (Konstantinov and Vandenberg 1996). The twelve described species in America north of Mexico have a combined range which includes all of the United States and southern Canada. Although most of the common species are identifiable, the genus in North America is in need of revision. A few of the described species are poorly known and four or five additional species, possibly undescribed, are present mostly in the southern states. This genus contains the potato flea beetle, *E. cucumeris* (Harris 1851), tobacco flea beetle, *E. hirtipennis* (Melsheimer 1847), tuber flea beetle, *E. tuberis* Gentner 1944, and other economically important species. Larvae are subterranean root and stem feeders. Food plants are Solanaceae, including *Datura*, *Lycopersicon*, *Nicotiana*, *Physalis*, and *Solanum*. Partial keys to species, Horn (1889), Gentner (1944), and Seeno and Andrews (1972). Larvae and biology, Chamberlin *et al.* (1924), Lawson (1991), Peterson (1951); biology, Gentner (1944), Martin and Herzog (1987).

Acallepitrix Bechyné 1959
This genus occurs throughout most of the Neotropical Region, represented by approximately 40 species (Scherer 1983). A single species, *A. nitens* (Horn 1889), occurs in America north of Mexico and ranges from Ohio to Florida west to Missouri and Texas. Adults feed on *Physalis* and *Solanum* (Solanaceae).

Hippuriphila Foudras 1860
Parachalcoides Chûjô 1959
Pseudocrepidosoma Medvedev 1966
Holarctic in distribution, this genus is comprised of four species, of which three are found in the Nearctic Region. Its distribution is northern, ranging from New York to northern California, northward into Canada and Alaska. Adults feed on *Equisetum* (Equisetaceae). Key to species, Brown (1942b).

Hornaltica Barber 1941
The single species, *H. bicolorata* (Horn 1889), is found in the eastern United States from New Jersey to Florida west to Michigan and Texas. Adults feed on *Acalypha* (Euphorbiaceae).

Margaridisa Bechyné 1958
About 15 Neotropical species are currently assigned to this genus (Scherer 1983), but the true number is probably much greater. A single described North American species, *M. atriventris*

(Melsheimer 1847), belongs here. It occurs from Ontario and Quebec, south to Florida, and west to Kansas and Texas. At least three additional undescribed species occur across the southern United States. Our species feed on *Acalypha* and *Tragia* (Euphorbiaceae). Larvae are root feeders. Biology, Balduf (1926), Chittenden (1924b).

Crepidodera Chevrolat 1836
Chalcoides Foudras 1860
Foudrasia Des Gozis 1881
This genus includes about 40 species found mostly in the Holarctic Region (Konstantinov and Vandenberg 1996); 16 species are widely distributed in America north of Mexico. Food plants in our region include *Populus* and *Salix* (Salicaceae), as well as *Crataegus* and *Prunus* (Rosaceae). Larvae are subterranean. Key to species, larval description, and notes on biology, Parry (1986).

Derocrepis Weise 1886
Aeschrocnemis Weise 1888
About 20 species comprise this genus that is found in Africa and the Holarctic Region (Konstantinov and Vandenberg 1996); three species occur in North America ranging from Quebec to North Carolina west to Colorado. Food plant genera in North America include *Cardamine* (=*Dentaria*) (Brassicaceae), *Robinia* (Fabaceae), *Aesculus* (Hippocastanaceae), and *Prunus* (Rosaceae). Key to species, Downie and Arnett (1996).

Neocrepidodera Heikertinger 1911
Asiorestia Jacobson 1926
Orestioides Hatch 1935
This genus is represented by two species in the Nearctic Region and is northern in distribution, found from Newfoundland and New Hampshire west to California and British Columbia. Our species are known from high elevations. In northern California, one species has been associated with *Carex* (Cyperaceae). About 100 other species occur in the Old World (Konstantinov and Vandenberg 1996).

Trichaltica Harold 1876
Hoplacerus Jacoby 1892
More than 30 species of this genus occur in the Neotropical Region, excluding the West Indies (Scherer 1983), and one additional species, *T. scabricula* (Crotch 1874), occurs in the eastern United States ranging from Ohio to Texas. A Mexican species, *T. tibialis* (Jacoby 1892), is recorded from Arizona. The North American species are on *Chionanthus*, *Fraxinus*, and *Ligustrum* (Oleaceae).

Hemiglyptus Horn 1889
When described, this genus was monotypic with *H. basalis* Crotch 1874, which is found in California and Oregon. A few additional species from Chile were subsequently added to this genus. Adults of *H. basalis* are taken on *Eriodictyon* and *Hydrophyllum* (Hydrophyllaceae).

Mantura Stephens 1831
Cardiapus Curtis 1833
Balanomorpha Chevrolat 1836
Stenomantura Heikertinger 1909
About 20 species are known, found mostly in the Palearctic Region (Konstantinov and Vandenberg 1996). One species, *M. floridana* Crotch 1873, is native to North America and is found in most of the United States and in most of adjacent Canada. *Mantura chrysanthemi* (Koch 1803), a Palearctic species previously recorded from Newfoundland, is now also known from Quebec, Maryland, New Hampshire, New Jersey, and Ohio. Both species feed on *Rumex* (Polygonaceae). Larvae are leaf miners. Larva, Bøving and Craighead (1931), Lawson (1991).

Chaetocnema Stephens 1831
Odontocnema: Stephens 1831 [error]
Plectroscelis Chevrolat 1836
Udorpes Motschulsky 1845 [error]
Ydorpes Motschulsky 1845
Tlanoma Motschulsky ,1845
Hydropus Agassiz 1846
Exorhina Weise 1886
Carcharodis Weise 1910
Brinckaltica Bechyné 1959
Chaetochema: Scherer 1983 [error]
This genus includes approximately 440 species and is found in all major zoogeographic regions of the world (White 1996a). Fifty-nine species are recognized from America north of Mexico, with a combined range throughout the United States and southern Canada. The genus includes some economically important species, namely the corn flea beetle, *C. pulicaria* Melsheimer 1847; desert corn flea beetle, *C. ectypa* Horn 1889; and the sweetpotato flea beetle, *C. confinis* (Crotch 1873). Some species serve as vectors for bacterial wilt. Larvae are subterranean. Food plant associations in America north of Mexico involve the genera *Artemisia* (Asteraceae); *Atriplex* (Chenopodiaceae); *Trianthema* (Aizoaceae); *Conocarpus* (Combretaceae); *Convolvulus*, *Dichondra*, *Ipomoea* (Convolvulaceae); *Cyperus*, *Fimbristylus* (Cyperaceae); *Abutilon*, *Hibiscus*, *Kosteletzkya*, *Malvastrum*, *Sphaeralcea* (Malvaceae); *Agrostis*, *Avena*, *Cynodon*, *Dactylis*, *Digitaria*, *Echinochloa*, *Eleusine*, *Elymus*, *Eragrostis*, *Hordeum*, *Lolium*, *Panicum*, *Phleum*, *Secale*, *Setaria*, *Sorghum*, *Triticum*, *Zea* (Poaceae); and *Polygonum* (Polygonaceae). There are several uncommonly collected species present in our region, and undoubtedly some species remain to be recognized and described. Many species are associated with grassy communities of open canopy habitats such as prairies, glades, and open pinelands. Several others are associated with sedges growing in wet places. Key to North American species, White (1996a). Larva, Anderson (1938); biology, Poos (1955).

Disonycha Chevrolat 1836
Orchestris Kirby 1837
Balzanica Bechyné 1959
This genus is composed of approximately 145 described species and is restricted to the New World. The 36 species recognized in America north of Mexico have a combined range throughout the

United States and southern Canada. They feed on a diverse array of plants, including *Phaulothamnus* (Achatocarpaceae); *Aster*, *Boltonia* (Asteraceae); *Amaranthus*, *Iresine* (Amaranthaceae); *Opuntia* (Cactaceae); *Stellaria* (Caryophyllaceae); *Beta*, *Chenopodium* (Chenopodiaceae); *Cassia*, *Coronilla*, *Schrankia* (Fabaceae); *Hypericum* (=*Ascyrum*) (Clusiaceae); *Passiflora* (Passifloraceae); *Phlox* (Polemoniaceae); *Polygonum* (Polygonaceae); *Salix* (Salicaceae); and *Itea* (Saxifragaceae). Larvae are folivorous and frequently found in association with adults on their food plants. Incomplete key to species, Blake (1933); also see Blake (1930, 1951, 1957, and 1970a). Larval description and biology, Chittenden (1899), Lawson (1991), Peterson (1951); biology, DeSwarte and Balsbaugh (1973), Hemenway and Whitcomb (1968), Watts (1990) (Volume 1, Color Figs. 8, 16).

Agasicles Jacoby 1905
Five species are native to South America. One species, the alligatorweed flea beetle, *A. hygrophila* Selman and Vogt 1971, was imported to Australia, China, New Zealand, Thailand, and the United States for the suppression of the introduced aquatic weed, *Alternanthera philoxeroides* (Mart.) Griseb. (Amaranthaceae). In the United States this beetle is now found from South Carolina to Texas. It was also released in California but failed to establish there. Larvae are folivorous. Identification of adults, Selman and Vogt (1971); biology and evolution, Vogt *et al.* (1979)(Volume 1, Color Fig. 13).

Parchicola Bechyné and Špringlová de Bechyné 1975
Two species found in the southeastern United States and long assigned to *Monomacra* Chevrolat 1836, belong to this genus (Riley *et al.*, in press). Both species feed on *Passiflora* (Passifloraceae).

Monomacra Chevrolat 1836
Lactica Erichson 1847
Camoena Baly 1882
Medonia Baly 1862
Monomacra is a large and diverse "catch-all" genus found throughout the Neotropical Region. Several genera, including *Parchicola* Bechyné and Špringlová de Bechyné, were split from this group. A complete revision of the species catalogued in *Monomacra* is needed since many of them probably belong to other genera. Two species found in the United States, namely *M. bumeliae* (Schaeffer 1905) from extreme southern Texas, and *M. opaca* Wilcox 1953, from Arizona, are still cataloged in *Monomacra*, although they do not belong here. Recorded plant associations for both species involve *Bumelia* (Sapotaceae). Taxonomic notes on this and related genera, Bechyné and Špringlová de Bechyné (1975, 1977), Duckett (1999).

Omophoita Chevrolat 1836
Ptena Chevrolat 1836
Homophoeta Erichson 1847
Homophoita Weise 1921 [error]
This genus includes about 50 species (Scherer 1983) and is found throughout the Neotropical Region. One wide-ranging subspecies, *O. cyanipennis octomaculata* Crotch 1873, is found in Texas and Florida. Taxonomic notes, Blake (1931b).

Asphaera Chevrolat 1836
Litosonycha H. Clark 1865
Two species of this large Neotropical genus are recorded from America north of Mexico. *Asphaera lustrans* (Crotch 1873) is common and is found from Arizona to Texas, north to Kansas, and in Florida. Larvae and adults feed on *Scutellaria* (Lamiaceae). *Asphaera abdominalis* (Chevrolat 1833) is common from northern South America to Mexico. It is recorded from Texas and Arizona, but these records need confirmation. In Central America and Mexico, adults and folivorous larvae feed on a common pasture weed belonging to the Loganiaceae. Key to species, Horn (1889).

[*Oedionychis* Latreille 1829, not in America north of Mexico, see *Capraita* and *Kuschelina*.]

Pachyonychis H. Clark 1860
Hamletia Crotch 1873
This genus is monotypic, with *P. paradoxus* H. Clark 1860 found from Pennsylvania to Florida to Alabama. This uncommon species has been collected from bogs in South Carolina. Nomenclature, Mignot (1969).

Capraita Bechyné 1957
Chloephaga Weise 1899 [not Eyton 1838 (Aves)]
About 60 species comprise this genus which occurs in North, Central, and South America (Scherer 1983). Our species were part of the former "catch-all" genus *Oedionychis* Latreille 1829, which is now considered to be restricted to the Old World. The New World "*Oedionychis*" were divided by Jan Bechyné over a period of years to form several new genera. Some of these are monotypic and the limits of the larger genera remain poorly defined. *Capraita* was proposed as a replacement for *Chloephaga* Weise 1899, a preoccupied name. Weise based *Chloephaga* on *Oedionychis cretica* Jacoby 1886, described from the island of Crete in the Mediterranean. It was later determined by Heikertinger (1922) not to be from Crete, but rather a synonym of *Oedionychis sexmaculata* (Illiger 1807), a species widely distributed in eastern North America. Balsbaugh and Hays (1972) and Wilcox (1975a) placed the species of Blake's (1927) *Oedionychis* "series B" into this genus. There is still confusion about the status and limits of the genus.

North American species are found mostly in the eastern United States and adjacent Canada, but two are found in the southwestern United States from western Texas to Arizona. Food plants include a diverse selection of plants, including *Ilex* (Aquifoliaceae); *Aster* (Asteraceae); *Campsis*, *Chilopsis* (Bignoniaceae); *Symphoricarpos*, *Triosteum*, *Viburnum* (Caprifoliaceae); *Cornus* (Cornaceae); *Vaccinium* (Ericaceae); *Cunila*, *Teucrium* (Lamiaceae); *Chionanthus*, *Fraxinus* (Oleaceae); *Plantago* (Plantaginaceae); *Scrophularia*, *Verbscum*, and *Veronica* (Scrophulariaceae). Adults of *Capraita circumdata* (Randall 1838) feed on many of the plant genera and families listed above; *C. sexmaculata* (Illiger 1807) confines its feeding to *Chionanthus* and

Fraxinus (Oleaceae); *C. obsidiana* (Fabricius 1801) prefers *Ilex* species; and *C. flavida* (Horn 1889) and *C. durangoensis* (Jacoby 1892) are on *Chilopsis linearis* (Cav.) Sweet (Bignoniaceae) in the southwestern United States. Larvae and their feeding habits are unknown. Key to species, Blake (1927). Notes on biology, Sholes (1987).

Kuschelina Bechyné 1951
Like *Capraita* above, this genus was created out of the "catch-all" genus *Oedionychis* Latreille. It includes about 30 species and is found throughout the New World (Scherer 1983) but was originally based on a few species from South America. Balsbaugh and Hays (1972) and Wilcox (1975a) placed the species from Blake's (1927) *Oedionychis* "series A" into this genus, but there is still some question about the morphological limits of the genus. Some of our species have been listed in *Alagoasa* Bechyné 1955 (Furth and Savini 1996), a morphologically close and perhaps inseparable genus found throughout the Neotropical Region. A conservative approach is followed here, retaining all the "series A" species in *Kuschelina* pending the publication of better morphological definitions for these and related genera.

Twenty-eight species are recorded from America north of Mexico, with a combined distribution throughout the United States and southern Canada. Food plants for our species include the genera *Chilopsis* (Bignoniaceae); *Physostegia*, *Teucrium* (Lamiaceae); *Agalinis*, *Penstemon*, *Seymeria*, and *Trichostema* (Scrophulariaceae). Keys to species, Blake (1927, 1954b), and see Blake (1954a, 1965). Larval description, Bøving (in Blake 1927), Lawson (1991), Peterson (1951).

Distigmoptera Blake 1943
Twelve species are recognized in this genus which is found in Central America, Mexico, and the Nearctic Region. In America north of Mexico, it is represented by nine species and is found across the southern provinces of Canada from British Columbia to New Brunswick, and it is generally distributed in the United States except the Pacific coast states. Key to species, Blake (1943), and see Balsbaugh and Kirk (1968).

[*Hypolampsis* H. Clark 1860, not in America north of Mexico; record from Kansas is doubtful.]

Pachyonychus Melsheimer 1847
Clarkaltica Weise 1921
This genus is monotypic, with *P. paradoxus* Melsheimer 1847, found from Missouri to Pennsylvania south to Alabama and Florida. This beetle is more commonly taken in the northern portions of its range. Adults feed on *Smilax* (Liliaceae). Nomenclature, Mignot (1969).

Pseudolampsis Horn 1889
This genus contains two species, one in South America and *P. guttata* (LeConte 1884) in the southeastern United States from Florida and Maryland west to Missouri and Texas. Adults and larvae feed on *Azolla caroliniana* Willd. (Azollaceae). Taxonomic notes and larvae, Casari and Duckett (1998). Biological notes, Buckingham and Buckingham (1981), Habeck (1979).

[*Phaedromus* H. Clark 1860, probably not in America north of Mexico; record from South Carolina is doubtful.]

Dibolia Latreille 1829
Petalopus Motschulsky 1845
Pseudodibolia Jablokoff-Khnzorian 1968, not Jacoby 1891 (Insecta)
Eudibolia Jablokoff-Khnzorian 1968
More than 60 species found mostly in the Holarctic Region comprise this genus (Konstantinov and Vandenberg 1996). It is widely distributed across the United States and southern Canada, where it is represented by thirteen species. Larvae mine the leaves of their host plants. Food plants in America north of Mexico include *Plantago* (Plantaginaceae); *Castilleja*, *Chelone*, *Dasistoma*, *Melampyrum*, and *Penstemon* (Scrophulariaceae). Key to species, larval description, and biological notes, Parry (1974); larva, Lawson (1991), Peterson (1951). Biology, Hawthorn (1978), Reed (1927).

Argopistes Motschulsky 1860
Sphaerophyma Baly 1878
Over 40 species comprise this genus which is found in most zoogeographic regions (Konstantinov and Vandenberg 1996). One species, *A. scyrtoides* LeConte 1878, occurs in south Florida where its larvae mine the leaves of *Forestiera* (Oleaceae). Larva, Bøving and Craighead (1931).

Psylliodes Latreille 1829
Macronema Stephens 1829
Macrocnema Stephens 1831
Eupus Wollaston 1854
Macrocnema Weise 1888, not Hubner 1816 (Insecta); not Stephens 1831 (Insecta)
Semicnema Weise 1888
Psyllomima Bedel 1898
Phyllomima: Waterhouse 1902 [error]
Psyllobactra Lopatin 1958
This genus includes about 200 species from all major zoogeographic regions but is especially well represented in the Palearctic (Konstantinov and Vandenberg 1996). Thirteen species are recorded from America north of Mexico, but nearly half (six) are introduced from the Palearctic Region, including one released in 1997 for suppression of *Carduus* thistles. The genus is found throughout the United States and southern Canada; most introduced species are confined to the northeastern United States and adjacent Canada. Our native species need taxonomic revision. Two are reported as common and widespread, while most others are known only from their original descriptions. A few undescribed species also appear to be present. Food plants in America north of Mexico include *Carduus* (Asteraceae); *Barbarea*, *Descurainia*, *Lepidium* (Brassicaceae); *Humulus* (Cannabaceae); and *Solanum* (Solanaceae). Larvae are subterranean on roots, or they mine in

flower and leaf buds. Larva, Bøving and Craighead (1931); biology, Dunn and Rizza (1976), Parker (1910).

Eumolpinae Hope 1840

Megascelidinae Chapuis 1854 (Megascelinae, of authors)
Synetinae Edwards 1953

Characteristics: Body generally robust, sometimes elongate and subparallel, glabrous or covered with hairs or scales. Antenna filiform to subclavate; antennal insertions separated by width of frons; anteclypeus and coronal suture usually absent; eyes entire or emarginate. Prothorax with marginal bead complete, incomplete or lacking; procoxae globose; tarsi 5-5-5, pseudotetramerous, without bifid setae. The only significant apomorphic character state shared by all members of this newly enlarged subfamily is the presence of vaginal glands (Reid 1995).

Known larvae of Eumolpinae are subterranean root feeders. The immature stages and biology of the group are poorly known, except for economically important species which have been well studied by comparison.

Recent phylogenetic work on Chrysomeloidea has united *Syneta* and *Megascelis* with the Eumolpinae. The tribal classification used here follows that of Seeno and Wilcox (1982) but with the addition of these two groups as tribes. There is little reason to believe that the current system of higher classification within the Eumolpinae will hold up under modern phylogenetic analysis. The key to genera given above proceeds directly to genus, since several genera are heterogeneous with respect to characters commonly employed to diagnose tribes. The tribal diagnoses provided below will not necessarily apply to the fauna outside our area. The use of sections ("-ites"endings) within tribes appears to have some utility, but some such groupings are probably highly artificial.

The Eumolpinae is one of the larger subfamilies of Chrysomelidae. There are approximately 420 genera worldwide assigned to 16 poorly defined tribes. Five tribes, 25 genera, and approximately 145 species occur in America north of Mexico.

Synetini Edwards 1953

Characteristics: Body elongate, subparallel-sided; dorsum either with sparse erect hairs or dense recumbent pubescence; head exserted; pronotum narrow, with incomplete marginal bead or bead absent; prosternal process narrow between coxae; procoxal cavities open; last ventrite of male with small lobe, that of female with deep, fimbriate, semicircular fossa; outer apical margins of meso- and metatibia entire; tarsal claws bifid.

Now recognized as belonging to the Eumolpinae, this group was placed in the Orsodacninae (now Orsodacnidae) in early literature and later recognized as a subfamily of Chrysomelidae. In the recent phylogenetic analysis of chrysomeloids, Reid (1995) united the group with Eumolpinae arguing that larval *Syneta* are very close to larvae of Eumolpinae and that the adult characteristics (deep abdominal fossa in the female, reductions in wing venation, procoxal process, metendosternite, tegmen, ovipositor, form of the prothorax) are autapomorphies or reduction characters that should not be used to infer relationship to other chrysomeloid groups.

This group is Holarctic in distribution with two genera and nine species in North America, and two additional species (of *Syneta*) in the Palearctic Region.

Syneta Dejean 1835
This genus includes eight species and two subspecies in North America, and is primarily northern in distribution, found from Alaska to Newfoundland, and south to the high elevations of North Carolina in the East, and California and Colorado in the West. Adults are associated with many trees, especially conifers, including *Abies*, *Larix*, *Picea*, *Pinus*, and *Tsuga* (Pinaceae). One of the eastern species, *S. ferruginea* (Germar 1811), prefers Betulaceae, including *Alnus*, *Betula*, *Corylus*, and *Ostrya*. One western species, *S. albida* LeConte 1860, is called the western fruit beetle and attacks various deciduous trees including some fruit trees, *Cydonia*, *Prunus*, and *Pyrus* (Rosaceae); and *Ribes* (Grossulariaceae). Key to species, Edwards (1953); key to eastern species, Brown (1961). Notes on classification, Mann and Crowson (1981). Larva, Kurcheva (1967); biology and immature stages, Yu et al. (1996).

Thricolema Crotch 1874
Tricolema: Brisley 1927 [error]
Thrichema: Yu 1996 [error]
This genus is monotypic, with *T. anomala* (Crotch 1874) known from California and Oregon. Adults are associated with *Calocedrus decurrens* (Torr.) Florin. (Cupressaceae); larvae are unknown.

Typophorini Chapuis 1874

Nodinini Chen 1940
Nodini Selman 1965

Characteristics: Body variable, but not especially elongate; dorsum glabrous; head frequently retracted into prothorax; pronotum usually transverse and with distinct marginal bead; pygidium without median longitudinal groove; outer apical margins of meso- and metatibiae with distinct emargination which contains a brush of strong setae; tarsal claws bifid.

About 100 genera are placed in this tribe, the vast majority being Old World tropical; three of the New World genera are known from north of Mexico.

Metachromites Chapuis 1874

Characteristics: Anterior margin of proepisternum straight without arcuate lobe, lower lateral margin of head exposed.

Metachroma Chevrolat 1836
Atrachela Gistel 1847
This genus contains about 138 species, mostly Caribbean and Nearctic. About 40 species are documented from America north

of Mexico, mostly from the southern United States. Two species extend as far north as Canada where they have a combined range from Ontario to Alberta. The adults have been collected from a variety of plants. Some of the more significant plant associations in America north of Mexico involve *Atriplex* (Chenopodiaceae), *Quercus* (Fagaceae), *Pinus* (Pinaceae), *Salix* (Salicaceae), and various grasses (Poaceae). In the southeastern United States, there are several species that share a clear association with oaks (*Quercus*, Fagaceae). A few other southeastern species show a coastal distribution pattern and are associated with vegetation along the margins of mud flats and behind beach dunes, including *Borrichia* (Asteraceae), *Salicornia* (Chenopodiaceae), and *Conocarpus* (Combretaceae). Key to species, Blake (1970b).

Typophorites Chapuis 1874

Characteristics: Anterior margin of proepisternum with arcuate lobe which covers lower lateral margin of head.

Paria LeConte 1858
This genus contains about 40 New World species. North of Mexico it is represented by 19 described species and found throughout the eastern United States and in parts of southern Canada, west to Arizona. Adults feed on a variety of plants, and one species, *Paria fragariae* Wilcox 1954, has been recognized as an occasional pest of roses and strawberries. Some notable food plants in the United States include *Iva*, *Solidago* (Asteraceae); *Cornus* (Cornaceae); *Hypericum* (Clusiaceae); *Chamaecyparis*, *Juniperus* (Cupressaceae); *Quercus* (Fagaceae); *Carya*, *Juglans* (Juglandceae); *Polygonum* (Polygonaceae); *Fragaria*, *Rosa*, *Rubus* (Rosaceae); *Salix* (Salicaceae); and *Taxodium* (Taxodiaceae). The taxonomy of the genus is difficult; some of our "species" may actually be complexes of sibling species. Keys to species, Balsbaugh (1970), Wilcox (1957). Larva Bøving and Craighead (1931), Lawson (1991), Peterson (1921, 1951); biology Weigel (1926).

Typophorus Chevrolat 1836
Hydrotica Gistel 1847
This is a large Neotropical genus of over 50 described species, of which one, *T. nigritus viridicyaneus* (Crotch 1873), is widely distributed in the eastern states from Pennsylvania to Florida, west to the Great Plains. The numerous other subspecies of *T. nigritus* (Fabricius 1801) range throughout most of the Neotropical Region. This beetle seems to be associated with Convolvulaceae everywhere it occurs. In the eastern United States it eats *Ipomoea* spp., including sweetpotato, *I. batatas* (L.) Lam., the roots of which are fed upon by the larvae. In Central and South America there appear to be two distinct species groups in *Typophorus*: one composed of medium-sized species (including *T. nigritus*) feeds on Convolvulaceae; the second group, composed of small species, has at least some members that feed on the tropical plant family Melastomataceae. The subspecies of *T. nigritus* need re-evaluation, and the genus as a whole is in need of revision. A second species, *T. pumilus* LeConte 1859, is recorded from Kansas, but this locality is most likely an error. Larva, Peterson (1951); biology, Brannon (1938).

Eumolpini Hope 1840

Colaspini Chapuis 1874
Iphimeini Chapuis 1874
Corynodini Marshall 1865
Chrysodinini Lefèvre 1885

Characteristics: Body variable, but not especially elongate and subparallel; dorsum glabrous; head free or sometimes retracted into prothorax; pronotum usually transverse with distinct marginal bead; pygidium with median longitudinal groove which is sometimes confined to base or rarely absent; outer apical margins of meso- and metatibiae entire or rarely with subapical emargination; tarsal claws appendiculate, each with broad basal tooth.

This is the largest eumolpine tribe, with approximately 170 genera worldwide; the majority of New World Eumolpinae are placed here. With the addition of many new genera over the years, the "sections" presently referred to this tribe and originally based on Chapuis (1874) seem to defy characterization.

Iphimeites Chapuis 1874

Spintherophyta Dejean 1836
Chrysodina Baly 1864
Chalcoparia Crotch 1873
This largely Neotropical genus is comprised of over 70 species. Three species penetrate into the southwestern United States from Texas to Arizona, and one additional species, *S. globosa* (Olivier 1808), is widespread east of the Rocky Mountains. The latter is apparently a general feeder, having been taken on many unrelated plants. Key to species, Schultz (1976).

Metaparia Crotch 1873
Phytospinthera Monrós and Bechyné 1956
Spintherophyta Lefèvre 1875, not Dejean 1836
This genus is composed of nine North and Central American species. The marked sexual dimorphism in the development of the head and mandibles separates it from similar genera. Three species are found in the southwestern United States, with a combined distribution from Texas to Arizona. Adults are abundant on trees and shrubs but seem to prefer *Prosopis* (Fabaceae).

Chrysodinopsis Bechyné 1950
Two Mexican species comprise this genus which is very close to, and possibly synonymous with, *Brachypnoea* Gistel. The primary distinguishing character is the curious shape of the male's body which is globose with horizontal elytral epipleura and a markedly flattened and steeply sloped elytral declivity. Females look much like typical *Brachypnoea*, but with multiple, well-developed elytral costae. One Mexican species, *C. basalis* (Jacoby 1890), is also found in southern Arizona.

Brachypnoea Gistel 1847
Noda Chevrolat 1836, not Shellenberg 1803 (Insecta)
Nodonota Lefèvre 1885
Colaspomorpha Weise 1921
Containing about 125 described Neotropical species and another seven species in America north of Mexico, this is among the largest New World eumolpine genera. Our species have long been treated under the generic name *Nodonota* Lefèvre 1885, but *Brachypnoea* has priority (Flowers *et al.* 1994). Our species have a combined distribution throughout the eastern United States and adjacent Canada, west to the Great Plains. In south Texas, *B. rotundicollis* (Schaeffer 1906) has a clear association with *Baccharis* (Asteraceae). The plant genus *Ambrosia* (Asteraceae) has been noted multiple times as a food plant for two wide-ranging species. However, these as well as some other species of the genus appear to be polyphagous, found on many unrelated plants. Key to species, Schultz (1980). Larva, Peterson (1951); larva and biology, Wood (1940).

Euphrytus Jacoby 1881
The center of diversity for this genus is in Mexico with three of the 24 species reaching Arizona. Further study is needed to adequately define the limits of this genus. The species originally described by Jacoby and later by Schaeffer form a more or less distinct group distinguished by characters given in the key. However, Bechyné (1957) transferred to *Euphrytus* several Mexican species of *Coytiera* Lefèvre 1875, which lack the dimorphism of the maxillary palpi typical of the genus.

[*Metaxyonycha* Chevrolat 1836, not in America north of Mexico.]

Promecosoma Lefèvre 1877
This genus of 26 species is almost entirely restricted to Mexico, but one species has a disjunct distribution, being found in northern South America. Two species, *P. arizonae* (Crotch 1873) and *P. inflatum* Lefèvre 1877, are recorded from southern Arizona (Riley *et al.*, in press).

Colaspis Fabricius 1801
Maecolaspis Bechyné 1950
This genus is found throughout our area except California and the Pacific Northwest. It is the largest New World genus in the Eumolpinae with over 200 species, of which 28 are found in our area. Despite the transfer of many species to other genera, *Colaspis* is still a heterogeneous assemblage which probably needs further subdivision. Adults of some species seem to be polyphagous, attacking a wide variety of plants, including some crops such as corn, beans and rice. Others seem to be rather specific in their food plant choices, including *Colaspis pini* Barber 1937, which causes serious damage to the growing shoots of pine trees in the southeastern United States. The species of the *nigrocyanea* group are associated with milkweeds of the genus *Cynanchum* (Asclepiadaceae) in Texas and northeastern Mexico. In south Texas, *C. subtropica* Schaeffer 1906, restricts its feeding to *Eupatorium* (Asteraceae). Members of the *favosa* group seem to be polyphagous in the Great Plains region, but in the southeastern United States they show a marked preference for *Myrica* (Myricaceae), while also feeding on *Rhododendron* (Ericaceae) and *Juglans* (Juglandaceae).

Barber (1937) and Blake (1974, 1976a, 1976b, 1977a, and 1977b) treated the taxonomy of most of the Nearctic species of *Colaspis*, but additional refinement is needed for some species covered, and others not covered, in these works. Additional taxonomic notes, Chapin (1979). Biology, Rolston and Rouse (1965).

Percolaspis Bechyné 1957
An unidentified species of this small Neotropical genus has recently been taken in southern Florida. It is undoubtedly an immigrant, but multiple collections in Florida strongly suggest that it is established there. Central American species of this genus have been associated with Fabaceae and Rubiaceae (Flowers 1996).

Rhabdopterus Lefèvre 1885
Rhabdophorus Lefèvre 1878 [not Swainson 1839 (Pisces)]
This genus contains about 67 species and ranges from South America to the eastern United States. The eight United States species are a homogeneous group, but the same cannot be said for the 59 described Central and South American species grouped under this name. Complicating matters further is the fact that the Nearctic species may not be congeneric with the South American *Colaspis hypochalceus* Harold 1875, type species of *Rhabdopterus*. Food plant choice by our species involves several unrelated plants, and some species are apparently polyphagous. Notable plant genera fed upon by adults incude *Vaccinium* (Ericaceae); *Quercus* (Fagaceae); *Smilax* (Liliaceae); *Celtis* (Ulmaceae); *Ampelopsis* and *Vitis* (Vitaceae). Adult feeding produces very characteristic perforations in the leaves of the host plant. Identification of species, Barber (1943), Schultz (1977). Immature stages and biology, Scammell (1915).

Tymnes Chapuis 1874
This genus includes seven United States and three Mexican species. *Tymnes* are forest-inhabiting beetles, found on many trees and shrubs. In California, *T. oregonensis* (Crotch 1873) is taken on *Calocedrus decurrens* (Torr.) Florin. (Cupressaceae). Identification of eastern species, Downie and Arnett (1996); also see Blake (1977b) and Riley *et al.* (in press).

Eumolpites Hope 1840

Eumolpus Weber 1801
Eudoxus Kirby 1837
Alphites Chapuis 1874
The 39 species in this genus are restricted to the New World tropics and include the largest members of the Eumolpinae. *Eumolpus robustus* (Horn 1885) is a wide-ranging Mexican and Central American species that was recorded from Arizona many years ago, but recent records from there are lacking. Adults of this species feed on Asclepiadaceae in Central America, and other

Eumolpus species are also known to feed on plants of this family. Revision of genus, Špringlová (1960).

Corynodites Marshall 1865

Chrysochus Chevrolat 1836
Atymius Gistel 1847
This genus includes 19 species and is Holarctic in distribution. Two species occur in North America: *C. auratus* (Fabricius 1775), found throughout the eastern areas west to the Rocky Mountains, feeds on *Apocynum* (Apocynaceae), and *C. cobaltinus* LeConte 1857, from the High Plains west to California and British Columbia feeds on *Asclepias* (Asclepiadaceae). Larva, Peterson (1951); biology, Weiss and West (1921)

Megascelidini Chapuis 1874

Megascelinae, of authors

Characteristics: Body elongate, narrow, subparallel-sided; dorsum more or less pubescent; head wider than pronotum; pronotum cylindrical, narrow, without marginal bead; pygidium broadly exposed, without median groove; outer apical margins of meso- and metatibiae entire; tarsal claws simple, contiguous at base, weakly divergent.

This tribe is classified as a separate subfamily in most of the recent literature, and its implied relationship has long been with the Criocerinae. Similarities to Eumolpinae have been noted by earlier workers, and its placement in this subfamily was confirmed by the recent analysis of Reid (1995). There are two genera, both confined to the New World tropics.

Megascelis Sturm 1826
This genus contains about 140 described species, of which only one, *M. texana* Linell 1893, is found in America north of Mexico. This species is found at the extreme southern tip of Texas near Brownsville, where adults are common and collected from the foliage of *Leucaena pulverulenta* (Schltdl.) Benth. (Fabaceae). Systematic position, Bechyné and Špringlová de Bechyné (1969). Larva, Cox (1998).

Adoxini Baly 1865

Characteristics: Body variable, but not especially elongate; dorsum usually covered with setae or scales; pronotum subcylindrical to cylindrical, usually without marginal bead; pygidium usually lacking longitudinal median groove; outer apical margins of meso- and metatibiae either entire or with variably developed subapical emargination; tarsal claws usually bifid, sometimes appendiculate, or rarely simple. This tribe as presently composed is an artificial assemblage of unrelated genera.

Scelodontites Chapuis 1874

Characteristics: Dark bronzed or metallic green or blue; head retracted, with deep supraorbital sulcus; pronotum without marginal bead; proepisternum with anterior margin straight, not arcuate; outer apical margins of meso- and metatibiae with indistinct subapical emargination; tarsal claws bifid.

Graphops LeConte 1884
Heteraspis LeConte 1859, not Blanchard 1845 (Insecta)
Phortus Weise 1899
This genus is generally distributed throughout the United States and southern Canada. All of its 19 species are found in America north of Mexico. Although the genus has not been reported from Mexico, a few of our species should be expected to occur in the northernmost areas. The larva of *G. marcassita* (Crotch 1873) feeds on the roots of strawberry, *Fragaria* (Rosaceae); adults of some other species show clear associations with *Oenothera*, *Calylophus*, and *Gaura* (Onagraceae), and with *Hypericum* (=*Ascyrum*) (Clusiaceae). Most species appear to have limited flight capabilities. Key to species, Blake (1955). Biology, Forbes (1884).

Leprotites Chapuis 1874

Characteristics: Body non-metallic; head without supraorbital sulcus; pronotum without marginal bead, though sometimes weakly serrate; proepisternum with anterior margin straight, not expanded to partially cover head; subapical emargination of outer margins of meso- and metatibiae absent or weakly expressed; tarsal claws bifid.

Xanthonia Baly 1863
Microlypesthes Pic 1936
This genus is found from southern Canada southward into Central America, and other species occur in Oriental Region. It contains a great many species, especially in Mexico where most remain undescribed. In America north of Mexico, there are at least 18 recognizable species, and several of the western species remain undescribed. The Oriental-Palearctic genus *Demotina* (see below) is closely related to the North American *Xanthonia* and is probably confused with it in many collections.

In North America, *Xanthonia* is found throughout the eastern half of the United States, adjacent Canada, across the southern states to Arizona, and south through the higher elevations of Mexico and northern Central America. These are forest-inhabiting beetles, and most species show a clear association with oaks (*Quercus*, Fagaceae), but are occasionally found on other trees. One undescribed species from Texas has a strict association with *Juniperus* (Cupressaceae). Key to eastern North American species, Staines and Weisman (2001).

Demotina Baly 1863
Twenty-six Asian species belong to this genus; one of these, *D. modesta* Baly 1874, is a naturalized introduction to North America that is native to Japan and Korea. This species is now widespread in the southeastern United States, recorded from North Carolina and Tennessee to Florida and Texas. It is abundant on oaks (*Quercus*, Fagaceae), an association also reported in its native range. Taxonomic notes, Isono (1990a, 1990b), Riley *et al.* (in press).

Fidia Walsh 1867

Atonia Gistel 1847 [unavailable]

This genus contains 15 described species from North and Central America, with six of these found in the United States. Additional undescribed species are also present in our region. Our species occur mostly in the eastern United States, but one is found in Arizona southward into Mexico. Known host plants are members of the Vitaceae including *Ampelopsis*, *Cissus*, *Parthenocissus*, and *Vitis*. One or more species are occasionally destructive to cultivated grapes; their larvae, called grape rootworms, damage the roots. Larva, Peterson (1951). Biology, Isley (1942).

Adoxites Baly 1865

Characteristics: Body non-metallic; head without supraorbital sulcus; proepisternum with anterior margin arcuate, partially covering lower lateral margin of head; pronotum without marginal bead; outer apical margins of meso- and metatibiae without subapical emargination; tarsal claws bifid.

Bromius Chevrolat 1836

Adoxus Kirby 1837

This is a monotypic genus containing *B. obscurus* (Linnaeus 1758) which is Holarctic in distribution. In North America it is found from Alaska and the Northwest Territories to Quebec, south to North Carolina and California. *Vitis* (Vitaceae) is generally cited as the host, but this species is also found on *Epilobium* (Onagraceae) in North America.

Myochroites Chapuis 1874

Characteristics: Body metallic or not; head without supraorbital sulcus; proepisternum with anterior margin arcuate, partially covering lower lateral margin of head; outer apical margins of meso- and metatibiae without subapical emargination; tarsal claws appendiculate, bifid, or rarely simple.

Myochrous Erichson 1847

Fifty-two species are assigned to this genus which is found throughout the New World. Thirteen species are found north of Mexico, with a combined range throughout the United States and parts of adjacent Canada. At least some species in this genus are unusual in that they are seldom found on plants during daylight hours, apparently spending much of their lives near the soil line. At night, however, they can be abundant on the upper portions of plants. One species, *M. denticollis* (Say 1824), sometimes called the southern corn leaf-beetle, has occasionally caused economic losses by damaging corn seedlings. Other species are suspected of living among reeds and grasses in marshy areas. In south Texas, *M. magnus* Schaeffer 1904, feeds on *Salix* (Salicaceae). In California, *M. whitei* Blake 1950, is on *Juncus* (Juncaceae), and *M. longulus* LeConte 1858, is on *Pluchea* (Asteraceae) and *Salicornia* (Chenopodiaceae). Key to species, Blake (1950b). Larva and biology, Kelly (1915).

Glyptoscelis Chevrolat 1836

Thirty-five species comprise this genus which ranges from Argentina and Chile north to southern Canada. Most of the 26 species found in America north of Mexico occur in the western United States. A few others are found in the eastern and central United States and adjacent Canada.

Adults of our *Glyptoscelis* species are found on woody shrubs and trees. One eastern species, *G. albicans* Baly 1865, has been taken on several different woody plants growing in bottomland habitats, without showing a particular preference. *Glyptoscelis squamulata* Crotch 1873, causes early-season damage to the leaf buds of grapes in California, and the adults of *G. pubescens* (Fabricius 1776) damage the new growth on pines in the eastern United States. Some other notable plant associations involve *Artemisia*, *Grindelia* (Asteraceae); *Calocedrus*, *Juniperus* (Cupressaceae); *Prosopis* (Fabaceae); *Carya* (Juglandaceae); *Ceanothus* (Rhamnaceae); *Salix* (Salicaceae); and *Sequoia* (Taxodiaceae). Key to species, Blake (1967). Biology, Ebeling (1939), Klein and Coppel (1969).

Colaspidea Laporte 1833

This genus has a disjunct distribution with seven species occurring in southern Europe and North Africa, and three species in California. Adults of the Californian species are associated with many trees and shrubs, including *Calocedrus* (Cupressaceae); *Eriodictyon* (Hydrophyllaceae); *Pinus* (Pinaceae); *Adenostoma* and *Purshia* (Rosaceae). Key to North American species, Fall (1933). This genus needs revision.

Lamprosomatinae Lacordaire 1848

Lamprosominae, of authors

Sphaerocharitinae Chapuis 1874

Characteristics: Body compact, round to nearly round, strongly convex; surface nearly smooth; head not visible from above, inserted into prothorax to the eyes; antenna short, somewhat serrate; antennal insertions separated by width of frons; pronotum convex, nearly as wide as elytra and deeply sinuate at base, fitted closely against elytral base; antennal groove present on each side of prosternal process. Scutellum large and triangular, or small and elongate-triangulate, or hidden; elytra covering pygidium; epipleuron excavated and sharply angled for reception of apex of metafemur; tibiae usually distinctly dilated; tarsi 5-5-5, pseudotetramerous, without bifid setae.

Larvae are casebearers, constructing cases with their fecal material and sometimes small wood particles. They feed on green plant material or graze the bark of woody plants.

This group is nearly worldwide, but its distribution is spotty. By far, most species are found in the Neotropical Region, and only one is confirmed for America north of Mexico. With three tribes, 12 genera, and approximately 190 known species, the Lamprosomatinae is one of the smallest chrysomelid subfamilies.

Lamprosomatini Lacordaire 1848

Characteristics: scutellum small, elongate-triangulate, or hidden.

Oomorphus Curtis 1831
subgenus *Oomorphus* Curtis 1831
Delphastobia Casey 1924
Of the 17 species belonging to this genus, three are assigned to the nominotypical subgenus, one each in Europe and Japan, and *O. floridanus* (Horn 1893) from south Florida and the Bahamas. Our species is common in the Florida Keys and adjacent mainland areas of south Florida. Adults are taken by beating many different trees and shrubs which are characteristic of West Indian hardwood hammock. Series of adults have been collected from *Ficus* (Moraceae) and *Metopium* (Anacardiaceae). Specimens have also been recovered from the nest of a packrat (*Neotoma*). Larva, Kasap and Crowson (1976).

subgenus *Histerogaster* Monrós 1948 [Neotropical]

[*Lamprosoma* Kirby 1818. Lacordaire (1848) gave "Californie" as a locality for his *L. opulentum*, but this record is highly doubtful. Presently, this genus is not known from America north of Mexico nor from the Baja California peninsula.]

Cryptocephalinae Gyllenhal 1814
The casebearers

Characteristics: Body robust, cylindrical, usually compact; head usually retracted into pronotum to eyes; antennal insertions separated by width of frons; lateral margin of elytron more or less lobed below humerus; ventrites 2-4 variably narrowed medially; mesal region of last ventrite of female usually with deep fossa, or at least shallow impression; pygidium broadly exposed, subvertical. Tarsi 5-5-5, without bifid adhesive setae; median lobe of male genitalia with lower margin of apex usually bearing setae; kotpresse (sclerotized rectal plates) well developed in females.

Larvae are casebearers, mostly detritivorous and living in litter on the soil surface; a small number consume green leaves.

The subfamily Cryptocephalinae is a moderate-sized group with at least 345 species in 22 genera in America north of Mexico. In most previous references, this group was considered three separate subfamilies, collectively recognized by the informal name "Camptosomata", but they are here recognized as tribes of a single subfamily following Reid (1995). Members are present in every region of the world, although the Chlamisini are largely tropical and the Clytrini are absent from Australia. As far as known, the larval stages are all casebearers, living in and protected by a case constructed of their fecal matter and sometimes plant debris. The larvae occupy a variety of niches, and numerous species of the Clytrini have been shown to be associated with ants (Selman 1988). General biology of the group, Erber (1988).

Cryptocephalini Gyllenhal 1813

Characteristics: Head deeply recessed into pronotum; antennae usually long, filiform, and reaching to elytral umbone or beyond, but sometimes shorter; eye nearly entire to distinctly emarginate; pronotum normal to swollen, surface not tuberculate; procoxae globose, distinctly separated by prosternum; prosternal process without antennal groove on each side. Elytra smooth to deeply punctate, but not tuberculate, usually with rows of punctures; sutural margin entire, not serrate. Pygidium exposed to varying degrees, usually punctured, smooth to alutaceous, not roughly sculptured. Legs normal, not contractile.

Larvae are mostly detritivorous, living in leaf litter and similar debris on the soil surface.

Pachybrachina Chapuis 1874

Characteristics: Antennae long, filiform; base of pronotum with marginal bead, not crenulate; prosternum subquadrate to longer than wide; profemur usually enlarged; tarsal claws simple.

Griburius Haldeman 1849
Scolochrus Suffrian 1852
As with most of the genera in the tribe, this genus is poorly known and needs revision. There are four described species from the eastern and southwestern United States and many more from throughout the New World. Adults are associated with many woody plants, including *Conocarpus* (Combretaceae), *Quercus* (Fagaceae), and *Salix* (Salicaceae). Larvae are undescribed but have been found in buzzards' nests (Beamer 1926) and in a packrat (*Neotoma*) nest. Key to species, Crotch (1873); taxonomic notes, Riley *et al.* (in press) (Volume 1, Color Fig. 4).

Pachybrachis Chevrolat 1836
subgenus *Pachybrachis* Chevrolat 1836
Pachybrachys: Mannerheim 1843 [error]
Pachystylus Rey 1883
Homoeostigmus Jacobson 1917
The North American species have not been assigned to subgenera. This is a very large and complex genus with over 150 species recognized from America north of Mexico and many more in Central and South America and the Palearctic Region. The genus is found throughout the United States and adjacent Canada, but most species occur in the southern areas, or the drier western and southwestern portions of the United States. Adults have been found on many different plants; the more notable associations involve *Rhus*, *Toxicodendron* (Anacardiaceae); *Artemisia*, *Hymenoclea* (Asteraceae); *Atriplex* (Chenopodiaceae); *Juniperus* (Cupressaceae); *Acacia*, *Baptisia*, *Desmanthus*, *Gleditsia*, *Leucaena*, *Mimosa*, *Pithecellobium*, *Prosopis*, *Robinia* (Fabaceae); *Quercus* (Fagaceae); *Juglans* (Juglandaceae); *Poplus*, *Salix* (Salicaceae); and *Larrea* (Zygophyllaceae). In the eastern United States, many species are taken without clear plant associations by sweeping in open-canopy habitats, such as glades, prairies, and open pine woodlands.

The genus in America north of Mexico was revised by Fall (1915) who did a remarkably thorough job given the limited amount of information available at the time and that only external characters were used. Since then, Balsbaugh and co-workers (see below) have reviewed two species complexes, making use of the male genitalia to help define species. In general, species of *Pachybrachis* are difficult to separate with external characters alone, owing to a generally high level of intraspecific variation in color, and to some extent, in sculpture. This is an especially acute problem in *Pachybrachis* since many species descriptions are based on few specimens, and most species show strong sexual dimorphism, with the most important taxonomic characters present only in males.

Revision, Fall (1915); select taxonomic changes, Balsbaugh and Hays (1972); *P. othonus* complex, Balsbaugh (1973); *P. nigricornis* complex and key to vittate species, Balsbaugh and Tucker (1976). Larval descriptions and biology, Lawson (1976, 1991), LeSage (1985), Stiefel (1993).

subgenus *Chloropachys* Rey 1883 [Palearctic]

Monachulina Leng 1920

Monachini Chapuis 1874

Characteristics: Antenna short, subserrate; prosternum distinctly wider than long; base of pronotum crenulate; profemur not enlarged; tarsal claws appendiculate.

About 15 genera are assigned to this group, most occurring in the Old World.

Lexiphanes Gistel 1847
Monachus Chevrolat 1836, not Fleming 1822 (Mammalia)
Monachulus Leng 1918

This genus is restricted to the New World and includes many species; five species are recorded from America north of Mexico. *Lexiphanes saponatus* (Fabricius 1801) is widespread over the eastern half of the United States and southern Canada. Two other species are limited to the southeastern states, and two more are found in the southwestern states, south into Mexico. There are many reported plant associations, but most are probably incidental. LeSage (1984b) collected larvae and adults of *L. saponatus* from *Cassandra* (Ericaceae), and adults from *Salix* (Salicaceae) and *Alnus* (Betulaceae). Species of the Southwest have been reported on *Prosopis* (Fabaceae) and from grasses (Poaceae). Key to species, Balsbaugh (1966), and see Riley and Balsbaugh (1988). Larva and biology, LeSage (1984b).

Cryptocephalina Gyllenhal 1813

Characteristics: Antennae usually long and filiform, sometimes short and clavate to subserrate; prosternum subquadrate to longer than wide; base of pronotum unmargined, crenulate; profemur usually not enlarged; tarsal claws simple or appendiculate.

This group is found worldwide and includes about 20 genera. A thorough review of the genera is needed.

Cryptocephalus Geoffroy 1762 [conserved name, ICZN 1994, Opinion 1754]
Cryptocephalus: Müller 1764 [a subsequent usage]
subgenus *Cryptocephalus* Geoffroy 1762
Physicerus Chevrolat 1836
Homalopus Chevrolat 1836
Strigophorus Chevrolat 1836 [*nomen nudum*]
Dicenopsis Saunders 1842
Mitocera Saunders 1842
Ochrosopsis Saunders 1843
Anodonta Saunders 1845, not Lamarck 1799 (Mollusca)
Idiocephala Saunders 1845
Ochrosopsus: Saunders 1845 [error]
Canthostethus Haldeman 1849
Mecostethus Stål 1857
Euphyma Baly 1877

This genus is worldwide in distribution and includes a great many species. Our species have not been assigned to subgenera which, thus far, apply to Palearctic species groups only. Seventy-seven species occur in America north of Mexico and have a combined distribution throughout the United States and adjacent Canada. Plant associations for adults are numerous, and many are probably incidental. Some of the most reliable include *Rhus* (Anacardiaceae); *Baccharis*, *Chrysothamnus*, *Gutierrezia*, *Hymenoclea*, *Isocoma* (Asteraceae); *Cornus* (Cornaceae); *Dudleya* (Crassulaceae); *Arctostaphylos*, *Cassandra*, *Kalmia*, *Vaccinium* (Ericaceae); *Acacia*, *Amorpha*, *Desmodium*, *Dalea*, *Leucaena*, *Prosopis* (Fabaceae); *Quercus* (Fagaceae); *Salvia* (Lamiaceae); *Comptonia* (Myricaceae); *Pinus* (Pinaceae); *Eriogonum* (Polygonaceae); *Ceanothus* (Rhamnaceae); *Cowania*, *Prunus*, *Rosa*, *Rubus* (Rosaceae); *Salix* (Salicaceae); *Anemopsis* (Saururaceae); and *Larrea* (Zygophyllaceae). Most larvae are probably associated with plant litter since they are not found on living plants with adults. Key to species, White (1968), and see Riley and Gilbert (2000). Larva, Lawson (1991), LeSage (1986a); biological notes, Stiefel (1993), Tilden (1949).

subgenus *Disopus* Chevrolat 1836 [Palearctic]
Taxaris Gistel 1847

subgenus *Protophysus* Chevrolat 1836 [Palearctic]
Proctophysus: Redtenbacher 1845 [error]

subgenus *Burlinius* Lopatin 1965 [Palearctic]

subgenus *Cerodens* Burlini 1969 [Palearctic]
Ceropachys Burlini 1953, not Costa 1847 (Insecta)

subgenus *Cryptodontus* Burlini 1969 [Palearctic]

subgenus *Bertiellus* Lopatin 1977 [Palearctic]

subgenus *Lamellosus* Tomov 1979 [Palearctic]

subgenus *Asionus* Lopatin 1988 [Palearctic]
Asiopus Lopatin 1965, not Sharp 1892 (Insecta)
Aryana Berti and Rapilly 1973, not Pate 1943 (Insecta)

subgenus *Heterichnus* Warchalowski 1991 [Palearctic]
Heterodactylus Medvedev 1963, not Spix 1825 (Reptilia)

Bassareus Haldeman 1849
This genus is composed of eight Nearctic species from the eastern half of the United States and adjacent Canada. Except for the characters given in the key, these beetles are very similar to *Cryptocephalus*. Some notable plant associations for adults involve *Baccharis* (Asteraceae), *Alnus* (Betulaceae), *Conocarpus* (Combretaceae), *Diospyros* (Ebenaceae), *Amorpha* (Fabaceae), *Cephalanthus* (Rubiaceae), and *Salix* (Salicaceae). Partial keys to species, LeConte (1880), Downie and Arnett (1996). This genus needs revision. Many of the available species-group names have been variously applied as either species, subspecies, or color forms (Vol. 1, Color Fig. 7).

Diachus LeConte 1880
Fabricianus Weise 1895
The species of this genus are poorly known and have not been revised since the genus was first established by LeConte (1880). One species, *D. auratus* (Fabricius 1801), is widely distributed in the New World, including most of our area, and has found its way to Australia, Hawaii, Japan, and islands of the South Pacific (Kimoto 1993). The remaining nine species found in America north of Mexico are known from limited areas, mainly in the eastern half of the United States, Arizona, California, or southeastern Canada. A few species are described from the West Indies, and there are many apparently undescribed species in Central America and at least two in Arizona. Adults seem to prefer plants that are in bloom, and can be found on many unrelated plants including, but not limited to, *Rhus* (Anacardiaceae); *Baccharis* (Asteraceae); *Acacia*, *Leucaena* (Fabaceae); *Ceanothus* (Rhamnaceae); *Rubus* (Rosaceae); and *Salix* (Salicaceae). Key to most species, LeConte (1880).

Triachus LeConte 1880
Parephistemus Casey 1924
Like *Diachus*, this genus has not been revised since it was established by LeConte (1880). Presently, there are five recognized Nearctic species. With the exception of *T. peninsularis* Schaeffer 1906, from Baja California Sur, the species of this genus are distributed throughout the eastern half of the United States and adjacent Canada. There is very little known about the biology of this group other than plant associations for adults which include *Rhus* (Anacardiaceae), *Baccharis* (Asteraceae), and *Myrica* (Myricaceae). Key to species, LeConte (1880).

Clytrini Lacordaire 1848

Characteristics: Head small to large, deeply recessed into pronotum to broadly exposed; antenna serrate beyond antennomere 4, short, usually not reaching beyond prothorax; eye entire to distinctly emarginate; pronotum normal; mesoscutellum not exposed; procoxae transverse to transversely conical, nearly contiguous to narrowly separated by prosternum; legs usually short and stout (except prolegs of *Anomoea*), not contractile; elytra smooth with rows of punctures, to irregularly punctate; sutural margin entire, not serrate; pygidium exposed to varying degrees. Larvae are mostly detritivorous, many are associated with ants. Key to New World genera and subgenera (except Clytrina), Moldenke (1981). Revision of the Clytrini north of Panama, Moldenke (1970).

Clytrina Lacordaire 1848

Characteristics: Head broadly exposed, not recessed into pronotum; mandibles large, usually sexually dimorphic; eyes with broad, shallow to distinct emargination; pronotum flattened, not convex; procoxae conical, contiguous; scutellum inclined, raised above plane of elytra; elytron usually thin, pliable, with weakly developed postbasal lobe; tarsal claws simple.

This is a fairly diverse group of about 40 genera. Almost all Old World Clytrini belong here. In contrast, the group is poorly represented in the New World with only two genera, both occurring in our area.

Anomoea Agassiz 1846
Anomoia Chevrolat 1836, not Walker 1835 (Insecta)
Euthyna Gistel 1847
This genus has four species which occur in America north of Mexico. Three of these are subdivided into subspecies. *Anomoea laticlavia laticlavia* (Forster 1771), is found over the eastern two-thirds of the United States and adjacent Canada. The other species are limited to the midwestern states or to the southern states from Florida to California, and south into Mexico. Six more species and many subspecies occur from Mexico to South America. Adult plant associations include *Baccharis* (Asteraceae); *Alnus*, *Betula* (Betulaceae); *Diospyros* (Ebenaceae); *Acacia*, *Desmanthus*, *Gleditsia*, *Mimosa*, *Robinia* (Fabaceae); *Quercus* (Fagaceae); and *Salix* (Salicaceae). Larva and biology, LeSage and Stiefel (1996); biology, Stiefel et al. (1995). Key to species, Moldenke (1970).

Smaragdina Chevrolat 1836
subgenus *Smaragdina* Chevrolat 1836
Cyaniris Chevrolat 1836
Carmentis Gistel 1847
Necyomantes Gistel 1847
Calyptorhina Lacordaire 1848
Gynandrophthalma Lacordaire 1848
Exomis Weise 1889
Smaragdinella Medvedev 1971
Monrosia Medvedev 1971

The genus is nearly worldwide in distribution, with seven species in the New World extending from Mexico to Brazil. The single North American species is not assigned to a subgenus. It is divided into two subspecies: *S. militaris militaris* LeConte 1858, ranging from Missouri to New Mexico, and *S. m. arizonica* Schaeffer 1919, ranging from Texas to Arizona. One European species is associated with ants (Jolivet 1988), but very little is known about the biology of the genus in the Nearctic Region. Moldenke (1970) indicated that adults are associated with mimosaceous shrubs (Fabaceae), but in Texas, *S. militaris* is taken on oaks (*Quercus*, Fagaceae). Key to species, Moldenke (1970).

subgenus *Otiocephala* Lefèvre 1872 [Palearctic]

Megalostomina Chapuis 1874

Characteristics: Head moderate to large, partially recessed into pronotum; mandibles large, usually sexually dimorphic; eyes entire to distinctly emarginate; lateral margin of prothorax without antennal groove; pronotum transverse, convex; procoxae globose, separated by prosternum; scutellum not inclined, level with plane of elytra; elytra rigid, usually with weakly developed postbasal lobe; tarsal claws simple. There are seven New World genera.

Megalostomis Chevrolat 1836

subgenus *Megalostomis* Chevrolat 1836 [Neotropical]

subgenus *Pygidiocarina* Moldenke 1970
At least three species of this subgenus occur in Arizona, California, or Texas, with their combined distribution extending south into Mexico, including Baja California Sur. Many other species occur from northern Mexico to South America. In the larval stage, this subgenus is associated with ants of the genus *Atta*, and adults are evidently specific to the Fabaceae (=Mimosaceae) (Moldenke 1970). Specific plant records include *Acacia*, *Prosopis*, and *Senna* (Fabaceae). Adults of an undetermined species from the cape region of Baja California Sur are found on *Jatropha* (Euphorbiaceae) and *Quercus* (Fagaceae). Key to species, Moldenke (1970).

subgenus *Scaphigenia* Lacordaire 1848 [Neotropical]

subgenus *Heterostomis* Lacordaire 1848 [Neotropical]

subgenus *Minturnia* Lacordaire 1848 [Neotropical]

subgenus *Coleobyersa* Moldenke 1981 [Neotropical]

subgenus *Snellingia* Moldenke 1981 [Neotropical]

Coscinoptera Lacordaire 1848
As presently defined, this genus is represented in America north of Mexico by a single species, *C. aeneipennis* (LeConte 1858), which ranges from southern California to southern Kansas and Arkansas, and south into northern Mexico and the Baja California peninsula. The genus includes 17 additional New World species (Moldenke 1981). Plant associations include *Acacia* and *Prosopis* (Fabaceae), as well as *Eriogonum* (Polygonaceae).

Coleothorpa Moldenke 1981
Moldenke (1981) considered most of the Nearctic species previously assigned to *Coscinoptera* sufficiently distinct as to warrant separation into this genus. Seven species are included in this genus which is found throughout the United States, southern Canada, and northern Mexico. Adults are associated with many plants, the more notable include *Rhus* (Anacardiaceae); *Viguiera* (Asteraceae); *Bursera* (Burseraceae); *Ephedra* (Ephedraceae); *Acacia*, *Cercidium*, *Prosopis* (Fabaceae); *Quercus* (Fagaceae); *Pinus* (Pinaceae); *Eriogonum* (Polygonaceae); and *Larrea* (Zygophyllaceae). Some Nearctic species are myrmecophiles (Jolivet 1988), but the immature stages of most are unknown. Key to species (as *Coscinoptera*), Moldenke (1970). Taxonomic and biological notes, Gilbert (1981).

[*Euryscopa* Lacordaire 1848, not in America north of Mexico.]

Coleorozena Moldenke 1981
This genus, as established by Moldenke (1981), contains 22 species, including the Nearctic species formerly assigned to *Euryscopa* Lacordaire. It is found throughout much of the New World, with seven species distributed in the southwestern United States and northern Mexico. Plant associations involve *Acacia*, *Cercidium*, *Prosopis* (Fabaceae); *Eriogonum* (Polygonaceae); and *Larrea* (Zygophyllaceae). Key to species (as *Euryscopa*), Moldenke (1970).

Babiina Chapuis 1874

Characteristics: Head small, recessed into pronotum; mandibles small, not sexually dimorphic; eyes emarginate; lateral margin of prothorax without antennal groove; pronotum transverse, convex; procoxae transversely conical, contiguous, usually partially hidden by retracted head; scutellum level with plane of elytra; elytra more or less rigid, usually with weakly developed postbasal lobe; tarsal claws appendiculate or bifid. This group includes thirteen New World genera.

Babia Chevrolat 1836
Harpasta Gistel 1847
subgenus *Babia* Chevrolat 1836
This subgenus is represented in America north of Mexico by a single species, *B. quadriguttata* Olivier 1796, which is divided into three subspecies. The combined range of the subspecies covers the central section of the United States from Massachusetts south to Florida, west to Arizona and northern Mexico. Ten additional species are known from Central and South America. In our area, adults appear to be associated with various trees and shrubs, including *Rhus* (Anacardiaceae), *Quercus* (Fagaceae), and *Carya* (Juglandaceae). Key to species, Moldenke (1970).

subgenus *Heterobabia* Monrós 1951 (Neotropical)

subgenus *Archaebabia* Moldenke 1981
Generally, the members of this subgenus are the smallest members of the tribe Clytrini. A single species, *B. tetraspilota* LeConte 1858, which is divided into four dubious subspecies, is present in America north of Mexico. It ranges from northern Mexico to southern California and Texas, north to Wyoming (uncommon in the northern limits of the range). This species is commonly collected on *Prosopis* (Fabaceae). A second species, *B. costalisdebaja* Moldenke 1970, occurs throughout the Baja California peninsula on *Cercidium* and *Prosopis* (Fabaceae), and on *Quercus* (Fagaceae). Four additional species occur in Central and South America. Key to species, Moldenke (1970).

subgenus *Coleolacordairei* Moldenke 1981 [Neotropical]

subgenus *Megababia* Moldenke 1981 [Neotropical]

Urodera Lacordaire 1848
subgenus *Boreurodera* Moldenke 1981
Two species are found in the Nearctic Region: *U. crucifera texana* Schaeffer 1919, from south Texas, and *U. dilaticollis* Jacoby 1889, from northern Mexico, Arizona, and Texas. Two additional species and five subspecies are present in Mexico and Central America. Adults are associated with plants in the family Fabaceae (=Mimosaceae). Key to species, Moldenke (1970).

subgenus *Urodera* Lacordaire 1848 [Neotropical]

subgenus *Austrurodera* Moldenke 1981 [Neotropical]

subgenus *Stereomoides* Moldenke 1981 [Neotropical]

subgenus *Familiurodera* Moldenke 1981 [Neotropical]

Saxinis Lacordaire 1848
subgenus *Boreosaxinis* Moldenke 1981
Twenty-two species make up this subgenus; ten occur in America north of Mexico. The three subspecies of *S. omogera* Lacordaire 1848, together occupy most of the area in the eastern two-thirds of the United States, south into Mexico. *Saxinis saucia* LeConte 1857, with nine subspecies, is found in the western one-third of the United States, extending south into Baja California, Mexico, and north into British Columbia, Canada. The remaining eight species from our area are restricted to Arizona, California, New Mexico, Texas, and northern Mexico. Plant associations include *Chrysothamnus* (Asteraceae); *Atriplex* (Chenopodiaceae); *Arctostaphylos* (Ericaceae); *Acacia*, *Prosopis* (Fabaceae); *Quercus* (Fagaceae); *Eriodictyon* (Hydrophyllaceae); *Eriogonum* (Polygonaceae); *Ceanothus* (Rhamnaceae); and *Adenostoma* (Rosaceae). Larvae of *S. saucia* have been found in association with ants (Moldenke 1970). Key to species, Moldenke (1970). Notes on biology, Spruyt (1925).

subgenus *Saxinis* Lacordaire 1848 [Neotropical]

Ischiopachina Chapuis 1874

Characteristics: Head small and recessed into pronotum; eye with distinct emargination; mandibles small, not sexually dimorphic; lateral margin of prothorax grooved just below marginal bead for reception of antenna; pronotum strongly convex medially; procoxae transversely conical, narrowly separated by prosternum; scutellum inclined but level with inclined plane of anterior portion of elytra; elytra rigid with pronounced, strongly angled, postbasal lobe; tarsal claws appendiculate.

This subtribe includes a single Neotropical genus.

[*Ischiopachys* Chevrolat 1836. Lacordaire (1848) gave "Californie" as a locality for his *I. proteus*, but this record is highly doubtful. Presently, this genus is not known from America north of Mexico nor from the Baja California peninsula.]

Chlamisini Gressitt 1946

Characteristics: Head deeply recessed into pronotum; antenna short, serrate from fifth or sixth antennomere; eyes distinctly emarginate. Pronotum faintly to highly gibbous, usually tuberculate; groove for reception of antenna present on each side of prosternum; procoxae transverse, widely separated by prosternal process. Elytra usually with tubercles and carinae; suture serrate or partially serrate. Pygidium broadly exposed, usually foveolate or otherwise coarsely sculptured. Legs contractile. Adults resemble caterpillar droppings.

Larvae are mostly folivorous. This tribe is worldwide in distribution, with the greatest diversity found in the Neotropical Region. Key to genera, Karren (1972). Key to genera of known larvae, LeSage (1984a).

Pseudochlamys Lacordaire 1848
This genus is represented in the Nearctic fauna by a single species, *P. semirufescens* Karren 1972, from Arizona. Adults have been collected on *Mimosa* (Fabaceae). Four other described species in this genus are Neotropical in distribution. Additional species from Mexico probably belong to this genus, but the limits of the genus are not clearly understood (Karren 1972).

Chlamisus Rafinesque 1815
Chlamys Knoch 1801, not Bolten 1798 (Mollusca)
Arthrochlamys H. Ihering 1905
Boloschesis Jacobson 1924
Bolochesis: Leonard 1928 [error]
This worldwide genus is the largest of the tribe and contains a few hundred species, mostly Neotropical. Presently, there are eight species known from America north of Mexico. One species, *C. foveolatus* (Knoch 1801), is widely distributed in the eastern half of the United States and south into Mexico. The remaining species are restricted to the southwestern United States from Texas and Kansas west to Arizona and south into Mexico. The morphological limits of the genus can not be clearly defined until the Neotropical fauna is revised (Karren 1972). Larval food is un-

known for the Nearctic species. Adults of *C. foveolatus* have been taken repeatedly on *Quercus* (Fagaceae); those of *C. quadrilobatus* (Schaeffer 1926) have been taken in series on *Waltheria* (Sterculiaceae), and on *Hyptis* and *Salvia* (Lamiaceae). Key to species, Karren (1972).

Diplacaspis Jacobson 1924
Diaspis Lacordaire 1848, not Costa 1835 (Insecta)
Diaspsis: Jacoby 1889 [error]
Skwarraia van Emden 1932
Diaclaspis: Monrós 1960 [error]

This genus is represented in the Nearctic fauna by a single species, *D. prosternalis* (Schaeffer 1906), found from Arizona, Texas and south into Mexico and Central America. The genus contains eight other species, all of which are Neotropical. Adults have been collected on *Acacia* and *Prosopis* (Fabaceae).

Neochlamisus Karren 1972

This genus was established by Karren (1972) to include species formerly included in *Diplacaspis* and *Chlamisus*, with similarity of the male and female genitalia as the defining morphological characters. Currently, there are 17 species recognized in America north of Mexico, with a few more found in Central America. Northern distribution is primarily in the United States east of the Rocky Mountains and adjacent Canada. Larvae are found on foliage of their food plants. Some species apparently develop on plants of more than one family, while others seem fairly restrictive in their food choice. Food plant genera include *Acer* (Aceraceae); *Alnus, Betula, Corylus* (Betulaceae); *Azalea, Chamaedaphne, Kalmia, Vaccinium* (Ericaceae); *Prosopis* (Fabaceae); *Quercus* (Fagaceae); *Juglans* (Juglandaceae); *Comptonia, Myrica* (Myricaceae); *Platanus* (Platanaceae); *Eriogonum, Polygonum* (Polygonaceae); *Ceanothus* (Rhamnaceae); *Geum, Fragaria, Potentilla, Rubus* (Rosaceae); *Salix* (Salicaceae); *Ulmus* (Ulmaceae); and *Larrea* (Zygophyllaceae). Key to species, Karren (1972). Larvae, Bøving and Craighead (1931), Lawson (1991), LeSage (1984a). Biology, Wallace (1970).

Exema Lacordaire 1848

As presently defined by Karren (1966), this genus is generally distributed throughout the Nearctic Region as far north as southern Canada. There are nine species found in America north of Mexico, some of which extend into northern Mexico or further south. Additional *Exema* species probably remain confused with *Chlamisus s. lat.* in other regions of the world. Most of our species are clearly associated with plants of the family Asteraceae, including *Ambrosia, Ageratina, Artemisia, Baccharis, Bebbia, Brickellia, Chrysothamnus, Encelia, Euthamia, Flourensia, Gnaphalium, Gutierrezia, Helianthus, Hymenoclea, Iva, Lepidospartum, Pluchea*, and *Solidago*. Larvae are found on the foliage of their hosts; however, *Exema gibber* (Fabricius 1798) of the southeastern United States is atypical. Its larva has been found grazing the bark on twigs of lychee, *Litchi chinensis* Sonn. (Sapindaceae), in Florida (Dekle 1957). Adults of this species have been taken mostly on *Quercus* (Fagaceae), but also from other woody plants. Key to species, Karren (1966). Biology and larvae, Boldt and White (1992), LeSage (1982).

Bibliography

ABDULLAH, M. and A. ABDULLAH. 1968. *Phyllobrotica decorata duportei*, a new sub-species of the Galerucinae (Coleoptera:Chrysomelidae) with a review of the species of *Phyllobrotica* in the Lyman Museum Collection. Entomologist's Monthly Magazine, 104: 4-9.

ANDERSON, L. D. and C. S. PAPP. 1961. The larger elm leaf beetle, *Monocesta coryli* (Say). Proceedings of the Entomological Society of Washington, 63: 203-207.

ANDERSON, W. H. 1938. Description of the larvae of *Chaetocnema denticulata* (Illiger) and *Chaetocnema pulicaria* Melsheimer (Coleoptera: Chrysomelidae). Proceedings of the Entomological Society of Washington, 40: 161-169.

ANDREWS, F. G. and A. J. GILBERT. 1992. *Cadiz hardyi*, a new genus and species of leaf beetle from a western North American sand dune (Coleoptera: Chrysomelidae). Coleopterists Bulletin, 46: 4-14.

ANDREWS, F. G. and A. J. GILBERT. 1993. Studies on the Chrysomelidae (Coleoptera) of the Baja California Peninsula: a new species of *Orthaltica* (Alticinae), with notes on the genus in Baja California. Pan-Pacific Entomologist, 69: 277-280.

ASKEVOLD, I. S. 1987a. The identity of *Donacia cuprea* Kirby, 1837, and *Donacia quadricollis* Say, 1827, with a taxonomic revision of members of the *Donacia subtilis* Kunze-Group (Coleoptera: Chrysomelidae: Donaciinae). Canadian Entomologist, 119: 629-645.

ASKEVOLD, I. S. 1987b. The identity of *Donacia caerulea* Olivier 1795 (Coleoptera: Chrysomelidae: Donaciinae). Coleopterists Bulletin, 41: 345-349.

ASKEVOLD, I. S. 1988. The genus *Neohaemonia* Székessy in North America (Coleoptera: Chrysomelidae: Donaciinae): systematics, reconstructed phylogeny, and geographic history. Transactions of the American Entomological Society, 113: 360-430.

ASKEVOLD, I. S. 1990. Reconstructed phylogeny and reclassification of the genera of Donaciinae (Coleoptera: Chrysomelidae). Quaestiones Entomologicae, 26: 601-664.

ASKEVOLD, I. S. 1991a. An annotated list of Nearctic Donaciinae (Coleoptera: Chrysomelidae): the generic classification and type specimens of the New World species. Psyche, 98(2-3): 165-192.

ASKEVOLD, I. S. 1991b. Classification, reconstructed phylogeny, and geographic history of the New World members of *Plateumaris* Thomson, 1859 (Coleoptera: Chrysomelidae: Donaciinae). Memoirs of the Entomological Society of Canada no. 157: 175 pp.

BALDUF, W. V. 1922. The striped cucumber beetle (*Diabrotica vittata*) its biology. Bulletin of the Ohio Agricultural Experiment Station, 388: 241-311.

BALDUF, W. V. 1926. The acalypha flea beetle (*Crepidodera atriventris* Melsh.). Journal of Economic Entomology, 19: 624-632.

BALSBAUGH, E. U., Jr. 1966. Genus *Lexiphanes* of America north of Mexico (Coleoptera: Chrysomelidae). Proceedings of the United States National Museum, 117: 655-680.

BALSBAUGH, E. U., Jr. 1969. *Pseudolampsis* (Coleoptera, Chrysomelidae, Alticinae): distribution and synonymy. Coleopterists Bulletin, 23: 16-23.

BALSBAUGH, E. U., Jr. 1970. Review of the genus *Paria* (Coleoptera: Chrysomelidae) of North America. Annals of the Entomological Society of America, 63: 453-460.

BALSBAUGH, E. U., Jr. 1973. Geographical variation in *Pachybrachis othonus* (Coleoptera: Chrysomelidae) with descriptions of a new subspecies. Annals of the Entomological Society of America, 66: 252-261.

BALSBAUGH, E. U., Jr. 1978. A second species of *Microtheca* Stål (Coleoptera: Chrysomelidae) found in North America. Coleopterists Bulletin, 32: 219-222.

BALSBAUGH, E. U., Jr. 1983. A taxonomic revision of the genus *Phaedon* north of Mexico (Coleoptera: Chrysomelidae). North Dakota Insects, Schafer-Post Series no. 15: 1-73.

BALSBAUGH, E. U., Jr. and K. L. HAYS. 1972. The leaf beetles of Alabama (Coleoptera: Chrysomelidae). Alabama Agricultural Experiment Station Bulletin no. 441: 1-223.

BALSBAUGH, E. U., Jr. and V. M. KIRK. 1968. Distributional and ecological notes on *Distigmoptera* and *Pseudolampsis* with a description of a new species of *Distigmoptera* (Coleoptera: Chrysomelidae). Journal of the Kansas Entomological Society, 41: 243-247.

BALSBAUGH, E. U., Jr. and W. L. TUCKER. 1976. Geographical variation in *Pachybrachis nigricornis* (Coleoptera: Chrysomelidae). Coleopterists Bulletin, 30: 117-131.

BARBER, H. S. 1916. A review of North American tortoise beetles (Chrysomelidae; Cassidinae). Proceedings of the Entomological Society of Washington, 18: 113-127.

BARBER, H. S. 1937. Some species of *Colaspis* from the *brunnea* confusion (Coleoptera: Chrysomelidae). Proceedings of the Entomological Society of Washington, 38: 198-204. (1936).

BARBER, H. S. 1943. Notes on *Rhabdopterus* in the United States (Coleoptera: Chrysomelidae). Bulletin of the Brooklyn Entomological Society, 38: 111-120.

BARBER, H. S. 1947. *Diabrotica* and two new genera (Coleoptera, Chrysomelidae). Proceedings of the Entomological Society of Washington, 49: 151-161.

BARROWS, E. M. 1979. Life cycles, mating, and color change in tortoise beetles (Coleoptera: Chrysomelidae: Cassidinae). Coleopterists Bulletin, 33: 9-16.

BARSTOW, D. L. and A. R. GITTINS. 1971. Life history studies on a willow leaf beetle *Altica bimarginata* Say in north Idaho (Coleoptera: Chrysomelidae). University of Idaho, Agriculture Research Bulletin no. 80: 1-20.

BARSTOW, D. L. and A. R. GITTINS. 1973. Descriptions of the life stages of *Altica bimarginata* (Coleoptera: Chrysomelidae). Journal of the Kansas Entomological Society, 46: 500-510.

BEAMER, R. H. 1926. Notes on *Griburius montezuma* (Suffrian) (Coleoptera-Chrysomelidae). Pan-Pacific Entomologist, 11:209-210.

BECHYNÉ, J. 1950. Les générotypes des Eumolpides de l'Amérique du Sud et du Centre avec les diagnoses des formes nouvelles (Col. Phytoph. Chrysomeloidea). Mitteilungen der Münchner Entomologische Gesellschaft (e. V.) 40: 264-292.

BECHYNÉ, J. 1952. Nachträge zu den Katalogen von Blackwelder und Junk-Schenkling de neotropischen echten Chrysomeliden (Col. Phytophaga). Entomologische Arbeiten aus dem Museum G. Frey, 3: 1-62 + 1 pl.

BECHYNÉ, J. 1953. Katalog der neotropischen Eumolpiden (Col. Phytoph. Chrysomeloidea). Entomologische Arbeiten aus dem Museum G. Frey, 4: 26-303.

BECHYNÉ, J. 1957. Eumolpides neotropicaux de la collection de Museo Civico di Storia Naturale "Giacomo Doria" di Genova (Col. Phytophaga). Annali del Museo Civico di Storia Naturale di Genova, 69: 226-247.

BECHYNÉ, J. and B. ŠPRINGLOVÁ DE BECHYNÉ. 1969. La posición sistemática de *Megascelis* Chevrolat (Col. Phytophaga). Revista de la Facultad de Agronomía (Maracay), 5(3): 65-76.

BECHYNÉ, J. and B. ŠPRINGLOVÁ DE BECHYNÉ. 1975. Notas sobre la serie filética de *Monomacra* y sus formas convergentes (Col. Phytophaga, Alticidae). Revista de la Facultad de Agronomía (Maracay), 8(4): 25-140.

BECHYNÉ, J. and B. ŠPRINGLOVA DE BECHYNÉ. 1977. Zur phylogenesis einiger neotropischen Alticiden (Col. Phytophaga). Studies on the Neotropical Fauna, 12: 81-145.

BECKER, E. C. 1979. *Pyrrhalta viburni* (Coleoptera: Chrysomelidae), a Eurasian pest of *Viburnum* recently established in Canada. Canadian Entomologist, 111: 417-419.

BELLER, S. and M. H. HATCH 1932. Coleoptera of Washington: Chrysomelidae. University of Washington Publications in Biology 1(2): 65-144.

BESHEAR, R. J. 1969. Observations on the life history of *Hemisphaerota cyanea* in Georgia (Coleoptera: Chrysomelidae). Journal of the Georgia Entomological Society, 4(4): 168-170.

BETHKE, J. A. and R. A. REDAK. 1996. Seasonal occurrence of the herbivore *Trirhabda geminata* (Coleoptera: Chrysomelidae) on *Encelia farinosa* in southern California. Annals of the Entomological Society of America, 89: 843-848.

BLAISDELL, F. E. 1939. A study of the species of Hispinae belonging to the genus *Stenopodius* with descriptions of new species. Transactions of the American Entomological Society, 64: 421-447 + pls. XVII-XIX.

BLAKE, D. H. 1927. Revision of the beetles of the genus *Oedionychis* occurring in America north of Mexico. Proceedings of the United States National Museum, 70(23): 1-44, pls. 1-2.

BLAKE, D. H. 1930. Synonymies of Antillean Chrysomelidae, with descriptions of new species. Bulletin of the Brooklyn Entomological Society, 25: 209-224.

BLAKE, D. H. 1931a. Revision of the beetles of the genus *Trirhabda* north of Mexico. Proceedings of the United States National Museum, 79(2): 1-36, pls. 1-2.

BLAKE, D. H. 1931b. Note on West Indian and Central American flea-beetles (Halticinae). Bulletin of the Brooklyn Entomological Society, 26: 76-83.

BLAKE, D. H. 1933. Revision of the beetles of the genus *Disonycha* occurring in America north of Mexico. Proceedings of the United States National Museum, 82(28): 1-66, pls.1-8.

BLAKE, D. H. 1935. Notes on *Systena*. Bulletin of the Brooklyn Entomological Society, 30(3): 89-109, pl. 3.

BLAKE, D. H. 1936. A redisposition of *Monoxia puncticollis* and allied species. Journal of the Washington Academy of Sciences, 26(10): 423-430.

BLAKE, D. H. 1939. A study of Leconte's types of the beetles of the genus *Monoxia*, with descriptions of new species. Proceedings of the United States National Museum, 87(3072): 145-171, pls. 18-19.

BLAKE, D. H. 1942. The chrysomelid beetles *Luperodes bivittatus* (Leconte) and *varicornis* (Leconte) and some allied species. Proceedings of the United States National Museum, 92(3129): 57-74, pls. 5-6.

BLAKE, D. H. 1943. The generic position of *Hypolampsis pilosa* (Illiger) and some related new species (Coleoptera, Halticidae). Proceedings of the Entomological Society of Washington, 45: 207-221.

BLAKE, D. H. 1945. The genus *Galeruca* in North America (Coleoptera: Galerucinae). Proceedings of the Entomological Society of Washington, 47: 53-63.

BLAKE, D. H. 1950a. The generic status of *Altica picta* Say with descriptions of four closely related species (Coleoptera, Chrysomelidae). Proceedings of the Entomological Society of Washington, 52: 178-183.

BLAKE, D. H. 1950b. A revision of the beetles of the genus *Myochrous*. Proceedings of the United States National Museum, 101(3271): 1-64, pls. 1-8.

BLAKE, D. H. 1951. New species of chrysomelid beetles of the genera *Trirhabda* and *Disonycha* (Coleoptera, Chrysomelidae). Journal of the Washington Academy of Sciences, 41(10): 324-328.

BLAKE, D. H. 1953. The chrysomelid beetles of the genus *Strabala* Chevrolat. Proceedings of the United States National Museum, 103(3319): 121-134.

BLAKE, D. H. 1954a. Five new species of chrysomelid beetles. Journal of the Washington Academy of Sciences, 44(8): 246-250.

BLAKE, D. H. 1954b. Chrysomelid beetles of the *Oedionychus miniatus* complex. Proceedings of the Entomological Society of Washington, 56: 139-147.

BLAKE, D. H. 1955. A study of LeConte's species of the chrysomelid genus *Graphops* with descriptions of some new species. Bulletin of the Museum of Comparative Zoology, 113(4): 263-301, pls.1-6.

BLAKE, D. H. 1956. Species of *Phyllobrotica* occurring in the Pacific Coast states (Coleoptera, Chrysomelidae). Proceedings of the Entomological Society of Washington, 58: 259-263.

BLAKE, D. H. 1957. A note on two chrysomelid beetles. Proceedings of the Entomological Society of Washington, 59: 278.

BLAKE, D. H. 1958. A review of some galerucine beetles with excised middle tibiae in the male. Proceedings of the United States National Museum, 108: 59-101.

BLAKE, D. H. 1965. Twelve new species of chrysomelid beetles from the West Indies (Coleoptera, Chrysomelidae). American Museum Novitates, no. 2217: 1-13.

BLAKE, D. H. 1966a. More new galerucine beetles with excised middle tibiae in the male. Proceedings of the United States National Museum, 118(3528): 233-266.

BLAKE, D. H. 1966b. A review of the beetles of the genus *Neobrotica* and some closely related genera. Proceedings of the United States National Museum, 118: 267-372.

BLAKE, D. H. 1967. A revision of the chrysomelid genus *Glyptoscelis* (Coleoptera, Chrysomelidae). Proceedings of the United States National Museum, 123(3604): 1-53.

BLAKE, D. H. 1970a. Notes on some chrysomelid beetles from the United States and Argentina. Proceedings of the Entomological Society of Washington, 72: 320-324.

BLAKE, D. H. 1970b. A review of the beetles of the genus *Metachroma* Chevrolat (Coleoptera: Chrysomelidae). Smithsonian Contributions to Zoology, no. 57: 1-111.

BLAKE, D. H. 1974. The costate species of *Colaspis* in the United States (Coleoptera: Chrysomelidae). Smithsonian Contributions to Zoology, 181: i-iii, 1-24.

BLAKE, D. H. 1976a. *Colaspis melancholica* Jacoby and its close relatives (Coleoptera: Chrysomelidae). Journal of the Washington Academy of Sciences, 65(4): 158-162. (1975).

BLAKE, D. H. 1976b. The brown semicostate and costate species of *Colaspis* in Mexico and Central America (Coleoptera: Chrysomelidae). United States Department of Agriculture, Technical Bulletin no. 1534: 1-40.

BLAKE, D. H. 1977a. *Colaspis favosa* Say and its close relatives (Coleoptera: Chrysomelidae). Proceedings of the Entomological Society of Washington, 79: 209-215.

BLAKE, D. H. 1977b. *Colaspis chrysis* Olivier and its close relatives (Coleoptera: Chrysomelidae). Proceedings of the Entomological Society of Washington, 79: 417-421.

BLATCHLEY, W. S. 1910. An illustrated descriptive catalogue of the Coleoptera or beetles (exclusive of the Rhynchophora) known to occur in Indiana - with bibliography and descriptions of new species. The Nature Publishing Co. Indianapolis, Indiana. 1386 pp. [This book was also issued simultaneously as Bulletin No. 1 of the Indiana Department of Geology and Natural Resources with the title "The Coleoptera or beetles of Indiana"].

BLATCHLEY, W. S. 1921. Notes on Indiana Halticini with characterization of a new genus and descriptions of new species. Journal of the New York Entomological Society, 24: 16-27.

BLUM, M. S. 1994. Antipredator devices in larvae of the Chrysomelidae: a unified synthesis for defensive eclecticism, Pp. 277-288 *In*: P. Jolivet, M. L. Cox and E. Petitpierre, eds., Novel aspects of the biology of Chrysomelidae. Kluwer Academic Publishers. Dordrecht, i-xxiii + pp. 1-582.

BOLDT, P. E. 1989. Biology and host specificity of *Trirhabda bacharidis* (Coleoptera: Chrysomelidae) on *Baccharis* (Asteraceae: Astereae). Environmental Entomology, 18: 78-84.

BOLDT, P. E. and C. L. STAINES, Jr. 1993. Biology and description of immature stages of *Pentispa suturalis* (Baly) (Coleoptera: Chrysomelidae) on *Baccharis bigelovii* (Asteraceae). Coleopterists Bulletin, 47: 215-220.

BOLDT, P. E. and R. E. WHITE. 1992. Life history and larval description of *Exema elliptica* Karren (Coleoptera: Chrysomelidae) on *Baccharis hamilifolia* L. (Asteraceae) in Texas. Proceedings of the Entomological Society of Washington, 94: 93-90.

BOROWIEC, L. 1984. Zoogeographical study on Donaciinae of the world (Coleoptera: Chrysomelidae). Polskie Pismo Entomologiczne, 53: 433-518.

BOROWIEC, L. 1989. Three new species of *Charidotella* Weise (Coleoptera, Chrysomelidae, Cassidinae), with checklist of the genus. Polskie Pismo Entomologiczne, 59: 203-222.

BOROWIEC, L. 1990. A review of the genus *Cassida* L. of the Australian Region and Papuan Subregion (Coleoptera: Chrysomelidae: Cassidinae). Genus, 1: 1-51.

BOROWIEC, L. 1994. A monograph of the Afrotropical Cassidinae (Coleoptera: Chrysomelidae). Part I. Introduction, key to the genera, and reviews of the tribes Epistictini, Basiprionotini and Aspidimorphini (except the genus *Aspidimorpha*). Genus (Supplement): pp. 1-276.

BOROWIEC, L. 1995. Tribal classification of the cassidoid Hispinae (Coleoptera: Chrysomelidae), Pp. 541-558, vol. 1. *In*: J. Pakaluk and S. A. Slipinskí, eds., Biology, Phylogeny, and Classification of Coleoptera: papers celebrating the 80th birthday of Roy A. Crowson. Muzeum i Instytut Zoologii PAN. Warsaw, Pp. i-xii + 1-558.

BOROWIEC, L. 1999. A world catalogue of the Cassidinae (Coleoptera: Chrysomelidae). Biologica Silesiae. Wroclaw. 476 pp.

BØVING, A. G. 1910. Natural history of the larvae of Donaciinae. Internationale Revue der gesamten Hydrobiologie und Hydrographie 3, Biol. Suppl. 1: 1-108, 7 pls.

BØVING, A. G. 1927. Descriptions of larvae of the genera *Diabrotica* and *Phyllobrotica*, with a discussion of the taxonomic validity of the subfamilies Galerucinae and Halticinae (Coleoptera: Chrysomelidae). Proceedings of the Entomological Society of Washington, 29: 193-205 + 1 pl.

BØVING, A. G. 1929. Beetle larvae of the subfamily Galerucinae. Proceedings of the United States National Museum, 75: 1-48, pls. 1-5.

BØVING, A. G. 1930. Description of the larva of *Cerotoma trifurcata* Forster (Coleoptera: Chrysomelidae). Proceedings of the Entomological Society of Washington, 32: 51-58, 1 pl.

BØVING, A. G. and F. C. CRAIGHEAD. 1931. An illustrated synopsis of the principal larval forms of the order Coleoptera. Entomologica Americana (new series), 11: i-viii + 1-351 [1930].

BRANNON, L. W. 1938. The sweet potato leaf beetle. United States Department of Agriculture Circular no. 495: 1-9.

BRITTON, W. E. 1907. The elm leaf beetle. Connecticut Agricultural Experiment Station Bulletin no. 155: 1-14.

BROWN, W. J. 1942a. The American species of *Phytodecta* Kby. (Coleoptera, Chrysomelidae). Canadian Entomologist, 74: 99-105.

BROWN, W. J. 1942b. The American species of *Entomoscelis* and *Hippuriphila* (Coleoptera, Chrysomelidae). Canadian Entomologist, 74: 172-176.

BROWN, W. J. 1945. Food-plants and distribution of the species of *Calligrapha* in Canada, with descriptions of new species (Coleoptera, Chrysomelidae). Canadian Entomologist, 77: 117-133.

BROWN, W. J. 1951. The American species of *Phratora* Chev. (Coleoptera: Chrysomelidae). Canadian Entomologist, 83: 121-130.

BROWN, W. J. 1956. The New World species of *Chrysomela* L. (Coleoptera: Chrysomelidae). Canadian Entomologist, 88 (supplement no. 3): 1-54.

BROWN, W. J. 1961. Notes on North American Chrysomelidae (Coleoptera). Canadian Entomologist, 93: 967-977.

BROWN, W. J. 1962. The American species of *Chrysolina* Mots. (Coleoptera: Chrysomelidae). Canadian Entomologist, 94: 58-74.

BROWN, W. J. 1969. A new species of *Pyrrhalta* from Sable Island (Coleoptera: Chrysomelidae). Canadian Entomologist, 101: 109.

BUCKINGHAM, G. R. and M. BUCKINGHAM. 1981. A laboratory biology of *Pseudolampsis guttata* (LeConte) (Coleoptera: Chrysomelidae) on waterfern, *Azolla caroliniana* Willd. (Pteridophyta: Azollaceae). Coleopterists Bulletin, 35: 181-188.

BUTTE, J. G. 1968a. The revision of the tribe Chalepini of America north of Mexico. I. Genus *Xenochalepus* Weise (Coleoptera: Chrysomelidae). Coleopterists Bulletin, 22: 45-62.

BUTTE, J. G. 1968b. The revision of the tribe Chalepini of America north of Mexico. II. Genus *Chalepus* Thunberg (Coleoptera: Chrysomelidae). Journal of the New York Entomological Society, 76: 117-133.

BUTTE, J. G. 1968c. The revision of the tribe Chalepini of America north of Mexico. III. Genus *Odontota* Chevrolat (Coleoptera: Chrysomelidae). Coleopterists Bulletin, 22: 101-124.

BUTTE, J. G. 1969. The revision of the tribe Chalepini of America north of Mexico. IV. Genus *Sumitrosis* Butte (Coleoptera: Chrysomelidae). Journal of the New York Entomological Society, 77: 12-30.

CAMPBELL, J. M. and W. J. CLARK. 1983. Observations on host selection by *Lysathia ludoviciana* (Chrysomelidae), a beetle with potential for biological control of certain aquatic weeds. Texas Journal of Science, 35: 165-167.

CASARI, S. A. and C. N. DUCKETT. 1998. Description of immature stages of two species of *Pseudolampsis* (Coleoptera: Chrysomelidae) and the establishment of a new combination in the genus. Journal of the New York Entomological Society, 105: 50-67. (1997).

CHAMBERLIN, F. S., J. N. TENHET, and A. G.BORING. 1924. Life-history studies of the tobacco flea-beetle in the southern cigar-wrapper district. Journal of Agricultural Research, 29: 573-584.

CHAPIN, J. B. 1979. A review of the Louisiana species of *Colaspis* (Coleoptera: Chrysomelidae). Coleopterists Bulletin, 33: 445-450.

CHAPUIS, F. 1874. Famille LXIX. Phytophages, vol. 10, Pp. 1-455. *In*: T. Lacordaire and F. Chapuis, Histoire naturelle des insectes. Genera des Coléoptères. La Librairie encylopédique de Roret. Paris. i-iv + 455 pp.

CHAPUIS, F. 1875. Famille LXIX. Phytophages, vol. 11, Pp. 1-420. *In*: T. Lacordaire and F. Chapuis, Histoire naturelle des insectes. Genera des Coléoptères. La Librairie encylopédique de Roret. Paris. 420 pp.

CHITTENDEN, F. H. 1899. The spinach flea-beetle. United States Department of Agriculture, Division of Entomology Bulletin no. 19 (n. s.): 80-85.

CHITTENDEN, F. H. 1902. The leaf-mining locust beetle, with notes on related species. United States Department of Agriculture, Division of Entomology Bulletin no. 38: 70-89.

CHITTENDEN, F. H. 1907. The water-cress leaf-beetle. United States Department of Agriculture, Bureau of Entomology, Bulletin no. 66, part II: 16-20.

CHITTENDEN, F. H. 1917. The horse-radish flea-beetle: its life history and distribution. United States Department of Agriculture Bulletin no. 535: 1-16.

CHITTENDEN, F. H. 1920. The beet leaf beetle. United States Department of Agriculture Bulletin no. 892: 1-24.

CHITTENDEN, F. H. 1924a. The argus tortoise beetle. Journal of Agricultural Research, 27: 43-51.

CHITTENDEN, F. H. 1924b. The acalypha flea-beetle (*Crepidodera atriventris* Melsh.). Canadian Entomologist, 56: 286.

CHITTENDEN, F. H. 1927. The species of *Phyllotreta* north of Mexico. Entomologica Americana, 8: 1-59.

CHITTENDEN, F. H. and H. O. MARSH. 1920. The western cabbage flea-beetle. United States Department of Agriculture Bulletin no. 902: 1-21.

CLARK, S. M. 1983. A revision of the genus *Microrhopala* (Coleoptera: Chrysomelidae) in America north of Mexico. Great Basin Naturalist, 43: 597-618.

CLARK, S. M. 1996. The genus *Scelolyperus* Crotch in North America (Coleoptera: Chrysomelidae: Galerucinae). Insecta Mundi, 10: 261-280.

CLARK, S. M. 1999. Descriptions of new luperine genera and species from Mexico, with keys to related taxa (Coleoptera: Chrysomelidae: Galerucinae). Insecta Mundi, 12: 189-206 (1998).

CLARK, S. M. 2001. The western North American genus *Androlyperus* Crotch, 1873 (Coleoptera: Chrysomelidae: Galerucinae). Insecta Mundi, 13: 217-227 (1999).

CLARK, S. M. and J. F. CAVEY. 1995. A new species of *Calligrapha* (Coleoptera: Chrysomelidae) from eastern North America. Insecta Mundi, 9: 329-333.

CLAVAREAU, H. 1913a. Chrysomelidae: 1. Sagrinae, 2. Donaciinae, 3. Orsodacninae, 4. Criocerinae. vol. 24 (Pars 51). *In*: W. Junk and S. Schenkling, eds. Coleopterorum Catalogus. W. Junk. Berlin, 103 pp.

CLAVAREAU, H. 1913b. Chrysomelidae: 5. Megascelinae, 6. Megalopodinae, 7. Clytrinae, 8. Cryptocephalinae, 9. Chamydinae, 10. Lamprosomatinae, vol. 24 (Pars 53). *In*: W. Junk and S. Schenkling, eds. Coleopterorum Catalogus. W. Junk. Berlin, 278 pp.

CLAVAREAU, H. 1914. Chrysomelidae: 11. Eumolpinae, vol. 24 (Pars 59). *In*: W. Junk and S. Schenkling, eds. Coleopterorum Catalogus. W. Junk. Berlin, 215 pp.

COX, M. L. 1994. Eggs bursters in the Chrysomelidae, with a review of their occurrence in the Chrysomeloidea, Pp. 75-110. *In*: P. Jolivet, M. L. Cox and E. Petitpierre, eds., Novel aspects of the biology of Chrysomelidae. Kluwer Academic Publishers. Dordrecht. xxiii + 582 pp.

COX, M. L. 1996. The pupae of Chrysomeloidea, Pp. 119-265. *In*: P. H. A. Jolivet and M. L. Cox, eds., Chrysomelidae Biology, vol. 1: The classification, phylogeny and genetics. SPB Academic Publishing. Amsterdam. 443 pp.

COX, M. L. 1998. The first instar larva of *Megascelis puella* Lacordaire (Coleoptera: Chrysomelidae: Megascelinae) and its value in the placement of the Megascelinae. Journal of Natural History, 32: 387-403.

CROTCH, G. R. 1873. Materials for the study of the Phytophaga of the United States. Proceedings of the Academy of Natural Sciences of Philadelphia, 25: 19-83.

CROWSON, R. A. 1946. A revision of the genera of the chrysomelid group Sagrinae (Coleoptera). Transactions of the Royal Entomological Society of London, 97(4): 75-115.

CUSHMAN, R. A. 1916. The cherry leaf-beetle, a periodically important enemy of cherries. United States Department of Agriculture Bulletin no. 352: 1-25.

DACCORDI, M. 1986. Nuove *Plagiodera* della regione Neotropicale (Coleoptera Chrysomelidae, Chrysomelinae). Bollettino della Museo Civico di Storia Naturale, Verona, 13: 443-454.

DACCORDI, M. 1994. Notes for phylogenetic study of Chrysomelinae, with descriptions of new taxa and a list of all known genera (Coleoptera: Chrysomelidae, Chrysomelinae), Pp. 60-84. *In*: D. G. Furth, ed., Proceedings of the Third International Symposium on the Chrysomelidae, Beijing, 1992. Backhuys. Leiden. 150 pp.

DACCORDI, M. 1996. Notes on the distribution of the Chrysomelinae and their possible origin, Pp. 399-412. *In*: P. H. A. Jolivet and M. L. Cox, eds., Chrysomelidae Biology, vol. 1: The classification, phylogeny and genetics. SPB Academic Publishing. Amsterdam. 443 pp.

DACCORDI, M. and L. LESAGE. 1999. Revision of the genus *Labidomera* Dejean with a description of two new species (Coleoptera: Chrysomelidae: Chrysomelinae), Pp. 437-461. *In*: M. L. Cox, ed. Advances in Chrysomelidae Biology I. Backhuys. Leiden. xii + 671 pp.

DAVIS, J. J. 1907. Life-history and habits of *Galeruca pomonae* Scopoli, in Illinois. Entomological News, 18: 269-275.

DEKLE, G. W. 1957. A leaf beetle feeding on the stems of Lychee. Proceedings of the Florida State Horticultural Society, 70: 331-333.

DESWARTE, D. H. and E. U. BALSBAUGH, Jr. 1973. Biologies of *Altica subplicata* and *Disonycha alternata* (Coleoptera: Chrysomelidae), two flea beetles that feed on sandbar willow. Annals of the Entomological Society of America, 66: 1349-1353.

DICKINSON, J. L. 1996. The behavior and ecology of *Labidomera* Chevrolat (Chrysomelidae: Chrysomelinae), Pp. 323-335. *In*: P. H. A. Jolivet and M. L. Cox, eds., Chrysomelidae Biology, vol. 2: Ecological studies. SPB Academic Publishing. Amsterdam. 465 pp.

DODGE, K. L. and P. W. PRICE. 1991. Life history of *Disonycha pluriligata* (Coleoptera: Chrysomelidae) and host plant relationships with *Salix exigua* (Salicaceae). Annals of the Entomological Society of America, 84: 248-254.

DOWNIE, N. and R. H. ARNETT, Jr. 1996. The beetles of northeastern North America. The Sandhill Crane Press. Gainesville, FL. x + 1721 pp. (Chrysomelidae, Volume 2, Pp. 1280-1410)

DUCKETT, C. N. 1999. A preliminary cladistic analysis of the subtribe Disonychina with special emphasis on the series *Paralactica* (Chrysomelidae: Galerucinae, Alticini). Pp 105-135. *In*: M. L. Cox, ed. Advances in Chrysomelidae Biology I. Backhuys Leiden. xii + 671 pp.

DUNN, P. H. and A. RIZZA. 1976. Bionomics of *Psylliodes chalcomera*, a candidate for biological control of musk thistle. Annals of the Entomological Society of America, 69: 395-398.

EBELING, W. 1939. The grape bud beetle, *Glyptoscelis squamulata* Crotch. Bulletin of the Department of Agriculture, State of California, 28: 459-465.

ECKBERG, T. B. and W. S. CRANSHAW. 1994. Larval biology and control of the rabbitbrush beetle, *Trirhabda nitidicollis* LeConte (Coleoptera: Chrysomelidae). Southwestern Entomologist, 19: 249-256.

EDWARDS, J. G. 1953. Species of the genus *Syneta* of the world (Coleoptera: Chrysomelidae). Wasmann Journal of Biology, 11: 23-82.

ERBER, D. 1988. Biology of Camptosomata Clytrinae - Cryptocephalinae - Chlamisinae - Lamprosomatinae, Pp. 513-552. *In*: P. Jolivet, E. Petitpierre and T. H. Hsiao, eds., Biology of Chrysomelidae. Kluwer Academic Publishers. Dordrecht. xxiv + 615 pp..

FALL, H. C. 1915. A revision of the North American species of *Pachybrachys*. Transactions of the American Entomological Society, 41: 291-486.

FALL, H. C. 1924. Part one. Systematic. The New England species of *Galerucella*, Pp. 81-91. *In*: H. C. Fall and W. C. Woods. The Blueberry leaf-beetle and some of its relatives. Maine Agricultural Experiment Station Bulletin no. 319: 81-140.

FALL, H. C. 1933. New Coleoptera XVI. Canadian Entomologist, 65: 229-234.

FARRELL, B. D. 1998. "Inordinate fondness" explained: why are there so many beetles. Science, 281(5376): 555-559.

FARRELL, B. and C. MITTER. 1990. Phylogenesis of insect/plant interactions: have *Phyllobrotica* leaf beetles (Chrysomelidae) and the Lamiales diversified in parallel? Evolution, 44: 1389-1403.

FINK, D. E. 1913. The twelve-spotted asparagus beetle (*Crioceris duodecimpunctata* L.). Cornell University Agricultural Experiment Station Bulletin no. 331: 422-435.

FLOWERS, R. W. 1996. La subfamilia Eumolpinae (Coleoptera: Chrysomelidae) en América Central. INBio Special Publications no.2: 1-59.

FLOWERS, R. W., D. G. FURTH and M. C. THOMAS. 1994. Notes on the distribution and biology of some Florida leaf beetles (Coleoptera: Chrysomelidae). Coleopterists Bulletin, 48: 79-89.

FORBES, S. A. 1884. On the life-histories and immature stages of three Eumolpinae. Psyche, 4: 123-130 + 1 pl.

FORD, E. J. and J. F. CAVEY. 1982. Biology and immature stages of the hispine beetle *Anisostena ariadne* (Coleoptera: Chrysomelidae). Coleopterists Bulletin, 36: 189-192.

FORD, E. J. and J. F. CAVEY. 1985. Biology and larval descriptions of some Maryland Hispinae. Coleopterists Bulletin, 39: 36-59.

FROST, S. W. 1972. Notes on *Blepharida dorothea* Mignot (Coleoptera: Chrysomelidae). Entomological News 83: 45-47.

FROST, S. W. 1973. Hosts and eggs of *Blepharida dorothea* (Coleoptera: Chrysomelidae). Florida Entomologist, 56: 121-122.

FUNK, D. J., D. J. FUTUYMA, G. ORTI and A. MEYER. 1995. Mitochondrial DNA sequences and multiple data sets: a phylogenetic study of phytophagous beetles (Chrysomelidae: *Ophraella*). Molecular Biology and Evolution, 12: 627-640.

FURTH, D. G. 1988. The jumping apparatus of flea beetles (Alticinae)-the metafemoral spring, Pp. 285-297. *In*: P. Jolivet, E. Petitpierre and T. H. Hsiao, eds., Biology of Chrysomelidae. Kluwer Academic Publishers. Dordrecht. xxiv + 615 pp.

FURTH, D. G. 1998. New World *Blepharida* Chevrolat 1836 (Coleoptera: Chrysomelidae: Alticinae). Memoirs of the Entomological Society of Washington, no. 21: 1-109.

FURTH, D. G. 1992. The New World *Blepharida* group, with a key to genera and description of a new species (Coleoptera: Chrysomelidae). Journal of the New York Entomological Society, 100: 399-414.

FURTH, D. G. and J. E. LEE. 2000. Similarity of the *Blepharida*-group genera using larval and adult characters (Coleoptera: Chrysomelidae: Alticinae). Journal of the New York Entomological Society, 108: 26-51.

FURTH, D. G. and V. SAVINI. 1996. Checklist of the Alticinae of Central America, including Mexico (Coleoptera: Chrysomelidae). Insecta Mundi, 10: 45-68.

FURTH, D. G. and K. SUZUKI. 1994. Character correlation studies of problematic genera of Alticinae in relation to Galerucinae (Coleoptera: Chrysomelidae), Pp. 116-135. *In*: D. G. Furth, ed., Proceedings of the Third International Symposium on the Chrysomelidae, Beijing, 1992. Backhuys. Leiden. 150 pp.

FUTUYMA, D. J. 1990. Observations on the taxonomy and natural history of *Ophraella* Wilcox (Coleoptera: Chrysomelidae), with a description of a new species. Journal of the New York Entomological Society, 98: 163-186.

FUTUYMA, D. J. 1991. A new species of *Ophraella* Wilcox (Coleoptera: Chrysomelidae) from the southeastern United States. Journal of the New York Entomological Society, 99: 643-653.

FUTUYMA, D. J. and S. S. MCCAFFERTY. 1990. Phylogeny and the evolution of host plant associations in the leaf beetle genus *Ophraella* (Coleoptera: Chrysomelidae). Evolution, 44: 1885-1913.

GENTNER, L. G. 1926. The mint flea-beetle. Special Bulletin of the Agricultural Experiment Station, Michigan State College, 155: 1-13.

GENTNER, L. G. 1944. The black flea beetles of the genus *Epitrix* commonly identified as *cucumeris* (Harris). Proceedings of the Entomological Society of Washington, 46: 137-149.

GERBER, G. H. 1994. Biology of *Entomoscelis* Chevrolat, Pp. 550-553. *In*: P. Jolivet, M. L. Cox and E. Petitpierre, eds., Novel aspects of the biology of Chrysomelidae. Kluwer Academic Publishers. Dordrecht. xxiii + 582 pp.

GILBERT, A. J. 1981. A new species of *Coscinoptera* Lacordaire from California (Coleoptera: Chrysomelidae). Pan-Pacific Entomologist, 57: 364-370.

GIRAULT, A. A. 1908. Outline life-history of the chrysomelid *Gastroidea cyanea* Melsheimer. Psyche, 15: 6-9.

GOE, M. T. 1918. Life history and habits of *Gastroidea caesia* Rog. (Col.). Entomological News, 29: 224-226.

GOEDEN, R. D. and D. W. RICKER. 1979. Life history of *Zygogramma tortuosa* Rogers on the ragweed, *Ambrosia eriocentra* (Gray) Payne, in southern California (Coleoptera: Chrysomelidae). Pan-Pacific Entomologist, 55: 261-266.

GOULD, G. E. 1944. The biology and control of the striped cucumber beetle. Purdue University Agricultural Experiment Station Bulletin no. 490: 1-28.

GREEN, G. 1939. The biology of *Lema sexpunctata* Oliv. Journal of the Kansas Entomological Society, 12: 128-132.

HABECK, D. H. 1979. Host plant of *Pseudolampsis guttata* (LeConte) (Coleoptera: Chrysomelidae). Coleopterists Bulletin, 33: 150.

HABECK, D. H. and R. WILKERSON. 1980. The life cycle of *Lysathia ludoviciana* (Fall) (Coleoptera: Chrysomelidae) on parrotfeather, *Myriophyllum aquaticum* (Velloso) Verde. Coleopterists Bulletin, 34: 167-170.

HARTZELL, F. Z. 1917. The cherry leaf-beetle. New York Agricultural Experiment Station Bulletin no. 444: 747-820.

HATCH, M. H. 1971. The beetles of the Pacific Northwest. Part V: Rhipiceroidea, Sternoxi, Phytophaga, Rhynchophora, and Lamellicornia. University of Washington Publications in Biology 16: 662 pp. (Chrysomelidae, Pp. 157-243)

HAWTHORN, W. R. 1978. Some effects of different *Plantago* species on feeding preference and egg laying in the flea beetle *Dibolia borealis* Chev. (Chrysomelidae). Canadian Journal of Zoology, 56: 1507-1513.

HEIKERTINGER, F. 1922. Die mit *Oedionychis* verwandten paläarktischen Halticinen (Coleopt.) (*Oedionychis, Lithonoma, Chloëphaga, Philopona, Hyphasoma*). Wiener Entomologische Zeitung, 39: 45-64.

HEIKERTINGER, F. and E. CSIKI. 1939. Chrysomelidae: 11. Halticinae I, vol. 25 (Pars 166). Pp. 1-336. *In*: W. Junk and S. Schenkling, eds., Coleopterorum Catalogus. W. Junk, 's-Gravenhage.

HEIKERTINGER, F. and E. CSIKI. 1940. Chrysomelidae: 11. Halticinae II, vol. 25 (Pars 166). Pp. 337-635. *In*: W. Junk and S. Schenkling, eds., Coleopterorum Catalogus. W. Junk, 's-Gravenhage.

HEMENWAY, R. and W. H. WHITCOMB. 1968. The life history of *Disonycha glabrata* (Coleoptera: Chrysomelidae). Journal of the Kansas Entomological Society, 41: 174-178.

HERRICK, F. Z. 1916. Observations on the life history of the cherry leaf beetle. Journal of Agricultural Research, 5: 943-950.

HILKER, M. 1994. Egg deposition and protection of eggs in Chrysomelidae, Pp. 263-276. *In*: P. Jolivet, M. L. Cox and E. Petitpierre, eds., Novel aspects of the biology of Chrysomelidae. Kluwer Academic Publishers. Dordrecht, xxiii + 582 pp.

HINCKS, W. D. 1952. The genera of the Cassidinae (Coleoptera: Chrysomelidae). Transactions of the Royal Entomological Society of London, 103(10): 327-355.

HODSON, A. C. 1942. Biological notes on the basswood leaf-miner, *Baliosus ruber* (Weber). Journal of Economic Entomology, 35: 570-573.

HOFFMAN, C. E. 1940. Morphology of the immature stages of some northern Michigan Donaciini (Chrysomelidae; Coleoptera). Papers of the Michigan Academy of Science Arts and Letters, 25: 243-290 + pls. i-x. (1939).

HOOD, C. E. 1940. Life history and control of the imported willow leaf beetle. United States Department of Agriculture Circular 572: 1-9.

HORN, G. H. 1889. A synopsis of the Halticinae of boreal America. Transactions of the American Entomological Society, 16: 163-320.

HORN, G. H. 1892. The Eumolpinae of boreal America. Transactions of the American Entomological Society, 19: 195-234.

HORN, G. H. 1893. The Galerucinae of boreal America. Transactions of the American Entomological Society, 20: 57-136.

HORN, G. H. 1895. Coleoptera of Baja California (supplement I.). Proceedings of the California Academy of Sciences (ser. 2), 4: 225-259 + 1 pl.

HORN, G. H. 1896. Coleoptera of Baja California (supplement II.). Proceedings of the California Academy of Sciences (ser. 2), 6: 367-381.

HSIAO, T. H. 1988. Host specificity, seasonality and bionomics of *Leptinotarsa* beetles, Pp. 581-599. *In*: P. Jolivet, E. Petitpierre and T. H. Hsiao. eds., Biology of Chrysomelidae. Kluwer Academic Publishers. Dordrecht. xxiv + 615 pp.

ISELY, D. 1920. Grapevine flea-beetles. United States Department of Agriculture Bulletin no. 901: 1-27.

ISELY, D. 1930. The biology of the bean leaf-beetle. Arkansas Agricultural Experiment Station Bulletin, 248: 1-20.

ISELY, D. 1942. The grape rootworm. Arkansas Agricultural Experiment Station Bulletin no. 426: 1-26.

ISONO, M. 1990a. A revision of the genus *Demotina* (Coleoptera, Chrysomelidae) from Japan, the Ryukyus, Taiwan and Korea, I. Japanese Journal of Entomology, 58: 375-382.

ISONO, M. 1990b. A revision of the genus *Demotina* (Coleoptera, Chrysomelidae) from Japan, the Ryukyus, Taiwan and Korea, II. Japanese Journal of Entomology, 58: 541-554.

JACKMAN, J. A. 1976. A tortoise beetle, *Hemisphaerota cyanea*, on palms in Texas. Southwestern Entomologist, 1: 181-183.

JACKSON, J. J. 1997. Biology of *Aphthona nigriscutis* (Coleoptera: Chrysomelidae) in the laboratory. Annals of the Entomological Society of America, 90: 433-437.

JACQUES, R. L., Jr. 1988. The potato beetles. The genus *Leptinotarsa* in North America (Coleoptera: Chrysomelidae). Flora and Fauna Handbook no. 3. J. E. Brill. Leiden. 144 pp.

JOLIVET, P. 1948. Contribution a l'étude des *Americanotimarcha* n. subg. (Col. Chrysomelidae). Bulletin du Musée Royal d'Histoire Naturelle de Belgique, 24(43): 1-11.

JOLIVET, P. 1951a. Contribution a l'étude du genre *Gastrophysa* Chevrolat (Coleoptera Chrysomelidae) (1re note). Bulletin du Musée Royal d'Histoire Naturelle de Belgique, 27(9): 1-11.

JOLIVET, P. 1951b. Contribution a l'étude du genre *Gastrophysa* Chevrolat (Coleoptera Chrysomelidae) (3me note) (1). Bulletin du Musée Royal d'Histoire Naturelle de Belgique, 27(21): 1-47.

JOLIVET, P. 1957. Chrysomelidae: Orsodacninae (Pars 51: Fasc. 3). *In*: W. D. Hincks, ed., Coleopterorum Catalogus, Supplementa. W. Junk, 's-Gravenhage. 16 pp.

JOLIVET, P. 1970. Chrysomelidae: Donaciinae (Pars 51: Fasc. 2). *In*: W. O. Steel, ed., Coleopterorum Catalogus, Supplementa. W. Junk, 's-Gravenhage. 71 pp.

JOLIVET, P. 1976. Notes préliminaires sur la biologia des *Timarcha* du Pacifique Nord Occidental Américain (*Americanotimarcha*) (Coleoptera Chrysomelidae). Cahiers du Pacifique, 19: 153-165.

JOLIVET, P. 1988. Food habits and food selection of Chrysomelidae. Bionomic and evolutionary perspectives, Pp. 1-24. *In*: P. Jolivet, E. Petitpierre and T. H. Hsiao, eds., Biology of Chrysomelidae. Kluwer Academic Publishers. Dordrecht. xxiv + 615 pp.

JOLIVET, P. 1989. A propos des *Timarcha* Nord-Américains (Col., Chrysomelidae). L'Entomologiste, 45: 27-34.

JOLIVET, P. 1992. Contribution à la taxonomie, la distribution et la biologie des *Chrysolina* nord-américains (Col. Chrysomelidae). L'Entomologiste, 48: 29-51.

JOLIVET, P. and T. HAWKESWOOD. 1995. Host-plants of Chrysomelidae of the world. Backhuys. Leiden. 281 pp.

JOLIVET, P. and E. PETITPIERRE. 1981. Biology of Chrysomelidae (Coleoptera). Butlletí de la Institució catalana d'historia Natural, 47 (Sec. Zool. 4): 105-138.

JONES, T. H. 1916. The eggplant tortoise beetle. United States Department of Agriculture Bulletin no. 422: 1-7.

KARREN, J. B. 1966. A revision of the genus *Exema* of America, north of Mexico (Chrysomelidae, Coleoptera). University of Kansas Science Bulletin, 46: 647-695.

KARREN, J. B. 1972. A revision of the subfamily Chlamisinae of America north of Mexico (Coleoptera: Chrysomelidae). University of Kansas Science Bulletin, 49: 875-98.

KASAP, H. and R. A. CROWSON. 1976. On the systematic relations of *Oomorphus concolor* (Sturm) (Col., Chrysomelidae), with descriptions of its larva and of an aberrant cryptocephaline larva from Australia. Journal of Natural History, 10: 99-112.

KAUFMANN, D. L. 1967. Notes on the biology of three species of *Lema* (Coleoptera: Chrysomelidae) with larval descriptions and key to described United States species. Journal of the Kansas Entomological Society, 40: 361-372.

KELLY, E. O. 1915. The southern corn leaf-beetle. United States Department of Agriculture Bulletin no. 221: 1-11.

KHRULEVA, O. A. 1996. Biology of arctic leaf beetle *Chrysolina cavigera* on Wrangle Island, Pp. 259-270. *In*: P. H. A. Jolivet and M. L. Cox, eds., Chrysomelidae Biology, vol. 3: General studies. SPB Academic Publishing, Amsterdam. 365 pp.

KIMOTO, S. 1993. New or little known Chrysomelidae (Coleoptera) from Japan and its adjacent regions, VI Entomological Review of Japan, 48: 93-101.

KLEIN, M. C. and H. C. COPPEL. 1969. The pine chrysomelid, *Glyptoscelis pubescens*, in northwestern Wisconsin. Annals of the Entomological Society of America, 62: 1-7.

KOGAN, M. and R. D. GOEDEN. 1970. The biology of *Lema trilineata daturaphila*, (Coleoptera: Chrysomelidae) with notes on efficiency of food utilization by larvae. Annals of the Entomological Society of America, 63: 537-546.

KOGAN, M. and D. D. KOGAN. 1979. *Odontota horni*, a hispine leaf miner adapted to soybean feeding in Illinois. Annals of the Entomological Society of America, 72: 456-461.

KONSTANTINOV, A. 1998a. Revision of the Palearctic species of *Aphthona* Chevrolat and cladistic classification of the Aphthonini (Coleoptera: Chrysomelidae: Alticinae). Memoirs on Entomology International, vol. 11. viii + 429 pp.

KONSTANTINOV, A. S. 1998b. On the structure and function of the female genitalia in flea beetles (Coleoptera: Chrysomelidae: Alticinae). Proceedings of the Entomological Society of Washington, 100: 353-360.

KONSTANTINOV, A. S. and N. J. VANDENBERG. 1996. Handbook of Palearctic flea beetles (Coleoptera: Chrysomelidae: Alticinae). Contributions on Entomology, International, 1(3): 237-439.

KRYSAN, J. L. and R. F. SMITH. 1987. Systematics of the *virgifera* species group of *Diabrotica* (Coleoptera: Chrysomelidae: Galerucinae). Entomography, 5: 375-464.

KRYSAN, J. L., R. F. SMITH, T. F. BRANSON and P. L. GUSS. 1980. A new subspecies of *Diabrotica virgifera* (Coleoptera: Chrysomelidae): description, distribution, and sexual compatibility. Annals of the Entomological Society of America, 73: 123-130.

KRYSAN, J. L., R. F. SMITH and P. L. GUSS. 1983. *Diabrotica barberi* (Coleoptera: Chrysomelidae) elevated to species rank based on behavior, habitat choice, morphometrics, and geographical variation in color. Annals of the Entomological Society of America, 76: 197-204.

KUKALOVÁ-PECK, J. and J. F. LAWRENCE. 1993. Evolution of the hind wing in Coleoptera. Canadian Entomologist, 125: 181-258.

KURCHEVA, G. F. 1967. Leaf beetle larvae of the subfamily Eumolpinae and of *Syneta betulae* F. (Coleoptera, Chrysomelidae). Entomological Review, 46: 132-137.

LACORDAIRE, J. T. 1845. Monographie des coléoptères subpentamères de la famille des phytophages, vol. 1, pt. 1. Mémoires de la Société Royale des Sciences de Liège, 3: 1-740.

LACORDAIRE, J. T. 1848. Monographie des coléoptères subpentamères de la famille des phytophages, vol. 2. Mémoires de la Société Royale des Sciences de Liège, 5: 1-890.

LAWSON, F. 1950. Biology of *Gastrophysa cyanea* Melsh. (Coleoptera: Chrysomelidae). Ohio Journal of Science, 50: 221-228.

LAWSON, F. A. 1976. Egg and larval case formation by *Pachybrachis bivittatus*. Annals of the Entomological Society of America, 69: 942-944.

LAWSON, F. A. 1991. Chrysomelidae (Chrysomeloidea) (=Cassididae, Cryptocephalidae, Megalopodidae, Sagridae, etc.). Pp. 568-585. *In*: F. W. Stehr, ed., Immature Insects, Vol. 2. Kendall/Hunt. Dubuque. xiv + 975 pp.

LECONTE, J. L. 1880. Short studies of North American Coleoptera. Transactions of the American Entomological Society, 8: 163-218.

LEE, J. E., S. W. LINGAFELTER and A. S. KONSTANTINOV. 1998. Larval morphology of *Systena blanda* Melsheimer (Coleoptera: Chrysomelidae: Alticinae). Proceedings of the Entomological Society of Washington, 100: 484-488.

LESAGE, L. 1982. The immature stages of *Exema canadensis* Pierce (Coleoptera: Chrysomelidae). Coleopterists Bulletin, 36: 318-327.

LESAGE, L. 1984a. Immature stages of Canadian *Neochlamisus* Karren (Coleoptera: Chrysomelidae). Canadian Entomologist, 116: 383-409.

LESAGE, L. 1984b. Egg, larva, and pupa of *Lexiphanes saponatus* (Coleoptera: Chrysomelidae: Cryptocephalinae). Canadian Entomologist, 116: 537-548.

LESAGE, L. 1985. The eggs and larvae of *Pachybrachis peccans* and *P. bivittatus*, with a key to the known immature stages of the Nearctic genera of Cryptocephalinae (Coleoptera: Chrysomelidae). Canadian Entomologist, 117: 203-220.

LESAGE, L. 1986a. The eggs and larvae of *Cryptocephalus quadruplex* Newman and *C. venustus* Fabricius, with a key to the known immature stages of the Nearctic genera of cryptocephaline leaf beetles (Coleoptera: Chrysomelidae). Canadian Entomologist, 118: 97-111.

LESAGE, L. 1986b. A taxonomic monograph of the Nearctic galerucine genus *Ophraella* Wilcox (Coleoptera: Chrysomelidae). Memoirs of the Entomological Society of Canada no. 133: 1-75.

LESAGE, L. 1988a. *Longitarsus huberi* n. sp. and the related species *L. mancus* LeConte (Coleoptera: Chrysomelidae: Alticinae). Coleopterists Bulletin, 42: 167-172.

LESAGE, L. 1988b. On the identity of *Longitarsus suspectus* Blatchley (Coleoptera: Chrysomelidae: Alticinae). Coleopterists Bulletin, 42: 253-254.

LESAGE, L. 1988c. Notes on European *Longitarsus* species introduced in North America (Coleoptera: Chrysomelidae: Alticinae). Canadian Entomologist, 120: 1133-1145.

LESAGE, L. 1989. Nomenclatorial changes in *Chrysolina* species described by Rogers (1896). Coleopterists Bulletin, 43: 74.

LESAGE, L. 1991. Family Chrysomelidae leaf beetles, Pp. 301-323. *In*: Y. Bousquet, ed., Checklist of beetles of Canada and Alaska. Canada Communication Group. Ottawa, Ontario. vi + 430 pp.

LESAGE, L. 1995. Revision of the costate species of *Altica* Müller of North America north of Mexico (Coleoptera: Chrysomelidae). Canadian Entomologist, 127: 295-411.

LESAGE, L. 1999. À propos de la série type d'*Altica ignita* Illiger, 1807 (Coleoptera, Chrysomelidae). Nouvelle Revue d'Entomologie (n. s.), 16: 373-376.

LESAGE, L. 2000. On the type series of *Altica chalybaea* (Coleoptera: Chrysomelidae). Entomological News, 111: 233-237.

LESAGE, L. and J. DENIS. 1999. The flea-beetle *Altica corni* Woods in North America (Coleoptera: Chrysomelidae, Alticinae), Pp. 533-544. *In*: M. L. Cox, ed. Advances in Chrysomelidae Biology I. Backhuys. Leiden. xii + 671 pp.

LESAGE, L. and P. PAQUIN. 1996. Identification keys for *Aphthona* flea beetles (Coleoptera: Chrysomelidae) introduced in Canada for the control of spurge (*Euphorbia* spp., Euphorbiaceae). Canadian Entomologist, 126: 593-603.

LESAGE, L. and V. L. STIEFEL. 1996. Biology and immature stages of the North American clytrines *Anomoea laticlavia* (Forster) and *A. flavokansiensis* Moldenke (Coleoptera: Chrysomelidae: Clytrinae), Pp. 217-238. *In*: P. H. A. Jolivet and M. L. Cox, eds., Chrysomelidae Biology, vol. 3: General Studies. SPB Academic Publishing. Amsterdam. vii + 365 pp.

LINDQUIST, O. H. and C. N. DAVIS. 1971. The biology of a birch leaf beetle, *Phratora hudsonia* (Coleoptera: Chrysomelidae), with a larval key to forest Chrysomelinae in Ontario. Canadian Entomologist, 103: 622-626.

LINGAFELTER, S. W. and A. S. KONSTANTINOV. 1998. *Systena* Chevrolat (Coleoptera: Chrysomelidae: Alticinae): notes on nomenclature, redescription of the genus, and a preliminary discussion of characters and phylogenetic relationships. Proceedings of the Entomological Society of Washington, 100: 467-483.

LINGAFELTER, S. W. and A. S. KONSTANTINOV. 2000. The monophyly and relative rank of alticine and galerucine leaf beetles: A cladistic analysis using adult morphological characters (Coleoptera: Chrysomelidae). Entomologica Scandinavica, 30: 397-416 (1999).

LOWE, V. H. 1898. The cottonwood leaf-beetle. New York Agricultural Experiment Station Bulletin no. 143: 3-24.

MACGILLIVRAY, A. D. 1903. Aquatic Chrysomelidae and a table of the families of coleopterous larvae. New York State Museum Bulletin, 68: 288-331.

MANGUIN, S., R. WHITE, B. BLOSSEY and S. D. HIGHT. 1993. Genetics, taxonomy, and ecology of certain species of *Galerucella* (Coleoptera: Chrysomelidae). Annals of the Entomological Society of America, 86: 397-410.

MANN, J. S. and R. A. CROWSON. 1981. The systematic positions of *Orsodacne* Latr. and *Syneta* Lac. (Coleoptera Chrysomelidae), in relation to characters of the larvae, internal anatomy and tarsal vestiture. Journal of Natural History, 15: 727-749.

MANN, J. S. and R. A. CROWSON. 1983. On the internal male reproductive organs and their taxonomic significance in the leaf beetles (Coleoptera: Chrysomelidae). Entomologia Generalis, 9: 75-99.

MARSHALL, J. E. 1979. The larvae of the British species of *Chrysolina* (Chrysomelidae). Systematic Entomology, 4: 409-417.

MARTIN, W. D. and G. A. HERZOG. 1987. Life history studies on the tobacco flea beetle, *Epitrix hirtipennis* (Melsheimer) (Coleoptera: Chrysomelidae). Journal of Entomological Science, 22: 237-244.

MARX, E. J. F. 1957. A review of the subgenus *Donacia* in the Western Hemisphere (Coleoptera, Donaciidae). Bulletin of the American Museum of Natural History, 112: 191-278.

MASON, M. L. and F. A. LAWSON. 1982. Biology of the American aspen beetle (Coleoptera: Chrysomelidae: *Gonioctena americana* (Schaeffer)) in the Medicine Bow National Forest, Wyoming. Journal of the Kansas Entomological Society, 55: 779-788.

MAW, M. G. 1976. Biology of the tortoise beetle, *Cassida hemisphaerica* (Coleoptera: Chrysomelidae), a possible biological control agent for bladder campion, *Silene cucubalus* (Caryophyllaceae), in Canada. Canadian Entomologist, 108: 945-954.

MCCAULEY, R. H., Jr. 1938. A revision of the genus *Microrhopala* in North America, north of Mexico. Bulletin of the Brooklyn Entomological Society, 33: 145-168 + 1 pl.

MESSINA, F. J. 1982. Comparative biology of the goldenrod leaf beetles, *Trirhabda virgata* LeConte and *T. borealis* Blake (Coleoptera: Chrysomelidae). Coleopterists Bulletin, 36: 255-269.

MIGNOT, E. C. 1969. *Pachyonychis* Clark and *Pachyonychus* Melsheimer: nomenclatural confusion. Coleopterists Bulletin, 23: 99-100.

MOLDENKE, A. R. 1970. A revision of the Clytrinae of North America north of the Isthmus of Panama. Stanford University. Stanford. 310 pp.

MOLDENKE, A. R. 1981. A generic reclassification of the New World Clytrinae (Coleoptera: Chrysomelidae) with a description of new species. Entomologische Arbeiten aus dem Museum G. Frey, 29: 75-116.

MONRÓS, F. 1951. Notas sobre Criocerinae. Acta Zoológica Lilloana, 11: 467-482.

MONRÓS, F. 1955. On some new genera of Nearctic Chrysomelinae (Chrysomelidae). Coleopterists Bulletin, 9: 53-60.

MONRÓS, F. 1958. Chrysomelidae: Sagrinae (Pars 51: Fasc. 1). *In*: W. D. Hincks, ed., Coleopterorum Catalogus, Supplementa. W. Junk. 's-Gravenhage. 19 pp.

MONRÓS, F. 1960a. Notas sobre Chrysomelidae (Coleoptera). Acta Zoológica Lilloana, 17: 1-24. (1959).

MONRÓS, F. 1960b. Los generos de Chrysomelidae. Opera Lilloana, 3: 1-337. (1959).

MONRÓS, F. 1960c. Chrysomelidae: Lamprosomatinae (Pars 53: Fasc. 10). *In*: W. D. Hincks, ed., Coleopterorum Catalogus, Supplementa. W. Junk. 's-Gravenhage. 16 pp.

MOURA, L. de A. 1998. Revisao de genero *Neolochmaea* (Coleoptera, Chrysomelidae, Galerucinae, Galerucini). Iheringia, Série Zoolgia, 85: 168-188.

MUNROE, D. D. and R. F. SMITH. 1980. A revision of the systematics of *Acalymma sensu stricto* Barber (Coleoptera: Chrysomelidae) from North America including Mexico. Memoirs of the Entomological Society of Canada no. 112: 1-92.

NEAL, J. W., Jr. 1989. Bionomics of immature stages and ethology of *Neochlamisus platani* (Coleoptera: Chrysomelidae) on American sycamore. Annals of the Entomological Society of America, 81: 64-72.

O'BRIEN, P. Y. and P. R. ATSATT. 1982. Life history and general bionomics of *Trirhabda sericotrachyla* Blake (Coleoptera: Chrysomelidae) in southern California. Pan-Pacific Entomologist, 58: 139-152.

ORTON, W. A. and F. H. CHITTENDEN. 1917. Control of diseases and insect enemies of the home vegetable garden. United States Department of Agriculture, Farmers' Bulletin no. 856: 70 pp.

PALMER, J. O. 1985. Phenology and dormancy in the milkweed leaf beetle *Labidomera clivicollis* (Kirby). American Midland Naturalist, 114: 13-18.

PALMER, W. A. 1986. The host range of *Trirhabda flavolimbata* (Mannerheim) (Coleoptera: Chrysomelidae) and its suitability as a biological control agent for *Baccharis* spp. (Asteraceae: Astereae). Coleopterists Bulletin, 40: 149-153.

PALMER, W. A. and R. D. GOEDEN. 1991. The host range of *Ophraella communa* LeSage (Coleoptera: Chrysomelidae). Coleopterists Bulletin, 45: 115-120.

PALMER, W. A. and W. H. HASELER. 1992. The host specificity and biology of *Trirhabda bacharidis* (Weber) (Coleoptera: Chrysomelidae), a species introduced into Australia for the biological control of *Baccharis halimifolia* L. Coleopterists Bulletin, 46: 61-66.

PAPP, C. S. 1959. Discusion of synonymy and first illustrations of larva and pupa of *Coreopsomela elegans* (Olivier 1807) from California (Notes on North American Coleoptera, No. 6). Journal of the Kansas Entomological Society, 32: 137-141.

PARKER, W. B. 1910. The life history and control of the hop flea-beetle (*Psylliodes punctulata* Melsh.). United States Department of Agriculture, Bureau of Entomology, Bulletin 82, part IV: 33-58.

PARRY, R. H. 1974. Revision of the genus *Dibolia* Latreille in America north of Mexico (Coleoptera: Chrysomelidae). Canadian Journal of Zoology, 52: 1317-1354.

PARRY, R. H. 1986. The systematics and biology of the flea beetle genus *Crepidodera* Chevrolat (Coleoptera: Chrysomelidae) in America north of Mexico. Insecta Mundi, 1: 156-195.

PATERSON, N. F. 1931. Studies on the Chrysomelidae. Part II. The bionomics and comparative morphology of the early stages of certain Chrysomelidae (Coleoptera, Phytophaga). Proceedings of the Zoological Society of London, 2: 879-949.

PETERSON, A 1921. The strawberry root-worm, a serious pest on roses in the greenhouse. New Jersey Agricultural Experiment Station Annual Report 41: 468-493.

PETERSON, A. 1951. Larvae of insects. An introduction to Nearctic species. Part II. Coleoptera, Diptera, Neuroptera, Siphonaptera, Mecoptera, Trichoptera. A. Peterson, Colombus, Ohio. 416 pp.

PIPER, G. L. 1978. Life history of *Zygogramma disrupta* in southeast Texas. Pan-Pacific Entomologist, 54: 226-230.

POOS, F. W. 1955. Studies of certain species of *Chaetocnema*. Journal of Economic Entomology, 48: 555-563.

POWELL, E. F. 1941. Relationships within the family Chrysomelidae (Coleoptera) as indicated by the male genitalia of certain species. American Midland Naturalist, 25: 148-195.

REDAK, R. A., J. A. BETHKE, T. D. PAINE and J. T. TRUMBLE. 1995. Biology and laboratory development of *Trirhabda geminata* (Coleoptera: Chrysomelidae) on the composite, *Encelia farinosa*. Annals of the Entomological Society of America, 88: 196-200.

REED, H. 1927. Some observations on the leaf-mining flea-beetle *Dibolia borealis* Chevrolat. Annals of the Entomological Society of America, 20: 540-549.

REID, C. A. M. 1995. A cladistic analysis of subfamilial relationships in the Chrysomelidae *sensu lato* (Chrysomeloidea), Pp. 559-631, vol. 2. *In*: J. Pakaluk and S. A. Slipinski, eds., Biology, Phylogeny, and Classification of Coleoptera: papers celebrating the 80th birthday of Roy A. Crowson. Muzeum i Instytut Zoologii PAN. Warsaw. pp. i-vi + 559-1092.

REID, C. A. M. 2000. Spilopyrinae Chapuis: a new subfamily in the Chrysomelidae and its systematic placement (Coleoptera). Invertebrate Taxonomy, 14: 837-862.

RILEY, E. G. 1979. A new species of *Phyllobrotica* Chevrolat (Coleoptera; Chrysomelidae) from the prairies of southwestern Missouri. Coleopterists Bulletin, 33: 331-335.

RILEY, E. G. 1982. *Erepsocassis* Spaeth, 1936: A valid genus (Coleoptera: Chrysomelidae: Cassidinae). Journal of the Kansas Entomological Society, 55: 651-657.

RILEY, E. G. 1985a. Identification of *Cassida atripes* LeConte, 1859, and *Coptocycla bisignata* Boheman, 1855, two North American tortoise beetles (Coleoptera: Chrysomelidae: Cassidinae). Journal of the Kansas Entomological Society, 58: 53-61.

RILEY, E. G. 1985b. Review of the North America species of *Glyphuroplata* Uhmann, 1940 (Coleoptera: Chrysomelidae: Hispinae). Journal of the Kansas Entomological Society, 58: 428-436.

RILEY, E. G. 1986a. Review of the tortoise beetle genera of the tribe Cassidini occurring in America north of Mexico (Coleoptera: Chrysomelidae: Cassidinae). Journal of the New York Entomological Society, 94: 98-114.

RILEY, E. G. 1986b. Notes on *Cassida relicta*, a tortoise beetle endemic to North America, with a key to the Nearctic species of *Cassida* (Coleoptera: Chrysomelidae). Entomological News, 97: 141-146.

RILEY, E. G. and E. U. BALSBAUGH, Jr. 1988. Two Middle American leaf beetles (Coleoptera: Chrysomelidae) newly recorded from the United States. Entomological News, 99: 143-152.

RILEY, E. G., S. M. CLARK and A. J. GILBERT. (in press). New records, nomenclatural changes and taxonomic notes for select North American leaf beetles (Coleoptera: Chrysomelidae). Insecta Mundi.

RILEY, E. G. and A. J. GILBERT. 2000. Three new species of *Cryptocephalus* Geoffroy from the United States and a new United States record (Coleoptera: Chrysomelidae, Cryptocephalinae). Occasional Papers of the Consortium Coleopterorum, 3: 30-35 (1999).

ROLSON, L. H., R. MAYES, P. EDWARDS and M. WINGFIELD. 1965. Biology of the eggplant tortoise beetle (Coleoptera: Chrysomelidae). Journal of the Kansas Entomological Society, 38: 361-366.

ROLSTON, L. H. and P. ROUSE. 1965. The biology and ecology of the grape colaspis, *Colaspis flavida* Say, in relation to rice production in the Arkansas Grand Prairie. Arkansas Agricultural Experiment Station Bulletin no. 694: 1-31.

SAILSBURY, M. B. 1943. The comparative morphology and taxonomy of some larval Criocerinae (Coleoptera: Chrysomelidae). Bulletin of the Brooklyn Entomological Society, 38: 59-74, 128-139.

SAMUELSON, G. A. 1994a. An elytron to body meshing mechanism of possible significance in the higher classification of Chrysomelidae, Pp. 136-147. *In*: D. G. Furth, ed. Proceedings of the Third International Symposium on the Chrysomelidae, Beijing, 1992. Backhuys. Leiden. 150 pp.

SAMUELSON, G. A. 1994b. Pollen consumption and digestion by leaf beetles, Pp. 179-183. *In*: P. Jolivet, M. L. Cox and E. Petitpierre, eds., Novel aspects of the biology of Chrysomelidae. Kluwer Academic Publishers. Dordrecht. xxiii + 582 pp.

SAMUELSON, G. A. 1996. Binding sites: elytron-to-body meshing structures of possible significance in the higher classification of Chrysomeloidea, Pp. 267-290. *In*: P. H. A. Jolivet and M. L. Cox, eds., Chrysomelidae Biology, vol. 1: The classification, phylogeny and genetics. SPB Academic Publishing. Amsterdam. 443 pp.

SANDERSON, M. W. 1948. Larval, pupal, and adult stages of North American *Physonota* (Chrysomelidae). Annals of the Entomological Society of America, 41: 468-477.

SANDERSON, M. W. 1951. A new record and two new species of North America Hispinae. Proceedings of the Entomological Society of Washington, 53: 160-163.

SCAMMELL, H. B. 1915. The cranberry rootworm. United States Department of Agriculture Bulletin no. 263: 1-8.

SCHAEFFER, C. F. A. 1925. Revision of the New World species of the tribe Donaciini of the coleopterous family Chrysomelidae. Brooklyn Museum Science Bulletin, 3: 45-165 + 1 pl.

SCHAEFFER, C. F. A. 1928. The North American species of *Hydrothassa* with notes on other Chrysomelidae and a description of new species and a variety (Col.). Journal of the New York Entomological Society, 36: 287-291.

SCHERER, G. 1962. Bestimmungsschlüssel der neotropischen Alticinen-Genera (Coleoptera: Chrysomelidae: Alticinae). Entomologische Arbeiten aus dem Museum G. Frey, 13: 497-607.

SCHERER, G. 1974. Review of North American species of *Orthaltica* with new generic synonymy (Coleoptera: Chrysomelidae: Alticinae). Coleopterists Bulletin, 28: 65-72.

SCHERER, G. 1983. Diagnostic key for the Neotropical Alticine genera (Coleoptera: Chrysomelidae: Alticinae). Entomologische Arbeiten aus dem Museum G. Frey, 31/32: 1-89.

SCHERER, G. 1988. The origins of the Alticinae, Pp. 115-130. *In*: P. Jolivet, E. Petitpierre and T. H. Hsiao, eds., Biology of Chrysomelidae. Kluwer Academic Publishers. Dordrecht. xxiv + 615 pp.

SCHMITT, M. 1988. The Criocerinae: biology, phylogeny and evolution, Pp.475-495, *In*: P. Jolivet, E. Petitpierre and T. H. Hsiao, eds., Biology of Chrysomelidae. Kluwer Academic Publishers. Dordrecht. xxiv + 615 pp.

SCHMITT, M. 1996. The phylogenetic system of the Chrysomelidae-history of ideas and present state of knowledge, Pp. 57-96. *In*: P. H. A. Jolivet and M. L. Cox, eds., Chrysomelidae Biology, vol. 1: The classification, phylogeny and genetics. SPB Academic Publishing. Amsterdam. 443 pp.

SCHULTZ, W. T. 1976. Review of the genus *Spintherophyta* (Coleoptera: Chrysomelidae) in North America north of Mexico. Annals of the Entomological Society of America, 69: 877-881.

SCHULTZ, W. T. 1977. Review of the genus *Rhabdopterus* (Coleoptera: Chrysomelidae) in America north of Mexico. Annals of the Entomological Society of America, 70: 968-974.

SCHULTZ, W. T. 1980. A new species of *Nodonota* (Coleoptera: Chrysomelidae) with a review of the United States species. Annals of the Entomological Society of America, 73: 200-203.

SEENO, T. N. and F. G. ANDREWS. 1972. Alticinae of California, part 1: *Epitrix* spp. (Coleoptera: Chrysomelidae). Coleopterists Bulletin, 26: 53-61.

SEENO, T. N. and J. A. WILCOX. 1982. Leaf beetle genera (Coleoptera: Chrysomelidae). Entomography, 1: 1-221.

SELMAN, B. J. 1988. Chrysomelids and ants, Pp. 463-473. *In*: P. Jolivet, E. Petitpierre and T. H. Hsiao, eds., Biology of Chrysomelidae. Kluwer Academic Publishers. Dordrecht. xxiv + 615.

SELMAN, B. J. 1994. Eggs and oviposition in chrysomelid beetles, Pp. 69-74. *In*: P. Jolivet, M. L. Cox and E. Petitpierre, eds., Novel aspects of the biology of Chrysomelidae. Kluwer Academic Publishers. Dordrecht. xxiii + 582 pp.

SELMAN, B. J. and G. B. VOGT. 1971. Lectotype designations in the South American genus *Agasicles* (Coleoptera: Chrysomelidae), with description of a new species important as a suppressant of alligatorweed. Annals of the Entomological Society of America, 64: 1016-1020.

SHOLES, O. D. V. 1987. Host plants and seasonal abundance of adult *Capraita subvittata* (Coleoptera: Chrysomelidae: Alticinae). Proceedings of the Entomological Society of Washington, 89: 818-820.

SILFVERBERG, H. 1994. Chrysomelidae in the Arctic, Pp. 503-510. *In*: P. Jolivet, M. L. Cox and E. Petitpierre, eds., Novel aspects of the biology of Chrysomelidae. Kluwer Academic Publishers. Dordrecht. xxiii + 582 pp.

SMEREKA, E. P. 1965. The life history and habits of *Chrysomela crotchi* Brown (Coleoptera: Chrysomelidae) in northwest Ontario. Canadian Entomologist, 97: 541-549.

SMITH, E. H. 1979. Genus *Tanygaster* Blatchley, a new synonym of *Phyllotreta* Chevrolat (Coleoptera: Chrysomelidae: Alticinae). Coleopterists Bulletin, 33: 359-362.

SMITH, E. H. 1985. Revision of the genus *Phyllotreta* Chevrolat of America north of Mexico part 1. The maculate species. Fieldiana, Zoology (new series) no. 28 : v + 168 pp.

SMITH, R. F. and J. F. LAWRENCE. 1967. Clarification of the status of the type specimens of *Diabroticites* (Coleoptera, Chrysomelidae, Galerucinae). University of California Publications in Entomology, 45: 1-174.

SPAETH, F. 1914. Chrysomelidae: 16. Cassidinae, vol. 25 (Pars 62). *In*: W. Junk and S. Schenkling, eds., Coleopterorum Catalogus. W. Junk, Berlin. 182 pp.

ŠPRINGLOVÁ, B. 1960. Essai monographique du genre *Eumolpus* (Coleoptera Phytophaga). Mémoires de Institut Royal des Sciences Naturelles de Belgique, 2nd ser. (fasc. 60): 1-79.

SPRUYT, F. J. 1925. Observations on the egg-laying habits of *Saxinis saucia* LeC. (Coleoptera-Chrysomelidae). Pan-Pacific Entomologist, 1: 176-178.

STAINES, C. L., Jr. 1986a. New combination and new synonymy in North American *Stenopodius* (Coleoptera: Chrysomelidae: Hispinae) with a taxonomic note on Uroplatini. Proceedings of the Entomological Society of Washington, 88: 192.

STAINES, C. L., Jr. 1986b. A revision of the genus *Brachycoryna* (Coleoptera: Chrysomelidae: Hispinae). Insecta Mundi, 1: 231-241.

STAINES, C. L., Jr. 1989. A revision of the genus *Octotoma* (Coleoptera: Chrysomelidae, Hispinae). Insecta Mundi, 3: 41-56.

STAINES, C. L., Jr. 1993. A revision of the genus *Anisostena* Weise (Coleoptera: Chrysomelidae, Hispinae). Part I. Introduction and the subgenera *Neostena* and *Apostena*. Insecta Mundi, 7: 183-190.

STAINES, C. L., Jr. 1994a. A revision of the genus *Anisostena* Weise (Coleoptera: Chrysomelidae, Hispinae). Part II. The subgenus *Anisostena*: key to the species groups and the *ariadne* species group. Insecta Mundi, 8: 125-135.

STAINES, C. L., Jr. 1994b. A revision of the genus *Anisostena* Weise (Coleoptera: Chrysomelidae, Hispinae). Part III. The *pilatei* species group. Insecta Mundi, 8: 213-226.

STAINES, C. L. Jr. 1994c. A revision of the genus *Anisostena* Weise (Coleoptera: Chrysomelidae, Hispinae). Part IV. The *nigrita* species group. Insecta Mundi, 8: 251-264.

STAINES, C. L., Jr. 1997a. Transfer of *Sumitrosis lateritia* (Smith) to *Platocthispa* Uhmann and a key to the genera of Hispinae known to occur in America north of Mexico (Coleoptera: Chrysomelidae, Hispinae). Coleopterists Bulletin, 51: 193-196.

STAINES, C. L., Jr. 1997b. Cereal leaf beetle (Coleoptera: Chrysomelidae) as a pest of ornamental grasses. Proceedings of the Entomological Society of Washington, 99: 376.

STAINES, C. L., Jr. and E. G. RILEY. 1994. Nomenclature and status of *Xenochalepus* and *Hemichalepus* (Coleoptera: Chrysomelidae, Hispinae). Journal of the Kansas Entomological Society, 67: 218-220.

STAINES, C. L. and D. M. WEISMAN. 2001. The species of *Xanthonia* Baly 1863 (Coleoptera: Chrysomelidae: Eumolpinae) in North America east of the Mississippi River. Proceedings of the Entomological Society of Washington, 103: 157-183.

STIEFEL, V. L. 1993. The larval habitat of *Pachybrachis pectoralis* (Melsheimer) and *Cryptocephalus fulgaratus* LeConte (Coleoptera: Chrysomelidae). Journal of the Kansas Entomological Society, 66: 450-453.

STIEFEL, V. L., J. R. NECHOLS and D. C. MARGOLIES. 1995. Overwintering biology of *Anomoea flavokansiensis* (Coleoptera: Chrysomelidae). Annals of the Entomological Society of America, 88: 342-347.

SUZUKI, K. 1988. Comparative morphology of the internal reproductive system of the Chrysomelidae, Pp. 316-355. *In*: P. Jolivet, E. Petitpierre and T. H. Hsiao, eds., Biology of Chrysomelidae. Kluwer Academic Publishers. Dordrecht. xxiv + 615 pp.

SUZUKI, K. 1994. Comparative morphology of the hindwing venation of the Chrysomelidae (Coleoptera), Pp. 337-354. *In*: P. Jolivet, M. L. Cox and E. Petitpierre, eds., Novel aspects of the biology of Chrysomelidae. Kluwer Academic Publishers. Dordrecht. xxiii + 582 pp.

TAKIZAWA, H. 1989. Notes on larvae of the subfamily Chrysomelinae (Coleoptera, Chrysomelidae), part 1. Kanagawa-Chuho, Yokohama, 90: 243-256.

THOMAS, M. C. 1994. *Chelymorpha cribraria* (Fabricius), a Neotropical tortoise beetle new to Florida (Coleoptera: Chrysomelidae). Florida Department of Agriculture and Consumer Services, Entomology Circular no. 363: 2 pp.

TILDEN, J. W. 1949. Oviposition of *Cryptocephalus confluens* Say (Coleoptera, Chrysomelidae). Entomological News, 60: 151-154.

UHMANN, E. 1957. Chrysomelidae: Hispinae (Hispinae Americanae) (Pars 35, Fasc. 1), Pp. 1-153. *In*: D. W. Hincks, ed., Coleopterorum Catalogus Supplementa. W. Junk. 's-Gravenhage.

UHMANN, E. 1958. Chrysomelidae: Hispinae (Hispinae Africanae, Eurasiaticae, Australicae) (Pars 35, Fasc. 2), Pp. i - ix, 156-398. *In*: D. W. Hincks, ed., Coleopterorum Catalogus Supplementa. W. Junk. 's-Gravenhage.

UHMANN, E. 1964. Chrysomelidae: Hispinae (Corrigenda et addenda) (Pars 35, Fasc. 3), Pp. i-v, 399-490. *In*: W. O. Steel, ed., Coleopterorum Catalogus Supplementa. W. Junk. 's-Gravenhage.

UNDERHILL, G. W. 1928. Life history and control of the pale-striped and banded flea beetles. Virginia Agricultural Experiment Station Bulletin no. 264: 1-20.

VOGT, G. M., J. U. MCGUIRE, Jr. and A. D. CUSHMAN. 1979. Probable evolution and morphological variation in South American disonychine flea beetles (Coleoptera: Chrysomelidae) and their amaranthaceous hosts. United States Department of Agriculture, Technical Bulletin no. 1593: xiv + 148 pp.

WADE, M. J. 1994. The biology of the imported willow leaf beetle, *Plagiodera versicolora* (Laicharting), Pp. 541-547. *In*: P. Jolivet, M. L. Cox and E. Petitpierre (eds.), Novel aspects of the biology of Chrysomelidae. Kluwer Academic Publishers, Dordrecht. xxiii + 582 pp.

WADE, M. J. and F. BREDEN. 1986. Life history of natural populations of the imported willow leaf beetle, *Plagiodera versicolora* (Coleoptera: Chrysomelidae). Annals of the Entomological Society of America, 79: 73-79.

WALLACE, J. B. 1970. The defensive function of a case on a chrysomelid beetle. Journal of the Georgia Entomological Society, 5: 19-24.

WARD, R. H. and R. L. PIENKOWSKI. 1978. Biology of *Cassida rubiginosa*, a thistle-feeding shield beetle. Annals of the Entomological Society of America, 71: 585-591.

WATTS, J. R. 1990. Eggs, larvae and biological notes on *Disonycha leptolineata* Blatchley (Coleoptera: Chrysomelidae). Insecta Mundi, 4: 93-97.

WEIGEL, C. A. 1926. The strawberry rootworm, a new pest on greenhouse roses. United States Department of Agriculture Bulletin no. 1357: 1-48.

WEISE, J. 1911. Chrysomelidae: 12. Hispinae, vol. 25 (Pars 35). *In*: W. Junk and S. Schenkling, eds., Coleopterorum Catalogus. W. Junk. Berlin. 94 pp.

WEISE, J. 1916. Chrysomelidae: 12. Chrysomelinae (Pars 68). *In*: W. Junk and S. Schenkling, eds., Coleopterorum Catalogus. W. Junk. Berlin. 255 pp.

WEISE, J. 1924. Chrysomelidae: 13. Galerucinae, vol. 25 (Pars 78). *In*: W. Junk and S. Schenkling, eds. Coleopterorum Catalogus. W. Junk. Berlin. 225 pp.

WEISS, H. B. and E. L. DICKERSON. 1917. *Plagiodera versicolora* Laich. - an imported poplar and willow pest. Canadian Entomologist, 49: 104-109.

WEISS, H. B. and E. WEST. 1921. Notes on the insects of the spreading dogbane, *Apocynum androsaemifolium* L., with a description of a new dogbane midge by Dr. E. P. Felt. Canadian Entomologist, 53: 146-152.

WELCH, K. A. 1978. Biology of *Ophraella notulata* (Coleoptera: Chrysomelidae). Annals of the Entomological Society of America, 71: 134-136.

WELLSO, S. G. and R. P. HOXIE. 1988. Biology of *Oulema*, Pp. 497-511. *In*: P. Jolivet, E. Petitpierre and T. H. Hsiao, eds., Biology of Chrysomelidae. Kluwer Academic Publishers. Dordrecht. xxiv + 615 pp.

WERNER, F. G. 1960. *Hilarocassis exclamationis* (L.), a tortoise beetle not previously reported from the United States (Chrysomelidae: Cassidinae). Coleopterists Bulletin, 14: 94.

WESTCOTT, R. L., R. E. BROWN, D. B. SHARRATT and R. E. WHITE. 1985. *Longitarsus*: a new species from Oregon and a new record for North America (Coleoptera: Chrysomelidae). Pan-Pacific Entomologist, 61: 323-330.

WHEELER, A. G., Jr. and E. R. HOEBEKE. 1979. Biology and seasonal history of *Calligrapha spiraeae* (Say) (Coleoptera: Chrysomelidae), with descriptions of the immature stages. Coleopterists Bulletin, 33: 257-268.

WHEELER, A. G., Jr. and W. A. SNOOK, II. 1986. Biology of *Sumitrosis rosea* (Coleoptera: Chrysomelidae), a leafminer of black locust, *Robinia pseudoacacia* (Leguminosae). Proceedings of the Entomological Society of Washington, 88: 521-530.

WHITE, R. E. 1968. A review of the genus *Cryptocephalus* in America north of Mexico (Chrysomelidae: Coleoptera). United States National Museum Bulletin no. 290: 1-124.

WHITE, R. E. 1979. A Neotropical leaf beetle established in the United States (Chrysomelidae). Annals of the Entomological Society of America, 72: 269-270.

WHITE, R. E. 1981. Homonymy in World species-group names of Criocerinae (Coleoptera: Chrysomelidae). United States Department of Agriculture, Technical Bulletin no. 1629: 1-69.

WHITE, R. E. 1993. A revision of the subfamily Criocerinae (Chrysomelidae) of North America north of Mexico. United States Department of Agriculture, Technical Bulletin no. 1805: 1-158.

WHITE, R. E. 1996a. A revision of the genus *Chaetocnema* of America north of Mexico (Coleoptera: Chrysomelidae). Contributions of the American Entomological Institute, 29: 1-158.

WHITE, R. E. 1996b. Leaf beetles as biological control agents against injurious plants in North America. Pp. 373-399, *In*: P. H. A. Jolivet and M. L. Cox, eds., Chrysomelidae Biology, vol. 2: Ecological studies. SPB Academic Publishing. Amsterdam. 465 pp.

WHITE, R. E. and H. S. BARBER. 1974. Nomenclature and definition of the tobacco flea beetle, *Epitrix hirtipennis* (Melsh.), and of *E. fasciata* Blatchley, (Coleoptera: Chrysomelidae). Proceedings of the Entomological Society of Washington, 76: 397-400.

WHITE, R. E. and W. H. DAY. 1979. Taxonomy and biology of *Lema trivittata* Say, a valid species with notes on *L. trilineata* (Oliv.) (Coleoptera: Chrysomelidae). Entomological News, 90: 209-217.

WILCOX, J. A. 1951. A new species and new genus of Galerucinae (Chrysomelidae: Coleoptera). Ohio Journal of Science, 51: 90-94.

WILCOX, J. A. 1953. New species of Galerucinae and Alticinae with notes on other species (Coleoptera: Chrysomelidae). Ohio Journal of Science, 53: 51-58.

WILCOX, J. A. 1954. Leaf beetles of Ohio (Chrysomelidae: Coleoptera). Ohio Biological Survey, Bulletin no. 43: 353-506.

WILCOX, J. A. 1957. A revision of the North American species of *Paria* Lec. (Coleoptera: Chrysomelidae). New York State Museum and Science Service Bulletin no. 365: 1-45.

WILCOX, J. A. 1965. A synopsis of the North American Galerucinae (Coleoptera: Chrysomelinae). New York State Museum and Science Service Bulletin no. 400: i-iv + 1-226.

WILCOX, J. A. 1971. Chrysomelidae: Galerucinae (Oidini, Galerucini, Metacyclini, Sermylini) (Pars 78, Fasc. 1). *In*: J. A. Wilcox, ed., Coleopterorum Catalogus Supplementa. W. Junk. 's-Gravenhage. 220 pp.

WILCOX, J. A. 1972a. A review of the North American Chrysomelinae leaf beetles (Coleoptera: Chrysomelidae). New York State Museum and Science Service Bulletin no. 421: 1-37.

WILCOX, J. A. 1972b. Chrysomelidae: Galerucinae (Luperini: Aulacophorina, Diabroticina) (Pars 78, Fasc. 2). *In*: J. A. Wilcox, ed., Coleopterorum Catalogus Supplementa. W. Junk. 's-Gravenhage. Pp. 221-431.

WILCOX, J. A. 1973. Chrysomelidae: Galerucinae (Luperini: Luperina) (Pars 78, Fasc. 3). *In*: J. A. Wilcox, ed., Coleopterorum catalogus supplementa. W. Junk. 's-Gravenhage. Pp. 432-664.

WILCOX, J. A. 1975a. Checklist of the beetles of Canada, United States, Mexico, Central America and the West Indies. Vol. 1, pt. 7. The leaf beetles (red version). [inside title: Checklist of the Chrysomelidae of Canada, United States, Mexico, Central America and the West Indies, family no. 104]. Biological Research Institute of America. New York. 166 pp. [The date "VIII-1-74" appears on the header of each page of the body of this work. This has led some authors to cite 1974 as the year of publication. The cover bears the date "January 1975".]

WILCOX, J. A. 1975b. Chrysomelidae: Galerucinae (addenda et index) (Pars 78, Fasc. 4). *In*: J. A. Wilcox, ed., Coleopterorum Catalogus Supplementa. W. Junk. 's-Gravenhage. Pp. 665-770.

WILCOX, J. A. 1979. Leaf beetle host plants in northeastern North America (Coleoptera: Chrysomelidae). The Biological Research Institute of America. Kinderhook. 30 pp.

WILSON, F. 1943. The entomological control of St. John's wort (*Hypericum perforatum* L.) with particular reference to the insect enemies of the weed in southern France. Commonwealth of Australia Council for Scientific and Industrial Research Bulletin 169: 1-87.

WOOD, M. 1940. The rose leaf beetle in Pennsylvania. Pennsylvania Agricultural Experiment Station Bulletin no. 387: 1-22.

WOODRUFF, R. E. 1965. A tortoise beetle (*Hemisphaerota cyanea* (Say)) on palms in Florida (Coleoptera: Chrysomelidae). Florida Department of Agriculture and Consumer Services, Entomology Circular no. 35: 1-2.

WOODRUFF, R. E. 1975. The tortoise beetles of Florida II, *Plagiometriona clavata* (Fabricius) (Coleoptera: Chrysomelidae). Florida Department of Agriculture and Consumer Services, Entomology Circular no. 155: 1-2.

WOODRUFF, R. E. 1976a. The tortoise beetles of Florida III, *Eurypepla calochroma floridensis* Blake (Coleoptera: Chrysomelidae). Florida Department of Agriculture and Consumer Services, Entomology Circular no. 163: 1-2.

WOODRUFF, R. E. 1976b. The tortoise beetles of Florida IV, *Metriona bicolor* (Fab.) (Coleoptera: Chrysomelidae). Florida Department of Agriculture and Consumer Services, Entomology Circular no. 164: 1- 2.

WOODS, W. C. 1918. The biology of Maine species of *Altica*. Maine Agricultural Experiment Station Bulletin no. 273: 149-204.

WOODS, W. C. 1924. Part two. Economic and biological, Pp. 92-141. *In*: H. C. Fall and W. C. Woods. The Blueberry leaf-beetle and some of its relatives. Maine Agricultural Experiment Station Bulletin no. 319: 80-141.

YU, P., X. YANG and S. WANG. 1996. Biology of *Syneta adamsi* Baly and its phylogenetic implication, Pp. 201-216. *In*: P. H. A. Jolivet and M. L. Cox, eds., Chrysomelidae Biology, vol. 3: General studies. SPB Academic Publishing. Amsterdam. 365 pp.

Superfamily CURCULIONOIDEA

125. NEMONYCHIDAE Bedel 1882

by Robert S. Anderson

Family common name: The pine flower snout beetles

Among the weevils, these rarely collected beetles are easily recognized by their straight antennae, and elongate rostrum combined with the presence of a distinct labrum. Adults are found in association with the male pollen-bearing flowers of *Pinus* species.

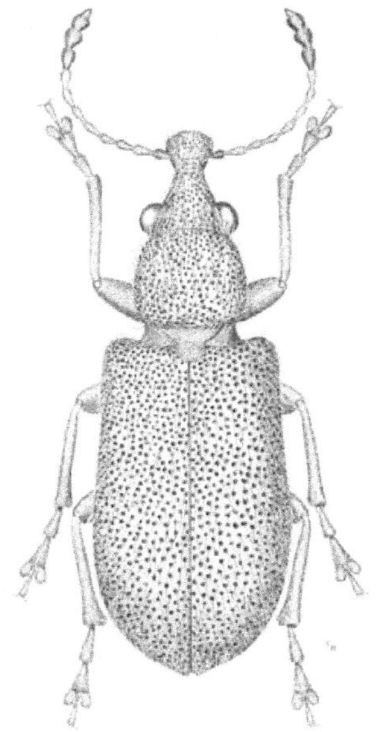

FIGURE 1.125. *Cimberis compta* (LeConte) (from Bright 1993, reproduced with the permission of the Minister of Public Works and Government Services, 2001)

Description (based on Lawrence 1982). Shape elongate, slightly convex; length 3.0-5.5 mm; color pale brown to black; vestiture of fine short to moderately long appressed or suberect pubescence. Rostrum moderately to very long and mostly narrow. Antennae straight, ending in a weak, loose club of three articles; antennal insertions lateral at the middle or near the apex of the rostrum. Labrum distinct, not fused with clypeus. Mandibles with a small but distinct mola. Maxillae with separate galea and lacinia and maxillary palps flexible. Labial palps attached ventrally near the base of the prementum. Gular sutures well-developed and separate. Proventriculus lacks sclerotized plates. Procoxae contiguous and the procoxal cavities narrowly closed posteriorly. Mesocoxal cavities either not closed laterally or narrowly so. Elytra without an inner subcostal flange. Hind wing mostly with four anal veins or fewer. Tarsal claws of some cleft. Visible sternites of the abdomen are free; pygydium concealed by the elytra. Tegmen simple or bilobed apically and the median lobe with a distinct dorsal plate.

Eggs are undescribed.

Larvae (based on Anderson 1991) when mature about 4.0-4.8 mm in length, of moderate thickness throughout length, strongly "C"-shaped. Body white, covered with mixture of long and short setae. Minute legs present on thorax. Head hypognathous, rounded at sides, pigmented, with few to many setae on frons and epicranium. Frontal sutures complete, reaching articulating membrane of mandible. Clypeus not distinguishable from frons and incompletely separated from labrum. One pair of anterior stemmata. Labrum short, the anterior margin rounded, bearing four pairs of setae. Antenna of a single membranous article bearing an accessory appendage. Mandible with two apical teeth, an obtuse protuberance on cutting edge, a distinctly produced molar area with a flattened grinding surface, and one pair of setae. Hypopharyngeal bracon present. Maxillary palp with three articles, palpiger present or absent. Labial palp of two articles. Premental sclerite present, may be divided medially. Thorax with pronotal sclerite transverse, lightly pigmented or unpigmented, sparsely covered with setae. Legs very small, subconical, of two or three segments, with or without a terminal claw. Abdomen with first eight segments with two dorsal folds and bearing annular or bicameral spiracles. Anal opening terminal.

Pupae are undescribed.

Habits and habitats. These beetles are rarely collected, likely because of their specialized habits and life history. In North America, adults are found on male flowers of several pine species very early in the season, often while snow is still on the ground. Adults feed on pollen. After mating, females lay eggs in the flowers where the larvae also feed on pollen. Mature larvae drop from the flowers to the ground and pupate in the soil. Pupation can take from a few months to two years. In some instances larvae have been observed feeding on plant parts other than pollen (Thomas and Herdy 1961). In other areas of the world gymnosperms such as Araucariaceae and Podocarpaceae, and some primitive angiosperms (Fagaceae and Ranunculacaeae) also serve as host plants. Whereas pines are the only documented hosts in North America, other genera of conifers may serve as hosts.

Status of the classification. This family is bipolar in distribution with approximately similar numbers of taxa found in the southern temperate zones of South America, New Zealand and Australia, and in the northern Holarctic Region. Whereas the family was once considered to be absent in tropical areas, a few species have recently been collected in Panama and Venezuela on *Podocarpus* and appear to represent an undescribed genus. The North American fauna has recently been revised (Kuschel 1989) and is well-known. A catalog of the North American species was prepared by Hamilton (1994).

Distribution. There are 5 genera and 15 species in North America. Two additional species in the genus *Atopomacer* are known from *Pinus* at high elevations in far northern Mexico. North American nemonychid species are generally distributed in the

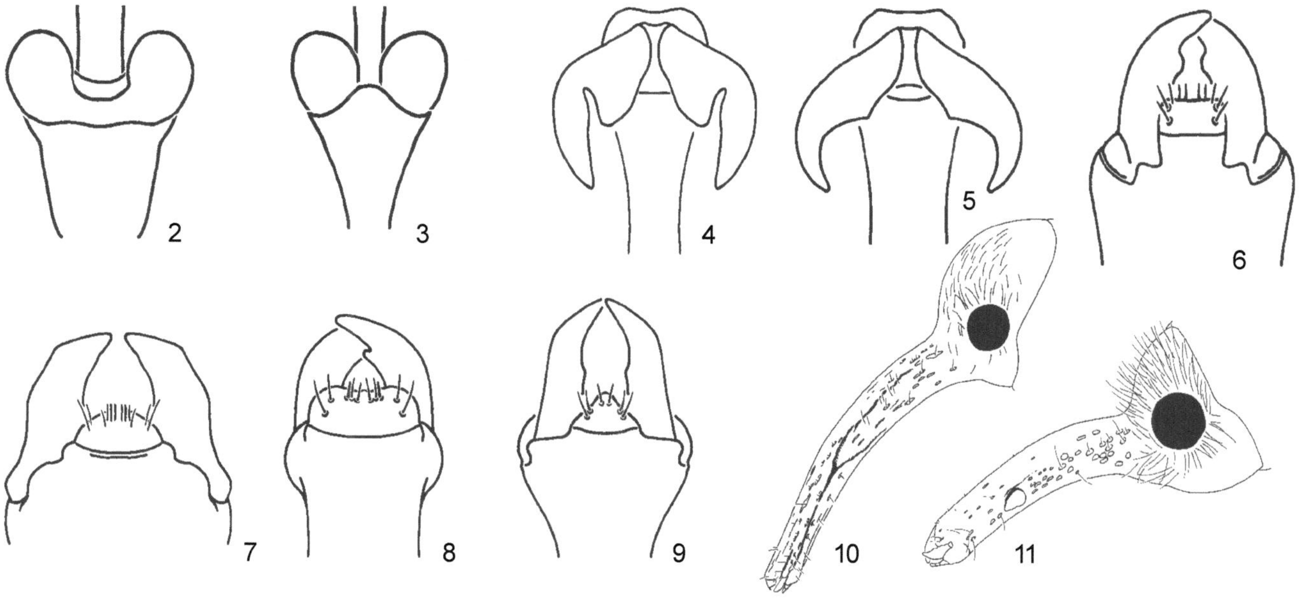

FIGURES 2.125-11.125. 2. *Atopomacer orites* Kuschel 1989, tarsus; 3. *Cimberis elongata* (LeConte 1876), tarsus; 4. *Atopomacer orites* Kuschel 1989, tarsal claw; 5. *Cimberis turbans* Kuschel 1989, tarsal claw; 6. *Lecontellus pinicola* Kuschel 1989, rostrum apex; 7. *Pityomacer carmelites* Kuschel 1989, rostrum apex; 8. *Cimberis decipiens* Kuschel 1989, rostrum apex; 9. *Acromacer bombifrons* (LeConte 1876), rostrum apex; 10. *Pityomacer pix* Kuschel 1989, head; 11. *Acromacer bombifrons* (LeConte 1876), head.

western montane and boreal regions and likely occur anywhere where pines are present.

Key to the Nearctic Genera

1. Second tarsomere truncate at middle, not projected over base of third (Fig. 2); hind tibia with single apical spur; tarsal claw with broad basal flange (Fig. 4); elytra with punctures arranged into indistinct striae (Rhinorhynchinae) .... *Atopomacer*

— Second tarsomere lobed at middle, projected over base of third (Fig. 3); hind tibia with two apical spurs; tarsal claw simple, lacking broad basal flange (Fig. 5); elytra with punctures not arranged into striae (Cimberidinae) 2

2(1). Mandibles inserted dorsally, the sockets fully exposed in dorsal view (Fig. 6); in lateral view, mandibles directed obliquely downwards in relation to rostral plane; antennae situated at middle or after middle of rostrum, as distant from mandibular sockets as combined length of first three articles (Doydirhynchini) .... *Lecontellus*

— Mandibles inserted laterally, the sockets only partially exposed in dorsal view (Figs. 7-9); in lateral view, mandibles continuous with rostral plane; antennae situated at middle or before middle of rostrum, distinctly closer to mandibular sockets than combined length of first three articles (Cimberidini) 3

3(2). Mandibles evenly curved on outer margin, each armed with well-developed tooth on inner margin (Fig. 8) .... *Cimberis*

— Mandibles angulate on outer margin, unarmed on inner margin (Figs. 7, 9) 4

4(3). Labrum trapezoidal or broadly rounded at apex, with six peg-like apical setae in addition to three pairs of dorsal setae (Fig. 7); base of rostrum moderately saddled against obliquely rising frons (Fig. 10) in dorsal view, with 1-3 low carinae flanked with rows of confluent punctures; females with setiferous patches mostly present on one or two ventrites .... *Pityomacer*

— Labrum triangular, pointed at apex, without any peg-like setae in addition to three pairs of dorsal setae (Fig. 9); base of rostrum very deeply saddled against vertically rising frons (Fig. 11) in dorsal view, smooth or sparsely punctate, not carinate; females without setiferous patches on ventrites .... *Acromacer*

Classification of the Nearctic Genera

Nemonychidae Bedel 1882

Cimberidinae Gozis 1882

Cimberidini Gozis 1882

Acromacer Kuschel 1989, 1 sp., *A. bombifrons* (LeConte 1876), British Columbia, Alberta, Washington, Oregon, California, Idaho and Nevada. Adults collected on *Pinus* species.

Cimberis Gozis 1881, 7 spp., generally distributed. Adults collected on *Pinus* species.

Pityomacer Kuschel 1989, 3 spp., British Columbia, Alberta, Washington, Oregon, California and Montana. Adults collected on *Pinus* species.

Doydirhynchini Pierce 1916

Lecontellus Kuschel 1989, 3 spp., Washington, Oregon, California and Nevada. Adults collected on *Pinus* species.

Rhinorhynchinae Voss 1922

Rhinorhynchini Voss 1922

Atopomacer Kuschel 1989, 1 sp., *A. ites* Kuschel 1989, Colorado and Arizona. Adults collected on *Pinus* species apparently at high elevations.

Bibliography

ANDERSON, D. M. 1991. Nemonychidae (Curculionoidea). Pp. 585-586. *In*: F. W. Stehr, ed. Immature Insects. Volume 2. Kendall/Hunt. Dubuque, Iowa.

BRIGHT, D. E. 1993. The Insects and Arachnids of Canada, Part. 21. The Weevils of Canada and Alaska: Volume 1. Centre for Land and Biological Research. Ottawa, 217 pp.

HAMILTON, R. W. 1994. A catalog of the Coleoptera of America north of Mexico. Family: Nemonychidae. USDA Agriculture Handbook 529-134, x + 8 pp.

KUSCHEL, G. 1989. The Nearctic Nemonychidae (Coleoptera: Curculionoidea). Entomologica Scandinavica, 20: 121-171.

LAWRENCE, J. L. 1982. Coleoptera. Pp. 482-553. *In*: S. P. Parker, ed. Synopsis and Classification of Living Organisms. Volume 2. McGraw Hill. New York.

THOMAS, J. B. and H. HERDY. 1961. A note on the life history of *Cimberis elongatus* (LeC.) (Coleoptera: Anthribidae). Canadian Entomologist, 93: 406-408.

126. ANTHRIBIDAE Billberg 1820

by Barry D. Valentine

Family common name: The fungus weevils

Family synonyms: Anthotribidae Gemminger and Harold 1872; Choragidae Kirby 1819; Platyrrhinidae Everts 1903; Platystomidae Pierce 1916; Platystomoidea Pierce 1916.

These primitive weevils greatly resemble the snout beetles but the beak is broad, the antennae are not geniculate, the pygidium is exposed, only the third tarsomere is spongy-pubescent beneath, the pronotal pubescence is directed anteriad, and the elytra usually have an abbreviated scutellar stria.

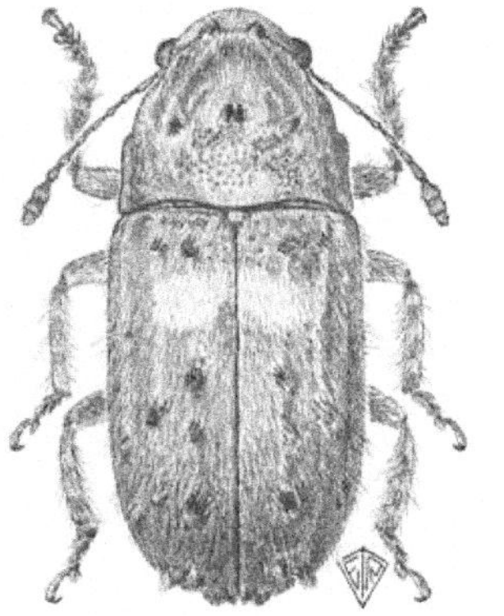

FIGURE 1.126. *Toxonotus cornutus* (Say)

Description: Shape unusually diverse; elongate and depressed, to oval, convex, and mite-like, mostly more or less elongate and convex above; length 0.4 to 40 mm, in U.S. 0.4 to 16 mm; vestiture rarely absent, of hair-like scales mostly in mixed shades of white, gray, straw, brown or black, in the tropics some with bright pink, red, yellow, or green. In collections, species are often confused with cerambycids, chrysomelids, ciids, scolytids, and mites.

Head large, rarely retractile into prothorax; if rostrum present mostly broad and flattened; surface mostly smooth on vertex, sculpture often progressively heavier on frons and rostral area. Antennae with eleven antennomeres (rarely nine or ten), not geniculate, most with an apical club of three antennomeres, club faint or absent in species with elongate antennae, and involving from two to eight antennomeres in some tropical species; inserted either on the lateral or ventral surface of the rostrum, or on the face between or below the eyes. Labrum distinct, mostly semicircular, setose; mandibles unusually large for a weevil, stout, curved, upper surface flattened and with a setose groove, the apices acute, mostly with a postmedian tooth; maxillary palpi slender, with four palpomeres, the first very short, the fourth long; gular sclerite and sutures invisible; postmentum (fused mentum and submentum) large, deeply emarginate in front, heavily sclerotized; ligula large, corneous, mostly notched to deeply emarginate in front, rarely entire; labial palpi with three palpomeres, the second shortest, both pairs of palpi elongate, cylindrical, flexible, and with acute (very rarely truncate) apices. Eyes dorsal, dorsolateral, or lateral, mostly large, ranging from entire to deeply emarginate, the facets very variable in size and number, and without intermixed setae.

Pronotum apex narrower than the maximum head width (except those choragine genera with retractile heads); some exotic genera with the eyes out on broad stalks; base mostly nearly as wide as the elytra, constricted or produced laterad in some; shape mostly trapezoidal, apex obliquely truncate, rarely somewhat produced over the head, base truncate with an antebasal transverse ridge, this mostly turned forward at the sides forming a short or more rarely complete side margin (called the lateral carina), the transverse ridge is always antebasal but it is referred to as basal in those species in which the surface behind it is more or less vertical; surface smooth, punctate, reticulate, rugose, or tuberculate; pleural region broad, supra-coxal sutures well developed; prosternum short to long in front of coxae, the intercoxal process narrow (rarely broader than a coxa); the coxal cavities closed behind. Mesosternum short, the process separating the mesocoxae extremely variable in shape and proportions. Metasternum mostly long, in some so short that meso- and metacoxae are barely separate, a transverse groove or suture near posterior margin, and in many a midventral longitudinal depression or groove. Legs robust to slender; trochantins not exposed; procoxae globular, mostly protruding, not touching or only barely so, and grooved on their inner faces for reception of the intercoxal process of prosternum; mesocoxae globular, separate; metacoxae elongate and transverse (except *Sicanthus* Valentine and *Cisanthribus* Zimmerman where they are globular or broadly tear-drop shaped), separate, not reaching elytral margin; trochanters moderate with the apices oblique; femora swollen postmedially, mostly without teeth or spines; tibiae slender, without movable spurs, but with an apical hook or tooth in males of some genera; tarsal formula 5-5-5, apparently 4-4-4, the second apically emarginate, the third deeply bilobed and tomentose beneath, partially embraced by the second, and rarely with the lobes fused, the fourth very small and invisible in dorsal view, the fifth elongate, with claws mostly toothed or cleft. Scutellum small, varied in shape, hidden in some flightless species. Elytral apices always locking into a longitudinal groove in pygidial base, pygidial disc always partially exposed in posterior view but may be invisible from above; normally 10 elytral striae or groups of punctures plus an abbreviated scutellar row, rarely 12, 14 or 19-21 rows of punctures; epipleural fold present, narrowed or obsolete apically. Wing venation and folding pattern approach that of the Chrysomelidae.

Abdomen with five visible sterna, the four sutures entire, fused, and immovable (rarely the fourth suture flexible); surface without coarse sculpturing. Male genitalia with the penis mostly slender, curved, with a movable dorsal plate which when raised gives an open "bird's beak" appearance, with paired slender basal struts; parameres and pars basalis fused, the latter with apex mostly hooded, rarely flattened, the ventral struts fused into a single median structure. Female genitalia sclerotized, the valvifers with long apodemes, the coxites apically toothed or ridged, the styli lateral, the spermatheca C-shaped.

Larvae are crescent-shaped, subcylindrical, fleshy, widest in the mid-abdominal region; size 4 to 12 mm in length (the majority still undescribed); body with lateral fleshy protuberances and a few to many short or long setae scattered over various segments and the head; color near white. Head exserted, hypognathous, setiferous, or rarely retracted into the prothorax, narrower than prothorax, with epicranial suture present. Antennae reduced, one-segmented or absent. Clypeus transverse; labrum setiferous; mandibles robust, with or without molar areas, bi- or tri-dentate; maxillae with cardo, stipes, three-segmented palpi, the galea setiferous, lacinia acute and inconspicuous; labium with submentum, mentum, ligula, and one-segmented palpi. Ocelli absent. Thorax with legs absent or if present, 1-, 2-, or 3-segmented, without tarsunguli. Abdomen nine-segmented, with two or three plicae per segment, the ninth segment smaller than the eighth. Spiracles annular or annular-biforous. Urogomphi absent. Anderson (1947) provides a partial key.

Habits and habitats. Distributions and biologies of Nearctic species are summarized by B. D. Valentine (1999). In general, adults may be found on the larval food plants. Those whose larvae feed in the stems or receptacles of various weeds appear to feed upon the pollen of the same plants; species with fungivorous larvae mostly are found feeding on the surface of the same fungi; those with wood-boring larvae occur on dead or dying tree trunks or branches, and in some cases at least feed upon bark. Adults of the majority of anthribid species in the United States are rare in museum collections. The best all-around collecting technique is beating dead or diseased branches, clumps of dead twigs or leaves, or tangles of dead vines. The beating cloth should be as close under the plants as possible and must be examined at once, for anthribids recover and fly off much more rapidly than other weevils. Some of the choragine genera complicate things by being active jumpers, and are as a result exasperatingly difficult to catch. Another technique is to range the woods examining the trunks of trees, especially the thin-barked ones, for signs of infirmity, and also all dead branches both above and on the ground. Look in particular for smooth gray or black patches flush with the wood surface, or small, protruding, black, brown or reddish lumps. These are pyrenomycete fungi, often in the genera *Hypoxylon* and *Biscogniauxia*, which are ignored by most collectors; they are very hard and when cut with a knife have a dry crumbly or charcoal-like consistency. All our species of *Eurymycter*, *Piesocorynus*, *Choragus*, and *Euxenus* are associated with these growths. Sweeping weedy fields usually produces only *Trigonorhinus* spp., the species *T. tomentosus* (Say) on common ragweed, *T. limbatus* (Say) and *T. griseus* (LeConte) on sneeze- or bitter-weed and other composites, *T. alternatus* (Say) on fungus-infested morning glory, *T. sticticus* (Boheman) on smutty grasses and corn, *T. rotundatus* (LeConte) on smutty grass (*Andropogon* sp.) and ferns, and the western *T. annulatus* (Carr) and *T. lepidus* Valentine on fungus-infested sagebrush. The hosts of *T. ornatus* (Schaeffer), *T. strigosus* (Jordan), and *T. nigromaculatus* (Schaeffer) are not known. Sweeping bushy, overgrown areas will sometimes reveal species of *Ormiscus* or *Eusphyrus*, especially in early spring, but beating is better for these and all other members of our fauna except species of *Euparius*, which occur on polypore fungi. Larvae of all our species feed on vegetation; they can be found in the twigs and branches of trees, in hard or polypore fungi, or under bark of dead or dying trees. One species, the coffee bean weevil, *Araecerus fasciculatus* (DeGeer 1775), lives in seeds and all sorts of dried plant materials from banana flour to strychnine. A related species in Australia and New Zealand occurs on dried fruit, especially apples. Our introduced species of the European genus *Anthribus* (formerly *Brachytarsus*) have larvae that feed on scale insects of the subfamily Lecaniinae; the female lays each of her eggs under a female scale that has laid her eggs, the weevil larvae then develop exclusively at the expense of the eggs of the scale insect. This is a startling deviation from the phytophagous habits of the Rhynchophora.

Status of the classification. This family has been subject to considerable nomenclatorial confusion and lack of studies of higher taxa beyond all proportions consistent with the rest of the order. The genera need considerable study. The world catalogues of Wolfrum (1929, 1953) need many additions and revisions. The Nearctic fauna is summarized by Valentine (1999) who provides a synonymic checklist and keys to most genera. Other basic references are LeConte (1876), Blatchley and Leng (1916), Pierce (1930), Ting (1936), and Valentine (1960, 1971, 1972, 1991).

Distribution. There are about 360 genera and 4,000 species known from all regions; 88 described and 32 undescribed species occur in North America. Of 30 North American genera, *Araecerus* is cosmopolitan; *Choragus*, *Eurymycter*, *Gonotropis*, *Allandrus*, and *Trigonorhinus* are holarctic (although the latter extends south to Argentina); *Sicanthus* is endemic but probably neotropical; *Euxenulus*, *Phoenicobiella*, and *Araeoderes* appear to be North American but are obviously related to Latin American groups; and all remaining genera are Neotropical, forming the northern fringe of an extensive Antillean, Central, or South American fauna.

Key to the Nearctic Tribes and Genera

1. Antennae inserted on anterior surface of rostrum or head (Fig. 2); antennal club never with four antennomeres (Choraginae) 2
— Antennae inserted on lateral or ventral surface of rostrum (Fig. 3); or antennal club sometimes with four antennomeres. (Anthribinae) 10

2(1). Eyes rounded, upper edges not closer together than lower 3
— Eyes elongate-oval, upper ends closer to each other than lower (Choragini) 7

3(2). Hind coxae elongate-transverse, almost reaching elytral edge (Araecerini) 4
— Hind coxae globular or short tear-drop shaped (Cisanthribini) .... *Sicanthus*

4(3). Head not retractile into prothorax, the eyes too wide; elytra with a scutellar plus 10 striae or rows of punctures 5
— Head capable of being retracted into prothorax past the eyes (Fig. 2); elytra with 12 or more rows of punctures 6

5(4). Lateral prothoracic carina present; transverse pronotal carina basal .... *Araecerus*
— Lateral prothoracic carina absent; transverse pronotal carina antebasal .... *Neoxenus*

6(4). Lateral prothoracic carina double, not upturned at apex; elytra with 12-14 rows of punctures .... *Habroxenus*
— Lateral prothoracic carina single, mostly with apex upturned; elytra with 19-21 rows of punctures . ... *Acaromimus*

7(2). Head not retractile into prothorax, the eyes too wide; pronotum with a raised reticulum forming a honeycomb pattern 8
— Head retractile into prothorax past the eyes; pronotum punctate, not reticulate 9

8(7). Antennae with 11 antennomeres .... *Choragus*
— Antennae with 9 or 10 antennomeres .... *Pseudochoragus*

9(7). Elytra with 10 rows or fields of punctures, plus a short scutellar row .... *Euxenus*
— Elytra with 11 to 15 rows of punctures, one or more mostly incomplete .... *Euxenulus*

10(1). Each elytron partially enclosing anterior part of scutellum, resulting scutellar notch shaped like an inverted omega (Fig. 4); eyes large, entire, very finely faceted (Gymnognathini) .... *Gymnognathus*
— Scutellar notch forming a simple V or U; eyes as above, or small, or notched, or coarsely faceted 11

11(10). Rostrum with a dorsal, median carina interrupted at base by an abrupt, small, very deep pit or transverse groove (Platystomini) 12
— Rostrum carinate or not, at most with a basal puncture 13

12(11). Lobes of all third tarsomeres separate .. *Toxonotus*
— Lobes of third tarsomeres fused..... *Phoenicobiella*

13(11). Mandibles with strongly toothed ventral cutting edge (Fig. 3), as well as normal dorsal edge (Cratoparini) .... *Euparius*
— Mandibles without a toothed ventral cutting edge 14

14(13). Eye entire, or truncate, or faintly sinuate on anterior face 15
— Eye strongly notched or emarginate on anterior face (nearest insertions of antennae) 24

15(14). Entire face with conspicuous white pubescence; center of pronotum with a pit or groove; our species with a conspicuous patch or band of white crossing the suture (Tropiderini) 16

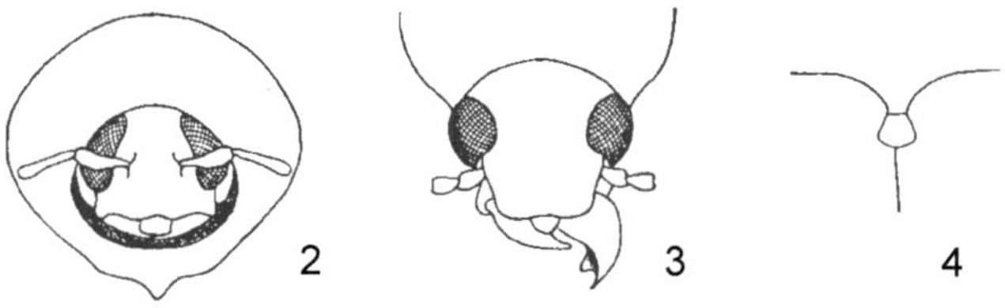

FIGURES 2.126-4.126. 2. *Euxenus* sp., head, anterior view; 3. *Euparius marmoreus* (Olivier), head, dorsal view; 4. *Gymnognathus* sp., scutellum, dorsal view.

— White pubescence, if present, confined to spots, if conspicuous, pronotum will have the disc concave with a central swelling, and elytra will have multiple tufts of erect pubescence 17

16(15). Pronotum with a smooth, longitudinal, shallow groove from central pit to antebasal carina; elytral pale patch antemedian and not reaching side margins .... *Gonotropis*
— Pronotum with a sinuous, transverse, median groove; elytral pale band postmedian and reaching the side margins .... *Eurymycter*

17(15). Antennal club with 4 antennomeres; rostrum thick, width at apex less than twice depth; scrobes dorso-lateral, interscrobal distance mostly less than interoccular distance (Discotenini) .... *Discotenes*
— Antennal club with 1 or 3 antennomeres; rostrum depressed apically, width at apex more than twice depth; interscrobal distance more than interoccular distance 18

18(17). Eyes finely faceted, more than 26 rows across maximum width 19
— Eyes coarsely faceted, 26 or fewer rows across maximum width (Piesocorynini) 23

19(18). Sides of rostral apex abruptly widened to cover the laterally protruding mandibular bases; our species with a small post-ocular tooth on apex of prothorax (Ischnocerini) .... *Ischnocerus*
— Sides of rostral apex with mandibular sheaths weakly or not wider than rostral dorsum; no small post-ocular teeth on prothoracic apex 20

20(19). Merger of rostrum with venter of head capsule forming a broad curve; rostrum long, flattened, and apically flared 21
— Merger of rostrum with venter of head capsule indicated by a transverse groove or abrupt angle; rostrum shorter, not apically flared (Platyrhinini) 22

21(20). Antennae with whorls of long, erect setae; length (head excluded) more than 5 mm (Stenocerini) . ... *Stenocerus*
— Antennae without whorls of long erect setae; length (head excluded) less than 4 mm (Allandrini) .... *Allandrus*

22(20). Face with a pair of abrupt depressions between the upper ends of the eyes; lateral prothoracic carina with an apical tubercle............ *Trachytropis*
— Face with at most a weak concave area at rostral base; lateral prothoracic carina not toothed at apex .... *Goniocloeus*

23(18). Eyes with 14 or more rows of facets across maximum width .... *Piesocorynus*
— Eyes with 12 or fewer rows of facets across maximum width .... *Brachycorynus*

24(14). All third tarsomeres with the lobes fused down the midline (Anthribini) .... *Anthribus*
— All third tarsomeres with the lobes separate 25

25(24). Rostrum quadrate, or with rounded apical angles 26
— Rostrum, excluding mandibles, narrowed from base to apex; rostral apex with central third longer than sides (Trigonorhinini) .... *Trigonorhinus*

26(25). Intercoxal process of mesosternum laterally angulate, or swollen, or bent; scrobes sulciform, continued across the rostral sides and ending below the eyes (Basitropidini) 27
— Intercoxal process of mesosternum simple; scrobes foveiform, more or less rounded and not continued transversly or ventrally around and under the rostrum (Zygaenodini) 28

27(26). Lateral prothoracic carina extending to apex where it is toothed; antennal scrobes strongly produced toward ventral midline .... *Eugonus*
— Lateral prothoracic carina not reaching anterior margin, nor apically toothed, antennal scrobes widely separated on venter .... *Phaenithon*

28(26). Lateral prothoracic carina absent or if present not reaching anterior margin 29
— Lateral prothoracic carina extending to apex .... *Araeoderes*

29(28). Transverse pronotal carina clearly antebasal, incapable of contacting the elytral base; pronotal hind angles not projecting laterad of the humerae .... *Ormiscus*
— Transverse pronotal carina subbasal or basal, capable of contacting elytral base at some point; pronotal hind angles often projecting laterad of the humerae .... *Eusphyrus*

Classification of the Nearctic Genera

Suprageneric taxa are diagnosed in the generic key. NOTE: Many undescribed species are included in the comments on distribution. As used below, the designation Neotropical includes Antillean species; Central and South America do not.

Anthribidae Billberg 1820

Choraginae Kirby 1819

Araecerini Lacordaire 1876

Araecerus Schoenherr 1823, 2 introduced spp., including the coffee bean weevil, *A. fasciculatus* (DeGeer 1775): Worldwide. Also about 75 Indopacific species.
Araeocerus Schoenherr 1839
Araeocorynus Jekel 1855
Araeosarus Walker 1859
Doticus Pascoe 1882
Metadoticus Olliff 1890

Neoxenus Valentine 1999, 1 sp., *N. versicolor* Valentine, 1999, Texas to Panama; four others in Central America and Antilles.

Habroxenus Valentine 1999, 1 sp., *H. politus* Valentine, 1999, Maryland, Texas; four others in Central America and Antilles.

Acaromimus Jordan 1907, 1 sp., *A. americanus* (Motschoulsky 1873), Florida, Alabama, and Texas; six others in Central America and Antilles.
Acaropsis Jordan 1907
Xenorchestes Motschulsky 1873, not Wollaston 1854
Euxenus Blatchley 1920, not LeConte 1876

Cisanthribini Zimmerman 1994

Sicanthus Valentine 1999, 1 sp., *S. rhizophorae* Valentine 1999, Florida keys.

Choragini Kirby 1819

Choragus Kirby 1819, 6 spp., eastern United States; 45 others almost worldwide.
Alticopus Villa and Villa 1833

Pseudochoragus Petri 1912, 1 sp., *P. nitens* (LeConte 1884), Massachusetts and Oklahoma; another species in Europe.
Choragus, of American authors, in part, not Kirby 1819
Holostilpna, of American authors, not Jordan 1907

Euxenus LeConte 1876, 3 spp., eastern United States; at least six more in Central America and Antilles.
Holostilpna Jordan 1907

Euxenulus Valentine 1960, 1 sp., *E. piceus* (LeConte 1878), south Florida; also three undescribed in Antilles.

Anthribinae Billberg 1820

Discotenini Lacordaire 1866

Discotenes Labram and Imhoff 1841, 2 spp., Texas; Arizona about 10 more south to Brazil.
Phanosolena Schaeffer 1904

Ischnocerini Lacordaire 1866

Ischnocerus Schoenherr 1839, 3 spp., Maryland to Texas; Arizona; 10 more in Neotropics.
Meconemus Labram and Imhoff 1839

Allandrini Pierce 1930

Allandrus LeConte 1876, 3 spp., United States except the desert southwest; also several in Palaearctic.
Tropiderinus Reitter 1916

Stenocerini Kolbe 1897

Stenocerus Schoenherr 1826, 1 sp., *S. longulus* Jekel 1855, south Texas to Brazil; seven more Mexico to Argentina.

Gymnognathini Valentine 1960

Gymnognathus Schoenherr 1826, 2 spp., south Texas; south Arizona; about 90 described and many undescribed Neotropical species.
Analotes Schoenherr 1839

Tropiderini Lacordaire 1866

Gonotropis Leconte 1876, 1 sp., *G. gibbosus* LeConte 1876, Canada south to Pennsylvania and Colorado; plus three Holarctic.
Tropideres, of European authors and Valentine 1960, in part, not Schoenherr 1823

Eurymycter LeConte 1876, 3 spp., United States except the desert southwest.
Tropideres, of Valentine 1960, not Schoenherr 1923

Piesocorynini Valentine 1960

Piesocorynus Dejean 1834, 5 spp., eastern United States; Arizona; about 30 others in Neotropics and Antilles.
Piezocorynus Schoenherr 1839
Camptotropis Jekel 1855

Brachycorynus Valentine 1999, 3 spp., eastern United States to Texas; a few others in Central America and Antilles.
Tropideres, of LeConte 1876 and others, not Schoenherr1823
Brevibarra, of Valentine 1960, not Jordan 1906

Platyrhinini Imhoff 1856

Goniocloeus Jordan 1904, 1 sp., *G. bimaculatus* (Olivier 1795), eastern United States to Texas; 36 others south to Brazil.
Tropideres, of various authors, in part, not Schoenherr 1823
Strabus Jekel 1860, not Gerstaecker 1860 [Curculionidae]

Trachytropis Jordan 1904, 1 sp., *T. arizonicus* (Sleeper 1954), southeastern Arizona; two others in Central America.

Platystomini Pierce 1916

Phoenicobiella Cockerell 1906, 2 spp., South Carolina to Lousiana, Florida and Bahamas; south Texas; a third species in Cuba.
Phoenicobius LeConte 1876, not Moerch 1852 [Mollusca]

Toxonotus Lacordaire 1866, 6 spp., eastern United States to New Mexico and Arizona; about 35 others throughout the Neotropics.
Anthribus, of authors, in part, not Geoffroy 1762
Neanthribus Jordan 1906
Pseudanthribus Pierce 1930

Basitropidini Lacordaire 1866

Eugonus Schoenherr 1833, 1 sp., *E. bicolor* Valentine 1972, southeastern Arizona; about 30 more in Central and South America.
Schimatocheilus Fahraeus 1839
Phaenisor Motschoulsky 1874

Phaenithon Schoenherr 1823, 1 sp., *P. platanum* (Schaeffer 1906), southeastern Arizona; about 65 others in Central and South America.
Camaroderes Jekel 1855
Griburiosoma Schaeffer 1906

Zygaenodini Lacordaire 1866

Ormiscus G. R. Waterhouse 1845, 14 described and about 30 undescribed spp., eastern United States and desert southwest; probably 200 more throughout the Neotropics, mostly undescribed.
Entomops Lacordaire 1866
Hormiscus Gemminger and Harold 1872
Toxotropis LeConte 1876
Gonops LeConte 1876

Eusphyrus LeConte 1876, 6 spp., eastern United States and desert southwest; 30 more in Central and South America.
Opisthotropis Hoffmann and Tempere 1954

Araeoderes Schaeffer, 1906, 1 sp., *A. texanus* Schaeffer, 1906, southern Texas and southern Alabama.

Trigonorhinini Valentine 1999

Trigonorhinus Wollaston 1861, 11 spp., southern Canada, entire United States; seven others south to Argentina, and in the Palearctic Region.
Brachytarsus, of many authors, not Schoenherr 1823
Trigonorrhinus Gemminger and Harold 1872
Anthribulus LeConte 1876
Brachytarsoides Pierce 1930

Cratoparini LeConte 1876

Euparius Schoenherr 1823, 5 spp., eastern United States to Montana and Arizona; at least 70 others in Central and South America, and Japan to Australia.
Cratoparis Dejean 1834
Caccorhinus Sharp 1891

Anthribini Billberg 1820

Anthribus Geoffroy 1762, 2 adventive spp. from Europe, Atlantic coast and ?California.

Brachytarsus Schoenherr 1823

Anthotribus Gemminger and Harold 1872

Pseudobrachytarsus Pierce 1930

BIBLIOGRAPHY

ANDERSON, W. H. 1947. Larvae of some genera of Anthribidae (Coleoptera). Annals of the Entomological Society of America, 40: 489-517, pl. I-IV.

BLATCHLEY, W. S. and C. W. LENG. 1916. Rhynchophora or Weevils of North Eastern America. The Nature Company. Indianapolis, 682 pp., fig. 1-155.

LECONTE, J. L. 1876. The Rhynchophora of America, North of Mexico. Proceedings of the American Philosophical Society, XV (96): i-xvi + 1-455.

PIERCE, W. D. 1930. Studies of the North American weevils of the superfamily Platystomoidea. Proceedings of the United States National Museum, 77(no. 2840): 1-34, pls. 1-5.

TING, P. C. 1936. The mouthparts of the coleopterous group Rhynchophora. Microentomology, 1: 93-114.

VALENTINE, B. D. 1960. The genera of the weevil family Anthribidae north of Mexico (Coleoptera). Transactions of the American Entomological Society, 86: 41-85.

VALENTINE, B. D. 1971. Family Anthribidae. Pp. 243-248. *In:* M. H. Hatch. Beetles of the Pacific Northwest, Part V. University of Washington Publications in Biology, 16: i-xvi + 1-662. pl. I-LV.

VALENTINE, B. D. 1972. Notes on anthribid weevils. III. New species and records primarily from Arizona. Coleopterists Bulletin, 26: 1-11, fig. 1-4.

VALENTINE, B. D. 1991. The *Choragus-Holostilpna* problem (Coleoptera: Anthribidae). Coleopterists Bulletin, 45: 301-307.

VALENTINE, B. D. 1998(1999). A review of Nearctic and some related Anthribidae (Coleoptera). Insecta Mundi, 12: 251-296.

WOLFRUM, P. 1929. Anthribidae. Coleopterorum Catalogus, 26(102): 1-145.

WOLFRUM, P. 1953. Anthribidae. Coleopterorum Catalogus, supplementum, 26(102): 1-63.

127. BELIDAE Schönherr 1826

by Robert S. Anderson

Family common name: The cycad weevils

The only North American members of this family are odd weevils, recognized by their straight antennae, somewhat truncated elytra mostly exposing the last one or two tergites, and the short stout legs with femora which are expanded in all legs, especially so in males. These weevils are only found in southern Florida where they are associated with native and adventive *Zamia* cycads.

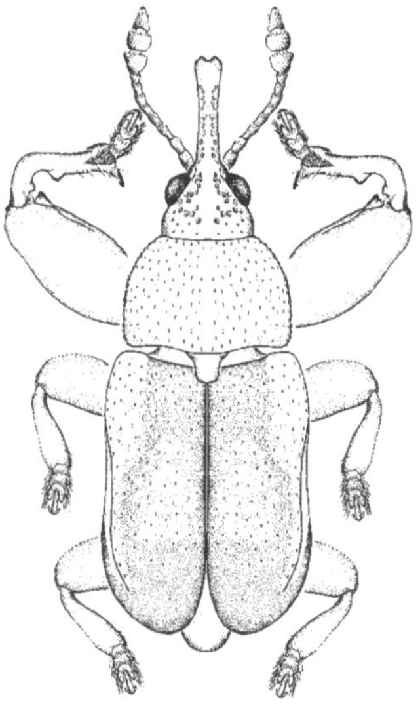

FIGURE 1.127. *Rhopalotria slossonae* (Schaeffer) (from Bright 1993, reproduced with the permission of the Minister of Public Works and Government Services, 2001)

Description. (based on Lawrence 1982). Shape elongate, slightly convex; length 3-6 mm; color pale to dark brown or black; vestiture of very fine short appressed pubescence. Rostrum moderately to very long and mostly narrow. Antennae straight, ending in a weak, loose club of three articles; articles 9 and 10 with deep apical pockets; antennal insertions ventral at or near the base of the rostrum. Labrum fused with clypeus. Labial palps with two articles and attached dorsally near the apex of the prementum. Gular sutures fused. Proventriculus lacks sclerotized plates. Pronotum with complete lateral edges. Procoxae contiguous and the procoxal cavities narrowly closed posteriorly. Elytra without an inner subcostal flange. Hindwing with fewer than five anal veins. Tarsal article 2 broadly lobed as article 3; tarsal claws simple. Visible sternites of the abdomen free; pygydium exposed by the elytra. Tegmen truncate or slightly emarginate apically and the median lobe with a distinct dorsal plate.

Eggs are undescribed.

Larvae (based in part on van Emden 1938, Lawrence 1982, Crowson 1986) short, broad, ventrally curved and subglabrous. Body lightly sclerotized. Legs absent. Head strongly retracted, endocarina indistinct, with very short antennae of one article and no epipharyngeal rods. Frontal sutures incomplete, not reaching articulating membrane of mandible. Maxilla with palp with two articles, palpiger absent. Thoracic spiracles on mesothorax. Abdomen with segments with two dorsal folds. Anal opening ventral.

Pupae are undescribed.

Habits and habitats. Within this family, only the genus *Rhopalotria* occurs in North America. It is restricted to south Florida where two species are associated with native and adventive species of *Zamia* (Cycadaceae). The biology has been well-studied by Norstog and Fawcett (1989). Adult weevils swarm on male cones of the cycads, where mating, feeding and oviposition occur. Larvae feed within the male cones. Weevils also visit female cones but do not feed. Pollen transport to the female cones occurs during these visits and these beetles (and their relatives elsewhere) may be obligate pollinators of the cycads.

Status of the classification. The family-level classification of these beetles is somewhat controversial. Crowson (1986) considered them as a distinct family, the Allocorynidae. Kuschel (1995) places them as Belidae, subfamily Oxycoryninae. The most recent catalog (Alonso-Zarazaga and Lyal 1999), probably following Lawrence (1982) and Thompson (1992), places them as the subfamily Allocoryninae of the Oxycorynidae. Here they are placed as Belidae, subfamily Allocoryninae.

Distribution. There is one genus with two species in North America. Both species occur only in Florida; one is adventive and one native.

Classification of the Nearctic Genera

Belidae Schönherr 1826

Allocoryninae Sharp 1890

Rhopalotria Chevrolat 1878, 2 species, *R. mollis* (Sharp 1890), Florida (adventive on *Zamia furfuracea* L.) and *R. slossonae* (Schaeffer 1905), Florida (native on *Zamia integrifolia* L., coontie).

Allocorynus Sharp 1890

Bibliography

ALONSO-ZARAZAGA, M. A. and C. H. C. LYAL. 1999. A world catalogue of families and genera of Curculionoidea (Insecta: Coleoptera) (Excepting Scolytidae and Platypodidae). Entomopraxis. Barcelona, Spain.

CROWSON, R. A. 1986. On the systematic position of Allocoryninae. Coleopterists Bulletin, 40: 243-244.

EMDEN, F. VAN. 1938. On the taxonomy of *Rhynchophora* larvae (Coleoptera). Transactions of the Royal Entomological Society of London, 87: 1-37.

KUSCHEL, G. 1995. A phylogenetic classification of Curculionoidea to families and subfamilies. Memoirs of the Entomological Society of Washington, 14: 5-33.

LAWRENCE, J. L. 1982. Coleoptera. Pp. 482-553. *In:* S. P. Parker, ed. Synopsis and Classification of Living Organisms. Volume 2. McGraw Hill. New York.

NORSTOG, K. J. and P. K. S. FAWCETT. 1989. Insect-cycad symbiosis and its relation to the pollination of *Zamia furfuracea* (Zamiaceae) by *Rhopalotria mollis* (Curculionidae). American Journal of Botany, 76: 1380-1394.

THOMPSON, R. T. 1992. Observations on the morphology and classification of weevils (Coleoptera, Curculionoidea) with a key to major groups. Journal of Natural History, 26: 835-891.

128. ATTELABIDAE Billberg 1820

by Robert W. Hamilton

Family common names: The leaf rolling weevils, tooth-nose snout beetles, and thief weevils.

Family synonyms: Rhynchitidae Gistel 1856

The members of this family are considered to be primitive weevils based on the straight antennae (non-geniculate), mouthpart structure, wing venation, etc. The family in the inclusive sense contains three subfamilies in America north of Mexico, Attelabinae, Rhynchitinae and Pterocolinae. In the Attelabinae (leaf rolling weevils), the body is more or less stout, the rostrum short and apically widened, the mandibles robust and toothed only on the inner margin, the front legs enlarged and the tarsal claws connate. In the Rhynchitinae (tooth nose snout beetles), the body is more or less elongate, the rostrum mostly narrow and elongate, the mandibles flat and toothed on the inner and outer margins, the legs subequal in size and the tarsal claws appendiculate. In the Pterocolinae (thief weevils), the body is short and compact, the short rostrum apically depressed and parallel-sided, the mandibles flat and toothed on the inner and outer margins (as in the Rhynchitinae), the middle and hind femora larger than the front femora, and the tarsal claws have a broad basal tooth. The Pterocolinae are a unique New World group that are clearly related to the Rhynchitinae although the rhynchitine link has not yet been determined.

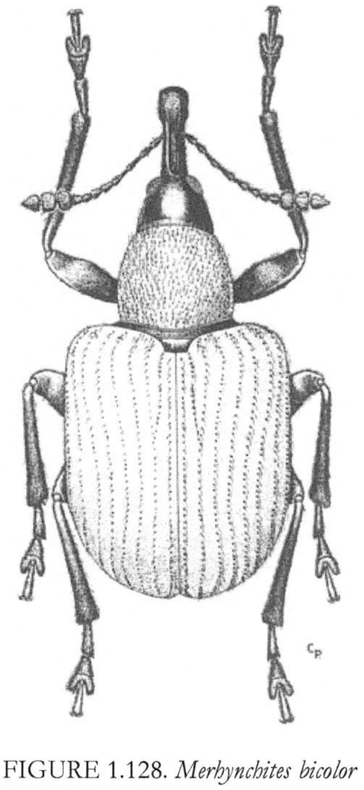

FIGURE 1.128. *Merhynchites bicolor* (Fabricius) (from Bright 1993, reproduced with the permission of the Minister of Public Works and Government Services, 2001)

Description: The following adult characters have been given by Thompson (1992) and Kuschel (1995) to support the single family status of Attelabidae: maxillary palps 3 or 4 segmented (if 3 segmented, elytral punctuation confused); tergite 8 concealed in both sexes; elytra separately and broadly rounded at apices, more or less exposing a pygidium (tergites 6 and 7); body setose, without broad scales; cuticle metallic or otherwise brightly colored in many; elytra with scutellary striole in many; abdominal ventrites decreasing in length, first 3 or 4 fused; pygidium visible in both sexes; male genitalia consisting of a free pedon and tectum and a large tegmen (as in all groups with non-geniculate antennae); intersegmental membrane between sternites 8 and 9 partly or entirely sclerotized.

The subfamily Attelabinae is characterized by the following: Adult with body more or less short, stout, glabrous or with appressed pubescence above, color variable; head rectangular to weakly triangular, elongate in males of some; eyes large, mostly reniform, longest diameter more or less vertical; rostrum short to moderate in length, distinctly widened apically, weakly arcuate in lateral view; lateral apical angle in some with tooth-like projection; postmentum in some males with pair of acuminate ventral projections; antennae inserted dorsolaterally, club 3-segmented, more or less compact, scrobe short, distinct, more or less vertical; mandibles robust, toothed on inner apical margin, outer margin more or less rounded; maxillary palps 4-segmented; labial palps indistinctly 1 or 2 segmented, located ventrally on prementum; elytra more or less quadrate, not much longer than wide, humeri simple to strongly protuberant, scutellary striole present or absent, striae becoming less distinct posteriorly in many; ventral abdominal sutures 1-3 more or less rigid; procoxae prominent, conical; prothoracic legs enlarged, profemora distinctly swollen, armed or unarmed; protibia longer, narrower and more arcuate in males; tibial apices uni-uncinate in males and bi-uncinate in females; tarsal claws connate.

Larva with thoracic spiracle located in mesothorax or in intersegmental fold between prothorax and mesothorax; prementum and mentum fused, equally sclerotized; maxillary palps 2-segmented; prodorsal fold more convex than postdorsal fold, postdorsal fold sloping posteriorly; abdominal tergum 9 at least twice as long as sternum 9; anus subterminal or ventral; cuticular asperities more or less inconspicuous.

The subfamily Rhynchitinae is characterized by the following: Adult with body more or less elongate, mostly with distinct semi-erect to erect setae above, color variable; head mostly triangular, widest at the base; eyes small to medium in size, rounded, longest diameter more or less horizonal; rostrum mostly slender, parallel-sided and longer than the head, more or less weakly arcuate in lateral view; antennae inserted laterally, club 3-segmented, more or less loosely united, scrobes shallow, longitudinally oblique, not well defined; mandibles flat, toothed on inner and outer margins; maxillary palps 4 segmented; labial palps 2 or 3 segmented, inserted laterally or apically on prementum; elytra elongate, mostly distinctly longer than wide, humeri simple, scutellary striole present or absent, striae mostly distinct throughout; ventral abdominal sutures distinct, suture between 1 and 2 rigid; all legs more or less equally developed; tibial apices mostly with one or two small straight mucros; tarsal claws with long clawlike inner processes; in some inner processes shorter, wider and more toothlike.

Larva with thoracic spiracle located on postero-lateral portion of prothorax; prementum and mentum separate, mostly unequally sclerotized; maxillary palp 2 or 3 segmented; prodorsal and postdorsal folds (segments 6-8) more or less subequal in convexity; tergum and sternum of abdominal segment 9 subequal in length; anus terminal or subventral, transverse x-shaped or simple transverse cleft; cuticular asperities more or less conspicuous.

The subfamily Pterocolinae is characterized by the following: Adult with body robust and dorsally convex, with appressed to semi-erect inconspicuous fine setae above, coloration metallic bluish-green to bluish-black; head weakly triangular; eyes large, oval, posteriorly more protuberant, anteriorly emarginate; rostrum short, as long as or shorter than the head, straight, in dorsal view more or less parallel-sided throughout, apically depressed, with beard like ventral setosity (more setose in males); antennae short, subequal to combined length of head and rostrum, inserted near basal 1/4 under basirostral ridge; scrobe fossa like; club large, strongly abrupt, compact; mandibles flat, toothed on inner and outer margins; maxillary palps 4-segmented, labial palps 3-segmented, inserted apically on prementum; pronotum with distinct lateral carinae; elytra short, apices individually rounded; pygidium and two complete abdominal terga visible; humeri simple; scutellary striole present; striae mostly distinct throughout; intervals more or less convex; all ventral abdominal sutures entire; mesepimeral "side pieces" strongly developed, visible in dorsal view between pronotum and elytra; front coxae small, globose, distinctly separated; middle and hind coxae widely separated; middle and hind femora larger than front femora; tibial apices with two blunt-tipped mucros; tarsal claws with broad basal tooth.

Larva is undescribed.

Attelabid eggs that have been examined are oval, creamy white to yellowish-white, shiny and without surface sculpture (Balduf 1959; Hamilton 1980, 1983, 1994).

Only a few North American pupae in the subfamily Rhynchitinae have been described (Hamilton 1980, 1983; Hamilton and Kuritsky 1981). They possess taxonomically significant tuberculate setae and posterior processes associated with abdominal segment 9.

Habits and habitats. All known members of the subfamily Attelabinae are leaf rollers and apparently cause no serious damage to their hosts although one Central American species of *Hybolabus* has been reported as a serious defoliator of its host tree, *Cariniana pyriformis* Miers (Lecythidaceae). All known females lay their egg(s) on leaves that they prepare by biting and cutting with their mandibles. The leaves are then rolled into a more or less barrel-shaped structure (nidus) that nourishes and protects the developing larvae. The enlarged front legs and the uncinate tibial apices manipulate the leaf tissue during nidus formation. Pupation occurs in the ground in the species that have been studied. Comments on the biology of three North American species are available. Packard (1890), Frost (1908), Blatchley and Leng (1916), and Lutz (1935) commented on the biology of *Himatolabus pubescens* (Say). Frost (1908) recorded a eulophid parasite bred from *Attelabus rhois* Boheman (=*H. pubescens*). Loding (1945) commented on the biology of *Homoeolabus analis* (Illiger). Murtfeldt (1872), Packard (1890), Girault (1904), and Edwards (1949) treated aspects of the biology of *Attelabus bipustulatus* Fabricius. Most of the information in these papers deals with taxonomy, host plants, nidus formation and distributions. Hopkins (1905) reported a trichogrammid egg parasite of *A. bipustulatus*. Van Emden (1938) included the larvae of *H. pubescens*, *Homoeolabus analis* and *Attelabus nigripes* LeConte in a key to the genera of Attelabini. Also, he associated the niduses of these species, and *A. bipustulatus* as well, with specific host plants. Vogt (1992) discussed the leaf rolling behavior, host plants and associated rhynchitid weevils (Pterocolinae) of attelabine weevils occurring from Canada to the Republic of Panama. Hamilton(1998), in a revision of the New World Pterocolinae, described 15 new species from Central and South America and clarified many attelabid-pterocoline-host plant associations initiated by Vogt.

Hamilton (1994) provided a summary of known North American rhynchitine biology. The larvae of these weevils develop in living or dead leaves as leaf miners, in fruits, in cut flower heads, in cut terminal shoots, in terminal buds, in flower buds, and in cut leaf primordia. The cuts are made by the adult female with her mandibles after oviposition in the specific plant part. The flower heads, buds or leaf primordia are either completely cut from the plant or are partially cut and eventually drop to the ground. Information has been published on the life stages or life cycles of only eight North American rhynchitine weevils - *Haplorhynchites aeneus* (Boheman), *Merhynchites bicolor* (Fabricius), *Merhynchites wickhami* (Cockerell), *Eugnamptus angustatus* (Herbst), *Auletobius cassandrae* (LeConte), *Temnocerus perplexus* (Blatchley), *Temnocerus naso* (Casey) and *Deporaus glastinus* (LeConte). Boving and Craighead (1931) provided the first basic illustrations of *H. aeneus* larvae. Hamilton (1973, 1981) provided detailed illustrations of life stages, host plants, and behavior for *H. aeneus*. Chittenden (1901), Dickerson (1910), Blatchley and Leng (1916), Ewing (1915), and Essig (1958) comment on the biology of the eastern rose curculio, *M. bicolor*. Balduf (1959) published the most comprehensive study on the biology of *M. bicolor* to date. Hamilton and Kuritsky (1981) commented on the life cycle of *M. bicolor* and provided detailed descriptions of the larva and pupa. Cooley (1903), Lovett (1915), Robertson (1923), and Hoerner (1936) comment on the life cycle of the western rose curculio, *M. wickhami*. Boving and Craighead (1931) illustrated what appears to be the larva of *E. angustatus* but they did not rear it and they determined it as *Orsodacne* sp. by the "method of elimination and locality." Hamilton (1980) provided notes on the biology of *E. angustatus* (Herbst) and described and illustrated the larva and pupa. Hamilton (1983) discussed the life cycles of *A. cassandrae* and *T. perplexus* and provided detailed descriptions and illustrations of their immature stages. Hamilton (1994) provided new life cycle data for *T. naso* and *D. glastinus* on *Quercus wislizenii* A.D.&C. in southern California.

Published biological information on other rhynchitid species is lacking except for brief comments by Kissinger (1964), given him by Vogt, stating that *Eugnamptus* spp. mine the dead leaves of various hardwoods and that females of *Temnocerus aeratus* (Say) lay eggs in and cut terminal oak twigs in which the larvae subsequently develop. Kissinger also indicated that, again according to Vogt, *Involvulus hirtus* (Fabricius) has habits similar to *T. aeratus*. Van Emden (1938) included three North American species in a paper on the taxonomy of Rhynchophora larvae. He provided key characters for the larvae of *H. aeneus*, *M. bicolor* and *Rhynchites velatus* LeConte and included them in a key to select species of Rhynchitini. He also associated these three species with host plants and localities. Based on his host plant data, he examined the larva of *H. aeneus* not *Rhynchites (Involvulus) hirtus* (Fabricius) as listed.

Haplorhynchites aeneus, *M. bicolor* and *M. wickhami* are the only North American rhynchitine species of known economic importance. *Haplorhynchites aeneus* attacks commercially grown sunflowers while *M. bicolor* and *M. wickhami* attack cultivated roses and sometimes damage blackberry and raspberry. Comments on the economic importance of *M. bicolor* have been made by Harris (1862), Chittenden (1901), Gates (1909), Dickerson (1910), Blatchley and Leng (1916) and Essig (1958). Cooley (1903), Lovett (1915), Robertson (1923), and Hoerner (1936), give information on the economic importance and control of *M. wickhami*. Schulz and Lipp (1969) and McBride and Oseto (1978) commented on the status of damage to sunflowers by *H. aeneus*.

Vogt (1992) discussed the biology of the Pterocolinae and their attelabid hosts and coined the term "thief weevils" for their habit of taking over the niduses of their attelabid hosts. Although not specifically stated, his comments on the biology of these weevils were probably based mainly on observations of *P. ovatus* in Maryland. According to Vogt, female pterocolines force their way into freshly made attelabid leaf rolls where they eat or destroy the host egg and oviposit their own. The pterocoline larva develops rapidly feeding on the decaying leaf tissue and leaves the hollowed out leaf roll to pupate in the ground.

Status of the classification. The subfamilies Rhynchitinae and Pterocolinae, here considered in the family Attelabidae, have been placed in the family Rhynchitidae by other workers. The family name Rhynchitidae is credited to Gistel 1856 (Alonso-Zarazaga and Lyal 1999). John L. LeConte (1876) was the first American entomologist to use the family name Rhynchitidae for weevils in the subfamilies Rhynchitinae and Pterocolinae. Subsequent workers (Sharp 1889, Pierce1909, Blatchley and Leng 1916, Ting 1936, Anderson 1991) grouped these weevils in the family Curculionidae. Pierce (1913) recognized the family Attelabidae including the Rhynchitinae and the new tribes Rhynchitini and Auletini. Voss (1922-1969), in a worldwide taxonomic monograph series on the subfamilies Attelabinae, Rhynchitinae and Pterocolinae, consistently listed these groups in the family Curculionidae. Boving and Craighead (1931), in a synopsis of the larval forms of Coleoptera placed the Rhynchitinae and Attelabinae in the family Attelabidae. Crowson (1955) grouped these weevils in the family Attelabidae but suggested a possible alternative would be to establish a family Rhynchitidae for the Rhynchitinae and Pterocolinae. Lawrence (1982) also placed these weevils in the family Attelabidae. O'Brien and Wibmer (1982), in an annotated checklist of North American weevils, recognized the families Rhynchitidae and Attelabidae but Wibmer and O'Brien (1986), in their annotated checklist of South American weevils, regrouped these weevils into the single family Attelabidae. Thompson (1992) using abdominal characters, etc. lumped the three subfamilies in the family Attelabidae. More recently Kuschel (1995), in a phylogenetic approach, recognized only the family Attelabidae and Farrell (1998), in a molecular approach, also grouped the rhynchitines and attelabines in the family Attelabidae. Hamilton (1969 to 1998) has consistently used the family Rhynchitidae for the Rhynchitinae and Pterocolinae. Most recently, Alonso-Zarazaga and Lyal (1999) have recognized the family Rhynchitidae in their world catalogue of families and genera of Curculionoidea.

The family name Attelabidae is credited to Billberg (1820). The genus *Attelabus*, on which the family name is based, was originally described by Linnaeus (1758). Linnaeus included only *A. coryli* Linnaeus in the genus and it was therefore the type by monotypy. Since the time of Linnaeus, many species from all over the world have been added to the genus *Attelabus*. Olivier (1807) decided that *A. coryli* was generically different from the other species that had been placed in *Attelabus* and described the new genus *Apoderus* with 13 species including *A. coryli*. *Attelabus coryli* is technically the type of *Attelabus* but was not recognized by Olivier, or anyone else, until Bedel (1888) pointed out that *Attelabus* should be used for those species placed by Olivier in *Apoderus*. Bedel proposed the name *Cyphus* for the species that were left in *Attelabus* by Olivier as well as many other species placed there by subsequent authors. Bedel's proprosal followed the International Code but subsequent workers, unaware of the change or unwilling to change the names of numerous species, have not followed Bedel. Silfverberg (1977) successfully appealed to the International Commission on Zoological Nomenclature to accept common usage of *Attelabus* in the interest of stability and to confirm the designation by Schoenherr (1823) of *Attelabus curculionoides* Linnaeus 1767 (= *Curculio nitens* Scopoli 1763) as type-species. Two major groups of attelabids are now officially recognized - *Apoderus* with *A. coryli* as its type and *Attelabus* with *Attelabus nitens* Scopoli as its type. Jekel (1860) divided the genus *Attelabus* into 16 subgenera and Voss (1925) elevated most of them to the generic level. The genus *Attelabus* is applicable to only two species of attelabids in America, north of Mexico - *A. bipustulatus* Fabricius and *A. nigripes* LeConte.

Distribution. The family includes 1,914 world species in 97 genera (Kuschel 1995). These weevils occur throughout the world but Kuschel points out that they do not occur in New Zealand, New Caledonia and the Pacific Islands. In the New World, 362 species have been described including 180 species of leaf rolling weevils in four subfamilies (Attelabinae, Euscelinae, Hybolabinae and Pilolabinae), 162 species of Rhynchitinae and 20 species of Pterocolinae. In America north of Mexico, there are 51 total species in 11 genera including 6 species of Attelabinae, 44 species of Rhynchitinae and one species of Pterocolinae.

KEY TO NEARCTIC SUBFAMILIES AND GENERA

1. Tarsal claws appendiculate (Figs. 7 and 8); mandibles depressed, toothed on inner and outer margin; front legs subequal to middle and hind legs or middle and hind femora larger than front femora; tibial apices unarmed or with small spurs or mucros 2

— Tarsal claws connate (Fig. 9); mandibles robust, not toothed on outer margin; front legs enlarged, distinctly larger than middle and hind legs; profemora distinctly swollen (Fig. 2); tibial apices uni-uncinate (male) or bi-uncinate (female) (Figs. 5 and 6) (Attelabinae) 10

2(1). Prothorax with distinct lateral carina (Fig. 3); propleura strongly excavated beneath carina; body small, robust; metallic bluish-green to bluish-black (Pterocolinae) .... *Pterocolus*

— Prothorax not laterally carinate (Fig. 4); propleura not strongly excavated; body size, shape and color variable (Rhynchitinae) 3

3(2). Scutellary striole present 4

— Scutellary striole absent 6

4(3). Pygidium completely or almost completely covered by elytra; elytra with some erect setae; males with one tooth on outer edge of mandibles and females with two .... *Eugnamptus*

— Pygidium mainly exposed, not completely or almost completely covered by elytra; elytra without erect setae; mandibular teeth similar in both sexes 5

5(4). Elytral striae distinct, more or less quadrate, moderately to deeply impressed; intervals narrower than width of striae, convex, smooth; interval punctures much smaller than striae; pubescence inconspicuous, more or less appressed; body dark colored, in some feebly metallic bronze or blue, less than 4 mm in length .... *Temnocerus*

— Elytral striae more or less indistinct (more distinct in *M. bicolor*), weakly impressed, especially posteriorly; discernable intervals wider than width of striae, more or less flat, in some minutely rugose; interval punctures numerous, as large or nearly as large as striae; pubescence conspicuous, fine, semi-erect; body variable reddish-orange and black to brownish-black with faint bluish metallic luster, greater than 4 mm in length. .... *Merhynchites*

6(3). Elytral striae distinctly rowed; intervals distinct, with punctures more or less smaller than striae; punctures not masked by pubescence 7

— Elytral striae not distinctly rowed; intervals indistinct, punctures as large or nearly as large as striae; punctures in some masked by pubescence 9

7(6). Elytra short, exposing pygidium and two abdominal terga; hind basitarsal segment longer than combined length of hind tarsal segments 3 and 4 ... . *Deporaus*

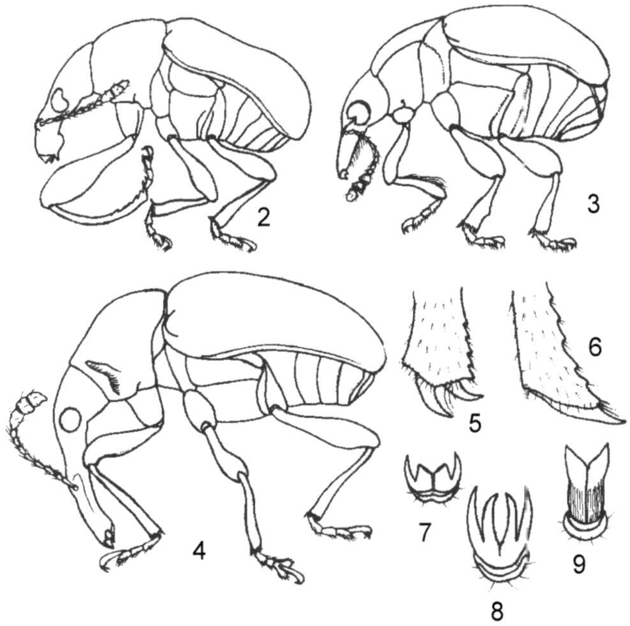

FIGURES 2.128-9.128. 2. *Himatolabus pubescens* (Say), male, lateral habitus; 3. *Pterocolus ovatus* (Fabricius), male, lateral habitus; 4. *Rhynchites velatus* LeConte, male, lateral habitus; 5. *Himatolabus pubescens* (Say), female, protibial apex; 6. *Himatolabus pubescens* (Say), male, protibial apex; 7. *Pterocolus ovatus* (Fabricius), tarsal claw; 8. *Haplorhynchites aeneus* (Boheman), tarsal claw; 9. *Himatolabus pubescens* (Say), tarsal claw.

— Elytra not short, covering all abdominal terga and part of the pygidium; hind basitarsal segment shorter or subequal to hind tarsal segments 3 and 4 combined 8

8(7). Elytral intervals wide, more or less flat, with numerous punctures, punctures in some as large or nearly as large as striae; males mostly more setose along elytral suture at declivity; antennal club symmetrical in both sexes .... *Haplorhynchites*

— Elytral intervals narrow, more or less convex, moderately punctured; punctures much smaller than striae; males not more setose along elytral suture at declivity; males with asymmetrical antennal club .... *Involvulus*

9(6). Body with violaceous to greenish metallic luster; elytral pubescence of appressed whitish setae and widely rowed, dark, erect setae; males with anterolateral spine on each side of pronotum(Fig. 4); length mostly greater than 4 mm .... *Rhynchites*

— Body without violaceous or greenish metallic luster; elytral pubescence not as above; males without anterolateral spines on each side of pronotum; length mostly less than 4 mm .... *Auletobius*

10(1). Upper surface with appressed pubescence......... .... *Himatolabus*

— Upper surface without appressed pubescence, glabrous except for a few scattered fine erect setae 11

11(10). Profemora unarmed in both sexes; submentum in males with pair of ventrally projecting acuminate spines; ventral rostral apex without median conical prominence 12

— Profemora in males armed with one or two blunt spine like projections; profemora in females unarmed or armed with a single peg like projection; submentum in males without pair of ventral projecting accuminate spines; ventral rostral apex in lateral view with median conical prominence (more pronounced in females) .... *Attelabus*

12(11). Body uniformly shiny black to brownish-black; head in both sexes subequal in size and shape; surface of abdominal sternites smooth, without tubercles; pronotum with pair of basilateral pitlike depressions .... *Xestolabus*

— Body bicolored; pronotum, elytra and abdomen red to reddish-orange; legs, sterna and head black; head elongate in males; females with abdominal sternites 1-3 with pair of acute tubercles; pronotum without pair of pitlike pronotal depressions .... *Homoeolabus*

Classification of Nearctic Genera

Attelabidae Billberg 1820

Attelabinae Billberg 1820

Attelabus Linnaeus 1758, 2 spp., *A. bipustulatus* Fabricius and *A. nigripes* LeConte; females are leaf rollers; larvae develop in leaf rolls; *A. bipustulatus* is associated with *Quercus* spp. and *Carpinus* sp. and ranges from SE Canada throughout the eastern U.S. and southwest to OK and TX; *A. nigripes* is associated with *Rhus copallina* L. and other *Rhus* sp. and ranges throughout the eastern half of the U.S. and south into Mexico; the two species can be distinguished by color. *A. bipustulatus* is black with reddish humeral maculae while *A. nigripes* is reddish throughout with variable darker areas (western form of *A. nigripes* is darker overall with pale areas in basal half of elytra).

Homoeolabus Jekel 1860, 1 sp., *H. analis* (Illiger); females are leaf rollers; larvae develop in leaf rolls; associated with many *Quercus* sp.; ranges from SE Canada throughout the eastern half of the U.S. (Volume 2, Color Fig. 28)

Himatolabus Jekel 1860, 2 spp., *H. pubescens* (Say), Fig. 2, and *H. axillaris* (Gyllenhal); females are leaf rollers; larvae develop in leaf rolls; both species are associated with *Quercus* spp.; *H. pubescens* also rolls the leaves of *Alnus incana* (L.) and *Corylus americana* Walt. in northeastern North America; *H. pubescens* ranges from SE Canada and the northeastern U.S. southwest into TX, NM and AZ and south through Mexico; *H. axillaris* is recorded in the southwestern states of AZ and UT and ranges south into Mexico; The two species can easily be separated by color. *H. pubescens* is reddish- brown throughout (with blackish-brown areas and darker extremities in some) and *H. axillaris* is black to dilute black with reddish humeral maculae; (key to New World species Hamilton 1992).

Xestolabus Jekel 1860, 1 sp., *X. constrictipennis* (Chittenden); females are leaf rollers; larvae develop in leaf rolls; occurs in the southwestern states of AZ and NM and south into Mexico; associated with *Rhus toxicodendron* L. and *Rhus* spp. in the southwestern U.S.

Rhynchitinae Gistel 1856

Rhynchitini Gistel 1856

Eugnamptus Schoenherr 1839, 9 spp.; known larvae are leaf miners; associated with species of *Quercus*, *Sassafras*, *Juglans*, *Carya*, *Cornus*, *Liquidambar*, etc.; distributed in the eastern half of the U.S. and southwestern states of AZ and NM; *Eugnamptus angustatus* (Herbst) is the most abundant species in North America and occurs in the eastern half of the U. S. in 4 sympatric color forms (key to Nearctic species, Hamilton 1990).

Haplorhynchites Voss 1938, 6 spp., larvae develop in flower heads cut by females; associated with various composite species of *Helianthus*, *Silphium*, *Coreopsis*, *Viquiera*, etc.; *H. aeneus* is the most common North American species and is widely distributed from south central Canada throughout the middle and eastern U.S.; the other species occur in the southwestern states (key to Nearctic species, Hamilton 1974).

Involvulus Schrank 1798, 1 sp., *I. hirtus* (Fabricius); larvae develop in cut terminal shoots of *Quercus* spp.according to Vogt in Kissinger (1964); ranges throughout northeastern U.S. and southwest to AZ.

Euvolvulus Reitter 1916

Merhynchites Sharp 1889, 4 spp.; larvae develop in buds or fruit (hips) of *Rosa* spp.; *M. bicolor* (Fabricius), the eastern rose curculio, develops in the hips and is distributed across the northern U.S. and southern Canada from coast to coast. The western rose curculio, *Merhynchites wickhami* (Cockerell), develops in the buds and ranges throughout the western half of the U.S. and southwestern Canada; the other 2 species occur locally in the southwestern U.S. (key to Nearctic species, Hamilton 1985).

Temnocerus Thunberg 1815, 14 spp.; known larvae develop in cut terminal shoots, buds and leaf primordia; associated with a wide variety of hosts including species of *Quercus*, *Comptonia*, *Acacia*, *Manzanita*, etc.; regionally distributed in U.S. (key to Nearctic species, Hamilton 1971).

Pselaphorhynchites Schilsky 1903

Rhynchites Schneider 1791, 1 sp., *R. velatus* LeConte (Fig. 4); larvae develop in fruits of the desert plum, *Prunus andersonii* Gray; recorded only from CA and NV.

Auletini Desbrochers 1908

Auletobius Desbrochers 1869, 9 spp.; known larvae develop in cut terminal leaf primordia; associated with a wide variety of hosts including species of *Quercus*, *Comptonia*, *Potentilla*, *Eriogonum*, etc.; regionally distributed in U.S. (key to Nearctic species, Pierce 1909; key to World species, Voss 1933-37).

Deporaini Voss 1929

Deporaus Samouelle 1819, 1 sp., *D. glastinus* LeConte; larvae are leaf miners; associated with *Q. wislizenii* in southern CA and other *Quercus* spp. throughout its range; ranges throughout western U.S., east to CO and southeast to NM and west TX.

Platyrhynchus Thunberg (1815)

Pterocolinae Lacordaire 1866

Pterocolus Say 1831, 1 sp., *P. ovatus* (Fabricius) (Fig. 3); leaf roll thief, north of Mexico the larvae develop in leaf rolls prepared by three species of attelabine weevils (*H. pubescens*, *A. bipustulatus* and *H. analis*); distributed throughout eastern half of U.S., southwest to TX and AZ and south into Mexico (key to New World species, Hamilton 1998).

Bibliography

ALONSO-ZARAZAGA, M. A. and C. H. C. LYAL. 1999. A world catalogue of families and genera of Curculionoidea (Insecta: Coleoptera) (Excepting Scolytidae and Platypodidae). Entomopraxis. Spain, 350 pp.

ANDERSON, D. M. 1991. Curculionidae (broad sense) (Curculionoidea). Pp. 594-612. *In:* F. W. Stehr, ed. Immature Insects. volume 2. Kendall/Hunt. Dubuque, IA.

BALDUF, W. V. 1959. Obligatory and facultative insects in rose hips, their recognition and bionomics. University of Illinois Press. Urbana, IL., 194 pp.

BEDEL, L. 1882-1888. Faune des Coleopteres du Bassin de la Seine. Rhynchophora. Annales del la Société Entomologique de France, 6 (hors. Serie), 442 pp.

BILLBERG, G. J. 1820. Enumeratio Insectorum in Museo G. J. Billberg. Gadel, Stockholm, 138 pp.

BLATCHLEY, W. S. and C. W. LENG. 1916. Rhynchophora or weevils of North Eastern America. The Nature Publishing Company. Indianapolis, IN, 682 pp.

BØVING, A. G. and F. C. CRAIGHEAD. 1931. An illustrated synopsis of the principal larval forms of the order Coleoptera. Entomologica Americana, 11: 1-351.

BRIGHT, D. E. 1993. The Insects and Arachnids of Canada. Part 21. The weevils of Canada and Alaska: Volume 1. Coleoptera: Curculionoidea, excluding Scolytidae and Curculionidae. Publication 1882. Research Branch, Agriculture Canada. Ottawa, Canada.

CHITTENDEN, F. H. 1901. Some insects injurious to violet, rose and other ornamental plants. United States Department of Agriculture Bulletin, 27: 1-114.

COOLEY, R. A. 1903. Two insect pests: the rosebud curculio, *Rhynchites bicolor* Fabr.- the poplar leaf-folding sawfly, *Pontania bozemani* Cooley. Montana Experiment Station. Bulletin, 46: 107-118.

CROWSON, R. A. 1955. The natural classification of the families of Coleoptera. E. W. Classey. Middlesex, England, 214 pp.

DICKERSON, E. L. 1910. Notes on *Rhynchites bicolor* Fabr. Journal of Economic Entomology, 3: 316-17.

EDWARDS, J. C. 1949. Coleoptera or beetles east of the Great Plains. J. W. Edwards. Ann Arbor, MI, 181 pp.

ESSIG, E. O. 1958. Insects and mites of western North America. MacMillan. New York, NY, ix-xiii + 1050 pp.

EWING, H. E. 1915. A case of persistent melanism. Biological Bulletin, 28: 224-228.

FARRELL, B.D. 1998. "Inordinate fondness" explained: Why are there so many beetles? Science, 281: 555-559.

FROST, C.A. 1908. Notes on *Attelabus rhois* and parasite. Psyche, 15: 26-32.

GATES, B. N. 1909. The rose curculio (*Rhynchites bicolor* Fabr.) in Massachusetts. Journal of Economic Entomology, 2: 465-66.

GIRAULT, A. A. 1904. *Attelabus bipustulatus* Fabr., the theory of oviposition and construction of nidus; misc. notes. Entomological News, 15: 189-193.

GISTEL, J. 1856. Die Mysterien der Europaischen Insectenwelt. Dannheimer. Kempten, 12 + 532 pp.

HAMILTON, R.W. 1969. Studies of the Rhynchophorous families Nemonychidae, Attelabidae and Rhynchitidae, with a revision of North American species of *Attelabus* Linnaeus, *Rhynchites* Schneider and *Eugnamptus* Schoenherr (Coleoptera: Curculionoidea). Unpublished Ph.D. dissertation, The Ohio State University, Columbus, Ohio, microfilm, 492 pp.

HAMILTON, R.W. 1971. The genus *Pselaphorhynchites* in America, north of Mexico (Coleoptera: Rhynchitidae). Annals of the Entomological Society of America, 64: 982-996.

HAMILTON, R.W. 1973. Observations on the biology of *Haplorhynchites aeneus* (Boheman) (Coleoptera: Rhynchitidae). Coleopterists Bulletin, 27: 83-86.

HAMILTON, R.W. 1974. The genus *Haplorhynchites* in America north of Mexico (Coleoptera: Rhynchitidae). Annals of the Entomological Society of America, 67: 787-794.

HAMILTON, R.W. 1979. Taxonomic use of endophallic structures in some Attelabidae and Rhynchitidae of America north of Mexico with notes on nomenclature. Annals of the Entomological Society of America, 72: 29-34.

HAMILTON, R.W. 1980. Notes on the biology of *Eugnamptus collaris* (Fabricius) (Coleoptera: Rhynchitidae) with descriptions of the larva and pupa. Coleopterists Bulletin, 34: 227-236.

HAMILTON, R. W. 1981. Description of the larva and pupa of *Haplorhynchites aeneus* (Boheman), (Coleoptera: Curculionoidea: Rhynchitidae). Journal of the Kansas Entomological Society, 54: 616-624.

HAMILTON, R. W. 1983. Biological data on two North American rhynchitids (Coleoptera: Rhynchitidae) associated with sweet fern, *Comptonia peregrina* (Linnaeus): with descriptions, illustrations and comparisons of their immature stages. Journal of the Kansas Entomological Society, 56: 511-522.

HAMILTON, R. W. 1985. The genus *Merhynchites* Sharp in America north of Mexico. Southwestern Entomologist, 10: 49-64.

HAMILTON, R. W. 1990. A revision of the weevil genus *Eugnamptus* Schoenherr (Coleoptera: Rhynchitidae) in America north of Mexico. Transactions of the American Entomological Society, 115: 475-502.

HAMILTON, R. W. 1992. Revision of the New World Genus *Himatolabus* Jekel (Coleoptera: Attelabidae) in North America. Transactions of the American Entomological Society, 118:197-226.

HAMILTON, R. W. 1994. New life cycle data for two western North American weevils (Coleoptera: Rhynchitidae) with a summary of North American rhynchitid biology. Coleopterists Bulletin, 48: 331-343.

HAMILTON, R. W. 1998. Taxonomic revision of the New World Pterocolinae (Coleoptera: Rhynchitidae. Transactions of the American Entomological Society, 124: 203-269.

HAMILTON, R. W. and S. S. KURITSKY. 1981. Description of the larva and pupa of *Merhynchites bicolor* (Fabricius). Coleopterists Bulletin, 35: 189-195.

HARRIS, T. W. 1862. A treatise on some of the insects injurious to vegetation. Crosby and Nichols. Boston, MA, 640 pp.

HOERNER, J. L. 1936. Western Rose Curculio, *Rhynchites bicolor wickhami* Cockerell. Colorado Agricultural Experiment Station Bulletin, 432: 1-19.

HOPKINS, A. A. 1905. (Secretarial Notes). Proceedings of the Entomological Society of Washington, 7: 130-133.

JEKEL, H. 1855-1860. Insecta saundersiana: or characters of undescribed insects in the collection of William Wilson Saunders, Esq. E. Newman. London, 242 pp.

KISSINGER, D. G. 1964. Curculionidae of America North of Mexico. A key to the genera. Taxonomic Publications. South Lancaster, MA, 143 pp.

KUSCHEL, G. 1995. A phylogenetic classification of Curculionoidea to families and subfamilies. Memoirs of the Entomological Society of Washington, 14: 5-33.

LAWRENCE, J. L. 1982. Coleoptera. Pp. 482-553. *In:* S. P. Parker, ed. Synopsis and classification of living organisms. McGraw/Hill. NY.

LECONTE, J. L. 1876. *In* J. L. LeConte and G. H. Horn, The Rhynchophora of America, north of Mexico. Proceedings of the American Philosophical Society, 15: 1-455.

LINNAEUS, C. V. 1758. Systema naturae per regna tria naturae secundum classes, ordines, genera, species, cum characteribus, differentis, synonymis, locis, ed. 10. Salvius. Holmiae, vol. 1, 823 pp.

LOVETT, A. L. 1915. The rose curculio, *Rhynchites bicolor* Fabr. injures blackberry buds. Reports of the Department of Entomology, Oregon Agricultural Experiment Station Bulletin, pp. 150-153.

LODING, H. P. 1945. Catalogue of the beetles of Alabama. Wetumpka Printing. Wetumpka, AL, 172 pp.

LUTZ, F. E. 1935. Field book of insects, 3rd ed. G. P. Putnam and Sons. New York, NY, 510 pp.

MCBRIDE, D. K. and C. Y. OSETO. 1978. Sunflower insect pests. Cooperative Extension Service. North Dakota State University, Fargo, ND, Circular E-623, pp. 1-6.

MURTFELDT, M. E. 1872. Notes on *Attelabus bipustulatus* Fabr. Canadian Entomologist, 4: 143-145.

O'BRIEN, C. W. and G. J. WIBMER. 1982. Annotated checklist of the weevils (Curculionidae, *sensu lato*) of North America, Central America and the West Indies (Coleoptera: Curculionoidea). Memoirs of the American Entomological Institute, 34: ii-ix, 1- 382.

OLIVIER, A. G. 1807. Entomologie, ou histoire naturelle des insectes, avec leurs caractères génériques et spécifiques, leur description, leur synonymie et leur figure enluminée. Desray. Paris, Coleopteres, vol. 5, 612 pp.

PACKARD, A. S. 1890. Fifth report of the U.S. Entomological Commission on insects injurious to forest and shade trees. Government Printing Office. Washington, DC, 957 pp.

PIERCE, W. D. 1909. Studies of North American weevils. Proceedings of the United States National Museum, 37(1708): 325-364.

PIERCE, W. D. 1913. Miscellaneous contributions to the knowledge of the weevils of the families Attelabidae and Brachyrhinidae. Proceedings of the United States National Museum, 45 (1988): 365-426.

ROBERTSON, H. A.1923. The rose curculio in Manitoba with notes on other insects affecting roses. Annual Reports of the Entomological Society of Ontario, 54: 12-16.

SCHOENHERR, C. J. 1823. Tabula synoptica familiae curculionidum. Isis Oken, heft X, columns 1132-1146.

SCHULZ, T. J. and W. V. LIPP. 1969. The status of the sunflower insect complex in the Red River Valley of North Dakota. Proceedings of the North Central Branch, Entomological Society of America, 24: 99-100.

SHARP, D. E. 1889. Biologia Centrali Americana, Insecta, Coleoptera, Curculionidae: Attelabinae. 4: 1-45.

SILFVERBERG, H. 1977. *Attelabus* Linnaeus, 1758 (Insecta, Coleoptera): Request for confirmation of designation of type-species. Z. N. (S.) 2209. Bulletin of Zoological Nomenclature, 34: 189-90.

THOMPSON, R. T. 1992. Observations on the morphology and classification of weevils (Coleoptera, Curculionoidea) with a key to major groups. Journal of Natural History, 26: 835-891.

TING, P.C. 1936. The mouthparts of the coleopterous group Rhynchophora. Microentomology, 1: 93-114.

VAN EMDEN, F. 1938. On the taxonomy of Rhynchophora larvae (Coleoptera). Transactions of the Royal Entomological Society of London, 87: 1-37.

VOGT, G. 1992. Leaf rolling weevils (Coleoptera: Attelabidae) their host plants, and associated rhynchitid weevils in North America (Canada through the Republic of Panama): Summary of a long term field study. Pp. 392-420. *In:* D. Quainter and A. Aiello, eds. Insects of Panama and Mesoamerica: Selected Studies. Oxford University Press, vxii + 692 pp.

VOSS, E. 1922. Monographische Bearbeitung der Unterfamilie Rhynchitinae (Curc.) I, Teil, Nemonychini-Auletini. Archiv für Natrurgeschichte, 88: 1-113.

VOSS, E. 1925. Die Unterfamilien Attelabinae und Apoderinae (Col. Curc.) (18. Beitrag zur Kenntnis der Curculioniden). Stettiner Entomologische Zeitung, 85: 1-78, 191-304.

VOSS, E. 1932-33. Monographie der Rhynchitinen-Tribus Rhynchitini. 2. Gattungsgruppe: Rhynchitina. V. 1. Teil der Monographie der Rhynchitinae-Pterocolinae. (41. Beitrag zur Kenntnis der Curculioniden.) Koleopterologische Rundschau, 18: 153-189 (1932); 19: 25-56 (1933).

VOSS, E. 1933-37. Monographie der Rhynchitinen-Tribus Auletini. III. Teil der Monographie der Rhynchitinae-Pterocolinae. (37. Beitrag zur Kenntnis der Curculioniden.) Stettiner Entomologische Zeitung. 94: 108-136, 273-286 (1933); 95: 109-135, 330-344 (1934); 96: 91-105, 229-241 (1935); 97: 279-289 (1936); 98: 101-108 (1937).

VOSS, E. 1938-1943. Monographie der Rhynchitinen-Tribus Deporaini sowie der Unterfamilien Pterocolinae- Oxycoryninae (Allocorynini). VII. Teil der Monographie der Rhynchitinae-Pterocolinae. (73. Beitrag zur Kenntnis der Curculioniden) Stettiner Entomologische Zeitung, 99: 59-117, 302-363 (1938); 103: 129-155 (1942); 104: 46-63 (1943).

VOSS, E. 1938. Monographie der Rhynchitinen-Tribus Rhynchitini. 2. Gattungsgruppe: Rhynchitina. V. 2. Teil der Monographie der Rhynchitinae-Pterocolinae. (45. Beitrag zur Kenntnis der Curculioniden), Koleopterologische Rundschau, 24: 129-171.

VOSS, E. 1941. Monographie der Rhynchitinen-Tribus Rhinocartini sowie der Gattungsgruppe Eugnamptina der Tribus Rhynchitini. IV. Teil der Monographie der Rhynchitinae-Pterocolinae. (32. Beitrag zur Kenntnis der Curculioniden). Deutche Entomologische Zeitschrift, 1941: 113-215.

VOSS, E. 1941. Monographie der Rhynchitinen-Tribus Rhynchitini. 2. Gattungsgruppe: Rhynchitina. V. 2. Teil der Monographie der Rhynchitinae-Pterocolinae. (45. Beitrag zur Kenntnis der Curculioniden). Mittheilungen Muenchener Entomologische Gesellschaft, 31: 628-680.

VOSS, E. 1965. Die Unterfamilie Camarotinae, ihre Beziehungen zur Familie Attelabidae sowie ein Versuch, die phylogenetischen Zusammenhange innerhalb der letzteren zur Darstellung zu bringen (Coleoptera, Curculionidae) (191. Beitrag zur Kenntnis der Curculioniden). Entomologische Abhandlungen Staatliches Museum fur Tierkunde in Dresden, 32 : 221-244.

VOSS, E. 1969. Monographie der Rhynchitinen-Tribus Rhynchitini. 2. Gattungsgruppe: Rhynchitina. V. 2. Teil der Monographie der Rhynchitinae-Pterocolinae. Entomologische Arbeiten, aus dem Museum G. Frey, 20: 117-375.

WIBMER, G. J. and C. W. O'BRIEN. 1986. Annotated checklist of the weevils (Curculionidae, *sensu lato*) of South America (Coleoptera: Curculionoidea). Memoirs of the American Entomological Institute, 39: i-xvi, 1- 563.

129. BRENTIDAE Billberg 1820

by Robert S. Anderson and David G. Kissinger

Family common name: The straight-snouted weevils; pear shaped weevils

Family synonyms: Apionidae Schoenherr 1823; Cycladidae Schoenherr 1823; Eurhynchidae Lacordaire 1863.

As used here, Brentidae is a composite of primitive weevil forms, not recognizable by any one or a few characters. Nearly all of them, with the exception of only the Nanophyinae, have straight, non-geniculate antennae. Apioninae and Nanophyinae are small 'pear-shaped' weevils immediately recognized by a long cylindrical trochanter with the femur attached at its apex, a character state that is known in no other beetles. Brentinae, Cyphagoginae and Trachelizinae have elongated, parallel-sided bodies and a long, generally straight rostrum that often exhibits marked sexual dimorphism. Cyladinae are the sweet potato weevils, recognizable by their unique habitus.

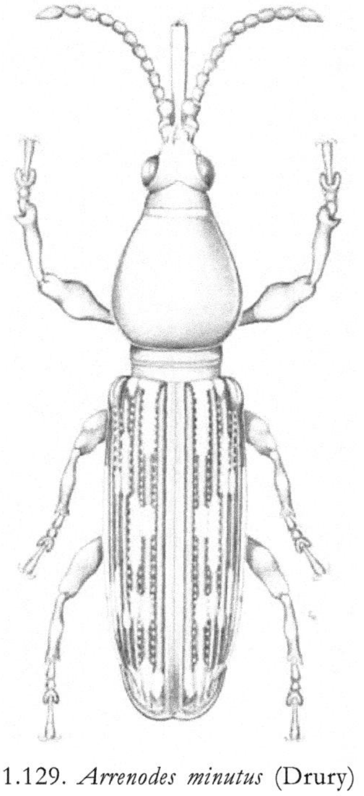

1.129. *Arrenodes minutus* (Drury) (from Bright 1993, reproduced with the permission of the Minister of Public Works and Government Services, 2001)

Description (based on Lawrence 1982). Shape variable elongate, narrow and parallel-sided (Brentinae, Cyphagoginae, Trachelizinae), or stouter and more robust with expanded oval elytra and a distinctive "pear-like" shape (Apioninae, Nanophyinae); flat to convex; length 1.0-40.0 mm; color pale brown to black, very rarely red or bicolored or with contrasting colors or markings; vestiture lacking, or of fine short to moderately long appressed or suberect pubescence; rarely with scales.

Rostrum moderately to very long and usually narrow and cylindrical; sexual dimorphism in rostrum form (Brentinae, Cyphagoginae, Trachelizinae) or length (Apioninae) generally evident. Antennae straight or rarely geniculate (Nano-phyinae), funicle very narrow and slender, rarely filiform or moniliform (Brentinae, Cyphagoginae, Trachelizinae), ending in a weak, loose club (Nanophyinae) or a compact club (Apioninae, Cyladinae) of three articles. Antennal insertions lateral at the middle of the rostrum or rarely dorsal near the base. Maxillary palps of two or three articles. Labial palps small, of one or two articles, attached dorsally near the apex of the prementum, rarely in cavities. Proventriculus lacks sclerotized plates. Front coxae vary from contiguous to widely separated. Middle coxae and hind coxae narrowly to widely separated. Front tibiae rarely equipped with an antenna-cleaning organ. Hind wing usually with reduced anal venation. Tarsal claws variable. First two visible sternites of the abdomen connate and much longer than following two; pygydium concealed by the elytra. Tegmen bilobed or notched apically and the median lobe with a "trough-like" ventral plate and a narrow dorsal plate.

Larvae (based on Lawrence 1982) when mature about 2.0-18.0 mm in length, slender and subcylindrical (Brentinae) or broad and strongly curved (Apioninae). Small legs rarely present on thorax (Brentinae), usually absent, but with well-developed pedal lobes (Apioninae). Frontal sutures complete although rarely indistinct, reaching articulating membrane of mandible. Antenna of a single membranous article. Hypopharyngeal bracon present. Abdomen with first eight segments with two, three or four dorsal folds and bearing annular or rarely biforous spiracles.

Habits and habitats. As noted, Brentidae is a composite of taxa of subfamily rank each of which has a rather different body form. Natural history aspects of Brentidae similarly are correlated with, and best discussed, at the subfamily level (Thomas 1996).

Brentinae, Cyphagoginae and Trachelizinae are the traditional brentids. They are small to large, elongate weevils that are usually found under bark as both adults and larvae. Where known, females lay their eggs in holes they have excavated with their rostrum in living, dying or recently felled hardwood trees. Larvae tunnel deep into the heartwood and appear to feed on wood and fungus mycelia. *Brentus anchorago* Linnaeus is one of North America's largest weevils, with males reaching a body length of up to 40 mm. In some species, adult males guard females during egg-laying.

Cyladinae includes only the adventive *Cylas formicarius* (Fabricius), the sweet potato weevil. Adults and larvae are associated with various Convolvulaceae (especially *Ipomoea pescapreae* L.) and are pests of sweet potatoes, *Ipomoea batatas* L. Larvae mine the stems.

Little is known about Nearctic Nanophyinae. Most Palearctic Nanophyinae are associated with plants of the family Lythraceae, especially the genus *Lythrum*, but also Crassulaceae, Tamaricaceae and Ericaceae. Larvae are associated with the fruits, leaves and stems and many appear to cause galls. One species of *Nanophyes* Schoenherr has been introduced into North America for biological control of *Lythrum salicaria* L. (purple loosestrife; Lythraceae)

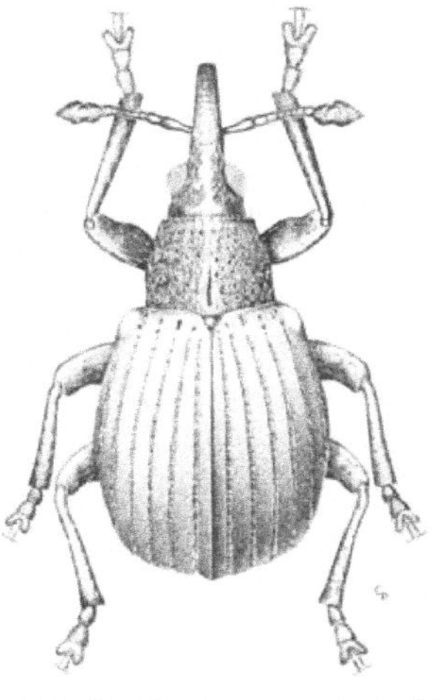

2.129. *Trichapion centrale* (Fall) (from Bright 1993, reproduced with the permission of the Minister of Public Works and Government Services, 2001)

(Harris 2001) and a second is under consideration. *Zeugonyx sabinae* Notman has been reared from galls on *Juniperus ashei* Buchh. in Texas, but otherwise no information is available on the native North American fauna.

Apioninae are associated with a wide variety of plants, mostly Fabaceae and Asteraceae, but also Malvaceae, Polygonaceae, Caprifoliaceae, and Umbelliferae. Many of the newly elevated genera, formerly subgenera of *Apion sensu lato*, are host specific on certain families of plants (*e.g.*, *Alocentron* on Malvaceae; *Perapion* on Polygonaceae). Larvae mine stems, feed on developing pericarp of fruit, or develop in the seeds. A few species are adventive. *Rhopalapion longirostre* (Olivier) is associated with the ornamental hollyhock, *Althea rosea* Cav. (Tuttle 1954) and *Ischnopterapion virens* (Herbst) with clover, *Trifolium* (Hoebeke *et al.* 2000). Two species of *Exapion* have been introduced into the western United States as biological control agents for gorse, *Ulex europaeus* L. and scotch broom, *Cytisus scoparius* (L.) (both Fabaceae) (O'Brien 1995) and *Omphalapion hookerorum* (Kirby) has been introduced for biological control of scentless chamomile, *Matricaria perforata* Mérat (Harris and McClay 2001). *Podapion gallicola* Riley 1883 has been associated with *Pinus*, larvae occur in galls on twigs (Bright 1993).

Status of the classification. At present there is continuing controversy over the classification of the members of this family. Alonso-Zarazaga and Lyal (1999) recognize Brentidae (including Brentinae, Cyphagoginae, Trachelizinae and Cyladinae) as distinct from Apionidae and Nanophyidae and accord each family level status. This system is different from others such as Lawrence (1982) which recognizes Brentidae (including Brentinae, Cyphagoginae and Trachelizinae) and Apionidae (including Apioninae, Nanophyinae and Cyladinae). The family level classification used here follows Lawrence and Newton (1995) and groups all of these taxa within Brentidae; however, at the subfamilial level, within Brentidae, the classification follows Alonso-Zarazaga and Lyal (1999). O'Brien and Wibmer (1982) provide a checklist of the North American species of Nanophyinae and species of Apioninae (all as *Apion sensu lato*).

Brentinae, Cyphagoginae, Trachelizinae and Cyladinae are well-known in North America and their classification remains unchanged (Warner 1960; Kissinger 1968). Brentinae have 3 species in 3 genera, and Cyphagoginae, Trachelizinae and Cyladinae each have only 1 genus and 1 species represented. Apioninae and Nanophyinae on the other hand have undergone substantial changes at the generic level over the last few years. In Nanophyinae,

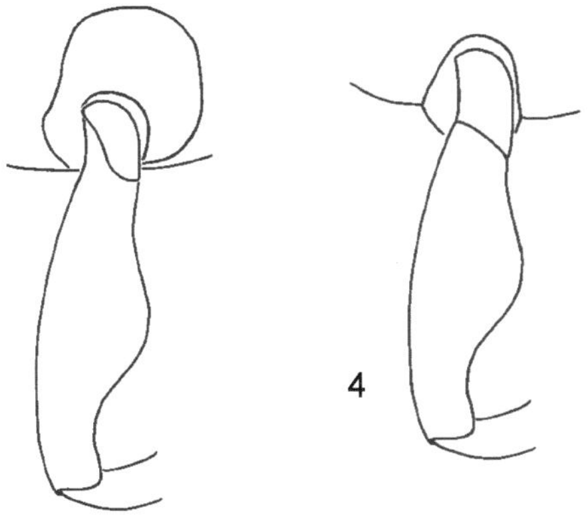

FIGURES 3.129-4.129. 3. Front leg, Brentinae (schematic); 4. Front leg, Apioninae (schematic).

there are now 5 genera and 5 species recognized in North America. This is based on phylogenetic evidence and follows the division of the once larger widespread genus *Nanophyes* Schoenherr by Alonso-Zarazaga (1989) and followed by Alonso-Zarazaga and Lyal (1999). The situation with Apioninae is even more complex with many of the various subgenera (*e.g.*, Kissinger 1968) now elevated to generic rank, many generic concepts have been narrowed with some North American species transferred into previously Palearctic genera, and a number of new genera have been proposed (Kissinger 1989, 1990, 1992, 1998, 1999; Alonso-Zarazaga 1990). Alonso-Zarazaga (1990) treats the Palearctic classification of Apioninae in detail and is the basis for the development and understanding of the new system as applied in North America. At present in North America there are approximately

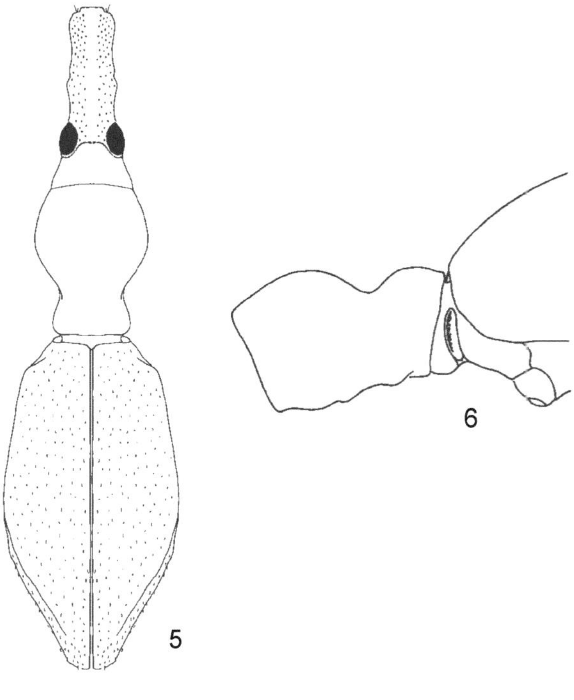

FIGURES 5.129-6.129. 5. Dorsal habitus, *Cylas formicarius* (Fabricius 1798); 6. Lateral view, pronotum and base of elytra, *Cylas formicarius*.

140 species placed in the 19 genera. A small number of North American species have not yet been assigned to a genus in this new system. Identifying Apioninae is no easy task and the key to the genera of Apioninae formerly placed in *Apion sensu lato* (starting at couplet 15) is based extensively on male characters only. Also, many of the characters used in establishing the new generic classification are based on male genitalia and require dissections and detailed study.

Distribution. Collectively this family is widely distributed in North America, but some of the constituent subfamilies have very restricted distributions. Apioninae are widespread in North America and found in most habitat types from sea level to the alpine zone. Nanophyinae are also widespread but do not extend as far north as do Apioninae, nor do they include the diversity of habitats of the latter group, but then there are also many fewer species. Excluding Cyladinae, which has only the one adventive species, *Cylas formicarius* Fabricius, Brentinae, Cyphagoginae and Trachelizinae are generally tropical in their distributions and 4 of the 5 species that occur in North America only do so in extreme southern Florida (*Brentus anchorago* (Linnaeus), *Stereodermus exilis* Suffrian, and *Paratrachelizus uncimanus* (Boheman)) or southern Texas (*Heterobrenthus texanus* Schaeffer). The remaining species, *Arrenodes minutus* (Drury), is widespread in eastern North America.

Key to the Nearctic Subfamilies and Genera

1. Trochanter short and triangular, femur attached to the side of trochanter (Fig. 3); body more elongate and narrow, length greater than 3.0 mm, most greater than 10.0 mm (Figs. 1, 5, 14) 2
— Trochanter long and somewhat cylindrical, femur attached to the apex of trochanter (Fig. 4); body "pear-shaped", length less than 5.0 mm, most less than 3.0 mm (Figs. 2, 16, 31) 7

2(1). Pronotum with a broad constricted collar (that is almost as long as the rest of the pronotum) before base (Fig. 5); mesothorax with a deep vertical sulcus near the posterior margin of the pronotum (Fig. 6); tarsal claws connate at base; elytra elongate-oval in dorsal view, lacking impressed striae (Fig. 5) (Cyladinae) .... *Cylas*
— Pronotum with at most a very short constriction before base (Figs. 1, 7, 9, 10, 14); mesothorax lacking a deep lateral pit near the posterior margin of the pronotum; tarsal claws free; elytra elongate and parallel-sided in dorsal view, with distinctly or indistinctly impressed striae 3

3(2). All femora lacking distinct tooth (although slight angulation may be present on front femur); rostrum with ventral surface punctate and with dense microsetae or sparse elongate recumbent setae 4
— At least front femur with obvious tooth; rostrum with ventral surface smooth or coarsely punctate, not setose (Brentinae) 5

4(3). Elytra with stria 3 bent inwards at middle towards elytral suture and coalescent with stria 2 (Fig. 7); front tibia with inner edge incised in apical one-

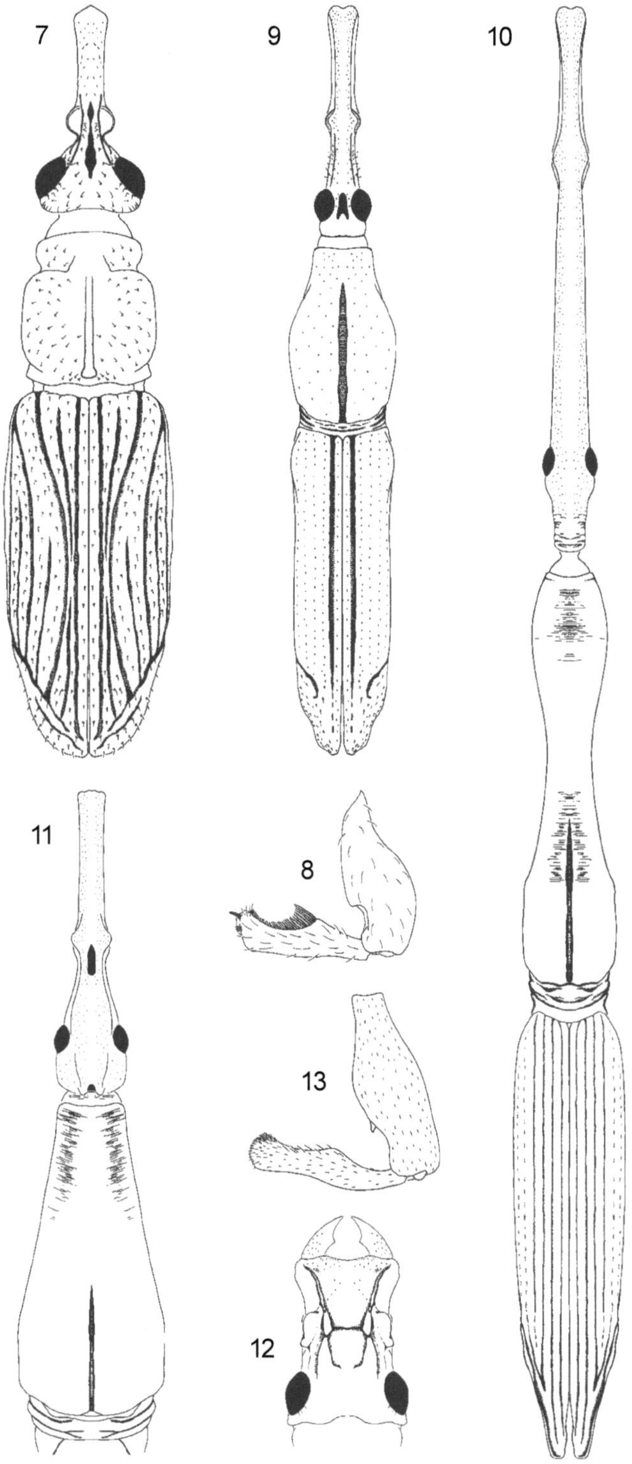

FIGURES 7.129-13.129. 7. Dorsal habitus, *Stereodermus exilis* Suffrian; 8. Front leg, *Stereodermus exilis;* 9. Dorsal habitus, *Paratrachelizus uncimanus* (Boheman); 10. Dorsal habitus, *Brentus anchorago* (Linnaeus), male; 11. Dorsal view, head and pronotum, *Brentus anchorago*, female; 12. Dorsal view, head, *Arrenodes minutus* (Drury), male; 13. Front leg, *Arrenodes minutus.*

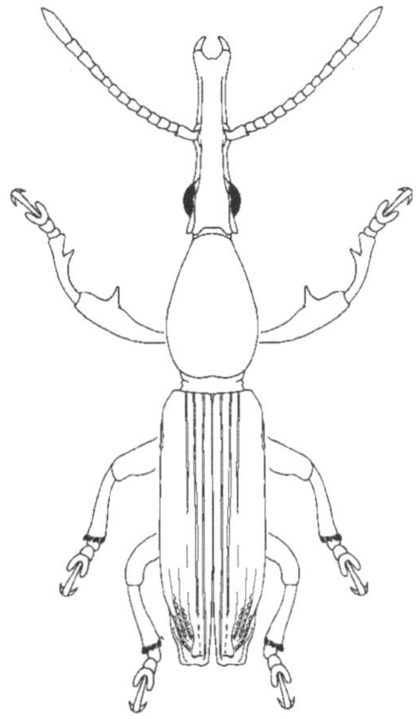

FIGURE 14.129. Dorsal habitus, *Heterobrenthus texanus* Schaeffer, male.

half, incision surrounded by very dense setae (Fig. 8); rostrum with ventral surface with sparse elongate recumbent setae (Cyphagoginae) .... *Stereodermus*

— Elytra with stria 3 evident only as a row of shallow punctures, straight throughout length, not coalescent with stria 2 (Fig. 9); front tibia with inner edge straight, not incised; rostrum with ventral surface with dense microsetae in basal portion (Trachelizinae) .... *Paratrachelizus*

5(4). Head longer than broad, not distinctly constricted behind eyes, lateral margins continuous, prolonged and convergent behind eyes (Figs. 10, 11); pronotum medially deeply sulcate in basal one-half (Figs. 10, 11); male with pronotum extremely elongated, lateral margins constricted near middle (Fig. 10) .... *Brentus*

— Head short, width more or less equal to length, distinctly constricted behind eyes, lateral margins at most slightly prolonged but interrupted by constriction, not convergent behind eyes (Figs. 1, 12, 14); pronotum uniformly smooth dorsally, not sulcate (Figs. 1, 14); male with pronotum not elongated, and lateral margins not constricted near middle 6

6(5). Antenna with articles 2 to 11 more or less equal in length and width towards apex (Fig. 1); head constricted behind eyes, hind angles not projecting (Figs. 1, 12); front tibia with at most a swelling on inner margin, no distinct tooth evident (Fig. 13); rostrum very dissimilar in male (short and broad with large mandibles, Fig. 12) and female (narrow and cylindrical, Fig. 1) .... *Arrenodes*

— Antenna with articles 2 to 11 increasing in length and width towards apex (Fig. 14); head strongly constricted behind eyes, hind angles projecting; front tibia with distinct tooth on inner margin (Fig. 14); rostrum dissimilar in male (dilated at apex) and female (long and cylindrical anterior to point of antennal insertion) .... *Heterobrenthus*

7(1). Antenna geniculate, scape distinctly longer than combined length of first 3 articles of funicle; funicle apparently with only 5 articles, articles 6 and 7 as wide as apical antennal article and forming a loose club (Nanophyinae) 8

— Antenna straight, not geniculate, scape shorter than combined length of first 3 articles of funicle; funicle with 7 articles, articles 6 and 7 narrow, not as wide as apical antennal article and distinct from compact apical club (Apioninae)............ 12

8(7). Tarsal claws equal in length and form 9

— Inner tarsal claw shorter and finer than outer claw 11

9(8). Elytra with interstria 8 with a row of small, dark tubercles (Fig. 15); elytra largely pale, with dark, oblique fascia extended from basal one-third at middle to humeri; vestiture consisting of sparse, erect setae and long, decumbent, moderately dense scales intermixed; length 2.0-2.3 mm .... *Pseudotychius*

— Elytra with interstria 8 smooth in basal one-quarter; elytral vestiture various; length 1.2-1.9 mm .. 10

10(9). Elytra lacking fascia, setae appressed and uniformly arranged, each seta short, not or hardly exceeding the base of the following one (Fig. 16); pronotum and tibia lacking specialized erect setae, only one subhumeral seta on elytral interstria 9 (Fig. 16) .... *Microon*

— Elytra with fascia, setae arcuate and disordered, each seta long, usually exceeding the base of the following one (Fig. 17); pronotum, tibia and odd elytral interstriae with specialized erect setae (Fig. 17) .... *Nanophyes*

11(8). Antenna with basal 5 articles of funicle slender; elytra largely pale, with basal dark area vestiture fine and short, consisting of whitish and dark scales .... *Nanodactylus*

— Antenna with basal 4 articles of funicle slender; elytra uniformly testaceous; vestiture of long whitish setae .... *Zeugonyx*

12(7). Middle coxae contiguous, not separated by junction between mesosternal process and intercoxal process of metasternum 13

— Middle coxae separated by junction between mesosternal process and intercoxal process of metasternum 15

13(12). Tarsal claws with acute basal tooth .... *Chrysapion*

— Tarsal claws simple, at most swollen as base, not toothed 14

14(13). Rostrum straight, in side view not forming angle at junction with frons, apical half of rostrum usually tapered to apex (Fig. 18); eyes not or scarcely prominent; antennae inserted at basal 0.29-0.47 of rostrum; scape at most as long as width of rostrum at midlength; prescutellar fovea conspicuous, lanceolate to sulciform; tarsi normal, front tarsus with first article 1.25-1.60 X as long as wide; internal sac of aedeagus without baculi, other

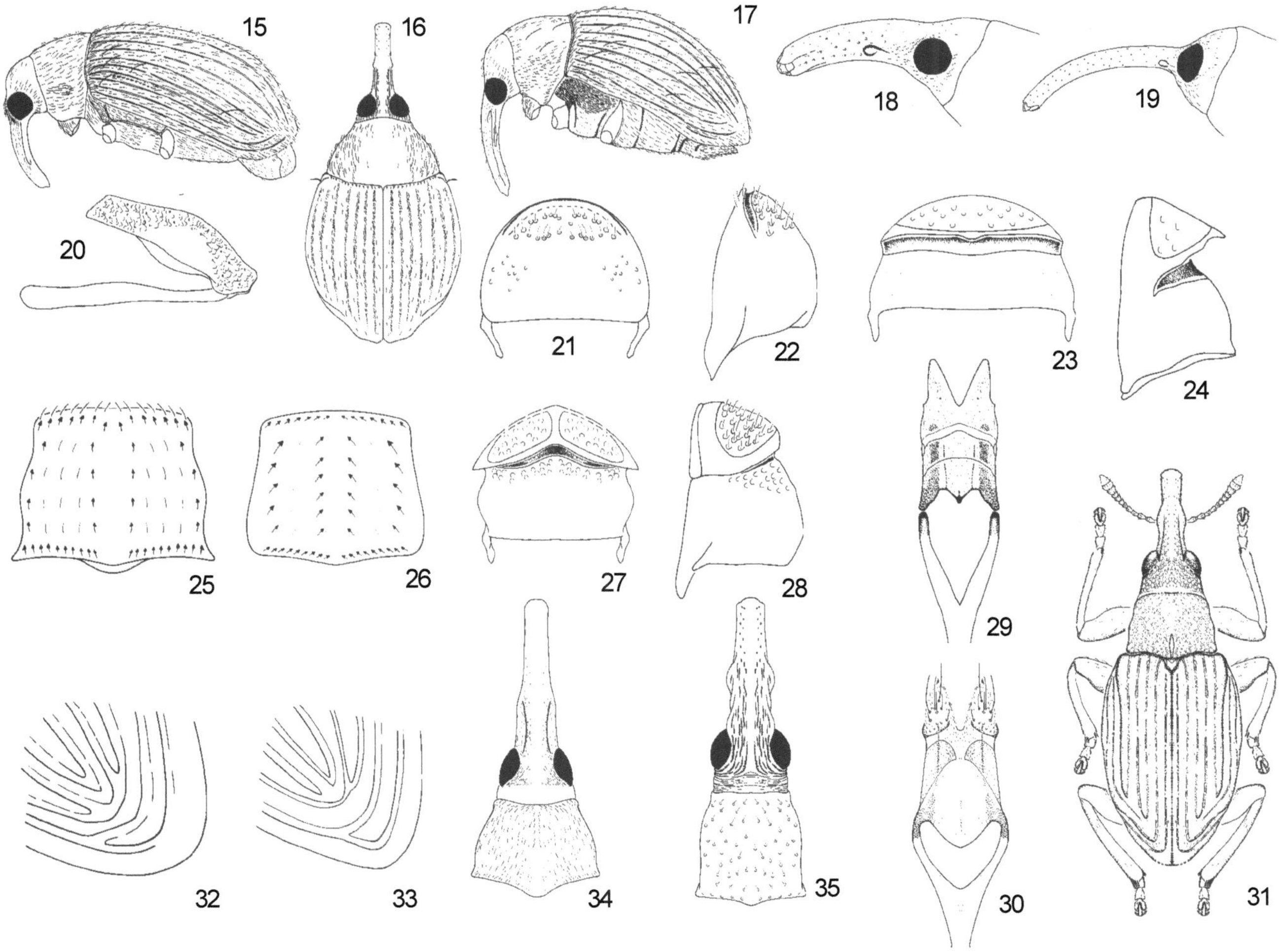

FIGURES 15.129-35.129. 15. Lateral habitus, *Pseudotychius watsoni* Blatchley; 16. Dorsal habitus, *Microon canadense* (Brown); 17. Lateral habitus, *Nanophyes marmoratus* (Goeze); 18. Lateral view, head, *Perapion*; 19. Lateral view, head, *Podapion*; 20.Front leg, *Fallapion*, male; 21. Dorsal view, pygydium, Aspidapiini; 22. Lateral view, pygydium, Aspidapiini; 23. Dorsal view, pygydium, Ixapiini; 24. Lateral view, pygydium, Ixapiini; 25. Dorsal view, pronotum, *Alocentron* (schematic); 26. Dorsal view, pronotum, *Rhopalapion* (schematic); 27. Dorsal view, pygydium, Oxystomatini; 28. Lateral view, pygydium, Oxystomatini; 29. Dorsal view, tegmen, *Betulapion simile* (Kirby), male; 30. Dorsal view, tegmen, *Mesotrichapion*, male; 31. Habitus, *Apion*, undetermined species, Alberta, Canada; 32. Elytral apex, *Apion;* 33. Elytral apex, *Ischnopterapion;* 34. Dorsal view, head, *Sayapion;* 35. Dorsal view, head, *Mesotrichapion*, male (Figs. 21-30 redrawn after Alonso-Zarazaga 1990).

structures may be present; associated with Polygonaceae .... *Perapion*

— Rostrum strongly curved, in side view forming a conspicuous angle at junction with frons, apical half of rostrum parallel-sided to slightly widened to apex (Fig. 19); eyes prominent; antennae inserted at basal 0.10-0.19 of rostrum; scape longer than width of rostrum at midlength; prescutellar fovea inconspicuous, similar to pronotal punctures; tarsi robust, front tarsus with first article 0.85-1.00 X as long as wide; internal sac of aedeagus with two baculi and denticles; associated with *Pinus* (Pinaceae) .... *Podapion*

15(12).* Front femur of male with polished (sometimes striate) area on ventral surface, the area generally limited posteriorly by a prominent longitudinal carina (Fig. 20); metasternum generally with a pair of spicules near the middle of the posterior margin .... *Fallapion*

— Front femur of male without polished area on ventral surface 16

16(15). Pygydium of male lacking distinct, deep, transverse preapical sulcus, profile of pygydium not interrupted (Figs. 21, 22); abdominal ventrite 5 of male rounded apically; elytra with specialized seta on interstria 7 lacking 17

— Pygydium of male with distinct, deep, transverse preapical sulcus, profile of pygydium distinctly interrupted by sulcus (Figs. 23, 24, 27, 28); abdominal ventrite 5 of male truncate apically; elytra with specialized seta on interstria 7 at middle or near apical 1/3 present or lacking 19

* From this point on the key is complex and relies extensively on characters of males. Males can be recognized by the shorter, more coarsely sculptured rostrum and (sometimes) by an upward deflection of the last abdominal ventrite. These are the taxa traditionally treated as *Apion sensu lato* and the reader may wish to consult Bright (1993) and Kissinger (1968) for species level identifications.

17(16). Antennal club with sutures obsolete or absent; tarsal claws simple, not toothed; on *Matricaria perforata* (Asteraceae) .... *Omphalapion*
— Antennal club with sutures distinct; tarsal claws toothed at base; on Malvaceae 18

18(17). Pronotum with vestiture directed away from the midline at the base, more or less parallel to the midline along the lateral margins, and perpendicular to the apical margin along the apical margin (Fig. 25); pronotum with basal flange well-developed (Fig. 25); on Malvaceae .... *Alocentron*
— Pronotum with vestiture directed towards the midline, in some cases, totally transverse at the basal or apical margins (Fig. 26); pronotum with basal flange obsolete (Fig. 26); on *Althea rosea* (Malvaceae) .... *Rhopalapion*

19(16). Elytra with transverse pattern of contrasting light and dark scales; pygydium of male with sulcus incomplete laterally, not reaching side margins of pygydium (Fig. 24); front tarsus with article 2 stout, about as wide as long; femora and tibiae robust; middle coxae widely separated (0.25-0.30 x diameter of middle coxa); aedeagus with internal sac with large apically bifurcate structure; on Rutaceae, Caprifoliaceae .... *Neapion*
— Elytra with scales more uniformly colored, not contrasting light and dark; pygydium of male with sulcus complete laterally, reaching side margins of pygydium (Figs. 27, 28); front tarsus with article 2 usually longer than wide; femora and tibiae slender; middle coxae moderately widely separated (<0.25 x diameter of middle coxa); aegeagus with internal sac without large structure; on Fabaceae and other plants 20

20(19). Genitalia of male with tegmen with prostegium articulated with free ring (Fig. 29) 21
— Genitalia of male with tegmen with prostegium fused with free ring (Fig. 30) 28

21(20). Tibiae of male simple, not mucronate 22
— Middle tibia (at least) of male mucronate 27

22(21). Scrobe with dorsal margin produced into a long, slender acute process projected ventrally; elytra with interstria 7 with specialized seta inserted near midlength of stria; metasternum very convex in lateral view; meso- and metasternal processes directed inwardly where meeting between hind coxae; on Fabaceae .... *Exapion*
— Scrobe with dorsal margin simple, not produced into a long, slender acute process; elytra with interstria 7 or 9, or 7 and 9, with or without specialized seta inserted at or behind apical 1/3 of elytron; metasternum slightly convex in lateral view; meso- and metasternal processes either flat or with metasternal process more prominent where meeting between hind coxae; on various plants .. 23

23(22). Body color red throughout; lateral margins of pronotum parallel in basal one-half (Fig. 31); male genitalia with prostegium with high median crista projected basally; elytral interstria 2 prolonged outward at apex (Fig. 32); on Polygonaceae .... *Apion*
— Body color black or with metallic sheen, occasionally with light colored appendages; lateral margins of pronotum various; male genitalia with prostegium with at most moderately high basal median carina not projected basally; elytral interstria 2 only slightly prolonged outward at apex (Fig. 33); on Fabaceae 24

24(23). Elytra blue-green; elytral interstria 7 with one specialized seta in apical one-third; legs dark; lateral margins of pronotum parallel in basal one-half; rostrum not prominently expanded at point of insertion of antennae; eyes prominent .... *Ischnopterapion*
— Elytra black; elytral interstriae 7 and 9 with one specialized seta near apex; legs various, may be light in color; lateral margins of pronotum various in basal one-half; rostrum variously expanded at point of insertion of antennae; eyes prominent or not 25

25(24). Pronotum without basal flange, with lateral margins nearly evenly convergent from base to apex, subconical in dorsal form; rostrum prominently expanded at point of insertion of antennae; legs and article 1 of antenna dark .... *Apionion*
— Pronotum with or without basal flange, with lateral margins various; rostrum smooth at point of insertion of antennae, not expanded; legs and article 1 of antenna dark or light 26

26(25). Pronotum with basal flange indistinct, subconical in dorsal form (Fig. 34); legs and article 1 of antenna light; aedeagus with median lobe and tegmen narrow, elongate, subcylindrical; internal sac with large teeth (>0.030 mm long) ...... *Sayapion*
— Pronotum without basal flange, lateral margins subparallel in basal one-half; legs and article 1 of antenna light or dark; aedeagus with median lobe slightly depressed, broader at base; internal sac with largest teeth (<0.020 mm long) .... *Kissingeria*

27(21). Head with ventral surface with subcephalic ridges lacking or low, not extended basally on head to point equivalent to middle of eye; vestiture of interstriae 2-5 sparse, generally inconspicuous, consisting of one row of scales; one specialized seta on interstria 9; aedeagus with parameroid lobe with one or more macrochaetae; on *Betula papyrifera* (Betulaceae) .... *Betulapion*
— Head with ventral surface with subcephalic ridges moderately to well developed, extended to or beyond the middle of the eye; vestiture of interstriae 2-5 relatively dense, consisting of two or more rows of scales; one specialized seta on each of interstriae 7 and 9; aedeagus with parameroid lobe generally lacking macrochaetae; associated mainly with Fabaceae and Asteraceae .... *Trichapion*

28(20). Elytra blue in color, legs dark; tarsus with claw simple; rostrum of male markedly expanded laterally at point of antennal insertion (Fig. 35); frons flat, striate .... *Mesotrichapion*
— Elytra not blue in color; rostrum of male smooth, not expanded laterally at point of antennal insertion 29

29(28). Male with metasternum tuberculate on median posterior margin and apical area of parameroid lobe

bearing four or more macrochaetae more than 0.055mm long .... *Eutrichapion*
— Male metasternum not tuberculate or parameroid lobe lacking macrochaetae more than 0.010mm long .... *Coelocephalapion*

Classification of the Nearctic Genera

Brentidae Billberg 1820

Brentinae Billberg 1820

Brentini Billberg 1820

Brentus Fabricius 1787, 1 sp., *B. anchorago* (Linnaeus 1758), Florida. Adults are often found under bark of *Bursera simaruba* (L.) Sarg.
Brenthus Illiger 1801

Arrhenodini Lacordaire 1866

Arrenodes Schoenherr 1823, 1 sp., *A. minutus* (Drury 1770), generally distributed in eastern United States, into extreme southern Canada. Adults are found under bark especially of oaks; larvae bore into the wood (Buchanan 1960). Males are territorial and guard females during egg-laying (Sanborne 1983).
Brentus Panzer 1788, not Fabricius 1787
Arrhenodes Schoenherr 1826
Eupsalis Lacordaire 1866
Platysystrophus Kleine 1917

Heterobrenthus Sharp 1895, 1 sp., *H. texanus* Schaeffer 1915, southern Texas.

Cyphagoginae Kolbe 1892

Stereodermini Sharp 1895

Stereodermus Lacordaire 1866, 1 sp., *S. exilis* Suffrian 1870, southern Florida.

Trachelizinae Lacordaire 1866

Trachelizini Lacordaire 1866

Paratrachelizus Kleine 1921, 1 sp., *P. uncimanus* (Boheman 1839), southern Florida.

Cyladinae Schoenherr 1823

Cylas Latreille 1802, 1 sp., *C. formicarius* (Fabricius 1798), generally distributed in southern United States. Adventive on Convolvulaceae including *Ipomoea batata* (L.) Lam. (sweet potato).
Cylanus Rafinesque 1815
Protocylas Pierce 1941

Apioninae Schoenherr 1823

Apionini Schoenherr 1823

Apion Herbst 1797, 1 sp., identity uncertain, Alberta. Adults have been collected around edges of high elevation snowfields. Associated with Polygonaceae in Palearctic Region.
Apiolum Kirby 1808
Apionus Rafinesque 1814
Apius Billberg 1820, not Panzer 1806
Apium Agassiz 1846
Oxystomum Gistel 1856
Oxeostomum Gistel 1856
Erythrapion Schilsky 1906

Aplemonini Kissinger 1968

Perapion Wagner 1907, 4 spp., generally distributed. Associated with Polygonaceae.
Eroosapion Ehret 1994 (valid subgenus)
Hemiperapion Wagner 1930 (valid subgenus)
Rhaphidoplectron Alonso-Zarazaga 1990 (valid subgenus)

Podapion Riley 1883, 1 sp., *P. gallicola* Riley 1883, generally distributed in eastern United States into extreme southern Canada; also in California and Oregon. Associated with *Pinus* spp.

Aspidapiini Alonso-Zarazaga 1990

Alocentron Schilsky 1901, 6 spp., generally distributed. Associated with Malvaceae.
Bulborhinapion Schatzmayr 1926 (valid subgenus)
Nearctalox Alonso-Zarazaga 1990 (valid subgenus)

Ceratapiini Alonso-Zarazaga 1990

Omphalapion Schilsky 1901, 1 sp., *O. hookerorum* (Kirby 1808), British Columbia, Alberta, Saskatchewan, Manitoba and Nova Scotia; adventive. This species was first discovered in Nova Scotia (Peschken *et al.* 1993) but has recently been introduced into western Canada for biological control of *Matricaria perforata* Mérat (scentless chamomile; Asteraceae) (Harris and McClay 2001).

Exapiini Alonso-Zarazaga 1990

Exapion Bedel 1997, 2 spp., California, Oregon and Washington. Adventive; introduced for biological control of gorse, *Ulex europaeus* L. and scotch broom, *Cytisus scoparius* (L.) (Fabaceae) (O'Brien 1995).
Ulapion Ehret 1997 (valid subgenus)

Ixapiini Alonso-Zarazaga 1990

Neapion Alonso-Zarazaga 1990, 6 spp., generally distributed in eastern North America west to southern Texas, north into extreme southern Canada. Subgenus *Neotropion*, 1 sp., *N. xanthoxyli* (Fall), associated with Rutaceae (genus *Zanthoxylum*); subgenus *Neapion*, 5 spp., associated with Caprifoliaceae (genus *Viburnum*).

Xixias Kissinger 1990

Neotropion Alonso-Zarazaga 1990 (valid subgenus)

Malvapiini Alonso-Zarazaga 1990

Rhopalapion Schilsky 1906, 1 sp., *R. longirostre* (Olivier 1807), generally distributed. Adventive on *Althea rosea* Cav. (hollyhock; Malvaceae).

Oxystomatini Alonso-Zarazaga 1990

Oxystomatina Alonso-Zarazaga 1990

Eutrichapion Reitter 1916, 3 spp., generally distributed. Associated with Fabaceae.

Cnemapion Bokor

Leconteapion Alonso-Zarazaga 1990 (valid subgenus)

Phalcrolobus Alonso-Zarazaga 1990 (valid subgenus)

Psilocalymma Alonso-Zarazaga 1990 (valid subgenus)

Mesotrichapion Györffy 1956, 1 sp., *M. cyanitinctum* (Fall 1927), Alaska to northern Quebec, south to southern Manitoba. Association with Fabaceae (genus *Astragalus*).

Trichapiina Alonso-Zarazaga 1990

Betulapion Ehret 1994, 1 sp., *B. simile* (Kirby 1811), generally distributed. Adventive on *Betula papyrifera* Marsh. (paper birch; Salicaceae); larvae develop in flowers.

Kissingeria Alonso-Zarazaga 1990, 6 spp., generally distributed. Associated with Fabaceae.

Trichapion Wagner 1912, 45 spp., generally distributed. Associated mostly with Fabaceae and Asteraceae.

Synapiina Alonso-Zarazaga 1990

Ischnopterapion Bokor 1923, 1 sp., *I. virens* (Herbst 1797), New York, Pennsylvania, Maryland, New Jersey, Connecticut, Delaware and Virginia. Adventive on various species of clover, *Trifolium* (Fabaceae) (Hoebeke *et al.* 2000).

Piezotrachelini Voss 1959

Chrysapion Kissinger 1968, 2 spp., Arizona, California and Texas.

Fallapion Kissinger 1968, 30 spp., generally distributed. Associated mostly with Asteraceae and Umbelliferae.

Incertae sedis

Apionion Kissinger 1998, 2 spp., *A. crassum* (Fall 1898), eastern United States; *A. dilatatum* (Smith 1884), Arizona. Associated with Fabaceae.

Coelocephalapion Wagner 1914, 22 spp., generally distributed. Associated mostly with Fabaceae and Asteraceae.

Sayapion Kissinger 1999, 5 spp., eastern United States, Texas and Arizona. One species has been associated with Fabaceae.

Nanophyinae Gistel 1856

Nanophyini Gistel 1856

Microon Alonso-Zarazaga 1989, 1 sp., *M. canadense* (Brown 1944), western United States south to Arizona and western Canada east to Manitoba. At least one Palearctic species is associated with the genus *Lythrum* (Lythraceae).

Nanodactylus Blatchely 1922, 1 sp., *N. obesulus* Blatchley 1922, Illinois, Indiana, Texas.

Nanophyes Schoenherr 1838, 1 sp., *N. marmoratus* (Goeze 1777), Manitoba, New York; introduced into Canada in 1997 for the biological control of *Lythrum salicaria* L. (Lythraceae) (Harris 2001). This weevil species is also approved for introduction into the United States and is established in the vicinity of Ithaca, New York. A second species, *N. brevis* Boheman 1845, also is under consideration for introduction.

Pseudotychius Blatchley 1922, 1 sp., *P. watsoni* Blatchley 1922, eastern United States, north into Ontario.

Zeugonyx Notman 1922, 1 sp., *Z. sabinae* Notman 1922, Texas. This species was reared from galls on twigs of *Juniperus ashei* Buchh. (Cupressaceae).

Bibliography

ALONSO-ZARAZAGA, M. A. 1989. Revision of the supraspecific taxa in the Palaearctic Apionidae Schoenherr, 1823. 1. Introduction and subfamily Nanophyinae Seidlitz, 1891 (Coleoptera, Curculionoidea). Fragmenta Entomologica, 21: 205-262.

ALONSO-ZARAZAGA, M. A. 1990. Revision of the supraspecific taxa in the Palaearctic Apionidae Schoenherr, 1823 (Coleoptera, Curculionoidea). 2. Subfamily Apioninae Schoenherr, 1823: Introduction, keys and descriptions. Graellsia, 46: 19-156.

ALONSO-ZARAZAGA, M. A. and C. H. C. LYAL. 1999. A world catalogue of families and genera of Curculionoidea (Insecta: Coleoptera) (Excepting Scolytidae and Platypodidae). Entomopraxis. Barcelona, Spain.

BRIGHT, D. E. 1993. The Insects and Arachnids of Canada. Part 21. The weevils of Canada and Alaska: Volume 1. Coleoptera: Curculionoidea, excluding Scolytidae and Curculionidae. Publication 1882. Research Branch, Agriculture Canada. Ottawa, Canada.

BUCHANAN, W. D. 1960. Biology of the oak timberworm, *Arrhenodes minutus.* Journal of Economic Entomology, 53: 510-513.

HARRIS, P. 2001. *Nanophyes marmoratus* (Goeze). Flower-feeding weevil. http://res2.agr.ca/lethbridge/weedbio/agents/ananmar.htm

HARRIS, P. and A. MCCLAY. 2001. *Omphalapion hookeri* Kirby. Seed-head weevil. http://res2.agr.ca/lethbridge/weedbio/agents/aomphook.htm

HOEBEKE, E. R., R. A. BYERS, M. A. ALONSO-ZARAZAGA, and J. F. STIMMEL. 2000. *Ischnopterapion (Chlorapion) virens* (Herbst) (Coleoptera: Curculionoidea: Brentidae: Apioninae), a Palearctic clover pest new to North America: recognition features, distribution, and bionomics. Proceedings of the Entomological Society of Washington, 102: 151-161.

KISSINGER, D. G. 1968. Curculionidae subfamily Apioninae of North and Central America with reviews of the world genera of Apioninae and world subgenera of *Apion* Herbst (Coleoptera). Taxonomic Publications. South Lancaster, Massachusetts.

KISSINGER, D. G. 1989. Apionidae from North and Central America. Part 1. Notes on the classification of the *Apion* subgenus *Trichapion* Wagner with description of new species from the United States of America (Coleoptera). Insecta Mundi, 3: 271-227.

KISSINGER, D. G. 1990. Apionidae from North and Central America. Part 2. Description of a new subgenus and two new species of *Apion* from Mexico. (Coleoptera). Insecta Mundi, 4: 33-40.

KISSINGER, D. G. 1992. Apionidae from North and Central America. Part 4. Generic classification and introduction to the genus *Coelocephalapion* Wagner, with new species from Mexico and Venezuela (Coleoptera). Insecta Mundi, 6: 65-77.

KISSINGER, D. G. 1998. Apionidae from North and Central America. Part 5. Description of the genus *Apionion* and 4 new species (Coleoptera). Insecta Mundi, 12: 93-102.

KISSINGER, D. G. 1999. Description of a new genus, *Sayapion*, from North and Central America (Coleoptera: Apionidae). Insecta Mundi, 13: 72.

LAWRENCE, J.L. 1982. Coleoptera, Pp. 482-553. *In*: S. P. Parker, ed. Synopsis and Classification of Living Organisms. Volume 2. McGraw Hill. New York.

LAWRENCE. J. F. and A. F. NEWTON, Jr. 1995. Families and subfamilies of Coleoptera (with selected genera, notes, references and data on family-group names). Pp 779-1006. *In*: J. Pakaluk and S. A. Slipinski, eds., Biology, phylogeny and classification of Coleoptera: Papers celebrating the 80th birthday of Roy A. Crowson. Muzeum i Instytut Zoologii PAN, Warsaw.

O'BRIEN, C. W. 1995. Curculionidae, premiere biological control agents. Memoirs of the Entomological Society of Washington, 14: 129-136.

O'BRIEN, C. W. and G. J. WIBMER. 1982. Annotated checklist of the weevils (Curculionidae *sensu lato*) of North America, Central America, and the West Indies (Coleoptera: Curculionoidea). Memoirs of the American Entomological Institute, 34: i-ix, 1-382.

PESCHKEN, D. P., K. C. SAWCHYN and D. E. BRIGHT. 1993. First record of *Apion hookeri* Kirby (Coleoptera: Curculionidae) in North America. Canadian Entomologist, 125: 629-631.

SANBORNE, M. 1983. Some observations on the behaviour of *Arrhenodes minutus* (Drury) (Coleoptera: Brentidae). Coleopterists Bulletin, 37: 106-113.

THOMAS, M. C. 1996. The primitive weevils of Florida (Coleoptera: Brentidae: Brentinae). Florida Department of Agriculture. Division of Plant Industry. Entomology Circular No. 375. 3 pp.

TUTTLE, D. M. 1954. Notes on the bionomics of six species of *Apion* (Curculionidae, Coleoptera). Annals of the Entomological Society of America, 47: 301-307.

WARNER, R. E. 1960. The genus *Stereodermus* new to America north of Mexico with a revised key to the genera of Brentidae. Coleopterists Bulletin, 14: 29.

130. ITHYCERIDAE Schönherr 1823

by Robert S. Anderson

Family common name: The New York weevil

Among the primitive weevils with straight, non-geniculate antennae, this enigmatic family contains only one species, whose adults can be recognized by their large size (12-18 mm long), stout form, and distinct pubescence.

FIGURE 1.130. *Ithycerus noveboracensis* (Forster) (from Bright 1993, reproduced with the permission of the Minister of Public Works and Government Services, 2001)

Description (based on Lawrence 1982): Body large, stout. Pubescence of scale-like bristles, distinct.

Head with single gular suture; pregular sutures absent. Rostrum broad, stout, not sexually dimorphic; antennal insertions lateral. Maxilla without lacinia; maxillary palp of three articles, rigid, partially retracted into large palpifer; labial palp of three articles; labrum absent; mandible stout. Antennae straight, moniliform; antennal club compact, of three articles. Proventriculus without sclerotized plates.

Notosternal suture complete. Procoxae contiguous, mesocoxae narrowly separated and metacoxae moderately widely separated. Hind wing with four anal veins. Legs with trochanter short; triangular; femur attached to side of trochanter. Tarsal claws toothed. Abdomen with ventrites 1 and 2 fused together but separated by distinct suture. Cap piece of tegmen strongly bilobed; the median lobe with a broad ventral plate and a narrow dorsal plate.

Egg (based on Sanborne 1981) with length 1.2-2.2 mm. Subspherical, longer than wide. External surface of chorion of hexagonal facets with impressed borders. Surface strongly punctate, with numerous aeropyles.

Larva (based in part on Sanborne 1981, Lawrence 1991) when mature about 20-25 mm in length. Body relatively short and broad, subcylindrical, strongly c-shaped. Body yellowish, with heavily sclerotized mouth frame and mandibles. Vestiture of moderately long, scattered setae. Head protracted and hypognathous, about as long as wide. Frontal sutures complete, reaching articulating membrane of mandible. Clypeus distinguishable from frons and completely separated from labrum. Three pairs of stemmata on each side. Labrum free, bearing one sensillum and four pairs of setae. Antenna of a single dome-like article bearing a sensorium, a bifurcate appendage and several setae. Mandible with two apical teeth, and three blunt teeth on inner edge, accessory ventral process and mola absent. Hypopharyngeal bracon present. Maxilla with palp with 2 articles, palpiger present. Labial palp of two articles; palps widely separated. Premental sclerite subtriangular. Thorax with pronotal sclerite moderately pigmented. Legs very small, widely separated, subconical, two or three jointed. Abdomen with first four segments with three dorsal folds, segments 5-8 with two dorsal folds. Spiracles annular.

Pupa (based on Sanborne 1981) with distinct labrum present; mandibles with one pair of short setae; setosity extensive.

Habits and habitats. Adult *I. noveboracensis* are associated with various species of Fagaceae, Betulacae and Juglandaceae. Adults appear to prefer white oak (*Quercus alba* L.) and American beech (*Fagus grandifolia* Ehrh.). They feed mainly on the bark of shoots, leaf petioles, leaf buds and acorn buds. Eggs are laid in the ground and larvae feed on the roots of the same host plants. Pupation takes place in the soil. A detailed study of the biology has been published by Sanborne (1981).

Status of the classification. Once enigmatic in its placement there now seems to be a consensus that this species belongs in a separate and distinct family (Lawrence 1982, Thompson 1992, Alonso-Zarazaga and Lyal 1999). Kuschel (1995) however, placed it in Curculionidae. Relationships to other Curculionoidea are still disputed.

Distribution. This family contains only the species *Ithycerus noveboracensis* (Forster). It is found throughout eastern North America.

Classification of the Nearctic Genera

Ithyceridae Schönherr 1823

Ithycerus Schönherr 1823, 1 sp., *I. noveboracensis* (Forster 1771), eastern United States into extreme southern Canada. Associated with various species of Fagaceae, Betulacae and Juglandaceae.

Pachyrhynchus Kirby 1837

Bibliography

ALONSO-ZARAZAGA, M. A. and C. H. C. LYAL. 1999. A world catalogue of families and genera of Curculionoidea (Insecta: Coleoptera) (Excepting Scolytidae and Platypodidae). Entomopraxis. Barcelona, Spain.

KUSCHEL, G. 1995. A phylogenetic classification of Curculionoidea to families and subfamilies. Memoirs of the Entomological Society of Washington, 14: 5-33.

LAWRENCE, J.L. 1982. Coleoptera. Pp. 482-553. *In:* S. P. Parker, ed. Synopsis and Classification of Living Organisms. Volume 2. McGraw Hill. New York.

LAWRENCE, J. L. 1991. Ithyceridae (Curculionoidea). Pp. 590-591. *In:* F. W. Stehr, ed. Immature Insects. Volume 2. Kendall/Hunt. Dubuque, Iowa.

SANBORNE, M. 1981. Biology of *Ithycerus noveboracensis* (Forster) (Coleoptera) and weevil phylogeny. Evolutionary Monographs, 4: 1-80.

THOMPSON, R.T. 1992. Observations on the morphology and classification of weevils (Coleoptera, Curculionoidea) with a key to major groups. Journal of Natural History, 26: 835-891.

131. CURCULIONIDAE Latreille 1802

by Robert S. Anderson

Family common name: The weevils or snout beetles

Weevils are one of the most diverse groups of organisms. Over 60,000 species have been described world wide and their diversity in North America is challenged among beetles perhaps only by Staphylinidae. Weevils are associated with virtually all kinds of plants and plant parts. Most feed on living plants but some are saprophagous. Weevils are immediately recognizable by their elongate rostrum (or snout), with mouthparts situated at the apex, geniculate antennae and compact antennal club. Some weevils in the subfamilies Entiminae, Cossoninae and Scolytinae have the rostrum reduced in form and not markedly produced anteriorly. Traditional considerations of the weevils do not include Scolytinae and Platypodinae but increasing evidence suggests these beetles are derived from within Curculionidae.

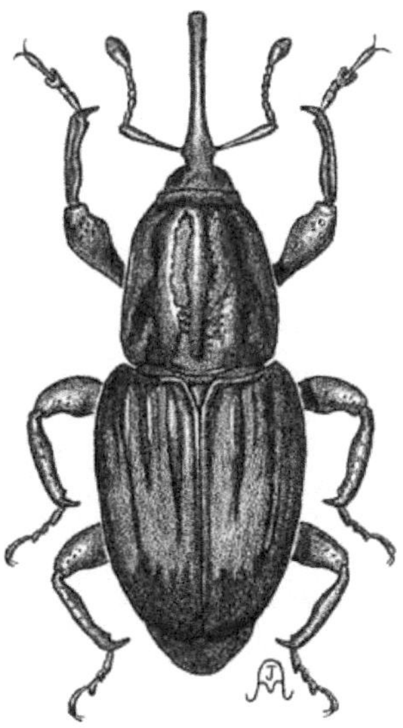

FIGURE 1.131. *Sphenophorus pertinax* (Olivier)

Description (based on Lawrence 1982): Shape very variable, broadly oval to elongate, slightly flattened to markedly convex, most covered with recumbent or appressed vestiture of scales, some with metallic sheen or forming contrasting patterns, some subglabrous or with erect or suberect hairs only; length from 1-40 mm (most 2-20 mm); color variable, typically black or dark brown, more rarely of other colors.

Eyes present, may be reduced or absent. Rostrum varying from very short and indistinct to very long and narrow; most sexually dimorphic with female rostrum longer, finer and with position of antennal insertion more basal. Antennae geniculate (very few exceptions where scape is very short and position of antennal insertion on rostrum is basal); club of three articles (sometimes with one), compact, in some the apical articles recessed in glabrous basal article; funicle of 5-7 articles, slender; point of antennal insertion on rostrum is various, mostly between midlength and apex, mostly lateral but in some, dorsal. Mandibles of some bearing a scar at apex or deciduous process. Maxillae in some concealed by expanded mentum, a few with distinct galea and lacinia. Labial palpi of one or two articles, rarely absent; in some weevils palpi inserted in cavities on the ventral surface of the prementum. Proventriculus of some lacking sclerotized plates. Front coxae contiguous or separated, middle and hind coxae variable. Tarsi of 5 articles but article 4 very small and hidden between lobes of article 3 (exception, Raymondionyminae with only 4 articles); tarsal claws of some connate and simple or with a basal process or tooth. Abdomen with first two ventrites connate, very rarely free. Pygydium formed by tergite VII or VIII, in most concealed beneath apex of elytra, exposed and/or sulcate in some. Cap piece of tegmen may be reduced and may or may not be bilobed, occasionally absent; aedeagus with a trough-like ventral plate and membranous dorsally; in some aedeagus with separate pedon and tectum.

Larvae (based on Lawrence 1982) subcylindrical, slightly curved; lightly sclerotized and grublike; usually with very fine hairs. Head hypognathous and free, rarely retracted into prothorax. Frontal arms "v-shaped" and not reaching mandibular articulations, endocarina usually present. Stemmata absent in most. Antennae of 1 or 2 articles and apical article sometimes a conical sensorium. Frontoclypeal suture present. Labrum free, usually with 4 pairs of setae. Maxillae with galea and lacinia fused to form mala, maxillary palpi usually of 2 articles. Labial palpi of 1 or rarely, and indistinctly, of 2 articles. Abdominal tergites usually with 3 or 4 transverse plicae. Thoracic spiracles found on the prothorax or between prothorax and mesothorax. Legs absent.

Habits and habitats. The habits and habitats of Scolytinae, long treated as a separate family, are summarized under that subfamily heading.

Weevils can be found associated with just about any kind of plant in any terrestrial or freshwater habitat. Most species are strictly phytophagous as adults and larvae and usually have a narrow range of suitable host plants. Most species are associated with angiosperms but a few are associated with gymnosperms, mainly the various conifers in the Pinaceae. Adult and larval feeding habits vary extensively but can loosely be classified into two groups: one in which both adults and larvae are polyphagous

Acknowledgments: This chapter includes major contributions by Robert J. Rabaglia (Scolytinae), Henry A. Hespenheide (Conoderinae), Boris A. Korotyaev (Ceutorhynchinae), and Anne T. Howden (Entiminae), to whom I am deeply grateful. I also thank the many people who contributed to this chapter in other ways, by providing specimens, literature or advice, by checking text or keys, or by answering my many questions. These people are Charles W. O'Brien, Horace R. Burke, Boris A. Korotyaev, Miguel Alonso-Zarazaga, Anne T. Howden, Donald E. Bright, Chris Lyal, Enzo Colonnelli, and Steve Lingafelter. Line illustrations are by Nadine Dupérré of Laval, Quebec. Paul Skelley and Mike Thomas provided the needed push to get it done.

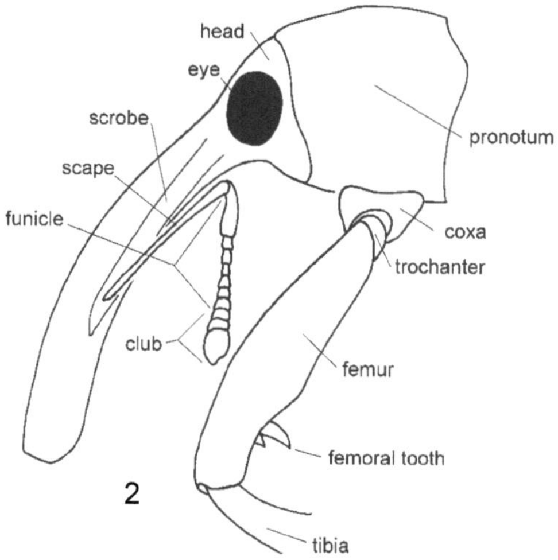

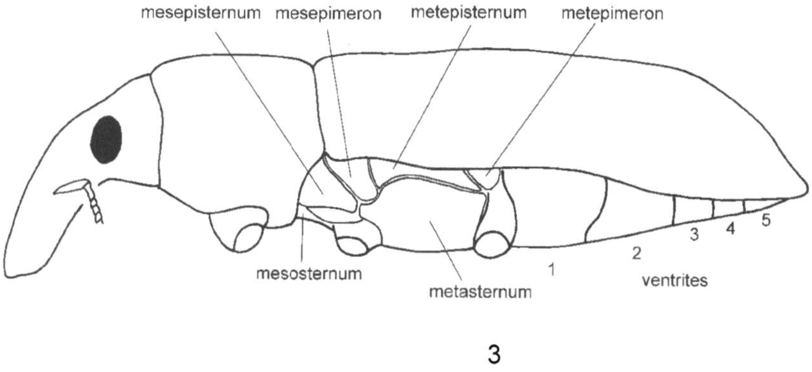

FIGURES 2.131-3.131. 2. Lateral view of a generalized curculionid head; schematic; 3. Lateral habitus of a generalized curculionid; schematic (both after Kissinger 1964)

(Entiminae), and one in which adults and larvae have a more restricted range of host plants (other subfamilies). Among the polyphagous species, the larvae feed externally in the soil on roots whereas the adults feed generally on foliage. Species with more restricted ranges of hosts usually feed little as adults (often visiting flowers) or feed on foliage or reproductive structures, and their larvae feed internally in the stems, roots, leaves or reproductive structures of a few congeric or confamilial plant taxa. Some weevil larvae in the Hyperinae and Ceutorhynchinae feed externally on foliage and reproductive structures. Pupation usually takes place in the host plant or in the soil but species of *Hypera* and *Cionus* construct a loosely woven cocoon that is attached to the host plant.

Adults of some weevil species (Raymondionyminae and Molytinae) have reduced eyes or are eyeless and live in the soil or leaf litter. Some weevil species in the Conoderinae, Cossoninae, Cryptorhynchinae and Molytinae feed in dead plant material, usually wood. Some species appear to live in association with ants, although this appears an obligate relationship only for *Liometophilus* (Cryptorhynchinae). Some species of Entiminae are parthenogenetic. Most Entiminae as well as some Cryptorhynchinae and Molytinae are flightless.

Curculionidae are a very important group economically. Some species are serious pests of ornamental, agricultural and forestry plants and have well-known common names (e.g., boll weevil, white pine weevil, strawberry root weevil, black vine weevil, etc.). Recently, species have become increasingly used in the biological control of introduced pest plants (e.g., *Neochetina*, *Hylobius*, *Cyphocleonus*, *Eustenopus*, etc.) particularly in western North American grasslands and southeastern aquatic habitats (O'Brien 1995). An excellent review of the biology of Anthonomini is by Burke (1976).

Some subfamilies as Erirhininae, Bagoinae, Cyclominae and Ceutorhynchinae have a number of genera and species associated with freshwater macrophytes. Many of these species are very good swimmers (Morris 1995) and adults spend most of their time in or near water. Most of these taxa are best found at night when adults come up onto the plants to feed. A few weevils are found in intertidal situations (e.g., many Cossoninae, *Emphyastes*, and *Thalasselephas*) where they develop in driftwood or seaweed. There are many weevils in arid habitats such as deserts (Entiminae) and grasslands (Baridinae and Ceutorhynchinae), likely because of their associations with the plants that dominate those habitats. A great number of weevils that have immigrated to North America from Europe are likely associated with imported ornamental plants or amongst ballast brought by ships at the turn of the 19th century.

Various groups of weevils are also common as Quaternary fossils in northern North America and are important in reconstructing the late Cenozoic history of northern habitats (Matthews 1982).

Obviously a more complete summary of the natural history of Curculionidae is beyond the scope of these few notes.

Status of the classification. The classification of the weevils was regarded by Crowson in 1955 as the last great problem to be clarified within the Coleoptera. While there have been many advances in the classification, much still remains to be resolved. The classification used herein largely is that of Alonso-Zarazaga and Lyal (1999) with a few changes in placement and ranking of certain taxa. A total of 18 subfamilies are recognized. Lawrence and Newton (1995), the classification at the family level adopted for this book, recognize only 6 subfamilies within Curculionidae, demoting many subfamilies to tribes within their Curculioninae. They also refer to the Entiminae as Brachycerinae although the constitution remains basically the same. They recognize Dryophthorinae as a separate family but not Raymondionyminae and Erirhininae, all three of which are recognized as families in the classifications of Thompson (1992) and Alonso-Zarazaga and Lyal (1999). These authors consider these as having family level status because they do not share the same derived male genitalic structure as the Curculionidae *sensu stricto*. Herein all are considered subfamilies within Curculionidae. The classification of Kuschel (1995) is very similar to that of Lawrence and Newton (1995) but includes Ithyceridae within the subfamily Brachycerinae of Curculionidae.

Catalogs are available for some groups of Curculionidae in North America (*e.g.*, Howden 1993; O'Brien 1986, 1989, 1996, 1997) and an annotated checklist (and supplements) with full synonyms, information about keys, and distributions has been published (O'Brien and Wibmer 1982, 1984; Wibmer and O'Brien 1989). A review of the state of knowledge about immatures is by Burke and Anderson (1976). Excellent (but outdated) regional works to the species level are those of Hatch (1971) for the Pacific Northwest and Downie and Arnett (1996) for northeastern North America. Blatchley and Leng (1916) remains an old standard. Many of the keys used herein are modified from Kissinger (1964).

Distribution. Curculionids are found just about everywhere in North America. Diversity is greatest in the southern United States but no recent regional counts are available. The last tabulation for the Nearctic Region as a whole was in 1978 by O'Brien and Wibmer who counted 239 genera and 2388 species. Bousquet (1991) recorded almost 700 species in Canada and Alaska. Anderson (1993a) counted 249 species in 115 genera in extreme southern Florida alone. Many recent additions to the fauna are the result of deliberate introductions for biological control purposes but also, a number of taxa recently added to the North American fauna are from extreme southern Florida or Texas and are recent discoveries. The species *Isochnus arcticus* (Korotyaev 1976) is found as far north as Ellesmere Island at almost 82 degrees north latitude.

Some weevil species are routinely intercepted at ports of entry of foreign materials (especially agricultural products) into the United States and Canada. Some of these taxa have traditionally or occasionally been considered as part of the North American fauna and included in keys and faunal lists. At present, there is no evidence to suggest they are established in North America and they are not included in the key. These genera are: *Diocalandra* Faust 1894; *Dynatopechus* Marshall 1931; *Sternochetus* Pierce 1917; *Liophloeus* Germar 1817; *Euophryum* Broun 1909.

Terminology. In general, standard terms for beetle anatomy are used in the keys and text (see Figs. 2, 3). Generally known and readily visible characters are used where possible but in some instances specialized characters requiring high magnification or dissections are required. Simply put, some weevil groups are difficult to identify. Measurements of body length are taken from the anterior margin of the eyes to the apex of the elytra; the snout is not included. On the elytra, intervals are numbered with the sutural interval being interval 1. Tarsal articles are numbered from 1 through 5, with 5 being the terminal or apical article bearing the claws; article 4 is very small and recessed between the lobes of article 3. I use the term ventrite to apply to the visible abdominal sternites and they are numbered from 1 through 5, the latter being terminal.

In older literature the terms 'uncus' and 'mucro' are used to describe the structure of the apical tooth on the hind tibia. Following Thompson (1992), I have chosen not to use these terms as comparative study shows them to refer to the same structure, the different names being used for different degrees of development and positioning of the apical tooth. Associated with this is the use of the term 'apical comb of setae' which I use to apply to the row of setae that may be across the apex of the hind tibia or in some weevils is displaced by a change in position of the apical tooth to be oriented longitudinally to the main axis of the tibia. We do use 'corbel' and related terms in the keys to Entiminae, contrary to the recommendations of Thompson (1992). See Thompson (1992) for details.

Classification of the Nearctic Subfamilies and Tribes

Curculionidae Latreille 1802
- I. Dryophthorinae
  - 1. Dryophthorini
  - 2. Orthognathini
  - 3. Rhynchophorini
- II. Erirhininae
  - 4. Erirhinini
- III. Raymondionyminae
  - 5. Raymondionymini
- IV. Curculioninae
  - 6. Curculionini
  - 7. Acalyptini
  - 8. Anthonomini
  - 9. Cionini
  - 10. Derelomini
  - 11. Ellescini
  - 12. Mecinini
  - 13. Otidocephalini
  - 14. Rhamphini
  - 15. Smicronychini
  - 16. Storeini
  - 17. Tychiini
- V. Bagoinae
- VI. Baridinae
  - 18. Baridini
  - 19. Madarini
  - 20. Madopterini
  - 21. Nertinini
- VII. Ceutorhynchinae
  - 22. Ceutorhynchini
  - 23. Cnemogonini
  - 24. Hypurini
  - 25. Mononychini
  - 26. Phytobiini
  - 27. Scleropterini
- VIII. Conoderinae
  - 28. Lechriopini
  - 29. Zygopini
  - 30. Tachygonini
- IX. Cossoninae
  - 31. Cossonini
  - 32. Acamptini
  - 33. Dryotribini

34. Onycholipini
35. Pentarthrini
36. Proecini
37. Rhyncolini

X. Cryptorhynchinae
38. Cryptorhynchini
39. Gasterocercini

XI. Cyclominae
40. Rhythirrinini

XII. Entiminae
41. Agraphini
42. Alophini
43. Anypotactini
44. Brachyderini
45. Cneorhinini
46. Cyphicerini
47. Eudiagogini
48. Eustylini
49. Geonemini
50. Hormorini
51. Naupactini
52. Omiini
53. Ophryastini
54. Otiorhynchini
55. Peritelini
56. Phyllobiini
57. Polydrusini
58. Sciaphilini
59. Sitonini
60. Tanymecini
61. Thecesternini
62. Trachyphloeini
63. Tropiphorini

XIII. Hyperinae
64. Hyperini

XIV. Lixinae
65. Lixini
66. Cleonini
67. Rhinocyllini

XV. Mesoptiliinae
68. Laemosaccini
69. Magdalidini

XVI. Molytinae
70. Molytini
71. Trachodini
72. Anchonini
73. Camarotini
74. Cholini
75. Cleogonini
76. Conotrachelini
77. Cycloterini
78. Erodiscini
79. Hylobiini
80. Lepyrini
81. Lymantini
82. Petalochilini
83. Piazorhinini
84. Pissodini
85. Sternechini
86. Thalasselephantini
87. Trypetidini

XVII. Scolytinae
88. Hylesinini
89. Scolytini

XVIII. Platypodinae
90. Platypodini

Key to the Nearctic Subfamilies of Curculionidae

1. Pregular sutures present; pregular sclerite distinct, located between median gular suture and labial articulation; head with rostrum virtually absent; at least one pair of tibiae with denticles or stout socketed setae along the dorsal (outer) margin 2

— Pregular sutures absent; pregular sclerite not evident; head with rostrum variable from very long and cylindrical to short and broad, or (rarely) nearly absent; tibiae lacking denticles or stout socketed setae along the dorsal (outer) margin 3

2(1). Tarsus with article 1 as long as articles 2-5 combined; head as wide as pronotum; pronotum usually with lateral constriction near middle; antennal club without sutures; lateral denticles on front tibia not socketed .... XVIII. Platypodinae (p. 805)

— Tarsus with article 1 not longer than articles 2 or 3; head narrower than pronotum, often concealed by pronotum when viewed dorsally; pronotum not constricted laterally; antennal club with sutures; lateral denticles on front tibia socketed or (rarely) not .... XVII. Scolytinae (p. 792)

3(1). Tarsus of 4 subequal articles (Fig. 12); eyes absent (Fig. 11); body size small (<5mm); body color generally pale orange-red or pale brown; tibia at inner apical angle with small tooth much shorter than a tarsal claw .... III. Raymondionyminae (p. 732)

— Tarsus of 5 articles, but with article 4 very small and difficult to see between lobes of article 3 (Fig. 88); eyes absent or present, well-developed, or reduced in size and represented by only from 1 to a few facets (Fig. 51); body size various; body color various; tibia at apex various but if eyes are lacking or almost so, tibia with large tooth arising from outer apical angle 4

4(3).[1] Tarsus with claws widely separated by dermal lobes extended between them from both dorsal and ventral surfaces at apex of article 5; mouthparts with prementum withdrawn into oral cavity, palpi mostly or entirely concealed; antenna inserted near base of rostrum, with scape long, projected some distance beyond the hind margin of the eye and not fitting into antennal scrobe (Fig. 5) (exceptions; *Dryophthorus* [Fig. 4], *Orthognathus* [Fig. 7], *Yuccaborus* [Fig. 6] have a more distal insertion of the antennae, possess a scrobe and the scape does not pass, or only slightly passes, beyond hind margin of eye); antenna with club of two basic parts, with basal glabrous and glossy portion, and apical uniformly pilose portion (Figs. 4-7); funicle with 4, 5 or 6 articles; body surface lacking broad flat scales; pygydium formed of tergite 7 in male .... I. Dryophthorinae (p. 728)

— Tarsus with claws single, connate at base or separate, but with dorsal and ventral surfaces at apex of article 5 not extended between bases of tarsal claws; mouthparts with prementum visible, not withdrawn, palpi mostly visible; antenna inserted variously along length of rostrum, usually some distance from base, with scape short or long, and fitting into antennal scrobe, but at most only slightly projected beyond the hind margin of the eye (Figs. 8, 13, 27, 70); antenna with club various, but mostly with three articles, each pilose to some extent, basal article not or rarely glossy, subequal in length to other articles or rarely variously longer than other 2 articles combined, sutures evident between all articles; funicle with 5, 6 or 7 articles; body surface mostly with some broad flat scales or fine hair-like scales; pygydium formed of tergite 8 in male 5

5(4). Male with aedeagus with tectum and pedon separate (dissection necessary), tegmen as long as or longer than aedeagus (including the apodemes); species associated with freshwater aquatic habitats, many with dense varnish-like coating over scales or with dense hydrofuge scales............ .... II. Erirhininae (p. 730)

— Male with aedeagus with tectum and pedon fused (dissection necessary), tegmen shorter than aedeagus (including the apodemes); species associated with various habitats, most with scales present, various in density, but lacking varnish-like coating (exception; *Bagous* recognized by presence of prosternal channel) or with scales lacking entirely 6

6(5).[2] Legs with well-developed, usually large hook-like tooth at apex of front, middle and hind tibiae: tooth arising from one of, a) outer apical angle (Fig. 57), b) from middle of apical margin (Fig. 93), or c) at inner apical angle, but if at inner apical angle, tooth on hind tibia is more or less as long as or longer than tarsal claw (Fig. 89) and outer curved face of tooth is continuous with apex of outer tibial margin or is connected to it by a distinct, continuous sharp carina which traverses the apical face of the tibia; apical comb of setae present or absent, if present, oriented either transversely, obliquely or subparallel to the length of the tibia 7

— Legs with apex of front, middle and hind tibiae with tooth, if present, small to moderately large (usually larger on front or middle tibiae), usually smaller than tarsal claw, arising from inner apical angle and with outer curved face distinctly separated from, and not continuous with, outer tibial margin or with carina traversing the apical face of the tibia; apical comb of setae oriented transversely to length of tibia 18

7(6). Mesepimeron strongly ascended, truncated by elytral humeri and visible (or nearly so) in dorsal view between pronotum and elytra (Figs. 23-26); tarsus with 1 (rarely) or 2 claws .... VI. Baridinae (most) (p. 740)

— Mesepimeron not ascended, not visible in dorsal view between pronotum and elytra (exception; *Laemosaccus* [Fig. 92] recognized by short, straight rostrum, basal margin of elytra extended over base of pronotum, exposed pygydium, and small, acute tooth on the inner margin of the front femur); tarsus with 2 claws 8

8(7). Rostrum in repose received into ventral channel which may be limited to prosternum or extended beyond into meso- or metasternum (Figs. 21, 59) 9

— Rostrum in repose not received into ventral channel, but may rest between front, middle and/or hind coxae 14

9(8). Eyes large, elongate-oval, subcontiguous (or nearly so) dorsally, frons very narrow (Fig. 45); eyes situated towards top front of head, in lateral view with lower margin of eye clearly situated above level of dorsum of base of rostrum (Fig 46) .... VIII. Conoderinae (p. 754)

— Eyes small to moderate in size, more or less rounded, more widely separated dorsally, frons broad; eyes situated towards sides of head, in lateral view with lower margin of eye situated near or below level of dorsum of base of rostrum 10

10(9). Rostrum very short, not much longer than wide, broad and flat dorsally, subquadrate in form (Fig. 82); dorsal vestiture of pronotum and elytra in part bifid (*Bangasternus*).................................... .... XIV. Lixinae (part) (p. 783)

— Rostrum moderately long, many times longer than wide, elongate and narrow; dorsal vestiture, if present, simple 11

11(10). Ventral channel extended beyond prosternum into meso- or metasternum (Fig. 59) .... X. Cryptorhynchinae (p. 761)

— Ventral channel limited to prosternum (Fig. 21); even though rostrum in repose may overlie meso-, metasternum and some abdominal ventrites) 12

12(11). Hind tibia with outer face at apex lacking apical comb of setae lateral to base of apical tooth (as in Fig. 57); body with distinct and dense suberect or erect broad scales, body of some specimens

[1] In small specimens it may be difficult to see the states of the tarsal claws and the mouthparts. There are only two genera of small-sized Dryophthorinae included here. *Dryophthorus* (Fig. 4) can be recognized by an antennal funicle of 4 articles in combination with the antennal club character, whereas *Sitophilus* may be recognized by the form of the apex of the hind tibia which has a small preapical tooth on the inner margin in addition to the larger hook-like tooth at the inner apical angle (the tibiae appearing "pincer-like"), in combination with the antennal club.

[2] This is often a difficult character to see clearly and to assess. Some groups (*e.g.*, many Baridinae and some Curculioninae) are equivocal and are thus considered in both halves of this couplet. In general, taxa associated with woody plants tend to develop a larger and curved apical tooth whereas those associated with herbaceous plants have a less developed tooth or apical spine, or none at all.

with crustose coating (Acamptini, *Acamptus*) ... . IX. Cossoninae (part) (p. 756)

— Hind tibia with outer face at apex with apical comb of setae lateral to base of apical tooth (Fig. 99); body vestiture various but surface not with crustose coating 13

13(12). Body lacking distinct vestiture, with smooth varnish-like coating over scales; elytra tuberculate or not; legs elongate, slender; commonly associated with aquatic habitats ... V. Bagoinae (p. 740)

— Body with vestiture of appressed scales or suberect or erect hair-like scales, lacking smooth varnish-like coating over scales, or obvious vestiture lacking; elytra tuberculate or not; legs more robust; rarely associated with aquatic habitats .... XVI. Molytinae (p. 786)

14(8). Mouthparts with labial palpi of 3 articles but short, globular, telescoping and appearing composed of 1 article, ventrally situated at apex of large prementum (Fig. 90); female with large paired symbiont sacs attached to vagina near base of gonocoxites; body size mostly medium to large (>5 mm) (exception; *Microlarinus*) .... XIV. Lixinae (p. 783)

— Mouthparts with labial palpi of 3 distinct articles but elongate, not telescoping, dorsally situated at apex of variously sized prementum; female lacking large paired symbiont sacs attached to vagina near base of gonocoxites (dissection necessary); body size mostly small to medium (<10 mm) 15

15(14). One or more of mesepisternum, mesepimeron, metepisternum and metepimeron with vestiture in form of dense plumose (pectinate) hairs (Fig. 15), rarely hairs may be sparse, fine and at most bifid only in anterior portion of metepisternum 16

— Mesepisternum, mesepimeron, metepisternum and metepimeron with vestiture, if present, simple not plumose or bifid 17

16(15). Tooth at apex of tibia, large and hook-like, larger than tarsal claw (Fig. 93); pronotum only slightly narrower than base of elytra in dorsal view (Fig. 91); elytra with basal margin at intervals 2-4 extended anteriorly overlapping base of pronotum (Fig. 91) .... XV. Mesoptilinae (p. 786)

— Tooth at apex of tibia, small, at most subequal in length to tarsal claw; pronotum distinctly narrower than base of elytra in dorsal view (Fig. 14); elytra with basal margin at intervals 2-4 straight, not overlapping base of pronotum (Fig. 14) (Otidocephalini) .... IV. Curculioninae (part) (p. 732)

17(15). Hind tibia with outer face at apex with apical comb of setae lateral to base of apical tooth, oriented either transversely, obliquely or subparallel to the length of the tibia (Figs. 99-101) .... XVI. Molytinae (most) (p. 786)

— Hind tibia with outer face at apex lacking apical comb of setae lateral to base of apical tooth (Fig. 57) .... IX. Cossoninae (most) (p. 756)

18(6). Mandible with prominent scar on outer apical face indicating point of attachment of deciduous process (Fig. 68), or else clothed on outer apical face with many fine scales and/or setae, mandibles generally robust and thick; rostrum short and broad, usually quadrate or subquadrate in form, often expanded laterally towards apex, not different in males and females in length or form (Figs. 70-77) .... XII. Entiminae (most) (p. 766)

— Mandible lacking scar and therefore lacking deciduous process, either glabrous or with a few small setae on outer apical face, mandibles generally less robust, smaller and thinner; rostrum more elongate and cylindrical, usually as long as or longer than pronotum, or (rarely) shorter than pronotum, different in males and females in length and/or form or not (Figs. 13, 17, 19) 19

19(18). Rostrum in repose received into distinct ventral channel in prosternum (rarely into mesosternum) 20

— Rostrum in repose not received into ventral channel, but may rest between front, middle and/or hind coxae 23

20(19). Rostrum very broad, more or less triangular in dorsal view, fitting into large, deep emargination in front of front coxae; emargination limited posteriorly by small, triangular prosternum (Fig. 69) (Thecesternini, *Thecesternus*)........................... .... XII. Entiminae (part) (p. 766)

— Rostrum more elongate and cylindrical in form, the prosternal channel extended behind the front coxae (rarely onto mesosternum) and the rostrum (when in repose) extended between and/or beyond front coxae 21

21(20). Antenna with funicle with 5 articles; prothorax lacking postocular lobes; claws free, simple; dorsum covered with fine, erect hair-like vestiture (Mecinini, *Cleopomiarus*) .... IV. Curculioninae (part) (p. 732)

— Antenna with funicle with 6 or 7 articles; other characters various 22

22(21). Pygydium covered by elytra; rostrum longer than pronotum, straight and slender, abruptly attenuate immediately beyond antennal insertion (Fig. 28); antenna with article 2 of funicle long, more or less one-half length of scape (Madarini, Zygobaridina, *Amercedes*) .... VI. Baridinae (part) (p. 740)

— Pygydium not covered by elytra; rostrum various in length, straight or slightly curved, more or less of uniform width throughout length, not abruptly attenuate (Fig. 34); antenna with article 2 of funicle short, much less than one-half length of scape .... VII. Ceutorhynchinae (part) (p. 747)

23(19). Mesepimeron strongly ascended, truncated by elytral humeri and visible in dorsal view between pronotum and elytra (Figs. 31-33); pygydium not covered by elytra (Figs. 31-33) .... VII. Ceutorhynchinae (part) (p. 747)

— Mesepimeron not ascended, not visible in dorsal view between pronotum and elytra; pygydium mostly covered by elytra 24

24(23). Tarsus with claws separate, each with basal process .... IV. Curculioninae (part) (p. 732)

— Tarsus with claws separate, simple 25

25(24). Eyes rounded, rostrum mostly very elongate, slender and cylindrical in cross section (Figs. 13, 15-19); antenna with scape not or just reaching anterior margin of eye (Figs. 13) .... IV. Curculioninae (part) (p. 732)

— Eyes more or less elongate-oval, rostrum shorter, more robust and subquadrate in cross section (Figs. 64, 81); antenna with scape just reaching or passing anterior margin of eye (Fig. 64) 26

26(25). Pronotum with anterolateral margin with distinct postocular lobe present (Fig. 64) .... XI. Cyclominae (p. 765)

— Pronotum with anterolateral margin straight, simple or postocular lobe at most very slightly developed (Fig. 81) 27

27(26). Vestiture with at least some bifid scales (limited on some specimens to thoracic sterna), if bifid scales appear absent, humeri obviously quadrate; humeri quadrate to subquadrate, rarely rounded, if humeri rounded, bifid scales are distinct on dorsum .... XIII. Hyperinae (p. 782)

— Vestiture simple, lacking bifid scales; humeri rounded.................. XII. Entiminae (part) (p. 766)

Classification of the Nearctic Curculionidae

I. Dryophthorinae Schoenherr 1825

by Robert S. Anderson

This group of weevils is characterized by the form of the antennal club with the basal article glabrous and glossy, the presence of what Zimmerman (1993) called 'dermal lobes' extended between the tarsal claws from both dorsal and ventral surfaces of the apex of tarsal article 5, the antenna (usually) with the scape long and extended far beyond the posterior margin of the eye, and male genitalia with a distinct lateral line dividing the aedeagus into upper (tectum) and lower (pedon) parts. This primitive form of genitalia is shared with Raymondionyminae and Erirhininae and is the basis for some authors removing these three subfamilies from Curculionidae and giving them each separate family status. By removing these three groups, the hypothesis of monophyly of Curculionidae is strengthened based on their unique derived form of genitalia not shared with other Curculionoidea.

Dryophthorinae are a tropical group, and few species occur in North America. Except for the diverse genus *Sphenophorus*, of the North American genera each is represented by but one or a few species. Most dryophthorines are associated with monocots, including Poaceae, Cyperaceae, Liliaceae and Arecaceae. Some species are serious pests of bananas, bromeliads, corn, turfgrass and stored products. Larvae generally mine stems or roots, some in semiaquatic habitats. The odd genus *Dryophthorus* is associated with moist dead wood.

Key to the Nearctic Genera of Dryophthorinae

1. Antenna with funicle of 4 articles (Fig. 4); tarsus with 5 distinct articles; body usually covered with a crusty deposit; size small, less than 4.0 mm in body length .... *Dryophthorus*

— Antenna with funicle of 6 articles (Figs. 5-7); tarsus with 5 articles but with article 4 small and difficult to see at base of article 5; body lacking surface deposit; size various 2

2(1). Pygydium covered by apex of elytra; antenna with scape not reaching anterior margin of eye (Figs. 6-7); metepimeron not visible 3

— Pygydium exposed at apex of elytra; antenna with scape projected at least past anterior margin of eye (Fig. 5); metepimeron visible (obscure in *Sitophilus*) 4

3(2). Front coxae contiguous; hind tibia expanded apically and with broad wide apical bevel; pronotum with postocular lobes; mandible large, lacking teeth on exterior face .... *Orthognathus*

— Front coxae separated by prosternum; hind tibia linear, not expanded apically and with narrow apical bevel; pronotum lacking postocular lobes; mandible small, with 3 teeth on exterior face ... . *Yuccaborus*

4(2). Size small, total body length less than 5 mm; tibiae (especially front) with distinct subapical tooth on inner margin in addition to larger apical tooth... .... *Sitophilus*

— Size moderate to large, total body length greater than 5 mm; tibia with at most a rounded subapical swelling on inner margin in addition to larger apical tooth 5

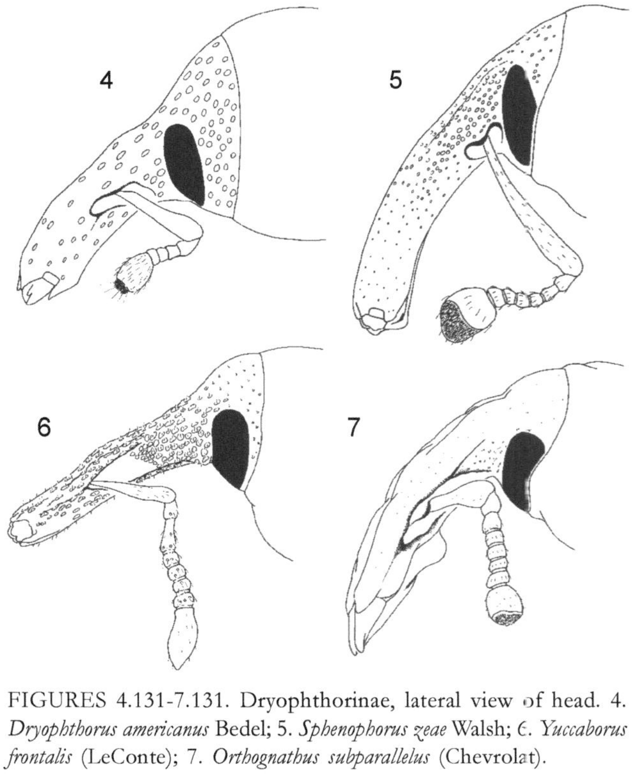

FIGURES 4.131-7.131. Dryophthorinae, lateral view of head. 4. *Dryophthorus americanus* Bedel; 5. *Sphenophorus zeae* Walsh; 6. *Yuccaborus frontalis* (LeConte); 7. *Orthognathus subparallelus* (Chevrolat).

5(4). Metepisternum very broad, length more or less 2 times width; antenna with club transverse, wider than long, lateral margins at base widely divergent, shape sub-triangular; body size very large, total body length greater than 25 mm .... *Rhynchophorus*
— Metepisternum narrow, length 3 or more times width; antenna with club elongate, longer than wide, lateral margins at base sub-parallel to slightly divergent, shape sub-quadrate or sub-oval; body size moderate to large, total body length greater than 5 mm but less than 25 mm 6

6(5). Scutellum (exposed portion) widest at or near middle, shape rhomboidal or sub-circular; more or less as long as wide .... *Cosmopolites*
— Scutellum (exposed portion) widest at or near base, shape triangular or sub-triangular; generally longer than wide 7

7(6). Tarsus with article 3 with ventral pilosity long, confined to apical margin as a continuous fringe, ventral surface otherwise glabrous; antenna with club obliquely truncate at apex with apical pilose part very short, appearing recessed within glabrous part, visible only as a narrow line in lateral view .... *Scyphophorus*
— Tarsus with article 3 with ventral pilosity long or short, uniformly covering 1/3 or more of ventral surface, or with pilosity sparse and confined to anterolateral angle or lateral margins, ventral surface otherwise glabrous; antenna with apex evenly rounded or truncate, with apical pilose part long, distinctly visible as more than a narrow line in lateral view 8

8(7). Tarsus with article 5 ventrally excavated and bilamellate at middle of apex; rostrum hump-like at base, directed posteroventrally; associated with Asteraceae, Asclepiadaceae .... *Rhodobaenus*
— Tarsus with article 5 ventrally evenly rounded at middle of apex; rostrum straight (few) or evenly rounded at base (many), directed anteroventrally; associated with monocotyledons 9

9(8). Tarsus with article 3 with ventral pilosity restricted to anterolateral areas, median area largely glabrous, article 3 narrow, subequal in width to article 2 (many) or broad, wider than article 2 (few) .... *Sphenophorus*
— Tarsus with article 3 with ventral pilosity extensive covering nearly all of ventral surface except near base at middle, article 3 broad, wider than article 2 10

10(9). Front coxae widely separated by width of antennal club; middle coxae widely separated by width of a coxa; prementum toothed ventrally or slightly emarginate at apex; Florida; on Arececeae, Bromeliaceae .... *Metamasius*
— Front coxae narrowly separated by one-half width of antennal club; middle coxae narrowly separated by one-half width of a coxa; prementum broadly sulcate throughout length; Arizona, California; on Cactaceae .... *Cactophagus*

Classification of the Nearctic Dryophthorinae

1. Dryophthorini Schoenherr 1825

Dryophthorus Germar 1824, 1 sp., *D. americanus* Bedel 1885, generally distributed in eastern North America. Adults are found under bark, in association with old rotten logs or in forest litter.
Bulbifer Dejean 1821
Dryophora Berthold 1827
Tetratemnus Wollaston 1873
Tetraspartus Pascoe 1885

2. Orthognathini Lacordaire 1866

Orthognathina Lacordaire 1866

Orthognathus Schoenherr 1838, 1 sp., *O. subparallelus* (Chevrolat 1880), Arizona. Adults have been collected at lights.
Sphenognathus Schoenherr 1840

Rhinostomina Kuschel 1995

Yuccaborus LeConte 1876, 1 sp., *Y. frontalis* (LeConte 1876), generally distributed in southwestern United States. Two subspecies are recognized. Adults and larvae are associated with *Yucca* (Liliaceae); adults come to lights.

3. Rhynchophorini Schoenherr 1833

Rhynchophorina Schoenherr 1833

Rhynchophorus Herbst 1795, 2 spp., *R. palmarum* (Linnaeus 1758) and *R. cruentatus* (Fabricius 1775). Extreme southeastern United States, Texas and California. Adults and larvae are associated with various species of palms (Arecaceae). See Wattanapongsiri (1966) to separate the species. (Volume 1, Color Fig. 14)
Cordyle Thunberg 1797

Litosomina Lacordaire 1866

Sitophilus Schoenherr 1838, 5 spp., generally distributed; adventive. Three species, *S. granarius* (Linnaeus 1758), *S. zeamais* Motschulsky 1855, and *S. oryzae* (Linnaeus 1763) are serious pests of stored grain products. See Kuschel (1961) for a partial key to species.

Sphenophorina Lacordaire 1866

Cactophagus LeConte 1876, 1 sp., *C. spinolae* (Gyllenhal 1838), Arizona and California, adults and larvae are associated with *Carnegiea gigantea* (Engelm.) and other cacti (Cactaceae) (Anderson 1948). *Cactophagus graphipterus* (Champion 1910) has been found in orchid houses in Connecticut, Washington DC, and New Jersey (Barber 1917). It is not known if this species is established there. See Vaurie (1967) to separate the species.

Eucactophagus Champion 1910
Phyllerythrurus Chevrolat 1885

Cosmopolites Chevrolat 1885, 1 sp., *C. sordidus* (Germar 1824), Florida, adventive. This species is associated with banana trees (*Musa sapientum* L.); larvae mine stem and corm (Woodruff 1969).

Metamasius Horn 1873, 3 spp., *M. hemipterus* (Linnaeus 1758) and *M. callizona* (Chevrolat 1883), adventive; *M. mosieri* Barber 1920, native; Florida. *Metamasius hemipterus* is associated with palms, sugar cane, and bananas (Woodruff and Baranowski 1985), whereas, *M. callizona* is a serious pest in *Tillandsia* (O'Brien and Thomas 1990, Frank and Thomas 2000, Larson and Frank 2000); *M. mosieri* is also associated with bromeliads (Larson *et al.* 2001). See Vaurie (1966) to separate the species.
Odontorhynchus Chevrolat 1880
Odontorrhynchus Kirby 1881
Metmasiopsis Champion 1910
Subphyllerythrurus Voss 1954

Rhodobaenus LeConte 1876, 2 spp., *R. tredecimpunctatus* (Illiger 1794) and *R. quinquepunctatus* (Say 1824), generally distributed in United States and southeastern Canada. Species are associated with various Asteraceae and Asclepiadaceae; larvae are in stems (Vaurie 1981). See Vaurie (1981) to separate the species.
Homalostylus Chevrolat 1885

Scyphophorus Schoenherr 1838, 2 spp., *S. acupunctatus* Gyllenhal 1838 and *S. yuccae* Horn 1873, generally distributed in extreme southern United States. Species are associated with *Agave* and *Yucca* (Liliaceae); larvae mine the roots and stems. See Vaurie (1971) to separate the species.

Sphenophorus Schoenherr 1838, 65 spp., generally distributed. Species are associated with various monocots including grasses (Poaceae) and sedges (Cyperaceae) (Vaurie 1951). Some species are pests of turfgrass or corn. See Vaurie (1951) to separate the species. (Volume 2, Color Fig. 30)
Sitonobia Gistel 1856
Merothricus Chevrolat 1885
Trichischius LeConte 1876
Nesorthognathus Voss 1943

Diocalandrina Zimmerman 1993

[*Diocalandra* Faust 1894, 3 spp., intercepted in quarantine; British Columbia, Washington, California and Arizona. Not established in North America.]

II. Erirhininae Schoenherr 1825

by Robert S. Anderson

This group of weevils is unfortunately very difficult to characterize based solely on external characters. Like Dryophthorinae and Raymondionyminae, they possess male genitalia that are primitive in structure with the aedeagus with separate tectum and pedon, and the tegmen as long as or longer than the aedeagus. Most species are associated with aquatic or semi-aquatic habitats and the members of the subtribe Stenopelmina possess a dense, varnish-like coating over the scales or have dense hydrofuge scales. Many species are active swimmers.

Most species mine the stems or other parts of aquatic macrophytes. Species in the genera *Cyrtobagous*, *Neochetina* and *Neohydronomus* have been introduced for biological control of aquatic weeds, mainly in Florida. *Grypus equiseti* (Fabricius 1775) is associated with primitive horsetails of the genus *Equisetum*.

Key to the Nearctic Genera of Erirhininae

1. Antenna with funicle of 6 articles 2
— Antenna with funicle of 7 articles 11

2(1). Tarsus with single claw .... *Brachybamus*
— Tarsus with two claws 3

3(2). Antenna with club with basal article glabrous and glossy and almost as long as rest of club (Fig. 8); tarsus with article 3 not emarginate, usually not wider than article 2 4
— Antenna with club uniformly pubescent (Fig. 9); tarsus with article 3 various 6

4(3). Pronotum with anterolateral margin straight, postocular lobe absent; tarsus with article 5 longer than four other articles combined; dorsal vestiture of only isolated appressed, rounded scales, no obvious varnish-like coating overlying scales.. .... *Cyrtobagous*
— Pronotum with anterolateral margin with well-developed postocular lobe; tarsus with article 5 shorter than four other articles combined; dorsal vestiture of dense appressed scales, with varnish-like coating overlying scales 5

5(4). Rostrum short, stout, nearly straight (Fig. 8); middle tibia flattened, with outer margin evenly curved, and with both inner and outer margins with numerous long, dense, fine hairs..... *Lissorhoptrus*
— Rostrum slender, elongate, evenly curved; middle tibia not flattened, with outer margin more or less

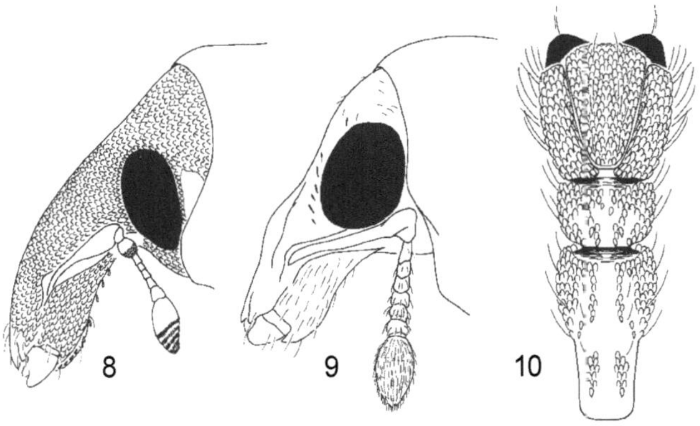

FIGURES 8.131-10.131. Erirhininae. 8-9. Lateral view of head. 8. *Lissorhoptrus oryzophilus* Kuschel; 9. *Stenopelmus rufinasus* Gyllenhal; 10. *Notiodes setosus* (LeConte), tarsus, dorsal view.

straight, and with both inner and outer margins with short, stout scales and at most a few scattered, fine longer hairs .... *Neobagoidus*

6(3). Tarsus with article 3 very broad, apex of article 5 not or very slightly projected beyond lobes of article 3 (Fig. 10) 7

— Tarsus with article 3 emarginate or bilobed, apex of article 5 distinctly projected beyond lobes of article 3 by at least one-half length article 5 9

7(6). Body size less than 1.5 mm; frons about half as wide as rostrum in dorsal view at point of antennal insertion; pronotum lacking postocular lobes .... *Tanysphyrus*

— Body size distinctly greater than 1.5 mm; frons wider than rostrum in dorsal view at point of antennal insertion; pronotum with postocular lobes present, slightly to well developed 8

8(7). Tarsus with article 5 very slightly projected beyond apices of lobes of article 3 .... *Neochetina*

— Tarsus with article 5 not projected beyond apices of lobes of article 3 .... *Notiodes*

9(6). Rostrum very short, subequal in length to scape (Fig. 9); pronotum with anterolateral margin straight, postocular lobe absent (Fig. 9) .... *Stenopelmus*

— Rostrum more elongate, from 1.5 to 2.0 times length of scape; pronotum with anterolateral margin with postocular lobe present, slightly to well developed 10

10(9). Rostrum straight, robust; eyes large, narrowly separated ventrally by less than the width of rostrum; pronotum with anterolateral margin with postocular lobe slightly developed ...... *Neohydronomus*

— Rostrum evenly curved, slender; eyes moderate, separated ventrally by about the width of rostrum; pronotum with anterolateral margin with postocular lobe well developed .... *Onychylis*

11(1). Each tibiae with small spur(s) in addition to small tooth at inner apical angle 12

— Tibiae all lacking spurs 13

12(11). Each tibia with 2 spurs .... *Procas*

— Front tibia with 1 spur, middle and hind tibiae each with 2 spurs .... *Notaris*

13(11). Antenna with funicle with fine pubescence; elytra with stria 10 not margined along last interval; body densely covered with broad scales .... *Grypus*

— Antenna with funicle with distinct setae; elytra with stria 10 finely margined along last interval; body with fine setae or elongate-linear scales .... *Tournotaris*

Classification of the Nearctic Erirhininae

4. Erirhinini Schoenherr 1825

Erirhinina Schoenherr 1825

Grypus Germar 1917, 3 spp., generally distributed in Canada and northern United States, south in West to Colorado. At least one species, *G. equiseti* (Fabricius 1775), is associated with *Equisetum* (Equisetaceae) in wetlands. See Cawthra (1957) to separate the species.

Aplopus Dejean 1821
Grypidius Schoenherr 1826

Notaris Germar 1817, 2 spp., *N. puncticollis* (LeConte 1876) and *N. aethiops* (Fabricius 1792), generally distributed in Canada and northern United States. *Notaris aethiops* is associated with *Sparganium ramosum* Curt. (Sparganiaceae) in Europe and *Typha* (Typhaceae) in wetlands in North America (Anderson 1997). See Buchanan (1927) to separate the species.

Pilumnus Dejean 1821
Erirhinus Schoenherr 1825
Erycus Tournier 1874

Procas Stephens 1831, 1 sp., *P. lecontei* Bedel 1879, Michigan, Ontario, Quebec and Yukon Territory.

Apachiscelus Desbrochers 1875
Notodermus Desbrochers 1875
Pseudypera Voss 1936

Tournotaris Alonso-Zarazaga and Lyal 1999, 2 spp., generally distributed in Canada, Alaska, and northern United States south into Nevada and California. At least one species, *T. bimaculata* (Fabricius 1787), is associated with *Typha* (Typhaceae) in wetlands (Anderson 1997). See Buchanan (1927) to separate some of the species.

Stenopelmina LeConte 1876

Brachybamus Germar 1835, 1 sp., *B. electus* Germar 1835, generally distributed in eastern North America. Adults have been associated with *Eleocharis* (Cyperaceae) in wetlands.

Cyrtobagous Hustache 1929, 1 sp., *C. salviniae* Calder and Sands 1985, Florida. This species has been introduced for biological control of *Salvinia molesta* Mitchell (Salviniaceae) (O'Brien 1995).

Lissorhoptrus LeConte 1876, 6 spp., generally distributed. Species are associated with wetlands; *L. oryzophilus* Kuschel 1952 is a pest of cultivated rice; larvae feed externally on roots (Anderson 1993a). See Kuschel (1952) to separate the species.

Lissocordylus Kuschel 1952

Neobagoidus O'Brien 1990, 1 sp., *N. carlsoni* O'Brien 1990, Florida. This species is associated with *Lachnanthes caroliniana* (Lamarck) Dandy (Haemodoraceae) in wetlands (O'Brien 1990).

Neochetina Hustache 1926, 2 spp., *N. bruchi* Hustache 1926 and *N. eichhorniae* Warner 1970, Florida, Louisiana and Texas. These species have been introduced for control of *Eichhornia crassipes* (Mart.) Solms. (water hyacinth; Pontederiaceae) (O'Brien 1995). See O'Brien (1976) or DeLoach (1975) to separate the species.

Neohydronomus Hustache 1926, 1 sp., *N. affinis* Hustache, Florida. This species has been introduced for control of *Pistia stratiotes* L. (water lettuce; Araceae) (O'Brien 1995).

Notiodes Schoenherr 1838, 12 spp., generally distributed. Associated with wetlands. At least three species of *Notiodes* have been associated with Cyperaceae but *Notiodes celatus* (Burke 1961) is associated with the fern *Marsilea mucronata* A. Br. (Marsileaceae) (Burke 1971). See Tanner (1943) and Burke (1961a, 1965) to separate the species.

Notiophilus Schoenherr 1835; not Duméril 1805
Endalus Laporte 1840
Notionomus Erichson 1842

Onychylis LeConte 1876, 6 spp., generally distributed in eastern North America. Species are associated with *Pontederia cordata* L. (Pontederiaceae) and *Nuphar luteum* (L.) Sibhorn and Smith (Nymphaeaceae) in wetlands (Burke 1961b, Anderson 1993a). See Burke (1961b) to separate the species. This genus is composite and is being subdivided by Charles O'Brien and Guillermo Wibmer.

Stenopelmus Schoenherr 1835, 1 sp., *S. rufinasus* Gyllenhal 1836, generally distributed in the United States and southern Canada. This species is associated with *Azolla* (Salviniaceae) in wetland habitats (Scherf 1964).

Panscopus Schoenherr 1843; not Schoenherr 1842
Monius Schoenherr 1845
Degorsia Bedel 1902

Tanysphyrina Gistel 1856

Tanysphyrus Germar 1817, 2 spp., generally distributed in the eastern United States and Canada west across the north to British Columbia and south to Utah. *Tanysphyrus lemnae* (Fabricius 1792) is a widespread Holarctic species associated with *Lemna* (duckweed; Lemnaceae) whereas *T. ater* Blatchley 1928 is associated with *Ricciocarpus natans* (L.) Corda (Bryophyta: Ricciaceae); larvae mine the leaves.

Tanysphyroides Egorov 1996 (valid subgenus)

III. Raymondionyminae Reitter 1913

by Robert S. Anderson

This is a small group of three genera of eyeless weevils found in North America only in California and adjacent Oregon. They are easily recognized by their eyeless condition (Fig. 11) but also by the tarsi, which have only 4 articles (Fig. 12). Like Dryophthorinae and Erirhininae they possess primitive male genitalia and have recently been given family status by Thompson (1992) and Alonso-Zarazaga and Lyal (1999). Adults are collected in various kinds of leaf litter. Nothing is known of larval biology.

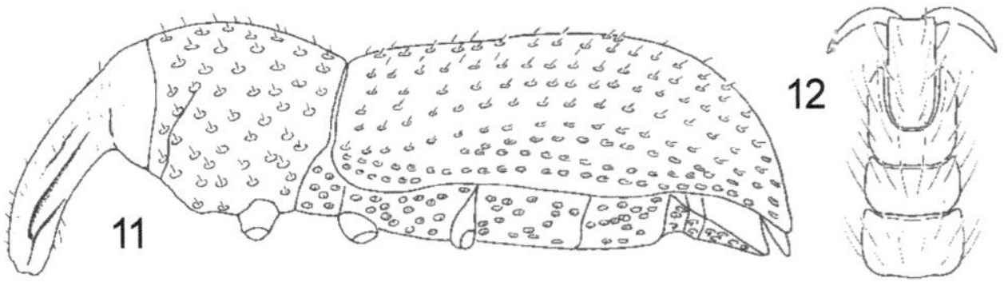

FIGURES 11.131-12.131. Raymondionyminae, *Alaocybites californica* Gilbert, 11. Lateral habitus; 12. Tarsus, dorsal view.

KEY TO THE NEARCTIC GENERA OF RAYMONDIONYMINAE

1. Front coxae not separated by prosternum; prosternum lacking lateral ridges in front of coxae; abdomen with ventrite 4 separated from 5 by a deep suture similar to suture between ventrites 3 and 4; antenna with funicle with 7 articles .... *Alaocybites*
— Front coxae narrowly separated by prosternum; prosternum with lateral ridge in front of each coxa slightly to well developed; abdomen with ventrite 4 separated from 5 by a shallow suture; antenna with funicle with 5 or 7 articles 2

2(1). Antenna with funicle with 5 articles; hind tibia linear or triangular in form; prosternum with ridges in front of coxae low .... *Gilbertiola*
— Antenna with funicle with 7 articles; hind tibia markedly expanded towards apex, subtriangular in form; prosternum with ridges in front of coxae well developed .... *Schizomicrus*

CLASSIFICATION OF THE NEARCTIC RAYMONDIONYMINAE

5. Raymondionymini Reitter 1913

Alaocybites Gilbert 1956, 2 spp., California. Adults have been collected in coniferous leaf litter. See Gilbert (1956) to separate the species.

Gilbertiola Osella 1982, 2 spp., California and Oregon. Adults have been collected in redwood leaf litter. See Gilbert (1956) to separate the species.

Gilbertia Osella 1977; not Cossman 1889; not Jordan and Eigenmann 1890; not Walsingham 1891

Schizomicrus Casey 1905, 1 sp., *S. caecus* (Casey 1892), California. Adults have been collected in leaf litter.

Schizonotus Casey 1892; not Ratzeburg 1852; not Thorell 1888; not Reuter 1892

IV. Curculioninae Latreille 1802

By Robert S. Anderson

Traditionally this subfamily has been restricted to members of the genus *Curculio* and some close relatives but it is now a large conglomerate of taxa of questionable relationships. Members have a small or no tooth on the inner angle at the apex of the hind tibia, eyes are rounded, the rostrum mostly elongate to very elongate and cylindrical in cross section, and the antenna with the scape not or just reaching the anterior margin of the eye. They may be confused with Baridinae or Ceutorhynchinae but members of these latter two subfamilies have the mesepimeron strongly ascended, truncated by elytral humeri and visible in dorsal view between the pronotum and elytra. Sexual dimorphism in rostral form in Curculioninae is extreme in some taxa; generally, the female rostrum in longer and finer and the antennae are inserted more basally than in males. This dimorphism appears to be related to oviposition and may be a key adaptation in explaining weevil diversity (Anderson 1995).

Curculionines tend to be associated with many herbaceous as well as some woody plants. Most have larvae that develop in reproductive structures such as fruits, seeds or flower buds; some also mine stems. Many plant families serve as hosts and knowledge of the host plant can facilitate identifications. Larvae of Rhamphini are leaf miners. Most species in Cionini and Mecinini are adventive. Anthonomini are the most diverse group, especially the genus *Anthonomus*. An excellent review of the natural history of Anthonomini is by Burke (1976).

Key to the Nearctic genera of Curculioninae

1. Rostrum in repose received into distinct ventral channel in prosternum; antennae with 5 funicle articles .... *Cleopomiarus*
— Rostrum in repose not received into ventral channel, but may rest between front, middle and hind coxae; antennae with 5-7 funicle articles 2

2(1). Tarsus with claws connate at base 3
— Tarsus with claws free at base, simple or with basal process 7

3(2). Antenna with funicle with 5 articles 4
— Antenna with funicle with 6 or 7 articles 6

4(3). Pygydium covered by elytral apices .... *Cionus*
— Pygydium exposed beyond elytral apices 5

5(4). Body oval, length less than twice greatest width; pronotum with lateral margins markedly arcuate from base to apex .... *Gymnetron*
— Body elongate and cylindrical, length more than twice greatest width; pronotum with lateral margins more or less subparallel in basal half .... *Mecinus*

6(3). Tarsus with article 5 shorter than articles 1 to 3 combined .... *Smicronyx*
— Tarsus with article 5 about as long as articles 1 to 3 combined .... *Promecotarsus*

7(2). Tarsus with claw simple, lacking basal process or tooth 8
— Tarsus with claw with basal tooth or process[3] .. 16

8(7). Front femur with ventral margin simple, lacking tooth 9
— Front femur with ventral margin with slightly to well-developed tooth 12

9(8). Pronotum with anterolateral margin with postocular lobe present; hind femur with ventral margin with large broad tooth .... *Pachytychius*
— Pronotum with anterolateral margin straight, postocular lobe absent; hind femur with ventral margin simple, lacking tooth 10

10(9). Pronotum with distinct lateral margin defined by low carina, apically with carina slightly produced laterally, denticulate or serrate .... *Elaeidobius*
— Pronotum with lateral margin rounded, not defined by carina, no lateral protrusions, denticulations or serrations 11

11(10). Abdomen with suture between ventrites 2 and 3 straight laterally; rostrum longer than pronotum; antenna with funicle from article 2 to apex, long and slender, about as long as club .... *Acalyptus*
— Abdomen with suture between ventrites 2 and 3 angulate posteriorly at lateral margin; rostrum shorter than pronotum; antenna with funicle from article 2 to apex, very short and stout, shorter than length of club .... *Phyllotrox*

12(8). Body greater than 2.3 mm in length; tarsal claws widely divergent, tooth on claw extended from underside of claw .... *Dorytomus*
— Body less than 2.3 mm in length; tarsal claws not widely divergent, tooth on claw extended from inside face of claw 13

13(12). Elytra nearly glabrous except for group of white scales near middle of interval 4; scutellum with dense white scales; middle coxae separated by distance nearly equal to width of a coxa .... *Ephelops*
— Elytra with more or less uniformly distributed scales or vestiture; middle coxae separated by distance distinctly less than width of a coxa 14

14(13). Body with sparse fine pubescence; hind tibia with apical tooth minute; body color light brown .... *Dietzianus*
— Body with sparse to dense scales; hind tibia with apical tooth about half as long as tarsal claw; body color darker reddish brown to black 15

15(14). Rostrum with lateral groove defined to anterior margin of eye, with at most a few scattered scales adjacent to eye; body elongate-oval; scales generally of one color; apical third of elytra in lateral view markedly rounded to apex; associated with Asteraceae .... *Epimechus*
— Rostrum with lateral groove not defined immediately anterior to eye, obliterated by dense scales adjacent to eye; body stout; scales ornate, of more than one color; apical half of elytra in lateral view sloped gradually to apex; associated with Solanaceae .... *Brachyogmus*

16(7). Abdomen with suture between ventrites 2 and 3 markedly extended posteriorly towards lateral margins, extended to or beyond suture between ventrites 3 and 4 (Fig. 16) 17

[3] *Nanops* has a minute tooth that is difficult to see at high magnification and it may appear absent. It can be recognized by its small size (1.4-1.5 mm) and front femur lacking a ventral tooth. Species are associated with *Hypericum* (Hypericaceae).

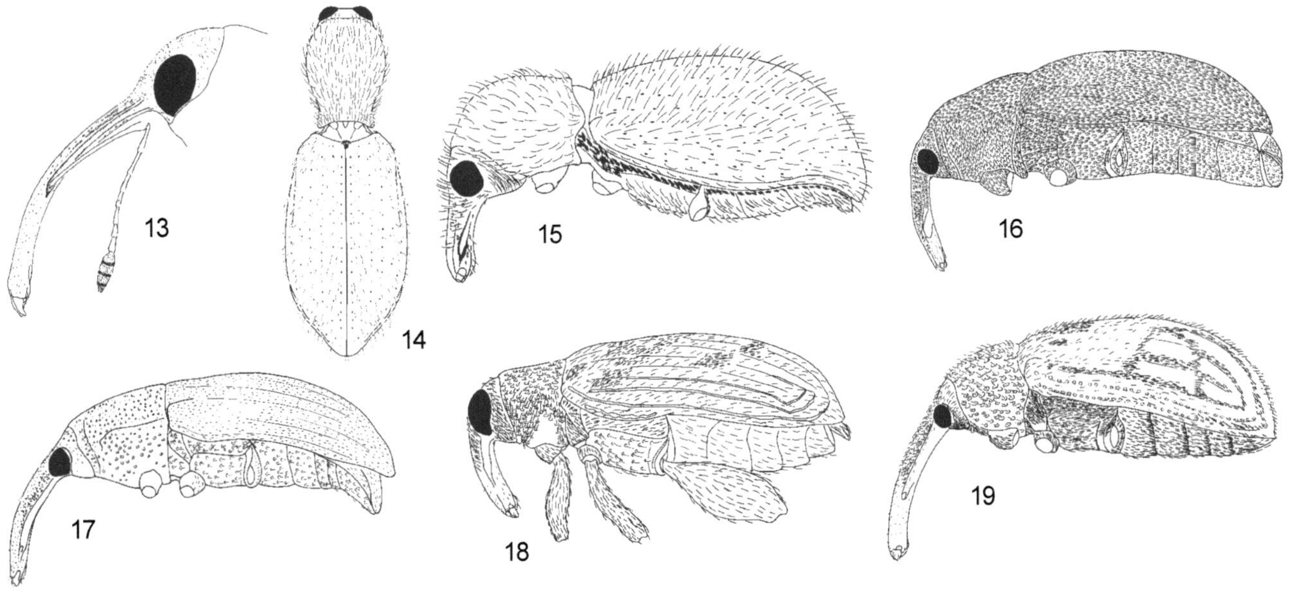

FIGURES 13.131-19.131. Curculioninae. 13. *Curculio monticola* (Casey), head, lateral view; 14. *Myrmex arizonicus* (Schaeffer), dorsal habitus. 15-19. Lateral habitus, 15. *Myrmex arizonicus* (Schaeffer); 16. *Tychius tectus* LeConte; 17. *Notolomus bicolor* LeConte; 18. *Tachyerges ephippiatus* (Say); 19. *Anthonomus fulvus* LeConte.

— Abdomen with suture between ventrites 2 and 3 more or less straight, if extended posteriorly, not extended to suture between ventrites 3 and 4 (Fig. 19) 18

17(16). Pygydium covered by elytra; antenna with funicle with 6 or 7 articles; associated with subfamily Papilionoideae (Fabaceae) .... *Tychius*

— Pygydium exposed beyond elytra apex (especially so in male); antenna with funicle with 5 or 6 articles; associated with subfamily Mimosoideae (Fabaceae) .... *Sibinia*

18(16). Front coxae positioned much closer to posterior margin of prosternum than to anterior margin, distance to anterior margin greater than twice distance to posterior margin (Fig. 17) 19

— Front coxae positioned near middle of prosternum, coxae more or less equidistant from anterior and posterior margins of prosternum (Figs. 15, 18-19) 22

19(18). Eye partly covered by anterior margin of pronotum (Fig. 13); mandible prominent, slender, triangular in outline, inner face simple, not dentate; rostrum very long and slender (Fig. 13) 20

— Eye distant from anterior margin of pronotum (Fig. 17); mandible not prominent, inner face dentate; rostrum moderately long and slender (Fig. 17) 21

20(19). Antenna with club longer than wide; tarsus with claw with distinct and long basal tooth; body with more or less uniform vestiture of brown or grey appressed scales; body size 4.2-13.0 mm; associated with Fagaceae, Juglandaceae and Betulaceae .... *Curculio*

— Antenna with club as wide as long; tarsus with claw with short and fine basal tooth; body with vestiture of scattered white appressed scales; body size 2.0-3.0 mm; associated with Salicaceae .... *Archarius*

21(19). Rostrum longer than head and pronotum combined; elytra black with sparse, recumbent vestiture; pronotum slightly constricted toward apex; southern Rocky Mountain United States; associated with Geraniaceae .... *Hypoleschus*

— Rostrum shorter than head and pronotum combined; elytra yellowish or light reddish brown, lacking obvious vestiture; pronotum rather markedly constricted toward apex; southeastern United States west into Texas; associated with Arecaceae .... *Notolomus*

22(18). Pronotum longer than wide, distinctly constricted at base such that width at midlength much greater than at base (Fig. 14); black, or black and red, glossy and ant-like in form 23

— Pronotum wider than long, base not distinctly constricted such that width at midlength is at most slightly greater than at base; form various 25

23(22). Head with supraocular sulcus present and angulate dorsolaterally; front femur lacking tooth on ventral margin; extreme southern Florida .... *Micromyrmex*

— Head with supraocular sulcus lacking, no obvious sulcus or impression above eye; front femur with tooth on ventral margin, tooth may be obsolete in some specimens; widespread 24

24(23). Elytra oval, humeri rounded, flight wings absent; eyes slightly reduced in size and number of facets .... *Oopterinus*

— Elytra elongate-oval, humeri quadrate (Fig. 14), flight wings present; eyes well-developed (Fig. 15) .... *Myrmex*

25(22). Hind tibia with apical comb of setae oblique, set at an angle to long axis of tibia; hind tibia narrowed apically; hind femur stouter than middle femur, slightly so in some specimens; jumping forms (Fig. 18) 26

— Hind tibia with apical comb of setae transverse, perpendicular to long axis of tibia; hind tibia not narrowed apically; hind femur not distinctly stouter than middle femur 28

26(25). Antenna with funicle of 7 articles; eyes subcontiguous to contiguous in anterior view; elytra with or without distinct pattern of contrasting pale vestiture .... *Tachyerges*

— Antenna with funicle of 6 articles; eyes distinctly separated at point of closest approach by a distance greater than 0.10 X width of an eye in anterior view; elytra without distinct pattern of contrasting pale vestiture 27

27(26). Metasternum (lateral portion), mesepisternum and metepisternum with short, dense, plumose white scales which contrast markedly with the rest of body vestiture; hind femur slightly expanded, length greater than 3.10 X maximum width, ventral margin simple; body size small, 1.0-1.8 mm . ... *Isochnus*

— Metasternum, mesepisternum and metepisternum with vestiture as on rest of body, mesepisternum and metepisternum in some speicmens with broad bifurcate (but not plumose) scales; hind femur slightly to markedly expanded, length less than 3.20 X maximum width, ventral margin with various spines and setae set in denticles; body size moderate, 1.6-2.5 mm .... *Orchestes*

28(25). Front coxae distinctly separated by process of prosternum; middle coxae widely separated by distance nearly equal to width of a coxa; body size 1.1-1.4 mm; extreme southern Florida .... *Huaca*

— Front coxae contiguous; middle coxae separated by distance less than width of a coxa; body size greater than 1.3 mm; widespread 29

29(28). Hind tibia with distinct apical tooth, tooth large and curved, subequal in size to tarsal claw 30

— Hind tibia with at most only small, usually straight apical tooth, tooth much smaller than tarsal claw, or tibial apex simple, lacking tooth 40

30(29). Antenna with funicle with coarse, elongate, erect scales; antenna with club compact, glossy and nearly glabrous; tarsus with claws usually with a long, slender tooth extended on inside of claw well distad of base .... *Magdalinops*

— Antenna with funicle with very fine setae; antenna with club various, usually less compact and with distinct pubescence; tarsus with claws various 31

31(30). Abdomen with sutures between ventrites angled posteriorly at lateral margins 32

— Abdomen with sutures between ventrites straight, not angled posteriorly at lateral margins 36

32(31). Front femur with large, broad, triangular tooth, tooth longer than tarsal claw .... *Ochyromera*

— Front femur lacking tooth or with at most a short tooth, not longer than tarsal claw 33

33(32). Tarsus with claw with short, broad, blunt basal process 34

— Tarsus with claw with long, fine, acute basal tooth 35

34(33). Front femur simple, lacking tooth; rostrum shorter than pronotum .... *Ellescus*

— Front femur with minute tooth on ventral margin; rostrum longer than pronotum .... *Proctorus*

35(33). Rostrum with scrobe descended, antenna with scape rested below lateral rostral groove and below ventral margin of rostrum; antenna with funicle with 6 or 7 articles; associated with Rubiaceae .... *Plocetes*

— Rostrum with scrobe not descended, rostrum without lateral grooves, antenna with scape parallel to long axis of rostrum; antenna with funicle with 7 articles; associated with Oleaceae .... *Lignyodes*

36(31). Front femur with ventral margin simple, lacking tooth; tooth on tarsal claw minute (may appear absent); body size small, 1.4-1.5 mm ..... *Nanops*

— Front femur with ventral margin with tooth; tooth on tarsal claw distinct; body size various, most greater than 1.5 mm 37

37(36). Front femur with large, broad, triangular tooth, middle and hind femora simple, lacking tooth; associated with Malvaceae .... *Macrorhoptus*

— Front, middle and hind femora each with tooth; associated with various plants (including Malvaceae) 38

38(37). Antenna with funicle with 6 articles; antenna with club with basal article glossy, almost glabrous, remaining articles densely pubescent; dorsal margin of eye elevated above level of interocular area .... *Anthonomopsis*

— Antenna with funicle with 7 articles; antenna with club various; eyes various 39

39(38). Elytra with surface even, not tuberculate; pygydium exposed beyond apices of elytra; scales of elytra evenly distributed, without contrasting pattern; body stout .... *Chelonychus*

— Elytra with serrate tubercle at base of interval 3; pygydium covered by elytra; elytra with broad, conspicuous band of white scales across elytra near base; body more elongate ...... *Smicraulax*

40(29). Rostrum with lateral groove short, apex of groove not extended to anterior margin of eye (short by distance at least equal to diameter of eye); associated with Viscaceae (mistletoe) ... *Cionomimus*

— Rostrum with lateral groove long, apex of groove extended to anterior margin of eye (if short, by much less than diameter of eye); associated with various plants 41

41(40). Antenna with funicle of 5 articles; antenna with club with basal article glossy, almost glabrous; front coxae of some slightly separated; middle coxae widely separated; femora simple, lacking tooth

on ventral margin; body size 1.3-1.5 mm .... *Neomastix*
— Antenna with funicle of 6 or 7 articles; antenna with club various; front coxae various; middle coxae various; femora with tooth or simple, lacking tooth on ventral margin; body size greater than 1.3 mm, most greater than 1.5 mm 42

42(41). Rostrum with lateral groove descended, directed slightly to well below middle of eye; elytra with base of interval 3 elevated; antenna with funicle of 6 articles .... *Pseudanthonomus*
— Rostrum with lateral groove not descended, directed to middle of eye; elytra with base of interval 3 various, flat to elevated; antenna with funicle of 6 or 7 articles 43

43(42). Rostrum with dense scales throughout almost entire length, scales obscuring underlying cuticle; head constricted behind eyes, hind margin of eye markedly produced; associated with *Bernardia* (Euphorbiaceae); Texas .... *Narberdia*
— Rostrum with scales if present, not dense and limited to basal half of length, scales not obscuring underlying cuticle; head not constricted behind eyes, eye produced but hind margin flat against cuticle; associated with various plants; widely distributed 44

44(43). Front femur markedly expanded, width about twice that of middle or hind femur, with large biserrate tooth on ventral margin; head subconical; associated with *Serjania* (Sapindaceae); southern Texas .... *Cionopsis*
— Front femur at most slightly expanded, width less than twice that of middle or hind femur, tooth on ventral margin various; head subconical or not, if subconical, then front femur not expanded; associated with various plants; widely distributed 45

45(44). Ventrite 5 of male very short at middle, deeply and broadly emarginate; pronotum with low median carina in basal third to one-half; rostrum short, subequal in length to pronotum or slightly longer, and straight; abdomen with ventrites flat .... *Coccotorus*
— Ventrite 5 of male longer at middle, at most slightly and shallowly emarginate; pronotum lacking median carina; rostrum moderate to long, slightly to distinctly longer than pronotum, and slightly to markedly curved; abdomen with ventrites convex 46

46(45). Front tibia moderately curved, with apical half of inner margin expanded and carinate; elytra with interval 2 descended lateral to scutellum, interval 3 with prominent swelling at base; mesosternum markedly declivious; body size 4.0 - 5.8 mm; southern Florida .... *Atractomerus*
— Front tibia only slightly curved, with apical half of inner margin simple; elytra with interval 2 flat lateral to scutellum, interval 3 with slight to prominent swelling at base; mesosternum at most slightly declivious; body size various, most less than 4.0 mm; widely distributed ..... *Anthonomus*

Classification of the Nearctic Curculioninae

6. Curculionini Latreille 1802
Curculionina Latreille 1802

Archarius Gistel 1856, 1 sp., *A. salicivorus* (Paykull 1792), Quebec; adventive. This species is associated with galls on *Salix* (Salicaceae). Recently confirmed as established in Quebec by Sylvain Côté (pers. comm.).

Archarias Villa and Villa 1833; not Dejean 1821
Balanobius Jekel 1861
Longifistulia Hong and Wang 1987
Toptaria Kwon and Lee 1990 (valid subgenus)

Curculio Linnaeus 1758, 27 spp., generally distributed. Species are associated with various Fagaceae, Juglandaceae and Betulaceae. See Gibson (1969) to separate the species.

Balaninus Germar 1817
Pelecinus Wiedemann 1823; not Latreille 1800
Tropibalaninus Heller 1927 (valid subgenus)
Carponinophilus Voss 1962 (valid subgenus)

7. Acalyptini Thomson 1859

Acalyptus Schoenherr 1833, 1 sp., *A. carpini* (Herbst 1795), generally distributed in Alaska, Canada and northern United States. This species is associated with *Salix* (Salicaceae) (Anderson 1997).

Orsophagus Roelofs 1874

8. Anthonomini Thomson 1859

Anthonomopsis Dietz 1891, 1 sp., *A. mixta* (LeConte 1876), generally distributed in eastern and central United States and Canada. This species is associated with *Prunus* (Rosaceae) (Ahmad and Burke 1972).

Anthonomus Germar 1817, 110 spp., generally distributed. Species are associated with various families of plants including Asteraceae, Caprifoliaceae, Cistaceae, Cupressaceae, Euphorbiaceae, Fabaceae, Juglandaceae, Krameriaceae, Malpighiaceae, Malvaceae, Myrtaceae, Rosaceae, Rutaceae, Salicaceae, Solanaceae, and Vitaceae; larvae mostly develop in reproductive structures or in galls (Burke 1976). See Dietz (1891), Hatch (1971), Blatchley and Leng (1916) to separate some of the species. The genus presently is being revised in the New World by Wayne Clark and Horace Burke; some of their papers include North American species (Clark 1987a, b, 1988, 1990, 1991a, b; Clark and Burke 1985, 1986, 1996).

Pallene Dejean 1821
Furcipus Desbrochers 1868 (valid subgenus)
Toplithus Gozis 1882
Anthomorphus Weise 1883 (valid subgenus)
Furcipes Bedel 1884
Toplethus Bedel 1884
Anthonomochaeta Dietz 1891 (valid subgenus)
Anthonomocyllus Dietz 1891 (valid subgenus)

Anthonomorphus Dietz 1891 (valid subgenus)
Cnemocyllus Dietz 1891 (valid subgenus)
Paranthonomus Dietz 1891
Tachypterus Dietz 1891; not Guérin-Méneville 1838
Trichobaropsis Dietz 1891
Listrorrhynchus Champion 1903
Tachypterellus Fall and Cockerell 1907 (valid subgenus)
Anthonomidius Reitter 1915 (valid subgenus)
Sexarthrus Blatchley 1916
Pterochalybs Ter-Minasian 1936 (valid subgenus)
Persexarthrus Voss 1944 (valid subgenus)
Parafurcipes Voss 1956 (valid subgenus)
Exanthonomus Voss 1960
Neobradybatus Hoffmann 1963

Atractomerus Duponchel and Chevrolat 1842, 1 sp., *A. punctipennis* (Gyllenhal 1836), southern Florida. This species is associated with *Eugenia* (Myrtaceae) (Anderson 1993a).
Leptarthrus Dietz 1891; not Stephens 1829
Cissoanthonomus Hustache 1939
Arthleptrus Burke 1982

Brachyogmus Linell 1897, 1 sp., *B. ornatus* Linell 1897, southwestern United States. This species is associated with *Lycium* (Solanaceae) (Burke 1968).

Chelonychus Dietz 1891, 2 spp., generally distributed in Western United States and Canada. See Clark and Burke (in press b) to separate the species.

Cionomimus Marshall 1939, 2 spp., southwestern and western United States. Species are associated with *Phoradendron* (mistletoe; Viscaceae) (Burke 1981). See Burke (1981) or Anderson (1994) to separate the species.
Cionistes Dietz 1891; not Wright 1861

Cionopsis Champion 1903, 2 spp., southern Texas. Species are associated with *Serjania* (Sapindaceae); larvae in fruits (Anderson and Burke 1990). See Burke (1982) to separate the species.

Coccotorus LeConte 1876, 4 spp., generally distributed in eastern and central United States and Canada. Species are associated with *Prunus* (Rosaceae) (Brown 1966a). See Brown (1966a) to separate the species.

Dietzianus Sleeper 1953, 2 spp., generally distributed in eastern United States. See Blatchley and Leng (1916) to separate the species.
Xanthus Dietz 1891; not Gistl 1834; not Agassiz 1843

Ephelops Dietz 1891, 1 sp., *E. triguttatus* Dietz 1891, southern Florida. This species may be associated with *Piscidia* (Fabaceae) (Anderson 1993a).

Epimechus Dietz 1891, 11 spp., generally distributed in western United States. Species are associated with various Asteraceae. See Clark and Burke (in press a) to separate the species.

Huaca Clark 1993, 2 spp., southern Florida. *Huaca apian* Clark 1993 has been associated with *Zanthoxylum flavum* Vahl. (Rutaceae) (as Anthonomini new genus 1, new species 1; Anderson 1993a). See Clark (1993a) to separate the species.

Magdalinops Dietz 1891, 4 spp., generally distributed in western United States and Canada. Species are associated with Asteraceae. See Clark and Burke (in press b) to separate the species.

Nanops Dietz 1891, 1 sp., *N. schwarzii* Dietz 1891, southeastern United States. This species is associated with *Hypericum* (Hypericaceae).

Narberdia Burke 1976, 1 sp., *N. aridulus* Burke 1976, Texas. This species is associated with *Bernardia myricaefolia* (Scheele) Wats. (Euphorbiaceae); larvae in fruits (Burke and Rector 1976).

Neomastix Dietz 1891, 1 sp., *N. solidaginis* Dietz 1891, southeastern United States. Adults have been associated with various plants (Clark 1993b).

Pseudanthonomus Dietz 1891, 7 spp., generally distributed in eastern and central United States and Canada extending as far west as Arizona and Colorado, and as far north as Yukon Territory. Species are associated with various Rosaceae, Ericaceae, Betulaceae, Saxifragaceae, Hamamelidaceae and Krameraceae; larvae in flower buds and fruits (Clark 1987c). See Clark (1987c) to separate the species.

Smicraulax Pierce 1908, 2 spp., Arizona and Texas. Species are associated with *Phoradendron* (mistletoe; Viscaceae); larvae mine stems. See Burke (1975) to separate the species.

9. Cionini Schoenherr 1825

Cionus Clairville 1798, 1 sp., *C. scrophulariae* (Linnaeus 1758), New York; adventive. This species is associated with *Scrophularia* and *Verbascum* (Scrophulariaceae); larvae feed externally on the leaves and pupate in round translucent cocoons among flowers and seed-capsules. Recently confirmed as established by Hoebeke (pers. comm.).

10. Derelomini Lacordaire 1866

Elaeidobius Kuschel 1952, 1 sp., *E. subvittatus* (Faust 1898), Florida; adventive. This species is associated with the male flowers of *Elaeis guineensis* Jacquin (African oil palm; Arecaceae) (O'Brien and Woodruff 1986).

Hypoleschus Fall 1907, 1 sp., *H. atratus* Fall 1907, New Mexico and Colorado. This species is associated with *Geranium* sp. (cranesbill; Geraniaceae) (C.W. O'Brien, pers. comm.).

Notolomus LeConte 1876, 2 spp., southeastern United States west to southern Texas. Species are associated with flowers of *Serenoa repens* (Bartr.) Small and *Sabal palmetto* (Walt.) Lodd (saw palmetto and cabbage palm; Arecaceae); larvae develop in male flowers (Anderson 1993a). See Blatchley and Leng (1916) to separate the species.

Phyllotrox Schoenherr 1843, 7 spp., generally distributed in the United States. *Phyllotrox canyonacerensis* Warner 1976 is associated with fruits of *Acer grandidentatum* Nutt. (maple; Aceraceae) (Warner 1976). The genus needs revision.

Euclyptus Dietz 1891

11. Ellescini Thomson 1859

Ellescina Thomson 1859

Ellescus Dejean 1821, 4 spp., generally distributed. Species are associated with *Salix* and *Populus* (willow, poplar and aspen; Salicaceae); larvae mine the central axis of female catkins (Scherf 1964). The genus needs revision.

Sarapus Villa and Villa 1833; not Fischer von Waldheim 1821
Elleschus Schoenherr 1838
Alyca LeConte 1876
Anisarctus Desbrochers 1907

Proctorus LeConte 1876, 2 spp., generally distributed in northern United States, Canada and Alaska. Associated with *Salix* (willow; Salicaceae). See LeConte and Horn (1876) to separate the species.

Encalus LeConte 1876

Dorytomina Bedel 1886

Dorytomus Germar 1817, 21 spp., generally distributed. Species are associated with *Salix* and *Populus* (willow, poplar and aspen; Salicaceae); larvae feed in catkins and one develops in sawfly galls in the stems of *Salix*. See O'Brien (1970a) to separate the species but note subsequent synonymy as summarized in O'Brien and Wibmer (1982).

Solenorhinus Motschulsky 1860
Doratotomus Gistel 1886
Eteophilus Bedel 1886
Alycodes Dietz 1891
Euolamus Reitter 1916 (valid subgenus)
Olamus Reitter 1916 (valid subgenus)
Praeolamus Zumpt 1932
Paradorytomus Zumpt 1932
Chaetodorytomus Iablokov-Khnzorian 1970 (valid subgenus)

12. Mecinini Gistel 1856

Cleopomiarus Pierce 1919, 1 sp., *C. hispidulus* (LeConte 1876), generally distributed in eastern United States. This species is associated with *Lobelia* (Campanulaceae); larvae in seed capsules (Anderson 1973).

Miaromimus Solari 1947
Hemimiarus Franz 1947

Gymnetron Schoenherr 1825, 4 spp., generally distributed; adventive. Species are associated with *Verbascum thapsis* Linnaeus, *Linaria vulgaris* Miller (both Scrophulariaceae) and *Plantago lanceolata* Linnaeus (Plantaginaceae); larvae in seed capsules (Anderson 1973). See Buchanan (1937) to separate three of the four species; Sleeper (1954a) presents notes on the fourth. Downie and Arnett (1996) provide a brief key to the four species.

Gymnetrum Agassiz 1846
Carpolinus Gistel 1848
Aprinus Desbrochers 1893
Eutemnoscelus Desbrochers 1893 (valid subgenus)

Mecinus Germar 1821, 2 spp., *M. pyraster* (Herbst 1795) and *M. janthinus* (Germar 1817), eastern and western United States and Canada (disjunct); adventive. *Mecinus pyraster* is associated with *Plantago lanceolata* Linnaeus (Plantaginaceae); larvae are in seed capsules (Anderson 1973). *Mecinus janthinus* has been introduced into Montana, Wyoming, Washington, British Columbia, Alberta and Nova Scotia (Harris *et al.* 2001; DeClerk-Floate and Harris in press) for the biological control of *Linaria vulgaris* Miller (yellow toad-flax) and *L. dalmatica* (L.) Miller (Dalmation toad-flax) (Scrophularaceae). There is no key to separate the two species in North America.

Hexaphyllus Dejean 1821
Macipus Fischer de Waldheim 1829
Mecinopsis Escalera 1914

13. Otidocephalini Lacordaire 1863

Micromyrmex Sleeper 1953, 2 spp., *M. cavirostris* (Casey 1892) and *M. poeyi* (Chevrolat 1832), southern Florida. See Blatchley and Leng (1916; as *Otidocephalus*) to separate the species.

Myrmex Sturm 1826, 31 spp., generally distributed in the United States and southeastern Canada; most species in southwestern United States. Species are associated mainly with various Asteraceae, also Fagaceae, Ulmaceae, Arecaceae, Smilacaceae, Viscaceae and Sapotaceae (Anderson 1993b). Larvae mostly mine stems. The genus needs revision. See Horn (1873) and Schaeffer (1907) to separate most of the species.

Otidocephalus Chevrolat 1832
Cycotida Pascoe 1872

Oopterinus Casey 1892, 2 spp., eastern United States. Larvae of *O. perforatus* develop in cynipid galls on the roots of *Quercus*. See O'Brien (1985) to separate the species.

14. Rhamphini Rafinesque 1815

Rhamphina Rafinesque 1815

Isochnus Thomson 1859, 5 spp., generally distributed in North America, including far northern Canada and Alaska; not in southeastern United States. Species are associated with *Salix* and *Populus* (willow, poplar and aspen; Salicaceae); larvae mine leaves (Anderson 1989a). See Anderson (1989a) to separate the species.

Orchestes Illiger 1798, 5 spp., generally distributed. Species are associated with Betulaceae, Rosaceae and Ulmaceae; larvae mine leaves (Anderson 1989a). See Anderson (1989a) to separate the species.

Salius Schrank 1798 (valid subgenus)
Alyctus Thomson 1859
Threcticus Thomson 1859
Euthoron Thomson 1859
Nomizo Morimoto 1984 (valid subgenus)

Tachyerges Schoenherr 1825, 3 spp., generally distributed. Species are associated with *Salix* and *Populus* (willow, poplar and aspen; Salicaceae); larvae mine leaves (Anderson 1989a). See Anderson (1989a) to separate the species.

15. Smicronychini Seidlitz 1891

Promecotarsus Casey 1892, 3 spp., generally distributed in western United States and Canada. See Casey (1892) to separate the species.

Smicronyx Schoenherr 1843, 70 spp., generally distributed. Species are associated with various plants, mostly Asteraceae and Convolvulaceae (*Cuscuta*; dodder); larvae are in seeds or may cause galls (Anderson 1962). See Anderson (1962) to separate the species.

Micronyx Schoenherr 1835; not Boisduval 1835
Desmoris LeConte 1876 (valid subgenus)
Pachyphanes Dietz 1894 (valid subgenus)
Pseudromicronyx Dietz 1894 (valid subgenus)
Synertha Dietz 1894
Chalybodontus Desbrochers 1897 (valid subgenus)
Oligocaricis Lea 1926

16. Storeini Lacordaire 1863

Pachytychius Jekel 1861, 1 sp., *P. haematocephalus* (Gyllenhal 1836), New York; adventive. This species is associated in Europe with *Lotus corniculatus* L. (Fabaceae) (Hoffmann 1958).

Styphlotychius Jekel 1861
Barytychius Jekel 1861
Scyphotychius Desbrochers 1875
Rabdotorhinus Desbrochers 1894
Fogatianus Caldara 1978

17. Tychiini Thomson 1859

Tychiina Thomson 1859

Sibinia Germar 1817, 22 spp., generally distributed in western United States. Species are associated with various Fabaceae (subfamily Mimosoideae); larvae in reproductive structures (Clark 1978). See Clark (1978) to separate the species.

Sibynes Schoenherr 1825
Campipterus Motschulsky 1845
Campopterus Agassiz 1846
Sibynia Agassiz 1846
Aocnus Schoenherr 1859
Sibynia Wollaston 1865; not Agassiz 1846
Paragoges LeConte 1876
Dichotychius Bedel 1885 (valid subgenus)
Mecynopyga Pierce 1908
Microtychius Casey 1910 (valid subgenus)
Teratonychus Bondar 1949
Itychus Kissinger 1962

Tychius Germar 1817, 16 spp., generally distributed; four species adventive (Anderson and Howden 1994). Species are associated with various native and adventive Fabaceae (subfamily Papilionoideae); larvae in reproductive structures (Clark 1971; Clark and Burke 1977). See Clark (1971, 1977) and Anderson and Howden (1994) to separate the species.

Miccotrogus Schoenherr 1825
Apeltarius Desbrochers 1873 (valid subgenus)
Ectatotychius Tournier 1874
Hypactus Marseul 1888
Henonia Pic 1897
Xenotychius Reitter 1897
Pseudolignyodes Pic 1899
Paratychius Casey 1910
Aoromius Desbrochers 1907
Lepidotychius Penecke 1922
Elleschidius Penecke 1938
Heliotychius Franz 1943
Neotychius Hustache 1945
Mongolotychius Korotyaev 1990

Lignyodina Bedel 1884

Lignyodes Dejean 1835, 17 spp., generally distributed. Species are associated with Oleaceae; larvae are in reproductive structures (Clark 1980a, 1980b, 1981). Subgenus *Lignyodes* are associated with *Fraxinus* (ash), subgenus *Chionanthobius* with *Chionanthus*, *Forestiera* and *Osmanthus*, and subgenus *Neotylopterus* with *Forestiera*. See Clark (1980a, 1980b, 1981) to separate the species.

Lignyodes Schoenherr 1835; not Dejean 1835
Stenorhynchus Villa and Villa 1835; not Lamarck 1818; not Hemprich 1820; not Berthold 1827
Rhaestes Gistel 1856
Thysanocnemis LeConte 1876
Tylopterus LeConte 1876; not Capiomont 1868
Chionanthobius Pierce 1912 (valid subgenus)
Lignyodius Dieckmann 1970
Neotylopterus Clark, Whitehead and Warner 1977 (valid subgenus)

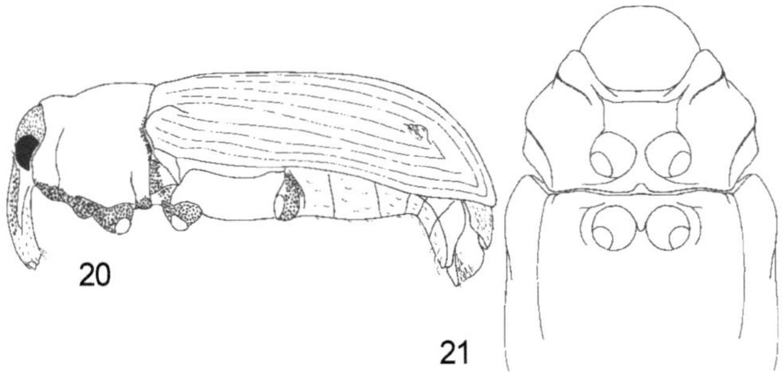

FIGURES 20.131-21.131. Bagoinae, *Bagous americanus* LeConte, 20. Lateral habitus; 21. Thoracic sterna, ventral view.

Plocetes LeConte 1876, 4 spp., generally distributed in southeastern United States west to southern Texas (two species are restricted to extreme southern Florida; one to extreme southern Texas). *Plocetes ulmi* LeConte 1876 is widespread in the southeastern United States and is associated with *Cephalanthus occidentalis* L. Species are all associated with Rubiaceae; larvae likely in reproductive structures (Clark 1982; Anderson 1991). See Clark (1982) and Anderson (1991) to separate the species.
Dietzia Champion 1903
Hamaba Casey 1910
Rosella Whitehead 1977

Ochyromerina Voss 1935

Ochyromera Pascoe 1874, 1 sp., *O. ligustri* Warner 1961, southeastern United States; adventive. This species is associated with *Ligustrum* (adventive; privet; Oleaceae) (Warner 1961).
Exochyromera Voss 1937

Incertae sedis (Curculioninae)

Macrorhoptus LeConte 1876, 6 spp., generally distributed in central and western United States and Canada. Species are associated with *Sphaeralcea*, *Sidalcea* and *Callirhoe* (Malvaceae); larvae are in reproductive structures (Burke 1973). The genus needs revision. See Sleeper (1957a) to separate the species.
Paraceratopus Brèthes 1910

V. Bagoinae Thomson 1859

by Robert S. Anderson

Only the genera *Bagous* and *Pnigodes* constitute Bagoinae in North America and the status of the latter as distinct is questionable. Most are found in aquatic or semi-aquatic habitats where larvae are associated with a variety of plant families. Members are easily recognized by the median prosternal channel (Fig. 21), the smooth varnish-like coating over the scales, the mostly tuberculate elytra, and the elongate and slender legs. They are very similar in appearance to the Stenopelmina (Erirhininae) but the members of that subtribe do not possess a sternal channel for reception of the rostrum and have different male genitalia.

Key to the Nearctic genera of Bagoinae

1. Pronotum slightly constricted behind apex (Fig. 20) .... *Bagous*
— Pronotum markedly constricted behind apex .... *Pnigodes*

Classification of the Nearctic Bagoinae

Bagous Germar 1817, 33 spp., generally distributed. Species are associated with various wetland plants such as *Limnobium spongia* (Bosc) Steud. (Hydrocharitaceae), *Brasenia schreberi* Gmel. and *Nymphaea* (Nymphaeaceae), *Eleocharis* and *Carex* (Cyperaceae), and *Potamogeton* (Potamogetonaceae) (O'Brien and Marshall 1979). *Bagous pictus* Blatchley 1920 is associated with *Sesuvium portulacastrum* (L.) L. (Aizoaceae). See Tanner (1943) to separate the species. The genus *Pnigodes* is questionably distinct from *Bagous*. The genus is being revised by Charles O'Brien.
Macropelmus Dejean 1821
Hydronomus Schoenherr 1825
Cyprus Schoenherr 1825
Lyprus Schoenherr 1826
Dicranthus Motschulsky 1845
Ephimeropus Hochhuth 1847
Elmidomorphus Cussac 1851
Bagoas Gistel 1856
Anactodes Brisout 1863
Helminthimorphus Bedel 1884
Bagoimorphus Desbrochers 1884
Parabagous Schilsky 1907
Abagous Sharp 1916
Parabagous Sharp 1916; not Schilsky 1907
Probagous Sharp 1916
Heterobagous Solari 1930
Himeniphades Kôno 1934
Memptorrhynchus Iablokov-Khnzorian 1960
Fontenelleus Hoffmann 1962

Pnigodes LeConte 1876, 1 sp., *P. setosus* LeConte 1876, generally distributed in central and southwestern United States. This genus is questionably distinct from *Bagous*.

VI. Baridinae Schoenherr 1836

by Robert S. Anderson

Among all weevils, those in the Baridinae are in need of the most study. The group as a whole is difficult to characterize and generic concepts and definitions need much refinement. There are many genera and some are of questionable validity. Some genera (*e.g.*, *Baris*, *Pseudobaris*, *Onychobaris*, *Sibariops*, etc.) have numerous included species but these have not been studied since their original descriptions and many of them are still known only from type series and localities. Thomas Lincoln Casey was the last person to seriously study this subfamily and is responsible for most of the generic and species concepts and names in use today. His types are all located at the Smithsonian Institution in Washington D.C. and a critical study of this collection is central to resolving the state of taxonomy in this subfamily.

Most baridines are glossy and black, with few (usually white) or no scales on the body, and are most readily recognized by an ascended mesepisternum that is visible between the hind angle of the pronotum and the elytral humerus. They share this latter feature with Ceutorhynchinae but the latter have an exposed pygydium (shared with some Baridinae) and have a very small or no apical tooth on the hind tibia. This tooth is generally well-developed in Baridinae or the outer curved face of the tooth is continuous with the apex of the outer tibial margin or is connected to it by a distinct, continuous sharp carina that traverses the apical face of the tibia.

The natural history of baridines is poorly known. Some species are associated with monocots such as various grasses, sedges, and palms. Larvae mostly mine stems. Some species in the genera *Buchananius* and *Plocamus* appear to be associated with fungi on dead wood. Adults, especially of the tribe Madopterini, frequently visit flowers. Many baridines can also be found in semi-aquatic habitats.

Key to the Nearctic genera of Baridinae
(slightly modified from Kissinger 1964)

1. Tarsus with a single claw 2
— Tarsus with two claws (may be connate at base) 3

2(1). Body subcylindrical in form, elytra about twice as long as wide; middle coxae separated by a distance less than the width of a coxa .. *Barilepton*
— Body elongate-oval in form, elytra about 1.5 times as long as wide; middle coxae separated by a distance about equal to the diameter of a coxa .... *Eisonyx*

3(1). Tarsus with claws connate at base 4
— Tarsus with claws separate at base 20

4(3). Pygydium not covered by elytra, broadly exposed, punctate, nearly vertical 5
— Pygydium more or less covered by elytra, mostly smooth, lacking obvious punctures, oblique 11

5(4). Front coxae widely separated by a distance greater than the width of a coxa 6
— Front coxae narrowly separated by a distance less than the width of a coxa 9

6(5). Prosternum unimpressed in front of coxae; antenna with club more or less subcylindrical in shape, about as wide as article 7 of funicle, basal article of club about half as long as club .... *Orchidophilus*
— Prosternum with median sulcus or apical impression in front of coxae; antenna with club more or less oval in shape, distinctly wider than article 7 of funicle, basal article of club about a third as long as club 7

7(6). Prosternum with apical excavation but lacking sulcus immediately anterior to coxae; elytra with intervals flat; body nearly glabrous; femora not toothed .... *Ampeloglypter*
— Prosternum with median sulcus extended from coxae to near apex; elytra with intervals various; elytra often with scattered white scales or with a patch of white scales at base of interval 3; femora with or without tooth 8

8(7). Elytra with intervals rather wide and flat on disk (Fig. 25); body color black or dark piceous; elytra often with scattered white scales or with a patch of white scales at base of interval 3; femora with or without tooth .... *Pseudobaris*
— Elytra with intervals narrow and convex on disk (Fig. 24); body color pale reddish brown; elytra nearly glabrous; femora without tooth ... *Desmoglyptus*

9(5). Antenna with club about as long as preceding six articles of funicle .... *Hesperobaris*
— Antenna with club shorter than preceding five articles of funicle 10

10(9). Rostrum distinctly separated from head by marked constriction at base of rostrum; body with dense scales; size greater than 3.0 mm ..... *Trichobaris*
— Rostrum at most slightly separated from head by slight constriction at base of rostrum; body nearly glabrous; size less than 2.0 mm .... *Microbaris*

11(4). Prosternum not sulcate in front of coxae and/or coxae separated by distance greater than width of a coxa 12
— Prosternum with deep, narrow sulcus in front of coxae and/or coxae separated by distance less than diameter of a coxa 14

12(11). Rostrum short and stout, shorter than pronotum; body nearly glabrous, with sparse, minute hair-like scales, elytra blue in color, intervals nearly impunctate .... *Zygobarinus*
— Rostrum long and slender, longer than pronotum; body with some sparse, broad scales, elytra black or piceous in color, intervals with deep, coarse punctures 13

13(12). Elytra with striae narrow, punctures wider than striae; prosternum in front of coxae with a pair of low ridges which are divergent posteriorly; southern Florida .... *Zygobaris*
— Elytra with striae broad, punctures not as wide as striae; prosternum in front of coxae with ridges developed only near apex; southern Texas .... *Zygobarella*

14(11). Prosternum behind coxae with deep narrow sulcus; rostrum longer than pronotum, abruptly attenuate immediately beyond antennal insertion, antennal insertion sub-basal (Fig. 28); antenna with article 2 of funicle long, more or less one-half length of scape .... *Amercedes*
— Prosternum behind coxae lacking sulcus; rostrum various in length, of more or less subequal width throughout length, not attenuate beyond antennal insertion, antennal insertion near or in front

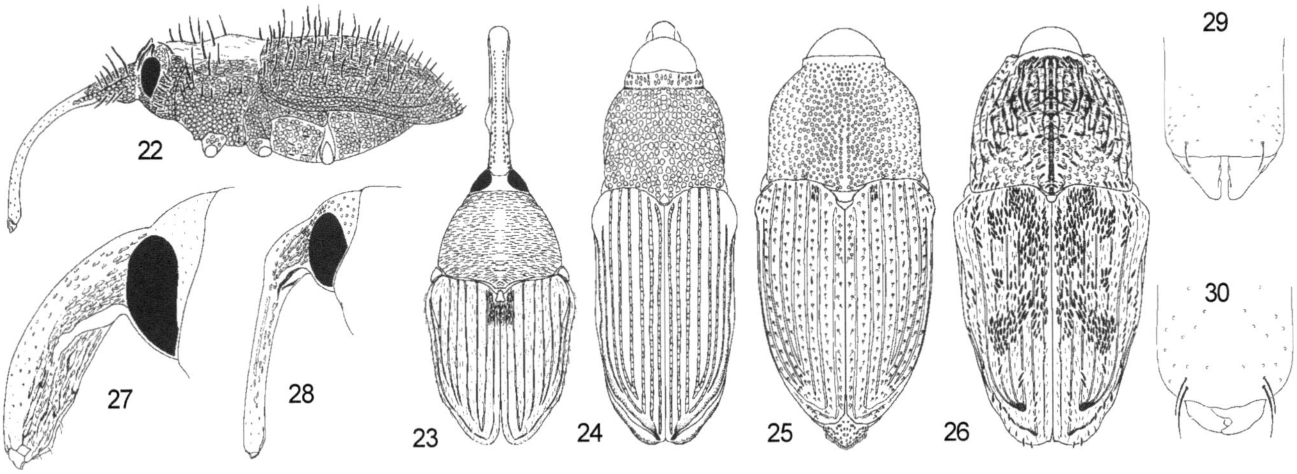

FIGURES 22.131-30.131. Baridinae. 22. *Plocamus echidna* (LeConte), lateral habitus. 23-26. Dorsal habitus, 23. *Geraeus patagoniensis* (Sleeper); 24. *Desmoglyptus arizonicus* Casey 1920; 25. *Pseudobaris nigrina* (Say); 26. *Glyptobaris lecontei* Champion. 27-28. Lateral view of head, 27. *Baris* sp.; 28. *Amercedes subulirostris* Casey. 29-30. Dorsal view of apex of rostrum and mandibles, 29. *Odontocorynus salebrosus* (Casey); 30. *Haplostethops* sp.

of middle of rostrum; antenna with article 2 of funicle short, much less than length of scape 15

15(14). Prosternum lacking sulcus medially in front of coxae; rostrum about as long as pronotum, slender, cylindrical, slightly curved .... *Strongylotes*

— Prosternum with median sulcus in front of coxae; rostrum shorter than pronotum, stout, distinctly curved 16

16(15). Body nearly glabrous; prosternum in front of coxae with glabrous, median, broad sulcus with acute lateral margins; pronotum with fine, shallow punctures .... *Stethobaris* (part)

— Body with narrow or broad scales; prosternum in front of coxae with median suclus with scales and with low, rounded lateral margins; pronotum with deep punctures 17

17(16). Mandibles prominent, not overlapped when closed; rostrum abruptly separated from head by deep constriction; hind tarsus with article 5 longer than articles 1 and 2 combined .... *Acentrinops*

— Mandibles small, not prominent, overlapped when closed; rostrum only slightly separated from head by slight constriction; hind tarsus with article 5 subequal in length to shorter than articles 1 and 2 combined 18

18(17). Antenna with basal article of funicle elongate, slender, longer than articles 2 to 5 combined; elytra with broad scales arranged in groups; body length greater than 3.0 mm; body form subcylindrical; tarsal claws connate to near apex .... *Barinus*

— Antenna with basal article of funicle stout, shorter than articles 2 to 4 combined; elytra with some solitary white scales; body length less than 2.5 mm; body form oval; tarsal claws connate only at base 19

19(18). Prosternum with median impression indistinctly defined laterally, wider posteriorly ...... *Catapastus*

— Prosternum with median impression distinctly defined and ridged laterally, wider anteriorly .... *Catapastinus*

20(3). Pygydium more or less completely exposed beyond elytral apex, punctate, nearly vertical 21

— Pygydium covered by elytra, mostly smooth, lacking obvious punctures, oblique 36

21(20). Hind tibia lacking tooth at apical margin or with tooth or process shorter than tarsal claw 22

— Hind tibia with tooth at apical margin about as long as tarsal claw 25

22(21). Mandible prominent, triangular, inner face smooth and straight .... *Pseudocentrinus* (part; male)

— Mandible with distinct teeth on inner face 23

23(22). Antenna with club shorter than articles 2-7 of funicle, article 2 of funicle longer than 3; abdominal ventrite 5 distinctly longer than 3 and 4 combined .... *Centrinogyna* (part; male)

— Antenna with club longer than articles 2-7 of funicle, articles 2 and 3 of funicle subequal in length; abdominal ventrite 5 shorter than 3 and 4 combined 24

24(23). Prosternum with shallow median sulcus in front of coxae; rostrum in lateral view distinctly separated from head by a marked transverse impression at base; scutellum quadrate .... *Orthoris*

— Prosternum flat in front of coxae, lacking median sulcus; rostrum in lateral view at most slightly separated from head by a slight transverse impression at base; scutellum triangular .... *Rhoptobaris*

25(21). Front coxae narrowly separated by a distance much less than the width of a coxa 26

— Front coxae widely separated by a distance greater than the width of a coxa 34

26(25). Prosternum with deep, narrow median sulcus in front of coxae 27

— Prosternum flat in front of coxae or with at most a slight median longitudinal impression 28

27(26). Antenna with article 2 of funicle more than twice as long as wide, as long as articles 3 and 4 combined; body form elliptical in dorsal view .... *Aulobaris*

— Antenna with article 2 of funicle less than twice as long as wide, shorter than articles 3 and 4 combined; body form elongate, subparallel in dorsal view .... *Trepobaris*

28(26). Elytra with striae 1and 2 deeply linearly punctate in basal one-third, striae deeply continuously impressed in apical two-thirds only; male with rostrum with ventral surface with long dense pilosity .... *Myctides*

— Elytra with striae 1and 2 deeply continuously impressed throughout entire length; male with rostrum with ventral surface glabrous or with a few short setae 29

29(28). Rostrum in lateral view continuous with head, not separated from head by transverse impression; eyes large, extended onto dorsal surface of head; frons about one-half as wide as rostrum at apex; body subcylindrical in form .... *Stenobaris*

— Rostrum in lateral view distinctly separated from head by transverse impression; eyes smaller, lateral, not extended onto dorsum of head; frons about as wide as rostrum at apex; body various in form 30

30(29). Pronotum with sides covered with broad, round scales; body with white and tan colored scales intermixed .... *Cosmobaris*

— Pronotum with sides lacking broad, round scales; body either subglabrous or with only white scales 31

31(30). Rostrum in lateral view separated from head by shallow, broad impression (Fig. 27) 32

— Rostrum in lateral view separated from head by a deep groove or dorsal constriction of the base of the rostrum 33

32(31). Elytra subglabrous, vestiture fine, minute, of uniform length and form; antenna with basal article of club less pubescent and more glossy than other articles of club; body size 2.8-6.0 mm .... *Baris*

— Elytra fine minute hair-like vestiture as well as broad, white scales which form a spot at the base of elytral interval 3 (other spots may also be present); antenna with club uniformly pubescent; body size 1.8-3.0 mm .... *Plesiobaris*

33(31). Pronotum broadly constricted at apex, not tubulate; vestiture of fine scales, recumbent .... *Pycnobaris*

— Pronotum sharply constricted at apex, tubulate; vestiture of fine scales and setae intermixed, suberect .... *Stictobaris*

34(25). Surface of pronotum distinctly rugose (Fig. 26); elytra with coarse elongate yellow scales and fine setae; prosternum produced posteriorly only slightly over mesosternum; mandible with inner face nearly smooth .... *Glyptobaris*

— Surface of pronotum punctate; elytra glabrous or nearly so, with at most fine short, hair-like scales; prosternum produced posteriorly over mesosternum; mandible with inner face deeply notched 35

35(34). Femora with ventral margin with tooth; pronotum and elytra with fine, sparse punctures; body nearly glabrous; elytra with surface uneven..... .... *Madarellus*

— Femora with ventral margin simple, lacking tooth; pronotum and elytra with deep, uniform punctures; body with fine short, suberect hair-like scales; elytra with surface more or less even... .... *Onychobaris*

36(20). Body with sparse, erect, coarse setae (Fig. 22); body size small, less than 3.0 mm 37

— Body with at most fine setae or appressed scales; body size various 38

37(36). Front coxae widely separated by distance much greater than width of a coxa; prosternum flat in front of coxae; body densely covered with broad, appressed scales in addition to erect setae (Fig. 22) .... *Plocamus*

— Front coxae narrowly separated by distance less than width of a coxa; prosternum medially shallowly, broadly impressed in front of coxae; body with at most some fine hair-like scales in addition to erect setae .... *Buchananius*

38(36). Mandibles prominent, triangular in form when viewed dorsally, not or only slightly overlapped or crossed when closed (Fig. 29) 39

— Mandibles not prominent, less obviously triangular in form when viewed dorsally, overlapped or crossed when closed (Fig. 30) 52

39(38). Mandible with inner face smooth (Fig. 29), not dentate or emarginate, usually straight but divergent in some specimens 40

— Mandible with inner face dentate or crenulate, straight 45

40(39). Abdomen with ventrite 5 tumid medially, more than three times as long as ventrites 3 and 4 combined; prosternum deeply, narrowly sulcate in front of coxae..... *Pseudocentrinus* (part; female)

— Abdomen with ventrite 5 flat medially, at most slightly longer than ventrites 3 and 4 combined; prosternum various 41

41(40). Antenna with distal articles of funicle obliquely truncate, club with basal article with glabrous area with blunt or dentiform process on inner face .. .. *Odontocorynus* (part; male)

— Antenna with distal articles of funicle and basal article of club simple, not modified 42

42(41). Male with prosternum flat in front of front coxae.. 43

— Male with prosternum deeply excavated anterior to front coxae.. 44

43(42)[4]. Pronotum with anterior tubulate portion with a longitudinal fold on each side; prosternum with vestiture on median line not radiating from a central point, but directed backwards (Fig. 23) .... *Geraeus*

— Pronotum with anterior tubulate portion lacking a longitudinal fold on each side; prosternum with vestiture on median line radiating from a central point, lying either behind, on, or before the posterior line of the tubulate portion ... *Linogeraeus*

44(42). Prosternum of male with large, deep pit between prosternal spines .... *Pachygeraeus*

— Prosternum of male flat between spines but with transverse, deep, median anterior fossa .... *Pycnogeraeus*

45(39). Front coxae widely separated by distance equal to width of a coxa; hind tibia with large tooth at apical margin at least as long as a tarsal claw 46

— Front coxae narrowly separated by distance obviously less than width of a coxa; hind tibia lacking tooth at apical margin or tooth obviously shorter than a tarsal claw 48

46(45). Rostrum with point of antennal insertion medial; prosternum flat, lacking sulcus; body elongate oval in form .... *Calandrinus*

— Rostrum with point of antennal insertion in basal one-half; prosternum with median sulcus in front of coxae; body broadly oval in form 47

47(46). Elytra, except at base, nearly glabrous; pronotum with basal margin distinctly produced posteriorly and emarginate immediately in front of scutellum; front tibia simple, not excavated to receive base of tarsus .... *Pachybaris*

— Elytra with scattered broad white scales; pronotum with basal margin distinctly produced posteriorly but not emarginate immediately in front of scutellum; front tibia with outer surface at apex deeply excavated to receive base of tarsus .... *Diorymeropsis*

48(45). Rostrum with point of antennal insertion behind midlength; prosternum of female with median longitudinal sulcus in front of coxae, in male, with two erect processes in front of which is an elongate-oval impression .... *Centrinopus*

— Rostrum with point of antennal insertion at or in front of midlength; prosternum of male unarmed in front of coxae 49

49(48). Metasternum short, middle and hind coxae separated by less than length of abdominal ventrite 1 behind hind coxa; body broadly oval in dorsal form .... *Microcholus*

— Metasternum longer, middle and hind coxae separated by a distance at least equal to length of abdominal ventrite 1 behind hind coxa; body elongate in dorsal form 50

50(49). Front coxae separated by more than one-half width of a coxa; mandible with outer face denticulate .... *Centrinogyna*

— Front coxae narrowly separated by distance less than one-half width of a coxa (exception some *Nicentrus*); mandible with outer margin not denticulate 51

[4] The genus *Centrinus* with only the species *C. pistor* (Germar 1824) should key here (not seen by me). Its relationship to *Geraeus* and *Linogeraeus* needs to be reassessed (see text).

51(50). Pronotum constricted apically, markedly tubulate .... *Centrinites*

— Pronotum not constricted or tubulate at apex .... *Nicentrus*

52(38). Lateral profile with dorsal surface markedly, evenly convex; elytra with striae obsolete, indicated only by rows on punctures 53

— Lateral profile with dorsal surface flattened near middle, not evenly convex; elytra with striae distinct, moderately deep 54

53(52). Prosternum with median longitudinal impression in front of coxae; pronotum markedly tubulate at apex .... *Oomorphidius*

— Prosternum flat in front of coxae; pronotum not tubulate at apex .... *Cholinobaris*

54(52). Shortest distance between middle and hind coxae less than, or equal to, one-half length of metepisternum 55

— Shortest distance between middle and hind coxae usually greater than one-half length of metepisternum 58

55(54). Antenna with article 2 of funicle about as wide as long, stout; antennal club large, elongate, about as long as funicle; body nearly glabrous, with minute hair-like scales .... *Stethobaris*

— Antenna with article 2 of funicle distinctly longer than wide, slender; antennal club moderately long, not as long as funicle; body with fine, but distinct, long hair-like scales 56

56(55). Pronotum constricted apically but not tubulate ... . *Oligolochus*

— Pronotum tubulate at apex 57

57(56). Prosternum deeply sulcate in front of coxae; body lacking dense white scales on venter .... *Idiostethus*

— Prosternum shallowly sulcate in front of coxae; lateral margins of thoracic sterna with dense, white, overlapped scales .... *Haplostethops*

58(54). Prosternum of male with a pair of erect, slender processes in front of coxae or with low transverse carina 59

— Prosternum of male unarmed in front of coxae 60

59(58). Body form in dorsal view elongate-oval; prosternum of male with short, slender spines ...... *Sibariops*

— Body form in dorsal view slender, subcylindrical; prosternum of male with long, slender spines and a deep rounded fossa anterior to spines .... *Cylindridia*

60(58). Front coxae narrowly separated by distance distinctly less than width of a coxa 61

— Front coxae widely separated by distance greater than width of a coxa 62

61(60). Elytra with fine, inconspicuous hair-like scales; pronotum strongly constricted at apex and tubulate .... *Dirabius*

— Elytra with moderately coarse, elongate, white conspicuous scales; pronotum not constricted or tubulate at apex .... *Trichodirabius*

62(60). Body with fine, elongate scales; antenna with funicle with very fine, long setae .... *Apinocis*
— Body with coarse, elongate scales; antenna with coarse, elongate scales .... *Barilepsis*

CLASSIFICATION OF THE NEARCTIC BARIDINAE

18. Baridini Schoenherr 1836

Baridina Schoenherr 1836

Aulobaris LeConte 1876, 10 spp., generally distributed in eastern United States and Canada, also California. Some species are associated with wetlands. See Casey (1892, 1920) to separate the species. The genus needs revision.

Baris Germar 1817, 92 spp., generally distributed. Species are associated with various plants, mostly Asteraceae. See Casey (1892) and Gilbert (1964) to separate some of the species. The genus needs revision; many species are of questionable validity.
Baridius Schoenherr 1825
Cyphirhinus Schoenherr 1826
Aegyptobaris Pic 1889 (valid subgenus)
Turkmenobaris Zaslavskij 1956 (valid subgenus)

Cosmobaris Casey 1920, 1 sp., *C. americana* Casey 1920, generally distributed. This species is associated with *Chenopodium* (Chenopodiaceae); larvae mine stems (Kissinger 1964).

Desmoglyptus Casey 1892, 2 spp., Maryland, Pennsylvania, Virginia, District of Columbia, and Arizona. Species are associated with *Vitis* (grape; Vitaceae) (Kissinger 1964). See Casey (1920) to separate the species.

Hesperobaris Casey 1892, 2 spp., Missouri, Kansas and Texas. See Casey (1892) to separate the species.

Microbaris Casey 1892, 1 sp., *M. galvestonica* Casey 1892, Texas.

Orthoris LeConte 1876, 7 spp., generally distributed in the western United States and Canada. Species are associated with *Mentzelia* (Loasaceae); larvae are in pods, stems and roots (Pierce 1907). See Casey (1892, 1920) to separate the species. The genus needs revision.

Plesiobaris Casey 1892, 6 spp., generally distributed in eastern United States. Species are associated with *Hypericum* (Hypericaceae) in wetlands. See Casey (1892, 1920) to separate the species.

Pseudobaris LeConte 1876, 31 spp., generally distributed in eastern United States and Canada, west to California, Colorado and Utah. At least one species is associated with *Lycopus* (Labiatae) (Kissinger 1963). See Casey (1892) to separate the species. The genus needs revision; many species are of questionable validity.
Pseudobaridia Casey 1920 (valid subgenus)

Pycnobaris Casey 1892, 4 spp., Kansas, Colorado, Texas and California. See Casey (1892, 1920) to separate the species.

Rhoptobaris LeConte 1876, 1 sp., *R. canescens* LeConte 1876, Kansas, Colorado, Texas and Oregon.

Stenobaris Linell 1897, 1 sp., *S. avicenniae* Linell 1897, Florida. This species is associated with *Avicennia germinans* (L.) L. (black mangrove; Avicenniaceae) (Linell 1897).

Trepobaris Casey 1892, 1 sp., *T. elongata* Casey 1892, Arizona and Texas.

Trichobaris LeConte 1876, 9 spp., generally distributed in eastern and southern United States and southern Canada. Species are associated with various Solanaceae; larvae are in stems (Barber 1935; Cuda and Burke 1985). See Barber (1935) to separate the species.

19. Madarini Jekel 1865

Madarina Jekel 1865

Ampeloglypter LeConte 1876, 3 spp., generally distributed in eastern United States and southern Canada. Species are associated with *Vitis* (grape; Vitaceae); larvae make galls on stems (Kissinger 1964). See Blatchley and Leng (1916) to separate the species.

Glyptobaris Casey 1892, 1 sp., *G. lecontei* Champion 1909, generally distributed in the eastern United States.

Madarellus Casey 1892, 5 spp., generally distributed in the eastern United States and Canada west to Texas. Species are associated with *Vitis* (Vitaceae) (Blatchley and Leng 1916). See Casey (1892, 1920) and Blatchley and Leng (1916) to separate the species. The genus needs revision.
Willinkia Bondar 1949

Onychobaris LeConte 1876, 33 spp., generally distributed in the United States, but especially in the southwest including California. See Casey (1892, 1920) to separate some of the species. The genus needs revision; many species are of questionable validity.

Orchidophilus Buchanan 1935, 3 spp., New Jersey, District of Columbia, and California; adventive in orchid houses, likely not established in the wild. See Buchanan (1935) to separate the species.

Stictobaris Casey 1892, 4 spp., southwestern and central United States. See Sleeper (1957b) to separate the species.

Tonesiina Alonso-Zarazaga and Lyal 1999

Myctides Pascoe 1874, 1 sp., *M. imberbis* Lea 1906, Florida; adventive. This species is associated with *Syzygium jambos* Alston (Myrtaceae); larvae in seeds (Woodruff 1977; Anderson 1993a).

20. Madopterini Lacordaire 1866

Torcina Bondar 1943

Sibariops Casey 1920, 41 spp., generally distributed in eastern United States and Canada. Species are associated with sedges in wetlands. See Casey (1920) to separate the species. The genus needs revision; many species are of questionable validity.

Zygobaridina Pierce 1907

Acentrinops Casey 1920, 1 sp., *A. brevicollis* Casey 1920, New Mexico and Texas.

Amercedes Casey 1894, 1 sp., *A. subulirostris* Casey 1893, Louisiana and Texas. This species is associated with *Zanthoxylum* (Rutaceae) (Pierce 1907).

Zygobaroides Pierce 1907

Apinocis Lea 1927, 15 spp., generally distributed. Larvae of at least one species, *A. saccharinus* (Marshall 1952), have been associated with grasses (Poacaeae) (Woodruff 1972). See Buchanan (1932) to separate most of the species. The genus needs revision.

Prosaldius Ogloblin 1930
Anacentrinus Buchanan 1932

Barilepis Casey 1920, 3 spp., generally distributed in the eastern United States, Arizona and Texas. See Casey (1920) to separate the species.

Barilepton LeConte 1876, 4 spp., generally distributed in eastern and southern United States. Species are associated with wetlands. See Casey (1892) to separate the species.

Barinus Casey 1892, 14 spp., generally distributed in eastern United States, also California. Species are associated with sedges in wetlands. See Sleeper (1956a) to separate the species.

Buchananius Kissinger 1957, 2 spp., generally distributed in eastern United States. Species are associated with various dead tree limbs on the ground (Kissinger 1964). See Kissinger (1958) or Blatchley and Leng (1916) to separate the species.

Zaglyptus LeConte 1876; not Foerster 1868

Calandrinus LeConte 1876, 2 spp., New Mexico, Texas, Colorado and Alberta. See Casey (1892, 1920) to separate the species.

Catapastinus Champion 1908, 1 sp., *C. caseyi* Champion 1908, southern Texas. This species is associated with *Zanthoxylum fagara* (L.) Sarg. (Rutaceae).

Catapastus Casey 1892, 6 spp., generally distributed in the eastern United States, Florida and Texas. Species are associated with various species of *Zanthoxylum* (Rutaceae). See Casey (1892, 1920) to separate the species.

Centrinites Casey 1892, 2 spp., generally distributed in the eastern United States. Species are associated with *Melanthium virginicum* L. (Liliaceae) (Blatchley and Leng 1916). See Casey (1920) to separate the species.

Leptosaldius Casey 1922

Centrinogyna Casey 1892, 5 spp., generally distributed in the western United States and central Canada. See Casey (1892, 1920) to separate the species. The genus needs revision.

Centrinopus Casey 1892, 6 spp., generally distributed in the eastern United States. Adults are found on flowers of Asteraceae (Kissinger 1964). See Casey (1920) to separate the species. The genus needs revision.

Centrinus Schoenherr 1825, 1 sp., *C. pistor* (Germar 1824), Kentucky. This species is of uncertain affinity. It has not been seen by me and its relationships to *Geraeus* and *Linogeraeus* need to be reassessed. Champion (1908: 261-261) limited *Centrinus* to large South American species with the mandibles strongly denticulate along the inner margins.

Toxeres Germar 1829
Toxeres Schoenherr 1833; not Germar 1829
Telephus Gistel 1848

Cholinobaris Casey 1920, 1 sp., *C. rhomboidea* Casey 1920, North Carolina.

Cylindridia Casey 1920, 4 spp., generally distributed in the eastern United States and southern Canada, west to Texas and Colorado. Species are associated with sedges (Cyperaceae). See Casey (1920) to separate the species.

Diorymeropsis Champion 1908, 1 sp., *D. xanthoxyli* (Linell 1897), Texas. This species is associated with *Zanthoxylum* (Rutaceae).

Pseudogarnia Casey 1920

Dirabius Casey 1920, 9 spp., generally distributed in the eastern United States and southern Canada, also one species in California. One species has been associated with *Scirpus cyperinus* L. (Cyperaceae); larvae are in the stems (Blatchley and Leng 1916).

Limnobaropsis Casey 1920 (valid subgenus)

Eisonyx LeConte 1880, 3 spp., Texas, New Mexico, Kansas, Oklahoma, Missouri, Iowa and Tennessee. Species are associated with *Senecio* and *Hymenoxys* (Asteraceae); larvae are in stems, crowns and roots (Pakaluk and Carlow 1994). See Pakaluk and Carlow (1994) to separate the species.

Eumononycha Casey 1893 (valid subgenus)

Geraeus Pascoe 1889, 21 spp., generally distributed in the eastern and southwestern United States and southeastern Canada. At least some species are associated with grasses (Poaceae) (Kissinger 1964); adults frequently visit flowers. See Casey (1892, 1920) and Blatchley and Leng (1916) to separate some of the species; consult O'Brien and Wibmer (1984) for a listing of species included in the genus. The genus needs revision; many species are of questionable validity. Generic definitions in this part of the Baridinae need much study.

Centrinaspis Casey 1920

Haplostethops Casey 1920, 7 spp., central United States, four species known only from Missouri. Species are associated with wetlands. See Casey (1920) to separate the species. The genus needs revision; some species are of questionable validity.

Idiostethus Casey 1892, 16 spp., generally distributed in the eastern United States and southern Canada. Adults have been associated with various flowers (Blatchley and Leng 1916); larvae may be associated with orchids. See Casey (1892, 1920) to separate the species. The genus needs revision; many species are of questionable validity.

Linogeraeus Casey 1920, 15 spp., generally distributed in the southeastern and southwestern United States. Adults frequently visit flowers. See Casey (1920) to separate some of the species; consult O'Brien and Wibmer (1984) for listing of species included in the genus. The genus needs revision; many species are of questionable validity.

Stereogeraeus Casey 1920
Conocentrinus Casey 1920
Glyptogeraeus Casey 1920
Brachygeraeus Casey 1920
Centrinaspidia Casey 1920
Lepidobaris Lea 1927; not Champion 1909

Microcholus LeConte 1876, 2 spp., New Jersey, Georgia and Florida. Species are associated with wetlands. See Casey (1892) and Blatchley and Leng (1916) to separate the species.

Nicentrus Casey 1892, 20 spp., generally distributed in the eastern United States west to Texas and Arizona. Adults are frequently found on flowers in various habitats. See Casey (1892, 1920) and Blatchley and Leng (1916) to separate some of the species. The genus needs revision.

Nicentrites Casey 1922
Eunicentrus Casey 1922

Odontocorynus Schoenherr 1844, 51 spp., generally distributed in the eastern United States and southern Canada, west to Texas and Colorado. Adults are frequently found on flowers (mostly Asteraceae) in various habitats. See Casey (1920) and Blatchley and Leng (1916) to separate some of the species. The genus needs revision; many species are of questionable validity.

Oligolochus Casey 1892, 7 spp., generally distributed in the eastern United States, Arizona and California. See Buchanan (1932) to separate the species.

Anacentrus Casey 1920

Oomorphidius Casey 1892, 2 spp., southeastern and central United States. See Casey (1892) to separate the species.

Pachybaris LeConte 1876, 1 sp., *P. porosa* LeConte 1876, Florida and Louisiana. Adults have been associated with flowers of *Serenoa repens* (Bartr.) Small and *Sabal palmetto* (Walt.) Lodd (saw palmetto and cabbage palm; Arecaceae) (Anderson 1993a).

Pachygeraeus Casey 1920, 3 spp., generally distributed in the central United States. See Casey (1920) to separate the species.

Plocamus LeConte 1876, 2 spp., generally distributed in the eastern United States and southern Canada. Adults have been associated with hickory, beech and maple (Blatchley and Leng 1916). See Blatchley and Leng (1916) to separate the species.

Euchaetes LeConte 1876; not Harris 1841; not Philippi 1843; not Sclater 1858
Eunyssobia Casey 1892
Epeuchaetes Lyman 1902

Pseudocentrinus Champion 1908, 1 sp., *P. ochraceus* (Boheman 1844), Texas.

Pycnogeraeus Casey 1920, 3 spp., Pennsylvania, Florida, Texas, New Mexico and Arizona. See Casey (1892) to separate the species.

Stethobaris LeConte 1876, 11 spp., generally distributed in the eastern United States and southern Canada, west to Texas and Arizona. A number of the species are associated with orchids (Brown 1966a; Hull Sieg and O'Brien 1993; Howden 1995). See Casey (1892) to separate some of the species. The genus needs revision.

Diorymerellus Champion 1908

Trichodirabius Casey 1920, 2 spp., Florida, Louisiana and Texas. See Casey (1920) to separate the species.

Zygobarella Casey 1920, 1 sp., *Z. xanthoxyli* (Pierce 1907), Texas. This species is associated with *Zanthoxylum* (Rutaceae); larvae develop in fruits (Pierce 1907).

Zygobarinus Pierce 1907, 1 sp., *Z. coelestinus* (Linell 1897), Florida.

Zygobaris LeConte 1876, 1 sp., *Z. nitens* LeConte 1876, Florida. Adults have been collected on *Zanthoxylum flavum* Vahl. (Rutaceae) (Anderson 1993a).

21. Nertinini Voss 1954

Strongylotes Schoenherr 1836, 1 sp., *S. parallelus* Champion 1907, Texas and Arizona.

VII. Ceutorhynchinae Gistel 1856

by Boris A. Korotyaev and Robert S. Anderson

Ceutorhynchinae are a relatively well-known group of small weevils found in both terrestrial and freshwater aquatic habitats throughout North America. They are readily recognized by the ascended mesepimeron (as in Baridinae), an exposed pygydium, and presence of a small or no tooth at the apex of the hind tibia. Many possess a prosternal channel for the reception of the rostrum and have pronotal postocular lobes that cover the eyes when the rostrum is in repose, but some do not. Adults of some species have expanded hind femora and are good jumpers.

Species of Ceutorhynchinae are associated with a variety of plant families. In terrestrial habitats the Cruciferae are a common host, whereas in aquatic habitats the most common host would appear to be Polygonaceae or semi-aquatic Cruciferae. Larvae of terrestrial species usually mine the stems or crowns of the plants but some aquatic taxa in the Phytobiini such as *Phytobius* have larvae that live and feed externally on plant reproductive structures. Some species in the genus *Ceutorhynchus* are adventive and serious pests of cultivated Cruciferae (especially rapeseed or canola) in western North America. Species in the genera *Phrydiuchus*, *Microplontus*, *Mogulones* and *Trichosirocalus* have been deliberately introduced for the biological control of pest weeds.

KEY TO THE NEARCTIC GENERA OF CEUTORHYNCHINAE

1. Tarsus with single claw; body size 3.6-5.5 mm; associated with *Iris versicolor* L. (Iridaceae) .... *Mononychus*

— Tarsus with 2 claws; body size 1.5-5.0 mm (most less than 3.5 mm); associated with various plants, not Iridaceae 2

2(1). Rostrum shorter than pronotum, thick, not more than 3 times as long as wide, weakly and unevenly curved; femora with tooth on ventral margin; hind femur rather strongly widened, 1.5-2.0 times as wide as middle femur; associated with *Portulaca oleracea* L. (Portulaceae) .... *Hypurus*

— Rostrum much longer than pronotum, more than 3 times as long as wide; if short and thick, then femora simple, unarmed, and hind femur weakly widened, less than 1.3 times as wide as middle one; associated with various plants, not Portulaceae 3

3(2). Rostrum no more than 3 times as long as wide (Fig. 35), wider than front femur; femora unarmed; anterior margin of pronotum not raised, often with 2 more or less acute denticles or with emargination limited by angular prominences....................... 4

— Rostrum more than 3 times as long as wide (Fig. 34); if less than 3 times, then femora dentate, anterior margin of pronotum strongly raised and size over 3.5 mm (*Phrydiuchus*), or metasternum between middle coxae depressed 12

4(3). Antenna with funicle of 7 articles; tarsal claws dentate; anterior margin of pronotum simple, without sharp denticles; prosternum anterior to front coxae long, with high keels; distance between front coxae usually not less than one-half width of rostrum .... *Rhinoncus*

— Antenna with funicle of 6 articles; tarsal claws dentate or simple; anterior margin of pronotum often with 2 sharp denticles; prosternum anterior to front coxae sometimes short, with low, sometimes obsolete keels; distance between front coxae less than one-half width of rostrum 5

5(4). Tarsal claw with well-developed tooth at base; prosternum in front of coxae deeply excavated with its anterior margin deeply angularly emarginate, the emargination extending behind the level of anterior margins of coxal cavities; inner, usually also posterior margins of eyes sharply raised 6

— Tarsal claw simple, if dentate (in *Neophytobius*), then front coxae separated at most by width of antennal club and apical margin of pronotum narrowly

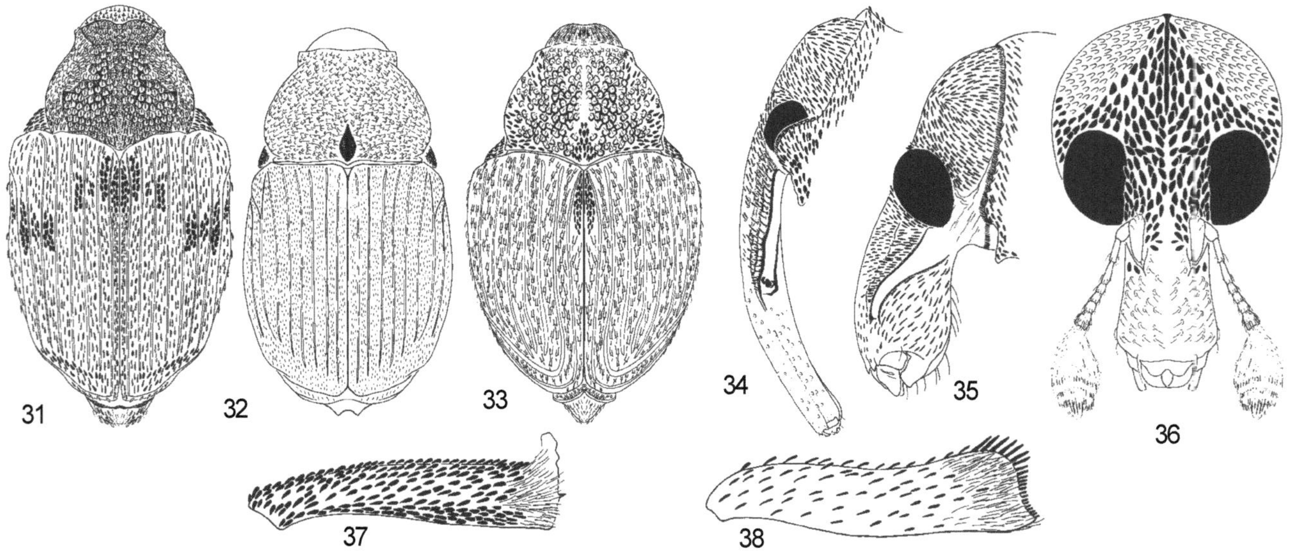

FIGURES 31.131-38.131. Ceutorhynchinae. 31-33. Dorsal habitus, 31. *Allosirocalus* sp.; 32. *Glocianus punctiger* (Gyllenhal); 33. *Homorosoma sulcipennis* (LeConte). 34-35. Lateral view of head, 34. *Ceutorhynchus rapae* Gyllenhal; 35. *Rhinoncus pericarpius* (Linnaeus). 36. *Pergasteromimus tetracanthus* (Champion), head and rostrum, dorsal view. 37-38. Front tibia, 37. *Acanthoscelis acephalus* (Say); 38. *Glocianus punctiger* (Gyllenhal).

excised; prosternum in front of coxae very weakly depressed and shallowly arcuately emarginate, the emargination not reaching the level of anterior margin of coxal cavities; inner margins of eyes slightly, if at all, raised 8

6(5). Rostrum about 3 times as long as wide, scarcely widened apically; antennal scrobe well developed, the dorsal margin reaching eye; elytra with alternate intervals moderately to rather strongly convex, more strongly so on apical prominences .... *Dietzella*

— Rostrum usually not more than twice as long as wide, strongly widened apically; antennal scrobe foveiform or very weakly developed, vanishing at most halfway to eye; elytra with alternate intervals not conspicuously convex 7

7(6). Rostrum less than twice as long as wide; antennae inserted on dorsal surface of rostrum, scape very short, shorter than two basal articles of funicle combined (Fig. 36); Florida ... *Perigasteromimus*

— Rostrum twice or more as long as wide, antennae inserted laterally; scape longer than two basal articles of funicle combined; widespread .... *Perigaster*

8(5). Body with very dense vestiture of short, recumbent or subrecumbent matte hydrophobous scales; narrow metallic-glossy scales may be present only on apical part of rostrum; tarsi narrow, 3rd article often scarcely wider than article 2; if (in *Parenthis*) 1.5 times as wide as the latter, then anterior margin of pronotum without denticles and inconspicuously emarginate medially; tarsal claw simple 9

— Body with sparse covering of scales composed partly (at least on rostrum) or mostly from metallic-glossy narrow scales; tarsi wider, 3rd article 1.5 to (mostly) 2.0 times as wide as article 2; tarsal claw simple or dentate 11

9(8). Body size larger, 2.6-3.0 mm long; tarsus very long and narrow, 3rd tarsal article slightly wider than article 2; dorsal and lateral surface of tarsal articles lacking long swimming hairs; pronotum with acute lateral tubercles; its rounded apical margin with shallow median emargination limited by acute angulations; elytra with interval 5 keel-shaped and finely muricate in basal one-third .... *Phytobius*

— Body size smaller, 2.2-2.7 mm long; 3rd tarsal article 1.6-1.7 times as wide as article 2; if scarcely wider, then tarsus dorsally with a few long and very fine swimming hairs; pronotum with obsolete obtuse lateral tubercles and apical margin lacking any trace of median emargination or acute angulations; elytra with interval 5 not carinate in basal part 10

10(9). Tarsus with article 3 less than 1.4 times as wide and about one-half as long as article 2; dorsal and lateral surface of tarsal articles covered with long very fine semi-erect swimming hairs; middle coxae separated by about one-half their width .... *Euhrychiopsis*

— Tarsus with 3rd article 1.6-1.7 times as wide and almost as long as article 2; tarsi lacking long swimming hairs; middle coxae separated by about their width .... *Parenthis*

11(8). Tarsus with claws dentate; apical margin of pronotum at middle produced anteriorly and not raised, with shallow emargination narrower than base of rostrum, margin lateral to emargination finely serrate; elytra with alternate intervals rather strongly convex, with rows of large sharp granules .... *Neophytobius*

— Tarsus with claws simple; apical margin of pronotum in the middle not produced anteriorly, with straight part limited by 2 sharp tubercles, distance between them not less than width of rostrum, margin lateral to tubercles smooth, not serrate; elytra with alternate intervals not conspicuously convex .... *Pelenomus*

12(3). Prosternum anterior to front coxae short, without traces of keels; distance between front coxae equal to width of antennal funicle; antenna with funicle with 6 articles .... *Amalus*

— Prosternum anterior to front coxae longer, with more or less developed keels; front coxae usually widely separated by distance greater than width of antennal funicle; antenna with funicle with 7 articles, if of 6 articles, then claws simple, or femora dentate, or rostrum wider than front femur 13

13(12). Rostrum usually wider than width of front femur; if about as wide, then at least mesosternum distinctly depressed (flat in *Phrydiuchus*, recognized by large body size of 4.0-5.0 mm); antenna with scape with elongate lamelliform translucent projection and/or 1-3 setae at apex 14

— Rostrum narrower than width of front femur; mesosternum mostly flat (only in *Nedyus* deeply depressed); antenna with scape lacking apical projections or setae 27

14(13). Body globose, elytra slightly longer than wide, with completely rounded shoulders and 7th stria almost reaching basal margin of elytron; meso- and metasterna very short, not more than half length of respective coxae .... *Acallodes*

— Body less convex and rounded, elytra usually rounded-triangular or with staightened sides; meso- and metasterna longer than one-half of respective coxae 15

15(14). Meso- and metasterna flat; body size large, 4.0-5.0 mm; associated with *Salvia* (Labiatae) .... *Phrydiuchus*

— Meso- and often metasternum more or less deeply depressed for reception of rostrum; body size smaller, less than 3.5 mm; associated with various plants, not Labiatae 16

16(15). Mesosternum more or less deeply depressed but depression not limited by keels at sides; if rostral sulcus extends onto metasternum, its sides gently sloping; femora with ventral tooth 17

— Mesosternum with depression limited by keels at sides, often extended onto metasternum and walls very steep or abrupt; femora simple, lacking ventral tooth 19

17(16). Antenna with funicle of 7 articles; meso- and metasterna very shallowly depressed; elytra rounded-triangular, with moderately large sharp granules densely arranged along intervals and provided with short scale-like subrecumbent seta apically (Fig. 33); elytral disc lacking scale pattern other than short postscutellar spot on sutural interval (Fig. 33); body size larger, 2.1-3.2 mm .... *Homorosoma*

— Antenna with funicle of 6 articles; meso- and metasterna moderately deeply depressed; elytra either with straightened sides in basal one-half and vague transverse band of white scales immediately behind middle, or with very large acute granules on intervals and entire body with very long erect hairs; body size smaller, 1.9-2.4 mm 18

18(17). Elytra with sides in basal one-half parallel or weakly rounded; body lacking erect hairy pubescence; elytral intervals with small rounded granules; associated with Salicaceae .... *Rutidosoma*

— Elytra with sides rounded evenly from base; body with long and fine, erect pubescence; elytral intervals with rows of sparsely arranged, very large, acute piliferous granules; associated with *Heuchera richardsoni* (Saxifragiaceae) .... *Asperosoma*

19(16). Rostrum dilated or, rarely, parallel-sided in female (in *Auleutes donaldi* Colonnelli) in apical part; rostral sulcus ending between middle coxae on metasternum, very deep, its walls abrupt; antenna with club with dense, short, very fine erect pubescence (may be absent on large basal segment); associated with Rubiaceae (*A. whiteheadi* Colonnelli, *A. tachygonoides* Dietz, *A. subfasciatus* Dietz) .... *Auleutes* (part)

— Rostrum somewhat tapered apically, not conspicuously dilated in apical part; rostral sulcus usually becoming less deep posteriorly; antenna with club lacking dense short, erect pubescence; associated with Vitaceae and Onagraceae 20

20(19). Pronotum and elytra with basal margins straight, noticeably raised at junction and crenulate; pronotum usually with a pair of discal prominences, lateral tubercles acute, well developed, located close to pronotal base; rostral sulcus extended onto metasternum (which is sometimes strongly convex longitudinally), often 1st ventrite is also deeply depressed medially; rostrum moderately to strongly curved; associated with Vitaceae 21

— Pronotum and elytra with basal margins neither raised nor crenulate; pronotum lacking discal prominences (but present in *Pelenosomus*, then rostral sulcus limited to mesosternum, very shallow, margined by fine low keels); rostral sulcus often limited to mesosternum, not extended onto 1st abdominal ventrite; metasternum not strongly convex longitudinally; rostrum moderately curved to straight; associated with Onagraceae 22

21(20). Outer margin of middle and hind (in *C. inaequalis* LeConte, also of front) tibiae emarginate, with well-developed tarsal grooves; pronotum with obtuse rounded or elongate prominences, not forming acute divergent ridges; elytra sparsely clothed with narrow, lanceolate scales arranged in narrow broken transverse bands; body size 2.3-3.6 mm .... *Craponius*

— Outer margin of middle and hind tibiae sometimes more or less distinctly grooved but not conspicuously emarginate; pronotum with 2 acute ridge-shaped discal prominences divergent anteriorly; elytra with moderately dense vestiture partly formed by lanceolate to oval scales; body size 2.1-2.3 mm .... *Orchestomerus*

22(20). Outer margin of front and middle tibiae very deeply emarginate and compressed, almost blade-shaped, lacking tarsal grooves, emarginations limited by large acute dentiform prominences .... *Cnemogonus*

— Outer margin of all tibiae straight or inconspicuously emarginate, sometimes with more or less developed tarsal grooves 23

23(22). Front tibia with outer margin with acute dentiform prominence apically (Fig. 37); body size large, 2.6-3.7 mm .... *Acanthoscelidius*

— Front tibia with outer apical angle not produced into acute prominence (Fig. 38); if weakly produced (in *Auleutes isolatus* Sleeper) or with 2 spines larger than other setae in apical comb (in *Auleutes asper* LeConte), then dorsal surface with moderately dense vestiture of white scales and more or less distinct scutellar spot, elytral intervals with one row of sharp granules each; body size usually smaller, 2.1-3.5 mm 24

24(23). Pronotum with a pair of rather high discal prominences and sharp lateral tubercles; elytra with dense semi-erect vestiture of varied brown to black very broadly lanceolate scales; long scutellar spot on 1st interval black, velvety .... *Pelenosomus*

— Pronotum without discal prominences; elytra with recumbent, usually sparse to moderately dense pubescence formed mostly by white or metallic-glossy narrow scales 25

25(24). Dorsal surface with sparse or moderately dense vestiture of white scales, lacking metallic-glossy scales, elytra with more or less distinct scutellar spot; rostral sulcus extended onto metasternum, on mesosternum sulcus deep, with abrupt walls; elytra with intervals often with sharp granules (antennal funicle then may be 6-segmented) (*A. epilobii* Paykull and several species from Canada and the U.S. except extreme southwestern and southeastern coastal regions) .... *Auleutes* (part)

— Dorsal surface either with a mixture of narrow dark scales with metallic sheen or with narrow broken bands of narrow white scales (then body chestnut-brown), scutellar spot poorly defined; rostral sulcus not extended onto metasternum, on mesosternum sulcus shallow, its sides gentle, or the sulcus limited by keels projected behind middle coxae; elytra with intervals lacking conspicuous granules 26

26(25). Depression on mesosternum shallow, broad, its margins very finely keel-shaped raised; suture between meso- and metasterna not raised; body black, legs often brown; elytra with more or less distinct scutellar spot and transverse bands, with moderately dense dark, narrow, metallic-glossy scales between bands; rostrum at least weakly curved (*A. nebulosus* LeConte) .... *Auleutes* (part)

— Depression on mesosternum margined by high lamelliform keels projecting behind middle coxae; body dark chestnut-brown; elytra with narrow broken bands of narrow white scales, lacking metallic glossy scales in between; rostrum nearly straight (*A. inspersus* Champion) .... *Auleutes* (part)

27(13). Body narrow, strongly elongate, elytra 1.5-1.6 times as long as wide; legs long, slender; tarsus with claws simple .... *Poophagus*

— Body less elongate-narrow, elytra less than 1.5 times as long as wide; legs various, but not long and slender; tarsus with claws various 28

28(27). Mesosternum shallowly, metasternum deeply depressed between coxae for reception of rostrum, the depression slightly extending beyond margins of middle coxae; associated with *Urtica dioica* L. (Urticaceae) .... *Nedyus*

— Meso- and metasterna lacking sulcus for reception of rostrum, not at all depressed medially; associated with plants other than Urticaceae 29

29(28). Antenna with funicle of 6 articles; body size large, 3.5-4.5 mm long, reddish brown with darker underside and rostrum; dorsal surface glossy, with erect or semi-erect long parallel-sided white and brown scales; basal margins of pronotum and elytra raised at junction and coarsely crenulate .... *Trichosirocalus*

— Antennal funicle mostly of 7 articles; if of 6 articles, then body size smaller (less than 3.5 mm), basal margins of pronotum and elytra not raised at junction and not crenulate, body lacking coarse erect vestiture, not glossy chestnut-brown 30

30(29). Tarsus with claw simple 31

— Tarsus with claw dentate 33

31(30). Antenna with funicle of 7 articles; femora with large tooth, all tibiae in male with large apical tooth, female tibiae simple; associated with Asteraceae, mostly *Cirsium* and *Carduus* .... *Hadroplontus*

— Antenna with funicle of 6 articles; femora with slightly defined tooth or angular prominence; front tibia of male simple; associated with Liliaceae, Scrophulariaceae, Fumariaceae or Papaveraceae 32

32(31). All tibiae in both sexes simple; base of pronotum distinctly bisinuate; elytra with pattern consisting of scutellar spot with lateral arms stretching obliquely back from sutural strip and separated by dark 5th interval from oblique bands on 6-9th intervals running anteriorly to sides behind humeral prominences and not touching them (Fig. 31), sometimes pale pattern reduced to inconspicuous macula in basal third of 6th interval on dull dark brown background; associated with Liliaceae and Scrophulariaceae .... *Allosirocalus*

— Middle and hind tibiae in male with small apical tooth; base of pronotum straight; elytra with pattern consisting only of scutellar spot on sutural interval; if this is more or less distinctly T-shaped and oblique bands on 6-8(9)th intervals also present, then bands touch humeral prominence; associated with Fumariaceae and Papaveraceae .... *Sirocalodes*

33(30). Elytra with pale pattern, consisting of scutellar spot with lateral branches oblique or perpendicular to the suture, and lateral bands on 6-8th intervals, running to sides of elytra behind humeral prominences 34

— Elytra without lateral bands, scutellar spot, if distinct, limited to sutural interval 35

34(33). Scutellar spot with oblique extensions from its transverse part, directed to lateral bands and either connected with them or separated by dark area; lateral bands running to humeral prominences or somewhat behind them; femora with tooth simple, medium-sized; body size small 2.2-2.8 mm; associated with *Matricaria perforata* Mérat and *Chrysanthemum leucanthemum* L. (Asteraceae) .... *Microplontus*

— Elytra with wide white cruciform scutellar spot and transverse, more or less oblique bands in middle of 6-9th intervals almost always separated from the scutellar spot by at least one dark interval; femora with tooth on front and middle femora with truncate apical (facing apex of femur) slope; body size large, 3.7-4.6 mm; associated with *Cynoglossum officinale* L. (Boraginaceae) .... *Mogulones*

35(33). Base of pronotum straight or slightly angularly produced posteriorly in the middle, not distinctly bisinuate; apical margin of pygidium, at least in male, deeply excised in the middle; associated with *Taraxacum* (Asteraceae) or (possibly) Liliaceae 36

— Base of pronotum more or less distinctly bisinuate; apical margin of pygidium entire; if sulcate and with combs of yellow hairs on sides of excision, then male hind tibia with usual small pointed apical tooth, female middle tibia with small fine apical tooth; associated with Cruciferae 37

36(35). Pronotum wide, with strongly rounded sides; anterior margin strongly bent; disc convex, without median sulcus, but with deep prescutellar fovea (Fig. 32); lateral tubercles fold-shaped; punctation dense, uniform and rather fine; elytra with short scutellar spot of dense white or yellowish scales on 1st interval, without oval white scales on base of other intervals and small pale spot at the end of basal third of 6th interval (Fig. 32); associated with *Taraxacum officinale* Weber (Asteraceae); southeastern Canada and northeastern United States .... *Glocianus*

— Pronotum wide at base, but with more or less concave sides and sharp, but not fold-shaped lateral tubercles; elytra with small, sometimes indistinct spot at the end of basal third of 6th interval; associated (possibly) with Liliaceae; Yukon Territory .... *Prisistus*

37(35). Antenna with funicle of 6 articles; anterior margin of pronotum not raised, sides without any trace of tubercles, in basal one-half weakly rounded; base distinctly bisinuate, disk without medial sulcus or shortly and narrowly sulcate only at base, strongly convex, densely and finely punctate, intervals between punctures with matte reticulate microsculpture; body and legs black; associated with *Nasturtium officinale* R. Br. (Cruciferae) .... *Amalorrhynchus*

— Antenna with funicle of 6 or 7 articles; if 6-segmented, then anterior margin of pronotum noticeably raised, sides usually with small but distinct tubercles, moderately rounded in basal one-half, disk with median sulcus, glossy, more or less coarsely punctate; elytral striae rather wide, legs often pale; associated with various Cruciferae. 38

38(37). Mesosternum not depressed; scales not concealing integument completely, not imbricate (except in *C. opertus* Brown) .... *Ceutorhynchus*

— Mesosternum between middle coxae moderately deeply depressed, sides of the depression gentle; scales brownish-grey and imbricate, concealing integument completely; semi-erect coarse setae or narrow scales also present...... .... *Rileyonymus*

Classification of the Nearctic Ceutorhynchinae

22. Ceutorhynchini Gistel 1856

Allosirocalus Colonnelli 1983, 5 spp., generally distributed in the central and western United States and southern Canada. Species may be associated with *Allium* (wild onion; Liliaceae) in Texas and *Mimulus* and *Pedicularis* (Scrophulariaceae) in the western United States. See Hatch (1971) to separate some of the species.

Amalorrhynchus Reitter 1913, 1 sp., *A. melanarius* (Stephens 1831), Quebec, Connecticut, Massachusetts and West Virginia; adventive. This species is associated with *Nasturtium officinale* R. Br. (watercress; Cruciferae).

Amalus Schoenherr 1825, 1 sp., *A. scortillum* (Herbst 1795), generally distributed in the northern United States and Canada; adventive. This species is associated with *Polygonum* (Polygonaceae); larvae feed in the crown (Hoffmann 1954).
Leptocaryurgus Gistel 1856

Ceutorhynchus Germar 1824, 68 spp., generally distributed; some adventive and of pest status. Species are associated with Cruciferae; larvae often mine in collars of roots or stems (Anderson 1993b). See Dietz (1896), Blatchley and Leng (1916) and Hatch (1971) to separate some of the species. An unpublished 1963 Ph.D. thesis by Rudolph Scheibner from Michigan State University allows for the identification of most species. The genus needs revision. *Calosirus* Thompson is given distinct generic status by Wibmer and O'Brien (1989) and Alonso-Zarazaga and Lyal (1999); 8 species could be placed provisionally in this taxon in North America.
Falciger Dejean 1821
Ceuthorhynchus Schoenherr 1837
Ceuthorhynchidius Jacquelin du Val 1855
Calosirus Thompson 1859
Ceuthorrhynchus Gemminger and Harold 1871
Ceuthorrhynchidius Gemminger and Harold 1871
Sirocalus Heyden 1906
Dionorenus Reitter 1913
Marklissus Reitter 1916
Heterosirocalus Wagner 1944
Neosirocalus Wagner 1944
Persirocalus Wagner 1944
Ceuthamiocolus Colonnelli 1983
Nipporhynchus Korotyaev 1996, not Chandler 1934
Heorhynchus Korotyaev 1999

Glocianus Reitter 1916, 1 sp., *G. punctiger* (Sahlberg 1835), generally distributed; adventive. This species is associated with *Taraxacum officinale* Weber (Asteraceae); larvae feed on seeds in flower heads (McAvoy *et al.* 1983).
Prenesdus Reitter 1916

Hadroplontus Thomson 1859, 1 sp., *H. litura* (Fabricius 1775), Nova Scotia, Ontario, Saskatchewan, Alberta, British Columbia, South Dakota, Montana, Idaho, Oregon and Washington. Introduced for biological control of *Carduus* and *Cirsium* thistles (Asteraceae) (Peschken and Wilkinson 1981); larvae mine the stems and crown (Hoffmann 1954).

Microplontus Wagner 1944, 2 spp., *M. edentulus* (Schultze 1896), Alberta, Saskatchewan, and *M. campestris* (Gyllenhal 1837), Ontario; both exotic. *Microplontus edentulus* has been released locally at two sites for biological control of scentless chamomile, *Matricaria perforata* Mérat (Asteraceae) (A. S. McClay, pers. comm.). *Microplontus campestris* has only recently been documented as present in North America; it is associated with *Chrysanthemum leucanthemum* L. (Asteraceae).

Mogulones Reitter 1916, 1 sp., *M. cruciger* (Herbst 1784), British Columbia and Alberta. This species has been introduced for the biological control of *Cynoglossum officinale* L. (hound's-tongue; Boraginaceae) (DeClerk-Floate and Schwarzländer in press).
Boraginobius Wagner 1944

Nedyus Schoenherr 1825, 2 spp., generally distributed in the eastern United States west to Texas and Canada west to Alberta. Species are associated with *Urtica dioica* L. (nettle; Urticaceae) (Blatchley and Leng 1916). See Blatchley and Leng (1916) to separate the species.
Cidnorhinus Thomson 1859

Phrydiuchus Gozis 1885, 2 spp., *P. tau* Warner and *P. spilmani* Warner (latter may not be established), California, Oregon, Washington; adventive. Introduced for the biological control of *Salvia aethiops* L. (Mediterranean sage; Labiatae) (Warner 1969). See Warner (1969) to separate the species.

Poophagus Schoenherr 1837, 1 sp., *P. sisymbrii* (Fabricius 1776), Quebec; adventive. This species is associated with *Nasturtium* (Cruciferae); larvae mine stems and roots (Hoffmann 1954).
Poephagus Gistel 1856
Acnemiscelis Desbrochers 1896

Prisistus Reitter 1916, 1 sp., *P. olgae* Korotyaev 1988, Yukon Territory. This species may be associated with Liliaceae.
Austroceutorhynchus Korotyaev 1980 (valid subgenus)
Ranunculiphilus Dieckmann 1970 (valid subgenus)
Svetlaniolus Korotyaev 1997 (valid subgenus)

Rileyonymus Dietz 1896, 1 sp., *R. relictus* Dietz 1896, Arizona, California. This genus is questionably distinct from *Ceutorhynchus*.

Sirocalodes Voss 1958, 3 spp., *S. tescorum* (Fall 1907), *S. sericans* (LeConte 1876), and *S. siculus* (Dietz 1896), generally distributed in the western and southern United States and Manitoba. *Sirocalodes tescorum* (and *S. wickhami* (Champion 1907) from Mexico) have been associated with *Argemone* (Papaveraceae) and *S. siculus* with *Corydalis* (Fumariaceae).

Trichosirocalus Colonnelli 1979, 1 sp., *T. horridus* (Panzer 1801), Virginia. Introduced for the biological control of *Carduus* (thistles; Asteraceae); larvae mine in the crown and stem (Trumble and Kok 1979).

23. Cnemogonini Colonnelli 1979

Acanthoscelidius Hustache 1930, 13 spp., generally distributed. Species are associated with *Oenothera*, *Gaura* and perhaps other Onagraceae (Anderson 1993b). See Dietz (1896) to separate most of the species. The genus needs revision and its relationships with *Auleutes* reassessed.
Acanthoscelis Dietz 1896; not Dejean 1825
Acantharthrus Marshall 1939

Auleutes Dietz 1896, 12 spp., generally distributed. Species are associated with *Ludwigia*, *Calylophus*, *Oenothera*, *Gaura* and perhaps other Onagraceae (Blatchley and Leng 1916; Anderson 1993b). *Auleutes donaldi* Colonnelli 1991 has been associated with *Bouvardia glaberrima* Engelm. (Rubiaceae). See Dietz (1896) to separate most of the species. The genus needs revision and its relationships with *Acanthoscelidius* and neotropical taxa placed as *Auleutes* reassessed. The key presented here recognizes three distinct groups of *Auleutes* likely warranting separate generic status.

Cnemogonus LeConte 1876, 1 sp., *C. lecontei* Dietz 1896, generally distributed in the northern United States and Canada. This species may be associated with Onagraceae.

Craponius LeConte 1876, 1 sp., *C. inaequalis* (Say 1831), generally distributed in the eastern and central United States and southern Canada. This species is associated with *Vitis* (grapes; Vitaceae); larvae feed within fruits on seeds (Blatchley and Leng 1916).

Dietzella Champion 1907, 2 spp., *D. zimmermanni* (Gyllenhal 1837), generally distributed in the eastern United States, southern Canada and western United States, and *D. sextuberculata* (Boheman 1845), Colorado. Species are associated with *Epilobium* (Onagraceae).

Orchestomerus Dietz 1896, 3 spp., generally distributed in the eastern United States, Texas and Arizona. At least one species, *O. whiteheadi* Colonnelli 1991, is associated with *Vitis* (wild grape; Vitaceae) in Arizona.
Platymeristes Dietz 1896

Pelenosomus Dietz 1896, 1 sp., *P. cristatus* Dietz 1896, southeastern United States.

Perigaster Dietz 1896, 4 spp., generally distributed in the eastern United States, southern Canada and western United States. Species are associated with *Ludwigia* (Onagraceae) (Anderson 1993a). See Buchanan (1931) to separate the species.

Perigasteromimus Colonnelli 1999, 1 sp., *P. tetracanthus* (Champion 1907), Florida. This species is associated with *Ludwigia* spp. (Onagraceae) (C. W. O'Brien, pers. comm.).

24. Hypurini Schultze 1902

Hypurus Rey 1882, 1 sp., *H. bertrandi* (Perris 1852), California and Florida; adventive. This species is associated with *Portulaca oleracea* L. (Portulacaceae) in Europe but other Portulacaceae are suitable hosts (Zimmerman 1957; Anderson 1993a).

25. Mononychini LeConte 1876

Mononychus Germar 1824, 1 sp., *M. vulpeculus* (Fabricius 1801), generally distributed in the eastern United States and southern Canada. This species is associated with *Iris versicolor* L. (Iridaceae); larvae are in seed pods (Blatchley and Leng 1916).

26. Phytobiini Gistel 1848

[*Eubrychius* Thomson 1859, 1 sp., *E. velutus* (Beck 1817). This species has been recorded from the eastern United States west through the north to the western United States and British Columbia. It does not occur in North America; all records are misidentifications of *Euhrychiopsis lecontei* (Dietz 1896), see Tamayo *et al.* 1999.]

Euhrychiopsis Dietz 1896, 1 sp., *E. lecontei* (Dietz 1896), eastern to central United States and western Canada. This species is associated with *Potamogeton* (Potamogetonaceae) and *Myriophyllum* (Haloragaceae) (Kissinger 1964; Hatch 1971). See Tamayo *et al.* (1999) for information about this species. All records of *Eubrychius velutus* (Beck 1817) are misidentifications of *E. lecontei* (Tamayo *et al.* 1999).

Neophytobius Wagner 1936, 1 sp., *N. cavifrons* (LeConte 1876), generally distributed in the western United States and Canada. This species is associated with *Polygonum* (Polygonaceae).

Nemophytobius Voss 1952

Parenthis Dietz, 1896, 1 sp., *P. vestitus* Dietz 1896, southeastern United States. This genus is considered as a junior synonym of *Phytobius* Schoenherr by Alonso-Zarazaga and Lyal (1999).

Pelenomus Thomson 1859, 13 spp., generally distributed in the more northerly United States and Canada. Most Palearctic species are associated with *Polygonum* (Polygonaceae) (Hoffmann 1954). See Dietz (1896), Blatchley and Leng (1916) and Hatch (1971) to separate some of the species. The genus needs revision.

Pachyrhinus Stephens 1829; not Schoenherr 1825
Phytobius Dejean 1835; not Schoenherr 1833
Mecopeltus Dietz 1896
Paraphytobius Wagner 1936

Phytobius Schoenherr 1833, 1 sp., *P. leucogaster* (Marsham 1802), generally distributed in the eastern United States west through the north to the western United States. This species is associated with *Myriophyllum* (Haloragaceae); larvae feed externally on flowers (Buckingham and Bennett 1981).

Hydaticus Schoenherr 1825; not Leach 1817
Litodactylus Redtenbacher 1849

Rhinoncus Schoenherr 1825, 7 spp., generally distributed; three species adventive. Species are associated with *Polygonum* (Polygonaceae) (Hoebeke and Whitehead 1980). See Hoebeke and Whitehead (1980) to separate six of the species. *Rhinoncus perpendicularis* (Reiche 1797) recently has been collected in Ontario.

Cryptorhis Billberg 1820
Campylirhynchus Dejean 1821
Camplirhynchus Gistl 1834
Campylorhynchus Agassiz 1846; not Spix 1824

27. Scleropterini Schultze 1902

Acallodes LeConte 1876, 3 spp., generally distributed in the eastern United States and southern Canada. One species is associated with *Lysimachia terrestris* L. (Primulaceae) (Blatchley and Leng 1916). See Blatchley and Leng (1916) to separate the species.

Asperosoma Korotyaev 1999, 1 sp., *A. echinatum* (Fall 1917), Manitoba. This odd species is associated with *Heuchera richardsoni* R. Br. (Saxifragiaceae) (Fall 1917).

Homorosoma Frivaldszky 1894, 1 sp., *H. sulcipenne* (LeConte 1876), generally distributed. This species is associated with *Polygonum* (Polygonaceae).

Rutidosoma Stephens 1831, 1 sp., *R. decipiens* (LeConte 1876), generally distributed in the eastern and western United States, Canada and Alaska. This species is associated with *Populus* (Salicaceae) (Anderson 1997).

Rhytidosomus Schoenherr 1837
Rhytidosoma Agassiz 1846
Oligodites Gistel 1856
Rhytidosomus Gemminger and Harold 1871; not Schoenherr 1837
Scleropteridius Otto 1897 (valid subgenus)
Prorutidosoma Korotyaev 1999 (valid subgenus)
Victorinus Korotyaev 1999 (valid subgenus)

VIII. Conoderinae Schoenherr 1833

by Henry A. Hespenheide

The Conoderinae have been defined by the combination of a prosternal channel for the reception of the rostrum, large approximate eyes, and the absence of postocular lobes on the anterolateral margin of the pronotum. They are usually placed between the Cryptorhynchinae and the Ceutorhynchinae and are probably more closely related to the former. Adults are typically diurnal and very wary and active fliers. There is considerable structural diversity within the subfamily, even among the North American forms, but the group is much more diverse in the Neotropical Region.

Most conoderine larvae are borers of wood or herbaceous stems although a few feed on seeds. The genus *Tachygonus*, provisionally placed here as a highly derived subgroup, has larvae which mine leaves. A few North American species such as *Cylindrocopturus adspersus* (LeConte 1876), the sunflower stem borer, are economically significant pests.

Key to the Nearctic Genera of Conoderinae

1. Antenna straight, not geniculate; hind coxae widely separated by a distance 4 or 5 times greater than width of a coxa; hind femur long (much longer than front or middle femora), spinose ventrally (Fig. 47); body form broadly ovate, flattened; with appressed scales and tufts of erect setae (Fig. 47) .... *Tachygonus*
— Antenna geniculate; hind coxae narrowly separated by a distance less than twice width of a coxa; hind femur short (subequal in length to front or middle femora), simple or with single ventral tooth; body form elongate ovate, subtriangular in lateral form, not flattened (Figs. 39-44); vestiture various, mostly with only appressed scales or scattered setae 2

2(1). Pygydium exposed dorsally (Fig. 41); body form more or less flattened dorsally and ventrally (Fig. 42); body length greater than 6.0 mm .... *Peltophorus*
— Pygydium covered by elytra (Figs. 39, 43); body form with ventral or dorsal surface convex (Figs. 40, 44); body length smaller than 6.0 mm 3

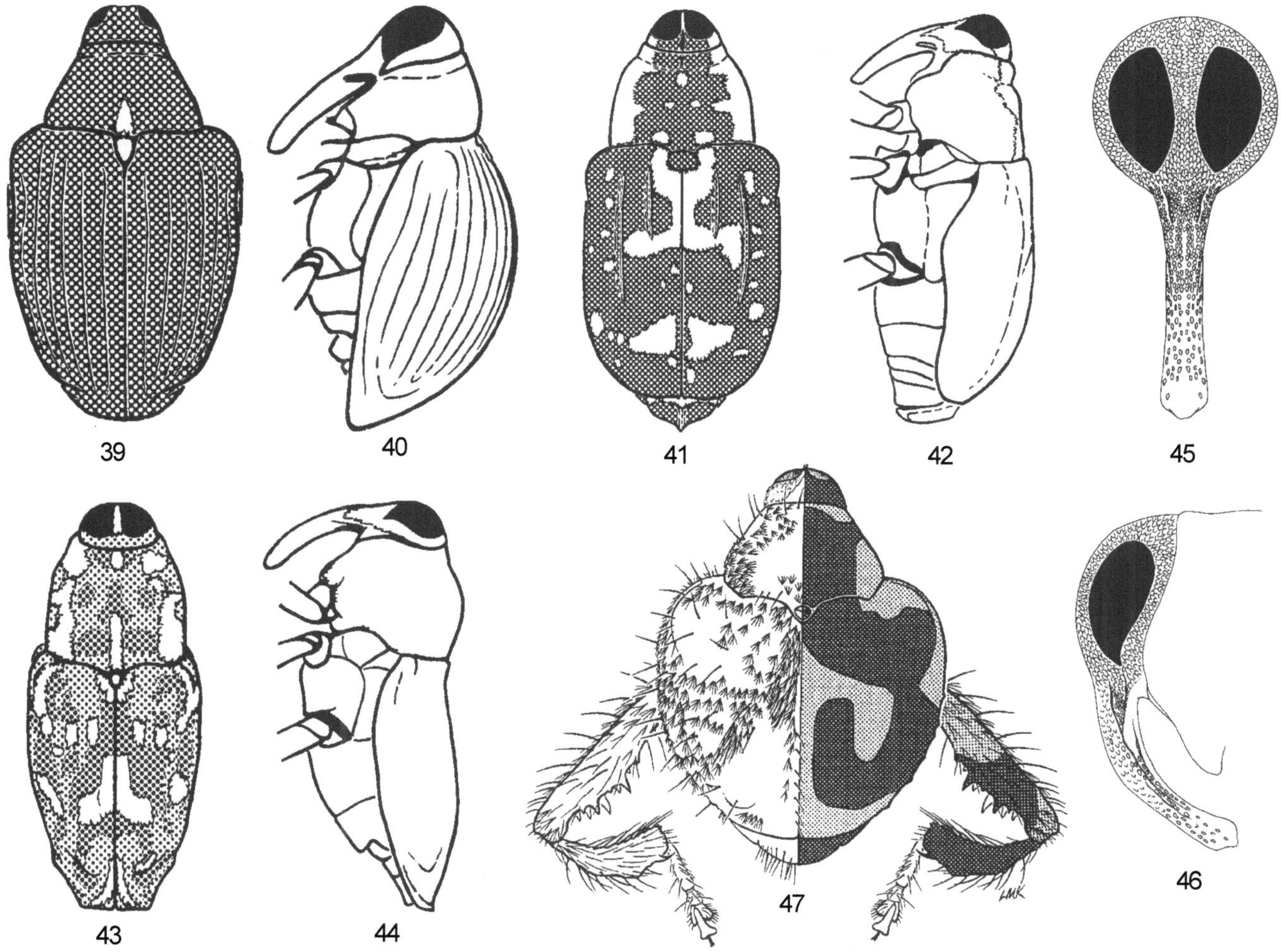

FIGURES 39.131-47.131. Conoderinae 39-40. *Psomus armatus* (Dietz), habitus, 39. Dorsal; 40. Lateral. 41-42. *Peltophorus polymitus seminiveus* (LeConte), habitus, 41. Dorsal; 42. Lateral. 43-44. *Cylindrocopturus adspersus* (LeConte), habitus, 43. Dorsal; 44. Lateral. 45-46. *Cylindrocopturus adspersus* (LeConte), head, 45. Anterior view; 46. Lateral view. 47. *Tachygonus lecontei* Gyllenhal (Conoderinae), dorsal habitus (left side show scale pattern, right side shows color).

3(2). Abdomen with ventrites in lateral view in same plane, horizontal; elytra in lateral view more or less continuously convex (Fig. 40) 4

— Abdomen with ventrites in lateral view ascended; elytra in lateral view flattened at base (Fig. 44) 5

4(3). Elytra with distinct setae; femora each with ventral tooth; body length greater than 3.0 mm .... *Acoptus*

— Elytra lacking distinct setae; femora simple, lacking ventral tooth; body length less than 2.0 mm .... *Psomus*

5(3). Mesosternum simple, unmodified, apex of rostrum free 6

— Mesosternum excavated, with lateral margins carinate; anterior margin of metasternum excavated to receive the tip of the rostrum or not 7

6(5). Antenna with articles 1 and 2 of funicle subequal in length; femora simple, lacking ventral tooth .... *Cylindrocopturus*

— Antenna with article 2 of funicle about twice length of article 1; femora with large ventral tooth .... *Copturus*

7(5). Femora carinate on outer face and with ventral tooth .... *Lechriops*

— Femora not carinate on outer face, lacking ventral tooth 8

8(7). Metasternum with anterior margin excavated for reception of rostrum; body length less than 3.0 mm .... *Eulechriops*

— Metasternum with anterior margin simple, not modified for reception of rostrum; body length greater than 3.0 mm .... *Cylindrocopturinus*

CLASSIFICATION OF THE NEARCTIC CONODERINAE

28. Lechriopini Lacordaire 1866

Acoptus LeConte 1876, 1 sp., *A. suturalis* LeConte 1876, generally distributed in eastern United States and southern Canada. This species is associated with dead wood of beech trees, also with hop-hornbeam and hickory (Blatchley and Leng 1916).

Copturus Schoenherr 1825, 1 sp., *C. floridanus* (Fall 1906), Florida. This species is associated with *Swietenia mahagoni* (L.) Jacq. (Meliaceae); larvae bore under bark of living branches (Anderson 1993a).

Zurus Heller 1895; not Amyot 1846
Neozurus O'Brien and Wibmer 1982

Cylindrocopturinus Sleeper 1963, 1 sp., *C. pictus* (Schaeffer 1908), Arizona. This species is associated with *Phoradendron* (mistletoe; Viscaceae) (Anderson 1994).

Eulechriops Faust 1896, 2 spp., generally distributed in the eastern United States and Arizona. Species are associated with *Quercus* (oak; Fagaceae).

Zygomicrus Casey 1897

Lechriops Schoenherr 1825, 4 spp., generally distributed. See Blatchley and Leng (1916) to separate some of the species.

Gelus Casey 1897

Psomus Casey 1892, 1 sp., *P. armatus* (Dietz 1891), northeastern United States and southern Canada. This species is associated with *Fraxinus americanus* L. (white ash; Oleacae) (Blatchley and Leng 1916).

29. Zygopini Lacordaire 1866

Cylindrocopturus Heller 1895, 29 spp., generally distributed. Many species are associated with various Asteraceae and some with Pinaceae. See Casey (1897), Fall (1906) and Hatch (1971) to separate some of the species. The genus needs revision.

Paratimorus Heller 1895
Gyrotus Casey 1897
Copturodes Casey 1897

Peltophorus Schoenherr 1845, 3 spp., Arizona, New Mexico and Texas. Species are associated with *Agave* (Amaryllidaceae); larvae mine the stalks or in the seeds (Kissinger 1964). See Sleeper (1963) to separate the species.

Apatorhynchus Desbrochers 1891
Opalocetus Desbrochers 1910

30. Tachygonini Lacordaire 1866

Alonso-Zarazaga and Lyal (1999) place Tachygonina as a subtribe within Curculioninae; Rhamphini. This distictive genus is placed here as a tribe within Conoderinae based upon the form of the unci at the apex of the tibiae, large eyes and form of vestiture. Similarity in the pectinate form of the scales suggests a relationship with the Neotropical genera *Philinna* Champion 1906 and *Philides* Champion 1906.

Tachygonus Guérin-Méneville 1833, 5 spp., generally distributed in eastern United States and southeastern Canada, west to Texas, Arizona, New Mexico and Colorado. Species are associated with *Quercus* (Fagaceae), *Ulmus* (Ulmaceae), *Robinia* (Fabaceae), *Coursetia* (Fabaceae)) and *Berchemia* (Rhamnaceae); larvae mine leaves (Hespenheide 1992). See Hespenheide (1992) to separate the species.

Tachygonus Schoenherr 1833; not Guérin-Méneville 1833
Tachyopus Zimmermann 1840

IX. Cossoninae Schoenherr 1825

by Robert S. Anderson

Cossonine weevils are easily recognized by the large, hook-like tooth at the apex of the hind tibia and the lack of an apical comb of setae. They are usually black or brown, lack scales but have appressed or erect hairs, and are generally long, slender and dorsoventrally compressed. Most species are associated with dead plant material of some sort, usually of woody angiosperms (where they live under bark), but some are also found in dead fern fronds, palm fronds, agave leaves, yucca stalks, etc. A number of taxa are found on sandy beaches in association with driftwood. Species of *Acamptus* and at least some *Pseudopentarthrum* are found in tree holes or rotten hollowed out trees. Most of the genera in North America are represented by only one or a few species. Some species, such as those in the genus *Hexarthrum*, are difficult to separate from Scolytinae.

Key to the Nearctic genera of Cossoninae

1. Prosternum with distinct channel for reception of rostrum when in repose; pronotum and elytra with dense, erect or suberect, broad scales .... *Acamptus*
— Prosternum simple, smooth, lacking channel for reception of rostrum; pronotum and elytra with vestiture various, if erect or suberect, composed of finer, hair-like scales or sparse, scattered broad scales 2

2(1). Eyes obviously located on base of rostrum, head distinctly constricted and globular behind eyes (Fig. 51); Florida 3
— Eyes located on head or at junction of head and rostrum, head not distinctly constricted dorsally behind eyes (Fig. 52); various locations 4

3(2). Eye reduced in size to less than 10 facets (Fig. 51); antenna with basal article of club subglabrous, glossy; rostrum in lateral view with ventral margin straight (Fig. 51) .... *Paralicus*
— Eye with more than 10 facets; antenna with basal article of club setose, not glossy; rostrum in lateral view with ventral margin curved ventrally towards apex .... *Dryotribus*

4(2). Eyes markedly reduced in size to 3 or 4 facets.... .... *Amaurorhinus*
— Eye with more than 10 facets 5

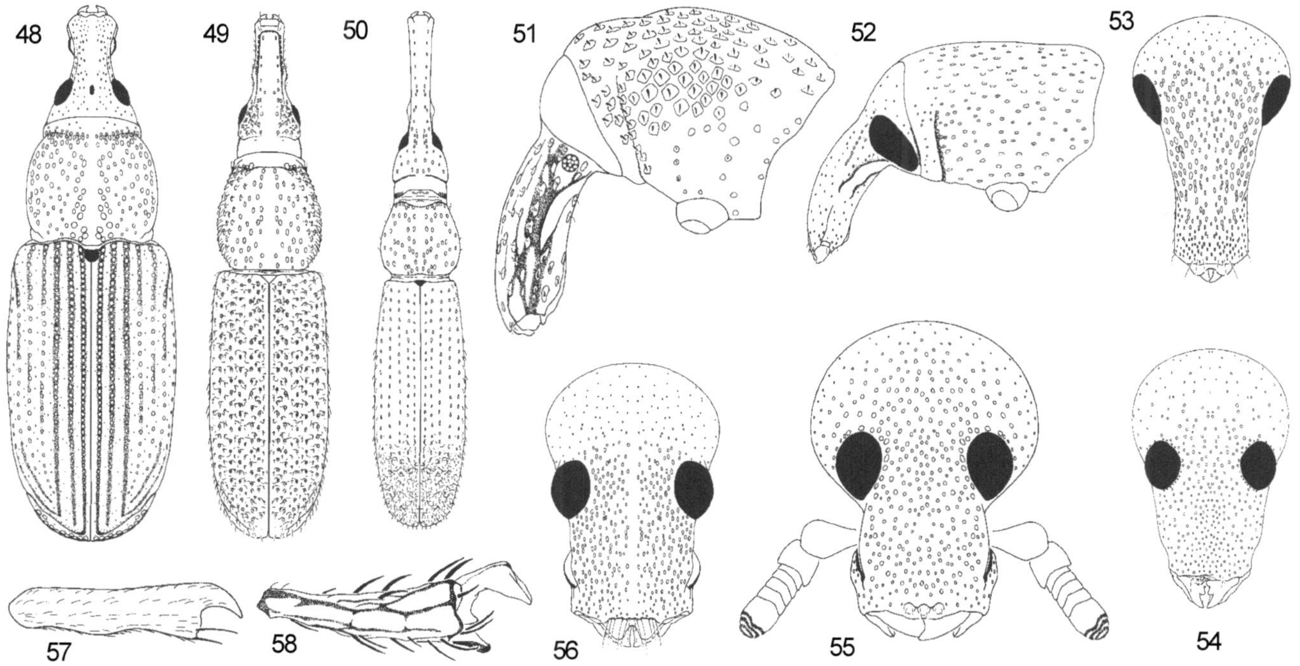

FIGURES 48.131-58.131. Cossoninae. 48-50. Dorsal habitus, 48. *Cossonus piniphilus* Boheman; 49. *Macroscytalus chisosensis* (O'Brien); 50. *Stenotrupis acicula* Wollaston. 51-52. Lateral view of head, 51. *Paralicus minyops* O'Brien; 52. *Micromimus minimus* (Boheman). 53-56. Dorsal view of head, 53. *Aphanommata tenuis* (Casey); 54. *Tomolips quercicola* (Boheman); 55. *Pseudopentarthrum robustum* Casey; 56. *Rhyncolus brunneus* Mannerheim. 57-58. Hind tibia, 57. *Cossonus piniphilus* Boheman; 58. *Elassoptes marinus* Horn.

5(4). Elytra with intervals with obvious vestiture of elongate-narrow, recurved hair-like scales, alternate intervals also with row of suberect, broad, truncate scales; eyes situated low on head, in lateral view with the ventral margin of rostrum directed to middle of eye .... *Himatium*

— Elytra with intervals with obvious vestiture absent, or with obvious vestiture of elongate, fine, hair-like setae only, no broad scales present; eyes situated higher on head, in lateral view with ventral margin of rostrum directed to lower one-half of eye or obviously below eye 6

6(5). Antenna with funicle of 5 or 6 articles 7

— Antenna with funicle of 7 articles 13

7(6). Antenna with funicle of 6 articles .... *Hexarthrum*

— Antenna with funicle of 5 articles 8

8(7). Dorsal vestiture of obvious, long, very fine hair-like setae, each at least as long as a strial puncture or longer 9

— Dorsal vestiture of at most indistinct, short, fine hair-like setae, each shorter than a strial puncture, or obvious vestiture absent 11

9(8). Body size greater than 2.0 mm, black in color; body form subcylindrical, more or less as wide as high; head not constricted behind eyes .... *Nyssonotus*

— Body size less than 2.0 mm, pale or dark brown in color; body form markedly dorsoventrally compressed, much wider than high; head constricted behind eyes, area posterior to constriction impunctate and glossy 10

10(9). Elytra three to four times as long as pronotum (Fig. 50); pronotum about as wide as long (Fig. 50); antenna with scape extended slightly beyond hind margin of eye; Florida .... *Stenotrupis*

— Elytra two to three times as long as pronotum (Fig. 49); pronotum longer than wide (Fig. 49); antenna with scape extended to about middle of eye; southwestern Texas .... *Macroscytalus*

11(8). Rostrum distinctly tapered apically in dorsal view (Fig. 54); antenna with club truncate at apex .... *Tomolips*

— Rostrum more or less subparallel or slightly expanded towards apex in dorsal view (Fig. 55); antenna with club rounded at apex 12

12(11). Antenna with club not expanded, article 1 of funicle about as wide as club (Fig. 55); widespread .... *Pseudopentarthrum*

— Antenna with club expanded, article 1 of funicle distinctly narrower than club; Quebec .... *Pentarthrum*

13(6). Dorsal vestiture of obvious, long, fine, hair-like setae, each at least as long as a strial puncture or longer 14

— Dorsal vestiture of at most indistinct, short, fine hair-like setae, each shorter than a strial puncture, or obvious vestiture absent 16

14(13). Front coxae very narrowly separated by much less than one-half width of a coxa; elytra with vestiture of long, erect, fine, hair-like setae, each much longer than a strial puncture .... *Pselactus*

— Front coxae more widely separated by at least one-half width of a coxa or more; elytra with vestiture of shorter hair-like or scale-like setae, each about as long as a strial puncture 15

15(14). Front coxae separated by more or less width of a coxa; elytra with vestiture of erect, stout, scale-like setae; eyes situated at junction between rostrum and head .... *Apotrepus*

— Front coxae separated by about one-half width of a coxa; elytra with vestiture of appressed, hair-like setae; eyes situated on head adjacent to base of rostrum .... *Carphonotus*

16(13). Hind tibia expanded somewhat laterally at apex, hook-like tooth at outer angle large and stout, tooth at inner angle, stout, spatulate, much longer than tarsal claw (Fig. 58) .... *Elassoptes*

— Hind tibia simple, not expanded laterally at apex, hook-like tooth at outer angle various (Fig. 57), tooth at inner angle (if present) not spatulate, smaller than tarsal claw 17

17(16). Rostrum in dorsal view with apical one-half more or less abruptly dilated beyond point of antennal insertion, wider than basal one-half, ventral margin of scrobe visible in dorsal view (Fig. 48); antenna inserted beyond midlength of rostrum .... *Cossonus*

— Rostrum subequal in width throughout length (Fig. 56) or tapered towards apex (Fig. 53), scrobe not visible in dorsal view; antenna with point of insertion in basal one-half of rostrum 18

18(17). Rostrum in dorsal view at least twice as long as width of frons between eyes 19

— Rostrum in dorsal view less than twice as long as width of frons between eyes (Figs. 53, 56) ... 24

19(18). Eyes large and elongate-oval in shape, height more or less twice maximum width (Fig. 52) .... *Micromimus*

— Eyes moderate in size, round or sub-oval in shape, height much less than twice width 20

20(19). Metepisternum wide throughout length, subequal in width to width of antennal club, with 2 or 3 rows of large, distinct, deep punctures; rostrum of female with point of antennal insertion basal, rostrum long, glabrous, cylindrical and slender, rostrum of male with point of antennal insertion at basal one-third, flat and deeply punctate dorsally, more quadrate in cross-section ... *Mesites*

— Metepisternum narrow throughout length, width much less than width of antennal club, with at most one indistinct row of shallow punctures; rostrum various in form, similar in both female and male 21

21(20). Front coxae very narrowly separated by much less than one-half width of a coxa, anterior and posterior prosternal processes acuminate apically ... . *Phloeophagus*

— Front coxae moderately to widely separated by at least one-half width of a coxa, anterior and posterior prosternal processes broadly truncate apically 22

22(21). Front coxae situated distant from posterior margin of prosternum, separated from margin by obviously more than the length of a coxa; body form markedly dorsoventrally compressed; body size less than 2.0 mm; Florida .... *Proeces*

— Front coxae situated close to posterior margin of prosternum, separated from margin by slightly less than the length of a coxa; body form subcylindrical; body size various; various locations, including Florida 23

23(22). Middle and hind femora very short, distinctly expanded apically and subtriangular in shape, length about twice width at apex; body size small, less than 2.0 mm; body color testaceous .... *Stenomimus*

— Middle and hind femora moderate in length, slightly expanded apically but more elongate in shape, length greater than twice width at apex; body size various, less than 5.0 mm; body color black, dark brown or testaceous .... *Caulophilus*

24(18). Front coxae very narrowly separated by less than one-half width of a coxa; posterior and anterior prosternal processes acuminate or subacuminate apically 25

— Front coxae moderately to widely separated by more than one-half width of a coxa; posterior and anterior prosternal processes truncate apically .. 27

25(24). Elytra at base with intervals 2 to 4 swollen and crenulate or minutely dentate; elytral declivity with numerous small denticles; rostrum very short, wider than long, in dorsal view with lateral margins subparallel .... *Stenoscelis*

— Elytra at base with intervals smooth, not swollen, dentate or crenulate; elytral declivity smooth, lacking denticles; rostrum as long as wide to slightly longer than wide; rostrum with lateral margins (excluding scrobes if visible in dorsal view) convergent apically 26

26(25). Eyes at most only slightly visible in dorsal view, flat, situated low on head (Fig. 53), in lateral view with ventral margin of rostrum directed towards middle or lower one-half of eye .... *Aphanommata*

— Eyes clearly visible in dorsal view, slightly convex, situated higher on head (Fig. 56), in lateral view with ventral margin of rostrum directed below eye (includes *Phloeophagus minor*, *P. californicus*) .. .. *Rhyncolus* (part)

27(24). Rostrum very short, wider than long; pronotum about as wide as long .... *Rhyncolus* (part)

— Rostrum longer than wide; pronotum longer than wide 28

28(27). Rostrum in dorsal view tapered towards apex, with apex narrower than width of rostrum at position of antennal insertion .... *Macrancylus*

— Rostrum in dorsal view subparallel or slightly wider towards apex, with apex subequal in width or slightly wider than width of rostrum at position of antennal insertion 29

29(28). Antenna with apex of scape extended well beyond the hind margin of the eye; eyes flat, or very slightly convex, slightly visible in dorsal view . ... *Macrorhyncolus*

— Antenna with apex of scape extended at most to the posterior margin of the eye; eyes convex, obviously visible in dorsal view 30

30(29). Antenna with scape extended to the hind margin of the eye; head slightly constricted behind eyes; elytral declivity lacking fine setae; Florida........ .... *Stenancylus*

— Antenna with scape extended to the midlength of the eye; head not constricted behind eyes; elytral declivity with numerous fine setae; California .... *Trichacorynus*

CLASSIFICATION OF THE NEARCTIC COSSONINAE

31. Cossonini Schoenherr 1825

Cossonus Clairville 1798, 20 spp., generally distributed. Adults are found under bark of various tree species, mostly conifers but also hardwoods. See Van Dyke (1915, 1916), Blatchley and Leng (1916) and Hatch (1971) to separate the species. The genus needs revision.
Borophloeus Wollaston 1873
Isotrogus Wollaston 1873
Hyponotus Wollaston 1873
Heterophasis Wollaston 1873 (valid subgenus)
Drepocossonus Voss 1939 (valid subgenus)
Caenocossonus Voss 1955 (valid subgenus)
Odontocossonus Voss 1956 (valid subgenus)
Otiorcossonus Voss 1956 (valid subgenus)

[*Dynatopechus* Marshall 1931, 1 sp., *D. aureopilosus* (Fairmaire 1849), intercepted in quarantine; California, Oregon and Washington. Not established in North America.]

Mesites Schoenherr 1838, 2 spp., likely adventive; eastern United States. Species are associated with driftwood on Atlantic and Gulf Coast beaches. See Blatchley and Leng (1916) to separate the species.
Odontomesites Wollaston 1873 (valid subgenus)

Stenotrupis Wollaston 1873, 1 sp., *S. acicula* Wollaston 1873, Florida. This species is associated with dead fronds of *Thrinax parviflora* Sw. (Arecaceae) (Anderson 1993a).
Diodimorpha Broun 1883
Pseudaphioda Voss 1956 (valid subgenus)

32. Acamptini LeConte 1876

Acamptus LeConte 1876, 3 spp., generally distributed in the eastern United States and southern Canada, west to Texas and Arizona. Species are associated with dead limbs and injured spots or areas of dead rotten wood such as tree holes or hollowed out trunks of various trees (Kissinger 1964). See Casey (1895) and Sleeper (1954b) to separate the species. The genus needs revision.
Pseudacamptus Champion 1909
Glyphostethus Marshall 1921

33. Dryotribini LeConte 1876

Amaurorhinus Fairmaire 1860, 1 sp., *A. beckwickianus* (Wollaston 1860), South Carolina; adventive.
Mesoxenus Wollaston 1861
Mazagranus Pic 1905 (valid subgenus)

Caulophilus Wollaston 1854, 4 spp., generally distributed in the eastern United States west to Texas, also California; one adventive species, *C. oryzae* (Gyllenhal 1838). Native species are associated with various species of dead trees and grapevines (Blatchley and Leng 1916). *Caulophilus oryzae* is the 'broad-nosed grain weevil' and is found in stored products as well as in avocado seeds and fruits (Anderson 1993a). See Blatchley and Leng (1916) to separate some of the species.
Allomimus LeConte 1876
Tytthomimus Champion 1909

Dryotribus Horn 1873, 1 sp., *D. mimeticus* Horn 1873; Florida and South Carolina. This species is associated with old boards and driftwood washed up along the coast (Anderson 1993a).
Thalattodora Perkins 1900
Pentacotaster Chûjô and Voss 1960

Micromimus Wollaston 1873, 2 spp., Florida. Adults have been collected under bark of various trees, especially *Bursera simaruba* (L.) Sarg. (Burseraceae) (Anderson 1993a).

Paralicus O'Brien 1984, 1 sp., *P. minyops* O'Brien 1984, southern Florida. Adults are found under driftwood and in litter along beaches (Anderson 1993a).

Stenomimus Wollaston 1873, 1 sp., *S. pallidus* (Boheman 1845), generally distributed in the eastern United States. Larvae have been found under bark of *Juglans nigra* L. (black walnut; Juglandaceae) (Anderson 1952).

34. Onycholipini Wollaston 1873

Hexarthrum Wollaston 1860, 3 spp., generally distributed in the eastern United States and southern Canada, disjunct to British Columbia and Idaho; one adventive. Adults occur in woodwork of buildings (Blatchley and Leng 1916). See Brown (1966a) to separate the species.

Pselactus Broun 1886, 1 sp., *P. spadix* (Herbst 1795), eastern United States and California; adventive. This species is associated with driftwood on coastal beaches (Blatchley and Leng 1916).
Codiosoma Bedel 1885; not Kirby 1874
Phloeophagia Aurivillius 1924

Pseudopentarthrum Wollaston 1873, 11 spp., generally distributed in the eastern United States west to Texas and Arizona. Species are associated with dead limbs and injured spots or areas of dead rotten wood such as tree holes or hollowed out trunks of vari-

ous trees (Kissinger 1964; Anderson 1993a). See Blatchley and Leng (1916), Blatchley 1922, 1925, 1928) and Sleeper (1954b) to separate the species. The genus needs revision.
Phloeophagomorphus Wollaston 1873
Pentarthrinus Casey 1892
Micropentarthrum Champion 1909
Neopentarthrum Mutchler 1925
Stenotylus Marshall 1933

Stenoscelis Wollaston 1861, 2 spp., generally distributed in the eastern United States and southern Canada. Species are associated with the dead wood of various trees (Kissinger 1964). See Buchanan (1948) to separate the species.
Dendroctonomorphus Wollaston 1873
Astenoscelis Hustache 1956 (valid subgenus)
Hexastenoscelis Voss 1964 (valid subgenus)

Trichacorynus Blatchley 1916, 2 spp., Indiana, New Jersey, Pennsylvania and California. Adults have been reared from *Yucca* stalks (Liliaceae) in California. See Sleeper (1957b) to separate the species.

35. Pentarthrini Lacordaire 1866

Macroscytalus Broun 1881, 1 sp., *M. chisosensis* (O'Brien 1973), southwestern Texas. This species is associated with the dead, dry leaves of *Agave havardiana* Trel. (Amaryllidaceae) (O'Brien 1973).
Rhinanisus Broun 1883
Baeorhopalus Broun 1883

Pentarthrum Wollaston 1854, 1 sp., *P. huttoni* Wollaston 1854, Quebec; adventive. Adults have been found in floor boards of houses (Warner 1952).
Attarus Broun 1909
Belka Broun 1909
Gaurocryphus Broun 1909
Trachyglyphus Broun 1909

36. Proecini Voss 1956

Proeces Schoenherr 1838, 1 sp., *P. depressus* (Boheman 1838), Florida. Adults have been collected in curled leaf sheaths of *Roystonea elata* Bartr. (F. Harper) (royal palm; Arecaceae) (Anderson 1993a).
Stenotis Wollaston 1854
Eucoptus Wollaston 1873
Borborhynchus Richard 1957

37. Rhyncolini Gistel 1856

Rhyncolina Gistel 1856

Aphanommata Wollaston 1873, 1 sp., *A. tenuis* (Casey 1892), southeastern United States west to Texas. Adults have been collected in tree hollow debris.
Rhamphocolus Casey 1892
Macrancyloides Champion 1909
Oocorynus Champion 1909
Brachytemnoides Folwaczny 1973

Apotrepus Casey 1892, 1 sp., *A. densicollis* Casey 1892, Arizona. This species is associated with scar tissue on wounds of saguaro cactus (Cactaceae) (Kissinger 1964).

Carphonotus Casey 1892, 1 sp., *C. testaceus* Casey 1892, generally distributed in the northern United States and across Canada. Adults have been associated with spruce (Blatchley and Leng 1916).

Elassoptes Horn 1873, 1 sp., *E. marinus* Horn 1873, western United States and Canada. Adults are associated with driftwood on beaches (Kissinger 1964).

Himatium Wollaston 1873, 2 spp., generally distributed in the eastern United States and southern Canada. Adults have been reared from dead branches of *Acer saccharum* L. (Aceraceae) and have been collected in leaf litter (Anderson 1993a). Adults have also been associated with the galleries of *Ips* bark beetles under pine bark. See Blatchley and Leng (1916) to separate the species.
Pholidonotus Wollaston 1873
Choerodemas Faust 1898 (valid subgenus)
Himatinum Cockerell 1906
Macrohimatium Konishi 1962

Macrancylus LeConte 1876, 1 spp., *M. linearis* LeConte 1876, southeastern United States west to Texas, Adults are associated with driftwood on coastal beaches (Blatchley and Leng 1916).
Haloxenus Perkins 1900

Macrorhyncolus Wollaston 1873, 1 sp., *M. littoralis* (Broun 1880), California; adventive. Adults are associated with driftwood on coastal beaches.

Nyssonotus Casey 1892, 1 sp., *N. seriatus* Casey 1892, Arizona, California and Texas. This species is associated with *Yucca* (Amaryllidaceae); larvae are in stalks (Anderson 1952).

Rhyncolus Germar 1817, 15 spp., generally distributed in the eastern United States west to Texas and north to Canada, then across Canada and south into the western United States. Species are associated with dead wood of various types of trees, mostly conifers but also willows, aspens and poplars (Salicaceae). *Rhyncolus pallens* Casey 1892 is associated with *Lupinus arboreus* Sims. (Fabaceae) in California. See Casey (1892), Blatchley and Leng (1916) and Buchanan (1946) to separate the species. Some species in *Phloeophagus* may be better placed as *Rhyncolus* (see note under *Phloeophagus*). The genus needs revision.
Rhyncholus Gistl 1834
Eremotes Wollaston 1861
Syntomocerus Wollaston 1865
Hyperemotes Voss 1934 (valid subgenus)
Xylocomesus Thatcher 1940

Axenomimetes Voss 1955 (valid subgenus)

Stenancylus Casey 1892, 2 spp., Florida. *Stenancylus colomboi* Casey 1892 is associated with *Acrostichum* (Pteridaceae) in Florida (Anderson 1993a). See Blatchley and Leng (1916) to separate the species.
Liolepta Blatchley 1916
Rhinonus Kuschel 1959

Tomolips Wollaston 1873, 1 sp., *T. quercicola* (Boheman 1845); generally distributed in the eastern United States. Larvae develop in dead wood of various trees (Blatchley and Leng 1916).
Wollastonia Horn 1873; not Heer 1852
Wollastoniella Cockerell 1906; not Reuter 1884
Parahornia Cockerell 1906

Phloeophagina Voss 1955

Phloeophagus Schoenherr 1838, 5 spp., generally distributed in the eastern United States north into Canada, west through the north, then south into the western United States. *Phloeophagus minor* Horn 1873 and *P. californicus* Van Dyke 1927 appear not to be congeneric with the remaining species and appear better placed as *Rhyncolus,* which is where they are treated in the key. See Blatchley and Leng (1916) and Van Dyke (1927) to separate the species.

X. Cryptorhynchinae Schoenherr 1825

by Robert S. Anderson

Members of this subfamily are easily recognized by the ventral channel on the sternum (Fig. 59), in which the rostrum lies in repose, being extended beyond the prosternum onto the mesosternum or even the metasternum. Usually, the eyes are covered by the anterolateral margins of the pronotum when the rostrum is in respose, and the tibia possess a large, curved apical tooth. The numbers of genera and species present in North America are very small compared to the taxonomic and structural diversity of cryptorhynchines in the Neotropical Region. A detailed study of the genera is much needed.

Cryptorhynchine larvae generally are borers in dead wood although some mine inside living plants and some species are found in seeds. Most genera are found in terrestrial habitats although species of the genus *Tyloderma* are associated with aquatic habitats. Many species are found in leaf litter and the odd southwestern species *Liometophilus manni* Fall 1912 is associated with ants. Many species are flightless. No species are serious economic pests although the mango weevil, *Sternochetus mangiferae* (Fabricius 1775), is frequently intercepted in quarantine at United States border inspections.

Key to the Nearctic genera of Cryptorhynchinae

1. Tarsus with claws separate, with basal tooth .... *Phyrdenus*
— Tarsus with claws separate or connate basally, lacking basal tooth 2

2(1). Tarsus with claws connate at base .... *Faustinus*
— Tarsus with claws separate, not connate at base. 3

3(2). Antenna with funicle of 5 or 6 articles 4
— Antenna with funicle of 7 articles 6

4(3). Body nearly glabrous or with sparse, narrow appressed scales; elytra with stria 10 ended above the hind coxa; frequently collected in wetlands .... *Tyloderma*
— Body with dense, often erect, scales; elytra with stria 10 ended anterior to, or posterior to hind coxa; not in wetlands 5

5(4). Pronotum with anterior portion with two subparallel costae, markedly declivitous posteriorly (Fig. 60); southwestern United States, in association with ants .... *Liometophilus*
— Pronotum with anterior portion simple, lacking costae, dorsal surface evenly rounded or on same plane; southern Florida .... *Neoulosomus*

6(3). Metasternum short, the distance between middle and hind coxae distinctly shorter than length of antennal club; generally, the metepisternal suture is not distinct and wings are absent or reduced; scutellum usually not visible, if visible, clothed with scales 7
— Metasternum long, the distance between middle and hind coxae subequal to or distinctly longer than length of antennal club; generally, the metepisternal suture is distinct and wings are present; scutellum usually visible 15

7(6). Abdomen with ventrite 1 about as long as or longer than length of ventrites 2-5 combined, with large, deep transverse depression or smaller lateral impressions near posterolateral margins .... *Eurhoptus*
— Abdomen with ventrite 1 shorter than length of ventrites 2-5 combined, simple, lacking any depressions 8

8(7). Hind tibia abruptly expanded in basal one-third, about twice as wide as at apex .... *Canistes*
— Hind tibia subequal in width throughout length or slightly expanded towards apex 9

9(8). Ventrite 2 about as long as ventrites 3-5 combined; pronotum markedly produced anteriorly over head, head not visible in dorsal view; elytra with apices produced, apex broadly truncate .... *Lembodes*
— Ventrite 2 distinctly shorter than ventrites 3-5 combined; pronotum at most only slightly produced anteriorly over head, head visible in dorsal view; elytra with apices not produced, apex evenly rounded 10

10(9). Abdomen with suture separating ventrites 1 and 2 deep, nearly straight 11
— Abdomen with suture separating ventrites 1 and 2 shallow, indistinct 14

11(10). Frons wider than apex of rostrum in dorsal view; ventrite 2 slightly longer than ventrites 3 and 4 combined .... *Calles*

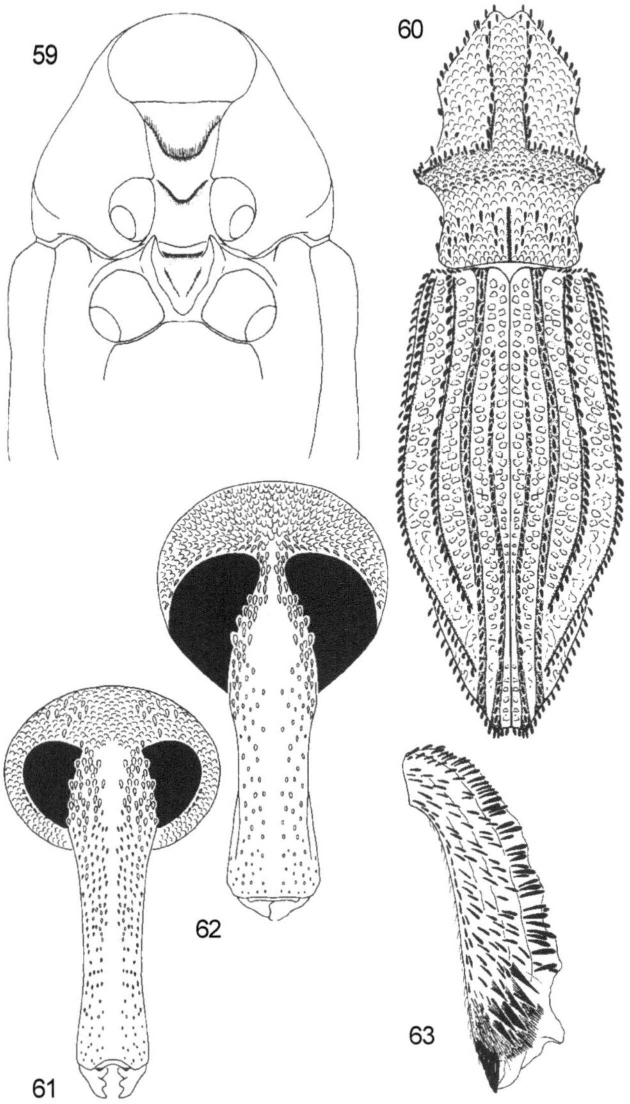

FIGURES 59.131-63.131. Cryptorhynchinae. 59. *Cryptorhynchus lapathi* (Linnaeus), thoracic sterna, ventral view; 60. *Liometophilus manni* Fall, dorsal habitus; 61. *Cryptorhynchus lapathi* (Linnaeus), head, anterior view; 62. *Cophes oblongus* (LeConte); head, anterior view; 63. *Zascelis irrorata* LeConte, middle tibia.

— Frons narrower than apex of rostrum in dorsal view; ventrite 2 subequal to or shorter than length of ventrites 3 and 4 combined 12

12(11). Sternal channel with posterior margin not extended posterior to level of hind margin of front coxae; mesosternum posterior to channel with median longitudinal carina; frons impressed between eyes; pronotum with deep, median longitudinal sulcus .... *Peracalles*

— Sternal channel with posterior margin extended posterior to level of hind margin of front coxae; mesosternum posterior to channel simple, lacking carina; frons mostly flat between eyes; pronotum simple, lacking sulcus 13

13(12). Elytra with stria 10 complete, extended to apex of elytra; scutellum visible, clothed with scales; pronotum in dorsal view widest at base .... *Pseudomus*

— Elytra with stria 10 absent or indicated at most only by a few small punctures in basal one-third of length; scutellum not visible; pronotum in dorsal view widest near middle .... *Gerstaeckeria*

14(10). Abdomen with ventrite 2 obviously longer than length of ventrites 3 and 4 combined, not shorter than ventrite 5; frons wider than apex of rostrum in dorsal view; sternal channel with posterior margin anterior to level of anterior margin of middle coxae .... *Pseudoacalles*

— Abdomen with ventrite 2 at most slightly longer than length of ventrites 3 and 4 combined, shorter than ventrite 5; frons as wide as or narrower than apex of rostrum in dorsal view; sternal channel with posterior margin at or posterior to level of anterior margin of middle coxae .... *Acalles*

15(6). Hind tibia lacking apical comb of stout setae; pronotum markedly produced over head, head not visible in dorsal view; tibia very short, less than one-half as long as femur ..... *Paracamptus*

— Hind tibia with apical comb of stout setae (usually arranged subparallel to long axis of tibia); pronotum not produced over head, head visible in dorsal view; tibia longer than one-half length of femur 16

16(15). Middle coxae separated by about two-thirds width of apex of rostrum 17

— Middle coxae separated by at least width of apex of rostrum 19

17(16). Abdomen with ventrite 2 longer than length of ventrites 3 and 4 combined; hind tibia with apical comb of stout setae longer than width of tibia at apex; metepisternal suture not visible .... *Maemactes*

— Abdomen with ventrite 2 shorter than length of ventrites 3 and 4 combined; hind tibia lacking apical comb of stout setae; metepisternal suture visible 18

18(17). Frons much wider than narrowest part of rostrum in dorsal view; elytra with stria 10 complete; hind tibia with apical comb composed of a complete apical row of setae and a confused row above that; femur simple, not sulcate ventrally .... *Euscepes*

— Frons about one-half as wide as narrowest part of rostrum in dorsal view; elytra with stria 10 ended anterior to hind coxa; hind tibia with short apical comb running perpendicular to long axis and another short row at 45 degree angle to apical comb and located above it; femur sulcate ventrally for reception of tibia................ *Apteromechus* (part)

19(16). Body with suberect or erect, very fine, long, hair-like vestiture 20

— Body with at most some scattered suberect or erect short, broad scales or scales arranged in tufts, no hair-like vestiture evident......................... 21

20(19). Legs with tibiae with outer margin with large serrations or denticles; antenna with funicle with ar-

ticle 2 long and slender, more or less twice as long as article 3 and about one and one-half times longer than article 1; body length greater than 4.0 mm; Arizona .... *Cnemidoprion*
— Legs with tibiae with outer margin simple; antenna with funicle with article 2 shorter and stouter, less than twice as long as article 3 and subequal to length of article 1; body length less than 4.0 mm; southern Florida .... *Eutinobothrus*

21(19). Leg with tibia with outer margin with acute serrate or dentate carina, especially in apical one-third (Fig. 63) .... *Zascelis*
— Leg with tibia with evenly rounded outer margin, lacking carina 22

22(21). Mandible prominent when closed, with obvious acute tooth on inner margin, angle between basal and apical cusps acute (Fig. 61).................... 23
— Mandible not prominent when closed, lacking tooth on inner margin or if tooth present, angle between basal and apical cusps obtuse (Fig. 62) 28

23(22). Frons not or only slightly narrower than narrowest part of rostrum in dorsal view (Fig. 61) 24
— Frons about one-half width of narrowest part of rostrum in dorsal view 26

24(23). Pronotum distinctly produced over head, head not visible in dorsal view; eyes almost completely covered by postocular lobes when rostrum is in repose; eyes widely separated by slightly more than width of rostrum at base; rostrum more or less straight .... *Troezon*
— Pronotum at most slightly produced over head, head partially visible in dorsal view; eyes only partially covered by postocular lobes when rostrum is in repose; eyes moderately separated by slightly less than width of rostrum at base; rostrum more or less evenly curved 25

25(24). Elytra with alternate intervals 3, 5 and 7 variously elevated, sharply carinate or lower and rounded; hind femur with ventral margin with single acute tooth (species with alternate elytral intervals carinate) or with two teeth that are broadly joined at their bases (species with alternate intervals convex) .... *Eubulus*
— Elytra with intervals of more or less equal height, although alternate intervals 3, 5 and 7 may possess patches of suberect or erect broad scales giving an elevated appearance; hind femur with ventral margin simple or with one or two separate acute teeth .... *Cryptorhynchus*

26(23). Elytra with stria 10 complete, indicated beyond level of hind coxa; hind tibia with apical comb arranged in two discrete parts .... *Apteromechus* (part)
— Elytra with stria 10 not extended beyond level of hind coxa; hind tibia with apical comb continuous, not separated into two parts 27

27(26). Abdomen with ventrite 2 markedly narrowed laterally, suture between ventrites 2 and 3 markedly angulate posteriorly at lateral margin for distance about one-half width of ventrite 3; leg with femur simple; rostrum with antenna inserted anterior to midlength; elytra with prominent humerus, obviously wider at base than pronotum at base .... *Sudus*
— Abdomen with ventrite 2 not narrowed laterally, suture between ventrites 2 and 3 straight lateral margin; leg with femur with ventral tooth; rostrum with antenna inserted posterior to midlength; elytra with humerus rounded, only slightly wider at base than pronotum at base .... *Pseudomopsis*

28(22). Hind tibia with apical comb composed of a double row of setae (may be irregular); elytra either with intervals 3 and 5 prominent or with no intervals prominent and no erect setae or nodules present .... *Cophes*
— Hind tibia with apical comb composed of a single row of setae; elytra either with erect short stout setae, nodules, or with sutural interval prominent immediately behind scutellum 29

29(28). Elytra with large, polished, prominent nodules; antenna with club lacking distinct sutures; pronotum with dense, round overlapping scales and elytra with small, narrow scales .... *Hohonus*
— Elytra with short, stout setae or with sutural interval prominent immediately behind scutellum; antenna with club with distinct sutures; pronotum and elytra with similar scales 30

30(29). Elytra laterally at base emarginate to receive prominent hind angle of pronotum; elytra with sutural interval prominent immediately behind scutellum, stria 10 indicated behind level of hind coxa as a fine line; pronotum with postocular lobes prominent and rounded; front femur with ventral margin with small tooth; front tarsus of male elongate and with long, fine setae .... *Episcirrus*
— Elytra laterally at base simple, not modified to receive hind angle of pronotum; elytra with sutural interval flat immediately behind scutellum, stria 10 not extended behind level of hind coxa; pronotum with postocular lobes slightly developed; front femur simple, lacking tooth; front tarsus of male normal .... *Rhynchus*

Classification of the Nearctic Cryptorhynchinae

38. Cryptorhynchini Schoenherr 1825

Cryptorhynchina Schoenherr 1825

Apteromechus Faust 1896, 4 spp., generally distributed in the eastern United States and Canada west to Texas and Arizona. Adults come to lights and are associated with various trees such as beech, oak and red bay; larvae have been found mining a dead sassafras limb (Kissinger 1964). See Whitehead (1979) to separate the species.

Acarlosia Hustache 1940

Cnemidoprion Marshall 1933, 1 sp., *C. oblongus* (Horn 1895), Arizona. Adults come to lights and have been collected on low roadside vegetation. A Brazilian species was reared from twigs of *Cienfugosia* (Malvaceae) (Anderson 1998).

Cryptorhynchus Illiger 1807, 4 spp., generally distributed in the eastern United States north into Canada, west across Canada and the northern United States to Oregon, Washington and British Columbia. Adults of *C. fuscatus* LeConte 1876, *C. helvus* LeConte 1878 and *C. minutissimus* LeConte 1876 come to lights and are associated with various species of dead or dying trees. *Cryptorhynchus lapathi* (Linnaeus 1758) is associated with living poplar and willow (Salicaceae). See Blatchley and Leng (1916) to separate the species. The genus in North America likely is composite and needs redefinition.

Arachnipes Villa and Villa 1833
Cryptorrhynchus Bedel 1884
Cryptorrhynchus Champion 1906
Cryptorhynchidius Pierce 1919
Atrichopsis Voss 1954 (valid subgenus)
Cryptorrhynchobius Voss 1965

Eubulus Kirsch 1870, 3 spp., generally distributed in the eastern United States into southern Canada, west in the south to Texas. Arizona and California. Adults of *E. bisignatus* (Say 1831) and *E. parochus* (Herbst 1797) have been associated with dead limbs of chestnut, beech, oak and birch (Blatchley and Leng 1916); larvae likely mine dead branches of various trees. A single adult of *E. obliquus* (Say 1831) has been associated with *Myrica cerifera* L. (Myricaceae) (Anderson 1993a). See Blatchley and Leng (1916) to separate the species.

Eubulosoma Voss 1954 (valid subgenus)

Eutinobothrus Faust 1896, 1 sp., *E. pilosellus* (Boheman 1844), southern Florida. This species is associated with *Ipomoea* (Convolvulaceae) (Anderson 1993a).

Gasterocercodes Pierce 1915

Liometophilus Fall 1912, 1 sp., *L. manni* Fall 1912, Arizona, New Mexico and Texas. This species is associated with nests of the ant *Liometopum apiculatum* Mayr (Formicidae). This is perhaps the oddest-looking weevil in North America.

Maemactes Schoenherr 1837, 1 sp., *M. cribratus* (LeConte 1876), Texas and Kansas.

Baropsis LeConte 1876
Baridopsis Rye 1878

Neoulosomus O'Brien and Wibmer 1982, 1 sp., *N. laticaudis* (Suffrian 1872), southern Florida. Adults have been collected on various dead limbs and vines (Anderson 1993a).

Ulosomus Schoenherr 1826; not Schoenherr 1825

Phyrdenus LeConte 1876, 2 spp., *P. divergens* (Germar 1824) generally distributed in the eastern United States, and *P. muriceus* (Germar 1824), Florida and Arizona. These species are associated with *Solanum* (Solanaceae) (Blatchley and Leng 1916; O'Brien 1961).

[*Sternochetus* Pierce 1917, 1 sp., *S. mangiferae* (Fabricius 1775), intercepted in quarantine; Florida and California. This species is a pest of mango but to date has not become established in North America.]

Sudus Kissinger 1964, 1 sp., *S. floridanus* Kissinger 1964, southeastern United States west to Texas.

Troezon Champion 1906, 1 sp., *T. lutosus* (LeConte 1884), Florida and Louisiana. This species is associated with *Dahlbergia ecastophyllum* (L.) Benth. (Fabaceae); larvae feed in the disc-like fruits (Blatchley and Leng 1916; Anderson 1993a).

Tyloderma Say 1831, 30 spp., generally distributed in the United States (most species in the southeast) and southern Canada. Species are associated with various plants, many in wetlands, in the families Onagraceae, Polygonaceae, Urticaceae, Haloragaceae, Rosaceae, Melostomataceae and Saururaceae (Wibmer 1981). Adults come to lights. See Wibmer (1981) to separate the species.

Analcis Say 1831; not Wagler 1830
Analcis Schoenherr 1833; not Wagler 1830; not Say 1831

Zascelis LeConte 1876, 1 sp., *Z. irrorata* LeConte 1876, southwestern United States. Adults come to lights. A second unnamed (and likely adventive) species is present in southern Florida (Anderson 1993a).

Tylodina Lacordaire 1866

Acalles Schoenherr 1825, 12 spp., generally distributed in the eastern United States into southern Canada, west to Texas, Arizona and New Mexico in the south. Species are associated with various dead branches or palm fronds on the ground, on dead vines and other hanging dead vegetation, and generally in leaf litter (Kissinger 1964; Anderson 1993a). Adults were also found sweeping *Borrichia* (Asteraceae), *Sesuvium* (Aizoaceae) and *Salicornia* and *Suaeda* (Chenopodiaceae) on beaches in southern Florida at night (Anderson 1993a). See Blatchley and Leng (1916) to separate some of the species. The genus needs revision and redefinition. Anderson (1993a) lists six undescribed species as present in southern Florida.

Ulosomus Schoenherr 1825
Microdalotes Gistel 1856
Trachodius Weise 1891
Milichacalles Voss 1960 (valid subgenus)
Trichacalles Voss 1960 (valid subgenus)

Calles Kissinger 1964, 1 sp., *C. cladotrichis* (Pierce 1912), Arizona, New Mexico and Texas. Adults were reared from roots of *Tidestromia lanuginosa* (Nutt.) Standl. (Amaranthaceae) (Pierce 1912). A second undescribed species has been collected in southern Florida on *Salicornia* and *Suaeda* (Chenopodiaceae) at night in coastal areas (Anderson 1993a) and another is known from Arizona.

Canistes Casey 1892, 1 sp., *C. schusteri* Casey 1892, eastern United States west to Texas. Adults have been collected in leaf litter.

Eurhoptus LeConte 1876, 2 spp., *E. pyriformis* LeConte 1876 and *E. sordidus* (LeConte 1876), generally distributed in the eastern and southcentral United States west to Texas and Oklahoma. Adults have been commonly collected in leaf litter. At least three undescribed species are known from Texas; the genus needs revision.

Eurrhoptus Rye 1878

Euscepes Schoenherr 1844, 1 sp., *E. porcellus* Boheman 1844, southern Florida. Adults have been collected on *Ipomoea* (Convolvulaceae) (Anderson 1993a); larvae are likely in the roots or stems.

Hyperomorpha Blackburn 1885
Batatarhynchus Hustache 1933

Faustinus Berg 1898, 1 sp., *F. cubae* (Boheman 1844), southern Florida. Adults and larvae are associated with various Solanaceae (Anderson 1993a).

Euxenus Faust 1896; not Gistel 1856; not LeConte 1876

Gerstaeckeria Champion 1905, 18 spp., southeastern and western United States, north into western Canada. Adults and larvae are associated with various Cactaceae. Adults are flightless and noctural. Larvae mine the pads of *Opuntia* and hollow out smaller pincushion cacti such as *Mamillaria* and *Coryphanta*. See O'Brien (1970b) to separate the species.

Opuntiaphila Pierce 1912
Philopuntia Pierce 1912

Lembodes Schoenherr 1844, 1 sp., *L. solitarius* Boheman 1844, southern Florida. Adults are collected on various types of dead vegetation (Anderson 1993a).

Paracamptus Casey 1895, 2 spp., *P. floridanus* Sleeper 1954 and *P. subtropicus* Casey 1895, southern Florida. Adults of *P. subtropicus* have been collected commonly on dead *Rhizophora mangle* L. (red mangrove; Rhizophoraceae) branches (Anderson 1993a). See Sleeper (1954b) to separate the species.

Peracalles Kissinger 1964, 2 spp., *P. pectoralis* (Leconte 1876), Illinois, Indiana, Ohio, Kentucky and Missouri, and *P. ventrosus* (LeConte 1878), Florida. Adults occur in leaf litter and *P. ventrosus* has been collected from emergent aquatic vegetation at night (C. W. O'Brien, pers. comm.). See Blatchley and Leng (1916; as *Acalles*) to separate the species.

Pseudoacalles Blatchley 1916, 1 sp., *P. nuchalis* (LeConte 1876), Florida and South Carolina. Adults are found in leaf litter (Anderson 1993a).

Pseudomopsis Champion 1905, 1 sp., *P. inflata* (LeConte 1876), southern Florida. Adults are commonly found on *Coccoloba uvifera* L. and *C. diversifolia* Jacq. (Polygonaceae); larvae feed in fruits (Anderson 1993a).

Pseudomus Schoenherr 1837, 2 spp., *P. sedentarius* (Say 1831), Florida and *P. truncatus* LeConte 1876, Georgia and South Carolina. See Blatchley and Leng (1916) to separate the species.

39. Gasterocercini Zherichin 1991

Cophes Champion 1905, 5 spp., generally distributed in eastern United States and southern Canada west into Texas. Adults of most species are associated with dead wood and come to lights. *Cophes texanus* Sleeper 1955 has been reared from dead *Baccharis neglecta* Britt. (Asteraceae). The genus needs revision and redefinition. See Sleeper (1955a) and Blatchley and Leng (1916; as *Cryptorhynchus*) to separate some of the species.

Coelosternus Schoenherr 1835; not Sahlberg 1823
Sternocoelus Kuschel 1955

Episcirrus Kuschel 1958, 1 sp., *E. brachialis* (LeConte 1884), Texas and Arizona. Adults are associated with *Bumelia lanuginosa* (Michx.) Pers. (Sapotaceae); larvae appear to mine dead branches.

Hohonus Kissinger 1964, 1 sp., *H. lacteicollis* (Champion 1906), Texas and Arizona. Adults are associated with *Phoradendron* (mistletoe; Viscaceae); larvae mine the stems (Anderson 1994).

Rhynchus Kissinger 1964, 1 sp., R. *apiculatus* (Gyllenhal 1837), southeastern United States. Adults are associated with *Myrica cerifera* L. (Myricaceae); larvae mine in dead trunks and larger branches (Ford 1985).

XI. Cyclominae Schoenherr 1826

by Robert S. Anderson

This is a small group of three genera of weevils, one of which appears not closely related to the other two. All members have a relatively short snout (but lack any deciduous processes and associated scars) and have well-developed postocular lobes (Fig. 66). The genus *Listroderes*, represented by only 3 species, is introduced from South America, a region of much greater cyclomine diversity. *Listronotus* is a large genus of over 80 species, most of which are associated with semi-aquatic and aquatic habitats. *Emphyastes* is an odd, unrelated genus found associated with seaweed along Pacific coastal beaches.

Key to the Nearctic Genera of Cyclominae

1. Front tibia prolonged beyond articulation of tarsus into long, flattened paddle; hind tibia markedly expanded at apex, wider than maximum width of hind femur (Fig. 67) .... *Emphyastes*
— Front tibia not prolonged beyond articulation of tarsus; hind tibia not expanded at apex, not as wide as maximum width of hind femur 2

2(1). Pronotum widest subapically, lateral margins straight and divergent from base to widest point, transversely quadrate in form (Fig. 66); pronotal disk

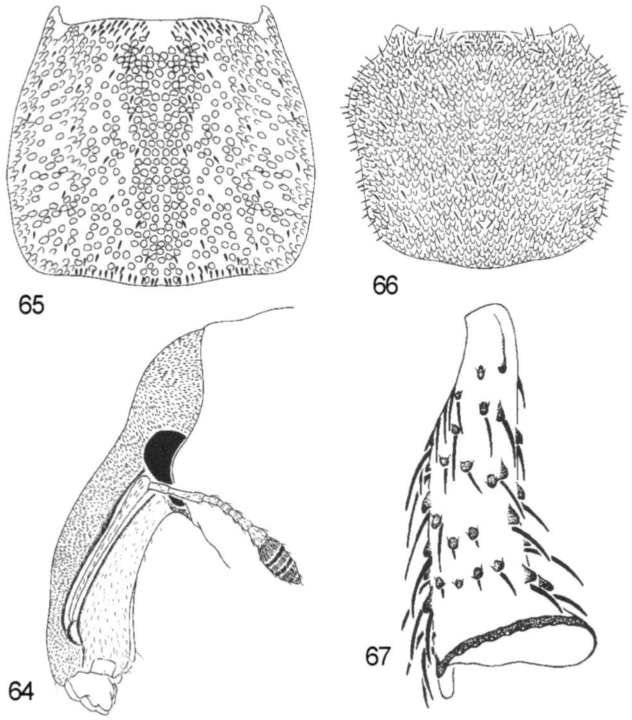

FIGURES 64.131-67.131. Cyclominae. 64. *Listronotus caudatus* (Say), head, lateral view; 65. *Listronotus oregonensis* (LeConte), pronotum, dorsal view; 66. *Listroderes costirostris* Schoenherr, pronotum, dorsal view; 67. *Emphyastes fucicola* Mannerheim, hind tibia.

broad and flat, lateral profile with anterior margin at middle elevated above level of pronotal disk; body size greater than 5 mm; elytra with erect, stout seta-like scales at least 2 to 3 times longer than diameter of adjacent rounded, flat scales . ... *Listroderes*

— Pronotum widest near base or at or before midlength, lateral margins rounded from base to widest point (Fig. 65); subcylindrical in form; pronotal disk more or less evenly convex, lateral profile evenly rounded from basal to anterior margin; body size various; elytra with or without erect, stout seta-like scales, but if seta-like scales as long as 2 to 3 times diameter of adjacent, rounded scales, body size is less than 5 mm .... *Listronotus*

CLASSIFICATION OF THE NEARCTIC CYCLOMINAE

40. Rhythirrinini Lacordaire 1863

Emphyastina Lacordaire 1863

Emphyastes Mannerheim 1852, 1 sp., *E. fucicola* Mannerheim 1852, western coastal United States and Canada north into southern Alaska. Adults and larvae are associated with decaying seaweed washed up and buried on sandy beaches (Anderson 1988b). Korotyaev and Egorov (1975) have suggested that this genus is related to *Thalasselephus* (Molytinae).

Listroderina LeConte 1876

Listroderes Schoenherr 1826, 3 spp., *L. costirostris* Schoenherr 1826, *L. difficilis* Germain 1895, and *L. apicalis* Waterhouse 1841, southeastern United States, Texas, Arizona and California; adventive. See Morrone (1993) to separate the species.

Listronotus Jekel 1865, 81 spp., generally distributed in the United States and Canada. Adults are found in wetlands and appear to be associated with a variety of plants (O'Brien 1981, Anderson 1993a). Adults of a few species in coastal Florida and Texas have been collected on *Borrichia frutescens* (L.) DC. (Asteraceae) and *Salicornia* (Chenopodiaceae) (Anderson 1993a). The genus, especially the smaller species originally placed in *Hyperodes*, needs revision. See O'Brien (1981) and Stockton (1963) to separate most of the species.

Macrops Kirby 1837; not Wagler 1830; not Burmeister 1835
Hyperodes Jekel 1865
Anchodemus LeConte 1876
Lixellus LeConte 1876
Mascarauxia Desbrochers 1898
Relistrodes Brèthes 1910
Aulametopiellus Brèthes 1926
Pseudohyperodes Hustache 1939

XII. Entiminae Schoenherr 1823

by Robert S. Anderson and Anne T. Howden

Entiminae are generally called the 'broad-nosed' weevils because this is the group of curculionids in which the snouts are the least developed. Aside from the possession of a shorter, broader snout, the best way to recognize them is that nearly all Entiminae have a mandible that bears a deciduous process that breaks off soon after emergence of the adult leaving a definite scar at the point of attachment on the outer face of the mandible. However, not all Entiminae possess this feature (*Thecesternus*, *Sitona*, and members of the tribe Alophini) and one must rely on other features in recognizing their inclusion in the subfamily. Entimines also possess only a short tooth or spine on the inner angle at the apex of the hind tibia and sexual dimorphism in the form of the rostrum is generally not as evident as in other weevils. The antennal scape of some species also extends to or beyond the anterior margin of the eye, a feature otherwise only found in Dryophthorinae.

Most entimines have larvae that feed externally in the soil on roots whereas the adults tend to feed on fresh foliage or reproductive structures such as flowers or buds. Many species are generalists and feed on a very broad range of plant taxa both as adults and larvae (*e.g.*, *Otiorhynchus ovatus* (Linnaeus 1758)) whereas others can be very host specific, feeding on a few closely related species or genera. Oviposition usually takes place in the soil or rarely on the foliage of the host plant, larvae then dropping to the ground to feed in the soil. Adults of many species of Entiminae are flightless and some are parthenogenetic.

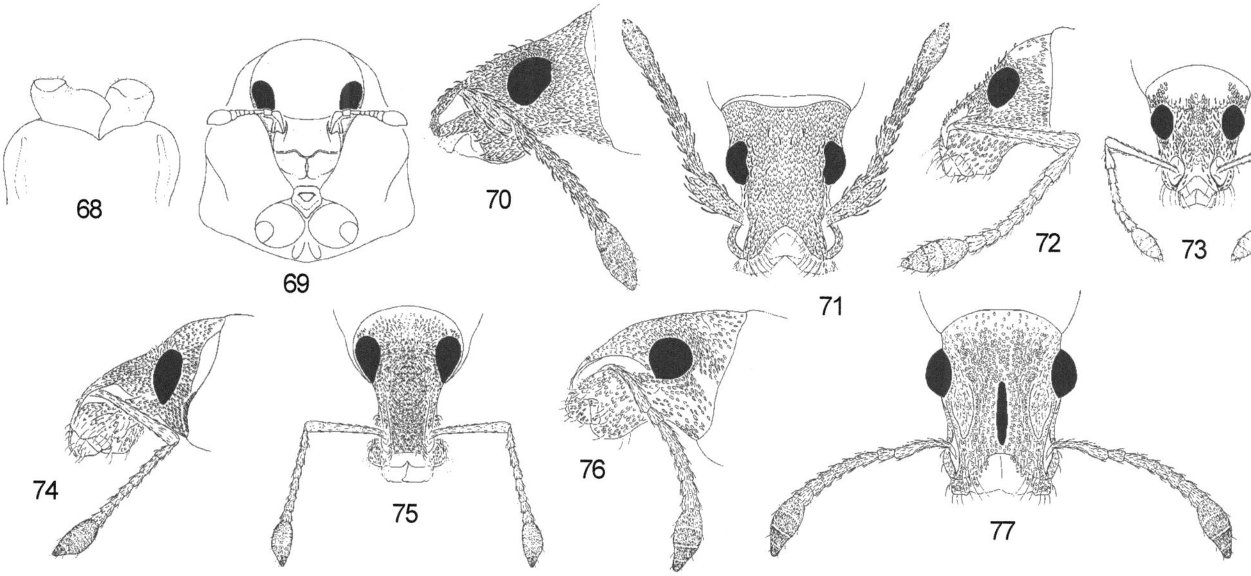

FIGURES 68.131-77.131. Entiminae. 68. Mandibles of Entiminae, schematic (after Kissinger 1964). 69. *Thecesternus* sp., head and thorax, ventral view (after Kissinger 1964). 70-71. *Brachystylus sayi* (Alonso-Zarazaga), head, 70. Lateral view; 71. Dorsal view. 72-73. *Sciopithes obscurus* Horn, head 72. Lateral view; 73. Dorsal view. 74-75. *Dyslobus lecontei* Casey, head, 74. Lateral view; 75. Dorsal view. 76-77. *Ericydeus lautus* (LeConte), head, 76. Lateral view; 77. Dorsal view.

Entiminae appear to be very adundant and diverse in arid habitats, particularly in the deserts of the southwestern United States. Some have developed adaptations for sand dwelling that include dense long hairs over the body and fossorial legs. Entimines are also the weevils most often found at higher elevations.

A number of species are pests of ornamental plants and of agricultural produce including citrus and other fruits. These include *Otiorhynchus ovatus* (strawberry root weevil), *O. sulcatus* (Fabricius 1775) (black vine weevil), *Cyrtepistomus castaneus* (Roelofs 1873) (Asiatic oak weevil), *Artipus floridanus* Horn 1876, and species of *Naupactus* and *Sitona*.

Entiminae are the most diverse subfamily in North America with 124 genera in 23 tribes recognized. In some tribes such as Peritelini, identification of the genera is very difficult and generic definitions need to be reassessed.

The portions of the key from couplet 68 to 84 and 105 to 122 are slightly modified from Kissinger (1964). The portion of the key from couplet 85 to 104 was adapted from Sleeper (1955b) and O'Brien (1984). An application is before the International Commission on Zoological Nomenclature to maintain Trachyphloeini (Alonso-Zarazaga and Lyal 1999, p. 8).

Key to the Nearctic Genera of Entiminae

1. Rostrum and sides of head recessed in deep cavity formed by the sides of the prothorax and by prosternum in the form of a small triangular plate originating between fore coxae (Fig. 69); when positioned within the cavity, only dorsal surfaces of rostrum and frons visible; mandible without deciduous process; tarsi without pads on ventral surface; tarsal claws free; body length 6.5-9.0 mm .... *Thecesternus*

— Rostrum and sides of head not recessed, prosternum not forming such a cavity; mandible with or without deciduous process; tarsi with or without pads on ventral surface, tarsal claws free or connate; body length various 2

2(1). Mandible large, hemispherical externally, inner surface slightly cupped; surface of mandible densely squamate except narrow median edge glabrous; mandible without deciduous process; bucal cavity large, maxillary palpus fully or mostly exposed; scrobes lateral; rostrum similar to head in length and width; anterior edge of prothorax straight laterally, not lobed beneath eye .... *Sitona*

— Mandible various; if large and densely squamate, then with postocular lobe and/or with scrobe dorsal and/or with deciduous process; other characters various 3

3(2). Mandible prognathous, like a thin roof extending over mouth parts; mandible without a deciduous process, or process very small and inconspicuous; all or most of mouthparts exposed in ventral view; elytra with humeri rounded; tarsal claws free 4

— Mandible not prognathous; mandible with deciduous process or its scar (Fig. 68); elytra with humeri various; tarsal claws free or not 6

4(3). Mentum as wide as bucal cavity, maxillae only briefly exposed basally; mandible with deciduous process or its scar present on extreme outer angle; elytra flat basally in vicinity of scutellum; pronotum with glabrous median carina; body length up to 9.0 mm .... *Byrsopages*

— Mentum as wide as one-third width of bucal cavity; mouthparts completely exposed in ventral view; deciduous process of mandible lacking or very small; elytra not flat basally; body length up to 6.0 mm 5

5(4). Dorsal surface clothed with appressed oval or fan-shaped scales with fine ribs radiating from point of attachment; body surface not encrusted with dirt; scrobe evanescent dorsally, not reaching anterior edge of eye; scape reaching posterior edge of eye; body length 3.5-5.6 mm .... *Dirotognathus* and *Lepidophorus*

— Dorsal surface dirt-encrusted, with papillae; scales not readily visible; scrobe well-defined dorsally, continuing above dorsal edge of eye; scape abruptly thickened distally, reaching anterior edge of eye; body length 3.0-3.3 mm .... *Vitavitus*

6(3). Dorsal surface dirt-encrusted, with papillae, without scales; mentum almost completely covering bucal cavity; scrobe ending well before eye; tarsal claws free; body length 3.5-6.3 mm .... *Leptopinara*

— Dorsal surface squamate, without papillae, with or without crust of dirt; other characters various 7

7(6). Bucal cavity deeply recessed, occupying approximately one-third width of rostrum, lateral walls of cavity perpendicular, posterior wall formed by distal edge of gula, anterior wall formed by ventral surface of mandibles when closed; mentum distally occupying full width of bucal cavity; maxillae slightly exposed proximally; densely covered with shiny, appressed, whitish, slightly opalescent scales; tarsal claws connate; body length 2.6-3.8 mm .... *Connatichela*

— Bucal cavity not so deeply recessed, if at all; mentum of various widths; vestiture dense or not; tarsal claws connate or not; other characters various 8

8(7). Side of prothorax with anterior margin produced into slight to very large rounded postocular lobe (Fig. 74); eye flat, tear-drop shaped (Fig. 74) 9

— Side of prothorax with anterior margin straight, not produced (Figs. 70, 72, 76); eye various 44

9(8). Mandible with four or more large setae; femur various 10

— Mandible with three large setae; femur with a tooth on inner edge distally 39

10(9). Ventral edge of postocular lobe very abrupt, angulate 11

— Ventral edge of postocular lobe gradual, more rounded 12

11(10). Elytra oval, scutellum seldom visible dorsally; body length 2.7-3.5 mm .... *Aracanthus*

— Elytra with humeral angle slanted from stria 7 outwards; scutellum very wide; prothorax and elytra patterned with lines and other markings; body length 4.1-8.0 mm .... *Eudiagogus*

12(10). Head with a deep sulcus across ventral surface of head-rostrum junction, sulcus continuous with scrobe; humeri quadrate .... *Colecerus*

— Head without a sulcus across ventral surface of head-rostrum junction; humeri rounded..... 13

13(12). Base of elytra carinate; rostrum dorsally with fine, deep median sulcus and shorter lateral sulcus at base that is abruptly turned laterally toward antennal scrobe; body length 4.3-5.6 mm .... *Sapotes*

— Base of elytra not carinate, sloping to mesothorax; rostrum with or without sulci, not with right-angled dorsolateral sulcus; postocular lobe with row of setae of graduated lengths; body length 5.0-25.0 mm 14

14(13). Rostrum with dorsal margin of scrobe well-defined, with a sharp upper angle; antenna with scape and funicle, and dorsal surfaces of tarsal articles all with round overlapping scales; body lacking long, fine, erect hairs .... *Ophryastes*

— Rostrum with dorsal margin of scrobe less well-defined, with a rounded upper angle with one or more of antennal scape and funicle, and dorsal surfaces of tarsal articles, lacking round overlapping scales; body with or without long, fine, erect hairs 15

15(14). Mandible with process or its scar linear, very small, rarely observed; sides of mandible large, flattened or slightly concave, irregularly sculptured; scape and funicle without broad scales; rostrum longer than head; eye teardrop shape or oval; size large, length 10.0-17.0 mm 16

— Mandibular process or its scar usually much larger and distinct (Fig. 68); sides of mandible not as above; antennal scape and funicle various; size various 19

16(15). Scutellum triangular, not conspicuous in dorsal view; integument shiny through sparse, elongate scales and stiff dark setae. *Acmaegenius*

— Scutellum rectangular, readily visible in dorsal view, conspicuously densely clothed with appressed setae/scales 17

17(16). Rostrum with three very deep sulci; one median dorsal sulcus from interantennal line to frons; one lateral sulcus on each side from near antennal insertion to upper edge of eye, thence following anterior edge of eye to scrobe .... *Triglyphulus*

— Rostrum with sulci, if present, not present along edge of eye 18

18(17). Elytra with slight humeral angle; striae with deep elongate punctures .... *Plinthodes*

— Elytra with humeral angle rounded; striae with shallow punctures .... *Trichalophus*

19(15). All surfaces densely squamate with smooth, shiny scales; with long fine erect hairs as much as 1.0 mm long on dorsal surface including legs and antennal funicle (Fig. 79); front tibia with outer angle expanded, middle tibia less so, hind tibia with outer and sometimes inner angle greatly expanded; tibiae edged distally with row of stout spines; many species with tarsal articles spinose ventrally, particularly article

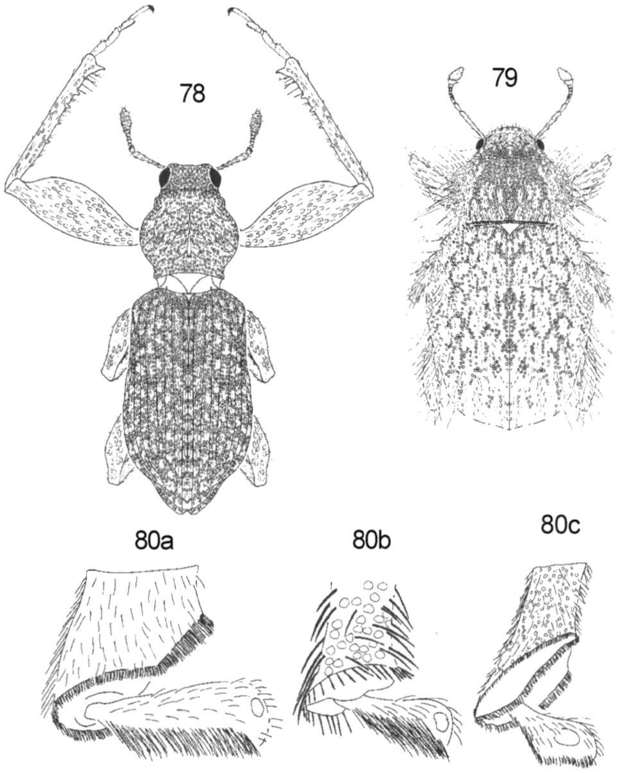

FIGURES 78.131-80.131. Entiminae. 78-79. Dorsal habitus, 78. *Pandeleteius rotundicollis* (Fall); 79. *Miloderes nelsoni* Kissinger. 80. Hind tibiae of various Entiminae, schematic. a) open corbel, b) semi-closed corbel, c) closed corbel.

1; lacking complete pads on all tarsal articles; associated with sandy environments 20

— Vestiture dense or not; without long fine erect hairs and expanded tibial apices (if long erect hairs and distinct postocular lobes present, see *Paracimbocera*); other characters various . 24

20(19). Epistoma abruptly perpendicular, posterior margin carinate .... *Trigonoscutoides*

— Epistoma on same plane as remainder of rostrum 21

21(20). Eye convex .... *Trigonoscuta*

— Eye flat 22

22(21). Tarsal claws connate basally or with a single claw .... *Eucilinus* (part)

— Tarsal claws free 23

23(22). Antennal scrobe shallow and greatly widened posteriorly .... *Cimbocera*

— Antennal scrobe deep and only slightly widened posteriorly .... *Miloderes*

24(19). Body with long fine erect hairs and distinct postocular lobes (as in *Cimbocera*, but tibial apices not expanded); tarsal article 3 with apical tufts (except females of *P. robusta*) .... *Paracimbocera*

— Body without long fine erect hairs; tarsal article 3 various 25

25(24). Dorsum very irregularly, coarsely sculptured; rostrum with three broad sulci; head with swelling above eye; elytra with costae on intervals 3, 5, 7; with large knob on apical umbone and large knob on declivity .... *Rhigopsis*

— Surface not irregularly sculptured as above; other characters various 26

26(25). Metepisternal suture present, well defined at least in basal half, usually complete, rarely obliterated; if suture is obliterated, then metepisternum with finer, smaller, and sparser scales than metasternum 27

— Metepisternal suture obliterated entirely or in large part at least in basal half, metepisternum clothed with scales similar in coarseness, size, and density 28

27(26). Dorsal margin of scrobe indistinct posteriorly; scape rests on eye when retracted next to head; rostrum separated from frons by distinct transverse impression; abdominal ventrite 2 longer than 3 and 4 united, suture between ventrites l and 2 deep, straight; article 1 of front tarsus dorsally clothed with sparse round scales .... *Adaleres*

— Abdominal ventrite 2 not longer than 3 and 4 united, suture between ventrites 1 and 2 deep, straight; article 1 of front tarsus lacking round scales on dorsal surface .... *Dyslobus*

28(26). Each puncture of elytral stria covered by round scale; elytra usually with sparse, coarse, erect setosity .... *Panscopus*

— Each puncture of elytral stria either with a fine seta, apparently glabrous, or puncture not distinct; elytral vestiture various 29

29(28). Scape of antenna clothed with round, flat scales 30

— Scape of antenna not clothed with round scales, only suberect, fine setae 35

30(29). Tibia with corbel open (Fig. 80a); body usually dirt-encrusted; with long, erect, sparse, spatulate setae; dorsal margin of scrobe poorly defined posteriorly; scape extended over middle of eye .... *Phyxelis*

— Tibia with corbel closed (Fig. 80c); other characters various 31

31(30). Elytra clothed with long, fine, erect setae, each seta about five times as long as diameter of an adjacent scale; hind tibia with corbel closed, corbel plate clothed with flat, round scales .. .. *Diamimus*

— Elytra either lacking erect setosity (*Orimodema*) or with stout bristles not more than three times as long as diameter of scales (*Dichoxenus*); hind tibia with corbel closed, corbel plate lacking scales 32

32(31). Body elongate in form; elytra lacking erect setosity, scales of elytra not at all overlapping; dorsal margin of scrobe poorly defined posteriorly; in dorsal view lateral margins of rostrum distinctly converging from base at anterior margin of eye to about the middle, thence nearly straight to point of insertion of antenna .... *Orimodema*

— Body shorter and stouter in form; elytra with sparse, erect bristles; in dorsal view lateral margins of rostrum slightly, evenly convergent from base at anterior margin of eye to point of insertion of antenna 33

33(32). Sternite 8 of female compressed distally, the vertical dimension much greater than horizontal dimension; elytra with scales not overlapping, on dorsal surface with short, erect, clavate setae; scape densely squamate .... *Paranametis*

— Sternite 8 of female shaped like a horizontal shovel, with broad, laterally expanded process distally; other characters various 34

34(33). Scape densely scaled; elytral setae blunt apically; ventral margin of scrobe clearly defined .... *Dichoxenus*

— Scape mostly setose with a few broad scales; elytral setae fine, acute; ventral margin of scrobe poorly defined .... *Anametis*

35(29). Basal margin of elytra produced strongly and abruptly perpendicularly before merging with sclerites covered by prothorax 36

— Elytra lacking distinct basal margin, evenly rounded to sclerites covered by prothorax (or elytra lacking erect setosity, *Melanolemma*); dorsal margin of scrobe poorly defined posteriorly 37

36(35). Dorsal margin of scrobe not defined posteriorly, scape passing over middle of eye; corbel of hind tibia open; body lacking erect setae or scales; tarsal claws connate .... *Tropiphorus*

— Dorsal margin of scrobe well defined posteriorly, scape passing over bottom of eye; corbel of hind tibia narrowly closed; body with erect fine setae usually numerous; tarsal claws free .... *Peritaxia*

37(35). Elytra clothed with long, fine, erect setae; elytral stria 10 extending posteriorly to above margin of hind coxa, there joining stria 9 .... *Crocidema*

— Elytra lacking erect setae; elytral stria 10 obscure, not joining stria 9 38

38(37). Rostrum flat on dorsal surface, there with fine, longitudinal, glabrous median line; frons two-fifths wider than distance between lateral margins of rostrum at point of insertion of antenna in dorsal view; rostrum lacking deep, fine sulcus distad of eye .... *Pseudorimus*

— Rostrum on dorsal surface convex, there with very vague, low, median longitudinal carina; frons much less than one-fifth wider than distance between lateral margins of rostrum at point of insertion of antenna in dorsal view; rostrum at base slightly distad of anterior margin of eye with short, deep, fine sulcus perpendicular to longitudinal axis of rostrum .... *Melanolemma*

39(9). Elytra with humeral angle 40

— Elytra with humerus rounded 42

40(39). Femora each with very large tooth bearing one or two smaller teeth on distal edge; scape reaching anterior third of prothorax; base of elytron forming a large lobe between scutellum and interval 5; dorsal elytral setae minute; color dark brown-black with whitish scales, with irregular pattern created by areas of very small scales allowing integument to show through; body length 6.0 -7.0 mm .... *Myllocerus*

— Femora each with single small tooth; dorsal elytral setae long, slender, arcuate; base of elytra straight; color and body length various 41

41(40). Eye small, separated from anterior margin of prothorax by three or more scales; elytral intervals flat; pronotum and elytra evenly, densely squamose; body length 3.3-4.0 mm .... *Cedophrys*

— Eye large, separated from anterior margin of prothorax by single row of scales; elytral intervals convex; pronotum and disc of elytra with scales very sparse or absent, replaced with minute setae; scales becoming more numerous laterally; body length 4.5-5.8 mm .... *Cyrtepistomus*

42(39). Pterygium open apically (Fig. 73); eye with approximately 14 facets along longest axis; prothorax with distinct lobe on anterior edge below eye, prothorax here 1.2x longer than length of pronotum; body length 3.5-3.8 mm .... *Calcmycterus*

— Pterygium various; prothorax straight on anterior edge, longer dorsally than ventrally; other characters various 43

43(42). Pterygium closed apically (Fig. 71); eye small, flattened, with approximately 20 facets along longest axis; prothorax 1.4x longer dorsally than ventrally; elytral scales not sculptured; body length 3.0-4.0 mm .... *Neoptochus*

— Pterygium open apically; eye with approximately 10 facets along longest axis; elytral scales sculptured in a fan shape; body length 2.4-4.0 mm .... *Myosides*

44(8). Anterior edge of prothorax laterally with postocular vibrissae in a cluster or tuft, or lacking; without lobe (except *Pachnaeus*); with or without tooth or knob 45

— Anterior edge of prothorax without postocular vibrissae 55

45(44). Elytra with humeral angle well-developed, distinct (Fig. 78; less so in *Pandeleteius simplarius*) 46

— Elytra with humerus rounded 52

46(45). Eye large, flattened; rostrum thick; front coxae contiguous or apparently so; body length 5.0-12.0 mm 47

— Eye smaller; front coxae distinctly separated by continuous prosternal integument; body length less than 5.0 mm 49

47(46). Hind tibia with straight comb of setae on outer edge, comb at least as long as width of tibia at apex; postocular vibrissae set on edge of prothorax; color dorsally dark, vaguely patterned

at most; corbel open; front coxae contiguous; body length 5.5-10.0 mm .... *Tanymecus*
— Hind tibia without straight comb of setae; postocular vibrissae set on knob or rounded tooth on edge of prothorax; color pale or brilliant metallic; corbel various; front coxae contiguous or apparently so; body length 6.4-12.0 mm 48

48(47). Front leg not larger, front femur not more swollen than middle and hind femora; eye large, flattened, oval, slightly diagonal; color pastel gray, green, or bluish; many with pollinosity; postocular vibrissae set on a prominent knob .... *Pachnaeus*
— Front leg larger than middle and hind legs, front femur enlarged; eye moderately convex, slightly transverse; postocular vibrissae lacking (or rudimentary in *H. opalina*); color glossy white or metallic blue, blue-green, green or copper; without pollinosity ... *Hadromeropsis*

49(46). Tarsal claws fused; rostrum extremely short; postocular vibrissae vestigial .... *Isodrusus*
— Tarsal claws free; rostrum short; postocular vibrissae various 50

50(49). Anterior margin of abdominal ventrites 3, 4, and 5 without modification; contour and vestiture more or less uniform; mandible without scales; postocular vibrissae various, well-developed in most species; front legs distinctly to greatly larger than middle and hind legs (Fig. 78) .... *Pandeleteius*
— Anterior margin of abdominal ventrites 3, 4 and 5 conspicuously modified; postocular vibrissae present or absent; front legs slightly larger than middle and hind legs 51

51(50). Anterior margin of abdominal ventrites 3, 4, and 5 deeply, narrowly sulcate across width of abdomen; posterior margin of sulcus carinate, right-angled in female, more rounded in male .... *Scalaventer*
— Anterior margin of abdominal ventrites 3, 4, and 5 shallowly sulcate across width of abdomen or medially only; posterior edge rounded .... *Pandeleteinus*

52(45). Tarsal articles 1, 2, and 3 on ventral surface with complete pad of dense fine setae; scrobe right-angled, reaching ventral surface, glabrous ... . *Isodacrys*
— Tarsal articles 1, 2, and 3 on ventral surface lacking complete pads; at most with small pad at apex of tarsal article 3; pads replaced with spines or long recumbent setae or neither; scrobe not right-angled, squamate 53

53(52). Dorsal surface with very fine long setae, recumbent or semi-erect, setae longer on sides and ventral surface; prothorax distinctly wider than long .... *Miloderoides*
— Dorsal surface with inconspicuous short, recumbent setae, similar on sides and ventral surface; prothorax distinctly longer than wide... 54

54(53). Side of rostrum with tooth-like extension over scrobe just apicad of eye .... *Piscatopus*
— Side of rostrum without extension over scrobe .... *Minyomerus*

55(44). Scrobe dorsal or dorsolateral, indefinite caudad of antennal insertion; scape in repose not situated in scrobe, usually passing over eye .. 56
— Scrobe lateral; scape in repose situated in scrobe 96

56(55). Corbel closed (Fig. 80c); tarsal claws free; humeral angle well-developed 57
— Corbel open or not distinctly closed (Fig. 80a, b); tarsal claws free or connate; humeral angle rounded 61

57(56). Rostrum twice as long as head, abruptly narrowed immediately caudad of antennal insertion where it is one-half as wide as dorsally .... *Evotus*
— Rostrum shorter, not shaped as above 58

58(57). Scrobe completely dorsal; apex of rostrum with approximately 16 long setae of graduated lengths; scape moderately thick, reaching to a point midway between eye and prothorax; dorsal outline of rostrum and head continuously flat; humeri obsolete; corbel not abruptly differentiated proximally, enclosed space filled with long scaly setae; body length 4.4-5.6 mm .... *Achrastenus*
— Scrobe lateral or dorsolateral; corbel closed (Fig. 80c), corbel plate various 59

59(58). Scape very thick, short, no longer than thickness of rostrum (Fig. 70, 71); apex of rostrum with 20-30 long setae (Fig. 71); mandible with many very long setae directed to mandibular scar (Fig. 71); body length 5.5-7.5 mm .... *Brachystylus*
— Scape longer, extended to or beyond eye ... 60

60(59). Eye lateral; anterior margin of prothorax without modified setae; humeri oblique, not prominent; base of each elytron arcuately produced between scutellum and stria 6; corresponding area on prothorax depressed, with different vestiture; body length 7.5-12.5 mm .... *Compsus*
— Eye slightly encroaching on dorsum; anterior margin of prothorax with a row of 20 or more very fine long setae of graduated lengths directed toward lower edge of eye; humeri quadrate, prominent; base of elytra very slightly produced; body length 13.5-18.0 mm .... *Diaprepes*

61(56). Funicle with six articles; tarsal claws free..... 62
— Funicle with seven articles; tarsal claws various 63

62(61). Prothorax lacking median sulcus; surface punctate; elytral intervals flat; corbel plate large, glabrous, oval; body length 5.8-6.8 mm .... *Agraphus*
— Prothorax with median longitudinal sulcus; elytral intervals 3, 5, and 7 more prominent; corbel plate indeterminate; body length 6.2 mm .... *Paragraphus*

63(61). Tarsal claws connate 64

— Tarsal claws free 84

64(63). Eye large, almost touching prothorax; anterior margin of prothorax with postocular lobe; corbel narrowly closed; elytra very convex, sides greatly rounded .... *Pseudocneorhinus*

— Eye smaller, prothorax without postocular lobe; corbel various; elytra various 65

65(64). Femora with large tooth on inner surface; dorsum with slender, decumbent setae only, or with very small metallic scales; scrobe more lateral than dorsal .... *Phyllobius*

— Femora without a tooth; other characters various 66

66(65). Rostrum in dorsal view more or less rectangular in outline; entire body and appendages densely scaled; scrobe completely dorsal .... *Aphrastus*

— Rostrum in dorsal view not rectangular 67

67(66). Mandible with more than five setae; elytra globose; body shiny black with white setae....... .... *Omias*

— Mandible with three setae; elytra more elongate; body color and vestiture various 68

68(67). Tarsal claws free at base; femora with minute tooth on inner edge distally; vestiture of sparse, erect, fine setae; without scales .... *Stomodes*

— Tarsal claws connate at base 69

69(68). Inner apical surface of hind tibia adjacent to tarsal insertion clothed with dense, round scales; funicle and dorsal surface of tarsi clothed with round scales; tarsi on ventral surface with coarse setae 70

— Inner apical surface of hind tibia adjacent to tarsal insertion glabrous; articles 4 to 7 of funicle and dorsal surface of tarsi clothed with elongate, rather fine, hairlike scales; pubescence on ventral surface of tarsi fine 74

70(69). Elytra and prothorax clothed with long, very fine, erect pile, pile longer on lateral portion of body, at least twice as long as width of an interval; dorsal surface of prothorax in lateral view slightly gibbous .... *Eucilinus* (part)

— Elytra and prothorax with short, stout setae, not longer than width of an interval; dorsal surface of prothorax nearly flat in lateral view 71

71(70). Dorsal comb of setae on apex of hind tibia about twice as long as apical comb and situated almost at 45 degree angle to longitudinal axis of tibia; abdominal ventrite 2 as long as ventrites 3 and 4 combined; ventral and dorsal margin of scrobe not distinct at distance one-half length of eye distad of front margin of eye .... *Rhypodilius*

— Dorsal comb of setae on apex of hind tibia about as long as apical comb and almost parallel to longitudinal axis of tibia; other characters various 72

72(71). Apical comb on hind tibia with anterior spines very short and coarse and posterior spines on ascending portion distinctly finer and almost twice as long; scrobe not reaching eye; scape slightly arcuate; abdominal ventrite 2 as long as 3 plus 4 combined.................... *Dysticheus*

— Apical comb on hind tibia made up of spines almost uniform in length and coarseness; other characters various 73

73(72). Dorsal and ventral margins of scrobe distinctly defined up to eye, scrobe a completely flat bottomed channel; elytra with erect, stout, moderately long, acute setae distinctly longer than scales composing decumbent vestiture; dorsal surface of head and rostrum in lateral view interrupted by erect, stout, acute setae; abdominal ventrite 2 shorter than ventrites 3 and 4 combined; scape strongly arcuate....... .... *Eucyllus*

— Dorsal margin of scrobe indicated up to slightly distad of eye, ventral margin either merging with dorsal margin distad of eye or else separated from dorsal margin by raised, convex area; elytra with suberect, short, wide, rounded scales slightly longer than those composing decumbent vestiture; dorsal surface of head and rostrum in lateral view slightly interrupted by suberect, short, rounded scales; abdominal ventrite 2 as long as ventrites 3 and 4 combined; scape slightly arcuate ...... *Thinoxenus*

74(69). Female with hind tibia unarmed at apex; scrobe situated more dorsally on rostrum, short, somewhat convergent posteriorly 75

— Female having hind tibia armed with one or more small spines or teeth at apex 76

75(74). Rostrum longer than head, antenna inserted at distance from anterior margin of eye about twice diameter of eye .... *Thricolepis*

— Rostrum about as long as head, antenna inserted at distance from anterior margin of eye less than the diameter of eye .... *Peritelinus*

76(74). Female with hind tibia armed with one spine or small tooth at apex 77

— Female with hind tibia armed with two or more spines or small teeth at apex 80

77(76). Scrobes lateral, not convergent posteriorly on dorsal surface of rostrum 78

— Scrobes dorsal, more or less convergent posteriorly on dorsal surface of rostrum 79

78(77). Antenna with scape much shorter than funicle .... *Aragnomus*

— Antenna with scape as long as funicle .... *Geodercodes*

79(77). Hind tibia at narrowest point in apical third at least two-thirds as wide as widest point of tibia at apex; epistoma separated from rostrum by fine, moderately raised, acute carina; in dorsal anterior view dorsal margins of rostrum distinctly, broadly emarginate above point of antennal insertion, greatly expanded toward apical region, in apical region subparallel sided, subparallel region about one-third wider than narrowest point between points of antennal insertions .... *Anchitelus*

— Hind tibia at narrowest point in apical third less than one-half as wide as tibia at widest point at apex; epistoma indistinctly separated from rostrum; in dorsal anterior view dorsal margins of rostrum straight, slightly diverging from behind point of antennal insertions to apex of rostrum, subparallel region distad of points of antennal insertions only slightly wider than narrowest point between points of antennal insertions. .... *Orthoptochus*

80(76). Female with hind tibia armed with three spines or small teeth at apex; funicle with article 7 distinctly longer than wide 81

— Female with hind tibia armed with two small spines or teeth at apex 82

81(80). Suture between abdominal ventrites 1 and 2 almost straight; scape with coarse, elongate scales; elytra with obvious (at 56X) erect, spatulate scales .... *Stenoptochus*

— Suture between abdominal ventrites 1 and 2 arcuate; scape with fine setae; elytra (at 56X) lacking obvious erect scales, at most with very fine, short, subdecumbent setae .... *Peritelopsis*

82(80). Antennal funicle with article 7 distinctly longer than wide .... *Nemocestes*

— Antennal funicle with article 7 about as wide as long 83

83(82). Suture between abdominal ventrites 1 and 2 almost straight; prothorax with coarse, erect setiform scales more or less uniformly distributed over dorsal surface, lacking a distinct clump at anterior margin .... *Paraptochus*

— Suture between abdominal ventrites 1 and 2 arcuate; prothorax with coarse, setiform scales subdecumbent on dorsal surface except at anterior margin with elongate clump of about 10 erect scales on either side of middle line .... *Peritelodes*

84(63). Antenna with scape with vestiture of fine setae and round flat scales; body size small, 2.3-4.5 mm 85

— Antenna with scape with vestiture of fine setae only or with at most a few scattered elongate, recumbent scales intermixed; body size larger, 3.2-13.0 mm, most larger than 5.0 mm 91

85(84). Epistoma large, distinct, occupying approximately half the anterior margin of rostrum, triangular, limited by distinct carina 86

— Epistoma very small, indistinct 87

86(85). Scrobe dorsolateral, usually reaching and often enclosing eye; all tibiae with single strong, almost horizontal apical tooth ... *Trachyphloeus*

— Scrobe dorsal, very short and deep, not reaching eye; front and middle tibiae with single, strong, almost horizontal apical tooth; hind tibia with pair of short vertical apical spines .... *Cercopeus*

87(85). Scrobe lateral, long, passing backward and below the lower angle of eye; all tibiae with apical spine or tooth 88

— Scrobe superior, very short and deep, not reaching eye, not directed below; only front and middle tibiae with apical spine or tooth 90

88(87). Appressed vestiture of dense radiate-pectinate scales; surface encrusted........ *Chaetechidius*

— Appressed vestiture of imbricate, oval scales, or scales concealed by encrustation 89

89(88). Vestiture short, suberect, broad, spatulate scale-like setae; eye with more than 5 facets across greatest width; elytra more elongate, parallel-sided; Oregon .... *Cathormiocerus*

— Vestiture of longer, suberect, fine hair-like setae; eye with less than 5 facets across greatest width; elytra more rounded, subglobose; southern United States .... *Trachyphloeosoma*

90(87). Rostrum short and broad; prementum not emarginate on apical margin; only labial palpi visible .... *Pseudocercopeus*

— Rostrum long, narrow, almost cylindrical; prementum broadly emarginate; both labial and maxillary palpi visible .... *Cercopedius*

91(84). Antennal scape stout, short, not exceeding eye 92

— Antennal scape reaching beyond anterior margin of prothorax 94

92(91). Eye not prominent; pronotum smooth, polished; with moderate punctures becoming larger and closer laterad; body length 4.0-4.5mm .... *Lupinocolus*

— Eye very prominent; pronotum with pronounced sculpture 93

93(92). Pronotum with large, irregularly spaced, flat-topped cylindrical protuberances with central depression set with a seta; integument and vestiture black; body length 8.3-9.6 mm .... *Agasphaerops*

— Pronotum with contiguous nodules each slightly depressed caudally; vestiture of slender, elongate coppery, green, or reddish metallic scales/setae on dorsum; oval, opalescent or metallic scales in groups forming pattern on sides of prothorax, elytra; elytra with intervals 3 and 5 slightly elevated; body length 7.5-9.0 mm .... *Hormorus*

94(91). Eye small, round, prominent, almost touching prothorax; rostrum greatly narrowed between antennal insertions, less than half the width of rostrum between outer edge of pterygia; scape arcuate; body length 3.2-4.0 mm ...... *Agronus*

— Eye moderate; rostrum wider between pterygia; scape arcuate or straight; body length various 95

95(94). Elytra of most species with few sparse scales or no scales; if densely squamose, pronotum similarly squamose or not and nodulate; femora with or without a tooth on inner surface; body length 4.0-13.0 mm .... *Otiorhynchus*

— Elytra and prothorax densely squamose; femora not dentate; body length 4.9-6.6 mm (Figs. 72-73) .... *Sciopithes*

96(55). Tarsal claws connate 97
— Tarsal claws free 105

97(96). Corbel closed (Fig. 80c), corbel plate covered with shiny round scales; epistoma concave, steeply angled, posterior margin conspicuously carinate; body length 4.5-8.0 mm .... *Philopedon*
— Corbel open (Fig. 80a) 98

98(97). Elytra with humerus prominent 99
— Elytra with humerus rounded 100

99(98). Rostrum with a conspicuous curved, glabrous callosity extending between the antennal insertions and paralleling the glabrous epistoma, surface squamose between the glabrous areas; elytral scales elongate .... *Pachyrhinus*
— Rostrum without a glabrous callosity; elytral scales round; some species with a minute tooth on fore femur .... *Polydrusus*

100(98). Head constricted dorsally behind eyes; eyes small, prominent; epistoma marked by fine carina; scape extended to posterior margin of eye .... *Strophosoma*
— Head not constricted behind eyes; other characters various 101

101(100). Epistoma not defined; body form narrow elongate .... *Brachyderes*
— Epistoma carinate or not (*Barypeithes*); eye flattened; humeri rounded 102

102(101). Vestiture of several sizes of very fine setae, lacking scales; color castaneous; epistoma minute, almost undetectable .... *Barypeithes*
— Vestiture of scales with or without setae; epistoma distinct 103

103(102). Femora with short, broad tooth .... *Sciaphilus*
— Femora without tooth 104

104(103). Striae on disc of elytra almost as wide as intervals; striae composed of large foveae; each interval with row of long stiff, erect setae; also with small appressed elongate scales, 2 to 4 abreast; elytra rotund; length less than 3.0 mm .... *Brachysomus*
— Elytral striae fine; stria 10 distinct; elytra narrow, elongate; length more than 3.0 mm .... *Mitostylus*

105(96). Apex of rostrum with keel across entire width of rostrum, keel forming posterior edge of epistoma; humeri distinct, quadrate; corbel semi-closed (Fig. 80b); mandibular cusp situated at the apex of an anterior projection of the mandible; front coxae very narrowly separated .... *Polydacrys*
— Epistoma indistinct, without carinate or keeled margin (except *Glaphyrometopus*, recognized by frons with deep, broad transverse concavity between eyes); humeri mostly rounded; other characters various 106

106(105). Eyes partly encroaching on head; epistoma poorly defined; humeri rounded (except *Lachnopus*, southern Florida only) 107
— Eyes lateral; humeri rounded (except *Ericydeus*, southwestern United States only); rostrum with longitudinal sulcus or impressed line reaching from interantennal line to head, continuing or not with fine impressed line reaching beyond eyes, in some genera to occiput............. 116

107(106). Rostrum separated from frons by distinct transverse sulcus or depression; dorsal aspect of tarsal articles with scales; tarsi clothed ventrally with "coarse setae" (stiff decumbent setae) 108
— Rostrum not separated from frons when viewed laterally, frons lacking tubercle above eye; maxillae concealed by mentum 110

108(107). Hind tarsus with article 3 bilobed and wider than article 2 .... *Stereogaster*
— Hind tarsus with article 3 not wider than article 2 109

109(108). Mentum not concealing maxillae; frons above each eye with distinct tubercle which conceals eye in dorsal view; elytra with inconspicuous, sparse, short, suberect spatulate scales or papillae; body length 2.5-3.0 mm ...... *Calyptillus*
— Mentum concealing maxillae; frons lacking distinct tubercle above eye; eye visible in dorsal view; elytra with suberect, fine, acute setae; body length greater than 3.0 mm .... *Cryptolepidus*

110(107). Head conspicuously constricted behind eye; eyes very prominent .... *Bradyrhynchoides*
— Head not constricted behind eye; eyes moderately convex but not protuberant............ 111

111(110). Mesepimeron triangular, anterior margin running straight to angle between elytron and peduncle of mesothorax, mesepisternum not touching side margin of elytron; scutellum well developed; metepisternal suture complete 112
— Mesepimeron short trapezoidal, anterior margin running to side margin of elytron, mesepisternum touching elytron on broad contact 113

112(111). Humeri well-developed, quadrate; scales on body sparse, not imbricate, prothorax and elytra lacking erect scales or setae; article 7 of funicle distinctly longer than wide .... *Lachnopus*
— Humeri rounded; scales on body dense, imbricate; prothorax and elytra with short suberect scales and setae; article 7 of funicle more or less wider than long .... *Omileus*

113(111). Metepisternal suture complete; rostrum lacking impressions on dorsal surface; eye distant from anterior margin of prothorax by more than half its greatest diameter; base of elytra as wide as base of prothorax ... *Stamoderes* and *Amotus*
— Metepisternal suture obliterated in basal half; rostrum on dorsal surface with distinct, lateral, longitudinal impression; eye separated from anterior margin of prothorax by half its diameter or less; base of elytra distinctly wider than base of prothorax 114

114(113). Dorsal surface of tarsi clothed with broad, rounded scales; tarsi on ventral surface with coarse setae on anterior part of articles 1 to 3 .... *Graphorhinus*
— Dorsal surface of tarsi clothed with fine setae or elongated scales; tarsi on ventral surface clothed with fine, dense setae................. 115

115(114). Rostrum (excluding mandibles) in dorsal view from anterior margin of eye to apex slightly longer than greatest width in apical region; antenna with short, narrow scales and sparse, long, fine setae; prosternum lacking two close adjacent tubercles behind front coxae .... *Epicaerus*
— Rostrum (excluding mandibles) in dorsal view from anterior margin of eye to apex distinctly longer than greatest width in apical region; antenna with both short and long, fine setae; prosternum with two closely adjacent tubercles behind front coxae .... *Barynotus*

116(106). Rostrum trisulcate (Fig. 77).......................... 117
— Rostrum with median sulcus only 118

117(116). Frons with deep, broad transverse concavity between eyes; rostrum with all three longitudinal sulci reaching transverse concavity; humerus absent; length 4.5-5.0 mm .... *Glaphyrometopus*
— Frons without transverse impression; rostrum with longitudinal sulci extending as far as between eyes (Fig. 77); humerus prominent; length 10.0-18.0 mm .... *Ericydeus*

118(116). Rostrum with very deep, wide median longitudinal "crater" from apex to at least middle; scape very thick, slightly curved; head with fine longitudinal impressed line; funicle with articles 1 and 2 equal in size; elytral stria 10 evanescent or absent beyond level of hind coxa .... *Platyomus*
— Rostrum not sulcate from apex; scape not thick; straight; other characters various............ 119

119(118). Epistoma conspicuous, very wide, occupying most of anterior edge of rostrum; prothorax and elytra with very irregularly shaped, randomly situated large foveae .... *Artipus*
— Epistoma inconspicuous, occupying half or less of anterior edge of rostrum; prothorax and elytra with only regular sculpture............ 120

120(119). Head and rostrum in dorsal view forming a flat triangle two times wider than apex of rostrum; epistoma small, inconspicuous; rostral longitudinal sulcus fine, reaching pronotum; corbel plate narrow.................................... *Aramigus*
— Head and rostrum less triangular, not flat.... 121

121(120). Elytral stria 10 complete, but sometimes concealed by vestiture, separate from stria 9; corbel open; some species resembling a *Sitona* or *Tanymecus*, others with more slender rostrum and tuberculate prothorax .... *Mesagroicus*
— Elytral stria 10 evanescent or absent beyond hind coxae 122

122(121). Funicle with article 2, 1.0-1.6x longer than article 1; scales of pronotum in clusters surrounding setae (except some *P. viridis*); corbel plate small, squamate .... *Phacepholis*
— Funicle with articles 2 and 1 with various ratios; other characters various 123

123(122). Scutellum glabrous, glossy; antennal funicle with article 2 approximately as long as article 1; corbel plate narrow .... *Atrichonotus*
— Scutellum squamate; antennal funicle with article 2 approximately 1.5 to 2.0x longer than article 1; corbel plate absent or present .... *Naupactus* and *Pantomorus*

Classification of the Nearctic Entiminae

41. Agraphini Horn 1876

Agraphus Say 1831, 1 sp., *A. bellicus* (Say 1831), eastern coastal United States from New York south to Florida. Adults are associated with sandy habitats along the Atlantic coast and in central Florida.

Agraphus Schoenherr 1834; not Say 1831

Paragraphus Blatchley 1916, 1 sp., *P. setosus* Blatchley 1916, Florida.

42. Alophini LeConte 1876

Acmaegenius LeConte 1876, 2 spp., *A. granicollis* Van Dyke 1927, Wyoming, and *A. hylobinus* LeConte 1876, Idaho and Oregon.

Lepidophorus Kirby 1837, 10 spp., generally distributed in the western United States and Canada, north into Alaska, Yukon Territory and the Northwest Territories, including *L. setiger* Hamilton 1895, generally distributed in the eastern United States from New York and Ohio south into Virginia and Tennessee. Adults of all species are flightless and collected in leaf litter or under rocks. The western species are often collected at high elevations. See Buchanan (1936a) to separate the species. This genus is questionably distinct from *Dirotognathus* (Tropiphorini), we could find no character to reliably distinguish the two.

Lophalophus LeConte 1876

Plinthodes LeConte 1876, 2 spp., *P. foveirostris* (Chittenden 1925), Ohio, North Carolina, Tennessee, and Virginia, and *P. taeniatus* (LeConte 1857), British Columbia, Oregon and Washington.

Trichalophus LeConte 1876, 8 spp., generally distributed in the western United States north into Canada and Alaska, then east across the north to Manitoba and Ontario. Adults are collected on several different kinds of plants. The genus needs revision. See Hatch (1971) to separate some of the species.

Triglyphulus Cockerell 1906, 2 spp., *T. ater* (LeConte 1876) and *T. nevadensis* Van Dyke 1938, California, Oregon, Washington and Nevada.

Triglyphus LeConte 1876; not Loew 1840; not Fraas 1866

43. Anypotactini Champion 1911

Polydacrys Schoenherr 1834, 1 sp., *P. depressifrons* Boheman 1840, southern Texas.

44. Brachyderini Schoenherr 1826

Brachyderes Schoenherr 1823, 1 sp., *B. incanus* (Linnaeus 1758), northeastern United States; adventive.
Thylacites Germar 1817
Brachylophus Fischer von Waldheim 1829 (valid subgenus)
Poloposes Gistel 1848
Eumonima Gistel 1856
Gastraspis Flach 1907 (valid subgenus)
Sulciurus Flach 1907
Echopus Desbrochers 1909

Strophosoma Billberg 1820, 1 sp., *S. melanogrammum* (Forster 1771), northeastern United States and adjacent southern Canada, also Washington and British Columbia; adventive. Adults feed generally on foliage.
Strophosomus Schoenherr 1823
Strophosomum Gistel 1856
Leucostrophus Flach 1907; not Rothschild and Jordan 1903
Morphostrophus Flach 1907 (valid subgenus)

45. Cneorhinini Lacordaire 1863

Philopedon Schoenherr 1826, 1 sp., *P. plagiatum* (Schaller 1783), Newfoundland, New Brunswick, Nova Scotia and Prince Edward Island; adventive. This species appears to be found in sandy habitats.
Philopedum Agassiz 1846
Dactylorhinus Tournier 1876
Dactylorrhinus Rye 1878

46. Cyphicerini Lacordaire 1863

Cyphicerina Lacordaire 1863

Cyrtepistomus Marshall 1913, 1 sp., *C. castaneus* (Roelofs 1873), generally distributed in the eastern and southeastern United States west to Texas; adventive. Adults generally feed on foliage. This species is known as the "Asiatic oak weevil".

Myosides Roelofs 1873, 1 sp., *M. seriehispidus* Roelofs 1873, Maryland, Massachusetts and Connecticut; adventive.

Oedophrys Marshall 1941, 1 sp., *O. hilleri* (Faust 1889), eastern United States from Connecticut and Pennsylvania south into Virginia; adventive. Adults feed generally on foliage.

Acanthotrachelina Marshall 1944

Calomycterus Roelofs 1873, 1 sp., *C. setarius* Roelofs 1873, generally distributed in the eastern United States west to Iowa and Nebraska; adventive. Adults generally feed on foliage.
Synolobus Faust 1886

Myllocerina Pierce 1913

Myllocerus Schoenherr 1823, 1 sp., *M. undatus* Marshall 1916, Florida; adventive.
Macrocorynus Schoenherr 1823
Hyperstylus Roelofs 1873
Exmyllocerus Voss 1937 (valid subgenus)
Pachymyllocerus Voss 1937 (valid subgenus)
Isomyllocerus Marshall 1954 (valid subgenus)
Pseudocanoixus Voss 1958 (valid subgenus)
Allomycterops Voss 1959 (valid subgenus)
Calomyllocerus Voss 1959 (valid subgenus)
Mylloceroversus Hoffmann 1961 (valid subgenus)
Corigetellus Hoffmann 1964 (valid subgenus)

Neoptochus Horn 1876, 1 sp., *N. adspersus* (Boheman 1834), Florida, Georgia and South Carolina.

47. Eudiagogini LeConte 1874

Aracanthus Say 1831, 1 sp., *A. pallidus* Say 1831, generally distributed in the southeastern United States west to Iowa and Texas.

Colecerus Schoenherr 1840, 2 spp., *C. dispar* (LeConte 1874), Arizona and Texas, and *C. marmoratus* (Horn 1876), Texas. Adults feed on foliage of various Fabaceae, especially *Prosopis* (mesquite), *Mimosa* and *Acacia*. See LeConte and Horn (1876) to separate the species.
Coleocerus Agassiz 1846
Coleocerus Gemminger and Harold 1871; not Agassiz 1846
Bathyris LeConte 1874

Eudiagogus Schoenherr 1840, 3 spp., generally distributed in the southeastern United States west to Texas, also Arizona and California. Adults feed on foliage of species of *Sesbania* (Fabaceae); larvae feed on nitrogen-fixing root nodules in the soil (Kovarik and Burke 1989). See Warner (1979) to separate the species.

48. Eustylini Lacrodaire 1863

Achrastenus Horn 1876, 1 sp., *A. griseus* Horn 1876, Texas.

Brachystylus Schoenherr 1824, 2 spp., *B. sayi* Alonso-Zarazaga 1994, southeastern United States, west to Texas, and *B. microphthalmus* Champion 1911, southern Texas. Adults of *B. sayi* feed on foliage of *Diospyros* (persimmon; Ebenaceae).
Syntomostylus Scudder 1893

Compsus Schoenherr 1823, 1 sp., *C. auricephalus* (Say 1824), southeastern United States west to Texas and Colorado. Adults generally feed on foliage.

Callopistus Say 1831

Diaprepes Schoenherr 1823, 1 sp., *D. abbreviatus* (Linnaeus 1758), adventive, Florida. Adults feed on foliage and are pests of citrus (Woodruff 1968, 1979).

49. Geonemini Gistel 1856

Barynotus Germar 1817, 3 spp., northeastern United States and eastern Canada, also British Columbia; adventive. Adults generally feed on foliage. See Brown (1950) to separate the species.

Merionus Dejean 1821
Kissodontus Desbrochers 1909

Bradyrhynchoides Pierce 1913, 1 sp., *B. constrictus* Pierce 1913, Texas.

Calyptillus Horn 1876, 1 sp., *C. cryptops* Horn 1876, New Mexico, Colorado, Kansas and Nebraska.

Cryptolepidus Van Dyke 1936, 7 spp., Arizona, California and Nevada. Adults are associated with *Artemisia* (sagebrush; Asteraceae) and other shrubs. See Ting (1940) to separate the species.

Lepidopus Van Dyke 1936; not Gouan 1770; not Dana 1852
Pseudoeucyllus Tanner 1950

Epicaerus Schoenherr 1834, 11 spp., generally distributed in the United States. Adults generally feed on foliage. The genus needs revision. See Pierce (1913) to separate most of the species.

Epagrius Schoenherr 1840
Diorynotus Sharp 1891 (valid subgenus)
Cacochromus Sharp 1891
Bradyrhynchus Sharp 1891
Melbonus Casey 1895
Epagriopsis Champion 1911

Graphorhinus Say 1831, 1 sp., *G. vadosus* Say 1831, Texas, Colorado, Kansas, Wyoming and Missouri.

Graphorhinus Schoenherr 1833; not Say 1831

Lachnopus Schoenherr 1840, 3 spp., southern Florida. *Lachnopus floridanus* Horn 1876 is native and feeds on foliage of various plants. *Lachnopus argus* (Reiche 1840) and *L. hispidus* (Gyllenhal 1834) are adventive species which doubtfully are established in Florida (Anderson 1993a).

Menoetius Dejean 1821
Ptilopus Schoenherr 1823

Omileus Horn 1876, 1 sp., *O. epicaeroides* Horn 1876, Texas.

Stamoderes Casey 1888, 2 spp., *S. lanei* (Van Dyke 1935), British Columbia, Oregon and Washington, and *S. uniformis* Casey 1888, California. Adults of *S. lanei* are found on *Artemisia* (sagebrush; Asteraceae). See Van Dyke (1935b) to separate the species. This genus is questionably distinct from *Amotus* (Tanymecini). We can find no characters to reliably distinguish these two genera.

Stereogaster Van Dyke 1936, 1 sp., *S. globosa* Van Dyke 1936, California.

Trigonoscuta Motschulsky 1853, 64 spp., western United States and adjacent southern Canada. Adults are associated with sandy habitats and feed on various plants. In a posthumous publication, Pierce (1975) described all but a few of the species as well as numerous subspecies. Most of these taxa are of questionable validity and need reassessment. See Pierce (1975) to separate the species and subspecies.

Panormus Casey 1888 (valid subgenus)
Eremocatoecus Pierce 1975 (valid subgenus)
Nesocatoecus Pierce 1975 (valid subgenus)

50. Hormorini Horn 1876

Agasphaerops Horn 1876, 1 sp., *A. niger* Horn 1876, California, Oregon, Washington and British Columbia. Adults are associated with lilies (Liliaceae).

Hormorus Horn 1876, 1 sp., *H. undulatus* (Uhler 1856), generally distributed in the northeastern United States and southern Canada.

Lupinocolus Van Dyke 1936, 1 sp., *L. blaisdelli* Van Dyke 1936, California and Nevada.

51. Naupactini Gistel 1856

Aramigus Horn 1876, 1 sp., *A. tesselatus* (Say 1824), central United States; adventive. Adults are pests on alfalfa and some other crops in Argentina (Lanteri and Díaz 1994). Lanteri and Díaz (1994) describe a number of morphotypes of *A. tesselatus*.

Aomopactus Jekel 1876

Artipus Sahlberg 1823, 1 sp., *A. floridanus* Horn 1876, Florida. This species is a pest of the Florida citrus industry (McCoy *et al.* 1985). (Volume 1, Color Fig. 2)

Artipus Schoenherr 1823; not Sahlberg 1823

Atrichonotus Buchanan 1939, 1 sp., *A. taeniatulus* (Berg 1881), southeastern United States west to Texas; adventive. Adults feed on foliage of various plants, but most frequently Fabaceae (Lanteri and O'Brien 1990).

Floresianus Hustache 1939
Floresianellus Lanteri 1981

Ericydeus Pascoe 1880, 2 spp., *E. lautus* (LeConte 1856) Arizona, California, Colorado, Utah and New Mexico, and *E. placidus* (Horn

1876), Arizona and California. Adults appear to be associated with Fabaceae. See Lanteri (1995) to separate the species.

Glaphyrometopus Pierce 1913, 1 sp., *G. ornithodorus* Pierce 1913, Texas.

Mesagroicus Schoenherr 1840, 9 spp., generally distributed. Adults generally feed on foliage. See Buchanan (1929a) and Burke (1960) to separate the species.

Mesagroecus Agassiz 1846
Lepidocricus Pierce 1910

Naupactus Dejean 1821, 4 spp., *N. godmanni* (Crotch 1867), *N. leucoloma* Boheman 1840, *N. minor* (Buchanan 1942) and *N. peregrinus* (Buchanan 1939), generally distributed in the southeastern United States; adventive. Adults are considered pests and feed on foliage of various plants. See Lanteri and Marvaldi (1995) and Lanteri (1986) to separate the species. We have not seen specimens of *Pantomorus pallidulus* Emden 1936 and thus cannot separate it from *Naupactus*.

Asynonychus Crotch 1867
Mimopactus Jekel 1875
Archopactus Heller 1921
Graphognathus Buchanan 1939

Pactorrhinus Ancey 1881, 1 sp., *P. grisescens* Ancey 1881, Arizona. This genus and species are unknown to us but likely refer to *Ericydeus lautus* (LeConte 1856). It does not appear in the key.

Pantomorus Schoenherr 1840, 1 sp., *P. pallidulus* Emden 1936, Texas. We have not seen specimens of *Pantomorus pallidulus* Emden 1936 and thus cannot separate it from *Naupactus*.

Pantoplanes Schoenherr 1840
Symmathetes Schoenherr 1847
Pantopactus Jekel 1876
Athetetes Pascoe 1886
Antelmia Hustache 1919
Pseudeudius Voss 1934

Phacepholis Horn 1876, 5 spp., central United States. See Lanteri (1990) to separate the species.

Platyomus Sahlberg 1823, 1 sp., *P. flexicaulis* (Schaeffer 1905), southern Texas.

Platyomus Schoenherr 1823; not Sahlberg 1823
Pseudocyphus Schaeffer 1905
Eustylomorphus Pierce 1915
Pachyus Kuschel 1955

52. Omiini Shuckard 1840

Omias Germar 1817, 6 spp., California, Oregon, Washington, Idaho and British Columbia. Adults generally feed on foliage. See Hatch (1971) to separate the species.

Mylacus Boheman 1843
Omorus Gistel 1856

53. Ophryastini Lacordaire 1863

Ophryastes Germar 1829, 35 spp., generally distributed in western United States and adjacent southern Canada. Adults are flightless and are associated with various arid habitat shrubs, mostly in the family Asteraceae, but also *Larrea tridentata* (DC.) Cov. (creosote bush; Zygophyllaceae) and *Atriplex* (saltbush; Chenopodiaceae). See Kissinger (1970) to separate the species.

Ophryastes Say 1831; not Germar 1929
Dystirus Pascoe 1872
Eupagoderes Horn 1876
Caccophryastes Sharp 1891
Tosastes Sharp 1891
Amydrogmus Pierce 1913

Sapotes Casey 1888, 2 spp., *S. longipilis* Van Dyke 1934 and *S. puncticollis* Casey 1888, Arizona, New Mexico and Texas. Adults are associated with arid habitat shrubs, especially *Larrea tridentata* (DC.) Cov. (creosote bush; Zygophyllaceae). See Van Dyke (1934) to separate the species.

54. Otiorhynchini Schoenherr 1826

Agronus Horn 1876, 3 spp., California, Oregon, Montana, Alberta and British Columbia. See Buchanan (1929b) to separate the species.

Otiorhynchus Germar 1822, 14 spp., generally distributed; all adventive. Adults and larvae generally feed on a variety of plants. This genus includes a number of common pest species; *O. ovatus* (Linnaeus 1758), the strawberry root weevil, and *O. sulcatus* (Fabricius 1775), the black vine weevil. Adults are flightless and a number of species have very restricted distributions in eastern coastal North America. See Warner and Negley (1976) to separate the species. Alonso-Zarazaga and Lyal (1999:168-170) list 105 valid subgeneric names, not including synonyms. For brevity, these are not repeated here.

Sciopithes Horn 1876, 6 spp., California, Oregon, Washington and British Columbia. Adults feed generally on foliage. See Van Dyke (1935b) to separate the species.

55. Peritelini Lacordaire 1863

Anchitelus Van Dyke 1936, 1 sp., *A. alboviridis* Van Dyke 1936, California.

Aragnomus Horn 1876, 3 spp., western United States.

Dysticheus Horn 1876, 2 spp., *D. insignis* Horn 1876 and *D. rotundicollis* Van Dyke 1953, California. See Van Dyke (1953) to separate the species.

Eucilinus Buchanan 1926, 3 spp., California, Utah, Idaho and Arizona. See Kissinger (1960) to separate the species.

Thysanocorynus Van Dyke 1938

Eucyllus Horn 1876, 8 spp., generally distributed in the far western United States. Adults are flightless and mostly nocturnal and are associated with various shrubs in arid habitats. See Pelsue and Sleeper (1972) to separate the species.

Geodercodes Casey 1888, 1 sp., *G. latipennis* Casey 1888, California, Oregon, Washington, Idaho, Montana and British Columbia.

Nemocestes Van Dyke 1936, 9 spp., generally distributed in the far western United States and adjacent British Columbia, including *N. horni* (Van Dyke 1936) also in Michigan, New York, Wisconsin, Ontario, Quebec, New Brunswick and Nova Scotia. Adults are flightless and mostly nocturnal and feed on foliage of various plants. See Van Dyke (1936a) and Hatch (1971) to separate the species.

Orthoptochus Casey 1888, 1 sp., *O. squamiger* Casey 1888, California.

Paraptochus Seidlitz 1868, 3 spp., California, Oregon and British Columbia. See Van Dyke (1935b) to separate the species.

Peritelinus Casey 1888, 3 spp., far western United States and adjacent British Columbia. See Van Dyke (1936a) to separate the species.

Peritelodes Casey 1888, 1 sp., *P. obtectus* Casey 1888, California.

Peritelopsis Horn 1876, 1 sp., *P. globiventris* (LeConte 1857), California.

Rhypodillus Cockerell 1906, 2 spp., *R. brevicollis* (Horn 1876), Arizona, New Mexico, Texas and Colorado, and *R. dilatatus* (Horn 1876), California. See LeConte and Horn (1876) to separate the species.

Rhypodes Horn 1876

Stenoptochus Casey 1888, 2 spp., *S. inconstans* Casey 1888 and *S. vanduzeei* Van Dyke 1935, California. See Van Dyke (1935b) to separate the species.

Stomodes Schoenherr 1826, 1 sp., *S. gyrosicollis* Boheman 1843, Maine; adventive.

Thinoxenus Horn 1876, 1 sp., *T. squalens* Horn 1876, California.

Thricolepis Horn 1876, 2 spp., *T. inornata* Horn 1876, generally distributed in the western United States, and *T. simulator* Horn 1876, California. See LeConte and Horn (1876) to separate the species.

56. Phyllobiini Schoenherr 1826

Aphrastus Say 1831, 3 spp., generally distributed in the eastern United States west to Texas, and southern Canada. Adults feed on the foliage of various plants. See Blatchley and Leng (1916) to separate the species.

Aphrastus Schoenherr 1833; not Say 1831

Micronychus Provancher 1877; not Motschulsky 1861

Evotus LeConte 1874, 1 sp., *E. naso* (LeConte 1857), northwestern United States and adjacent Canada. Adults feed on foliage of various plants.

Phyllobius Germar 1824, 3 spp., northeastern United States and adjacent Canada; adventive. Adults feed on foliage of various trees. No key to the three species in North American species is available but the status of *P. glaucus* as established is undetermined. See Côté and Bright (1995) to separate two of the species. Alonso-Zarazaga and Lyal (1999:173-174) list 19 valid subgeneric names, not including synonyms. For brevity, these are not repeated here.

57. Polydrusini Schoenherr 1823

[*Liophloeus* Germar 1817, 1 sp., *L. tessulatus* (Mueller 1776), intercepted in quarantine; New York. Not established in North America.]

Pachyrhinus Schoenherr 1823, 8 spp., generally distributed in the western United States and Canada east to Nova Scotia and south to Pennsylvania and Connecticut. Adults are associated with *Pinus* (pine; Pinaceae). The genus needs revision. See Fall (1901) to separate the species.

Scythropus Schoenherr 1826

Carpomanes Gistel 1856

Polydrusus Germar 1817, 7 spp., generally distributed in the northeastern United States and adjacent southern Canada, also in Arizona and New Mexico; includes 3 adventive species, *P. cervinus* (Linnaeus 1758), *P. impressifrons* (Gyllenhal 1834), and *P. sericeus* (Schaller 1783). Adults generally feed on foliage of various trees. The genus needs revision and the relationships between the native and adventive species reassessed. See Sleeper (1957c) to separate most of the species. Alonso-Zarazaga and Lyal (1999:175) list 24 valid subgeneric names, not including synonyms. For brevity, these are not repeated here.

58. Sciaphilini Sharp 1891

Barypeithes Jacquelin du Val 1854, 1 sp., *B. pellucidus* (Boheman 1834), generally distributed in the northeastern United States and adjacent southern Canada, also in California, Oregon, Washington and British Columbia; adventive. Adults feed on foliage of a wide variety of plants.

Barypithes Gemminger and Harold 1871

Exomias Bedel 1883 (valid subgenus)

Moroderia Reitter 1915

Brachysomus Schoenherr 1823, 1 sp., *B. echinatus* (Bonsdorff 1785), Massachusetts, Minnesota, Quebec and Newfoundland; adventive.
Pavrosomus Fischer de Waldheim 1829
Platytarsus Schoenherr 1840
Thricolepoides O'Brien 1979

Mitostylus Horn 1876, 3 spp., southwestern United States including Texas and Oklahoma. Adults are found on various kinds of low vegetation. The genus needs revision. See Van Dyke (1936b) and Burke (1963) to separate the species.
Derosomus Sharp 1891

Sciaphilus Schoenherr 1823, 1 sp., *S. asperatus* (Bonsdorff 1785), generally distributed in the northeastern United States and adjacent southern Canada, also Idaho, South Dakota and British Columbia; adventive. Adults generally feed on foliage.
Lygophilus Fischer von Waldheim 1829; not Rafinesque 1815
Sphaerilethmus Gistel 1856

59. Sitonini Gistel 1856

Sitona Germar 1817, 11 spp., generally distributed; includes 5 adventive species. Adults are associated with various herbaceous species of Fabaceae. Larvae feed in the soil on roots. Some species such as the pea leaf weevil, *S. lineatus* (Linnaeus 1758), the sweetclover weevil, *S. cylindricollis* (Fahraeus 1840), and the clover root curculio, *S. hispidulus* (Fabricius 1776), are agricultural pests. See Bright (1994) to separate the species.
Charagmus Schoenherr 1826 (valid subgenus)
Clyptus Villa and Villa 1833
Sitones Schoenherr 1840
Parasitones Sharp 1896
Sitonidius Mueller 1913
Coelositona González 1971 (valid subgenus)

60. Tanymecini Lacordaire 1863

Tanymecina Lacordaire 1863

Amotus Casey 1888, 3 spp., California. Adults are associated with *Artemisia* (Asteraceae) and perhaps other shrubs and trees. See Van Dyke (1935b) to separate the species. This genus is questionably distinct from *Stamoderes* (Geonemini). We can find no characters to reliably distiunguish these two genera.
Mimetes Schoenherr 1847; not Eschscholtz 1818; not Leach 1820; not Huebner 1821; not Vigors 1827; not Gloger 1841

Hadromeropsis Pierce 1913, 1 sp., *H. opalinus* (Horn 1876), Arizona. Adults have been collected on *Calliandra eriophylla* Benth. and *Acacia* (Fabaceae) (Howden 1982).
Hadromerus Schoenherr 1834; not Schoenherr 1823
Hadrorestes Howden 1982 (valid subgenus)

Isodacrys Sharp 1911, 2 spp., *I. ovipennis* (Schaeffer 1908) and *I. burkei* Howden 1961, Texas. Adults have been collected sweeping herbaceous Asteraceae. See Howden (1961) to separate the species.

Isodrusus Sharp 1911, 1 sp., *I. debilis* Sharp 1911, Texas.

Miloderoides Van Dyke 1936, 3 spp., Nevada, Colorado, Utah, Wyoming and Idaho. See Tanner (1942) to separate the species.

Minyomerus Horn 1876, 6 spp., southwestern United States, north into Kansas, Wyoming, Montana and Alberta. Adults have been found on various kinds of low vegetation. The genus needs revision and the relationships of this genus to *Piscatopus* need reassessment. No key to species exists.
Elissa Casey 1888
Pseudelissa Casey 1888

Pachnaeus Schoenherr 1826, 2 spp., *P. litus* (Germar 1824), southern Florida, and *P. opalus* (Olivier 1807), coastal southeastern United States from northern Florida north to New Jersey. Adults are general foliage feeders and are citrus pests. The larvae feed on roots. See Woodruff (1981) to separate the species.
Docorhinus Schoenherr 1823
Pachneus Gemminger and Harold 1871

Pandeleteinus Champion 1911, 3 spp., generally distributed in the southwestern United States. Adults have been collected on various species of trees. See Howden (1959) to separate the species.

Pandeleteius Schoenherr 1834, 13 spp., generally distributed in the southwestern and eastern United States and adjacent southern Canada. Adults are frequently found on *Quercus* (oak; Fagaceae) or on various trees in the family Fabaceae. See Howden (1959) to separate the species.
Pandeletius Agassiz 1846
Menetypus Kirsch 1868
Pandeletius Gemminger and Harold 1871; not Agassiz 1846
Pandeletejus Horn 1876
Exmenetypus Voss 1954 (valid subgenus)

Piscatopus Sleeper 1960, 1 sp., *P. griseus* Sleeper 1960. Adults have been found on *Larrea tridentata* (Zygophyllaceae). The relationships of this genus to *Minyomerus* need reassessment.

Scalaventer Howden 1970, 1 sp., *S. subtropicus* (Fall 1907), southern Florida. Adults have been collected on *Bumelia celastrina* H.B.K. (Sapotaceae) and various other trees (Anderson 1993a).

Tanymecus Germar 1817, 3 spp., generally distributed in the eastern, central and southwestern United States north into the prairie provinces of Canada. Adults are usually collected on low herbaceous plants. Adults of *T. lacaena* (Herbst 1797) have been collected commonly on *Sesuvium portulacastrum* (L.) L. (Aizoaceae) in

southern Florida (Anderson 1993a). See Van Dyke (1935b) and Blatchley and Leng (1916) to separate the species.
Hynnulus Villa and Villa 1833
Episomechus Reitter 1903 (valid subgenus)
Geomecus Reitter 1903 (valid subgenus)
Indomecus Pajni and Gandhi 1987

Trigonoscutoides O'Brien 1977, 1 sp., *T. texanus* O'Brien 1977, Texas. Adults are collected on and under *Quercus havardii* Rydb. (O'Brien 1977a).

61. Thecesternini Lacordaire 1863

Thecesternus Say 1831, 7 spp., eastern, central and southwestern United States north into Alberta. Adults are flightless and found under rocks and cow dung (Kissinger 1964). Larvae of *T. hirsutus* Pierce 1909 feed on the roots of *Parthenium hysterophorus* L. (Asteaceae) (McClay and Anderson 1985). The genus needs revision. See Pierce (1909) to separate the species.
Lithodus Germar 1834
Thicosternus Gemminger and Harold 1871

62. Trachyphloeini Lacordaire 1863

Trachyphloeina Lacordaire 1863

Cathormiocerus Schoenherr 1842, 1 sp., *C. curvipes* Wollaston 1854, Oregon; adventive. This genus is questionably distinct from *Trachyphloeus*. We can find no characters to reliably distinguish these two genera.
Scoliocerus Wollaston 1854
Mitomermus Jacquelin du Val 1854 (valid subgenus)
Schaumius Brisout 1866 (valid subgenus)
Cathormiocerinus Escalera 1918 (valid subgenus)

Cercopedius Sleeper 1955, 1 sp., *C. artemisiae* (Pierce 1910), western United States from Nevada, Utah and Colorado north, into British Columbia. Adults are found on *Artemisia* (sagebrush; Asteraceae).

Cercopeus Schoenherr 1842, 11 spp., generally distributed in the eastern United States west into Texas. Adults are flightless and found in leaf litter. The genus needs revision. See Sleeper (1955b), Burke (1963) and O'Brien (1977b) to separate the species.
Cercopius Gemminger and Harold 1871

Chaetechidius Sleeper 1955, 1 sp., *C. speciosus* Sleeper 1955, Colorado. Adults were found under a stone.

Pseudocercopeus Sleeper 1955, 1 sp., *P. setosus* Sleeper 1955, Arizona.

Pseudocneorhinus Roelofs 1873, 1 sp., *P. bifasciatus* (Roelofs 1880), eastern United States; adventive. Adults and larvae feed on a wide variety of plants (Maier 1983).

Trachyphloeosoma Wollaston 1869, 1 sp., *T. advena* Zimmerman 1956, Alabama, Florida, Georgia, Mississippi, North Carolina and South Carolina; adventive.
Trachyphloeops Roelofs 1873

Trachyphloeus Germar 1817, 4 spp., *T. aristatus* (Gyllenhal 1827), *T. asperatus* Boheman 1843, *T. angustisetulus* Hansen 1915, and *T. bifoveolatus* (Beck 1817), northeastern United States and adjacent Canada, also northwestern United States and adjacent Canada; adventive. See Brown (1965) and Borovec (1989) to separate the species. This genus is questionably distinct from *Cathormiocerus*. We can find no characters to reliably distinguish these two genera.
Phyllastolus Gistel 1856
Lacordairius Brisout 1866 (valid subgenus)
Chaetechus Horn 1876
Paratrachyphloeus Desbrochers 1895
Pseudolacordairius Escalera 1923 (valid subgenus)

63. Tropiphorini Marseul 1863

Adaleres Casey 1895, 3 spp., California. See Casey (1895) to separate the species.

Anametis Horn 1876, 2 spp., *A. granulata* (Say 1831), generally distributed in the eastern United States and adjacent southern Canada, and *A. subfusca* Fall 1907, Arizona, Texas, New Mexico and Colorado. The relationships of this genus to *Dichoxenus* Horn 1876, *Peritaxia* Horn 1876 and some species of the Mexican genus *Amphidees* Schoenherr 1842 need to be reassessed. See Fall and Cockerell (1907) to separate the species.

Byrsopages Schoenherr 1842, 1 sp., *B. villosus* Boheman 1842, Alaska.
Strongylophthalmus Motschulsky 1860
Strongylophthalmus Faust 1894; not Motschulsky 1860
Kurilonus Sharp 1896

Cimbocera Horn 1876, 4 spp., generally distributed in the western inland United States and Canada. Adults are found on various woody shrubs at night. See Van Dyke (1935a) and Tanner (1941) to separate the species.

Connatichela Anderson 1984, 1 sp., *C. artemisiae* Anderson 1984, Yukon Territory and Alaska. Adults are associated with *Artemisia frigida* Willd. (Asteraceae) (Anderson 1984).

Crocidema Van Dyke 1934, 5 spp., Arizona, Texas, Utah and California. Adults are found on various woody shrubs at night. The genus needs revision and its relationships to *Pseudorimus* Van Dyke 1934 and *Melanolemma* Van Dyke 1935 reassessed. See Van Dyke (1934, 1951) to separate the species.

Diamimus Horn 1876, 1 sp., *D. subsericeus* Horn 1876, western inland United States north to Montana.

Dichoxenus Horn 1876, 4 spp., eastern and central United States from Illinois and Arkansas west to Texas, Colorado and Wyoming. The relationships of this genus to *Anametis* Horn 1876, *Peritaxia* Horn 1876 and some species of the Mexican genus *Amphidees* Schoenherr 1842 need to be reassessed. See Sleeper (1956b) to separate the species.

Dirotognathus Horn 1876, 2 spp., *D. punctatus* Hatch 1971, Oregon, and *D. sordidus* Horn Arizona, California, Nevada and Colorado. See Hatch (1971) to separate the species. This genus is questionably distinct from *Lepidophorus* (Alophini).

Dyslobus LeConte 1869, 34 spp., generally distributed in the western United States and adjacent Canada. Adults are flightless and noctural and are found on foliage of various plants. The genus needs revision. See Van Dyke (1933) and Hatch (1971) to separate some of the species.

Lepesoma Motschulsky 1845; not Spix 1825
Lepidosoma Agassiz 1846; not Wagler 1830; not Swainson 1839
Ledidosoma Gemminger and Harold 1871; not Wagler 1830; not Swainson 1839; not Agassiz 1846
Melamomphus Horn 1876
Amnesia Horn 1876
Thricomigus Horn 1876

Leptopinara O'Brien 1981, 2 spp., *L. papillata* O'Brien 1981, New Mexico and Texas, and *L. flemingi* Anderson 1993, Texas. See Anderson (1993c) to separate the species.

Melanolemma Van Dyke 1935, 1 sp., *M. montana* Van Dyke 1935, Colorado. The relationships of this genus to *Pseudorimus* Van Dyke 1934 and *Crocidema* Van Dyke 1934 need to be reassessed.

Miloderes Casey 1888, 6 spp., Utah, Nevada, Arizona and California. See Kissinger (1960) to separate some of the species.

Orimodema Horn 1876, *O. protracta* Horn 1876, generally distributed in the southwestern United States.

Panscopus Schoenherr, 1842, 28 spp., generally distributed throughout the United States and southern Canada. Adults are mostly noctural and are associated with various types of plants. Most species are in forests but a few are found in more arid habitats. The genus needs revision. See Buchanan (1936b) and Hatch (1971) to separate most of the species.

Nocheles LeConte 1874 (valid subgenus)
Phymatinus LeConte 1876 (valid subgenus)
Nomidus Casey 1895 (valid subgenus)
Neopanscopus Pierce 1913 (valid subgenus)
Panscopidius Pierce 1913
Pseudopanscopus Buchanan 1927 (valid subgenus)
Dolichonotus Buchanan 1936 (valid subgenus)
Parapanscopus Buchanan 1936 (valid subgenus)

Paracimbocera Van Dyke 1938, 3 spp., Wyoming, Idaho, Colorado, Nevada and California. *Paracimbocera robusta* (Van Dyke 1935) has been recorded from *Ephedra nevadensis* S. Wats. (Ephedraceae) (Sleeper and Jenkins 1967) and *P. artemisiae* Ting 1940 from *Artemisia* (Asteraceae) (Ting 1940). See Ting (1940) to separate the species.

Paranametis Burke 1960, 1 sp., *P. distincta* Burke 1960, Texas.

Peritaxia Horn 1876, 7 spp., southwestern United States north to Wyoming and Montana. Adults are noctural and collected on various kinds of plants. The genus needs revision and the relationships of this genus to *Dichoxenus* Horn 1876, *Anametis* Horn 1876 and some species of the Mexican genus *Amphidees* Schoenherr 1842 need to be reassessed. There is no key to species.

Phyxelis Schoenherr 1842, 2 spp., generally distributed in the eastern United States and adjacent southern Canada. The genus needs revision since at least two undescribed species are known. See Blatchley and Leng (1916) to separate the species.

Geoderces Horn 1876

Pseudorimus Van Dyke 1934, 2 spp., Arizona and New Mexico. The relationships of this genus to *Melanolemma* Van Dyke 1935 and *Crocidema* Van Dyke 1934 need to be reassessed. See Van Dyke (1934) to separate the species.

Rhigopsis LeConte 1874, 1 sp., *R. effracta* LeConte 1874, California.

Tropiphorus Schoenherr 1842, 3 spp, Newfoundland, Nova Scotia and Quebec; adventive. See Brown (1967) to separate the species.

Brius Dejean 1821
Tropidophorus Gistel 1856; not Duméril and Bibron 1839
Tropidophorus Gemminger and Harold 1871; not Duméril and Bibron 1839; not Gistel 1856; not Jan 1865
Synirmus Bedel 1883
Dochorhynchus Desbrochers 1897

Vitavitus Kissinger 1974, 1 sp., *V. thulius* Kissinger 1974, Alaska, Yukon Territory, Northwest Territories and Nunavut. Adults are flightless and are collected in tundra and dry south-facing slopes (Anderson 1997).

XIII. Hyperinae Marseul 1863

by Robert S. Anderson

Only the genus *Hypera* occurs in North America. It is recognized by a short snout (Fig. 81) (lacking deciduous processes and associated scars), the pronotum lacks a postocular lobe, and at least some of the scales of the body are bifid. In Mexico, Central and South America several related genera are found. Larvae of all hyperines feed externally on plant foliage and make loosely woven cocoons, which they attach to the host plants, in which they

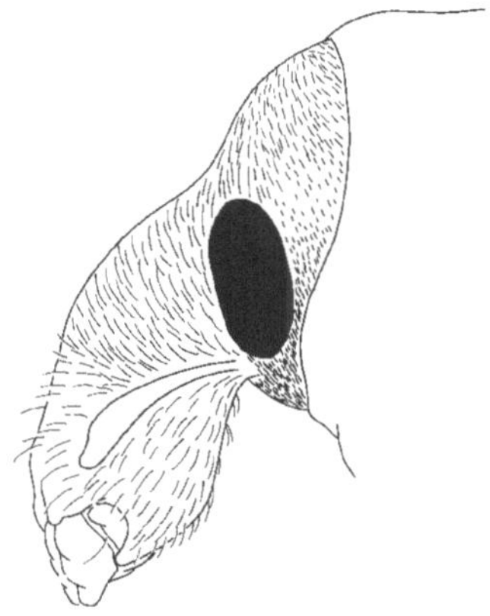

FIGURE 81.131. Hyperinae. 81. *Hypera punctata* (Fabricius), head, lateral view.

pupate. Plants in the Polygonaceae and Fabaceae appear to be the primary hosts.

CLASSIFICATION OF THE NEARCTIC HYPERINAE

64. Hyperini Marseul 1863

Hypera Germar 1817, 17 species, generally distributed throughout the United States and Canada, north into Alaska; 6 species are adventive. Adults and larvae feed externally on foliage of various Fabaceae and Polygonaceae (Titus 1911, Puttler *et al.* 1973). A number of pest species are included in the genus, namely, *H. postica* (Gyllenhal 1813), the alfalfa weevil; *H. brunneipennis* (Boheman 1834), the Egyptian alfalfa weevil; *H. nigrirostris* (Fabricius 1775), the lesser clover weevil; *H. punctata* (Fabricius 1775), the clover leaf weevil; *H. meles* (Fabricius 1792), the clover head weevil; and *H. rumicis* (Linnaeus 1758). The genus needs revision. See Titus (1911) to separate most of the species.

Phytonomus Schoenherr 1823
Dapalinus Capiomont 1868 (valid subgenus)
Eririnomorphus Capiomont 1868 (valid subgenus)
Tigrinellus Capiomont 1868 (valid subgenus)
Phytonomidius Capiomont 1868
Antidonus Bedel 1886 (valid subgenus)
Spongifer Petri 1901
Heteromorphus Petri 1901
Boreohypera Korotyaev 1999 (valid subgenus)

XIV. Lixinae Schoenherr 1823

by Robert S. Anderson

This is a relatively small subfamily of generally large-sized weevils associated mostly with arid habitats. Lixines are readily recognized by the large tooth at the apex of the hind tibia, their larger size, and short, globular and telescoping labial palpi of 3 articles (but appearing composed of 1 article) (Fig. 90), ventrally situated at the apex of the large prementum. Females possess large paired symbiont sacs attached to the vagina near the base of the gonocoxites but this can only be seen in dissections. The rostrum of lixines can be short and wide or long and slender and most species are grey or otherwise dull in color and scale pattern.

Larvae of most species mine in the roots and stems of various plants but a few such as *Rhinocyllus* and *Larinus* have larvae that feed in reproductive structures on seeds. Many species are associated with Asteraceae and Fabaceae but members of the genus *Lixus* are also found in semi-aquatic and aquatic habitats. Several genera have been introduced into North America for biological control of noxious or pest weeds especially in rangelands of western North America.

KEY TO THE NEARCTIC GENERA OF LIXINAE

1. Rostrum short and broad in dorsal view, from apex of epistoma to anterior margin of eye, more or less as long as greatest width (Fig. 82); body size small, at most slightly greater than 5 mm 2
— Rostrum more elongate, in dorsal view, from apex of epistoma to anterior margin of eye, much longer than greatest width; body size moderate to large, subequal to or greater than 5 mm 4

2(1). Elytron with intervals each with a row of conspicuous, erect setae in addition to appressed hair-like scales; anterolateral margin of pronotum straight, with long postocular vibrissae immediately behind eye; eye more or less round .... *Microlarinus*
— Elytron with intervals lacking erect setae, with only appressed to suberect hair-like scales; anterolateral margin of pronotum with rounded postocular lobe and short postocular vibrissae; eye distinctly elongate-oval 3

3(2). Prosternum with a pair of prominent lateral ridges which form a deep ventral channel; dorsal vestiture of pronotum and elytra with scales in part bifid .... *Bangasternus*
— Prosternum lacking ridges, no ventral channel evident; dorsal vestiture of pronotum and elytra with scales simple .... *Rhinocyllus*

4(1). Ventral and dorsal surfaces of body with numerous long, fine, erect hairs, some hairs about as long as antennal scape; white vestiture of both dorsal and ventral surfaces, especially prosternum and abdomen, composed of numerous pectinate suberect or appressed scales .... *Eustenopus*
— Ventral and dorsal surfaces of body with at most suberect short hair-like scales, or erect hairs if present, sparse and many times shorter than antennal scape; white vestiture of both dorsal and ventral surfaces composed of simple appressed scales, some scales of mesosternum and coxae bifid or pectinate 5

5(4). Pronotum dorsally and laterally with numerous shiny, glabrous tubercles, lacking distinct punctures (Figs. 83-84) 6
— Pronotum dorsally and laterally with distinct punctures, with at most the outer margins of punctures at lateral margins swollen, glabrous and shiny 7

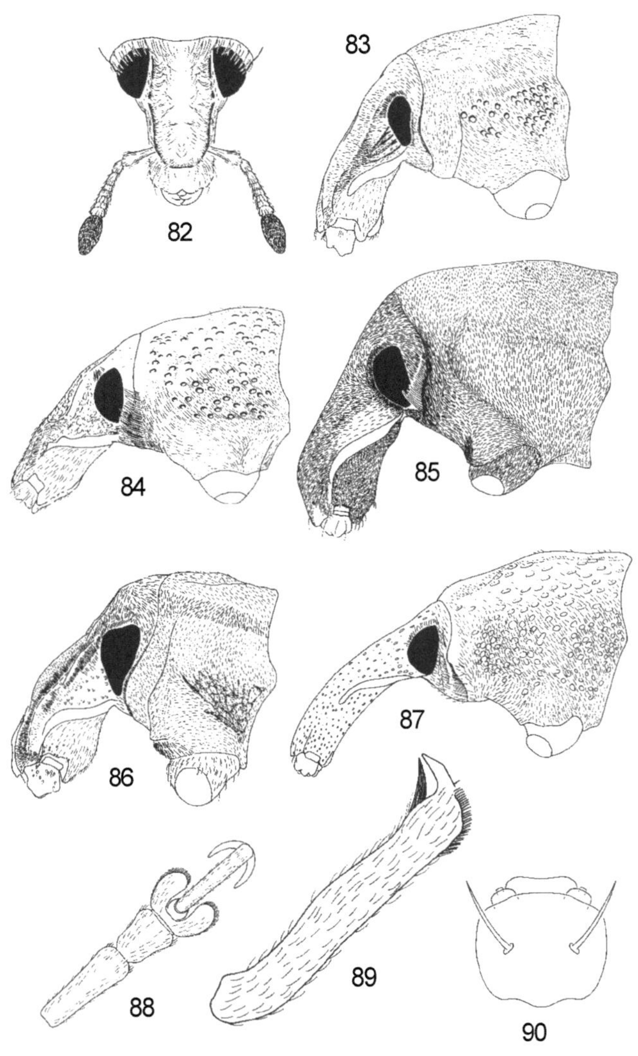

FIGURES 82.131-90.131. Lixinae. 82. *Bangasternus orientalis* (Capiomont), head, dorsal view. 83-87. Head and pronotum, lateral view, 83. *Cleonis pigra* (Scopoli); 84. *Cyphocleonus achates* (Fahraeus); 85. *Apleurus albovestitus* (Casey); 86. *Stephanocleonus plumbeus* LeConte; 87. *Lixus scrobicollis* Boheman. 88-89. *Lixus scrobicollis* Boheman, 88. Fore tarsus; 89. Fore tibia. 90. Labial palpi, Lixinae, schematic (after Anderson 1988a)

6(5). Rostrum with single median sulcus; pronotum with anterolateral margin with rounded postocular lobe (Fig. 83) .... *Cleonis*

— Rostrum carinate medially throughout length, lateral margins raised towards base, appearing as lateral carinae; pronotum with anterolateral margin straight behind eye, lacking postocular lobe (Fig. 84) .... *Cyphocleonus*

7(5). Mesosternum with mesosternal process markedly tumescent; male with aedeagus expanded laterally from midlength to apical one-third .... *Apleurus* (part)

— Mesosternum with mesosternal process flat or at most slightly convex, not tumescent; male with aedeagus more or less uniform in width throughout median portion of length 8

8(7). Tibia with apical flange rounded (Fig. 89); elytron with all intervals equally flat or with at most only humerus and very base of interval 3 swollen and convex; pronotum with disk with scale pattern various; prosternum with or without swellings, swellings, if present, situated immediately anterior to each front coxal cavity and lateral to each prosternal impression 9

— Tibia with apical flange sharp, carinate; elytron with all intervals equally flat or with humerus and variously sutural interval, intervals 3, 5, 7 and 9 elevated and convex throughout the greater part of their length; pronotum with disk with large white scales in a lateral stripe of various width, small and fine in moderately broad to very broad apically narrowed median stripe, median area largely black in color, underlying dark cuticle not obscured by overlying scales; prosternum with or without swellings, swellings, if present, situated immediately anterior to each prosternal impression 11

9(8). Antenna with article 2 of funicle distinctly longer than wide; distinctly longer than each of articles 3 to 6, slightly shorter than to distinctly longer than article 1; pronotum with anterolateral margin straight, slightly sinuate, or with at most variously developed (usually small) acute, postocular projection, postocular vibrissae unequal in length, greatest length (more or less equal to one-half width of eye) behind base of each eye (Fig. 87); femur with ventral surface dentate or not .... *Lixus*

— Antenna with article 2 of funicle more or less as long as wide; more or less subequal in length to each of articles 3 to 6, shorter than article 1; pronotum with anterolateral margin straight or with slightly to well-developed rounded postocular lobe, postocular vibrissae uniformly short (less than one-half width of eye in length) to unequal in length, greatest length (more or less equal to or greater than one-half width of eye) behind base of each eye; femur with ventral surface not dentate 10

10(9). Elytra elongate-narrow (width at midlength less than 0.65 times length); pronotal disk with distinct white scales of various sizes; elytra with white scales, various in size, but more or less obscuring view of underlying cuticle over large part of elytral surface .... *Scaphomorphus*

— Elytra more robust (width at midlength greater than 0.65 times length); pronotal disk with at most only very short fine setae, distinct scales absent; elytra with scattered patches of elongate fine hair-like scales in addition to very short setae, underlying cuticle not obscured by scales .... *Larinus*

11(8). Epistoma with anterior margin emarginate; pronotum with well-developed, rounded postocular lobes, postocular vibrissae indistinct, uniformly short (Fig. 86); eyes widest near upper margin, flat or only slightly convex in dorsal view .... *Stephanocleonus*

— Epistoma with anterior margin rounded; pronotum with anterior margin straight behind eyes or with small acute projection immediately behind base of eye, postocular vibrissae distinct and long, longest immediately behind base of eye (Fig. 85);

eyes widest near midheight, protruding and moderately to markedly convex in dorsal view .... *Apleurus* (part)

CLASSIFICATION OF THE NEARCTIC LIXINAE

65. Lixini Schoenherr 1823

Eustenopus Petri 1907, 1 sp., *E. villosus* (Boheman 1843), locally distributed in the western United States. Introduced for the biological control of *Centaurea solstitialis* L. (yellow star-thistle) (Asteraceae).

Larinus Dejean 1821, 3 spp., locally distributed in the northeastern and northwestern United States, British Columbia, Manitoba and Nova Scotia, adventive. In North America, the species *L. planus* (Fabricius 1792), *L. obtusus* Gyllenhal 1836, and *L. minutus* Gyllenhal 1836 have been introduced for the biological control of *Cirsium arvense* (L.) Scop. (Canada thistle), and *Centaurea solstitialis* L. (yellow star-thistle), *C. maculosa* Lam. (spotted knapweed) and *C. diffusa* Lam. (diffuse knapweed) (all Asteraceae) (Lang 1997a, b). There is no key to the North American species.

Rhinobatus Germar 1817; not Walbaum 1792; not Schneider 1801
Larinus Germar 1824; not Dejean 1821
Phyllonomeus Gistel 1856 (valid subgenus)
Larinodontes Faust 1898
Cryphopus Petri 1907 (valid subgenus)
Lariniorhynchus Reitter 1924
Larinomesius Reitter 1924 (valid subgenus)

Lixus Fabricius 1801, 69 species, generally distributed throughout the United States and Canada. Adults are associated with various plants in the Asteraceae and Polygonaceae. The genus needs revision. See Blatchley and Leng (1916) to separate some of the species. Alonso-Zarazaga and Lyal (1999:190) list 18 valid subgeneric names, not including synonyms. For brevity, these are not repeated here.

Microlarinus Hockhuth 1847, 2 spp., *M. lareynii* (Jacquelin du Val 1852) and *M. lypriformis* (Wollaston 1861), southwestern United States and Washington. Adults have been introduced for the biological control of *Tribulus terrestris* L. (puncturevine; Zygophyllaceae) (Kirkland and Goeden 1977, 1978a, b). See Hatch (1971) to separate the species.

66. Cleonini Schoenherr 1826

Apleurus Chevrolat 1873, 6 spp., southwestern United States east to Texas, north to Idaho. Adults are associated with various arid habitat plants (Anderson 1988a). See Anderson (1988a) to separate the species.

Cleonopsis LeConte 1876
Cleonaspis LeConte 1876
Centrocleonus LeConte 1876; not Chevrolat 1873
Dinocleus Casey 1891
Gibbostethus Anderson 1988 (valid subgenus)

Cleonis Dejean 1821, 1 sp., *C. pigra* (Scopoli 1763), far northeastern United States and adjacent southern Canada, adventive. Adults are associated with *Cirsium arvense* (L.) Scop. (Canada thistle) and *C. vulgare* (Savi) Tenore (bull thistle) (Anderson 1988a).

Geomorphus Schoenherr 1823
Cleonus Schoenherr
Xerobia Gistel 1856

Cyphocleonus Motschulsky 1860, 1 sp., *C. achates* (Fahraeus 1842), Colorado, Wyoming, Montana, Oregon and British Columbia. This species was introduced as a biological control agent for *Centaurea maculosa* Lam. (spotted knapweed) and *C. diffusa* Lam. (diffuse knapweed) (Asteraceae) (Lang 1997c). It has been introduced in other states but does not appear to be established.

Scaphomorphus Motschulsky 1860, 19 spp., generally distributed in the western and central United States and adjacent southern Canada, also along the eastern coastal United States from Florida north into New York. Adults mostly are associated with various arid habitat Fabaceae and Asteraceae (Anderson 1988a). See Anderson (1988a; as *Cleonidius*) to separate the species.

Scaphidomorphus Lacordaire 1863; not Hope 1841
Cleonidius Casey 1891
Lixestus Reitter 1916

Stephanocleonus Motschulsky 1860, 6 spp., western montane United States north into the Yukon Territory and east across Canada to Newfoundland. See Anderson (1988a, 1989b) to separate the species.

Eucleonus Faust 1904; not Gistel 1856
Deracanthopsis Voss 1967 (valid subgenus)
Eremocleonus Ter-Minasian 1974 (valid subgenus)
Taeniocleonus Ter-Minasian 1974 (valid subgenus)
Sanzia Alonso-Zarazaga and Lyal 1999 (valid subgenus)

67. Rhinocyllini Lacordaire 1863

Bangasternus Gozis 1882, 2 spp., *B. orientalis* (Capiomont 1873) and *B. fausti* Reitter 1890, California, Montana, Nebraska and Oregon. These species have been introduced for the control of *Centaurea solstitialis* L. (yellow star-thistle), *C. diffusa* Lam. (diffuse knapweed) and *C. maculosa* Lam. (spotted knapweed) (Asteraceae) (Lang 1997d). There is no key to separate the two species in North America.

Coelostethus Capiomont 1873; not LeConte 1861

Rhinocyllus Germar 1817, 1 sp., *R. conicus* (Froelich 1792), locally distributed throughout most of the United States and adjacent southern Canada. This species was introduced for the biological control of *Carduus nutans* L. (nodding or musk thistle; Asteraceae); larvae feed in flowerheads (Kok 1998). Louda *et al.* (1997) report the species has apparently expanded its host range and is now a threat to native species of *Cirsium* at various locations in the United States.

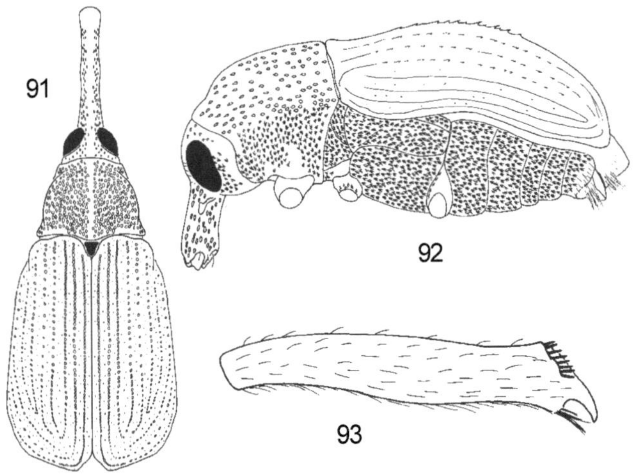

FIGURES 91.131-93.131. Mesoptiliinae. 91. *Magdalis lecontei* Horn, dorsal habitus; 92. *Laemosaccus nephele* complex, lateral habitus; 93. *Magdalis lecontei* Horn, hind tibia.

XV. Mesoptiliinae Lacordaire 1863

by Robert S. Anderson

This is a very small subfamily of only three genera in North America. They are recognized by the presence of a large hook-like apical tooth on the hind tibia (Fig. 93), the pronotum is only slightly narrower than the base of the elytra (Fig. 91), and the elytra have the basal margin from intervals 2-4 extended anteriorly and overlapping the base of the pronotum (Fig. 91). The genus *Laemosaccus* are compact cylindrical beetles with a short, straight rostrum (Fig. 92), whereas *Magdalis* and *Trichomagdalis* have a very different, anteriorly tapered form, with the width across the apices of the elytra generally the widest part of the beetle.

Larvae of all species mine in wood (both hardwoods and conifers) or in stems of herbaceous plants. The genus *Magdalis* does not extend far into Mexico, but the genus *Laemosaccus* has many Neotropical species, most undescribed. The center of diversity for Mesoptilinae appears to be in Chile.

Key to the Nearctic Genera of Mesoptiliinae

1. Front coxae separated by prosternum; rostrum stout, short and straight (Fig. 92); elytra black with red markings or entirely black .... *Laemosaccus*
— Front coxae contiguous; rostrum elongate, cylindrical, curved ventrally or straight; elytra unicolorous, black or reddish 2

2(1). Elytra more or less parallel sided in dorsal view, with numerous fine, elongate scales; tarsus with claws simple; pronotum with anterolateral angles smooth, not serrate .... *Trichomagdalis*
— Elytra usually widened posteriorly in dorsal view (Fig. 91), nearly glabrous; tarsus with claws simple or toothed basally; pronotum with anterolateral angles serrate or smooth .... *Magdalis*

Classification of the Nearctic Mesoptiliinae

68. Laemosaccini Lacordaire 1866

Laemosaccus Schoenherr 1823, 2 spp., *L. nephele* (Herbst 1797), generally distributed in the eastern United States and adjacent southern Canada west into Texas, New Mexico and Arizona; and, *L. texanus* Champion 1903, southern Texas and Arizona. Adults of *L. nephele* are associated with species of *Quercus* (Fagaceae) and *Prosopis* (Fabaceae), and *L. texanus* with various Malvaceae (Blatchley and Leng 1916); larvae mine twigs or stems. The genus needs revision; *L. nephele* is a composite of a number of distinct host-specific species. See Champion (1903) to separate the species.

69. Magdalidini Pascoe 1870

Magdalis Germar 1817, 25 spp., generally distributed, *M. barbicornis* (Latreille 1804) is adventive. Adults are associated with various trees; larvae mine in bark of dead or dying trees (Blatchley and Leng 1916). The genus needs revision. See Fall (1913), Blatchley and Leng (1916) and Hatch (1971) to separate some of the species.

Rhina Latreille 1802; not Schneider 1801
Edo Germar 1819 (valid subgenus)
Porrothus Dejean 1821 (valid subgenus)
Rhinodes Dejean 1821
Panus Schoenherr 1823 (valid subgenus)
Thamnophilus Schoenherr 1823; not Vieillot 1816
Magdalinus Germar 1843
Porrhothus Agassiz 1846
Scardamyctes Gistel 1848
Panopsis Daniel 1903 (valid subgenus)
Neopanus Reitter 1916
Laemosaccidius Smreczynski 1972 (valid subgenus)
Odontomagdalis Barrios 1984 (valid subgenus)

Trichomagdalis Fall 1913, 3 spp., California. See Fall (1913) to separate the species.

XVI. Molytinae Schoenherr 1823

by Robert S. Anderson

Like Curculioninae, this subfamily is also somewhat of a conglomerate of likely unrelated forms. They are grouped together here primarily because they all share a large, hook-like apical tooth on the hind tibia, or have various modifications to the apex of the hind tibia related to the development of the tooth (Figs. 99-101).

Most taxa of molytines are associated with woody plants and have larvae that feed in dead wood or other dead and decaying plant material. Some taxa such as *Lepilius*, *Epacalles* and *Lymantes* are associated with leaf litter and likely have larvae that develop in fallen plant debris. Larvae of others such as *Cholus* mine stems, or some such as *Conotrachelus*, feed in the seeds,

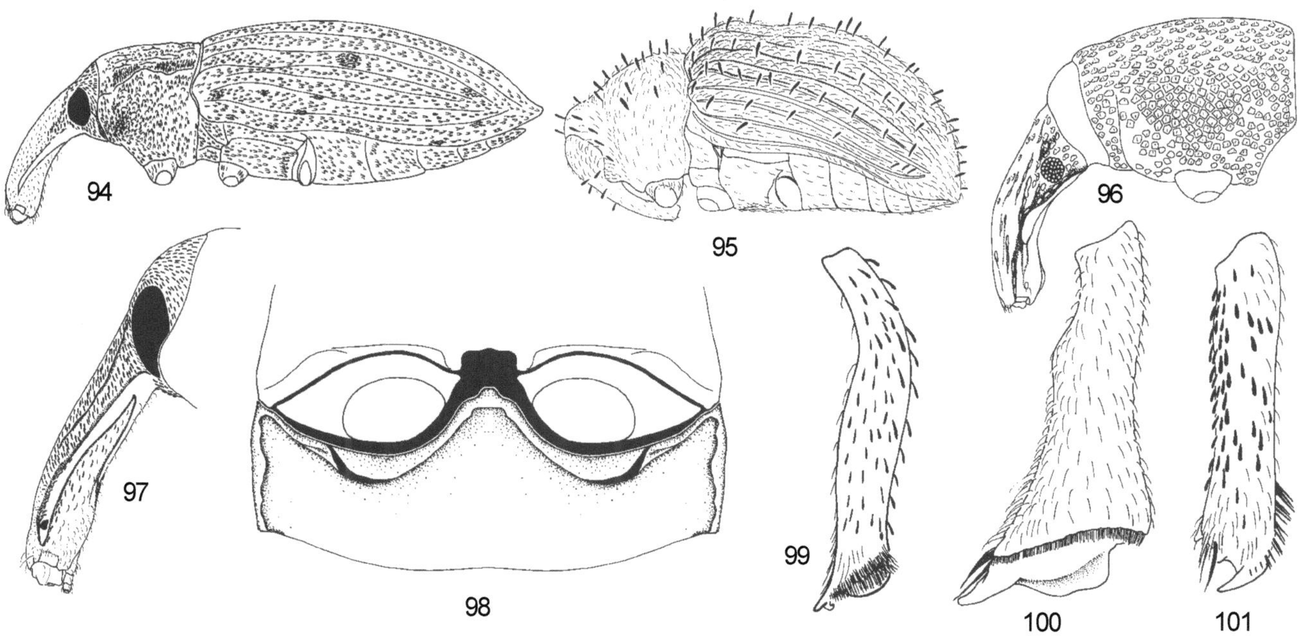

FIGURES 94.131-101.131. Molytinae. 94-95. Lateral habitus, 94. *Lepyrus* sp.; 95. *Microhyus setiger* LeConte. 96. *Dioptrophorus repens* (Casey), head and pronotum, lateral view. 97. *Hilipinus nearcticus* O'Brien, head, lateral view. 98. *Heilus bioculatus* (Boheman), hind coxae and first abdominal ventrite. 99-101. Hind tibia, 99. *Conotrachelus posticatus* Boheman; 100. *Pachylobius picivorus* (Germar); 101. *Pissodes strobi* (Peck).

fruits or reproductive structures of living plants. Several taxa represented in the United States by only one species, such as *Hilipinus*, *Heilus* and *Heilipus*, are significantly more diverse in the Neotropical Region with hundreds of species found there. *Odontopus* and *Piazorhinus* have larvae that mine leaves.

With over 500 species, the genus *Conotrachelus* may prove to be the most diverse genus of weevil in the Americas. The odd genera *Thalasselephas* and *Hormops* are associated with seaweed and tree squirrel nests, respectively.

KEY TO THE NEARCTIC GENERA OF MOLYTINAE

1. Eyes absent or reduced in size to less than 12 facets 2
— Eyes present, well-developed, composed of more than 12 facets 3

2(1). Antenna with funicle of 8 articles; eyes absent; southern Florida ..... *Caecossonus*
— Antenna with funicle of 7 articles; eyes absent (but indicated by setose swelling on basal portion of rostrum) or present, composed of up to 11 facets; eastern United States (not southern Florida) west into southwestern Texas ..... *Lymantes*

3(2). Head with eyes obviously situated on basal portion of rostrum, head distinctly constricted and globular behind eyes (Fig. 96) ..... *Dioptrophorus*
— Head with eyes obviously situated on head, not on rostrum (Fig. 97); head may be constricted and globular behind base of rostrum 4

4(3). Rostrum in repose received into ventral channel on prosternum 5
— Rostrum in repose not received into ventral channel, but may rest between front, middle and/or hind coxae 13

5(4). Tarsus with claw simple, free or connate, lacking basal tooth 6
— Tarsus with claw with basal tooth 10

6(5). Tarsus with claws connate at base; pronotum coarsely strigose or punctate; some with metallic sheen ..... *Chalcodermus*
— Tarsus with claws free; pronotum finely and shallowly punctate; not with metallic sheen 7

7(6). Pronotum and elytra with sparse, long, erect stout seta-like scales (Fig. 95); body length less than 2.5 mm ..... *Microhyus*
— Pronotum and elytra lacking long, erect vestiture, with only appressed scales; length various 8

8(7). Body form elongate; California ..... *Micromastus*
— Body form oval; eastern United States west into Texas 9

9(8). Pronotum and elytra with scattered short, clavate, recurved setae and appressed scales; eyes lateral in placement, separated dorsally by a distance slightly greater than width of rostrum at base; metasternum steeply sloped between hind coxae; extreme southwestern Texas ... *Lepilius*
— Pronotum and elytra with only appressed scales; eyes lateral in placement, but with upper portion encroached on dorsal surface of head, separated dorsally by a distance slightly less than width of rostrum at base; metasternum gradually sloped between hind coxae; eastern United States into central Texas ..... *Epacalles*

10(5). Pronotum with only slightly developed postocular lobe; dorsum with at most a few scattered, unpigmented scales; pronotum coarsely strigose or rugulose or regularly punctate ... *Rhyssomatus*

— Pronotum with distinct postocular lobe; dorsum with more or less dense, appressed, pigmented scales; pronotum regularly finely or coarsely punctate, median impunctate line or carina present or not 11

11(10). Elytra with all intervals carinate or alternate intervals carinate or at least swollen throughout most of their length, in some specimens carina or swelling only or most evident on declivity; femora, especially middle and hind with distinct tooth on ventral margin .... *Conotrachelus*

— Elytra with all intervals flat or slightly but evenly swollen throughout most of their length; femora with or without tooth of ventral margin 12

12(11). Dorsum of pronotum and elytra with long, erect setae and appressed scales; rostrum long and slender .... *Pheleconus*

— Dorsum of pronotum and elytra with only appressed scales; rostrum short and stout .... *Micralcinus*

13(4). Front coxae very narrowly to widely separated by prosternal processes 14

— Front coxae contiguous, not separated by prosternal processes 24

14(13). Metepisternal suture absent; pronotum deeply, coarsely punctate, distance between punctures less than the diameter of a puncture; elytra with numerous, low, setiferous tubercles .... *Anchonus*

— Metepisternal suture present, although may be defined only in anterior one-half; pronotum various, but not deeply, coarsely punctate; elytra smooth or with two large tubercles at about midlength on interval 2 15

15(14). Metepisternal suture visible and subcarinate in only anterior one-half; elytra with scattered tufts of suberect broad scales .... *Trachodes*

— Metepisternal suture visible throughout length; elytra with appressed scales or fine vestiture 16

16(15). Hind tibia with equally large tooth at both outer and inner apical angles; mandibles with apices divergent, not overlapping 17

— Hind tibia with large tooth at outer apical angle only (Fig. 101), or also with much smaller tooth at inner apical angle; mandibles convergent, overlapping 18

17(16). Pronotum with anterolateral margin with distinct postocular lobe; adventive, in greenhouses .... *Cholus*

— Pronotum with anterolateral margin almost straight, lacking distinct postocular lobes; native, extreme southern Arizona .... *Neoerethistes*

18(16). Eyes narrowly separated by less than width of antennal club; front femur with obvious tooth; tarsal claw with basal process 19

— Eyes separated by more than width of antennal club; front femur lacking obvious tooth; tarsal claw simple 20

19(18). Front femur with large, serrate tooth; rostrum about as long as pronotum, subcylindrical; elytra lacking scales .... *Odontopus*

— Front femur with small, simple tooth; rostrum shorter than length of pronotum, dorsoventrally compressed, spatulate, especially towards apex; elytra with narrow scales .... *Piazorhinus*

20(18). Front coxae widely separated by the width of a coxa; body form markedly dorsoventrally compressed, upper contour flat .... *Nanus*

— Front coxae very narrowly separated by much less than one-third the width of a coxa; body form more subcylindrical, upper contour rounded 21

21(20). Hind tibia with apex with apical comb of stout setae oriented longitudinally along axis of tibia on outer margin (Fig. 101) .... *Pissodes*

— Hind tibia with apex with apical comb transverse or obliquely oriented to long axis of tibia across outer apical margin 22

22(21). Body size moderate, greater than 3.0 mm; scutellum large and distinct; California, adventive .... *Tranes*

— Body size small, less than 3.0 mm; scutellum minute or not visible; coastal beaches of Pacific states, British Columbia, and Florida 23

23(22). Pronotum sculptured with paramedian broad impression, lateral impressions and low lateral tubercles at anterior one-third; elytra with alternate intervals elevated especially so on declivity; body color dark brown or black; Florida .... *Gononotus*

— Pronotum smooth and virtually impunctate, lateral margins evenly rounded; elytra with intervals flat; body color pale brown; Pacific coastal states and British Columbia .... *Thalasselephas*

24(13). Tarsus with claws connate at base 25

— Tarsus with claws free at base 26

25(24). Elytra with acute lateral subhumeral tubercle; eyes moderate in size, widely separated ventrally; front tibia with tooth on inner margin ...... *Sternechus*

— Elytra with lateral margins simple, lacking tubercle; eyes very large and elongate, subcontiguous ventrally; front tibia with inner margin simple .... *Hormops*

26(24). Metepisternal suture absent; body size small to moderate, length less than 6.0 mm 27

— Metepisternal suture present; body size moderate to large, length greater than 6.0 mm 28

27(26). Metasternum and abdominal ventrites 1 and 2 with large, deep excavations .... *Gastrotaphrus*

— Metasternum and abdominal ventrites 1 and 2 smooth, lacking large, deep excavations .... *Sthereus*

28(26). Pronotum markedly constricted and tubulate at base; rostrum very long and fine, about twice as long as length of pronotum; body black and glossy, lacking distinct broad scales .... *Sicoderus*

— Pronotum not constricted or tubulate at base, broad; rostrum moderate in length and somewhat stout, shorter than length of pronotum; body various, not black and glossy, with at least some broad appressed scales 29

29(28). Hind tibia markedly expanded at apex, wider at apex than apex of femur; area adjacent to tarsal articulation on hind femur with large flat flange (Fig. 100); front tibia with outer angle produced and spatulate .... *Pachylobius*

— Hind tibia not significantly expanded at apex, width at apex narrower than width of femur at apex; area adjacent to tarsal articulation on hind femur with at most a low cariniform extension; front tibia with outer angle rounded, not produced or spatulate 30

30(29). Femur with inner margin lacking tooth, evenly rounded at apical one-third; tibia with inner margin simple, not cariniform or expanded, outer margin more or less straight 31

— Femur with inner margin with variously developed tooth at apical one-third; tibia with inner margin cariniform and slightly to markedly expanded at point of occlusion with femoral tooth, outer margin arcuate 33

31(30). Pronotum with anterolateral margins lacking distinct postocular lobes; eyes rounded, distinctly convex; metepimeron visible, with vestiture and sculpture similar to metepisternum (Fig. 94) .... *Lepyrus*

— Pronotum with anterolateral margins with distinct postocular lobes; eyes elongate-oval, flat or slightly convex; metepimeron not visible, concealed by elytra, if metepimeron visible because of displaced elytra, vestiture and sculpture finer than on metepisternum 32

32(31). Antenna with funicle with article 2 longer than 1; elytra with pattern of brown scales with patches of paler scales surrounded by black scales along length of interval 4 and at apical callus, scales about twice as long as wide, striae with small, rounded punctures, each with a broad flat scale; associated with Taxodiaceae .... *Eudociminus*

— Antenna with funicle with article 2 shorter than 1; elytra with pattern of scattered, white or cream colored scales, scales many times longer than wide, striae with large, deep, elongate punctures, each with a fine hair-like seta; associated with Pinaceae .... *Hylobius* (part)

33(30). Metasternum between middle and hind coxae shorter than length of a middle coxa; pronotum and elytra with numerous, small, glossy, round nodules; surface sculpture coarse and irregular .... *Steremnius*

— Metasternum between middle and hind coxae longer than length of a middle coxa; pronotum and elytra punctate or rugose (pronotum); surface sculpture regular and more or less smooth 34

34(33). Abdomen with ventrite 1 with raised anterior margin with posteriorly expanded, slightly crenulate area immediately behind hind coxa (Fig. 98); elytra with single, rounded patch of black scales forming an 'eyespot' at posterior two-fifths .... *Heilus*

— Abdomen with ventrite 1 with anterior margin evenly developed immediately behind hind coxa, no expanded area evident; elytra with scale pattern various, lacking 'eyespots' 35

35(34). Rostrum above scrobe coarsely punctate, not at all sulcate; pronotum and elytra with contrasting pattern of black cuticle and bright white, glossy scales arranged around periphery; eyes separated dorsally by less than one-half width of the rostrum at base .... *Heilipus*

— Rostrum above scrobe with shallow, longitudinal sulcus oriented parallel to dorsal margin of scrobe (Fig. 97); pronotum and elytra scale pattern various, not contrasting; eyes separated dorsally by a distance greater than or subequal to width of the rostrum at base 36

36(35). Elytra with punctures each with a fine hair-like scale; rostrum with a slight longitudinal impression immediately above scrobe, impression not defined dorsally but more or less continuous with dorsum of rostrum; hind tibia with apical comb composed of a single row of setae ... *Hylobius* (part)

— Elytra with punctures each with a broad scale; rostrum with a distinct, moderately deep longitudinal impression immediately above scrobe, impression defined dorsally by a low but distinct carina, not continuous with dorsum of rostrum; hind tibia with apical comb composed of a long apical row of setae and a second short preapical row towards the dorsal end of the apical row .... *Hilipinus*

Classification of the Nearctic Molytinae

70. Molytini Schoenherr 1823

Plinthina Lacordaire 1863

Gastrotaphrus Buchanan 1936, 1 sp., *G. barberi* Buchanan 1936, far western United States and British Columbia. Adults have been collected in moss and leaf litter (Anderson 1988b).

Steremnius Schoenherr 1835, 3 spp., *S. carinatus* (Boheman 1842) and *S. tuberosus* Gyllenhal 1836, far western United States, British Columbia and Alaska, and *S. shermani* (Fiske 1906), North Carolina, Tennessee and Virginia (at high elevations). Adults have been collected in leaf litter; larvae feed in phloem of slash or roots of dead conifers (Anderson 1988b). See Brown (1966b) to separate the species.

Paraplinthus Faust 1892

Sthereus Motschulsky 1845, 4 spp., far western United States, British Columbia, Alaska, Nova Scotia and Newfoundland. Adults of *S. multituberculatus* Buchanan 1936, *S. quadrituberculatus* Motschulsky 1845, and *S. horridus* (Mannerheim 1952) have been associated with various conifers and collected in leaf litter; adults of *S. ptinoides* have been collected under driftwood on beaches

(Anderson 1988b). See Hatch (1971) or Zimmerman (1964) to separate the species.

Stereus Mannerheim 1846
Lobosoma Buchanan 1936
Philostratus Zimmerman 1964
Lobosoma Zimmerman 1964

71. Trachodini Gistel 1848

Trachodes Germar 1824, 1 sp., *T. hispidus* (Linnaeus 1758), Newfoundland; adventive.

Blastophila Gistel 1856
Metrachodes Marshall 1948
Atrachodes Morimoto 1962 (valid subgenus)

72. Anchonini Imhoff 1856

Anchonus Schoenherr 1825, 4 spp., Florida. Adults frequently are collected in association with driftwood and in the litter of coastal hardwood hammocks (Thomas and O'Brien 1999). See Thomas and O'Brien (1999) to separate the species.

Choristorhinus Fairmaire 1899

73. Camarotini Schoenherr 1833

Alonso-Zarazaga and Lyal (1999) place these weevils as Curculioninae but the structure of the uncus at the tibial apex suggests they are Molytinae or related. Here they are placed as a tribe within Molytinae.

Prionomerina Lacordaire 1863

Odontopus Say 1831, 1 sp., *O. calceatus* (Say 1831), generally distributed in eastern United States. This species is associated with *Sassafras* (Lauraceae) and *Liriodendron* (Magnoliaceae); larvae mine leaves.

Prionomerus Schoenherr 1835

74. Cholini Schoenherr 1825

Cholina Schoenherr 1825

Cholus Germar 1824, 1 sp., *C. cattleyae* (Champion 1916), Washington, DC, New Jersey and Wisconsin; adventive. This species has been found in greenhouses; it is not established in the wild in North America.

Archarias Dejean 1821
Dionychus Germar 1824
Litomerus Schoenherr 1833
Polyderces Schoenherr 1833
Aphyoramphus Guérin-Méneville 1844 (valid subgenus)
Lonchocerus Chevrolat 1879
Sternoxus Chevrolat 1879
Platypachys Chevrolar 1879
Gymnonotus Chevrolat 1879
Ardoleucus Checrolat 1881
Atroniscus Desbrochers 1906

Rhinastina Vaurie 1973

Neoerethistes O'Brien and Wibmer 1982, 1 sp., *N. arizonicus* (Sleeper 1954), Arizona.

75. Cleogonini Gistel 1856

Rhyssomatus Schoenherr 1837, 17 spp., generally distributed with the exception of the northwestern United States and adjacent Canada. Adults are associated with various plants in the families Asclepiadaceae, Asteraceae, Convolvulaceae and Fabaceae (Blatchley and Leng 1916; Kissinger 1964; Anderson 1993a). The genus needs revision. See Casey (1895) and Blatchley and Leng (1916) to separate some of the species.

Polydus Pascoe 1872
Sermysatus Casey 1895 (valid subgenus)

76. Conotrachelini Jekel 1865

[*Chaleponotus* Casey 1892, 1 sp., *C. elusus* Casey 1892, Indiana. This genus and species are known only from the type specimen, labelled from "Indiana". There is considerable doubt that this is a North American taxon. At the time of its description Casey was studying Brazilian Baridinae and there is the possibility that the locality reference is to Indiana, Brazil and not the state of Indiana in the United States.]

Conotrachelus Dejean 1835, 63 spp., generally distributed throughout the United States and Canada. Adults are associated with various plants; many come to lights. Larvae feed in developing fruits and in injured or dying wood (Kissinger 1964). A number of species are associated with *Quercus* (Fagaceae) and other hardwoods. *Conotrachelus nenuphar* (Herbst 1797) is the plum curculio and *C. crataegi* Walsh 1863 is the quince curculio. The genus needs revision; a number of undescribed species are known from Florida (Anderson 1993a). See Schoof (1942) and Blatchley and Leng (1916) to separate some of the species. This treatment of *Conotrachelus* includes *C. parvulus* Champion 1904, and *Pheleconus cribricollis* (Say 1831) and *P. infector* (Boheman 1845); see also notes about *Micralcinus* and *Pheleconus*. Relationships of these genera need to be reassessed.

Cyphorhynchus Schoenherr 1837
Glycaria Pascoe 1880
Edesius Pascoe 1881
Enops Pascoe 1889
Loceptes Casey 1910
Pseudocomus Varga 1951 (valid subgenus)

Epacalles Kissinger 1964, 1 sp., *E. inflatus* Blatchley 1916, eastern United States west into central Texas. Adults are collected in leaf litter.

Lepilius Champion 1905, 1 sp., undescribed, extreme southwestern Texas. Adults of an undescribed species have been collected in leaf litter in Big Bend National Park, Texas.

Micralcinus LeConte 1876, 3 spp., southeastern United States west into Texas. Adults of *M. cribratus* LeConte 1876 have been associated with *Amaranthus* (Amaranthaceae) and adults of *M. maculatus* (Blatchley 1916) with *Sesuvium portulacastrum* (L.) L. Aizoaceae (Anderson 1993a). I have not seen specimens of *M. kalmbachi* Buchanan 1927. See Sleeper (1955c) to separate the species (note that *M. stehri* Sleeper 1955 is a junior synonym of *Conotrachelus parvulus* Champion 1904; Wibmer and O'Brien 1989).

Microhyus LeConte 1876, 1 sp., *M. setiger* LeConte 1876, eastern United States and adjacent southern Canada. Adults have been associated with dead *Fagus* (beech; Fagaceae).
Echinaspis Blatchley 1922; not Haeckel 1881

Micromastus LeConte 1876, 1 sp., *M. gracilis* (Boheman 1859), California.

Pheloconus Roelofs 1875, 1 sp., *P. hispidus* (LeConte 1876) generally distributed in the eastern United States west to Louisiana. Adults of *P. hispidus* (LeConte 1876) have been associated with *Malvastrum corchorifolium* (Desc.) Britt. (Malvaceae). Two additional species, *P. infector* (Boheman 1845) and *P. cribricollis* (Say 1831) have been considered as *Pheleconus* but do not fit the generic definition and are likely *Conotrachelus*. See Blatchley and Leng (1916; as *Conotrachelus* groups III and VI) to separate this complex of species.

77. Cycloterini Lacordaire 1863

Cycloterina Lacordaire 1863

Gononotus LeConte 1876, 1 sp., *G. angulicollis* (Suffrian 1871), Florida. Adults are common under debris on beaches (Anderson 1993a).
Nemosinus Faust 1892

78. Erodiscini Lacordaire 1863

Alonso-Zarazaga and Lyal (1999) place these weevils as Curculioninae but the structure of the uncus at the tibial apex suggests they are Molytinae or related. Here they are placed as a tribe within Molytinae.

Sicoderus Vanin 1986, 1 sp., *S. tinamus* (LeConte 1884), Florida. This species appears to be associated with *Bumelia celastrina* (Nutt.) R. W. Long (Sapotaceae) (Anderson 1993a).

79. Hylobiini Kirby 1837

Hylobiina Kirby 1837

Eudociminus Leng 1918, 1 sp., *E. mannerheimi* (Boheman 1836), southeastern United States west to Louisiana. Adults are associated with *Taxodium distichum* (L.) Rich. (bald cypress; Taxodiaceae).
Eudocimus Boheman 1836; not Wagler 1832

Heilipus Germar 1824, 1 sp., *H. apiatus* (Olivier 1807), Florida, Georgia, South Carolina, North Carolina and Tennessee. Adults have been associated with various plants. In Florida the larvae bore into the cambium at the base of *Persea americana* Mill. (American avocado; Lauraceae) (Woodruff 1963). Two additional species, *H. lauri* Boheman 1845 and *H. pittieri* Barber 1919 have been found in greenhouses.
Hilipus Agassiz 1846
Hilipus Gemminger and Harold 1871; not Agassiz 1846

Heilus Kuschel 1955, 1 sp., *H. bioculatus* (Boheman 1843), southern Florida, adventive. Adults and larvae have been associated with *Bursera simaruba* (L.) Sarg. (Burseraceae) (Anderson 1993a).

Hilipinus Champion 1902, 1 sp., *H. nearcticus* O'Brien 1982, Florida, Louisiana and Mississippi. Adults come to lights.

Hylobius Germar 1817, 8 spp., generally distributed throughout the eastern and central United States and all of Canada; one adventive species. Adults and larvae of the native species are associated with conifers (Warner 1966). *Hylobius transversovittatus* (Goeze 1777) has been introduced from Europe for the biological control of *Lythrum salicaria* (purple loosestrife; Lythraceae) and is now established in New York, Pennsylvania, Maryland, Virginia, Ohio, Indiana, Illinois, Iowa, Michigan, Wisconsin, Minnesota, South Dakota, Colorado, Montana, Oregon and Washington in the United States (Weeden 2000), and in British Columbia, Alberta, Manitoba and Nova Scotia in Canada (Harris 2001). See Warner (1966) to separate the seven native species.
Callirus Dejean 1821 (valid subgenus)
Hypomolyx LeConte 1876
Hylobitelus Reitter 1923
Poiyaunbus Kôno 1934

Pachylobius LeConte 1876, 1 sp., *P. picivorus* (Germar 1824), generally distributed in the eastern United States and adjacent southern Canada. Adults are associated with *Pinus* (Pinaceae); larvae mine the inner bark of roots and stumps of dying or injured trees (Franklin and Taylor 1970).

80. Lepyrini Kirby 1837

Lepyrus Germar 1817, 6 spp., generally distributed in the northern and western montane United States and across Canada including the far north and Alaska. Adults are often associated with *Salix* (willow; Salicaceae) but larvae likely feed on the roots of other plants (Anderson 1997). The genus needs revision. Several subspecies of questionable status are recognized in North America and no attempt has been made to compare the North American fauna to those of Asia. See Van Dyke (1928) to separate the forms.

Dirus Dejean 1821

81. Lymantini Lacordaire 1866

Caecossonus Gilbert 1955, 1 sp., *C. dentipes* Gilbert 1955, southern Florida. Adults are frequently collected in soil and leaf litter (Howden 1992).

Dioptrophorus Faust 1892, 1 sp., *D. repens* (Casey 1892), California, Oregon and Washington. Adults have been collected in leaf litter.

Metopotoma Casey 1892

Anculopus Van Dyke 1927

Lymantes Schoenherr 1838, 4 spp., southeastern United States north to Ohio and west to western Texas and Oklahoma. Adults have been collected in leaf litter. The genus needs revision. See Sleeper (1965) to separate the species.

Typhloglymma Dury 1901

Stewpeckia Osella 1980

82. Petalochilini Lacordaire 1863

Hormops LeConte 1876, 1 sp., *H. abducens* LeConte 1876, southeastern United States north to Ohio and west to Texas. Adults are found in the nests of fox and grey squirrels (Blatchley 1918).

83. Piazorhinini Lacordaire 1863

Alonso-Zarazaga and Lyal (1999) place these weevils as Curculioninae but the structure of the uncus at the tibial apex suggests they are Molytinae or related. Here they are placed as a tribe within Molytinae.

Piazorhinus Schoenherr 1835, 4 spp., generally distributed in eastern United States and southeastern Canada. Species are associated with *Quercus* (Fagaceae) and *Coccoloba diversifolia* Jacq. (Polygonaceae) (Anderson 1993a, b). See Blatchley and Leng (1916) to separate the species.

Acamatus Schoenherr 1833

Polyponus Kirsch 1875

Piazorrhinus Champion 1903

84. Pissodini Gistel 1856

Pissodina Gistel 1856

Pissodes Germar 1817, 22 spp., generally distributed throughout the United States and Canada. Adults and larvae are associated with various conifers. Some species are of economic importance. The genus needs revision. See Hopkins (1911) to separate the species.

Piniphilus Dejean 1821

Epipissodes Voss 1956 (valid subgenus)

85. Sternechini Lacordaire 1863

Chalcodermus Dejean 1835, 7 spp., generally distributed in the eastern and central United States west into Texas and Arizona. Adults of *C. aeneus* Boheman 1837 have been associated with *Vigna luteola* (Jacq.) Benth. (Fabaceae); larvae develop in seed pods (Ainslie 1910). Adults of *C. collaris* Horn 1873 have been reared from seed pods of *Cassia chamaechrista* L. (Fabaceae) (Alsterlund 1937a, b). Adults of *C. martini* Van Dyke 1930 have been collected from two species of *Brickellia* (Asteraceae) in Arizona and Texas. The genus needs revision. See Blatchley and Leng (1916) to separate some of the species.

Anthobates Gistel 1848

Sternechus Schoenherr 1826, 2 spp., *S. armatus* (Casey 1895), southeastern United States north to Illinois and New Jersey, and *S. paludatus* (Casey 1895), Arizona. Some tropical species are associated with Fabaceae (Anderson 1993b).

Sternuchus Gemminger and Harold 1871; not LeConte 1850

Sternuchus Suffrian 1871; not LeConte 1850; not Gemminger and Harold 1871

Plectromodes Casey 1895

86. Thalasselephantini Alonso-Zarazaga and Lyal 1999

Thalasselephas Egorov and Korotyaev 1976, 1 sp., *T. testaceus* (LeConte 1876), California, Oregon and British Columbia. Adults are found under seaweed on sandy coastal beaches (Anderson 1988b). Korotyaev and Egorov (1975) have suggested that this genus is related to *Emphyastes* (Cyclominae).

Phycocoetes LeConte 1876; not Agassiz 1846

Neophycocoetes O'Brien and Wibmer 1982

87. Trypetidini Lacordaire 1866

Nanus Schoenherr 1844, 1 sp., *N. uniformis* Boheman 1844, southern Florida. Adults are associated with palms and banana trees.

Homaloxenus Wollaston 1873

Incertae sedis

Tranes Schoenherr 1843, 1 sp., *T. internatus* Pascoe 1870, California, adventive. Adults have been collected in association with introduced *Encephalartos* (Cycadaceae) from Australia. It is not known whether the genus is established in North America.

Platyphaeus Pascoe 1877

XVII. Scolytinae Latreille 1807

by Robert J. Rabaglia

Subfamily common name: The bark and ambrosia beetles

Subfamily synonyms: Hylurgidae Zimmerman 1868; Ipidae Latreille 1804

The general body shape of these small beetles ranges from very stout to moderately elongate and cylindrical. Typically the body is brownish with moderate pubescence. The geniculate antennae have a distinct club.

Description: (modified from Wood 1982) Very small to small in size, 1-9 mm, mostly 1-3 mm; shape stout to cylindrical; color brownish or piceous; pubescence sparse to abundant, mostly consisting of very fine, short setae or stout, flat setae.

Head prominent, or withdrawn into pronotum; surface punctate to granulate. Antennal scape well developed, funicle one to seven segmented, club large, either solid, annulated or rarely pseudolamellate; inserted on the sides of head between eyes and mandibles. Labrum absent; mandibles short, curved, the apices blunt, dentate; maxillary palpi three segmented, segments short and stout. Gular region reduced to a small pregula, gular sutures confluent; mentum moderate, variable; labial palpi three segmented, small, stout, apically acute. Eyes lateral, moderate, flat, transverse.

Pronotum slightly broader than head; shape truncate anteroventrally, quadrate to subcircular, borders margined or not; surface punctate, asperate, rugose or striate; pleural region broad; prosternum short in front of coxae, some with a small median process projecting posteriorly; procoxal cavities closed behind. Legs moderate in length; trochantins not exposed; anterior coxae globular, contiguous to widely separated; middle coxae round, flat, separate; hind coxae subtriangular, separate; trochanters small, triangular; femora swollen, short; tibiae compressed, mostly toothed with apical hooks or, with marginal teeth or denticles; tarsal formula 5-5-5, apparently 4-4-4, slender, third segment narrow or dilated, fourth segment minute; claws large, simple divergent. Scutellum small, quadrate, triangular or absent. Elytra entire, apically rounded, mostly declivous and often with tubercles, denticles or spines apically; striae mostly distinct, punctate; epipleural fold obscure. Wing venation and folding pattern not described.

Abdomen with five visible sterna, sutures entire; surface microrugose to punctate. Male genitalia with penis stout, apically blunt, basally with a pair of slender, articulating struts; parameres absent; pars basalis reduced to a slender complete or incomplete ring and a curved, slender basal strut of variable length. Female genitalia undescribed.

Larvae C-shaped, subcylindrical, fleshy; size 2 mm - 10 mm in length; vestiture ranges from absent to a few, simple setae; color near white. Head partly retracted or distinctly exserted, mouthparts hypognathous or nearly prognathous with a faint epicranial suture surrounding the frons. Antennae very small to absent. Mandible mostly short, stout, gouge-shaped, subtriangular without mola or retinaculum; maxilla with cardo, fused stipes and mola; maxillary palpi one or two segmented. Stemmata absent in most. Thorax frequently broader than abdomen; legs absent, but with fleshy lobes ventrally. Abdomen with three or more plicae on each segment; nine or ten segmented, segments 8 -10 in some with pigmented tubercles dorsally. Spiracles on mesothorax and abdominal segments one to eight, annular, annular-biforous or biforous, or inconspicuous.

Habits and habitats. Most bark and ambrosia beetles live in injured, weakened or dying woody plants. Hosts must contain sufficient moisture for development and most species complete only one generation in a host. A few species breed in roots and stems of non-woody plants, others breed in seed or cones, but the majority of species are considered bark beetles or ambrosia beetles. Bark beetles feed on the phloem of the inner bark of their woody host plant. Fewer than half the species in the family are bark beetles, but they are the majority of species in the temperate regions. Ambrosia beetles cultivate and feed on symbiotic ambrosia fungi in the xylem of the host plant. Most tropical species exhibit this habit.

Typically, adult bark and ambrosia beetles bore through the outer bark and construct an egg gallery either in the phloem-cambial region (bark beetles) or in the xylem (ambrosia beetles). Females lay eggs at regular intervals on either side of the gallery. Among bark beetles, larval feeding mines radiate out from the egg gallery, and engrave the inner bark or wood or both. These characteristic engravings can often be found under the bark of dead or dying trees. Ambrosia beetle larvae feed on the ambrosia fungus in small cradles off of the egg gallery. After pupation, the next generation of bark beetles emerges through individual exit holes in the bark, giving it a characteristic "shot hole" appearance. Ambrosia beetle adults usually emerge through the parental entrance hole.

Most of the life stages of these beetles occur within the host plant, however, upon emergence adults must find suitable host material in which to feed and breed. They are often among the first insects to colonize a dying tree; therefore, rapid location of hosts is an important part of their biology. In many species, host location is mediated by olfactory responses to host odors (e.g., terpene hydrocarbons), tree degradation products (e.g., alcohols) or conspecific semiochemicals (pheromones). Several species utilize pheromones not only for attraction of potential mates, but also for mass aggregation to overcome resistance of the host tree. The pheromone biology of species of *Dendroctonus*, *Ips* and *Scolytus*, among others, has been well studied, and the complex inter- and intraspecific interactions elucidated (Wood, D.L. 1982, Borden 1982, Raffa *et al.* 1993).

Many bark and ambrosia beetle species have distinctive, subsocial behaviors. Social organization associated with reproductive behavior ranges from simple monogamy to heterosanguineous polygyny to consanguineous polygyny. Division of labor in gallery construction and maintenance is marked by sexual dimorphism, especially in structures on the head and elytral declivity.

Ecologically and economically this is a very important group of beetles. Members of *Dendroctonus* and *Ips* kill or degrade vast expanses of forest each year. Species of *Scolytus* are well known as vectors of the Dutch elm disease fungus. In the tropics, ambrosia beetles stain and degrade valuable wood products. In North America, several species of exotic xyleborines cause damage to young, stressed trees in the landscape and nurseries, and species of *Gnathotrichus, Monarthrum* and *Trypodendron* degrade wood products in the Pacific Northwest (Furniss and Carolin 1977).

There have been numerous studies on the biology, chemical ecology and control of many of the economically important genera.

Status of the classification. This book treats bark and ambrosia beetles as a subfamily of Curculionidae following Crowson (1967); however, the following Key and Classification of Tribes and Genera follow Wood (1973) and a family catalog by Wood and Bright (1992), but with the status of the subfamilies and tribes reduced to tribes and subtribes. See Wood (1973 and 1986) for a discussion of this issue.

Bark and ambrosia beetles occur on all continents except Antarctica. In North America, the fauna has been well studied within the past century, and is now well known. Wood (1982) published a monograph on the bark and ambrosia beetles of North and Central America, including a key to all genera and species in the region (at the time about 1430 species were recognized). Wood and Bright (1992) published a catalog of the worldwide Scolytidae, followed by a recent update (Bright and Skidmore 1997). The taxonomic status of tribes and genera in the Key and Classification sections below follows these catalogs. The two exceptions are the new genera *Dryoxylon* Bright and Rabaglia (1999) and *Pseudips* Cognato (2000) which have been added to the key.

Distribution. There are approximately 5,800 species worldwide, with about 525 species and subspecies described from the United States and Canada. Bark beetles can be found from the subalpine forests of the north to the subtropical forests of Florida. Distinctive faunas exist in the desert plateau of the southwest, the deciduous forests of the southeast, the northern coniferous forests, the Pacific Coast and southern Florida. Bark beetles tend to be more restricted by host than ambrosia beetles. Within a bark beetle genus, most species are restricted to a limited number of host species; for example, *Phloeosinus* are found mostly in Cupressaceae and *Pseudopityophthorus* are found almost exclusively in *Quercus*.

Wood (1977) estimated that there were 37 Old World species established in North and Central America. Since then approximately 10 additional species new to North America have been reported (Atkinson *et al.* 1990, Hoebeke 1991, Haack and Kucera 1993, Rabaglia and Cavey 1994, Bright and Rabaglia 1999, Vandenberg *et al.* 2000, Hoebeke 2001 and Mudge *et al.* 2001).

Key to the Nearctic Genera of Scolytinae
(Modified from D. E. Bright, unpublished 2000)

1. Anterior margins of elytra procurved and bearing a series of crenulations; pronotum unarmed in most; head visible from above (Fig. 102) (Hylesinini) 2

— Anterior margins of elytra forming a straight line across body, unarmed, smooth and either rounded or with a fine raised line; pronotum, in most, armed by granules or asperities on at least anterior third; head concealed from above (Fig. 103) (Scolytini) 24

2(1). Prothorax longitudinally strigose; prothoracic tibia with a curved bifid process, meso- and metathoracic tibiae with a single curved spine extending beyond spine of inner apical angle; antennal funicle 7-segmented; lateral prosternal area bearing a sharply elevated ridge from coxa to anterior margin; crenulations on elytral bases rather small (Bothrosternina) 3

— Prothorax punctate or asperate, never longitudinally strigose; all tibiae bearing several teeth, none extending beyond tarsal insertion; antennal funicle and prosternal area variable 4

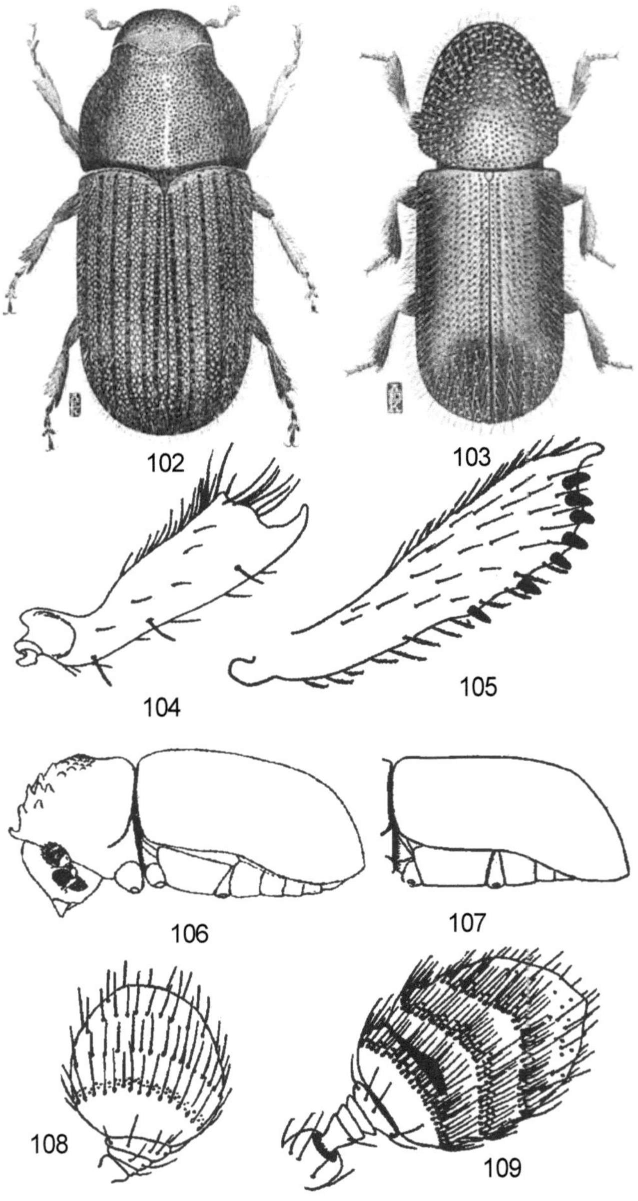

FIGURES 102.131-109.131. Scolytinae. 102-103. Dorsal habitus, 102. *Dendroctonus pseudotsugae* Hopkins; 103. *Dryocoetes affaber* Mannerheim. 104-105. Protibia, 104. *Scolytus* sp.; 105. *Procryphalus* sp. 106-107. Lateral habitus, 106. *Hypothenemus* sp.; 107. *Pityophthorus* sp. 108-109. Antennal club, 108. *Cryptocarenus* sp.; 109. *Hypothenemus* sp. (Figures 102-103 after Swaine 1918; Figure 104 after Chamberlin 1958; Figures 105-109 after Wood 1982.)

3(2). Sutures of antennal club straight; rostrum distinctly wider than distance between eyes; body and frons not as below .... *Cnesinus*
— Sutures of antennal club procurved; rostrum width at tip equal to distance between eyes; body oval; frons excavated, with median tubercle just above epistoma .... *Pagiocerus*

4(2). Prothoracic precoxal area rather large, lateral margin strongly elevated from anterior margin to coxa 5
— Prothoracic precoxal area short, lateral prosternal ridge poorly developed or absent 8

5(4). Crenulations on elytral bases forming a single row of teeth; first and second segments of antennal club subequal in length; body rather stout, length less than 2.5 mm; in roots of herbaceous legumes (Hylesinina, part) .... *Hylastinus*
— Crenulations on elytral bases obsolete, if visible, then irregularly placed, not forming a definite single row; first segment of antennal club distinctly longer than second; body mostly larger than 3 mm, very slender if smaller; not in herbaceous legumes (Hylastina) 6

6(5). Anterior coxae widely separated; surface of elytra and between punctures on pronotum dull; vestiture sparse, recumbent, yellow; body color dull reddish brown .... *Scierus*
— Anterior coxae narrowly separated, almost contiguous; surface between punctures on pronotum and elytra smooth and glossy; the longer hairlike vestiture erect; mature color glossy, dark brown or black 7

7(6). Pronotum, in most, constricted anteriorly, discal surface with about equal numbers of small and large punctures intermixed; third tarsomere broad, bilobed .... *Hylurgops*
— Pronotum not noticeably constricted anteriorly, discal surface with punctures uniformly large, with very few small punctures; third tarsomere narrower, emarginate .... *Hylastes*

8(4). Scutellum visible, elytral bases notched for its reception 9
— Scutellum not visible, elytral bases straight 20

9(8). Antennal club symmetrical, sutures transverse 10
— Antennal club with sutures oblique, pseudolamellate or absent 18

10(9). Pronotum asperate on anterolateral areas (Hylesinina, part) 11
— Anterolateral areas of pronotum unarmed 12

11(10). Eye entire; vestiture scalelike; costal margins of elytra ascending slightly at apex, abdomen ascending to meet them; hosts *Fraxinus* species .... *Hylesinus*
— Eye shallowly emarginate; vestiture hairlike; costal margins of elytra descending to apex, abdomen horizontal; hosts *Alnus* species ...... *Alniphagus*

12(10). Scutellar notch between elytra very deep, acute; elytra extended anteriorly over pronotum, posterolateral area of pronotum abruptly grooved to accommodate elytral margins; xylophagous species (Phloeosinina, part) .... *Dendrosinus*
— Scutellar notch between elytra emarginate, but not deeply grooved; elytra not extended anteriorly, pronotum not grooved; phloeophagous species (Tomicina) 13

13(12). Fore coxae widely separated 14
— Fore coxae contiguous, or at most very narrowly separated 16

14(13). Elytral vestiture hair-like; antennal club slightly flattened, segment 1 occupying one-fourth of club length; in *Ulmus* .... *Hylurgopinus*
— Elytral vestiture scale-like; antennal club conical, segment 1 occupying less than one-fourth of club length; in conifers 15

15(14). Each elytral interspace bearing a row of erect, flattened scales in addition to recumbent ground cover; antennal funicle 5-segmented .... *Xylechinus*
— Elytral interspaces bearing a row of erect, hairlike setae, ground cover scale-like or stout setae; antennal funicle 7-segmented .... *Pseudohylesinus*

16(13). Antennal funicle 5-segmented; antennal club with sutures slightly procurved; anterior margin of pronotum distinctly emarginate; 2.5-9.0 mm in length .... *Dendroctonus*
— Antennal funicle 6-segmented 17

17(16). Elytra with erect interstrial setae abundant, randomly placed; a short median carina on frons extending from epistomal margin to level of antennal insertion, ending dorsally in an acute elevation; elytra densely rugose .... *Hylurgus*
— Elytra with erect interstrial setae in uniseriate rows; a fine median carina on frons extending from epistoma to middle of frons, of equal height throughout; elytra smooth .... *Tomicus*

18(9). Antennal club pseudolamellate, constricted at sutures and movable at intersegmental lines (Phloeotribina) .... *Phloeotribus*
— Antennal club fused at sutures, sutures oblique or partly to entirely obsolete (Phloeosinina, part).. 19

19(18). Antennal club with three oblique sutures; funicle attached to base of club; pronotum unarmed; eye deeply emarginate; hosts Cupressinine trees, rarely other conifers .... *Phloeosinus*
— Antennal club solid and unmarked by sutures; funicle attached to side of club; pronotum, in most, armed by a few asperities in anterolateral areas; eye entire; hosts mostly hardwoods .... *Chramesus*

20(8). Eye emarginate or completely divided; pronotum never armed by asperities; crenulations at bases of elytra widely distributed, extending laterally beyond interstriae 5; antennal funicle 5- or 6-segmented (Polygraphina) 21
— Eye sinuate or entire; pronotum armed by a few scattered or clustered asperities; crenulations at bases of elytra restricted to area between suture and interstriae 5; antennal funicle 4- or 5-segmented (Hypoborina) 23

21(20). Eye completely divided into two parts; antennal club solid, unmarked by sutures ... *Polygraphus*
— Eye less than half divided by an emargination; antennal club marked by sutures 22

22(21). Antennal funicle 5-segmented .... *Carphoborus*
— Antennal funicle 6-segmented .... *Carphobius*

23(20). Antennal funicle 4-segmented, sutures of club indicated only by marginal notches; elytra with uniseriate rows of erect, broad interstrial scales and recumbent strial hair of equal length; pronotum armed by 3 or 4 pairs of median tubercles, the anterior pair marginal .. *Liparthrum*
— Antennal funicle 5-segmented, sutures of club transverse, distinct; elytral vestiture without conspicuous recumbent hair; pronotum armed by 2 or 3 widely separated paired clusters of lateral teeth .... *Chaetophloeus*

24(1). Lateral margin of anterior and posterior tibia unarmed except for a single curved process at outer apical angle that curves toward and extends beyond process of inner apical angle (Fig. 104); lateral line of pronotum sharply elevated; antennal club flattened, the sutures strongly procurved; antennal funicle 7-segmented (Scolytina) 25
— Lateral margin of anterior tibia armed by several toothlike processes, none of which curve toward the inner process (Fig. 105); lateral line of pronotum raised or not; antennal club and funicle variable.. 26

25(24). Elytra slightly if at all declivous behind, the abdomen ascending abruptly behind to meet them; scutellum depressed; antennal scape distinctly shorter than funicle .... *Scolytus*
— Elytral declivity rather steep, descending to meet the horizontal abdomen; scutellum small, flush with surface of elytra; antennal scape at least as long as funicle .... *Cnemonyx*

26(24). Metepisternum visible to posterior extremity (Fig. 106); antennal club varying from flat to obliquely truncate; tibia and antennal funicle variable 27
— Metepisternum largely covered by elytra, visible only in front (Fig. 107); antennal club strongly flattened with sutures on both sides, those on posterior surface not strongly displaced apically; tibia slender, in most, bearing about three teeth on apical portion; antennal funicle 1- to 5-segmented 66

27(26). Lateral margins of prothorax subacutely elevated; procoxae widely separated (Ctenophorina) .. 28
— Lateral margins of prothorax rounded; procoxae subcontiguous 29

28(27). Anterior area of pronotum transversely rugose; pronotum and elytra subglabrous .... *Scolytodes*
— Pronotum uniformly punctured, unarmed; vestiture of pronotum and elytra abundant, consisting of erect, stout, almost scalelike bristles .... *Pycnarthrum*

29(27). Fore tibia with sides parallel, in most, armed only on apical margin by small teeth never with process on outer apical angle exceeding tarsal insertion; procoxae separated; (Micracina) 30
— Fore tibia much wider apically, armed on lateral margin by several denticles; procoxae contiguous 36

30(29). Antennal club small, greatest width through basal half, apex narrowly rounded, sutures straight, transverse 31
— Antennal club larger, greatest width through apical half, apex broadly rounded, sutures procurved 32

31(30). Elytral declivity subvertical, bisulcate, obtusely angulate behind; sutures of antennal club distinctly marked by rows of setae; antennal pedicle and scape about equal in length ... *Stenocleptus*
— Elytral declivity more gradual, evenly convex, rather narrowly rounded behind; sutures of antennal club indicated only by marginal notches; scape distinctly longer than pedicle .... *Pseudothysanoes* (part)

32(30). Elytra broadly rounded behind; margins of antennal club, in most, constricted at first suture 33
— Elytra acuminate behind; antennal club without sutural constrictions at sides 34

33(32). Pronotum wider than long, widest near base, summit more prominent; fore tibia more slender, apically obliquely truncate, mucro often bifurcate .... *Psuedothysanoes* (part)
— Pronotum longer than wide, widest near middle, summit less prominent; fore tibia rather broad, more nearly truncate apically, mucro undivided .... *Thysanoes*

34(32). Sutures of antennal club broadly procurved, the first appearing bisinuate and extending less than one-third length of club; scape club-shaped, with few setae; eye oval, rather small; fore tibia more slender, slightly wider apically, with supplemental tubercles on posterior face .... *Hylocurus*
— Sutures of club very strongly, narrowly procurved, the first most often reaching middle of club; scape compressed, subtriangular, with numerous long setae; eye elongate, large; fore tibia broad, sides subparallel, posterior surface devoid of tubercles except for teeth on apical margin 35

35(34). Eyes moderately separated beneath, entire; fore tibia with all five teeth on distal margin, mucro broad .... *Micracis*
— Eyes subcontiguous beneath, emarginate; fore tibia with at least one of the five teeth on outer margin, mucro more slender .... *Micracisella*

36(29). Male frons bearing a very large, long, partly double process which may curve upward and backward over prothorax, in some, reaching its posterior margin; pronotum asperate to base in median area, summit on basal third, in most, extending behind its basal margin and over scutellum; body usually covered by an incrustation (Cactopinina) .... *Cactopinus*
— Male frons not armed by a large median process; pronotal summit at or slightly behind middle of prothorax, basal third devoid of asperities.... 37

37(36). Antennal club more strongly flattened, with sutures on both faces, those on posterior face strongly procurved and limited to apical half; costal margins of elytra slightly ascending posteriorly; vestiture scale-like (Cryphalina) 38

— Antennal club obliquely truncate or at least with sutures of posterior face restricted to less than apical one-fourth; costal margins of elytra descending posteriorly; vestiture hairlike setae... 46

38(37). Pronotum without a fine, raised lateral line; eye, in some, sinuate, never emarginate; costal margins of elytra ascending only slightly posteriorly 39

— Pronotum acutely margined at sides, and with a fine, raised line at least on basal one-third; eye emarginate or entire; costal margins of elytra distinctly ascending posteriorly 42

39(38). Antennal funicle 5-segmented; antennal club narrow, pointed at tip, sutures straight, not septate; basal half of pronotum without scale-like setae .... *Trypophloeus*

— Antennal funicle 4-segmented; antennal club broadly rounded at tip, sutures curved, partly septate or not septate; basal half of pronotum with scalelike setae 40

40(39). Antennal club not septate, sutures indicated by 3 strongly procurved rows of setae (Fig. 108) .... *Ernoporicus*

— Antennal club with at least part of first suture septate, none of sutures indicated by strongly procurved rows of setae (Fig. 109) 41

41(40). Sutures of antennal club straight, the first septate; anterior margin of pronotum slightly produced; pronotum with no indication of a fine raised lateral margin .... *Procryphalus*

— Antennal club with a strongly oblique septum on one side, no other sutures indicated; anterior margin of pronotum broadly rounded; pronotum with an indistinct, fine, raised lateral line .... *Scolytogenes*

42(38). Antennal club with sutures indicated by rather strongly recurved rows of setae; third tarsomere broad and emarginate .... *Cryphalus*

— Sutures of antennal club straight or procurved; third tarsomere cylindrical 43

43(42). Eye entire; antennal club large, aseptate, funicle normally 3-segmented, rarely 4-segmented; body stout, less than 2.3 times longer than wide; body shorter than 1.1 mm .... *Trischidias*

— Eye emarginate; antennal funicle 5-segmented, rarely 4- or 3-segmented; in most, body longer than 1.1 mm 44

44(43). Strial punctures obscure, not impressed; posterior half of pronotum finely granulate; antennal club large, not septate; male and female similar in size and appearance......................... *Hypocryphalus*

— Strial punctures distinct; posterior half of pronotum not closely granulate, in most, punctate; male much smaller than female 45

45(44). Antennal club not septate; raised lateral margin of pronotum extending two-thirds of distance from basal margin; elytra glabrous except for a few subcapitate interstrial bristles ... *Cryptocarenus*

— Antennal club with suture 1 partly septate; raised lateral margin extending only one-third of distance from basal to anterior lateral margin; elytra clothed by rows of strial and interstrial setae ... . *Hypothenemus*

46(37). Antennal funicle 2- or 3-segmented; pronotum unarmed, punctured over entire surface, lateral line not sharply raised; length 2.0 mm or less (Crypturgina) 47

— Antennal funicle 4- or 5-segmented; pronotum mostly armed anteriorly by granules or asperities, if unarmed, lateral line sharply raised; length mostly over 2.0 mm 48

47(46). Antennal funicle 2-segmented, club with 1 obscure suture indicated at tip .... *Crypturgus*

— Antennal funicle 3-segmented, club with 3 sutures .... *Dolurgus*

48(46). Eye completely divided by an emargination; antennal funicle 4-segmented, club without distinct sutures (Xyloterina) 49

— Anterior margin of eye sinuate or emarginate, never completely divided; antennal funicle 4- or 5-segmented, club, in most, with evident sutures 50

49(48). Antennal club with subcorneous basal area strongly, rather narrowly procurved; protibia of female thickened and tuberculate on posterior face, flattened and finely tuberculate in male; male head deeply, broadly excavated, the prothorax sub-quadrate; female frons convex, anterior margin of female pronotum rounded .... *Trypodendron*

— Antennal club with subcorneous basal area broadly procurved; protibia flattened and devoid of tubercles on posterior face; frons not excavated in either sex; anterior margin of prothorax rounded in both sexes .... *Xyloterinus*

50(48). Pronotum either punctate or else finely granulate over almost entire surface, dorsal profile evenly convex, not strongly declivous anteriorly, anterior margin never armed; tibia rather slender and armed by few, coarse teeth; declivity unarmed (Dryocoetina) 51

— Pronotum coarsely asperate and strongly declivous anteriorly, in most, punctate at least on posterior third, in some, anterior margin armed; tibia variable; declivity frequently armed by spinous processes 55

51(50). Antennal club compressed or with membranous apical portion extended beyond corneous portion, sutures procurved; scutellum very small 52

— Antennal club subtruncate, sutures transverse or recurved; scutellum moderate to large 53

52(51). Antennal funicle 4-segmented; club compressed, sutures strongly arcuate; pronotum granulate on anterior half, punctate behind; host *Acer* .... *Lymantor*

— Antennal funicle 5-segmented; club less strongly compressed, sutures rather broadly procurved; pronotum granulate to base; host *Cucurbita* .... *Dendrocranulus*

53(51). Frons convergently aciculate; elytral declivity evenly convex, extending over at least posterior one-third of elytra, granules absent; protibia armed on lateral margin by 2-4 socketed teeth; posterior face of antennal club with 2 sutures . ... *Coccotrypes*

— Frons never convergently aciculate; elytral declivity flattened or impressed, confined to posterior one-fourth of elytra, granules mostly present; protibia armed on lateral margin by 5 or more socketed teeth; posterior face of antennal club without sutures or with 1 suture 54

54(53). Pronotum 1.4 times longer than wide, anterior margin slightly notched or emarginate; elytral declivity moderately deeply, evenly sulcate .... *Dryoxylon*

— Pronotum 1.0-1.2 times longer than wide, anterior margin evenly rounded; elytral declivity evenly convex to slightly flattened, may have second interspace impressed .... *Dryocoetes*

55(50). Meso- and metathoracic tibia rather slender, abruptly narrowed apically, armed by a few rather widely spaced coarse teeth; males and females similar in size and general shape (Ipina) 56

— Meso- and metathoracic tibia rather broadly dilated to a point slightly beyond middle then gradually narrowed to apex, and armed by a series of small closely set teeth of more or less uniform size and shape; males rare, in most, smaller and radically different in shape (Xyleborina) 61

56(55). Elytral declivity rather narrowly bisulcate, margins moderately elevated, rounded and armed by not more than 3 teeth; lower margin of declivity rounded; in most, body shorter than 3.0 mm 57

— Elytral declivity broadly, rather deeply excavated, margins acutely elevated and armed by 3 or more tubercles or teeth; lower margin of declivity with an acutely elevated transverse ridge separating declivital excavation from apical margin; body mostly longer than 3.0 mm 58

57(56). Female frons deeply, rather narrowly excavated; male declivity with 2 or 3 pairs of enlarged teeth; antennal club compressed, 2 sutures visible on distal third of posterior face .... *Pityogenes*

— Female frons convex; male declivity more narrowly impressed with 2 or 3 pairs of very small teeth or granules; antennal club obliquely truncate, without sutures on posterior face .... *Pityokteines*

58(56). Antennal club obliquely truncate, with sutures recurved; elytral declivity less strongly excavated, the third tooth displaced mesally, not on summit of declivital margin .... *Orthotomicus*

— Antennal club flattened, with sutures procurved or strongly bisinuate; elytral declivity broadly excavated, armed by 3 to 6 major denticles, all denticles on summit of lateral margin 59

59(58). Lateral margins of elytral declivity armed by 4 to 6 pairs of spinelike denticles; ventrolateral margin of elytral declivity very strongly produced, circumscribing an arc much less than one-third of a circle, its lateral extremities ending a long distance from largest denticle; sutures 1 and 2 of antennal club weakly bisinuate to strongly angulate .... *Ips* (part)

— Lateral margins of elytral declivity armed by 3 pairs of spinelike denticles; ventrolateral margin of elytral declivity only slightly to moderately produced, circumscribing an arc at least one-third of a circle, its lateral extremities ending near third (last and largest) denticle; sutures 1 and 2 of antennal club weakly to very strongly, broadly procurved 60

60(59). Sutures on antennal club weakly procurved, almost straight; strial punctures at least twice as large as those of interstriae, in clearly defined rows; spine 3 on elytral declivity cylindrical or conical, not constricted before apex; body length 2.3-3.6 mm .... *Ips* (part, *latidens* group)

— Sutures on antennal club very strongly procurved; strial and interstrial punctures subequal in size, not always in clearly definable rows; spine 3 on declivity subcapitate, distinctly constricted before apex; body length 3.5-5.0 mm ...... *Pseudips*

61(55). Antennal club more strongly compressed, corneous area small, near base, its distal margin strongly procurved, distal pubescent portion reaching basal one-fifth at sides; pregula not impressed; elytra obliquely truncate behind, declivity broadly, concavely excavated and acutely margined on a complete circle at periphery .... *Premnobius*

— Antennal club thickened basally, corneous area larger with its distal margin recurved, pubescent area not reaching basal third; pregula depressed; elytral declivity convex, not acutely margined on upper half 62

62(61). Procoxae widely separated; body stout, elytra less than 1.3 times as long as pronotum .... *Xylosandrus*

— Procoxae contiguous; body elongate, often slender, elytra at least 1.5 times as long as pronotum 63

63(62). Pronotum wider than long, subquadrate, anterior margin unarmed 64

— Pronotum longer than wide, subcircular, anterior margin armed by a series of median serrations 65

64(63). Pronotum asperate to base; declivity steep, bearing several granules or rather large denticles, strial and interstrial punctures small .... *Ambrosiodmus*

— Pronotum asperate only on anterior half, punctate on basal half; declivity more sloping, bearing small tubercles, strial and interstrial punctures larger .... *Euwallacea*

65(63). Scutellum conical; lower margin of declivity, beginning about interspace 7, bearing a series of pointed tubercles, the one nearest suture (at end of interspace 2) largest .... *Xyleborinus*

— Scutellum flat; lower margin of declivity acute or rounded, unarmed .... *Xyleborus*

66(26). Antennal funicle 5-segmented (3-segmented in *Dendroterus*) club mostly small, symmetrical; pubescence more abundant; bark or twig beetles (Pityophthorina) 67

— Antennal funicle 1-, 2-, or 5-segmented, club much larger, asymmetrical in most; pubescence less abundant; ambrosia beetles (Corthylina) 73

67(66). Basal and lateral margins of prothorax rounded, without a fine raised line; antennal club somewhat large in size; vestiture shorter and more uniform in length 68

— Basal and posterior portion of lateral margins of prothorax with an obvious, fine, raised line; antennal club proportionately smaller; most with vestiture longer on declivity than on disc 69

68(67). Antennal funicle 3-segmented; club less than twice as long as funicle; female pronotum without patches of pilose pubescence; elytral pubescence abundant .... *Dendroterus*

— Antennal funicle 5-segmented, club at least twice as long as funicle; female prothorax with a pair of pilose pubescent areas on middle third of lateral areas; elytral pubescence sparse ... *Pityoborus*

69(67). Antennal club devoid of sutures except for one strongly oblique septum on anterior half of club only; prothorax evenly rounded in dorsal profile, summit inconspicuous, asperities fine, transition from asperate to punctate area gradual .... *Araptus*

— Antennal club with at least two complete sutures indicated at least by setae; prothorax more strongly declivous anteriorly, summit and arrangement of asperities variable 70

70(69). Sutures of antennal club not septate; in most, pronotal asperities extending behind middle at sides, the transition from asperate to punctate area gradual; body moderately to very stout .... *Conophthorus*

— First and second sutures of antennal club septate; pronotal asperities mostly not reaching middle, the transition from asperate to punctate area usually abrupt, summit usually well developed; body slender to moderately stout 71

71(70). Pronotum and elytra minutely densely punctured; vestiture very short, mostly dense, almost always scalelike; antennal club with first segment shorter than others; greater development of frontal vestiture a male character; hosts *Quercus*, rarely other broadleaf trees ...... *Pseudopityophthorus*

— Pronotum and elytra more coarsely, less densely punctured; vestiture usually longer, less abundant, always hairlike; greater development of frontal vestiture a female character; hosts usually conifers, but also broadleaf trees and shrubs ... 72

72(71). Pregular area greatly enlarged and ornamented by a beard-like brush of exceedingly long hair...... .... *Pityotrichus*

— Pregular area small, without conspicuous vestiture .... *Pityophthorus*

73(66). Antennal funicle 5-segmented, club smaller, less than twice as long as funicle ..... *Gnathotrichus*

— Antennal funicle 1- or 2-segmented; club very large, more than three times as long as funicle 74

74(73). Antennal funicle 2-segmented; posterior surface of fore tibia tuberculate; elytra emarginate or divaricate at sutural apex .... *Monarthrum*

— Antennal funicle 1-segmented; posterior surface of fore tibia smooth; elytra evenly rounded behind, without a sutural notch at apex .... *Corthylus*

Classification of the Nearctic Scolytinae

88. Hylesinini Erichson 1836

Hylastina Leconte 1876

Scierus LeConte 1876, 2 spp., northern and western North America in *Picea*; usually found in the phloem of roots and stumps of standing dead trees or next to the ground in boles of downed trees.

Hylurgops LeConte 1876, 6 spp., 2 with subspecies, throughout coniferous forests of North America; all species breed in the phloem of stumps, roots and souring logs. The genus is closely related to *Hylastes* from which some species are distinguished with difficulty.

Hylesinites Germar 1813
Hylastities Hagedorn 1906
Myelophites Hagedorn 1906
Hylescierites Schedl 1947

Hylastes Erichson 1836, 14 spp. in Pinaceae throughout America north of Mexico, *H. opacus* Erichson 1836 is an adventive from Europe. All species breed in the phloem of stumps and roots.

Hylesinina Erichson 1836

Hylastinus Bedel 1888, 1 sp., *H. obscurus* (Marsham 1802), native to Palearctic, now found throughout North America. Breeds in roots of legumes, especially *Trifolium* species.

Alniphagus Swaine 1918, 2 spp. in western North America (an additional species occurs in east Asia). All species breed in phloem of *Alnus* species.

Hylastinoides Spessivtev 1919

Hylesinus Fabricius 1801, 7 spp. throughout America north of Mexico in mostly *Fraxinus* hosts. Adults and larvae deeply mine the wood in the phloem-cambial area. Adults construct biramous galleries and larvae mine parallel to the grain of the wood.

Leperisinus Reitter 1913
Apidocephalus Wickham 1916

Tomicina Thomson 1859

Hylurgopinus Swaine 1918, 1 sp., *H. rufipes* (Eichhoff 1868) occurs east of the Rocky Mountains. This phloeophagous species breeds in large branches and boles of *Ulmus,* and is a vector of the Dutch elm disease fungus.

Pseudohylesinus Swaine 1917, 9 spp., 2 with subspecies, occur in western North America (2 additional species occur in Mexico). They breed in the phloem of limbs, boles and roots of weakened conifers.

Xylechinus Chapuis 1869, 2 spp. occur in northern and western North America coincident with their *Picea* hosts. They are phloeophagous in small, weakened trees.

Pruniphagus Murayama 1958
Squamosinus Nunberg 1964
Xylechinops Browne 1973

Hylurgus Latreille 1807, 1 sp., *H. ligniperda* (Fabricius 1787), native to Europe, was recently found in cut pine stumps in New York State (Hoebeke 2001). All species are native to Palearctic.

Tomicus Latreillle 1802, 1 sp., *T. piniperda* (L. 1758), native to Palearctic, was first found in North America in 1992. It is now recorded from the Lake States, Maine, Maryland, New Hampshire, New York, Pennsylvania, Vermont, West Virginia, Ontario and Quebec. Adults feed in shoots of *Pinus* and breed in boles of weakened or downed trees.

Blastophagus Eichhoff 1864
Myelophilus Eichhoff 1878

Dendroctonus Erichson 1836, 13 spp. found throughout America north of Mexico. Most species breed in the boles of conifers and some are capable of killing healthy hosts. Species in this genus are among the most economically important bark beetles.

Bothrosternina Blandford 1896

Cnesinus LeConte 1868, 1 sp., *C. strigicollis* LeConte 1868, in southeast United States and Mexico (an additional 100 species occur from Mexico to Argentina). Twigs and small woody stems are selected for attack. Adults bore through the bark and into the wood, normally reaching the pith. Larvae feed in the center of twigs extending the parental gallery.

Nemophilus Chapuis 1869

Pagiocerus Eichhoff 1868, 1 sp., *P. frontalis* (Fabricius 1801) occurs north of South America from North Carolina to Mexico (additional species occur in South America.). This species infests large seeds of trees and other plants, especially corn.

Phloeotribina Chapuis 1869

Phloeotribus Latreille 1796, 9 spp. occur north of Mexico; 2 in the west and 7 in the east, especially in the southeast. Adults breed in the phloem-cambial region of hosts. *P. liminarus* (Harris 1852) occasionally is a pest of *Prunus.*

Phloeophthorus Wollaston 1854
Dryotomus Chapuis 1869
Phthorophloeus Rey 1885
Elzearius Guillebeau 1893
Eulytocerus Blandford 1897
Comesiella DelGuercio 1925
Neophleotribus Eggers 1943
Dryotomicus Wood 1962

Phloeosinina Nusslin 1912

Dendrosinus Chapuis 1869, 1 sp., *D. bourreriae* Schwarz 1920, in the Florida Keys (nine additional species occur in Central and South America). Adults and larvae feed in the wood of small woody plants.

Phloeosinus Chapuis 1869, 25 spp., 3 of which occur in the east and the remainder in the west. All species, except *P. pini* Swaine 1915, attack Cupressaceae and Taxodiaceae. Adults construct longitudinal galleries under the bark that usually deeply engrave the wood.

Phloeosinites Hagedorn 1906

Chramesus LeConte 1868, 9 spp. are found north of Mexico. These small beetles are phloeophagous in twigs and small branches of hardwood trees and shrubs.

Rhopalopleurus Chapuis 1869
Thaumasinulus Reitter 1913
Prochramesus Wood 1956

Hypoborina Nusslin 1911

Chaetophloeus LeConte 1876, 9 spp., eight in western North America and one in the Florida Keys and adjacent islands. All species attack branches and twigs. Long larval mines radiate from the parental gallery, deeply engraving the xylem and phloem.

Renocis Casey 1886
Pseudocryphalus Swaine 1917

Liparthrum Wollaston 1854, 2 spp., one in Arizona and one in Mississippi and Indiana. These phloeophagous species are very small and attack small twigs of woody plants.

Erineosinus Blackman 1920
Phloeochilus Schedl 1953
Phloeotrypetus Wood 1960
Dacryophthous Schedl 1971
Trypanophellos Bright 1982

Polygraphina Chapuis 1869

Polygraphus Erichson 1836, 3 spp. of the 60 worldwide species occur in North America. They are phloeophagous in recently broken, cut or fallen *Picea.*

Lepisomus Kirby 1837

Spongotarsus Hagedorn 1908
Pseudopolygraphus Seitner 1911
Ozophagus Eggers 1919
Nipponopolygraphus Nobuchi 1981

Carphoborus Eichhoff 1864, 9 spp. in the 48 states and one additional species in northern Canada and Alaska. All are phloeophagous in small or broken branches of Pinaceae.
Estenoborus Reitter 1913

Carphobius Blackman 1943; 1 sp., *C. arizonicus* Blackman 1943, in Arizona, extends north from Central America. Two additional species occur in Central America. They are phloeophagous in small broken branches of conifers.

89. Scolytini Latreille 1807

Scolytina Latreille 1807

Cnemonyx Eichhoff 1868, 2 spp. of this Neotropical genus are found in the Florida Keys. They are phloeophagous in woody hosts.
Ceratolepis Chapuis 1869
Loganius Chapuis 1869
Minulus Eggers 1912
Coptodryas Schedl 1948
Coptosomus Schedl 1952

Scolytus Geoffroy 1762, 20 spp. found throughout America north of Mexico. Native western species are found in conifers, while most eastern species are in hardwoods. Three Palearctic species are established in North America, most notably, *S. multistriatus* (Marsham 1802), which transmits the Dutch elm disease fungus. All are phloeophagous and construct characteristic galleries under the bark.
Ekkoptogaster Herbst 1793
Coptogaster Illiger 1807
Eccoptogaster Gyllenhal 1813
Scolytochelus Reitter 1913
Ruguloscolytus Butovitsch 1929
Archaeoscolytus Butovitsch 1929
Spinuloscolytus Butovitsch 1929
Tubuloscolytus Butovitsch 1929
Pygmaeoscolytus Butovitsch 1929
Pinetoscolytus Butovitsch 1929
Confusoscolytus Tsai and Huang 1962

Ctenophorina Chapuis 1869

Pycnarthrum Eichhoff 1878, 1 sp., *P. hispidum* (Ferrari 1867), infests *Ficus* limbs and boles in south Florida and Texas.
Nemobius Chapuis 1869
Monebius Hopkins 1914
Nomebius Navas 1915

Scolytodes Ferrari 1867, 1 sp., *S. schwarzi* (Hopkins 1902) infests *Ficus* in south Florida. Approximately 100 spp. occur in Central and South America.
Hexacolus Eichhoff 1868
Ctenophrus Chapuis 1869
Prionosceles Blandford 1897
Epomadius Blandford 1897
Erinophlius Hopkins 1902
Hylocurosoma Eggers 1940
Hexacolinus Schedl 1963

Micracina LeConte 1876

Pseudothysanoes Blackman 1920, 19 spp. throughout the United States, one of which extends into Canada; most inhabit arid areas in the western states, three species are found in the east and southeast (approximately 60 additional species occur in Central America). Within this genus a wide variety of hosts are attacked. Several western species breed in the phloem of dying mistletoe (*Phoradendron*), other species occur in the twigs of hardwood trees. This is a very diverse genus, and several species groups were previously treated as distinct genera. Species keying out to couplet 31 in this section's key were previously recognized as the genus *Cryptocleptus*. Species with the antennal scape short and broadly expanded are placed in the subgenus *Aphanocleptus*, and those with an elongate and slender antennal scape are placed in the subgenus *Psuedothysanoes*.
Cryptocleptes Blackman 1920
Chalcohyus Blackman 1943
Bostrichips Schedla 1951
Gretschkinia Sokanovskii 1959
Aphanocleptus Wood 1960
Cryptulocleptus Wood 1967
Neoglostatus Schedl 1978

Stenoclyptus Blackman 1943, 1 sp. in U.S., *S. sulcatus* (Bruck 1936). Two species in the genus, one in California and one in Mexico. The genus is closely related to *Pseudothysanoes*. They are phloeophagous in small branches of woody plants.

Thysanoes LeConte 1876, 7 spp. across the southern United States, 1 species extends north to Illinois and Pennsylvania. Apparently they are xylophagous in small branches of trees.

Hylocurus Eichhoff 1872, 15 spp. north of Mexico, most of which occur in the southeast (more than 40 additional species occur in Central and South America). The *rudis* group needs further study; Atkinson (1989) suggests the synonymy of some species. All species are xylophagous in small branches.
Micracisoides Blackman 1920

Micracisella Blackman 1928, 5 spp. in eastern and southern United States. These small (1.0-2.5 mm) beetles breed in the pith of damaged, small twigs.
Pseudomicracis Blackman 1920

Micracis LeConte 1868, 4 spp. in the United States, 1 extends to Canada, 2 are known only from Arizona. They are xylophagous in twigs.

Cactopinina Chamberlin 1939

Cactopinus Schwarz 1899, 5 spp. in southwestern United States, additional species occur in Mexico. The unique, paired epistomal male horns distinguish this genus. They are phloeophagous in woody plants, but more commonly feed subepidermally in *Cereus* and related cacti.

Cactopinorus Bright 1967

Ipina Bedel 1888

Pityogenes Bedel 1888, 7 spp. across the United States and Canada. One species, *P. bidentatus* (Herbst 1784), is native to the Palearctic. The North American species of this primarily Eurasian genus are phloeophagous in branches, limbs and boles of *Pinus*.

Eggersia Lebedev 1926
Pityoceragenes Balachowsky 1947

Pityokteines Fuchs 1911, 6 spp. in North America, one of which, *P. sparsus* (LeConte 1868) occurs in the east. They often construct star-shaped galleries in the phloem of limbs and boles of dying trees. This genus is closely related to *Orthotomicus*. Various Pinaceae serve as hosts.

Othotomides Wood 1951

Orthotomicus Ferrari 1867, 1 sp. found across North America, *Orthotomicus caelatus* (Eichhoff 1868), is phloeophagous in *Pinus, Picea and Larix* (about 10 species are known from the Palearctic).

Neotomicus Fuchs 1911

Ips DeGeer 1775, 23 spp. plus subspecies are currently recognized from across North America. Some species placed in synonomy by Wood (1982) are recognized as valid species (Lanier 1987, Lanier *et al.* 1991). Species in this relatively large genus have been put into various species groups by several workers (Hopping 1963, Lanier 1970a, 1970b, 1972, Wood 1982, Cognato and Sperling 2000). Cognato and Vogler (2001) recently revised *Ips* as monophyletic with the removal of the *latidens* group and their tentative placement in *Orthotomicus*. In addition, they also named four subgenera for monophyletic groups of *Ips* species. This well known and important genus is phloeophagous in *Pinus* and *Picea*. Most breed in dying trees and slash, but some may attack the boles and tops of healthy trees. Characteristic egg galleries engrave the phloem-cambial area.

Cumatomicus Ferrari 1867
Cyrtotomicus Ferrari 1868

Pseudips Cognato 2000, 2 spp. in North America and 1 species in Asia. Cognato (2000) used molecular, morphological and behavioral characters to separate these species from *Ips*. The two North American species occur in the west where they are phloeophagous on *Picea* (*Pseudips concinnus* (Mannerheim 1852)) and *Pinus* (*Pseudips mexicanus* (Hopkins 1905)).

Dryocoetina Lindemann 1876

Dendrocranulus Schedl 1937, 3 spp. in southern and western United States. All species infest stems of Cucurbitaceae. The genus is closely related to the Old World *Xylocleptes* Ferrari.

Lymantor Lovendal 1889, 1 sp. in eastern United States and Canada and 1 species in Alaska. These beetles are phloeophagous in small, dry, often dead, branches of *Acer* and, rarely, other hosts.

Dryocoetes Eichhoff 1864, 7 spp. in United States and Canada. They are phloeophagous in the boles of mostly conifers, except *D. betulae* Hopkins 1915, which infests the bole of *Betula*.

Anodius Motschulsky 1860
Dryocoetinus Balachowsky 1949

Dryoxylon Bright and Rabaglia 1999, 1 sp., *D. onoharaensum* (Murayama 1934), native to Japan, originally described as a *Xyleborus*, is established in southeastern United States. Little is known about the biology, but it appears to feed in the xylem (Bright and Rabaglia 1999). Normark *et al.* (1999) discussed the genetic affinities of this genus and other Dryocoetini to Xyleborini.

Coccotrypes Eichhoff 1878, 9 spp. are known from United States, mostly Florida and California. This genus contains many species, mostly from southeast Asia and Africa, and species found in most other areas, including the United States, have arrived through commerce (Wood 1986). Females mate with dwarfed siblings before they emerge to seek a new host. They most often infest large seeds; however, a few species are phloeophagous. Wood (1986) stated that this genus is "in a state of taxonomic chaos". Jordal *et al.* (2000) showed the genetic relatedness of the genus to Xyleborini.

Poecilips Schaufuss 1897
Cryphaloides Formanek 1908
Thamnurgides Hopkins 1915
Spermatoplex Hopkins 1915
Dendrurgus Eggers 1923

Crypturgina LeConte 1876

Dolurgus Eichhoff 1868, 1 sp. is known from western North America. *Dolurgus pumilus* (Mannerheim 1843) occurs from Alaska to California where it breeds in dying *Picea*. It utilizes the entrance holes of larger bark beetles, and its galleries are often wholly in the bark.

Crypturgus Erichson 1836, 3 spp. occur in America north of Mexico, one of which, *C. pusillus* (Gyllenhal 1813), is native to Europe and Asia. They utilize the entrance holes of other beetles to gain access to the phloem in the boles of conifers.

Xyloterina Lindemann 1876

Trypodendron Stephens 1830, 5 spp. in North America, additional species occur in Europe and Asia. These are monogamous ambrosia beetles that breed in either conifers or hardwoods. *Trypodendron lineatum* (Olivier 1795), which occurs across North America and into northern Europe and Asia, is often a pest of conifer logs in processing yards.
Xyloterus Erichson 1836

Xyloterinus Swaine 1918, 1 sp., *Xyloterinus politus* (Say 1826), is recognized in the genus, which is found throughout eastern North America. This monogamous ambrosia beetle is commonly found attacking weakened hardwood trees.

Xyleborina LeConte, 1876

Premnobius Eichhoff 1878, 1 sp., *P. cavipennis* Eichhoff 1878, from Africa is found in Florida. This genus is unique within the Xyleborini. Browne (1961) treated it as a distinct tribe, and Normark *et al.* (1999), using DNA, showed a separate origin from Xyleborini and a closer relationship to Ipini. Males of these ambrosia beetles are flightless and mate with siblings (consanguineous polygyny) before the females leave the brood gallery.
Premnophilus Browne 1962

Ambrosiodmus Hopkins 1915, 7 spp. occur in the eastern United States, mostly in the southeast. They are consanguineously polygynous in a wide variety of hosts. Most attacks occur in the lower bole and stumps of trees.
Phloeotrogus Motschulsky 1863
Brownia Nunberg 1963

Euwallacea Hopkins 1915, 1 sp., *E. validus* (Eichhoff 1875), native to Asia, is now established in the eastern United States. It is a consanguineously polygynous ambrosia beetle that breeds in the stumps and boles of hardwoods and conifers.

Xyleborus Eichhoff 1864, (Vandenberg *et al.* 2000, key to eastern United States species); 17 spp. are identified from America north of Mexico, 5 of which are native to Europe and Asia. Most United States species are found in the east. More than 500 species are described from the neotropics, Africa and Asia. Representatives of this large and important genus attack almost all parts of woody plants. Most of these ambrosia beetles attack declining trees, but some may attack apparently healthy plants. Flightless, haploid males mate with sibling or parental females within the brood galleries before emergence. Extreme inbreeding and partial parthenogenesis may be the cause of the many morphological races and species. In addition, this mating system has allowed for new founder populations to be easily distributed through commerce (Atkinson *et al.* 1990). The generic and tribal limits of these rapidly radiating species need taxonomic revision. Jordal *et al.* (2000) and Normark *et al.* (1999) showed genetic relatedness to Dryocoetini and Wood (1986) suggests a relationship with Xyloterini.
Anisandrus Ferrari 1867
Anaertus Duges 1887
Progenius Blandford 1896
Heteroborips Reitter 1913
Xyleborips Reitter 1913
Boroxylon Hopkins 1915
Notoxyleborus Schedl 1934

Xylosandrus Reitter 1913, 4 spp. in eastern North America, 3 of which are native to Asia. The three exotic species are becoming very common, and occasionally aggressively attack apparently healthy, small trees. *Xylosandrus compactus* (Eichhoff 1875) often attacks healthy, vigorous twigs of living trees. All species cultivate ambrosia fungi and are consanguineously polygynous.
Apoxyleborus Wood 1980

Xyleborinus Reitter 1913, 3 spp. occur in America north of Mexico, 2 are exotic. *Xyleborinus saxeseni* (Ratzburg 1837), native to Europe, is found across the United States; *X. alni* (Niisima), from Europe and Asia, has recently been found on the west coast of North America (Mudge *et al.* 2001). The genus was often treated as a subgenus or synonym of *Xyleborus,* but it is morphologically distinct. Their biology is similar to *Xyleborus*, attacking limbs and boles of weakened trees.

Cryphalina Lindemann 1876

Trypophloeus Fairmaire 1868, 4 spp. in northern and western North America. These small, less than 2 mm, beetles are monogamous and phloeophagous in the bark of thin-barked limbs and boles of *Alnus*, *Salix* and *Populus*.
Glyptoderes Eichhoff 1878

Procryphalus Hopkins 1915, 2 spp. in western North America, one additional species in Asia. Biology is similar to *Trypophloeus*.

Ernoporicus Berger 1917, 1 sp., *E. kanawhae* Hopkins 1915, known only from the type series taken in flight in West Virginia.
Eocryphalus Kurenzov 1941
Ernopocerus Balachowsky 1949

Scolytogenes Eichhoff 1878, 1 sp., *S. knabi* (Hopkins 1915), occurs in vines in south Florida. Many other species are found in subtropical and tropical areas of the world.
Lepicerus Eichhoff 1878
Cryphalomorhpus Schaufuss 1891
Letznerella Reitter 1913
Hypothenoides Hopkins 1915
Neocryphalus Eggers 1922
Negritus Eggers 1923
Cylindrotomicus Eggers 1936
Lepicerinus Hinton 1936
Cryphalophilus Schedl 1970
Xylocryptus Schedl 1975

Hypocryphalus Hopkins 1915, 1 sp., *H. mangiferae* (Stebbing 1914), native to Asia, occurs in mango, *Mangifera*, in south Florida. They are phloeophagous in branches of their host.

Dacryphalus Hopkins 1915

Cryphalus Erichson 1836, 3 spp. occur in conifers in northern and western North America. They are generally less than 2 mm and infest declining branches and small trees. Adults construct cave-type galleries in the phloem. Several hundred nominate species occur in Asia to Australia, and a worldwide taxonomic revision is needed (Wood 1986).

Pseudocryphalus Ferrari 1868
Taenioglyptes Bedel 1888
Cryptarthrum Blandford 1896
Allarthrum Hagedorn 1912
Ericryphalus Hopkins 1915
Piperius Hopkins 1915
Ernocryphalus Murayama 1958
Acryphalus Tsai and Li 1963
Jugocryphalus Tsai and Li 1963

Cryptocarenus Eggers 1937, 2 spp. are found in south Texas and Florida and extend through Central and South America. Males are flightless in these consanguineous polygynous pith borers of small twigs.

Tachyderes Blackman 1943

Hypothenemus Westwood 1836, 21 spp. have been recorded from the United States, many of which are native to Asia or Africa. Most United States species occur in the southern half of the country. These small, less than 2 mm, beetles infest twigs, vines, pith, seeds and other plant material. They are consanguineously polygynous and have been widely distributed through commerce. Over 200 species have been assigned to this genus, and species identification is often difficult.

Stephanoderes Eichhoff 1872
Homoeocryphalus Lindemann 1876
Triarmocerus Eichhoff 1878
Adiaeretus Hagedorn 1909
Stylotentus Schedl 1939
Chondronoderes Schedl 1940
Archeophalus Schedl 1941
Pachynoderus Schedl 1941
Lepiceroides Schedl 1957
Ernophloeus Nunberg 1958
Epsips Beeson 1941
Macrocryphalus Nobuchi 1981

Trischidias Hopkins 1915, 5 spp. occur in the southeastern United States. These very small, less than 1 mm, beetles are relatively rare. One species feeds on fungus pustules under the bark of mangrove, and others are phloeophagous in injured, often fungus-infested twigs.

Pityophthorina Eichhoff 1878

(This group has been treated as a subytribe of Corthylina by Wood and Bright (1992).)

Dendroterus Blandford 1904, 2 spp. in United States, one in Texas in *Jatropha* and one in California in *Bursera*. They are phloeophagous in the bark of declining branches.

Plesiophthorus Schedl 1940
Xylochilus Schedl 1956

Araptus Eichhoff 1872, 1 sp., *A. dentifrons* Wood 1974, occurs in south Florida (Atkinson and Peck 1994) and possibly Texas. This Neotropical species breeds in the pith of vines. *Araptus politus* (Blandford 1904) has been intercepted in large seeds in the port of Miami, but it is not known to be established.

Neodryocoetes Eggers 1933
Thamnophthorus Schedl 1938
Neopityophthorus Schedl 1938
Sphenoceros Schedl 1939
Hypertensus Hagedorn 1950
Brachydendrulus Schedl 1951
Gnathocranus Schedl 1951
Gnathoborus Schedl 1970

Conophthorus Hopkins 1915, 8 spp. are currently recognized from America north of Mexico, 2 spp. occur in the east and 6 in the west. All species breed in the cones of *Pinus*.

Pityoborus Blackman 1922, 2 spp. in United States, one in southeast and one in southwest. They are phloeophagous in dying branches of *Pinus*. Their galleries in the cambium deeply score the xylem.

Pityotrichus Wood 1962, 2 spp. in southwest United States (Arizona and New Mexico). These species are distinguished from closely related *Pityophthorus* by the unique pregula referenced in the key. They are monogamous and feed in the phloem of small branches.

Pityophilus Blackman 1928

Pseudopityophthorus Swaine 1918, 11 spp. across America north of Mexico. All species breed in branches or boles of *Quercus*, except *P. fagi* Blackman 1931 which is found in *Fagus*.

Xenophthorus Wood and Yin 1986

Pityophthorus Eichhoff 1864, 104 spp. are recognized north of Mexico, more than 200 additional species occur in Central and South America and more than 50 in Europe, Asia and Africa. This large and diverse genus is found throughout the United States in many different hosts. Representatives may be found breeding in twigs, seedlings, boles or pith. Most are heterosanguinously polygynous and some are monogamous.

Trigonogenius Hagedorn 1912
Hagedornus Lucus 1920

Myeloborus Blackman 1928
Gnathophorus Schedl 1935
Conophthocranulus Schedl 1935
Breviophthorus Schedl 1938
Pityophthoroides Blackman 1942
Cladoborus Sawamoto 1942
Neomips Schedl 1954
Ctenyophthorus Schedl 1955
Gnathophthorus Wood 1962
Hypopityophthorus Bright 1981

Corthylina LeConte 1876

Gnathotrichus Eichhoff 1869, 7 spp. north of Mexico. *G. materiarius* (Fitch 1858) occurs in *Pinus* throughout eastern North America, the remaining species are in the west in oaks and conifers. They are monogamous ambrosia beetles breeding in dying or fallen trees or logs. Some species are pests in wood processing yards, especially in the Pacific Northwest.
Gnathotrichoides Blackman 1931
Ancyloderes Blackman 1938
Paraxyleborus Hoffman 1942
Prognathotrichus Bright 1972

Monarthrum Kirsch 1866, 5 spp. north of Mexico, 2 spp. throughout the east and 3 spp. in the west. More than 100 additional species are found in Central and South America. These ambrosia beetles attack logs and boles of dying hardwoods, especially oaks.
Corthylomimus Ferrari 1867
Cosmocorynus Ferrari 1867
Pterocyclon Eichhoff 1869
Anchonocerus Eichhoff 1878
Phthorius Eichhoff 1878
Trypocranus Eichhoff 1878
Eupteroxylon Eggers 1936

Corthylus Erichson 1836, 3 spp. in United States, one of which occurs in Canada; all are found east of the Rocky Mountains. Approximately 100 additional species occur in Central and South America. These ambrosia beetles breed in a variety of locations on a tree. *Corthylus papulans* Eichhoff 1869, in Florida, breeds in small branches; *C. punctatissimus* (Zimmermann 1868) breeds in sapling trees, especially *Acer*, near ground level and *C. columbianus* Hopkins 1895 breeds in the xylem of living trees, usually *Acer*, which survive after the brood emerges.
Morizus Ferrari 1867
Pseudocorthylus Ferrari 1867
Corthylomimus Schedl 1972

XVIII. Platypodinae Shuckard 1840

by Robert S. Anderson

Platypodinae are an enigmatic group that have been recognized either as a distinct family or a subfamily within Curculionidae. There are 4 genera in North America based on the recent division of the genus *Platypus* into a variety of smaller genera (Wood 1993). Traditionally they have been closely allied with Scolytinae, but Lyal (1995) could not find support for a monophyletic group comprised only of scolytines and platypodines nor could he find support for them having a separate ancestry from Curculionidae. Similarly, Thompson (1992) chose to give Platypodidae family level status while at the same time considering Scolytinae as a subfamily within Curculionidae. A review of their phylogenetic position is given by Kuschel *et al.* (2000).

Platypodinae are easily recognized by the lack of a rostrum, presence of pregular sutures, pregular sclerite distinct, located between median gular suture and labial articulation, at least one pair of tibiae with denticles or stout socketed setae along the dorsal (outer) margin, tarsus with article 1 as long as articles 2-5 combined, pronotum usually with a lateral constriction near the middle and the antennal club without sutures (Fig. 131).

Where known, adults and larvae infest the wood of dead or recently cut or dying trees. Larvae mine galleries deep into the wood which become stained black by ambrosia fungi which grow on the walls of the tunnels and serve as the larval food (Bright 1993).

Key to the Nearctic genera of Platypodinae

1. Metasternum and metepisternum near hind coxa weakly or not impressed for reception of femur, anterior margin of impressed area not continuously carinate or with a row of small spines, surface of impressed area with at least some setae .... *Treptoplatypus*
— Metasternum and metepisternum near hind coxa impressed for reception of femur, anterior margin of impressed area either continuously carinate or with a series of small spines, surface of impressed area glabrous 2

2(1). Male with ventrite 3, 4 or 5 simple, not armed with spines .... *Euplatypus*
— Male with ventrite 3, 4 or 5 with a pair of widely separated coarse spines 3

3(2). Male with ventrite 3 with a pair of spines; female with mycetangia pores moderate in size .... *Myoplatypus*
— Male with ventrite 4 with a pair of spines; female with mycetangia pores unusually large in size . ... *Oxoplatypus*

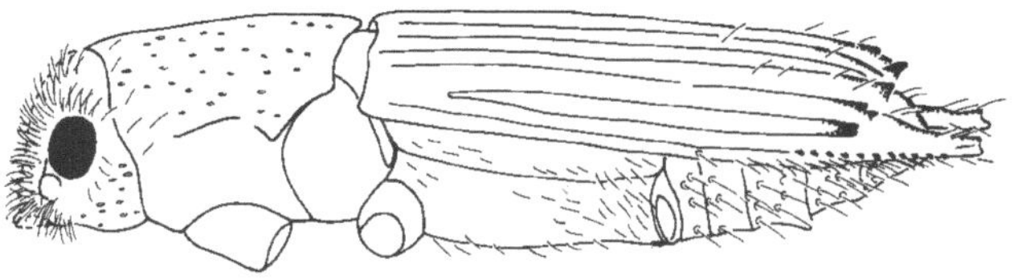

FIGURE 110.131. Platypodinae. 110. *Myoplatypus flavicornis* (Fabricius), lateral habitus.

Classification of the Nearctic Platypodinae

90. Platypodini Shuckard 1840

Treptoplatypus Schedl 1972, 2 spp., *T. abietis* (Wood 1958) and *T. wilsoni* (Swaine 1916), northwestern United States and British Columbia. Species are associated with timber of conifers (Bright 1993).

Myoplatypus Wood 1993, 1 sp., *M. flavicornis* (Fabricius 1776), southern Florida.

Oxoplatypus Wood 1993, 1 sp., *O. quadridentatus* (Olivier 1795), southeastern United States. This species is associated with various species of *Quercus* (oak; Fagaceae) (Wood 1993).

Euplatypus Wood 1993, 3 spp., *E. parallelus* (Fabricius 1801), *E. compositus* (Say 1824) and *E. pini* (Hopkins 1905), southern United States; one species adventive.

Bibliography

AHMAD, M. and H. R. BURKE. 1972. Larvae of the weevil tribe Anthonomini (Coleoptera: Curculionidae). Miscellaneous Publications of the Entomological Society of America, 8(2): 33-81.

AINSLIE, G. G. 1910. The cowpea curculio. United States Department of Agriculture Bureau of Entomology, Bulletin, 85: 133-146.

ALONSO-ZARAZAGA, M. A. and C. H. C. LYAL. 1999. A world catalogue of families and genera of Curculionoidea (Insecta: Coleoptera) (Excepting Scolytidae and Platypodidae). Entomopraxis. Barcelona, Spain. 315 pp.

ALSTERLUND, J. F. 1937a. Notes on the biology of the partridge-pea weevil, *Chalcodermus collaris* Horn (Coleop., Curculionidae). Entomological News, 48: 31-35.

ALSTERLUND, J. F. 1937b. The larva of *Chalcodermus collaris* Horn with key to related species. Proceedings of the Entomological Society of Washington, 39: 216-222.

ANDERSON, D. M. 1962. The weevil genus *Smicronyx* in America north of Mexico (Coleoptera: Curculionidae). Proceedings of the United States National Museum, 113: 185-372.

ANDERSON, D. M. 1973. Keys to larvae and pupae of the Gymnetrinae of America north of Mexico (Coleoptera: Curculionidae). Proceedings of the Entomological Society of Washington, 75: 133-140.

ANDERSON, R. S. 1984. *Connatichela artemisiae*, a new genus and species of weevil from the Yukon Territory (Coleoptera: Curculionidae; Leptopiinae): Taxonomy, paleontology and biogeography. Canadian Entomologist, 116: 1571-1580.

ANDERSON, R. S. 1988a. Systematics, phylogeny and biogeography of New World weevils of the tribe Cleonini (Coleoptera: Curculionidae). Quaestiones Entomologicae, 23: 431-709.

ANDERSON, R. S. 1988b. The Curculionidae of the Queen Charlotte Islands, British Columbia (Coleoptera) Canadian Journal of Zoology, 66: 2406-2414.

ANDERSON, R. S. 1989a. A revision of the subfamily Rhynchaeninae in the New World (Coleoptera: Curculionidae). Transactions of the American Entomological Society, 115: 207-312.

ANDERSON, R. S. 1989b. New synonymy in North American *Stephanocleonus* Motschulsky. Coleopterists Bulletin, 43: 93.

ANDERSON, R. S. 1991. A new species of *Plocetes* from the Florida Keys with notes on other species occurring in the United States (Curculionidae; Curculioninae; Tychiini). Florida Entomologist, 74: 105-110.

ANDERSON, R. S. 1993a. The Curculionoidea of southern Florida: An annotated checklist (Coleoptera: Curculionoidea [excluding Curculionidae; Scolytinae, Platypodinae]). Insecta Mundi, 6: 193-248.

ANDERSON, R. S. 1993b. Weevils and plants: Phylogenetic versus ecological mediation of evolution of host plant associations in Curculionidae (Curculioninae). Memoirs of the Entomological Society of Canada, 165: 197-232.

ANDERSON, R. S. 1993c. A new species of *Leptopinara* from west Texas (Coleoptera: Curculionidae; Brachycerinae; Entimini). Coleopterists Bulletin, 47: 203-206.

ANDERSON, R. S. 1994. A review of New World weevils associated with Viscaceae (mistletoes [in part]) including descriptions of new genera and new species (Coleoptera: Curculionidae). Journal of Natural History, 28: 435-492.

ANDERSON, R. S. 1995. An evolutionary perspective on diversity in Curculionoidea. Memoirs of the Entomological Society of Washington 14, 103-114.

ANDERSON, R. S. 1997. Weevils (Coleoptera: Curculionoidea, excluding Scolytinae and Platypodinae) of the Yukon. Pp. 523-562. *In*: H. V. Danks and J. A. Downes, eds. Insects of the Yukon. Biological Survey of Canada (Terrestrial Arthropods) Ottawa. 1034 pp.

ANDERSON, R. S. 1998. Nomenclatural changes in New World Curculionidae (Curculioninae [Cryptorhynchini, Molytini], Cossoninae). Coleopterists Bulletin, 52: 285-290.

ANDERSON, R. S. and H. R. BURKE. 1990. Natural history of weevils of the genus *Cionopsis* Champion, with description of the larva of *Cionopsis maculata* Burke. Southwestern Entomologist, 15: 101-107.

ANDERSON, R. S. and A. T. HOWDEN. 1994. *Tychius meliloti* Stephens new to Canada with a brief review of the species of *Tychius* Germar introduced into North America (Coleoptera: Curculionidae). Canadian Entomologist, 126: 1363-1368.

ANDERSON, W. H. 1948. Larvae of some genera of Calendrinae (=Rhynchophorinae) and Stromboscerinae (Coleoptera: Curculionidae). Annals of the Entomological Society of America, 41: 413-437.

ANDERSON, W. H. 1952. Larvae of some genera of Cossoninae (Coleoptera: Curculionidae). Annals of the Entomological Society of America, 45: 281-309.

ATKINSON, T.H. 1989. New synonomy, new species and notes on Scolytidae (Coleoptera) from southeastern United States. Coleopterists Bulletin, 43:325-337.

ATKINSON, T.H. and S.B. PECK. 1994. Annotated checklist of the bark and ambrosia beetles (Coleoptera: Platypodidae and Scolytidae) of tropical southern Florida. Florida Entomologist, 77:313-329.

ATKINSON, T.H., R.J. RABAGLIA and D.E. BRIGHT. 1990. Newly detected exotic species of *Xyleborus* (Coleoptera: Scolytidae) with a revised key to species in eastern North America. Canadian Entomologist, 122:92-104.

BARBER, H. S. 1917. Notes and descriptions of some orchid weevils. Proceedings of the Entomological Society of Washington, 19: 12-22.

BARBER, H. S. 1935. The tobacco and solanum weevils of the genus *Trichobaris*. United States Department of Agriculture. Miscellaneous Publication No. 226. pp. 1-28.

BLATCHLEY, W. S. 1918. The home of *Hormops* and its proper position among other Rhynchophora. Journal of the New York Entomological Society, 26: 155-161.

BLATCHLEY, W. S. 1922. Notes on the Rhynchophora of eastern North America, with characterizations of new genera and descriptions of new species. Journal of the New York Entomological Society, 30: 95-106, 113-127.

BLATCHLEY, W. S. 1925. Notes on the Rhynchophora of eastern North America with descriptions of new species, III. Journal of the New York Entomological Society, 33: 87-113.

BLATCHLEY, W. S. 1928. Notes on the Rhynchophora of eastern North America with descriptions of new species, IV. Journal of the New York Entomological Society, 36: 235-262.

BLATCHLEY, W. S. and C. W. LENG. 1916. Rhynchophora or weevils of North Eastern America. The Nature Publishing Company. Indianapolis, IN. 682 pp.

BORDEN, J.H. 1982. Aggregation pheromones. Pp 74-139. *In*: J. B. Mitton and K. B. Sturgeon, eds. Bark Beetles in North American Conifers. University of Texas Press. Austin.

BOROVEC, R. 1989. Les espèces françaises du genre *Trachyphloeus* Germar 1824. Pp. 383-416. *In*: G. Tempère and J. Péricart, eds. Coléoptères. Curculionides (Quatrième Partie). Faune de France. Vol. 74. Fédération Française des Sociétés de Sciences Naturelles, Paris. pp. 1-534.

BRIGHT, D. E. 1993. The Insects and Arachnids of Canada. Part 21. The weevils of Canada and Alaska: Volume 1. Coleoptera: Curculionoidea, excluding Scolytidae and Curculionidae. Publication 1882. Research Branch, Agriculture Canada. Ottawa, Canada.

BRIGHT, D. E. 1994. Revision of the genus *Sitona* (Coleoptera: Curculionidae) of North America. Annals of the Entomological Society of America, 87: 277-306.

BRIGHT, D.E. and R. J. RABAGLIA. 1999. *Dryoxylon*, a new genus for *Xyleborus onoharaensis* Murayama, recently established in the southeastern United States (Coleoptera: Scolytidae). Coleopterists Bulletin, 53: 333-337.

BRIGHT, D. E. and R. E. SKIDMORE. 1997. A catalog of Scolytidae and Platypodidae (Coleoptera), Supplement 1 (1990-1994). NRC Research Press. Ottawa, Ontario Canada. 368p.

BROWN, W. J. 1932. New species of Coleoptera III. Canadian Entomologist, 64: 3-12.

BROWN, W. J. 1950. The extralimital distribution of some species of Coleoptera. Canadian Entomologist, 82: 197-205.

BROWN, W. J. 1965. *Trachyphloeus* Germar (Coleoptera: Curculionidae) in North America. Canadian Entomologist, 97: 189-192.

BROWN, W. J. 1966a. Chrysomelinae and Curculionidae (Coleoptera); descriptions and notes. Canadian Entomologist, 98: 855-859.

BROWN, W. J. 1966b. The species of *Steremnius* Schoenherr (Coleoptera: Curculionidae). Canadian Entomologist, 98: 586-587.

BROWN, W. J. 1967. Notes on the extralimital distribution of some species of Coleoptera. Canadian Entomologist, 99: 85-93.

BROWNE, F. G. 1961. The generic characters, habits and taxonomic status of *Premnobius* Eichh. (Coleoptera, Scolytidae). West African Timber Borer Research Unit Report, 4:45-51.

BUCHANAN, L. L. 1927. A short review of *Notaris* (Coleoptera: Curculionidae) Bulletin of the Brooklyn Entomological Society, 22: 36-40.

BUCHANAN, L. L. 1929a. North American species of the weevils of the otiorhynchid genus *Mesagroicus*. Proceedings of the United States National Museum, 76: 1-14.

BUCHANAN, L. L. 1929b. A new *Agronus* from Canada (Coleoptera: Otiorhynchidae). Proceedings of the Entomological Society of Washington, 31: 102-104.

BUCHANAN, L. L. 1931. Synopsis of *Perigaster* (Coleoptera: Curculionidae). Journal of the Washington Academy of Sciences, 21: 320-325.

BUCHANAN, L. L. 1932. A new barine curculionid injurious to sugarcane in Louisiana with synopses of *Anacentrinus* and *Oligolochus*. Annals of the Entomological Society of America, 25: 328-336.

BUCHANAN, L. L. 1935. A new genus and species of orchid weevils (Coleoptera, Curculionidae, Barinae). Proceedings of the Hawaiian Entomological Society, 9: 45-48.

BUCHANAN, L. L. 1936a. Synopsis of *Lepidophorus* (Coleoptera, Curculionidae). Bulletin of the Brooklyn Entomological Society, 31: 1-10.

BUCHANAN, L. L. 1936b. The genus *Panscopus* Schoenherr (Coleoptera: Curculionidae). Smithsonian Miscellaneous Collections, 94(16): 1-18.

BUCHANAN, L. L. 1937. Notes on Curculionidae (Coleoptera). Journal of the Washington Academy of Sciences, 27: 312-316.

BUCHANAN, L. L. 1946. Notes on American *Rhyncolus*, with description of a new species (Coleoptera, Curculionidae). Bulletin of the Brooklyn Entomological Society, 41: 129-136.

BUCHANAN, L. L. 1948. A new species of *Stenoscelis*, and notes on other Curculionidae (Coleoptera). Bulletin of the Brooklyn Entomological Society, 43: 61-66.

BUCKINGHAM, G. R. and C. A. BENNETT. 1981. Laboratory biology and behavior of *Litodactylus leucogaster*, a ceutorhynchine weevil that feeds on watermilfoils. Annals of the Entomological Society of America, 74: 451-458.

BURKE, H. R. 1960. A new genus and two new species of weevils from Texas with notes on others (Curculionidae). Coleopterists Bulletin, 14: 121-127.

BURKE, H. R. 1961a. Review of the *limatulus-setosus* group of the genus *Endalus* in America north of Mexico (Coleoptera: Curculionidae). Bulletin of the Brooklyn Entomological Society, 56: 9-19.

BURKE, H. R. 1961b. Notes on *Onychylus* LeConte with descriptions of two new species (Curculionidae). Coleopterists Bulletin, 15: 1-7.

BURKE, H. R. 1963. New species of Texas weevils with notes on others (Coleoptera: Curculionidae). Southwestern Naturalist, 8: 162-172.

BURKE, H. R. 1965. A new species of *Endalus* in the *limatulus-setosus* group with notes on *Endalus celatus* Burke (Coleoptera, Curculionidae). Southwestern Naturalist, 10: 9-13.

BURKE, H. R. 1968. Biological and taxonomic notes on *Brachyogmus ornatus*, with descriptions of larval and pupal stages (Coleoptera: Curculionidae). Coleopterists Bulletin, 22: 126-132.

BURKE, H. R. 1971. Observations on the life history and habits of *Endalus celatus* Burke (Coleoptera: Curculionidae). Coleopterists Bulletin, 25: 63-65.

BURKE, H. R. 1973. Taxonomic relationships and biology of *Macrorhoptus* (Curculionidae). Coleopterist Bulletin, 27: 175-181.

BURKE, H. R. 1975. A new species of *Smicraulax* from Mexico, with key to species of the genus (Coleoptera: Curculionidae). Entomological News, 86: 167-171.

BURKE, H. R. 1976. Bionomics of the anthonomine weevils. Annual Review of Entomology, 21: 283-303.

BURKE, H. R. 1981. Review of the genus *Cionomimus* Marshall with descriptions of two new species (Coleoptera: Curculionidae). Southwestern Entomologist, 6: 174-183.

BURKE, H. R. 1982. The genus *Cionopsis* Champion: New species, key and taxonomic notes (Coleoptera: Curculionidae). Southwestern Entomologist, 6: 288-297.

BURKE, H.R. and D. M. ANDERSON. 1976. Systematics of larvae and pupae of American Curculionoidea: Status report, historical review and bibliography. Southwestern Entomologist, 1: 56-73.

BURKE, H. R. and B. S. RECTOR. 1976. A new anthonomine genus and species from Texas and Mexico, with larval and pupal descriptions and biological notes (Coleoptera: Curculionidae). Journal of the Kansas Entomological Society, 49: 541-550.

CASEY, T. L. 1892. Coleopterological notices. IV. Annals of the New York Academy of Sciences, 6: 359-712.

CASEY, T. L. 1895. Coleopterological notices. VI. Annals of the New York Academy of Sciences, 8: 435-838.

CASEY, T. L. 1897. Coleopterological notices. VII. Annals of the New York Academy of Sciences, 9: 285-681.

CASEY, T. L. 1920. Some descriptive studies among the American Barinae. Memoirs on the Coleoptera, 9: 300-516.

CAWTHRA, E. M. 1957. Description of a new species of *Grypidius* Schoenherr (Col.: Curculionidae) with a key to the genus. Proceedings of the Royal Entomological Society of London, Series B, 26: 127-130.

CHAMBERLIN, W. J. 1958. Scolytoidea of the Northwest: Oregon, Washington, Idaho and British Columbia. Oregon State Monograph 2:1-208. Oregon State College, Corvallis, OR.

CHAMPION, G. C. 1903. Biologia Centrali-Americana. Insecta. Coleoptera. Rhynchophora. Curculionidae. Curculioninae (part), volume 4, part 4, pp. 145-312.

CHAMPION, G. C. 1908. Biologia Centrali-Americana. Insecta. Coleoptera. Rhynchophora. Curculionidae. Curculioninae (part), volume 4, part 5, pp. 241-400.

CLARK, W. E. 1971. A taxonomic revision of the weevil genus *Tychius* Germar in America north of Mexico (Coleoptera: Curculionidae). Brigham Young University Science Bulletin (Biological Series), 13(3): 1-39.

CLARK, W. E. 1977. North American *Tychius*: New synonymy and observations of phylogeny and zoogeography (Coleoptera: Curculionidae). Entomologica Scandinavica, 8: 287-300.

CLARK, W. E. 1978. The weevil genus *Sibinia* Germar: Natural history, taxonomy, phylogeny and zoogeography, with revision of the New World species (Coleoptera: Curculionidae). Quaestiones Entomologicae, 14: 91-387.

CLARK, W. E. 1980a. Revision of the weevil genus *Neotylopterus* Hustache (Coleoptera: Curculionidae). Annals of the Entomological Society of America, 73: 216-230.

CLARK, W. E. 1980b. Revision of Nearctic weevils of the genus *Lignyodes* Dejean (Coleoptera: Curculionidae). Transactions of the American Entomological Society, 106: 273-326

CLARK, W. E. 1981. The genus *Chionanthobius* Pierce (Coleoptera: Curculionidae): Descriptions of a new species on *Forestiera* (Oleaceae) and of the larva and pupa of *C. schwarzi* Pierce. Proceedings of the Entomological Society of Washington, 83: 185-197.

CLARK, W. E. 1982. Classification of the weevil tribe Lignyodini (Coleoptera, Curculionidae, Tychiinae), with a revision of the genus *Plocetes*. Transactions of the American Entomological Society, 108: 11-151.

CLARK, W. E. 1987a. Revision of the *Anthonomus* subgenus *Anthomorphus* Weise (Coleoptera: Curculionidae). Quaestiones Entomologicae, 23: 317-364.

CLARK, W. E. 1987b. Revision of the *unipustulatus* group of the weevil genus *Anthonomus* Germar (Coleoptera: Curculionidae). Coleopterists Bulletin, 41: 73-88.

CLARK, W. E. 1987c. Revision of the Nearctic species of *Pseudanthonomus* Dietz (Coleoptera: Curculionidae). Coleopterists Bulletin, 41: 263-285.

CLARK, W. E. 1988. The species of *Anthonomus* in the *albolineatus* group (Coleoptera: Curculionidae). Transactions of the American Entomological Society, 113: 309-359.

CLARK, W. E. 1990. Revision of the *Anthonomus* subgenus *Anthonomocyllus* Dietz (Coleoptera: Curculionidae). Quaestiones Entomologicae, 26: 559-600.

CLARK, W. E. 1991a. Revision of the *Anthonomus alboannulatus* and *Anthonomus triensis* species groups (Coleoptera: Curculionidae). Coleopterists Bulletin, 45: 206-226.

CLARK, W. E. 1991b. The *Anthonomus curvirostris* species group (Coleoptera: Curculionidae). Transactions of the American Entomological Society, 117: 39-66.

CLARK, W. E. 1993a. *Huaca* Clark, a new genus of Neotropical Anthonomini (Coleoptera: Curculionidae). Transactions of the American Entomological Society, 119: 1-45.

CLARK, W. E. 1993b. The weevil genus *Neomastix* Dietz (Coleoptera: Curculionidae, Anthonomini). Coleopterists Bulletin, 47: 1-19.

CLARK, W. E. and H. R. BURKE. 1985. Revision of the *venustus* species group of the weevil genus *Anthonomus* Germar (Coleoptera: Curculionidae). Transactions of the American Entomological Society, 111: 103-170.

CLARK, W. E. and H. R. BURKE. 1986. Revision of the *gularis* species group of the genus *Anthonomus* Germar (Coleoptera: Curculionidae). Coleopterists Bulletin, 40: 1-26.

CLARK, W. E. and H. R. BURKE. 1996. The species of *Anthonomus* Germar (Coleoptera: Curculionidae) associated with plants in the family Solanaceae. Southwestern Entomologist, Supplement No. 19: 1-114.

CLARK, W. E. and H. R. BURKE. (In Press a). Revision of the weevil genus *Epimechus* Dietz (Coleoptera: Curculionidae: Anthonomini). Insecta Mundi.

CLARK, W. E. and H. R. BURKE. (In Press b). Revision of the weevil genera *MagdalinopsDietz* and *Chelonychus* Dietz (Coleoptera: Curculionidae: Anthonomini). Coleopterists Bulletin.

COGNATO, A. I. 2000. Phylogenetic analysis reveals new genus of Ipini bark beetle (Scolytidae). Annals of the Entomological Society of America, 93:362-366.

COGNATO, A. I. and F. A. H. SPERLING. 2000. Phylogeny of *Ips* DeGerr species inferred from mitochondrial cytochrome oxidase I sequence. Molecular Phylogenetics and Evolution, 14:445-460.

COGNATO, A. I. and A. P. VOGLER. 2001. Exploring data interaction and nucleotide alignment in a multiple gene analysis of *Ips* (Coleoptera: Scolytidae). Systematic Biology, in press.

CÔTÉ, S. and D. E. BRIGHT. 1995. Premières mentions canadiennes de *Phyllobius intrusus* Kôno (Coleoptera: Curculionidae) et tableaux de détermination des espèces de *Phyllobius* et *Polydrusus* au Canada. Fabreries, 20: 81-89.

CROWSON, R. A. 1967 [1955]. The natural classification of the families of Coleoptera. Classey. Hampton, England. [Reprinted from Entomological Monograph Magazine 1950-1954 and from Lloyd, London 1955, 214 pp.]

CUDA, J. P. and H. R. BURKE. 1985. Biology and impact of *Trichobaris texana* (Coleoptera: Curculionidae) on silverleaf nightshade, *Solanum elaeagnifolium* in central Texas, Pp. 721-734. *In*: E. S. Delfosse, ed. Proceedings VI International Symposium on Biological Control of Weeds, 19-25 August 1984, Vancouver, Canada.

DeCLERK-FLOATE, R. and P. HARRIS. (In Press). *Linaria dalmatica* (L.) Miller, Dalmatian Toadflax (Scrophulariaceae). *In*: P. G. Mason and J. T. Huber, eds, Biological Control Programmes in Canada, 1981-2000. CABI Publishing, Wallingford, Oxon, UK.

DeCLERK-FLOATE, R. and M. SCHWARZLÄNDER. (In Press). *Cynoglossum officinale* (L.), Houndstongue (Boraginaceae). *In*: P. G. Mason and J. T. Huber, eds, Biological Control Programmes in Canada, 1981-2000. CABI Publishing, Wallingford, Oxon, UK.

DeLOACH, C. J. 1975. Identification and biological notes on the species of *Neochetina* that attack Pontederiaceae in Argentina (Coleoptera: Curculionidae: Bagoini). Coleopterists Bulletin, 29: 257-265.

DIETZ, W. G. 1891. Revision of the genera and species of Anthonomini inhabiting North America. Transactions of the American Entomological Society, 18: 177-276.

DIETZ, W. G. 1896. Revision of the genera and species of Ceutorhynchini inhabiting North America. Transactions of the American Entomological Society, 23: 387-480.

DOWNIE, N. M. and R. H. ARNETT, Jr. 1996. The Beetles of Northeastern North America. Volume II. Polyphaga: Series Bostrichiformia through Curculionoidea. The Sandhill Crane Press. Gainesville, FL. i-ix, pp. 891-1721.

FALL, H. C. 1901. Notes on *Dichelonycha* and *Cantharis*, with descriptions of new species in other genera. Transactions of the American Entomological Society, 27: 277-310.

FALL, H. C. 1907. New species of American Coleoptera of the tribe Zygopini. Transactions of the American Entomological Society, 32: 53-56, 57-61. [1906]

FALL, H. C. 1913. A brief review of our species of *Magdalis*, with notes and descriptions of other North American Rhynchophora. Transactions of the American Entomological Society, 39: 23-72.

FALL, H. C. 1917. New Coleoptera. VII. Canadian Entomologist, 49: 385-391.

FALL, H. C. and T. D. A. COCKERELL. 1907. The Coleoptera of New Mexico. Transactions of the American Entomological Society, 33: 145-272.

FORD, E. J. 1985. Descriptions of the larva and pupa of *Rhynchus apiculatus*, with biological notes (Coleoptera: Curculionidae). Coleopterists Bulletin, 39: 29-32.

FRANK, J. H. and M. C. THOMAS. 2000. Weevils that eat bromeliads. http://bromeliadbiota.ifas.ufl.edu/wvbrom.htm

FRANKLIN, R. T. and J. W. TAYLOR, Jr. 1970. Biology of *Pachylobius picivorus* (Coleoptera: Curculionidae) in the Georgia Piedmont. Canadian Entomologist, 102: 962-968.

FURNISS, R.L. and V.M. CAROLIN. 1977. Western Forest Insects. Miscellaneous Publication No. 1339, USDA Forest Service.

GIBSON, L. P. 1969. Monograph of the genus *Curculio* in the New World (Coleoptera: Curculionidae). Part I. United States and Canada. Miscellaneous Publications of the Entomological Society of America, 6(5): 239-285.

GILBERT. E. E. 1956. The raymondionymine weevils of California, with a description of a new genus, and several new species (Coleoptera: Curculionidae). Pan-Pacific Entomologist, 32: 55-72.

GILBERT. E. E. 1964. The genus *Baris* Germar in California (Coleoptera: Curculionidae). University of California Publications in Entomology, 34: 1-153.

HAACK, R. and D. KUCERA. 1993. Common pine shoot beetle. USDA Forest Service, Northeastern Area, Pest Alert. NA-TP-05-93.

HARRIS, P. 2001. *Hylobius transversovittatus* Goeze. Root-feeding weevil. http://res2.agr.ca/lethbridge/weedbio/agents/ahyltra.htm.

HARRIS, P., R. DeCLERCK-FLOATE and A. McCLAY. 2001. *Mecinus janthinus* Germar. Stem boring weevil. http://res2.agr.ca/lethbridge/weedbio/agents/amecinus.htm.

HATCH, M. H. 1971. The Beetles of the Pacific Northwest. Part V. Rhipiceroidea, Sternoxi, Phytophaga, Rhynchophora, and Lamellicornia. University of Washington Publications in Biology, 16: xiv + 662 pp.

HESPENHEIDE, H. A. 1992. A review of the genus *Tachygonus* (Coleoptera: Curculionidae) north of Mexico. Proceedings of the Entomological Society of Washington, 94: 1-11.

HOEBEKE, E.R. 1991. An Asian ambrosia beetle, *Ambrosiodmus lewisi*, new to North America (Coleoptera: Scolytidae). Proceedings of the Entomological Society of Washington, 93:420-424.

HOEBEKE, E.R. 2001. *Hylurgus ligniperda*: A new exotic pine bark beetle in the United States. Newsletter of the Michigan Entomological Society, 46: 1-2.

HOEBEKE, E. R. and D. R. WHITEHEAD. 1980. New records of *Rhinoncus bruchoides* (Herbst) for the Western Hemisphere and a revised key to the North American species of the genus *Rhinoncus* (Coleoptera: Curculionidae: Ceutorhynchinae). Proceedings of the Entomological Society of Washington, 82: 556-561.

HOFFMANN, A. 1954. Coléoptères. Curculionides (Deuxième Partie). Faune de France. Vol. 59. Fédération Française des Sociétés de Sciences Naturelles, Paris. Pp. 487-1208.

HOFFMANN, A. 1958. Coléoptères. Curculionides (Troisième Partie). Faune de France. Vol. 62. Fédération Française des Sociétés de Sciences Naturelles, Paris. Pp. 1209-1839.

HOPKINS, A. D. 1911. Contributions toward a monograph of the bark-weevils of the genus *Pissodes*. United States Department of Agriculture, Bureau of Entomology, Technical Series, 20: i-x + 1-68.

HOPPING, G.R. 1963. The natural groups of *Ips* DeGeer (Coleoptera: Scolytidae). Canadian Entomologist, 95: 508-516.

HORN, G. H. 1873. Contributions to a knowledge of the Curculionidae of the United States. Proceedings of the American Philosophical Society, 13: 407-489.

HOWDEN, A. T. 1959. A revision of the species of *Pandeleteius* Schoenherr and *Pandeleteinus* Champion of America north of Mexico (Coleoptera: Curculionidae). Proceedings of the California Academy of Sciences, Series 4, 29(10): 361-421.

HOWDEN, A. T. 1961. A revision of the genus *Isodacrys* Sharp (Curculionidae, Tanymecini). Coleopterists Bulletin, 15: 75-95.

HOWDEN, A. T. 1982. Revision of the New World genus *Hadromeropsis* Pierce (Coleoptera, Curculionidae, Tanymecini). Contributions of the American Entomological Institute, 19(6): i-iii, 1-180.

HOWDEN, A. T. 1992. Review of the New World eyeless weevils with uncinate tibiae (Coleoptera, Curculionidae; Molytinae, Cryptorhynchinae, Cossoninae). Memoirs of the Entomological Society of Canada, 162: 1-76.

HOWDEN, A. T. 1993. A catalog of the Coleoptera of America north of Mexico. Family Curculionidae. Subfamily Polydrosinae. Tribe Tanymecini. United States Department of Agriculture, Agriculture Handbook Number 529-143e: x + 13 pp.

HOWDEN, A. T. 1995. Structures related to oviposition in Curculionoidea. Memoirs of the Entomological Society of Washington, 14: 53-100.

HULL SIEG, C. and C. W. O'BRIEN. 1993. *Stethobaris commixta* Blatchley (Coleoptera: Curculionidae) collected from a species of orchid, *Platanthera praeclara* Sheviak and Bowles, in North Dakota tall-grass prairie. Prairie Naturalist, 25: 81.

JORDAL, B.H., B.B. NORMARK and B.D. FARRELL. 2000. Evolutionary radiation of an inbreeding haplodiploid beetle lineage (Curculionidae, Scolytinae). Biological journal of the Linnean Society, 71: 483-499.

KIRKLAND, R. L. and R. D. GOEDEN. 1977. Descriptions of the immature stages of imported puncturevine weevils, *Microlarinus lareynii* and *M. lypriformis*. Annals of the Entomological Society of America, 70: 583-587.

KIRKLAND, R. L. and R. D. GOEDEN. 1978a. Biology of *Microlarinus lareynii* (Col.: Curculionidae) on puncturevine in southern California. Annals of the Entomological Society of America, 71: 13-18.

KIRKLAND, R. L. and R. D. GOEDEN. 1978b. Biology of *Microlarinus lypriformis* (Col.: Curculionidae) on puncturevine in southern California. Annals of the Entomological Society of America, 71: 65-69.

KISSINGER, D. G. 1958. A new name for *Zaglyptus* LeConte, 1876 Forster, 1868) and a review of the North American species (Curculionidae, Baridinae). Coleopterists Bulletin, 11: 47-52.

KISSINGER, D. G. 1960. Description of a new species of *Miloderes* Casey with notes on some broad-nosed weevils (Curculionidae). Coleopterists Bulletin, 14: 25-28.

KISSINGER, D. G. 1963. Notes on the habits of some Curculionidae. Coleopterists Bulletin, 17: 53-57.

KISSINGER, D. G. 1964. Curculionidae of America north of Mexico. A key to genera. Taxonomic Publications, South Lancaster, MA. vi + 143 pp.

KISSINGER, D. G. 1970. Curculionidae tribe Ophryastini of North America (Coleoptera). Taxonomic Publications, South Lancaster, MA. v + 238 pp.

KOK, L. T. 1998. *Rhinocyllus conicus* (Coleoptera: Curculionidae). Biological Control: A guide to Natural Enemies in North America. Cornell University. http://www.nysaes.cornell.edu/ent/biocontrol/weedfeeders/rhinocyllus_c.html

KOROTYAEV, B. A. and A.B. EGOROV. 1976. Review of the weevil tribe Emphyastini (Coleoptera, Curculionidae) habitants on supralittoral of the Sea of Japan, Ochotian and Bering Seas. Proceedings of the Zoological Institute of the Academy of Sciences of the USSR, 77: 43-55 (in Russian).

KOVARIK, P. W. and H. R. BURKE. 1989. Observations on the biology and ecology of two species of *Eudiagogus* (Coleoptera: Curculionidae). Southwestern Naturalist, 34: 196-212.

KUSCHEL, G. 1952. Revisión de *Lissorhoptrus* LeConte y géneros vecinos de América (Ap. 11 de Coleoptera Curculionidae). Revista Chilena de Entomología, 1: 23-74.

KUSCHEL, G. 1961. On problems of synonymy in the *Sitophilus oryzae* complex (30th contribution, Col. Curculionoidea). Annals and Magazine of Natural History, Series 13, 4: 241-244.

KUSCHEL, G. 1995. A phylogenetic classification of Curculionoidea to families and subfamilies. Memoirs of the Entomological Society of Washington, 14: 5-33.

KUSCHEL, G., R. A. B. LESCHEN and E. C. ZIMMERMAN. 2000. Platypodidae under scrutiny. Invertebrate Taxonomy, 14: 771-805.

LANG, R. F. 1997a. *Larinus minutus* (Coleoptera: Curculionidae). Biological Control: A guide to Natural Enemies in North America. Cornell University. http://www.nysaes.cornell.edu/ent/biocontrol/weedfeeders/larinus_minutus.html.

LANG, R. F. 1997b. *Larinus obtusus* (Coleoptera: Curculionidae). Biological Control: A guide to Natural Enemies in North America. Cornell University. http://www.nysaes.cornell.edu/ent/biocontrol/weedfeeders/larinus_obtusus.html

LANG, R. F. 1997c. *Cyphocleonus achates* (Coleoptera: Curculionidae). Biological Control: A guide to Natural Enemies in North America. Cornell University. http://www.nysaes.cornell.edu/ent/biocontrol/weedfeeders/cyphocleonus.html

LANG, R. F. 1997d. *Bangasternus fausti* (Coleoptera: Curculionidae). Biological Control: A guide to Natural Enemies in North America. Cornell University. http://www.nysaes.cornell.edu/ent/biocontrol/weedfeeders/bangasternus_fausti.html

LANIER, G.N. 1970a. Biosystematics of the genus *Ips* (Coleoptera: Scolytidae) in North America: Hopping's group IX. Canadian Entomologist, 102:1139-1163.

LANIER, G.N. 1970b. Biosystematics of the genus *Ips* (Coleoptera: Scolytidae) in North America: Hopping's group III. Canadian Entomologist, 102:1404-1423.

LANIER, G.N. 1972. Biosystematics of the genus *Ips* (Coleoptera: Scolytidae) in North America: Hopping's groups IV and X. Canadian Entomologist, 104:361-388.

LANIER, G.N. 1987. The validity of *Ips cribicollis* (Eichhoff) as distinct from *I. grandicollis* (Eichhoff) and occurrence of both species in Central America. Canadian Entomologist, 119:179-187.

LANIER, G.N., S.A. TEALE. and J.A. PAJARES. 1991. Biosystematics of the genus *Ips* (Coleoptera: Scolytidae) in North America: review of *Ips calligraphus* group. Canadian Entomologist, 123:1103-1124.

LANTERI, A. A. 1986. Revision del genero *Asynonychus* Crotch (Coleoptera: Curculionidae). Revista de la Asociación de Ciencias Naturales del Litoral, 17: 161-174.

LANTERI, A. A. 1990. Systematic revision and cladistic analysis of *Phacepholis* Horn. Southwestern Entomologist, 15: 179-204.

LANTERI, A. A. 1995. Systematic revision of *Ericydeus* Pascoe (Coleoptera: Curculionidae). Entomologica Scandinavica, 26: 393-424.

LANTERI, A. A. and N. B. DÍAZ. 1994. Systematic study and cladistic analysis of the genus *Aramigus* Horn (Coleoptera: Curculionidae). Transactions of the American Entomological Society, 120: 113-144.

LANTERI, A. A. and A. E. MARVALDI. 1995. *Graphognathus* Buchanan a new synonym of *Naupactus* Dejean and systematics of the *N. leucoloma* species group (Coleoptera: Curculionidae). Coleopterists Bulletin, 49: 206-228.

LANTERI, A. A. and C. W. O'BRIEN. 1990. Taxonomic revision and cladistic analysis of *Atrichonotus* Buchanan (Coleoptera: Curculionidae). Transactions of the American Entomological Society, 116: 697-725.

LARSON, B. and J. H. FRANK. 2000. Mexican bromeliad weevil, *Metamasius callizona* (Chevrolat). Featured Creatures. Department of Entomology and Nematology, University of Florida. http://creatures.ifas.ufl.edu/orn/m_callizona.htm

LARSON, B., J. H. FRANK and O. R. CREEL. 2001. Florida bromeliad weevil, *Metamasius mosieri* Barber. Featured Creatures. Department of Entomology and Nematology, University of Florida. http://creatures.ifas.ufl.edu/orn/m_mosieri.htm

LAWRENCE, J. F. 1982. Coleoptera, Pp. 482-553. *In*: S. P. Parker, ed. Synopsis and Classification of Living Organisms. Volume 2. McGraw Hill. New York.

LAWRENCE, J. F. and A. F. NEWTON, Jr. 1995. Families and subfamilies of Coleoptera (with selected genera, notes, references and data on family-group names), Pp. 779-1006. *In*: J. Pakaluk and S. A. Slipinski, eds., Biology, Phylogeny, and Classification of Coleoptera: Papers celebrating the 80th birthday of Roy A. Crowson. Museum i Instytut Zoologii PAN, Warsaw.

LECONTE, J. L. and G. H. HORN. 1876. The Rhynchophora of America, north of Mexico. Proceedings of the American Philosophical Society, 15(96): i-xvi, 1-455.

LINELL, M. L. 1897. New genera and species of North American Curculionidae. Journal of the New York Entomological Society, 5: 49-56.

LOUDA, S. M., D. KENDALL, J. CONNOR and D. SIMBERLOFF. 1997. Ecological effects of an insect introduced for the biological control of weeds. Science, 277: 1088-1090.

LYAL, C. H. C. 1995. The ventral structures of the weevil head (Coleoptera: Curculionoidea). Memoirs of the Entomological Society of Washington, 14: 35-51.

MAIER, C. T. 1983. Influence of host plants on the reproductive success of the parthenogenetic two-banded Japanese weevil, *Callirhopalus bifasciatus* (Roelofs) (Coleoptera: Curculionidae). Environmental Entomology, 12: 1197-1203.

MATTHEWS, J. V. Jr. 1982. East Beringia during late Wisconsin time: A review of the biotic evidence, Pp. 127-150. *In*: D.M. Hopkins, J.V. Matthews, Jr., C. E. Schweger and S.B. Young, eds., Paleoecology of Beringia, Academic Press, New York. 489 pp.

McAVOY, T. J., L. T. KOK and J. T. TRUMBLE. 1983. Biological studies of *Ceutorhynchus punctiger* (Coleoptera: Curculionidae) on dandelion in Virginia. Annals of the Entomological Society of America, 76: 671-674.

McCLAY, A. S. and D. M. ANDERSON. 1985. Biology and immature stages of *Thecesternus hirsutus* Pierce (Coleoptera, Curculionidae) in north-eastern Mexico. Proceedings of the Entomological Society of Washington, 87: 207-215.

McCOY, C. W., C. SEGRETAIN, G. M. BEAVERS and C. TARRANT. 1985. Laboratory rearing and some aspects of the biology of *Artipus floridanus* Horn (Coleoptera: Curculionidae). Florida Entomologist, 68: 379-385.

MORRIS, M. G. 1995. Surface swimming in some Curculionidae. Memoirs of the Entomological Society of Washington, 14: 129-136.

MORRONE, J. J. 1993. Systematic revision of the *costirostris* species group of the weevil genus *Listroderes* Schoenherr (Coleoptera: Curculionidae). Transactions of the American Entomological Society, 119(4): 271-315.

MUDGE, A.D., J.R. LABONTE, K.J.R. JOHNSON and E. LAGASA. 2001. Exotic woodboring Coleoptera (Micromalthidae, Scolytidae) and Hymenoptera (Xiphydriidae) new to Oregon and Washington. Proceedings of the Entomological Society of Washington, 103: 1011-1019.

NORMARK, B.B., B.H. JORDAL and B.D. FARRELL. 1999. Origin of a haplodiploid beetle lineage. Proceedings of the Royal Society of London, B. 266:2253-2259.

O'BRIEN, C. W. 1961. *Phyrdenus muriceus* (Germar) attacking tomatoes in Arizona (Coleoptera: Curculionidae). Pan-Pacific Entomologist, 37: 185-186.

O'BRIEN, C. W. 1970a. A taxonomic revision of the genus *Dorytomus* in North America (Coleoptera: Curculionidae). University of California Publications in Entomology, 60: 1-80.

O'BRIEN, C. W. 1970b. A taxonomic revision of the genus *Gerstaeckeria* north of Mexico (Coleoptera: Curculionidae). Annals of the Entomological Society of America, 63: 255-272.

O'BRIEN, C. W. 1973. *Rhinanisus chisosensis*, a new species of cossonine weevil in a genus new to the United States (Coleoptera: Curculionidae). Coleopterists Bulletin, 27: 7-9.

O'BRIEN, C. W. 1976. A taxonomic revision of the New World subaquatic genus *Neochetina* (Coleoptera: Curculionidae; Bagoini). Annals of the Entomological Society of America, 69: 165-174.

O'BRIEN, C. W. 1977a. *Trigonoscutoides texanus* (new genus, new species) from Texas sand dunes (Tanymecinae: Curculionidae: Coleoptera). Coleopterists Bulletin, 31: 155-158.

O'BRIEN, C. W. 1977b. *Cercopeus komareki*, new species from Florida and Georgia (Coleoptera: Curculionidae: Otiorhynchinae). Florida Entomologist, 60: 257-261.

O'BRIEN, C. W. 1981. The larger (4.5 + mm.) *Listronotus* of America, north of Mexico (Cylindrorhininae, Curculionidae, Coleoptera). Transactions of the American Entomological Society, 107: 69-123.

O'BRIEN, C. W. 1984a. *Paralicus minyops* O'Brien, new genus and new species of Cossoninae from Florida and the Bahama Islands (Curculionidae: Coleoptera). Southwestern Entomologist, 9: 346-349.

O'BRIEN, C. W. 1984b. *Trachyphloeosoma advena* Zimmerman, new to the continental United States, with a key to U. S. genera of Trachyphloeini (Coleoptera: Curculionidae: Otiorhynchinae). Coleopterists Bulletin, 38: 181-184.

O'BRIEN, C. W. 1985. A new *Oopterinus* from Arkansas (Coleoptera: Curculionidae). Entomological News, 96: 101-104.

O'BRIEN, C.W. 1986. A catalog of the Coleoptera of America north of Mexico. Family Curculionidae. Subfamily Hylobiinae. United States Department of Agriculture, Agriculture Handbook Number 529-143c. viii + 11 pp.

O'BRIEN, C.W. 1989. A catalog of the Coleoptera of America north of Mexico. Family Curculionidae. Subfamily Pissodinae. United States Department of Agriculture, Agriculture Handbook Number 529-143d. x + 8 pp.

O'BRIEN, C. W. 1990. *Neobagoidus carlsoni*, new genus, new species of aquatic weevil from Florida. Southwestern Entomologist, 15: 71-76.

O'BRIEN, C. W. 1995. Curculionidae, premiere biological control agents. Memoirs of the Entomological Society of Washington, 14: 129-136.

O'BRIEN, C.W. 1996. A catalog of the Coleoptera of America north of Mexico. Family Curculionidae. Subfamily Erirhininae. United States Department of Agriculture, Agriculture Handbook Number 529-143f. x + 39 pp.

O'BRIEN, C.W. 1997. A catalog of the Coleoptera of America north of Mexico. Family Curculionidae. Subfamilies Acicnemidinae, Cossoninae, Rhytirrhininae, Molytinae, Petalochilinae, Trypetidinae, Dryophthorinae, Tachygoninae, Thecesterninae. United States Department of Agriculture, Agriculture Handbook Number 529-143g. x + 48 pp.

O'BRIEN, C. W. and G. B. MARSHALL. 1979. U. S. *Bagous*, bionomic notes, a new species, and a new name (Bagoini, Erirhininae, Curculionidae, Coleoptera). Southwestern Entomologist, 4: 141-149.

O'BRIEN, C. W. and M. C. THOMAS. 1990. The species of *Metamasius* in Florida (Coleoptera: Curculionidae). Florida Department of Agriculture. Division of Plant Industry. Entomology Circular No. 330. 4 pp.

O'BRIEN, C. W. and G. J. WIBMER. 1978. Numbers of genera and species of Curculionidae (Coleoptera). Entomological News, 89: 89-92.

O'BRIEN, C. W. and G. J. WIBMER. 1982. Annotated checklist of the weevils (Curculionidae *sensu lato*) of North America, Central America, and the West Indies (Coleoptera: Curculionoidea). Memoirs of the American Entomological Institute, 34: i-ix, 1-382.

O'BRIEN, C. W. and G. J. WIBMER. 1984. Annotated checklist of the weevils (Curculionidae *sensu lato*) of North America, Central America, and the West Indies (Coleoptera: Curculionoidea) - Supplement 1. Southwestern Entomologist, 9: 286-307.

O'BRIEN, C. W and R. E. WOODRUFF. 1986. First records in the United States and South America of the African oil palm weevils, *Elaeidobius subvittatus* (Faust) and *E. kamerunicus* (Faust) (Coleoptera: Curculionidae). Florida Department of Agriculture. Division of Plant Industry. Entomology Circular No. 284. 2 pp.

PAKALUK, J. and T. A. CARLOW. 1994. Revision of the genus *Eisonyx* LeConte (Coleoptera: Curculionidae: Baridinae). Coleopterists Bulletin, 48: 153-169.

PELSUE, F. W. and E. L. SLEEPER. 1972. A review of *Eucyllus* Horn (Coleoptera: Curculionidae, Brachyrhininae, Peritelini). Bulletin of the Southern California Academy of Sciences, 71: 80-91.

PESHKEN, D. P. and A. T. S. WILKINSON. 1981. Biocontrol of Canada thistle (*Cirsium arvense*): Releases and effectiveness of *Ceutorhynchus litura* (Coleoptera: Curculionidae) in Canada. Canadian Entomologist, 113: 777-785.

PIERCE, W. D. 1907. On the biologies of the Rhynchophora of North America. Studies from the Zoological Laboratory, the University of Nebraska, Lincoln, 1907: 249-319.

PIERCE, W. D. 1909. Studies of North American weevils. Proceedings of the United States National Museum, 37: 325-364.

PIERCE, W. D. 1912. Systematic notes and descriptions of some weevils of economic or biological importance. Proceedings of the United States National Museum, 42: 155-170.

PIERCE, W. D. 1913. Miscellaneous contributions to the knowledge of the weevils of the families Attelabidae and Brachyrhinidae. Proceedings of the United States National Museum, 45: 365-426.

PIERCE, W. D. 1975. The sand dune weevils of the genus *Trigonoscuta* with a correlation of their anatomy to the geological history of our coast lines. Privately published, Orange, CA. iii + 162 pp.

PUTTLER, B., S. E. THEWKE and R. E. WARNER. 1973. Bionomics of three Nearctic species, one new, of *Hypera* (Coleoptera: Curculionidae), and their parasites. Annals of the Entomological Society of America, 66: 1299-1306.

RABAGLIA, R. J. and J.F. CAVEY. 1994. Note on the distribution of the immigrant bark beetle, *Hylastes opacus* Erichson, in North America (Coleoptera: Scolytidae). Entomological News, 105: 277-279.

RAFFA, K.F., T.W.PHILLIPS and S.M. SALOM. 1993. Strategies and mechanisms of host colonization by bark beetles. *In*: T. D. Schowalter and G.M. Filip, eds. Beetle-Pathogen Interactions in Conifer Forests. Academic Press. London.

SCHAEFFER, C. F. A. 1907. New Rhynchophora II. Journal of the New York Entomological Society, 15: 75-80.

SCHERF, H. 1964. Die Entwicklundsstadien der mitteleuropäischen Curculioniden (Morphologie, Bionomie, Ökologie). Abhandlungen der Senckenbergischen Naturforschenden Gesellschaft, 506: 1-335.

SCHOOF, H. F. 1942. The genus *Conotrachelus* Dejean (Coleoptera, Curculionidae) in the north central United States. Illinois Biological Monographs, 19(3): 1-170.

SLEEPER, E. L. 1954a. A European weevil in North America. Entomological News, 65: 129-130.

SLEEPER, E. L. 1954b. New Rhynchophora II (Coleoptera, Curculionidae). Ohio Journal of Science, 54: 180-186.

SLEEPER, E. L. 1955a. A synopsis of the genus *Cophes* in the United States and Mexico (Coleoptera, Curculionidae). Ohio Journal of Science, 55: 188-191.

SLEEPER, E. L. 1955b. A review of the Trachy[ph]loeini of America north of Mexico (Coleoptera, Curculionidae). Ohio Journal of Science, 55: 279-292.

SLEEPER, E. L. 1955c. A synopsis of *Micralcinus* LeConte (Coleoptera, Curculionidae). Ohio Journal of Science, 55: 95-98.

SLEEPER, E. L. 1956a. A synopsis of the genus *Barinus* Casey in North America (Coleoptera Curculionidae). Ohio Journal of Science, 56: 76-86.

SLEEPER, E. L. 1956b. The genus *Dichoxenus* Horn (Coleoptera, Curculionidae). Ohio Journal of Science, 56: 165-169.

SLEEPER, E. L. 1957a. Notes on the genus *Macrorhoptus* LeConte (Coleoptera: Curculionidae, Anthonominae). Ohio Journal of Science, 57: 70-74.

SLEEPER, E. L. 1957b. Notes on the Curculionoidea. Ohio Journal of Science, 57: 38-42.

SLEEPER, E. L. 1957c. Notes on North American species of *Polydrusus* Germar (Coleoptera: Curculionidae, Brachyderinae). Ohio Journal of Science, 57: 129-134.

SLEEPER, E. L. 1963. A study of the Zygopinae (Coleoptera: Curculionidae) of America north of Mexico, I. Bulletin of the Southern California Academy of Sciences, 62: 209-220.

SLEEPER, E. L. 1965. On *Lymantes* Schoenherr (Coleoptera, Curculionidae). Bulletin of the Southern California Academy of Sciences, 64: 144-152.

SLEEPER, E. L. and S. L. JENKINS. 1967. Notes on *Paracimbocera robusta* Van Dyke (Coleoptera: Curculionidae). Bulletin of the Southern California Academy of Sciences, 66: 46-48.

STOCKTON, W. D. 1963. New species of *Hyperodes* Jekel and a key to the Nearctic species of the genus (Coleoptera: Curculionidae). Bulletin of the Southern California Academy of Sciences, 62: 140-149.

SWAINE, J.M. 1918. Canadian bark-beetles. Part II. Dominion of Canada Department of Agriculture, Entomological Branch Technical Bulletin 14(2):1-143.

TAMAYO, M., C. W. O'BRIEN, R. P. CREED, C. E. GRUE and K. HAMEL. 1999. Distribution and classification of aquatic weevils (Coleoptera: Curculionidae) in the genus *Euhrychiopsis* in Washington State. Entomological News, 110: 103-112.

TANNER, V. M. 1941. Studies in the weevils of the western United States, No. IV: A new species of *Cimbocera*. Great Basin Naturalist, 2: 29-32.

TANNER, V.M. 1942. Studies in the weevils of the western United States, No. V: A new species of *Miloderoides*. Great Basin Naturalist, 3: 23-26.

TANNER, V. M. 1943. A study of the subtribe Hydronomi with a description of new species, (Curculionidae) study No. VI. Great Basin Naturalist, 4: 1-38.

THOMAS, M. C. and C. W. O'BRIEN. 1999. On the genus *Anchonus* Schönherr in Florida (Coleoptera: Curculionidae). Insecta Mundi, 13: 229-233.

THOMPSON, R.T. 1992. Observations on the morphology and classification of weevils (Coleoptera, Curculionoidea) with a key to major groups. Journal of Natural History, 26: 835-891.

TING, P. C. 1940. Revisional notes concerned with *Cimbocera* and related genera (Coleoptera: Curculionidae). Bulletin of the Southern California Academy of Sciences, 39: 128-157.

TITUS, E. G. 1911. The genera *Hypera* and *Phytonomus* (Coleoptera, Family Curculionidae) in America, north of Mexico. Annals of the Entomological Society of America, 4: 383-473.

TRUMBLE, J. T. and L. T. KOK. 1979. *Ceuthorhynchidius horridus* (Coleoptera: Curculionidae): Life cycle and development on *Carduus* thistles in Virginia. Annals of the Entomological Society of America, 72: 563-564.

VANDENBERG, N.J., R.J. RABAGLIA and D.E. BRIGHT. 2000. New records of two *Xyleborus* (Coleoptera: Scolytidae) in North America. Proceedings of the Entomological Society of Washington, 102:62-68.

VAN DYKE, E. C. 1915. The species of *Cossonus* Clairv. (Coleoptera) of America north of Mexico. Bulletin of the Brooklyn Entomological Society, 10: 1-23.

VAN DYKE, E. C. 1916. Supplementary notes and descriptions of North American Ostomidae, Cleridae, and *Cossonus* (Col.). Bulletin of the Brooklyn Entomological Society, 11: 71-79.

VAN DYKE, E. C. 1927. New species of North American Rhynchophora (Coleoptera). Pan-Pacific Entomologist, 5: 53-58.

VAN DYKE, E. C. 1928. The species of the genus *Lepyrus* Germ. (Coleoptera-Curculionidae) in North America. Pan-Pacific Entomologist, 5: 53-58.

VAN DYKE, E. C. 1933. A short review of *Dyslobus* LeConte, a genus of broad-nosed weevils of the subfamily Otiorhynchinae with descriptions of new species. Pan-Pacific Entomologist, 9: 31-47.

VAN DYKE, E C. 1934. New species of North American weevils in the family Curculionidae, subfamily Brachyrhininae. Pan-Pacific Entomologist, 10: 175-191.

VAN DYKE, E. C. 1935a. New species of North American weevils in the family Curculionidae, subfamily Brachyrhininae, II. Pan-Pacific Entomologist, 11: 1-13.

VAN DYKE, E. C. 1935b. New species of North American weevils in the family Curculionidae, subfamily Brachyrhininae, III. Pan-Pacific Entomologist, 11: 83-96.

VAN DYKE, E. C. 1936a. New species of North American weevils in the family Curculionidae, subfamily Brachyrhininae, IV. Pan-Pacific Entomologist, 12: 19-32.

VAN DYKE, E. C. 1936b. New species of North American weevils in the family Curculionidae, subfamily Brachyrhininae, V. Pan-Pacific Entomologist, 12: 73-85.

VAN DYKE, E. C. 1951. New species of Coleoptera from North America. Pan-Pacific Entomologist, 27: 27-35.

VAN DYKE, E. C. 1953. New Coleoptera from western North America (Carabidae, Melasidae, Buprestidae, Curculionidae). Pan-Pacific Entomologist, 29: 102-107.

VAURIE, P. 1951. Revision of the genus *Calendra* (formerly *Sphenophorus*) in the United States and Mexico (Coleoptera, Curculionidae). Bulletin of the American Museum of Natural History, 98(2): 33-186.

VAURIE, P. 1966. A revision of the Neotropical genus *Metamasius* (Coleoptera, Curculionidae, Rhynchophorinae). Species groups I and II. Bulletin of the American Museum of Natural History, 131(3): 211-337.

VAURIE, P. 1967. A revision of the Neotropical genus *Metamasius* (Coleoptera, Curculionidae, Rhynchophorinae). Species group III. Bulletin of the American Museum of Natural History, 136(4): 175-268.

VAURIE, P. 1971. Review of *Scyphophorus* (Curculionidae: Rhynchophorinae). Coleopterists Bulletin, 25: 1-8.

VAURIE, P. 1981. Revision of *Rhodobaenus*. Part 2. Species in North America (Canada to Panamá) (Coleoptera, Curculionidae, Rhynchophorinae). Bulletin of the American Museum of Natural History, 171(2): 117-198.

WARNER, R. E. 1952. Another European weevil, *Pentarthrum huttoni* Woll. in North America. Coleopterists Bulletin, 6: 51-52.

WARNER, R. E. 1961. The genus *Ochyromera* new to the Western Hemisphere, with a new species and additions to the Junk-Schenkling Coleopterorum Catalogus (Curculionidae: Prionomerinae, Endaeini). Coleopterists Bulletin, 15: 121-124.

WARNER, R. E. 1966. A review of the *Hylobius* of North America, with a new species injurious to slash pine (Coleoptera: Curculionidae). Coleopterists Bulletin, 20: 65-81.

WARNER, R. E. 1969. The genus *Phrydiuchus*, with the description of two new species (Coleoptera: Curculionidae). Annals of the Entomological Society of America, 62: 1293-1302.

WARNER, R. E. 1976. Description of a new species of *Phyllotrox* infesting seeds of *Acer grandidentatum*, with new synonymy and a note on *Euclyptus* (Coleoptera: Curculionidae: Erirhininae). Great Basin Naturalist, 36: 463-466.

WARNER, R. E. 1979. The genus *Eudiagogus* (Coleoptera: Curculionidae: Leptopiinae), with two new species on the weed *Sesbania* (Leguminosae). Proceedings of the Entomological Society of Washington, 81: 304-320.

WARNER, R. E. and F. B. NEGLEY. 1976. The genus *Otiorhynchus* in America north of Mexico (Coleoptera: Curculionidae). Proceedings of the Entomological Society of Washington, 78: 240-262.

WATTANAPONGSIRI, A. 1966. A revision of the genera *Rhynchophorus* and *Dynamis* (Coleoptera: Curculionidae). Department of Agriculture Science Bulletin, Bangkok, 1:1-328.

WEEDEN, C. 2000. *Hylobius transversovittatus* (Coleoptera: Curculionidae). Biological Control: A guide to Natural Enemies in North America. Cornell University. http://www.nysaes.cornell.edu/ent/biocontrol/weedfeeders/hylobius.html

WHITEHEAD, D. R. 1979. Notes on *Apteromechus* Faust of America north of Mexico (Coleoptera: Curculionidae: Cryptorhynchinae). Proceedings of the Entomological Society of Washington, 81: 230-233.

WIBMER, G. J. 1981. Revision of the New World weevil genus *Tyloderma* in America north of Mexico (Coleoptera: Curculionidae: Cryptorhynchinae). Southwestern Entomologist, Supplement 3, 1-95 pp.

WIBMER, G.J. and C. W. O'BRIEN. 1989. Additions and corrections to annotated checklist of the weevils of North America, Central America, and the West Indies, and of South America (Coleoptera: Curculionoidea). Southwestern Entomologist, Supplement 13, 1-49 pp.

WOOD, D.L. 1982. The role of pheromones, kairomones and allomones in the host selection and colonization behavior of bark beetles. Annual Review of Entomology, 27: 411-446.

WOOD, S.L. 1973. On the taxonomic status of Platypodidae and Scolytidae (Coleoptera). Great Basin Naturalist, 33:77-90.

WOOD, S.L. 1977. Introduced and exported American Scolytidae (Coleoptera). Great Basin Naturalist, 37:67-74.

WOOD, S.L. 1982. The bark and ambrosia beetles of North America (Coleoptera: Scolytidae), a taxonomic monograph. Great Basin Naturalist Memoirs 6. 1359p.

WOOD, S.L. 1986. A reclassification of the genera of Scolytidae (Coleoptera). Great Basin Naturalist Memoirs 10. 126p

WOOD, S. L. 1993. Revision of the genera of Platypodidae (Coleoptera). Great Basin Naturalist, 53: 259-281.

WOOD, S.L. and D.E. BRIGHT. 1992. A catalog of Scolytidae and Platypodidae (Coleoptera), Part 2: Taxonomic Index, Volumes A and B. Great Basin Naturalist Memoirs 13. 1553p.

WOODRUFF, R. E. 1963. An avocado weevil (*Heilipus apiatus* Oliv.) (Coleoptera: Curculionidae). Florida Department of Agriculture. Division of Plant Industry. Entomology Circular No. 1. 1 p.

WOODRUFF, R. E. 1968. The present status of a West Indian weevil (*Diaprepes abbreviata* (L.)) in Florida (Coleoptera: Curculionidae). Florida Department of Agriculture. Division of Plant Industry. Entomology Circular No. 77. 4 pp.

WOODRUFF, R. E. 1969. The banana root borer (*Cosmopolites sordidus* (Germar)), in Florida (Coleoptera: Curculionidae). Florida Department of Agriculture. Division of Plant Industry. Entomology Circular No. 88. 2 pp.

WOODRUFF, R. E. 1972. *Nicentrus saccharinus* Marshall, a potential sugarcane pest recently introduced into Florida (Coleoptera: Curculionidae). Florida Department of Agriculture. Division of Plant Industry. Entomology Circular No. 124. 2 pp.

WOODRUFF, R. E. 1973. An orchid weevil new to the United States (Coleoptera: Curculionidae). Florida Department of Agriculture. Division of Plant Industry. Entomology Circular No. 129. 2 pp.

WOODRUFF, R. E. 1977. A weevil pest of rose-apple (*Syzygium jambos*), apparently new to the United States (Coleoptera: Curculionidae). Florida Department of Agriculture. Division of Plant Industry. Entomology Circular No. 183. 2 pp.

WOODRUFF, R. E. 1979. Florida citrus weevils (Coleoptera: Curculionidae). Florida Department of Agriculture. Division of Plant Industry. Entomology Circular No. 202. 4 pp.

WOODRUFF, R. E. 1981. Florida citrus weevils of the genus *Pachnaeus* in Florida. Florida Department of Agriculture. Division of Plant Industry. Entomology Circular No. 231. 4 pp.

WOODRUFF, R. E. and R. M. BARANOWSKI. 1985. *Metamasius hemipterus* (Linnaeus) recently established in Florida (Coleoptera: Curculionidae). Florida Department of Agriculture. Division of Plant Industry. Entomology Circular No. 272. 4 pp.

ZIMMERMAN, E. C. 1957. The *Portulaca* leafmining weevil, *Hypurus bertrandi*, in Hawaii (Coleoptera: Curculionidae: Ceutorhynchinae). Annals of the Entomological Society of America, 50: 221-222.

ZIMMERMAN, E. C. 1964. On the supposed North American Trachodinae (Coleoptera: Curculionidae). Coleopterists Bulletin, 18: 25-32.

ZIMMERMAN, E. C. 1993. Australian Weevils (Coleoptera: Curculionoidea). Volume III. CSIRO Press. Melbourne.

Keys to Families of Beetles in America North of Mexico

by Michael A. Ivie

These keys are specifically designed for North American taxa and may lead to incorrect identifications of many taxa from outside this region. They are aimed at the successful family placement of all beetles in North America north of Mexico, and as such will not always be simple to use. A key to the most common 50% of species in North America would be short and simple to use. However, after an initial learning period, most coleopterists recognize those groups on sight, and never again key them out. It is the odd, the rare and the exceptional that make a complex key necessary, and it is in its ability to correctly place those taxa that a key is eventually judged. Although these keys build on many previous successful efforts, especially those of Crowson (1955), Arnett (1973) and Borror *et al.* (1989), in many ways I have taken a new approach that owes more to Lawrence and Britton (1994) and Lawrence *et al.* (1999).

Nearly 60% of the families in *American Beetles* have had their membership redefined or ranks changed since the last comprehensive treatment of North American beetles (Arnett 1973). Over 35% of the families included in the last credible linear key to North American beetle families (Borror *et al.* 1989) have been redefined. Several taxa have even been moved since the most recent key, the Herculean computerized key to world taxa by Lawrence *et al.* (1999). So, this is the first attempt at a conventional family key to these particular taxa. Instead of simply modifying existing keys by adding couplets to further divide the old endpoints, I have started from the base and rebuilt the key structure around the new family concepts. The add-on approach would have resulted in keys totaling nearly 400 couplets that would have greatly obscured family concepts instead of the <200 couplet keys provided here. The success of this approach will be determined after testing by you, the users.

An effort has been made to key out a specific family in a single couplet, but when this was in conflict with making reliable identifications, it was overridden by practicality. This interface of goal and practicality sometimes causes considerable use of "and," "or," "if...then," and "usually." The keys are deliberately not phylogenetic, but when possible and practical I used shared derived features to facilitate the one-family, one-couplet goal. I hope the user will supply the patience needed to deal with the magnificent complexity of beetles as represented in the longer couplets presented below.

Several characters that have been traditionally favored for use in North American beetle keys are not emphasized here – especially tarsal formulae – and other characters that have rarely been used previously have been placed in critical couplets. These unfamiliar characters will undoubtedly cause some initial discomfort for experienced users until, through practice, they become familiar and proficient with them. However, I hope new users will find these characters an aid in understanding today's definitions of families. I have attempted to avoid characters requiring dissection and, where possible, overly long lists of options, but when necessary, I have erred on the side of directing the user to a correct identification.

No key will work on all specimens because of abnormalities of development, poor preservation, previously unknown species, sexes or variation, or simple errors in characterization. Furthermore, with more than 30,000 species to be considered, there are undoubtedly rare forms that escaped my notice and even possibly some common and easily collected species with exceptional characters that I overlooked. While this key should work for at least 95% of specimens collected and 90% of North American species, the specialized collector who delves into unique habitats or uses specialized methods may find a higher percentage of problems. Even in North America there are still many taxa to be discovered in specialized niches like deep soils, unique unsampled habitats, or with specialized techniques like flight intercept traps, soil washing, or Berlese funnels. These taxa will undoubtedly stretch our understanding of family characterizations in the North American fauna. Invasions by exotic species also will continue to introduce exceptions.

A high-quality microscope (at least 40X, preferably 60X) with good illumination is required to see many characters, especially those of small specimens. Specimens must be clean and properly mounted so that dorsal, ventral, and lateral surfaces are visible (see Borror *et al.* [1989] for a discussion of mounting). Card-mounted specimens with the venter obscured will prove mostly frustrating and should be remounted on points before attempting the key. In many cases, characters will be easier to see in dry specimens than in those preserved in liquid.

Dirty specimens should be cleaned before identification is attempted. First, place the specimen in hot, but not boiling, water for a minute or two, and then vibrate in an ammonia bath (household cleaning ammonia, available at grocery stores) using an ultrasonic cleaner (inexpensive to very expensive options are available from entomological suppliers, jewelry stores, and architectural supply houses) for 5-20 minutes works well for cleaning and degreasing. Resistant coatings of foreign material, particularly those encrustations secreted by the beetle itself, may require gentle abrasion under the microscope with a pin or camel's-hair brush between ammonia baths. To neutralize the base in the ammonia before mounting and drying for storage, the specimen should be rinsed in several baths of clean water in the ultrasonic cleaner, then in a final wash of high percentage alcohol to help dry the specimen (this last should NOT be done in the ultrasonic cleaner because of fire hazard).

Pubescent specimens may require further work to keep setae from matting after this treatment. Critical-point drying is the best method, but requires an expensive and complex critical-point drier. As an alternative for species with sparse, stiff setae, a dip in 100% ETOH and gentle blowing will usually return the setae to their normal positions. If this does not work adequately, chemical drying is another option. Run the cleaned specimens through a dehydrating wash of 100% ETOH, one or two 30-minute soaks in hexamethyldisilazane (HMDS, available from chemical suppliers), and then place them in a shallow dish of HMDS, and allow it to evaporate (Brown 1993). HMDS must be used with adequate ventilation, see the material safety data sheet that comes with the chemical.

Legs and antennae should be positioned to allow for clear viewing of the total length of the structure itself, as well as coxal cavities and sternal surfaces. Critical characters of the pro- and mesocoxal cavities and thoracic sterna will prove difficult to see in specimens with the legs retracted. This instruction is easy to make, but frustrating to follow, as the need to see a particular structure is not obvious until after the specimen is mounted. In many cases, especially for small species, when at a critical juncture in the key, a specimen will be found to have the leg in just the wrong position to see a particular structure. A tiny amount of Barber's Relaxing Fluid (Peterson 1964) applied with a fine brush to the offending joint will usually allow enough movement of the structure in a dried specimen to avoid the need for relaxing.

Barber's Relaxing Fluid (Peterson 1964) is made with the following formula (parts by volume):

53 parts 95% Ethanol (ETOH)
49 parts distilled water
19 parts ethyl acetate (acetic ether)
7 parts benzol (benzene)

This solution must be used with adequate ventilation because of the presence of benzene, known to be a hazardous substance that may cause acute and chronic health effects, including cancer, in humans. Consult the material safety data sheet that accompanies the benzene.

Conventions. Figures are cited from the text throughout Volumes 1 and 2 with the convention of "x.Y" where "x" is the figure number and "Y" the chapter number, *e.g.*, "4.22" is Figure 4 of chapter 22. The introductory chapter is denoted "I," *i.e.*, "2.I" is the second figure in the Introduction. Figures in this Family Key Chapter are denoted "K", *i.e.*, "3.K" is the third figure in this chapter. A number before the family name at the end of a couplet refers to the chapter number for that family. Chapters I and 1-22 are in Volume 1, Chapters 23-131 and K are in Volume 2.

Qualifiers

Especially – Most strongly or often expressed.

IF...THEN – In the case denoted by IF, accompanied by the condition indicated by THEN.

Rarely – Character that occurs as an exception in a group, but may be encountered in less than ca. 2% of specimens seen by the normal user.

Seldom – Character that occurs as an exception in a group, but may be seen in a distinct minority of cases, expected to be in the 10% or less range.

Often – Character expected to be present in a large proportion of the specimens seen by the average collector or identifier, but may be absent in any given, or even a majority of, species or specimens encountered.

Usually – Character present in a majority (>51%) of species and specimens of the group, but exceptions occur.

Variable – Used when a character state in the opposing couplet occurs, along with other states of that character.

AND – When a capitalized AND is present, it means that all of the characteristics before and all of the characteristics after it must occur together.

AND/OR – meets either all the conditions before AND all of the characteristics after it; OR that either the characteristics before it or the characteristics after it apply.

OR – When a capitalized OR is present, it means that either the characteristics before it or the characteristics after it apply, but not necessarily both.

Morphological Terms

Most terms used can be interpreted using the introductory discussion of characters on pages 2 through 9 of Volume 1, or by reference to cited figures. It is assumed that the user is either familiar with basic insect morphology or has access to a general textbook such as Borror *et al.* (1989) and an entomological glossary such as Nichols and Schuh (1989). More detailed beetle-specific terminology follows Lawrence and Britton (1994) and/or Lawrence *et al.* (1999). A few specific and important terms are defined here because the user may not have access to these latter works. Some important usages unique to this key are also defined to avoid ambiguity.

Use of "Segment". The difference between true anatomical segments and apparent articulated joints in insect appendages causes considerable confusion. In an anatomical sense, the term "segment" should only be used in insects for the homologue of either a body metamere or the true segments of the primitive arthropod appendage (Nichols and Schuh 1989, Chapman 1998). However, the term is often misused in reference to antennae and tarsi (Chapman 1998), and its correct use for the palpi is uncertain in some cases. Therefore, the following terms are used in these keys (following Lawrence *et al.* 1999):

Antennomere vs. Antennal Segment. There are only 3 true segments in the beetle antenna (scape, pedicel, and flagellum) (Chapman 1998). The annulate sections of the flagellum are not true segments and should not be referred to as antennal segments. The technically correct use of scape, pedicel, and flagellomeres is unwieldy, so I use antennomere for the visible articulated parts of the antenna, numbered from proximal to distal. The scape is always antennomere 1, the pedicel is antennomere 2, etc.

Palpomere vs. Palpal Segment. The joints of the maxillary and labial palpi may be properly referred to as segments, being anatomically homologous to appendage segments (Chapman 1998). However, it is unclear if all of the situations involving articulated pieces of beetle palpi may be correctly considered true segments, so the term palpomere is employed here.

Tarsomere vs. Tarsal Segment. There are 2 true segments involved in what is called the beetle tarsus: the true tarsus, usually subdivided in beetles into 2-4 pieces, and the apical claw-bearing pretarsus (Chapman 1998). Many authors refer to each of the articulated pieces of these 2 true segments as tarsal segments, but I prefer the term tarsomere to distinguish them from the correct use of segment.

Clicking Mechanism. This mechanism consists of a long prosternal intercoxal process with the dorsal or dorso-apical surface of the apex notched to fit against a slight projection on the anterior margin of the relatively large, deep mesocoxal cavity. In some compact species there is a plate-like, margined ventral face to the postcoxal portion of the intercoxal process that is tightly received by the deeply emarginate mesosternum. In these cases the true apex of the prosternum is hidden in a deep cavity between the mesocoxae. This later condition co-occurs with prosternal or hypomeral antennal grooves.

Connate Ventrites. These are abdominal ventrites that are fused and immovable relative to each other. This condition can often be detected as a difference in the quality of the suture between those ventrites that are connate and those that are not, by the absence of a membrane between the connate sternites (Figs. 4-5.106) vs. a distinct membrane clearly visible at the other sutures, or by a reduction in the depth of the suture itself, especially medially (Fig. 58.41). However, the easiest and most certain way to tell is to view the upturned lateral portion of the ventrite that is held against the elytron in repose. Connate ventrites will be obviously nonmovable in this view, and lack the hinged form of the free, movable state.

Mesocoxal Cavities Open or Closed. Closed mesocoxal cavities are defined as having the meso- and metasterna in contact laterad the mesocoxa (Fig. 12.58). Open mesocoxal cavities are defined as having the mesepimeron and/or the mesepisternum separating the meso- and metasterna and reaching the mesocoxa (Figs. 23.I, 13-14.58, 4.K).

Metacoxa Reaching the Elytra or Lateral Portion of the Body. This character is defined as having the metacoxa completely separating the visible portions of the metathorax and first abdominal ventrite all the way to the elytral epipleuron (Figs. 23.I, 17.70, 19.70, 1-2.K), or to the point where the side of the body is indicated in cases where the elytra are not tightly held against the body at this level. If the metasternite, metepisternum, or metepimeron is in visible contact with the anterolateral angle of the first ventrite (Fig. 3.K), even as a narrow lamina, the metacoxa is deemed to not reach the elytra or lateral portion of the body.

Posterior Face of Metacoxa. A distinct posterior face on the metacoxa can be detected by viewing the specimen from the side. If there is a "step" between the level of the metasternum and the first ventrite formed by an angle at a carina running across the coxa (or the presence of a distinct metacoxal flange that extends posteriorly from the coxa) and another angle where the coxa contacts the ventrite in a ‾|_ or ‾_ manner, it has a posterior face (Figs. 1.K, 3.K). If there is a distinct excavation on the posterior face overhung by a ventral flange, it is also considered to have a posterior face (Fig. 2.K). If the level of the anterior edge of the coxa is on a different level from the abdomen and the ventral face of the coxa (viewed half-way between median and lateral ends of the coxa) is S-shaped in longitudinal (antero-posterior) section, *i.e.*, evenly curved rather than angulate, it does not have a distinct posterior face.

Pro-, Meso-, and Meta-. These are used to denote anterior, middle, and hind segments, respectively, when referring to the thoracic segments and their appendages.

Ventrite. This term refers to a normally visible abdominal sternite numbered from the base of the abdomen irrespective of the true morphological segment it represents.

Acknowledgments

This chapter is dedicated to Charles A. Triplehorn, John F. Lawrence, Richard S. Miller and LaDonna L. Ivie. They have provided me with the inspiration, education, information, motivation, confidence and support needed to undertake this project. I want to thank all the authors of the family chapters of *American Beetles* for their contributions. Without them, this key would not have been possible. Many of them checked characters for me, or sent specimens on short notice. The full list of people deserving thanks is too long to reproduce here. I had no idea what a huge task I was taking on when I agreed to write this key, and without the support of these people, I would never have made it even to this imperfect stage. Particular thanks go to Robert S. Anderson, George E. Ball, Cheryl B. Barr, Charles L. Bellamy, Donald S. Chandler, Shawn M. Clark, Zachary H. Falin, J. Joseph Giersch, Richard L. Hurley, LaDonna L. Ivie, Paul J. Johnson, John F. Lawrence, Richard A. B. Leschen, Katharine A. Marske, Joseph V. McHugh, Kelly B. Miller, Richard S. Miller, Darren A. Pollock, T. Keith Philips, Alistair S. Ramsdale, Brett C. Ratcliffe, Robert E. Roughley, Paul E. Skelley, Kipling Will, and Daniel K. Young, who discussed characters, checked couplets, reviewed all or part of the manuscript, discussed the project and/or listened when I needed to rant. They have provided information to correct many errors in previous drafts, but responsibility for the remaining errors remains solely with me. Thanks also for the help of Richard S. Miller, Charles O'Brien, Alistair S. Ramsdale, and Paul E. Skelley, each of whom provided critical specimens of rare taxa on short notice. A special thanks to John Sulzycki of CRC Press for resisting the urge to do something drastic during the frustration of my missing deadlines for completion of this key. Lastly, I want to acknowledge the Montana Agricultural Experiment Station and Montana State University for giving me the freedom to spend so much of my career looking at beetles.

I. Key to the Suborders of North American Coleoptera

1. Notopleural sutures present (Figs. 19.I, 11.6, 2.8, 4-5.10, 3.12); OR, abdomen with only 3 ventrites; body form hemispherical, minute beetles (length <1.3 mm) (Fig. 1.3); OR, small (length <2.6 mm), soft-bodied beetles (Fig. 1.2) with wings rolled in a spiral "cigar" manner (*i.e.*, not folded) 2
— Notopleural sutures absent; abdomen with 4 or more ventrites; wings folded or not, not rolled .... **Polyphaga (Key D)**

2(1). Hind coxae immovably fused to metasternum, completely dividing first ventrite (Figs. 33.I, 11.6, 2.8, 4-5.10, 3.12, 13-17.12) .... **Adephaga (Key C)**
— Hind coxae free, first visible abdominal sternite extending entirely across venter behind them 3

3(2). Minute beetles, length 2 mm or less in length (Figs. 1.3, 1.4); antenna with short club of 1 to 3 antennomeres (Figs. 1.3, 1.4); wing folded in repose .... **Myxophaga (Key B)**
— Minute to moderately large beetles, length 1.5 - 25 mm; antennae either filiform (Fig. 1-2.1), or sub-moniliform and gradually widening from 4th segment (Fig. 1.2); hind wings in repose spirally rolled in a spiral "cigar" manner .... **Archostemata (Key A)**

A. Key to the Families of North American Archostemata

1. Notopleural sutures present; elytra reticulate, long, covering pygidium (Figs. 1-2.1); body covered in scales; antennae filiform to subserrate (Figs. 1-2.1), length >4 mm .... 1. Cupedidae
— Notopleural sutures absent, elytra smooth, short, leaving at least pygidium exposed (Fig. 1.2); body without scales; antennae sub-moniliform and gradually widened from 4th segment (Fig. 1.2); length <2.6 mm .... 2. Micromalthidae

B. Key to the Families of North American Myxophaga

1. Body hemispherical (Fig. 1.3); elytra covering all terga; abdomen with 3 ventrites; antenna with 11 antennomeres, 9-11forming club .... 3. Microsporidae
— Body more elongate-oval and depressed dorsoventrally (Fig. 1.4); elytra short, 3-4 tergites exposed; abdomen with 6 or 7 ventrites; antenna with 9 antennomeres, antennomere 9 forming narrow club .... 4. Hydroscaphidae

C. Key to the Families of North American Adephaga

1. Metacoxa greatly enlarged, a ventral plate concealing trochanter and basal half of femur, covering most of 3 basal ventrites (Fig. 2.8) .... 8. Haliplidae
— Metacoxa greatly enlarged or not (Figs. 11.6), IF metacoxa greatly enlarged, THEN all ventrites visible laterally, coxa not concealing trochanter, basal half of femur or first 3 ventrites (Figs. 4-5.10, 3.12, 17.12) 2

2(1). Metacoxa not reaching elytron laterally, metepimeron and first ventrite in contact laterad of metacoxa and mesad of elytral margin (Figs. 6.6, 12.6); antenna usually at least partly pubescent (in addition to scattered long sensory setae); procoxal cavities usually closed behind; IF metacoxa reaches elytron and procoxal cavities open (one tiny species, length 2 mm or less), THEN second ventrite (first behind metacoxa) 3 times as long as metacoxa at insertion of leg and last maxillary palpomere distinctly narrower than penultimate (aciculate) 3
— Metacoxa reaching elytron laterally, junction of metepimeron and first ventrite not visible with elytron in place; antenna not pubescent, with only scattered long sensory setae; procoxal cavities open behind; second ventrite less than 3 times as long as metacoxa; last maxillary palpomere not distinctly narrower than penultimate 4

3(2). Mentum expanded, fused laterally to head capsule, covering ventral mouthparts completely when mandibles closed, mentum extending anteriorly beyond other mouthparts to form cutting edge; outer angle of protibia with large inwardly curved uncus (Fig. 1.5); body cylindrical; antenna moniliform; head, pronotum, and elytra with deep canaliculate grooves (Fig. 1.5) .... 5. Rhysodidae
— Mentum not fused laterally to head capsule or extending beyond other mouthparts, maxilla and labium with at least palpi visible (Figs. 5.I, 4.6, 45-48.6); outer angle of protibia with straight or outwardly curved teeth or spines (Figs. 10.6, 13.6, 33.6, 38.6, 43.6); head pronotum and elytra without deep canaliculate grooves; body form and antennae variable ... . 6. Carabidae

4(2). Protibia with antenna cleaner on inner apical angle (*cf.* Fig. 13.6); head with supraorbital setae (*cf.* Fig. 52.6) 9.Trachypachidae
— Protibia without antenna cleaner on inner apical angle; head lacking supraorbital setae 5

5(4). Pedicel of antenna greatly enlarged, offset from main line of antenna, flagellum very short and compact, not extended beyond hind margin of head; mid and hind legs very short; eyes usually divided into 2 isolated parts on each side, rarely with only a very narrow canthus extending between upper and lower portions .... 7. Gyrinidae
— Pedicel of antenna normal, antenna extended beyond hind margin of head; mid and hind legs not especially short; eyes not divided 6

6(5). Metafemur and metatibia narrow and subcylindrical in cross section; metatarsus shorter than metatibia and not tapered distally (Fig. 1.11); body not streamlined, outline of thorax and elytron discontinuous, base of pronotum distinctly narrower than elytra (Fig. 1.11); length 11-16 mm .... 11. Amphizoidae

— Metafemur and metatibia more or less distinctly compressed, especially so in larger species (length 6 mm or greater); metatarsus usually as long or longer than metatibia (Fig. 3.12), distinctly tapering distally (Figs. 3-4.12); body streamlined, outline of pronotum and elytron usually conjointly rounded (Figs. 1.10, 4-5.10, 32-47.12); length 1-40 mm 7

7(6). EITHER scutellum not visible; protarsus with 5 distinct tarsomeres; eyes distinct and length 1.0-1.6 mm; OR scutellum not visible; protarsus with 5 distinct tarsomeres; length 1.9-5 mm; outer margin of protibia evenly curved and bearing a distinct comb of stout parallel and contiguous setae (Fig. 3.10); AND inner apical angle with large, curved protibial spur (Fig. 3.10) .... 10. Noteridae

— Scutellum visible or not; protibia less evenly rounded on outer apical angle (Fig. 2.12), outer margin lacking setal comb (Figs. 2-4.12); inner angle without large inner protibia spur (Fig. 4.12); length 1.5-40 mm; IF less than 2 mm, THEN protarsus either pseudotetramerous (Figs. 7.12, 39.12) or eyes absent or greatly reduced and indistinct .... 12. Dytiscidae

D. Key to the Families of North American Polyphaga (by Michael A. Ivie, couplets 3-13 by Mary Liz Jameson and Brett Ratcliffe).

[Does not include unrecognized females of the Telegeusidae. Expected to be anelytrous or larviform, they will probably key out to couplet 183.]

1. Elytra present, complete, short, or reduced to flap-like stubs on the mesothorax 2

— Elytra totally absent 182

2(1). Antenna with strongly asymmetrical, usually lamellate club of 3-8 antennomeres (Figs. 17.I, 2.23, 3.23, 2.31, 2.34, 56-57.34, etc.); procoxae large, strongly transverse or conical and projecting below prosternum; procoxal cavities closed; trochantins concealed (except in Diphyllostomatidae); protibiae flattened with one or more teeth on outer edge; tarsi with 5 distinct tarsomeres, none of which are lobed or densely pubescent 3

— Antenna not lamellate, or coxae, tibiae or tarsi not as above 14

3(2). Antennae with 11 antennomeres 4

— Antennae with fewer than 11 antennomeres ... 5

4(3). Antennal club with 4-7 elongate antennomeres (Fig. 1.28) .... 28. Pleocomidae

— Antennal club with 3 circular or oval antennomeres (Fig. 2.29) ...... 29. Geotrupidae

5(3). Body capable of being rolled into contractile ball (Fig. 2.32); middle and posterior tibiae flattened and dilated .... 32. Ceratocanthidae

— Body oblong, not capable of being rolled into ball; middle and posterior tibiae not significantly flattened and dilated 6

6(5). Longer apical spur of mesotibia pectinate along one edge (*cf.* Fig. 2.30) .... 30. Ochodaeidae

— Longer apical spur of mesotibia simple, not pectinate (*cf.* Fig. 3.30) 7

7(6). Antennomeres of antennal club not capable of being tightly closed together (Figs. 1-3.23, 1.24, 1.25) 8

— Antennomeres of antennal club capable of being closed together (Figs. 13.I, 2.31, 1.33, 2.34, 56.34, etc.) 10

8(7). Abdomen with 7 ventrites, first divided by metacoxa; head strongly constricted behind eyes; protibia lacking apical spurs; trochantin exposed; mesocoxae conical and projecting; length 5-9 mm .... 24. Diphyllostomatidae

— Abdomen with 5-6 ventrites, first not divided; head not strongly constricted behind eyes; protibia with one or 2 apical spurs; trochantin not visible; mesocoxae not projecting; length 8-60 mm 9

9(8). Mentum with apex deeply emarginate; mesocoxal cavities closed laterally; body distinctly flattened dorsally (Fig. 1.25) .... 25. Passalidae

— Mentum with apex simple, not deeply emarginate; mesocoxal cavities open laterally; body evenly convex dorsally (Fig. 1.23) .... 23. Lucanidae

10(7). Antennal club with 3 antennomeres, first hollowed out to receive second (Fig. 2.31) .... 31. Hybosoridae

— Antennal club with 3-7 antennomeres, first simple, not hollowed out to receive second (*e.g.*, Fig. 2.34) 11

11(10). Abdomen with 5 ventrites; dorsal surface roughened or tuberculate, not shining (Figs. 1.26, 1-3.27) 12

— Abdomen with 6 ventrites (Fig. 5.34); dorsal surface variably sculptured, shining or not 13

12(11). Eyes not divided by canthus; clypeus with sides narrowing apically; color brown, gray, or black; metafemora and metatibia not enlarged, not covering abdomen .... 27. Trogidae

— Eyes divided by prominent canthus; clypeus with sides subparallel to divergent anteriorly; color testaceous to light reddish brown; metafemora and metatibia enlarged, covering most of abdomen .... 26. Glaresidae

13(11). Elytra shortened and widely divergent at apex (except in *Lichnanthe lupina*), not covering pygidium (Fig. 1.33); eighth morphological abdominal segment with spiracle ..... 33. Glaphyridae

— Elytra not shortened or widely divergent at apex, pygidium exposed or not; eighth morphological abdominal segment lacking spiracle (Fig. 90.34) .... 34. Scarabaeidae

14(2). Tarsi with 2-5 tarsomeres, not pseudotetramerous on ALL legs (*i.e.*, third of 5 tarsomeres on hind leg not lobed and enclosing small fourth, any other configuration possible); antennae, mouthparts, femora, and metacoxae variable; OR tarsi pseudotetramerous and metacoxa with distinct posterior face (at least medially) set off from ventral surface by a carina or flange; OR tarsi pseudotetramerous, head not at all rostrate, and antennae strongly or weakly clubbed but not geniculate 15

— Tarsi pseudotetramerous on all legs, with apparent penultimate tarsomere lobed below, enclosing and nearly hiding true fourth tarsomere (Figs. 31.I, 3.120, 36-37.124, 57-59.124, etc.); often with long antennae (Fig. 1.120), rostrate head (Figs. 1.125, 1.127, 1.128, 1.129, 1.130), or enlarged hind femora (Fig. 54-56.124); metacoxa without exposed posterior face 16

15(14). Palps very short, usually immovably fixed and not visible; head rostrate, prolonged into a variously developed beak (Figs. 1.125, 1.127, 1.128, 1.129, 1.130) AND/OR antennae geniculate with compact club (Fig. 18.I) 16

— Palps longer, flexible, and usually evident (*e.g.*, Figs. 6-7.I); head usually not prolonged into a beak but if rostrate or antennae elbowed and club compact, then palps longer and flexible 26

16(14,15). Antenna usually without distinct club, filiform, moniliform, serrate or pectinate (Figs. 8-12.I); head not rostrate; if antenna distinctly clubbed, club of 5 or more antennomeres and length of head from top to clypeal margin less than or equal to width of head just behind eyes.... 17

— Antenna distinctly clubbed with 4 or fewer antennomeres in club (Figs. 15-16.I, 18.I); OR if antennae moniliform, head distinctly rostrate (Fig. 1.129); OR if club with 5 or more antennomeres, length of head from vertex to clypeal margin greater than width of head just behind eyes 20

17(16). Antenna usually more than half length of body, often inserted on prominence, capable of being reflexed backward over body (Fig. 1.120); tibiae with 2 obvious apical spurs (Fig. 5.120); first antennomere usually several times longer that second; pygidium never sclerotized and exposed; length 3-75 mm .... 120. Cerambycidae

— Antenna usually less than half length of body, seldom inserted on prominence, not reflexible back over body; tibiae without or with one or two apical spurs; first antennomere seldom more than 2-3 times length of second; pygidium of some species sclerotized and exposed; length usually less than 12 mm 18

18(17). All tibiae with 2 distinct apical spurs AND front without "X" grooves; mesonotum with or without stridulatory file; ligula large, membranous and bilobed; aedeagus with median struts and tegmen bilobed 19

— At least one tibia without 2 apical spurs OR front with deep "X" grooves (Figs. 5-6.124); mesonotum without stridulatory file; ligula normal; aedeagus without median struts .... 124. Chrysomelidae

[NOTE: The Bruchinae, treated in Chapter 121, key out here. See Status of Classification and subfamily key in Chapter 124]

19(18). Head with short but distinct temple behind eye, set off from narrowed neck (Fig. 1.122); apex of mandible bidentate; ligula with a single lobe; mesonotum with stridulatory file .... 122. Megalopodidae

— Head lacking temples, evenly narrowed from behind eyes to neck (Fig. 1.123); apex of mandible unidentate or bidentate; ligula bilobed; mesonotum without stridulatory file .... 123. Orsodacnidae

20(16). Antenna geniculate (rarely appearing straight or nearly so), club compact (Figs. 1-2.131, 69-77.131); metatrochanter not cylindrical, femur attached obliquely (Fig. 3.129) .... 131. Curculionidae

— Antenna straight (very rarely geniculate), club loose or not evident; metatrochanter variable but if antenna geniculate, trochanter cylindrical and squarely attached to femur (Fig. 4.129) 21

21(20). Labrum visible and free (Fig. 6-9.125, 2-3.126); second tarsomere not spongy beneath (Figs 2-3.125); maxillary palpi normal 22

— Labrum never free; tarsi variable; maxillary palpi rigid 23

22(21). Antenna situated adjacent to eye or laterally near base of short dorsoventrally flat rostrum; apex of third antennomere reaching well beyond front margin of eye; all tibiae lacking spurs or spurs vestigial; notosternal sutures indistinct to obsolete .... 126. Anthribidae

— Antenna situated distally on long cylindrical rostrum; apex of third antennomere not or barely reaching front margin of eye; all tibiae with spurs; notosternal sutures distinct .... 125. Nemonychidae

23(21). Antenna either moniliform and body elongate (Fig. 1.129, 7.129, 9-11.129, 14.129) (Brentinae, Cyphagoginae, Trachelizinae); OR antenna straight and clubbed, body pear-shaped (Fig. 2.129, 15-17.129, 31.129) and metatrochanter cylindrical, squarely joined to femur (Fig. 4.129) (Apioninae, Nanophyinae); OR antenna geniculate, body pear-shaped and metatrochanter cylindrical, squarely joined to femur (Fig. 4.129) (Nanophyinae) OR antenna with 9-10 antennomeres and body elongate-cylindrical (Fig. 5.129) (Cyladinae, Nanophyinae) .... 129. Brentidae

— Antenna straight, not geniculate, with 11 antennomeres, club distinct; metatrochanter triangular or diamond-shaped, obliquely joined to femur (Fig. 3.129); body form variable 24

24(23). Gena produced anteriorly on each side, visible in frontal view as large tooth on each side of apex of rostrum, laterad mandible; dorsal surface with obvious, recumbent, scale-like setae; body surface lacking metallic sheen; length 12 mm or more .... 130. Ithyceridae

— Gena not produced anteriorly; upper surface glabrous or with fine hair-like setae; body surface often with distinct metallic sheen; length variable, mostly less than 10 mm 25

25(24). Antenna situated at least length of antennomere 1 from eye, positioned laterally on long quadrate rostrum (Fig. 1.128) OR very close to eye at base of short, robust rostrum (Fig. 3.128); protibia with anterior face apically flat, simple, not distinct from rest of surface (Figs. 5-6.128); metafemur with dorsal margin slightly to moderately arched (Figs. 2-4.128); pygidium oblique to vertical (Figs. 2-4.128); elytron often with a scutellary striole (Fig. 1.128); body surface often with distinct metallic sheen .... 128. Attelabidae

— Antenna situated immediately in front of eye at base of long cylindrical rostrum (Fig. 1.127); protibia with front face at apex with shallow grooved area filled with short, fine pilosity (Fig. 1.127); metafemur with dorsal margin markedly arched, paddle-like in shape, femur almost as wide as long; pygidium nearly horizontal; elytron lacking scutellary striole; body surface lacking metallic sheen .... 127. Belidae

26(15). Length 1.2 mm or less; antenna long, thin, with loose to indistinct club (Figs. 1.17, 77.17); antennomeres each with a whorl of long setae at apex; wing fringed with long setae that are longer than width of wing (Figs. 2-4.17, 77.17), or wing absent .... 17. Ptiliidae

— Length variable, antenna not as above, wings rarely with fringe longer than width of wing 27

27(26). Head with paired ocelli (Figs. 89-92.22) 28

— Head without paired ocelli (a single median ocellus may be present) 30

28(27). Anterior edge of scutellum abruptly and sharply elevated above mesoscutum; metepisternum reaching mesocoxal cavity and contacting first ventrite to separate metacoxa from elytral edge .... 66. Derodontidae

— Anterior edge of scutellum not abruptly elevated, continuous with mesoscutum; metepisternum variable 29

29(28). Elytra completely covering abdomen; antenna short, not reaching middle of pronotum, antenna with 9 antennomeres, club of 5 pubescent antennomeres (Figs. 3-4.16); ventral surface with hydrofuge pubescence (Ochthebiinae) .... 16. Hydraenidae

— Elytra usually exposing 1 or more abdominal terga (Fig. 6.22); antenna short to long, reaching beyond middle of pronotum in species with long elytra (Fig. 2.22); antennal club, if present, not involving 5 antennomeres; underside of body without hydrofuge surface (Omaliinae) .... 22. Staphylinidae

30(27). Elytra very short, leaving 3 or more abdominal tergites exposed (Figs. 1.22, 3.22, 3.102, 7.102, etc.) 31

— Elytra longer, leaving no more than 1 or 2 abdominal tergites exposed 51

31(30). Metatarsus with 1 fewer tarsomere than mesotarsus 32

— Metatarsus and mesotarsus with same number of tarsomeres 35

32(31). Body greatly flattened dorsoventrally, abdomen with 5 ventrites (Inopeplinae)........................ .... 116. Salpingidae

— Body not greatly flattened, abdomen with 6-7 ventrites 33

33(32). Antenna strongly serrate to pectinate, flabellate, bipectinate or biflabellate (Figs. 8-18.102)..... .... 102. Ripiphoridae

— Antenna, at most, very weakly serrate........... 34

34(33). Tarsal claw with long, acute process or blade arising from base, usually more than half as long as claw (Figs. 13-16.111), rarely (*Hornia*) reduced to hyaline spine; antenna filiform; body corpulent and soft............................. 111. Meloidae

— Tarsal claw simple; antenna weakly clubbed; body cylindrical (Fig. 31.22) (Euaesthetinae) .... 22. Staphylinidae

35(31). Eyes large, separated frontally by less than diameter of third antennomere; wings well developed, folded longitudinally at rest; maxillary palp complex; antenna with antennomeres 9-11 less than half the width of antennomeres 3-5 (*Atractocerus*) .... 71. Lymexylidae

— Eyes separated by more than diameter of third antennomere; wings, if well developed, usually folded transversely; maxillary palpi simple; antenna not as above 36

36(35). Scutellary striole present; 2 basal ventrites connate, suture not diminished medially; antennae of males pectinate to flabellate or plumose; serrate in females (Xenorhipidina) .... 41. Buprestidae

— Scutellary striole absent; ventrites all free or 4 ventrites connate; antennae variable 37

37(36). Antenna with distinct club (Figs. 13.I, 15-18.I) 38

— Antenna not clubbed (Figs. 8-12.I, 14.I) 43

[NOTE: The myxophagan family Hydroscaphidae will key out here if an easily made mistake is made in the suborder key (above). These tiny (length 1.0-1.2 mm) beetles can be recognized by the elongate, narrow last antennomere (Fig. 1.4) which does not fit either the "distinct club" or "not clubbed" choice, as well as by the presence of notopleural sutures.]

38(37). Mesotarsus with 2, 3 or 4 tarsomeres .... 22. Staphylinidae

— Mesotarsus with 5 tarsomeres 39

39(38). Antenna with 4 apical antennomeres expanded into asymmetrical club, first antennomere shining, other 3 tomentose (Fig. 1.21); elytra usually some combination of black and orange but occasionally all black; fifth tergite with pair of longitudinal carinae topped by stridulatory files; 12 mm or greater in length, usually more than 15 mm (Nicrophorinae)................21. Silphidae

[*Thanatophilus* (Silphidae) may key here for individuals with an extended abdomen; it lacks the stridulatory files of the fifth tergite and is 8-14 mm in length, but otherwise fits here because of antennal configuration.]

— Antenna not as above; fifth tergite without stridulatory files; color variable; length 13 mm or less, usually less than 10 mm 40

40(39). Antenna with 3 antennomeres; pronotum with antennal pockets anterolaterally above lateral margins; dorsoventrally flattened, louse-like parasites of beaver (Fig. 23.19) (*Platypsyllus*) .... 19. Leiodidae

— Antenna with 9-11 antennomeres; pronotum without antennal pockets 41

41(40). Procoxal cavities open .... 22. Staphylinidae

— Procoxal cavities closed 42

42(41). Lateral margins of pronotum complete; 5 ventrites .... 77. Nitidulidae

— Lateral margins of prontum incomplete; 6 ventrites (*Cylidrella*) .... 72. Trogossitidae

43(37). Mesotarsus with 4 or fewer tarsomeres .... 22. Staphylinidae

— Mesotarsus with 5 tarsomeres 44

44(43). Antenna with 12 antennomeres; antenna biserrate, bipectinate or biramose (Fig. 1.61). .... 61. Phengodidae

— Antenna with fewer than 12 antennomeres; antennal type variable 45

45(44). Last maxillary and labial palpomere long, nearly as long as, or longer than, antenna (Fig. 1.60) .... 60. Telegeusidae

— Last maxillary and labial palpomeres much shorter than antenna 46

46(45). Head covered above by pronotum (Fig. 9.62); often with luminous organs on abdomen (Fig. 19.62) .... 62. Lampyridae

— Head visible from above; never with luminous organs 47

47(46). Anterior edge of scutellum abruptly elevated, with distinct step to mesoscutum (female *Anorus*) .... 38. Dascillidae

— Anterior edge of scutellum in same plane as mesoscutum 48

48(47). Pronotum with lateral eversible vesicles (Fig. 2.74) (Malachiinae) .... 74. Melyridae

— Pronotum without eversible vesicles 49

49(48). Mesosternum medially excavated, forming a cavity to receive extended prosternal process; Southwestern USA (female Cebrioninae)........ .... 58. Elateridae

— Mesosternum not excavated to receive extended prosternal process; widespread 50

50(49). Elytra individually rounded, not meeting apically at suture (Fig. 4.64); mandible long and narrow (Figs. 25-26.64) .... 64. Cantharidae

— Elytra truncate, meeting at suture apically (Figs. 5-40.22); mandible often short and broad .... 22. Staphylinidae

[NOTE: The archostematan family Micromalthidae will key out to this couplet if an easily made mistake is made in the suborder key. They are exceptional for non-polyphagans in lacking notopleural sutures. At this couplet they will match neither choice because of the individually rounded elytra character of the cantharids, and the short mandible of the staphylinids. The combination of a concealed trochantin and posteriorly emarginate scutellum will further distinguish this family at this point. The rolled wing exhibited by the micromalthids is unique to the Archostemata. See Chapter 2.]

51(30). Apices of penultimate 2 or 3 antennomeres each completely ringed with microsetose groove (periarticular gutters) (Fig. 9.18) (must be viewed distally, difficult to see in very small specimens or in those with very compact antennal club); antenna with distinct to indistinct loose club; prothorax with sharp lateral margins; 5 or 6 ventrites; protrochantin exposed or hidden, if hidden and antenna with 11 antennomeres, antennomere 8 smaller than 7 or 9 52

— Antennae usually lacking periarticular gutters on antennal club; other characters variable; if complete periarticular gutters present, protrochantin hidden, antenna with 11 antennomeres AND antennomere 8 not smaller than 7 and 9 53

52(51). Metatibial spurs subequal in length (Figs. 16-18.19); small (1-6 mm), round to elongate oval, shining, granulate or transversely strigulate beetles; elytra glabrous or pubescent, striate or not; prothorax as broad as elytra (Figs. 1-5.19); procoxae strongly projecting and constricted by procoxal cavity; often capable of retracting into a ball-shape by curling head and prothorax under body; antenna distinctly clubbed, often with 11 antennomeres, 5 of which are involved in club and antennomere 8 smaller than 7 or 9. Some genera with 10 or 11 antennomeres and with distinct club of 3 or 4 antennomeres (Fig. 12.19); these latter with flattened, externally flanged hind femora, apical portion of which are excavate to receive tibiae; tarsal formula highly variable, 3-3-3, 4-4-4, 5-4-4, 5-5-4 or 5-5-5; one genus (*Colon*) with 11 antennomeres and somewhat gradually clubbed antenna that lacks small eighth antennomere (Fig. 7.19) has elytra pubescent, with characteristic shape and sutural stria (Fig. 2.19) (see also, couplet 112) ..... 19. Leiodidae

[NOTE: Three very aberrant and ecologically restricted genera that lack distinctly clubbed antennae belong here. *Glacicavicolla* is restricted to ice caves in Idaho and Wyoming and characterized by elongate head, pronotum and elytra, each separately constricted; cuticle translucent, shining; eyes absent, and with elongate, slender legs and antennae. Two genera of Platypsyllinae are associated with mammal nests or mammals and are characterized by oval, strongly dorsoventrally flattened body (Fig. 5.19), recumbent pubescence, an occipital crest overlapping anterior margin of pronotum (Fig. 5.19) and eyes absent or barely indicated.]

— Metatibial spurs distinctly unequal; moderately sized (4-14 mm), somewhat flattened shining beetles; elytra striate and glabrous; pronotum somewhat narrowed relative to elytra (Figs. 2-3.18); procoxae strongly projecting or transverse; body not retractile; antenna long, club loose and indistinct, eighth antennomere never smaller than 7 and 9; femora simple; tarsi 5-5-5 .... 18. Agyrtidae

53(51). Mesotarsus with 3 apparent tarsomeres, either clearly with 3 tarsomeres, or second tarsomere strongly lobed and hiding small penultimate (third) tarsomere (Figs. 5.92, 40.93, 43-44.93) 54

— Mesotarsus with 4 or 5 distinct tarsomeres OR first tarsomere distinctly lobed, engulfing very small second and small third of four, appearing to have 2 or 3 tarsomeres 62

54(53). Mesotarsus pseudotrimerous, with second tarsomere strongly lobed, hiding small penultimate (third) tarsomere (Figs. 5. 92, 40.93, 43-44.93) 55

— Mesotarsus truly with 3 tarsomeres, second tarsomere not greatly lobed 57

55(54). Procoxal cavities closed (except in *Holopsis*); head small, usually covered by hood-like pronotum (Figs. 1-9.94); if head exposed from above (Figs. 10-11.94), procoxal cavities closed; mostly tiny beetles less than 2 mm ... . 94. Corylophidae

— Head visible from above in front of pronotum; procoxal cavities open; size variable, up to 11 mm 56

56(55). Frontoclypeal suture distinctly impressed; all ventrites free; first ventrite without postcoxal lines; pronotum often with sublateral lines (Figs. 6-8.92) .... 92. Endomychidae

— Frontoclypeal suture absent; 2 basal ventrites connate, first ventrite with postcoxal lines (Figs. 49-55.93); pronotum lacking sublateral lines .... 93. Coccinellidae

57(54). Eyes absent (Fig. 8.90) (*Anommatus*) .... 90. Bothrideridae

— Eyes present 58

58(57). Head gradually narrowed behind eyes, without distinct temples or neck; procoxal cavities open; oval or elongate oval with base of pronotum subequal to elytral base 59

— Head sharply narrowed behind eyes or temples, with distinct neck; procoxal cavities open or closed, elongate or elongate oval, with base of pronotum distinctly narrower than elytra... 60

59(58). Antennal scape normal, shorter than club; funicle longer than entire club; posterior edge of last ventrite crenulate (*Ostomopsis*) .... 91. Cerylonidae

[NOTE: The myxophagan family Microsporidae will key out here if an easily made mistake is made in the suborder key (above). They will match the antennal characters, but lack the crenulation on the last ventrite. These tiny (length 0.5-1.2 mm) beetles can be easily recognized, having only 3 ventrites (5 in *Ostomopsis*). See Chapter 3.]

— Antennal scape large, subequal to length of club; funicle with 3 antennomeres, shorter than first antennomere of club (*Micropsephodes*) .... 92. Endomychidae

60(58). Abdomen with six ventrites, head narrowed immediately behind eyes (Fig. 10.22), lacking temples; procoxal cavities open; lateral margin of pronotum coarsely dentate; trochantin exposed; mesocoxal cavities open (*Dasycerus*) .... 22. Staphylinidae

— Abdomen with five ventrites; head behind eyes with distinct temples; procoxal cavities open or closed; lateral margin of pronotum simple to finely dentate or absent; trochantin concealed; mesocoxal cavities variable 61

61(60). Abdomen very short, half length of metasternum; pronotum not margined laterally; mesocoxal cavities unstudied in North American species; scutellum not visible; elytron at base with pit at end of impressed groove (Fig. 1.65); 2 rare species known from Florida.......................... .... 65. Jacobsoniidae

— Abdomen longer than metasternum (except *Akalyptoischion*, California); lateral margin of pronotum absent to finely dentate; mesocoxal cavities closed; scutellum small but visible; elytra usually striate; common and widespread .... 95. Latridiidae

62(53). Antenna with 9 antennomeres, last 5 involved in club (Figs. 3-4.16); 6 or 7 ventrites; tiny intercoxal sclerite between metacoxae; maxillary palp long relative to antenna (Fig. 4.16); ventral surface with hydrofuge pubescence; 3.0 mm or less .... 16. Hydraenidae

— Antenna not as above; other characters not in combination above 63

63(62). Antenna with 7-9 antennomeres, antennomeres 7-9 usually forming loose, tomentose club (Figs. 25-26.13), antennomere 6 often forming a cupule at base of club (Figs. 8.13, 10.13); maxillary palp often as long or longer than antenna (Figs. 1.13, 16.13, 19.13), always more than 1/2 antennal length (Figs. 27.13, 38.13); metacoxa with ventro-posterior carina setting off convex posterior face (Fig. 1.K) that rotates against anterior excavation of first ventrite; planes of ventral surface of metacoxa and first ventrite discontinuous; metatrochanter inserted on ventral (not posterior) surface of metacoxa (Figs. 40.13, 42.13), femur held against ventral face of coxa, not against posterior face of coxa or flat to abdominal surface when fully retracted .... 13. Hydrophilidae

— Antenna variable but not as above; maxillary palp usually much shorter than antenna; metacoxa configured differently 64

64(63). Metacoxa with distinct posterior face (at least medially) set off from ventral surface by carina or flange (Fig. 3.K), posterior face often excavated (Fig. 2.K); ventral surface of metacoxa not co-planar with first ventrite; metafemur inserted on posterior face of metacoxa and femur held posterior to coxa when retracted (Figs. 2-3.K); procoxal cavities open; meso and metatarsi with equal number of tarsomeres 65

— Metacoxa without distinct posterior face; metatrochanter often inserted on ventral surface or on small medial projection of coxa, never received in coxal excavation and resting ventrad of metacoxa in retracted position; ventral surface of metacoxa more-or-less continuous with first ventrite OR metatarsus with one fewer tarsomere than mesotarsus; procoxal cavities open or closed 106

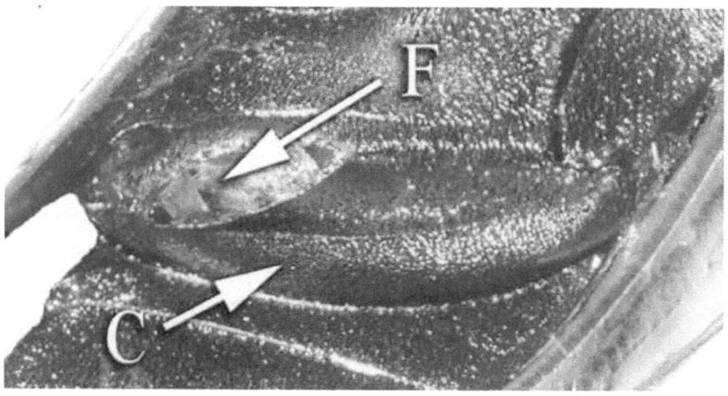

FIGURE 1.K. *Tropisternus* sp. (13. Hydrophilidae) metacoxa, oblique posterior view. C – posterior face of metacoxa. F – base of metacoxa (remainder removed).

65(64). Abdomen with 7-8 ventrites, metatarsus with 5 tarsomeres 66

— Abdomen with 6 or fewer ventrites, metatarsus with 5 or 4 tarsomeres 71

66(65). Head with median ocellus (male *Thylodrias*)...... .... 68. Dermestidae

— Head without median ocellus 67

67(66). Antenna with 12 antennomeres, biramose (Fig. 2.61) (male *Zarhipis*)............ 61. Phengodidae

— Antenna with 11 antennomeres, simple to uniramose or biramose 68

68(67). Mesothoracic coxae distinctly separated; elytra often reticulate (Fig. 1.59, 5.59), at least feebly costate; femur and/or tibia compressed; pronotum with distinct longitudinal median carina (Fig. 1.59), groove (Fig. 4.59) or cell (Fig. 5.59), occasionally restricted to base or disc .... 59. Lycidae

— Mesocoxae contiguous or nearly so; elytra not reticulate; femur and tibia seldom compressed; pronotum rarely with distinct longitudinal median carina, groove or cell 69

69(68). Pronotum extended forward, covering head in dorsal view (Figs. 1.62, 8-9.62 18.62, 20.62, 23-36.62); 1 or more ventrites often with luminous organs (most obvious in males) (Fig. 19.62); separation of antennal insertions equal to or less than diameter of antennal fossa (Fig. 22.62) .. .. 62. Lampyridae

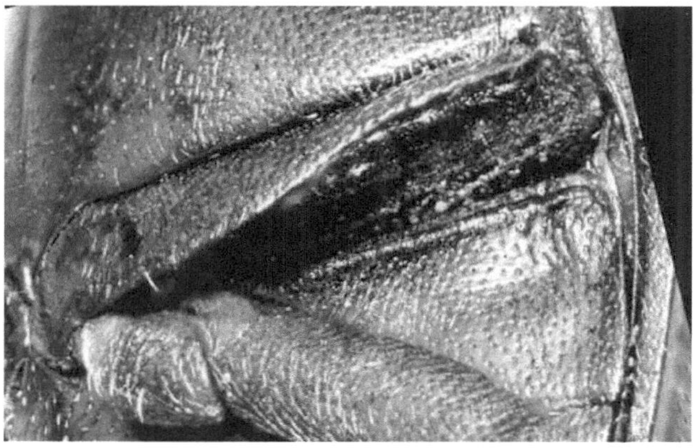

FIGURE 2.K. Elateridae metacoxa, oblique posterior view.

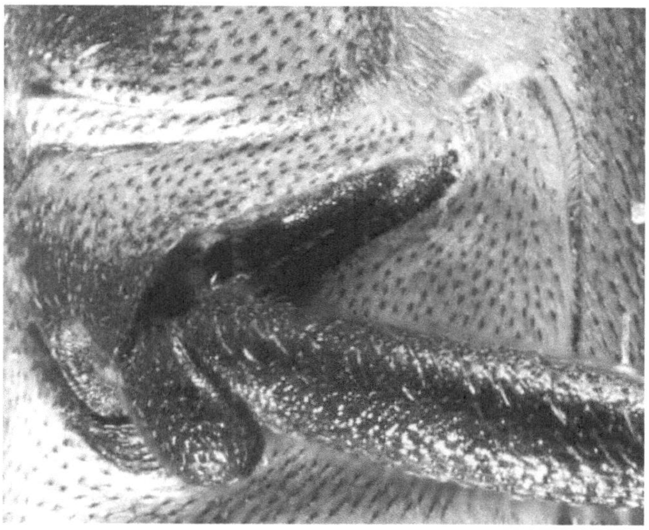

FIGURE 3.K. *Helichus immsi* Hinton (44. Dryopidae) metacoxa, oblique posterior view.

— Head exposed in dorsal view when extended, OR if covered by pronotum, antennae separated by nearly twice diameter of antennal fossa; abdomen lacking luminous organs .. 70

70(69). Labrum not distinct, membranous and often hidden beneath clypeus (Figs. 8-9.64); abdomen with paired glandular openings on lateral edge of tergites (Fig. 24.64); tarsomere 4 with bifid ventral lobe (Figs. 27-28.64) .. 64. Cantharidae

— Labrum distinct and sclerotized; abdomen lacking paired glandular openings on tergites; tarsomeres 3 and 4 with bifid ventral lobes... .... 63. Omethidae

71(65). Posterior angles of prothorax acute, embracing elytral humeri (Figs. 1-2.56,1.57, 1-3.58); metatarsus with 5 tarsomeres; 3 or more ventrites connate; prothorax dorsoventrally mobile relative to mesothorax; intercoxal process of prosternum long, notched dorsally, received in deep mesocoxal cavity as a clicking mechanism; IF clicking mechanism cannot be seen because visible portion of intercoxal process is flat ventrally and received tightly in deeply emarginate mesosternum, THEN sternopleural suture or hypomeron grooved to receive antennae 72

— Posterior angles of prothorax not acute and embracing elytral humeri, or rarely somewhat acute and weakly embracing humeri (Fig. 1.54, etc.); metatarsus with 5 or 4 tarsomeres; ventrites variable; prosternal process variable, but if large and received in deeply emarginate mesosternum, apex of prosternal process not notched dorsally nor capable of clicking; if large prosternal process received tightly in deep mesocoxal cavity AND underside of prothorax grooved to receive antennae, then metatarsus with 4 tarsomeres 74

72(71). Labrum not externally visible; abdomen with 5 connate ventrites .... 35. Eucnemidae

— Labrum free and visible; abdomen with 3, 4 or 5 connate ventrites 73

73(72). Antenna indistinctly to distinctly clubbed (Fig. 1.57), apex received in margined cavity on posterolateral portion of hypomeron, just anterior to retracted foreleg; metasternum with or without oblique margined groove for mesotarsus; prosternum with click mechanism hidden by plate-like ventral surface of postcoxal intercoxal process which fits tightly against exposed portion of mesosternal cavity; elytra strongly striate and covered with silky, subrecumbent setae; abdomen with 5 connate ventrites; length 1-5 mm .... 57. Throscidae

— Antenna variable (filiform, serrate, pectinate, etc.), but not clubbed; antennal groove, if present, at or near sternopleural suture; metasternum without margined groove for mesotarsus; IF click mechanism hidden as above, THEN elytra not strongly striate and setae suberect; abdomen with 3 or 4 connate ventrites; length 1-60 mm .... 58. Elateridae

74(71). Mesocoxal cavities closed laterally, the mesosternum and metasternum meeting laterad mesocoxa OR antenna elongate, antennomeres 3-8 with long rami, 9-11 flattened, elongate-serrate (Fig. 27.70); pronotum often hood-like, covering head from above (Figs. 1.69, 4.69, 11-13.69) 75

— Mesocoxal cavities open laterally, the mesosternum and metasternum separated laterad mesocoxa by the mesepimera or mesepimera and mesepisternum; antennae not as above; pronotum variable 76

75(74). Metatrochanter cylindrical, short to long, squarely attached to femur, distinctly separating coxa and tibia (Fig. 29.I) ...... 70. Anobiidae

— Metatrochanter short, triangular, obliquely attached to femur so that femur and coxa are adjacent to narrowly separated, on one side (Figs. 5-7.69) .... 69. Bostrichidae

76(74). Anterior margin of scutellum with abrupt, carinate elevation that fits against posterior margin of pronotum, or scutellum absent or not visible 77

— Anterior margin of scutellum not abruptly elevated, fitting under overlapping posterior margin of pronotum 101

77(76). Procoxae strongly and distinctly projecting ventrad of prosternum, 1/3 or more of dorsoventral length ventrad of intercoxal process (Figs. 2-3.49, 3.67), procoxae usually conical or transversely conical 78

— Procoxae not or weakly projecting ventrad of prosternum; if procoxae conical, then lying longitudinally and not or weakly projecting ventrally ventrad of intercoxal process 85

78(77). Tarsi with 4 distinct tarsomeres; metacoxal plates greatly expanded, hiding most of first ventrite; hind wing, when developed, often fringed with long setae; length 0.7-2 mm ..... 36. Clambidae

— Tarsi with 5 distinct tarsomeres; metacoxal plates distinct but not hinding most of first ventrite; wing not fringed; size variable 79

79(78). Antenna with distinct, simple club of 3 antennomeres (Figs. 1.14, 1.67, 4.68, 7.68) 80

— Antenna variously constructed, but without a simple club of 3 compact antennomeres ... 82

80(79). Elytra truncate; pygidium sclerotized and completely or nearly completely exposed (Fig. 1.14) .... 14. Sphaeritidae

— Elytra complete; pygidium not sclerotized, completely covered or with only small portion exposed 81

81(80). Upper surface of body glabrous; body contractile; protibia held anterior to profemur and covering antenna in hypomeral cavity when contracted (Fig. 3.67) (Orphilinae) .... 67. Nosodendridae

— Upper surface of body variously pubescent, setose or scaled (Fig. 1.68); body not strongly contractile; protibia held posterior to profemur and antennal club not covered by leg when contracted (Fig. 4.68) .... 68. Dermestidae

82(79). Base of pronotum crenulate; scutellum usually medially notched on anterior margin; antennal insertions not elevated; mandibles moderate and evenly curved; labrum large, sclerotized and dorsal to mandibles .... 49. Ptilodactylidae

— Base of pronotum simple; anterior margin of scutellum not notched; dorsal margin of antennal insertions elevated and protuberent; mandibles large, abruptly curved mesad at nearly right angle; labrum either short and membraneous or extending between and below mandibles 83

83(82). Empodium not obvious, hidden between bases of claws or absent; base of pronotum nearly straight (Fig. 1.38) .... 38. Dascillidae

— Empodium large, 1/3 length of claw, obviously plurisetose; base of pronotum strongly trisinuate around scutellum (Fig. 1.39, 1.52) 84

84(83). Tarsomeres 1-4 with large, membranous, divided lobes; antenna lamellate (males) or increasingly serrate apically (females) ...... 39. Rhipiceridae

— Tarsi simple, without ventral lobes; antennae serrate to pectinate .... 52. Callirhipidae

85(77). Head with single median ocellus (Fig. 4.68) .... 68. Dermestidae

— Head without ocellus 86

86(85). Antenna short, not reaching middle of pronotum, scape and pedicel (antennomeres 1-2) relatively large, together 1/3 or more of total length; antennomeres 3 to last transverse; body covered in dense tomentum 87

— Antenna short to long, scape and pedicel (antennomeres 1-2) not 1/3 of total length; antennomeres 3 to last variable; body vestiture variable 89

87(86). Head distinctly prognathous, mandibles strongly projecting forward (Fig. 1.47); profemur widened medially and armed externally with strong spines (Fig. 1-2.47); mesotarsus with 4 tarsomeres .... 47. Heteroceridae

— Head distinctly hypognathous, mandibles either directed ventrad or hidden (Figs. 1.44, 1.45); profemur simple, neither widened medially nor armed with large spines; mesotarsus with 5 tarsomeres 88

88(87). Metasternite with postcoxal lines delimiting retractile position of mesotibia; antenna hidden in subocular groove and cavity between head and pronotum; body oval (Fig. 1.45) .... 45. Lutrochidae

— Metasternite without postcoxal lines; subocular groove absent or very weakly developed, antenna not hidden in pronotum; body nearly parallel-sided (Fig. 1.44) .... 44. Dryopidae

89(86). Scape and pedicel received in deeply excavate pro- and mesosterna between pro- and mesocoxae (Fig. 2.50); pedicel longer than scape, scape and pedicel together more than 2/3 length of serrate flagellum (Fig. 2-3.50); body strongly contractile, all legs received in cavities (Fig. 2.50); mesotarsus with 5 tarsomeres, with long lobe on third tarsomere, fourth small and sometimes difficult to see (pseudotetramerous) .... 50. Chelonariidae

— Antennae not received in excavations between pro- and mesocoxae; antennae not as above; mesotarsus usually not pseudotetramerous 90

90(89). Head with subgenal ridges that fit against procoxae when head deflexed .. 37. Scirtidae

— Head without subgenal ridges, genae not in contact with procoxae 91

91(90). Two basal ventrites connate, either with suture between them partially obliterated medially OR if suture between ventrites 1 and 2 not medially indistinct, sternopleural sutures at least moderately grooved to receive antennae 92

— Ventrites all free, OR 3 or 5 ventrites connate; ventrital and sternopleural sutures variable 94

92(91). Suture between 2 basal sternites distinct medially; mesotarsus with small, bisetose empodium; antenna filiform to distinctly clubbed; body strongly convex .... 42. Byrrhidae

— Suture between 2 basal sternites weak to absent medially (Figs. 9.41, 58.41, etc.); mesotarsus lacking visible empodium; antenna usually serrate, pectinate or flabellate (Figs. 7-8.41, 28.41, 46-48.41); body weakly dorsoventrally flattened 93

93(92). Fourth tarsomere with long lobe beneath, completely divided into 2 parts; metepisternum broad, approximately twice as long as wide .. .. 40. Schizopodidae

— Fourth tarsomere with variable entire, undivided lobe beneath (Figs. 4-6.41, 54-55.41); metepisternum narrow, at least 3 times as long as wide (Figs. 9.41, 11.49, 18-19.41, etc.) or almost completely concealed under elytra (Fig. 15.41) .... 41. Buprestidae

94(91). Legs retractile, rotated forward in repose, with tibia held anterior to femur; profemur with flange on posterior face covering tibial excavation, protibia grooved to receive tarsus; usually with margined excavations on propleuron, mesosternite, and ventrites to receive legs.. 95

— If legs retractile, protibia held posterior to or ventral to femur; profemoral flange, if present, located on anterior face 96

95(94). Mentum strongly sclerotized, expanded, covering labium and maxillae (Fig. 2.67); head not deflexible; antenna covered by prolegs in broad sternopleural pocket (Fig. 2.67); ventrites 1 and 2 excavate for metathoracic leg; mesotibia with marginal spines; length 4-9 mm .... 67. Nosodendridae

— Mentum normal, head usually retractable into pronotum to anterior margin of eyes (one exception) (Fig. 1.46); antennae received in internal pronotal cavities beneath head, external anterior pronotal cavities or partly in sternopleural grooves and partly under legs against hypomeron; excavation for metathoracic leg, if present, limited to ventrite 1; margin of mesotiba not spinose; length 1-2 mm .. .. 46. Limnichidae

96(94). Elytra with thumb-like process on inner lateral surface near subapical curve, locking into ventrite 5 .... 53. Artematopodidae

[NOTE: Elytra must be separated from side of abdomen to see this character.]

— Elytra without such a locking device 97

97(96). Posterior angles of pronotum with short discal carinae (Fig. 1.54); procoxal cavity with narrow lateral extension at pleurosternal suture .... 54. Brachypsectridae

— Posterior angles of prontum without short discal carinae; procoxal cavity broad at pleurosternal suture 98

98(97). Propleuron extended mesad behind procoxa for approximately half length of trochantin; length 10-15 mm 99

— Margin of propleuron curved laterad posteriorly, not extended mesad posterior to procoxa; length 1-8 mm 100

99(98). Posterior margin of pronotum crenulate; mesotibial spines subequal in size, smooth; antenna compressed serrate (Fig. 1.51); tarsomeres simple; empodium large and setose .... 51. Eulichadidae

— Posterior margin of pronotum simple; mesotibial spines unequal in size, finely serrate; antenna cylindrical-serrate (Fig. 1.38); tarsomeres 1-4 with large, divided membranous lobes; empodium absent .... 38. Dascillidae

100(98). Posterior edge of pronotum simple; last tarsomere much longer than others, usually half or more total length of tarsus .... 43. Elmidae
— Posterior edge of pronotum crenulate; last tarsomere subequal in length to first .... 48. Psephenidae

101(76). Head with subgenal ridges that fit against procoxae when head deflexed; prosternum in front of coxa narrow, shorter than intercoxal process .... 37. Scirtidae
— Head without subgenal ridges, genae usually not in contact with procoxae; prosternum in front of coxae nearly as long as or longer than intercoxal process 102

102(101). Metacoxal plates large, plate-like, longer medially than metasternite, hiding most of metafemur, even when fully extended (Figs. 8-9.35) .... 35. Eucinetidae
— Metacoxal plates narrow, forming either a parallel plate or simple carina; metafemur fully visible 103

103(102). Length of body 4 or more times maximum width (Fig. 1-2.71); male maxillary palp complex, multilobate .... 71. Lymexylidae
— Length of body 2.5 or less times maximum width; maxillary palp not branched 104

104(103). Prosternal intercoxal process complete, reaching behind procoxa to level of mesosternum; posterior portion of hypomeron not extending behind procoxa; elytral epipleuron with an internally carinate edge complete to suture; head with face narrowed; clypeal margin straight; 3 basal ventrites connate .... 48. Psephenidae
— Prosternal intercoxal process incomplete, not reaching beyond midpoint of procoxa; posterior portion of hypomeron variable behind procoxa; elytral epipleuron narrowed before reaching suture (complete in one genus); head with face not greatly narrowed; clypeal margin emarginate (Figs. 5-6.21, 11.21); all ventrites free 105

105(104). Elytra with 9 or 10 punctate striae (Figs. 1-4.18); posterior portion of hypomeron extending up to half the distance to mesal edge of procoxa; length 7-14 mm .... 18. Agyrtidae
— Elytra without punctate striae, otherwise variable, irregularly punctate (Fig. 1.21), with complex low sculpture (Fig. 2.21) or up to 3 carinate costae (Fig. 3.21); posterior portion of hypomeron not extending behind procoxa or extending only a short distance mesad of lateral edge of procoxa (Figs. 9-10.21); length 7-45 mm .... 21. Silphidae

106(64). Hind coxae widely separated by broad, truncate intercoxal process of first ventrite 107
— Intercoxal process of first ventrite absent, acute or rounded 109

107(106). Mesocoxal cavities open laterally, closure involving mesepisterna (Georissinae) .... 13. Hydrophilidae
— Mesocoxal cavities open or not; if open, closure solely involving mesepimeron 108

108(107). Antenna geniculate, club usually of 3 antennomeres; elytra short and truncate, exposing 2 non-flexing terga; body compact .... 15. Histeridae
— Antenna not obviously geniculate, clubbed or not; elytra rarely exposing 2 terga, IF 2 terga exposed, THEN exposed abdominal segments flexible, body not oval or body cylindrical and compact 109

109(106,108). Procoxae with exposed trochantin 110
— Trochantin concealed or absent 128

110(109). Metacoxa extending laterally to reach elytral epipleuron or side of body, no visible contact between metathorax and first ventrite 111
— Metacoxa not reaching elytron, first ventrite and metathorax visibly in contact laterad coxa 116

111(110). Hind tarsus with 5 tarsomeres 112
— Hind tarsus with 4 tarsomeres 157

112(111). Head with temples and occipital ridge distinct, occipital ridge closely fitting against pronotum, constricted behind to a distinct neck (difficult to see when head is retracted with ridge and temples against pronotum); elytra with strong characteristic sutural stria, no other striae evident (Fig. 2.19); 11 antennomeres, gradual club of 3-4 antennomeres (Fig. 7.19); 4 (females) or 5 (males) ventrites (*Colon*, see couplet 52) .... 19. Leiodidae
— Head without ridge and constricted neck that fits against pronotum; elytra striate or not, but not as above; antenna variable; at least 5 ventrites 113

113(112). Prosternal process between coxae distinctly elevated above level of prosternum, apex strongly curved dorsally, reaching level of postcoxal extensions of hypomeron; cervical sclerites absent; antenna not clubbed; elytra glabrous or subglabrous; length 8-20 mm .... 120. Cerambycidae
— Prosternal process not elevated between coxae nor with apex strongly curved dorsad; cervical sclerites present; antenna clubbed or not; elytra densely to sparsely setose, subglabrous or glabrous; length 1-24 mm 114

114(113). Procoxae not projecting distinctly below intercoxal process, large and transverse; antenna distinctly clubbed; prothorax with sharp lateral margins; IF procoxae slightly projecting, THEN antenna distinctly clubbed and tarsi not lobed beneath; not bright red .... 72. Trogossitidae
— Procoxae projecting distinctly below intercoxal process, conical or transverse (Figs. 21.73, 116-117.73); antennae variable; margins of prothorax variable; IF procoxae are only slightly projecting, THEN antennae feebly clubbed (Fig. 6.73), tarsi lobed beneath (Fig. 8.73) AND color bright red 115

115(114). Tarsi not lobed beneath; procoxal cavity strongly transverse; labrum subtruncate to convex, rounded or acute (Figs. 18-19.74); eye not emarginate (Figs. 18-19.74); antenna rarely with distinct apical club, and if so, club of 5 or more antennomeres; elytra usually confusedly punctate; pronotum and abdomen sometimes with eversible glands (Fig. 2.74) .... 74. Melyridae

— Tarsi with lobes on multiple tarsomeres (Figs. 8-9.73, 50.73, 87.73); procoxal cavity circular (Figs. 21.73, 54.73), elongate or slightly transverse (Fig. 13.73); labrum subtruncate to concave or deeply emarginate (Fig. 77.73); eye often emarginate (Figs. 19-20.73, 77.73, 114-115.73); antenna usually apically clubbed, club of 1 or more antennomeres (Figs. 5-7.73, 30-37.73, 42-43.73, 88-97.73, 104-112.73); elytra often punctate-striate (Fig. 1.73); pronotum and abdomen never with eversible glands .... 73. Cleridae

116(110). Elytra short, completely exposing 1 or more tergites (Figs. 1.76, 5-6.77, 1.78, 2.79) 117

— Elytra covering all of abdomen or exposing apex of 1 tergite 120

117(116). Procoxal cavities broadly open (by more than half width of coxa); labium with 2 palpomeres; abdominal intercoxal process truncate; pygidium and last ventrite longer than preceding 4 combined (Fig. 1.78) .... 78. Smicripidae

— Procoxal cavities closed or narrowly open (by less than half width of coxa); labium with 3 palpomeres or non-articulated; abdominal intercoxal process acute to broadly rounded or absent; pygidium variable 118

118(117). Labial palps non-articulated; prosternal process elevated between procoxae and strongly curved dorsally behind .... 76. Brachypteridae

— Labium with 3 palpomeres; prosternal process flat or elevated between procoxae, but not strongly curved dorsally behind 119

119(118). Antenna with 10 antennomeres, club of only 1 antennomere; elytra more than twice as long as wide (Rhizophaginae) ..... 79. Monotomidae

— Antenna with 10 or 11 antennomeres, club of 3 or more antennomeres; elytra less that twice as long as wide .... 77. Nitidulidae

120(116). Mesotarsus with 4 tarsomeres; tarsal lobes, if present, small, not obscuring penultimate tarsomere 121

— Mesotarsus with 5 tarsomeres, fourth possibly obscured by enlarged lobe of third (pseudotetramerous) 123

121(120). Lateral margin of pronotum crenulate (Fig. 4.98), antennal insertions concealed from above (*Sphindocis*) .... 98. Ciidae

— Lateral margins of pronotum smooth or minutely denticulate, antennal insertions visible from above 122

122(121). Body nearly spherical, capable of being rolled into a ball; mandibles resting against metasternum in retracted position (*Cybocephalus*) .... 77. Nitidulidae

— Body flattened-cylindrical, not at all spherical (Fig. 1.96) (Mycetophaginae) .... 96. Mycetophagidae

123(120). Antenna with 10 antennomeres, one involved in club (Fig. 1.79) (Rhizophaginae) .... 79. Monotomidae

— Antenna with 10 or 11 antennomeres, if clubbed, club of 2 or more antennomeres 124

124(123). Body extremely flattened; elytra nearly parallel-sided, disc almost perfectly flat between rounded lateral carinae running from humeri to near apex, setting off vertical sides and guttered epipleural margin (Fig. 1.82); either large (>10 mm) and red with expanded temples (Fig. 1.82) or small (<5 mm) and dull brown without temples .... 82. Cucujidae

— Body not so distinctly flattened; elytra distinctly transversely arched, not fitting other combinations above 125

125(124). Dorsal face of mandible with tubercle that fits into cavity on clypeus, setose cavity at base, hidden when mandibles are closed (mycangium) (Fig. 3.75); elytra with scutellary striole (Figs. 8-11.80, 1.75); antenna with 2 or 3 antennomeres forming club (Figs. 1.75, 4-5.75, 8-11.75); body oval to cylindrical (Figs. 1.75, 8-11.75) .... 75. Sphindidae

— Mandible without dorsal mycangium; elytra without scutellary striole; antenna and body shape variable 126

126(125). Antenna with a distinct club AND meso- and metatarsi with equal numbers of tarsomeres ... 127

— Metatarsus with one fewer tarsomere than mesotarsus; antenna distinctly clubbed or not 157

127(126). Pygidium at least partially exposed, strongly sclerotized, punctate, distinctly different from other tergites (Figs. 1.77, 7-8.77); tibiae usually spinose or denticulate on external margin .... 77. Nitidulidae

— Pygidium not exposed, not strongly sclerotized, similar to other tergites (Fig. 1.88); tibiae smooth on external margin .... 88. Byturidae

128(109). Antennal insertions concealed from above by lateral expansion of frons (Figs. 7-10.106, 53-54.106, 65-67.106, etc.); AND 3 basal ventrites connate (Figs. 4-5.106), fourth and fifth movable (Figs. 138-139.106); AND procoxal cavities closed by the mesad extension of the posterior portion of the hypomeron; AND procoxal process not expanded laterally at apex to close procoxal cavities (Figs. 131-132.106, 135-137.106); antenna usually with 11 antennomeres (rarely with 9 or 10 antennomeres) . ... 106. Tenebrionidae

— Without this combination of characters 129

129(128). Abdomen with first 4 ventrites connate 130

— Abdomen with fewer than 4 ventrites connate 131

130(129). Antenna serrate or pectinate (Fig. 1.55); antennal insertions exposed from above (Fig. 1.55); metacoxa laterally reaching epipleuron; intercoxal process of prosternum with long, notched, apical projection, received in deep mesosternal cavity to form clicking mechanism; last ventrite without submarginal groove; mentum without setose pit .... 55. Cerophytidae

— Antenna moniliform, clavate or capitate (Figs. 3.104, 5.104, 1-2.105, 8.105); antennal insertions concealed from above (Fig. 5.104); metacoxa not reaching elytron, first ventrite and metepimeron in contact laterad coxa and mesad epipleuron; prosternal process broad, widened apically (Figs. 2.104, 4.104, 5-7.105); last ventrite usually with submarginal groove; males often with median setose pit on mentum .... 105. Zopheridae

[NOTE: The Monommatini, treated in Chapter 104, key out here. See Status of Classification and the key in Chapter 105]

131(129). Metatarsus with 5 tarsomeres, first reduced, often difficult to see; metatarsomere 1 either hidden in apical excavation of metatibia OR, IF metatarsal insertion fully exposed, THEN first metatarsomere less than 1/4 length of second and obliquely attached under second (may only be visible from below in oblique distal angle); elytra covering pygidium; antenna with distinct club of 2-4 capitate to elongate-loose antennomeres (Figs. 14-16.69, 21-22.69, 31-33.69); AND one of the following combinations: 1) head somewhat to distinctly hypognathous; pronotum hood-like, projecting anteriorly (or ventrally in some fully hypognathous species) beyond anterolateral angles above head (Figs. 1.69, 2.69, 4.69,11-13.69,17-20.69) (Bostrichinae, Dinoderinae, Endecatominae); 2) head prognathous (Figs. 27-30.69); intercoxal process of first ventrite truncate, metacoxae widely separated, metacoxa reaching elytra laterally, separating metathorax and first ventrite (Lyctinae); OR 3) head prognathous (Fig. 3.69); procoxae transversly cylindrical, projecting at sides, proleg attached and directed laterally, femur and trochanter large (Polycaoninae) .... 69. Bostrichidae

— First tarsomere not so reduced; other characters variable; if first metatarsomere is reduced relative to second, pronotum not hood-like, head not hypognathous AND/OR metatarsus of 4 tarsomeres, pygidium exposed, first ventrite not widely truncate between metacoxae; procoxae not transversely cylindrical and projecting at sides 132

132(131). Mesotarsus with 4 distinct tarsomeres 133

— Mesotarsus with 5 tarsomeres, or tarsi pseudotetramerous 144

133(132). Mesocoxal cavities closed laterally 134

— Mesocoxal cavities open laterally 140

134(133). Antennal insertions concealed from above 135

— Antennal insertions exposed from above.... 136

135(134). Eyes usually present; IF eyes absent, THEN elytra with flat tubercles (Fig. 11.103) .... 103. Colydiidae

— Eyes absent; elytra smooth (*Aglenus*) .... 116. Salpingidae

136(134). Genae with pair of anteriorly directed horns extending beyond labium, visible from above... .... 107. Prostomidae

— Genae lacking gular horns 137

137(136). Abdomen with 6 ventrites; pronotum usually large, hood-like, covering or nearly covering head (Figs. 1-9.94); pygidium usually exposed (Figs. 6-7.94, 10-11.94); epipleuron incomplete; frontoclypeal suture absent; length less than 2 mm .... 94. Corylophidae

— Abdomen with 5 or 6 ventrites; pronotum never hood-like, head visible from above; pygidium, epipleuron and frontoclypeal suture variable; IF with 6 ventrites, THEN length 4 mm or greater and frontoclypeal suture present 138

138(137). Antenna longer, reaching to or beyond middle of pronotum, club loose; pronotum usually with pair of sublateral discal carinae or grooves, running from base laterad of basal pits (Fig. 1.92); body usually round to ovoid .... 92. Endomychidae

— Antenna shorter, not reaching beyond middle of pronotum, club compact; IF pronotum with discal carinae or grooves, THEN usually a median groove or pit and body elongate 139

139(138). Posterior margin of last ventrite crenulate OR body distinctly oval, length no more than twice maximum width; antenna with 8, 9, or 10 antennomeres; hind trochanter obliquely attached to femur, but distinctly separating coxa from femur .... 91. Cerylonidae

— Posterior margin of last ventrite never crenulate, antenna with 10-11 antennomeres; body elongate, at least 2.75 times maximum width; hind trochanter offset so that femur and coxa are in contact or nearly so .... 90. Bothrideridae

140(133). Metacoxae separated by more than 1/2 transverse coxal diameter 141

— Metacoxae separated by less than 1/2 transverse coxal diameter 142

141(140). Procoxal cavities narrrowly closed; pro- and mesocoxae strongly transverse; mandible tucked into cavity when closed, not visible from side; antenna with 9 antennomeres, last 5 forming club (Fig. 3.94); pronotum not grooved or carinate on disc (Fig. 3.94); small beetles, less than 2 mm in length (*Orthoperus*) .... 94. Corylophidae

— Procoxal cavities narrowly to widely open; pro- and mesocoxae circular to slightly transverse; mandible visible from side; antenna with 8-11 antennomeres, if clubbed, club of 1, 2 or 3 antennomeres (Figs. 1-3.92); pronotum usually with submarginal grooves or carinae, especially basally (Figs. 6-8.92); size 1-10 mm; IF less than 2 mm and lacking grooves or carinae on pronotum (*Eidoreus*), THEN antenna with 10-11 antennomeres, 1 or 2 of which form a distinct club .... 92. Endomychidae

142(140). Intercoxal process of first ventrite absent, no part of ventrite extending between coxae to contact metaventrite; first ventrite lacking margined metacoxal cavitites; metacoxae conical and projecting; body soft; small triangular part of morphological abdominal sternite 2 usually visible laterad of metacoxa (*i.e.*, ventrite 1 small, divided); often colorful, with red, yellow or metallic blue/green markings (Fig. 10.69); length 5-12 mm (Psoinae)...... 69. Bostrichidae

— Intercoxal process of first ventrite complete; first ventrite with margined coxal cavities; metacoxae transverse; body fully sclerotized; ventrite 1 closing anterolateral angle between metacoxa and abdomen, not divided by metacoxae; never metallic; length 0.5-6.5 mm 143

143(142). Body elongate-oval and somewhat cylindrical; pronotum usually very convex in transverse section, edges often directed ventrally (Figs. 1.98, 6.98, 10.98, 20.98, 36-42.98); pronotum without basal pits or impressions; head or pronotum of male often with horns or tubercles (Figs. 13-14.98, 33.98, 36.98, 39-40.98, 42.98, 43-45.98); antenna with 8-10 antennomeres and club of 2-3 antennomeres; males often with pubescent median fova on first ventrites; head without distinct temples or neck. One species from California has relatively flat pronotum with crenulate margins directed laterally (Fig. 4.98), antenna with 11 antennomeres and 3 antennomeres in club, but 2 basal sternites are connate .... 98. Ciidae

— Body oval to elongate oval, usually somewhat dorsoventally depressed; pronotum usually weakly convex transversely, edges directed laterally; pronotum with 2 basal pits or impressions laterad scutellum (Fig. 1.96) (sometimes in posterior marginal groove and difficult to discern); head and pronotum without horns or tubercles; antenna with 11 antennomeres, last 2-5 forming club; all ventrites free, without median fovea. One genus (length less than 2 mm) somewhat cylindrical, with very convex pronotum in transverse section, with head abruptly constricted behind short temples to form distinct neck.......... 96. Mycetophagidae

144(132). Abdomen with six ventrites AND metatarsi with five tarsomeres; terminal maxillary palpomere (4) shorter and narrower than penultimate (Figs. 2-8.20); shape rather characteristic (Figs. 25-46); length 0.6-2-7 mm ...... 20. Scydmaenidae

— Abdomen with four or five ventrites; tarsi variable; terminal maxillary palpomere (3 or 4) as wide or wider AND/OR as long or longer than penultimate; size variable 145

145(144). Pregular area on each side with a laterally facing surface bearing setose pit or cavity near end of distinct antennal groove; first ventrite with postcoxal lines .... 89. Biphyllidae

— Pregular area without laterally facing setose pit; antennal grooves and postcoxal lines variable 146

146(145). First ventrite much longer than second (measured behind coxa); elytra without punctate or impressed striae (traces of striae occasionally visible through cuticle, but not expressed on the surface); epipleuron distinct in basal half, not reaching apex (usually narrowed at level of third ventrite); genae carinate and projecting ventrally between eye and mentum; apex of elytra with double suture or "subapical gap" caused by wide flange of elytral coupling system; elytra complete, exposing at most tip of last tergite....................... 85. Cryptophagidae

[NOTE: Two genera of tiny (<1.3 mm) cryptophagids (*Amydropa*, Baja California, and *Hypocoprus*, Rocky Mountain region) lack the subgenal carinae. *Hypocoprus* has the first 2 ventrites subequal and the pygidium exposed while *Amydropa* lacks the double suture on the elytra. The other characters fit these 2 rare genera. *Ambydropa* has greatly reduced eyes (10 facets or fewer) and *Hypocoprus* has distinct temples.]

— Not fitting this combination of characters, EITHER with the first ventrite short, elytra striate, epipleuron complete to apex, gena flat between eye and mentum, OR elytra not covering most of pygidium 147

147(146). Metatrochanter transversely or obliquely attached to femur, distinctly separating femur from coxa (Figs. 28-29.I) 148

— Metatrochanter obliquely attached to femur, offset so that femur abuts coxa (Fig. 26-27.I) 157

148(147). Antennal insertions approximate or separated by less than 1/2 width of head behind eyes AND pronotum without lateral carinae; metatarsus with 5 tarsomeres; metatrochanter elongate, cylindrical (Fig. 29.I, 69.70, 74.70) (Ptininae).. .... 70. Anobiidae

— Without combination of narrowly separated antennal insertions and no lateral carina on pronotum; other characters variable........ 149

149(148). Pronotum with sublateral lines or grooves that extend from base anterad midpoint, often to anterior margin (Figs. 1-11.83); head usually with sublateral lines from median margin of eye to pronotum; lateral margins of pronotum smooth or wavy or with few obtuse angles (Figs.1-11.83), not acutely denticulate or serrate; head not sharply constricted to a distinct neck; body oval to elongate, subcylindrical to strongly dorsoventrally flattened (Figs. 1-11.83) .... 83. Laemophloeidae

— Pronotum usually without sublateral lines that extend from base anterad midpoint; head variable; IF pronotum with sublateral lines that extend from base to or beyond midpoint, THEN lateral margins of pronotum sharply denticulate, anterior angles acutely projecting AND/OR head sharply constricted behind small temples (Figs. 17.80, 19.80, 22.80); body variable 150

150(149). Mesocoxal cavities open laterally............... 151

— Mesocoxal cavities closed laterally 153

151(150). Antenna with 10 antennomeres, distinctly clubbed; elytra shortened, exposing all of pygidium (Fig. 2.79); head abruptly constricted to form neck; 1-4 mm .... 79. Monotomidae

— Not fitting one or more of above characters 152

152(151). Body elongate, flattened (Figs. 1.80, 16-25.80); meso- and metatarsi with same number of tarsomeres; head usually with distinct temples before abrubtly constricted neck (Figs. 1.80, 4.80, 16-25.80); procoxae either closed behind (Fig. 3.80) or, if open (Brontinae, Fig. 2.80), elytra transversely flat or slightly concave between slightly to distinctly raised interstria between stria 6 and 7; elytron with scutellary striole; base of mandible with dorsal setose pit (mycangium) hidden beneath clypeus when closed; antenna filiform, with scape more than 3 times length of pedicel .... 80. Silvanidae

— Metatarsus with one tarsomere fewer than mesotarsus; other characters variable..... 157

153(150). Body shining, oval and strongly convex; pronotum tightly embracing elytra (Fig. 1.84), pronotum laterobasally with a vaguely transparent, thin flange which slides over a smooth area on base of humeral angle of elytron, this area on elytron delimited posteriorly by a thin carina; pronotum and elytra with wide propleura and epipleura, lateral margins sharp, explanate, strongly directed ventrally so that lateral margins are far below level of procoxa and mesad epipleural margin, dorsal surface forming an inverted "U" in transverse section; tarsal claw toothed or appendiculate ...... 84. Phalacridae

— Body usually not so evenly oval, pronotum not coadapted to pronotum in the above manner, without described flanged basal angles of pronotum or associated elytral area; lateral margins of pronotum and elytra laterad, rather than ventrad, to procoxa and eplipleura; tarsal claws toothed only in groups with pronotum narrowed behind 154

154(153). Meso- and metatarsi with same number of tarsomeres; face often with beaded lateral margins (Figs. 1.81, 3-4.86, 9-10.87) 155

— Mesotarsus with one more tarsomere than metatarsus; face without beaded lateral margins 157

155(154). Gular sutures confluent; genae expanded anteriorly, plate-like, concealing maxillae (Fig. 4.81) .... 81. Passandridae

— Gular sutures separate or absent; genae not so expanded 156

156(155). Procoxal cavities usually open behind (Fig. 8.86); terminal maxillary palpomere narrow, elongate; IF procoxal cavities closed behind (Fig. 7.86), THEN closure by mesad extension of hypomeron, length less than 3 mm and pronotum somewhat narrowed near base (*Cryptophilus*) .... 86. Languriidae

— Procoxal cavities closed behind by laterad expansion of the prosternal process (Figs. 5-6.87); terminal maxillary palpomere often securiform (Figs. 5-6.87), or narrow and elongate (Fig. 4.87); length 3-22 mm .... 87. Erotylidae

157(111,126,147,152,154).Last visible segment of abdomen forming a terminal spine (Fig. 1.101); body wedge-shaped, humpbacked; head retracted to hypognathous position (Fig. 1.101); metatibia and metatarsus usually with oblique or transverse, comb-like serrate ridges subapically on lateral faces (Figs. 2-7.101).. 101. Mordellidae

— Abdomen not prolonged into a terminal spine; body otherwise variable; metatibia and metatarsus without comb-like serrate ridges as above, IF similar combs are present, THEN they are apical 158

158(157). Tarsal claw with a ventral blade or elongate lobe beneath (Figs. 13-16.111) (reduced to a large fused tooth ending about 2/3 length of upper blade in *Phodaga* [Fig. 15.111] SW-USA); head sharply or gradually constricted behind eyes to distinct neck 159

— Tarsal claw without ventral blade or elongate lobe beneath, if claw toothed or apendiculate, not as in Fig. 15.111; head constricted or not 160

159(158). Ventral appendage of tarsal claw usually lobe-like, membranous, occasionally blade-like and sclerotized; elytra usually meeting along suture to very near apex, which may be narrowly separately rounded (Fig. 1.110); lateral margin of pronotum absent, complete (Fig. 2.110), or indicated only at base (Fig. 3.110); mesocoxal cavities usually narrowly separated, occasionally contiguous; maxillae not forming sucking tube; antenna without club or with vague to distinct club of 3 antennomeres; hind wing with well-developed radial cell; if pronotal margin completely absent, antenna with at least vague indication of club in last 3 antennomeres and mesocoxal cavities narrowly separated; if elytra broadly separately rounded, pronotum with lateral carina at base (Fig. 3.110) ....110. Stenotrachelidae

— Ventral appendage of tarsal claw blade-like and sclerotized (Figs. 13-16.111); elytra usually diverging along suture before apex, broadly separately rounded (Fig. 1.111); pronotum lacking marginal carina laterally; antenna without club of 3 antennomeres; mesocoxal cavities contiguous; maxillae usually normal; radial cell absent in hind wing; if elytra meeting on suture to very near apex, maxillae modified into sucking tube that extends beyond mandibular apices .... 111. Meloidae

160(158). Base of pronotum with marginal groove (Fig. 2.117, 6.117) that extends laterally onto hypomeron, ending in a pit near posterior margin of coxa (Fig. 3.117); pronotum narrowed posteriorly, not margined laterally (Figs. 2.115, 1-3.117); head sharply narrowed behind distinct temples to form narrow neck (Figs. 2.115, 1-3.117, 18-20.117); elytra sparsely to densely setose 161

— Basal groove of pronotum, if present, not ending in pit on hypomeron; pronotum margined laterally or not; elytra with or without setae 162

161(160). Antenna with last 3 antennomeres forming elongate apical club, club more than 1/2 total length of antenna (Fig. 2.115) (*Anisotria*) .... 115. Pyrochroidae

[Note: *Anisotria* was moved to the Anthicidae by Lawrence *et al.* 1999.]

— Antenna not clubbed, with weak, short club, with last 3 antennomeres shorter than 1/2 total length of antenna (Fig. 1.117) OR with only last antennomere long (subequal to antennomeres 7-10) .... 117. Anthicidae

162(160). Mesocoxal cavities closed laterally 163

— Mesocoxal cavities open laterally 165

163(162). Basal 3 ventrites connate; antenna with 11 antennomeres, submoniliform/triangular, filiform, serrate to subflabellate (Figs. 1-7.112, 9-10.112); cervical sclerites present .... 112. Mycteridae

— Two or no ventrites connate; antenna with 10 to 11 antennomeres, moniliform to capitate (Figs. 1-5.116, 12-13.116); cervical sclerites absent 164

164(163). Prothorax with pleurosternal suture ending in a large setose pit at antero-lateral margin of procoxal cavity; 2 basal ventrites connate; 11 antennomeres; 1.5-3.8 mm; deserts of western USA from Idaho to Mexican border (*Cononotus*) .... 115. Pyrochroidae

— Prothorax with or without pleurosternal suture, lacking large setose pit on anterior margin of procoxal cavity; all ventrites usually free, or 2 basal ventrites connate (*Aegialites*); 10-11 antennomeres; length 1.5-7 mm; widespread in forests and Pacific beaches (*Aegialites*); if in deserts, antenna with 10 antennomeres (*Dacoderus*) .... 116. Salpingidae

165(162). Body deep, mildly to distinctly wedge-shaped (Figs. 1-2.102, 4-6.102); antenna serrate, pectinate or flabellate, often bipectinate or biflabellate (Figs. 8-18.102); vertex often inflated and narrowed above eyes in frontal view; vertex usually extending dorsally above plane of pronotum in lateral view, vertex and pronotum at least coplanar; tarsi toothed, bifid or pectinate; maxillary lobes sometimes stylet-like, extending beyond tips of mandibles .... 102. Ripiphoridae

— Body usually not deep and wedge-shaped, IF body deep and wedge-shaped, THEN antenna simple and head coplanar with or slipping under front margin of pronotum; tarsi variable; maxillary lobes not stylet-like 166

166(165). Pronotum lacking lateral carina 167

— Pronotum with complete or incomplete lateral carina 171

167(166). Metacoxa extending laterally to elytron or side of body, completely separating metepisternum and first ventrite 168

— Metacoxa not reaching elytron or side of body, metepisternum and first ventrite in contact laterad metacoxa 170

168(167). Tarsi appearing 4-4-3 (actually 5-5-4, pseudotetramerous/pseudotrimerous); eyes coarsely faceted, appearing hairy, interfacetal setae as coarse, long and dense as those on front and sides of head adjacent to eyes; 1-4 mm .... 118. Aderidae

— Tarsi distinctly 5-5-4; eyes with or without interfacetal setae, IF interfacetal setae present, THEN setae not as coarse, long or obvious as on front and sides of head adjacent to the eyes; 4-21 mm 169

169(168). Head prognathous, not abruptly constricted to a narrow neck, lacking distinct temples (Fig. 1.109); anterior portion of prosternum as long or longer than prosternal process; first 2 ventrites connate .... 109. Oedemeridae

— Head distinctly declined, abruptly constricted to form narrow neck behind distinct temples; anterior portion of prosternum shorter than prosternal process (Eurygeniinae) .... 117. Anthicidae

170(167). Elytra distinctly setose; eye emarginate anteriorly; penultimate tarsomere with large lobe beneath (Figs. 1.115, 3.115) .... 115. Pyrochroidae

— Elytra glabrous; eye not emarginate; penultimate tarsomere simple (Fig. 1.114) (*Pytho* and *Priognathus*) .... 114. Pythidae

171(166). Metacoxa extending laterally to elytra or side of body, completely separating metepisternum and first ventrite; mesotibial spurs serrate, pectinate or pubescent 172

— Metacoxa not reaching elytra or side of body, metathorax and first ventrite at least narrowly closing metacoxal cavity laterally; mesotibial spurs variable 174

172(171). Head vertically narrowed behind eyes to form narrow neck, head not received into prothorax, either bulging beyond pronotal margin, or fitting closely against pronotal margin so that head in lateral view has a posterior carina or crest meeting anterior margin of pronotum (Fig. 5.K) .... 119. Scraptiidae

— Head gradually narrowed behind eyes, fitting into pronotum in a telescoping manner 173

173(172). Tarsus without lobes on penultimate tarsomere; sutural stria deeply impressed near apex of elytra, distinctly more so than in basal half; 2 basal ventrites connate; metatibia longer than first metatarsomere; prosternal intercoxal process level with ventral surface of non-projecting procoxae; EITHER intercoxal process long, parallel-sided, reaching behind procoxae OR prosternal process incomplete, narrowed apically between coxae; length 7-13 mm .... 108. Synchroidae

— Tarsus with penultimate tarsomere lobed beneath (Figs. 1.100, 21-22.100) OR metatibia shorter than first metatarsomere; IF sutural stria deeply impressed near apex, THEN also impressed on

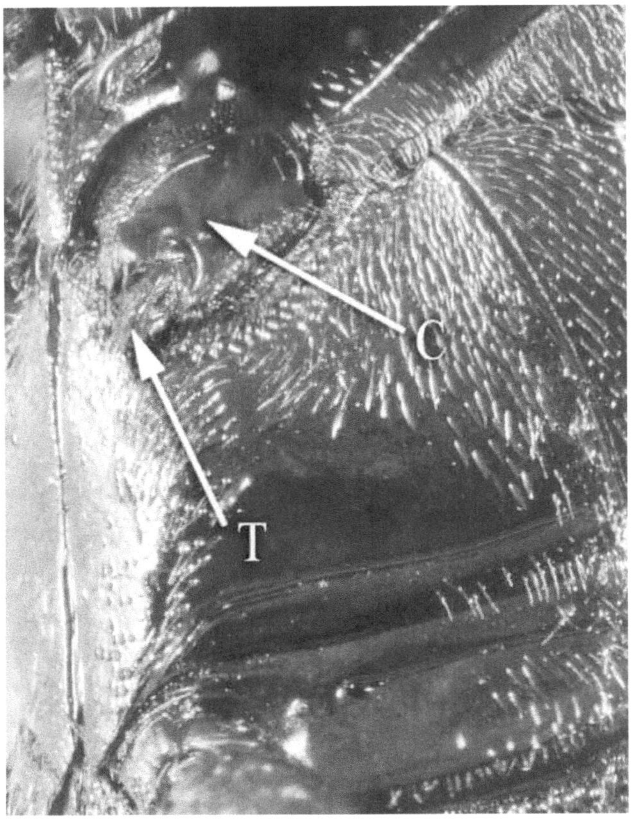

FIGURE 4.K. *Eustrophinus arizonensis* (Horn) (99. Tetratomidae) meso- and metacoxa, ventral view. C – polished ventral face of mesocoxa. T – mesotrochanter, femur removed.

basal half (Fig. 1.100); 2 basal ventrites connate or all free; prosternal process incomplete, not reaching behind procoxae, narrowed apically, resting below ventral surface of usually projecting procoxae; length 2-20 mm .... 100. Melandryidae

174(171). Procoxal cavities closed behind; first 2 ventrites connate; body strongly rounded (Fig. 1.97)... .... 97. Archeocrypticidae

— Procoxal cavities open behind; ventrites connate or free; body form variable, often elongate 175

175(174). Elytra with sutural and epipleural margins elevated; strongly elevated carina running from humeral angle to near apex resulting in distinctly concave elytral disc; pronotum with median longitudinal elevated carina on basal 1/3, deep transverse grooves with pits at each end on either side of carina (*Ischalia*)........... .... 117. Anthicidae

— Elytra and pronotum without strongly elevated carinae, elytral disc convex 176

176(175). Prosternal intercoxal process incomplete or absent, not separating procoxae 177

— Prosternal intercoxal process complete, fully separating procoxae 178

177(176). Antenna filiform; terminal labial and maxillary palpomeres expanded apically; prosternum shorter than diameter of procoxae; mesotibial spurs pubescent or serrate; tarsi lobed on penultimate tarsomeres (Fig. 22.100) (Osphyinae)...................... 100. Melandryidae

— Antenna with long, serrate club of last 3 antennomeres (Figs. 7-8.114); terminal labial and maxillary palpomeres cylindrical; prosternum as long as procoxal diameter; mesotibial spurs smooth; tarsi not lobed (*Trimitomerus*) .... 114. Pythidae

178(176). Tarsi simple, lacking lobes below 179

— At least some tarsomeres distinctly lobed below 181

179(178). Median longitudinal line (discrimen) of metasternum short, extended from hind margin less than 1/2 total length of ventrite; mesocoxa normal, convex and punctate anterior to trochanteral insertion... 180

— Median longitudinal line (discrimen) of metaventrite longer, extended from hind margin more than 1/2 total length of ventrite (Fig. 4.K); mesocoxa with unique, polished, ventral face anterior to trochanteral insertion (Fig. 4.K) (this polished area is rubbed by flat opposing surface of trochanter and base of femur when leg rotated is forward).......... 99. Tetratomidae

180(179). Antenna short, not reaching middle of pronotum; apical 3 antennomeres forming a distinct, rather abrupt club .... 113. Boridae

— Antenna longer, reaching base of elytron; apical antennomeres somewhat wider than basal antennomeres, not forming abrupt club (*Sphalma*) .... 114. Pythidae

181(178). Antenna filiform; setae on elytra very short and indistinct, shorter than diameter of punctures; elytra uniform in color; California and Nevada (*Tydessa*)........................... 115. Pyrochroidae

— Antenna strongly serrate; setae on elytra conspicuous, several times longer than diameter

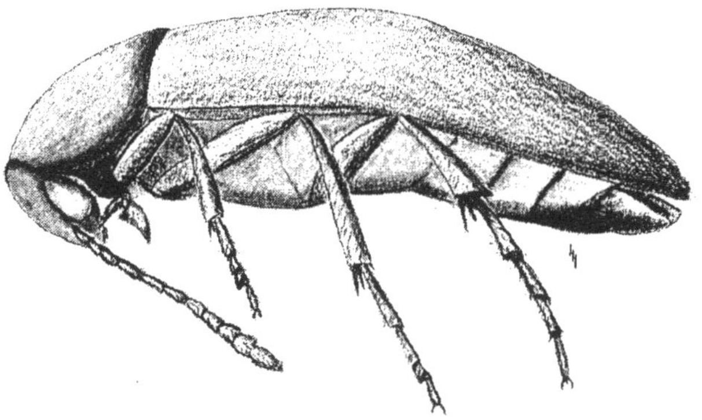

FIGURE 5.K. *Anaspis atrata* Champion (119. Scraptiidae), lateral view (modified from Hatch 1965).

of punctures; elytra reddish with dark markings, a macula around scutellum and transverse band at apical 1/3, usually joined by line along suture; south Texas .... *Polypria cruxrufa* Chevrolat

[NOTE: Family placement *incertae sedis*, see end of Chapter 100.]

182(1). One tarsal claw; eye reduced to a single ommatidium 183

— Two tarsal claws; compound eye normal, reduced or with single ommatidium 184

183(182). Gonopore present (females) ... 61. Phengodidae

— Gonopore not present....................................... .... Larvae (not further keyed)

184(182). Head with median ocellus (female *Thylodrius*)... .... 68. Dermestidae

— Head without ocellus 185

185(184). Head prognathous; pronotum expanded anteriorly, extending over head in retracted position (*Phausis*, *Microphotus*) or head retractile into tubular prothorax (*Pterotus*); distinctly to slightly dorso-ventrally flattened; antenna with 9 or fewer antennomeres; some, possibly all, bioluminescent; widespread (females).......... .... 62. Lampyridae

— Head hypognathous, not retractile into prothorax; body globular-cylindrical; antenna with 11 antennomeres; not bioluminescent; Florida or near ports of entry (female Ripidiinae, North American females unknown)........................... .... 102. Ripiphoridae

Literature cited

ARNETT, R. H., Jr. 1973. Beetles of the United States (a Manual for Identification). American Entomological Institute. Ann Arbor, MI. 1112 pp.

BORROR, D. J., C. A. TRIPLEHORN and N. F. JOHNSON. 1989. An introduction to the study of insects (sixth ed.). Saunders College Publishing. Philadelphia, PA. 875 pp.

BROWN, B. V. 1993. A further chemical alternative to critical-point-drying for preparing small (or large) flies. Fly Times 11: 10.

CHAPMAN, R. A. 1998. The Insects. Structure and Function, 4th ed. Cambridge University Press. Cambridge, MA. 770 pp.

CROWSON, R. A. 1955. The Natural Classification of the Families of Coleoptera. Nathaniel Lloyd. London. 187 pp.

HATCH, M. H. 1965. The Beetles of the Pacific Northwest. Part IV: Macrodactyles, Palpicornes, and Heteromera. University of Washington Publications in Biology 16(4):1-268.

LAWRENCE, J. F. and E. B. BRITTON. 1994. Australian Beetles. Melbourne University Press. Carlton, Victoria. x + 192 pp., 16 pls.

LAWRENCE, J. F., A. M. HASTINGS, M. J. DALLWITZ, T. A. PAINE and E. J. ZURCHER. 1999. Beetles of the World: A Key and Information System for Families and Subfamilies. CD-ROM, Version 1.0 for MS-Windows. CSIRO Publishing. Melbourne.

NICHOLS, S. W. and R. T. SCHUH 1989. The Torre-Bueno Glossary of Entomology. New York Entomological Society and American Museum of Natural History, New York. 840 pp.

PETERSON, A. 1964. Entomological Techniques, How to Work with Insects (tenth ed.). A. Peterson. Columbus, OH. 435 pp.

Taxonomic Index

(**Note:** Taxon names of genus rank and above, both valid and synonyms, are indexed in the keys and classification sections.)

A

Ababa 269, 273
Abagous 740
Abcrana 58
Abdera 420, 421
Abstrulia 415
Abutiloneus 605, 606
Acallepitrix 635
Acalles 762, 764
Acalligrapha 650
Acallodes 749, 754
Acalymma 631, 657
Acalyptini 724, 736
Acalyptus 733, 736
Acamatus 792
Acamptini 724, 727, 759
Acamptus 727, 756, 759
Acantharthrus 753
Acanthinus 553, 555
Acanthocerus 35
Acanthocinini 584, 599
Acanthocinus 585, 599
Acanthoderes 598
Acanthoderini 584, 598
Acanthogethes 315
Acanthoptera 596
Acanthopterus 596
Acanthoscelidae 602
Acanthoscelides 606
Acanthoscelidius 750, 753
Acanthoscelis 753
Acanthotrachelina 776
Acanthurus 70
Acarlosia 763
Acaromimus 697, 698
Acaropsis 698
Acentrinops 742, 746
Achrastenus 771, 776
Achryson 576, 591
Achrysonini 576
Acletus 289
Aclinidia 65
Acmaegenius 768, 775
Acmaeodera 102, 106
Acmaeoderina 102, 105
Acmaeoderini 102, 105
Acmaeoderoides 102, 106
Acmaeoderoidina 102, 106
Acmaeoderopsis 102, 106
Acmaeops 573, 587,
588, 589
Acmaeopsilla 588
Acmaeopsoides 574, 587
Acnemiscelis 753
Acneus 134
Acolpus 231
Acoma 53, 60
Aconobius 476, 493
Acoptus 755
Acratus 57
Acrepis 241
Acribus 315
Acrocyum 634, 660
Acromacer 693
Acropteroxys 344, 345
Acroschatia 488
Acryphalus 804
Actenicerus 166, 169
Actenobium 255
Actenobius 255
Actenodes 104, 108
Actenodina 104
Acylomus 336
Adaleres 769, 781
Adalia 381, 386
Adasytes 286, 296
Adelina 475, 491
Adelinina 483, 491
Adelonia 466, 484
Adelothyreus 154, 155
Adephaga 819
Aderidae 559, 560, 833
Aderina 562
Aderini 561
Aderus 561, 562
Adetus 583, 597
Adiaeretus 804
Adicolytta 527
Adimonia 656
Adistemia 396, 397
Adoceta 176, 177
Adonia 386
Adoniates 385
Adoxini 636, 637, 639, 670
Adoxites 639, 671
Adoxus 671
Adrastites 160
Aedilis 599
Aegialia 43, 45
Aegialiini 42, 45
Aegialites 545, 547, 833
Aegialitinae 545, 547
Aegilopsis 597
Aegomorphus 584, 598
Aegoschema 598
Aegyptobaris 745
Aenigmaticini 392, 393
Aenigmaticum 393
Aeolus 162, 169
Aethecerinus 580, 596
Aethecerus 596
Aethina 312, 314
Aethiopoctines 598
Aethriosia 230
Afissa 388
Afreminae 553
Africaraea 314
Afrocolotes 288
Afroglaresis 15
Agaeocera 107
Agaeocerina 103, 107
Agalissus 595
Agallissini 579, 595
Agallissus 579, 595
Agapanthiini 583, 597
Agasicles 634, 665
Agasphaerops 773, 777
Agathengis 341
Agelastica 656
Agennopsis 597
Ageonoma 461
Agleninae 545, 546
Aglenus 545, 546, 830
Aglycoptera 63
Agnathinae 541, 542
Agnathus 542
Agrabia 386
Agraphini 725, 775
Agraphus 771, 775
Agrilaxia 104, 108
Agrilina 105, 108
Agrilinae 102, 104, 108
Agrilini 104, 105, 108
Agrilus 105, 108
Agriotella 167
Agriotes 164, 167
Agriotina 167
Agriotini 167
Agriotites 160
Agroiconota 624, 647
Agronus 773, 778
Agrostithassa 652
Agrypnides 160
Agrypninae 168
Agrypnini 168
Agrypnus 163, 168
Agyrtidae 823, 828
Ahasverus 324, 326
Airora 265, 266
Akalyptoischion 396, 397
Alabameubria 134
Alaephus 473, 488
Alaites 160
Alamona 62
Alaocybites 732
Alaudes 471, 485
Alaus 163, 169
Albridarollia 65
Alcidion 585, 599
Alesiaires 385
Alethia 480, 499
Algarobius 605, 606
Alindria 266
Alitargus 400
Alitene 652
Aliziarii 387
Allandrini 697, 699
Allandrus 697, 699
Allarthrum 804
Allecula 499
Alleculidae 463
Alleculina 479, 483, 499
Alleculinae 466, 483, 498
Alleculini 466, 479 483, 499
Alledoya 647
Allobregmus 256
Allocoryninae 701
Allocorynus 701
Alloeocnemis 239
Alloeoscelis 598
Allomimus 759
Allomycterops 776
Allonyx 294
Alloparmulus 393
Allophaedon 627, 652
Allophasia 492
Allopoda 565, 566
Allopodini 566
Allopogon 147
Allopogonia 147
Allorhagium 589
Allosirocalus 751, 752
Alloxacis 518
Alniphagus 795, 799
Alobates 482, 500
Alocentron 716, 717
Alonsina 652
Alophini 725, 775
Alosternida 573, 587
Alphites 669
Alphitobiini 468, 479,
483, 497
Alphitobius 479, 492, 497
Alphitophagus 475, 491
Althaeus 605, 606
Altica 636, 662

Alticidae 617
Alticini619, 628, 629, 639, 660
Alticopus 698
Alyca 738
Alycodes 738
Alyctus 739
Alymeris 292
Amalorrhynchus 752
Amalus 749, 752
Amannus 596
Amarantha 497
Amartus 310
Amarygmimi 479
Amarygmini 468, 483, 498
Amaurorhinus 756, 759
Amaurostoma 600
Amblis 108, 649
Amblycerinae 604, 606
Amblycerini 606
Amblycerus 604, 606
Amblyctis 420
Amblycyphrus 486
Amblyderus 553, 555
Amblyophileurus 67
Ambrosiodmus 798, 803
Amechanus 25
Amecocerus 295
Amercedes 727, 741, 746
Americanotimarcha 649
Amilia 600
Ammanus 580
Ammidaenemia 494
Ammidium 494
Ammodonus 475, 493
Amnesia 782
Amniscus 599
Amobium 255
Amotus 774, 780
Ampedidae 160
Ampedina 168
Ampedini 168
Ampedus 164, 168
Ampeloglypter 741, 745
Amphelasma 631, 657
Ampheremus 103, 107
Amphibolus 254
Amphicerus 236, 240
Amphicrossus 312, 314
Amphicyrta 114, 115
Amphicyrtinae 115
Amphidora 495
Amphimallon 54, 58
Amphionycha 600
Amphivectura 284, 294
Amphizoidae 819
Amphotis 313, 314
Ampumixis 118, 119
Amydrogmus 778
Amydropa 340, 341
Anabium 255
Anacentrinus 746
Anacentrus 747
Anachilus 167
Anacomis 594
Anactodes 740
Anaedus 471, 484
Anaemia 494
Anaerea 600
Anaertus 803
Anaglyptini 582, 595
Anagrylius 66
Analampis 107
Analcis 764
Analeptura 573, 587
Analotes 699
Anambodera 102, 106
Anamesus 167
Anametis 770, 781
Anamorphinae 369
Anamorphus 369
Anamphidora 480, 500
Anaspididae 564
Anaspidinae 565, 566
Anaspis 565, 567
Anastrangalia 574, 587
Anastrategus 66
Anataxis 107
Anatis 381, 386
Anatrinodia 70
Anatropis 69
Anbium 255
Anchastus 164, 168
Anchicera 341
Anchitelus 778
Anchodemus 766
Anchomma 465, 473, 486
Anchonini 725, 790
Anchonocerus 805
Anchonus 788, 790
Anchorius 356, 357
Anchycteis 137
Anchytarsinae 137
Anchytarsus 137
Ancistronycha 207
Ancognatha 64, 65
Anculopus 792
Ancyclonycha 58
Ancylocera 582, 595
Ancylocerina 582, 595
Ancylocheira 107
Ancyloderes 805
Ancylotela 106
Ancyronychini 119
Ancyronyx 118, 119
Andrector 657
Andrimus 481, 499
Androchirus 481, 499
Androlyperini 658
Androlyperus 632, 659
Aneflomorpha 578, 591
Aneflus 578, 579, 591
Anelaphus 578, 591
Anelastes 153, 154
Anelastini 154
Anelpistus 521
Anemia 494
Anemiadena 494
Anepsiini 465, 470, 473, 483, 486
Anepsius 473, 486
Anespyra 591
Aneurops 320, 321
Aniara 496
Aniarus 496
Anisandrus 803
Anisarctus 738
Anischinae 160
Anisocalvia 387
Anisocera 310
Anisocerini 584, 598
Anisomeristes 393
Anisophileurus 67
Anisorus 589
Anisostena 623, 643, 644
Anisostica 386
Anisosticta 381, 386
Anisostictini 386
Anisotria 541, 542, 833
Anisoxya 419, 421
Annegialia 45, 46
Anobiidae 235, 245, 253, 254, 826, 831
Anobiinae 235
Anobiopsis 250, 256
Anobium 251, 255
Anocomis 594
Anodius 802
Anodonta 673
Anomaearthrum 501
Anomala 61, 62
Anomalacra 61, 62
Anomalaegialia 45
Anomalepta 62
Anomalini 60, 62
Anomalonychus 527
Anomalonyx 527
Anomalopides 62
Anomalopus 62
Anomatocera 310
Anommatidae 358
Anommatus 360, 361, 824
Anomoea 638, 674
Anomoia 674
Anomotheca 257
Anophorus 314
Anopliomorpha 577, 591
Anoplis 107
Anoplitis 644
Anoplium 591, 592
Anoplocephalus 66
Anoplocurius 577, 591
Anoplodera 587, 588, 589, 590
Anoplognatho 66
Anoplophora 584, 597
Anoplotrupes 26
Anorus 91, 823
Anotheca 257
Anovia 377, 385
Antelmia 778
Antennalia 571, 587
Anteros 496
Anthaxia 104, 108
Anthaxiina 104, 108
Anthaxini 104
Antherophagi 340
Antherophagus 340, 341
Anthicidae 549, 551, 553, 833, 834
Anthicinae 551, 552, 555
Anthicini 552, 555
Anthicus 553, 555
Anthobates 566, 792
Anthobatula 566
Anthocomus 287, 290, 291
Anthomorphus 736
Anthonaeus 310
Anthonomidius 737
Anthonomini 724, 736
Anthonomochaeta 736
Anthonomocyllus 736
Anthonomopsis 735, 736
Anthonomorphus 737
Anthonomus 736
Anthophilax 588
Anthophylax 588, 589
Anthophyllax 572
Anthotribidae 695
Anthotribus 700
Anthracalaus 162, 169
Anthracopteryx 166, 170
Anthrenini 229
Anthrenus 230
Anthribidae 695, 698, 821
Anthribinae 696, 698
Anthribini 698, 700
Anthribulus 699
Anthribus 698, 699, 700
Antidonus 783
Anypotactini 725, 776
Aocnus 739
Aomopactus 777
Aoromius 739
Apachiscelus 731
Apatidae 233
Apatides 237, 240
Apatini 236, 239
Apatophysis 588
Apatorhynchus 756
Apatura 107
Apelocera 595
Apeltarius 739
Aphanisticina 105, 108
Aphanisticini 105, 108
Aphanisticus 105, 108
Aphanocleptus 801
Aphanommata 758, 760
Aphanotus 497
Aphenolia 312, 314
Aphidecta 375, 381, 387
Aphideita 387
Aphidiphages 385
Aphodiinae 42, 45
Aphodiini 42, 45
Aphodius 44, 45
Aphonides 64, 65, 66
Aphonus 65, 66
Aphorista 369
Aphotaenius 45, 46

Aphotistus 170
Aphrastus 772, 779
Aphricus 167, 168
Aphthona 634, 661
Aphthonia 661
Aphtona 661
Aphyoramphus 790
Apidocephalus 799
Apilocera 595
Apinocis 746
Apiolum 717
Apion 716, 717
Apionidae 711
Apioninae 714, 717
Apionini 717
Apionion 716, 718
Apionus 717
Apium 717
Apius 717
Aplastinae 160
Aplastini 167
Aplastus 167
Aplectrus 595
Aplemonini 717
Apleurus 784, 785
Aplopus 731
Apocrypha 466, 498
Apocryphini 466, 483, 498
Apomecynini 583, 597
Aporataxia 597
Aposacium 393
Apostena 644
Apotrepus 758, 760
Apoxyleborus 803
Apria 314
Aprinus 738
Apsectus 229, 230
Apsena 478, 495
Apsida 481, 500
Apterius 661
Apteromechus 762, 763
Apterospasta 527
Apterulomoides 496
Aptopus 163, 168
Aracanthus 768, 776
Arachnipes 764
Araecerini 696, 698
Araecerus 697, 698
Araeocerus 698
Araeocorynus 698
Araeoderes 698, 699
Araeodontia 269, 275
Araeopidius 137
Araeosarus 698
Araeoschizina 483, 485
Araeoschizus 471, 486
Aragnomus 772, 778
Aramigus 775, 777
Araptus 799, 804
Arawana 375, 384
Archaebabia 638, 676
Archaeoscolytus 801
Archarias 736, 790
Archarius 734, 736
Archeocrypticidae 401, 402, 834
Archeophalus 804
Archimedii 70
Archodontes 570, 586
Archopactus 778
Archophileurus 64, 67
Archostemata 819
Arctobyrrhus 115
Arctolina 626, 650
Ardoleucus 790
Areopidiinae 137
Argaleus 589
Argalia 600
Argopistes 634, 666
Argoporis 478, 496
Arhagus 155
Arhina 315
Arhopalus 571, 587, 594
Aridionomus 397
Aridius 396, 397
Ariotus 561, 562
Armalia 475, 489
Arrenodes 714, 717
Arrhabaeus 485
Arrhenodes 717
Arrhenodini 717
Arrhenoplita 492
Arrhipis 154, 155
Arrowella 386
Artematopodidae 146, 147, 827
Artematopodinae 147
Arthleptrus 737
Arthrochlamys 676
Arthrolipinae 393
Arthrolips 392, 393
Arthromacra 471, 484
Artipus 775, 777
Artobium 256
Aryana 674
Asaphes 169
Asbolus 471, 486
Asclera 516, 517
Asclerini 517
Ascoliocerus 165, 170
Aseminae 570, 571, 587
Asemini 571, 587
Asemum 571, 587
Aserica 57
Asidina 473, 488
Asidini 470, 471, 472, 483, 487
Asidopsis 473, 488
Asidora 231
Asiocnemis 154, 156
Asionus 674
Asiopus 674
Asiorestia 664
Asiphytodecta 649
Aspathines 455
Asperosoma 750, 754
Asphaera 633, 634, 665
Aspidapiini 717
Aspidimorpha 646
Aspidimorphini 646
Aspidiphoridae 305
Aspidius 493
Aspidocha 393
Aspidodella 65
Aspidomorpha 646
Aspidotites 65
Astenoscelis 760
Asterizini 646
Astromula 591
Astrotus 488
Astyleiopus 599
Astylidius 586, 599
Astylogelhes 315
Astylopsis 585, 586, 599
Astynomus 599
Asydates 284, 295
Asynonychus 778
Ataenius 45, 47
Atahualpina 491
Ataxia 583, 598
Ataxiini 583
Atelestus 289
Ateuchus 49, 50
Athetetes 778
Athoini 169
Athoites 160
Athoomorphinae 160
Athous 164, 169
Atimia 571, 587
Atimiini 571, 587
Atimuropsis 597
Atomaria 340, 341
Atomariinae 341
Atomariini 340, 341
Atopa 91
Atopidae 90
Atopomacer 694
Atrachela 667
Atrachodes 790
Atractelmis 118, 119
Atractoceridae 261
Atractocerus 262, 822
Atractomerus 737
Atractopterus 167
Atrichonotus 775, 777
Atrichopsis 764
Atroniscus 790
Attagenini 229
Attagenus 230
Attalomimidae 281
Attalus 286, 289
Attalusinus 286, 289
Attarus 760
Attelabidae 703, 707, 822
Attelabinae 706, 707
Attelabus 707
Atylostagma 578, 591
Atymius 670
Atysites 628, 639, 654
Auchenia 611
Auchmobius 474, 489
Augyles 130
Augyliini 129, 130
Aulacon 168
Aulacoscelidinae 614, 615
Aulacoscelinae 615
Aulacoscelis 614, 615
Aulacostenus 155
Aulacosternus 155
Aulametopiellus 765
Auletini 708
Auletobius 706
Auleutes 750, 751, 753
Aulicus 270, 277
Aulobaris 743, 745
Aulonothroscus 159
Austroceutorhynchus 753
Austrolema 611
Austrurodera 676
Autobium 256
Autoserica 57
Axenomimetes 761
Axestinus 591
Axinocerus 230
Axion 375, 384
Axylophilus 561
Aztecalius 66
Azya 377, 385
Azyae 385
Azyaires 385
Azyinae 385
Azyini 377, 385

B

Babia 638, 675
Babiina 638, 639, 675
Bactridium 320, 321
Bactrocerus 551, 554
Baeorhopalus 760
Bagoas 740
Bagoimorphus 740
Bagoinae 724, 727, 740
Bagous 740
Balaninus 736
Balanobius 736
Balanomorpha 664
Baliosus 623, 644
Balophila 661
Bangasternus 726, 783, 785
Barberia 106
Baridina 745
Baridinae 724, 726, 727, 740, 741, 745
Baridini 724, 745
Baridius 745
Baridopsis 764
Barilepis 746
Barilepsis 745
Barilepton 741, 746
Barinus 746
Baris 743, 745
Baropsis 764
Barrellus 103, 107
Barymorpha 63
Barynotus 775, 777

Barypeithes 774, 779
Barypithes 779
Barytychius 739
Basiptini 646
Basitropidini 698, 699
Bassareus 638, 674
Batatarhynchus 765
Bathona 393
Bathyris 776
Batuliodes 486
Batuliomorpha 473, 486
Batulius 473, 486
Batyle 580, 596
Batyleoma 596
Baulius 553, 555
Beerellus 102, 105
Belidae 701, 822
Belka 760
Bellamira 573, 588
Belonia 397
Belopini 466, 483, 484
Belotus 209
Benedictia 55, 60
Benibotarus 175, 176
Bergininae 400
Berginus 400
Berninelsonius 165, 170
Bertiellus 673
Betarmon 167
Betulapion 716, 718
Bidensomela 626, 650
Biomorphus 496
Biphyllidae 356, 831
Birrhii 113
Bius 479, 497
Bladus 166, 168
Blaisdelliana 379, 383
Blanchardia 199
Blaps 468, 496
Blapsidae 463
Blapstinus 476, 493
Blaptini 468, 483, 496
Blapylis 477, 495
Blastophagus 800
Blastophila 790
Blatchleya 198, 199
Blauta 164, 168
Blepharida 619, 634, 660
Blepharidella 660
Blepharidina 660
Blitopertha 62
Bogcia 269, 274
Bolbelasmus 24, 25
Bolboceras 24, 25
Bolbocerastes 25
Bolboceratinae 24, 25
Bolbocerosoma 24, 25
Bolborhombus 25
Boliographa 651
Bolitophagina 475, 483, 490
Bolitophaginae 465, 466, 483, 490
Bolitophagini 465, 466, 475, 483, 490
Bolitophagus 475, 490
Bolitotherus 475, 490
Bolochesis 676
Boloschesis 676
Boraginobius 752
Borborhynchus 760
Boreades 309, 310
Boreocanthon 49
Boreohypera 783
Boreosaxinis 676
Boreurodera 676
Boridae 534, 535
Borinae 535
Borophloeus 759
Boros 535
Borowiecius 605, 606
Boroxylon 803
Boschella 271, 279
Bostrichidae 233, 238, 826, 830, 831
Bostrichinae 235, 236, 239
Bostrichini 236, 239
Bostrichips 801
Bostrichoclerus 269, 274
Bostrichoidea 224
Bostrichus 236, 240
Bostrychidae 233
Bostrychoplites 236, 240
Bostrychopsis 237, 240
Bostrychulus 240
Bostrychus 240
Bothrideres 360, 361
Bothrideridae 358, 361, 824, 830
Bothriderinae 361
Bothriophorini 126
Bothriospilina 575, 595
Bothroscelis 615
Bothrosternina 794, 800
Bothrotes 474, 489
Botrodus 364
Boucardius 48
Brachiacantha 379, 384
Brachiacanthini 379, 383
Brachyacantha 384
Brachyacanthadini 383
Brachyacanthaires 383
Brachybamus 730, 731
Brachycilibe 485
Brachycopris 50
Brachycoryna 623, 644
Brachycorynus 698, 699
Brachycrepis 168
Brachydendrulus 804
Brachyderes 774, 776
Brachyderini 725, 776
Brachygeraeus 747
Brachyina 105, 109
Brachyleptura 575, 588
Brachyleptus 310
Brachylon 365
Brachylophus 776
Brachymera 113
Brachyogmus 737
Brachyontis 487
Brachypeplus 312, 313
Brachyphytodecta 649
Brachypnoea 636, 668, 669
Brachypsectra 148
Brachypsectridae 148, 827
Brachypteridae 309, 310, 829
Brachypterolus 309, 310
Brachypterus 309, 310
Brachys 105, 109
Brachysomida 573, 588
Brachysomus 774, 780
Brachyspasta 526
Brachystylus 771, 776
Brachyta 588
Brachytarsoides 699
Brachytarsus 699, 700
Brachytemnoides 760
Bradyceris 641
Bradycinetulus 25
Bradycinetus 25
Bradylema 642
Bradylemoides 642
Bradyrhynchoides 774, 777
Bradyrhynchus 777
Branchini 471, 483, 487
Branchus 471, 487
Brenthus 717
Brentidae 711, 717, 821
Brentinae 713, 717
Brentini 717
Brentus 714, 717
Brevibarra 699
Breviophthorus 805
Brinkaltica 664
Brius 782
Bromius 637, 671
Brontes 325
Brontinae 324, 325
Brontini 324, 325
Brontophloeus 333
Brothylus 577, 591
Brounia 140
Brownia 803
Bruchidae 602, 606
Bruchidini 606
Bruchidius 605, 606
Bruchinae 606, 621, 639
Bruchini 606
Bruchus 604, 606
Brucita 630, 655
Brumoides 375, 384
Brumus 375, 384
Brydaeon 592
Buchananius 743, 746
Bulbifer 729
Bulborhinapion 717
Buprestidae 98, 105, 822, 827
Buprestina 103, 107
Buprestinae 102, 106
Buprestini 103, 104
Buprestis 104, 107
Burlinius 673
Byrrhida 113
Byrrhidae 113, 115, 827
Byrrhinae 115
Byrrhini 115
Byrrhocerus 257
Byrrhodes 251, 252, 257
Byrrhoidea 113
Byrrhoides 257
Byrrhus 115, 255
Byrroidea 113
Byrsopages 767, 781
Byrsopolis 63
Bystus 369
Byturellus 355
Byturidae 829
Byturosoma 297
Byturosomus 285, 297
Byturus 355

C

Caccodes 209
Caccophryastes 778
Caccorhinus 699
Cacicula 385
Cacidula 385
Cacochromus 777
Cacophrissus 576, 591
Cacoplia 597
Cacostola 583, 597
Cacoteminus 255
Cacotemnus 255
Cactophagus 729
Cactopinina 796, 802
Cactopinorus 802
Cactopinus 796, 802
Cadiz 625, 653
Caecossonus 787, 792
Caelius 45
Caenia 176, 177
Caeniella 177
Caenocara 251, 257
Caenocorse 497
Caenocossonus 759
Caenomychus 368
Caenoptera 593
Caenoscelini 340, 341
Caenoscelis 340, 341
Calandrinus 744, 746
Calitinae 265
Calitys 265
Callergates 586
Calles 761, 764
Callichroma 593
Callichromatini 579, 593
Callidiellum 581, 593
Callidiini 581, 593
Callidium 582, 593
Callidostola 593
Calligramma 626, 651
Calligrapha 626, 650, 651
Callimellum 593
Callimoxys 581, 593
Callimus 593
Callipogonini 570, 586
Callipogonius 584, 598
Callirhipidae 144, 145, 826

Callirus 791
Callistethus 62
Calloides 583, 594
Callona 580, 596
Callopistus 777
Callosobruchus 604, 606
Callotillus 269, 275
Calocerus 169
Calochrominae 175, 177
Calochromus 176, 177
Calomacraspis 62, 63
Calomycterus 770, 776
Calomyllocerus 776
Calopodinae 516, 517
Calopterini 176, 177
Calopteron 176, 177
Caloptognatha 176, 177
Calopus 516, 517
Calosirus 752
Calospasta 526, 527
Calotheca 660
Calvia 387
Calymaderus 257
Calymmaderini 257
Calymmaderus 248, 251, 257
Calymmoderus 257
Calypterus 256
Calyptillus 774, 777
Calyptobium 368
Calyptomerus 85
Calyptorhina 674
Calytheca 251, 257
Camaroderes 699
Camarotini 725, 790
Camoena 665
Campipterus 739
Camplirhynchus 754
Campopterus 739
Campsosterninae 160
Campta 314
Camptodes 313, 314
Camptorhina 57
Camptosomes 617
Camptotropis 699
Campylidae 160
Campylirhynchus 754
Campylorhynchus 754
Campyloxeninae 160
Campylus 169
Canifa 565, 566
Canistes 761, 764
Canonura 599
Cantharidae 202, 825
Cantharinae 206, 207, 208
Cantharini 207
Cantharis 208
Cantharolethrus 7
Canthidium 49, 50
Canthon 49
Canthonini 49
Canthostethus 673
Capnochroa 481, 499
Capraita 633, 665
Carabidae 819
Carcharodis 664
Cardiapus 664
Cardiophorinae 168
Cardiophorites 160
Cardiophorus 164, 168
Cardiorhinites 160
Carebara 420
Cariaires 385
Cariites 386
Carinisphindus 308
Carmentis 674
Carphobius 796, 801
Carphoborus 796, 801
Carphonotus 758, 760
Carphuridae 281
Carphurini 286, 287
Carphuroides 286, 287
Carpolinus 738
Carpomanes 779
Carponinophilus 736
Carpophilinae 312, 314
Carpophilus 312, 314
Cartodere 396, 397
Cartoderema 397
Caryedes 605
Caryedon 604, 606
Caryobruchus 604, 606
Caspyria 440
Cassida 624, 646
Cassididae 617
Cassidini 623, 639, 646
Cassidula 647
Cassidulella 647
Catapastinus 742, 746
Catapastus 742, 746
Cataphagus 167
Cateretes 310
Cateretidae 309, 310
Cathartosilvanus 324, 325
Cathartus 324, 326
Catheretes 310
Cathetopteron 586, 600
Cathorama 258
Cathormiocerinus 781
Cathormiocerus 781
Catopochrotidae 338
Catoptyx 393
Catorama 258
Caulophilus 758, 759
Causima 527
Caverneleodes 478, 495
Cavicoxumidae 160
Cebriognathinae 160
Cebrionates 160
Cebrioninae 167, 823
Cebrionini 167
Celetes 177
Cenophengus 184
Cenoscelis 496
Centrinaspidia 747
Centrinaspis 747
Centrinites 744, 746
Centrinogyna 742, 744, 746
Centrinopus 744
Centrinus 744, 746
Centrioptera 486
Centrocerum 592
Centrochilus 68, 70
Centrocleonus 785
Centrodera 572, 588
Centronopini 468, 481, 482, 483, 498
Centronopus 482, 498
Centuriatus 129, 130
Cephalaon 521
Cephalips 315
Cephalistes 289
Cephalobyrrhinae 126
Cephaloidae 520
Cephaloinae 521
Cephaloleiini 622, 639, 643
Cephaloon 521
Cephalopycnus 361
Cephaloscymnini 377, 382
Cephaloscymnus 377, 382
Ceracis 406, 409
Cerambycidae 568, 586, 828
Cerambycinae 570, 575, 590
Cerambycoidea 568
Cerandria 491
Cerapheles 287, 291
Cerasphorus 595
Ceraspis 54
Ceratapiini 717
Ceratobothris 108
Ceratocanthidae 34, 35, 820
Ceratocanthus 35
Ceratoderus 556
Ceratographis 599
Ceratolepis 801
Ceratoma 657
Ceratomagilla 386
Ceratomegilla 381, 386
Ceratophyus 25
Ceratopidonia 589
Ceratotaxia 155
Cercidiestes 607
Cercolia 590
Cercometes 310
Cercopedius 781
Cercopeus 781
Cercopius 781
Cercus 310
Cercyonops 649
Cerenopini 469, 478, 483, 496
Cerenopus 478, 496
Cerodens 673
Ceropachys 673
Cerophorus 314
Cerophytidae 150, 151, 830
Cerophytum 151
Cerotana 657
Cerotoma 631, 657
Cerotomites 629, 639, 657
Ceruchini 7
Ceruchus 7, 8
Cerylcautomus 364
Cerylon 364
Cerylonidae 363, 824, 830
Ceryloninae 364
Cetoniinae 42, 67, 68
Cetoniini 68, 69
Ceuthamiocolus 752
Ceuthorhynchidius 752
Ceuthorhynchus 752
Ceuthorrhynchidius 752
Ceuthorrhynchus 752
Ceutorhynchinae724, 727, 747, 748, 752
Ceutorhynchini 724, 752
Ceutorhynchus 752
Chaerocassis 624, 648
Chaeroceroides 652
Chaeroceta 652
Chaetechidius 781
Chaetechus 781
Chaetocanthinae 29
Chaetocera 652
Chaetochema 664
Chaetocnema 634, 664
Chaetocoelus 286, 287
Chaetodorytomus 738
Chaetophloeus 796, 800
Chaetophora 114, 116
Chalcodermus 787, 792
Chalcoderus 51
Chalcohyus 801
Chalcoidea 626, 650
Chalcoides 664
Chalcolepidiides 160
Chalcolepidius 163, 169
Chalcoparia 668
Chalcophora 104, 106
Chalcophorella 106
Chalcophorina 104, 106
Chalcophorini 103, 104, 106
Chalepini 623, 639, 643
Chaleponotus 790
Chalepus 66, 623, 644
Chalybodontus 739
Chanchamayia 657
Charagmus 780
Charaphloeus 332, 334
Charidotella 624, 648
Charidotini 646
Chariessa 271, 278
Charisalia 573, 588
Charopus 286, 289
Chasmatopterini 53, 57
Chauliognathidae 202
Chauliognathinae 206, 209
Chauliognathini 206, 209
Chaunocolus 53, 57
Cheirodes 471, 494
Chelonariidae 139, 141, 827
Chelonarium 140, 141
Chelonychus 735, 737
Chelymorpha 623, 645
Cherostus 490
Chesas 18
Chilocoriens 384
Chilocorinae 375, 384
Chilocorini 375, 384
Chilocorus 375, 384
Chilometopon 489

Chilostetha 107
Chion 595
Chionanthobius 739
Chirida 647
Chirodines 54, 58
Chlaenobia 55, 58
Chlamisini 619, 638, 639, 676
Chlamisus 639, 676, 677
Chlamydidae 617
Chlamys 676
Chloephaga 665
Chlorida 575, 595
Chlorixanthe 68, 69
Chloropachys 673
Chlorophorus 594
Chnaunanthus 53, 57
Chnoodiaires 385
Chnoodiens 385
Choeridium 50
Choerodemas 760
Cholina 790
Cholini 725, 790
Cholinobaris 744, 746
Cholus 788, 790
Chondronoderes 804
Choragidae 695
Choraginae 696, 698
Choragini 696, 698
Choragus 697, 698
Chorea 151
Choriolaus 573, 588
Choristorhinus 790
Chramesus 795, 800
Chromatia 499
Chromogethes 315
Chrotoma 575, 595
Chrysapion 714, 718
Chrysina 62, 63
Chrysobothrina 104, 108
Chrysobothrini 103, 108
Chrysobothris 104, 108
Chrysochus 636, 670
Chrysodina 668
Chrysodinini 668
Chrysodinopsis 636, 668
Chrysolina 619, 626, 650
Chrysolinina 650
Chrysomela 627, 652
Chrysomelidae 617, 619, 639
Chrysomelina 625, 626, 639, 651
Chrysomelinae 619, 624, 639, 648
Chrysomelini 625, 639, 649, 653
Chrysomeloidea 568
Chrysophana 102, 105
Chujoita 641
Cibdelis 482, 500
Cicanyssodrys 599
Cidnorhinus 752
Ciidae 403, 408, 829, 831
Ciinae 405, 408
Cillaeinae 312, 313
Cimberidinae 693
Cimberidini 693
Cimberis 693
Cimbocera 769, 781
Cinyra 107
Ciocotalpa 63
Cioidae 403
Cionini 724, 737
Cionistes 737
Cionomimus 735, 737
Cionopsis 736, 737
Cionus 737
Circornus 497
Cis 407, 409
Cisanthribini 696, 698
Cissides 403
Cissites 527
Cissoanthonomus 737
Cistela 499
Cladoborus 805
Cladotominae 137
Clambidae 85, 826
Clambus 85
Clamoris 471, 485
Clanini 382, 384
Clanoptilus 287, 290
Clarkaltica 666
Claudius 254
Clavicornia 305
Cleis 387
Clemmus 369
Clemmys 369
Clemnus 369
Clemora 58
Cleniocerus 170
Cleogonini 725, 790
Cleomenini 582, 595
Cleonaspis 785
Cleonidius 785
Cleonini 725, 785
Cleonis 784, 785
Cleonopsis 785
Cleonus 785
Cleopomiarus 727, 738
Cleptelmis 118, 119
Cleridae 267, 273, 829
Clerinae 270, 277
Clinocera 108
Clitostethus 383
Clypastraea 392, 393
Clypeastodes 393
Clypeogethes 315
Clyptus 780
Clythridae 617
Clytini 582, 594
Clytoleptus 583, 594
Clytridae 617
Clytrina 638, 639, 674
Clytrini 638, 639, 674
Clytus 583, 594
Cnecus 255
Cnemapion 718
Cnemarachis 54, 58
Cnemeplatiini 470, 471, 483
Cnemidoprion 763
Cnemocyllus 737
Cnemodini 469, 483, 488
Cnemodinus 469, 489
Cnemodus 489
Cnemogonini 724, 753
Cnemogonus 750, 753
Cnemonyx 796, 801
Cnemotrupes 26
Cneorane 659
Cneorhinini 725, 776
Cnesinus 795, 800
Cnestocera 527
Cnopina 562
Cnopus 560, 562
Coccidophilus 377, 382
Coccidula 377, 385
Coccidulides 385
Cocciduliens 384, 385
Coccidulina 385
Coccidulinae 384
Coccidulini 377, 385
Coccinella 381, 386
Coccinellaires 385
Coccinellates 385
Coccinellidae 371, 381, 385, 386, 824
Coccinellides 385, 386
Coccinelliens 385
Coccinellina 386, 387
Coccinellinae 375, 385
Coccinellini 375, 385, 386
Coccinellites 386
Coccinelliti 385
Coccinelloidea 305
Coccotorus 736, 737
Coccotrypes 798, 802
Codiosoma 759
Coeleburia 591
Coelocephalapion 717, 718
Coelocnemis 482, 500
Coelomerites 628, 639, 654
Coelometopinae 483, 500
Coelometopini 468, 481, 483, 500
Coelomorpha 487
Coelopalorus 478, 497
Coelophora 381, 386
Coelophoraires 386
Coelopterini 382
Coelosattus 487
Coelositona 780
Coelosternus 765
Coelostethus 255, 785
Coelus 473, 487
Coenocara 257
Coenonycha 54
Coenopoeus 585, 599
Colaspidea 637, 671
Colaspini 668
Colaspis 637, 669
Colaspomorpha 669
Colastus 314
Colaulon 168
Colecerus 768, 776
Coleobyersa 675
Coleocerus 776
Coleolacordairei 676
Coleomegilla 381, 386
Coleomethia 575, 590
Coleorozena 638, 675
Coleothorpa 638, 675
Collapterix 597
Collapteryx 597
Collops 286, 291
Colon 828
Colopterus 312, 314
Colposternini 255
Colposternus 255
Colydiidae 830
Colymmaderus 257
Comacmaeops 572, 588
Comazus 85
Comesiella 800
Compsidia 600
Compsomorphus 498
Compsus 771, 777
Conalia 427
Conaliini 427
Conchopterus 593
Condylops 286, 288
Confusoscolytus 801
Conibiosoma 476, 493
Conibius 476, 493
Conionitini 487
Coniontellus 487
Coniontides 487
Coniontini 471, 473, 483
Coniontis 473, 487
Coniophthalma 341
Coniopoda 369
Conipinus 487
Coniporidae 305
Conisattus 473, 487
Conithassa 397
Connatichela 768, 781
Conocentrinus 747
Conoderes 169
Conoderinae 724, 726, 754, 755
Conoderus 162, 169
Conoecus 473, 489
Cononotus 541, 542, 833
Conopheribium 254
Conophoribium 254
Conophthocranulus 805
Conophthorus 799, 804
Conoploribium 254
Conotelus 312, 313
Conotrachelini 725, 790
Conotrachelus 788, 790
Conradsia 642
Cophes 763, 765
Copidita 517, 518
Copobaeninae 553
Coprini 49, 50
Copris 49, 50
Coprophanaeus 49, 51
Coptocycla 647
Coptodryas 801

Coptogaster 801
Coptosomus 801
Copturodes 756
Copturus 755, 756
Coraebina 108
Coraebini 105, 108
Coraebus 108
Coraia 630, 654
Cordyle 729
Cordylospasta 525, 526
Coreopsomela 650
Corigetellus 776
Corphyra 542
Corthylina 799, 805
Corthylomimus 805
Corthylus 799, 805
Corticaria 397
Corticariinae 397
Corticariini 396
Corticarina 397
Corticeus 466, 493
Corticotomus 265, 266
Cortilena 396, 397
Cortinicara 397
Cortodera 573, 588
Corylophi 393
Corylophidae 390, 393, 824, 830
Corylophinae 392, 393
Corylophini 393
Corylophodes 393
Corylophodini 393
Corymbites 170
Corymbitodes 166, 170
Corynetidae 267
Corynodini 668
Corynodites 639, 670
Coscinocephalus 65, 66
Coscinopter 496
Coscinoptera 638, 675
Coscinoptilix 496
Cosmobaris 743, 745
Cosmocorynus 805
Cosmopolites 729, 730
Cosmosalia 575, 588
Cossoninae 724, 727, 756, 759
Cossonini 724, 759
Cossonus 758, 759
Cossyphinae 483
Cossyphodidae 463
Cotalpa 61, 63
Cotinis 68, 69
Cotinorrhina 69
Cradytes 285, 296
Craniotus 472, 488
Craponius 750, 753
Cratidus 477, 495
Cratonychus 168
Cratoparini 699
Cratoparis 699
Cregya 271, 279
Cremastocheilini 68, 70
Cremastocheilus 68, 70
Crepidaspis 647
Crepidodera 635, 664
Crepidomenites 160
Crepidotritus 168
Crigmus 167
Criniflavia 69
Criocephalum 587
Criocephalus 587
Crioceridae 617
Criocerinae 619, 621, 622, 639, 641
Criocerini 622, 639, 641
Crioceris 622, 641
Criomorphus 587
Crioproposus 580
Crioprosopus 596
Crioscapha 420
Crocidema 781
Crosimus 341
Crossidius 580, 596
Crymodes 535
Cryphalina 797, 803
Cryphaloides 802
Cryphalomorhpus 803
Cryphalophilus 803
Cryphalus 797, 804
Cryphopus 785
Cryptadius 474, 489
Cryptafricini 341
Cryptamorpha 324, 325
Cryptarcha 313, 315
Cryptarchina 315
Cryptarchinae 312, 315
Cryptarchips 315
Cryptarchula 315
Cryptarchus 315
Cryptarthrum 804
Crypticini 468, 475, 483
Crypticomorpha 487
Cryptocarenus 797, 804
Cryptocephalidae 617
Cryptocephalina 638, 639, 673
Cryptocephalinae 618, 619, 620, 621, 638, 639, 672
Cryptocephalini 619, 638, 639, 672
Cryptocephalus 638, 673
Cryptocleptes 801
Cryptoglossa 471, 486
Cryptoglossini 471, 483, 486
Cryptognatha 377, 382, 384
Cryptognathaires 384
Cryptognathini 375, 377, 384
Cryptolaemus 377, 383
Cryptolepidus 774, 777
Cryptolestes 333
Cryptophagidae 338, 831
Cryptophaginae 340
Cryptophagini 340
Cryptophagistes 341
Cryptophagus 340, 341
Cryptophilinae 345
Cryptophilini 345
Cryptophilus 344, 345, 832
Cryptopleura 595
Cryptops 497
Cryptorama 252, 258
Cryptoramorphini 257
Cryptoramorphus 252, 257
Cryptorhis 754
Cryptorhopalum 230
Cryptorhynchidius 764
Cryptorhynchina 763
Cryptorhynchinae 725, 726, 761, 763
Cryptorhynchini 725, 763
Cryptorhynchus 763, 764
Cryptorrhynchobius 764
Cryptorrhynchus 764
Cryptosomatulini 341
Cryptostoma 154
Cryptostomes 617
Cryptoweisea 382
Cryptulocleptus 801
Crypturgina 797, 802
Crypturgus 797, 802
Cryticini 492
Crytorama 258
Crytoramorphus 257
Ctenicera 165, 166, 170
Ctenicerinae 160
Ctenicerus 170
Cteniopus 499
Ctenobium 255
Ctenochira 648
Ctenonychus 168
Ctenophorina 796, 801
Ctenophrus 801
Ctenyophthorus 805
Ctesias 230
Cucujidae 329, 829
Cucujoidea 305
Cucujus 329
Cultellunguis 208
Cumatomicus 802
Cupedidae 819
Curculio 734, 736
Curculionidae 722, 724
Curculionina 736
Curculioninae 724, 727, 728, 732, 736
Curculionini 724, 736
Curelius 340, 341
Curiini 577, 592
Curimopsis 114, 116
Curinus 375, 384
Curius 577, 592
Curtomerus 579, 591
Cyanauges 368
Cyaniris 674
Cybocephalidae 311
Cybocephalinae 312, 315
Cybocephalus 312, 315, 829
Cybotus 476, 493
Cychramus 313, 314
Cycladidae 711
Cyclica 617
Cycliopleurus 592
Cyclocassida 647
Cyclocephala 64, 65
Cyclocephalini 64, 65
Cyclodera 596
Cyclodinus 553, 555
Cyclominae 725, 728, 765, 766
Cycloneda 381, 386
Cycloterina 791
Cycloterini 725, 791
Cyclotrypema 631, 657
Cycotida 738
Cydoniaires 386
Cyladinae 713, 717, 821
Cylanus 717
Cylas 713, 717
Cylidrella 265, 266, 823
Cylindera 591
Cylindrataxia 597
Cylindridia 744, 746
Cylindrocopturinus 755, 756
Cylindrocopturus 755, 756
Cylindrodera 591
Cylindrotomicus 803
Cyllene 594
Cyllodes 313, 314
Cylloepus 119
Cyllopodes 314
Cymatodera 269, 275
Cymatoderella 269, 276
Cymatothes 479, 498
Cymbolus 291
Cynaeus 491
Cynauges 368
Cyphagoginae 714, 717
Cyphanobium 257
Cypherotylus 352
Cyphicerina 776
Cyphicerini 725, 776
Cyphirhinus 745
Cyphocleonus 784, 785
Cyphogaster 640
Cyphomorpha 645
Cyphon 88
Cyphonidae 87
Cyphonotida 574, 588
Cyphorhynchus 790
Cypriacis 104, 107
Cyprus 740
Cyrtepistomus 770, 776
Cyrtinini 583, 599
Cyrtinus 583, 599
Cyrtobagous 730, 731
Cyrtomerus 591
Cyrtomoptera 207
Cyrtomoptila 208
Cyrtophorus 582, 595
Cyrtotomicus 802
Cyrtotriplax 352
Cysteodemus 525, 526
Cytilus 115

D

Dacne 350
Dacnidae 348
Dacninae 350

Dacoderinae 545, 547
Dacoderus 545, 547, 833
Dacryophthous 800
Dacryphalus 804
Dactyliini 46
Dactylorhinus 776
Dactylorrhinus 776
Dadopora 314
Dalopius 164, 167
Dampfius 129, 131
Danae 369
Danosoma 163, 168
Dapalinus 783
Dapsiloderus 420
Dascillida 90
Dascillidae 90, 91, 823, 826, 827
Dascillinae 91
Dascilloidea 90
Dascillus 91
Dasycerus 824
Dasydactylus 344, 345
Dasypogon 147
Dasyrhadus 292
Dasytastes 284, 293
Dasytellus 293
Dasytes 284, 293
Dasytidae 281
Dasytinae 284, 292
Dasytini 292
Daulis 387
Dearthrus 230, 231
Decadiomus 379, 383
Decamerinae 265
Dechomus 460
Dectes 585, 599
Degorsia 732
Deilelater 162, 169
Deinophloeus 333, 334
Deloyala 624, 647
Delphastobia 672
Delphastus 377, 382
Delpygus 496
Deltaspis 596
Deltochilum 49
Deltometopus 154, 156
Demotina 637, 670
Dendrobias 596
Dendrobiella 237, 240
Dendrobium 255
Dendrocharini 155
Dendrocharis 153, 155
Dendrocranulus 798, 802
Dendroctonomorphus 760
Dendroctonus 795, 800
Dendroides 542
Dendrophagus 324, 325
Dendrosinus 800
Dendroterus 799, 804
Dendrurgus 802
Denierota 526
Denticollis 164, 169
Deporaini 708
Deporaus 706, 708
Depressoscymnus 383
Deracanthopsis 785
Derancistrus 586, 587
Derelomini 724, 737
Deretaphrini 361
Deretaphrus 360, 361
Dermestes 229, 230
Dermestidae 228, 826, 835
Dermestini 229, 230
Derobrachus 571, 587
Deroconus 169
Derocrepis 635, 664
Derodontidae 221, 222, 822
Derodontinae 222
Derodontus 222
Derolathrus 220
Deronocus 162, 169
Derosomus 780
Derospidea 630, 654
Descarpentriesiola 107
Desmatogaster 251, 255
Desmiphora 584, 598
Desmiphorini 584, 598
Desmocerini 571, 590
Desmocerus 571, 590
Desmoglyptus 745
Desmophora 598
Desmoris 739
Desolakerrus 165
Desolakkerus 170
Deuterobrotica 658
Deuteroleptidea 593
Diabrotica 631, 656
Diabroticina 628, 639, 656
Diabroticites 628, 639, 656
Diacanthus 170
Diachus 638, 674
Diaclaspis 677
Dialytellus 44, 46
Dialytes 43, 46
Diamimus 781
Diaperidae 463
Diaperina 483, 491
Diaperinae 483, 490
Diaperini 466, 475, 483, 491
Diaperis 475, 492
Diaprepes 777
Diaptalia 65
Diaspis 677
Diaspsis 677
Dibolia 633, 666
Dicenopsis 673
Dicentrus 575, 590
Dicerca 104, 107
Dicercina 104, 107
Dicercini 103, 104, 107
Dichelonyx 54
Dichelotarsus 208
Dichotomiini 49, 50
Dichotomius 49, 50
Dichotychius 739
Dichoxenus 770, 782
Dichromina 65
Diclidia 565, 566
Dicranopselaphus 134
Dicranthus 740
Dicrepidiina 168
Dicrepidiites 160
Dicrepidius 164, 168
Dicronychidae 160
Dictyalotus 238
Dictyoptera 175, 176
Dicytopterus 176
Didactylia 43, 46
Didactyliini 43, 46
Didion 379, 383
Dienerella 396, 397
Dietzella 749, 753
Dietzia 740
Dietzianus 733, 737
Digitonthophagus 51
Dignamptus 501
Digrapha 177
Dihammaphora 582, 595
Dilandius 555
Dimalia 45
Dimites 160
Dinacoma 55, 58
Dinapate 236, 239
Dinapatini 236, 239
Dinocleus 785
Dinoderinae 235, 241
Dinoderus 236, 241
Diocalandra 730
Diocalandrina 730
Diodimorpha 759
Dioedus 471, 485
Diomini 379, 383
Diomus 379, 383
Dionorenus 752
Dionychus 790
Dioptrophorus 787, 792
Diorhabda 629, 654
Diorymerellus 747
Diorymeropsis 744, 746
Diorynotus 777
Diozodes 592
Diphyaulon 168
Diphyllidae 356
Diphyllocis 410
Diphyllostoma 10
Diphyllostomatidae 10, 820
Diphytaxis 154, 156
Diplacaspis 639, 677
Diplocoelus 357
Diplostethus 164, 167
Diplotaxini 53
Diplotaxis 53
Dipropus 164, 168
Dirabius 744, 746
Dircaea 419, 421
Dirhagus 155
Dirotognathus 768, 782
Dirrhagofarsus 155
Dirus 792
Discodemus 487
Discodon 209
Discogenia 478, 495
Discotenes 697, 698
Discotenini 697, 698
Discrepidius 168
Disonycha 634, 664
Disonychina 660
Disopus 673
Dissacantha 591
Dissacanthus 591
Distenia 569, 590
Disteniidae 568
Disteniinae 569, 590
Distigmoptera 633, 666
Distremocephalus 184
Ditaphrus 126
Ditemnus 209
Ditylini 517
Ditylus 516, 517
Dobzhanskia 386
Dochorhynchus 782
Docorhinus 780
Doichopterum 501
Dolerosomus 165, 167
Dolichocassida 647
Dolichocis 407, 409
Dolichonotus 782
Dolichosoma 293
Doliema 491
Doliodesmus 475, 491
Dolopiosomus 167
Dolopius 167
Dolurgus 797, 802
Donacia 622, 640
Donaciella 622, 640
Donaciidae 617
Donaciinae 619, 621, 622, 639, 640
Donaciini 622, 639, 640
Donaciomima 622, 640
Donacocia 640
Doratoma 257
Doratotomus 738
Dorcadoma 257
Dorcaschema 583, 597
Dorcaschematini 583, 597
Dorcasina 574, 588
Dorcasta 583, 597
Dorcatoma 252, 257
Dorcatominae 257
Dorcatomini 257
Dorcatorna 257
Dorchaschema 597
Dorcini 7, 8
Dorcotoma 257
Dorcus 7, 8
Dorkatoma 257
Dorthesia 440
Doryphorina 625, 639, 650
Dorytomina 738
Dorytomus 733, 738
Doticus 698
Doydirhynchini 693, 694
Drachylis 521
Drapetes 164, 168
Drapetini 160
Drepocossonus 759
Dromaeolus 154, 156
Drymo 591
Dryobiini 581, 593

Dryobius 581, 593
Dryocoetes 798, 802
Dryocoetina 797, 802
Dryocoetinus 802
Dryophilinae 254
Dryophora 729
Dryophthorinae 724, 726, 728, 729
Dryophthorini 724, 729
Dryophthorus 728, 729
Dryopidae 121, 122, 827
Dryops 122
Dryotomicus 800
Dryotomus 800
Dryotribini 724, 759
Dryotribus 756, 759
Dryoxylon 798, 802
Dubiraphia 118, 119
Duboisius 552, 554
Durangoum 255
Dylobolus 600
Dynastes 65, 67
Dynastinae 42, 64, 65
Dynastini 64, 67
Dynatopechus 759
Dyscinetus 64, 66
Dysides 235, 239
Dysidinae 235, 238
Dyslobus 769, 782
Dysmerus 333
Dysphaga 590
Dysphenges 634, 662
Dystaxia 96
Dystaxiella 96
Dystaxiina 96
Dystaxiini 96
Dysticheus 772, 778
Dystirus 778
Dytiscidae 820

E

Eanoides 170
Eanus 165, 170
Earophilidae 329
Eba 497
Ebaeus 288
Eburia 576, 591
Eburiini 591
Eccoptogaster 801
Echinaspis 791
Echmatophorus 63
Echocerus 491
Echopus 776
Echtrogasterini 155
Ecnomorphus 314
Ectamenogonus 168
Ectatotychius 739
Ectinoplectron 62, 63
Ectonia 661
Ectopria 134
Ectrephidae 245
Ecyrus 585, 598
Edesius 790
Edo 786
Edrotes 474, 489
Efflagitatus 130, 131
Eggersia 802
Egoliinae 266
Eidocolastus 314
Eidoreus 368
Eisonyx 746
Ekkoptogaster 801
Elacatis 545, 546
Elaeidobius 737
Elaphidiini 577, 591
Elaphidion 577, 578, 592
Elaphidionoides 591, 592
Elaphidionopsis 577, 592
Elaphidium 592
Elassoptes 758, 760
Elater 165, 167, 168
Elateridae 160, 167, 823, 826
Elaterinae 167
Elaterini 167
Elateroidea 146
Elateroides 262
Elateropsis 570, 586
Elathous 164, 169
Elatocoelus 155
Elatotrypes 582, 593
Elatrinus 168
Eleates 475, 490
Eleodes 476, 477, 495
Eleodimorpha 476, 495
Eleodini 469, 471, 476, 483, 494
Eleodopsis 495
Eleusinus 547
Eleutho 591
Elianus 199
Elissa 780
Elleschidius 739
Elleschus 738
Ellescina 738
Ellescini 724, 738
Ellescus 738
Elmidae 117, 119, 828
Elmidomorphus 740
Elminae 119
Elmini 119
Elodes 88
Elodidae 87
Elonus 562
Elytroleptus 579, 596
Elzearius 800
Emarhopa 656
Embaphion 476, 495
Emeax 487
Emelinini 561
Emelinus 560, 561
Emenadia 440
Emmenotarsis 297
Emmenotarsus 297
Emmesa 419, 421
Emmetrus 652
Emperochela 645
Emphyastes 765, 766
Emphyastina 766
Emphyli 340
Emporius 326
Empysara 491
Emyirhagus 155
Enagria 646
Enallonyx 294
Enaphalodes 579, 592
Enbrachys 108
Encalus 738
Enchodes 420
Encyclops 588
Encylops 572
Endalus 732
Endecatomidae 233
Endecatominae 234, 238
Endecatomus 234
Endectomus 238
Endectus 460
Endeodes 286, 289
Endomerus 314
Endomiini 555
Endomychidae 366, 824, 830, 831
Endomychinae 368
Endomychus 368
Endrosa 58
Engidae 348
Engididae 348
Engis 350
Engyaulus 108
Enicmoderes 397
Enicmus 396, 397
Eniconyx 168
Enneacoides 487
Ennearthron 407, 409
Enneatoma 257
Enneboeus 401, 402
Enoclerus 270, 277
Enocys 108
Enopliinae 270, 278
Enoplolema 642
Enops 790
Entiminae 725, 727, 728, 766, 767, 775
Entomophthalmus 153, 155
Entomops 699
Entomoscelina 625, 639, 649
Entomoscelis 626, 649
Entomosterna 596
Entomotrogus 231
Entypus 410
Eocaria 387
Eocryphalus 803
Eoglaresis 15
Epacalles 787, 790
Epagriopsis 777
Epagrius 777
Epailaeus 254
Epantius 478, 496
Epauloecus 254
Epeuchaetes 747
Ephalus 475, 493
Ephamillus 521
Ephelops 737
Ephimeropus 740
Ephisteminі 341
Ephistemus 340, 341
Epicaerus 775, 777
Epicauta 525, 526
Epicautini 526
Epilachna 375, 387
Epilachnidae 371
Epilachniens 387
Epilachninae 375, 387
Epilachnini 375, 387
Epimechus 737
Epiphaleria 491
Epiphanini 155
Epiphanis 154, 155
Epiphileurus 67
Epiphloeinae 270, 278
Epipissodes 792
Epipocus 369
Episcirrus 763, 765
Episernus 249, 254
Epismilia 382
Episomechus 781
Epithrix 663
Epitragini 470, 473, 483, 489
Epitragodes 474, 489
Epitragosoma 474, 489
Epitrix 635, 663
Eplophorus 595
Epomadius 801
Eponomastus 369
Epopterus 369
Epsips 804
Epuraea 312, 314
Epuraeanella 314
Eremocatoecus 777
Eremocleonus 785
Eremonomus 494
Eremosis 652
Eremotes 760
Erepsocassis 624, 648
Ergates 570, 586
Ericryphalus 804
Ericydeus 775, 777
Eridaulus 409
Erineosinus 800
Erinophlius 801
Eriopsilus 598
Erirhinina 731
Erirhininae 724, 726, 730, 731
Erirhinini 724, 731
Erirhinus 731
Erirhipidia 69
Erirhipis 69
Eririnomorphus 783
Ernobiinae 254
Ernobiini 254
Ernobius 249, 254
Ernocharis 500
Ernocryphalus 804
Ernophloeus 804
Ernopocerus 803
Ernoporicus 797, 803
Erodiscini 725, 791
Eronyxa 265
Eroosapion 717

Eropterus 175, 176
Eros 175, 176
Erotides 177
Erotinae 175, 176
Erotylidae 348, 350, 832
Erotylinae 350, 352
Erycus 731
Erynephala 630, 655
Erythrapion 717
Erythrolitus 337
Eschatocrepis 284, 293
Eschatomoxys 474, 489
Eschatoporis 471, 484
Eschscholtz 154
Estenoborus 801
Esthesopus 164, 168
Esthlogena 598
Estoloides 585, 598
Eteophilus 738
Euaesthetinae 822
Eubrianacinae 134
Eubrianax 134
Eubriinae 134
Eubrychius 753
Eubulosoma 764
Eubulus 763, 764
Eucactophagus 730
Eucamptodes 314
Eucanthus 24, 25
Euceratocerini 255
Euceratocerus 255
Euchaestes 598
Euchaetes 747
Euchrysolina 650
Eucilinus 769, 772, 778
Eucinetidae 82, 83, 828
Eucinetus 83
Eucleonus 785
Euclidii 70
Euclyptus 738
Eucnemidae 152, 825
Eucnemididae 152
Eucneminae 155
Eucnemini 155
Eucnemis 153, 155
Eucnocerus 231
Euconibius 493
Eucoptus 760
Eucrada 249, 254
Eucradinae 254
Eucradini 254
Eucrossus 577, 591
Euctenochira 648
Eucyllus 772, 779
Eucymbolus 286, 292
Eucyphus 115
Eudasytes 296
Euderces 582, 595
Euderiinae 235, 242
Eudiagogini 725, 776
Eudiagogus 768, 776
Eudibolia 666
Eudichronychinae 160
Eudistenia 575, 590
Eudociminus 789, 791
Eudocimus 791
Eudomychus 368
Eudoxilus 596
Eudoxus 669
Euetheola 64, 66
Eufallia 396, 397
Eugastra 58
Eugenysini 645
Euglenesidae 559, 560
Euglenesina 561
Euglenesini 561
Euglenidae 559, 560
Eugnamptus 706, 707
Eugoniopus 315
Eugonus 698, 699
Euhrychiopsis 749, 753
Eulabini 469, 478, 483, 495
Eulabis 496
Eulechriops 755, 756
Eulichadidae 142, 827
Euligyrus 66
Eulimnichus 126
Eulytocerus 800
Eumecomera 517, 518
Eumichthini 580, 593
Eumichthus 581, 593
Eummenotarsus 297
Eumolpidae 617
Eumolpinae 619, 621, 636, 639, 667
Eumolpini 636, 637, 639, 668
Eumolpites 639, 669
Eumolpus 636, 669
Eumonima 776
Eumononycha 746
Eunausibius 324, 325
Eunicentrus 747
Eunyssobia 747
Euolamus 738
Euoniticellus 49, 50
Eupactidius 257
Eupactus 257
Eupagoderes 778
Euparia 45, 47
Eupariini 43, 46
Euparius 697, 699
Euparixia 45, 47
Euphorhipis 69
Euphoria 68, 69
Euphoriaspis 69
Euphrytus 637, 669
Euphyma 673
Eupisenus 415
Euplastius 166, 169
Euplateumaris 640
Euplatypus 805, 806
Euplectroscelis 634, 660
Eupleurida 553
Eupodes 617
Eupogonius 584, 598
Eupogonocherus 598
Eupompha 525, 526
Eupomphini 526
Eupristocerus 105, 108
Eupsalis 717
Eupsilobiinae 368
Eupsilobius 368
Eupsophulus 473, 488
Eupsophus 488
Eupteroxylon 805
Eupus 666
Eurelymis 291
Eurhopalus 231
Eurhoptus 761, 765
Eurhynchidae 711
Europini 321
Europs 320, 321
Eurostus 254
Eurrhoptus 765
Eurygeniinae 551, 554, 833
Eurygeniini 551, 554
Eurygenius 552, 554
Eurymeloe 527
Eurymetopini 470, 474, 483, 489
Eurymetopon 474, 490
Eurymycter 697, 699
Euryomia 69
Euryopa 184
Euryoptera 594
Eurypalpus 134
Eurypepla 624, 646
Euryplatus 547
Eurypogon 147
Euryptera 588
Euryptychini 156
Euryptychus 153, 156
Euryscelis 583, 594
Euryscopa 675
Eurysphindus 308
Eurystethes 547
Eurystethinae 547
Eusattodera 631, 632, 659
Eusattodes 487
Eusattus 473, 487
Euscaphurus 83
Euscepes 762, 765
Euschaefferia 265, 266
Euschides 488
Eusemostene 497
Eusomalia 115
Eusphyrus 698, 699
Eustenopus 783, 785
Eustilbus 336
Eustroma 592
Eustromula 578, 592
Eustrongylium 501
Eustrophinae 414, 415
Eustrophini 416
Eustrophinus 416
Eustrophopsis 416
Eustrophus 416
Eustylini 725, 776
Eustylomorphus 778
Eutemnoscelus 738
Eutheca 257, 660
Eutheia 597
Euthoron 739
Euthuorus 597
Euthyna 674
Euthysanius 167
Eutinobothrus 763, 764
Eutochia 496
Eutomus 490
Eutoxotus 589
Eutrichapion 717, 718
Eutrichillus 585, 599
Eutricholistra 285, 295
Eutrichopleurus 285, 296
Eutrilia 393
Eutriorophus 490
Eutylistes 257
Eutylistus 257
Euvacusus 553, 555
Euvolvulus 707
Euvrilleta 256
Euvrilletta 252, 256
Euwallacea 798, 803
Euxenulus 697, 698
Euxenus 697, 698, 765
Euxestinae 364
Euxestus 364
Evander 596
Evodinus 572, 588
Evoplus 492
Evotus 771, 779
Exanthonomus 737
Exapiini 717
Exapion 716, 717
Exema 639, 677
Exerestus 498
Exiguipenna 652
Exilia 591
Exmenetypus 780
Exmyllocerus 776
Exocentrus 598
Exochomaires 384
Exochomus 375, 384
Exochyromera 740
Exoma 115
Exomala 62
Exomella 115
Exomias 779
Exomis 674
Exopioides 239
Exoplectra 377, 385
Exoplectrae 385
Exoplectrides 385
Exoplectrinae 385
Exoplectrini 375, 377, 385
Exops 239
Exorhina 664
Exorini 656
Exosomites 628, 639, 659
Explorator 129, 130

F

Fabricianus 674
Fachus 257
Falacer 498
Falciger 752
Fallapion 715, 718

Falsogastrallus 255
Falsomordellistena 428
Falsocalleros 176, 177
Familiurodera 676
Fassatia 656
Fastuolina 650
Faustinus 761, 765
Fidia 637, 671
Fleutiauxellus 165, 170
Floresianellus 777
Floresianus 777
Floridocassis 624, 647
Florilinus 230
Fluvicola 134
Foadia 393
Fogatianus 739
Fontenelleus 740
Formicilla 553, 555
Formicomini 555
Fornax 154, 156
Fossanobium 257
Fossipedes 90
Fossocarus 54, 58
Foudrasia 664
Frostia 209
Fuchsina 396, 397
Fulcidacidae 617
Furcipes 736
Furcipus 736

G

Galba 155
Galerima 656
Galerotoma 656
Galeruca 629, 656
Galerucella 630, 654
Galerucidae 617
Galerucinae 619, 621, 628, 629, 639, 653
Galerucini 619, 628, 639, 654
Galerucites 628, 639, 656
Galleruca 656
Gambrinus 169
Ganascus 561
Ganimus 590
Gasterocercini 725, 765
Gasterocercodes 764
Gastragallus 255
Gastrallini 255
Gastrallomimum 255
Gastrallus 255
Gastraspis 776
Gastroeidea 651
Gastroidea 651
Gastrophysa 628, 651
Gastrotaphrus 788, 789
Gaurambe 265
Gaurocryphus 760
Gaurotes 572, 588, 589
Gelus 756
Genattus 230
Genuchinus 68, 70
Geocanthon 49
Geoderces 782
Geodercodes 772, 779
Geohowdenius 26
Geomecus 781
Geomorphus 785
Geonemini 725, 777
Geopsammodius 44, 47
Geotrupes 25, 26
Geotrupidae 23, 25, 820
Geotrupinae 24, 25
Geraeus 743, 746
Germarostes 35
Geropa 576, 591
Gerstaeckeria 762, 765
Gibbifer 350, 352
Gibbiinae 253
Gibbium 254
Gibbobruchus 604, 607
Gibbostethus 785
Gietellidae 281
Gilbertia 732
Gilbertiola 732
Gillaspytes 65, 66
Ginglymocladus 198
Glaphyra 593
Glaphyridae 37, 38, 820
Glaphyrinae 38
Glaphyrocanthon 49
Glaphyrometopus 775, 778
Glaresidae 15, 820
Glaresis 15
Glaucotes 585, 599
Glenidion 635, 662
Gleosoma 393
Glipodes 427
Glipostenoda 428
Glischrochilus 313, 315
Glisonotha 346
Globicornis 231
Glocianus 752
Gloeosoma 393
Gloeosomatini 393
Glycaria 790
Glycobius 582, 594
Glyphonyx 163, 168
Glyphostethus 759
Glyphuroplata 623, 645
Glyploscelimorpha 96
Glyptasida 472, 488
Glyptina 634
Glyptobaris 743, 745
Glyptoderes 803
Glyptogeraeus 747
Glyptophorus 341
Glyptoscelimorpha 96
Glyptoscelis 636, 637, 671
Glyptotus 482, 500
Gnaphalodes 575, 595
Gnathacmaeops 573, 588
Gnathium 524, 527
Gnathoborus 804
Gnathocranus 804
Gnathophorus 805
Gnathophthorus 805
Gnathospasta 527
Gnathotrichoides 805
Gnathotrichus 799, 805
Gnathoweisea 377, 382
Gnatocerus 491
Gnorimella 68, 70
Gnostidae 245
Gnostus 254
Goes 584, 597
Golbachia 154, 155
Gomya 220
Gonasida 472, 488
Gondwanocrypticus 475, 492
Goniaderini 467, 471, 483, 484
Gonielmis 119
Gonioctoeus 699
Gonioctena 625, 649
Gonioctenini 649
Goniomena 649
Gonocallus 594
Gonocyphus 51
Gonoderina 480, 483, 499
Gononotus 788, 791
Gonops 699
Gonotropis 697, 699
Gorginus 336, 337
Gracilema 642
Gracilia 576, 592
Graciliini 576, 592
Grammoptera 573, 588, 589
Graphalia 65
Graphicallo 626, 651
Graphisurus 599
Graphognathus 778
Graphops 637, 670
Graphorhinus 777
Graptodera 662
Gratiana 624, 647
Gratusus 400
Gretschkinia 801
Griburiosoma 699
Griburius 638, 672
Grobbenia 341
Gronevus 394
Gronocarus 54, 58
Grylius 66
Grynocharis 265
Gryphinus 393
Grypidius 731
Grypus 731
Gyascutus 103, 107
Gymnetina 68, 69
Gymnetini 68
Gymnetis 68, 69
Gymnetoides 69
Gymnetron 738
Gymnetrum 738
Gymnodi 70
Gymnodus 70
Gymnoganascus 560, 562
Gymnognathini 697, 699
Gymnognathus 697, 699
Gymnonotus 790
Gymnoparomius 315
Gymnopsyra 578, 592
Gymnopyge 54
Gymnosomides 385
Gymnota 107, 652
Gynaecomeloe 526
Gynandrophthalma 674
Gynnis 58
Gynopterus 254
Gyrinidae 819
Gyrotus 756

H

Habrobregmus 255
Habroxenus 697, 698
Hadraule 405, 410
Hadrobegmus 256
Hadrobregmini 255
Hadrobregmus 251 255
Hadrobresmus 255
Hadrobrogmus 255
Hadromeropsis 771 780
Hadromerus 780
Hadromorphus 166, 170
Hadroplontus 751, 752
Hadrorestes 780
Haematochiton 350, 352
Haemoniini 619, 622, 639, 641
Hagedornus 804
Haliplidae 819
Hallomeninae 414, 415
Hallomenus 414, 415
Halmus 375, 384
Halotosia 65
Haloxenus 760
Haltica 662
Halticidae 617
Halticidea 654
Halyziaires 387
Halyziates 387
Halyziides 386, 387
Halyziini 375, 387
Hamaba 740
Hamaxobium 83
Hamletia 665
Hapalips 344, 346
Hapalorhinus 290
Hapalosalia 589
Haplandrus 482, 500
Haplanthaxia 108
Haplidoeme 590
Haplidus 577, 591
Haplophoria 69
Haplorhynchites 706, 707
Haplostethops 744, 747
Haplostethus 106
Hapsidolema 642
Hapsidolemoides 622, 642
Haptoncus 314
Haptoscelis 656
Hargium 589
Harmonia 375, 381, 387
Harmoniaspis 386
Harpasta 675

Harpium 589
Hatchiana 209
Hebestola 584, 597
Hedobia 249, 254
Hedobiinae 254
Hedobiini 254
Heilipus 789, 791
Heilus 789, 791
Helesius 379, 383
Helichus 122
Heliocis 517, 518
Heliomanes 593
Heliotychius 739
Hellodes 652
Helminthidae 117
Helminthimorphus 740
Helocerus 230
Helodes 652
Helodidae 87
Helopidae 463
Helopini 468, 478, 483, 496
Helops 478, 496
Hemantus 555
Hemicallidium 594
Hemichalepus 644
Hemicoelinum 255
Hemicoelum 255
Hemicoelus 251, 255
Hemicrepidiini 160
Hemicrepidius 164, 169
Hemierana 586, 600
Hemiglyptus 636, 664
Hemilophini 586, 600
Hemimiarus 738
Hemiopsida 154, 155
Hemipeplinae 532
Hemipeplus 532
Hemiperapion 717
Hemiphaedon 652
Hemiphileurus 64, 67
Hemiphrynus 634
Hemipodistra 208
Hemiptychus 258
Hemirhipides 160
Hemirhipini 169
Hemirhipus 169
Hemisphaerica 386
Hemisphaerota 623
Hemisphaerotini 623, 639, 645
Hemispilota 62
Hemitychus 258
Hendecatomus 238
Henonia 739
Henoticoides 341
Henoticus 340, 341
Henotiderus 341
Henous 527
Heorhynchus 752
Heptaphylla 490
Hesperandra 570, 586
Hesperanoplium 577, 591
Hesperobaenus 320, 321
Hesperobaris 745
Hesperophanes 577, 591
Hesperophanini 576, 591
Hesperorhipis 104, 108
Heterachtes 592
Heterachthes 579, 592
Heterachthon 592
Heterarthron 239
Heteraspis 670
Heterelmis 119
Heterhelus 309, 310
Heterichnus 674
Heterlimnius 119
Heterobabia 675
Heterobagous 740
Heteroborips 803
Heterobostrychus 237, 240
Heterobrachium 310
Heterobrenthus 714, 717
Heteroceridae 127, 130, 827
Heterocerinae 130
Heterocerini 129, 130
Heterocerus 131
Heterodactylus 674
Heteroderes 162, 169
Heterodontus 314
Heterodromia 325
Heteromorphus 783
Heteromus 310
Heterophagus 497
Heterophasis 759
Heteropromus 477, 495
Heterops 582, 595
Heteropsini 582, 595
Heteropus 493
Heterosirocalus 752
Heterostomis 675
Hetoemis 597
Hexacolinus 801
Hexacolus 801
Hexacylloepus 119
Hexaphyllus 8, 738
Hexarthrum 757, 759
Hexastenoscelis 760
Hilarocassis 623, 645
Hilipinus 789, 791
Hilipus 791
Himatinum 760
Himatium 757, 760
Himatolabus 706, 707
Himeniphades 740
Hippodamia 381, 386
Hippodamiaires 385
Hippodamiidae 386
Hippodamiina 386
Hippodamiini 385
Hippodamiites 386
Hippomelanina 103, 107
Hippomelanini 103, 107
Hippomelas 103, 107
Hippopsis 583, 597
Hippuriphila 635, 663
Hirsutographis 599
Hirsutotriplax 350, 352
Hispidae 617
Hispinae 619, 620, 622, 639, 643
Histeridae 828
Histerogaster 672
Histeropsis 492
Histiaea 494
Hohonus 763, 765
Holaniara 496
Holeleodes 477, 495
Hologymnetis 68, 69
Holomallus 297
Holoparamecus 368
Holopleura 579, 595
Holopleurini 595
Holopsis 392, 393
Holostilpna 698
Holostrophini 414, 416
Holostrophinus 416
Holostrophus 414, 416
Homaesthesis 571, 587
Homalostylus 730
Homaloxenus 792
Homochromina 65
Homoeocryphalus 804
Homoeolabus 707
Homoeostigmus 672
Homoiosternus 61, 63
Homophoeta 665
Homophoita 665
Homorosoma 750, 754
Homovalgus 70
Hopatrinus 494
Hoplacerus 664
Hoplia 53, 56
Hoplicnema 393
Hopliini 53, 56
Hoplocephala 492
Hoplopteryx 587
Hoplosia 598
Hoppingiana 284, 293
Horiidae 522
Horiini 527
Horistonotus 164, 168
Hormiscus 699
Hormops 788, 792
Hormorini 725, 777
Hormorus 773, 777
Hornaltica 635, 663
Hornia 526, 527
Hornietus 43, 46
Horticola 355
Hospitopterus 230
Huaca 735, 737
Huleechius 119
Hyagniellus 597
Hybodera 580, 593
Hyboderini 580, 593
Hybosoridae 32, 33, 820
Hybosorus 33
Hydaticus 754
Hydnocerinae 269, 276
Hydraenidae 822, 824
Hydrogaleruca 654
Hydronomus 740
Hydrophilidae 824, 828
Hydrophytophagus 341
Hydrophytophilus 341
Hydropus 664
Hydroscaphidae 819
Hydrothassa 627, 652
Hydrotica 668
Hyerisus 255
Hylaspini 639, 656
Hylastes 795, 799
Hylastina 795, 799
Hylastinoides 799
Hylastinus 795, 799
Hylastities 799
Hylecerus 262
Hylecoetinae 262
Hylecoetus 262
Hylescierites 799
Hylesinina 795, 799
Hylesinini 725, 794, 799
Hylesinites 799
Hylesinus 795, 799
Hyliota 325
Hylis 154, 155
Hylobiina 791
Hylobiini 725, 791
Hylobitelus 791
Hylobius 789, 791
Hyloceotus 262
Hylochares 154, 155
Hylocharini 155
Hylocrinus 490
Hylocrius 594
Hylocurosoma 801
Hylocurus 796, 801
Hyloecotus 262
Hylophilidae 559
Hylophilus 562
Hylotrupes 581, 593, 594
Hylurgidae 792
Hylurgopinus 795, 800
Hylurgops 795, 799
Hylurgus 795, 800
Hylus 155
Hymenochara 481, 500
Hymenorus 480, 499
Hynnulus 781
Hypactus 739
Hypasclera 516, 518
Hypebaeus 287, 288
Hypera 783
Hyperaspidae 383
Hyperaspides 383
Hyperaspidina 383
Hyperaspidini 383
Hyperaspidius 379, 383
Hyperaspiens 383
Hyperaspini 379, 383
Hyperaspis 379, 383
Hyperaspites 383
Hyperemotes 760
Hypericia 626, 650
Hyperinae 725, 728, 782, 783
Hyperini 725, 783
Hyperisus 255
Hyperius 255
Hypermallus 592

Hyperodes 766
Hyperomorpha 765
Hyperplatys 585, 599
Hyperstylus 776
Hypertensus 804
Hypexilis 576, 592
Hypnoidini 160, 170
Hypnoidus 165, 170
Hypoborina 795, 800
Hypocaelus 155
Hypocephalidae 568
Hypocopridae 338
Hypocopriini 341
Hypocoprini 340
Hypocoprus 341
Hypocryphalus 797, 804
Hypodacne 364
Hypodesites 160
Hypoganus 166, 170
Hypogena 478, 497
Hypohylis 155
Hypolampsis 666
Hypoleschus 737
Hypolithus 165, 170
Hypomolyx 791
Hyponotus 759
Hypophloeinae 483, 492
Hypophloeini 466, 483, 493
Hypophloeus 493
Hypopityophthorus 805
Hypora 256
Hyporhagus 455
Hypothenemus 797, 804
Hypothenoides 803
Hypothyce 55, 58
Hypotrichia 55, 58
Hypulini 421
Hypulus 419, 421
Hypurini 724, 753
Hypurus 748, 753

I

Ibidion 592
Ibidionini 579, 592
Iccius 475, 491
Ichnea 271, 278
Ichnodes 521
Ichthyurini 206, 209
Ichthyurus 209
Ictistygnini 554
Idalia 386
Idiobates 497
Idiocephala 673
Idiogethes 315
Idiopidonia 574, 588
Idiopteron 177
Idiostethus 744, 747
Idiotarsus 153, 155
Idocolastus 314
Idoemea 590
Ignelater 162, 169
Ignotus 230
Ilendus 400
Inamblyderus 555
Inca 68, 70
Incisophthalma 642
Incolia 415
Indomecus 781
Ino 547
Inopelonia 661
Inopeplinae 545, 547, 822
Inopeplus 545, 547
Involvulus 706, 707
Iodema 527
Iperaspini 383
Iphicorynus 492
Iphimeini 668
Iphimeites 639, 668
Iphthiminus 500
Iphthimus 500
Ipidae 792
Ipina 798, 802
Ipinae 315
Ipochus 584, 596
Ips 315, 798, 802
Isarthron 587
Isarthrum 587
Isarthrus 154, 156
Ischalia 551, 553, 834
Ischaliidae 549
Ischaliinae 551, 553
Ischiodontus 168
Ischiopachina 638, 639, 676
Ischiopachys 638, 676
Ischnocerini 697, 698
Ischnocerus 697, 698
Ischnocnemis 596
Ischnopterapion 716, 718
Ischyropalpus 552, 556
Ischyrosonychini 623, 639, 646
Ischyrus 350
Isochnus 735, 738
Isocoryna 65
Isodacrys 771, 780
Isodrusus 771, 780
Isohydnocera 270, 276
Isomira 481, 499
Isomyllocerus 776
Isonychus 54
Isopentra 527
Isorhipina 69
Isorhipis 155
Isoriphis 153
Isosceles 600
Isostrabala 663
Isotrilophus 427
Isotrogus 759
Ithyceridae 821
Itomarus 345
Itychus 739
Ixapiini 717

J

Jacobsoniidae 219, 220, 824
Janbechynea 614, 615
Janessa 345
Jentozkus 83
Jodema 527
Jonthonota 624, 648
Judolia 573, 588
Jugocryphalus 804
Juliusiana 640
Juliusina 640
Juniperella 104, 107

K

Karumiinae 91
Kateretes 310
Keitheatus 658
Kissingeria 716, 718
Kissodontus 777
Knausia 480, 499
Knowltonia 108
Knullanoplium 591
Knulliana 575, 595
Knulliella 106
Knulliobuprestis 107
Kolbeus 25
Korynetidae 267
Korynetinae 272, 279
Kurilonus 781
Kuschelina 633, 665, 666
Kytorhininae 606
Kytorhinus 604, 606

L

Labidomera 626, 651
Lacconotinae 532
Lacconotus 532
Lachnodactyla 137
Lachnodere 493
Lachnopus 774, 777
Lachnosterna 58
Lacon 163, 168
Lacordairius 781
Lactica 665
Ladioderma 257
Laemophloeidae 331
Laemophloeus 332, 334
Laemosaccidius 786
Laemosaccini 725, 786
Laemosaccus 786
Lagocheirus 586, 599
Lagochirus 599
Lagopelas 18
Lagriidae 463
Lagriiformes 483
Lagriinae 483
Lagriini 466, 471, 483, 484
Lagrioidinae 553
Lamellicornia 1
Lamellosus 674
Lamiidae 568
Lamiinae 569, 583, 596
Lamiini 583, 597
Lamperos 496
Lampetis 104, 107
Lamprocleptes 600
Lampromerus 591
Lampropterus 581, 593
Lamprosoma 637, 672
Lamprosomatidae 617
Lamprosomatinae 619, 620, 637, 639, 671
Lamprosomatini 639, 672
Lamprosomidae 617
Lamprosominae 671
Lampyridae 823, 825, 835
Lanelater 162, 169
Languiria 345
Langura 345
Languria 344, 345
Languriidae 832
Languriinae 344
Languriini 345
Langurites 344, 345
Lanternarius 130, 131
Lapecautomus 365
Lapethus 365
Laphyra 593
Lappus 556
Lapsus 130, 131
Lar 397
Lara 118, 120
Larainae 120
Laria 606
Laricobiinae 222
Laricobius 222
Lariidae 602
Lariniorhynchus 785
Larinodontes 785
Larinomesius 785
Larinotinae 265
Larinus 784, 785
Lariopsis 315
Larisia 567
Lariversius 476, 495
Lasiderma 257
Lasiocassis 647
Lasioderma 252, 256
Lasiodermini 256
Lasiopus 57
Lassioderna 257
Latemnis 69
Latheticus 478, 497
Lathridiidae 395
Lathridiomimus 220
Lathridiomorphus 220
Lathridius 397
Lathropus 333
Latometini 460
Latridiidae 395, 397, 824
Latridiinae 396, 397
Latridius 396, 397
Leanum 497
Lebasiella 272, 279
Lechriopini 724, 755
Lechriops 756
Leconteapion 718
Lecontella 269, 275
Lecontellus 693, 694
Lecontia 535
Lederia 418, 420
Ledidosoma 782

Leichenini 471, 483, 494
Leichenum 471, 494
Leiestinae 368
Leiodidae 823, 828
Leiopleura 105, 108
Leiopleurella 108
Leiopleurina 105, 108
Leiopsammodius 44, 47
Leiopus 599
Lema 622, 641, 642
Lembodes 761, 765
Lemiini 622, 639, 641
Lemodinae 553
Leonia 527
Leonidia 527
Lepadodes 393
Lepargus 598
Leperisinus 799
Lepesoma 782
Lepiarcha 315
Lepicerinus 803
Lepiceroides 804
Lepicerus 803
Lepidelater 168
Lepidobaris 747
Lepidocnemeplatia 471, 485
Lepidocricus 778
Lepidophorus 768, 775
Lepidopus 777
Lepidosoma 782
Lepidotus 168
Lepidotychius 739
Lepilius 787, 791
Lepismadera 105, 108
Lepisomus 800
Leprotites 639, 670
Leptacmaeops 588
Leptaegialia 45
Leptalia 573, 588
Leptanthicus 552, 555
Leptarthrus 737
Leptideella 593
Leptides 593
Leptinotarsa 626, 651
Leptipsius 320, 321
Leptocaryurgus 752
Leptoceletes 176, 177
Leptohoplia 60, 62
Leptophloeus 333
Leptopinara 768, 782
Leptoremus 551, 554
Leptorhabdium 572, 588
Leptosaldius 746
Leptoschema 164, 167
Leptostilbus 336
Leptostylis 599
Leptostylopsis 586, 599
Leptostylus 586, 599
Leptotrichaltica 663
Leptotrix 663
Leptovectura 284, 294
Leptura 574, 588, 590
Lepturges 585, 599
Lepturgoides 599
Lepturidae 568
Lepturinae 570, 571, 587
Lepturini 571, 587
Lepturoides 169
Lepturopsis 574, 588
Lepyrini 725, 791
Lepyrus 789, 791
Leropeplus 314
Letznerella 803
Leucostrophus 776
Lewisium 394
Lexiphanes 638, 673
Liagrica 591
Liasemum 587
Liatongus 49, 50
Liberocera 69
Librodor 315
Lichadidae 142
Lichenophanes 236, 240
Lichminus 126
Lichnanthe 38
Ligmargus 165, 170
Ligniperda 8
Lignyodes 739
Lignyodina 739
Lignyodius 739
Ligyrellus 66
Ligyrodes 66
Ligyrus 66
Lilioceris 622, 641
Limnichidae 125, 126, 827
Limnichinae 126
Limnichini 126
Limnichites 126
Limnichoderus 125, 126
Limniidae 117
Limnobaropsis 746
Limomius 593
Limonius 166, 169
Lina 652
Linamorpha 653
Lindia 497
Lindorus 385
Linogeraeus 744, 747
Linsleya 526, 527
Linsleyana 589
Linsleyonides 577, 592
Liobaulius 555
Liodema 475, 492
Liolepta 761
Lioligus 115
Liometophilus 761, 764
Lioolius 257
Lioon 115
Liophloeus 779
Liopinus 585, 599
Liopus 599
Lioscymnus 382
Liotrichus 166, 170
Liozoum 254
Liozum 254
Liozuom 254
Liparopeplus 314
Liparthrum 796, 800
Lissocordylus 731
Lissomelas 68, 70
Lissomidae 160
Lissominae 168
Lissomini 168
Lissonotini 579, 595
Lissonotus 579, 595
Lissorhoptrus 730, 731
Listemus 115
Listrimorpha 284, 298
Listrini 295
Listrochelus 54, 58
Listroderes 766
Listroderina 766
Listromimus 284, 297
Listronotus 766
Listropsis 286, 297
Listrorrhynchus 737
Listrus 295
Litargellus 400
Litargus 400
Litasida 472, 488
Litheleodes 477, 495
Lithodus 781
Lithophorus 360, 361
Litochropus 336, 337
Litochrus 336
Litodactylus 754
Litolibrus 336
Litomerus 790
Litosomina 729
Litosonycha 665
Litropeda 256
Lixellus 766
Lixestus 785
Lixinae 725, 726, 727, 783, 785
Lixini 725, 785
Lixus 784, 785
Loberina 346
Loberinae 345
Loberus 344, 346
Lobiopa 313, 314
Lobometopon 474, 489
Lobopoda 479, 499
Lobosoma 790
Lobostoma 314
Loceptes 790
Lochmaea 655
Lochmaeocles 583, 597
Lochmaeodes 597
Lochmalodes 597
Lodinus 493
Loedelia 272, 280
Loganius 801
Lonchocerus 790
Lonchophorus 51
Longifistulia 736
Longitarsus 634, 661
Lophalia 580, 596
Lophalophus 775
Lopheros 175, 176
Lophocateres 265
Lophocateridae 263
Lophocaterinae 265
Lophopogonius 585, 598
Lordicassis 647
Lordiconia 647
Loriarulus 314
Loricaster 85, 86
Lucaina 176, 177
Lucanidae 6, 7, 820
Lucaninae 7
Lucanini 7
Lucanus 7, 8
Ludiides 160
Ludius 167, 170
Luperaltica 634, 661
Luperina 628, 639, 658
Luperini 639, 656
Luperites 660
Luperodes 660
Luperosoma 631, 658
Lupinocolus 773, 777
Lupraea 634, 661
Luprea 661
Lutrochidae 123, 124, 827
Lutrochus 124
Lycaina 177
Lychochoriolaus 573, 588
Lycidae 174, 176, 825
Lycinae 177
Lyconotus 176, 177
Lycoperdina 369
Lycoperdininae 369
Lycoptis 265
Lycostomus 177
Lyctidae 233
Lyctinae 235, 238, 242
Lyctini 238, 242
Lyctoxylon 238, 242
Lyctus 238, 242
Lycus 176, 177
Lygistopterini 177
Lygistopterus 175, 177
Lygistus 658
Lygophilus 780
Lymantes 787, 792
Lymantini 725, 792
Lymantor 797, 802
Lymexylidae 261, 262, 822, 828
Lymexyloidea 261
Lymexylonidae 261
Lyphia 478, 497
Lyphontes 67
Lyprus 740
Lypsimena 584, 598
Lysathia 635, 662
Lystronychina 480, 483, 500
Lystronychus 480, 500
Lytopeplus 365
Lytta 525, 526, 527
Lyttidae 522
Lyttini 527
Lyttomeloe 527

M

Machlotes 361
Macipus 738
Macrancyloides 760

Macrancylus 758, 760
Macranoxia 58
Macrarthria 554
Macrarthrius 554
Macratria 551, 554
Macratriinae 551, 554
Macraulacinae 155
Macraulacini 156
Macrelmis 118, 119
Macreurops 320, 321
Macrobasis 527
Macrocis 409
Macrocnema 666
Macrocorynus 776
Macrocryphalus 804
Macrodactylini 53
Macrodactylus 54
Macrodea 341
Macrohaltica 662
Macrohimatium 760
Macrolina 627, 652
Macronaemia 381, 386
Macronovius 385
Macronychini 120
Macronychus 118, 120
Macropelmus 740
Macropodina 70
Macropogon 147
Macroprion 231
Macrops 766
Macrorhoptus 735, 740
Macrorhyncolus 758, 760
Macroscytalus 757, 760
Macrosiagon 439, 440
Macrosiagonini 439, 440
Macrotomini 570, 586
Macrozeugophora 611
Macrozophobas 498
Maculicauta 527
Maculurges 599
Madarellus 743, 745
Madarina 745
Madarini 724, 727, 745
Madoniella 270, 278
Madopterini 724, 746
Maecolaspis 669
Maemactes 762, 764
Magdalidini 725, 786
Magdalinops 737
Magdalinus 786
Magdalis 786
Malacamerus 659
Malachiidae 281
Malachiinae 284, 287, 823
Malachiini 286, 288
Malachius 287, 290
Malacocis 406
Malacomacrus 590
Malacopterus 575, 590
Malacorhinus 632, 656
Maladera 54, 57
Malagoniella 49
Mallodon 570, 586
Mallodrya 513
Malloea 169
Malobidion 591
Malporus 553, 555
Malthininae 207, 209
Malthinini 207, 209
Malthinus 209
Malthodes 209
Malthodini 207, 209
Malthomethes 198, 199
Malthophia 590
Malthopia 590
Malvapiini 718
Mannophorus 580, 596
Mantura 635, 664
Maphoca 410
Margaiostus 165, 170
Margaridisa 635, 663
Marginus 357
Margus 497
Marinexa 314
Marklissus 752
Martianus 492
Martinezia 47
Martineziella 45, 47
Mascarauxia 766
Mastinocerini 184
Mastinocerus 184
Mastogeniina 102, 106
Mastogeniini 102, 106
Mastogenius 102, 106
Mastoremus 551, 554
Matheteinae 198, 200
Matheteini 200
Matheteus 198, 200
Mauroniscidae 281
Mazagranus 759
Mecas 586, 600
Mecinini 724, 727, 738
Mecinopsis 738
Mecinus 733, 738
Mecomycter 294
Meconemus 698
Mecopeltus 754
Mecostethus 673
Mectemycor 286, 294
Mecynopyga 739
Mecynotarsus 552, 556
Mecysmus 476, 493
Medonia 665
Megababia 676
Megabregmus 256
Megacerus 605, 606
Megacheuma 582, 594
Megachoriolaus 573, 588
Megacyllene 582, 594
Megaderus 579, 596
Megagarpolus 314
Megaleptura 588
Megalodacne 350
Megalodacninae 350
Megalopodidae 609, 611
Megalosoma 67
Megalostomina
638, 639, 675
Megalostomis 638, 675
Megalotrox 18
Megaltica 662
Meganemognatha 527
Meganoplium 577, 591
Megapenthes 164, 167
Megapenthini 167
Megasattus 487
Megascelidae 617
Megascelididae 617
Megascelidinae 667
Megascelidini 637, 639, 670
Megascelinae 667, 670
Megascelis 637, 667, 670
Megasemum 571, 587
Megasida 473, 488
Megasoma 65, 67
Megasominus 67
Megatoillidae 228
Megatoma 230, 231
Megatrupes 26
Megeleates 475, 490
Megetra 525, 526
Megilla 386
Megobrium 581, 593
Megorama 252, 257
Meibomeus 605, 607
Meioderus 393
Meizoderus 393
Melalgus 235, 239
Melamomphus 782
Melanactes 165, 169
Melanactides 160
Melanastus 475, 490
Melandrya 419, 421
Melandryidae 417, 420, 834
Melandryinae 418, 420
Melandryini 421
Melaneleodes 477, 495
Melaneros 177
Melanimini 471, 483, 494
Melanocanthon 49
Melanolemma 770, 782
Melanophila 104, 107
Melanophilina 104, 107
Melanophilini 104, 107
Melanophthalma 396, 397
Melanotina 168
Melanotites 160
Melanotus 163, 168
Melanthaxia 108
Melasia 496
Melasidae 152
Melasina 155
Melasinae 155
Melasini 155
Melasis 153, 155
Melasmetus 594
Melasoma 652
Melasomida 653
Melbonus 777
Meligethes 312, 315
Meligethinae 312, 315
Melittomma 262
Melittommatinae 262
Melittomminae 262
Melittophagus 527
Mellieicis 410
Meloe 526, 527
Meloidae 522, 526, 822, 832
Meloinae 526
Meloini 527
Melolonthinae 42, 51, 56
Melolonthini 53, 57
Melosoma 652
Melyridae 281, 823, 829
Melyrinae 284, 292
Melyris 284, 292
Melyrodes 284, 292
Memptorrhynchus 740
Menechides 498
Menedrio 498
Menetypus 780
Menoeceus 479, 499
Menoetius 777
Menovectura 294
Meracantha 479, 498
Merhynchites 706, 707
Meridion 593
Meriellum 582, 594
Merinus 482, 500
Merionus 777
Meristhus 163, 168
Merium 594
Merobruchus 606, 607
Merophysiidae 366
Meroscelisini 570, 587
Merotemnus 484
Merothricus 730
Mesagroecus 778
Mesagroicus 775, 778
Mesites 758, 759
Mesocoelopini 258
Mesocoelopodinae 258
Mesocoelopus 252, 258
Mesogenini 155
Mesohelops 496
Mesomphaliidae 617
Mesomphaliini 623, 639, 645
Mesoptiliinae 725, 786
Mesoptilinae 727
Mesosa 597
Mesosini 583, 597
Mesotrichapion 716, 718
Mesoxenus 759
Metablapylis 478, 495
Metachroma 637, 667
Metachromites 639, 667
Metaclisa 479, 497
Metacmaeops 573, 589
Metacycla 632, 656
Metacyclini 628, 639, 656
Metadoticus 698
Metaleptus 580, 596
Metallographa 651
Metallotimarcha 649
Metamasius 729
Metaparia 636, 668
Metapetalium 257
Metatypus 397
Metaxyonycha 669
Metaxyphloeus 332, 334

Methia 576, 590
Methiini 575, 590
Metmasiopsis 730
Metophthalmus 396, 397
Metopoloba 474, 489
Metoponium 474, 490
Metopotoma 792
Metrachodes 790
Metrioidea 631, 632, 660
Metriona 647
Metrionaspis 648
Metrionella 624, 648
Metrobrotica 657
Mezium 254
Miaromimus 738
Miccotrogus 739
Michthisoma 571, 587
Michthisomini 571, 587
Michthysoma 587
Micracina 796, 801
Micracis 796, 802
Micracisella 796, 801
Micracisoides 801
Micraegialia 43, 45
Micralcinus 788, 791
Micraltica 662
Micrambinus 341
Micraneflus 579, 592
Micranobium 255
Micranoplium 579, 592
Micrapate 237, 240
Micraspaires 385
Micraspidarii 385
Micrasta 102, 106
Micrathous 164, 169
Microanobium 255
Microaugyles 129, 130
Microbaris 741, 745
Microbregma 251, 255
Microbregmum 255
Microcallidium 594
Microcholus 744, 747
Microchondras 369
Microclytus 582, 595
Microctenochira 648
Microcylloepus 119
Microdalotes 764
Microdera 652
Microgoes 584, 597
Microhoriini 555
Microhypnus 165, 170
Microhyus 787, 791
Microlarinus 727, 783, 785
Microlema 642
Microlina 652
Microlipus 287, 290
Microlypesthes 670
Micromalthidae 819
Micromastus 787, 791
Micromes 474, 490
Micromimus 758, 759
Micromyrmex 734, 738
Micronaemia 386
Micronychus 779
Micronyx 739
Microon 714, 718
Micropentarthrum 760
Microplontus 751, 752
Micropsephodes 824
Microrhagus 154, 155
Microrhopala 623, 645
Microscapha 418, 420
Microschatia 472, 488
Microsphera 393
Microsporidae 819
Microstagetus 392, 393
Microsternus 350
Microtheca 626, 649
Microtonus 419, 421
Microtrichodesma 256
Microtychius 739
Microum 341
Microweisea 377, 382
Microweiseini 382
Microweisini 377, 382
Microxanthus 314
Microzogus 255
Micruria 314
Micrurula 314
Migiwa 165, 170
Migneauxia 396, 397
Milichacalles 764
Miloderes 769, 782
Miloderoides 771, 780
Mimetes 780
Mimopactus 778
Mimosestes 605, 607
Minaderus 589
Minthea 238, 242
Minturnia 675
Minulus 801
Minyomerus 771, 780
Mionycha 624, 647
Mionychella 647
Miraces 630, 654
Mitocera 673
Mitomermus 781
Mitostylus 780
Mitraelabrini 554
Mnionomus 341
Mogulones 751, 752
Molamba 393
Molorchini 581, 593
Molorchus 581, 593
Molytinae 725, 727, 786, 787, 789
Molytini 725, 789
Monachammus 597
Monachini 673
Monachulina 638, 639, 673
Monachulus 673
Monachus 673
Monanus 324
Monapus 51
Monarthrum 799, 805
Monebius 801
Moneilema 584, 597
Moneilemini 584, 597
Mongolotychius 739
Monilema 597
Monius 732
Monocerus 556
Monocesta 630, 654
Monochamus 584, 597
Monocrepidiites 160
Monocrepidius 169
Monohammus 597
Monohamus 597
Monoleptites 628, 639, 659
Monomacra 636, 665
Monommatidae 454, 455, 457
Monommatini 454, 455, 459, 460
Monommidae 454, 457, 460
Mononidia 65
Mononychini 724, 753
Mononychus 748, 753
Monophylla 269, 274
Monoplatini 660
Monoplesa 597
Monotoma 320
Monotomidae 319, 320, 829, 832
Monotominae 320
Monotomini 320
Monoxia 619, 630, 655
Monrosia 674
Mordella 427
Mordellidae 423, 426, 832
Mordellinae 426
Mordellini 426
Mordellistena 427
Mordellistenini 427
Mordellochroa 428
Morizus 805
Moroderia 779
Moronillus 394
Morphostrophus 776
Morychus 115
Mulsantina 375, 381, 387
Murmidiinae 364
Murmidius 364
Myacetochares 500
Mycetaea 369
Mycetaeidae 366
Mycetaeinae 369
Mycetina 369
Mycetochara 481, 500
Mycetocharina 480, 483, 500
Mycetocharis 500
Mycetocharus 500
Mycetophagidae 399, 400, 829, 831
Mycetophaginae 400, 829
Mycetophagus 400
Mycetophila 500
Mychocerus 364
Mychothenidae 366
Mychthisoma 587
Mycotretus 350, 352
Mycotrogus 478, 497
Mycotrupes 25, 26
Mycteridae 530, 532, 833
Mycterinae 532
Mycterus 532
Myctides 743, 745
Myeloborus 805
Myelophilus 800
Myelophites 799
Mylabridae 602
Mylabris 606
Mylacus 778
Myllocerina 776
Mylloceroversus 776
Myllocerus 770, 776
Mymechixenini 492
Myochroites 639, 671
Myochrous 636, 671
Myocoryna 651
Myodes 440
Myodites 440
Myoplatypus 805, 806
Myosides 770, 776
Myothorax 314
Myrmeceicon 70
Myrmechixenini 465, 483
Myrmechixenis 492
Myrmechixenus 465
Myrmecotonus 70
Myrmedophila 340, 341
Myrmex 734, 738
Mysia 387
Mysiates 385
Myxophaga 819
Myzia 379, 386, 387

N

Nacerdes 516, 517
Nacerdinae 516, 517
Naemia 381, 386
Nakanea 588
Nanodactylus 714, 718
Nanophyes 714, 718
Nanophyinae 714, 718, 821
Nanophyini 718
Nanops 735, 737
Nanularia 103, 107
Nanus 788, 792
Narberdia 736, 737
Narpus 118, 119
Narthecius 333, 334
Nassipa 567
Nathicus 555
Nathrenus 230
Nathriini 581, 593
Nathriobrium 581, 593
Nathrius 581, 593
Naucles 565, 567
Naupactini 725, 777
Naupactus 775, 778
Nausibius 324, 325
Nautes 478, 496
Neacanthocinus 599
Neadasytes 284, 297
Neandra 586
Neaneflus 578, 592
Neanthophylax 572, 589
Neanthribus 699

Neapion 716, 717
Nearctalox 717
Neatus 479, 497
Necrobia 272, 280
Necydalini 571, 590
Necydalis 571, 590
Necyomantes 674
Nedyus 751, 752
Nefoncerus 53, 56
Negalius 526
Negastriinae 160, 170
Negastrius 165, 170
Negritus 803
Nelites 492
Nelsonocheira 107
Neltumius 605, 607
Nematodes 154, 156
Nematodini 156
Nematodinus 155
Nematoplinae 521
Nematoplus 521
Nemicelus 532
Nemobius 801
Nemocestes 773, 779
Nemognatha 525, 527
Nemognathinae 522, 527
Nemognathini 527
Nemonychidae 692, 693, 821
Nemophilus 800
Nemophytobius 754
Nemosinus 791
Nemosoma 265, 266
Nemosomidae 263
Neoalosterna 574, 589
Neoathyreus 24, 25
Neobagoidus 731
Neobaphion 476, 495
Neobellamira 573, 589
Neobium 256
Neobradybatus 737
Neobrotica 657
Neoceletes 177
Neochalepus 644
Neochetina 731
Neochlamisus 639, 677
Neochodaeus 29
Neoclytus 583, 594
Neococcinella 386
Neocompsa 579, 592
Neocondylops 286, 288
Neocrepidodera 635, 664
Neocryphalus 803
Neocylloepus 119
Neodryocoetes 804
Neoelmis 119, 120
Neoerethistes 788, 790
Neoeurygenius 552, 554
Neogalerucella 655
Neogastrallus 255
Neoglostatus 801
Neohaemonia 622, 641
Neoharmonia 375, 381, 386, 387
Neohedobia 249, 254
Neoheterocerus 129, 131
Neohydronomus 732
Neohylecoetus 262
Neohypdonus 165, 170
Neolema 622, 642
Neolochmaea 629, 655
Neolycus 177
Neomallodon 570, 586
Neomastix 736, 737
Neomida 475, 492
Neomips 805
Neomysia 387
Neopachyta 589
Neopalla 386
Neopallodes 314
Neopanscopus 782
Neopanus 786
Neopedilus 542
Neopentarthrum 760
Neophleotribus 800
Neophycocoetes 792
Neophytobius 749, 754
Neopityophthorus 804
Neopolyarthron 571, 587
Neopristilophus 166, 170
Neopsammodius 44, 47
Neoptochus 770, 776
Neoptychodes 584, 597
Neopyrochroa 542
Neorthopleura 271, 279
Neoscraptia 565, 566
Neosirocalus 752
Neosothes 252, 258
Neostena 644
Neotomicus 802
Neotrichophorus 167
Neotropion 718
Neotychius 739
Neotylopterus 739
Neoulosomus 761, 764
Neoxenus 697, 698
Neozonitis 528
Neozurus 756
Nephasis 383
Nephaspis 379, 383
Nephus 379, 383
Nerrocerus 257
Nertinini 724, 747
Nesaecrepida 636
Nesocatoecus 777
Nesorthognathus 730
Nesostes 487
Nesostizocera 592
Nevermannia 257
Nevermannus 257
Nicaginae 7
Nicagus 7, 8
Nicentrites 747
Nicentrus 744, 747
Nicobiini 256
Nicobium 250, 256
Nicrophorinae 822
Nilionidae 463
Nipponopolygraphus 801
Nipponoserica 54, 57
Nipporhynchus 752
Niptinus 254
Niptus 254
Nipus 377, 382
Nitidula 313, 314
Nitidulidae 311, 313, 823, 829
Nitidulinae 312, 314
Nitidulopsis 314
Nitops 314
Nocheles 782
Nocibiotes 476, 494
Noda 669
Nodini 667
Nodinini 667
Nodonota 669
Nodopus 286, 288
Nomaspis 527
Nomebius 801
Nomidus 782
Nomizo 739
Nomophloeus 321
Noserosus 461
Noserus 461
Nosodendridae 224, 226, 826, 827
Nosodendrinae 226
Nosodendron 226
Nosodermini 461
Nosodes 266
Notaris 731
Noteridae 820
Nothopleurus 570, 586
Nothorhina 587
Nothrus 592
Notibius 476, 494
Notiodes 732
Notiomimus 256
Notionomus 732
Notiophilus 732
Notiotheca 257
Notodermus 731
Notolomus 734, 737
Notorhina 587
Notoxidae 549
Notoxini 552, 556
Notoxus 552, 556
Notoxyleborus 803
Novelsis 229, 230
Noviares 385
Noviini 377, 385
Nuzonia 647
Nycteus 83
Nyctobates 500
Nyctoporini 470, 483, 486
Nyctoporis 470, 487
Nyssodrys 599
Nyssodrysina 585, 599
Nyssodrystes 599
Nyssonotus 757, 760

O

Oberea 586, 600
Oberonus 122
Oberopa 600
Obriini 592
Obrium 576, 592
Ochodaeidae 28, 29, 820
Ochodaeinae 29
Ochodaeus 29
Ochraesthes 594
Ochraethes 583, 594
Ochrestes 594
Ochresthes 594
Ochroesthes 594
Ochrolitus 336, 337
Ochrosanis 532
Ochrosidia 65
Ochrosopsis 673
Ochrosopsus 673
Ochthebiinae 822
Ochyromera 735, 740
Ochyromerina 740
Octinodes 167
Octocryptites 160
Octodesmus 237, 240
Octotemnidae 403
Octotemnus 405, 410
Octotoma 623, 644
Odonotomus 108
Odontaeus 25
Odonteus 25
Odonthogethes 315
Odontionycha 647
Odontocera 579, 593
Odontocnema 664
Odontocorynus 743, 747
Odontocossonus 759
Odontomagdalis 786
Odontomesites 759
Odontonyx 137
Odontopsammodius 44, 47
Odontopus 788, 790
Odontorhynchus 730
Odontorrhynchus 730
Odontosphindinae 307
Odontosphindus 308
Odontota 644
Odontotaenius 13
Oebaceres 598
Oedemeridae 514, 517, 833
Oedemerinae 516, 517
Oediarthrus 369
Oedionychina 660
Oedionychis 665
Oedophrys 770, 776
Oedostethus 165, 170
Oeme 575, 590
Oemini 590
Oeneini 384
Oeneis 382
Oeniini 384
Oenopion 482, 500
Oestodes 166, 168
Oestodini 168
Oesyophila 592
Ohkubous 63
Oistus 165, 169
Olamus 738
Olibroporus 336

Olibrus 336
Oliganomala 62
Oligocaricis 739
Oligodites 754
Oligolochus 744, 747
Oligomerinus 256
Oligomerodes 256
Oligomeroides 256
Oligomerus 251, 256
Olistherarthrus 288
Olistherus 336
Olla 381, 387
Ologlyptus 488
Olotelina 562
Omaliinae 822
Omethes 198, 199
Omethidae 197, 199, 825
Omethinae 198, 199
Omethini 199
Omias 772, 778
Omiini 725, 778
Omileus 774, 777
Omonadus 553, 555
Omophoita 633, 665
Omorgus 18
Omorus 778
Omoscylon 597
Omosiphora 314
Omosita 313, 314
Omphalapion 716, 717
Oncerini 53, 56
Oncerus 53, 56
Oncideres 583, 598
Onciderini 583, 597
Onichodon 154, 156
Onicotis 314
Oniticellini 49, 50
Onitini 49, 50
Onitis 49, 50
Ontherus 49, 50
Onthophagini 49, 50
Onthophagus 49, 51
Onthurgus 51
Onychobaris 743, 745
Onychocassis 647
Onycholipini 725, 759
Onychomira 480, 499
Onychotrupes 26
Onychylis 732
Oochila 486
Ooconibius 493
Oocorynus 760
Oomorphidius 744, 747
Oomorphus 637, 672
Oophorini 169
Oophorus 169
Oopterinus 734, 738
Opacinota 624, 648
Opalocetus 756
Opatrinae 468, 471, 483, 493
Opatrini 471, 475, 483, 493
Opatrinus 471, 494
Ophiostomis 589
Ophistomis 588, 589
Ophraea 630, 655
Ophraella 619, 630, 655
Ophryastes 768, 778
Ophryastini 725, 778
Opilo 270, 277
Opiocanthon 49
Opisthotropis 699
Oplosia 584, 598
Opsimini 575, 590
Opsimus 575, 590
Optioservus 120
Opuntiaphila 765
Ora 88
Orchesia 418, 420
Orchesiini 418, 420
Orchestes 739
Orchestomerus 750, 753
Orchestris 661
Orchidophilus 741, 745
Orcus 384
Ordobrevia 118, 120
Orectis 647
Orestioides 664
Orimodema 782
Orizabus 65, 66
Ormiscus 698, 699
Ornithia 581, 593
Orophiina 408
Orophius 410
Orophyidae 403, 408
Orphilinae 226, 826
Orphilini 229
Orphilus 226, 230
Orphinus 230
Orsodachna 614
Orsodacne 614
Orsodacnidae 613, 614, 667
Orsodacninae 614, 667
Orsodanca 614
Orsophagus 736
Orthaltica 635, 663
Orthochoriolaus 573, 589
Orthocis 410
Orthognathina 729
Orthognathini 724, 729
Orthognathus 728, 729
Ortholeptura 574, 589
Orthopeplus 313, 314
Orthoperidae 390, 393
Orthoperinae 393
Orthopérites 393
Orthoperus 392, 393, 830
Orthoptochus 779
Orthoris 742, 745
Orthosoma 571, 587
Orthostethus 164, 167
Orthosticha 652
Orthotomicus 798, 802
Orthygia 660
Orwellion 578, 592
Oryctini 64, 66
Oryzaephilus 325
Osmidus 576, 591
Osmoderma 68, 70
Osmopleura 579, 595
Osphya 418, 421
Osphyinae 418, 421, 834
Ostoma 265
Ostomatidae 263
Ostomidae 263
Ostomodes 265
Ostomopsinae 364
Ostomopsis 364, 824
Othniinae 545, 546
Othnius 546
Othotomides 802
Otidocephalini 724, 727, 738
Otidocephalus 738
Otiocephala 675
Otiorcossonus 759
Otiorhynchini 725, 778
Otiorhynchus 773, 778
Oulema 622, 641, 642
Oulimnius 119, 120
Oxacis 517, 518
Oxeostomum 717
Oxoplatypus 805, 806
Oxoplus 596
Oxycnemus 315
Oxycopis 517, 518
Oxygonodera 474, 490
Oxygonus 166, 170
Oxygrylius 64, 66
Oxylaemus 361
Oxynchus 383
Oxynopterides 160
Oxynopterini 169
Oxyomus 43, 46
Oxypteris 107
Oxystomatina 718
Oxystomatini 718
Oxystomum 717
Ozognathini 255
Ozognathus 255
Ozophagus 801

P

Pachnaeus 771, 780
Pachneus 780
Pachybaris 744, 747
Pachybrachina 638, 639, 672
Pachybrachis 638, 672
Pachybrachys 672
Pachybruchus 606
Pachydemini 53
Pachygeraeus 744, 747
Pachylema 642
Pachylina 628, 652
Pachylobius 789, 791
Pachymerinae 604, 606
Pachymyllocerus 776
Pachynoderus 804
Pachyonychis 633, 665
Pachyonychus 633, 666
Pachyphanes 739
Pachyplectrus 33
Pachyrhinus 754, 774, 779
Pachyschelina 105, 108
Pachyschelus 105, 108
Pachystethus 62
Pachystylus 672
Pachyta 572, 589
Pachytychius 733, 739
Pachyurgus 498
Pachyus 778
Pactopus 159
Pactorrhinus 778
Pactostoma 488
Pagiocerus 795, 800
Palacoxeninae 154
Palaeothona 661
Palaeothroscus 159
Palaeoxenus 153, 154
Palechus 66
Palembus 492
Paleocopris 51
Palla 386
Pallene 736
Pallodes 313, 314
Palorus 478, 497
Pandeleteinus 771, 780
Pandeleteius 771, 780
Pandeletejus 780
Pandeletius 780
Pangauridae 160
Panopsis 786
Panormus 777
Panscopidius 782
Panscopus 732, 769, 782
Pantomallus 591
Pantomorus 775, 778
Pantopactus 778
Pantoplanes 778
Panus 786
Parabagous 740
Parabaliosus 644
Parabrachypterus 310
Parabyrsopolis 61, 63
Paracamptus 762, 765
Paraceratopus 740
Parachalcoides 663
Parachalepus 644
Parachelepus 644
Parachrysina 61, 63
Paracimbocera 782
Paraclinidia 65
Paraclytus 594
Paracotalpa 61, 63
Paradasytes 284, 297
Paradonus 165, 170
Paradorytomus 738
Parafilumis 266
Parafurcipes 737
Paragoges 739
Paragraphus 771, 775
Paragrilus 105, 108
Paragymnetis 69
Parahornia 761
Paralichus 137
Paralicus 756, 759
Paralitargus 400
Parallelina 588
Parallelostethus 164, 167
Paralobium 249, 254
Paralytta 527

Paramallus 586
Paramordellaria 426
Paranaemia 381, 386
Paranametis 782
Paranapiacaba 631, 657
Parandra 570, 586
Parandrinae 569, 570, 586
Parandrita 332, 334
Paranomala 62
Paranomus 170
Paranoplium 575, 590
Paranovelsis 230
Paranthonomus 737
Paraopsimus 590
Parapachyta 588
Parapanscopus 782
Parapelonides 271, 279
Paraphaedon 652
Paraphloeus 334, 493
Paraphymatodes 594
Paraphytobius 754
Paraplagiomorpha 653
Paraplinthus 789
Paraptochus 773, 779
Paraptorthodius 185
Parareoda 63
Parasida 488
Parasitones 780
Parastasia 61, 63
Paratenetus 471, 484
Parathyce 55, 58
Paratillus 272, 279
Paratimia 571, 587
Paratimorus 756
Paratrachelizus 717
Paratrachyphloeus 781
Paratriarius 630, 657
Paratucumaniella 597
Paratychius 739
Paratyndaris 102
Paraxyleborus 805
Parazonitis 528
Parchicola 636, 665
Parelaphidion 578, 579, 592
Parenthis 749, 754
Parepectasis 598
Parephistemus 674
Parevander 579, 596
Parhapsidolema 642
Paria 637, 668
Parilendus 400
Parmenini 584, 596
Parmenonta 583, 597
Parmenosoma 584, 596
Parmulini 392, 393
Parmulus 393
Parnus 122
Parochodaeus 29
Paroedostethus 170
Paropsimena 651
Paropsina 625, 639, 649
Parorectis 624, 647
Paroxacis 517, 518
Parysatis 598
Passalidae 12, 13, 820
Passalinae 13
Passalini 13
Passalus 13
Passandridae 327, 832
Pauronemognatha 527
Pavrosomus 780
Pechalius 474, 489
Pectotoma 565, 566
Pedetes 167, 169
Pediacus 329
Pedilinae 542
Pedilophorini 115
Pedilus 541, 542
Peditatus 129, 131
Pedonoeces 493
Pedrillia 611
Pedrillimorpha 611
Pedrilliomorpha 611
Pedrilonga 611
Pedrinella 611
Pelecinus 736
Pelecotoides 440
Pelecotoma 438, 439
Pelecotominae 438, 439
Pelecotomoides 440
Pelecyphorus 488
Pelenomus 749, 754
Pelenosomus 750, 753
Pelidnota 62, 63
Pelidnotidia 63
Pelidnotopsis 63
Pelina 386
Pelonides 279
Pelonium 271, 279
Pelonomus 122
Peltastica 222
Peltasticinae 222
Peltidae 263
Peltidea 266
Peltinae 265
Peltinodinae 392, 393
Peltinoditae 393
Peltis 265, 266
Peltophorus 754, 756
Peltotrupes 25, 26
Pelurosticha 650
Penetini 467, 471, 483, 485
Penichroa 576, 591
Pensus 324, 325
Pentacotaster 759
Pentanodes 582, 595
Pentaphyllus 475, 492
Pentaria 565, 566
Pentariini 566
Pentarthrini 725, 760
Pentarthrinus 760
Pentarthrum 757, 760
Penthe 414, 415
Penthelispa 460
Penthinae 414, 415
Penthispa 645
Pentiliaires 384
Pentispa 623, 645
Pentodontini 64, 66
Peracalles 762, 765
Peranoplium 591
Perapion 715, 717
Perarthrus 580, 596
Percolaspis 637, 669
Pergetus 551, 554
Peridorcatoma 258
Perigaster 749, 753
Perigasteromimus 749, 753
Perigracilia 576, 592
Perilypus 270, 277
Periphileurus 67
Perissarthron 169
Peritapnia 584, 598
Peritaxia 770, 782
Peritelini 725, 778
Peritelinus 772, 779
Peritelodes 779
Peritelopsis 773, 779
Perothopidae 152
Perothopinae 154
Perothops 153, 154
Persexarthrus 737
Persirocalus 752
Perthalycra 314
Petaliini 257
Petalium 251, 257
Petalochilini 725, 792
Petalon 91
Petalopus 666
Petriidae 463
Pezocrosita 626, 650
Phacepholis 775, 778
Phaea 586, 600
Phaedon 627, 652
Phaedromus 666
Phaenisor 699
Phaenithon 698, 699
Phaenocerus 155
Phaenops 104
Phaeton 652
Phalacridae 335, 336, 832
Phalacropsis 336
Phalacrus 336
Phalcrolobus 718
Phaleria 475, 491
Phaleriini 467, 475, 483, 491
Phalerisida 491
Phaleromela 475, 491
Phanaeini 49, 51
Phanaeus 49, 51
Phanocerus 118, 120
Phanosolena 698
Phantozomerus 315
Pharaxonotha 344, 345
Pharaxonothini 345
Pharini 382
Phedius 479, 499
Phegoneus 473, 489
Pheleconus 788
Pheletes 169
Phellidius 490
Phellopsini 459, 460
Phellopsis 459, 460
Pheloconus 791
Phengodella 184
Phengodes 184
Phengodidae 181, 184, 823, 835
Phengodinae 184
Phengodini 184
Phenolia 313, 314
Pherhimius 163, 169
Phidola 598
Philagathes 596
Philaspis 648
Phileurini 64, 67
Phileurus 64, 67
Philolithus 472, 488
Philopedon 774, 775
Philopedum 776
Philopuntia 765
Philostratus 790
Philothermus 364, 365
Philoxylon 254
Phloeochilus 800
Phloeodes 460, 461
Phloeolaemus 333, 334
Phloeophagia 759
Phloeophagina 761
Phloeophagomorphus 760
Phloeophagus 758, 761
Phloeophthorus 800
Phloeosinina 795, 800
Phloeosinites 800
Phloeosinus 795, 800
Phloeotribina 795, 800
Phloeotribus 795, 800
Phloeotrogus 803
Phloeotrya 420, 421
Phloeotrypetus 800
Phobetus 55, 60
Phodaga 525, 526
Phoenicobiella 697 699
Phoenicobius 699
Phoeodes 461
Pholidonotus 760
Phomalus 562
Phoracantha 578, 592
Photheca 257
Phratora 626, 652
Phrenapatinae 483, 484
Phrixiina 103, 107
Phrydiuchus 749, 752
Phryganophilus 419, 421
Phthora 485
Phthorius 805
Phthorophloeus 800
Phycocoetes 792
Phyconomus 320, 321
Phydanis 634, 660
Phyletes 491
Phylethus 491
Phyllastolus 781
Phyllechthrus 658
Phyllecthris 631
Phyllecthrites 628, 639, 657, 658
Phyllecthrus 658
Phyllerythrurus 730
Phyllis 651

Phyllobaenus 269, 276
Phyllobiini 725, 779
Phyllobius 772, 779
Phyllobrotica 631, 659
Phyllobroticites 628, 639, 659
Phylloceridae 152
Phyllocerinae 154
Phyllodecta 652
Phyllomima 666
Phyllonomeus 785
Phyllopertha 62
Phyllophaga 54, 55, 58
Phyllophoridae 160
Phyllostoma 10
Phyllotreta 634, 661
Phyllotrox 733, 738
Phymaphora 368
Phymatinus 782
Phymatodellus 594
Phymatoderus 594
Phymatodes 582, 594
Phyrdenus 761, 764
Physemus 125, 126
Physicerus 673
Physocnemum 582, 594
Physocoelus 498
Physodactylides 160
Physonaspis 646
Physonota 624, 646
Physonotini 646
Physorhinina 168
Physorhinites 160
Physorhinus 164, 168
Phytalus 55, 58
Phytobaenini 561
Phytobiini 724, 753
Phytobius 749, 754
Phytodecta 649
Phytodectella 649
Phytoeciini 586, 600
Phyton 592
Phytonomidius 783
Phytonomus 783
Phytophaga 568, 617
Phytophagoidea 568
Phytospinthera 668
Phyxelis 782
Piazorhinini 725, 792
Piazorhinus 788, 792
Piazorrhinus 792
Picatoma 257
Picrotini 341
Pidnota 589
Pidonia 574, 589
Piesocorynini 697, 699
Piesocorynus 698, 699
Piezocera 579, 592
Piezocerini 579, 592
Piezocorynus 699
Piezotrachelini 718
Pilema 593
Pilipalpinae 542
Pilumnus 731
Pimeliidae 463
Pimeliiformes 483
Pimeliinae 466, 469, 483, 485
Pinetoscolytus 801
Piniphilus 792
Pinotus 50
Piodes 572, 589
Piperius 804
Pirhidius 439, 441
Piscatopus 771, 780
Piseninae 414, 415
Pisenus 414, 415
Pissodes 788, 792
Pissodina 792
Pissodini 725, 792
Pithophilus 393
Pitnus 254
Pityobius 164, 169
Pityoborus 799, 804
Pityoceragenes 802
Pityocis 410
Pityogenes 798, 802
Pityokteines 798, 802
Pityomacer 694
Pityophagini 315
Pityophagus 313, 315
Pityophilus 598, 804
Pityophthorina 799, 804
Pityophthoroides 805
Pityophthorus 799, 804
Pityotrichus 799, 804
Pityphilus 598
Placonotus 332, 334
Placopterus 270, 278
Placosternus 583, 594
Plagiodera 628, 652
Plagiometriona 624, 647
Plagiomorpha 628, 653
Plagionotus 594
Plagiosalia 65
Plagioschema 628, 653
Plagiosterna 653
Plagithmysus 594
Planismus 345
Plastoceridae 160
Plastocerus 167
Platerodinae 177
Platerodini 176
Plateros 176, 177
Plateumarini 622, 639, 640
Plateumaris 622, 640
Platocthispa 623, 645
Platybregmus 251, 255
Platycerini 7
Platyceroides 7, 8
Platyceropsis 7, 8
Platycerus 7, 8
Platycis 175, 177
Platycycla 646
Platydema 475, 492
Platymeristes 753
Platynotini 471, 483, 494
Platyomus 775, 778
Platyotus 497
Platypachys 790
Platyphaeus 792
Platyphytodecta 649
Platypodinae 725, 805, 806
Platypodini 725, 806
Platypsyllus 823
Platyrhinini 697, 699
Platyrrhinidae 695
Platystomidae 695
Platystomini 697, 699
Platystomoidea 695
Platysystrophus 717
Platytarsus 780
Platytomus 45, 48
Plectonotum 209
Plectris 54
Plectrodera 584, 597
Plectrodes 55, 58
Plectromerus 577, 592
Plectromodes 792
Plectroscelis 664
Plectrura 584, 596
Pleganophorinae 369
Pleocoma 21
Pleocomidae 20, 21, 820
Pleonides 271
Plesiobaris 743, 745
Plesiocis 407, 410
Plesioclytus 583, 594
Plesiophthorus 804
Plesiosternus 61, 63
Pleuridium 361
Pleuropasta 525, 526
Pleurophorus 45, 48
Pleuropompha 527
Pleurosticha 626, 650
Plinthina 789
Plinthocoalium 593
Plinthocoelium 579, 593
Plinthodes 768, 775
Plionoma 580, 596
Plocamus 743, 747
Plocetes 740
Plusiotina 63
Plusiotis 63
Pnigodes 740
Pocadius 313, 314
Pocalta 63
Podabrini 206, 208
Podabrus 209
Podapion 715, 717
Podistra 208
Podolasia 53, 57
Podolasiini 53, 56
Podostena 54, 57
Poecilips 802
Poecilium 594
Poecilobrium 581, 593
Poecilocera 622, 640
Poecilocrypticus 475, 492
Poecilonota 104, 107
Poephagus 753
Pogonochaerus 598
Pogonocherini 584, 598
Pogonocherus 585, 598
Poiyaunbus 791
Polemiotus 473, 489
Polemius 209
Poliaenus 584, 598
Polinexa 314
Polopinus 482, 501
Poloposes 776
Polycaon 235, 239
Polycaoninae 235, 239, 830
Polycesta 102, 105
Polycestina 105
Polycestinae 101, 105
Polycestini 102, 105
Polyctesina 102, 105
Polyctesini 102
Polydacrys 774, 776
Polyderces 790
Polydora 107
Polydrusini 725, 779
Polydrusus 774, 779
Polydus 790
Polygramma 651
Polygraphina 795, 800
Polygraphus 796, 800
Polylamina 58
Polymoechus 63
Polyopsia 597
Polyphaga 819, 820
Polyphylla 55, 58
Polypleurus 482, 501
Polyponus 792
Polypria 421, 835
Polyspila 651
Pomachiliina 167
Pomachilites 160
Pomphopoea 527
Poophagus 751, 753
Popillia 61, 62
Porcinolus 115
Poreospasta 527
Porphyraspis 645
Porrhothus 786
Porrothus 786
Postelichus 122
Potergini 159
Potocallidium 594
Praeolamus 738
Prasinalia 103, 107
Prasocuris 627, 652
Prataeus 484
Prateus 471
Premnobius 798, 803
Premnophilus 803
Prenesdus 752
Priocera 270, 277
Priognathus 539, 833
Prionellus 587
Prionidae 568
Prionina 587
Prioninae 569, 570, 586
Prionini 570, 587
Prionocyphon 88
Prionomerina 790
Prionomerus 790
Prionosceles 801
Prionus 571, 587
Prioscelida 496

Priotoma 257
Prisistus 751, 753
Pristilophus 170
Pristoscelis 284, 295
Probagous 740
Probatius 599
Procas 731
Prochramesus 800
Procryphalus 797, 803
Proctophysus 673
Proctorus 735, 738
Proculini 13
Proderops 498
Proeces 758, 760
Proecha 598
Proecini 725, 760
Progenius 803
Prognathotrichus 805
Proludius 166, 170
Prolyctus 360, 361
Promecosoma 636
Promecotarsus 733, 739
Prometopia 313, 314
Prometopion 489
Promoresia 120
Promus 477, 495
Pronemognatha 525, 527
Pronocera 581, 594
Pronocerus 594
Prophyllis 654
Propylaea 387
Propylea 381, 387
Prorutidosoma 754
Prosaldius 746
Proscarabaeus 527
Prosopotropis 155
Prostephanus 235, 241
Prosternidae 160
Prosterninae 169
Prosternini 169
Prosternon 166, 170
Prosternoptinus 254
Prostomidae 510, 511, 830
Prostominiinae 546
Prostomis 511
Protaetia 68, 69
Protaneflus 591
Protelateridae 160
Prothalpia 419, 421
Protheca 252, 257
Prothecini 257
Protocallidium 594
Protocylas 717
Protofarsus 154, 155
Protopeltinae 265
Protophysus 673
Proutianini 155
Proutianus 153, 155
Pruniphagus 800
Psacus 231
Psalidonota 624, 647
Psammobius 48
Psammodiini 43, 47
Psammodius 44, 48
Psammoporus 45
Psapharochrus 598
Psebiini 581, 593
Pselactus 757, 759
Pselaphorhynchites 707
Psenocerus 584, 598
Psephenidae 133, 134, 828
Psepheninae 134
Psephenus 134, 643
Pseudacamptus 759
Pseudacratus 57
Pseudalexia 368
Pseudalindria 266
Pseudallonyx 294
Pseudanemia 494
Pseudanidorus 562
Pseudanostirus 166, 170
Pseudanthonomus 736, 737
Pseudanthribus 699
Pseudaphioda 759
Pseudaphonus 66
Pseudariotus 561, 562
Pseudastylopsis 585, 599
Pseudasydates 286, 295
Pseudataenius 45, 47
Pseudatrox 18
Pseudebaeus 288
Pseudeleodes 478, 495
Pseudelissa 780
Pseudephalus 476, 494
Pseudeudius 778
Pseudeurostus 254
Pseudibidion 592
Pseudino 547
Pseudips 798, 802
Pseudischyrus 350, 352
Pseudoabsidia 208
Pseudoacalles 762, 765
Pseudoazya 377, 385
Pseudobaridia 745
Pseudobaris 741, 745
Pseudobrachytarsus 700
Pseudocamptodes 314
Pseudocanoixus 776
Pseudocanthon 49
Pseudocassida 646
Pseudocentrinus 742, 743, 747
Pseudocercopeus 773, 781
Pseudochelonarium 140
Pseudochina 256
Pseudochinus 257
Pseudochlamys 639, 676
Pseudochodaeus 29, 30
Pseudochoragus 698
Pseudocistela 481, 499
Pseudocleis 387
Pseudocneorhinus 772, 781
Pseudocoelus 487
Pseudocomus 790
Pseudocorthylus 805
Pseudocotalpa 61, 63
Pseudocotomus 266
Pseudocrepidosoma 663
Pseudocryphalus 800, 804
Pseudocyphus 778
Pseudodibolia 634, 661, 666
Pseudodinoptera 587
Pseudoeucyllus 777
Pseudogarnia 746
Pseudogaurotina 573, 589
Pseudoholostrophus 414, 416
Pseudohylesinus 795, 800
Pseudohyperodes 766
Pseudolacordairius 781
Pseudolampsis 633, 666
Pseudolignyodes 739
Pseudolina 653
Pseudolucanus 8
Pseudoluperus 632, 658
Pseudomelanactes 169
Pseudomelanactini 169
Pseudomeninae 154
Pseudomethia 576, 590
Pseudomicracis 801
Pseudomopsis 763, 765
Pseudomus 762, 765
Pseudonomus 493
Pseudonosoderma 460
Pseudopachyta 589
Pseudopanscopus 782
Pseudoparmulus 393
Pseudoparopsis 653
Pseudopentarthrum 757, 759
Pseudophanus 325
Pseudophymatodes 594
Pseudopidonia 589
Pseudopilema 581, 593
Pseudopityophthorus 799, 804
Pseudoplateros 177
Pseudopoecilium 594
Pseudopolygraphus 801
Pseudorimus 782
Pseudorthygia 634, 660
Pseudoscymnus 379, 383
Pseudostrangalia 574, 589
Pseudostylopsis 599
Pseudotelegeusis 179, 180
Pseudotetramera 568
Pseudothalycra 314
Pseudothysanoes 796, 801
Pseudotolida 428
Pseudotychius 714, 718
Pseudoweisea 382
Pseudoxyletinus 256
Pseudozonitis 525, 527
Pseudromicronyx 739
Pseudypera 731
Psilax 51
Psilocalymma 718
Psilocnemis 68, 70
Psiloptera 107
Psilopterina 104, 107
Psilopterini 104, 107
Psilopyga 313, 315
Psoa 235, 239, 241
Psoidae 233
Psoinae 235, 241, 831
Psomus 755, 756
Psuedothysanoes 796
Psylliodes 633, 666
Psyllobactra 666
Psyllobora 375, 387
Psylloborini 375, 387
Psyllomima 666
Psyrassa 577, 578, 592
Pteleon 632, 659
Ptena 665
Pterichthya 597
Pterichtya 597
Pterichyta 597
Pterna 500
Pterochalybs 737
Pterocolinae 706, 708
Pterocolus 706, 708
Pterocyclon 805
Pteroplatini 579
Pteropliini 598
Pteryngium 340, 341
Ptichopus 13
Ptiliidae 822
Ptilineurini 254
Ptilineurus 249, 254
Ptilininae 256
Ptilinini 256
Ptilinobium 249, 256
Ptilinus 249, 256
Ptilodactyla 137, 138
Ptilodactylidae 135 137, 826
Ptilodactylinae 137
Ptilophorinae 438, 440
Ptilopus 777
Ptinidae 245
Ptininae 253, 831
Ptinodes 256
Ptinomorpha 662
Ptinus 254
Ptosima 102, 106
Ptosimina 102, 106
Pulion 310
Pullus 383
Purpuricenus 580, 596
Pyanisia 498
Pycnarthrum 796, 801
Pycnobaris 745
Pycnocephalus 361
Pycnogeraeus 744, 747
Pycnomerinae 457
Pycnomerini 459, 460
Pycnomerus 459, 460
Pycnomorhus 460
Pycnomorpha 472, 488
Pycnonotida 488
Pycnotomina 320, 321
Pygidiocarina 675
Pygmaeopsis 584, 598
Pygmaeoscolytus 801
Pygoleptura 574, 589
Pyres 498
Pyrochroa 542
Pyrochroidae 540, 542, 833, 834
Pyrochroinae 541, 542
Pyrophorini 169
Pyrophorites 160
Pyrophorus 169

Pyrota 526
Pyrotini 526
Pyrotrichus 572, 589
Pyrracita 598
Pyrrhacita 598
Pyrrhalta 630, 655
Pythidae 537, 539, 834
Pytho 539, 833
Pyticeroides 271, 278

Q

Qadrius 552, 554
Quadrifrons 314
Quadrius 554
Quasilema 622, 642
Quirinus 369

R

Rabdotorhinus 739
Racenisa 658
Radine 257
Radymna 654
Rahdine 257
Ranunculiphilus 753
Raymondionyminae 724, 725, 732
Raymondionymini 724
Redistes 361
Relistrodes 766
Reminius 501
Renocis 800
Rentoniinae 265
Retocomus 551, 554
Revelieria 396, 397
Rhabdophloeus 333
Rhabdopterus 636, 669
Rhabduchus 369
Rhaciaspis 168
Rhacius 484
Rhacostoma 314
Rhadalidae 281
Rhadalinae 284, 291
Rhadalus 286, 291
Rhadine 257
Rhaeboscelina 105, 108
Rhaeboscelis 108
Rhaestes 739
Rhagium 572, 589
Rhagomicrus 153, 155
Rhagonycha 207
Rhamnusium 588
Rhamphina 738
Rhamphini 724, 738
Rhamphocolus 760
Rhanidea 368
Rhanis 368
Rhaphidoplectron 717
Rhaxonycha 207
Rhigopsis 782
Rhina 786
Rhinandrus 479, 498
Rhinanisus 760
Rhinastina 790
Rhinobatus 785
Rhinocyllini 725, 785
Rhinocyllus 783, 785
Rhinodes 786
Rhinoncus 748, 754
Rhinonus 761
Rhinoplatia 516, 518
Rhinorhynchinae 693, 694
Rhinorhynchini 694
Rhinosimus 546, 547
Rhinostomina 729
Rhinotragini 579, 593
Rhipiceridae 92, 94, 826
Rhipidandrina 475, 483, 490
Rhipidandrus 475, 490
Rhipidophorus 440
Rhipiphorina 69
Rhizelmis 118, 120
Rhizobiares 385
Rhizobiellus 385
Rhizobiides 385
Rhizobiinae 385
Rhizobiini 385
Rhizobius 385
Rhizopertha 242
Rhizophagidae 319
Rhizophaginae 320, 829
Rhizophagus 319, 320
Rhodobaenus 729, 730
Rhodoleptus 580, 596
Rhombonalia 62
Rhopalapion 716, 718
Rhopalodontus 405, 410
Rhopalomerus 594
Rhopalopachys 594
Rhopalophora 582, 595
Rhopalophorini 582, 595
Rhopalophorus 595
Rhopalopleurus 800
Rhopalopus 594
Rhopalotria 701
Rhoptobaris 742, 745
Rhymbomicrus 369
Rhymbus 369
Rhynchaenus 735
Rhyncheros 177
Rhynchites 706, 707
Rhynchitidae 703
Rhynchitinae 706, 707
Rhynchitini 707
Rhyncholus 760
Rhynchophorina 729
Rhynchophorini 724, 729
Rhynchophorus 729
Rhynchus 763, 765
Rhyncolina 760
Rhyncolini 725, 760
Rhyncolus 758, 760
Rhyphonemognatha 525, 527
Rhypodes 779
Rhypodilius 772
Rhypodillus 779
Rhysodidae 819
Rhysopaussidae 463
Rhysothorax 45
Rhyssemus 44, 48
Rhyssomatus 788, 790
Rhythirrinini 725, 766
Rhytidosoma 754
Rhytidosomus 754
Rhyzobius 377, 385
Rhyzopertha 236, 242
Rilettius 552, 554
Rileyonymus 752, 753
Ripidiinae 439, 441, 835
Ripidius 441
Ripiphoridae 431, 833, 835
Ripiphorinae 439, 440
Ripiphorini 439, 440
Ripiphorus 439, 440
Riponus 587
Rismethus 163, 168
Rodolia 377, 385
Rodoliaires 385
Romaleum 592
Romulus 578, 592
Ropalodontus 410
Ropalopus 582, 594
Roplisa 70
Rosalia 581, 593
Rosaliini 581, 593
Rosella 740
Rostricephalinae 160
Ruguloscolytus 801
Rushia 420
Rustleria 361
Rutela 62, 64
Rutelinae 42, 60, 62
Rutelini 60, 62
Rutelisca 61, 64
Rutidosoma 750, 754
Rybakowia 662
Ryletinus 256
Rypobiinae 392, 393
Rypobiini 393
Rypobius 393

S

Sachalinobia 572, 589
Sacina 393
Sacinops 393
Sacium 393
Sacodes 88
Sagra 618
Sagridae 617
Sagrinae 618, 639
Sahlberginus 394
Salebius 340, 341
Salius 739
Salltiini 341
Salpingidae 544, 546, 822, 830, 833
Salpinginae 545, 547
Salpingus 547
Sandalus 94
Sandytes 561
Sangaria 662
Sanzia 785
Saperda 584, 600
Saperdini 584, 600
Sapintus 552, 555
Sapotes 768, 778
Saprobia 314
Sarabandus 88
Sarapus 738
Sarosesthes 583, 594
Sarpedon 153, 155
Satorystia 355
Saxinis 638, 676
Sayapion 716, 718
Scaeother 352
Scalaventer 771, 780
Scalidia 328
Scalidiidae 327
Scalopterus 289
Scanylus 562
Scaphidema 466, 492
Scaphidemini 466, 483, 492
Scaphidomorphus 785
Scaphigenia 675
Scaphinus 571, 587
Scaphomorphus 784, 785
Scaptes 493
Scaptolenus 166, 167
Scarabaeidae 39, 820
Scarabaeiformia 1
Scarabaeinae 42, 48, 49
Scarabaeoidea 1, 42
Scardamyctes 786
Scelida 632, 659
Scelidites 628, 639, 658
Scelodontites 639, 670
Scelolyperus 659
Schaumius 781
Schedarosus 491
Schematizites 628, 639
Schimatocheilus 699
Schistoceros 240
Schizax 580, 596
Schizillus 471, 486
Schizomicrus 732
Schizonotus 732
Schizophilini 154
Schizophilus 153, 154
Schizophthalmotribolium 497
Schizopodidae 95, 96, 827
Schizopodina 96
Schizopodinae 96
Schizopodini 96
Schizopus 96
Schizotus 542
Schoenicus 473, 489
Sciaphilini 725, 779
Sciaphilus 774, 780
Scierus 795, 799
Sciopithes 773, 778
Scirtes 88
Scirtidae 87, 828
Sclerocerus 590
Sclerodes 147
Scleropteridius 754
Scleropterini 724, 754
Scobicia 237, 240

Scoliocerus 781
Scolochrus 672
Scolytina 796, 801
Scolytinae 725, 792, 794, 799
Scolytini 725, 794, 801
Scolytochelus 801
Scolytodes 796, 801
Scolytogenes 797, 803
Scolytus 796, 801
Scotias 254
Scotobaenus 482, 498
Scotobates 498
Scotochroa 420
Scotochroides 420
Scraptia 565, 566
Scraptiidae 564, 566
Scraptiinae 565, 566
Scraptiini 566
Sculptotheca 252, 257
Scuromanius 294
Scymniaires 383
Scymniens 382
Scymnillini 377, 382
Scymnillodes 382
Scymnillus 382
Scymninae 382
Scymnini 377, 379, 383
Scymnobius 383
Scymnus 379, 382, 383
Scyphophorus 729, 730
Scyphotychius 739
Scythropus 779
Securipalpes 371
Selatosomus 166, 170
Selenedon 167
Selis 314
Selonodon 166, 167
Selvadiini 379, 383
Selvadius 379, 383
Semanotus 581, 594
Semicnema 666
Semijulistus 291
Semiotina 160
Sennius 605, 607
Senodoniinae 160
Serangiini 377, 382
Serica 54, 57
Sericini 53, 57
Sericoderini 393
Sericoderistes 393
Sericoderus 392, 393
Sericosomus 167
Sericus 164, 167
Sermylassa 656
Sermylini 656
Sermysatus 790
Serratibruchus 606
Serraticollis 663
Serrifornax 154, 156
Serrocerus 257
Serropalpidae 417
Serropalpini 420
Serropalpus 420
Setasomus 166, 170
Sexarthrus 737
Sibariops 744, 746
Sibinia 734, 739
Sibynes 739
Sibynia 739
Sicanthus 696, 698
Sicinus 491
Sicoderus 789, 791
Sicyobius 597
Sidis 383
Sidrotepa 256
Sierraclava 114, 116
Sigerpes 361
Silaria 567
Silenus 154
Silinae 206, 207, 209
Silini 206, 207, 209
Silphidae 822, 828
Silvanidae 322, 832
Silvaninae 323, 325
Silvanophloeus 334
Silvanoprus 324, 326
Silvanus 324
Simplocaria 115
Simplocariini 115
Sinodendrini 7, 8
Sinodendron 7, 8
Sinolus 593
Sinomela 649
Sinoxylini 236, 241
Sinoxylon 236, 241
Siolaires 385
Sirocalodes 751, 753
Sirocalus 752
Sisenes 517, 518
Sisyphini 49, 51
Sisyphus 49, 51
Sitodrepa 256
Sitona 767, 780
Sitones 780
Sitonidius 780
Sitonini 725, 780
Sitonobia 730
Sitophagus 491
Sitophilus 728, 729
Sitoprepa 256
Skwarraia 677
Smaragdina 638, 674
Smaragdinella 674
Smicraulax 735, 737
Smicripidae 316, 317, 829
Smicrips 317
Smicronychini 724, 739
Smicronyx 733, 739
Smileceras 596
Smilia 382
Smodicini 575, 590
Smodicum 575, 590
Snellingia 675
Solanophila 388
Solenopterini 570, 586
Solenorhinus 738
Soronia 313, 314
Sosylus 361
Sotenus 591
Spacalopsis 597
Spalacopsis 583, 597
Spaniophaeni 340
Sparedrus 516, 517
Sparteus 606
Spartiophila 649
Spartomena 649
Spartophila 649
Spartoxena 649
Spastonyx 525, 527
Spectralia 103, 107
Spermatoplex 802
Spermophagidae 602
Spermophagini 606
Spermophagus 606
Sphaenothecus 580, 596
Sphaericus 254
Sphaeriestes 546, 547
Sphaerilethmus 780
Sphaeriontis 487
Sphaeritidae 826
Sphaerobothris 104, 108
Sphaerocharitinae 671
Sphaeromela 626, 650
Sphaeromorphus 35
Sphalma 539, 834
Sphenoceros 804
Sphenognathus 729
Sphenophorina 729
Sphenophorus 729, 730
Sphenoptera 104, 107
Sphenopterina 104, 107
Sphenostethus 570, 587
Sphenothecus 596
Sphindidae 305, 307, 829
Sphindinae 308
Sphindociinae 405
Sphindocis 405, 408, 829
Sphindus 308
Sphingocephalus 565, 567
Sphoenothecus 596
Spiladelphia 386
Spilosota 65
Spilota 62, 386
Spilotus 419, 420
Spinhyporhagus 455, 456
Spinianemia 494
Spintherophyta 636, 668
Spinthoptera 107
Spinuloscolytus 801
Spondylidae 568
Spondylidinae 569, 571, 587
Spondylis 571, 587
Spongifer 783
Spongotarsus 801
Sponidium 521
Squamanotoxus 552, 556
Squamodera 102, 106
Squamosinus 800
Stachysivora 659
Stagetus 251, 257
Stamoderes 774, 777
Staphylinidae 822, 823, 824
Staphylinoeme 590
Statira 471, 484
Stator 607
Stauroglossicus 314
Steatoderus 167
Stegnomorpha 315
Stegobiini 256
Stegobium 256
Stelidota 313, 314
Stenancylus 759, 761
Stenaspis 580
Stenaspsis 596
Stencorus 589
Stene 497
Stenelaphus 578, 592
Steneleodes 478, 495
Stenelmis 118, 120
Stenidea 598
Stenispa 622, 643
Stenobaris 745
Stenocerini 697, 699
Stenocerus 697, 699
Stenochia 501
Stenochidus 479, 499
Stenochorus 589
Stenocleptus 796
Stenoclyptus 801
Stenocolus 142
Stenocorus 572, 589
Stenodema 266
Stenodontes 570, 586
Stenomantura 664
Stenomimus 758, 759
Stenomorpha 472, 473, 488
Stenopelmina 731
Stenopelmus 731, 732
Stenophrixothrix 184
Stenopodius 623, 644
Stenopterini 581, 593
Stenoptochus 773, 779
Stenorhynchus 739
Stenoscelis 758, 760
Stenosides 472, 488
Stenosini 470, 471, 483, 485
Stenosoma 598
Stenosphenopsis 592
Stenosphenus 577, 592
Stenostrophia 574, 589
Stenotarsoides 369
Stenotarsus 369
Stenotis 760
Stenotrachelidae 520, 521, 832
Stenotrachelinae 521
Stenotrachelus 521
Stenotrichus 496
Stenotrupis 757, 759
Stenotylus 760
Stenura 588
Stenuris 107
Stephanocleonus 784, 785
Stephanoderes 804
Stephanopachys 236, 242
Stephanucha 68, 69

Stephostethus 396, 397
Steremnius 789
Stereodermini 717
Stereodermus 714, 717
Stereogaster 774, 777
Stereogeraeus 747
Stereomoides 676
Stereopalpus 552, 554
Stereopselaphus 554
Stereosa 107
Stereus 790
Steriphanus 490
Sternechini 725, 792
Sternechus 788, 792
Sternidius 585, 599
Sternidocinus 585, 599
Sternochetus 764
Sternocoelus 765
Sternodea 340, 341
Sternoxus 790
Sternuchus 85, 792
Steropinae 553
Stethasida 473, 488
Stethobaris 742, 744, 747
Stethon 155
Stethorhanis 368
Stethorini 377, 383
Stethorus 377, 383
Steupeckia 792
Sthereus 788, 789
Stibia 474, 490
Sticholotidinae 375, 382
Sticholotini 382
Stichosoma 650
Stichtoplychus 257
Stichtoptychus 252, 257
Sticticomus 555
Stictobaris 743, 745
Stictocera 107
Stictoleptura 574, 589
Stigmalia 65
Stigmatoma 500
Stilbolemma 57
Stilbus 336
Stilpnus 169
Stizocera 577, 592
Stolaini 645
Stomodes 772, 779
Storeini 724, 739
Storthephora 484
Strabala 634, 663
Strabus 699
Strangalepta 574, 589
Strangalia 573, 588, 589, 590
Strangalina 589
Strategodes 66
Strategopsis 66
Strategus 65, 66
Striatheca 251, 257
Strickerus 652
Stricticomus 553, 555
Strigocis 406, 410
Strigoderma 61, 62
Strigodermella 62
Strigophorus 673
Stromatium 577, 591
Strongylaspis 570, 586, 648
Strongyliini 468, 483, 501
Strongylium 468, 501
Strongylocassis 624, 648
Strongylophthalmus 781
Strongylotes 747
Strongylus 314, 385
Strophiona 575, 590
Strophosoma 774, 776
Strophosomum 776
Strophosomus 776
Stygnesis 597
Stygoparnus 122
Stylantheus 606, 607
Styloleptus 585, 599
Stylotentus 804
Styloxus 576, 590
Styphlotychius 739
Subcoccinella 375, 388
Subphyllerythrurus 730
Sudus 763, 764
Sulcacis 406, 410
Sulcatolema 642
Sulciurus 776
Sumitrosis 623, 644
Svetlaniolus 753
Sybra 583, 597
Sydates 296
Sydatopsis 285, 297
Symbiotes 369
Symmathetes 778
Symphora 419, 421
Symphyomethes 198, 200
Sympiezocera 594
Synanobium 257
Synaphaeta 583, 597
Synaphe 597
Synapiina 718
Synaptidae 160
Synaptina 168
Syncalypta 116
Syncalyptidae 113
Syncalyptinae 116
Synchroa 513
Synchroidae 512, 513, 833
Synclytus 594
Syndesinae 7
Synertha 739
Syneta 637, 667
Synetidae 617
Synetinae 667
Synetini 637, 639, 667
Synetocephala 658
Synetocephalus 632, 658
Synirmus 782
Synolobus 776
Synonychina 386
Synonychinae 386
Synonychini 386
Synstrophus 414, 416
Syntomocerus 760
Syntomostylus 776
Syphraea 662
Syphrea 636, 662
Sysphenges 662
Syssitos 328
Systaechea 369
Systena 634, 662
Systene 597
Systenocerus 8
Systenus 8
Syzeton 561, 562
Syzetoninina 562

T

Tachyderes 804
Tachyerges 735, 739
Tachygonini 724, 756
Tachygonus 754, 756
Tachyopus 756
Tachypterellus 737
Tachypterus 737
Taeniocleonus 785
Taenioglyptes 804
Taenobates 501
Tagalus 485
Taiwania 647
Talaepora 597
Talanini 468, 483, 501
Talanus 501
Tanaops 286, 289
Tanarthropsis 555
Tanarthrus 552, 555
Tanilotes 562
Tanygaster 661
Tanymecina 780
Tanymecini 725, 780
Tanymecus 771, 780
Tanyochraethes 583, 594
Tanysphryina 732
Tanysphyroides 732
Tanysphyrus 731, 732
Taphrocerus 105, 109
Taphroscelidia 328
Taranomis 596
Taraxis 611
Tarpela 478, 496
Tarsosteninae 272
Tarsostenninae 279
Tarsostenus 272, 279
Tasioderma 257
Tasmus 314
Tauronthophagus 51
Tautoclines 597
Taxaris 673
Tedinus 481, 499
Tegrodera 525, 526
Teinodactyla 661
Telabis 474, 490
Telegeusidae 179, 823
Telegeusis 179, 180
Telephanini 324
Telephanus 324
Telephoridae 202
Telephus 746
Telesicles 480, 499
Telmatophilides 340
Telmatophilus 340, 341
Temnocerus 706, 707
Temnochila 266
Temnochilidae 263
Temnopsophus 286, 288
Temnoscheila 265, 266
Tenebrio 479, 498
Tenebrioloma 497
Tenebriomimus 492
Tenebrionellus 498
Tenebrionidae 463, 483
Tenebrioniformes 483
Tenebrioninae 483, 495
Tenebrionini 469, 479, 483, 497
Tenebroides 265, 266
Tentyridae 483
Tentyriidae 463
Teplinini 392, 393
Teratonychus 739
Terchaetes 598
Teredinae 361
Terobyturus 355
Tesarius 44, 48
Tessaroma 493
Tessaropa 576, 590
Tessaroptes 590
Testergus 661
Tetracanthus 595
Tetralobites 160
Tetramera 568
Tetranodes 595
Tetranodus 582, 595
Tetraonycidae 522
Tetraonycinae 527
Tetraonycini 527
Tetraonyx 525, 527
Tetraopes 586, 600
Tetraopini 586, 600
Tetrapriocera 237, 240
Tetraspartus 729
Tetratemnus 729
Tetratoma 414, 415
Tetratomidae 413, 414, 834
Tetratominae 414
Tetropium 571, 587
Tetrops 586, 600
Texania 104, 106
Texaponium 474, 490
Thalassa 379, 383
Thalasselephantini 725, 792
Thalasselephas 788, 792
Thalattodora 759
Thalycra 313, 314
Thambospasta 551, 554
Thambus 154, 156
Thamnophilus 786
Thamnophthorus 804
Thamnurgides 802
Thanasimus 270, 277
Thaneroclerinae 268, 273

Thaneroclerus 268, 273
Thapter 257
Thaptor 257
Tharops 155
Tharsus 478, 497
Thaumaglossa 230
Thaumaphrastus 230
Thaumasinulus 800
Thea 387
Theca 257
Thecesternini 725, 727, 781
Thecesternus 727, 767, 781
Thecha 257
Thelydrias 230
Theogenes 67
Therchaetes 598
Thersalus 592
Thes 397
Thesalia 589
Thia 590
Thicanus 555
Thicosternus 781
Thinalmus 209
Thinoxenus 772, 779
Thione 319, 321
Thionini 320
Thlibocoryna 651
Thorictidae 228
Thorictini 229, 230
Thorictodes 230
Threcticus 739
Thricolema 637, 667
Thricolepis 772, 779
Thricolepoides 780
Thricomigus 782
Thrimolus 400
Thrincopyge 101, 105
Thrincopygina 101, 105
Thrincopygini 101, 105
Throscidae 158, 159, 826
Throscinae 159
Throscinus 125, 126
Throscites 159
Throscus 159
Thryallis 584, 598
Thyamis 661
Thyce 55, 58
Thylacites 776
Thylacosterninae 160
Thylodrias 229, 230
Thylodriini 229
Thylodrius 835
Thymalops 265
Thymalus 265
Thysanocnemis 739
Thysanocorynus 779
Thysanoes 796, 801
Tigrinellus 783
Tigrinestola 585, 598
Tilargus 400
Tillinae 269, 274
Tilloclytus 582, 595
Tillomorpha 595
Tillomorphini 582, 595
Timarcha 625, 649
Timarchini 639, 648
Timarchostoma 649
Tinodemus 336
Tinopus 595
Tipnus 254
Tiresias 230
Tisactia 340, 341
Tisiphone 317
Tlanoma 664
Toblezkus 82, 83
Tomarus 65, 66, 345
Tomicina 795, 800
Tomicus 795, 800
Tomoderinae 551, 556
Tomoderus 551, 556
Tomolips 757, 761
Tomoxia 426
Tonesiina 745
Tonibius 476, 494
Toplethus 736
Toplithus 736
Toposcopus 438, 440
Toptaria 736
Toraminae 345
Toramus 344, 345
Torcina 746
Torneutina 575, 595
Torneutini 595
Tosastes 778
Tostegoptera 55, 58
Totropiidae 568
Tournotaris 731
Toxeres 746
Toxognathinae 160
Toxonotus 697, 699
Toxotochorus 589
Toxotopsis 589
Toxotropis 699
Toxotus 589
Trachelizinae 714, 717
Trachelizini 717
Trachodes 788, 790
Trachodini 725, 790
Trachodius 764
Trachyderes 579, 596
Trachyderina 579, 596
Trachyderini 579, 582, 595
Trachyglyphus 760
Trachyina 105, 108
Trachyini 105, 108
Trachykele 104, 107
Trachykelina 104, 107
Trachymela 625, 649
Trachymetopa 661
Trachypachidae 819
Trachyphloeina 781
Trachyphloeini 725, 781
Trachyphloeops 781
Trachyphloeosoma 781
Trachyphloeus 773, 781
Trachys 105, 108
Trachyscelida 632, 658
Trachyscelidites 629, 639, 658
Trachyscelini 467, 483, 492
Trachyscelis 492
Trachysida 574, 590
Trachytropis 697, 699
Tragidion 580, 596
Tragosoma 570, 587
Tranes 788, 792
Trapezicomus 555
Treiodous 527
Trepobaris 743, 745
Treptoplatypus 805, 806
Triachus 638, 674
Triarius 632, 658
Triarmocerus 804
Triboliini 467, 478, 483, 497
Tribolium 478, 497
Tribrachys 314
Trichacalles 764
Trichacorynus 759, 760
Trichalophus 768, 775
Trichaltica 635, 664
Trichanemia 494
Trichapiina 718
Trichapion 716, 718
Trichastylopsis 586, 599
Tricheleodes 477, 495
Trichesthes 58
Trichiasida 472, 473, 488
Trichiini 68, 70
Trichinorhipis 104, 108
Trichiorhyssemus 44, 48
Trichiotes 474, 490
Trichiotinus 68, 70
Trichischius 730
Trichobaris 741, 745
Trichobaropsis 737
Trichocanonura 585, 599
Trichochroides 297
Trichochronellus 285, 297
Trichochrous 285, 296
Trichocnemis 586
Trichoderulus 495
Trichodes 270, 277
Trichodesma 256
Trichodirabius 744, 747
Trichoferus 591
Tricholochmaea 630, 655
Trichomagdalis 786
Trichopsephus 369
Trichoscelis 596
Trichosirocalus 751, 753
Trichosomides 384
Trichoton 476, 494
Tricoryninae 258
Tricorynini 258
Tricorynus 252, 258
Tricrania 523, 527
Tricranioides 527
Tricrepidius 168
Triglyphulus 768, 775
Triglyphus 775
Trigonarthris 574, 590
Trigonodera 438, 440
Trigonogenius 254, 804
Trigonogya 102, 106
Trigonopeltastes 68, 70
Trigonorhinini 698, 699
Trigonorhinus 698, 699
Trigonorrhinus 699
Trigonoscuta 769, 777
Trigonoscutoides 769, 781
Trihabda 654
Trimera 305
Trimitomerus 539, 834
Trimytis 474, 490
Trinodia 70
Trinodini 229
Triodoclytus 583, 594
Triodonyx 54, 58
Triorophus 474, 490
Triphalopsis 474, 490
Triphyllia 414, 415
Triplax 350, 352
Tripopithys 256
Tripopitys 256
Trirhabda 630, 654
Trirrhabda 654
Trischidias 797, 804
Tritaenia 651
Tritoma 350, 352, 400
Tritominae 350
Triungulinus 527
Trixagus 159, 355
Trochoideus 369
Troezon 763, 764
Trogidae 17, 18, 820
Trogloderus 495
Troglomethes 198, 200
Troglops 286, 288
Trogoderma 231
Trogosita 266
Trogositidae 263
Trogossita 266
Trogossitidae 263, 265, 823, 828
Trogossitinae 265, 266
Trogoxylini 238, 242
Trogoxylon 238, 242
Trophimus 286, 290
Tropibalaninus 736
Tropicini 129, 131
Tropicus 129, 131
Tropideres 699
Tropiderini 697, 699
Tropiderinus 699
Tropidophorus 782
Tropiphorini 725, 781
Tropiphorus 782
Trox 18
Trpopitys 256
Truncatophloeus 333
Truncatus 661
Trypanophellos 800
Trypetidini 725, 792
Trypherus 209
Trypocranus 805
Trypodendron 797, 803
Trypophloeus 797, 803
Trypopithys 256
Trypopitys 256
Tubuloscolytus 801

Tucsonia 106
Turboscymnus 383
Turkmenobaris 745
Tychiina 739
Tychiini 724, 739
Tychius 739
Tydessa 541, 542, 834
Tydessinae 541, 542
Tylcus 582, 595
Tylistis 257
Tylocentra 647
Tylocerina 599
Tyloderma 761, 764
Tylodina 764
Tylonotus 576, 591
Tylopterus 739
Tylosis 580, 596
Tymnes 636, 637, 669
Tyndarina 102, 106
Tyndarini 102, 106
Typhaea 400
Typhloglymma 792
Typhlusechina 483, 486
Typhlusechus 471, 486
Typhobia 492
Typitium 521
Typocerus 574, 590
Typophorini 639, 667
Typophorites 639, 668
Typophorus 637, 668
Tytthaspides 386
Tytthomimus 759
Tytthonyx 209
Tytthonyxini 207, 209

U

Udorpes 664
Ulapion 717
Uleiota 324, 325
Ulema 642
Ulochaetes 571, 590
Uloma 478, 491, 496
Ulomini 466, 478, 483, 496
Ulomoides 475, 492
Uloporus 402
Ulosomus 764
Ulosonia 497
Ulus 476, 494
Upis 482, 501
Urgleptes 585, 599
Urilletta 256
Urleta 63
Urodera 638, 676
Urographis 585, 599
Urophorus 312, 314
Uroplatini 643
Usechimorpha 459, 460
Usechini 459, 460
Usechus 459, 460
Utobium 255

V

Vacronini 470, 473, 483, 488
Vacronus 488
Vacusus 553, 555
Valenus 585, 599
Valgini 68, 70
Valgus 68, 70
Vandykea 575, 590
Vanonus 561, 562
Vasaces 517, 518
Vectura 284, 294
Vecturoides 284, 294
Vesperelater 162, 169
Victorinus 754
Vincenzellus 546, 547
Virbius 310
Vitavitus 782
Vitellius 155
Volvoxis 314
Vriletta 256
Vrilleta 256
Vrilletta 252, 256

W

Willinkia 745
Wolcottia 270, 276
Wollastonia 761
Wollastoniella 761

X

Xanthochroa 516, 517
Xanthochroina 517, 518
Xanthogaleruca 630, 655
Xanthonia 637, 670
Xanthus 737
Xarifa 248, 255
Xenelmis 120
Xenochalepus 623, 644
Xenodacne 350
Xenodorum 594
Xenoheptaulacus 46
Xenomelanophila 104, 107
Xenomycetes 369
Xenomycetinae 369
Xenophthorus 804
Xenorchestes 698
Xenorhipidina 822
Xenorhipina 104, 108
Xenorhipis 104, 108
Xenoscelinae 345
Xenoscelini 346
Xenotychius 739
Xeranobium 249
Xeranoplium 577, 591
Xerasia 355
Xerobia 785
Xeronthobius 256
Xeropsamobeus 43, 46
Xesfobium 255
Xestobiini 255
Xestobium 249, 255
Xestocis 409
Xestolabus 707
Xestoleptura 574, 590
Xeuxes 415
Xiletinus 256
Xixias 718
Xoidolema 642
Xyleborina 798, 803
Xyleborinus 798, 803
Xyleborips 803
Xyleborus 799, 803
Xylechinops 800
Xylechinus 795, 800
Xylecoethus 262
Xylentinus 256
Xyletininae 256
Xyletinini 256
Xyletomeridius 256
Xyletomerus 252, 256
Xyletomimus 256
Xylion 237, 240
Xylita 420
Xylobiini 155
Xylobiops 237, 241
Xylobius 155
Xyloblaptus 237, 241
Xylochilus 804
Xylocomesus 760
Xylocrius 582, 594
Xylocryptus 803
Xyloecus 155
Xylomeira 237, 241
Xylopertha 240, 241
Xyloperthella 237, 241
Xyloperthini 236, 237, 240
Xylophilidae 559
Xylophilus 154, 155, 562
Xylopinus 482, 501
Xyloprista 237, 241
Xylopsocus 238, 241
Xyloryctes 65, 67
Xylosandrus 798, 803
Xylosteus 572, 590
Xyloterina 797, 803
Xyloterinus 797, 803
Xyloterus 803
Xylotheca 258
Xylothrips 238, 241
Xylotinus 256
Xylotrechus 582, 595
Xytelinus 256

Y

Yakuhananomia 427
Ydorpes 664
Yermoella 96
Yuccaborus 728, 729
Yunocassis 647

Z

Zabrotes 604, 606
Zagloba 377, 383
Zaglyptus 746
Zagymnus 579, 595
Zaitzevia 118, 120
Zalepia 168
Zamodes 576, 591
Zaplous 584, 598
Zarhipis 184, 825
Zascelis 763, 764
Zenoa 145
Zenodosus 268, 273
Zestobium 255
Zeugonyx 714, 718
Zeugophora 611
Zeugophorinae 611
Zidalus 494
Zilini 382
Zilora 419, 421
Zilus 377, 382
Zodinus 494
Zonantes 560, 561
Zonitis 525, 528
Zopheridae 457, 460, 830
Zopherini 459, 460
Zopherodes 461
Zopherosini 461
Zopherus 459, 461
Zophobas 479, 498
Zorochros 165, 170
Zorochrus 170
Zurecheria 501
Zurus 756
Zygaenodini 698, 699
Zygobarella 741, 747
Zygobaridina 727, 746
Zygobarinus 741, 747
Zygobaris 747
Zygobaroides 746
Zygogramma 626, 651
Zygogrammini 650
Zygomicrus 756
Zygopini 724, 756
Zygospila 651